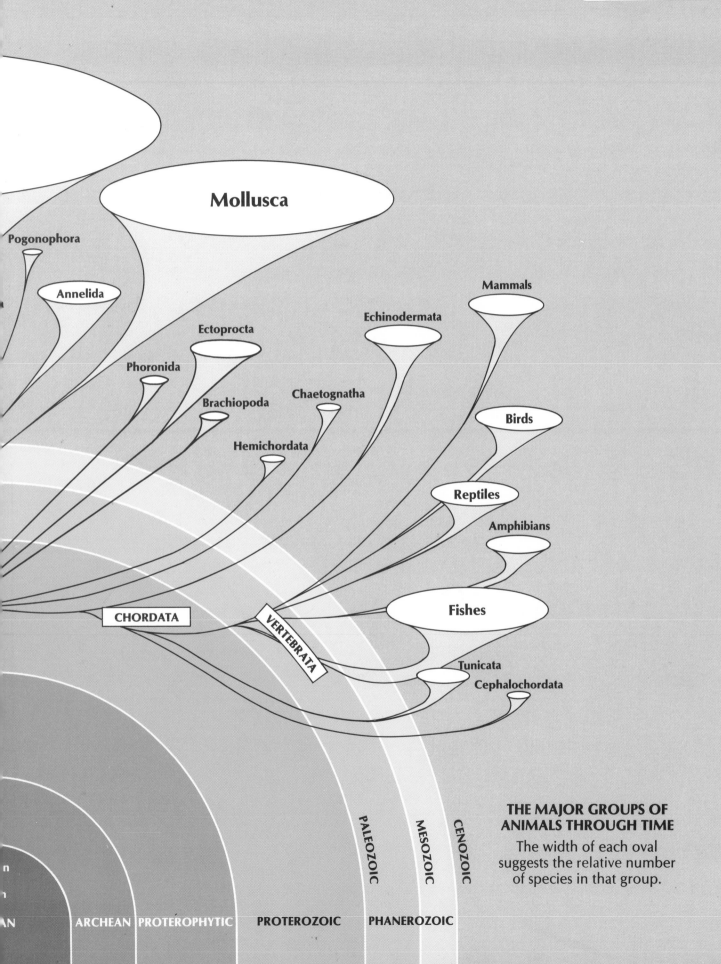

Pogonophora

Mollusca

Annelida

Ectoprocta

Phoronida

Echinodermata

Brachiopoda

Chaetognatha

Mammals

Hemichordata

Birds

Reptiles

Amphibians

CHORDATA

VERTEBRATA

Fishes

Tunicata

Cephalochordata

PALEOZOIC

MESOZOIC

CENOZOIC

**THE MAJOR GROUPS OF
ANIMALS THROUGH TIME**

The width of each oval
suggests the relative number
of species in that group.

ARCHEAN PROTEROPHYTIC   PROTEROZOIC   PHANEROZOIC

INTEGRATED PRINCIPLES *of*

ZOOLOGY

INTEGRATED PRINCIPLES *of*

ZOOLOGY

TENTH EDITION

10

Cleveland P. Hickman, Jr.
Washington and Lee University

Larry S. Roberts
University of Miami

Allan Larson
Washington University

Original Artwork by
William C. Ober *and* **Claire Garrison**

WCB **Wm. C. Brown Publishers**

Dubuque, IA   Bogotá   Buenos Aires   Caracas   Chicago   Guilford, CT   London
Madrid   Mexico City   Seoul   Singapore   Sydney   Taipei   Tokyo   Toronto

Project Team

Editor *Margaret J. Kemp*
Developmental Editor *Kathleen R. Loewenberg*
Production Editor *Jane E. Matthews*
Marketing Manager *Thomas C. Lyon*
Designer *K. Wayne Harms*
Art Editor *Jodi K. Banowetz*
Photo Editor *John C. Leland*
Permissions Coordinator *Karen L. Storlie*
Advertising Coordinator *Heather Wagner*

 Wm. C. Brown Publishers

President and Chief Executive Officer *Beverly Kolz*
Vice President, Director of Editorial *Kevin Kane*
Vice President, Sales and Market Expansion *Virginia S. Moffat*
Vice President, Director of Production *Colleen A. Yonda*
Director of Marketing *Craig S. Marty*
National Sales Manager *Douglas J. DiNardo*
Executive Editor *Michael Lange*
Advertising Manager *Janelle Keeffer*
Production Editorial Manager *Renée Menne*
Publishing Services Manager *Karen J. Slaght*
Royalty/Permissions Manager *Connie Allendorf*

 A Times Mirror Company

Copyedited by Cathy DiPasquale Conroy

Cover credit © 1996 Tom & Pat Leeson/DRK Photo

Freelance permissions editor *Karen Dorman*

The credits section for this book begins on page 871 and
is considered an extension of the copyright page.

Copyright © 1997 Times Mirror Higher Education Group, Inc.
All rights reserved

Library of Congress Catalog Card Number: 95–83331

ISBN 0–697–24366–4

Printed in the United States of America by Times Mirror Higher Education Group, Inc.,
2460 Kerper Boulevard, Dubuque, IA 52001

BRIEF CONTENTS

CONTENTS

PART III

The Diversity of Animal Life

PART IV

Activity of Life

PREFACE

Forty-two years ago, when the senior author's father completed the first edition of this textbook, zoology was already a well-established discipline, having emerged in the 19th century from the enormously popular study of natural history. The principles that form the conceptual framework of professional zoology today were formalized largely before the first edition appeared, and zoologists increasingly were embracing and profiting from the methods and technologies developed in other branches of science. Yet in the relatively short time separating the first edition of this text from the present one, the acceleration of scientific discovery has produced dramatic conceptual changes in zoological science. Forty-two years ago, zoologists still adhered to the two-kingdom, plant and animal system of classification. Populational thinking was relatively new to the evolutionary sciences, and cladistic methodology, introduced by Willi Hennig in 1950, was to remain largely ignored for two more decades. Only two years before the first edition of this book appeared, James Watson and Francis Crick published the much sought-after chemical structure of DNA. Molecular genetics was in its infancy, and genetic engineering with its enormous benefits to humanity was not even a promise. The first reported case of AIDS was 26 years in the future. The principles of ethology were unknown in the United States, and it would be six years before Jane Goodall began her studies on the social life of chimpanzees. The world's population 42 years ago was less than half today's population, but already there was rising concern over serious environmental problems. It would be eight more years before Rachael Carson would publish her famous polemic, *Silent Spring,* that documented the widespread destruction of songbirds and other wildlife from the indiscriminate spraying of DDT.

In responding to the growth of zoological discovery, *Integrated Principles of Zoology* evolved gradually as the authors of each edition strived to reflect the state of understanding at that time. Like other modern science textbooks with lavish art programs and extensive pedagogical aids, this tenth edition bears little resemblance to the first. Nevertheless, certain features that distinguished the first edition, and that were unique to zoology textbooks of the time, survive to the present. The statement and integration of evolutionary principles and emphasis on zoological investigation as a process for answering questions have distinguished all editions. Other features that originated in the first edition include an introductory statement in each of the survey chapters that explains the position of the group in the animal kingdom and the characteristics that distinguish the group; an historical appendix listing key discoveries and publications that have greatly influenced the development of zoology; the annotation of references; and the derivation of technical terms.

NEW TO THE TENTH EDITION

Although this edition of *Integrated Principles of Zoology,* with its three-column design and many new color illustrations, will look different from the previous edition, the chapter organization remains mostly unchanged. In addition to the numerous updatings and changes in individual chapters, we made the cladograms in all survey chapters more user friendly by adding color icons representing specific animal groups. Introductory statements covering "Position in the Animal Kingdom" and "Biological Contributions" were added to all chordate chapters. We have increased coverage of environmental issues in this edition. Today, concern with the animal world takes on new urgency as we are forced to reappraise our position in nature and decide what we wish to make of this world and the creatures with whom we share it.

The principal revisions made in each of the five parts of the text are explained below. We reworked all chapters to make the text current while eliminating excessive detail and placing greater emphasis on basic concepts and the role of experimentation and comparative studies in zoology. We retained and strengthened pedagogical features introduced in previous editions. These include the opening prologues for each chapter that relate a theme or topic drawn from the chapter; chapter summaries and review questions to aid student comprehension and study; in-text derivations of animal generic names; chapter notes (previously marginal notes) that enhance and enlarge in-text material; and an extensive glossary providing pronunciation, derivation, and definition of zoological terms.

PART ONE: INTRODUCTION TO THE LIVING ANIMAL

The five chapters of this unit explain the general features of living systems, the chemistry and origin of life, and the structure and function of animal cells. Chapter 1, completely revised in the ninth edition, introduces evolution, genetics and hypothetico-deductivism, and principles that underlie zoological study. Chapter 2, on the chemistry of life, is designed to help students with limited backgrounds in chemistry. Additional information on the basic structure of matter is found in Appendix B. In Chapter 3, on the origin of life, we incorporated new information on the earth's early atmosphere, the idea that methane and ammonia could have originated from planetisimals or hydrothermal vents, and Knoll's hypothesis to explain the Cambrian explosion. Among additions to the last two chapters of this unit are the concept of potocytosis, the remarkable behavior of the spindle fibers "feeling out" the kinetochores in mitosis, and a completely revised and re-illustrated explanation of the chemiosmotic hypothesis for generating ATP. We also reworked the treatment of glycolysis and revised several of the existing illustrations.

PART TWO: CONTINUITY AND EVOLUTION OF ANIMAL LIFE

The four chapters of Part Two focus on widely shared properties of living systems: reproduction, development, possession of a genetic program, and possession of an evolutionary history. We revised and reorganized Chapter 6 on reproduction to provide a more comprehensive and comparative treatment. We broadened the explanation of the proposed benefits of sex, reworked the section on the formation of reproductive cells to include the concept of germ cell line, explained nongenetic sex determination, and inserted a new section on invertebrate reproductive systems. In Chapter 7, we rewrote and expanded the section on gene expression to include the recent exciting work on homologies among the homeobox genes that encode positional information during the development of metazoans. Because most students are introduced to genetics in some course other than introductory zoology, we condensed

Chapters 8 and 9 of the previous edition into a new Chapter 8, now entitled, "The Principles of Genetics: A Review." We substantially rewrote the section on genetics of cancer and introduced Ras and p53 proteins as products of oncogenes and tumor suppressor genes, respectively. Chapter 9, on organic evolution, begins with an historical account of Charles Darwin's life, then develops the five fundamental components of Darwin's evolutionary theory, an approach introduced successfully in the ninth edition of this text.

PART THREE: THE DIVERSITY OF ANIMAL LIFE

The 22 chapters of this unit form the core of most zoology courses. The 20 chapters that survey the animal phyla and the animal-like protista (protozoa) are preceded by two introductory chapters. The first of these is a short but important chapter on animal architecture, which describes the components of the metazoan body and defines the organization of animal body plans. The second chapter in this unit explains the principles of animal taxonomy and how they are applied by the competing schools of evolutionary taxonomy and cladistics. The material in both of these chapters prepares students for the organizational structure of the remainder of this book.

Of the many changes made in the 14 invertebrate chapters (Chapters 12 to 25), we mention here only a few: an updated discussion of ameboid movement, new material on ammonoids, expanded coverage of the Polychaeta, greater emphasis on ticks as vectors, inclusion of a table of classification for Crustacea, and a discussion of the evolution of wings in insects. Among several recent changes in classification that have been incorporated is recent molecular evidence that allows us to place the Ctenophora phylogenetically outside a group containing the cnidarians and placozoans. The Turbellaria are now considered to be paraphyletic, and the name is retained only because of its familiarity to

zoologists. The quite remarkable subversion of host gene expression by *Trichinella* spp. is emphasized further, and we note that five species are now recognized in the genus. In reorganizing the discussions of protostome phylogeny, we note that evidence points to the existence of two groups of lesser protostomes, one with affinities to the annelids, the other to arthropods. The section on echinoderm phylogeny is expanded to include the carpoids and helicoplacoids, and we note that molecular evidence provides serious concerns as to whether the chaetognaths are deuterostomes, the group to which they traditionally have been assigned.

Among numerous changes to the six chordate chapters (Chapters 26 to 31) are new sections that include implications of research on homeobox genes in amphioxus, the discovery of the conodont animal, the evolution of the tetrapod leg, and a box treatment of the evolution of dinosaurs. We also revised and updated the classifications of birds and mammals. The appearance and usefulness of these chapters has been enhanced with several new illustrations. The treatment of human evolution, which in previous editions had been included in the chapter on organic evolution, was moved to the end of the mammal chapter (Chapter 31). Many instructors treat human evolution and mammals together in their classes, and we believe that the human evolutionary story illustrates mammalian adaptation better than it illustrates general evolutionary principles.

PART FOUR: ACTIVITY OF LIFE

This section consists of six chapters that take a comparative approach to the functional systems of animals, and a closing chapter on animal behavior. For this edition we added a new section on the control of muscle contraction (Chapter 32), completely revised the treatment of vertebrate kidney function (Chapter 33), and made numerous revisions and updatings in Chapter 37 on the endocrine system.

As with previous editions, rapid advances in the field of immunology required us to rewrite this section completely. Because research on cytokines in the 1980s virtually revolutionized immunology, we mention specifically the most important cytokines, including their functions. The evolutionary focus of the closing updated chapter on animal behavior, with its many examples of instinctive and social behavior, reinforces for students the evolutionary theme woven throughout the text.

PART FIVE: THE ANIMAL AND ITS ENVIRONMENT

Many instructors use the chapter on the biosphere and animal distribution (Chapter 39) as an introduction to Chapter 40, which covers the principles of ecology. For this edition we expanded the treatment of the Great American Interchange in Chapter 39 and provided a new accompanying illustration. A section entitled "Major Faunal Realms" was deleted because the concept is considered now of marginal significance by many biogeographers. The realms were erected mainly on the basis of disjunct mammal distribution, which is explained more suitably in the final section on continental drift theory. Chapter 40 (Animal Ecology) examines some of the fundamental principles of ecology. Among changes for this edition is the addition of an explanation and accompanying illustration of the guild concept, exemplified by Robert MacArthur's classic study of wood warblers.

TEACHING AND LEARNING AIDS

To help students in **vocabulary development,** we have boldfaced key words, and provided the derivations of technical and zoological terms, and generic names of animals where they first appear in the text. In this way students gradually become familiar with the more common roots that comprise many technical terms. An extensive glossary of almost 1100 terms provides pronunciation, derivation, and definition of each term. More than 160 new terms were added to the glossary or rewritten for this edition.

A distinctive feature of this text is a **chapter prologue** for each chapter that draws out some theme or fact relating to the subject of the chapter. Some present biological, particularly evolutionary, principles; others (especially those in the survey sections) illuminate distinguishing characteristics of the group treated in the chapter. Each is intended to present an important concept drawn from the chapter in an interesting manner that will facilitate learning by students, as well as engage their interest and pique their curiosity.

Chapter notes, which appear throughout the book, augment the text material and offer interesting sidelights without interrupting the narrative. We prepared many new notes for this edition and revised several of the existing notes.

To assist students in chapter review, each **chapter** ends with a concise **summary,** a list of **review questions,** and annotated **selected references.** The review questions enable the student to self-test retention and understanding of the more important chapter material.

The **historical appendix,** unique to this textbook, lists key discoveries in zoology, and separately describes books and publications that have greatly influenced the development of zoology. Many readers have found this appendix an invaluable reference to be consulted long after their formal training in zoology.

Again, William C. Ober and Claire W. Garrison have **enhanced** the **art program** for this text with many new full color paintings that replace older art, or that illustrate new material. Bill's artistic skills, knowledge of biology, and experience gained from an earlier career as a practicing physician, have enriched this text through six of its editions. Claire practiced pediatric and obstetric nursing before turning to scientific illustration as a full-time career. Texts illustrated by Bill and Claire have received national recognition and won awards from the Association of Medical Illustrators, American Institute of Graphic Arts, Chicago Book Clinic, Printing Industries of America, and Bookbuilders West. They are also recipients of the Art Directors Award.

SUPPLEMENTS

The **Instructor's Manual and Test Item File** provide chapter outlines, commentaries, lesson plans prepared by the authors, and a listing of resource references for each chapter. Also included is a listing of the transparencies and slides available with the book, and a comprehensive test bank offering 35 to 40 objective questions per chapter. We trust this will be of particular value to first-time users of the text, although experienced teachers may also find the ancillary helpful.

The **Laboratory Manual** by Cleveland P. Hickman, Jr., Frances M. Hickman, and Lee Kats, *Laboratory Studies in Integrated Zoology,* now in its ninth edition, has been extensively rewritten and reillustrated in full color. It can be adapted conveniently for two-semester, one-semester, or term courses by judicious selection of exercises. The popular wall chart, "Chief taxonomic subdivisions and organ systems of animals," is available to users of the laboratory manual at no additional charge.

Test questions contained in the **Instructor's Manual and Test File** are also available as a **Computerized Test Bank**, a test generation system for IBM and Macintosh computers. Using this system, instructors can create tests or quizzes quickly and easily. Questions can be sorted by type or level of difficulty, and instructors can also add their own material to the bank of questions provided.

A set of 150 full color **transparency acetates** of important textual illustrations are available with this edition of *Integrated Principles of Zoology*. Labeling is clear, dark, and bold for easy reading.

A set of 150 **Animal Diversity Slides,** photographed by the authors and Bill Ober on their various excursions, are offered in this unique textbook supplement. Both invertebrates and vertebrates are represented. Descriptions, including specific names of each animal and brief overview of the animal's ecology and/or behavior, accompany the slides.

Tapes 3, 4, and 5 in the **Life Science Animations Video Series** will provide you with more than 30 animations of the most difficult-to-learn concepts found in a zoology course.

The **Life Sciences Living Lexicon** is an interactive CD-ROM that includes complete glossaries for all life science disciplines, a section describing the classification system, an overview of word construction, and more than 500 vivid illustrations and animations of key processes.

Approximately 2,000 still images and 15 minutes of moving video of invertebrates in their own environment are found on the new **Invertebrate Zoology Videodisc.** This visual forum for discovery provides a wide variety of images from the West and East coasts, as well as the Caribbean and other regions of the world.

ACKNOWLEDGMENTS

We wish to thank the following reviewers who suggested numerous improvements and whose collective wisdom was of the greatest assistance to us as we approached this edition. Their experience with students of varying backgrounds, and their interest in, and knowledge of, zoology helped to shape the text into its final form.

The following individuals reviewed specific chapters, or entire sections, of the tenth edition manuscript:

Barbara J. Abraham, Hampton University
Sylvester Allred, Northern Arizona University
Helen I'Anson, Washington and Lee University

Dean C. T. Bratis, Delaware Community College
D. Charles Dailey, Sierra College
Larry Hurd, Washington and Lee University
Bretton W. Kent, University of Maryland–College Park
Josef Kren, University of Nebraska–Lincoln
Catherine O'Brien, San Jacinto College North
Donna Bruns Stockrahm, Moorhead State University
Kathy Thompson, Louisiana State University
Scott Turner, SUNY College of Environmental Science and Forestry
C. David Vanicek, California State University–Sacramento
John J. Weilgus, Washington and Lee University
A. Quinton White, Jacksonville University
Samuel Zeveloff, Weber State University

The following individuals provided valuable assistance by responding to a questionnaire concerning their experiences and opinions after teaching with the ninth edition.

Sylvester Allred, Northern Arizona University
Carol Armstrong, Iowa Central Community College
Edmond J. Bacon, University of Arkansas at Monticello
Linda Barham, Meridian Community College
Francis F. Belcik, East Carolina University
Latsy Best, Palm Beach Community College
Allen D. Bidol, Oakland Community College–Orchard Ridge
Carol Bixler, Louisburg College
Howard D. Booth, Eastern Michigan University
Marian Borgmann-Ingwersen, Wayne State College
Kimberly R. Brown, Jackson County Community College
Willis A. Brown, Jr., Mount Olive College

John M. Chapin, St. Petersburg Junior College
Suzette F. Chopin, Texas A & M University–Corpus Christi
Mariette Cole, Concordia College St. Paul
Melinda F. Davis, Fort Valley State College
Kathryn Dickson, California State University–Fullerton
Walter J. Diehl, Mississippi State University
J. Roger Eagen, Adirondack Community College
David A. Easterla, Northwest Missouri State University
Peter Elliott, Okanagan University College
Joseph G. Engemann, Western Michigan University
Robert E. Espinoza, Colorado State University
Edward J. Greding, Jr., Del Mar College
Peggy Green, Broward Community College
Thomas F. Grittinger, University of Wisconsin Center–Sheboygan
Richard J. Haeusler, Bay College
Leon E. Hallacher, University of Hawaii at Hilo
Mary F. Haskins, Rockhurst College
Ben Hawkins, College of the Redwoods
Kenneth L. Heacock, Oklahoma State University
Scott J. Herrmann, University of Southern Colorado
Galen R. Hunsicker, Southern California College
Edward C. Hurlbut, Mesa State College
Jude Johnson, College of New Caledonia
Charles C. Jones, Miles College
Frank Jordan, Jacksonville University
Eunice R. Knouse, Spartanburg Methodist College
Martin Kopenski, Northern Michigan University
Gene Kritsky, College of Mount St. Joseph
Lorraine Larison, Pasadena City College
Kevin Lyon, Jones County Junior College

Christine R. Maher, Montana State University-Billings

Wendy H. McCullen, Columbus State Community College

Alison M. Mostrom, Philadelphia College of Pharmacy and Science

Thomas A. Nelson, Eastern Illinois University

Lloyd M. Pederson, San Joaquin Delta College

Dan F. Penney, San Jacinto College District

Susan E. Peters, University of North Carolina at Charlotte

Gary S. Phillips, Iowa Lakes Community College

Terry L. Phipps, Cedarville College

William H. Pritchett, Fairmont State College

M. E. Pryor, Morehead State University

James A. Raines, North Harris College

Donald C. Rizzo, Marygrove College

Vaughn M. Rundquist, Montana State University

Dennis J. Russell, University of Alaska Southeast

Thomas J. Saleska, Concordia University of Wisconsin

Judith Salley, South Carolina State

Donald O. Santana, Gavilan College

A. Jewell Schock, Wayne State College

Sherry Schmidt, Mt. San Antonio College

Barbara A. Sheilds, Eastern Michigan University

Amy Sheldon, Faulkner State Community College

David L. Smith, Valencia Community College

Richard L. Stoffer, Ashland University

Ed Story, Maysville Community College

Robert M. Sullivan, Texas A & M University–Kingsville

Sarah H. Swain, Middle Tennessee State University

LeLeng To, Goucher College

Eddie Trevino, Marymount College

Mario A. Vecchiarelli, Housatonic Community Technical College

Steven H. Vee, Rock Valley College

Gary O. Wallace, Milligan College

Judith K. Weyer, Iowa Wesleyan College

Eugene A. Young, Southwestern College

The authors express their appreciation to the editors and support staff at William C. Brown Publishers who made this project possible. Special thanks are due Marge Kemp, Acquisitions Editor, and Kathy Loewenberg, Developmental Editor, who were the driving forces in piloting this text throughout its production. Jane Matthews, Production Editor, somehow kept authors, text, art, and production programs on schedule. Others who played key roles and to whom we express our gratitude are Cathy DiPasquale Conroy, who copy edited the manuscript; John Leland and Jodi Banowetz who oversaw the extensive photographic program and art program respectively; and Karen Dorman, who handled permissions. The text was designed by Wayne Harms. We are indebted to them for their talents and dedication. They were all a pleasure to work with.

Cleveland P. Hickman, Jr.
Larry S. Roberts
Allan Larson

Introduction to the Living Animal

1

Life: Biological Principles and the Science of Zoology

The Uses of Principles

We gain knowledge of the animal world not in a passive or haphazard manner but by actively applying important guiding principles to our investigations. Just as the exploration of outer space is both guided and limited by available technologies, exploration of the animal world depends critically on our questions, methods, and principles. The body of knowledge that we call zoology makes sense only when the principles that we use to construct it are clear.

The principles of modern zoology have a long history and many sources. Some principles derive from the laws of physics and chemistry, which all living systems obey. Others derive from the scientific method, which tells us that our hypotheses regarding the animal world are useless unless they guide us to gather data that potentially can refute them. Many important principles derive from previous studies of the living world, of which animals are one part. The principles of heredity, variation, and organic evolution guide the study of life from the simplest unicellular forms to the most complex animals, fungi, and plants. Because all of life shares a common evolutionary origin, principles learned from the study of one group often may be applied to other groups as well. By tracing the origins of our operating principles, we see that zoologists are not an island unto themselves but form an integrated part of the scientific community.

We begin our study of zoology not by focusing narrowly within the animal world, but by searching broadly for our most basic principles and their diverse sources. These principles simultaneously guide our studies of animals and integrate those studies into the broader context of human knowledge. ■

Zoology, the scientific study of animal life, builds on centuries of human inquiry into the animal world. The mythologies of nearly every human culture document attempts to solve the mysteries of animal life and its origin. Zoologists now confront these same mysteries with the most advanced methods and technologies developed throughout all branches of science. We start by documenting the diversity of animal life and organizing it in a systematic way. This complex and exciting process builds on the contributions of thousands of zoologists working in all dimensions of the biosphere (Figure 1-1). We strive

through this work to understand how animal diversity originated and how animals perform the basic processes of life that permit them to thrive in many diverse environments.

This chapter introduces the fundamental properties of animal life, the methodological principles on which their study is based, and two important theories that guide our research: (1) the theory of evolution, which is the central organizing principle of biology, and (2) the chromosomal theory of inheritance, which guides our study of heredity and variation in animals. These theories unify our knowledge of the animal world.

FUNDAMENTAL PROPERTIES OF LIFE

CAN LIFE BE DEFINED?

Our discussion begins with the difficult question, What is life? Although many attempts have been made throughout the years to define life, it is now clear that simple definitions are doomed to failure. When we try to give life a simple definition, we assume that life has maintained certain fixed properties throughout all of its history. However, the properties that life exhibits today (pp. 4–9) are very different from those that were present at its origin. Rather

A

B

C

D

Figure 1-1

A few of the many dimensions of zoological research: **A**, Observing moray eels in Maui, Hawaii; **B**, Working with tranquilized polar bears; **C**, Banding mallard ducks; **D**, observing *Daphnia pulex* (×150) microscopically.

D, SIU/Visuals Unlimited, **inset,** T. E. Adams/Visuals Unlimited.

than demonstrating fixed properties, the history of life shows perpetual change, which we call *evolution*. As the genealogy of life progressed and branched from the earliest living form to the millions of species living today, new properties evolved and passed from parents to their offspring. Through this process, living systems have generated many rare and spectacular features that are without counterparts in the nonliving world. Unexpected properties emerge on many different lineages in life's evolutionary history and give us the great organismal diversity that we observe today.

We might be tempted to define life on the basis of those most widely shared properties that were evident at life's origin. The replication of molecules, for example, can be traced to life's origin and represents one of life's most universal properties. An attempt to define life based on properties present at its origin faces a major problem, however. These are the very properties that are most likely to be shared by some nonliving forms. To study the origin of life, we must ask how organic molecules acquire the ability for precise replication. But where do we draw the line between those replicative processes that characterize life and those that are merely general chemical features of the matter from which life arose? Replication of complex crystalline structures in nonliving chemical assemblages might be confused, for example, with the replicative molecular properties that we associate with life. We could try alternatively to define life using only the most advanced properties that characterize the highly evolved living systems that we observe today. The nonliving world would not intrude on such a definition, but we would be eliminating those early forms of life from which all others descended and which give life its historical unity. Ultimately our definition must be based on the common history of life on earth that gives it an identity and separates it from the nonliving world. We can trace this common history back-

ward through time from the diverse forms that we observe today and in the fossil record to their common ancestor that arose in the atmosphere of the primitive earth (see Chapter 3). All organisms forming part of this long history of hereditary descent from life's common ancestor are included in our concept of life even if they are no longer alive today.

Although we do not force life into a simple definition, we can readily identify the living world and separate it from the nonliving. Many remarkable properties have arisen during life's history and are observed in living forms in various combinations. They clearly identify their possessors as part of the unified historical entity that we call life. All such features are present in the most highly evolved forms of life, such as those that compose the animal kingdom. We consider it highly unlikely that any of these properties will be lost during life's future evolution because they are so important to the maintenance and functioning of the living forms that possess them. By choosing not to define life narrowly on the basis of these properties, however, we do not exclude this possibility.

GENERAL PROPERTIES OF LIVING SYSTEMS

The most outstanding general features that have arisen during life's history include chemical uniqueness; complexity and hierarchical organization; reproduction (heredity and variation); possession of a genetic program; metabolism; development; and environmental interaction.

1. **Chemical uniqueness.** *Living systems demonstrate a unique and complex molecular organization.* The history of life has featured the assembly of large molecules, known as macromolecules, that are far more complex than the small molecules that constitute nonliving matter. These macromolecules are composed of the same kinds of atoms and chemical bonds that occur in nonliving matter and they obey all fundamental laws of

chemistry (Appendix B); it is only the complex organizational structure of these macromolecules that makes them unique. We recognize four major categories of biological macromolecules: nucleic acids, proteins, carbohydrates, and lipids. These categories differ in the structures of their component parts, the kinds of chemical bonds that link their subunits together, and their functions in the living system.

The general structures of these macromolecules evolved and stabilized early in the history of life. With some modifications, these same general structures are found in every form of life that we observe today. Proteins, for example, contain about 20 specific kinds of amino acid subunits linked together by peptide bonds in a linear sequence (Figure 1-2). Additional bonds occurring between amino acids that are not adjacent to each other in the protein chain give the protein a complex, three-dimensional structure (see Figures 1-2 and 2-14). A typical protein contains several hundred amino acid subunits. Despite the stability of this basic protein structure, the ordering of the different amino acids in the protein molecule is subject to enormous variation. This variation underlies much of the diversity that we observe among different kinds of living forms. The nucleic acids, carbohydrates, and lipids likewise contain characteristic bonds that link variable subunits. This gives living systems both a biochemical unity and a great potential for diversity.

2. **Complexity and hierarchical organization.** *Living systems demonstrate a unique and complex hierarchical organization.* Nonliving matter is organized at least into atoms and molecules and often has a higher degree of organization as well. However, atoms and molecules are combined into patterns in the living world that do not exist in the nonliving world. In living systems, we find a hierarchy of levels that includes, in ascending order of complexity, macromolecules, cells, organisms, populations, and species

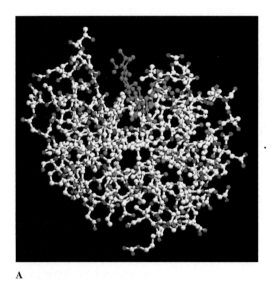

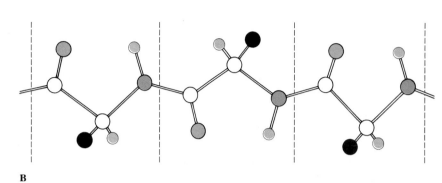

A   **B**

Figure 1-2

A computer simulation of the three-dimensional structure of the lysozyme protein **(A),** which is used by animals to destroy bacteria. The protein is a linear string of molecular subunits called amino acids, connected as shown in **B,** that fold in a three-dimensional pattern to form the active protein. The white balls correspond to carbon atoms, the red balls to oxygen, the blue balls to nitrogen, the yellow balls to sulfur, the green balls to hydrogen, and the black balls **(B)** to molecular groups formed by various combinations of carbon, oxygen, nitrogen, hydrogen, and sulfur atoms that differ among amino acids. Hydrogen atoms are not shown in **A.** The purple molecule in **A** is a structure from the bacterial cell wall that is broken by lysozyme.

A, Courtesy of IBM UK Scientific Centre.

(Figure 1-3). Each level builds on the level below it and has its own internal structure, which is also often hierarchical. Within the cell, for example, macromolecules are compounded into structures such as ribosomes, chromosomes, and membranes, and these are likewise combined in various ways to form even more complex subcellular structures called organelles, such as mitochondria (see Chapters 4 and 5). The organismal level also has a hierarchical substructure; cells are combined into tissues, which are combined into organs, which likewise are combined into organ systems (see Chapter 10).

The cell (Figure 1-4) is the smallest unit of the biological hierarchy that is semiautonomous in its ability to conduct its basic functions, including reproduction. The replication of molecules and subcellular components occurs only within the cellular context, not independently. The cell is therefore viewed as the basic unit of living systems (Chapter 4). We can isolate cells from an organism and cause them to grow and multiply under laboratory conditions in the presence of nutrients alone. This is not possible for any individual molecules or subcellular components, which require additional cellular constituents for their reproduction.

Figure 1-3

Volvox globator (see p. 224) is a multicellular phytoflagellate that illustrates three different levels of the biological hierarchy: cellular, organismal, and populational. Each individual spheroid (organism) contains cells embedded in a gelatinous matrix. The larger cells function in reproduction, and the smaller ones perform the general metabolic functions of the organism. The individual spheroids together form a population.

Each successively higher level of the biological hierarchy is composed of units of the preceding lower level in the hierarchy. An important characteristic of this hierarchy is that the prop-

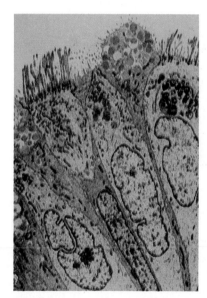

Figure 1-4

Electron micrograph of ciliated epithelial cells and mucus-secreting cells (see pp. 49 and 187–188). The cell is the basic building block of living organisms.

erties of any given level cannot be obtained from even the most complete knowledge of the properties of its component parts. A physiological feature, such as blood pressure, is a property of the organismal level; it is impossible to predict someone's blood pressure simply by knowing the physical characteristics of the individual

Table 1-1	Different Hierarchical Levels of Biological Complexity that Display Reproduction, Variation, and Heredity			
Level	**Timescale of Reproduction**	**Fields of Study**	**Methods of Study**	**Some Emergent Properties**
Cell	Hours (mammalian cell = ~16 hours)	Cell biology	Microscopy (light, electron), biochemistry	Chromosomal replication (meiosis, mitosis), synthesis of macromolecules (DNA, RNA, proteins, lipids, polysaccharides)
Organism	Hours to days (unicellular); days to years (multicellular)	Organismal anatomy, physiology, genetics	Dissection, genetic crosses, clinical studies	Structure, functions and coordination of tissues, organs and organ systems (blood pressure, body temperature, sensory perception, feeding)
Population	Up to thousands of years	Population biology, population genetics, ecology	Statistical analysis of variation; abundance, geographical distribution	Social structures, systems of mating, age distribution of organisms, levels of variation, action of natural selection
Species	Thousands to millions of years	Systematics and evolutionary biology, community ecology	Study of reproductive barriers, phylogeny, paleontology, ecological interactions	Method of reproduction, reproductive barriers

cells of the body. Likewise, systems of social interaction, as observed in bees, occur at the level of the population; it would not be possible to infer the properties of this social system by knowing only the properties of individual bees.

The appearance of new characteristics at a given level of organization is called **emergence,** and these characteristics are known as **emergent properties.** These properties arise from the interactions that occur among the component parts of the system. For this reason, we must study all levels directly, and the subdivisions of the field of biology (molecular biology; cell biology; organismal anatomy, physiology and genetics; population biology) reflect this fact (Table 1-1). We find that the emergent properties expressed at a particular level of the biological hierarchy are certainly influenced and restricted by the properties of the lower-level components. For example, it would be impossible for a population of organisms that lack

hearing to develop a spoken language. Nonetheless, the properties of the parts of a living system do not rigidly determine the properties of the whole. Many different spoken languages have emerged in human culture from the same basic anatomical structures that permit hearing and speech. The freedom of the parts to interact in different ways makes possible a great diversity of potential emergent properties at each level of the biological hierarchy.

The different levels of the biological hierarchy and their particular emergent properties are products of evolution. Before multicellular organisms evolved, there was no distinction between the organismal and cellular levels, and it is still absent from the single-celled organisms (Chapter 12). The diversity of emergent properties that we see at all levels of the biological hierarchy contributes to the difficulty of giving life a simple definition or description.

3. **Reproduction.** *Living systems can reproduce themselves.* Life does

not arise spontaneously but comes only from prior life, through a process of reproduction. Although life certainly originated from nonliving matter at least once (Chapter 3), this required enormously long periods of time and conditions very different from those of the modern biosphere. At each level of the biological hierarchy, living forms reproduce to generate others like themselves (Figure 1-5). Genes are replicated to produce new genes. Cells divide to produce new cells. Organisms reproduce, sexually or asexually, to produce new organisms (Chapter 6). Populations can become fragmented to give rise to new populations, and species can give rise to new species through a process known as speciation. Reproduction at any level of the hierarchy usually features an increase in numbers. Individual genes, cells, organisms, populations, or species may fail to reproduce themselves, but reproduction is nonetheless an expected property of these individuals.

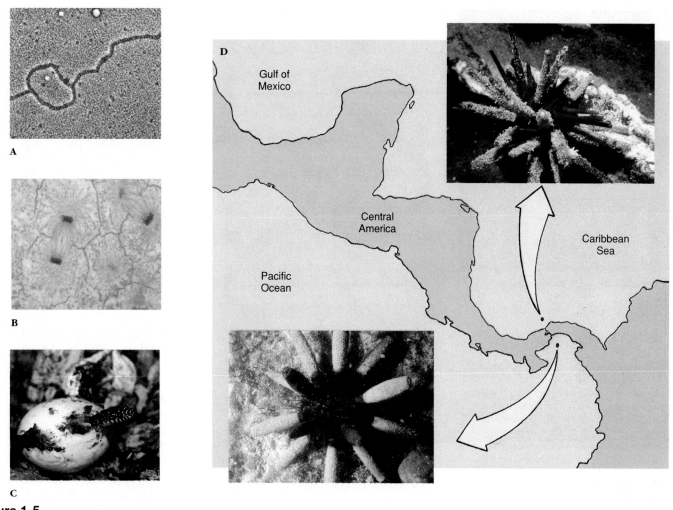

Figure 1-5

Reproductive processes observed at four different levels of biological complexity: **A,** Molecular level—electron micrograph of a replicating DNA molecule; **B,** Cellular level—micrograph of cell division at mitotic telophase; **C,** Organismal level—a kingsnake hatching; **D,** Species level—formation of new species in the sea urchin (*Eucidaris*) after geographic separation of Caribbean (*E. tribuloides*) and Pacific (*E. thouarsi*) populations by the formation of a land bridge.

Reproduction at each of these levels features the complementary, and yet apparently contradictory, phenomena of **heredity** and **variation.** Heredity is the faithful transmission of traits from parents to offspring, usually (but not necessarily) observed at the organismal level. Variation is the production of *differences* among the traits of different individuals. In the reproductive process, the properties of descendants resemble those of their parents to varying degrees but are usually not identical to them. Replication of deoxyribonucleic acid (DNA) occurs with high fidelity, but errors occur at repeatable rates. Cell division is an exceptionally precise process, especially with regard to the nuclear material, but chromosomal changes occur nonethe-

less at measurable rates. Organismal reproduction likewise demonstrates both heredity and variation, the latter being obvious especially in sexually reproducing forms. The production of new populations and species also demonstrates conservation of some properties and changes of others. Two closely related frog species may have similar mating calls but differ in the rhythm of repeated sounds.

We will see later in this book that the interaction of heredity and variation in the reproductive process is the basis for organic evolution (Chapter 9). If heredity were perfect, living systems would never change; if variation were uncontrolled by heredity, biological systems would lack the stability that allows them to persist through time.

4. **Possession of a genetic program.** *A genetic program provides fidelity of inheritance* (Figure 1-6). The structures of the protein molecules needed for organismal development and functioning are encoded in nucleic acids (Chapter 8). For animals and most other organisms, the genetic information is contained in DNA. DNA is a very long, linear chain of subunits called nucleotides, each of which contains a sugar phosphate (deoxyribose phosphate) and one of four nitrogenous bases (adenine, cytosine, guanine, or thymine, abbreviated A, C, G, and T, respectively). The sequence of nucleotide bases represents a code for the order of amino acids in the protein specified by the DNA molecule. The correspondence between the sequence

A

Figure 1-6

James Watson and Francis Crick with a model of the DNA double helix **(A).** Genetic information is coded in the nucleotide base sequence inside the DNA molecule. Genetic variation is shown **(B)** in DNA molecules that are similar in base sequence but differ from each other at four positions. Such differences can encode alternative traits, such as different eye colors.

A, © A. C. Barrington Brown/Photo Researchers, Inc.

B

synthesis of molecules and structures. Metabolism is often viewed as an interaction of destructive (catabolic) and constructive (anabolic) reactions. The most fundamental anabolic and catabolic chemical processes used by living systems arose early in the evolutionary history of life and are shared by all living forms. These include synthesis of carbohydrates, lipids, nucleic acids, and proteins and their constituent parts and the cleavage of chemical bonds to recover energy stored in them. In animals, many fundamental metabolic reactions occur at the cellular level, often in specific organelles that are found throughout the animal kingdom. Cellular respiration occurs, for example, in the mitochondria. The cellular and nuclear membranes regulate metabolism by controlling the movement of molecules across the cellular and nuclear boundaries, respectively. The study of the performance of complex metabolic functions is known as **physiology.** We will devote a large portion of this book to describing and comparing the diverse tissues, organs, and organ systems that different groups of animals have evolved to perform the basic physiological functions of life (Chapters 12 through 31).

6. **Development.** *All organisms pass through a characteristic life cycle.* Development describes the characteristic changes that an organism undergoes from its origin (usually the fertilization of the egg by sperm) to its final adult form (Chapter 7). Development usually features changes in size and shape, and the differentiation of structures within the organism. Even the simplest one-celled organisms grow in size and replicate their component parts until they divide into two or more cells. Multicellular organisms undergo more dramatic changes during their lives. In some multicellular forms, different stages of the life cycle are so dissimilar that they are hardly recognizable as part of the same species. Embryos are distinctly different from the juvenile and adult forms into which they will develop. Even the postembryonic development of some

of bases in DNA and the sequence of amino acids in a protein is known as the genetic code.

The genetic code was established early in the evolutionary history of life, and the same code is present in bacteria and in the nuclear genomes of almost all animals and plants. The near constancy of this code among living forms provides strong evidence for a single origin of life. The genetic code has undergone very little evolutionary change since its origin because an alteration would disrupt the structure of nearly every protein, which would in turn severely disrupt cellular functions that require very specific protein structures. Only in the rare instance in which the altered protein structures are still compatible with their cellular functions would such a change have the chance to survive and be reproduced. This has hap-

pened in the DNA contained in animal mitochondria, the organelles that regulate cellular energy. The genetic code in animal mitochondrial DNA is slightly different from the standard code of nuclear and bacterial DNA. Because the mitochondrial DNA specifies far fewer proteins than the nuclear DNA, the likelihood of getting a change in the code that does not disrupt cellular functions is greater there than in the nucleus.

5. **Metabolism.** *Living organisms maintain themselves by obtaining nutrients from their environments* (Figure 1-7). The nutrients are broken down to obtain chemical energy and molecular components for use in building and maintaining the living system (Chapter 5). We call these essential chemical processes **metabolism.** They include digestion, production of energy (respiration), and

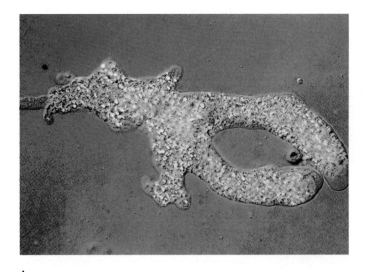

A

B

Figure 1-7

Feeding processes illustrated by (A) an ameba surrounding food and (B) a chameleon capturing insect prey with its projectile tongue.

organisms includes stages that are dramatically different from each other. The transformation that occurs from one stage to the other is called **metamorphosis.** There is little resemblance, for example, among the egg, larval, pupal, and adult stages of metamorphic insects (Figure 1-8). Among animals, the early stages of development are often more similar among organisms of related species than are later developmental stages. In our survey of animal diversity, we will describe all stages of the observed life histories, but we will concentrate on the adult stages in which diversity both within and between different animal groups tends to be greatest.

7. **Environmental interaction.** *All animals interact with their environments.* The study of organismal interaction with the environment is known as **ecology.** Of special interest are the factors that affect the geographic distribution and abundance of animals (Chapters 39 and 40). The science of ecology permits us to understand how an organism can perceive environmental stimuli and respond in appropriate ways by adjusting its metabolism and physiology (Figure 1-9). All organisms respond to stimuli in their environment, and this property is called **irritability.** The stimulus and response may be simple, such as a unicellular organism moving from

A

B

Figure 1-8

Pupal and adult stages of the insect life cycle: **A,** Adult monarch butterfly emerging from the pupa; **B,** Fully formed adult monarch butterfly.

or toward a light source or away from a noxious substance, or it may be quite complex, such as a bird responding to a complicated series of signals in a mating ritual. Life and the environment are inseparable. We cannot isolate the evolutionary history of a lineage of organisms from the environments in which it occurred.

LIFE OBEYS PHYSICAL LAWS

To untrained observers, these seven properties of life may appear to violate the basic laws of physics. Vitalism, the idea that life is endowed with a mystical vital force that violates physical and chemical laws, was once widely advocated. Biological research

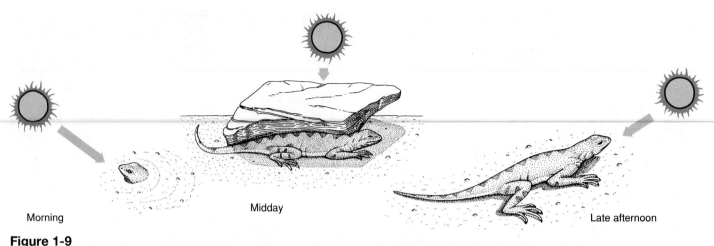

Figure 1-9
A lizard regulates its body temperature by choosing different locations (microhabitats) at different times of day.

has consistently rejected vitalism, showing instead that all living systems operate and evolve within the constraints of the basic laws of physics and chemistry. The laws governing energy and its transformations (thermodynamics) are particularly important for understanding life (Chapter 5). The **first law of thermodynamics** is the law of conservation of energy. Energy is neither created nor destroyed, but it can be transformed from one form to another. All aspects of life require energy and its transformation. The energy to support life on earth flows from the fusion reactions in our sun and reaches the earth in the form of light and heat. Sunlight is captured by green plants and cyanobacteria and transformed by photosynthesis into chemical bonds. The energy in chemical bonds is a form of potential energy that can be released when the bond is broken; the energy is used to perform numerous cellular tasks. Energy transformed and stored in plants is then used by the animals that eat the plants, and these animals may in turn provide energy for other animals that eat them.

The **second law of thermodynamics** states that physical systems tend to proceed toward a state of greater disorder, or **entropy.** The energy obtained and stored by plants is subsequently released by a variety of mechanisms and finally dissipated as heat. The high degree of molecular organization found in living cells is attained and maintained only as long as

energy fuels the organization. The ultimate fate of materials in the cells is degradation and dissipation of their chemical bond energy as heat. The process of evolution whereby organismal complexity can increase over time may appear at first to violate the second law of thermodynamics, but it does not. Organismal complexity is achieved and maintained only by the constant use and dissipation of energy flowing into the biosphere from the sun. The survival, growth, and reproduction of animals requires energy that comes from breaking complex food molecules into simple organic waste products. The processes by which animals acquire energy through nutrition and respiration command the attention of the many physiological sciences.

ZOOLOGY AS A PART OF BIOLOGY

Animals form a distinct branch on the evolutionary tree of life (Figure 1-10). It is a large and old branch that originated in the Precambrian seas over 600 million years ago. The animals form part of an even larger limb known as the **eukaryotes,** organisms whose cells contain membrane-enclosed nuclei. This larger limb includes the plants and fungi. Perhaps the most distinctive characteristic of the animals as a group is their means of nutrition, which consists of eating other organisms. This basic way of life has led to the evolution of

many diverse systems for locomotion and for capturing and processing a wide array of food items.

Animals can be distinguished also by the absence of properties that have evolved in other eukaryotes. Plants, for example, have evolved the ability to use light energy to produce organic compounds (photosynthesis), and they have evolved rigid cell walls that surround their cell membranes; photosynthesis and cell walls are absent from animals. Fungi have evolved the ability to acquire nutrition by absorption of small organic molecules from their environment, and they have a body plan consisting of tubular filaments called *hyphae;* structures of this kind are absent from the animal kingdom.

Some organisms combine the properties of animals and plants. For example, *Euglena* (Figure 1-11) is a motile, single-celled organism that resembles plants in being photosynthetic, but it resembles animals in its ability to eat food particles. *Euglena* is part of a separate eukaryotic lineage that diverged from those of plants and animals early in the evolutionary history of eukaryotes. *Euglena* and other unicellular eukaryotes are sometimes grouped into the kingdom Protista, although this kingdom may be an arbitrary grouping of unrelated lineages, in which case it would violate taxonomic principles (see Chapter 11).

The fundamental structural and developmental features evolved by the animal kingdom are presented in detail in Chapters 7 and 10.

Figure 1-10

An abbreviated evolutionary tree of life, showing the evolutionary position of multicellular animals relative to several other major branches (see Chapter 11 for further details).

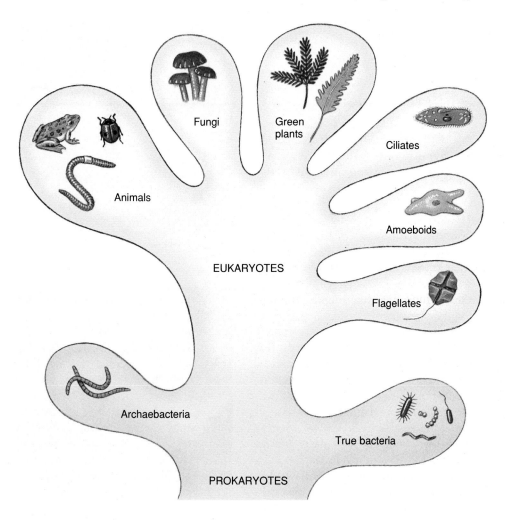

Principles of Science

Nature of Science

We stated in the first sentence of this chapter that zoology is the scientific study of animals. A basic understanding of zoology therefore requires an understanding of what science is, what it is not, and how knowledge is gained by using the scientific method.

Science is a way of asking questions about the natural world and obtaining precise answers to them. Although science, in the modern sense, has arisen recently in human history (within the last 200 years or so), the tradition of asking questions about the natural world is an ancient one. In this section we examine the methodology that zoology shares with science as a whole. These features distinguish the sciences from those activities that we exclude from the realm of science, such as art and religion.

Despite the enormous impact that science has had on our lives, many people have only a minimal understanding of the real nature of science. For example, on March 19, 1981, the governor of Arkansas signed into law the Balanced Treatment for Creation-Science and Evolution-Science Act (Act 590 of 1981). This act falsely presented "creation-science" as a valid scientific endeavor. "Creation-science" is actually a religious position advocated by a minority of the American religious community, and it does not qualify as science. The enactment of this law led to a historic lawsuit tried in December 1981 in the court of Judge William R. Overton, U.S. District Court, Eastern District of Arkansas. The suit was brought by the American Civil Liberties Union on behalf of 23

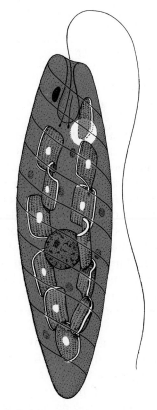

Figure 1-11

Some organisms, such as the flagellate *Euglena* (shown here) and *Volvox* (see Figure 1-3) combine properties that are normally associated with both animals (motility) and plants (photosynthetic ability).

plaintiffs, including a number of religious leaders and groups representing several denominations, individual parents, and educational associations. The plaintiffs contended that the law was a violation of the First Amendment to the U.S. Constitution, which prohibits "establishment of religion" by the government. This prohibition includes passing a law that would aid one religion or prefer one religion over another. On January 5, 1982, Judge Overton permanently enjoined the State of Arkansas from enforcing Act 590.

Considerable testimony during the trial dealt with the nature of science. Some witnesses defined science simply, if not very informatively, as "what is accepted by the scientific community" and "what scientists do." However, on the basis of other testimony by scientists, Judge Overton was able to state explicitly these essential characteristics of science:

1. It is guided by natural law.
2. It has to be explanatory by reference to natural law.
3. It is testable against the observable world.
4. Its conclusions are tentative, that is, are not necessarily the final word.
5. It is falsifiable.

The pursuit of scientific knowledge must be guided by the physical and chemical laws that govern the state of existence. Scientific knowledge must explain what is observed by reference to natural law without requiring the intervention of a supernatural being or force. We must be able to observe events in the real world, directly or indirectly, to test hypotheses about nature. If we draw a conclusion relative to some event, we must be ready always to discard or to modify our conclusion if further observations contradict it. As Judge Overton stated, "While anybody is free to approach a scientific inquiry in any fashion they choose, they cannot properly describe the methodology used as scientific if they start with a conclusion and refuse to change it regardless of the evidence developed during the course of the investigation." Science is neutral on the question of religion, and the results of science do not favor one religious position over another.

SCIENTIFIC METHOD

These essential criteria of science form the basis for an approach known as the **hypothetico-deductive method.** The first step of this method is the generation of hypotheses or potential answers to the question being asked. These hypotheses are usually based on prior observations of nature, or they are derived from theories based on such observations. Scientific hypotheses often constitute general statements about nature that may explain a large number of diverse observations. Darwin's hypothesis of natural selection, for example, explains the observations that many different species have properties that adapt them to their environments. On the basis of the hypothesis, the scientist must make a prediction about future observations. The scientist must say, "If my hypothesis is a valid explanation of past observations, then future observations ought to have certain characteristics." The best hypotheses are those that make many predictions which, if found erroneous, will lead to rejection, or falsification, of the hypothesis.

The hypothesis of natural selection was invoked to explain variation observed in British moth populations (Figure 1-12). In industrial areas of England having heavy air pollution, many populations of moths contain primarily darkly pigmented (melanic) individuals, whereas moth populations inhabiting clean forests show a much higher frequency of lightly pigmented individuals. The hypothesis suggests that moths can survive most effectively by matching their surroundings, thereby remaining invisible to birds that seek to eat them. Experimental studies have shown that, consistent with this hypothesis, birds are able to locate and then to eat moths that do not match their surroundings, but that birds in the same area frequently fail to find moths that match their surroundings. Another testable prediction of the hypothesis of natural selection

A

B

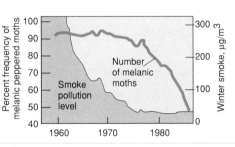

C

Figure 1-12

Light and melanic forms of the peppered moth, *Biston betularia* on, **A,** a lichen-covered tree in unpolluted countryside and, **B,** a soot-covered tree near industrial Birmingham, England. These color variants have a simple genetic basis. **C,** Recent decline in the frequency of the melanic form of the peppered moth with falling air pollution in industrial areas of England. The frequency of the melanic form still exceeded 90% in 1960, when smoke and sulfur dioxide emissions were still high. Later, as emissions fell and light-colored lichens began to grow again on the tree trunks, the melanic form became more conspicuous to predators. By 1986, only 50% of the moths were still of the melanic form, the rest having been replaced by the light form.

is that when polluted areas are cleaned, the moth populations should demonstrate an increase in the frequency of lightly pigmented individuals. Observations of such populations confirmed the result predicted by natural selection.

If a hypothesis is very powerful in explaining a wide variety of related phenomena, it attains the status of a theory. Natural selection is a good example. Our example of the use of natural selection to explain observed pigmentation patterns in moth populations is only one of many phenomena to which natural selection applies. Natural selection provides a potential explanation for the occurrence of many different traits distributed among virtually all animal species. Each of these instances constitutes a specific hypothesis generated from the theory of natural selection. Note, however, that falsification of a specific hypothesis does not necessarily lead to rejection of the theory as a whole. Natural selection may fail to explain the origins of human behavior, for example, but it provides an excellent explanation for many structural modifications of the pentadactyl (five-fingered) vertebrate limb for diverse functions. Scientists test many subsidiary hypotheses of their major theories to ask whether their theories are generally applicable. The most useful theories are those that can explain the largest array of different natural phenomena.

We emphasize that the meaning of the word "theory," when used by scientists, is not "speculation" as it is in ordinary English usage. Failure to make this distinction has been prominent in creationist challenges to evolution. The creationists have spoken of evolution as "only a theory," as if it were little better than a guess. In fact, the theory of evolution is supported by such massive evidence that most biologists view repudiation of evolution as tantamount to repudiation of reason. Nonetheless, evolution, along with all other theories in science, is not proven in a mathematical sense, but it is testable, tentative, and falsifiable. Powerful theories that guide extensive research are called **paradigms.** The history of science has

shown that even major paradigms are subject to refutation and replacement when they fail to account for our observations of the natural world. They are then replaced by new paradigms in a process known as a **scientific revolution.** For example, prior to the 1800s, animal species were studied as if they were specially created entities whose essential properties remained unchanged through time. Darwin's theories led to a scientific revolution that replaced these views with the evolutionary paradigm. The evolutionary paradigm has guided biological research for more than 130 years, and to date there is no scientific evidence that falsifies it; it continues to guide active inquiry into the natural world, and it is generally accepted as the cornerstone of biology.

PHYSIOLOGICAL VERSUS EVOLUTIONARY SCIENCES

The many questions that people have asked about the animal world since the time of Aristotle can be grouped into two major categories.* The first category seeks to understand the **proximate** or **immediate causes** that underlie the functioning of biological systems at a particular time and place. These include the problems of explaining how animals perform their metabolic, physiological, and behavioral functions at the molecular, cellular, organismal, and even populational levels. For example, how is genetic information expressed to guide the synthesis of proteins? What causes cells to divide to produce new cells? How does population density affect the physiology and behavior of organisms?

The biological sciences that address proximate causes are known as **physiological sciences,** and they proceed using the experimental method. This method consists of three steps: (1) predicting how a system being studied will respond to a disturbance, (2) making the disturbance, and then (3) comparing the observed

*Mayr, E. 1982. *The Growth of Biological Thought.* Cambridge, Harvard University Press, pp. 67–71.

results with the predicted ones. Experimental conditions are repeated to eliminate chance occurrences that might produce erroneous conclusions. **Controls**—repetitions of the experimental procedure that lack the disturbance—are established to protect against any unperceived factors that may bias the outcome of the experiment. The processes by which animals maintain a body temperature under different environmental conditions, digest their food, migrate to new habitats, or store energy are some additional examples of physiological phenomena that are studied by experiment (Chapters 32 through 38). Subfields of biology that constitute physiological sciences include molecular biology, cell biology, endocrinology, developmental biology, and community ecology.

In contrast to questions concerning the proximate causes of biological systems are questions of the **ultimate causes** that have produced these systems and their distinctive characteristics through evolutionary time. For example, what are the evolutionary factors that caused some birds to acquire complex patterns of seasonal migration between North and South America? Why do different species of animals have different numbers of chromosomes in their cells? Why do some animal species maintain complex social systems, whereas the animals of other species are largely solitary?

The biological sciences that address questions of ultimate cause are known as **evolutionary sciences,** and they proceed largely using the **comparative method** rather than experimentation. Characteristics of molecular biology, cell biology, organismal structure, development, and ecology are compared among related species to identify their patterns of variation. The patterns of similarity and dissimilarity are then used to test hypothesis of relatedness, and thereby to reconstruct the evolutionary tree that relates the species being studied. The evolutionary tree is then used to examine hypotheses of the evolutionary origins of the diverse molecular, cellular, organismal, and populational properties observed in the animal world. Clearly, the evolutionary sciences rely on results of the physiological sciences as a starting point. Evolutionary sciences include comparative biochemistry, molecular evolution, comparative cell biology, comparative anatomy, comparative physiology, and phylogenetic systematics.

THEORIES OF EVOLUTION AND HEREDITY

We turn now to a specific consideration of the two major paradigms that guide zoological research today: Darwin's theory of evolution and the chromosomal theory of inheritance.

DARWIN'S THEORY OF EVOLUTION

Darwin's theory of evolution is now over 130 years old (Chapter 9). Darwin articulated the complete theory when he published his famous book *On the Origin of Species by Means of Natural Selection* in England in 1859 (Figure 1-13). Biologists today are frequently asked, "What is Darwinism?" and "Do biologists still accept Darwin's theory of evolution?" These questions cannot be given simple answers, because Darwinism encompasses several different, although mutually compatible, theories. Professor Ernst Mayr of Harvard University has argued that Darwinism should be viewed as five major theories.* These five theories have somewhat different origins and different fates and cannot be discussed accurately as if they were only a single statement. The theories are (1) perpetual change, (2) common descent, (3) multiplication of species, (4) gradualism, and (5) natural selection. The first three theories are generally accepted as having universal application throughout the living world. The theories of gradualism and natural selection are controversial among evolutionists, although both are strongly advocated

*Mayr, E. 1985. Chapter 25 in D. Kohn, ed. *The Darwinian Heritage*. Princeton, Princeton University Press.

Figure 1-13
Modern evolutionary theory is strongly identified with Charles Robert Darwin who, with Alfred Russel Wallace, provided the first credible explanation of evolution. This photograph of Darwin was taken in 1854 when he was 45 years old. His most famous book, *On the Origin of Species,* appeared five years later.

Courtesy American Museum of Natural History. Neg. #326668

by a large portion of the evolutionary community and are important components of the Darwinian evolutionary paradigm. Gradualism and natural selection are clearly part of the evolutionary process, but their explanatory power might not be as widespread as Darwin intended. Legitimate controversies regarding gradualism and natural selection are often misrepresented by creationists as challenges to the first three theories presented above, although the validity of those first three theories is strongly supported by all relevant facts.

1. **Perpetual change.** This is the basic theory of evolution on which the others are based. It states that the living world is neither constant nor perpetually cycling, but is always changing. The properties of organisms undergo transformation across generations throughout time. This theory originated in antiquity but did not gain widespread acceptance until Darwin advocated it in the context of his other four theories. "Perpetual change" is documented by the fossil record, which clearly refutes creationists' claims for a recent origin of all

living forms. Because it has withstood repeated testing and is supported by an overwhelming number of observations, we now regard "perpetual change" as a scientific fact.

2. **Common descent.** The second Darwinian theory, "common descent," states that all forms of life descended from a common ancestor through a branching of lineages (Figure 1-14). The opposing argument, that the different forms of life arose independently and descended to the present in linear, unbranched genealogies, has been refuted by comparative studies of organismal form, cell structure, and macromolecular structures (including those of the genetic material, DNA). All of these studies confirm the theory that life's history has the structure of a branching evolutionary tree, known as a **phylogeny.** Species that share relatively recent common ancestry have more similar features at all levels than do species that have only an ancient common ancestry. Much current research is guided by Darwin's theory of common descent toward reconstructing life's phylogeny using the patterns of similarity and dissimilarity observed among species. The resulting phylogeny serves as the basis for our taxonomic classification of animals (Chapter 11).

3. **Multiplication of species.** Darwin's third theory states that the evolutionary process produces new species by the splitting and transformation of older ones. Species are now generally viewed as reproductively distinct populations of organisms that usually but not always differ from each other in organismal form. Once species are fully formed, interbreeding among members of different species does not occur. Evolutionists generally agree that the splitting and transformation of lineages produces new species, although there is still much controversy concerning the details of this process (Chapter 9) and the precise meaning of the term "species" (Chapter 11). The study of the historical processes that generate new species guides much active scientific research.

4. **Gradualism.** Gradualism states that the large differences in anatomi-

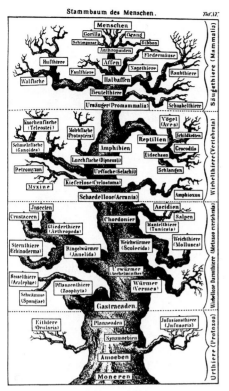

Figure 1-14

An early tree of life drawn in 1874 by the German biologist, Ernst Haeckel, who was strongly influenced by Darwin's theory of common descent. Many of the phylogenetic hypotheses shown in this tree, including the unilateral progression of evolution toward humans (= Menschen, *top*), have since been refuted.

cal traits that characterize different species originate through the accumulation of many small incremental changes over very long periods of time. This theory is important because genetic changes having very large effects on organismal form are usually harmful to the organism. It is possible, however, that some genetic variants that have large effects on the organism are nonetheless sufficiently beneficial to be favored by natural selection. Therefore, although gradual evolution is known to occur, it may not explain the origin of all structural differences that we observe among species (Figure 1-15). Scientists are still actively studying this question.

5. **Natural selection.** Natural selection, Darwin's most famous theory, rests on three propositions. First, there is variation among organisms (within populations) for anatomical, behavioral, and physiological traits. Second, the variation is at least partly heritable so that offspring tend to resemble their parents. Third, organisms with different variant forms leave different numbers of offspring to future generations. Variants that permit their

possessors most effectively to exploit their environments will preferentially survive and be transmitted to future generations. Over many generations, favorable new traits will spread throughout the population. Accumulation of such changes leads, over long periods of time, to the production of new organismal features and new species. Natural selection is therefore a creative process that generates novel features from the small individual variations that occur among organisms within a population.

Natural selection explains why organisms are constructed to meet the demands of their environments, a phenomenon called **adaptation** (Figure 1-16). Adaptation is the expected result of a process that accumulates the most favorable variants occurring in a population throughout long periods of evolutionary time. Adaptation was viewed previously as strong evidence against evolution, and Darwin's theory of natural selection was therefore important for convincing people that a natural process, capable of being studied scientifically, could produce new species. The

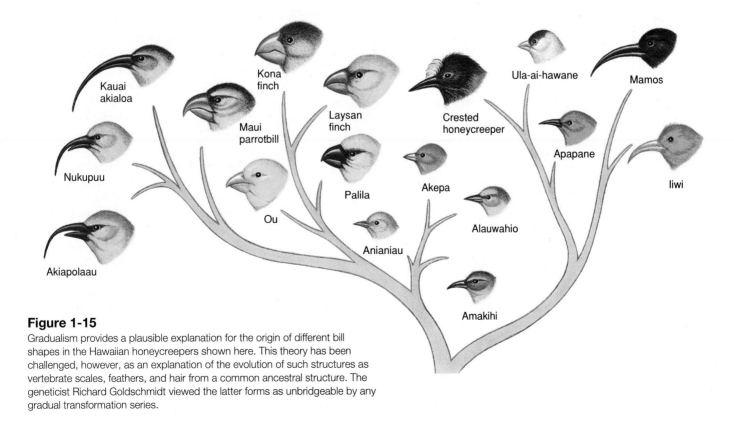

Figure 1-15

Gradualism provides a plausible explanation for the origin of different bill shapes in the Hawaiian honeycreepers shown here. This theory has been challenged, however, as an explanation of the evolution of such structures as vertebrate scales, feathers, and hair from a common ancestral structure. The geneticist Richard Goldschmidt viewed the latter forms as unbridgeable by any gradual transformation series.

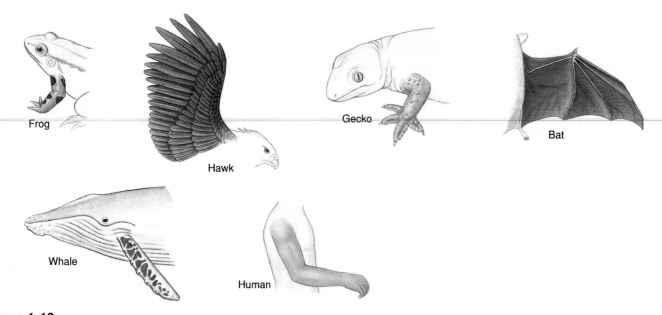

Figure 1-16

According to Darwinian evolutionary theory, the different forms of these vertebrate forelimbs were molded by natural selection to adapt them for different functions. We will see in later chapters that, despite these adaptive differences, these limbs share basic structural similarities.

demonstration that natural processes could produce adaptation was important to the eventual acceptance of all five Darwinian theories.

Darwin's theory of natural selection faced a major obstacle when it was first proposed: it lacked a theory of heredity. People assumed incorrectly that heredity was a blending process, and that any favorable new variant appearing in a population therefore would be lost. The new variant arises initially in a single organism, and that organism therefore must mate with one lacking the favorable new trait. Under blending inheritance, the organism's offspring would then have only a diluted form of the favorable trait. These offspring likewise would mate with others that lack the favorable trait. With its effects diluted by half each generation, the trait eventually would cease to exist. Natural selection would be completely ineffective in this situation.

Darwin was never able to counter this criticism successfully. It did not occur to Darwin that hereditary factors could be discrete and nonblending and that a new genetic variant therefore could persist unaltered from one generation to the next. This principle is known as **particulate inheritance.** It was established after 1900 with the discovery of Gregor Mendel's genetic experiments, and it was even-

tually incorporated into what we now call the **chromosomal theory of inheritance.** We use the term **neo-Darwinism** to describe Darwin's theories as modified by incorporating this theory of inheritance.

MENDELIAN HEREDITY AND THE CHROMOSOMAL THEORY OF INHERITANCE

The chromosomal theory of inheritance is the foundation for current studies of genetics and evolution in animals (Chapters 8 and 9). This theory comes from the consolidation of research done in the fields of genetics, which was founded by the experimental work of Gregor Mendel (Figure 1-17), and cell biology.

Genetic Approach

The genetic approach consists of mating or "crossing" populations of organisms that are true-breeding for contrasting traits, and then following the hereditary transmission of those traits through subsequent generations. "True-breeding" means that a population maintains across generations only one of the contrasting states of a particular feature when propagated in isolation from other populations.

Gregor Mendel studied the transmission of seven variable features in garden peas, crossing populations that were true-breeding for alternative traits (for example, tall versus short plants). In the first generation (called the F_1 generation, for "filial"), only one of the alternative parental traits was observed; there was no indication of blending of the parental traits. In the example, the offspring (called F_1 hybrids) formed by crossing the tall and short plants were tall, regardless of whether the tall trait was transmitted through the male or the female parent. These F_1 hybrids were allowed to self-pollinate, and both parental traits were found among their offspring (called the F_2 generation), although the trait observed in the F_1 hybrids (tall plants in this example) was three times more common than the other trait. Again, there was no indication of blending of the parental traits (Figure 1-18).

Mendel's experiments showed that the effects of a genetic factor can be masked in a hybrid individual, but that these factors were not physically altered during the transmission process. He postulated that variable traits are specified by paired hereditary factors, which we now call "genes." When germ cells (eggs or sperm) are produced, the two genes controlling a particular feature are segregated from each other and each germ cell receives only one of them. Fertilization restores the paired condition. If an organism possesses different forms of the paired genes for a feature, only one of them is expressed in its appearance, but both genes nonetheless will be transmitted unaltered in equal numbers to the gametes produced. Transmission of these genes is particulate, not blending. Mendel observed that the inheritance of one pair of traits is independent of the inheritance of other paired traits. We now know, however, that not all pairs of traits are inherited independently of each other. Numerous studies, particularly of the fruit fly, *Drosophila melanogaster,* have shown that the principles of inheritance discovered initially in plants apply also to animals.

Contributions of Cell Biology

Improvements in microscopes during the 1800s permitted cytologists to study the production of germ cells by direct observation of reproductive tissues. Interpreting the observations was initially difficult, however. Some prominent biologists hypothesized, for example, that sperm were parasitic worms in the semen (Figure 1-19). This hypothesis was soon falsified, and the true nature of germ cells was clarified. As the precursors of germ cells prepare to divide in the early stages of germ cell production, the nuclear material condenses to reveal discrete, elongate structures called chromosomes. Chromosomes

A

B

Figure 1-17

A, Gregor Johann Mendel. **B,** The monastery in Brno, Czech Republic, now a museum, where Mendel carried out his experiments with garden peas.

Courtesy Gregor Mendel Museum, Brno, Czechoslovakia.

PARTICULATE INHERITANCE (observed)

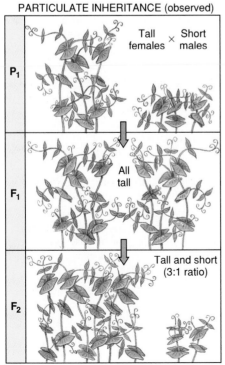

BLENDING INHERITANCE (not observed)

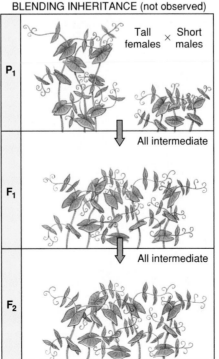

Figure 1-18

Different predictions of particulate versus blending inheritance regarding the outcome of Mendel's crosses of tall and short plants. The prediction of particulate inheritance is upheld and the prediction of blending inheritance is falsified by the results of the experiments. The reciprocal experiments (crossing short female parents with tall male parents) produced similar results. (P_1 = parental generation; F_1 = first filial generation; F_2 = second filial generation.)

The Animal Rights Controversy

In recent years, the debate surrounding the use of animals to serve human needs has intensified. Most controversial of all is the issue of animal use in biomedical and behavioral research and in the testing of commercial products.

A few years ago, Congress passed a series of amendments to the Federal Animal Welfare Act, a body of laws covering animal care in laboratories and other facilities. These amendments have become known as the three R's: **Reduction** in the number of animals needed for research; **Refinement** of techniques that might cause stress or suffering; **Replacement** of live animals with simulations or cell cultures whenever possible. As a result, the total number of animals used each year in research and in testing of commercial products has declined. Developments in cellular and molecular biology also have contributed to a decreased use of animals for research and testing. The animal rights movement, composed largely of vocal antivivisectionists, has created an awareness of the needs of animals used in research and has stimulated researchers to discover cheaper, more efficient, and more humane alternatives.

However, computers and culturing of cells can simulate the effects on organismal systems of, for instance, drugs, only when the basic principles involved are well known. When the principles themselves are being scrutinized and tested, computer modeling is not sufficient. A recent report by the National Research Council concedes that although the search for alternatives to the use of animals in research and testing will continue, "the chance that alternatives will completely replace animals in the foreseeable future is nil." Realistic immediate goals, however, are reduction in number of animals used, replacement of mammals with other vertebrates, and refinement of experimental procedures to reduce discomfort of the animals being tested.

Medical and veterinary progress depends on research using animals. Every drug and every vaccine developed to improve the human condition has

According to the U.S. Department of Health and Human Services, animal research has helped extend our life expectancy by 20.8 years.

been tested first on animals. Research using animals has enabled medical science to eliminate smallpox and polio, and to immunize against diseases previously common and often deadly, including diphtheria, mumps, and rubella. It also has helped to create treatments for cancer, diabetes, heart disease, and manic-depressive psychoses, and to develop surgical procedures including heart surgery, blood transfusions, and cataract removal. AIDS research is wholly dependent on studies using animals. The similarity of simian AIDS, identified in rhesus monkeys, to human AIDS has permitted the disease in monkeys to serve as a model for the human disease. Recent work indicates that cats, too, may prove to be useful models for the development of an AIDS vaccine. Skin grafting experiments, first done with cattle and later with other animals, opened a new era in immunological research with vast ramifications for treatment of disease in humans and other animals.

Research using animals also has benefited *other animals* through the

development of veterinary cures. The vaccines for feline leukemia and canine parvovirus were first introduced to other cats and dogs. Many other vaccinations for serious diseases of animals were developed through research on animals: for example, rabies, distemper, anthrax, hepatitis, and tetanus. No endangered species are used in general research (except to protect that species from total extinction). Thus, research using animals has provided enormous benefits to humans and other animals. Still, much remains to be learned about treatment of diseases such as cancer, AIDS, diabetes, and heart disease, and research with animals will be required for this purpose.

Despite the remarkable benefits produced by research on animals, advocates of animal rights often present an inaccurate and emotionally distorted picture of this research. The ultimate goal of most animal rights activists, who have focused specifically on the use of animals in science rather than on the treatment of animals in all contexts,

continued

THE ANIMAL RIGHTS CONTROVERSY—CONTINUED.

remains the total abolition of all forms of research using animals. The scientific community is deeply concerned about the impact of these attacks on the ability of scientists to conduct important experiments that will benefit people and animals. They argue that if we are justified to use animals for food and fiber and as pets, we are justified in experimentation to benefit human welfare when these studies are conducted humanely and ethically.

REFERENCES ON ANIMAL RIGHTS CONTROVERSY

Commission on Life Sciences, National Research Council. 1988. Use of laboratory animals in biomedical and behavioral research. Washington, D.C., National Academy Press. *Statement of national policy on guidelines for the use of animals in biomedical research. Includes a chapter on the benefits derived from the use of animals.*

Goldberg, A. M., and J. M. Frazier. 1989. Alternatives to animals in toxicity testing. Sci. Am. **261**:24–30 (Aug.). *Describes alternatives that are being developed for the costly and time-consuming use of animals in the testing of thousands of chemicals that each year must be evaluated for potential toxicity to humans.*

Pringle, L. 1989. The animal rights controversy. San Diego, California, Harcourt Brace Jovanovich, Publishers. *Although no one writing about the animal rights movement can honestly claim to be totally objective and impartial on such an emotionally charged issue, this book comes as close as any to presenting a balanced treatment.*

Rowan, A. N. 1984. Of mice, models, and men: a critical evaluation of animal research. Albany, New York, State University of New York Press. *Good review of the issues. Chapter 7 deals with the use of animals in education, and notes that our educational system provides little help in resolving the contradiction of teaching kindness to animals while using animals in experimentation in biology classes.*

Sperling, S. 1988. Animal liberators: research and morality. Berkeley, University of California Press. *Thoughtful and carefully researched study of the animal rights movement, its ideological roots, and the passionate idealism of animal rights activists.*

Figure 1-19

An early nineteenth-century micrographic drawing of spermatozoa from (1) guinea pig, (2) white mouse, (3) hedgehog, (4) horse, (5) cat, (6) ram, and (7) dog (Prévost and Dumas, 1821). Some biologists initially interpreted these as parasitic worms in the semen, but on further examination found them to be male reproductive cells.

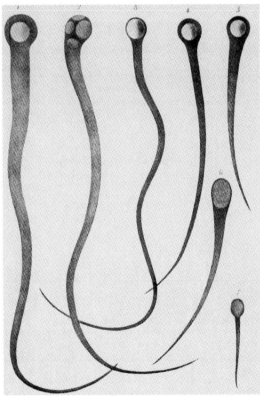

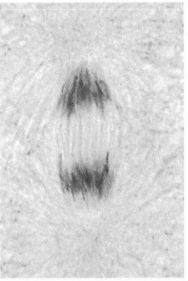

Figure 1-20

Paired chromosomes being separated before nuclear division in the process of forming reproductive cells.

occur in pairs that are usually similar but not identical in appearance and informational content. The number of chromosomal pairs varies among species. One member of each pair is derived from the female parent and the other from the male parent. Paired chromosomes are physically associated and then segregated into different daughter cells during cell division prior to gamete formation (Figure 1-20). Each resulting gamete receives one chromosome from each pair. Different pairs of chromosomes are sorted into gametes independently of each other. Because the behavior of the chromosomal material during germ cell formation parallels that postulated for Mendel's genes, Sutton and Boveri in 1903 through 1904 hypothesized that chromosomes were the physical bearers of the genetic material. This hypothesis met with extreme skepticism when first proposed. A long series of tests designed to falsify it nonetheless showed that its predictions were upheld. The chromosomal theory of inheritance is now well established.

Summary

Zoology is the scientific study of animals, and it is part of biology, the scientific study of life. Animals and life in general can be identified by attributes that they have acquired over their long evolutionary histories. The most outstanding attributes of life include chemical uniqueness, complexity and hierarchical organization, reproduction, possession of a genetic program, metabolism, development, and interaction with the environment. Biological systems comprise a hierarchy of integrative levels (molecular, cellular, organismal, populational, and species levels), each of which demonstrates a number of specific emergent properties.

Science is characterized by the acquisition of knowledge by constructing and then testing hypotheses through observations of the natural world. Science is guided by natural law, and its hypotheses are testable, tentative, and falsifiable. Zoological sciences can be subdivided into two categories, the physiological sciences and the evolutionary sciences. The physiological sciences use the experimental method to ask how animals perform their basic metabolic, developmental, behavioral, and reproductive functions, including investigations of their molecular, cellular, and populational systems. The evolutionary sciences use the comparative method to reconstruct the history of life, and then use that history to understand how diverse species and their molecular, cellular, organismal, and populational properties arose through evolutionary time. Hypotheses that withstand repeated testing and therefore explain many diverse phenomena gain the status of a theory.

Powerful theories that guide extensive research are called "paradigms." The major paradigms that guide the study of zoology are Darwin's theory of evolution and the chromosomal theory of inheritance.

The principles given in this chapter illustrate the unity of biological science. All components of biological systems are guided by natural laws and are constrained by those laws. Living organisms can come only from other living organisms, just as new cells can be produced only from pre-existing cells. Reproductive processes occur at all levels of the biological hierarchy and demonstrate both heredity and variation. The interaction of heredity and variation at all levels of the biological hierarchy produces evolutionary change and has generated the great diversity of animal life documented throughout this book.

Review Questions

1. Why is it difficult to define life?
2. What are the basic chemical differences that distinguish living from nonliving systems?
3. Describe the hierarchical organization of life. How does this organization lead to the emergence of new properties at different levels of biological complexity?
4. What is the relationship between heredity and variation in reproducing biological systems?
5. Describe how the evolution of complex organisms is compatible with the second law of thermodynamics.
6. What are the essential characteristics of science? Describe how evolutionary studies fit these characteristics whereas "scientific creationism" does not.
7. Use studies of natural selection in British moth populations to illustrate the hypothetico-deductive method of science.
8. How do we distinguish the terms hypothesis, theory, paradigm, and scientific fact?
9. How do biologists distinguish physiological and evolutionary sciences?
10. What are Darwin's five theories of evolution (as identified by Ernst Mayr)? Which are accepted as fact and which continue to stir controversy among biologists?
11. What major obstacle confronted Darwin's theory of natural selection when it was first proposed? How was this obstacle overcome?
12. How does neo-Darwinism differ from Darwinism?
13. Describe the respective contributions of the genetic approach and cell biology to formulating the chromosomal theory of inheritance.

Selected References

Futuyma, D. J. 1995. Science on trial: the case for evolution. Sunderland, Massachusetts, Sinauer Associates, Inc. *A defense of evolutionary biology as the exclusive scientific approach to the study of life's diversity.*

Kitcher, P. 1982. Abusing science: the case against creationism. Cambridge, Massachusetts, MIT Press. *A treatise on how knowledge is gained in science and why creationism does not qualify as science.*

Kuhn, T. S. 1970. The structure of scientific revolutions. ed. 2, enlarged. Chicago, University of Chicago Press. *An influential and controversial commentary on the process of science.*

Mayr, E. 1982. The growth of biological thought: diversity, evolution and inheritance. Cambridge, Massachusetts, The Belknap Press of Harvard University Press. *An interpretive history of biology with special reference to genetics and evolution.*

Medawar, P. B. 1989. Induction and intuition in scientific thought. London, Methuen & Company. *A commentary on the basic philosophy and methodology of science.*

Moore, J. A. 1993. Science as a way of knowing: the foundations of modern biology. Cambridge, Massachusetts, Harvard University Press. *A lively, wide-ranging account of the history of biological thought and the workings of life.*

Perutz, M. F. 1989. Is science necessary? Essays on science and scientists. New York, E. P. Dutton. *A general discussion of the utility of science.*

2

Chemistry of Life

Water, Water, Everywhere

As land animals, we may not appreciate just how much of earth is covered with water—over 70%. We are composed mostly of water, too, as are other animals. Water is so common, we take it for granted. But water is an extraordinary substance and is by no means common in the universe. Earth is the only planet (as far as we know) that has liquid water on its surface. Water is the only substance that occurs in nature in the three phases of solid, liquid, and vapor within the ordinary range of the earth's temperatures. The significance of this and other unique properties of water are discussed in this chapter.

All the chemical reactions that enable our cells to derive energy, reproduce, and perform other vital functions take place in an aqueous medium, and these reactions quite often consume or produce water molecules. Water is within our cells and bathes the cells within our bodies. Fans of televised science fiction, such as the series *Star Trek, the Next Generation* are sometimes treated to dramatized encounters with "life forms" composed only of energy or crystals or some other product of the writer's imagination. Even Data the android is "alive" for dramatic purposes. However, none of these entities is life as we know it, because life, conceived as it was in water, requires water to continue as life. The life that exists on earth could not have originated or evolved in the absence of that unique compound. Water is not everywhere, as it seemed to the Ancient Mariner, but wherever there is life there is water. ∎

Living systems and their constituents obey physical and chemical laws. Within the cells of any organism, the living substance is composed of a multitude of nonliving constituents: proteins, nucleic acids, fats, carbohydrates, waste metabolites, crystalline aggregates, pigments, and many others. Physical and chemical interactions of such substances account for the many processes essential to life, including digestion and absorption of nutrients, release of energy, removal of waste, communication of cells with each other, conduction of nerve impulses, and transmission of genetic information from one generation to the next. Because these phenomena will be discussed in later pages, we must present some information on biochemistry here. Basic information on atoms, elements, and molecules; bonds; acids, bases, and salts; and buffers is given at the end of the book in Appendix B.

WATER AND LIFE

Water is the most abundant of all compounds in cells, making up about 60% to 90% of most living organisms. The maintenance of a constant aqueous internal environment is a major physiological task for all organisms, both terrestrial and aquatic.

Water has several extraordinary properties that make it especially fit for its essential role in living systems. We now know that the remarkable properties of water can be explained in large part on the basis of the hydrogen bonds that form between its molecules (Figure 2-1). Although hydrogen bonds are much weaker than the covalent bonds within a water molecule, they require significant energy for breakage.

Water has a **high specific heat capacity:** 1 calorie* is required to elevate the temperature of 1 g of water 1° C. Every other liquid but ammonia

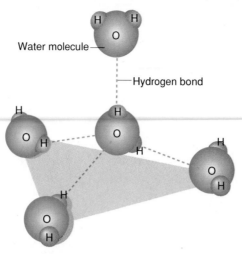

Figure 2-1

Geometry of water molecules. Each water molecule is linked by hydrogen bonds (*dashed lines*) to four other molecules. If imaginary lines are used to connect the divergent oxygen atoms, a tetrahedron is obtained.

requires less heat to accomplish the same temperature increase. When water is heated, much of the heat energy is used to rupture some of the hydrogen bonds, leaving less heat to increase the kinetic energy (molecular movement), and thus the temperature, of the water. The high thermal capacity of water has a great moderating effect on environmental temperature changes and is a great protective agent for all life.

Water also has a **high heat of vaporization.** More than 500 calories is required to change 1 g of liquid water into water vapor. This is so because all of the hydrogen bonds between a water molecule and its neighbors must be ruptured before that water molecule can escape the surface and enter the air. For terrestrial animals (and plants), cooling produced by the evaporation of water is an important means of getting rid of excess heat.

Another important property of water from a biological standpoint is its **unique density behavior** during changes of temperature. Most liquids become denser with decreasing temperature. Water, however, reaches its maximum density at 4° C *while still a liquid,* then becomes less dense with further cooling. Therefore ice *floats* rather than forming on the bottom of lakes and ponds. If it were not for this

*A calorie is defined as the amount of heat required to heat 1 g of water from 14.5° C to 15.5° C. While the calorie is the traditional unit of heat widely used in publications and tables, it is not part of the International System of Units (the SI system) which uses the joule (J) as the energy unit (1 cal = 4.184 J).

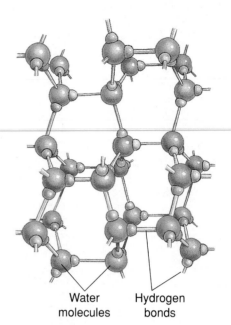

Water molecules **Hydrogen bonds**

Figure 2-2

When water freezes at 0° C, the four partial charges of each atom in the molecule interact with the opposite charges of atoms in other water molecules. The hydrogen bonds between all the molecules form a crystal-like lattice structure, and the molecules are farther apart (and thus less dense) than when some of the molecules have not formed hydrogen bonds at 4° C.

property, bodies of water would freeze solid from the bottom up in winter, and, except in warmer climates, would not necessarily completely melt in summer. Under these conditions, aquatic life would be severely limited. In ice, all molecules form hydrogen bonds with others (Figure 2-2). The molecules form an extensive, open, crystal-like network held together by hydrogen bonds. The molecules in this latticelike form are farther apart, and thus less dense, than when some of the molecules have not formed hydrogen bonds at 4° C.

Water has a **high surface tension,** greater than that of any other liquid but mercury. This property is an aspect of the great cohesiveness of water molecules: their tendency to be held together by hydrogen bonds between them. Cohesiveness is important in the maintenance of protoplasmic form and movement, and the high surface tension creates a unique ecological niche (see p. 803) for insect forms, such as water striders (Figure 2-3) and whirligig beetles, that

Figure 2-3
Because of hydrogen bonds between water molecules at the water-air interface, the water molecules cling together and create a high surface tension. Thus some insects, such as this water strider, can literally walk on water.

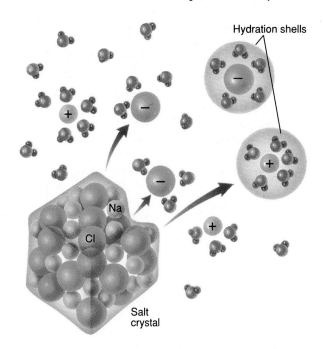

Figure 2-4
When a crystal of sodium chloride dissolves in water, the negative ends of the dipolar molecules of water surround the Na⁺ ions, while the positive ends of water molecules face the Cl⁻ ions. The ions are thus separated and do not reenter the salt lattice.

skate on the surfaces of ponds. Despite its high surface tension, water has **low viscosity,** a property that favors the movement of blood through minute capillaries and of cytoplasm inside cellular boundaries.

Water is an excellent **solvent.** Salts dissolve to a much greater degree in water than they do in any other solvent. This results from the dipolar nature of water, which causes it to orient itself around charged particles dissolved in it. When, for example, NaCl is dissolved in water, the Na⁺ and Cl⁻ ions present in the solid salt rapidly separate into independent Na⁺ and Cl⁻ ions. The negative zones of the water dipoles align themselves around the Na⁺ ions while the positive zones align themselves around the Cl⁻ ions (Figure 2-4). This keeps the ions separated, promoting a high degree of dissociation. Other solvents having less or no dipolar character are less able to align effectively around such ions and are therefore less able to dissolve the salt. Water binding to dissolved protein molecules may be essential to the proper functioning of the protein.

Water is a participant in many of the chemical reactions that take place in living organisms. Many compounds are split into smaller pieces by the addition of a molecule of water, a process called **hydrolysis.** Likewise, larger compounds are often synthesized from smaller components by the reverse of hydrolysis, called **condensation** reactions.

$$R{-}R + H_2O \xrightarrow[\text{Hydrolysis}]{} R{-}OH + H{-}R$$

$$R{-}OH + H{-}R \xrightarrow[\text{Condensation}]{} R{-}R + H_2O$$

It is ironic that, despite the crucial role of water in living systems, the world's water supply is in grave danger because of human activities, population growth, and poverty (la Rivière, 1989). Humans are polluting the world's surface fresh water, groundwater, and oceans with both natural and industrial wastes. Only small numbers of water management projects are in progress, and these are limited to the more heavily industrialized countries. In newly industrializing countries, pollution control is neglected in favor of industrial development, and in some regions (East Asia, for example) the most serious environmental problem may be degradation of water resources. In less developed countries, pollution from organic wastes, responsible for much human disease, is widespread, and financial support of waste treatment does not exist. Humanity desperately needs water management on a worldwide scale if we are to avoid dire consequences.

ORGANIC MOLECULES

The term "organic compounds" was formerly applied to substances derived from living matter, but many organic compounds can be synthesized from inorganic chemicals. Thus the term "organic" is now applied broadly to compounds that contain carbon. Many also contain hydrogen, oxygen, nitrogen, sulfur, phosphorus, salts, and other elements.

Carbon has a great ability to bond with other carbon atoms in chains of varying lengths and configurations. More than a million organic compounds have been identified; more are being added daily. Carbon-to-carbon combinations introduce the possibility of enormous complexity and variety into molecular structure. Examples are found in the pages that follow.

CARBOHYDRATES: NATURE'S MOST ABUNDANT ORGANIC SUBSTANCE

Carbohydrates are compounds of carbon, hydrogen, and oxygen. They are usually present in the ratio of 1 C: 2 H: 1 O and are grouped as H—C—OH. Familiar examples of carbohydrates are sugars, starches, and cellulose (the woody structure of plants). There is more cellulose on earth than all other organic materials combined. Carbohydrates are synthesized by green plants from water and carbon dioxide, with the aid of the sun's energy. This process, called **photosynthesis,** is a reaction upon which all life depends, for it is the starting point in the formation of food.

Carbohydrates are usually divided into the following three classes: (1) **monosaccharides,** or simple sugars; (2) **disaccharides,** or double sugars; (3) **polysaccharides,** or complex sugars. Simple sugars are composed of carbon chains containing 4 carbons (tetroses), 5 carbons (pentoses), or 6 carbons (hexoses). Other simple sugars have up to 10 carbons, but these are not biologically important. Simple sugars, such as glucose, galactose, and fructose, all contain a free sugar group,

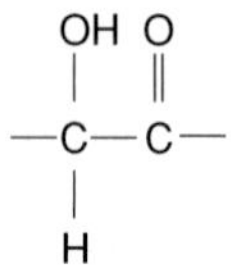

$$-\overset{\displaystyle \overset{OH}{|}}{C}-\overset{\displaystyle \overset{O}{\|}}{C}-$$
$$\underset{\displaystyle \underset{H}{|}}{}$$

in which the double-bonded O may be attached to the terminal C of a chain or to a nonterminal C. The hexose **glucose** (also called dextrose) is the most important carbohydrate in the living world. Glucose is often shown as a straight chain (Figure 2-5A), but in water it tends to form a cyclic compound (Figure 2-5B). The "chair" diagram (Figure 2-6) of glucose best represents its true configuration, but we must remember that all forms of glucose, however represented, are the same molecule.

Other hexoses of biological significance are galactose and fructose. Their straight-chain structures are compared with that of glucose in Figure 2-7.

Disaccharides are double sugars formed by the bonding of two simple sugars. An example is maltose (malt sugar), composed of two glucose molecules. As shown in Figure 2-8, the two glucose molecules are condensed

together by the removal of a molecule of water. This condensation reaction, with the sharing of an oxygen atom by the two sugars, characterizes the formation of all disaccharides. Two other common disaccharides are sucrose (ordinary cane, or table, sugar), formed by the linkage of glucose and fructose, and lactose (milk sugar), composed of glucose and galactose.

Polysaccharides are made up of many molecules of simple sugars (usually glucose) linked together in long chains and are referred to by the chemist as polymers. Their empirical formula is usually written $(C_6H_{10}O_5)_n$, where n stands for the unknown number of simple sugar molecules of which they are composed. Starch is the common storage form of sugar in most plants and is an important food constituent for animals. **Glycogen** is an important storage form for sugar in animals. It is found mainly in liver and muscle cells in vertebrates. When needed, glycogen is converted into glucose and is delivered by the blood to the tissues. Another polymer is **cellulose,** which is the principal structural carbohydrate of plants.

The main role of carbohydrates in protoplasm is to serve as a source of chemical energy. Glucose is the most important of these energy carbohydrates. Some carbohydrates become basic components of protoplasmic structure, such as the pentoses that form constituent groups of nucleic acids and of nucleotides.

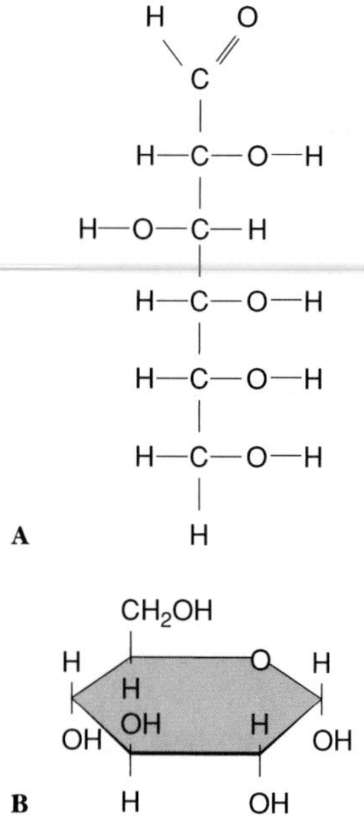

Figure 2-5

Two ways of depicting the structural formula of the simple sugar glucose. In **A,** the carbon atoms are shown in open-chain form. When dissolved in water, glucose tends to assume a ring form as in **B.** In this ring model the carbon atoms located at each turn in the ring are usually not shown.

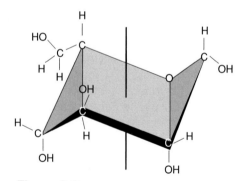

Figure 2-6

"Chair" representation of a glucose molecule.

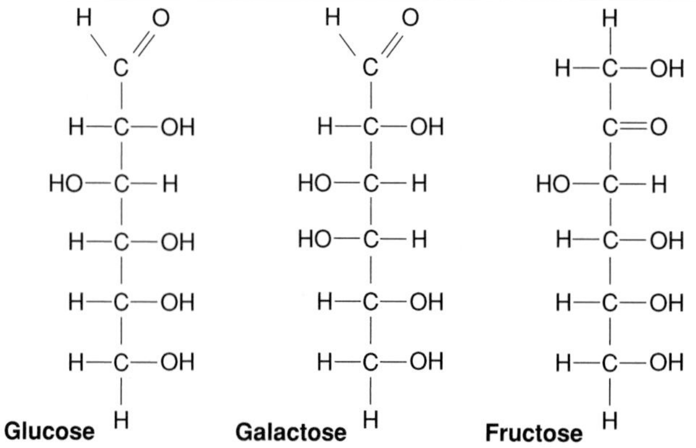

Glucose   Galactose   Fructose

Figure 2-7

These three hexoses are the most common monosaccharides.

LIPIDS: FUEL STORAGE AND BUILDING MATERIAL

Lipids are fats and fatlike substances. They are composed of molecules of low polarity; consequently, they are virtually insoluble in water but are soluble in organic solvents such as acetone and ether. Three principal groups of lipids are neutral fats, phospholipids, and steroids.

Neutral Fats

The neutral or "true" fats are major fuels of animals. Stored fat may be derived directly from dietary fat or indirectly from dietary carbohydrates that are converted to fat for storage. Fats are oxidized and released into the bloodstream as needed to meet tissue demands, especially the demands of active muscle.

Neutral fats are triglycerides, which are molecules consisting of glycerol and three molecules of fatty acids. Neutral fats are therefore esters, that is, a combination of an alcohol (glycerol) and an acid. Fatty acids in triglycerides are simply long-chain monocarboxylic acids; they vary in size but are commonly 14 to 24 carbons long. The production of a typical fat by the union of glycerol and stearic acid is shown in Figure 2-9A. In this reaction it can be seen that the three fatty acid molecules have united with the OH group of the glycerol to form stearin (a neutral fat), with the production of three molecules of water.

Most triglycerides contain two or three different fatty acids attached to glycerol, bearing ponderous names such as myristoyl stearoyl glycerol (Figure 2-9B). The fatty acids in this triglyceride are **saturated;** that is, every carbon within the chain holds two hydrogen atoms. Saturated fats, more common in animals than in plants, are usually solid at room temperature. **Unsaturated** fatty acids, typical of plant oils, have two or more carbon atoms joined by double bonds; that is, the carbons are not "saturated" with hydrogen atoms and are able to form bonds with other atoms. Two common unsaturated fatty acids are oleic acid and linoleic acid (Figure 2-10). Plant fats such as peanut oil and corn oil tend to be liquid at room temperature.

Phospholipids

Unlike the fats that are fuels and serve no structural roles in the cell, phospholipids are important components of the molecular organization of tissues, especially membranes. They resemble triglycerides in structure, except that one of the three fatty acids is replaced by phosphoric acid and an organic base. An example is lecithin, an important phospholipid of nerve membrane (Figure 2-11). Because the

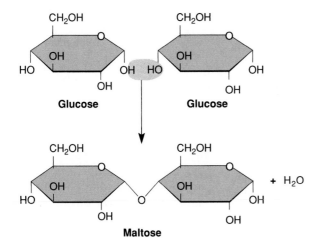

Figure 2-8
Formation of a double sugar (disaccharide maltose) from two glucose molecules with the removal of one molecule of water.

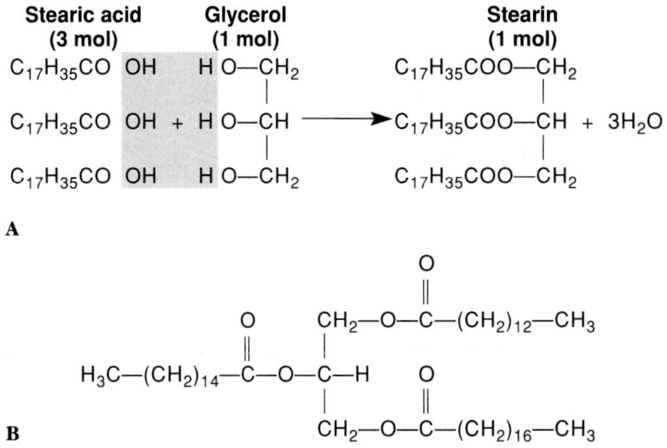

Figure 2-9
Neutral fats. **A,** Formation of a neutral fat from three molecules of stearic acid (a fatty acid) and glycerol. **B,** A neutral fat bearing three different fatty acids.

$$CH_3-(CH_2)_7-CH=CH-(CH_2)_7-COOH$$
Oleic acid

$$CH_3-(CH_2)_4-CH=CH-CH_2-CH=CH-(CH_2)_7-COOH$$
Linoleic acid

Figure 2-10
Unsaturated fatty acids: oleic acid having one double bond and linoleic acid having two double bonds. The remainder of the hydrocarbon chains of both acids is saturated.

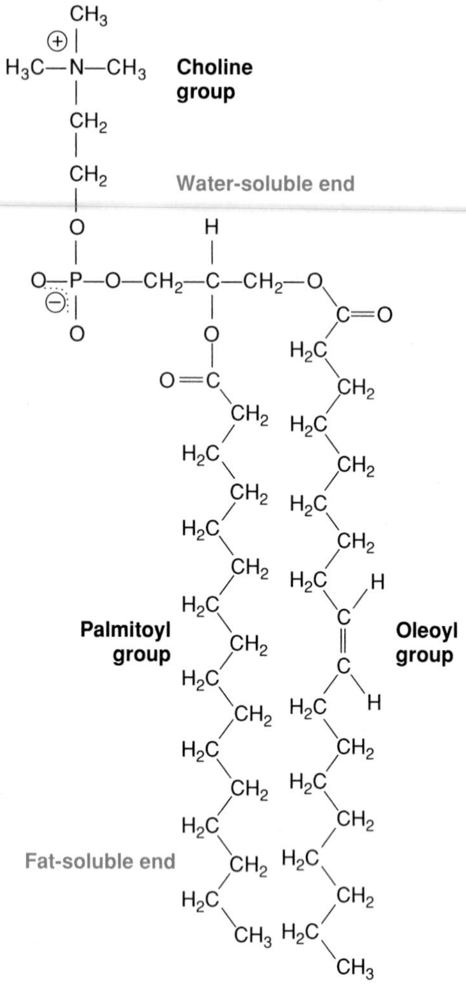

Figure 2-11

Lecithin (phosphatidyl choline), an important phospholipid of nerve membranes.

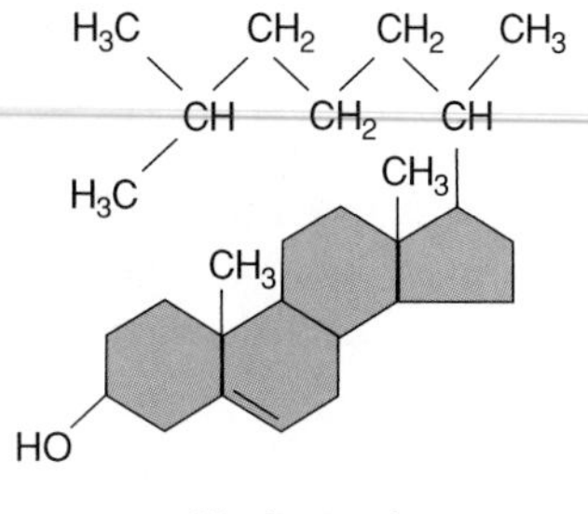

Figure 2-12

Cholesterol, a steroid. All steroids have a basic skeleton of four rings (three 6-carbon rings and one 5-carbon ring) with various side groups attached.

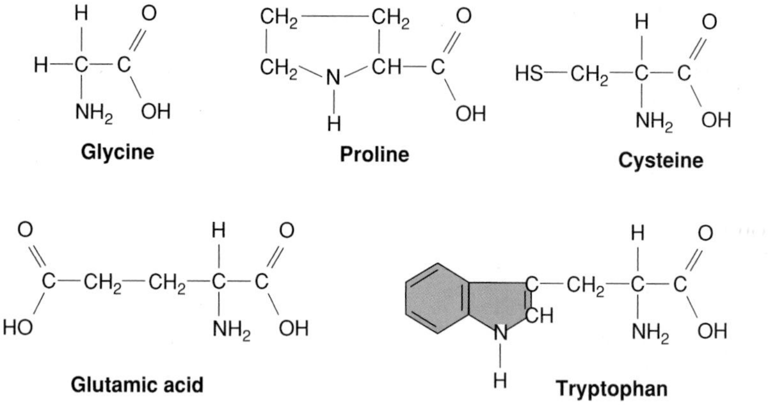

Cholesterol

Figure 2-13

Five of the twenty naturally occurring amino acids.

phosphate group on phospholipids is charged and polar and therefore soluble in water and the remainder of the molecule is nonpolar, phospholipids can bridge two environments and bind water-soluble molecules such as proteins to water-insoluble materials.

Steroids

Steroids are complex alcohols; although they are structurally unlike fats, they have fatlike properties. The steroids are a large group of biologically important molecules, including cholesterol (Figure 2-12), vitamin D, many adrenocortical hormones, and the sex hormones.

AMINO ACIDS AND PROTEINS

Proteins are large, complex molecules composed of 20 commonly occurring amino acids (Figure 2-13). The amino acids are linked together by **peptide bonds** to form long, chainlike polymers. In the formation of a peptide bond, the carboxyl group of one amino acid is linked by a covalent bond to the amino group of another, with the elimination of water, as follows:

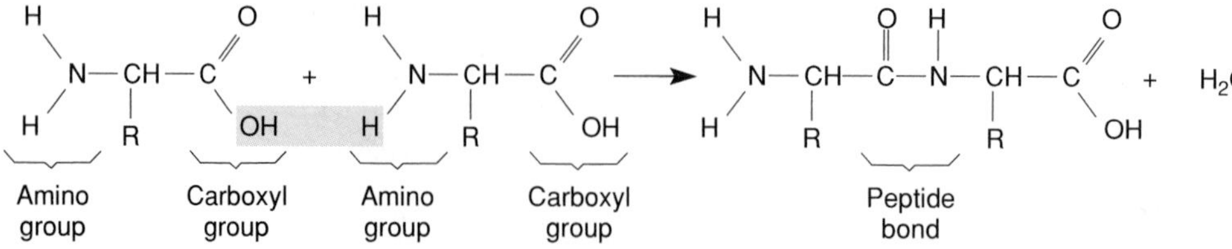

The combination of two amino acids by a peptide bond forms a dipeptide, and, as is evident, there is still a free amino group on one end and a free carboxyl group on the other; therefore additional amino acids can be joined to both ends until a long chain is produced. The 20 different kinds of amino acids can be arranged in an enormous variety of sequences of up to several hundred amino acid units;

therefore it is not difficult to account for practically countless varieties of proteins among living organisms.

A protein is not just a long string of amino acids; it is a highly organized molecule. For convenience, biochemists have recognized four levels of protein organization called primary, secondary, tertiary, and quaternary.

The **primary structure** of a protein is determined by the kind of sequence of amino acids making up the polypeptide chain. Because the bonds between the amino acids in the chain are characterized by a limited number of stable angles, certain recurrent structural patterns are assumed by the chain. This is called the **secondary structure,** and it is often that of an **alpha-helix,** that is, helical turns in a clockwise direction like a screw (Figure 2-14). The spirals of the chains are stabilized by hydrogen bonds, usually between a hydrogen atom of one amino acid and the peptide-bond oxygen of another in an adjacent turn of the helix.

Not only does the polypeptide chain (primary structure) spiral into helical configurations (secondary structure) but also the helices themselves bend and fold, giving the protein its complex, yet stable, three-dimensional **tertiary structure** (Figure 2-14). The folded chains are stabilized by the interactions between side groups of amino acids. One of these interactions is the **disulfide bond,** a covalent bond between the sulfur atoms in pairs of cysteine (sis'tee-in) units that are brought together by folds in the polypeptide chain. Other kinds of bonds that help stabilize the tertiary structure of proteins are hydrogen bonds, ionic bonds, and hydrophobic bonds.

The term **quaternary structure** describes proteins that contain more than one polypeptide chain unit. For example, hemoglobin (the oxygen-carrying substance in blood) of higher vertebrates is composed of four polypeptide subunits nested together into a single protein molecule (Figure 2-14).

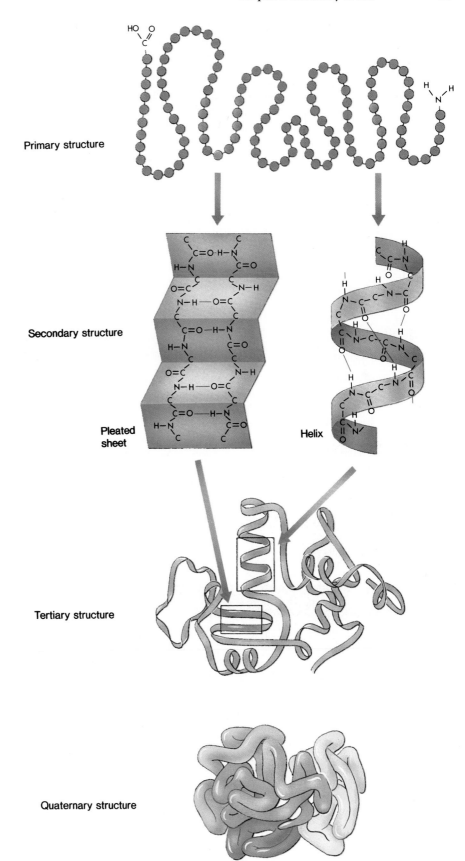

Figure 2-14

Structure of proteins. The amino acid sequence of a protein (*primary structure*) encourages the formation of hydrogen bonds between nearby amino acids, producing coils and foldbacks (the *secondary structure*). Bends and helices cause the chain to fold back on itself in a complex manner (*tertiary structure*). Individual polypeptide chains of some proteins aggregate together to form the functional molecule composed of several subunits (*quaternary structure*).

Proteins as Enzymes

Proteins perform many functions in living things. They serve as the structural framework of protoplasm and form many cell components. However, the most important roles of proteins by far are as **enzymes,** the biological catalysts required for almost every reaction in the body.

Enzymes lower the activation energy required for specific reactions and enable life processes to proceed at moderate temperatures. They control the reactions by which food is digested, absorbed, and metabolized. They promote the synthesis of structural materials for growth and to replace the wear and tear on the body. They determine the release of energy used in respiration, growth, muscle contraction, physical and mental activities, and many other activities. Enzyme action is described in Chapter 5 (p. 62).

NUCLEIC ACIDS

Nucleic acids are complex substances of high molecular weight that represent a basic manifestation of life. The sequence of nitrogenous bases in these polymeric molecules encodes the genetic information necessary for all aspects of biological inheritance. They not only direct the synthesis of enzymes and other proteins, but they are also the only molecules that have the power (with the help of the right enzymes) to replicate themselves. The two kinds of nucleic acids in cells are **deoxyribose nucleic acid (DNA) and ribose nucleic acid (RNA).** They are polymers of repeated units called **nucleotides,** each containing a sugar, a nitrogenous base, and a phosphate group. The structure of nucleic acids is crucial to the mechanism of inheritance and protein synthesis, so this subject is discussed further in Chapter 8 (p. 125).

Summary

The unique structure of water and its ability to form hydrogen bonds between adjacent water molecules are the characteristics responsible for its special properties: solvency; high heat capacity, boiling point, and surface tension; and lower density as a solid than as a liquid. Life on earth could not have appeared without water.

Carbon is especially versatile in bonding with itself or with other atoms, and it is the only element capable of forming the variety of molecules found in living things.

Carbohydrates are composed primarily of carbon, hydrogen, and oxygen grouped as H—C—OH. Sugars serve as immediate sources of energy in living systems. Monosaccharides, or simple sugars, may bond together to form disaccharides or polysaccharides, which serve as storage forms of sugar or perform structural roles. Lipids exist principally as fats, phospholipids, and steroids.

Proteins are large molecules composed of amino acids linked together by peptide bonds. Proteins have a primary, secondary, tertiary, and often, quaternary structure. Proteins perform many functions, especially as enzymes (biological catalysts).

Nucleic acids are polymers of nucleotide units, each composed of a sugar, a nitrogenous base, and a phosphate group. They contain the material of inheritance and function in protein synthesis.

Review Questions

1. Explain why water molecules tend to form hydrogen bonds with other water molecules.
2. Explain each of the following properties of water, and tell how each is conferred by the dipolar nature of the water molecule: high specific heat capacity; high heat of vaporization; unique density behavior; high surface tension; good solvent for ions of salts.
3. Name two simple carbohydrates, two storage carbohydrates, and a structural carbohydrate.
4. What are characteristic differences in molecular structure between lipids and carbohydrates?
5. Explain the difference between the primary, secondary, tertiary, and quaternary structures of a protein.
6. What are the important nucleic acids in a cell, and of what units are they constructed?

Selected References

Standard texts in freshman chemistry and in biochemistry and cellular biology are good sources of additional information on the topics in this chapter. The following selection is by no means exhaustive.

Karplus, M., and J. A. McCammon. 1986. The dynamics of proteins, Sci. Am. **254:**42–51 (Apr.). *Atoms in a protein molecule are in constant motion and would not be able to function if they were rigid.*

la Rivière, J. W. M. 1989. Threats to the world's water. Sci. Am. **261:**80–94 (Sept.). *We must take appropriate steps soon to avoid severe shortages of this vital resource.*

Lehninger, A. L. 1993. Principles of biochemistry, ed. 2. New York, Worth Publishers, Inc. *Clearly presented advanced text.*

Lodish, H., D. Baltimore, A. Berk, S. L. Zipursky, P. Matsudira, and J. Darnell. 1995. Molecular cell biology, ed. 2. New York, Scientific American Books, Inc. *Thorough treatment; begins with fundamentals such as energy, chemical reactions, bonds, pH, and biomolecules, then proceeds to advanced molecular biology.*

Rand, R. P. 1992. Raising water to new heights. Science **256:**618. *Cites some ways that water can affect function of protein molecules.*

Weinberg, R. A. 1985. The molecules of life. Sci. Am. **253:**48–57 (Oct.). *This entire issue of Scientific American is devoted to the modern findings of molecular biology.*

3

Origin of Life

That Mystery of Mysteries . . .

Cosmologists endeavor excitedly to understand the origins of the universe in toto and our own little solar system in particular. With backgrounds variously in physics, chemistry, astronomy, and mathematics, these scientists agree that the universe, including the planet Earth, is about 4.5 billion years old. Difficult as the question of cosmic origin is, more challenging still is the issue of the origin of life itself. The most intriguing of recent findings is the consistent picture of very early life on earth as revealed by the fossil record. The oldest fossils yet discovered are filamentous and sheath-enclosed colonial microorganisms from Western Australia that have been firmly dated radioactively at 3.3 to 3.5 billion years old. These are the earliest known life forms, among the first so far as we know to have closed themselves off from the rest of the universe as living, self-reproducing units. Responding

then to the forces of natural selection, protocells such as these began to evolve, eventually yielding the web of life on earth today. Throughout time, life and environment have evolved together, each deeply marking the other. The primitive earth, with its reducing atmosphere of ammonia, methane, and water, was superbly fit for the prebiotic synthesis that led to life's beginnings. Yet, it was totally unsuited, indeed lethal, for the kinds of organisms that inhabit the earth today, just as early forms of life could not survive in our present environment.

Charles Darwin pondered the origins of life and species, the "mystery of mysteries," as he called it. Since Darwin's day, we have learned more and more about the history of living forms that have evolved or are evolving on this earth. ■

All organisms, from humans to the smallest microbes that transcend the rather arbitrary boundary between life and nonlife, share a remarkable uniformity in cellular building blocks and cell function. Some of the similarities, such as the genetic code in DNA, the same 20 amino acids assembled to form all their proteins, and common metabolic pathways, were mentioned in Chapter 1. These conformities, along with other examples of molecular and functional identity, suggest that all life must have had a common beginning.

Although we acknowledge the kinship of living things, we must admit at the beginning that we do not know how life on earth originated. For many years the study of life's origins was not considered worthy of serious speculation by biologists because, it was argued, the absence of a geological record made the course of events resulting in the appearance of life unknowable. This situation has changed.

Since 1950 several laboratories around the world have been devoting full-time research to origin-of-life studies. It is a multidisciplinary effort that requires the contributions of scientists of several specialties: biologists, chemists, physicists, geologists, and astronomers. From such studies it has been possible to reconstruct a scenario of ancient events in which the earliest life form could have evolved more than 4 billion years BP (before the present) from inorganic constituents present on the surface of the earth. These studies are not attempts to prove or disprove any religious or philosophical belief, but rather they are endeavors to provide an intellectually satisfying account of how life on earth could have arisen by natural means.

HISTORICAL PERSPECTIVE

From ancient times it was commonly believed that life could arise by spontaneous generation from nonliving material, in addition to arising from parental organisms by reproduction (biogenesis). Frogs appeared to arise from damp earth, mice from putrefied matter, insects from dew, maggots from decaying meat, and so on. Warmth, moisture, sunlight, and even starlight were often mentioned as beneficial factors that encouraged spontaneous generation.

Among the accounts of early efforts to produce organisms spontaneously in the laboratory is a recipe for making mice, given by the Belgian plant nutritionist Jean Baptiste van Helmont (1648). "If you press a piece of underwear soiled with sweat together with some wheat in an open jar, after about 21 days the odor changes and the ferment . . . changes the wheat into mice. But what is more remarkable is that the mice which came out of the wheat and underwear were not small mice, not even miniature adults or aborted mice, but adult mice emerge!"

The first attack on the doctrine of spontaneous generation occurred in 1668 when the Italian physician Francesco Redi exposed meat in jars, some uncovered and some covered with parchment or wire gauze. The meat in all the vessels spoiled, but only the open vessels had maggots, and he noticed that flies were constantly entering and leaving these vessels. He concluded that, if flies had no access to the meat, no worms would be found.

Although Redi's refutation of spontaneous generation became widely known, the doctrine was too firmly entrenched to be abandoned. In 1748 the English Jesuit priest John T. Needham boiled mutton broth and put it in corked containers. After a few days the medium was swarming with microscopic organisms. He concluded that spontaneous generation was real, because he believed that he had killed all living organisms by boiling the broth and that he had excluded the access of others by sealing the tubes.

Figure 3-1
Louis Pasteur, who refuted the idea of spontaneous generation.

However, an Italian investigator, Abbé Lazzaro Spallanzani (1767), was critical of Needham's experiments and conducted experiments that dealt another blow against the notion of spontaneous generation. He thoroughly boiled extracts of vegetables and meat, placed these extracts in clean vessels, and sealed the necks of the flasks hermetically in flame. He then immersed the sealed flasks in boiling water for several minutes to make sure that all germs were destroyed. As controls, he left some tubes open to the air. At the end of 2 days, he found the open flasks swarming with organisms; the others contained none.

This experiment still did not settle the issue, for the advocates of spontaneous generation maintained either that air, which Spallanzani had excluded, was necessary for the production of new organisms or that the method he used destroyed the vegetative power of the medium. When oxygen was discovered (1774), opponents of Spallanzani seized on this as the vital element that he had destroyed in his experiments.

It remained for the great French scientist Louis Pasteur (Figure 3-1) to silence all but the most stubborn proponents of spontaneous generation with an elegant series of experiments with his famous "swan-neck" flasks.

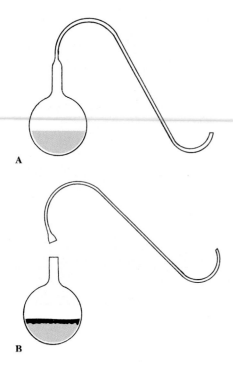

Figure 3-2
Louis Pasteur's swan-neck flask experiment.
A, Sugared yeast water boiled in a swan-neck
flask remains sterile until the neck is broken.
B, Within 48 hours, the flask is swarming with life.

Pasteur (1861) answered the objection to the lack of air by introducing fermentable material into a flask with a long S-shaped neck that was open to air (Figure 3-2). The flask and its contents were then boiled for a long time. Afterward the flask was cooled and left undisturbed. No fermentation occurred because all organisms that entered the open end were deposited on the floor of the neck and did not reach the flask contents. When the neck of the flask was cut off, the organisms in the air could fall directly on the fermentable mass and fermentation occurred within it in a short time. Pasteur concluded that, if suitable precautions were taken to keep out the germs and their reproductive elements, such as eggs and spores, no fermentation or putrefaction could take place.

Pasteur brought an end to the long and tenacious career of the concept of spontaneous generation. Pasteur's work showed that no living organisms come into existence except as descendants of similar organisms. In announcing his results before the French Academy, Pasteur proclaimed, "Never will the doctrine of spontaneous generation arise from this mortal blow." Paradoxically, in showing that spontaneous generation did not occur as previously claimed (production of mice, maggots, frogs, and others), Pasteur also ended for a time further inquiry into the spontaneous origins of life. A lengthy period of philosophical speculation followed, but virtually no experimentation on life's origins was performed for 60 years.

RENEWAL OF INQUIRY: OPARIN-HALDANE HYPOTHESIS

The rebirth of interest in the origins of life occurred in the 1920s. In that decade the Russian biochemist Alexander I. Oparin and the British biologist J. B. S. Haldane independently proposed that life originated on earth after an inconceivably long period of "abiogenic molecular evolution." Rather than arguing that the first living organisms miraculously originated all at once, a notion that had constrained fresh thinking for so long, Oparin and Haldane suggested that the simplest living units (for example, bacteria) came into being gradually by the progressive assembly of inorganic molecules into more complex organic molecules. These molecules would react with each other to form living microorganisms.

The hypotheses of Haldane and Oparin differed in detail, but both proposed that the earth's primitive atmosphere consisted of simple compounds such as water, carbon dioxide, methane, and ammonia, but lacked oxygen. When such a gas mixture is exposed to ultraviolet radiation, many organic substances such as sugars and amino acids are formed. Ultraviolet light must have been very intense on the primitive earth before the production of oxygen (by photosynthetic organisms), which reacted with ultraviolet rays to form ozone, a three-atom form of oxygen. Today ozone serves as a protective screen to prevent such intense ultraviolet radiation from reaching the earth's surface. Haldane believed that the early organic molecules could accumulate in the primitive oceans to form a "hot dilute soup." In this primordial broth carbohydrates, fats, proteins, and nucleic acids might have been assembled to form the earliest microorganisms.

The Oparin-Haldane hypothesis greatly influenced theoretical speculation on the origins of life during the 1930s and 1940s. Finally in 1953 Stanley Miller, working with Harold Urey in Chicago, made the first successful attempt to simulate with laboratory apparatus the conditions thought to prevail on the primitive earth. This experiment, described in more detail later, demonstrated that important biomolecules are formed in surprisingly large amounts when an electrical discharge is passed through a reducing atmosphere of the kind proposed by Haldane and Oparin. The realization that it was possible to simulate a prebiotic milieu in the laboratory ushered in a new era in origin-of-life studies. It coincided with the dawn of the space age and a new public interest in the question of life's origins.

PRIMITIVE EARTH

According to the big bang model, the universe originated from a primeval fireball and has been expanding and cooling since its inception 10 to 20 billion years ago. The sun and the planets formed approximately 4.6 billion years ago out of a spherical cloud of cosmic dust and gases that had some angular momentum. The cloud collapsed under the influence of its own gravity into a rotating disc. As the material in the central part of the disc condensed to form the sun, a substantial amount of gravitational energy was released as radiation. The pressure of this outwardly directed radiation prevented the complete collapse of the nebula into the sun. The material left behind began to cool and eventually gave rise to the planets (Figure 3-3).

Figure 3-3
Solar system showing narrow range of conditions suitable for life.

ORIGIN OF EARTH'S ATMOSPHERE

Primeval Atmosphere

Present evidence indicates that after the earth was formed, it was subjected to a period of heavy bombardment with large (100 km diameter) comets and meteorites. During this time, the heat generated by such impacts would have continuously or periodically vaporized the oceans. At the time heavy bombardment ended (about 3.8 billion years ago), we now believe the atmosphere was composed mostly of carbon dioxide and molecular nitrogen, with traces of carbon monoxide, molecular hydrogen, and reduced sulfur gases. It is generally agreed that the atmosphere at that time contained no more than a trace of oxygen, if any. Miller's experiments presume that the primeval atmosphere was strongly reducing, that is, it contained a predominance of molecules having less oxygen than hydrogen. Methane (CH_4) and ammonia (NH_3) are examples of fully reduced compounds. Current opinion, however, is that the atmosphere at the time of life's origin was only mildly reducing.

The character of the primeval atmosphere is important in any discussion of the origins of life, because the organic compounds of which living organisms are made are not stable if they are free in an oxidizing atmosphere. Organic compounds are not synthesized nonbiologically in our oxidizing atmosphere today, and they are not stable if they are introduced into it.

Appearance of Oxygen

Our atmosphere today is strongly oxidizing. It contains 78% molecular nitrogen, approximately 21% free oxygen, 1% argon, and 0.03% carbon dioxide. Although the time course for its development is much debated, at some point oxygen began to appear in significant amounts in the atmosphere.

The most important source of oxygen is photosynthesis. Almost all oxygen produced at the present time is produced by cyanobacteria (blue-green algae), eukaryotic algae, and plants. Each day these organisms combine approximately 400 million tons of carbon with 70 million tons of hydrogen to set free 1.1 billion tons of oxygen. Oceans are a major source of

oxygen. Almost all oxygen produced today is consumed by organisms for respiration; if this did not occur, the amount of oxygen in the atmosphere would double in approximately 3000 years. Because Precambrian fossil cyanobacteria resemble modern cyanobacteria, it is probable that most of the oxygen in the early atmosphere was produced by photosynthesis.

CHEMICAL EVOLUTION

SOURCES OF ENERGY

All of the current hypotheses for the origin of life require that a variety of carbon compounds be accumulated on earth during a period of prebiotic chemical evolution. If the simple gaseous compounds composing the early atmosphere, plus methane and ammonia, are mixed together in a closed glass system and allowed to stand at room temperature, they never chemically react with each other. To promote a chemical reaction, a continuous source of **free energy** sufficient to overcome reaction-activation barriers must be supplied (we discuss the concept of free energy on p. 63).

Table 3-1	Present Sources of Energy Averaged over the Earth
Source	**Energy (cal/cm²/year)**
Total radiation from sun	260,000
Infrared (above 700 nm)	143,000
Visible (350–700 nm)	113,600
Ultraviolet	
250–350 nm	2837
200–250 nm	522
150–200 nm	39
<150 nm	1.7
Electrical discharges	4
Shock waves	1.1
Radioactivity (to 1 km depth)	0.8
Volcanoes	0.13
Cosmic rays	0.0015

Source: Data from S. L. Miller and L. E. Orgel. *The origins of life on the earth.* Prentice-Hall, Inc., Englewood Cliffs, NJ, 1974.

The sun is by far the most powerful source of free energy for the earth. Each year each square centimeter of the earth receives an average 260,000 calories of radiant energy. The way in which this energy is distributed over the infrared, visible, and ultraviolet regions of the spectrum is shown in Table 3-1. Some investigators have suggested that life arose in vast clouds of water and dust particles surrounding the primitive earth. Such particles could have absorbed considerable energy in the visible and near infrared wavelengths. Since the solar energy available falls off rapidly in the ultraviolet region, only approximately 0.2% of the total energy is at wavelengths shorter than 200 nm,* where it can be absorbed by molecules such as methane, water, and ammonia. Nevertheless, ultraviolet radiation could have been an important source of energy for photochemical reactions in the primitive atmosphere.

Electrical discharges could have provided another source of energy for chemical evolution. Although the total amount of electrical energy released by lightning is small compared with solar energy, nearly all of the energy of lightning is effective in synthesizing organic compounds in a reducing atmosphere. A single flash of lightning through a reducing atmosphere generates a large amount of organic matter. Thunderstorms may have been one of the most important sources of energy for organic synthesis.

The widespread volcanic activity on the primitive earth also could have been a source of energy. One hypothesis maintains, for example, that life did not originate on the surface of the earth, but deep beneath the sea in or around **hydrothermal vents** (p. 797). Hydrothermal vents are submarine hot springs, in which seawater seeps through cracks in the bottom until the water comes close to hot magma. The water is superheated and expelled forcibly, carrying a variety of molecules it has dissolved out of the superheated rocks. These include hydrogen sulfide, methane, iron ions, and sulfide ions. Hydrothermal vents have been discovered in several locations beneath the deep sea, but they would have been much more widely prevalent on the early earth. If life originated there, they could have provided refuges while the comet and asteroid bombardment tapered off and the oceans were only partially evaporated. Interestingly, there are now many heat- and sulfur-loving bacteria that grow in hot springs.

PREBIOTIC SYNTHESIS OF SMALL ORGANIC MOLECULES

Earlier we referred to Stanley Miller's pioneering simulation of primitive earth conditions in the laboratory. Miller built an apparatus designed to circulate a mixture of methane, hydrogen, ammonia, and water past an electric spark (Figure 3-4). Water in the flask was boiled to produce steam that helped to circulate the gases. The products formed in the electrical discharge (representing lightning) were condensed in the condenser and collected in the U-tube and small flask (representing ocean).

After a week of continuous sparking, the water containing the products was analyzed. The results were surprising. Approximately 15% of the carbon that was originally in the reducing "atmosphere" had been converted into organic compounds that collected in the "ocean." The most striking finding was that many compounds related to life were synthesized. These included four amino acids commonly found in proteins, urea, and several simple fatty acids.

We can appreciate the astonishing nature of this synthesis when we consider that there are thousands of known organic compounds with structures no more complex than those of the amino acids formed. Yet in Miller's synthesis most of the relatively few substances formed were compounds found in living organisms. This was surely no coincidence, and it suggests that prebiotic synthesis on the primitive earth may have occurred under conditions that were not greatly different from those that Miller chose to simulate.

Miller's experiments have been criticized in light of current opinion that the early atmosphere of the earth was quite different from Miller's strongly reducing simulated atmosphere. Nevertheless, Miller's work

*Units of measurement commonly used in microscopic study are micrometers, nanometers, and angstroms: 1 micrometer (μm) = 0.000001 meter (about 1/25,000 inch); 1 nanometer (nm) = 0.000000001 meter; 1 angstrom (Å) = 0.0000000001 meter. Thus, 1 m = 10^3 mm = 10^6 μm = 10^9 nm = 10^{10} Å. Use of the angstrom is being discontinued.

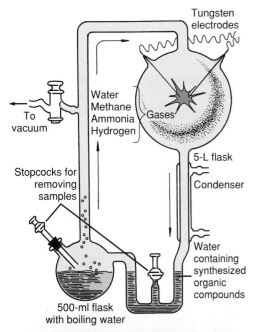

Figure 3-4

Dr. S. L. Miller with a replica of the apparatus used in his 1953 experiment on the synthesis of amino acids with an electric spark in a strongly reducing atmosphere.

stimulated many other investigators to repeat and extend his experiment. It was soon found that amino acids could be synthesized in many different kinds of gas mixtures that were heated (volcanic heat), irradiated with ultraviolet light (solar radiation), or subjected to electrical discharge (lightning). All that was required to produce amino acids was that the gas mixture be reducing and that it be subjected violently to some energy source. In other experiments, electrical discharges were passed through mixtures of carbon monoxide, nitrogen, and water, yielding amino acids and nitrogenous bases. Although reaction rates were much slower than in atmospheres containing methane and ammonia, and yields were poor in comparison, these experiments sup-

port the hypothesis that the chemical beginnings of life can occur in atmospheres that are only mildly reducing. The need for methane and ammonia, however, led to proposals that these substances might have been brought in by comets or meteorites, or that they were synthesized near the hydrothermal vents.

Thus the experiments of many scientists have shown that highly reactive intermediate molecules such as hydrogen cyanide, formaldehyde, and cyanoacetylene are formed when a reducing mixture of gases is subjected to a violent energy source. These react with water and ammonia or nitrogen to form more complex organic molecules, including amino acids, fatty acids, urea, aldehydes, sugars, and purine and pyrimidine bases—indeed all the building blocks required for the synthesis of the most complex organic compounds of living matter.

FORMATION OF POLYMERS

Need for Concentration

The next stage in chemical evolution involved the condensation of amino acids, purines, pyrimidines, and sugars to yield larger molecules that resulted in proteins and nucleic acids. Such condensations do not occur easily in dilute solutions, because the presence of excess water tends to drive reactions toward decomposition (hydrolysis). Although the primitive ocean might have been called a primordial soup, it was probably a rather dilute one containing organic material that was approximately one-tenth to one-third as concentrated as chicken bouillon.

Prebiotic synthesis must have occurred in restricted regions where concentrations were higher. Violent weather on the primitive earth would have created enormous dust storms, and impacts of meteorites would have lofted great amounts of dust into the atmosphere. The dust particles could have become foci of water droplets. Salt concentration in the particles could have been high and provided a

concentrated medium for chemical reactions. Alternatively, perhaps the surface of the earth was too warm to have oceans but not too hot for a damp surface. This would have resulted from constant rain and rapid evaporation. Thus the earth's surface could have become coated with organic molecules, an "incredible scum." Prebiotic molecules might have been concentrated by adsorption on the surface of clay and other minerals. Clay has the capacity to concentrate and condense large amounts of organic molecules. The surface of iron pyrite (FeS_2) has also been suggested as a site for the evolution of biochemical pathways. The positively charged surface of pyrite would attract a wide variety of negative ions, which would become bound to its surface. Further, there is an abundance of pyrite around hydrothermal vents, thus supporting the hydrothermal vent hypothesis.

Thermal Condensations

Most biological polymerizations are condensation (dehydration) reactions; that is, monomers are linked together by the removal of water (p. 23). In living systems condensation reactions always take place in an aqueous (cellular) environment in the presence of the appropriate enzymes. Without enzymes and energy supplied by ATP, the macromolecules (proteins and nucleic acids) of living systems soon break down into their constituent monomers.

One of the ways in which dehydration reactions could have occurred in primitive earth conditions without enzymes is by thermal condensation. The simplest dehydration is accomplished by driving off water from solids by direct heating. For example, if a mixture of all 20 amino acids is heated to 180° C, a good yield of polypeptides is obtained.

The thermal synthesis of polypeptides to form "proteinoids" has been studied extensively by the American scientist Sidney Fox. He showed that heating dry mixtures of amino acids

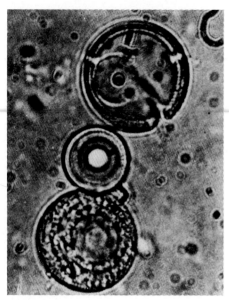

Figure 3-5
Electron micrograph of proteinoid microspheres. These proteinlike bodies can be produced in the laboratory from polyamino acids and may represent precellular forms. They have definite internal ultrastructure. (×1700)

and then mixing the resulting polymers with water will form small spherical bodies. These proteinoid microspheres (Figure 3-5) possess certain characteristics of living systems. Each is not more than 2 µm in diameter and is comparable in size and shape to spherical bacteria. Their outer walls appear to have a double layer, and they show osmotic and selective diffusion properties. They may grow by accretion or proliferate by budding like bacteria. There is no way to know whether proteinoids may have been the ancestors of the first cells or whether they are just interesting creations of the chemist's laboratory. They must be formed under conditions that would have been found only in volcanoes. Possibly organic polymers might have condensed on or in volcanoes and then, wetted by rain or dew, reacted further in solution to form polypeptides or polynucleotides.

ORIGIN OF LIVING SYSTEMS

We now have evidence from the fossil record that life existed by 3.8 billion years ago; therefore the origin of

the earliest life form can be estimated at 4 billion years BP. The first living organisms were protocells: autonomous membrane-bound units with a complex functional organization that permitted the essential activity of self-reproduction. The primitive chemical systems we have described lack this essential property. The principal problem in understanding the origin of life is explaining how primitive chemical systems could have become organized into living, autonomous, self-reproducing cells.

As we have seen, a lengthy chemical evolution on the primitive earth produced several molecular components of living forms. In a later stage of evolution, nucleic acids (DNA and RNA) began to behave as simple genetic systems that directed the synthesis of proteins, especially enzymes. However, this has led to a troublesome chicken-egg paradox: (1) How could nucleic acids have appeared without enzymes to synthesize them? (2) How could enzymes have evolved without nucleic acids to direct their synthesis? These questions are predicated on a long-accepted dogma that only proteins could act as enzymes. Startling evidence presented in the 1980s indicates that RNA in some instances has catalytic activity. Catalytic RNA (ribozymes) can mediate processing of messenger RNA, splicing out introns (p. 141), and can catalyze formation of peptide bonds. Evidence is strong that translation of mRNA by ribosomes (p. 142) is catalyzed by their RNA, not protein, content.

Therefore the earliest enzymes could have been RNA, and the earliest self-replicating molecules could have been RNA. Investigators are now calling this stage the "RNA world." Nevertheless, proteins have several important advantages over RNA as catalysts, and DNA is a more stable carrier of genetic information. The first protocells with protein enzymes and DNA would have had a powerful selective advantage over those with only RNA.

Once this stage of organization was reached, natural selection (p. 15) began acting on these primitive self-replicating systems. This was a critical

point. Before this stage, biogenesis was shaped by the favorable environmental conditions on the primitive earth and by the nature of the reacting elements themselves. When self-replicating systems became responsive to the forces of natural selection, they began to evolve. The more rapidly replicating and more successful systems were favored, and they replicated even faster. In short, the most efficient forms survived. From this evolved the genetic code and fully directed protein synthesis. The system now could be called a living organism, the common ancestor of all living organisms.

ORIGIN OF METABOLISM

Living cells today are organized systems that possess complex and highly ordered sequences of enzyme-mediated reactions. How did such vastly complex metabolic schemes develop?

Organisms that depend on nutrient molecules they have not synthesized for their food supplies are known as **heterotrophs** (Gr., *heteros,* another, + *trophos,* feeder), whereas organisms that can synthesize their food from inorganic sources using light or another source of energy are called **autotrophs** (Gr., *autos,* self, + *trophos,* feeder) (Figure 3-6). The earliest microorganisms are sometimes referred to as **primary heterotrophs** because they existed before there were any autotrophs. They were probably anaerobic, bacterium-like organisms similar to modern *Clostridium,* and they obtained all their nutrients directly from the environment. Chemical evolution had already supplied generous stores of nutrients in the prebiotic soup. There would be neither advantage nor need for the earliest organisms to synthesize their own compounds, as long as they were freely available from the environment.

Once the supply of a required compound was exhausted or became precarious, perhaps because of an increase in the numbers of organisms using it, those protocells able to convert a precursor to the required compound would have had a

Figure 3-6
Koala, a heterotroph, feeding on a eucalyptus tree, an autotroph. All heterotrophs depend for their nutrients directly or indirectly on autotrophs that capture the sun's energy to synthesize their own nutrients.

In photosynthesis water is the source of the hydrogen that is used to reduce carbon dioxide to sugars (that is, to add hydrogen) and molecular oxygen is liberated:

$$6\,CO_2 + 6\,H_2O \xrightarrow{\text{light}} C_6H_{12}O_6 + O_2$$

This equation summarizes the many reactions now known to take place in the process of photosynthesis. Undoubtedly these reactions did not appear all at once, and other reduced compounds, such as hydrogen sulfide (H_2S), were probably the early electron donors, rather than H_2O.

$$6\,CO_2 + 12\,H_2S \xrightarrow{\text{light}} C_6H_{12}O_6 + 12S + H_2O$$

As hydrogen sulfide and other reducing agents except water were used up, oxygen-evolving photosynthesis appeared. This may have occurred as early as 3.5 billion years BP. Gradually oxygen began to accumulate in the atmosphere. When atmospheric oxygen reached approximately 1% of its present level, ozone began to accumulate and ultraviolet radiation was screened out. Now land and surface waters could be occupied, and oxygen production probably increased sharply.

At this important juncture, accumulating atmospheric oxygen began to interfere with cellular metabolism, which up to this point had evolved under strictly reducing conditions. As the atmosphere slowly changed from a somewhat reducing to a highly oxidizing one, a new and highly efficient kind of metabolism appeared: **oxidative (aerobic) metabolism.** By using the available oxygen as a terminal electron acceptor (p. 72) and completely oxidizing glucose to carbon dioxide and water, much of the bond energy stored by photosynthesis could be recovered. Most living forms became completely dependent on oxidative metabolism.

tremendous advantage over those that lacked this capability. They would be selected for survival and would thrive. As the situation was repeated over and over again, long reaction sequences could develop.

We present here the traditional view that the first organisms were primary heterotrophs. Carl Woese finds it easier to visualize membrane-associated molecular aggregates that absorbed visible light and converted it with some efficiency into chemical energy. Thus the first organisms would have been autotrophs. Woese has also challenged the notion that metabolism has evolved step by step backward, suggesting instead that the earliest "metabolism" may have consisted of numerous chemical reactions catalyzed by nonprotein cofactors (substances necessary for the function of many of the protein enzymes in living cells). These cofactors would also have been associated with membranes.

An enzyme is normally required to catalyze each of these reactions.

So when we say that early protocells developed a reaction sequence as we have described—*A* made from *B*, *B* from *C*, and so on—we are really assuming that the appropriate enzymes appeared to catalyze these reactions. The numerous enzymes of cellular metabolism appeared when cells became able to utilize proteins for catalytic functions and thereby gained a selective advantage. No planning was required; the results were achieved through natural selection.

APPEARANCE OF PHOTOSYNTHESIS AND OXIDATIVE METABOLISM

Eventually, almost all usable energy-rich nutrients of the prebiotic soup were consumed. This ushered in the next stage of biochemical evolution: the use of readily available solar radiation to provide metabolic energy. It is easy to appreciate what an advantage the autotrophs had over the primary heterotrophs in an age of increasing scarcity of nutrient molecules.

PRECAMBRIAN LIFE

As depicted on the inside back cover of this book, the Precambrian period spanned the geological time before the beginning of the Cambrian period some 570 to 600 million years BP. At the beginning of the Cambrian period, most of the major phyla of invertebrate animals made their appearance within a few million years.

This has been called the "Cambrian explosion" because before this time, fossil deposits were rare and almost devoid of anything more complex than single-celled bacteria. We now recognize that the apparent rarity of Precambrian fossils was because they escaped notice owing to their microscopic size, although some soft-bodied macroscopic animals may have been present. What forms of life existed on earth before the burst of evolutionary activity in the early Cambrian world, and what organisms were responsible for the momentous change from a reducing to an oxidizing atmosphere?

PROKARYOTES AND THE AGE OF CYANOBACTERIA (BLUE-GREEN ALGAE)

The earliest bacterium-like organisms proliferated, giving rise to a great variety of bacterial forms, some of which were capable of photosynthesis. From these arose the oxygen-producing **cyanobacteria** some 3 billion years ago.

The name "algae" is misleading because it suggests a relationship to the eukaryotic algae, and many scientists prefer the alternative name "cyanobacteria" rather than "blue-green algae." These were the organisms responsible for producing oxygen initially released into the atmosphere. Study of the biochemical reactions in present cyanobacteria suggests that they evolved in a time of fluctuating oxygen concentration. For example, although they can tolerate atmospheric concentrations of oxygen (21%), the optimum concentration for many of their metabolic reactions is only 10%.

Bacteria are called **prokaryotes,** meaning literally "before the nucleus." They contain a single, large molecule of DNA not located in a membrane-bound nucleus, but found in a nuclear region, or **nucleoid.** The DNA is not complexed with histone proteins, and prokaryotes lack membranous organelles such as mitochondria, plastids, Golgi apparatus, and endoplasmic reticulum (Chapter 4). During cell division, the nucleoid divides and replicates of the cell's DNA are distributed to the daughter cells. Prokaryotes lack the chromosomal organization and chromosomal (mitotic) division seen in animals, fungi, and plants.

Bacteria and especially cyanobacteria ruled the earth's oceans unchallenged for some 1½ to 2 billion years. The cyanobacteria reached the zenith of their success approximately 1 billion years BP, when filamentous forms produced great floating mats on the ocean surface. This long period of cyanobacterial dominance, encompassing approximately two-thirds of the history of life, has been called with justification the "age of blue-green algae." Bacteria and cyanobacteria are so completely different from forms of life that evolved later that they have been placed in a separate kingdom, Monera.

More recently, however, Carl Woese and his colleagues at the University of Illinois have discovered that the prokaryotes actually comprise at least two distinct lines of descent: the Eubacteria ("true" bacteria) and the Archaebacteria. Although these two groups of bacteria look very much alike when viewed with the electron microscope, they are biochemically distinct. The cell walls of the archaebacteria do not contain muramic acid, as do those of all other bacteria, and there are fundamental differences in their metabolism. But the most compelling evidence for differentiating these two groups comes from the use of one of the newest and most powerful tools at the disposal of the evolutionist, the molecular sequencing technique (see note). Woese found that the sequence of bases in one kind of

RNA, ribosomal RNA, is sharply different from that of all other bacteria as well as from those found in the eukaryotes (see the following discussion). Woese believes that the archaebacteria are so distinctly different from the true bacteria that they should be considered as a separate kingdom, Archaebacteria. The Monera would then comprise only the true bacteria.

Molecular sequencing has emerged as a very successful approach to unraveling the genealogies of ancient forms of life. The sequences of nucleotides in the DNA of an organism's genes are a record of evolutionary relationship, because every gene that exists today is an evolved copy of a gene that existed millions, even billions, of years ago. Genes become altered by mutations through the course of time, but vestiges of the original gene usually persist. With modern techniques, one can determine the sequence of nucleotides in an entire molecule of DNA or in short segments of the molecule. When genes for the same function in two different organisms are compared, the extent to which they differ can be correlated with the time elapsed since the two organisms diverged from a common ancestor. Similar comparisons can be accomplished with some types of RNA and proteins.

Finally, Rivera and Lake (1992) have reported evidence that one group of archaebacteria, the eocytes, share a common ancestor with the eukaryotes (see following discussion). If this finding is supported by further work, the eocytes are a third prokaryote group and cannot be retained within the Archaebacteria.

APPEARANCE OF THE EUKARYOTES

The **eukaryotes** ("true nucleus"; Figure 3-7) have cells with membrane-bound nuclei containing **chromosomes** composed of **chromatin.** In

Figure 3-7

Comparison of prokaryotic and eukaryotic cells. The prokaryotic cell is about one-tenth the size of the eukaryotic cell.

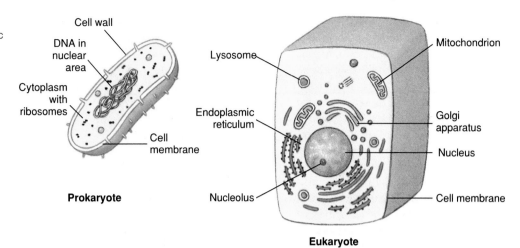

Prokaryote

Eukaryote

contrast to the prokaryote chromosome, constituents of chromatin include proteins called **histones** and RNA, in addition to the DNA. Some nonhistone proteins are found associated with both prokaryotic DNA and eukaryotic chromosomes. Eukaryotes are generally larger than prokaryotes, contain much more DNA, and usually divide by some form of mitosis. Within their cells are numerous membranous organelles, including mitochondria in which the enzymes for oxidative metabolism are packaged. Protistans (protozoa and algae), fungi, plants, and multicellular animals are composed of eukaryotic cells. There is fossil evidence that single-celled eukaryotes arose at least 1.5 million years ago (Figure 3-8).

Prokaryotes and eukaryotes are profoundly different from each other (Figure 3-7) and clearly represent a marked dichotomy in the evolution of life. The ascendancy of the eukaryotes resulted in a rapid decline in the dominance of cyanobacteria as the eukaryotes proliferated and fed on them.

Why were the eukaryotes so successful? Probably because they developed an important process facilitating rapid evolution: sex. Sex promotes genetic variability in populations by mixing the genes of each two individuals that mate. By preserving favorable genetic variants, natural selection encourages rapid evolutionary change (p. 167). Prokaryotes propagate effectively and efficiently, but their mechanisms for interchange of genes, which

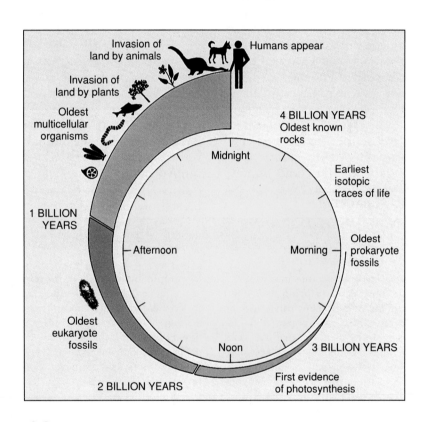

Figure 3-8

The clock of biological time. A billion seconds ago it was 1961, and most students using this text had not yet been born. A billion minutes ago the Roman empire was at its zenith. A billion hours ago Neanderthals were alive. A billion days ago the first bipedal hominids walked the earth. A billion months ago the dinosaurs were at the climax of their radiation. A billion years ago no creature had ever walked on the surface of the earth.

does occur in some cases, lack the systematic genetic recombination characteristic of sexual reproduction.

The organizational complexity of the eukaryotes is so much greater than that of the prokaryotes that it is difficult to visualize how a eukaryote could have arisen from any known prokaryote. The American biologist

Lynn Margulis and others have proposed that eukaryotes did not in fact arise from any single prokaryote but were derived from a symbiosis ("life together") of two or more types. Mitochondria and plastids, for example, each contain their own complement of DNA (apart from the nucleus of the cell), which has some prokaryote characteristics. Nuclei, plastids, and mitochondria all contain genes encoding ribosomal RNA, and comparisons of the sequence of bases of these genes show that the nuclear, plastid, and mitochondrial DNAs represent distinct evolutionary lineages. Plastids are closest evolutionarily to cyanobacteria, and mitochondria are closest to another group of bacteria (purple bacteria), consistent with the symbiotic hypothesis of eukaryotic origins. Mitochondria contain the enzymes of oxidative metabolism, and plastids (a plastid with chlorophyll is a chloroplast) carry out photosynthesis. It is easy to see how a host cell that was able to accommodate such guests in its cytoplasm would have had enormous competitive advantages.

Eukaryotes may have originated more than once. The first eukaryotes were undoubtedly unicellular, and many were photosynthetic autotrophs. Some of these lost their photosynthetic ability and became

In addition to maintaining that mitochondria and plastids originated as bacterial symbionts, Lynn Margulis argues that eukaryote flagella, cilia (locomotory structures), and even the spindle of mitosis came from a kind of bacterium like a spirochete. Indeed, she suggests that this association (the spirochete with its new host cell) was what made the evolution of mitosis possible. Margulis's evidence that the organelles are former partners of the ancestral cell is now accepted by most biologists.

heterotrophs, feeding on the autotrophs and the prokaryotes. As the cyanobacteria were cropped, their dense filamentous mats began to thin, providing space for other species. Carnivores appeared and fed on herbivores. Soon a balanced ecosystem of carnivores, herbivores, and primary producers appeared. This was ideal for evolutionary diversity. By freeing space, cropping herbivores encouraged a greater diversity of producers, which in turn promoted the evolution of new and more specialized croppers. An ecological pyramid developed with carnivores at the top.

The burst of evolutionary activity that followed at the end of the Precambrian period and beginning of the Cambrian period was unprecedented, and nothing approaching it has occurred since. Nearly all animal and plant phyla appeared and established themselves within a relatively brief period of a few million years. Some investigators hypothesize that the explanation for the "Cambrian explosion" lies in the accumulation of oxygen in the atmosphere up to a threshold level. Larger, multicellular animals required the increased efficiency of oxidative metabolism; these pathways could not be supported under conditions of limiting oxygen concentration.

Summary

There is a remarkable uniformity in the chemical constituents of living things and their cellular metabolism; this suggests that life on earth had a common origin. Historically, it was believed that life could arise anytime, spontaneously, as long as the conditions were right. Some 60 years after Louis Pasteur disproved spontaneous generation, A. I. Oparin and J. B. S. Haldane proposed a long "abiogenic molecular evolution" on earth in which organic molecules slowly accumulated in a "primordial soup." The atmosphere of the primitive earth was reducing, and little or no free oxygen was present. Ultraviolet radiation, electrical discharges of lightning, or energy from hydrothermal vents could have provided energy for the formation of organic molecules. Stanley Miller and Harold Urey showed the plausibility of the Oparin-Haldane hypothesis

by simple but ingenious experiments. Subsequent findings have shown that the atmosphere was probably only mildly reducing. Concentration necessary for condensation reactions to produce larger molecules might have been provided by damp surfaces, clay particles, iron pyrite, or other conditions. RNA may have been the primordial biomolecule, performing the functions of both genetic coding and catalysis. When self-replicating systems became responsive to the forces of natural selection, evolution proceeded more rapidly.

The first organisms were the primary heterotrophs, living on the energy stored in molecules dissolved in the primordial soup. As such molecules were used up, autotrophs had a great selective advantage. Molecular oxygen began to accumulate in the atmosphere as an end product of photosynthesis, and finally the atmosphere became oxidizing. The organisms responsible for generation of atmospheric oxygen were apparently cyanobacteria. All bacteria are prokaryotes, organisms that lack a membrane-bound nucleus and other organelles in their cytoplasm. The prokaryotes consist of two genetically distinct groups that some believe should be considered separate kingdoms, Archaebacteria and Monera.

The eukaryotes apparently arose from symbiotic unions of two or more types of prokaryotes. Eukaryotes have most of their genetic material (DNA) borne in a membrane-bound nucleus and have mitochondria and sometimes plastids. They include the eukaryotic algae, fungi, plants, and animals. Their evolutionary success results in great degree from the variability conferred by sexual reproduction.

Review Questions

1. In regard to the experiments of Louis Pasteur and Stanley Miller described in this chapter, explain what constituted the following in each case: observations, hypothesis, deduction, prediction, data, control. (The scientific method was described on p. 12.)
2. What was the composition of the earth's atmosphere at the time of the origin of life, and how did it differ from the atmosphere of today?

3. Name three different sources of energy that could have powered reactions on early earth to form organic compounds.
4. Explain the significance of the Miller-Urey experiments.
5. What are several mechanisms by which organic molecules in the prebiotic world could have been concentrated so that further reactions could occur?
6. Distinguish among the following: primary heterotroph, autotroph, secondary heterotroph.

7. What is the origin of the oxygen in the present-day atmosphere, and what is its metabolic significance to most organisms living today?
8. Distinguish between prokaryotes and eukaryotes as completely as you can.
9. Describe Margulis' view on the origin of eukaryotes from prokaryotes.
10. What was the "Cambrian explosion" and how might you account for it?

Selected References

Conway, Morris, S. 1993. The fossil record and the early evolution of the Metazoa. Nature **361**:219–225. *An important summary correlating fossil and molecular evidence.*

Gesteland, R. F., and J. F. Atkins, editors. 1993. The RNA world. Cold Spring Harbor, New York, Cold Spring Harbor Laboratory Press. *Evidence that there was a period when RNA served in both catalysis and transmission of genetic information.*

Kasting, J. F. 1993. Earth's early atmosphere. Science **259**:920–926. *Most investigators agree that there was little or no oxygen in the atmosphere of early earth and that there was a significant increase about 2 billion years ago.*

Knoll, A. H. 1991. End of the Proterozoic Eon. Sci. Am. **265**:64–73 (Oct.). *Multicellular animals probably originated only after oxygen in the atmosphere accumulated to a critical level.*

Margulis, L. 1993. Symbiosis in cell evolution, ed. 2. New York, W. H. Freeman. *An important updating of the author's 1981 book on this topic.*

Orgel, L. E. 1994. The origin of life on the earth. Sci. Am. **271**:77–83 (Oct.). *There is growing evidence for an RNA world, but difficult questions are unanswered.*

Rivera, M. C., and J. A. Lake. 1992. Evidence that eukaryotes and eocyte prokaryotes are immediate relatives. Science **257**:74–76. *A controversial paper indicating that eukaryotes may share a common ancestor with a prokaryote group classified among the Archaebacteria.*

Wainright, P. O., G. Hinkle, M. L. Sogin, and S. L. Stickel. 1993. Monophyletic origins of the Metazoa: an evolutionary link with fungi. Science **260**:940–942. *Molecular evidence that multicellular animals share*

a closer common ancestor with Fungi than with plants or other eukaryotes.

Waldrop, M. M. 1992. Finding RNA makes proteins gives 'RNA world' a big boost. Science **256**:1396–1397. *Reports on the significance of Noller et al. (1992, Science **256:**1416), who found that ribosomal RNA can catalyze the formation of peptide bonds and hence proteins.*

Woese, C. R. 1984. The origin of life. Carolina Biology Readers, no. 13. Burlington, North Carolina, Carolina Biological Supply Company. *Thought-provoking account of the author's views, critique of traditional concepts, focus on problems.*

4

The Cell as the Unit of Life

The Fabric of Life

It is a remarkable fact that living forms, from amebas and unicellular algae to whales and giant redwood trees, are formed from a single type of building unit: the cell. All animals and plants are composed of cells and cell products. Thus the cell theory is another of the great unifying concepts of biology.

New cells come from division of preexisting cells, and the activity of a multicellular organism as a whole is the sum of the activities of its constituent cells and their interactions. The energy to support virtually all of life's activities flows from sunlight that is captured by green plants and algae and transformed by photosynthesis into chemical bond energy. Chemical bond energy is a form of potential energy that can be released when the bond is broken; the energy is used to perform electrical, mechanical, and osmotic tasks in the cell. Ultimately, all energy is dissipated, little by little, into heat. This is in accord with the second law of thermodynamics, which states that there is a tendency in nature to proceed toward a state of greater molecular disorder, or entropy. Thus the high degree of molecular organization in living cells is attained and maintained only as long as energy fuels the organization. ∎

Cell Concept

More than 300 years ago the English scientist and inventor Robert Hooke, using a primitive compound microscope, observed boxlike cavities in slices of cork and leaves. He called these compartments "little boxes or cells." In the years that followed Hooke's first demonstration of the remarkable powers of the microscope before the Royal Society of London in 1663, biologists gradually began to realize that cells were far more than simple containers filled with "juices."

Cells are the fabric of life. Even the most primitive cells are enormously complex structures that form the basic units of all living matter. All tissues and organs are composed of cells. In a human an estimated 60 trillion cells interact, each performing its specialized role in an organized community. In single-celled organisms all the functions of life are performed within the confines of one microscopic package. There is no life without cells. The idea that the cell represents the basic structural and functional unit of life is an important unifying concept of biology.

With the exception of some eggs, which are the largest cells (in volume) known, cells are small and mostly invisible to the unaided eye. Consequently, our understanding of cells paralleled technical advances in the resolving power of microscopes. The Dutch microscopist A. van Leeuwenhoek sent letters to the Royal Society of London containing detailed descriptions of the numerous organisms he had observed using high-quality single lenses that he had made (1673 to 1723). In the early nineteenth century, the improved design of the microscope permitted biologists to see separate objects only one μm apart. This advance was quickly followed by new discoveries that laid the groundwork for the **cell theory**—a theory stating that all living organisms are composed of cells.

In 1838 Matthias Schleiden, a German botanist, announced that all plant tissue was composed of cells. A

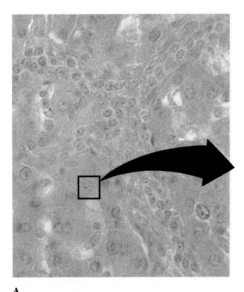
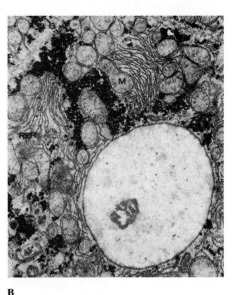

A   **B**

Figure 4-1

Liver cells. **A,** Magnified approximately 400 times through light microscopy. Note the prominently stained nucleus in each polyhedral cell. **B,** Portion of single liver cell, magnified approximately 5000 times by electron microscopy. A single large nucleus dominates the field; mitochondria **(M)**, rough endoplasmic reticulum **(RER)**, and glycogen granules **(G)**, are also seen.

year later one of his countrymen, Theodor Schwann, described animal cells as being similar to plant cells, an understanding that had been long delayed because the animal cell is bounded only by a nearly invisible plasma membrane rather than the distinct cell wall characteristic of the plant cell. Schleiden and Schwann are thus credited with the unifying cell theory that ushered in a new era of productive exploration in cell biology.

In 1840 J. Purkinje introduced the term **protoplasm** to describe the cell contents. Protoplasm was at first thought to be a granular, gel-like mixture with special and elusive life properties of its own; the cell was viewed as a bag of thick soup containing a nucleus. Later the interior of the cell became increasingly visible as microscopes were improved and better tissue-sectioning and staining techniques were introduced. Rather than being a uniform granular soup, the cell's interior is composed of numerous **cellular organelles,** each performing a specific function in the life of the cell. Today we realize that the components of a cell are so highly organized, structurally and functionally, that describing its contents as "proto-

plasm" is a bit like describing the contents of an automobile engine as "autoplasm."

How Cells Are Studied

The light microscope, with all its variations and modifications, has contributed more to biological investigation than any other instrument developed by humans. It has been a powerful exploratory tool for 300 years, and it continues to be so more than 50 years after the invention of the electron microscope. However, the electron microscope has vastly enhanced our understanding of the delicate internal organization of cells, and modern biochemical, immunological, physical, and molecular techniques have contributed enormously to our understanding of cell structure and function.

The electron microscope employs high voltages to direct a beam of electrons through the object examined. The wavelength of the electron beam is approximately 0.00001 that of ordinary white light, thus permitting far greater magnification and resolution (compare A and B of Figure 4-1). In preparation for viewing, specimens

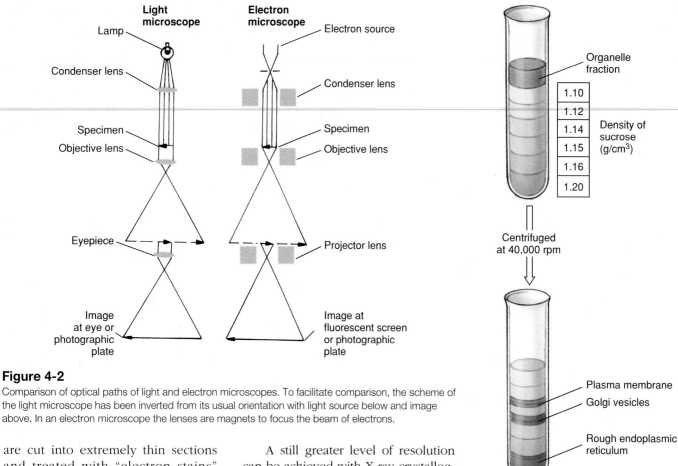

Figure 4-2

Comparison of optical paths of light and electron microscopes. To facilitate comparison, the scheme of the light microscope has been inverted from its usual orientation with light source below and image above. In an electron microscope the lenses are magnets to focus the beam of electrons.

Figure 4-3

Separation of cell organelles in a density gradient by ultracentrifugation. The gradient is formed by layering sucrose solutions in a centrifuge tube, then carefully placing a preparation of mixed organelles on top. The tube is centrifuged at about 40,000 revolutions per minute for several hours, and the organelles become separated down the tube according to their density.

are cut into extremely thin sections and treated with "electron stains" (ions of elements such as osmium, lead, and uranium) to increase contrast between different structures. Images are seen on a fluorescent screen and photographed (Figure 4-2). Because electrons pass through the specimen to the photographic plate, the instrument is called a transmission electron microscope.

In contrast, specimens prepared for the scanning electron microscope are not sectioned, and electrons do not pass through them. The whole specimen is bombarded with electrons, causing secondary electrons to be emitted. An apparent three-dimensional image is recorded in the photograph. Although the magnification capability of the scanning instrument is not as great as the transmission microscope, much has been learned about the surface features of organisms and cells. Examples of scanning electron micrographs are shown on pp. 105, 659, and 670.

A still greater level of resolution can be achieved with X-ray crystallography and nuclear magnetic resonance (NMR) spectroscopy. These techniques reveal a great deal about the shape of biomolecules and the relationship of the atoms within them to each other. Both techniques are laborious, but NMR spectroscopy does not require purification and crystallization of a substance, and the molecule can be observed in solution.

Advances in the techniques of cell study (cytology) are not limited to improvements in microscopes but include new methods of tissue preparation, staining for microscopic study, and the great contributions of modern biochemistry and molecular biology. For example, the various organelles of cells have differing, characteristic densities. Cells can be broken up with most of the organelles remaining intact, then centrifuged in a density gradient (Figure 4-3), and relatively pure preparations of each organelle may be recovered. Thus the biochemi-

cal functions of the various organelles may be studied separately. The DNA and various types of RNA can be extracted and studied. Many enzymes can be purified and their characteristics determined. The use of radioactive isotopes has allowed elucidation of many metabolic reactions and pathways in the cell. Modern chromatographic techniques can separate chemically similar intermediates and products. A particular protein in cells can be extracted and purified, and specific antibodies (see p. 675) against the protein can be prepared. When the antibody is complexed with a fluorescent substance and the complex is used to stain cells, the complex binds to the protein of interest, and its precise location in cells can be determined. Many more examples could be cited, and these have contributed enormously to our present understanding of cell structure and function.

ORGANIZATION OF CELLS

If we were to restrict our study of cells to fixed and sectioned tissues, we would be left with the erroneous impression that cells are static, quiescent, rigid structures. In fact, the cell interior is in a constant state of upheaval. Most cells are continually changing shape, pulsing, and heaving; their organelles twist and regroup in a cytoplasm teeming with starch granules, fat globules, and vesicles of various sorts. This description is derived from studies of living cell cultures with time-lapse photography and video. If we could see the swift shuttling of molecular traffic through gates in the cell membrane and the metabolic energy transformations within cell organelles, we would have an even stronger impression of internal turmoil. However, the cell is

anything but a bundle of disorganized activity. There is order and harmony in the cell's functioning that represents the elusive phenomenon we call life. Studying this dynamic miracle of evolution through the microscope, we realize that, as we gradually comprehend more and more about this unit of life and how it operates, we are gaining a greater understanding of the nature of life itself.

PROKARYOTIC AND EUKARYOTIC CELLS

We already described the radically different cell plan of prokaryotes and eukaryotes (p. 39). A fundamental distinction, expressed in their names, is that prokaryotes lack the membrane-bound nucleus present in all eukaryotic cells. Other major differences are summarized in Table 4-1.

Despite these differences, which are of paramount importance in cell studies, prokaryotes and eukaryotes have much in common. Both have DNA, use the same genetic code, and synthesize proteins. Many specific molecules such as ATP perform similar roles in both. These fundamental similarities imply common ancestry.

Prokaryotic organisms are bacteria separated into the kingdoms Archaebacteria and Monera. The most complex of these are the filamentous forms of cyanobacteria and certain other monerans. All other organisms are eukaryotes that, according to our present taxonomic system, are distributed among four kingdoms: the kingdom Protista (protozoa and eukaryotic algae), Plantae (green plants), Fungi (true fungi), and Animalia (multicellular animals). We will discuss the kingdom classifications in Chapter 11 (p. 208). The following discussion is restricted to eukaryotic cells, of which all animals are composed.

COMPONENTS OF EUKARYOTIC CELLS AND THEIR FUNCTIONS

Typically, the eukaryotic cell is enclosed within a thin, sturdy, selectively permeable **plasma membrane** (Figures 4-4 and 4-5). This structure regulates the flow of materials between the cell and its surroundings. In some cells, such as nerve cells, the plasma membrane also is involved in intercellular communication. In other cells, such as intestinal epithelium, the plasma membrane is modified into numerous, small, fingerlike projections

Table 4-1	Comparison of Prokaryotic and Eukaryotic Cells	
Characteristic	**Prokaryotic Cell**	**Eukaryotic Cell**
Cell size	Mostly small (1–10 μm)	Mostly large (10–100 μm)
Genetic system	DNA with some nonhistone protein; simple, circular DNA molecule in nucleoid; nucleoid is not membrane bound	DNA complexed with histone and nonhistone proteins in complex chromosomes within nucleus with membranous envelope
Cell division	Direct by binary fission or budding; no mitosis	Some form of mitosis; centrioles in many; mitotic spindle present
Sexual system	Absent in most; highly modified if present	Present in most; male and female partners; gametes that fuse
Nutrition	Absorption by most; photosynthesis by some	Absorption, ingestion, photosynthesis by some
Energy metabolism	No mitochondria; oxidative enzymes bound to cell membrane, not packaged separately; great variation in metabolic pattern	Mitochondria present; oxidative enzymes packaged therein; more unified pattern of oxidative metabolism
Intracellular movement	None	Cytoplasmic streaming, phagocytosis, pinocytosis
Flagella/cilia	Not with "9 + 2" microtubular pattern	With "9 + 2" microtubular pattern
Cell wall	Contains disaccharide chains cross-linked with peptides	If present, not with disaccharide polymers linked with peptides

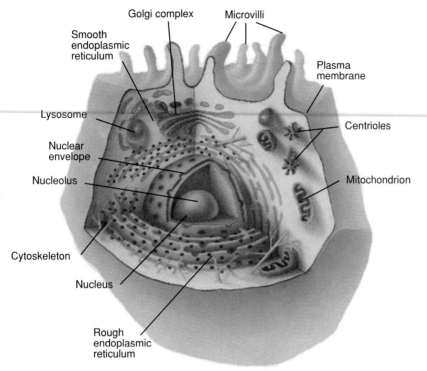

Figure 4-4
Generalized cell with principal organelles, as might be seen with the electron microscope. No single cell contains all these organelles, but many cells contain a large number of them.

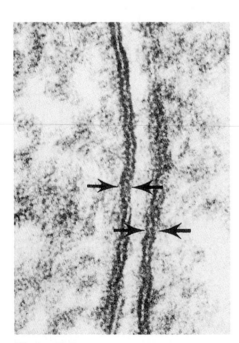

Figure 4-5
Plasma membranes of two adjacent cells. Each membrane (*between arrows*) shows a typical dark-light-dark staining pattern. (× 325,000)

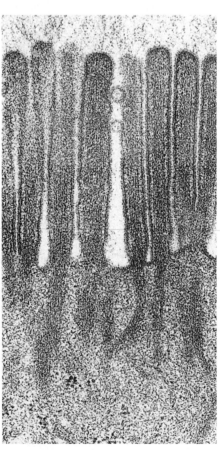

Figure 4-6
Electron micrograph of microvilli. (× 59,000)

Courtesy S. Ito.

called **microvilli** (sing., **microvillus**) that increase the surface area of the cell (Figure 4-6).

The most prominent organelle is the spherical or ovoid nucleus enclosed within *two* membranes to form a double-layered **nuclear envelope** (Figure 4-7). At intervals the nuclear envelope is perforated by pores, permitting some continuity between the nuclear contents and the cytoplasm surrounding the nucleus. The nucleus contains chromatin, one or more dense, granular structures called **nucleoli** (sing., **nucleolus**) (Figure 4-7), and a complicated network of protein strands, the **nuclear matrix.** The chromatin is a complex of DNA and histone and nonhistone protein, and it carries the genetic information of the cell. Nucleoli are specialized parts of certain chromosomes that carry multiple copies of the DNA information to synthesize ribosomal RNA. After transcription from the nucleolar DNA, the ribosomal RNA combines with several different proteins to form a **ribosome** (Figure 4-8), detaches from the nucleolus, and passes through nuclear pores to the cytoplasm. The nuclear matrix organizes the various components in the nucleus necessary for DNA and RNA synthesis.

The cytoplasm contains many organelles such as mitochondria, Golgi complexes, and centrioles. Plant cells typically contain **plastids,** the photosynthetic organelles, and bear a cell wall containing cellulose outside the plasma membrane. Animal cells lack plastids and a cell wall.

The space between the membranes comprising the nuclear envelope connects at some points with the space (channels, or **cisternae** [sing., **cisterna**]) within the membranes of the **endoplasmic reticulum (ER).** The ER (Figures 4-4 and 4-8) is a complex of membranes that separates some of the products of the cell from the synthetic machinery that produces them. The cisternae of the ER apparently function as routes for transport of certain substances within the cell. Often the membranes of the ER are lined on their outer surfaces with ribosomes and are thus designated

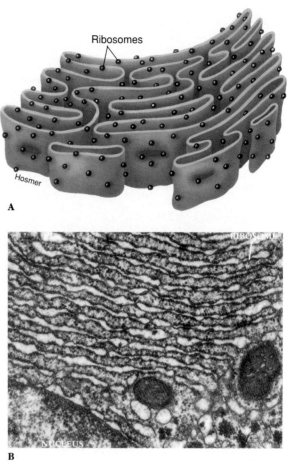

Ribosomes

Hosmer

A

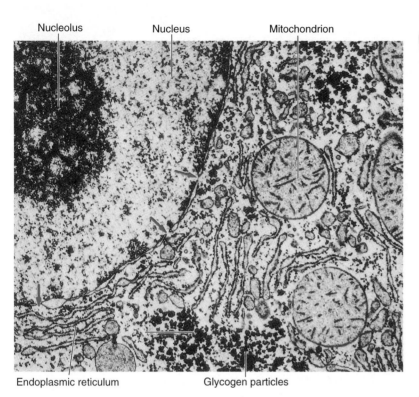

Nucleolus   Nucleus   Mitochondrion

Endoplasmic reticulum   Glycogen particles

Figure 4-7

Electron micrograph of part of hepatic cell of rat showing portion of nucleus (*left*) and surrounding cytoplasm. Endoplasmic reticulum and mitochondria are visible in cytoplasm, and pores (*arrows*) can be seen in nuclear envelope. (× 14,000)

B

Figure 4-8

Endoplasmic reticulum. **A,** Endoplasmic reticulum is continuous with the nuclear envelope. It may have associated ribosomes (rough endoplasmic reticulum) or not (smooth endoplasmic reticulum). **B,** Electron micrograph showing rough endoplasmic reticulum. (× 28,000)

B, Courtesy Richard Rodewald.

Figure 4-9

Golgi complex (=Golgi body, Golgi apparatus). **A,** The smooth cisternae of the Golgi complex have enzymes that modify proteins synthesized by the rough endoplasmic reticulum. **B,** Electron micrograph of a Golgi complex. (× 46,000)

B, Courtesy Charles Flickinger.

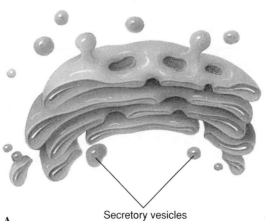

Secretory vesicles

A

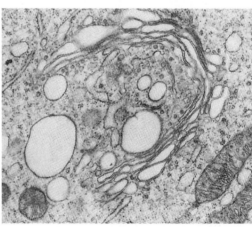

B

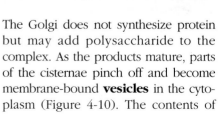

rough ER (contrasted with **smooth ER,** without ribosomes) (Figure 4-8). In some instances it has been shown that the protein synthesized by the ribosomes on the rough ER enters the cisternae and from there is transported to the Golgi apparatus or complex (Figure 4-9). The **Golgi complex** is a stack of smooth, membranous cisternae bearing enzymes that function in the storage, modification, and packaging of protein products, especially secretory products. The Golgi does not synthesize protein but may add polysaccharide to the complex. As the products mature, parts of the cisternae pinch off and become membrane-bound **vesicles** in the cytoplasm (Figure 4-10). The contents of

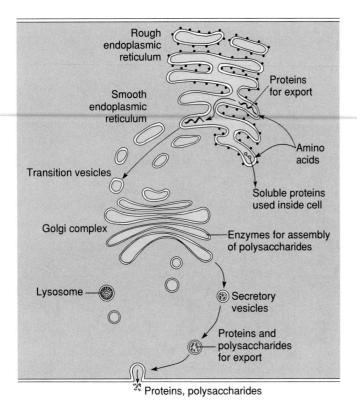

Figure 4-10

System for assembling, isolating, and secreting proteins for export in a eukaryotic cell.

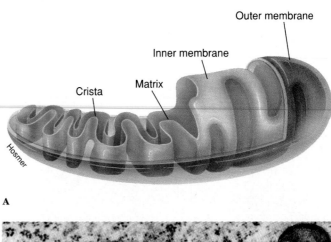

A

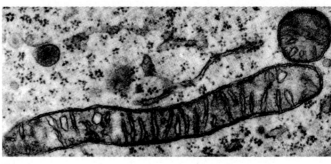

B

Figure 4-11

Mitochondria. **A,** Structure of a typical mitochondrion. **B,** Electron micrograph of mitochondria in cross and longitudinal section. (× 30,000)

B, Courtesy Charles Flickinger.

some of these vesicles may be expelled to the outside of the cell as secretory products destined to be exported from the glandular cell. Others may contain digestive enzymes that remain in the cell that produces them. Such vesicles are called **lysosomes** (literally, "releasing body," a body capable of causing lysis, or disintegration). The enzymes they contain are involved in the breakdown of foreign material, including bacteria engulfed by the cell. Lysosomes also are capable of breaking down injured or diseased cells and worn-out cellular components, since the enzymes they contain are so powerful that they kill the cell that formed them if the lysosome membrane ruptures. In normal cells the enzymes remain safely enclosed within the protective membrane.

Mitochondria (sing., **mitochondrion**) (Figure 4-11) are conspicuous organelles present in nearly all eukaryotic cells. They are diverse in size, number, and shape; some are rodlike, and others are more or less spherical. They may be scattered uniformly

through the cytoplasm, or they may be localized near cell surfaces and other regions where there is high metabolic activity. A mitochondrion is composed of a double membrane. The outer membrane is smooth, whereas the inner membrane is folded into numerous platelike or fingerlike projections called **cristae** (Figure 4-11). These characteristic features make mitochondria easy to identify among the organelles. Mitochondria are often called "powerhouses of the cell," because enzymes located on the cristae carry out the energy-yielding steps of aerobic metabolism. ATP (adenosine triphosphate), the most important energy-transfer molecule of all cells, is produced in this organelle. Mitochondria are self-replicating. They have a tiny, circular genome (complete set of all the genes), resembling those of prokaryotes but much smaller. The genome contains DNA that specifies some, but not all, of the proteins of the mitochondrion, plus the tRNAs and rRNAs (p. 142) that function in the mitochondrion.

Eukaryotic cells characteristically have a system of tubules and filaments that form the **cytoskeleton** (Figures 4-12 and 4-13). These provide support and maintain the form of the cell, and in many cells, they provide a means of locomotion and translocation of organelles within the cell. **Microfilaments** are thin, linear structures, first observed distinctly in muscle cells, where they are responsible for the ability of the cell to contract. They are made of a protein called **actin.** Several dozen other proteins are known that bind with actin and determine its configuration and behavior in particular cells. One of these is **myosin,** whose interaction with actin causes contraction in muscle and other cells (p. 643). Actin microfilaments also provide a means for moving messenger RNA (p. 142) from the nucleus to particular positions within the cell. **Microtubules,** somewhat larger than microfilaments, are tubular structures composed of a protein called **tubulin** (Figure 4-13). They play a vital role in moving the chromosomes toward the daughter cells during cell division as will be seen later, and they are important in intracellular architecture, organization, and transport. In addition, microtubules form essential parts of the structures of cilia and flagella. Microtubules radiate out from a microtubule organizing center near the nucleus, the **centrosome.** Within the centrosome are found a pair of **centrioles** (Figures 4-4 and 4-14), which are themselves composed of microtubules. Each centriole of a pair lies at right angles to the other and is a short cylinder of nine triplets of microtubules. They replicate before cell division. Although cells of higher plants do not have centrioles, a microtubule organizing center is present. **Intermediate filaments** are larger than microfilaments and smaller than microtubules. There are five biochemically distinct types of intermediate filaments, and their composition and arrangement depend on the cell type in which they are found.

SURFACES OF CELLS AND THEIR SPECIALIZATIONS

The free surface of epithelial cells (cells that cover the surface of a structure or line a tube or cavity) sometimes bears either **cilia** or **flagella** (sing., **cilium, flagellum**). These are motile extensions of the cell surface that sweep materials past the cell. In single-celled organisms and some small multicellular forms, they propel the entire organism through a liquid medium. Flagella provide the means of locomotion for the sperm of most animals and many plants.

Cilia and flagella have different beating patterns (see p. 640), but their internal structure is the same. With few exceptions, the internal structures of locomotory cilia and flagella are composed of a long cylinder of nine pairs of microtubules enclosing a central pair. At the base of each cilium or flagellum is a **basal body (kinetosome),** which is identical in structure to a centriole.

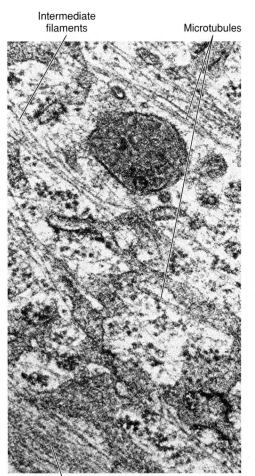

Intermediate filaments   Microtubules

Microfilaments

Figure 4-12
Cytoskeleton of a cell, showing its complex nature. Three visible cytoskeletal elements, in order of increasing diameter, are microfilaments, intermediate filaments, and microtubules. (× 66,600)

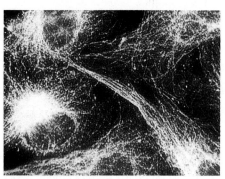

Figure 4-13
The microtubules in kidney cells of a baby hamster have been rendered visible by treatment with a preparation of fluorescent proteins that specifically bind to tubulin.

Figure 4-14
Centrioles. **A,** Each centriole is composed of nine triplets of microtubules arranged as a cylinder. **B,** Electron micrograph of a pair of centrioles, one in longitudinal (*right*) and one in cross section (*left*). The normal orientation of centrioles is at right angles to each other.

Microtubule triplet

A

B

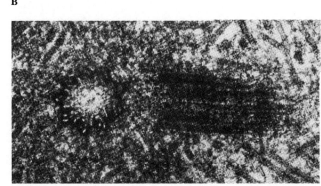

Indeed, cilia and flagella are so alike in the details of their structure that it seems highly likely that they had a common evolutionary origin. Whether their origin was the symbiosis of a spirochete-like bacterium and host cell, as suggested by Margulis, is more conjectural. Margulis and others prefer the term undulipodia to include both cilia and flagella, and it is less awkward to use one word for structures that are alike in structure and origin. However, the terms "cilia" and "flagella" are so common and widely used that the student should be familiar with them.

Many cells move neither by cilia nor flagella but by **ameboid movement** using **pseudopodia.** Some groups of protozoa (p. 216), migrating cells in embryos of multicellular animals, and some cells of adult multicellular animals, such as white blood cells, show ameboid movement. Cytoplasmic streaming through the action of actin microfilaments extends a lobe (pseudopodium) outward from the surface of the cell. Continued streaming in the direction of the pseudopodium brings the cytoplasmic organelles into the lobe and accomplishes movement of the entire cell. Some specialized pseudopodia have cores of microtubules (p. 218), and movement is effected by assembly and disassembly of the tubular rods.

Cells covering the surface of a structure or cells packed together in a tissue may have specialized junctional complexes between them (Figure 4-15). Nearest the free surface, the two apposing cell membranes appear to fuse, forming a **tight junction.** At various points small ellipsoid discs occur, just beneath the cell membrane in each cell. These discs are known as **desmosomes.** Specialized proteins extend through the cell membrane of each cell and project a **glycoprotein** (a protein with a sugar molecule attached) that binds with the glycoprotein of the other cell. Intermediate filaments are attached to the desmosome within each cell and

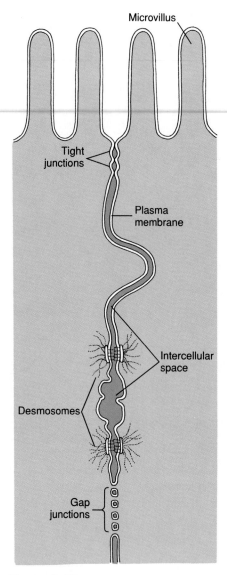

Figure 4-15

Two apposing plasma membranes forming the boundary between two epithelial cells. Various kinds of junctional complexes are found. The tight junction is a firm, adhesive band completely encircling the cell. Desmosomes are isolated "spot-welds" between cells. Gap junctions serve as sites of intercellular communication. Intercellular space may be greatly expanded in cells of some tissues.

extend into the cytoplasm. **Gap junctions,** rather than serving as points of attachment, provide a means of intercellular communication. They form tiny canals between cells so that the cytoplasm becomes continuous, and small molecules can pass from one cell to another.

Another specialization of the cell surface is the lacing together of adjacent cell surfaces where the plasma

membranes of the cells infold and interdigitate much like a zipper. They are especially common in the epithelium of kidney tubules. Microvilli, mentioned earlier, are another type of cell surface specialization (Figures 4-6 and 4-15). With the electron microscope they can be seen clearly in the lining of the intestine where they greatly increase the absorptive and digestive surface. Such specializations appear as brush borders when viewed with the light microscope.

MEMBRANE STRUCTURE AND FUNCTION

STRUCTURE OF THE CELL MEMBRANE

The incredibly thin, yet sturdy, plasma membrane that encloses every cell is vitally important in maintaining cellular integrity. Once believed to be a rather static entity that defined cell boundaries and kept cell contents from spilling out, the plasma membrane (also called the plasmalemma) is a dynamic structure having remarkable activity and selectivity. It is a permeability barrier that separates the interior from the external environment of the cell, regulates the vital flow of molecular traffic into and out of the cell, and provides many of the unique functional properties of specialized cells.

Membranes inside the cell surround a variety of organelles. Indeed, the cell is a system of membranes that divide it into numerous compartments. Someone has estimated that if all the membranes present in one g of liver tissue were spread out flat, they would cover 30 square meters! Internal membranes share many of the structural features of the plasma membrane and are the site for many, perhaps most, of the cell's enzymatic reactions.

The basic structure of all biological membranes is a phospholipid bilayer, which forms a matrix in which other molecules are attached. (The structure of phospholipids is given on p. 26.) One end of each molecule of phospholipid is water soluble

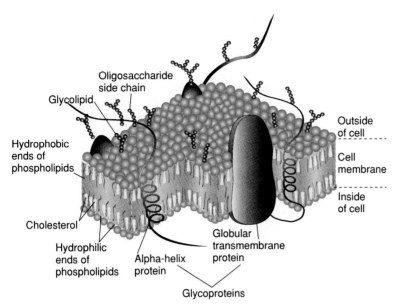

Figure 4-16
Diagram illustrating fluid-mosaic model of a cell membrane.

(hydrophilic, or "water loving"), whereas the other end, consisting of hydrocarbon chains of fatty acids that are insoluble in water, is hydrophobic ("water fearing"). In membranes the hydrophobic ends of the phospholipids in each layer point toward each other, and the hydrophilic ends are directed toward the water phase (Figure 4-16). On mixing phospholipids with water, the phospholipid molecules spontaneously arrange themselves in a bilayer, which is the state of lowest free energy for these molecules in water. Because ions and most biological molecules are water soluble, the hydrocarbon forms a barrier between the cell's interior and its environment. An important characteristic of the phospholipid bilayer is that it is a liquid. This gives the membrane flexibility and allows the phospholipid molecules to diffuse sideways freely within their own monolayer.

Another important lipid in eukaryotic membranes is cholesterol (see p. 26). There are about equal numbers of cholesterol and phospholipid molecules in a cell membrane. The cholesterol in the membrane makes it even less permeable to small molecules and decreases the flexibility. Cholesterol is synthesized in the endoplasmic reticulum, whose membranes, however, contain far less cholesterol than the plasma membrane.

Glycoproteins are essential components of plasma membranes. The glycoproteins fall into two broad categories according to their shape within the hydrocarbon region of the membrane (Figure 4-16). The hydrophilic, carbohydrate portions of both types are located on the extracellular ends of the protein. One type has a substantial globular portion within the hydrocarbon core. Some of these proteins catalyze the transport of substances such as negatively charged ions across the membrane (permeases, p. 53). The other type has a tightly coiled alpha-helix (p. 27) spanning the hydrophobic core of the membrane, and the hydrophobic side chains of the amino acids project outward from the helix. A hydrophilic end anchors the protein within the cytoplasm. Important functions of this type of protein include action as specific receptors for various molecules or as highly specific markings. For example, the self-nonself recognition that enables the immune system to react to invaders (Chapter 34) is based on proteins of this type.

Like the phospholipid molecules, most of the glycoproteins can move laterally in the membrane, although more slowly. This view of the membrane, in which a variety of protein molecules are embedded in a phospholipid bilayer, and in which they

can move about, is known as the **fluid-mosaic model** and is now widely accepted.

FUNCTION OF THE CELL MEMBRANE

The plasma membrane acts as a gatekeeper for the entrance and exit of the many substances involved in cell metabolism. Some substances can pass through with ease, others enter slowly and with difficulty, and still others cannot enter at all. This is called the **selective behavior** of the cell membrane. Because conditions outside the cell are different from and more variable than conditions within the cell, it is necessary that the passage of substances across the membrane be rigorously controlled.

We recognize three principal ways that a substance may traverse the cell membrane: (1) by **diffusion** along a concentration gradient; (2) by a **mediated transport system,** in which the substance binds to a specific site that in some way assists it across the membrane; and (3) by **endocytosis,** in which the substance is enclosed within a vesicle that forms on and detaches from the membrane surface to enter the cell.

Diffusion and Osmosis

If a living cell surrounded by a membrane is immersed in a solution having more solute molecules than the fluid inside the cell, a **concentration gradient** instantly exists between the two fluids. Assuming that the membrane is **permeable** to the solute, there is a net movement of solute toward the inside, the side having the lower concentration. The solute diffuses "downhill" across the membrane until its concentrations on each side are equal.

Most cell membranes are **selectively permeable,** that is, permeable to water but variably permeable or impermeable to solutes. In free diffusion it is this selectiveness that regulates molecular traffic. As a rule, gases (such as oxygen and carbon dioxide), urea, and lipid-soluble solutes (such as hydrocarbons and

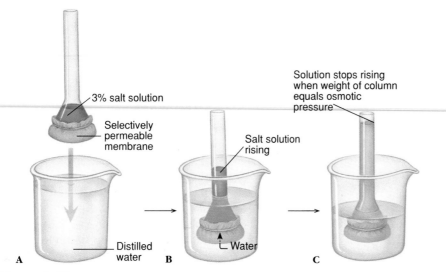

Figure 4-17

Simple membrane osmometer. **A,** The end of a tube containing a salt solution is closed at one end by a selectively permeable membrane. The membrane is permeable to water but not to salt. **B,** When the tube is immersed in pure water, water molecules diffuse through the membrane into the tube. Water molecules are in higher concentration in the beaker because they are diluted inside the tube by salt ions. Because the salt cannot diffuse out through the membrane, the volume of fluid inside the tube increases, and the level rises. **C,** When the weight of the column of water inside the tube exerts a downward force (hydrostatic pressure) causing water molecules to leave through the membrane in equal number to those that enter, the volume of fluid inside the tube stops rising. At this point the hydrostatic pressure is equivalent to the osmotic pressure.

alcohol) are the only solutes that can diffuse through biological membranes with any degree of freedom. Because many water-soluble molecules readily pass through membranes, such movements cannot be explained by simple diffusion. Sugars, as well as many electrolytes and macromolecules, are moved across membranes by carrier-mediated processes, which are described in the next section.

Diffusion of a solvent (usually water) across a selectively permeable membrane from a region of lower solute concentration to a region of higher solute concentration is **osmosis.** We can demonstrate osmosis by a simple experiment in which we tie a selectively permeable membrane such as cellophane tightly over the end of a funnel. We fill the funnel with a salt solution and place it in a beaker of pure water so that the water levels inside and outside the funnel are equal. In a short time the water level in the glass tube of the funnel rises, indicating a net movement of water through the cellophane membrane into the salt solution (Figure 4-17).

Inside the funnel are salt molecules, as well as water molecules. In the beaker outside the funnel are only water molecules. Thus the concentration of water is less on the inside because some of the available space is occupied by the larger, nondiffusible salt molecules. A concentration gradient exists for water molecules in the system. Water diffuses from the region of greater concentration of water (pure water outside) to the region of lesser concentration (salt solution inside).

As water enters the salt solution, the fluid level in the funnel rises. Gravity creates a hydrostatic pressure inside the osmometer. Eventually the pressure produced by the increasing weight of solution in the funnel pushes water molecules out as fast as they enter. The level in the funnel becomes stationary and the system is in equilibrium. The **osmotic pressure** of the solution is equivalent to the **hydrostatic pressure** necessary to prevent further net entry of water.

The concept of osmotic pressure is not without problems. A solution reveals an osmotic "pressure" only when it is separated from solvent by a selectively permeable membrane. It can be disconcerting to think of an isolated bottle of salt solution as having "pressure" much as compressed gas in a bottle (*hydrostatic* pressure) would have. Furthermore, the osmotic pressure is really the hydrostatic pressure that must be applied to a solution to keep it from gaining water *if* the solution were separated from pure water by a selectively permeable membrane. Consequently, biologists frequently use the term **osmotic potential** rather than osmotic pressure. However, since the term "osmotic pressure" is so firmly fixed in our vocabulary, it is necessary to understand the usage despite its potential confusion.

The *direct* measurement of osmotic pressure in biological solutions is seldom done today because the osmotic pressures of most biological solutions are so great that it would be impractical if not impossible to measure them with the simple membrane osmometer described. The osmotic pressure of human blood plasma would lift a fluid column more than 250 feet—if we could construct such a long, vertical tube and find a membrane that would not rupture from the pressure.

Indirect methods of measuring osmotic pressure are more practical. By far the most widely used measurement is **freezing point depression.** This is a much faster and more accurate determination than is the direct measurement of osmotic pressure. Pure water freezes at exactly 0° C. As solutes are added, the freezing point is lowered; the greater the concentration of solutes, the lower the freezing point. Human blood plasma freezes at approximately –0.56° C; seawater freezes at approximately –1.80° C. Although the lowering of the freezing point of water by the presence of solutes is small, great accuracy of measurement is possible because the instruments used by biologists can detect differences of as little as 0.001° C.

Mediated Transport

We have seen that the cell membrane is an effective barrier to the free diffusion of most molecules of biological significance. Yet it is essential that such materials enter and leave the cell. Nutrients such as sugars and materials for growth such as amino acids must enter the cell, and the wastes of metabolism must leave. Such molecules are moved across the membrane by special proteins called **transporters** or **permeases**. Permeases form a small passageway through the membrane, enabling the solute molecule to cross the phospholipid bilayer (Figure 4-18A). Permeases are usually quite specific, recognizing and transporting only a limited group of chemical substances or perhaps even a single substance.

At high concentrations of solute, mediated transport systems show a saturation effect. This means simply that the rate of influx reaches a plateau beyond which increasing the solute concentration has no further effect on influx rate (Figure 4-18B). This is evidence that the number of transporters available in the membrane is limited. When all become occupied by solutes, the rate of transport is at a maximum and it cannot be increased. Simple diffusion shows no such limitation; the greater the difference in solute concentrations on the two sides of the membrane, the faster the influx.

Two distinctly different kinds of mediated transport mechanisms are recognized: (1) **facilitated diffusion,** in which the permease assists a molecule to diffuse through the membrane that it cannot otherwise penetrate, and (2) **active transport,** in which energy is supplied to the system to transport molecules in the direction opposite a concentration gradient. Facilitated diffusion therefore differs from active transport in that it sponsors movement in a downhill direction (in the direction of the concentration gradient) only and requires no metabolic energy to drive the transport system.

In many animals facilitated diffusion aids in the transport of glucose (blood sugar) into body cells that oxidize it as a principal energy source for the synthesis of ATP. The concentration of glucose is greater in the blood than in the cells that consume it,

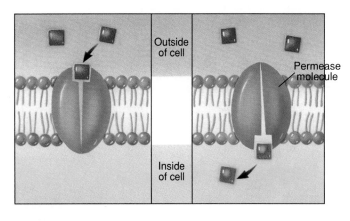

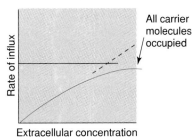

Figure 4-18

Facilitated transport. **A,** The permease molecule binds with a molecule to be transported (substrate) on one side of the plasma membrane, changes shape, and releases the molecule on the other side. Facilitated transport takes place in the direction of a concentration gradient. **B,** Rate of transport increases with increasing substrate concentration until all permease molecules are occupied.

favoring inward diffusion, but glucose is a water-soluble molecule that does not, by itself, penetrate the membrane rapidly enough to support the metabolism of many cells; the carrier system increases the inward flow of glucose.

In active transport, molecules are moved uphill against the forces of passive diffusion. Active transport always involves the expenditure of energy (from ATP) because materials are pumped against a concentration gradient. Among the most important active transport systems in all animals are those that maintain sodium and potassium ion gradients between cells and the surrounding extracellular fluid or external environment. Most animal cells require a high internal concentration of potassium ions for protein synthesis at the ribosome and for certain enzymatic functions. The potassium ion concentration may be 20 to 50 times greater inside the cell than outside. Sodium ions, on the other hand, may be 10 times more concentrated outside the cell than inside. Both of these ionic gradients are maintained by the active transport of potassium ions into and sodium ions out of the cell. In many cells one kind of molecule on one side of the membrane is transported simultaneously with another kind of molecule or ion

on the opposite side. The **sodium-potassium pump** that maintains the sodium and potassium ion gradients across nerve cell membranes is an example (Figure 4-19). There is evidence that 10% to 40% of all energy produced by some cells is used to power the sodium-potassium pump.

Endocytosis

Endocytosis, the ingestion of material by cells, is a collective term that describes three similar processes, **phagocytosis, potocytosis,** and **receptor-mediated endocytosis** (Figure 4-20). They are pathways for specifically internalizing solid particles, small molecules and ions, and macromolecules, respectively. All require energy and thus may be considered forms of active transport.

Phagocytosis, which literally means "cell eating," is a common method of feeding among the protozoa and lower metazoa. It is also the way in which white blood cells (leukocytes) engulf cellular debris and uninvited microbes in the blood. By phagocytosis, an area of the cell membrane, coated internally with actin-myosin, forms a pocket that engulfs the solid material. The membrane-enclosed vesicle then detaches from the cell surface and

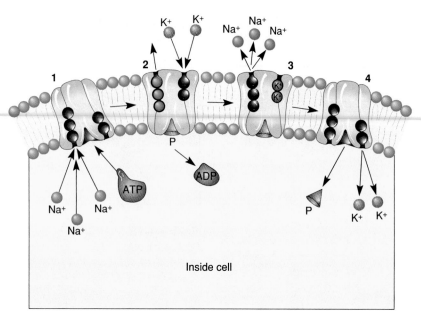

moves into the cytoplasm where its contents are digested by intracellular enzymes (Figure 34-7, p. 675).

Potocytosis is similar to phagocytosis except that small areas of the surface membrane are invaginated into cells to form tiny vesicles. The invaginated pits and vesicles are called **caveolae** (ka-vee'o-lee). Specific binding receptors for the molecule or ion to be internalized are concentrated on the cell surface of caveolae. Potocytosis apparently functions for intake of at least some vitamins, and similar mechanisms may be important in translocating substances from one side of a cell to the other (see "exocytosis," following) and internalizing signal molecules, such as some hormones or growth factors.

Figure 4-19

Sodium-potassium pump, powered by bond energy of ATP, maintains the normal gradients of these ions across the cell membrane. The pump works by a series of conformational changes in the permease: *Step 1.* Three ions of Na⁺ bind to the interior end of the permease, producing a conformational (shape) change in the protein complex. *Step 2.* The complex binds a molecule of ATP and cleaves it. *Step 3.* The binding of the phosphate group to the complex induces a second conformational change, passing the three Na⁺ ions across the membrane, where they are now positioned facing the exterior. This new conformation has a very low affinity for the Na⁺ ions, which dissociate and diffuse away, but it has a high affinity for K⁺ ions and binds two of them as soon as it is free of the Na⁺ ions. *Step 4.* Binding of the K⁺ ions leads to another conformational change in the complex, this time leading to dissociation of the bound phosphate. Freed of the phosphate, the complex reverts to its original conformation, with the two K⁺ ions exposed on the interior side of the membrane. This conformation has a low affinity for K⁺ ions so that they are now released, and the complex has the conformation it started with, having a high affinity for Na⁺ ions.

In phagocytosis, potocytosis, and receptor-mediated endocytosis some amount of extracellular fluid is necessarily trapped in the vesicle and nonspecifically brought within the cell. We describe this as *bulk-phase endocytosis,* and because it is nonspecific, the process corresponds roughly to what we have called traditionally *pinocytosis,* or "cell drinking." Actually, potocytosis also means "cell-drinking" but was coined to distinguish internalization of specific small molecules or ions.

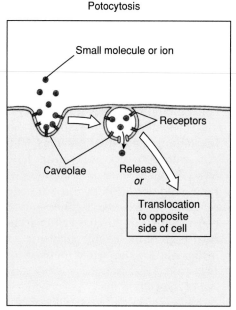

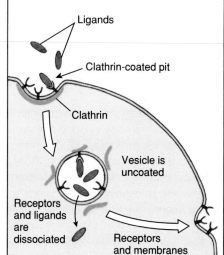

Figure 4-20

Three types of endocytosis. In phagocytosis the cell membrane binds to a large particle and extends to engulf it. In potocytosis small areas of cell membrane, bearing specific receptors for a small molecule or ion, invaginate to form caveolae. Receptor-mediated endocytosis is a mechanism for selective uptake of large molecules in clathrin-coated pits. Binding of the ligand to the receptor on the surface membrane stimulates invagination of pits.

Receptor-mediated endocytosis is a specific mechanism for bringing large molecules within the cell. Proteins of the plasma membrane specifically bind particular molecules (referred to as **ligands** in this process), which may be present in the extracellular fluid in very low concentrations. The invaginations of the cell surface that bear the receptors are coated within the cell with a protein called **clathrin,** hence they are described as **clathrin-coated pits.** As a clathrin-coated pit with its receptor-bound ligand invaginates and is brought within the cell, it is uncoated, the receptor and the ligand are dissociated, and the receptor and membrane material are recycled back to the surface membrane. Some important proteins and peptide hormones are brought into cells in this manner.

Exocytosis

Just as materials can be brought into the cell by invagination and formation of a vesicle, the membrane of a vesicle can fuse with the plasma membrane and extrude its contents to the surrounding medium. This is the process of **exocytosis.** This process occurs in various cells to remove undigestible residues of substances brought in by endocytosis, to secrete substances such as hormones (Figure 4-10), and to transport a substance completely across a cellular barrier, as we just mentioned. For example, a substance may be picked up on one side of the wall of a blood vessel by potocytosis, moved across the cell, and released by exocytosis.

MITOSIS AND CELL DIVISION

All cells arise from the division of preexisting cells. All the cells found in most multicellular organisms originated from the division of a single cell, the **zygote,** which is formed from the union (fertilization) of an **egg** and a **sperm.** Cell division provides the basis for one form of growth, for both

sexual and asexual reproduction, and for the transmission of hereditary qualities from one cell generation to another cell generation.

In the formation of **body cells** (somatic cells) the process of nuclear division is referred to as **mitosis.** By mitosis each "daughter cell" is ensured of receiving a complete set of genetic instructions. Mitosis is a delivery system for distributing the chromosomes and the DNA they contain to continuing cell generations. Thus all somatic cells, which number hundreds of billions in large animals, have the same genetic content, because all are descended by reproduction of the original fertilized egg. As an animal grows, its somatic cells differentiate and assume different functions and appearances because of differential gene action. Even though most of the genes in specialized cells remain silent and unexpressed throughout the lives of those cells, every cell possesses a complete genetic complement. Mitosis ensures equality of genetic potential; later, other processes direct the orderly expression of genes during embryonic development by selecting from the genetic instructions that each cell contains. (These fundamental properties of cells of multicellular organisms are discussed further in Chapter 7.)

In animals that reproduce **asexually,** mitosis is the only mechanism for the transfer of genetic information from parent to progeny. In animals that reproduce **sexually,** the parents must produce **sex cells** (gametes or germ cells) that contain only half the usual number of chromosomes, so that the offspring formed by the union of the gametes will not contain double the number of parental chromosomes. This requires a special type of *reductional* division called **meiosis,** described in Chapter 6 (p. 86).

CELL CYCLE

Cycles are conspicuous attributes of life. The descent of a species through time is in a very real sense a sequence of life cycles. Similarly, cells undergo

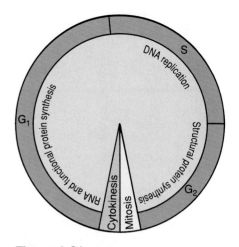

Figure 4-21

Cell cycle, showing relative duration of recognized periods. S, G_1, and G_2 are periods within interphase; S, synthesis of DNA; G_1, presynthetic period; G_2, postsynthetic period. Actual duration of the cycle and the different periods varies considerably in different cell types. After mitosis and cytokinesis the cell may go into an arrested, quiescent stage known as G_0.

cycles of growth and replication as they repeatedly divide. A cell cycle is a mitosis-to-mitosis cycle, that is, the interval between one cell generation and the next (Figure 4-21).

Actual nuclear division occupies only about 5% to 10% of the cell cycle; the rest of the cell's time is spent in **interphase,** the stage between nuclear divisions. For many years it was thought that interphase was a period of rest, because the nucleus appeared inactive when observed with the ordinary light microscope. In the early 1950s new techniques for revealing DNA replication in nuclei were introduced at the same time that biologists came to appreciate fully the significance of DNA as the genetic material. It was then discovered that DNA replication occurred during the interphase stage. Further studies revealed that many other protein and nucleic acid components essential to normal cell growth and division were synthesized during the seemingly quiescent interphase period.

Replication of DNA occurs during a phase called the S period (period of synthesis). In mammalian cells in tissue

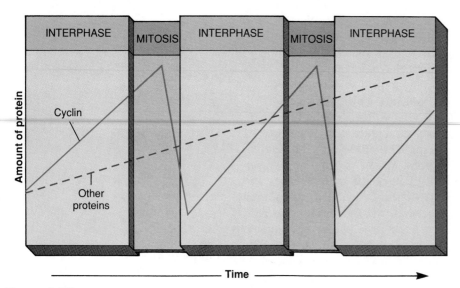

Figure 4-22

Variations in the level of cyclin in dividing cells of early sea urchin embryos. Cyclin binds with its cyclin-dependent kinase to activate the enzyme.

culture, the S period lasts about six of the 18 to 24 hours required to complete one cell cycle. In this period both strands of DNA must replicate; that is, new complementary partners are synthesized for both strands so that two identical molecules are produced from the original strand.

The S period is preceded and succeeded by G_1 and G_2 periods, respectively (G stands for "gap"), during which no DNA synthesis is occurring. For most cells, G_1 is an important preparatory stage for the replication of DNA that follows. During G_1, transfer RNA, ribosomes, messenger RNA, and several enzymes are synthesized. During G_2, spindle and aster proteins are synthesized in preparation for chromosome separation during mitosis. G_1 is typically of longer duration than G_2, although there is much variation in different cell types. Embryonic cells divide very rapidly because there is no cell growth between divisions, only subdivision of mass. DNA synthesis may proceed a hundred times more rapidly in embryonic cells than in adult cells, and the G_1 period is very shortened. As the organism develops, the cycle of most of its cells lengthens, and many cells may be arrested for long periods in a nonproliferative or quiescent phase called G_0. Neurons, for example, divide no further and are essentially in a permanent G_0.

Recent results have yielded much information on the exquisite regulation of events in the cell cycle. The transitions of the cell cycle are mediated by **cyclin-dependent kinases (cdk's)** and activating subunits of the cdk's called **cyclins.** Kinases are enzymes that add phosphate groups to other proteins to activate or inactivate them, and kinases themselves may require activation. The cdk's become active only when they are bound with the appropriate cyclin, and cyclins are synthesized and degraded during the cell cycle (Figure 4-22). The mechanisms involved in cdk regulation of cell cycles are mostly not known.

STRUCTURE OF CHROMOSOMES

As mentioned earlier, DNA in the eukaryotic cell occurs in chromatin, a complex of DNA with histone and nonhistone protein. The chromatin is organized into a number of discrete bodies called **chromosomes** (color bodies), so named because they stain deeply with certain biological dyes. In cells that are not dividing, the chromatin is loosely organized and dispersed, so that the individual chromosomes cannot be distinguished (Chapter 8, p. 125). Before division the chromatin condenses, and the chromosomes can be recognized and their individual morphological characteristics

determined. They are of varied lengths and shapes, some bent and some rod-like. Their number is constant for the species, and every body cell (but not the germ cells) has the same number of chromosomes regardless of the cell's function. A human, for example, has 46 chromosomes in each somatic cell.

During mitosis (nuclear division) the chromosomes shorten and become increasingly condensed and distinct, and each assumes a shape partly characterized by the position of a constriction, the **centromere** (Figure 4-23). The centromere is the location of the **kinetochore,** a disc of proteins specialized to bind with the microtubules of the spindle fibers during mitosis.

The problem of packaging the cell's DNA so that the genetic instructions are accessible during the transcription process is formidable. Transcription is the formation of messenger RNA from nuclear DNA (Chapter 8, p. 125).

PHASES IN MITOSIS

There are two distinct stages of cell division: the division of the nuclear chromosomes **(mitosis)** and the division of the cytoplasm **(cytokinesis).** Mitosis, the division of the nucleus (that is, chromosomal segregation), is certainly the most obvious and complex part of cell division and that of greatest interest to the cytologist. Cytokinesis normally immediately follows mitosis, although occasionally the nucleus may divide a number of times without a corresponding division of the cytoplasm. In such a case the resulting mass of protoplasm containing many nuclei is referred to as a **multinucleate cell.** An example is the giant resorptive cell type of bone (osteoclast), which may contain 15 to 20 nuclei. Sometimes a multinucleate mass is formed by cell fusion rather than nuclear proliferation. This arrangement is called a **syncytium.** An example is vertebrate skeletal muscle, which is composed of multinucleate fibers formed by the fusion of numerous embryonic cells.

The process of mitosis is divided into four successive stages or phases, although one stage merges into the next without sharp lines of transition. These phases are prophase, metaphase, anaphase, and telophase (Figure 4-24). When the cell is not actively dividing, it is in interphase. The DNA already has been replicating during the S phase of interphase.

Prophase

At the beginning of prophase, the centrosomes (along with their centrioles) replicate, and the two centrosomes migrate to opposite sides of the nucleus (Figure 4-24). At the same time, microtubules appear between the two centrosomes to form a football-shaped **spindle,** so named because of its resemblance to nineteenth-century wooden spindles, used to twist thread together in spinning. Other microtubules radiate outward from each centrosome to form **asters.**

At this time the diffuse nuclear chromatin condenses to form visible

Figure 4-23

Structure of a metaphase chromosome. The sister chromatids are still attached at their centromere. Each chromatid has a kinetochore, to which the kinetochore fibers are attached. Kinetochore microtubules from each chromatid run to one of the centrosomes, which are located at opposite poles.

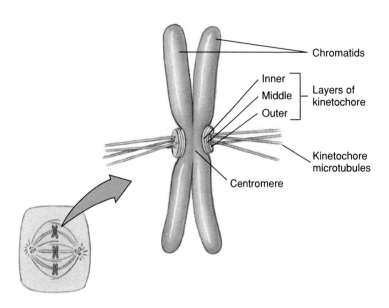

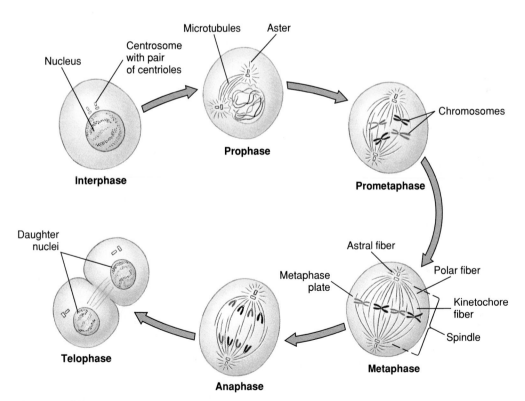

Figure 4-24

Stages of mitosis, showing division of a cell with two pairs of chromosomes. One chromosome of each pair is shown in red.

chromosomes. These actually consist of two identical sister **chromatids** formed during interphase. The sister chromatids are joined together at their centromere. Dynamic spindle fibers repeatedly extend and retract from the centrosome. When a fiber encounters a kinetochore, it binds to the kinetochore, ceases extending and retracting, and is now called a **kinetochore fiber.** It is as if the centrosome is sending out "feelers" to find the chromosome.

Microtubules are long, hollow, inelastic cylinders composed of the protein tubulin. Each tubulin molecule is actually a doublet composed of two globular proteins. The molecules are attached head-to-tail to form a strand, and 13 strands aggregate to form a microtubule. Because the tubulin subunits in a microtubule are always attached head-to-tail, the ends of the microtubule differ chemically and functionally. One end (called the plus end) both adds and deletes tubulin subunits more rapidly than the other end (the minus end). In a mitotic spindle, the plus ends of the kinetochore and polar fibers are away from the centrosome, and the minus ends are at the centrosome. The microtubule grows when the rate of adding subunits exceeds that of removing them, and it becomes shorter when the rate of removal exceeds that of addition.

Metaphase

Each centromere has two kinetochores, and each of the kinetochores is attached to one of the centrosomes by a kinetochore fiber. By a kind of tug-of-war during metaphase, the condensed sister chromatids are moved to the middle of the nuclear region to form a **metaphasic plate** (Figure 4-25). The centromeres line up precisely on the plate with the arms of the chromatids trailing off randomly in various directions.

Anaphase

The single centromere that has held the two chromatids together now splits so that two independent chromosomes, each with its own centromere, are formed. The chromosomes move toward their respective poles, pulled by the kinetochore fibers. This phase is often called **anaphase A.** The arms of each chromosome trail along behind as the microtubules shorten to drag the chromosomes along. Present evidence indicates that the force moving the chromosomes is disassembly of the tubulin subunits at the kinetochore end of the microtubules (see the boxed note).

As the chromosomes approach their respective centrosomes, the spindle lengthens, and the centrosomes move farther apart. This is **anaphase B.** The mechanism of this movement appears to involve the interdigitating free ends of the polar fibers. The tubulin in these microtubules has other protein molecules associated with it that serve as "motor molecules." These motor molecules interact with the adjacent fiber (or motor molecules on the adjacent fiber) and push the two halves of the spindle away from each other.

Telophase

When the daughter chromosomes reach their respective poles, telophase has begun. The daughter chromosomes are crowded together and stain intensely with histological stains. The spindle fibers disappear and the chromosomes lose their identity, reverting to the diffuse chromatin network characteristic of the interphase nucleus. Finally, the nuclear membranes reappear around the two daughter nuclei.

CYTOKINESIS: CYTOPLASMIC DIVISION

During the final stages of nuclear division a **cleavage furrow** appears on the surface of the dividing cell and encircles it at the midline of the spindle. The cleavage furrow deepens and pinches the plasma membrane as though it were being tightened by an invisible rubber band. Microfilaments of actin are present just beneath the surface in the furrow between the cells. Interaction with myosin, similar to that which occurs when muscle cells contract (p. 643), draws the furrow inward. Finally, the infolding edges of the plasma membrane meet and fuse, completing the cell division. As with other aspects of the cytoskeleton, such as the spindle, the centrosomes are responsible for locating and contracting microfilaments equidistant between them and at right angles to the spindle.

FLUX OF CELLS

Cell division is important for growth, for replacement of cells lost to natural attrition and wear and tear, and for wound healing. Cell division is especially rapid during the early development of the organism. At birth the human infant has about 2 trillion cells from repeated division of a single fertilized egg. This immense number could be attained by just 42 cell divisions, with each generation dividing once every six to seven days. With only five more cell divisions, the cell number would increase to approximately 60 trillion, the number of cells in a mature man weighing 75 kg. But of course no organism develops in this machinelike manner. Cell division is rapid during early embryonic development, then slows with age. Furthermore, different cell populations divide at widely different rates. In some the average period between divisions is measured in hours, whereas in others it is measured in days, months, or even years. Cells in the central nervous system stop dividing altogether after the early months of fetal development and persist without further division for the life of the individual. Muscle cells also stop dividing during the third month of fetal development, and future growth depends on enlargement of fibers already present.

In other tissues that are subject to wear and tear, lost cells must be constantly replaced. It has been estimated that in humans about 1% to 2% of all body cells—a total of 100 billion—are shed daily. Mechanical rubbing wears

Figure 4-25
Stages of mitosis in whitefish.

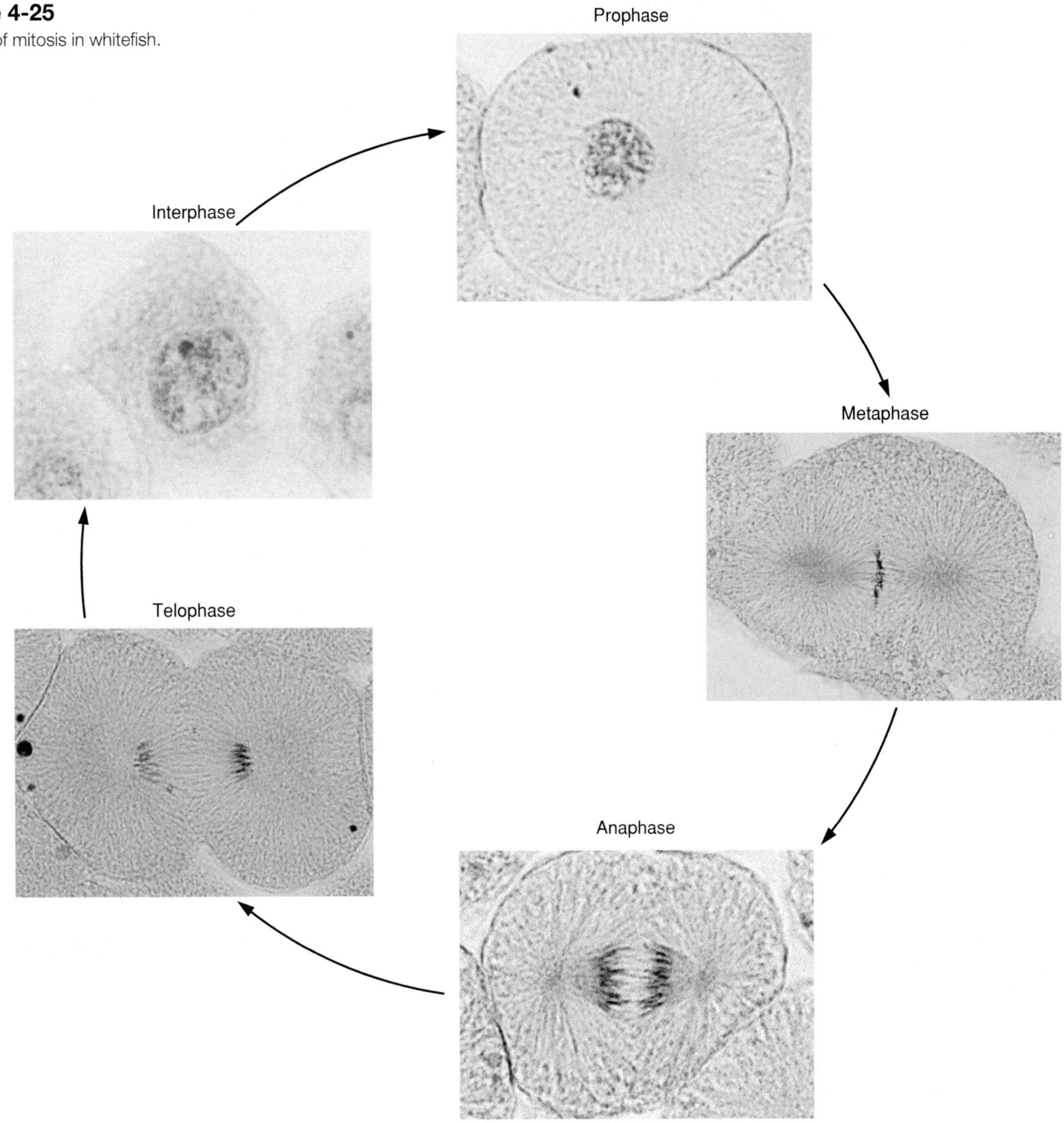

Interphase

Prophase

Metaphase

Telophase

Anaphase

away the outer cells of the skin, and food in the alimentary canal rubs off lining cells. The restricted life cycle of blood corpuscles involves enormous numbers of replacements, and during active sex life of males many millions of sperm are produced each day. Such losses of cells are made up by mitosis.

Normal development, however, does entail cell death in which the cells are not replaced. They may become senescent, accumulating damage from destructive oxidizing agents and eventually dying. Other cells undergo a programmed cell death, or **apoptosis** (Gr. *apo-,* from, away from; + *ptosis,* a falling), which is in many cases necessary for the continued health and development of the organism. For example, during embryonic development of vertebrates, excess nerve cells and immune cells that would attack the body's own tissues "commit suicide" in this manner. Apoptosis consists of a well-coordinated and predictable series of events: The cells round up and form bulges from the cytoplasm, the nuclear membrane and other organelles break down, and the DNA is broken up by enzymes.

Apoptosis currently is receiving a great deal of attention from researchers. One of the most valuable laboratory models is a tiny free-living nematode, *Caenorhabditis elegans* (see p. 308). The effects of apoptosis are not always beneficial to the organism. For example, an important disease mechanism in AIDS (acquired immune deficiency syndrome) seems to be an inappropriate triggering of programmed cell death among important cells of the immune system.

Summary

Cells are the basic structural and functional units of all living organisms. Eukaryotic cells differ from the prokaryotic cells of bacteria and archaebacteria in several respects, the most distinctive of which is the presence of a membrane-bound nucleus containing chromosomes that carry the hereditary material.

Cells are surrounded by a plasma membrane that regulates the flow of molecular traffic between the cell and its surroundings. The nucleus, enclosed by a double membrane, contains chromatin and one or more nucleoli. Outside the nuclear envelope is the cell cytoplasm, subdivided by a membranous network, the endoplasmic reticulum. Among the organelles within the cell are the Golgi complex, mitochondria, lysosomes, and other membrane-bound vesicles. The cytoskeleton is composed of microfilaments (actin), microtubules (tubulin), and intermediate filaments (several types). Cilia and flagella are hairlike, motile appendages that contain microtubules. Ameboid movement by pseudopodia operates by means of actin microfilaments. Tight junctions, desmosomes, and gap junctions are structurally and functionally distinct connections between cells.

Membranes in the cell are composed of a phospholipid bilayer and other materials including cholesterol and proteins. Hydrophilic ends of the phospholipid molecules are on the outer and inner surface of membranes, and the fatty acid portions are directed inward, toward each other, to form a hydrophobic core.

Substances can enter cells by diffusion, mediated transport, and endocytosis. Osmosis is diffusion of water through a selectively permeable membrane as a result of osmotic pressure. Solutes to which the membrane is impermeable require a transporter or permease molecule to traverse the membrane. Permease-mediated systems include facilitated diffusion (in the direction of a concentration gradient) and active transport (against a concentration gradient, which requires energy). Endocytosis includes bringing droplets (pinocytosis, potocytosis) or particles (phagocytosis) into the cell. In exocytosis the process of endocytosis is reversed.

Cell division in eukaryotes includes mitosis, the division of the nuclear chromosomes, and cytokinesis, the division of the cytoplasm. Mitosis itself is only a small part of the total cell cycle. In interphase, G_1, S, and G_2 periods are recognized, and the S period is the time when DNA is synthesized (the chromosomes are replicated).

The replicated chromosomes are each held together by a centromere. In prophase, the replicated chromosomes condense into recognizable bodies. A spindle forms between the centrosomes as they separate to opposite poles of the cell. At the end of prophase the nuclear envelope disintegrates, and the kinetochores of each chromosome become attached to both centrosomes by microtubules (kinetochore fibers). At metaphase the sister chromatids are moved to the center of the cell. At anaphase the centromeres divide, and one of each kind of chromosome is pulled toward the centrosome by the attached kinetochore fiber. At telophase the chromosomes gather in the position of the nucleus in each cell and revert to a diffuse chromatin network. The nuclear membrane reappears, and cytokinesis occurs.

Cells divide rapidly during embryonic development, then more slowly with age. Some cells continue to divide throughout the life of the animal to replace cells lost by attrition and wear, whereas others, such as nerve and muscle cells, complete their division during early development and never divide again. Some cells undergo a programmed cell death, or apoptosis.

Review Questions

1. Explain the difference (in principle) between a light microscope and an electron microscope.
2. Give a one-sentence definition of each of the following: plasma membrane, chromatin, nucleus, nucleolus, rough endoplasmic reticulum (rough ER), Golgi complex, lysosomes, mitochondria, microfilaments, microtubules, intermediate filaments, centrioles, basal body (kinetosome), tight junction, gap junction, desmosome, glycoprotein, microvilli.
3. Name two functions each for actin and for tubulin.
4. Distinguish between cilia, flagella, and pseudopodia.
5. What are the functions of each of the main constituents of the plasma membrane?
6. Our current concept of the plasma membrane is known as the fluid-mosaic model. Why?
7. You place some red blood cells in a solution and observe that they swell and burst. You place some cells in another solution, and they shrink and become wrinkled. Explain what has happened in each case.
8. Explain why a beaker containing a salt solution, placed on a table in your classroom, can have a high osmotic pressure, yet be subjected to a hydrostatic pressure of only one atmosphere.
9. The cell membrane is an effective barrier to molecular movement across it, yet many substances do enter and leave the cell. Explain the mechanisms through which this is accomplished and comment on the energy requirements of these mechanisms.
10. Distinguish between phagocytosis, potocytosis, receptor-mediated endocytosis, and exocytosis.
11. Define the following: chromosome, centromere, centrosome, kinetochore, mitosis, cytokinesis, syncytium.
12. Explain the phases of the cell cycle, and comment on important cellular processes that take place during each phase. What is G_0?
13. Name the stages of mitosis in order, and describe the behavior of the chromosomes at each stage.
14. Briefly describe ways that cells may die during the normal life of a multicellular organism.

Selected References

Anderson, R. G. W., B. A. Kamen, K. G. Rothberg, and S. W. Lacey. 1992. Potocytosis: sequestration and transport of small molecules by caveolae. Science **255:**410–413. *Describes the mechanism of cell internalization of small molecules.*

Bretscher, M. S. 1985. The molecules of the cell membrane. Sci. Am. **253:**100–108 (Oct.). *Good presentation of molecular structure of cell membranes, junctions, and mechanism of receptor-mediated endocytosis.*

Bretscher, M. S., and S. Munro. 1993. Cholesterol and the Golgi apparatus. Science **261:**1280–1281. *Good description of behavior of cholesterol in the cell and how it is concentrated in the plasma membrane by the Golgi.*

Dautry-Varsat, A., and H. F. Lodish. 1984. How receptors bring proteins and particles into cells. Sci. Am. **250:**52–58 (May). *Good coverage of receptor-mediated endocytosis.*

Glover, D. M., C. Gonzalez, and J. W. Raff. 1993. The centrosome. Sci. Am. **268:**62–68 (June). *The centrosome of animal cells serves as an organizing center for the cytoskeleton.*

Hartwell, L. H., and M. B. Kastan. 1994. Cell cycle control and cancer. Science **266:**1821–1828. *Genetic changes in the coordination of cyclin-dependent kinases, checkpoint controls, and repair pathways can lead to uncontrolled cell division.*

Lodish, H., D. Baltimore, A. Berk, S. L. Zipursky, P. Matsudira, and J. Darnell. 1995. Molecular biology, ed. 2. New York, Scientific American Books, W.H. Freeman & Company. *Up-to-date, thorough, and readable. Includes both cell biology and molecular biology. Advanced, but highly recommended.*

McIntosh, J. R., and K. L. McDonald. 1989. The mitotic spindle. Sci. Am. **261:**48–56 (Oct.). *Current knowledge and hypotheses on the function of the microtubules of mitosis.*

Murray, A., and T. Hunt. 1993. The cell cycle. An introduction. New York, Oxford University Press. *A good review of our present understanding of the cell cycle.*

Murray, A. W., and M. W. Kirschner. 1991. What controls the cell cycle. Sci. Am. **264:**56–63 (Mar.). *Presents the evidence for the fascinating role of cdc2 kinase and cyclin in the cell cycle.*

5

Physiology of the Cell

Deferring the Second Law

Living systems appear to contradict the second law of thermodynamics, which states that energy in the universe has direction and that it has been, and always will be, running down. In effect all forms of energy inevitably will be degraded to heat. This increase in disorder, or randomness, in any closed system is termed entropy. Living systems, however, *decrease* their entropy by *increasing* the molecular orderliness of their structure. Certainly an organism becomes vastly more complex during its development from fertilized egg to adult. The second law of thermodynamics, however, applies to closed systems, and living organisms are not closed systems. Animals grow and maintain themselves by borrowing free energy from the environment. When a deer feasts on the acorns and beechnuts of summer, it transfers potential energy, stored as chemical bond energy in the nuts' tissues, to its own body. Then, in step-by-step sequences called biochemical pathways, this energy is gradually released to fuel the deer's many activities. In effect, the deer decreases its own internal entropy by increasing the entropy of its food. The orderly structure of the deer is not permanent, however, but will be dissipated when it dies.

The ultimate source of this energy for the deer—and for almost all life on earth—is the sun. Sunlight is captured by green plants, which fortunately accumulate enough chemical bond energy to sustain both themselves and the animals that feed on them. Thus the second law is not violated, it is simply held at bay by life on earth, which uses the continuous flow of solar energy to maintain a biosphere of high internal order, at least for the period of time that life exists on earth. ■

In protozoa all life processes occur within the borders of a single cell. In metazoa they are distributed among groups of specialized cells that have specific functions, such as nervous system coordination, digestion, and excretion, that benefit the whole organism—a "division of labor" that has contributed to the great evolutionary diversity of multicellular animals. Nevertheless, certain activities are performed by every animal cell, whether a liver cell, a nerve cell, or a muscle cell. All cells must produce energy, synthesize their own internal structure, control much of their own activity, and guard their boundaries. In this chapter we consider these shared activities.

ENERGY AND THE LAWS OF THERMODYNAMICS

The concept of energy is fundamental to all life processes. We usually express energy as the capacity to do work, that is, to bring about change. Yet energy is a somewhat abstract quantity that is difficult to define and elusive to measure. Energy cannot be seen, it can only be defined and described by how it affects matter.

Energy can exist in either of two states: kinetic or potential. **Kinetic energy** is the energy of motion. **Potential energy** is stored energy, energy that is not doing work but has the capacity to do so. Energy can be transformed from one state to another.

Especially important for living organisms is chemical energy, a form of potential energy that is stored in chemical bonds of molecules. Chemical energy can be tapped when bonds are rearranged to release kinetic energy. Much of the work done by living organisms involves the conversion of potential energy to kinetic energy.

The conversion of one form of energy to another is governed by the two laws of thermodynamics. The **first law of thermodynamics** states that energy cannot be created or destroyed. It can change from one form to another, but the total amount of energy in a system remains the same. In short, energy is conserved. If we burn gasoline in an engine, we do not create new energy but merely convert the chemical energy in gasoline to another form, in this example, mechanical energy and heat. The **second law of thermodynamics,** introduced in the prologue to this chapter, concerns the transformation of energy. This fundamental law states that a closed system moves toward increasing disorder, or entropy, as energy is dissipated from the system (Figure 5-1). Living systems, however, are open systems that not only maintain their organization but also increase it, as during the development of an animal from egg to adult.

FREE ENERGY

To describe the energy changes that take place in chemical reactions,

biochemists use the concept of **free energy.** Free energy is simply the energy in a system available for doing work. In a molecule, free energy equals the energy present in chemical bonds minus the energy that cannot be used. The majority of reactions in cells release free energy and are said to be **exergonic** (Gr. *ex,* out; + *ergon,* work). Such reactions are spontaneous and always proceed "downhill" since free energy is lost from the system. Thus:

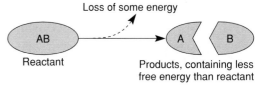

However, many important reactions in cells require the addition of free energy and are said to be **endergonic** (Gr. *endon,* within; + *ergon,* work). Such reactions have to be "pushed uphill" because they end up with more energy than they started with:

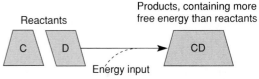

As we will see in a later section, ATP is the ubiquitous, energy-rich intermediate used by organisms to power important uphill reactions such as those required for active transport of molecules across membranes and cellular synthesis.

Figure 5-1

Diffusion of a solute through a solution, an example of entropy. When the solute (sugar molecules) is first introduced into a solution, the system is ordered and unstable **(B).** Without energy to maintain this order, the solute particles become distributed into solution, reaching a state of disorder (equilibrium) **(D).** Entropy has increased from left diagram to right diagram.

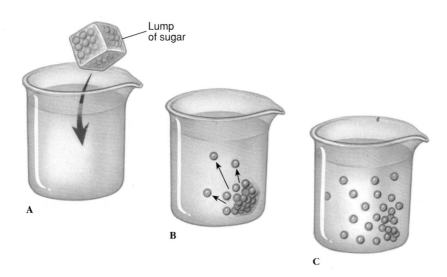

THE ROLE OF ENZYMES

ENZYMES AND ACTIVATION ENERGY

For any reaction to occur, even exergonic ones that tend to proceed spontaneously, the chemical bonds first must be destabilized. For example, if the reaction involves splitting a covalent bond, the atoms forming the bond must first be stretched apart to make them less stable. Some energy, termed the **activation energy,** must be supplied before the bond will be stressed enough to break. Only then will an overall loss of free energy and formation of reaction products occur. This requirement can be likened to the energy needed to push a cart over the crest of a hill before it will roll spontaneously down the other side, the cart liberating its potential energy as it descends.

One way to activate chemical reactants is to raise the temperature. By increasing the rate of molecular collisions and pushing chemical bonds apart, heat can impart the necessary activation energy to make a reaction proceed. However metabolic reactions must occur at biologically tolerable temperatures, temperatures too low to allow reactions to proceed beyond imperceptible rates. Instead, living systems have evolved a different strategy: they employ **catalysts.**

Catalysts are chemical substances that accelerate reaction rates without affecting the products of the reaction and without being altered or destroyed as a result of the reaction. A catalyst cannot make an energetically impossible reaction happen; it simply accelerates a reaction that would have proceeded at a very slow rate otherwise.

Enzymes are the catalysts of the living world. The special catalytic talent of an enzyme is its power to reduce the amount of activation energy required for a reaction. In effect, an enzyme steers the reaction through one or more intermediate steps, each of which requires much less activation energy than that required for a single-step reaction (Figure 5-2). Note

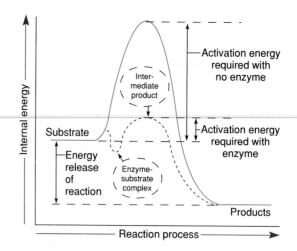

Figure 5-2

Energy changes during enzyme catalysis of a substrate. The overall reaction proceeds with a net release of energy (exergonic). In the absence of an enzyme, substrate is stable because of the large amount of activation energy needed to disrupt strong chemical bonds. The enzyme reduces the energy barrier by forming a chemical intermediate with a much lower internal energy state.

that enzymes do not supply the activation energy. Instead they lower the activation energy barrier, making a reaction more likely to proceed. Enzymes affect only the reaction rate. They do not in any way alter the free energy change of a reaction, nor do they change the proportions of reactants and products in a reaction.

NATURE OF ENZYMES

Enzymes are complex molecules that vary in size from small, simple proteins with a molecular weight of 10,000 to highly complex molecules with molecular weights up to 1 million. Many enzymes are pure proteins—delicately folded and interlinked chains of amino acids. Other enzymes require the participation of small nonprotein groups called **cofactors** to perform their enzymatic function. In some cases these cofactors are metallic ions (such as ions of iron, copper, zinc, magnesium, potassium, and calcium) that form a functional part of the enzyme. Examples are carbonic anhydrase, which contains zinc; the cytochromes, which contain iron; and troponin (a muscle contraction enzyme), which contains calcium. Another class of cofactors, called **coenzymes,** is organic. All coenzymes contain groups derived from vitamins, compounds that must be supplied in

the diet. All of the B complex vitamins are coenzymatic compounds. Since animals have lost the ability to synthesize the vitamin components of coenzymes, it is obvious that a vitamin deficiency can be serious. However, unlike dietary fuels and nutrients that must be replaced after they are burned or assembled into structural materials, vitamins are recovered in their original form and are used repeatedly. Examples of coenzymes that contain vitamins are nicotinamide adenine dinucleotide (NAD), which contains the vitamin nicotinic acid (niacin); coenzyme A, which contains the vitamin pantothenic acid; and flavin adenine dinucleotide (FAD), which contains riboflavin (vitamin B_2).

ACTION OF ENZYMES

An enzyme functions by associating in a highly specific way with its **substrate,** the molecule whose reaction it catalyzes. The enzyme bears an active site located within a cleft or pocket and contains a unique molecular configuration. The active site has a flexible surface that enfolds and conforms to the substrate (Figure 5-3). The binding of enzyme to substrate forms an **enzyme-substrate complex (ES complex),** in which the substrate is secured by covalent bonds to one or more points in the active site of the

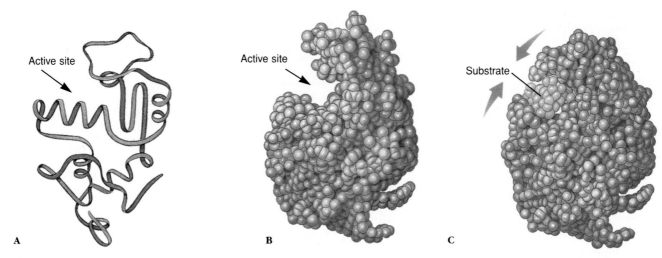

Figure 5-3

How an enzyme works. This space-filling model shows that the enzyme lysozyme bears a pocket containing the active site. When a chain of sugars (substrate) enters the pocket, the protein enzyme changes shape slightly so that the pocket enfolds the substrate and conforms to its shape. This positions the active site (an amino acid in the protein) next to a bond between adjacent sugars in the chain, causing the sugar chain to break.

enzyme. The ES complex is not strong and will quickly dissociate, but during this fleeting moment the enzyme provides a unique chemical environment that stresses certain chemical bonds in the substrate so that much less energy is required to complete the reaction.

How can biochemists be certain that an enzyme-substrate (ES) complex exists? The original evidence offered by Leonor Michaelis in 1913 is that, when the substrate concentration is increased while the enzyme concentration is held constant, the reaction rate reaches a maximum velocity. This *saturation effect* is interpreted to mean that all catalytic sites become filled at high substrate concentration. It is not seen in uncatalyzed reactions. Other evidence includes the observation that the ES complex displays unique spectroscopic characteristics not displayed by either the enzyme or the substrate alone. Furthermore, some ES complexes can be isolated in pure form, and at least one kind (nucleic acids and their polymerase enzymes) has been directly visualized with the electron microscope.

Enzymes that engage in important main-line sequences—such as the crucial energy-providing reactions of the cell that proceed constantly—seem to operate in sets rather than in isolation. For example, the conversion of glucose to carbon dioxide and water proceeds through 19 reactions, each requiring a specific enzyme. Main-line enzymes are found in relatively high concentrations in the cell, and they may implement quite complex and highly integrated enzymatic sequences. One enzyme carries out the first step, then passes the product to another enzyme that catalyzes another step, and this process continues until the end of the enzymatic pathway is reached. The reactions may be said to be coupled. Coupled reactions will be explained in a following section on chemical energy transfer by ATP.

SPECIFICITY OF ENZYMES

One of the most distinctive attributes of enzymes is their high specificity. This is a consequence of the exact molecular fit that is required between enzyme and substrate. Furthermore, an enzyme catalyzes only one reaction. Unlike reactions carried out in the organic chemist's laboratory, no

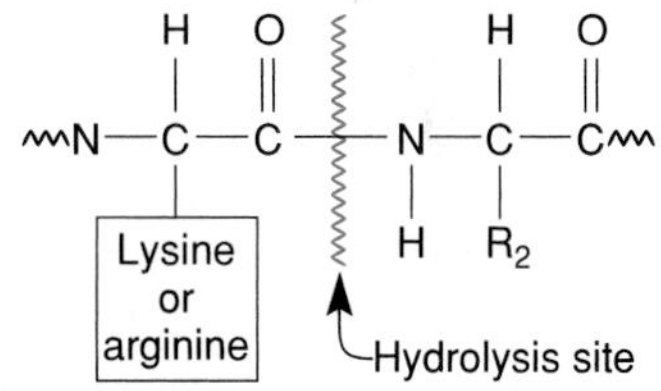

Figure 5-4

High specificity of trypsin. It splits only peptide bonds adjacent to lysine or arginine.

side reactions or by-products result. Specificity of both substrate and reaction is obviously essential to prevent a cell from being swamped with useless by-products.

However, there is some variation in degree of specificity. Some enzymes catalyze the oxidation (dehydrogenation) of only one substrate. For example, succinic dehydrogenase catalyzes the oxidation of succinic acid only. Others, such as proteases (for example, pepsin and trypsin), will act on almost any protein, although each protease has its particular point of attack in the protein (Figure 5-4). Usually an enzyme will take on one substrate molecule at a time, catalyze its chemical change, release the product, and then repeat the process with another substrate molecule. The enzyme may repeat this process billions of times until it is finally worn out (after a few hours to several years) and is broken down by scavenger enzymes in the cell. Some enzymes undergo successive catalytic cycles at speeds of up to a million cycles per minute, but most operate at slower rates.

ENZYME-CATALYZED REACTIONS

Enzyme-catalyzed reactions are reversible. This is signified by the double arrows between substrate and products. For example:

Fumaric acid + H_2O ⇌ Malic acid

However, for various reasons the reactions catalyzed by most enzymes tend to go predominantly in one direction. For example, the proteolytic enzyme pepsin degrades proteins into amino acids (a **catabolic** reaction), but it does not accelerate the rebuilding of amino acids into any significant amount of protein (an **anabolic** reaction). The same is true of most enzymes that catalyze the cleavage of large molecules such as nucleic acids, polysaccharides, lipids, and proteins. There is usually one set of reactions and enzymes that break them down (catabolism), but they must be resynthesized by a different set of reactions that are catalyzed by different enzymes (anabolism). This apparent irreversibility exists because the chemical equilibrium usually favors the formation of the smaller degradation products.

The building of complex molecules from simpler ones by living systems is called **anabolism** (Gr. *ana*, up; + *bole*, throw); the breaking down of complex molecules is called **catabolism** (Gr. *kata*, down; + *bole*, throw).

The net **direction** of any chemical reaction depends on the relative energy contents of the substances involved. If there is little change in the chemical bond energy of the substrate and the products, the reaction is more easily reversible. However, if large quantities of energy are released as the reaction proceeds in one direction,

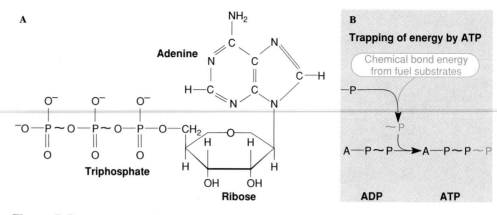

Figure 5-5
A, Structure of ATP. **B,** ATP formation from ADP.

more energy must be provided in some way to drive the reaction in the reverse direction. For this reason many if not most enzyme-catalyzed reactions are in practice irreversible unless the reaction is coupled to another that makes energy available. In the cell both reversible and irreversible reactions are combined in complex ways to make possible both synthesis and degradation.

Hydrolysis literally means "breaking with water." In hydrolysis reactions, a molecule is broken down by the addition of water. A hydrogen is attached to one subunit and a hydroxyl (—OH) unit is attached to another. This breaks the covalent bond between subunits. Hydrolysis is the opposite of condensation (water-losing) reactions in which the subunits of molecules are linked together by the removal of water. Macromolecules are built by condensation reactions.

CHEMICAL ENERGY TRANSFER BY ATP

We have seen that endergonic reactions are those that will not proceed spontaneously by themselves because the products require an input of free energy. However, an endergonic reaction may be driven by coupling the energy-requiring reaction with an energy-yielding reaction. ATP is the most common intermediate in **coupled**

reactions, and because it can drive such energetically unfavorable reactions, it is of central importance in metabolic processes.

The ATP molecule consists of adenosine (the purine adenine and the 5-carbon sugar ribose) and a triphosphate group (Figures 5-5 and 5-6). Most of the free energy in ATP resides in the triphosphate group, especially in two **phosphoanhydride bonds** between the three phosphate groups. These two bonds are called **high-energy bonds** because a great deal of free energy in the bonds is liberated when ATP is hydrolyzed to adenosine diphosphate (ADP) and inorganic phosphate.

$$ATP + H_2O \rightarrow ADP + P_i$$

where P_i represents inorganic phosphate (i = inorganic). The high-energy groups in ATP are designated by the "tilde" symbol ~. A high-energy phosphate bond is shown as ~P and a low-energy bond (such as the bond linking the triphosphate group to adenosine) as —P. ATP may be symbolized as A—P~P~P and ADP as A—P~P.

The way that ATP can act to drive a coupled reaction is shown in Figure 5-7. A coupled reaction is really a system involving two reactions linked by an energy shuttle (ATP). The conversion of substrate A to product A is endergonic because the product contains more free energy than the substrate. Therefore energy must be supplied by coupling the reaction to one that is exergonic, the conversion of substrate B to product B. Substrate B in this

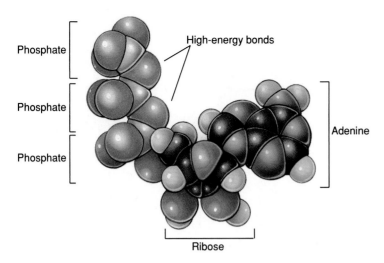

Figure 5-6

Space-filling model of ATP. In this model, carbon is shown in black; nitrogen in blue; oxygen in red; and phosphorus in yellow.

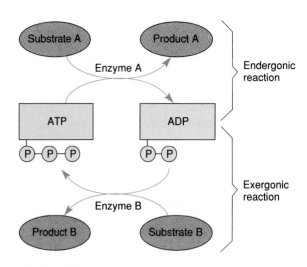

Figure 5-7

A coupled reaction. The endergonic conversion of substrate A to product A will not occur spontaneously but requires an input of energy from another reaction involving a large release of energy. ATP is the intermediate through which the energy is shuttled.

reaction is commonly called a **fuel** (for example, glucose or a lipid). The bond energy that is released in reaction B is transferred to ADP, which in turn is converted to ATP. ATP now contributes its phosphate-bond energy to reaction A, and ADP is produced again.

The high-energy bonds of ATP are actually rather weak, unstable bonds. Because they are unstable, the energy of ATP is readily released when ATP is hydrolyzed in cellular reactions. Note that ATP is an **energy-coupling agent** and *not* a fuel. It is not a storehouse of energy set aside for some future need. Rather it is produced by one set of reactions and is almost immediately consumed by another. ATP is formed as it is needed, primarily by oxidative processes in the mitochondria. Oxygen is not consumed unless ADP and phosphate molecules are available, and these do not become available until ATP is hydrolyzed by some energy-consuming process. *Metabolism is therefore mostly self-regulating.*

CELLULAR RESPIRATION

HOW ELECTRON TRANSPORT IS USED TO TRAP CHEMICAL BOND ENERGY

Having seen that ATP is the one common energy denominator by which all cellular machines are powered, we are in a position to ask how this energy is captured from fuel substrates. This question directs us to an important generalization: *all cells obtain their chemical energy requirements from oxidation-reduction reactions.* This means that in the degradation of fuel molecules, hydrogen atoms (electrons and protons) are passed from electron donors to electron acceptors with a release of energy. A portion of this energy is trapped and used to form the high-energy bonds in ATP.

Because they are so important, let us review what we mean by oxidation-reduction ("redox") reactions. In these reactions there is a transfer of electrons from an electron donor (the reducing agent) to an electron acceptor (the oxidizing agent). As soon as the electron donor loses its electrons, it becomes oxidized. As soon as the electron acceptor accepts electrons, it becomes reduced (Figure 5-8). In other words, a reducing agent becomes oxidized when it reduces another compound, and an oxidizing agent becomes reduced when it oxidizes another compound. Thus for every oxidation there must be a corresponding reduction.

In an oxidation-reduction reaction the electron donor and electron acceptor form a redox pair:

Electron donor $\rightleftharpoons$ e⁻ + Electron acceptor
+ Energy
(reducing agent;   (oxidizing agent;
becomes oxidized)   becomes reduced)

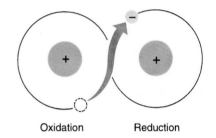

Oxidation   Reduction

Figure 5-8

A redox pair. The molecule at left is oxidized by the loss of an electron. The molecule at right is reduced by gaining an electron.

When electrons are accepted by the oxidizing agent, energy is liberated because the electrons move to a more stable position. In the cell, the electrons flow through a series of carriers. Each carrier is reduced by accepting electrons and then is reoxidized by passing electrons to the next carrier in the series. By transferring electrons stepwise in this manner, energy is gradually released, and a maximum yield of ATP is realized.

AEROBIC VERSUS ANAEROBIC METABOLISM

Ultimately, the electrons are transferred to a **final electron acceptor.** The nature of this final acceptor is the key that determines the overall efficiency of cellular metabolism. Heterotrophs (organisms that cannot synthesize their own food but must obtain nutrients from the environment, that is, animals, fungi, and most protists and bacteria) can be divided into two groups: **aerobes,** those which use molecular oxygen as the final electron acceptor, and **anaerobes,** those which employ some other molecule as the final electron acceptor.

As discussed in Chapter 3, life originated in the absence of oxygen, and the abundance of atmospheric oxygen was produced after photosynthetic organisms (autotrophs) evolved. Some strictly anaerobic organisms still exist and indeed play some important roles in specialized habitats. However, evolution has favored aerobic metabolism, not only because oxygen became available, but also because aerobic metabolism is vastly more efficient than anaerobic metabolism. In the absence of oxygen, only a very small fraction of the bond energy present in foodstuffs can be released. For example, when an anaerobic microorganism degrades glucose, the final electron acceptor (such as pyruvic acid) still contains most of the energy of the original glucose molecule. An aerobic organism on the other hand, using oxygen as the final electron acceptor, can completely oxidize glucose to carbon dioxide and water. Almost 20 times as much energy is released when glucose is completely oxidized as when it is degraded only to the stage of pyruvic acid. An obvious advantage of aerobic metabolism is that a much smaller quantity of foodstuffs is required to maintain a given rate of metabolism.

Overview of Respiration

Aerobic metabolism is more familiarly known as true **cellular respiration,** defined as the oxidation of fuel molecules with molecular oxygen as the final electron acceptor. As mentioned previously, the oxidation of fuel molecules describes the *removal of electrons* and *not* the direct combination of molecular oxygen with fuel molecules. Let us look at this process in general before considering it in more detail.

Hans Krebs, the British biochemist who contributed so much to our understanding of respiration, described three stages in the complete

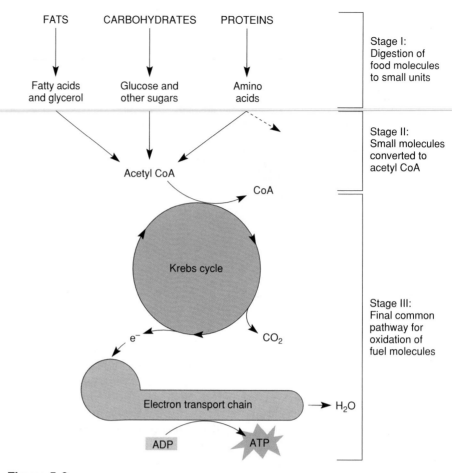

Figure 5-9
Overview of cellular respiration, showing the three stages in the complete oxidation of food molecules to carbon dioxide and water.

oxidation of fuel molecules to carbon dioxide and water (Figure 5-9). In stage I, foodstuffs passing through the intestinal tract are digested into small molecules that can be absorbed into the circulation. There is no useful energy yield during digestion, which is discussed in Chapter 35. In stage II, most of the degraded foodstuffs are converted into two 3-carbon units (pyruvic acid). This occurs in the cell cytoplasm. The pyruvic acid molecules then enter the mitochondria, where in another reaction they join with a coenzyme (coenzyme A) to form acetyl-CoA. Some ATP is generated in stage II, but the yield is small compared with that obtained in stage III of respiration. In stage III the final oxidation of fuel molecules occurs, with a large yield of ATP. This stage takes place entirely in the mitochondria. Acetyl coenzyme A is channeled into the Krebs cycle where the acetyl group is completely oxidized to car-

bon dioxide. Electrons released from the acetyl groups are transferred to special carriers that pass them to electron acceptor compounds in the electron transport chain. At the end of the chain the electrons (and the protons accompanying them) are accepted by molecular oxygen to form water.

Glycolysis

We begin our journey through the stages of respiration with glycolysis, a nearly universal pathway in living organisms that converts glucose into pyruvic acid. In a series of reactions, glucose and other 6-carbon monosaccharides are split into 3-carbon fragments, **pyruvic acid** (Figure 5-10). A single oxidation occurs during glycolysis, and each molecule of glucose yields two molecules of ATP. In this pathway the carbohydrate molecule is phosphorylated twice by ATP, first to glucose-6-phosphate (not shown in

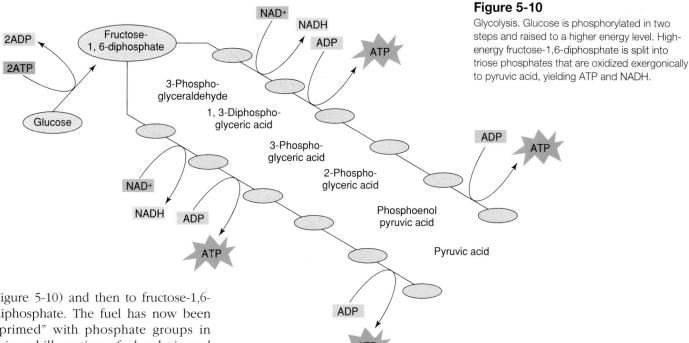

Figure 5-10
Glycolysis. Glucose is phosphorylated in two steps and raised to a higher energy level. High-energy fructose-1,6-diphosphate is split into triose phosphates that are oxidized exergonically to pyruvic acid, yielding ATP and NADH.

Figure 5-10) and then to fructose-1,6-diphosphate. The fuel has now been "primed" with phosphate groups in this uphill portion of glycolysis and is sufficiently reactive to enable subsequent reactions to proceed. This is a kind of deficit financing that is required for an ultimate energy return many times greater than the original energy investment.

The term "glycolysis" ("sweet releasing") was coined to describe the anaerobic conversion of glucose to lactic acid. However, many biochemists today use "glycolysis" to refer to the breakdown of sugar to pyruvic acid, the initial sequence of reactions leading into the Krebs cycle. This is the sense in which we use the term here. The anaerobic breakdown of sugar to lactic acid is referred to as anaerobic glycolysis. The distinction between "glycolysis" and "anaerobic glycolysis" is one of convenience, because the enzyme sequence from glucose to pyruvic acid is the same under both aerobic and anaerobic conditions.

In the downhill portion of glycolysis, fructose-1,6-diphosphate is cleaved into two 3-carbon sugars, which undergo an oxidation (electrons are removed), with the electrons and one of the hydrogen ions being accepted by nicotinamide adenine dinucleotide (NAD, a derivative of the vitamin niacin) to produce a reduced form called NADH. NADH serves as a carrier molecule to convey high-energy elec-

trons to the final electron transport chain, where ATP will be produced.

The two 3-carbon sugars next undergo a series of reactions, ending with the formation of two molecules of pyruvic acid (Figure 5-10). In two of these steps, a molecule of ATP is produced. In other words, each 3-carbon sugar yields two ATP molecules, and since there are two 3-carbon sugars, four ATP molecules are generated. Recalling that two ATP molecules were used to prime the glucose initially, the net yield up to this point is two ATP molecules. The 10 enzymatically catalyzed reactions in glycolysis can be summarized as:

$$\text{glucose} + 2ADP + 2P_i + 2NAD^+ \rightarrow$$
$$2 \text{ pyruvic acid} + 2NADH + 2ATP$$

ACETYL COENZYME A: STRATEGIC INTERMEDIATE IN RESPIRATION

In aerobic metabolism the two molecules of pyruvic acid formed during glycolysis enter the mitochondrion. There, each molecule is oxidized, and one of the carbons is released as carbon dioxide (Figure 5-11). The 2-carbon residue condenses with **coenzyme A** to form **acetyl coenzyme A** (acetyl-CoA).

Pyruvic acid is the undissociated form of the acid $CH_3—\overset{\overset{\text{O}}{\|}}{C}—COOH$. Under physiological conditions pyruvic acid typically dissociates into pyruvate $(CH_3—\overset{\overset{\text{O}}{\|}}{C}—COO^-)$ and H^+. It is correct to use either term in describing this and other organic acids (such as lactic acid, or lactate) in metabolism.

Acetyl coenzyme A is a critically important compound. Its oxidation in the Krebs cycle (following) provides energized electrons to generate ATP, and it is a crucial intermediate in lipid metabolism (p. 74).

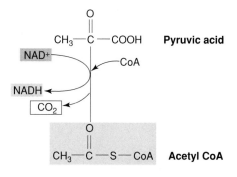

Figure 5-11
Formation of acetyl coenzyme A from pyruvic acid.

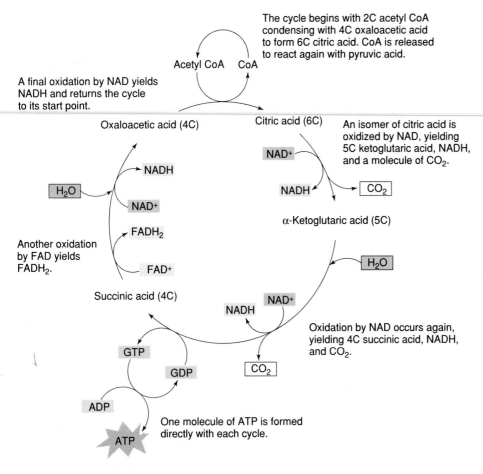

The cycle begins with 2C acetyl CoA condensing with 4C oxaloacetic acid to form 6C citric acid. CoA is released to react again with pyruvic acid.

A final oxidation by NAD yields NADH and returns the cycle to its start point.

An isomer of citric acid is oxidized by NAD, yielding 5C ketoglutaric acid, NADH, and a molecule of CO_2.

Another oxidation by FAD yields $FADH_2$.

Oxidation by NAD occurs again, yielding 4C succinic acid, NADH, and CO_2.

One molecule of ATP is formed directly with each cycle.

Figure 5-12
Krebs cycle in outline form, showing the production of three molecules of reduced NAD, one molecule of reduced FAD, one molecule of ATP, and two molecules of carbon dioxide. The molecules of NADH and $FADH_2$ will yield 11 molecules of ATP when oxidized in the electron transport system.

KREBS CYCLE: OXIDATION OF ACETYL COENZYME A

The degradation (oxidation) of the 2-carbon acetyl group of acetyl coenzyme A occurs in a cyclic sequence called the **Krebs cycle** (also called the citric acid cycle and the tricarboxylic acid cycle [TCA cycle]) (Figure 5-12). The acetyl coenzyme A condenses with a 4-carbon acid (oxaloacetic acid), releasing the coenzyme A to react again with pyruvic acid. Through a series of reactions the two carbons from the acetyl group are released as carbon dioxide, and the oxaloacetic acid is regenerated. Hydrogen ions and electrons in the oxidations are transferred to NAD and to FAD (flavine adenine dinucleotide, another electron acceptor), and a

pyrophosphate bond is generated in the form of guanosine triphosphate (GTP). This high-energy phosphate is readily transferred to ADP to form ATP. The overall products of the Krebs cycle are CO_2, ATP, NADH, and $FADH_2$:

acetyl unit + $3NAD^+$ + FAD + ADP
+ $P_i \rightarrow 2CO_2$ + 3NADH + $FADH_2$ + ATP

The molecules of NADH and $FADH_2$ formed will yield 11 molecules of ATP when oxidized by O_2 in the electron transport chain. The other molecules in the cycle behave as intermediate reactants and products which are continuously regenerated as the cycle turns.

ELECTRON TRANSPORT CHAIN

The transfer of hydrogen ions and electrons from reduced NAD and

FAD to the final electron acceptor, molecular oxygen, is accomplished in an elaborate electron transport chain embedded in the inner membrane of the mitochondria (Figure 5-13, see also p. 48). Each carrier molecule in the chain (labeled I to IV in Figure 5-13) is a large protein-based complex that accepts and releases electrons at lower energy levels than the carrier preceding it in the chain. As electrons pass from one carrier molecule to the next, free energy is released. Some of this energy is used to drive the synthesis of ATP by setting up a H^+ gradient across the mitochondrial membrane. At three points along the electron transport system, ATP production occurs by the phosphorylation of ADP. By this means, the oxidation of one NADH yields three ATP molecules. The reduced FAD from the Krebs cycle enters the electron transport chain at a lower level than NADH and so yields two ATP molecules. This method of energy capture is called **oxidative phosphorylation** because the formation of high-energy phosphate is coupled to oxygen consumption, and this depends on the demand for ATP by other metabolic activities within the cell.

How is ATP actually generated during oxidative phosphorylation? The most widely accepted explanation is a process called chemiosmotic coupling (Figure 5-13). According to this model, as the electrons contributed by NADH and $FADH_2$ are carried down the electron transport chain, they activate proton pumping channels which pump protons (hydrogen ions) outward and into the space between the two mitochondrial membranes. This causes the proton concentration outside to rise, producing a diffusion pressure that drives the protons back into the mitochondrion through special proton channels. These channels are ATP-forming protein complexes that use the inward passage of protons to induce the formation of ATP. Exactly how proton movement is coupled to ATP synthesis is not yet understood.

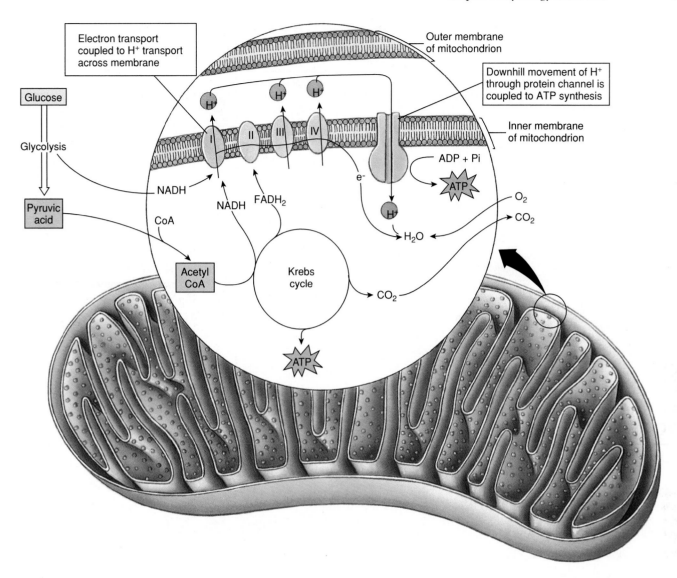

Figure 5-13

Oxidative phosphorylation. Most of the ATP in living organisms is produced in the electron transport chain. Electrons removed from fuel molecules in cellular oxidations (glycolysis and the Krebs cycle) flow through the electron transport chain, the major components of which are four protein complexes (I, II, III, and IV). Electron energy is tapped by the major complexes and used to push H^+ outward across the inner mitochondrial membrane. The H^+ gradient created drives H^+ inward through proton channels which couple H^+ movement to ATP synthesis.

EFFICIENCY OF OXIDATIVE PHOSPHORYLATION

We are now in a position to calculate the ATP yield from the complete oxidation of glucose (Figure 5-14). The overall reaction is:

$$\text{Glucose} + 2\ \text{ATP} + 36\ \text{ADP} + 36\ \text{P} + 6\ O_2 \rightarrow 6\ CO_2 + 2\ \text{ADP} + 36\ \text{ATP} + 6\ H_2O$$

ATP has been generated at several points along the way (Table 5-1). The cytoplasmic NADH generated in glycolysis requires a molecule of ATP to fuel transport of each molecule of NADH into the mitochondrion; therefore each NADH from glycolysis results in only two ATP (total of four), compared with the three ATP per NADH (total of six) formed within the mitochondria. Accounting for the two ATP used in the priming reactions in glycolysis, the net yield may be as high as 36 molecules of ATP per molecule of glucose. (The yield of 36 ATP is a theoretical maximum because some of the H^+ gradient produced by electron transport may be used for other functions, such as transporting substances in and out of the mitochondrion.) The overall efficiency of aerobic oxidation of glucose is about 38%, comparing very favorably with human-designed energy conversion systems, which seldom exceed 5% to 10% efficiency.

ANAEROBIC GLYCOLYSIS: GENERATING ATP WITHOUT OXYGEN

Up to this point we have been describing aerobic cellular respiration.

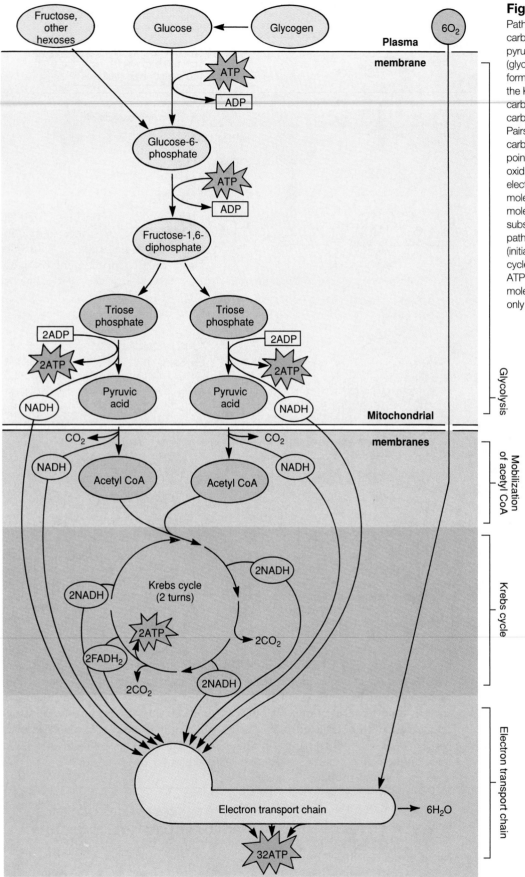

Figure 5-14

Pathway for oxidation of glucose and other carbohydrates. Glucose is degraded to pyruvic acid by cytoplasmic enzymes (glycolytic pathway). Acetyl coenzyme A is formed from pyruvic acid and is fed into the Krebs cycle. An acetyl group (two carbons) is oxidized to two molecules of carbon dioxide with each turn of the cycle. Pairs of electrons are removed from the carbon skeleton of the substrate at several points in the pathway and are carried by oxidizing agents NADH or $FADH_2$ to the electron transport chain where 32 molecules of ATP are generated. Four molecules of ATP are also generated by substrate phosphorylation in the glycolytic pathway, and two molecules of ATP (initially GTP) are formed in the Krebs cycle. This yields a total of 38 molecules of ATP (36 molecules net) per glucose molecule. Molecular oxygen is involved only at the very end of the pathway.

Table 5-1	Calculation of Total ATP Molecules Generated in Respiration	
ATP Generated	**Source**
4	Directly in glycolysis
2	As GTP ($\rightarrow$ ATP) in Krebs cycle
4	From NADH in glycolysis
6	From NADH produced in pyruvic acid to acetyl coenzyme A reaction
4	From reduced FAD in Krebs cycle
18	From NADH produced in Krebs cycle
38 Total	
-2	Used in priming reactions in glycolysis
36 Net	

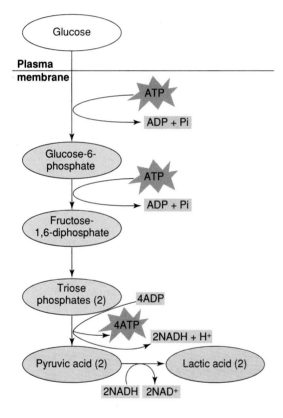

Figure 5-15

Anaerobic glycolysis, a process that proceeds in the absence of oxygen. Glucose is broken down to two molecules of pyruvic acid, generating four molecules of ATP and yielding two, since two molecules of ATP are used to produce fructose-1,6-diphosphate. Pyruvic acid, the final electron acceptor for the hydrogen ions and electrons released during pyruvic acid formation, is converted to lactic acid.

We will now consider how animals generate ATP without oxygen, that is, anaerobically.

Under anaerobic conditions, glucose and other 6-carbon sugars are first broken down stepwise to a pair of 3-carbon pyruvic acid molecules, yielding two molecules of ATP and four atoms of hydrogen (four reducing equivalents, represented by 2 NADH + H$^+$). But in the absence of molecular oxygen, further oxidation of pyruvic acid cannot occur because the Krebs cycle and electron transport chain cannot operate and cannot, therefore, provide a mechanism for reoxidizing the NADH produced in glycolysis. The problem is neatly solved in most animal cells by reducing pyruvic acid to lactic acid (Figure 5-15). Pyruvic acid becomes the final electron acceptor and lactic acid the end product of anaerobic glycolysis. This frees the hydrogen-bound carrier to recycle and pick up more H$^+$. In **alcoholic fermentation** (as in yeast, for example) the steps are identical to glycolysis down to pyruvic acid. One of its carbons is then released as carbon dioxide, and the resulting 2-carbon compound is reduced to ethanol, thus regenerating the NAD.

Anaerobic glycolysis is only one-eighteenth as efficient as the complete oxidation of glucose to carbon dioxide and water, but its key virtue is that it provides *some* high-energy phosphate in situations in which oxygen is absent or in short supply. Many microorganisms live in places where oxygen is severely depleted, such as waterlogged soil, in mud of lake or sea bottom, or within a decaying carcass. Vertebrate skeletal muscle may rely heavily on glycolysis during short bursts of activity when contraction is so rapid and powerful that oxygen delivery to the tissues is not sufficient to supply energy demands by oxidative phosphorylation alone. At such times the animal has no choice but to supplement oxidative phosphorylation with anaerobic glycolysis. Intense activity is followed by a period of increased oxygen consumption as lactic acid diffuses from muscle to the liver where it is metabolized. Because oxygen consumption increases following heavy activity, the animal is said to have acquired an **oxygen debt** during activity which is repaid when activity ceases and the accumulated lactic acid is metabolized.

Some animals rely heavily on anaerobic glycolysis during normal activities. For example, diving birds and mammals fall back on glycolysis almost entirely to give them the energy needed to sustain long dives. Salmon would never reach their spawning grounds were it not for anaerobic glycolysis providing almost all of the ATP used in the powerful muscular bursts needed to carry them up rapids and falls. Many parasitic animals have dispensed with oxidative phosphorylation entirely. They secrete relatively reduced end products of their energy metabolism, such as succinic acid, acetic acid, and propionic acid. These compounds are produced in mitochondrial reactions that derive several more molecules of ATP than does the cycle from glycolysis to lactic acid, although these sequences are still far less efficient than the classical electron transport chain.

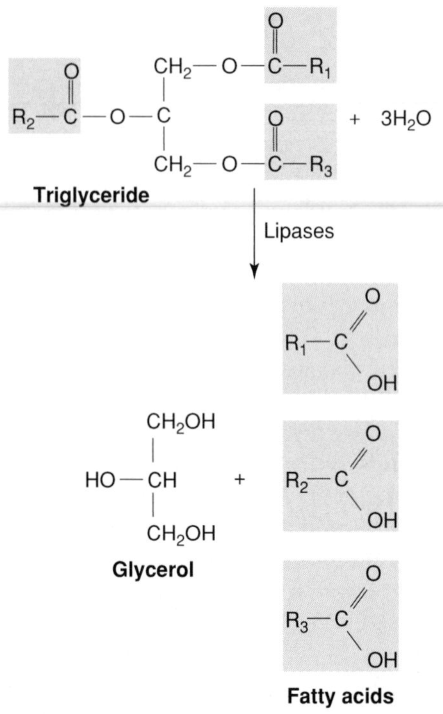

Figure 5-16
Hydrolysis of a triglyceride (neutral fat) by intracellular lipase. The R groups of each fatty acid represent a hydrocarbon chain.

METABOLISM OF LIPIDS

The first step in the breakdown of a triglyceride is its hydrolysis to glycerol and the three fatty acid molecules (Figure 5-16). Glycerol is phosphorylated and enters the glycolytic pathway.

The remainder of the triglyceride molecule consists of fatty acids. One of the abundant naturally occurring fatty acids is **stearic acid.**

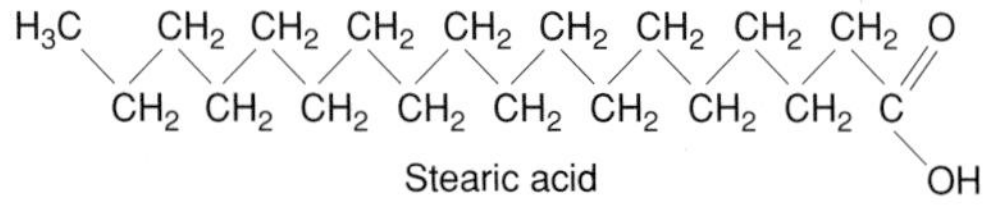

The long hydrocarbon chain of a fatty acid is sliced up by oxidation, two carbons at a time; these are released from the end of the molecule as acetyl coenzyme A. Although two high-energy phosphate bonds are required to prime each 2-carbon fragment, energy is derived both from the reduction of NAD and FAD in the oxidations and from the acetyl group as it is degraded in the Krebs cycle. It can be calculated that the complete oxidation of one molecule of 18-carbon

stearic acid will net 146 ATP molecules. By comparison, three molecules of glucose (also totaling 18 carbons) yield 108 ATP molecules. Since there are three fatty acids in each triglyceride molecule, a total of 440 ATP molecules are formed. An additional 22 molecules of ATP are generated in the breakdown of glycerol, giving a grand total of 462 molecules of ATP. Little wonder that fat is considered the king of animal fuels! Fats are more concentrated fuels than carbohydrates, because fats are almost pure hydrocarbons; they contain more hydrogen per carbon atom than sugars do, and it is the energized electrons of hydrogen that generate high-energy bonds, when they are carried through the mitochondrial electron transport chain.

Fat stores are derived principally from surplus fats and carbohydrates in the diet. Acetyl coenzyme A is the source of carbon atoms used to build fatty acids. Since all major classes of organic molecules (carbohydrates, fats, and proteins) can be degraded to acetyl coenzyme A, all can be converted into stored fat. The biosynthetic pathway for fatty acids resembles a reversal of the catabolic pathway already described, but it requires an entirely different set of enzymes. From acetyl coenzyme A, the fatty acid chain is assembled two carbons at a time. Because fatty acids release energy when they are oxidized, they obviously require an input of energy for their synthesis. This is provided principally by electron energy from glucose degradation. Thus the total ATP derived from oxidation of a molecule of triglyceride is not as great as calculated, because varying amounts of energy are required for synthesis and storage.

Stored fats are the greatest reserve fuel in the body. Most of the usable fat resides in adipose tissue that is composed of specialized cells packed with globules of triglycerides. Adipose tissue is widely distributed in the abdominal cavity, in muscles, around deep blood vessels, and especially under the skin. Women average about 30% more fat than men, and

this is largely responsible for differences in shape between males and females. Humans can only too easily deposit large quantities of fat, generating personal unhappiness and hazards to health.

The physiological and psychological aspects of obesity are now being investigated by many researchers. There is increasing evidence that body fat deposition is regulated by a feeding control center located in the lateral and ventral regions of the hypothalamus, an area in the floor of the forebrain. The set point of this regulator determines the normal weight for the individual, which may be rather persistently maintained above or below what is considered normal for the human population. Thus obesity is not always caused by overindulgence and lack of self-control, despite popular notions to the contrary.

METABOLISM OF PROTEINS

Since proteins are composed of amino acids, of which 20 kinds commonly occur (p. 26), the central topic of our consideration is amino acid metabolism. Amino acid metabolism is complex. For one thing, each of the 20 amino acids requires a separate pathway for biosynthesis and degradation. For another, amino acids are precursors to tissue proteins, enzymes, nucleic acids, and other nitrogenous constituents that form the fabric of the cell. The central purpose of carbohydrate and fat oxidation is to provide energy needed to construct and maintain these vital macromolecules.

Let us begin with the **amino acid pool** in the blood and extracellular fluid from which the tissues draw their requirements. When animals eat proteins, most are digested in the gut, releasing the constituent amino acids, which are then absorbed (Figure 5-17). Tissue proteins also are hydrolyzed during normal growth, repair, and tissue restructuring; their amino acids

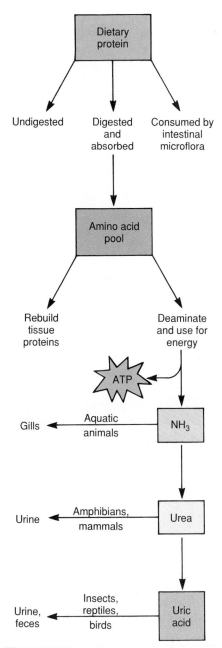

Figure 5-17

Fate of dietary protein.

join those derived from protein foodstuffs to enter the amino acid pool. A portion of the amino acid pool is used to rebuild tissue proteins, but most animals ingest a surplus of protein. Since amino acids are not excreted as such in any significant amounts, they must be disposed of in some other way. In fact, amino acids can be and are metabolized through oxidative pathways to yield high-energy phosphate. In short, excess proteins serve as fuel as do carbohydrates and fats. Their importance as fuel obviously

depends on the nature of the diet. In carnivores that ingest a diet of almost pure protein and fat, nearly half of their high-energy phosphate is derived from amino acid oxidation.

Before entering the fuel depot, nitrogen must be removed from the amino acid molecule. This can be by deamination (the amino group splits off to form ammonia and a keto acid) or by transamination (the amino group is transferred to a keto acid to yield a new amino acid). Thus amino acid degradation yields two main products, carbon skeletons and ammonia, which are handled in different ways. Once the nitrogen atoms are removed, the carbon skeletons of amino acids can be completely oxidized, usually by way of pyruvic acid or acetic acid. These residues then enter regular routes used by carbohydrate and fat metabolism.

The other product of amino acid degradation is ammonia. Ammonia is highly toxic because it reacts with α-ketoglutaric acid to form glutamic acid (an amino acid). Any accumulation of ammonia effectively removes this important intermediate from the Krebs cycle (see Figure 5-12) and inhibits respiration. Disposal of ammonia offers little problem to aquatic animals because it is soluble and readily diffuses into the surrounding medium through the respiratory surfaces. Terrestrial forms cannot get rid of ammonia so conveniently and must detoxify it by converting it to a relatively nontoxic compound. The two principal compounds formed are **urea** and **uric acid,** although a variety of other detoxified forms of ammonia are excreted by different invertebrate and vertebrate groups. Among vertebrates, amphibians and especially mammals produce urea. Reptiles and birds, as well as many terrestrial invertebrates, produce uric acid (the excretion of uric acid by insects and birds is described on pages 415 and 581, respectively).

The key feature that seems to determine the choice of nitrogenous waste is the availability of water in the environment. When water is abundant, the chief nitrogenous waste is ammonia. When water is restricted, it

is urea. And for animals living in truly arid habitats, it is uric acid. Uric acid is highly insoluble and easily precipitates from solution, allowing its removal in solid form. The embryos of birds and reptiles benefit greatly from excretion of nitrogenous waste as uric acid, because the waste cannot be eliminated through their shells. During embryonic development, the harmless, solid uric acid is retained in one of the extraembryonic membranes. When the hatchling emerges into its new world, the accumulated uric acid, along with the shell and membranes that supported development, is discarded.

MANAGEMENT OF METABOLISM

The complex pattern of enzymatic reactions that constitutes metabolism cannot be explained entirely in terms of physicochemical laws or chance happenings. Although some enzymes do indeed "flow with the tide," the activity of others is rigidly controlled. In the former case, suppose the function of an enzyme is to convert A to B. If B is removed by conversion into another compound, the enzyme will tend to restore the original ratio of B to A. Since many enzymes act reversibly, this can result, according to the metabolic situation prevailing, in synthesis or degradation. For example, an excess of an intermediate in the Krebs cycle would result in its contribution to glycogen synthesis; a depletion of such a metabolite would lead to glycogen breakdown. This automatic compensation (equilibration) is not, however, sufficient to explain all that actually takes place in an organism, as for example, what happens at branch points in a metabolic pathway.

Mechanisms exist for critically regulating enzymes in both *quantity* and *activity*. In bacteria, genes leading to synthesis of an enzyme are switched on or off, depending on the presence or absence of a substrate molecule. In this way the *quantity* of an enzyme is controlled. It is a relatively slow process.

Mechanisms that alter activity of enzymes can quickly and finely adjust metabolic pathways to changing conditions in a cell. The presence or increase in concentration of some molecules can alter the shape (conformation) of particular enzymes, thus activating or inhibiting the enzyme (Figure 5-18). For example, phosphofructokinase, which catalyzes the phosphorylation of glucose-6-phosphate to fructose-1,6-diphosphate (Figure 5-14), is inhibited by high concentrations of ATP or citric acid. Their presence means that a sufficient amount of precursors has reached the Krebs cycle and additional glucose is not needed.

As well as being subject to alteration in physical shape, some enzymes exist in both an active and an inactive form. These may be chemically different. Enzymes that degrade glycogen (phosphorylase) and synthesize it (synthase) are examples. Conditions that activate the phosphorylase tend to inactivate synthase and vice versa.

Many cases of enzyme regulation are known, but these selected examples must suffice to illustrate the importance of enzyme regulation in the integration of metabolism.

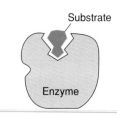

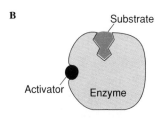

Figure 5-18
Enzyme regulation. **A,** The active site of an enzyme may only loosely fit its substrate in the absence of an activator. **B,** With the regulatory site of the enzyme occupied by an activator, the enzyme binds the substrate, and the site becomes catalytically active.

Summary

Living systems are subject to the same laws of thermodynamics that govern non-living systems. The first law states that energy cannot be destroyed, although it may change form. The second law states that the structure of systems proceeds toward total randomness, or increasing entropy, as energy is dissipated. Solar energy trapped by photosynthesis as chemical bond energy is passed through the food chain where it is used for biosynthesis, active transport, and motion, before finally being dissipated as heat. Living organisms are able to decrease their entropy and maintain high internal order because the biosphere is an open system from which energy can be captured and used. Energy available for use in biochemical reactions is termed "free energy."

Enzymes are proteins, associated with nonprotein cofactors, that vastly accelerate rates of chemical reactions in living systems. An enzyme does this by temporarily binding its reactant (substrate) onto an active site in a highly specific fit. In this configuration, internal activation energy barriers are lowered enough to modify the substrate, and the enzyme is restored to its original form.

Cells use the energy stored in chemical bonds of organic fuels by breaking the fuels down through a series of enzymatically controlled steps. This bond energy is transferred to ATP and packaged in the form of "high-energy" phosphate bonds. ATP is produced as it is required in cells to power various synthetic, secretory, and mechanical processes.

Glucose is an important source of energy for cells. In aerobic metabolism (respiration), the 6-carbon glucose is split into two 3-carbon molecules of pyruvic acid. Pyruvic acid is decarboxylated to form 2-carbon acetyl coenzyme A, a strategic intermediate that leads to the Krebs cycle. Acetyl coenzyme A can also be derived from breakdown of fat. In the Krebs cycle, acetyl coenzyme A is oxidized in a series of reactions to carbon dioxide, yielding, in the course of the reactions, energized electrons that are passed to electron acceptor molecules (NAD and FAD). In the final stage, the energized electrons are passed along an electron transport chain consisting of a series of electron carriers located in the inner membranes of the mitochondrion. ATP is generated at three points along the chain as the electrons are passed from carrier to carrier and finally to oxygen. A net total of 36 molecules of ATP may be generated from one molecule of glucose.

In the absence of oxygen (anaerobic glycolysis), glucose is degraded to two 3-carbon molecules of lactic acid, yielding two molecules of ATP. Although anaerobic glycolysis is vastly less efficient than respiration, it provides essential energy for muscle contraction when heavy energy expenditure outstrips the oxygen-delivery system of an animal; it also is the only source for microorganisms living in oxygen-free environments.

Triglycerides (neutral fats) are especially rich depots of metabolic energy because the fatty acids of which they are composed are highly reduced and free of water. Fatty acids are degraded by sequential removal of 2-carbon units, which enter the Krebs cycle through acetyl-CoA.

Amino acids in excess of requirements for synthesis of proteins and other biomolecules are used as fuel. They are degraded by deamination or transamination to yield ammonia and carbon skeletons. The latter enter the Krebs cycle to be oxidized. Ammonia is a highly toxic waste product that aquatic animals quickly dispose through respiratory surfaces. Terrestrial animals, however, convert ammonia into much less toxic compounds, urea or uric acid, for disposal.

The integration of metabolic pathways is finely regulated by mechanisms that control both the amount and activity of enzymes. The quantity of some enzymes is regulated by certain molecules that switch on or off enzyme synthesis. Enzyme activity may be altered by the presence or absence of metabolites that cause conformational changes in enzymes and thus improve or diminish their effectiveness as catalysts.

Review Questions

1. State the first and second laws of thermodynamics. Living systems may appear to violate the second law of thermodynamics because living things maintain a high degree of organization despite a universal trend toward increasing disorganization. What is the explanation for this paradox?

2. Explain what is meant by "free energy" in a system. Will a reaction that proceeds spontaneously have a positive or negative change in free energy?

3. Many biochemical reactions proceed slowly unless the energy barrier to the reaction is lowered. How is this accomplished in living systems?

4. What happens in the formation of an enzyme-substrate complex that favors the disruption of substrate bonds?

5. What is meant by a "high-energy bond"?

6. Although ATP supplies energy to an endergonic reaction, why is it not considered a fuel?

7. What is an oxidation-reduction reaction and why are such reactions considered so important in cellular metabolism?

8. Give an example of the final electron acceptor found in aerobic and anaerobic organisms. Why is aerobic metabolism more efficient than anaerobic metabolism?

9. Why is it necessary for glucose to be "primed" with a high-energy phosphate bond before it can be degraded in the glycolytic pathway?

10. What happens to the electrons that are removed during the oxidation of triose phosphates during glycolysis?

11. Why is acetyl coenzyme A considered a "strategic intermediate" in respiration?

12. Why are oxygen atoms important in oxidative phosphorylation?

13. Explain how animals can generate ATP *without* oxygen. Given that anaerobic glycolysis is much less efficient than oxidative phosphorylation, why has anaerobic glycolysis not been discarded during animal evolution?

14. Why are animal fats sometimes called "the king of fuels"? What is the significance of acetyl coenzyme A to lipid metabolism?

15. The breakdown of amino acids yields two products: ammonia and carbon skeletons. What happens to these products?

16. Explain the relationship between the amount of water in an animal's environment and the kind of nitrogenous waste it produces.

17. Explain three ways that enzymes may be regulated in cells.

Selected References

Dickerson, R. E. 1980. Cytochrome c and the evolution of energy metabolism. Sci. Am. **242**:136–153 (Mar.). *How the metabolism of modern organisms evolved.*

Hinkle, P., and R. McCarty. 1978. How cells make ATP. Sci. Am. **238**:104–123 (Mar.). *Good description of the chemiosmotic hypothesis which explains how ATP is made from electrons removed from foodstuffs.*

Lodish, H., D. Baltimore, A. Berk, S. L. Zipursky, P. Matsudira, and J. Darnell. 1995. Molecular cell biology, ed. 2, New York, Scientific American Books, Inc. *Chapter 15 is a comprehensive, well illustrated treatment of intermediary metabolism.*

Stryer, L. 1988. Biochemistry, ed. 3. San Francisco, W.H. Freeman & Company Publishers. *One of the best of undergraduate biochemistry texts.*

Wolfe, S. L. 1993. Molecular and cellular biology. Belmont, California, Wadsworth Publishing Company. *Chapter 9 provides a well organized explanation of energy metabolism.*

II

Continuity and Evolution of Animal Life

6

The Reproductive Process

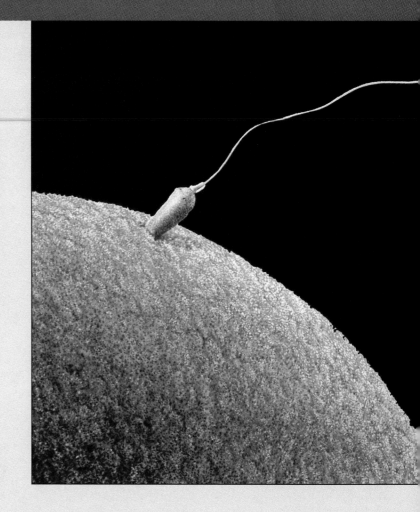

Omne vivum ex ovo

In 1651, late in a long life, William Harvey, the English physiologist who earlier had ushered in experimental physiology by explaining the circuit of the blood, published a treatise on reproduction. He asserted that all life developed from the egg—*omne vivum ex ovo*. This was curiously insightful, since Harvey had no means for visualizing the eggs of many animals, in particular the microscopic mammalian egg, which is no larger than a speck of dust to the unaided eye. Further, argued Harvey, the egg is launched into its developmental course by some influence from the semen, a conclusion that was either remarkably perceptive or a lucky guess, since sperm also were invisible to Harvey. Such ideas differed sharply from existing notions of biogenesis, which saw life springing from many sources of which eggs were but one.

Harvey was describing characteristics of sexual reproduction in which two parents, male and female, must come together physically to ensure fusion of gametes from each.

Despite the importance of Harvey's aphorism that all life arises from eggs, it was too sweeping to be wholly correct. Life springs from the reproduction of preexisting life, and reproduction may not be bound up in eggs and sperm. Nonsexual reproduction, the creation of new, genetically identical individuals by budding or fragmentation or fission from a single parent, is common, indeed characteristic, among some phyla. Most animals have found sex to be the winning strategy, probably because sexual reproduction promotes diversity, enhancing long-term survival of the lineage in a world of perpetual change. ■

Reproduction is one of the ubiquitous miracles of life. Evolution is inextricably linked to reproduction, because the ceaseless replacement of aging predecessors with new life gives animals the means to respond and evolve in a changing environment as the earth itself has changed over the ages. In this chapter we will distinguish asexual and sexual reproduction and explore the reasons why, for multicellular animals at least, sexual reproduction appears to offer important advantages over asexual. We will then take up, in turn, the origin and maturation of germ cells; the plan of reproductive systems; the reproductive patterns in animals; and, finally, the endocrine events that orchestrate reproduction.

Nature of the Reproductive Process

The two modes of reproduction are asexual and sexual. In **asexual** reproduction (Figure 6-1A and B) there is only one parent and there are no special reproductive organs or cells. Each organism is capable of producing genetically identical copies of itself as soon as it becomes an adult. The production of copies is marvelously simple, direct, and typically rapid. **Sexual** reproduction (Figure 6-1C and D) as a rule involves two parents, each of which contributes special **sex cells** (also called germ cells, or **gametes**) that in union (fertilization) develop into a new individual. The **zygote** formed from this union receives genetic material from *both* parents and accordingly is different from both. The combination of genes produces a genetically unique individual, still bearing the characteristics of the species but also bearing traits that make it different from its parents. Sexual reproduction, by recombining the parental characters, tends to multiply variations and makes possible a richer and more diversified evolution.

Mechanisms for interchange of genes between individuals are more limited in organisms with only asexual reproduction. Of course, in asexual organisms that are haploid (bear only one set of genes), mutations are immediately expressed and evolution can proceed quickly. In sexual animals, on the other hand, a gene mutation is often not expressed immediately, since it may be masked by its normal partner on the homologous chromosome. (Homologous chromosomes, discussed on p. 87, are those that pair during

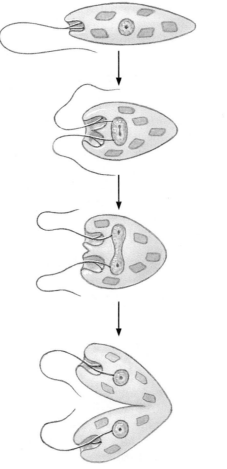

A   Binary fission in *Euglena*

Figure 6-1
Asexual and sexual reproduction in animals.
A, Binary fission in *Euglena,* a flagellate protozoan, results in two individuals. **B,** Budding, a simple form of asexual reproduction as shown in a hydra, a radiate animal. The buds eventually detach themselves and grow into fully formed individuals. **C,** Barnacles reproduce sexually, but are hermaphroditic, with each individual bearing both male and female organs. Each barnacle possesses a pair of enormously elongated penises—an obvious advantage to a sessile animal—that can be extended many times the length of the body to inseminate another barnacle some distance away. The partner may reciprocate with its own penises. **D,** Frogs, here in mating position (amplexus), represent bisexual reproduction, the most common form of sexual reproduction involving separate male and female individuals.

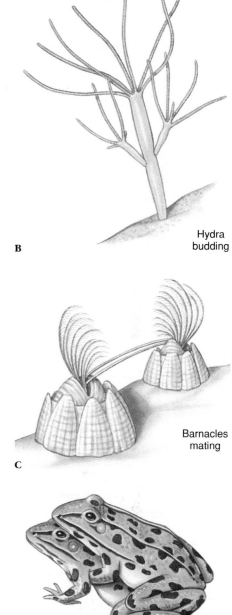

B   Hydra budding

C   Barnacles mating

D   Frogs in amplexus

meiosis and have genes controlling the same characteristics.) There is only a remote chance that both members of a gene pair will mutate in the same way at the same moment.

Asexual Reproduction: Reproduction without Gametes

Asexual reproduction (see Figure 6-1A and B) is the production of individuals without gametes, that is, eggs or sperm. It includes a number of distinct processes to be described below, all without involving sex or a second parent. The offspring produced by asexual reproduction of an individual all have the same genotype and are referred to as **clones.**

It would be a mistake to conclude that asexual reproduction is in any way a "defective" form of reproduction relegated to the minute forms of life that have not yet discovered the joys of sex. Given the facts of their abundance, that they have persisted on earth for 3.5 billion years, and that they form the roots of the food chain on which all higher forms depend, the single-celled asexual organisms are both resoundingly abundant and supremely important. For these forms the advantages of asexual reproduction are its rapidity (many bacteria divide every half hour) and simplicity (no sex cells to produce and no time and energy expended in finding a mate).

Asexual reproduction appears in bacteria and protists and in many invertebrate phyla, such as cnidarians, bryozoans, annelids, echinoderms, and hemichordates. In animal phyla in which asexual reproduction occurs, most members employ sexual reproduction as well. In these groups, asexual reproduction ensures rapid increase in numbers when the differentiation of the organism has not advanced to the point of forming gametes. Asexual reproduction is absent among the vertebrates (although some forms of parthenogenesis have been interpreted as asexual by some authors; see p. 83).

The basic forms of asexual reproduction are fission (binary and multiple), budding, gemmulation, and fragmentation. **Binary fission** is common among bacteria and protozoa (Figure 6-1A). In binary fission the body of the parent divides by mitosis into two approximately equal parts, each of which grows into an individual similar to the parent. Binary fission may be lengthwise, as in flagellate protozoa, or transverse, as in ciliate protozoa. In **multiple fission** the nucleus divides repeatedly before division of the cytoplasm, giving rise to many daughter cells simultaneously. Spore formation, called sporogony, is a form of multiple fission that is common among some parasitic protozoa, for example, the malaria parasite.

Budding is an unequal division of the organism. The new individual arises as an outgrowth (bud) from the parent, develops organs like those of the parent, and then detaches itself. Budding occurs in several animal phyla and is especially prominent in the cnidarians (Figure 6-1B).

Gemmulation is the formation of a new individual from an aggregation of cells surrounded by a resistant capsule, called a gemmule. In many freshwater sponges gemmules develop in the fall and survive the winter in the dried or frozen body of the parent. In the spring, the enclosed cells become active, emerge from the capsule, and grow into a new sponge.

In **fragmentation** a multicellular animal breaks into two or more parts, with each fragment capable of becoming a complete individual. Many invertebrates can reproduce asexually by simply breaking in two and then regenerating the missing parts of the fragments.

SEXUAL REPRODUCTION: REPRODUCTION WITH GAMETES

The essential feature of sexual reproduction is the *production of offspring formed by the union of gametes from two genetically different parents* (Figures 6-1C and D, and 6-2). The offspring will thus have a new genotype different from either of the parents. The individuals sharing parenthood are characteristically of different **sexes,** male and female (there are exceptions among sexually reproducing organisms, such as bacteria and some protozoa in which sexes are lacking). The distinction between male and female is based, not on any differences in parental size or appearance, but on the size and mobility of the gametes (sex cells) they produce. The **ovum** (egg) is produced by the female. Ova are large (because of stored yolk to sustain early development), nonmotile, and produced in relatively small numbers. The **spermatozoon** (sperm) is produced by the male. Sperm are small, motile, and produced in enormous numbers. Each is a stripped-down package of highly-condensed genetic material designed for the single mission of finding and fertilizing the egg.

There is another crucial event that distinguishes sexual from asexual reproduction: **meiosis,** a distinctive type of gamete-producing nuclear division. As will be described later (p. 86), meiosis differs from ordinary cell division (mitosis) in being a double division. The chromosomes split once, but the cell divides *twice,* producing four cells, each with half the original number of chromosomes (the haploid number). Meiosis is followed by **fertilization** in which two haploid gametes are combined to restore the normal (diploid) chromosomal number of the species.

The new cell (zygote), which now begins to divide by mitosis, has equal numbers of chromosomes from each parent and accordingly is different from each. It is a unique individual bearing a recombination of parental characteristics. This is the great strength of sexual reproduction that keeps feeding new genetic combinations into the population.

Many protozoa reproduce both sexually and asexually. When sexual reproduction does occur, it may or may not involve male and female gametes. Sometimes two mature sexual parents merely join together to exchange nuclear material or merge cytoplasm (conjugation, p. 233). It is not possible in these cases to distinguish sexes.

The male-female distinction is more clearly evident in the metazoa. Organs that produce the germ cells are known as **gonads.** The gonad that produces the sperm is called the **testis** (see Figure 6-13) and that which forms the egg, the **ovary** (see Figure 6-14). The

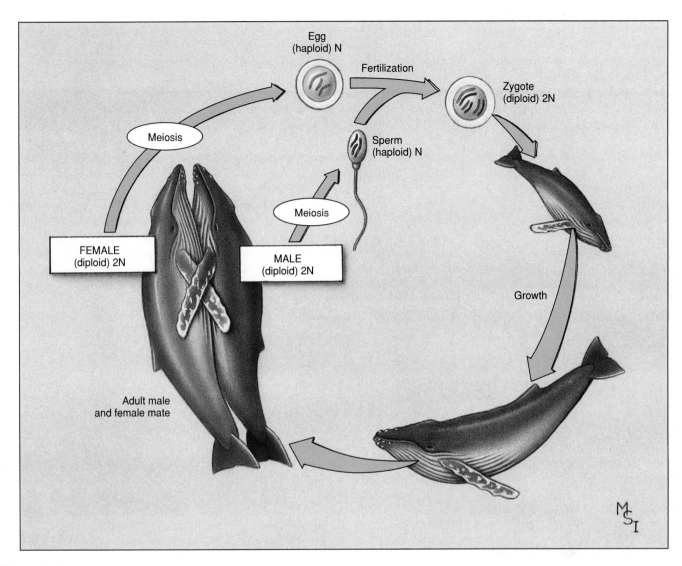

Figure 6-2
A sexual life cycle. The life cycle begins with haploid germ cells, formed by meiosis, combining to form a diploid zygote, which grows by mitosis to an adult. Most of the life cycle is spent as a diploid organism.

gonads represent the **primary sex organs,** the only sex organs found in certain groups of animals. Most metazoa, however, have various **accessory sex organs** (such as penis, vagina, oviducts, and uterus) that transfer and receive sex cells. In the primary sex organs the sex cells undergo many complicated changes during their development, the details of which are described later. In our present discussion we will distinguish bisexual reproduction from two alternatives: parthenogenesis and hermaphroditism.

Bisexual Reproduction

Bisexual reproduction is the common and familiar method of sexual reproduction involving separate and distinct male and female individuals (Figure 6-2). Each has its own reproductive system and produces only one kind of sex cell, spermatozoon or ovum, but never both. Nearly all vertebrates and many invertebrates have separate sexes, and such a condition is called **dioecious** (Gr. *di-*, two; + *oikos,* house).

Parthenogenesis

Parthenogenesis ("virgin origin") is the development of an embryo from an unfertilized egg or, if a spermatozoon does penetrate the egg, there is no union of male and female pronuclei. There are many patterns of parthenogenesis. In one type, called **ameiotic parthenogenesis,** there is no meiosis, and the egg is formed by mitotic cell division. This "asexual" form of parthenogenesis is known to occur in some species of flatworms, rotifers, crustaceans, insects, and probably others. In these cases, the offspring are clones of the parent because, without meiosis, there is no reshuffling of genes between chromosomes (this important event during meiosis, called crossing over, is described on p. 87).

In **meiotic parthenogenesis** a haploid ovum is formed by meiosis, and it may or may not be activated by the influence of a male. For example, in some species of fishes, the female may be inseminated by the male of the same or related species, but the sperm serves only to activate the egg; the male's genome is rejected before it can penetrate the egg. In several species of flatworms, rotifers, annelids, mites, and insects, the haploid egg begins development spontaneously; no males are required to stimulate activation of the ovum. The diploid condition is restored by chromosomal duplication. A variant of this type of parthenogenesis occurs in many bees, wasps, and ants. In honey bees, for example, the queen bee can either fertilize the eggs as she lays them or allow them to pass unfertilized. The fertilized eggs become diploid females (queens or workers), and the unfertilized eggs develop parthenogenetically to become haploid males (drones); this type of sex determination is known as **haplodiploidy.** In some animals meiosis may be so severely modified that the offspring are clones of the parent. This happens in certain populations of whiptail lizards of the American southwest, which are clones consisting solely of females (Cole, 1984).

From time to time claims arise that spontaneous parthenogenetic development to term has occurred in humans. A British investigation of about 100 cases in which the mother denied having had intercourse revealed that in nearly every case the child possessed characteristics not present in the mother, and consequently must have had a father. Nevertheless, mammalian eggs very rarely will spontaneously start developing into embryos without fertilization. In certain strains of mice, such embryos will develop into fetuses and then die. The most remarkable instance of parthenogenetic development among the higher vertebrates has been found in turkeys in which certain strains, selected for their ability to develop without sperm, grow to reproducing adults.

Parthenogenesis is surprisingly widespread in animals. It is an abbreviation of the usual steps required of bisexual reproduction. It may have evolved to avoid the problem—which may be great in some animals—of bringing together males and females at the right moment for successful fertilization. The disadvantage of parthenogenesis is that if the environment should suddenly change, as it often does, parthenogenetic species have limited capacity to shift gene combinations to adapt to the new conditions. Bisexual species, by recombining parental characteristics, have a better chance of producing variant offspring that can utilize new environments.

Hermaphroditism

Animals that have both male and female organs in the same individual are called hermaphrodites, and the condition is called hermaphroditism (from a combination of the names of the Greek god Hermes and goddess Aphrodite). In contrast to the dioecious state of separate sexes, hermaphrodites are **monoecious** (Gr. *monos,* single; + *oikos,* house), meaning that both male and female organs are in the same organism. Many sessile, burrowing, or endoparasitic invertebrate animals (for example, most flatworms, some hydroids and annelids, and all barnacles and pulmonate snails) and a few vertebrates (some fishes), are hermaphroditic. Some hermaphrodites fertilize themselves, but most avoid self-fertilization by exchanging germ cells with another member of the same species (Figure 6-3). An advantage is that with every individual producing eggs, a hermaphroditic species could potentially produce twice as many offspring as could a bisexual species in which half the individuals are nonproductive males. In some fishes, called sequential hermaphrodites, the animal experiences a genetically programmed sex change during its life. In many species of reef fishes, for example, the wrasses, the animal begins life as either a female or a male (depending on the species) but later becomes the opposite sex.

WHAT GOOD IS SEX?

The question "What good is sex?" appears to have an easy answer: it serves the purpose of reproduction. But if we rephrase the question to ask, "Why do so many animals reproduce sexually rather than asexually?" the answer is not so apparent. Because sexual reproduction is so nearly universal among animals, it might be inferred that it must be highly advantageous. Yet it is easier to list disadvantages to sex than advantages. Sexual reproduction is

Figure 6-3
Hermaphroditic snails mating. Pulmonate snails are "simultaneous" hermaphrodites; during mating each partner inserts its penis into the female opening of the other.

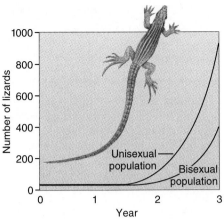

Figure 6-4

Comparison of the growth of a population of unisexual whiptail lizards with a population of bisexual lizards. Because all individuals of the unisexual population are females, all produce eggs, whereas only half the bisexual population are egg-producing females. By the end of the third year the unisexual lizards are more than twice as numerous as the bisexual ones.

complicated, requires more time, and uses much more energy than asexual reproduction. Mating partners must come together and coordinate their activities to produce young. Many biologists believe that an even more troublesome problem is the "cost of meiosis." A female that reproduces asexually passes all of her genes to her offspring. But when she reproduces sexually the genome is divided during meiosis and only half her genes flow to the next generation. Another cost is wastage in the production of males, many of whom fail to reproduce and thus consume resources that could be applied to the production of females. Whiptail lizards of the American southwest offer a fascinating example of the potential advantage of parthenogenesis. When unisexual and bisexual species of the same genus are reared under similar conditions in the laboratory, the population of the unisexual species grows more quickly. This happens because all of the unisexual lizards (all females) deposit eggs, whereas only 50% of the bisexual lizards do so (Figure 6-4).

Clearly, the costs of sexual reproduction are substantial. How are they offset? Biologists have disputed this question for years without producing an answer that satisfies everyone.

Many biologists believe that sexual reproduction, with its breakup and recombination of genomes, keeps producing novel genotypes that *in times of environmental change* may survive and reproduce, whereas most others die. Variability, advocates of this viewpoint argue, is sexual reproduction's trump card.

Variety may make sexual reproduction a winning strategy for the unstable environment, but some biologists believe that for many vertebrates sexual reproduction is unnecessary and may even be maladaptive. In animals in which most of the young survive to reproductive age (humans, for example), there is no demand for novel recombinations to cope with changing habitats. One offspring appears as successful as the next in each habitat. Significantly, parthenogenesis has evolved in several species of fish and in a few amphibians and reptiles. Such species are exclusively parthenogenetic, suggesting that where it has been possible to overcome the numerous constraints to making the transition, bisexual reproduction loses out.

But is variability worth the biological costs of sexual reproduction? The underlying problem keeps coming back: asexual organisms, because they can have more offspring in a given time, appear to be more fit in Darwinian terms. And yet most metazoan animals are determinedly committed to sexuality. There is considerable evidence that asexual reproduction is most successful in colonizing new environments. When habitats are empty what matters most is rapid reproduction; variability matters little. But as habitats become more crowded, competition between species for resources increases. Selection becomes more intense, and genetic variability—new genotypes produced by recombination in sexual reproduction—furnishes the diversity that permits the population to resist extinction. Therefore, on a geological timescale, asexual lineages, because

of the lack of genetic flexibility, may be more prone to extinction than sexual lineages. Sexual reproduction is therefore favored by species selection (species selection is described on p. 175).

There are many invertebrates that use both sexual and asexual reproduction, thus enjoying the advantages each has to offer.

FORMATION OF REPRODUCTIVE CELLS

The vertebrate body has two basically different types of cells: the **somatic cells,** which are differentiated for specialized functions and die with the individual, and the **germ cells,** which form the gametes: eggs and sperm. Germ cells provide the continuity of life between generations and ensure the species' survival. The germ cells, or their precursors, the **primordial germ cells,** are set aside at the beginning of embryonic development, usually in the endoderm, and migrate to the gonads. Here they develop into eggs and sperm—nothing else. The continuity of germ cells from one generation to the next is called the **germ cell line.** The other cells of the gonads are somatic cells. They cannot form eggs or sperm, but they are necessary for the support, protection and nourishment of the germ cells during their development (gametogenesis).

A traceable germ cell line as present in the vertebrates is not distinguishable in most other animals. In many invertebrates the germ cells develop directly from somatic cells at some period in the life of the individual.

ORIGIN AND MIGRATION OF GERM CELLS

In vertebrates, the actual tissue from which the gonads arise appears in early development as a pair of **genital ridges,** growing into the coelom from the dorsal coelomic lining on each side of the gut near the anterior end of the kidney (mesonephros).

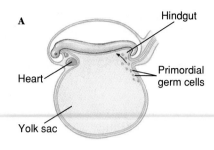

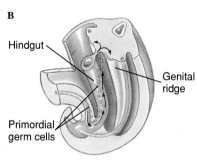

Figure 6-5
Migration of mammalian primordial germ cells. **A,** From the yolk sac the primordial germ cells migrate through the region of the hindgut into the genital ridges **(B).** In human embryos, the migration is complete by the end of the fifth week of gestation.

Surprisingly perhaps, the primordial germ cells do not arise in the developing gonad but in the yolk-sac endoderm (p. 117). From studies with frogs and toads, it has been possible to trace the germ cell line back to the fertilized egg, in which a localized area of germinal cytoplasm (called **germ plasm**) can be identified in the vegetal pole of the uncleaved egg mass. This material can be followed through subsequent cell divisions of the embryo until it becomes situated in primordial germ cells in gut endoderm. From here they migrate by ameboid movement to the genital ridges. A similar migration of the primordial germ cells occurs in mammals (Figure 6-5). The primordial germ cells are the future stock of gametes for the animal. Once in the gonad they begin to divide by mitosis, increasing their numbers from a few dozen to several thousand.

At first the gonad is sexually indifferent. In mammals the indifferent gonad has an inherent tendency to become an ovary. In rabbits, for example, the removal of the fetal gonads before they have differentiated will invariably produce a female with oviducts, uterus, and vagina, even if the rabbit is a genetic male. In the normal male, however, a male-determining gene on the Y chromosome regulates the formation of a testis-determining factor that organizes the developing gonad into a testis instead of an ovary. Once formed, the testis secretes the steroid **testosterone.** This hormone masculinizes the fetus, causing the differentiation of penis, scrotum, and the male ducts and glands. It also destroys the incipient breast primordia, but leaves behind the nipples as a reminder of the indifferent ground plan from which both sexes develop. Testosterone is also responsible for the masculinization of the brain, but it does so indirectly. Surprisingly, testosterone is enzymatically converted to estrogen in the brain, and it is estrogen that determines the organization of the brain for male-typical behavior.

If the animal is a genetic female, no special stimulus is required for female development. The *absence* of male-determining genes allows the gonad to follow its inherent tendency to become an ovary. Then, in the absence of testosterone, the embryo develops female sexual organs: vagina, clitoris, and uterus. This arrangement makes sense for mammals, in which the fetus develops within a female's body (the mother) and is constantly exposed to the mother's sex hormones. If female hormones fostered female development, male embryos would not develop normally unless specially protected from the mother's hormones. In fact, female brain development requires special protection from the effects of estrogen because, as mentioned previously, estrogen causes masculinization of the brain. To prevent this from happening in females, a blood protein (alpha-fetoprotein) binds to estrogen and keeps the hormone from reaching the brain.

The genetics of sex determination are treated in Chapter 8 (p. 132). Sex determination is strictly chromosomal in mammals, birds, amphibians, many reptiles, and probably most fishes.

For every structure in the reproductive system of the male or female, there is a homologous structure in the other. This happens because during early development male and female characteristics begin to differentiate from the embryonic genital ridge and two duct systems that at first are identical in both sexes. Under the influence of the sex hormones, the genital ridge develops into the testes of the male and the ovaries of the female. One duct system (Wolffian) becomes ducts of the testes in the male and a vestigial structure adjacent to the ovaries in the female. The other duct (Müllerian) develops into the oviducts, uterus, and vagina of the female and into the small, vestigial appendix of the testes in the male. Similarly, the clitoris and labia of the female are homologous to the penis and scrotum of the male, since they develop from the same embryonic structures.

However, many fishes and reptiles lack sex chromosomes altogether; in these, gender is determined by nongenetic factors such as temperature or behavior. In crocodilians, many turtles, and some lizards the incubation temperature of the nest determines the sex ratio by some as yet unknown sex-determining mechanism. Alligator eggs, for example, incubated at low temperature become all females; those incubated at higher temperature become all males (Figure 6-6). Sex determination of many fishes is behavior dependent. Most of these species are hermaphroditic, possessing both male and female gonads. Sensory stimuli from the animal's social environment determine whether it will be male or female.

MEIOSIS: NUCLEAR DIVISION OF GERM CELLS TO PRODUCE GAMETES

In sexually reproducing organisms there are two sets of chromosomes in every somatic cell; this is the **diploid** (Gr. *diploos,* double) number of chromosomes. In humans the diploid number is 46 (abbreviated 2*N*). These can

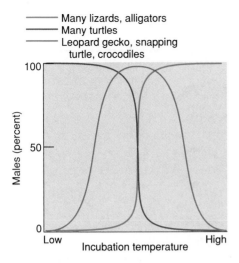

Many lizards, alligators
Many turtles
Leopard gecko, snapping turtle, crocodiles

Figure 6-6

Temperature-dependent sex determination. In many reptiles that lack sex chromosomes the incubation ~~temperature~~ gender. The graph shows that the embryos of many turtles develop into males at low temperature, whereas the embryos of many lizards and alligators become males at high temperatures. The embryos of crocodiles become males at intermediate temperatures, and become females at higher or lower temperatures.

From David Crews, "Animal Sexuality," Scientific American 270(1):108–114, January 1994. Copyright ©1994 Scientific American, Inc. All rights reserved. Reprinted by permission.

be matched in pairs and the members of each pair are called **homologous** chromosomes. Therefore every body cell contains *two* chromosomes bearing genes coding for the same set of characteristics, one on each of the homologs. These may be alternative forms of the same gene, and if so they are **allelic genes,** or **alleles** (multiple alleles are described on p. 132). Sometimes only one of the alleles has an effect on the organism, although both are present in each cell, and either may be passed on to the progeny as a result of meiosis and subsequent fertilization.

During an individual's growth, all the chromosomes of the mitotically dividing cells are replicated during the S period of each cell cycle (Figure 4-21, p. 55), so that each new cell contains the double set of chromosomes. Such cells, containing the diploid number of chromosomes, are called **diploid cells.** In the reproductive organs the gametes (eggs and sperm) are formed by a kind of maturation division, called meiosis, which *separates* the

two sets of chromosomes. The gametes formed in meiosis contain a single set of chromosomes, the **haploid** (Gr. *haploos,* single) number (abbreviated *N*). In humans the haploid number for gametes is 23.

The principle difference between mitosis (described on p. 56) and meiosis is that the chromosome number is reduced to half in meiosis but not in mitosis. Mitosis produces diploid cells that are genetically identical to each other and to their parent cells. Meiosis produces haploid germ cells (gametes) that differ from each other and from their parent cells. If it were not for this reductional division of meiosis, the union of egg and sperm would produce an individual with twice as many chromosomes as the parents. Continuation of this process in just a few generations could yield body cells with astronomical numbers of chromosomes.

Meiosis consists of *two* nuclear divisions in which the chromosomes divide only once (Figure 6-7). Most of the unique features of meiosis occur during the prophase of the first meiotic division. The two members of each pair of homologous chromosomes come into side-by-side contact **(synapsis)** to form a **bivalent.** Each chromosome of the bivalent has already replicated to form two chromatids, each of which will become a new chromosome. The two chromatids within each homologous chromosome are joined at one point, the centromere. Because each bivalent is made up of two pairs of chromatids, or *four* future chromosomes, it is also called a **tetrad** (Gr. *tetra,* four). The position or location of any gene on a chromosome is the gene **locus** (pl., **loci**), and in synapsis all gene loci on a chromosome normally lie exactly opposite the corresponding loci on the homologous chromosome. Toward the end of prophase, the chromosomes shorten and thicken and are ready to enter into the first meiotic division. In contrast to mitosis, the centromeres holding the chromatids together *do not divide* at the beginning of anaphase. As a result, one of each pair of double-stranded chromosomes **(dyads)** is pulled toward each pole by

the microtubules of the division spindle. Therefore at the end of the first meiotic division, also called the reduction division, the daughter cells contain *one of each* of the homologous chromosomes, so the total chromosome number has been reduced to the haploid number (*N*). However, because the chromatids are still joined by the centromeres, each cell contains the 2*N* amount of DNA.

The second meiotic division more closely resembles the events in mitosis. The dyads are split at the beginning of anaphase by division of the centromeres, and single-stranded chromosomes move toward each pole. By the end of the second meiotic division, the cells have the haploid number of chromosomes and the *N* amount of DNA. Each chromatid of the original tetrad exists in a separate nucleus. Four cells (gametes) are formed from the original germ cell, each containing one complete haploid set of chromosomes and only one allele of each gene.

Crossing Over

An important event occurs when a chromatid of one chromosome exchanges parts with an adjacent chromatid of the paired homologous chromosome. This phenomenon, called **crossing over,** is clearly shown in Figure 6-7. Crossing over is important because the hereditary material is redistributed between chromatids of different homologous chromosomes. The chromosomes exchange equivalent sections bearing alleles for the same genes, so that each chromatid contains a full set of genes, but the alleles are in new combinations on the homologous chromosomes.

While the chromosomes are in synapsis, a strand of one chromatid becomes joined with the homologous chromatid. As prophase continues, the homologs begin to move apart, revealing **chiasmata** (sing., **chiasma**), the connection points between *nonsister* chromatids and the probable points of crossover (refer to prophase I in Figure 6-7; also the discussion of crossing over in Chapter 8, p. 136). There may be one

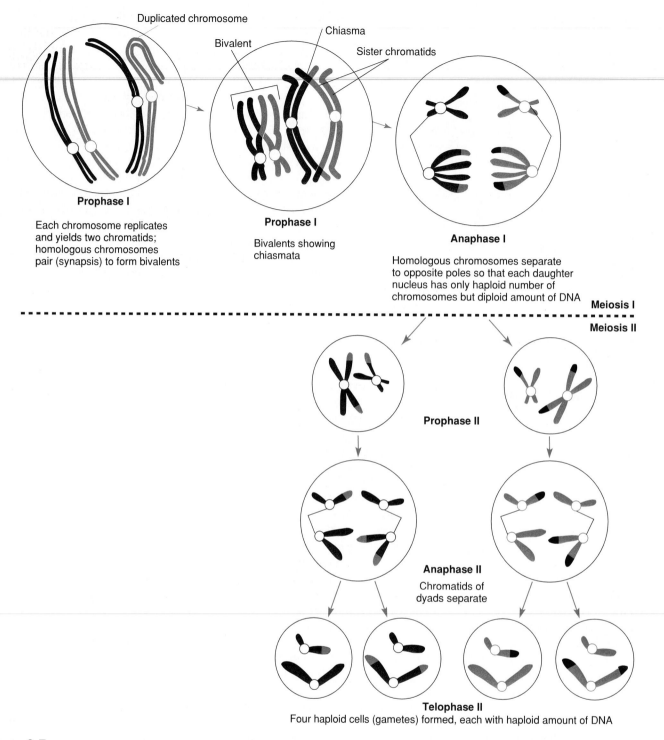

Prophase I

Each chromosome replicates and yields two chromatids; homologous chromosomes pair (synapsis) to form bivalents

Prophase I

Bivalents showing chiasmata

Anaphase I

Homologous chromosomes separate to opposite poles so that each daughter nucleus has only haploid number of chromosomes but diploid amount of DNA

Meiosis I

Meiosis II

Prophase II

Anaphase II

Chromatids of dyads separate

Telophase II

Four haploid cells (gametes) formed, each with haploid amount of DNA

Figure 6-7

Meiosis in a germ cell with two pairs of chromosomes ($2N = 4$, that is, a diploid number of 4) showing the behavior of the chromosomes in two successive cell divisions. Compare this figure with Figure 4-24, p. 57, showing mitosis.

chiasma or more present in each bivalent, depending on the number of times the adjacent homologs have joined. When the chiasmata pull apart, the exchange is complete. The resulting four gametes at the end of meiosis are all genetically different (Figure 6-7).

GAMETOGENESIS

The series of transformations that results in the formation of mature gametes is called gametogenesis. Although the same essential processes are involved in the maturation of both sperm and eggs, there are some important differences. Gametogenesis in the testis is called **spermatogenesis,** and in the ovary it is called **oogenesis.**

Spermatogenesis

The walls of the seminiferous tubules contain the differentiating sex cells arranged in a stratified layer five to eight cells deep (Figure 6-8). The outermost layers contain **spermatogonia** (Figure 6-9), diploid cells which have increased in number by ordinary mitosis. Each spermatogonium increases in size and becomes a **primary spermatocyte.** Each primary spermatocyte then undergoes the first meiotic division, as described previously, to become two **secondary spermatocytes.**

Each secondary spermatocyte enters the second meiotic division without the intervention of a resting period. In the two steps of meiosis each primary spermatocyte gives rise to four **spermatids,** each containing the haploid number (23 in humans) of chromosomes. A spermatid may contain all chromosomes that the male inherited from his mother, those he inherited from his father, or a combination of his parents' chromosomes. Without further divisions the spermatids are

Figure 6-8

Section of a seminiferous tubule containing male germ cells. More than 200 long, highly coiled seminiferous tubules are packed in each human testis. This scanning electron micrograph reveals, in the tubule's central cavity, numerous tails of mature spermatozoa that have differentiated from germ cells in the periphery of the tubule. (×525)

From Tissues and Organs: A Text-Atlas of Scanning Electron Microscopy, *by Richard G. Kessel and Randy H. Kardon, W. H. Freeman and Co., © 1979.*

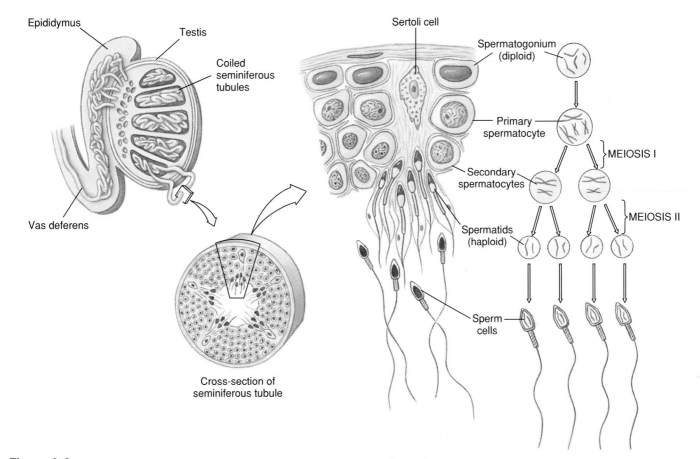

Figure 6-9

Spermatogenesis. Section of seminiferous tubule showing spermatogenesis. The germ cells develop within the recesses of large Sertoli cells that extend from the periphery of the seminiferous tubule to the lumen and that provide nourishment to the germ cells. Stem germ cells from which the sperm differentiate are the spermatogonia, diploid cells located peripherally in the tubule. These divide by mitosis to produce either more spermatogonia or primary spermatocytes. Meiosis begins when the primary spermatocytes divide to produce haploid secondary spermatocytes with double-stranded chromosomes. The second meiotic division forms four haploid spermatids with single-stranded chromosomes. As the sperm develop they are gradually pushed toward the lumen of the seminiferous tubule.

transformed into mature spermatozoa by losing a great deal of cytoplasm, condensing the nucleus into a head, and forming a middle piece and a whiplike, flagellar tail for locomotion (Figure 6-9). The head consists of the nucleus containing the chromosomes for heredity and an **acrosome,** a distinctive feature of nearly all the metazoa (exceptions are the teleost fishes and certain invertebrates). In many species, both invertebrate and vertebrate, the acrosome contains egg-membrane lysins that serve to clear an entrance path through the membranes that form a barrier around the egg. In mammals at least, the lysin is the enzyme hyaluronidase. A striking feature of many invertebrate spermatozoa is the acrosome filament, an extension of varying length in different species that projects suddenly from the sperm head when the latter first contacts the egg membrane. Its fusion with the egg's plasma membrane is the initial event of fertilization (see Contact and Recognition between Egg and Sperm, p. 104).

The total length of the human sperm is 50 to 70 μm. Some toads have sperm that exceed 2 mm (2000 μm) in length (Figure 6-10) and are easily visible to the unaided eye. Most sperm, however, are microscopic in size (see p. 19 for an early nineteenth-century drawing of mammalian sperm, interpreted by biologists of the time as parasitic worms in the semen). In all sexually reproducing animals the number of sperm in males is far greater than the number of eggs in corresponding females. The number of eggs produced is related to the chances of the young to hatch and reach maturity. This explains the enormous number of eggs produced by certain fishes, as compared with the small number produced by mammals.

Oogenesis

The early germ cells in the ovary, called **oogonia,** increase in number by ordinary mitosis. Each oogonium contains the diploid number of chromosomes. In human females after puberty, one of these oogonia typically

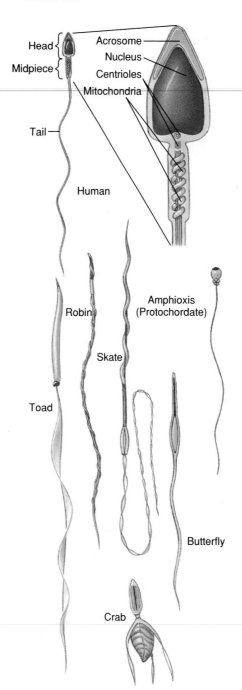

Figure 6-10
Types of vertebrate and invertebrate sperm.

develops each menstrual month into a functional egg. After the oogonia cease to increase in number, they grow in size and become **primary oocytes** (Figure 6-11). Before the first meiotic division, the chromosomes in each primary oocyte meet in pairs, paternal and maternal homologs, just as in spermatogenesis. When the first maturation (reduction) division occurs, the cytoplasm is divided unequally.

One of the two daughter cells, the **secondary oocyte,** is large and receives most of the cytoplasm; the other is very small and is called the **first polar body** (Figure 6-11). Each of these daughter cells, however, has received half of the chromosomes.

In the second meiotic division, the secondary oocyte divides into a large **ootid** and a small polar body. If the first polar body also divides in this division, which sometimes happens, there are three polar bodies and one ootid (Figure 6-11). The ootid develops into a functional ovum. The polar bodies are nonfunctional, and they disintegrate. The formation of the nonfunctional polar bodies is necessary to enable the egg to get rid of excess chromosomes, and the unequal cytoplasmic division makes possible a large cell with sufficient yolk for the development of the young. Thus the mature ovum has the N (haploid) number of chromosomes, the same as the sperm. However, each primary oocyte gives rise to only *one* functional gamete instead of four as in spermatogenesis.

In most vertebrates and many invertebrates the egg does not actually complete all the meiotic divisions before fertilization occurs. The general rule is that development is arrested during prophase of meiosis I. Meiosis is resumed and completed either at the time of ovulation (birds and most mammals) or shortly after fertilization (many invertebrates, teleost fishes, amphibians, and reptiles). In humans, the ova begin meiosis I at about the thirteenth week of fetal development, then their development arrests in prophase. They remain in this condition for up to 20 to 30 years. Meiosis I is completed, and the first polar body is ejected, just before ovulation. Meiosis II is completed when the ovum is penetrated by a spermatozoon.

The most obvious feature of egg maturation is the deposition of yolk. Yolk, usually stored as granules or more organized platelets, is not a definite chemical substance but may be lipid or protein or both. In insects and vertebrates, all having more or less

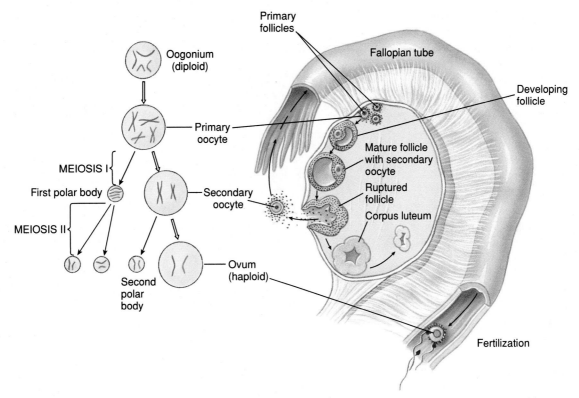

Figure 6-11

Oogenesis. Early germ cells (oogonia) increase by mitosis during embryonic development to form diploid primary oocytes. After puberty, each menstrual month a diploid primary oocyte is divided in the first meiotic division into a haploid secondary oocyte and a haploid polar body. If the secondary oocyte is fertilized, it enters the second meiotic division. The double-stranded chromosomes separate into a large ootid and small second polar body. Both ootid and second polar body now contain the N amount of DNA. Fusion of the haploid egg nucleus with a haploid sperm nucleus produces a diploid (2N) zygote.

yolky eggs, the yolk may be synthesized within the egg from raw materials supplied by the surrounding follicle cells, or preformed lipid or protein yolk may be transferred by pinocytosis from follicle cells to the oocyte.

The result of the enormous accumulation of yolk granules and other nutrients (glycogen and lipid droplets) is that an egg grows well beyond the normal limits that force ordinary body (somatic) cells to divide. A young frog oocyte 50 μm in diameter, for example, grows to 1500 μm in diameter when mature after 3 years of growth in the ovary, and the volume has increased by a factor of 27,000. Bird eggs attain even greater absolute size; a hen egg will increase 200 times in volume in only the last 6 to 14 days of rapid growth preceding ovulation.

Thus eggs are remarkable exceptions to the otherwise universal rule that organisms are composed of relatively minute cellular units. This cre-

ates a surface area-to-cell volume ratio problem, since everything that enters and leaves the ovum (nutrients, respiratory gases, wastes, and so on) must pass through the cell membrane. As the egg becomes larger, the available surface per unit of cytoplasmic volume (mass) becomes smaller. As we would anticipate, the metabolic rate of the egg gradually diminishes until, when mature, the ovum is in suspended animation awaiting fertilization.

MATERNAL SUPPORT OF THE EMBRYO

The great majority of invertebrates, as well as many vertebrates, lay their eggs in the environment for development; these animals are called **oviparous** ("egg-birth"). Fertilization may be either internal (the eggs are fertilized inside the body of the female before she lays them) or external (the eggs are fertilized by the male after the female lays them).

While many oviparous animals simply abandon their eggs rather indiscriminately, others display extreme care in finding places that will provide immediate and suitable sources of food for the young when they hatch.

Some animals retain their eggs in the body (usually the oviduct) while they develop, with the embryo deriving all of its nourishment from yolk stored within the egg. These animals are called **ovoviviparous** ("egg-live-birth"). Ovoviviparity occurs in several invertebrate groups (for example, various annelids, brachiopods, insects, and gastropod molluscs) and is common among certain fishes and reptiles.

In the third pattern, **viviparous** ("live-birth"), the egg develops in the oviduct or uterus with the embryo deriving its nourishment directly from the mother. Usually some kind of intimate anatomical relationship is established between the developing embryo and the mother. In both ovoviviparity and viviparity, fertilization must be internal (that is, within the body of the female) and the mother gives birth to young in an advanced stage of development. Viviparity is confined mostly to the mammals and elasmobranch fishes, although viviparous invertebrates (some scorpions, for example), amphibians, and reptiles are known. Development of the embryos within the mother's body, whether ovoviviparous or viviparous, obviously affords more protection to the offspring than egg-laying.

PLAN OF REPRODUCTIVE SYSTEMS

The basic components of the reproductive systems are similar in sexual animals, although differences in reproductive habits, methods of fertilization, and so on have produced many variations. Sexual systems consist of two components: (1) **primary organs,** which are the gonads that produce the sperm and eggs and the steroid sex hormones; and (2) **accessory organs,** which assist the gonads in the formation and delivery of the gametes, and may also serve to support the embryo. They are of great variety, and include the gonoducts (sperm ducts and oviducts), accessory organs for transferring spermatozoa into the female, storage organs for spermatozoa or yolk, packaging systems for eggs, and nutritional organs such as yolk glands and the placenta.

INVERTEBRATE REPRODUCTIVE SYSTEMS

Invertebrates that transfer sperm from male to female for internal fertilization require organs and plumbing to facilitate this function that may be as complex as that of any vertebrate. In contrast, the reproductive systems of invertebrates that simply release their gametes into the water for external fertilization may be little more than centers for gametogenesis. Polychaete annelids, for example, have no permanent reproductive organs. Gametes arise by proliferation of cells lining the body cavity. When mature

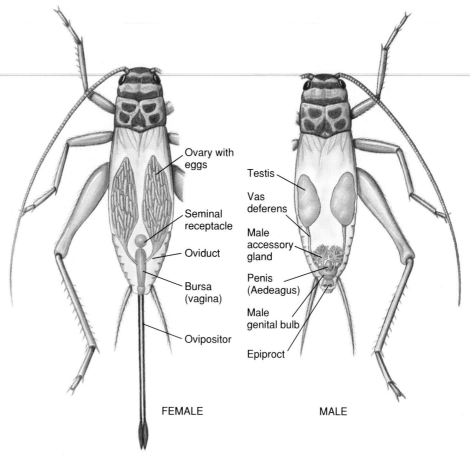

FEMALE   **MALE**

Figure 6-12

Reproductive system of crickets. Sperm from the paired testes of the male pass through sperm tubes (vas deferens) to an ejaculatory duct housed in the penis. In females, eggs from the ovaries pass through oviducts to the genital bursa. At mating, sperm enclosed in a membranous sac (spermatophore) formed by the secretions of the accessory gland are deposited in the genital bursa of the female, then migrate to the seminal receptacle where they are stored. The female controls the release of a few sperm to fertilize her eggs at the moment they are laid, using the needlelike ovipositor to deposit the eggs in the soil.

the gametes are released through coelomic or nephridial ducts or, in some species, they may spill out through ruptures in the body wall.

Insects have separate sexes (dioecious), practice internal fertilization by copulation and insemination, and consequently have complex reproductive systems (Figure 6-12). Sperm from the testes of males pass through sperm ducts to seminal vesicles (where the sperm are stored) and then through a single ejaculatory duct to a penis. Seminal fluid from one or more accessory glands is added to the semen in the ejaculatory duct. Females have a pair of ovaries formed from a series of egg tubes (ovarioles). Mature ova pass through oviducts to a common genital chamber and then to a short copula-

tory bursa (vagina). In most insects, the male transfers sperm by inserting the penis directly into the female system where they are stored in a seminal vesicle. Often a single mating provides sufficient sperm to last the reproductive life of the female.

VERTEBRATE REPRODUCTIVE SYSTEMS

In vertebrates the reproductive and excretory systems are often referred to as the **urogenital system** because of their close anatomical connection, especially in the male. This association is very striking during embryonic development. In male fishes and amphibians the duct that drains the kidney (**Wolffian duct**) also serves as the sperm duct. In male reptiles, birds, and mammals in which the kidney develops its own independent duct (**ureter**) to carry away waste, the old Wolffian duct becomes exclusively a sperm duct (**vas deferens**). In all these forms, with the exception of most mammals, the ducts open into a **cloaca** (derived, appropriately, from the Latin meaning "sewer"), a common chamber into which the intestinal, reproductive, and excretory canals empty. Almost all placental mammals have no cloaca; instead the urogenital system has its own opening separate from the anal opening. The **oviduct** of the female is an independent duct that does, however, open into the cloaca in forms that have a cloaca.

Male Reproductive System

The male reproductive system of vertebrates, such as that of the human male (Figure 6-13) includes testes, vasa efferentia, vasa deferentia, glands, and (in some birds, some reptiles, and all mammals) a penis.

The paired **testes** are the sites of sperm production. Each testis is made up of numerous **seminiferous tubules,** in which the sperm develop (Figure 6-9), and the **interstitial tissue** lying along the tubules, which produces the male sex hormone (testosterone). In most mammals the two testes are housed permanently in a sac-like scrotum suspended outside the abdominal cavity, or the testes descend into the scrotum during the breeding season. This strange and seemingly insecure arrangement provides an environment of slightly lower temperature, since in at least some mammals (including humans) sperm apparently do not form at temperatures maintained within the body. In the marine mammals and all other vertebrates the testes are positioned permanently within the abdomen.

The sperm are conveyed from the seminiferous tubules to the **vasa efferentia,** small tubes passing to a coiled **epididymis** (one for each testis) (Figure 6-9) and then to a **vas deferens,** the ejaculatory duct. In mammals the

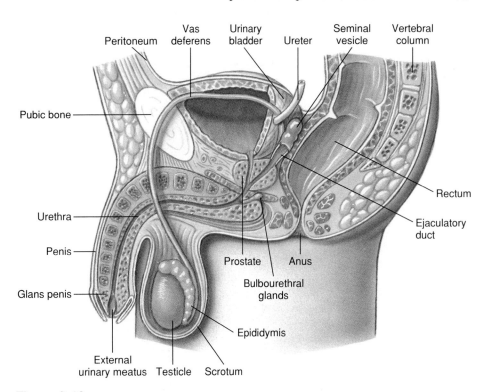

Figure 6-13
Human male reproductive system showing the reproductive structures in sagittal view.

vas deferens joins the **urethra,** a duct that serves to carry both sperm and urinary products through the penis, or external intromittent organ.

Most aquatic vertebrates have no need for a penis, since sperm and eggs are liberated into the water in close proximity to each other. However, in terrestrial (and some aquatic) vertebrates that bear their young alive or enclose the egg within a shell, sperm must be transferred to the female. In most birds, this is a rather haphazard process of simply presenting cloaca to cloaca. Only reptiles and mammals have a true penis. In mammals the normally flaccid organ is erected when engorged with blood. Many mammals, although not humans, possess a bone in the penis (baculum), which presumably helps with rigidity.

In most mammals three pairs of glands open into the reproductive channels: **seminal vesicles, prostate glands,** and the **bulbourethral glands** (Figure 6-13). Fluid secreted by these glands furnishes food to the sperm, lubricates the

passageways for the sperm, and counteracts the acidity of the urine so that the sperm are not harmed.

Female Reproductive System

The ovaries of female vertebrates produce both ova and the female sex hormones (estrogens and progesterone). In all jawed vertebrates, mature ova from each ovary enter the funnel-like opening of an **oviduct** which typically has a fringed margin that embraces the ovary. The posterior end of the oviduct is unspecialized in most fishes and amphibians, but in the cartilaginous fishes, reptiles, and birds that produce a large, shelled egg, special regions have developed for the production of albumin and shell. In the amniotes (reptiles, birds, and mammals; see Amniotes and the Amniotic Egg, p. 117) the terminal portion of the oviduct is expanded into a muscular **uterus** in which shelled eggs are held before laying or embryos complete their development. In placental mammals, the walls of the uterus establish a close vascular association with the embryonic membranes through a **placenta** (see p. 118).

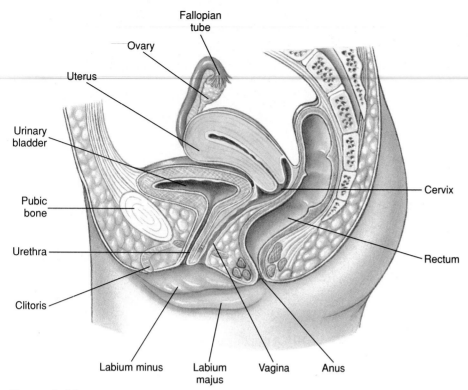

Figure 6-14
Human female reproductive system showing the pelvis in sagittal section.

The paired ovaries of the human female (Figure 6-14), slightly smaller than the male testes, contain many thousands of eggs (ova). Each egg develops within a **follicle** that enlarges and finally ruptures to release the mature egg (Figure 6-11). During the fertile period of the woman, approximately 13 eggs mature each year, and usually the ovaries alternate in releasing an egg. Since the woman is fertile for only some 30 years, of the half a million primary oocytes in her ovaries at birth, only 300 to 400 have a chance to reach maturity; the others degenerate and are absorbed.

The **oviducts,** or fallopian tubes, are lined with cilia for propelling the egg in its course. The two ducts open into the upper corners of the uterus, or womb, which is specialized for housing the embryo during the 9 months of its intrauterine existence. It is provided with thick muscular walls, many blood vessels, and a specialized lining: the **endometrium.** The uterus varies with different mammals. It was originally paired but tends to fuse in many eutherian mammals.

The **vagina** is a muscular tube adapted for receiving the male's penis and for serving as the birth canal during expulsion of the fetus from the uterus. Where the vagina and the uterus meet, the uterus projects down into the vagina to form the **cervix.**

The external genitalia of the human female, or vulva, include folds of skin, the **labia majora** and **labia minora,** and a small erectile organ, the **clitoris** (the female homolog of the glans penis of the male). The opening into the vagina is normally reduced in size in the virgin state by a membrane of no known function, the **hymen.**

HORMONES OF VERTEBRATE REPRODUCTION

HORMONAL CONTROL OF THE TIMING OF REPRODUCTIVE CYCLES

From fish to mammals, reproduction in vertebrates is usually a seasonal or cyclic activity. Timing is crucial, because the young should appear when food is available and other environmental conditions are optimal for survival. Once set in motion by some environmental cue, such as seasonal change in temperature or photoperiod, or some social force, the sexual reproductive process is controlled by hormones. Hormones of the anterior pituitary gland link the neurosecretory centers of the brain to the endocrine tissues of the gonads (neurosecretion and the pituitary gland are described in Chapter 37, beginning on p. 743). This delicately balanced hormonal system controls the development of the gonads, the accessory sex structures, and the secondary sexual characteristics (see the following text) as well as the timing of reproduction.

The cyclic reproductive patterns of mammals are of two types: the **estrous cycle,** characteristic of most mammals, and the **menstrual cycle,** characteristic only of the anthropoid primates (monkeys, apes, and humans). These two cycles differ in two important ways. First, in the estrous cycle the female is receptive to the male only during brief periods of **estrus,** or "heat," whereas in the menstrual cycle receptivity may occur throughout the cycle. Second, the menstrual cycle, but not the estrous cycle, ends with the collapse and discharge of the inner portion of the endometrium (uterine lining). In the estrous animal each cycle ends with the uterine lining simply reverting to its original state, without the discharge characteristic of the menstrual cycle.

THE GONADAL STEROIDS AND THEIR CONTROL

The ovaries produce two kinds of steroid sex hormones—**estrogens and progesterone** (Figure 6-15). Estrogens are responsible for the development of the female accessory sex structures (oviducts, uterus, and vagina) and for stimulating female reproductive activity. The secondary sex characters, that is, characteristics that are not primarily involved in the formation and delivery of ova or sperm but that are essential for the behavioral and functional success of reproduction, are also controlled or maintained by estrogens. These include characteristics such as distinctive skin or feather coloration, bone development, body size and, in mammals, the initial development of the mammary glands. In mammals, progesterone is responsible for preparing the uterus to receive the developing embryo. These hormones are controlled by the **pituitary gonadotropins,** follicle-stimulating hormone (FSH), and luteinizing hormone (LH) (Figure 6-16). The gonadotropins are in turn governed by the **gonadotropin releasing hormones (GnRH)** produced by neurosecretory centers in the hypothalamus (p. 746 and Table 37-1). Through this control system environmental factors such as light, nutrition, and stress may influence reproductive cycles.

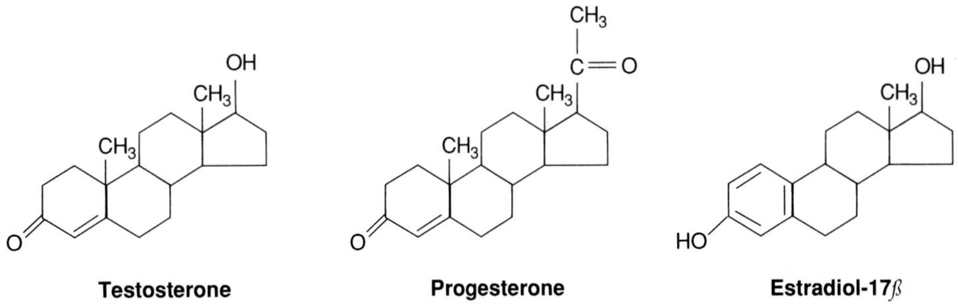

Oral contraceptives (the "pill") are usually combined preparations of estrogen and progesterone that act to decrease the output of the pituitary gonadotropins FSH and LH. This prevents the ovarian follicles from ripening, and ovulation from occurring. Oral contraceptives are highly effective, with a failure rate of less than 1%, if the treatment procedure is followed properly.

The male sex hormone **testosterone** (Figure 6-15) is manufactured by the **interstitial cells** of the testes. Testosterone is necessary for the growth and development of the male accessory sex structures (penis, sperm ducts, and glands), development of secondary male sex characteristics (such as bone and muscle growth, male plumage or pelage coloration, antlers in deer, and, in humans, voice quality), and male sexual behavior. The development of the testes and the secretion of testosterone is controlled by FSH and LH, the same pituitary hormones that regulate the female reproductive cycle.

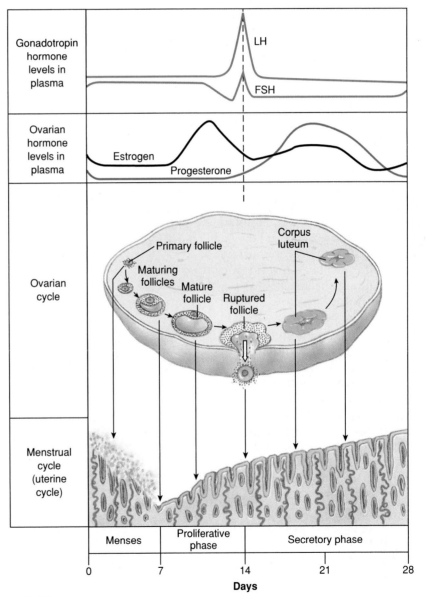

Figure 6-15

Sex hormones. These three sex hormones show the basic four-ring steroid structure. The female sex hormone estradiol (an estrogen) is a C_{18} (18-carbon) steroid with an aromatic A ring (first ring to left). The male sex hormone testosterone is a C_{19} steroid with a carbonyl group (C=O) on the A ring. The female pregnancy hormone progesterone is a C_{21} steroid, also bearing a carbonyl group on the A ring.

Figure 6-16

Human menstrual cycle, showing changes in blood hormone levels and uterine endometrium during the 28-day cycle. FSH promotes maturation of the ovarian egg follicles, which secrete estrogen. Estrogen prepares the uterine endometrium and causes a surge in LH, which in turn stimulates the corpus luteum to secrete progesterone. Progesterone production will persist only if the egg is fertilized; without pregnancy, progesterone declines and menstruation follows.

THE MENSTRUAL CYCLE

The human menstrual cycle (L. *mensis,* month) consists of three distinct phases: menstrual phase, follicular phase, and luteal phase (Figure 6-16). Menstruation (the "period") signals the menstrual phase, when part of the lining of the uterus (endometrium) degenerates and sloughs off, producing the menstrual discharge. By day 3 of the cycle the blood levels of FSH and LH begin to rise slowly, prompting some of the ovarian follicles to begin growing and to secrete estrogen. As estrogen levels in the blood increase, the uterine endometrium begins to thicken and uterine glands within the endometrium enlarge. By day 10 most of the ovarian follicles that began to develop at day 3 now degenerate, leaving only one (sometimes two or three) to continue ripening until it appears like a blister on the surface of the ovary.

At day 13 or 14 in the cycle, a surge of LH from the pituitary induces the largest follicle to rupture **(ovulation),** releasing the egg onto the ovarian surface. Now follows a critical period, for unless the mature egg is fertilized within a few hours it will die. During the luteal phase, a **corpus luteum** ("yellow body") forms from the wall of the follicle that ovulated (Figures 6-11 and 6-16). The corpus luteum, responding to the continued stimulation of LH, becomes a transitory endocrine gland that secretes progesterone in addition to estrogen. Progesterone ("before carrying [gestation]"), as its name implies, stimulates the uterus to undergo the final maturational changes that prepare it for gestation. The uterus is now fully ready to house and nourish the embryo. If fertilization has *not* occurred, the corpus luteum disappears, and its hormones are no longer secreted. Since the uterine lining (endometrium) depends on progesterone and estrogen for its maintenance, their disappearance causes the uterine lining to deteriorate, leading to the menstrual discharge.

HORMONES OF HUMAN PREGNANCY AND BIRTH

If fertilization occurs, the developing blastocyst will contact the uterine surface after about 6 days and bury itself in the endometrium. This process is called implantation. Growth of the embryo continues, producing a spherically shaped trophoblast. This embryonic stage contains three distinct tissue layers, the amnion, chorion, and the embryo proper, the inner cell mass (p. 109). The chorion becomes the source of the **human chorionic gonadotropin (HCG),** which appears in the bloodstream soon after implantation. HCG stimulates the corpus luteum to synthesize and release both estrogen and progesterone (Figure 6-17). The point of attachment between the trophoblast and the uterus becomes the placenta (the evolution and development of the placenta is described in the next chapter, p. 118). Besides serving as a medium for the transfer of materials between the maternal and fetal bloodstreams, the placenta also serves as an endocrine gland. After about the third month of pregnancy, the corpus luteum degenerates, but by then the

While women in more than 90 other countries benefit from safe, recently-developed, easier-to-use contraceptives, American couples are limited to the standby contraceptives developed more than 30 years ago: the Pill, condom, IUD, diaphragm, and surgical sterilization. Without a drastic change in federal policy, which is unlikely, serious economic, political, and social problems will continue to mire reproductive research in controversy. Because of high development costs, an inhospitable regulatory climate, the growing number of costly lawsuits, and threats of boycotts of drug companies by anti-abortion activists, only one U.S. company is now actively pursuing contraceptive research, while at least eight American companies did so in the early 1970s. The unfortunate consequence is that contraceptive failures account for some 2 million unwanted pregnancies each year in the United States and for about half the 1.5 million abortions, one of the highest abortion rates in the industrialized world. Without a change in the present adverse political climate, there is little hope of reducing unwanted pregnancies.

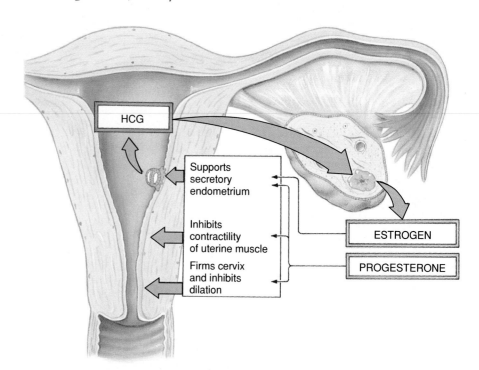

Figure 6-17

The multiple roles of progesterone and estrogen in normal human pregnancy. After implantation of the embryo in the uterus, the trophoblast (the future placenta) secretes human chorionic gonadotropin (HCG) which maintains the corpus luteum until the placenta, at about the seventh week of pregnancy, begins producing the sex hormones progesterone and estrogen.

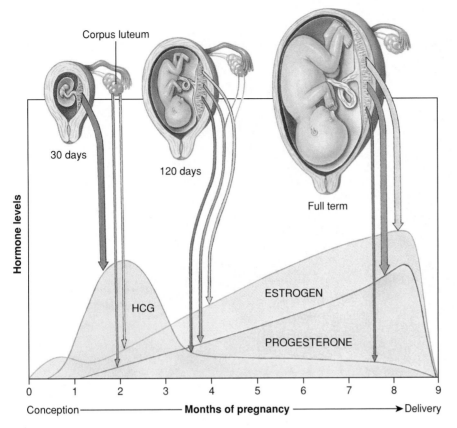

Corpus luteum

30 days

120 days

Full term

Hormone levels

HCG

ESTROGEN

PROGESTERONE

0   1   2   3   4   5   6   7   8   9

Conception ——————— **Months of pregnancy** ——————→ Delivery

Figure 6-18

Hormone levels released from the corpus luteum and placenta during pregnancy. The width of the arrows suggests the amount of hormones released. HCG (human chorionic gonadotropin) is produced solely by the placenta. The synthesis of progesterone and estrogen shifts during pregnancy from the corpus luteum to the placenta.

placenta itself is secreting both progesterone and estrogen (Figure 6-18). Later the placenta begins to synthesize **relaxin;** this hormone allows some expansion of the pelvis by increasing the flexibility of the pubic symphysis, and also dilates the cervix in preparation for delivery.

The preparation of the mammary glands for secretion of milk requires two additional hormones, **prolactin** and **human placental lactogen.** Prolactin is produced by the pituitary, but in the nonpregnant woman its secretion is inhibited. During pregnancy, the elevated levels of progesterone and estrogen depress the inhibitory signal, and prolactin begins to appear in the blood. Prolactin, in combination with human placental lactogen, prepare the mammary glands for secretion. After birth, the actual secretion of milk is triggered when the infant sucks on the nipple. This leads to the reflex release of **oxytocin** from

the pituitary; when oxytocin reaches the mammary gland it causes contraction of smooth muscles lining the ducts and sinuses of the mammary glands and the ejection of milk.

Birth, or parturition, begins with a series of strong, rhythmic contractions of the uterine musculature, called **labor.** The exact signal that triggers birth is not fully understood, but several important factors have been identified. Just before birth the placenta begins to die and two essential hormonal changes happen. The secretion of estrogen, which stimulates uterine contractions, rises sharply, while the level of progesterone, which inhibits uterine contractions, declines (Figure 6-18). This removes the "progesterone block" that keeps the uterus quiescent throughout pregnancy. Prostaglandins, a group of hormone-like long-chain fatty acids, also increase at this time, making the uterus more irritable. Finally, stretching of the uterus

sets in motion neural reflexes which stimulate the secretion of oxytocin from the pituitary. Oxytocin also stimulates uterine smooth muscle, leading to stronger and more frequent labor contractions.

Given the intricacy of pregnancy it may seem remarkable that healthy babies are ever born! In fact we are the lucky survivors of pregnancy, for miscarriages are quite common and serve as a mechanism to reject prenatal abnormalities such as chromosomal damage and other genetic errors, exposure to drugs or toxins, immune irregularities, or improper hormonal priming of the uterus. Modern hormonal tests show that about 30% of fertile zygotes are spontaneously aborted before or right after implantation; such miscarriages are unknown to the mother or are expressed as a slightly late menstrual period. Another 20% of established pregnancies end in miscarriage (those known to the mother), giving a spontaneous abortion rate of about 50%.

Childbirth occurs in three stages. In the first stage the neck (cervix), or opening of the uterus into the vagina, is enlarged by the pressure of the baby in its bag of amniotic fluid, which may be ruptured at this time. In the second stage the baby is forced out of the uterus and through the vagina to the outside (Figure 6-19). In the third stage the placenta, or afterbirth, is expelled from the mother's body, usually within 10 minutes after the baby is born.

Multiple Births

Many mammals give birth to more than one offspring at a time or to a litter, each member of which has come from a separate egg. There are many mammals, however, that have only one offspring at a time, although occasionally they may have plural young. The armadillo (*Dasypus*) is almost unique among mammals in giving birth to four young at one time—all of the same sex, either male or female, and all derived from one zygote.

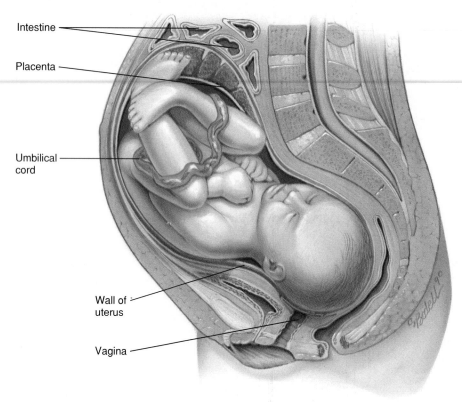

Intestine

Placenta

Umbilical cord

Wall of uterus

Vagina

Figure 6-19
Position of human fetus just before birth.

Human twins may come from one zygote (identical, or monozygotic, twins; Figure 6-20B) or two zygotes (nonidentical, dizygotic, or fraternal, twins; Figure 6-20A). Triplets, quadruplets, and quintuplets may include a pair of identical twins. The other babies in such multiple births usually come from separate zygotes. Fraternal twins do not resemble each other more than other children born separately in the same family, but identical twins are, of course, strikingly alike and always of the same sex. Embryologically, each member of fraternal twins has its own placenta and amnion (Figure 6-20A). About 33% of identical twins have separate placentas, indicating that the blastomeres separated at an early, possibly the two-cell, stage (Figure 6-20B, *top*). All other identical twins share a common placenta, indicating that splitting occurred after formation of the inner cell mass (see Figure 7-24 on p. 119). If the splitting were to happen after placenta formation but before the amnion forms, the twins would have individual amniotic sacs (Figure 6-20B, *middle*). This happens in the great majority of identical twins. Finally, a very small percentage of identical twins share one amniotic sac and a single placenta (Figure 6-20B, *bottom*). For this to happen, separation must have occurred after day 9 of pregnancy, since the amnion has formed by this time. In these cases, the twins are at risk of becoming conjoined, a condition known as Siamese twinning.

The frequency of twin births in comparison to single births is approximately 1 in 86, that of triplets 1 in 86^2, and that of quadruplets approximately 1 in 86^3. The frequency of identical twin births to all births is about the same the world over, whereas the frequency of fraternal births varies with race and country. In the United States, three-fourths of all twin births are dizygotic (fraternal), whereas in Japan only a little more than one-fourth are dizygotic. The tendency for fraternal twinning (but apparently not identical twinning) seems to run in family lines; fraternal twinning (but not identical twinning) also increases in frequency as mothers get older.

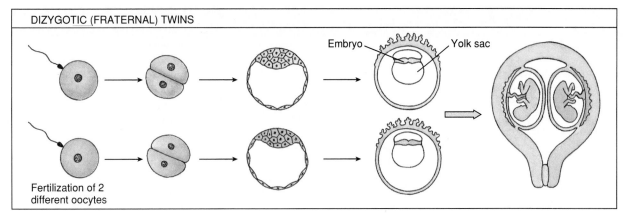

A

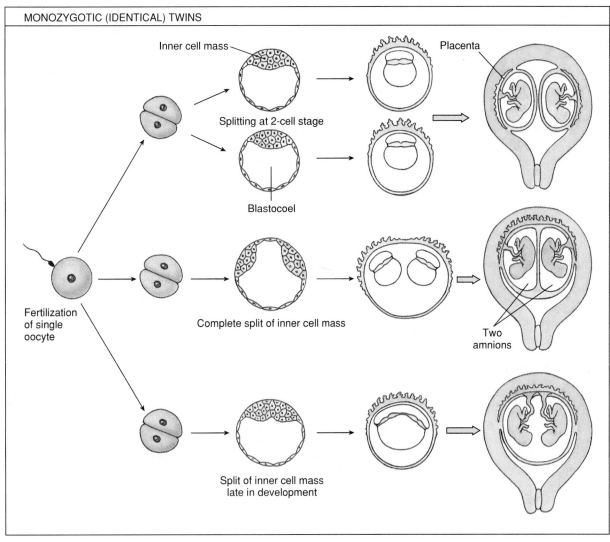

B

Figure 6-20

Formation of human identical twins. See text for explanation.

Summary

Reproduction is a universal property of all living organisms. Asexual reproduction is a rapid and direct process by which a single organism produces genetically identical copies of itself. It may occur by fission, budding, gemmulation, and fragmentation. Sexual reproduction involves the production of sex cells (gametes), usually by two parents, which combine by fertilization to form a zygote that develops into a new individual. The sex cells are formed by meiosis, reducing the number of chromosomes to haploid, and the diploid chromosome number is restored at fertilization. Sexual reproduction recombines parental characters and thus reshuffles and amplifies genetic diversity. This is important for evolution. Two alternatives to typical bisexual reproduction are parthenogenesis, the development of an unfertilized egg, and hermaphroditism, the presence of both male and female organs in the same individual.

Sexual reproduction exacts heavy costs in time and energy, requires cooperative investments in mating, and results in a 50% loss of genetic representation of each parent in the offspring. The classical view of why sex is needed is that it maintains variable offspring within the population having superior fitness for environmental change.

In vertebrates the primordial germ cells arise in the yolk sac endoderm, then migrate to the gonad. In mammals the gonad will become a testis in response to masculinizing signals from the Y chromosome of the male, and an ovary in the absence of such signals in the female.

In bisexual animals the genetic material is distributed to the offspring in the gametes (eggs and sperm), produced by meiosis. Each somatic cell in an organism has two chromosomes of each kind (homologous chromosomes) and is thus diploid. Meiosis separates the homologous chromosomes, so that each gamete has half the somatic chromosome number (haploid). In the first meiotic division the centromeres do not divide, and each daughter cell receives one of each of the replicated homologous chromosomes with the chromatids still attached to the centromere. At the beginning of the first meiotic division, the replicated homologous chromosomes come to lie alongside each other (synapsis), forming a bivalent. The gene loci on one set of chromatids lie opposite the corresponding loci on the chromatids of the homologous chromosomes. Chromatids can exchange material with nonsister chromatids (crossing over) to produce new genetic combinations. At the second meiotic division, the centromeres divide, separating each replicated chromosome into two unreplicated chromosomes that go to different daughter cells. The diploid number is restored when the male and female gametes fuse to form the zygote.

Germ cells mature in the gonads by a process called gametogenesis (spermatogenesis in the male and oogenesis in the female), involving both mitosis and meiosis. In spermatogenesis, each primary spermatocyte gives rise by meiosis and growth to four motile sperm, each bearing the haploid number of chromosomes. In oogenesis, each primary oocyte gives rise to only one mature, nonmotile, haploid ovum. The remaining nuclear material is discarded in polar bodies. During oogenesis the egg accumulates large food reserves.

Sexual reproductive systems vary enormously in complexity, ranging from some invertebrates, such as polychaete worms that lack any permanent reproductive structures to the complex systems of vertebrates and many invertebrates consisting of permanent gonads and various accessory structures for transferring, packaging, and nourishing the gametes and the embryo.

The male reproductive system of humans includes the testes, composed of seminiferous tubules in which millions of sperm develop, and a duct system (vasa efferentia and vas deferens) that joins the urethra, glands (seminal vesicles, prostate, bulbourethral), and the penis. The human female system includes the ovaries, containing thousands of eggs within follicles; egg-carrying oviduct; the uterus; and the vagina.

The seasonal or cyclic nature of reproduction in vertebrates has required the evolution of precise hormonal mechanisms that control the production of sex cells, signal readiness for mating, and prepare ducts and glands for successful fertilization of eggs. The neurosecretory centers of the brain are linked to the endocrine cells of the gonads by the gonadotropic hormones: follicle-stimulating hormone (FSH) and luteinizing hormone (LH). Estrogens in the female and testosterone in the male control the growth of the accessory sex structures and the secondary sex characteristics.

In the human menstrual cycle estrogen induces the initial proliferation of the uterine endometrium. A surge in LH midway in the cycle induces ovulation and causes the corpus luteum to secrete progesterone which completes preparation of the uterus for implantation. If the egg is fertilized, pregnancy is maintained by sex hormones produced by the placenta.

Human pregnancy is divided into three periods: the germinal period (first 2 weeks) when the major body organs begin formation; the embryonic period (2 to 8 weeks) when organs develop rapidly and body shape is established; and the fetal period (3 months to term) when final growth occurs. At term, the sudden decline of progesterone secretion together with the action of other hormones produces labor contractions. Birth follows.

Multiple births in mammals may result from the division of one zygote, producing identical twins, or from separate zygotes, producing fraternal twins. Identical twins in humans may have separate placentas, or (most commonly) they may share a common placenta but have individual amniotic sacs.

Review Questions

1. Define asexual reproduction, and describe four forms of asexual reproduction in invertebrates.

2. Define sexual reproduction and explain why meiosis contributes to one of its great strengths.

3. Explain why genetic mutations in asexual organisms lead to much more rapid evolutionary change than do genetic mutations in sexual forms.

4. Define two alternatives to bisexual reproduction—parthenogenesis and hermaphroditism—and offer a specific example of each from the animal kingdom. What is the difference between ameiotic and meiotic parthenogenesis?

5. Define the terms dioecious and monoecious. Can either of these terms be used to describe a hermaphrodite?

6. A paradox of sexual reproduction is that despite being widespread in nature, the question of why it exists at all is still unresolved. What are some disadvantages of sex? What are some consequences of sex that make it so important?

7. What is a germ cell line? How do the germ cells (or the germ plasm) pass from one generation to the next?

8. Name the principal phases of meiosis. What events in meiosis clearly distinguish meiosis from mitosis? Which of the two meiotic divisions most closely resembles mitosis?

9. Explain how a spermatogonium, containing a diploid number of chromosomes, develops into four functional sperm, each containing a haploid number of chromosomes. In what significant way(s) does oogenesis differ from spermatogenesis?

10. Define, and distinguish between, the terms oviparous, ovoviviparous, and viviparous.

11. Name the general location and give the function of the following reproductive structures: seminiferous tubules, vas deferens, urethra, seminal vesicles, mature follicle, oviducts, endometrium.

12. How do the two kinds of mammalian reproductive cycles—estrous and menstrual—differ from each other?

13. What is the male sex hormone and what is its function?

14. Explain how the female hormones FSH, LH, and estrogen interact during the menstrual cycle to bring about ovulation and, subsequently, formation of the corpus luteum.

15. Explain the function of the corpus luteum in the menstrual cycle. If fertilization of the ovulated egg happens, what endocrine events occur to support pregnancy?

16. Describe the principal developmental events that occur during germinal, embryonic, and fetal periods of human pregnancy.

17. If identical human twins develop from separate placentas, when would this indicate the embryo must have separated? When must separation have occurred if the twins share a common placenta but develop within separate amnions?

Selected References

Bell, G. 1982. The masterpiece of nature: the evolution and genetics of sexuality. Berkeley, University of California Press. *Scholarly synthesis of ideas on the meaning of sex. Advanced treatment.*

Blüm, V. 1986. Vertebrate reproduction: a textbook. New York, Springer-Verlag (translated from the German by A. C. Whittle). *Broad, readable treatment of structure and function of vertebrate reproductive systems.*

Cole, C. J. 1984. Unisexual lizards. Sci. Am. **250:**94–100 (Jan.). *Some populations of whiptail lizards from the American southwest consist only of females that reproduce by virgin birth.*

Crews, D. 1994. Animal sexuality. Sci. Am. **270:**108–114 (Jan.). *Sex is determined genetically in mammals and most other vertebrates, but not in many reptiles and fishes which lack sex chromosomes*

altogether. *The author describes nongenetic sex determination and suggests a new framework for understanding the origin of sexuality.*

Forsyth, A. 1986. A natural history of sex: the ecology and evolution of sexual behavior. New York, Charles Scribner's Sons. *Engagingly written, factually accurate account of the sex lives of animals from protozoa to humans, abounding in imagery and analogy. Highly recommended.*

Halliday, T. 1982. Sexual strategy. Survival in the wild. Chicago, University of Chicago Press. *Semipopular treatment of sexual strategies, especially vertebrate mating systems, rested in a framework of natural selection. Well-chosen illustrations.*

Jameson, E. W. 1988. Vertebrate reproduction. New York, John Wiley & Sons. *Comparative treatment of diversity of*

reproductive patterns in vertebrates; includes parental investment and environmental responses.

Maxwell, Kenneth. 1994. The sex imperative: an evolutionary tale of sexual survival. New York, Plenum Press. *Witty survey of sex in the animal kingdom.*

Michod, Richard E. 1995, Eros and evolution: a natural philosophy of sex. Reading, Massachusetts, Addison-Wesley Publishing Company. *In this engaging book, the author argues that sex evolved as a way of coping with genetic errors and avoiding homozygosity.*

Pollard, Irina. 1994. A guide to reproduction: social issues and human concerns. Cambridge, Cambridge University Press. *This comprehensive treatment of human reproduction extends biology to the social and environmental consequences of human reproductive potential.*

7

Principles of Development

The Primary Organizer

In the 1920s and 1930s, research in embryology was dominated by one issue: embryonic induction, the capacity of one tissue to influence the developmental fate of another. The new paradigm of induction began with the work of the German embryologist Hans Spemann (1869 to 1941) who set out to discover how different parts of an embryo influence one another. In experiments carried out in 1916, Spemann had noted the capacity of tissue transplanted from the dorsal lip of the salamander gastrula to transform the tissue it touched. These delicate experiments were repeated in 1921 and 1922 by his student Hilde Pröscholdt, who, despite great difficulties with the amphibian material used, produced six successful embryos in which the transplanted tissue had induced the host embryo to form a secondary embryo (the results are described in more detail on p. 114). Spemann designated the dorsal lip tissue the **primary organizer** because it was the only tissue that had the capacity to organize, by induction, the principal axis of a secondary embryo. The classic experiments were published in 1924 but Hilde, who in the meantime had married the embryologist Otto Mangold, had already died as the result of a household accident. Spemann (above, photographed in his laboratory) was awarded the Nobel Prize in 1935, the only biologist ever cited purely for research in embryology. By demonstrating the central importance of induction, Spemann had ushered in the golden age of embryology, which continued until after World War II when induction research began to yield to studies of genetic control of body form. ■

How is it possible that a tiny, spherical fertilized human egg, scarcely visible to the naked eye, can unfold into a fully formed, unique person, consisting of thousands of billions of cells, each cell performing a predestined functional or structural role? How is this marvelous unfolding controlled? Clearly all the information needed must originate from the nucleus and in the surrounding cytoplasm. But knowing where the control system lies is very different from understanding how it guides the conversion of a fertilized egg into a fully differentiated animal. Despite intense scrutiny by thousands of scientists over many decades, it seemed until very recently that developmental biology, almost alone among the biological sciences, lacked a satisfactory conceptual coherence. This has now changed. During the last two decades the combination of genetics with modern techniques of cellular and molecular biology has produced an avalanche of information that has solved many questions. The causal relationships between development and evolution have also become the focus of research. We do at last appear to have a conceptual framework to account for development.

Early Concepts: Preformation Versus Epigenesis

Early scientists and lay people alike speculated at length about the mystery of development long before the process was submitted to modern techniques of biochemistry, molecular biology, tissue culture, and electron microscopy. An early and persistent idea was that the young animal was preformed in the egg and that development was simply a matter of unfolding what was already there. Some claimed they could actually see a miniature of the adult in the egg or the sperm (Figure 7-1). Even the more cautious argued that all the parts of the embryo were in the egg, needing only to unfold, but so small and transparent they could not be

Figure 7-1
Preformed human infant in sperm as imagined by seventeenth-century Dutch histologist Niklass Hartsoeker, one of the first to observe sperm with a microscope of his own construction. Other remarkable pictures published during this period depicted the figure sometimes wearing a nightcap!

seen. The concept of **preformation** was strongly advocated by most seventeenth- and eighteenth-century naturalist-philosophers.

William Harvey, the great physiologist of blood circulation fame, published a book on embryology in his old age (1651) that contained the conclusion "Omne vivum ex ova" (All life from the egg, see p. 80). Harvey could not accept the naive doctrine of preformation but was unable to offer experimental evidence to refute it.

In 1759 German embryologist Kaspar Friedrich Wolff clearly showed that in the earliest developmental stages of the chick, there was no preformed individual, only undifferentiated granular material that became arranged into layers. These continued to thicken in some areas, to become thinner in others, to fold, and to segment, until the body of the embryo appeared. Wolff called this **epigenesis** ("origin upon or after"), an idea that the fertilized egg contains building material only, somehow assembled by an unknown directing force. Current ideas of development are essentially epigenetic in concept, although we know far more about what directs growth and differentiation.

Development describes the progressive changes in an individual from its beginning to maturity (Figure 7-2). In sexual multicellular organisms,

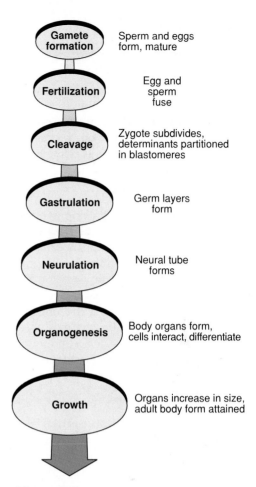

Figure 7-2
Key events in animal development.

development usually begins with a fertilized egg that divides mitotically to produce the animal's body plan and all of the many kinds of cells in the body. This generation of cellular diversity is not defined all at once but is formed as the result of a **hierarchy of developmental decisions.** The many familiar cell types that make up the body do not simply "unfold" at some point, but arise from conditions created in preceding stages. At each stage of development new structures arise from the subdivision of less committed rudiments. Each subdivision is increasingly restrictive, and the decision made at each stage in the hierarchy further limits developmental fate. Once cells embark on a course of differentiation, they become irrevocably committed to that course. They no longer depend on the stage that preceded them, nor do they have the option of becoming something different. Once a structure becomes committed it is said to be

determined. Thus the hierarchy of commitment is progressive, it is irreversible, and the final differentiated tissue is relatively stable. The two basic processes that are responsible for this progressive subdivision are **cytoplasmic localization** and **induction.** We will discuss both processes as we proceed through this chapter.

FERTILIZATION

The initial event in development in sexual reproduction is **fertilization,** the union of male and female gametes to form a **zygote.** Fertilization accomplishes two things: it provides for the recombination of paternal and maternal genes, thus restoring the original diploid number of the chromosomes characteristic of the species, and it activates the egg to begin development. However, while normally it is sperm contact that activates the egg, the sperm is not always required for development. Some animals, such as certain species of rotifers, crustaceans, insects, fishes, and desert lizards, are naturally parthenogenetic (p. 83), and the eggs of many species can be artificially induced to develop without sperm fertilization. Thus neither sperm contact nor the paternal genome is *always* essential for egg activation, although, in the great majority of species, artificial parthenogenesis will not take an embryo very far down the developmental path.

OOCYTE MATURATION

During oogenesis, described in the preceding chapter, the egg prepares itself for fertilization, and for the beginning of development. Whereas the sperm eliminates all of its cytoplasm and condenses its nucleus to the smallest possible dimensions, the egg grows in size by accumulating yolk reserves to support future growth. The egg cytoplasm also contains vast amounts of messenger RNA, ribosomes, transfer RNA, and other elements that will be required for protein synthesis. In addition, the eggs of most species contain **morphological**

determinants that will direct the activation and repression of specific genes later in postfertilization development. The nucleus also grows rapidly in size during egg maturation, becoming bloated with nuclear sap and so changed in appearance that it is given a special name, the **germinal vesicle.**

Most of this intense preparation occurs during the prolonged prophase of the first meiotic division. The oocyte is now poised to resume the meiotic divisions that are essential to produce the haploid female pronucleus that will join the male haploid pronucleus at fertilization. After resumption of meiosis, the egg rids itself of excess chromosomal material in the form of polar bodies (described in Chapter 6, p. 90). A vast amount of synthetic activity has preceded this stage. The oocyte is now a highly structured system, provided with a dowry which, after fertilization, will support the nutritional requirements of the embryo and direct its development through cleavage.

FERTILIZATION AND ACTIVATION

Our current understanding of fertilization and activation derives in large part from more than a century of research on marine invertebrates, especially the sea urchin. Sea urchins produce large numbers of eggs and sperm, which can be combined in the laboratory for study. Fertilization has also been studied in many vertebrates and, more recently, in mammals, using the sperm and eggs of mouse, hamster, and rabbit.

Contact and Recognition between Egg and Sperm

Most marine invertebrates and many marine fishes simply release their gametes into the ocean. Even though an egg is a large target for a sperm, the enormous dispersing effect of the ocean and the limited swimming range of a spermatozoon conspire against an egg and a sperm coming together by chance encounter. To improve likelihood of contact, the eggs of numerous marine species release a chemotactic factor that attracts sperm to the egg. The chemotactic molecule is species-specific, attracting to the egg only sperm of the same species.

In sea urchin eggs, the sperm first penetrates a jelly layer surrounding the egg, then contacts the egg's vitelline envelope, a thin membrane lying just above the egg plasma membrane (Figure 7-3). At this point the acrosomal process of the sperm (Figure 7-4) releases an egg recognition protein that binds to specific sperm receptors on the vitelline envelope. This ensures that the egg will recognize only sperm of the same species; all others are screened out. This is important in the marine environment

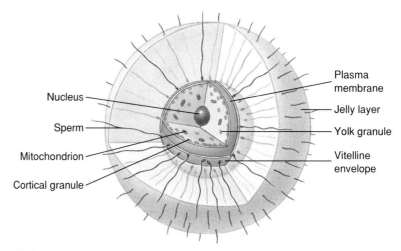

Figure 7-3
Structure of sea urchin egg at the moment of fertilization.

Nucleus

Sperm

Mitochondrion

Cortical granule

Plasma membrane

Jelly layer

Yolk granule

Vitelline envelope

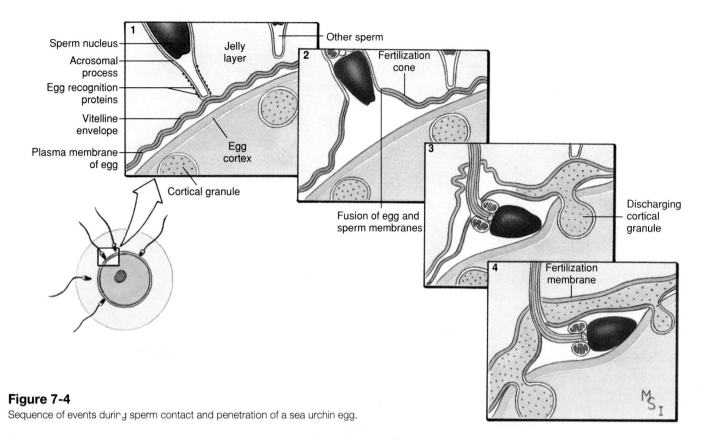

Figure 7-4
Sequence of events during sperm contact and penetration of a sea urchin egg.

where many closely related species may be spawning at the same time. Similar recognition proteins have been found on the sperm of vertebrate species (including mammals) and presumably are a universal property of all animals.

Prevention of Polyspermy

At the point of sperm contact with the egg vitelline envelope a **fertilization cone** appears into which the sperm head is later drawn (see Figure 7-4). This is followed immediately by important changes in the egg surface that block the entrance of additional sperm, which, in marine eggs especially, may quickly surround the egg in swarming numbers (Figure 7-5). The entrance of more than one sperm, called **polyspermy,** must be prevented because the union of more than two haploid nuclei would be ruinous for normal development. The egg membrane must be ready to bind with the membrane of the first sperm, but must immediately lose this ability after the first sperm has entered. In the sea urchin egg, contact of the first sperm with the egg membrane is instantly followed by

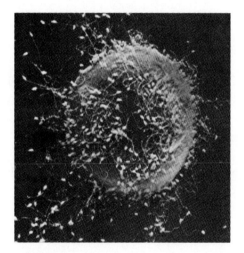

Figure 7-5
Binding of sperm to the surface of a sea urchin egg. Only one sperm penetrates the egg surface, the others being blocked from entrance by rapid changes in the egg membranes. Unsuccessful sperm are soon lifted away from the egg surface by a newly-formed fertilization membrane.

an electrical potential change in the egg membrane that prevents additional sperm from fusing with the membrane. This is followed by the **cortical reaction,** in which thousands of enzyme-rich cortical granules, located just beneath the egg membrane, fuse with the

membrane and release their contents into the space between the egg membrane and the overlying vitelline envelope (see Figure 7-4). Water immediately flows into this space, elevating the envelope and lifting away all sperm bound to it, except the one sperm that has successfully fused with the egg membrane. The hardened vitelline envelope is now called a **fertilization membrane,** and the block to polyspermy is complete. The timing sequence of these early events is summarized in Figure 7-6. Mammals have a similar security system that is erected within seconds after the sperm fuses with the egg membrane.

Fusion of Pronuclei and Egg Activation

Once the sperm and egg membranes have fused, the sperm loses its flagellum, which disintegrates. The nuclear envelope then breaks apart, allowing the sperm chromatin to expand from its extremely condensed state. The enlarged sperm nucleus, now called a **pronucleus,** migrates inward to contact the female pronucleus. Their fusion forms the diploid **zygote nucleus.**

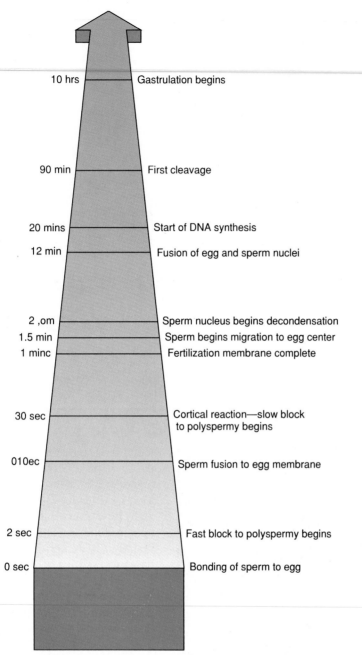

10 hrs	Gastrulation begins
90 min	First cleavage
20 mins	Start of DNA synthesis
12 min	Fusion of egg and sperm nuclei
2 ,om	Sperm nucleus begins decondensation
1.5 min	Sperm begins migration to egg center
1 minc	Fertilization membrane complete
30 sec	Cortical reaction—slow block to polyspermy begins
010ec	Sperm fusion to egg membrane
2 sec	Fast block to polyspermy begins
0 sec	Bonding of sperm to egg

Figure 7-6
Timing of events during fertilization and early development in the sea urchin.

Nuclear fusion takes much less than an hour in sea urchin eggs (Figure 7-6), but requires about 12 hours in mammals.

Fertilization sets in motion several important changes in the cytoplasm of the egg, which prepare it for cleavage. It serves to remove one or more inhibitors that have blocked metabolism and kept the egg in its quiescent, suspended-animation state. This is immediately followed by a burst of DNA and protein synthesis, the latter utilizing the abundant supply of messenger RNA previously stored in the egg cytoplasm. Fertilization also initiates an almost complete reorganization of the egg cytoplasm within which are the morphogenetic determinants that will activate or repress specific genes as development proceeds. The movement of cytoplasm repositions the determinants into new and correct spatial arrangements that are essential for proper development. The zygote now enters cleavage.

CLEAVAGE AND EARLY DEVELOPMENT

During cleavage the zygote divides repeatedly to convert the large, unwieldy cytoplasmic mass into a large number of small, maneuverable cells (called **blastomeres**) clustered together like a mass of soap bubbles. There is no growth during this period, only subdivision of mass, which continues until normal somatic cell size and nucleocytoplasmic ratios are attained. At the end of cleavage the zygote has been divided into many hundreds or thousands of cells (about 1000 in polychaete worms, 9000 in amphioxus, and 700,000 in frogs). There is a rapid increase in DNA content during cleavage as the number of nuclei and the amount of DNA are doubled with each division. Apart from this, there is little change in chemical composition or displacement of constituent parts of the egg cytoplasm during cleavage. **Polarity,** that is, a polar axis, is present in the egg, and this establishes the direction of cleavage and subsequent differentiation of the embryo.

PATTERNS OF CLEAVAGE

Although cleavage is usually very regular, the pattern of cleavage is greatly affected (1) by the quantity and distribution of yolk present and (2) by the genes controlling the symmetry of cleavage. Four principal types of cleavage are shown in Figures 7-7 and 7-8.

How the Amount and Distribution of Yolk Affects Cleavage

Eggs with very little yolk that is evenly distributed in the egg are called **isolecithal** (Gr. *isos,* equal; + *lekithos,* yolk). In such eggs, which are typical of echinoderms, tunicates, cephalochordates, nemerteans, most molluscs, and several other invertebrate groups as well as marsupial and placental mammals (including humans), cleavage is **holoblastic** (Gr. *holo,* whole;

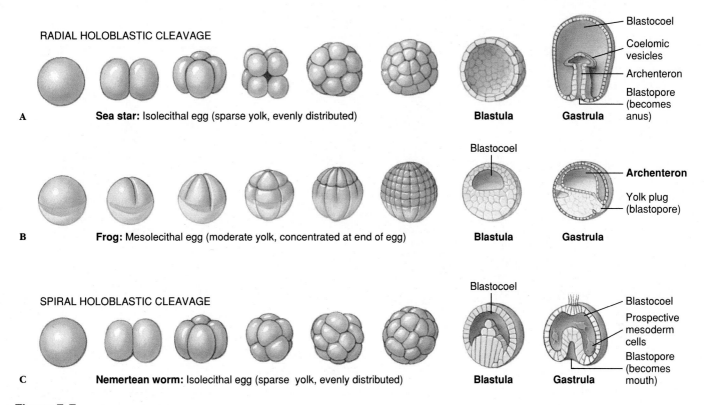

Figure 7-7

Early development of sea star, frog, and nemertean.

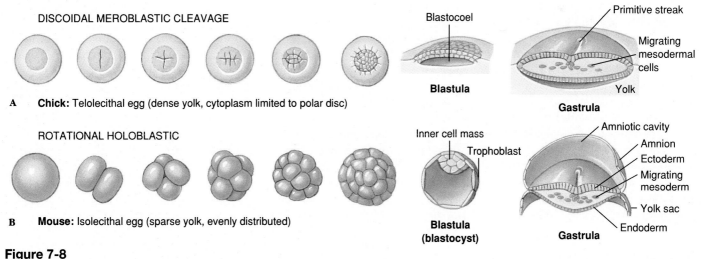

Figure 7-8

Early development of chick and mouse.

+ *blastos,* germ), meaning that the cleavage furrow extends completely through the egg (see sea star and nemertean worm development in Figure 7-7; and mouse development in Figure 7-8). The zygotes of most aquatic invertebrates contain limited yolk for growth and typically transform rapidly into a free-swimming, larval stage that must feed itself to sustain further development. This is called **indirect development** because the larval stage is interposed in the developmental sequence between embryo and adult. The larva will later undergo a **metamorphosis** into the juvenile-adult body form. Mammalian zygotes, such as those of the mouse (Figure 7-8B), also contain little yolk but have evolved a strategy that allows them to bypass the larval stage. They develop a placental attachment to the mother through which they are nourished during the long gestation. This is an example of **direct development.**

Amphibian eggs (Figure 7-7B) are called **mesolecithal** (Gr. *mesos,* middle; + *lekithos,* yolk) because they have a moderate amount of yolk concentrated in the **vegetal pole.** The opposite **animal pole** contains mostly cytoplasm and very little yolk. They also cleave holoblastically, but cleavage is somewhat retarded in the yolk-rich vegetal pole. Most amphibians develop rapidly into a feeding larval stage, which will later metamorphose into a juvenile.

Birds, reptiles, most fishes, a few tropical amphibians, cephalopod molluscs, and monotreme mammals produce the largest eggs of all animals, called **telolecithal** (Gr. *telos,* end; + *lekithos,* yolk) because they contain an abundance of yolk that is densely concentrated at one end of the egg (refer to chick development in Figure 7-8). The actively dividing cytoplasm is confined to a narrow disc-shaped mass lying on top of the yolk. Cleavage is partial, or **meroblastic** (Gr. *meros,* part; + *blastos,* germ), because the cleavage furrows cannot cut through the heavy yolk concentration, but instead stop at the border between the cytoplasm and the yolk below. Reptiles and birds have no larval stage or placental attachment but are provisioned with enough yolk to support growth until they hatch as juveniles.

We see then that the amount and distribution of yolk is an evolutionarily plastic adaptation that crosses phyletic lines. Yolk is simply an adaptation that enables embryos to develop without an external food supply. Embryos having little yolk either develop rapidly into a larval stage that seeks its own food (for example, sea star embryos) or develop a nourishing placenta (placental mammals).

How Cleavage is Affected by Different Inherited Patterns

Another important influence on a species' cleavage pattern is its inherited symmetry of cell division. Zygotes of the great majority of invertebrates cleave by one of two patterns: radial or spiral. In **radial cleavage** (represented by the sea star in Figure 7-7A), the cleavage planes produce symmetrical tiers, or layers, of cells on top of each other. Radial cleavage is

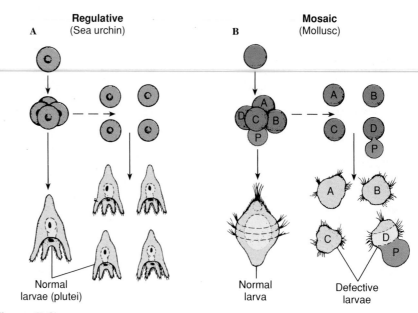

Figure 7-9

Regulative and mosaic cleavage. **A,** Regulative cleavage. Each of the early blastomeres (such as that of a sea urchin) when separated from the others develops into a small pluteus larva. **B,** Mosaic cleavage. In the mollusc, when the blastomeres are separated, each gives rise to only a part of an embryo. The larger size of one of the defective larvae is the result of the formation of a polar lobe (P) composed of clear cytoplasm of the vegetal pole, which this blastomere alone receives.

also said to be **regulative** because each blastomere of the early embryo, if separated from the others, can adjust or "regulate" its development into a complete and well-proportioned (though possibly smaller) embryo. For example, a blastomere from a four-cell embryo that would normally give rise only to ectoderm components of the organism, would, if separated from contact with its neighbor cells, form a complete, integrated organism. This happens because the blastomeres are at first equipotential in their developmental fate, that is, there is no definite relation between the position of any of the *early* blastomeres and the specific tissue it will form in the developing embryo.

Spiral cleavage (represented by nemertean worm development in Figure 7-7C) is different from radial in several ways. Rather than the eggs dividing parallel or perpendicular to the animal-vegetal axis, they cleave oblique to this axis and typically produce quartets of cells that come to lie, not on top of each other, but in the furrows between the cells. In addition, spirally cleaving eggs pack themselves tightly together much like

a group of soap bubbles, rather than just lightly contacting each other as do many radially cleaving embryos. Spirally cleaving embryos also differ from radial embryos in having a **mosaic** form of development. This means that the organ-forming determinants in the egg cytoplasm become strictly localized in the egg, even before the first cleavage division. The result is that if the early blastomeres are separated, each will continue to develop for a time as though it were still part of the whole. Each forms a defective, partial embryo (Figure 7-9). A curious feature of most spirally cleaving embryos (but not in nemerteans) is that at about the 29-cell stage a blastomere called the 4d cell is formed, and this cell will give rise to all the mesoderm of the embryo.

The importance of these two cleavage patterns extends well beyond the differences we have described. They are signals of a fundamental dichotomy, the early evolutionary divergence of bilateral metazoan animals into two separate lineages. Spiral cleavage is found in the annelids, nemerteans, turbellarian flatworms, all molluscs except the cephalopods, some brachiopods, and the echiurans. These (except the brachiopods) and several other invertebrate phyla are included in the **Protostomia** division of the animal kingdom (see the illustration inside the front cover of this book). The name Protostomia, meaning "mouth first," refers to the formation of the mouth from the first embryological opening, the blastopore. Radial cleavage is characteristic of the **Deuterostomia** division of the animal kingdom, a grouping that includes the echinoderms (sea stars and their kin), three lophophorate phyla, hemichordates, and chordates. In the Deuterostomia ("mouth second") the blastopore usually becomes the anus, while the mouth forms secondarily. Other distinguishing developmental hallmarks of these two divisions are summarized in Figure 7-10.

Radial cleavage is typical for the deuterostomes, but it is strikingly modified in many chordate groups. The zygotes of reptiles, birds, and most fish divide by **discoidal** cleavage. Because of the great mass of yolk in these meroblastic eggs, cleavage is confined to a small disc of cytoplasm lying atop a mound of yolk (see chick development in Figure 7-8A). A single layer of blastoderm is formed and later divides with equatorial cleavages to form several layers of cells. At first, the developing embryo is nourished by direct absorption of food material from the yolk; later a circulation becomes established that delivers food from the yolk mass to the growing embryo.

Cleavage in mammals is called **rotational** because of the absence of any obvious polarity, and the random orientation of asynchronous divisions (mouse development in Figure 7-8B). Cleavage in mammals is slower than in any other animal group. In humans, the first division is completed about 36 hours after fertilization, and the next divisions follow at 12- to 24-hour intervals. After the third division, the cells suddenly close into a tightly packed configuration, which is stabilized by tight junctions that form between outermost cells of the embryo. These

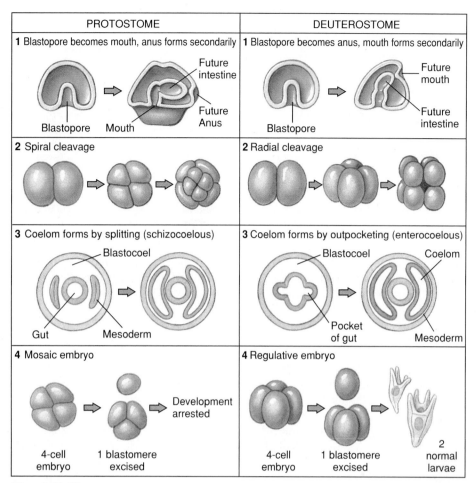

Figure 7-10

Developmental tendencies of protostomes and deuterostomes. These tendencies are much modified in some groups, for example, the vertebrates. Cleavage in mammals is rotational rather than radial; in reptiles, birds, and many fishes cleavage is discoidal. Vertebrates have also evolved a derived form of coelom formation that is basically schizocoelous.

outer cells form the **trophoblast.** The trophoblast is not part of the embryo proper but will form the embryonic portion of the placenta when the embryo implants in the uterine wall. The cells that actually give rise to the embryo proper form from the inner cells, called the **inner cell mass** (see blastula stage in Figure 7-8B).

BLASTULATION

Cleavage, however modified by different cleavage patterns and by the presence of varying amounts of yolk, results in a cluster of cells called a **blastula** (commonly called a blastocyst in mammals) (see Figures 7-7 and 7-8). In many animals the cells arrange themselves around

a central fluid-filled cavity called the **blastocoel.** At this point the embryo consists of a few hundred to several thousand cells poised for further development. There has been a great increase in total DNA content, since each of the many daughter cell nuclei, by chromosomal replication at mitosis, contains as much DNA as the original zygote nucleus. The whole embryo, however, has not increased in size above the zygote; it has been subdivided into smaller and smaller cells.

Polarity of the Embryo

Following the formation of the blastula, the blastomeres become progressively rearranged into the body plan of the

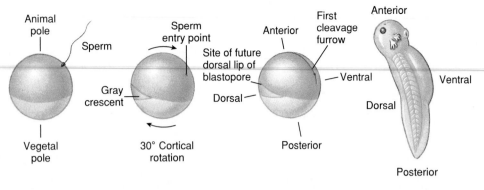

Figure 7-11
Diagram of early development of a frog embryo, showing how the site of sperm entry determines the polarity of the embryo. **A,** Sperm enters the radially symmetrical egg. **B,** Cortex of pigmented animal hemisphere rotates in direction of sperm entry to reveal the gray crescent on the opposite side of the egg. **C,** The first cleavage furrow forms, delineating the right and left sides of the embryo. **D,** Future larva.

larva or adult animal. This occurs in discrete steps. An important early step in establishing the body plan of a metazoan animal is setting up the main axis of the embryo, an event called **polarization.** Surprisingly, in some animals it is the entrance of the sperm that superimposes bilateral symmetry on a radial egg and determines the body axes (Figure 7-11). In amphibian eggs, the entrance of the sperm causes the egg cortex to rotate so that the pigmented animal hemisphere shifts toward the site of sperm entry (in many animals, the sperm may enter anywhere on the egg surface, but in amphibians sperm entrance is restricted to the pigmented animal hemisphere). The shifting of the dark animal hemisphere exposes a lightly pigmented area of cytoplasm, called the **gray crescent,** a half moon shaped area on the side of the egg *opposite* the site of sperm entry. All three axes of the embryo are now defined: the **anteroposterior axis,** the **dorsoventral axis,** and the **left-right axis** (see Figure 7-11).

The timing of polarization varies widely in different animal groups. It may occur as early as during oogenesis or as late as the beginning of gastrulation.

Gastrulation and the Formation of Germ Layers

The second major step in establishing the body plan is gastrulation. Up to this point the embryo has divided itself into a multicellular complex in which the cytoplasm of these cells is nearly in the same position as in the original undivided egg. During gastrulation, cells from the surface of the blastula are displaced to the interior of the embryo. At this time new and important cell associations are formed, and they will establish the embryonic body plan. The early cell movements in gastrulation are remarkably similar in most protostomes and deuterostomes but the mechanisms of gastrulation depend very much on the amount and distribution of yolk. Gastrulation is relatively simple in most non-yolky embryos, but it is much more complex in embryos developing from yolk-laden eggs.

In the sea star, a deuterostome (see Figure 7-7A), cells at the vegetal pole indent, then migrate inward by a process called **invagination.** In a way similar to pushing in the side of a soft tennis ball, an internal cavity is created called the **archenteron.** The archenteron is the primitive gut and its opening to the outside, the **blastopore,** marks the position of the future anus of the sea star larva. As the archenteron continues to invaginate inward, its anterior end expands into two pouchlike **coelomic vesicles,** which pinch off to form left and right **coelomic compartments** (see Figure 7-7A). The gastrula is now an embryo of three **germ layers.** The outer layer

is the **ectoderm;** it will give rise to the epithelium of the body surface and to the nervous system. The inner layer that forms the archenteron is the **endoderm;** it will give rise to the epithelial lining of the digestive tube. The outpocketing of the archenteron is the origin of the **mesoderm.** This third germ layer will form the muscular system, reproductive system, peritoneum (lining of the coelomic compartments), and the calcareous plates of the sea star's endoskeleton. The mesoderm is also the origin of the water vascular system of the sea star, a system unique to echinoderms.

Gastrulation in the nemertean worm, a protostome, resembles gastrulation in the sea star (see Figure 7-7C). As in the sea star, an archenteron is formed by invagination, creating an embryo with two germ layers of outer ectoderm and inner endoderm. The mesoderm, however, forms in a different way. Cells destined to become mesoderm arise ventrally at the lip of the blastopore and proliferate between the walls of the archenteron (endoderm) and outer body wall (ectoderm). Meticulous cell lineage studies by early embryologists established that in many protostomes (for example, flatworms, annelids, and molluscs) these mesodermal precursors arise from a single large blastomere at the 29- to 64-cell stage embryo called the 4d cell (see Figure 11-13, p. 210). In most nemerteans, the precise origin of the mesoderm is not yet known; in some it is probably the 4d cell, but in others it apparently derives from an earlier blastomere.

In nemerteans, as in other protostomes, the blastopore becomes the mouth of the future worm. This is another significant developmental characteristic that distinguishes protostomes from deuterostomes. Other important differences in the developmental tendencies of these two great divisions of the metazoa are summarized in Figure 7-10.

In the frog, a deuterostome with radial cleavage (see Figure 7-7B), the morphogenetic movements of gastrulation are greatly influenced by the mass of inert yolk in the vegetal half of the embryo. Cleavage divisions are slowed in this half so that the resulting blastula consists of many small cells in the animal half and a few large cells in the vegetal half (see Figure 7-7B). Cells on the surface begin to sink inward (invaginate) at the blastopore and continue to move inward as a sheet to form a flattened cavity, the archenteron. Although the cell movements are different, the result of gastrulation is the same in the frog and the sea star: an embryo of two germ layers (endoderm and ectoderm) is formed. As invagination proceeds, the third germ layer, the mesoderm, rolls over the lateral and ventral lip of the blastopore and, once inside, penetrates between the endoderm on the inside and the ectoderm on the outside. The three germ layers now formed are the primary structural layers that play crucial roles in the further differentiation of the embryo.

In the bird and reptile embryos (see Figure 7-8A), gastrulation begins with a thickening of the blastoderm at the caudal end of the embryo that migrates forward to form the **primitive streak** (Figure 7-12). This becomes the anteroposterior axis of the embryo and the center of early growth. The primitive streak is homologous to the blastopore of the frog embryo, but in the chick it does not open into the gut cavity because of the impeding effect of the mass of yolk. Cells of the upper layer migrate toward the primitive streak, then roll over the edge and into the blastocoel. The migrating cells form the mesoderm and contribute to the endoderm. The cells on the surface of the embryo compose the ectoderm. The embryo now has three germ layers, at this point arranged as sheetlike layers with ectoderm on top and endoderm at the bottom. This changes, however, when all three germ layers lift from the underlying yolk (Figure 7-12), then fold under to form a three-layered embryo that is pinched off from the yolk except for a stalk attachment to the yolk at midbody.

Gastrulation in mammals is remarkably similar to gastrulation in reptiles and birds (see Figure 7-8B). Gastrulation movements in the inner cell mass produce a primitive streak, much as they do in yolky bird eggs,

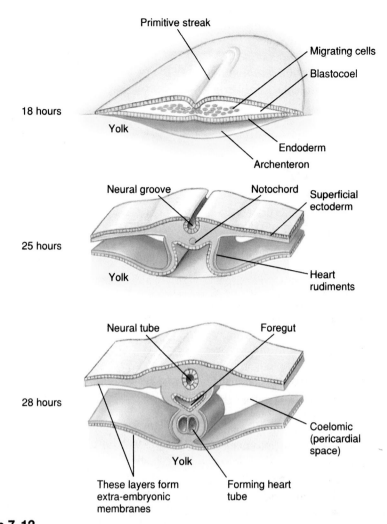

Figure 7-12
Gastrulation in the chick. Transverse sections through the heart-forming region of the chick show development at 18, 25, and 28 hours of incubation.

even though the mammalian embryo is nearly devoid of yolk. Migrating cells sink inward to form a mesoderm layer between the ectoderm and endoderm, and an empty yolk sac is formed from the endoderm.

In the least derived of the true metazoa, the Cnidaria and the Ctenophora, only two germ layers are formed, the endoderm and ectoderm. These animals are **diploblastic.** In all more derived metazoa the mesoderm also appears, either from pouches of the archenteron or from other cells associated with endoderm formation. This three-layered condition is called **triploblastic.**

FORMATION OF THE COELOM

The coelom, or true body cavity that contains the viscera, may be formed by one of two methods—**schizocoelous** or **enterocoelous** (see Figure 7-10)—or by modification of these methods. (The two terms are descriptive, for *schizo* comes from the Greek *schizein,* to split; *entero* is a Greek form from *enteron,* gut; *coelous* comes from the Greek *koilos,* hollow or cavity.) In schizocoelous formation the coelom arises, as the word implies, from the splitting of mesodermal bands that originate from the blastopore region and grow between the ectoderm and endoderm; in enterocoelous formation the coelom comes from pouches of the archenteron, or primitive gut.

These two quite different origins for the coelom are another expression of the deuterostome-protostome dichotomy of bilateral animals. The coelom of protostome animals develops by the schizocoelous method. The deuterostomes primitively follow the enterocoelous plan. Advanced chordates, however, are exceptions to this distinction because their coelom is formed by mesodermal splitting (schizocoelous), although in other respects they develop as deuterostomes, the division to which they are assigned.

MECHANISMS OF DEVELOPMENT

NUCLEAR EQUIVALENCE

How does the developing embryo generate the multitude of many cell types of a complete multicellular organism from the starting point of the single diploid nucleus of the zygote? To many nineteenth-century embryologists there seemed only one acceptable answer: as cell division ensued, the hereditary material had to be parceled out unequally to daughter cells. In this view, the genome gradually became broken up and isolated into smaller and smaller units until finally only the information required to impart the characteristics of a single cell type remained. This became known as the Roux-Weismann hypothesis, after the two German embryologists who developed the concept.

The efforts of Hans Driesch to disrupt egg development are poetically described by Peattie: "Behold Driesch grinding the eggs of Loeb's favorite sea urchin up between plates of glass, pounding and breaking and deforming them in every way. And when he ceased from thus abusing them, they proceeded with their orderly and normal development. Is any machine conceivable, Driesch asks, which could thus be torn down . . . have its parts all disarranged and transposed, and still have them act normally? One cannot imagine it. But of the living egg, fertilized or not, we can say that there lie latent within it all the potentialities presumed by Aristotle, and all of the sculptor's dream of form, yes, and the very power in the sculptor's arm."

From Peattie, D. C. 1935. *An Almanac for Moderns*. New York, G. P. Putnams Sons.

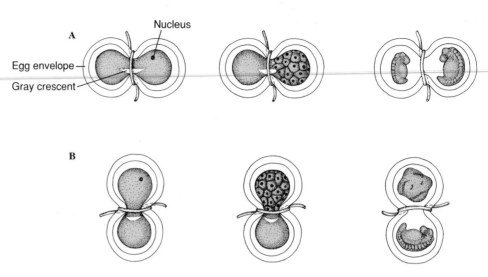

Figure 7-13

Spemann's delayed nucleation experiments. Two kinds of experiments were performed. **A,** Hair ligature was used to constrict an uncleaved fertilized newt egg. Both sides contained part of the gray crescent. The nucleated side alone cleaved until a descendant nucleus crossed over the cytoplasmic bridge. Then both sides completed cleavage and formed two complete embryos. **B,** Hair ligature was placed so that the nucleus and gray crescent were completely separated. The side lacking the gray crescent became an unorganized piece of belly tissue; the other side developed normally.

However, in 1892 Hans Driesch discovered that if he mechanically shook apart a two-celled sea urchin into separate cells, both half-embryos developed into normal larvae. Driesch concluded that both cells contained all the genetic information of the original zygote. Still, this did not settle the argument, because many embryologists believed that even if all cells contained complete genomes, the nuclei might become progressively modified in some way to dispense with the information they do not use in forming differentiated cells.

Around the turn of the century Hans Spemann introduced a new approach to testing the Roux-Weismann hypothesis. Spemann placed minute ligatures of human hair around salamander eggs just as they were about to divide, constricting them until they were almost, but not quite, separated into two halves (Figure 7-13). The nucleus lay in one half of the partially divided egg; the other side was anucleate, containing only cytoplasm. The egg then completed its first cleavage division on the side containing the nucleus; the anucleate side remained undivided. Eventually, when the nucleated side had divided into about 16 cells, one of the cleavage nuclei would wander across the narrow cytoplasmic bridge to the anucleate side. Immediately this side began to divide.

With both halves of the embryo containing nuclei, Spemann drew the ligature tight, separating the two halves of the embryo. He then watched their development. Usually two complete embryos resulted. Although the one embryo would have possessed only one-sixteenth the original nuclear material (according to the Roux-Weismann hypothesis), and the other contained fifteen-sixteenths, they both developed normally. The one-sixteenth embryo was initially smaller, but it caught up in size in about 140 days. This showed that a single nucleus selected from the 16-cell embryo contained a complete set of genes; all were equivalent.

Sometimes, however, Spemann observed that the nucleated half of the embryo developed only into an abnormal ball of "belly" tissue, although the half that received the delayed nucleus developed normally. Why should the more generously endowed fifteen-sixteenths embryo fail to develop and the small one-sixteenth embryo live? The explanation, Spemann discovered, depended on the position of the **gray crescent,** the pigment-free area that appears at the moment of fertilization (see Figure 7-11). If one half of the constricted embryo lacked a part of the gray crescent, it would not develop (Figure 7-13).

CYTOPLASMIC LOCALIZATION: SIGNIFICANCE OF THE CORTEX

Spemann's delayed nucleation experiments served as compelling evidence for two important conclusions: (1) all cells contained the same nuclear information (thus disproving the Roux-Weismann hypothesis), and (2) the cytoplasm in the area of the gray crescent must contain information essential for normal development.

During oogenesis the molecules responsible for organizing the future embryo are deposited mainly in the outer layer, or **cortex,** of the egg. It was once believed that the visible particulate matter in the cytoplasm had determinative properties. However, in the 1930s it was discovered that if the egg was centrifuged until the cytoplasm was thoroughly displaced into layers—yolk granules at the bottom and all the oil droplets at the top—the embryo would still develop normally, or almost so. Even when up to 50% of the egg's cytoplasm was removed with a micropipette, the embryo would develop perfectly. Only if great centrifugal force was applied, enough to disrupt the pattern of the cortex, were abnormalities observed.

These surprising results showed conclusively that the cortex, and not the central cytoplasm, contained an invisible but dynamic organization that determined the pattern of cleavage. Cortical organization is at first labile (especially so in regulative eggs) but soon becomes fixed and irreversible. As cleavage progresses, the cortex and the enclosed cytoplasm become segregated into territories with specific determinative properties.

NUCLEAR TRANSPLANTATION EXPERIMENTS

It was now clear that the egg cortex contained the primary guiding force

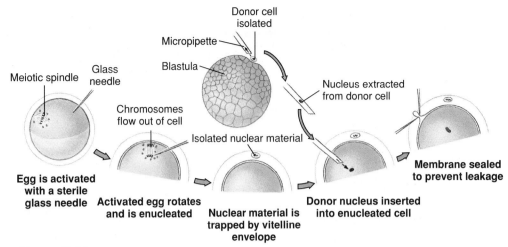

Figure 7-14

Procedure used to transplant blastula nuclei into activated enucleated eggs of the frog *Rana pipiens.* The egg is activated with a glass needle causing the mitotic spindle to move to the animal pole in preparation for forming a polar body. The chromosomes are removed with a second glass needle. A donor nucleus (and surrounding cytoplasm) from a frog embryo is then inserted into the enucleated egg.

for development through cleavage and blastula formation and that synthetic activities during this period were supported by RNA and proteins laid down in the egg cytoplasm *before* fertilization. However, continued differentiation past cleavage or the blastula stage required the action of genetic information in the nuclear chromosomes of the developing embryo. Dividing nuclei were equivalent up to this point, but did they remain equivalent after the initiation of gene action?

This problem was approached with a new technique: **nuclear transplantation,** the transfer of nuclei from one cell to another, or to an enucleated cell. In 1952 R. Briggs and T. J. King at Indiana University pioneered this technique with a two-step procedure (Figure 7-14). They first removed the nucleus from an activated egg with a glass needle. Then they drew out the nucleus from a donor cell with a micropipette and introduced it into the enucleated egg. The results were of great interest. Donor nuclei from blastula cells supported normal development of the recipient embryos to the tadpole stage. Later experiments by other workers showed many of the embryos would develop into normal postmetamorphic frogs. Since the donor cells came from embryos with 1000 to 3000 cells, this was unequivocal proof that nuclear equivalence remained far beyond Spemann's 16-cell stage. However, Briggs and King

found that normal development declined as the developmental stage of the donor nuclei increased. Nuclei from late gastrula or postgastrula stages showed increased incidence of developmental arrest at the gastrula stage (Figure 7-15).

These and other experiments showed that as cells become more highly differentiated, their nuclei become increasingly restricted in ability to support development of enucleated eggs. Developmental biologists describe this restriction as a progressive loss of nuclear **potency.** (Completely competent nuclei, such as the zygote nucleus, are said to be **totipotent.**) But if nuclei do not dispense with unneeded genetic information, how do they select from the genome the specific information needed for cell differentiation? To answer this question, we must now consider the way cells interact to determine which part of the genome will be expressed and when.

EMBRYONIC INDUCTION

We have learned that early development of the embryo is guided by cytoplasmic localization in which determinants in specialized regions of the egg cytoplasm direct the pattern of cell division. At the beginning of gastrulation, cells become committed to a particular fate through a new process, their interactions with neighboring cells. This process is called **induction.**

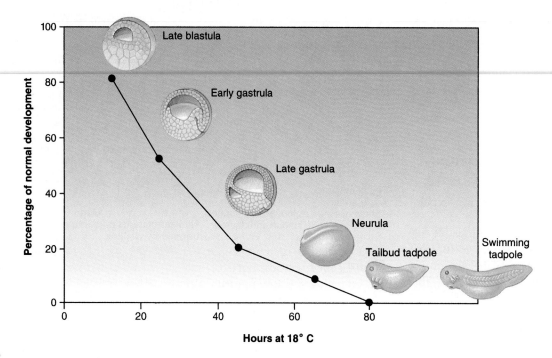

Figure 7-15

Graph showing how the percentage of successful nuclear transplants declines as the age of the donor nucleus increases. The horizontal axis shows the developmental stage of the embryo from which the donor nucleus was isolated and inserted into the enucleated egg. The vertical axis shows the percentage of transplants capable of developing to the swimming tadpole stage. Note that no somatic nuclei from embryos later than the neurula stage could support development to normal swimming tadpoles.

Induction, the capacity of some cells to evoke a specific developmental response in others, is a widespread phenomenon in development. The classic experiments, cited in the opening essay on p. 102, were reported by Hans Spemann and Hilde Mangold in 1924. When a piece of dorsal blastopore lip from a salamander gastrula was transplanted into a ventral or lateral position of another salamander gastrula, it invaginated and developed a notochord and muscle somites. It also induced the *host* ectoderm to form a neural tube. Eventually a whole system of organs developed where the graft was placed, and this grew into a nearly complete secondary embryo (Figure 7-16). This creature was composed partly of grafted tissue and partly of induced host tissue.

It was soon found that *only* grafts from the dorsal lip of the blastopore were capable of inducing the formation of a complete or nearly complete secondary embryo. This area corresponds to the presumptive areas of notochord, muscle somites, and prechordal plate. It was also found that only ectoderm of the host would develop a nervous

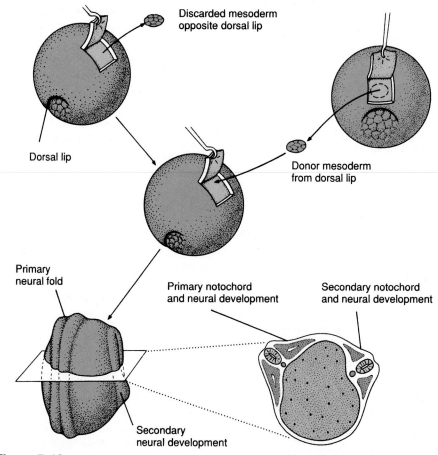

Figure 7-16

The Spemann-Mangold primary organizer experiment.

system in the graft and that the reactive ability was greatest at the early gastrula stage and declined as the recipient embryo got older.

Spemann designated the dorsal lip area the **primary organizer** because it was the only tissue capable of inducing the development of a secondary embryo in the host. He also termed this inductive event **primary induction** because he believed it to be the first inductive event in development. Subsequent studies showed that many other cell types originate by inductive interactions, a process called **secondary induction.**

Usually cells that have differentiated act as inductors for adjacent undifferentiated cells. Timing is important. Once a primary inductor sets in motion a specific developmental pattern in some cells, numerous secondary inductions follow. What emerges is a sequential pattern of development involving not only inductors but cell movement, changes in adhesive properties of cells, and cell proliferation. There is no "hard-wired" master control panel that directs development, but rather a sequence of local patterns in which one step in development is a subunit of another. In showing that each step in the developmental hierarchy is a necessary preliminary for the next, Hans Spemann's induction experiments were among the most significant events in experimental embryology.

GENE EXPRESSION DURING DEVELOPMENT

Early embryonic development is directed by products synthesized during oogenesis and stored in the egg. After fertilization, protein synthesis is translated from stored mRNA which was transcribed from the maternal genome (transcription and translation are described on pp. 140 through 143). In many animals, maternal mRNA directs protein synthesis through cleavage and to the early or mid-blastula stage. After this stage, protein synthesis gradually switches from maternal to zygotic control as the descendants of the zygotic nucleus begin to transcribe their own mRNA.

The Homeotic Genes

As development proceeds, gene expression must be regulated to ensure the orderly development of the embryo. Among the earliest and most important genes to be expressed are those that control the overall body plan of the embryo. Recently much attention has been focused on the **homeotic genes** (Gr. *homoios,* to be like, or resembling) of fruit flies, which specify the identity of segments and assure that structures such as legs and wings and antennae develop in the right place. As mutations, homeotic genes cause a body part to be replaced with a structure normally found elsewhere in the body (Figure 7-17). It soon became evident that such mutations were affecting master genes that controlled the expression of many subordinate genes. During the cloning and sequencing of homeotic genes, a 180-nucleotide DNA sequence was discovered, called the **homeobox.** Homeoboxes were soon found in the genomes of other animals, including other arthropods, nematodes, annelids, sea urchins, amphibians, and mammals. An important characteristic of the homeobox is that the 180-nucleotide sequence is highly conserved, that is, the sequence is remarkably similar in homeobox genes in different species, even those widely separated in evolutionary origin. For example, the homeobox in a mouse homeotic gene shares two-thirds of its base pairs with a homeobox in one of the fruit fly homeotic genes. The genes carrying the homeobox sequence are all expressed during development, suggesting that the homeobox carries out a broadly essential function.

All of the proteins coded by the homeobox genes contain a highly conserved 60-amino acid sequence called the **homeodomain.** All the evidence suggests that the homeodomain proteins studied are all regulatory proteins that recognize and bind to specific sequences of DNA in the genes regulated by the homeotic genes. In this way the homeodomain proteins switch subordinate genes on or off at specific times during development.

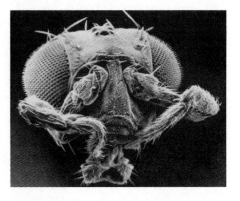

Figure 7-17
Head of a fruit fly with a pair of legs growing out of head sockets where antennae normally grow. The *Antennapedia* homeotic gene normally specifies the second thoracic segment (with legs), but the dominant mutation of this gene leads to this bizarre phenotype.

Much of our understanding of homeoboxes comes from studies of homeobox control of segmentation in insects, especially fruit flies. It was discovered that the homeobox-containing genes were lined up along the fly's third chromosome in precisely the same order as the parts of the fly's body that they controlled. Genes at the beginning of the cluster produce proteins that control the formation of the upper body; those farther along the cluster control development of the upper abdomen; and those at the end of the cluster control development of the lower abdomen (Figure 7-18). Mice and humans have four clusters of homeobox-containing genes, each cluster located on a separate chromosome. Researchers who first revealed the order of these genes in mice discovered that they show clear homology to the fruit fly's homeotic genes: they are structurally similar, they match each other in order, and they obey the same rule of order of expression. That is, genes located near one end of the cluster are expressed in the upper half of the mouse body while those at the other end of the cluster are expressed in the lower half of the body (Figure 7-18).

Amphibian development provides an excellent example of how homeotic genes control development. In amphibians, one homeotic gene produces a homeobox protein that controls the

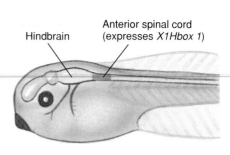

Control tadpole

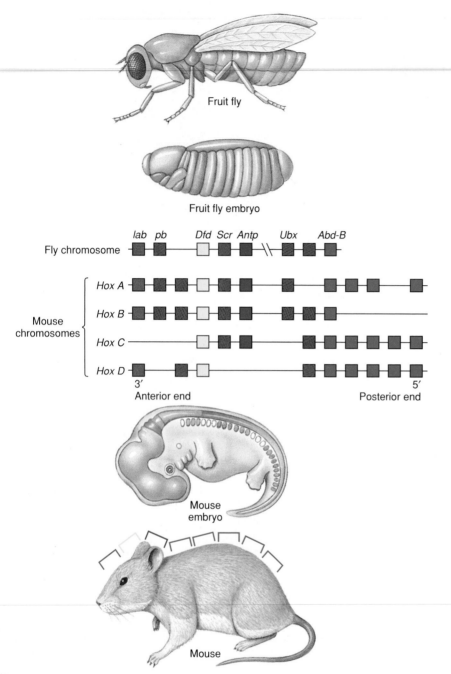

Figure 7-18

Homology of homeobox genes in insects and mammals. These genes in both insects (fruit fly) and mammals (mouse) control the subdivision of the embryo into regions of different developmental fates along the anterior-posterior axis. The homeobox-containing genes lie on a single chromosome of the fruit fly and on four separate chromosomes in the mouse. Clearly defined homologies between the two, and the parts of the body in which they are expressed, are shown in color. The black boxes denote areas where it is difficult to identify specific homologies between the two.

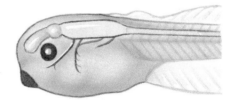

Tadpole injected with antibodies to *X1Hbox 1* protein

Figure 7-19

How the inhibition of a homeodomain regulatory protein alters normal development of the central nervous system of a frog tadpole. When the protein (encoded by a homeobox DNA sequence known as *X1Hbox 1*) was inactivated by antibodies directed against it, the area that should have become anterior spinal cord transformed into hindbrain instead.

expression of target genes that direct the formation of the anterior spinal cord. When researchers injected antibodies directed against the homeobox protein, the structure that should have become spinal cord became hindbrain instead. The portion of spinal cord that should have formed was missing altogether (Figure 7-19), because the genes that directed its development were not turned on by the homeobox regulatory protein.

The amazing similarity of the homeobox complexes in animals as phylogenetically distant as nematodes and mammals suggests that the cluster arose very early in the history of life and was in place in the common ancestor of all the Metazoa. The homeobox-containing genes may be considered a defining character, or, in the language of cladistics, a synapomorphy of the animal kingdom. Their function was to specify the fundamental anteroposterior axis of an early metazoan. Once such a complex had evolved, it could be modified to produce new body shapes for the different animal phyla.

VERTEBRATE DEVELOPMENT

THE COMMON VERTEBRATE HERITAGE

A prominent outcome of the shared ancestry of vertebrates is their common pattern of development. This is best seen in the remarkable similarity of postgastrula vertebrate embryos (Figure 7-20). The likeness occurs at a brief moment in the development of

FISH   SALAMANDER   TORTOISE   CHICK   HUMAN

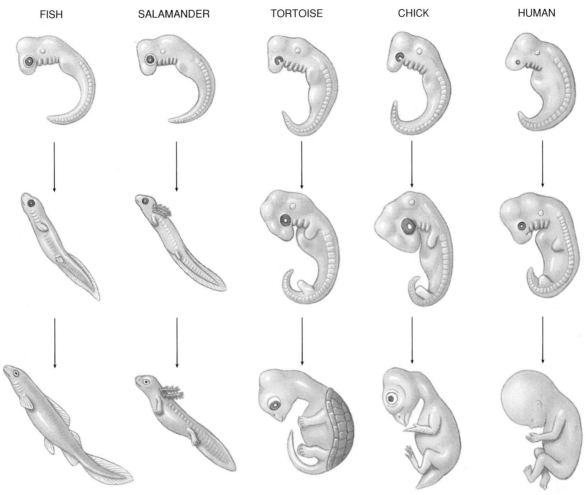

Figure 7-20

Early vertebrate embryos. Embryos as diverse as fish, salamander, tortoise, bird, and human show remarkable similarity following gastrulation (extraembryonic structures such as yolk sac and umbilical cord are omitted from the drawings to emphasize similarities). At this stage (top row) they reveal features common to the entire subphylum Vertebrata. As development proceeds they diverge, each becoming increasingly recognizable as belonging to a specific class, order, family, and finally, species.

vertebrates when the shared chordate hallmarks of dorsal neural tube, notochord, pharyngeal gill pouches with aortic arches, ventral heart, and postanal tail are present at about the same stage of development. Their moment of similarity—when the embryos seem almost interchangeable—is all the more extraordinary considering the great variety of eggs and widely different types of early development that have converged toward a common design. Then, as development continues, the embryos diverge in pace and direction, becoming recognizable as members of their class, then their order, then family, and finally their species. The important contribution of early vertebrate development to our understanding of homology and evolutionary

common descent is described in Chapter 9 in the section on Ontogeny, Phylogeny, and Recapitulation, p. 161.

AMNIOTES AND THE AMNIOTIC EGG

Reptiles, birds, and mammals form a monophyletic grouping of vertebrates called **amniotes,** so named because their embryos develop within a membranous sac, the **amnion.** The amnion is one of four extraembryonic membranes that compose a sophisticated support system within the **amniotic egg** (Figure 7-21) that evolved when the first amniotes appeared in the late Paleozoic era. The shelled, amniotic egg could be buried in nests on land, thus freeing the early amniotes from

the aquatic environment and making possible the unfettered conquest of land by the vertebrates.

The evolution of the first extraembryonic membrane, the **yolk sac,** actually predates the appearance of the amniotes many millions of years. The yolk sac with its enclosed yolk is a conspicuous feature of all fish embryos. After hatching, the growing fish larva depends on the remaining yolk provisions to sustain it until it can begin to feed itself (Figure 7-22). The mass of yolk is an extraembryonic structure because it is not really a part of the embryo proper, and the yolk sac is an **extraembryonic membrane** because it is an accessory structure that develops beyond the embryonic body and is discarded after the yolk is consumed.

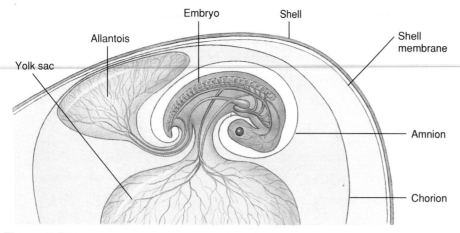

Figure 7-21

Amniotic egg at an early stage of development showing a chick embryo and its extraembryonic membranes. The sites of some of the extraembryonic interchanges are indicated by the labeling.

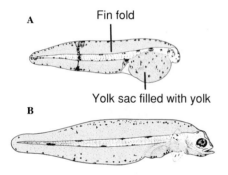

Figure 7-22

Fish larvae showing yolk sac. **A,** The one-day-old larva of a marine flounder has a large yolk sac. **B,** After 10 days of growth the larva has developed mouth, sensory organs, and a primitive digestive tract. With its yolk supply now exhausted, it must capture food to grow and survive.

The amniotic egg that evolved with the earliest Paleozoic amniotes contained three extraembryonic membranes in addition to the yolk sac: the **amnion,** a fluid-filled sac that encloses the embryo and provides an aqueous environment in which the embryo floats, protected from shocks and adhesions; the **allantois,** a bag that grows out of the hindgut to serve as repository for metabolic wastes that accumulate during development and as a respiratory surface for the exchange of oxygen and carbon dioxide; and the **chorion,** the outermost membrane that completely encloses the rest of the embryonic system. The chorion lies just beneath the egg shell (see Figure 7-21). As the embryo grows and its need for oxygen increases, the allan-

tois and chorion fuse to form the **chorioallantoic membrane.** This double membrane is provided with a rich vascular network, connected to the embryonic circulation. Lying just beneath the porous shell, the vascular chorioallantois serves as a provisional "lung" across which oxygen and carbon dioxide can freely exchange. Thus the amniotic egg provides a complete life-support system for the embryo, enclosed by a tough outer shell. The amniotic egg is one of the most important adaptations to have evolved in vertebrate ancestry.

The evolution of a shelled amniotic egg made internal fertilization a reproductive requirement. The male must introduce sperm directly into the female reproductive tract, since sperm must reach and fertilize the egg before the egg shell is wrapped around it.

THE MAMMALIAN PLACENTA AND EARLY MAMMALIAN DEVELOPMENT

Because mammals are descendants of one of the three lineages that originated with a common amniote ancestor, they inherited the amniotic egg. However, rather than developing within the egg shell like other amniotes, most mammalian embryos evolved the propitious strategy of developing within the mother's body. We have already seen that early mammalian development closely parallels

that of the egg-laying amniotes. The earliest mammals were egg layers, and even today the most primitive mammals, the **monotremes** (duck-billed platypus and spiny anteater), lay large yolky eggs that closely resemble bird eggs. In the **marsupials** (pouched mammals such as opossums and kangaroos), the embryos develop for a time within the mother's uterus. But the embryo does not "take root" in the uterine wall, and consequently it receives little nourishment from the mother before birth. The young of marsupials are therefore born immature and are sheltered in a pouch in the mother's abdominal wall and nourished with milk (reproduction in marsupials is described on p. 612).

All other mammals, composing 94% of the Class Mammalia, are **placental mammals.** These have evolved a **placenta,** a remarkable fetal structure through which the embryo is nourished. The evolution of this fetal organ required substantial restructuring, not only of the extraembryonic membranes to form the placenta but also of the maternal oviduct, part of which had to expand into a long-term housing for the embryo, the **uterus.** Despite these modifications, the development of the extraembryonic membranes in placental mammals is remarkably similar to their development in the egg-laying amniotes (compare Figures 7-21 and 7-23).

The early stages of mammalian cleavage, shown in Figure 7-8, occur while the blastocyst is traveling down the oviduct toward the uterus, propelled by ciliary action and muscular peristalsis. When the human blastocyst is about six days old and composed of about 100 cells, it contacts the uterine endometrium (uterine lining) and implants (Figure 7-24). On contact, the trophoblast cells proliferate rapidly and produce enzymes that break down the epithelium of the uterine endometrium. This allows the blastocyst to sink into the endometrium. By the eleventh or twelfth day the blastocyst is totally buried and surrounded by a pool of maternal blood. The trophoblast thickens, sending out thousands of tiny, finger-like projections, the **chorionic villi.** These projections sink like roots into the uterine endometrium after the embryo implants. As development proceeds and embryonic demands for food and gas exchange increase, the great proliferation of villi in the placenta vastly increases its total surface area. Although the human placenta at term measures only 18 cm (7 inches) across, its total absorbing surface is approximately 13 square meters—50 times the surface area of the skin of the newborn infant.

Since the mammalian embryo is protected and nourished through the placenta rather than with stored yolk, what becomes of the four extraembryonic membranes it has inherited from the early amniotes? The amnion remains unchanged, a protective water jacket in which the embryo

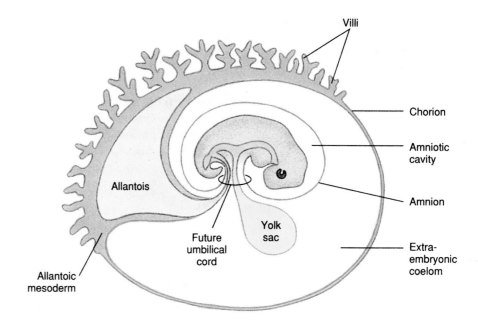

Figure 7-23

Generalized diagram of the extraembryonic membranes of a mammal, showing how their development parallels that of the chick (compare with Figure 7-21). Most of the extraembryonic membranes of the mammal have been redirected to new functions.

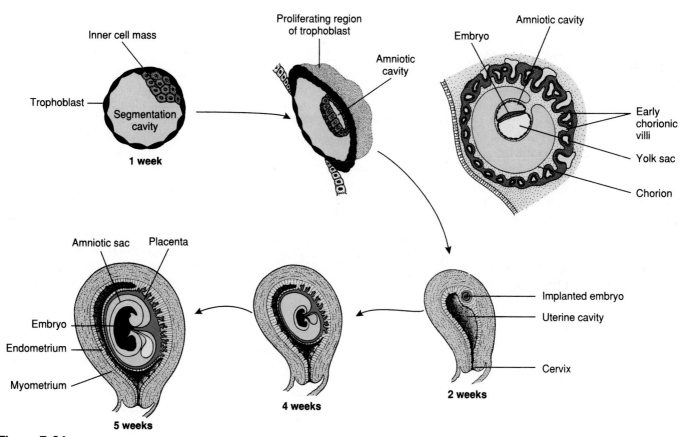

Figure 7-24

Early development of the human embryo and its extraembryonic membranes.

One of the most intriguing questions the placenta presents is, why is it not immunologically rejected by the mother? Both placenta and embryo are genetically alien to the mother because they contain proteins (called major histocompatibility proteins, p. 675) that differ from those of the mother. We would expect uterine tissues to reject the embryo just as the mother would reject an organ transplanted from her own child. The placenta is a uniquely successful foreign transplant, or **allograft,** because it has evolved measures for suppressing the immune response that normally would be mounted against it and the fetus by the mother. Recent experiments suggest that the chorion produces proteins and lymphocytes that block the normal immune response by suppressing the formation of specific antibodies by the mother.

floats. A fluid-filled yolk sac is also retained, although it contains no yolk. It has acquired a new function: during early development it is the source of stem cells that give rise to blood and lymphoid cells. These stem cells later migrate into the developing embryo. The two remaining extraembryonic membranes, the allantois and the chorion, have been completely recommitted to new functions. The allantois is no longer needed for the storage of metabolic wastes. Instead it contributes to the **umbilical cord** that links the embryo physically and functionally with the placenta. The chorion, the outermost membrane, forms most of the placenta itself.

The embryo grows rapidly, and all of the major organs of the body have begun their formation by the end of the fourth week of development. The embryo is now about 5 mm in length and weighs approximately 0.02 g. During the first two weeks of development (the **germinal period**) the embryo is quite resistant to outside influences. However, during the next eight weeks, when all of the major organs are being established and body shape is forming (the **embryonic period**), the embryo is more sensitive to disturbances that might cause malformations (such as exposure to alcohol or drugs taken by the mother) than at any other time in its development. The embryo becomes a **fetus** at approximately two months after fertilization. This ushers in the **fetal period,** which is primarily a growth phase although some of the organ systems (especially the nervous and endocrine systems) will continue to develop. The fetus grows from approximately 28 mm and 2.7 g at 60 days to approximately 350 mm and 3000 g at term (nine months).

DEVELOPMENT OF SYSTEMS AND ORGANS

DERIVATIVES OF ECTODERM: NERVOUS SYSTEM AND NERVE GROWTH

During vertebrate gastrulation the three germ layers are formed. These differentiate, as we have seen, first into primordial cell masses and then into specific organs and tissues. During this process, cells become increasingly committed to specific directions of differentiation. The derivatives of the three germ layers are diagrammed in Figure 7-25.

The assignment of early embryonic layers to specific "germ layers" (not to be confused with "germ cells," which are the eggs and sperm) is for the convenience of embryologists and of no concern to the embryo. Whereas the three germ layers normally differentiate into the tissue and organs described here, it is not the germ layer itself that determines differentiation but rather the precise position of an embryonic cell with relation to other cells.

The brain, spinal cord, and nearly all the outer epithelial structures of the body develop from the primitive ectoderm. They are among the earliest organs to appear. Just above the notochord, the ectoderm thickens to form a **neural plate.** The edges of this plate rise up, fold, and join together at the top to create an elongated, hollow **neural tube.** The neural tube gives rise to most of the nervous system: anteriorly it enlarges and differentiates into the brain and cranial nerves; posteriorly it forms the spinal cord and spinal motor nerves. Much of the rest of the peripheral nervous system is derived from the **neural crest cells,** which pinch off from the neural tube before it closes (Figure 7-26). Among the multitude of different cell types and structures that originate with the neural crest are portions of the cranial nerves, pigment cells, cartilage, and bone of most of the skull, including the jaws, ganglia of the autonomic nervous system, medulla of the adrenal gland, and contributions to several other endocrine glands. Neural crest tissue is unique to vertebrates and was probably of prime importance in the evolution of the vertebrate head and jaws.

How are the billions of nerve axons in the body formed? What directs their growth? Biologists were intrigued with these questions that seemed to have no easy solutions. Since a single nerve axon may be more than a meter in length (for example, motor nerves running from the spinal cord to the toes), it seemed impossible that a single cell could spin out so far. It was suggested that a nerve fiber grew from a series of preformed protoplasmic bridges along its route. The answer had to await the development of one of the most powerful tools available to biologists, the cell culture technique.

In 1907 embryologist Ross G. Harrison discovered that he could culture living neuroblasts (embryonic nerve cells) for weeks outside the body by placing them in a drop of frog lymph hung from the underside of a cover slip. Watching nerves grow for periods of days, he saw that each nerve fiber was the outgrowth of a single cell. As the fibers extended outward, materials for growth flowed down the axon center to the growing tip where they were incorporated into new protoplasm.

The second question—what directs nerve growth—has taken longer to unravel. An idea held well into the 1940s was that nerve growth is a

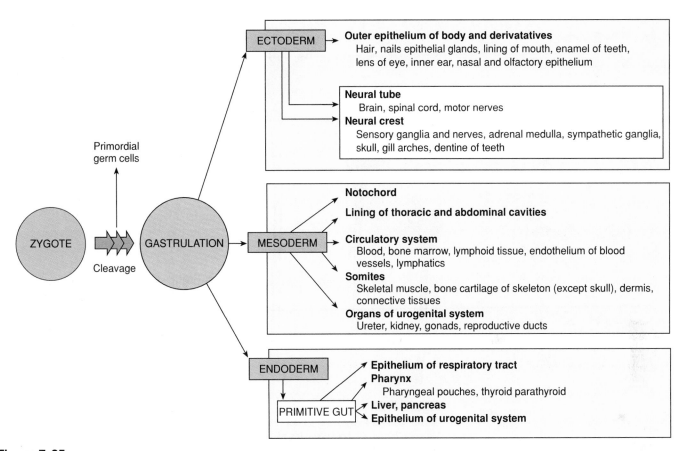

Figure 7-25
Derivatives of the primary germ layers in mammals.

random, diffuse process. It was believed that the nervous system developed as an equipotential network, or blank slate, that later would be shaped by usage into a functional system. The nervous system just seemed too incredibly complex for us to imagine that nerve fibers could find their way selectively to so many predetermined destinations. Yet it appears that this is exactly what they do! Research with invertebrate nervous systems indicated that each of the billions of nerve cell axons acquires a distinct identity that in some way directs it along a specific pathway to its destination. Many years ago Harrison observed that a growing nerve axon terminated in a "growth cone," from which extend numerous tiny thread-like pseudopodial processes (filopodia) (Figure 7-27). Recent research has shown that the growth cone is steered by an array of guidance molecules secreted along the pathway and by the axon's target. This chemical guidance system, which must, of

course, be genetically directed, is just one example of the amazing precision that characterizes the entire process of differentiation.

The tissue culture technique developed by Ross G. Harrison is now used extensively by scientists in all fields of active biomedical research, not just by developmental biologists. The great impact of the technique has been felt only in recent years. Harrison was twice considered for the Nobel Prize (1917 and 1933), but he failed ever to receive the award because, ironically, the tissue culture method was then believed to be "of rather limited value."

DERIVATIVES OF ENDODERM: DIGESTIVE TUBE AND SURVIVAL OF GILL ARCHES

In the frog embryo the primitive gut makes its appearance during gastrulation with the formation of an internal cavity, the **archenteron.** From this

simple endodermal cavity develop the lining of the digestive tract, the lining of the pharynx and lungs, most of the liver and pancreas, the thyroid and parathyroid glands, and the thymus (Figure 7-25).

The **alimentary canal** develops from the primitive gut and is folded off from the yolk sac by the growth and folding of the body wall (Figure 7-28). The ends of the tube open to the exterior and are lined with ectoderm, whereas the rest of the tube is lined with endoderm. The **lungs, liver,** and **pancreas** arise from the foregut.

Among the most intriguing derivatives of the digestive tract are the pharyngeal (gill) arches and pouches, which make their appearance in the early embryonic stages of all vertebrates (see Figure 7-20). In fishes, the gill arches develop into gills and supportive structures and serve as respiratory organs. When the early vertebrates moved onto land, gills were unsuitable for aerial respiration and were replaced by lungs.

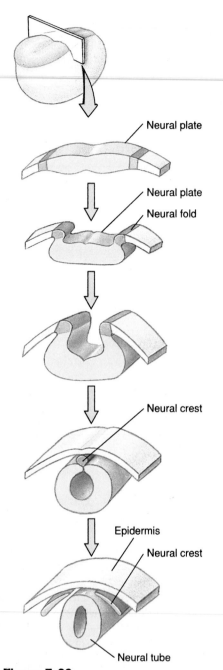

Figure 7-26
Development of the neural tube and neural crest cells from the neural plate ectoderm.

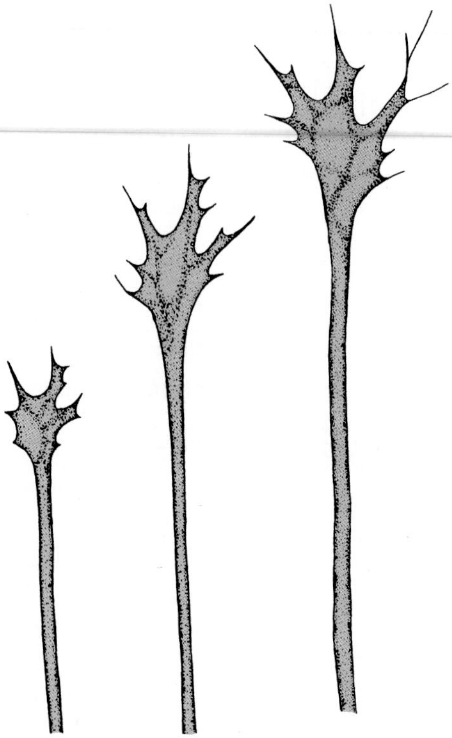

Why then do gill arches persist in the embryos of terrestrial vertebrates? Certainly not for the convenience of biologists who use these and other embryonic structures to reconstruct lines of vertebrate descent. Even though the gill arches serve no respiratory function in either the embryos or adults of terrestrial vertebrates, they remain as necessary primordia for a great variety of other structures. For example, the first arch and its endoderm-lined pouch (the space between adjacent arches) form

Figure 7-27
Growth cone at the growing tip of a nerve axon. Materials for growth flow down the axon to the growth cone from which numerous threadlike filopodia extend. These serve as a pioneering guidance system for the developing axon.

the upper and lower jaws and inner ear of vertebrates. The second, third, and fourth gill pouches contribute to the tonsils, parathyroid gland, and thymus. We can understand then why gill arches and other fishlike structures appear in early mammalian embryos. Their original function has been abandoned, but the structures are retained for new purposes. It is the great conservatism of early embryonic development that has so conveniently provided us with a telescoped evolutionary history.

DERIVATIVES OF MESODERM: SUPPORT, MOVEMENT, AND BEATING HEART

The intermediate germ layer, the mesoderm, forms the vertebrate skeletal, muscular, and circulatory structures and the kidney (Figure 7-25). As vertebrates have increased in size and complexity, the mesodermally derived supportive, movement, and transport structures make up an even greater proportion of the body bulk.

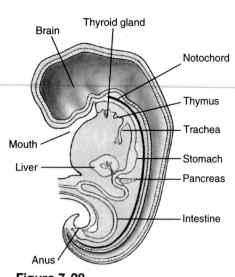

Figure 7-28
Derivatives of the alimentary canal of a human embryo.

Most **muscles** arise from the mesoderm along each side of the spinal cord (Figure 7-29). This mesoderm divides into a linear series of somites (38 in humans), which by splitting, fusion, and migration become the muscles of the body and axial parts of the skeleton. The **limbs** begin as buds from the side of the body. Projections of the limb buds develop into fingers and toes.

Although the primitive mesoderm appears after the ectoderm and endoderm, it gives rise to the first functional organ, the embryonic heart. Guided by the underlying endoderm, clusters of precardiac mesodermal cells move ameba-like into a central position between the underlying primitive gut and the overlying neural tube (see Figure 7-12, p. 111). Here the heart is established, first as a single, thin tube.

Even while the cells group together, the first twitchings are evident. In the chick embryo, a favorite and nearly ideal animal for experimental embryology studies, the primitive heart begins to beat on the second day of the 21-day incubation period; it begins beating before any true blood vessels have formed and before there is any blood to pump. As the ventricle primordium develops, the spontaneous cellular twitchings become coordinated into a feeble but rhythmical beat. Then, as the atrium develops behind the ventricle, followed by the development of the sinus venosus behind the

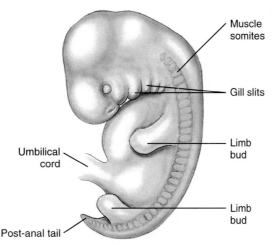

Figure 7-29

Human embryo showing muscle somites, which differentiate into skeletal muscles and axial skeleton.

atrium, the heart rate quickens. Each new heart chamber has an intrinsic beat that is faster than its predecessor.

Finally a specialized area of heart muscle called the **sinus** (or sinoatrial) node develops in the sinus venosus and takes command of the entire heartbeat (the role of the sinus node in the excitation of the heart is described on p. 684). This becomes the heart's **pacemaker.** As the heart builds up a strong and efficient beat, vascular channels open within the embryo and across the yolk. Within the vessels are the first primitive blood cells suspended in plasma.

The early development of the heart and circulation is crucial to continued embryonic development, because without a circulation the embryo could not obtain materials for growth. Food is absorbed from the yolk and carried to the embryonic body; oxygen is delivered to all the tissues, and carbon dioxide and other wastes are carried away. The embryo is totally dependent on these extraembryonic support systems, and the circulation is the vital link between them.

Summary

Developmental biology is concerned with the emergence of order and complexity during the development of a new individual from a fertilized egg, and with the control of this process. The early preformation concept of development gave way in the eighteenth century to the theory of epigenesis, which holds that development is the progressive appearance of new structures that arise as the products of antecedent development. Fertilization of an egg by a sperm restores the diploid number of chromosomes and activates the egg for development. Both sperm and egg have evolved devices to promote efficient fertilization. The sperm is a highly condensed haploid nucleus provided with a locomotory flagellum. Many eggs release chemical sperm attractants, most have surface receptors that recognize and bind only with sperm of their own species, and all have developed devices to prevent polyspermy.

During cleavage the zygote divides rapidly and usually synchronously, producing a multicellular blastula. Cleavage is greatly influenced by the quantity and distribution of yolk in the egg. Eggs with little yolk, such as those of most marine invertebrates, divide completely (holoblastic) and usually have indirect development with a larval stage interposed between the embryo and adult. Eggs having an abundance of yolk, such as those of birds, divide only partially (meroblastic) and often have no larval stage.

Based on several developmental characteristics, the bilateral metazoan animals are divided into two great lineages. The Protostomia are characterized by spiral cleavage, mosaic cleavage, and the mouth forming at or near the embryonic blastopore. The Deuterostomia are characterized by radial cleavage, regulative cleavage, and the mouth forming secondarily to the anus and not from the blastopore.

At gastrulation, cells on the embryo's surface move inward to form the germ layers (endoderm, ectoderm, mesoderm) and the embryonic body plan. Like cleavage, gastrulation is much influenced by the quantity of yolk.

Despite the different developmental fates of embryonic cells, every cell contains a complete genome and thus the same nuclear information. Early development is governed by the products of the maternal genome because the cortex of the egg contains cytoplasmic determinants, deposited during oogenesis, that guide development through cleavage. With the approach of gastrulation, control gradually shifts from maternal to zygotic as the embryo's own nuclear genes begin transcribing mRNA.

The harmonious differentiation of tissues depends in large part on induction, the ability of one tissue to produce a specific developmental response in another. In vertebrates, cell movements that establish the body plan are coordinated by the primary organizer; in amphibians the primary organizer is centered in the dorsal lip of the blastopore. Induction guides a sequence of local events, with each step serving as a preliminary for the next step in a developmental hierarchy.

During development, certain parts of each cell's genome are expressed while the remainder are switched off. Genes expressed early in development produce proteins that regulate the expression of subordinate genes in the developmental hierarchy. One group of control genes, called homeobox genes, encodes regulatory proteins that contain highly conserved DNA-binding regions called homeodomains. Homeobox genes control the subdivision of the embryo into different developmental fates along the anterior-posterior axis.

The postgastrula stage of vertebrate development represents a remarkable convergence of morphology when jawed vertebrates from fish to humans exhibit features common to all. As development proceeds, these common features become more and more characteristic of the species.

The amniotes are terrestrial vertebrates that develop extraembryonic membranes during embryonic life. The four membranes are the amnion, allantois, chorion, and yolk sac, each serving a specific life-support function for the embryo that develops within a self-contained egg (as in birds and reptiles) or within the maternal uterus (mammals).

The mammalian embryo is nourished by the placenta, a complex fetal-maternal structure that develops in the uterine wall. During pregnancy the placenta becomes an independent nutritive, endocrine, and regulatory organ for the embryo.

The germ layers formed at gastrulation differentiate into tissues and organs. The ectoderm gives rise to the skin and nervous system; the endoderm gives rise to the alimentary canal, pharynx, lungs, and certain glands; and the mesoderm forms the muscular, skeletal, circulatory, and excretory systems.

Review Questions

1. What is meant by epigenesis? How did Kaspar Friedrich Wolff's concept of epigenesis differ from the early notion of preformation?
2. How is the egg (oocyte) prepared during oogenesis for fertilization? Why is preparation essential to development?
3. Describe the events that follow contact of a spermatozoon with an egg. What is polyspermy and how is it prevented?
4. What is meant by the term "activation"?
5. How does the amount of yolk affect cleavage? Compare cleavage in a sea star with that in a bird. What is indirect development?
6. What is the difference between radial and spiral cleavage?
7. What are distinguishing developmental hallmarks of the two great lineages of bilateral metazoans, the Protostomia and the Deuterostomia?
8. When and how is polarity of the body plan established in amphibian eggs?
9. Using the sea star embryo as an example, describe gastrulation. Explain how the mass of inert yolk affects gastrulation in frog and bird embryos.
10. What is the difference between schizocoelous and enterocoelous origins of the body cavity?
11. Describe two different experimental approaches that serve as evidence for nuclear equivalence in animal embryos.
12. What is meant by "induction" as the term is used in embryology? Describe the famous organizer experiment of Spemann and Mangold and explain its significance.
13. What are homeotic genes and what is the "homeobox" contained in such genes? What is the function of the homeobox in animal development? What is unique about the regulatory proteins encoded by homeobox genes? Why are such genes and the proteins they encode said to be "strongly conserved"?
14. What is the embryonic evidence that the vertebrates share a common evolutionary ancestor?
15. What are the four extraembryonic membranes of the amniotic egg of a bird or reptile and what is the function of each membrane?
16. What is the fate of the four extraembryonic membranes of the amniotic egg of placental mammals?
17. Explain what the "growth cone" that Ross Harrison observed at the ends of growing nerve fibers has to do with the direction of nerve growth.
18. Name two organ system derivatives of each of the three germ layers.

Selected References

Browder, L. W., C. A. Erickson, and W. R. Jeffery. 1991. Developmental biology, ed. 3. Philadelphia, Saunders College Publishing. *Comprehensive description of development and mechanisms of the developmental process. Well-structured account and one of the most readable of the developmental texts.*

De Robertis, E. M., O. Guillermo, and C. V. E. Wright. 1990. Homeobox genes and the vertebrate body plan. Sci. Am. **263:**46–52 (July). *How a family of regulatory genes, first discovered in fruit flies, determines the shape of the vertebrate body.*

Gehring, W. J. 1985. The molecular basis of development. Sci. Am. **253:**153–162 (Oct.). *The role of homeotic genes in establishing the body plan of the fruit fly is explained.*

Gilbert, S. F. 1994. Developmental biology, ed. 4. Sunderland, Massachusetts, Sinauer Associates. *Combines descriptive and mechanistic aspects; good selection of examples from many animal groups.*

Goodman, C. S., and M. J. Bastiani. 1984. How embryonic nerve cells recognize one another. Sci. Am. **251:**58–66 (Dec.). *Research with insect larvae shows that developing neurons follow pathways having specific molecular labels.*

McGinnis, W., and M. Kuziora. 1994. The molecular architects of body design. Sci. Am. **270:**58–66 (Feb.). *Describes the nearly identical molecular mechanisms that define the body shapes in all animals.*

Slack, J. M. W. 1991. From egg to embryo: regional specification in early development, ed. 2. New York, Cambridge University Press. *Emphasis on pattern formation; comparative approach.*

Wolpert, L. 1991. The triumph of the embryo. Oxford, Oxford University Press. *Written for the nonspecialist, this engaging book is rich in detail and insight for all biologists interested in the development of life.*

8

Principles of Genetics: A Review

A Code for All Life

The principle of hereditary transmission is a central tenet of life on earth: all organisms inherit a structural and functional organization from their progenitors. What is inherited by an offspring is not necessarily an exact copy of the parent but a set of coded instructions that gives rise to a certain expressed organization. These instructions are in the form of genes, the fundamental units of inheritance. One of the great triumphs of modern biology was the discovery in 1953 by James Watson and Francis Crick of the nature of the coded instructions in genes. This was followed by the discovery of the way in which the code is translated into the expression of characteristics. The genetic material (deoxyribonucleic acid, DNA) is composed of nitrogenous bases arranged on a backbone of sugar phosphate units. The genetic code lies in the linear order or sequence of bases in the DNA strand.

Because the DNA molecules replicate themselves in their passage from generation to generation, genetic variations can persist once they have happened. Such molecular alterations, called mutations, are the ultimate source of biological variation and the raw material of evolution. ■

A basic principle of modern evolutionary theory is that organisms attain their diversity of form, function, and behavior through hereditary modifications of preexisting lines of ancestors. It means that all known lineages of plants and animals are related by descent from simpler, common ancestral groups.

Heredity establishes the continuity of life forms. Although offspring and parents in a particular generation may look different, there is nonetheless a basic sameness that runs from generation to generation for any species of plant or animal. In other words, "like begets like." Yet children are not precise replicas of their parents. Some of their characteristics show resemblances to one or both parents, but they also demonstrate many traits not found in either parent. What is actually inherited by an offspring from its parents is a certain type of germinal organization **(genes)** that, under the influence of environmental factors, guides the orderly sequence of differentiation of the fertilized egg into a human being, bearing the unique physical characteristics as we see them. Each generation hands on to the next the instructions required for maintaining continuity of life.

The gene is the unit entity of inheritance, the germinal basis for every characteristic that appears in an organism. The study of what genes are and how they work is the science of genetics. It is a science that deals with the underlying causes of *resemblance,* as seen in the remarkable fidelity of reproduction, and of *variation,* which is the working material for organic evolution. Genetics has shown that all living forms use the same information storage, transfer, and translation system, and thus it has provided an explanation for both the stability of all life and its probable descent from a common ancestral form. This is one of the most important unifying concepts of biology.

MENDEL'S INVESTIGATIONS

The first man to formulate the cardinal principles of heredity was Gregor Johann Mendel (1822–1884) (Figure 8-1 and p. 17), who lived with the Augustinian monks at Brünn (Brno),

Purple vs. white flowers
F1 = all purple
F2 = 705 purple
224 white
Ratio: 3.15:1

Round vs. wrinkled seeds
F1 = all round
F2 = 5474 round
1850 wrinkled
Ratio: 2.96:1

Yellow vs. green seeds
F1 = all yellow
F2 = 6022 yellow
2001 green
Ratio: 3.01:1

Green vs. yellow pods
F1 = all green
F2 = 428 green
152 yellow
Ratio: 2.82:1

Inflated vs. constricted pods
F1 = all inflated
F2 = 882 inflated
299 constricted
Ratio: 2.95:1

Long vs. short stems
F1 = all long
F2 = 787 long
277 short
Ratio: 2.84:1

Axial vs. terminal flowers
F1 = all axial
F2 = 651 axial
207 terminal
Ratio: 3.14:1

Figure 8-1

Seven experiments on which Gregor Mendel based his postulates. These are the results of monohybrid crosses for first and second generations.

Moravia. At that time Brünn was a part of Austria, but now it is in the eastern part of the Czech Republic. While conducting breeding experiments in a small monastery garden from 1856 to 1864, Mendel examined with great care the progeny of many thousands of plants. He worked out in elegant simplicity the laws governing the transmission of characters from parent to offspring. His discoveries, published in 1866, were of great significance, coming just after Darwin's publication of *The Origin of Species.* Yet these discoveries remained unappreciated and forgotten

until 1900—some 35 years after the completion of the work and 16 years after Mendel's death.

Mendel's classic observations were based on the garden pea because it had been produced in pure strains by gardeners over a long period of time by careful selection. For example, some varieties were definitely dwarf and others were tall. A second reason for selecting peas was that they were self-fertilizing but also capable of cross-fertilization. To simplify his problem Mendel chose single characters that displayed sharply contrasting traits. He carefully avoided mere quantitative and intermediate characteristics. Mendel selected pairs of contrasting traits, such as tall plants versus dwarf plants and smooth seeds versus wrinkled seeds (Figure 8-1).

By not reporting conflicting findings, which must surely have arisen as they do in any original research, Mendel has been accused of "cooking" his results. The chances are, however, that he carefully avoided ambiguous material to strengthen his central message, which we still regard as an exemplary achievement in experimental analysis.

Mendel crossed plants having one of these traits with others having the contrasting trait. He did this by removing the stamens from a flower to prevent self-fertilization and then placing on the stigma (female part of flower) pollen from the flower of the plant that had the contrasting character. He also prevented the experimental flowers from being pollinated from other sources such as wind and insects. When the cross-fertilized flower bore seeds, he noted the kind of plants (hybrids) that were produced from the planted seeds. Subsequently he crossed these hybrids among themselves to see what would happen.

Mendel knew nothing of the cytological basis of heredity, since chromosomes and genes were unknown to him. Although we can admire Mendel's power of intellect in his discovery of the principles of inheritance without knowledge of chromosomes, the principles are certainly easier to understand if we first review chromosomal behavior, especially in meiosis (p. 86).

In sexually reproducing animals gametes (ova and sperm, p. 89) are responsible for providing the genetic information to the offspring. Scientific explanation for genetic principles required a study of germ cells and their behavior, which meant working backward from certain visible results of inheritance to the mechanism responsible for such results. The nuclei of sex cells were early suspected of furnishing the real answer to the mechanism. This applied especially to the chromosomes

(p. 56), since they appeared to be the only entities passed on in equal quantities from both parents to offspring.

When Mendel's laws were rediscovered in 1900, their parallelism with the cytological behavior of the chromosomes was obvious. Later experiments showed that the mechanism of heredity could be definitively assigned to the chromosomes. The next problem was to find out how chromosomes affected the hereditary pattern.

MENDELIAN LAWS OF INHERITANCE

MENDEL'S FIRST LAW

Mendel's **law of segregation** states that *in the formation of gametes, paired factors affecting a phenotype (apparent traits) segregate independently of one another.* In one of Mendel's original experiments, he pollinated pure-line tall plants with the pollen of pure-line dwarf plants. Thus the visible characteristics, or **phenotypes,** were tall and dwarf. Mendel found that all the progeny in the first generation (F_1) were tall, just as tall as the tall parents of the cross. The reciprocal cross—dwarf plants pollinated with tall plants—gave the same result. This always happened no matter which way the cross was made. Obviously, this kind of inheritance was not a blending of two traits, since none of the progeny was of intermediate size.

Next Mendel selfed (self-fertilized) the tall F_1 plants and raised several hundred progeny, the second (F_2) generation. This time, *both* tall and dwarf plants appeared. Again, there was no blending (no plants of intermediate size), but the appearance of dwarf plants from all tall parental plants was surprising. The dwarf trait, present in the grandparents but not in the parents, had reappeared. When he counted the actual number of tall and dwarf plants in the F_2 generation, he discovered that there were almost exactly three times more tall plants than dwarf ones.

Mendel then repeated this experiment for the six other contrasting traits that he had chosen, and in every case he obtained ratios very close to 3:1

(see Figure 8-1). At this point it must have been clear to Mendel that he was dealing with hereditary determinants for the contrasting traits that did not blend when brought together. Even though the dwarf trait disappeared in the F_1 generation, it reappeared fully expressed in the F_2 generation. He realized that the F_1 generation plants carried determinants (which he called "factors") of both tall and dwarf parents, even though only the tall trait was expressed in the F_1 generation.

Mendel called the tall factor **dominant** and the short **recessive.** Similarly, the other pairs of traits that he studied showed dominance and recessiveness. Whenever a dominant factor is present, the recessive one cannot produce its effect. The recessive trait will appear **only** when both factors are recessive, or in other words, in a pure condition.

In representing his crosses, Mendel used letters as symbols; dominant traits were represented by capital letters, and for recessive traits he used the corresponding lowercase letters. Modern geneticists still follow this custom. Thus the factors for pure tall plants might be represented by *T/T,* the pure recessive by *t/t,* and the mix, or hybrid, of the two plants by *T/t.* The slash mark is to indicate that the alleles are on homologous chromosomes. (Alleles are alternative forms of genes for the same trait; Chapter 6, p. 87.) When the gametes unite in any fertilization, a zygote is formed. The zygote bears the complete genetic constitution of the organism. All the gametes produced by *T/T* must necessarily be *T,* whereas those produced by *t/t* must be *t.* Therefore a zygote produced by union of the two must be *T/t,* or a **heterozygote.** On the other hand, the pure tall plants (*T/T*) and pure dwarf plants (*t/t*) are **homozygotes,** meaning that the paired factors (alleles) are alike on the homologous chromosomes. A cross involving only one pair of contrasting traits is called a **monohybrid cross.**

As we mentioned before, in the cross between tall and dwarf plants there were two phenotypes: tall and dwarf. On the basis of genetic formulas there are three *hereditary* types: *T/T, T/t,* and *t/t.* These are called

genotypes. A genotype is an allelic combination (*T/T*, *T/t*, or *t/t*), and the phenotype is the corresponding appearance of the organism (tall or dwarf).

In diagram form, one of Mendel's original crosses (tall plant and dwarf plant) could be represented as follows:

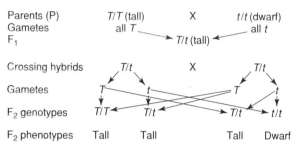

In other words, all possible combinations of F_1 gametes in the F_2 zygotes will yield a 3:1 phenotypic ratio and a 1:2:1 genotypic ratio. It is convenient in such crosses to use the checkerboard method devised by Punnett (Punnett square) for representing the various combinations resulting from a cross. In the F_2 cross the following scheme would apply:

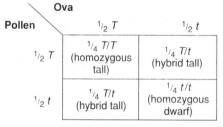

Ratio: 3 tall to 1 dwarf

The next step was an important one because it enabled Mendel to test his hypothesis that every plant contained nonblending factors from both parents. He self-fertilized the plants in the F_2 generation; that is, the stigma of a flower was fertilized by the pollen of the same flower. The results showed that self-pollinated F_2 dwarf plants produced only dwarf plants, whereas one-third of the F_2 tall plants produced tall and the other two-thirds produced both tall and dwarf in the ratio of 3:1, just as the F_1 plants had done. Genotypes and phenotypes were as follows:

This experiment showed that the dwarf plants were pure because they at all times gave rise to short plants when self-pollinated; the tall plants contained both pure tall and hybrid tall. It also demonstrated that, although the dwarf trait disappeared in the F_1 plants, which were all tall, dwarfness appeared in the F_2 plants.

Mendel reasoned that the factors for tallness and dwarfness were units that did not blend when they were together. The F_1 generation (the first generation of hybrids, or first filial generation) contained both of these units or factors, but when these plants formed their germ cells, the factors separated so that each germ cell had only one factor. In a pure plant both factors were alike; in a hybrid they were different. He concluded that individual germ cells were always pure with respect to a pair of contrasting factors, even though the germ cells were formed from hybrid individuals possessing both contrasting factors.

This idea formed the basis for his law of segregation, that is, that whenever two factors are brought together in a hybrid, they segregate into separate gametes that are produced by the hybrid. Both factors or alleles of the parent pass with equal frequency to the gametes. We now understand that the factors segregate because there are two alleles for the character, one on each chromosome of a homologous pair, but the gametes receive only one of each in meiosis.

Mendel's great contribution was his quantitative approach to inheritance. This really marks the birth of genetics, because before Mendel, people believed that traits were blended like mixing together two colors of paint, a notion that unfortunately still lingers in the minds of many and was a problem for Darwin's theory of natural selection when he first proposed it (p. 16). If traits were blended, variability would be lost in hybridization between individuals. With particulate inheritance, on the other hand, different variations are retained and can be shuffled about and re-sorted like blocks.

Testcross

When one of the alleles is dominant, heterozygous individuals are identical in phenotype to individuals homozygous for the dominant allele. Therefore you cannot determine the genotypes of these individuals just by observing their phenotypes. For instance, in Mendel's experiment of tall and dwarf traits, it is impossible to determine the genetic constitution of the tall plants of the F_2 generation by mere inspection of the tall plants. Three-fourths of this generation are tall, but which of them are heterozygotes?

As Mendel reasoned, the test is to cross the questionable individuals with pure recessives. If the tall plant is homozygous, all the offspring in such a testcross are tall, thus:

Parents   *T/T* (tall) x *t/t* (dwarf)

Pollen \ Ova	T	T
t	*T/t* (hybrid tall)	*T/t* (hybrid tall)
t	*T/t* (hybrid tall)	*T/t* (hybrid tall)

All of the offspring are *T/t* (hybrid tall). If, on the other hand, the tall plant is heterozygous, half of the offspring are tall and half dwarf, thus:

Parents   *T/t* (hybrid tall) x *t/t* (dwarf)

Pollen \ Ova	T	t
t	*T/t* (hybrid tall)	*t/t* (homozygous dwarf)
t	*T/t* (hybrid tall)	*t/t* (homozygous dwarf)

F_2 plants: Tall $\begin{cases} \frac{1}{4}\ T/T \xrightarrow{\text{Selfed}} \text{all } T/T \text{ (homozygous tall)} \\ \frac{1}{2}\ T/t \xrightarrow{\text{Selfed}} 1\ T/T : 2\ T/t : 1\ t/t \text{ (3 tall: 1 dwarf)} \end{cases}$

Dwarf $\frac{1}{4}\ t/t \xrightarrow{\text{Selfed}}$ all t/t (homozygous dwarf)

The **testcross** is often used in modern genetics for the analysis of the genetic constitution of the offspring, as well as for a quick way to make desirable homozygous stocks of animals and plants.

Intermediate Inheritance

In some cases neither allele is completely dominant over the other, and the heterozygous phenotype appears either intermediate between or even quite distinct from those of the parents. This is called **intermediate inheritance,** or **incomplete dominance.** In the four-o'clock flower (*Mirabilis*), two allelic variants determine red versus pink or white flowers; homozygotes are red or white flowered, but heterozygotes have pink flowers. In a certain strain of chickens, a cross between those with black and splashed white feathers produces offspring that are not gray but a distinctive color called Andalusian blue (Figure 8-2). In each case, if the F_1s are crossed, the F_2s have a ratio of 1:2:1 in colors, or 1 red: 2 pink: 1 white in four-o'clock flowers and 1 black: 2 blue: 1 white for Andalusian chickens. This can be illustrated for the chickens as follows (when neither of the alleles is recessive, we can represent both by capital letters and distinguish them by the addition of a "prime" sign (B') or by superscript letters, for example, B^b for black and B^w for white):

Parents	B/B	(black feathers)	X	B'/B'	(white feathers)
Gametes	all B			all B'	
F_1			B/B' (all blue)		
Crossing hybrids		B/B'	X	B/B'	
Gametes		B, B'		B, B'	
F_2 genotypes	B/B		B/B'	B/B'	B'/B'
F_2 phenotypes	Black		Blue	Blue	White

In this kind of a cross, the heterozygous phenotype is indeed a blending of both parental types. It is easy to see how such observations

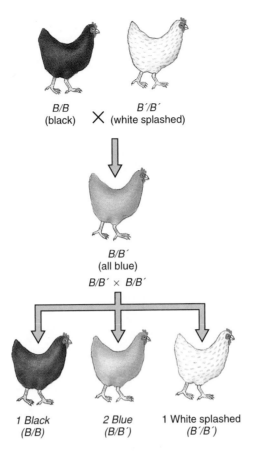

B/B (black) X B'/B' (white splashed)

B/B' (all blue)

$B/B' \times B/B'$

1 Black (B/B)   2 Blue (B/B')   1 White splashed (B'/B')

Figure 8-2
Cross between chickens with black and splashed white feathers. Black and white are homozygous; Andalusian blue is heterozygous.

would encourage the notion of the blending concept of inheritance. However, in the cross of black and white chickens or red and white flowers, there is no blending of genetic factors; the homozygous offspring of phenotypically intermediate hybrid individuals breed true to the original parental phenotypes.

MENDEL'S SECOND LAW

According to Mendel's **law of independent assortment,** *genes located on different pairs of homologous chromosomes assort independently during meiosis.* Thus the law deals with genes for two different characters that are borne on two different pairs of chromosomes. Mendel carried out experiments on peas that differed from each other at two or more genes, that is, experiments involving two or more phenotypic characters.

Mendel had already established that tall plants were dominant to dwarf. He also noted that crosses between plants bearing yellow seeds and plants bearing green seeds produced plants with yellow seeds in the F_1 generation; therefore yellow was dominant to green. The next step was to make a cross between plants differing in these two characteristics. When a tall plant with yellow seeds ($T/T\ Y/Y$) was crossed with a dwarf plant with green seeds ($t/t\ y/y$), the F_1 plants were tall and yellow as expected ($T/t\ Y/y$).

Parents	$T/T\ Y/Y$	x	$t/t\ y/y$
	(tall, yellow)		(dwarf, green)
Gametes	all TY		all ty
F_1		$T/t\ Y/y$	
		(tall, yellow)	

The F_1 hybrids were then crossed with each other, and the F_2 results are shown in Figure 8-3.

Mendel already knew that a cross between two plants bearing a single pair of alleles of the genotype T/t would yield a 3:1 ratio. Similarly, a cross between two plants with the genotypes Y/y would yield the same 3:1 ratio. If we examine *only* the tall and dwarf phenotypes expected in the outcome of the dihybrid experiment, they produce a ratio of 12 tall to 4 dwarf, which reduces to a ratio of 3:1. Likewise, a total of 12 plants have yellow seeds for every 4 plants that have green—again a 3:1 ratio. Thus the monohybrid ratio prevails for both traits when they are considered independently. The 9:3:3:1 ratio is nothing more than a combination of the two 3:1 ratios.

$$3:1 \times 3:1 = 9:3:3:1$$

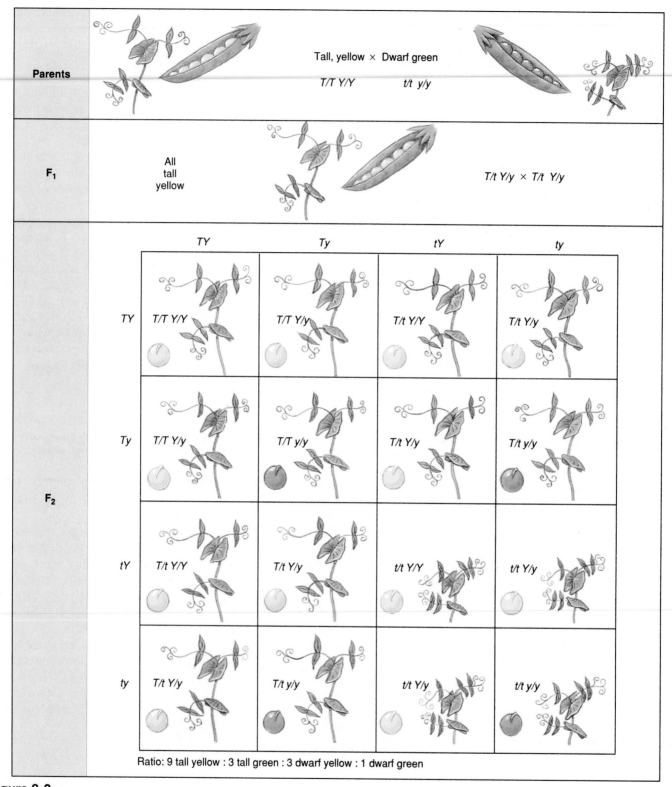

Ratio: 9 tall yellow : 3 tall green : 3 dwarf yellow : 1 dwarf green

Figure 8-3
Punnett square method for determination of genotypes and phenotypes expected in a dihybrid cross for independently assorting genes.

When one of the alleles is unknown, it can be designated by a dash (T/—). This designation can also be used when it is immaterial whether the genotype is heterozygote or homozygote, as when we total all of a certain phenotype. The dash could represent either T or t.

The F_2 genotypes and phenotypes are as follows:

$$
\begin{array}{l}
\left.\begin{array}{ll}
1 & T/T\ Y/Y \\
2 & T/t\ Y/Y \\
2 & T/T\ Y/y \\
4 & T/t\ Y/y
\end{array}\right\} \quad 9\ T/{-}Y/{-} \quad\quad \text{9 Tall yellow} \\[2em]
\left.\begin{array}{ll}
1 & T/T\ y/y \\
2 & T/t\ y/y
\end{array}\right\} \quad 3\ T/{-}y/y \quad\quad \text{3 Tall green} \\[1.5em]
\left.\begin{array}{ll}
1 & t/t\ Y/Y \\
2 & t/t\ Y/y
\end{array}\right\} \quad 3\ t/t{-}Y/{-} \quad\quad \text{3 Dwarf yellow} \\[1.5em]
1\quad t/t\ y/y \quad\quad\quad 1\ t/t\ y/y \quad\quad \text{1 Dwarf green}
\end{array}
$$

The results of this experiment show that the segregation of alleles for plant height is entirely independent of the segregation of alleles for seed color. Neither has any influence on the other. Thus another way to state Mendel's law of independent assortment is that *allelic variants of different genes on different chromosomes segregate independently of one another*. The reason is that during meiosis the member of any pair of homologous chromosomes received by a gamete is independent of which member of any other pair of chromosomes it receives. Of course, independent assortment assumes that the genes are on different chromosomes. If they were on the same chromosome, they would assort together unless crossing over occurred (p. 136).

One way to estimate genotypic or phenotypic ratios of progeny from a cross of particular genotypes is to construct a Punnett square. With a monohybrid cross, this is easy; with a dihybrid cross, a Punnett square is rather laborious; and with a trihybrid cross, it is very tedious. We can make such estimates much more easily by taking advantage of simple probability calculations. The basic assumption is that all the genotypes of gametes of one sex have an equal chance of uniting with all the genotypes of gametes of the other sex, in proportion to the numbers of each present. This is generally true when the sample size is large enough, and the actual numbers observed come close to those predicted by the laws of probability.

We may define probability as follows:

$$
\text{Probability (p)} = \frac{\text{Number of times an event happens}}{\begin{array}{c}\text{Total number of trials or possibilities}\\ \text{for the event to happen}\end{array}}
$$

For example, the probability (p) of a coin falling heads when tossed is 1/2, because the coin has two sides. The probability of rolling a three on a die is 1/6, because the die has six sides.

The probability of independent events occurring together (ordered events) involves the **product rule,** which is simply the product of their individual probabilities. When two coins are tossed together, the probability of getting two heads is $1/2 \times 1/2 = 1/4$, or 1 chance in 4. The probability of rolling two threes simultaneously with two dice is as follows:

$$
\text{Probability of two threes} = 1/6 \times 1/6 = 1/36
$$

Note, however, that a small sample size may give a result quite different from that predicted. Thus if we tossed the coin three times and it fell heads each time, we would not be much surprised. But if we tossed the coin 1000 times, and the number of times it fell heads diverged very much from 500, we would strongly suspect that there was something wrong with the coin or with the way we were tossing it.

We can use the product rule to predict the ratios of inheritance in monohybrid or dihybrid (or larger) crosses if the genes sort independently in the gametes (as they did in all of Mendel's experiments) (Table 8-1).

Table 8-1	**Use of Product Rule for Determination of Genotype and Phenotype Ratios in a Dihybrid Cross for Independently Assorting Genes**		
Parents' genotypes	$T/t\ Y/y$	$\times$	$T/t\ Y/y$
Equivalent monohybrid crosses	$T/t \times T/t$	and	$Y/y \times Y/y$
Genotype ratios in F_1s of monohybrid crosses	1/4 T/T		1/4 Y/Y
	2/4 T/t		2/4 Y/y
	1/4 t/t		1/4 y/y
Combine two monohybrid ratios to determine dihybrid genotype ratios	1/4 T/T　$\times$	1/4 Y/Y = 1/16 $T/T\ Y/Y$	
		2/4 Y/y = 2/16 $T/T\ Y/y$	
		1/4 y/y = 1/16 $T/T\ y/y$	
	2/4 T/t　$\times$	1/4 Y/Y = 2/16 $T/t\ Y/Y$	
		2/4 Y/y = 4/16 $T/t\ Y/y$	
		1/4 y/y = 2/16 $T/t\ y/y$	
	1/4 t/t　$\times$	1/4 Y/Y = 1/16 $t/t\ Y/Y$	
		2/4 Y/y = 2/16 $t/t\ Y/y$	
		1/4 y/y = 1/16 $t/t\ y/y$	
Phenotype ratios in F_1s of monohybrid crosses		3/4 $T/{-}$(tall), 1/4 t/t (dwarf)	
		3/4 $Y/{-}$(yellow), 1/4 y/y (green)	
Combine two monohybrid ratios to determine phenotype ratios	3/4 $T/{-}$　$\times$	3/4 $Y/{-}$ = 9/16 $T/{-}Y/{-}$ (tall, yellow)	
		1/4 y/y = 3/16 $T/{-}y/y$ (tall, green)	
	1/4 t/t　$\times$	3/4 $Y/{-}$ = 3/16 $t/t\ Y/{-}$ (dwarf, yellow)	
		1/4 y/y = 1/16 $t/t\ y/y$ (dwarf, green)	

Therefore phenotype ratios = 9 tall, yellow:3 tall, green:3 dwarf, yellow:1 dwarf, green

MULTIPLE ALLELES

Previously we defined alleles as the alternate forms of a gene. Whereas an individual can have no more than two alleles at a given locus (one each on each chromosome of the homologous pair, p. 87), many more dissimilar alleles can exist in the population. An example is the set of multiple alleles that affects coat color in rabbits. The different alleles are C (normal color), c^{ch} (chinchilla color), c^h (Himalayan color), and c (albino). The four alleles fall into a dominance series with C dominant over everything. The dominant allele is always written to the left and the recessive to the right:

C/c^h	=	Normal color
c^{ch}/c^h	=	Chinchilla color
c^h/c	=	Himalayan color

Multiple alleles arise through mutations at the same gene locus over periods of time. Any gene may mutate (p. 146) if given time and thus can give rise to slightly different alleles at the same locus.

GENE INTERACTION

The types of crosses previously described are simple in that the character variation involved results from the action of a single gene, but many cases are known in which the variation of a character is the result of two or more genes. Mendel probably did not appreciate the real significance of the genotype, as contrasted with the visible character—the phenotype. We now know that many different genotypes may be expressed as a single phenotype.

Also, many genes have more than a single effect on organismal phenotypes. A gene for eye color, for instance, may be the ultimate cause of eye color, yet at the same time it may be responsible for influencing the development of other characters as well. An allele at one locus may mask or prevent the expression of an allele at another locus acting on the same trait, a phenomenon called **epistasis.** Another case of gene interaction is that in which several sets of alleles may produce a cumulative effect on the same character, or **polygenic inheritance.**

Several characters in humans are polygenic. In such cases the characters, instead of having discrete alternative phenotypes, show continuous variation between two extremes. This is sometimes called **blending,** or **quantitative inheritance.** In this kind of inheritance the children are often more or less intermediate between the two parents.

One illustration of such a type is the degree of pigmentation in matings between the black and white human races. The cumulative genes in such matings have a quantitative expression. Three or four genes are probably involved in skin pigmentation, but we will simplify our explanation by assuming that there are only two pairs of independently assorting genes. Thus a person with very dark pigment has two genes for pigmentation on separate chromosomes (A/A B/B). Each dominant allele contributes one unit of pigment. A person with very light pigment has alleles (a/a b/b) that contribute no color. (Freckles that commonly appear in the skin of very light people represent pigment contributed by entirely separate genes.) The offspring of very dark and very light parents would have an intermediate skin color (A/a B/b).

The children of parents having intermediate skin color show a range of skin color, depending on the number of genes for pigmentation that they inherit. Their skin color ranges from very dark (A/A B/B), to dark (A/A B/b or A/a B/B), intermediate (A/A b/b or A/a B/b or a/a B/B), light (A/a b/b or a/a B/b), to very light (a/a b/b). It is thus possible for parents heterozygous for skin color to produce children with darker or lighter colors than themselves.

SEX DETERMINATION AND SEX-LINKED INHERITANCE

Before the importance of the chromosomes in heredity was realized in the early 1900s, how gender was determined was totally unknown. The first really scientific clue to the determination of sex came in 1902 when C. McClung observed that bugs (Hemiptera) produced two kinds of sperm in approximately equal numbers. One kind contained among its regular set of chromosomes a so-called accessory chromosome that was lacking in the other kind of sperm. Since all the eggs of these species had the same number of haploid chromosomes, half the sperm would have the same number of chromosomes as the eggs, and half of them would have one chromosome less. When an egg was fertilized by a spermatozoon carrying the accessory (sex) chromosome, the resulting offspring was a female; when fertilized by the spermatozoon without an accessory chromosome, the offspring was a male. There are therefore two kinds of chromosomes: **sex chromosomes,** which determine sex (and sex-linked traits), and **autosomes,** which determine the other body traits. The particular type of sex determination just described is often called the XX-XO type, which indicates that the females have two X chromosomes and the males only one X chromosome (the O indicates absence of the chromosome). The XX-XO method of sex determination is depicted in Figure 8-4.

Later, other types of sex determination were discovered. In humans

The inheritance of eye color in humans is another example of gene interaction. One allele (B) determines whether pigment is present in the front layer of the iris. This allele is dominant over the allele for the absence of pigment (b). The genotypes B/B and B/b pigment generally produce brown eyes, and b/b produces blue eyes. However, these phenotypes are greatly affected by many modifier genes influencing, for example, the amount of pigment present, the tone of the pigment, and its distribution. Thus a person with B/b may even have blue eyes if modifier genes determine a lack of pigment, thus explaining the rare instances of a brown-eyed child of blue-eyed parents.

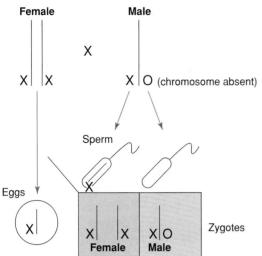

Figure 8-4
XX-XO sex determination.

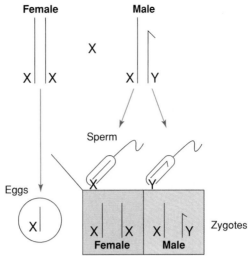

Figure 8-5
XX-XY sex determination.

and many other animals each sex has the same number of chromosomes; however, the sex chromosomes (XX) are alike in the female but unlike (XY) in the male. Hence the human egg contains 22 autosomes + 1 X chromosome; the sperm are of two kinds; half carry 22 autosomes + 1 X and half bear 22 autosomes + 1 Y. The Y chromosome is much smaller than the X. At fertilization, when two X chromosomes come together, the offspring are female; when X and Y chromosomes come together, the offspring are male. The XX-XY kind of determination is shown in Figure 8-5.

Speculation on how sex was determined in animals produced several incredible beliefs, for example, that the two testicles of the male contain different types of semen, one begetting males, the other females. It is not difficult to imagine the abuse and mutilation of domestic animals that occurred when attempts were made to alter the sex ratios of herds. Another idea asserted that sex of the offspring was determined by the more heavily sexed parent. An especially masculine father should produce sons, an effeminate father only daughters. Such ideas were not testable and have lingered until recently.

A third type of sex determination is found in birds, moths, and butterflies in which the male has two X (or sometimes called ZZ) chromosomes

and the female an X and a Y (or ZW). Finally, there are both invertebrates (p. 433) and vertebrates (p. 565) in which sex is determined by environmental or behavioral conditions rather than by sex chromosomes, or by genetic loci whose variation is not associated with visible difference in chromosomal structure.

Since the X chromosome carries several genes that are essential for normal cellular function and the Y chromosome carries none, every individual must have *at least* one X chromosome. And of course the XX-XY system ensures that they will: the female has two, the male one. To keep the balance between X-chromosomal genes and autosomal genes the same in the female as in the male, one of the female's X chromosomes is always inactivated in mammalian somatic cells. This happens early in development when the embryo consists of only a few cells, according to the widely accepted Lyon hypothesis (after Mary Lyon, who discovered this and other aspects of human sex determination). At this time it is strictly a matter of chance whether the maternal or paternal X chromosome is the one inactivated in each cell. All descendants of that cell, however, retain the same inactivated chromosome. Consequently, tissues of adult females are a patchwork of cells (mosaic) expressing the genes of one or the other X chromosome, never both. Interest-

ingly, the inactivated X chromosome condenses in the interphase nucleus of most somatic cells into a tiny, darkly staining object called a Barr body (Figure 8-6). This furnishes a simple method for determining whether an individual is a genetic male or female, as has been done, for example, in examination of entrants in women's athletic events.

Normally this X-chromosome mosaic is of no consequence in human females, but it has a dramatic effect in cats. Tortoise-shell (calico, tabby) cats, which are spotted black and yellow, are always females heterozygous for black and yellow alleles; the males having either allele are all black or all yellow. This happens because the genes for fur color are located on the X chromosome; that is, they are sex linked. A male, having only one X chromosome, will be either black or yellow. But a female has two X chromosomes and, if she is heterozygous for fur color, her fur will be a mosaic of yellow and black patches. Each patch represents the descendants of an early embryonic skin cell having one operating and one inactive X chromosome.

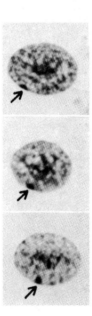

Figure 8-6
Nuclei from three epithelial cells taken from the mouth of a normal human female. The darkly staining Barr bodies (*arrows*) are the heterochromatinized (inactive) X chromosome.

Genetic sex is just one aspect of sex determination. Having an XX or an XY genetic constitution does not in itself produce a female or male. The embryo is at first totally indifferent sexually, despite its genetic sex. Even the gonad is sexually indifferent and bipotential, since it can differentiate into either a testis or an ovary. Gonad differentiation was discussed in Chapter 6 (p. 86), and we mentioned there that the primordial gonad has an inherent tendency to become an ovary, and the rest of the body to become female, *unless* a male-determining gene on the Y chromosomes redirects gonadal differentiation into a testis. As the presumptive gonad differentiates into an ovary or testis, it begins to secrete sex hormones (androgens by the testes, estrogens by the ovaries). These hormones influence the other developing organs to become male or female.

Sex-Linked Inheritance

It has long been known that the inheritance of some characters depends on the sex of the parent carrying the gene and the sex of the offspring. One of the best-known sex-linked traits of humans is hemophilia (Chapter 34, p. 674). Another example is red-green color blindness in which red and green colors are indistinguishable to varying degrees. Color-blind men greatly outnumber color-blind women. When color blindness does appear in women, their fathers are color blind. Furthermore, if a woman with normal vision who is a carrier of color blindness bears sons, half of them are likely to be color blind, regardless of whether the father had normal or affected vision. How are these observations explained?

The color blindness and hemophilia defects are recessive traits carried on the X chromosome that are phenotypically expressed either when both genes are defective in the female or when only one defective gene is present in the male. The inheritance pattern of these defects is shown for color blindness in Figure 8-7. When the mother is a carrier and the father is normal, half of the sons but none of the daughters are color blind. However, if the father is color blind and the mother is a carrier, half of the sons *and* half of the daughters are color blind (on the average and in a large sample). It is easy to understand then why such defects are much more prevalent in males: a single sex-linked recessive gene in the male has a visible effect. What would be the outcome of a mating between a homozygous normal woman and a color-blind man?

Another example of a sex-linked character was discovered by Thomas Hunt Morgan (1910) in *Drosophila*. The normal eye color of this fly is red, but mutations for white eyes do occur (Figure 8-8). A gene for eye color is carried on the X chromosome. If

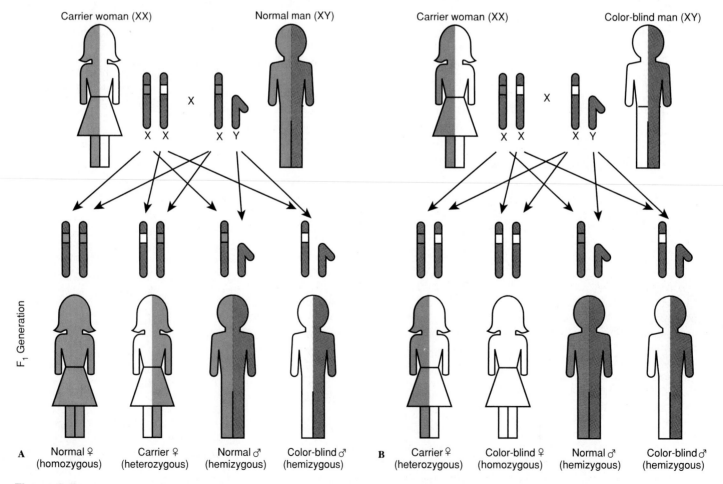

Figure 8-7

Sex-linked inheritance of red-green color blindness in humans. **A,** Carrier mother and normal father produce color-blindness in one-half of their sons but in none of their daughters. **B,** Half of both sons and daughters of carrier mother and color-blind father are color blind.

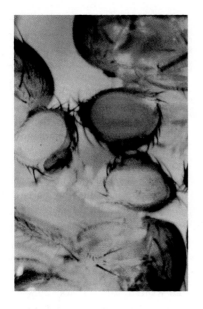

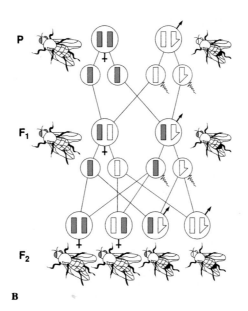

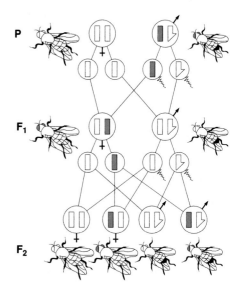

Figure 8-8

Sex-linked inheritance of eye color in fruit fly *Drosophila melanogaster*. **A,** White and red eyes of *D. melanogaster*. **B,** Genes for eye color are carried on X chromosome; Y carries no genes for eye color. Normal red is dominant to white. Homozygous red-eyed female mated with white-eyed male gives all red-eyed in F₁. F₂ ratios from F₁ cross are one homozygous red-eyed female and one heterozygous red-eyed female to one red-eyed male and one white-eyed male.

Figure 8-9

Reciprocal cross of Figure 8-8 (homozygous white-eyed female with red-eyed male) gives white-eyed males and red-eyed females in F₁. F₂ shows equal numbers of red-eyed and white-eyed females and red-eyed and white-eyed males.

truebreeding white-eyed males and red-eyed females are crossed, all the F₁ offspring are red eyed because this trait is dominant (Figure 8-8). If these F₁ offspring are interbred, all F₂ females have red eyes; half of the males have red eyes and the other half have white eyes. No white-eyed females are found in this generation; only the males show the recessive trait (white eyes). The allele for white eyes is recessive and should only appear in a homozygous condition. However, since the male has only one X chromosome (the Y does not carry a gene for eye color), white eyes appear whenever the X chromosome carries the gene for this trait. Males are said to be **hemizygous** for traits carried on the X chromosome.

If the reciprocal cross is made in which the females are white eyed and the males red eyed, all the F₁ females are red eyed and all the males are white eyed (Figure 8-9). If these F₁ offspring are interbred, the F₂ generation shows equal numbers of red-eyed and white-eyed males and females.

AUTOSOMAL LINKAGE AND CROSSING OVER

Linkage

Since Mendel's laws were rediscovered in 1900, it has become clear that, in seeming contradiction to Mendel's second law, not all factors segregate independently. Indeed, many traits are inherited together. Since the number of chromosomes in any organism is relatively small compared with the number of characters, each chromosome must contain many genes. All genes present on a chromosome are said to be **linked.** Linkage simply means that the genes are on the same chromosome, and all genes present on homologous chromosomes belong to the same linkage groups. Therefore there should be as many linkage groups as there are chromosome pairs.

Geneticists commonly use the word "linkage" in two somewhat different senses. Sex linkage refers to inheritance of a trait on the sex chromosomes, and thus its phenotypic expression depends on the sex of the organism and the factors already discussed. Autosomal linkage, or simply, linkage, refers to inheritance of the genes on a given autosomal chromosome. Letters used to represent such genes are normally written without a slash mark between them, indicating that they are on the same chromosome. For example, *AB/ab* shows that genes *A* and *B* are on the same chromosome, and *a* and *b* are on the homologous chromosome. Interestingly, Mendel studied seven characteristics of garden peas, which assorted independently because they were on seven different chromosomes. If he had studied eight characteristics, he might not have found independent assortment in two of the traits, because garden peas have only seven pairs of homologous chromosomes. If genes are very far apart on the same chromosome, however, they often assort independently because crossing over between them happens regularly.

How can we reconcile these observations on recessive sex-linked defects with the inactivation of one X chromosome in females? If the normal X chromosome was the one inactivated in cells of heterozygous females, would the person not show the recessive trait because the active chromosome was the one with the recessive allele? The answer is that which X chromosome is inactivated is entirely random in cells of the embryo, and there are almost always enough cells with normal X chromosomes that the recessive trait is not expressed.

In *Drosophila,* in which this principle has been worked out most extensively, there are four linkage groups that correspond to the four pairs of chromosomes found in these fruit flies. Usually, small chromosomes have small linkage groups, and large chromosomes have large groups.

Crossing Over

Linkage, however, is usually not complete. If we perform an experiment in which animals such as *Drosophila* are crossed, we find that linked traits separate in some percentage of the offspring. Separation of alleles located on the same chromosome occurs because of **crossing over.**

As described in Chapter 6 (p. 87), during the protracted prophase of the first meiotic division, paired homologous chromosomes break and exchange equivalent portions; genes "cross over" from one chromosome to its homolog, and vice versa (Figure 8-10). Each chromosome consists of two sister chromatids held together by means of a proteinaceous structure called a **synaptonemal complex.** Breaks and exchanges occur at corresponding points on nonsister chromatids. (Breaks and exchanges also occur between sister chromatids but usually have no genetic significance because sister chromatids are identical.) Crossing over is a means for exchanging genes between homologous chromosomes and as such greatly increases the amount of genetic recombination. The frequency of crossing over varies depending on the species, but usually at least one and often several crossovers occur each time chromosomes pair.

Because the frequency of recombination is proportional to the distance between loci, the comparative linear position of each locus can be determined. Laborious genetic experiments over many years have produced gene maps that indicate the position of more than 500 genes distributed on the four *Drosophila* chromosomes. Crossing over is so prevalent that genes located far apart on the same chromosome often will obey Mendel's second law, assorting independently in genetic crosses.

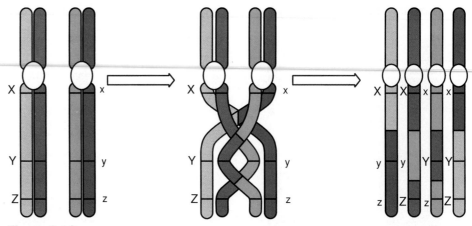

Figure 8-10

Crossing over during meiosis. Nonsister chromatids exchange portions, so that none of the resulting gametes is genetically the same as any other. Gene X is farther from gene Y than Y is from Z; therefore gene X is more frequently separated from Y in crossing over than Y is from Z.

CHROMOSOMAL ABERRATIONS

Structural and numerical deviations from the norm that affect many genes at once are called chromosomal aberrations. They are sometimes called chromosomal mutations, but most cytogeneticists prefer to use the term "mutation" to refer to qualitative changes within a gene; gene mutations are discussed on p. 146.

Despite the incredible precision of meiosis, chromosomal aberrations do occur, and they are more common than one might think. They have great economic benefit in agriculture. Unfortunately, they are also responsible for many human genetic malformations. It is estimated that five out of every 1000 humans are born with *serious* genetic defects attributable to chromosomal anomalies. An even greater number of embryos with chromosomal defects are aborted spontaneously, far more than ever reach term.

Changes in chromosome numbers are called **euploidy** when there is the addition or deletion of whole sets of chromosomes and **aneuploidy** when a single chromosome is added to or subtracted from a set. A "set" of chromosomes contains one member of each homologous pair as would be present in the nucleus of a gamete. The most common kind of euploidy is **polyploidy,** the carrying of one or more additional sets of chromosomes. Such aberrations are much more common in plants than in animals. Animals are much less tolerant of chromosomal aberrations, because sex determination requires a delicate balance between the numbers of sex chromosomes and autosomes. Many domestic plant species are polyploid (cotton, wheat, apples, oats, tobacco, and others), and perhaps 40% of flowering plants are believed to have originated in this manner. Horticulturists favor polyploids and often try to develop them because they have more intensely colored flowers and more vigorous vegetative growth.

Aneuploidy is usually caused by failure of chromosomes to separate during meiosis **(nondisjunction).** If a pair of chromosomes fails to separate during the first or second meiotic divisions, both members go to one pole and none to the other. This results in one gamete having $n - 1$ number of chromosomes and another having $n + 1$ number of chromosomes. If the $n - 1$ gamete is fertilized by a normal n gamete, the result is a **monosomic** animal. Survival is rare because the lack of one chromosome gives an uneven balance of genetic instructions. **Trisomy,** the result of the fusion of a normal n gamete and an $n + 1$ gamete, is much more common, and several kinds of trisomic conditions are known in humans. Perhaps the most familiar is **trisomy 21, or Down syndrome.** As the name indicates, it involves an extra chromosome 21 combined with the chromosome pair 21, and it is caused by nondisjunction of that pair during meiosis. It occurs spontaneously, and there is seldom any family history of the abnormality. However, the risk of its appearance rises dramatically with increasing age of the mother; it occurs

40 times as often in women over 40 years old as among women between the ages of 20 and 30.

A syndrome is a group of symptoms associated with a particular disease or abnormality, although not every symptom is necessarily shown by every patient with the condition. In 1866 an English physician, John Langdon Down, described the syndrome that we know is caused by trisomy 21 (also called Down syndrome and Down's syndrome). Among the numerous characteristics of the condition, the most disabling is a varying degree of mental retardation. This and other conditions caused by chromosomal aberrations can be diagnosed *prenatally* by amniocentesis. The physician inserts a hypodermic needle through the abdominal wall of the mother and into the fluids surrounding the fetus (*not into* the fetus) and withdraws some of the fluid, which contains some fetal cells. The cells are grown in culture, their chromosomes are examined, and other tests are done. If a severe birth defect is found, the mother has the option of having an abortion. As an extra "bonus," the sex of the fetus is learned after amniocentesis. How?

Structural aberrations involve whole sets of genes within a chromosome. A portion of a chromosome may be reversed, placing the linear arrangement of genes in reverse order (inversion); nonhomologous chromosomes may exchange sections (translocation); entire blocks of genes may be lost (deletion); or an extra section of chromosome may attach to a normal chromosome (duplication). These structural changes often produce phenotypic changes. Duplications, although rare, are important for evolution because they supply additional genetic information that may enable new functions.

GENE THEORY

GENE CONCEPT

The term "gene" (Gr. *genos,* descent) was coined by W. Johannsen in 1909 to refer to the hereditary factors of Mendel. Both cytological and genetic studies showed that genes, although of unknown chemical nature, were the fundamental units of inheritance. They were regarded as indivisible units of the chromosomes on which they were located. Studies with multiple mutant alleles demonstrated that alleles are in fact divisible by recombination; that is, *portions* of a gene are separable, and they have a fine structure. Furthermore, parts of many genes in eukaryotes are separated by sections of DNA, called **introns,** that do not specify a part of the finished product. Because of their ability to mutate, to be assorted and shuffled around in different combinations, genes have become the basis for our modern interpretation of evolution. Genes are molecular patterns that can maintain their identities for many generations and can be self-duplicated in each generation.

One Gene—One Polypeptide Hypothesis

Since genes act to produce different phenotypes, we may infer that their action follows the scheme: gene → gene product → phenotypic expression. The direct products of genes are RNA molecules that function in the synthesis of proteins. Different genes encode one of many messenger RNA molecules that specify the sequence of amino acids of a particular protein, ribosomal RNA molecules that form the structural site for protein synthesis, or one of the many transfer RNA molecules needed to transport amino acids to the site of protein synthesis. The proteins thus produced act as enzymes, antibodies, hormones, and structural elements throughout the body.

The first clear, well-documented study to link genes and enzymes was carried out on the common bread mold *Neurospora* by Beadle and Tatum in the early 1940s. This organism was ideally suited to a study of gene function for several reasons: these molds are much simpler to handle than fruit flies, they grow readily in well-defined chemical media, and they are haploid organisms (through most of their life cycle) that are consequently unencumbered with dominance relationships. Furthermore, mutations were readily induced by irradiation with ultraviolet light. Ultraviolet-light-induced mutants, grown and tested in specific nutrient media, had single-gene mutations that were inherited in accord with Mendelian principles of segregation. Each mutant strain was defective in one enzyme, which prevented that strain from synthesizing one or more complex molecules. Putting it another way, the ability to synthesize a particular molecule was controlled by a single gene.

From these experiments Beadle and Tatum set forth an important and exciting formulation: **one gene produces one enzyme.** For this work they were awarded the Nobel Prize in 1958. The new hypothesis was soon validated by the research of others who studied other biosynthetic pathways. Hundreds of inherited disorders, including dozens of human hereditary diseases, are caused by single mutations that result in the loss of a specific enzyme. We now know that a particular protein may be made of several chains of amino acids (polypeptides), each of which may be specified by a different gene, and not all proteins specified by genes are enzymes (for example, structural proteins, antibodies, transport proteins, and hormones). Furthermore, genes directing the synthesis of ribosomal and transfer RNA were not included in Beadle and Tatum's formulation. Therefore a gene now may be defined more inclusively as **a nucleic acid sequence** (usually DNA) that encodes a functional polypeptide or RNA sequence.

STORAGE AND TRANSFER OF GENETIC INFORMATION

NUCLEIC ACIDS: MOLECULAR BASIS OF INHERITANCE

As we noted in Chapter 2, cells contain two kinds of nucleic acids: deoxyribonucleic acid (DNA), which is the genetic material, and ribonucleic acid (RNA), which functions in protein synthesis. Both DNA and RNA are polymers built of repeated units called **nucleotides.** Each nucleotide contains three parts: a **sugar,** a **nitrogenous**

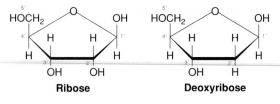

Figure 8-11

Ribose and deoxyribose, the pentose sugars of nucleic acids. A carbon atom lies in each of the four corners of the pentagon (labeled 1' to 4'). Ribose has a hydroxyl group (—OH) and a hydrogen on the number 2' carbon; deoxyribose has two hydrogens at this position.

Table 8-2	Chemical Components of DNA and RNA	
	DNA	**RNA**
Purines	Adenine	Adenine
	Guanine	Guanine
Pyrimidines	Cytosine	Cytosine
	Thymine	Uracil
Sugar	2-Deoxyribose	Ribose
Phosphate	Phosphoric acid	Phosphoric acid

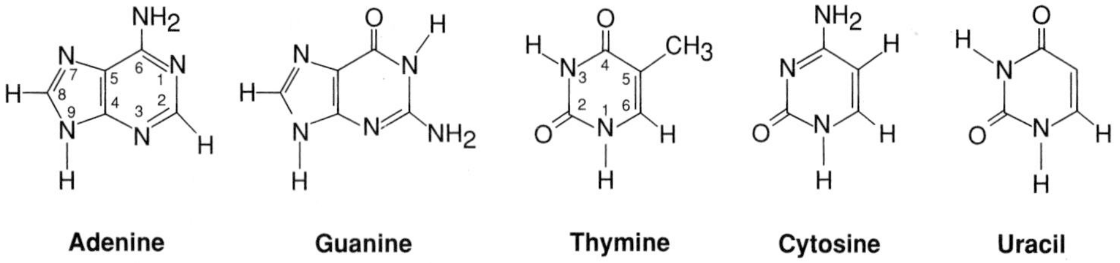

PURINES

Adenine   Guanine

PYRIMIDINES

Thymine   Cytosine   Uracil

Figure 8-12

Purines and pyrimidines of DNA and RNA.

base, and a **phosphate group.** The sugar is a pentose (5-carbon) sugar; in DNA it is **deoxyribose** and in RNA it is **ribose** (Figure 8-11).

The nitrogenous bases of nucleotides are also of two types: pyrimidines, which consist of a single, 6-membered ring, and purines, which are composed of two fused rings. Both of these types of compounds contain nitrogen as well as carbon in their rings, which is why they are called "nitrogenous" bases. The purines in both RNA and DNA are adenine and guanine (Table 8-2). The pyrimidines in DNA are thymine and cytosine, and in RNA they are uracil and cytosine. The carbon atoms in the bases are numbered (for identification) according to standard biochemical notation (Figure 8-12). The carbons in the ribose and deoxyribose are also numbered, but to distinguish them from the carbons in the bases, the numbers for the carbons in the sugars are given prime signs (see Figure 8-11).

The sugar, phosphate group, and nitrogenous base are linked as shown in the generalized scheme for a nucleotide:

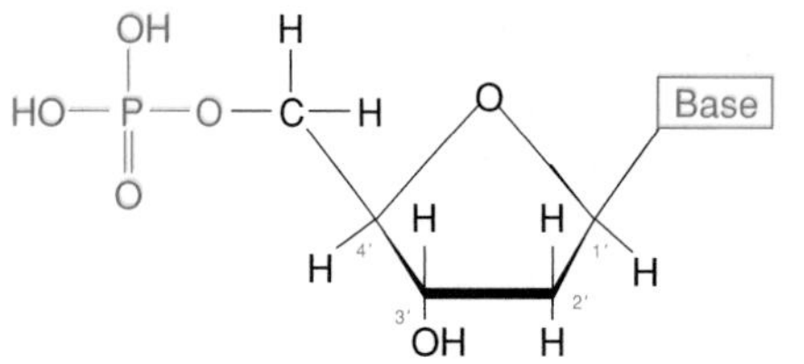

Phosphate   Sugar   Nitrogenous base

In DNA the backbone of the molecule is built of phosphoric acid and deoxyribose; to this backbone are attached the nitrogenous bases (Figure 8-13). The **5' end** of the backbone has a free phosphate group on the **5'** carbon of the ribose, and the **3' end** has a free hydroxyl group on the **3'** carbon. However, one of the most interesting and important discoveries about the nucleic acids is that DNA is not a single polynucleotide chain; rather it consists of *two* complementary chains that are precisely cross-linked by specific hydrogen bonding between purine and pyrimidine bases. The number of adenines is equal to the number of thymines, and the number of guanines equals the number of cytosines. This fact suggested a pairing of bases: adenine with thymine (AT) and guanine with cytosine (GC) (Figures 1-6 and 8-14).

The result is a ladder structure (Figure 8-15). The upright portions are the sugar phosphate backbones, and the connecting rungs are the paired nitrogenous bases, AT or GC. However, the ladder is twisted into a **double helix** with approximately 10 base pairs for each complete turn of the helix (Figure 8-16). The two DNA strands run in opposite directions, that is they are **antiparallel,** and the 5' end of one strand is the 3' end of the other. This is evident from an examination of Figure 8-16. The two strands are also **complementary—** the sequence of bases along one strand specifies the sequence of bases along the other strand.

The determination of the structure of DNA has been widely acclaimed as the single most important biological discovery of this century. It was based on the x-ray diffraction studies of Maurice H. F. Wilkins and Rosalind Franklin and on the ingenious proposals of Francis H. C. Crick and James D. Watson published in 1953. Watson, Crick, and Wilkins were later awarded the Nobel Prize for their momentous discovery.

RNA is similar to DNA in structure except that it consists of a *single* polynucleotide chain, has ribose instead of deoxyribose, and has uracil instead of thymine. The three kinds of RNA (ribosomal, transfer, and messenger) are described below.

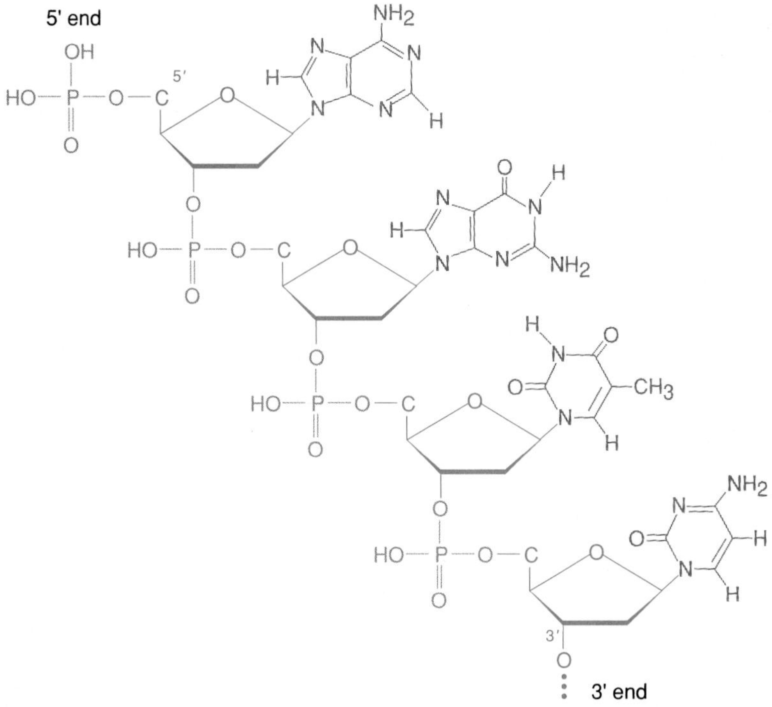

Figure 8-13

Section of a strand of DNA. Polynucleotide chain is built of a backbone of phosphoric acid and deoxyribose sugar molecules. Each sugar holds a nitrogenous base side arm. Shown from top to bottom are adenine, guanine, thymine, and cytosine.

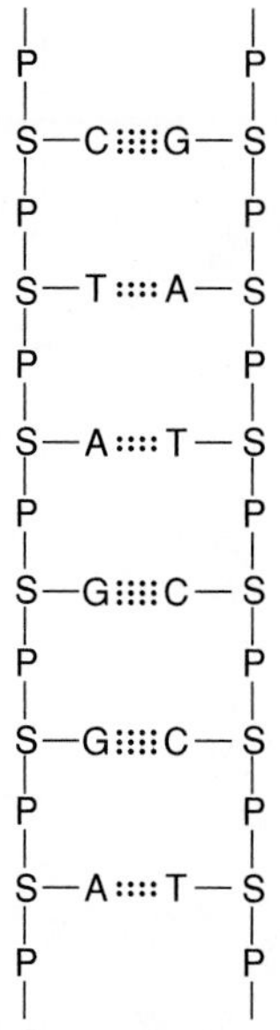

Figure 8-15

DNA, showing how the complementary pairing of bases between the sugar-phosphate "backbones" keeps the double helix at a constant diameter for the entire length of the molecule. Dotted lines represent the three hydrogen bonds between each cytosine and guanine and the two hydrogen bonds between each adenine and thymine.

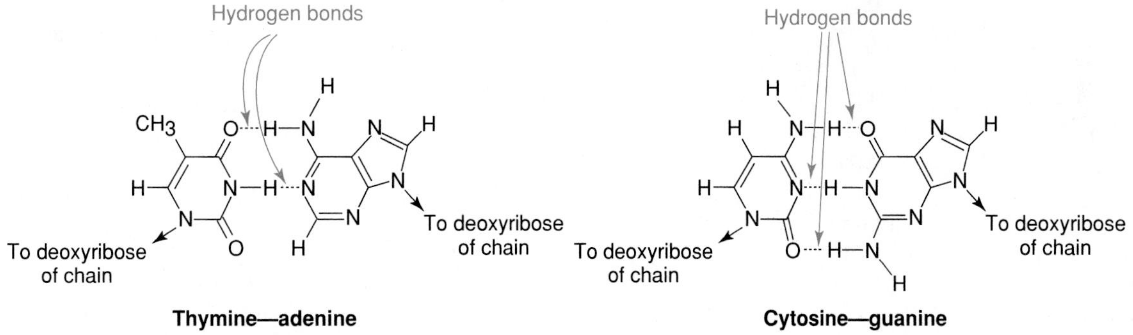

Figure 8-14

Positions of hydrogen bonds between thymine and adenine and between cytosine and guanine in DNA.

Every time a cell divides, the structure of DNA must be precisely copied in the daughter cells. This is called **replication** (Figure 8-17). During replication, the two strands of the double helix unwind, and each separated strand serves as a **template** against which a complementary strand is synthesized. That is, an enzyme (DNA polymerase) assembles a new strand of polynucleotides with a thymine group going next to the adenine group in the template strand, a guanine group next to the cytosine group, and so on.

DNA Coding by Base Sequence

Since DNA is the genetic material, and it is composed of a linear sequence of base pairs, an obvious extension of the Watson-Crick model is that the sequence of base pairs in DNA codes for, and is colinear with, the sequence of amino acids in a protein. The coding hypothesis had to account for the way a string of four different bases—a four-letter alphabet—could dictate the sequence of 20 different amino acids.

In the coding procedure, obviously there cannot be a 1:1 correlation between four bases and 20 amino acids. If the coding unit (often called a word, or **codon**) consists of two bases, only 16 words (4^2) can be formed, which cannot account for 20 amino acids. Therefore the codon must consist of at least three bases or three letters, because 64 possible words (4^3) can be formed by four bases when taken as triplets. This means that there could be a considerable redundancy of triplets (codons), since DNA codes for just 20 amino acids. Later work confirmed that nearly all of the amino acids are specified by more than one triplet code (Table 8-3).

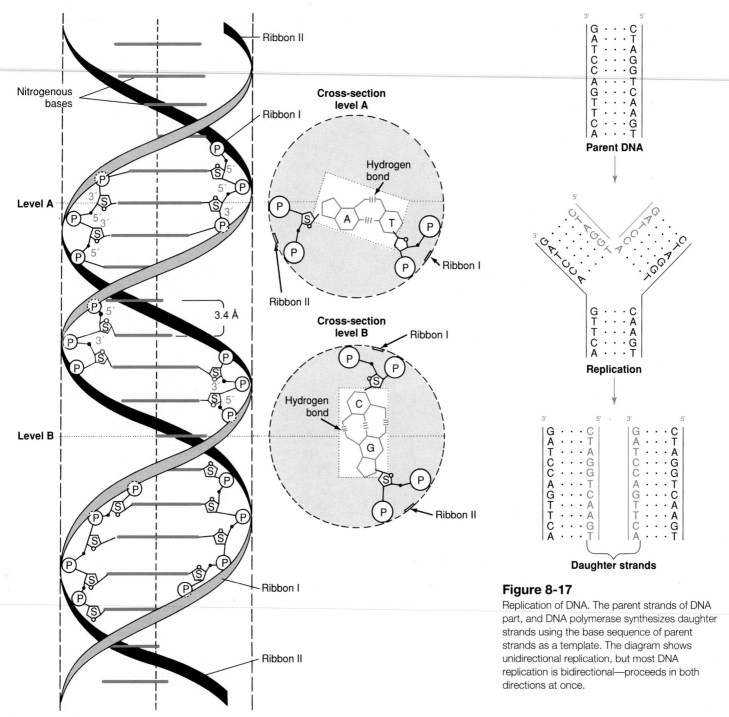

Figure 8-16
DNA molecule.

Figure 8-17
Replication of DNA. The parent strands of DNA part, and DNA polymerase synthesizes daughter strands using the base sequence of parent strands as a template. The diagram shows unidirectional replication, but most DNA replication is bidirectional—proceeds in both directions at once.

DNA shows a surprising stability, both in prokaryotes and in eukaryotes. Interestingly, it is susceptible to damage by harmful chemicals in the environment and radiation. Such damage is usually not permanent, because cells have an efficient repair system. Various types of damage and repair are known, one of which is called **excision repair.** Ultraviolet irradiation often causes adjacent pyrimidines to link together by covalent bonds (dimerize), preventing transcription and replication. A series of several enzymes "recognizes" the area of the damaged strand and excises the pair of dimerized pyrimidines and several bases following them. DNA polymerase then synthesizes the missing strand along the remaining one, according to the base-pairing rules, and the enzyme **DNA ligase** joins the end of the new strand to the old one.

TRANSCRIPTION AND THE ROLE OF MESSENGER RNA

Information is coded in DNA, but DNA does not participate directly in protein synthesis. It is obvious that an

Table 8-3	The Genetic Code: Amino Acids Specified by Codons of Messenger RNA

Second Letter

First Letter		U		C		A		G		Third Letter
		U		**C**		**A**		**G**		
U	UUU UUC	Phenylalanine	UCU UCC UCA UCG	Serine	UAU UAC	Tyrosine	UGU UGC	Cysteine	U C
	UUA UUG	Leucine			UAA UAG	End chain	UGA UGG	End chain Tryptophane	A G
C	CUU CUC CUA CUG	Leucine	CCU CCC CCA CCG	Proline	CAU CAC	Histidine	CGU CGC CGA CGG	Arginine	U C A G
					CAA CAG	Glutamine			
A	AUU AUC AUA	Isoleucine	ACU ACC ACA ACG	Threonine	AAU AAC	Asparagine	AGU AGC	Serine	U C A
	AUG	Methionine*			AAA AAG	Lysine	AGA AGG	Arginine	G
G	GUU GUC GUA GUG	Valine	GCU GCC GCA GCG	Alanine	GAU GAC	Aspartic acid	GGU GGC GGA GGG	Glycine	U C A G
					GAA GAG	Glutamic acid			

*Also, begin chain.

intermediary is required. This intermediary is another nucleic acid called **messenger RNA (mRNA).** The triplet codes in DNA are **transcribed** into mRNA, with uracil substituting for thymine (Table 8-3).

Ribosomal, transfer, and messenger RNAs are transcribed directly from DNA, each encoded by different sets of genes. In this process of making a complementary copy of one strand or gene of DNA in the formation of mRNA, an enzyme, **RNA polymerase,** is needed. (In fact, in eukaryotes each type of RNA [ribosomal, transfer, and messenger] is transcribed by a specific type of RNA polymerase.) The mRNA contains a sequence of bases that complements the bases in one of the two DNA strands just as the DNA strands complement each other. Thus A in the coding DNA strand is replaced by U in mRNA; C is replaced by G; G is replaced by C; and T is replaced by A. Only one of the two chains is used as the template for RNA synthesis because only one bears the AUG codon that initiates a message (Table 8-3). The reason why only one strand of the double-stranded DNA is a "coding strand" is that mRNA otherwise would always be formed in complementary pairs, and enzymes also would be synthesized in complementary pairs. In other words, two different enzymes would be produced for every DNA coding sequence instead of one. This would lead to metabolic chaos.

Genes on the DNA of prokaryotes are coded on a continuous stretch of DNA, which is transcribed into mRNA and then translated (see the following section). It was assumed that this was also the case for eukaryotic genes until the surprising discovery that stretches of DNA that are transcribed into RNA in the nucleus are not found in the corresponding mRNA in the cytoplasm. In other words, pieces of the nuclear mRNA were removed in the nucleus before the finished mRNA was transported to the cytoplasm (Figure 8-18). It was thus discovered that many genes are split, interrupted by sequences of bases that do not code for the final product, and the mRNA transcribed from them must be edited or matured before translation in the cytoplasm. The noncoding segments of nuclear mRNA are now known as **introns,** while the

coding sections that are translated into gene product are called **exons.** Before the mRNA leaves the nucleus, a methylated guanine "cap" is added at the 5' end, and a tail of adenine nucleotides (poly-*A*) is often added at the 3' end. The introns are then removed and the flanking pieces of exons are spliced together (Figure 8-18).

In mammals the genes coding for the histones and for interferon are on continuous stretches of DNA. However, we now know that genes coding for many proteins are split. In lymphocyte differentiation the parts of the split genes coding for immunoglobulins are actually *rearranged* during development, so that different proteins result from subsequent transcription and translation. This helps account for the enormous diversity of antibodies manufactured by the descendants of the lymphocytes (p. 676).

Scientists have learned further that the belief that introns are nonfunctional regions is not necessarily true. Although DNA of animal mitochondria has no introns, in mitochondrial mRNAs of some other organisms, the introns may code for proteins, and in one case for a

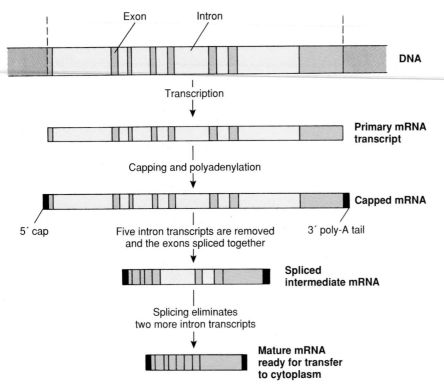

Figure 8-18

Transcription and maturation of ovalbumin gene of chicken. The entire gene of 7700 base pairs is transcribed to form the primary mRNA, then the 5′ cap of methyl guanine and the 3′ polyadenylate tail are added. After the introns are spliced out, the mature mRNA is transferred to the cytoplasm.

ribosomal component. In most of those studied so far, the proteins coded by the introns are "maturases," proteins that play some role in the splicing of the introns from which they came or introns from different genes. There are base sequences in some introns that are complementary to other base sequences in the intron, suggesting that the intron could fold so that the complementary sequences would pair. This may be necessary to control proper alignment of intron boundaries before splicing. Most surprising of all has been the discovery that, at least in some cases, the RNA can "self-catalyze" the excision of introns. The ends of the intron join; the intron thus becomes a small circle of RNA, and the exons are spliced together. This process does not fit the classical definition of an enzyme or other catalyst since the molecule itself is changed in the reaction.

TRANSLATION: FINAL STAGE IN INFORMATION TRANSFER

The **translation** process takes place on **ribosomes,** granular structures

composed of protein and **ribosomal RNA (rRNA).** Ribosomal RNA is composed of a large and a small subunit, and the small subunit comes to lie in a depression of the large subunit to form the functional ribosome (Figure 8-19). The mRNA molecules fix themselves to the ribosomes to form a messenger RNA-ribosome complex. Since only a short section of mRNA molecule is in contact with a single ribosome, the mRNA usually fixes itself to several ribosomes at once. The entire

complex, called a **polyribosome** or **polysome,** allows several molecules of the same kind of protein to be synthesized at once, one on each ribosome of the polysome (Figure 8-19).

The assembly of proteins on the mRNA-ribosome complex requires the action of another kind of RNA called **transfer RNA (tRNA).** The tRNAs are surprisingly large molecules that are folded in a complicated way in the form of a cloverleaf (Figure 8-20). The tRNA molecules collect the free amino acids from the cytoplasm and deliver them to the polysome, where they are assembled into a protein. There are special tRNA molecules for every amino acid. Furthermore, each tRNA is accompanied by a specific tRNA synthetase. The tRNA synthetases are enzymes that are necessary to sort out and attach the correct amino acid to a site on the end of each tRNA by a process called **charging.**

On the cloverleaf-shaped molecule of tRNA, a special sequence of three bases (the **anticodon**) is exposed in just the right way to form base pairs with complementary bases (the codon) in the mRNA. The codons are read and proteins assembled along the mRNA in a 5′ to 3′ direction. The anticodon of the tRNA is the key to the correct sequencing of amino acids in the protein being assembled.

For example, alanine is assembled into a protein when it is signaled by the codon GCG in an mRNA. The translation is accomplished by alanine tRNA in which the anticodon is CGC. The alanine tRNA is first charged with

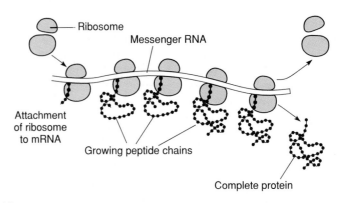

Figure 8-19

How the protein chain is formed. As ribosomes move along messenger RNA, the amino acids are added stepwise to form the polypeptide chain.

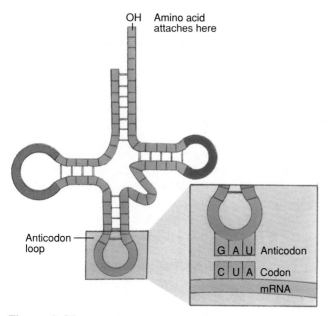

Figure 8-20

Structure of a tRNA molecule. The anticodon loop bears bases complementary to those in the mRNA codon. The other two loops function in binding to the ribosomes in protein synthesis. The amino acid is added to the free single-stranded —OH end by tRNA synthetase.

alanine by its tRNA synthetase. The alanine tRNA complex enters the ribosome where it fits precisely into the right place on the mRNA strand. Then the next charged tRNA specified by the mRNA code (glycine tRNA, for example) enters the ribosome and attaches itself beside the alanine tRNA. The two amino acids are united with a peptide bond (with the energy from a molecule of guanosine triphosphate), and the alanine tRNA falls off. The process continues stepwise as the protein chain is built (Figure 8-21). A protein of 500 amino acids can be assembled in less than 30 seconds.

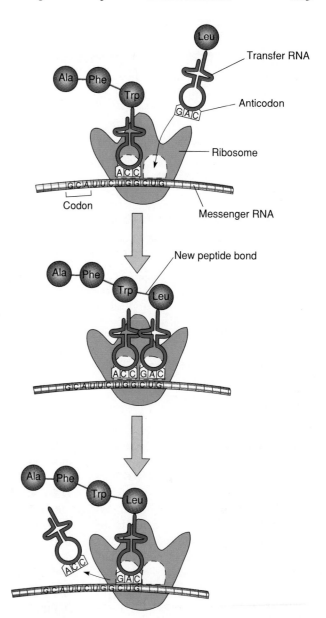

Figure 8-21

Formation of polypeptide chain on messenger RNA. As ribosome moves down messenger RNA molecule, transfer RNA molecules with attached amino acids enter ribosomes (*top*). Amino acids are joined together into polypeptide chain, and transfer RNA molecules leave ribosome (*bottom*).

REGULATION OF GENE EXPRESSION

In Chapter 7 we saw how the orderly differentiation of an organism from fertilized ovum to adult requires the involvement of genetic material at every stage of development. Developmental biologists have provided convincing evidence that every cell in a developing embryo is genetically equivalent. Thus it is clear that as tissues differentiate (change developmentally), they use only a part of the genetic instruction present in every cell. Certain genes express themselves only at certain times and not at others. Indeed, there is reason to believe that in a particular cell or tissue, most of the genes are inactive at any given moment. The problem in development is to explain how, if every cell has a full gene complement, certain genes are "turned on" and produce proteins that are required for a particular developmental stage while the other genes remain silent.

Actually, although the developmental process brings the question of gene activation clearly into focus, gene regulation is necessary throughout an organism's existence. The cellular enzyme systems that control all functional processes obviously require genetic regulation because enzymes have powerful effects even in minute amounts. Enzyme synthesis must be responsive to the influences of supply and demand.

The discoveries of two French scientists, J. Monod and F. Jacob, contributed greatly to our understanding of gene regulation. Using bacteria, they described the **operon model.** Knowledge of gene regulation in prokaryotes is important in recombinant DNA technology, but we now know that the operon concept has little if any applicability to gene regulation in eukaryotes.

Gene Regulation in Eukaryotes

There are a number of different phenomena in eukaryotic cells that can serve as control points, and the following are a few examples.

Transcriptional Control. This may be the most important mechanism. **Transcription factors** are molecules that may have a positive or a negative effect on transcription of RNA from the DNA of the target genes. The factors may act within the cells that produce them or they may be transported to different parts of the body prior to action. An example of a positive transcription factor is a steroid receptor. Steroid hormones produced by endocrine glands elsewhere in the body enter the cell and bind with a receptor protein in the nucleus. The steroid-receptor complex then binds with DNA near the target gene (p. 741). Progesterone, for example, binds with a nuclear receptor in cells of the chicken oviduct; the hormone-receptor complex then activates the transcription of genes encoding egg albumin and other substances.

Translational Control. Genes can be transcribed and the mRNA sequestered in some way so that translation is delayed. This commonly happens in the development of eggs of many animals. The oocyte accumulates large quantities of messenger RNA during its development, then fertilization activates metabolism and initiates translation of maternal mRNA.

Gene Rearrangement. Vertebrates contain cells called lymphocytes that bear genes coding for proteins called antibodies (p. 675). Each type of antibody has the capacity to bind specifically with a particular foreign substance (antigen). Because the number of different antigens is enormous, the genetic diversity of antibody genes must be equally great. One source of this diversity is rearrangement of DNA sequences coding for the antibodies during the development of lymphocytes. Rearrangements are also responsible for a switch in mating types in some fungi.

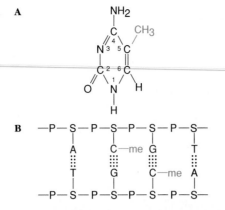

Figure 8-22

Some genes in eukaryotes are turned off by the methylation of some cytosine residues in the chain. **A,** Structure of 5-methyl cytosine. **B,** Cytosine residues next to guanine are those that are methylated in a strand, thus allowing both strands to be symmetrically methylated.

DNA Modification. An important mechanism for turning genes off appears to be methylation of cytosine residues in the 5 position, that is, adding a methyl group (CH_3—) to the carbon in the 5 position in the cytosine ring (Figure 8-22A). This usually happens when the cytosine is next to a guanine residue; thus, the bases in the complementary DNA strand would also be a cytosine and a guanine (Figure 8-22B). When the DNA is replicated, an enzyme recognizes the CG sequence and quickly methylates the daughter strand, maintaining the gene in an inactive state.

GENETIC ENGINEERING

Progress in our understanding of genetic mechanisms on the molecular level, as discussed in the last few pages, has been almost breathtaking in the last few years. We can expect many more discoveries in the near future. This progress has been due largely to the effectiveness of many biochemical techniques now used in molecular biology. We have space to describe only a few briefly.

One of the most important tools in this technology is a series of enzymes called **restriction endonucleases.** Each of these enzymes, derived from bacteria, cleaves double-stranded DNA at particular sites determined by the particular base sequences at that point. Many of these endonucleases cut the DNA strands so that one has several bases projecting farther than the other strand (Figure 8-23), leaving what are called "sticky ends." When these DNA fragments are mixed with others that have been cleaved by the same endonuclease, they tend to anneal (join) by the rules of complementary base pairing. They are sealed into their new position by the enzyme **DNA ligase.**

Besides their chromosomes, most prokaryote and at least some eukaryote cells have small circles of double-stranded DNA called *plasmids*. Though comprising only 1% to 3% of the bacterial genome, they may carry important genetic information, for example, resistance to an antibiotic. Plastids in plant cells (for example, chloroplasts) and mitochondria, found in most eukaryotic cells, are self-replicating and have their own complement of DNA in the form of small circles. The DNA of mitochondria and plastids codes for some of their proteins, and some of their proteins are specified by nuclear genes.

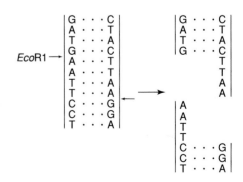

Figure 8-23

Action of restriction endonuclease, *Eco*R1. Such enzymes recognize specific base sequences that are palindromic (a palindrome is a word spelled the same backward and forward). *Eco*R1 leaves "sticky ends," which anneal to other DNA fragments cleaved by the same enzyme. The strands are joined by DNA ligase.

If the DNA annealed after cleavage by the endonuclease is from two different sources, for example, a plasmid (see note on page 144) and a mammal, the product is **recombinant DNA.** To make use of the recombinant DNA, the modified plasmid must be cloned in bacteria. The bacteria are treated with dilute calcium chloride to make them more susceptible to taking up the recombinant DNA, but the plasmids do not enter most of the cells present. The bacterial cells that have taken up the recombinant DNA can be identified if the plasmid has a marker, for example, resistance to an antibiotic. Then, only the bacteria that can grow in the presence of the antibiotic are those that have absorbed the recombinant DNA. Some bacteriophages (bacterial viruses) have also been used as carriers for recombinant DNA. Plasmids and bacteriophages that carry recombinant DNA are called **vectors.** The vectors retain the ability to replicate in the bacterial cells; therefore the recombinant insert is amplified.

A clone is a collection of individuals or cells all derived by asexual reproduction from a single individual. When we speak of cloning a gene or plasmid in bacteria, we mean that we isolate a colony or group of bacteria derived from a single ancestor into which the gene or plasmid was inserted.

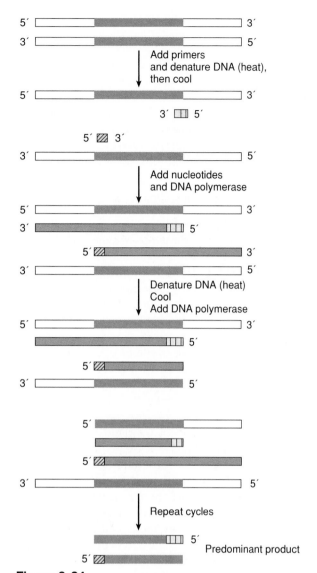

Figure 8-24

Steps in the polymerase chain reaction (PCR).

Recent advances have made it a simple task to clone a specific gene enzymatically from any organism as long as part of the sequence of that gene is known. The technique is called the **polymerase chain reaction (PCR).** Two short chains of nucleotides called primers are synthesized; primers are complementary to different DNA strands in the known sequence. A large excess of each primer is added to a sample of DNA from the organism, and the mixture is heated to separate the double helix into single strands. When the mixture is cooled, there is a much greater probability that each strand of the gene of interest will anneal to a primer than to the other strand of the gene—because there is so much more primer present. DNA polymerase is added along with the four deoxyribonucleotide triphosphates, and DNA synthesis proceeds from the 3′ end of each primer, extending the primer in the 5′ to 3′ direction. If the primers are chosen so that each anneals toward the 3′ end of each of the template strands, entire new complementary strands will be synthesized, and the number of copies of the gene has doubled (Figure 8-24). The reaction mixture is then reheated and cooled again to allow more primers to bind original and new copies of each strand. With each cycle of DNA synthesis, the number of copies of the gene doubles. Since each cycle can take less than five minutes, the number of copies of a gene can increase from one to over one million in less than two hours! The PCR allows cloning a known gene from an individual patient, identification of a drop of dried blood at a crime scene, or cloning the DNA of a 40,000-year-old woolly mammoth.

Recombinant DNA technology and the PCR are currently being used in many areas with great positive potential and many practical uses.

The techniques of molecular biology have allowed scientists to accomplish feats of which few could dream only a decade or so ago. These accomplishments will bring enormous benefits for humanity in the form of enhanced food production and treatment of disease. Progress with crop plants have been so rapid that we can expect genetically engineered soybean, cotton, rice, corn, sugarbeet, tomato, and alfalfa to reach the market before the year 2000. Development of transgenic animals of potential use has not progressed as far as development of such plants. Gene therapy for inherited diseases presents many difficulties, but research in this area is vigorous, and clinical trials for certain conditions are beginning.

SOURCES OF PHENOTYPIC VARIATION

The creative force of evolution is natural selection acting on biological variation. Without variability among individuals, there could be no continued adaptation to a changing environment and no evolution.

There are actually several sources of variability, some of which we have already described. The independent assortment of chromosomes during meiosis is a random process that creates new chromosomal recombinations in the gametes. In addition, chromosomal crossing over during meiosis allows recombination of linked genes between homologous chromosomes, further increasing variability. The random fusion of gametes from both parents produces still another source of variation.

Thus sexual reproduction multiplies variation and provides the diversity and plasticity necessary for a species to survive environmental change. Sexual reproduction with its sequence of gene segregation and recombination, generation after generation, is, as the geneticist T. Dobzhansky

has said, the "master adaptation which makes all other evolutionary adaptations more readily accessible."

Although sexual reproduction reshuffles and amplifies whatever genetic diversity exists in the population, there must be ways to generate *new* genetic variation. This happens through gene mutations and, sometimes, through chromosomal aberrations.

GENE MUTATIONS

Gene mutations are chemicophysical changes in genes resulting in an alteration of the sequence of bases in the DNA. These mutations can be studied directly by determining the DNA sequence and indirectly through their effects on organismal phenotype, if such effects are present. A mutation may result in a codon substitution as, for example, in the condition in humans known as **sickle cell anemia.** Homozygotes with sickle cell trait usually die before the age of 30 because the ability of their red blood cells to carry oxygen is greatly impaired, a result of the substitution of only a single amino acid in the amino acid sequence of their hemoglobin. Other mutations may involve the deletion of one or more bases or the insertion of additional bases into the DNA chain. The translation of mRNA will thus be shifted, leading to codons that specify incorrect amino acids.

Once a gene is mutated, it faithfully reproduces its new self just as it did before it was mutated. Many mutations are harmful, many are neither helpful nor harmful, and sometimes mutations are advantageous. Helpful mutations are of great significance to evolution because they furnish new possibilities on which natural selection works to build adaptations. Natural selection determines which new alleles merit survival; the environment imposes a screening process that passes the beneficial and eliminates the harmful.

When an allele of a gene is mutated to the new allele, it tends to be recessive and its effects are normally masked by its partner allele. Only in the homozygous condition can such mutant alleles be expressed. Thus a population carries a reservoir of mu-

tant recessive alleles, some of which are homozygous lethals but which are rarely present in the homozygous condition. Inbreeding encourages the formation of homozygotes and increases the probability of recessive mutants being expressed in the phenotype.

Most mutations are destined for a brief existence. There are cases, however, in which mutations may be harmful or neutral under one set of environmental conditions and helpful under a different set. Should the environment change, there could be a new adaptation beneficial to the species. The earth's changing environment has provided numerous opportunities for new gene combinations and mutations, as evidenced by the great diversity of animal life today.

Frequency of Mutations

Although mutation occurs randomly, different mutation rates prevail at different loci. Some *kinds* of mutations are more likely to occur than others, and individual genes differ considerably in length. A long gene (more base pairs) is more likely to have a mutation than a short gene. Nevertheless, it is possible to estimate average spontaneous rates for different organisms and traits.

Relatively speaking, genes are extremely stable. In the well-studied fruit fly *Drosophila* there is approximately one detectable mutation per 10,000 loci (rate of 0.01% per locus per generation). The rate for humans is one per 10,000 to one per 100,000 loci per generation. If we accept the latter, more conservative figure, then a single normal allele is expected to go through 100,000 generations before it is mutated. However, since human chromosomes contain 100,000 loci, every person carries approximately one new mutation. Similarly, each ovum or spermatozoon produced contains, on the average, one mutant allele.

Since most mutations are deleterious, these statistics are anything but cheerful. Fortunately, most mutant genes are recessive and are not expressed in heterozygotes. Only a few will by chance increase enough in frequency for homozygotes to be produced.

Molecular Genetics of Cancer

The crucial defect in cancer cells is that they proliferate in an unrestrained manner **(neoplastic growth).** The mechanism that controls the rate of division of normal cells has somehow broken down, and the cancer cells multiply much more rapidly, invading other tissues in the body. Cancer cells originate from normal cells that lose their constraint on division and become dedifferentiated (less specialized) to some degree. Thus there are many kinds of cancer, depending on the original founder cells of the tumor. In recent years mounting evidence has indicated that the change in many cancerous cells, perhaps all, has a genetic basis, and investigation of the genetic damage that causes cancer is now a major thrust of cancer research.

Oncogenes and Tumor Suppressor Genes

We now recognize that cancer is a result of a series of specific genetic changes that take place in a particular clone of cells. These include alterations in two types of genes: **oncogenes** and **tumor suppressor genes,** and there are numerous specific genes of each type now known.

Of the many ways that cellular DNA can sustain damage, the three most important are ionizing radiation, ultraviolet radiation, and chemical mutagens. The high energy of ionizing radiation (x rays and gamma rays) causes electrons to be ejected from the atoms it encounters, resulting in ionized atoms with unpaired electrons (free radicals). The free radicals (principally from water) are highly reactive chemically, and they react with molecules in the cell, including DNA. Some damaged DNA is repaired, but if the repair is inaccurate, a mutation results. Ultraviolet radiation is of much lower energy than ionizing radiation and does not produce free radicals. It is absorbed by pyrimidines in DNA and causes formation of a double covalent bond between the adjacent pyrimidines. UV repair mechanisms can also be inaccurate. Chemical mutagens react with the DNA bases and cause mispairing during replication.

Oncogenes (Gr. *onkos,* bulk, mass; + *genos,* descent) are genes whose activity has been associated for some time with the production of cancer. They are genes that are normally found in cells, and in their normal form they are called **proto-oncogenes.** One of these codes for a protein known as **Ras.** Ras protein is a guanosine triphosphatase (**GTP**ase) that is located just beneath the cell membrane. When a receptor on the cell surface binds a growth factor, Ras is activated and initiates a cascade of reactions, ultimately leading to cell division. The oncogene form codes for a protein that initiates the cell-division cascade even when the growth factor has not bound to the surface receptor, that is, the growth factor is absent.

Gene products of tumor suppressor genes act as a constraint on cell proliferation. One such product is called **p53** (for "53-kilodalton protein," a reference to its molecular weight). Mutations in the gene coding for p53 are present in about half of the 6.5 million cases of human cancer diagnosed each year. Normal p53 has a number of crucial functions, depending on the circumstances of the cell. It can trigger apoptosis (p. 59), act as a transcription activator or repressor (turning genes on or off), control progression from G_1 to S phase in the cell cycle, and promote repair of damaged DNA. Many of the mutations known in p53 interfere with its binding to DNA and thus its function.

Summary

Genes are the unit entities that determine all the characteristics of an organism and are inherited by offspring from their parents. Allelic variants of genes may be dominant, recessive, or intermediate; the recessive allele in the heterozygous genotype will not be expressed in the phenotype but requires the homozygous condition for overt expression. In a monohybrid cross involving a dominant allele and its recessive allele (both parents homozygous), the F_1 generation will be all heterozygous, whereas the F_2 genotypes will occur in a 1:2:1 ratio, and the phenotypes in a 3:1 ratio. This demonstrates Mendel's law of segregation. Heterozygotes in intermediate inheritance show phenotypes intermediate between the homozygous phenotypes, or sometimes they show a different phenotype altogether, with corresponding alterations in the phenotypic ratios. Dihybrid crosses (in which the genes for two different characteristics are carried on separate pairs of homologous chromosomes) demonstrate Mendel's law of independent assortment, and the phenotypic ratios will be 9:3:3:1 with dominant and recessive alleles for each gene. The ratios for monohybrid and dihybrid crosses can be determined by construction of a Punnett square, but the laws of probability allow calculation of the ratios in crosses of two or more characters much more easily.

Genes can have more than two alleles, and different combinations of alleles can produce different phenotypic effects. Alleles of different genes can interact in producing a phenotype, as in polygenic inheritance, in which one gene affects the expression of another gene.

Gender is determined in most animals by the sex chromosomes; in humans, fruit flies, and many other animals, females have two X chromosomes, and males have an X and a Y. A gene on the X chromosome shows sex-linked inheritance and will produce an effect in the male, even if a recessive allele is present, because the Y chromosome does not carry a corresponding allele. All genes on a given autosomal chromosome are linked, and their variants do not assort independently unless they are very far apart on the chromosome, so that crossing over occurs between them in nearly every meiosis. Crossing over increases the amount of genetic recombination in a population. Since the frequency of crossing over between two genes increases with the distance between the loci of the genes on the chromosome, gene maps of each chromosome can be constructed.

Occasionally, nondisjunction of one of the chromosomes occurs in meiosis, and one of the gametes gets one chromosome too many and the other gets n - 1 chromosomes. Resulting zygotes usually do not survive; humans with 2n + 1 chromosomes may live, but they are born with serious defects, such as Down syndrome.

One gene most commonly controls the production of one protein or polypeptide (one gene-one polypeptide hypothesis), but the various types of RNA are also encoded on the genes.

The nucleic acids in the cell are DNA and RNA, which are large polymers of nucleotides composed of a nitrogenous base, pentose sugar, and phosphate group. The nitrogenous bases in DNA are adenine (A), guanine (G), thymine (T), and cytosine (C), and those in RNA are the same except that uracil (U) is substituted for thymine. DNA is a double-stranded, helical molecule in which the bases extend toward each other from the sugar-phosphate backbone: A always pairs with T and G with C. Thus the strands are antiparallel and complementary, being held in place by hydrogen bonds between the paired bases. In DNA

replication the strands part, and the enzyme DNA polymerase synthesizes a new strand along each parental strand, using the parental strand as a template.

Many DNA sequences encode protein. The sequence of the bases in such DNA is a code for the amino acid sequence in the ultimate product protein. Each triplet of three bases specifies a particular amino acid.

Proteins are synthesized by transcription of DNA into the base sequence of a molecule of messenger RNA (mRNA), which functions in concert with ribosomes (containing ribosomal RNA [rRNA] and protein) and transfer RNA (tRNA). Ribosomes attach to the strand of mRNA and move along it, assembling the amino acid sequence of the protein. Each amino acid is brought into position for assembly by a molecule of tRNA, which itself bears a base sequence (anticodon) complementary to the respective codons of the mRNA. In eukaryotic nuclear DNA the sequences of bases in DNA coding for amino acids in a protein (exons) are interrupted by intervening sequences (introns). The introns are removed from the primary mRNA before it leaves the nucleus, and the protein is synthesized in the cytoplasm.

Genes, and the synthesis of the products for which they are responsible, must be regulated: turned on or off in response to varying environmental conditions or cell differentiation. Gene regulation in eukaryotes is complex, and a number of mechanisms

are known. One inactivation mechanism is the methylation of cytosine residues that lie next to guanine residues in the DNA strand, thus preventing transcription. Other mechanisms involve transcriptional control, translational control, and gene rearrangement.

Modern methods in molecular genetics have made spectacular advances possible. Restriction endonucleases cleave DNA at specific base sequences, and such cleaved DNA from different sources can be rejoined to form recombinant DNA. Combining mammalian with plasmid or viral DNA, a mammalian gene can be introduced into bacterial cells, which then multiply and express the mammalian gene. The polymerase chain reaction (PCR) makes it relatively simple to clone specific genes if only a small sequence of the gene is known.

A mutation is a physicochemical alteration in the bases of the DNA that may change the phenotypic effect of the gene. Although rare and usually detrimental to the survival and reproduction of the organism, mutations are occasionally beneficial and provide new genetic material on which natural selection can work.

Cancer (neoplastic growth) is associated with a series of genetic changes in a clone of cells that allow unrestrained proliferation of those cells. Oncogenes (such as the gene coding for Ras protein) and inactivation of tumor suppressor genes (such as that coding for p53 protein) have been implicated in many cancers.

Review Questions

1. What is the relationship between homologous chromosomes and alleles?
2. Define the following: dominant, recessive, zygote, heterozygote, homozygote, phenotype, genotype, monohybrid cross, dihybrid cross.
3. Diagram by Punnett square a cross between individuals with the following genotypes: $A/a \times A/a$; $A/a\ B/b \times A/a\ B/b$.
4. Concisely state Mendel's law of segregation and his law of independent assortment.
5. Assuming brown eyes (B) are dominant over blue eyes (b), determine the genotypes of all the following individuals. The blue-eyed son of two brown-eyed parents marries a brown-eyed woman whose

mother was brown eyed and whose father was blue eyed. Their child is blue eyed.
6. Recall that red color (R) in four-o'clock flowers is incompletely dominant over white (R'). In the following crosses, give the genotypes of the gametes produced by each parent and the flower color of the offspring: $R/R' \times R/R'$; $R'/R' \times R/R'$; $R/R \times R/R'$; $R/R \times R'/R'$.
7. A brown male mouse is mated with two female black mice. When each female has produced several litters of young, the first female has had 48 black and the second female has had 14 black and 11 brown young. Can you deduce the pattern of inheritance of coat color and the genotypes of the parents?

8. Rough coat (R) is dominant over smooth coat (r) in guinea pigs, and black coat (B) is dominant over white (b). If a homozygous rough black is mated with a homozygous smooth white, give the appearance of each of the following: F_1; F_2; offspring of F_1 mated with smooth, white parent; offspring of F_1 mated with rough, black parent.
9. Assume right-handedness (R) dominates over left-handedness (r) in humans, and that brown eyes (B) are dominant over blue (b). A right-handed, blue-eyed man marries a right-handed, brown-eyed woman. Their two children are right handed, blue eyed and left handed, brown eyed. The man marries again, and this time the woman is right handed and

brown eyed. They have 10 children, all right handed and brown eyed. What are the genotypes of the man and his two wives?

10. In *Drosophila,* red eyes are dominant to white and the recessive characteristic is on the X chromosome. Vestigial wings (*v*) are recessive to normal (*V*). What will be the appearance of the following crosses: $X^W/X^w \ V/v \times X^w/Y \ v/v$, $X^w/X^w \ V/v \times X^W/Y \ V/v$.

11. Assume that color blindness is a recessive character on the X chromosome. A man and woman with normal vision have the following offspring: daughter with normal vision who has one color-blind son and one normal son; daughter with normal vision who has six normal sons; and a color-blind son who has a daughter with normal vision. What are the probable genotypes of all the individuals?

12. Distinguish the following: euploidy, aneuploidy, and polyploidy; monosomy and trisomy.

13. Name the purines and pyrimidines in DNA and tell which pair with each other in the double helix. What are the purines and pyrimidines in RNA and to what are they complementary in DNA?

14. Explain how DNA is replicated.

15. Why is it not possible for a codon to consist of only two bases?

16. Explain the transcription and processing of mRNA in the nucleus.

17. Explain the role of mRNA, tRNA, and rRNA in protein synthesis.

18. What are four ways that genes can be regulated in eukaryotes?

19. In modern molecular genetics, what is recombinant DNA, and how is it prepared?

20. Name three sources of phenotypic variation.

21. Distinguish between proto-oncogene and oncogene. What are two mechanisms whereby cancer can be caused by genetic changes?

22. What are Ras protein and p53? How can mutations in the genes for these proteins contribute to cancer?

23. Outline the essential steps in the procedure for the polymerase chain reaction.

Selected References

Cavenee, W. K., and R. L. White. 1995. The genetic basis of cancer. Sci. Am. **272:**72–79 (Mar.). *Describes mutations in cells of colorectal cancer and brain tumors.*

Culotta, D., and D. E. Koshland, Jr. 1993. p53 sweeps through cancer research. Science **262:**1958–1961. *p53 was discovered in 1979, but it was 10 years before scientists began to uncover its importance.*

Erlich, H. A., D. Gelfand, and J. J. Sninsky. 1991. Recent advances in the polymerase chain reaction. Science **252:**1643–1651. *A review of recent developments in methods and applications of the PCR.*

Friend, S. 1994. p53: a glimpse at the puppet behind the shadow play. Science **265:**334–335. *A short summary of the crucial roles of p53 protein and how mutations in the gene coding for it lead to inactivation.*

Hall, A. 1994. A biochemical function for Ras— at last. Science **264:**1413–1414. *Ras protein is an enzyme in a signal transduction cascade stimulating a cell to divide.*

Klug, W. S. 1991. Concepts of genetics, ed. 3. New York, Macmillan Publishing Company. *A shorter text.*

Koshland, D. E., Jr. 1989. The engineering of species. Science **244:**1233. *This is the lead editorial in an issue of the journal containing several reviews on genetic engineering.*

Mange, E. J., and A. P. Mange. 1994. Basic human genetics. Sunderland, Massachusetts, Sinauer Associates. *A readable, introductory text concentrating on the genetics of the animal species of greatest concern to most of us.*

Marx, J. 1994. Oncogenes reach a milestone. Science **266:**1942–1944. *Research on oncogenes has helped us understand many normal cell processes.*

Mullis, K. B. 1990. The unusual origin of the polymerase chain reaction. Sci. Am. **262:**56–65 (Apr.). *How the author had the idea for the simple production of unlimited copies of DNA while driving through the mountains of California.*

Russell, P. J. 1992. Genetics, ed. 3. New York, HarperCollins Publishers. *Popular general genetics text.*

Verma, I. M. 1990. Gene therapy. Sci. Am. **263:**68–84 (Nov.). *A review of prospects for treating and preventing genetic diseases by putting healthy genes into the body.*

Weinberg, R. A. 1991. Tumor suppressor genes. Science **254:**1138–1146. *How inactivation of tumor suppressor genes is a step in production of cancer.*

9

Organic Evolution

A Legacy of Change

The predominant feature of life's history is the legacy of perpetual change. Despite the apparent permanence of the natural world, change characterizes all things on earth and in the universe. Countless kinds of animals and plants have flourished and disappeared, leaving behind an imperfect fossil record of their existence. Many, but not all, have left living descendants that bear a partial resemblance to them.

Life's changes are perceived and measured in many ways. On a short evolutionary timescale, we see changes in the frequencies of different genetic variants within populations. Evolutionary changes in the relative frequencies of light- and dark-colored moths were observed within a single human lifetime in the polluted countryside of industrial England. The fossil record reveals the formation of new species and dramatic changes in the appearances of organisms on longer timescales covering 100,000 to 1 million years. Major evolutionary trends and periodic mass extinctions occur on even larger timescales covering tens of millions of years. The fossil record of horses through the past 50 million years shows a series of different species replacing older ones through time and ending with the familiar horses that we know today. The fossil record of marine invertebrates shows us a series of mass extinctions separated by intervals of approximately 26 million years.

The earth bears its own record of the irreversible, historical change that we call organic evolution. Because every feature of life as we know it today is a product of the evolutionary process, biologists consider organic evolution the keystone of all biological knowledge. ■

In Chapter 1, we introduced Darwinian evolutionary theory as the dominant paradigm that guides biology. Charles Robert Darwin and Alfred Russel Wallace (Figure 9-1) were the first to establish this paradigm. Today the reality of organic evolution can be denied only by abandoning reason. As the noted English biologist Sir Julian Huxley wrote, "Charles Darwin effected the greatest of all revolutions in human thought, greater than Einstein's or Freud's or even Newton's, by simultaneously establishing the fact and discovering the mechanism of organic evolution." Darwinian theory allows us to understand both the genetics of populations and long-term trends in the fossil record. We review the historical development of thought about evolution as it led to Darwinism and then discuss current evidence supporting Darwin's five theories of evolution and modifications of them.

A   **B**

Figure 9-1

Founders of the theory of natural selection. **A,** Charles Robert Darwin (1809 to 1882), as he appeared in 1881, the year before his death. **B,** Alfred Russel Wallace (1823 to 1913) in 1895. Darwin and Wallace independently developed the same theory. A letter and essay from Wallace written to Darwin in 1858 spurred Darwin into writing *The Origin of Species,* published in 1859.

A, *Courtesy American Museum of Natural History, New York, Neg. # 32662.* **B,** *The Natural History Museum, London.*

HISTORICAL DEVELOPMENT OF THE IDEA OF ORGANIC EVOLUTION

PRE-DARWINIAN EVOLUTIONARY IDEAS

Before the eighteenth century, speculation on the origin of species rested on myth and superstition, not on anything resembling a testable scientific theory. Creation myths viewed the world as a constant entity that did not change after its creation. Nevertheless, some thinkers approached the idea that nature has a long history of perpetual and irreversible change.

Early Greek philosophers, notably Xenophanes, Empedocles, and Aristotle, developed a primitive idea of evolutionary change. They recognized fossils as evidence of a former life that they believed had been destroyed by natural catastrophe. Despite their spirit of intellectual inquiry, the Greeks failed to establish an evolutionary concept and the issue declined well before the rise of Christianity. The opportunity for evolutionary thinking became even more restricted as the biblical account of the earth's creation became accepted as a tenet of faith. The year 4004 B.C. was fixed by Archbishop James Ussher (mid-seventeenth century) as the time of life's creation. Evolutionary views were considered rebellious and heretical. Still, some speculation continued. The French naturalist Georges Louis Buffon (1707 to 1788) stressed the influence of environment on the modifications of animal type. He also extended the age of earth to 70,000 years.

Lamarckism: The First Scientific Explanation of Evolution

The first complete explanation of evolution was authored by the French biologist, Jean Baptiste de Lamarck (1744 to 1829) (Figure 9-2) in 1809, the year of Darwin's birth. He made the first convincing case for the idea that fossils were the remains of extinct animals. Lamarck's evolutionary mechanism, **inheritance of acquired characteristics,** was engagingly simple: organisms, by striving to meet the demands of their environments, acquire adaptations and pass them by heredity to their offspring. According to Lamarck, the giraffe evolved its long neck because its ancestors lengthened their necks by stretching to obtain food and then passed the lengthened neck to their offspring. Over many

Figure 9-2

Jean Baptiste de Lamarck (1744 to 1829), French naturalist who offered the first scientific explanation of evolution. Lamarck's hypothesis that evolution proceeds by the inheritance of acquired characteristics has been disproven.

The Natural History Museum, London.

generations, these changes accumulated to produce the long neck of the modern giraffe.

We call Lamarck's concept of evolution **transformational,** because it claims that individual organisms transform their appearance to produce evolution. We now reject transformational theories because genetic studies show that traits acquired by an organism during its lifetime, such as strengthened muscles, are not inherited by offspring (Chapter 8). Darwin's evolutionary theory differs from Lamarck's in being a **variational** theory. Evolutionary change is caused by the differential survival and reproduction of genetically variable organisms (Chapter 1), not by inheritance of acquired characteristics.

Charles Lyell and Uniformitarianism

The geologist Sir Charles Lyell (1797 to 1875) (Figure 9-3) established in his *Principles of Geology* (1830 to 1833) the principle of **uniformitarianism,** which influenced Darwin's theory of gradualism (Chapter 1). Uniformitarianism encompasses several different principles. Some are assumptions that guide the scientific study of nature and others are testable hypotheses about

Figure 9-3

Sir Charles Lyell (1797 to 1875), English geologist and friend of Darwin. His book *Principles of Geology* greatly influenced Darwin during Darwin's formative period. This photograph was made about 1856.

The Natural History Museum, London.

geological change. Lyell's assumptions are (1) that the laws of physics and chemistry remain the same throughout the history of the earth, and (2) that past geological events occurred by natural processes similar to those that we observe in action today. Lyell showed that natural forces, acting over long periods of time, could explain the formation of fossil-bearing rocks. His geological studies led him to conclude that the earth's age must be reckoned in millions of years.

Lyell's two testable claims about the natural world are "gradualism" and "nondirectionalism." Gradualism claims that large geological changes (such as the formation of mountains) occurred not in catastrophic events but by gradual accumulation of small geological changes over very long periods of time. Nondirectionalism claims that geological changes are not programmed toward any predetermined final condition. Darwin later applied both principles to the biological world, rejecting both catastrophism and the notion that evolution proceeds toward predetermined goals.

DARWIN'S GREAT VOYAGE OF DISCOVERY

"After having been twice driven back by heavy southwestern gales, Her Majesty's ship *Beagle,* a ten-gun brig, under the command of Captain FitzRoy, R.N., sailed from Devonport on the 27th of December, 1831." Thus began Charles Darwin's account of the historic five-year voyage of the *Beagle* around the world (Figure 9-4). Darwin, not quite 23 years old, had been asked to serve as naturalist without pay on the *Beagle,* a small vessel only 90 feet in length, which was about to depart on an extensive surveying voyage to South America and the Pacific. It was the beginning of one of the most important voyages of the nineteenth century.

During the voyage (1831 to 1836) (Figure 9-5), Darwin endured seasickness and the erratic companionship of the authoritarian Captain FitzRoy. But Darwin's youthful physical strength

and early training as a naturalist equipped him for his work. The *Beagle* made many stops along the harbors and coasts of South America and adjacent regions. Darwin made extensive collections and observations on the fauna and flora of these regions. He unearthed numerous fossils of animals long since extinct and noted the resemblance between fossils of the South American pampas and the known fossils of North America. In the Andes he encountered seashells embedded in rocks at 13,000 feet. He experienced a severe earthquake and watched the mountain torrents that relentlessly wore away the earth. These observations strengthened his conviction that natural forces were responsible for the geological features of the earth.

In mid-September of 1835, the *Beagle* arrived at the Galápagos Islands, a volcanic archipelago straddling the equator 600 miles west of Ecuador (Figure 9-6). The fame of the islands stems from their infinite strangeness. They are unlike any other islands on earth. Some visitors today are struck with awe and wonder; others with a sense of depression and dejection. Circled by capricious currents, surrounded by shores of twisted lava, bearing skeletal brushwood baked by the equatorial sun, almost devoid of vegetation, inhabited by strange reptiles and by convicts stranded by the Ecuadorian government, the islands indeed had few admirers among mariners. By the middle of the seventeenth century, the islands were already known to the Spaniards as "Las Islas Galápagos"—the tortoise islands. The giant tortoises, used for food first by buccaneers and later by American and British whalers, sealers, and ships of war, were the islands' principal attraction. At the time of Darwin's visit, the tortoises were already heavily exploited.

During the *Beagle's* five-week visit to the Galápagos, Darwin began to develop his views of the evolution of life on earth. His original observations of the giant tortoises, marine iguanas, mockingbirds, and ground finches, all contributed to the turning point in Darwin's thinking.

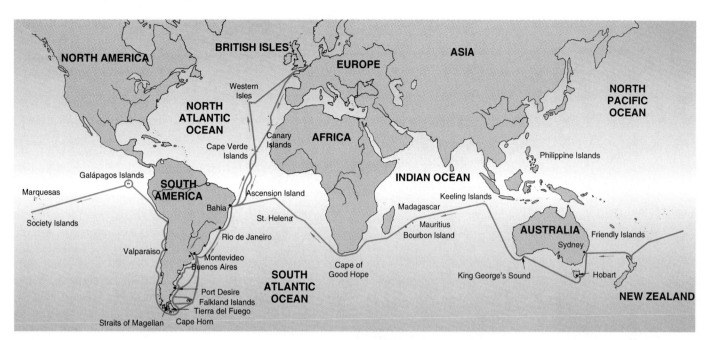

Figure 9-4
Five-year voyage of H.M.S. *Beagle.*

A   B

Figure 9-5
Charles Darwin and H.M.S. *Beagle.* **A,** Darwin in 1840, four years after the *Beagle* returned to England, and a year after his marriage to his cousin, Emma Wedgwood. **B,** The H.M.S. *Beagle* sails in Beagle Channel, Tierra del Fuego, on the southern tip of South America in 1833. The watercolor was painted by Conrad Martens, one of two official artists during the voyage of the *Beagle.*

Figure 9-6
The Galápagos Islands viewed from the rim of a volcano.

Figure 9-7
Darwin's study at Down House in Kent, England, is preserved today much as it was when Darwin wrote *The Origin of Species.*

Darwin was struck by the fact that, although the Galápagos Islands and the Cape Verde Islands (visited earlier in this voyage of the *Beagle*) were similar in climate and topography, their fauna and flora were altogether different. He recognized that Galápagos plants and animals were related to those of the South American mainland, yet differed from them in curious ways. Each island often contained a unique species of a particular group of animals that was nonetheless related to forms on other islands. In short, Galápagos life must have originated in continental South America and then undergone modification in the various environmental conditions of the different islands. He concluded that living forms were neither divinely created nor immutable; they were, in fact, the products of evolution. Although Darwin devoted only a few pages to Galápagos animals and plants in his monumental *On the Origin of Species,* published more than two decades later, his observations on the unique character of the animals and plants were, in his own words, the "origin of all my views."

On October 2, 1836, the *Beagle* returned to England, where Darwin conducted the remainder of his scientific work (Figure 9-7). Most of Darwin's extensive collections had preceded him there, as had most of his notebooks and diaries kept during the cruise. Darwin's journal was published three years after the *Beagle's* return to England. It was an instant success and required two additional printings within the first year. In later versions, Darwin made extensive changes and titled his book *The Voyage of the Beagle.* The fascinating account of his observations written in a simple, appealing style has made the book one of the most lasting and popular travel books of all time.

Curiously, the main product of Darwin's voyage, his theory of evolution, did not appear in print for more than 20 years after the *Beagle's* return. In 1838 he "happened to read for amusement" an essay on populations by T. R. Malthus (1766 to 1834), who stated that animal and plant populations, including human populations, tend to increase beyond the capacity of the environment to support them. Darwin had already been gathering information on the artificial selection of animals under domestication by humans. After reading Malthus's

article, Darwin realized that a process of selection in nature, a "struggle for existence" because of overpopulation, could be a powerful force for evolution of wild species.

He allowed the idea to develop in his own mind until it was presented in 1844 in a still unpublished essay. Finally in 1856 he began to pull together his voluminous data into a work on the origin of species. He expected to write four volumes, a very big book, "as perfect as I can make it." However, his plans were to take an unexpected turn.

In 1858, he received a manuscript from Alfred Russel Wallace (1823 to 1913), an English naturalist in Malaya with whom he was corresponding. Darwin was stunned to find that in a few pages, Wallace summarized the main points of the natural selection theory on which Darwin had been working for two decades. Rather than withhold his own work in favor of Wallace as he was inclined to do, Darwin was persuaded by two close friends, the geologist Lyell and the botanist Hooker, to publish his views in a brief statement that would appear together with Wallace's paper in the *Journal of the Linnaean Society.* Portions of both papers were read before an unimpressed audience on July 1, 1858.

"Whenever I have found that I have blundered, or that my work has been imperfect, and when I have been contemptuously criticized, and even when I have been overpraised, so that I have felt mortified, it has been my greatest comfort to say hundreds of times to myself that 'I have worked as hard and as well as I could, and no man can do more than this.' " *Charles Darwin, in his autobiography, 1876.*

For the next year, Darwin worked urgently to prepare an "abstract" of the planned four-volume work. This was published in November 1859, with the title *On the Origin of Species by Means of Natural Selection, or the Preservation of Favoured Races in the Struggle for Life.* The 1250 copies of the first printing were sold the first day! The book instantly generated a storm that has never completely abated. Darwin's views were to have extraordinary consequences on scientific and religious beliefs and remain among the greatest intellectual achievements of all time.

Once Darwin's caution had been swept away by the publication of *On the Origin of Species,* he entered an incredibly productive period of evolutionary thinking for the next 23 years, producing book after book. He died April 19, 1882, and was buried in Westminster Abbey. The little *Beagle* had already disappeared, having been retired in 1870 and presumably broken up for scrap.

DARWINIAN EVOLUTIONARY THEORY: THE EVIDENCE

PERPETUAL CHANGE

The main premise underlying Darwinian evolution is that the living world is neither constant nor perpetually cycling, but is always changing. Perpetual change in the form and diversity of animal life throughout its 600- to 700-million-year history is seen most directly in the fossil record. A **fossil** is a remnant of past life uncovered from the crust of the earth (Figure 9-8). Some fossils constitute complete remains (mammoths and amber insects), actual hard parts (teeth and bones), or petrified skeletal parts that are infiltrated with silica or other minerals (ostracoderms and molluscs). Other fossils include molds, casts, impressions, and fossil excrement (coprolites). In addition to documenting organismal evolution, fossils reveal profound changes in the earth's environment, including major changes in the distributions of lands and seas. Because many organisms left no fossils, a complete record of the past is always beyond our reach; nonetheless, discovery of new fossils and reinterpretation of existing ones expand our knowledge of how animal form and diversity changed through geological time.

A

B

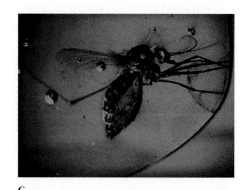

C

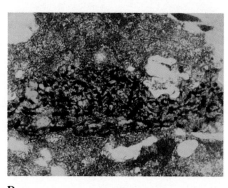

D

Figure 9-8

Four examples of fossil material. **A,** Fish fossil from rocks of the Green River Formation, Wyoming. Such fish swam here during the Eocene epoch of the Tertiary period, approximately 55 million years ago. **B,** Stalked crinoids (class Crinoidea, p. 464) from 85-million-year-old Cretaceous rocks. The fossil record of these echinoderms shows that they reached their peak millions of years earlier and began a slow decline to the present. **C,** An insect fossil that got stuck in the resin of a tree 40 million years ago and that has since hardened into amber. **D,** Electron micrograph of tissue from a fly fossilized as shown in **C;** the nucleus of a cell is marked in red.

*A, Photo by Ken Lucas/Biological Photo Service. **B,** Photo by A.J. Copley/Visuals Unlimited.*

Fossil remains may on rare occasions include soft tissues preserved so well that recognizable cellular organelles can be viewed with the electron microscope! Insects are frequently found entombed in amber, the fossilized resin of trees. One study of a fly entombed in 40-million-year-old amber revealed structures corresponding to muscle fibers, nuclei, ribosomes, lipid droplets, endoplasmic reticulum, and mitochondria (Figure 9-8D). This extreme case of mummification probably occurred because chemicals in the plant sap diffused into the embalmed insect's tissues.

Interpreting the Fossil Record

The fossil record is biased because preservation is selective. Vertebrate skeletal parts and invertebrates with shells and other hard structures left the best record (Figure 9-8). Soft-bodied animals, including the jellyfishes and most worms, are fossilized only under very unusual circumstances such as those that formed the Burgess shale deposits of British Columbia (Figure 9-9). Exceptionally favorable conditions for fossilization produced the Precambrian fossil bed of south Australia, the tar pits of Rancho La Brea (Hancock Park, Los Angeles), the great dinosaur beds

Figure 9-9

A, Fossil trilobites visible at the Burgess Shale Quarry, British Columbia. **B,** Animals of the Cambrian period, approximately 580 million years ago, as reconstructed from fossils preserved in the Burgess Shale of British Columbia, Canada. The main new body plans that appeared rather abruptly at this time established the body plans of animals familiar to us today. **C,** Key to Burgess Shale drawing. *Amiskwia* (1), from an extinct phylum; *Odontogriphus* (2), from an extinct phylum; *Eldonia* (3), a possible echinoderm; *Halichondrites* (4), a sponge; *Anomalocaris canadensis* (5), from an extinct phylum; *Pikaia* (6), an early chordate; *Canadia* (7), a polychaete; *Marrella splendens* (8), a unique arthropod; *Opabinia* (9), from an extinct phylum; *Ottoia* (10), a priapulid; *Wiwaxia* (11), from an extinct phylum, *Yohoia* (12), a unique arthropod; *Xianguangia* (13), an anemone-like animal; *Aysheaia* (14), an onychophoran or new phylum; *Sidneyia* (15), a unique arthropod; *Dinomischus* (16), from an extinct phylum; *Hallucigenia* (17), from an extinct phylum.

A

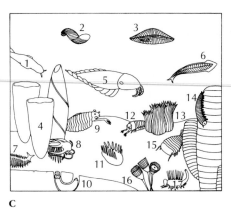

C

B

Figure 9-10
A hadrosaur skeleton from Dinosaur Provincial Park, Alberta.

(Alberta, Canada, and Jensen, Utah; Figure 9-10) and the Olduvai Gorge of Tanzania.

Fossils are laid down in stratified layers with new deposits forming on top of older ones. If left undisturbed, which is rare, a sequence is preserved with the ages of fossils being directly proportional to their depth in the stratified layers. Characteristic fossils often serve to identify particular layers. Certain widespread marine invertebrate fossils, including various foraminiferans and echinoderms, are such good indicators of specific geological periods that they are called "index," or "guide," fossils. Unfortunately, the layers are usually tilted or folded or show faults (cracks). Old deposits exposed by erosion may be covered with new deposits in a different plane. When exposed to tremendous pressures or heat, stratified sedimentary rock metamorphoses into crystalline quartzite, slate, or marble, which destroys fossils.

Geological Time

Long before the earth's age was known, geologists divided its history into a table of succeeding events based on the ordered layers of sedimentary rock. The "law of stratigra-phy" produced a relative dating with the oldest layers at the bottom and the youngest at the top of the sequence. Time was divided into eons, eras, periods, and epochs as shown on the endpaper inside the back cover of this book. Time during the last eon (Phanerozoic) is expressed in eras (for example, Cenozoic), periods (for example, Tertiary), epochs (for example, Paleocene), and sometimes smaller divisions of an epoch.

In the late 1940s, radiometric dating methods were developed for determining the absolute age in years of rock formations. Several independent methods are now used, all based on the radioactive decay of naturally occurring elements into other elements. These "radioactive clocks" are independent of pressure and temperature changes and therefore are not affected by often violent earth-building activities.

One method, potassium-argon dating, depends on the decay of potassium-40 (^{40}K) to argon-40 (^{40}Ar) (12%) and calcium-40 (^{40}Ca) (88%).

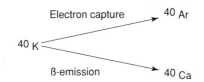

The half-life of potassium-40 is 1.3 billion years. This means that half of the original atoms will decay in 1.3 billion years, and half of the remaining atoms will be gone at the end of the next 1.3 billion years. This decay continues until all radioactive potassium-40 atoms are gone. To measure the age of the rock, one calculates the ratio of remaining potassium-40 atoms to the amount of potassium-40 originally there (the remaining potassium-40 atoms plus the argon-40 and calcium-40 into which they have decayed). Several such isotopes exist for dating purposes, some for dating the age of the earth itself. One of the most useful radioactive clocks depends on the decay of uranium into lead. With this method, rocks over 2 billion years old can be dated with a probable error of less than 1%.

The fossil record of macroscopic organisms begins near the base of the Cambrian period of the Paleozoic era, approximately 600 million years BP. The period before the Cambrian is called the Precambrian era. Although the Precambrian era occupies 85% of all geological time, it has received much less attention than later eras, partly because oil, which provides the commercial incentive for much geological work, seldom exists in Precambrian formations. There is, however, evidence for life in the Precambrian era: well-preserved fossils of bacteria and algae, and casts of jellyfishes, sponge spicules, soft corals, segmented flatworms, and worm trails. Most, but not all, are microscopic fossils.

Evolutionary Trends

The fossil record allows us to view evolutionary change across the broadest scale of time. Species arise and then become extinct repeatedly throughout the fossil record. Animal species typically survive approximately 1 million to 10 million years, although their duration is highly variable. When we study patterns of species or taxon replacement through time, we observe trends. Trends are directional changes in the characteristic features or patterns of diversity in a group of organisms. Fossil trends clearly demonstrate Darwin's principle of perpetual change. We must emphasize that trends are observed only "after the fact." We cannot predict from the early fossils of a group what the appearance or diversity of the later fossils will be. The evolutionary process has no predetermined directions built into it.

A well-studied fossil trend is the evolution of horses from the Eocene epoch to the present (Figure 9-11). George Gaylord Simpson (p. 202) showed that this trend is compatible with Darwinian evolutionary theory.

The three characteristics that show the clearest trends in horse evolution are body size, foot structure, and tooth structure. Looking back at the Eocene epoch, we see many different genera and species of horses that replaced each other through time (see Figure 9-11). Compared to modern horses, the horses in extinct genera were small, their teeth had a relatively small grinding surface and their feet had a relatively large number of toes (four). Throughout the subsequent Oligocene, Miocene, Pliocene, and Pleistocene epochs, there are continuing patterns of new genera arising and old ones becoming extinct. In each case, there is a net increase in body size, expansion of the grinding surface of the teeth, and reduction in the number of toes. As the number of toes is reduced, the central digit becomes increasingly more prominent in the foot and eventually only this central digit remains.

The fossil record shows a net change not only in the characteristics of horses but also variation in the numbers of different horse genera (and numbers of species) that exist through time. The many horse genera of past epochs have been lost to extinction, leaving only a single survivor. Evolutionary trends in diversity are observed in fossils of many different groups of animals (Figure 9-12).

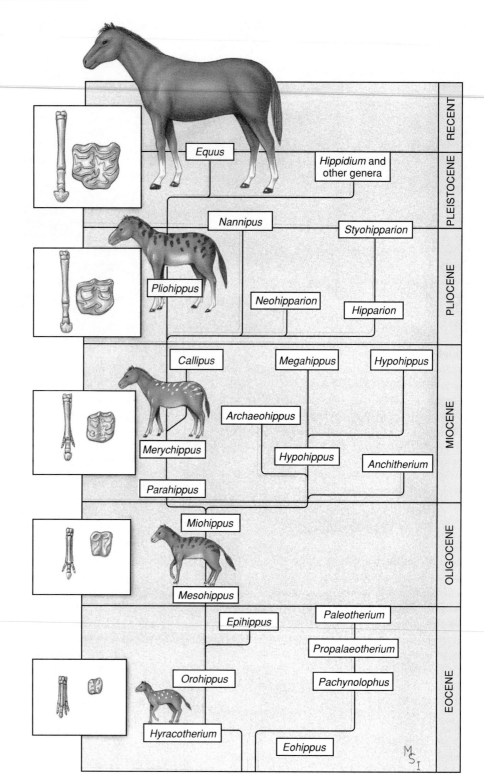

Figure 9-11

A reconstruction of the genera of horses from the Eocene to the present. Evolutionary trends toward increased size, elaboration of molars, and loss of toes are shown together with a hypothetical genealogy of extant and fossil genera.

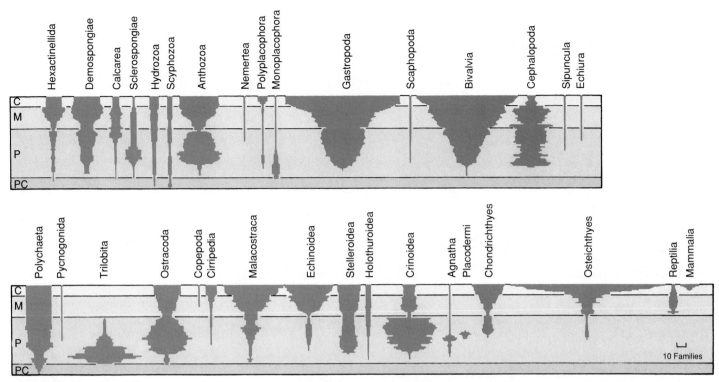

Figure 9-12
Diversity profiles of taxonomic families from different animal groups in the fossil record. The scale marks off the Precambrian (*PC*) and the Paleozoic (*P*), Mesozoic (*M*), and Cenozoic (*C*) eras. The number of families is indicated from the width of the profile.

Trends in fossil diversity through time are produced by different rates of species formation versus extinction through time. Why do some lineages generate large numbers of new species whereas others generate relatively few? Why do different lineages undergo higher or lower rates of extinction (of species, genera, or families) throughout evolutionary time? To answer these questions, we must turn to Darwin's other four theories of evolution (Chapter 1). Regardless of how we answer these questions, however, the observed trends in animal diversity clearly illustrate Darwin's principle of perpetual change. Because the remaining four theories of Darwinism rely on the theory of perpetual change, evidence supporting these theories strengthens Darwin's theory of perpetual change.

COMMON DESCENT

Darwin proposed that all plants and animals have descended from "some one form into which life was first breathed." Life's history is depicted as a branching tree, called a **phylogeny,** that gives all of life a unified evolutionary history. Pre-Darwinian evolutionists, including Lamarck, advocated multiple independent origins of life, each of which gave rise to lineages that changed through time without extensive branching. Like all good scientific theories, common descent makes several important predictions that can be tested and potentially used to reject it. According to this theory, we should be able to trace the genealogies of all modern species backward until they converge on ancestral lineages shared with other species, both living and extinct. We should be able to continue this process, moving farther backward through evolutionary time, until we reach the primordial ancestor of all life on earth. All living forms will connect to this tree somewhere, including many extinct forms that represent dead branches. Although reconstructing the history of life in this manner may seem almost impossible, it has in fact been extraordinarily successful. How has this difficult task been accomplished?

Homology and Phylogenetic Reconstruction

Darwin recognized the major source of evidence for common descent in the concept of **homology.** Darwin's contemporary, Richard Owen (1804 to 1892), used this term to denote "the same organ in different organisms under every variety of form and function." The classic example of homology is the limb skeleton of vertebrates. The bones of the vertebrate limb maintain characteristic structures and patterns of connection despite diverse modifications for different functions (Figure 9-13). According to Darwin's theory of common descent, the structures that we call homologies represent characteristics inherited with some modification from a corresponding feature in a common ancestor.

Darwin devoted an entire book, *The Descent of Man and Selection in Relation to Sex,* largely to the idea that humans share common descent with apes and other animals. This idea was repugnant to the Victorian world, which responded with predictable outrage (Figure 9-14). Darwin

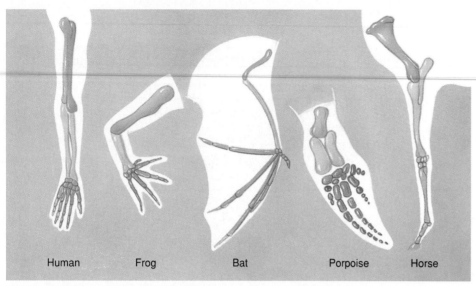

Figure 9-13

Forelimbs of five vertebrates show skeletal homologies: *green,* humerus; *yellow,* radius and ulna; *purple,* "hand" (carpals, metacarpals, and phalanges). Clear homologies of bones and patterns of connection are evident despite evolutionary modification for various particular functions.

Figure 9-14

This 1873 advertisement for Merchant's Gargling Oil ridicules Darwin's theory of the common descent of humans and apes, which hardly received universal acceptance during Darwin's lifetime.

built his case mostly on anatomical comparisons revealing homology between humans and apes. To Darwin, the close resemblances between apes and humans could be explained only by common descent.

Throughout the history of all forms of life, evolutionary processes generate new features that are transmitted across generations. Every time a new feature arises on a lineage destined to be ancestral to others, we see the origin of a new homology. The pattern formed by these homologies provides evidence for common descent and allows us to reconstruct the branching evolutionary history of life. We can illustrate such evidence using a phylogenetic tree of the ground-dwelling ratite birds (Figure 9-15). A new skeletal homology (see Figure 9-15) arises on each of the lineages shown (descriptions of the homologies are not included because they are highly technical). The different groups of species located at the tips of the tree contain different combinations of these homologies that reflect ancestry. For example, the ostriches show homologies 1 through 5 and 8, whereas the kiwis show homologies 1, 2, 13, and 15. The branches of the tree combine these species into a **nested hierarchy** of groups within groups (see Chapter 11). Smaller groups (species grouped near terminal branches) are contained within larger ones (species grouped by basal branches, including the trunk of the tree). If we erase the tree structure but retain the patterns of homology observed in the terminal groups of species, we will still be able to reconstruct the branching structure of the entire tree. Evolutionists test the theory of common descent by observing the patterns of homology present within all groups of organisms. The pattern formed by all homologies taken together should specify a single branching tree structure that represents the evolutionary genealogy of the group.

The nested hierarchical structure of homology is so pervasive in the living world that it forms the basis for our systematic classification of all forms of life (genera grouped into

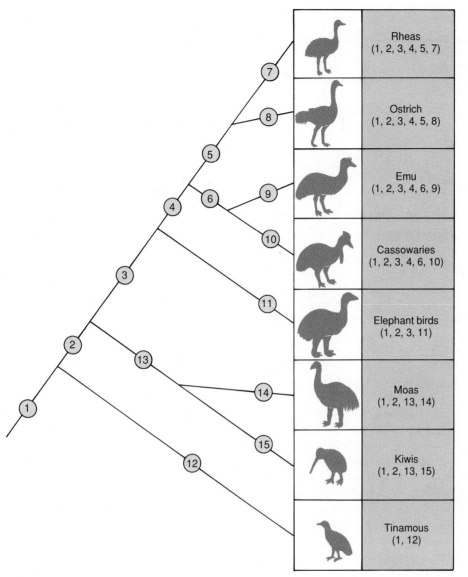

Figure 9-15

The phylogenetic pattern specified by twelve homologous structures in the skeletons of ratite birds. Homologous features are numbered 1 through 12 and are marked both on the branches of the tree on which they arose and on the birds that have them. If you were to erase the tree structure, you would be able to reconstruct it without error from the distributions of homologous features shown for the birds at the terminal branches.

Image labels: Rheas (1, 2, 3, 4, 5, 7); Ostrich (1, 2, 3, 4, 5, 8); Emu (1, 2, 3, 4, 6, 9); Cassowaries (1, 2, 3, 4, 6, 10); Elephant birds (1, 2, 3, 11); Moas (1, 2, 13, 14); Kiwis (1, 2, 13, 15); Tinamous (1, 12)

argument, which is not a scientific hypothesis, can make no testable predictions about any pattern of homology; it only devises supernatural excuses for whatever observations are made by evolutionists.

Ontogeny, Phylogeny, and Recapitulation

Ontogeny is the history of the development of an organism through its entire life. Early developmental and embryological features contribute greatly to our knowledge of homology and common descent. Comparative studies of ontogeny show how the evolutionary alteration of developmental timing generates new phenotypes, thereby causing evolutionary divergence among lineages.

The German zoologist Ernst Haeckel, a contemporary of Darwin, believed that each successive stage in the development of an individual represented one of the adult forms that appeared in its evolutionary history. The human embryo with gill depressions in the neck was believed, for example, to signify a fishlike ancestor. On this basis Haeckel gave his generalization: *ontogeny (individual development) recapitulates (repeats) phylogeny (evolutionary descent).* This notion later became known simply as **recapitulation** or the **biogenetic law.** Haeckel based his biogenetic law on the flawed premise that evolutionary change occurs by successively adding stages onto the end of an unaltered ancestral ontogeny, compressing the ancestral ontogeny into earlier developmental stages. This notion was based on Lamarck's concept of the inheritance of acquired characteristics (p. 151).

The nineteenth-century embryologist, K. E. von Baer, gave a more satisfactory explanation of the relationship between ontogeny and phylogeny. He argued that early developmental features were simply more widely shared among different animal groups than later ones. Figure 9-16 shows, for example, the early embryological similarities of organisms whose adult

families, families grouped into orders, and so on). Hierarchical classification even preceded Darwin's theory because this pattern is so evident, but it was not adequately explained before Darwin. Once the idea of common descent was accepted, biologists began investigating the structural, molecular, and/or chromosomal homologies of all animal groups. Taken together, the nested hierarchical patterns uncovered by these studies have permitted us to reconstruct the evolutionary trees of many groups and to

continue investigating others. The use of Darwin's theory of common descent to reconstruct the evolutionary history of life and to classify animals is the subject of Chapter 11.

Note that the earlier evolutionary hypothesis that life arose many times, forming unbranched lineages, predicts linear sequences of evolutionary change with no nested hierarchy of homologies among species. Because we do observe nested hierarchies of homologies, that hypothesis is rejected. Note also that the creationist

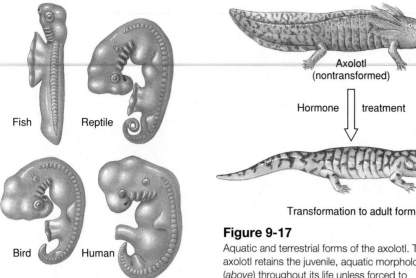

Figure 9-16
Comparison of gill arches of different embryos. All are shown separated from the yolk sac. Note the remarkable similarity of the four embryos at this early stage in development.

Axolotl
(nontransformed)

Hormone   treatment

Transformation to adult form

Figure 9-17
Aquatic and terrestrial forms of the axolotl. The axolotl retains the juvenile, aquatic morphology (*above*) throughout its life unless forced to metamorphose (*below*) by hormone treatment. The axolotl evolved from metamorphosing ancestors, an example of paedomorphosis.

forms are very different. The adults of animals with relatively short and simple ontogenies often resemble pre-adult stages of other animals whose ontogeny is more elaborate, but the embryos of descendants do not necessarily resemble the adults of their ancestors. Even early development undergoes evolutionary divergence among groups, however, and it is not quite as stable as von Baer believed.

We now know that there are many parallels between ontogeny and phylogeny, but the features of an ancestral ontogeny can be shifted either to earlier or later stages in descendant ontogenies. Evolutionary change in the timing of development is called **heterochrony,** a term initially used by Haeckel to denote exceptions to recapitulation. If the descendant's ontogeny extends beyond the ancestral one, new characteristics can be added late in development, beyond the point at which development would have terminated in the evolutionary ancestor. Features observed in the ancestor often are moved to earlier stages of development in this process, and ontogeny therefore does recapitulate phylogeny to some degree. Ontogeny also can be shortened during evolution, however. Terminal stages of the ancestor's ontogeny are deleted, and

the adults of descendants come to resemble pre-adult stages of their ancestors (Figure 9-17). This reverses the parallel between ontogeny and phylogeny (reverse recapitulation) producing **paedomorphosis** (the retention of ancestral juvenile characters by later ontogenetic stages of descendants). Because the lengthening or shortening of ontogeny can change different parts of the body independently, we often see a mosaic of different kinds of developmental evolutionary change in a single lineage. Therefore, cases in which an entire ontogeny recapitulates phylogeny are rare.

MULTIPLICATION OF SPECIES

The multiplication of species through time is a logical corollary to Darwin's theory of common descent. A branch point on the evolutionary tree means that an ancestral species has split into two different ones. Darwin's theory postulates that the variation present within a species, especially variation that occurs between geographically separated populations, provides the material from which new species are produced. Because evolution is a branching process, the total number of species produced by evolution increases through time, although most of these species eventually become extinct. A major challenge for evolution-

ists is to discover the process by which an ancestral species "branches" to form two or more descendant species.

Before we explore the multiplication of species, we must decide what we mean by "species." As we will see in Chapter 11, there is no consensus regarding the definition of species. Most biologists would agree, however, that important criteria for recognizing species include (1) descent from a common ancestral population, (2) reproductive compatibility (ability to interbreed) within and reproductive incompatibility between species, and (3) maintenance within species of genotypic and phenotypic cohesion (lack of abrupt differences among populations in allelic frequencies [see the following text] and organismal appearance). The criterion of reproductive compatibility has received the greatest attention in studies of species formation, also called **speciation.**

The biological factors that prevent different species from interbreeding are called **reproductive barriers.** The primary problem of speciation is to discover how two initially compatible populations evolve reproductive barriers that cause them to become distinct, separately-evolving lineages. How do populations diverge from each other in their reproductive properties while maintaining complete reproductive compatibility within each population?

Reproductive barriers between populations usually evolve gradually. Evolution of reproductive barriers requires that diverging populations must be kept physically separate for long periods of time. If the diverging populations were reunited before reproductive barriers were completely formed, interbreeding would occur between the populations and they would merge. Speciation by gradual divergence in animals may require extraordinarily long periods of time, perhaps 10,000 to 100,000 years or more. Geographical isolation followed by gradual divergence is the most effective way for reproductive barriers to evolve, and many evolutionists consider geographical separation a prerequisite for branching speciation.

Figure 9-18

Professor Ernst Mayr, a major contributor to our knowledge of speciation and of evolution in general.

Allopatric Speciation

Allopatric ("in another land") populations of a species are those that occupy separate geographical areas. They cannot interbreed, but would be expected to do so if the geographic barriers between them were removed. Speciation that results from the evolution of reproductive barriers between geographically separated populations is known as **allopatric speciation** or geographic speciation. The separated populations evolve independently and adapt to their different environments, generating reproductive barriers between them as a result of their separate evolutionary paths. Ernst Mayr (Figure 9-18) has contributed greatly to our knowledge of allopatric speciation through his studies of speciation in birds.

Allopatric speciation begins when a species splits into two or more geographically separated populations. This can happen in either of two ways: by **vicariant speciation** or by a **founder event.** Vicariant speciation is initiated when climatic or geological changes fragment a species' habitat, producing impenetrable barriers that separate different populations. For example, a mammalian species inhabiting a lowland forest could be divided by the uplifting of a mountain barrier, the sinking and flooding of a geological fault, or climatic changes that cause prairie or desert conditions to encroach on the forest.

Vicariant speciation has two important consequences. Although the ancestral population is fragmented, the individual fragments are usually left fairly intact. The vicariant process itself does not induce genetic change by reducing populations to small size or by transporting them to different environments. Another important consequence is that the same vicariant events may fragment several different species simultaneously. For example, fragmentation of the lowland forest described above most likely would disrupt numerous and diverse species, including salamanders, frogs, snails, and many other forest dwellers. Indeed, the same geographic patterns are observed among closely related species in different groups of organisms whose habitats are similar. Such patterns provide strong evidence for vicariant speciation.

The alternative means of initiating allopatric speciation is for a small number of individuals to disperse to a distant place where no other members of their species are present. They may establish a new population in what is called a founder event. Allopatric speciation caused by founder events has been observed, for example, in the native fruit flies of Hawaii. Hawaii contains numerous patches of forest separated by volcanic lava flows. On rare occasions, strong winds can transport a small group of flies from one forest to another, geographically isolated forest where the flies are able to start a new population. Sometimes, a single fertilized female may found a new population. Unlike what happens in vicariant speciation, the new population initially has a very small size, which can cause its genetic structure to change dramatically from that of its ancestral population (see p. 171). When this happens, phenotypic characteristics that were stable in the ancestral population often reveal unprecedented variation in the new population. As the newly expressed variation is sorted by natural selection, large changes in phenotype and reproductive properties occur, hastening the evolution of reproductive barriers between the ancestral and newly founded populations.

Surprisingly, we often learn most about the genetics of allopatric speciation from cases in which formerly separated populations regain geographic contact following the evolution of incipient reproductive barriers that are not absolute. The occurrence of mating between divergent populations is called **hybridization** and the offspring of these matings are called **hybrids** (Figure 9-19). By studying

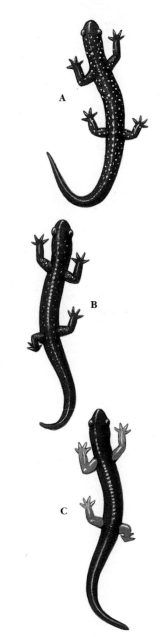

Figure 9-19

Pure and hybrid salamanders. The hybrid is intermediate in appearance between the parental populations. **A,** Pure white-spotted *Plethodon teyahalee;* **B,** a hybrid between white-spotted *P. teyahalee* and red-legged *P. jordani,* intermediate in appearance for both spotting and leg color; **C,** pure red-legged *P. jordani.*

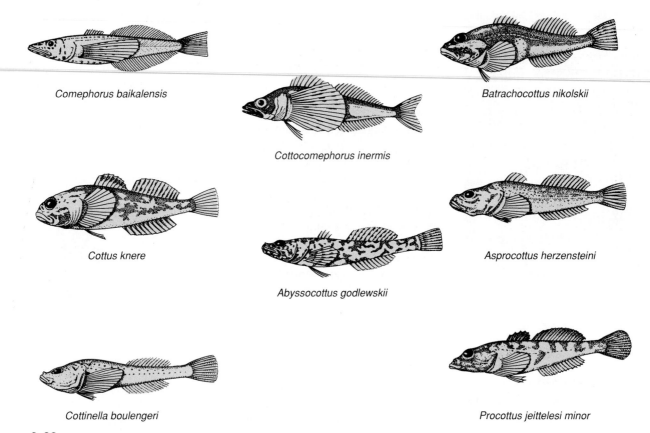

Comephorus baikalensis

Cottocomephorus inermis

Batrachocottus nikolskii

Cottus knere

Abyssocottus godlewskii

Asprocottus herzensteini

Cottinella boulengeri

Procottus jeittelesi minor

Figure 9-20

The sculpins of Lake Baikal, products of speciation that occurred within a single lake.

the genetics of hybrid populations, we can identify the genetic bases of reproductive barriers.

Biologists often distinguish between reproductive barriers that impair fertilization (premating barriers) and those that impair the growth and development, survival, or reproduction of hybrid individuals (postmating barriers). Premating barriers may cause members of divergent populations either not to recognize each other as potential mates or not to complete the mating ritual successfully. In some cases, the female and male genitalia of the different populations will be incompatible. In others, premating barriers may be strictly behavioral, with the members of different species being otherwise nearly identical in phenotype. Different species that are indistinguishable in organismal appearance are called **sibling species.** Sibling species arise when allopatric populations diverge in the seasonal timing of reproduction or in auditory, behavioral, or chemical signals required for mating. Evolutionary divergence in these fea-

tures can produce effective premating barriers without obvious changes in organismal appearance. Sibling species occur in groups as diverse as ciliates, flies, and salamanders.

Nonallopatric Speciation

Can speciation ever occur without geographic separation of populations? Allopatric speciation may seem an unlikely explanation for situations where many closely related species occur together in restricted areas that have no traces of physical barriers to animal dispersal. For example, several large lakes around the world contain very large numbers of closely related species of fish. The great lakes of Africa (Lake Malawi, Lake Tanganyika, and Lake Victoria) each contain many species of cichlid fishes that are found nowhere else. Likewise, Lake Baikal in Siberia contains many different species of sculpins that occur nowhere else in the world (Figure 9-20). It is difficult to conclude that these species arose anywhere other than in the lakes they

inhabit, and yet those lakes are young on an evolutionary timescale and have no obvious environmental barriers that would fragment fish populations.

To explain the speciation of fish in freshwater lakes and other examples like these, **sympatric** ("same land") **speciation** has been hypothesized. According to this hypothesis, different individuals within a species become specialized for occupying different subsets of the environment. By seeking out and using very specific habitats in a single geographic area, different populations achieve sufficient physical and adaptive separation to evolve reproductive barriers. For example, the cichlid species of the African lakes are very different from each other in their feeding specializations. In many parasitic organisms, particularly parasitic insects, different populations may use different host species, thereby providing the physical separation necessary for reproductive barriers to evolve. Supposed cases of sympatric speciation have been criticized, however, because the reproductive distinctness of the different populations often

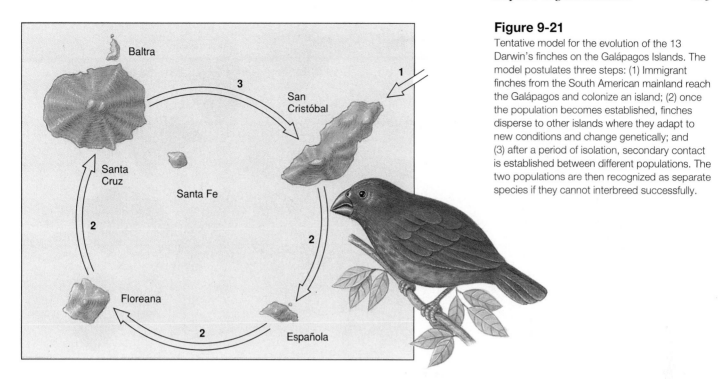

Figure 9-21

Tentative model for the evolution of the 13 Darwin's finches on the Galápagos Islands. The model postulates three steps: (1) Immigrant finches from the South American mainland reach the Galápagos and colonize an island; (2) once the population becomes established, finches disperse to other islands where they adapt to new conditions and change genetically; and (3) after a period of isolation, secondary contact is established between different populations. The two populations are then recognized as separate species if they cannot interbreed successfully.

is not well demonstrated, so that we may be observing fewer different species than we think.

The occurrence of sudden sympatric speciation is most likely among the higher plants. Between one-third and one-half of flowering plant species may have evolved by polyploidy (doubling of chromosome numbers), without prior geographic isolation of populations. In animals, however, speciation through polyploidy is an exceptional event.

Adaptive Radiation

The production of ecologically diverse species from a common ancestral stock is called adaptive radiation. Some of our best examples of **adaptive radiation** are associated with lakes and young islands, which are sources of new evolutionary opportunities for aquatic and terrestrial organisms, respectively. Oceanic islands that were formed by volcanoes are initially devoid of life. They are gradually colonized by plants and animals from a continent or from other islands in separate founder events. The founders encounter ideal situations for evolutionary diversification, because environmental resources that were heavily

exploited by other species on the mainland are free for colonization on the sparsely populated island. Archipelagoes, such as the Galápagos Islands, greatly increase the opportunities for both founder events and ecological diversification. The entire archipelago is isolated from the continent and each island is geographically isolated from the others by the sea; moreover, each island is different from every other in its physical, climatic, and biotic characteristics.

The Galápagos finches clearly illustrate adaptive radiation on an oceanic archipelago (Figures 9-21 and 9-22). The Galápagos finches (the name "Darwin's finches" was popularized in the 1940s by the British ornithologist David Lack) are close relatives, but each species differs from the others in the size and shape of the beak and in feeding habits. If the finches were specially created, it would require the strangest kind of coincidence for 13 similar kinds of finches to be created on the Galápagos Islands and nowhere else. Darwin's finches descended from a single ancestral population that arrived from the mainland and subsequently colonized the different islands of the Galápagos archipelago. The finches underwent

adaptive radiation, occupying habitats that on the mainland would have been denied to them by the presence of other species that are better able to exploit those habitats. The Galápagos finches thus assumed the characteristics of mainland families as diverse and unfinchlike as warblers and woodpeckers. A fourteenth Darwin's finch, found on isolated Cocos Island far north of the Galápagos archipelago, is similar in appearance to the Galápagos finches and almost certainly descended from the same ancestral stock.

GRADUALISM

Darwin's theory of gradualism opposed arguments for the sudden origin of species. Small differences, resembling those that we observe among organisms within populations today, are the raw material from which the different major forms of life evolved. This theory shares with Lyell's uniformitarianism the notion that we must not explain past changes by invoking unusual catastrophic events that are not observable today. If new species originated in single, catastrophic events, we should be able to see this happening today and we do not. What we observe instead are small, continuous

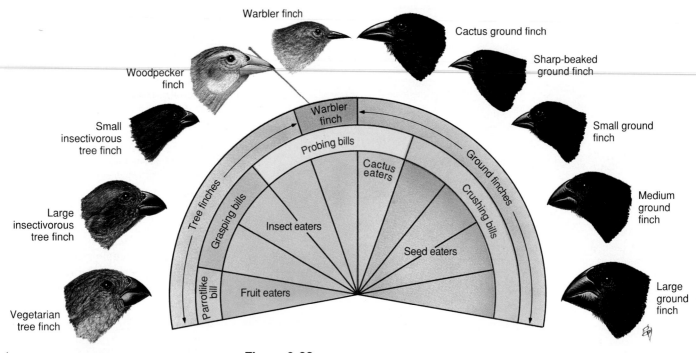

A

Figure 9-22

A, Adaptive radiation in ten species of Darwin's finches from Santa Cruz, one of the Galápagos Islands. Differences in bills and feeding habits are shown. All apparently descended from a single common ancestral finch from the South American continent. **B,** Woodpecker finch, one of the 13 species of Galápagos Islands finches, using a slender twig as a tool for feeding. This finch worked for about 15 minutes before spearing and removing a wood roach from a break in the tree.

B

changes in the phenotypes that are present in natural populations. Such continuous changes can produce major differences among species only by accumulating over many thousands to millions of years. A simple statement of Darwin's theory of gradualism is that accumulation of quantitative changes leads to qualitative change.

Mayr (see Figure 9-18) makes an important distinction between populational gradualism and phenotypic gradualism. **Populational gradualism** states that new traits become established in a population by increasing their frequency initially from a small fraction of the population to a majority of the population. Populational gradualism is well established and is not controversial. **Phenotypic gradualism** states that new traits,

even those that are strikingly different from ancestral ones, are produced in a series of small, incremental steps.

Phenotypic Gradualism

Phenotypic gradualism was controversial when Darwin first proposed it, and it is still controversial. Not all phenotypic changes are small, incremental ones. Some mutations that appear during artificial breeding, traditionally called "sports," change the phenotype substantially in a single mutational step. Sports that produce dwarfing are observed in many species, including humans, dogs, and sheep, and have been used by animal breeders to achieve desired results; for example, a sport that deforms the limbs was used to produce ancon sheep, which cannot jump hedges and are therefore easily contained (Figure 9-23). Many colleagues of Darwin who accepted his other theories considered phenotypic gradualism too extreme. If sporting mutations can be used in animal breeding, why must we exclude them from our

Figure 9-23

The ancon breed of sheep arose from a "sporting mutation" that caused dwarfing of the legs. Many of his contemporaries criticized Darwin for his claim that such mutations are not important in the process of evolution by natural selection.

evolutionary theory? In favor of gradualism, some have replied that sporting mutations always have negative side-effects that would prevent them from surviving in natural populations. Indeed, it is questionable whether the ancon sheep, despite their attractiveness to farmers, would propagate successfully in the presence of their long-legged relatives in nature.

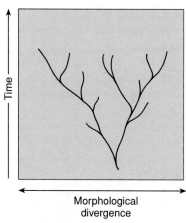

Figure 9-24
The gradualist model of evolutionary change in morphology, viewed as proceeding more or less steadily through geological time (millions of years). Bifurcations followed by gradual divergence led to speciation.

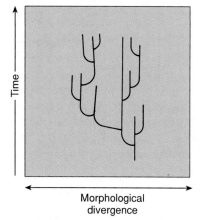

Figure 9-25
The punctuated equilibrium model sees evolutionary change being concentrated in relatively rapid bursts of branching speciation (lateral lines) followed by prolonged periods of no change throughout geological time (millions of years).

Punctuated Equilibrium

When we view Darwinian gradualism on a geological timescale, we may expect to find in the fossil record a long series of intermediate forms connecting the phenotypes of ancestral and descendant populations (Figure 9-24). This predicted pattern is called **phyletic gradualism.** Darwin recognized that phyletic gradualism is not often revealed by the fossil record. Studies conducted since Darwin's time likewise have failed to produce the predicted continuous series of fossils predicted by phyletic gradualism. Is the theory of gradualism therefore refuted by the fossil record? Darwin and others claim that it is not, because the fossil record is too imperfect to preserve transitional series. Although evolution is a slow process by our standards, it is rapid relative to the rate at which good fossil deposits accumulate. Others have argued, however, that the abrupt origins and extinctions of species in the fossil record force us to conclude that phyletic gradualism is rare.

Niles Eldredge and Stephen Jay Gould proposed **punctuated equilibrium** to explain the discontinuous evolutionary changes observed throughout geological time. Punctuated equilibrium states that phenotypic evolution is concentrated in relatively brief events of branching speciation, followed by much longer intervals of evolutionary stasis (Figure 9-25). Speciation is an episodic event, occurring over a period of approximately 10,000 to 100,000 years. Because species may survive for 5 million to 10 million years, the speciation event is a "geological instant," representing 1% or less of a species' life span. Ten thousand years is plenty of time, however, for Darwinian evolution to accomplish dramatic changes. A small fraction of the evolutionary history of a group therefore accounts for most of the morphological evolutionary change that we observe.

The process of allopatric speciation by founder events provides a possible explanation for punctuated equilibria. Remember that founder-induced speciation requires the breaking of genetic equilibrium in a small, geographically isolated population. Such small populations have very little chance of being preserved in the fossil record. After a new genetic equilibrium forms and stabilizes, the new population may increase in size, thereby increasing the likelihood that some of its members will be preserved as fossils.

Evolutionists who lamented the imperfect state of the fossil record were treated in 1981 to the opening of an uncensored page of fossil history in Africa. Peter Williamson, a British paleontologist working in fossil beds 400 m deep near Lake Turkana, documented a remarkably clear record of speciation in freshwater snails. The geology of the Lake Turkana basin reveals a history of instability. Earthquakes, volcanic eruptions, and climatic changes caused the waters to rise and fall periodically, sometimes by hundreds of feet. Thirteen lineages of snails show long periods of stability interrupted by relatively brief periods of rapid change in shell shape when snail populations were fragmented by receding waters. These populations diverged to produce new species that then remained unchanged through thick deposits before becoming extinct and being replaced by descendant species. The transitions occurred within 5000 to 50,000 years. In the few meters of sediment where speciation occurred, transitional forms were visible. Williamson's study conforms well to the punctuated equilibrium model of Eldredge and Gould.

NATURAL SELECTION

Natural selection is the centerpiece of Darwin's theory of evolution. It gives us a natural explanation for the origins of **adaptation,** including all developmental, behavioral, anatomical, and physiological attributes that enhance the organism's ability to use environmental resources to survive and to reproduce. Darwin developed his theory of natural selection as a series of five observations and three inferences drawn from them:

Observation 1—Organisms have great potential fertility. All populations produce large numbers of gametes and potentially large numbers of offspring each generation. Population size would increase exponentially at an enormous rate if all individuals that were produced each generation survived and reproduced. Darwin calculated that, even in a slow-breeding species such as the elephant, a single pair breeding from age 30 to 90 and having only six young could produce 19 million descendants in 750 years.

Observation 2—Natural populations normally remain constant in size, except for minor fluctuations. Natural populations fluctuate in size across generations and sometimes go extinct, but no natural populations show the continued exponential growth that their reproductive biology theoretically could sustain.

Observation 3—Natural resources are limited. Exponential growth of a natural population would require unlimited natural resources to provide food and habitat for the expanding population, but natural resources are finite.

Inference 1—There exists a continuing *struggle for existence* among members of a population. Survivors represent only a part, often a very small part, of the individuals produced each generation. Darwin wrote in *The Origin of Species* that "it is the doctrine of Malthus applied with manifold force to the whole animal and vegetable kingdoms." The struggle for food, shelter, and space becomes increasingly severe as overpopulation develops.

Observation 4—All organisms show *variation*. No two individuals are exactly alike. They differ in size, color, physiology, behavior, and many other ways.

Observation 5—Some variation is heritable. Darwin noted that offspring tend to resemble their parents, although he did not understand how. The hereditary mechanism discovered by Gregor Mendel would be applied to Darwin's theory many years later.

Inference 2—There is *differential survival and reproduction* among varying organisms in a population. Survival in the struggle for existence is not random with respect to hereditary variation present in the population. Some traits give their possessors an advantage in using the environment for effective survival and reproduction.

Inference 3—Over many generations, differential survival and reproduction generates new adaptations and new species. The differential reproduction of varying organisms gradually transforms species and results in the long-term "improvement" of types. Darwin knew that people often use hereditary variation to produce useful new breeds of livestock and plants. *Natural* selection acting over millions of years should be even more effective in producing new types than the *artificial* selection imposed during a human lifetime. Natural selection acting independently on geographically separated populations would cause them to diverge from each other, thereby generating the reproductive barriers that lead to speciation.

The popular phrase "survival of the fittest" was not originated by Darwin but was coined a few years earlier by the British philosopher Herbert Spencer, who anticipated some of Darwin's principles of evolution. Unfortunately the phrase later came to be coupled with unbridled aggression and violence in a bloody, competitive world. In fact, natural selection operates through many other characteristics of living things. The fittest animal may be one that enhances the living conditions of its population. Fighting prowess is only one of several means toward successful reproductive advantage.

Natural selection can be viewed as a two-step process with a random component and a nonrandom component. The production of variation among organisms is the random component. The mutational process does not preferentially generate traits that are favorable to the organism; if anything, the reverse is probably true. The nonrandom component is the survival of different traits. This is determined by the effectiveness of different traits in permitting their possessors to use environmental resources to survive and to reproduce. The phenomenon of differential survival and reproduction among varying organisms is now called **sorting** and should not be equated with natural selection. We now know that even random processes (genetic drift, p. 171) can produce sorting among varying organisms. Selection states that sorting occurs *because certain traits give their possessors advantages in survival and reproduction* relative to others that lack those traits. Selection is therefore a specific cause of sorting.

Darwin's theory of natural selection has been challenged repeatedly. One challenge claims that directed (nonrandom) variation governs evolutionary change. In the decades around 1900, diverse evolutionary hypotheses collectively called **orthogenesis** proposed that variation has momentum that forces a lineage to evolve in a particular direction that is not always adaptive. The extinct Irish elk was a popular example of orthogenesis. Newly produced variation was considered preferentially to increase the size of the antlers. Natural selection was considered ineffective at stopping the antlers eventually from becoming so large and cumbersome that they forced the Irish elk into extinction (Figure 9-26). Orthogenesis explained apparently nonadaptive evolutionary trends that led a species into decline. Subsequent genetic research on the nature of variation, however, has rejected the genetic predictions of orthogenesis.

Another recurring criticism of natural selection is that it cannot generate new structures or species but can only modify old ones. Most structures in their early evolutionary stages

Figure 9-26

The Irish elk, a fossil species that once was used to support the orthogenetic idea that momentum in variation caused the antlers to become so large that the species was forced into extinction.

could not have performed the biological roles that the fully formed structures perform, and it is therefore unclear how natural selection could have favored them. What use is half a wing or the rudiment of a feather for a flying bird? To answer this criticism, we propose that many structures evolved initially for purposes different from the ones they have today. Rudimentary feathers could have been useful in thermoregulation, for example. The feathers later became useful for flying after they incidentally acquired some aerodynamic properties. Natural selection then could act to improve the usefulness of feathers for flying. Because the structural changes that separate members of different species are similar in kind to those that we observe within species, it is unreasonable to propose that selection will never lead beyond the species boundary.

REVISIONS OF DARWIN'S THEORY

NEO-DARWINISM

The most serious weakness in Darwin's theory was his failure to identify correctly the mechanism of inheritance. Darwin saw heredity as a blending phenomenon in which the characteristics of the parents melded together in the offspring. Darwin also invoked the Lamarckian hypothesis that an organism could alter its heredity through the use and disuse of body parts and through the direct influence of the environment. August Weissmann rejected Lamarckian inheritance by showing experimentally that modifications of an organism during its lifetime do not change its heredity (see Chapter 7), and he revised Darwin's theory accordingly. We now use the term, **neo-Darwinism** to denote Darwin's theory as revised by Weissmann.

Mendelian genetics eventually clarified the particulate inheritance that Darwin's theory of natural selec-

tion required (p. 127). Ironically, when Mendel's work was rediscovered in 1900, it was viewed as antagonistic to Darwin's theory of natural selection. When mutations were discovered in the early 1900s, most geneticists thought that they produced new species in single large steps. They relegated natural selection to the role of executioner, a negative force that merely eliminated the obviously unfit.

EMERGENCE OF MODERN DARWINISM: THE SYNTHETIC THEORY

In the 1930s a new breed of geneticists began to reevaluate Darwin's theory from a different perspective. These were population geneticists, scientists who studied variation in natural populations of animals and plants and who had a sound knowledge of statistics and mathematics. Gradually, a new comprehensive theory emerged that brought together population genetics, paleontology, biogeography, embryology, systematics, and animal behavior in a Darwinian framework.

Population genetics studies evolution as a change in the genetic composition of populations. With the establishment of population genetics, evolutionary biology became divided into two different subfields. **Microevolution** is the field that studies evolutionary changes in the frequencies of different allelic forms of genes within populations. **Macroevolution** studies evolution on a grand scale, encompassing the origin of new organismal structures and designs, evolutionary trends, adaptive radiation, phylogenetic relationships of species, and mass extinction. Macroevolutionary research is based in systematics and the comparative method (p. 199). Following the evolutionary synthesis, both macroevolution and microevolution have operated firmly within the tradition of neo-Darwinism, and both have expanded Darwinian theory in important ways.

MICROEVOLUTION: GENETIC VARIATION AND CHANGE WITHIN SPECIES

Microevolution is the study of genetic change occurring within natural populations. The observation of different allelic forms of a gene in a population is called **polymorphism.** All of the alleles of all genes possessed by members of a population collectively form the **gene pool.** The amount of polymorphism present in large populations is potentially enormous, because at observed mutation rates, many different alleles are expected for all genes.

Population geneticists study polymorphism by identifying the different allelic forms of a gene that are present in a population and then measuring their relative frequencies in the population. The relative frequency of a particular allelic form of a gene in a population is known as its **allelic frequency.** For example, in the human population, there are three different allelic forms of the gene encoding the ABO blood types: I^A, I^B, and i (p. 680). Because each individual contains two copies of this gene, the total number of copies present in the population is twice the number of individuals. What fraction of this total is represented by each of the three different allelic forms? In the French population, we find the following allelic frequencies: I^A = .46, I^B = .14, and i = .40. In the Russian population, the corresponding allelic frequencies differ (I^A = .38, I^B = .28, and i = .34), demonstrating microevolutionary divergence between these populations (see Figure 9-27). Genetically, alleles I^A and I^B are dominant to i, but i is nearly as frequent as I^A and exceeds the frequency of I^B in both populations. Dominance describes the *phenotypic effect* of an allele in heterozygous individuals, not its relative abundance in a population of individuals. We will demonstrate that Mendelian inheritance and dominance do not alter allelic frequencies directly or produce evolutionary change in a population.

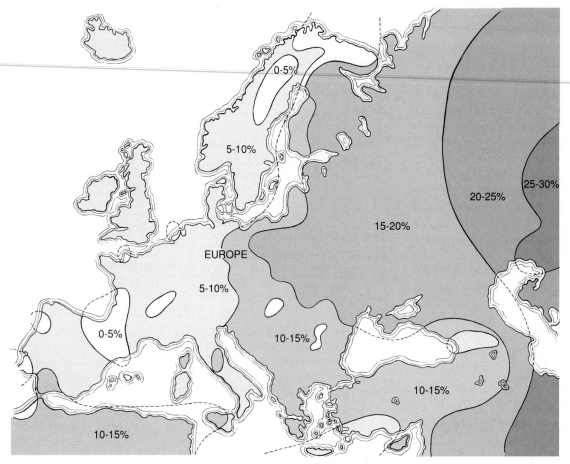

Figure 9-27
Frequencies of the blood-type B allele among humans in Europe. The allele is more common in the east and rarer in the west. The allele may have arisen in the east and gradually diffused westward through the genetic continuity of human populations. There is no known selective advantage of this allele, and its changing frequency probably represents the effects of random genetic drift.

GENETIC EQUILIBRIUM

In many human populations, genetically recessive traits, including the O blood type, blond hair, and blue eyes, are very common. Why have not the genetically dominant alternatives gradually supplanted these recessive traits? It is a common misconception that a characteristic associated with a dominant allele increases in proportion because of its genetic dominance. This is not the case, because there is a tendency in *large* populations for allelic frequencies to remain in equilibrium generation after generation. This principle is based on the **Hardy-Weinberg equilibrium** (see box), which forms the foundation for population genetics. According to this theorem, the hereditary process alone does not produce evolutionary change. In large biparental populations, allelic frequencies and

genotypic ratios attain an equilibrium in one generation and remain constant thereafter unless disturbed by recurring mutations, natural selection, migration, nonrandom mating, or genetic drift (random sorting). Such disturbances are the sources of microevolutionary change.

A rare allele, according to this principle, does not disappear from a large population merely because it is rare. That is why certain rare traits, such as albinism and cystic fibrosis, persist for endless generations. For example, albinism in humans is caused by a rare recessive allele *a*. Only one person in 20,000 is an albino, and this individual must be homozygous (*a/a*) for the recessive allele. Obviously there are many carriers in the population with normal pigmentation, that is, people who are heterozygous (*A/a*) for albinism. What is their frequency? A convenient way

to calculate the frequencies of genotypes in a population is with the binomial expansion of $(p + q)^2$ (see box). We will let p represent the allelic frequency of A and q the allelic frequency of a.

Assuming that mating is random (a questionable assumption, but one that we will accept for our example), the distribution of genotypic frequencies is $p^2 = A/A$, $2pq = A/a$, and $q^2 = a/a$. Only the frequency of genotype a/a is known with certainty, 1/20,000; therefore:

$$q^2 = 1/20,000$$

$$q = (1/20,000)^{1/2} = 1/141$$

$$p = 1 - q = 140/141$$

The frequency of carriers is as follows:

$$A/a = 2pq = 2 \times 140/141 \times 1/141$$

$$= 1/70$$

Hardy-Weinberg Equilibrium: Why the Hereditary Process Does Not Change Allelic Frequencies

The Hardy-Weinberg law is a logical consequence of Mendel's first law of segregation and expresses the tendency toward equilibrium inherent in Mendelian heredity.

Let us select for our example a population having a single locus bearing just two alleles T and t. The phenotypic expression of this gene might be, for example, the ability to taste a chemical compound called phenylthiocarbamide. Individuals in the population will be of three genotypes for this locus, T/T, T/t (both tasters), and t/t (nontasters). In a sample of 100 individuals, let us suppose we have determined that there are 20 of T/T genotype, 40 of T/t genotype, and 40 of t/t genotype. We could then set up a table showing the allelic frequencies as follows (remember that every individual has two copies of the gene):

Genotype	Number of Individuals	Copies of the T Allele	Copies of the t Allele
T/T	20	40	
T/t	40	40	40
t/t	40		80
TOTAL	100	80	120

Of the 200 copies, the proportion of the T allele is $80/200 = 0.4$ (40%); and the proportion of the t allele is $120/200 = 0.6$ (60%). It is customary in presenting this equilibrium to use "p" and "q" to represent the two allelic frequencies. The genetically dominant allele is represented by p, and the genetically recessive by q. Thus:

$$p = \text{frequency of } T = 0.4$$
$$q = \text{frequency of } t = 0.6$$
$$\text{Therefore } p + q = 1$$

Having calculated allelic frequencies in the sample, let us determine whether these frequencies will change spontaneously in a new generation of the population. Assuming the mating is random (and this is important; all mating combinations of genotypes must be equally probable), each individual will contribute an equal number of gametes to the "common pool" from which the next generation is formed. This being the case, the frequencies of gametes in the "pool" will be proportional to the allelic frequencies in the sample. That is, 40% of the gametes will be T, and 60% will be t (ratio of 0.4:0.6). Both ova and sperm will, of course, show the same frequencies. The next generation is formed as follows:

	Ova	
Sperm	$T = 0.4$	$t = 0.6$
$T = 0.4$	$T/T = 0.16$	$T/t = 0.24$
$t = 0.6$	$T/t = 0.24$	$t/t = 0.36$

Collecting the genotypes, we have:

frequency of $T/T = 0.16$
frequency of $T/t = 0.48$
frequency of $t/t = 0.36$

Next, we determine the values of p and q from the randomly mated populations. From the table above, we see that the frequency of T will be the sum of genotypes T/T, which is 0.16, and one-half of the genotype T/t, which is 0.24:

$$T(p) = 0.16 + .5(0.48) = 0.4$$

Similarly, the frequency of t will be the sum of genotypes t/t, which is 0.36, and one-half the genotype T/t, which is 0.24:

$$t(p) = 0.36 + .5(0.48) = 0.6$$

The new generation bears exactly the same allelic frequencies as the parent population! Note that there has been no increase in the frequency of the genetically dominant allele T. Thus *in a freely interbreeding, sexually reproducing population, the frequency of each allele would remain constant generation after generation in the absence of natural selection, migration, recurring mutation and genetic drift* (see text). The more mathematically minded reader will recognize that the genotype frequencies T/T, T/t, and t/t are actually a binomial expansion of $(p + q)^2$:

$$(p + q)^2 = p^2 + 2pq + q^2 = 1$$

One person in every 70 is a carrier! Although a recessive trait may be rare, it is amazing how common a recessive allele may be in a population. There is a message here for anyone proposing to eliminate a "bad" recessive allele from a population by controlling reproduction. It is practically impossible. Because only the homozygous recessive individuals reveal the phenotype against which artificial selection could act (by sterilization, for example), the allele would continue to surface from heterozygous carriers. For a recessive allele present in 2 of every 100 persons (but homozygous in only 1 in 10,000 persons), it would require 50 generations of complete selection against the homozygotes just to reduce its frequency to one in 100 persons.

How Genetic Equilibrium Is Upset

Genetic equilibrium is disturbed in natural populations by (1) random genetic drift, (2) nonrandom mating, (3) recurring mutation, (4) migration, (5) natural selection, and interactions among these factors. Recurring mutation is the ultimate source of variability in all populations, but it usually requires interaction with one or more of the other factors to upset genetic equilibrium. We will look at these other factors individually.

Genetic Drift

Some species, like the cheetah (Figure 9-28), contain very little genetic varia-

tion, probably because their ancestral lineages contained some very small populations. A small population clearly cannot contain large amounts of genetic variation. Each individual organism has at most two different allelic forms of each gene, and a single breeding pair contains at most four different allelic forms of a gene. Suppose that we have such a breeding pair. We know from Mendelian genetics (Chapter 8) that chance decides which of the different allelic forms of a gene gets passed to offspring. It is therefore possible by chance alone that one or two of the parental alleles in this example will not be passed to any offspring. It is highly unlikely that the different alleles present in a small ancestral population are all passed to descendants without any

Figure 9-28
The cheetah, a species whose genetic variability
has been depleted to very low levels because of
small population size in the past.

change of allelic frequency. This chance fluctuation in allele frequency from one generation to the next, including loss of alleles from the population, is called **genetic drift.**

Genetic drift occurs to some degree in all populations of finite size. Perfect constancy of allelic frequencies, as predicted by Hardy-Weinberg equilibrium, occurs only in infinitely large populations, and such populations occur only in mathematical models. All populations of animals are finite and therefore experience some effect of genetic drift, which becomes greater, on average, as population size declines. Genetic drift erodes the genetic variability of a population. If population size remains small for many generations in a row, genetic variation can be greatly depleted. This loss is harmful to a species' evolutionary success because it restricts potential genetic responses to environmental change. Indeed, biologists are concerned that cheetah populations may have insufficient variation for continued survival.

Nonrandom Mating

If mating is nonrandom, genotypic frequencies will deviate from the Hardy-Weinberg expectations. For example, if two different alleles of a gene are equally frequent ($p = q = .5$), we expect half of the genotypes to be heterozygous ($2pq = 2 [.5] [.5] = .5$) and

one-quarter to be homozygous for each of the respective alleles ($p^2 = q^2 = [.5]^2 = .25$). If we have **positive assortative mating,** individuals mate preferentially with others of the same genotype, such as albinos mating with other albinos. Matings among homozygous parents generate offspring that are homozygous like themselves. Matings among heterozygous parents produce on average 50% heterozygous offspring and 50% homozygous offspring (25% of each alternative type) each generation. This increases the frequency of homozygous genotypes and decreases the frequency of heterozygous genotypes in the population but does not change allelic frequencies.

Preferential mating among close relatives also increases homozygosity and is called **inbreeding.** Whereas positive assortative mating usually affects one or a few traits, inbreeding simultaneously affects all variable traits. Strong inbreeding greatly increases the chances that rare recessive alleles will become homozygous and be expressed.

Because inbreeding and genetic drift are both promoted by small population size, they are often confused with each other. Their effects are very different, however. Inbreeding alone cannot change allelic frequencies in the population, only the ways that alleles are combined into genotypes. Genetic drift changes allelic frequencies and consequently also changes genotypic frequencies. Even very large populations have the potential for being highly inbred if there is a behavioral preference for mating with close relatives, although this rarely occurs in nature. Genetic drift, however, will be relatively weak in very large populations.

Migration

Migration prevents different populations of a species from diverging. If a large species is divided into many small populations, genetic drift and selection acting separately in the different populations can produce evolutionary divergence between them. A small amount of migration each

generation keeps the different populations from becoming too different. For example, the French and Russian populations whose ABO allele frequencies were discussed previously show some genetic divergence, but continuing migration between them prevents them from becoming completely distinct.

Natural Selection

Natural selection can change both allelic frequencies and genotypic frequencies in a population. Although the effects of selection are often reported for particular polymorphic genes, we must stress that natural selection acts on the whole animal, not on isolated traits. The organism that possesses the superior combination of traits will be favored. An animal may have traits that confer no advantage or even a disadvantage, but it is successful overall if its combination of traits is favorable. When we claim that a genotype at a particular gene has a higher **relative fitness** than others, we state that on average that genotype confers an advantage in survival and reproduction in the population. If alternative genotypes have unequal probabilities of survival and reproduction, the Hardy-Weinberg equilibrium will be upset.

Some traits and combinations of traits are advantageous for certain aspects of the organism's survival or reproduction and disadvantageous for others. Darwin used the term, **sexual selection,** to denote the selection of traits that are advantageous for obtaining mates but may be harmful for survival. Bright colors and elaborate feathers may enhance a male bird's competitive ability in obtaining mates while simultaneously increasing his vulnerability to predators (Figure 9-29). Changes in the environment can alter the selective value of different traits. The action of selection on character variation is therefore very complex.

Interactions of Selection, Drift, and Migration

Subdivision of a species into small populations that exchange migrants is

Figure 9-29

A pair of wood ducks. Brightly-colored feathers of male birds probably confer no survival advantage and might even be harmful by alerting predators. Such colors nonetheless confer advantage in attracting mates, which overcomes, on average, the negative consequences of these colors for survival. Darwin used the term "sexual selection" to denote traits that give an individual an advantage in attracting mates, even if the traits are neutral or harmful for survival.

an optimal one for promoting rapid adaptive evolution of the species as a whole. The interaction of genetic drift and selection in the different populations permits many different genetic combinations of many polymorphic genes to be tested against natural selection. The migration among the populations permits particularly favorable new genetic combinations to spread throughout the species as a whole. The interaction of selection, genetic drift, and migration in this

example produces evolutionary change that is qualitatively different from what would result if any of these three factors acted alone. Natural selection, genetic drift, mutation, nonrandom mating, and migration interact in natural populations to create an enormous opportunity for evolutionary change; the perpetual stability predicted by Hardy-Weinberg equilibrium almost never lasts across any significant amount of evolutionary time.

MEASURING GENETIC VARIATION WITHIN POPULATIONS

Protein Polymorphism

How do we measure the genetic variation that occurs in natural populations? Genetic dominance, interactions between alleles of different genes, and environmental effects on the phenotype make it difficult to quantify genetic variation indirectly by observing organismal phenotypes. Variability can be quantified, however, at the molecular level. Different allelic forms of genes encode proteins that may differ slightly in their amino acid sequence. This is called **protein polymorphism.** If these differences affect the protein's net electric charge, the different allelic forms can be separated using protein electrophoresis (Figure 9-30). We can identify the genotypes of particular individuals for protein-coding genes and measure allelic frequencies in the population.

Over the last 25 years, geneticists using this approach have discovered far more variation than was previously expected. Despite the high levels of polymorphism discovered using protein electrophoresis (Table 9-1), these studies underestimate both protein polymorphism and the total genetic variation present in a population. For

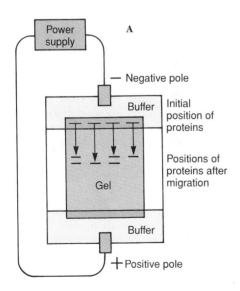

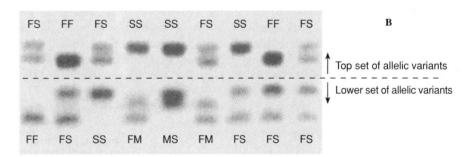

Figure 9-30

The study of genetic variation in proteins using gel electrophoresis. **A,** The electrophoretic apparatus separates allelic variants of proteins that differ in charge because of differences in their sequence of amino acids. **B,** Genetic variation in the protein leucine aminopeptidase for nine individuals of the brown snail, *Helix aspersa.* Two different sets of allelic variants are revealed. The top set contains two alleles [denoted fast (F) and slow (S) according to their relative movement in the electric field]. Individuals homozygous for the fast allele show only a single fast band on the gel (FF), those homozygous for the slow allele show only a single slow band (SS), and heterozygous individuals have both bands (FS). The lower set contains three different alleles denoted fast (F), medium (M), and slow (S). Note that no individuals shown are homozygous for the medium (M) allele.

Table 9-1	Values of Polymorphism (P) and Heterozygosity (H) for Various Animals and Plants as Measured Using Protein Electrophoresis		
(a) Species	**Number of Proteins**	**P**	**H**
Humans	71	0.28	0.067
Northern elephant seal	24	0.0	0.0
Horseshoe crab	25	0.25	0.057
Elephant	32	0.29	0.089
Drosophila pseudoobscura	24	0.42	0.12
Barley	28	0.30	0.003
Tree frog	27	0.41	0.074
(b) Taxa	**Number of Species**	**P**	**H**
Plants	—	0.31	0.10
Insects (excluding *Drosophila*)	23	0.33	0.074
Drosophila	43	0.43	0.14
Amphibians	13	0.27	0.079
Reptiles	17	0.22	0.047
Birds	7	0.15	0.047
Mammals	46	0.15	0.036
Average		0.27	0.078

Source: Data from P. W. Hedrick, *Population biology.* Jones and Bartlett, Boston, 1984.
*P, the average number of alleles per gene; H, the proportion of heterozygous genes per individual.

example, protein polymorphism that does not involve charge differences is not detected. Furthermore, because the genetic code is degenerate (more than one codon for most amino acids, p. 139), protein polymorphism does not reveal all of the genetic variation present in protein-coding genes. Genetic changes that do not alter protein structure may alter patterns of protein synthesis during development and can be very important to the organism. When all kinds of variation are considered, it is evident that most species have an enormous potential for further evolutionary change.

QUANTITATIVE VARIATION

Quantitative traits are those that show continuous variation with no obvious pattern of Mendelian segregation in their inheritance. The values of the trait in offspring often are intermediate between the values in the parents. Such traits are influenced by variation at many genes, each of which follows Mendelian inheritance and contributes a small, incremental amount to the total phenotype. Traits that show

quantitative variation include tail length in mice, length of a leg segment in grasshoppers, number of gill rakers in sunfishes, number of peas in pods, and height of adult males of the human species. When the values are graphed with respect to frequency distribution, they often approximate a normal, or bell-shaped, probability curve (Figure 9-31A). Most individuals fall near the average, fewer fall well above or below the average, and the extremes make up the "tails" of the frequency curve with increasing rarity. Usually, the larger the population sample, the more closely the frequency distribution resembles a normal curve.

Selection can act on quantitative traits to produce three different kinds of evolutionary response (see Figure 9-31B, C, and D). One is to favor average values of the trait and to disfavor extreme ones; this is called **stabilizing selection** (Figure 9-31B). **Directional selection** favors an extreme value of the phenotype and causes the population average to shift toward it over time (Figure 9-31C). When we think about natural selection producing evo-

lutionary change, it is usually directional selection that we have in mind, although we must remember that this is not the only possibility. A third alternative is **disruptive selection** in which two different extreme phenotypes are simultaneously favored, but the average is disfavored (Figure 9-31D). The population will become bimodal, meaning that two very different phenotypes will predominate.

MACROEVOLUTION: MAJOR EVOLUTIONARY EVENTS

Macroevolution describes large-scale events in organic evolution. The process of speciation links macroevolution and microevolution. The major trends in the fossil record described earlier (see Figures 9-11 and 9-12) fall clearly within the realm of macroevolution. The patterns and processes of macroevolutionary change emerge from those of microevolution, but they acquire some degree of autonomy in doing so. The emergence of new adaptations and species, and the varying rates of speciation and extinction observed in the fossil record go beyond the fluctuations of allelic frequencies within populations.

Stephen Jay Gould recognizes three different "tiers" of time at which we observe distinct evolutionary processes. The first tier constitutes the timescale of population genetic processes, from tens to thousands of years. The second tier covers millions of years, the scale on which rates of speciation and extinction can be measured and compared among different groups of organisms. The third tier covers tens to hundreds of millions of years, and is marked by the occurrence of periodic mass extinctions. In the fossil record of marine organisms, mass extinctions recur at intervals of approximately 26 million years. Five of these mass extinctions have been particularly disastrous (Figure 9-32). The study of long-term changes in animal diversity focuses on the third-tier timescale (see Figures 9-12 and 9-32).

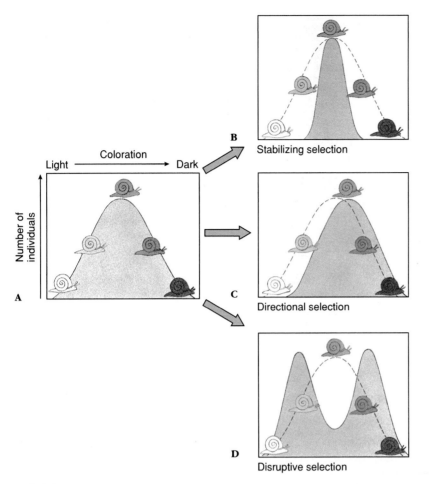

Figure 9-31

Responses to selection on a continuous (polygenic) character, coloration in a snail. **A,** The frequency distribution of coloration before selection. **B,** Stabilizing selection culls extreme variants from the population, in this case eliminating individuals that are unusually light or dark, thereby stabilizing the mean. **C,** Directional selection shifts the population mean, in this case by favoring darkly colored variants. **D,** Disruptive selection favors both extremes but not the mean; the mean is unchanged but the population no longer has a bell-shaped distribution of phenotypes.

SPECIATION AND EXTINCTION THROUGH GEOLOGICAL TIME

Evolutionary change at the second tier provides a new perspective on Darwin's theory of natural selection. A species has two possible evolutionary fates: it may give rise to new species or become extinct without leaving descendants. Rates of speciation and extinction vary among lineages, and the lineages that have the highest speciation rates and lowest extinction rates produce the greatest diversity of living forms. The characteristics of a species may make it more or less likely than others to undergo speciation or extinction events. Because many characteristics are passed from ancestral to descendant species (analogous to heredity at the organismal level), lineages whose properties enhance the probability of speciation and confer resistance to extinction should come to dominate the living world. This species-level process that produces differential rates of speciation and extinction among lineages is analogous in many ways to natural selection. It represents an expansion of Darwin's theory of natural selection.

Species selection is the differential survival and multiplication of species through geological time based on variation among lineages in emergent, species-level properties. These

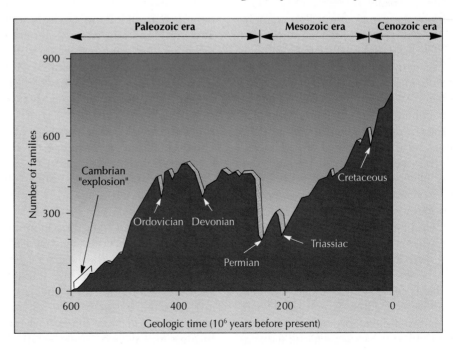

Figure 9-32

Changes in numbers of families of marine animals through time from the Cambrian period to the present. Sharp drops represent five major extinctions of skeletonized marine animals. Note that despite the extinctions, the overall number of marine families has increased to the present.

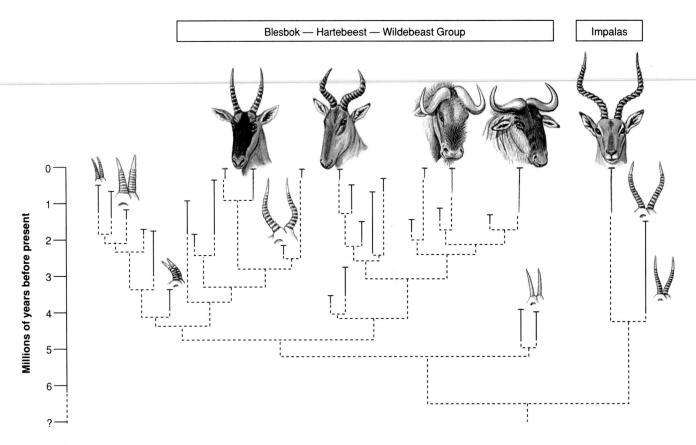

Figure 9-33

Contrasting diversity between two lineages of African antelopes. Higher speciation and extinction rates in the lineage containing the blesboks, hartebeests, and wildebeests is attributed to greater specialization in feeding relative to the impala lineage, an example of effect macroevolution.

species-level properties include mating rituals, social structuring, migration patterns, geographic distribution, and all other properties that emerge at the species level (see p. 6). Descendant species usually resemble their ancestors in these properties. For example, a "harem" system of mating in which a single male and several females compose a breeding unit characterizes some mammalian lineages but not others. We expect speciation rates to be enhanced by social systems that promote the founding of new populations by small numbers of individuals. Certain social systems may increase the likelihood that a species will survive environmental challenges through cooperative action. Such properties would be favored over geological time by species selection.

Effect macroevolution is similar to species selection except that differential speciation and extinction among lineages is caused by variation in organismal-level properties (such as specialized versus generalized feeding)

rather than species-level properties (see p. 6). Organisms that specialize in eating a restricted range of foods, for example, may be subjected more readily than generalized feeders to geographic fragmentation caused by spatial fluctuation of the environment. Such fragmentation could generate more frequent opportunities for speciation to occur throughout geological time. The fossil records of two major lineages of African antelopes demonstrate this result (Figure 9-33). A lineage of specialized grazers that contains the blesboks, hartebeests, and wildebeests shows high speciation and extinction rates; since the late Miocene, 33 extinct and 7 living species are found, representing at least 18 events of branching speciation and 12 terminal extinctions. In contrast, a lineage of generalist grazers and browsers that contains the impalas shows neither branching speciation nor terminal extinction during this same interval of time. Interestingly, although these two lineages differ

greatly in speciation rates, extinction rates, and species diversity, they do not differ significantly in the total number of individual animals alive today.

MASS EXTINCTIONS

When we study evolutionary change on an even larger timescale, we observe periodic events in which large numbers of taxa go extinct simultaneously. These events are called **mass extinctions** (see Figure 9-32). The most cataclysmic of these extinction episodes happened about 225 million years ago, when at least half of the families of shallow-water marine invertebrates, and fully 90% of marine invertebrate species disappeared within a few million years. This was the **Permian extinction.** The **Cretaceous extinction,** which occurred about 65 million years ago, marked the end of the dinosaurs, as well as numerous marine invertebrates and many small reptilian taxa.

The causes of mass extinctions and their occurrence at intervals of

Figure 9-34
Twin craters of Clearwater Lakes in Canada show that multiple impacts on the earth are not as unlikely as they might seem. Evidence suggests that at least two impacts within a short time were responsible for the Cretaceous mass extinction.

approximately 26 million years are difficult to explain. Some have proposed biological explanations for these periodic mass extinctions and others consider them artifacts of our statistical and taxonomic analyses. Walter Alvarez proposed that the earth was periodically bombarded by asteroids, causing these mass extinctions (Figure 9-34). The drastic effects of bombardment of a planet by asteroids were observed several years ago when a series of asteroids bombarded Jupiter. Such bombardments could change the earth's climate drastically, sending debris into the atmosphere and blocking sunlight. Temperature changes would have challenged the ecological tolerances of many species. This hypothe-

sis is being tested in several ways, including a search for impact craters left by the asteroids and for altered mineral content of the rock strata where mass extinctions occurred. Atypical concentrations of the rare earth element iridium in some strata imply that this element entered the earth's atmosphere through asteroid bombardment.

Sometimes, lineages favored by species selection or effect macroevolution are unusually susceptible to mass extinction. The climatic changes produced by the hypothesized asteroid bombardments could produce selective challenges very different from those encountered at other times in the earth's history. Selective discrimination of particular biological traits by

events of mass extinction is termed **catastrophic species selection.** For example, mammals survived the end-Cretaceous mass extinction that destroyed the dinosaurs and other prominent vertebrate and invertebrate groups. Following this event, the mammals were able to use environmental resources that previously had been denied them, leading to their adaptive radiation.

Natural selection, species selection, effect macroevolution, and catastrophic species selection interact to produce the macroevolutionary trends that we see in the fossil record. The study of these interacting causal processes has made modern paleontology an active and exciting field.

Summary

Organic evolution explains the diversity of living organisms as the historical outcome of gradual change from previously existing forms. Evolutionary theory is strongly identified with Charles Robert Darwin who presented the first credible explanation for evolutionary change. Darwin derived much of the material used to construct his theory from his experiences on a five-year voyage around the world aboard the H.M.S. *Beagle.*

Darwin's evolutionary theory has five major components. Its most basic proposition is *perpetual change,* the theory that

the world is neither constant nor perpetually cycling but is steadily undergoing irreversible change. The fossil record amply demonstrates perpetual change in the continuing fluctuation of animal form and diversity following the Cambrian explosion 600 million years ago. Darwin's theory of *common descent* states that all organisms descend from a common ancestor through a branching of genealogical lineages. This theory explains morphological homologies among organisms as characteristics inherited with modification from a corresponding feature in their common evolutionary

ancestor. Patterns of homology formed by common descent with modification permit us to classify organisms according to their evolutionary relationships.

A corollary of common descent is the *multiplication of species* through evolutionary time. Allopatric speciation describes the evolution of reproductive barriers between geographically separated populations to generate new species. In some animals, especially parasitic insects that specialize on different host species, speciation may occur without geographical isolation, which is known as sympatric speciation. Adaptive radiation is the proliferation of many adaptively diverse species from a single ancestral lineage. Oceanic archipelagoes, such as the Galápagos Islands, are particularly conducive to the adaptive radiation of terrestrial organisms.

Darwin's theory of *gradualism* states that large phenotypic differences between species are produced by the accumulation through evolutionary time of many individually small changes. Gradualism is still controversial. Mutations that have large effects on the phenotype have been useful in animal breeding, leading some to dispute Darwin's claim that such mutations are not important in evolution. On a macroevolutionary perspective, punctuated equilibrium states that most evolutionary change occurs in relatively brief events of branching speciation, separated by long intervals in which little phenotypic change accumulates.

Darwin's fifth major statement is that *natural selection* is the guiding force of evolution. This principle is founded on observations that all species overproduce their kind, causing a struggle for the limited resources that support existence. Because no two organisms are exactly alike, and because variable traits are at least partially heritable, those whose hereditary endowment enhances their use of resources for survival and reproduction contribute disproportionately to the next generation. Over many generations, the sorting of variation by selection produces new species and new adaptations.

Mutations are the ultimate source of all new variation on which selection acts. Darwin's theory emphasizes that variation is produced at random with respect to the organism's needs and that differential survival and reproduction provide the direction for evolutionary change. Darwin's theory of natural selection was modified in this century by correction of his genetic errors. This modified theory became known as neo-Darwinism.

Population geneticists discovered the principles by which genetic properties of populations change through time. A particularly important discovery, known as Hardy-Weinberg equilibrium, showed that the hereditary process itself does not change the genetic composition of populations. The important sources of evolutionary change include mutation, genetic drift, nonrandom mating, migration, natural selection, and their interactions.

Neo-Darwinism, as elaborated by population genetics, formed the basis for the Evolutionary Synthesis of the 1930s and 1940s. Genetics, natural history, paleobiology, and systematics were brought together under the common goal of expanding our knowledge of Darwinian evolution. Microevolution comprises the study of genetic change within contemporary populations. These studies show that most natural populations contain enormous amounts of variation. Macroevolution comprises the study of evolutionary change on a geological timescale. Macroevolutionary studies measure rates of speciation, extinction, and changes of diversity through time. These studies have expanded Darwinian evolutionary theory to include higher-level processes that regulate rates of speciation and extinction among lineages, including species selection, effect macroevolution, and catastrophic species selection.

Review Questions

1. Briefly summarize Lamarck's concept of the evolutionary process. What is wrong with this concept?
2. What is "uniformitarianism"? How did it influence Darwin's evolutionary theory?
3. Why was the *Beagle's* journey so important to Darwin's thinking?
4. What was the key idea contained in Malthus's essay on populations that was to help Darwin formulate his theory of natural selection?
5. Explain how each of the following contributes to Darwin's evolutionary theory: fossils; geographic distributions of closely related animals; homology; animal classification.
6. How do modern evolutionists view the relationship between ontogeny and phylogeny? Explain how the observation of paedomorphosis refutes Haeckel's biogenetic law.
7. What are the important differences between the vicariant and founder effect modes of allopatric speciation?
8. What are reproductive barriers? How do premating and postmating barriers differ?
9. Under what conditions is sympatric speciation proposed?
10. What is the main evolutionary lesson provided by Darwin's finches on the Galápagos Islands?
11. How is the observation of "sporting mutations" in animal breeding used to challenge Darwin's theory of gradualism? Why did Darwin reject such mutations as having little evolutionary importance?
12. What does the theory of punctuated equilibrium state about the occurrence of speciation throughout geological time? What observation led to this theory?
13. Describe the observations and inferences that compose Darwin's theory of natural selection.
14. Identify the random and nonrandom components of Darwin's theory of natural selection.
15. Describe some recurring criticisms of Darwin's theory of natural selection. How can these be refuted?
16. It is a common but mistaken belief that because some alleles are dominant and others are recessive, the dominants will eventually replace (drive out) all the recessives. How does the Hardy-Weinberg equilibrium answer this notion?
17. Assume you are sampling a trait in animal populations; the trait is controlled by a single allelic pair A and a, and you can distinguish all three phenotypes AA, Aa, and aa (intermediate inheritance). Your sample includes:

Population	AA	Aa	aa	TOTAL
I	300	500	200	1000
II	400	400	200	1000

Calculate the distribution of phenotypes in each population as expected under Hardy-Weinberg equilibrium. Is population I in equilibrium? Is population II in equilibrium?
18. If after studying a population for a trait determined by a single pair of alleles you find that the population is not in equilibrium, what possible reasons might explain the lack of equilibrium?
19. Explain why genetic drift is more powerful in small populations.
20. Describe how the effects of genetic drift and natural selection can interact in a subdivided species.
21. Is it easier for selection to remove a deleterious recessive allele from a randomly mating population or a highly inbred population? Why?
22. Distinguish between microevolution and macroevolution.

Selected References

Avise, J. C. 1994. Molecular markers, natural history and evolution. New York, Chapman and Hall. *An exciting and readable account of the evolutionary discoveries made using molecular studies, with particular attention to conservational issues.*

Bowlby, J. 1990. Charles Darwin: a new life. New York, W. W. Norton & Company. *An interpretive biography of Charles Darwin.*

Buss, L. W. 1987. The evolution of individuality. Princeton, New Jersey, Princeton University Press. *An original and provocative thesis on the relationship between development and evolution, with examples drawn from many different animal phyla.*

Darwin, C. 1859. On the origin of species by means of natural selection, or the preservation of favoured races in the struggle for life. London, John Murray. *There were five subsequent editions by the author.*

Endler, J. A. 1986. Natural selection in the wild. Princeton, New Jersey, Princeton University Press. *A review of what we have learned about selection from studying natural populations.*

Futuyma, D. J. 1986. Evolutionary biology. Sunderland, Massachusetts, Sinauer Associates. *A very thorough introductory textbook on evolution.*

Glen, W. 1994. The mass extinction debates: how science works in a crisis. Stanford, Stanford University Press. *A discussion of mass extinction presented in the form of a debate and panel discussion among concerned scientists.*

Gould, S. J. 1977. Ontogeny and phylogeny. Cambridge, Massachusetts, Harvard University Press. *An examination of relationship between ontogeny and phylogeny with a detailed history of this subject.*

Gould, S. J. 1989. Wonderful life: the Burgess Shale and the nature of history. New York, W. W. Norton & Company. *An insightful discussion of what fossils tell us about the nature of life's evolutionary history.*

Hall, B. K. 1992. Evolutionary developmental biology. New York, Chapman and Hall. *A review of the interaction of genetics and development in evolving lineages, with particular attention to issues of heterochrony, homology, and developmental constraints on evolution.*

Hartl, D. L., and A. G. Clark. 1989. Principles of population genetics. Sunderland, Massachusetts, Sinauer Associates. *A current textbook on population genetics.*

Keller, E. F., and E. A. Lloyd (eds.). 1992. Keywords in evolutionary biology. Cambridge, Massachusetts, Harvard University Press. *Many terms used in evolutionary biology are plagued by varying or multiple meanings. Forty keywords are defined in essays by 47 contributors.*

Kohn, D. 1985. The Darwinian heritage. Princeton, New Jersey, Princeton University Press. *An edited volume on Darwin and Darwinism with contributions from many leading historians of evolutionary theory.*

Levinton, J. 1988. Genetics, paleontology and macroevolution. Cambridge, England, Cambridge University Press. *A study of the relationship between microevolution and macroevolution.*

Mayr, E. 1988. Toward a new philosophy of biology. Cambridge, Massachusetts, Harvard University Press. *A collection of essays on many aspects of evolution by a leading evolutionary biologist.*

Otte, D., and J. A. Endler. 1989. Speciation and its consequences. Sunderland, Massachusetts, Sinauer Associates. *An edited volume covering recent issues in the study of speciation, with contributions from many active evolutionary biologists.*

Ross, R. M., and W. D. Allmon. 1990. Causes of evolution: a paleontological perspective. Chicago, University of Chicago Press. *An edited volume of studies on the causes of evolution as viewed from a primarily paleontological perspective.*

The Diversity of Animal Life

10

Architectural Pattern of an Animal

New Designs for Living

Zoologists today recognize 32 phyla of multicellular animals, each phylum characterized by a distinctive body plan and biological properties that set it apart from all other phyla. All are the survivors of perhaps 100 phyla that were generated 600 million years ago during the Cambrian explosion, the most important evolutionary event in the history of animal life. Within the space of a few million years virtually all of the major body plans that we see today, together with many other novel plans that we know only from the fossil record, were established. Entering a world sparse in species and mostly free of competition, these new life forms began widespread experimentation, producing new themes in animal architecture.

Nothing since has equaled the Cambrian explosion. Later bursts of speciation that followed major extinction events produced only variations on established themes.

Once forged, a major body plan becomes a limiting determinant of body form for descendants of that ancestral line. Molluscs beget only molluscs and birds beget birds, nothing else. Despite the appearance of structural and functional adaptations for distinctive ways of life, the evolution of new forms always develops within the architectural constraints of the phylum's ancestral pattern. This is why we shall never see molluscs that fly or birds confined within a protective shell. ∎

Table 10-1	**Levels of Organization in Organismal Complexity**

1. *Protoplasmic grade of organization.* Protoplasmic organization is found in protozoa and other unicellular organisms. All life functions are confined within the boundaries of a single cell, the fundamental unit of life. Within the cell, protoplasm is differentiated into organelles capable of carrying on specialized functions.

2. *Cellular grade of organization.* Cellular organization is an aggregation of cells that are functionally differentiated. A division of labor is evident, so that some cells are concerned with, for example, reproduction, others with nutrition. Such cells have little tendency to become organized into tissues (a tissue is a group of similar cells organized to perform a common function). Some protozoan forms, such as *Volvox,* that have distinct somatic and reproductive cells might be placed at the cellular level of organization. Many authorities also place sponges at this level.

3. *Cell-tissue grade of organization.* A step beyond the preceding is the aggregation of similar cells into definite patterns or layers, thus becoming a **tissue.** Sponges are considered by some authorities to belong to this grade, although the jellyfishes and their relatives (Cnidaria) more clearly demonstrate the tissue plan. Both groups are still largely of the cellular grade of organization because most of the cells are scattered and not organized into tissues. An excellent example of a tissue in cnidarians is the **nerve net,** in which the nerve cells and their processes form a definite tissue structure, with the function of coordination.

4. *Tissue-organ grade of organization.* The aggregation of tissues into organs is a further step in advancement. Organs are usually made up of more than one kind of tissue and have a more specialized function than tissues. The first appearance of this level is in the flatworms (Platyhelminthes), in which there are a number of well-defined organs such as eyespots, proboscis, and reproductive organs. In fact, the reproductive organs are well organized into a reproductive system.

5. *Organ-system grade of organization.* When organs work together to perform some function, we have the highest level of organization—the organ system. The systems are associated with the basic body functions—circulation, respiration, digestion, and the others. The simplest animals that show this type of organization are the nemertean worms, which have a complete digestive system distinct from the circulatory system. Most animal phyla demonstrate this type of organization.

The English satirist Samuel Butler proclaimed that the human body was merely "a pair of pincers set over a bellows and a stewpan and the whole thing fixed upon stilts." Although human attitudes toward the human body are distinctly ambivalent, most people less cynical than Butler would agree that the body is a triumph of intricate, living architecture. Less obvious, perhaps, is that the architecture of humans and most other animals conforms to the same set of rules of functional anatomy. The essential uniformity of biological organization derives from the common ancestry of animals and from their basic cellular construction. Despite the vast differences in structural complexity of organisms ranging from the simplest protozoa to humans, all share an intrinsic material design and fundamental functional plan. In this chapter we will consider the limited number of body plans that underlie the apparent diversity of animal form and examine some of the common architectural themes that animals share.

THE HIERARCHICAL ORGANIZATION OF ANIMAL COMPLEXITY

Among the different unicellular and metazoan groups, we can recognize five major grades of organization (Table 10-1). Each grade is more complex than the one before, and builds on it in a hierarchical manner.

The unicellular protozoa (Protista) are the simplest animal-like organisms. These unicellular forms are nonetheless complete organisms that perform all of the basic functions of life as seen in the more complex animals. Within the confines of their cell, they show remarkable organization and division of labor, possessing distinct supportive structures, locomotor devices, fibrils, and simple sensory structures. The diversity observed among unicellular organisms is achieved by varying the architectural patterns of subcellular structures, organelles, and the cell as a whole (Chapter 12).

The **metazoa,** or multicellular animals, evolved greater structural complexity by combining cells into larger units. The metazoan cell is a specialized part of the whole organism and, unlike the protozoan cell, it is not capable of independent existence. The cells of a multicellular organism are specialized for performing the various tasks accomplished by subcellular elements in the protozoa. The simplest metazoans show the *cellular* grade of organization in which cells demonstrate division of labor but are not strongly associated to perform a specific collective function (Table 10-1). In the more complex **tissue** grade, similar cells are grouped together and perform their common functions as a highly coordinated unit. In animals of the tissue-organ grade of organization, tissues are assembled into still larger functional units called **organs.** Usually one type of tissue carries the burden of an organ's chief function, as muscle tissue does in the heart; other tissues—epithelial, connective, and nervous—perform supportive roles. The chief functional cells of an organ are called the **parenchyma** (pa-ren′ka-ma; Gr. *para,* beside, + *enchyma,* infusion). The supportive tissues are its **stroma** (Gr. bedding). For instance, in the vertebrate pancreas, the secreting cells are the parenchyma; the capsule and connective tissue framework represent the stroma.

Most metazoa (nemerteans and all more structurally complex phyla) have an additional level of complexity in which different organs operate together as **organ systems.** Eleven different kinds of organ systems are ob-

served in metazoans: skeletal, muscular, integumentary, digestive, respiratory, circulatory, excretory, nervous, endocrine, immune, and reproductive. The great evolutionary diversity of these organ systems is covered in Chapters 15 through 31.

COMPLEXITY AND BODY SIZE

The most complex grades of metazoan organization permit and to some extent even promote the evolution of large body size (Figure 10-1). Large size has several important physical and ecological consequences for the organism. As animals become larger, the body surface increases much more slowly than body volume. This happens because surface area increases as the square of body length (length^2), whereas volume (and therefore mass) increases as the cube of body length (length^3). This disparity soon makes it difficult for activities served from the surface to provision the mass of cells within. In the course of animal evolution this problem was solved by developing internal transport systems to shuttle nutrients, gases, and waste products between the cells and the external environment. This arrangement also buffers the organism against environmental fluctuations.

The tendency for body size to increase within lines of descent is known as "Cope's law of phyletic increase," named after nineteenth-century American paleontologist and naturalist Edward Drinker Cope. Cope noted that lineages begin with small organisms that give rise to larger and ultimately to giant forms. These frequently become extinct, providing opportunities for new lineages which in turn evolve larger forms. Cope's rule holds well for nonflying vertebrates and many invertebrate groups, even though Cope's Lamarckian explanation for the trend—that organisms evolved from an inner urge to attain a higher state of being (and larger size)—was preposterous. Exceptions to Cope's rule are few (but the insects are a particularly large one).

Larger size permits a more efficient use of metabolic energy. A large mammal uses more oxygen than a small mammal, but the cost of maintaining body temperature is less per gram of weight for the large mammal than for a small one. Large animals also can move themselves at less energy cost than can small animals. A large mammal uses more oxygen in running than a small mammal, but the energy cost of moving 1 g of its body over a given distance is much less for a large mammal than for a small one (Figure 10-2). Another important consequence of large body size is greater protection from predation. For all of these reasons, ecological opportunities of larger animals are very different from those of small ones. The extensive adaptive radiations observed in the taxa of large animals are detailed in subsequent chapters.

EXTRACELLULAR COMPONENTS OF THE METAZOAN BODY

In addition to the hierarchically arranged cellular structures discussed in the preceding text, the metazoan animal contains two important noncellular components: the body fluids and the extracellular structural elements. In all eumetazoans, the body fluids are subdivided into two fluid "compartments": those that occupy the **intracellular space,** within the body's cells, and those that occupy the **extracellular space,** outside the cells. In animals with closed vascular systems (such as segmented worms and vertebrates), the extracellular fluids are subdivided further into the **blood plasma** (the fluid portion of the blood outside the cells; blood cells are really part of the intracellular compartment) and **interstitial fluid.** The interstitial fluid, also called tissue fluid, occupies the space surrounding the cells. Many invertebrates have open blood systems, however, with no true separation of blood plasma from interstitial fluid. These relationships are explored further in Chapter 34.

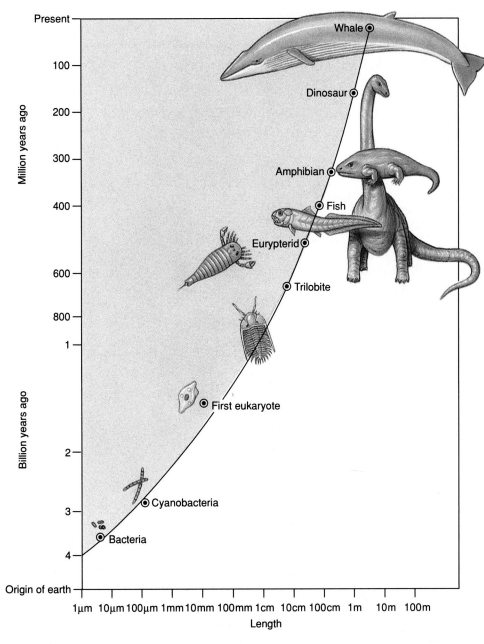

Figure 10-1
Graph showing the evolution of size (length) increase in organisms at different periods of life on earth. Note that both scales are logarithmic.

If we were to remove all the specialized cells and body fluids from the interior of the body, we would be left with the third element of the animal body: extracellular structural elements. This is the supportive material of the organism, including loose connective tissue (especially well developed in vertebrates but present in all metazoa), cartilage (molluscs and chordates), bone (vertebrates), and cuticle (arthropods, nematodes, annelids, and others). These elements provide mechanical stability and protection (Chapter 32). In some instances, they act also as a depot of materials for exchange, and serve as a medium for extracellular reactions. The diversity of extracellular skeletal elements characteristic of the different groups of animals is described in Chapters 16 through 31.

TYPES OF TISSUES

A **tissue** is a group of similar cells (together with associated cell products) specialized for the performance of a common function. The study of tissues is called **histology** (Gr. *histos,* tissue, + *logos,* discourse). All cells in metazoan animals take part in the formation of tissues. Sometimes the cells of a tissue may be of several kinds, and some tissues have a great many intercellular materials.

> The term "intercellular," meaning "between cells," should not be confused with the term "intracellular," meaning "within cells."

During embryonic development, the germ layers become differentiated into four kinds of tissues. These are epithelial, connective, muscular, and nervous tissues (Figure 10-3). This is a surprisingly short list of only four basic tissue types that are able to meet the diverse requirements of animal life.

EPITHELIAL TISSUE

An **epithelium** (pl., epithelia) is a sheet of cells that covers an external or internal surface. On the outside of

Figure 10-2
Net cost of running for mammals of various sizes. Each point represents the cost (measured in rate of oxygen consumption) of moving 1 g of body over 1 km. The cost decreases with increasing body size.

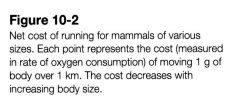

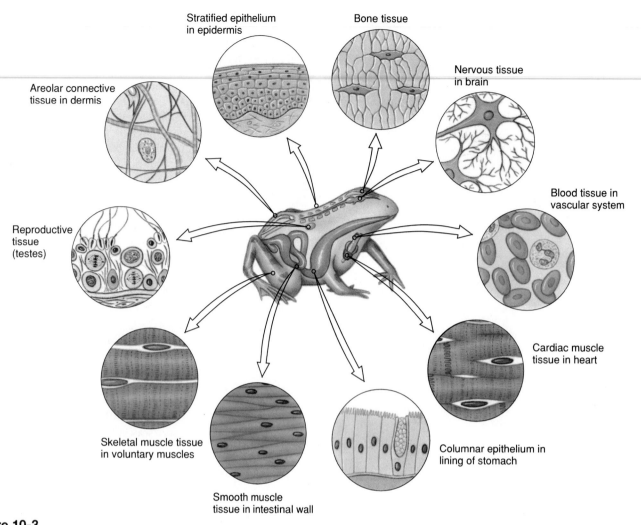

Stratified epithelium in epidermis

Bone tissue

Nervous tissue in brain

Areolar connective tissue in dermis

Blood tissue in vascular system

Reproductive tissue (testes)

Cardiac muscle tissue in heart

Skeletal muscle tissue in voluntary muscles

Columnar epithelium in lining of stomach

Smooth muscle tissue in intestinal wall

Figure 10-3

Types of tissues in a vertebrate, showing examples of where different tissues are located in a frog.

the body, the epithelium forms a protective covering. Inside, the epithelium lines all the organs of the body cavity, as well as ducts and passageways through which various materials and secretions move. On many surfaces the epithelial cells are often modified into glands that produce lubricating mucus or specialized products such as hormones or enzymes.

Epithelia are classified on the basis of cell form and number of cell layers. Simple epithelia (Figure 10-4) are found in all metazoan animals, while stratified epithelia (Figure 10-5) are mostly restricted to the vertebrates. All types of epithelia are supported by an underlying basement membrane, which is a condensation of the ground substance of connective tissue. Blood vessels never penetrate

into epithelial tissues, so they depend on the diffusion of oxygen and nutrients from the underlying tissues.

CONNECTIVE TISSUE

Connective tissues are a diverse group of tissues that serve various binding and supportive functions. They are so widespread in the body that the removal of other tissues would still leave the complete form of the body clearly apparent. Connective tissue is made up of relatively few cells, a great many extracellular fibers, and a **ground substance** (also called **matrix**), in which the fibers are embedded. We recognize several different types of connective tissue. **Connective tissue proper** includes both **loose connective tissue,** composed

of fibers and fixed and wandering cells suspended in a syrupy ground substance, and **dense connective tissue,** such as tendons and ligaments, that is composed largely of densely packed fibers (Figure 10-6). Much of the fibrous tissue of connective tissue is composed of **collagen** (Gr. *kolla,* glue, + *genos,* descent), a protein material of great tensile strength. Collagen is the most abundant protein in the animal kingdom, found in animal bodies wherever both flexibility and resistance to stretching are required. The connective tissue of invertebrates, as in vertebrates, consists of cells, fibers, and ground substance, but it is not as elaborately developed.

Other types of connective tissue include **blood, lymph,** and **tissue fluid** (collectively considered vascular

Figure 10-4
Types of simple epithelium.

Simple
squamous
epithelial   Basement   Free
cell   membrane   Nucleus   surface

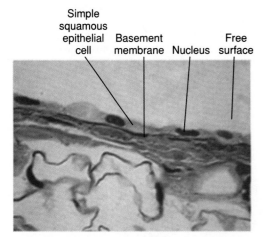

Simple squamous epithelium, composed of flattened cells that form a continuous delicate lining of blood capillaries, lungs, and other surfaces where it permits the passive diffusion of gases and tissue fluids into and out of cavities.

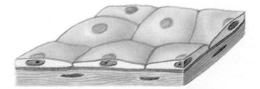

Simple squamous epithelium

Simple cuboidal   Basement   Lumen
epithelial cell   membrane   (free space)

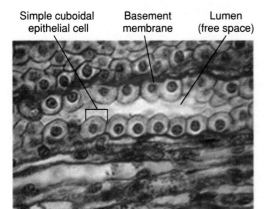

Simple cuboidal epithelium is composed of short, boxlike cells. Cuboidal epithelium usually lines small ducts and tubules, such as those of the kidney and salivary glands, and may have active secretory or absorptive functions.

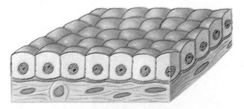

Simple cuboidal epithelium

Simple columnar epithelium resembles cuboidal epithelium but the cells are taller and usually have elongate nuclei. This type of epithelium is found in highly absorptive surfaces such as the intestinal tract of most animals. The cells often bear minute, fingerlike projections called microvilli that greatly increase the absorptive surface. In some organs, such as the female reproductive tract, the cells are ciliated.

Basement   Epithelial   Microvilli on
membrane   cells   cell surface   Nuclei

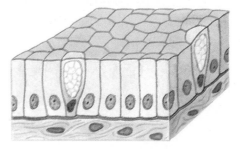

Simple columnar epithelium

Figure 10-5
Types of stratified epithelium.

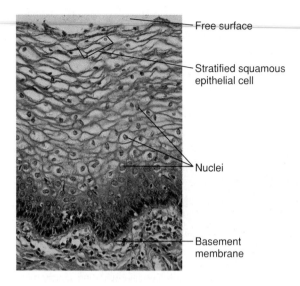

— Free surface

— Stratified squamous epithelial cell

— Nuclei

— Basement membrane

Stratified squamous epithelium consists of two to many layers of cells adapted to withstand mild mechanical abrasion. The basal layer of cells undergoes continuous mitotic divisions, producing cells that are pushed toward the surface where they are sloughed off and replaced by new cells from beneath. This type of epithelium lines the oral cavity, esophagus, and anal canal of many vertebrates, and the vagina of mammals.

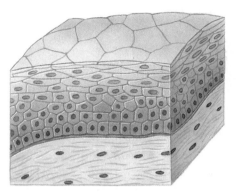

Stratified squamous epithelium

Free surface

Basement membrane

Transitional epithelium is a type of stratified epithelium specialized to accommodate great stretching. This type of epithelium is found in the urinary tract and bladder of vertebrates. In the relaxed state it appears to be four or five cell layers thick, but when stretched out it appears to have only two or three layers of extremely flattened cells.

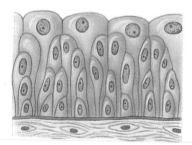

Connective   Nucleus   Transitional
tissue   epithelial cell

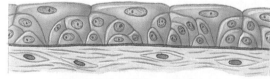

Transitional epithelium—unstretched

Transitional epithelium—Stretched

Figure 10-6
Types of connective tissue.

Nucleus   Collagen fiber   Elastic fiber

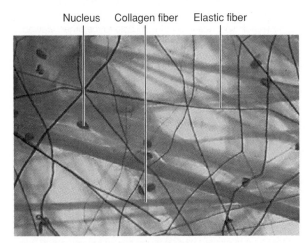

Loose connective tissue, also called areolar connective tissue, is the "packing material" of the body that anchors blood vessels, nerves, and body organs. It contains fibroblasts that synthesize the fibers and ground substance of connective tissue and wandering macrophages that phagocytize pathogens or damaged cells. The different fiber types include strong collagen fibers (thick and red in micrograph) and thin elastic fibers (black and branching in micrograph) formed of the protein elastin. Adipose (fat) tissue is considered a type of loose connective tissue.

Nucleus   Fibers

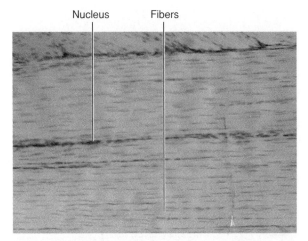

Dense connective tissue forms tendon, ligaments, and fasciae (fa'sha), the latter arranged as sheets or bands of tissue surrounding skeletal muscle. In tendon (shown here) the collagenous fibers are extremely long and tightly packed together.

Chondrocyte   Lacuna   Matrix

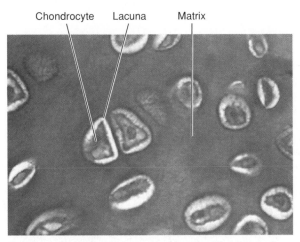

Cartilage is a vertebrate connective tissue composed of a firm gel ground substance (matrix) containing cells (chondrocytes) living in small pockets called lacunae, and collagen or elastic fibers (depending on the type of cartilage). In hyaline cartilage shown here, both collagen fibers and matrix are stained uniformly purple and cannot be distinguished one from the other. Because cartilage lacks a blood supply, all nutrients and waste materials must diffuse through the ground substance from surrounding tissues.

Central   Osteocytes   Mineralized
canal   in lacunae   matrix

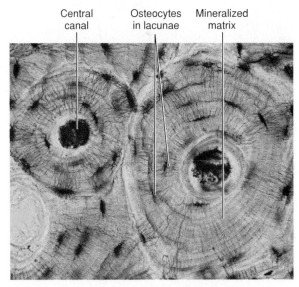

Bone, the strongest of vertebrate connective tissues, contains mineralized collagen fibers. Small pockets (lacunae) within the matrix contain bone cells, called osteocytes. The osteocytes communicate with blood vessels that penetrate into bone by means of a tiny network of channels called canaliculi. Unlike cartilage, bone undergoes remodeling during an animal's life, and can repair itself following even extensive damage.

Figure 10-7
Types of muscle tissue.

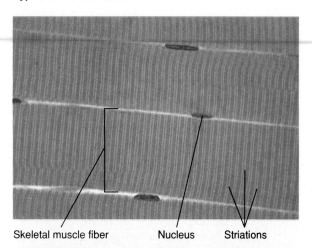

Skeletal muscle fiber   Nucleus   Striations

Skeletal muscle is a type of striated muscle found in both invertebrates and vertebrates. It is composed of extremely long, cylindrical fibers, which are multinucleate cells that may reach from one end of the muscle to the other. Viewed through the light microscope, the cells appear to have a series of stripes, called striations, running across them. Skeletal muscle is called voluntary muscle (in vertebrates) because it contracts when stimulated by nerves under conscious cerebral control.

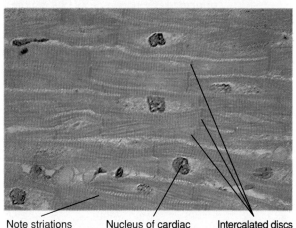

Note striations   Nucleus of cardiac muscle cell   Intercalated discs (special junctions between cells)

Cardiac muscle is another type of striated muscle found only in the vertebrate heart. The cells are much shorter than those of skeletal muscle and have only one nucleus per cell (uninucleate). Cardiac muscle tissue is a branching network of fibers with individual cells interconnected by junctional complexes called intercalated discs. Cardiac muscle is considered involuntary muscle because it does not require nerve activity to stimulate contraction. Instead, heart rate is controlled by specialized pacemaker cells located in the heart itself. However, autonomic nerves from the brain may alter pacemaker activity.

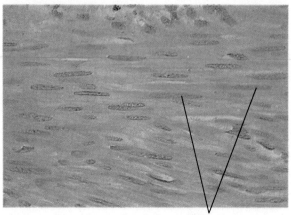

Nuclei of smooth muscle cells

Smooth muscle is nonstriated muscle found in both invertebrates and vertebrates. Smooth muscle cells are long, tapering strands, each containing a single nucleus. Smooth muscle is the most common type of muscle in invertebrates in which it serves as body wall musculature and lines ducts and sphincters. In vertebrates, smooth muscle lines the walls of blood vessels and surrounds internal organs such as the intestine and uterus. It is called involuntary muscle in vertebrates since its contraction is usually not consciously controlled.

tissue), composed of distinctive cells in a watery ground substance, the plasma. Vascular tissue lacks fibers under normal conditions. **Cartilage** is a semirigid form of connective tissue with closely packed fibers embedded in a gel-like ground substance (matrix). **Bone** is a calcified connective tissue containing calcium salts organized around collagen fibers (see Figure 10-6).

MUSCULAR TISSUE

Muscle is the most common tissue in the body of most animals. It is made up of elongated cells of fibers specialized for contraction. It originates (with few exceptions) from the mesoderm, and its unit is the cell or **muscle fiber.** The unspecialized cytoplasm of muscles is called **sarcoplasm,** and the contractile elements within the fiber are the **myofibrils.** Structurally, muscles are either **smooth** (fibers unstriped) or **striated** (fibers cross-striped) (Figure 10-7).

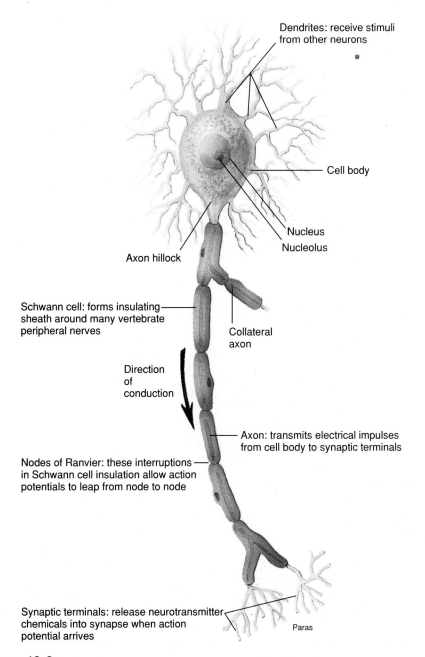

Dendrites: receive stimuli from other neurons

Cell body

Nucleus

Nucleolus

Axon hillock

Schwann cell: forms insulating sheath around many vertebrate peripheral nerves

Collateral axon

Direction of conduction

Axon: transmits electrical impulses from cell body to synaptic terminals

Nodes of Ranvier: these interruptions in Schwann cell insulation allow action potentials to leap from node to node

Synaptic terminals: release neurotransmitter chemicals into synapse when action potential arrives

Paras

Figure 10-8

Functional anatomy of a neuron. From the nucleated body, or **soma,** extend one or more **dendrites** (Gr. *dendron,* tree), which receive electrical impulses from receptors or other nerve cells, and a single **axon** that carries impulses away from the cell body to other nerve cells or to an effector organ. The axon is often called a **nerve fiber.** Nerves are separated from other nerves or from effector organs by specialized junctions called synapses.

NERVOUS TISSUE

Nervous tissue is specialized for the reception of stimuli and the conduction of impulses from one region to another. The two basic types of cells in nervous tissue are **neurons** (Gr. nerve), the basic functional unit of the nervous system, and **neuroglia** (nu-rog'le-a; Gr. nerve, + *glia,* glue), a variety of nonnervous cells that insulate neuron membranes and serve various supportive functions. The functional anatomy of a typical nerve cell is diagrammed in Figure 10-8.

ANIMAL BODY PLANS

As pointed out in the prologue to this chapter, while the diversity of animal body form is enormous, the possibilities for designs for living are constrained by ancestral history. Nevertheless, animals are shaped by their particular habitat and way of life. A worm that adopts a parasitic life in a vertebrate's intestine will look and function very differently from a free-living member of the same group. Yet both will share the distinguishing hallmarks of the phylum.

The major grades in body architecture are, in sequence, multicellularity, bilateral symmetry, the "tube-within-a-tube" plan, and, finally, the eucoelomate (true coelom) body plan. These advancements with their principal alternatives can be arranged in a branching pattern as shown in Figure 10-9.

ANIMAL SYMMETRY

Symmetry applies to the arrangement or correspondence of body parts with reference to some axis of the body. Most animals have some kind of symmetry. Those that do not, such as many sponges that are irregular in shape, are said to be asymmetrical.

Spherical symmetry means that any plane passing through the center divides the body into equivalent, or mirrored, halves (Figure 10-10, *top left*). This type of symmetry is found chiefly among some of the protozoa and is rare in animals. Spherical forms are best suited for floating and rolling.

Radial symmetry (Figure 10-10, *top right center*) applies to forms that can be divided into similar halves by more than two planes passing through one main axis. These are the tubular, vase, or bowl shapes found in some sponges and in the hydras, jellyfish, sea urchins, and the like, in which one end of the longitudinal axis is usually the mouth. A variant form is **biradial symmetry** in which, because of some part that is paired rather than radial, only two planes passing through the longitudinal axis will produce mirrored halves. Sea walnuts (phylum Ctenophora, p. 272), which are more or less globular in form but have a pair of tentacles, are an example. Radial

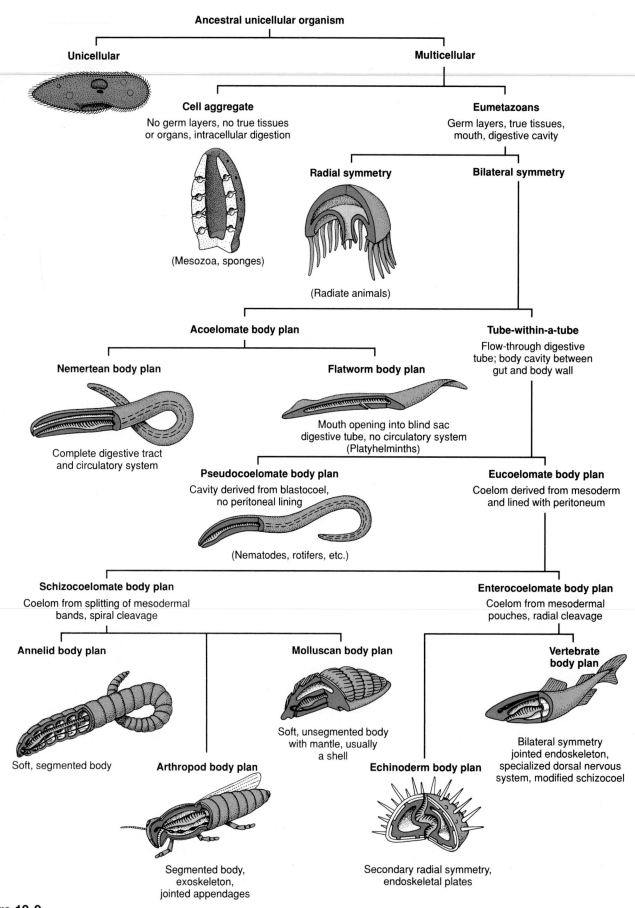

Ancestral unicellular organism

Unicellular

Multicellular

Cell aggregate
No germ layers, no true tissues
or organs, intracellular digestion

Eumetazoans
Germ layers, true tissues,
mouth, digestive cavity

(Mesozoa, sponges)

Radial symmetry

Bilateral symmetry

(Radiate animals)

Acoelomate body plan

Tube-within-a-tube
Flow-through digestive
tube; body cavity between
gut and body wall

Nemertean body plan

Flatworm body plan

Mouth opening into blind sac
digestive tube, no circulatory system
(Platyhelminths)

Complete digestive tract
and circulatory system

Pseudocoelomate body plan
Cavity derived from blastocoel,
no peritoneal lining

Eucoelomate body plan
Coelom derived from mesoderm
and lined with peritoneum

(Nematodes, rotifers, etc.)

Schizocoelomate body plan
Coelom from splitting of mesodermal
bands, spiral cleavage

Enterocoelomate body plan
Coelom from mesodermal
pouches, radial cleavage

Annelid body plan

Molluscan body plan

**Vertebrate
body plan**

Soft, segmented body

Arthropod body plan

Soft, unsegmented body
with mantle, usually
a shell

Bilateral symmetry
jointed endoskeleton,
specialized dorsal nervous
system, modified schizocoel

Echinoderm body plan

Segmented body,
exoskeleton,
jointed appendages

Secondary radial symmetry,
endoskeletal plates

Figure 10-9

Architectural patterns of animals. These basic body plans have been variously modified during evolutionary descent to fit animals to a great variety of habitats. Ectoderm is shown in gray, mesoderm in red, and endoderm in yellow.

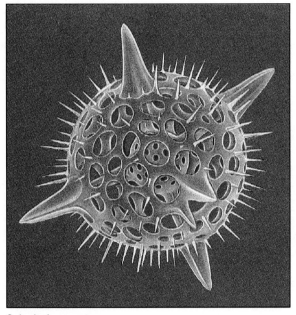

Spherical symmetry

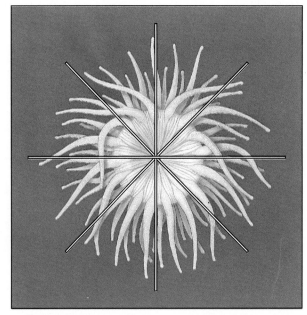

Radial symmetry

Bilateral symmetry

Figure 10-10

Animal symmetry. Illustrated are animals showing spherical, radial, and bilateral symmetry.

and biradial animals are usually sessile, freely floating, or weakly swimming. Radial animals, with no front or back end, can interact with their environment in all directions—an advantage to a sessile form with feeding structures arranged to snare prey approaching from any direction.

The two phyla that are primarily radial, Cnidaria and Ctenophora, are called the **Radiata.** The echinoderms (sea stars and their kin) are primarily bilateral animals (their larvae are bilateral) that have become secondarily radial as adults.

Bilateral symmetry applies to animals that can be divided along a sagittal plane into two mirrored portions— right and left halves (Figure 10-10, *bottom*). The appearance of bilateral symmetry in animal evolution was a major advancement, because bilateral animals are much better fitted for directional (forward) movement than are radially symmetrical animals. Bilateral animals are collectively called the **Bilateria.** Bilateral symmetry is strongly associated with **cephalization,** discussed in the following text.

Let us review some of the convenient terms used for locating regions of animal bodies (see Figure 10-11). **Anterior** is used to designate the head end; **posterior,** the opposite or tail end; **dorsal,** the back side; and **ventral,** the front or belly side. **Medial** refers to the midline of the body; **lateral,** to the sides. **Distal** parts are farther from the middle of the body than some point of reference; **proximal** parts are nearer. **Pectoral** refers to the chest region or the area supported by the forelegs, and **pelvic** refers to the hip region or the area supported by the hind legs. A **frontal plane** (sometimes called coronal plane) divides a bilateral body into dorsal and ventral halves by running through the anteroposterior axis and the right-left axis at right angles to the **sagittal plane,** the plane dividing an animal into right and left halves. A **transverse plane** (also called a cross section) would cut through a dorsoventral and a right-left axis at right angles to both the sagittal and frontal planes and would result in anterior and posterior portions (Figure 10-11).

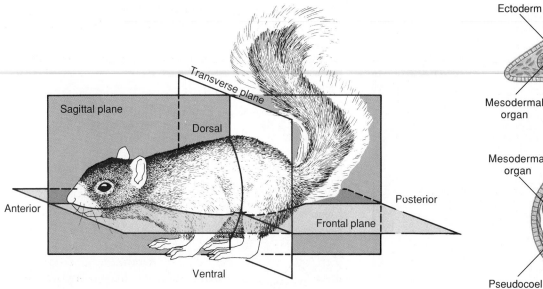

Figure 10-11

The planes of symmetry as illustrated by a bilaterally symmetrical animal.

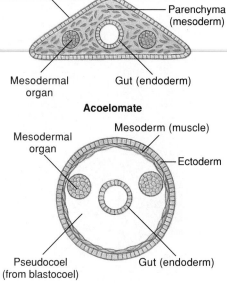

Figure 10-12

Acoelomate, pseudocoelomate, and eucoelomate body plans.

Body Cavities

The bilateral animals can be grouped according to their body-cavity type or lack of body cavity (see Figure 10-9). In higher animals the main body cavity is the **coelom,** a fluid-filled space that surrounds the gut. The evolution of the coelom, like bilateral symmetry, was a major advancement in animal architecture because it provides a "tube-within-a-tube" arrangement (Figure 10-9) that allows much greater body flexibility than is possible in animals lacking an internal body cavity. The coelom also provides space for visceral organs and permits greater size and complexity by exposing more cells to surface exchange. The fluid-filled coelom additionally serves as a hydrostatic skeleton in some forms, especially many worms, aiding in such activities as movement and burrowing.

As shown in Figure 10-9, the presence or absence of a coelom is a key determinant in the evolutionary advancement of the Bilateria.

Acoelomate Bilateria

The flatworms (phylum Platyhelminthes) and ribbon worms (phylum Nemertea) have *no body cavity* surrounding the gut (Figure 10-12, *top*); they are "acoelomate" (Gr. *a,* without, + *koilōma,* cavity). The region between the ectodermal epidermis and the endodermal digestive tract is completely filled with mesoderm in the form of a spongy mass of space-filling cells called **parenchyma.** The parenchyma is derived from an in-wandering of ectodermal cells from the general surface of the early embryo. In at least some acoelomates, the parenchymal cells are cell bodies of muscle cells (see p. 282).

Pseudocoelomate Bilateria

Nematodes and several other phyla have a cavity surrounding the gut called a **pseudocoel** ("false cavity"), and its possessors have a tube-within-a-tube arrangement (Figure 10-12, *center*). The pseudocoel is best defined by a negative attribute, the *absence* of a **peritoneum,** a thin cellular membrane derived from mesoderm that, in animals with a true coelom, lines the body cavity. The pseudocoel is derived from the blastocoel of the embryo and represents a persistent blastocoel.

Eucoelomate Bilateria

The remainder of the bilateral animals possess a **true coelom** lined with mesodermal peritoneum (Figure 10-12, *bottom*). The true coelom arises within the mesoderm itself and may be formed by one of two methods, **schizocoelous** or **enterocoelous** (Figure 10-13), or by modifying these methods. The two terms are descriptive, for *schizo* comes from the Greek *schizein,* to split; *entero* is derived from the Greek *enteron,* meaning gut; and *coelous* comes from the Greek *koilos,* meaning hollow or cavity. In schizocoelous formation the coelom arises, as the word implies, from the splitting of mesodermal bands that originate from cells in the blastopore region. (Mesoderm is one of the three primary germ layers that appear very early in the development of all bilateral animals, lying between the innermost endoderm and outermost ectoderm, p. 120 and Figure 7-25). In enterocoelous formation the coelom comes from pouches of the archenteron, or primitive gut.

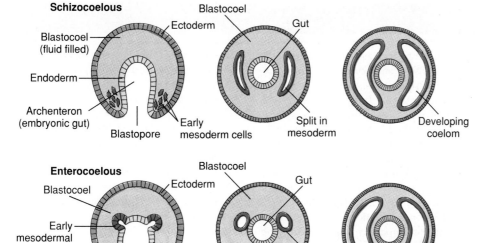

Figure 10-13

Types of mesoderm and coelom formation. In schizocoelous formation, the mesoderm originates from the wall of the archenteron near the blastopore and proliferates into a band of tissue that splits to form the coelom. In enterocoelous formation, most mesoderm originates as a series of pouches from the archenteron; these pinch off and enlarge to form the coelom. In both formations, the coeloms expand to obliterate the blastocoel.

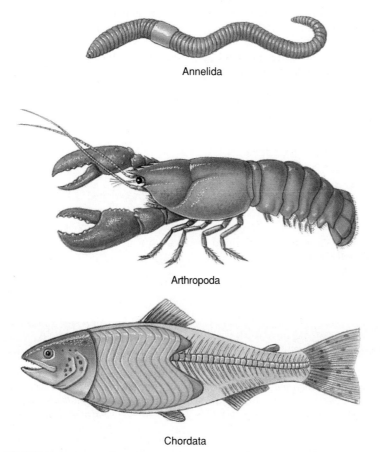

Figure 10-14

Segmented phyla. These three phyla have all made use of an important principle in nature: metamerism, or repetition of structural units. Segmentation in annelids and arthropods is homologous, but chordates may have derived their segmentation independently. Segmentation brings more varied specialization because segments, especially in arthropods, have become modified for different functions.

Once development is complete, the results of schizocoelous and enterocoelous formations are indistinguishable. Both processes give rise to a true coelom lined with a mesodermal peritoneum (Gr. *peritonaios,* stretched around) and having mesenteries in which the visceral organs are suspended.

METAMERISM (SEGMENTATION)

Metamerism is the serial repetition of similar body segments along the longitudinal axis of the body. Each segment is called a **metamere,** or **somite.** In forms such as the earthworm and other annelids, in which metamerism is most clearly represented, the segmental arrangement includes both external and internal structures of several systems. There is repetition of muscles, blood vessels, nerves, and the setae of locomotion. Some other organs, such as those of sex, may be repeated in only a few somites. In more evolved animals much of the segmental arrangement has become obscure.

True metamerism is found in only three phyla: Annelida, Arthropoda, and Chordata (Figure 10-14), although superficial segmentation of the ectoderm and the body wall may be found among many diverse groups of animals.

CEPHALIZATION

The differentiation of a head is called **cephalization** and is found chiefly in bilaterally symmetrical animals. The concentration of nervous tissue and sense organs in the head bestows obvious advantages to an animal moving through its environment head first. This is the most efficient positioning of instruments for sensing the environment and responding to it. Usually the mouth of the animal is located on the head as well, since so much of an animal's activity is concerned with procuring food. Cephalization is always accompanied by differentiation along an anteroposterior axis **(polarity).** Polarity usually involves gradients of activities between limits, such as between the anterior and posterior ends.

Summary

From the relatively simple organisms that made up the beginnings of life on earth, animal evolution has progressed through a history of ever more intricately organized forms. Organelles became integrated into cells, cells into tissues, tissues into organs, and organs into systems. Whereas a protozoan carries out all life functions within the confines of a single cell, an advanced multicellular animal is an organization of subordinate units that are united at successive levels.

During the course of evolution the upper limit of body size in animal lineages has tended to increase; this led inevitably to the division of labor among body parts and increased complexity.

The metazoan body consists of cells, most of which are functionally specialized; body fluids, divided into intracellular and extracellular fluid compartments; and extracellular structural elements, which are fibrous or formless elements that serve various structural functions in the extracellular space. The cells of metazoa develop into various tissues made up of similar cells performing common functions. The basic tissue types are epithelial, connective, muscular, and nervous. Tissues are organized into larger functional units called organs, and organs are associated to form systems.

Every organism has an inherited body plan that may be described in terms of broadly inclusive characteristics, such as symmetry, presence or absence of body cavities, partitioning of body fluids, presence or absence of segmentation, degree of cephalization, and type of nervous system.

Review Questions

1. Name the five levels of organization in organismal complexity and explain how each successive level is more complex than the one preceding it.
2. Can you suggest why, during the evolution of separate animal lineages, there has been a tendency for the maximum body size to increase? Why should it be inevitable that complexity tends to increase along with body size?
3. What is the meaning of the terms "parenchyma" and "stroma" as they relate to body organs?
4. Body fluids of eumetazoan animals are separated into fluid "compartments." Name these compartments and explain how compartmentalization may differ in animals with open and closed circulatory systems.
5. What are the four major types of tissues in the body of a metazoan?
6. How would you distinguish between simple and stratified epithelium? What characteristic of stratified epithelium

might explain why it, rather than simple epithelium, is found lining the oral cavity, esophagus, and vagina?
7. What are the three elements present in all connective tissue? Give some examples of the different types of connective tissue.
8. What are three different kinds of muscle found among animals? Explain how each is specialized for particular functions.
9. Describe the principal structural and functional features of a neuron.
10. Match the animal group with its body plan:

____ Unicellular	a. Nematode
____ Cell aggregate	b. Vertebrate
____ Blind sac, acoelomate	c. Protozoan
____ Tube-within-a-tube, pseudocoelomate	d. Flatworm
____ Tube-within-a-tube, eucoelomate	e. Sponge
	f. Arthropod
	g. Nemertean

11. Distinguish among spherical, radial, biradial, and bilateral symmetry.
12. Use the following terms to identify regions on your body and on the body of a frog: anterior, posterior, dorsal, ventral, lateral, distal, proximal.
13. How would frontal, sagittal, and transverse planes divide your body?
14. What is meant by metamerism? Name three phyla showing metamerism.

Selected References

Bonner, J. T. 1988. The evolution of complexity by means of natural selection. Princeton, New Jersey, Princeton University Press. *Levels of complexity in organisms and how size affects complexity.*

Caplan, A. J. 1984. Cartilage. Sci. Am. **251**:84–94 (Oct.). *Structure, aging, and development of vertebrate cartilage.*

Grene, M. 1987. Hierarchies in biology. Am. Sci. **75**:504–510 (Sept.–Oct.). *The term "hierarchy" is used in many different senses in biology. The author points out that current evolutionary theory carries the hierarchical concept beyond the Darwinian restriction to the two levels of gene and organism.*

Kessel, R. G., and R. H. Kardon. 1979. Tissues and organs: a text-atlas of scanning electron microscopy. San Francisco, W. H. Freeman & Co. *Collection of excellent scanning electron micrographs with text.*

McMahon, T. A., and J. T. Bonner. 1983. On size and life. New York, Scientific American Books, Inc. *A well-illustrated book about size and scale in the living world; clear examples and explanations.*

Welsch, U., and V. Storch. 1976. Comparative animal cytology and histology. London, Sidgwick & Jackson. *Comparative histology with good treatment of invertebrates.*

11

Classification and Phylogeny of Animals

Order in Diversity

Zoologists have named more than 1.5 million species of animals, and thousands more are described each year. Some zoologists estimate that the species named so far constitute less than 20% of all living animals and less than 1% of all those that have existed in the past.

Despite its magnitude, the diversity of animals is not without limits. There are many conceivable forms that do not exist in nature as our myths of minotaurs and winged horses demonstrate. The characteristic features of humans and cattle never occur together in nature as they do in the mythical minotaur. Nor do the characteristic wings of birds and bodies of horses occur naturally as they do in the mythical horse, Pegasus. Humans, cattle, birds, and horses are distinct groups of animals, yet they do share some important features, including vertebrae and homeothermy, that separate them from even more dissimilar forms such as insects and flatworms.

All human cultures classify their familiar animals according to various patterns in animal diversity. These classifications have many purposes. Animals may be classified in some societies according to their usefulness or destructiveness to human endeavors. Others may group animals according to their roles in mythology. Biologists group animals according to their evolutionary relationships as revealed by ordered patterns in their sharing of homologous features. This classification is called a "natural system" because it reflects relationships that exist among animals in nature, outside the context of human activity. The systematic zoologist has three major goals: to discover all species of animals, to reconstruct their evolutionary relationships, and then to classify them accordingly. ■

Darwin's theory of common descent (Chapter 1) is the underlying principle that guides our search for order in the diversity of animal life. Our science of taxonomy ("arrangement law") produces a formal system for naming and classifying species that reflects this order. Animals that have very recent common ancestry share many features in common and are grouped most closely in our taxonomic classification; dissimilar animals that share only very ancient common ancestry are placed in different taxonomic groups except at the "highest" or most inclusive levels of taxonomy. Taxonomy is part of the broader science of systematics, or comparative biology, in which everything that is known about animals is used to understand their evolutionary relationships. The study of taxonomy predates evolutionary biology, however, and many taxonomic practices are remnants of a pre-evolutionary world view. Adjusting our taxonomic system to accommodate evolution has produced many problems and controversies. Taxonomy has reached an unusually active and controversial point in its development in which several alternative taxonomic systems are competing for use. To understand this controversy, it is necessary first to review the history of animal taxonomy.

LINNAEUS AND THE DEVELOPMENT OF CLASSIFICATION

The Greek philosopher and biologist Aristotle was the first to classify organisms on the basis of their structural similarities. Following the Renaissance in Europe, the English naturalist John Ray (1627 to 1705) introduced a more comprehensive system of classification and a new concept of species. The flowering of systematics in the eighteenth century culminated in the work of Carolus Linnaeus (1707 to 1778; Figure 11-1), who gave us our current scheme of classification.

Figure 11-1
Carolus Linnaeus (1707 to 1778). This portrait was made of Linnaeus at age 68, three years before his death.

Linnaeus was a Swedish botanist at the University of Uppsala. He had a great talent for collecting and classifying objects, especially flowers. Linnaeus produced an extensive system of classification for both plants and animals. This scheme, published in his great work, *Systema Naturae,* used morphology (the comparative study of organismal form) for arranging specimens in collections. He divided the animal kingdom into species and gave each one a distinctive name. He grouped species into genera, genera into orders, and orders into classes. Because his knowledge of animals was limited, his lower categories, such as the genera, were very broad and included animals that are only distantly related. Much of his classification has been drastically altered, but the basic principle of his scheme is still followed.

Linnaeus's scheme of arranging organisms into an ascending series of groups of ever-increasing inclusiveness is the **hierarchical system** of classification. The major categories, or **taxa** (sing., **taxon**), into which organisms are grouped were given one of several standard **taxonomic ranks** to indicate the general degree of inclusiveness of the group. The hierarchy of taxonomic ranks has been expanded considerably since Linnaeus's time (Table 11-1). It now includes seven mandatory ranks for the animal kingdom, in descending series: kingdom, phylum, class, order, family, genus, and species. All organisms being classified must be placed into at least seven taxa, one at each of the mandatory ranks. Taxonomists have the option of subdividing these seven ranks even further to recognize more than seven taxa (superclass, subclass, infraclass, superorder, suborder, etc.) for any particular group of organisms. In all, more than 30 taxonomic ranks are recognized. For very large and complex groups, such as the fishes and insects, these additional ranks are needed to express different degrees of evolutionary divergence. Unfortunately, they also contribute complexity to the system.

Linnaeus's system for naming species is known as **binomial nomenclature.** Each species has a latinized name composed of two words (hence binomial) written in italics (underlined if handwritten or typed). The first word is the name of the **genus,** written with a capital initial letter; the second word is the **species epithet** which is peculiar to the species within the genus and is written with a small initial letter (see Table 11-1). The genus name is always a noun, and the species epithet is usually an adjective that must agree in gender with the genus. For instance, the scientific name of the common robin is *Turdus migratorius* (L. *turdus,* thrush; *migratorius,* of the migratory habit). The species epithet never stands alone; the complete binomial must be used to name a species. Names of genera must refer only to single groups of organisms; the same name cannot be given to two different genera of animals. The same species epithet may be used in different genera, however, to denote different and unrelated species. For example, the scientific name of the white-breasted nuthatch is *Sitta carolinensis.* The species epithet *"carolinensis"* is used in other genera, including *Parus carolinensis* (Carolina chickadee) and *Anolis carolinensis*

Table 11-1	Examples of Taxonomic Categories to Which Representative Animals Belong			
	Human	**Gorilla**	**Southern Leopard Frog**	**Katydid**
Kingdom	Animalia	Animalia	Animalia	Animalia
Phylum	Chordata	Chordata	Chordata	Arthropoda
Subphylum	Vertebrata	Vertebrata	Vertebrata	Uniramia
Class	Mammalia	Mammalia	Amphibia	Insecta
Subclass	Eutheria	Eutheria	—	Pterygota
Order	Primates	Primates	Anura	Orthoptera
Suborder	Anthropoidea	Anthropoidea	—	Ensifera
Family	Hominidae	Pongidae	Ranidae	Tettigoniidae
Subfamily	—	—	Raninae	Phaneropterinae
Genus	*Homo*	*Gorilla*	*Rana*	*Scudderia*
Species	*Homo sapiens*	*Gorilla gorilla*	*Rana sphenocephala*	*Scudderia furcata*
Subspecies	—	—	—	*Scudderia furcata furcata*

The hierarchical system of classification applied to four species (human, gorilla, Southern leopard frog, and katydid). Higher taxa generally are more inclusive than lower-level taxa, although taxa at two different levels may be equivalent in content (for example, the family Hominidae contains only the genus *Homo,* making the content of these taxa equivalent, whereas the family Pongidae contains the genera *Gorilla, Pan,* and *Pongo,* making it more inclusive than any of these genera). Closely-related species are united at a lower point in the hierarchy than are distantly related species. For example, humans and gorillas are united at the level of the suborder (Anthropoidea) and above; they are united with the Southern leopard frog at the subphylum level (Vertebrata) and with the katydid at the level of the kingdom (Animalia).

(green anole, a lizard) to mean "of Carolina." All ranks above the species are designated using uninomial nouns, written with a capital initial letter.

There are times when a species is divided into subspecies, in which case a trinomial nomenclature is employed (see katydid example, Table 11-1). Thus to distinguish the southern form of the robin from the eastern robin, the scientific term *Turdus migratorius achrustera* (duller color) is employed for the southern type. The generic, specific, and subspecific names are printed in italics (underlined if handwritten or typed). The subspecies name may be a repetition of the species epithet. Formal recognition of subspecies has lost popularity among taxonomists because the boundaries between subspecies are rarely distinct. Recognition of subspecies is usually based on one or a few superficial characters and does not denote an evolutionarily distinct unit. Subspecies designations, therefore, should not be taken too seriously.

TAXONOMIC CHARACTERS AND PHYLOGENETIC RECONSTRUCTION

A major goal of systematics is to reconstruct the evolutionary tree or **phylogeny** that relates all extant and extinct species. This is done by studying organismal features, formally called **characters,** that vary among species. A character is any feature that the taxonomist uses to study variation within and among species. We find potentially useful taxonomic characters in morphological, chromosomal, and molecular features (see p. 201). Taxonomists find characters by observing patterns of similarity among organisms. If two organisms possess similar features, this indicates that they may have inherited these features from a common ancestor. Character similarity that results from common ancestry is called **homology** (see Chapter 9). Similarity does not always reflect common ancestry, however. Independent evolutionary origin of similar features on different lineages produces patterns of similarity among organisms that do not reflect common descent; this occurrence complicates the work of taxonomists. Character similarity that misrepresents common descent is called nonhomologous similarity or **homoplasy.**

USING CHARACTER VARIATION TO RECONSTRUCT PHYLOGENY

To reconstruct the phylogeny of a group using characters that vary among its members, the first step is to determine which variant form of each character was present in the common ancestor of the entire group. This character state is called **ancestral** for the group as a whole. We presume that all other variant forms of the character arose later within the group, and these are called evolutionarily **derived character states.** The **polarity** of a character refers to the ancestral/descendant relationships among its different states. For example, if we consider as a character the dentition of amniotic vertebrates (reptiles, birds, and mammals), presence versus absence of teeth in the jaws constitute two different character states. Teeth are absent from birds but present in the other amniotes. To evaluate the polarity of this character, we must determine which character state, presence or absence of teeth, characterized the most recent common ancestor of amniotes and which state was derived subsequently within the amniotes.

The method that we use to examine the polarity of a variable character is called **outgroup comparison.** We consult an additional group of organisms, called the **outgroup,** that is phylogenetically close but not within the group being studied. We infer that any character state found both within the group being studied and in the outgroup is ancestral for the study group. The amphibians and different groups of bony fishes constitute appropriate outgroups to the amniotes for polarizing variation in the dentition of amniotes. Teeth are usually present in amphibians and bony fishes; therefore, we infer that presence of teeth is ancestral for amniotes and absence of teeth is derived. The polarity of this character indicates that teeth were lost in the ancestral lineage of all modern birds. Polarity of characters is evaluated most effectively when several different outgroups are used. All character states found in the study group that are absent from appropriate outgroups are considered derived.

The organisms or species that share derived character states form subsets within the group called **clades** (Gr. *klados,* branch). A derived character shared by the members of a clade is formally called a **synapomorphy** (Gr. *synapsis,* joining together, + *morphē,* form) of that clade. Taxonomists use synapomorphies as evidence of homology to infer that a particular group of organisms forms a clade. Within the amniotes, absence of teeth and presence of feathers are synapomorphies that identify the birds as a clade. A clade corresponds to a unit of evolutionary common descent; it includes all descendants of a particular ancestral lineage. The pattern formed

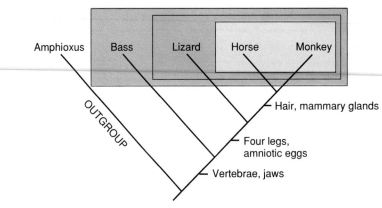

Figure 11-2

The cladogram as a nested hierarchy of taxa. Amphioxus is the outgroup, and the study group comprises four vertebrates (bass, lizard, horse, and monkey). Four characters that vary among vertebrates are used to generate a simple cladogram: presence versus absence of four legs, amniotic eggs, hair, and mammary glands. For all four characters, absence is the ancestral state in vertebrates because this is the condition found in the outgroup, Amphioxus; for each character, presence is the derived state in vertebrates. Because they share presence of four legs and amniotic eggs as synapomorphies, the lizard, horse, and monkey form a clade relative to the bass. This clade is subdivided further by two synapomorphies (presence of hair and mammary glands) that unite the horse and monkey relative to the lizard. We know from comparisons involving even more distantly related animals that presence of vertebrae and jaws constitute synapomorphies of the vertebrates and that Amphioxus, which lacks these features, falls outside the vertebrate clade.

by the derived states of all characters within our study group will take the form of a **nested hierarchy** of clades within clades. The goal is to identify all of the different clades nested within the study group, which would give a complete account of the patterns of common descent among the species in the group.

Character states ancestral for a taxon are often called **plesiomorphic** for that taxon and the sharing of ancestral states among organisms is termed **symplesiomorphy.** Unlike synapomorphies, however, symplesiomorphies do not provide useful information on the nesting of clades within clades. In the example given above, we found that the presence of teeth in jaws was plesiomorphic for amniotes. If we grouped together the mammalian and reptilian groups, which possess teeth, to the exclusion of birds, we would not obtain a valid clade. Birds also descend from all common ancestors of reptiles and mammals and must be included within any clade that includes all reptiles and mammals. Errors in determining polarity of characters therefore clearly can produce errors in inference of phylogeny. It is important to note, however, that character states

that are plesiomorphic at one taxonomic level can be apomorphic at a more inclusive level. For example, the presence of jaws bearing teeth is a synapomorphy of gnathostome vertebrates (p. 494), a group that includes the amniotes plus amphibians, bony fishes, and cartilaginous fishes (although teeth have been lost in birds and some other gnathostomes). The goal of phylogenetic analysis therefore can be restated as one of finding the appropriate taxonomic level at which any given character state is a synapomorphy. The character state is then used at that level to identify a clade.

The nested hierarchy of clades is presented as a branching diagram called a **cladogram** (Figure 11-2; see also Figure 9-15). Taxonomists often make a technical distinction between a cladogram and a **phylogenetic tree.** The branches of a cladogram are only a formal device for indicating the nested hierarchy of clades within clades. The cladogram is not strictly equivalent to a phylogenetic tree, whose branches represent real lineages that occurred in the evolutionary past. To obtain a phylogenetic tree, we must add to the cladogram important additional information concerning ancestors, the durations of evolutionary lineages, or the amounts of evolutionary change that occurred on the lineages. Because the structure of a cladogram is congruent with that of the corresponding phylogenetic tree, however, the cladogram is often used as a first approximation of the phylogenetic tree.

SOURCES OF PHYLOGENETIC INFORMATION

We find the characters used to construct cladograms in comparative morphology (including embryology), comparative cytology, and comparative biochemistry. **Comparative morphology** examines the varying shapes and sizes of organismal structures, including their developmental origins. As we will see in later chapters, the variable structures of skull bones, limb bones, and integument (scales, hair, feathers) are particularly important for reconstructing the phylogeny of vertebrates. Comparative morphology uses specimens obtained from both living organisms and fossilized remains. **Comparative biochemistry** uses the sequences of amino acids in proteins and the sequences of nucleotides in nucleic acids (see Chapter 8) to identify variable characters for constructing a cladogram (Figure 11-3). Recent studies have shown that comparative biochemistry can be applied to some fossils in addition to living organisms. **Comparative cytology** uses variation in the numbers, shapes, and sizes of chromosomes and their parts (Chapter 4) to obtain variable characters for constructing cladograms. Comparative cytology is used almost exclusively on living rather than fossilized organisms.

To add the evolutionary timescale necessary for producing a phylogenetic tree, we must consult the fossil record. We can look for the earliest appearance in fossils of derived morphological characters to estimate the ages of clades distinguished by those characters. The age of a fossil showing the derived characters of a particular clade is determined by radioactive dating (p. 157). An example of a phylogenetic tree constructed using these methods is Figure 27-1, p. 500.

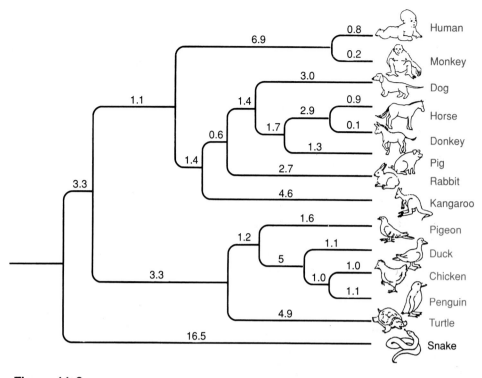

Figure 11-3

A phylogenetic tree of representative amniotes based on inferred base substitutions in the gene that encodes the respiratory protein, cytochrome c. Numbers on the branches are the estimated numbers of mutational changes that occurred in this gene along the different evolutionary lineages.

We can use comparative biochemical data to estimate the ages of different lineages on a phylogenetic tree. Some protein and DNA sequences undergo approximately linear rates of divergence through evolutionary time. The age of the most recent common ancestor of two species is therefore proportional to the differences measured between their proteins and DNA sequences. We calibrate the evolution of proteins and DNA sequences by measuring their divergence between species whose most recent common ancestor has been dated using fossils. We then use the molecular evolutionary calibration to estimate the ages of other branches on the phylogenetic tree.

THEORIES OF TAXONOMY

A theory of taxonomy establishes the principles that we use to recognize and to rank taxonomic groups. There are two currently popular theories of taxonomy, (1) traditional evolutionary taxonomy and (2) phylogenetic systematics (cladistics). Both are based on evolutionary principles. We will see, however, that these two theories differ on how evolutionary principles are used. These differences have important implications for how we use a taxonomy to study the evolutionary process.

The relationship between a taxonomic group and a phylogenetic tree or cladogram is important for both of these theories. This relationship can take one of three forms: **monophyly, paraphyly,** or **polyphyly** (Figure 11-4). A taxon is monophyletic if it includes the most recent common ancestor of the group and all descendants of that ancestor (see Figure 11-4A). A taxon is paraphyletic if it includes the most recent common ancestor of all members of a group and some but not all of the descendants of that ancestor (see Figure 11-4B). A taxon is polyphyletic if it does not include the most recent common ancestor of all members of a group; this requires that the group has had at least two separate evolutionary origins, usually requiring independent evolutionary acquisition

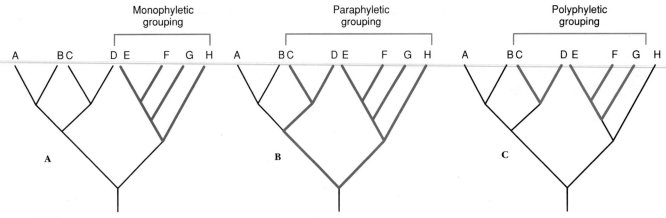

Figure 11-4

Relationships between phylogeny and taxonomic groups illustrated for a hypothetical phylogeny of eight species (A through H). **A,** *Monophyly*—a monophyletic group contains the most recent common ancestor of all members of the group and all of its descendants. **B,** *Paraphyly*—a paraphyletic group contains the most recent common ancestor of all members of the group and some but not all of its descendants. **C,** *Polyphyly*—a polyphyletic group does not contain the most recent common ancestor of all members of the group, thereby requiring that the group have at least two separate phylogenetic origins.

of similar features (see Figure 11-4C). Both evolutionary and cladistic taxonomy accept monophyletic groups and reject polyphyletic groups in their classifications. They differ on the acceptance of paraphyletic groups, however, and this difference has important evolutionary implications.

TRADITIONAL EVOLUTIONARY TAXONOMY

Traditional **evolutionary taxonomy** incorporates two different evolutionary principles for recognizing and ranking higher taxa: (1) common descent and (2) amount of adaptive evolutionary change, as shown on a phylogenetic tree. Evolutionary taxa must have a single evolutionary origin, and must show unique adaptive features.

The mammalian paleontologist George Gaylord Simpson (Figure 11-5) was highly influential in developing and formalizing the procedures of evolutionary taxonomy. According to Simpson, a particular branch on the evolutionary tree is given the status of a higher taxon if it represents a distinct **adaptive zone.** Simpson describes an adaptive zone as "a characteristic reaction and mutual relationship between environment and organism, a way of life and not a place where life is led." By entering a new adaptive zone through a fundamental change in organismal structure and

behavior, an evolving population can use environmental resources in a completely new way.

A taxon that comprises a distinct adaptive zone is termed a **grade.** Simpson gives the example of penguins as a distinct adaptive zone within birds. The lineage immediately ancestral to all penguins underwent fundamental changes in the form of the body and wings to permit a switch from aerial to aquatic locomotion (Figure 11-6). Aquatic birds that can fly both in the air and underwater are somewhat intermediate in habitat,

Figure 11-5

George Gaylord Simpson (1902 to 1984) formulated the principles of evolutionary taxonomy.

American Museum of Natural History, Neg. #334101.

morphology, and behavior between aerial and aquatic adaptive zones. Nonetheless, the obvious modifications of the wings and body of penguins for swimming represent a new grade of organization. The penguins are therefore recognized as a distinct taxon within the birds, the family Spheniscidae. The broader the adaptive zone when fully occupied by a group of organisms, the higher the rank that the corresponding taxon is given.

Evolutionary taxa may be either monophyletic or paraphyletic. Recognition of paraphyletic taxa requires, however, that our taxonomies distort patterns of common descent. An evolutionary taxonomy of the anthropoid primates provides a good example (Figure 11-7). This taxonomy places humans (genus *Homo*) and their immediate fossil ancestors in the family Hominidae, and it places the chimpanzees (genus *Pan*), gorillas (genus *Gorilla*), and orangutans (genus *Pongo*) in the family Pongidae. However, the pongid genera *Pan* and *Gorilla* share more recent common ancestry with the Hominidae than they do with the remaining pongid genus, *Pongo*. This makes the family Pongidae paraphyletic because it does not include humans, who also descend from the most recent common ancestor of all pongids (see Figure 11-7). Evolutionary taxonomists nonetheless recognize the pongid genera as a single, family-level

A   B

Figure 11-6

A, Penguin. **B,** Diving petrel. Penguins (avian family Spheniscidae) were recognized by George G. Simpson as a distinct adaptive zone within birds because of their adaptations for submarine flight. Simpson believed that the adaptive zone ancestral to the penguins resembled that of the diving petrels, which display adaptations for combined aerial and aquatic flight. The adaptive zones of penguins and diving petrels are distinct enough to be recognized taxonomically as different families within a common order (Ciconiiformes).

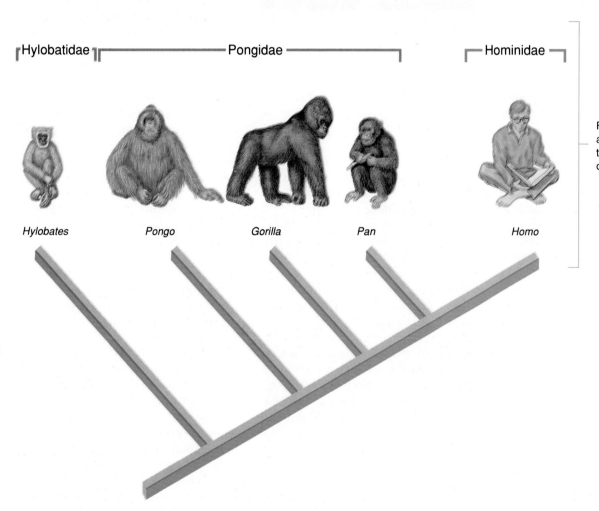

Family level classification according to evolutionary taxonomy, based principally on unique adaptive zones

Figure 11-7

Phylogeny and family-level classification of anthropoid primates. Evolutionary taxonomy groups the genera *Gorilla, Pan,* and *Pongo* into the paraphyletic family Pongidae because they share the same adaptive zone or grade of organization. Humans (genus *Homo*) are phylogenetically closer to *Gorilla* and *Pan* than any of these genera are to *Pongo,* but humans are placed in a separate family (Hominidae) because they represent a new grade of organization. Cladistic taxonomy requires either that the family Pongidae be split into three monophyletic family-level taxa or that *Homo* be included in the Pongidae and Hominidae discontinued as a family-level taxon. The gibbons (genus *Hylobates*) form the monophyletic family Hylobatidae, which is recognized in both evolutionary and cladistic classifications.

grade of arboreal, herbivorous primates having limited mental capacity; in other words, they show the same family-level adaptive zone. Humans are terrestrial, omnivorous primates who have greatly expanded mental and cultural attributes, thereby comprising a distinct adaptive zone at the taxonomic level of the family. Unfortunately, if we want our taxa to constitute adaptive zones, we compromise our ability to present common descent in the most straightforward taxonomic manner.

Traditional evolutionary taxonomy has been challenged from two opposite directions. One challenge states that because phylogenetic trees can be very difficult to obtain, it is impractical to base our taxonomic system on common descent and adaptive evolution. We are told that our taxonomy should represent a more easily measured feature, the overall similarity of organisms evaluated without regard to phylogeny. This principle is known as **phenetic taxonomy.** Phenetic taxonomy did not have a strong impact on animal classification, and scientific interest in this approach is in decline. Despite the difficulties of reconstructing phylogeny, zoologists still consider this endeavor a central goal of their systematic work, and they are unwilling to compromise this goal for purposes of methodological simplicity.

PHYLOGENETIC SYSTEMATICS/CLADISTICS

A second and stronger challenge to evolutionary taxonomy is one known as **phylogenetic systematics** or **cladistics.** As the first name implies, this approach emphasizes the criterion of common descent and, as the second name implies, it is based on the cladogram of the group being classified. This approach to taxonomy was first proposed in 1950 by the German entomologist, Willi Hennig (Figure 11-8) and therefore is sometimes called "Hennigian systematics." All taxa recognized by Hennig's cladistic system must be monophyletic. We saw above how the evolutionary taxonomists' recognition of the primate families Hominidae and

Figure 11-8
Willi Hennig (1913 to 1976), German entomologist who formulated the principles of phylogenetic systematics/cladistics.

Pongidae distorts genealogical relationships to emphasize the adaptive uniqueness of the Hominidae. Because the most recent common ancestor of the paraphyletic family Pongidae is also an ancestor of the Hominidae, recognition of the Pongidae is incompatible with cladistic taxonomy.

The disagreement on the validity of paraphyletic groups may seem trivial at first, but its important consequences become clear when we discuss evolution. For example, the claims that amphibians evolved from bony fish, that birds evolved from reptiles, or that humans evolved from apes may be made by an evolutionary taxonomist but are meaningless to a cladist. We imply by these statements that a descendant group (amphibians, birds, or humans) evolved from part of an ancestral group (bony fish, reptiles, and apes, respectively) to which the descendant does not belong. This usage automatically makes the ancestral group paraphyletic, and indeed bony fish, reptiles, and apes as traditionally recognized are paraphyletic groups. How are such paraphyletic groups recognized? Do they share distinguishing features that are not shared by the descendant group?

Paraphyletic groups are usually defined in a negative manner. They are distinguished only by features ab-

sent from a particular descendant group, because any traits that they share from their common ancestry are present also in the excluded descendants (unless secondarily lost). For example, apes are those "higher" primates that are not humans. Likewise, fish are those vertebrates that lack the distinguishing characteristics of tetrapods (amphibians and amniotes). What does it mean then to say that humans evolved from apes? To the evolutionary taxonomist, apes and humans are different adaptive zones or grades of organization; to say that humans evolved from apes states that bipedal, tailless organisms of large brain capacity evolved from arboreal, tailed organisms of smaller brain capacity. To the cladist, however, the statement that humans evolved from apes says essentially that humans evolved from something that they are not, a trivial statement that contains no useful information. To the cladist, any statement that a particular monophyletic group descends from a paraphyletic one is nothing more than a claim that the descendant group evolved from something that it is not. Extinct ancestral groups are always paraphyletic because they exclude a descendant that shares their most recent common ancestor. Although many such groups have been recognized by evolutionary taxonomists, none are recognized by cladists.

Zoologists often construct paraphyletic groups because they are interested in a terminal, monophyletic group (such as humans), and they want to ask questions about its ancestry. It is often convenient to lump together organisms whose features are considered approximately equally distant from the group of interest and to ignore their own unique features. It is significant in this regard that humans have never been placed in a paraphyletic group, whereas most other organisms have been. Apes, reptiles, fishes, and invertebrates are all terms that traditionally designate paraphyletic groups formed by combining various "side branches" that are found when human ancestry is traced backward through the tree of life. Such a taxonomy can give the erroneous impression that all of evolution is a progressive march toward humanity or, within other groups, a progressive march toward whatever species humans designate as being the most "advanced." Such thinking is a relic of pre-Darwinian views that there is a linear scale of nature having "primitive" creatures at the bottom and humans near the top just below angels. Darwin's theory of common descent states, however, that evolution is a branching process with no linear scale of increasing perfection along a single branch. Nearly every branch will contain its own combination of ancestral and derived features. In cladistics, this perspective is emphasized by recognizing taxa only by their own unique properties and not grouping organisms only because they lack the unique properties found in related groups.

Fortunately, there is a convenient way to express the common descent of groups without constructing paraphyletic taxa. It is done by finding what is called the **sister group** of the taxon of interest to us. Two different monophyletic taxa are termed sister groups if they share common ancestry with each other more recently than either one does with any other taxa. The sister group of humans appears to be the chimpanzees, with the gorillas forming a sister group to the humans and chimpanzees combined. The orangutans are the sister group of the clade that includes humans, chimpanzees, and gorillas, and the gibbons form the sister group of the clade that includes the orangutans, chimpanzees, gorillas, and humans (see Figure 11-7).

CURRENT STATE OF ANIMAL TAXONOMY

The formal taxonomy of animals that we use today was established using the principles of evolutionary systematics and has been revised recently in part using the principles of cladistics. Introduction of cladistic principles initially has the effect of replacing paraphyletic groups with monophyletic subgroups while leaving the remaining taxonomy mostly unchanged. A thorough revision of taxonomy along cladistic principles, however, will require profound changes, one of which almost certainly will be abandonment of the Linnaean ranks. In our coverage of animal taxonomy, we will try as much as possible to use taxa that are monophyletic and therefore consistent with the criteria of both evolutionary and cladistic taxonomy. We will continue, however, to use Linnaean ranks. In some cases in which commonly recognized taxa are clearly paraphyletic grades, we will note this fact and suggest alternative taxonomic schemes that contain only monophyletic taxa.

In discussing patterns of descent, we will avoid statements such as "mammals evolved from reptiles" that imply paraphyly and will instead specify appropriate sister-group relationships. We will avoid referring to groups of organisms as being primitive, advanced, specialized, or generalized because all groups of animals contain combinations of primitive, advanced, specialized, and generalized features; these terms are best restricted to describing specific characteristics and not the group as a whole.

Revision of taxonomy according to cladistic principles can cause confusion. In addition to new taxonomic names, we see old ones used in unfamiliar ways. For example, the cladistic use of "bony fishes" includes amphibians and amniotes (including reptilian groups, birds, and mammals) in addition to the finned, aquatic animals that we normally group under the term "fish." The cladistic use of "reptiles" includes birds in addition to the snakes, lizards, turtles, and crocodilians; however, it excludes some fossil forms, such as the synapsids, that were traditionally placed in the Reptilia (see Chapters 29 through 31). Taxonomists must be very careful to specify when using these seemingly familiar terms whether the traditional evolutionary taxa or newer cladistic taxa are being discussed.

SPECIES

While discussing Darwin's book, *The Origin of Species,* in 1859, Thomas Henry Huxley asked, "In the first place, what is a species? The question is a simple one, but the right answer to it is hard to find, even if we appeal to those who should know most about it." We have used the term "species" so far as if it had a simple and unambiguous meaning. Actually, Huxley's commentary is as valid today as it was 135 years ago. Our concepts of species have become more sophisticated, but the diversity of different concepts and the disagreements surrounding their use are as evident now as they were in Darwin's time.

CRITERIA FOR RECOGNITION OF SPECIES

Despite widespread disagreement about the nature of species, biologists have repeatedly designated certain criteria as being important to their identification of species. First, the criterion of **common descent** is central to nearly all modern concepts of species. The members of a species must trace their ancestry to a common ancestral population although not necessarily to a single pair of parents. Species are thus historical entities. A second criterion is that species must be the **smallest distinct groupings** of organisms sharing patterns of ancestry and descent; otherwise, it would be difficult to separate species from higher taxa whose members also share common descent. Morphological characters traditionally have been important in identifying such groupings, but chromosomal and molecular characters increasingly are being used for this purpose. A third important criterion is that of **reproductive community,** which applies only to sexually reproducing organisms; members of a species must form a reproductive community that excludes members of other species. This criterion is very important to many modern species concepts.

TYPOLOGICAL SPECIES CONCEPT

Before Darwin, a species was considered a distinct and immutable entity. Species were defined by fixed, essential features (usually morphological)

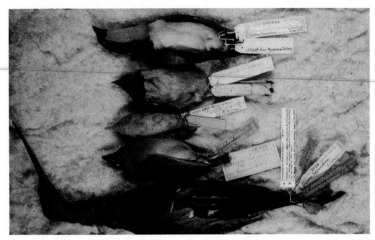

Figure 11-9
Specimens of birds from the Smithsonian Institution (Washington D.C.), including birds originally collected by John J. Audubon, Theodore Roosevelt, John Gould, and Charles Darwin.

that were interpreted as a divinely created pattern or archetype. This practice constitutes the **typological (or morphological) species concept.** Scientists recognized species formally by designating a **type specimen** that was labeled and deposited in a museum to represent the ideal form or morphology for the species (Figure 11-9). When scientists obtained additional specimens and wanted to assign them to a species, the type specimens of described species were consulted. The new specimens were assigned to a previously described species if they possessed the essential features of its type specimen. Small differences from the type specimen were considered accidental imperfections. Large differences from existing type specimens would lead the scientist to describe a new species with the designation of its own type specimen. In this manner, the living world was categorized into species.

Evolutionists discarded the typological species concept, but some of its traditions remain. Scientists still name species by describing type specimens deposited in museums. Organismal morphology is likewise still important in recognizing species; however, species are no longer viewed as classes defined by the possession of certain morphological features. The basis of the evolutionary world view is that species are historical entities whose properties are subject always to change. The variation that we observe among organisms within a species is

not the imperfect manifestation of an eternal "type"; the type itself is only an abstraction taken from the very real and important variation present within the species. The type is at best an average form that will change as organismal variation is sorted through time by natural selection. The type specimen serves only as a guide to the general kinds of morphological features that we may expect to find in the species as we observe it today.

The person who first describes a type specimen and publishes the name of a species is called the authority. This person's name and date of publication are often written after the species name. Thus, *Didelphis marsupialis* Linnaeus, 1758, tells us that Linnaeus was the first person to publish the species name of the opossum. The authority citation is not part of the scientific name but rather is an abbreviated bibliographical reference. Sometimes, the generic status of a species is revised following its initial description. In this case, the name of the authority is presented in parentheses.

BIOLOGICAL SPECIES CONCEPT

The most influential concept of species inspired by Darwinian evolutionary theory is the **biological species concept** formulated by Theodosius Dobzhansky and Ernst Mayr. This concept solidified during the evolutionary synthesis of the

1930s and 1940s from earlier ideas, and it has been refined and reworded several times since then. In 1983, Mayr stated the biological species concept as follows: *"A species is a reproductive community of populations (reproductively isolated from others) that occupies a specific niche in nature."* Note that the species is identified here according to the reproductive properties of populations, not according to possession of any specific organismal characteristics. The species is an **interbreeding population** of individuals having common descent and sharing intergrading characteristics. The study of populational variation in organismal morphology, chromosomal structure, and molecular genetic features will be very useful for evaluating the geographical boundaries of interbreeding populations in nature. The criterion of the "niche" (Chapter 40) recognizes that members of a reproductive community are expected also to have common ecological properties.

Because a reproductive community should maintain genetic cohesiveness, we expect that organismal variation will be relatively smooth and continuous within species and discontinuous between them. Although the biological species is based on reproductive properties of populations rather than organismal morphology, morphology nonetheless can help us to diagnose biological species. Sometimes species status can be evaluated directly by conducting breeding experiments. This is practical only in a minority of cases, however, and our decisions regarding species membership are usually made by studying character variation. Variation in molecular characters is very useful for identifying geographical boundaries of reproductive communities. Molecular studies have revealed the presence of cryptic or **sibling species** (p. 164) that are too similar in morphology to be diagnosed as separate species by morphological characters alone.

ALTERNATIVES TO THE BIOLOGICAL SPECIES CONCEPT

The biological species concept has received strong criticism. To understand why, we must keep in mind several important facts about species. First, a species has dimensions in space and time, and this usually creates problems for locating discrete boundaries between species. Second, we view the species both as a unit of evolution and as a rank in the taxonomic hierarchy. There are sometimes important conflicts between these roles, as we will see in the following text. A third problem is that according to the biological species concept, species do not exist in groups of organisms that reproduce only asexually. It is common systematic practice, however, to describe species in all groups of organisms, regardless of whether reproduction is sexual or asexual.

The Species in Space and Time

Any species has a distribution through space, known as its **geographic range,** and a distribution through time, known as its **evolutionary duration.** Species differ greatly from each other in both of these dimensions. Species having very large geographic ranges or worldwide distributions are called **cosmopolitan,** whereas those with very restricted geographic distributions are called **endemic.** If a species were restricted to a single point in space and time, we would have little difficulty recognizing it, and nearly every species concept would lead us to the same decision. We have little difficulty distinguishing from each other the different species of animals that we can find living in our local park or woods. However, when we compare the local population of a species to similar but not identical populations located hundreds of miles away, it may be hard to determine whether these populations represent parts of a single species or different species (Figure 11-10).

Throughout the evolutionary duration of a species, its geographic range may change many times. A geographic range may be either continuous or disjunct, the latter having breaks within it where the species is absent. Suppose that we find two similar but not identical populations living 300 miles apart with no related populations between

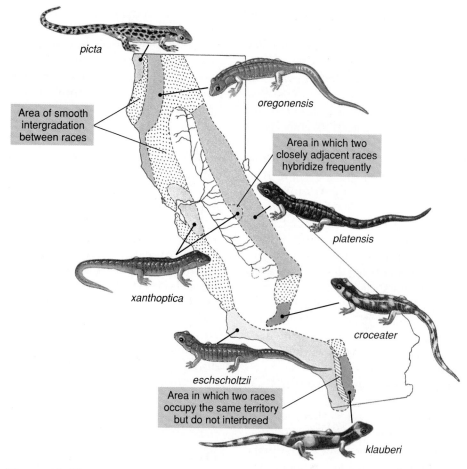

Figure 11-10

Geographic variation of color patterns in the salamander genus *Ensatina*. The species status of these populations has puzzled taxonomists for generations and continues to do so. Current taxonomy recognizes only a single species (*Ensatina eschscholtzii*) divided into subspecies as shown. Hybridization is evident between most adjacent populations, but studies of variation in proteins and DNA show large amounts of genetic divergence among populations. Furthermore, populations of the subspecies *eschscholtzii* and *klauberi* can overlap geographically without interbreeding.

them. Are we observing a single species with a disjunct distribution or two different but closely related species? Suppose that these populations have been separated historically for 50,000 years. Is this enough time for them to have evolved separate reproductive communities, or can we still view them as being part of the same reproductive community? The answers to such questions are very hard to find. Much of the disagreement among different species concepts relates to solving these problems.

Evolutionary Species Concept

The time dimension described above creates obvious problems for the biological species concept. How do we

assign fossil specimens to biological species that are recognized today? If we trace a lineage backward through time, how far must we go before we have crossed a species boundary? If we could follow the unbroken genealogical chain of populations backward through time to the point where two sister species converge on their common ancestor, we would need to cross at least one species boundary somewhere. It would be very hard to decide, however, where to draw a sharp line between the two species.

To address this problem, the **evolutionary species concept** was proposed by Simpson in the 1940s to add an evolutionary time dimension to the biological species concept. This concept persists in a modified form today.

A current definition of the evolutionary species is *a single lineage of ancestor-descendant populations that maintains its identity from other such lineages and that has its own evolutionary tendencies and historical fate.* Note that the criterion of common descent is retained here in the need for a lineage to have a distinct historical identity. The same kinds of diagnostic features discussed previously (p. 208) will be relevant for identifying evolutionary species, although in most cases only morphological features will be available from fossils. Unlike the biological species concept, the evolutionary species concept applies both to sexually and asexually reproducing forms. As long as continuity of diagnostic features is maintained by the evolving lineage, it will be recognized as a species. Abrupt changes in diagnostic features will mark the boundaries of different species in evolutionary time.

Phylogenetic Species Concept

The last concept that we present is the **phylogenetic species concept.** The phylogenetic species concept is defined as an *irreducible (basal) grouping of organisms diagnosably distinct from other such groupings and within which there is a parental pattern of ancestry and descent.* This concept emphasizes most strongly the criterion of common descent. Both asexual and sexual groups are covered.

The phylogenetic species is a strictly monophyletic unit. The main difference in practice between the evolutionary and phylogenetic species concepts is that the latter emphasizes recognizing as separate species the smallest groupings of organisms that have undergone independent evolutionary change. The evolutionary species concept would group into a single species geographically disjunct populations that demonstrate some genetic divergence but are judged similar in their "evolutionary tendencies," whereas the phylogenetic species concept would treat them as separate species. In general, a greater number of species would be described using the phylogenetic species concept than any other species concept, and many

taxonomists consider it impractical for this reason. For strict adherence to cladistic systematics, however, the phylogenetic species concept is ideal because only this concept generates strictly monophyletic units at the species level.

The phylogenetic species concept intentionally disregards details of evolutionary process and gives us a criterion that allows us to describe species without first needing to conduct detailed studies on evolutionary processes. Advocates of the phylogenetic species concept do not necessarily disregard the importance of studying evolutionary process. They argue, however, that the first step in studying evolutionary process is to have a clear picture of life's history. To accomplish this task, the pattern of common descent must be reconstructed in the greatest detail possible by starting with the smallest taxonomic units that have a history of common descent.

DYNAMISM OF SPECIES CONCEPTS

The current disagreements concerning concepts of species should not be considered discouraging. Whenever a field of scientific investigation enters a phase of dynamic growth, old concepts will be reevaluated and either refined or replaced with newer, more progressive ones. The active debate occurring within systematics shows that this field has acquired unprecedented activity and importance in biology. Just as Thomas Henry Huxley's time was one of enormous advances in biology, so is the present time. Both times are marked by fundamental reconsiderations of the meaning of species. We cannot predict which concept of species will be dominant 10 years from now, or even whether any of the concepts of species currently being advocated will survive until then. The conflicts between the current concepts, however, will lead us into the future. Understanding the conflicting perspectives, rather than learning a single species concept, is therefore of greatest importance for people now entering the study of zoology.

MAJOR DIVISIONS OF LIFE

From Aristotle's time to the late 1800s it was traditional to assign every living organism to one of two kingdoms: plant or animal. However, the two-kingdom system was problematic. Although it was easy to place rooted, photosynthetic organisms such as trees, flowers, mosses, and ferns among the plants and to place food-ingesting, motile forms such as insects, fishes, and mammals among the animals, unicellular organisms presented difficulties (Chapter 1). Some forms were claimed both for the plant kingdom by botanists and for the animal kingdom by zoologists. An example is *Euglena* (p. 223), which is motile, like animals, but has chlorophyll and photosynthesis, like plants. Other groups, such as bacteria, were rather arbitrarily assigned to the plant kingdom.

Several alternative systems have been proposed to solve the problem of classifying unicellular forms. In 1866 Haeckel proposed the new kingdom Protista to include all single-celled organisms. At first the bacteria and cyanobacteria (blue-green algae), forms that lack nuclei bounded by a membrane, were included with nucleated unicellular organisms. Finally, the important differences between the anucleate bacteria and cyanobacteria (prokaryotes) and all other organisms that have membrane-bound nuclei (eukaryotes) were recognized. In 1969 R. H. Whittaker proposed a five-kingdom system that incorporated the basic prokaryote-eukaryote distinction. The kingdom Monera contained the prokaryotes. The kingdom Protista contained the unicellular eukaryotic organisms (protozoa and unicellular eukaryotic algae). The multicellular organisms were split into three kingdoms on the basis of mode of nutrition and other fundamental differences in organization. The kingdom Plantae included multicellular photosynthesizing organisms, higher plants, and multicellular algae. Kingdom Fungi contained the molds, yeasts, and fungi that obtain their food by absorption. The invertebrates (except the protozoa) and the vertebrates make up the kingdom Animalia. Most of these forms ingest their

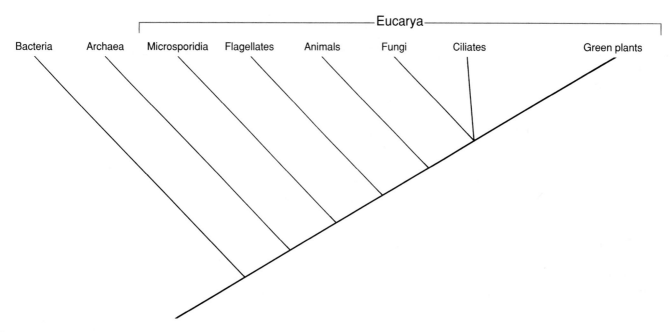

Figure 11-11
Evolutionary relationships among major groups of living organisms as inferred from ribosomal RNA sequence comparisons. Data are not yet available for all groups of organisms.

food and digest it internally, although some parasitic forms are absorptive.

All of these different systems were proposed without regard to the phylogenetic relationships that are needed to construct evolutionary or cladistic taxonomies. The oldest phylogenetic events in the history of life have been obscure, because the different forms of life share very few characters that can be compared among them to reconstruct phylogeny. Recently, however, a cladistic classification of all life forms has been proposed based on phylogenetic information obtained from molecular data (the nucleotide base sequence of ribosomal RNA, Figure 11-11). According to this tree, Woese, Kandler, and Wheelis (1990) recognized three monophyletic domains above the kingdom level: Eucarya (all eukaryotes), Bacteria (the true bacteria), and Archaea (prokaryotes differing from bacteria in membrane structure and ribosomal RNA sequences). They did not divide the Eucarya into kingdoms, although if we retain Whittaker's kingdoms Plantae, Animalia, and Fungi, the Protista become a paraphyletic group (Figure 11-12). To maintain a cladistic classification, the Protista must be broken up by recognizing as separate kingdoms the Ciliata, Flagellata, and Microsporidia as shown in Figure 11-11, and phylogenetic information must be gathered for additional

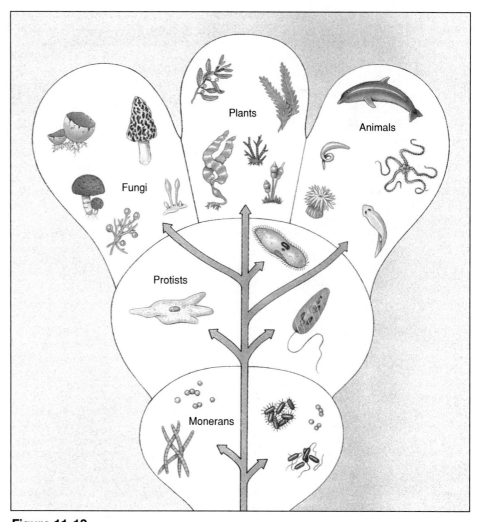

Figure 11-12
Whittaker's five-kingdom classification superimposed on a phylogenetic tree showing living representatives of these kingdoms. Note that the kingdoms Monera and Protista constitute paraphyletic groups and are therefore unacceptable to cladistic systematics.

protistan groups, including the amebas. This taxonomic revision has not been made; however, if the phylogenetic tree in Figure 11-11 is supported by further evidence, revision of the taxonomic kingdoms will be necessary.

Until a few years ago, the animal-like protistans were traditionally studied in zoology courses as the animal phylum Protozoa. Given current knowledge and the principles of phylogenetic systematics, this commits two taxonomic errors; "protozoa" are neither animals nor are they a valid monophyletic taxon at any level of the Linnaean hierarchy. The kingdom Protista is likewise invalid because it is not monophyletic. The animal-like protistans, now divided into seven separate phyla, are nonetheless of interest to students of zoology because of their animal-like properties.

MAJOR SUBDIVISIONS OF THE ANIMAL KINGDOM

The phylum is the largest formal taxonomic category in the Linnaean classification of the animal kingdom. Animal phyla are often grouped together to produce additional, informal taxa intermediate between the phylum and the animal kingdom. These taxa are based on embryological and anatomical characters that reveal the phylogenetic affinities of the different animal phyla. Zoologists in the past have recognized subkingdom Protozoa, which contains the primarily unicellular phyla, and the subkingdom Metazoa, which contains the multicellular phyla. As noted above, however, the Protozoa are not a valid taxonomic group and do not

belong within the animal kingdom, which is synonymous with the Metazoa. The higher-level groupings of the true animal phyla are as follows:

Branch A (Mesozoa): phylum Mesozoa, the mesozoa

Branch B (Parazoa): phylum Porifera, the sponges, and phylum Placozoa

Branch C (Eumetazoa): all other phyla

Grade I (Radiata): phyla Cnidaria, Ctenophora

Grade II (Bilateria): all other phyla

Division A (Protostomia): characteristics in Figure 11-13

Acoelomates: phyla Platyhelminthes, Gnathostomulida, Nemertea

Figure 11-13

Basis for the distinction between divisions of bilateral animals.

PROTOSTOMES		DEUTEROSTOMES	
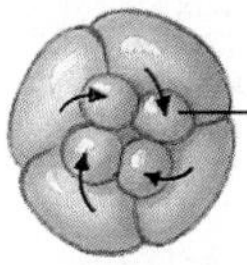 Spiral cleavage	Cleavage mostly spiral	Cleavage mostly radial	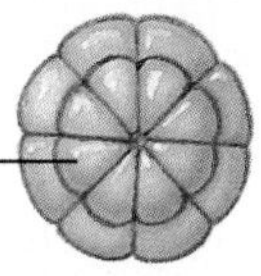 Radial cleavage
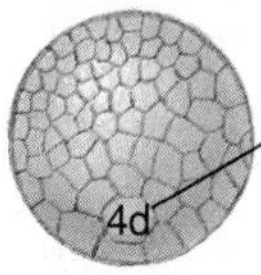 Cell from which mesoderm will derive 4d	Endomesoderm usually from a particular blastomere designated 4d	Endomesoderm from enterocoelous pouching (except chordates)	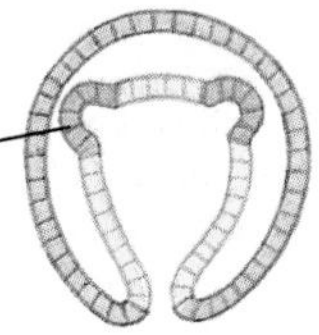 Endomesoderm from pouches from primitive gut
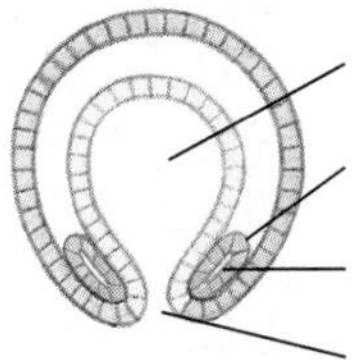 Primitive gut Mesoderm Coelom Blastopore	In coelomate protostomes the coelom forms as a split in mesodermal bands (schizocoelous)	All coelomate, coelom from fusion of enterocoelous pouches (except chordates, which are schizocoelous)	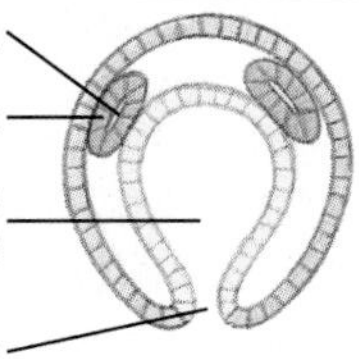 Coelom Mesoderm Primitive gut Blastopore
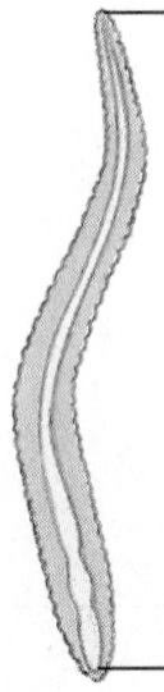 Anus Annelid (earthworm) Mouth	Mouth from, at, or near blastopore; anus a new formation Embryology mostly determinate (mosaic) Includes phyla Platyhelminthes, Nemertea, Annelida, Mollusca, Arthropoda, minor phyla	Anus from, at, or near blastopore, mouth a new formation Embryology usually indeterminate (regulative) Includes phyla Echinodermata, Hemichordata, Chaetognatha, Phoronida, Ectoprocta, Brachiopoda, Chordata	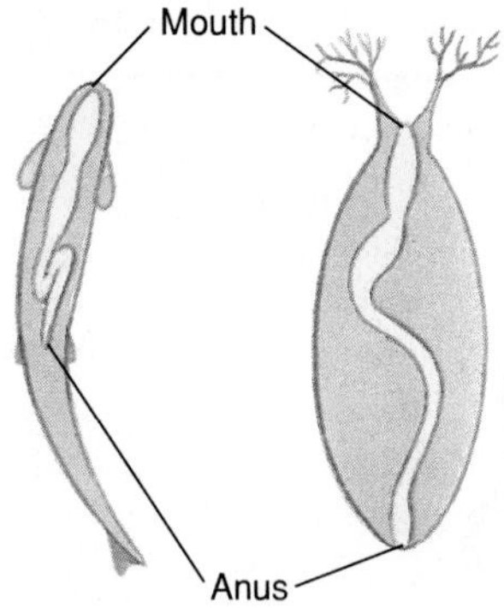 Mouth Anus

Pseudocoelomates: phyla
  Rotifera, Gastrotricha,
  Kinorhyncha, Nematoda,
  Nematomorpha,
  Acanthocephala,
  Entoprocta, Priapulida,
  Loricifera
Eucoelomates: phyla
  Mollusca, Annelida,
  Arthropoda, Echiurida,
  Sipunculida, Tardigrada,

Pentastomida,
  Onychophora,
  Pogonophora
Division B (Deuterostomia):
  characteristics in Figure
  11-13
  phyla Phoronida,
  Ectoprocta, Chaetognatha,
  Brachiopoda,
  Echinodermata,
  Hemichordata, Chordata

As in the outline, the bilateral animals are customarily divided into protostomes and deuterostomes on the basis of their embryological development (Figure 11-13). However, some of the phyla are difficult to place into one of these two categories because they possess some characteristics of each group (Chapters 23, 25).

Summary

Animal systematics has three major goals: (1) to identify all species of animals, (2) to evaluate the evolutionary relationships among animal species, and (3) to group animal species in a hierarchy of taxonomic groups (taxa) that conveys evolutionary relationships. The taxa are ranked to denote increasing inclusiveness as follows: species, genus, family, order, class, phylum, and kingdom. All of these ranks can be subdivided to signify taxa that are intermediate between them. The names of species are binomial, with the first name designating the genus to which the species belongs (capitalized) followed by a species epithet (lower case), both written in italics. Taxa at all other ranks are given single nonitalicized names.

There are currently two major schools of taxonomy. Traditional evolutionary taxonomy groups species into higher taxa according to the joint criteria of common descent and adaptive evolution; such taxa have a single evolutionary origin and occupy a distinctive adaptive zone. A second approach, known as phylogenetic systematics or cladistics, emphasizes common descent exclusively in grouping species into higher taxa. Only monophyletic taxa (those having a single evolutionary origin

and containing all descendants of the group's most recent common ancestor) are used in cladistics. In addition to monophyletic taxa, evolutionary taxonomy recognizes some taxa that are paraphyletic (having a single evolutionary origin but excluding some descendants of the most recent common ancestor of the group). Both schools of taxonomy exclude polyphyletic taxa (those having more than one evolutionary origin).

Both evolutionary taxonomy and cladistics require that patterns of common descent among species be assessed before higher taxa are recognized. Comparative morphology (including development), cytology, and biochemistry are used to reconstruct the nested hierarchical relationships among taxa that reflect the branching of evolutionary lineages through time. The fossil record provides estimates of the ages of evolutionary lineages. Comparative studies and the fossil record jointly permit us to reconstruct a phylogenetic tree representing the evolutionary history of the animal kingdom.

The biological species concept has guided the recognition of most animal species. A biological species is defined as a reproductive community of populations

(reproductively isolated from others) that occupies a specific niche in nature. It is not immutable through time but changes during the course of evolution. Because the biological species concept may be difficult to apply in spatial and temporal dimensions, and because it excludes asexually reproducing forms, alternative concepts have been proposed. These include the evolutionary species concept and the phylogenetic species concept. No single concept of species is universally accepted by all zoologists.

Traditionally, all living forms were placed into two kingdoms (animal and plant) but more recently, a five-kingdom system (animals, plants, fungi, protistans, and monerans) has been followed. Neither of these systems conforms to the principles of evolutionary or cladistic taxonomy because they place single-celled organisms into either paraphyletic or polyphyletic groups. Based on our current knowledge of the phylogenetic tree of life, "protozoa" do not form a monophyletic group and they do not belong within the animal kingdom. Because many unicellular forms share animal-like properties, however, they are of great interest to students of zoology.

Review Questions

1. List in order, from most inclusive to least inclusive, the principal categories (taxa) in Carolus Linnaeus's system of classification.
2. Explain why the system for naming species that originated with Linnaeus is "binomial."

3. How does the biological species concept differ from earlier typological concepts of a species? Why do evolutionary biologists prefer it to typological species concepts?
4. How do monophyletic, paraphyletic, and polyphyletic taxa differ? How do

these differences affect the validity of such taxa for both evolutionary and cladistic taxonomies?
5. How are taxonomic characters recognized? How are such characters used to construct a cladogram?

6. What is the difference between a cladogram and a phylogenetic tree? Given a cladogram for a group of species, what additional information is needed to obtain a phylogenetic tree?

7. How would cladists and evolutionary taxonomists differ in their interpretations of the statement that humans evolved from apes, which evolved from monkeys?

8. What taxonomic practices based on the typological species concept are retained in systematics today? How has their interpretation changed?

9. What problems have been identified with the biological species concept? How do other species concepts attempt to overcome these problems?

10. What are the five kingdoms distinguished by Whittaker? How does their recognition conflict with the principles of cladistic taxonomy?

Selected References

Ereshefsky, M. (ed.). 1992. The units of evolution. Cambridge, Massachusetts, MIT Press. *A thorough coverage of concepts of species, including reprints of important papers on the subject.*

Hall, B. K. 1994. Homology: the hierarchical basis of comparative biology. San Diego, Academic Press. *A collection of papers discussing the many dimensions of homology, the central concept of comparative biology and systematics.*

Hillis, D. M., C. Moritz and B. K. Mable (eds.). 1996. Molecular systematics, ed. 2. Sunderland, Massachusetts, Sinauer Associates, Inc. *A detailed coverage of the biochemical and analytical procedures of comparative biochemistry.*

Hull, D. L. 1988. Science as a process. Chicago, University of Chicago Press. *A study of the working methods and interactions of systematists, containing a thorough review of the principles of evolutionary, phenetic, and cladistic taxonomy.*

Jeffrey, C. 1973. Biological nomenclature. London, Edward Arnold, Ltd. *A concise, practical guide to the principles and practice of biological nomenclature and a useful interpretation of the Codes of Nomenclature.*

Maddison, W. P., and D. R. Maddison. 1992. MacClade version 3.01. Sunderland, Massachusetts, Sinauer Associates, Inc. *A computer program for the MacIntosh that conducts phylogenetic analyses of systematic characters. The instruction manual stands alone as an excellent introduction to phylogenetic procedures. The computer program is user-friendly and excellent for instruction in addition to serving as a tool for analyzing real data.*

Margulis, L., and K. V. Schwartz. 1987. Five kingdoms: an illustrated guide to the phyla of life on earth, ed. 2. San Francisco, W.H. Freeman & Co. *Illustrated catalog and descriptions of all major groups with bibliography and glossary.*

Mayr, E., and P. D. Ashlock. 1991. Principles of systematic zoology. New York, McGraw-Hill. *A detailed survey of systematic principles as applied to animals.*

Panchen, A. L. 1992. Classification, evolution, and the nature of biology. New York, Cambridge University Press. *Excellent explanations of the methods and philosophical foundations of biological classification.*

Wiley, E. O. 1981. Phylogenetics: the theory and practice of phylogenetic systematics. New York, John Wiley & Sons, Inc. *Excellent, thorough presentation of cladistic theory.*

Wiley, E. O., D. Siegel-Causey, D. R. Brooks, and V. A. Funk. 1991. The compleat cladist: a primer of phylogenetic procedures. Lawrence, University of Kansas Printing Service. *A workbook presenting detailed instruction in cladistic concepts and methods.*

Woese, C. R., O. Kandler, and M. L. Wheelis. 1990. Towards a natural system of organisms: proposal for the domains Archaea, Bacteria, and Eucarya. Proceedings of the National Academy of Sciences, USA, **87:**4576—4579. *Proposed cladistic classification for the major taxonomic divisions of life.*

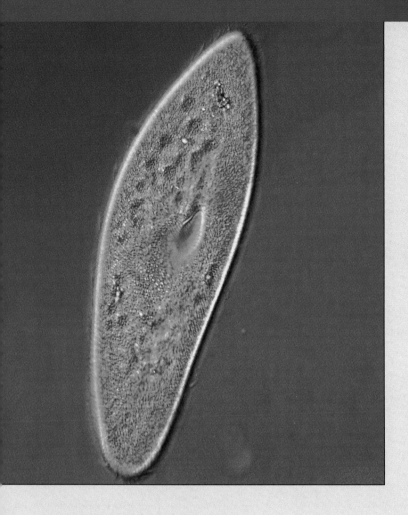

The Animal-Like Protista

Emergence of Eukaryotes

The first reasonable evidence for life on earth dates from approximately 3.5 billion years ago. These first cells were prokaryotic, bacteria-like organisms. After an enormous time span of evolutionary diversification at the prokaryotic level, unicellular eukaryotic organisms appeared. Although the origin of single-celled eukaryotes can never be known with certainty, it probably came about through a process of symbiosis. Certain aerobic bacteria may have been engulfed by other bacteria that were unable to cope with the increasing concentrations of oxygen in the atmosphere. The aerobic bacteria had the enzymes necessary for deriving energy in the presence of oxygen, and they would have become the ancestors of mitochondria. Most, but not all, of the genes of the mitochondria would come to reside in the host-cell nucleus. Almost all present-day eukaryotes have mitochondria and are aerobic.

Some ancestral eukaryotic cells engulfed photosynthetic bacteria, which evolved to become chloroplasts, and the eukaryotes thereby were able to manufacture their own food molecules using energy from sunlight. The descendants of one line, the green algae, eventually gave rise to the kingdom Plantae.

Some of the eukaryotes that did not become residences for chloroplasts, and even some that did, evolved animal-like characteristics and gave rise to the protistan subkingdom Protozoa. Protozoa are a diverse assemblage of unicellular organisms with puzzling affinities. They are distinctly animal-like in several respects: they lack a cell wall, have at least one motile stage in the life cycle, and most ingest their food. Throughout their long history, protozoa have radiated to generate a bewildering array of morphological forms within the constraints of a single cell. ■

POSITION RELATIVE TO THE ANIMAL KINGDOM

The protozoan is a complete organism in which all life activities are carried on within the limits of a single plasma membrane. Because their protoplasmic mass is not subdivided into cells, protozoa sometimes have been termed "acellular," but most people prefer "unicellular" to emphasize the many structural similarities to the cells of multicellular animals.

In the five-kingdom classification of living organisms, the protozoan phyla are considered members of the Protista. This group contains all the unicellular eukaryotes, as well as the multicellular algae, which are included because of their simple structure and clear relationship to unicellular algae. Because the word "protist" has been associated with small, unicellular organisms, some biologists object to using it for a group that contains multicellular algae, some of which are many meters long. They prefer **Protoctista** (Gr., *protos,* very first; + *ktistos,* to establish). Whether Protista or Protoctista, it is an extremely heterogeneous, paraphyletic group, some of whose members are more closely related to the multicellular kingdoms Plantae, Fungi, and Animalia than they are to each other. There can be little doubt, however, that the Animalia share common ancestry with one or more groups of animal-like protista, or protozoa.

BIOLOGICAL CONTRIBUTIONS

1. **Intracellular specialization** (division of labor within the cell) involves the organization of functional organelles in the cell.
2. The simplest example of **division of labor between cells** is seen in certain colonial protozoa that have both somatic and reproductive zooids (individuals) in the colony.
3. **Asexual reproduction** by mitotic division appears in the protists.
4. **True sexual reproduction** with zygote formation is found in some protozoa.
5. The responses (taxes) of protozoa to stimuli represent the **simplest reflexes and instincts** as we know them in metazoans.
6. The simplest animal-like organisms with **exoskeletons** are certain shelled protozoa.
7. **All types of nutrition** are developed in the protozoa; autotrophic, saprozoic, and holozoic. **Basic enzyme systems** to accomplish these types of nutrition are developed.
8. Means of **locomotion** in aqueous media are developed.

The organisms referred to as protozoa are united only on the basis of a single, negative characteristic: they are not multicellular. This concept was recognized, in a way, by the American zoologist Libbie Hyman (1940),[*] who preferred the term "acellular" rather than the traditional "unicellular" to describe protozoa. She distinguished them as "animals whose body substance is not partitioned into cells." Although most zoologists have returned to describing protozoa as unicellular because of electron microscope studies subsequent to Hyman's book, the concept of acellularity is still a valuable one. It reminds us that the traditionally recognized phylum Protozoa was not a natural phylogenetic grouping. An enormous amount of information on protozoan structure, life histories, and physiology has accumulated in recent years, and the Society of Protozoologists published a new classification of protozoa in 1980, recognizing *seven* separate phyla. We adopt the classification because it comes closer to reflecting real evolutionary relationships than the older, simpler systems, but there is no way to give adequate treatment to all groups, even all phyla, of protozoa in a book of this scope. We will introduce you here to the most important and largest phyla: Sarcomastigophora (containing the flagellates and amebas), Apicomplexa (important intracellular parasites, including the malarial organism), and Ciliophora (ciliates).

The protozoan phyla do demonstrate a basic body plan or grade—the single eukaryotic cell—and they amply demonstrate the enormous adaptive potential of that grade. Over 64,000 species have been named, and over half of these are fossils. Although they are unicellular, protozoa are functionally complete organisms with many complicated, microanatomical structures. Their various organelles tend to be more specialized than those of the average cell in a multicellular organism. Particular organelles may perform as skeletons, sensory structures, conducting mechanisms, and so on.

Protozoa are found wherever life exists. They are highly adaptable and easily distributed from place to place. They require moisture, whether they live in marine or freshwater habitats, soil, decaying organic matter, or plants and animals. They may be sessile or free swimming, and they form a large part of the floating plankton. The same species are often found widely separated in time as well as in space. Some species may have spanned geological eras of more than 100 million years.

Despite their wide distribution, many protozoa can live successfully only within narrow environmental ranges. Species adaptations vary greatly, and successions of species frequently occur as environmental conditions change. These changes may be brought about by physical factors, such as the drying up of a pond or seasonal changes in temperature, or by biological changes, such as predator pressure.

Protozoa play an enormous role in the economy of nature. Their fantastic numbers are attested by the gigantic ocean soil deposits formed by their skeletons. About 10,000 species of protozoa are symbiotic in or on animals or plants, sometimes even other protozoa. The relationship may be

---

[*]Hyman, L. H. 1940. The invertebrates: Protozoa through Ctenophora. New York, McGraw-Hill Book Company.

mutualistic (both partners benefit), commensalistic (one partner benefits, no effect on the other), or parasitic (one partner benefits at the expense of the other), depending on the species involved. Some of the most important diseases of humans and domestic animals are caused by parasitic protozoa.

A number of species are colonial and some have multicellular stages in their life cycles, which may lead one to wonder why such protozoa are not considered metazoa. The reasons are that they usually have clearly recognizable, noncolonial relatives and, more arbitrarily, that they do not have more than one kind of nonreproductive cell and they do not undergo embryonic development. By definition, metazoa have more than one kind of nonreproductive cell in their bodies and undergo embryogenesis.

FORM AND FUNCTION

Inasmuch as protozoa are cells, in many aspects their structures and physiology are the same as those of cells of multicellular organisms. However, because they must conduct all the functions of life as individual organisms, and because they show such enormous diversity in form, habitat, feeding, and so on, many features are unique to various protozoan cells.

CHARACTERISTICS OF PROTOZOAN PHYLA

1. **Unicellular;** some colonial, and some with multicellular stages in their life cycles
2. **Mostly microscopic,** although some are large enough to be seen with the unaided eye
3. All symmetries represented in the group; shape variable or constant (oval, spherical, or other)
4. **No germ layer present**
5. No organs or tissues, but **specialized organelles** are found; nucleus single or multiple
6. Free-living, mutualism, commensalism, parasitism all represented in the group
7. Locomotion by **pseudopodia, flagella, cilia,** and direct cell movements; some sessile
8. Some provided with a **simple endoskeleton** or **exoskeleton,** but mostly naked
9. **Nutrition of all types:** autotrophic (manufacturing own nutrients by photosynthesis), heterotrophic (depending on other plants or animals for food), saprozoic (using nutrients dissolved in the surrounding medium)
10. Aquatic or terrestrial habitat; free-living or symbiotic mode of life
11. Reproduction **asexually** by fission, budding, and cysts and **sexually** by conjugation or by syngamy (union of male and female gametes to form a zygote)

NUCLEUS AND CYTOPLASM

As in other eukaryotes, the nucleus is a membrane-bound structure whose interior communicates with the cytoplasm by small pores. Within the nucleus the genetic material (DNA) is borne on chromosomes. Except during cell division, the chromosomes are not usually condensed in a form that can be distinguished, although during fixation of the cells for light microscopy, the chromosomal material (chromatin) often clumps together irregularly, leaving some areas within the nucleus relatively clear. The appearance is described as **vesicular** and is characteristic of many protozoan nuclei (Figure 12-1). Condensations of chromatin may be distributed around the periphery of the nucleus or internally in distinct patterns. In some flagellates the chromosomes are visible through the interphase as they would appear during prophase of mitosis.

Also within the nucleus, one or more **nucleoli** are often present. **Endosomes** are nucleoli that remain as

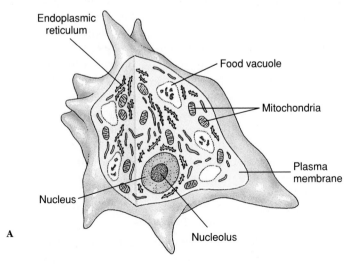

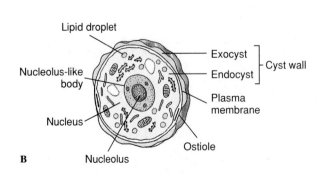

Figure 12-1
Structure of *Acanthamoeba palestinensis.* **A,** Active, feeding form. **B,** Cyst.

discrete bodies during mitosis; they are characteristic of phytoflagellates, parasitic amebas, and trypanosomes (see Figures 12-1, 12-11, and 12-14).

The **macronuclei** of ciliates are described as **compact** or **condensed** because the chromatin material is more finely dispersed and clear areas cannot be observed with the light microscope (see Figure 12-24).

Cellular organelles like those in the cells of multicellular animals can be distinguished in the cytoplasm of many protozoa. These include mitochondria, endoplasmic reticulum, Golgi apparatus, and various vesicles. Chloroplasts, the membrane-bound organelles in which photosynthesis takes place, are found in most phytoflagellates (see Figure 12-12).

Sometimes the peripheral and the central areas of the cytoplasm can be distinguished as **ectoplasm** and **endoplasm** (see Figure 12-4). The endoplasm appears more granular and contains the nucleus and cytoplasmic organelles. The ectoplasm appears more transparent (hyaline) under the light microscope, and it bears the bases of the cilia or flagella. The ectoplasm is often more rigid and is in the gel state of a colloid, whereas the more fluid endoplasm is in the sol state.

LOCOMOTOR ORGANELLES

The chief means by which protozoa move are by cilia and flagella and by pseudopodial movement. These mechanisms are extremely important in the biology of higher animals as well.

Cilia and Flagella

Many small metazoans use cilia not only for locomotion but also to create water currents for their feeding and respiration. Ciliary movement is vital to many species in such functions as handling food, reproduction, excretion, and osmoregulation (as in flame cells, p. 283).

There is no real morphological distinction between cilia and flagella (Figure 12-2), and some investigators have preferred to call them both undulipo-

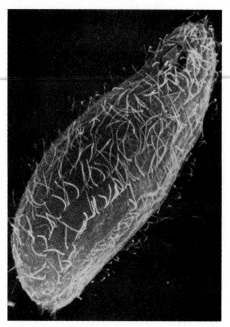

Figure 12-2

Scanning electron micrograph of the free-living ciliate *Tetrahymena thermophila* showing rows of cilia (× 2000). Beating of flagella either pushes or pulls the organism through its medium, while cilia propel the organism by a "rowing" mechanism. Their structure is similar, whether viewed by scanning or transmission electron microscopy.

dia (L. dim. of *unda,* a wave, + Gr. *podos,* a foot). However, a cilium propels water parallel to the surface to which the cilium is attached, whereas a flagellum propels water parallel to the main axis of the flagellum. Thus there are important effects on the mechanics and speed of propulsion. Each flagellum or cilium contains nine pairs of longitudinal microtubules arranged in a circle around a central pair (Figure 12-3), and this is true for all motile flagella and cilia in the animal kingdom, with a few notable exceptions. This "9 + 2" tube of microtubules in the flagellum or cilium is its **axoneme;** the axoneme is covered by a membrane continuous with the cell membrane covering the rest of the organism. At about the point where the axoneme enters the cell proper, the central pair of microtubules ends at a small plate within the circle of nine pairs (Figure 12-3A). Also at about that point, another microtubule joins each of the nine pairs, so that these form a short tube extending from the base of the flagellum into the cell

and consisting of nine *triplets* of microtubules. The short tube of nine triplets is the **kinetosome (or basal body)** and is exactly the same in structure as the **centriole** that organizes the mitotic spindle during cell division (see p. 49 and Figure 4-24, p. 57). The centrioles of some flagellates may give rise to the kinetosomes, or the kinetosomes may function as centrioles. All typical flagella and cilia have a kinetosome at their base, regardless of whether they are borne by a protozoan or metazoan cell.

> Description of the axoneme as "9 + 2" is traditional, but it is also misleading because there is only a single pair of microtubules in the center. If we were consistent, we would have to describe the axoneme as "9 + 1."

The current explanation for ciliary and flagellar movement is the **sliding microtubule hypothesis.** The movement is powered by the release of chemical bond energy in ATP (p. 66). Two little arms are visible in electron micrographs on each of the pairs of peripheral tubules in the axoneme (level *X* in Figure 12-3), and these bear the enzyme adenosine triphosphatase (ATPase), which cleaves the ATP. When the bond energy in ATP is released, the arms "walk along" one of the filaments in the adjacent pair, causing it to slide relative to the other filament in the pair. Shear resistance, causing the axoneme to bend when the filaments slide past each other, is provided by "spokes" from each doublet to the central pair of fibrils. These are also visible in electron micrographs. Direct evidence for the sliding microtubule hypothesis was obtained by attaching tiny gold beads to axonemal microtubules and observing their movement microscopically.

Pseudopodia

Although pseudopodia are the chief means of locomotion in the Sarcodina (see Classification, p. 236), they can be formed by a variety of flagellate

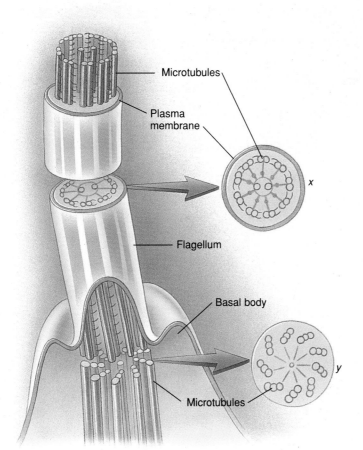

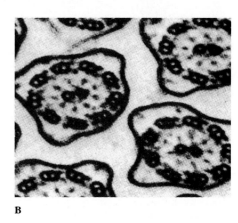

B

A

Figure 12-3

A, The axoneme is composed of nine pairs of microtubules plus a central pair, and it is enclosed within the cell membrane. The central pair ends at about the level of the cell surface in a basal plate (axosome). The peripheral microtubules continue inward for a short distance to compose two of each of the triplets in the kinetosome (at level y in **A**). **B,** Electron micrograph of section through several cilia, corresponding to section x in **A.** ($\times$ 133,000)

protozoa, as well as by ameboid cells of many invertebrates. In fact, much of the defense against disease in the human body depends on ameboid white blood cells, and ameboid cells in many other animals, vertebrate and invertebrate, play similar roles.

In the protozoa, pseudopodia exist in several forms. The most familiar are the **lobopodia** (Figures 12-4 and 12-5), which are rather large, blunt extensions of the cell body containing both endoplasm and ectoplasm. Some amebas characteristically do not extend individual pseudopodia, but the whole body moves with pseudopodial movement; this is known as the **limax** form (for a genus of slugs, *Limax*). **Filopodia** are thin extensions, usually branching, and containing only ectoplasm. They

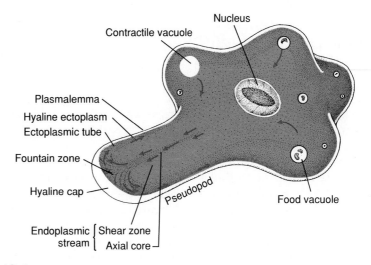

Figure 12-4

Ameba in active locomotion. Arrows indicate the direction of streaming protoplasm. The first sign of new pseudopodium is thickening of the ectoplasm to form a clear hyaline cap, into which the fluid endoplasm flows. As the endoplasm reaches the forward tip, it fountains out and is converted into ectoplasm, forming a stiff outer tube that lengthens as the forward flow continues. Posteriorly the ectoplasm is converted into fluid endoplasm, replenishing the flow. Substratum is necessary for ameboid movement.

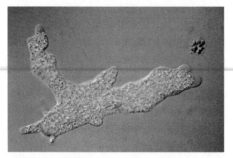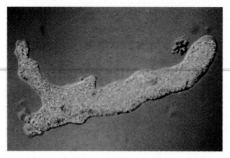

Figure 12-5

Ameboid movement. At left and center, the ameba extends a pseudopodium toward a *Pandorina* colony. At right, the ameba surrounds the *Pandorina* before engulfing it by phagocytosis.

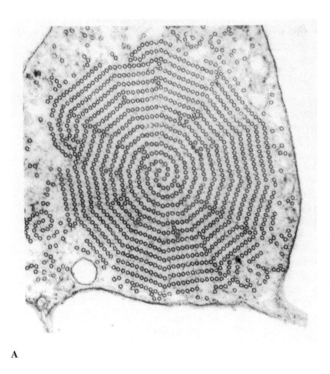

A

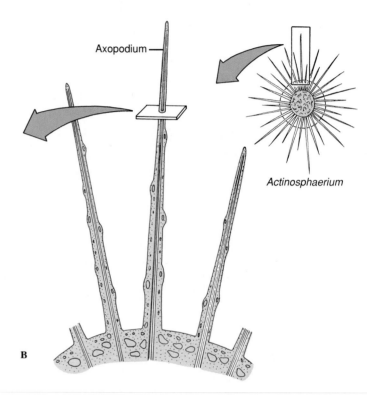

B

Figure 12-6

A, Diagram of axopodium to show orientation of **B,** electron micrograph of axopodium (from *Actinosphaerium nucleofilum*) in cross section. Protozoa with axopodia are shown in Figure 12-16 (*Actinophrys* and *Clathrulina*). The axoneme of the axopodium is composed of an array of microtubules, which may vary from three to many in number depending on the species. Some species can extend or retract their axopodia quite rapidly. (× 99,000)

are found in members of the sarcodine class Filosea, such as *Euglypha* (see Figure 12-9B). **Reticulopodia** (see Figure 12-16) are distinguished from filipodia in that reticulopodia repeatedly rejoin to form a netlike mesh, although some protozoologists believe that the distinction between filipodia and reticulopodia is artificial. Members of the superclass Actinopoda have **axopodia** (see Figure 12-16), which are long, thin pseudopodia supported by axial rods of microtubules (Figure 12-6). The microtubules are arranged in a definite spiral or geometrical array, depending on the species, and constitute the axoneme of the axopod. Axopodia can be extended or retracted, apparently by addition or removal of microtubular material. Since the tips can adhere to the substrate, the organism can progress by a rolling motion, shortening the axonemes in front and extending those in the rear. Cytoplasm can flow along the axonemes, toward the body on one side and in the reverse direction on the other.

How pseudopodia work has long attracted the interest of zoologists, but we have only recently gained some insight into the phenomenon. When a typical lobopodium begins to form, an extension of ectoplasm called the hyaline cap appears, and endoplasm begins to flow toward and into the hyaline cap (Figure 12-4). As the endoplasmic material flows into the hyaline cap, it fountains out to the periphery and changes from the sol to the gel state; that is, it becomes ectoplasm. Thus the ectoplasm is a tube through which the endoplasm flows as the pseudopodium extends. On the trailing side of the organism, ectoplasm becomes endoplasm. At some point the pseudopodium becomes anchored to the substrate, and the cell is drawn forward. The current hypothesis of pseudopodial movement involves the participation of actin, myosin, and other components. As endoplasm fountains out in the hyaline cap, actin subunits become polymerized into microfilaments, which in turn become cross-linked to each other by actin-binding protein (ABP) to form a gel, that is, the endoplasm becomes ectoplasm. At the "posterior" the ABP releases the microfilaments, which return to the sol state of endoplasm. Before the microfilaments are disassembled into actin subunits, they interact with myosin, contracting and causing the endoplasm to flow in the direction of the pseudopodium by hydrostatic pressure.

EXCRETION AND OSMOREGULATION

Vacuoles can be seen in the cytoplasm of many protozoa under the light microscope, and some of these vacuoles periodically fill with a fluid substance that is then expelled. Evidence is strong that these **contractile vacuoles** (see Figures 12-4, 12-12, and 12-24) function principally in osmoregulation. They are more prevalent and fill and empty more frequently in freshwater protozoa than in marine and endosymbiotic species, where the surrounding medium would be more nearly isosmotic (having the same osmotic pressure) to the cytoplasm. Smaller species, which have a greater surface-to-volume ratio, generally have more rapid filling and expulsion rates in their contractile vacuoles. Excretion of metabolic wastes, on the other hand, is almost entirely by diffusion. The main end product of nitrogen metabolism is ammonia, which readily diffuses out of the small bodies of protozoa.

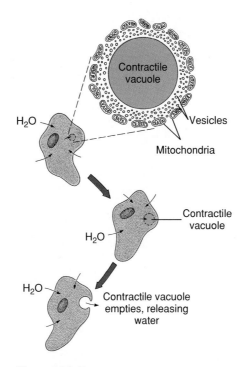

Figure 12-7

How an ameba pumps out water. Water enters the ameba's body by osmosis and is removed by the rhythmic filling and emptying of the contractile vacuole. The contractile vacuole of *Amoeba proteus* is surrounded by tiny vesicles that fill with fluid, which is then emptied into the vacuole. Note the numerous mitochondria that are believed to provide energy needed to adjust the salt content of the tiny vesicles.

The contractile vacuoles, sometimes called water expulsion vesicles, differ considerably in complexity among the various types of protozoa. In amebas the contractile vacuoles are not usually found at any particular site, being passively carried around in the endoplasm (Figure 12-7). Small vesicles join the contractile vacuole, emptying their contents into it as the vacuole fills. The vacuole finally joins its membrane to the surface membrane and empties its contents to the outside.

Some ciliates, such as *Blepharisma,* have contractile vacuoles with filling mechanisms similar to that described for amebas. Others, such as *Paramecium,* have more complex contractile vacuoles. In these the contractile vacuoles are located in a specific position beneath the cell membrane, with an "excretory" pore leading to the outside, and surrounded by the ampul-

lae of about six feeder canals (see Figure 12-24). The feeder canals, in turn, are surrounded by fine tubules about 20 nm in diameter, which connect with the canals during filling of the ampullae and at their lower ends connect with the tubular system of the endoplasmic reticulum. The ampullae and the contractile vacuole are surrounded by bundles of fibrils, which may play a role in the contraction of these structures. Contraction of the ampullae fills the vacuole. When the vacuole contracts to discharge its contents to the outside, the ampullae become disconnected from the vacuole, so that backflow is prevented.

NUTRITION

Protozoa can be categorized broadly into autotrophs (which synthesize their own organic constituents from inorganic substrates) and heterotrophs (which must obtain organic molecules synthesized by other organisms). Another kind of classification, usually applied to heterotrophs, involves those that ingest visible particles of food (**phagotrophs,** or **holozoic** feeders) as contrasted with those ingesting food in a soluble form (**osmotrophs,** or **saprozoic** feeders). However, reality is not so simple, even among the one-celled organisms. Autotrophic protozoa use light energy to synthesize their organic molecules (phototrophs), but they often practice phagotrophy and osmotrophy as well. Even among the heterotrophs, few are exclusively either phagotrophic or osmotrophic. The single order Euglenida (class Phytomastigophorea) contains some forms that are mainly phototrophs, some that are mainly osmotrophs, and some that are mainly phagotrophs. The species of *Euglena* show considerable variety in nutritional capability. Some require certain preformed organic molecules, even though they are autotrophs, and some lose their chloroplasts if maintained in darkness, thus becoming permanent osmotrophs.

Holozoic nutrition implies phagocytosis (Figure 12-8), in which there is an infolding or invagination of the cell

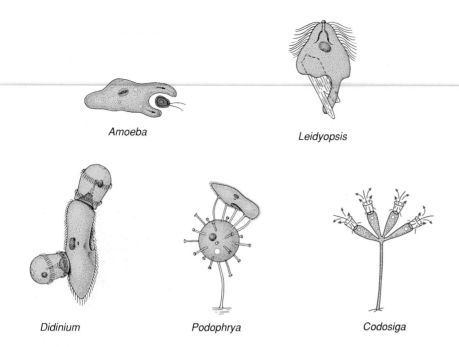

Amoeba

Leidyopsis

Didinium

Podophrya

Codosiga

Figure 12-8

Some feeding methods among protozoa. *Amoeba* surrounds a small flagellate with pseudopodia. *Leidyopsis*, a flagellate living in the intestine of termites, forms pseudopodia and ingests wood chips. *Didinium*, a ciliate, feeds only on *Paramecium*, which it swallows through a temporary cytostome in its anterior end. Sometimes more than one *Didinium* feed on the same *Paramecium*. *Podophrya* is a suctorian ciliophora. Its tentacles attach to its prey and suck prey cytoplasm into the body of the *Podophrya*, where it is pinched off to form food vacuoles. *Codosiga*, a sessile flagellate with a collar of microvilli, feeds on particles suspended in the water drawn through its collar by the beat of its flagellum. Technically, all of these methods are types of phagocytosis.

membrane around the food particle. As the invagination extends farther into the cell, it is pinched off at the surface (p. 53). The food particle is thus contained in an intracellular, membrane-bound vesicle, the **food vacuole** or **phagosome.** Lysosomes, small vesicles containing digestive enzymes, fuse with the phagosome and pour their contents into it, where digestion begins. As the digested products are absorbed across the vacuole membrane, the phagosome becomes smaller. Any undigestible material may be released to the outside by exocytosis, the vacuole again fusing with the cell surface membrane. In most ciliates, many flagellates, and many apicomplexans, the site of phagocytosis is a definite mouth structure, the **cytostome** (Figures 12-8 and 12-24). In amebas, phagocytosis can occur at almost any point by envelopment of the particle with pseudopodia. Particles must be ingested through the opening of the test, or shell, in amebas that

have tests. Flagellates may form a temporary cytostome, usually in a characteristic position, or they may have a permanent cytostome with specialized structure. Many ciliates have a characteristic structure for expulsion of waste matter, the **cytopyge** or **cytoproct,** found in a characteristic location. In some, the cytopyge also serves as the site for expulsion of the contents of the contractile vacuole.

Saprozoic feeding may be by pinocytosis or by transport of solutes directly across the outer cell membrane. Pinocytosis and transport across the cell membrane are discussed on p. 53. Direct transport across the membrane may be by diffusion, facilitated transport, or active transport. Diffusion is probably of little or no importance in the nutrition of protozoa, except possibly in some endosymbiotic species. Some important food molecules, such as glucose and amino acids, may be brought into the cell by facilitated diffusion and active transport.

It has been shown that a stimulatory substance, or "inducer," must be present in the surrounding medium for many protozoa to initiate pinocytosis. Several proteins act as inducers, but some salts and other substances can also stimulate pinocytosis; it appears that the inducer must be a positively charged molecule. Pinocytosis takes place at the inner end of the cytopharynx in the protozoa possessing that structure.

REPRODUCTION

Sexual phenomena occur widely among the protozoa, and sexual processes may precede certain phases of asexual reproduction, but embryonic development does not occur; protozoa do not have embryos. The essential features of sexual processes include a reduction division of the chromosome number to half (diploid number to haploid number), the development of sex cells (gametes) or at least gamete nuclei, and usually a fusion of the gamete nuclei (p. 233).

Fission

The cell multiplication process that produces more individuals in the protozoa is called fission. The most common type of fission is **binary,** in which two essentially identical individuals result (Figure 12-9). When the progeny cell is considerably smaller than the parent and then grows to adult size, the process is called **budding.** This occurs in some ciliates. In **multiple fission,** division of the cytoplasm (cytokinesis) is preceded by several nuclear divisions, so that a number of individuals are produced almost simultaneously (Figure 12-20). Multiple fission, or **schizogony,** is common among the Sporozoea and some Sarcodina. If the multiple fission is preceded by or associated with union of gametes, it is referred to as **sporogony.**

All of the foregoing types of division are accompanied by some form of mitosis (p. 55). However, the process is often somewhat unlike that found in metazoans. For example, the nuclear

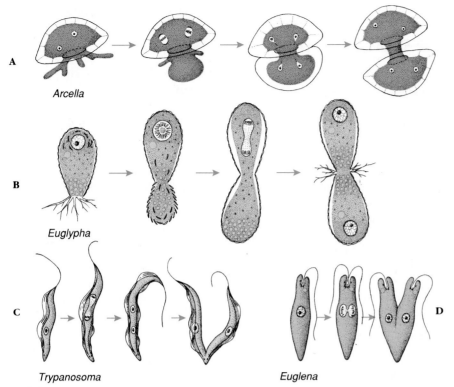

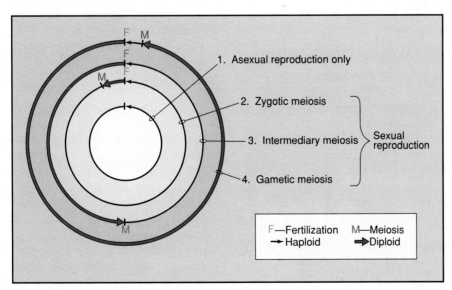

Figure 12-9

Binary fission in some sarcodines and flagellates. **A,** The two nuclei of *Arcella* divide as some of its cytoplasm is extruded and begins to secrete a new test for the daughter cell. **B,** The test of another sarcodine, *Euglypha,* is constructed of secreted platelets. Secretion of the platelets for the daughter cell is begun before the cytoplasm begins to move out of the aperture. As these are used to construct the test of the daughter cell, the nucleus divides. **C,** *Trypanosoma* has a kinetoplast (part of the mitochondrion) near the kinetosome of its flagellum close to its posterior end in the stage shown. All of these parts must be replicated before the cell divides. **D,** Division of *Euglena*. Compare **C** and **D** with Figure 12-26, fission in a ciliophoran.

Figure 12-10

Timing of fertilization and meiosis (reduction division) demonstrated among various protozoa. *1,* Asexual reproduction only, all individuals presumably haploid. *2,* Zygotic meiosis, in which reduction division occurs immediately after fertilization (zygote formation). *3,* Intermediary meiosis, in which asexually reproducing, diploid individuals undergo meiosis to produce asexually reproducing, haploid individuals. Fertilization restores diploid condition. *4,* Gametic meiosis, in which meiosis occurs during gamete formation, and all cells except gametes are diploid.

membrane often persists through mitosis, and the microtubular spindle may be formed within the nuclear membrane. Centrioles have not been observed in nuclear division of ciliates; the nuclear membrane persists in micronuclear mitosis, with the spindle within the nucleus. The macronucleus of ciliates seems simply to elongate, constrict, and divide without any recognizable mitotic phenomena **(amitosis).**

Sexual Processes

Although all protozoa reproduce asexually, and some are apparently exclusively asexual, the widespread occurrence of sex among the protozoa testifies to its importance as a means of genetic recombination. The gamete nuclei, or pronuclei, which fuse in fertilization to restore the diploid number of chromosomes, are usually borne in special gametic cells. When the gametes all look alike, they are called **isogametes,** but most species have two dissimilar types, or **anisogametes.**

We are accustomed to thinking of the timing of meiosis as it occurs in metazoa, during or just before gamete formation: **gametic meiosis** (Figure 12-10). This is indeed the case in some flagellates, the Heliozoea, and the Ciliophora. However, in other flagellates and in the Sporozoea, the first divisions *after* fertilization are meiotic **(zygotic meiosis),** and all the individuals produced asexually (mitotically) in the life cycle up to the next zygote are haploid. It is believed that most protozoa that do not reproduce sexually are haploid, but this is difficult to show because there is no meiosis. In some of the Granuloreticulosa (foraminiferans), there is an alternation of haploid and diploid generations **(intermediary meiosis),** a phenomenon widespread among plants.

The fertilization of an individual gamete by another is **syngamy,** but some sexual phenomena in protozoa do not involve that process. Examples are **autogamy,** in which gametic nuclei arise by meiosis and fuse to form a zygote within the same organism that produced them, and **conjugation,** in which there is an exchange of gametic nuclei between paired organisms (conjugants). We will describe conjugation further in the discussion of the paramecium.

ENCYSTMENT AND EXCYSTMENT

Separated as they are from their external environment only by their delicate external cell membrane, it seems astonishing that protozoa could be so successful in habitats frequently subjected to extremely harsh conditions. This is surely related to their ability to form cysts: dormant forms marked by the possession of resistant external coverings and a more or less complete shutdown of metabolic machinery. Cyst formation is also important to many parasitic forms that must survive a harsh environment between hosts (Figure 12-1). However, some parasites do not form cysts, apparently depending on direct transfer from one host to another. Reproductive phases such as fission, budding, and syngamy may occur in the cysts of some species. Encystment has not been found in *Paramecium,* and it is rare or absent in marine forms.

Cysts of some soil-inhabiting and freshwater protozoa have amazing durability. The cysts of the soil ciliate *Colpoda* can survive 7 days in liquid air and 3 hours at 100° C. Survival of *Colpoda* cysts in dried soil has been shown for up to 38 years, and those of a certain small flagellate (*Podo*) can survive up to 49 years! Not all cysts are so sturdy, however. Those of *Entamoeba histolytica* will tolerate gastric acidity but not desiccation, temperature above 50° C, or sunlight.

The conditions stimulating encystment are incompletely understood, although in some cases cyst formation is cyclic, occurring at a certain stage in the life cycle. In most free-living forms, adverse change in the environment favors encystment. Such conditions may include food deficiency,

desiccation, increased osmotic pressure of the environment, decreased oxygen concentration, or change in pH or temperature.

During encystment a number of organelles, such as cilia or flagella, are resorbed, and the Golgi apparatus secretes the cyst wall material, which is carried to the surface in vesicles and extruded.

Although the exact stimulus for excystation (escape from cysts) is usually unknown, a return of favorable conditions initiates excystment in those protozoa in which the cysts are a resistant stage. In parasitic forms the excystment stimulus may be more specific, requiring conditions similar to those found in the host.

REPRESENTATIVE TYPES

This section describes some representatives of each large group of protozoa to provide a basis for comparing the groups and an idea of the diversity of protozoa. Forms such as *Amoeba* and *Paramecium,* although large and easy to obtain for study, are not wholly representative because their life histories are somewhat simpler than those of other members of the respective groups.

PHYLUM SARCOMASTIGOPHORA

The Sarcomastigophora includes both protozoa that move by flagella (Mastigophora) and those that move by pseudopodia (Sarcodina). These characteristics are not mutually exclusive; some mastigophorans (flagellates) can form and use pseudopodia, and a number of sarcodines have flagellated stages in their life cycles.

Subphylum Mastigophora: The Flagellated Protozoa

Although some flagellates can form pseudopodia, their primary means of locomotion is by one or more flagella. The group is divided into the phytoflagellates (class Phytomastigophorea), which usually have chlorophyll and are thus plantlike, and the zooflagel-

lates (class Zoomastigophorea), which do not have chlorophyll, are either holozoic or saprozoic, and thus are animal-like.

Phytoflagellates. Phytoflagellates usually have one or two (sometimes four) flagella and **chloroplasts,** which contain the pigments used in photosynthesis. They are mostly free living and include such familiar forms as *Euglena, Chlamydomonas, Peranema, Volvox,* and the dinoflagellates. *Peranema* (Figure 12-11) is related to *Euglena* but is a colorless phytoflagellate with holozoic nutrition. *Chilomonas* is another common form that is an important food item for amebas. Some of the flagellates are colonial, living in groups of zooids (each individual in a colonial animal or protist is a zooid). In some species the number of zooids per colony is characteristic (Figure 12-11).

Traditionally, zoologists have considered phytoflagellates as protozoa, but botanists call them algae. As Phytomastigophorea they make up only one class of a single phylum, but as algae, they comprise six to nine divisions (a taxon of plants equivalent to a phylum). A curious anomaly: the same organisms are treated quite differently as taxa, depending on what course you take. We hope that further phylogenetic research clarifies these questions.

Among the most interesting of all the phytoflagellates are the dinoflagellates. They have a longitudinal and an equatorial flagellum, each borne at least partly in grooves on the body. The body may be naked or covered by cellulose plates or valves or by a cellulose membrane. Most dinoflagellates have brown or yellow chromatophores, although some are colorless. Many species, both colorless and pigmented, can ingest prey through a mouth region between the plates near the posterior area of the body. *Ceratium* (Figure 12-11), for example, has a thick covering with long spines, into which the body extends, but it can catch food with posterior pseudopodia and ingest it between the flexible plates in the posterior groove. *Noctiluca* (Figure 12-11), a colorless dinoflagellate, is a voracious predator and has a long, motile tentacle, near the base of which its single, short flagellum emerges. *Noctiluca* is one of many marine organisms that can produce light (bioluminescence). Several groups of phytoflagellates are planktonic primary producers (p. 796) in freshwater and marine environments; however, dinoflagellates are the most important, particularly in the sea. Zooxanthellae are dinoflagellates that live in mutualistic association in the tissues of certain invertebrates, including other protozoa, sea anemones, horny and stony corals, and clams. The association with stony corals is of ecological and economic importance because only corals with symbiotic zooxanthellae can form coral reefs (Chapter 14).

Dinoflagellates can also damage other organisms, such as when they produce a "red tide." Although this name was originally applied to situations in which the organisms reproduced in such profusion (producing a "bloom") that the water turned red from their color, any instance of a bloom producing detectable levels of toxic substances is now called a red tide. The water may be red, brown, yellow, or not remarkably colored at all. The toxic substances are apparently not harmful to the organisms that produce them, but they may be highly poisonous to fish and other marine life. Several different types of dinoflagellates and one species of cyanobacterium have been responsible for red tides. Red tides have resulted in considerable economic losses to the shellfish industry. Another flagellate produces a toxin that is concentrated in the food chain, especially in large, coral reef fishes. The illness produced in humans after eating such fish is known as ciguatera.

Euglena viridis. *Euglena viridis* (Figure 12-12) is a flagellate commonly studied in introductory zoology courses. Its natural habitat is freshwater streams and ponds where there is considerable vegetation. The organisms are spindle shaped and about 60 μm long, but some species of *Euglena* are smaller and some larger (*E. oxyuris* is 500 μm long). Just beneath the outer membrane of *Euglena* are proteinaceous strips and microtubules that form the **pellicle.** In *Euglena* the pellicle is flexible enough to permit bending, but in other euglenids it may be more rigid. A **flagellum** extends from a flask-shaped **reservoir** at the anterior end, and another, short flagellum ends within the reservoir. A **kinetosome** is found at the base of each flagellum, and a **contractile vacuole** empties into the reservoir. A red eyespot, or **stigma,** apparently functions in orientation to light. Within the cytoplasm are oval **chloroplasts** that bear chlorophyll and give the organism its greenish color. **Paramylon bodies** of various shapes are masses of a starchlike food storage material.

The nutrition of *Euglena* is normally autotrophic (holophytic), but if kept in the dark the organism makes use of saprozoic nutrition, absorbing nutrients through its body surface. Mutants of *Euglena* can be produced that have permanently lost their photosynthetic ability. Although *Euglena* does not ingest solid food, some euglenids are phagotrophic. *Peranema* has a cytostome that opens alongside the flagellar reservoir.

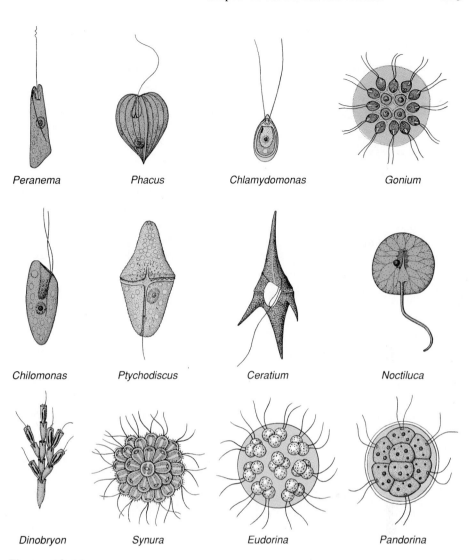

Figure 12-11

Diversity among the Phytomastigophorea. *Pandorina, Eudorina, Synura, Gonium,* and *Dinobryon* are colonial. *Ptychodiscus, Ceratium,* and *Noctiluca* are dinoflagellates. *Noctiluca, Peranema,* and *Chilomonas* have no pigments and are not photosynthetic. *Phacus* has two flagella, one of which is very short, as in *Euglena.*

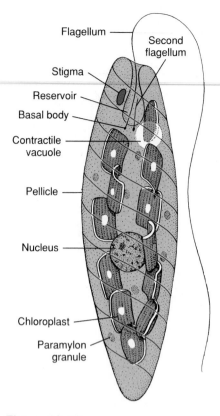

Figure 12-12

Euglena. Features shown are a combination of those visible in living and stained preparations.

Euglena reproduces by binary fission and can encyst to survive adverse environmental conditions.

Volvox globator. *Volvox* (Figure 12-13) is a colonial phytoflagellate (some biologists consider it multicellular because it contains separate somatic and reproductive cells, see p. 5) often studied in introductory courses because its mode of colony formation is somewhat similar to embryonic development of some metazoa. The order to which *Volvox* belongs (Volvocida) includes many freshwater flagellates, mostly green, with a cellulose cell wall through which two short flagella project. Many are colonial forms (Figure 12-11, *Pandorina, Eudorina, Gonium*).

Volvox (Figure 12-13) is a green, hollow sphere that may reach a diameter of 0.5 to 1 mm. It is a colony of many thousands of zooids (up to 50,000) embedded in the gelatinous surface of a jelly ball. Each cell is much like a euglenid, with a nucleus, a pair of flagella, a large chloroplast, and a red stigma. Adjacent cells are connected with each other by cytoplasmic strands. At one pole (usually in front as the colony moves), the stigmata are a little larger. Coordinated action of the flagella causes the colony to move by rolling over and over.

In *Volvox* we have a division of labor to the extent that most of the zooids are somatic cells concerned with nutrition and locomotion, and a few germ cells located in the posterior half are responsible for reproduction. Reproduction is asexual or sexual. In either case only certain zooids located around the equator or in the posterior half take part.

The original polarity of the zooids in the colony is such that the flagella are protruding into the interior cavity. To end up with the flagella on the outside, so that locomotion is possible, the entire spheroid must turn itself inside out. This process, called inversion, is *very unusual*. Of all other living organisms, only the sponges (phylum Porifera) have a comparable developmental process.

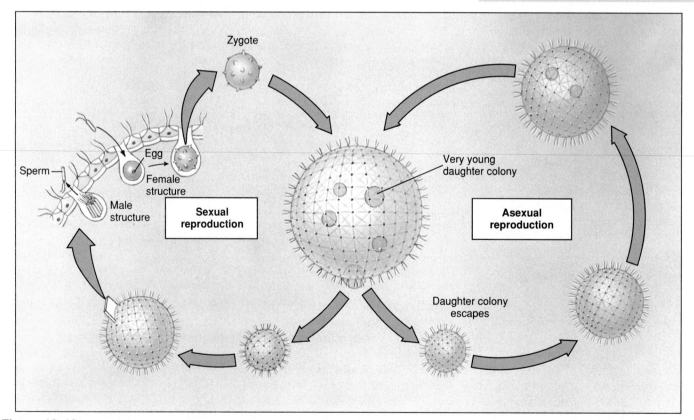

Figure 12-13

Life cycle of *Volvox*. Asexual reproduction occurs in spring and summer when specialized diploid reproductive cells divide to form young colonies that remain in the mother colony until large enough to escape. Sexual reproduction occurs largely in autumn when haploid sex cells develop. The fertilized ova may encyst and so survive the winter, developing into a mature asexual colony in the spring. In some species the colonies have separate sexes; in others both eggs and sperm are produced in the same colony.

Asexual reproduction in *Volvox* occurs by the repeated mitotic division of one of the germ cells to form a hollow sphere of cells, with the flagellated ends of the cells inside. The sphere then turns itself inside out to form a daughter colony similar to the parent colony. Several daughter colonies are formed inside the parent colony before they escape by rupture of the parent.

In **sexual reproduction** some of the zooids differentiate into **macrogametes** or **microgametes** (Figure 12-13). The macrogametes are fewer and larger and are loaded with food for nourishment of the young colony. The microgametes, by repeated division, form bundles or balls of small flagellated sperm that leave the mother colony when they mature and swim about to find a mature ovum. After fertilization, the zygote secretes a hard, spiny, protective shell around itself. When released by the breaking up of the parent colony, the zygote remains quiescent during the winter. Within the shell the zygote undergoes repeated division, producing a small colony that breaks out in the spring. A number of asexual generations may follow, during the summer, before sexual reproduction occurs again.

Zooflagellates. The zooflagellates are all colorless, lack chromoplasts, and have holozoic or saprozoic nutrition. Most are symbiotic.

Some of the most important protozoan parasites are zooflagellates. Many of them belong to the genus *Trypanosoma* (Figure 12-14) and live in the blood of fish, amphibians, reptiles, birds, and mammals. Some are nonpathogenic, but others produce severe diseases in humans and domestic animals. *Trypanosoma brucei gambiense* and *T. brucei rhodesiense* cause African sleeping sickness in humans, and *T. brucei brucei* causes a related disease in domestic animals. They are transmitted by the tsetse fly (*Glossina*). *Trypanosoma b. rhodesiense,* the more virulent of the sleeping sickness trypanosomes, and *T. b. brucei* have natural reservoirs (antelope and other wild mammals) that are apparently not harmed by

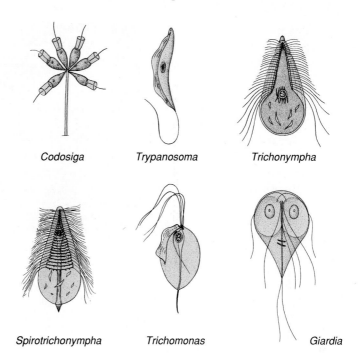

Figure 12-14

Some Zoomastigophorea. *Codosiga* is a colonial flagellate with cells similar to those found in sponges (phylum Porifera). The others are all symbiotic. *Trichonympha, Spirotrichonympha,* and *Trichomonas* are commonly found in the gut of termites and wood roaches, where they help digest cellulose from the wood eaten by the insects. Species of *Trichomonas* are also found in humans. *Trypanosoma* is a parasite of various animals, and some species cause serious disease in humans and domestic animals. *Giardia* is an intestinal parasite of mammals that causes diarrhea in humans.

the parasites. *Trypanosoma cruzi* causes Chagas' disease in humans in Central America and South America. It is transmitted by the bite of the "kissing bug" (Triatominae). Three species of *Leishmania* cause disease in humans. One is a serious visceral disease affecting especially the liver and spleen, one causes lesions in the mucous membranes of the nose and throat, and the least serious causes a type of skin lesion. They are transmitted by sand flies. Visceral leishmaniasis and cutaneous leishmaniasis are common in parts of Africa and Asia, and the mucocutaneous form is found in Central America and South America.

The several species of *Trichomonas* (Figure 12-14) are symbiotic. *Trichomonas hominis* is found in the cecum and colon of humans and apparently causes no disease. *T. vaginalis* inhabits the urogenital tract of humans, is transmitted venereally, and may cause vaginitis. Other species of *Trichomonas* are widely distributed through all classes of vertebrates and many invertebrates.

Giardia lamblia often causes no disease in the intestine of humans but may sometimes produce severe diarrhea. It is transmitted through fecal contamination.

Giardia lamblia is commonly transmitted through water supplies contaminated with sewage. The same species, however, lives in a variety of mammals other than humans. Beavers seem to be an important source of infection in the mountains of the western United States. When one has hiked for miles in the wild on a hot day, it can be very tempting to fill a canteen and drink from a crystal-clear beaver pond. Many cases of infection have been acquired that way.

Subphylum Sarcodina

Superclass Rhizopoda

Amoeba proteus. Of the amebas, the most commonly studied species is *Amoeba proteus.* They live in slow streams and ponds of clear water, often in shallow water on aquatic vegetation or on the sides of ledges. They are rarely found free in water, for they require a substratum on which to crawl. They have an irregular shape because of their power to thrust out lobopodia at any point on their bodies. They are colorless and about 250 to 600 µm in greatest diameter. Unlike *Euglena,* the **pellicle** consists only of the cell membrane. **Ectoplasm** and **endoplasm** are prominent. Organelles such as **nucleus, contractile vacuole, food vacuoles,** and small **vesicles** can be observed easily with the light microscope. Amebas live on algae, protozoa, rotifers, and even other amebas, upon which they feed by phagocytosis. An ameba can live for many days without food but decreases in volume during this process. The time necessary for the digestion by a food vacuole varies with the kind of food but is usually around 15 to 30 hours. When the ameba reaches full size, it divides by binary fission with typical mitosis.

Other Rhizopoda. There are many species of amebas; for example, *A. verrucosa* has short pseudopodia; *Chaos carolinense* (*Pelomyxa carolinensis*) is several times as large as *A. proteus;* and *A. radiosa* has many slender pseudopodia.

There are many entozoic amebas, most of which live in the intestines of humans or other animals. Two common genera are *Endamoeba* and *Entamoeba. Endamoeba blattae* is an endocommensal in the intestine of cockroaches, and related species are found in termites. *Entamoeba histolytica* is the most important rhizopod parasite of humans. It lives in the large intestine and on occasion can invade the intestinal wall by secreting enzymes that attack the intestinal lining. If this occurs, a serious and sometimes fatal amebic dysentery may result. The organisms may be carried by the blood to the liver and other organs and cause abscesses there. Many infected persons show few or no symptoms but are carriers, passing cysts in their feces. Infection is spread by contaminated water or food containing the cysts.

Figure 12-15

A, Living foraminiferan, showing thin pseudopodia extending from test. **B,** Test of foraminiferan, *Vertebralima striata.* Foraminiferans (class Granuloreticulosea) are ameboid marine protozoans that secrete a calcareous, many-chambered test in which to live and then extrude protoplasm through pores to form a layer over the outside. The animal begins with one chamber, and as it grows, it secretes a succession of new and larger chambers, continuing this process throughout life. Many foraminiferans are planktonic, and when they die, their shells are added to the ooze on the ocean's bottom.

Other species of *Entamoeba* found in humans are *E. coli* in the intestine and *E. gingivalis* in the mouth. Neither of these species is known to cause disease.

Not all rhizopods are "naked" as are the amebas. Some have their delicate plasma membrane covered with a protective **test** or shell. *Arcella* and *Difflugia* (Figure 12-16) are common sarcodines. They have a test of secreted siliceous or chitinoid material that may be reinforced with grains of sand. They move by means of pseudopodia that project from openings in the shell.

The **foraminiferans** (class Granuloreticulosea) are an ancient group of shelled rhizopods found in all oceans, with a few in fresh and brackish water. Most foraminiferans live on the ocean floor in incredible numbers, having perhaps the largest biomass of any animal group on earth. Their tests are of numerous types (Figures 12-15 and 12-16). Most tests are many chambered and are made of calcium carbonate, although they sometimes use silica, silt, and other foreign materials. Slender pseudopodia extend through openings in the test, then branch and run together to form a protoplasmic net (**reticulopodia**) in which they ensnare their prey. Here the captured prey is digested, and the digested products are carried into the interior by the flowing protoplasm. The life

cycles of foraminiferans are complex, for they have multiple fission and alternation of haploid and diploid generations (intermediary meiosis).

Some of the slime molds (class Eumycetozoa), especially *Dictyostelium discoideum,* have been studied intensively because of their fascinating developmental cycle. Under natural conditions this species lives in forest detritus throughout the world. It feeds on bacteria and reproduces by binary fission as long as the food supply is plentiful. When food runs short, however, the amebas are attracted to each other, streaming toward a central point to form a **pseudoplasmodium** (large mass of discrete cells). Under the same conditions, some species actually fuse to become a large multinucleate individual (**plasmodium**). The pseudoplasmodium of *Dictyostelium* may migrate some distance to a favorable location, where it forms a stalk with a fruiting body on top (Figure 12-17). It forms resistant cysts within the fruiting body, which are widely dispersed upon rupture of the fruiting body. Many details about the development, genetics, and biochemistry of these organisms are known.

Superclass Actinopoda. The Actinopoda is composed of the mostly freshwater class Heliozoea and the three marine classes Acantharea,

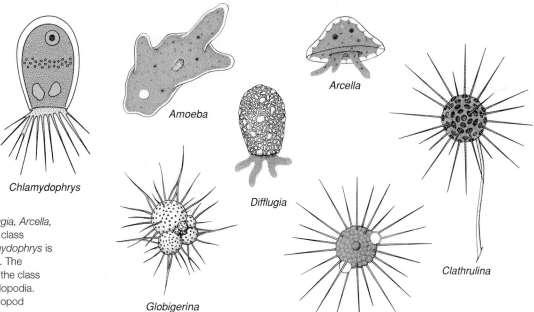

Figure 12-16

Diversity among the Sarcodina. *Difflugia, Arcella,* and *Amoeba* belong to the rhizopod class Lobosea and have lobopodia. *Chlamydophrys* is in the class Filosea and has filopodia. The foraminiferan *Globigerina* belongs to the class Granuloreticulosea and shows reticulopodia. *Actinophrys* and *Clathrulina* are actinopod heliozoeans. They have axopodia.

Phaeodarea, and Polycystinea. Members of the marine classes are commonly known as **radiolarians.** All have axopodia, and except for some of the heliozoeans, they have tests (Figure 12-18). These protozoa are beautiful little organisms.

The biological characteristics of the freshwater Heliozoea are somewhat better known than those of the other actinopods. Examples are *Actinosphaerium,* which is about 1 mm in diameter and can be seen with the unassisted eye, and *Actinophrys* (Figure 12-16), only 50 μm in diameter; neither has a test. *Clathrulina* (Figure 12-16) secretes a latticed test.

The oldest known protozoa are found among the radiolarians. Radiolarians are nearly all pelagic (live in open water). Most of them are planktonic in shallow water, although some live in deep water. Their highly specialized skeletons are intricate in form and of great beauty (Figure 12-18). The body is divided by a central capsule that separates inner and outer zones of cytoplasm. The central capsule, which may be spherical, ovoid, or branched, is perforated to allow cytoplasmic continuity. The skeleton is made of silica, strontium sulfate, or a combination of silica and organic matter and usually has a radial arrangement of spines that

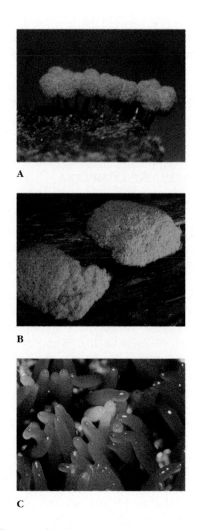

Figure 12-17

Fruiting bodies of three genera of plasmodial slime molds. **A,** *Arcyria.* **B,** *Fuligo.* **C,** *Tubifera.*

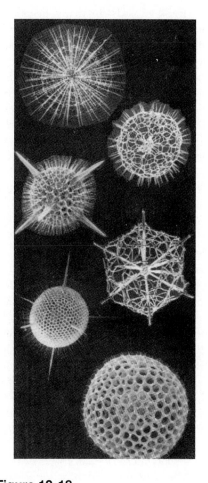

Figure 12-18

Types of radiolarian tests (class Polycystinea). In his study of these beautiful forms collected on the famous *Challenger* expedition of 1872 to 1876, Haeckel worked out our present concepts of symmetry.

extend through the capsule from the center of the body. At the surface a shell may be fused with the spines. Around the capsule is a frothy mass of cytoplasm from which axopodia arise (p. 218). These are sticky to catch the prey, which are carried by the streaming protoplasm to the central capsule to be digested. The ectoplasm on one side of the axial rod moves outward, or toward the tip, while on the other side it moves inward, or toward the test.

Radiolarians may have one or many nuclei. Their life history is not completely known, but binary fission, budding, and sporulation have been observed in them.

Role of Sarcodina in Building Earth Deposits. The foraminiferans and the radiolarians have existed since Precambrian times and have left excellent fossil records. In many instances their hard shells have been preserved unaltered. Many of the extinct species closely resemble present-day ones. They were especially abundant during the Cretaceous and Tertiary periods. Some of them were among the largest protozoa that have ever existed, measuring up to 100 mm (about 4 in) or more in diameter.

For untold millions of years the tests of dead foraminiferans have been sinking to the bottom of the ocean, building up a characteristic ooze rich in lime and silica. About one-third of the sea bottom is covered with ooze that is made up of the shells of the genus *Globigerina*. This ooze is especially abundant in the Atlantic Ocean.

Radiolarians (Figure 12-18), with their less soluble siliceous shells, are usually found at greater depths (4600 to 6100 meters), mainly in the Pacific and Indian oceans. Radiolarian ooze probably covers about 5 to 8 million square kilometers. Under certain conditions, radiolarian ooze forms rocks (chert). Many fossil radiolarians are found in the Tertiary rocks of California.

The thickness of these deep-sea sediments has been estimated at 700 to 4000 m. Although the average rate of sedimentation must vary greatly, it is always very slow. *Globigerina* ooze has

probably increased 1 to 12.5 mm in 1000 years. As many as 50,000 shells of foraminiferans may be found in a single gram of sediment, which gives some idea of the magnitude of numbers of these microorganisms and the length of time it has taken them to form the sediment carpet on the ocean floor.

Of equal interest and of greater practical importance are the limestone and chalk deposits that were laid down by the accumulation of these microorganisms when sea covered what is now land. Later, through a rise in the ocean floor and other geological changes, this sedimentary rock emerged as dry land. The chalk deposits of many areas of England, including the White Cliffs of Dover, were formed in this way. The great pyramids of Egypt were made from stone quarried from limestone beds that were formed by a very large foraminiferan population that flourished during the early Tertiary period.

Since fossil foraminiferans and radiolarians can be brought up in well drillings, their identification is often important to oil geologists for correlation of rock strata.

PHYLUM APICOMPLEXA

All apicomplexans are endoparasites, and their hosts are found in many animal phyla. The presence of a certain combination of organelles, the **apical complex,** distinguishes this subphylum (Figure 12-19A). The apical complex is usually present only in certain developmental stages of the organisms; for example, **merozoites** and **sporozoites** (Figure 12-20). Some of the structures, especially the **rhoptries** and **micronemes,** apparently aid in penetrating the host's cells or tissues.

Locomotor organelles are less obvious in this group than in other protozoa. Pseudopodia occur in some intracellular stages, and gametes of some species are flagellated. Tiny contractile fibrils can form waves of contraction across the body surfaces to propel the organism through a liquid medium.

The life cycle usually includes both asexual and sexual reproduction,

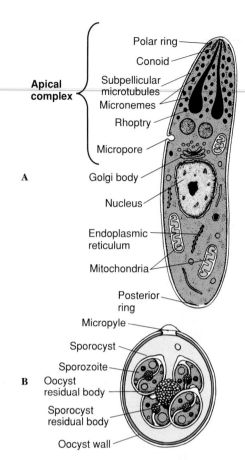

Figure 12-19

A, Diagram of an apicomplexan sporozoite or merozoite at the electron microscope level, illustrating the apical complex. The polar ring, conoid, micronemes, rhoptries, subpellicular microtubules, and micropore (cytostome) are all considered components of the apical complex. **B,** Infective oocyst of *Eimeria.* The oocyst is the resistant stage and has undergone multiple fission after zygote formation (sporogony).

and there is sometimes an invertebrate intermediate host. At some point in the life cycle, the organisms develop a **spore (oocyst),** which is infective for the next host and is often protected by a resistant coat.

Class Sporozoea

The most important class of the phylum Apicomplexa, the Sporozoea, contains three subclasses: the Gregarinia, the Coccidia, and the Piroplasmia. The gregarines are common parasites of invertebrates, but they are of little economic significance. Piroplasms are of some veterinary importance; for example, *Babesia bigemina* causes

Texas red-water fever in cattle. Humans are occasionally infected with species of *Babesia* normally parasitic in other animals.

Subclass Coccidia.   The Coccidia are intracellular parasites in invertebrates and vertebrates, and the group includes species of very great medical and veterinary importance.

Eimeria.   The name "coccidiosis" is generally applied only to infections with *Eimeria* or *Isospora.* Humans can be infected with species of *Isospora,* but there is usually little disease. However, *Isospora* infections can be very serious in AIDS patients. Some species of *Eimeria* may cause serious disease in some domestic animals. The symptoms usually include severe diarrhea or dysentery.

Eimeria tenella is often fatal to young fowl, producing severe pathogenesis in the intestine. The organisms undergo schizogony (p. 220) in the intestinal cells, finally producing gametes. After fertilization the zygote forms an oocyst that passes out of the host in the feces (Figure 12-19B). Sporogony occurs within the oocyst outside the host, producing eight sporozoites in each oocyst. Infection occurs when a new host accidentally ingests a sporulated oocyst and the sporozoites are released by digestive enzymes.

Toxoplasma gondii.   A similar life cycle occurs in *Toxoplasma gondii,* a parasite of cats, but this species produces extraintestinal stages as well. The extraintestinal stages can develop in a wide variety of animals other than cats—for example, rodents, cattle, and humans. Gametes and oocysts are not produced by the extraintestinal forms, but they can initiate the intestinal cycle in a cat that eats infected prey. In humans *Toxoplasma* causes little or no ill effects except in an AIDS patient or in a woman infected during pregnancy, particularly in the first trimester. Such infection greatly increases the chances of a birth defect in the baby; it is now believed that 2% of all mental retardation in the United States is the result of con-

genital toxoplasmosis. Toxoplasmosis can also become a serious disease in persons who are immunosuppressed, either with drugs or in AIDS. The normal route of infection for humans is apparently the consumption of infected meat that is insufficiently cooked.

> Another important source of infection with *Toxoplasma* is domestic cats. The oocysts are passed in the cat's feces, and a pregnant woman should not empty the litter box. If such a chore cannot be avoided, daily clean-up should be acceptable because it takes three days for the oocysts to sporulate and become infective.

Plasmodium: The Malarial Organism.   The best known of the coccidians is *Plasmodium,* the causative organism of the most important infectious disease of humans: **malaria.** Malaria is a very serious disease, difficult to control and widespread, particularly in tropical and subtropical countries. Four species of *Plasmodium* infect humans. Although each produces its own peculiar clinical picture, all four have similar cycles of development in their hosts (Figure 12-20).

The parasite is carried by mosquitoes (*Anopheles*), and sporozites are injected into the human with the insect's saliva during its bite. The sporozoites penetrate liver cells and initiate schizogony. The products of this division then enter other liver cells to repeat the schizogonous cycle, or in *P. falciparum* they penetrate the red blood cells after only one cycle in the liver. The period when the parasites are in the liver is the **incubation period,** and it lasts from 6 to 15 days, depending on the species of *Plasmodium.*

Merozoites released as a result of the liver schizogony enter red blood cells, where they begin a series of schizogonous cycles. When they enter the cells, they become ameboid **trophozoites,** feeding on hemoglobin. The end product of the parasite's digestion of hemoglobin is a dark, insoluble pigment: **hemozoin.** The hemozoin accumulates in the host cell, is

released when the next generation of merozoites is produced, and eventually accumulates in the liver, spleen, or other organs. The trophozoite within the cell grows and undergoes schizogony, producing 6 to 36 merozoites, depending on the species, which burst forth to infect new red cells. When the red blood cell containing the merozoites bursts, it releases the parasite's metabolic products, which have accumulated there. The release of these foreign substances into the patient's circulation results in the chills and fever characteristic of malaria.

Since the populations of schizonts maturing in the red blood cells are synchronized to some degree, the episodes of chills and fever have a periodicity characteristic of the particular species of *Plasmodium.* In *P. vivax* (benign tertian) malaria and *P. ovale* malaria, the episodes occur every 48 hours; in *P. malariae* (quartan) malaria, every 72 hours; and in *P. falciparum* (malignant tertian) malaria, about every 48 hours, although the synchrony is less well defined in this species. People usually recover from infections with the first three species, but mortality may be high in untreated cases of *P. falciparum* infection. Sometimes grave complications, such as **cerebral malaria,** occur. Unfortunately, *P. falciparum* is the most common species, accounting for 50% of all malaria in the world. Certain genes, for example the gene for sickle cell hemoglobin (p. 146 and p. 693), confer some resistance to malaria on people that carry them.

After some cycles of schizogony in the red blood cells, infection of new cells by some of the merozoites results in the production of **microgametocytes** and **macrogametocytes** rather than another generation of merozoites. When the gametocytes are ingested by a mosquito feeding on the patient's blood, they mature into **gametes,** and fertilization occurs. The zygote becomes a motile **ookinete,** which penetrates the stomach wall of the mosquito and becomes the **oocyst.** Within the oocyst, sporogony occurs, and thousands of **sporozoites** are produced. The oocyst ruptures, and the sporozoites migrate to

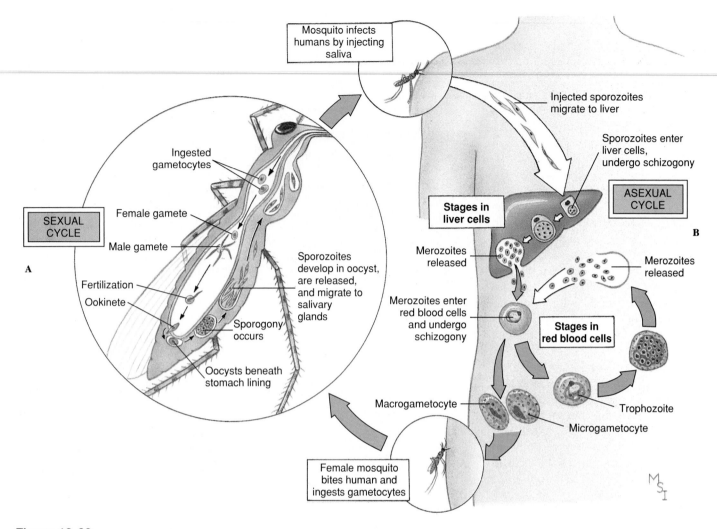

Figure 12-20

Life cycle of *Plasmodium vivax,* one of the protozoa (class Sporozoea) that causes malaria in humans. **A,** Sexual cycle produces sporozoites in body of mosquito. Meiosis occurs just after zygote formation (zygotic meiosis). **B,** Sporozoites infect a human and reproduce asexually, first in liver cells and then in red blood cells. Malaria is spread by *Anopheles* mosquito, which sucks up gametocytes along with human blood, then, when biting another victim, leaves sporozoites in new wound.

the salivary glands, from which they are transferred to a human by the bite of the mosquito. Development in the mosquito requires 7 to 18 days but may be longer in cool weather.

The elimination of mosquitoes and their breeding places by insecticides, drainage, and other methods has been effective in controlling malaria in some areas. However, the difficulties in carrying out such activities in remote areas and the acquisition of resistance to insecticides by mosquitoes and to antimalarial drugs by *Plasmodium* (especially *P. falciparum*) mean that malaria will be a serious disease of humans for a long time to come.

Other species of *Plasmodium* parasitize birds, reptiles, and mammals. Those of birds are transmitted chiefly by the *Culex* mosquito.

PHYLUM CILIOPHORA

The ciliates are a large and interesting group, with a great variety of forms living in all types of freshwater and marine habitats. They are the most structurally complex and diversely specialized of all the protozoa. The majority are free living, but some are commensal or parasitic. They are usually solitary and motile, but some are sessile and some colonial. There is great diversity of shape and size. In general,

they are larger than most other protozoa, but they range from very small (10 to 12 μm) up to 3 mm long. All have cilia that beat in a coordinated rhythmical manner, although the arrangement of the cilia may vary and some lack cilia as adults.

Ciliates are always multinucleate, possessing at least one **macronucleus** and one **micronucleus,** but varying from one to many of either type. The macronuclei are apparently responsible for metabolic and developmental functions and for maintaining all the visible traits, such as the pellicular apparatus. Macronuclei vary in shape among the different species (Figures 12-21 and 12-24). The mi-

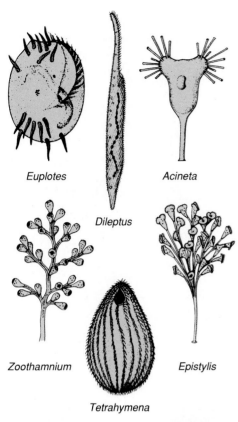

Euplotes

Acineta

Dileptus

Zoothamnium

Epistylis

Tetrahymena

Vorticella   *Stentor*

Figure 12-21
Some representative ciliates. *Euplotes* have stiff cirri used for crawling about. Contractile fibrils in ectoplasm of *Stentor* and in stalks of *Vorticella* allow great expansion and contraction. Note the macronuclei, long and curved in *Euplotes* and *Vorticella,* shaped like a string of beads in *Stentor.*

cronuclei participate in sexual reproduction and give rise to macronuclei after exchange of micronuclear material between individuals. The micronuclei divide mitotically, and the macronuclei divide amitotically (see p. 221).

The **pellicle** of ciliates may consist only of the cell membrane or in some species may form a thickened

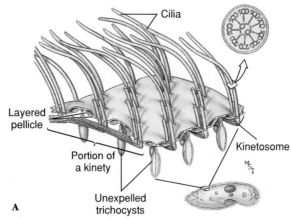

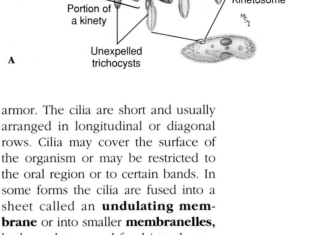

Cilia

Layered pellicle

Portion of a kinety

Kinetosome

Unexpelled trichocysts

A

armor. The cilia are short and usually arranged in longitudinal or diagonal rows. Cilia may cover the surface of the organism or may be restricted to the oral region or to certain bands. In some forms the cilia are fused into a sheet called an **undulating membrane** or into smaller **membranelles,** both used to propel food into the **cytopharynx** (gullet). In other forms there may be fused cilia forming stiffened tufts called **cirri,** often used in locomotion by the creeping ciliates (Figure 12-21).

An apparently structural system of fibers, in addition to the kinetosomes, makes up the **infraciliature,** just beneath the pellicle (Figure 12-22). Each cilium terminates beneath the pellicle in its kinetosome, and from each kinetosome a fibril arises and passes along beneath the row of cilia, joining with the other fibrils of that row. The cilia, kinetosomes, and other fibrils of that ciliary row make up what is known as a **kinety** (Figure 12-22). All ciliates seem to have kinety systems, even those that lack cilia at some stage. The infraciliature apparently does not coordinate the ciliary beat, as formerly thought. Coordination of the ciliary movement seems to be by waves of depolarization of the cell membrane moving down the animal, similar to the phenomenon in a nerve impulse.

Most ciliates are holozoic. Most of them possess a cytostome (mouth) that in some forms is a simple opening and in others is connected to a gullet or ciliated groove. The mouth in some is strengthened with stiff, rodlike

B

Figure 12-22
Infraciliature and associated structures in ciliates. **A,** Structure of the pellicle and its relation to the infraciliature system. **B,** Expelled trichocyst.

trichites for swallowing larger prey; in others, such as the paramecia, ciliary water currents carry microscopic food particles toward the mouth. *Didinium* has a proboscis for engulfing the paramecia on which it feeds (Figure 12-8). Suctorians paralyze their prey and then ingest the contents through tube-like tentacles by a complex feeding mechanism that apparently combines phagocytosis with a sliding filament action of microtubules in the tentacles (Figure 12-8).

Some ciliates have curious small bodies in their ectoplasm between the bases of the cilia. Examples are **trichocysts** (Figures 12-22 and 12-24) and **toxicysts.** Upon mechanical or chemical stimulation, these bodies explosively expel a long, thread-like structure. The mechanism of expulsion is unknown. The function of trichocysts is thought to be defensive, although this is unclear. When attacked by a *Didinium,* a paramecium expels its trichocysts but to no avail. Toxicysts, however, release a poison that paralyzes the prey of carnivorous ciliates. Toxicysts are structurally quite distinct from trichocysts. Many dinoflagellates have structures very similar to trichocysts.

Among the more striking and familiar of the ciliates are *Stentor* (Gr. herald with a loud voice), trumpet shaped and solitary, with a bead-shaped macronucleus (Figure 12-21); *Vorticella* (L. dim. of *vortex,* a whirlpool), bell shaped and attached by a contractile stalk (Figure 12-21); and *Euplotes* (Gr. *eu,* true, good, + *ploter,* swimmer) with a flattened body and groups of fused cilia (cirri) that function as legs.

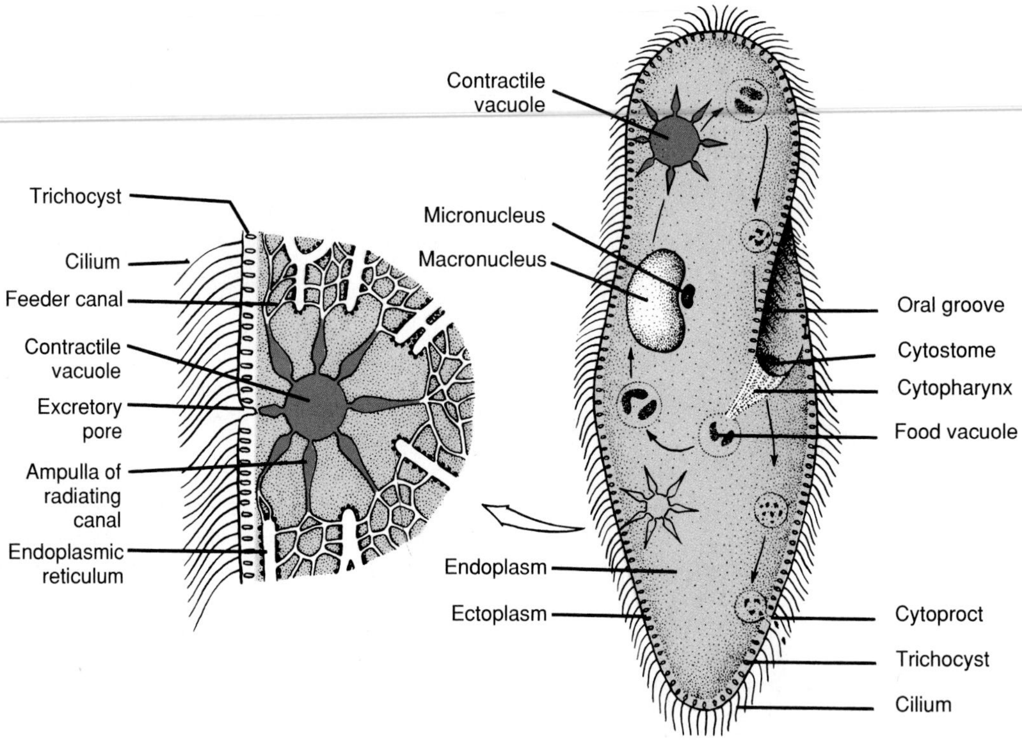

Paramecium: A Representative Ciliate

Paramecia are usually abundant in ponds or sluggish streams containing aquatic plants and decaying organic matter.

Form and Function. Paramecia are often described as slipper shaped. *Paramecium caudatum* is 150 to 300 μm in length and is blunt anteriorly and somewhat pointed posteriorly (Figure 12-23). The organism has an asymmetrical appearance because of the **oral groove,** a depression that runs obliquely backward on the ventral side.

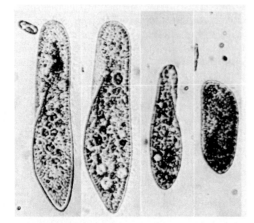

Figure 12-23

Comparison of four common species of *Paramecium* photographed at the same magnification. *Left to right: P. multimicronucleatum, P. caudatum, P. aurelia,* and *P. bursaria.*

Figure 12-24

Left, enlarged section of a contractile vacuole (water expulsion vesicle) of *Paramecium.* Water is apparently collected by endoplasmic reticulum, emptied into feeder canals and then into the vesicle. The vesicle contracts to empty its contents to the outside, thus serving as an osmoregulatory organelle. *Right, Paramecium,* showing cytopharynx, food vacuoles, and nuclei.

The **pellicle** is a clear, elastic membrane that may be ornamented by ridges or papillalike projections (Figure 12-22), and its entire surface is covered with cilia arranged in lengthwise rows. Just below the pellicle is the thin clear **ectoplasm** that surrounds the larger mass of granular **endoplasm** (Figure 12-24). Embedded in the ectoplasm just below the surface are the spindle-shaped **trichocysts,** which alternate with the bases of the cilia. The infraciliature can be seen only with special fixing and staining methods.

The **cytostome** at the end of the oral groove leads into a tubular **cytopharynx,** or **gullet.** Along the gullet an undulating membrane of modified cilia keeps food moving. Fecal material is discharged through a **cytoproct** posterior to the oral groove (Figure 12-24). Within the endoplasm are food vacuoles containing food in various stages of digestion. There are two **contractile vacuoles,** each consisting of a central space surrounded by several **radiating canals** (Figure 12-24) that collect fluid and empty it into the central vacuole. We described excretion and osmoregulation on p. 219.

Paramecium caudatum has two nuclei: a large kidney-shaped **macronucleus** and a smaller **micronucleus** fitted into the depression of the former. These can usually be seen only in stained specimens. The number of micronuclei varies in different species, for example, *P. multimicronucleatum* may have as many as seven.

Paramecia are holozoic, living on bacteria, algae, and other small organisms. The cilia in the oral groove sweep food particles in the water into the cytostome, from which point they are carried into the cytopharynx by the undulating membrane. From the cytopharynx the food is collected into a food vacuole that is constricted into the endoplasm. The food vacuoles circulate in a definite course through the cytoplasm while the food is being digested by enzymes from the endoplasm. The indigestible part of the food is ejected through the cytoproct.

The body is elastic, allowing it to bend and squeeze its way through narrow places. Its cilia can beat either forward or backward, so that the organism can swim in either direction. The cilia beat obliquely, causing the

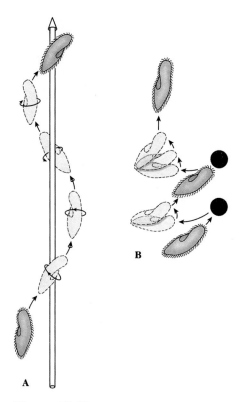

Figure 12-25
A, Spiral path of swimming *Paramecium*.
B, Avoidance reaction of *Paramecium*.

increases the rate of the forward ciliary beat, and depolarization results in ciliary reversal and backward swimming.

Locomotor responses, by which an organism more or less continuously orients itself with respect to a stimulus, are called *taxes* (sing., *taxis*). Movement toward the stimulus is a positive taxis; movement away is a negative taxis. Some examples are thermotaxis, response to heat; phototaxis, response to light; thigmotaxis, response to contact; chemotaxis, response to chemical substances; rheotaxis, response to currents of air or water; galvanotaxis, response to constant electric current; and geotaxis, response to gravity. Some stimuli do not cause an orienting response but simply a change in movement: more rapid movement, more frequent random turning, or slowing or cessation of movement. Such responses are known as kineses. Is the avoiding reaction of a paramecium a taxis or a kinesis?

organism to rotate on its long axis. In the oral groove the cilia are longer and beat more vigorously than the others so that the anterior end swerves aborally. As a result of these factors, the organism moves forward in a spiral path (Figure 12-25A).

When a ciliate, such as a paramecium, comes in contact with a barrier or a disturbing chemical stimulus, it reverses its cilia, backs up a short distance, and swerves the anterior end as it pivots on its posterior end. This is called an **avoiding reaction** (Figure 12-25B). It may continue to change its direction to keep itself some distance away from the noxious stimulus, and it may react in a similar fashion to keep itself within the zone of an attractant. The paramecium may also change its swimming speed. How does the paramecium "know" when to change directions or swimming speed? Interestingly, the reactions of the organism depend on the effects of the stimulus on the electrical potential difference across its cell membrane. The paramecia slightly hyperpolarize in attractants and depolarize in repellents that produce the avoiding reaction. Hyperpolarization

Reproduction. Paramecia reproduce only by binary fission across kineties (ciliary rows) but have certain forms of sexual phenomena called conjugation and autogamy.

In **binary fission** the micronucleus divides mitotically into two daughter micronuclei, which move to opposite ends of the cell (Figure 12-26). The macronucleus elongates and divides amitotically.

Conjugation occurs at intervals in ciliates. Conjugation is the temporary union of two individuals to exchange chromosomal material (Figure 12-27). During the union the macronucleus disintegrates and the micronucleus of each individual undergoes meiosis, giving rise to four haploid micronuclei, three of which degenerate (Figure 12-27A to D). The remaining micronucleus then divides into two haploid pronuclei, one of which is exchanged with the other conjugant. The pronuclei fuse to restore the diploid number of chromosomes, followed by several more nuclear events detailed in Figure 12-27. Following this complicated process, the protists may con-

tinue to reproduce by binary fission without the necessity of conjugation.

The result of conjugation is similar to that of zygote formation, for each exconjugant contains hereditary material from two individuals. The advantage of sexual reproduction is that it permits gene recombinations, thus increasing genetic variation in the population. Although ciliates in clone cultures can apparently reproduce repeatedly and indefinitely without conjugation, the stock seems eventually to lose vigor. Conjugation restores vitality to a stock. Seasonal changes or a deteriorating environment will usually stimulate sexual reproduction.

Autogamy is a process of self-fertilization that is similar to conjugation except that there is no exchange of nuclei. After the disintegration of the macronucleus and the meiotic divisions of the micronucleus, the two haploid pronuclei fuse to form a synkaryon that is homozygous (Chapter 8, p. 127), rather than heterozygous as in the case of exconjugants.

Symbiotic Ciliates

Many symbiotic ciliates live as commensals, but some can be harmful to their hosts. *Balantidium coli* lives in the large intestine of humans, pigs, rats, and many other mammals (Figure 12-28). There seem to be host-specific strains, and the organism is not easily transmitted from one species to another. Transmission is by fecal contamination of food or water. Usually the organisms are not pathogenic, but in humans they sometimes invade the intestinal lining and cause a dysentery similar to that caused by *Entamoeba histolytica*. The disease can be serious and even fatal. Infections are common in parts of Europe, Asia, and Africa but are rare in the United States.

Other species of ciliates live in other hosts. *Entodinium* (Figure 12-28) belongs to a group that has very complex structure and lives in the digestive tract of ruminants, where they may be very abundant. *Nyctotherus* live in the colon of frogs and toads. In aquarium and wild freshwater fish, *Ichthyophthirius* causes a disease known to many fish culturists as "ick." Untreated, it can cause much loss of exotic fish.

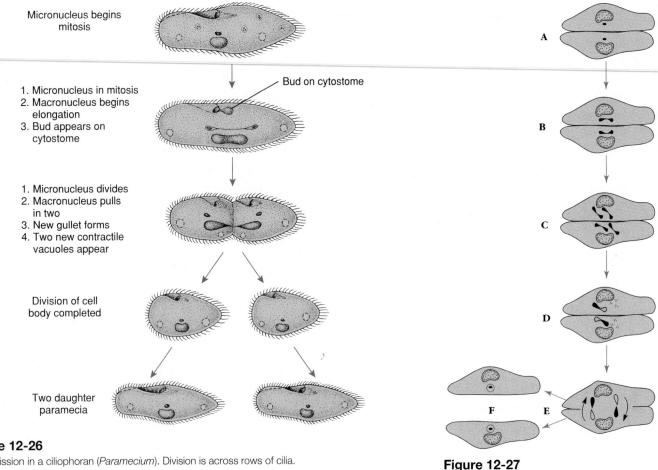

Micronucleus begins mitosis

Bud on cytostome

1. Micronucleus in mitosis
2. Macronucleus begins elongation
3. Bud appears on cytostome

1. Micronucleus divides
2. Macronucleus pulls in two
3. New gullet forms
4. Two new contractile vacuoles appear

Division of cell body completed

Two daughter paramecia

A

B

C

D

E

F

Figure 12-26
Binary fission in a ciliophoran (*Paramecium*). Division is across rows of cilia.

Figure 12-27
Scheme of conjugation in *Paramecium*. **A,** Two individuals come in contact on oral surface. **B** and **C,** Micronuclei divide twice (meiosis), resulting in four haploid nuclei in each partner. **D,** Three micronuclei degenerate; remaining one divides to form "male" and "female" pronuclei. **E,** Male pronuclei are exchanged between conjugants. **F,** Male and female pronuclei fuse, and individuals separate. Subsequently old macronuclei are absorbed and replaced by new macronuclei.

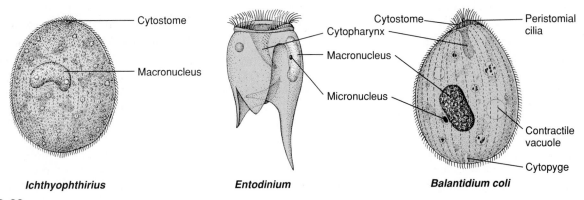

Cytostome

Macronucleus

Ichthyophthirius

Cytostome

Cytopharynx

Macronucleus

Micronucleus

Entodinium

Cytostome

Peristomial cilia

Contractile vacuole

Cytopyge

Balantidium coli

Figure 12-28
Some symbiotic ciliates. *Balantidium coli* is a parasite of humans and other mammals. *Ichthyophthirius* causes a common disease in aquarium and wild freshwater fish. *Entodinium* is found in the rumen of cows and sheep.

Suctorians

Suctorians are ciliates in which the young possess cilia and are free swimming, and the adults grow a stalk for attachment, become sessile, and lose their cilia. They have no cytostome but feed by long, slender, tubelike tentacles. The suctorian captures living prey, usually a ciliate, by the tip of one or more tentacles and paralyzes it. The cytoplasm of the prey then flows through the attached tentacles, forming food vacuoles in the feeding suctorian (Figure 12-8).

One of the best places to find freshwater suctorians is in the algae that grow on the carapace of turtles. Common genera of suctorians found there are *Anarma* (without stalk or test) and *Squalorophrya* (with stalk and test). Other freshwater representatives are *Podophrya* (Figure 12-8) and *Dendrosoma. Acinetopsis* and *Ephelota* are saltwater forms.

Suctorian parasites include *Trichophrya,* the species of which are found on a variety of invertebrates and freshwater fish; *Allantosoma,* which live in the intestine of certain mammals; and *Sphaerophrya,* which are found in *Stentor.*

PHYLOGENY AND ADAPTIVE RADIATION

PHYLOGENY

Protozoa represent early phylogenetic splits from all multicellular (metazoan) animals. The common ancestor of protozoan and metazoan animals was almost certainly unicellular, and the evolutionary derivation of multicellular forms has occurred more than once. Sponges, for example, may well have been derived separately from other metazoa. Some protozoa, particularly among the colonial and multicellular flagellates (Figure 12-11), show various degrees of cell aggregation and some differentiation that may parallel the body plans of early metazoa.

The mechanism of ciliary coordination and kineses shown by ciliates, depending on establishment of a membrane potential, polarization, and depolarization, suggests the antecedent among the protozoa of the mechanism of neural transmission, coordination, and cell-to-cell communication among the metazoa.

With the exception of certain shell-bearing Sarcodina, such as foraminiferans and radiolarians, protozoa have left no fossil records. Mastigophorans may be the oldest of all protozoa, perhaps having arisen from bacteria and spirochetes, but the group is probably polyphyletic. Some of the phytoflagellates are more closely related to the green algae than they are to zooflagellates. Some colorless phytoflagellates have chlorophyll-bearing relatives, and some autotrophic forms are facultatively saprophytic in darkness. Hence, the Phytomastigophorea may include the closest protistan relatives of animals and plants. Common ancestry of the amebas and the flagellates is indicated by the pseudopodia of some flagellates and the flagellated states of some amebas.

Apicomplexa, which are all specialized parasites, probably came from flagellated ancestors: they often have ameboid feeding stages and flagellated gametes. Aggregation and fruiting body formation in the Eumycetozoa suggest affinities to the fungi, as is indicated by their common name, slime molds, and some scientists do consider them fungi. The origin of the ciliates is somewhat obscure, but the basic structural similarity of the flagellum to the cilium seems to indicate that the ciliates and flagellates share common ancestry.

ADAPTIVE RADIATION

We have described some of the wide range of adaptations of protozoa in the preceding pages. The Sarcodina range from bottom-dwelling, naked species to planktonic forms such as the foraminiferans and radiolarians with beautiful, intricate tests. There are many symbiotic species of amebas. Flagellates likewise show adaptations for a similarly wide range of habitats, with the added variation of photosynthetic ability in many species of Phytomastigophorea. In fact, botanists usually refer to these organisms as unicellular algae and divide them into some six to nine divisions (botanical equivalents of phyla).

Within a single-cell body plan, the division of labor and specialization of organelles are carried furthest in the ciliates. These have become the most complex of all protozoa. Specializations for intracellular parasitism have been adopted by the Apicomplexa, Microspora, and Myxozoa.

CLASSIFICATION OF PROTOZOAN PHYLA

The four main groups of protozoa traditionally recognized were the flagellates, amebas, spore formers, and ciliates. The system that follows reflects a much more phylogenetic arrangement, including the recognition that amebas and flagellates are more closely related to each other than they are to other groups, and that the "spore formers" represent several completely unrelated forms. This taxonomy continues to follow the principles of evolutionary taxonomy rather than cladistic taxonomy because some clearly paraphyletic groups (marked by*) are included (see Chapter 11).

Kingdom Protista* (pro-tees´ta) (Gr. *protistos,* first of all). Single-celled eukaryotes and their immediate descendants (for example, multicellular algae).

Subkingdom Protozoa* (pro-to-zo´a) (Gr. *protos,* first, primary; + *zōon,* animal). Animal-like protistans.

Phylum Sarcomastigophora* (sar´ko-mas-ti-gof´o-ra) (Gr. *sarkos,* flesh, + *mastix,* whip, + *phora,* bearing). Flagella, pseudopodia, or both types of locomotory organelles; usually with only one type of nucleus; typically no spore formation; sexuality, when present, essentially syngamy.

    Subphylum Mastigophora* (mas-ti-gof´o-ra) (Gr. *mastix,* whip, + *phora,* bearing). One or more flagella typically present in adult stages; autotrophic or heterotrophic or both; reproduction usually asexual by fission.

        Class Phytomastigophorea* (fi´to-mas-ti-go-for´e-a) (Gr. *phyton,* plant, + *mastix,* whip, + *phora,* bearing). Plantlike flagellates, usually bearing chromoplasts (pigment-bearing bodies; chromoplasts with chlorophyll are chloroplasts), which contain chlorophyll. Examples: *Chilomonas, Euglena, Volvox, Ceratium, Peranema, Noctiluca.*

        Class Zoomastigophorea (zo´o-mas-ti-go-for´e-a) (Gr. *zōon,* animal, + *mastix,* whip, + *phora,* bearing). Flagellates without chromoplasts; one to many flagella; ameboid forms with or without flagella in some groups; species predominantly symbiotic. Examples: *Trichomonas, Trichonympha, Trypanosoma, Leishmania, Dientamoeba.*

    Subphylum Opalinata (o´pa-lin-a´ta) (N.F. *opaline,* like opal in appearance, + *ata,* group suffix). Body covered with longitudinal rows of cilium-like organelles; parasitic; cytostome (cell mouth) lacking; two to many nuclei of one type. Examples: *Opalina, Protoopalina.*

    Subphylum Sarcodina (sar-ko-di´na) (Gr. *sarkos,* flesh, + *ina,* belonging to). Pseudopodia typically present; flagella present in developmental stages of some; free living or parasitic.

        Superclass Rhizopoda (ri-zop´o-da) (Gr. *rhiza,* root, + *pous, podos,* foot). Locomotion by lobopodia, filopodia, or reticulopodia, or by cytoplasmic flow without production of discrete pseudopodia. Composed of eight classes, some of which are listed here.

           Class Lobosea (lo-bo´se-a) (Gr. *lobos,* lobe). Pseudopodia lobose or more or less filiform but produced from broader lobe; usually uninucleate; no fruiting bodies. Examples: *Amoeba, Entamoeba, Acanthamoeba, Naegleria, Chaos, Arcella, Difflugia.*

           Class Eumycetozoea (yu´mi-set-o-zo´e-a) (Gr. *eu,* good, true, + *mykes,* fungus, + *zōon,* animal). Ameboid feeding stage, flagellated stage present or absent; produce aerial fruiting bodies with one to thousands of spores. Examples: *Dictyostelium, Physarum.*

           Class Filosea (fi-los´e-a) (L. *filum,* thread). Hyaline, filiform pseudopodia, often branching, sometimes rejoining; no spores or flagellated stages known. Examples: *Vampyrella, Euglypha, Gromia.*

           Class Granuloreticulosea (gran´yu-lo-re-tik´yu-los´e-a) (L. *granulum,* dim. of *granum,* grain, + *reticulum,* dim. of *rete,* net). Delicate, finely granular or hyaline reticulopodia or, rarely, finely pointed, granular but nonrejoining pseudopodia. Examples: *Allogromia, Fusulina, Textularia, Elphidium, Globigerina,* other foraminiferans.

        Superclass Actinopoda (ak´ti-nop´o-da) (Gr. *aktis, aktinos,* ray, + *pous, podos,* foot). Often spherical, usually planktonic; pseudopodia in form of axopodia, with microtubular supporting structure.

           Class Acantharea (a´kan-thar´e-a) (Gr. *akantha,* spine or thorn). Strontium sulfate skeleton composed of 20 or more radiating spines more or less joined in cell center; marine, usually planktonic. Examples: *Acanthometra, Lithoptera.*

           Class Polycystinea (pol´e-sis-tin´e-a) (Gr. *polys,* many, + *kystis,* bladder). Siliceous skeleton in most species, usually of solid elements, consisting of one or more latticed shells with or without radial spines, or of spicules; capsular membrane usually of grossly polygonal plates with many more than three pores; marine, planktonic. Example: *Thalassicolla.*

           Class Phaeodarea (fe´o-dar´e-a) (Gr. *phaios,* dusky, + *daria,* suffix). Skeleton of mixed silica and organic matter, consisting of usually hollow spines and shells; very thick capsular membrane with three pores; marine, planktonic. Examples: *Aulacantha, Challengeron.*

           Class Heliozoea (he´le-o-zo´e-a) (Gr. *helios,* sun, + *zōon,* animal). Without central capsule; skeletal structures, if present, siliceous or organic; axopodia radiating on all sides; most species freshwater. Examples: *Clathrulina, Actinophrys, Actinosphaerium.*

Phylum Labyrinthomorpha (la´bi-rinth-o-morf´a) (Gr. *labyrinth,* maze, labyrinth, + *morph,* form; + *a,* suffix). Small group living on algae; mostly marine or estuarine. Example: *Labyrinthula.*

Phylum Apicomplexa (a´pi-com-plex´a) (L. *apex,* tip or summit, + *complex,* twisted around, + *a,* suffix). Characteristic set of organelles (apical complex) associated with anterior end present in some developmental stages; cilia and flagella absent except for flagellated microgametes in some groups; cysts often present; all species parasitic.

Class Perkinsea (per-kin´se-a). Small group parasitic in oysters.

Class Sporozoa (spor´o-zo´e-a) (Gr. *sporos*, seed, + *zōon*, animal). Spores or oocysts typically present that contain infective sporozoites; flagella present only in microgametes of some groups; pseudopods ordinarily absent, if present they are used for feeding, not locomotion; one or two host life cycles. Examples: *Monocystis, Gregarina, Eimeria, Plasmodium, Toxoplasma, Babesia, Pneumocystis.* [Note: taxonomic position of *Pneumocystis* not known with certainty.]

Phylum Myxozoa (mix-o-zo´a) (Gr. *myxa*, slime, mucus: + *zōon*, animal). Parasites of lower vertebrates, especially fishes, and invertebrates. (This phylum is regarded multicellular by some scientists.)

Phylum Microspora (mi-cros´por-a) (Gr. *micro*, small, + *sporos*, seed). Parasites of invertebrates, especially arthropods, and lower vertebrates.

Phylum Ascetospora (as-e-tos´por-a) (Gr. *asketos*, curiously wrought, + *sporos*, seed). Small group that is parasitic in invertebrates and a few vertebrates.

Phylum Ciliophora (sil-i-of´or-a) (L. *cilium*, eyelash, + Gr. *phora*, bearing). Cilia or ciliary organelles in at least one stage of life cycle; two types of nuclei, with rare exception; binary fission across rows of cilia, budding and multiple fission also occur; sexuality involving conjugation, autogamy, and cytogamy; nutrition heterotrophic; contractile vacuole typically present; most species free living, but many commensal, some parasitic. (This is a very large group, now divided by the Society of Protozoologists classification into three classes and numerous orders and suborders. The classes are separated on the basis of technical characteristics of the ciliary patterns, especially around the cytostome, the development of the cytostome, and other characteristics.) Examples: *Paramecium, Colpoda, Tetrahymena, Balantidium, Stentor, Blepharisma, Epidinium, Euplotes, Vorticella, Carchesium, Trichodina, Podophrya, Ephelota.*

Summary

The assemblage of protists known as protozoa is a large, heterogeneous group now recognized as being composed of seven phyla. The largest and most important of the phyla are the Sarcomastigophora (flagellates and amebas), the Apicomplexa (coccidians, malaria-causing organisms, and others), and the Ciliophora (ciliates). They demonstrate the great adaptive potential of the basic body plan, the single eukaryotic cell. They occupy a vast array of niches and habitats, and many species have complex and specialized organelles.

All protozoa have one or more nuclei, and these often appear vesicular under the light microscope. Macronuclei of ciliates are compact. Endosomes are often present in the nuclei. Many protozoa have organelles similar to those found in metazoan cells.

Pseudopodial or ameboid movement is a locomotory and food-gathering mechanism in protozoa and plays a vital role as a defense mechanism in metazoa. It is accomplished by microfilaments moving past each other, and it requires expenditure of energy from ATP. Ciliary movement is likewise important in both protozoa and metazoa. Currently, the most widely accepted mechanism to account for ciliary movement is the sliding microtubule hypothesis.

Various protozoa feed by holophytic, holozoic, or saprozoic means. The excess water that enters their bodies is expelled by contractile vacuoles (water-expulsion vesicles). Respiration and waste elimination are through the body surface. Protozoa can reproduce asexually by binary fission, multiple fission, and budding; sexual processes are common. Cyst formation to withstand adverse environmental conditions is an important adaptation in many protozoa.

Most phytoflagellates are photosynthetic, and many zooflagellates are important parasites. They move by beating one or more flagella. Sarcodines move by pseudopodia; many are important members of planktonic communities, and some are parasites. Many have a test, or shell. All apicomplexans are parasitic, and they include *Plasmodium,* which causes malaria. The Ciliophora move by means of cilia or ciliary organelles. They are a large and diverse group, and many are complex in structure.

The common ancestor of protozoa and metazoa was unicellular, and the structural similarity of flagella and cilia suggests common ancestry of groups with those organelles.

Review Questions

1. Explain why a protozoan may be very complex, even though it is composed of only one cell.
2. What are eight characteristics of protozoa?
3. Distinguish among the following protozoan phyla: Sarcomastigophora, Apicomplexa, Ciliophora.
4. Distinguish vesicular and compact nuclei.
5. Explain the transitions of endoplasm and ectoplasm in ameboid movement. What is a current hypothesis regarding the role of actin in ameboid movement?
6. Distinguish lobopodia, filipodia, reticulopodia, and axopodia.
7. Contrast the structure of an axoneme of a cilium with that of a kinetosome.
8. What is the sliding microtubule hypothesis?
9. Explain how protozoa eat, digest their food, osmoregulate, and respire.
10. Distinguish the following: binary fission, budding, multiple fission, and sexual and asexual reproduction.
11. Distinguish gametic meiosis, zygotic meiosis, and intermediary meiosis.

12. What is the survival value of encystment?

13. Contrast and give an example of phytoflagellates and zooflagellates.

14. Name three kinds of sarcodines, and tell where they are found (their habitats).

15. Outline the general life cycle of malaria organisms.

16. What is the public health importance of *Toxoplasma,* and how do humans become infected with it?

17. Define the following with reference to ciliates: macronucleus, micronucleus, pellicle, undulating membrane, cirri, infraciliature, trichocysts, conjugation.

18. Outline the steps in conjugation of ciliates.

19. What are indications that the Sarcodina, Apicomplexa, and Ciliophora may share common ancestry at some level with the Phytomastigophorea?

Selected References

See also general references for Part III, p. 626.

Allen, R. D. 1987. The microtubule as an intracellular engine. Sci. Am. **256:**42–49 (Feb.). *The action of microtubules accounts for the movement of chromosomes in mitosis and pseudopodial movement of filopodia and reticulopodia.*

Anderson, O. R. 1988. Comparative protozoology: ecology, physiology, life history. New York, Springer-Verlag. *Good treatment of the aspects mentioned in the subtitle.*

Fenchel, T. 1987. Ecology of protozoa: the biology of free-living phagotrophic protists. Madison, Wisconsin, Science Tech Publishers.

Harrison, G. 1978. Mosquitoes, malaria and man: a history of the hostilities since 1880. New York, E. P. Dutton. *A fascinating story, well told.*

Lee, J. 1993. "On a piece of chalk"—updated. J. Eukaryotic Microbiol. **40:**395–410. *Excellent summary of current knowledge of foraminiferans.*

Lee, J. J., S. H. Hutner, and E. C. Bovee (eds). 1985. An illustrated guide to the protozoa. Lawrence, Kansas, Society of Protozoologists. Allen Press. *A comprehensive guide and essential reference for students of the protozoa.*

Roberts, L. S., and J. J. Janovy, Jr. 1995. Foundations of parasitology, ed. 5. Dubuque, Iowa, William C. Brown Publishers. *Up-to-date and readable information on parasitic protozoa.*

Sleigh, M. A. 1989. Protozoa and other protists. London, Edward Arnold. *Extensively updated version of the author's* The biology of protozoa.

Stossel, T. P. 1994. The machinery of cell crawling. Sci. Am. **271:**54–63 (Sept.). *Ameboid movement—how cells crawl—is important throughout the animal kingdom, as well as in Protista. We now understand quite a bit about its mechanism.*

13

The Mesozoa and Parazoa

Phylum Mesozoa
Phylum Placozoa
Phylum Porifera: Sponges

The Advent of Multicellularity

Sponges are the simplest of multicellular animals. Because the cell is the elementary unit of life, the evolution of organisms larger than unicellular protistans arose as an aggregate of such building units. Nature has experimented with producing larger organisms without cellular differentiation—certain large, single-celled marine algae, for example—but such examples are rarities. Typically nature has held stubbornly to multicellular construction in the progress toward higher organization. There are many advantages to multicellularity as opposed to simply increasing the mass of a single cell. Since it is at cell surfaces that exchange takes place, dividing a mass into smaller units greatly increases the surface area available for metabolic activities. It is impossible to maintain a workable surface-to-mass ratio by simply increasing the size of a single-celled organism. Thus multicellularity is a highly adaptive path toward increasing body size.

Strangely, while sponges are multicellular, they are phylogenetically distinct from other metazoans. They certainly derive from unicellular ancestors, but from which group and in what manner is unknown. The sponge body is an assemblage of cells embedded in a gelatinous matrix and supported by a skeleton of minute needlelike spicules and protein. Because sponges neither look nor behave like other animals, it is understandable that they were not completely accepted as animals by zoologists until well into the nineteenth century. ∎

POSITION IN ANIMAL KINGDOM

The multicellular organisms of the kingdom Animalia (the metazoa) are typically divided into three grades: (1) Mesozoa (a single phylum), (2) Parazoa (phylum Porifera, the sponges; and phylum Placozoa), and (3) Eumetazoa (all other phyla). Although Mesozoa and Parazoa are multicellular, their plan of organization is distinct from that in the eumetazoan phyla. Such cellular layers as they possess are not homologous to the germ layers of the Eumetazoa, and neither group has developmental patterns in line with the other metazoa. The name Parazoa means the "beside-animals."

BIOLOGICAL CONTRIBUTIONS

1. Although the simplest in organization of all the metazoa, these groups do compose a higher level of morphological and physiological integration than that found in protozoan colonies. The Mesozoa and Parazoa may be said to belong to a **cellular level of organization.**
2. The mesozoans, although composed simply of an outer layer of somatic cells and an inner layer of reproductive cells, nevertheless have a very complex reproductive cycle somewhat suggestive of that of the trematodes (flukes). Mesozoans are entirely parasitic.
3. Placozoans are essentially composed of two epithelia with fluid and some fibrous cells between them.
4. The sponges (poriferans) are more complex, with several types of cells differentiated for various functions, some of which are organized into **incipient tissues** of a low level of integration.
5. The developmental patterns of these three phyla are different from those of other phyla, and their embryonic layers are not homologous to the germ layers of Eumetazoa.
6. The sponges have developed a unique system of **water currents** on which they depend for food and oxygen.

ORIGIN OF METAZOA

Unraveling the origin of the multicellular animals (metazoans) has presented many problems for zoologists. It is generally believed that they evolved from unicellular organisms, but there is much disagreement over which group of unicellular organisms gave rise to them, and how. Three prevalent hypotheses in current use are (1) that the metazoans arose from a syncytial (multinucleate) ciliated form in which cell boundaries later evolved, (2) that they arose from a colonial flagellated form in which the cells gradually became more specialized and interdependent, and (3) that the origin of metazoans was polyphyletic, or derived from more than one group of unicellular organisms.

Proponents of the **syncytial ciliate hypothesis** believe that metazoans arose from an ancestor shared with the single-celled ciliates. The common ancestor of metazoans acquired multiple nuclei within a single cell membrane and later became compartmentalized into the multicellular condition. It is assumed that the body form of the ancestor resembled that of modern ciliates and thus tended toward bilateral symmetry. Therefore the earliest metazoans would have been bilateral and similar to the present primitive flatworms. There are several objections to this hypothesis. It ignores the embryology of the flatworms in which nothing similar to cellularization occurs; it does not explain the presence of flagellated sperm in the metazoans; and, perhaps more important, it implies that the radial symmetry of cnidarians is derived from a primary bilateral symmetry, for which there is no evidence.

The **colonial flagellate hypothesis**—first proposed by Haeckel in 1874—is the classic scheme, which, with various revisions, still has many followers. According to this hypothesis, the metazoans descended from ancestors characterized by a hollow, spherical, colony of flagellated cells. The individual cells within the colony became differentiated for specific functional roles (reproductive cells, nerve cells, somatic cells, and so on), thus subordinating cellular independence to the welfare of the colony as a whole. The colonial ancestral form was at first radially symmetrical, similar perhaps to the free-swimming planula larvae of the cnidarians (jellyfishes and others, p. 252). This larva is radially symmetrical and has no mouth. The cnidarians with their radial symmetry could have evolved from this form.

Bilateral symmetry could have evolved later when some of these planula-like ancestors became adapted for a creeping form of locomotion on the ocean floor. Dorsal and ventral surfaces would have differentiated, a ventral mouth would have appeared, and a start would have been made toward cephalization (a concentration of neurons and sensory structures at the anterior). This would have led to primitive bilateral symmetry, resembling that of the flatworms.

Some zoologists prefer the idea that the metazoans had a **polyphyletic origin** and suggest that the sponges, cnidarians, ctenophores, and remaining eumetazoans evolved independently. Thus no single scheme might account for them all.

We now have evidence based on small subunit ribosomal RNA sequences and on similarities in complex biochemical pathways.[*] This evidence generally supports the colonial flagellate hypothesis, that is, that metazoans represent a monophyletic assemblage including choanoflagellates ("collared" flagellates such as *Codosiga,* see p. 225). The sister group of metazoans appears to be the fungi.

PHYLUM MESOZOA

The name Mesozoa (mes-o-zo´a) (Gr. *mesos,* in the middle, + *zōon,* animal) was coined by an early investigator (van Beneden, 1876) who believed that the group was a "missing link" between protozoa and metazoa. These minute, ciliated, wormlike animals represent an extremely simple level of organization. All mesozoans

---

*Wainright, P. O., et al. 1993. Science **260:**340–342.

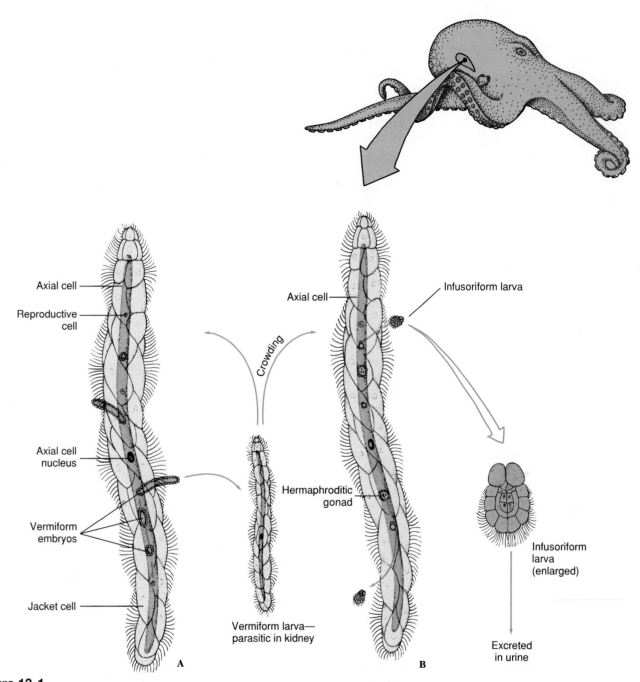

Figure 13-1

Two methods of reproduction by mesozoans. **A,** Asexual development of vermiform larvae from reproductive cells in the axial cell of the adult. **B,** Under crowded conditions in the host kidney, reproductive cells develop into gonads with gametes that produce infusoriform dispersal larvae that emerge in the host urine.

live as parasites in marine invertebrates, and the majority of them are only 0.5 to 7 mm in length. Most are made up of just 20 to 30 cells arranged basically in two layers. The layers are not homologous to the germ layers of higher metazoans.

There are two classes of mesozoans, the Rhombozoa and the Orthonectida, but they differ so much from each other that some authorities believe they should be placed in separate phyla.

The rhombozoans (Gr. *rhombos,* a spinning top, + *zōon,* animal) live in the kidneys of benthic cephalopods (bottom-dwelling octopuses, cuttlefishes, and squids). The adults, called **vermiforms** (or nematogens), are long and slender (Figure 13-1). Their inner, reproductive cells give rise to vermiform larvae that grow and then reproduce. When the population becomes crowded, the reproductive cells of some adults develop into gonadlike structures pro-

ducing male and female gametes. The zygotes grow into minute (0.04 mm) ciliated infusoriform larvae (Figure 13-1B), quite unlike the parent. These are shed with the host urine into the seawater. The next part of the life cycle is unknown because the infusoriform larvae are not immediately infective to a new host.

Orthonectids (Gr. *orthos,* straight, + *nektos,* swimming) (Figure 13-2) parasitize a variety of invertebrates, such as brittle stars, bivalve molluscs,

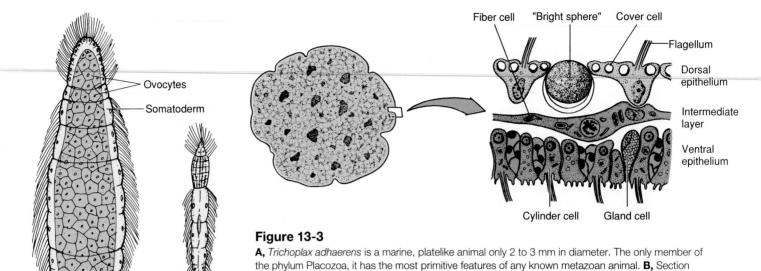

Figure 13-2

A, Female and, **B,** male orthonectid (*Rhopalura*). This mesozoan parasitizes such forms as flatworms, molluscs, annelids, and brittle stars. The structure consists of a single layer of ciliated epithelial cells surrounding an inner mass of sex cells.

Figure 13-3

A, *Trichoplax adhaerens* is a marine, platelike animal only 2 to 3 mm in diameter. The only member of the phylum Placozoa, it has the most primitive features of any known metazoan animal. **B,** Section through *Trichoplax adhaerens,* showing histological structure.

polychetes, and nemerteans. The life cycles involve sexual and asexual phases, and the asexual stage is quite different from that of the rhombozoans. It consists of a multinucleated mass called a **plasmodium,** which by division ultimately gives rise to males and females.

PHYLOGENY OF MESOZOANS

There is still much to learn about these mysterious little parasites, but probably one of the most intriguing questions is the place of mesozoans in the evolutionary picture. Some investigators believe they represent primitive or degenerate flatworms and even believe that they should be classed with the phylum Platyhelminthes. Others place them close to protozoa, possibly related to the ciliates. Whether metazoans and mesozoans derived independently from protozoan beginnings or whether mesozoans are indeed degenerate flatworms is still an enigma.

PHYLUM PLACOZOA

The phylum Placozoa (Gr. *plax, plakos,* tablet, plate, + *zōon,* animal) was proposed in 1971 by K. G. Grell to contain a single species, *Trichoplax adhaerens* (Figure 13-3A), a tiny (2 to 3 mm) marine form that had been considered either a mesozoan or a cnidarian larva by various workers in the past. The body is platelike and has no symmetry, no organs, and no muscular or nervous system. It is composed of a dorsal epithelium of cover cells and shiny spheres, a thick ventral epithelium containing monociliated cells, (cylinder cells) and nonciliated gland cells, and a space between the epithelia containing fluid and fibrous cells (Figure 13-3B). The organisms glide over their food, secrete digestive enzymes on it, and then absorb the products. Grell considers *Trichoplax* diploblastic (see p. 111), with the dorsal epithelium representing ectoderm and the ventral epithelium representing endoderm because of its nutritive function. The phylogenetic position of placozoans is uncertain, although recent molecular evidence places them as a sister group to the phylum Cnidaria (p. 252).

PHYLUM PORIFERA: SPONGES

Sponges belong to phylum Porifera (po-rif´-er-a) (L. *porus,* pore, + *fera,* bearing). The bodies of sponges bear myriads of tiny pores and canals that constitute a filter-feeding system adequate for their inactive life habit. They are sessile animals and depend on the water currents carried through their unique canal systems to bring them food and oxygen and to carry away their body wastes. Their bodies are little more than masses of cells embedded in a gelatinous matrix and stiffened by a skeleton of minute **spicules** of calcium carbonate or silica and collagen (p. 186). They have no organs or true tissues, and even their cells show a certain degree of independence. As sessile animals with only negligible body movement, they have not evolved a nervous system or sense organs and have only the simplest of contractile elements.

So, although they are multicellular, sponges share few of the characteristics of other metazoan phyla. They seem to be outside the line of evolution leading from the protozoa to the other metazoa: a dead-end branch. It is for this reason that they are often called the Parazoa (Gr. *para,* beside or alongside of, + *zōon,* animal).

Sponges vary in size from a few millimeters to the great loggerhead sponges, which may reach 2 m or more across. Many sponge species are brightly colored because of pigments in the dermal cells. Red, yellow, orange, green, and purple sponges are not uncommon. However, the color fades quickly when the sponges are removed from the water. Some sponges, including the simplest and

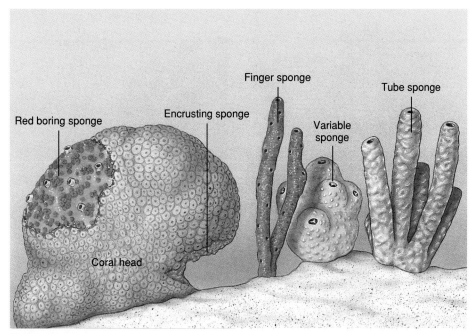

Figure 13-4
Some growth habits and forms of sponges.

most primitive, are radially symmetrical, but many are quite irregular in shape. Some stand erect, some are branched or lobed, and others are low, even encrusting, in form (Figure 13-4). Some bore holes into shells or rocks.

Most of the 5000 or more sponge species are marine, although some 150 species live in fresh water. Marine sponges are abundant in all seas and at all depths, and a few even exist in brackish water. Although the embryos are free swimming, the adults are always attached, usually to rocks, shells, corals, or other submerged objects. Some bottom-dwelling forms even grow on sand or mud bottoms. Their growth patterns often depend on the shape of the substratum, the direction and speed of the water currents, and the availability of space, so that the same species may differ markedly in appearance under different environmental circumstances. Sponges in calm waters may grow taller and straighter than those in rapidly moving waters.

Many animals (crabs, nudibranchs, mites, bryozoans, and fish) live as commensals or parasites in or on sponges. The larger sponges particularly tend to harbor a large variety of invertebrate commensals. On the other hand, sponges grow on many other living animals, such as molluscs, barnacles, brachiopods, corals, or hydroids. Some crabs attach pieces of sponge to their carapace for camouflage and for protection, since most predators seem to find sponges distasteful. Some reef fishes, however, graze on shallow-water sponges.

The sponges are an ancient group, with an abundant fossil record extending back to the early Cambrian period and even, according to some claims, the Precambrian. Living poriferans traditionally have been assigned to three classes: Calcarea (with calcareous spicules), Hexactinellida (six-rayed siliceous spicules), and Demospongiae (with a skeleton of siliceous spicules or **spongin** [a specialized collagen] or both). A fourth class (Sclerospongiae) was erected to contain sponges with a massive calcareous skeleton and siliceous spicules. Some zoologists (Wood, 1990) believe that the class Sclerospongiae is not needed, but we will retain it for the present.

Certainly one reason for the success of sponges as a group is that they have few enemies. Because of a sponge's elaborate skeletal framework and often noxious odor, most potential predators find sampling a sponge about as pleasant as eating a mouthful of glass splinters embedded in evil-smelling gristle.

CHARACTERISTICS OF PHYLUM PORIFERA

1. Multicellular; body a loose aggregation of cells of mesenchymal origin
2. Body with pores (ostia), canals, and chambers that serve for passage of water
3. Mostly marine; all aquatic
4. Radial symmetry or none
5. Epidermis of flat pinacocytes; most interior surfaces lined with flagellated collar cells (choanocytes) that create water currents; a gelatinous protein matrix called mesohyl (mesoglea) contains amebocytes of various types and skeletal elements
6. Skeletal structure of fibrillar collagen (a protein) and calcareous or siliceous crystalline spicules, often combined with variously modified collagen (spongin)
7. No organs or true tissues; digestion intracellular; excretion and respiration by diffusion
8. Reactions to stimuli apparently local and independent; nervous system probably absent
9. All adults sessile and attached to substratum
10. Asexual reproduction by buds or gemmules and sexual reproduction by eggs and sperm; free-swimming ciliated larvae

FORM AND FUNCTION

The only body openings of these unusual animals are pores, usually many tiny ones called **ostia** for incoming water, and a few large ones

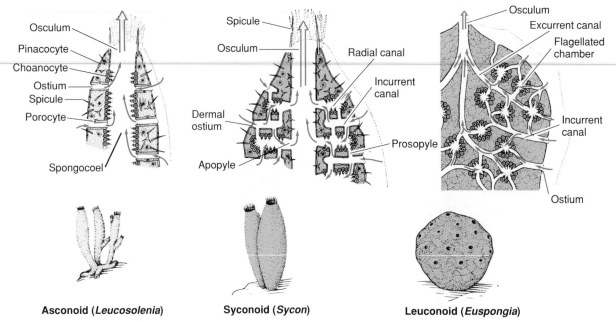

Asconoid (*Leucosolenia*)   Syconoid (*Sycon*)   Leuconoid (*Euspongia*)

Figure 13-5

Three types of sponge structure. The degree of complexity from simple asconoid to complex leuconoid type has involved mainly the water-canal and skeletal systems, accompanied by outfolding and branching of the collar cell layer. The leuconoid type is considered the major plan for sponges, for it permits greater size and more efficient water circulation.

called **oscula** (sing., **osculum**) for water outlet. These openings are connected by a system of canals, some of which are lined with peculiar flagellated collar cells called **choanocytes,** whose flagella maintain a current of environmental water through the canals. Water enters the canals through a multitude of tiny incurrent pores **(dermal ostia)** and leaves by way of one or more large oscula. The choanocytes not only keep the water moving but also trap and phagocytize food particles that are carried in the water. The cells lining the passageways are very loosely organized. Collapse of the canals is prevented by the skeleton, which, depending on the species, may be made up of needlelike calcareous or siliceous spicules, a meshwork of organic spongin fibers, or a combination of the two.

Sessile animals make few movements and therefore need little in the way of nervous, sensory, or locomotor parts. Sponges apparently have been sessile from their earliest appearance and have never acquired specialized nervous or sensory structures, and they have only the very simplest of contractile systems.

Figure 13-6

Clathrina canariensis (class Calcarea) is common on Caribbean reefs in caves and under ledges.

Types of Canal Systems

Most sponges fall into one of three types of canal systems: asconoid, syconoid, or leuconoid (Figure 13-5).

Asconoids: Flagellated Spongocoels.

The asconoid sponges have the simplest type of organization. They are small and tube shaped. Water enters through microscopic dermal pores into a large cavity called the **spongocoel,** which is lined with

choanocytes. The choanocyte flagella pull the water through the pores and expel it through a single large osculum (see Figure 13-5). *Leucosolenia* (Gr. *leukos,* white, + *solen,* pipe) is an asconoid type of sponge. Its slender, tubular individuals grow in groups attached by a common stolon, or stem, to objects in shallow seawater. *Clathrina* (L. *clathri,* lattice work) is an asconoid with bright yellow, intertwined tubes (Figure 13-6). Asconoids are found only in the Calcarea.

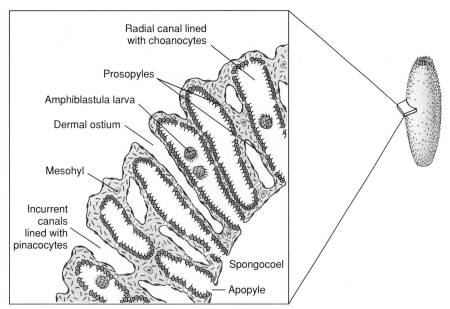

Figure 13-7

Cross section through wall of sponge *Sycon,* showing canal system.

These three types of canal systems—asconoid, syconoid, and leuconoid—demonstrate an increase in complexity and efficiency of the water pumping system, but they do not imply an evolutionary or developmental sequence. The leuconoid grade of construction has evolved independently many times in sponges. Possession of the leuconoid plan is of clear adaptive value; it increases the proportion of flagellated surfaces compared with the volume, thus providing more collar cells to meet food demands. It makes possible a much larger body size than asconoid or syconoid grades.

Types of Cells

Sponge cells are loosely arranged in a gelatinous matrix called **mesohyl** (mesoglea, mesenchyme) (Figures 13-7 and 13-9). The mesohyl is the "connective tissue" of the sponges; in it are found various ameboid cells, fibrils, and skeletal elements. There are several types of cells in sponges.

Pinacocytes. The nearest approach to a true tissue in sponges is found in the arrangement of the **pinacocyte** cells of the **pinacoderm** (Figure 13-9). These are thin, flat, epithelial-type cells

Syconoids: Flagellated Canals. Syconoid sponges look somewhat like larger editions of asconoids, from which they were derived. They have the tubular body and single osculum, but the body wall, which is thicker and more complex than that of asconoids, contains choanocyte-lined **radial canals** that empty into the spongocoel (see Figure 13-5). The spongocoel in syconoids is lined with epithelial-type cells rather than flagellated cells as in asconoids. Water enters through a large number of dermal ostia into **incurrent canals** and then filters through tiny openings called **prosopyles** into the radial canals (Figure 13-7). There food is ingested by the choanocytes, whose flagella force the water on through internal pores **(apopyles)** into the spongocoel. From there it emerges through the osculum. Syconoids do not usually form highly branched colonies as the asconoids do. During development, syconoid sponges pass through an asconoid stage; the flagellated canals form by evagination of the body wall. This is evidence that syconoid sponges were derived from asconoid ancestral stock. Syconoids are found in classes Calcarea and Hexactinellida. *Sycon* (Gr. *sykon,* a fig) is a commonly studied example of the syconoid type of sponge (see Figure 13-5).

Leuconoids: Flagellated Chambers. Leuconoid organization is the most complex of the sponge types and the best adapted for increase in sponge size. Most leuconoids form large masses with numerous oscula (Figure 13-8). Clusters of flagellated chambers are filled from incurrent canals and discharge water into excurrent canals that eventually lead to the osculum (Figure 13-5). Most sponges are of the leuconoid type, which occurs in most Calcarea and in all other classes.

Figure 13-8

This orange demosponge, *Mycale laevis,* often grows beneath platelike colonies of the stony coral, *Montastrea annularis.* The large oscula of the sponge are seen at the edges of the plates. Unlike some other sponges, *Mycale* does not burrow into the coral skeleton and may actually protect coral from invasion by more destructive species. Pinkish radioles of a Christmas tree worm, *Spirobranchus giganteus* (phylum Annelida, class Polychaeta) also project from the coral colony. An unidentified reddish sponge can be seen to the right of the Christmas tree worm.

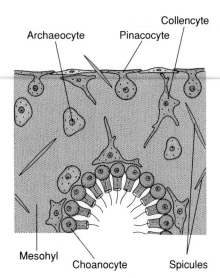

Figure 13-9

Small section through sponge wall, showing four types of sponge cells. Pinacocytes are protective and contractile; choanocytes create water currents and engulf food particles; archaeocytes have a variety of functions; collencytes secrete collagen.

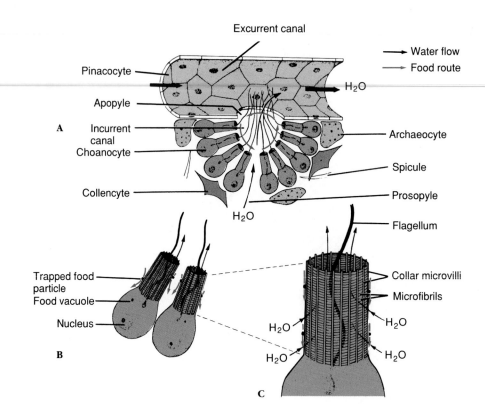

Figure 13-10

Food trapping by sponge cells. **A,** Cutaway section of canals showing cellular structure and direction of water flow. **B,** Two choanocytes and **C,** structure of the collar. Small red arrows indicate movement of food particles.

that cover the exterior surface and some interior surfaces. Some are T-shaped, with their cell bodies extending into the mesohyl. Pinacocytes are somewhat contractile and help regulate the surface area of the sponge. Some of the pinacocytes are modified as contractile **myocytes,** which are usually arranged in circular bands around the oscula or pores, where they help regulate the rate of water flow.

Choanocytes. The choanocytes, which line the flagellated canals and chambers, are ovoid cells with one end embedded in the mesohyl and the other exposed. The exposed end bears a flagellum surrounded by a collar (Figures 13-9 and 13-10). The electron microscope shows that the collar is made up of adjacent microvilli, connected to each other by delicate microfibrils, forming a fine filtering device for straining food particles from the water (Figure 13-10B and C). The beat of the flagellum pulls water through the sievelike collar and forces it out through the open top of the collar. Particles too large to enter the collar become trapped in secreted mucus and slide down the collar to the base where they are phagocytized by the cell body. Larger particles have al-

ready been screened out by the small size of the dermal pores and prosopyles. The food engulfed by the cells is passed on to a neighboring archaeocyte for digestion.

Archaeocytes. **Archaeocytes** are ameboid cells that move about in the mesohyl (Figure 13-9) and carry out a number of functions. They can phagocytize particles at the pinacoderm and receive particles for digestion from the choanocytes. Archaeocytes apparently can differentiate into any of the other types of more specialized cells in the sponge. Some, called **sclerocytes,** secrete spicules. Others, called **spongocytes,** secrete the spongin fibers of the skeleton, and **collencytes** secrete fibrillar collagen (p. 186). **Lophocytes** secrete large quantities of collagen but are distinguishable morphologically from collencytes.

Types of Skeletons

The skeleton gives support to the sponge, preventing collapse of the

canals and chambers. The major structural protein in the animal kingdom is collagen, and fibrils of collagen are found throughout the intercellular matrix of all sponges. In addition, various Demospongiae secrete a form of collagen traditionally known as spongin. There are several types of spongin, differing in chemical composition and form (fibers, spicules, filaments, spongin surrounding spicules, and so on) that are found in the various demosponges. Demospongiae also secrete siliceous spicules, as do Sclerospongiae. The siliceous spicules and organic collagen skeleton of the sclerosponges is limited to a thin layer of living tissue above their massive skeleton of calcium carbonate. Calcareous sponges secrete spicules composed mostly of crystalline calcium carbonate that have one, three, or four rays (Figure 13-11). Glass sponges have siliceous spicules with six rays arranged in three planes at right angles to each other. There are many variations in the shape of spicules, and these structural variations are of taxonomic importance.

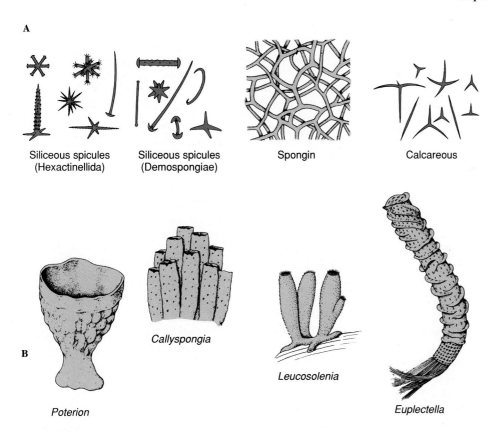

Figure 13-11

A, Types of spicules found in sponges. There is amazing diversity, complexity, and beauty of form among the many types of spicules. **B,** Some sponge body forms.

Sponge Physiology

All the life activities of the sponge depend on the current of water flowing through the body. A sponge pumps a remarkable amount of water. *Leuconia* (Gr. *leukos*, white), for example, is a small leuconoid sponge about 10 cm tall and 1 cm in diameter. It is estimated that water enters through some 81,000 incurrent canals at a velocity of 0.1 cm/second. However, *Leuconia* has more than 2 million flagellated chambers whose combined diameter is much greater than that of the canals, so that in the chambers the water slows down to 0.001 cm/second, allowing ample opportunity for food capture by the collar cells. All of the water is expelled through a single osculum at a velocity of 8.5 cm/second: a jet force capable of carrying waste products some distance away from the sponge. Some large sponges can filter 1500 liters of water a day.

Sponges feed primarily on particles suspended in the water pumped through their canal systems. Detritus particles, planktonic organisms, and bacteria are consumed nonselectively in the size range from 50 µm (average diameter of ostia) to 0.1 µm (width of spaces between microvilli of choanocyte collar).Pinacocytes may phagocytize particles at the surface, but most of the larger particles are consumed in the canals by archaeocytes that move close to the lining of the canals. The smallest particles, accounting for about 80% of the particulate organic carbon, are phagocytized by the choanocytes. Sponges can also absorb dissolved nutrients from the water passing through the system. Protein molecules are taken into the choanocytes by pinocytosis (p. 55).

Digestion is entirely **intracellular** (occurs within cells), and present evidence indicates that archaeocytes perform this chore. Choanocytes pass particles of food to the archaeocytes for digestion.

There are no respiratory or excretory organs; both functions are apparently carried out by diffusion in individual cells. Contractile vacuoles have been found in archaeocytes and choanocytes of freshwater sponges.

The only visible activities and responses in sponges, other than the propulsion of water, are slight alterations in shape and the closing and opening of the incurrent and excurrent pores, and these movements are very slow. The most common response is closure of the oscula. Apparently excitation spreads from cell to cell, although some zoologists point to the possibility of coordination by means of substances carried in the water currents, and some have tried, not very successfully, to demonstrate the presence of nerve cells.

Reproduction

Sponges reproduce both asexually and sexually. **Asexual reproduction** occurs by means of bud formation and by regeneration following fragmentation. **External buds,** after reaching a certain size, may become detached from the parent and float away to form new sponges, or they may remain to form colonies. **Internal buds,** or **gemmules** (Figure 13-12), are formed in freshwater sponges and some marine sponges. Here, archaeocytes collect together in the mesohyl and become surrounded by a tough spongin coat incorporating

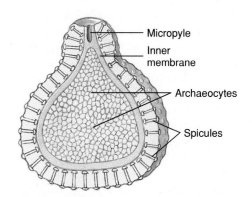

Figure 13-12

Section through a gemmule of a freshwater sponge (Spongillidae). Gemmules are a mechanism for survival of the harsh conditions of winter. On return of favorable conditions, the archaeocytes exit through the micropyle to form a new sponge. The archaeocytes of the gemmule give rise to all the cell types of the new sponge structure.

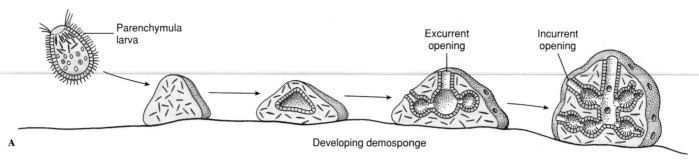

A

Developing demosponge

Figure 13-13

A, Development of demosponges. **B,** Development of the syconoid sponge *Sycon*.

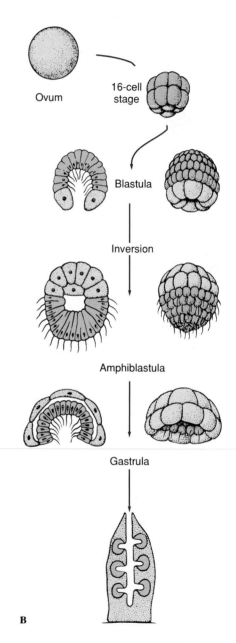

Ovum

16-cell stage

Blastula

Inversion

Amphiblastula

Gastrula

B

siliceous spicules. When the parent animal dies, the gemmules survive and remain dormant, preserving the life of the species during periods of freezing or severe drought. Later the cells in the gemmules escape through a special opening, the **micropyle,** and develop into new sponges. Gemmulation in freshwater sponges (Spongillidae) is thus an adaptation to the changing seasons. Gemmules are also a means of colonizing new habitats, since they can be spread by streams or by animal carriers. What prevents the gemmules from hatching during the season of formation rather than remaining dormant? Some species secrete a substance that inhibits early germination of the gemmules, and the gemmules do not germinate as long as they are held in the body of the parent. Other species undergo a period of maturation at low temperatures (as in winter) before they germinate. Gemmules in marine sponges also seem to be an adaptation to pass the cold of winter; they are the only form in which *Haliclona loosanoffi* exists during the colder parts of the year in the northern part of its range.

In **sexual reproduction** most sponges are **monoecious** (have both male and female sex cells in one individual). Sperm arise from transformation of choanocytes. In Calcarea and at least some Demospongiae, oocytes also develop from choanocytes; in other demosponges oocytes apparently are derived from archaeocytes. Most sponges are viviparous, that is, after fertilization the zygote is retained in and derives nourishment from the parent, and a ciliated larva is released. In such sponges, sperm are released into the water by one individual and

are taken into the canal system of another. There choanocytes phagocytize them, then transform into carrier cells and carry the sperm through the mesohyl to the oocytes. Other sponges are oviparous, and both oocytes and sperm are expelled free into the water. The free-swimming larva of most sponges is a solid-bodied **parenchymula** (Figure 13-13A). The outwardly directed, flagellated cells migrate to the interior after the larva settles and become the choanocytes in the flagellated chambers. The Calcarea and a few Demospongiae have a very strange developmental pattern. A hollow blastula, called an **amphiblastula** (Figure 13-13B), develops, with flagellated cells toward the interior. The blastula then turns *inside out* **(inversion),** the flagellated ends of the cells becoming directed to the outside! The flagellated cells **(micromeres)** of the larva are at one end, and the larger, nonflagellated cells **(macromeres)** are at the other. In contrast to other metazoan embryos, the micromeres invaginate into and are overgrown by the macromeres. The flagellated micromeres become the choanocytes, archeocytes, and collencytes of the new sponge, and the nonflagellated cells give rise to pinacoderm and sclerocytes.

Regeneration and Somatic Embryogenesis

Sponges have a tremendous ability to repair injuries and to restore lost parts, a process called **regeneration.** Regeneration does not imply a reorganization of the entire animal, but only of the wounded portion.

On the other hand, if a sponge is cut into small fragments, or if the cells of a sponge are entirely dissociated and are allowed to fall into small groups, or aggregates, entire new sponges can develop from these fragments or aggregates of cells. This process has been termed **somatic embryogenesis.** Somatic embryogenesis involves a complete reorganization of the structure and functions of the participating cells or bits of tissue. Isolated from the influence of adjoining cells, they can realize their own potential to change in shape or function as they develop into a new organism.

A great deal of experimental work has been done in this field. The process of reorganization appears to differ in sponges of differing complexity. There is still some controversy concerning just what mechanisms lead to the adhesion of the cells and the share that each type of cell plays in the formative process.

Class Calcarea (Calcispongiae)

The Calcarea (also called Calcispongiae) are the calcareous sponges, so called because their spicules are composed of calcium carbonate. The spicules are straight (monaxons) or have three or four rays. These sponges tend to be small—10 cm or less in height—and tubular or vase shaped. They may be asconoid, syconoid, or leuconoid in structure. Though many are drab in color, some are bright yellow, red, green, or lavender. *Leucosolenia* and *Sycon* (often called *Scypha* or *Grantia* by biological supply companies) are marine shallow-water forms commonly studied in the laboratory. *Leucosolenia* is a small asconoid sponge that grows in branching colonies, usually arising from a network of horizontal, stolonlike tubes (Figure 13-6). *Sycon* is a solitary sponge that may live singly or form clusters by budding. The vase-shaped, typically syconoid animal is 1 to 3 cm long, with a fringe of straight spicules around the osculum that discourages small animals from entering.

Class Hexactinellida (Hyalospongiae): Glass Sponges

The glass sponges make up the class Hexactinellida (or Hyalospongiae).

They are nearly all deep-sea forms that are collected by dredging. Most of them are radially symmetrical, with vase- or funnel-shaped bodies that are usually attached by stalks of root spicules to a substratum (Figure 13-11, *Euplectella*) (N. L. from Gr. *euplektos,* well plaited). They range from 7.5 cm to more than 1.3 m in length. Their distinguishing features are the skeleton of six-rayed siliceous spicules that are commonly bound together into a network forming a glasslike structure and the **trabecular net** of living tissue produced by the fusion of the pseudopodia of archaeocytes. Within the trabecular net are elongated, finger-shaped chambers lined with choanocytes and opening into the spongocoel. The osculum is unusually large and may be covered over by a sievelike plate of silica. There is no pinacoderm or gelatinous mesohyl, and both the external surface and the spongocoel are lined with trabecular net. The skeleton is rigid, and muscular elements (myocytes) appear to be absent. The general arrangement of the chambers fits glass sponges into both syconoid and leuconoid types. Their structure is adapted to the slow, constant currents of sea bottoms, for the channels and pores of the sponge wall are relatively large and uncomplicated and permit an easy flow of water. Little, however, is known about their physiology, doubtless because of their deep-water habitat.

The latticelike network of spicules found in many glass sponges is of exquisite beauty, such as that of *Euplectella,* or Venus' flower basket (Figure 13-11), a classic example of the Hexactinellida.

Class Demospongiae

Class Demospongiae contains 95% of living sponge species, including most of the larger sponges. The spicules are siliceous but are not six rayed, and they may be bound together by spongin or may be absent altogether. All members of the class are leuconoid, and all are marine except one family, the Spongillidae, or freshwater sponges.

Freshwater sponges are widely distributed in well-oxygenated ponds and streams, where they encrust plant stems and old pieces of submerged wood. They may resemble a bit of wrinkled scum, be pitted with pores, and be brownish or greenish in color. Common genera are *Spongilla* (L. *spongia,* from Gr. *spongos,* sponge) and *Myenia.* Freshwater sponges are most common in midsummer, although some are more easily found in the fall. They die and disintegrate in late autumn, leaving the gemmules (already described) to produce the next year's population. They also reproduce sexually.

The marine Demospongiae are quite varied and may be quite striking in color and shape (Figure 13-14). Some are encrusting; some are tall and fingerlike; some are low and spreading; some bore into shells; and some are shaped like fans, vases, cushions, or balls (Figure 13-14). Loggerhead sponges may grow several meters in diameter.

The so-called bath sponges (*Spongia, Hippospongia*) belong to the group called horny sponges, which have spongin skeletons and lack siliceous spicules entirely.

Class Sclerospongiae

The Sclerospongiae are a small group of sponges that secrete a massive calcareous skeleton and are thus often called coralline sponges. Living tissue extends from 1 mm to 3 cm or more into the skeleton but only 1 mm above it. Their organization is leuconoid, as in the Demospongiae, and siliceous spicules and spongin are present in three of the four orders. Sclerosponges live in cryptic habitats (deeply shaded or completely dark locations) on coral reefs, such as crevices, caves, undersurfaces of overhangs, and deep water. They are apparently relict representatives of ancient groups with a geological history extending from the Cambrian period. When modern corals and their symbiosis with algal cells (see p. 271) arose in the Mesozoic era, they were very successful and came to dominate the reefs. However, because the corals depend physiologically on the

A

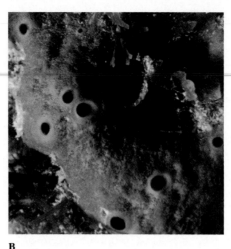

B

C

Figure 13-14

Marine Demospongiae on Caribbean coral reefs. **A,** *Pseudoceratina crassa* is a colorful sponge growing at moderate depths. **B,** *Ectyoplasia ferox* is irregular in shape and its oscula form small, volcano-like cones. It is toxic and may cause skin irritation if touched. **C,** *Monanchora unguifera* with commensal brittle star, *Ophiothrix suensoni* (phylum Echinodermata, class Ophiuroidea).

algal cells growing in their tissues, they must have adequate light. Thus sclerosponges and certain other organisms were able to exploit the unoccupied cryptic habitats.

PHYLOGENY AND ADAPTIVE RADIATION

Phylogeny

The sponges originated before the Cambrian period. Two groups of calcareous spongelike organisms occupied early Paleozoic reefs. The Devonian period saw rapid development of many glass sponges. The possibility that sponges arose from choanoflagellates (protozoa that bear collars and flagella) earned support for a time. However, many zoologists object to that hypothesis because sponges do not acquire collars until later in their embryological development. The outer cells of the larvae are flagellated but not collared, and they do not become collar cells until they become internal. Also, collar cells are found in certain corals and echinoderms, so they are not unique to the sponges.

However, these objections are countered by evidence based on the sequences of ribosomal RNA. This evidence supports the hypothesis of a common ancestor for choanoflagellates and metazoans. It suggests also that sponges and Eumetazoa are sister groups, with the Porifera having split off before the origin of the radiates and placozoans, but sharing a common ancestor.

Adaptive Radiation

The Porifera have been a highly successful group that has branched out into several thousand species and a variety of marine and freshwater habitats. Their diversification centers largely on their unique water-current system and its various degrees of complexity. The proliferation of the flagellated chambers in the leuconoid sponges was more favorable to an increase in body size than that of the asconoid and syconoid sponges because facilities for feeding and gaseous exchange were greatly enlarged.

CLASSIFICATION OF PHYLUM PORIFERA

Class Calcarea (cal-ca´re-a) (L. *calcis,* lime) **(Calcispongiae).** Have spicules of calcium carbonate that often form a fringe around the osculum (main water outlet); spicules needle shaped or three or four rayed; all three types of canal systems (asconoid, syconoid, leuconoid) represented; all marine. Examples: *Sycon, Leucosolenia.*

Class Hexactinellida (hex-ak-tin-el´i-da) (Gr. *hex, six,* + *aktis,* ray, + L. *-ellus,* dim. suffix). **(Hyalospongiae).** Have six-rayed, siliceous spicules extending at right angles from a central point; spicules often united to form network; body often cylindrical or funnel shaped; flagellated chambers in simple syconoid or leuconoid arrangement; habitat mostly deep water; all marine. Examples: Venus' flower basket (*Euplectella*), *Hyalonema.*

Class Demospongiae (de-mo-spun´je-e) (Gr. *demos,* people, + *spongos,* sponge). Have siliceous spicules that are not six rayed, or spongin, or both; leuconoid-type canal systems; one family found in fresh water; all others marine. Examples: *Thenea, Cliona, Spongilla, Myenia,* and all bath sponges.

Class Sclerospongiae (skler´o-spun´je-e) (Gr. *skleros,* hard, + *spongos,* sponge). Secrete massive basal skeleton of calcium carbonate, with living tissue extending into the skeleton from 1 mm to 3 cm or more, extending above skeleton less than 1 mm; have siliceous spicules similar to Demospongiae (sometimes absent), and spongin fibers; leuconoid organization; inhabit caves, crevices, tunnels, and deep water on coral reefs. Examples: *Astrosclera, Calcifibrospongia, Ceratoporella, Merlia.*

Summary

Members of the phylum Mesozoa are very simply organized animals that are parasitic in the kidneys of cephalopod molluscs (class Rhombozoa) and in several other invertebrate groups (class Orthonectida). They have only two cell layers, but these are not homologous to the germ layers of higher metazoans. They have a complicated life history that is still incompletely known. Whether their simple organization is primitive or derived from a more advanced group is unknown.

The phylum Placozoa has only one member, a small platelike marine organism. It too has only two cell layers, but some workers believe that these layers are homologous to ectoderm and endoderm of higher metazoans. The closest relatives of the placozoans seem to be the cnidarians.

The sponges (phylum Porifera) are an abundant marine group with some freshwater representatives. They have various specialized cells, but these are not organized into tissues or organs. They depend on the flagellar beat of their choanocytes to circulate water through their bodies for food gathering and respiratory gas exchange. They are supported by secreted skeletons of fibrillar collagen, collagen in the form of large fibers or filaments (spongin), calcareous or siliceous spicules, or a combination of spicules and spongin in most species.

Sponges reproduce asexually by budding, fragmentation, and gemmules (internal buds). Most sponges are monoecious but produce sperm and oocytes at different times. Embryogenesis is unusual, with a migration of flagellated cells at the surface to the interior (parenchymella) or the production of an amphiblastula with inversion and growth of macromeres over micromeres. Sponges have great regenerative abilities.

Sponges are an ancient group, remote phylogenetically from other metazoa, but some evidence suggests that they are a sister group to the Eumetazoa. Their adaptive radiation is centered on elaboration of the water circulation and filter feeding system.

Review Questions

1. Briefly describe and contrast the syncytial ciliate hypothesis, the colonial flagellate hypothesis, and the polyphyletic origin of the metazoa. Which hypothesis seems most compatible with available data?
2. Describe the body plan of the Mesozoa and Placozoa.
3. Give eight characteristics of sponges.
4. Briefly describe asconoid, syconoid, and leuconoid body types in sponges.
5. What sponge body type is most efficient and makes possible the largest body size?
6. Define the following: ostia, osculum, spongocoel, apopyles, prosopyles.
7. Define the following: pinacocytes, choanocytes, archaeocytes, sclerocytes, spongocytes, collencytes.
8. What material is found in the skeleton of all sponges?
9. Describe the skeletons of each of the classes of sponges.
10. Describe how sponges feed, respire, and excrete.
11. What is a gemmule?
12. Describe how gametes are produced and the process of fertilization in most sponges.
13. Contrast embryogenesis in most Demospongiae with that in the Calcarea.
14. What is the largest class of sponges, and what is its body type?
15. What are possible ancestors to sponges? Justify your answer.

Selected References

See also general references for Part III, p. 626.

Bergquist, P. R. 1978. Sponges. Berkeley, University of California Press. *Excellent monograph on sponge structure, classification, evolution, and general biology.*

Grell, K. G. 1982. Placozoa. In S. P. Parker (ed). Synopsis and classification of living organisms, vol. 1. New York, McGraw-Hill Book Company. *Synopsis of placozoan characteristics.*

Hartman, W. D. 1982. Porifera. In S. P. Parker (ed). Synopsis and classification of living organisms, vol. 1. New York, McGraw-Hill Book Company. *Review of sponge classification.*

Simpson, T. L. 1984. The cell biology of sponges. New York, Springer-Verlag. *A review and synthesis, points out many problems yet to be solved.*

Wainright, P. O., G. Hinkle, M. L. Sogin, S. K. Stickel. 1993. Monophyletic origins of the Metazoa: an evolutionary link with Fungi. Science **260:**340–342. *Reports molecular evidence that the sister group of metazoans is Fungi and that multicellular animals are monophyletic.*

Wood, R. 1990. Reef-building sponges. Am. Sci. **78:**224–235. *The author presents evidence that the known sclerosponges belong to either the Calcarea or the Demospongiae and that a separate class Sclerospongiae is not needed.*

14

The Radiate Animals

Phylum Cnidaria
Phylum Ctenophora

A Fearsome Tiny Weapon

Although members of the phylum Cnidaria are more highly organized than sponges, they are still relatively simple animals. Most are sessile; those that are unattached, such as jellyfish, can swim only feebly. None can chase their prey. Indeed, we might easily get the false impression that the cnidarians were placed on earth to provide easy meals for other animals. The truth is, however, many cnidarians are very effective predators that are able to kill and eat prey that are much more highly organized, swift, and intelligent. They manage these feats because they possess tentacles that bristle with tiny, remarkably sophisticated weapons called nematocysts.

As it is secreted within the cell that contains it, the nematocyst is endowed with potential energy to power its discharge. It is as though a factory manufactured a gun, cocked and ready with a bullet in its chamber, as it rolls off the assembly line. Like the cocked gun, the completed nematocyst requires only a small stimulus to make it fire. Rather than a bullet, a tiny thread bursts from the nematocyst. Achieving a velocity of 2 meters/sec and an acceleration of $40,000 \times$ gravity, it instantly penetrates its prey and injects a paralyzing toxin. A small animal unlucky enough to brush against one of the tentacles is suddenly speared with hundreds or even thousands of nematocysts and quickly immobilized. Some nematocyst threads can penetrate human skin, resulting in sensations ranging from minor irritation to great pain, even death, depending on the species. A fearsome, but wondrous, tiny weapon. ■

POSITION IN ANIMAL KINGDOM

The two phyla Cnidaria and Ctenophora make up the radiate animals, which are characterized by **primary radial** or **biradial symmetry,** which we believe is ancestral for the eumetazoans. Radial symmetry, in which the body parts are arranged concentrically around the oral-aboral axis, is particularly suitable for sessile or sedentary animals and for free-floating animals because they approach their environment (or it approaches them) from all sides equally. Biradial symmetry is basically a type of radial symmetry in which only two planes through the oral-aboral axis divide the animal into mirror images because of the presence of some part that is single or paired. All other eumetazoans have a primary bilateral symmetry; that is, they are bilateral or were derived from an ancestor that was bilateral. Neither phylum has advanced generally beyond the **tissue level of organization,** although a few organs occur. In general, the ctenophores' structure is more complex than that of the cnidarians.

BIOLOGICAL CONTRIBUTIONS

1. Both phyla have developed two well-defined **germ layers,** ectoderm and endoderm; a third, or mesodermal, layer, which is derived embryologically from the ectoderm, is present in some. The body plan is saclike, and the body wall is composed of two distinct layers, epidermis and gastrodermis, derived from the ectoderm and endoderm, respectively. The gelatinous matrix, mesoglea, between these layers may be structureless, may contain a few cells and fibers, or may be composed largely of mesodermal connective tissue and muscle fibers.

2. An internal body cavity, the **gastrovascular cavity,** is lined by the gastrodermis and has a single opening, the mouth, which also serves as the anus.

3. **Extracellular digestion** occurs in the gastrovascular cavity, and intracellular digestion takes place in the gastrodermal cells. Extracellular digestion allows ingestion of larger food particles.

4. Most radiates have **tentacles,** or extensible projections around the oral end, that aid in food capture.

5. Radiates are the simplest animals to possess true **nerve cells** (protoneurons), but the nerves are arranged as a nerve net, with no central nervous system.

6. Radiates are the simplest animals to possess sense organs, which include well-developed statocysts (organs of equilibrium) and ocelli (photosensitive organs).

7. Locomotion in the free-moving forms is achieved by either **muscular contractions** (cnidarians) or **ciliary comb plates** (ctenophores). However, both groups are still better adapted to floating or being carried by currents than to strong swimming.

8. **Polymorphism**[*] in the cnidarians has widened their ecological possibilities. In many species the presence of both a polyp (sessile and attached) stage and a medusa (free-swimming) stage permits occupation of a benthic (bottom) and a pelagic (open-water) habitat by the same species. Polymorphism also widens the possibilities of structural complexity.

9. Some unique features are found in these phyla, such as **nematocysts** (stinging organelles) in cnidarians and **colloblasts** (adhesive organelles) and **ciliary comb plates** in ctenophores.

[*]Note that polymorphism here refers to more than one structural form of individual within a species, as contrasted with the use of the word in genetics (p. 169), in which it refers to different allelic forms of a gene in a population.

PHYLUM CNIDARIA

The phylum Cnidaria (ny-dar´e-a) (Gr. *knide,* nettle, + L. *aria* [pl. suffix], like or connected with) is an interesting group of more than 9000 species. It takes its name from cells called **cnidocytes,** which contain the stinging organelles **(nematocysts)** characteristic of the phylum. Nematocysts are *formed and used* only by cnidarians. Another name for the phylum, Coelenterata (se-len´te-ra´ta) (Gr. *koilos,* hollow, + *enteron,* gut, + L. *ata* [pl. suffix], characterized by), is used less commonly than formerly, and it sometimes now refers to both radiate phyla, since its meaning is equally applicable to both.

The cnidarians are generally regarded as originating close to the basal stock of the metazoan line. They are an ancient group with the longest fossil history of any metazoan, reaching back more than 700 million years. Although their organization has a structural and functional simplicity not found in other metazoans, they form a significant proportion of the biomass in some locations. They are widespread in marine habitats, and there are a few in fresh water. Although they are mostly sessile, or at best, fairly slow moving or slow swimming, they are quite efficient predators of organisms that are much swifter and more complex. The phylum includes some of nature's strangest and loveliest creatures: the branching, plant-like hydroids; the flowerlike sea anemones; the jellyfishes; and those architects of the ocean floor, the horny corals (sea whips, sea fans, and others) and the stony corals whose thousands of years of calcareous house-building have produced great reefs and coral islands (p. 270).

Cnidarians are found most abundantly in shallow marine habitats, especially in warm temperatures and tropical regions. There are no terrestrial species. Colonial hydroids are usually found attached to mollusc shells, rocks, wharves, and other animals in shallow coastal water, but some species are found at great depths. Floating and free-swimming medusae are found in open seas and

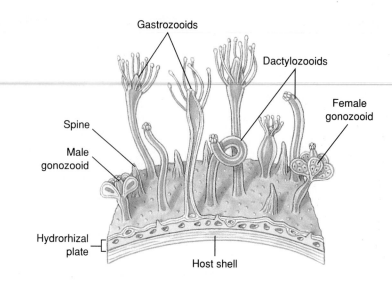

Figure 14-1

A, A hermit crab with its cnidarian mutuals. The shell is blanketed with polyps of the hydrozoan *Hydractinia milleri*. The crab gets some protection from predation by the cnidarians, and the cnidarians get a free ride and bits of food from their host's meals. **B,** Portion of a colony of *Hydractinia*, showing the types of zooids and the stolon (hydrorhiza) from which they grow.

lakes, often far from the shore. Floating colonies such as the Portuguese man-of-war and *Velella* (L. *velum,* veil; + *ellus,* dim. suffix) have floats or sails by which the wind carries them.

Some ctenophores, molluscs, and flatworms eat hydroids bearing nematocysts and use these stinging structures for their own defense. Some other animals feed on cnidarians, but cnidarians rarely serve as food for humans.

Cnidarians sometimes live symbiotically with other animals, often as commensals on the shell or other surface of their host. Certain hydroids (Figure 14-1) and sea anemones commonly live on snail shells inhabited by hermit crabs, providing the crabs some protection from predators. Algae frequently live as mutuals in the tissues of cnidarians, notably in some freshwater hydras and in reef-building corals. The presence of the algae in reef-building corals limits the occurrence of coral reefs to relatively shallow, clear water where there is sufficient light for the photosynthetic requirements of the algae. These kinds of corals are an essential component of coral reefs, and reefs are extremely important habitats in tropical waters. Coral reefs are discussed further later in the chapter.

Although many cnidarians have little economic importance, reef-building corals are an important exception. Fish and other animals associated with reefs provide substantial amounts of food for humans, and reefs are of

economic value as tourist attractions. Precious coral is used for jewelry and ornaments, and coral rock serves for building purposes.

Planktonic medusae may be of some importance as food for fish that are of commercial value; the reverse is also true—the young fish fall prey to cnidarians.

Four classes of Cnidaria are commonly recognized: Hydrozoa (the most variable class, including hydroids, fire corals, Portuguese man-of-war, and others), Scyphozoa (the "true" jellyfishes), Cubozoa (cube jellyfishes), and Anthozoa (the largest class, including sea anemones, stony corals, soft corals, and others).

CHARACTERISTICS OF PHYLUM CNIDARIA

1. Entirely aquatic, some in fresh water but mostly marine
2. **Radial symmetry** or biradial symmetry around a longitudinal axis with **oral** and **aboral** ends; no definite head
3. Two basic types of individuals: **polyps** and **medusae**
4. Exoskeleton or endoskeleton of chitinous, calcareous, or protein components in some
5. Body with two layers, epidermis and gastrodermis, with mesoglea **(diploblastic);** mesoglea with cells and connective tissue (ectomesoderm) in some **(triploblastic)**
6. **Gastrovascular cavity** (often branched or divided with septa) with a single opening that serves as both mouth and anus; extensible tentacles usually encircling the mouth or oral region
7. Special stinging cell organelles called **nematocysts** in either epidermis or gastrodermis or in both; nematocysts abundant on tentacles, where they may form batteries or rings
8. **Nerve net** with symmetrical and asymmetrical synapses; with some sensory organs; diffuse conduction
9. Muscular system (epitheliomuscular type) of an outer layer of longitudinal fibers at base of epidermis and an inner one of circular fibers at base of gastrodermis; modifications of this plan in higher cnidarians, such as separate bundles of independent fibers in the mesoglea
10. Asexual reproduction by budding (in polyps) or sexual reproduction by gametes (in all medusae and some polyps); sexual forms monoecious or dioecious; **planula larva;** holoblastic indeterminate cleavage
11. No excretory or respiratory system
12. No coelomic cavity

FORM AND FUNCTION

Dimorphism and Polymorphism in Cnidarians

One of the most interesting—and sometimes puzzling—aspects of this phylum is the dimorphism and often polymorphism displayed by many of its members. All cnidarian forms fit into one of two morphological types (dimorphism): the **polyp,** or hydroid form, which is adapted to a sedentary or sessile life, and the **medusa,** or jellyfish form, which is adapted for a floating or free-swimming existence (Figure 14-2).

Most polyps have tubular bodies with a mouth at one end surrounded by tentacles. The aboral end is usually attached to a substratum by a pedal disc or other device. Polyps may live singly or in colonies. Colonies of some species include morphologically differing individuals (polymorphism), each specialized for a certain function, such as feeding, reproduction, or defense (Figure 14-1).

Medusae are usually free swimming and have bell-shaped or umbrella-shaped bodies and tetramerous symmetry (body parts arranged in fours). The mouth is usually centered on the concave side, and tentacles extend from the rim of the umbrella.

The sea anemones and corals (class Anthozoa) are all polyps; hence, they are not dimorphic. The true jellyfishes (class Scyphozoa) have a conspicuous medusoid form, but many have a polypoid larval stage. The colonial hydroids of class Hydrozoa, however, sometimes have life histories that feature both the polyp, or hydroid, stage and the free-swimming medusa stage—rather like a Jekyll-and-Hyde existence. A species that has both the attached polyp and the floating medusa within its life history can take advantage of the feeding and distribution possibilities of both the pelagic (open-water) and the benthic (bottom) types of environment. Many hydrozoans are also polymorphic, with several distinct types of polyps in a colony.

Superficially the polyp and medusa seem very different. But actually each has retained the saclike

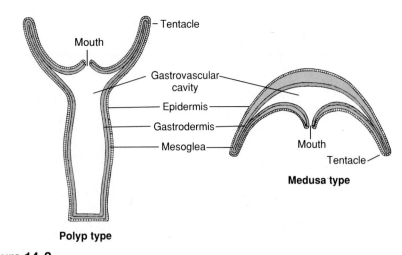

Figure 14-2

Comparison between the polyp and medusa types of individuals.

body plan that is basic to the phylum (Figure 14-2). The medusa is essentially an unattached polyp with the tubular portion widened and flattened into the bell shape.

Both the polyp and the medusa possess the three body wall layers typical of the cnidarians, but the jelly-like layer of mesoglea is much thicker in the medusa, constituting the bulk of the animal and making it more buoyant. It is because of this mass of mesoglea "jelly" that the medusae are commonly called jellyfishes.

Nematocysts: The Stinging Organelles

One of the most characteristic structures in the entire cnidarian group is the stinging organelle called the **nematocyst** (Figure 14-3). Over 20 different types of nematocysts (Figure 14-4) have been described in the cnidarians so far; they are important in taxonomic determinations. The nematocyst is a tiny capsule composed of material similar to chitin and containing a coiled tubular "thread" or filament, which is a continuation of the narrowed end of the capsule. This end of the capsule is covered by a little lid, or **operculum.** The inside of the undischarged thread may bear tiny barbs, or spines.

The nematocyst is enclosed in the cell that has produced it, the **cnidocyte** (during its development, the cnidocyte is properly called the

cnidoblast). Except in Anthozoa, cnidocytes are equipped with a triggerlike **cnidocil,** which is a modified cilium. Anthozoan cnidocytes have a somewhat different ciliary mechanoreceptor. In some sea anemones, and perhaps other cnidarians, small organic molecules from the prey "tune" the mechanoreceptors, sensitizing them to the frequency of vibration caused by the prey swimming. Tactile stimulation causes the nematocyst to discharge. Cnidocytes are borne in invaginations of ectodermal cells and, in some forms, in gastrodermal cells, and they are especially abundant on the tentacles. When a nematocyst is discharged, its cnidocyte is absorbed and a new one replaces it. Not all nematocysts have barbs or inject poison. Some, for example, do not penetrate the prey but rapidly recoil like a spring after discharge, grasping and holding any part of the prey caught in the coil (Figure 14-4). Adhesive nematocysts usually do not discharge in food capture.

The mechanism of nematocyst discharge is remarkable. Present evidence indicates that discharge is due to a combination of tensional forces generated during nematocyst formation and due to an astonishingly high osmotic pressure within the nematocyst: 140 atmospheres. When stimulated to discharge, the high internal osmotic pressure causes water to rush into the capsule. The operculum opens, and the rapidly increasing *hydrostatic pressure* within the capsule forces the

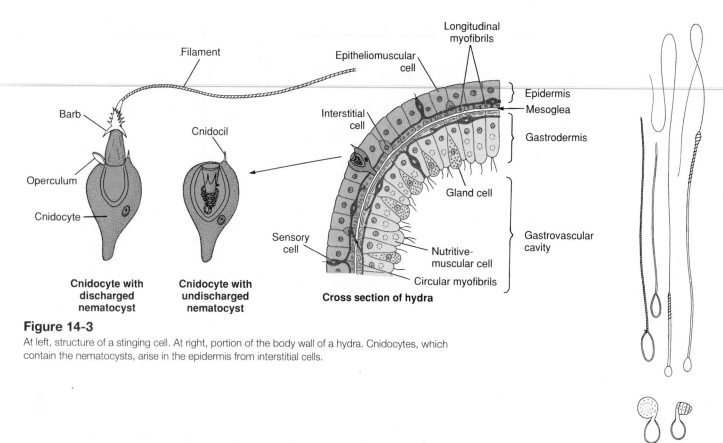

Figure 14-3

At left, structure of a stinging cell. At right, portion of the body wall of a hydra. Cnidocytes, which contain the nematocysts, arise in the epidermis from interstitial cells.

thread out with great force, the thread turning inside out as it goes. At the everting end of the thread, the barbs flick to the outside like tiny switch-blades. This minute but awesome weapon then injects poison when it penetrates the prey.

> Note again the distinction between osmotic and hydrostatic pressure (p. 52). The nematocyst is never required actually to contain 140 atmospheres of hydrostatic pressure within itself; such a hydrostatic pressure would doubtless cause it to explode. As the water rushes in during discharge, the osmotic pressure falls rapidly, while the hydrostatic pressure rapidly increases.

The nematocysts of most cnidarians are not harmful to humans and are a nuisance at worst. However, the stings of the Portuguese man-of-war (Figure 14-5) and certain jellyfish are quite painful and sometimes dangerous.

Nerve Net

The nerve net of the cnidarians is one of the best examples of a diffuse nervous system in the animal kingdom. This plexus of nerve cells is found both at the base of the epidermis and at the base of the gastrodermis, forming two interconnected nerve nets. Nerve processes (axons) end on other nerve cells at synapses or at junctions with sensory cells or effector organs (nematocysts or epitheliomuscular cells). Nerve impulses are transmitted from one cell to another by release of a neurotransmitter from small vesicles on one side of the synapse or junction (p. 718). One-way transmission between nerve cells in higher animals is ensured because the vesicles are located on only one side of the synapse. However, cnidarian nerve nets are peculiar in that many of the synapses have vesicles of neurotransmitters on both sides, allowing transmission across the synapse in either direction. Another peculiarity of cnidarian nerves is the absence of any sheathing material (myelin) on the axons.

Figure 14-4

Several types of nematocysts shown after discharge. At bottom are two views of a type that does not impale the prey, rather it recoils like a spring, catching any small part of the prey in the path of the recoiling thread.

> Note that there is little adaptive value for a radially symmetrical animal to have a central nervous system with a brain. The environment approaches from all sides equally, and there is no control over the direction of approach to a prey organism.

There is no concentrated grouping of nerve cells to suggest a "central nervous system." Nerves are grouped, however, in the "ring nerves" of hydrozoan medusae and in the marginal sense organs of scyphozoan medusae. In some cnidarians the nerve nets form two or more systems: in Scyphozoa there is a fast conducting system to coordinate swimming movements and a slower one to coordinate movements of tentacles.

Reduced medusae
(gonophores)

Figure 14-5

In some hydroids, such as this *Tubularia crocea,* medusae are reduced to gonadal tissue and do not detach. These reduced medusae are known as gonophores.

Figure 14-6

Hydra with developing bud and ovary.

The nerve cells of the net have synapses with slender sensory cells that receive external stimuli, and the nerve cells have junctions with epitheliomuscular cells and nematocysts. Together with the contractile fibers of the epitheliomuscular cells, the sensory-nerve cell net combination is often referred to as a **neuromuscular system,** an important landmark in the evolution of the nervous system. The nerve net arose early in metazoan evolution, and it has never been completely lost phylogenetically. Annelids have it in their digestive systems. In the human digestive system it is represented by nerve plexuses in the musculature. The rhythmical peristaltic movements of the stomach and intestine are coordinated by this counterpart of the cnidarian nerve net.

CLASS HYDROZOA

The majority of hydrozoa are marine and colonial in form, and the typical life cycle includes both the asexual polyp and the sexual medusa stages. Some, however, such as the freshwater hydra, have no medusa stage.

Some marine hydroids do not have free medusae (Figure 14-5), whereas some hydrozoans occur only as medusae and have no polyp.

The hydra, although not a typical hydrozoan, has become a favorite as an introduction to Cnidaria because of its size and ready availability. Combining its study with that of a representative colonial marine hydroid such as *Obelia* (Gr. *obelias,* round cake) gives an excellent idea of the class Hydrozoa.

Hydra: A Freshwater Hydrozoan

The common freshwater hydra (Figure 14-6) is a solitary polyp and one of the few cnidarians found in fresh water. Its normal habitat is the underside of aquatic leaves and lily pads in cool, clean fresh water of pools and streams. The hydra family is found throughout the world, with 16 species occurring in North America.

Body Plan. The body of the hydra can extend to a length of 25 to 30 mm or can contract to a tiny, gelatinous mass. It is a cylindrical tube with the aboral end drawn out into a slender stalk, ending in a **basal** (or pedal) **disc** for attachment. This basal disc is provided with gland cells to enable the hydra to adhere to a substratum and

also to secrete a gas bubble for floating. In the center of the disc there may be an excretory pore. The mouth, located on a conical elevation called the **hypostome,** is encircled by 6 to 10 hollow tentacles that, like the body, can be greatly extended when the animal is hungry.

The mouth opens into the **gastrovascular cavity** (also called a coelenteron) which communicates with the cavities in the tentacles. In some individuals **buds** may project from the sides, each with a mouth and tentacles like the parent. Testes or ovaries, when present, appear as rounded projections on the surface of the body (Figure 14-6).

Body Wall. The body wall surrounding the gastrovascular cavity consists of an outer **epidermis** (ectodermal) and an inner **gastrodermis** (endodermal) with **mesoglea** between them (Figure 14-3).

Epidermis. The epidermal layer contains epitheliomuscular, interstitial, gland, cnidocyte, and sensory and nerve cells.

Epitheliomuscular cells make up most of the epidermis and serve both for covering and for muscular contraction (Figure 14-7). The bases of most of these cells are extended parallel to the tentacle or body axis and

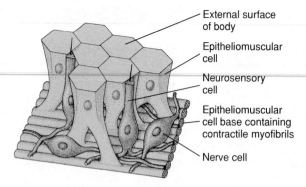

External surface
of body

Epitheliomuscular
cell

Neurosensory
cell

Epitheliomuscular
cell base containing
contractile myofibrils

Nerve cell

Figure 14-7
Epitheliomuscular and nerve cells in hydra.

contain myofibrils, thus forming a layer of longitudinal muscle next to the mesoglea. Contraction of these fibrils shortens the body or tentacles.

Interstitial cells are the undifferentiated stem cells found among the bases of the epitheliomuscular cells. Differentiation of the interstitial cells gives rise to cnidoblasts, sex cells, buds, nerve cells, and others, but generally not to epitheliomuscular cells (which reproduce themselves).

Gland cells are tall cells around the basal disc and mouth that secrete an adhesive substance for attachment and sometimes a gas bubble for floating.

Over 230 years ago, Abraham Trembley was astonished to discover that isolated sections of the stalk of hydra could regenerate and each become a complete animal. Since then, over 2000 investigations of hydra have been published, and the organism has become a classic model for the study of morphological differentiation. The mechanisms governing morphogenesis have great practical importance, and the simplicity of hydra lends itself to these investigations. Substances controlling development (morphogens), such as those determining which end of a cut stalk will develop a mouth and tentacles, have been discovered, and they may be present in the cells in extremely low concentrations (10^{-10} M).

Cnidocytes containing nematocysts occur throughout the epidermis. The hydra has three functional types of nematocysts: those that penetrate the prey and inject poison (penetrants, Figure 14-3); those that recoil and entangle the prey (volvents); and those that secrete an adhesive substance used in locomotion and attachment (glutinants).

Sensory cells are scattered among the other epidermal cells, especially near the mouth and tentacles and on the basal disc. The free end of each sensory cell bears a flagellum, which is the sensory receptor for chemical and tactile stimuli. The other end branches into fine processes, which synapse with the nerve cells.

Nerve cells of the epidermis are generally multipolar (have many processes), although in more highly organized cnidarians the cells may be bipolar (with two processes). Their processes (axons) form synapses with sensory cells and other nerve cells and junctions with epitheliomuscular cells and cnidocytes. There are both one-way (morphologically asymmetrical) and two-way synapses with other nerve cells.

Gastrodermis. The gastrodermis, a layer of cells lining the gastrovascular cavity, is chiefly large, ciliated, columnar epithelial cells with irregular flat bases. The cells of the gastrodermis include nutritive-muscular, interstitial, and gland cells.

Nutritive-muscular cells are usually tall columnar cells and have laterally extended bases containing myofibrils. The myofibrils run at right angles to the body or tentacle axis and so form a circular muscle layer. However, this muscle layer in hydras is very weak,

and longitudinal extension of the body and tentacles is brought about mostly by increasing the volume of water in the gastrovascular cavity. The water is brought in through the mouth by the beating of the cilia on the nutritive-muscular cells. Thus, the water in the gastrovascular cavity serves as a **hydrostatic skeleton.** The two cilia on the free end of each cell also serve to circulate food and fluids in the digestive cavity. The cells often contain large numbers of food vacuoles. Gastrodermal cells in the green hydra (*Chlorohydra*) (Gr. *chloros,* green; + *hydra,* a mythical nine-headed monster slain by Hercules) bear green algae (zoochlorellae), which give the hydras their color. This is probably a case of symbiotic mutualism, since the algae use the respiratory carbon dioxide from the hydra to form organic compounds useful to the host. They receive shelter and probably other physiological requirements in return.

Interstitial cells are scattered among the bases of the nutritive cells. They transform into other types of cells when the need arises.

Gland cells in the hypostome and in the column secrete digestive enzymes. Mucous glands about the mouth aid in ingestion.

Nematocysts are not found in the gastrodermis because cnidocytes are lacking in this layer.

Mesoglea. The mesoglea lies between the epidermis and gastrodermis and is attached to both layers. It is gelatinous, or jellylike, and both epidermal and gastrodermal cells send processes into it. It is a continuous layer that extends over both body and tentacles, thickest in the stalk portion and thinnest in the tentacles. This arrangement allows the pedal region to withstand great mechanical strain and gives the tentacles more flexibility. The mesoglea helps to support the body and acts as a type of elastic skeleton.

Locomotion. Unlike colonial polyps, which are permanently attached, the hydra can move about freely by gliding on its basal disc, aided by mucous secretions. Or using an "inch worm" movement, it can loop along by bending over and attaching its tentacles to

Figure 14-8
Hydra catches an unwary water flea with the nematocysts of its tentacles. This hydra already contains one water flea eaten previously.

the substratum. It may even turn end over end or detach itself and, by forming a gas bubble on its basal disc, float to the surface.

Feeding and Digestion. Hydras feed on a variety of small crustaceans, insect larvae, and annelid worms. The hydra awaits its prey with tentacles extended (Figure 14-8). The food organism that brushes against its tentacles may find itself harpooned by scores of nematocysts that render it helpless, even though it may be larger than the hydra. The tentacles move toward the mouth, which slowly widens. Well moistened with mucous secretions, the mouth glides over and around the prey, totally engulfing it.

The activator that actually causes the mouth to open is the reduced form of glutathione, which is found to some extent in all living cells. Glutathione escapes from the prey through the wounds made by the nematocysts, but only animals releasing enough of the chemical to activate the feeding response are eaten by the hydra. This explains how a hydra distinguishes between *Daphnia,* which it relishes, and some other forms that it refuses. If we place glutathione in water containing hydras, each hydra will go through the motions of feeding even though no prey is present.

Inside the gastrovascular cavity, gland cells discharge enzymes on the food. The digestion starts in the gastrovascular cavity (extracellular digestion), but many of the food particles are drawn by pseudopodia into the nutritive-muscular cells of the gastrodermis, where intracellular digestion occurs. Ameboid cells may carry undigested particles to the gastrovascular cavity, where they are eventually expelled with other indigestible matter.

Reproduction. The hydra reproduces sexually and asexually. In asexual reproduction, buds appear as outpocketings of the body wall and develop into young hydras that eventually detach from the parent. Most species are dioecious. Temporary gonads (Figure 14-6) usually appear in the autumn, stimulated by the lower temperatures and perhaps also by the reduced aeration of stagnant waters. Eggs in the ovary usually mature one at a time and are fertilized by sperm shed into the water.

The zygotes undergo holoblastic cleavage to form a hollow blastula. The inner part of the blastula delaminates to form the endoderm (gastrodermis), and the mesoglea is laid down between the ectoderm and endoderm. A cyst forms around the embryo before it breaks loose from the parent, enabling it to survive the winter. Young hydras hatch out in spring when the weather is favorable.

Hydroid Colonies

Far more representative of class Hydrozoa than the hydras are hydroids that have a medusa stage in their life cycle. We often use *Obelia* in laboratory exercises for beginning students to illustrate the hydroid type (Figure 14-9).

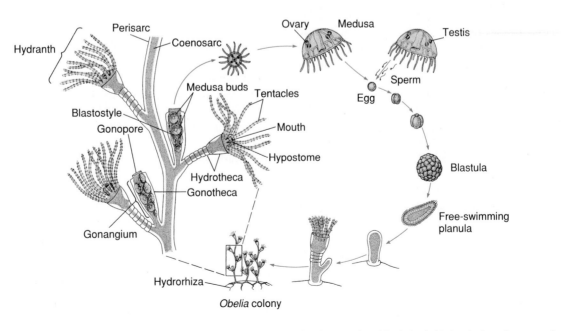

Obelia colony

Figure 14-9
Life cycle of *Obelia,* showing alternation of polyp (asexual) and medusa (sexual) stages. *Obelia* is a calyptoblastic hydroid; that is, its polyps as well as its stems are protected by continuations of the perisarc. Contrast with *Eudendrium* (Figure 14-10).

A

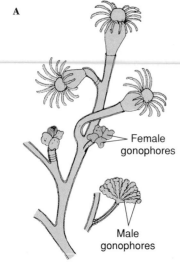

Female
gonophores

Male
gonophores

B

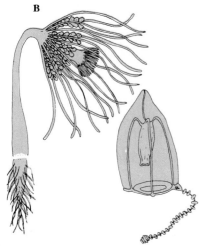

Figure 14-10
Gymnoblastean hydroids. **A,** *Eudendrium* forms
a bushy colony with naked hydranths and
gonophores. There are no free medusae.
B, *Corymorpha* is a solitary hydroid. It produces
free-swimming medusae, each with a single
trailing tentacle.

A typical hydroid has a base, a
stalk, and one or more terminal
zooids. The base by which the colonial hydroids attach to the substratum
is a rootlike stolon, or **hydrorhiza,**
which gives rise to one or more stalks
called **hydrocauli.** The living cellular
part of the hydrocaulus is the tubular
coenosarc, composed of the three
typical cnidarian layers surrounding
the coelenteron (gastrovascular cavity).
The protective covering of the hydrocaulus is a nonliving chitinous sheath,
or **perisarc.** Attached to the hydrocaulus are the individual polyp animals, or zooids. Most of the zooids are

Figure 14-11
Bell medusa, *Polyorchis penicillatus,* medusa
stage of an unknown attached polyp.

feeding polyps called **hydranths,** or
gastrozooids. They may be tubular,
bottle shaped, or vaselike, but all have
a terminal mouth and a circlet of tentacles. In some forms, such as *Obelia,*
the perisarc continues as a protective
cup around the polyp into which it
can withdraw for protection (Figure
14-9). In others the polyp is naked
(Figure 14-10). In some forms the
perisarc is an inconspicuous, thin film.

The hydranths, much like miniature hydras, capture and ingest prey,
such as tiny crustaceans, worms, and
larvae, thus providing nutrition for the
entire colony. After partial extracellular digestion in the hydranth, the digestive broth passes along the common gastrovascular cavity where it is
taken up by gastrodermal cells, and
intracellular digestion occurs.

Circulation within the gastrovascular cavity is a function of the ciliated
gastrodermis but is also aided by
rhythmical contractions and pulsations
of the body, which occur in hydroids.

Just as hydras reproduce asexually by budding, the colonial hydroids
bud off new individuals, thus increasing the size of the colony. New feeding polyps arise by budding, and
medusa buds also arise on the
colony. In *Obelia* these medusae bud

from a reproductive polyp called a
gonangium. The young medusae
leave the colony as free-swimming
individuals that mature and produce
gametes (eggs and sperm) (Figure 14-
9). In some species the medusae remain attached to the colony and shed
their gametes there. In other species
the medusae never develop and the
gametes are shed by male and female
gonophores (Figure 14-10). Embryonation of the zygote results in a ciliated planula larva that swims about
for a time. Then it settles down to a
substratum to develop into a minute
polyp that gives rise, by asexual budding, to the hydroid colony, thus
completing the life cycle.

Hydroid medusae are usually
smaller than scyphozoan medusae,
ranging from 2 to 3 mm to several centimeters in diameter (Figure 14-11). The
margin of the bell projects inward as a
shelflike **velum,** which partly closes
the open side of the bell and is used in
swimming (Figure 14-12). Muscular
pulsations that alternately fill and empty
the bell propel the animal forward, aboral side first, with a sort of weak "jet
propulsion." The tentacles attached to
the bell margin are rich in nematocysts.

The mouth opening at the end of
a suspended **manubrium** leads to a
stomach and four radial canals that
connect with a ring canal around the
margin. This in turn connects with the
hollow tentacles. Thus the gastrovascular cavity is continuous from mouth
to tentacles, and gastrodermis lines
the entire system. Nutrition is similar
to that of the hydranths.

The nerve net is usually concentrated into two nerve rings at the base
of the velum. The bell margin has a
liberal supply of sensory cells. It usually also bears two kinds of specialized sense organs: **statocysts,** which
are small organs of equilibrium (Figure 14-12B), and **ocelli,** which are
light-sensitive organs.

Freshwater Medusae

The freshwater medusa *Craspedacusta
sowerbyi* (Figure 14-13) (order Hydroida) may have evolved from marine
ancestors in the Yangtze River of China.

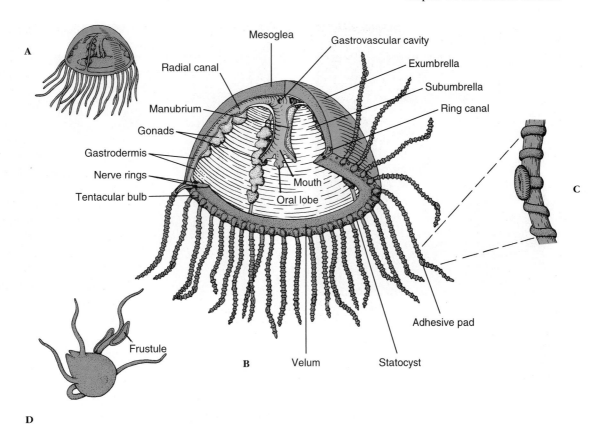

Figure 14-12

Structure of *Gonionemus.* **A,** Medusa with typical tetramerous arrangement. **B,** Cutaway view showing morphology. **C,** Portion of a tentacle with its adhesive pad and ridges of nematocysts. **D,** Tiny polyp, or hydroid stage, that develops from the planula larva. It can produce more polyps by budding (frustules) or produce medusa buds.

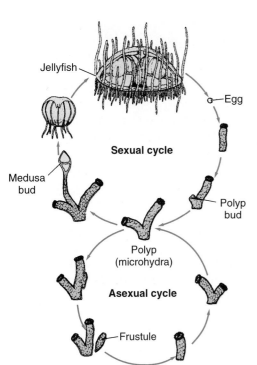

Figure 14-13

Life cycle of *Craspedacusta,* a freshwater hydrozoan. The polyp has three methods of asexual reproduction: by budding off new individuals, which may remain attached to the parent (colony formation); by constricting off nonciliated planula-like larvae (frustules), which can move around and give rise to new polyps; and by producing medusa buds, which develop into sexual jellyfish.

Probably introduced with shipments of aquatic plants, this interesting form has now been found in many parts of Europe, all over the United States, and in parts of Canada. The medusae may attain a diameter of 20 mm.

The polyp phase of this animal is tiny (2 mm) and has a very simple form with no perisarc and no tentacles. It occurs in colonies of a few polyps. For a long time its relation to the medusa was not recognized, and thus the polyp was given a name of its own, *Microhydra ryderi.* On the basis of its relationship to the jellyfish and the law of priority, both the polyp and the medusa should be called *Craspedacusta* (N.L. *craspedon,* velum; + Gr. *kystis,* bladder).

The polyp has three methods of asexual reproduction, as shown in Figure 14-13.

Floating Colonies

Members of the orders Siphonophora and Chondrophora are among the most specialized of the Hydrozoa. They form polymorphic swimming or floating colonies made up of several types of modified medusae and polyps.

Physalia (Gr. *physallis,* bladder), the Portuguese man-of-war (Figure 14-15), is one such colony with a rainbow-hued float of blues and pinks that carries it along on the surface

A   B

Figure 14-14

These hydrozoans form calcareous skeletons that resemble true coral. **A,** *Stylaster roseus* (order Stylasterina) occurs commonly in caves and crevices in coral reefs. These fragile colonies branch in only a single plane and may be white, pink, purple, red, or red with white tips. **B,** Species of *Millepora* (order Milleporina) form branching or platelike colonies and often grow over the horny skeleton of gorgonians (see p. 270), as is shown here. They have a generous supply of powerful nematocysts that produce a burning sensation on human skin, justly earning the common name fire coral.

waters of tropical seas. Many are blown to shore on the eastern coast of the United States. The long, graceful tentacles, actually zooids, are laden with nematocysts and are capable of inflicting painful stings. The float, called a **pneumatophore,** is believed to have expanded from the original larval polyp. It contains a sac arising from the body wall and filled with a gas similar to air. The float acts as a type of nurse-carrier for future generations of individuals that bud from it and hang suspended in the water. Some of the siphonophores, such as *Stephalia* and *Nectalia,* possess swimming bells as well as a float.

An interesting mutualistic relationship exists between *Physalia* and a small fish called *Nomeus* (Gr. herdsman) that swims among the tentacles with perfect safety. Why the fish is not stung to death by its host's nematocysts is unclear, but like the anemone fish to be discussed later, *Nomeus* is probably protected by a skin mucus that does not stimulate nematocyst discharge.

Figure 14-15

Three Portuguese man-of-war colonies, *Physalia physalis* (order Siphonophora, class Hydrozoa), lie stranded in shallow water. Colonies often drift onto southern ocean beaches, where they are a hazard to bathers. Each colony of medusa and polyp types is integrated to act as one individual. As many as a thousand zooids may be found in one colony. The nematocysts secrete a powerful neurotoxin.

There are several types of polyp individuals. The **gastrozooids** are feeding polyps with a single long tentacle arising from the base of each. Some of these long, stinging tentacles become separated from the feeding polyp and are called **dactylozooids,** or fishing tentacles. These sting the prey and lift them to the lips of the feeding polyps. Among the modified medusoid individuals are the **gonophores,** which are little more than sacs containing either ovaries or testes.

CLASS SCYPHOZOA

Class Scyphozoa (si-fo-zo´a) (Gr. *skyphos,* cup) includes most of the

Figure 14-16

Giant jellyfish, *Cyanea capillata* (order Semeaeostomeae, class Scyphozoa). A North Atlantic species of *Cyanea* reaches a bell diameter exceeding 2 m. It is known as the "sea blubber" by fishermen.

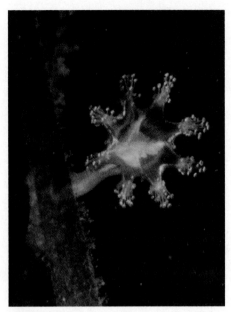

Figure 14-17

Thaumatoscyphus hexaradiatus (order Stauromedusae, class Scyphozoa). Members of this order are unusual scyphozoans in that the medusae are sessile and attached to seaweed or other objects.

larger jellyfishes, or "cup animals." A few, such as *Cyanea* (Gr. *kyanos,* dark-blue substance), may attain a bell diameter exceeding 2 m and tentacles 60 to 70 m long (Figure 14-16). Most scyphozoans, however, range from 2 to 40 cm in diameter. Most are found floating in the open sea, some even at depths of 3000 m, but one unusual order is sessile and attaches by a stalk to seaweeds and other objects on the sea bottom (Figure 14-17). Their coloring may range from colorless to striking orange and pink hues.

The bells of different species vary in depth from a shallow saucer shape to a deep helmet or goblet shape. The jelly (mesoglea) layer is unusually thick, giving the bell a fairly firm consistency. The jelly is 95% to 96% water. Unlike the hydromedusae, this layer in the scyphomedusae also contains ameboid cells and fibers. Movement is by rhythmical pulsations of the umbrella. There is no velum as in the hydromedusae. There may be many tentacles or few, and they may be short as in *Aurelia* (L. *aurum,* gold) or long as in *Cyanea. Aurelia aurita* (Figure 14-18) is a familiar species 7 to 10 cm in diameter, commonly found in the waters off both

the east and west coasts of the United States, and widely used for the study of Scyphozoa.

The margin of the umbrella is scalloped, usually with each indentation bearing a pair of **lappets,** and between them is a sense organ called a **rhopalium** (tentaculocyst). *Aurelia* has eight such notches. Some scyphozoans have 4, others 16. Each rhopalium is club shaped and contains a hollow statocyst for equilibrium and one or two pits lined with sensory epithelium. In some species the rhopalia also bear ocelli.

The mouth is centered on the subumbrella side. The manubrium is usually drawn out to form four frilly **oral arms** that are used in capturing and ingesting prey.

The tentacles, the manubrium, and often the entire body surface are well supplied with nematocysts that can give painful stings. However, the primary function of scyphozoan nematocysts is not to attack humans but to paralyze prey animals, which are conveyed to the mouth lobes with the help of the other tentacles or by the bending of the umbrella margin.

Aurelia, which has comparatively short tentacles, feeds on small plank-

tonic animals. These are caught in the mucus of the umbrella surface, are carried to "food pockets" on the umbrella margin by cilia, and are picked up from the pockets by the oral lobes whose cilia carry the food to the gastrovascular cavity. Cilia in the gastrodermis layer keep a current of water moving to bring food and oxygen into the stomach and carry out wastes.

Cassiopeia (L. mythical queen of Ethiopia), the "upside-down jellyfish" common to Florida waters, and *Rhizostoma* (Gr. *rhiza,* root, + *stoma,* mouth), which can be found in colder temperatures, belong to a group differing from that of *Aurelia* both in their lack of tentacles on the umbrella margin and in the structure of the oral arms. During development, the edges of the oral lobes fold over and fuse, forming canals (**arms** or **brachial canals**) that become highly branched. These open to the surface at frequent intervals by pores called "mouths"; the original mouth is obliterated in the fusion of the oral lobes. Planktonic organisms caught in the mucus of the frilly oral arms are transported to the mouths and then up the brachial canals to the gastric cavity by cilia. In contrast to the usual swimming habit of medusae, *Cassiopeia* is usually found lying on its "back" in shallow lagoons. Its umbrella margin contracts about 20 times a minute, creating water currents to bring plankton into contact with the mucus and nematocysts of its oral lobes. Its tissues are abundantly supplied with symbiotic algal cells (**zooxanthellae).** As they lie sunning themselves in shallow water, *Cassiopeia* are thus reminiscent of large flowers in more ways than one.

Internally, extending out from the stomach of scyphozoans are four **gastric pouches** in which gastrodermis extends down in little tentaclelike projections called **gastric filaments.** These are covered with nematocysts to quiet further any prey that may still be struggling. Gastric filaments are lacking in the hydromedusae. A complex system of **radial canals** branches out from the pouches to a **ring canal** in the margin and makes up a part of the gastrovascular cavity.

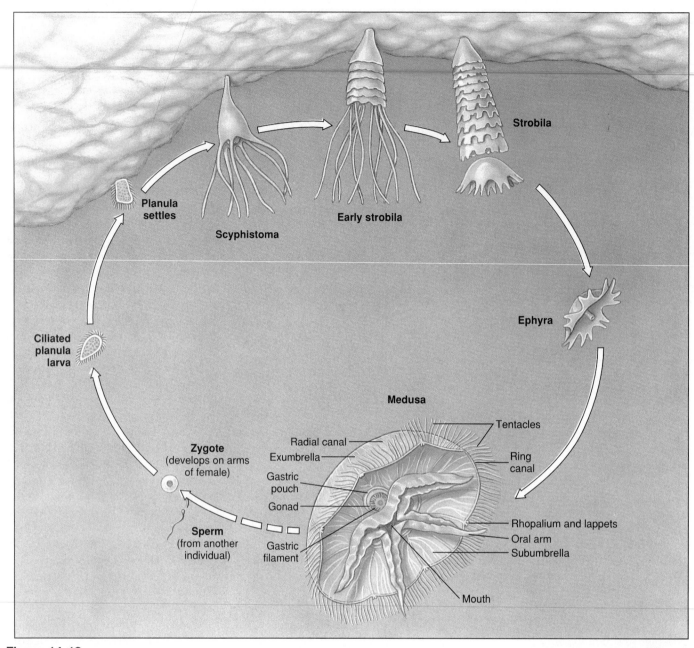

Figure 14-18

Life cycle of *Aurelia,* a marine scyphozoan medusa.

The **nervous system** in scypho-zoans is a nerve net, with a subum-brella net that controls bell pulsations and another, more diffuse net that con-trols local reactions such as feeding.

The sexes are separate, with the gonads located in the gastric pouches. Fertilization is internal, with the sperm being carried by ciliary cur-rents into the gastric pouch of the fe-male. The zygotes may develop in the seawater or may be brooded in folds of the oral arms. The ciliated planula larva becomes attached and develops into a **scyphistoma,** a hydralike form

(Figure 14-19). By a process of **stro-bilation** the scyphistoma of *Aurelia* forms a series of saucerlike buds, **ephyrae,** and is now called a **stro-bila** (Figures 14-18 and 14-19). When the ephyrae break loose, they grow into mature jellyfish.

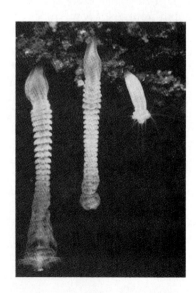

Figure 14-19

Scyphistoma (*right*) *and strobilas* (*left*) (polyp stages) of *Aurelia aurita* hanging from a rock. The strobila at left has an ephyra nearly ready to separate.

Photo by D. P. Wilson/Frank Lane Picture Agency Lmtd.

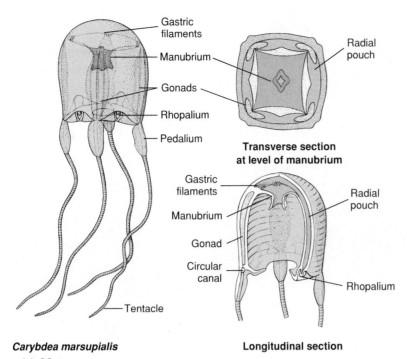

Gastric filaments
Manubrium
Gonads
Rhopalium
Pedalium
Tentacle

Carybdea marsupialis

Radial pouch

Transverse section at level of manubrium

Gastric filaments
Manubrium
Gonad
Circular canal
Radial pouch
Rhopalium

Longitudinal section

Figure 14-20
Carybdea, a cubozoan medusa.

CLASS CUBOZOA

The Cubozoa until recently were considered an order (Cubomedusae) of Scyphozoa. The medusoid is the predominant form (Figure 14-20); the polypoid is inconspicuous and in most cases unknown. Some cubozoan medusae may range up to 25 cm tall, but most are about 2 to 3 cm. In transverse section the bells are almost square. A tentacle or group of tentacles is found at each corner of the square at the umbrella margin. The base of each tentacle is differentiated into a flattened, tough blade called a **pedalium** (Figure 14-20). Rhopalia are present. The umbrella margin is not scalloped, and the subumbrella edge turns inward to form a **velarium.** The velarium functions as a velum does in hydrozoan medusae, increasing swimming efficiency, but it differs structurally. Cubomedusae are strong swimmers and voracious predators, feeding mostly on fish. Stings of some species can be fatal to humans.

The complete life cycle is known for only one species, *Tripedalia cystophora* (L. *tri,* three, + Gr. *pedalion,* rudder). The polyp is tiny (1 mm tall), solitary, and sessile. It buds off new polyps laterally, which detach and creep away. The polyps do not produce ephyrae but metamorphose directly into medusae.

Chironex fleckeri (Gr. *cheir,* hand; + *nexis,* swimming) is a large cubomedusa known as the sea wasp. Its stings are quite dangerous and sometimes fatal. Most of the fatal stings have been reported from tropical Australian waters, usually following quite massive stings. Witnesses have described victims as being covered with "yards and yards of sticky wet string." The stings are very painful, and death, if it is to occur, ensues within a matter of minutes. If death does not occur within 20 minutes after stinging, complete recovery is likely.

CLASS ANTHOZOA

The anthozoans, or "flower animals," are polyps with a flowerlike appearance (Figure 14-21). There is no medusa stage. Anthozoa are all marine and are found in both deep and shallow water and in polar seas as well as tropical seas. They vary greatly in size and may be solitary or colonial. Many of the forms are supported by skeletons.

The class has three subclasses: **Zoantharia** (or **Hexacorallia**), made up of the sea anemones, hard corals, and others; the **Ceriantipatharia,** which includes only the tube anemones and thorny corals; and the **Alcyonaria** (or **Octocorallia**), containing the soft and horny corals, such as sea fans, sea pens, sea pansies, and others. The zoantharians and ceriantipatharians have a **hexamerous** plan (of six or multiples of six) or polymerous symmetry and have simple tubular tentacles arranged in one or more circlets on the oral disc. The alcyonarians are **octomerous** (built on a plan of eight) and always have eight pinnate (featherlike) tentacles arranged around the margin of the oral disc (Figure 14-22).

The gastrovascular cavity is large and partitioned by septa, or mesenteries, that are inward extensions of the body wall. Where one septum extends into the gastrovascular cavity from the body wall, another extends from the diametrically opposite side; thus, they are said to be **coupled.** In the Zoantharia, the septa are not only coupled, they are also **paired** (Figure 14-23). The muscular arrangement varies among the different groups, but there are usually circular muscles in the body wall and longitudinal and transverse muscles in the septa.

Figure 14-21
Sea anemones are the familiar and colorful "flower animals" of tide pools, rocks, and pilings of the intertidal zone. Most, however, are subtidal, their beauty seldom revealed to human eyes. These are rose anemones, *Tealia piscivora*.

The mesoglea is a mesenchyme containing ameboid cells. There is a general tendency towards biradial symmetry in the septal arrangement and in the shape of the mouth and pharynx. There are no special organs for respiration or excretion.

Sea Anemones

Sea anemone (order Actiniaria) polyps are larger and heavier than hydrozoan polyps (Figure 14-21). Most of them range from 5 mm or less to 100 mm in diameter, and from 5 mm to 200 mm long, but some grow much larger.

Some of them are quite colorful. Anemones are found in coastal areas all over the world, especially in the warmer waters. They attach by means of their pedal discs to shells, rocks, timber, or whatever submerged substrata they can find. Some burrow in the bottom mud or sand.

 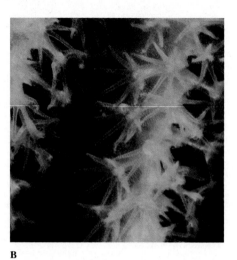

A   **B**

Figure 14-22

Orange sea pen *Ptilosarcus gurneyi* (order Pennatulacea, class Anthozoa). Sea pens are colonial forms that inhabit soft bottoms. **A,** Entire colony. The base of the fleshy body of the primary polyp is buried in the bottom. It gives rise to numerous secondary, branching polyps, shown enlarged in **B.** The pinnate tentacles characteristic of the subclass Alcyonaria are apparent.

Figure 14-23

Structure of the sea anemone. The free edges of the septa and the acontia threads are equipped with nematocysts to complete the paralyzation of prey begun by the tentacles.

Figure 14-24
A sea anemone that swims. When attacked by a predatory sea star *Dermasterias,* the anemone *Stomphia didemon* detaches from the bottom and rolls or swims spasmodically to a safer location.

Sea anemones are cylindrical in form with a crown of tentacles arranged in one or more circles around the mouth of the flat **oral disc** (Figure 14-23). The slit-shaped mouth leads into a **pharynx.** At one or both ends of the mouth is a ciliated groove called a **siphonoglyph,** which extends into the pharynx. The siphonoglyph creates a water current directed into the pharynx. The cilia elsewhere on the pharynx direct water outward. The currents thus created carry in oxygen and remove wastes. They also help maintain an internal fluid pressure, providing a hydrostatic skeleton that serves in lieu of a true skeleton as a support for opposing muscles.

The pharynx leads into a large **gastrovascular cavity** that is divided into six radial chambers by means of six pairs of **primary (complete) septa,** or **mesenteries,** extending vertically from the body wall to the pharynx (Figure 14-23). Openings between the chambers (septal perforations) in the upper part of the pharyngeal region help in water circulation. Smaller **(incomplete)** septa partially subdivide the large chambers and provide a means of increasing the surface area of the gastrovascular cavity. The free edge of each incomplete septum forms a type of sinuous cord called a **septal filament** that is provided with nematocysts and with gland cells for digestion. In some anemones (such as *Metridium*) the lower ends of the septal filaments are prolonged into **acontia threads,** also provided with nematocysts and gland cells, that can be protruded through the mouth or through pores in the body wall to help overcome prey or provide defense. The pores also aid in the rapid discharge of water from the body when the animal is endangered and contracts to a small size.

Sea anemones are carnivorous, feeding on fish or almost any live (and sometimes dead) animals of suitable size. Some species live on minute forms caught by ciliary currents.

Feeding behavior in many zoantharians is under chemical control. Some respond to reduced glutathione. In certain others two compounds are involved: asparagine, the feeding activator, causes a bending of the tentacles toward the mouth; then reduced glutathione induces swallowing of the food.

Muscles are well developed in sea anemones, but the arrangement is quite different from that in Hydrozoa. Longitudinal fibers of the epidermis occur only in the tentacles and oral disc of most species. The strong longitudinal muscles of the column are gastrodermal and are located in the septa (Figure 14-23). Gastrodermal circular muscles in the column are well developed.

Most anemones can glide along slowly on their pedal discs. They can expand and stretch out their tentacles in search of small vertebrates and invertebrates, which they overpower with tentacles and nematocysts and carry to the mouth. When disturbed, sea anemones contract and withdraw their tentacles and oral discs. Some anemones are able to swim, to a limited extent, by rhythmical bending movements, which may be a mechanism for escape from enemies such as sea stars and nudibranchs. *Stomphia,* for example, at the touch of a predatory sea star, will detach its pedal disc and make creeping or swimming movements to escape (Figure 14-24). This escape reaction is elicited not only by the touch of the star but also by exposure to drippings exuded by the star or to crude extracts made from its tissues. The sea drippings contain steroid saponins that are toxic and irritating to most invertebrates. Extracts given off by nudibranchs also can provoke this reaction in sea anemones.

Anemones form some interesting mutualistic relationships with other organisms. Many species harbor symbiotic algae (zooxanthellae) within their tissues, similar to the hard coral-zooxanthellae association (described later in the chapter), and the anemones profit from the product of algal photosynthesis (p. 271). Some anemones habitually attach to the

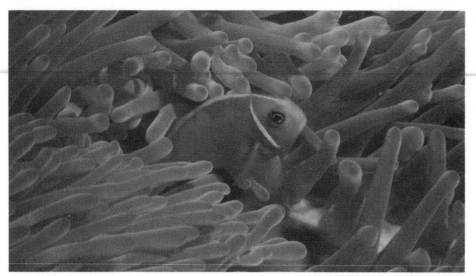

Figure 14-25

Pink clown fish (*Amphiprion perideraion*) nestles in the tentacles of its sea anemone host. Clown fishes do not elicit stings from their hosts but may lure unsuspecting other fish to become meals for the anemone.

A

B

C

Figure 14-26

A, Cup coral *Tubastrea* sp. The polyps form clumps resembling groups of sea anemones. Although often found on coral reefs, *Tubastrea* is not a reef-building coral (ahermatypic) and has no symbiotic zooxanthellae in its tissues. **B,** The polyps of *Montastrea cavernosa* are tightly withdrawn in the daytime but open to feed at night, as in **C.**

shells occupied by certain hermit crabs. The hermit encourages the relationship and, finding its favorite species, which it recognizes by touch, it massages the anemone until it detaches. The hermit crab holds the anemone against its own shell until the anemone is firmly attached. The crab derives some protection against predators by the anemone. The anemone gets free transportation and particles of food dropped by the hermit crab.

Certain damselfishes (clown fishes) (family Pomacentridae) form associations with large anemones, especially in tropical Indo-Pacific waters (Figure 14-25). An unknown property of the skin mucus of the fish causes the anemone's nematocysts not to discharge, but if some other fish is so unfortunate as to brush the anemone's tentacles, it is likely to become a meal. The anemone obviously provides shelter for the anemone fish, and the fish may help ventilate the anemone by its movements, keep the anemone free of sediment, and even lure an unwary victim to seek the same shelter.

The sexes are separate in some sea anemones, and some are hermaphroditic. Monoecious species are **protandrous** (produce sperm first, then eggs). Gonads are arranged on the margins of the septa, and fertilization takes place externally or in the gastrovascular cavity. The zygote develops

into a ciliated larva. Asexual reproduction commonly occurs by **pedal laceration** or by longitudinal fission, occasionally by transverse fission or by budding. In pedal laceration, small pieces of the pedal disc break off as the animal moves, and each of these regenerates a small anemone.

Zoantharian Corals

The zoantharian corals belong to the order Scleractinia, sometimes known as the true or stony corals. The stony corals might be described as miniature sea anemones that live in calcareous cups they themselves have secreted (Figures 14-26 and 14-27). Like that of the anemones, the coral polyp's gastrovascular cavity is subdivided by septa arranged in multiples of six (hexamerous) and its hollow tentacles surround the mouth, but there is no siphonoglyph.

Instead of a pedal disc, the epidermis at the base of the column secretes the limy skeletal cup, including the sclerosepta, which project up into the polyp between the true septa (Figure 14-27). The living polyp can retract into the safety of the cup when not feeding. Since the skeleton is secreted below the living tissue rather than within it, the calcareous material is an exoskeleton. In many colonial corals, the skeleton may become massive,

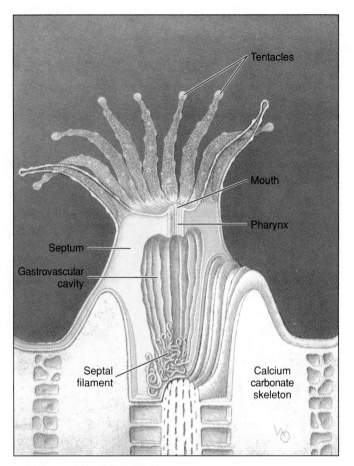

Figure 14-27
Polyp of a zoantharian coral (order Scleractinia) showing calcareous cup (exoskeleton), gastrovascular cavity, sclerosepta, septa, and septal filaments.

Figure 14-28
A tube anemone (subclass Ceriantipatharia, order Ceriantharia) extends from its tube at night. Its oral disc bears long tentacles around the margin and short tentacles around the mouth.

building up over many years, with the living coral forming a sheet of tissue over the surface. The gastrovascular cavities of the polyps are all connected through this sheet of tissue.

Three other small orders of Zoantharia are recognized.

Tube Anemones and Thorny Corals

Members of the subclass Ceriantipatharia have coupled but unpaired septa. The tube anemones (order Ceriantharia) (Figure 14-28) are solitary and live in soft bottom sediments, buried to the level of the oral disc. They occupy tubes constructed of secreted mucus and threads of nematocyst-like organelles, into which they can withdraw. The thorny or black corals (order Antipatharia) (Figure 14-29) are colonial and attached to a firm substratum. Their skeleton is of a horny material and has thorns. Both of these orders are small in numbers of species and are limited to warmer waters of the sea.

Alcyonarian Corals

Alcyonarians (Octocorallia) have strict octomerous symmetry, with eight pinnate tentacles and eight unpaired, complete septa (Figure 14-22). They are all colonial, and the gastrovascular cavities of the polyps communicate

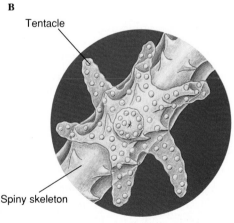

Enlargement of single polyp

Figure 14-29
A, Colony of *Antipathes,* a black or thorny coral (order Antipatharia, class Anthozoa). Most abundant in deep waters in the tropics, black corals secrete a tough, proteinaceous skeleton that can be worked into jewelry. **B,** The polyps of Antipatharia have six simple, nonretractile tentacles. The spiny processes in the skeleton are the origin of the common name thorny corals.

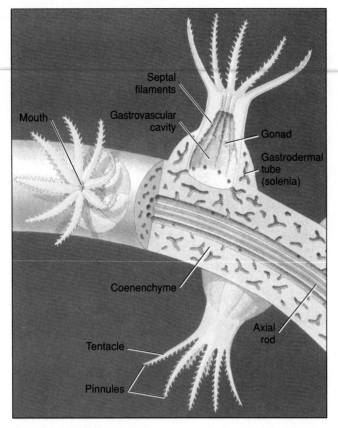

Figure 14-30

Polyps of an alcyonarian coral (octocoral). Note the eight pinnate tentacles, coenenchyme, and solenia. They have an endoskeleton of limy spicules often with a horny protein, which may be in the form of an axial rod.

through a system of gastrodermal tubes called **solenia** (Figure 14-30). The solenia run through an extensive mesoglea **(coenenchyme)** in most alcyonarians, and the surface of the colony is covered by epidermis. The skeleton is secreted in the coenenchyme and consists of limy spicules, fused spicules, or a horny protein, often in combination. Thus the skeletal support of most alcyonarians is an endoskeleton. The variation in pattern among the species of alcyonarians lends great variety to the form of the colonies: from the soft coral *Gersemia,* with its spicules scattered through the coenenchyme, to the tough, axial supports of the sea fans and other gorgonian corals (Figure 14-31), to the fused spicules of the organ-pipe coral. *Renilla* (L. *ren,* kidney, + *illa,* suffix), the sea pansy, is a colony reminiscent of a pansy flower. Its polyps are embedded in the fleshy upper side and a short stalk that supports the colony is embedded in the sea floor (Figure 14-32).

Ptilosarcus (Gr. *ptilon,* feather, + *sarkos,* flesh), a sea pen, is a member of the same order and may reach a length of 50 cm (Figure 14-22).

The graceful beauty of the alcyonarians—in hues of yellow, red, orange, and purple—helps create the "submarine gardens" of the coral reefs.

Coral Reefs

Most students will have seen photographs or movies giving a glimpse of the vibrant color and life found on coral reefs, and some may have been fortunate enough to visit a reef. Coral reefs are among the most productive of all ecosystems, and they have a diversity of life forms rivaled only by the tropical rain forest. They are large formations of calcium carbonate (limestone) in shallow tropical seas laid down by living organisms over thousands of years; living plants and animals are confined to the top layer of reefs where they add more and more calcium carbonate to that deposited by

Figure 14-31

Colonial gorgonian, or horny, corals (order Gorgonacea, class Anthozoa) are conspicuous components of reef faunas, especially of the West Indies. **A,** Red gorgonian *Melithaea* sp. on a Pacific coral reef. **B,** A sea plume, *Pseudopterogorgia* sp. and **C,** a sea fan *Gorgonia ventalina* in the Caribbean.

Figure 14-32
Sea pansy *Renilla reniformis* (order
Pennatulacea, class Anthozoa). The eight
pinnate tentacles characteristic of the subclass
Alcyonaria can be distinguished.

their predecessors. The most important
organisms that precipitate calcium carbonate from seawater to form reefs are
the scleractinian, **hermatypic** (reef-
building) **corals** (Figure 14-26) and
coralline algae. Not only do the
coralline algae contribute to the total
mass of calcium carbonate, but their
precipitation of the substance helps to
hold the reef together. Some alcyonarians and hydrozoa (especially *Millepora*
[L. *mille,* thousand, + *porus,* pore] spp.,
the "fire coral," Figure 14-14B) contribute in some measure to the calcareous material, and an enormous variety
of other organisms contributes small
amounts. However, hermatypic (Gr.
herma, support, mound, + *typos,* type)
corals seem essential to the formation
of large reefs, since such reefs do not
occur where these corals cannot live.

Hermatypic corals require warmth,
light, and the salinity of undiluted seawater. This limits coral reefs to shallow
waters between 30 degrees north and
30 degrees south latitude and excludes
them from areas with upwelling of
cold water or areas near major river
outflows with attendant low salinity
and high turbidity. These corals require light because they have mutualistic algae (zooxanthellae) living in their
tissues. The microscopic zooxanthellae
are very important to the corals; their
photosynthesis and fixation of carbon
dioxide furnish food molecules for
their hosts, they recycle phosphorus
and nitrogenous waste compounds
that otherwise would be lost, and they
enhance the ability of the coral to deposit calcium carbonate.

Because zooxanthellae are vital to
hermatypic corals, and water absorbs
light, hermatypic corals rarely live
below a depth of 100 feet (30 m).
Interestingly, some deposits of coral
reef limestone, particularly around
Pacific islands and atolls, reach great
thickness—even thousands of feet.
Clearly, the corals and other organisms
could not have grown from the bottom
in the abyssal blackness of the deep
sea and reached shallow water where
light could penetrate. Charles Darwin
was the first to realize that such reefs
began their growth in *shallow* water
around volcanic islands; then as the
islands slowly sank beneath the sea,
the growth of the reefs kept up with
the rate of sinking, thus accounting for
the depth of the deposits.

Several types of reefs are commonly recognized. The **fringing reef**
is close to a landmass with either no lagoon or a narrow lagoon between it
and the shore. A **barrier reef** (Figure
14-33) runs roughly parallel to shore
and has a wider and deeper lagoon
than does a fringing reef. **Atolls** are
reefs that encircle a lagoon but not an
island. These types of reefs typically
slope rather steeply into deep water at
their seaward edge. **Patch** or **bank
reefs** occur some distance back from
the steep, seaward slope in the lagoons
of barrier reefs or atolls. The so-called
Great Barrier Reef, extending 2027 km
long and up to 145 km from the shore
off the northeast coast of Australia, is
actually a complex of reef types.

A

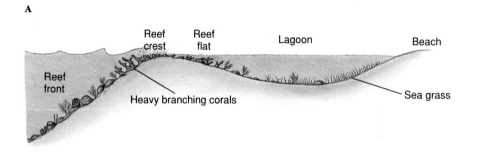

B

Figure 14-33
A, Profile of a barrier reef. **B,** Portion of an atoll from the air. Reef slope plunges into deep water at left
(dark blue), lagoon at right.

Fringing, barrier, and atoll reefs all have distinguishable zones that are characterized by different groups of corals and other animals. The side of the reef facing the sea is the **reef front** or **fore reef slope** (Figure 14-33). The reef front is more or less parallel to the shore and perpendicular to the predominant direction of wave travel. It slopes downward into deeper water, sometimes gently at first, then precipitously. Characteristic assemblages of scleractinian corals grow deep on the slope, high near the crest, and in zones between. In shallow water or slightly emergent at the top of the reef front is the **reef crest.** The upper front and the crest bear the greatest force of the waves and must absorb great energy during storms. Pieces of coral and other organisms are broken off at such times and thrown shoreward onto the **reef flat,** which slopes down into the lagoon. The reef flat thus receives a supply of calcareous material that is eventually broken down into coral sand. The sand is stabilized by the growth of plants such as turtle grass and coralline algae and ultimately becomes cemented into the mass of the reef by precipitation of carbonates. A reef is not an unbroken wall facing the sea but is highly irregular, with grooves, caves, crevices, channels through from the flat to the front, and deep, cup-shaped holes ("blue holes"). Alcyonarians tend to grow in these areas that are more protected from the full force of the waves, as well as on the flat and the deeper areas of the fore reef slope. Many other kinds of organisms inhabit the cryptic locations such as caves and crevices.

Enormous numbers of species and individuals of invertebrate groups and fishes populate the reef ecosystem. For example, there are 300 *common* species of fishes on Caribbean reefs and more than 1200 on the Great Barrier Reef complex of Australia. It is marvelous that such diversity and productivity can be maintained, since the reefs are washed by the nutrient-poor waves of the open ocean. Although relatively little nutrient enters the ecosystem, little is lost because the interacting organisms are

so efficient in recycling. The corals even feed on the feces of the fish swimming over them!

Despite their great intrinsic and economic value, coral reefs in many areas are threatened today by a variety of factors, mostly of human origin. These include nutrients from sewage and agricultural fertilizer that wash into the water from land. Agricultural pesticides, as well as sediment from tilled fields, also contribute to reef degradation. Corals in the Persian Gulf have withstood a surprising amount of pollution, high salinity, and temperature swings—much more than reefs in other parts of the world have been able to endure. However, whether they can survive the greatest oil slick ever created by humans (in the Gulf War of 1991) is a question yet to be answered.

Classification of Phylum Cnidaria

Class Hydrozoa (hi-dro-zo´a) (Gr. *hydra,* water serpent, + *zōon,* animal). Solitary or colonial; asexual polyps and sexual medusae, although one type may be suppressed; hydranths with no mesenteries; medusae (when present) with a velum; both freshwater and marine. Examples: *Hydra, Obelia, Physalia, Tubularia.*

Class Scyphozoa (si-fo-zo´a) (Gr. *skyphos,* cup, + *zōon,* animal). Solitary; polyp stage reduced or absent; bell-shaped medusae without velum; gelatinous mesoglea much enlarged; margin of bell or umbrella typically with eight notches that are provided with sense organs; all marine. Examples: *Aurelia, Cassiopeia, Rhizostoma.*

Class Cubozoa (ku´bo-zo´a) (Gr. *kybos,* a cube + *zōon,* animal). Solitary; polyp stage reduced; bell-shaped medusae square in cross section, with tentacle or group of tentacles hanging from a bladelike pedalium at each corner of the umbrella; margin of umbrella entire, without velum but with velarium; all marine. Examples: *Tripedalia, Carybdea, Chironex, Chiropsalmus.*

Class Anthozoa (an-tho-zo´a) (Gr. *anthos,* flower, + *zōon,* animal). All polyps; no medusae; solitary or colonial; enteron subdivided by at least eight mesenteries or septa bearing nematocysts; gonads endodermal; all marine.

Subclass Zoantharia (zo´an-tha´re-a) (N. L. from Gr. *zōon,* animal, + *anthos,* flower, + L. *aria,* like or connected with) **(Hexacorallia).** With simple unbranched tentacles; mesenteries in pairs; sea anemones, hard corals, and others. Examples: *Metridium, Anthopleura, Tealia, Astrangia, Acropora.*

Subclass Ceriantipatharia (se´re-ant-ip´a-tha´re-a) (N. L. combination of Ceriantharia and Antipatharia). With simple unbranched tentacles; mesenteries unpaired; tube anemones and black or thorny corals. Examples: *Cerianthus, Antipathes, Stichopathes.*

Subclass Alcyonaria (al´ce-o-na´re-a) (Gr. *alkonion,* kind of sponge resembling nest of kingfisher [*alkyon,* kingfisher], + L. *aria,* like or connected with) **(Octocorallia).** With eight pinnate tentacles; eight complete, unpaired mesenteries; soft and horny corals. Examples: *Tubipora, Alcyonium, Gorgonia, Plexaura, Renilla.*

Phylum Ctenophora

Ctenophora (te-nof´o-ra) (Gr. *kteis, ktenos,* comb, + *phora,* pl. of bearing) is composed of fewer than 100 species. All are marine forms occurring in all seas but especially in warm waters. They take their name from the eight rows of comblike plates they bear for locomotion. Common names for ctenophores are "sea walnuts" and "comb jellies." Ctenophores, along with cnidarians, represent the only two phyla having primary radial symmetry, in contrast to other metazoans, which have primary bilateral symmetry.

Ctenophores do not have nematocysts, except in one species (*Haeckelia rubra,* after Ernst Haeckel, nineteenth-century German zoologist) that carries nematocysts on certain regions of its tentacles but lacks colloblasts. These nematocysts are apparently appropriated from cnidarians on which it feeds.

In common with the cnidarians, ctenophores have not advanced beyond the tissue grade of organization. There are no definite organ systems in the strict meaning of the term.

Except for a few creeping and sessile forms, ctenophores are free-swimming. Although they are feeble swimmers and are more common in surface waters, ctenophores are sometimes found at considerable depths. They are often at the mercy of tides and strong currents, but they avoid storms by swimming downward in the water. In calm water they may rest vertically with little movement, but when moving they use their ciliated comb plates to propel themselves mouth-end forward. Highly modified forms such as *Cestum* (L. *cestus,* girdle) use sinuous body movements as well as their comb plates in locomotion.

The fragile, transparent bodies of ctenophores are easily seen at night when they emit light (luminesce).

CLASS TENTACULATA

Representative Type: Pleurobrachia

Pleurobrachia (Gr. *pleuron,* side, + L. *brachia,* arms) is a representative of this group of ctenophores. Its transparent body is about 1.5 to 2 cm in diameter (Figure 14-34A). The oral pole bears the mouth opening, and the aboral pole has a sensory organ, the **statocyst.**

Comb Plates. On the surface are eight equally spaced bands called **comb rows** that extend as meridians from the aboral pole and end before reaching the oral pole (Figure 14-35). Each band consists of transverse plates of long fused cilia called **comb plates** (Figure 14-36B). Ctenophores are propelled by the beating of the cilia on the comb plates. The beat in each row starts at the aboral end and proceeds successively along the combs to the oral end. All eight rows

CHARACTERISTICS OF PHYLUM CTENOPHORA

1. Symmetry **biradial;** arrangement of internal canals and the position of the paired tentacles change the radial symmetry into a combination of the two **(radial + bilateral)**
2. Usually ellipsoidal or spherical in shape, with **radially arranged rows of comb plates for swimming**
3. Ectoderm, endoderm, and a mesoglea (ectomesoderm) with scattered cells and muscle fibers; may be considered **triploblastic**
4. Nematocysts absent (except in one species), but **adhesive cells (colloblasts)** present
5. Digestive system consisting of mouth, pharynx, stomach, a series of canals, and anal pores
6. Nervous system consisting of a subepidermal plexus concentrated around the mouth and beneath the comb plate rows; an **aboral sense organ** (statocyst)
7. No polymorphism or dimorphism
8. Reproduction monoecious; gonads (endodermal origin) on the walls of the digestive canals, which are under the rows of comb plates; mosaic cleavage; cydippid larva
9. Luminescence common

COMPARISON WITH CNIDARIA

Ctenophores resemble the cnidarians in the following ways:
1. Form of radial symmetry; together with the cnidarians, they form the group Radiata
2. Aboral-oral axis around which the parts are arranged
3. Well-developed gelatinous ectomesoderm (collenchyme)
4. No coelomic cavity
5. Diffuse nerve plexus
6. Lack of organ systems

They differ from the cnidarians in the following ways:
1. They do not form nematocysts
2. Development of distinct muscle cells from mesenchyme
3. Presence of comb plates and colloblasts
4. Mosaic, or determinate type of development
5. Presence of pharynx generally
6. No polymorphism or dimorphism
7. Never colonial
8. Presence of anal openings

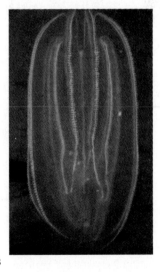

A   B

Figure 14-34

A, Comb jelly *Pleurobrachia* sp. (order Cydippida, class Tentaculata). Its fragile beauty is especially evident at night when it luminesces from its comb rows. **B,** *Mnemiopsis* sp. (order Lobata, class Tentaculata).

normally beat in unison. The animal is thus driven forward with the mouth in advance. The animal can swim backward by reversing the direction of the wave.

Tentacles. The two **tentacles** are long, solid and very extensible, and they can be retracted into a pair of **tentacle sheaths.** When completely extended, they may measure 15 cm in

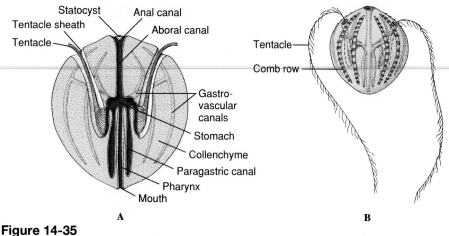

Figure 14-35

The comb jelly *Pleurobrachia,* a ctenophore. **A,** Hemisection. **B,** External view.

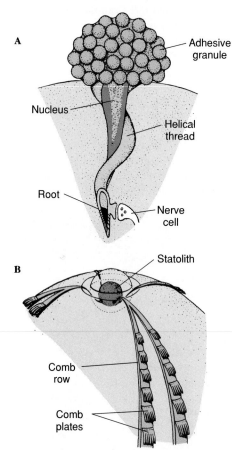

Figure 14-36

A, Colloblast, an adhesive cell characteristic of ctenophores. **B,** Portion of comb rows showing comb plates, each composed of transverse rows of long fused cilia.

length. The surface of the tentacles bears **colloblasts,** or glue cells (Figure 14-36A), which secrete a sticky substance that is used for catching and holding small animals.

Body Wall. The cellular layers of ctenophores are generally similar to those of cnidarians. Between the epidermis and gastrodermis is a gelatinous **collenchyme** that makes up most of the interior of the body and contains muscle fibers and ameboid cells. Although they are derived from ectodermal cells, the muscle cells are distinct and are not contractile portions of epitheliomuscular cells (in contrast to the Cnidaria).

Digestive System and Feeding. The **gastrovascular system** consists of a mouth, a pharynx, a stomach, and a system of gastrovascular canals that branch through the jelly to extend to the comb plates, tentacular sheaths, and elsewhere (Figure 14-35). There are two blind canals that terminate near the mouth, and an aboral canal that passes near the statocyst and then divides into two small **anal canals** through which undigested material is expelled.

Ctenophores prey on small planktonic organisms such as copepods. The glue cells on the tentacles stick to the small prey and enable the tentacles to carry the prey to the ctenophore's mouth. Digestion is both extracellular and intracellular.

Respiration and Excretion. Respiration and excretion occur through the body surface.

Nervous and Sensory Systems. Ctenophores have a nervous system similar to that of the cnidarians. It is made up of a subepidermal plexus, which is concentrated under each comb plate, but there is no central control as is found in more complex animals.

The sense organ at the aboral pole is a statocyst. Tufts of cilia support a calcareous statolith, with the whole being enclosed in a bell-like container. Alterations in the position of the animal change the pressure of the statolith on the tufts of cilia. The sense organ is also concerned in coordinating the beating of the comb rows but does not trigger their beat.

The epidermis of ctenophores bears abundant sensory cells, so the animals are sensitive to chemical and other forms of stimuli. When a ctenophore comes in contact with an unfavorable stimulus, it often reverses the beat of its comb plates and backs up. The comb plates are very sensitive to touch, which often causes them to be withdrawn into the jelly.

Reproduction and Development. *Pleurobrachia,* like other ctenophores, are monoecious. The gonads are located on the lining of the gastrovascular canals under the comb plates. Fertilized eggs are discharged through the epidermis into the water.

Cleavage in the ctenophores is determinate (mosaic), since the various parts of the animal that will be formed by each blastomere are determined early in embryogenesis. If one of the blastomeres is removed in the early stages, the resulting embryo will be deficient. This type of development differs from that of cnidarians, which is regulative (indeterminate). The free-swimming **cydippid larva** is superficially similar to the adult ctenophore and develops directly into an adult.

Some biologists have regarded the ctenophores and some more complex cnidarians (for example, some anthozoans) as triploblastic because the highly cellular nature of the mesoglea would constitute a mesoderm. However, others define meso-

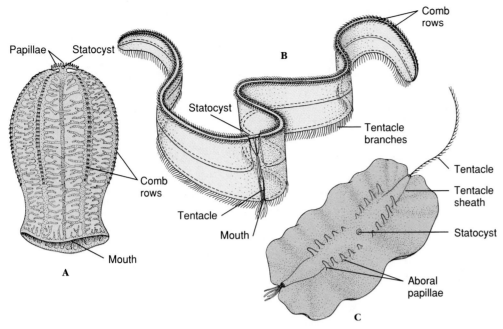

Figure 14-37
Diversity among the phylum Ctenophora. **A,** *Beroe* sp. (order Beroida, class Nuda). **B,** *Cestum* sp. (order Cestida, class Tentaculata). **C,** *Coeloplana* sp. (order Platyctenea, class Tentaculata).

derm strictly as a layer derived from endoderm; thus both cnidarians and ctenophores would be diploblastic.

OTHER CTENOPHORES

Ctenophores are fragile and beautiful creatures. Their transparent bodies glisten like fine glass, brilliantly iridescent during the day and luminescent at night.

One of the most striking ctenophores is *Beroe* (L. a nymph), which may be more than 100 mm in length and 50 mm in breadth (Figure 14-37A). It is conical or thimble shaped and is flattened in the tentacular plane. The tentacular plane in *Beroe* is defined as where the tentacles would have been, because it has a large mouth but no tentacles. The animal is pink or rusty brown. Its body wall is covered with an extensive network of canals formed by the union of the paragastric and meridional canals. Venus' girdle (*Cestum,* Figure 14-37B) is highly compressed in the tentacular plane. Bandlike, it may be more than 1 m long and presents a graceful appearance as it swims in the oral direction. The highly modified *Ctenoplana* (Gr. *ktenos,* comb, + L. *planus,* flat) and *Coeloplana* (Gr. *koilos,* hollow, + L. *planus,*

flat) (Figure 14-37C) are rare but are interesting because they have disc-shaped bodies flattened in the oral-aboral axis and are adapted for creeping rather than swimming. A common ctenophore along the Atlantic and Gulf coasts is *Mnemiopsis* (Gr. *mneme,* memory, + *opsis,* appearance) (Figure 14-34B), which has a laterally compressed body with two large oral lobes and unsheathed tentacles.

Nearly all ctenophores give off flashes of luminescence at night, especially such forms as *Mnemiopsis* (Figure 14-34B). The vivid flashes of light seen at night in southern seas are often caused by members of this phylum.

Since the 1980s population explosions of *Mnemiopsis leidyi* in the Black and Azov Seas have led to catastrophic declines in fisheries there. Inadvertently introduced from the coast of the Americas with ballast water of ships, the ctenophores feed on zooplankton, including small crustaceans and the eggs and larvae of fish. The normally inoffensive *M. leidyi* is kept in check in the Atlantic by certain specialized predators, but introduction of such predators into the Black Sea carries its own dangers.

PHYLOGENY AND ADAPTIVE RADIATION

PHYLOGENY

The origin of the cnidarians and ctenophores is obscure, although the most widely supported hypothesis today is that the radiate phyla arose from a radially symmetrical, planula-like ancestor. Such an ancestor could have been common to the radiates and to the higher metazoans, the latter having been derived from a branch whose members habitually crept about on the sea bottom. Such a habit would select for bilateral symmetry. Others became sessile or free floating, conditions for which radial symmetry is a selective advantage. A planula larva in which an invagination formed to become the gastrovascular cavity would correspond roughly to a cnidarian with an ectoderm and an endoderm.

Some researchers believe the trachyline medusae (an order of class Hydrozoa) resemble the ancestral cnidarian because of their direct development from the planula and actinula larvae to the medusa (Figure 14-38). The trachyline-like ancestor would have given rise to other cnidarian lines after the evolution of the polyp stage and alternation of sexual (medusa) and asexual (polyp) generations. Subsequently, the medusa

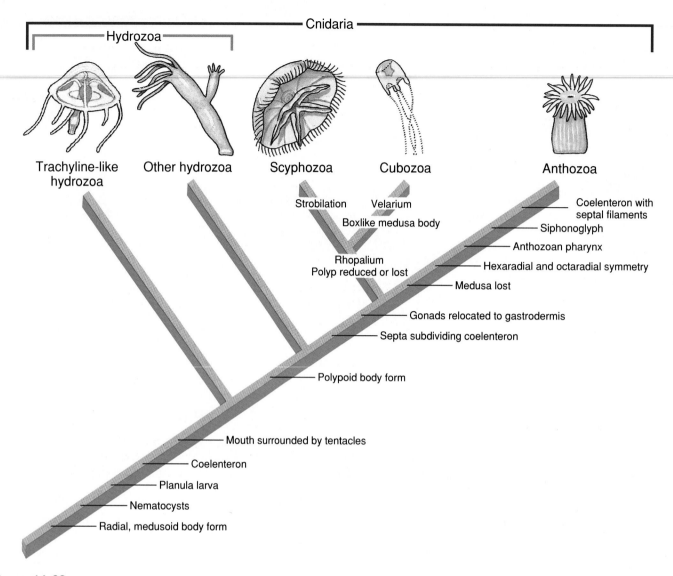

Figure 14-38

Cladogram showing hypothetical relationships of cnidarian classes with some shared derived characters indicated. This hypothesis suggests that the hydrozoan order Trachylina retains the ancestral cnidarian life cycle, having branched off before the evolution of the polyp stage. Note that this arrangement makes the Hydrozoa paraphyletic; the trachyline-like Hydrozoa is a sister group to all the other Cnidaria.

Source: Modified from R. C. Brusca and G. J. Brusca, Invertebrates. *Sinauer Associates, Inc., Sunderland, MA, 1990.*

was completely lost in the anthozoan line. If the order Trachylina is retained within the class Hydrozoa, however, then the Hydrozoa becomes paraphyletic. Future investigators may resolve this problem.

In the past it was assumed that the ctenophores arose from a medusoid cnidarian, but this assumption has been questioned recently. The similarities between the groups are mostly of a general nature and do not seem to indicate a close relationship. Some molecular evidence suggests that the ctenophores branched off the

metazoan line after the sponges but before the cnidarians and placozoans.

ADAPTIVE RADIATION

In their evolution neither phylum has deviated far from its basic plan of structure. In the Cnidaria, both the polyp and medusa are constructed on the same scheme. Likewise, the ctenophores have adhered to the arrangement of the comb plates and their biradial symmetry.

Nonetheless, the cnidarians have achieved large numbers of individuals

and species, demonstrating a surprising degree of diversity considering the simplicity of their basic body plan. They are efficient predators, many feeding on prey quite large in relation to themselves. Some are adapted for feeding on small particles. The colonial form of life is well explored, with some colonies growing to great size among the corals, and others, such as the siphonophores, showing astonishing polymorphism and specialization of individuals within the colony.

Summary

The phyla Cnidaria and Ctenophora have a primary radial symmetry; radial symmetry is an advantage for sessile or free-floating organisms because environmental stimuli come from all directions equally. The Cnidaria are surprisingly efficient predators because they possess stinging organelles called nematocysts. Both phyla are essentially diploblastic (some triploblastic, according to definition of mesoderm), with a body wall composed of epidermis and gastrodermis and a mesoglea between. The digestive-respiratory (gastrovascular) cavity has a mouth and no anus. Cnidarians are at the tissue level of organization. They have two basic body types (polypoid and medusoid), and in many hydrozoans and scyphozoans the life cycle involves both the asexually reproducing polyp and the sexually reproducing medusa.

That unique organelle, the nematocyst, is produced by a cnidoblast (which becomes the cnidocyte) and is coiled within a capsule. When discharged, some types of nematocysts penetrate the prey and inject poison. Discharge is effected by a change in permeability of the capsule and an increase in internal hydrostatic pressure because of the high osmotic pressure within the capsule.

Most hydrozoans are colonial and marine, but the freshwater hydras are commonly demonstrated in class laboratories. They have a typical polypoid form but are not colonial and have no medusoid stage. Most marine hydrozoans are in the form of a branching colony of many polyps (hydranths). The medusa may be free-swimming or remain attached to the colony.

The scyphozoans are typical jellyfishes, in which the medusoid is the dominant body form, and many have an inconspicuous polypoid stage. Cubozoans are predominantly medusoid. They include the dangerous sea wasps.

Anthozoans are all marine and are polypoid; there is no medusoid stage. The most important subclasses are the Zoantharia (with hexamerous or polymerous symmetry) and Alcyonaria (with octomerous symmetry). The largest zoantharian orders contain the sea anemones, which are solitary and do not have a skeleton, and the stony corals, which are mostly colonial and secrete a calcareous exoskeleton. Stony corals are the critical component in coral reefs, which are habitats of great beauty, productivity, and ecological and economic value. The Alcyonaria contain the soft and horny corals, many of which are important and beautiful components of coral reefs.

The Ctenophora are biradial and swim by means of eight comb rows. Colloblasts, with which they capture small prey, are characteristic of the phylum.

Cnidaria and Ctenophora are probably derived from an ancestor that resembled the planula larva of the cnidarians. Despite their relatively simple level of organization, the cnidarians are an important phylum.

Review Questions

1. Explain the selective advantage of radial symmetry for sessile and free-floating animals.
2. What characteristics of the phylum Cnidaria do you think are most important in distinguishing it from other phyla?
3. Name and distinguish the classes in the phylum Cnidaria.
4. Distinguish between the polyp and medusa forms.
5. Explain the mechanism of nematocyst discharge. How can a hydrostatic pressure of one atmosphere be maintained within the nematocyst until it receives an expulsion stimulus?
6. What is an unusual feature of the nervous system of cnidarians?
7. Diagram a hydra and label the main body parts.
8. Name and give the functions of the main cell types in the epidermis and in the gastrodermis of hydra.
9. What stimulates feeding behavior in hydras?
10. Define the following with regard to hydroids: hydrorhiza, hydrocaulus, coensosarc, perisarc, hydranth, gonangium, manubrium, statocyst, ocellus.
11. Give an example of a highly polymorphic, floating, colonial hydrozoan.
12. Distinguish the following from each other: statocyst and rhopalium; scyphomedusae and hydromedusae; scyphistoma, strobila, and ephyrae; velum, velarium, and pedalium; Zoantharia and Alcyonaria.
13. Define the following with regard to sea anemones: siphonoglyph; primary septa or mesenteries; incomplete septa; septal filaments; acontia threads; pedal laceration.
14. Describe three specific interactions of anemones with nonprey organisms.
15. Contrast the skeletons of zoantharian and alcyonarian corals.
16. Coral reefs generally are limited in geographic distribution to shallow marine waters. How do you account for this?
17. Specifically, what kinds of organisms are most important in deposition of calcium carbonate on coral reefs?
18. How do zooxanthellae contribute to the welfare of hermatypic corals?
19. Distinguish each of the following from each other: fringing reefs; barrier reefs; atolls; patch or bank reefs.
20. What characteristics of Ctenophora do you think are most important in distinguishing it from other phyla?
21. How do ctenophores swim, and how do they obtain food?
22. Compare cnidarians and ctenophores, giving five ways in which they resemble each other and five ways in which they differ.
23. What is a widely held hypothesis on the origin of the radiate phyla?

Selected References

See also general references for Part III, p. 626.

Brown, B. E., and J. C. Ogden. 1993. Coral bleaching. Sci. Am. **268:**64–70 (Jan.). *Abnormally warm water is apparently the cause of reef corals losing their zooxanthellae.*

Crossland, C. J., B. G. Hatcher, and S. V. Smith. 1991. Role of coral reefs in global ocean production. Coral Reefs **10:**55–64. *Because of extensive recycling of nutrients within reefs, their net energy production for export is relatively minor. However, they play a major role in inorganic carbon precipitation by biologically-mediated processes.*

Fishman, D. J. 1991. Corals in a troubled sea of crude: the harder marine ecosystem of the Persian Gulf. Ocean Realm Spring: pp. 9–11. *The corals in the Persian Gulf are tougher than corals in other areas, but they may not withstand the disastrous oil spill from the Persian Gulf War.*

Goreau, T. F., N. I. Goreau, and T. J. Goreau. 1979. Corals and coral reefs. Sci. Am. **2451:**124–135 (Aug.). *A good summary of the biology, ecology, and physiology of reef corals.*

Humann, P. 1992. Reef creature identification. Florida, Caribbean, Bahamas. Jacksonville, Florida, New World Publications, Inc. *This is the best field guide available for identification of "non-coral" cnidarians.*

Humann, P. 1993. Reef coral identification. Florida, Caribbean, Bahamas. Jacksonville, Florida, New World Publications, Inc. *Superb color photographs and accurate identifications make this by far the best field guide to corals now available; includes sea grasses and some algae.*

Kenchington, R., and G. Kelleher. 1992. Crown-of-thorns starfish management conundrums. Coral Reefs **11:**53–56. *The first article of an entire issue on the starfish:* Acanthaster planci, *a predator of corals. Another entire issue was devoted to this predator in 1990 (p. 456).*

Lenhoff, H. M., and S. G. Lenhoff. 1988. Trembley's polyps. Sci. Am. **258:**108–113 (Apr.). *Trembley's elegant experiments on hydras in the 1740s were the dawn of experimental zoology.*

Shick, J. M. 1991. A functional biology of sea anemones. New York, Chapman and Hall. *A comprehensive coverage for serious students of sea anemone physiology and ecology.*

Ward, F. 1990. Florida's coral reefs are imperiled. National Geographic **178**(1):115–132. *Describes the degradation suffered by reefs in the Florida Keys, along with the probable causes.*

15

The Acoelomate Animals

Phylum Platyhelminthes
Phylum Nemertea
Phylum Gnathostomulida

Getting Ahead

For animals that spend their lives sitting and waiting, as do most members of the two radiate phyla we considered in the preceding chapter, radial symmetry is ideal. One side of the animal is just as important as any other for snaring prey coming from any direction. But if an animal is active in seeking food, shelter, home sites, and reproductive mates, it requires a different set of strategies and a new body organization. Active, directed movement requires an elongated body form with head (anterior) and tail (posterior) ends. In addition, one side of the body is kept up (dorsal) and the other side, specialized for locomotion, is kept down (ventral). What results is a bilaterally symmetrical animal in which the body can be divided along only one plane of symmetry to yield two halves which are mirror images of each other. Furthermore, since it is better to determine where one is going than where one has been, sense organs and centers for nervous control have come to be located on the head. This is called cephalization. Thus cephalization and primary bilateral symmetry evolved together.

The three acoelomate phyla considered in this chapter are not greatly more complex in organization than the Radiata except in symmetry. The evolutionary consequence of that development alone was enormous, however, for it is the type of symmetry assumed by all more complex animals. ■

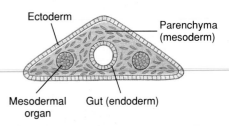

Figure 15-1
Acoelomate body plan.

Position in Animal Kingdom

1. The Platyhelminthes, or flatworms, the Nemertea, or ribbon worms, and the Gnathostomulida, or jaw worms, are the simplest animals to have **primary bilateral symmetry.**
2. These phyla have only one internal space, the digestive cavity, with the region between the ectoderm and endoderm filled with mesoderm in the form of muscle fibers and mesenchyme (parenchyma). Since they lack a coelom or a pseudocoelom, they are termed **acoelomate animals,** and because they have three well-defined germ layers, they are termed triploblastic.
3. Acoelomates show more specialization and division of labor among their organs than do the radiate animals because the mesoderm makes more elaborate organs possible. Thus the acoelomates are said to have reached the **organ-system level of organization.**
4. They belong to the protostome division of the Bilateria and have spiral cleavage, and at least the platyhelminths and nemerteans have determinate (mosaic) cleavage.

Biological Contributions

1. The acoelomates developed the basic **bilateral** plan of organization that has been widely exploited in the animal kingdom.
2. The **mesoderm** developed into a well-defined embryonic germ layer **(triploblastic),** making available a great source of tissues, organs, and systems.
3. Along with bilateral symmetry, **cephalization** was established. There is some centralization of the nervous system evident in the **ladder type of system** found in flatworms.
4. Along with the subepidermal musculature, there is also a mesenchymal system of muscle fibers.
5. They are the simplest animals with an **excretory system.**
6. The nemerteans are the simplest animals to have a **circulatory system** with blood and a **one-way alimentary canal.** Although not stressed by zoologists, the rhynchocoel cavity in ribbon worms is technically a true coelom, but because it is merely a part of the proboscis mechanism, it is probably not of evolutionary significance.
7. Unique and specialized structures occur in all three phyla. The parasitic habit of many flatworms has led to many specialized adaptations, such as organs of adhesion.

The three phyla considered in this chapter have the simplest organization within the Bilateria, a grouping of phyla that includes all the rest of the animal kingdom. These three are the Platyhelminthes (Gr. *platys,* flat, + *helmins,* worm), or flatworms; the Nemertea (Gr. *Nemertes,* one of the nereids, unerring one), or ribbon worms; and the Gnathostomulida (Gr. *gnathos,* jaw, + *stoma,* mouth, + L. *ulus,* dim.), or jaw worms. They have only one internal space, the digestive cavity, with the region between the ectoderm and endoderm filled with mesoderm in the form of muscle fibers and mesenchyme (parenchyma). Since they lack a coelom or a pseudocoel, they are termed **acoelomate** animals (Figure 15-1), and because they have three well-defined germ layers,

they are **triploblastic.** Acoelomates show more specialization and division of labor among their organs than do the radiate animals because the mesoderm makes more elaborate organs possible; thus, the acoelomates are said to have reached the organ-system level of organization.

These phyla belong to the protostome division of the Bilateria and typically have spiral cleavage. They have some centralization of the nervous system, with a concentration of nerves anteriorly and a ladder-type arrangement of trunks and connectives down the body. They have an excretory (or osmoregulatory) system, and the nemerteans also have a circulatory system. They also have a one-way digestive system, with an anus as well as a mouth.

Phylum Platyhelminthes

The word "worm" is loosely applied to elongated, bilateral invertebrate animals without appendages. At one time zoologists considered worms (Vermes) a group in their own right. Such a group included a highly diverse assortment of forms. This unnatural assemblage was reclassified into various phyla. By tradition, however, zoologists still refer to the various groups of these animals as flatworms, ribbon worms, roundworms, segmented worms, and the like.

The Platyhelminthes were derived from an ancestor that probably had many cnidarian-like characteristics, including a gelatinous mesoglea. Nonetheless, replacement of the gelatinous mesoglea with a cellular, mesodermal **parenchyma** laid the basis for a more complex organization. Parenchyma is a form of "packing" tissue containing more cells and fibers than the mesoglea of the cnidarians. In at least some platyhelminths, the parenchyma is made up of noncontractile cell bodies of muscle cells, that is, the cell body containing the nucleus and other organelles is connected to an elongated contractile portion in somewhat the same manner as the epitheliomuscular cells of the cnidarians (see Figure 14-7).

Flatworms range in size from a millimeter or less to some of the tapeworms that are many meters in length. Their flattened bodies may be slender, broadly leaflike, or long and ribbonlike.

The flatworms include both free-living and parasitic forms, but the free-living members are found exclusively in the Class Turbellaria. A few

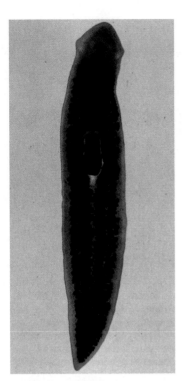

Figure 15-2
Stained planarian.

CHARACTERISTICS OF PHYLUM PLATYHELMINTHES

1. Three germ layers (**triploblastic**)
2. **Bilateral symmetry;** definite polarity of anterior and posterior ends
3. **Body flattened dorsoventrally;** oral and genital apertures mostly on ventral surface
4. Epidermis may be cellular or syncytial (ciliated in some); **rhabdites** in epidermis of most Turbellaria; epidermis a syncytial **tegument** in Monogenea, Trematoda, and Cestoda
5. Muscular system primarily of a sheath form and of mesodermal origin; layers of circular, longitudinal, and sometimes oblique fibers beneath the epidermis
6. No internal body space other than digestive tube (acoelomate); spaces between organs filled with parenchyma, a form of connective tissue or mesenchyme
7. Digestive system incomplete (gastrovascular type); absent in some
8. **Nervous system consisting of a pair of anterior ganglia with** longitudinal nerve cords connected by transverse nerves and located in the mesenchyme in most forms; similar to cnidarians in primitive forms
9. Simple sense organs; eyespots in some
10. Excretory system of two lateral canals with branches bearing **flame cells (protonephridia);** lacking in some primitive forms
11. Respiratory, circulatory, and skeletal systems lacking; lymph channels with free cells in some trematodes
12. Most forms monoecious; reproductive system complex, usually with well-developed gonads, ducts, and accessory organs; internal fertilization; development direct in free-swimming forms and those with a single host in the life cycle; usually indirect in internal parasites in which there may be a complicated life cycle often involving several hosts
13. Class Turbellaria mostly free living; classes Monogenea, Trematoda, and Cestoda entirely parasitic

---

turbellarians are symbiotic or parasitic, but the majority are adapted as bottom dwellers in marine or fresh water or live in moist places on land. Many, especially of the larger species, are found on the underside of stones and other hard objects in freshwater streams or in the littoral zones of the ocean.

The majority of species of turbellarians are marine, but there are many freshwater species. Planarians (Figure 15-2) and some others frequent streams and spring pools; others prefer flowing water of mountain streams. Some species occur in fairly hot springs. Terrestrial turbellarians are found in rather moist places under stones and logs. There are about six species of terrestrial turbellarians in the United States.

All members of the classes Monogenea and Trematoda (the flukes) and the Class Cestoda (the tapeworms) are parasitic. Most of the Monogenea are ectoparasites, but all the trematodes and cestodes are endoparasitic. Many species have indirect life cycles with more than one host; the first host is often an invertebrate, and the final host is usually a vertebrate. Humans serve as hosts for a number of species. Certain larval stages may be free living.

CLASS TURBELLARIA

Turbellarians are mostly free-living worms that range in length from 5 mm or less to 50 cm. Usually covered with ciliated epidermis, these are mostly creeping worms that combine muscular with ciliary movements to achieve locomotion. The mouth is on the ventral side. Unlike the trematodes and cestodes, they have simple life cycles.

As traditionally recognized, the turbellarians form a paraphyletic group. Several synapomorphies show that some turbellarians are phylogenetically closer to the Trematoda, Monogenea, and Cestoda than they are to other turbellarians. For example, in some turbellarians the yolk for nutrition of the developing embryo is contained within the egg cell itself **(endolecithal),** and the embryogenesis shows the spiral determinate cleavage typical of protostomes (p. 108). The endolecithal egg is considered ancestral for flatworms. The other turbellarians plus all trematodes, monogeneans, and cestodes share a derived condition in which the egg cell contains little or no yolk, and the yolk is contributed by cells released from organs called **vitellaria.** Usually a number of yolk cells surrounds the zygote within the eggshell **(ectolecithal),** affecting cleavage in such a way that the spiral pattern cannot be distinguished. The ectolecithal turbellarians therefore appear to form a clade with the Trematoda, Monogenea, and Cestoda to the exclusion of endolecithal turbellarians. Another character, the dual-gland adhesive system (described later), shows that the endolecithal turbellarians also are paraphyletic; presence of the dual-gland shows that some endolecithal turbellarians form a clade with the ectolecithal flatworms to the exclusion of other endolecithal turbellarian lineages. The term Turbellaria is therefore used here only for simplicity of organization and presentation because it describes an artificial group.

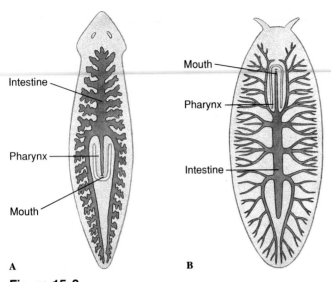

Figure 15-3
Intestinal pattern of two orders of turbellarians. **A,** Tricladida. **B,** Polycladida.

Figure 15-4
Pseudoceros hancockanum, a marine polyclad turbellarian. Marine polyclads are often large and beautifully colored. The orange polyps of *Tubastrea aurea,* an ahermatypic coral, and *Aplidium cratiferum,* a colonial tunicate (Chapter 26) that looks something like cartilage, are also in the photograph.

Other characteristics of value in distinguishing orders of Turbellaria are the form of the gut (present or absent; simple or branched; pattern of branching) and the pharynx (simple; folded; bulbous). Except for the order Polycladida (Gr. *poly,* many, + *klados,* branch), the turbellarians with endolecithal eggs have a simple gut or no gut and a simple pharynx. In a few there is no recognizable pharynx. The polyclads have a folded pharynx and a gut with many branches (Figure 15-3). The polyclads include many marine forms of moderate to large size (3 to more than 40 mm) (Figure 15-4), and a highly branched intestine is correlated with larger size in Turbellaria. Members of the order Tricladida (Gr. *treis,* three, + *klados,* branch), which are in the ectolecithal group and include the freshwater planaria, have a three-branched intestine (Figure 15-3).

Members of the order Acoela (Gr. *a,* without, + *koilos,* hollow) are often regarded as having changed least from the ancestral form. They are small and have a mouth but no gastrovascular cavity or excretory system. Food is merely passed through the mouth into temporary spaces that are surrounded by mesenchyme, where gastrodermal phagocytic cells digest the food intracellularly.

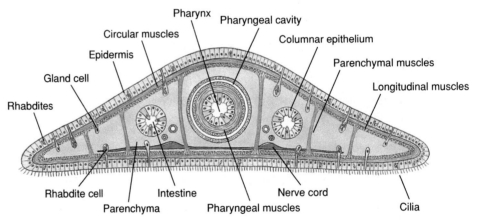

Figure 15-5
Cross section of planarian through pharyngeal region, showing relationships of body structures.

Form and Function

Epidermis, Muscles, Locomotion.
The freshwater planarians, such as *Dugesia* (formerly called *Euplanaria* but changed by priority to *Dugesia* after Dugès, who first described the form in 1830), belong to the Tricladida and are used extensively in introductory laboratory courses. The outer covering is a ciliated epidermis resting on a basement membrane. It contains rod-shaped **rhabdites** that, when discharged with water, swell and form a protective mucous sheath around the body. Single-cell mucous glands open on the surface of the epidermis (Figure 15-5). Most orders of turbellarians have **dual-gland** adhesive organs in the epidermis. These consist of three cell types: viscid and releasing gland cells and anchor cells (Figure 15-6). Secretions of the viscid gland cells apparently fasten the microvilli of the anchor cells to the substrate, and the secretions of the releasing gland cells provide a quick, chemical detaching mechanism.

In the body wall below the basement membrane are layers of **muscle fibers** that run circularly, longitudinally, and diagonally. A meshwork of **parenchyma** cells, developed from mesoderm, fills the spaces between muscles and visceral organs. Parenchyma cells in some, perhaps all, flatworms are not a separate cell type but are the noncontractile portions of muscle cells.

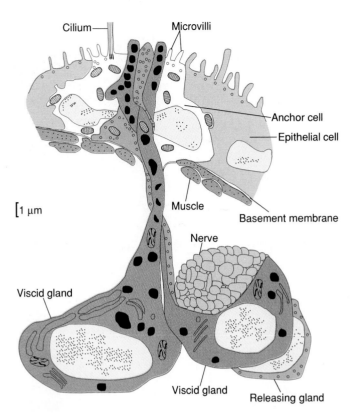

Figure 15-6
Reconstruction of dual-gland adhesive organ of the turbellarian *Haplopharynx* sp. There are two viscid glands and one releasing gland, which lie beneath the body wall. The anchor cell lies within the epidermis, and one of the viscid glands and the releasing gland are in contact with a nerve.

15-7A). The flame cell is cup shaped with a tuft of flagella extending from the inner face of the cup. In an evolutionarily derived condition that groups some turbellarians with the other classes of flatworms, the protonephridia form a **weir** (Old English *wer,* a fence placed in a stream to catch fish), that is, the rim of the cup is elongated into fingerlike projections that interdigitate with similar projections of a tubule cell. The space (lumen) enclosed by the tubule cell continues into collecting ducts that finally open to the outside by pores. The beat of the flagella (resembling a flickering flame) provides a negative pressure to draw fluid through the delicate interdigitations between the flame cell and the tubule cell. The wall of the duct beyond the flame cell commonly bears folds or microvilli that probably function in resorption of certain ions or molecules.

Osmoregulatory organs known as nephridia occur in many invertebrates. When they are closed at the inner end, they are called **protonephridia,** but when they open into a coelomic space at their inner end, they are **metanephridia.** There is a wide range from simple to complex in both types, but protonephridia are generally considered more primitive.

Among the various flatworms, there may be a single protonephridium or from one to four pairs. In planarians they join and rejoin into a network along each side of the animal (Figure 15-7) and may empty through many nephridiopores. This system is mainly osmoregulatory because it is reduced or absent in marine turbellarians, which do not have to expel excess water.

Metabolic wastes are removed largely by diffusion through the body wall.

Respiration.   There are no respiratory organs. Exchange of gases takes place through the body surface.

Very small planaria swim by means of their cilia. Others move by gliding, head slightly raised, over a slime track secreted by the marginal adhesive glands. The beating of the epidermal cilia in the slime track drives the animal along, while rhythmical muscular waves can be seen passing backward from the head. Large polyclads and terrestrial turbellarians crawl by muscular undulations, much in the manner of a snail.

Nutrition and Digestion.   The digestive system includes a mouth, a pharynx, and an intestine (Figure 15-3). The pharynx, enclosed in a **pharyngeal sheath** (Figure 15-7), opens posteriorly just inside the mouth, through which it can extend (Figure 15-7). The intestine has three many-branched trunks, one anterior and two posterior. The whole forms a **gastrovascular cavity** lined with columnar epithelium (Figure 15-7).

Planarians are mainly carnivorous, feeding largely on small crustaceans, nematodes, rotifers, and insects. They can detect food from some distance by means of chemoreceptors. They entangle their prey in mucous secretions from the mucous glands and rhabdites. The planarian grips its prey with its anterior end, wraps its body around the prey, extends its proboscis, and sucks up minute bits of the food. Intestinal secretions contain proteolytic enzymes for some **extracellular digestion.** Bits of food are sucked up into the intestine, where the phagocytic cells of the gastrodermis complete the digestion **(intracellular).** The gastrovascular cavity extends to most parts of the body, and food is absorbed through its walls into the body cells. Undigested food is egested through the pharynx.

Excretion and Osmoregulation. Except in the Acoela, the osmoregulatory system of turbellarians consists of **protonephridia** (excretory or osmoregulatory organs closed at the inner end) with **flame cells** (Figure

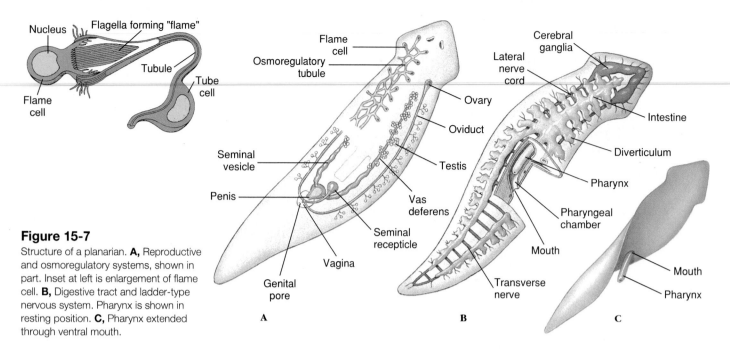

Figure 15-7
Structure of a planarian. **A,** Reproductive and osmoregulatory systems, shown in part. Inset at left is enlargement of flame cell. **B,** Digestive tract and ladder-type nervous system. Pharynx is shown in resting position. **C,** Pharynx extended through ventral mouth.

Nervous System.   The most primitive flatworm nervous system, found in some of the acoels, is a **subepidermal nerve plexus** resembling the nerve net of the cnidarians. Other flatworms have, in addition to a nerve plexus, one to five pairs of **longitudinal nerve cords** lying under the muscle layer. The more advanced flatworms tend to have the lesser number of nerve cords. Freshwater planarians have one ventral pair (Figure 15-7B). Connecting nerves form a "ladder-type" pattern. The brain is a bilobed mass of ganglion cells arising anteriorly from the ventral nerve cords. Except in the acoels, which have a diffuse system, the neurons are organized into sensory, motor, and association types—an important development in the evolution of the nervous system.

Sense Organs.   Active locomotion in flatworms has favored not only cephalization in the nervous system but also further evolution of sense organs. **Ocelli,** or light-sensitive eyespots, are common in turbellarians (Figure 15-2).

Tactile cells and chemoreceptive cells are abundant over the body, and in planarians they form definite organs on the auricles (the earlike lobes on the sides of the head). Some species

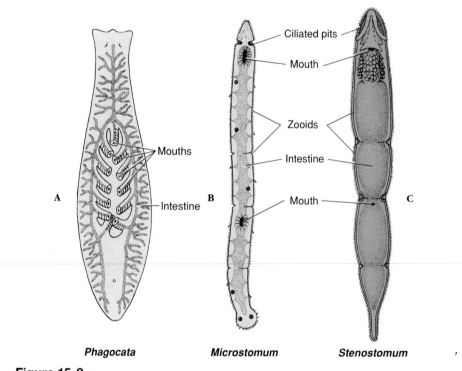

Phagocata   *Microstomum*   *Stenostomum*

Figure 15-8
Some small freshwater turbellarians. **A,** *Phagocata* has numerous pharynges. **B** and **C,** Incomplete fission results for a time in a series of attached zooids.

also have statocysts for equilibrium and rheoreceptors for sensing direction of the water current.

Reproduction and Regeneration.
Many turbellarians reproduce both asexually (by fission) and sexually. Asexually, the freshwater planarian merely constricts behind the pharynx and separates into two animals, each of which regenerates the missing

parts—a quick means of population increase. There is evidence that a reduced population density results in an increase in the rate of fissioning. In some forms, such as *Stenostomum* (Gr. *stenos,* narrow, + *stoma,* mouth) and *Microstomum* (Gr. *mikros,* small, + *stoma,* mouth), in which fissioning occurs, the individuals do not separate at once but remain attached, forming chains of zooids (Figure 15-8B and C).

The considerable powers of regeneration in planarians have provided an interesting system for experimental studies of development. For example, a piece excised from the middle of the planarian can regenerate both a new head and a new tail. However, the piece retains its original polarity: the head grows at the anterior end and the tail at the posterior end. An extract of heads added to a culture medium containing headless worms will prevent regeneration of new heads, suggesting that substances in one region will suppress the regeneration of the same region at another level of the body. Many other experiments could be cited.

Turbellarians are monoecious (hermaphroditic) but practice cross-fertilization. During the breeding season each individual develops both male and female organs, which usually open through a common genital pore (Figure 15-7A). After copulation one or more fertilized eggs and some yolk cells become enclosed in a small cocoon. The cocoons are attached by little stalks to the underside of stones or plants. Embryos emerge as juveniles that resemble mature adults. In some marine forms the egg develops into a ciliated free-swimming larva.

CLASS TREMATODA

Trematodes are all parasitic flukes, and as adults they are almost all found as endoparasites of vertebrates. They are chiefly leaflike in form and are structurally similar in many respects to the ectolecithal Turbellaria. A major difference is found in the body covering, or **tegument,** which does not bear cilia in the adult. Furthermore, in common with Monogenea and Cestoda, cell bodies are sunk beneath the outer layer (Figure 15-9) and superficial muscle layers, and they communicate with the outer layer (distal cytoplasm) by processes extending between the muscles. Because the distal cytoplasm is continuous, with no intervening cell membranes, the tegument is **syncytial.** This peculiar epidermal arrangement is probably related to adaptations for parasitism in ways that are still unclear.

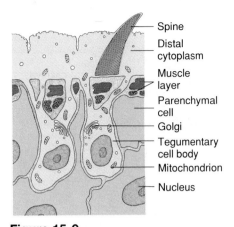

Figure 15-9
Diagrammatic drawing of the structure of the tegument of a trematode *Fasciola hepatica.*

Other structural adaptations for parasitism are more apparent: various penetration glands or glands to produce cyst material; organs for adhesion such as suckers and hooks; and increased reproductive capacity. Otherwise, trematodes retain several turbellarian characteristics, such as a well-developed alimentary canal (but with the mouth at the anterior, or cephalic, end) and similar reproductive, excretory, and nervous systems, as well as a musculature and parenchyma that are only slightly modified from those of the Turbellaria. Sense organs are poorly developed.

Of the subclasses of Trematoda, Aspidogastrea and Didymozoidea are small and poorly known groups, but Digenea (Gr. *dis,* double, + *genos,* race) is a large group with many species of medical and economic importance.

Subclass Digenea

With rare exceptions, digeneans have an indirect life cycle, the first **(intermediate)** host being a mollusc and the **definitive** host (the host in which sexual reproduction occurs, sometimes called the **final** host) being a vertebrate. In some species a second, and sometimes even a third, intermediate host intervenes. The group has radiated greatly, and its members parasitize almost all kinds of vertebrate hosts. Digeneans inhabit, according to species, a wide variety of sites in their hosts: all parts of the digestive tract, respiratory tract, circulatory system, urinary tract, and reproductive tract.

One of the world's most amazing biological phenomena is the digenean life cycle. Although the cycles of different species vary widely in detail, a typical example would include the adult, egg, miracidium, sporocyst, redia, cercaria, and metacercaria stages (Figure 15-11). The egg usually passes from the definitive host in the excreta and must reach water to develop further. There, it hatches to a free-swimming, ciliated larva, the **miracidium.** The miracidium penetrates the tissues of a snail, where it transforms into a **sporocyst.** The sporocyst reproduces asexually to yield either more sporocysts or a number of **rediae.** The rediae, in turn, reproduce asexually to produce more rediae or to produce **cercariae.** In this way a single egg can give rise to an enormous number of progeny. The cercariae emerge from the snail and penetrate a second intermediate host or encyst on vegetation or other objects to become **metacercariae,** which are juvenile flukes. The adult grows from the metacercaria when that stage is eaten by the definitive host.

Some of the most serious parasites of humans and domestic animals belong to the Digenea. The first digenean life cycle to be worked out was that of *Fasciola hepatica* (L. *fasciola,* a small bundle, band), which causes "liver rot" in sheep and other ruminants. The adult fluke lives in the bile passage of the liver, and the eggs are passed in the feces. After hatching, the miracidium penetrates a snail to become a sporocyst. There are two generations of rediae, and the cercaria encysts on vegetation. When the infested vegetation is eaten by the sheep or other ruminant (or sometimes humans), the metacercariae excyst and grow into young flukes.

Clonorchis sinensis: Liver Fluke in Humans

Clonorchis (Gr. *clon,* branch, + *orchis,* testis) is the most important liver fluke of humans and is common in many regions of eastern Asia, especially in China, Southeast Asia, and Japan. Cats, dogs, and pigs are also often infected.

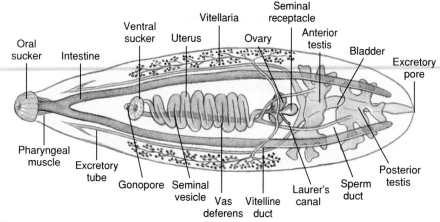

Figure 15-10

Structure of human liver fluke *Clonorchis sinenesis.*

Structure. The worms vary from 10 to 20 mm in length (Figure 15-10). Their structure is typical of many trematodes in most respects. They have an **oral sucker** and a **ventral sucker.** The **digestive system** consists of a pharynx, a muscular esophagus, and two long, unbranched intestinal ceca. The **excretory system** consists of two protonephridial tubules, with branches provided with flame cells or bulbs. The two tubules unite to form a single median bladder that opens to the outside. The **nervous system,** like that of turbellarians, is made up of two cerebral ganglia connected to longitudinal cords that have transverse connectives.

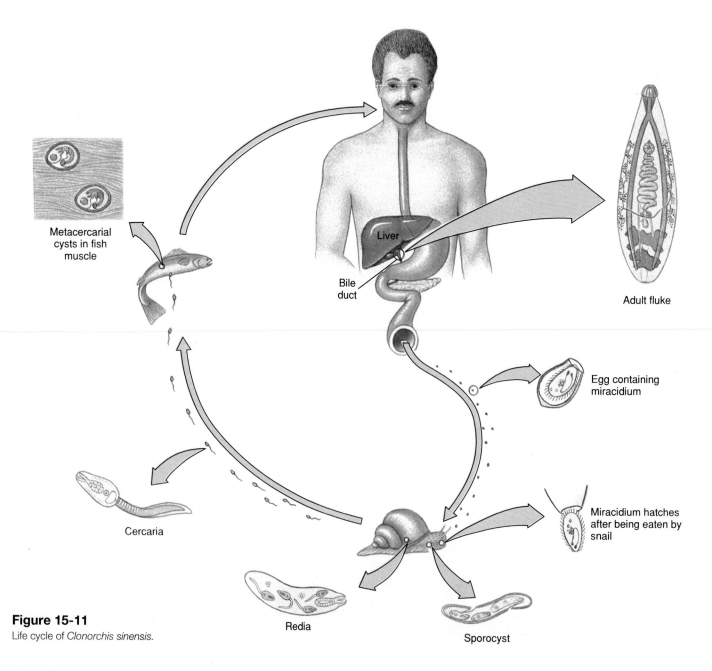

Figure 15-11

Life cycle of *Clonorchis sinensis.*

The **reproductive system** is hermaphroditic and complex. The male system is made up of two branched **testes** and two **vasa efferentia** that unite to form a single **vas deferens,** which widens into a **seminal vesicle.** The seminal vesicle leads into an **ejaculatory duct,** which terminates at the genital opening. Unlike most trematodes, *Clonorchis* does *not* have a protrusible copulatory organ, the cirrus. The female system contains a branched **ovary** with a short **oviduct,** which is joined by ducts from the **seminal receptacle** and the **vitellaria** at the **ootype.** The ootype is surrounded by a glandular mass, **Mehlis' gland,** of uncertain function. From Mehlis' gland the much-convoluted **uterus** runs to the genital pore. Cross-fertilization between individuals is usual, and sperm are stored in the seminal receptacle. When an oocyte is released from the ovary, it is joined by a sperm and a group of vitelline cells and is fertilized. The vitelline cells release a proteinaceous shell material, which is stabilized by a chemical reaction; the Mehlis' gland secretions are added; and the egg passes into the uterus.

Life Cycle. The normal habitat of the adults is in the bile passageways of humans and other fish-eating mammals (Figure 15-11). The eggs, each containing a complete miracidium, are shed into the water with the feces but do not hatch until they are ingested by the snail *Parafossarulus* or related genera. The eggs, however, may live for some weeks in water. In the snail the miracidium enters the tissues and transforms into the sporocyst (a baglike structure with embryonic germ cells), which produces one generation of rediae. The redia is elongated, with an alimentary canal, a nervous system, an excretory system, and many germ cells in the process of development. The rediae pass into the liver of the snail where the germ cells continue embryonation and give rise to the tadpolelike cercariae.

The cercariae escape into the water, swim about until they meet with a fish of the family Cyprinidae, and then bore into the muscles or under the scales. Here the cercariae lose their tails and encyst as metacercariae. If a mammal eats raw infected fish, the metacercarial cyst dissolves in the intestine, and the young flukes apparently migrate up the bile duct, where they become adults. There the flukes may live for 15 to 30 years.

The effect of the flukes on humans depends mainly on the extent of the infection. A heavy infection can cause a pronounced cirrhosis of the liver and can result in death. Cases are diagnosed through fecal examinations. To avoid infection, all fish used as food should be thoroughly cooked. Destruction of the snails that carry larval stages is a method of control.

Schistosoma: Blood Flukes

Schistosomiasis, infection with blood flukes of the genus *Schistosoma* (Gr. *schistos,* divided, + *soma,* body), ranks as one of the major infectious diseases in the world, with 200 million people infected. The disease is widely prevalent over much of Africa and parts of South America, the West Indies, the Middle East, and the Far East. The old generic name for the worms was *Bilharzia* (from Theodor Bilharz, German parasitologist who discovered *Schistosoma haematobium*), and the infection was called bilharziasis, a name still used in many areas.

Unfortunately, some projects intended to raise the standard of living in some tropical countries, such as the Aswan High Dam in Egypt, have increased the prevalence of schistosomiasis by creating more habitats for the snail intermediate hosts. Before the dam was constructed, the 500 miles of the Nile River between Aswan and Cairo was subjected to annual floods; alternate flooding and drying killed many snails. Four years after dam completion, the prevalence of schistosomiasis had increased sevenfold along that segment of the river. The prevalence in fishermen around the lake above the dam increased from a very low level to 76%.

The blood flukes differ from most other flukes in being dioecious and having the two branches of the digestive tube united into a single tube in the posterior part of the body. The male is broader and heavier and has a large, ventral groove, the **gynecophoric canal,** posterior to the ventral sucker. The gynecophoric canal embraces the long, slender female (Figure 15-12).

Three species account for most of the schistosomiasis in humans: *S. mansoni,* which lives primarily in the venules draining the large intestine; *S. japonicum,* which is found mostly in the venules of the small intestine; and *S. haematobium,* which lives in the venules of the urinary bladder. *S. mansoni* is common in parts of Africa, Brazil, northern South America, and the West Indies; species of *Biomphalaria* are the principal snail intermediate hosts. *S. haematobium* is widely prevalent in Africa, using snails of the genera *Bulinus* and *Physopsis* as the main intermediate hosts. *S. japonicum* is confined to the Far East, and its hosts are several species of *Oncomelania.*

The life cycle of blood flukes is similar in all species. Eggs are discharged in human feces or urine; if they get into water, they hatch out as ciliated miracidia, which must contact the required kind of snail within a few hours to survive. In the snail, they transform into sporocysts, which produce another generation of sporocysts. The daughter sporocysts give rise to cercariae directly, without the formation of rediae. The cercariae escape from the snail and swim about until they come in contact with the bare skin of a human. They penetrate the skin, shedding their tails in the process, and reach a blood vessel where they enter the circulatory system. There is no metacercarial stage. The young schistosomes make their way to the hepatic portal system of blood vessels and undergo a period of development in the liver before migrating to their characteristic sites. As eggs are released by the adult female, they are somehow extruded through

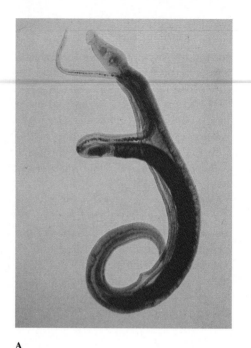

A

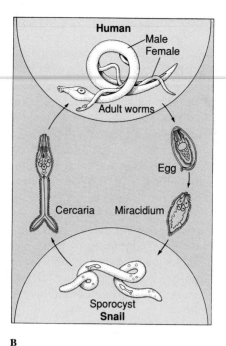

B

Figure 15-12

A, Adult male and female *Schistosoma mansoni* in copulation. The male has a long gynecophoric canal that holds the female (the darkly stained individual). Humans are usually hosts of adult parasites, found mainly in Africa but also in South America and elsewhere. Humans become infected by wading or bathing in cercaria-infested waters. **B,** Life cycle of *Schistosoma mansoni*.

Figure 15-13

Human abdomen, showing schistosome dermatitis caused by penetration of schistosome cercariae that are unable to complete development in humans. Sensitization to allergenic substances released by cercariae results in rash and itching.

the wall of the venule and through the gut or bladder lining, to be voided with the feces or urine, according to species. Many eggs do not make this difficult transit and are swept by the blood flow back to the liver or other areas, where they become centers of inflammation and tissue reaction.

The main ill effects of schistosomiasis result from the eggs. With *S. mansoni* and *S. japonicum,* eggs in the intestinal wall cause ulceration, abscesses, and bloody diarrhea with abdominal pain. Similarly, *S. haematobium* causes ulceration of the bladder wall with bloody urine and pain on urination. Eggs swept to the liver or other sites cause symptoms associated with the organs where they lodge. When they are caught in the capillary bed of the liver, they impede circulation and cause cirrhosis, a fibrotic reaction that interferes with liver function. Of the three species, *S. haematobium* is considered the least serious and *S. japonicum* the most severe. The prognosis is poor in heavy infections of *S. japonicum* without early treatment.

Control is best carried out by educating people to dispose of their body wastes hygienically, a difficult problem with poor people living under primitive conditions.

> Although proper disposal of body wastes is the best control for schistosomiasis, other strategies are being pursued with varying success: chemotherapy, vector control, and vaccination. Development of a vaccine is the subject of much research, but an effective vaccine is not yet available. Vector control by environmental management and by biological means appear promising. Biological controls include introduction of species of snails, crayfish, and fish that prey on the snail vectors.

Schistosome Dermatitis (Swimmer's Itch). Various species of schistosomes in several genera cause a rash or dermatitis when their cercariae penetrate hosts that are unsuitable for further development (Figure 15-13). The cercariae of several genera whose normal hosts are North American birds

cause dermatitis in bathers in northern lakes. The severity of the rash increases with an increasing number of contacts with the organisms, or sensitization. After penetration, the cercariae are attacked and killed by the host's immune mechanisms, and they release allergenic substances, causing itching. The condition is more an annoyance than a serious threat to health, but there may be economic losses to persons depending on vacation trade around infested lakes.

Paragonimus: Lung Flukes

Several species of *Paragonimus* (Gr. *para,* beside, + *gonimos,* generative), a fluke that lives in the lungs of its host, are known from a variety of mammals. *Paragonimus westermani* (Figure 15-14), found in east Asia, southwest Pacific, and some parts of South America, parasitizes a number of wild carnivores, humans, pigs, and rodents. Its eggs are coughed up in the sputum, swallowed, then eliminated with the feces. The metacercariae develop in freshwater crabs, and the infection is acquired by eating

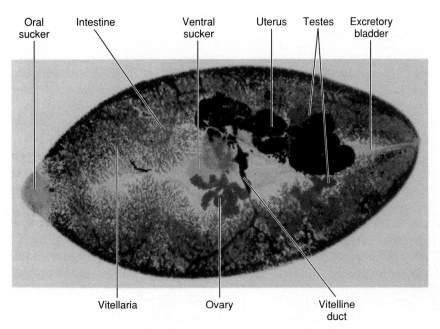

Oral sucker · Intestine · Ventral sucker · Uterus · Testes · Excretory bladder

Vitellaria · Ovary · Vitelline duct

Figure 15-14

Lung fluke *Paragonimus westermani.* Adults are up to 2 cm long. Eggs discharged in sputum or feces hatch into free-swimming miracidia that enter snails. Cercariae from snails enter freshwater crabs and encyst in soft tissues. Humans are infected by eating poorly cooked crabs or by drinking water containing larvae freed from dead crabs.

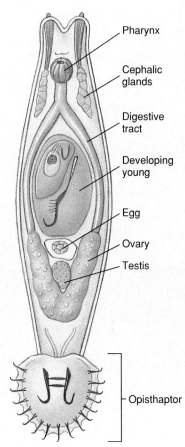

Pharynx · Cephalic glands · Digestive tract · Developing young · Egg · Ovary · Testis · Opisthaptor

Figure 15-15

A monogenetic fluke *Gyrodactylus cylindriformis,* ventral view.

uncooked crab meat. The infection causes respiratory symptoms, with breathing difficulties and chronic cough. Fatal cases are common. A closely related species, *P. kellicotti,* occurs in mink and similar animals in North America, but only one human case has been recorded. Its metacercariae are in crayfish.

Some Other Trematodes

Fasciolopsis buski (L. *fasciola,* small bundle, + Gr. *opsis,* appearance) parasitizes the intestine of humans and pigs in India and China. Larval stages occur in several species of planorbid snails, and the cercariae encyst on water chestnuts, an aquatic vegetation eaten raw by humans and pigs.

Leucochloridium is noted for its remarkable sporocysts. Snails (*Succinea*) eat vegetation infected with eggs from bird droppings. The sporocysts become much enlarged and branched, and the cercariae encyst within the sporocyst. The sporocysts enter the snail's head and tentacles, become brightly striped with orange and green bands, and pulsate at frequent intervals. Birds are attracted by

the enlarged and pulsating tentacles, eat the snails, and so complete the life cycle.

CLASS MONOGENEA

The monogenetic flukes traditionally have been placed as an order of the Trematoda, but they are sufficiently different to deserve a separate class. Cladistic analysis places them closer to the Cestoda. Monogeneans are all parasites, primarily of the gills and external surfaces of fish. A few are found in the urinary bladders of frogs and turtles, and one parasitizes the eye of a hippopotamus. Although widespread and common, monogeneans seem to cause little damage to their hosts under natural conditions. However, like numerous other fish pathogens, they become a serious threat when their hosts are crowded together, as, for example, in fish farming.

The life cycles of monogeneans are direct, with a single host. The egg hatches a ciliated larva, the **oncomiracidium,** that attaches to the host or swims around awhile before attachment. The oncomiracidium bears hooks on its posterior, which in

many species become the hooks on the large posterior attachment organ **(opisthaptor)** of the adult. Because the monogenean must cling to the host and withstand the force of water flow over the gills or skin, adaptive radiation has produced a wide array of opisthaptors in different species. Opisthaptors may bear large and small hooks, suckers, and clamps, often in combination with each other.

Common genera are *Gyrodactylus* (L. *gyro,* a circle, + Gr. *daktylos,* toe, finger) (Figure 15-15) and *Dactylogyrus* (Gr. *daktylos,* toe, finger, + L. *gyro,* a circle), both of economic importance to fish culturists, and *Polystoma* (Gr. *polys,* many, + *stoma,* mouth), found in the urinary bladder of frogs.

CLASS CESTODA

Cestoda, or tapeworms, differ in many respects from the preceding classes. They usually have long flat bodies in

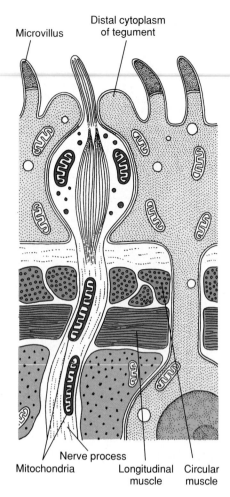

Figure 15-16
Schematic drawing of a longitudinal section through a sensory ending in the tegument of *Echinococcus granulosus*.

Table 15-1 | **Common Cestodes of Humans**

Common and Scientific Name	Means of Infection; Prevalence in Humans
Beef tapeworm (*Taeniarhynchus saginatus*)	Eating rare beef; most common of all tapeworms in humans
Pork tapeworm (*Taenia solium*)	Eating rare pork; less common than *T. saginatus*
Fish tapeworm (*Diphyllobothrium latum*)	Eating rare or poorly cooked fish; fairly common in Great Lakes region of United States, and other areas of world where raw fish is eaten
Dog tapeworm (*Dipylidium caninum*)	Unhygienic habits of children (juveniles in flea and louse); moderate frequency
Dwarf tapeworm (*Vampirolepis nana*)	Juveniles in flour beetles; common
Unilocular hydatid (*Echinococcus granulosus*)	Cysts of juveniles in humans; infection by contact with dogs; common wherever humans are in close relationship with dogs and ruminants
Multilocular hydatid (*Echinococcus multilocularis*)	Cysts of juveniles in humans; infection by contact with foxes; less common than unilocular hydatid

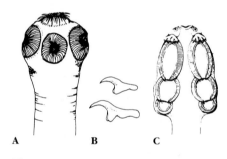

Figure 15-17
Two tapeworm scolices. **A,** Scolex of *Taenia solium* (pork tapeworm) with apical hooks and suckers. (Scolex of *Taeniarhynchus saginatus* is similar, but without hooks.) **B,** Hooks of *T. solium*. **C,** Scolex of *Acanthobothrium coronatum,* a tapeworm of sharks. This species has large leaflike sucker organs divided into chambers with apical suckers and hooks.

which there is a linear series of sets of reproductive organs. Each set is called a **proglottid** and is usually set off at its anterior and posterior ends by zones of muscle weakness, marked externally by grooves. There is a complete lack of a digestive system. As in Monogenea and Trematoda, there are no external, motile cilia in the adult, and the tegument is of a distal cytoplasm with sunken cell bodies beneath the superficial muscle layer (Figure 15-16). In contrast to the monogeneans and trematodes, however, the entire surface of cestodes is covered with minute projections similar to the microvilli of the vertebrate small intestine (p. 46). The microvilli greatly enlarge the surface area of the tegument, which is a vital adaptation

of the tapeworm since it must absorb all its nutrients across the tegument.

Tapeworms are nearly all monoecious. They have well-developed muscles, and their excretory system and nervous system are somewhat similar to those of other flatworms. They have no special sense organs but do have sensory endings in the tegument that are modified cilia (Figure 15-16). One of their most specialized structures is the **scolex,** or holdfast, which is the organ of attachment. It is usually

provided with suckers or suckerlike organs and often with hooks or spiny tentacles (Figure 15-17).

With rare exceptions, all cestodes require at least two hosts, and the adult is a parasite in the digestive tract of vertebrates. Often one of the intermediate hosts is an invertebrate.

The subclass Eucestoda contains the great majority of species in the class. With the exception of two small orders, the members of this subclass have the body divided into a series of proglottids and are thus referred to as **polyzoic.** Their larval forms all have six hooks. The main body of the worms, the chain of proglottids, is called the **strobila.** Typically, there is a **germinative zone** just behind the scolex where new proglottids are formed. As younger proglottids are differentiated in front of it, each individual proglottid moves posteriorly in the strobila, and its gonads mature. The proglottid is usually fertilized by another proglottid in the same or a different strobila. The shelled embryos form in the uterus of the proglottid, and either they are expelled through a uterine pore or the entire proglottid is shed from the worm as it reaches the posterior end.

Some zoologists have maintained that the proglottid formation of cestodes represents "true" segmentation (metamerism), but we do not support this view. Segmentation of tapeworms is best considered a replication of sex organs to increase reproductive capacity and is not related to the metamerism found in Annelida, Arthropoda, and Chordata (see pp. 195 and 365).

More than 1000 species of tapeworms are known to parasitologists. Almost all vertebrate species are infected. Normally, adult tapeworms do little harm to their hosts. The most common tapeworms found in humans are given in Table 15-1.

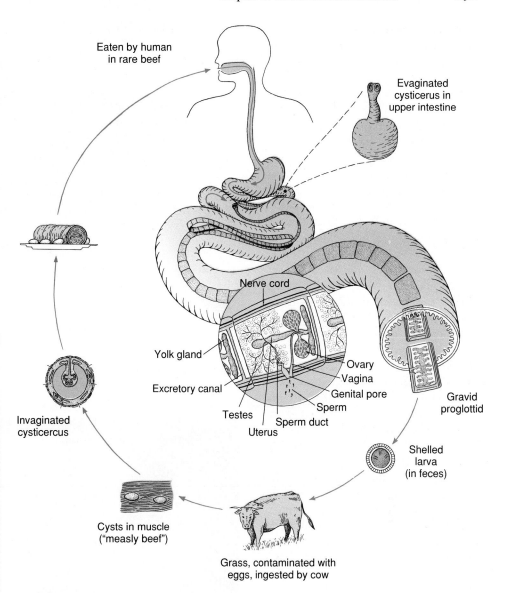

Figure 15-18

Life cycle of beef tapeworm, *Taeniarhynchus*. Ripe proglottids break off in the human intestine, leave the body in feces, crawl out of feces onto grass, and are ingested by cattle. Eggs hatch in the cow's intestine, freeing oncospheres, which penetrate into muscles and encyst, developing into "bladder worms." Human eats infected rare beef, and cysticercus is freed in intestine where it attaches to the intestinal wall, forms a strobila, and matures.

Taeniarhynchus saginatus: Beef Tapeworm

Structure. *Taeniarhynchus saginatus* (Gr. *tainia,* band, ribbon, + *rhynchos,* beak, snout) is called the beef tapeworm, but it lives as an adult in the alimentary canal of humans. The juvenile form is found primarily in the intermuscular tissue of cattle. The mature adult may reach a length of 10 m or more. Its scolex has four suckers for attachment to the intestinal wall, but no hooks. A short neck connects the scolex to the strobila, which may have as many as 2000 proglottids. The gravid proglottids bear shelled, infective larvae (Figure 15-18) and become detached and pass in the feces.

The tapeworm shows some unity in its organization, for **excretory canals** in the scolex are also connected to the canals, two on each side, in the proglottids, and two longitudinal **nerve cords** from a **nerve ring** in the scolex run back into the proglottids also (Figure 15-19). Attached to the excretory ducts are the flame cells. Each mature proglottid also contains muscles and parenchyma as well as a complete set of male and female organs similar to those of a trematode.

In the order to which this species belongs, however, the vitellaria are typically a single, compact **vitelline gland** located just posterior to the ovaries. When the gravid proglottids break off and pass out with the feces, they usually crawl out of the fecal mass and onto vegetation nearby. There they may be picked up by grazing cattle. The proglottid ruptures as it dries up, further scattering the embryos on soil and grass. The embryos may remain viable on grass for as long as 5 months.

Life Cycle. When cattle swallow the shelled larvae, they hatch, and the larvae (**oncospheres**) use their hooks to burrow through the intestinal wall into the blood or lymph vessels and finally reach voluntary muscle, where they encyst to become **bladder worms** (juveniles called **cysticerci**). There the juveniles develop an invaginated scolex but remain quiescent. When infected "measly" meat is eaten by a suitable host, the cyst wall dissolves, the scolex evaginates and attaches to the intestinal mucosa, and new proglottids begin to develop. It takes 2 to 3 weeks for a mature worm to form. When a person is infected with one of these tapeworms, numerous gravid proglottids are expelled daily, sometimes crawling out the anus by themselves. Humans become infected by eating rare roast beef, steaks, and barbecues. Considering that about 1% of American cattle are infected, that 20% of all cattle slaughtered are not federally inspected, and that even in inspected meat one-fourth of infections are missed, it is not surprising that tapeworm infection is fairly common. Infection is precluded when meat is thoroughly cooked.

Some Other Tapeworms

Taenia solium: Pork Tapeworm. The adult *Taenia solium* (Gr. *tainia*, band, ribbon) lives in the small intestine of humans, whereas the juveniles live in the muscles of pigs. The scolex has both suckers and hooks arranged on its tip (Figure 15-17), the **rostellum.** The life history of this worm is similar to that of the beef tapeworm, except that humans become infected by eating improperly cooked pork.

The pork tapeworm is much more dangerous than *T. saginatus* because the cysticerci, as well as the adults, can develop in humans. If eggs or proglottids are accidentally ingested by a human, the liberated em-

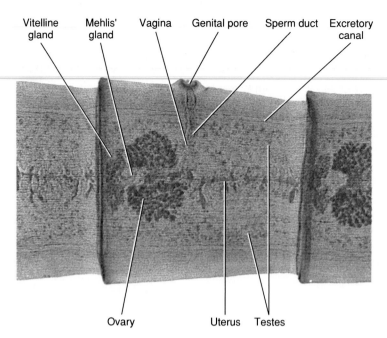

Figure 15-19
Mature proglottid of *Taenia pisiformis,* a dog tapeworm. Portions of two other proglottids also shown.

Labels: Vitelline gland, Mehlis' gland, Vagina, Genital pore, Sperm duct, Excretory canal, Ovary, Uterus, Testes

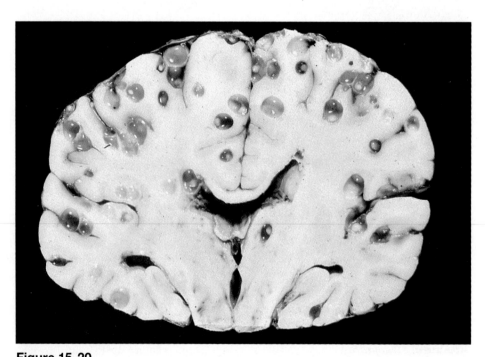

Figure 15-20
Section through the brain of a person who died of cerebral cysticercosis, an infection with the cysticerci of *Taenia solium.*

bryos migrate to any of several organs and form cysticerci (Figure 15-20). The condition is called **cysticercosis.** Common sites are the eye or brain, and infection in such locations can result in blindness, serious neurological symptoms, or death.

Diphyllobothrium latum: Fish Tapeworm. The adult *Diphyllobothrium* (Gr. *dis,* double, + *phyllon,* leaf, + *bothrion,* hole, trench) is found in the intestine of humans, dogs, cats, and other mammals; the immature stages are in crustaceans and fish. With

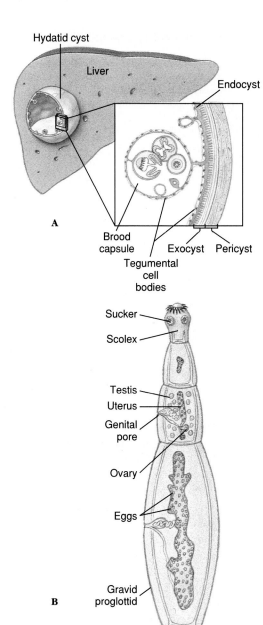

Figure 15-21
Echinococcus granulosus, a dog tapeworm, which may be dangerous to humans. **A,** Early hydatid cyst or bladder-worm stage found in cattle, sheep, hogs, and sometimes humans produces hydatid disease. Humans acquire disease by unsanitary habits in association with dogs. When eggs are ingested, liberated larvae usually encyst in the liver, lungs, or other organs. Brood capsules containing scolices are formed from the inner layer of each cyst. The cyst enlarges, developing other cysts with brood pouches. It may grow for years to the size of a basketball, necessitating surgery. **B,** The adult tapeworm lives in intestine of a dog or other carnivore.

CLASSIFICATION OF PHYLUM PLATYHELMINTHES

Class Turbellaria (tur´bel-lar´e-a) (L. *turbellae* [pl.], stir, bustle, + *aria,* like or connected with): the **turbellarians.** Usually free-living forms with soft, flattened bodies; covered with ciliated epidermis containing secreting cells and rodlike bodies (rhabdites); mouth usually on ventral surface sometimes near center of body; no body cavity except intercellular lacunae in parenchyma; mostly hermaphroditic, but some have asexual fission. A paraphyletic taxon. Examples: *Dugesia* (planaria), *Microstomum, Planocera.*

Class Trematoda (trem´a-to´da) (Gr. *trematodes,* with holes, + *eidos,* form): the **digenetic flukes.** Body of adults covered with a syncytial tegument without cilia; leaflike or cylindrical in shape; usually with oral and ventral suckers, no hooks; alimentary canal usually with two main branches; mostly monoecious; development indirect, with first host a mollusc, final host usually a vertebrate; parasitic in all classes of vertebrates. Examples: *Fasciola, Clonorchis, Schistosoma.*

Class Monogenea (mon´o-gen´e-a) (Gr. *mono,* single, + *gene,* origin, birth): the **monogenetic flukes.** Body of adults covered with a syncytial tegument without cilia; body usually leaflike to cylindrical in shape; posterior attachment organ with hooks, suckers, or clamps, usually in combination; monoecious; development direct, with single host and usually with free-swimming, ciliated larva; all parasitic, mostly on skin or gills of fish. Examples: *Dactylogyrus, Polystoma, Gyrodactylus.*

Cass Cestoda (ses-to´da) (Gr. *kestos,* girdle, + *eidos,* form): the **tapeworms.** Body of adults covered with nonciliated, syncytial tegument; general form of body tapelike; scolex with suckers or hooks, sometimes both, for attachment; body usually divided into series of proglottids; no digestive organs; usually monoecious; larva with hooks; parasitic in digestive tract of all classes of vertebrates; development indirect with two or more hosts; first host may be vertebrate or invertebrate. Examples: *Diphyllobothrium, Hymenolepis, Taenia.*

a length of up to 20 m, it is the largest of the cestodes that infect humans. Fish tapeworm infections can occur anywhere in the world where people commonly eat raw fish; in the United States infections are most common in the Great Lakes region. In Finland, but apparently not other areas, the worm can cause a serious anemia.

Echinococcus granulosus: Unilocular Hydatid.

Adult *E. granulosus* (Gr. *echinos,* hedgehog, + *kokkos,* kernel) (Figure 15-21B), parasitizes dogs and other canines; the juveniles develop in more than 40 species of mammals, including humans, monkeys, sheep, reindeer, and cattle. Thus humans may serve as an intermediate host in the case of this tapeworm. The juvenile stage is a special kind of cysticercus called a **hydatid cyst** (Gr. *hydatis,* watery vesicle). It grows slowly, but it can grow for a long time—up to 20 years— reaching the size of a basketball in an unrestricted site such as the liver. If the hydatid grows in a critical location, such as the heart or central nervous system, serious symptoms may appear in a much shorter time. The main cyst maintains a single or unilocular chamber, but within the main cyst, daughter cysts bud off, and each contains thousands of scolices. Each scolex will produce a worm when eaten by a canine. The only treatment is surgical removal of the hydatid.

PHYLUM NEMERTEA (RHYNCHOCOELA)

Nemerteans (nem-er´te-ans) (Gr. *Nemertes,* one of the Nereids, unerring one) are often called the ribbon worms. Their name refers to the unerring aim of the proboscis, a long muscular tube that can be thrust out swiftly to grasp the prey. The phylum is also

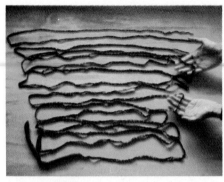

Figure 15-22

Baseodiscus is a genus of nemerteans whose members typically measure several meters in length. This *B. mexicanus* from the Galápagos Islands was over 10 m long.

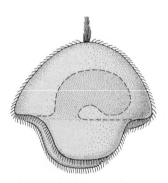

Figure 15-23

The pilidium larva, typical of most nemerteans, is ciliated and free swimming, and has lateral lobes.

CHARACTERISTICS OF PHYLUM NEMERTEA

1. Bilateral symmetry; highly contractile body that is cylindrical anteriorly and flattened posteriorly
2. Three germ layers
3. Epidermis with cilia and gland cells; rhabdites in some
4. Body spaces with parenchyma, which is partly gelatinous
5. An **eversible proboscis,** which lies free in a cavity (rhynchocoel) above the alimentary canal
6. **Complete digestive system** (mouth to anus)
7. Body-wall musculature of outer circular and inner longitudinal layers with diagonal fibers between the two; sometimes another circular layer inside the longitudinal layer
8. **Blood-vascular system with two or three longitudinal trunks**
9. Acoelomate, although the rhynchocoel technically may be considered a true coelom
10. Nervous system usually a four-lobed brain connected to paired longitudinal nerve trunks or, in some, middorsal and midventral trunks
11. Excretory system of two coiled canals, which are branched with **flame cells**
12. Sexes separate with simple gonads; asexual reproduction by fragmentation; few hermaphrodites; **pilidium larvae** in some
13. No respiratory system
14. Sensory **ciliated pits** or **head slits** on each side of head, which communicate between the outside and the brain; tactile organs and ocelli (in some)
15. In contrast to Platyhelminthes, there are few parasitic nemerteans

called Rhynchocoela (ring´ko-se´la) (Gr. *rhynchos,* beak, + *koilos,* hollow), which also refers to the proboscis. They are thread-shaped or ribbon-shaped worms; nearly all of them are marine. Some live in secreted gelatinous tubes. There are about 650 species in the group.

Nemertean worms are usually less than 20 cm long, although a few are several meters in length (Figure 15-22). *Lineus longissimus* (L. *linea,* line) is said to reach 30 m. Their colors are often bright, although most are dull or pallid. In the odd genus *Gorgonorhynchus* (Gr. *Gorgo,* name of a female monster of terrible aspect, + *rhynchos,* beak, snout) the proboscis is divided into many proboscides, which appear as a mass of wormlike structures when everted.

With a few exceptions, the general body plan of the nemerteans is similar to that of Turbellaria. Like the latter,

their epidermis is ciliated and has many gland cells. Another striking similarity is the presence of flame cells in the excretory system. Rhabdites have been found in several nemerteans, including *Lineus.* However, nemerteans differ from flatworms in their reproductive system. They are mostly dioecious. In the marine forms there is a ciliated **pilidium larva** (Gr. *pilidion,* a small felt nightcap) (Figure 15-23). This helmet-shaped larva has a ventral mouth but no anus—another flatworm characteristic. It also has some resemblance to the trochophore larva that is found in annelids and molluscs. Other flatworm characteristics are the presence of bilateral symmetry and a mesoderm and the lack of a coelom. All in all, the present evidence seems to indicate that the nemerteans came from an ancestral form similar in body plan to Platyhelminthes.

The nemerteans show some derived features absent from the flatworms. One of these is the eversible **proboscis** and its sheath, for which there are no counterparts among Platyhelminthes. Another difference is the presence of an **anus** in the adult, producing a **complete digestive system.** A digestive system with an anus

is more efficient because ejection of waste materials back through the mouth is not necessary. Nemerteans are also the simplest animals to have a **blood-vascular system.**

A few of the nemerteans occur in moist soil and fresh water, but by far the larger number are marine. At low tide they are often coiled under stones. It seems probable that they are active at high tide and quiescent at low tide. Some nemerteans such as *Cerebratulus* (L. *cerebrum,* brain, + *ulus,* dim. suffix) often live in empty mollusc shells. The small species live among seaweed, or they may be found swimming near the surface of the water. Nemerteans are often secured by dredging at depths of 5 to 8 m or deeper. A few are commensals or parasites. *Prostoma rubrum* (Gr. *pro,* before, in front of, + *stoma,* month) which is 20 mm or less in length, is a well-known freshwater species.

FORM AND FUNCTION

Nemerteans are slender worms and very fragile (Figure 15-22) with a great diversity in size. Longer ones are difficult to study in the laboratory. *Am-*

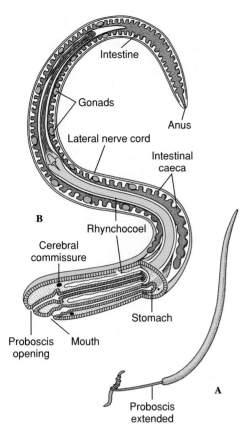

Figure 15-24

A, *Amphiporus,* with proboscis extended to catch prey. **B,** Structure of female nemertean worm *Amphiporus* (diagrammatic). Dorsal view to show proboscis.

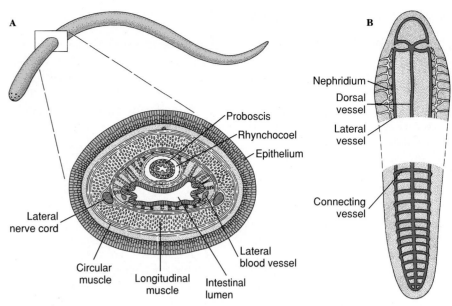

Figure 15-25

A, Diagrammatic cross section of female nemertean worm. **B,** Excretory and circulatory systems of nemertean worm. Flame bulbs along nephridial canal are closely associated with lateral blood vessels.

phiporus (Gr. *amphi,* both sides of, + *porus,* pore) (Figure 15-24), which is taken here as a representative type, is one of the smaller ones. It is from 20 to 80 mm long and about 2.5 mm wide. It is dorsoventrally flattened and has rounded ends. The body wall consists of an epidermis of ciliated columnar cells and layers of circular and longitudinal muscles (Figure 15-25A). A partly gelatinous parenchyma fills the space around the visceral organs. Ocelli are located at the anterior end. The thick-lipped mouth is anteroventral, with the opening of the proboscis just above it.

The proboscis is not connected with the digestive tract but is an eversible organ that can be protruded from its cavity, the **rhynchocoel,** and used for defense and catching prey (Figure 15-24). It lies within a sheath to which it is attached by muscles. The rhynchocoel is filled with fluid, and by muscular pressure on this fluid the an-

terior part of the tubular proboscis is everted, or turned inside out. The proboscis apparatus is an invagination of the anterior body wall, and its structure therefore duplicates that of the body wall. The retractor muscles attached at the end are used to retract the everted proboscis, much like inverting the tip of a finger of a glove by a string attached inside at its tip. The proboscis is armed with a sharp-pointed stylet. A frontal gland also opens at the anterior end by a pore.

Locomotion

Nemerteans can move with considerable speed by the combined action of their well-developed musculature and cilia. They glide mainly against a substratum; some species make use of muscular waves in crawling. Some nemerteans have the interesting method of protruding the proboscis, attaching themselves by means of the stylet, and then drawing the body up to the attached position.

Feeding and Digestion

The nemerteans are carnivorous and voracious, eating either dead or living prey. In seizing their prey they thrust out the slime-covered proboscis, which quickly ensnares the prey by

wrapping around it (Figure 15-24A). The stylet also pierces and holds the prey. Then retracting the proboscis, the nemertean draws the prey near the mouth, where it is engulfed by the esophagus that is thrust out to meet it.

The digestive system is complete and extends straight through the length of the body to the terminal anus, lying ventral to the proboscis sheath. The esophagus is straight and opens into a dilated part of the tract, the stomach. The blind anterior end of the intestine as well as the main intestine is provided with paired **lateral ceca.** The alimentary tract is lined with cilated epithelium, and in the wall of the esophagus there are glandular cells.

Digestion is largely extracellular in the intestinal tube, and when the food is ready for absorption, it passes through the cellular lining of the intestinal tract into the blood-vascular system. The indigestible material passes out the anus (Figure 15-24B), in contrast to Platyhelminthes in which it leaves by the mouth.

Circulation

The blood-vascular system is simple and enclosed with a single dorsal vessel and two lateral vessels (Figure 15-25B) connected by transverse vessels. All three longitudinal vessels join together anteriorly to form a type of collar. The blood is usually colorless, containing nucleated corpuscles. However, in some nemerteans the blood is red, green, yellow, or orange from the presence of pigments whose function is unknown. There is no heart, and the blood is propelled by the muscular walls of the blood vessels and by bodily movements.

Excretion and Respiration

The excretory system contains a pair of lateral tubes with many branches and flame cells (Figure 15-25B). Each lateral tube opens to the outside by one or more pores. Waste is picked up from the parenchymal spaces and blood by the flame cells and carried by the excretory ducts to the outside. Many of the protonephridia are so closely associated with the circulatory system that their function may be truly excretory, in contrast to their apparently osmoregulatory function in Platyhelminthes. Respiration occurs through the body surface.

Nervous System

The nervous system includes a brain composed of four fused ganglia, one pair dorsal and one pair ventral, united by commissures (connecting nerves). Five longitudinal nerves extend from the brain posteriorly—a large lateral trunk on each side of the body, paired dorsolateral trunks, and one middorsal trunk. These are connected by a network of nerve fibers. From the brain, nerves run to the proboscis, to the ocelli and other sense organs, and to the mouth and esophagus. In addition to the ocelli, there are other sense organs, such as tactile papillae, sensory pits and grooves, and probably auditory organs.

Reproduction and Development

The reproductive system in *Amphiporus* is dioecious. The gonads in either sex lie between the intestinal ceca (Figure 15-24). From each gonad a short duct (gonopore) runs to the

dorsolateral body surface. Eggs and sperm are discharged into the water, where fertilization occurs. Egg production in the females is usually accompanied by degeneration of the other visceral organs.

Nemerteans have a spiral, determinate cleavage (Figures 7-7C and 7-10). The mesoderm is derived partly from the endoderm and partly from the ectoderm. The rhyncocoel develops as a cavity in the mesoderm and is, therefore, technically a coelomic cavity, but it is not homologous to the coelom in higher forms.

A pilidium larva (Figure 15-23) develops, which bears a dorsal spike of fused cilia and a pair of lateral lobes. The entire larva is covered with cilia and has a mouth and alimentary canal but no anus. In some nemerteans the zygote develops directly without undergoing metamorphosis. The freshwater species, *Prostoma rubrum,* is hermaphroditic. A few nemerteans are viviparous.

Regeneration

Nemerteans have great powers of regeneration. At certain seasons some of them fragment by autotomy, and from each fragment a new individual develops. This is especially noteworthy in the genus *Lineus.* A fragment from the anterior region will produce a new individual more quickly than will one from the posterior part. Sometimes the proboscis is shot out with such force that it is broken off from the body. In such a case a new proboscis develops within a short time.

CLASSIFICATION OF PHYLUM NEMERTEA

Class Enopla (en´o-pla) (Gr. *enoplos,* armed). Proboscis usually armed with stylets; mouth opens in front of brain. Examples: *Amphiporus, Prostoma.*

Class Anopla (an´o-pla) (Gr. *anoplos,* unarmed). Proboscis lacks stylets; mouth opens below or posterior to brain. Examples: *Cerebratulus, Tubulanus, Lineus.*

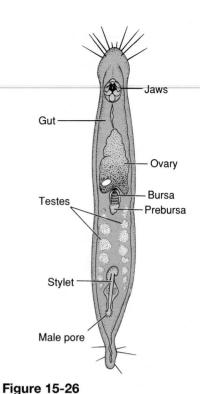

Figure 15-26

Gnathostomula jenneri (phylum Gnathostomulida) is a tiny member of the interstitial fauna between grains of sand or mud. Species in this family are among the most commonly encountered jaw worms, found in shallow water and down to depths of several hundred meters.

PHYLUM GNATHOSTOMULIDA

The first species of the Gnathostomulida (nath´o-sto-myu´lid-a) (Gr. *gnathos,* jaw, + *stoma,* mouth, + L. *ulus,* dim. suffix) was observed in 1928 in the Baltic, but its description was not published until 1956. Since then jaw worms have been found in many parts of the world, including the Atlantic coast of the United States, and over 80 species in 18 genera have been described.

Gnathostomulids are delicate wormlike animals and are 0.5 to 1 mm long (Figure 15-26). They live in the interstitial spaces of very fine sandy coastal sediments and silt and can endure conditions of very low oxygen. They often occur in large numbers and frequently in association with gastrotrichs, nematodes, ciliates, tardigrades, and other small forms.

Lacking a pseudocoel, a circulatory system, and an anus, the

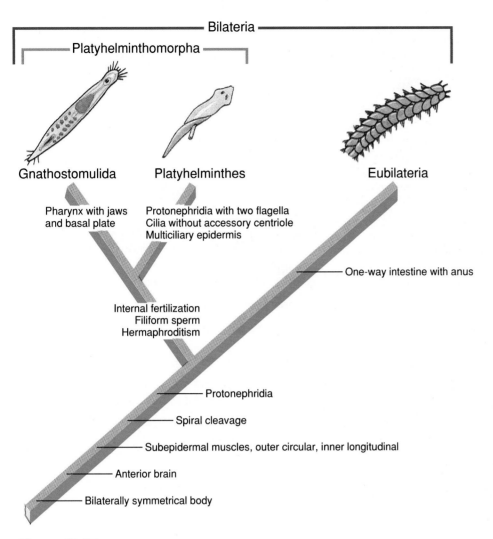

Figure 15-27

Cladogram showing relationships of acoelomates and other bilateral animals with some of the characters used to construct the hypothesis. Shared derived characters of the earliest bilateral ancestor are indicated at the base. The ancestor also had monociliary epidermal cells and protonephridia with a single flagellum, characters retained in the Gnathostomulida, while the Platyhelminthes developed multiciliated epidermal cells and multiflagellated protonephridia, first through a biflagellated stage. The specialized filiform sperm, shared by platyhelminths and gnathostomulids, is strong evidence of common ancestry of these groups.

Source: Based on P. Ax in The Origins and Relationships of Lower Invertebrates, *edited by M. S. Conway, J. D. George, R. Gibson, and H. M. Platt. Clarendon Press, Oxford, 1985.*

gnathostomulids show some similarities to the turbellarians and were at first included in that group. However, their parenchyma is poorly developed, and their pharynx is reminiscent of the rotifer mastax. The pharynx is armed with a pair of lateral jaws used to scrape fungi and bacteria off the substratum. And, although the epidermis is ciliated, each epidermal cell has but one cilium, a condition rarely found in the lower bilateral animals except in some gastrotrichs.

Gnathostomulids can glide, swim in loops and spirals, and bend the head from side to side. Sexual stages may include males, females, and hermaphrodites. Fertilization is internal.

PHYLOGENY AND ADAPTIVE RADIATION

PHYLOGENY

There can be little doubt that the bilaterally symmetrical animals were derived from a radial ancestor, perhaps one very similar to the planula larva of the cnidarians. Some investigators believe that this **planuloid ancestor** may

have given rise to one branch of descendants that were sessile or free floating and radial, which became the Cnidaria, and another branch that acquired a creeping habit and bilateral symmetry. Bilateral symmetry is a selective advantage for creeping or swimming animals because sensory structures are concentrated on the anterior (cephalization), which is the end that first encounters environmental stimuli.

According to Ax (1985), an early branch from the bilateral line would have given rise to the Platyhelminthes and Gnathostomulida (forming the Platyhelminthomorpha, Figure 15-27), which would become a sister group to all the other bilateral animals (Eubilateria). The scheme shown in Figure 15-27 would separate the Platyhelminthes and Nemertea, traditionally considered closely related, because the Nemertea have a flow-through gut. If this view is correct, the Nemertea must have branched off the Eubilateria line soon after the one-way intestine with anus was established.

Among the Platyhelminthes, it seems clear that the Turbellaria is paraphyletic (Ehlers, 1985), but we are retaining the taxon for the present because presentation based on thorough cladistic analysis would require introduction of many more taxa and characteristics beyond the scope of this book. For example, the ectolecithal turbellarians should be allied with the trematodes and cestodes in a sister group to the endolecithal turbellarians. Some of the ectolecithal turbellarians share a number of other derived characters with the trematodes and cestodes and have been placed by Brooks (1989) in a group designated Cercomeria (Gr. *kerkos,* tail, + *meros,* part) (Figure 15-28). The unique architecture of the tegument in cestodes and trematodes, designated a "neodermis" by Brooks and others, indicates with high probability that these groups share a common ancestor.

ADAPTIVE RADIATION

The flatworm body plan, with its creeping adaptation, placed a selective advantage on bilateral symmetry

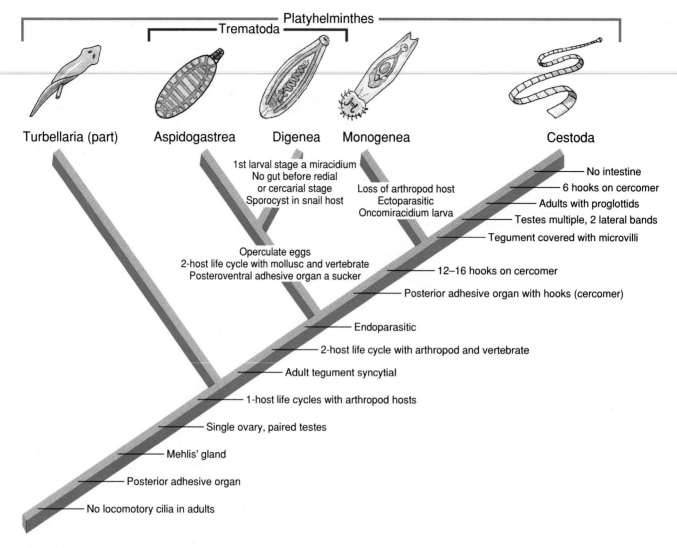

Figure 15-28

Hypothetical relationships among parasitic Platyhelminthes. The traditionally accepted class Turbellaria is paraphyletic. Some turbellarians have ectolecithal development and, together with the Trematoda, Monogenea, and Cestoda, form a monophyletic clade and a sister group of the endolecithal turbellarians. For the sake of simplicity, the synapomorphies of those turbellarians and of the Aspidogastrea, as well as many others given by Brooks (1989) are omitted. All of these organisms comprise a clade (called Cercomeria) with a posterior adhesive organ.

Source: Modified from D. R. Brooks, "The phylogeny of the Cercomeria (Platyhelminthes: Rhabdocoela) and general evolutionary principles" in Journal of Parasitology *75:606-616 (1989).*

and further development of cephalization, ventral and dorsal regions, and caudal differentiation. Because of their body shape and metabolic requirements, early flatworms must have been well preadapted for parasitism and gave rise to symbiotic lines on numerous occasions. These lines produced descendants that radiated abundantly as parasites, and many flatworms became very highly specialized for that mode of existence.

The ribbon worms have stressed the proboscis apparatus in their evolutionary diversity. Its use in capturing prey may have been secondarily evolved from its original function as a highly sensitive organ for exploring the environment. Although the ribbon worms have evolved beyond the flatworms in their complexity of organization, they have been dramatically less abundant as a group. Perhaps the proboscis was so efficient as a preda-

tor tool that there was little selective pressure to explore parasitism, or perhaps some critical preconditions were simply not present.

Likewise, the jaw worms have not radiated nor become nearly as abundant or diverse as the flatworms. However, they have exploited the marine interstitial environment, particularly zones of very low oxygen concentration.

Summary

The Platyhelminthes, the Nemertea, and the Gnathostomulida are the simplest phyla that are bilaterally symmetrical, a condition of adaptive value for actively crawling or swimming animals. They have neither a coelom nor a pseudocoel and are thus acoelomate. They are triploblastic and at the organ-system level of organization.

The body surface of turbellarian flatworms is a cellular epithelium, at least in part ciliated, containing mucous cells and rod-shaped rhabdites that function together in locomotion. Members of all other classes of flatworms are covered by a nonciliated, syncytial tegument with a vesicular distal cytoplasm and cell bodies beneath superficial muscle layers. Digestion is extracellular and intracellular in most; cestodes must absorb predigested nutrients across their tegument because they have no digestive tract. Osmoregulation is by flame-cell protonephridia, and removal of metabolic wastes and respiration occur across the body wall. Except for the primitive turbellarians, flatworms have a ladder-type nervous system with motor, sensory, and association neurons. Most flatworms are hermaphroditic, and asexual reproduction occurs in some groups.

The class Turbellaria is a paraphyletic group with mostly free living and carnivorous members. The digenetic trematodes have a mollusc intermediate host and almost always a vertebrate definitive host. The great amount of asexual reproduction that occurs in the intermediate host helps to increase the chances that some of the offspring will reach a definitive host. Aside from the tegument, digeneans share many basic structural characteristics with the Turbellaria. The Digenea includes a number of important parasites of humans and domestic animals. These contrast with the Monogenea, which are important ectoparasites of fishes and have a direct life cycle (without intermediate hosts).

Cestodes (tapeworms) generally have a scolex at their anterior end, followed by a long chain of proglottids, each of which contains a complete set of reproductive organs of both sexes. Cestodes live as adults in the digestive tract of vertebrates. They have microvilli on their tegument, which increase its surface area for absorption. The shelled larvae are passed in the feces, and the juveniles develop in a vertebrate or invertebrate intermediate host.

Members of the Nemertea have a complete digestive system with an anus and a true circulatory system. They are free living, mostly marine, and they capture their prey by ensnaring it with their long, eversible proboscis.

The Gnathostomulida are a curious phylum of little wormlike marine animals living among sand grains and silt. They have no anus, and they share certain characteristics with such widely diverse groups as turbellarians, rotifers, sponges, and cnidarians.

The flatworms and the cnidarians both probably evolved from a common ancestor (planuloid), some of whose descendants became sessile or free floating and radial, while others became creeping and bilateral.

Review Questions

1. Why is bilateral symmetry of adaptive value for actively motile animals?
2. Match the terms in the right column with the classes in the left column:

   _____ Turbellaria   a. Endoparasitic
   _____ Monogenea   b. Free-living and
   _____ Trematoda   commensal
   _____ Cestoda   c. Ectoparasitic

3. Give ten characteristics of the Platyhelminthes.
4. Distinguish two mechanisms by which flatworms supply yolk for their embryos. Which system is evolutionarily ancestral for flatworms and which one is derived?
5. Briefly describe the body plan of turbellarians.
6. What do planarians eat, and how do they digest it?
7. Briefly describe the osmoregulatory system, the nervous system, and the sense organs of planaria.
8. Contrast asexual reproduction in Turbellaria, Trematoda, and Cestoda.
9. Contrast the typical life cycle of Monogenea with that of a digenetic trematode.
10. Describe and contrast the tegument of turbellarians and the other classes of platyhelminths. Could this be evidence that the trematodes, monogeneans, and cestodes form a clade within the Platyhelminthes? Why?
11. Answer the following questions with respect to both *Clonorchis* and *Schistosoma:* (a) how do humans become infected? (b) what is the general geographical distribution? (c) what are the main disease conditions produced?
12. Why is *Taenia solium* a more dangerous infection than *Taeniarhynchus saginatus?*
13. What are two cestodes for which humans can serve as intermediate hosts?
14. Define each of the following with reference to cestodes: scolex, microvilli, proglottids, strobila.
15. Give three differences between nemerteans and platyhelminths.
16. Where do gnathostomulids live?
17. Explain how a planuloid ancestor could have given rise to both the Cnidaria and the Bilateria.
18. What is an important character of the Nemertea that might suggest that the phylum is closer to other bilateral protostomes than to the Platyhelminthes? What are some characteristics of nemerteans suggesting that they share an ancestor with platyhelminths?

Selected References

See also general references for Part III, p. 626.

Ax, P. 1985. The position of the Gnathostomulida and Platyhelminthes in the phylogenetic system of the Bilateria. In Conway Morris, S., J. D. George, R. Gibson, and H. M. Platt (eds). The origins and relationships of lower invertebrates. Oxford, Clarendon Press. *Cladistic analysis supporting Platyhelminthomorpha as a sister group of the Eubilateria.*

Brooks, D. R. 1989. The phylogeny of the Cercomeria (Platyhelminthes: Rhabdocoela) and general evolutionary principles. J. Parasitol. **75**:606–616. *Cladistic analysis of parasitic flatworms.*

Desowitz, R. S. 1981. New Guinea tapeworms and Jewish grandmothers. New York, W. W. Norton & Company. *Accounts of parasites and parasitic diseases of humans. Entertaining and instructive. Recommended for all students.*

Ehlers, U. 1985. Phylogenetic relationships within the Platyhelminthes. In Conway Morris, S., J. D. George, R. Gibson, and H. M. Platt (eds). The origins and relationships of lower invertebrates. Oxford, Clarendon Press. *Presents relationships of groups normally assigned to the paraphyletic Turbellaria.*

Schell, S. C. 1985. Handbook of trematodes of North America north of Mexico. Moscow, Idaho, University Press of Idaho. *Good for trematode identification.*

Schmidt, G. D. 1985. Handbook of tapeworm identification. Boca Raton, Florida, CRC Press. *Up-to-date manual for cestode identification.*

Smyth, J. D., and D. P. McManus. 1989. The physiology and biochemistry of cestodes. Cambridge, Cambridge University Press. *Good summary, covers much literature.*

Strickland, G. T. 1991. Hunter's tropical medicine, ed 7. Philadelphia, W. B. Saunders Company. *A valuable source of information on parasites of medical importance.*

16

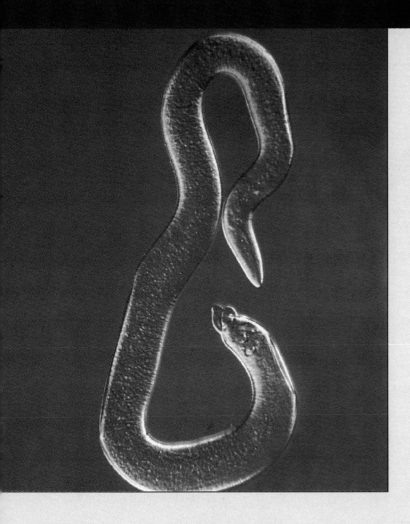

The Pseudocoelomate Animals

Phylum Rotifera
Phylum Gastrotricha
Phylum Kinorhyncha
Phylum Loricifera
Phylum Priapulida
Phylum Nematoda
Phylum Nematomorpha
Phylum Acanthocephala
Phylum Entoprocta

A World of Nematodes

Without any doubt, nematodes are the most important pseudocoelomate animals, in terms of both numbers and their impact on humans. Nematodes are abundant over most of the world, yet most people are only occasionally aware of them as parasites of humans or of their pets. We are not aware of the millions of these worms in the soil, in ocean and freshwater habitats, in plants, and in all kinds of other animals. Their dramatic abundance moved N. A. Cobb* to write in 1914:

If all the matter in the universe except the nematodes were swept away, our world would still be dimly recognizable,

and if, as disembodied spirits, we could then investigate it, we should find its mountains, hills, vales, rivers, lakes and oceans represented by a thin film of nematodes. The location of towns would be decipherable, since for every massing of human beings there would be a corresponding massing of certain nematodes. Trees would still stand in ghostly rows representing our streets and highways. The location of the various plants and animals would still be decipherable, and, had we sufficient knowledge, in many cases even their species could be determined by an examination of their erstwhile nematode parasites. ■

*From N. A. Cobb. 1914. Yearbook of the United States Department of Agriculture, p. 472.

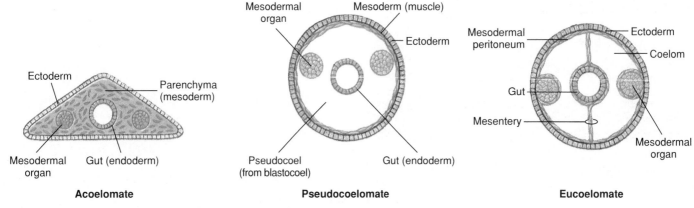

Figure 16-1
Acoelomate, pseudocoelomate, and eucoelomate body plans.

THE PSEUDOCOELOMATES

Vertebrates and more complex invertebrates have a true **coelom,** or peritoneal cavity, which is formed in the mesoderm during embryonic development and is therefore lined with a layer of mesodermal epithelium, the **peritoneum** (Figure 16-1). The pseudocoelomate phyla have a pseudocoel rather than a true coelom. It is derived from the embryonic blastocoel rather than from a secondary cavity within the mesoderm. It is a space between the gut and the mesodermal and ectodermal components of the body wall, and it is not lined with peritoneum.

Nine distinct groups of animals belong to the pseudocoelomate category: Rotifera, Gastrotricha, Kinorhyncha, Nematoda, Nematomorpha, Loricifera, Priapulida, Acanthocephala, and Entoprocta. Since the first five of these groups have certain similarities, some authorities place them as classes in a phylum called Aschelminthes (as´kel-min´theez) (Gr. *askos,* bladder, + *helmins,* worm). However, they differ so much that any phylogenetic relationship is highly debatable, and other authorities consider them separate phyla. Some group the five loosely as individual phyla under a superphylum Aschelminthes. The Entoprocta have sometimes been grouped with the Ectoprocta, together called the Bryozoa (moss animals). However, because the ectoprocts have a true coelom, they are usually considered a separate phylum, and the term "bryozoans" is currently taken to exclude the entoprocts.

However one classifies them, the pseudocoelomates are a heterogeneous assemblage of animals. Most of them are small; some are microscopic; some are fairly large. Some, such as the nematodes, are found in freshwater, marine, terrestrial, and parasitic habitats; others, such as the Acanthocephala, are strictly parasitic. Some have unique characteristics, such as the lacunar system of the acanthocephalans or the ciliated corona of the rotifers.

Even in such a diversified grouping, some characteristics are shared. All have a body wall of epidermis (often syncytial), a dermis, and muscles surrounding the pseudocoel. The digestive tract is complete (except in Acanthocephala), and it, along with the gonads and excretory organs, is within the pseudocoel and bathed in perivisceral fluid. The epidermis in many secretes a nonliving cuticle with some specializations such as bristles or spines.

CHARACTERISTICS OF THE PSEUDOCOELOMATE PHYLA

1. Symmetry bilateral; unsegmented; triploblastic (three germ layers)
2. Body cavity a **pseudocoel**
3. Size mostly small; some microscopic; a few a meter or more in length
4. Body vermiform; body wall a **syncytial** or cellular epidermis with thickened cuticle, sometimes molted; muscular layers mostly of **longitudinal fibers;** cilia absent in several phyla
5. Digestive system (lacking in acanthocephalans) complete with mouth, enteron, and anus; pharynx muscular and well developed: **tube-within-a-tube arrangement;** digestive tract usually only an epithelial tube with **no definite muscle layer**
6. Circulatory and respiratory organs lacking
7. Excretory system of canals and protonephridia in some; cloaca that receives excretory, reproductive, and digestive products may be present
8. Nervous system of cerebral ganglia or of a circumenteric nerve ring connected to anterior and posterior nerves; sense organs of ciliated pits, papillae, bristles, and some eyespots
9. Reproductive system of gonads and ducts that may be single or double; sexes nearly always separate, with the male usually smaller than the female; eggs microscopic with shell often containing chitin
10. Development may be direct or within a complicated life history; cleavage mostly mosaic; **constancy in number of cells or nuclei common**

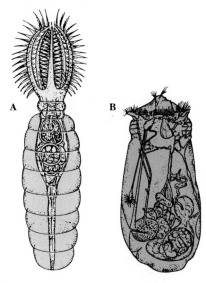

Stephanoceros fimbriatus   Asplanchna priodonta

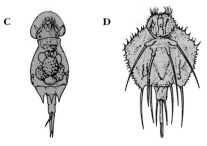

Squatinella rostrum   Macrochaetus longipes

Figure 16-2

Variety of form in rotifers. **A,** *Stephanoceros* has five long, fingerlike coronal lobes with whorls of short bristles. It catches its prey by closing its funnel when food organisms swim into it, and the bristly lobes prevent the prey from escaping. **B,** *Asplanchna* is a pelagic, predatory genus with no foot. **C,** *Squatinella* has a semicircular nonretractable, transparent hoodlike extension covering the head. **D,** *Machrochaetus* is dorsoventrally flattened.

A constant number of cells or nuclei in individuals of a species, or in parts of their bodies, is known as **eutely,** which is common to several of the groups. In most of them there is an emphasis on the longitudinal muscle layer.

PHYLUM ROTIFERA

Rotifera (ro-tif´e-ra) (L. *rota,* wheel, + *fera,* those that bear) derive their name from the characteristic ciliated crown, or **corona,** that, when beating, often gives the impression of rotating wheels. Rotifers range from 40 μm to 3 mm in length, but most are between 100 and 500 μm long. Some have beautiful colors, although most are transparent, and some have bizarre shapes (Figure 16-2). Their shapes are often correlated with their mode of life. The floaters are usually globular and saclike; the creepers and swimmers are somewhat elongated and wormlike; and the sessile types are commonly vaselike, with a thickened outer epidermis (lorica). Some are colonial. One of the best-known genera is *Philodina* (Gr. *philos,* fond of, + *dinos,* whirling) (Figure 16-3), which is often used for study.

Rotifers are a cosmopolitan group of about 1800 species, some of which are found throughout the world. Most of the species are freshwater inhabitants, a few are marine, some are terrestrial, and some are epizoic (live on the body of another animal) or parasitic.

Rotifers are adapted to many kinds of ecological conditions. Most species are benthic, living on the bottom or in vegetation of ponds or along the shores of freshwater lakes where they swim or creep about on the vegetation. A large proportion of the species that live in the water film between sand grains of beaches (meiofauna) are rotifers. Pelagic forms (Figure 16-2B) are common in the surface waters of freshwater lakes and ponds, and they may exhibit cyclomorphosis, that is, variations in body form resulting from seasonal or nutritional changes.

Many species of rotifers can endure long periods of desiccation, during which they resemble grains of sand. While in a desiccated condition, rotifers are very tolerant of temperature variations, especially those rotifers that dwell in mosses. True encystment occurs in only a few rotifers. On addition of water, desiccated rotifers resume their activity.

Strictly marine species are rather few in number. Some of the littoral (intertidal) species of the sea may be freshwater ones that are able to adapt to seawater.

FORM AND FUNCTION

External Features

The body of the rotifer is composed of a head bearing a ciliated corona, a trunk, and a posterior tail, or foot. It is covered with a cuticle and is nonciliated except for the corona.

The ciliated corona, or crown, surrounds a nonciliated central area of the head, which may bear sensory bristles or papillae. The appearance of the head end depends on which of the several types of corona it has— usually a circlet of some sort, or a pair of trochal (coronal) discs (the

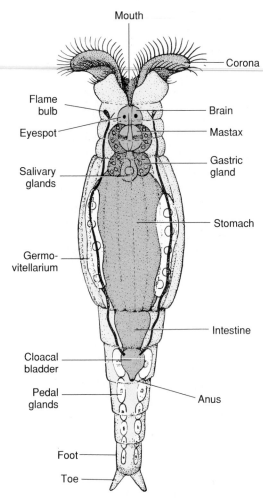

Figure 16-3

Structure of *Philodina* rotifer.

term *trochal* comes from a Greek word meaning wheel). The cilia on the corona beat in succession, giving the appearance of a revolving wheel or pair of wheels. The **mouth** is located in the corona on the midventral side. The coronal cilia are used in both locomotion and feeding.

The trunk may be elongated, as in *Philodina* (Figure 16-3), or saccular in shape (see Figure 16-2). It contains the visceral organs and often bears sensory antennae. The body wall of many species is superficially ringed so as to simulate segmentation. Though some rotifers have a true, secreted cuticle, all have a fibrous layer within their epidermis. The fibrous layer in some is quite thick and forms a case-like **lorica,** which is often arranged in plates or rings.

The **foot** is narrower and usually bears one to four **toes.** Its cuticle may

be ringed so that it is telescopically retractile. It is tapered gradually in some forms (Figure 16-3) and sharply set off in others (see Figure 16-2). The foot is an attachment organ and contains **pedal glands** that secrete an adhesive material used by both sessile and creeping forms. In swimming pelagic forms, the foot is usually reduced. Rotifers move by creeping with leechlike movements aided by the foot, or by swimming with the coronal cilia, or both.

Internal Features

Underneath the cuticle is the **syncytial epidermis,** which secretes the cuticle, and bands of **subepidermal muscles,** some circular, some longitudinal, and some running through the pseudocoel to the visceral organs. The **pseudocoel** is large, occupying the space between the body wall and the viscera. It is filled with fluid, some of the muscle bands, and a network of mesenchymal ameboid cells.

The digestive system is complete. Some rotifers feed by sweeping minute organic particles or algae toward the mouth by the beating of the coronal cilia. The cilia are able to sort out and dispose of the larger unsuitable particles. The pharynx **(mastax)** is fitted with a muscular portion that is equipped with hard jaws **(trophi)** for sucking in and grinding up the food particles. The constantly chewing pharynx is often a distinguishing feature of these tiny animals. Carnivorous species feed on protozoa and small metazoans, which they capture by trapping or grasping. The trappers have a funnel-shaped area around the mouth. When small prey swim into the funnel, the lobes fold inward to capture and hold them until they are drawn into the mouth and pharynx. The hunters have trophi that can be projected and used like forceps to seize the prey, bring it back into the pharynx, and then pierce it or break it up so that the edible parts can be sucked out and the rest discarded. The **salivary** and **gastric glands** are believed to secrete enzymes for extracellular digestion. Absorption occurs in the stomach.

The excretory system typically consists of a pair of **protonephridial tubules,** each with several **flame cells,** that empty into a common bladder. The bladder, by pulsating, empties into the **cloaca**—into which the intestine and oviducts also empty. The fairly rapid rate of pulsation of the protonephridia—one to four times per minute—would indicate that the protonephridia are important osmoregulatory organs. The water apparently enters through the mouth rather than across the epidermis; even marine species empty their bladder at frequent intervals.

The nervous system consists of a bilobed **brain,** dorsal to the mastax, that sends paired nerves to the sense organs, mastax, muscles, and viscera. Sensory organs include paired **eyespots** (in some species such as *Philodina*), sensory bristles and papillae, and ciliated pits and dorsal antennae.

Reproduction

Rotifers are dioecious, and the males are usually smaller than the females. In the class Bdelloidea males are entirely unknown, and in Monogononta they seem to occur only for a few weeks of the year.

The female reproductive system in the Bdelloidea and Monogononta consists of combined ovaries and yolk glands **(germovitellaria)** and oviducts that open into the cloaca. Yolk is supplied to the developing ova by way of flow-through cytoplasmic bridges, rather than as separate yolk cells as in many Platythelminthes.

Mictic (Gr., *miktos,* mixed, blended) refers to the capacity of the haploid eggs to be fertilized (that is, "mixed") with the male's sperm nucleus to form a diploid embryo. Amictic ("without mixing") eggs are already diploid and can only develop parthenogenetically.

In the Bdelloidea (*Philodina,* for example), all females are parthenogenetic and produce diploid eggs that hatch into diploid females. These reach

Figure labels

Mouth
Corona
Flame bulb
Brain
Eyespot
Mastax
Salivary glands
Gastric gland
Germo-vitellarium
Stomach
Cloacal bladder
Intestine
Pedal glands
Anus
Foot
Toe

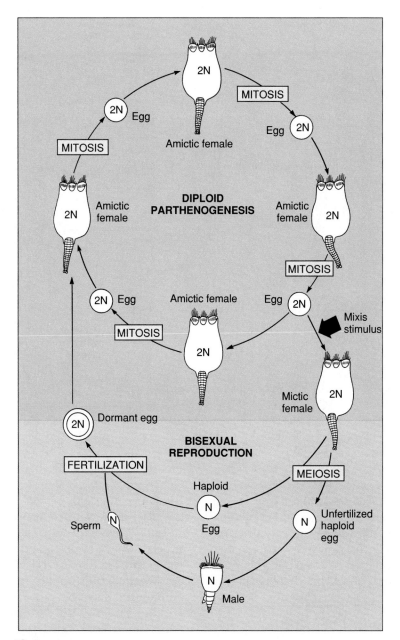

Figure 16-4
The reproduction of some rotifers (class Monogononta) is parthenogenetic during the part of the year when environmental conditions are suitable. In response to certain stimuli, the females begin to produce haploid (N) eggs. If the haploid eggs are not fertilized, they hatch into haploid males. The males provide sperm to fertilize other haploid eggs, which then develop into diploid (2N), dormant eggs that can resist the rigors of winter. When suitable conditions return, the dormant eggs resume development, and a female hatches.

CLASSIFICATION OF PHYLUM ROTIFERA

Class Seisonidea (sy´son-id´e-a) (Gr. *seison,* earthen vessel, + *eidos,* form). Marine; elongated form; corona vestigial; sexes similar in size and form; females with pair of ovaries and no vitellaria; single genus (*Seison*) with two species; epizoic on gills of a crustacean (*Nebalia*).

Class Bdelloidea (del-oid´e-a) (Gr. *bdella,* leech, + *eidos,* form). Swimming or creeping forms; anterior end retractile; corona usually with pair of trochal discs; males unknown; parthenogenetic; two germovitellaria. Examples: *Philodina* (Figure 16-3), *Rotaria*.

Class Monogononta (mon´o-go-non´ta) (Gr. *monos,* one, + *gonos,* primary sex gland). Swimming or sessile forms; single germo-vitellarium; males reduced in size; eggs of three types (amictic, mictic, dormant). Examples: *Asplanchna* (Figure 16-2B), *Epiphanes*.

maturity in a few days. In the class Seisonidea the females produce haploid eggs that must be fertilized and that develop into either males or females. In the Monogononta, however, females produce two kinds of eggs (Figure 16-4). During most of the year diploid females produce thin-shelled, **diploid amictic eggs.** These develop parthenogenetically into diploid amictic females. However, such rotifers often live in temporary ponds or streams and are cyclic in their reproductive patterns. Any one of several environmental factors—for example, crowding, diet, or photoperiod (according to species)—may induce the amictic eggs to develop into diploid mictic females that will produce thin-shelled **haploid mictic eggs.** If these eggs are not fertilized, they develop into haploid males. But if fertilized, the eggs develop a thick, resistant shell and become dormant. They survive over winter ("winter eggs") or until environmental conditions are again suitable, at which time they hatch into diploid females. Dormant eggs are often dispersed by winds or birds, which may account for the peculiar distribution patterns of rotifers.

The male reproductive system includes a single testis and a ciliated sperm duct that runs to a genital pore (males usually lack a cloaca). The end of the sperm duct is specialized as a copulatory organ. Copulation is usually by hypodermic impregnation; that is, the penis can penetrate any part of the female body wall and inject the sperm directly into the pseudocoel.

Females hatch with adult features, needing only a few days' growth to reach maturity. Males often do not grow and are sexually mature at hatching.

Nuclear Constancy

Most structures in rotifers are syncytial, but the nuclei in the various organs are said to show a remarkable constancy in numbers in any given

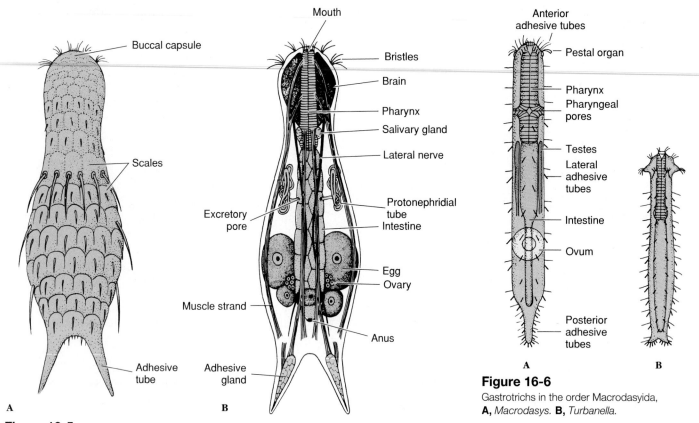

Figure 16-5

Chaetonotus, a gastrotrich. **A,** Dorsal surface. **B,** Internal structure, ventral view.

Figure 16-6

Gastrotrichs in the order Macrodasyida, **A,** *Macrodasys.* **B,** *Turbanella.*

species (eutely). For example, E. Martini (1912) reported that in one species of rotifer he always found 183 nuclei in the brain, 39 in the stomach, 172 in the corona epithelium, and so on. Organisms with eutely show a highly precise genetic control of nuclear division and differentiation. The nuclei are programmed to differentiate and divide an exact number of times, then halt when the appointed number is reached.

Phylum Gastrotricha

Gastrotricha (gas-tro-tri´ka) (N. L. fr. Gr., *gaster, gastros,* stomach or belly, + *thrix, trichos,* hair) includes small, ventrally flattened animals about 65 to 500 μm long, somewhat like rotifers but lacking the corona and mastax and having a characteristically bristly or scaly body. They are usually found gliding on the bottom, or on an aquatic plant or animal substrate, by means of their ventral cilia, or they compose part of the meiofauna in the interstitial spaces between bottom particles.

Gastrotrichs are found in both fresh and salt water. The 400 or so species are about equally divided between the two media. Many species are cosmopolitan, but only a few occur in both fresh water and the sea. Much is yet to be learned about their distribution.

Form and Function

The gastrotrich (Figures 16-5 and 16-6) is usually elongated, with a convex dorsal surface bearing a pattern of bristles, spines, or scales, and a flattened ciliated ventral surface. Cells on the ventral surface may be monociliated or multiciliated. The head is often lobed and ciliated, and the tail end may be forked.

A syncytial epidermis is found beneath the cuticle. Longitudinal muscles are better developed than are circular ones, and in most cases they are unstriated. Adhesive tubes secrete a substance for attachment. A dual-gland system for attachment and release is present, similar to that described for the Turbellaria (p. 282). The pseudocoel is somewhat reduced and contains no amebocytes.

The digestive system is complete and is made up of a mouth, a muscular pharynx, a stomach-intestine, and an anus (Figure 16-5B). The food is largely algae, protozoa, and detritus, which are directed to the mouth by the head cilia. Digestion appears to be extracellular. The protonephridia are equipped with **solenocytes** rather than flame cells. Solenocytes have a single flagellum enclosed in a cylinder of cytoplasmic rods.

The nervous system includes a brain near the pharynx and a pair of lateral nerve trunks. Sensory structures are similar to those in rotifers, except that the eyespots are generally lacking. Sensory bristles, often concentrated on the head, are modified from cilia.

Gastrotrichs are hermaphroditic, although the male system of some is so rudimentary that they are functionally parthenogenetic females. Like the rotifers, some gastrotrichs produce thin-walled, rapidly developing eggs and thick-shelled, dormant eggs. The thick-shelled eggs can withstand harsh environmental conditions and may survive dormancy for some years. Development is direct, and the juveniles have the same form as adults.

PHYLUM KINORHYNCHA

Kinorhyncha (kin´o-ring´ka) (Gr. *kinein,* to move, + *rhynchos,* beak) are marine worms a little larger than rotifers and gastrotrichs but usually not more than 1 mm long. The phylum has also been called Echinodera, meaning spiny necked. About 75 species have been described.

Kinorhynchs are cosmopolitan, living from pole to pole, from intertidal areas to 6000 m in depth. Most live in mud or sandy mud, but some have been found in algal holdfasts, sponges, or other invertebrates. They feed mainly on diatoms. About 100 species have been reported. Among the best-known genera of the Kinorhyncha are *Echinoderes, Pycnophyes,* and *Kinorhynchus.*

FORM AND FUNCTION

The body of the kinorhynch is divided into 13 segments, which bear spines but have no cilia (Figure 16-7). The retractile head has a circlet of spines with a small retractile proboscis. The body is flat ventrally and arched dorsally. The body wall is made up of a cuticle, a syncytial epidermis, and longitudinal epidermal cords, much like those of nematodes. The arrangement of the muscles is correlated with the segments, and circular, longitudinal, and diagonal muscle bands are all represented.

A kinorhynch cannot swim. In the silt and mud where it commonly lives, it burrows by extending the head into the mud and anchoring it with spines. It then draws the body forward until the head is retracted into the body. When disturbed, the kinorhynch draws in the head and protects it with a closing apparatus of cuticular plates (Figure 16-7).

The digestive system is complete, with a mouth at the tip of the proboscis, a pharynx, an esophagus, a stomach-intestine, and an anus.

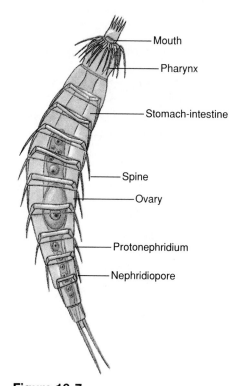

Figure 16-7

Echinoderes, a kinorhynch, is a minute marine worm. Segmentation is superficial. Head with its circle of spines is retractile.

Kinorhynchs feed on diatoms or on organic material in the mud where they burrow.

The pseudocoel is filled with amebocytes containing fluid. The excretory system is made up of a multinucleated solenocyte protonephridium on each side of the tenth and eleventh segments. Each solenocyte has one long and one short flagellum.

The nervous system is in contact with the epidermis, with a multilobed brain encircling the pharynx, and with a ventral ganglionated nerve cord extending throughout the body. Sense organs are represented by eyespots in some and by the sensory bristles.

Sexes are separate, with paired gonads and gonoducts. There is a series of about six juvenile stages and a definitive, nonmolting adult.

PHYLUM LORICIFERA

The Loricifera (L., *lorica,* corselet, + Gr., *phora,* bearing) are a very recently described phylum of animals (1983). The tiny animals (0.25 mm

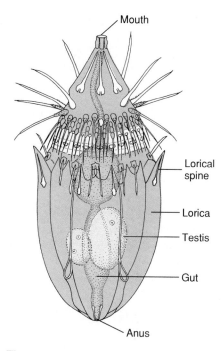

Figure 16-8

Dorsal view of adult loriciferan, *Nanoloricus mysticus.*

long) live in the spaces between the grains of marine gravel, to which they cling tightly. Though they were described from specimens collected off the coast of France, they are apparently widely distributed in the world.

FORM AND FUNCTION

Loriciferans have oral styles and scalids rather similar to those of the kinorhynchs, and the entire forepart of the body can be retracted into the circular lorica (Figure 16-8). The nature of their diet is unknown. Their brain fills most of the head, and the scalids are innervated by nerves from the brain and other ganglia. The sexes are separate, but details of reproduction are unknown. Juveniles resemble adults in several respects but have a pair of tapering toes that are believed to function in locomotion.

PHYLUM PRIAPULIDA

The Priapulida (pri´a-pyu´li-da) (Gr. *priapos,* phallus, + *ida,* pl. suffix) are a small group (only 15 species) of marine worms found chiefly in the colder waters of both hemispheres. They have been reported along the Atlantic coast from Massachusetts to Greenland and along the Pacific coast from California to Alaska. They live in the mud and sand of the sea floor and range from intertidal zones to depths of several thousand meters. *Tubiluchus* (L. *tubulus,* dim. of *tubus,* waterpipe) is a minute detritus feeder adapted to interstitial life in warm coralline sediments. *Maccabeus* (named for a Judean patriot who died in 160 B.C.) is a tiny tube-dweller discovered in muddy Mediterranean bottoms.

FORM AND FUNCTION

Priapulids have cylindrical bodies, rarely more than 12 to 15 cm long. Most of them are burrowing predaceous animals that usually orient themselves upright in the mud with the mouth at the surface. They are adapted for burrowing by body contractions.

The body includes a proboscis, trunk, and usually one or two caudal appendages (Figure 16-9). The eversible proboscis is ornamented with papillae and ends with rows of curved spines that surround the mouth. The proboscis is used in sampling the surroundings as well as for the capture of small, soft-bodied prey. *Maccabeus* has a crown of brachial tentacles around the mouth.

The trunk is not metameric but is superficially divided into 30 to 100 rings and is covered with tubercles and spines. The tubercles are probably sensory in function. The anus and urogenital pores are located at the posterior end of the trunk. The caudal appendages are hollow stems believed to be respiratory and probably chemoreceptive in function. A chitinous cuticle, molted periodically throughout life, covers the body.

The digestive system contains a muscular pharynx and a straight intestine and rectum (Figure 16-9). There is a nerve ring around the pharynx and a midventral nerve cord. The body cavity contains amebocytes and,

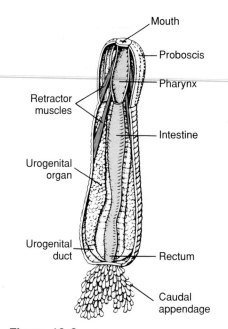

Figure 16-9
Major internal structures of *Priapulus.*

at least in *Priapulus caudatus,* corpuscles containing a respiratory pigment called hemerythrin.

Sexes are separate. The paired urogenital organs are each made up of a gonad and clusters of solenocytes, both connected to a protonephridial tubule that carries both gametes and excretory products to the outside. The embryology is poorly known. In some the egg undergoes radial cleavage and develops into a stereogastrula. The larvae of *Priapulus* dig into the mud and become detritus feeders.

Long considered pseudocoelomate, priapulids were erroneously judged coelomate when nuclei were found in the membranes lining the body cavity, the membranes thus interpreted as a peritoneum. However, electron microscopy showed that the nuclei of their muscle cells were peripheral, and the muscles secreted an extracellular membrane. The muscle nuclei and extracellular membrane gave the false appearance of an epithelial lining.

PHYLUM NEMATODA: ROUNDWORMS

Approximately 12,000 species of Nematoda (nem-a-to´da) (Gr., *nematos,* thread) have been named, but it has

been estimated that if all species were known, the number would be nearer 500,000. They live in the sea, in fresh water, and in the soil, from polar regions to the tropics, and from mountaintops to the depths of the sea. Good topsoil may contain billions of nematodes per acre. Nematodes also parasitize virtually every type of animal and many plants. The effects of nematode infestation on crops, domestic animals, and humans make this phylum one of the most important of all parasitic animal groups.

In 1963 Sydney Brenner started studying a free-living nematode, *Caenorhabditis elegans,* the beginning of some extremely fruitful research. Now this small worm has become one of the most important experimental models in biology. The origin and lineage of all the cells in its body (959) have been traced from the zygote to the adult, and the complete "wiring diagram" of its nervous system is known—all the neurons and all connections between them. The genome has been almost completely mapped, and scientists are working toward sequencing the entire genome of 3 million bases. Many basic discoveries have been made and will be made using *C. elegans.*

Free-living nematodes feed on bacteria, yeasts, fungal hyphae, and algae. They may be saprozoic or coprozoic (live in fecal material). Predatory species may eat rotifers, tardigrades, small annelids, and other nematodes. Many species feed on plant juices from higher plants, which they penetrate, sometimes causing agricultural damage of great proportions. Nematodes themselves may be prey for mites, insect larvae, and even nematode-capturing fungi. One of the free-living nematodes, *Caenorhabditis elegans,* is easy to culture in the laboratory and has become an invaluable model for studies of basic developmental biology.

Virtually every species of vertebrate and many invertebrates serve as hosts for one or more types of parasitic nematodes. Nematode parasites in humans cause much discomfort, disease, and death, and in domestic animals they are the source of great economic loss.

FORM AND FUNCTION

Distinguishing characteristics of this large group of animals are their cylindrical shape; their flexible, nonliving cuticle; their lack of motile cilia or flagella (except in one species); the muscles of their body wall, which have several unusual features, such as the fact that the muscles run in a longitudinal direction only, and eutely. Correlated with their lack of cilia, nematodes do not have protonephridia; their excretory system consists of one or more large gland cells opening by an excretory pore, or a canal system without gland cells, or both cells and canals together. Their pharynx is characteristically muscular with a triradiate lumen and resembles the pharynx of the gastrotrichs and of the kinorhynchs. Use of the pseudocoel as a hydrostatic organ is highly developed in the nematodes, and much of the functional morphology of the nematodes can be best understood in the context of the high **hydrostatic pressure** (turgor) in the pseudocoel.

Most nematode worms are less than 5 cm long, and many are microscopic, but some parasitic nematodes are more than 1 m in length.

The outer body covering is a relatively thick, noncellular **cuticle** secreted by the underlying epidermis **(hypodermis).** The hypodermis is syncytial, and its nuclei are located in four **hypodermal cords** that project inward (Figure 16-10). The dorsal and ventral hypodermal cords bear the longitudinal dorsal and ventral nerves, and the lateral cords bear excretory canals. The cuticle is of great functional importance to the worm, serving to contain the high hydrostatic pressure exerted by the fluid in the pseudocoel. The several layers of the cuticle are primarily of **collagen,** a structural protein also abundant in vertebrate connective tissue. Three of the layers are composed of crisscrossing fibers, which confer some longitudinal elasticity on the worm but severely limit its capacity for lateral expansion.

The body wall muscles of the nematodes are very unusual. They lie beneath the hypodermis and contract longitudinally only. There are no circular muscles in the body wall. The muscles are arranged in four bands, or quadrants, marked off by the four hypodermal cords (Figure 16-10). Each muscle cell has a contractile

fibrillar portion (or **spindle**) and a noncontractile **sarcoplasmic** portion (cell body). The spindle is distal and abuts the hypodermis, and the cell body projects into the pseudocoel. The spindle is striated with bands of actin and myosin, reminiscent of vertebrate skeletal muscle (see Figure 10-7, p. 190, and p. 642). The cell bodies contain the nuclei and are a major depot for glycogen storage in the worm. From each cell body a process or **muscle arm** extends either to the ventral or the dorsal nerve. Though not unique to nematodes, this arrangement is very curious; in most animals nerve processes (axons, p. 715) extend to the muscle, rather than the other way around.

The fluid-filled pseudocoel, in which the internal organs lie, constitutes a hydrostatic skeleton. Hydrostatic skeletons, found in many invertebrates, lend support by transmitting the force of muscle contraction to the enclosed, noncompressible fluid. Normally, muscles are arranged antagonistically, so that movement is effected in one direction by contraction of one group of muscles, and movement back in the opposite direction is effected by the antagonistic set of muscles. However, nematodes do not have circular body wall muscles to antagonize the longitudinal muscles; therefore the

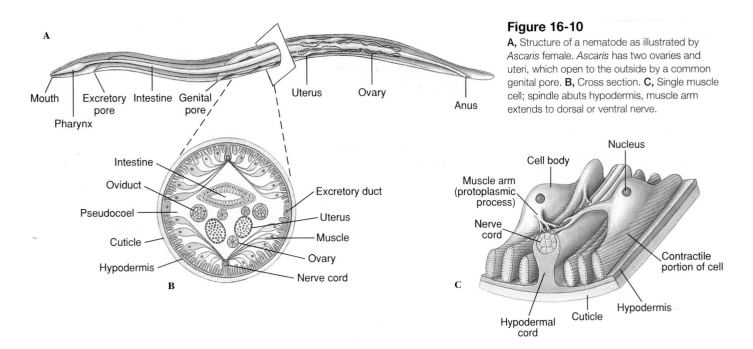

A, Structure of a nematode as illustrated by *Ascaris* female. *Ascaris* has two ovaries and uteri, which open to the outside by a common genital pore. **B,** Cross section. **C,** Single muscle cell; spindle abuts hypodermis, muscle arm extends to dorsal or ventral nerve.

Figure 16-10

A

Mouth   Pharynx   Excretory pore   Intestine   Genital pore   Uterus   Ovary   Anus

B

Intestine
Oviduct
Pseudocoel
Cuticle
Hypodermis

Excretory duct
Uterus
Muscle
Ovary
Nerve cord

C

Muscle arm (protoplasmic process)
Cell body
Nucleus
Nerve cord
Contractile portion of cell
Hypodermis
Cuticle
Hypodermal cord

cuticle must serve that function. As the muscles on one side of the body contract, they compress the cuticle on that side, and the force of the contraction is transmitted (by the fluid in the pseudocoel) to the other side of the nematode, stretching the cuticle on that side. This compression and stretching of the cuticle serve to antagonize the muscle and are the forces that return the body to resting position when the muscles relax; this produces the characteristic thrashing motion seen in nematode movement. An increase in efficiency of this system can be achieved only by an increase in hydrostatic pressure. Consequently, the hydrostatic pressure in the nematode pseudocoel is much higher than is usually found in other kinds of animals that have hydrostatic skeletons but that also have antagonistic muscle groups.

The alimentary canal of the nematode consists of a mouth (Figure 16-10), a muscular pharynx, a long nonmuscular intestine, a short rectum, and a terminal anus. Food is sucked into the pharynx when the muscles in its anterior portion contract rapidly and open the lumen. Relaxation of the muscles anterior to the food mass closes the lumen of the pharynx, forcing the food posteriorly toward the intestine. The intestine is one cell layer thick. Food matter is moved posteriorly by the body movements and by additional food being passed into the intestine from the pharynx. Defecation is accomplished by muscles that simply pull the anus open, and the expulsive force is provided by the high pseudocoelomic pressure that surrounds the gut.

The adults of many parasitic nematodes have an anaerobic energy metabolism; thus, a Krebs cycle and cytochrome system characteristic of aerobic metabolism are absent. They derive energy through glycolysis and probably through some incompletely known electron transport sequences. Interestingly, some free-living nematodes and free-living stages of parasitic nematodes are obligate aerobes and have a Krebs cycle and cytochrome system.

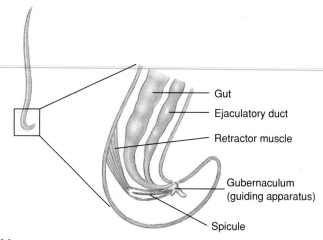

Figure 16-11
Posterior end of a male nematode.

A **ring of nerve tissue and ganglia** around the pharynx give rise to small nerves to the anterior end and to two **nerve cords,** one dorsal and one ventral. **Sensory papillae** are concentrated around the head and tail. The **amphids** are a pair of somewhat more complex sensory organs that open on each side of the head at about the same level as the cephalic circle of papillae. The amphidial opening leads into a deep cuticular pit with sensory endings of modified cilia. The amphids are usually reduced in nematode parasites of animals, but most parasitic nematodes bear a bilateral pair of **phasmids** near the posterior end. They are rather similar in structure to the amphids.

Most nematodes are dioecious. The male is smaller than the female, and its posterior end usually bears a pair of **copulatory spicules** (Figure 16-11). Fertilization is internal, and eggs are usually stored in the uterus until deposition. After embryonation a juvenile worm hatches from the egg. The four juvenile stages are separated by a molt, or shedding, of the cuticle. Many parasitic nematodes have free-living juvenile stages. Others require an intermediate host to complete their life cycles.

SOME NEMATODE PARASITES

As mentioned previously, nearly all vertebrates and many invertebrates

The copulatory spicules of male nematodes are not true intromittent organs, since they do not conduct the sperm, but are another adaptation to cope with the high internal hydrostatic pressure. The spicules must hold the vulva of the female open while the ejaculatory muscles overcome the hydrostatic pressure in the female and rapidly inject sperm into her reproductive tract. Furthermore, nematode spermatozoa are unique among those studied in the animal kingdom in that they lack a flagellum and acrosome. Within the female reproductive tract, the sperm become ameboid and move by pseudopods. Could this be another adaptation to the high hydrostatic pressure in the pseudocoel?

are parasitized by nematodes. A number of these are very important pathogens of humans and domestic animals. A few nematodes are common in humans in North America (Table 16-1), but they and many others usually abound in tropical countries. Space permits mention of only a few in this discussion.

Ascaris Lumbricoides: The Large Roundworm of Humans

Because of its size and availability, *Ascaris* (Gr., *askaris,* intestinal worm) is usually selected as a type for study in zoology, as well as in experimental work. Thus it is proba-

Table 16-1	Common Parasitic Nematodes of Humans in North America	
Common and Scientific Names	**Mode of Infection; Prevalence**	
Hookworm (*Ancylostoma duodenale* and *Necator americanus*)	Contact with juveniles in soil that burrow into skin; common in southern states	
Pinworm (*Enterobius vermicularis*)	Inhalation of dust with ova and by contamination with fingers; most common worm parasite in United States	
Intestinal roundworm (*Ascaris lumbricoides*)	Ingestion of embryonated ova in contaminated food; common in rural areas of Appalachia and southeastern states	
Trichina worm (*Trichinella spiralis*)	Ingestion of infected muscle; occasional in humans throughout North America	
Whipworm (*Trichuris trichiura*)	Ingestion of contaminated food or by unhygienic habits; usually common wherever *Ascaris* is found	

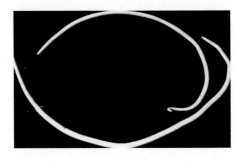

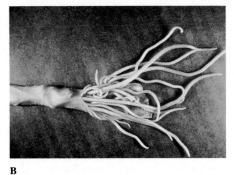

A   B

Figure 16-12

A, Intestinal roundworm *Ascaris lumbricoides,* male and female. Male, *top,* is smaller and has characteristic sharp kink in the end of the tail. The females of this large nematode may be over 30 cm long. **B,** Intestine of a pig, nearly completely blocked by *Ascaris suum.* Such heavy infections are also fairly common with *A. lumbricoides* in humans.

ble that parasitologists know more about the structure, physiology, and biochemistry of *Ascaris* than of any other nematode. This genus includes several species. One of the most common, *A. megalocephala,* is found in the intestine of horses. *A. lumbricoides* (Figure 16-12) is one of the most common parasites found in humans; recent surveys have shown a prevalence of up to 64% in some areas of the southeastern United States, and over a *billion* people are infected worldwide. The large roundworm of pigs, *A. suum,* is morphologically close to *A. lumbricoides,* and they were long considered the same species.

A female *Ascaris* may lay 200,000 eggs a day, passing out in the host's feces. Given suitable soil conditions, embryonation is complete within 2 weeks. Direct sunlight and high temperatures are rapidly lethal, but the eggs have an amazing tolerance to other adverse conditions, such as desiccation or lack of oxygen. The shelled juveniles can remain viable for many months or even years in the soil. Infection usually occurs when the eggs are ingested with uncooked vegetables or when children put soiled fingers or toys in their mouths. Unsanitary defecation habits "seed" the soil, and viable eggs remain long after all signs of the fecal matter have disappeared.

Other ascarids are common in wild and domestic animals. Species of *Toxocara,* for example, are found in dogs and cats. Their life cycle is generally similar to that of *Ascaris,* but the juveniles often do not complete their tissue migration in adult dogs, remaining in the host's body in a stage of arrested development. Pregnancy in the female, however, stimulates the juveniles to wander, and they infect the embryos in the uterus. The puppies are then born with worms. These ascarids also survive in humans but do not complete their development, leading to an occasionally serious condition in children known as *visceral larva migrans.* This is a good argument for pet owners to practice hygienic disposal of canine wastes!

When a host swallows embryonated eggs, the tiny juveniles hatch. They burrow through the intestinal wall into the veins or lymph vessels and are carried through the heart to the lungs. There they break out into the alveoli and are carried up to the tracheae. If the infection is large, they may cause a serious pneumonia at this stage. On reaching the pharynx, the juveniles are swallowed, passed through the stomach, and finally mature about 2 months after the eggs were ingested. In the intestine, where they feed on intestinal contents, the worms cause abdominal symptoms and allergic reactions, and in large numbers they may cause intestinal blockage. Perforation of the intestine with resultant peritonitis is not uncommon, and wandering worms may occasionally emerge from the anus or throat or may enter the trachea or eustachian tubes and middle ears.

Hookworms

Hookworms are so named because of the anterior end curves dorsally, suggesting a hook. The most common species is *Necator americanus* (L. *necator,* killer), whose females are up to 11 mm long. The males can reach 9 mm in length. Large plates in their mouths (Figure 16-13) cut into the intestinal mucosa of the host where they suck blood and pump it through

Plates

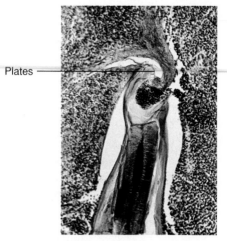

Figure 16-13
Section through anterior end of hookworm attached to dog intestine. Note cutting plates of mouth pinching off mucosa from which the thick muscular pharynx sucks blood. Esophageal glands secrete anticoagulant to prevent blood clotting.

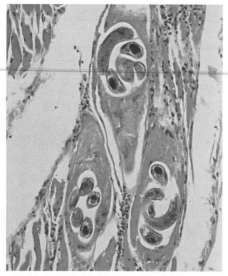

Figure 16-14
Section of muscle infected with trichina worm *Trichinella spiralis,* human case. The juveniles lie within muscle cells that the worms have induced to transform into nurse cells (commonly called cysts). An inflammatory reaction is evident around the nurse cells. The juveniles may live 10 to 20 years, and the nurse cells eventually may calcify.

A

B

Figure 16-15
Pinworms, *Enterobius vermicularis.* **A,** Female worm from human large intestine (slightly flattened in preparation), magnified about 20 times. **B,** Group of pinworm eggs, which are usually discharged at night around the anus of the host, who, by scratching during sleep, gets fingernails and clothing contaminated. This may be the most common and widespread of all human helminth parasites.

their intestines, partially digesting it and absorbing the nutrients. They suck much more blood than they need for food, and heavy infections cause anemia in the patient. Hookworm disease in children may result in retarded mental and physical growth and a general loss of energy.

Eggs pass in the feces, and the juveniles hatch in the soil, where they live on bacteria. When human skin comes in contact with infected soil, the juveniles burrow through the skin to the blood, and reach the lungs and finally the intestine in a manner similar to that described for *Ascaris.*

Trichina Worm

Trichinella spiralis (Gr. *trichinos,* of hair, + *-ella,* diminutive) is one of the species of tiny nematodes responsible for the potentially lethal disease trichinosis. Adult worms burrow in the mucosa of the small intestine where the female produces living young. The juveniles penetrate into blood vessels and are carried throughout the body, where they may be found in almost any tissue or body space. Eventually, they penetrate skeletal muscle cells, becoming one of the largest known intracellular parasites. The juvenile causes

astonishing redirection of gene expression in its host cell, which loses its striations and becomes a **nurse cell** that nourishes the worm (Figure 16-14). When meat containing live juveniles is swallowed, the worms are liberated into the intestine where they mature.

Trichinella spp. can infect a wide variety of mammals in addition to humans, including hogs, rats, cats, and dogs. Hogs become infected by eating garbage containing pork scraps with juveniles or by eating infected rats. In addition to *T. spiralis,* we now know there are four other sibling species in the genus. They differ in geographic distribution, infectivity to different host species, and freezing resistance.

Heavy infections may cause death, but lighter infections are much more common—about 2.4% of the population of the United States is infected.

Pinworms

The pinworm, *Enterobius vermicularis* (Gr. *enteron,* intestine, + *bios,* life), causes relatively little disease, but it is the most common helminth parasite in the United States, estimated at 30% in children and 16% in adults. The adult parasites (Figure 16-15) live in the large intestine and cecum. The fe-

males, up to about 12 mm in length, migrate to the anal region at night to lay their eggs (Figure 16-15). Scratching the resultant itch effectively contaminates the hands and bedclothes. Eggs develop rapidly and become infective within 6 hours at body temperature. When they are swallowed, they hatch in the duodenum, and the worms mature in the large intestine.

Diagnosis of most intestinal roundworms is usually made by examination of a small bit of feces under the microscope and finding characteristic eggs. However, pinworm eggs are not often found in the feces because the female deposits them on the skin around the anus. The "Scotch tape method" is more effective. The sticky side of cellulose tape is applied around the anus to collect the eggs, then the tape is placed on a glass slide and examined under the microscope. Several drugs are effective against this parasite, but all members of a family should be treated at the same time because the worm easily spreads through a household.

Members of this order of nematodes have **haplodiploidy,** a characteristic shared with a few other animal groups, notably many hymenopteran insects (p. 422). Males are haploid and are produced parthenogenetically; females are diploid and arise from fertilized eggs.

Filarial Worms

At least eight species of filarial nematodes infect humans, and some of these are major causes of disease. Some 250 million people in tropical countries are infected with *Wuchereria bancrofti* (named for Otto Wucherer) or *Brugia malayi* (named for S. L. Brug), which places these species among the scourges of humanity. The worms live in the lymphatic system, and the females are as long as 100 mm. The disease symptoms are associated with inflammation and obstruction of the lymphatic system. The females release live young, the tiny **microfilariae,** into the blood and lymphatic system. As they feed, mosquitos pick up the microfilariae, and they develop in the mosquitos to the infective stage. They escape from the mosquito when it is feeding again on a human and penetrate the wound made by the mosquito bite.

The dramatic manifestations of elephantiasis are produced occasionally after long and repeated exposure to the worms. The condition is marked by an excessive growth of connective tissue and enormous swelling of affected parts, such as the scrotum, legs, arms, and more rarely, the vulva and breasts (Figure 16-16).

Another filarial worm causes river blindness (onchocerciasis) and is carried by the blackfly. It infects more than 30 million people in parts of Africa, Arabia, Central America, and South America.

The most common filarial worm in the United States is probably the dog heartworm, *Dirofilaria immitis* (Figure 16-17). Carried by mosquitos, it also can infect other canids, cats, ferrets, sea lions, and occasionally hu-

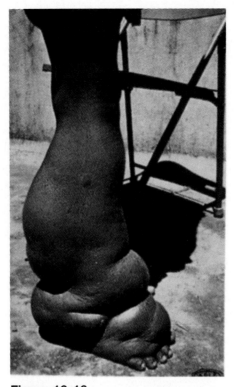

Figure 16-16
Elephantiasis of leg caused by adult filarial worms of *Wuchereria bancrofti,* which live in lymph passages and block the flow of lymph. Tiny juveniles, called microfilariae, are picked up with blood meal of mosquitos, where they develop to infective stage and are transmitted to a new host.

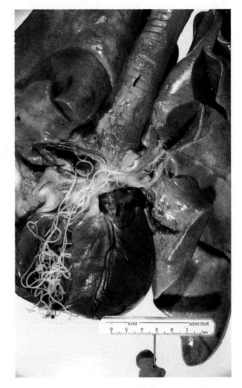

Figure 16-17
Dirofilaria immitis in right ventricle, extending up into right and left pulmonary arteries of an eight-year-old Irish setter.

mans. Along the Atlantic and Gulf Coast states and northward along the Mississippi River throughout the midwestern states, prevalence in dogs is up to 45%. It occurs in other states at a lower prevalence. This worm causes a very serious disease among dogs, and no responsible owner should fail to provide "heartworm pills" for a dog during mosquito season.

CLASSIFICATION OF PHYLUM NEMATODA

Classification of the nematodes is somewhat more satisfactory at the order and superfamily level; the division into classes relies on characteristics that are not striking and that are difficult for the novice to distinguish. The classification given here is that proposed by Adamson,* whose analysis indicated that the traditional class Aphasmidia was paraphyletic.

Class Rhabditea (rab-di´te-a) (Gr. *rhabdos,* a rod) Amphids ventrally coiled or derived therefrom; three esophageal glands; some with phasmids; both free-living and parasitic forms. Examples: *Caenorhabditis, Ascaris, Enterobius, Necator, Wuchereria.*

Class Enoplea (ee-no´ple-a) (Gr. *enoplos,* armed) Amphids generally well-developed, pocket-like; five or more esophageal glands; phasmids absent; excretory system lacking lateral canals, formed of single, ventral, glandular cells, or entirely absent; mostly free living, but includes some parasites. Examples: *Dioctophyme, Trichinella, Trichuris.*

*Adamson, M. 1987. Canad. J. Zool. **65:**1478–1482.

PHYLUM NEMATOMORPHA

The popular name for the Nematomorpha (nem´a-to-mor´fa) (Gr. *nema, nematos,* thread, + *morphe,* form) is "horsehair worms," based on an old superstition that the worms arise from horsehairs that happen to fall into the water, and they look something like hairs from a horse's tail. They were long included with the nematodes, with which they share the structure of the cuticle, presence of epidermal cords, longitudinal muscles only, and pattern of nervous system. However, since the early larval form of some species has a striking resemblance to the Priapulida, it is impossible to say to what group the nematomorphs are most closely related.

About 250 species of horsehair worms have been named. Worldwide in distribution, they are free living as adults and parasitic in arthropods as juveniles. Adults do not feed but will live almost anywhere in wet to moist surroundings if the oxygen is adequate. Some juveniles, such as *Gordius* (named for an ancient king who tied an intricate knot), a cosmopolitan genus, are believed to encyst on vegetation that may later serve as food for a grasshopper or other arthropod. In the marine form *Nectonema* (Gr. *nektos,* swimming, + *nema,* thread), juveniles occur in hermit crabs and other crabs.

FORM AND FUNCTION

Horsehair worms are extremely long and slender, with a cylindrical body. Their length ranges from 10 to 70 cm, but their diameter is only 0.3 to 2.5 mm. The anterior ends are usually rounded, and the posterior ends are rounded or have two or three caudal lobes (Figure 16-18).

The body wall is much like that of nematodes: a secreted cuticle, a hypodermis, and musculature of **longitudinal muscles** only. Ventral, or dorsal and ventral, but not lateral, hypodermal cords are present. In most nematomorphs the ventral nerve cord is connected to the ventral hypodermal cord by the **nervous lamella.**

The digestive system is vestigial. The pharynx is a solid cord of cells, and the intestine does not open to the cloaca. The larval forms absorb food from their arthropod hosts through the body wall, and the adults apparently live on stored nutrients.

Circulatory, respiratory, and excretory systems are lacking. There are a nerve ring around the pharynx and a midventral nerve cord.

Juveniles do not emerge from the arthropod host unless there is water nearby. Adults are often seen wriggling slowly about in ponds or streams, with males being more active than females. Each sex has a pair of gonads and a pair of gonoducts that empty into the cloaca. The female discharges her eggs into the water in long strings. The young hatch from the eggs and somehow gain entry into the arthropod host. After several months in the hemocoel of the host, the matured worm emerges into the water. Curiously, if the host is a terrestrial insect, the parasite stimulates the insect by an unknown mechanism to seek water.

PHYLUM ACANTHOCEPHALA

The members of the phylum Acanthocephala (a-kan´tho-sef´a-la) (Gr. *akantha,* spine or thorn, + *kephalē,* head) are commonly known as "spiny-headed worms." The phylum derives its name from one of its most distinctive features, a cylindrical, invaginable proboscis bearing rows of recurved spines, by which it attaches itself to the intestine of its host. All acanthocephalans are endoparasitic, living as adults in the intestine of vertebrates.

Various species range in size from less than 2 mm to more than 1 m in length, with the females of a species usually larger than the males. The body is usually bilaterally flattened, with numerous transverse wrinkles. The worms are typically cream color but may be yellowish or brown as a result of absorption of pigments from the intestinal contents.

Acanthocephalans inflict traumatic damage by penetrating the host's intestinal wall with the spiny proboscis. In many cases there is remarkably little inflammation, but in some species the inflammatory response of the host

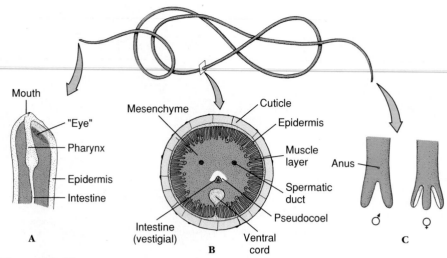

Figure 16-18

Structure of *Paragordius,* a nematomorph. **A,** Longitudinal section through the anterior end. **B,** Transverse section. **C,** Posterior end of male and female worms. Nematomorphs, or "horsehair worms," are very long and very thin. Their pharynx is usually a solid cord of cells and is nonfunctional. *Paragordius,* whose pharynx opens through to the intestine, is unusual in this respect and also in the possession of a photosensory organ ("eye").

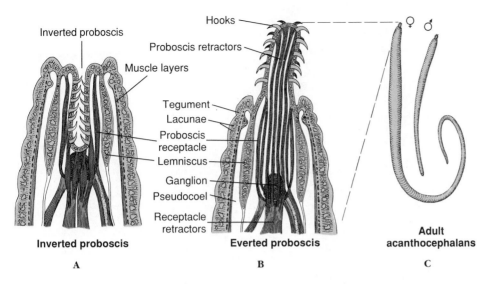

Figure 16-19

Structure of a spiny-headed worm (Phylum Acanthocephala). **A** and **B,** Eversible spiny proboscis by which the parasite attaches to the intestine of the host, often doing great damage. Because they lack a digestive tract, food is absorbed through the tegument. **C,** Male is typically smaller than female.

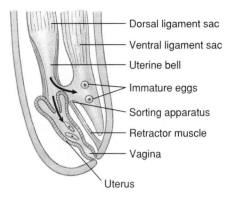

Figure 16-20

Scheme of the genital selective apparatus of a female acanthocephalan. It is a unique device for separating immature from mature fertilized eggs. Eggs containing larvae enter the uterine bell and pass on to the uterus and exterior. Immature eggs are shunted into the ventral ligament sac or into the pseudocoel to undergo further development.

is intense. Infection with these worms can cause great pain, particularly if the gut wall is completely perforated.

The phylum is cosmopolitan, and more than 500 species are known, most of which parasitize fish, birds, and mammals. However, no species is normally a parasite of humans, although species that usually occur in other hosts occasionally infect humans. *Macracanthorhynchus hirudinaceus* (Gr. *makros,* long, large, + *akantha,* spine, thorn, + *rhynchos,* beak) occurs throughout the world in the small intestine of pigs and sometimes in other mammals.

Larvae of spiny-headed worms develop in arthropods, either crustaceans or insects, depending on the species.

FORM AND FUNCTION

In life the body is somewhat flattened, although it is usual for specimens to be treated with tap water before fixation so that fixed specimens are turgid and cylindrical (Figure 16-19C).

The body wall is syncytial, and its surface is punctured by minute crypts 4 to 6 μm deep, which greatly increase the surface area of the tegument. About 80% of the thickness of the tegument is the radial fiber zone, which contains a **lacunar system** of ramifying fluid-filled canals (Figure 16-19A and B). Curiously, the body-wall muscles are tubelike and filled

with fluid. The tubes in the muscles are continuous with the lacunar system; therefore circulation of the lacunar fluid may well bring nutrients to and remove wastes from the muscles. There is no heart or other circulatory system, and contraction of the muscles would serve to move the lacunar fluid through the canals and muscles. Both longitudinal and circular body-wall muscles are present.

The proboscis, which bears rows of recurved hooks, is attached to the neck region (Figure 16-19) and can be inverted into a **proboscis receptacle** by retractor muscles. Attached to the neck region (but not within the proboscis) are two elongated **lemnisci** (extensions of the tegument and lacunar system) that may serve as reservoirs of the lacunar fluid from the proboscis when that organ is invaginated.

There is no respiratory system. When present, the excretory system consists of a pair of **protonephridia** with flame cells. These unite to form a common tube opening into the sperm duct or uterus.

The nervous system has a central ganglion within the proboscis receptacle and nerves to the proboscis and body. There are sensory endings on the proboscis and genital bursa.

Acanthocephalans have no digestive tract, and they must absorb all nutrients through their tegument. They can absorb various molecules by spe-

cific membrane transport mechanisms, and other substances can cross the tegumental membrane by pinocytosis (probably potocytosis). The tegument bears some enzymes, such as peptidases, which can cleave several dipeptides, and the amino acids are then absorbed by the worm. Like cestodes, acanthocephalans require host dietary carbohydrate, but their mechanism for absorption of glucose is different. As glucose is absorbed, it is rapidly phosphorylated and compartmentalized, so that a metabolic "sink" is created into which glucose in the surrounding medium can flow. Glucose diffuses down the concentration gradient into the worm because it is constantly removed as soon as it enters.

Acanthocephalans are dioecious. A pair of tubular **genital ligaments,** or **ligament sacs,** extends posteriorly from the end of the proboscis receptacle. The male has a pair of testes, each with a vas deferens, and a common ejaculatory duct that ends in a small penis. During copulation sperm are ejected into the vagina, travel up the genital duct, and escape into the pseudocoel.

In the female the ovarian tissue in the ligament sac breaks up into **ovarian balls** that rupture the ligament sacs and float free in the pseudocoel. One of the ligament sacs leads to a funnel-shaped **uterine bell** that receives the developing shelled embryos and passes them on to the uterus (Figure 16-20).

An interesting and unique selective apparatus operates here. Fully developed embryos are slightly longer than immature ones, and they are passed on into the uterus, while immature eggs are retained for further maturation.

The shelled embryos, which are discharged in the feces of the vertebrate host, do not hatch until eaten by the intermediate host. For *M. hirudinaceus* this is any of several species of soil-inhabiting beetle larvae, especially scarabeids. Grubs of the June beetle (*Phyllophaga*) are frequent hosts. Here the larva **(acanthor)** burrows through the intestine and develops into the juvenile **(cystacanth).** Pigs become infected by eating the grubs. Multiple infections may do considerable damage to the pig's intestine, and perforations can occur.

Phylum Entoprocta

Entoprocta (en´to-prok´ta) (Gr. *entos,* within, + *proktos,* anus) is a small phylum of about 150 species of tiny, sessile animals that superficially resemble hydroid cnidarians but have ciliated tentacles that tend to roll inward (Figure 16-21). Most entoprocts are microscopic, and none is more than 5 mm long. They are all stalked and sessile forms; some are colonial, and some are solitary. All are ciliary feeders.

With the exception of the genus *Urnatella* (L. *urna,* urn, + *ellus,* dim. suffix), all entoprocts are marine forms that have a wide distribution from the polar regions to the tropics. Most marine species are restricted to coastal and brackish waters and often grow on shells and algae. Some are commensals on marine annelid worms. Freshwater entoprocts occur on the underside of rocks in running water. *U. gracilis* is the only common freshwater species in North America (Figure 16-21A).

Form and Function

The body, or **calyx,** of the entoproct is cup shaped, bears a crown, or circle, of ciliated tentacles, and may be attached to a substratum by a single

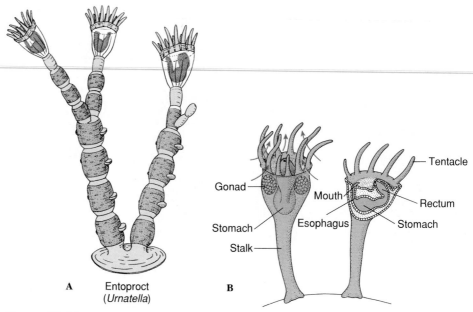

Figure 16-21
A, *Urnatella,* a freshwater entoproct, forms small colonies of two or three stalks from a basal plate.
B, *Loxosomella,* a solitary entoproct. Both solitary and colonial entoprocts can reproduce asexually by budding, as well as sexually.

stalk and an attachment disc with adhesive glands, as in the solitary *Loxosoma* and *Loxosomella* (Gr. *loxos,* crooked, + *soma,* body) (Figure 16-21B), or by two or more stalks in colonial forms. Both tentacles and stalk are continuations of the body wall. The 8 to 30 tentacles making up the crown are ciliated on their lateral and inner surfaces, and each can move individually. The tentacles can roll inward to cover and protect the mouth and anus but cannot be retracted into the calyx.

Movement is usually restricted in entoprocts, but *Loxosoma,* which lives in the tubes of marine annelids, is quite active, moving over the annelid and its tube freely.

The gut is U-shaped and ciliated, and both the mouth and the anus open within the circle of tentacles. Entoprocts are **ciliary filter feeders.** Long cilia on the sides of the tentacles keep a current of water containing protozoa, diatoms, and particles of detritus moving in between the tentacles. Short cilia on the inner surfaces of the tentacles capture the food and direct it downward toward the mouth.

The body wall consists of a cuticle, cellular epidermis, and longitudinal muscles. The pseudocoel is largely filled with a gelatinous parenchyma in which is embedded a pair of protonephridia and their ducts, which unite and empty near the mouth. There is a well-developed **nerve ganglion** on the ventral side of the stomach, and the body surface bears sensory bristles and pits. Circulatory and respiratory organs are absent. Exchange of gases occurs through the body surface, probably much of it through the tentacles.

Some species are monoecious, some dioecious, and some appear to be protandrous; that is, the gonad at first produces sperm and later eggs. The gonoducts open within the circle of tentacles.

Fertilized eggs develop in a depression, or brood pouch, between the gonopore and the anus. Entoprocts have a modified spiral cleavage pattern with mosaic blastomeres. The embryo gastrulates by invagination. The trochophore-like larva (see p. 326) is ciliated and free swimming. It has an apical tuft of cilia at the anterior end and a ciliated girdle around the ventral margin of the body. Eventually the larva settles to the substratum and inverts to form the adult.

PHYLOGENY AND ADAPTIVE RADIATION

PHYLOGENY

Hyman (1951) grouped the Rotifera, Gastrotricha, Kinorhyncha, Nematoda, and Nematomorpha into a single phylum (Aschelminthes). All of these phyla share a certain combination of characteristics, including their usually wormlike bodies, a cuticle secreted by an epidermis that is underlain by muscles not arranged in regular circular and longitudinal layers, a brain that is a circumenteric nerve ring, determinate cleavage and eutely, and absence of a muscle layer in the intestine. Hyman contended that the evidences of relationships were so concrete and specific that they could not be disregarded. Nevertheless, most authors now consider that differences between the groups are sufficient to merit phylum status for each, although some accept the concept of the Aschelminthes as a superphylum. These phyla may well have been derived originally from the protostome line via an acoelomate common ancestor (Figure 16-22).

Loriciferans bear some similarity to kinorhynchs, larval Priapulida, larval nematomorphs, rotifers, and tardigrades (p. 438). Although the loriciferans are poorly known, cladistic analysis suggests that they form a sister group to the kinorhynchs and that these two phyla together are a sister group of the priapulids (Figure 16-22).

Acanthocephalans are highly specialized parasites with a unique structure and have doubtless been so for millions of years. Any ancestral or other related group that would shed a clue to the phyletic relationships of the Acanthocephala is probably long since extinct. Like the cestodes, acanthocephalans have no digestive tract and must absorb all nutrients across the tegument, but the tegument of the two groups is quite different in structure. Also, acanthocephalans are pseudocoelomate and show eutely, as in the nematodes, although here, too, the structural and developmental differences are great.

The entoprocts were once included with the phylum Ectoprocta in a phylum called Bryozoa, but the ectoprocts are true coelomate animals, and many zoologists prefer to place them in a separate group. Ectoprocts are still often referred to as bryozoans. The Entoprocta may be distantly related to the Ectoprocta, but there is little evidence of close relationship. The entoprocts may have arisen as an early offshoot of the same line that led to the ectoprocts. These relationships remain controversial.

ADAPTIVE RADIATION

Certainly the most impressive adaptive radiation in this group of phyla is shown by the nematodes. They are by far the most numerous in terms of both individuals and species, and they have been able to adapt to almost every habitat available to animal life. Their basic pseudocoelomate body plan, with the cuticle, hydrostatic skeleton, and longitudinal muscles, has proved generalized and plastic enough to adapt to an enormous variety of physical conditions. Free-living lines gave rise to parasitic forms on at least several occasions, and virtually all potential hosts have been exploited. All types of life cycle occur: from the simple and direct to the complex, with intermediate hosts; from normal dioecious reproduction to parthenogenesis, hermaphroditism, and alternation of free-living and parasitic generations. A major factor contributing to the evolutionary opportunism of the nematodes has been their extraordinary capacity to survive suboptimal conditions, for example, the developmental arrests in many free-living and animal parasitic species and the ability to undergo cryptobiosis (survival in harsh conditions by assuming a very low metabolic rate) in many free-living and plant parasitic species.

In December 1995, P. Funch and R. M. Kristensen reported that they had found some very strange little creatures clinging to the mouthparts of the Norway lobster (*Nephrops norvegicus*), so strange that they did not fit into any known phylum (Nature **378**:711–714). Funch and Kristensen concluded that the organisms, only 0.35 mm long, represented a new phylum, which was named **Cycliophora.** The name refers to a crown of compound cilia, reminiscent of rotifers, with which the organisms feed. They are described as "acoelomate," although whether they might have a pseudocoel is unclear, and they have a cuticle. Their life cycle seems bizarre. The sessile feeding stages on the lobster's mouthparts undergo internal budding to produce motile stages: (1) larvae containing new feeding stages; (2) dwarf males, which become attached to feeding stages that contain developing females; and (3) females, which also attach to the lobster's mouthparts, then produce dispersive larvae and degenerate.

Whether the proposed new phylum will withstand the scrutiny of further research is unknown, and its possible relationships to other phyla are quite unclear. Funch and Kristensen think that the organisms are protostomes and see affinities with the Entoprocta and Ectoprocta. Little short of astonishing, however, is their abundance on the mouthparts of a host as well known as the Norway lobster. How could biologists have failed to notice them before? At a time when habitat destruction drives many species to extinction every year, we wonder if there are *phyla* suffering the same fate. S. Conway Morris ponders the possibility of further undiscovered phyla (Nature **378**:661–662), suggesting you may need a couple of zoology textbooks and a decent microscope when you next dine at your favorite seafood restaurant: "Who knows what might be found lurking under the lettuce?"

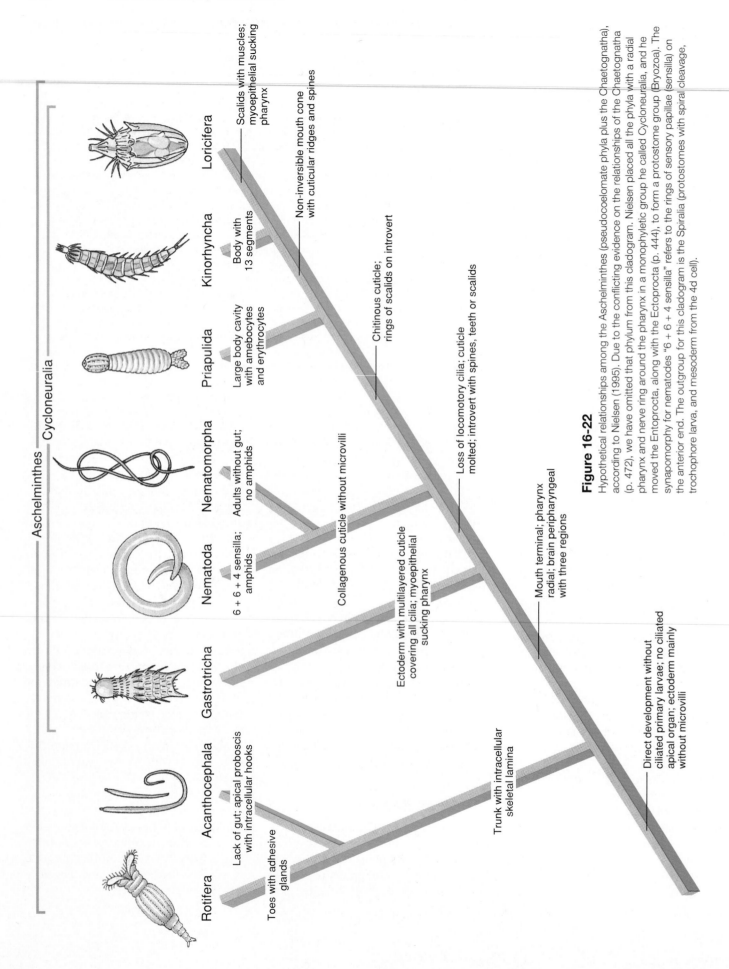

Figure 16-22

Hypothetical relationships among the Aschelminthes (pseudocoelomate phyla plus the Chaetognatha), according to Nielsen (1995). Due to the conflicting evidence on the relationships of the Chaetognatha (p. 472), we have omitted that phylum from this cladogram. Nielsen placed all the phyla with a radial pharynx and nerve ring around the pharynx in a monophyletic group he called Cycloneuralia, and he moved the Entoprocta, along with the Ectoprocta (p. 444), to form a protostome group (Bryozoa). The synapomorphy for nematodes "6 + 6 + 4 sensilla" refers to the rings of sensory papillae (sensilla) on the anterior end. The outgroup for this cladogram is the Spiralia (protostomes with spiral cleavage, trochophore larva, and mesoderm from the 4d cell).

Loricifera — Scalids with muscles; myoepithelial sucking pharynx

Non-inversible mouth cone with cuticular ridges and spines

Kinorhyncha — Body with 13 segments

Chitinous cuticle; rings of scalids on introvert

Priapulida — Large body cavity with amebocytes and erythrocytes

Loss of locomotory cilia; cuticle molted; introvert with spines, teeth or scalids

Nematomorpha — Adults without gut; no amphids

Collagenous cuticle without microvilli

Nematoda — 6 + 6 + 4 sensilla; amphids

Ectoderm with multilayered cuticle covering all cilia; myoepithelial sucking pharynx

Mouth terminal; pharynx radial; brain peripharyngeal with three regions

Gastrotricha

Acanthocephala — Lack of gut; apical proboscis with intracellular hooks

Trunk with intracellular skeletal lamina

Direct development without ciliated primary larvae; no ciliated apical organ; ectoderm mainly without microvilli

Rotifera — Toes with adhesive glands

Aschelminthes — Cycloneuralia

Summary

The phyla covered in this chapter possess a body cavity called a pseudocoel, which is derived from the embryonic blastocoel, rather than a secondary cavity in the mesoderm (coelom). Several of the groups exhibit eutely, a constant number of cells or nuclei in adult individuals of a given species.

The phylum Rotifera is composed of small, mostly freshwater organisms with a ciliated corona, which creates currents of water to draw planktonic food toward the mouth. The mouth opens into a muscular pharynx, or mastax, that is equipped with jaws.

The Gastrotricha, Kinorhyncha, and Loricifera are small phyla of tiny, aquatic pseudocoelomates. Gastrotrichs move by cilia or adhesive glands, and kinorhynchs anchor and then pull themselves by the spines on their head. Loriciferans can withdraw their bodies into the lorica. Priapulids are marine burrowing worms.

By far the largest and most important of this group of phyla are the nematodes, of which there may be as many as 500,000 species in the world. They are more or less cylindrical, tapering at the ends, and covered with a tough, secreted cuticle. The body-wall muscles are longitudinal only, and to function well in locomotion, such an arrangement must enclose a volume of fluid in the pseudocoel at high hydrostatic pressure. This fact of nematode life has a profound effect on most of their other physiological functions, for example, ingestion of food, egestion of feces, excretion, copulation, and others. Most nematodes are dioecious, and there are four juvenile stages, each separated by a molt of the cuticle. Almost all invertebrate and vertebrate animals and many plants have nematode parasites, and many other nematodes are free living in soil and aquatic habitats. Some parasitic nematodes have part of their life cycle free living, some undergo a tissue migration in their host, and some have an intermediate host in their life cycle. Some of the parasitic nematodes cause severe diseases in humans and other animals.

The Nematomorpha or horsehair worms are related to the nematodes and have parasitic juvenile stages in arthropods, followed by a free-living, aquatic, nonfeeding adult stage.

Acanthocephalans are all parasitic in the intestines of vertebrates as adults, and their juvenile stages develop in arthropods. They have an anterior, invaginable proboscis armed with spines, which they embed in the intestinal wall of their host. They do not have a digestive tract and so must absorb all nutrients across their tegument.

The Entoprocta are small, sessile, aquatic animals with a crown of ciliated tentacles encircling both the mouth and anus.

The Rotifera, Gastrotricha, Kinorhyncha, Nematoda, and Nematomorpha have been included by some workers in one Phylum Aschelminthes, but most biologists believe that the groups are not sufficiently related to be encompassed by a single phylum. It is possible that they are derived from a common ancestor in the protostome line. Of all these phyla, the Nematoda have achieved enormous evolutionary success and undergone great adaptive radiation.

Review Questions

1. Give seven characteristics of pseudocoelomate animals.
2. Explain the difference between a true coelom and a pseudocoel.
3. What is the normal size of a rotifer; where is it found; and what are its major features?
4. Explain the difference between mictic and amictic eggs of rotifers. What is the adaptive value of each?
5. What is eutely?
6. About how big are loriciferans, priapulids, gastrotrichs, and kinorhynchs? Where are each of them found?
7. What is a hydrostatic skeleton?
8. Distinguish a solenocyte from a flame cell protonephridium.
9. Explain two peculiar features of the body-wall muscles in nematodes.
10. What feature of body-wall muscles in nematodes requires a high hydrostatic pressure in the pseudocoelomic fluid for efficient function?
11. Explain the interaction of the cuticle, body-wall muscles, and pseudocoelomic fluid in the locomotion of nematodes.
12. Explain how the high pseudocoelomic pressure affects feeding and defecation in nematodes.
13. Outline the life cycle of each of the following: *Ascaris lumbricoides,* hookworm, *Enterobius vermicularis, Trichinella spiralis, Wuchereria bancrofti.*
14. Where in the human body are the adults of each species in question 13 found?
15. Outline the life cycle of a typical nematomorph.
16. How are nematodes and nematomorphs alike, and how are they different?
17. Describe the major features of the acanthocephalan body.
18. How do acanthocephalans get food?
19. What are distinguishing characteristics of entoprocts?
20. What are the "aschelminth" phyla, and what do they have in common?
21. What phylum covered in this chapter has radiated into the most diversity? How do the members of this phylum impact humans?

Selected References

See also general references for Part III, p. 626.

Bird, A. F., and J. Bird. 1991. The structure of nematodes, ed. 2. New York, Academic Press. *The most authoritative reference available on nematode morphology. Highly recommended.*

Bundy, D. A. P., and E. S. Cooper. 1989. *Trichuris* and trichuriasis in humans. In J. R. Baker and R. Muller (eds), Advances in parasitology, vol. 28. London, Academic Press. *A recent estimate of people in the world infected with whipworm (Trichuris) is 687 million. This is exceeded among nematodes only by hookworms (932 million) and Ascaris (1.26 billion).*

Despommier, D. D. 1990. *Trichinella spiralis:* the worm that would be virus. Parasitol. Today **6:**193–196. *The first stage juveniles of* Trichinella *are among the largest of all intracellular parasites.*

Duke, B. O. L. 1990. Onchocerciasis (river blindness)—can it be eradicated? Parasitol. Today **6:**82–84. *Despite the introduction of a very effective drug, the author predicts that this parasite will not be eradicated in the foreseeable future.*

Ogilvie, B. M., M. E. Selikirk, and R. M. Maizels. 1990. The molecular revolution and nematode parasitology: yesterday, today, and tomorrow. J. Parasitol. **76:**607–618.

Modern molecular biology has wrought enormous changes in investigations on nematodes.

Poinar, G. O., Jr. 1983. The natural history of nematodes. Englewood Cliffs, New Jersey, Prentice-Hall, Inc. *Contains a great deal of information about these fascinating creatures, including free-living and plant and animal parasites.*

Roberts, L. 1990. The worm project. Science **248:**1310–1313. *A nematode,* Caenorhabditis elegans, *has been of great value in studies of development and genetics.*

17

The Molluscs
Phylum Mollusca

A Significant Space

Long ago in the Precambrian era, the most complex animals populating the seas were acoelomate. They must have been inefficient burrowers, and they were unable to exploit the rich subsurface ooze. Any that developed fluid-filled spaces within the body would have had a substantial selective advantage because these spaces could serve as a hydrostatic skeleton and improve burrowing efficiency.

The simplest, and probably the first, mode of achieving a fluid-filled space within the body was retention of the embryonic blastocoel, as in the pseudocoelomates. This was not the best evolutionary solution because, for example, the organs lay loose in the body cavity.

Some descendants of Precambrian acoelomate organisms evolved a more elegant arrangement: a fluid-filled space

within the mesoderm, the *coelom*. This meant that the space was lined with mesoderm and the organs were suspended by mesodermal membranes, the *mesenteries*. Not only could the coelom serve as an efficient hydrostatic skeleton, with circular and longitudinal body-wall muscles acting as antagonists, but a more stable arrangement of organs with less crowding resulted. The mesenteries provided an ideal location for networks of blood vessels, and the alimentary canal could become more muscular, more highly specialized, and more diversified without interfering with other organs.

Development of the coelom was a major step in the evolution of larger and more complex forms. All the major groups in the chapters to follow are coelomates. ■

POSITION IN ANIMAL KINGDOM

1. The molluscs are one of the major groups of true **coelomate** animals.
2. They belong to the **protostome** branch, or schizocoelous coelomates, and have spiral cleavage and determinate (mosaic) development.
3. Many molluscs have a **trochophore larva** similar to the trochophore larva of marine annelids and other marine protostomes. Developmental evidence thus indicates that molluscs and annelids share a common ancestor.
4. Because molluscs are not metameric, they must have diverged from their common ancestor with the annelids before the advent of metamerism.
5. All the **organ systems** are present and well developed.

BIOLOGICAL CONTRIBUTIONS

1. In molluscs gaseous exchange occurs not only through the body surface as in phyla discussed previously, but also in specialized **respiratory organs** in the form of **gills** or **lungs.**
2. Most classes have an **open circulatory system** with pumping **heart,** vessels, and blood sinuses. In most cephalopods the circulatory system is closed.
3. The efficiency of the respiratory and circulatory systems in the cephalopods has made greater body size possible. Invertebrates reach their largest size in some of the cephalopods.
4. They have a fleshy **mantle** that in most cases secretes a shell and is variously modified for a number of functions.
5. Features unique to the phylum are the **radula** and the muscular **foot.**
6. The highly developed direct **eye** of cephalopods is similar to the indirect eye of the vertebrates but arises as a skin derivative in contrast to the brain eye of vertebrates.

THE MOLLUSCS

The Mollusca (mol-lus´ka) (L. *molluscus,* soft) is one of the largest animal phyla after the Arthropoda. There are nearly 50,000 living species and some 35,000 fossil species. The name Mollusca indicates one of their distinctive characteristics, a soft body. This very diverse group (Figure 17-1) includes the chitons, tooth shells, snails, slugs, nudibranchs, sea butterflies, clams, mussels, oysters, squids, octopuses, and nautiluses. The group ranges from fairly simple organisms to some of the most complex of invertebrates, and in size from almost microscopic to the giant squid *Architeuthis.* These huge molluscs may grow to 18 m long, including their tentacles. They may weigh 450 kg (1000 pounds). The shells of some of the giant clams, *Tridacna gigas,* which inhabit the Indo-Pacific coral reefs, reach 1.5 m in length and weigh more than 225 kg. These are extremes, however, for probably 80% of all molluscs are less than 5 cm in maximum shell size. The phylum includes some of the most sluggish and some of the swiftest and most active of the invertebrates. It includes herbivorous grazers, predaceous carnivores, filter feeders, detritus feeders, and parasites.

Molluscs are found in a great range of habitats, from the tropics to polar seas, at altitudes exceeding 7000 m, in ponds, lakes, and streams, on mud flats, in pounding surf, and in open ocean from the surface to the abyssal depths. Most of them live in the sea, and they represent a variety of life-styles, including bottom feeders, burrowers, borers, and pelagic forms.

A

B

C

D

Figure 17-1

Molluscs: a diversity of life forms. The basic body plan of this ancient group has become variously adapted for different habitats. **A,** A chiton (*Tonicella lineata*), Class Polyplacophora. **B,** A marine snail (*Calliostoma annulata*), Class Gastropoda. **C,** A nudibranch (*Chromodoris* sp.) Class Gastropoda. **D,** Pacific giant clam (*Panope abrupta*), with siphons to the left, Class Bivalvia. **E,** An octopus (*Octopus dofleini*), Class Cephalopoda, foraging in an eel grass bed.

E

According to the fossil evidence, the molluscs originated in the sea, and most of them have remained there. Much of their evolution occurred along the shores, where food was abundant and habitats were varied. Only the bivalves and gastropods moved on to brackish and freshwater habitats. As filter feeders, the bivalves were unable to leave aquatic surroundings. Only the snails (gastropods) actually invaded the land. Terrestrial snails are limited in their range by their need for humidity, shelter, and the presence of calcium in the soil.

A wide variety of molluscs is used as food. Pearl buttons are obtained from shells of bivalves. The Mississippi and Missouri river basins have furnished material for most of this industry in the United States; however, supplies are becoming so depleted that attempts are being made to propagate bivalves artificially. Pearls, both natural and cultured, are produced in the shells of clams and oysters, most of them in a marine oyster, *Meleagrina,* found around eastern Asia.

Some molluscs are destructive. The burrowing shipworms, which are bivalves of several species (Figure 17-27), do great damage to wooden ships and wharves. To prevent the ravages of shipworms, wharves must be either creosoted or built of concrete (unfortunately, some ignore the creosote, and some bivalves bore into concrete). Snails and slugs frequently damage garden and other vegetation. In addition, snails often serve as intermediate hosts for serious parasites. The boring snail *Urosalpinx* rivals the sea star in destroying oysters.

In this chapter we explore the various major groups of molluscs, including those that apparently met with little evolutionary success (classes Caudofoveata, Solenogastres, Monoplacophora, and Scaphopoda). Members of the class Polyplacophora (chitons) are common to abundant marine animals, especially in the intertidal zone. Bivalves (class Bivalvia) have evolved many species, both marine and freshwater. Largest and most

CHARACTERISTICS OF PHYLUM MOLLUSCA

1. Body bilaterally symmetrical (bilateral asymmetry in some); unsegmented; often with definite head
2. Ventral body wall specialized as a muscular **foot,** variously modified but used chiefly for locomotion
3. Dorsal body wall forms pair of folds called the **mantle,** which encloses the **mantle cavity,** is modified into **gills** or **lungs,** and secretes the **shell** (shell absent in some)
4. Surface epithelium usually ciliated and bearing mucous glands and sensory nerve endings
5. **Coelom** limited mainly to area around heart, and perhaps lumen of gonads and part of kidneys
6. Complex digestive system; rasping organ **(radula)** usually present; anus usually emptying into mantle cavity
7. **Open circulatory system** (mostly closed in cephalopods) of heart (usually three chambered), blood vessels, and sinuses; respiratory pigments in blood
8. Gaseous exchange by **gills, lungs, mantle,** or **body surface**
9. One or two kidneys **(metanephridia)** opening into the pericardial cavity and usually emptying into the mantle cavity
10. Nervous system of paired cerebral, pleural, pedal, and visceral ganglia, with nerve cords and subepidermal plexus; ganglia centralized in nerve ring in gastropods and cephalopods
11. Sensory organs of touch, smell, taste, equilibrium, and vision (in some); eyes highly developed in cephalopods
12. Internal and external **ciliary tracts** often of great functional importance
13. Both **monoecious** and **dioecious** forms; **spiral cleaveage;** primitive larva a **trochophore,** many with a **veliger** larva, some with direct development

intelligent of all invertebrates are in the class Cephalopoda (squids, octopuses, and others). Most abundant and widespread of molluscs, however, are the snails and their relatives (class Gastropoda). Although enormously diverse, molluscs have in common a basic body plan, which is described later in the chapter. It seems peculiar, though, that the molluscs have failed to exploit the coelom. The coelom in molluscs is limited to a space around the heart, and perhaps around the gonads and part of the kidneys. Although it develops embryonically in a manner similar to the coelom of annelids, the functional consequences of the space are quite different. Some zoologists believe that the molluscs arose from a flatworm-type ancestor separately from the annelids and that their coeloms are not homologous.

FORM AND FUNCTION

The enormous variety, great beauty, and easy availability of the shells of molluscs have made shell collecting a popular pastime. However, many amateur shell collectors, even though able to name hundreds of the shells that grace our beaches, know very little about the living animals that created those shells and once lived in them. Reduced to its simplest dimensions, the mollusc body plan may be said to consist of a **head-foot** portion and a **visceral mass** portion (Figure 17-2). The head-foot is the more active area, containing the feeding, cephalic sensory, and locomotor organs. It depends primarily on muscular action for its function. The visceral mass is the portion containing digestive, circulatory, respiratory, and reproductive organs, and it depends primarily on ciliary tracts for its functioning. Two folds of skin, outgrowths of the dorsal body wall, make up a protective **mantle,** or **pallium,** which encloses a space between the mantle and body wall called the **mantle cavity (pallial cavity).** The mantle cavity houses the **gills (ctenidia)** or lung, and in some molluscs the mantle secretes a protective **shell** over the visceral mass. Modifications of the structures that make up the head-foot and the visceral mass produce the

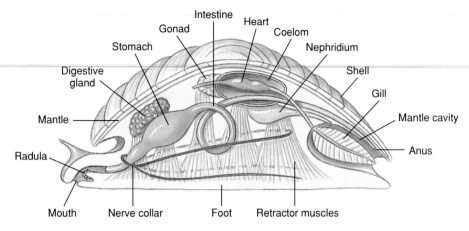

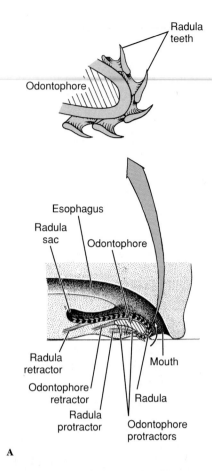

Figure 17-2

Generalized mollusc. Although this construct is often presented as a "hypothetical ancestral mollusc (HAM)," most experts now reject this interpretation. For example, the molluscan ancestor probably was covered with calcareous spicules, rather than a univalve shell. Such a diagram is useful, however, to facilitate description of the general body plan of molluscs.

great confusion of different patterns making up this major group of animals. Greater emphasis on either the head-foot portion or the visceral mass portion can be observed in the various classes of molluscs.

HEAD-FOOT

Most molluscs have well-developed heads, which bear the mouth and some specialized sensory organs. Photosensory receptors range from fairly simple up to the complex eyes of the cephalopods. Tentacles are often present. Within the mouth is a structure unique to molluscs, the radula, and usually posterior to the mouth is the chief locomotor organ, or foot.

Radula

The radula is a rasping, protrusible, tonguelike organ found in all molluscs except the bivalves and most solenogasters. It is a ribbonlike membrane on which are mounted rows of tiny teeth that point backward (Figure 17-3). Complex muscles move the radula and its supporting cartilages (**odontophore**) in and out while the membrane is partly rotated over the tips of the cartilages. There may be a few or as many as 250,000 teeth, which, when protruded, can scrape, pierce, tear, or cut. The usual function

of the radula is twofold: to rasp off fine particles of food material and to serve as a conveyor belt for carrying the particles in a continuous stream toward the digestive tract. As the radula wears away anteriorly, new rows of teeth are continuously replaced by secretion at its posterior end. The pattern and number of teeth in a row are specific for each species and are used in the classification of molluscs. Very interesting radular specializations, such as for boring through hard materials or for harpooning prey, are found in some forms.

Foot

The molluscan foot (see Figure 17-2) may be variously adapted for locomotion, for attachment to a substratum, or for a combination of functions. It is usually a ventral, solelike structure in which waves of muscular contraction effect a creeping locomotion. However, there are many modifications, such as the attachment disc of limpets, the laterally compressed "hatchet foot" of bivalves, or the siphon for jet propulsion in the squids and octopuses. Secreted mucus is often used as an aid to adhesion or as a slime tract by small molluscs that glide on cilia.

In snails and bivalves the foot is extended from the body hydraulically,

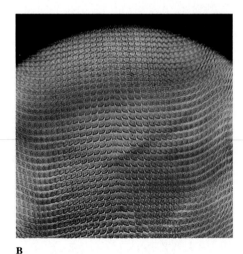

Figure 17-3

A, Diagrammatic longitudinal section of a gastropod head showing the radula and radula sac. The radula moves back and forth over the odontophore cartilage. As the animal grazes, the mouth opens, the odontophore is thrust forward, the radula gives a strong scrape backward bringing food into the pharynx, and the mouth closes. The sequence is repeated rhythmically. As the radula ribbon wears out anteriorly, it is continually replaced posteriorly. **B,** Radula of a snail prepared for microscopic examination.

by engorgement with blood. Burrowing forms can extend the foot into the mud or sand, enlarge it with blood pressure, then use the engorged foot as an anchor to draw the body forward. In pelagic (free-swimming) forms the foot may be modified into winglike parapodia, or thin, mobile fins for swimming.

VISCERAL MASS

Mantle and Mantle Cavity

The mantle is a sheath of skin extending from the visceral mass that hangs down on each side of the body, protecting the soft parts and creating between itself and the visceral mass the space called the mantle cavity. The outer surface of the mantle secretes the shell.

The mantle cavity (Figure 17-2) plays an enormous role in the life of the mollusc. It usually houses the respiratory organs (gills or lung), which develop from the mantle, and the mantle's own exposed surface serves also for gaseous exchange. The products from the digestive, excretory, and reproductive systems are emptied into the mantle cavity. In aquatic molluscs a continuous current of water, kept moving by surface cilia or by muscular pumping, brings in oxygen and, in some forms, food; flushes out wastes; and carries reproductive products out to the environment. In aquatic forms the mantle is usually equipped with sensory receptors for sampling the environmental water. In cephalopods (squids and octopuses) the muscular mantle and its cavity create the jet propulsion used in locomotion. Many molluscs can withdraw the head or foot into the mantle cavity, which is surrounded by the shell, for protection.

In primitive form, the mollusc ctenidium (gill) consists of a long, flattened axis extending from the wall of the mantle cavity (Figure 17-4). Many leaflike gill filaments project from the central axis. Water is propelled by cilia between the gill filaments, and blood diffuses from an afferent vessel in the central axis through the filament to an efferent vessel. The direction of blood movement is opposite to the direction of water movement, thus establishing a countercurrent exchange mechanism (see p. 516). The two ctenidia are located on opposite sides of the mantle cavity and are arranged so that the cavity is functionally divided into an incurrent chamber and an excurrent chamber. Such gills are found in the less derived gastropods, but they are variously modified in many molluscs.

Shell

The shell of the mollusc, when present, is secreted by the mantle and is lined by it. Typically there are three layers (Figure 17-5A). The **periostracum** is the outer horny layer, composed of an organic substance called conchiolin, which consists of quinone-tanned protein. It helps to protect the underlying calcareous layers from erosion by boring organisms. It is secreted by a fold of the mantle edge, and growth occurs only at the margin of the shell. On the older parts of the shell the periostracum often becomes worn away. The middle **prismatic layer** is composed of densely packed prisms of calcium carbonate laid down in a protein matrix. It is secreted by the glandular margin

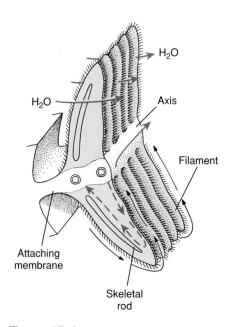

Figure 17-4

Primitive condition of mollusc ctenidium. Circulation of water between the gill filaments is by cilia, and blood diffuses through the filament from the afferent vessel to the efferent vessel. Black arrows are ciliary cleansing currents.

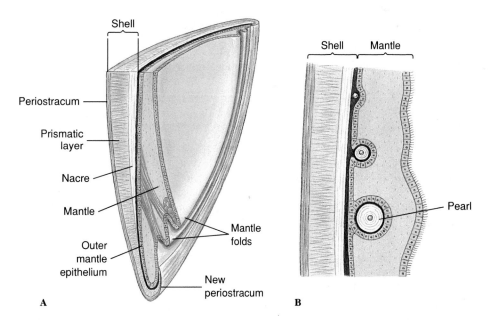

Figure 17-5

A, Diagrammatic vertical section of shell and mantle of a bivalve. The outer mantle epithelium secretes the shell; the inner epithelium is usually ciliated. **B,** Formation of pearl between mantle and shell as a parasite or bit of sand under the mantle becomes covered with nacre.

of the mantle, and increase in shell size occurs at the shell margin as the animal grows. The inner **nacreous layer** of the shell lies next to the mantle and is secreted continuously by the mantle surface, so that it increases in thickness during the life of the animal. The calcareous nacre is laid down in thin layers. Very thin and wavy layers produce the iridescent mother-of-pearl found in the abalones (*Haliotis*), the chambered nautilus (*Nautilus*), and many bivalves. Such shells may have 450 to 5000 fine parallel layers of crystalline calcium carbonate (aragonite) for each centimeter of thickness.

Freshwater molluscs usually have a thick periostracum that gives some protection against the acids produced in the water by the decay of leaf litter. In many marine molluscs the periostracum is relatively thin, and in some it is absent. There is great variation in shell structure. Calcium for the shell comes from the environmental water or soil or from food. The first shell appears during the larval period and grows continuously throughout life.

Internal Structure and Function

Gaseous exchange occurs through the body surface, particularly the mantle, and in specialized respiratory organs such as ctenidia, secondary gills, and lungs. There is an **open circulatory system** with a pumping heart, blood vessels, and blood sinuses. Most cephalopods have a closed blood system with heart, vessels, and capillaries. The digestive tract is complex and highly specialized, according to the feeding habits of the various molluscs, and is usually provided with extensive ciliary tracts. Most molluscs have a pair of kidneys (**metanephridia,** a type of nephridium in which the inner end opens into the coelom by a **nephrostome**); the ducts of the kidneys in many forms also serve for the discharge of eggs and sperm.

The **nervous system** consists of several pairs of ganglia with connecting nerve cords, and it is generally

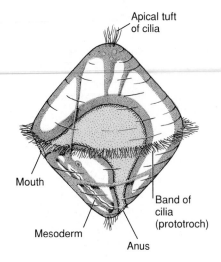

Figure 17-6

Generalized trochophore larva. Molluscs and annelids with primitive embryonic development have trochophore larvae, as do several other phyla.

simpler than that of the annelids and arthropods. Neurosecretory cells have been identified in the nervous system that, at least in certain airbreathing snails, produce a growth hormone and function in osmoregulation. There are various types of highly specialized sense organs.

Reproduction and Life History

Most molluscs are dioecious, although some are hermaphroditic. The free-swimming larva that emerges from the egg in primitive molluscs is the **trochophore,** which is also the primitive larval type of the annelids (Figure 17-6). Direct metamorphosis of the trochophore into a small juvenile, as in chitons, is viewed as a primitive character, and the intervention of another free-swimming larval stage, the **veliger,** as in many gastropods and bivalves, is a derived character. The veliger (Figure 17-7) has the beginnings of a foot, shell, and mantle. In many molluscs the trochophore is passed in the egg, and a veliger hatches to become the only free-swimming stage. Cephalopods, freshwater and some marine snails, and some freshwater bivalves have no free-swimming larvae, and a juvenile hatches from the egg.

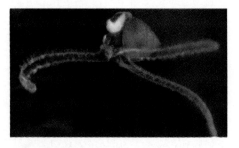

Figure 17-7

Veliger of a snail, *Pedicularia,* swimming. The adults are parasitic on corals. The ciliated process (velum) develops from the prototroch of the trochophore (Figure 17-6).

The trochophore larva (Figure 17-6) is minute, translucent, and more or less pear shaped and has a prominent circlet of cilia (prototroch) and sometimes one or two accessory circlets. It is found in molluscs and annelids with primitive embryonic development and is considered one of the evidences for common phylogenetic origin of the two phyla. Some form of trochophore-like larva is also found in marine turbellarians, nemertines, brachiopods, phoronids, sipunculids, and echiurids, and it probably reflects some phylogenetic relationship among all these phyla.

Classes of Molluscs

For more than 50 years five classes of living molluscs were recognized: Amphineura, Gastropoda, Scaphopoda, Bivalvia (also called Pelecypoda), and Cephalopoda. The discovery of *Neopilina* in the 1950s added another class (Monoplacophora), and Hyman[1] contended that solenogasters and chitons make up separate classes (Polyplacophora and Aplacophora), lapsing the name Amphineura. Recognition of important differences between organisms such as *Chaetoderma* and the

[1]Hyman, L. H. 1967. The invertebrates, vol. VI. New York, McGraw-Hill Book Company.

other solenogasters has led to the separation of the Aplacophora into the sister groups Caudofoveata and the Solenogastres.[2]

CLASS CAUDOFOVEATA

Members of the Class Caudofoveata are wormlike, marine organisms ranging from 2 to 140 mm in length (Figure 17-41). They are mostly burrowers and orient themselves vertically, with the terminal mantle cavity and gills at the entrance of the burrow. They feed on microorganisms and detritus. They have no shell, but their bodies are covered with calcareous scales. There are no spicules or scales on the oral pedal shield, an organ apparently associated with food selection and intake. A radula is present, although reduced in some, and the sexes are separate. This little group has fewer than 70 species; however, its features may be closer to those of the common ancestor of molluscs than any other living molluscs.

CLASS SOLENOGASTRES

The solenogasters (see Figure 17-41) and the caudofoveates were formerly united in the Class Aplacophora, and some zoologists retain the name Aplacophora for the solenogasters. Both the caudofoveates and the solenogasters are marine, wormlike, shell-less, with calcareous scales or spicules in their integument, with reduced head, and without nephridia. The solenogasters, however, usually have no radula and no gills (although secondary respiratory structures may be present). Their foot is represented by a midventral, narrow furrow, the pedal groove. They are hermaphroditic. Rather than burrowing, solenogasters live free on the bottom, and they often live and feed on cnidarians. The solenogasters are also a small group, numbering about 250 species.

[2]Boss, K. J. 1982. Mollusca. In Parker, S. P. ed., Synopsis and classification of living organisms, vol. 1. New York, McGraw-Hill Book Company.

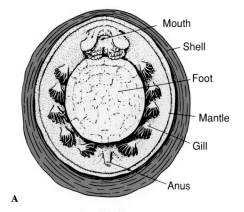

A

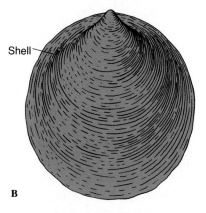

B

Figure 17-8

Neopilina, Class Monoplacophora. Living specimens range from 3 mm to about 3 cm in length. **A,** Ventral view. **B,** Dorsal view.

CLASS MONOPLACOPHORA

Until 1952 it was thought that the Monoplacophora were extinct; they were known only from Paleozoic shells. However, in that year living specimens of *Neopilina* (Gr. *neo,* new, + *pilos,* felt cap) were dredged up from the ocean bottom near the west coast of Costa Rica. Nearly a dozen species of monoplacophorans are now known. These molluscs are small and have a low, rounded shell and a creeping foot (Figure 17-8). They have superficial resemblance to limpets, but unlike most other molluscs, a number of organs are serially repeated. Such serial repetition occurs to a more limited extent in the chitons. *Neopilina* has five pairs of gills, two pairs of auricles, six pairs of nephridia, one or two pairs of go-

nads, and a ladderlike nervous system with 10 pairs of pedal nerves. The mouth bears the characteristic radula.

CLASS POLYPLACOPHORA: CHITONS

The chitons (Gr. coat of mail, tunic) (Figures 17-9 and 17-10) represent a somewhat more diverse molluscan group. They are rather flattened dorsoventrally and have a convex dorsal surface that bears eight articulating limy plates, or valves, hence their name Polyplacophora ("many plate bearers"). The plates overlap posteriorly and are usually dull colored to match the rocks to which the chitons cling. Their head and cephalic sensory organs are reduced, but photosensitive structures **(esthetes),** which have the form of eyes in some chitons, pierce the plates.

Most chitons are small (2 to 5 cm); the largest, *Cryptochiton* (Gr. *crypto,* hidden, + *chiton,* coat of mail), rarely exceeds 30 cm. They prefer rocky surfaces in intertidal regions, although some live at great depths. Most chitons are stay-at-home organisms, straying only very short distances for feeding. In feeding, the radula projects from the mouth to scrape algae from the rocks. The radula is reinforced with the iron-containing mineral, magnetite. The chiton clings tenaciously to its rock with the broad, flat foot. If detached, it can roll up like an armadillo for protection.

The mantle forms a girdle around the margin of the plates, and in some species mantle folds cover part or all of the plates. Compared with the primitive condition, the mantle cavity has been extended along the side of the foot, and the gills have been increased in number. Thus the gills are suspended from the roof of the mantle cavity along each side of the broad ventral foot. With the foot and the mantle margin adhering tightly to the substrate, these grooves become closed chambers, open only at the ends. Water enters the grooves anteriorly,

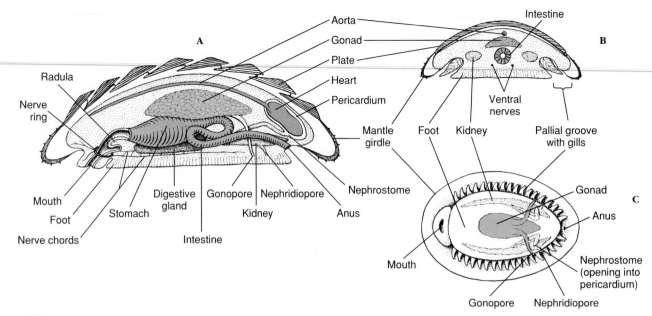

Figure 17-9
Anatomy of a chiton (Class Polyplacophora). **A,** Longitudinal section. **B,** Transverse section. **C,** External ventral view.

Figure 17-10
Mossy chiton, *Mopalia muscosa*. The upper surface of the mantle, or "girdle," is covered with hairs and bristles, an adaptation for defense.

flows across the gills, and leaves posteriorly, bringing a continuous supply of oxygen to the gills. At low tide the margins of the mantle can be tightly pressed to the substratum to diminish water loss, but in some circumstances, the mantle margins can be held open for limited air breathing. A pair of **osphradia** (sense organs for sampling water) are found in the mantle grooves near the anus of many chitons.

Blood pumped by the three-chambered heart reaches the gills by way of an aorta and sinuses. A pair of kidneys (metanephridia) carries waste from the pericardial cavity to the exterior. Two pairs of longitudinal nerve cords are connected in the buccal region.

Sexes are separate in most chitons, and the trochophore larva metamorphoses directly into a juvenile, without an intervening veliger stage.

CLASS SCAPHOPODA

The Scaphopoda, commonly called the tusk shells or tooth shells, are benthic marine molluscs found from the subtidal zone to over 6000 m depth. They have a slender body covered with a mantle and a tubular shell open at both ends. In the scaphopods the molluscan body plan has taken a new direction, with the mantle wrapped around the viscera and fused to form a tube. Most scaphopods are 2.5 to 5 cm long, although they range from 4 mm to 25 cm long. *Dentalium* (L. *dentis,* tooth) is a common Atlantic genus.

The foot, which protrudes through the larger end of the shell, is used to burrow into mud or sand, always leaving the small end of the shell exposed to the water above (Figure 17-11). Respiratory water circulates through the mantle cavity both

by movements of the foot and ciliary action (Figure 17-11). Gaseous exchange occurs in the mantle, for gills are absent. Most of the food is detritus and protozoa from the substratum. It is caught on the cilia of the foot or on the mucus-covered, ciliated knobs of the long tentacles extending from the head (**captacula**) and is conveyed to the nearby mouth. The radula carries the food to a crushing gizzard. The captacula may serve some sensory function, but the eyes, tentacles, and osphradia typical of many other molluscs are lacking.

The sexes are separate, and the larva is a trochophore.

CLASS GASTROPODA

Among the molluscs the Class Gastropoda is by far the largest and most diverse, containing about 40,000 living and 15,000 fossil species. It is made up of members of such diversity that there is no single general term in our language that can apply to them as a group. They include snails, limpets, slugs, whelks, conchs, periwinkles, sea slugs, sea hares, and sea butterflies. They range from marine molluscs with many primitive characters to the highly evolved terrestrial, air-breathing snails and slugs. These animals are basically bilaterally

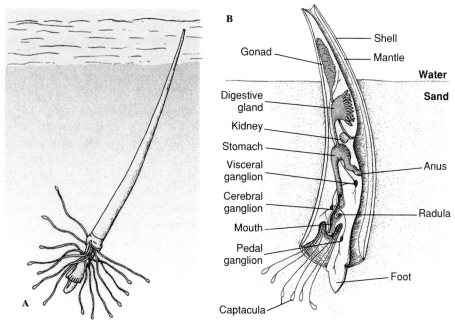

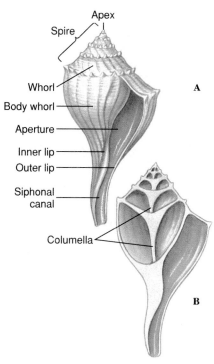

Figure 17-11

Scaphopoda. **A,** *Dentalium* (class Scaphopoda), with its tubular shell, burrows into soft mud or sand and feeds by means of its prehensile tentacles (captacula). Water enters and leaves by way of the posterior aperture. **B,** Internal anatomy of *Dentalium*.

Busycon carica
(knobbed whelk)

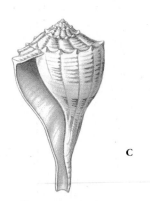

Busycon contrarium
(lightning whelk)

Figure 17-12

Shell of the whelk *Busycon*. **A** and **B,** *Busycon carica,* a dextral, or right-handed, shell. A dextral shell has the aperture on the right side when the shell is held with the apex up and the aperture facing the observer. **C,** *B. contrarium,* a sinistral, or left-handed, shell.

symmetrical, but because of **torsion,** a twisting process that occurs in the veliger stage, the visceral mass has become asymmetrical.

The shell, when present, is always of one piece **(univalve)** and may be coiled or uncoiled. Starting at the **apex,** which contains the oldest and smallest **whorl,** the whorls become successively larger and spiral about the central axis, or **columella** (Figure 17-12). The shell may be right handed **(dextral)** or left handed **(sinistral),** depending on the direction of coiling. Dextral shells are far more common. The direction of coiling is genetically controlled.

Gastropods range from microscopic forms to giant marine forms such as *Pleuroploca gigantea,* a snail with a shell up to 60 cm long, and the sea hare *Aplysia* (see Figure 17-21), some species of which reach 1 m in length. Most of them, however, are between 1 and 8 cm in length. Some fossil gastropods were as much as 2 m long.

The range of gastropod habitats is large. In the sea, gastropods are common both in the littoral zones and at great depths, and some are even pelagic. Some are adapted to brackish water and others to fresh water. On land they are restricted by such factors as the mineral content of the soil and extremes of temperature, dryness, and acidity. Even so, they are widespread, and some have been found at great altitudes and some even in polar regions. Snails occupy all kinds of habitats: in small pools or large bodies of water, in woodlands, in pastures, under rocks, in mosses, on cliffs, in trees, underground, and on the bodies of other animals. They have successfully undertaken every mode of life except aerial locomotion.

Gastropods are usually sluggish, sedentary animals because most of them have heavy shells and slow locomotion. Some are specialized for climbing, swimming, or burrowing. Shells are their chief defense, although they are also protected by coloration and by secretive habits. Many snails have an **operculum,** a horny plate that covers the shell **aperture** when the body is withdrawn into the shell. Others lack shells altogether. Some are distasteful to other animals, and a few such as *Strombus* can deal an active blow with the foot, which bears a sharp operculum. Nevertheless, they are eaten by birds, beetles, small mammals, fish, and other predators. Serving as intermediate hosts for many kinds of parasites, especially trematodes, snails are often harmed by the larval stages of parasites.

Torsion

Of all the molluscs, only gastropods undergo torsion. Torsion is a peculiar phenomenon that moves the mantle

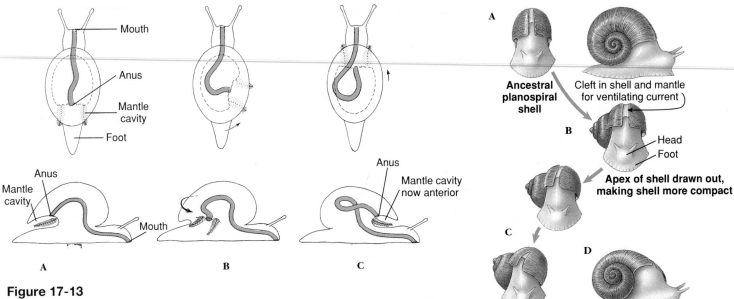

Figure 17-13

Torsion in gastropods. **A,** Ancestral condition before torsion. **B,** Hypothetical intermediate condition. **C,** Early gastropod, torsion complete; direction of crawling now tends to carry waste products back into mantle cavity, resulting in fouling.

Figure 17-14

Evolution of shell in gastropods. **A,** Earliest coiled shells were planospiral, each whorl lying completely outside the preceding whorl. **B,** Better compactness was achieved by snails in which each whorl lay partially to the side of the preceding whorl. **C** and **D,** Better weight distribution resulted when shell was moved upward and posteriorly.

cavity, which was originally (primitively) posterior, to the front of the body, thus twisting the visceral organs as well through a 90- to 180-degree rotation. It occurs during the veliger stage, and in some species the first part may take only a few minutes. The second 90 degrees typically takes a longer period. Before torsion occurs, the embryo's mouth is anterior and the anus and mantle cavity are posterior (Figure 17-13). The change is brought about by an uneven growth of the right and left muscles that attach the shell to the head-foot.

After torsion, the anus and mantle cavity become anterior and open above the mouth and head. The left gill, kidney, and heart auricle are now on the right side, whereas the original right gill, kidney, and heart auricle are now on the left, and the nerve cords have been twisted into a figure eight. Because of the space available in the mantle cavity, the animal's sensitive head end can now be withdrawn into the protection of the shell, with the tougher foot forming a barrier to the outside.

Varying degrees of **detorsion** are seen in the opisthobranchs and pulmonates, and the anus opens to the right side or even to the posterior. However, both of these groups were derived from torted ancestors.

The curious arrangement that results from torsion poses a serious sanitation problem by creating the possibility of wastes being washed back over the gills **(fouling)** and causes us to wonder what strong evolutionary pressures selected for such a strange realignment of body structures. Several explanations have been proposed, none entirely satisfying. For example, sense organs of the mantle cavity (osphradia) would better sample water when turned in the direction of travel. Certainly the consequences of torsion and the resulting need to avoid fouling have been very important in the subsequent evolution of gastropods. These consequences cannot be explored, however, until another unusual feature of gastropods—coiling—has been described.

Coiling

The coiling, or spiral winding, of the shell and visceral mass is not the same as torsion. Coiling may occur in the larval stage at the same time as torsion, but the fossil record shows that coiling was a separate evolutionary event and originated in gastropods earlier than torsion did. Nevertheless, all living gastropods have descended from coiled, torted ancestors, whether or not they now show these characteristics.

Early gastropods had a bilaterally symmetrical **planospiral** shell; that is, all the whorls lay in a single plane (Figure 17-14A). Such a shell was not very compact, since each whorl had to lie completely outside the preceding one. Curiously, a few modern species have secondarily returned to the planospiral form. The compactness problem of the planospiral shell was solved by the **conispiral** shape, in which each succeeding whorl is at the side of the preceding one (Figure 17-14B). However, this shape was clearly unbalanced, hanging as it was with much weight over to one side. Better weight distribution was achieved by shifting the shell upward and posteriorly, with the shell axis oblique to the longitudinal axis of the foot (Figure 17-14C). The weight and bulk of the main body whorl, the largest whorl of the shell, pressed on the right side of the mantle cavity, however, and apparently interfered with the organs on that side. Accordingly, the gill, auricle, and kidney of the right side have been lost in all except primitive living gastropods, leading to a condition of *bilateral asymmetry.*

Although the loss of the right gill was probably an adaptation to the mechanics of carrying the coiled shell, that condition made possible a way to avoid the problem of torsion—fouling—that is displayed in most modern prosobranchs. Water is brought into the left side of the mantle cavity and out the right side, carrying with it the wastes from the anus and nephridiopore, which lie near the right side. Ways in which fouling is avoided in other gastropods are mentioned later in the chapter.

Feeding Habits

Feeding habits of gastropods are as varied as their shapes and habitats, but all include the use of some adaptation of the radula. The majority of gastropods are herbivorous, rasping off particles of algae. Some herbivores are grazers, some are browsers, and some are planktonic feeders. *Haliotis,* the abalone (Figure 17-15A), holds seaweed with the foot and breaks off pieces with the radula. Land snails forage at night for green vegetation.

Some snails, such as *Bullia* and *Buccinum,* are scavengers living on dead and decaying flesh; others are carnivores that tear their prey with the radular teeth. *Melongena* feeds on clams, especially *Tagelus,* the razor clam, thrusting its proboscis between the gaping shell valves. *Fasciolaria* and *Polinices* (Figure 17-15B) feed on a variety of molluscs, preferably bivalves. *Urosalpinx cinerea,* the oyster borer, or tingle, drills holes through the shell of the oyster. Its radula, bearing three longitudinal rows of teeth, is used first to begin the drilling action, then the animal glides forward, everts an accessory boring organ through a pore in the anterior sole of its foot, and holds it against the shell, using a chemical agent to soften the shell. Short periods of rasping alternate with long periods of chemical activity until a neat round hole is completed. With its proboscis inserted through the hole, the snail may feed continuously for hours or

A   **B**

Figure 17-15

A, Red abalone, *Haliotus rufescens.* This huge, limpetlike snail is prized as food and extensively marketed. Abalones are strict vegetarians, feeding especially on sea lettuce and kelp. **B,** Moon snail, *Polinices lewisii.* A common inhabitant of West Coast sand flats, the moon snail is a predator of clams and mussels. It uses its radula to drill neat holes through its victim's shell, through which the proboscis is then extended to eat the bivalve's fleshy body.

Figure 17-16

Conus extends its long, wormlike proboscis. When the fish attempts to consume this tasty morsel, the *Conus* stings it in the mouth and kills it. The snail engulfs the fish with its distensible stomach, then regurgitates the scales and bones some hours later.

days, using the radula to tear away the soft flesh. *Urosalpinx* is attracted to its prey at some distance by sensing some chemical, probably one released in the metabolic wastes of the prey.

Cyphoma gibbosum and related species live and feed on gorgonians (Phylum Cnidaria, Chapter 14) in shallow, tropical coral reefs. This snail is commonly known as the flamingo tongue. During normal activity the brightly colored mantle entirely envelops the shell, but it can be quickly withdrawn into the shell aperture when the animal is disturbed.

Members of the genus *Conus* (Figure 17-16) feed on fish, worms, and molluscs. Their radula is highly modi-

fied for prey capture. A gland charges the radular teeth with a highly toxic venom. When *Conus* senses the presence of its prey, a single radular tooth slides into position at the tip of the proboscis. Upon striking the prey, the proboscis expels a tooth like a harpoon, and the poison quiets the prey at once. This is an effective adaptation for a slowly moving predator to prevent the escape of a swiftly moving prey. Some species of *Conus* can deliver very painful stings, and in several species the sting is lethal to humans. The venom consists of a series of toxic peptides, and each *Conus* species carries peptides (**conotoxins**) that are specific for the neuroreceptors of its preferred prey. Conotoxins have become valuable tools in research on the various receptors and ion channels of nerve cells.

Some gastropods feed on organic deposits on the sand or mud. Others collect the same sort of organic debris but can digest only the microorganisms contained in it. Some sessile gastropods, such as some limpets, are ciliary feeders that use the gill cilia to draw in particulate matter, roll it into a mucous ball, and carry it to the mouth. Some sea butterflies secrete a mucous net to catch small planktonic forms; then they draw the web into the mouth.

After maceration by the radula or by some grinding device, such as the gizzard in the sea hare *Aplysia,* digestion is usually extracellular in the lumen of the stomach or digestive glands. In ciliary feeders the stomachs are sorting regions, and most of the digestion is intracellular in the digestive glands.

Internal Form and Function

Respiration in most gastropods is carried out by a **ctenidium** (two ctenidia is the primitive condition, found in some prosobranchs) located in the mantle cavity, though some aquatic forms, lacking gills, depend on the mantle and skin. After the more derived prosobranchs lost one of the gills, most of them lost half of the remaining one, and the central axis became attached to the wall of the mantle cavity (Figure 17-17). Thus they attained the most efficient gill arrangement for the way the water circulated through the mantle cavity (in one side and out the other).

The pulmonates have a highly vascular area in the mantle that serves as a **lung** (Figure 17-18). Most of the mantle margin seals to the back of the animal, and the lung opens to the outside by a small opening called the **pneumostome.** Many aquatic pulmonates must surface to expel a bubble of gas from the lung. To take in air, they curl the edge of the mantle around the pneumostome to form a siphon.

Most gastropods have a single nephridium (kidney). The circulatory and nervous systems are well developed (Figure 17-18). The latter incorporates three pairs of ganglia connected by nerves. Sense organs include eyes or simple photoreceptors, stato-

cysts, tactile organs, and chemoreceptors. The simplest type of gastropod eye is simply a cuplike indentation in the skin lined with pigmented photoreceptor cells. In many gastropods the eyecup contains a lens and is covered with a cornea. A sensory area called the **osphradium,** located at the base of the incurrent siphon of most gastropods, is chemosensory in some forms, although its function may be mechanoreceptive in some and is still unknown in others.

There are both dioecious and monoecious gastropods. Many gastropods perform courtship ceremonies. During copulation in monoecious species there is an exchange of spermatozoa or spermatophores (bundles of sperm). Many terrestrial pulmonates eject a dart from a dart sac (Figure 17-18) into the partner's body to heighten excitement before copulation. After copulation each partner deposits its eggs in shallow burrows in the ground. Gastropods with the most primitive characteristics discharge ova and sperm into the seawater where fertilization occurs, and the embryos soon hatch as free-swimming trochophore larvae. In most gastropods fertilization is internal.

Fertilized eggs encased in transparent shells may be emitted singly to float among the plankton or may be laid in gelatinous layers attached to

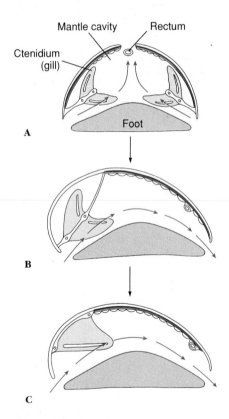

Figure 17-17

Evolution of the ctenidia in gastropods.
A, Primitive prosobranchs with two ctenidia and excurrent water leaving the mantle cavity by a dorsal slit or hole. **B,** Condition after one ctenidium had been lost. **C,** Derived condition found in majority of prosobranchs, in which filaments on one side of remaining gill are lost, and axis is attached to mantle wall.

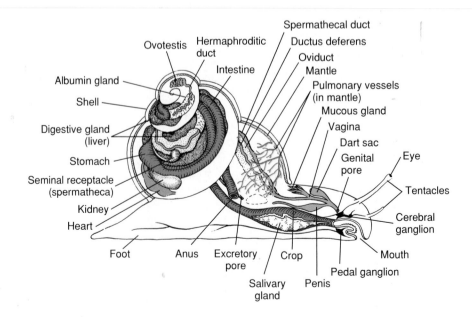

Figure 17-18

Anatomy of a pulmonate snail.

the substratum. Some marine forms enclose the eggs, either in small groups or in large numbers, in tough egg capsules, in a wide variety of egg cases (Figure 17-19). The young generally emerge as veliger larvae (Figure 17-7), or they may spend the veliger stage in the case or capsule and emerge as young snails. Some species, including many freshwater snails, are ovoviviparous, brooding their eggs and young in the pallial oviduct.

Major Groups of Gastropods

There are three subclasses of gastropods: the Prosobranchia, Opisthobranchia, and Pulmonata.

Prosobranchia. This group contains most of the marine snails and some freshwater and terrestrial gastropods. The mantle cavity is anterior as a result of torsion, with the gill or gills lying in front of the heart. Water enters the left side and exits from the right side, and the edge of the mantle often extends into a long siphon to separate incurrent from excurrent flow. In prosobranchs with two gills (for example, the abalone *Haliotis* and the keyhole limpet *Diodora,* Figures 17-15A and 17-20B), fouling is avoided by having the excurrent water go up and out through one or more holes in the shell above the mantle cavity.

Prosobranchs have one pair of tentacles. The sexes are usually separate. An operculum is often present.

They range in size from the periwinkles and small limpets (*Patella* and *Diodora*) (Figure 17-20B) to the horse conch (*Pleuroploca*), the largest gastropod in the Atlantic Ocean. Familiar examples of prosobranchs are the abalone (*Haliotis*), which has an ear-shaped shell; the whelk (*Busycon*), which lays its eggs in double-edged, disc-shaped capsules attached to a cord a meter long; the common periwinkle (*Littorina*); the moon snail (*Polinices,* Figure 17-15B); the oyster borer (*Urosalpinx*), which bores into oysters and sucks out their juices; the rock shell (*Murex*), a European

species that was used to make the royal purple of the ancient Romans; and some freshwater forms (*Goniobasis* and *Viviparus*).

Opisthobranchia. The opisthobranchs are an odd assemblage of molluscs that include sea slugs, sea hares, sea butterflies, and canoe shells. They are nearly all marine; most of them are shallow-water forms, hiding under stones and seaweed; a few are pelagic. Currently nine or more orders of opisthobranchs are recognized, but for convenience they can be divided into two classical groups: tectibranchs, with gill and shell usually present,

and nudibranchs, in which there is no shell or true gill, but in which secondary gills are present along the dorsal side or around the anus. The opisthobranchs show partial or complete detorsion; thus the anus and gill (if present) are displaced to the right side or rear of the body. Clearly, the fouling problem is obviated if the anus is moved away from the head toward the posterior. Two pairs of tentacles are usually found, and the second pair is often further modified (**rhinophores,** Figure 17-21), with platelike folds that apparently increase the area for chemoreception. The shell is typically reduced or absent. All are monoecious.

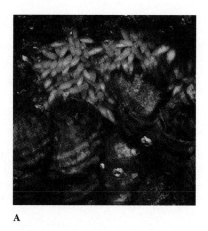

Figure 17-19
Eggs of marine gastropods. **A,** The wrinkled whelk, *Thais lamelosa,* lays egg cases resembling grains of wheat; each contains hundreds of eggs. **B,** Egg ribbon of a dorid nudibranch.

Figure 17-20
A, Cowry, *Jenneria pustulata,* crawls over zoanthid cnidarians. The brightly patterned and polished shells of cowries have been used as ornaments for thousands of years. **B,** *Diodora aspera,* a prosobranch gastropod with a hole in the apex through which the water leaves the mantle cavity.

Rhinophore   Oral tentacle

A   **B**

Figure 17-21
A, The sea hare, *Aplysia dactylomela,* crawls and swims across a tropical seagrass bed, assisted by large, winglike parapodia, here curled above the body.
B, When attacked, the sea hare squirts a copious protective secretion from its "purple gland" in the mantle cavity.

Figure 17-22
An aeolid nudibranch, *Flabellina iodinea.* Its long, dorsal cerata contain nematocysts, which the animal obtains from its cnidarian diet.

Among the tectibranchs are the large sea hare *Aplysia* (Figure 17-21), which has large, earlike anterior tentacles and a vestigial shell, and the pteropods, or sea butterflies (*Cavolina* and *Clione*). In pteropods the foot is modified into fins for swimming; thus, they are pelagic and form a part of the plankton fauna.

The nudibranchs are represented by the sea slugs, which are often brightly colored and carnivorous (Figure 17-22). The plumed sea slug *Aeolis,* which lives on sea anemones and hy-

droids, often draws the color of its prey into the elongated papillae (cerata) that cover its back. It and some other nudibranchs also salvage the nematocysts of cnidarian prey for their own use. The frilled sea slug *Tridachia* is a lovely little green or blue and white form common in Florida waters. *Hermissenda* is one of the more common West Coast nudibranchs.

Pulmonata.   The pulmonates show some detorsion and include the land and most freshwater snails and slugs

(and a few brackish and saltwater forms). They have lost their ancestral ctenidia, but the vascularized mantle wall has become a lung, which fills with air by contraction of the mantle floor (some aquatic species have developed secondary gills in the mantle cavity). The anus and nephridiopore open near the pneumostome, and waste is expelled forcibly with air or water from the lung. They are monoecious. The aquatic species have one pair of nonretractile tentacles, at the base of which are the eyes; land forms have two pairs of tentacles, with the posterior pair bearing the eyes (Figure 17-23). Among the thousands of land species, some of the most familiar American forms are *Helix, Polygyra, Succinea, Anguispira, Zonitoides, Limax,* and *Agriolimax.* Aquatic forms are represented by *Helisoma, Lymnaea,* and *Physa. Physa* is a left-handed (sinistral) snail.

CLASS BIVALVIA (PELECYPODA)

The Bivalvia are also known as Pelecypoda (pel-e-sip´o-da), or "hatchet-footed" animals, as their name implies (Gr. *pelekys,* hatchet, + *pous, podos,* foot). They are the bivalved molluscs that include the mussels, clams, scallops, oysters, and shipworms (Figures 17-24 to 17-27) and they range in size

Pneumostome

A

B

Figure 17-23
A, Pulmonate land snail. Note two pairs of tentacles; the second, larger pair bears the eyes. **B,** Banana slug, *Ariolimax columbianus.* Note pneumostome.

A

B

Figure 17-24
Bivalve molluscs. **A,** Mussels, *Mytilus edulis,* occur in northern oceans around the world; they form dense beds in the intertidal zone. A host of marine creatures live protected beneath attached mussels. **B,** Scallops (*Chlamys opercularis*) swim to escape the attack by starfish (*Asterias rubens*). When alarmed, these most agile of bivalves swim by clapping the two shell valves together.

B, Photo by D. P. Wilson/Frank Lane Picture Agency Lmtd.

Figure 17-25
Representing a group that has evolved from burrowing ancestors, the surface-dwelling bivalve *Pecten* sp. has developed sensory organs along its mantle edges (tentacles and a series of blue eyes).

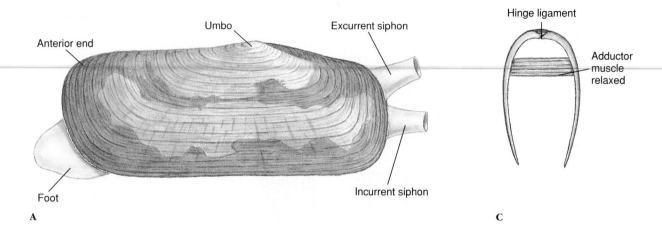

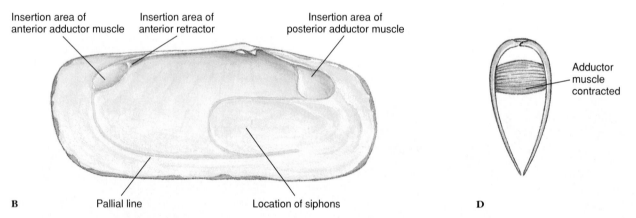

Figure 17-26

Tagelus plebius, the stubby razor clam (class Bivalvia). **A,** External view of right valve. **B,** Inside of left shell showing scars where muscles were attached. The mantle was attached at the pallial line. **C** and **D,** Sections showing function of adductor muscles and hinge ligament. In **C** the adductor muscle is relaxed, allowing the hinge ligament to pull the valves apart. In **D** the adductor muscle is contracted, pulling the valves together.

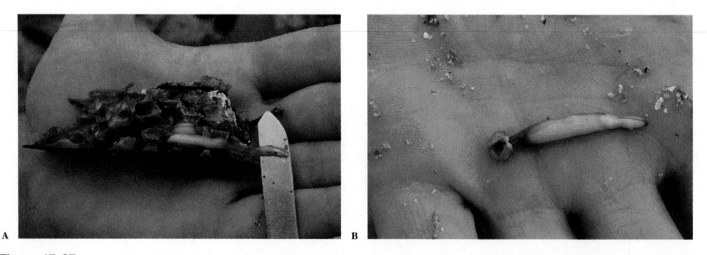

Figure 17-27

A, Shipworms are bivalves that burrow in wood, causing great damage to unprotected wooden hulls and piers. **B,** The two small, anterior valves, seen at left, are used as rasping organs to extend the burrow.

from tiny seed shells 1 to 2 mm in length to giant South Pacific clams *Tridacna,* which may reach more than 1 m in length and as much as 225 kg (500 pounds) in weight (see Figure 17-35). Most bivalves are sedentary **filter feeders** that depend on ciliary currents produced by the gills to bring in food materials. Unlike the gastropods, they have no head, no radula, and very little cephalization.

Most bivalves are marine, but many live in brackish water and in streams, ponds, and lakes.

Freshwater clams were once abundant and diverse in streams throughout the eastern United States, but they are now easily the most jeopardized group of animals in the country. Of the more than 300 species once present, 12 are extinct, 42 are listed as threatened or endangered, and as many as 88 more may be listed soon. A combination of causes is responsible, of which a decline in water quality is among the most important. Pollution and sedimentation from mining, industry, and agriculture are among the culprits. Poaching to supply the Japanese cultured pearl industry is partially to blame. And in addition to everything else, the prolific zebra mussels (see next note) attach in great numbers to the native clams, exhausting food supplies (phytoplankton) in the surrounding water.

A

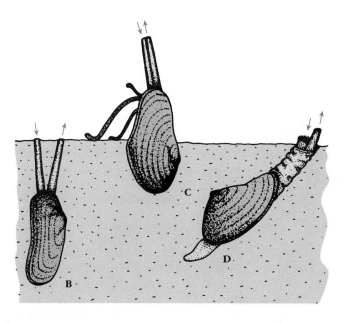

Figure 17-28
Adaptations of siphons in bivalves. **A,** In the northwest ugly clam, *Entodesma saxicola,* the incurrent and excurrent siphons are clearly visible. **B** to **D,** In many marine forms the mantle is drawn out into long siphons. In **A, B,** and **D,** the incurrent siphon brings in both food and oxygen. In **C,** *Yoldia,* the siphons are respiratory; long ciliated palps feel about over the mud surface and convey food to the mouth.

Form and Function

Shell. Bivalves are laterally compressed, and their two shells **(valves)** are held together dorsally by a hinge ligament that causes the valves to gape ventrally. The valves are drawn together by adductor muscles that work in opposition to the hinge ligament (Figure 17-26C and D). The **umbo** is the oldest part of the shell, and growth occurs in concentric lines around it (Figure 17-26A).

Pearl production is the by-product of a protective device used by the animal when a foreign object (grain of sand, parasite, or other) becomes lodged between the shell and mantle. The mantle secretes many layers of nacre around the irritating object (Figure 17-5). Pearls are cultured by inserting particles of nacre, usually taken from the shells of freshwater clams, between the shell and mantle of a certain species of oyster and by keeping the oysters in enclosures for several years. *Meleagrina* is an oyster used extensively by the Japanese for pearl culture.

Body and mantle. The **visceral mass** is suspended from the dorsal midline, and the muscular foot is attached to the visceral mass anteroventrally. The ctenidia hang down on each side, each covered by a fold of the mantle. The posterior edges of the mantle folds are modified to form dorsal excurrent and ventral incurrent openings (Figure 17-28A). In some marine bivalves the mantle is drawn out into long muscular siphons that allow the clam to burrow into the mud or sand and extend the siphons to the water above (Figure 17-28B to D).

Locomotion. Pelecypods initiate movement by extending a slender muscular foot between the valves (Figure 17-28D). Blood is pumped into the foot, causing it to swell and to act as an anchor in the mud or sand, then longitudinal muscles contract to shorten the foot and pull the animal forward.

Scallops and file shells are able to swim with a jerky motion by clapping their valves together to create a sort of jet propulsion. The mantle edges can direct the stream of expelled water, so that the animals can swim in virtually any direction (Figure 17-24).

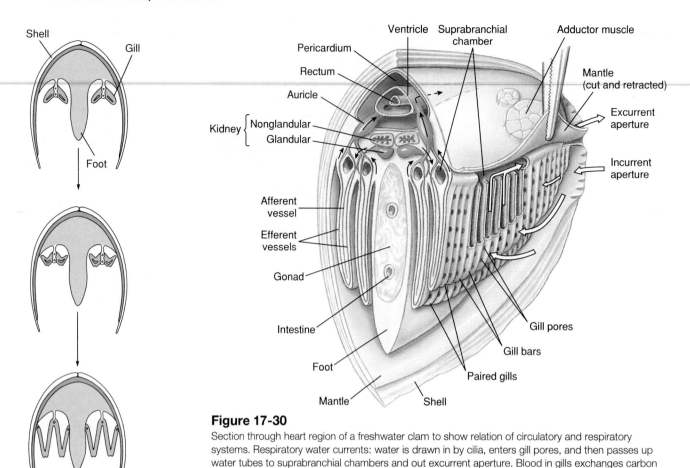

Shell
Gill
Foot

Figure 17-29

Evolution of bivalve ctenidia. By a great lengthening of individual filaments, the ctenidia became adapted for filter feeding and separated the incurrent chamber from the excurrent, suprabranchial chamber.

Ventricle   Suprabranchial chamber   Adductor muscle
Pericardium
Rectum   Mantle (cut and retracted)
Auricle   Excurrent aperture
Kidney {
Nonglandular
Glandular   Incurrent aperture
Afferent vessel
Efferent vessels   Gill pores
Gonad   Gill bars
Intestine   Paired gills
Foot
Mantle   Shell

Figure 17-30

Section through heart region of a freshwater clam to show relation of circulatory and respiratory systems. Respiratory water currents: water is drawn in by cilia, enters gill pores, and then passes up water tubes to suprabranchial chambers and out excurrent aperture. Blood in gills exchanges carbon dioxide for oxygen. Blood circulation: ventricle pumps blood forward to sinuses of foot and viscera, and posteriorly to mantle sinuses. Blood returns from mantle to auricles; it returns from viscera to the kidney, and then goes to the gills, and finally to the auricles.

Gills. Gaseous exchange occurs through both the mantle and the gills. The gills of most bivalves are highly modified for filter feeding; they are derived from the primitive ctenidia by a great lengthening of the filaments on each side of the central axis (Figure 17-29). As the ends of the long filaments became folded back toward the central axis, the ctenidial filaments took the shape of a long, slender W. The filaments lying beside each other became joined by ciliary junctions or tissue fusions, forming platelike **lamellae** with many vertical water tubes inside. Thus water enters the incurrent siphon, propelled by ciliary action, then enters the water tubes through pores between the filaments in the lamellae, proceeds dorsally into a common **suprabranchial chamber** (Figure 17-30), and then out the excurrent aperture.

Feeding. Most bivalves are filter feeders. The respiratory currents bring both oxygen and organic materials to the gills where ciliary tracts direct them to the tiny pores of the gills. Gland cells on the gills and labial palps secrete copious amounts of mucus, which entangles particles suspended in the water going through gill pores. These mucous masses slide down the outside of the gills toward food grooves at the lower edge of the gills (Figure 17-31). Heavier particles of sediment drop off the gills as a result of gravitational pull, but smaller particles travel along the food grooves toward the labial palps. The palps, being also grooved and ciliated, direct the mucous mass into the mouth.

Some bivalves, such as *Nucula* and *Yoldia,* are deposit feeders and have long proboscides attached to the labial palps (Figure 17-28C). These can be protruded onto the sand or mud to collect food particles, in addition to the particles attracted by the gill currents.

Shipworms (Figure 17-27) burrow in wood and feed on the particles they excavate. Symbiotic bacteria live in a special organ in the bivalve and produce cellulase to digest the wood.

Septibranchs, another group of bivalves, draw small crustaceans or bits of organic debris into the mantle cavity by sudden inflow of water created by the pumping action of a muscular septum in the mantle cavity.

Internal structure and function. The floor of the stomach of filter-feeding bivalves is folded into ciliary tracts for sorting the continuous stream of particles. A cylindrical **style**

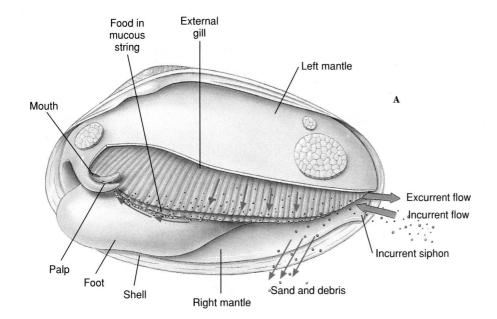

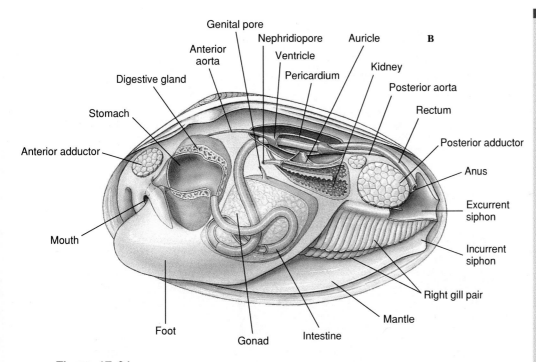

Figure 17-31

A, Feeding mechanism of freshwater clam. Left valve and mantle are removed. Water enters the mantle cavity posteriorly and is drawn forward by ciliary action to the gills and palps. As water enters the tiny openings of the gills, food particles are sieved out and caught up in strings of mucus that are carried by cilia to the palps and directed to the mouth. Sand and debris drop into the mantle cavity and are removed by cilia. **B,** Clam anatomy.

The three-chambered heart, which lies in the pericardial cavity (Figure 17-31), has two auricles and a ventricle and beats slowly, ranging from 0.2 to 30 times per minute. Part of the blood is oxygenated in the mantle and returns to the ventricle through the auricles; the rest circulates through sinuses and passes in a vein to the kidneys, from there to the gills for oxygenation, and back to the auricles.

A pair of U-shaped kidneys (nephridial tubules) lies just ventral and posterior to the heart (Figure 17-31B). The glandular portion of each tubule opens into the pericardium; the bladder portion empties into the suprabranchial chamber.

Zebra mussels, *Dreissena polymorpha*, are a recent and potentially disastrous biological introduction into North America. They were apparently picked up as veligers with ballast water by one or more ships in freshwater ports in northern Europe and then expelled between Lake Huron and Lake Erie in 1986. This 4 cm bivalve spread throughout the Great Lakes by 1990, and by 1994 it was as far south on the Mississippi as New Orleans, as far north as Duluth, Minnesota, and as far east as the Hudson River in New York. It attaches to any firm surface and filter feeds on phytoplankton. Large numbers build up rapidly. They foul water intake pipes of municipal and industrial plants, impede intake of water for municipal supplies, and have far-reaching effects on the ecosystem (see preceding note). Zebra mussels may cost $5 billion or more to control by the end of the century.

The nervous system consists of three pairs of widely separated ganglia connected by commissures and a system of nerves. Sense organs are poorly developed. They include a pair of statocysts in the foot, a pair of osphradia of uncertain function in the mantle cavity, tactile cells, and sometimes simple pigment cells on the mantle. Scallops (*Pecten, Chlamys*) have a row of small blue eyes along

sac opening into the stomach secretes a gelatinous rod called the **crystalline style,** which projects into the stomach and is kept whirling by means of cilia in the style sac (Figure 17-32). Rotation of the style helps to dissolve its surface layers, freeing digestive enzymes (especially amylase) that it contains, and to roll the mucous food mass. Dislodged particles are sorted, and suitable ones are directed to the digestive gland or picked up by amebocytes. Further digestion is intracellular.

each mantle edge (Figure 17-25). Each eye has a cornea, lens, retina, and pigmented layer. Tentacles on the margin of the mantle of *Pecten* (Figure 17-25) and *Lima* (p. 321) have tactile and chemoreceptor cells.

Reproduction and development. Sexes are usually separate. Gametes are discharged into the suprabranchial chamber to be carried out with the excurrent flow. An oyster may produce 50 million eggs in a single season. In most bivalves fertilization is external. The embryo develops into trochophore, veliger, and spat stages (Figure 17-33).

In most freshwater clams fertilization is internal. Eggs drop into the water tubes of the gills where they are fertilized by sperm entering with the incurrent flow. They develop there into a bivalved **glochidium larva** stage, which is a specialized veliger (Figure 17-34). When discharged, the glochidia are carried by water currents, and if they come in contact with a passing fish, they attach to its gills or skin and live as parasites for several weeks. Then they sink to the bottom to begin independent lives. Larval "hitchhiking" helps distribute a form whose locomotion is very limited.

Boring. Many pelecypods can burrow into mud or sand, but some have evolved a mechanism for burrowing into much harder substances, such as wood or stone.

Teredo, Bankia, and some other genera are called shipworms. They can be very destructive to wooden ships and wharves. These strange little clams have a long, wormlike appearance, with a pair of slender siphons on the posterior end that keep water flowing over the gills, and a pair of small globular valves on the anterior end with which they burrow (Figure 17-27). The valves have microscopic teeth that function as very effective wood rasps. The animals extend their burrows with an unceasing rasping motion of the valves. This sends a continuous flow of fine wood particles into the digestive tract where they are attacked by cellulase produced by symbiotic bacteria. Interestingly, these

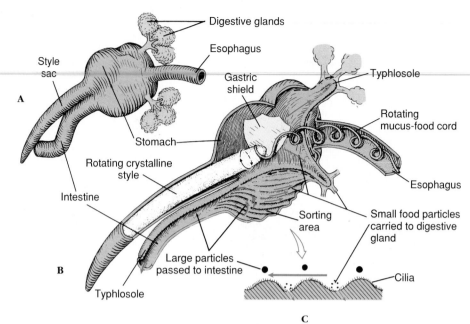

Figure 17-32
Stomach and crystalline style of ciliary-feeding clam. **A,** External view of stomach and style sac. **B,** Transverse section showing direction of food movements. Food particles in incoming water are caught in a cord of mucus that is kept rotating by the crystalline style. Ridged sorting areas direct large particles to the intestine and small food particles to digestive glands. **C,** Sorting action of cilia.

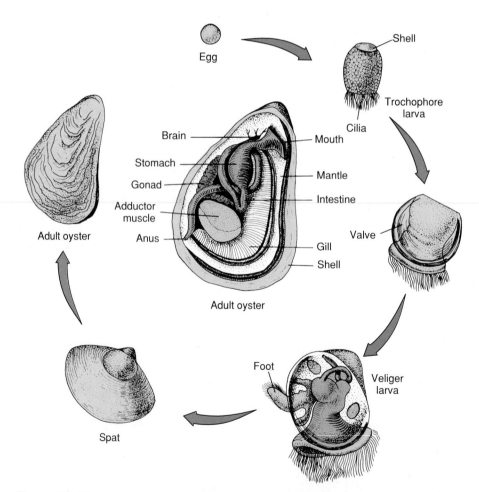

Figure 17-33
Life cycle of the oyster. Oyster larvae swim about for approximately 2 weeks before settling down for attachment to become spats. Oysters take about 4 years to grow to commercial size.

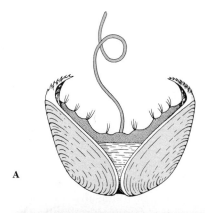

A

B

Figure 17-34

A, Glochidium, or larval form, for some freshwater clams. When the larva is released from brood pouch of mother, it may become attached to a fish's gill by clamping its valves closed. It remains as a parasite on the fish for several weeks. Its size is approximately 0.3 mm. **B,** Some clams have adaptations that help their glochidia find a host. The mantle edge of this female pocketbook mussel (*Lampsilis ovata*) mimics a small minnow, complete with eye. When a smallmouth bass comes to dine, it gets doused with glochidia.

Figure 17-35

Clam (*Tridacna gigas*) lies buried in coral rock with greatly enlarged siphonal area visible. These tissues are richly colored and bear enormous numbers of symbiotic single-celled algae (zooxanthellae) that provide much of the clam's nutriment.

bacteria also fix nitrogen, an important property for their hosts, which live on a diet (wood) high in carbon but deficient in nitrogen.

Some clams bore into rock. The piddock (*Pholas*) bores into limestone, shale, sandstone, and sometimes wood or peat. It has strong valves that bear spines, which it uses to cut away the rock gradually while anchoring itself with its foot. *Pholas* may grow to 15 cm long and make rock burrows up to 30 cm long.

CLASS CEPHALOPODA

The Cephalopoda (Gr. *kephalē,* head, + *pous, podos,* foot) include the squids, octopuses, nautiluses, devil-fish, and cuttlefish. All are marine, and all are active predators.

The modified foot is concentrated in the head region. It is in the form of a funnel for expelling water from the mantle cavity. The anterior margin of the head is drawn out into a circle or crown of arms or tentacles.

Cephalopods range upward in size from 2 or 3 cm. The common squid of markets, *Loligo,* is about 30 cm long. The giant squid *Architeuthis* is the largest invertebrate known.

Fossil records of cephalopods go back to Cambrian times. The earliest shells were straight cones; others were curved or coiled, culminating in the coiled shell similar to that of the modern *Nautilus,* the only remaining member of the once flourishing nautiloids (Figure 17-36). Cephalopods without shells or with internal shells (such as octopuses and squids) apparently evolved from some early straight-shelled ancestor. Many ammonoids, which are extinct, had quite elaborate shells (Figure 17-36C).

The enormous giant squid, *Architeuthis,* is very poorly known because no one has ever been able to study a living specimen. The anatomy has been studied from stranded animals, from those captured in the nets of fishermen, and from specimens found in the stomach of sperm whales. The mantle length is 5 to 6 m, and the head is up to one meter. They have the largest eyes in the animal kingdom: up to 25 cm (10 inches) in diameter. They apparently eat fish and other squids, and they are an important food item for sperm whales. They are thought to live on or near the sea bottom at a depth of 1000 m, but some have been observed swimming at the surface.

Figure 17-36

Nautilus, a cephalopod. **A,** Live *Nautilus,* feeding on a fish. **B,** Longitudinal section, showing gas-filled chambers of shell, and diagram of body structure. **C,** Longitudinal section through shell of an ammonoid.

A

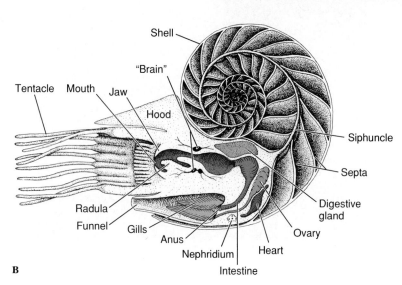

B

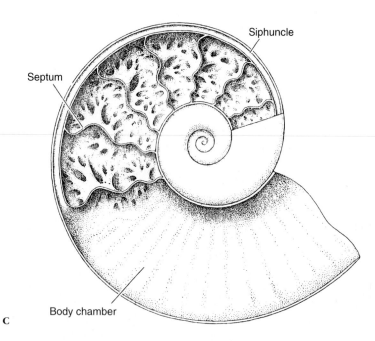

C

taken at depths of 5000 m. *Nautilus* is usually found near the bottom in water 50 to 560 m deep, near islands in the southwestern Pacific.

Form and Function

Shell. Although early nautiloid and ammonoid shells were heavy, they were made buoyant by a series of **gas chambers,** as is that of *Nautilus* (Figure 17-36B), enabling the animal to swim while carrying its shell. The shell of *Nautilus,* although coiled, is quite different from that of a gastropod. The shell is divided by transverse septa into internal chambers (Figure 17-36B). The living animal inhabits only the last chamber. As it grows, it moves forward, secreting behind a new septum. The chambers are connected by a cord of living tissue called the **siphuncle,** which extends from the visceral mass. Cuttlefishes (Figure 17-37) also have a small, curved shell, but it is entirely enclosed by the mantle. In the squids most of the shell has disappeared, leaving only a thin, horny strip called a pen, which is enclosed by the mantle. In *Octopus* (Gr. *oktos,* eight, + *pous, podos,* foot) the shell has disappeared entirely.

Locomotion. Cephalopods swim by forcefully expelling water from the mantle cavity through a ventral **funnel** (or **siphon**)—a sort of jet propulsion method. The funnel is mobile and can be pointed forward or backward to control direction; the force of water expulsion controls speed.

Squids and cuttlefishes are excellent swimmers. The squid body is streamlined and built for speed (Figure 17-38). Cuttlefishes swim more slowly. The lateral fins of squids and cuttlefishes serve as stabilizers, but they are held close to the body for rapid swimming.

Nautilus is active at night; its gas-filled chambers keep the shell upright. Although not as fast as the squid, it moves surprisingly well.

Octopus has a rather globular body and no fins (Figure 17-1E). The octo-

The natural history of some cephalopods is fairly well known. They are marine animals and appear sensitive to the degree of salinity. Few are found in the Baltic Sea, where the water has a low salt content. Cephalopods are found at various depths. The octopus is often seen in the intertidal zone, lurking among rocks and crevices, but occasionally it is found at great depths. The more active squids are rarely found in shallow water, and some have been

Figure 17-37
The cuttlefish, *Sepia* sp., has an internal shell familiar to keepers of caged birds as "cuttlebone."

A

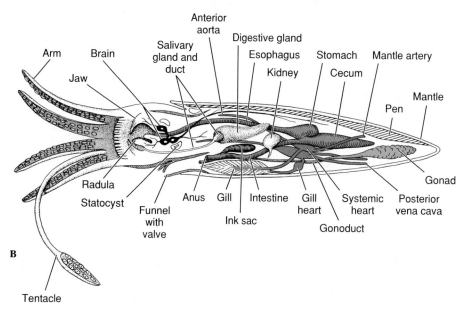

Figure 17-38
A, Reef squid *Sepioteuthis lessoniana*. **B,** Lateral view of squid anatomy, with the left half of the mantle removed.

pus can swim backward by spurting jets of water from its funnel, but it is better adapted to crawling about over the rocks and coral, using the suction discs on its arms to pull or to anchor itself. Some deep-water octopods have the arms webbed like an umbrella and swim in a sort of medusa fashion.

Internal features. The active habits of cephalopods are reflected in their internal anatomy, particularly their respiratory, circulatory, and nervous systems.

Respiration and Circulation.

Except for the nautiloids, cephalopods have one pair of gills. Because ciliary propulsion would not circulate enough water for their high oxygen requirements, there are no cilia on the gills. Instead, radial muscles in the mantle wall thin the wall and enlarge the mantle cavity, drawing water in. Strong circular muscles contract and expel the water forcibly through the funnel. A system of one-way valves prevents the water from being taken in through the funnel and expelled around the mantle margin.

Likewise, the open circulatory system inherited from their ancestral molluscs would be inadequate for cephalopods. Their circulatory system consists of a closed network of vessels, and capillaries conduct the blood through the gill filaments. Furthermore, the molluscan plan of circulation places the entire systemic circulation before the blood reaches the gills (in contrast

to vertebrates, in which the blood leaves the heart and goes directly to the gills or lungs). This functional problem was solved by the development of **accessory** or **branchial hearts** (Figure 17-38B) at the base of each gill to increase the pressure of the blood going through the capillaries there.

After *Nautilus* secretes a new septum, the new chamber is filled with fluid similar in ionic composition to that of the *Nautilus'* blood (and of seawater). Fluid removal involves the active secretion of ions into tiny intercellular spaces in the siphuncular epithelium, so that a very high local osmotic pressure is produced, and the water is drawn out of the chamber by osmosis. The gas in the chamber is just the respiratory gas from the siphuncle tissue that diffuses into the chamber as fluid is removed. Thus the gas pressure in the chamber is 1 atmosphere or less because it is in equilibrium with the gases dissolved in the seawater surrounding the *Nautilus,* which are in turn in equilibrium with air at the surface of the sea, despite the fact that the *Nautilus* may be swimming at 400 m beneath the surface. That the shell can withstand implosion by the surrounding 41 atmospheres (about 600 pounds per square inch), and that the siphuncle can remove water against this pressure are marvelous feats of natural engineering!

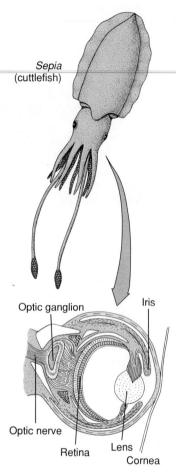

Sepia
(cuttlefish)

Optic ganglion

Iris

Optic nerve

Retina   Lens

Cornea

Figure 17-39
Eye of a cuttlefish (*Sepia*). The structure of
cephalopod eyes shows a high degree of
convergent evolution with the eyes of vertebrates.

Nervous and Sensory Systems.

The nervous and sensory systems are
more elaborate in cephalopods than
in other molluscs. The brain, the
largest in any invertebrate, consists of
several lobes with millions of nerve
cells. Squids have giant nerve fibers
(among the largest known in the animal kingdom) that are activated when
the animal is alarmed and that initiate
maximal contractions of the mantle
muscles for a speedy escape.

Sense organs are well developed.
Except for *Nautilus,* which has relatively simple eyes, cephalopods have
highly complex eyes with cornea, lens,
chambers, and retina (Figure 17-39).
Orientation of the eyes is controlled by
the statocysts, which are larger and
more complex than in other molluscs.
The eyes are held in a constant relation

to gravity, so that the slit-shaped pupils
are always in a horizontal position. Octopods are apparently color-blind but
can be taught to discriminate between
shapes—for example, a square and a
rectangle—and to remember such a
discrimination for a considerable time.
Experimenters find it easy to modify
their behavior patterns by devices of
reward and punishment. They are capable of observational learning, that is,
when one octopus observes another
being rewarded by making a correct
choice, the observer learns which
choice is rewarded and consistently
makes the same selection when given
the opportunity.

Octopods use their arms for tactile
exploration and can discriminate between textures by feel but apparently
not between shapes. The arms are
well supplied with both tactile and
chemoreceptor cells. Cephalopods
seem to lack a sense of hearing.

Communication.   Little is known of
the social behavior of nautiloids or
deep-water cephalopods, but inshore
and littoral forms such as *Sepia, Sepioteuthis, Loligo,* and *Octopus* have been
studied extensively. Although their
tactile sense is well developed and
they have some chemical sensitivity,
visual signals are the predominant
means of communication. These consist of a host of movements of the
arms, fins, and body, as well as many
color changes. The movements may
range from minor body motions to exaggerated spreading, curling, raising,
or lowering of some or all of the arms.
Color changes are effected by chromatophores, cells in the skin that contain pigment granules. Tiny muscle
cells surround each elastic chromatophore, whose contractions pull
the cell boundary of the chromatophore outward, causing it to expand greatly. As the cell expands, the
pigment becomes dispersed, changing
the color pattern of the animal. When
the muscles relax, the chromatophores
return to their original size and the
pigment becomes concentrated again.
By means of the chromatophores,
which are under nervous and proba-

bly hormonal control, an elaborate
system of changes in color and pattern
is possible, including general darkening or lightening; flushes of pink, yellow, or lavender; and the formation of
bars, stripes, spots, or irregular
blotches. These may be used variously
as danger signals, as protective coloring, in courtship rituals, and probably
in other ways.

By assuming different color patterns of different parts of the body, a
squid can transmit three or four different messages *simultaneously* to different individuals and in different directions, and it can instantaneously
change any or all of the messages.
Probably no other system of communication in invertebrates can convey
so much information so rapidly.

Deep-water cephalopods may
have to depend more on chemical or
tactile senses than their littoral or surface cousins, but they also produce
their own type of visual signals, for
they have evolved many elaborate luminescent organs.

Most cephalopods other than nautiloids have another protective device.
An ink sac that empties into the rectum contains an **ink gland** that secretes **sepia,** a dark fluid containing
the pigment melanin, into the sac.
When the animal is alarmed, it releases a cloud of ink, which may hang
in the water as a blob or be contorted
by water currents. The animal quickly
departs from the scene, leaving the
ink as a decoy to the predator.

Reproduction.   Sexes are separate
in cephalopods. In the male seminal
vesicle the spermatozoa are encased
in spermatophores and stored in a sac
that opens into the mantle cavity. One
arm of the adult male is modified as
an intromittent organ, called a **hectocotylus,** that he uses to pluck a spermatophore from his own mantle cavity and insert it into the mantle cavity
of the female near the oviduct opening (Figure 17-40). Before copulation
males often undergo color displays,
apparently directed against rival
males. Eggs are fertilized as they
leave the oviduct and are then usually

Figure 17-40
Copulation in cephalopods. **A,** Mating cuttlefishes. **B,** Male octopus uses modified arm to deposit spermatophores in female mantle cavity to fertilize her eggs. Octopuses often tend their eggs during development.

attached to stones or other objects. Some octopods tend their eggs. *Argonauta,* the paper nautilus, secretes a fluted "shell," or capsule, in which she broods her eggs.

The large yolky eggs undergo meroblastic cleavage. During the embryonic development, the head and foot become indistinguishable. The ring around the mouth, which bears the arms, or tentacles, may be derived from the anterior part of the foot. A juvenile hatches from the egg; there is no free-swimming larva in cephalopods.

Major Groups of Cephalopods

There are three subclasses of cephalopods: the Nautiloidea, which have two pairs of gills; the entirely extinct Ammonoidea; and the Coleoidea, which have one pair of gills. The Nautiloidea populated the Paleozoic and Mesozoic seas, but there survives only one genus, *Nautilus* (see Figure 17-36), of which there are five or six species. The *Nautilus'* head, with its 60 to 90 or more tentacles, can be extruded from the opening of the body compartment of the shell. Its tentacles have no suckers but are made adhesive by secretions. They are used in searching for, sensing, and grasping food. Beneath the head is the funnel. The mantle, mantle cavity, and visceral mass are sheltered by the shell.

The Ammonoidea were widely prevalent in the Mesozoic era but became extinct by the end of the Creta-

ceous period. They had chambered shells analogous to nautiloids, but the septa were more complex, and the septal sutures (where the septa contact the inside of the shell) were frilled (compare shells in Figure 17-36B and C). The reasons for their extinction remain a mystery. Present evidence suggests that they were gone before the asteroid bombardment at the end of the Cretaceous period (p. 345), and some nautiloids, which some ammonoids closely resembled, survive to the present.

The subclass Coleoidea includes all living cephalopods except *Nautilus.* There are four orders of coleoids. Members of the order Sepioidea (cuttlefishes and their relatives) have a rounded or compressed, bulky body bearing fins (Figure 17-37). They have eight arms and two tentacles. Both the arms and the tentacles have suckers, but the tentacles bear suckers only at their ends (Figure 17-37). Members of the order Teuthoidea (squids, Figure 17-38) have a more cylindrical body but also have eight arms and two tentacles. The order Vampyromorpha (vampire squid) contains only a single, deepwater species. Members of the order Octopoda have eight arms and no tentacles (see Figure 17-1E). Their bodies are short and saclike, with no fins. The suckers in squids are stalked (pedunculated), with horny rims bearing teeth; in octopuses the suckers are sessile and have no horny rims.

PHYLOGENY AND ADAPTIVE RADIATION

The first molluscs probably arose during Precambrian times because fossils attributed to the Mollusca have been found in geological strata as old as the early Cambrian period. On the basis of such shared features as spiral cleavage, mesoderm from the 4d blastomere, and trochophore larva, most zoologists have accepted the Mollusca as protostomes, allied with the annelids and arthropods. Opinions differ, however, as to whether molluscs were derived from a flatwormlike ancestor independent of the annelids, share an ancestor with the annelids after the advent of the coelom, or share a metameric common ancestor with the annelids. This last hypothesis is strengthened if *Neopilina* (Class Monoplacophora) can be considered metameric, as some scientists have contended. However, it is unlikely that such a successful adaptation as metamerism would have been lost in all later molluscs, and there is no trace of metamerism in the development of any known molluscan larva. Therefore most zoologists now suggest that the replication of body parts found in the monoplacophorans is pseudometamerism. The most reasonable hypothesis is that the molluscs branched off from the annelid line after the coelom arose but before the advent of metamerism. Some analyses suggest that molluscs and annelids are more closely related to each other than either is to the arthropods.

Fossils are remains of past life uncovered from the crust of the earth (Chapter 9). They can be actual parts or products of animals (teeth, bones, shells, and so on), petrified skeletal parts, molds, casts, impressions, footprints, and others. Soft and fleshy parts rarely leave recognizable fossils. Therefore we have no record of molluscs before they had shells, and there can be some doubt that certain early fossil shells are really remains of molluscs, particularly if the group they represent is now extinct. The issue of how to define a mollusc from hard parts alone was emphasized by Yochelson (1978, Malacologia **17**:165), who said, "If scaphopods were extinct and soft parts were unknown, would they be called mollusks? I think not."

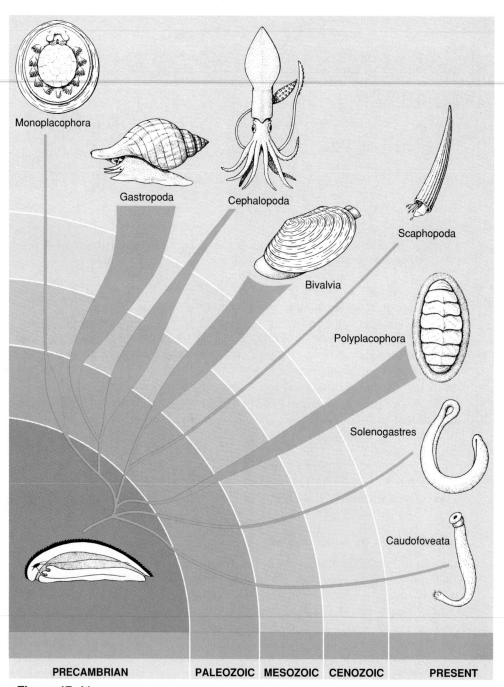

Figure 17-41
The classes of Mollusca, showing their derivations and relative abundance.

The "hypothetical ancestral mollusc" (see Figure 17-2) was long viewed as representing the original mollusc ancestor, but neither a solid shell nor a broad, crawling foot are now considered universal characters for the Mollusca. The primitive ancestral mollusc was probably a more or less wormlike organism with a ventral gliding surface and a dorsal mantle with a chitinous cuticle and calcareous scales (Figure 17-41). It had a posterior mantle cavity with two gills, a radula, a ladderlike nervous system, and an open circulatory system with a heart. Among living molluscs the primitive condition is most nearly approached by the caudofoveates, although the foot is reduced to an oral shield in members of this class. The solenogasters have lost the gills, and the foot is represented by the ventral groove. Both these classes probably branched off from the primitive ancestor before the development of a solid shell, a distinct head with sensory organs, and the ventral muscularized foot. The polyplacophorans probably also branched off early from the main lines of molluscan evolution before the veliger was established as the larva. Some workers believe that the shells of polyplacophorans are not homologous to the shells of other molluscs because they differ structurally and developmentally. The Polyplacophora and the remaining classes are sister groups (Figure 17-42).

Some investigators believe that the Gastropoda are polyphyletic, perhaps being composed of several groups independently derived from an ancestor shared with the monoplacophorans, but cladistic analysis suggests that the Gastropoda and Cephalopoda form a sister group to the Monoplacophora (see Figure 17-42). Both the gastropods and cephalopods have a greatly expanded visceral mass. The mantle cavity was brought toward the head by torsion in gastropods, but in cephalopods the mantle cavity was extended ventrally. The evolution of the chambered shell in cephalopods was a very important contribution to their freedom from the substratum and their ability to swim. The development of their respiratory, circulatory, and nervous systems is correlated with their predatory and swimming habits.

Scaphopods and bivalves have an expanded mantle cavity that essentially envelops the body. Adaptations for burrowing characterize this

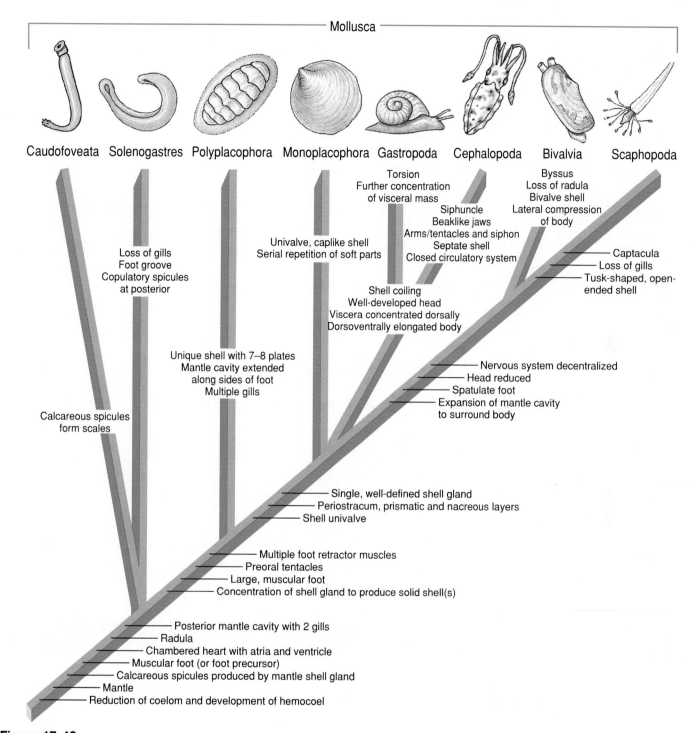

Figure 17-42

Cladogram showing hypothetical relationships among the classes of Mollusca. The synapomorphies that identify the various clades are shown, although a number of these have been modified or lost in some descendants. For example, the univalve shell (as well as shell coiling) has been reduced or lost in many gastropods and cephalopods, and many gastropods have undergone detorsion. The bivalve shell of the Bivalvia was derived from the ancestral univalve shell. The byssus is not present in most adult bivalves but functions in larval attachment in many; therefore the byssus is considered a synapomorphy of Bivalvia.

Source: Modified from R. C. Brusca and G. J. Brusca, Invertebrates. Sinauer Associates, Inc., Sunderland, MA, 1990.

clade: the spatulate foot and reduction of the head and sense organs.

Most of the diversity among molluscs is related to their adaptation to different habitats and modes of life and to a wide variety of feeding methods, ranging from sedentary filter feeding to active predation. There are many adaptations for food gathering within the phylum and an enormous variety in radular structure and function, particularly among the gastropods.

The versatile glandular mantle has probably shown more plastic adaptive capacity than any other molluscan structure. Besides secreting the shell and forming the mantle cavity, it is variously modified into gills, lungs, siphons, and apertures, and it sometimes functions in locomotion, in the feeding processes, or in a sensory capacity. The shell, too, has undergone a variety of evolutionary adaptations.

CLASSIFICATION OF PHYLUM MOLLUSCA

Useful characteristics for distinction of classes of molluscs are the type of foot and the type of shell. Several other characteristics are important in particular classes.

Class Caudofoveata (kaw´do-fo-ve-at´a) (L. *cauda*, tail, + *fovea*, small pit). Wormlike; shell, head, and excretory organs absent; radula usually present; mantle with chitinous cuticle and calcareous scales; oral pedal shield near anterior mouth; mantle cavity at posterior end with pair of gills; sexes separate; formerly united with solenogasters in Class Aplacophora. Examples: *Chaetoderma, Limifossor.*

Class Solenogastres (so-len´o-gas´trez) (Gr. *solen*, pipe, + *gaster*, stomach): **solenogasters.** Wormlike; shell, head, and excretory organs absent; radula present or absent; mantle usually covered with scales or spicules; mantle cavity posterior, without true gills, but sometimes with secondary respiratory structures; foot represented by long, narrow, ventral pedal groove; hermaphroditic. Example: *Neomenia.*

Class Monoplacophora (mon´o-pla-kof´o-ra) (Gr. *monos*, one, + *plax*, plate, + *phora*, bearing). Body bilaterally symmetrical with a broad flat foot; a single limpetlike shell; mantle cavity with five or six pairs of gills; large coelomic cavities; radula present; six pairs of nephridia, two of which are gonoducts; separate sexes. Example: *Neopilina* (see Figure 17-8).

Class Polyplacophora (pol´y-pla-kof´o-ra) (Gr. *polys*, many, several, + *plax*, plate, + *phora*, bearing): **chitons.** Elongated, dorsoventrally flattened body with reduced head; bilaterally symmetrical; radula present; shell of eight dorsal plates; foot broad and flat; gills multiple along sides of body between foot and mantle edge; sexes usually separate, with a trochophore but no veliger larva. Examples: *Mopalia* (see Figure 17-10), *Tonicella* (see Figure 17-1A).

Class Scaphopoda (ska-fop´o-da) (Gr. *skaphe*, trough, boat, + *pous, podos*, foot): **tusk shells.** Body enclosed in a one-piece tubular shell open at both ends; conical foot; mouth with radula and tentacles; head absent; mantle for respiration; sexes separate; trochophore larva. Example: *Dentalium* (see Figure 17-11).

Class Gastropoda (gas-trop´o-da) (Gr. *gaster*, stomach, + *pous, podos*, foot): **snails and slugs.** Body asymmetrical and shows effects of torsion; body usually in a coiled shell (shell uncoiled or absent in some); head well developed, with radula; foot large and flat; one or two gills, or with mantle modified into secondary gills or a lung; most with single auricle and single nephridium; nervous system with cerebral, pleural, pedal, and visceral ganglia; dioecious or monoecious, some with trochophore, typically with veliger, some without pelagic larva. Examples: *Busycon, Polinices* (see Figure 17-15B), *Physa, Helix, Aplysia* (see Figure 17-21).

Class Bivalvia (bi-val´ve-a) (L. *bi*, two, + *valva*, folding door, valve) **(Pelecypoda): bivalves.** Body enclosed in a two-lobed mantle; shell of two lateral valves of variable size and form, with dorsal hinge; head greatly reduced, but mouth with labial palps; no radula; no cephalic eyes, a few with eyes on mantle margin; foot usually wedge shaped; gills platelike; sexes usually separate, typically with trochophore and veliger larvae. Examples: *Anodonta, Venus, Tagelus* (see Figure 17-26), *Teredo* (see Figure 17-27).

Class Cephalopoda (sef´a-lop´o-da) (Gr. *kephalē*, head, + *pous, podos*, foot): **squids and octopuses.** Shell often reduced or absent; head well developed with eyes and a radula; head with arms or tentacles; foot modified into siphon; nervous system of well-developed ganglia, centralized to form a brain; sexes separate, with direct development. Examples: *Loligo* (see Figure 17-38), *Octopus* (see Figure 17-1E), *Sepia* (see Figure 17-37).

Summary

The Mollusca is one of the largest and most diverse phyla, its members ranging in size from very small organisms to the largest of invertebrates. Their basic body divisions are the head, the foot, and the visceral mass, which is usually covered by a shell. The majority are marine, but some are freshwater, and a few are terrestrial. They occupy a wide variety of niches. A number are economically important, and a few are medically important as hosts of parasites.

The molluscs are coelomate (have a coelom), although their coelom is limited to the area around the heart and gonads. The evolutionary development of a coelom was important because it enabled better organization of visceral organs and, in many of the animals that have it, an efficient hydrostatic skeleton.

The mantle and mantle cavity are important characteristics of molluscs. The mantle secretes the shell and overlies a part of the visceral mass to form a cavity housing the gills. The mantle cavity has been modified into a lung in some molluscs. The foot is usually a ventral, sole-like, locomotory organ, but it may be variously modified, as in the cephalopods, where it has become arms and a funnel. The radula is found in all molluscs except bivalves and solenogasters and is a protrusible, tonguelike organ with teeth used in feeding. Except in the cephalopods, which have a closed circulatory system, the circulatory system of molluscs is open, with a heart and blood sinuses. Molluscs usually have a pair of nephridia connecting with the coelom and a complex nervous system with a variety of sense organs. The primitive larva of molluscs is the trochophore, and most marine molluscs have a more advanced larva, the veliger.

The classes Caudofoveata and Solenogastres are small groups of wormlike molluscs with no shell. The Scaphopoda is a slightly larger class with a tubular shell, open at both ends, and the mantle wrapped around the body.

The Class Monoplacophora is a tiny, univalve marine group showing pseudometamerism. The Polyplacophora are more common, marine organisms with shells in the form of a series of eight plates. They are rather sedentary animals with a row of gills along each side of their foot.

The Gastropoda are the most successful and largest class of molluscs. Their interesting evolutionary history includes torsion, or the twisting of the posterior end to the anterior, so that the anus and head are at the same end, and coiling, an elongation and spiraling of the visceral mass. Torsion has led to the problem of fouling, which is the release of excreta over the head and in front of the gills, and this has been solved in various ways among different gastropods. Among the solutions to fouling are bringing water into one side of the mantle cavity and out the other (many prosobranchs), some degree of detorsion (opisthobranchs), and conversion of the mantle cavity into a lung (pulmonates).

The Class Bivalvia are marine and freshwater, and they have their shell divided into two valves joined by a dorsal ligament and held together by an adductor muscle. Most of them are filter feeders, drawing water through their gills by ciliary action.

The members of the Class Cephalopoda are the most advanced molluscs; they are all predators and many can swim rapidly. Their tentacles capture prey by adhesive secretions or by suckers. They swim by forcefully expelling water from their mantle cavity through a funnel, which was derived from the foot.

There is strong embryological evidence that the molluscs share a common ancestor with the annelids, although the molluscs are not metameric.

Review Questions

1. Members of such a large and diverse phylum as Mollusca impact humans in many ways. Discuss this statement.
2. How does the coelom develop embryologically? Why was the evolutionary development of the coelom important?
3. What are characteristics of Mollusca that distinguish it from other phyla?
4. Briefly describe the characteristics of the primitive ancestral mollusc, and tell how each class of molluscs (Caudofoveata, Solenogastres, Monoplacophora, Polyplacophora, Scaphopoda, Gastropoda, Bivalvia, Cephalopoda) differs from the primitive condition with respect to each of the following: shell, radula, foot, mantle cavity and gills, circulatory system, and head.
5. Define the following: ctenidia, odontophore, periostracum, prismatic layer, nacreous layer, metanephridia, nephrostome, trochophore, veliger, glochidium, osphradium.
6. Briefly describe the habitat and habits of a typical chiton.
7. Define the following with respect to gastropods: operculum, columella, torsion, fouling, bilateral asymmetry, rhinophore, pneumostome.
8. What was the survival problem that was created by torsion? What ways have evolved in gastropods to avoid this problem?
9. The gastropods have radiated enormously. Illustrate this statement by describing variations in feeding habits found in gastropods.
10. Distinguish among prosobranchs, opisthobranchs, and pulmonates.
11. Briefly describe how a typical bivalve feeds and how it burrows.
12. How is the ctenidium modified from the ancestral form in a typical bivalve?
13. What is the function of the siphuncle of cephalopods?
14. Describe how cephalopods swim and how they eat.
15. Describe adaptations in the circulatory and neurosensory systems of cephalopods that are particularly valuable for actively swimming, predaceous animals.
16. Distinguish between ammonoids and nautiloids.
17. To what other major invertebrate groups are molluscs related, and what is the nature of the evidence for the relationship?

Selected References

See also general references for Part III, p. 626.

Abbott, R. T. 1974. American seashells, ed. 2 New York, Van Nostrand Reinhold Company, Inc. *Identification guide to 1500 Atlantic and Pacific species.*

Barinaga, M. 1990. Science digests the secrets of voracious killer snails. Science **249:**250–251. *Describes current research on the toxins produced by cone snails.*

Boss, K. J. 1982. Mollusca. In Parker, S. P. (ed.). Synopsis and classification of living organisms, vol. 1. New York, McGraw-Hill Book Company. *Includes an account of the hypothetical ancestral mollusc and relation to Caudofoveata.*

Gosline, J. M., and M. D. DeMont. 1985. Jet-propelled swimming in squids. Sci. Am. **252:**96–103 (Jan.). *Mechanics of swimming in squid are analyzed; elasticity of collagen in mantle increases efficiency.*

Kuznik, F. 1993. America's aching mussels. National Wildlife (Oct.–Nov.) pp. 34–39. *Details the miserable status of freshwater clams (or mussels) in the United States.*

Morris, P. A. (W. J. Clench [editor]). 1973. A field guide to shells of the Atlantic and Gulf coasts and the West Indies, ed. 3. Boston, Houghton Mifflin Company. *An excellent revision of a popular handbook.*

Morse, A. N. C. 1991. How do planktonic larvae know where to settle? Am. Sci. **79:**154–167. *Abalone larvae recognize a chemical cue from certain red algae, which causes them to settle and transform into juveniles.*

Moynihan, M. 1985. Communication and noncommunication by cephalopods. Bloomington, Indiana University Press. *Readable summarization of our understanding of communication in this remarkable group of molluscs.*

Roper, C. R. E., and K. J. Boss. 1982. The giant squid. Sci. Am. **246:**96–105 (April). *Many mysteries remain about the deep-sea squid, Architeuthis, because it has never been studied alive. It can reach a weight of 1000 pounds and a length of 18 m, and its eyes are as large as automobile headlights.*

Ross, J. 1994. An aquatic invader is running amok in U.S. waterways. Smithsonian **24**(11):40–50 (Feb.). *A small bivalve, apparently introduced into the Great Lakes with ballast water from ships, is clogging up intake pipes and municipal water supplies. It will take billions of dollars to control.*

Ward, P. 1983. The extinction of the ammonites. Sci. Am. **249:**136–147 (Oct.). *Like the nautiloids, the ammonoids arose in the Paleozoic. Subsequently, they underwent several explosive radiations, the last of which was in the late Mesozoic, and then became extinct.*

Ward, P., L. Greenwald, and O. E. Greenwald. 1980. The buoyancy of the chambered nautilus. Sci. Am. **243:**190–203 (Oct.). *Reviews discoveries on how the nautilus removes the water from a chamber after secreting a new septum.*

18

The Segmented Worms

Phylum Annelida

Dividing Up the Body

Although a spacious, fluid-filled coelom provided an efficient hydrostatic skeleton for burrowing, precise control of body movements was not possible in the earliest coelomates. The force of muscle contraction in one area was carried throughout the body by the fluid in the undivided coelom. This defect was remedied when a series of partitions (septa) evolved in the common ancestor of annelids and arthropods. When the septa divided the coelom into a series of compartments, components of most other body systems, such as circulatory, nervous, and excretory, were repeated in each segment. This body plan is known as *metamerism*.

The evolutionary advent of metamerism was highly significant because it made possible development of much greater complexity in structure and function. Metamerism not only increased the efficiency of burrowing but also made possible independent and separate movements by the separate segments. The need for fine control of movements led, in turn, to the evolution of a more sophisticated nervous system. Moreover, repetition of body parts gave the organisms a built-in redundancy, as in some human-made systems. This provided a safety factor: if one segment should fail, the others could still function. Thus an injury to one part would not necessarily be fatal.

The evolutionary potential of the metameric body plan is amply demonstrated by the large and diverse phylum Arthropoda. Metamerism also arose independently in the deuterostome line, which includes the numerous and adaptively diverse vertebrates. ■

POSITION IN ANIMAL KINGDOM

1. Annelids belong to the **protostome** branch of the animal kingdom and have **spiral cleavage** and **determinate (mosaic) development,** characters in common with and that indicate relationship with the molluscs and primitive arthropods.
2. Together with the molluscs and arthropods, they form a sister group with the flatworms.
3. Annelids as a group show a primitive metamerism with comparatively few differences between the different somites.
4. Characters shared with arthropods include an outer secreted cuticle and a similar nervous system.

BIOLOGICAL CONTRIBUTIONS

1. The introduction of **metamerism** by the group represents the greatest innovation seen in this phylum. An homologous but more highly specialized metamerism is seen in the arthropods.
2. A true coelomic cavity reaches a high stage of development in this group.
3. Specialization of the head region into differentiated organs, such as the tentacles, palps, and eyespots of the polychaetes, is carried further in some annelids than in other invertebrates so far considered.
4. There are modifications of the **nervous system,** with cerebral ganglia (brain), two closely fused ventral nerve cords with giant fibers running the length of the body, and various ganglia with their lateral branches.
5. The circulatory system is much more complex than any we have so far considered. It is a closed system with muscular blood vessels and aortic arches ("hearts") for propelling the blood.
6. The appearance of the fleshy **parapodia,** with their respiratory and locomotor functions, introduces a suggestion of the paired appendages and specialized gills found in the more highly organized arthropods.
7. The well-developed **nephridia** in most of the somites have reached a differentiation that involves removal of waste from the blood as well as from the coelom.
8. Annelids are the most highly organized animals capable of complete regeneration. However, this ability varies greatly within the group.

The Phylum Annelida (an-nel´i-da) (L. *annelus,* little ring, + *ida,* pl. suffix) consists of the segmented worms. It is a large phylum, numbering approximately 15,000 species, the most familiar of which are the earthworms and freshwater worms (oligochaetes) and the leeches (hirudineans). However, approximately two-thirds of the phylum is composed of the marine worms (polychaetes), which are less familiar to most people. Among the latter are many curious members; some are strange, even grotesque, whereas others are graceful and beautiful. They include the clam-worms, plumed worms, parchment worms, scaleworms, lugworms, and many others. The annelids are true coelomates and belong to the protostome branch, with spiral cleavage and mosaic development. They are a highly developed group in which the nervous system is more centralized and the circulatory system more complex than those of the phyla we have studied thus far.

The Annelida are worms whose bodies are divided into similar rings, or **segments,** arranged in linear series and externally marked by circular grooves called **annuli;** the name of the phylum is descriptive of this characteristic. Body segmentation, or **metamerism,** in the annelids is not merely an external feature but is also seen internally in the repetitive arrangement of organs and systems and in the delimiting of segments (also called metameres or somites) by septa. Metamerism is not limited to annelids; it is shared by the arthropods (insects, crustaceans, and others), whose metamerism is homologous to the annelids, and also by the vertebrates, in which it evolved independently.

Annelids are sometimes called "bristle worms" because, with the exception of the leeches, most annelids bear tiny chitinous bristles called **setae** (L. *seta,* hair or bristle). Short needlelike setae help anchor the somites during locomotion to prevent backward slipping; long, hairlike setae aid aquatic forms in swimming. Since many annelids either are burrowers or live in secreted tubes, the stiff setae also aid in preventing the worm from being pulled out or washed out of its home. Robins know from experience how effective the earthworms' setae are.

Annelids have a worldwide distribution, and a few species are cosmopolitan. Polychaetes are chiefly marine forms. Most are benthic, but some live free in the open sea. Oligochaetes and leeches occur predominantly in fresh water or terrestrial soils. Some freshwater species burrow in the bottom mud and sand and others among submerged vegetation. Many of the leeches are predators, and many are specialized for piercing their prey and feeding on blood or soft tissues. A few leeches are marine, but most of them live in fresh water or in damp regions. Suckers are typically found at both ends of the body for attachment to the substratum or to their prey.

CHARACTERISTICS OF PHYLUM ANNELIDA

1. Body **metameric;** symmetry bilateral
2. Body wall with outer circular and inner longitudinal muscle layers; outer transparent moist cuticle secreted by epithelium
3. **Chitinous setae** often present; setae absent in leeches
4. Coelom (schizocoel) well developed and divided by septa, except in leeches; coelomic fluid supplies turgidity and functions as hydrostatic skeleton
5. **Circulatory system closed** and segmentally arranged; respiratory pigments (hemoglobin, hemerythrin, or chlorocruorin) often present; amebocytes in blood plasma
6. Digestive system complete and not metamerically arranged
7. Respiratory gas exchange through skin, **gills,** or **parapodia**
8. Excretory system typically a **pair of nephridia for each metamere**
9. Nervous system with a double ventral nerve cord and a pair of ganglia with lateral nerves in each metamere; brain a pair of dorsal cerebral ganglia with connectives to cord
10. Sensory system of tactile organs, taste buds, statocysts (in some), photoreceptor cells, and eyes with lenses (in some)
11. Hermaphroditic or separate sexes; larvae, if present, are trochophore type; asexual reproduction by budding in some; spiral cleavage and mosaic development

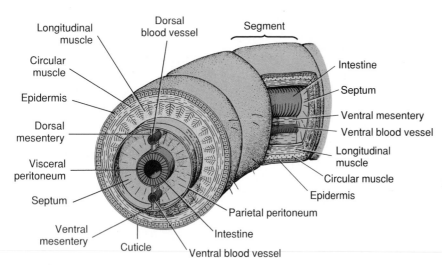

Figure 18-1
Annelid body plan.

BODY PLAN

The annelid body typically has an anterior **prostomium,** a segmented body, and a terminal portion bearing the anus **(pygidium).** The prostomium and the pygidium are not considered metameres, but anterior segments often fuse with the prostomium to make up the head. New metameres form during development just in front of the pygidium; thus the oldest segments are at the anterior end and the youngest segments are at the posterior.

The body wall has strong circular and longitudinal muscles adapted for swimming, crawling, and burrowing and is covered with epidermis and a thin, outer layer of nonchitinous cuticle (Figure 18-1).

In most annelids the coelom develops embryonically as a split in the mesoderm on each side of the gut **(schizocoel),** forming a pair of coelomic compartments in each segment. **Peritoneum** (a layer of mesodermal epithelium) lines the body wall of each compartment, forms dorsal and ventral **mesenteries,** and covers all the organs (Figure 18-1). The peritonea of adjacent segments meet to form the **septa.** The septa are perforated by the gut and longitudinal blood vessels. Not only is the coelom metamerically arranged, but practically every body system is affected in some way by this segmental arrangement.

Except in the leeches, the coelom is filled with fluid and serves as a **hydrostatic skeleton.** Because the volume of the fluid is essentially constant, contraction of the longitudinal bodywall muscles causes the body to shorten and become larger in diameter, whereas contraction of the circular muscles causes it to lengthen and become thinner. Separation of the hydrostatic skeleton into a metameric series of coelomic cavities increases its efficiency greatly, because the force of local muscle contraction is not transferred throughout the length of the worm. Widening and elongation can occur in restricted areas. Crawling motions are effected by alternating waves of contraction of longitudinal and circular muscles (peristaltic contraction) passing down the body. Segments in which longitudinal muscles are contracted widen and anchor themselves against burrow walls or other substratum while other segments, in which circular muscles are contracted, elongate and stretch forward. Forces powerful enough for burrowing as well as locomotion can thus be generated. Swimming forms use undulatory rather than peristaltic movements in locomotion.

CLASS POLYCHAETA

The polychaetes form the largest class of annelids with more than 10,000 species, most of them marine. Although the majority of them are 5 to 10 cm long, some are less than 1 mm, and others may be as long as 3 m. Some are brightly colored in reds and greens; others are dull or iridescent. Some are picturesque, such as the "featherduster" worms (Figure 18-2).

Polychaetes differ from other annelids in having a well-differentiated head with specialized sense organs; paired appendages, called **parapodia,** on most segments; and no clitellum (see p. 361) (Figure 18-3). As their name implies, they have many setae, usually arranged in bundles on the parapodia. They show a pronounced

A

B

Figure 18-2

Tube-dwelling sedentary polychaetes. **A,** One of the featherduster worms (called a Christmas-tree worm), *Spirobranchus giganteus,* has a double crown of radioles and lives in a calcareous tube. **B,** Sabellid polychaetes, *Bispira brunnea,* live in leathery tubes.

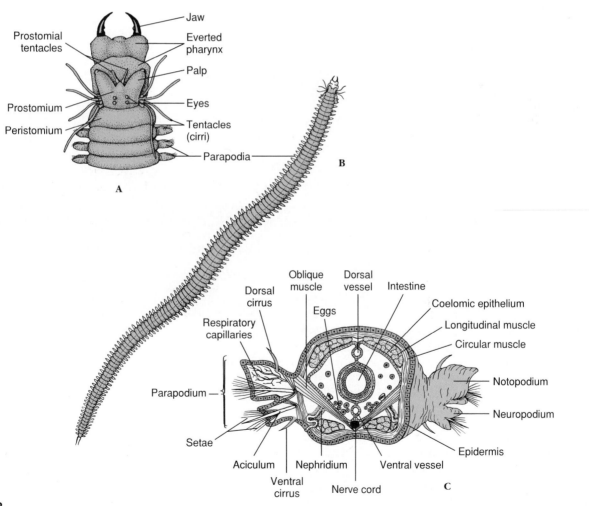

Figure 18-3

Nereis virens, an errant polychaete. **A,** Anterior end, with pharynx everted. **B,** External structure. **C,** Generalized transverse section through region of the intestine.

differentiation of some body somites and a specialization of sensory organs practically unknown among clitellates (see p. 367).

Many polychaetes are euryhaline (can tolerate a wide range of environmental salinity) and occur in brackish water. The freshwater polychaete fauna is more diversified in warmer regions than in the temperate zones.

Polychaetes live under rocks, in coral crevices, or in abandoned shells, or they burrow into mud or sand; some build their own tubes on submerged objects or in bottom material; some adopt the tubes or homes of other animals; some are pelagic, making up a part of the planktonic population. They are extremely abundant in some areas; for example, a square meter of mud flat may contain thousands of polychaetes. They play a significant part in marine food chains because they are eaten by fish, crustaceans, hydroids, and many others.

They are often divided for convenience into two groups (formerly the basis of subclasses): the sedentary polychaetes, and the errant or free-moving polychaetes. Sedentary polychaetes are mainly tubicolous; that is, they spend all or much of their time in tubes or permanent burrows. Many of them, especially those that live in tubes, have elaborate devices for feeding and respiration. Errant polychaetes (L. *errare,* to wander), include the free-moving pelagic forms, active burrowers, crawlers, and the tube worms that leave their tubes for feeding or breeding. Most of these, like the clam worm *Nereis* (Gr. name of a sea nymph) (Figure 18-3), are predatory forms equipped with jaws or teeth. They have a muscular eversible pharynx armed with teeth that can be thrust out with surprising speed and dexterity for capturing prey.

FORM AND FUNCTION

The polychaete typically has a head, or **prostomium,** which may or may not be retractile and which often bears eyes, tentacles, and sensory palps (Figures 18-3 and 18-7). The first segment **(peristomium)** surrounds the mouth and may bear setae, palps, or, in predatory forms, chitinous jaws. Cil-

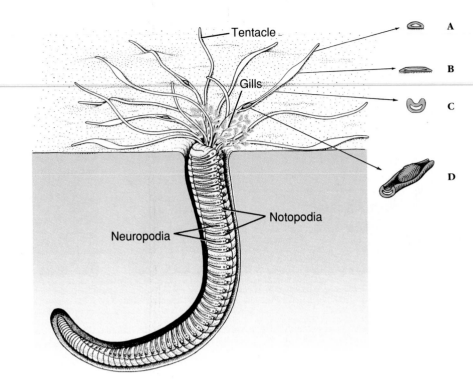

Figure 18-4

Amphitrite, which builds its tubes in mud or sand, extends long grooved tentacles out over the mud to pick up bits of organic matter. The smallest particles are moved along food grooves by cilia, larger particles by peristaltic movement. Its plumelike gills are blood red. **A,** Section through exploratory end of tentacle. **B,** Section through tentacle in area adhering to substratum. **C,** Section showing ciliary groove. **D,** Particle being carried toward mouth.

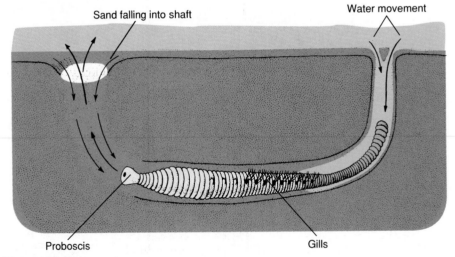

Figure 18-5

Arenicola, the lugworm, lives in an L-shaped burrow in intertidal mud flats. It burrows by successive eversions and retractions of its proboscis. By peristaltic movements it keeps water filtering through the sand. The worm then ingests the food-laden sand.

iary feeders may bear a tentacular crown that can be opened like a fan or withdrawn into the tube.

The trunk is segmented, and most segments bear parapodia, which may have lobes, cirri, setae, and other parts on them (see Figure 18-3). The parapodia are used in crawling, swimming, or anchoring in tubes. They usually serve

as the chief respiratory organs, although some polychaetes may also have gills. *Amphitrite,* for example, has three pairs of branched gills and long extensible tentacles (Figure 18-4). *Arenicola,* the lugworm (Figure 18-5), which burrows through the sand leaving characteristic castings at the entrance to its burrow, has paired gills on certain somites.

Nutrition

The polychaete digestive system consists of a foregut, midgut, and hindgut. The foregut includes the stomodeum, pharynx, and anterior esophagus. It is lined with cuticle, and the jaws, where present, are constructed of cuticular protein. The midgut is derived from endoderm. The more anterior portions secrete digestive enzymes, and absorption takes place toward the posterior. The short, ectodermally-derived hindgut connects the midgut to the exterior via the anus, which is on the pygidium.

Errant polychaetes are mostly predators and scavengers. Sedentary polychaetes feed on suspended particles, or they are deposit feeders, consuming particles on or in the bottom. We will discuss food habits of some specific polychaetes in the following text.

Circulation and Respiration

Polychaetes show considerable diversity in both circulatory and respiratory structure and function. As mentioned before, parapodia and gills serve for gaseous exchange in various species. In some polychaetes there are no special organs for respiration, and gaseous exchange takes place across the general body surface.

The circulatory pattern varies greatly. In *Nereis* a dorsal longitudinal vessel carries blood anteriorly, and a ventral longitudinal vessel conducts it posteriorly (Figure 18-3C). Blood flows between these two vessels via segmental networks in the parapodia, septa, and around the intestine. In *Glycera* the circulatory system is reduced and joins with the coelom. Septa are incomplete, and thus the coelomic fluid assumes the function of circulation.

Many polychaetes have respiratory pigments such as hemoglobin, chlorocruorin, or hemerythrin.

Excretion

Although there is some variety in excretory organs, including possession of protonephridia and mixed proto- and metanephridia in some, most polychaetes have metanephridia (Figure 18-3). There is one pair per metamere, with the inner end of each **(nephrostome)** opening into a coelomic compartment. Coelomic fluid passes into the nephrostome, and selective resorption occurs along the nephridial duct, as in oligochaetes (see Figure 18-14).

Nervous System and Sense Organs

The organization of the central nervous system in polychaetes follows the basic annelid plan (see Figure 18-15). Dorsal cerebral ganglia connect with a subpharyngeal ganglion via a circumpharyngeal commissure. A double ventral nerve cord courses the length of the worm, with metamerically arranged ganglia.

Sense organs are more highly developed in polychaetes than in oligochaetes and include eyes, nuchal organs (see the following), and statocysts. Eyes, when present, may range from simple eyespots to well-developed organs. They are most conspicuous in errant worms. Usually the eyes are retinal cups, with rodlike photoreceptor cells lining the cup wall and directed toward the lumen of the cup. The highest degree of development occurs in the family Alciopidae, which has large, image-resolving eyes similar in structure to those of some cephalopod molluscs (Figure 17-39, p. 344), with cornea, lens, retina, and retinal pigment. The alciopid eye also has accessory retinas, a characteristic shared by deep-sea fishes and some deep-sea cephalopods. Different wavelengths of light penetrate to different depths in water, and the accessory retinas of alciopids are sensitive to different wavelengths. The eyes of these pelagic animals may be well adapted to function as the light varies with depth. Studies with electroencephalograms show that they are sensitive to the dim light of the deep sea. Nuchal organs are ciliated sensory pits or slits that appear to be chemoreceptive, an important factor in food gathering. Some burrowing and tube-building polychaetes have statocysts that function in body orientation.

Reproduction and Development

In contrast to clitellates, polychaetes have no permanent sex organs, and they usually have separate sexes. Reproductive systems are simple. Gonads appear as temporary swellings of the peritoneum and shed their gametes into the coelom. They are carried outside through gonoducts, through the metanephridia, or by rupture of the body wall. Fertilization is external, and the early larva is a trochophore.

Some polychaetes live most of the year as sexually immature animals called atokes, but during the breeding season a portion of the body develops into a sexually mature worm called an epitoke, which is swollen with gametes (Figure 18-6). An example is the palolo worm, which lives in burrows among coral reefs. During the swarming period, the epitokes break off and swim to the surface. Just before sunrise, the sea is literally covered with them, and at sunrise they burst, freeing the eggs and sperm for fertilization. The anterior portions of the worms regenerate new posterior sections. Swarming is of great adaptive value because the synchronous maturation of all the epitokes ensures the maximum number of fertilized eggs. However, it is very hazardous; many types of predators have a feast. In the meantime, the atoke remains safely in its burrow to produce another epitoke at the next cycle.

CLAM WORMS: *NEREIS*

The clam worms (Figure 18-7), or sand worms as they are sometimes called, are errant polychaetes that live in mucus-lined burrows in or near low tide. Sometimes they are found in temporary hiding places, such as under stones, where they stay with their bodies covered and their heads protruding. They are most active at night, when they wiggle out of their hiding places and swim about or crawl over the sand in search of food.

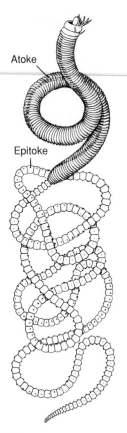

Figure 18-6
Eunice viridis, the Samoan palolo worm. The posterior segments make up the epitokal region, consisting of segments packed with gametes. Each segment has an eyespot on the ventral side. Once a year the worms swarm, and the epitokes detach, rise to the surface, and discharge their ripe gametes, leaving the water milky. By the next breeding season, the epitokes are regenerated.

The body, containing about 200 somites, may grow to 30 or 40 cm in length. The head is made up of a prostomium and a peristomium. The prostomium bears a pair of stubby palps, sensitive to touch and taste; a pair of short sensory tentacles; and two pairs of small dorsal eyes that are light sensitive. The peristomium bears the ventral mouth, a pair of chitinous jaws, and four pairs of sensory tentacles (Figure 18-3A).

Each parapodium has two lobes: a dorsal **notopodium** and a ventral **neuropodium** (Figure 18-3C). One or more chitinous spines **(acicula)** supports each lobe. The parapodia bear setae and are abundantly supplied with blood vessels. The parapodia are used for both creeping and swimming and

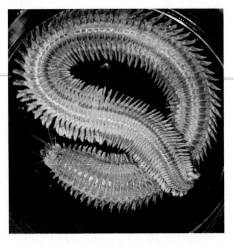

Figure 18-7
Nereis diversicolor. Note the well-defined segments, the lobed parapodia, and the prostomium with tentacles.

are manipulated by oblique muscles that run from the midventral line to the parapodia in each somite. The worm swims by lateral undulatory wriggling of the body—unlike the peristaltic movement of the earthworms. It can dart through the water with considerable speed. These undulatory movements can also be used to suck water into or pump it out of the burrow. The worm will usually adapt some kind of burrow if it can find one. When a worm is placed near a glass tube, it will wriggle in without hesitation.

The clam worm feeds on small animals, other worms, and larval forms. It seizes them with its chitinous jaws, which it protrudes through the mouth when it everts its pharynx. As it withdraws its pharynx, it swallows the food. Movement of the food through the alimentary canal is by peristalsis.

OTHER INTERESTING POLYCHAETES

Scale worms (Figure 18-8) are members of the family Polynoidae (Gr. *Polynoe,* the daughter of Nereus and Doris, a sea god and goddess), one of the most abundant and widespread of polychaete families. Their rather flattened bodies are covered with broad scales, modified from dorsal parts of the parapodia. Most are of modest size, but some are enormous (up to

Figure 18-8
The scale worm *Hesperonoe adventor* normally lives as a commensal in the tubes of *Urechis* (Phylum Echiura, p. 433).

Figure 18-9
The fireworm *Hermodice carunculata* feeds on gorgonians and stony corals. Its setae are like tiny glass fibers and serve to ward off predators.

190 mm long and 100 mm wide). They are carnivorous and feed on a wide variety of animals. Many are commensal, living in burrows of other polychaetes or in association with cnidarians, molluscs, or echinoderms.

Hermodice carunculata (Gr. *berma,* reef, + *dex,* a worm found in wood) (Figure 18-9) and related species are called fireworms. Their setae are hollow, brittle, and contain a poisonous secretion. When touched, the setae break off in the wound and cause skin irritation. They feed on corals, gorgonians, and other cnidarians.

Tube dwellers secrete many types of tubes. Some are parchmentlike or leathery (Figure 18-2B); some are firm, calcareous tubes attached to

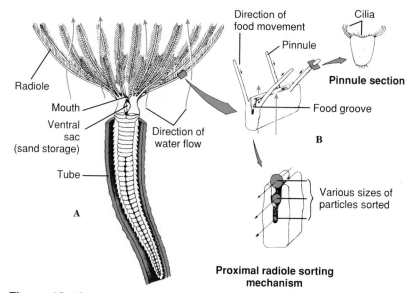

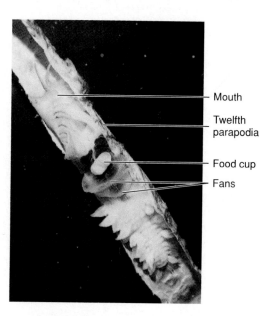

Figure 18-10

Sabella, a polychaete ciliary feeder, extends its crown of feeding radioles from its leathery secreted tube, reinforced with sand and debris. **A,** Anterior view of the crown. Cilia direct small food particles along grooved radioles to mouth and discard larger particles. Sand grains are directed to storage sacs and later are used in tube building. **B,** Distal portion of radiole showing ciliary tracts of pinnules and food grooves.

Figure 18-11

Chaetopterus, a sedentary polychaete, lives in a U-shaped tube in the sea bottom. It pumps water through the parchmentlike tube (of which one-half has been cut away here) with its three pistonlike fans. The fans beat 60 times per minute to keep water currents moving. The winglike notopodia of the twelfth segment continuously secrete a mucous net that strains out food particles. As the net fills with food, the food cup rolls it into a ball, and when the ball is large enough (about 3 mm), the food cup bends forward and deposits the ball in a ciliated groove to be carried to the mouth and swallowed.

rocks or other surfaces (Figure 18-2A); and some are simply grains of sand or bits of shell or seaweed cemented together with mucous secretions. Many burrowers in sand and mud flats simply line their burrows with mucus (Figure 18-5).

Most sedentary tube and burrow dwellers are particle feeders, using ciliary or mucoid methods of obtaining food. The principal food source is plankton and detritus. Some, like *Amphitrite* (Gr. a mythical sea nymph) (Figure 18-4), with head peeping out of the mud, send out long extensible tentacles over the surface to deposit feed. Cilia and mucus on the tentacles entrap particles found on the sea bottom and move them toward the mouth. The lugworm *Arenicola* (L. *arena,* sand, + *colo,* inhabit) employs an interesting combination of suspension and deposit feeding. It lives in an L-shaped burrow in which, by peristaltic movements, it causes water to flow. Food particles are filtered out by the sand at the front of its burrow, and it ingests the food-laden sand (Figure 18-5).

The fanworms, or "featherduster" worms, are beautiful tubeworms, fascinating to watch as they emerge from their secreted tubes and unfurl their lovely tentacular crowns to feed (Fig-

ure 18-2). A slight disturbance, sometimes even a passing shadow, causes them to duck quickly into the safety of the homes they have built. Food attracted to the feathery arms, or **radioles,** by ciliary action is trapped in mucus and is carried down ciliated food grooves to the mouth (Figure 18-10). Particles too large for the food grooves pass along the margins and drop off. Further sorting may occur near the mouth where only the small particles of food enter the mouth, and sand grains are stored in a sac to be used later in enlarging the tube.

The parchment worm *Chaetopterus* (Gr. *chaitē,* long hair, + *pteron,* wing) feeds on suspended particles by an entirely different mechanism (Figure 18-11). It lives in a U-shaped, parchmentlike tube buried, except for the tapered ends, in sand or mud along the shore. The worm attaches to the side of the tube by ventral suckers. Fans (modified parapodia) on segments 14 to 16 pump water through the tube by rhythmical movements. A pair of enlarged parapodia in the twelfth segment secretes a long mucous bag that reaches back to a small food cup just in front of the fans. All the water passing through the tube is filtered through this mucous bag, the end of which is rolled up into

a ball by cilia in the cup. When the ball is about the size of a BB shot, the fans stop beating and the ball of food and mucus is rolled forward by ciliary action to the mouth and is swallowed.

CLASS OLIGOCHAETA

The more than 3000 species of oligochaetes are found in a great variety of sizes and habitats. They include the familiar earthworms and many species that live in fresh water. Most are terrestrial or freshwater forms, but some are parasitic, and a few live in marine or brackish water.

With few exceptions, oligochaetes bear setae, which may be long or short, straight or curved, blunt or needlelike, or arranged singly or in bundles. Whatever the type, they are less numerous in oligochaetes than in polychaetes, as is implied by the class name, which means "few long hairs." Aquatic forms usually have longer setae than do earthworms.

EARTHWORMS

The most familiar of the oligochaetes are the earthworms ("night crawlers"), which burrow in moist, rich soil, emerging at night to explore their surroundings. In damp, rainy weather they stay near the surface, often with mouth or anus protruding from the burrow. In very dry weather they may burrow several feet underground, coil up in a slime chamber, and become dormant. *Lumbricus terrestris* (L. *lumbricum,* earthworm), the form commonly studied in school laboratories, is approximately 12 to 30 cm long (Figure 18-12). Giant tropical earthworms may

Aristotle called earthworms the "intestines of the soil." Some 22 centuries later Charles Darwin published his observations in his classic *The Formation of Vegetable Mould Through the Action of Worms*. He showed how worms enrich the soil by bringing subsoil to the surface and mixing it with the topsoil. An earthworm can ingest its own weight in soil every 24 hours, and Darwin estimated that from 10 to 18 tons of dry earth per acre pass through their intestines annually, thus bringing up potassium and phosphorus from the

subsoil and also adding to the soil nitrogenous products from their own metabolism. They expose the mold to the air and sift it into small particles. They also drag leaves, twigs, and organic substances into their burrows closer to the roots of plants. Their activities are important in aerating the soil. Darwin's views were at odds with his contemporaries, who thought earthworms were harmful to plants. But recent research has amply confirmed Darwin's findings, and earthworm management is now practiced in many countries.

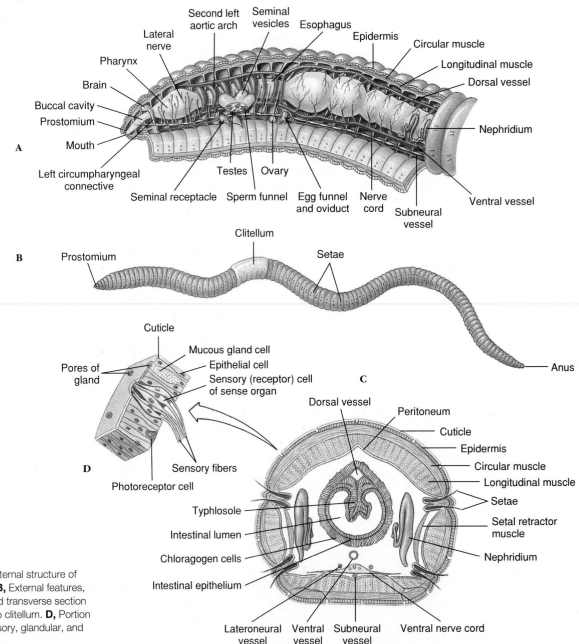

Figure 18-12

Earthworm anatomy. **A,** Internal structure of anterior portion of worm. **B,** External features, lateral view. **C,** Generalized transverse section through region posterior to clitellum. **D,** Portion of epidermis showing sensory, glandular, and epithelial cells.

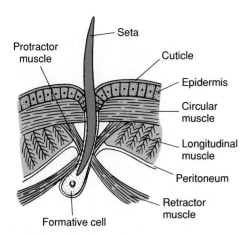

Figure 18-13
Seta with its muscle attachments showing relation to adjacent structures. Setae lost by wear and tear are replaced by new ones, which develop from formative cells.

have from 150 to 250 or more segments and may grow to as much as 4 m in length. They usually live in branched and interconnected tunnels.

Form and Function

In earthworms the mouth is overhung by a fleshy prostomium at the anterior end, and the anus is on the posterior end (Figure 18-12B). In most earthworms each segment bears four pairs of chitinous setae (Figure 18-12C), although in some oligochaetes each segment may have up to 100 or more. Each seta is a bristlelike rod set in a sac within the body wall and moved by tiny muscles (Figure 18-13). The setae project through small pores in the cuticle to the outside. In locomotion and burrowing, setae anchor parts of the body to prevent slipping. Earthworms move by peristaltic movement. Contractions of circular muscles in the anterior end lengthen the body, pushing the anterior end forward where it is anchored by setae; contractions of longitudinal muscles then shorten the body, pulling the posterior end forward. As these waves of contraction pass along the entire body, it is gradually moved forward.

Nutrition.   Most oligochaetes are scavengers. Earthworms feed mainly on decayed organic matter, bits of leaves and vegetation, refuse, and ani-mal matter. After being moistened by secretions from the mouth, food is drawn in by the sucking action of the muscular pharynx. The liplike prostomium aids in manipulating the food into position. The calcium from the soil swallowed with the food tends to produce a high blood calcium level. **Calciferous glands** along the esophagus secrete calcium ions into the gut and so reduce the calcium ion concentration of the blood. Calciferous glands are really ionoregulatory, rather than digestive, organs. They also function in regulating the acid-base balance of the body fluids, maintaining the pH at a fairly stable value.

Leaving the esophagus, food is stored temporarily in the thin-walled **crop** before being passed on into the **gizzard,** which grinds the food into small pieces. Digestion and absorption take place in the **intestine.** Along the dorsal side, the wall of the intestine is infolded to form a **typhlosole,** which greatly increases the absorptive and digestive surface (Figure 18-12C). The digestive system secretes various enzymes to break down the food.

Surrounding the intestine and dorsal vessel and filling much of the typhlosole is a layer of yellowish **chloragogen tissue** derived from the peritoneum. This tissue serves as a center for the synthesis of glycogen and fat, a function roughly equivalent to that of liver cells. The chloragogen cells when ripe (full of fat) are released into the coelom where they float free as cells called **eleocytes** (Gr. *elaio,* oil, + *kytos,* hollow vessel [cell]), which transport materials to the body tissues. They apparently can pass from segment to segment and have been found to accumulate around wounds and regenerating areas, where they break down and release their contents into the coelom. Chloragogen cells also function in excretion.

Circulation and Respiration.   Annelids have a double transport system: the coelomic fluid and the circulatory system. Food, wastes, and respiratory gases are carried by both coelomic fluid and blood in varying degrees. Blood circulates in a closed system of vessels, including capillary systems in the tissues. There are five main blood trunks, all running lengthwise through the body.

The **dorsal vessel** (single) runs above the alimentary canal from the pharynx to the anus. It is a pumping organ, provided with valves, and it functions as the true heart. This vessel receives blood from vessels of the body wall and digestive tract and pumps it anteriorly into the five pairs of **aortic arches.** The function of the aortic arches is to maintain a steady pressure of blood into the ventral vessel.

The **ventral vessel** (single) serves as the aorta. It receives blood from the aortic arches and delivers it to the brain and rest of the body, giving off segmental vessels to the walls, nephridia, and digestive tract.

The blood contains colorless ameboid cells and a dissolved respiratory pigment, hemoglobin (p. 693). The blood of some annelids may have respiratory pigments other than hemoglobin, as noted above.

Earthworms have no special respiratory organs, but gaseous exchange takes place across their moist skin.

Excretion.   Each somite except the first three and the last one bears a pair of **metanephridia.** Each nephridium occupies parts of two successive somites (Figure 18-14). A ciliated funnel, the **nephrostome,** lies just anterior to an intersegmental septum and leads by a small ciliated tubule through the septum into the somite behind, where it connects with the main part of the nephridium. Several complex loops of increasing size compose the nephridial duct, which terminates in a bladderlike structure leading to an aperture, the **nephridiopore.** The nephridiopore opens to the outside near the ventral row of setae. By means of cilia, wastes from the coelom are drawn into the nephrostome and tubule, where they are joined by salts and organic wastes transported from blood capillaries in the glandular part of the nephridium. The waste is discharged to the outside through the nephridiopore.

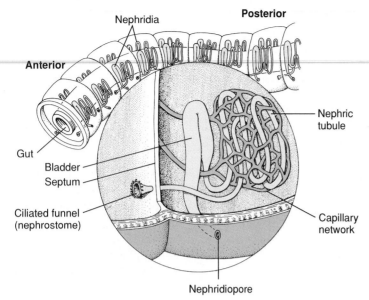

Figure 18-14

Nephridium of earthworm. Wastes are drawn into the ciliated nephrostome in one segment, then passed through the loops of the nephridium, and expelled through the nephridiopore of the next segment.

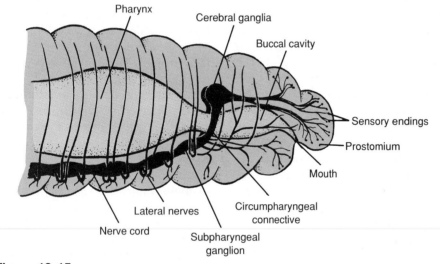

Figure 18-15

Anterior portion of earthworm and its nervous system. Note concentration of sensory endings in this region.

Aquatic oligochaetes excrete ammonia; terrestrial oligochaetes excrete the much less toxic urea. *Lumbricus* produces both, the level of urea depending somewhat on environmental conditions. Both urea and ammonia are produced by chloragogen cells, which may break off and enter the nephridia directly, or their products may be carried by the blood. Some nitrogenous waste is also eliminated through the body surface.

Oligochaetes are largely freshwater animals, and even such terrestrial forms as earthworms must exist in a moist environment. Osmoregulation is a function of the body surface and the nephridia, as well as the gut and the dorsal pores. *Lumbricus* will gain weight when placed in tap water and lose it when returned to the soil. Salts as well as water can pass across the integument, salts apparently being carried by active transport.

Nervous System and Sense Organs. The nervous system in earthworms (Figure 18-15) consists of a central system and peripheral nerves. The central system reflects the typical an-

nelid pattern: a pair of **cerebral ganglia** (the brain) above the pharynx, a pair of **connectives** passing around the pharynx connecting the brain with the first pair of ganglia in the nerve cord; a **ventral nerve cord,** really double, running along the floor of the coelom to the last somite; and a pair of fused ganglia on the nerve cord in each somite. Each pair of fused ganglia gives off nerves to the body structures, which contain both sensory and motor fibers.

Neurosecretory cells have been found in the brain and ganglia of annelids, both oligochaetes and polychaetes. They are endocrine in function and secrete neurohormones concerned with the regulation of reproduction, secondary sex characteristics, and regeneration.

For rapid escape movements most annelids have from one to several very large axons commonly called **giant axons** (Figure 18-16), or giant fibers, located in the ventral nerve cord. Their large diameter increases the rate of conduction (see p. 717) and makes possible simultaneous contractions of muscles in many segments.

In the dorsal median giant fiber of *Lumbricus,* which is 90 to 160 μm in diameter, the speed of conduction has been estimated at 20 to 45 m/second, several times faster than in ordinary neurons of this species. This is also much faster than in polychaete giant fibers, probably because in the earthworms the giant fibers are enclosed in myelinated sheaths. The speed of conduction may be altered by changes in temperature.

Simple sense organs are distributed all over the body. Earthworms have no eyes but do have many lens-shaped photoreceptors in the epidermis. Most oligochaetes are negatively phototactic to strong light but positively phototactic to weak light. Many single-celled sense organs are widely distributed in the epidermis. What are presumably chemoreceptors are most numerous on the prostomium. There are many free nerve endings in the integument, which are probably tactile.

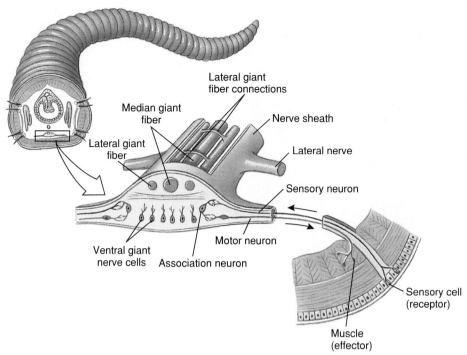

Figure 18-16
Portion of nerve cord of earthworm showing arrangement of simple reflex arc (*in foreground*) and the three dorsal giant fibers that are adapted for rapid reflexes and escape movements. Ordinary crawling involves a succession of reflex acts, the stretching of one somite stimulating the next to stretch, and so on. Impulses are transmitted much faster in giant fibers than in regular nerves so that all segments can contract simultaneously when quick withdrawal into a burrow is necessary.

General Behavior. Earthworms are among the most defenseless of creatures, yet their abundance and wide distribution indicate their ability to survive. Although they have no specialized sense organs, they are sensitive to many stimuli. They react positively to mechanical stimuli when they are moderate and negatively to a strong vibration (such as footfall near them), which causes them to retire quickly into their burrows. They react to light, which they avoid unless it is very weak. Chemical responses aid them in the choice of food.

Chemical as well as tactile responses are very important to the worm. It not only must be able to sample the organic content of the soil to find food, but also must sense its texture, acidity, and calcium content.

Experiments show that earthworms have some learning ability. They can be taught to avoid an electric shock, and thus an association reflex can be built up in them. Darwin credited earthworms with a great deal of intelligence in pulling leaves into

their burrows: he observed that they seized the leaves by the narrow end, the easiest way for drawing a leaf-shaped object into a small hole. Darwin assumed that the seizure of the leaves by the worms did not result from random handling or from chance but was purposeful in its mechanism. However, investigations since Darwin's time have shown that the process is mainly one of trial and error, for they often seize a leaf several times before getting it right.

Reproduction and Development.
Earthworms are monoecious (hermaphroditic); that is, both male and female organs are found in the same animal (Figure 18-12A). In *Lumbricus* the reproductive systems are found in somites 9 to 15. Two pairs of small testes and two pairs of sperm funnels are surrounded by three pairs of large seminal vesicles. Immature sperm from the testes mature in the seminal vesicles, then pass into the sperm funnels and down sperm ducts to the male genital pores in somite 15,

where they are expelled during copulation. Eggs are discharged by a pair of small ovaries into the coelomic cavity, where the ciliated funnels of the oviducts pick them up and carry them to the outside through female genital pores on somite 14. Two pairs of seminal receptacles in somites 9 and 10 receive and store sperm from the mate during copulation.

Reproduction in earthworms may occur at any season, but they usually copulate at night during warm, moist weather (Figure 18-17). When mating, the worms extend their anterior ends from their burrows and bring their ventral surfaces together (Figure 18-18). They are held together by mucus secreted by the **clitellum** (L. *clitellae,* packsaddle) and by special ventral setae, which penetrate each other's bodies in the regions of contact. After discharge, sperm travel to the seminal receptacles of the other worm in its seminal grooves. After copulation each worm secretes first a mucous tube and then a tough, chitinlike band that forms a **cocoon** around its clitellum. As the cocoon passes forward, eggs from the oviducts, albumin from the skin glands, and sperm from the mate (stored in the seminal receptacles) pour into it. Fertilization of the eggs then takes place within the cocoon. When the cocoon leaves the worm, its ends close, producing a lemon-shaped body. Embryogenesis occurs within the cocoon, and the form that hatches from the egg is a young worm similar to the adult. Thus development is direct with no metamorphosis. The juvenile does not develop a clitellum until it is sexually mature.

FRESHWATER OLIGOCHAETES

Freshwater oligochaetes usually are smaller and have more conspicuous setae than do earthworms. They are more mobile than earthworms and tend to have better-developed sense organs. They are generally benthic forms that creep about on the bottom or burrow in the soft mud. Aquatic oligochaetes are an important food source for fishes. A few are ectoparasitic.

Figure 18-17
Two earthworms in copulation. Their anterior ends point in opposite directions as their ventral surfaces are held together by mucous bands secreted by the clitella. Mutual insemination occurs during copulation. After separation each worm secretes a cocoon to receive its eggs and sperm.

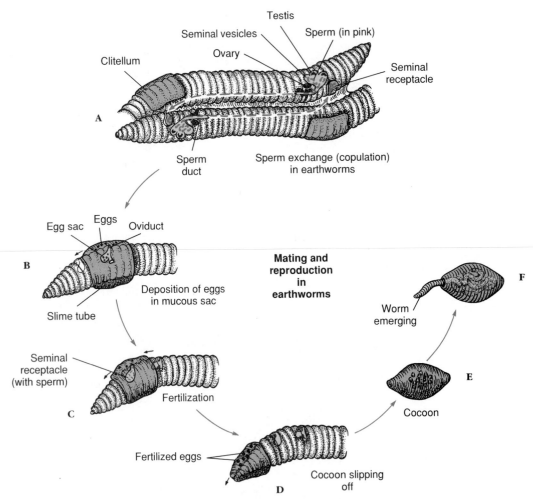

Figure 18-18
Earthworm copulation and formation of egg cocoons. **A,** Mutual insemination; sperm from genital pore (somite 15) pass along seminal grooves to seminal receptacles (somites 9 and 10) of each mate. **B** and **C,** After worms separate, a slime tube formed over the clitellum passes forward to receive eggs from oviducts and sperm from seminal receptacles. **D,** As cocoon slips off over anterior end, its ends close and seal. **E,** Cocoon is deposited near burrow entrance. **F,** Young worms emerge in 2 to 3 weeks.

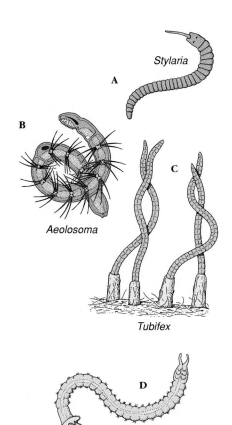

Figure 18-19
Some freshwater oligochaetes. **A,** *Stylaria* has the prostomium drawn out into a long snout. **B,** *Aeolosoma* uses cilia around the mouth to sweep in food particles, and it buds off new individuals asexually. **C,** *Tubifex* lives head down in long tubes. **D,** *Dero* has ciliated anal gills.

Some of the more common freshwater oligochaetes are the 1 mm long *Aeolosoma* (Gr. *aiolos,* quick-moving, + *soma,* body) (Figure 18-19B), which contains red or green pigments, has bundles of setae, and is often found in hay cultures; the 2 to 4 mm long *Nais* (L. *nais,* water nymph), which is brownish and has two bundles of setae on anterior segments and four bundles of setae on each posterior segment; the 10 to 25 mm long *Stylaria* (Gr. *stylos,* pillar) (Figure 18-19A), with setae arranged like those of *Nais,* a prostomium extended into a long process, and black eyespots; the 5 to 10 mm long *Dero,* (Gr. *dere,* neck or throat), which is reddish, lives in tubes, and usually has 3 to 4 pairs of tail gills (Figure 18-19D); the 30 to 40 mm long *Tubifex* (L. *tubus,* tube, + *faciens,* to make or do) (Figure 18-19C), which is reddish and lives with its head in mud

at the bottom of ponds and its tail waving in the water; the 10 to 15 mm long *Chaetogaster* (N.L. *chaeta,* bristle, + *gastrula,* belly), which has only ventral bundles of setae; and *Enchytraeus* (Gr. *enchytraeus,* living in an earthen pot), small whitish worms that live both in moist soil and in water. Some oligochaetes, such as *Aeolosoma,* may form chains of zooids asexually by transverse fission (Figure 18-19B).

CLASS HIRUDINEA: THE LEECHES

Leeches occur predominantly in freshwater habitats, but a few are marine, and some have even adapted to terrestrial life in warm, moist places. They are more abundant in tropical countries than in temperate zones. Some leeches attack human beings and are a nuisance.

Most leeches are between 2 and 6 cm in length, but some are smaller; some, including the "medicinal" leech, reach 20 cm, but the giant of all is the Amazonian *Haementeria* (Gr. *haimateros,* bloody) (Figure 18-20), which reaches 30 cm.

Leeches occur in a variety of patterns and colors: black, brown, red or olive green. They are usually flattened dorsoventrally. Some are adapted for forcing their pharynx or proboscis into soft tissues such as the gills of fish. The most specialized leeches, however, have sawlike chitinous jaws with which they can cut through tough skin. Many leeches live as carnivores on small invertebrates; some are temporary parasites; and some are permanent parasites, never leaving their host.

Like the oligochaetes, leeches are hermaphroditic and have a clitellum, but this appears only during the breeding season. The clitellum secretes a cocoon for the reception of eggs. Leeches are more highly specialized than the oligochaetes. As fluid feeders and bloodsuckers, they have lost the setae used by the oligochaetes in locomotion and have developed suckers for attachment while sucking blood; their gut is specialized for storage of large quantities of blood.

Figure 18-20
The world's largest leech, *Haementeria ghilianii,* on the arm of Dr. Roy K. Sawyer, who found it in French Guiana, South America.

FORM AND FUNCTION

Unlike other annelids, leeches have a fixed number of somites (usually 34; 17 or 31 in some groups), but they appear to have many more because each somite is marked by transverse grooves to form from two to 16 superficial rings **(annuli)** (Figure 18-21).

The coelom represents another difference between leeches and other annelids; leeches lack distinct coelomic compartments. In all but one species the septa have disappeared, and the coelomic cavity is filled with connective tissue and a system of spaces called **lacunae.** The coelomic lacunae form a regular system of channels filled with coelomic fluid, which in some leeches serves as an auxiliary circulatory system.

Most leeches creep with looping movements of the body, by attaching first one sucker and then the other and pulling up the body. Aquatic leeches can also swim with a graceful undulatory movement.

Nutrition

Leeches are popularly considered parasitic, but many are predaceous. Even the true bloodsuckers rarely remain on

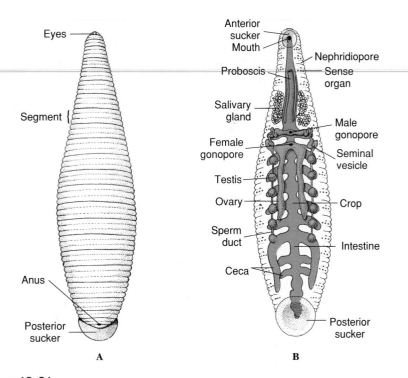

Figure 18-21
Structure of a leech, *Placobdella*. **A,** External appearance, dorsal view. **B,** Internal structure, ventral view.

the host for a long period of time. Most freshwater leeches are active predators or scavengers equipped with a proboscis that can be extended to draw in small invertebrates or to take blood from cold-blooded vertebrates. Some freshwater leeches are true bloodsuckers, preying on cattle, horses, humans, and others. Some terrestrial leeches feed on insect larvae, earthworms, and slugs, which they hold by an oral sucker while using a strong sucking pharynx to draw in the food. Other terrestrial forms climb bushes or trees to reach warm-blooded vertebrates such as birds or mammals.

Most leeches are fluid feeders. Many prefer to feed on tissue fluids and blood pumped from wounds already open. The true bloodsuckers, which include the so-called medicinal leech *Hirudo medicinalis* (L. *hirudo*, a leech) (Figure 18-22), have cutting plates, or "jaws," for cutting tissues. Some parasitic leeches leave their hosts only during the breeding season, and certain fish parasites are permanently parasitic, depositing their cocoons on the host fish.

For centuries the "medicinal leech" (*Hirudo medicinalis*) was used for bloodletting because of the mistaken idea that bodily disorders and fevers were caused by an excess of blood. A 10- to 12-cm-long leech can extend to a much greater length when distended with blood, and the amount of blood it can suck is considerable. Leech collecting and leech culture in ponds were practiced in Europe on a commercial scale during the nineteenth century. Wordsworth's poem "The Leech-Gatherer" was based on this use of the leech.

Leeches are once again being used medically. When fingers, toes, or ears are severed, microsurgeons can reconnect arteries but not all the more delicate veins. Leeches are used to relieve congestion until the veins can grow back into the healing digit.

Respiration and Excretion

Gas exchange occurs only through the skin except in some of the fish leeches, which have gills. There are 10 to 17 pairs of nephridia, in addition to coelomocytes and certain other specialized cells that also may be involved in excretory functions.

Nervous and Sensory Systems

Leeches have two "brains"; one is in the anterior and is composed of six pairs of fused ganglia forming a ring around the pharynx, and one is in the posterior and is composed of seven pairs of fused ganglia. There are an additional 21 pairs of segmental ganglia along the double nerve cord. In addition to free sensory nerve endings and photoreceptor cells in the epidermis, there is a row of sense organs, called sensillae, in the central annulus of each segment; there are also a number of pigment-cup ocelli.

Leeches are highly sensitive to stimuli associated with the presence of a prey or host. They are attracted by and will attempt to attach to an object smeared with appropriate host substances, such as fish scales, oil secretions, or sweat. Those that feed on the blood of mammals are attracted by warmth, and the terrestrial haemadipsids of the tropics will converge on a person standing in one place.

Reproduction

Leeches are hermaphroditic but practice cross-fertilization during copulation. Sperm are transferred by a penis or by hypodermic impregnation (a spermatophore is expelled from one worm and penetrates the integument of the other). After copulation the clitellum secretes a cocoon that receives the eggs and sperm. They bury their cocoons in bottom mud, attach them to submerged objects, or, in terrestrial species, place them in damp soil. Development is similar to that of oligochaetes.

Circulation

In leeches the coelom has been reduced by the invasion of connective tissue

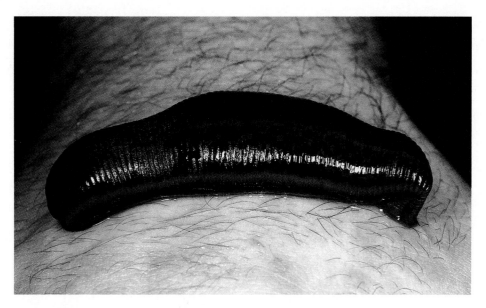

Figure 18-22
Hirudo medicinalis feeding on blood from human arm.

and, in some, by a proliferation of chloragogen tissue, to a system of coelomic sinuses and channels. Some orders of leeches retain the typical oligochaete circulatory system, and in these the coelomic sinuses act as an auxiliary blood-vascular system. In other orders the traditional blood vessels are lacking and the system of coelomic sinuses forms the only blood-vascular system. In those orders contractions of certain longitudinal channels provide propulsion for the blood (the equivalent of coelomic fluid).

Evolutionary Significance of Metamerism

No truly satisfactory explanation has yet been given for the origins of metamerism and the coelom, although the subject has stimulated much speculation and debate over the years. All of the classical explanations of the origin of metamerism and the coelom have had important arguments leveled against them, and more than one may be correct, or none, as suggested by R. B. Clark.* The coelom

*Clark, R. B. 1964. Dynamics in metazoan evolution. The origin of the coelom and segments. Oxford, England, Clarendon Press.

and metamerism may have evolved independently in more than one group of animals, as, for example, in the chordates and in the protostomes. Clark stressed the functional and evolutionary significance of these features to the earliest animals that possessed them. He argued forcefully that the adaptive value of the coelom in the protostomes, at least, was as a **hydrostatic skeleton** in a burrowing animal. Thus contraction of muscles in one part of the animal could act antagonistically on muscles in another part by transmission of the force of contraction through the enclosed constant volume of fluid in the coelom.

Although the original function of the coelom may have been burrowing in the substrate, certain other advantages accrued to its possessors. Some of these were mentioned in the prologue to Chapter 17. In addition, the coelomic fluid would have acted as a circulatory fluid for nutrients and wastes, making large numbers of flame cells distributed throughout the tissues unnecessary. Gametes could be stored in the spacious coelom for release simultaneously with other individuals in the population, thus enhancing chances of fertilization, and this would have selected for greater nervous and endocrine control. Finally, separation of the coelom into a series of compart-

ments by septa (metamerism) would have increased burrowing efficiency and made possible independent and separate movements by separate metameres, as mentioned in the prologue to this chapter. Independent movements of metameres in different parts of the body would have placed selective value on a more sophisticated nervous system for control of the movements and led to elaboration of the central nervous system.

Phylogeny and Adaptive Radiation

Phylogeny

There are so many similarities in the early development of the molluscs, annelids, and primitive arthropods that there seems little doubt about their close relationship. These three phyla apparently form a sister group to the flatworms. Many marine annelids and molluscs have an early embryogenesis typical of protostomes, in common with some marine flatworms, which is probably a shared plesiomorphic trait (p. 200). Annelids share with the arthropods an outer secreted cuticle and have a similar nervous system, and the lateral appendages (parapodia) of many marine annelids are similar to the appendages of certain arthropods with other primitive characters. The most important resemblance, however, probably lies in the metameric plan of the annelid and the arthropod body structure.

What can we infer about the common ancestor of the annelids? This has been the subject of a long and continuing debate. Most hypotheses of annelid origin have assumed that metamerism arose in connection with the development of lateral appendages (parapodia) resembling those of the polychaetes. However, the oligochaete body is adapted to vagrant burrowing in the substratum with a peristaltic movement that is highly benefited by a metameric coelom. On the other hand, polychaetes with well-developed parapodia are generally adapted to swimming and crawling in a medium too fluid for effective peristaltic locomotion. Although

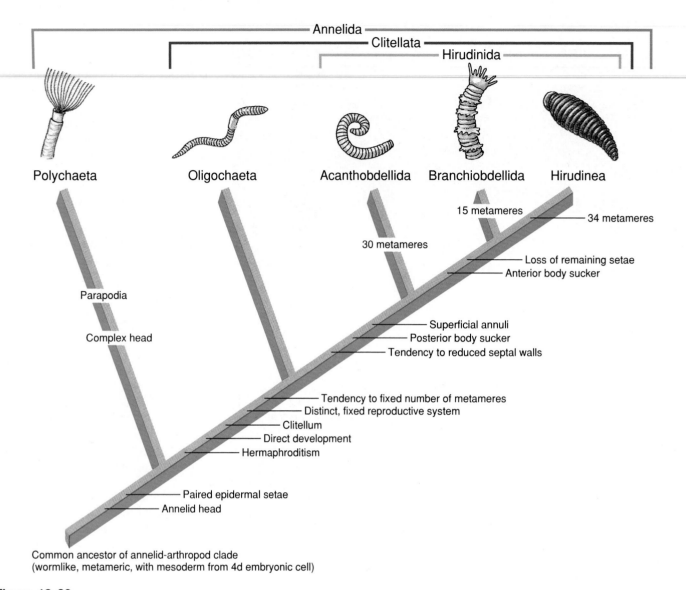

Figure 18-23

Cladogram of the annelids, showing the appearance of shared derived characters that specify the five monophyletic groups (based on Brusca and Brusca, 1990). The Acanthobdellida and the Branchiobdellida are two small groups discussed briefly in the marginal note on p. 367. Brusca and Brusca place both groups, together with the Hirudinea ("true" leeches), within a single taxon, the Hirudinida. This clade has several synapomorphies: tendency toward reduction of septal walls, the appearance of a posterior sucker, and the subdivision of body segments by superficial annuli. Note also that, according to this scheme, the Oligochaeta have no defining synapomorphies, that is, they are defined solely by retention of plesiomorphies (retained primitive characters, p. 200), and thus might be paraphyletic.

Source: Modified from R. C. Brusca and G. J. Brusca, Invertebrates. *Sinauer Associates, Inc., Sunderland, MA, 1990.*

the parapodia do not prevent such locomotion, they do little to further it, and it seems likely that they evolved as an adaptation for swimming. Although the polychaetes are most primitive in some respects, such as in their reproductive system, some authorities have argued that the ancestral annelids were more similar to the oligochaetes in overall body plan and that those of the polychaetes and leeches are more evolutionarily derived. The leeches are closely related to the oligochaetes but have diverged from them in connection with a swimming existence and the abandonment of a burrowing mode of life. This relationship is shown by the cladogram in Figure 18-23.

ADAPTIVE RADIATION

Annelids are an ancient group that has undergone extensive adaptive radiation. The basic body structure, particularly of the polychaetes, lends itself to almost endless modification. As marine worms, polychaetes have a wide range of habitats in an environment that is not physically or physiologically demanding. Unlike the earthworms, whose environment imposes strict physical and physiological demands, the polychaetes have been free to experiment and thus have achieved a wide range of adaptive features.

A basic adaptive feature in evolution of annelids is their septal arrangement, resulting in fluid-filled coelomic compartments. Fluid pressure in these compartments is used as a hydrostatic skeleton in precise movements such as burrowing and swimming. Powerful circular and longitudinal muscles can flex, shorten, and lengthen the body.

There is a wide variation in feeding adaptations, from the sucking pharynx of the oligochaetes and the chitinous jaws of carnivorous polychaetes to the specialized tentacles and radioles of the particle feeders.

In polychaetes the parapodia have been adapted in many ways and for a variety of functions, chiefly locomotion and respiration.

In leeches many adaptations, such as suckers, cutting jaws, pumping pharynx, distensible gut, and the production of hirudin, relate to their predatory and bloodsucking habits.

Classification of Phylum Annelida

Higher classification of the annelids is based primarily on the presence or absence of parapodia, setae, and other morphological features. Because both the oligochaetes and the hirudineans (leeches) bear a clitellum, these two groups are often placed under the heading Clitellata (cli-tel-la´ta) and members are called clitellates. On the other hand, because both the Oligochaeta and the Polychaeta possess setae, some authorities place them together in a group called Chaetopoda (ke-top´o-da) (N.L. *chaeta*, bristle, from Gr. *chaitē*, long hair, + *pous, podos,* foot).

Class Polychaeta (pol´e-ke´ta) (Gr. *polys,* many, + *chaitē,* long hair). Mostly marine; head distinct and bearing eyes and tentacles; most segments with parapodia (lateral appendages) bearing tufts of many setae; clitellum absent; sexes usually separate; gonads transitory; asexual budding in some; trochophore larva usually present; mostly marine. Examples: *Nereis, Aphrodita, Glycera, Arenicola, Chaetopterus, Amphitrite.*

Class Oligochaeta (ol´i-go-ke´ta) (Gr. *oligos,* few, + *chaitē,* long hair). Body with conspicuous segmentation; number of segments variable; setae few per metamere; no parapodia; head absent; coelom spacious and usually divided by intersegmental septa; hermaphroditic; development direct, no larva; chiefly terrestrial and freshwater. Examples: *Lumbricus, Stylaria, Aeolosoma, Tubifex.*

Class Hirudinea (hir´u-din´e-a) (L. *hirudo,* leech, + *ea,* characterized by): **leeches.** Body with fixed number of segments (normally 34; 17 or 31 in some groups) with many annuli; oral and posterior suckers usually present; clitellum present; no parapodia; setae absent (except in *Acanthobdella*); coelom closely packed with connective tissue and muscle; development direct; hermaphroditic; terrestrial, freshwater, and marine. Examples: *Hirudo, Placobdella, Macrobdella.*

The Branchiobdellida, a group of small annelids that are parasitic or commensal on crayfish and show similarities to both oligochaetes and leeches, are here placed with the oligochaetes, but they are considered a separate class by some authorities. They have 14 or 15 segments and bear a head sucker.

One genus of leech, *Acanthobdella,* has some characteristics of leeches and some of oligochaetes; it is sometimes separated from the other leeches into a special class, Acanthobdellida, that characteristically has 27 somites, setae on the first five segments, and no anterior sucker.

Summary

The phylum Annelida is a large, cosmopolitan group containing the marine polychaetes, the earthworms and freshwater oligochaetes, and the leeches. Certainly the most important structural innovation underlying the diversification of this group is metamerism, the division of the body into a series of similar segments, each of which contains a repeated arrangement of many organs and systems. The coelom is also highly developed in the annelids, and this, together with the septal arrangement of fluid-filled compartments and a well-developed body-wall musculature, is an effective hydrostatic skeleton for precise burrowing and swimming movements. Further metameric specialization occurs in the arthropods, the subjects of the next three chapters.

Polychaetes are the largest class of annelids and are mostly marine. On each somite they have many setae, which are borne on paired parapodia. Parapodia show a wide variety of adaptations among polychaetes, including specialization for swimming, respiration, crawling, maintaining position in a burrow, pumping water through a burrow, and accessory feeding. Some polychaetes are mostly predaceous and have an eversible pharynx with jaws. Other polychaetes rarely leave the burrows or tubes in which they live. Several styles of deposit and filter feeding are shown among the members of this group. Polychaetes are dioecious, have a primitive reproductive system, no clitellum, external fertilization, and a trochophore larva.

The Class Oligochaeta contains the earthworms and many freshwater forms;

they have a small number of setae per segment (compared to the Polychaeta) and no parapodia. They have a closed circulatory system, and the dorsal blood vessel is the main pumping organ. There is a pair of nephridia in most somites. Earthworms contain the typical annelid nervous system: dorsal cerebral ganglia connected to a double, ventral nerve cord with segmental ganglia, running the length of the worm. Oligochaetes are hermaphroditic and practice cross-fertilization. The clitellum plays an important role in reproduction, including secretion of mucus to surround the worms during copulation and secretion of a cocoon to receive the eggs and sperm and in which embryonation occurs. A small, juvenile worm hatches from the cocoon.

The leeches (Class Hirudinea) are mostly freshwater, although a few are marine and a few are terrestrial. They feed mostly on fluids; many are predators, some are temporary parasites, and a few are permanent parasites. The hermaphroditic leeches reproduce in a fashion similar to oligochaetes, with cross-fertilization and cocoon formation by the clitellum.

Embryological evidence supports a phylogenetic relationship of the annelids with the molluscs and arthropods.

Review Questions

1. What are characteristics of the Phylum Annelida that distinguish it from other phyla?
2. Distinguish among the classes of the Phylum Annelida.
3. Describe the annelid body plan, including the body wall, segments, coelom and its compartments, and coelomic lining.
4. Explain how the hydrostatic skeleton of the annelids helps them to burrow. How is the efficiency for burrowing increased by metamerism?
5. Describe three ways that various polychaetes obtain food.
6. Define each of the following: prostomium, peristomium, pygidium, radioles, parapodium, neuropodium, notopodium.
7. Explain the function of each of the following in earthworms: pharynx, calciferous glands, crop, gizzard, typhlosole, chloragogen tissue.
8. Describe the main features of each of the following in each class of annelids: circulatory system, nervous system, excretory system.
9. Describe the function of the clitellum and the cocoon.
10. How are freshwater oligochaetes generally different from earthworms?
11. Describe the ways in which leeches obtain food.
12. What are the main differences in reproduction and development among the three classes of annelids?
13. What was the evolutionary significance of metamerism and the coelom to its earliest possessors?
14. What are the phylogenetic relationships between the molluscs, annelids, and arthropods? What is the evidence for these relationships?

Selected References

See also general references to Part III, p. 626.

Conniff, R. 1987. The little suckers have made a comeback. Discover **8**:84–94 (Aug.). *Describes the medical uses for leeches in microsurgery.*

Dales, R. P. 1967. Annelids, ed. 2. London, The Hutchinson Publishing Group, Ltd. *A concise account of the annelids.*

Hartman, O. 1968. Atlas of the errantiate polychaetous annelids from California. Los Angeles, University of Southern California, Allan Hancock Foundation. *This and the following reference have extensive keys for identification.*

Hartman, O. 1969. Atlas of the sedentariate polychaetous annelids from California. Los Angeles, University of Southern California, Allan Hancock Foundation.

Kingman, J., and P. Kingman. 1993. The dance of the luminescent threadworms. Underwater Naturalist **22**(2):36 *Describes the spectacular swarming of the epitokes of Odontosyllis enopla off Belize during July and August, the third and fourth nights after a full moon.*

Lent, C. M., and M. H. Dickinson. 1988. The neurobiology of feeding in leeches. Sci. Am. **258**:98–103 (June). *Feeding behavior in leeches is controlled by a single neurotransmitter (serotonin).*

19

The Arthropods

Phylum Arthropoda
Subphylum Trilobita
Subphylum Chelicerata

A Suit of Armor

Sometime, somewhere in the Precambrian era, a major milestone in the evolution of life on earth was passed. The soft cuticle in an ancestor of the animals we now call arthropods was stiffened by deposition of additional amounts of protein and an inert polysaccharide called chitin. The cuticular exoskeleton was some protection against predators and other environmental hazards, and it conferred on its possessors a formidable array of other selective advantages. Of course, the suit of armor could not be uniformly stiff; the animal would be as unable to move as the rusted tin woodsman in the *Wizard of Oz*. Stiff sections of cuticle were separated from each other by thin, flexible sections, which formed joints. The cuticular exoskeleton had enormous evolutionary potential. Jointed extensions on each metamere became appendages.

Once the stiffened cuticle evolved, or perhaps concurrently with it, many other changes were necessary in the bodies of the proto-arthropods. To allow for growth, a sequence of cuticular molts was required. This mandated hormonal control. The hydrostatic skeletal function of the coelomic compartments was lost, leading to a regression of the coelom and its replacement with an open system of sinuses (hemocoel). Motile cilia were lost. These changes and others are called "arthropodization." Some zoologists argue that all the changes in arthropodization follow from the development of a cuticular exoskeleton. If several different annelid-like ancestors had independently evolved a cuticular exoskeleton, then they independently would have evolved the identical suite of characters we associate with arthropodization. The huge phylum we call Arthropoda would be in reality polyphyletic. However, we agree with other zoologists who feel that the weight of evidence still supports single-phylum status. ■

POSITION IN ANIMAL KINGDOM

1. Shared derived characters indicate that both annelids and arthropods evolved from a line of coelomates, segmented protostomes with spiral cleavage and mosaic development.
2. Evolution of the hard cuticular exoskeleton was followed or accompanied by arthropodization, which included loss of intersegmental septa; development of hemocoel and loss of closed circulatory system; jointed appendages; conversion of body-wall muscles to insert on cuticle.
3. Like the annelids, the arthropods have conspicuous metamerism, but their somites have greater variety and more grouping for specialized purposes; specialization of appendages, with pronounced division of labor, results in greater variety of action.

BIOLOGICAL CONTRIBUTIONS

1. Cephalization becomes more pronounced, with centralization of fused ganglia and sensory organs in the head.
2. Compared with the annelids, the **somites** are more **specialized** for a variety of purposes, forming functional groups **(tagmosis)**.
3. The presence of paired **jointed appendages** diversified for numerous uses produces greater adaptability.
4. Locomotion is by extrinsic limb muscles, in contrast to the body musculature of annelids. Striated muscles confer rapidity of movement.
5. Although **chitin** is found in a few groups other than arthropods, its use is better developed in the arthropods. The **cuticular exoskeleton,** containing chitin, is a great innovation, making possible a wide range of adaptations.
6. The **tracheae** represent a breathing mechanism more efficient than that of most invertebrates.
7. The alimentary canal shows greater specialization by having, in various arthropods, chitinous teeth, compartments, and gastric ossicles.
8. Behavioral patterns are much more complex than those of most invertebrates, with a wider occurrence of **social** organization.
9. Many arthropods have well-developed protective coloration and protective resemblances.

PHYLUM ARTHROPODA

Phylum Arthropoda (ar-throp´o-da) (Gr. *arthron,* joint, + *pous, podos,* foot) is the most extensive phylum in the animal kingdom, composed of more than three-fourths of all known species. Approximately 900,000 species of arthropods have been recorded, and probably at least as many more remain to be classified. However, based on surveys of insect fauna in the canopy of rain forests, many estimates of yet undescribed species are much higher. Arthropods include the spiders, scorpions, ticks, mites, crustaceans, millipedes, centipedes, insects, and some others. In addition, there is a rich fossil record extending to the very late Precambrian period.

Arthropods are eucoelomate protostomes with well-developed organ systems, and they share with the annelids the property of conspicuous metamerism.

Arthropods have an exoskeleton containing chitin, and their primitive pattern is that of a linear series of similar somites, each with a pair of jointed appendages. However, the pattern of somites and appendages varies greatly in the phylum. There is a tendency for the somites to be combined or fused into functional groups, called **tagmata** (sing., **tagma**), for specialized purposes; the appendages are frequently differentiated and specialized for pronounced division of labor.

Few arthropods exceed 60 cm in length, and most are far below this size. The largest is the Japanese crab *Macrocheira* (Gr. *makros,* large, + *cheir,* hand), which has approximately a 4 m span; the smallest is the parasitic mite *Demodex* (Gr. *dēmos,* body, frame, + *dex,* a wood worm), which is less than 0.1 mm long.

Arthropods are usually active, energetic animals. Judging by their great abundance and their wide ecological distribution, as well as by the vast number of species, their diversity is surpassed by no other group of animals.

Although arthropods compete with humans for food and spread serious diseases, they are essential in pollination of many food plants, and they also serve as food, yield drugs and dyes, and produce products such as silk, honey, and beeswax.

The arthropods are more widely and more densely distributed throughout all regions of the earth than are members of any other phylum. They are found in all types of environment from low ocean depths to very high altitudes, and from the tropics far into both north and south polar regions. Different species are adapted for life in the air; on land; in fresh, brackish, and marine waters; and in or on the bodies of plants and other animals. Some species live in places where no other animal could survive.

Although all modes of feeding—carnivorous, herbivorous, and omnivorous—occur in this vast group, the majority are herbivorous. Most aquatic arthropods depend on algae for their nourishment, and the majority of land forms live chiefly on plants. In diversity of ecological distribution, the arthropods have no rivals.

We will cover the subphyla Trilobita (all extinct) and Chelicerata in this chapter, and successive chapters will be devoted to each of the subphyla Crustacea and Uniramia (classification of the Arthropoda on p. 381).

Characteristics of Phylum Arthropoda

1. Bilateral symmetry; **metameric body** divided into **tagmata** consisting of head and trunk; head, thorax, and abdomen; or cephalothorax and abdomen
2. **Jointed appendages;** primitively, one pair to each somite, but number often reduced; appendages often modified for specialized functions
3. **Exoskeleton of cuticle** containing protein, lipid, chitin, and often calcium carbonate secreted by underlying epidermis and shed (molted) at intervals
4. **Complex muscular system,** with exoskeleton for attachment, **striated muscles** for rapid actions, smooth muscles for visceral organs; no cilia
5. **Reduced coelom** in adult; most of body cavity consisting of hemocoel (sinuses, or spaces, in the tissues) filled with blood
6. **Complete digestive system;** mouthparts modified from appendages and adapted for different methods of feeding
7. **Open circulatory system,** with dorsal **contractile heart,** arteries, and hemocoel (blood sinuses)
8. Respiration by **body surface, gills, tracheae** (air tubes), or **book lungs**
9. Paired excretory glands called **coxal, antennal,** or **maxillary glands** present in some, homologous to metameric nephridial system of annelids; some with other excretory organs, called **malpighian tubules**
10. **Nervous system** of **annelid plan,** with dorsal brain connected by a ring around the gullet to a double nerve chain of ventral ganglia; fusion of ganglia in some species; well-developed sensory organs
11. **Sexes usually separate,** with paired reproductive organs and ducts; usually internal fertilization; oviparous or ovoviviparous; often with **metamorphosis;** parthenogenesis in some

Comparison of Arthropoda with Annelida

Similarities between Arthropoda and Annelida are as follows:

1. External segmentation marked
2. Segmental arrangement of muscles
3. Ventral nerve cord with metamerically arranged ganglia and dorsal cerebral ganglia
4. Spiral cleavage (found in some arthropods)

Arthropods differ from annelids in having the following:

1. Fixed number of segments (in adults)
2. Usually lack intersegmental septa
3. Pronounced tagmatization (compared with limited tagmatization in annelids)
4. Coelomic cavity reduced; main body cavity a hemocoel
5. Open (lacunar) circulatory system
6. Special mechanisms (gills, tracheae, book lungs) for respiration
7. Exoskeleton containing chitin
8. Jointed appendages
9. Compound eyes (also present in a few annelids) and other well-developed sense organs
10. Absence of cilia

Why Have Arthropods Achieved Such Great Diversity and Abundance?

The arthropods have achieved a great diversity, number of species, wide distribution, variety of habitats and feeding habits, and power of adaptation to changing conditions. In the following discussion we briefly summarize some of the structural and physiological patterns that have been helpful to them.

1. A versatile exoskeleton. The arthropods possess an exoskeleton that is highly protective without sacrificing mobility. This skeleton is the **cuticle,** an outer covering secreted by the underlying epidermis. The cuticle is made up of an inner and thicker **procuticle** and an outer, relatively thin **epicuticle.** The procuticle is divided into the **exocuticle,** which is secreted before a molt, and **endocuticle,** which is secreted after molting. Both layers of the procuticle contain **chitin** bound with protein. Chitin is a tough, resistant, nitrogenous polysaccharide that is insoluble in water, alkalis, and weak acids. Thus the procuticle not only is flexible and lightweight but also affords protection, particularly against dehydration. In some crustaceans the chitin may make up as much as 60% to 80% of the procuticle, but in insects it is probably not more than 40% (the remainder being protein). In most crustaceans the procuticle is also impregnated with **calcium salts,** which reduce its flexibility. In the hard shells of lobsters and crabs, for instance, this calcification is extreme. The outer epicuticle is composed of protein and lipid. The protein is stabilized and hardened by tanning, adding further protection. Both the procuticle and epicuticle are laminated, that is, composed of several layers each (see Figure 32-1, p. 632).

The cuticle may be soft and permeable or may form a veritable coat of armor. Between body segments and between the segments of appendages it is thin and flexible, creating movable joints and permitting free movements. In crustaceans and insects the cuticle forms ingrowths (apodemes) that serve for muscle attachment. It may also line the foregut and hindgut, line and support the trachea, and be adapted for biting mouthparts, sensory organs, copulatory organs, and ornamental purposes. It is indeed a versatile material.

The nonexpansible cuticular exoskeleton does, however, impose important restrictions on growth. To grow, an arthropod must shed its outer covering at intervals and grow a larger one—a process called **ecdysis,** or **molting.** Arthropods molt four to seven times before reaching adulthood, and some continue to molt after that. An exoskeleton is also relatively heavy and becomes proportionately heavier with increasing size. This tends to limit the ultimate body size.

2. Segmentation and appendages for more efficient locomotion. Typically each somite bears a pair of jointed appendages, but this arrangement is often modified, with both segments and appendages specialized for adaptive functions. The limb segments are essentially hollow levers that are moved by internal muscles, most of which are striated for rapid action. The jointed appendages have sensory hairs and may be modified and adapted for sensory functions, food handling, swift and efficient walking, and swimming.

3. Air piped directly to cells. Most terrestrial arthropods have the highly efficient tracheal system of air tubes, which delivers oxygen directly to the tissues and cells and makes a high metabolic rate possible. This system also tends to limit body size. Aquatic arthropods breathe mainly by some form of gill that is quite efficient.

4. Highly developed sensory organs. Sensory organs are found in great variety, from the compound (mosaic) eye to those accomplishing touch, smell, hearing, balancing, chemical reception. Arthropods are keenly alert to what happens in their environment.

5. Complex behavior patterns. Arthropods exceed most other invertebrates in the complexity and organization of their activities. Innate (unlearned) behavior unquestionably controls much of what they do, but learning also plays an important part in the lives of many of them.

6. Use of diverse resources through metamorphosis. Many arthropods pass through metamorphic changes, including a larval form quite different from the adult in structure. The larval form often is adapted for eating food different from that of the adult and occupies a different space, resulting in less competition within a species.

SUBPHYLUM TRILOBITA

The trilobites probably had their beginnings before the Cambrian period, in which they flourished. They have been extinct for 200 million years, but were

A

B

Figure 19-1
Fossils of early arthropods. **A,** Trilobite fossils, dorsal view. These animals were abundant in mid-Cambrian period. **B,** Eurypterid fossil. Eurypterids flourished in Europe and North America from Ordovician to Permian periods.

abundant during the Cambrian and Ordovician periods. Their name refers to the trilobed shape of the body, caused by a pair of longitudinal grooves. They were bottom dwellers and probably scavengers (Figure 19-1A). Most of them could roll up like pill bugs, and they ranged from 2 to 67 cm in length.

The exoskeleton contained chitin, strengthened in some areas by calcium carbonate. There were three tagmata in the body: head, thorax, and pygidium. The head was one piece but showed signs of former segmentation; the thorax had a variable number of somites; and the somites of the pygidium, at the posterior end, were fused into a plate. The head bore a pair of antennae, compound eyes, mouth, and four pairs of jointed appendages. Each body somite except the last also bore a pair of biramous (two-branched) appendages. One of the branches had a fringe of filaments that may have served as gills.

SUBPHYLUM CHELICERATA

The chelicerate arthropods are an ancient group that includes the eurypterids (extinct), horseshoe crabs, spiders, ticks and mites, scorpions, and sea spiders. They are characterized by having six pairs of appendages that include a pair of chelicerae, a pair of pedipalps, and four pairs of walking legs (a pair of chelicerae and five pairs of walking legs in horseshoe crabs). They have no mandibles and no antennae. Most chelicerates suck liquid food from their prey.

CLASS MEROSTOMATA

Class Merostomata is represented by the eurypterids, all now extinct, and the xiphosurids, or horseshoe crabs, an ancient group sometimes called "living fossils."

Subclass Eurypterida

The eurypterids, or giant water scorpions (Figure 19-1B) were the largest of all fossil arthropods, some reaching a length of 3 m. Their fossils occur in rocks from the Ordovician to the Permian periods. They had many resemblances to the marine horseshoe crabs (Figure 19-2) and also to the scorpions, their land counterparts. The head had six fused segments and bore both simple and compound eyes and six pairs of appendages. The abdomen had 12 segments and a spikelike telson.

Ideas regarding their early habitats differ. Some believe the eurypterids evolved mainly in fresh water; others hold that they arose in brackish lagoons.

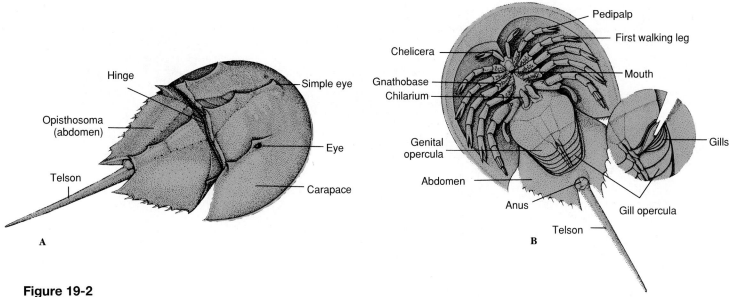

Figure 19-2
A, Dorsal view of horseshoe crab *Limulus* (class Merostomata). They grow to 0.5 m in length. **B,** Ventral view.

Subclass Xiphosurida: Horseshoe Crabs

The xiphosurids are an ancient marine group that dates from the Cambrian period. Our common horseshoe crab *Limulus* (L. *limus,* sidelong, askew) (Figure 19-2) goes back practically unchanged to the Triassic period. Only three genera (five species) survive today: *Limulus,* which lives in shallow water along the North American Atlantic coast; *Carcinoscorpius* (Gr. *karkinos,* crab, + *skorpiōn,* scorpion), along the southern shore of Japan; and *Tachypleus,* (Gr. *tachys,* swift, + *pleutēs,* sailor), in the East Indies and along the coast of southern Asia. They usually live in shallow water.

Xiphosurids have an unsegmented, horseshoe-shaped **carapace** (hard dorsal shield) and a broad abdomen, which has a long **telson,** or tailpiece. The cephalothorax bears five pairs of walking legs and a pair of chelicerae, whereas the abdomen has six pairs of broad, thin appendages that are fused in the median line (Figure 19-2). On some of the abdominal appendages, **book gills** (flat, leaflike gills) are exposed. There are two compound and two simple eyes on the carapace. The horseshoe crab swims by means of its abdominal plates and can walk with its walking legs. It feeds at night on worms and small molluscs, which it seizes with its chelicerae.

During the mating season the horseshoe crabs come to shore at high tide to mate. The female burrows into the sand where she lays her eggs, with one or more smaller males following her closely to add their sperm to the nest before she covers the eggs with sand. The eggs are warmed by the sun and protected from the waves until the young larvae hatch and return to the sea by another high tide. The larvae are segmented and are often called "trilobite larvae" because they resemble the trilobites, to which the xiphosurids may be related.

CLASS PYCNOGONIDA: SEA SPIDERS

Some sea spiders are only a few millimeters long, but others are much larger. They have small, thin bodies and usually four pairs of long, thin walking legs. In addition, they have a feature unique among arthropods: somites are duplicated in some groups, so that they possess five or six pairs of legs instead of the four pairs normally characteristic of arachnids. The males of many species bear a subsidiary pair of legs **(ovigers)** (Figure 19-3) on which they carry developing eggs, and ovigers are often absent in females. Many species also are equipped with chelicerae and palps.

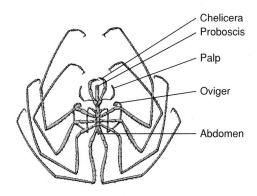

Figure 19-3
Pycnogonid, *Nymphon* sp. In this genus all the anterior appendages (chelicerae, palps, and ovigers) are present in both sexes, although ovigers are often not present in females of other genera.

The mouth is at the tip of a long **proboscis,** which sucks juices from cnidarians and soft-bodied animals. Most pycnogonids have four simple eyes. The circulatory system is limited to a simple dorsal heart, and excretory and respiratory systems are absent. The long, thin body and legs provide a large surface, in proportion to volume, that is evidently sufficient for diffusion of gases and wastes. Because of the small size of the body, the digestive system sends branches into the legs, and most of the gonads are also in the legs.

Sea spiders are found in all oceans, but they are most abundant in polar waters. *Pycnogonum* (Figure 19-4) is a

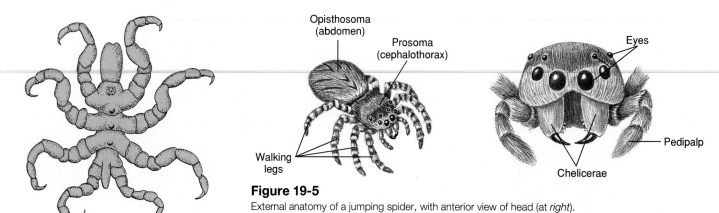

Figure 19-4

Pycnogonum, a pycnogonid with relatively short legs. Females of this genus have none of the anterior appendages, and only males have ovigers.

Figure 19-5

External anatomy of a jumping spider, with anterior view of head (at *right*).

common intertidal genus found on both the Atlantic and Pacific coasts of the United States; it has relatively short, heavy legs. *Nymphon* (Figure 19-3) is the largest genus of pycnogonids, with over 200 species. It occurs from subtidal depths to 6800 m in all seas except the Black and Baltic seas.

Some authorities believe that the pycnogonids are more closely related to the crustaceans than to other arthropods (their larva is rather similar in appearance to the nauplius larva of crustaceans); others place them closer to the arachnids.

CLASS ARACHNIDA

The arachnids (Gr. *arachnē,* spider) show wider anatomical divergence than the insects. In addition to spiders, the group includes scorpions, pseudoscorpions, whip scorpions, ticks, mites, daddy longlegs (harvestmen), and others. There are many differences among these with respect to form and appendages. They are mostly free living and are far more common in warm, dry regions than elsewhere.

The arachnid tagmata are a cephalothorax and abdomen, and the cephalothorax usually bears a pair of chelicerae, a pair of pedipalps, and four pairs of walking legs (Figure 19-5). Antennae and mandibles are lacking. Most arachnids are predaceous and have claws, fangs (claws and fangs are

modified pedipalps and chelicerae), poison glands, or stingers. They usually have sucking mouthparts or a strong sucking pharynx with which they ingest the fluids and soft tissues from the bodies of their prey. Among their interesting adaptations are the spinning glands of the spiders.

Arachnids have become extremely diverse. More than 70,000 species have been described so far. They were the first of the arthropods to move into terrestrial habitats. Scorpions are among Silurian fossils, and by the end of the Paleozoic period mites and spiders had appeared.

Most arachnids are harmless to humans and actually do much good by destroying injurious insects. A few, such as the black widow and brown recluse spiders, can give painful or even dangerous bites. The sting of the scorpion may be quite painful. Some ticks and mites are carriers of diseases as well as causes of annoyance and painful irritations. Certain mites damage a number of important food and ornamental plants by sucking their juices. Several smaller orders are not included in the following discussion.

Order Araneae: Spiders

The spiders are a large group of 35,000 species, distributed throughout the world. The spider body is compact: a **cephalothorax (prosoma)** and **abdomen (opisthosoma),** both unsegmented and joined by a slender pedicel.

The anterior appendages are a pair of **chelicerae** (Figure 19-5), which have terminal **fangs** through

which run ducts from poison glands, and a pair of **pedipalps** having basal parts with which they chew (Figure 19-5). Four pairs of **walking legs** terminate in claws.

All spiders are predaceous, feeding largely on insects. They effectively dispatch their prey with their fangs and poison. Some spiders chase their prey, others ambush them, and many trap them in a net of silk. After the spider seizes its prey with its chelicerae and injects venom, it liquefies the tissues with a digestive fluid and sucks the resulting broth into the stomach. Spiders with teeth at the bases of the chelicerae crush or chew the prey, aiding digestion by enzymes from the mouth.

Spiders breathe by means of book lungs or tracheae or both. Book lungs, which are unique in spiders, consist of many parallel air pockets extending into a blood-filled chamber (Figure 19-6). Air enters the chamber by a slit in the body wall. The tracheae make up a system of air tubes that carry air directly to the tissues from openings called spiracles. The tracheae are similar to those in insects (p. 414) but are much less extensive.

Spiders and insects have a unique **excretory system of malpighian tubules** (Figure 19-6), which work in conjunction with specialized rectal glands. Potassium and other solutes and waste materials are secreted into the tubules, which drain the fluid, or "urine," into the intestine. The rectal glands reabsorb most of the potassium and water, leaving behind such wastes as uric acid. By this cycling of water and potassium, species living in dry

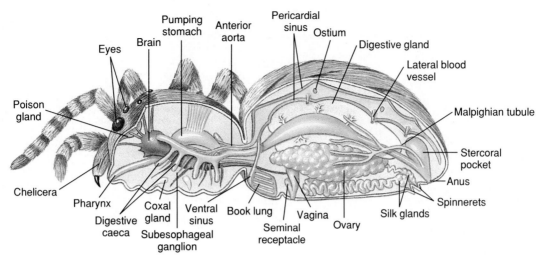

Figure 19-6
Spider, internal anatomy.

environments may conserve body fluids, producing a nearly dry mixture of urine and feces. Many spiders also have **coxal glands,** which are modified nephridia that open at the coxa, or base, of the first and third walking legs.

Spiders usually have eight **simple eyes,** each with a lens, optic rods, and a retina (Figure 19-6). They are used chiefly for perception of moving objects, but some, such as those of the hunting and jumping spiders, may form images. Since a spider's vision is usually poor, its awareness of its environment depends largely on its hairlike **sensory setae.** Every seta on its surface, regardless of whether it is actually connected to receptor cells, is useful in communicating some information about the surroundings, air currents, or changing tensions in the spider's web. By sensing the vibrations of its web, the spider can judge the size and activity of its entangled prey or can receive the message tapped out by a prospective mate.

Web-Spinning Habits. The ability to spin silk is central to a spider's life, as it is in some other arachnids. Two or three pairs of spinnerets containing hundreds of microscopic tubes run to special abdominal **silk glands** (Figure 19-6). A scleroprotein secretion emitted as a liquid apparently hardens as a result of being pulled from the spinnerets and forms the silk thread. Spiders' silk threads are stronger than

steel threads of the same diameter and are said to be second in strength only to fused quartz fibers. The threads will stretch one-fifth of their length before breaking.

The web used for trapping insects is the use of silk familiar to most people. The kind of net varies with the species. Some are simple and consist merely of a few strands of silk radiating out from a spider's burrow or place of retreat. Others spin the beautiful, geometrical orb webs. However, spiders use silk threads for many purposes besides web making. They use them to line their nests; form sperm webs or egg sacs; build draglines; make bridge lines, warning threads, molting threads, attachment discs, or nursery webs; or wrap up their prey securely (Figure 19-7). Not all spiders spin webs for traps. Some, such as the wolf spiders, jumping spiders (Figure 19-5), and fisher spiders (Figure 19-8), simply chase and catch their prey.

Reproduction. Before mating, the male spins a small web, deposits a drop of sperm on it, and then picks the sperm up and stores it in the special cavities of his pedipalps. When he mates, he inserts the pedipalps into the female genital opening to store the sperm in his mate's seminal receptacles. Before mating, there is usually a courtship ritual. The female lays her eggs in a silken net, which

Figure 19-7
Grasshopper, snared and helpless in the web of a golden garden spider (*Argiope aurantia*), is wrapped in silk while still alive. If the spider is not hungry, the prize will be saved for a later meal.

she may carry about or may attach to a web or plant. A cocoon may contain hundreds of eggs, which hatch in approximately two weeks. The young usually remain in the egg sac for a few weeks and molt once before leaving it. Several molts occur before adulthood.

Are Spiders Really Dangerous? It is truly amazing that such small and helpless creatures as the spiders have generated so much unreasoning fear in the human mind. Spiders are timid creatures that, rather than being dangerous enemies to humans, are actually allies in the continuing battle with insects. The venom produced to kill

A   B

Figure 19-8

Fisher spider, *Dolomedes triton,* feeds on a minnow. This handsome spider feeds mostly on aquatic and terrestrial insects but occasionally captures small fishes and tadpoles. It pulls its paralyzed victim from the water, pumps in digestive enzymes, then sucks out the predigested contents.

Figure 19-10

A, Black widow spider, *Latrodectus mactans,* suspended on her web. Note the red "hourglass" on the ventral side of her abdomen. **B,** Brown recluse spider, *Loxosceles reclusa,* is a small venomous spider. Note the small violin-shaped marking on its cephalothorax. The venom is hemolytic and dangerous.

Figure 19-9

Tarantula, *Rhechostica hentzi.*

the prey is usually harmless to humans. Even the most poisonous spiders bite only when threatened or when defending their eggs or young. The American tarantulas (Figure 19-9), despite their fearsome size, are *not* dangerous. They rarely bite, and their bite is not serious.

There are, however, two genera in the United States that can give severe or even fatal bites: *Latrodectus* (L. *latro,* robber, + *dektēs,* biter), and *Loxosceles* (Gr. *loxos,* crooked, + *skelos,* leg). The most important species are *Latrodectus mactans,* the **black**

widow, and *Loxosceles reclusa,* the **brown recluse.** The black widow is moderate to small in size and shiny black, with a bright orange or red "hourglass" on the underside of the abdomen (Figure 19-10A). The venom is neurotoxic; that is, it acts on the nervous system. About four or five out of each 1000 bites reported have proved fatal.

The brown recluse is brown and bears a violin-shaped dorsal stripe on its back (Figure 19-10B). Its venom is hemolytic rather than neurotoxic, producing death of the tissues and skin surrounding the bite. Its bite can be mild to serious and occasionally fatal.

Some spiders in other parts of the world are dangerous, for example, the funnelweb spider *Atrax robustus* in Australia. Most dangerous of all are certain ctenid spiders in South America, for example, *Phoneutria fera.* In contrast to most spiders, these are quite aggressive.

Order Scorpionida: Scorpions

Although scorpions are more common in tropical and subtropical regions, some occur in temperate zones. Scorpions are generally secretive, hiding in burrows or under ob-

jects by day and feeding at night. They feed largely on insects and spiders, which they seize with the pedipalps and tear up with the chelicerae.

Sand-dwelling scorpions apparently locate their prey by sensing surface waves generated by movements of insects on or in the sand. These waves are picked up by compound slit sensilla located on the basitarsal segments of the legs. The scorpion can locate a burrowing cockroach 50 cm away and reach it in three or four quick orientation movements.

The scorpion tagmata are a rather short **cephalothorax,** which bears the appendages, a pair of large median eyes, and two to five pairs of small lateral eyes; a **preabdomen** of seven segments; and a long slender **postabdomen,** or tail, of five segments, which ends in a stinging apparatus (Figure 19-11A). The chelicerae are small and three jointed; the pedipalps are large, chelate (pincerlike), and six jointed; and the four pairs of walking legs are eight jointed.

On the ventral side of the abdomen are curious comblike **pectines,** which are tactile organs used for exploring the ground and for sex recognition. The stinger on the last segment consists of a bulbous

A

B

Figure 19-11

A, Scorpion (order Scorpionida) with young, which stay with the mother until the first molt. **B,** Harvestmen, *Mitopus* sp. (order Opiliones). Harvestmen run rapidly on their stiltlike legs. They are especially noticeable during the harvesting season, hence the common name.

base and a curved barb that injects the venom. The venom of most species is not harmful to humans but may produce a painful swelling. However, the sting of certain species of *Androctonus* in Africa and *Centruroides* (Gr. *kenteō,* to prick, + *oura,* tail, + *oides,* form) in Mexico can be fatal unless antivenin is administered.

Scorpions perform a complex mating dance, the male holding the female's chelae and stepping back and forth. He taps her genital area with his forelegs and stings her pedipalp. Finally, he deposits a spermatophore and pulls the female over it until the sperm mass is taken up in the female orifice. Scorpions are either ovoviviparous or truly viviparous; that is, the females brood their young within the female reproductive tract. After several months or a year of development, anywhere from 6 to 90 young are produced, depending on the species. The young, only a few millimeters long,

crawl up on the mother's back until after the first molt (Figure 19-11). They mature in about a year.

Order Opiliones: Harvestmen

Harvestmen, often known as "daddy longlegs," are common in the United States and other parts of the world (Figure 19-11B). These curious creatures are easily distinguished from spiders by the fact that their abdomen and cephalothorax are broadly joined, without the constriction of the pedicel, and their abdomen shows external segmentation. They have four pairs of usually long, spindly legs, and they can cast off one or more of these without apparent ill effect if they are grasped by a predator (or human hand). The ends of their chelicerae are pincerlike, and they feed much more as scavengers than do spiders.

Order Acari: Ticks and Mites

Members of the order Acari are without doubt the most medically and economically important group of arachnids. They far exceed the other orders in number of individuals and species. Although about 30,000 species have been described, some authorities estimate that from 500,000 to 1 million species exist. Hundreds of individuals of several species of mites may be found in a small portion of leaf mold in forests. They are found throughout the world in both terrestrial and aquatic habitats, even extending into such inhospitable regions as deserts, polar areas, and hot springs. Many acarines are parasitic during one or more stages of their life cycle.

Most mites are 1 mm or less in length. Ticks, which make up only one suborder of the Acari, range from a few millimeters to occasionally 3 cm. A tick may become enormously distended with blood after feeding on its host.

Acarines differ from all other arachnids in having complete fusion of the cephalothorax and abdomen, with no sign of external division or segmentation (Figure 19-12). They carry their mouthparts on a little anterior projection, the **capitulum.** The capitulum mainly consists of the feed-

A

B

Figure 19-12

A, Wood tick, *Dermacentor* variabilis (order Acari). Larvae, nymphs, and adults are all parasitic but drop off their hosts to molt to the next stage. **B,** Red velvet (harvest) mite, *Trombidium* sp. As with the chigger (*Trombicula*), only the larvae of *Trombidium* are parasitic. Nymphs and adults are free living and feed on insect eggs and small invertebrates.

ing appendages surrounding the mouth. On each side of the mouth is a chelicera, which functions in piercing, tearing, or gripping food. The form of the chelicerae varies greatly in different families. Lateral to the chelicerae is a pair of segmental pedipalps, which also vary greatly in form and function related to feeding. Ventrally the bases of the pedipalps are fused to form a **hypostome,** whereas a **rostrum,** or **tectum,** extends dorsally over the mouth. Adult mites and ticks usually have four pairs of legs, although there may be only one to three in some specialized forms.

Most acarines transfer sperm directly, but many species transfer sperm by means of a spermatophore. A larva with six legs hatches from the egg, and one or more eight-legged nymphal stages follow before the adult stage is reached.

Many species of mites are entirely free living. *Dermatophagoides farinae*

Figure 19-13
Scanning electron micrograph of house dust mite, *Dermatophagoides farinae.*

The inflamed welt and intense itching that follows a chigger bite is not the result of the chigger burrowing into the skin, as is popularly believed. Rather the chigger bites through the skin with its chelicerae and injects a salivary secretion containing powerful enzymes that liquefy skin cells. Human skin responds defensively by forming a hardened tube that the larva uses as a sort of drinking straw and through which it gorges itself with host cells and fluid. Scratching usually removes the chigger but leaves the tube, which is a source of irritation for several days.

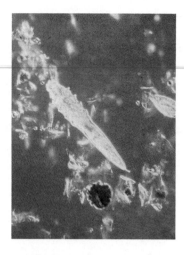

Figure 19-15
Demodex folliculorum, the human follicle mite.

(Gr. *dermatos,* skin, + *phagō,* to eat, + *eidos,* likeness of form) (Figure 19-13) and related species are denizens of house dust all over the world, sometimes causing allergies and dermatoses. There are some marine mites, but most aquatic species are found in fresh water. They have long, hairlike setae on their legs for swimming, and their larvae may be parasitic on aquatic invertebrates. Such abundant organisms must be important ecologically, but many acarines have more direct effects on our food supply and health. The spider mites (family Tetranychidae) are serious agricultural pests on fruit trees, cotton, clover, and many other plants. They suck out the contents of plant cells, causing a mottled appearance to the leaves (Figure 19-14), and construct a protective web from silk glands opening near the base of the chelicerae. Larvae of the genus *Trombicula* are called chiggers or redbugs. They feed on the dermal tissues of terrestrial vertebrates, including humans, and may cause an irritating dermatitis; some species of chiggers transmit a disease called Asiatic scrub typhus. The hair follicle mite, *Demodex* (Figure 19-15), is apparently nonpathogenic in humans; it infects most of us although we are unaware of it. Other species of *Demodex* and other genera of mites cause mange in domestic animals. The human itch mite, *Sarcoptes scabiei* (Figure 19-16), causes intense itching as it burrows beneath the skin.

Figure 19-14
Damage to *Chamaedorea* sp. palm caused by mites of the family Tetranychidae (order Acari). Over 130 species of this family occur in North America, and a number of them are serious agricultural pests. Mites pierce plant cells and suck out contents, resulting in mottled appearance shown here.

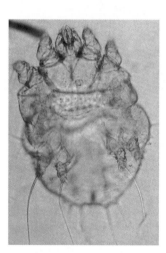

Figure 19-16
Sarcoptes scabiei, the itch mite.

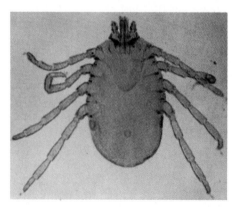

Figure 19-17
Boophilus annulatus, a tick that carries Texas cattle fever.

In addition to disease conditions that they themselves cause, ticks are among the world's premier disease vectors, ranking second only to mosquitos. They carry a greater variety of infectious agents than any other arthropods, including protozoan, rickettsial, viral, bacterial, and fungal organisms. Species of *Ixodes* carry the most common arthropod-borne infection in the United States, Lyme disease (see note). Species of *Dermacentor* (Figure 19-12A) and other ticks transmit Rocky Mountain spotted fever, a poorly named disease because most cases occur in the eastern United States. *Dermacentor* also transmits tularemia and the agents of several other diseases. Texas cattle fever, also called red-water fever, is caused by a protozoan parasite transmitted by the cattle tick *Boophilus annulatus* (Figure 19-17). Many more examples could be cited.

An epidemic of arthritis occurred in the 1970s in the town of Lyme, Connecticut. Subsequently known as Lyme disease, it is caused by a bacterium and carried by ticks of the genus *Ixodes*. There are now thousands of cases a year in Europe and North America, and other cases have been reported from Japan, Australia, and South Africa. Many people bitten by infected ticks recover spontaneously or do not get the disease. Others, if not treated at an early stage, develop a chronic, disabling disease.

PHYLOGENY AND ADAPTIVE RADIATION

PHYLOGENY

The sharing of derived characters between annelids and arthropods gives strong support to the hypothesis that both phyla originated from a line of coelomate segmented protostomes, which in time diverged to form a protoannelid line with laterally located parapodia and one or more protoarthropod lines with more ventrally located parapodia. Some authors have contended that the Arthropoda is polyphyletic and that some or all of the present subphyla are derived from different annelid-like ancestors that underwent "arthropodization." The crucial development is the hardening of the cuticle to form an arthropod exoskeleton, and most of the features that distinguish arthropods from annelids (p. 371) result from the stiffened exoskeleton (see Prologue for this chapter). For example, the vital role of the coelomic compartments as a hydrostatic skeleton was gone; therefore intersegmental septa were unnecessary, as was a closed circulatory system. Jointed appendages, of course, are necessary if the external surface is hard, and the body-wall muscles of the annelid could be converted and inserted on the considerable inner surfaces of the cuticle for efficient movement of body parts. Compared with annelids, there was a great restriction in permeable surfaces for respiration and excretion. Thus arthropods *could* have evolved more than once. However, other zoologists argue strongly that the derived similarities of the arthropod subphyla strongly support monophyly of the phylum. The phyla Onychophora and Tardigrada (Chapter 22) may be sister taxa to the arthropods. A cladogram depicting possible relationships is presented in Chapter 22 (p. 440).

Some evidence based on ribosomal RNA sequences supports monophyly of the Arthropoda and the inclusion of the Onychophora in the phylum.* These data also suggest that the myriapods (millipedes and centipedes) are a sister group to all other arthropods and that the crustaceans and insects form a monophyletic group! If these conclusions are supported by further investigations, our concepts of arthropod phylogeny and classification are subject to major revision.

Controversy on phylogeny within the Chelicerata also exists, especially on the relationship of the Pycnogonida (Figure 19-18). Some workers place the pycnogonids as a sister group with the chelicerates in a larger grouping called Cheliceriformes.

ADAPTIVE RADIATION

Annelids show limited tagmatization and little differentiation of appendages. However, in arthropods the adaptive trend has been toward pronounced tagmatization by differentiation or fusion of somites, giving rise in more derived groups to such tagmata as head and trunk; head, thorax, and abdomen; or cephalothorax (fused head and thorax) and abdomen. Arthropods with primitive characters tend to have similar appendages on each somite, and each somite bears a pair of appendages. The more derived forms have appendages specialized for specific functions, or some somites lack appendages entirely.

Much of the amazing diversity in arthropods seems to have developed because of modification and specialization of their cuticular exoskeleton and their jointed appendages, resulting in a wide variety of locomotor and feeding adaptations.

W. S. Bristowe (1971) estimated that at certain seasons a Sussex field that had been undisturbed for several years had a population of 2 million spiders to the acre. He concluded that so many could not successfully compete except for the many specialized adaptations they had evolved. These include adaptations to cold and heat, wet and dry conditions, and light and darkness.

Some spiders capture large insects, some only small ones; web-builders snare mostly flying insects, whereas hunters seek those that live on the ground. Some lay eggs in the spring, others in the late summer. Some feed by day, others by night, and some have developed flavors that are distasteful to birds or to certain predatory insects. As it is with the spiders, so has it been with other arthropods; their adaptations are many and diverse and contribute in no small way to their long success.

*Ballard, J. W. O., et al. 1992. Science **258**:1345–1348.

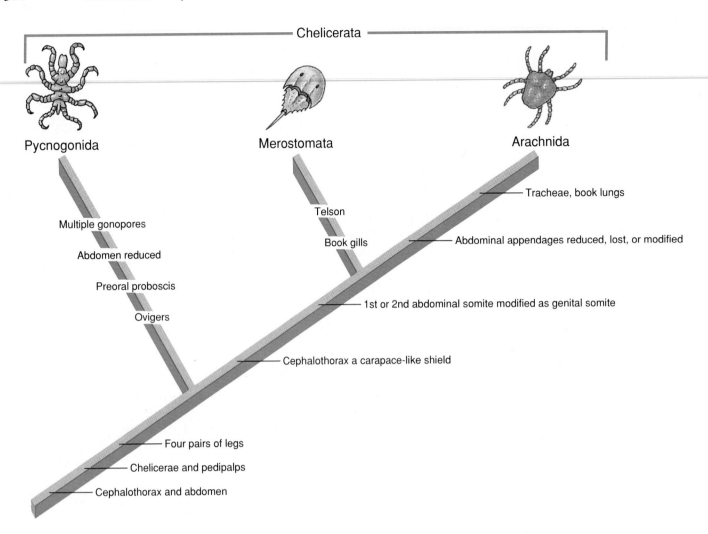

Figure 19-18

Cladogram of the chelicerates, showing one proposed ancestor-descendant relationship within the chelicerate clade. Shared derived characters used to construct the cladogram are shown adjacent to the branch lines (based on Brusca and Brusca, 1990). Some workers separate the Pycnogonida from the Chelicerata, placing it as a sister group to the chelicerates in a larger Cheliceriformes grouping.

Source: Modified from R. C. Brusca and G. J. Brusca, Invertebrates, *Sinauer Associates, Inc., Sunderland, MA, 1990.*

CLASSIFICATION OF PHYLUM ARTHROPODA

Subphylum Trilobita (tri´lo-bi´ta) (Gr. *tri,* three, + *lobos,* lobe): **trilobites.** All extinct forms; Cambrian to Carboniferous; body divided by two longitudinal furrows into three lobes; distinct head, thorax, and abdomen, biramous (two-branched) appendages.

Subphylum Chelicerata (ke-lis´e-ra´ta) (Gr. *chēlē,* claw, + *keras,* horn, + *ata,* group suffix): **eurypterids, horseshoe crabs, spiders, ticks.** First pair of appendages modified to form chelicerae; pair of pedipalps and four pairs of legs; no antennae, no mandibles; cephalothorax and abdomen usually unsegmented.

    Class Merostomata (mer´o-sto´ma-ta) (Gr. *mēros,* thigh, + *stoma,* mouth, + *ata,* group suffix): **aquatic chelicerates.** Cephalothorax and abdomen; compound lateral eyes; appendages with gills; sharp telson; subclasses Eurypterida (all extinct) and Xiphosurida, the horseshoe crabs. Example: *Limulus.*

    Class Pycnogonida (pik´no-gon´i-da) (Gr. *pyknos,* compact, + *gony,* knee, angle): **sea spiders.** Small (3 to 4 mm), but some reach 500 mm; body chiefly cephalothorax; tiny abdomen; usually four pairs of long walking legs (some with five or six pairs); mouth on long proboscis; four simple eyes; no respiratory or excretory system. Example: *Pycnogonum.*

    Class Arachnida (ar-ack´ni-da) (Gr. *arachnē,* spider): **scorpions, spiders, mites, ticks, harvestmen.** Four pairs of legs; segmented or unsegmented abdomen with or without appendages and generally distinct from cephalothorax; respiration by gills, tracheae, or book lungs; excretion by malpighian tubules or coxal glands; dorsal bilobed brain connected to ventral ganglionic mass with nerves; simple eyes; chiefly oviparous; no true metamorphosis. Examples: *Argiope, Centruroides.*

Subphylum Crustacea (crus-ta´she-a) (L. *crusta,* shell, + *acea,* group suffix): **crustaceans.** Mostly aquatic, with gills; cephalothorax usually with dorsal carapace; biramous appendages, modified for various functions. Head appendages consisting of two pairs of antennae, one pair of mandibles, and two pairs of maxillae. Development primitively with nauplius stage (see classification of crustaceans, p. 402).

Subphylum Uniramia (yu-ni-ra´me-a) (L. *unus,* one, + *ramus,* a branch): **insects and myriapods.** All appendages uniramous; head appendages consisting of one pair of antennae, one pair of mandibles, and one or two pairs of maxillae.

    Class Diplopoda (di-plop´o-da) (Gr. *diploos,* double, + *pous, podos,* foot): **millipedes.** Body almost cylindrical; head with short antennae and simple eyes; body with variable number of somites; short legs, usually two pairs of legs to a somite; oviparous. Examples: *Julus, Spirobolus.*

    Class Chilopoda (ki-lop´o-da) (Gr. *cheilos,* lip, + *pous, podos,* foot): **centipedes.** Dorsoventrally flattened body; variable number of somites, each with one pair of legs; one pair of long antennae; oviparous. Examples: *Cermatia, Lithobius, Geophilus.*

    Class Pauropoda (pau-ro´po-da) (Gr. *pauros,* small, + *pous, podos,* foot) **pauropods.** Minute (1 to 1.5 mm); cylindrical body consisting of double segments and bearing 9 or 10 pairs of legs; no eyes. Example: *Pauropus.*

    Class Symphyla (sym´fy-la) (Gr. *syn,* together, + *phylē,* tribe): **garden centipedes.** Slender (1 to 8 mm) with long, filiform antennae; body consisting of 15 to 22 segments with 10 to 12 pairs of legs; no eyes. Example: *Scutigerella.*

    Class Insecta (in-sek´ta) (L. *insectus,* cut into): **insects.** Body with distinct head, thorax, and abdomen; pair of antennae; mouthparts modified for different food habits; head of six fused somites; thorax of three somites; abdomen with variable number, usually 11 somites; thorax with two pairs of wings (sometimes one pair or none) and three pairs of jointed legs; usually oviparous; gradual or abrupt metamorphosis. (Brief description of insect orders: pp. 425 to 426.)

Summary

The Arthropoda is the largest, most abundant and diverse phylum in the world. They are metameric, coelomate protostomes with well-developed organ systems. Most show marked tagmatization. They are extremely diverse and occur in all habitats capable of supporting life. Perhaps more than any other single factor, the prevalence of the arthropods is accounted for by adaptations made possible by their cuticular exoskeleton. Other important elements are jointed appendages, tracheal respiration, efficient sensory organs, complex behavior, and metamorphosis.

The trilobites were a dominant Paleozoic subphylum, now extinct. Members of the subphylum Chelicerata have no antennae, and their main feeding appendages are chelicerae. In addition, they have a pair of pedipalps (which may be similar to the walking legs) and four pairs of walking legs. The class Merostomata includes the extinct eurypterids and the ancient, although still extant, horseshoe crabs. The class Pycnogonida contains the sea spiders, which are odd little animals with a large suctorial proboscis and vestigial abdomen. The great majority of living chelicerates are in the class Arachnida: spiders (order Araneae), scorpions (order Scorpionida), harvestmen (order Opiliones), ticks and mites (order Acari), and others.

The tagmata of spiders (cephalothorax and abdomen) show no external segmentation and are joined by a waistlike pedicel. Spiders are predaceous, and their chelicerae are provided with poison glands for paralyzing or killing their prey.

They breathe by book lungs, tracheae, or both. Spiders can spin silk, which they use for a variety of purposes, including webs for trapping prey in some cases.

Distinctive characters of scorpions are their large, clawlike pedipalps and their clearly segmented abdomen, which bears a terminal stinging apparatus. Harvestmen have small, ovoid bodies with very long, slender legs. Their abdomen is segmented and broadly joined to the cephalothorax.

The cephalothorax and abdomen of ticks and mites are completely fused and the mouthparts are borne on the anterior capitulum. They are the most numerous of any arachnids; some are important carriers of disease, and others are serious plant pests.

Review Questions

1. What are important distinguishing features of arthropods?
2. Name the subphyla of arthropods, and give a few examples of each.
3. How do arthropods differ from annelids, and how are they alike?
4. Briefly discuss the contribution of the cuticle to the success of arthropods, and name some other factors that have contributed to their success.
5. What is a trilobite?
6. What appendages are characteristic of chelicerates?
7. Briefly describe the appearance of each of the following: eurypterids, horseshoe crabs, pycnogonids.
8. What are the tagmata of arachnids, and which of the tagmata bears appendages?
9. Describe the mechanism of each of the following with respect to spiders: feeding, excretion, sensory reception, web-spinning, reproduction.
10. What are the most important spiders in the United States that are dangerous to humans?
11. Distinguish each of the following orders from each other: Araneae, Scorpionida, Opiliones, Acari.
12. Discuss the importance of members of the Acari to human well-being.
13. Some biologists suggest that the Arthropoda is polyphyletic. Explain why this could be so despite the characteristics shared by all arthropods.

Selected References

See also general references for Part III, p. 626.

Foelix, R. F. 1982. Biology of spiders. Cambridge, Massachusetts, Harvard University Press. *Attractive, comprehensive book with extensive references; of interest to both amateurs and professionals.*

Hadley, N. F. 1986. The arthropod cuticle. Sci. Am. **255**:104–112 (July). *Modern studies on the chemistry and structure of arthropod cuticle help to explain its remarkable properties.*

Jackson, R. R. 1985. A web-building jumping spider. Sci. Am. **253**:102–115 (Sept.). *This unusual jumping spider eats mostly other spiders, rather than insects. It often fastens its web to that of another species to invade the alien web and prey on the spider that built it.*

Jaenson, T. G. T. 1991. The epidemiology of Lyme borreliosis. Parasitology Today **7**:39–45. *Excellent summary of clinical and ecological aspects of Lyme disease.*

Kaston, B. J. 1978. How to know the spiders, ed. 3. Dubuque, Iowa, William C. Brown Publishers. *Spiral-bound identification manual.*

Lane, R. P., and R. W. Crosskey, (eds). 1993. Medical insects and arachnids. London, Chapman and Hall. *This is the best book currently available on medical entomology.*

McDaniel, B. 1979. How to know the ticks and mites. Dubuque, Iowa, William C. Brown Publishers. *Useful, well-illustrated keys to genera and higher categories of ticks and mites in the United States.*

Polis, G. A., (ed). 1990. The biology of scorpions. Stanford, California, Stanford University Press. *The editor brings together a readable summary of what is known about scorpions.*

Shear, W. A. 1994. Untangling the evolution of the web. Amer. Sci. **82**:256–266. *Fossil spider webs are nonexistent. Evolution of the web must be studied by comparing modern spider webs to each other and correlating studies of spider anatomy.*

Vollrath, F. 1992. Spider webs and silks. Sci. Am. **266**:70–76 (Mar.). *Spider web design and silk must obey the same constraints as materials used in human structural engineering; we can learn useful lessons from spiders.*

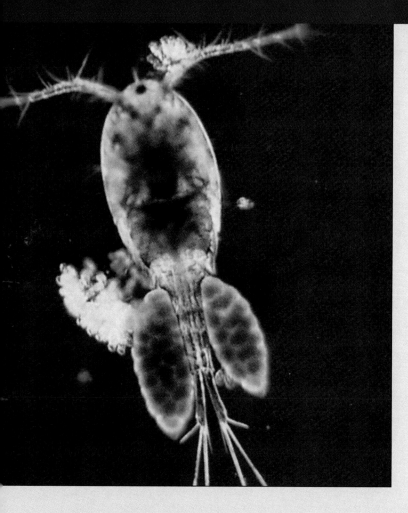

20

The Aquatic Mandibulates

Phylum Arthropoda
Subphylum Crustacea

Insects of the Sea

The subphylum Crustacea (L. *crusta,* shell) gets its name from the hard shell that most crustaceans bear. Over 30,000 species have been described, and several times that number probably exist. Most familiar to people are the edible ones, for example, lobsters, crayfishes, shrimps, and crabs. In addition to these "crusty" crustaceans, there is an astonishing array of less familiar forms such as copepods, ostracods, water fleas, whale lice, tadpole shrimp, and krill. They fill a wide variety of ecological roles and show enormous variation in morphological characteristics, making a satisfactory definition of the group singularly hard to frame.

We live in the age of arthropods, notwithstanding our anthropocentric attachment to the tradition of calling the current era the age of mammals. Together, insects and crustaceans compose some 80% of all named animal species. Just as insects pervade the terrestrial habitat (at least a million named species and countless billions of individuals), crustaceans abound in oceans, lakes, and rivers. Some walk or creep on the bottom, some burrow, and some (such as barnacles) are sessile. Some swim upright, others swim upside down, and many are delicate microscopic forms that drift as plankton in the oceans or in lakes. Indeed, it is probable that the *most abundant animals in the world* are members of the copepod genus *Calanus*. In recognition of their dominance of marine habitats, it is understandable that crustaceans have been called "insects of the sea." ■

Arthropods that possess mandibles (jawlike appendages) are known as mandibulates and traditionally have been united in the subphylum Mandibulata. As we noted in the previous chapter, some authors think that the phylum Arthropoda is polyphyletic and that arthropodization occurred more than once. In addition, many investigators now believe that there are sufficient differences between the crustaceans and the uniramians (insects, millipedes, centipedes, pauropods, and symphylans) to justify separation at least to subphylum level. Both the Crustacea and the Uniramia have, at least, a pair of **antennae,** a pair of **mandibles,** and a pair of **maxillae** on the head. These appendages perform sensory, masticatory, and food-handling functions, respectively. The body may consist of a head and trunk, but in the more derived forms, a high degree of tagmatization (p. 370) has occurred so that there is a well-defined head, thorax, and abdomen. In most Crustacea one or more thoracic segments have become fused with the head to form a **cephalothorax.** The thoracic and abdominal appendages are mainly for walking or swimming, but in some groups they are highly specialized in function. The Crustacea are mainly marine; however, there are many freshwater and a few terrestrial species, whereas the uniramians are mainly terrestrial. There are numerous species of insects in freshwater habitats, but only a few in marine.

SUBPHYLUM CRUSTACEA

GENERAL NATURE OF A CRUSTACEAN

Although crustaceans differ from other arthropods in a variety of ways, the only truly distinguishing characteristic is that crustaceans are the only arthropods with **two pairs of antennae.** In addition to the two pairs of antennae and a pair of mandibles, crustaceans have two pairs of maxillae on the head, followed by a pair of appendages on each body segment

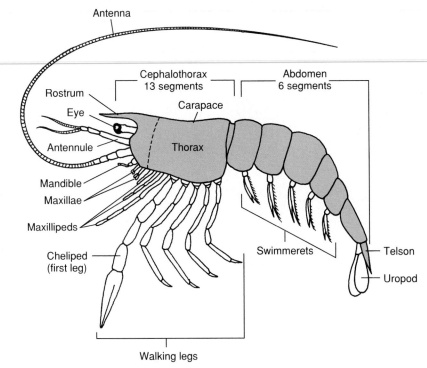

Figure 20-1
Archetypical plan of the Malacostraca. The two maxillae and three maxillipeds have been separated diagrammatically to illustrate the general plan.

(in some crustaceans not all somites bear appendages). All appendages, except perhaps the first antennae, are primitively **biramous** (two main branches), and at least some of the appendages of present-day adults show that condition. Organs specialized for respiration, if present, are in the form of **gills.**

Most crustaceans have between 16 and 20 somites, but some forms have 60 somites or more. A larger number of somites is a primitive feature. The more derived condition is to have fewer segments and increased tagmatization (see p. 370). The major tagmata are the head, thorax, and abdomen, but these are not homologous throughout the class (or even within some subclasses) because of varying degrees of fusion of somites, for example, as in the cephalothorax.

By far the largest group of crustaceans is the class Malacostraca, which includes the lobsters, crabs, shrimps, beach hoppers, sow bugs, and many others. These show a surprisingly constant arrangement of body segments and tagmata that are often called the

caridoid facies* and is considered the ancestral plan of the class (Figure 20-1). This typical body plan has a head of five (six embryonically) fused somites, a thorax of eight somites, and an abdomen of six somites (seven in a few species). At the anterior end is the nonsegmented **rostrum** and at the posterior end is the nonsegmented **telson,** which with the last abdominal somite and its **uropods** forms the tail fan in many forms.

In many crustaceans the dorsal cuticle of the head may extend posteriorly and around the sides of the animal to cover or be fused with some or all of the thoracic and abdominal somites. This covering is called the **carapace.** In some groups the carapace forms clamshell-like valves that cover most or all of the body. In the decapods (including lobsters, shrimp, crabs, and others), the carapace covers the entire cephalothorax but not the abdomen.

---

*"Caridoid" derives from the scientific name of a group of crustaceans; "facies" means face or general appearance.

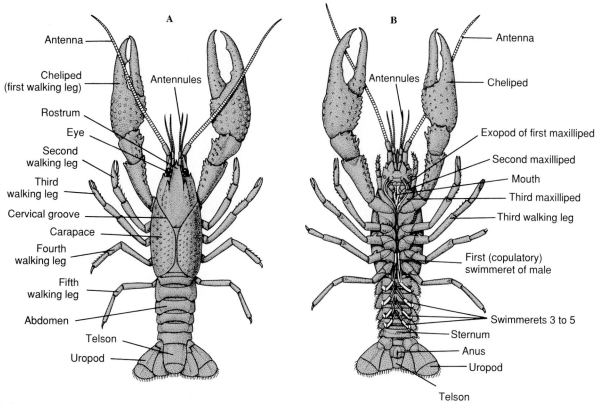

Figure 20-2
External structure of crayfishes. **A,** Dorsal view. **B,** Ventral view.

FORM AND FUNCTION

Because of their size and easy availability, large crustaceans such as crayfishes have been studied more than other groups. They are also commonly studied in introductory laboratory courses. Therefore many of the comments that follow apply specifically to crayfishes and their relatives.

External Features

The bodies of crustaceans are covered with a secreted cuticle composed of chitin, protein, and limy, calcareous material. The harder, heavy plates of the larger crustaceans are particularly high in calcareous deposits. The hard protective covering is soft and thin at the joints between the somites, allowing flexibility of movement. The carapace, if present, covers much or all of the cephalothorax; in decapods such as crayfishes, all of the head and thoracic segments are enclosed dorsally by the carapace. Each somite not en-

closed by the carapace is covered by a dorsal cuticular plate, or **tergum** (Figure 20-2A), and a ventral transverse bar, the **sternum,** lies between the segmental appendages (Figure 20-2B). The abdomen terminates in a telson, which is not considered a somite and bears the anus. (The telson may be homologous to the annelid pygidium.) In several groups the telson bears a pair of processes, forming the **caudal furca.**

The position of the **gonopores** varies according to sex and group of crustaceans. They may be on or at the base of a pair of appendages, at the terminal end of the body, or on somites without legs. In crayfishes the openings of the vasa deferentia are on the median side at the base of the fifth pair of walking legs, and those of the oviducts are at the base of the third pair. In the female the opening to the seminal receptacle is usually located in the midventral line between the fourth and fifth pairs of walking legs.

Appendages. Members of the classes Malacostraca (including crayfishes) and Remipedia typically have a pair of jointed appendages on each somite (Figure 20-3), although the abdominal somites in the other classes do not bear appendages. Considerable specialization is evident in the appendages of the derived crustaceans such as the crayfishes. However, all are variations of the basic, biramous plan, illustrated by a crayfish appendage such as a maxilliped (a thoracic limb modified to become a head appendage) (Figures 20-3 and 20-4). The basal portion, the **protopod,** bears a lateral **exopod** and a medial **endopod.** The protopod is made up of one or two joints **(basis and coxa),** whereas the exopod and endopod have from one to several joints each. Some appendages, such as the walking legs of crayfishes, have become secondarily uniramous. Medial or lateral processes sometimes occur on crustacean limbs, called **endites** and **exites,** respectively, and

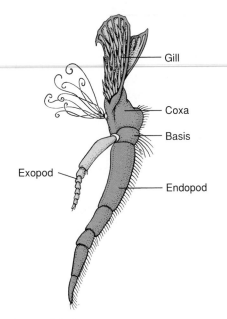

Figure 20-3

Parts of a biramous crustacean appendage (third maxilliped of a crayfish).

an exite on the protopod is called an **epipod.** Epipods are often modified as gills. Table 20-1 shows how the various appendages have become modified from the biramous plan to fit specific functions.

The terminology applied by various workers to crustacean appendages has not been blessed with uniformity. At least two systems are in wide use. Alternative terms to those we have used, for example, are protopodite, exopodite, endopodite, basipodite, coxopodite, and epipodite. The first and second pairs of antennae may be referred to as the antennules and antennae, and the first and second maxillae are often called maxillules and maxillae. A rose by any other name. . . .

Structures that have a similar basic plan and have descended from a common form are said to be homologous, whether they have the same function or not. Since the specialized walking legs, mouthparts, chelipeds, and swimmerets have all developed from a common biramous type but have become modified to perform different functions, they are all homologous to each other, a condition known as **serial homology.** Primitively they were all very similar, but during the evolution of structural modifications, some branches have been reduced, some lost, some greatly altered, and some new parts added. The crayfishes and their allies possess the best examples of serial homology in the animal kingdom.

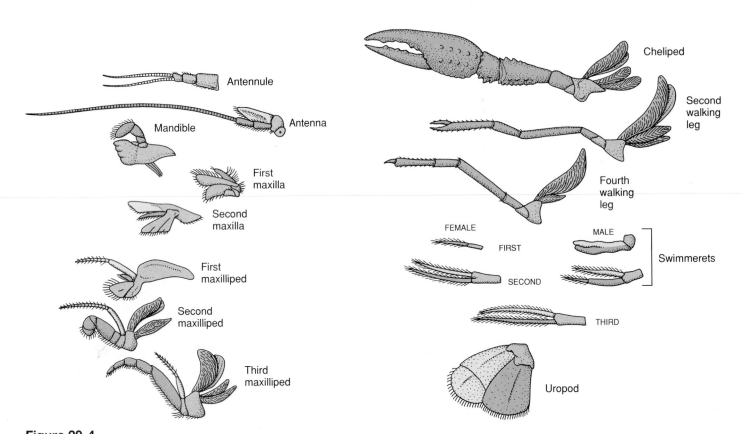

Figure 20-4

Appendages of a crayfish showing how they have become modified from the basic biramous plan, as found in a swimmeret. Protopod, pink; endopod, purple; exopod, yellow.

Table 20-1	Crayfish Appendages			
Appendage	**Protopod**	**Endopod**	**Exopod**	**Function**
First antenna (antennule)	3 segments, statocyst in base	Many-jointed feeler	Many-jointed feeler	Touch, taste, equilibrium
Second antenna (antenna)	2 segments, excretory pore in base	Long, many-jointed feeler	Thin, pointed blade	Touch, taste
Mandible	2 segments, heavy jaw and base of palp	2 distal segments of palp	Absent	Crushing food
First maxilla (maxillule)	2 segments with 2 thin endites	Small unjointed lamella	Absent	Food handling
Second maxilla (maxilla)	2 segments, with 2 endites and 1 scaphognathite (epipod)	1 small pointed segment	Part of scaphognathite (bailer)	Drawing currents of water into gills
First maxilliped	2 medial plates and epipod	2 small segments	1 basal segment, plus many-jointed filament	Touch, taste, food handling
Second maxilliped	2 segments plus gill (epipod)	5 short segments	2 slender segments	Touch, taste, food handling
Third maxilliped	2 segments plus gill (epipod)	5 larger segments	2 slender segments	Touch, taste, food handling
First walking leg (cheliped)	2 segments plus gill (epipod)	5 segments with heavy pincer	Absent	Offense and defense
Second walking leg	2 segments plus gill (epipod)	5 segments plus small pincer	Absent	Walking and prehension
Third walking leg	2 segments plus gill (epipod); genital pore in female	5 segments plus small pincer	Absent	Walking and prehension
Fourth walking leg	2 segments plus gill (epipod)	5 segments, no pincer	Absent	Walking
Fifth walking leg	2 segments; genital pore in male; no gill	5 segments, no pincer	Absent	Walking
First swimmeret	In female reduced or absent; in male fused with endopod to form tube			In male, transferring sperm to female
Second swimmeret Male	Structure modified for transfer of sperm to female	Structure modified for transfer of sperm to female		
Female	2 segments	Jointed filament	Jointed filament	Creating water currents; carrying eggs and young
Third, fourth, and fifth swimmerets	2 short segments	Jointed filament	Jointed filament	Creating water currents; in female carrying eggs and young
Uropod	1 short, broad segment	Flat, oval plate	Flat, oval plate; divided into 2 parts with hinge	Swimming; egg protection in female

Internal Features

The muscular and nervous systems and segmentation in the thorax and abdomen clearly show the metamerism inherited from the annelid-like ancestors, but there are marked modifications in other systems. Most of the changes involve concentration of parts in a particular region or else reduction or complete loss of parts, such as the intersepta.

Hemocoel.   The major body space in the arthropods is not the coelom but a blood-filled **hemocoel.** During the embryonic development of most arthropods, vestigial coelomic cavities open within the mesoderm of at least some somites. These are soon obliterated or become continuous with the space between the developing mesodermal and ectodermal structures and the yolk. This space becomes the hemocoel and is thus not lined by a mesodermal

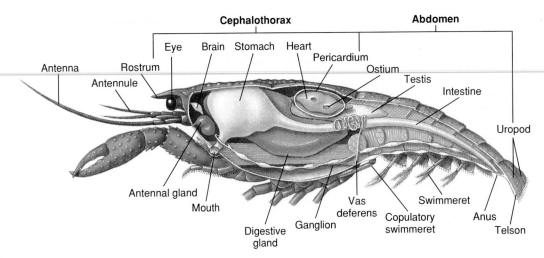

Figure 20-5

Internal structure of a male crayfish.

peritoneum. In the crustaceans the only coelomic compartments remaining are the end sacs of the excretory organs and the space around the gonads.

Muscular System. Striated muscles make up a considerable part of the body of most Crustacea. The muscles are usually arranged in antagonistic groups: **flexors,** which draw a part toward the body, and **extensors,** which straighten it out. The abdomen of a crayfish has powerful flexors (Figure 20-5), which are used when the animal swims backward—its best means of escape. Strong muscles on either side of the stomach control the mandibles.

Respiratory System. Respiratory gas exchange in the smaller crustaceans takes place over thinner areas of the cuticle (for example, in the legs) or over the entire body, and specialized structures may be absent. The larger crustaceans have gills, which are delicate, featherlike projections with very thin cuticle. In the decapods the sides of the carapace enclose the gill cavity, which is open anteriorly and ventrally (Figure 20-6). Gills may project from the pleural wall into the gill cavity, from the articulation of the thoracic legs with the body, or from the thoracic coxae. The latter two types are typical of crayfishes. The "bailer," a part of the second maxilla, draws water over the gill filaments, into the gill cavity at the bases of the legs, and out of the gill cavity at the anterior.

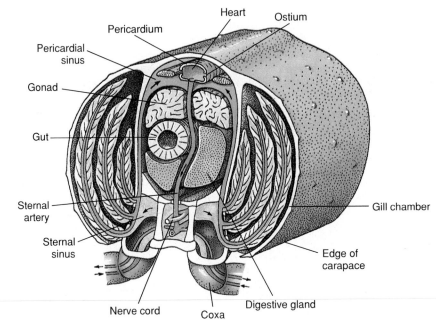

Figure 20-6

Diagrammatical cross section through heart region of a crayfish showing direction of blood flow in this "open" blood system. Heart pumps blood to body tissues through arteries, which empty into tissue sinuses. Returning blood enters sternal sinus, then goes through gills for gas exchange, and finally back to pericardial sinus by efferent channels. Note absence of veins.

Circulatory System. Crustaceans and other arthropods have an "open" or lacunar type of circulatory system. This means that there are no veins and no separation of blood from interstitial fluid, as there is in animals with closed systems (see p. 681). The hemolymph (blood) leaves the heart by way of arteries, circulates through the hemocoel, and returns to venous sinuses, or spaces, instead of veins before it reenters the heart. The annelids have a closed system, as do the vertebrates.

A dorsal heart is the chief propulsive organ. It is a single-chambered sac of striated muscle. Hemolymph enters the heart from the surrounding **pericardial sinus** through paired ostia, with valves that prevent backflow into the sinus (Figure 20-6). From the heart the hemolymph enters one or more arteries. Valves in the arteries prevent a backflow of hemolymph. Small arteries empty into the tissue sinuses, which in turn often discharge into the large **sternal sinus** (Figure 20-6).

From there, afferent sinus channels carry hemolymph to the gills, if any, for oxygen and carbon dioxide exchange. The hemolymph then returns to the pericardial sinus by efferent channels (Figure 20-6).

Hemolymph in arthropods is largely colorless. It includes ameboid cells of at least two types. Hemocyanin, a copper-containing respiratory pigment, or hemoglobin, an iron-containing pigment, may be carried in solution. Hemolymph has the property of clotting, which prevents its loss in minor injuries. Some ameboid cells release a thrombinlike coagulant that precipitates the clotting.

Excretory System. The excretory organs of adult crustaceans are a pair of tubular structures located in the ventral part of the head anterior to the esophagus (Figure 20-5). They are called **antennal glands** or **maxillary glands,** depending on whether they open at the base of the antennae or of the second maxillae. A few adult crustaceans have both. The excretory organs of the decapods are antennal glands, also called **green glands** in this group. Crustaceans do not have malpighian tubules, the excretory organs of spiders and insects.

The **end sac** of the antennal gland, which is derived from an embryonic coelomic compartment, consists of a small vesicle **(saccule)** and a spongy mass called a **labyrinth.** The labyrinth connects by an **excretory tubule** to a dorsal **bladder,** which opens to the exterior by a pore on the ventral surface of the basal antennal segment (Figure 20-7). Hydrostatic pressure within the hemocoel provides the force for filtration of fluid into the end sac, and as the filtrate passes through the excretory tubule and bladder, it is modified by resorption of salts, amino acids, glucose, and some water and is finally excreted as urine.

Excretion of nitrogenous wastes (mostly ammonia) takes place by diffusion across thin areas of cuticle, especially the gills, and the so-called excretory organs function principally to

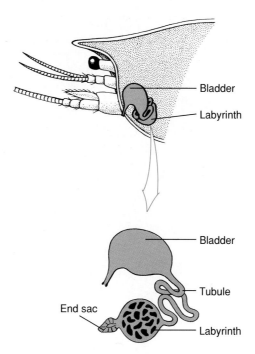

Figure 20-7
Scheme of antennal gland (green gland) of crayfishes. (In natural position organ is much folded.) Some crustaceans lack a labyrinth, and the excretory tubule (nephridial canal) is a much-coiled tube.

regulate the ionic and osmotic composition of the body fluids. Freshwater crustaceans, such as crayfishes, are constantly threatened with overdilution of the blood by water, which diffuses across the gills and other water-permeable surfaces. The green glands, by forming a dilute, low-salt urine, act as an effective "flood-control" device. Some Na^+ and Cl^- are lost in the urine, however, but this is made up by active absorption of salt from the water by the gills. In marine crustaceans, such as lobsters and crabs, the kidney functions to adjust the salt composition of the hemolymph by selective modification of the salt content of the tubular urine. In these forms the urine remains isosmotic to the blood.

Nervous and Sensory Systems. The nervous systems of crustaceans and annelids have much in common, although those of crustaceans have more fusion of ganglia (Figure 20-5). The brain is a pair of **supraesophageal ganglia** that supplies

nerves to the eyes and the two pairs of antennae. It is joined by connectives to the **subesophageal ganglion,** a fusion of at least five pairs of ganglia that supply nerves to the mouth, appendages, esophagus, and antennal glands. The double ventral nerve cord has a pair of ganglia for each somite and gives off nerves to the appendages, muscles, and other parts. In addition to this central system, there may be a sympathetic nervous system associated with the digestive tract.

Crustaceans have better-developed sense organs than do the annelids. The largest sense organs of crayfishes are the eyes and the statocysts. Tactile organs are widely distributed over the body in the form of **tactile hairs,** delicate projections of the cuticle that are especially abundant on the chelae, mouthparts, and telson. The chemical senses of taste and smell are found in hairs on the antennae, mouthparts, and other places.

A saclike **statocyst,** opening to the surface by a dorsal pore, is found on the basal segment of each first antenna of crayfishes. The statocyst contains a ridge that bears sensory hairs formed from the chitinous lining and grains of sand that serve as **statoliths.** Whenever the animal changes its position, corresponding changes in the position of the grains on the sensory hairs are relayed as stimuli to the brain, and the animal can adjust itself accordingly. Each molt (ecdysis) of the cuticle results in loss of the cuticular lining of the statocyst and with it the sand grains. New grains are picked up through the dorsal pore after ecdysis.

The eyes in many crustaceans are compound, that is, made up of many photoreceptor units called **ommatidia** (Figure 20-8). Covering the rounded surface of each eye is a transparent area of the cuticle, the **cornea,** which is divided into many small squares or hexagons known as facets. These are the outer ends of the ommatidia. Each ommatidium behaves like a tiny eye and contains several kinds of cells arranged in a columnar fashion (Figure 20-8). Black pigment cells are found between adjacent ommatidia.

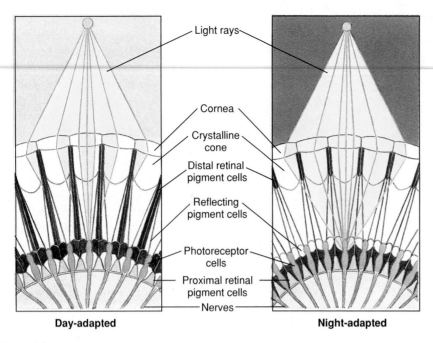

Day-adapted   **Night-adapted**

Figure 20-8

Portion of compound eye of arthropod showing migration of pigment in ommatidia for day and night vision. Five ommatidia represented in each diagram. In daytime each ommatidium is surrounded by a dark pigment collar so that each ommatidium is stimulated only by light rays that enter its own cornea (mosaic vision); in nighttime, pigment forms incomplete collars and light rays can spread to adjacent ommatidia (continuous, or superposition, image).

Movement of pigment in the arthropod compound eye permits it to adjust for different amounts of light. There are three sets of pigment cells in each ommatidium: distal retinal, proximal retinal, and reflecting; these are so arranged that they can form a more or less complete collar or sleeve around each ommatidium. For strong light or day adaptation the distal retinal pigment moves inward and meets the outward-moving proximal retinal pigment so that a complete pigment sleeve forms around the ommatidium (Figure 20-8). In this condition only rays that strike the cornea directly will reach the photoreceptor (retinular) cells, for each ommatidium is shielded from the others. Thus each ommatidium will see only a limited area of the field of vision (a mosaic, or apposition, image). In dim light the distal and proximal pigments separate so that the light rays, with the aid of the reflecting pigment cells, have a chance to spread to adjacent ommatidia and to form a continuous, or superposition, image. This second type of vision is less precise but takes maximum advantage of the limited amount of light received.

Reproduction, Life Cycles, and Endocrine Function

Most crustaceans have separate sexes, and there are various specializations for copulation among the different groups. Barnacles are monoecious but generally practice cross-fertilization. In some ostracods males are scarce, and reproduction is usually parthenogenetic. Most crustaceans brood their eggs in some manner: branchiopods and barnacles have special brood chambers, copepods have brood sacs attached to the sides of the abdomen (Figure 20-19), and many malacostracans carry eggs and young attached to their abdominal appendages.

Crayfishes have direct development: there is no larval form. A tiny juvenile with the same form as the adult and a complete set of appendages and somites hatches from the egg. However, development is indirect in the majority of crustaceans, and a larva quite unlike the adult in structure and appearance hatches from the egg. Change from larva ultimately to an adult is **metamorphosis**. The primitive and most widely occurring larva in the Crustacea is the

nauplius (Figures 20-9 and 20-23). The nauplius bears only three pairs of appendages: uniramous first antennules, biramous antennae, and biramous mandibles. They all function as swimming appendages at this stage. Subsequent development may involve a gradual change to the adult body form, and appendages and somites are added through a series of molts, or assumption of the adult form may involve more abrupt changes. For example, the metamorphosis of a barnacle proceeds from a free-swimming nauplius to a larva with a bivalve carapace called a cyprid and finally to the sessile adult with calcareous plates.

Ecdysis. Ecdysis (ek-die´sis) (Gr. *ekdyein,* to strip off), or molting, is necessary for the body to increase in size because the exoskeleton is nonliving and does not grow as the animal grows. Much of the crustacean's functioning, including its reproduction, behavior, and many metabolic processes, is directly affected by the physiology of the molting cycle.

The **cuticle,** which is secreted by the underlying epidermis, has several layers (Figure 20-10). The outermost is the **epicuticle,** a very thin layer of lipid-impregnated protein. The bulk of the cuticle is the several layers of **procuticle:** (1) the **exocuticle,** which is just beneath the epicuticle and contains protein, calcium salts, and chitin; (2) the **endocuticle,** which itself is composed of (3) a **principal layer,** which contains more chitin and less protein and is heavily calcified, and (4) an uncalcified **membranous layer,** a relatively thin layer of chitin and protein.

Some time before the actual ecdysis, the epidermal cells enlarge considerably. They separate from the membranous layer, secrete a new epicuticle, and begin secreting a new exocuticle (Figure 20-11). Enzymes are released into the area above the new epicuticle. These enzymes begin to dissolve the old endocuticle, and the soluble products are resorbed and stored within the body of the crustacean. Some of the calcium salts are stored as **gastroliths** (mineral accretions) in the walls of the stomach. Finally, only the exocuticle

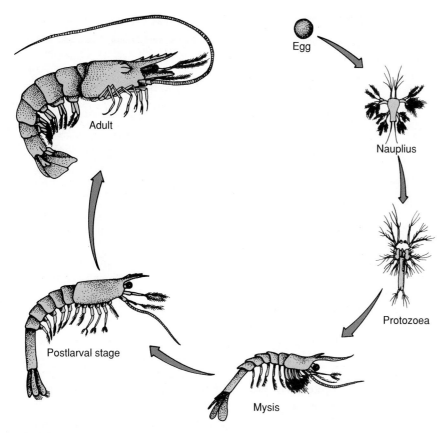

Figure 20-9

Life cycle of the Gulf shrimp *Penaeus*. Penaeids spawn at depths of 40 to 90 m. The young larval forms make up part of the plankton fauna and work their way inshore to water of lower salinity to develop as juveniles. Older shrimp return to deeper water offshore.

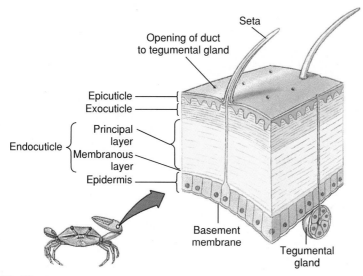

Figure 20-10

Structure of crustacean cuticle.

and epicuticle of the old cuticle remain, underlain by the new epicuticle and new exocuticle. The animal swallows water, which it absorbs through its gut, and its blood volume increases greatly. The internal pressure causes the cuticle to split, and the animal pulls itself out of its old exoskeleton (Figure 20-12). Then follow a stretching of the still soft new cuticle, deposition of the new endocuticle, redeposition of the salvaged inorganic salts and other constituents, and hardening of the new cuticle. During the period of molting, the animal is defenseless and remains hidden away.

When the crustacean is young, ecdysis must occur frequently to allow growth, and the molting cycle is relatively short. As the animal approaches maturity, intermolt periods become progressively longer, and in some species molting ceases altogether. During intermolt periods, increase in tissue mass occurs as living tissue replaces water.

Hormonal Control of the Ecdysis Cycle. Although ecdysis is hormonally controlled, the cycle is often initiated by an environmental stimulus perceived by the central nervous system. Such stimuli may include temperature, day length, and humidity (in the case of land crabs). The action of the signal from the central nervous system is to decrease the production of a **molt-inhibiting hormone** by the **X-organ.** The X-organ is a group of neurosecretory cells in the medulla terminalis of the brain. In crayfishes and other decapods, the medulla terminalis is found in the eyestalk. The hormone is carried in the axons of the X-organ to the **sinus gland** (which is probably not glandular in function), also in the eyestalk, where it is released into the hemolymph.

A drop in the level of molt-inhibiting hormone, promotes release of a **molting hormone** from the **Y-organs.** The Y-organs are beneath the epidermis near the adductor muscles of the mandibles, and they are homologus to the prothoracic glands of insects, which produce the hormone ecdysone. The action of the molting hormone is to initiate the processes leading to ecdysis (proecdysis). Once

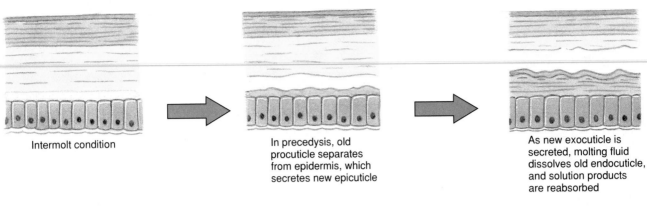

Intermolt condition

In precedysis, old procuticle separates from epidermis, which secretes new epicuticle

As new exocuticle is secreted, molting fluid dissolves old endocuticle, and solution products are reabsorbed

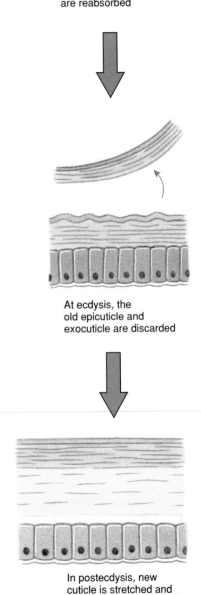

At ecdysis, the old epicuticle and exocuticle are discarded

In postecdysis, new cuticle is stretched and unfolded, and endocuticle is secreted

Figure 20-11
Cuticle secretion and resorption in ecdysis.

initiated, the cycle proceeds automatically without further action of hormones from either the X- or Y-organs.

Other Endocrine Functions. Not only does removal of eyestalks accelerate molting, it was also found over 100 years ago that crustaceans whose eyestalks have been removed can no longer adjust body coloration to background conditions. About 50 years later it was discovered that the defect was caused not by loss of vision but by loss of hormones in the eyestalks. Body color of crustaceans is largely a result of pigments in special branched cells (chromatophores) in the epidermis. Concentration of the pigment granules in the center of the cells causes a lightening effect, and dispersal of the pigment throughout the cells causes a darkening effect. The pigment behavior is controlled by hormones from neurosecretory cells in the eyestalk, as is migration of retinal pigment for light and dark adaptation in the eyes (Figure 20-8).

Neurosecretory cells are nerve cells that are modified for secretion of hormones. They are widespread in invertebrates and also occur in vertebrates. Cells in the vertebrate hypothalamus and in the posterior pituitary are good examples (see p. 746).

Release of neurosecretory material from the pericardial organs in the wall of the pericardium causes an increase in the rate and amplitude of the heartbeat.

Androgenic glands, first found in an amphipod (*Orchestia*, a common beach hopper), occur in male malacostracans. Unlike most other endocrine organs in the crustaceans, these are not neurosecretory organs. Their secretion stimulates the expression of male sexual characteristics. Young malacostracans have rudimentary androgenic glands, but in females these glands fail to develop. If they are artificially implanted in a female, her ovaries transform to testes and begin to produce sperm, and her appendages begin to take on male characteristics at the next molt. In isopods the androgenic glands are found in the testes; in all other malacostracans they are between the muscles of the coxopods of the last thoracic legs and partly attached near the ends of the vasa deferentia. Although females do not possess organs similar to androgenic glands, their ovaries produce one or two hormones that influence secondary sexual characteristics.

Hormones that influence other body processes in Crustacea may be present, and evidence suggests that a neurosecretory substance produced in the eyestalk regulates the level of blood sugar.

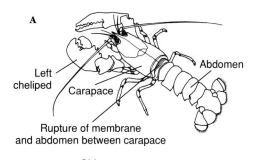

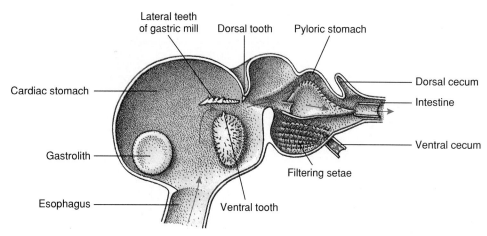

Figure 20-13

Malacostracan stomach showing gastric "mill" and directions of food movements. Mill has chitinous ridges, or teeth, for mastication, and setae for straining the food before it passes into the pyloric stomach.

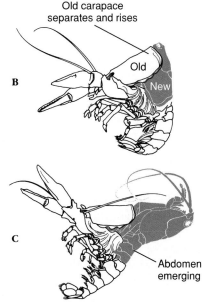

Figure 20-12

Molting sequence in the lobster *Homarus americanus*. **A,** Membrane between carapace and abdomen ruptures, and carapace begins slow elevation. This step may take up to 2 hours. **B** and **C,** Head, thorax, and finally abdomen withdraw. This process usually takes no more than 15 minutes. Immediately after ecdysis, chelipeds are desiccated and body is very soft. Lobster continues rapid absorption of water so that within 12 hours the body increases about 20% in length and 50% in weight. Tissue water will be replaced by protein in succeeding weeks.

Feeding Habits

Feeding habits and adaptations for feeding vary greatly among crustaceans. Many forms can shift from one type of feeding to another depending on environment and food availability, but all use the same fundamental set of mouthparts. The mandibles and maxillae are involved in the actual ingestion; maxillipeds hold and crush food. In predators the walking legs, particularly the chelipeds, serve in food capture.

Many crustaceans, both large and small, are predatory, and some have interesting adaptations for killing their prey. One shrimplike form, *Lygiosquilla*, has on one of its walking legs a specialized digit that can be drawn into a groove and released suddenly to pierce a passing prey. The pistol shrimp *Alpheus* has one enormously enlarged chela that can be cocked like the hammer of a gun and snapped with a force that stuns its prey.

The food of **suspension feeders** ranges from plankton, detritus, and bacteria, and **predators** consume larvae, worms, crustaceans, snails, and fishes. **Scavengers** eat dead animal and plant matter. Suspension feeders, such as the fairy shrimps, water fleas, and barnacles, use their legs, which bear a thick fringe of setae, to create water currents that sweep food particles through the setae. The mud shrimp *Upogebia* uses long setae on its first two pairs of thoracic appendages to strain food material from water circulated through its burrow by movements of its swimmerets.

Crayfishes have a two-part stomach (Figure 20-13). The first part contains a **gastric mill** in which food, already torn up by the mandibles, can be further ground up by three calcareous teeth into particles fine enough to pass through a setose filter in the second part; the food particles then pass into the intestine for chemical digestion.

A BRIEF SURVEY OF CRUSTACEANS

The crustaceans are an extensive group with many subdivisions. They have many patterns of structure, habitat, and mode of living. Some are much larger than crayfishes; others are smaller, even microscopic. Some are highly developed and specialized; others have simpler organization.

You should realize that the following summary of crustaceans and the classification on p. 402 are misleadingly brief. Although we mention all classes, a complete presentation of taxa in the hierarchy below the class level would require coverage beyond the scope of this textbook.

CLASS REMIPEDIA

Remipedia (Figure 20-14) is a very small, recently described class of Crustacea. The 10 species described so far have come from caves with connections to the sea. Remipedes have some very primitive features. There are 25 to 38 trunk segments (thorax and abdomen), all bearing paired, biramous, swimming appendages that are all essentially alike. The antennules are biramous. Both pairs of maxillae and a pair of maxillipeds, however, are prehensile and apparently adapted for feeding. The shape of the swimming appendages is similar to that found in the Copepoda,

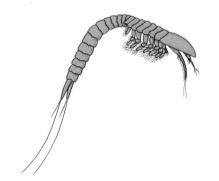

Figure 20-14

A remiped crustacean of the class Remipedia.

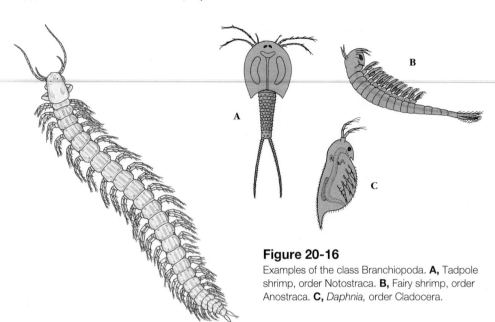

Figure 20-16

Examples of the class Branchiopoda. **A,** Tadpole shrimp, order Notostraca. **B,** Fairy shrimp, order Anostraca. **C,** *Daphnia,* order Cladocera.

Figure 20-17

An ostracod of the subclass Ostracoda, class Maxillopoda.

Figure 20-15

A cephalocarid crustacean of the class Cephalocarida.

but unlike copepods and cephalocarids, the swimming legs are directed laterally rather than ventrally.

CLASS CEPHALOCARIDA

Cephalocarida (Figure 20-15) is also a small group, with only nine species known. Cephalocarids occur along both coasts of the United States, in the West Indies, and in Japan. They are 2 to 3 mm long and have been found in bottom sediments from the intertidal zone to a depth of 300 m. Some of their features are quite primitive. The thoracic limbs are very similar to each other, and the second maxillae are similar to the thoracic limbs. The second maxillae and the first seven tho-

racic legs have a large epipod on their protopod, and the protopod is a single joint. Cephalocarids do not have eyes, a carapace, or abdominal appendages. True hermaphrodites, they are unique among the Arthropoda in discharging both eggs and sperm through a common duct.

CLASS BRANCHIOPODA

Branchiopoda also represents a crustacean type with some primitive characters. Four orders are recognized: **Anostraca** (fairy shrimp and brine shrimp, Figure 20-16B), with no carapace; **Notostraca** (tadpole shrimp, Figure 20-16A), whose carapace forms a large dorsal shield; **Conchostraca** (clam shrimp), with a bivalve carapace usually enclosing the entire body; and **Cladocera** (water fleas, Figure 20-16C), typically with a carapace that encloses the body but not the head. Branchiopods have reduced first antennae and second maxillae. Their legs are flattened and leaflike **(phyllopodia)** and are the chief respiratory organs (hence the name branchiopods). Most branchiopods also use their legs for suspension feeding, and in groups other than the cladocerans, they use their legs for locomotion as well.

Most branchiopods are freshwater forms. The most important and abundant order is the Cladocera, which often forms a large segment of the

freshwater zooplankton. Their reproduction is very interesting and is reminiscent of that occurring in some rotifers (Chapter 16). During the summer they often produce only females, by parthenogenesis, rapidly increasing the population. With the onset of unfavorable conditions, some males are produced, and eggs that must be fertilized are produced by normal meiosis. The fertilized eggs are highly resistant to cold and desiccation, and they are very important for survival of the species over the winter and for passive transfer to new habitats. Cladocera have mostly direct development, whereas other branchiopods have gradual metamorphosis.

CLASS MAXILLOPODA

The class Maxillopoda includes a number of crustacean groups traditionally considered classes themselves. Specialists have recognized evidence that these groups descended from a common ancestor and thus form a monophyletic group within the Crustacea. They basically have five cephalic, six thoracic, and usually four abdominal somites plus a telson, but reductions are common. There are no typical appendages on the abdomen. The eye of the nauplius (when present) has a unique structure and is referred to as a **maxillopodan eye.**

Subclass Ostracoda

Members of Ostracoda are, like the conchostracans, enclosed in a bivalve carapace and resemble tiny clams, 0.25 to 8 mm long (Figure 20-17). Ostracods show considerable fusion of trunk somites, and numbers of thoracic appendages are reduced to two or none. Feeding and locomotion are principally

Figure 20-18
A mystacocarid crustacean of the subclass Mystacocarida, subclass Maxillopoda.

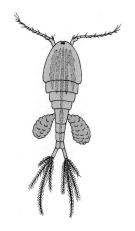

Figure 20-19
A copepod with attached ovisacs; subclass Copepoda, class Maxillopoda.

by use of the head appendages. Most ostracods live on the bottom or climb on plants, but some are planktonic, burrowing, or parasitic. Feeding habits are diverse; there are particle, plant, and carrion feeders and predators. They are widespread in both marine and freshwater habitats. Development is gradual metamorphosis.

Subclass Mystacocarida

The Mystacocarida is a class of tiny crustaceans (less than 0.5 mm long) that live in the interstitial water between sand grains of marine beaches (psammolittoral habitat) (Figure 20-18). Only 10 species have been described, but mystacocarids are widely distributed through many parts of the world. They are primitive in several characteristics.

Subclass Copepoda

The group is second only to the Malacostraca in numbers of species. The copepods are small (usually a few millimeters or less in length) and rather elongate, tapering toward the posterior. They lack a carapace and retain the simple, median, nauplius (maxillopodan) eye in the adult (Figure 20-19). They have a single pair of uniramous maxillipeds and four pairs of rather flattened, biramous, thoracic swimming appendages. The fifth pair of legs is reduced. The posterior part of the body is usually separated from the anterior, appendage-bearing portion by a major articulation. The antennules are often longer than the other appendages. The Copepoda have become very diverse and evolutionarily enterprising, with large numbers of symbiotic as well as free-living species. Many of the parasites are highly modified, and the adults may be so highly modified (and may depart so far from the description just given) that they can hardly be recognized as arthropods, let alone crustaceans.

Ecologically, the free-living copepods are of extreme importance, often dominating the primary consumer level (p. 799) in aquatic communities. In many marine localities the copepod *Calanus* is the most abundant organism in the zooplankton and has the greatest proportion of the total biomass (p. 799). In other localities it may be surpassed in the biomass only by euphausids (p. 398). *Calanus* forms a major portion of the diet of such economically and ecologically important fish as herring, menhaden, sardines, and the larvae of larger fish and (along with euphausids) is an important food item for some whales and sharks. Other genera commonly occur in the marine zooplankton, and some forms such as *Cyclops* and *Diaptomus* may form an important segment of the freshwater plankton. Many species of copepods are parasites of a wide variety of other marine invertebrates and marine and freshwater fish, and some of the latter are of economic importance. Some species of free-living copepods serve as intermediate hosts of parasites of humans, such as *Diphyllobothrium* (a tapeworm) and *Dracunculus* (a nematode), and of other animals.

Development in the copepods is indirect, and some of the highly modified parasites show striking metamorphoses.

Subclass Tantulocarida

The Tantulocarida (Figure 20-20) is the most recently described class (here considered a subclass) of crustaceans (1983). Only about 12 species are known so far. They are tiny (0.15 to 0.2 mm) copepod-like ectoparasites of other deep-sea benthic crustaceans. They have no recognizable head appendages except one pair of antennae on the sexual female. The life cycle is not known with certainty, but present evidence suggests that there is a parthenogenetic cycle and a bisexual cycle with fertilization. The **tantulus** larvae penetrate the cuticle of their hosts by a mouth tube. Then their abdomen and all thoracic limbs are lost during metamorphosis to the adult. Alone among the maxillopodans, the juveniles bear six to seven abdominal somites, but other evidence supports inclusion in this class.

Subclass Branchiura

Branchiurans are a small group of primarily fish parasites, which, despite their name, have no gills (Figure 20-21). Members of this group are usually between 5 and 10 mm long and may be found on marine or freshwater fish. They typically have a broad, shieldlike carapace, compound eyes, four biramous thoracic appendages for swimming, and a short, unsegmented abdomen. The second maxillae have become modified as suction cups, enabling the parasites to move about on their fish host or even from fish to fish. Development is almost direct: there is no nauplius, and the young resemble the adults except in size and degree of development of the appendages.

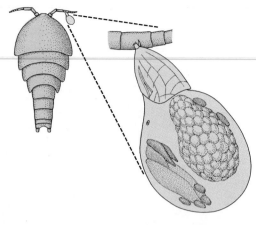

Figure 20-20

A tantulocarid. This curious little parasite is shown attached to the first antenna of its copepod host at left; subclass Tantulocarida, class Maxillopoda.

Figure 20-21

Fish louse; subclass Branchiura, class Maxillopoda.

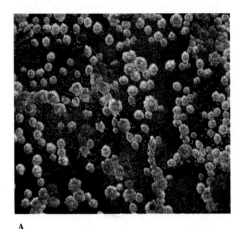

A

B

Figure 20-22

Barnacles; order Thoracica, subclass Cirripedia, class Maxillopoda. **A,** Acorn barnacles, *Balanus balanoides,* on an intertidal rock await the return of the tide. **B,** Common gooseneck barnacles, *Lepas anatifera.* Note the feeding legs, or cirri, on *Lepas.* Barnacles attach themselves to a variety of firm substrates, including rocks, pilings, and boat bottoms.

Subclass Cirripedia

The Cirripedia includes the barnacles (order Thoracica), which are usually enclosed in a shell of calcareous plates, as well as three smaller orders of burrowing or parasitic forms. Barnacles are sessile as adults and may be attached to the substrate by a stalk (gooseneck barnacles) (Figure 20-22B) or directly (acorn barnacles) (Figure 20-22A). Typically the carapace (mantle) surrounds the body and secretes a shell of calcareous plates. The head is reduced, the abdomen is absent, and the thoracic legs are long, many-jointed cirri with hairlike setae. The cirri are extended through an opening between the calcareous plates to filter from the water the small particles on which the animal feeds (Figure 20-22). Although all barnacles are marine, they are often found in the intertidal zone and are therefore exposed to drying and sometimes fresh water for some periods of time. During these periods the aperture between the plates closes to a very narrow slit.

Barnacles frequently foul ship bottoms by settling and growing there. So great may be their number that the speed of the ship may be reduced 30% to 40%, necessitating drydocking the ship to clean them off.

Barnacles are hermaphroditic and undergo a striking metamorphosis during development. Most hatch as nauplii, which soon become cyprid larvae, so called because of their resemblance to the ostracod genus *Cypris.* They have a bivalve carapace and compound eyes. The cyprid attaches to the substrate by means of its first antennae, which have adhesive glands, and begins its metamorphosis. This involves several dramatic changes, including secretion of the calcareous plates, loss of the eyes, and transformation of the swimming appendages to cirri.

Members of the cirripede order Rhizocephala, such as *Sacculina,* are highly modified parasites of crabs. They start life as cyprid larvae, just as other cirripedes, but when they find a host, most species metamorphose into a **kentrogon** (Gr. *kentron,* a point, spine, + *gonos,* progeny) which injects cells of the parasite into the hemocoel of the crab (Figure 20-23). Eventually, rootlike absorptive processes grow throughout the crab's body, and the parasite's reproductive structures become externalized between the cephalothorax and the reflexed abdomen of the crab.

The exact position at which the reproductive structures become externalized from the crab's body is of great adaptive value for the rhizocephalan parasite. Because the crab's egg mass (if it had one) would be borne in this position, the crab treats the parasite as if it were a mass of the crab's own eggs. It protects, ventilates, and grooms the parasite and actually assists in the parasite reproduction by performing spawning behavior at the appropriate time. The crab's grooming is necessary to the continued good health of the parasite. But what if the rhizocephalan's larva is so unlucky as to infect a male crab? No problem. During the parasite's internal growth in the male crab, it castrates its host, and the crab becomes structurally and behaviorally like a female!

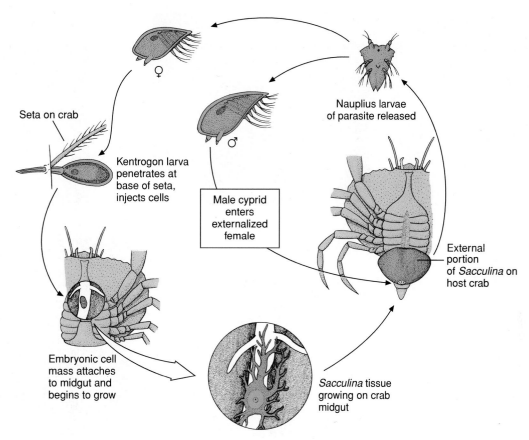

Figure 20-23
Life cycle of *Sacculina* (order Rhizocephala, subclass Cirripedia; class Maxillopoda), parasite of crabs (*Carcinus*).

CLASS MALACOSTRACA

The Malacostraca is the largest class of Crustacea and shows great diversity. The diversity is indicated by the higher classification of the group, which includes three subclasses, 14 orders, and many suborders, infraorders, and superfamilies. We confine our coverage to mentioning a few of the most important orders. We described the characteristic caridoid facies of the malacostracans on p. 384.

Order Isopoda

Isopods are one of the few crustacean groups to have successfully invaded terrestrial habitats in addition to freshwater and seawater habitats and the only crustaceans to have become truly terrestrial.

They are commonly dorsoventrally flattened, lack a carapace, and have sessile compound eyes; their first pair of thoracic limbs are maxillipeds. The remaining thoracic limbs lack exopods and are similar, while the abdominal appendages bear the gills and, except the uropods, also are similar to each other (hence the name isopods).

Common land forms are the sow bugs, or pill bugs (*Porcellio* and *Armadillidium,* Figure 20-24A), which live under stones and in damp places. Although they are terrestrial, they do not have the efficient cuticular covering and other adaptations possessed by insects to conserve water; therefore they must live in moist conditions. *Caecidotea* (Figure 20-24B) is a common freshwater form found under rocks and among aquatic plants. *Ligia* is a common marine form that scurries

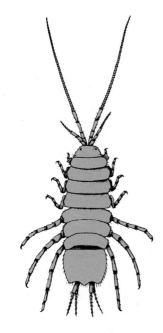

Figure 20-24
A, Four pill bugs, *Armadillidium vulgare* (order Isopoda), common terrestrial forms. **B,** Freshwater sow bug, *Caecidotea* sp., an aquatic isopod.

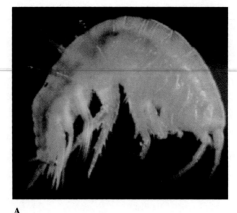

A

B

Figure 20-25

An isopod parasite (*Anilocra* sp.) on a coney (*Cephalopholis fulvus*) inhabiting a Caribbean coral reef.

Figure 20-26

Marine amphipods. **A,** Free-swimming amphipod, *Anisogammarus* sp. **B,** Skeleton shrimp, *Caprella* sp., shown on a bryozoan colony, resemble praying mantids. **C,** *Phronima,* a marine pelagic amphipod, takes over the tunic of a salp (subphylum Urochordata, Chapter 26). Swimming by means of its abdominal swimmerets, which protrude from the opening of the barrel-shaped tunic, the amphipod maneuvers to catch its prey. The tunic is not seen.

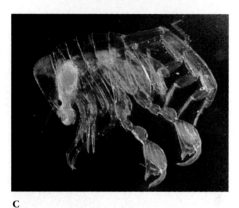

C

about on the beach or rocky shore. Some isopods are parasites of fish (Figure 20-25) or crustaceans and some are highly modified.

Development is essentially direct but may be highly metamorphic in the specialized parasites.

Order Amphipoda

Amphipods resemble isopods in that the members lack a carapace and have sessile compound eyes and one pair of maxillipeds (Figure 20-26). However, they are usually compressed laterally, and their gills are in the typical thoracic position. Furthermore, their thoracic and abdominal limbs are each arranged in two or more groups that differ in form and function. For example, one group of abdominal legs may be for swimming and another for jumping. There are many marine amphipods, including some beach-dwelling forms (for example, *Orchestia,* one of the beach hoppers), numerous freshwater genera (*Hyalella* and *Gammarus*), and a few parasites. Development is direct.

Order Euphausiacea

The Euphausiacea is a group of only about 90 species, but they are important as the oceanic plankton known as "krill." They are about 3 to 6 cm long,

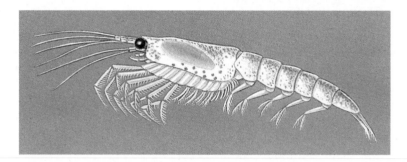

Figure 20-27

Meganyctiphanes (order Euphausiacea) "northern krill."

have a carapace that is fused with all the thoracic segments but does not entirely enclose the gills, have no maxillipeds, and have all thoracic limbs with exopods (Figure 20-27). Most are bioluminescent, with a light-producing substance in an organ called a **photophore.** Some species may occur in enormous swarms, covering up to 45 m² and extending 100 to 500 m in one direction. They form a major portion of the diet of baleen whales and many fishes. Eggs hatch as nauplii, and development is indirect.

Order Decapoda

Decapods have three pairs of maxillipeds and five pairs of walking legs, of which the first is modified in many to form pincers (chelae). They range in size from a few millimeters to the largest of all arthropods, the Japanese spider crab, whose chelae span 4 m. The crayfishes, lobsters, crabs, and "true" shrimp belong in this group (Figures 20-28 and 20-29). There are about 10,000 species of decapods, and the order is extremely diverse. They are

Figure 20-28

Decapod crustaceans. **A,** The bright orange tropical rock crab, *Grapsus grapsus,* is a conspicuous exception to the rule that most crabs bear cryptic coloration. **B,** The hermit crab, *Elassochirus gilli,* which has a soft abdominal exoskeleton, lives in a snail shell that it carries about and into which it can withdraw for protection. **C,** The male fiddler crab, *Uca* sp., uses its enlarged cheliped to wave territorial displays and in threat and combat. **D,** The red night shrimp, *Rhynchocinetes rigens,* prowls caves and overhangs of coral reefs, but only at night. **E,** The spiny lobster, *Panulirus argus,* (shown here) and the northern lobster, *Homarus americanus,* are consumed with gusto by many people.

Figure 20-29

Sponge crab, *Dromidia antillensis.* This crab is one of several species that deliberately mask themselves with material from their immediate environment.

very important ecologically and economically, and numerous species are relished as items of food for humans.

The crabs, especially, exist in a great variety of forms. Although resembling the pattern of crayfishes, they differ from the latter in having a broader cephalothorax and a reduced abdomen. Familiar examples along the seashore are the hermit crabs (Figure 20-28B), which live in snail shells because their abdomens are not protected by the same heavy exoskeleton as the anterior parts are; the fiddler crabs, *Uca* (Figure 20-28C), which burrow in the sand just below the high-tide level and come out to run about over the sand while the tide is out; and the spider crabs such as *Libinia* and the interesting decorator crabs *Oregonia* and others, which cover their carapaces with sponges and sea anemones for protective camouflage (Figure 20-29).

PHYLOGENY AND ADAPTIVE RADIATION

PHYLOGENY

The relationship of the crustaceans to other arthropods has long been a puzzle. The controversy over whether the Arthropoda is polyphyletic was mentioned in Chapter 19. The crustaceans have traditionally been allied with the uniramians (insects, myriapods, see Chapter 21) in a group known as Mandibulata because they both have mandibles, as contrasted with chelicerae. Critics of this traditional grouping have argued that the mandibles in each group were so different that they could not have been inherited from a common ancestor. In addition to some differences in the muscles, the mandibles of the crustaceans are multijointed, and the chewing and biting surfaces are at the bases "gnathobasic mandible." Uniramian mandibles, on the other hand, are of a single joint, and the biting surface is on the distal portion ("entire limb mandible"). However, advocates of the "mandibulate hypothesis" maintain that these differences are not so fundamental that they could not have arisen during the 550-million-year history of the mandibulate taxa. They also point out the numerous other similarities between crustaceans and uniramians, such as the basic structure of the ommatidia, the tripartite brain, and the head primitively of five somites, each with a pair of appendages. This mandibulate hypothesis can be depicted in a cladogram (Figure 20-30).

Among the Crustacea, the Remipedia seem to be the most primitive of all known forms (Figure 20-30). They have a long body, with no tagmatization behind the head, a double ventral nerve cord, and serially arranged digestive ceca. Fossils of a puzzling arthropod from the Mississippian period seem to be a sister group of the remipedians and may shed light on the origin of biramous appendages. They have *two pairs* of uniramous limbs on each somite. It is possible that each somite actually represents two ancestral somites that fused together (diplopodous condition).* We see such a condition in the Diplopoda (p. 406), a uniramian class with two pairs of legs on each somite. In the ancestral crustacean, the bases of the pairs of limbs on the diplopodous somites would have fused together to become a biramous appendage, with two branches on a common protopod.

ADAPTIVE RADIATION

The adaptive radiation demonstrated by the crustaceans is great, with all manner of aquatic niches exploited. They are unquestionably the dominant arthropod group in the marine environment, and they share dominance of freshwater habitats with insects. Invasions of terrestrial environments have been much more limited, with isopods being the only notable success. There are a few other terrestrial examples, such as land crabs. The most diverse class is the Malacostraca, and the most abundant group is the Copepoda. Members of both taxa include planktonic suspension feeders and numerous scavengers. Copepods have been particularly successful as parasites of both vertebrates and invertebrates, and it is clear that the present parasitic copepods are the products of numerous invasions of such niches.

*Emerson, J. J., and F. R. Schram. 1990. Science **250:**667–669.

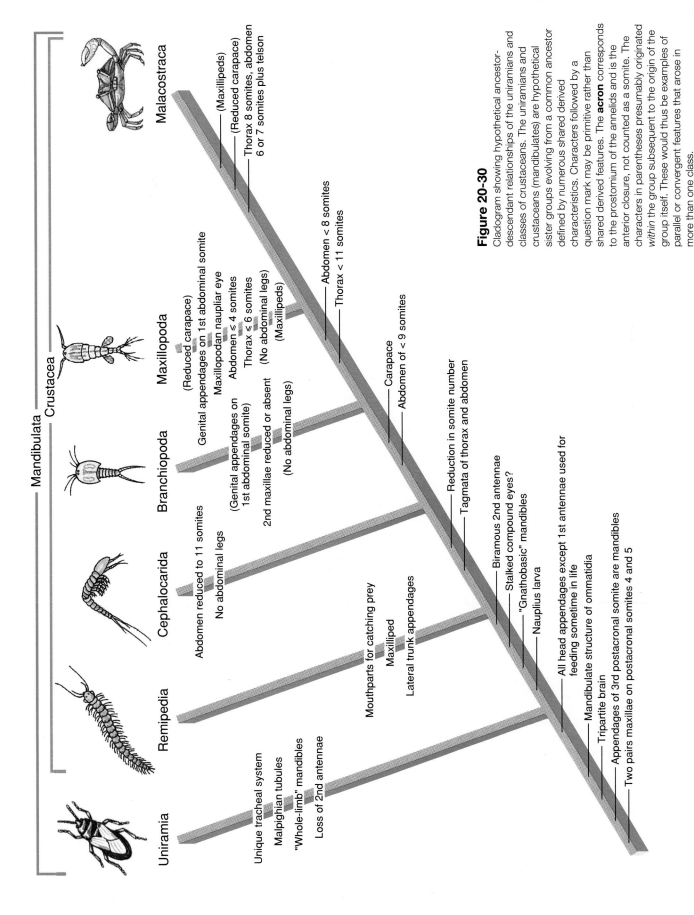

Figure 20-30

Cladogram showing hypothetical ancestor-descendant relationships of the uniramians and classes of crustaceans. The uniramians and crustaceans (mandibulates) are hypothetical sister groups evolving from a common ancestor defined by numerous shared derived characteristics. Characters followed by a question mark may be primitive rather than shared derived features. The **acron** corresponds to the prostomium of the annelids and is the anterior closure, not counted as a somite. The characters in parentheses presumably originated *within* the group subsequent to the origin of the group itself. These would thus be examples of parallel or convergent features that arose in more than one class.

CLASSIFICATION OF SUBPHYLUM CRUSTACEA

Higher classification of the Crustacea is complex and subject to change as new data become available. The following relies on several sources. We are omitting many smaller taxa.

Class Remipedia (ri-mi-pee´dee-a) (L. *remipedes,* oar-footed). No carapace; one-segmented protopods; biramous antennules and antennae; all trunk appendages similar; cephalic appendages large and raptorial; maxilliped somite fused to head; trunk unregionalized. Example: *Speleonectes.*

Class Cephalocarida (sef´a-lo-kar´i-da) (Gr. *kephalē,* head, + *karis,* shrimp, + *ida,* pl. suffix). No carapace; phyllopodia, one-segmented protopods; uniramous antennules and biramous antennae; compound eyes lacking; no abdominal appendages; maxilliped similar to thoracic leg. Example: *Hutchinsoniella.*

Class Branchiopoda (bran´kee-op´o-da) (Gr. *branchia,* gills, + *pous, podos,* foot). Phyllopodia; carapace present or absent; no maxillipeds; antennules reduced; compound eyes present; no abdominal appendages; maxillae reduced.

Order Anostraca (an-os´tra-ka) (Gr. *an-,* prefix meaning without, + *ostrakon,* shell): **fairy shrimp and brine shrimp.** No carapace; no abdominal appendages; uniramous antennae. Examples: *Artemia, Branchinecta.*

Order Notostraca (no-tos´tra-ka) (Gr. *nōtos,* the back, + *ostrakon,* shell): **tadpole shrimp.** Carapace forming large dorsal shield; abdominal appendages present, reduced posteriorly; antennae vestigial. Examples: *Triops, Lepidurus.*

Order Cladocera (kla-dah´se-ra) (Gr. *klados,* a branch, + *keras,* a horn): **water fleas.** Carapace folded, usually enclosing trunk but not head; biramous antennae; abdominal appendages absent. Examples: *Daphnia, Leptodora.*

Order Conchostraca (kon-kos´tra-ka) (Gr. *konche,* + *ostrakon,* shell): **clam shrimp.** Bivalved carapace enclosing entire body; biramous antennae; all trunk appendages similar. Example: *Lynceus.*

Class Maxillopoda (maks-i-lah´po-da) (L. *maxilla,* the jawbone, + *pous, podos,* a foot). Usually five cephalic, six thoracic, and four abdominal somites plus a telson, but reductions common; no typical appendages on abdomen; naupliar eye of unique structure (maxillopodan eye); carapace present or absent.

Subclass Ostracoda (os-rak´o-da) (Gr. *ostrakodes,* having a shell): **ostracods.** Bivalve carapace entirely encloses body; body unsegmented or indistinctly segmented; no more than two pairs of trunk appendages. Examples: *Cypris, Cypridina, Gigantocypris.*

Subclass Mystacocarida (mis-tak´o-kar´i-da) (Gr. *mystax,* mustache, + *karis,* shrimp, + *ida,* pl. suffix): **mustache shrimps.** No carapace; body of cephalon and ten-segmented trunk; telson with clawlike caudal rami; cephalic appendages nearly identical, but antennae and mandibles biramous, other head appendages uniramous; second through fifth trunk somites with short, single-segment appendages. Example: *Derocheilocaris.*

Subclass Copepoda (ko-pep´o-da) (Gr. *kōpē,* oar, + *pous, podos,* foot): **copepods.** No carapace; thorax typically of seven somites, of which first and sometimes second fuse with head to form cephalothorax; antennules uniramous; antennae bi- or uniramous; four to five pairs swimming legs; parasitic forms often highly modified. Examples: *Cyclops, Diaptomus, Calanus, Ergasilus, Lernaea, Salmincola, Caligus.*

Subclass Tantulocarida (tan´tu-lo-kar´i-da) (L. *tantulus,* so small, + *caris,* shrimp). No recognizable cephalic appendages except antennae on sexual female; solid median cephalic stylet; six free thoracic somites, each with pair of appendages, anterior five biramous; six abdominal somites; minute copepod-like ectoparasites. Examples: *Basipodella, Deoterthron.*

Subclass Branchiura (bran-ki-ur´a) (Gr. *branchia,* gills, + *ura,* tail): **fish lice.** Body oval, head and most of trunk covered by flattened carapace, incompletely fused to first thoracic somite; thorax with four pairs of appendages, biramous; abdomen unsegmented, bilobed; eyes compound; antennae and antennules reduced; maxillules often forming suctoral discs. Examples: *Argulus, Chonopeltis.*

Subclass Cirripedia (sir-i-ped´i-a) (L. *cirrus,* curl of hair, + *pes, pedis,* foot): **barnacles.** Sessile or parasitic as adults; head reduced and abdomen rudimentary; paired compound eyes absent; body segmentation indistinct; usually hermaphroditic; in free-living forms carapace becomes mantle, which secretes calcareous plates; antennules become organs of attachment, then disappear. Examples: *Balanus, Policipes, Sacculina.*

Class Malacostraca (mal-a-kos´tra-ka) (Gr. *malakos,* soft, + *ostrakon,* shell). Usually with eight somites in thorax and six plus telson in abdomen; all segments with appendages; antennules often biramous; first one to three thoracic appendages often maxillipeds; carapace covering head and part or all of thorax, sometimes absent; gills usually thoracic epipods.

Order Isopoda (i-sop´o-da) (Gr. *isos,* equal, + *pous, podos,* foot): **isopods.** No carapace; antennules usually uniramous, sometimes vestigial; eyes sessile (not stalked); gills on abdominal appendages; body commonly dorso-ventrally flattened; second thoracic appendages usually not prehensile. Examples: *Armadillidium, Caecidotea, Ligia, Porcellio.*

Order Amphipoda (am-fip´o-da) (Gr. *amphis,* on both sides, + *pous, podos,* foot): **amphipods.** No carapace; antennules often biramous; eyes usually sessile; gills on thoracic coxae; second and third thoracic limbs usually prehensile; typically bilaterally compressed body form. Examples: *Orchestia, Hyalella, Gammarus.*

Order Euphausiacea (yu-faws-i-a´si-a) (Gr. *eu,* well, + *phausi,* shining bright, + L. *acea,* suffix, pertaining to): **krill.** Carapace fused to all thoracic segments but not entirely enclosing gills, no maxillipeds; all thoracic limbs with exopods. Example: *Meganyctiphanes.*

Order Decapoda (de-kap´o-da) (Gr. *deka,* ten, *pous, podos,* foot): **shrimps, crabs, lobsters.** All thoracic segments fused with and covered by carapace; eyes on stalks; first three pairs of thoracic appendages modified to maxillipeds. Examples: *Penaeus, Cancer, Pagurus, Grapsus, Homarus, Panulirus.*

Summary

In addition to a pair of mandibles, the Crustacea and the Uniramia have in common at least one pair of antennae, and a pair of maxillae. Their tagmata are a head and trunk or a head, thorax, and abdomen.

The Crustacea is a large, primarily aquatic subphylum. Crustaceans have two pairs of antennae, their appendages are primitively biramous, and many have a carapace.

All arthropods must periodically cast off their cuticle (ecdysis) and grow in dimensional size before the newly secreted cuticle hardens. Premolt and postmolt periods are hormonally controlled, as are several other processes, such as change in body color and expression of sexual characteristics.

Feeding habits vary greatly in Crustacea, and there are many predators, scavengers, suspension feeders, and parasites. Respiration is through the body surface or by gills, and excretory organs take the form of maxillary or antennal glands. Circulation, as in other arthropods, is through an open system of sinuses (hemocoel), and a dorsal, tubular heart is the chief pumping organ. Most crustaceans have compound eyes composed of units called ommatidia. Sexes are usually separate.

The crustacean class Branchiopoda is characterized by the possession of phyllopodia and contains, among others, the order Cladocera, which is ecologically important as zooplankton. Members of the maxillopodan subclass Copepoda lack a carapace and abdominal appendages. They are abundant and are among the most important of the primary consumers in many freshwater and marine ecosystems. Many are parasitic. Most members of the subclass Cirripedia (barnacles) are sessile as adults, secrete a shell of calcareous plates, and filter feed by means of their thoracic appendages.

The Malacostraca is the largest crustacean class, and the most important orders are the Isopoda, Amphipoda, Euphausiacea, and Decapoda. All have both abdominal and thoracic appendages. Isopods lack a carapace and are usually dorsoventrally flattened. Amphipods also lack a carapace but are usually laterally flattened. Euphausiaceans are important oceanic plankton called krill. Decapods include crabs, shrimps, lobsters, crayfishes, and others; they have five pairs of walking legs (including the chelipeds) on their thorax.

Review Questions

1. What are the tagmata and the appendages on the head of crustaceans? What other important characteristics of Crustacea distinguish them from other arthropods?
2. Define each of the following: tergum, sternum, caudal furca, telson, protopod, exopod, endopod, epipod, endite.
3. What is meant by homologous structures? What is meant by serial homology, and how do crustaceans show serial homology?
4. Distinguish a hemocoel from a coelom.
5. Briefly describe respiration and circulation in crayfishes.
6. Briefly describe the function of antennal and maxillary glands in Crustacea.
7. How does a crayfish detect changes in position?
8. What is the photoreceptor unit of a compound eye? How does this unit adjust to varying amounts of light?
9. What is a nauplius? What is the difference between direct and indirect development in the Crustacea?
10. Describe the molting process in Crustacea, including the action of the hormones.
11. Which of the classes and subclasses of Crustacea, the Branchiopoda, Ostracoda, Copepoda, Cirripedia, and Malacostraca are the most diverse? Most numerous? Distinguish them from each other.
12. Compare and contrast the Isopoda, Amphipoda, Euphausiacea, and Decapoda.
13. What is the significance of the Remipedia to hypotheses concerning the origin of crustaceans?

Selected References

See also general references for Part III, p. 626.

Bliss, D. E. (editor in chief). 1982–1985. The biology of Crustacea, vols. 1–10. New York, Academic Press, Inc. *This series is a standard reference for all aspects of crustacean biology.*

Cameron, J. N. 1985. Molting in the blue crab. Sci. Am. **252:**102–109 (May). *The life cycle and development of the commercially valuable blue crab* Callinectes sapidus *is described. Studies on the chemistry of the molting process may have important economic benefits.*

Cronin, T. W., N. J. Marshall, and M. F. Land. 1994. The unique visual system of the mantis shrimp. Amer. Sci. **82:**356–365. *The ancestors of mantis shrimps diverged from other crustaceans about 400 million years ago. Accuracy in the raptorial strike of these aggressive predators requires a highly refined visual system.*

Emerson, M. J., and F. R. Schram. 1990. The origin of crustacean biramous appendages and the evolution of the Arthropoda. Science **250:**667–669. *Gives evidence from the Mississippian fossil* Tesnusocaris *that the biramous limbs of crustaceans may have originated by fusion of legs on a diplopodous somite.*

Huys, R., G. A. Boxshall, and R. J. Lincoln. 1993. The tantulocaridan life cycle: the circle closed? J. Crust. Biol. **13:**432–442. *The current hypothesis of a parthenogenetic cycle alternating with a cycle that includes fertilization in these bizarre little creatures.*

Ritchie, L. E., and J. T. Høeg. 1981. The life history of *Lernaeodiscus porcellanae* (Cirripedia: Rhizocephala) and co-evolution with its porcellanid host. J. Crust. Biol. **1:**334–347. *The life history and description of the maternal care given the parasite by its host. A fascinating story.*

Schmitt, W. L. 1965. Crustaceans. Ann Arbor, The University of Michigan Press. *A good little book by one of the most eminent students of the group; very interesting reading and easy to understand.*

Schram, F. R. 1986. Crustacea. New York, Oxford University Press. *The most recent comprehensive account.*

21

The Terrestrial Mandibulates

Phylum Arthropoda
Uniramians: Classes Chilopoda,
Diplopoda, Pauropoda,
Symphyla, and Insecta

A Winning Combination

Tunis, Algeria—Treating it as an invading army, Tunisia,
Algeria, and Morocco have mobilized to fight the most
serious infestation of locusts in over 30 years. Billions of the
insects have already caused extensive damage to crops and
are threatening to inflict great harm to the delicate
*economies of North Africa.**

The staggering losses occasionally inflicted by the billions
of locusts in Africa serve as only one reminder of our
ceaseless struggle with the dominant group of animals on
earth today: the insects. Insects far outnumber all the other
species of animals in the world combined; and numbers of
individuals are equally enormous. Some scientists have
estimated that there are 200 million insects for every single
human alive today! Insects have an unmatched ability to adapt

to all land environments and to virtually all climates. Having
originally evolved as land animals, insects developed wings
and invaded the air 150 million years before flying reptiles,
birds, or mammals. Many have exploited freshwater habitats,
where they are now widely prevalent; only in the sea are the
numbers of insects more limited.

How can we account for the enormous numbers of these
creatures? In common with other arthropods, insects have a
combination of valuable structural and physiological
adaptations, including a versatile exoskeleton, segmentation,
an efficient respiratory system, and highly developed sensory
organs. In addition, insects have a waterproofed cuticle, and
many have extraordinary abilities to survive adverse
environmental conditions. ■

*From the *New York Times*, 20 April 1988.

The members of the subphylum Uniramia are primarily terrestrial arthropods. Only a few of them have returned to aquatic life, usually in fresh water.

The term "myriapod," meaning "many footed," is commonly used for a group of four classes of uniramians that have evolved a pattern of two tagmata—head and trunk—with paired appendages on most or all trunk somites. The myriapods include the Chilopoda (centipedes), Diplopoda (millipedes), Pauropoda (pauropods), and Symphyla (symphylans).

The insects have evolved a pattern of three tagmata—head, thorax, and abdomen—with appendages on the head and thorax but greatly reduced on or absent from the abdomen. The common ancestor of insects probably resembled the myriapods in general body form.

The uniramians have only one pair of antennae, and their appendages are always uniramous, never biramous like those of the crustaceans. Although some insect young are aquatic and have gills, the gills are not homologous with those of the crustaceans.

The insects and myriapods use tracheae to carry the respiratory gases directly to and from all body cells in a manner similar to the onychophorans and some of the arachnids.

Excretion is usually by malpighian tubules.

CLASS CHILOPODA

The Chilopoda (ki-lop´o-da) (Gr. *cheilos,* margin, lip, + *pous, podos,* foot), or centipedes, are land forms with somewhat flattened bodies that may contain from a few to 177 somites (Figure 21-1). Each somite, except the one behind the head and the last two in the body, bears a pair of jointed legs. The appendages of the first body segment are modified to form poison claws.

The head appendages are similar to those of an insect. There are a pair of antennae, a pair of mandibles, and one or two pairs of maxillae. A pair of eyes on the dorsal side of the head consists of groups of ocelli.

The digestive system is a straight tube into which salivary glands empty at the anterior end. Two pairs of malpighian tubules empty into the hind part of the intestine. There is an elongated heart with a pair of arteries to each somite. The heart has a series of ostia to provide for the return of the blood to the heart from the hemocoel. Respiration is by means of a tracheal system of branched air tubes that come from a pair of spiracles in each somite. The nervous system is typically arthropod, and there is also a visceral nervous system.

Sexes are separate, with unpaired gonads and paired ducts. Some centipedes lay eggs and others are viviparous. The young are similar to the adults.

Centipedes prefer moist places such as under logs, bark, and stones. They are very agile and are carnivorous in their eating habits, living on earthworms, cockroaches, and other insects. They kill their prey with their poison claws and then chew it with their mandibles. The common house centipede, *Scutigera* (L. *scutum,* shield, + *gera,* bearing), which has 15 pairs of legs, is often seen scurrying around bathrooms and damp cellars, where it catches insects. Most species are harmless to humans, although many of the tropical centipedes, some of which may reach a length of 30 cm, are dangerous.

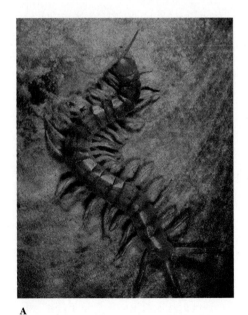

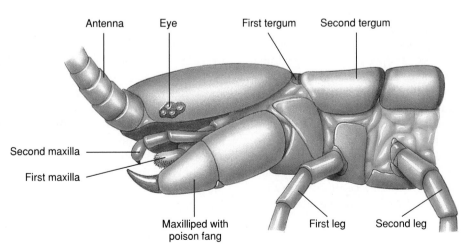

A   **B**

Figure 21-1

A, Centipede, *Scolopendra* (class Chilopoda). Most segments have one pair of appendages each. First segment bears a pair of poison claws, which in some species can inflict serious wounds. Centipedes are carnivorous. **B,** Head of centipede.

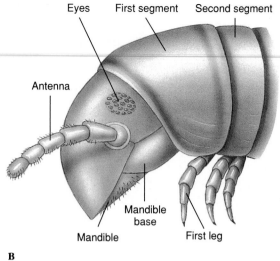

A   **B**

Figure 21-2
A, Millipede *Narceus americanus*. Note the typical doubling of appendages on most segments, hence diplosegments. **B,** Head of millipede.

CLASS DIPLOPODA

The Diplopoda (Gr. *diplos,* double, two + *pous, podos,* foot) are commonly called millipedes, which literally means "thousand feet" (Figure 21-2). Even though they do not have that many legs, they do have a large number of appendages, since each abdominal somite has two pairs, a condition that may have arisen from the fusion of pairs of somites (p. 400). Their cylindrical bodies are made up of 25 to 100 somites. The short thorax consists of four somites, each bearing one pair of legs.

The head bears two clumps of simple eyes and a pair each of antennae, mandibles, and maxillae. The general body structures are similar to those of centipedes, with a few variations here and there. Two pairs of spiracles on each abdominal somite open into air chambers that give off the tracheal air tubes. There are two genital apertures toward the anterior end.

In most millipedes the appendages of the seventh somite are specialized for copulatory organs. After millipedes copulate, the female lays the eggs in a nest and guards them carefully. The larval forms have only one pair of legs to each somite.

Millipedes are not as active as centipedes. They walk with a slow, graceful motion, not wriggling as the centipedes do. They prefer dark, moist places under logs or stones. Most are herbivorous, feeding on decayed plant matter, although sometimes they eat living plants. Because they are slow-moving animals, many millipedes roll up into a coil when disturbed. Many millipedes also protect themselves from predation by secreting toxic or repellent fluids from special glands **(repugnatorial glands)** positioned along the sides of the body. Common examples of this class are *Spirobolus* and *Julus,* both of which have wide distribution.

CLASS PAUROPODA

The Pauropoda (Gr. *pauros,* small, + *pous, podos,* foot) are a group of minute (2 mm or less), soft-bodied myriapods, numbering almost 500 species. They have a small head with branched antennae and no eyes, but they have a pair of sense organs that have the appearance of eyes (Figure 21-3A). Their 12 trunk segments usually bear nine pairs of legs (none on the first or the last two segments). They have only one tergal plate covering each two segments.

Tracheae, spiracles, and circulatory system are lacking. Pauropods are probably most closely related to the diplopods, but they have more primitive characteristics.

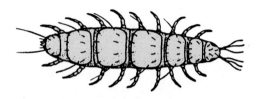

A

B

Figure 21-3
A, Pauropod. Pauropods are minute, whitish myriapods with three-branched antennae and nine pairs of legs. They live in leaf litter and under stones. They are eyeless but have sense organs that resemble eyes. **B,** *Scutigerella,* a symphylan, is a minute whitish myriapod that is sometimes a greenhouse pest.

Although widely distributed, the pauropods are the least well known of the myriapods. They live in moist soil, leaf litter, or decaying vegetation and under bark and debris. Representative genera are *Pauropus* and *Allopauropus.*

CLASS SYMPHYLA

The Symphyla (Gr. *sym*, together, + *phylon*, tribe) are small (2 to 10 mm) and have centipede-like bodies (Figure 21-3B). They live in humus, leaf mold, and debris. *Scutigerella* (L. dim. of *Scutigera*) are often pests on vegetables and flowers, particularly in greenhouses. They are soft bodied, with 14 segments, 12 of which bear legs and one a pair of spinnerets. The antennae are long and unbranched. Only 160 species have been described.

> The mating behavior of *Scutigerella* is unusual. The male places a spermatophore at the end of a stalk. When the female finds it, she takes it into her mouth, storing the sperm in special buccal pouches. Then she removes the eggs from her gonopore with her mouth and attaches them to moss or lichen, or to the walls of crevices, smearing them during the handling with some of the semen and so fertilizing them. The young at first have only six or seven pairs of legs.

Symphylans are eyeless but have sensory pits at the bases of the antennae. The tracheal system is limited to a pair of spiracles on the head and tracheal tubes to the anterior segments only.

CLASS INSECTA

The Insecta (L. *insectus,* cut into) are the most diverse and abundant of all the groups of arthropods. There are more species of insects than species of all the other classes of animals combined. The recorded number of insect species has been estimated at close to 1 million, but this figure may represent only a fraction of the species that exist. There is also striking evidence of continuing evolution among insects at the present time, even though the fossil record indicates that the group as a whole is stable.

It is difficult to appreciate fully the significance of this extensive group and its role in the biological pattern of animal life. The study of insects **(entomology)** occupies the time and resources of skilled men and women all over the world. The struggle between humans and their insect competitors seems to be endless, yet paradoxically insects have so interwoven themselves into the economy of nature in so many useful roles that we would have a difficult time without them.

Insects differ from other arthropods in having **three pairs of legs** and usually **two pairs of wings** on the thoracic region of the body, although some have one pair of wings or none. Insects range in size from less than 1 mm to 20 cm in length, the majority being less than 2.5 cm long. Generally, the largest insects live in tropical areas.

DISTRIBUTION

Insects are among the most abundant and widespread of all land animals. They have spread into practically all habitats that will support life except the deeper waters of the sea. Relatively few are marine. The marine water striders (*Halobates*), which live on the surface of the ocean, are the only marine invertebrates that live on the sea-air interface. Insects are common in brackish water, in salt marshes, and on sandy beaches. They are abundant in fresh water, in soil, in forests (especially the tropical forest canopy), and in plants, and they are found even in deserts and wastelands, on mountaintops, and as parasites in and on the bodies of plants and animals.

Their wide distribution is made possible by their powers of flight and their highly adaptable nature. In most cases they can easily surmount barriers that are virtually impassable to many other animals. Their small size allows them to be carried by currents of both wind and water to far regions. Their well-protected eggs can withstand rigorous conditions and can be carried long distances by birds and other animals. Their agility and aggressiveness enable them to fight for every possible niche in a habitat. No single pattern of biological adaptation can be applied to them.

ADAPTABILITY

Insects, during their evolution, have shown an amazing adaptability, as evidenced by their wide distribution and enormous diversity of species. Most of their structural modifications have taken place in the wings, legs, antennae, mouthparts, and alimentary canal. Such wide diversity enables this vigorous group to take advantage of all available resources of food and shelter. Some are parasitic, some suck the sap of plants, some chew up the foliage of plants, some are predaceous, and some live on the blood of various animals. Within these different groups, specialization occurs, so that a particular kind of insect will eat, for instance, the leaves of only one kind of plant. This specificity of eating habits lessens competition with other species and to a great extent accounts for their biological diversity.

Insects are well adapted to dry and desert regions. The hard and protective exoskeleton helps prevent evaporation, but some insects also extract the utmost in fluid from food and fecal material, as well as moisture from the water by-product of bodily metabolism.

As in other arthropods, the exoskeleton is made up of a complex system of plates known as **sclerites,** connected to one another by concealed, flexible hinge joints. The muscles between the sclerites enable the insect to make precise movements. The rigidity of its exoskeleton is attributable to the unique scleroproteins and not to its chitin component, and its lightness makes flying possible. By contrast, the cuticle of crustaceans is stiffened mostly by mineral matter.

EXTERNAL FORM AND FUNCTION

Insects show a remarkable variety of morphological characteristics, but as a group they are much more homogeneous in tagmatization than are the Crustacea. Some insects are fairly generalized in body structure; some are highly specialized. The grasshopper,

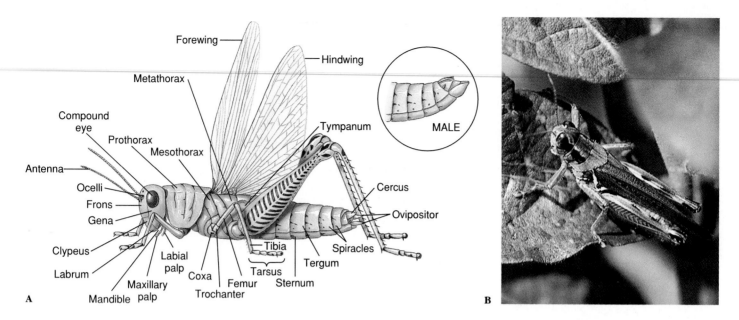

Figure 21-4

A, External features of a female grasshopper. The terminal segment of a male with external genitalia is shown in inset. **B,** A spur-throated grasshopper, *Melanoplus* sp. (order Orthoptera). This large genus contains many agricultural pest species.

or locust, is a generalized type that is usually used in laboratories to demonstrate the general features of insects (Figure 21-4).

The insect tagmata are the head, thorax, and abdomen. The cuticle of each body somite typically is composed of four plates (sclerites), a dorsal notum (tergum), a ventral sternum, and a pair of lateral pleura. The pleura of abdominal segments are membranous rather than sclerotized.

The head usually bears a pair of relatively large compound eyes, a pair of antennae, and usually three ocelli. The antennae, which vary greatly in size and form (Figure 21-5), act as tactile organs, olfactory organs, and in some cases as auditory organs. Mouthparts, formed from specially hardened cuticle, typically consist of a labrum, a pair each of mandibles and maxillae, a labium, and a tonguelike hypopharynx. The type of mouthparts an insect possesses determines how it feeds. We will discuss some of these modifications later.

The thorax is composed of three somites: prothorax, mesothorax, and metathorax, each bearing a pair of

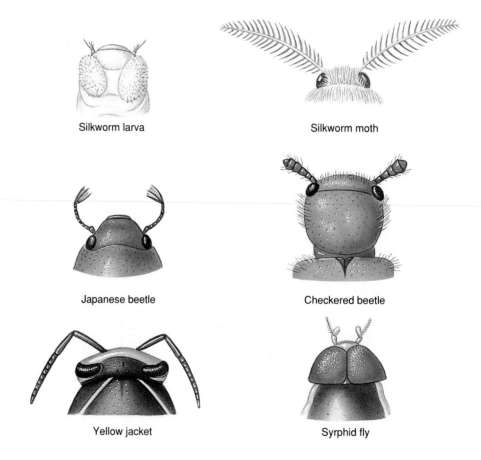

Figure 21-5

A few of the various types of insect antennae.

legs (Figure 21-4). In most insects the mesothorax and metathorax each bear a pair of wings. The wings are cuticular extensions formed by the epidermis. They consist of a double membrane containing veins of thicker cuticle that serve to strengthen the wing. Although these veins vary in their patterns among the different species, they are constant within a species and serve as one means of classification and identification.

Legs of insects often are modified for special purposes. Terrestrial forms have walking legs with terminal pads and claws as in beetles. These pads may be sticky for walking upside down, as in houseflies. The hindlegs of the grasshoppers and crickets are adapted for jumping (Figure 21-6). The mole cricket has the first pair of legs modified for burrowing in the ground. Water bugs and many beetles have paddle-shaped appendages for swimming. For grasping its prey, the forelegs of the praying mantis are long and strong (Figure 21-7). The legs of honeybees show complex adaptations for collecting pollen (Figure 21-8).

The abdomen of insects is composed of 9 to 11 segments; the eleventh, when present, is reduced to a pair of cerci (appendages at the posterior end). Larval or nymphal forms have a variety of abdominal appendages, but these are lacking in the adults. The end of the abdomen bears the external genitalia (Figure 21-4A).

There are innumerable variations in body form among the insects. Beetles are usually thick and plump (Figure 21-9A); damselflies, ant lions, and walking sticks are long and slender (Figure 21-9B); many aquatic beetles are streamlined; and cockroaches are flat, adapted to living in crevices. The ovipositor of the female ichneumon wasp is extremely long (Figure 21-10). The cerci form horny forceps in the earwigs and are long and many jointed in stoneflies and mayflies. Antennae are long in cockroaches and katydids, short in dragonflies and most beetles, knobbed in butterflies, and plumed in most moths. Other variations exist (Figure 21-5).

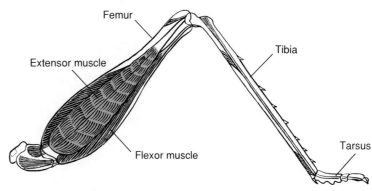

Figure 21-6

Hindleg of grasshopper. Muscles that operate the leg are found within a hollow cylinder of exoskeleton. Here they are attached to the internal wall, from which they manipulate segments of limb on the principle of a lever. Note pivot joint and attachment of tendons of extensor and flexor muscles, which act reciprocally to extend and flex the limb.

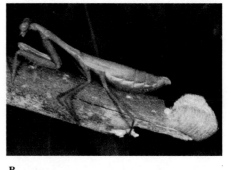

Figure 21-7

A, Praying mantis (order Orthoptera) feeding on an insect. **B,** Praying mantis laying eggs.

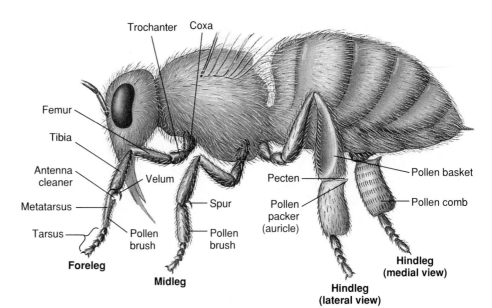

Figure 21-8

Adaptive legs of worker honeybee. In the foreleg, the toothed indentation covered with the velum combs out the antennae. The spur on the middle leg removes wax from wax glands on the abdomen. Pollen brushes on the front and middle legs comb off pollen picked up on body hairs and deposit it on the pollen brushes of the hindlegs. Long hairs of the pecten on the hindleg remove pollen from the brush of the opposite leg; then the auricle (pollen packer) presses it into a pollen basket when the leg joint is flexed back. A bee carries her load in both baskets to the hive and pushes pollen into a cell, to be cared for by other workers.

A

B

Figure 21-9
A, A giant horned beetle *Diloboderus abderus* (order Coleoptera) from Uruguay. Though the ferocious-looking processes from the head and thorax might appear to be for pinching or stabbing an opponent, they actually are used to lift or pry up a rival of the same species away from resources. **B,** Walking sticks *Diapheromera femorata* (order Orthoptera), mating. The species is common in much of North America. It is wingless, and despite its camouflage as a twig, it is fed upon by numerous predators.

Locomotion

Walking. When walking, most insects use a triangle of legs involving the first and last leg of one side together with the middle leg of the opposite side. In this way, insects keep three of their six legs on the ground, a tripod arrangement for stability.

Some insects, such as the water strider *Gerris* (L. *gero,* to carry), are able to walk on the surface of water. The water strider has on its footpads nonwetting hairs that do not break the surface film of water but merely indent it. As it skates along, *Gerris* uses only the two posterior pairs of legs and steers with the anterior pair (Figure 21-11). The body of the marine water

Figure 21-10
An ichneumon wasp with the end of the abdomen raised to thrust her long ovipositor into wood to find a tunnel made by the larva of a wood wasp or wood-boring beetle. She can bore 13 mm or more into the wood to lay her eggs in the larva of the wood-boring beetle, which will become host for the ichneumon larvae. Other ichneumon species attack spiders, moths, flies, crickets, caterpillars, and other insects.

strider *Halobates* (Gr. *halos,* the sea, + *bātes,* one that treads), an excellent surfer on rough ocean waves, is further protected by a water-repellent coat of close-set hairs shaped like thick hooks.

Power of Flight. Insects share the power of flight with birds and flying mammals. However, their wings evolved in a different manner from that of the limb buds of birds and mammals and are not homologous to them. Insect wings are formed by outgrowths from the body wall of the mesothoracic and metathoracic segments and are composed of cuticle.

Most insects have two pairs of wings, but the Diptera (true flies) have only one pair, the hindwings being represented by a pair of tiny **halteres** (balancers) that vibrate and are responsible for equilibrium during the flight. Males of the order Strepsiptera have only the hind pair of wings and an anterior pair of halteres. Males of the scale insects also have one pair of wings but no halteres. Some insects are wingless. Ants and termites, for example, have wings only on males, and on females during certain periods; workers are always wingless. Lice and fleas are always wingless.

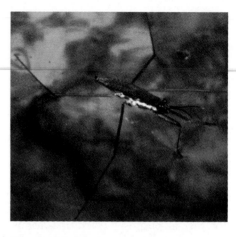

Figure 21-11
Water strider, *Gerris* sp. (order Hemiptera). The animal is supported on its long, slender legs by the water's surface tension.

Wings may be thin and membranous, as in flies and many others (Figure 21-10); thick and horny, such as the front wings of beetles (Figure 21-9A); parchmentlike, such as the front wings of grasshoppers; covered with fine scales, as in butterflies and moths; or covered with hairs, as in caddis flies.

Wing movements are controlled by a complex of muscles in the thorax. **Direct flight muscles** are attached to a part of the wing itself. **Indirect flight muscles** are not attached to the wing and cause wing movement by altering the shape of the thorax. The wing is hinged at the thoracic tergum and also slightly laterally on a pleural process, which acts as a fulcrum (Figure 21-12). In all insects, the upstroke of the wing is effected by contracting indirect muscles that pull the tergum down toward the sternum (Figure 21-12A). Dragonflies and cockroaches accomplish the downstroke by contracting direct muscles attached to the wings lateral to the pleural fulcrum. In Hymenoptera and Diptera all the flight muscles are indirect. The downstroke occurs when the sternotergal muscles relax and longitudinal muscles of the thorax arch the tergum (Figure 21-12B), pulling up the tergal articulations relative to the pleura. The downstroke in beetles and grasshoppers involves both direct and indirect muscles.

Flight muscle contraction has two basic types of neural control: **synchro-**

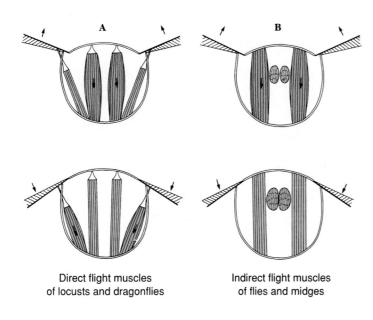

Direct flight muscles
of locusts and dragonflies

Indirect flight muscles
of flies and midges

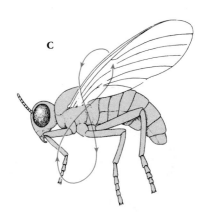

Figure 21-12
A, Flight muscles of insects such as
cockroaches, in which upstroke is by indirect
muscles and downstroke is by direct muscles.
B, In insects such as flies and bees, both
upstroke and downstroke are by indirect
muscles. **C,** The figure-eight path followed by
the wing of a flying insect during the upstroke
and downstroke.

nous and **asynchronous.** Larger insects such as dragonflies and butterflies have synchronous muscles, in which a single volley of nerve impulses stimulates a muscle contraction and thus one wing stroke. Asynchronous muscles are found in the more specialized insects. Their mechanism of action is complex and depends on the storage of potential energy in resilient parts of the thoracic cuticle. As one set of muscles contracts (moving the wing in one direction), they stretch the antagonistic set of muscles, causing them to contract (and move the wing in the other direction). Because the muscle contractions are not phase-related to nervous stimulation, only occasional nerve impulses are necessary to keep the muscles responsive to alternating stretch activation. Thus extremely rapid wing beats are possible. For example, butterflies (with synchronous muscles) may beat as few as four times per second. Insects with asynchronous muscles, such as flies and bees, may vibrate at 100 beats per second or more. The fruit fly *Drosophila* (Gr. *drosos,* dew, + *philos,* loving) can fly at 300 beats per second, and midges have been clocked at more than 1000 beats per second.

Obviously flying entails more than a simple flapping of wings; a for-ward thrust is necessary. As the indirect flight muscles alternate rhythmically to raise and lower the wings, the direct flight muscles alter the angle of the wings so that they act as lifting airfoils during both the upstroke and the downstroke, twisting the leading edge of the wings downward during the downstroke and upward during the upstroke. This produces a figure-eight movement (Figure 21-12C) that aids in spilling air from the trailing edges of the wings. The quality of the forward thrust depends, of course, on several factors, such as variations in wing venation, how much the wings are tilted, and how they are feathered.

Flight speeds vary. The fastest flyers usually have narrow, fast-moving wings with a strong tilt and a strong figure-eight component. Sphinx moths and horseflies are said to achieve approximately 48 km (30 miles) per hour and dragonflies approximately 40 km (25 miles) per hour. Some insects are capable of long continuous flights. The migrating monarch butterfly *Danaus plexippus* (Gr. after Danaus, mythical king of Arabia) (Figure 21-26) travels south for hundreds of miles in the fall, flying at a speed of approximately 10 km (6 miles) per hour.

INTERNAL FORM AND FUNCTION

Nutrition

The digestive system (Figure 21-13) consists of a foregut (mouth with salivary glands, esophagus, crop for storage, and gizzard for grinding in some); a midgut (stomach and gastric ceca); and a hindgut (intestine, rectum, and anus). Some digestion may take place in the crop as the food mixes with enzymes from the saliva, but no absorption takes place there. The main site for digestion and absorption is the midgut, and the ceca may increase the digestive and absorptive area. Little absorption of nutrients occurs in the hindgut (with certain exceptions, such as wood-eating termites), but this is a major area for resorption of water and some ions (see p. 415).

The majority of insects feed on plant juices and plant tissues **(phytophagous** or **herbivorous).** Some insects feed on specific plants; others, such as grasshoppers, will eat almost any plant. The caterpillars of many moths and butterflies eat the foliage of only certain plants. Certain species of ants and termites cultivate fungus gardens as a source of food.

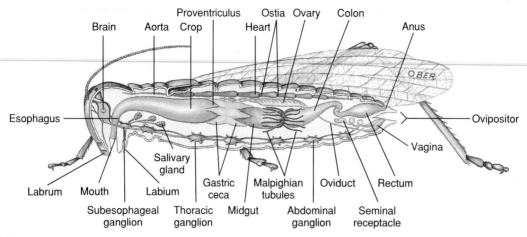

Figure 21-13
Internal structure of female grasshopper.

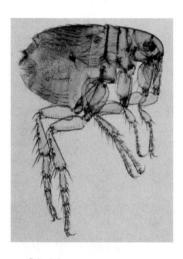

Figure 21-14
Female human flea, *Pulex irritans*.

A   B

Figure 21-15
A, Hornworm, larval stage of a sphinx moth (order Lepidoptera). The more than 100 species of North American sphinx moths are strong fliers and mostly nocturnal feeders. Their larvae, called hornworms because of the large, fleshy posterior spine, are often pests of tomatoes, tobacco, and other plants. **B,** Hornworm parasitized by a tiny wasp *Apanteles,* which laid its eggs inside the caterpillar. The wasp larvae have emerged, and their pupae are on the caterpillar's skin. Young wasps emerge in 5 to 10 days, but the caterpillar usually dies.

Many beetles and the larvae of many insects live on dead animals **(saprophagous).** A number of insects are **predaceous,** catching and eating other insects as well as other types of animals (Figure 21-7). However, the so-called predaceous diving beetle *Cybister fimbriolatus* (Gr. *kybistēr,* diver) is not as predaceous as once supposed, but is largely a scavenger.

Many insects, adults as well as larvae, are **parasitic.** For instance, fleas (Figure 21-14) live on the blood of mammals, and the larvae of many varieties of wasps live on spiders and caterpillars (Figure 21-15). In turn, many are parasitized by other insects. Some of the latter are beneficial to humans by controlling the numbers of injurious insects. Parasitism of parasitic insects by other insects, a condition known as **hyperparasitism,** often becomes quite complex.

For each type of feeding, the mouthparts are adapted in a specialized way. The **sucking mouthparts** are usually arranged in the form of a tube and can pierce the tissues of plants or animals. The water scorpion (*Ranatra fusca* order Hemiptera) demonstrates this arrangement well. The water scorpion is a sticklike aquatic insect with a slender caudal respiratory tube. It has a beak in which there are four piercing, needle-like stylets made up of two mandibles and two maxillae. These parts fit together to form two tubes, a salivary tube for injecting saliva into the prey and a food tube for drawing out the body fluid of the prey. The mosquito also combines piercing with needle-like stylets and sucking through a food channel (Figure 21-16B). In honeybees the labium forms a flexible and contractile "tongue" covered with many hairs. When the bee plunges its proboscis into nectar, the tip of the tongue bends upward and moves

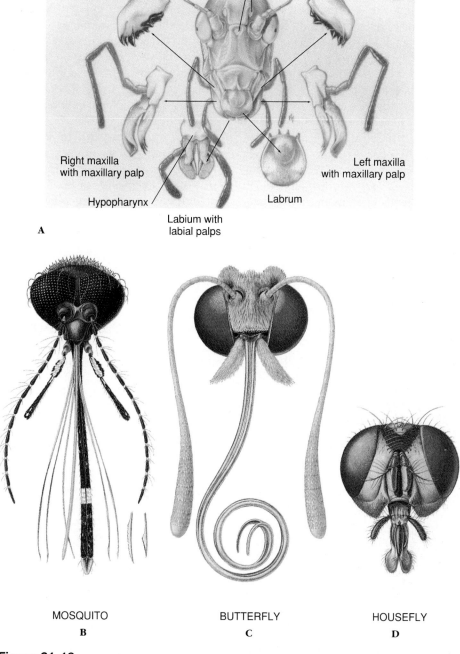

Figure 21-16
Four types of insect mouthparts. (See text for description of types and examples.)

A — Right mandible, Ocelli, Left mandible, Right maxilla with maxillary palp, Hypopharynx, Labium with labial palps, Labrum, Left maxilla with maxillary palp

B — MOSQUITO

C — BUTTERFLY

D — HOUSEFLY

Biting mouthparts such as those of the grasshopper and many other herbivorous insects are adapted for seizing and crushing food (Figure 21-16A); those of most carnivorous insects are sharp and pointed for piercing their prey. The mandibles of chewing insects are strong, toothed plates whose edges can bite or tear while the maxillae hold the food and pass it toward the mouth. Enzymes secreted by the salivar glands add chemical action to the chewing process.

Circulation

A tubular heart in the pericardial cavity (Figure 21-13) moves hemolymph (blood) forward through the only blood vessel, a dorsal aorta. The heartbeat is a peristaltic wave. Accessory pulsatory organs help move the hemolymph into the wings and legs, and flow is also facilitated by various body movements. The hemolymph consists of plasma and amebocytes and apparently has little to do with oxygen transport.

Gas Exchange

Terrestrial animals require efficient respiratory systems that permit rapid oxygen and carbon dioxide exchange but at the same time restrict water loss. In insects this is the function of the **tracheal system,** an extensive network of thin-walled tubes that branch into every part of the body (Figure 21-17). The tracheal trunks open to the outside by paired **spiracles,** usually two on the thorax and seven or eight on the abdomen. A spiracle may be merely a hole in the integument, as in primary wingless insects, but there is usually a valve or some sort of closing mechanism that cuts down water loss. The evolution of such a device must have been very important in enabling insects to move into drier habitats. The spiracle may also possess a filtering device such as a sieve plate or a set of interlocking bristles that may prevent entrance of water, parasites, or dust into the tracheae.

The **tracheae** are composed of a single layer of cells and are lined with cuticle that is shed, along with the outer cuticle, during the molt. Spiral thickenings of the cuticle (called taenidia)

back and forth rapidly. Liquid enters the tube by capillarity and is drawn up continuously by a pumping pharynx. In butterflies and moths, mandibles are usually absent, and the maxillae form a long sucking proboscis (Figure 21-16C) for drawing nectar from flowers. At rest the proboscis coils into a flat spiral. In feeding it extends, and fluid is pumped up by pharyngeal muscles.

Houseflies, blowflies, and fruit flies have **sponging** and **lapping mouthparts** (Figure 21-16D). At the apex of the labium is a pair of large, soft lobes with grooves on the lower surface that serve as food channels. These flies lap up liquid food or liquefy food first with salivary secretions. Horseflies not only sponge up surface liquids but bite into the skin with slender, tapering mandibles and then sponge up blood.

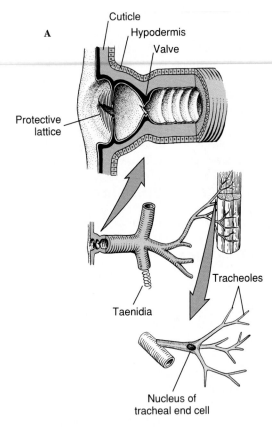

A

Cuticle
Hypodermis
Valve
Protective lattice
Tracheoles
Taenidia
Nucleus of tracheal end cell

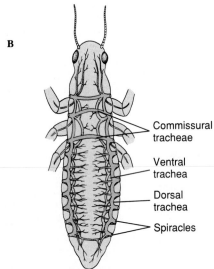

B

Commissural tracheae
Ventral trachea
Dorsal trachea
Spiracles

Figure 21-17

A, Relationship of spiracle, tracheae, taenidia (chitinous bands that strengthen the tracheae), and tracheoles (diagrammatic). **B,** Generalized arrangement of insect tracheal system (diagrammatic). Air sacs and tracheoles not shown.

support the tracheae and prevent their collapse. The tracheae branch out into smaller tubes, ending in very fine, fluid-filled tubules called **tracheoles** (lined with cuticle, but not shed at ecdysis), which branch into a fine network over the cells. In large insects the largest tracheae may be several millimeters in diameter but taper down to 1 to 2 μm. The tracheoles then taper to 0.5 to 0.1 μm in diameter. In one of the stages of the silkworm larva, it is estimated that there are 1.5 million tracheoles! Scarcely any living cell is more than a few micrometers away from a tracheole. In fact, the ends of some tracheoles actually indent the membranes of the cells they supply, so that they terminate close to the mitochondria. The tracheal system affords an efficient system of transport without the use of oxygen-carrying pigments in the hemolymph.

Although the diving beetle *Dytiscus* (Gr. *dytikos*, able to swim) can fly, it spends most of its life in the water as an excellent swimmer. It uses an "artificial gill" in the form of a bubble of air held under its wing covers. The bubble is kept stable by a layer of hairs on top of the abdomen and is in contact with the spiracles on the abdomen. Oxygen from the bubble diffuses into the tracheae and is replaced by diffusion of oxygen from the water. Thus the bubble can last for several hours before the beetle must surface to replace it. Mosquito larvae are not good swimmers but live just below the surface, putting out short breathing tubes like snorkels to the surface for air (Figure 21-22B). Spreading oil on the water, a favorite method of mosquito control, clogs the tracheae with oil and so suffocates the larvae. "Rattailed maggots" of the syrphid flies have an extensible tail that can stretch as much as 15 cm to the water surface.

The tracheal system may also include **air sacs,** which are apparently dilated tracheae without taenidia (Figure 21-18A). They are thin walled and flexi-

ble and are mostly in the body cavity but also in appendages. In many insects the air sacs increase the volume of air inspired and expired. Muscular movements in the abdomen draw air into the tracheae and expand the sacs, which collapse on expiration. In some insects—locusts, for example—additional pumping is provided by telescoping the abdomen, pumping with the prothorax, or thrusting the head forward and backward. In some insects, the air sacs have functions other than respiratory. For example, they may allow internal organs to change in volume during growth without changing the shape of the insect, and they reduce the weight of large insects.

In some very small insects, gas transport occurs entirely by diffusion along a concentration gradient. Consumption of oxygen causes a reduced pressure in the tracheae that sucks air in through the spiracles.

The tracheal system is an adaptation for air breathing, but many insects (nymphs, larvae, and adults) live in water. In small, soft-bodied aquatic nymphs, the gaseous exchange may occur by diffusion through the body wall, usually into and out of a tracheal network just under the integument. The aquatic nymphs of stoneflies and mayflies have **tracheal gills,** which are thin extensions of the body wall containing a rich tracheal supply. The gills of dragonfly nymphs are ridges in the rectum (rectal gills) where gas exchange occurs as water moves in and out.

Excretion and Water Balance

Insects and spiders have a unique excretory system consisting of **malpighian tubules** that operate in conjunction with specialized glands in the wall of the rectum. The malpighian tubules, variable in number, are thin, elastic, blind tubules attached to the juncture between the midgut and hindgut (Figures 21-13 and 21-18A). The free ends of the tubules lie free in the hemocoel and are bathed in hemolymph.

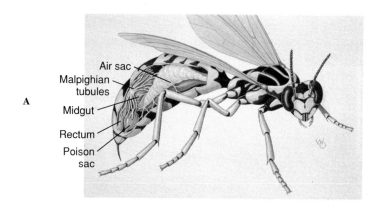

A

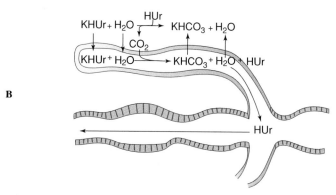

B

Figure 21-18

Malpighian tubules of insect. **A,** Malpighian tubules are located at the juncture of the midgut and hindgut (rectum) as shown in the cutaway view of a wasp. **B,** Function of malpighian tubules. Solutes, especially potassium, are actively secreted into the tubules. Water and potassium acid urate (KHUr) follow. This fluid moves into the rectum where solutes and water are actively resorbed, leaving uric acid (HUr) to be excreted.

The mechanism of urine formation in the malpighian tubules of herbivorous insects appears to depend on the active secretion of potassium into the tubules (Figure 21-18B). This primary secretion of ions pulls water along with it by osmosis to produce a potassium-rich fluid. Other solutes and waste materials also are secreted or diffuse into the tubule. The predominant waste product of nitrogen metabolism in most insects is uric acid, which is virtually insoluble in water (see p. 657). This enters the upper end of the tubule, where the pH is slightly alkaline, as relatively soluble potassium and urate (abbreviated KHUr in Figure 21-18). As the formative urine passes into the lower end of the tubule, the potassium combines with carbon dioxide, is reabsorbed as potassium bicarbonate ($KHCO_3$), the pH changes to acidic (pH 6.6), and the insoluble uric acid

(HUr) precipitates out. As the urine drains into the intestine and passes through the hindgut, specialized rectal glands absorb chloride, sodium (and in some cases potassium), and water.

Since water requirements vary among different types of insects, this ability to cycle water and salts is very important. Insects living in dry environments may resorb nearly all water from the rectum, producing a nearly dry mixture of urine and feces. Leaf-feeding insects take in and excrete quantities of fluid. Freshwater larvae need to excrete water and conserve salts. Insects that feed on dry grains need to conserve water and excrete salt.

Nervous System

The nervous system in general resembles that of the larger crustaceans, with a similar tendency toward fusion of ganglia (Figure 21-13). A number

of insects have a giant fiber system. There is also a stomodeal nervous system that corresponds in function to the autonomic nervous system of vertebrates. Neurosecretory cells located in various parts of the brain have an endocrine function, but, except for their role in molting and metamorphosis, little is known of their activity.

Sense Organs

Along with neuromuscular coordination, insects have unusually keen sensory perception. Their sense organs are mostly microscopic and are located chiefly in the body wall. Each type usually responds to a specific stimulus. The various organs are receptive to mechanical, auditory, chemical, visual, and other stimuli.

Mechanoreception. Mechanical stimuli, or those dealing with touch, pressure, vibration, and the like, are picked up by **sensilla.** A sensillum may be simply a seta, or hairlike process, connected with a nerve cell, a nerve ending just under the cuticle and lacking a seta, or a more complex organ (scolopophorous organ) consisting of sensory cells with their endings attached to the body wall. Such organs are widely distributed over the antennae, legs, and body.

Auditory Reception. Very sensitive setae (hair sensilla) or tympanal organs may detect airborne sounds. In tympanal organs a number of sensory cells (ranging from a few to hundreds) extend to a very thin tympanic membrane that encloses an air space in which vibrations can be detected. Tympanal organs occur in certain Orthoptera (Figure 21-4), Homoptera, and Lepidoptera. Some insects are fairly insensitive to airborne sounds but can detect vibrations reaching them through the substrate. Organs on the legs usually detect vibrations of the substrate.

Chemoreception. Chemoreceptors (for taste or smell) are usually bundles of sensory cell processes that are

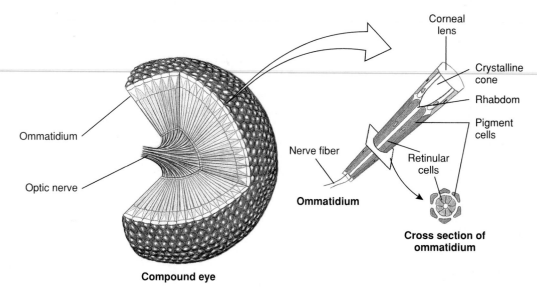

Figure 21-19
Compound eye of an insect. A single ommatidium is shown enlarged to the right.

often located in sensory pits. These are often on mouthparts, but in ants, bees, and wasps they are also on the antennae, and butterflies, moths, and flies also have them on the legs. The chemical sense is generally keen, and some insects can detect certain odors for several kilometers. Many of the patterns of insect behavior such as feeding, mating, habitat selection, and host-parasite relations are mediated through the chemical senses. These senses play a crucial role in the responses of insects to artificial repellents and attractants.

Visual Reception.   Insect eyes are of two types, simple and compound. Simple eyes are found in some nymphs and larvae and in many adults. Most insects have three ocelli on the head. Evidence indicates that honeybees use ocelli to monitor light intensity but not to form images.

Most adult insects have compound eyes, which may cover much of the head. They consist of thousands of ommatidia—6300 in the eye of a honeybee, for example. The structure of the compound eye is similar to that of crustaceans (Figure 21-19). An insect such as a honeybee can see simultaneously in almost all directions around its body, but it is more myopic than the human, and images, even of nearby objects, are

fuzzy. However, most flying insects rate much higher than humans in flicker-fusion tests. Flickers of light become fused in the human eye at a frequency of 45 to 55 per second, but bees and blowflies can distinguish as many as 200 to 300 separate flashes of light per second. This is probably an advantage in analyzing a fast-changing landscape during flight.

A bee can distinguish colors, but its sensitivity begins in the ultraviolet range, which the human eye cannot see, and extends into the orange; honeybees cannot distinguish shades of red from shades of gray.

Other Senses.   Insects also have well-developed senses for temperature, especially on the antennae and legs, and for humidity, proprioception (sensation of muscle stretch and body position), gravity, and other physical properties.

Neuromuscular Coordination

Insects are active creatures with excellent neuromuscular coordination. Arthropod muscles are typically cross-striated, just as vertebrate skeletal muscles are. A flea can leap a distance of 100 times its own length, and an ant can carry in its jaws a load greater than its own weight. This sounds as though insect muscle were

stronger than that of other animals. Actually, however, the force a particular muscle can exert is related directly to its cross-sectional area, not its length. Based on maximum load moved per square centimeter of cross section, the strength of insect muscle is relatively the same as that of vertebrate muscle. The illusion of great strength of insects (and other small animals) is simply a consequence of a small body size.

In terms of proportionate body length, the flea's jump would be the equivalent of a 6-foot human executing a standing high jump of 600 feet! Actually, the insect's muscles are not entirely responsible for its jump; they cannot contract rapidly enough to reach the required acceleration. The flea depends on pads of *resilin,* a protein with unusual elastic properties, that is also found in the wing-hinge ligaments of many other insects. Resilin releases 97% of its stored energy on returning from a stretched position, compared with only 85% in most commercial rubber. When the flea prepares to jump, it rotates its hind femurs and compresses the resilin pads, then engages a "catch" mechanism. In effect, it has cocked itself. To take off, the flea must exert the relatively small muscular action to unhook the catches, allowing the resilin to expand.

A

B

Figure 21-20

Copulation in insects (see also Figure 21-9B). **A,** *Omura congrua* (order Orthoptera) are a kind of grasshopper found in Brazil. **B,** Bluet damselflies *Enallagma* sp. (order Odonata) are common throughout North America. Here, the male still grasps the female after copulation as the female (white abdomen) lays eggs.

Reproduction

Sexes are separate in insects, and fertilization is usually internal. Insects have various means of attracting mates. The female moth gives off a powerful pheromone that males can detect for a great distance. Fireflies use flashes of light; some insects find each other by means of sounds or color signals and by various kinds of courtship behavior.

Sperm are usually deposited in the vagina of the female at the time of copulation (Figures 21-13 and 21-20). In some orders the sperm are encased in spermatophores that may be transferred at copulation or deposited on the substratum to be picked up by the female. The male silverfish deposits a

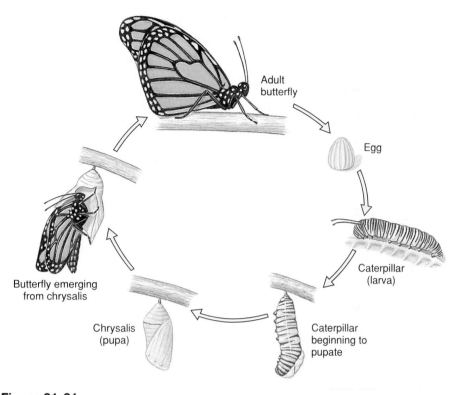

Figure 21-21

Complete (holometabolous) metamorphosis in a butterfly, *Danaus plexippus.* Eggs hatch to produce first of several larval instars. Last larval instar molts to become a pupa. Adult emerges at pupal molt.

spermatophore on the ground, then spins signal threads to guide the female to it. During the evolutionary transition of ancestral insects from aquatic to terrestrial life, spermatophores were used widely and copulation evolved much later.

Usually the sperm are stored in the spermatheca of the female in numbers sufficient to fertilize more than one batch of eggs. Many insects mate only once during their lifetime, but male damselflies copulate several times per day.

Insects usually lay a great many eggs. A queen honeybee, for example, may lay more than 1 million eggs during her lifetime. On the other hand, some flies are viviparous and bring forth only a single offspring at a time. Insects that make no provision for the care of their young may lay many more eggs than do insects that provide for their young or those that have a very short life cycle.

Most species lay their eggs in a particular type of place to which visual, chemical, or other cues guide them. Butterflies and moths lay their

eggs on the specific kind of plant on which the caterpillar must feed. The tiger moth may look for a pigweed, the sphinx moth for a tomato or tobacco plant, and the monarch butterfly for a milkweed plant (Figure 21-21). Insects whose immature stages are aquatic characteristically lay their eggs in water (Figure 21-22A). A tiny braconid wasp lays her eggs on the caterpillar of the sphinx moth where they will feed and pupate in tiny white cocoons (Figure 21-15). The ichneumon wasp, with unerring accuracy, seeks out a certain kind of larva in which her young will live as internal parasites. Her long ovipositors may have to penetrate 1 to 2 cm of wood to find the larva of a wood wasp or a wood-boring beetle in which she will deposit her eggs (Figure 21-10).

METAMORPHOSIS AND GROWTH

Early development occurs within the egg, and the hatching young escape from the egg in various ways. During the postembryonic development most insects change in form; that is, they

A

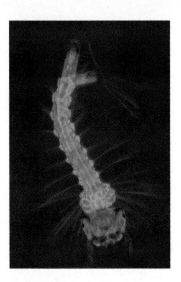

B

Figure 21-22

A, Mosquito *Culex* (order Diptera) lays her eggs in small packets or rafts on the surface of standing or slowly moving water. **B,** Mosquito larvae are the familiar wrigglers of ponds and ditches. To breathe, they hang head down, with respiratory tubes projecting through the surface film of water. Motion of vibratile tufts of fine hairs on the head brings a constant supply of food.

undergo **metamorphosis** (Figure 21-21). During this period they must undergo a number of molts to grow, and each stage of the insect between molts is called an **instar.**

Although metamorphosis occurs in many animals, insects illustrate it more dramatically than any other group. The transformation, for instance, of the hickory horned devil caterpillar into the beautiful royal walnut moth represents an astonishing morphological change. In insects metamorphosis is associated with the evolution of wings, which are restricted to the reproductive stage where they can be of the most benefit.

Holometabolous Metamorphosis

Approximately 88% of insects undergo a **holometabolous** (Gr. *holo,* complete, + *metabole-* change) metamorphosis, which separates the physiological processes of growth (larva) from those of differentiation (pupa) and reproduction (adult) (Figure 21-21). Each stage functions efficiently without competition with the other stages, for the larvae often live in entirely different surroundings and eat different foods from the adults. The wormlike larvae, which usually have chewing mouthparts, are known as caterpillars, maggots, bagworms, fuzzy worms, grubs, and so on. After a series of instars during which the wings develop internally, the larva forms a case or cocoon about itself and becomes a pupa, or chrysalis, a nonfeeding stage in which many insects pass the winter. When the final molt occurs over winter, the full-grown adult emerges, pale and with wings wrinkled. In a short time the wings expand and harden, and the insect is on its way. The stages, then, are egg, larva (several instars), pupa, and adult (Figure 21-21). The adult undergoes no further molting.

Hemimetabolous Metamorphosis

Some insects undergo a **hemimetabolous** (Gr. *hemi,* half, + *metabole-,* change), or gradual (incomplete), metamorphosis. These include the grasshoppers, cicadas, mantids, and terrestrial bugs, which have terrestrial young, and mayflies, stoneflies, dragonflies, and aquatic bugs, whose young are aquatic and lay their eggs in water. The young are called **nymphs,** and their wings develop externally as budlike outgrowths in the early instars and increase in size as the animal grows by successive molts and becomes a winged adult (Figures 21-23 and 21-24). Aquatic nymphs in some orders have tracheal gills or other modifications for aquatic life (Figure 21-25). The stages are egg, nymph (several instars), and adult (Figure 21-24).

Figure 21-23

Ecdysis in the dog-day cicada, *Tibicen pruinosa* (order Homoptera). The old cuticle splits along a dorsal midline as a result of increased blood pressure and of air forced into the thorax by muscle contraction. The emerging insect is pale, and its new cuticle is soft. The wings will be expanded by blood pumped into veins, and the insect enlarges by taking in air.

The biological meaning of the word "bug" is a great deal more restrictive than in common English usage. People often refer to all insects as "bugs," even extending the word to include such non-animals as bacteria, viruses, and glitches in computer programs. Strictly speaking, however, a bug is a member of the order Hemiptera and nothing else.

Direct Development

A few insects, such as silverfish and springtails, undergo direct development. The young, or juveniles, are similar to the adults except in size and sexual maturation. The stages are egg, juvenile, and adult. Such insects include the primitively wingless insects.

Physiology of Metamorphosis

Hormones regulate metamorphosis in insects. The major endocrine organs that are involved in development are the **brain,** the **prothoracic (ecdysial) glands,** the **corpora cardiaca,** and the **corpora allata** (Figure 37-4, p. 743).

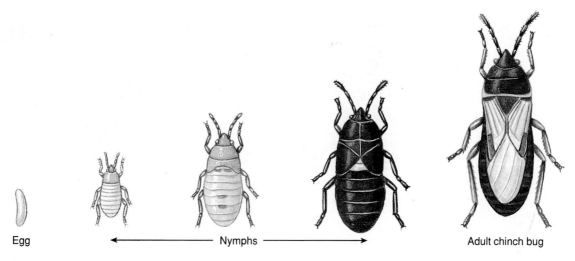

Egg   ←———— Nymphs ————→   Adult chinch bug

Figure 21-24
Life history of a hemimetabolous insect.

A   B   C

Figure 21-25
A, Stonefly, *Perla* sp. (order Plecoptera). **B,** Ten-spot dragonfly, *Libellula pulchella* (order Odonata). **C,** Nymph (larva) of dragonfly. Both stoneflies and dragonflies have aquatic larvae that undergo gradual metamorphosis.

C, Carolina Biological Supply/Phototake.

The intercerebral part of the brain and the ganglia of the nerve cord contain several groups of neurosecretory cells that produce an endocrine substance called **brain hormone (ecdysiotropin** [ek-die´ze-o-tro´pin]). These neurosecretory cells send their axons to paired organs behind the brain, the corpora cardiaca, which serve as storage and release organs for the brain hormone (and also produce other hormones). The brain hormone is carried in the hemolymph to the prothoracic gland, a glandular organ, that in the head or the prothorax produces **molting hormone,** or **ecdysone** (ek-die´sone) in response to brain hormone. Ecdysone sets in motion certain processes that lead to the casting off of the old cuticle (ecdysis).

Simple molting persists as long as **juvenile hormone (neotenine)** is present in sufficient amounts, along with the molting hormone in the hemolymph, and each molt produces a larger larva. The corpora allata produce juvenile hormone (Figure 37-3).

In later instars, the corpora allata release progressively less juvenile hormone. When juvenile hormone is at a very low level, the larva molts to become a pupa, and cessation of juvenile hormone production in the pupa leads to an adult at the next molt (metamorphosis). Control of development is the same in hemimetabolous insects, except that there is no pupa, and cessation of juvenile hormone production occurs in the final nymphal instar. The corpora allata again become active in adult insects, in which juvenile hormone is important in normal egg production. The prothoracic glands degenerate in the adults of most insects, and the adult does not molt.

Insect hormones have been the subject of much interesting experimental work. For example, if the corpora allata (and thus juvenile hormone) are removed surgically from the larva, the following molt will result in metamorphosis. Conversely, if the corpora allata from a young larva are transplanted into a final larval instar, the latter can be converted into a giant larva, because metamorphosis to the pupa cannot occur.

DIAPAUSE

Many animals, including many types of insects, undergo a period of dormancy in their annual life cycle. In temperate zones there may be a period of winter dormancy, called hibernation, or a period of summer dormancy, called estivation, or both. There are periods in the life cycle of many insects when eggs, larvae, pupae, or even adults remain dormant for a long time because external conditions of climate, moisture, and the like are too harsh or unfavorable for survival in states of normal activity. Thus the life cycle is synchronized with periods of suitable environmental conditions and abundance of food. Most insects enter a dormant state when some factor of the environment, such as temperature, becomes unfavorable, and the dormancy continues until conditions again become favorable.

However, some species have a prolonged arrest of growth that occurs regardless of the environment, that is, whether or not favorable conditions prevail. This type of dormancy is called **diapause** (dī´a-poz) (Gr. *dia,* through, dividing into two parts, + *pausis,* a stopping), and it is an important adaptation to survive adverse environmental conditions. Diapause is genetically determined in each species and sometimes varies between strains within a species, but it is usually set off by some certain signal. In the environment of the insect, such signals forecast adverse conditions to come, for example, the lengthening or shortening of the days. Thus photoperiod, or day length, is often the signal that initiates diapause. The arrival of a critical length of day starts the physiological machinery for establishing diapause, which continues until the proper day length or other signal is received.

Diapause always occurs at the end of an active growth stage of the molting cycle so that, when the diapause period is over, the insect is ready for another molt. One species of the ant *Myrmica* reaches the third instar stage in late summer. Many of the larvae do not develop beyond this point until the following spring, even if temperatures are mild or if the larvae are kept in a warm laboratory.

DEFENSE

Insects as a group display many colors. This is especially true of butterflies, moths, and beetles. Even in the same species the color pattern may

A

B

Figure 21-26

Mimicry in butterflies. **A,** Monarch butterfly is distasteful to, and avoided by, birds because as a caterpillar it fed on the acrid milkweed. **B,** The monarch is mimicked by the smaller viceroy butterfly, *Limenitis archippus,* which feeds on willows and is presumably tasteful to birds, but is not eaten because it so closely resembles the monarch in color and markings. This kind of mimicry is called Batesian mimicry.

vary in a seasonal way, and there also may be color differences between males and females. Some of the color patterns in insects are probably highly adaptive, such as those for **protective coloration, warning coloration,** and **mimicry** (Figures 21-26 and 21-27).

Besides color, insects have other methods of protecting themselves. The cuticular exoskeleton affords good protection for many of them; some, such as stinkbugs, have repulsive odors and tastes; others protect themselves by a good offense, for many are very aggressive and can put up a good fight (for example, bees and ants); and still others are swift in running for cover when danger threatens.

Many insects practice chemical warfare in a variety of ingenious ways. Some repel an assault by virtue of their bad taste, odor, or poisonous properties, others use chemical exudates that mechanically prevent a predator from attacking. The caterpillars of some monarch butterflies (Figure 21-26) assimilate cardiac glycosides from certain species of milkweed (family Asclepiadaceae); this substance confers unpalatability on the butterflies after metamorphosis and induces vomiting in their predators. The bombardier beetle, on the other hand, produces an irritating spray that it aims accurately at attacking ants or other enemies.

A

B

C

Figure 21-27

Camouflage in insects. **A,** *Estigena pardalis* (order Lepidoptera) in Java resembles a dead leaf. **B,** Bizarre processes from the thorax of a treehopper from Mexico, *Sphongophorus* sp. (order Homoptera), masquerade as parts of the twig on which it feeds. **C,** Broken outlines and color of a katydid (*Dysonia* sp., order Orthoptera) in Costa Rica give it the appearance of the leaves on which it has been feeding.

Figure 21-28
Tumble bugs, or dung beetles, *Canthon pilularis* (order Coleoptera), chew off a bit of dung, roll it into a ball, and then roll it to where they will bury it in soil. One beetle pushes while the other pulls. Eggs are laid in the ball, and the larvae feed on the dung. Tumble bugs are black, an inch or less in length, and common in pasture fields.

BEHAVIOR AND COMMUNICATION

The keen sensory perceptions of insects make them extremely responsive to many stimuli. The stimuli may be internal (physiological) or external (environmental), and the responses are governed by both the physiological state of the animal and the pattern of nerve pathways traveled by the impulses. Many of the responses are simple, such as orientation toward or away from the stimulus, for example, attraction of a moth to light, avoidance of light by a cockroach, or attraction of carrion flies to the odor of dead flesh.

Much of the behavior of insects, however, is not a simple matter of orientation but involves a complex series of responses. A pair of tumble bugs, or dung beetles, chew off a bit of dung, roll it into a ball, and roll the ball laboriously to where they intend to bury it, after laying their eggs in it (Figure 21-28). The cicada slits the bark of a twig and then lays an egg in each of the slits. The female potter wasp *Eumenes* scoops up clay into pellets, carries them one by one to her building site, and fashions them into dainty little narrow-necked clay pots, into each of which she lays an egg. Then she hunts and paralyzes a number of caterpillars, pokes them into the opening of a pot, and closes up the opening with clay. Each egg, in its own protective pot, hatches to find a well-stocked larder of food.

Much of such behavior is innate; however, a great deal more learning is involved than was once believed. The potter wasp, for example, must learn where she has left her pots if she is to return to fill them with caterpillars one at a time. Social insects, which have been studied extensively, are capable of most of the basic forms of learning used by mammals. The exception is insight learning. Apparently insects, when faced with a new problem, cannot reorganize their memories to construct a new response.

Some insects can memorize and perform in sequence tasks involving multiple signals in various sensory areas. Worker honeybees have been trained to walk through mazes that involved five turns in sequence, using such clues as the color of a marker, the distance between two spots, or the angle of a turn. The same is true of ants. Workers of one species of *Formica* learned a six-point maze at a rate only two or three times slower than that of laboratory rats. The foraging trips of ants and bees often wind and loop about in a circuitous route, but once the forager has found food, the return trip is relatively direct. One investigator suggests that the continuous series of calculations necessary to figure the angles, directions, distance, and speed of the trip and to convert it into a direct return could involve a stopwatch, a compass, and integral vector calculus. How the insect does it is unknown.

Insects communicate with other members of their species by means of chemical, visual, auditory, and tactile signals. **Chemical signals** take the form of **pheromones,** which are substances secreted by one individual that affect the behavior or physiological processes of another individual. Pheromones include sex attractants, releasers of certain behavior patterns, trail markers, alarm signals, territorial markers, and the like. Like hormones, pheromones are effective in minute quantities. Social insects, such as bees, ants, wasps, and termites, can recognize a nestmate—or an alien in the nest—by means of identification pheromones. Pheromones determine caste in termites and to some extent in ants and bees. In fact, pheromones are probably a primary integrating force in populations of social insects. Many insect pheromones have been extracted and chemically identified.

Sound production and **reception** (phonoproduction and phonoreception) in insects have been studied extensively, and although a sense of hearing is not present in all insects, this means of communication is meaningful to those insects that use it. Sounds serve as warning devices, advertisement of territorial claims, or courtship songs. The sounds of crickets and grasshoppers seem to be concerned with courtship and aggression. Male crickets scrape the modified edges of the forewings together to produce their characteristic chirping. The long, drawn-out sound of the male cicada, a call to attract females, is produced by the vibrating membranes in a pair of organs on the ventral side of the basal abdominal segment.

There are many forms of **tactile communication,** such as tapping, stroking, grasping, and antennae touching, which evoke responses varying from recognition to recruitment and alarm. Certain kinds of flies, springtails, and beetles manufacture their own **visual signals** in the form of **bioluminescence.** The best known of the luminescent beetles are the fireflies, or lightning bugs (which are neither flies nor bugs, but beetles), in which the flash of light helps to locate a prospective mate. Each species has its own characteristic flashing rhythm produced on the ventral side of the last abdominal segments. The females flash an answer to the species-specific pattern to attract the males. This interesting "love call" has been adopted by species of *Photuris,* which prey on the male fireflies of other species they attract (Figure 21-29).

Figure 21-29
Firefly femme fatale, *Photuris versicolor,* eating a male *Photinus tanytoxus,* which she attracted with false mating signals.

Social Behavior

Insects rank very high in the animal kingdom in their organization of social groups, and cooperation within the more complex groups depends heavily on chemical and tactile communication. Social communities are not all complex, however. Some community groups are temporary and uncoordinated, as are the hibernating associations of carpenter bees or the feeding gatherings of aphids. Some are coordinated for only brief periods, and some cooperate more fully, such as the tent caterpillars *Malacosoma,* that join in building a home web and feeding net. However, all of these are open communities with limited social behavior.

In the true societies of some orders, such as the Hymenoptera (honeybees and ants) and Isoptera (termites), a complex social life is necessary for the perpetuation of the species. They involve all stages of the life cycle, the communities are usually permanent, all activities are collective, and there is reciprocal communication and division of labor. The society usually demonstrates polymorphism, or **caste** differentiation.

The honeybees have one of the most complex organizations in the insect world. Instead of lasting one season, their organization continues for a more or less indefinite period. As

Figure 21-30
Queen bee surrounded by her court. The queen is the only egg layer in the colony. The attendants, attracted by her pheromones, constantly lick her body. As food is transferred from these bees to others, the queen's presence is communicated throughout the colony.

many as 60,000 to 70,000 honeybees may live in a single hive. Of these, there are three castes: a single sexually mature female, or **queen;** a few hundred **drones,** which are sexually mature males; and the rest are **workers,** which are sexually inactive genetic females (Figure 21-30).

The workers take care of the young, secrete wax with which they build the six-sided cells of the honeycomb, gather the nectar from flowers, manufacture honey, collect pollen, and ventilate and guard the hive. One drone, sometimes more, fertilizes the queen during the mating flight, at which time enough sperm is stored in her spermatheca to last her a lifetime.

Castes are determined partly by fertilization and partly by what is fed to the larvae. Drones develop parthenogenetically from unfertilized eggs (and consequently are haploid); queens and workers develop from fertilized eggs (and thus are diploid; see haplodiploidy, p. 84). Female larvae that will become queens are fed royal jelly, a secretion from the salivary glands of the nurse workers. Royal jelly differs from the "worker jelly" fed

to ordinary larvae, but the components in it that are essential for queen determination have not yet been identified. Honey and pollen are added to the worker diet about the third day of larval life. Pheromones in the "queen substance," which is produced by the queen's mandibular glands, prevent the female workers from maturing sexually. Workers produce royal jelly only when the level of "queen substance" pheromone in the colony drops. This occurs when the queen becomes too old, dies, or is removed. Then the workers' ovaries develop, and they start enlarging a larval cell and feeding the larva the royal jelly that produces a new queen.

Honeybees have evolved an efficient system of communication by which, through certain body movements, their scouts inform the workers of the location and quantity of food sources (Figure 38-19, p. 769).

Termite colonies contain several castes, consisting of fertile individuals, both males and females, and sterile individuals (Figure 21-31). Some of the fertile individuals may have wings and may leave the colony, mate, lose their wings, and as **king** and **queen** start a new colony. Wingless fertile individuals may under certain conditions substitute for the king or queen. Sterile members are wingless and become **workers** and **soldiers.** Soldiers have large heads and mandibles and serve for the defense of the colony. As in bees and ants, extrinsic factors cause caste differentiation. Reproductive individuals and soldiers secrete inhibiting pheromones that pass throughout the colony to the nymphs through a mutual feeding process, called **trophallaxis,** so that they become sterile workers. Workers also produce pheromones, and if the level of "worker substance" or "soldier substance" falls, as might happen after an attack by marauding predators, for example, the next generation produces compensating proportions of the appropriate caste.

Ants also have highly organized societies. Superficially, they resemble termites, but they are quite different

Figure 21-31

A, Termite workers, *Reticulitermes flavipes* (order Isoptera), eating yellow pine. Workers are wingless sterile adults that tend the nest, care for the young, and so forth. **B,** Termite queen (*Macrotermes bellicosus* from Ghana) becomes a distended egg-laying machine. The queen and several workers and soldiers are shown here.

(belong to a different order) and can be distinguished easily. In contrast to termites, ants are usually dark in color, are hard bodied, and have a constriction between the thorax and abdomen.

In ant colonies the males die soon after mating and the queen either starts her own new colony or joins some established colony and does the egg laying. The sterile females are wingless workers and soldiers that do the work of the colony: gather food, care for the young, and protect the colony. In many larger colonies there may be two or three types of individuals within each caste.

Ants have evolved some striking patterns of "economic" behavior, such as making slaves, farming fungi, herd-

Figure 21-32

A, Ants attending treehopper nymphs on a jackfruit in Brazil. **B,** A weaver ant nest in Australia.

ing "ant cows" (aphids or other homopterans) (Figure 21-32A), sewing their nests together with silk (Figure 21-32B), and using tools.

INSECTS AND HUMAN WELFARE

BENEFICIAL INSECTS

Although most of us think of insects primarily as pests, humanity would have great difficulty in surviving if all insects were suddenly to disappear. Some of them produce useful materi-

als: honey and beeswax from bees, silk from silkworms, and shellac from a wax secreted by the lac insects. More important, however, insects are necessary for the cross-fertilization of many crops. Bees pollinate almost $10 billion worth of food crops per year in the United States alone, and this does not include pollination of forage crops for livestock or pollination by other insects.

Very early in their evolution, insects and flowering plants formed a relationship of mutual adaptations that have been to each other's advantage. Insects exploit flowers for food, and flowers exploit insects for pollination. Each floral development of petal and sepal arrangement is correlated with the sensory adjustment of certain pollinating insects. Among these mutual adaptations are amazing devices of allurements, traps, specialized structures, and precise timing.

Many predaceous insects, such as tiger beetles, aphid lions, ant lions, praying mantids, and lady bird beetles, destroy harmful insects. Some insects control harmful ones by parasitizing them or by laying their eggs where their young, when hatched, may devour the host. Dead animals are quickly consumed by maggots hatched from eggs laid in carcasses (Figure 21-33).

Insects serve as an important source of food for many birds, fish, and other animals.

HARMFUL INSECTS

Harmful insects include those that eat and destroy plants and fruits, such as grasshoppers, chinch bugs, corn borers, boll weevils, grain weevils, San Jose scale, and scores of others (Figure 21-34). Practically every cultivated crop has some insect pest. Lice, bloodsucking flies, warble flies, botflies, and many others attack humans or domestic animals or both. Malaria, carried by the *Anopheles* mosquito, is still one of the world's killers; mosquitos also transmit yellow fever and filariasis. Fleas carry plague, which at many times in history has almost wiped out whole human populations.

Figure 21-33
Fly maggots (order Diptera) feeding on a deer carcass.

A

B

C

Figure 21-34
Insect pests. **A,** Japanese beetles, *Popillia japonica* (order Coleoptera), are serious pests of fruit trees and ornamental shrubs. They were introduced into the United States from Japan in 1917. **B,** Walnut caterpillars, *Datana ministra* (order Lepidoptera), defoliating a hickory tree. **C,** Corn ear worms, *Heliothis zea* (order Lepidoptera). An even more serious pest of corn is the infamous corn borer, an import from Europe in 1908 or 1909.

The housefly is the vector of typhoid, as is the louse for typhus fever; the tsetse fly carries African sleeping sickness; and a bloodsucking bug, *Rhodnius,* is a carrier of Chagas' disease. In addition there is tremendous destruction of food, clothing, and property by weevils, cockroaches, ants, clothes moths, termites, and carpet beetles. Not the least of the insect pests is the bedbug, *Cimex,* a bloodsucking hemipterous insect that humans may have contracted, probably early in their evolution, from bats that shared their caves.

The gypsy moth, introduced into the United States in 1869 in an ill-advised attempt to breed a better silkworm, has spread throughout the northeast as far south as Virginia. It defoliates oak forests in years when there are outbreaks. In 1981, it defoliated 13 million acres in 17 northeastern states.

CONTROL OF INSECTS

Because all insects are an integral part of the ecological communities to which they belong, their total destruction would probably do more harm than good. Food chains would be disturbed, some of our favorite birds would disappear, and the biological cycles by which dead animal and plant matter disintegrates and returns to enrich the soil would be seriously impeded. The beneficial role of insects in our environment has often been overlooked, and in our zeal to control the pests we have sprayed the landscape indiscriminately with extremely effective "broad-spectrum" insecticides that eradicate the good, as well as the harmful, insects. We have also found, to our dismay, that many of the chemical insecticides persist in the environment and accumulate as residues in the bodies of animals higher up in the food chains. Furthermore, many strains of insects have developed a resistance to the insecticides in common use.

In recent years, methods of control other than chemical insecticides have been under intense investigation and experimentation. Economics, concern for the environment, and consumer demand are causing thousands of farmers across the United States to use alternatives to strict dependence on chemicals.

Several types of biological controls have been developed and are under investigation. All of these areas present problems but show great possibilities. One is the use of bacterial, viral, and fungal pathogens. A bacterium, *Bacillus thuringiensis,* is quite effective in control of lepidopteran pests (cabbage looper, imported cabbage worm, tomato worm, gypsy moth). Other strains of *B. thuringiensis* attack insects in other orders, and the species range of target insects is being widened by techniques of genetic engineering. Genes coding for the toxin produced by *B. thuringiensis* are also being introduced into other bacteria and even into the plants themselves, which makes the plants resistant to insect attack.

A number of viruses and fungi that have potential as insecticides have been isolated. Difficulties and expense in rearing these agents are being overcome in certain cases, and some are nearing commercial production.

Introduction of natural predators or parasites of the insect pests has met with some success. In the United States the vedalia beetle from Australia helps control the cottony-cushion scale on citrus plants, and numerous instances of control by use of insect parasites have been recorded.

Another approach to biological control is to interfere with the reproduction or behavior of insect pests with sterile males or with naturally occurring organic compounds that act as hormones or pheromones. Such research, although very promising, is slow because of our limited understanding of insect behavior and the problems of isolating and identifying complex compounds that are produced in such minute amounts. Nevertheless, pheromones will probably play an important role in biological pest control in the future.

A systems approach referred to as **integrated pest management** is being increasingly practiced. This involves integrated utilization of all possible, practical techniques to contain pest infestations at a tolerable level,

The sterile male approach has been used effectively in eradicating screwworm flies, a livestock pest. Large numbers of male insects, sterilized by irradiation, are introduced into the natural population; females that mate with the sterile flies lay infertile eggs.

for example, cultural techniques (resistant plant varieties, crop rotation, tillage techniques, timing of sowing, planting or harvesting, and others), use of biological controls, and sparing use of insecticides.

Classification of Class Insecta

Insects are divided into orders mainly on the basis of wing structure, mouthparts, and metamorphosis. Entomologists do not all agree on the names of the orders or on the limits of each order. Some choose to combine and others to divide the groups. However, the following synopsis of the orders is one that is rather widely accepted.

Order Protura (pro-tu´ra) (Gr. *protos,* first, + *oura,* tail). Minute (1 to 1.5 mm); no eyes or antennae; appendages on abdomen as well as thorax; live in soil and dark, humid places; slight, gradual metamorphosis.

Order Diplura (dip-lu´ra) (Gr. *diploos,* double, + *oura,* tail): **japygids.** Usually less than 10 mm; pale, eyeless; a pair of long terminal filaments or pair of caudal forceps; live in damp humus or rotting logs; development direct.

Order Collembola (col-lem´bo-la) (Gr. *kolla,* glue, + *embolon,* peg, wedge): **springtails** and **snow fleas.** Small (5 mm or less); no eyes; respiration by trachea or body surface; a springing organ folded under the abdomen for leaping; abundant in soil; sometimes swarm on pond surface film or on snowbanks in spring; development direct.

Order Thysanura (thy-sa-nu´ra) (Gr. *thysanos,* tassel, + *oura,* tail): **silverfish** and **bristletails.** Small to medium size; large eyes; long antennae; three long terminal cerci; live under stones and leaves and around human habitations; development direct.

Order Ephemeroptera (e-fem-er-op´ter-a) (Gr. *ephēmeros,* lasting but a day, + *pteron,* wing): **mayflies.** Wings membranous; forewings larger than hindwings; adult mouthparts vestigial; nymphs aquatic, with lateral tracheal gills.

Order Odonata (o-do-na´ta) (Gr. *odontos,* tooth, + *ata,* characterized by): **dragonflies, damselflies** (Figures 21-20B, and 21-25B). Large; membranous wings are long, narrow, net veined, and similar in size; long and slender body; aquatic nymphs with gills and prehensile labium for capture of prey.

Order Orthoptera (or-thop´ter-a) (Gr. *orthos,* straight, + *pteron,* wing): **grasshoppers** (Figure 21-4), **locusts, crickets, cockroaches, walking sticks** (Figure 21-9B), **praying mantids** (Figure 21-7). Wings, when present, with forewings thickened and hindwings folded like a fan under forewings; chewing mouthparts.

Order Dermaptera (der-map´ter-a) (Gr. *derma,* skin, + *pteron,* wing): **earwigs.** Very short forewings; large and membranous hindwings folded under forewings when at rest; biting mouthparts; forcepslike cerci.

Order Plecoptera (ple-kop´ter-a) (Gr. *plekein,* to twist, + *pteron,* wing): **stoneflies** (Figure 21-25A). Membranous wings; larger and fanlike hindwings; aquatic nymph with tufts of tracheal gills.

Order Isoptera (i-sop´ter-a) (Gr. *isos,* equal, + *pteron,* wing): **termites** (Figure 21-31). Small; membranous, narrow wings similar in size with few veins; wings shed at maturity; erroneously called "white ants"; distinguishable from true ants by broad union of thorax and abdomen; complex social organization.

Order Embioptera (em-bi-op´ter-a) (Gr. *embios,* lively, + *pteron,* wing): **webspinners.** Small; male wings membranous, narrow, and similar in size; wingless females; chewing mouthparts; colonial; make silk-lined channels in tropical soil.

Order Psocoptera (so-cop´ter-a) (Gr. *psoco,* rub away, + *pteron,* wing) **(Corrodentia): psocids, book lice, bark lice.** Body usually small, may be as large as 10 mm; membranous, narrow wings with few veins, usually held rooflike over abdomen when at rest; some wingless species; found in books, bark, bird nests, on foliage.

Order Zoraptera (zo-rap´ter-a) (Gr. *zōros,* pure, + *apterygos,* wingless): **zorapterans.** As large as 2.5 mm; membranous, narrow wings usually shed at maturity; colonial and termitelike.

Order Mallophaga (mal-lof´a-ga) (Gr. *mallos,* wool, + *phagein,* to eat): **biting lice.** As large as 6 mm; wingless; chewing mouthparts; legs adapted for clinging to host; live on birds and mammals.

Order Anoplura (an-o-plu´ra) (Gr. *anoplos,* unarmed, + *oura,* tail): **sucking lice**. Depressed body; as large as 6 mm; wingless; mouthparts for piercing and sucking; adapted for clinging to warm-blooded host; includes the head louse, body louse, crab louse, others.

Order Thysanoptera (thy-sa-nop´ter-a) (Gr. *thysanos,* tassel, + *pteron,* wing): **thrips**. Length 0.5 to 5 mm (a few longer); wings, if present, long, very narrow, with few veins, and fringed with long hairs; sucking mouthparts; destructive plant-eaters, but some feed on insects.

Order Hemiptera (he-mip´ter-a) (Gr. *hemi,* half, + *pteron,* wing) **(Heteroptera): true bugs**. Size 2 to 100 mm; wings present or absent; forewings with basal portion leathery, apical portion membranous; hindwings membranous; at rest, wings held flat over abdomen; piercing-sucking mouthparts; many with odorous scent glands; includes water scorpions, water striders (Figure 21-11), bedbugs, squash bugs, assassin bugs, chinch bugs, stinkbugs, plant bugs, lace bugs, others.

Order Homoptera (ho-mop´ter-a) (Gr. *homos,* same, + *pteron,* wing): **cicadas, aphids, scale insects, leafhoppers, treehoppers** (Figure 21-27B). (Often included as suborder under Hemiptera.) If winged, either membranous or thickened front wings and membranous hindwings; wings held rooflike over body; piercing-sucking mouthparts; all plant-eaters; some destructive; a few serving as source of shellac, dyes, and so on; some with complex life histories.

Order Neuroptera (neu-rop´ter-a) (Gr. *neuron,* nerve, + *pteron,* wing): **dobsonflies, ant lions, lacewings**. Medium to large size; similar, membranous wings with many cross veins; chewing mouthparts; dobsonflies with greatly enlarged mandibles in males, and with aquatic larvae; ant lion larvae (doodlebugs) make craters in sand to trap ants.

Order Coleoptera (ko-le-op´ter-a) (Gr. *koleos,* sheath, + *pteron,* wing): **beetles** (Figures 21-9A, 21-28, and 21-34A), **fireflies** (Figure 21-29), **weevils** (Figure 21-35D). The largest order of animals in the world; front wings (elytra) thick, hard, opaque; membranous hindwings folded under front wings at rest; mouthparts for biting and chewing; includes ground beetles, carrion beetles, whirligig beetles, darkling beetles, stag beetles, dung beetles, diving beetles, boll weevils, others.

Order Strepsiptera (strep-sip´ter-a) (Gr. *strepsis,* a turning, + *pteron,* wing): **stylops**. Females with no wings, eyes, or antennae; males with vestigial forewings and fan-shaped hindwings; females and larvae parasitic in bees, wasps, and other insects.

Order Mecoptera (me-kop´ter-a) (Gr. *mekos,* length, + *pteron,* wing): **scorpionflies**. Small to medium size; wings long, slender, with many veins; at rest, wings held rooflike over back; scorpion-like male clasping organ at end of abdomen; carnivorous; live in moist woodlands.

Order Lepidoptera (lep-i-dop´ter-a) (Gr. *lepidos,* scale, + *pteron,* wing): **butterflies and moths**. Membranous wings covered with overlapping scales, wings coupled at base; mouthparts a sucking tube, coiled when not in use; larvae (caterpillars) with chewing mandibles for plant eating, stubby prolegs on the abdomen, and silk glands for spinning cocoons; antennae knobbed in butterflies and usually plumed in moths (Figure 21.35).

Order Diptera (dip´ter-a) (Gr. *dis,* two, + *pteron,* wing): **true flies**. Single pair of wings, membranous and narrow; hindwings reduced to inconspicuous balancers (halteres); sucking mouthparts or adapted for sponging, lapping, or piercing; legless larvae (maggots); includes crane flies, mosquitos, moth flies, midges, fruit flies, flesh flies, houseflies, horseflies, botflies, blowflies, and many others.

Order Trichoptera (tri-kop´ter-a) (Gr. *trichos,* hair, + *pteron,* wing): **caddis flies**. Small, soft bodies; wings well veined and hairy, folded rooflike over hairy body; chewing mouthparts; aquatic larvae of many species construct cases of leaves, sand, gravel, bits of shell, or plant matter, bound together with secreted silk or cement; some make silk feeding nets attached to rocks in stream.

Order Siphonaptera (si-fon-ap´ter-a) (Gr. *siphon,* a siphon, + *apteros,* wingless): **fleas** (Figure 21-14). Small; wingless; bodies laterally compressed; legs adapted for leaping; no eyes; ectoparasitic on birds and mammals; larvae legless and scavengers.

Order Hymenoptera (hi-men-op´ter-a) (Gr. *hymen,* membrane, + *pteron,* wing): **ants, bees, wasps** (Figure 21-35C). Very small to large; membranous, narrow wings coupled distally; subordinate hindwings; mouthparts for biting and lapping up liquids; ovipositor sometimes modified into stinger, piercer, or saw (Figure 21-10); both social and solitary species, most larvae legless, blind, and maggotlike.

PHYLOGENY AND ADAPTIVE RADIATION

Insect fossils, although not abundant, have been found in numbers sufficient to give a general idea of the evolutionary history of insects. Although several groups of marine arthropods, such as trilobites, crustaceans, and xiphosurans, were present in the Cambrian period, the first terrestrial arthropods—the scorpions and millipedes—did not appear until the Silurian period. The first insects, which were wingless, date from the Devonian period. By the Carboniferous period, several orders of winged insects, most of which are now extinct, had appeared.

Most zoologists agree that the insects and myriapods share a number of important characteristics and that they probably evolved from a common ancestor (Figure 21-36). The ancestor probably had a head and trunk of many similar somites, a primitive character retained by the myriapods. Evolution of the insects involved specialization of the first three post-cephalic somites to become the locomotor segments (thorax) and a loss or reduction of appendages on the rest of the body (abdomen). The wingless apterygotes have traditionally been regarded as having the most primitive characteristics, but the subclass

Figure 21-35

A, *Papilio krishna* (order Lepidoptera) is a beautiful swallowtail butterfly from India. Members of the Papilionidae grace many areas of the world, both tropical and temperate, including North America. Compare the knobbed antennae with the plumed antennae in **B,** *Rothschildia jacobaea,* a saturniid moth from Brazil. *Hyalophora cecropia* is a common saturniid in North America. **C,** Paper wasp (order Hymenoptera) attending her pupae and larvae. **D,** *Curculio proboscideus,* the chestnut weevil, is a member of the largest family (Curculionidae) of the largest insect order (Coleoptera). This family includes many serious agricultural pests.

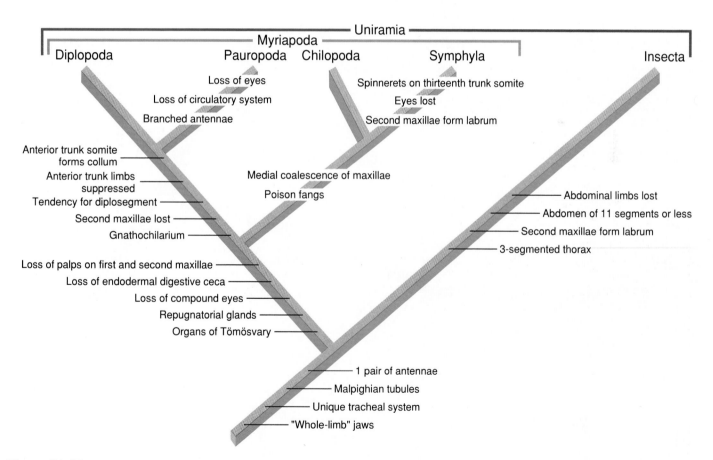

Figure 21-36

Cladogram showing hypothetical relationships of uniramians. Here the myriapods and insects are sister groups; therefore Diplopoda, Pauropoda, Chilopoda, and Symphyla would become subclasses under the Class Myriapoda. Organs of Tömösvary are unique sensory organs opening at the bases of the antennae, and repugnatorial glands, located on certain somites or legs, secrete an obnoxious substance for defense. The gnathochilarium is formed in diplopods and pauropods by fusion of the first maxillae, and the collum is the collarlike tergite of the first trunk segment. Formation of a labrum from the second maxillae has been sometimes considered evidence of relationship of symphylans and insects; it is viewed here as convergence. Outgroups for this cladogram would be the non-uniramian arthropod lineages.

Source: Modified from R. C. Brusca and G. J. Brusca, Invertebrates. *Sinauer Associates, Inc., Sunderland, MA, 1990.*

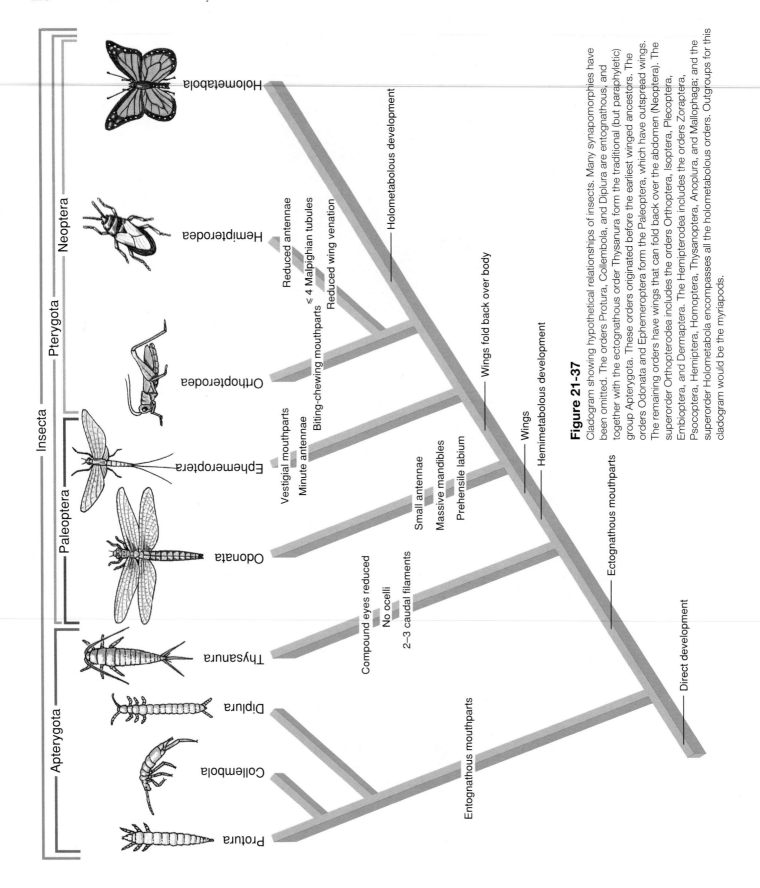

Figure 21-37

Cladogram showing hypothetical relationships of insects. Many synapomorphies have been omitted. The orders Protura, Collembola, and Diplura are entognathous, and together with the ectognathous order Thysanura form the traditional (but paraphyletic) group Apterygota. These orders originated before the earliest winged ancestors. The orders Odonata and Ephemeroptera form the Paleoptera, which have outspread wings. The remaining orders have wings that can fold back over the abdomen (Neoptera). The superorder Orthopterodea includes the orders Orthoptera, Isoptera, Plecoptera, Embioptera, and Dermaptera. The Hemipterodea includes the orders Zoraptera, Psocoptera, Hemiptera, Homoptera, Thysanoptera, Anoplura, and Mallophaga; and the superorder Holometabola encompasses all the holometabolous orders. Outgroups for this cladogram would be the myriapods.

Apterygota is apparently paraphyletic (Figure 21-37). Three of the apterygote orders (Diplura, Collembola, Protura) have their mandibles and first maxillae located deeply in pouches in the head, a condition known as **endognathy.** They share other primitive and derived characters, and there are many similarities between the endognathous insects and the myriapods. All other insects are **ectognathous,** including the wingless order Thysanura. Ectognathous insects do not have their mandibles and maxillae in pouches, and they share other synapomorphies. Endognathous and ectognathous insects form sister groups, and the Thysanura diverged from the common ancestor of ectognathous insects before the advent of flight,* which unites the remaining ectognathous orders.

The evolutionary origin of insect wings has long been a puzzle. The adaptive value of wings for flight is clear, but such structures do not spring into existence fully developed. They had to have evolved from earlier structures that would have been too small to support flight, but what possible value to their owners were "wings" too small to fly? One plausible suggestion is that proto-wings were used by their possessors for skimming across the surface of water, much as present-day stoneflies do.

The ancestral winged insect gave rise to three lines, which differed in their ability to flex their wings. Two of these (Odonata and Ephemeroptera) have outspread wings. The other line has wings that can fold back over the abdomen. It branched into three groups, all of which were present by the Permian period. One group with hemimetabolous metamorphosis, chewing mouthparts, and cerci includes the Orthoptera, Dermaptera, Isoptera, and Embioptera; another group with hemimetabolous metamorphosis and a tendency toward sucking mouthparts includes the Thysanoptera, Hemiptera, and Homoptera and perhaps also the Psocoptera, Zoraptera, Mallophaga, and Anoplura, although there is some disagreement among authorities about the last group. Insects with holometabolous metamorphosis have the most specialized life history, and these apparently form a clade.

The adaptive properties of the insects have been stressed throughout this chapter. The directions and ranges of their adaptive radiation, both structurally and physiologically, have been amazingly varied. Whether it be in the area of habitat, feeding adaptations, means of locomotion, reproduction, or general mode of living, the adaptive achievements of the insects are truly remarkable.

Summary

Members of the subphylum Uniramia have uniramous appendages and bear one pair of antennae, a pair of mandibles, and two pairs of maxillae (one pair of maxillae in millipedes) on the head. The tagmata are the head and trunk in the myriapods and head, thorax, and abdomen in the insects.

The Insecta is the largest class of the world's largest phylum. Insects are easily recognized by the combination of their tagmata and the possession of three pairs of thoracic legs.

The evolutionary success of insects is largely explained by several features allowing them to exploit terrestrial habitats, such as waterproofing their cuticle and other mechanisms to minimize water loss and the ability to become dormant during adverse conditions.

Most insects bear two pairs of wings on their thorax, although some have one pair and some are wingless. Wing movements in some insects are controlled by synchronous, direct flight muscles, which insert directly on the base of the wings in the thorax, whereas others have asynchronous, indirect flight muscles, which move the wings by changing the shape of the thorax.

Feeding habits vary greatly among insects, and there is an enormous variety of specialization of mouthparts reflecting the particular feeding habits of a given insect. They breathe by means of a tracheal system, which is a system of tubes that opens by spiracles on the thorax and abdomen. Excretory organs are malpighian tubules.

Sexes are separate in insects, and fertilization is usually internal. Almost all insects undergo metamorphosis during development. In hemimetabolous (gradual) metamorphosis, the larval instars are called nymphs, and the adult emerges at the last nymphal molt. In holometabolous (complete) metamorphosis, the last larval molt gives rise to a nonfeeding stage (pupa). A winged adult emerges at the final, pupal, molt. Both types of metamorphosis are hormonally controlled.

Insects are important to human welfare, particularly because they pollinate food and forage crop plants, control populations of other, harmful insects by predation and parasitism, and serve as food for other animals. Many insects are harmful to human interests because they feed on crop plants, and many are carriers of important diseases affecting humans and domestic animals.

Modern insects and myriapods show certain basic similarities, and the insects probably descended from a common ancestor resembling the myriapods in body form. The endognathous insects retain many primitive characters and perhaps most closely resemble the ancestral insect.

Adaptive diversity and the numbers of both species and individuals in the Insecta are enormous.

*Brusca, R. C., and G. J. Brusca. 1990. Invertebrates. Sunderland, Massachusetts, Sinauer Associates, Inc.

Review Questions

1. Distinguish the following from each other: Diplopoda, Chilopoda, Insecta.
2. What characteristics of insects distinguish them from *all* other arthropods?
3. Explain why indirect flight muscles can beat much more rapidly than direct flight muscles.
4. How do insects walk?
5. What are the parts of the insect gut, and what are the functions of each?
6. Describe three different types of mouthparts found in insects, and tell how they are adapted for feeding on different foods.
7. Describe the tracheal system of a typical insect and explain why it is able to function efficiently without the use of oxygen-carrying pigments in the hemolymph. Why would a tracheal system not be suitable for humans?

8. Describe the unique excretory system of insects. How is uric acid formed?
9. Describe the sensory receptors on insects for the various stimuli.
10. Explain the difference between holometabolous and hemimetabolous metamorphosis in insects, including the stages of each.
11. Describe the hormonal control of metamorphosis in insects, including the action of each hormone and where each is produced.
12. What is diapause, and what is its adaptive value?
13. Briefly describe three features that insects have evolved to avoid predation.
14. Describe and give an example of each of four ways insects can communicate with each other.

15. What are the castes found in honeybees and in termites, and what is the function of each?
16. What are the mechanisms of caste determination in honeybees and termites?
17. What is trophallaxis? What function(s) does it serve in termites?
18. Name several ways in which insects are beneficial to humans and several ways they are detrimental.
19. What are ways in which detrimental insects can be controlled? What is integrated pest management?
20. What are the most probable characteristics of the most recent common ancestor of the insects? What major lineages descended from it?

Selected References

See also general references for Part III, p. 626.

Berenbaum, M. R. 1995. Bugs in the system. Reading, Massachusetts, Addison-Wesley Publishing Company. *How insects impact human affairs. Well written for a wide audience, highly recommended.*

Blum, M. S. (ed). 1985. Fundamentals of insect physiology. New York, John Wiley & Sons. *Good, multi-authored text on insect physiology. Recommended.*

Borror, D. J., D. M. Delong, and C. A. Triplehorn. 1989. An introduction to the study of insects, ed. 6. Philadelphia, Saunders College Publishing. *A good entomology text.*

Chapman, R. F. 1982. The insects: structure and function, ed. 3. Cambridge, Massachusetts, Harvard University Press. *Comprehensive text on morphology and physiology of insects.*

Heinrich, B., and H. Esch. 1994. Thermo-regulation in bees. Amer. Sci. **82:**164–170. *Fascinating behavioral and physiological adaptations for increasing and decreasing body temperature allow bees to function in a surprisingly wide range of environmental temperatures.*

Hölldobler, B. H., and E. O. Wilson. 1990. The ants. Cambridge, Massachusetts, Harvard University Press. *The fascinating story of social organization in ants.*

Huber, F., and J. Thorson. 1985. Cricket auditory communication. Sci. Am. **253:**60–68 (Dec.). *Responses of female crickets to male songs using a spherical treadmill were studied. Insight has been gained into the auditory mechanism of crickets and the elements in the male's song that are recognized by the female.*

McMasters, J. H. 1989. The flight of the bumblebee and related myths of entomological engineering. Am. Sci. **77:**164–169. *There is a popular myth about an aerodynamicist who "proved" that a bumblebee cannot fly—but his assumptions were wildly wrong.*

Moffat, A. S. 1991. Research on biological pest control moves ahead. Science **252:**211–212. *A report on the current status of biological pest control, including the contributions of genetic engineering.*

Topoff, H. 1990. Slave-making ants. Am. Sci. **78:**520–528. *An amazing type of social parasitism in which certain species of ants raid the colonies of related species, abduct their pupae, then exploit them to do all the work in the host colony.*

Wootton, R. J. 1990. The mechanical design of insect wings. Sci. Am. **263:**114–120 (Nov.). *The ingenious architecture of insect wings and how they are adapted to flight.*

22

The Lesser Protostomes

Phylum Sipuncula, Phylum Echiura, Phylum Pogonophora, Phylum Pentastomida, Phylum Onychophora, Phylum Tardigrada

Some Evolutionary Experiments

During the Cambrian Period, about 535 to 530 million years ago, a most fertile time occurred in evolutionary history. For over 3 billion years before this time, evolution had forged little more than prokaryotes and protistans, and a vast number of potential environments awaited occupants. Then, within the space of a few million years, all of the major phyla of macroscopic invertebrates, and probably all of the smaller phyla, became established. This was the Cambrian explosion, the greatest evolutionary "bang" the world has known. In fact, the fossil record suggests that more phyla existed in the Paleozoic Era than exist now, but some disappeared during major extinction events that punctuated the evolution of life on earth. The greatest of these disruptions was the Permian extinction about 230 million years ago. Thus evolution has led

to many "experimental models." Some of these models failed because they were unable to survive in changing conditions. Others gave rise to abundant and dominant species and individuals that inhabit the world today. Still others radiated but little, with small numbers of species persisting, while others were formerly more abundant but are now in decline.

The great evolutionary flow that began with the appearance of the coelom and led to the three huge phyla of molluscs, annelids, and arthropods produced other lines as well. Those that have survived are small and lack great economic and ecological importance; they are usually grouped together as "lesser protostomes." They probably diverged at different times from different ancestors, but in all likelihood each is phylogenetically close to the annelids or arthropods or both. ■

THE LESSER PROTOSTOMES

This chapter includes a brief discussion of six phyla whose positions in the phylogenetic lines of the animal kingdom are somewhat problematical, as are their relationships to each other. The coelomate, protostome ancestors that eventually produced the three major phyla—Mollusca, Annelida, and Arthropoda—also produced a number of other lines. Some are now extinct, whereas others, although small in number of species and marked by very little evolutionary divergence within each phylum, have survived.

Three of the phyla, Sipuncula, Echiura, and Pogonophora, are benthic (bottom-dwelling) marine worms that may be phylogenetically close to the annelids. The first two have a variety of proboscis devices used in burrowing and food gathering. The pogonophores live in tubes, mostly in deep-sea mud, have long anterior tentacles, and lack a digestive tract. The Pentastomida, Onychophora, and Tardigrada have sometimes been grouped together and called the pararthropods because they have unjointed limbs with claws (at some stage) and a cuticle that undergoes molting, suggesting that they share an ancestor with the arthropods. The Pentastomida are entirely parasitic; the Onychophora are terrestrial but are limited to damp areas; the Tardigrada are found in marine, freshwater, and terrestrial habitats.

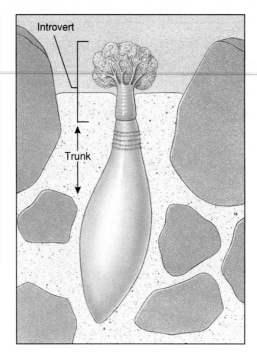

Figure 22-1
Themiste, a sipunculan.

PHYLUM SIPUNCULA

The phylum Sipuncula (sigh-pun´kyu-la) (L. *sipunculus,* little siphon) consists of benthic marine worms, predominantly littoral or sublittoral. They live sedentary lives in burrows in mud or sand, occupy borrowed snail shells, or live in coral crevices or among vegetation. Some species construct their own rock burrows by chemical and perhaps mechanical means. More than half the species are restricted to tropical zones. Some are tiny, slender worms, but the majority range from 15 to 30 cm in length. Some of them are commonly known as "peanut worms" because, when disturbed, they can contract to a peanut shape (Figure 22-1).

Sipunculans have no segmentation or setae. They are most easily recognized by a slender retractile **introvert,** or **proboscis,** that is continually and rapidly being run in and out of the anterior end. The walls of the **trunk** are muscular. When the introvert is everted, the mouth can be seen at its tip surrounded by a crown of ciliated tentacles. Undisturbed sipunculans usually extend the anterior end from the burrow or hiding place and

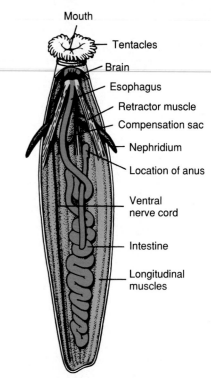

Figure 22-2
Internal structure of *Sipunculus.*

stretch out the tentacles to explore and feed. They are largely deposit feeders living on organic matter collected in mucus on the tentacles and moved to the mouth by ciliary action. The introvert is extended by hydrostatic pressure produced by contraction of the body-wall muscles against the coelomic fluid. The lumen of the hollow tentacles is not connected to the coelom but rather to one or two blind, tubular compensation sacs that lie along the esophagus (Figure 22-2). The sacs receive the fluid from the tentacles when the introvert is retracted. Retraction is effected by special retractor muscles. The surface of the introvert is often rough because of surface spines, hooks, or papillae.

There is a large, fluid-filled coelom traversed by muscle and connective tissue fibers. The digestive tract is a long tube that doubles back on itself to end in the anus near the base of the introvert (Figure 22-2). A pair of large nephridia opens to the outside to expel waste-filled coelomic amebocytes; the nephridia also serve as gonoducts. Circulatory and respiratory systems are lacking, but the

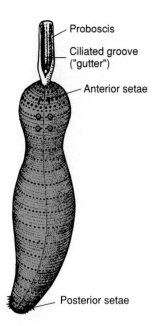

Figure 22-3
Echiurus, an echiurian common on both Atlantic and Pacific coasts of North America. The shape of the proboscis lends them the common name of "spoon worms."

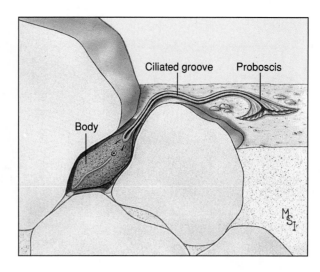

Figure 22-4
Bonellia (phylum Echiura) is a detritus feeder. Lying in its burrow, it explores the surface with its long proboscis, which picks up organic particles and carries them along a ciliated groove to the mouth.

coelomic fluid contains red corpuscles that contain a respiratory pigment, hemerythrin, used in the transportation of oxygen. The nervous system has a bilobed cerebral ganglion just behind the tentacles and a ventral nerve cord extending the length of the body. The sexes are separate. Permanent gonads are lacking, and ovaries or testes develop seasonally in the connective tissue covering the origins of one or more of the retractor muscles. Sex cells are released through the nephridia. The larval form is usually a trochophore. Asexual reproduction also occurs by transverse fission, the posterior one-fifth of the parent constricting off to become the new individual.

There are approximately 330 species and 16 genera, which are placed by some authorities into four families. The best-known genera are probably *Sipunculus, Phascolosoma* (Gr. *phaskōlos,* leather bag, pouch, + *sōma,* body), *Aspidosiphon* (Gr. *aspidos,* shield, + *siphōn,* siphon), and *Golfingia* (named by E. R. Lankester in honor of an afternoon of golfing at St. Andrews, Scotland).

PHYLUM ECHIURA

The phylum Echiura (ek-ee-yur´a) (Gr. *echis,* viper, serpent, + *oura,* tail, + *ida,* pl. suffix) consists of marine worms that burrow into mud or sand or live in empty snail shells or sand dollar tests, rocky crevices, and so on. They are found in all oceans—most commonly in littoral zones of warm waters—but some have been found in polar waters and some have been dredged from depths of 2000 m. They vary in length from a few millimeters to 40 or 50 cm.

The echiurans have only about one-third as many species (140) as the sipunculans. There are two classes: Echiurida and Sactosomatida. Echiurida is much larger and includes two orders and five families.

The body of echiurans is cylindrical and somewhat sausage shaped (Figure 22-3). Anterior to the mouth is a flattened, extensible proboscis which, unlike that of the sipunculids, cannot be retracted into the trunk. Echiurids are often called "spoonworms" because of the shape of the contracted proboscis in some species. The proboscis, which contains the brain, is actually a cephalic lobe, probably homologous to the annelid prostomium. The proboscis has a ciliated groove leading to the mouth.

While the animal lies buried, the proboscis can extend out over the mud for exploration and deposit feeding (Figure 22-4). *Bonellia viridis* picks up very small particles and moves them along the proboscis by cilia; larger particles are moved by a combination of cilia and muscular action or by muscular action alone. Unwanted particles can be rejected along the route to the mouth. The proboscis is short in some forms and long in others. *Bonellia,* which is only 8 cm long, can extend its proboscis up to 2 m.

In some species sexual dimorphism is pronounced, with the female being much the larger of the two. *Bonellia* has an extreme sexual dimorphism, and the tiny male lives on the body of the female or in her nephridia. Determination of sex in *Bonellia* is very interesting. The free-swimming larvae are sexually undifferentiated. Those that settle on the proboscis of a female become males (1 to 3 mm long). About 20 males are usually found in a single female. Larvae that do not contact a female proboscis metamorphose into females. The stimulus for development into males is apparently a hormone produced by the female proboscis.

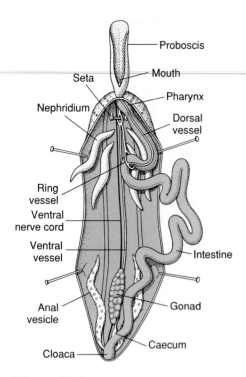

Figure 22-5
Internal anatomy of an echiuran.

One common form, *Urechis* (Gr. *oura,* tail, + *echis,* viper, serpent), lives in a U-shaped burrow in which it secretes a funnel-shaped mucous net. It pumps water through the net, capturing bacteria and fine particulate material in it. *Urechis* periodically swallows the food-laden net. *Lissomyema* (Gr. *lissos,* smooth, + *mys,* muscle) lives in empty gastropod shells in which it constructs galleries irrigated by rhythmical pumping of water and feeds on sand and mud drawn in by the irrigation process.

Cuticle and epithelium, which may be smooth or ornamented with papillae (Figure 22-3), cover the muscular body wall. There may be a pair of anterior setae or a row of bristles around the posterior end. The coelom is large. The digestive tract is long and coiled and terminates at the posterior end (Figure 22-5). A pair of anal sacs may have an excretory and osmoregulatory function. Most echiurans have a closed circulatory system with colorless blood but contain hemoglobin in coelomic corpuscles and certain body cells. Two to many nephridia serve mainly as gonoducts. A nerve ring runs around the pharynx and forward into the proboscis, and

Figure 22-6
A colony of giant beardworms at great depth near a hydrothermal vent along the Galápagos Trench, eastern Pacific Ocean.

Photo by J. F. Grassle, Woods Hole Oceanographic Institute.

there is a ventral nerve cord. There are no specialized sense organs.

The sexes are separate, with a single gonad in each sex. The mature sex cells break loose from the gonads and leave the body cavity by way of the nephridia, and fertilization is usually external.

Early cleavage and trochophore stages are very similar to those of annelids and sipunculans. The trochophore stage, which may last from a few days to 3 months, according to the species, is followed by gradual metamorphosis to the wormlike adult.

PHYLUM POGONOPHORA

The phylum Pogonophora (po´go-nof´e-ra) (Gr. *pogon,* beard, + *phora,* bearing), or beardworms, was entirely unknown before the twentieth century. The first specimens to be described were collected from deep-sea dredgings off the coast of Indonesia in 1900. They have since been discovered in several seas, including the western Atlantic off the U.S. eastern coast. Some 145 species have been described so far. We recognize two classes, Perviata and Vestimentifera, but some authorities consider the Vestimentifera a separate phylum. The usual length of perviatans is from 5 to 85 cm, with a diameter usually of less than a millimeter. Vestimentiferans, however, live around deepwater hydrothermal vents and grow much larger: up to 1.5 m long and 5 cm in diameter (Figure 22-6).

Among the most amazing animals found in the deep-water, Pacific rift communities (Chapter 40, p. 797) are the giant pogonophorans, *Riftia pachyptila,* much larger than any pogonophores reported before. The trophosome of other pogonophores is confined to the posterior part of the trunk, which is buried in sulfide-rich sediments, but the trophosome of *Riftia* occupies most of its large trunk. It has a much larger supply of hydrogen sulfide, enough to nourish its large body, in the effluent of the hydrothermal vents.

These elongated tube-dwelling forms have left no known fossil record. Their closest affinity seems to be to the annelids.

Most pogonophores live in the bottom ooze on the ocean floor, always below the intertidal zone and usually at depths of more than 200 m. This accounts for their delayed discovery, for they are obtained only by dredging. They are sessile animals that secrete very long chitinous tubes in which they live, probably extending the anterior end of the body only for feeding. The tubes are generally oriented upright in the bottom ooze. The tube is usually about the same length as the animal, which can move up or down inside the tube but cannot turn around.

The beardworm has a long, cylindrical body covered with cuticle. The body is divided into a short anterior **forepart;** a long, very slender **trunk;**

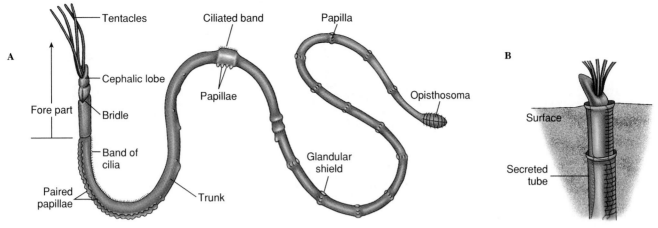

Figure 22-7

Diagram of a typical pogonophoran. **A,** External features. In life, the body is much more elongated than shown in this diagram. **B,** Position in tube.

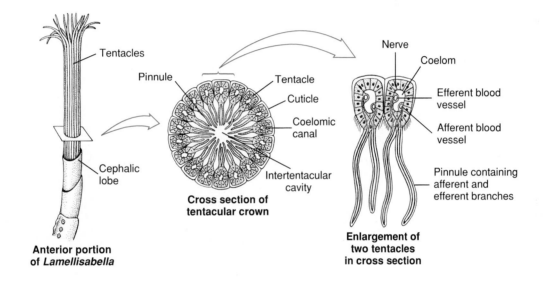

Figure 22-8

Cross section of tentacular crown of pogonophore *Lamellisabella*. Tentacles arise from ventral side of forepart at base of cephalic lobe. Tentacles (which vary in number in different species) enclose a cylindrical space, with the pinnules forming a kind of food-catching network. Food molecules may be absorbed into the blood supply of tentacles and pinnules.

and a small, segmented **opisthosoma** (Figure 22-7). At its anterior, the cephalic lobe bears from 1 to 260 long tentacles (the "beard" that gives the phylum its name), depending on the species. The tentacles are hollow extensions of the coelom and bear minute pinnules. For a part or all of their length the tentacles lie parallel with each other, enclosing a cylindrical intertentacular space into which the pinnules project (Figure 22-8).

The long trunk bears papillae and, about midway back, two rings of short toothed setae called **girdles,** which grip the tube wall, allowing the two halves of the body to contract or

extend independently in the tube. Posterior to the girdles, the trunk is very thin and easily broken when the animals are collected. In fact, the segmented tail end of the animal, or opisthosoma, was not found and described until after 1963! It is thicker than the trunk and has 5 to 23 short segments that bear setae.

Cuticle, epidermis, and circular and longitudinal muscles compose the body wall. The cuticle is similar in structure to that of annelids and sipunculans.

Pogonophores are remarkable in having no mouth or digestive tract, making their mode of nutrition a puzzling matter. They absorb some nutri-

ents dissolved in the seawater, such as glucose, amino acids, and fatty acids, through the pinnules and microvilli of their tentacles. Most of their energy, however, apparently is derived from a mutualistic association with chemoautotrophic bacteria. These bacteria oxidize hydrogen sulfide to provide the energy to produce organic compounds from carbon dioxide. The pogonophore bears the bacteria in an organ called the **trophosome,** which is derived embryonically from the midgut (all traces of the foregut and hindgut are absent in the adult).

Sexes are separate, with a pair of gonads and a pair of gonoducts in the trunk section. The cleavage is un-equal and atypical. It seems to be closer to radial than to spiral. The development of the apparent coelom is schizocoelic, not enterocoelic as was originally described. The worm-shaped embryo is ciliated but a poor swimmer. It is probably swept along by water currents until it settles.

Because the first specimens of Pogonophora that were dredged up lacked the segmented opisthosoma, Ivanov and other early workers, who believed they were working with whole specimens, described the coelom as trimeric (composed of three parts), like that of the hemichordates, and assumed that the organisms were deuterostomes. Ivanov also described the larval coelom as being trimeric. The later discovery of the segmented posterior end brought about some revision of the hypothesis. The adult coelom has proved to be polymeric, not trimeric. That fact and the schizocoelic development of the larva point toward an affinity with protostomes rather than deuterostomes. The pogonophore tubes were originally thought to resemble those of the hemichordate pterobranchs, but analysis of their amino acid and chitin content shows no relationship to the pterobranchs. Pogonophores have photoreceptor cells very similar to those of annelids (oligochaetes and leeches), and the structure of the cuticle, the makeup of the setae, and the segmentation of the opisthosoma all strongly suggest a relationship with the annelids. However, the phylogenetic position of the Pogonophora must be considered still unsettled until the embryology of more species of more than one family is studied.

Adaptive radiation has not been extensive. The chief areas of diversity are in the structure of the tentacular crown and the tube.

PHYLUM PENTASTOMIDA

The Pentastomida (pen-ta-stom´i-da) (Gr. *pente,* five, + *stoma,* mouth), or tongue worms, are a phylum of about 90 species of wormlike parasites of the respiratory system of vertebrates. The

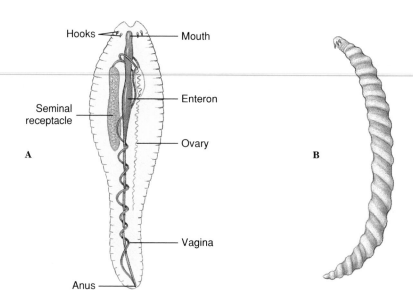

Figure 22-9

Two pentastomids. **A,** *Linguatula,* found in nasal passages of carvivorous mammals. Female is shown with some internal structures. **B,** Female *Armillifer,* a pentastomid with pronounced body rings. In parts of Africa and Asia, humans are parasitized by immature stages; adults (10 cm long or more) live in lungs of snakes. Human infection may occur from eating snakes or from contaminated food or water.

adults live mostly in the lungs of reptiles, such as snakes, lizards, and crocodiles, but one species, *Reighardia sternae,* lives in the air sacs of terns and gulls, and another, *Linguatula serrata* (Gr. *lingua,* tongue), lives in the nasopharynx of canines and felines (and occasionally humans). Although more common in tropical areas, they also occur in North America, Europe, and Australia.

The adults range from 1 to 13 cm in length. Transverse rings give their bodies a segmented appearance (Figure 22-9). The body is covered with a chitinous cuticle that is molted periodically during larval stages. The anterior end may bear five short protuberances (hence the name Pentastomida). Four of these bear claws. The fifth bears the mouth and two pairs of sclerotized hooks for attachment to the host tissues (Figure 22-10). There is a simple straight digestive system, adapted for sucking. The nervous system, similar to that of annelids and arthropods, has paired ganglia along the ventral nerve cord. The only sense organs appear to be papillae. There are no circulatory, excretory, or respiratory organs.

Sexes are separate, and the females are usually larger than the males. A female may produce several million

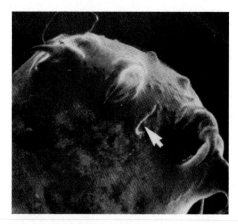

Figure 22-10

Anterior end of a pentastome. Note both the mouth (*arrow*), between the middle hooks, and the apical sensory papillae.

Courtesy J. Ubelaker.

eggs, which pass up the trachea of the host, are swallowed, and pass out with the feces. The larvae hatch out as oval, tailed creatures with four stumpy legs. Most pentastomid life cycles require an intermediate vertebrate host such as a fish, a reptile, or, rarely, a mammal, that is eaten by the definitive vertebrate host. After ingestion by the intermediate host, the larva penetrates the intestine, migrates randomly in the body, and finally metamorphoses into a nymph. After growth and several molts, the nymph finally becomes

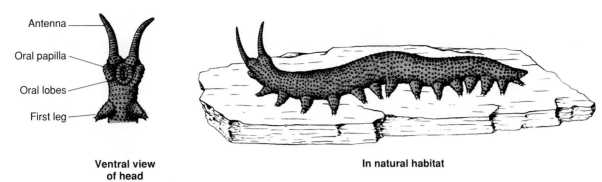

Antenna
Oral papilla
Oral lobes
First leg

**Ventral view
of head**

In natural habitat

Figure 22-11
Peripatus, a caterpillar-like onychophoran that has characteristics in common with both annelids and arthropods.

encapsulated and dormant. When eaten by the final host, the juvenile finds its way to the lung, feeds on blood and tissue, and matures.

Several species have been found encysted in humans, the most common being *Armillifer armillatus* (L. *armilla,* ring, bracelet, + *fero,* to bear), but usually they cause few symptoms. *Linguatula serrata* is a cause of nasopharyngeal pentastomiasis, or "halzoun," a disease of humans in the Middle East and India.

PHYLUM ONYCHOPHORA

Members of the phylum Onychophora (on-y-kof´o-ra) (Gr. *onyx,* claw, + *pherein,* to bear) are commonly called the "velvet worms," or "walking worms." They compose approximately 70 species of caterpillar-like animals, ranging from 1.4 to 15 cm in length. They live in rain forests and other moist, leafy habitats in tropical and subtropical regions and in some temperate regions of the Southern Hemisphere.

The fossil record of the onychophorans shows that they have changed little in their 500-million-year history. A fossil form, *Aysheaia,* discovered in the Burgess shale deposit of British Columbia and dating back to mid-Cambrian times, is very much like the modern onychophorans. Onychophorans have been of unusual interest to zoologists because they share so many characteristics with both the annelids and the arthropods. They have been called, a bit too hopefully perhaps, the "missing link" between the two

phyla. Onychophorans were probably far more common at one time than they are now. Today they are terrestrial and extremely retiring, coming out only at night or when the air is nearly saturated with moisture.

FORM AND FUNCTION

External Features

The onychophoran body is more or less cylindrical and shows no external segmentation except for the paired appendages (Figure 22-11). The skin is soft and velvety and is covered with a thin, flexible cuticle that contains protein and chitin. In structure and chemical composition it resembles arthropod cuticle; however, it never hardens like arthropod cuticle, and it is molted in patches rather than all at one time. The body is studded with tiny **tubercles,** some of which bear sensory bristles. The color may be green, blue, orange, dark gray, or black, and minute scales on the tubercles give the body an iridescent and velvety appearance. The head bears a pair of large **antennae,** each with an annelid-like eye at the base (Figure 22-11). The ventral mouth has a pair of clawlike **mandibles** and is flanked by a pair of **oral papillae** which can expel a defensive secretion.

The **unjointed legs** are short, stubby, and clawed. Onychophorans crawl by passing waves of contraction from anterior to posterior. When a segment extends, the legs lift up and move forward. The legs are more ventrally located than are the parapodia of annelids.

Internal Features

The body wall is muscular like that of the annelids. The body cavity is a **hemocoel,** imperfectly divided into compartments, or sinuses, much like those of the arthropods. **Slime glands** on each side of the body cavity open on the oral papillae. When disturbed by a predator, the animal can eject from the slime glands two streams of a sticky substance that rapidly hardens.

The mouth, surrounded by lobes of skin, contains a dorsal tooth and a pair of lateral mandibles for grasping and cutting prey. There is a muscular pharynx and a straight digestive tract (Figure 22-12). Most velvet worms are predaceous, feeding on caterpillars, insects, snails, worms, and the like. Some onychophorans live in termite nests and feed on termites.

Each segment contains a pair of **nephridia,** each nephridium with a vesicle, ciliated funnel and duct, and nephridiopore opening at the base of a leg. Absorptive cells in the midgut excrete crystalline uric acid, and certain pericardial cells function as nephrocytes, storing excretory products taken from the blood.

For respiration there is a **tracheal system** that ramifies to all parts of the body and communicates with the outside by many openings, or **spiracles,** scattered all over the body. They cannot close their spiracles to prevent water loss, so although the tracheae are efficient, the animals are restricted to moist habitats. The tracheal system is somewhat different from that of arthropods and probably has evolved independently.

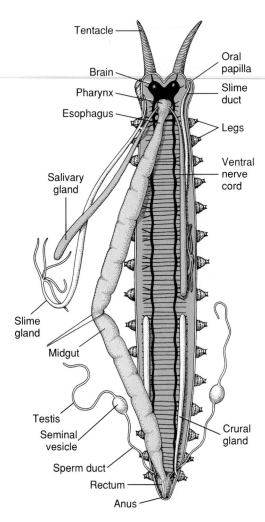

Figure 22-12
Internal anatomy of an onychophoran.

Figure 22-13
Scanning electron micrograph of an aquatic tardigrade, *Pseudobiotus*.

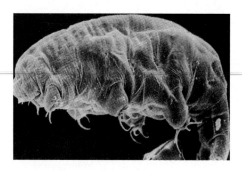

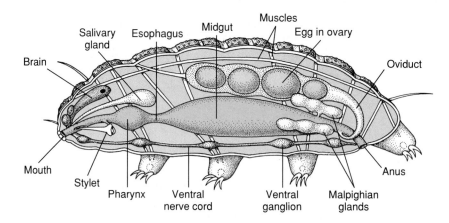

Figure 22-14
Internal anatomy of a tardigrade.

The open circulatory system has, in the pericardial sinus, a dorsal, tubular heart with a pair of ostia in each segment.

There are a pair of cerebral ganglia with connectives and a pair of widely separated nerve cords with connecting commissures. The brain gives off nerves to the antennae and head region, and the cords send nerves to the legs and body wall. Sense organs include pigment cup ocelli, taste spines around the mouth, tactile papillae on the integument, and hygroscopic receptors that orient the animal toward water vapor.

Onychophorans are dioecious, with paired reproductive organs. The males usually deposit their sperm in spermatophores in the female seminal receptacle. The male deposits the spermatophores on the female's back,

which may accumulate a number of them. White blood cells dissolve the skin beneath the spermatophores. The sperm can then enter the body cavity and migrate in the blood to the ovaries to fertilize the eggs. Onychophorans may be oviparous, ovoviviparous, or viviparous. Only two Australian genera are oviparous, laying shell-covered eggs in moist places. In all other onychophorans the eggs develop in the uterus, and living young are produced. In some species there is a placental attachment between mother and young (viviparous); in others the young develop in the uterus without attachment (ovoviviparous).

Phylum Tardigrada

Tardigrada (tar-di-gray´da) (L. *tardus,* slow, + *gradus,* step), or "water bears," are minute organisms usually less than a millimeter in length. Most of the 300 to 400 species are terrestrial forms that live in the water film surrounding mosses and lichens. Some live in freshwater algae or mosses or in the bottom debris, and a few are marine, inhabit-

ing the interstitial spaces between sand grains, in both deep and shallow seawater. They share many characteristics with the arthropods.

They have an elongated, cylindrical, or a long oval body that is unsegmented. The head is merely the anterior part of the trunk. The trunk bears four pairs of short, stubby, unjointed legs, each armed with four to eight claws (Figure 22-13). The body is covered by a nonchitinous cuticle that is molted along with the claws and buccal apparatus four or more times in the life history. Cilia are absent. Common American genera are *Macrobiotus* (Gr. *makros,* large, + *biotos,* life), *Echiniscus* (Gr. *echinos,* hedgehog, + *iskos,* dim. suffix), and *Hypsibius* (Gr. *hypsos,* high height, + *bios,* life).

The mouth opens into a buccal tube that empties into a muscular pharynx that is adapted for sucking (Figure 22-14). Two needlelike stylets flanking the buccal tube can be protruded through the mouth. The stylets pierce the cellulose walls of plant cells, and the pharynx sucks in the liquid contents. Some tardigrades

Figure 22-15
Scanning electron micrograph of the highly ornate egg of the tardigrade, *Macrobiotus hufelandii.*

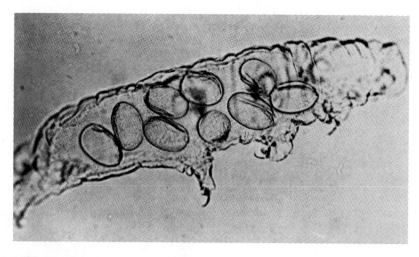

Figure 22-16
Molted cuticle of a tardigrade, containing a number of fertilized eggs.

suck the body juices of nematodes, rotifers, and other small animals. Some, such as *Echiniscus,* expel feces when molting, leaving the feces in the discarded cuticle. At the junction of the stomach and rectum, three glands, thought to be excretory and often called malpighian tubules, empty into the digestive system.

Most of the body cavity is a hemocoel, with the true coelom restricted to the gonadal cavity. There are no circulatory or respiratory systems, gaseous exchange occurring through the body surface.

The muscular system consists of a number of long muscle bands. Circular muscles are absent, but the hydrostatic pressure of the body fluid may act as a skeleton. Being unable to swim, the water bear creeps about with apparent awkwardness, clinging to the substrate with its claws.

The brain is large and covers most of the dorsal surface of the pharynx. Circumpharyngeal connectives link it to the subpharyngeal ganglion, from which the double ventral nerve cord extends posteriorly as a chain of four ganglia.

Sexes are separate in tardigrades. In some freshwater and moss-dwelling species, males are unknown and parthenogenesis seems to be the rule. In marine species, however, males and females occur with approximately equal frequency. Eggs of some species are highly ornate (Figure 22-15). Egg laying, like defecation, apparently occurs only at molting, when the volume of coelomic fluid is reduced. Females of some species deposit the eggs in the molted cuticle (Figure 22-16). Males gather around the old cuticle and shed sperm into it. Fertilization in other species is internal but only at the time of molting.

Cleavage is holoblastic but atypical, and a stereogastrula is formed. Six pairs of coelomic pouches arise from the gut, but all except the last pair disaggregate to form the buccal apparatus, pharynx, and body musculature. The last pair fuses to form the gonad. Thus the gonocoel (which is enterocoelic) is the only true coelom left in the adult. Development is direct.

One of the most intriguing features of terrestrial tardigrades is their capacity to enter a state of suspended animation, called cryptobiosis (formerly called anabiosis), during which metabolism is virtually imperceptible; the organism can withstand harsh environmental conditions. Under gradual drying conditions, the water content of the body decreases from 85% to only 3%, movement ceases, and the body becomes barrel shaped. In a cryptobiotic state tardigrades can resist temperature extremes, ionizing radiation, oxygen deficiency, and other adverse conditions and may survive for years. Activity resumes when moisture is again available.

PHYLOGENY

The early embryological development of sipunculans, echiurans, and annelids is almost identical, showing a very close relationship among the three. It is also similar to molluscan development. Some authors group the four phyla into a supraphyletic assemblage called the "Trochozoa" because of the common possession of a trochophore larva. Other similarities, too, point to close relationship of the sipunculans to the echiurans and annelids, such as the nature of the nervous system and body wall. The sipunculans and echiurans are not metameric and thus are more primitive in that characteristic than annelids. They probably represent collateral evolutionary lines that branched from protoannelid stock before the origin of metamerism.

Several characters suggest relationship of the Pogonophora to the Annelida, as we noted previously.

The phylogenetic affinities of the Pentastomida are uncertain. They have some similarities to the Annelida. Their larval appendages and molting cuticle, however, are arthropod characteristics. Their larvae resemble tardigrade larvae. Most modern taxonomists align

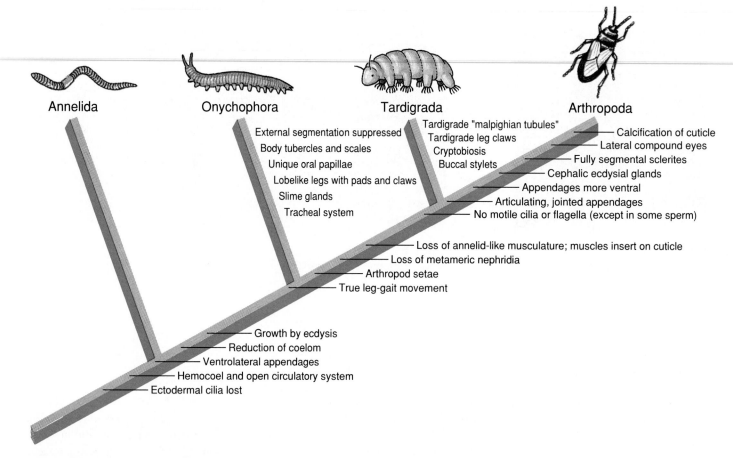

Figure 22-17

Cladogram depicting hypothetical relationships of Onychophora and Tardigrada to annelids and arthropods. The onychophorans diverged from the arthropod line after the development of such synapomorphies as hemocoel and growth by ecdysis, but they share several primitive characters with the annelids, such as the metameric arrangement of the nephridia. Note that the tracheal system of the onychophorans is not homologous to that of the arthropods but represents a convergence. The phylogenetic relationships of the other phyla covered in this chapter are too difficult to evaluate to permit construction of a cladogram that includes them. The Echiura and Sipuncula probably arose from the protoannelid line but diverged before the origin of metamerism.

Source: Modified from R. C. Brusca and G. J. Brusca, Invertebrates. *Sinauer Associates, Inc., Sunderland, MA, 1990.*

them with the arthropods, however, and evidence is accumulating that they are most closely related to the crustacean subclass Branchiura (p. 395). This evidence includes similarities in morphology of their sperm and in base sequences of ribosomal RNA. If the pentastomids really are close to the branchiurans, then their status as a phylum should be revoked, and they should be classified as crustacean arthropods.

Onychophorans share a number of characteristics with the annelids: metamerically arranged nephridia, muscular body wall, pigment cup ocelli, and ciliated reproductive ducts. Characteristics shared with arthropods include the cuticle, tubular heart and hemocoel with open circulatory system, presence of tracheae (probably not homologous), and large size of the brain. Unique characteristics include oral papillae, slime glands, body tubercles, and suppression of external segmentation.

Some authors believe the onychophorans should be included with the arthropods, but that would involve redefining the phylum Arthropoda. Manton* recommended placing the Onychophora with the myriapods and insects in the phylum Uniramia. However, most authors believe that the differences seem to warrant keeping them in a separate phylum (Figure 22-17).

The affinities of tardigrades are among the most puzzling of all animal groups. They have some similarities to rotifers, particularly in their reproduction and their cryptobiotic tendencies, and some authors have called them pseudocoelomates. Their embryogenesis, however, would seem to put them among the coelomates. The enterocoelic origin of the meso-

derm is a deuterostome characteristic. Other authors identify several important synapomorphies that suggest an ancestor in common with the arthropods (Figure 22-17).

Recent discoveries of Cambrian fossil pentastomids and tardigrades and additional fossil onychophorans lend strong support that these small phyla arose during the Cambrian explosion, just as did the major phyla. Because this period was long before terrestrial vertebrates evolved, the identity of the hosts for Cambrian pentastomids remains enigmatic; some authors have suggested that they might have been conodonts (see p. 494).

*Manton, S. M. 1977. The Arthropoda: habits, functional morphology, and evolution. Oxford, England, Clarendon Press.

Summary

The six small phyla covered in this chapter are grouped together here for convenience. The Sipuncula and the Echiura apparently diverged from the protostome line before the advent of metamerism, but the Pogonophora and the Annelida may share a metameric ancestor. The Onychophora and Tardigrada probably share an ancestor with the Arthropoda (Figure 22-17). The Pentastomida have certain arthropod-like characteristics, and available evidence indicates an ancestor shared with the crustacean subclass Branchiura. If so, they should lose their status as a phylum and be classified with the Crustacea.

Sipunculans are small, burrowing marine worms with an eversible introvert at their anterior end. The introvert bears tentacles, which they use for deposit-feeding. They are not metameric.

Echiurans are more diverse than sipunculans, but there are a smaller number of species. They are also burrowing marine worms, and most are deposit-feeders, with a proboscis anterior to their mouth. They also are not metameric.

Pogonophorans live in tubes on the deep ocean floor, and they are metameric. They have no mouth or digestive tract but apparently absorb some nutrient by the crown of tentacles at their anterior end. Much of their energy is due to chemoautotrophy of bacteria in their trophosome.

The Pentastomida are wormlike parasites in the lungs and nasal passages of carnivorous vertebrates. They are related to the arthropods.

The Onychophora are caterpillar-like animals found in humid, mostly tropical habitats. They are metameric and crawl by means of a series of unjointed, clawed appendages. They share characteristics with both annelids and arthropods.

Tardigrades are minute animals, mostly terrestrial, living in the water film that surrounds mosses and lichens. They have eight unjointed legs and a nonchitinous cuticle. Their chief body cavity is a hemocoel, as in arthropods. They can undergo cryptobiosis, withstanding adverse conditions for long periods.

Review Questions

1. Distinguish the following phyla from each other, and describe each one's habitat: Sipuncula, Echiura, Pogonophora, Pentastomida, Onychophora, Tardigrada.
2. What do the members of each of the aforementioned groups feed upon?
3. What is the evidence that the Sipuncula and Echiura diverged from the protostome line before the origin of the annelids? Why are these phyla considered closely related?
4. What is the largest pogonophoran known? Where is it found, and how is it nourished?
5. Briefly describe the life cycle of a typical pentastomid.
6. The onychophorans have been regarded by some biologists as a "missing link" between annelids and arthropods. What is evidence for this hypothesis? What is evidence that onychophorans form a distinct phylum in their own right?
7. What is the survival value of cryptobiosis in tardigrades?
8. How does the introvert of sipunculans differ from the proboscis of echiurans?
9. The Onychophora and Tardigrada seem to be related to Arthropoda, but some workers regard the Pentastomida as arthropods in fact. Why?

Selected References

See also general references for Part III, p. 626.

Childress, J. J., H. Felbeck, and G. N. Somero. 1987. Symbiosis in the deep sea. Sci. Am. **256**:114–120 (May). *The amazing story of how the animals around deep-sea vents, including* Riftia pachyptila, *manage to absorb hydrogen sulfide and transport it to their mutualistic bacteria. For most animals, hydrogen sulfide is highly toxic.*

Crowe, J. H., and A. F. Cooper, Jr. 1971. Cryptobiosis. Sci. Am. **225**:30–36 (Dec.). *Cryptobiotic nematodes, rotifers, and tardigrades can withstand adverse conditions of astonishing rigor, yet perceptible metabolism continues in their state of suspended animation.*

Gould, S. J. 1995. Of tongue worms, velvet worms, and water bears. Natural History **104**(1):6–15. *Intriguing essay on affinities of Pentastomida, Onychophora, and Tardigrada and how they, along with larger phyla, were products of the Cambrian explosion.*

Haugerud, R. E. 1989. Evolution in the pentastomids. Parasitol. Today **5**:126–132. *Much remains to be learned of this puzzling group, but there is strong evidence of its crustacean affinities.*

Rice, M. E., and M. Todorovic (eds). 1975. Proceedings of the International Symposium on the biology of the Sipuncula and Echiura, 2 vols. Washington,

D.C., National Museum of Natural History. *A series of technical articles, but much of interest for further reading on these two phyla.*

Southward, E. C. 1975. Fine structure and phylogeny of the Pogonophora. In E. J. W. Barrington and R. P. S. Jefferies, (eds). Protochordates. London, Zoological Society of London, no. 36.

23

The Lophophorate Animals

Phylum Phoronida
Phylum Ectoprocta (Bryozoa)
Phylum Brachiopoda

Animal Weeds

When a plant grows somewhere that humans do not want it to grow, we call it a "weed." Sessile organisms that settle and grow on pilings, boat hulls, pipes, cables, and other structures placed there by humans are referred to as "fouling organisms." Since we do not want them there, we might call them the marine equivalent of terrestrial weeds. And like terrestrial weeds, they are very persistent.

Members of one of the phyla covered in this chapter, the Ectoprocta, are among the most important fouling organisms, especially on ship and boat hulls. Fouling of boat hulls causes turbulence as the vessel proceeds through the water, and the increased resistance decreases the speed of the vessel and increases its fuel consumption. It is costly to scrape the organisms from the boat hull either in dry dock or in the water. Consequently, boat hulls have often been painted with paints containing a toxic antifouling agent. One of the most effective of

these is a substance called tributyl tin (TBT). After application in the paint, TBT is released at a low rate over a long period of time, making repainting necessary less frequently. Unfortunately, release of the toxin into the seawater, especially in harbors and basins where many boats were concentrated, has catastrophic effects on many organisms, particularly bivalves, which concentrate the compound in their tissues.

In 1988 the U.S. Congress passed a law severely restricting use of TBT in antifouling paints, and the problem should be considerably alleviated by this law.

Ironically, eighteenth-century naturalists included ectoprocts (along with cnidarians and some others) in a group designated "zoophytes," meaning "animal plants." These early workers thought that zoophytes were akin to both animals and plants. Comparing ectoprocts with plant weeds gives a new meaning to the term zoophytes. ■

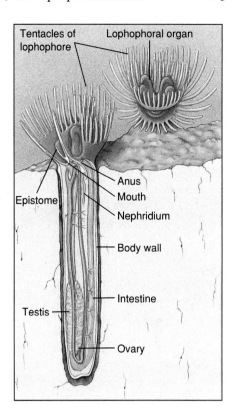

Figure 23-1
Internal structure of *Phoronis* (phylum Phoronida), in diagrammatic vertical section.

THE LOPHOPHORATES

The Phoronida are wormlike marine forms that live in secreted tubes in sand or mud or attached to rocks or shells. The Ectoprocta are minute forms, mostly colonial, whose protective cases often form encrusting masses on rocks, shells, or plants. The Brachiopoda are bottom-dwelling marine forms that superficially resemble molluscs because of their bivalved shells.

One might wonder why these three apparently different types of animals are lumped together in a group called lophophorates. Actually they have more in common than first appears. They are all coelomate; all have some deuterostome characteristics; all are sessile; and none has a distinct head. But other phyla share these characteristics. What really sets this group apart from other phyla is the common possession of a ciliary feeding device called a **lophophore** (Gr. *lophos,* crest or tuft, + *phorein,* to bear).

A lophophore is a unique arrangement of ciliated tentacles borne on a ridge (a fold of the body wall), which surrounds the mouth but not the anus. The lophophore with its crown of tentacles contains within it an extension of the coelom, and the thin, ciliated walls of the tentacles are not only an efficient feeding device but also serve as a respiratory surface for exchange of gases between the environmental water and the coelomic fluid. The lophophore can usually be extended for feeding or withdrawn for protection.

In addition, all three phyla have a U-shaped alimentary canal, with the anus placed near the mouth but outside the lophophore. The coelom is primitively divided into three compartments, the **protocoel,** the **mesocoel** and the **metacoel,** and the mesocoel extends into the hollow tentacles of the lophophore. The protocoel, where present, forms a cavity in a flap over the mouth, the **epistome.** The portion of the body that contains the mesocoel is known as the **mesosome,** and that containing the metacoel is the **metasome.** Members of all three phyla have a free-swimming larval stage but are sessile as adults.

PHYLUM PHORONIDA

The phylum Phoronida (fo-ron'i-da) (L. *Phoronis,* in mythology, surname of Io, who was turned into a white heifer) contains approximately 10 species of small, wormlike animals that live on the bottom of shallow coastal waters, especially in temperate seas. They range from a few millimeters to 30 cm in length. Each worm secretes a leathery or chitinous tube in which it lies free, but which it never leaves. The tubes may be anchored singly or in a tangled mass on rocks, shells, or pilings or buried in the sand. They thrust out the tentacles on the lophophore for feeding, but if the animal is disturbed, it can withdraw completely into its tube.

The lophophore consists of two parallel ridges curved in a horseshoe shape, the bend located ventrally and the mouth lying between the two ridges (Figure 23-1). The horns of the ridges often coil into twin spirals. Each ridge carries hollow ciliated tentacles, which, like the ridges themselves, are extensions of the body wall.

The cilia on the tentacles direct a water current toward a groove between the two ridges, which leads toward the mouth. Plankton and detritus caught in this current are carried by the cilia to the mouth. The anus lies dorsal to the mouth, outside the lophophore, flanked on each side by a nephridiopore (Figure 23-1). Water leaving the lophophore passes over the anus and nephridiopores, carrying away the wastes. Cilia in the stomach area of the U-shaped gut aid in food movement.

The body wall consists of cuticle, epidermis, and both longitudinal and circular muscles. The protocoel is present as a small cavity in the epistome; it connects to the mesocoel along the lateral aspects of the epistome. A septum separates the metacoel from the mesocoel. The phoronids have a closed system of contractile blood vessels but no heart; the red blood contains hemoglobin within nucleated cells. There is a pair of metanephridia. A nerve ring sends nerves to the tentacles and body wall; a single giant motor fiber lies in the epidermis; and an epidermal nerve plexus supplies the body wall and epidermis.

There are both monoecious (the majority) and dioecious species of Phoronida, and at least one species reproduces asexually. The cleavage pattern is radial, and development is regulative. Coelom formation is by a highly modified enterocoelous route, but the blastopore becomes the mouth. The free-swimming, ciliated larva, called an actinotroch, metamorphoses into the adult, which sinks to the bottom, secretes a tube, and becomes sessile.

Phoronopsis californica is a large, orange form about 30 cm long found along the west coast of the United States. *Phoronis architecta* is a smaller (approximately 12 cm long) Atlantic coast species that has a very wide distribution.

PHYLUM ECTOPROCTA (BRYOZOA)

The Ectoprocta (ek-to-prok'ta) (Gr. *ektos,* outside, + *proktos,* anus) have long been called bryozoans, or moss animals (Gr. *bryon,* moss, + *zōon,* animal), a term that originally included the Entoprocta also. However, because the entoprocts are pseudocoelomates and have the anus located within the tentacular crown, they are no longer classified with the ectoprocts, which, like the other lophophorates, are eucoelomate and have the anus outside the circle of tentacles. Some authors continue to use the name "Bryozoa" but now exclude the entoprocts from the group.

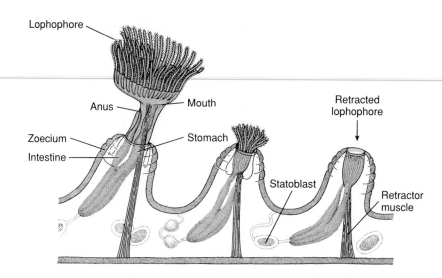

Figure 23-2

Small portion of freshwater colony of *Plumatella* (phylum Ectoprocta), which grows on the underside of rocks. These tiny individuals disappear into their chitinous zoecia when disturbed. Statoblasts are resistant capsules containing germinative cells.

Of the 4000 or so species of ectoprocts, few are more than 0.5 mm long. All are aquatic, both freshwater and marine, but are found largely in shallow waters. With very few exceptions they are colony builders. Ectoprocts have become diverse and abundant. They have left a rich fossil record since the Ordovician period. Marine forms today exploit all kinds of firm surfaces, such as shells, rocks, large brown algae, mangrove roots, and ship bottoms.

Each member of a colony lives in a tiny chamber, called a **zoecium,** which is secreted by its epidermis (Figure 23.2). Each individual, or **zooid,** consists of a feeding polypide and a case-forming cystid. The **polypide** includes the lophophore, digestive tract, muscles, and nerve centers. The **cystid** is the body wall of the animal, together with its secreted exoskeleton. The exoskeleton, or zoecium, may, according to the species, be gelatinous, chitinous, or stiffened with calcium and possibly also impregnated with sand. The shape may be boxlike, vaselike, oval (Figure 23-3A), or tubular.

Some colonies form limy encrustations on seaweed, shells, and rocks; others form fuzzy or shrubby growths on erect, branching colonies that look like seaweed (Figure 23-3B). Some ectoprocts might easily be mistaken for hydroids but can be distinguished under a microscope by the fact that they have an anus (Figure 23-2). In some freshwater forms the individuals are borne on finely branching stolons that form delicate tracings on the underside of rocks or plants. Other freshwater ectoprocts are embedded in large masses of gelatinous material. Although the zooids are minute, the colonies may be several centimeters in diameter, some encrusting colonies may be a meter or more in width (Figure 23-4), and erect forms may reach 30 cm or more in height. Freshwater ectoprocts may form mosslike colonies on the stems of plants or on rocks, usually in shallow ponds or pools. They may be able to slide along slowly on the object that supports them.

The polypide lives a type of jack-in-the-box existence, popping up to feed and then quickly withdrawing into its little chamber, which often has a tiny trapdoor (operculum) that shuts to conceal its inhabitant. To extend the tentacular crown, certain muscles contract, which increases the hydrostatic pressure within the body cavity and pushes the lophophore out by a hydraulic mechanism. Other muscles can contract to withdraw the crown to safety with great speed.

The lophophore ridge tends to be circular in marine ectoprocts (Figure

A

B

Figure 23-3

Colonies of marine ectoprocts. **A,** The zooids are extended in this lacy colony of *Triphyllozoon* sp. **B,** *Reteporella graffei* has upright, branching colonies.

B

Figure 23-5

A, Ciliated lophophore of *Electra pilosa,* a marine ectoproct. **B,** *Plumatella repens,* a freshwater bryozoan (phylum Ectoprocta). It grows on the underside of rocks and vegetation in lakes, ponds, and streams.

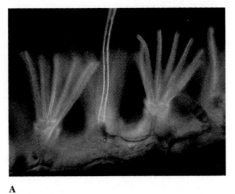

A

Figure 23-4

Skeletal remains of a colony of *Membranipora,* a marine encrusting form of Ectoprocta. Each little oblong zoecium is the calcareous former home of a tiny ectoproct.

23-5A) and U-shaped in freshwater species (Figure 23-5B). When feeding, the animal extends the lophophore and spreads the tentacles out into a funnel. Cilia on the tentacles draw water into the funnel and out between the tentacles. Food particles caught by cilia in the funnel are drawn into the mouth, both by the pumping action of the muscular pharynx and by the action of cilia in the pharynx. Undesirable particles can be rejected by reversing the ciliary action, by drawing the tentacles close together, or by retracting the whole lophophore into the zoecium. Digestion in the ciliated, U-shaped digestive tract appears to be extracellular for protein and starches and intracellular for fats.

Respiratory, vascular, and excretory organs are absent. Gaseous exchange is through the body surface, and since the ectoprocts are small, coelomic fluid is adequate for internal transport. Coelomocytes engulf and store waste materials. There are a ganglionic mass and a nerve ring around the pharynx, but no sense organs are present. A septum divides the mesocoel in the lophophore from the larger posterior metacoel. The protocoel and epistome are present only in the freshwater ectoprocts.

Pores in the walls between adjoining zooids permit exchange of materials by way of the coelomic fluid.

Most colonies contain only feeding individuals, but polymorphism occurs in some species. One type of modified zooid resembles a bird beak that snaps at small invading organisms that might foul a colony. Another type has a long bristle that sweeps away foreign particles.

Most ectoprocts are hermaphroditic. Some species shed eggs into the seawater, but most brood their eggs, some within the coelom and some externally in a special ovicell, which is a modified zoecium in which the embryo develops. Cleavage is radial but apparently mosaic. Little is known of mesoderm derivation, but there is no evidence of protostome characteristics. Larvae of nonbrooding species have a functional gut and swim about for a few months; larvae of brooding species do not feed and settle after a brief free-swimming existence. They attach to the substratum by mucopolysaccharide and protein secretions from an **adhesive sac,** then metamorphose to the adult form.

Ectoprocts reproduce asexually by budding and form colonies. Freshwater ectoprocts have another type of

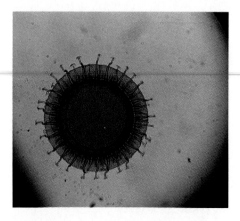

Figure 23-6
Statoblast of a freshwater ectoproct *Cristatella.*
This statoblast is about 1 mm in diameter and
bears hooked spines.

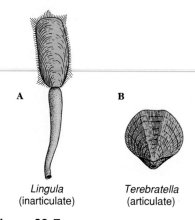

Lingula
(inarticulate)

Terebratella
(articulate)

Figure 23-7
Brachiopods. **A,** *Lingula,* an inarticulate
brachiopod that normally occupies a burrow.
The contractile pedicel can withdraw the body
into the burrow. **B,** An articulate brachiopod,
Terebratella. The valves have a tooth-and-socket
articulation, and a short pedicel projects through
the pedicel valve to attach to the substratum.

budding that produces **statoblasts**
(Figure 23-6), which are hard, resistant capsules containing a mass of
germinative cells. Statoblasts are
formed during the summer and fall.
When the colony dies in late autumn,
the statoblasts remain, and in spring
they give rise to new polypides and
eventually to new colonies.

PHYLUM BRACHIOPODA

The Brachiopoda (brak-i-op′o-da)
(Gr. *brachĭon,* arm, + *pous, podos,*
foot), or lamp shells, are an ancient
group. Although about 325 species
are now living, some 12,000 fossil
species, which once flourished in the
Paleozoic and Mesozoic seas, have
been described. Modern forms have
changed little from the early ones.
The genus *Lingula* (L. tongue) (Figure
23-7A) is probably the most ancient of
these "living fossils," having existed
virtually unchanged since Ordovician
times. Most modern brachiopod shells
range between 5 to 80 mm in length,
but some fossil forms reached 30 cm.

Brachiopods are attached, bottom-
dwelling, marine forms that mostly
prefer shallow water. Externally brachiopods resemble the bivalved molluscs in having two calcareous shell
valves secreted by the mantle. They
were, in fact, classed with the molluscs until the middle of the nineteenth century, and their name refers

to the arms of the **lophophore,**
which were thought homologous to
the mollusc foot. Brachiopods, however, have dorsal and ventral valves
instead of right and left lateral valves
as do the bivalve molluscs and, unlike
the bivalves, most of them are attached to a substrate either directly or
by means of a fleshy stalk called a
pedicel (or pedicle). Some, such as
Lingula, live in vertical burrows in
sand or mud. Muscles open and close
the valves and provide movement for
the stalk and tentacles.

In most brachiopods the ventral
(pedicel) valve is slightly larger than
the dorsal (brachial) valve, and one
end projects in the form of a short,
pointed beak that is perforated where
the fleshy stalk passes through (Figure 23-7B). In many the shape of the
pedicel valve is like that of the classic
oil lamp of ancient Greece and Rome,
so that the brachiopods came to be
known as the "lamp shells."

There are two classes of brachiopods, based on shell structure. The
shell valves of Articulata have a connecting hinge with an interlocking
tooth-and-socket arrangement, as in
Terebratella (L. *terebratus,* a boring, +
ella, dim. suffix); those of Inarticulata
lack the hinge and are held together by
muscles only, as in *Lingula* and *Glot-
tidia* (Gr. *glōttidos,* mouth of windpipe).

The body occupies only the posterior part of the space between the

valves (Figure 23-8), and extensions
of the body wall form mantle lobes
that line and secrete the shell. The
large horseshoe-shaped lophophore
in the anterior mantle cavity bears
long, ciliated tentacles used in respiration and feeding. Ciliary water currents carry food particles between the
gaping valves and over the
lophophore. The tentacles catch food
particles, and ciliated grooves carry
the particles along the arm of the
lophophore to the mouth. Rejection
tracts carry unwanted particles to the
mantle lobe where they are swept out
in ciliary currents. Organic detritus
and some algae are apparently the
primary food sources. The brachiopod lophophore not only can create
food currents, as do other lophophorates, but also seems able to absorb
dissolved nutrients directly from the
environmental seawater.

There is no cavity in the epistome
of articulates, but in inarticulates there
is a protocoel in the epistome that
opens into the mesocoel. As in the
other lophophorates, the posterior
metacoel bears the viscera. One or two
pairs of nephridia open into the
coelom and empty into the mantle
cavity. Coelomocytes, which ingest
particulate wastes, are expelled by the
nephridia. There is an open circulatory
system with a contractile heart. The
lophophore and mantle are probably
the chief site of gaseous exchange.
There is a nerve ring with a small dorsal and a larger ventral ganglion.

Sexes are separate, and paired
gonads discharge gametes through
the nephridia. Most fertilization is external, but a few species brood their
eggs and young.

Cleavage is radial, and coelom
and mesoderm formation in at least
some brachiopods is enterocoelic.
The blastopore closes, but its relationship to the mouth is uncertain. In
the articulates, metamorphosis of the
larva occurs after it has attached by a
pedicel. In the inarticulates, the juvenile resembles a minute brachiopod
with a coiled pedicel in the mantle
cavity. There is no metamorphosis.
As the larva settles, the pedicel attaches to the substratum, and adult
existence begins.

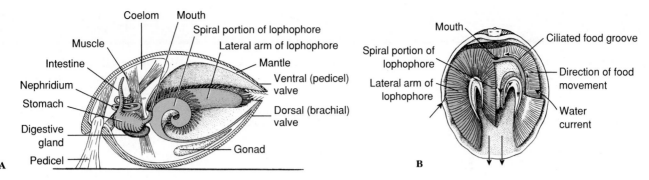

Figure 23-8
Phylum Brachiopoda. **A,** An articulate brachiopod (longitudinal section). **B,** Feeding and respiratory currents. Large arrows show water flow over lophophore; small arrows indicate food movement toward mouth in ciliated food groove.

PHYLOGENY AND ADAPTIVE RADIATION

The phylogenetic position of the lophophorates has been the subject of much controversy and debate. Sometimes they have been considered protostomes with some deuterostome characters, and at other times deuterostomes with some protostome characters. Brusca and Brusca* contend that there is overwhelming evidence that they are a monophyletic clade and are deuterostomes. Their common possession of a lophophore is a unique synapomorphy. Other features, such as the U-shaped digestive tract, metanephridia (except in ectoprocts), and tendency to secrete outer casings may be homologous within the clade, but they are convergent with many other taxa.

The division of the coelom into three parts (**trimerous,** or **tripartite)**

is a feature shared with other deuterostomes, but some recent authors question the trimerous nature and homologies of the coelom in some lophophorates (for example, whether the space in the epistome of inarticulate brachiopods is a protocoel, whether the mesocoel and metacoel in brachiopods are homologous to those spaces in other lophophorates, and whether the body coelom of ectoprocts is homologous to that of brachiopods and phoronids). The blastopore becomes the mouth in the phoronids, and development in ectoprocts is mosaic. Their larvae have been called trochophore-type in the past, although the resemblance to trochophores of annelids and molluscs is not close. Thus they show some protostome characters, and evidence for others is incomplete. Hyman suggested that the deuterostomes may have branched from the protostome line by way of a lophophore type of

ancestor. Brusca and Brusca concluded that, although the lophophorates were derived from a common stock, their relationships were too uncertain to construct a cladogram.

All lophophorates are filter feeders, and most of their evolutionary diversification has been constrained by this function. The tubes of phoronids vary according to their habitats. Many of the adaptations of ectoprocts seem related to the miniaturization of individual zooids and colony formation. Various ectoprocts tend to build their protective exoskeletons of chitin or gelatin, which may or may not be impregnated with calcium and sand. Brachiopod variations occur largely in their shells and lophophores.

*Brusca, R. C., and G. J. Brusca. 1990. Invertebrates. Sunderland, Massachusetts, Sinauer Associates, Inc.

Summary

The Phoronida, Ectoprocta, and Brachiopoda all bear a lophophore, which is a crown of ciliated tentacles surrounding the mouth but not the anus and containing an extension of the mesocoel. They are also sessile as adults, have a U-shaped digestive tract, and have a free-swimming larva. The lophophore functions as both a respiratory and a feeding structure, its cilia creating water currents from which food particles are filtered.

Phoronida are the least common of the lophophorates, living in tubes mostly

in shallow coastal waters. They thrust the lophophore out of the tube for feeding.

Ectoprocts are abundant in marine habitats, living on a variety of submerged substrata, and a number of species are common in fresh water. Ectoprocts are colonial, and although each individual is quite small, the colonies are commonly several centimeters or more in width or height. Each individual lives in a chamber (zoecium), which is a secreted exoskeleton of chitinous, calcium carbonate, or gelatinous material.

Brachiopods were very abundant in the Paleozoic era but have been declining in numbers and species since the early Mesozoic era. Their bodies and lophophores are covered by a mantle, which secretes a dorsal and a ventral valve (shell). They are usually attached to the substrate directly or by means of a pedicel.

The lophophorates have coelomic compartments that apparently correspond to the three compartments protocoel, mesocoel, and metacoel, found in many deuterostomes. Their embryogenesis shows both protostome and deuterostome characteristics.

Review Questions

1. What characters do the three lophophorate phyla have in common? What characters distinguish them from each other?
2. Define each of the following: lophophore, zoecium, zooid, polypide, cystid, statoblasts.
3. What are some protostome characters found among the lophophorates? What are their deuterostome characters?
4. What are the coelomic compartments found in the lophophorates?
5. What is the difference in orientation of the valves of brachiopods compared with bivalve molluscs?
6. How is the lophophore of ectoprocts extended?

Selected References

See also general references for Part III, p. 626.

American Society of Zoologists. 1977. Biology of lophophorates. Am. Zool. **17**(1):3–150. *A collection of 13 papers.*

Leffler, M. 1988. TBT on its way out. Oceans **21**:56–57 (Sept.–Oct.). *News of the highly toxic, antifouling agent, tributyl tin.*

Nielson, C. 1977. The relationships of the Entoprocta, Ectoprocta and Phoronida. Am. Zool. **17**:149–150. *On the basis of their development, the author considers phoronids deuterostomes and reunites Entoprocta with Ectoprocta.*

Richardson, J. R. 1986. Brachiopods. Sci. Am. **255**:100–106. (Sept.). *Reviews brachiopod biology and adaptations and contends that in the next few million years there may be an increase in the number of species, rather than a decline.*

Woollacott, R. M., and R. C. Zimmer (eds). 1977. Biology of bryozoans. New York, Academic Press, Inc. *Contains 15 articles on ectoprocts. Advanced.*

The Echinoderms

Phylum Echinodermata

A Design to Puzzle the Zoologist

The distinguished American zoologist Libbie Hyman once described the echinoderms as a "noble group especially designed to puzzle the zoologist." With a combination of characteristics that should delight the most avid reader of science fiction, the echinoderms would seem to confirm Lord Byron's observation that

> *Tis strange—but true;*
> *for truth is always strange;*
> *Stranger than fiction.*

Despite the adaptive value of bilaterality for free-moving animals, and the merits of radial symmetry for sessile animals, echinoderms confounded the rules by becoming free moving but radial. That they evolved from a bilateral ancestor there can be no doubt, for their larvae are bilateral. They undergo a bizarre metamorphosis to a radial adult in which there is a 90° reorientation in body axis.

A compartment of the coelom has been transformed in echinoderms into a unique water-vascular system that uses hydraulic power to operate a multitude of tiny tube feet used in food gathering and locomotion. An endoskeleton of dermal ossicles may fuse together to invest the echinoderm in armor, or it may be reduced in some to microscopic bodies. Many echinoderms have miniature jawlike pincers (pedicellariae) scattered on their body surface, often stalked and sometimes equipped with poison glands.

This constellation of characteristics is unique in the animal kingdom. It has both defined and limited the evolutionary potential of the echinoderms. Despite the vast amount of research that has been devoted to them, we are still far from understanding many aspects of echinoderm biology. ■

POSITION IN ANIMAL KINGDOM

1. Phylum Echinodermata (e-ki′no-der′ma-ta) (Gr. *echinos*, sea urchin, hedgehog, + *derma*, skin, + *ata*, characterized by) belongs to the **Deuterostomia** branch of the animal kingdom, the members of which are enterocoelous coelomates. The other phyla traditionally assigned to this group are Chaetognatha, Hemichordata, and Chordata. We are also placing the lophophorate phyla (Phoronida, Ectoprocta, and Brachiopoda) in the Deuterostomia.
2. Primitively, deuterostomes have the following embryological features in common: anus developing from or near the blastopore, and mouth developing elsewhere; coelom budded off from the archenteron (enterocoel); radial and regulative (indeterminate) cleavage; and endomesoderm (mesoderm derived from or with the endoderm) from enterocoelic pouches.
3. Thus the echinoderms, the chordates, and the smaller deuterostome phyla are presumably derived from a common ancestor. Nevertheless, their evolutionary history has taken the echinoderms to the point where they are very much unlike any other animal group.

BIOLOGICAL CONTRIBUTIONS

1. There is one word that best describes the echinoderms: strange. They have a unique constellation of characteristics found in no other phylum. Among the more striking of the features shown by the echinoderms are as follows:

   a. The system of channels composing the **water-vascular system,** derived from a coelomic compartment.
   b. The **dermal endoskeleton** composed of calcareous ossicles.
   c. The **hemal system,** whose function remains mysterious, also enclosed in a coelomic compartment.
   d. Their **metamorphosis,** which changes a bilateral larva to a radial adult.

THE ECHINODERMS

The echinoderms are marine forms and include the sea stars (also called starfishes), brittle stars, sea urchins, sea cucumbers, and sea lilies. They represent a bizarre group sharply distinguished from all other members of the animal kingdom. Their name is derived from their external spines or protuberances. A calcareous endoskeleton is found in all members of the phylum, either in the form of plates or represented by scattered tiny ossicles.

The most noticeable characteristics of the echinoderms are (1) the spiny endoskeleton of plates, (2) the water-vascular system, (3) the pedicellariae, (4) the dermal branchiae, and (5) radial or biradial symmetry. Radial symmetry is not limited to echinoderms, but no other group with such complex organ systems has radial symmetry.

Echinoderms are an ancient group of animals extending back to the Cambrian period. Despite the excellent fossil record, the origin and early evolution of the echinoderms are still obscure. It seems clear that they descended from bilateral ancestors because their larvae are bilateral but become radially symmetrical later in their development. Many zoologists believe that early echinoderms were sessile and evolved radiality as an adaptation to the sessile existence. Bilaterality is of adaptive value to animals that travel through their environment, while radiality is of value to animals whose environment meets them on all sides equally. Hence, the body plan of present-day echinoderms seems to have been derived from one that was attached to the bottom by a stalk, had radial symmetry and radiating grooves (ambulacra) for food gathering, and had an upward-facing oral side. Attached forms were once plentiful, but only about 80 species, all in the class Crinoidea, still survive. Oddly, conditions have favored the survival of their free-moving descendants, although they are still quite radial, and among them are some of the most abundant marine animals. Nevertheless, in the exception that proves the rule (that bilaterality is adaptive for free-moving animals), at least three groups of echinoderms (two groups of sea urchins and the sea cucumbers) have evolved back toward bilaterality.

Echinoderms have no ability to osmoregulate and thus rarely venture into brackish waters. They occur in all oceans of the world and at all depths, from the intertidal to the abyssal regions. Often the most common animals in the deep ocean are echinoderms. The most abundant species found in the Philippine Trench (10,540 m) was a holothurian. Echinoderms are virtually all bottom dwellers, although there are a few pelagic species.

No parasitic echinoderms are known, but a few are commensals. On the other hand, a wide variety of other animals make their homes in or on echinoderms, including parasitic or commensal algae, protozoa, ctenophores, turbellarians, cirripedes, copepods, decapods, snails, clams, polychaetes, fish, and other echinoderms.

The asteroids, or sea stars (Figure 24-1), are commonly found in various types of bottom habitats, often on hard, rocky surfaces, but numerous species are at home on sandy or soft bottoms. Some species are particle feeders, but many are predators, feeding particularly on sedentary or sessile prey, since the sea stars themselves are relatively slow moving.

Ophiuroids—brittle stars, or serpent stars (Figure 24-11)—are by far the most active echinoderms, moving by their arms rather than by tube feet. A few species are reported to have swimming ability, and some burrow. They may be scavengers, browsers, or deposit or filter feeders. Some are commensal in large sponges, in whose water canals they may live in great numbers.

Holothurians, or sea cucumbers (Figure 24-21), are widely prevalent in all seas. Many are found on sandy or mucky bottoms, where they lie concealed. Compared with other echinoderms, holothurians are greatly

A

B

C

D

Figure 24-1

Some sea stars (class Asteroidea) from the Pacific. **A,** Cushion star *Pteraster tessellatus* can secrete incredible quantities of mucus when disturbed, presumably as a defense reaction. **B,** Leather star *Dermasterias imbricata* lacks spines and feeds on sea anemones. **C,** *Penta-gonaster duebeni* from the Great Barrier Reef is brilliant red and orange. **D,** *Crossaster papposus,* one of the sun stars, feeds on other sea stars.

CHARACTERISTICS OF PHYLUM ECHINODERMATA

1. Body unsegmented (nonmetameric) with **radial, pentamerous symmetry;** body rounded, cylindrical, or star shaped, with five or more radiating areas, or **ambulacra,** alternating with interambulacral areas
2. **No head or brain;** few specialized sensory organs; sensory system of tactile and chemoreceptors, podia, terminal tentacles, photoreceptors, and statocysts
3. Nervous system with circumoral ring and radial nerves; usually two or three systems of networks located at different levels in the body, varying in degree of development according to group
4. **Endoskeleton of dermal calcareous ossicles** with **spines** or of calcareous **spicules** in dermis; covered by an epidermis (ciliated in most); **pedicellariae** (in some)
5. A unique **water-vascular system** of coelomic origin that extends from the body surface as a series of tentacle-like projections **(podia,** or **tube feet)** that are protracted by increase of fluid pressure within them; an opening to the exterior **(madreporite** or **hydropore)** usually present
6. Locomotion by tube feet, which project from the ambulacral areas, by movement of spines, or by movement of arms, which project from central disc of body
7. Digestive system usually complete; axial or coiled; anus absent in ophiuroids
8. Coelom extensive, forming the perivisceral cavity and the cavity of the water-vascular system; coelom of enterocoelous type; coelomic fluid with amebocytes
9. Blood-vascular system **(hemal system)** much reduced, playing little if any role in circulation, and surrounded by extensions of coelom **(perihemal sinuses);** main circulation of body fluids (coelomic fluids) by peritoneal cilia
10. Respiration by **dermal branchiae, tube feet, respiratory tree** (holothuroids), and **bursae** (ophiuroids)
11. **Excretory organs absent**
12. Sexes separate (except a few hermaphroditic) with large gonads, single in holothuroids but multiple in most; simple ducts, with no elaborate copulatory apparatus or secondary sexual structures; fertilization usually external; eggs brooded in some
13. Development through **free-swimming, bilateral, larval stages** (some with direct development); metamorphosis to radial adult or subadult form
14. Autotomy and regeneration of lost parts conspicuous

extended in the oral-aboral axis. They are oriented with that axis more or less parallel to the substrate and lying on one side. Most are suspension or deposit feeders.

Echinoids, or sea urchins (Figure 24-16), are adapted for living on the ocean bottom and always keep their oral surface in contact with the substratum. The "regular" sea urchins prefer hard bottoms, but the sand dollars and heart urchins ("irregular" urchins) are usually found on sand. The regular urchins, which are radially symmetrical, feed chiefly on algae or detritus, while the irregulars, which are secondarily bilateral, feed on small particles.

Crinoids (Figure 24-26) stretch their arms out and up like a flower's petals and feed on plankton and suspended particles. Most living species become detached from their stems as adults, but they nevertheless spend most of their time on the substrate, holding on by means of aboral appendages called cirri.

The zoologist who admires the fascinating structure and function of echinoderms can share with the layperson an admiration of the beauty of their symmetry, often enhanced by bright colors. Many species are rather drab, but others may be orange, red, purple, blue, and often multicolor.

Because of the spiny nature of their structure, echinoderms are not often the prey of other animals—except other echinoderms (sea stars). Some fishes have strong teeth and other adaptations that enable them to feed on echinoderms. A few mammals, such as sea otters, feed on sea urchins. In scattered parts of the world, humans relish sea urchin gonads, either raw or roasted on the half shell. Trepang, the cured body wall of certain large holothurians, is a delicacy in some east Asian countries. It is highly nutritious, since more than 50% is easily digestible protein, and it is said to impart a delicate flavor to soups.

Sea stars feed on a variety of molluscs, crustaceans, and other invertebrates. In some areas they may perform an important ecological role as a top carnivore in the habitat. Their chief economic impact is on clams and oysters. A single starfish may eat as many as a dozen oysters or clams in a day. To rid shellfish beds of these pests, quicklime is sometimes spread over areas where they abound. The quicklime damages the delicate epidermal membrane, destroying the dermal branchiae and ultimately the animal itself. Unfortunately, other soft-bodied invertebrates are also damaged. However, the oysters remain with their shells tightly closed until the quicklime is degraded.

Echinoderms have been widely used in developmental studies, for their gametes are usually abundant and easy to collect and handle in the laboratory. The investigator can follow the embryonic developmental stages with great accuracy. We know more about the molecular biology of sea urchin development than that of almost any other embryonic system. Artificial parthenogenesis was first discovered in sea urchin eggs, when it was found that, by treating the eggs with hypertonic seawater or subjecting them to a variety of other stimuli, development would proceed without the presence of sperm.

CLASS ASTEROIDEA

Sea stars, often called starfishes, demonstrate the basic features of echinoderm structure and function very well, and they are easily obtainable. Thus we will consider them first, then comment on the major differences shown by the other groups.

Sea stars are familiar along the shoreline where large numbers of them may aggregate on the rocks. Sometimes they cling so tenaciously that they are difficult to dislodge without tearing off some of the tube feet. They also live in muddy or sandy bottoms and among coral reefs. They are often brightly colored and range in size from a centimeter in greatest diameter to about a meter across. *Asterias* (Gr. *asteros,* a star) is one of the common genera of the east coast of the United States and is commonly studied in zoology laboratories. *Pisaster* (Gr. *pisos,* a pea, + *asteros,* a star) is common on the west coast of the United States, as is *Dermasterias* (Gr. *dermatos,* skin, leather, + *asteros,* a star), the leather star (Figure 24-1B).

FORM AND FUNCTION

External Features

Sea stars are composed of a central disc that merges gradually with the tapering arms (rays). The body is somewhat flattened, flexible, and covered with a ciliated, pigmented epidermis. The mouth is centered on the under, or oral, side, surrounded by a soft peristomial membrane. An **ambulacrum** (pl., **ambulacra,** L. *ambulacrum,* a covered way, an alley, a walk planted with trees) or **ambulacral area,** runs from the mouth on the oral side of each arm to the tip of the arm. Sea stars typically have five arms, but they may have more (Figure 24-1D), and there are as many ambulacral areas as there are arms. An **ambulacral groove** is found along the middle of each ambulacral area, and the groove is bordered by rows of **tube feet (podia)** (Figure 24-2). These in turn are usually protected by movable

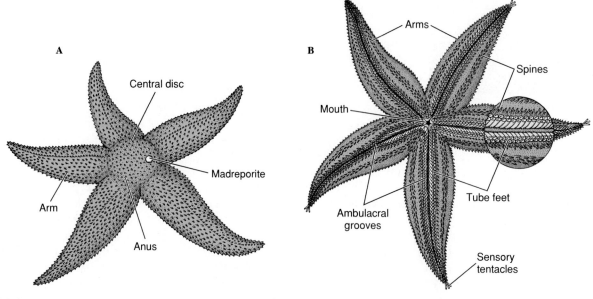

Figure 24-2
External anatomy of asteroid. **A,** Aboral view. **B,** Oral view.

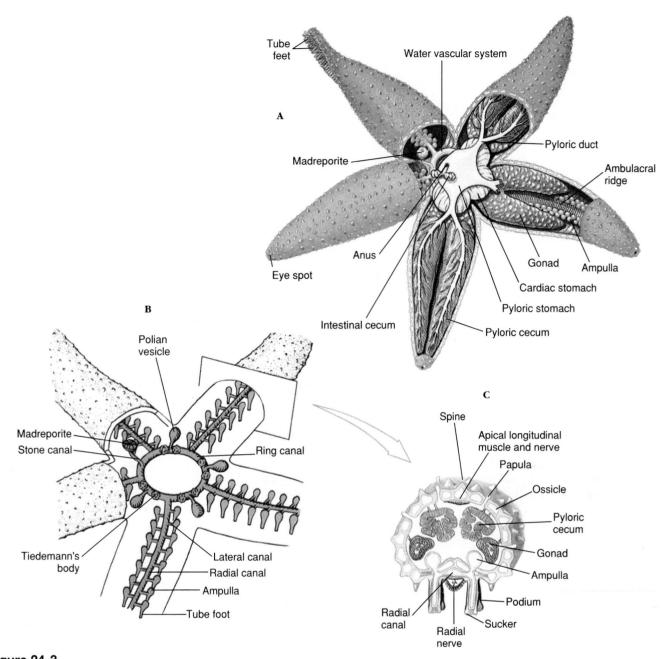

Figure 24-3

A, Internal anatomy of a sea star. **B,** Water-vascular system. Podia penetrate between ossicles. (Polian vesicles are not present in *Asterias*.) **C,** Cross section of arm at level of gonads, illustrating open ambulacral grooves.

spines. The large **radial nerve** can be seen in the center of each ambulacral groove (Figure 24-3C), between the rows of tube feet. The nerve is very superficially located, covered only by thin epidermis. Under the nerve is an extension of the coelom and the radial canal of the water-vascular system, all of which are external to the underlying ossicles (Figure 24-3C). In all other classes of living echinoderms except the crinoids, these structures are covered over by ossicles or other dermal tissue;

thus the ambulacral grooves in asteroids and crinoids are said to be *open,* and those of the other groups are *closed.*

The aboral surface is usually rough and spiny, although the spines of many species are flattened, so that the surface appears smooth (Figure 24-1C). Around the bases of the spine are groups of minute, pincerlike **pedicellariae,** bearing tiny jaws manipulated by muscles (Figure 24-4). These help keep the body surface free of debris, protect the papulae, and some-

times aid in food capture. The **papulae (dermal branchiae** or **skin gills)** are soft delicate projections of the coelomic cavity, covered only with epidermis and lined internally with peritoneum; they extend out through spaces between the ossicles and are concerned with respiration (Figures 24-3C, and 24-4F). Also on the aboral side are the inconspicuous anus and the circular **madreporite** (Figure 24-2A), a calcareous sieve leading to the water-vascular system.

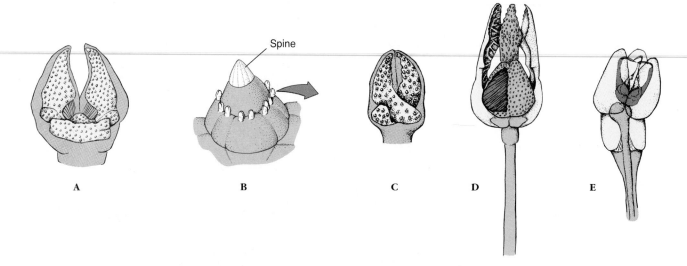

Spine

A   B   C   D   E

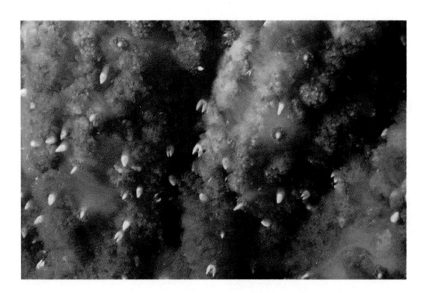

F

Figure 24-4

Pedicellariae of sea stars and sea urchins.
A, Forceps-type pedicellaria of *Asterias*. **B,** and
C, Scissors-type pedicellariae of *Asterias;* size
relative to spine is shown in **B. D,** Tridactyl
pedicellaria of *Strongylocentrotus.* **E,** Globiferous
pedicellaria of *Strongylocentrotus.* **F,** Close-up
view of the aboral surface of the sea star
Pycnopodia helianthoides. Note the large
pedicellariae, as well as the groups of small
pedicellariae around the spines. Many thin-walled
papulae can be seen.

Endoskeleton

Beneath the epidermis of sea stars is a
mesodermal endoskeleton of small
calcareous plates, or **ossicles,** bound
together with connective tissue. From
these ossicles project the spines and
tubercles that make up the spiny surface. The ossicles are penetrated by a
meshwork of spaces, usually filled
with fibers and dermal cells. This internal meshwork structure is described as **stereom** (Figure 24-23)
and is unique to the echinoderms.

Muscles in the body wall move
the rays and can partially close the
ambulacral grooves by drawing their
margins together.

Coelom, Excretion, and Respiration

The coelomic compartments of larval
echinoderms give rise to several structures in the adult, one of which is a
spacious body coelom filled with
fluid. The fluid contains amebocytes
(coelomocytes), bathes the internal
organs, and projects into the papulae.
The ciliated peritoneal lining of the
coelom circulates the fluid around the
body cavity and into the papulae. Exchange of respiratory gases and excretion of nitrogenous waste, principally ammonia, take place by
diffusion through the thin walls of the
papulae and tube feet. Some wastes

may be picked up by the coelomocytes, which migrate through the epithelium of the papulae or tube feet to
the exterior, or the tips of papulae
containing the waste-laden coelomocytes may pinch off.

Water-Vascular System

The water-vascular system is another
coelomic compartment and is unique
to the echinoderms. Showing exploitation of hydraulic mechanisms to
a greater degree than in any other animal group, it is a system of canals
and specialized tube feet that, together with the dermal ossicles, has

determined the evolutionary potential and limitations of this phylum. In sea stars the primary functions of the water-vascular system are locomotion and food gathering, in addition to respiration and excretion.

Structurally, the water-vascular system opens to the outside through small pores in the madreporite. The madreporite of asteroids is on the aboral surface (Figure 24-2A) and leads into the **stone canal,** which descends toward the **ring canal** around the mouth (Figure 24-3B). **Radial canals** diverge from the ring canal, one into the ambulacral groove of each ray. Also attached to the ring canal are four or five pairs of folded, pouchlike **Tiedemann's bodies** and from one to five **polian vesicles** (polian vesicles are absent in some sea stars, such as *Asterias*). The Tiedemann's bodies may produce coelomocytes, and the polian vesicles are apparently for fluid storage.

A series of small **lateral canals,** each with a one-way valve, connects the radial canal to the cylindrical podia, or tube feet, along the sides of the ambulacral groove in each ray. Each podium is a hollow, muscular tube, the inner end of which is a muscular sac, the **ampulla,** that lies within the body coelom (Figure 24-3A and C), and the outer end of which usually bears a **sucker.** Some species lack the suckers. The podia pass to the outside between the ossicles in the ambulacral groove.

The water-vascular system operates hydraulically and is an effective locomotor mechanism. The valves in the lateral canals prevent backflow of fluid into the radial canals. Each tube foot has in its walls connective tissue that maintains the cylinder at a relatively constant diameter. Contraction of muscles in the ampulla forces fluid into the podium, extending it. Conversely, contraction of the longitudinal muscles in the tube foot retracts the podium, forcing fluid back into the ampulla. Contraction of muscles in one side of the podium bends the organ toward that side. Small muscles at the end of the tube foot can raise the middle of the disclike end, creating suction when the end is applied to

A   B

Figure 24-5
A, *Orthasterias koehleri* eating a clam. **B,** This *Pycnopodia helianthoides* has been overturned while eating a large sea urchin *Strongylocentrotus franciscanus*. This sea star has 20 to 24 arms and can range up to 1 m in diameter (arm tip to arm tip).

the substrate. It has been estimated that by combining mucous adhesion with suction, a single podium can exert a pull equal to 25 to 30 g. Coordinated action of all or many of the tube feet is sufficient to draw the animal up a vertical surface or over rocks. The ability to move while firmly adhering to the substrate is a clear advantage to an animal living in a sometimes wave-churned environment.

On a soft surface, such as muck or sand, the suckers are ineffective (numerous sand-dwelling species have no suckers), so the tube feet are employed as legs. Locomotion becomes mainly a stepping process. Most sea stars can move only a few centimeters per minute, but some very active ones can move 75 to 100 cm per minute; for example, *Pycnopodia* (Gr. *pyknos,* compact, dense, + *pous, podos,* foot) (Figure 24-5B). When inverted, the sea star bends its rays until some of the tubes reach the substratum and attach as an anchor; then the sea star slowly rolls over.

Tube feet are innervated by the central nervous system (ectoneural and hyponeural systems, see following text). Nervous coordination enables the tube feet to move in a single direction, although not in unison, so that the sea star may progress. If the radial nerve in an arm is cut, the podia in that arm lose coordination,

although they can still function. If the circumoral nerve ring is cut, the podia in all arms become uncoordinated, and movement ceases.

Feeding and Digestive System

The mouth on the oral side leads through a short esophagus to a large stomach in the central disc. The lower (cardiac) part of the stomach can be everted through the mouth during feeding (Figure 24-2B), and excessive eversion is prevented by gastric ligaments. The upper (pyloric) part is smaller and connects by ducts to a pair of large **pyloric ceca (digestive glands)** in each arm (Figure 24-3A). Digestion is mostly extracellular, although some intracellular digestion may occur in the ceca. A short intestine leads aborally from the pyloric stomach, and there are usually a few small, saclike **intestinal ceca** (Figure 24-3A). The anus is inconspicuous, and some sea stars lack an intestine and anus.

Many sea stars are carnivorous and feed on molluscs, crustaceans, polychaetes, echinoderms, other invertebrates, and sometimes small fish. Sea stars consume a wide range of food items, but many show particular preferences (Figures 24-5 and 24-6). Some select brittle stars, sea urchins, or sand dollars, swallowing them whole and later regurgitating undigestible ossicles

A   **B**

Figure 24-6

A, Crown-of-thorns star *Acanthaster planci* feeding on coral. **B,** Close-up view of arm of *A. planci.* Puncture wounds from these spines are painful; the spines are equipped with poison glands.

and spines (Figure 24-5B). Some attack other sea stars, and if they are small compared with their prey, they may attack and begin eating at the end of one arm.

Since 1963 there have been numerous reports of increasing numbers of the crown-of-thorns starfish (*Acanthaster planci* [Gr. *akantha,* thorn, + *asteros,* star]) (Figure 24-6) that were damaging large areas of coral reef in the Pacific Ocean. The crown-of-thorns star feeds on coral polyps, and it sometimes occurs in large aggregations, or "herds." There is some evidence that outbreaks have occurred in the past, but an increase in frequency during the past 30 years suggests that some human activity may be affecting the starfish. Efforts to control the organism are very expensive and of questionable effectiveness. The controversy continues, especially in Australia, where it is exacerbated by extensive media coverage.

Some asteroids feed heavily on molluscs (Figure 24-5A), and *Asterias* is a significant predator on commercially important clams and oysters. When feeding on a bivalve, a sea star will hump over its prey, attaching its podia to the valves, and then exert a steady pull, using its feet in relays. A force of some 1300 g can thus be exerted. In half an hour or so the adductor muscles of the bivalve fatigue and relax. With a very small gap available, the star inserts its soft everted stomach into the space between the valves and wraps it around the soft parts of the shellfish. After feeding, the sea star draws in its stomach by contraction of the stomach muscles and relaxation of body-wall muscles.

Some sea stars feed on small particles, either entirely or in addition to carnivorous feeding. Plankton and other organic particles coming in contact with the animal's surface are carried by the epidermal cilia to the ambulacral grooves and then to the mouth.

Hemal System

The so-called hemal system is not very well developed in asteroids, and its function in all echinoderms is unclear. The hemal system has little or nothing to do with the circulation of body fluids. It is a system of tissue strands enclosing unlined sinuses and is itself enclosed in another coelomic compartment, the **perihemal channels** (Figure 24-7). The hemal system may be useful in distributing digested products, but its specific functions are not really known.

Nervous System

The nervous system consists of three units at different levels in the disc and arm. The chief of these systems is the **oral (ectoneural)** system composed of a **nerve ring** around the mouth and a main **radial nerve** into each arm. It appears to coordinate the tube feet. A **deep (hyponeural)** system lies aboral to the oral system, and an **aboral** system consists of a ring around the anus and radial nerves along the roof of the rays. An **epidermal nerve plexus** or nerve net freely connects these systems with the body wall and related structures. The epidermal plexus coordinates the responses of the dermal branchiae to tactile stimulation—the only instance known in echinoderms in which coordination occurs through a nerve net.

The sense organs are not well developed. Tactile organs and other sensory cells are scattered over the surface, and an ocellus is at the tip of each arm. Their reactions are mainly to touch, temperature, chemicals, and differences in light intensity. Sea stars are usually more active at night.

Reproductive System, Regeneration, and Autotomy

Most sea stars have separate sexes. A pair of gonads lies in each interradial space (Figure 24-3A). Fertilization is external and occurs in early summer when the eggs and sperm are shed into the water. A secretion from neurosecretory cells located on the radial nerves stimulates maturation and shedding of asteroid eggs.

Echinoderms can regenerate lost parts. Sea star arms can regenerate readily, even if all are lost. Sea stars also have the power of autotomy and can cast off an injured arm near the base. It may take months to regenerate a new arm.

Some species can regenerate a complete new sea star (Figure 24-8) from an arm that was broken off or removed if it contains a part (about one-fifth) of the central disc. In former times fishermen used to dispatch sea

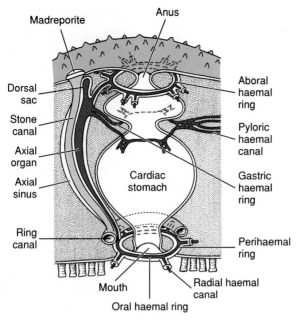

Figure 24-7
Hemal system of asteroids. The main perihemal channel is the thin-walled axial sinus, which encloses both the axial organ and the stone canal. Other features of the hemal system are shown.

Figure 24-8
Pacific sea star *Echinaster luzonicus* can reproduce itself by splitting across the disc, then regenerating missing arms. The one shown here has evidently regenerated six arms from the longer one at top left.

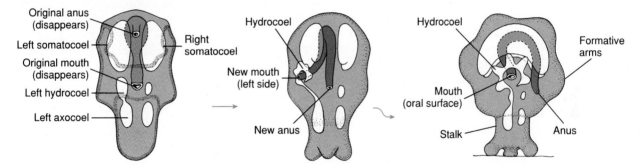

Figure 24-9
Asteroid metamorphosis. The left somatocoel becomes the oral coelom, and the right somatocoel becomes the aboral coelom. The left hydrocoel becomes the water-vascular system and the left axocoel the stone canal and perihemal channels. The right axocoel and hydrocoel are lost.

stars they collected from their oyster beds by chopping them in half with a hatchet—a worse than futile activity. Some sea stars reproduce asexually under normal conditions by cleaving the central disc, each part regenerating the rest of the disc and missing arms.

Development

In some species the liberated eggs are brooded, either under the oral side of the animal or in specialized aboral structures, and development is direct, but in most species the embryonating eggs are free in the water and hatch to free-swimming larvae.

Early embryogenesis shows the typical primitive deuterostome pattern.

Gastrulation is by invagination, and the anterior end of the archenteron pinches off to become the coelomic cavity, which expands in a U shape to fill the blastocoel. Each of the legs of the U, at the posterior, constricts to become a separate vesicle, and these eventually give rise to the main coelomic compartments of the body (metacoels, called **somatocoels** in echinoderms). The anterior portion of the U undergoes subdivision to form the protocoels and mesocoels (called **axocoels** and **hydrocoels** in echinoderms) (Figure 24-9). The left hydrocoel will become the water-vascular system, and the left axocoel will give rise to the stone canal and perihemal channels. The right axocoel and

hydrocoel will disappear. The free-swimming larva has cilia arranged in bands and is called a **bipinnaria** (Figure 24-10A). The ciliated tracts become extended into larval arms. Soon the larva grows three adhesive arms and a sucker at its anterior end and is then called a **brachiolaria.** At that time it attaches to the substratum, forms a temporary attachment stalk, and undergoes metamorphosis.

Metamorphosis involves a dramatic reorganization of a bilateral larva into a radial juvenile. The anteroposterior axis of the larva is lost, and *what was the left side becomes the oral surface, and the larval right side becomes the aboral surface* (Figure 24-9). Correspondingly, the larval

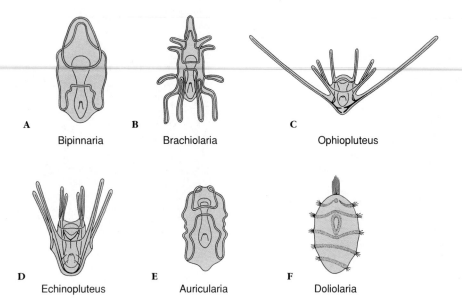

A

Figure 24-10
Larvae of echinoderms. **A,** Bipinnaria of asteroids. **B,** Brachiolaria of asteroids. **C,** Ophiopluteus of ophiuroids. **D,** Echinopluteus of echinoids. **E,** Auricularia of holothuroids. **F,** Doliolaria of crinoids.

B

Figure 24-11
A, Brittle star *Ophiura lutkeni* (class Ophiuroidea). Brittle stars do not use their tube feet for locomotion but can move rapidly (for an echinoderm) by means of their arms. **B,** Basket star *Astrophyton muricatum* (class Ophiuroidea). Basket stars extend their many-branched arms to filter feed, usually at night.

mouth and anus disappear, and a new mouth and anus form on what were originally the left and right sides, respectively. The portion of the anterior coelomic compartment from the left side expands to form the ring canal of the water-vascular system around the mouth, and then it grows branches to form the radial canals. As the short, stubby arms and the first podia appear, the animal detaches from its stalk and begins life as a young sea star.

CLASS OPHIUROIDEA

The brittle stars are the largest of the major groups of echinoderms in numbers of species, and they are probably the most abundant also. They abound in all types of benthic marine habitats, even carpeting the abyssal sea bottom in many areas.

FORM AND FUNCTION

Apart from the typical possession of five arms, the brittle stars are surprisingly different from the asteroids. The arms of brittle stars are slender and sharply set off from the central disc (Figure 24-11). They have no pedicellariae or papulae, and their ambulacral grooves are closed and covered with arm ossicles. The tube feet are

without suckers; they aid in feeding but are of limited use in locomotion. In contrast to that in the asteroids, the madreporite of the ophiuroids is located on the oral surface, on one of the oral shield ossicles (Figure 24-12). Ampullae on the podia are absent, and force for protrusion of the podium is generated by a proximal muscular portion of the podium.

Each of the jointed arms consists of a column of articulated ossicles (the so-called **vertebrae**), connected by muscles and covered by plates. Locomotion is by arm movement. The arms are moved forward in pairs and are placed against the substratum, while one (any one) is extended forward or trailed behind, and the animal is pulled or pushed along in a jerky fashion.

Five movable plates that serve as jaws surround the mouth (Figure 24-12). There is no anus. The skin is leathery, with dermal plates and spines arranged in characteristic patterns. Surface cilia are mostly lacking.

The visceral organs are confined to the central disc, since the rays are too slender to contain them (Figure 24-13). The stomach is saclike, and there is no intestine. Indigestible material is cast out of the mouth.

Five pairs of invaginations called **bursae** open toward the oral surface by genital slits at the bases of the

arms. Water circulates in and out of these sacs for exchange of gases. On the coelomic wall of each bursa are small gonads that discharge into the bursa their ripe sex cells, which pass through the genital slits into the water for fertilization (Figure 24-14A). Sexes are usually separate; a few ophiuroids are hermaphroditic. Some brood their young in the bursae; the young escape through the genital slits or by rupturing the aboral disc. The larva is called the ophiopluteus, and its ciliated bands extend onto delicate, beautiful larval arms (Figure 24-10C). During the metamorphosis to the juvenile, there is no temporarily attached phase, as there is in asteroids.

Water-vascular, nervous, and hemal systems are similar to those of

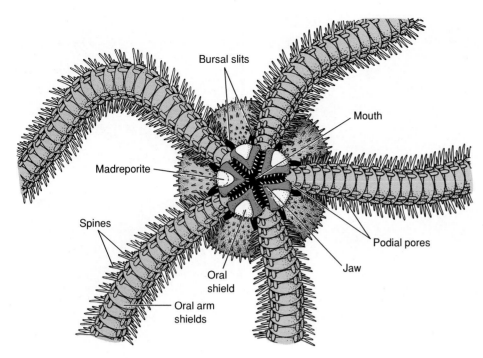

Figure 24-12
Oral view of spiny brittle star *Ophiothrix.*

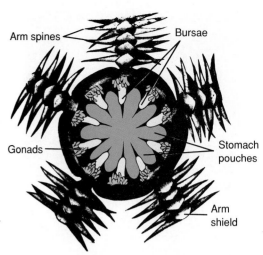

Figure 24-13
Ophiuroid with aboral disc wall cut away to show principal internal structures. The bursae are fluid-filled sacs in which water constantly circulates for respiration. They also serve as brood chambers. Only bases of arms are shown.

the sea stars. Each arm contains a small coelom, a radial nerve, and a radial canal of the water-vascular system.

BIOLOGY

Brittle stars tend to be secretive, living on hard bottoms where little or no light penetrates. They are generally negatively phototropic and insinuate themselves into small crevices between rocks, becoming more active at night. They are commonly fully exposed on the bottom in the permanent darkness of the deep sea. Ophiuroids feed on a variety of small particles, either browsing food from the bottom or suspension feeding. Podia are important in transferring food to the mouth. Some brittle stars extend arms into the water and catch suspended particles in mucous strands between the arm spines.

Regeneration and autotomy are even more pronounced in brittle stars than in sea stars. Many seem very fragile, releasing an arm or even part of the disc at the slightest provocation. Some can reproduce asexually by cleaving the disc; each progeny then regenerates the missing parts.

Some common ophiuroids along the coast of the United States are *Amphipholis* (Gr. *amphi,* both sides of, + *pholis,* horny scale) (viviparous and hermaphroditic), *Ophioderma* (Gr. *ophis,* snake, + *dermatos,* skin), *Ophiothrix* (Gr. *ophis,* snake, + *thrix,* hair), and *Ophiura* (Gr. *ophis,* snake, + *oura,* tail) (Figure 24-11). The basket stars *Gorgonocephalus* (Gr. *Gorgo,* name of a female monster of terrible aspect, + *kephalē,* a head) (Figure 24-14B) and *Astrophyton* (Gr. *asteros,* star, + *phyton,* creature, animal) (Figure 24-11B) have arms that branch repeatedly. Most ophiuroids are drab, but some are attractive, with bright color patterns (Figure 24-14A).

CLASS ECHINOIDEA

The echinoids have a compact body enclosed in an endoskeletal test, or shell. The dermal ossicles, which have become closely fitting plates, make up the test. Echinoids lack arms, but their tests reflect the typical pentamerous plan of the echinoderms in their five ambulacral areas. The most notable modification of the ancestral body plan is that the oral surface has

A

B

Figure 24-14
A, This brittle star *Ophiopholis aculeata* has its bursae swollen with eggs, which it is ready to expel. The arms have been broken and are regenerating. **B,** Oral view of a basket star *Gorgonocephalus eucnemis,* showing pentaradial symmetry.

Figure 24-15
Purple sea urchin *Strongylocentrotus purpuratus* is common along the Pacific coast of North America where there is heavy wave action.

expanded around to the aboral side, so that the ambulacral areas extend up to the area around the anus **(periproct).** The majority of living species of sea urchins are referred to as "regular"; they have a hemispherical shape, radial symmetry, and medium to long spines (Figures 24-15 and 24-16). Sand dollars (Figure 24-17) and heart urchins (Figure 24-18) are "irregular" because the orders to which they belong have become secondarily bilateral; their spines are usually very short. Regular urchins move by means of their tube feet, with some assistance from their spines, and irregular urchins move chiefly by their spines (Figure 24-17). Some echinoids are quite colorful.

Echinoids have a wide distribution in all seas, from the intertidal regions to the deep oceans. Regular urchins often prefer rocky or hard bottoms, whereas sand dollars and heart urchins like to burrow into a sandy substrate. Distributed along one or both coasts of North America are common genera of regular urchins (*Arbacia* [Gr. Arbakēs, first king of Media], *Strongylocentrotus* [Gr. strongylos, round, compact, + kentron, point, spine] [Figure 24-15], and *Lytechinus* [Gr. lytos, dissolvable, broken, + echinos, sea urchin]) and sand dollars (*Dendraster* [Gr. dendron, tree, stick, + asteros, star] and *Echinarachnius* [Gr. echinos, sea urchin, + arachnē, spider]). The West Indies–Florida region is rich in echinoderms, including echinoids, of which *Diadema* (Gr. diadeō, to bind around), with its long, needle-sharp spines, is a notable example (Figure 24-16D).

Figure 24-16
Diversity among regular sea urchins (class Echinoidea). **A,** Pencil urchin *Eucidaris tribuloides.* Members of this order have many primitive characters and have survived since the Paleozoic era. They may be closest to the common ancestor to all other extant echinoids. **B,** Slate-pencil urchin *Heterocentrotus mammilatus.* The large, triangular spines of this urchin were formerly used for writing on slates. **C,** Aboral spines of the intertidal urchin *Colobocentrotus atratus* are flattened and mushroom shaped, while the marginal spines are wedge shaped, giving the animal a streamlined form to withstand pounding surf. **D,** *Diadema antillarum* is a common species in the West Indies and Florida. **E,** *Astropyga magnifica* is one of the most spectacularly colored sea urchins, with bright-blue spots along its interambulacral areas.

FORM AND FUNCTION

The echinoid test is a compact skeleton of 10 double rows of plates that bear movable, stiff spines (Figure 24-19). The plates are sutured firmly. The five pairs of ambulacral rows are homologous to the five arms of the sea star and have pores (Figure 24-19B) through which the long tube feet extend. The plates bear small tubercles on which the round ends of the spines

articulate as ball-and-socket joints. The spines are moved by small muscles around the bases.

There are several kinds of pedicellariae, the most common of which are three jawed and are mounted on long stalks (Figure 24-4D and E). Pedicellariae help keep the body clean and capture small organisms. The pedicellariae of many species bear poison glands, and the toxin paralyzes small prey.

Diadema antillarum is not nearly as prominent as it once was. In January 1983, an epidemic swept through the Caribbean and along the Florida Keys. Its cause has never been determined, but it decimated the *Diadema* population, leaving less than 5% of the original numbers. Other species of sea urchins were unaffected. However, various types of algae, formerly grazed heavily by the *Diadema* have increased greatly on the reefs, and the *Diadema* populations have not recovered. This has had a disastrous effect on coral reefs around Jamaica. Herbivorous fish around that island had been chronically overharvested, and then, after the *Diadema* epidemic, there was nothing left to control algal overgrowth. Coral reefs around Jamaica have been largely destroyed.

A   B

Figure 24-17

Two sand dollar species. **A,** *Encope grandis* as they are normally found burrowing near the surface on a sandy bottom. **B,** Removed from the sand. The short spines and petaloids on the aboral surface of this *Encope micropora* are easily seen.

Figure 24-18

An irregular echinoid *Meoma,* one of the largest heart urchins (test up to 18 cm). *Meoma* occurs in the West Indies and from the Gulf of California to the Galápagos Islands. **A,** Aboral view. Anterior ambulacral area is not modified as a petaloid in the heart urchins, although it is in the sand dollars. **B,** Oral view. Note curved mouth at anterior end and periproct at posterior end.

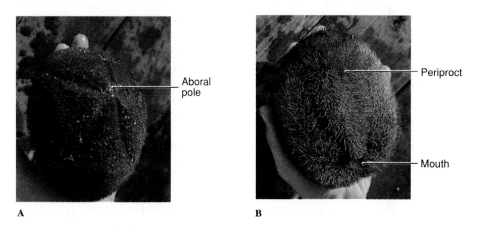

A   B

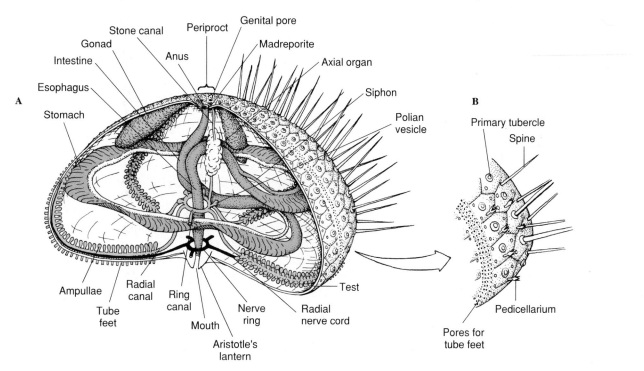

Figure 24-19

A, Internal structure of the sea urchin; water-vascular system in tan. **B,** Detail of portion of endoskeleton.

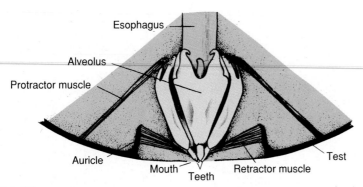

Figure 24-20
Aristotle's lantern, the complex mechanism used by the sea urchin for masticating its food. Five pairs of retractor muscles draw the lantern and teeth up into the test; five pairs of protractors push the lantern down and expose the teeth. Other muscles produce a variety of movements. Only major skeletal parts and muscles are shown in this diagram.

Five converging teeth surround the mouth of regular urchins. In some sea urchins branched gills (modified podia) encircle the peristome. The anus, genital pores, and madreporite are located aborally in the periproct region (Figure 24-19). The sand dollars also have teeth, and the mouth is located at about the center of the oral side, but the anus has shifted to the posterior margin or even the oral side of the disc, so that an anteroposterior axis and bilateral symmetry can be recognized. Bilateral symmetry is even more accentuated in the heart urchins, with the anus near the posterior on the oral side and the mouth moved away from the oral pole toward the anterior (Figure 24-18).

Inside the test (Figure 24-19) are the coiled digestive system and a complex chewing mechanism (in the regular urchins and in sand dollars), called **Aristotle's lantern** (Figure 24-20), to which the teeth are attached. A ciliated **siphon** connects the esophagus to the intestine and enables the water to bypass the stomach to concentrate the food for digestion in the intestine. Sea urchins eat algae and other organic material, which they graze with their teeth. Sand dollars have short club-shaped spines that move the sand and its organic contents over the aboral surface and down the sides. Fine food particles drop between the spines, and ciliated tracts on the oral side carry the particles to the mouth.

The hemal and nervous systems are basically similar to those of the as-teroids. The ambulacral grooves are closed, and the radial canals of the water-vascular system run just beneath the test, one in each of the ambulacral radii (Figure 24-19). The ampullae for the podia are within the test, and each ampulla usually communicates with its podium by *two* canals through pores in the ambulacral plate; consequently, such pores in the plates are in pairs. The peristomial gills, where present, are of little or no importance in respiratory gas exchange, this function being carried out principally by the other podia. In the irregular urchins the respiratory podia are thin walled, flattened, or lobulate and are arranged in ambulacral fields called **petaloids** on the aboral surface. The irregular urchins also have short, suckered, single-pored podia in the ambulacral and sometimes interambulacral areas; these function in food handling.

Sexes are separate, and both eggs and sperm are shed into the sea for external fertilization. Some, such as certain of the pencil urchins, brood their young in depressions between the spines. The **echinopluteus larvae** (Figure 24-10D) of nonbrooding echinoids may live a planktonic existence for several months and then metamorphose quickly into young urchins.

CLASS HOLOTHUROIDEA

In a phylum characterized by odd animals, class Holothuroidea (sea cucumbers) contains members that both struc-turally and physiologically are among the strangest. These animals have a remarkable resemblance to the vegetable after which they are named (Figure 24-21). Compared with the other echinoderms, the holothurians are greatly elongated in the oral-aboral axis, and the ossicles are much reduced in most, so that the animals are soft bodied. Some species crawl on the surface of the sea bottom, others are found beneath rocks, and some are burrowers.

Common species along the east coast of North America are *Cucumaria frondosa* (L. *cucumis,* cucumber), *Sclerodactyla briareus* (Gr. *skleros,* hard, + *daktylos,* finger) (Figure 24-22), and the translucent, burrowing *Leptosynapta* (Gr. *leptos,* slender, + *synapsis,* joining together). Along the Pacific coast there are several species of *Cucumaria* (Figure 24-21C) and the striking reddish brown *Parastichopus* (Gr. *para,* beside, + *stichos,* line or row, + *pous, podos,* foot) (Figure 24-21A), with very large papillae.

FORM AND FUNCTION

The body wall is usually leathery, with tiny ossicles embedded in it (Figure 24-23), although a few species have large ossicles forming a dermal armor (Figure 24-21B). Because of the elongate body form of the sea cucumbers, they characteristically lie on one side. In some species the locomotor tube feet are equally distributed to the five ambulacral areas (Figure 24-21C) or all over the body, but most have well-developed tube feet only in the ambulacra normally applied to the substratum (Figure 24-21A and B). Thus a secondary bilaterality is present, albeit of quite different origin from that of irregular urchins. The side applied to the substratum has three ambulacra and is called the sole; the tube feet in the dorsal ambulacral areas, if present, are usually without suckers and may be modified as sensory papillae. All tube feet, except oral tentacles, may be absent in burrowing forms.

The oral tentacles are 10 to 30 retractile, modified tube feet around the

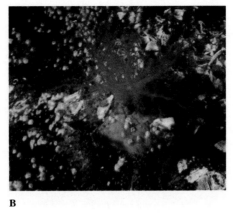

A   **B**   **C**

Figure 24-21

Sea cucumbers (class Holothuroidea). **A,** Common along the Pacific coast of North America, *Parastichopus californicus* grows to 50 cm in length. Its tube feet on the dorsal side are reduced to papillae and warts. **B,** In sharp contrast to most sea cucumbers, the surface ossicles of *Psolus chitonoides* are developed into a platelike armor. The ventral surface is a flat, soft, creeping sole, and the mouth (surrounded by tentacles) and anus are turned dorsally. **C,** Tube feet are found in all ambulacral areas of *Cucumaria miniata* but are better developed on its ventral side, shown here.

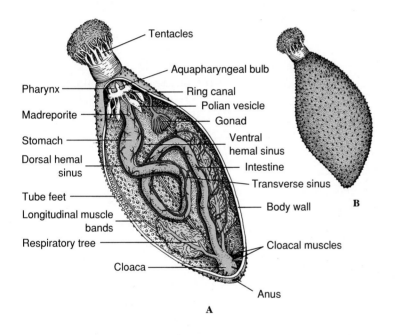
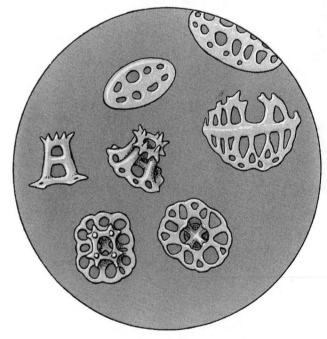

Figure 24-22

Anatomy of the sea cucumber *Sclerodactyla*. **A,** Internal. **B,** External. *Red,* hemal system.

Figure 24-23

Ossicles of sea cucumbers are usually microscopic bodies buried in the leathery dermis. They can be extracted from the tissue with commercial bleach and are important taxonomic characteristics. The ossicles shown here, called tables, buttons, and plates, are from the sea cucumber *Holothuria difficilis.* They illustrate the meshwork (stereom) structure observed in ossicles of all echinoderms at some stage in their development (× 250).

mouth. The body wall contains circular and longitudinal muscles along the ambulacra.

The coelomic cavity is spacious and fluid filled and has many coelomocytes. Because of the reduction in the dermal ossicles, they no longer function as an endoskeleton, and the fluid-filled coelom now serves as a hydrostatic skeleton.

The digestive system empties posteriorly into a muscular **cloaca** (Figure 24-22). A **respiratory tree** composed of two long, many-branched tubes also empties into the cloaca, which pumps seawater into it. The respiratory tree serves for both respiration and excretion and is not present in any other group of living echinoderms. Gas exchange also occurs through the skin and tube feet.

The hemal system is more well developed in holothurians than in other echinoderms. The water-vascular system is peculiar in that the madreporite lies free in the coelom.

The sexes are usually separate, but some holothurians are hermaphroditic. Among the echinoderms, only the sea cucumbers have a single gonad, and this is considered a primitive character.

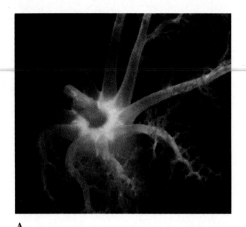

A   B   C

Figure 24-24

A, *Eupentacta quinquesemita* extends its tentacles to collect particulate matter in the water, then puts them one by one into its mouth and cleans the food from them. **B,** Moplike tentacles of *Parastichopus californicus* are used for deposit feeding on the bottom. **C,** *Bohadschia argus* expels its cuvierian tubules, modified parts of its respiratory tree, when it is disturbed. These sticky strands, containing a toxin, discourage potential predators.

The gonad is usually in the form of one or two clusters of tubules that join at the gonoduct. Fertilization is external, and the free-swimming larva is called an **auricularia** (Figure 24-10E). Some species brood the young either inside the body or somewhere on the body surface.

BIOLOGY

Sea cucumbers are sluggish, moving partly by means of their ventral tube feet and partly by waves of contraction in the muscular body wall. The more sedentary species trap suspended food particles in the mucus of their outstretched oral tentacles or pick up particles from the surrounding bottom. They then stuff the tentacles into the pharynx, one by one, sucking off the food material (Figure 24-24A). Others crawl along, grazing the bottom with their tentacles (Figure 24-24B).

Sea cucumbers have a peculiar power of what appears to be self-mutilation but may be a mode of defense. When irritated or when subjected to unfavorable conditions, many species can cast out a part of their viscera by a strong muscular contraction that may either rupture the body wall or evert its contents through the anus. The lost parts are soon regenerated. Certain species have organs of Cuvier (cuvierian tubules) attached to the posterior part of the respiratory tree, which can be expelled in the direction of an enemy

(Figure 24-24C). These can become long and sticky after expulsion, and some contain toxins.

There is an interesting commensal relationship between some sea cucumbers and a small fish, *Carapus,* that uses the cloaca and respiratory tree of the sea cucumber as shelter.

CLASS CRINOIDEA

The crinoids include the sea lilies and feather stars. They have several primitive characters. As fossil records reveal, crinoids were once far more numerous than they are now. They differ from other echinoderms by being attached during a substantial part of their lives. Sea lilies have a flower-shaped body that is placed at the tip of an attached stalk (Figure 24-25). Feather stars have long, many-branched arms, and the adults are free moving, though they may remain in the same spot for long periods (Figure 24-26). During metamorphosis feather stars become sessile and stalked, but after several months they detach and become free moving. Many crinoids are deep-water forms, but feather stars may inhabit shallow waters, especially in the Indo-Pacific and West-Indian–Caribbean regions, where the largest numbers of species are found.

FORM AND FUNCTION

The body disc, or **calyx,** is covered with a leathery skin **(tegmen)** con-

taining calcareous plates. The epidermis is poorly developed. Five flexible arms branch to form many more arms, each with many lateral **pinnules** arranged like barbs on a feather (Figure 24-25). The calyx and arms together are called the **crown.** Sessile forms have a long, jointed **stalk** attached to the aboral side of the body. This stalk is composed of plates, appears jointed, and may bear **cirri.** Madreporite, spines, and pedicellariae are absent.

The upper (oral) surface bears the mouth, which opens into a short esophagus, from which the long intestine with diverticula proceeds aborally for a distance and then makes a complete turn to the **anus,** which may be on a raised cone (Figure 24-25B). With the aid of tube feet and mucous nets, crinoids feed on small organisms that are caught in the ambulacral grooves. The **ambulacral grooves** are open and ciliated and serve to carry food to the mouth (Figure 24-25B). Tube feet in the form of tentacles are also found in the grooves.

The water-vascular system has the basic echinoderm plan. The nervous system has an oral ring and a radial nerve that runs to each arm. The aboral or entoneural system is more highly developed in crinoids than in most other echinoderms. Sense organs are scanty and primitive.

The sexes are separate. The gonads are simply masses of cells in the genital cavity of the arms and pinnules.

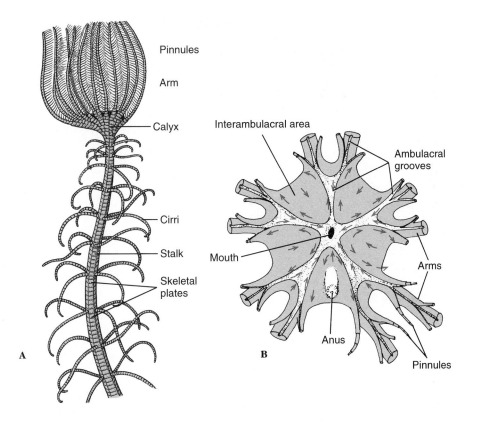

Figure 24-25

Crinoid structure. **A,** Sea lily (stalked crinoid) with portion of stalk. Modern crinoid stalks rarely exceed 60 cm, but fossil forms were as much as 20 m long. **B,** Oral view of calyx of the crinoid *Antedon,* showing direction of ciliary food currents. Ambulacral grooves with podia extend from mouth along arms and branching pinnules. Food particles touching podia are tossed into ambulacral grooves and carried, tangled in mucus, by strong ciliary currents toward mouth. Particles falling on interambulacral areas are carried by cilia first toward mouth and then outward and finally dropped off the edge, thus keeping the oral disc clean.

Figure 24-26

Davidaster spp. are crinoids found on Caribbean coral reefs. They extend their arms into the water to catch food particles while maintaining their body in a crevice.

The gametes escape without ducts through a rupture in the pinnule wall. Brooding occurs in some forms. The **doliolaria** larvae (Figure 24-10F) are free swimming for a time before they become attached and metamorphose.

Most living crinoids are from 15 to 30 cm long, but some fossil species had stalks 20 m in length.

CLASS CONCENTRICYCLOIDEA

Strange little (less than 1 cm diameter), disc-shaped animals (Figure 24-27) were discovered in water over 1000 m deep off New Zealand. Sometimes called sea daisies, they are the most recently described (1986) class of echinoderms, and only two species are known so far. They are pentaradial in symmetry but have no arms. Their tube feet are located around the periphery of the disc, rather than along ambulacral areas, as in other echinoderms. Their water-vascular system includes two concentric ring canals; the outer ring may represent the radial canals since the podia arise from it. A hydropore, homologous to the madreporite, connects the inner

ring canal to the aboral surface. One species has no digestive tract; its oral surface is covered by a membranous **velum,** by which it apparently absorbs nutrients. The other species has a shallow, sac-like stomach but no intestine or anus.

PHYLOGENY AND ADAPTIVE RADIATION

PHYLOGENY

Despite the existence of an extensive fossil record, there have been numerous contesting hypotheses on echinoderm phylogeny. Based on the embryological evidence of the bilateral larvae, there can be little doubt that their ancestors were bilateral and that their coelom had three pairs of spaces (trimeric). Some investigators have held that the radial symmetry arose in a free-moving echinoderm ancestor and that sessile groups were derived several times independently from the free-moving ancestors. However, this view does not account for the adaptive significance of the radial symmetry, that is, as an adaptation for the sessile existence. The more traditional view is that the first echinoderms were sessile, became radial as an adaptation to that existence, and then gave rise to the free-moving groups. Figure 24-29 is consistent with this hypothesis. It views the evolution of endoskeletal plates with stereom structure and of external ciliary grooves for feeding as early echinoderm (or pre-echinoderm) developments. The extinct carpoids (Figures 24-28A, 24-29) had stereom ossicles but were not radially symmetrical, and the status of their water-vascular system, if any, is uncertain. Some investigators regard carpoids as a separate subphylum of echinoderms (Homalozoa), and others believe they represent a group of pre-echinoderms that shows affinities to the chordates (Calcichordata, p. 484). The fossil helicoplacoids (Figures 24-28B, 24-29) show evidence of three, true ambulacral grooves, and their mouth was on the side of the body.

Attachment to the substratum by the aboral surface would have led to radial symmetry and the origin of the Pelmatozoa. Both the Cystoidea

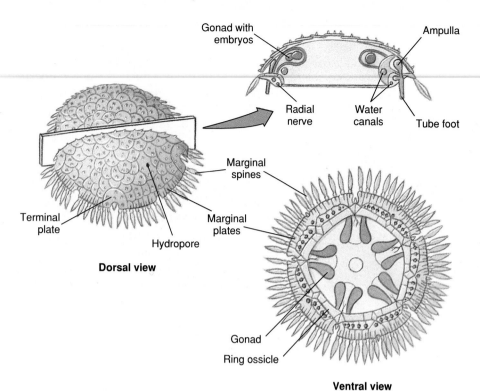

Gonad with embryos

Ampulla

Radial nerve

Water canals

Tube foot

Terminal plate

Marginal spines

Marginal plates

Hydropore

Dorsal view

Gonad

Ring ossicle

Ventral view

Figure 24-27

Xyloplax spp. (class Concentricycloidea) are bizarre little disc-shaped echinoderms. With their podia around the margin, they are the only echinoderms not having podia distributed along ambulacral areas.

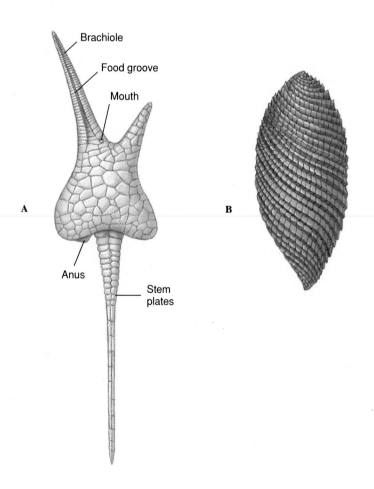

Brachiole

Food groove

Mouth

Anus

Stem plates

A

B

Figure 24-28

A, *Dendrocystites,* a carpoid (subphylum Homalozoa) with one brachiole. Brachioles are so called to distinguish them from the heavier arms of asteroids, ophiuroids and crinoids. This group bore some characters interpreted as chordate in nature. It is called Calcichordata by some investigators (p. 485). **B,** *Helicoplacus,* a helicoplacoid, had three ambulacral areas and apparently a water-vascular system. It represents a sister group to the modern echinoderms.

(extinct) and the Crinoidea primitively were attached to the substratum by an aboral stalk. An ancestor that became free-moving and applied its oral surface to the substratum would have given rise to the Eleutherozoa. Phylogeny within the Eleutherozoa is controversial. Most investigators agree that the echinoids and holothuroids are related and form a single clade, but opinions diverge on the relationship of the ophiuroids and asteroids. Figure 24-29 illustrates the view that the ophiuroids arose after the closure of ambulacral grooves, but this scheme treats the evolution of five ambulacral rays (arms) in the ophiuroids and asteroids as independently evolved. Alternatively, if the ophiuroids and asteroids are a single clade, then closed ambulacral grooves must have evolved separately in ophiuroids and in the common ancestor of echinoids and holothuroids.

Data on the Concentricycloidea are insufficient to place this group on a cladogram.

ADAPTIVE RADIATION

The radiation of the echinoderms has been determined by the limitations and potentials inherent in their most important characters: radial symmetry, water-vascular system, and dermal endoskeleton. If their ancestors had a brain and specialized sense organs, these were lost in the adoption of radial symmetry. Thus it is not surprising that there are large numbers of creeping, benthic forms with filter-feeding, deposit-feeding, scavenging, and herbivorous habits, comparatively few predators, and very rare pelagic forms. In this light the relative success of the asteroids as predators is impressive and probably attributable to the extent to which they have exploited the hydraulic mechanism of the tube feet.

The basic body plan of echinoderms has severely limited their evolutionary opportunities to become parasites. Indeed, the most mobile of the echinoderms, the ophiuroids, which are also the ones most able to insert their bodies into small spaces, are the only group with significant numbers of commensal species.

CLASSIFICATION OF PHYLUM ECHINODERMATA

There are about 6000 living and 20,000 extinct or fossil species of Echinodermata. The traditional classification placed all the free-moving forms that were oriented with oral side down in the subphylum Eleutherozoa, containing most of the living species. The other subphylum, Pelmatozoa, contained mostly forms with stems and oral side up; most of the extinct classes and the living Crinoidea belong to this group. Although alternative schemes have strong supporters, cladistic analysis provides evidence that the two traditional subphyla are monophyletic clades.* The following includes only groups with living members.

Subphylum Pelmatozoa (pel-ma′to-zo′a) (Gr. *pelmatos;* a stalk, + *zōon,* animal). Body in form of cup or calyx, borne on aboral stalk during part or all of life; oral surface directed upward; open ambulacral grooves; madreporite absent; both mouth and anus on oral surface; several fossil classes plus living Crinoidea.

Class Crinoidea (krin-oi′de-a) (Gr. *krinon,* lily; + *eidos,* form; + *ea,* characterized by): **sea lilies** and **feather stars.** Five arms branching at base and bearing pinnules; ciliated ambulacral grooves on oral surface with tentacle-like tube feet for food gathering; spines, madreporite, and pedicellariae absent. Examples: *Antedon, Davidaster* (Figure 24-26).

Subphylum Eleutherozoa (e-lu′ther-o-zo′a) (Gr. *eleutheros,* free, not bound, + *zōon,* animal). Body form star-shaped, globular, discoidal, or cucumber-shaped; oral surface directed toward substratum or oral-aboral axis parallel to substratum; body with or without arms; ambulacral grooves open or closed.

Class Concentricycloidea (kon-sen′tri-sy-kloy′de-a) (L. *cum,* together, + *centrum,* center [having a common center], + Gr. *kyklos,* circle, + *eidos,* form, + *ea,* characterized by): **sea daisies.** Disc-shaped body, with marginal spines but no arms; concentrically arranged skeletal plates; ring of suckerless podia near body margin; hydropore present; gut present or absent, no anus. Example: *Xyloplax* (Figure 24-27).

Class Asteroidea (as′ter-oy′de-a) (Gr. *aster,* star, + *eidos,* form, + *ea,* characterized by): **sea stars** and **starfish.** Star-shaped, with arms not sharply marked off from central disc; ambulacral grooves open, with tube feet on oral side; tube feet often with suckers; anus and madreporite aboral; pedicellariae present. Examples: *Asterias, Pisaster* (p. 449).

Class Ophiuroidea (o′fe-u-roy′de-a) (Gr. *ophis,* snake, + *oura,* tail, + *eidos,* form, + *ea,* characterized by): **brittle stars** and **basket stars.** Star shaped, with arms sharply marked off from central disc; ambulacral grooves closed, covered by ossicles; tube feet without suckers and not used for locomotion; pedicellariae absent; anus absent. Examples: *Ophiura* (Figure 24-11A), *Gorgonocephalus* (Figure 24-14B).

Class Echinoidea (ek′i-noy′de-a) (Gr. *echinos,* sea urchin, hedgehog, + *eidos,* form, + ea, characterized by): **sea urchins, sea biscuits,** and **sand dollars.** More or less globular or disc-shaped, with no arms; compact skeleton or test with closely fitting plates; movable spines; ambulacral grooves closed; tube feet with suckers; pedicellariae present. Examples: *Arbacia, Strongylocentrotus* (Figure 24-15), *Lytechinus, Mellita.*

Class Holothuroidea (hol′o-thu-roy′de-a) (Gr. *holothourion,* sea cucumber, + *eidos,* form, + *ea,* characterized by): **sea cucumbers.** Cucumber-shaped, with no arms; spines absent; microscopic ossicles embedded in thick muscular wall; anus present; ambulacral grooves closed; tube feet with suckers; circumoral tentacles (modified tube feet); pedicellariae absent; madreporite internal. Examples: *Sclerodactyla, Parastichopus, Cucumaria* (Figure 24-21C).

*Brusca, R. C., and G. J. Brusca. 1990. Invertebrates. Sunderland, Massachusetts, Sinauer Associates; Meglitsch, P. A., and F. R. Schram. 1991. Invertebrate zoology, ed. 3. New York, Oxford University Press; Paul, C. R. S., and A. B. Smith. 1984. Biol. Rev. **59:**443–481.

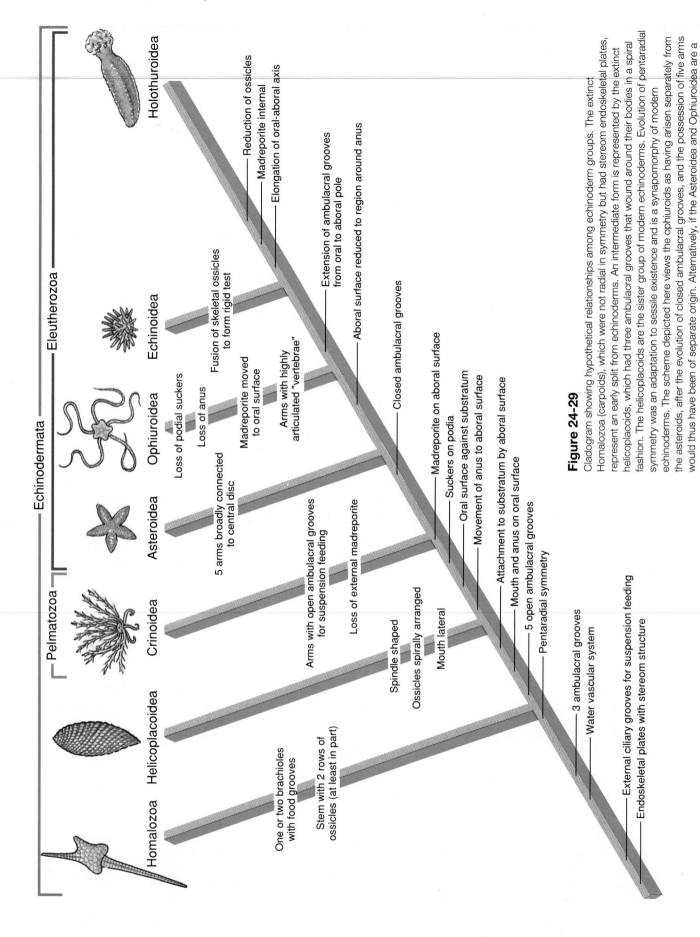

Figure 24-29

Cladogram showing hypothetical relationships among echinoderm groups. The extinct Homalozoa (carpoids), which were not radial in symmetry but had stereom endoskeletal plates, represent an early split from echinoderms. An intermediate form is represented by the extinct helicoplacoids, which had three ambulacral grooves that wound around their bodies in a spiral fashion. The helicoplacoids are the sister group of modern echinoderms. Evolution of pentaradial symmetry was an adaptation to sessile existence and is a synapomorphy of modern echinoderms. The scheme depicted here views the ophiuroids as having arisen separately from the asteroids, after the evolution of closed ambulacral grooves, and the possession of five arms would thus have been of separate origin. Alternatively, if the Asteroidea and Ophiuroidea are a monophyletic clade, with five arms being synapomorphic, then closed ambulacral grooves in the ophiuroids would have evolved separately from that character in the echinoids and holothuroids.

Summary

The phylum Echinodermata shows the characteristics of the Deuterostomia division of the animal kingdom. They are an important marine group sharply distinguished from other phyla of animals. They have a radial symmetry but were derived from bilateral ancestors.

The sea stars (class Asteroidea) can be used to illustrate the echinoderms. Sea stars usually have five arms, which merge gradually with a central disc. Like other echinoderms, they have no head and few specialized sensory organs. The mouth is directed toward the substratum. They have stereom dermal ossicles, respiratory papulae, and open ambulacral grooves. Many sea stars have pedicellariae. Their water-vascular system is an elaborate hydraulic system derived embryonically from one of their coelomic compartments. Along the ambulacral areas, branches of the water-vascular system (tube feet) are important in locomotion, food gathering, respiration, and excretion. Many sea stars are predators, whereas others feed on small particles. Sexes are separate, and reproductive systems are very simple. The bilateral, free-swimming larva becomes attached, transforms to a radial juvenile, then detaches and becomes a motile sea star.

The arms of brittle stars (class Ophiuroidea) are slender and sharply set off from the central disc. They have no pedicellariae or ampullae and their ambulacral grooves are closed. Their tube feet have no suckers, and their madreporite is on the oral side. They crawl by means of their arms, and their tube feet function in food gathering.

The dermal ossicles of sea urchins (class Echinoidea) are closely fitting plates, the body is compact, and there are no arms. The ambulacral areas are closed and extend around their body toward the aboral pole. Sea urchins move by means of tube feet or by their spines. Some urchins (sand dollars and heart urchins) have returned to bilateral symmetry.

The dermal ossicles in sea cucumbers (class Holothuroidea) are very small; therefore the body wall is soft. Their ambulacral areas also are closed and extend toward the aboral pole. Holothuroids are greatly elongated in the oral-aboral axis and lie on their side. Because certain of the ambulacral areas are characteristically against the substratum, sea cucumbers have also undergone some return to bilateral symmetry. The tube feet around the mouth are modified into tentacles, with which they feed. They have an internal respiratory tree, and the madreporite hangs free in the coelom.

Sea lilies and feather stars (class Crinoidea) are the only group of living echinoderms, other than the asteroids, with open ambulacral grooves. They are mucociliary particle feeders and lie with their oral side up.

Sea daisies (class Concentricycloidea) are a newly discovered class of very small echinoderms that are circular in shape, have marginal tube feet, and two concentric ring canals in their water-vascular system.

The ancestors of echinoderms were bilaterally symmetrical, and they probably evolved through a sessile stage that became radially symmetrical and then gave rise to the free-moving forms.

Review Questions

1. What is the constellation of characteristics possessed by echinoderms that is found in no other phylum?
2. How do we know that echinoderms were derived from an ancestor with bilateral symmetry?
3. Distinguish the following groups of echinoderms from each other: Crinoidea, Asteroidea, Ophiuroidea, Echinoidea, Holothuroidea, Concentricycloidea.
4. What is an ambulacrum, and what is the difference between open and closed ambulacral grooves?
5. Trace or make a rough copy of Figure 24-3B without labels; then from memory label the parts of the water-vascular system of the sea star.
6. Briefly explain the mechanism of action of a sea star's tube foot.
7. What are the structures involved in the following functions in sea stars? Briefly describe the action of each: respiration, feeding and digestion, excretion, reproduction.
8. Compare the structures and functions in question 7 as they are found in brittle stars, sea urchins, sea cucumbers, and crinoids.
9. Briefly describe development in sea stars, including metamorphosis.
10. Match the groups in the left column with *all* correct answers in the right column.

    ____ Crinoidea
    ____ Asteroidea
    ____ Ophiuroidea
    ____ Echinoidea
    ____ Holothuroidea
    ____ Concentri-
         cycloidea

    a. Closed ambulacral grooves
    b. Oral surface generally upward
    c. With arms
    d. Without arms
    e. Approximately globular or disc-shaped
    f. Elongated in oral-aboral axis
    g. With pedicellariae
    h. Madreporite internal
    i. Madreporite on oral plate

11. Define the following: pedicellariae, madreporite, respiratory tree, Aristotle's lantern.
12. What is some evidence that the ancestral echinoderm was sessile?
13. Give four examples of how echinoderms are important to humans.
14. What is a major difference in the function of the coelom in the holothurians compared with the other echinoderms?
15. Describe a reason for the hypothesis that the ancestor of the eleutherozoan groups was a radial, sessile organism.

Selected References

See also general references for Part III, p. 626.

Baker, A. N., F. W. E. Rowe, and H. E. S. Clark. 1986. A new class of Echinodermata from New Zealand. Nature **321:**862–864. *Describes the strange class Concentricycloidea.*

Birkeland, C. 1989. The Faustian traits of the crown-of-thorns starfish. Am. Sci. **77:**154–163. *The fast growth in the early years of the life of an* Acanthaster planci *results in loss of body integrity in later life.*

Davidson, E. H., B. R. Hough-Evans, and R. J. Britten. 1982. Molecular biology of the sea urchin embryo. Science **217:**17–26. *Many fundamental insights into the process of embryogenesis have been revealed through studies of sea urchins.*

Fell, H. B. 1982. Echinodermata. In Parker, S. P. (ed). Synopsis and classification of living organisms, vol. 2. New York, McGraw-Hill Book Company. *Presents an alternative classification, not recognizing Pelmatozoa and Eleutherozoa and placing Asteroidea and Ophiuroidea in subclasses of the class Stelleroidea, subphylum Asterozoa.*

Gilbert, S. F. 1994. Developmental biology, ed. 4. Sunderland, Massachusetts, Sinauer Associates. *Any modern text in developmental biology, such as this one, provides a multitude of examples in which studies on echinoderms have contributed (and continue to contribute) to our knowledge of development.*

Hughes, T. P. 1994. Catastrophes, phase shifts and large-scale degradation of a Caribbean coral reef. Science **265:**1547–1551. *Describes the sequence of events, including the die-off of sea urchins, leading to the destruction of the coral reefs around Jamaica.*

Lawrence, J. 1989. A functional biology of echinoderms. Baltimore, The Johns Hopkins Press. *Well-researched book on echinoderm biology with emphasis on feeding, maintenance, and reproduction.*

Moran, P. J. 1990. *Acanthaster planci* (L.): biographical data. Coral Reefs **9:**95–96. *Presents a summary of essential biological data on* A. planci. *This entire issue of* Coral Reefs *is devoted to* A. planci.

25

Chaetognaths and Hemichordates

Phylum Chaetognatha
Phylum Hemichordata

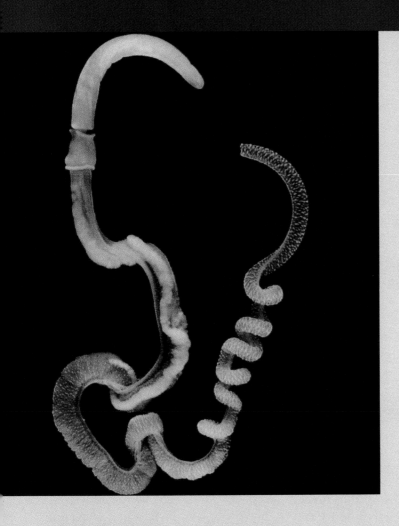

Part Chordates

In the mid-nineteenth century, with interest in the origin of the chordates running high, a group of wormlike marine invertebrates of unknown relationship began to attract attention with their chordatelike characteristics. In 1885 W. Bateson named them Hemichordata and forcefully argued that these organisms should be included in the phylum Chordata. Bateson pointed to several structures found in hemichordates that he believed were homologous with comparable features in chordates: a dorsal nerve cord, gill slits, and, most importantly, a sac-like evagination of the mouth region that he interpreted as a notochord. The notochord, a rodlike, supportive structure lying dorsal to the gut of early growth stages of all chordates, is a key distinguishing feature of the phylum Chordata. If the hemichordates possessed a notochord—even half a notochord—they had to be chordates.

Unfortunately, the structure that Bateson interpreted as a notochord neither looks like a notochord nor develops like a notochord. These and other problems with giving the hemichordates membership in the chordate club were noted in the 1930s, but by this time the concept had become firmly established in textbooks and began to assume a life of its own. Some zoologists and texts doggedly continued to assign the subphylum Hemichordata to the phylum Chordata for 25 or more years. Eventually most zoologists agreed that the hemichordates should be viewed as a distinct phylum of "lesser" deuterostomes. Bateson's name—Hemichordata—has stuck, however, and oddly enough seems appropriate for a group of animals that, although lacking a notochord, does bear certain characters in common with the true chordates. As the likely sister group of the chordates, the hemichordates are indeed half (or part) chordates. ∎

The deuterostomes include, along with the Echinodermata and the lophophorates, two other phyla: Hemichordata and Chordata. Two of the chordate subphyla—Urochordata and Cephalochordata—are also invertebrate groups. The phylum Chaetognatha traditionally has been included among the deuterostomes, but this arrangement is not supported by recent molecular evidence.* The chaetognaths do bear a number of deuterostome characters, however, and we will continue to include them in this chapter for the present. These phyla have enterocoelous development of the coelom and some form of radial cleavage.

PHYLUM CHAETOGNATHA

A common name for the chaetognaths is arrowworms. They are all marine animals and are highly specialized for their planktonic existence. Their relationship to other groups is obscure, although embryological characters indicate deuterostome affinities.

The name Chaetognatha (ketog'na-tha) (Gr. *chaite*, long flowing hair, + *gnathos,* jaw) refers to the sickle-shaped bristles on each side of the mouth. This is not a large group, for only some 65 species are known. Their small, straight bodies resemble miniature torpedoes, or darts, ranging from 2.5 to 10 cm in length.

The arrowworms are all adapted for a planktonic existence, except for *Spadella* (Gr. *spadix,* palm frond, + *ella,* dim. suffix), a benthic genus. They usually swim to the surface at night and descend during the day. Much of the time they drift passively, but they can dart forward in swift spurts, using the caudal fin and longitudinal muscles—a fact that no doubt contributes to their success as planktonic predators. Horizontal fins bordering the trunk are used in flotation rather than in active swimming.

*Telford, M. J., and P. W. H. Holland. 1993. Mol. Biol. Evol. **10:**660–676; Wada, H., and N. Satoh. 1994. Proc. Natl. Acad. Sci. **91:**1801–1804.

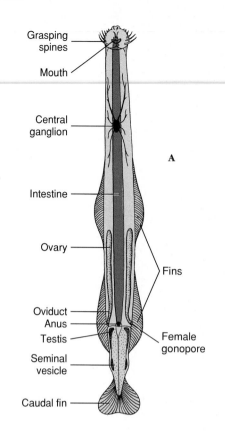

Grasping spines

Mouth

Central ganglion

A

Intestine

Ovary

Fins

Oviduct
Anus
Testis

Female gonopore

Seminal vesicle

Caudal fin

FORM AND FUNCTION

The body of the arrowworm is unsegmented and includes a head, trunk, and postanal tail (Figure 25-1A). On the underside of the head is a large vestibule leading to the mouth. The vestibule contains teeth and is flanked on both sides by curved chitinous spines used in seizing the prey. A pair of eyes is on the dorsal side. A peculiar hood formed from a fold of the neck can be drawn forward over the head and spines. When the animal captures prey, it retracts the hood, and the teeth and raptorial spines spread apart and then snap shut with startling speed. Arrowworms are voracious feeders, living on planktonic forms, especially copepods, and even small fish (Figure 25-1B). When they are abundant, as they often are, they may have a substantial ecological impact. They are nearly transparent, a characteristic of adaptive value in their role as planktonic predators.

A thin cuticle covers the body, and the epidermis is single layered except along the sides of the body, where it is stratified in a thick layer. These are the only invertebrates with a many-layered epidermis.

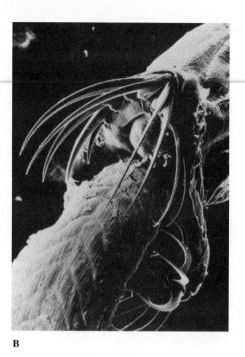

B

Figure 25-1
Arrowworms. **A,** Internal structure of *Sagitta.* **B,** Scanning electron micrograph of a juvenile arrowworm, *Flaccisagitta hexaptera* (35 mm length) eating a larval fish.

Arrowworms have a complete digestive system, a well-developed coelom, and a nervous system with a nerve ring containing large dorsal and ventral ganglia and a number of lateral ganglia. Sense organs include the eyes, sensory bristles, and a unique U-shaped ciliary loop that extends over the neck from the back of the head. The ciliary loop may detect water currents or may be chemosensory. However, vascular, respiratory, and excretory systems are entirely lacking.

Arrowworms are hermaphroditic with either cross- or self-fertilization. The eggs of *Sagitta* (L. arrow) bear a coat of jelly and are planktonic. Eggs of other arrowworms may be attached to the body and carried about for a time. The juveniles develop directly without metamorphosis. Chaetognath embryogenesis differs from that of other deuterostomes in that the coelom is formed by a backward extension from the archenteron rather than by pinched-off coelomic sacs. There is no true peritoneum lining the coelom. Cleavage is radial, complete, and equal.

A common arrowworm is *Sagitta* (Figure 25-1A).

PHYLUM HEMICHORDATA

The Hemichordata (hem′i-kor-da′ta) (Gr. *hemi,* half, + *chorda,* string, cord) are marine animals that were formerly considered a subphylum of the chordates, based on their possession of gill slits and a rudimentary notochord. However, the so-called hemichordate notochord is really a buccal diverticulum (called a stomochord, meaning "mouth-cord") and not homologous with the chordate notochord, so the hemichordates are given the rank of a separate phylum.

Hemichordates are vermiform bottom dwellers, living usually in shallow waters. Some are colonial and live in secreted tubes. Most are sedentary or sessile. Their distribution is almost cosmopolitan, but their secretive habits and fragile bodies make collecting them difficult.

Members of class Enteropneusta (Gr. *enteron,* intestine, + *pneustikos,* of, or for, breathing) (acorn worms) range from 20 mm to 2.5 m in length. Members of class Pterobranchia (Gr. *pteron,* wing, + *branchia,* gills) are smaller, usually 1 to 12 mm, not including the stalk. About 70 species of enteropneusts and two small genera of pterobranchs are recognized.

Hemichordates have the typical tricoelomate structure of deuterostomes.

CLASS ENTEROPNEUSTA

The enteropneusts, or acorn worms, are sluggish, wormlike animals that live in burrows or under stones, usually in mud or sand flats of intertidal zones. *Balanoglossus* (Gr. *balanos,* acorn, + *glōssa,* tongue) and *Saccoglossus* (Gr. *sakkos,* sac, strainer, + *glōssa,* tongue) (Figure 25-2) are common genera.

Form and Function

The mucus-covered body is divided into a tonguelike proboscis, a short collar, and a long trunk (protosome, mesosome, and metasome).

Proboscis. The proboscis is the active part of the animal. It probes about in the mud, examining its surroundings and collecting food in mucous strands on its surface. Cilia carry the particles to the groove at the edge of the collar, direct them to the mouth on the underside, and then the particles are swallowed. Large particles can be rejected by covering the mouth with the edge of the collar (Figure 25-3).

Burrow dwellers use the proboscis to excavate, thrusting it into the mud or sand and allowing cilia and mucus to move the sand backward. Or they may eat the sand and mud as they go, extracting its organic contents. They build U-shaped mucus-lined burrows, usually with two openings 10 to 30 cm apart and with the base of the U 50 to 75 cm below the surface. They can thrust the proboscis out the front opening for feeding. Defecation at the back opening builds characteristic spiral mounds of feces that leave a telltale clue to the location of the burrows.

In the posterior end of the proboscis is a small coelomic sac (protocoel) into which extends the **buccal diverticulum,** a slender, blindly ending pouch of the gut that reaches forward into the buccal region and was formerly considered a notochord. A

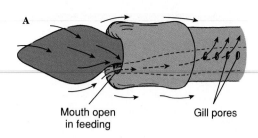

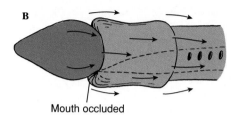

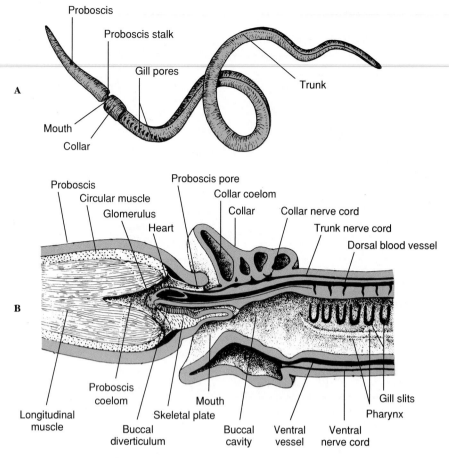

Figure 25-2
Acorn worm *Saccoglossus* (Hemichordata, class Enteropneusta). **A,** External lateral view.
B, Longitudinal section through anterior end.

Figure 25-3
Food currents of enteropneust hemichordate.
A, Side view of acorn worm with mouth open,
showing direction of currents created by cilia
on proboscis and collar. Food particles are
directed toward mouth and digestive tract.
Rejected particles move toward outside of
collar. Water leaves through gill pores.
B, When mouth is occluded, all particles are
rejected and passed onto the collar.
Nonburrowing and some burrowing
hemichordates use this feeding method.

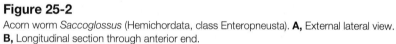

slender canal connects the protocoel with a **proboscis pore** to the outside (Figure 25-2B). The paired coelomic cavities in the collar also open by pores. By taking in water through the pores into the coelomic sacs, the proboscis and collar can be stiffened to aid in burrowing. Contraction of the body musculature then forces the excess water out through the gill slits, reducing the hydrostatic pressure and allowing the animal to move forward.

Branchial System.   A row of **gill pores** is located dorsolaterally on each side of the trunk just behind the collar (Figure 25-3A). These open from a series of gill chambers that in turn connect with a series of **gill slits** in the sides of the pharynx. There are no gills on the gill slits, but some respiratory gaseous exchange occurs in the vascular branchial epithelium, as well as in the body surface. Ciliary currents keep a fresh supply of water moving from the mouth through the pharynx and out the gill slits and branchial chambers to the outside.

Feeding and the Digestive System.
Hemichordates are largely ciliary-mucus feeders. Behind the buccal cavity lies the large pharynx containing in its dorsal part the U-shaped gill slits (Figure 25-2B). Since there are no gills, the primary function of the branchial mechanism of the pharynx is presumably food gathering. Having been caught in mucus and brought to the mouth by ciliary action on the proboscis and collar, food particles are strained out of the branchial water that leaves through the gill slits. The food then passes to the ventral part of the pharynx and esophagus to the intestine, where digestion and absorption occur (Figure 25-3).

Circulatory and Excretory Systems.
A middorsal vessel carries the colorless blood forward above the gut. In the collar the vessel expands into a sinus and a heart vesicle above the buccal diverticulum. Blood then enters a network of blood sinuses called the **glomerulus,** which partially surrounds these structures. The glomerulus is assumed to have an excretory function (Figure 25-2B). Blood travels posteriorly through a ventral vessel below the gut, passing through extensive sinuses to the gut and body wall.

Nervous and Sensory Systems.
The nervous system consists mostly of a subepithelial network, or plexus, of nerve cells and fibers to which processes of epithelial cells are attached. Thickenings of this net form dorsal and ventral nerve cords that are united posterior to the collar by a ring connective. The dorsal cord continues into the collar and furnishes many fibers to the plexus of the proboscis. The collar cord is hollow in some species and contains giant

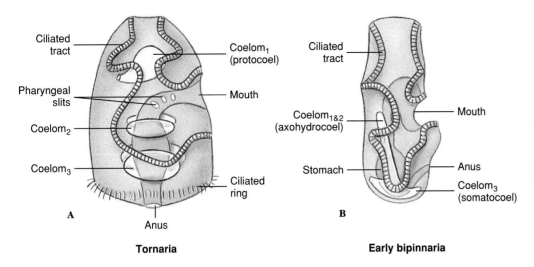

Ciliated tract
Coelom₁ (protocoel)
Pharyngeal slits
Mouth
Coelom₂
Coelom₃
Ciliated ring

A

Anus

Tornaria

Ciliated tract
Coelom₁&₂ (axohydrocoel)
Mouth
Stomach
Anus
Coelom₃ (somatocoel)

B

Early bipinnaria

Figure 25-4
Comparison of a hemichordate tornaria **(A)** to an echinoderm bipinnaria **(B)**.

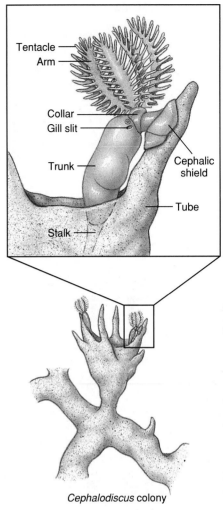

Cephalodiscus colony

Figure 25-5
Cephalodiscus, a pterobranch hemichordate. These tiny (5 to 7 mm) forms live in tubes in which they can move freely. Ciliated tentacles and arms direct currents of food and water toward mouth.

nerve cells with processes running to the nerve trunks. This nerve plexus system is quite reminiscent of that of the cnidarians and echinoderms.

Sensory receptors include neurosensory cells throughout the epidermis (especially in the proboscis, a preoral ciliary organ that may be chemoreceptive) and photoreceptor cells.

Reproductive System and Development. Sexes are separate in enteropneusts. A dorsolateral row of gonads runs along each side of the anterior part of the trunk. Fertilization is external, and in some species a ciliated **tornaria** larva develops that at certain stages is so similar to the echinoderm bipinnaria that it was once believed to be an echinoderm larva (Figure 25-4). The familiar *Saccoglossus* of American waters has direct development without a tornaria stage.

CLASS PTEROBRANCHIA

The basic plan of the class Pterobranchia is similar to that of the Enteropneusta, but certain structural differences are correlated with the sedentary life-style of pterobranchs. The first pterobranch ever reported was obtained by the famed *Challenger* expedition of 1872 to 1876. Although first placed among the Polyzoa (Entoprocta and Ectoprocta), its affinities to the hemichordates were

later recognized. Only two genera (*Cephalodiscus* and *Rhabdopleura*) are known in any detail.

Pterobranchs are small animals, usually within the range of 1 to 7 mm in length, although the stalk may be longer. Many individuals of *Cephalodiscus* (Gr. *kephalē,* head, + *diskos,* disc) (Figure 25-5) live together in collagenous tubes, which often form an anastomosing system. The zooids are not connected, however, and live independently in the tubes. Through apertures in these tubes, they extend their crown of tentacles. They are attached to the walls of the tubes by extensible stalks that can jerk the owners back into the tubes when necessary.

The body of *Cephalodiscus* is divided into the three regions—proboscis, collar, and trunk—characteristic of the hemichordates. There is only one pair of gill slits, and the alimentary canal is U-shaped, with the anus near the mouth. The proboscis is shield shaped. At the base of the proboscis are five to nine pairs of branching arms with tentacles containing an extension of the coelomic compartment of the mesosome, as in a lophophore. Ciliated grooves on the tentacles and arms collect food. Some species are dioecious, and others are monoecious. Asexual reproduction by budding may also occur.

In *Rhabdopleura* (Gr. *rhabdos,* rod, + *pleura,* a rib, the side), which

is smaller than *Cephalodiscus,* the members remain together to form a colony of zooids connected by a stolon and enclosed in secreted tubes (Figure 25-6). The collar in these forms bears two branching arms. No gill clefts or glomeruli are present. New individuals are reproduced by budding from a creeping basal stolon, which branches on a substratum. None of the pterobranchs has a tubular nerve cord in the collar, but otherwise their nervous system is similar to that of the Enteropneusta.

The fossil graptolites of the middle Paleozoic era often are placed as an extinct class under Hemichordata. They are important index fossils of

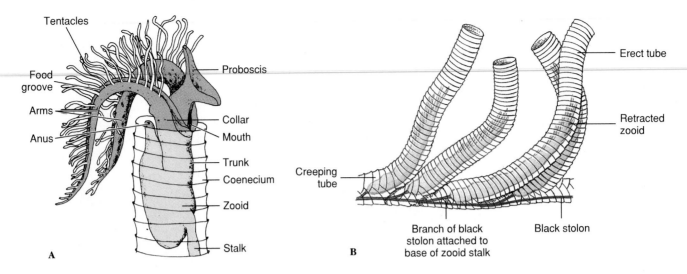

Figure 25-6

A, *Rhabdopleura*, a pterobranch hemichordate in its tube. Individuals live in branching tubes connected by stolons, and protrude the ciliated tentacles for feeding. **B,** Portion of a colony.

the Ordovician and Silurian geological strata. Alignment of the graptolites with the hemichordates has been very controversial, but discovery of an organism that seems to be a living graptolite lends strong support to the hypothesis. It has been described as a new species of pterobranch, called *Cephalodiscus graptolitoides.*

PHYLOGENY AND ADAPTIVE RADIATION

PHYLOGENY

Hemichordate phylogeny has long been puzzling. Hemichordates share characters with both the echinoderms and the chordates. With chordates they share gill slits, which serve primarily for filter feeding and secondarily for breathing, as they do in some of the protochordates. In addition, a short dorsal, somewhat hollow nerve cord in the collar zone may be homologous to the nerve cord of the chordates (Figure 25-7). The buccal diverticulum in the hemichordate

mouth cavity, long thought homologous to the notochord of chordates, is now considered a synapomorphy of hemichordates themselves. The early embryogenesis of the hemichordates is remarkably like that of echinoderms, and the early tornaria larva is almost identical to the bipinnaria larva of asteroids. However, the hypothetical relationships shown in Figure 25-7 place the lophophorates as sister groups of the hemichordates and chordates, required by the proposed synapomorphy for all these groups of a crown of ciliated tentacles containing extensions of the mesocoel.*

Other than their shared deuterostome characters, the relationship of the chaetognaths to the other deuterostome phyla is enigmatic. Studies on base sequences of rRNA indicate that chaetognaths may be closer to the protostomes than we heretofore thought. Some investigators suggest, however, that chaetognaths are neither protostomes nor deuterostomes but originated independently from an early coelomate lineage.

ADAPTIVE RADIATION

Because of their sessile lives and their habitat in secreted tubes in ocean bottoms, where conditions are fairly stable, the pterobranchs have undergone little adaptive divergence. They have retained a tentacular type of ciliary feeding. The enteropneusts, on the other hand, although sluggish, are more active than the pterobranchs. Having lost the tentaculated arms, they use a proboscis to trap small organisms in mucus, or they eat sand as they burrow and digest organic sediments from the sand. Their evolutionary divergence, although greater than that of the pterobranchs, is still modest.

*Brusca, R. C., and G. J. Brusca. 1990. Invertebrates. Sunderland, Massachusetts, Sinauer Associates.

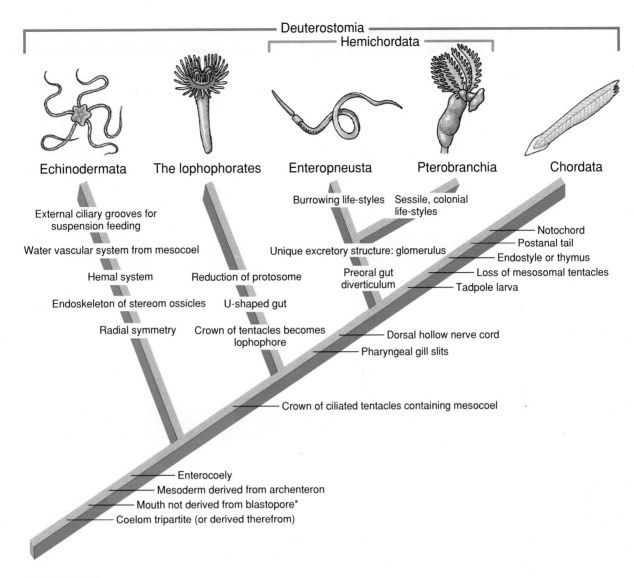

*Except in phoronids.

Figure 25-7

Cladogram showing hypothetical relationships among deuterostome phyla. The crown of ciliated tentacles (containing extensions of the mesocoel) is here considered a character borne by the ancestors of the lophophorates, hemichordates, and chordates. The tentacular crown would have been lost in the line leading to the chordates, and among the hemichordates, the enteropneusts. The pterobranchs retain the primitive character, while in the lophophorate phyla, it is modified into a lophophore. Uncertainties regarding the relationships of the Chaetognatha preclude placement in this cladogram. The Protostomia serves as outgroup.

Source: Modified from R. C. Brusca and G. J. Brusca, Invertebrates. *Sinauer Associates, Inc., Sunderland, MA, 1990.*

Summary

The arrowworms (phylum Chaetognatha) are a small group but are important as a component of marine plankton. They have a well-developed coelom and are effective predators, catching other planktonic organisms with the teeth and chitinous spines around their mouth.

Members of the phylum Hemichordata are marine worms that were formerly considered chordates because their buccal diverticulum was considered a notochord. However, like the chordates, some of them do have gill slits and a hollow, dorsal nerve cord. The divisions of their body (proboscis, collar, trunk) contain the typical deuterostome coelomic compartments (protocoel, mesocoel, metacoel). The hemichordate class Enteropneusta contains burrowing worms that feed on particles strained out of the water by their gill slits. Members of the class Pterobranchia are tube dwellers, filter feeding with tentacles. The hemichordates are important phylogenetically because they show affinities with the chordates, echinoderms, and lophophorates, and they are the likely sister group of the chordates.

Review Questions

1. What is evidence that Chaetognatha are deuterostomes? What is evidence that conflicts with this hypothesis?
2. What is the ecological importance of arrowworms?
3. What characteristics do the Hemichordata have in common with the Chordata, and how do the two phyla differ?
4. Distinguish the Enteropneusta from the Pterobranchia.
5. What is the evidence that the Hemichordata are related both to the echinoderms and the lophophorate phyla?

Selected References

See also general references for Part III, p. 626.

Barrington, E. J. W. 1965. The biology of Hemichordata and Protochordata. San Francisco, W. H. Freeman & Company. *Concise account of behavior, physiology, and reproduction of hemichordates, urochordates, and cephalochordates.*

Bieri, R., and E. V. Thuesen. 1990. The strange worm *Bathybelos*. Am. Sci. **78:**542–549. *Bathybelos is a peculiar chaetognath with a dorsal nervous system, a characteristic shared in the animal kingdom only with the Hemichordata and Chordata. The authors contend that the character in chaetognaths is convergent with that in the hemichordates and chordates.*

Svitii, K. A. 1993. It's alive, and it's a graptolite. Discover **14**(7):18–19. *Short account of the discovery of the "living fossil,"* Cephalodiscus graptolitoides.

Thuesen, E. V., and K. Kogure. 1989. Bacterial production of tetrodotoxin in four species of Chaetognatha. Biol. Bull. **176:**191–194. *Chaetognaths use venom to enhance prey capture, and the venom (tetrodotoxin) is produced by bacteria* (Vibrio alginolyticus).

26

The Chordates

*General Characteristics,
Protochordates, and Ancestry
of the Earliest Vertebrates*

It's a Long Way From Amphioxus

Along the more southern coasts of North America, half
buried in sand on the sea floor, lives a small fishlike
translucent animal quietly filtering organic particles from
seawater. Inconspicuous, of no commercial value and
largely unknown, this creature is nonetheless one of the
famous animals of classical zoology. It is amphioxus, an
animal that wonderfully exhibits the four distinctive
hallmarks of the phylum Chordata—(1) dorsal, tubular nerve
cord overlying (2) a supportive notochord, (3) gill slits for
filter feeding, and (4) a postanal tail for propulsion—all
wrapped up in one creature with textbook simplicity.
Amphioxus is an animal that might have been designed by a
zoologist for the classroom. During the nineteenth century,
with interest in vertebrate ancestry running high, amphioxus
was considered by many to resemble very closely the direct
ancestor of the vertebrates. Its exalted position was later
acknowledged by Philip Pope in a poem sung to the tune of
"Tipperary." It ends with the refrain:

It's a long way from amphioxus
　　It's a long way to us.
It's a long way from amphioxus
　　To the meanest human cuss.
Well, it's good-bye to fins and gill slits
　　And its welcome lungs and hair,
It's a long, long way from amphioxus
　　But we all came from there.

But amphioxus' place in the sun was not to endure. For one
thing, amphioxus lacks one of the most important of vertebrate
characteristics, a distinct head with special sense organs and
the equipment for shifting to an active predatory mode of life.
This, together with several specialized features, suggests to
zoologists today that amphioxus represents an early departure
from the main line of chordate descent. It seems that we are a
very long way indeed from amphioxus. Nevertheless, while
amphioxus is denied the vertebrate ancestral award, we
believe that it more closely resembles the earliest prevertebrate
than any other animal we know. ■

POSITION IN THE ANIMAL KINGDOM

Phylum Chordata (kor-da′ta) (L. *chorda,* cord) belongs to the Deuterostomia branch of the animal kingdom that includes the phyla Echinodermata, Hemichordata, and the three lophophorate phyla—Phoronida, Ectoprocta, and Brachiopoda. These six phyla share many embryological features and are probably descended from an ancient common ancestor. From humble beginnings, the chordates have evolved a vertebrate body plan of enormous adaptability that always remains distinctive, while it provides almost unlimited scope for specialization in life habitat, form, and function.

BIOLOGICAL CONTRIBUTIONS

1. The **endoskeleton** of the vertebrates permits continuous growth without molting and the attainment of large body size, and it provides an efficient framework for muscle attachment.
2. The **perforated pharynx** of protochordates that originated as a suspension-feeding device served as the framework for the subsequent evolution of true internal gills with pharyngeal muscular pump, and jaws.
3. The adoption of a **predatory habit** by the early vertebrates and the accompanying evolution of a **highly differentiated brain** and **paired special sense organs** contributed in large measure to the successful adaptive radiation of the vertebrates.
4. The **paired appendages** that appeared in the aquatic vertebrates were successfully adapted later as jointed limbs for efficient locomotion on land or as wings for flight.

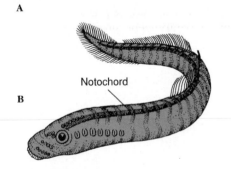

Figure 26-1

A, Structure of the notochord and its surrounding sheaths. Cells of the notochord proper are thick walled, pressed together closely, and filled with semifluid. Stiffness is caused mainly by turgidity of fluid-filled cells and surrounding connective tissue sheaths. This primitive type of endo-skeleton is characteristic of all chordates at some stage of the life cycle. The notochord provides longitudinal stiffening of the main body axis, a base for trunk muscles, and an axis around which the vertebral column develops. **B,** In hagfishes and lampreys it persists throughout life, but in higher vertebrates it is largely replaced by the vertebrae. In mammals slight remnants are found in nuclei pulposi of intervertebral discs. The method of notochord formation is different in the various groups of animals. In amphioxus it originates from the endoderm; in birds and mammals it arises as an anterior outgrowth of the embryonic primitive streak.

THE CHORDATES

The animals most familiar to most people belong to the phylum Chordata (kor-da′ta) (L. *chorda,* cord). Humans are members and share with other chordates the characteristic from which the phylum derives its name—the **notochord** (Gr. *nōton,* back, + L. *chorda,* cord) (Figure 26-1). This structure is possessed by all members of the phylum, in either the larval or the embryonic stages or throughout life. The notochord is a rodlike, semirigid body of cells enclosed by a fibrous sheath, which extends, in most cases, the length of the body between the gut tract and the central nervous system. Its primary purpose is to support and stiffen the body, that is, to act as a skeletal axis.

The structural plan of chordates shares features of many nonchordate invertebrates, such as bilateral symmetry, anteroposterior axis, coelom, tube-within-a-tube arrangement, metamerism, and cephalization. However, the exact phylogenetic position of the chordates within the animal kingdom is unclear.

Two possible lines of descent have been proposed. Earlier speculations that focused on the arthropod-annelid-mollusc group (Protostomia branch) of the invertebrates have fallen from favor. It is now believed that only members of the echinoderm-hemichordate assemblage (Deuterostomia branch) deserves serious consideration as a chordate sister group. The chordates share with the other Deuterostomes several important characteristics: radial cleavage (p. 108), an anus derived from the first embryonic opening (blastopore) and a mouth derived from an opening of secondary origin, and a coelom primitively formed by fusion of enterocoelous pouches (except in vertebrates in which the coelom is basically schizocoelous). These common characteristics indicate a natural unity among the Deuterostomia.

As a whole, there is more fundamental unity of plan throughout all the organs and systems of this phylum than there is in many of the invertebrate phyla. Ecologically the chordates are among the most adapt-

CLASSIFICATION OF PHYLUM CHORDATA

Phylum Chordata

Group Protochordata (Acrania)

Subphylum Urochordata (u′ro-kor-da′ta) (Gr. *oura,* tail, + L. *chorda,* cord, + *ata,* characterized by) **(Tunicata): tunicates.** Notochord and nerve cord in free-swimming larva only; ascidian adults sessile, encased in tunic.

Subphylum Cephalochordata (sef′a-lo-kor-da′ta) (Gr. *kephale̅,* head, + L. *chorda,* cord): **lancelets (amphioxus).** Notochord and nerve cord found along entire length of body and persist throughout life; fishlike in form.

Group Craniata

Subphylum Vertebrata (ver′te-bra′ta) (L. *vertebratus,* backboned). Bony or cartilaginous vertebrae surrounding spinal cord; notochord only in embryonic stages, persisting in some fishes; also may be divided into two groups (superclasses) according to presence of jaws.

Superclass Agnatha (ag′na-tha) (Gr. *a,* without, + *gnathos,* jaw) **(Cyclostomata): hagfishes, lampreys.** Without true jaws or paired appendages. (Probably a paraphyletic group.)

Class Myxini (mik-sin′y) (Gr. *myxa,* slime): **hagfishes.** Terminal mouth with four pairs of tentacles; buccal funnel absent; nasal sac with duct to pharynx; 5 to 15 pairs of gill pouches; partially hermaphroditic.

Class Cephalaspidomorphi (sef-a-lass′pe-do-morf′e) (Gr. *kephale̅,* head, + *aspidos,* shield, *morphe̅,* form) **(Petromyzones): lampreys.** Suctorial mouth with horny teeth; nasal sac not connected to mouth; seven pairs of gill pouches.

Superclass Gnathostomata (na′tho-sto′ma-ta) (Gr. *gnathos,* jaw, + *stoma,* mouth): **jawed fishes, all tetrapods.** With jaws and (usually) paired appendages.

Class Chondrichthyes (kon-drik′thee-eez) (Gr. *chondros,* cartilage, + *ichthys,* a fish): **sharks, skates, rays, chimaeras.** Streamlined body with heterocercal tail; cartilaginous skeleton; five to seven gills with separate openings, no operculum, no swim bladder.

Class Osteichthyes (ost′e-ik′thee-eez) (Gr. *osteon,* bone, + *ichthys,* a fish): **bony fishes.** Primitively fusiform body but variously modified; mostly ossified skeleton; single gill opening on each side covered with operculum; usually swim bladder or lung.

Class Amphibia (am-fib′e-a) (Gr. *amphi,* both or double, + *bios,* life): **amphibians.** Ectothermic tetrapods; respiration by lungs, gills, or skin; development through larval stage; skin moist, containing mucous glands, and lacking scales.

Class Reptilia (rep-til′e-a) (L. *repere,* to creep): **reptiles.** Ectothermic tetrapods possessing lungs; embryo develops within shelled egg; no larval stage; skin dry, lacking mucous glands, and covered by epidermal scales. (A paraphyletic group.)

Class Aves (ay′veez) (L. pl. of *avis,* bird): **birds.** Endothermic vertebrates with front limbs modified for flight; body covered with feathers; scales on feet.

Class Mammalia (ma-may′lee-a) (L. *mamma,* breast): **mammals.** Endothermic vertebrates possessing mammary glands; body more or less covered with hair; well-developed neocerebrum.

able of organic forms and are able to occupy most kinds of habitat. They illustrate perhaps better than any other animal group the basic evolutionary processes of the origin of new structures, adaptive strategies, and adaptive radiation.

TRADITIONAL AND CLADISTIC CLASSIFICATION OF THE CHORDATES

The traditional Linnaean classification of the chordates provides a simple and convenient way to indicate the taxa included in each major group. However, in cladistic usage, some of the traditional taxa, such as Agnatha and Reptilia, are no longer recognized. This happens because such taxa do not satisfy the requirement of cladistics that only **monophyletic** groups are valid taxonomic entities, that is, groups that contain all known descendants of a single common ancestor. The reptiles, for example, are considered a **paraphyletic** grouping because this group does not contain all of the descendants of their most recent common ancestor. The common ancestor of reptiles is also an ancestor of birds and mammals. As shown in the cladogram (Figure 26-3), reptiles, birds, and mammals compose a monophyletic clade called the Amniota, so named because all develop from an egg having special extraembryonic membranes, one of which is the amnion. Therefore according to cladistics, the reptiles can be grouped only in a negative manner as amniotes that are not birds or mammals; there are no positive or novel features that unite the reptiles to the exclusion of birds and mammals. Similarly, the agnathans (hagfishes and lampreys) are a paraphyletic grouping because the most recent common ancestor of agnathans is also an ancestor of all the remaining vertebrates (the gnathostomes). The reasons why paraphyletic groups are not used in cladistic taxonomy are explained in more detail in Chapter 11 (p. 204).

It is important to recognize that the phylogenetic tree of the chordates (Figure 26-2) and the cladogram of the chordates (Figure 26-3) provide different kinds of information. The cladogram shows a nested hierarchy of taxa grouped by their sharing of derived characters. These characters may be morphological, physiological, embryological, behavioral, chromosomal, or molecular in nature. Although the cladogram shows the *relative* time of origin of the novel properties of taxonomic groups and their specific position in the hierarchical system of evolutionary common descent, it contains no timescale or information on ancestral lineages. By contrast, the branches

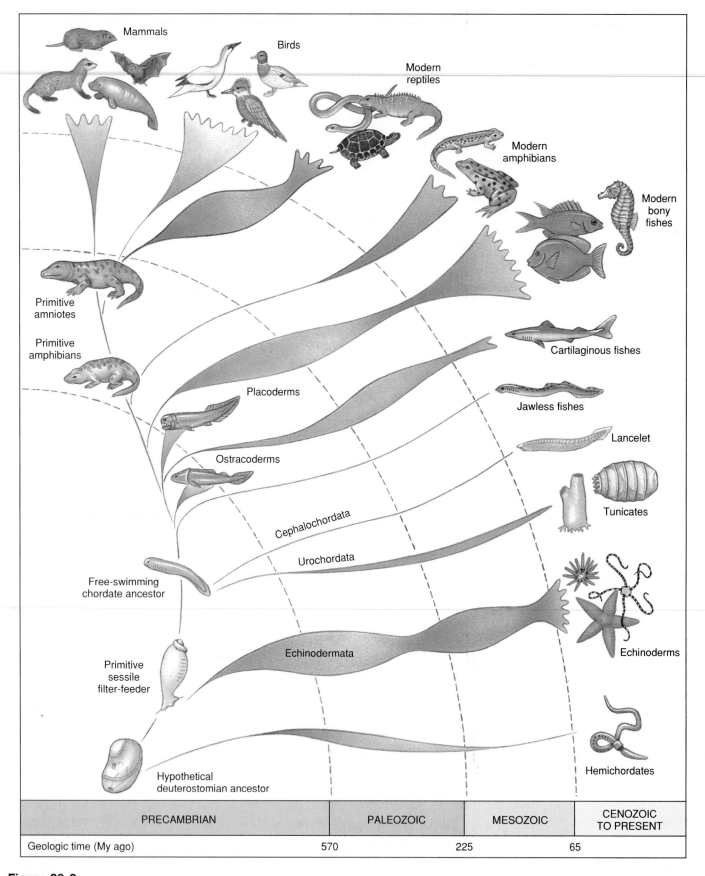

Figure 26-2
Phylogenetic tree of the chordates, suggesting probable origin and relationships. Other schemes have been suggested and are possible. The relative abundance in numbers of species of each group through geological time, as indicated by the fossil record, is suggested by the bulging and thinning of that group's line of descent. Dashed lines indicate a poor or nonexistent fossil record.

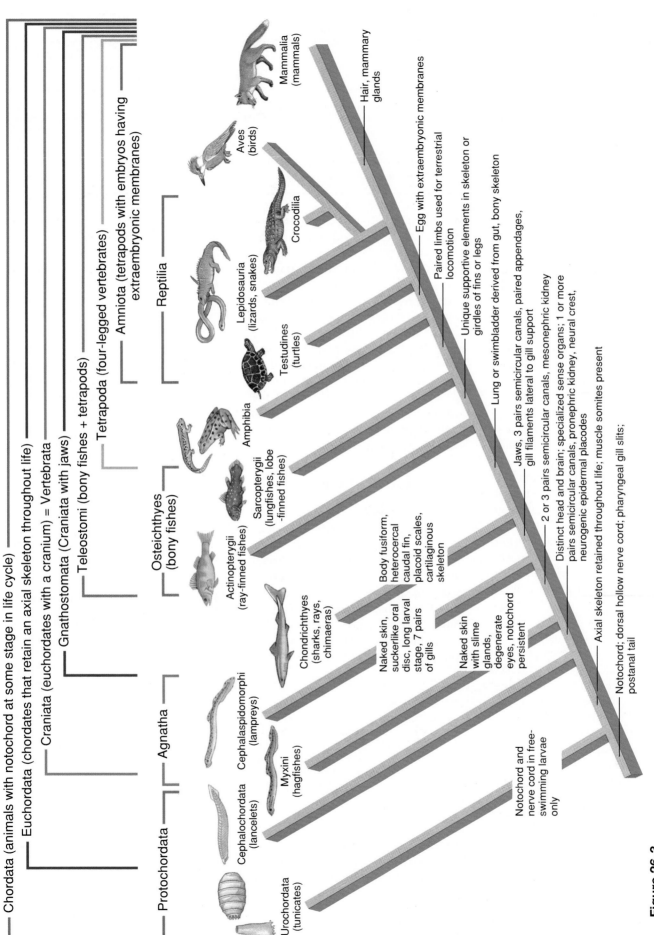

Figure 26-3

Cladogram of the Phylum Chordata showing the probable relationships of monophyletic groups composing the phylum. Each branch in the cladogram represents a monophyletic group. Some of the derived character states that identify the branchings are shown to the right of the branch points. Across the top of the cladogram are shown the progressive nestings of monophyletic groupings within the phylum. The term Craniata, although commonly equated with Vertebrata, is preferred by many authorities because it recognizes that the jawless vertebrates (Agnatha) have a cranium but no vertebrae. The traditional groupings Protochordata, Agnatha, Osteichthyes, and Reptilia, although paraphyletic and not recognized in cladistic usage, are retained because of their conceptual usefulness.

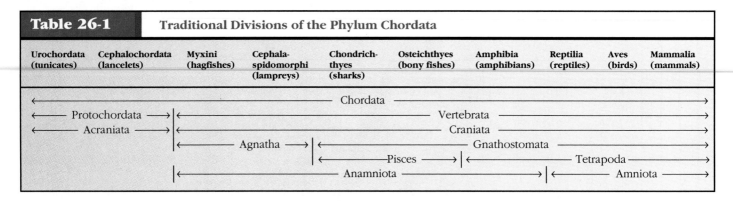

Table 26-1		**Traditional Divisions of the Phylum Chordata**							
Urochordata (tunicates)	Cephalochordata (lancelets)	Myxini (hagfishes)	Cephalaspidomorphi (lampreys)	Chondrichthyes (sharks)	Osteichthyes (bony fishes)	Amphibia (amphibians)	Reptilia (reptiles)	Aves (birds)	Mammalia (mammals)

of a phylogenetic tree are intended to represent real lineages that occurred in the evolutionary past. Geological information regarding ages of lineages is added to the information from the cladogram to generate a phylogenetic tree for the same taxa.

In our treatment of the chordates, we have retained the traditional Linnaean classification (p. 481) because of its conceptual usefulness and because the alternative—thorough revision following cladistic principles—would require extensive change and the virtual abandonment of familiar rankings. However, we have tried to use monophyletic taxa as much as possible, because such usage is consistent with both evolutionary and cladistic taxonomy (see p. 204).

Several of the traditional divisions of the phylum Chordata used in Linnaean classifications are shown in Table 26-1. A fundamental separation is the Protochordata from the Vertebrata. Since the former lack a well-developed head, they are also called Acraniata. All vertebrates have a well-developed skull case enclosing the brain and are called Craniata. The vertebrates (craniates) may be variously subdivided into groups based on shared possession of characteristics. Two such subdivisions shown in Table 26-1 are: (1) Agnatha, vertebrates lacking jaws (hagfishes and lampreys), and Gnathostomata, vertebrates having jaws (all other vertebrates) and (2) Amniota, vertebrates whose embryos develop within a fluid-filled sac, the amnion (reptiles, birds, and mammals), and Anamniota, vertebrates lacking this adaptation (fishes

and amphibians). The Gnathostomata in turn can be subdivided into Pisces, jawed vertebrates with limbs (if any) in the shape of fins; and the Tetrapoda (Gr. *tetras,* four, + *podos,* foot), jawed vertebrates with two pairs of limbs. Note that several of these groupings are paraphyletic (Protochordata, Acraniata, Agnatha, Anamniota, Pisces) and consequently are not accepted in cladistic classifications. Accepted monophyletic taxa are shown at the top of the cladogram in Figure 26-3 as a nested hierarchy of increasingly more inclusive groupings.

FOUR CHORDATE HALLMARKS

The four distinctive characteristics that, taken together, set chordates apart from all other phyla are the **notochord, dorsal tubular nerve cord, pharyngeal pouches,** and **postanal tail.** These characteristics are always found at some embryonic stage, although they may be altered or may disappear in later stages of the life cycle.

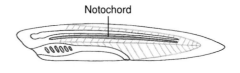

Notochord

NOTOCHORD

The notochord is a flexible, rodlike structure, extending the length of the body. It is the first part of the endoskeleton to appear in the embryo.

The notochord is an axis for muscle attachment, and because it can bend without shortening, it permits undulatory movements of the body. In most of the protochordates and in some vertebrates, the notochord persists throughout life (Figure 26-1). In all vertebrates a series of cartilaginous or bony vertebrae are formed from mesenchymal cells derived from blocks of mesodermal cells (somites) lateral to the notochord. In most vertebrates, the notochord is entirely displaced by the vertebrae, although remains of the notochord usually persist between or within the vertebrae.

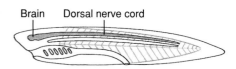

Brain   Dorsal nerve cord

DORSAL TUBULAR NERVE CORD

In most invertebrate phyla that have a nerve cord, it is ventral to the alimentary canal and is solid, but in the chordates the single cord is dorsal to the alimentary canal and is a tube (although the hollow center may be nearly obliterated during growth). The anterior end becomes enlarged to form the brain. The hollow cord is produced in the embryo by the infolding of ectodermal cells on the dorsal side of the body above the notochord. Among the vertebrates, the nerve cord passes through the protective neural arches of the vertebrae, and the anterior brain is surrounded by a bony or cartilaginous cranium.

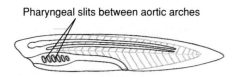

Pharyngeal slits between aortic arches

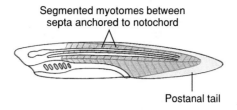

Segmented myotomes between septa anchored to notochord

Postanal tail

PHARYNGEAL POUCHES AND GILL SLITS

Pharyngeal gill slits are perforated slitlike openings that lead from the pharyngeal cavity to the outside. They are formed by the inpocketing of the outside ectoderm (pharyngeal grooves) and the evagination, or outpocketing, of the endodermal lining of the pharynx (pharyngeal pouches). In aquatic chordates, the two pockets break through when they meet, forming the slit. In amniotes these pockets may not break through and only grooves are formed instead of slits. In tetrapod vertebrates the pharyngeal pouches give rise to several different structures, including the Eustachian tube, middle ear cavity, tonsils, and parathyroid glands (see pp. 121–122).

The perforated pharynx originated as a suspension-feeding apparatus and is used as such in the protochordates. Water with suspended food particles is drawn by ciliary action through the mouth and flows out through the gill slits where food is trapped in mucus. Later, in the vertebrates, ciliary action was replaced by a muscular pump that drives water through the pharynx by expanding and contracting the pharyngeal cavity. Also modified were the aortic arches that carry blood through the gill bars. In the protochordates these are simple vessels surrounded by connective tissue. The early fishes added a capillary network having only thin, gas-permeable walls separating the water outside from the blood inside. This improved the efficiency of gas transfer. These adaptations led to the evolution of **internal gills,** completing the conversion of the pharynx from a suspension-feeding apparatus in protochordates to a respiratory organ in aquatic vertebrates.

POSTANAL TAIL

The postanal tail, together with somatic musculature and the stiffening notochord, provides the motility that larval tunicates and amphioxus need for their free-swimming existence. As a structure added to the body behind the end of the digestive tract, it clearly has evolved specifically for propulsion in water. Its efficiency is later increased in fishes with the addition of fins. The tail is evident in humans only as a vestige (the coccyx, a series of small vertebrae at the end of the spinal column) but most other mammals have a waggable tail as adults.

ANCESTRY AND EVOLUTION

Since the mid-nineteenth century when the theory of organic evolution became the focal point for ferreting out relationships between groups of living organisms, zoologists have debated the question of vertebrate origins. It has been very difficult to reconstruct lines of descent because the earliest protochordates were in all probability soft-bodied creatures that stood little chance of being preserved as fossils even under the most ideal conditions. Consequently, such reconstructions largely come from the study of living organisms, especially from an analysis of early developmental stages that tend to be more evolutionarily conserved than the differentiated adult forms that they become.

Zoologists at first speculated that the chordates evolved within the protostome lineage (annelids and arthropods) but discarded such ideas when they realized that supposed morphological similarities had no developmental basis. Early in this century when further theorizing became rooted in developmental patterns of animals, it

Most of the early efforts to identify kinship of chordates to other phyla are now recognized as based on similarities related to analogy rather than homology. Analogous structures are those that perform similar functions but have altogether different origins (such as wings of birds and butterflies). Homologous structures, on the other hand, share a common origin but may look quite different (at least superficially) and perform quite different functions. For example, all vertebrate forelimbs are homologous because they are derived from a pentadactyl limb, even though they may be modified as differently as the human arm and a bird's wing. Homologous structures share a genetic heritage; analogous structures do not. Obviously, only homologous similarities have any bearing in ancestral connections.

soon became apparent that the chordates must have originated within the deuterostome branch of the animal kingdom. As explained earlier (p. 109 and Figure 7-10), the Deuterostomia, a grouping that includes the echinoderms, hemichordates, lophophorates, and chordates, has several important embryological features that clearly separate it from the Protostomia and establish its monophyly. Thus the deuterostomes are almost certainly a natural grouping of interrelated animals that have their common origin in ancient Precambrian seas. There are several lines of anatomical, developmental, and molecular evidences suggesting that somewhat later, at the base of the Cambrian period some 570 million years ago, the first distinctive chordates appeared, evolved from an early echinoderm, early hemichordate, or a common ancestor of both (Figure 26-2; see also Figure 25-7, p. 477).

Although modern echinoderms look nothing at all like modern chordates, there is one curious group of fossil echinoderms, the "Calcichordata," that had gill slits and possibly other chordate attributes (Figure 26-4). These small nonsymmetrical forms had a head resembling a long-toed medieval boot,

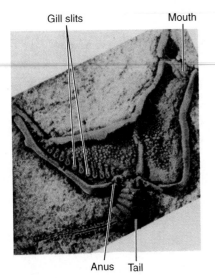

Figure 26-4

Fossil of a primitive echinoderm, a calcichordate, that lived during the Ordovician period (450 million years BP). It shows affinities with both echinoderms and chordates and may belong to a lineage that was ancestral to the chordates.

Courtesy of R. P. S. Jeffries, The Natural History Museum, London.

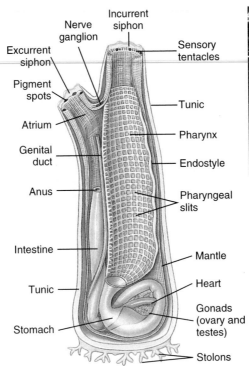

Figure 26-5

Structure of a common tunicate, *Ciona* sp.

Figure 26-6

Compound sea squirt *Botryllus* sp., common in shallow coastal waters and rock tide pools. Each of the star-shaped patterns represents a colonial arrangement in which the arms of the star are individual organisms, each with its own incurrent siphon at the end of the arm. All are united centrally where they share a common test, forming a compound tunicate.

a series of gill slits covered with flaps much like the gill openings of sharks, a flexible "arm" that could be interpreted as a postanal tail, and structures that are tentatively interpreted as a notochord and muscle blocks. They apparently used their gill slits for suspension feeding, as do the primitive chordates today. Although the calcichordates apparently had some of the right chordate characters based on soft anatomy, there is no convincing similarity between the hard skeleton of calcichordates (which was calcium carbonate) and that of vertebrates (which is composed of a hydrated complex of calcium and phosphate called hydroxyapatite). While such speculations bring us closer to an understanding of chordate origins, we are not yet in a position to know the precise characteristics of the long-sought chordate ancestor. However, we do know a great deal about the two living protochordate groups that descended from it. These we will now consider.

SUBPHYLUM UROCHORDATA (TUNICATA)

The urochordates ("tail-chordates"), more commonly called tunicates, in-clude about 3000 species. They are found in all seas from near the shoreline to great depths. Most of them are sessile as adults, although some are free living. The name "tunicate" is suggested by the usually tough, nonliving **tunic,** or test, that surrounds the animal and contains cellulose (Figure 26-5). As adults, tunicates are highly specialized chordates, for in most species only the larval form, which resembles a microscopic tadpole, bears all the chordate hallmarks. During adult metamorphosis, the notochord (which, in the larva, is restricted to the tail, hence the group name Urochordata) and the tail disappear altogether, while the dorsal nerve cord becomes reduced to a single ganglion.

Urochordata is divided into three classes: **Ascidiacea** (Gr. *askiolion,* little bag, + *acea,* suffix), **Larvacea** (L. *larva,* ghost, + *acea,* suffix), and **Thaliacea** (Gr. *thalia,* luxuriance, + *acea,* suffix). Of these the members of Ascidiacea are by far the most common, diverse, and best known. They are often called "sea squirts" because some species forcefully discharge a jet of water from the excurrent siphon when irritated. All but a few ascidian species are sessile animals, attached to rocks or other hard substrates such as pilings or the bottoms of ships. In many areas, they are among the most abundant of intertidal animals.

Ascidians may be solitary, colonial, or compound. Each of the solitary and colonial forms has its own test, but among the compound forms many individuals may share the same test (Figure 26-6). In some of these compound ascidians each member has its own incurrent siphon, but the excurrent opening is common to the group.

Solitary ascidians (Figure 26-5) are usually spherical or cylindrical forms. Lining the tunic is an inner membrane, the **mantle.** On the outside are two projections: the **incurrent siphon,** or oral siphon, which corresponds to the anterior end of the body, and the **excurrent siphon,** or atrial siphon, that marks the dorsal side. When the sea squirt is expanded, water enters the incurrent siphon and passes into a capacious ciliated **pharynx** that is minutely subdivided by gill slits to form an elaborate basketwork. Water passes through the gill slits into an **atrial cavity** and out through the excurrent siphon.

Feeding depends on the formation of a mucous net that is secreted by a glandular groove, the **endostyle,** located along the midventral side of the pharynx. Cilia on gill bars of the pharynx pull the mucus into a sheet that spreads dorsally across the inner

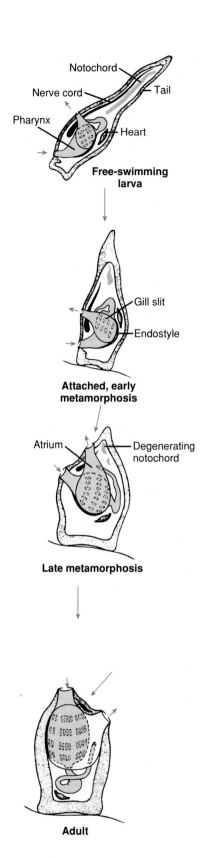

Figure 26-7

Metamorphosis of a solitary ascidian from a free-swimming tadpole larva stage.

face of the pharynx. Food particles brought in the incurrent opening are trapped on the mucous net, which is then worked into a rope and carried posteriorly by cilia into the esophagus and stomach. Nutrients are absorbed in the midgut and indigestible wastes are discharged from the anus, located near the excurrent siphon.

The circulatory system consists of a ventral heart and two large vessels, one on either side of the heart; these vessels connect to a diffuse system of smaller vessels and spaces serving the pharyngeal basket (where respiratory exchange occurs), the digestive organs, gonads, and other structures. An odd feature found in no other chordate is that the heart drives the blood first in one direction for a few beats, then pauses, reverses its action, and drives the blood in the opposite direction for a few beats. Another remarkable feature is the presence of strikingly high amounts of rare elements in the blood, such as vanadium and niobium. The vanadium concentration in the sea squirt *Ciona* may reach 2 million times its concentration in seawater. The function of these rare metals in the blood is a mystery.

The nervous system is restricted to a **nerve ganglion** and a plexus of nerves that lie on the dorsal side of the pharynx. Beneath the nerve ganglion is located the **subneural gland,** which is connected by a duct to the pharynx. Apparently this gland samples the water coming into the pharynx and may additionally perform an endocrine function concerned with reproduction. A notochord is lacking in adult sea squirts.

Sea squirts are hermaphroditic, with usually a single ovary and a single testis in the same animal. Germ cells are carried by ducts into the atrial cavity, and then into the surrounding water where fertilization occurs.

Of the four chief characteristics of chordates, adult sea squirts have only one: pharyngeal gill slits. However, the larval form gives away the secret of their true relationship. The tadpole larva (Figure 26-7) is an elongate,

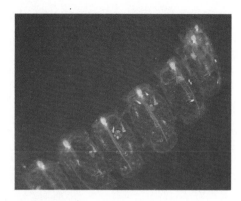

Figure 26-8

Colonial thaliacean. The transparent individuals of this delicate, planktonic species are grouped in a chain. Visible within each individual is an orange gonad, an opaque gut, and a long serrated gill bar.

transparent form with all four chordate characteristics: notochord, hollow dorsal nerve cord, propulsive postanal tail, and a large pharynx with endostyle and gill slits. The larva does not feed but swims about for some hours before fastening itself vertically by its adhesive papillae to some solid object. It then undergoes a dramatic metamorphosis (Figure 26-7) to become the sessile adult, so modified as to become almost unrecognizable as a chordate.

Tunicates of the class Thaliacea, known as thaliaceans or salps, are barrel- or lemon-shaped pelagic forms with transparent, gelatinous bodies that, despite the considerable size that some species reach, are nearly invisible in sunlit surface waters. They occur singly or in colonial chains that may reach several meters in length (Figure 26-8). The cylindrical thaliacean body is typically surrounded by bands of circular muscle, with incurrent and excurrent siphons at opposite ends. Water pumped through the body by muscular contraction (rather than by cilia as in ascidians) is used for locomotion by a sort of jet propulsion, for respiration, and as a source of particulate food that is filtered on mucous surfaces. Many are provided with luminous organs and give a brilliant light at night. Most of the body is hollow, with the viscera forming a compact mass on the ventral side.

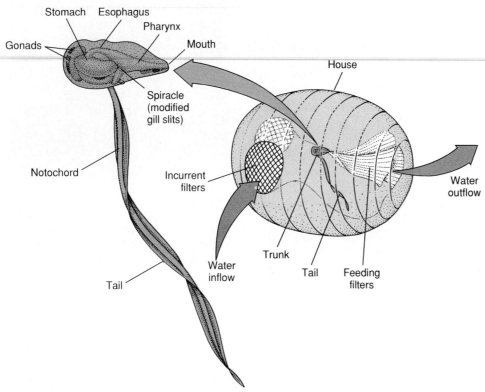

Figure 26-9

Larvacean adult (*left*) and as it appears within its transparent house (*right*), which is about the size of a walnut. When the feeding filters become clogged with food, the tunicate abandons its house and builds a new one.

The life histories of thaliaceans are often complex and are adapted to respond to sudden increases in their food supply. The appearance of a phytoplankton bloom, for example, is met by an explosive population increase leading to an extremely high density of thaliaceans. Common forms include *Doliolum* and *Salpa*, both of which reproduce by an alternation of sexual and asexual generations. Thaliaceans are believed to have evolved from sessile ancestors as did the ascidians.

The third tunicate class, the Larvacea (Appendicularia in some classifications) are curious larva-like pelagic creatures shaped like a bent tadpole. In fact their resemblance to the larval stages of other tunicates has given them their class name of Larvacea. They feed by a method unique in the animal world. Each builds a delicate house, a transparent hollow sphere of mucus interlaced with filters and passages through which the water enters (Figure 26-9). Particulate food trapped on a feeding filter inside the house is drawn into the animal's mouth through a strawlike tube. When the filters become clogged with waste, which happens about every 4 hours, the larvacean abandons its house and builds a new house, a process that takes only a few minutes. Like the thaliaceans, the larvaceans can quickly build up dense populations when food is abundant. Scuba diving among the houses, which are about the size of walnuts, is likened to swimming through a snowstorm! Larvaceans are paedomorphic, that is, they are sexually mature animals that have retained the larval body form of their evolutionary ancestors (see the boxed note explaining paedomorphosis on p. 492).

SUBPHYLUM CEPHALOCHORDATA

The cephalochordates are the marine lancelets: slender, laterally compressed, translucent animals about 5 to 7 cm in length (Figure 26-10) that inhabit the sandy bottoms of coastal waters around the world. Lancelets originally bore the generic name *Amphioxus* (Gr. *amphi*, both ends, + *oxys*, sharp), later surrendered by priority to *Branchiostoma* (Gr. *branchia*, gills, + *stoma*, mouth). This left amphioxus as a convenient common name for all of the approximately 25 species in this diminutive subphylum. Four species of amphioxus are found in North American coastal waters.

Amphioxus is especially interesting because it has the four distinctive characteristics of chordates in simple form. Water enters the mouth, driven by cilia in the buccal cavity, then passes through numerous gill slits in the pharynx where food is trapped in mucus, which is then moved by cilia into the intestine. Here the smallest food particles are separated from the mucus and passed into the hepatic cecum (diverticulum) where they are phagocytized and digested intracellularly. As in the tunicates, the filtered water passes first into an atrium, then leaves the body by an atriopore (equivalent to the excurrent siphon of tunicates).

The closed circulatory system is complex for so simple a chordate. The flow pattern is remarkably similar to that of the primitive fishes, although there is no heart. Blood is pumped forward in the ventral aorta by peristaltic-like contractions of the vessel wall, then passes upward through the branchial arteries (aortic arches) in the gill bars to the paired dorsal aortas, which join to become a single dorsal aorta. From here the blood is distributed to the body tissues by microcirculation and then is collected in veins, which return it to the ventral aorta. Lacking both erythrocytes and hemoglobin, the blood is thought to transport nutrients but play little role in gas exchange.

The nervous system is centered around a hollow nerve cord lying above the notochord. Pairs of spinal nerve roots emerge at each trunk myomeric (muscle) segment. Sense organs are simple, unpaired bipolar receptors located in various parts of the body. The "brain" is a simple vesicle at the anterior end of the nerve cord.

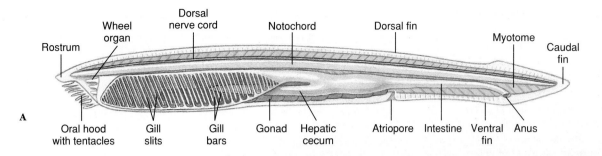

Figure 26-10

Amphioxus. This interesting bottom-dwelling cephalochordate illustrates the four distinctive chordate characteristics (notochord, dorsal nerve cord, pharyngeal gill slits, and postanal tail). The vertebrate ancestor is thought to have had a similar body plan. **A,** Internal structure. **B,** Living amphioxus in typical position for suspension feeding. Note the oral hood with tentacles surrounding the mouth.

Sexes are separate. The sex cells are set free in the atrial cavity, then pass out the atriopore to the outside where fertilization occurs. Cleavage is total (holoblastic) and a gastrula is formed by invagination. The larvae hatch soon after deposition and gradually assume the shape of adults.

No other chordate shows the basic diagnostic chordate characteristics as clearly as the amphioxus. In addition to the four chordate anatomical hallmarks, amphioxus possesses several structural features that suggest the vertebrate plan. Among these are a liver cecum, a diverticulum that resembles the vertebrate pancreas in secreting digestive enzymes, segmented trunk musculature, and the basic circulatory plan of more advanced chordates. As discussed later (p. 492), many zoologists consider amphioxus to be a living descendant of an ancestor that gave rise to both the cephalochordates and the vertebrates. Therefore the cephalochordates are, in cladistic terms, the sister group of the vertebrates (Figure 26-3).

SUBPHYLUM VERTEBRATA (CRANIATA)

The third subphylum of the chordates is the large and eminently diverse Vertebrata. This monophyletic group shares the basic chordate characteristics with the other two subphyla, but in addition it demonstrates a number of novel homologies that the others do not share. The alternative name of the subphylum, Craniata, more accurately describes the group since all have a cranium (bony or cartilaginous brain case) but some, the jawless fishes, lack vertebrae.

ADAPTATIONS THAT HAVE GUIDED VERTEBRATE EVOLUTION

From the earliest fishes to the mammals, the evolution of the vertebrates has been guided by the specialized basic adaptations of the living endoskeleton, pharynx and efficient respiration, advanced nervous system, and paired limbs.

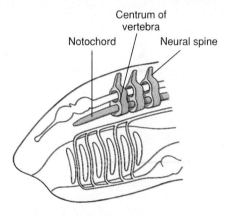

Living Endoskeleton

The endoskeleton of vertebrates, as in the echinoderms, is an internal supportive structure and framework for the body. This is a departure in animal architecture, since invertebrate skeletons generally enfold the body. Exoskeletons and endoskeletons have their own particular sets of advantages and limitations that are related to size (see note on p. 635). For vertebrates, the living endoskeleton possesses an overriding advantage over the dead

CHARACTERISTICS OF SUBPHYLUM VERTEBRATA

1. Chief diagnostic features of chordates—**notochord, dorsal nerve cord, pharyngeal gill pouches,** and **postanal tail**—all present at some stage of the life cycle

2. **Integument** basically of two divisions, an outer **epidermis** of stratified epithelium from the ectoderm and an inner **dermis** of connective tissue derived from the mesoderm; many modifications of skin among the various classes, such as glands, scales, feathers, claws, horns, and hair

3. Distinctive **endoskeleton** consisting of vertebral column (notochord persistent in jawless fishes which lack vertebrae), limb girdles, and two pairs of jointed appendages derived from somatic mesoderm, and a head skeleton (cranium and pharyngeal skeleton) derived largely from neural crest cells

4. Muscular, perforated pharynx; in fishes pharyngeal slits possess gills and muscular aortic arches; in tetrapods the much reduced pharynx is embryonic source of glandular tissue

5. **Many muscles** attached to the skeleton to provide for movement

6. Complete digestive system ventral to the spinal column and provided with large digestive glands, liver, and pancreas

7. Circulatory system consisting of the **ventral heart** of two to four chambers; a closed blood vessel system of arteries, veins, and capillaries; blood fluid containing red blood corpuscles with hemoglobin and white corpuscles; paired aortic arches connecting the ventral and dorsal aortas and giving off branches to the gills among the gill-breathing vertebrates; in the terrestrial types modification of the aortic arch plan into pulmonary and systemic systems

8. Well-developed **coelom** largely filled with the visceral systems

9. Excretory system consisting of **paired kidneys** (mesonephric or metanephric types in adults) provided with ducts to drain the waste to cloaca or anal region

10. Highly differentiated **brain;** 10 or 12 pairs of **cranial nerves** with both motor and sensory functions usually; a pair of spinal nerves for each primitive myotome; an **autonomic nervous system** in control of involuntary functions of internal organs; **paired special sense organs** derived from **epidermal placodes**

11. **Endocrine system** of ductless glands scattered through the body

12. Nearly always separate sexes; each sex containing paired gonads with ducts that discharge their products either into the cloaca or into special openings near the anus

13. **Body plan** consisting typically of **head, trunk,** and **postanal tail; neck** present in some, especially terrestrial forms; two pairs of appendages usually, although entirely absent in some; coelom divided into a pericardial space and a general body cavity; mammals with a thoracic cavity

---

exoskeleton of arthropods. Growing with the body as it does, the endoskeleton permits almost unlimited body size with much greater economy of building materials. Some vertebrates have become the most massive animals on earth. The endoskeleton forms an excellent jointed scaffolding for muscles and the muscles in turn protect the skeleton and cushion it from potentially damaging impact.

We should note that the vertebrates have not wholly lost the protective function of a firm external covering. The skull and the thoracic rib cage enclose and protect vulnerable organs. Most vertebrates are further protected with a tough integument, often bearing nonliving structures such as scales, hair, or feathers that may provide insulation as well as physical security.

The endoskeleton was probably composed initially of cartilage that later gave way to bone. Cartilage forms a perfectly suitable endoskeleton for aquatic animals. Cartilage is superior to bone for fast growth and

is therefore ideal for constructing the first skeletal framework of all vertebrate embryos. In the agnathans (hagfish and lampreys), the sharks and their kin, and even in some bony fishes such as sturgeons, the adult endoskeleton is composed mostly or entirely of cartilage. Bone appears in the endoskeleton of more advanced vertebrates, perhaps because it offers two clear advantages to cartilage. First, it serves as a reservoir for phosphate, an indispensable component of compounds with high-energy bonds, of membranes, and of nucleic acids. Second, only bone could provide the structural strength required for life on land, where mechanical stresses on the endoskeleton are far greater than they are in water.

Pharynx and Efficient Respiration

The perforated pharynx (gill slits), present as pharyngeal pouches in all chordates at some stage in their life cycle, evolved as a suspension-

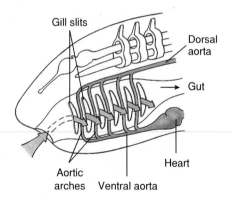

feeding apparatus. In primitive chordates (such as amphioxus), water with suspended food particles is drawn through the mouth by ciliary action and flows out through the gill slits where the food is trapped in mucus. As the protovertebrates shifted from suspension-feeding to a predatory life habit, the pharynx became modified into a muscular feeding apparatus through which water could be pumped by expanding and contracting the pharyngeal cavity. Circulation to the internal gills was improved by the addition of capillary beds (lacking in protochordates) and the development of a ventral heart and muscular aortic arches. All of these changes supported an increased metabolic rate that would have to accompany the switch to an active life of selective predation.

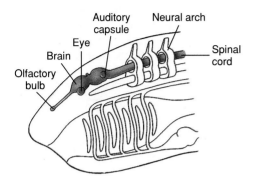

Advanced Nervous System

No single system in the body is more strongly associated with functional and structural advancement than is the nervous system. When the vertebrate ancestors shifted from suspension feeding to active predation, new sensory, motor, and integrative controls became essential for the location and capture of larger prey items. Paired special sense organs designed for distance reception evolved. These included paired eyes with lenses and inverted retinas; pressure receptors, such as paired ears designed for equilibrium and later redesigned to include sound reception; electroreceptors that could signal the direction of potential prey; and chemical receptors, including taste receptors and exquisitely sensitive olfactory organs.

The development of the vertebrate head and paired sense organs was largely the result of two embryonic innovations present only in vertebrates: the **neural crest** and **epidermal placodes.** The neural crest, a population of ectodermal cells lying along the length of the embryonic neural tube, contributes to the formation of many different structures, among them the cranium, cranial nerves, branchial skeleton, and the aortic arches. The epidermal placodes are plate-like ectodermal thickenings (the term "placode" derives from a

Greek word meaning "plate") that appear anteriorly on either side of the neural tube. These give rise to the nose, eyes, ears, taste receptors, and lateral line mechanoreceptors and electroreceptors. Thus the vertebrate head with its sensory structures located adjacent to the mouth (later equipped with prey-capturing jaws), stemmed from the creation of completely new cell types—a rare event in animal evolution.

Paired Limbs

Pectoral and pelvic appendages are present in most vertebrates in the form of paired fins or jointed legs. These originated as swimming stabilizers and later became prominently developed into legs for locomotion on land. Jointed limbs are especially suited for life on land because they permit finely graded leveling motions against a substrate.

THE SEARCH FOR THE VERTEBRATE ANCESTRAL STOCK

The earliest vertebrate Paleozoic fossils, the jawless ostracoderm fishes to be considered at the end of this chapter, share many novel features or organ system development with living vertebrates. These organ systems must have originated either in an early vertebrate lineage, possibly at the origin of vertebrates, or in an invertebrate chordate lineage. With one exception, hardly any invertebrate chordates are known as fossils. The exception is *Pikaia gracilens,* a ribbon-shaped, somewhat fishlike, creature about 5 cm in length discovered in the famous Burgess Shale of British Columbia (Figure 26-11). *Pikaia* is a mid-Cambrian form that precedes the earliest vertebrate fossils in many millions of years. Although this fossil has not yet been described in detail, we know that it possessed both a notochord and the characteristic chordate >-shaped muscle bands (myotomes). Without question *Pikaia* is a chordate. It shows a remarkable resemblance to living amphioxus, at least in overall

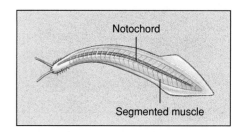

Figure 26-11
Pikaia, the earliest known chordate, from the Burgess Shale of British Columbia, Canada.

body organization, and may in fact be an early cephalochordate. *Pikaia* is a provocative fossil but, until other Cambrian chordate fossils are discovered, it is not possible to assess its relationship to the earliest vertebrates. In the absence of additional fossil evidence, most speculations on vertebrate ancestry have focused on the living cephalochordates and tunicates, since it is widely believed that the vertebrates must have emerged from a lineage related to one or the other of these two protochordate groups.

Garstang's Hypothesis of Chordate Larval Evolution

At first glance, more unlikely candidates for vertebrate ancestry than the tunicates can hardly be imagined. Adult tunicates, which spend their lives anchored to some marine surface, lack a notochord, tubular nerve cord, postanal tail, sense organs, and segmented musculature. Their larvae, however, bear all the right qualifications for chordate membership. Called "tadpole" larvae because of their superficial resemblance to larval frogs, these tiny, site-seeking forms have a notochord, hollow dorsal nerve cord, pharyngeal gill slits, and postanal tail, as well as a brain and sense organs.

At the time of its discovery in 1869, the tadpole larva was considered the descendant of an ancient free-swimming chordate ancestor of the tunicates. The adults then came to be regarded as sessile descendants of the free-swimming form. In 1928, Walter Garstang in England introduced fresh thinking into the vertebrate ancestor debate by turning this sequence

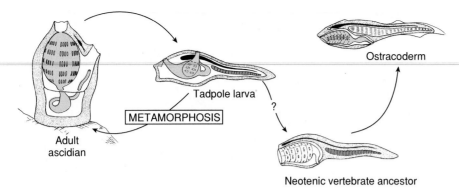

Figure 26-12

Garstang's hypothesis of larval evolution. Adult tunicates live on the sea floor but reproduce through a free-swimming tadpole larva. More than 500 million years ago, some larvae began to reproduce in the swimming stage. These evolved into the ostracoderms, the first known vertebrates.

around; rather than the ancestral tadpole larva giving rise to a tunicate sessile adult, he suggested that the sessile adults *were* the ancestral stock. The tadpole larva then evolved as an adaptation for spreading to new habitats. Next, Garstang suggested that at some point the larva failed to metamorphose into an adult, but developed gonads and reproduced in the larval stage. With continued larval evolution, a new group of free-swimming animals appeared (Figure 26-12).

Garstang called this process **paedomorphosis** (Gr. *pais,* child + *morphē,* form), a term that describes the presence of evolutionary juvenile or larval traits in the adult body. Garstang departed from previous thinking by suggesting that evolution may occur in the larval stages of animals—and in this case, lead to the vertebrate lineage. Paedomorphosis is a well-known phenomenon in several different animal groups (paedomorphosis in amphibians is described on p. 536). Furthermore, Garstang's hypothesis agrees with the embryological evidence. Nevertheless, it remains untested and thus speculative.

Position of Amphioxus

It has long been hypothesized that the cephalochordate amphioxus reveals many morphological characteristics of the ancestral vertebrate. No other protochordate shows the basic diagnostic characteristics of the chor-

Paedomorphosis, the displacement of ancestral larval or juvenile features into a descendant adult, can be produced by three different evolutionary-development processes: neoteny, progenesis, and post-displacement. In neoteny, the growth rate of body form is slowed so that the animal does not attain the ancestral adult form at the time it reaches reproductive maturity. Progenesis is the precocious maturation of gonads in a larval (or juvenile) body that then stops growing and never attains the adult body form. In post-displacement, the onset of a developmental process is delayed relative to reproductive maturation, so that the ancestral adult form is not attained at the time of reproductive maturation. Neoteny, progenesis and post-displacement thus describe different ways in which paedomorphosis can happen. Biologists use the inclusive term paedomorphosis to describe the results of these evolutionary-developmental processes.

dates so well. However, as pointed out in the prologue to this chapter (p. 479), amphioxus is no longer considered to lie in the main line of chordate descent. It lacks a brain and all of the specialized sensory equipment that characterizes the vertebrates. There are no gills in the pharynx and no mouth or pharyngeal musculature for pumping water through the gill slits; movement of water is entirely by the action of cilia.

Recent studies of the expression of homeobox-containing genes which control the body plan of chordate embryos (homeobox genes are described on p. 115) suggest that the ancestor of both amphioxus and vertebrates was cephalized, that is, it had a head region with a brain and sense organs. In amphioxus and other cephalochordates the notochord grows forward to the anterior tip of the animal, obliterating most traces of the primitive head region. Despite these specializations and others peculiar to modern cephalochordates, many zoologists believe that amphioxus has largely retained the primitive pattern of the immediate prevertebrate condition.

THE AMMOCOETE LARVA OF THE LAMPREY AS A MODEL OF THE PRIMITIVE VERTEBRATE BODY PLAN

The lampreys (jawless fishes of the class Cephalaspidomorphi, discussed in the next chapter) have a freshwater larval stage known as the ammocoete (Figure 26-13). In body form, appearance, life habit, and most anatomical details, the ammocoete larva resembles amphioxus. In fact, the lamprey larva was given the genus name *Ammocoetes* (Gr. *ammos,* sand, + *koitē,* bed, referring to the preferred larval habitat) in the nineteenth century when it was erroneously thought to be an adult cephalochordate, closely allied with amphioxus. The ammocoete larva is so different from the adult lamprey that the mistake is understandable; the exact relationship was not explained until metamorphosis into the adult lamprey was observed.

The ammocoete larva has a long, slender body with an oral hood surrounding the mouth much like amphioxus (Figure 26-13). The ammocoete is a suspension feeder, but instead of drawing water by ciliary action into the pharynx as amphioxus does, the ammocoete produces the feeding current by muscular pumping action much like modern fishes. In the floor of the pharynx is an endostyle, as

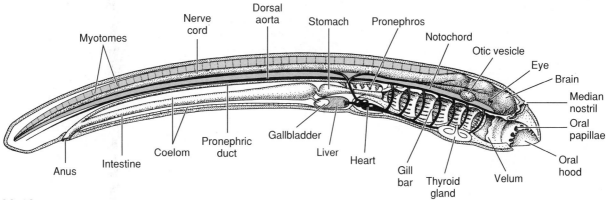

Figure 26-13

Ammocoete larva, freshwater larval stage of a sea lamprey. Although it resembles amphioxus in many ways, the ammocoete has a well-developed brain, paired eyes, pronephric kidney, and other features lacking in amphioxus but representative of the vertebrate body plan.

in amphioxus. The endostyle produces a food-ensnaring mucus that is passed directly to the intestine. The arrangement of body muscle into myotomes, the presence of a notochord serving as the chief skeletal axis, and the plan of the circulatory system all closely resemble these features in amphioxus.

The ammocoete does have several characteristics lacking in amphioxus that are homologous to those of vertebrates. These include a two-chambered heart (atrium and ventricle), a three-part brain (forebrain, midbrain, hindbrain), special sense organs derived from epidermal placodes (two eyes, one on each side of the midbrain; a median nostril; and auditory vesicles located lateral to the midbrain), a thyroid gland, and a pituitary gland. The kidney is pronephric (p. 657) and conforms to the basic vertebrate plan. Instead of the numerous gill slits of amphioxus, there are only seven pairs of gill pouches and slits in ammocoetes. From gill bars separating the gill slits project gill filaments bearing secondary lamellae much like the more extensive gills of modern fishes (see Figure 27-27, p. 517). The ammocoete also has a true liver replacing the hepatic cecum of amphioxus, a gallbladder, and pancreatic tissue (but no distinct pancreatic gland).

The ammocoete larva displays the most primitive condition for these characteristics of any living vertebrate. It clearly illustrates many shared derived characters of vertebrates that are obscured in the development of other vertebrates. It may approach most closely the supposed body plan of the ancestral vertebrate.

THE EARLIEST VERTEBRATES: JAWLESS OSTRACODERMS

The earliest vertebrate fossils are late Cambrian articulated skeletons from the United States, Bolivia, and Australia. They were small, jawless creatures collectively called ostracoderms (os-trak'o-derm) (Gr. *ostrakon*, shell, + *derma*, skin), which belong to the Agnatha division of the vertebrates. These earliest ostracoderms lacked paired fins that later fishes found so important for stability (Figure 26-14). The swimming movements of one of the early groups, the **heterostracans** (Gr. *heteros*, different, + *ostrakon*, shell) (also called pteraspidiforms), must have been clumsy, although sufficient to propel them along the ocean bottom where they searched for food. With fixed circular or slitlike mouth openings they probably filtered small food particles from the water or ocean bottom. However, unlike the ciliary suspension-feeding protochordates, the ostracoderms sucked water into the pharynx by muscular pumping, an important innovation that suggests to some authorities that the ostracoderms may have been mobile predators that fed on soft-bodied animals.

The term "ostracoderm" does not denote a natural evolutionary assemblage but rather is a term of convenience for describing several groups of heavily armored extinct jawless fishes.

During the Devonian period, the heterostracans underwent a major radiation, resulting in the appearance of several peculiar-looking forms varying in shape and length of the snout, dorsal spines, and dermal plates. Without ever evolving paired fins or jaws, these earliest vertebrates flourished for 150 million years until becoming extinct near the end of the Devonian period.

Coexisting with the heterostracans throughout much of the Devonian period were the **osteostracans** (Gr. *osteon*, bone, + *ostrakon*, shell) (also called cephalaspidiforms). The osteostracans improved the efficiency of their benthic life by evolving paired pectoral fins that provided control over pitch and yaw. This ensured well-directed forward movement. A typical osteostracan, such as *Cephalaspis* (Gr. *kephalē*, head, + *aspis*, shield) (Figure 26-14), was a small animal, seldom exceeding 30 cm in length. It was covered by a well-developed armor—the head by a solid shield and the body by bony plates—but it had no axial skeleton or vertebrae. The jawless mouth was toothless. Other distinctive features included a sensory lateral line system, paired eyes with complex eye muscle patterns, and inner ears with semicircular canals. The evolution of the basic vertebrate head pattern in the ostracoderms, although lacking jaws, was an advance of great significance in vertebrate history. As a group the bottom-feeding ostracoderms enjoyed a respectable radiation in the Silurian and Devonian periods.

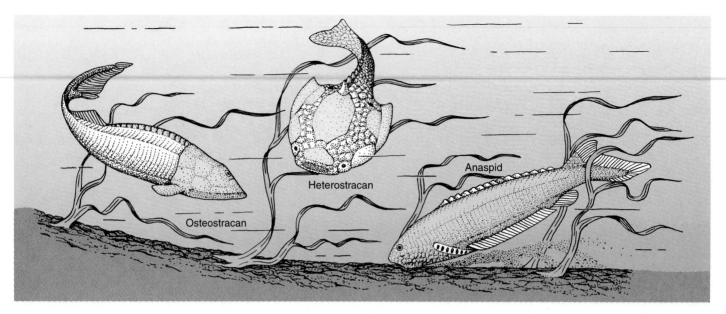

Figure 26-14

Three ostracoderms, jawless fishes of Silurian and Devonian times. They are shown as they might have appeared while searching for food on the floor of a Devonian sea. All were probably suspension-feeders, but employed a strong pharyngeal pump to circulate water rather than the much more limiting mode of ciliary feeding used by their protovertebrate ancestors (presumably resembling amphioxus for this feature). Modern lampreys are believed to be derived from the anaspid group.

The Swedish paleozoologist Erik Stensiö was the first to approach fossil anatomy with the same painstaking attention to minute detail that morphologists have long applied to the anatomical study of living fishes. He developed novel and exacting methods for gradually grinding away a fossil, a few micrometers at a time, to reveal internal features. He was able to reconstruct not only bone anatomy, but nerves, blood vessels, and muscles in numerous groups of Paleozoic and early Mesozoic fishes. His innovative methods are widely used today by paleozoologists.

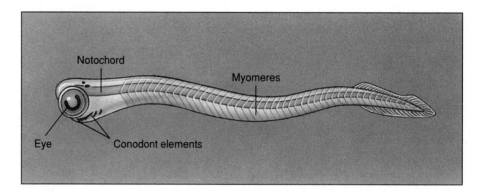

Figure 26-15

Restoration of a living conodont animal. The conodont superficially resembled amphioxus, but it possessed a much greater degree of encephalization (large, paired eyes, possible auditory capsules) and bone-like mineralized elements—all indicating that the conodont animal was a vertebrate. The conodont elements are believed to be gill-supporting structures or part of a suspension-feeding apparatus.

For decades, geologists have used strange microscopic, toothlike fossils called **conodonts** (Gr. *kōnos,* cone, + *odontos,* tooth) to date Paleozoic marine sediments without having any idea what kind of creature originally possessed these elements. The discovery in the early 1980s of the fossils of complete conodont animals has changed all this: conodont elements belonged to a small early marine vertebrate (Figure 26-15). It is widely believed that as more is learned about conodont animals they will play an important role in understanding the origin of vertebrates. At present, however, their position in the vertebrate phylogeny is a matter of debate.

EARLY JAWED VERTEBRATES

All jawed vertebrates, whether extinct or living, are collectively called **gnathostomes** ("jaw mouth") in contrast to the jawless vertebrates, the **agnathans** ("without jaw"). The living agnathans, the naked hagfishes and lampreys, are also often referred to as cyclostomes ("circle mouth"). The gnathostomes are a monophyletic group since the presence of jaws is a derived character state shared by all jawed fishes and tetrapods. The agnathans, however, are characterized by the absence of jaws—a primitive feature—and are therefore almost certainly paraphyletic.

The origin of jaws was one of the most important events in vertebrate evolution. The utility of jaws is obvious: they allow predation on large and active forms of food not available to the jawless vertebrates. There is ample evidence that jaws arose through modifications of the first two of the serially repeated cartilaginous gill arches. The beginnings of this trend can, in fact, be seen in some of the jawless ostracoderms where the mouth becomes bordered by strong dermal plates that could be manipulated somewhat like jaws with the gill arch musculature. Later, the anterior gill arches became hinged and bent forward into the characteristic position of vertebrate jaws. The evidence for this remarkable transformation is threefold. First, both gill arches and jaws form from upper and lower bars that bend forward and are hinged in the middle (Figure 26-16). Second, both gill arches and jaws are derived from neural crest cells rather than from mesodermal tissue, the source of most bones. Third, the jaw musculature is homologous to the original gill support musculature. Nearly as remarkable as this drastic morphological remodeling is the subsequent evolutionary fate of jawbone elements—their transformation into the ear ossicles of the mammalian middle ear (see the note on p. 731).

Among the first jawed vertebrates were the heavily armored **placoderms** (plak'o-derm) (Gr. *plax*, plate, + *derma*, skin). These first appear in the fossil record in the early Devonian period (Figure 26-17). Placoderms evolved a great variety of forms, some very large (one was 10 m in length!) and grotesque in appearance. They were armored fish covered with diamond-shaped scales or with large plates of bone. All became extinct by the end of the Paleozoic era and appear to have left no descendants. However, the **acanthodians** (Figure 26-17), a group of early jawed fishes that were contemporary with the placoderms, are believed by many authorities to have given rise to the great radiation of bony fishes that dominate the waters of the world today.

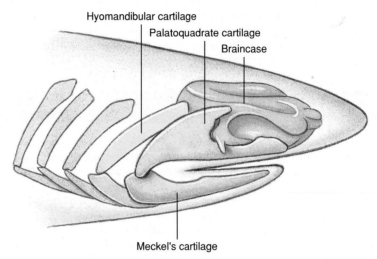

Figure 26-16

How the vertebrates got their jaw. The resemblance between jaws and the gill supports of the primitive fishes such as this carboniferous shark suggests that the upper jaw (palatoquadrate) and lower jaw (Meckel's cartilage) evolved from structures that originally functioned as gill supports. The gill supports immediately behind the jaws are hinged like the jaws and served to link the jaws to the brain case. Relics of this transformation are seen during the development of modern sharks.

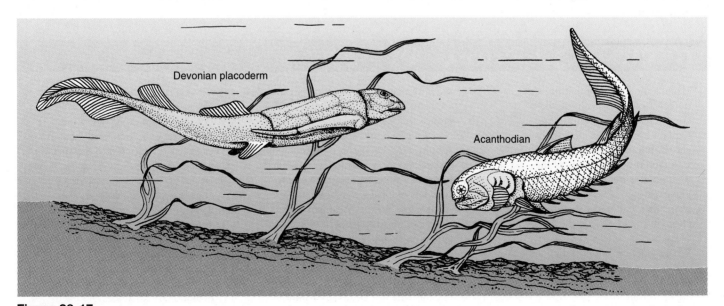

Figure 26-17

Early jawed fishes of the Devonian period, 400 million years ago. Shown are a placoderm (*left*) and a related acanthodian (*right*). The jaws and the gill supports from which the jaws evolved develop from neural crest cells, a diagnostic character of vertebrates. Most placoderms were bottom dwellers that fed on detritus although some were active predators. The acanthodians, the earliest-known true jawed fishes, carried less armor than the placoderms. Most were marine but several species entered fresh water.

EVOLUTION OF MODERN FISHES AND TETRAPODS

Reconstruction of the origins of the vast and varied assemblage of modern living vertebrates is, as we have seen, based largely on fossil evidence. Unfortunately the fossil evidence for the earliest vertebrates is often incomplete and tells us much less than we would like to know about subsequent trends in evolution. Affinities become much easier to establish as the fossil record improves. For instance, the descent of birds and mammals from early tetrapod ancestors has been worked out in a highly convincing manner from the relatively abundant fossil record available. By contrast, the ancestry of modern fishes is shrouded in uncertainty.

Despite the difficulty of clarifying early lines of descent for the vertebrates, they are clearly a natural, monophyletic group, distinguished by a large number of shared characteristics. They have almost certainly descended from a common ancestor, although we still do not know from which chordate group the vertebrate lineage originated. Very early in their evolution, the vertebrates divided into the agnathans and the gnathostomes. These two groups differ from each other in many fundamental ways, in addition to the absence of jaws in the former group and their presence in the latter. The appearance of both jaws and paired fins were major innovations in vertebrate evolution, among the most important reasons for the subsequent major radiations of the vertebrates that produced the modern fishes and all of the tetrapods, including the reader of this book.

Summary

The phylum Chordata is named for the rodlike notochord that forms a stiffening body axis at some stage in the life cycle of every chordate. All chordates share four distinctive hallmarks that set them apart from all other phyla: notochord, dorsal tubular nerve cord, pharyngeal pouches, and postanal tail. Two of the three chordate subphyla are invertebrates and lack a well-developed head. They are the Urochordata (tunicates), most of which are sessile as adults but all of which have a free-swimming larval stage, and the Cephalochordata (lancelets), fishlike forms that include the famous amphioxus.

The chordates may have descended from echinoderm-like ancestors, probably in the Precambrian period, but the true origin of the chordates is not yet, and may never be, known with certainty. Taken as a whole, the chordates have a greater fundamental unity of organ systems and body plan than have many of the invertebrate phyla.

The subphylum Vertebrata includes the backboned members of the animal kingdom (the living jawless vertebrates, the hagfishes and lampreys, actually lack vertebrae but are included with the Vertebrata by tradition because they share numerous homologies with vertebrates). As a group the vertebrates are characterized by having a well-developed head, and by their comparatively large size, high degree of motility, and a distinctive body plan that embodies several distinguishing features that permitted the exceptional adaptive radiation of the group. Most important of these are the living endoskeleton that allows continuous growth and provides a sturdy framework for efficient muscle attachment and action, a pharynx perforated with gill slits (lost or greatly modified in higher vertebrates) with vastly increased respiratory efficiency, advanced nervous system with clear separation of the brain and spinal cord, and paired limbs.

Review Questions

1. What is the evidence that the six deuterostome phyla should be considered a natural grouping?
2. Explain how the use of a cladistic classification for the vertebrates results in important regroupings of the traditional vertebrate taxa (refer to Figure 26-3).
3. Name four hallmarks shared by all chordates, and explain the function of each.
4. What evidences support the idea that chordates evolved within the deuterostome assemblage? What characteristics does the fossil echinoderm group Calcichordata possess that suggest it might closely resemble the ancestor of the chordates?

Why are neither the annelids nor the arthropods considered possible sister groups to the chordates?

5. Name the two invertebrate chordate subphyla, and give three distinguishing characteristics of each.
6. Both sea squirts (urochordates) and lancelets (cephalochordates) are suspension-feeding organisms. Describe the suspension-feeding apparatus of a sea squirt and explain in what ways its mode of feeding is similar to, and different from, that of amphioxus.
7. Explain why it is necessary to know the life history of a tunicate to understand why tunicates are chordates.

8. What is distinctive about the way a larvacean tunicate feeds?
9. List four adaptations that guided vertebrate evolution, and explain how each has contributed to the success of vertebrates.
10. Discuss Garstang's hypothesis of chordate larval evolution.
11. Explain why the body plan of the ammocoete larva of the lamprey can be considered a model of the primitive vertebrate body plan.
12. Distinguish between ostracoderms and placoderms. What important evolutionary advances did each contribute to vertebrate evolution? What are conodonts?

Selected References

See also general references for Part III, p. 626.

Alldredge, A. 1976. Appendicularians. Sci. Am. **235:**94–102 (July). *Describes the biology of larvaceans, which build delicate houses for trapping food.*

Bone, Q. 1979. The origin of chordates. Oxford Biology Readers, No. 18, New York, Oxford University Press. *Synthesis of hypotheses and range of disagreements bearing on an unsolved riddle.*

Carroll, R. L. 1988. Vertebrate paleontology and evolution. New York, W.H. Freeman & Company. *Authoritative treatment of the vertebrate fossil record. The first two chapters contain discussions of cladistic classification of the vertebrates, the vertebrate body plan, and the origin of vertebrate characters.*

Gans, C. 1989. Stages in the origin of vertebrates: analysis by means of scenarios. Biol. Rev. **64:**221–268. *Reviews the diagnostic characters of protochordates and ancestral vertebrates and presents a scenario for the protochordate-vertebrate transition.*

Gould, S. J. (ed.) 1993. The book of life. New York, W.W. Norton & Company. *A sweeping, handsomely illustrated view of (almost entirely) vertebrate life.*

Jeffries, R. P. S. 1986. The ancestry of the vertebrates. Cambridge, Cambridge University Press. *Jeffries argues that the Calcichordata are the direct ancestors of the vertebrates, a view that most zoologists are not willing to accept. Still, this book is an excellent summary of the deuterostome groups and of the various competing hypotheses of vertebrate ancestry.*

Long, J. A. 1995. The rise of fishes: 500 million years of evolution. Baltimore, The Johns Hopkins University Press. *An authoritative, liberally illustrated evolutionary history of fishes.*

Pough, F. H., J. B. Heiser, and W. N. McFarland. 1989. Vertebrate life, ed. 3. New York, Macmillan Publishing Company. *Vertebrate morphology, physiology, ecology, and behavior cast in a cladistic framework.*

27

The Fishes

*Phylum Chordata
Classes Myxini,
Cephalaspidomorphi,
Chondrichthyes, and
Osteichthyes*

What Is a Fish?

In common (and especially older) usage, the term fish has often been used to describe a mixed assortment of water-dwelling animals. We speak of jellyfish, cuttlefish, starfish, and shellfish, knowing full well that when we use the word "fish" in such combinations, we are not referring to a true fish. In earlier times, even biologists did not make such a distinction. Sixteenth century natural historians classified seals, whales, amphibians, crocodiles, even hippopotamuses, as well as a host of aquatic invertebrates, as fish. Later biologists were more discriminating, eliminating first the invertebrates and then the amphibians, reptiles, and mammals from their narrowing concept of a fish. Today we recognize a fish as a gill-breathing, ectothermic, aquatic vertebrate that possesses fins, and skin that is usually covered with scales. Even this

modern concept of the term "fish" is controversial, at least as a taxonomic unit, because fishes do not compose a monophyletic group. The common ancestor of the fishes is also an ancestor to the land vertebrates, which we exclude from the term "fish," unless we use the term in an exceedingly nontraditional way. Because fishes live in a habitat that is basically alien to humans, people have rarely appreciated the remarkable diversity of these vertebrates. Nevertheless, whether appreciated by humans or not, the world's fishes have enjoyed an effusive proliferation that has produced an estimated 24,600 living species—more than all other species of vertebrates combined—with adaptations that have fitted them to almost every conceivable aquatic environment. No other animal group threatens their domination of the seas. ■

POSITION IN THE ANIMAL KINGDOM

The fishes are a vast array of distantly related gill-breathing aquatic vertebrates with fins. Fishes are the most ancient and the most diverse of the monophyletic subphylum Vertebrata within the phylum Chordata, constituting four of the eight vertebrate classes and one-half of the approximately 48,000 recognized vertebrate species. Although they are a heterogeneous assemblage, they exhibit phylogenetic continuity within the group and with the tetrapod vertebrates. The jawless fishes, hagfishes and lampreys, are the living forms that resemble most closely the armored ostracoderms that appeared in the Cambrian period of the Paleozoic. The living jawed fishes, cartilaginous and bony fishes, are related phylogenetically to the acanthodians, a group of jawed fishes that were contemporary with the placoderms of the Silurian and Devonian periods of the Paleozoic. The tetrapod vertebrates, the amphibians, reptiles, birds, and mammals, arose from one lineage of bony fishes, the sarcopterygians (fleshy-finned fishes). The evolution of fishes paralled the appearance of numerous advances in vertebrate history.

BIOLOGICAL CONTRIBUTIONS

1. The basic vertebrate body plan was established in the common ancestor of all vertebrates. Foremost was the evolution of **cellular bone** and the **first endoskeleton.** The **vertebral column** replaced the notochord as the main stiffening axis in adult vertebrates and provided attachment for the skull, many muscles, and the appendages.
2. With the **brain and spinal cord enclosed** and protected within the cranium and vertebral column, the early fishes were the first animals to house the central nervous system separate from the rest of the body. **Specialized sense organs** for taste, smell, and hearing evolved with a tripartite brain. Other sensory innovations include an inner ear with semicircular canals, intricate lateral line sensory systems, and complex eye muscle patterns.
3. The development of **jaws with teeth** permitted predation of large and active foods. This gave rise to a predator-prey arms race that became a major shaping element in vertebrate evolution through the ages.
4. The evolution of **paired pectoral and pelvic fins** supported by shoulder and hip girdles provided greatly improved maneuverability and became the precursors of arms and legs of tetrapod vertebrates.
5. Fishes developed the appropriate physiological adaptations that enabled them to invade every conceivable type of aquatic habitat. The origin of lungs and air gulping in early fleshy-finned fishes permitted limited penetration of semiterrestrial habitats and prepared for the invasion of land with the evolution of tetrapods.

The life of a fish is bound to its body form. Their mastery of river, lake, and ocean is revealed in the many ways that fishes have harmonized their life design to the physical properties of their aquatic surroundings. Suspended in a medium that is 800 times more dense than air, a trout or pike can remain motionless, varying its neutral buoyancy by adding or removing air from the swim bladder. Or it may dart forward or at angles, using its fins as brakes and tilting rudders. With excellent organs for salt and water exchange, bony fishes can steady and finely tune their body fluid composition in their chosen freshwater or seawater environment. Their gills are the most effective respiratory devices in the animal kingdom for extracting oxygen from a medium that contains less than $\frac{1}{20}$ as much oxygen as air. Fishes have excellent olfactory and visual senses and a unique lateral line system, which with its exquisite sensitivity to water currents and vibrations provides a "distance touch" in water.

Thus in mastering the physical problems of their element, fishes evolved a basic body plan and set of physiological strategies that both shaped and constrained the evolution of the vertebrate groups that descended from them.

ANCESTRY AND RELATIONSHIPS OF MAJOR GROUPS OF FISHES

The fishes are of ancient ancestry, having descended from an unknown free-swimming protochordate ancestor (hypotheses of chordate and vertebrate origins were discussed in Chapter 26). Whatever their origin, during the Cambrian period, or perhaps even in the Precambrian, the earliest fishlike vertebrates branched into the jawless **agnathans** and the jawed **gnathostomes** (Figure 27-1). All vertebrates have descended from one or the other of these two ancestral groups.

The use of *fishes* as the plural form of *fish* may sound odd to most people accustomed to using *fish* in both the singular and the plural. Both plural forms are correct but the zoologist uses *fishes* to mean more than one kind of fish.

The jawless agnathans, the more primitive of the two groups, include the extinct ostracoderms and the living **hagfishes** and **lampreys,** fishes adapted as scavengers or parasites. The agnathans have no vertebrae but are nevertheless included within the subphylum Vertebrata because they have a cranium and many other vertebrate homologies. The ancestry of hagfishes and lampreys is uncertain; they bear little resemblance to the extinct ostracoderms. Although the hagfishes and the more derived lampreys superficially look much alike, they are in fact so different from each other that they have been assigned to separate classes by ichthyologists.

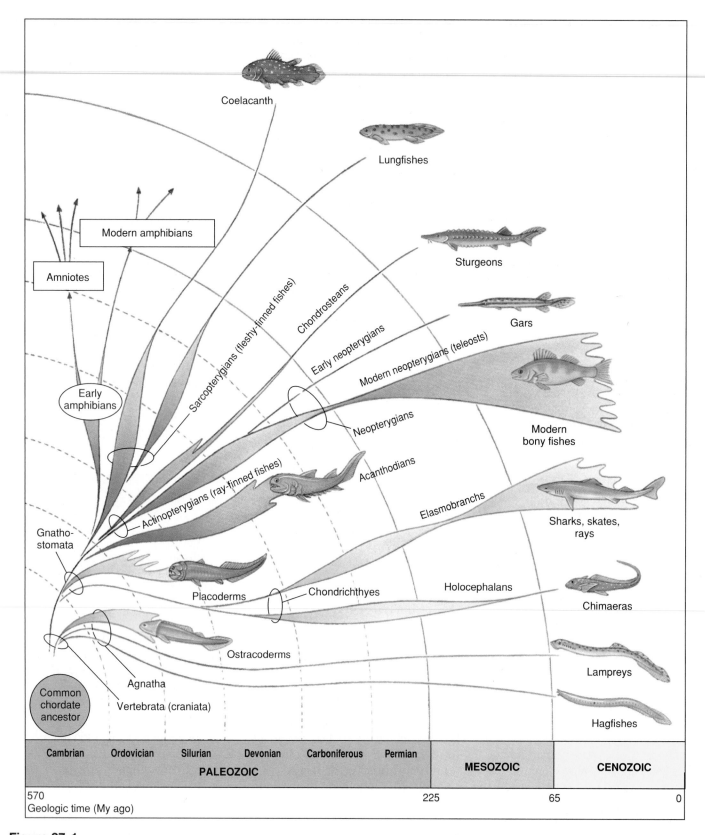

Figure 27-1

Graphic representation of the family tree of fishes, showing the evolution of major groups through geological time. Numerous lineages of extinct fishes are not shown. Widened areas in the lines of descent indicate periods of adaptive radiation and the relative number of species in each group. The fleshy-finned fishes (sarcopterygians), for example, flourished in the Devonian period, but declined and are today represented by only four surviving genera (lungfishes and coelacanth). Homologies shared by the sarcopterygians and tetrapods suggest that they are sister groups. The sharks and rays radiated during the Carboniferous period. They came dangerously close to extinction during the Permian period but staged a recovery in the Mesozoic era and are a secure group today. Johnny-come-latelies in fish evolution are the spectacularly diverse modern fishes, or teleosts, which make up most living fishes.

All remaining fishes have paired appendages and jaws and are included, along with the tetrapods (land vertebrates) in the monophyletic lineage of gnathostomes. They appear in the fossil record in the Silurian period with fully formed jaws, and no forms intermediate between agnathans and gnathostomes are known. By the Devonian period, the Age of Fishes, several distinct groups of jawed fishes were well represented. One of these, the placoderms (p. 495), became extinct, leaving no descendants, in the Carboniferous period, which followed the Devonian. A second group, the **cartilaginous fishes** of the class Chondrichthyes (sharks, rays, and chimaeras), lost the heavy dermal armor of the early jawed fishes and adopted cartilage rather than bone for the skeleton. Most are active predators with a sharklike body form that has undergone only minor changes over the ages. As a group, the sharks and their kin flourished during the Devonian and Carboniferous periods of the Paleozoic era but declined dangerously close to extinction at the end of the Paleozoic. They staged a recovery in the early Mesozoic and radiated to form the modest but thoroughly successful assemblage of modern sharks (Figure 27-1).

The other two groups of gnathostome fishes, the **acanthodians** (p. 495) and the **bony fishes,** were well represented in the Devonian period. The acanthodians somewhat resembled bony fishes but were distinguished by having heavy spines on all the fins except the caudal fin. They became extinct in the lower Permian period. Although the affinities of the acanthodians are much debated, many authors believe that they are the sister group of the bony fishes. The **bony fishes** (class Osteichthyes, Figure 27-2) are the dominant fishes today. We can recognize two distinct lineages of bony fishes. Of these two, by far the most diverse are the **ray-finned fishes** (subclass Actinopterygii), which radiated to form the modern bony fishes. The other lineage, the **fleshy-finned fishes** (subclass Sarcopterygii), though a relic group today, carry the distinction of being the sister group of the tetrapods.

The fleshy-finned fishes are represented today by the **lungfishes** and the **coelacanth**—meager remnants of important stocks that flourished in the Devonian period (Figure 27-1). A classification of the major fish taxa is on p. 524.

SUPERCLASS AGNATHA: JAWLESS FISHES

The living members of the Agnatha are represented by approximately 84 species divided between two classes: Myxini (hagfishes) with about 43 species and Cephalaspidomorphi (lampreys) with 41 species (Figures 27-3 and 27-4). Members of both groups lack jaws, internal ossification, scales, and paired fins, and both groups share porelike gill openings and an eel-like body form. In other respects, however, the two groups are morphologically very different. The hagfishes are certainly the least derived of the two, while the lampreys bear many derived morphological characters that place them phylogenetically much closer to the jawed bony fishes than to the hagfishes. Because of these differences, hagfishes and lampreys have been assigned to separate vertebrate classes, leaving the grouping "agnatha" as a paraphyletic assemblage of jawless fishes of uncertain relationship.

CLASS MYXINI: HAGFISHES

The hagfishes are an entirely marine group that feeds on dead or dying fishes, annelids, molluscs, and crustaceans. Thus they are neither parasitic like lampreys nor predaceous, but are scavengers. There are 43 described species of hagfishes, of which the best known in North America are the Atlantic hagfish *Myxine glutinosa* (Gr. *myxa,* slime) (Figure 27-3) and the Pacific hagfish *Eptatretus stouti* (N. L. *ept,* Gr. *hepta,* seven + *tretos,* perforated). Although almost completely blind, the hagfish is quickly attracted to food, especially dead or dying fishes, by its keenly developed senses of smell and touch. It attaches itself to its prey by means of two toothed, horny plates that fold

While the unique anatomical and physiological features of the strange hagfishes are of interest to biologists, hagfishes have not endeared themselves to either sports or commercial fishermen. In earlier days of commercial fishing mainly by gill nets and set lines, hagfish often bit into the bodies of captured fish and ate out the contents, leaving behind a useless sack of skin and bones. But as large and efficient otter trawls came into use, hagfishes ceased to be an important pest.

CHARACTERISTICS OF CLASS MYXINI

1. Body slender, eel-like, rounded, with **naked skin containing slime glands**
2. **No paired appendages,** no dorsal fin (the caudal fin extends anteriorly along the dorsal surface)
3. **Fibrous and cartilaginous skeleton;** notochord persistent
4. Biting mouth with two rows of eversible teeth
5. Heart with one atrium and one ventricle; **accessory hearts** in caudal region; aortic arches in gill region
6. Five to 16 pairs of gills with a variable number of gill openings
7. **Pronephric kidney** anteriorly and independent **mesonephric kidney** posteriorly; marine, **body fluids isosmotic with seawater**
8. Digestive system **without stomach;** no spiral valve or cilia in intestinal tract
9. Dorsal nerve cord with differentiated brain; **no cerebellum;** 10 pairs of cranial nerves; dorsal and ventral nerve roots united
10. Sense organs of taste, smell, and hearing; **eyes degenerate; one pair semicircular canals**
11. Sexes separate (ovaries and testes in same individual but only one is functional); external fertilization; large yolky eggs, **no larval stage**

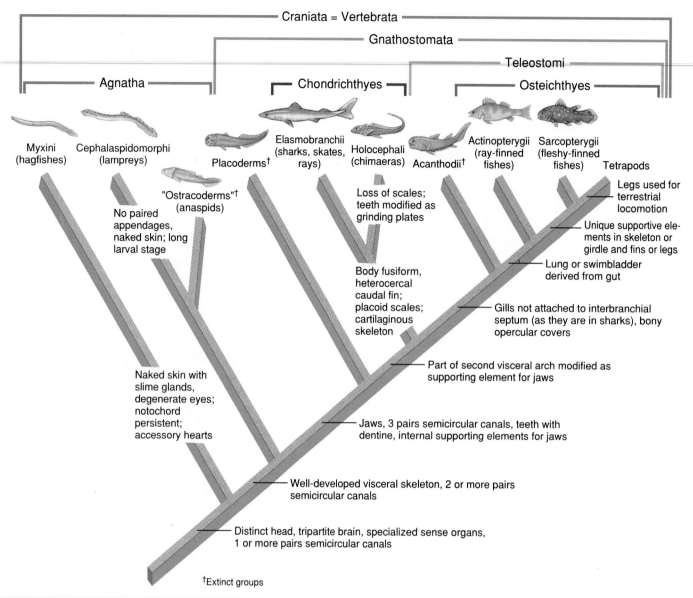

Figure 27-2

Cladogram of the fishes, showing the probable relationships of major monophyletic fish taxa. *Several alternative relationships have been proposed.* Extinct groups are designated by a dagger (†). Some of the shared derived characters that mark the branchings are shown to the right of the branch points. The groups Agnatha and Osteichthyes, although paraphyletic structural grades considered undesirable in cladistic classification, are commonly and conveniently recognized in evolutionary systematics because they share broad structural and functional patterns of organization.

together in a pincerlike action. Then the tongue is thrust forward to rasp off pieces of tissue. For extra leverage, the hagfish often ties a knot in its tail, then passes it forward along the body until it is pressed securely against the side of its prey.

Hagfishes are renowned for their ability to generate enormous quantities of slime. If disturbed or roughly handled, the hagfish exudes a milky fluid from special glands positioned along the body. On contact with sea-

water, the fluid forms a slime so slippery that the animal is almost impossible to grasp.

Unlike any other vertebrate, the body fluids of hagfishes are in osmotic equilibrium with seawater, as in most marine invertebrates. Hagfishes have several other anatomical and physiological peculiarities, including a low-pressure circulatory system served by three accessory hearts in addition to the main heart positioned behind the gills.

The reproductive biology of hagfishes remains largely a mystery, despite a still unclaimed prize offered more than 100 years ago by the Copenhagen Academy of Science for information on the animal's breeding habits. It is known that although both male and female gonads are found in the same animal, only one gonad becomes functional. The females produce small numbers of surprisingly large, yolky eggs up to 3 cm in diameter. There is no larval stage and growth is direct.

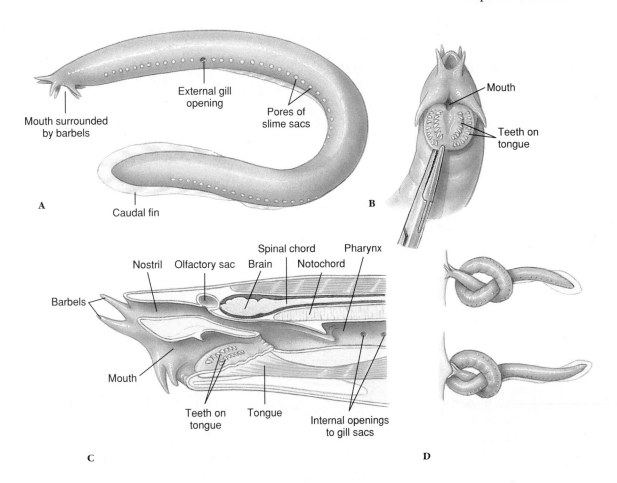

Figure 27-3

The Atlantic hagfish *Myxine glutinosa* (Class Myxini). **A,** External anatomy; **B,** Ventral view of head, showing horny plates used to grasp food during feeding; **C,** Sagittal section of head region (note retracted position of rasping tongue and internal openings into a row of gill sacs); **D,** Hagfish knotting, showing how it obtains leverage to tear flesh from prey.

CLASS CEPHALASPIDOMORPHI (PETROMYZONTES): LAMPREYS

All the lampreys of the Northern Hemisphere belong to the family Petromyzontidae (Gr. *petros,* stone, + *myzon,* sucking). The group name refers to the lamprey's habit of grasping a stone with its mouth to hold position in a current. The destructive marine lamprey *Petromyzon marinus* is found on both sides of the Atlantic Ocean (in America and Europe) and may attain a length of 1 m (Figure 27-4). *Lampetra* (L. *lambo,* to lick or lap up) also has a wide distribution in North America and Eurasia and ranges from 15 to 60 cm long. There

CHARACTERISTICS OF CLASS CEPHALASPIDOMORPHI

1. Body slender, eel-like, rounded with naked skin
2. One or two **median fins, no paired appendages**
3. **Fibrous and cartilaginous skeleton;** notochord persistent
4. Suckerlike oral disc and tongue with well-developed teeth
5. Heart with one atrium and one ventricle; aortic arches in gill region
6. Seven pairs of gills each with external gill opening
7. **Mesonephric kidney;** anadromous and fresh water; **body fluids osmotically and ionically regulated**
8. Dorsal nerve cord with differentiated brain, **small cerebellum present;** 10 pairs cranial nerves; dorsal and ventral nerve roots separated
9. Digestive system without stomach; intestine with **spiral fold and cilia**
10. Sense organs of taste, smell, hearing; **eyes well developed** in adult; **two pairs semicircular canals**
11. Sexes separate; single gonad without duct; external fertilization; **long larval stage** (ammocoete)

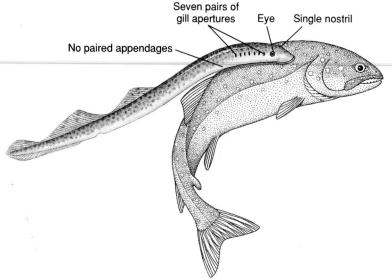

Seven pairs of
gill apertures
Eye
Single nostril

No paired appendages

Figure 27-4

External anatomy of the sea lamprey *Petromyzon marinus.*

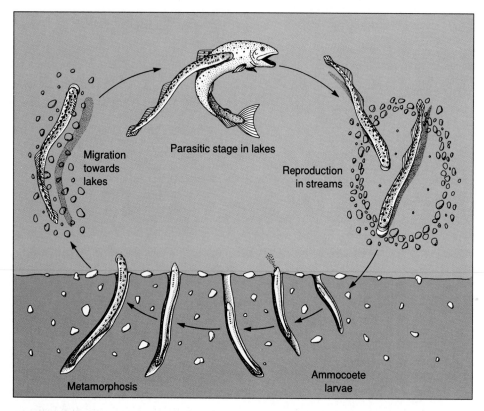

Migration
towards
lakes

Parasitic stage in lakes

Reproduction
in streams

Metamorphosis

Ammocoete
larvae

Figure 27-5

Life cycle of the "landlocked" form of the sea lamprey *Petromyzon marinus.*

are 17 species of lampreys in North America. About half of these belong to the nonparasitic brook type; the others are parasitic. The genus *Ichthyomyzon* (Gr. *ichthyos,* fish, + *myzon,* sucking), which includes three parasitic and three nonparasitic species, is restricted to eastern North America. On the west coast of North America the chief marine form is *Lampetra tridentatus.*

All lampreys ascend freshwater rivers or streams to breed. The marine forms are anadromous (Gr. *anadromos,* running upward) that is, they leave the sea where they spend their adult lives to swim up rivers and streams to spawn. In North America all lampreys spawn in the winter or spring. The males begin nest building and are joined later by females. Using their oral discs to lift stones and pebbles and vigorous body vibrations to sweep away light debris, they form an oval depression (Figure 27-5). At spawning, with the female attached to a rock to maintain her position over the nest, the male attaches to the dorsal side of her head. As the eggs are shed into the nest, they are fertilized by the male. The sticky eggs adhere to pebbles in the nest and quickly become covered with sand. The adults die soon after spawning.

The eggs hatch in about 2 weeks, releasing small larvae (ammocoetes), which are so unlike their parents that early biologists thought they were a separate species. The larva bears a remarkable resemblance to amphioxus and possesses the basic chordate characteristics in such simplified and easily visualized form that it has been considered a chordate archetype (p. 492). After absorbing the remainder of its yolk supply, the young ammocoete, now about 7 mm long, leaves the nest gravel and drifts downstream to burrow in some suitable sandy, low-current area. The larva takes up a suspension-feeding existence while growing slowly for 3 to 7 or more years, then rapidly metamorphoses into an adult. This change involves the development of larger eyes, the replacement of the hood by the oral disc with teeth, a shifting of the nostril to the top of the head, and the development of a rounder but shorter body.

Parasitic lampreys either migrate to the sea, if marine, or remain in fresh water, where they attach themselves by their suckerlike mouth to a fish and, with their sharp horny teeth, rasp away the flesh and suck out body fluids (Figure 27-6). To promote the flow of blood, the lamprey injects an anticoagulant into the wound. When the lamprey is gorged, it releases its hold but leaves the fish with a large, gaping wound that may prove fatal. The parasitic fresh-

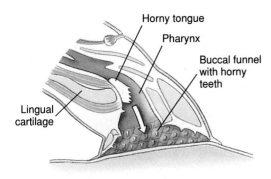

Attachment to fish with horny teeth and suction

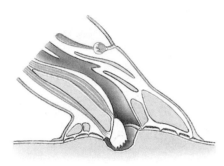

Tongue protruded for rasping flesh

Figure 27-6

How the lamprey uses its horny tongue to feed. After firmly attaching to a fish by its sucker, the protrusible tongue rapidly rasps an opening through the fish's integument. Body fluid, abraded skin, and muscle are eaten.

CHARACTERISTICS OF CLASS CHONDRICHTHYES

1. **Body fusiform,** with a **heterocercal** caudal fin (diphycercal in chimaeras) (Figure 27-16); paired pectoral and pelvic fins, two dorsal median fins; pelvic fins in male modified for **"claspers"**

2. **Mouth ventral;** two olfactory sacs that do not open into the mouth cavity in elasmobranchs; nostrils open into mouth cavity in chimaeras; jaws present

3. Skin with **placoid scales** and mucous glands in elasmobranchs (Figure 27-18); skin naked in chimaeras; modified placoid scales for teeth in elasmobranchs; teeth modified as grinding plates in chimaeras

4. **Endoskeleton entirely cartilaginous;** notochord persistent; vertebrae complete and separate in elasmobranchs; vertebrae absent in chimaeras; appendicular, girdle, and visceral skeletons present

5. Digestive system with J-shaped stomach (stomach absent in chimaeras) and **intestine with spiral valve;** liver, gallbladder, and pancreas present

6. Circulatory system of several pairs of aortic arches; dorsal and ventral aorta, capillary and venous systems, hepatic portal and renal portal systems; two-chambered heart

7. Respiration by means of five to seven pairs of gills; separate and exposed gill slits in elasmobranchs; operculum over four gill openings in chimaeras

8. No swim bladder or lung

9. Mesonephric kidney and rectal gland; blood isosmotic or slightly hyperosmotic to seawater; **high concentrations of urea and trimethylamine oxide in blood**

10. Brain of two olfactory lobes, two cerebral hemispheres, two optic lobes, cerebellum, medulla oblongata; 10 pairs of cranial nerves; **three pairs of semicircular canals**

11. Senses of smell, vibration reception (lateral line system), and electroreception well developed, vision moderately developed

12. Sexes separate; gonads paired; reproductive ducts open into cloaca (separate urogenital and anal openings in chimaeras); oviparous, ovoviviparous, or viviparous; direct development; **fertilization internal**

water adults live a year or more before spawning and then die; the anadromous forms live longer.

The nonparasitic lampreys do not feed after emerging as adults and their alimentary canal degenerates to a nonfunctional strand of tissue. Within a few months they also spawn and die.

The invasion of the Great Lakes by the landlocked sea lamprey *Petromyzon marinus* in this century has had a devastating effect on the fisheries. No lampreys were present in the Great Lakes west of Niagara Falls until the Welland Ship Canal was built in 1829. Even then nearly 100 years elapsed before sea lampreys were first seen in Lake Erie. After that the spread was rapid, and the sea lamprey was causing extraordinary damage in all the Great Lakes by the late 1940s. No fish species was immune from attack, but the lampreys preferred lake trout, and this multimillion dollar fishing in-

dustry was brought to total collapse in the early 1950s. Lampreys then turned to rainbow trout, whitefish, turbot, yellow perch, and lake herring, all important commercial species. These stocks were decimated in turn. The lampreys then began attacking chubs and suckers. Coincident with the decline in attacked species, the sea lampreys themselves began to decline after reaching a peak abundance in 1951 in Lakes Huron and Michigan and in 1961 in Lake Superior. The fall has been attributed both to depletion of food and to the effectiveness of control measures (mainly chemical larvicides in selected spawning streams). Lake trout, aided by a restocking program, are now recovering. Wounding rates are low in Lake Michigan but still high in Lake Huron. Fishery organizations are now experimenting with the release of sterilized male lampreys into spawning streams; these mate with fertile females, and the eggs fail to develop.

CLASS CHONDRICHTHYES: CARTILAGINOUS FISHES

There are nearly 850 living species in the class Chondrichthyes, an ancient, compact, and highly developed group. Although a much smaller and less diverse assemblage than the bony fishes, their impressive combination of well-developed sense organs, powerful jaws and swimming musculature, and predaceous habits ensures them a secure and lasting niche in the aquatic community. One of their distinctive features is their cartilaginous skeleton. Although there is some calcification here and there, bone is entirely absent throughout the class—a curious evolutionary feature, since the Chondrichthyes are derived from ancestors having well-developed bone. Almost all cartilaginous fishes are marine; only about 25 species live primarily in fresh water.

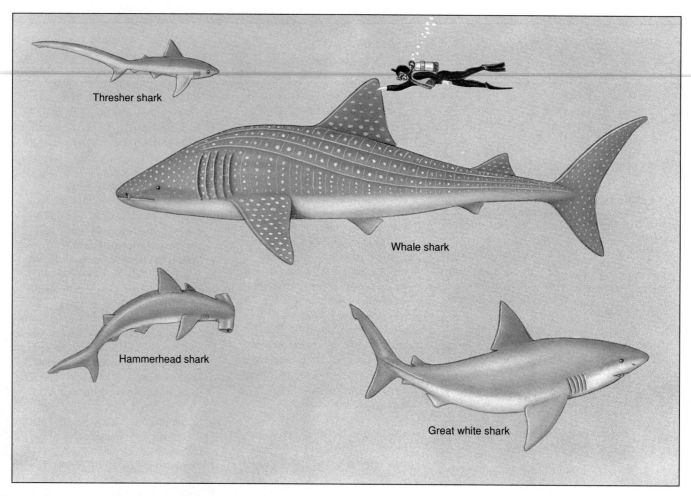

Figure 27-7

Diversity in sharks of the subclass Elasmobranchii. The thresher shark *Alopias vulpinus,* exceptional because of its long upper tail lobe, may exceed 4 m in length. The great white shark *Carcharodon carcharias,* largest and most notorious of dangerous sharks, is a heavy-bodied, spindle-shaped shark that may reach 6 m in length. The nine species of hammerheads (genus *Sphyrna*) are distinguished from all other sharks by the flattened head with hammerlike lobes bearing eyes and nostrils on the ends. The whale shark *Rhincodon typus* is the world's largest fish, reaching 12 m in length. It is a suspension feeder that feeds on plankton collected on a sievelike mesh over its gills.

With the exception of whales, sharks include the largest living vertebrates. The larger sharks may reach 15 m in length. The dogfish sharks so widely studied in zoological laboratories rarely exceed 1 m.

SUBCLASS ELASMOBRANCHII: SHARKS, SKATES, AND RAYS

The elasmobranchs are carnivores that track their prey using their lateral line system and large olfactory organs. Their vision is not well developed.

Fertilization is internal (a curiously advanced feature in an ancient group), and many sharks have evolved elaborate reproductive modes; some are livebearers with gestation periods up to 2 years—the longest of any vertebrate.

There are five living orders of elasmobranchs, numbering about 815 species. All of the more notorious sharks belong to the order Lamniformes, including the requiem sharks, hammerhead sharks, whale sharks, mackerel sharks, nurse sharks, and sand sharks (Figure 27-7). The order Squaliformes includes the dogfish shark, familiar to generations of comparative anatomy students. The skates and several groups of rays (sawfish rays, electric rays, stingrays, eagle rays, manta rays, and devil rays) belong to the order Rajiformes.

Much has been written about the propensities of sharks to attack humans, both by those exaggerating their ferocious nature and by those seeking to write them off as harmless. It is true,

as the latter group of writers argues, that sharks are by nature timid and cautious. But it also is a fact that certain of them are dangerous to humans. There are numerous authenticated cases of shark attacks by *Carcharodon* (Gr. *karcharos,* sharp, + *odous,* tooth), the great white shark (commonly reaching 6 m and often larger); the mako shark *Isurus* (Gr. *is,* equal, + *ouros,* tail); the tiger shark *Galeocerdo* (Gr. *galeos,* shark, + *kerdō,* fox); and the hammerhead shark *Sphyrna* (Gr. *sphyra,* hammer). More shark casualties have been reported from the tropical and temperate waters of the Australian region than from any other. During World War II there were several reports of mass shark attacks on the victims of ship sinkings in tropical waters.

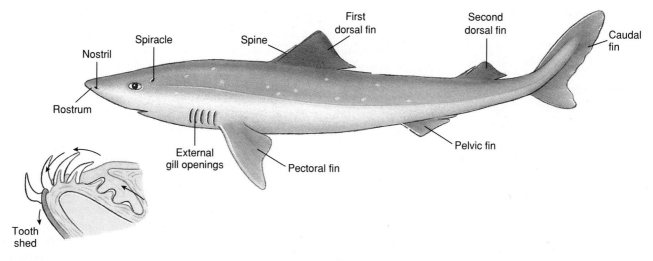

Figure 27-8
Dogfish shark, *Squalus acanthias*. Inset: section of lower jaw shows new teeth developing inside the jaw. These move forward to replace lost teeth. Rate of replacement varies in different species.

The worldwide shark fishery is experiencing unprecedented pressure, driven by the high price of shark fins for shark-fin soup, an oriental delicacy (which commonly sells for $50.00 per bowl). Coastal shark populations in general have declined so rapidly that "finning" is to be outlawed in the United States; other countries, too, are setting quotas to protect threatened shark populations. Even in the Marine Resources Reserve of the Galápagos Islands, one of the world's exceptional wild places, tens of thousands of sharks have been killed illegally for the Asian shark-fin market. That illegal fishery continues at this writing. Contributing to the threatened collapse of shark fisheries worldwide is the long time required by most sharks to reach sexual maturity; some species take as long as 15 years.

Form and Function

Although to most people sharks have a sinister appearance and fearsome reputation, they are at the same time among the most gracefully streamlined of all fishes. The body of a dogfish shark (Figure 27-8) is fusiform (spindle shaped). In front of the ventral mouth is a pointed **rostrum;** at the posterior end the vertebral column turns up to end in the longer upper lobe of the tail. This type of tail is called **heterocercal.** The fins consist of the paired **pectoral** and **pelvic** fins supported by appendicular skeletons, two median **dorsal** fins (each with a spine in *Squalus* [L. a kind of sea fish]), and a median **caudal** fin. A median **anal** fin is present in the smooth dogfish (*Mustelus* [L. *mustela,* weasel]). In the male, the medial part of the pelvic fin is modified to form a **clasper,** which is used in copulation. The paired **nostrils** (blind pouches) are ventral and anterior to the mouth (Figure 27-9). The lateral eyes are lidless, and behind each eye is a spiracle (remnant of the first gill slit). Five gill slits are found anterior to each pectoral fin. The tough, leathery skin is covered with toothlike, dermal **placoid scales** arranged to reduce the turbulence of water flowing along the body surface during swimming.

Sharks are well equipped for their predatory life. Their vision is less acute than that of most bony fishes, but this is more than compensated by a keen sense of smell used to guide them to food. A well-developed **lateral line system** is used for detecting and locating objects and moving animals (predators, prey, and social partners). It is composed of a canal system extending along the side of the body and over the head (Figure 27-10). Inside are special receptor organs (**neuro-**

Figure 27-9
Head of sand tiger shark *Carcharias* sp. Note the series of successive teeth. Also visible in a row below the eye are the ampullae of Lorenzini.

masts) which are extremely sensitive to vibrations and currents in the water. Sharks also can detect and aim attacks at prey buried in the sand by sensing the bioelectric fields that surround all animals. The receptors, the **ampullary organs of Lorenzini,** are located on the shark's head.

Both the upper and lower jaws of sharks are provided with many sharp, triangular teeth. The front row of functional teeth on the edge of the jaw is backed by rows of developing teeth that replace worn teeth throughout the life of the shark (Figures 27-8 and 27-9). The mouth cavity opens into the large **pharynx,** which contains openings to the separate gill slits and spiracles. A short, wide esophagus

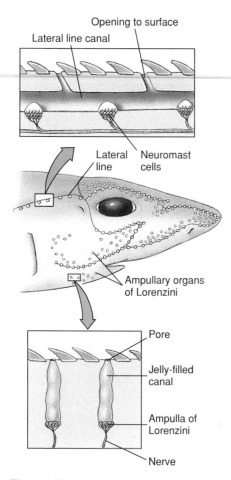

Figure 27-10

Sensory canals and receptors in a shark. The ampullae of Lorenzini respond to weak electric fields, and possibly to temperature, water pressure, and salinity. The lateral line sensors, called neuromasts, are sensitive to disturbances in the water, enabling the shark to detect nearby objects by reflected waves in the water.

runs to the J-shaped stomach. A **liver** and **pancreas** open into the short, straight **intestine,** which contains the unique **spiral valve** that delays the passage of food and increases the absorptive surface (Figure 27-11). Attached to the short rectum is the **rectal gland,** also peculiar to the cartilaginous fishes, which secretes a colorless fluid containing a high concentration of sodium chloride. It assists the **mesonephric kidney** in regulating the salt concentration of the blood. The chambers of the **heart** are arranged in tandem formation (Figure 27-11), and the flow pattern of the circulatory system is basically the same as that of other gill-breathing vertebrates.

All cartilaginous fishes have internal fertilization, but maternal support of the embryo is highly variable. Many elasmobranchs lay large, yolky eggs immediately after fertilization; these are known as **oviparous.** Some oviparous sharks and rays deposit their eggs in a horny capsule called a "mermaid's purse," which often is provided with tendrils that wrap around the first firm object it contacts, much like the tendrils of grape vines. The embryo is nourished from the yolk for a prolonged period—6 to 9

months in some, as much as 2 years in one species—before hatching as a miniature replica of the adult. Many sharks, however, retain the embryos in the reproductive tract for prolonged periods. Some are **ovoviviparous** species, which retain the developing young in the oviduct while they are nourished by the contents of their yolk sac until born. Still other species have true **viviparous** reproduction. In these, a form of **placenta** develops through which each embryo receives nourishment from the maternal bloodstream. The evolution of prolonged retention of embryos by many elasmobranchs was an important innovation that contributed to the success of these fish. However, regardless of the form of maternal support, once the eggs are laid, or the young born, all parental care ends.

The marine elasmobranchs have developed an interesting solution to the physiological problem of living in a salty medium. To prevent water from being drawn out of the body osmotically, the elasmobranchs have retained nitrogenous compounds, especially urea and trimethylamine oxide, in the blood. These solutes, combined with the blood salts, raise the blood solute concentration to exceed slightly

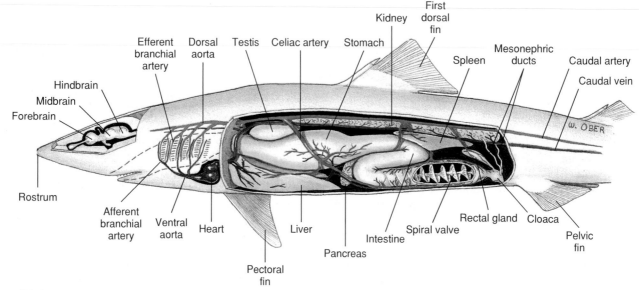

Figure 27-11

Internal anatomy of dogfish shark *Squalus acanthias.*

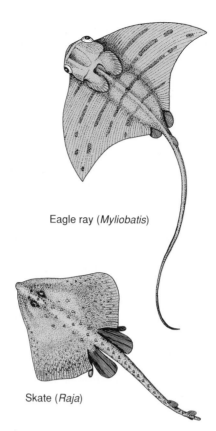

Eagle ray (*Myliobatis*)

Skate (*Raja*)

Figure 27-12
Skates and rays are specialized for life on the sea floor. They are flattened dorsoventrally and move by undulations of greatly expanded winglike pectoral fins.

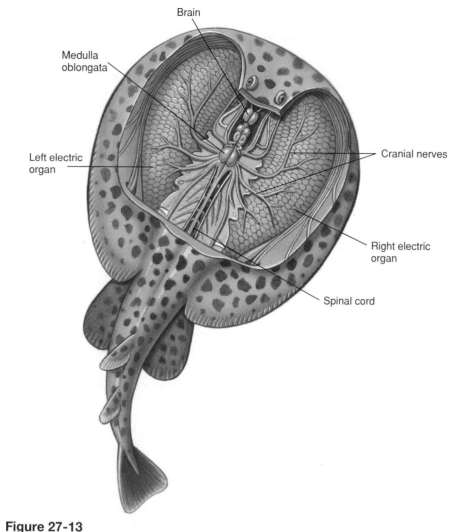

Brain

Medulla oblongata

Left electric organ

Cranial nerves

Right electric organ

Spinal cord

Figure 27-13
Electric ray *Torpedo* with electric organs exposed. Organs are built up of disclike, multinucleated cells called electrocytes. When all cells are discharged simultaneously, a high-amperage current flows into the surrounding water to stun prey or discourage predators. Power output may be several kilowatts.

that of seawater, eliminating an osmotic inequality between their bodies and the surrounding seawater.

A little more than half of all elasmobranchs are rays, a group that includes skates, electric rays, sawfishes, stingrays, eagle rays, and manta rays. Most are specialized for bottom dwelling, with greatly enlarged pectoral fins that are fused to the head and used like wings in swimming (Figure 27-12). The gill openings are on the underside of the head, but the large spiracles are on top. Water for breathing is taken in through these spiracles to prevent clogging the gills, for the mouth is often buried in sand. Their teeth are adapted for crushing their prey: molluscs, crustaceans, and an occasional small fish.

In the order containing the skates and rays (Rajiformes), we commonly refer to members of only one family (Rajidae) as skates. Alone among members of the Rajiformes, skates do not bear living young but lay large, yolky eggs enclosed within a horny covering (the "mermaid's purse") that often washes up on beaches. Although the tail is slender, skates have a somewhat more muscular tail than most rays, and they usually have two dorsal fins and sometimes a caudal fin.

The stingrays have a slender and whiplike tail that is armed with one or more saw-edged spines with venom glands at the base. Wounds from the spines are excruciatingly painful, and may heal slowly and with complications. Electric rays are sluggish fish with large electric organs on each side of the head (Figure 27-13). Each organ is made up of numerous vertical stacks of disclike cells connected in parallel so that when all the cells are discharged simultaneously, a high-amperage current is produced that flows out into the surrounding water. The voltage produced is relatively low but the power output may be several kilowatts—quite sufficient to stun prey or discourage predators. Electric rays were used by the ancient Egyptians for a form of electrotherapy in the treatment of afflictions such as arthritis and gout.

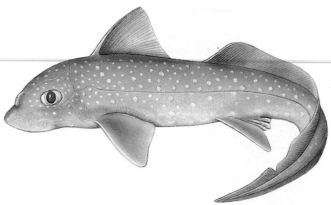

Figure 27-14
Chimaera, or ratfish, of North American west coast. This species is one of the most handsome of chimaeras, which tend toward bizarre appearances.

SUBCLASS HOLOCEPHALI: CHIMAERAS

The members of the small subclass Holocephali, distinguished by such suggestive names as ratfish (Figure 27-14), rabbitfish, spookfish, and ghostfish, are remnants of a line that diverged from the earliest shark lineage which originated at least 350 million years ago (Devonian period of the Paleozoic). Fossil chimaeras (ky-meer'uz) first occurred in the Jurassic period, reached their zenith in the Cretaceous and early Tertiary periods (120 million to 50 million years ago), and have declined ever since. Today there are only about 30 species extant.

Anatomically the chimaeras present an odd mixture of sharklike and bony fish-like features. Instead of a toothed mouth, their jaws bear large flat plates. The upper jaw is completely fused to the cranium, a most unusual development in fishes. Their food is seaweed, molluscs, echinoderms, crustaceans, and fishes—a surprisingly mixed diet for such a specialized grinding dentition. Chimaeras are not commercial species and are seldom caught. Despite their grotesque shape, they are beautifully colored with a pearly iridescence.

CLASS OSTEICHTHYES: BONY FISHES

ORIGIN, EVOLUTION, AND DIVERSITY

The bony fishes, the largest and most diverse taxon of all vertebrates, origi-nated in the late Silurian, approximately 410 million years ago. Details of head structure of the earliest complete bony fish fossils indicate that they probably descended from an ancestor shared with the acanthodians (p. 495). By the middle of the Devonian the bony fishes already had radiated extensively, with adaptations that fitted them for every aquatic habitat except the most inhospitable.

The early bony fishes had an **operculum** over the gill slits, composed of bony plates attached to the first gill arch. This feature increased respiratory efficiency because the outward rotation of the operculum created a negative pressure so that water would be drawn across the gills, as well as pushed across by the mouth pump. These earliest bony fishes also had a pair of lungs that served as accessory breathing structures. The fin pattern of bony fishes became established at this time with two sets of ventral paired fins: the **pectoral fins** positioned anteriorly and a smaller set of **pelvic fins** positioned posteriorly. These were supported by bony pectoral and pelvic girdles embedded in the body musculature. They also had one or two median dorsal fins above and a median anal fin below. This fin pattern, which provided excellent directional control in movement, persists in the advanced bony fishes today (Figure 27-15).

Several key adaptations of bony fishes contributed to their radiation. An obvious manifestation of their adaptability is the great diversity of body form among the living bony fishes. Progressive specialization of jaw structure and feeding mechanisms is another key feature in their evolution. Bony fishes have high levels of activity, supported by efficient gill design for gas exchange, rapid metabolic oxidation of food, and an effective form of undulatory locomotion that persists in many tetrapods (for example, salamanders, snakes, and many lizards).

By the middle Devonian, the Osteichthyes had divided into two distinct lineages. One lineage, the **ray-finned**

CHARACTERISTICS OF CLASS OSTEICHTHYES

1. **Skeleton more or less bony,** representing the primitive skeleton; vertebrae numerous; notochord may persist in part; **tail usually homocercal** (Figure 27-16)
2. Skin with mucous glands and with embedded dermal scales (Figure 27-17) of three types: **ganoid, cycloid,** or **ctenoid; some without scales; no placoid scales (Figure 27-18)
3. Fins both median and paired, with **fin rays of cartilage or bone**
4. **Mouth terminal** with many teeth (some toothless); jaws present; olfactory sacs paired and may or may not open into mouth
5. Respiration by gills supported by bony gill arches and covered by a **common operculum**
6. **Swim bladder** often present with or without duct connected to pharynx
7. Circulation consisting of a two-chambered heart, arterial and venous systems, and characteristically four pairs of aortic arches; blood containing nucleated red cells
8. Nervous system of a brain with small olfactory lobes and cerebrum; large optic lobes and cerebellum; 10 pairs of cranial nerves; three pairs of semicircular canals
9. Sexes separate (sex reversal in some), gonads paired; fertilization usually external; larval forms may differ greatly from adults

Figure 27-15

Internal anatomy of the yellow perch *Perca flavescens,* a freshwater teleost fish.

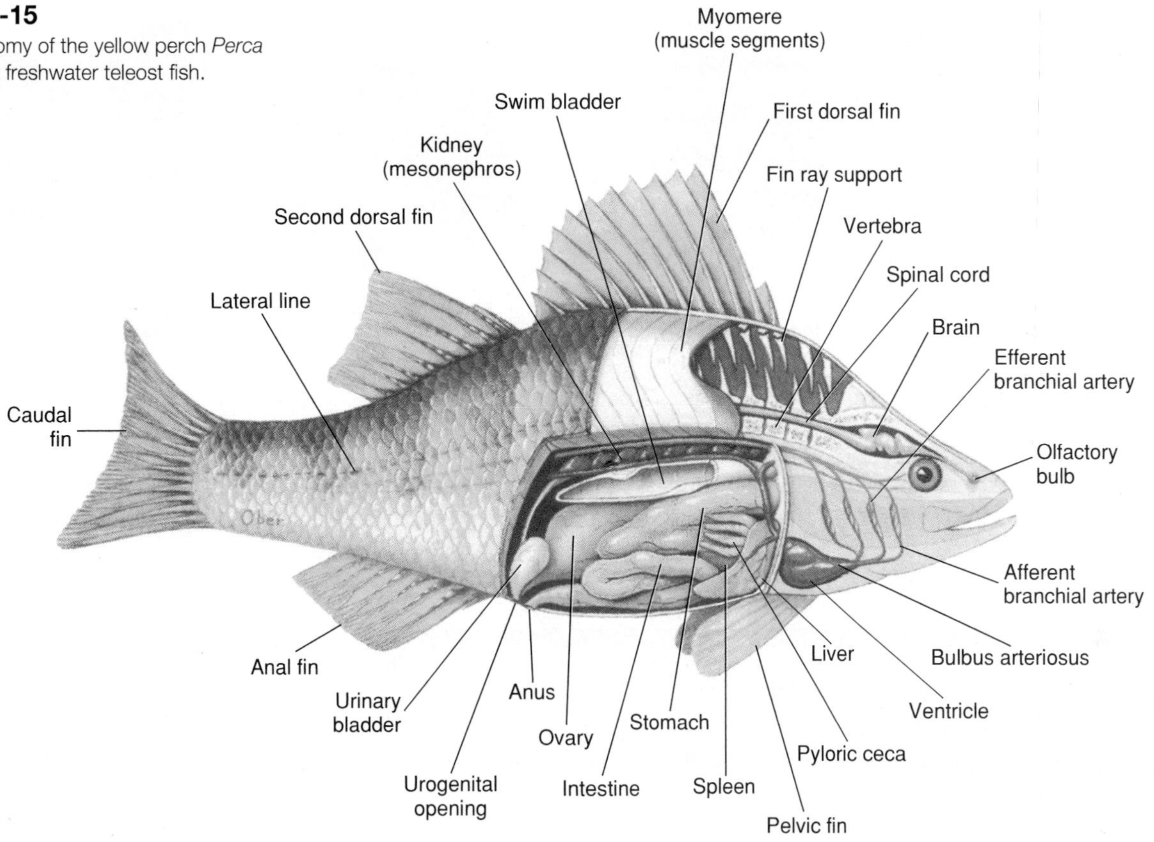

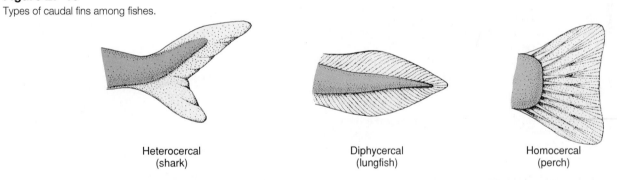

Figure 27-16

Types of caudal fins among fishes.

Heterocercal
(shark)

Diphycercal
(lungfish)

Homocercal
(perch)

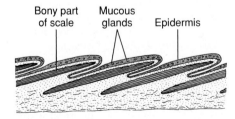

Figure 27-17

Section through the skin of a bony fish, showing the overlapping scales (*red*). The scales lie in the dermis and are covered by epidermis.

fishes (Actinopterygii), includes the modern bony fishes, the largest of all vertebrate radiations. The other lineage is the **fleshy-finned fishes** (Sarcopterygii), a group represented today by the **lungfishes** and the **coelacanth** (see Figures 27-21 and 27-22). Their evolutionary history is of great interest because they are the sister group of all land vertebrates (tetrapods). Although the name "Osteichthyes" means "bony fish," the presence of bone is not an exclusive characteristic of the group. Many of the Paleozoic fishes that preceded the Osteichthyes possessed bone, while some living bony fishes (sturgeons, for example) have much more cartilage than bone in their skeletons. Most bony fishes, however, have exclusively bony skeletons.

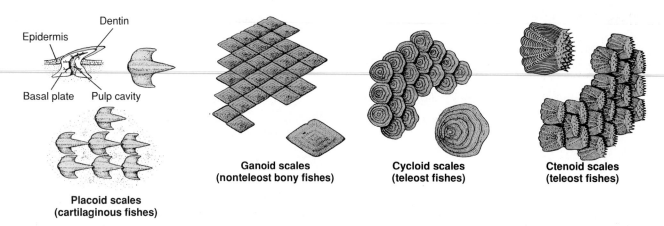

Epidermis
Dentin
Basal plate
Pulp cavity

**Placoid scales
(cartilaginous fishes)**

**Ganoid scales
(nonteleost bony fishes)**

**Cycloid scales
(teleost fishes)**

**Ctenoid scales
(teleost fishes)**

Figure 27-18

Types of fish scales. Placoid scales are small, conical toothlike structures characteristic of Chondrichthyes. Diamond-shaped ganoid scales, present in early bony fishes such as the gar, are composed of layers of silvery enamel (ganoin) on the upper surface and bone on the lower. Teleosts have either cycloid or ctenoid scales. These are thin and flexible and are arranged in overlapping rows.

CLASSIFICATION OF CLASS OSTEICHTHYES

Class Osteichthyes: bony fishes. Two subclasses, 45 living orders.

Subclass Actinopterygii: ray-finned fishes. Two superorders, 42 living orders.

Superorder Chondrostei (kon-dros'tee-i) (Gr. *chondros,* cartilage, + *osteon,* bone): **early ray-finned fishes.** Ten extinct orders; two living orders containing the bichir (*Polypterus*), sturgeons, and paddlefish. The fossil record extends from the Devonian to the present.

Superorder Neopterygii (nee-op-te-rij'ee-i) (Gr. *neos,* new, + *pteryx,* fin, wing): **modern bony fishes.** The early neopterygians include five extinct orders and two living orders, the latter represented by the gars (*Lepisosteus*) and the bowfin (*Amia*). The fossil record extends from the late Permian to the present. Modern neopterygians are placed in the Division Teleostei represented by 38 living orders. Body covered with thin scales without bony layer (cycloid or ctenoid) or scaleless; jaw suspension and cheek bone arrangement allows great expansion of feeding apparatus; mouth terminal; caudal fin mostly homocercal and symmetrical; notochord a mere vestige; swim bladder mainly a hydrostatic organ and usually not opened to the esophagus; endoskeleton mostly bone. The 40 living orders embrace 428 families, and approximately 23,640 living species, representing 96% of all living fishes.

Subclass Sarcopterygii: fleshy-finned fishes. Ten extinct orders and three living orders containing the coelacanth, *Latimeria chalumnae,* and three genera of lungfishes: *Neoceratodus, Lepidosiren,* and *Protopterus.*

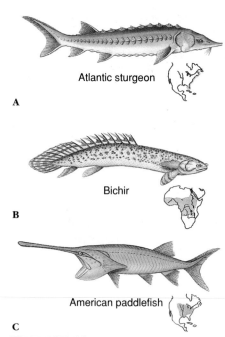

Atlantic sturgeon

A

Bichir

B

American paddlefish

C

Figure 27-19

Chondrostean ray-finned fishes of the subclass Actinopterygii. **A,** Atlantic sturgeon, *Acipenser oxyrhynchus* (now uncommon), of Atlantic coastal rivers. **B,** Bichir *Polypterus bichir* of equatorial west Africa. It is a nocturnal predator. **C,** Paddlefish *Polyodon spathula* of the Mississippi River reaches a length of 2 m and a weight of 90 kg.

SUBCLASS ACTINOPTERYGII: RAY-FINNED FISHES

The ray-finned fishes are an enormous assemblage containing all of our familiar bony fishes—more than 24,600 species. The group had its beginnings in Devonian freshwater lakes and streams. They were small fish with large eyes, extended mouths, and a **heterocercal** tail (Figure 27-16). These earliest ray-finned fishes, known from the fossil record as palaeoniscids (pay'lee-o-nis'ids), were heavily armored with bony **ganoid scales** (Figure 27-18), had fins supported by numerous skeletal rays that were moved as a unit by body-wall musculature, and had functional lungs as well as gills. They looked distinctly different from the lungfishes and lobe-finned fishes with which they shared the Devonian swamps and rivers.

From these earliest ray-finned fishes, two major groups emerged. Those with the most primitive charac-

teristics are the chondrosteans (Gr. *chondros,* cartilage, + *osteon,* bone), represented today by the freshwater and anadromous sturgeons, paddle-fishes, and the bichir *Polypterus* (Gr. *poly,* many, + *pteros,* winged) of African rivers (Figure 27-19). *Polypterus* is an interesting relic with a lunglike swim bladder and many other primitive characteristics; it resembles an ancient

A

B

Figure 27-20

Nonteleost neopterygian fishes. **A,** Bowfin *Amia calva*. **B,** Longnose gar *Lepisosteus osseus*. The bowfin lives in the Great Lakes region and Mississippi basin. Gars are common fishes of eastern and southern North America. They frequent slow-moving streams where they may hang motionless in the water, ready to snatch passing fish.

palaeoniscid more than any other living descendant. There is no satisfactory explanation for the survival to the present of certain fishes such as this one and the coelacanth *Latimeria* when all of their kin perished millions of years ago.

The second major group of ray-finned fishes to emerge from the early palaeoniscid stock were **neopterygians** (Gr. *neos,* new, + *pteryx,* fin). The neopterygians appeared in the late Permian and radiated extensively during the Mesozoic era (Figure 27-1). During the Mesozoic one lineage gave rise to a secondary radiation that led to the modern bony fishes, the teleosts. There are, however, two surviving genera of early neopterygians, the bowfin *Amia* (Gr. tunalike fish) of shallow, weedy waters of the Great Lakes and Mississippi Valley, and the gars *Lepisosteus* (Gr. *lepidos,* scale, + *osteon,* bone)

of eastern and southern North America (Figure 27-20). The seven species of gars are large predators with elongate bodies and jaws. Gars belie their lethargic appearance by suddenly dashing alongside their prey to grasp them in needle-sharp teeth.

The major lineage of neopterygians are the **teleosts** (Gr. *teleos,* perfect, + *osteon,* bone), the modern bony fishes (Figure 27-15). Diversity appeared early in teleost evolution, foreshadowing the truly incredible variety of body forms among teleosts today. The heavy armorlike scales of the more primitive ray-finned fishes have been replaced in the teleosts by light, thin, and flexible **cycloid** and **ctenoid** scales. These look much alike (Figure 27-18) except that ctenoid scales have comblike ridges on the exposed edge that may be an adaptation for reducing frictional drag. Some teleosts, such as some catfishes and sculpins, lack scales altogether. Nearly all teleosts have a **homocercal tail,** with the upper and lower lobes of about equal size (Figure 27-16). Several changes in the jaw suspension and cheek bones provided the teleosts with an expandable and highly mobile oral cavity, and allowed a great variety of feeding mechanisms to evolve. The lungs of early forms have been transformed in the teleosts to a **swim bladder** with a buoyancy function. Teleosts have highly maneuverable fins for control of body movement. In some teleosts the fins are provided with stout, sharp spines, thus making themselves prickly mouthfuls for would-be predators. With these adaptations (and many others), teleosts have become the most diverse of fishes.

SUBCLASS SARCOPTERYGII: FLESHY-FINNED FISHES

The fleshy-finned fishes are today represented by only seven species: six species (three genera) of lungfishes and a single lobe-finned fish, the coelacanth—survivors of a group once abundant during the Devonian period of the Paleozoic (Figures 27-21 and 27-22). The relationships within the sarcopterygians are problematical; some authorities divide the lungfishes

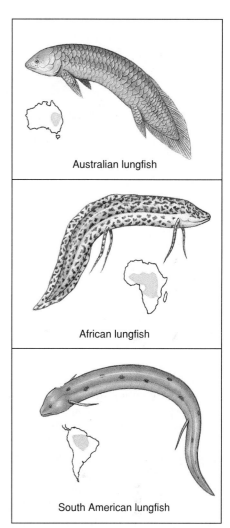

Figure 27-21

Lungfishes are fleshy-finned fishes of the subclass Sarcopterygii. The Australian lungfish *Neoceratodus forsteri* is the least specialized of three lungfish genera. The African lungfish *Protopterus* sp. is best adapted of the three for remaining dormant in mucous-lined cocoons breathing air during prolonged periods of drought.

and lobe-finned fishes into separate subclasses of bony fishes, but the distinction between the two groups may not be as great as once thought.

All of the early sarcopterygians had lungs as well as gills, and a tail of the **heterocercal** type. However, during the Paleozoic the orientation of the vertebral column changed so that the tail became symmetrical, with the median dorsal and ventral fins displaced posteriorly to form one continuous, flexible fin around the tail. This type of tail is called **diphycercal** (Figure 27-16). The strong, fleshy, paired lobed fins of the sarcopterygians (pectoral and pelvic)

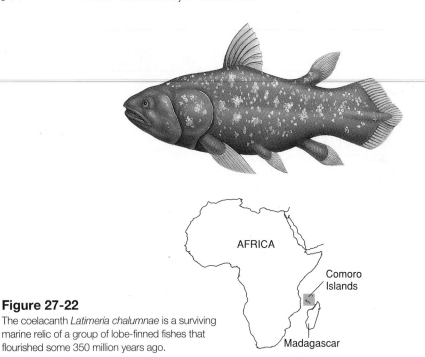

Figure 27-22
The coelacanth *Latimeria chalumnae* is a surviving marine relic of a group of lobe-finned fishes that flourished some 350 million years ago.

may have been used much like four legs to scuttle along the bottom. They had powerful jaws and their skin was covered with heavy scales that consisted of a dentine-like material called **cosmine** overlaid by a thin enamel.

Of the three surviving genera of lungfishes, the least specialized is *Neoceratodus* (Gr. *neos,* new, + *keratos,* horn, + *odes,* form), the living Australian lungfish, which may attain a length of 1.5 m (Figure 27-21). This lungfish is able to survive in stagnant, oxygen-poor water by coming to the surface and gulping air into its single lung, but it cannot live out of water. The South American lungfish *Lepidosiren* (L. *lepidus,* pretty, + *siren,* mythical mermaid) and the African lungfish *Protopterus* (Gr. *prōtos,* first, + *pteron,* wing) are separate evolutionary lineages, and they can live out of water for long periods of time. *Protopterus* lives in African streams and rivers that run completely dry during the dry season, with their mud beds baked hard by the hot tropical sun. The fish burrows down at the approach of the dry season and secretes a copious slime that is mixed with mud to form a hard cocoon in which it estivates until the rains return. Surprisingly little is known about the ecology of the South American lungfish *Lepidosiren.*

The lobe-finned fishes (called crossopterygians in some classifications)

consist of two groups. One group known as the **rhipidistians** flourished in the late Paleozoic era and then became extinct. The rhipidistians are of special importance because they include the ancestors of the tetrapods (and, in cladistic terms, are therefore a paraphyletic group). The second group was the **coelacanths.** These also arose in the Devonian period, radiated somewhat, and reached their evolutionary peak in the Mesozoic era. At the end of the Mesozoic era they nearly disappeared but left one remarkable surviving species, *Latimeria chalumnae* (named for M. Courtenay-Latimer, South African museum director) (Figure 27-22). Since the last coelacanths were believed to have become extinct 70 million years ago, the astonishment of the scientific world can be imagined when the remains of a coelacanth were found on a dredge off the coast of South Africa in 1938. An intensive search was begun in the Comoro Islands area near Madagascar where, it was learned, native Comoran fishermen occasionally caught them with hand lines at great depths. Numerous specimens have now been caught, many in excellent condition, although none has been kept alive beyond a few hours after capture.

The "modern" marine coelacanth is a descendant of the Devonian freshwater stock. The tail is of the diphycercal type (Figure 27-16) but possesses a

small lobe between the upper and lower caudal lobes, producing a three-pronged structure (Figure 27-22). Coelacanths also show some degenerative features, such as more cartilaginous parts and a swim bladder that was either calcified or else persisted as a mere vestige. They also lack the internal nostril so characteristic of the early lobe-finned fishes, but this is probably a secondary loss after the adoption of a deep-water existence; obviously neither internal nostrils nor functional lungs have any relevance for such a life habit.

STRUCTURAL AND FUNCTIONAL ADAPTATIONS OF FISHES

LOCOMOTION IN WATER

To the human eye, some fishes appear capable of swimming at extremely high speeds. But our judgment is unconsciously tempered by our own experience that water is a highly resistant medium through which to move. Most fishes, such as a trout or a minnow, can swim maximally about 10 body lengths per second, obviously an impressive performance by human standards. Yet when these speeds are translated into kilometers per hour it means that a 30 cm (1 foot) trout can swim only about 10.4 km (6.5 miles) per hour. As a general rule, the larger the fish the faster it can swim.

Measuring fish cruising speeds accurately is best done in a "fish wheel," a large ring-shaped channel filled with water that is turned at a speed equal and opposite to that of the fish. Much more difficult to measure are the sudden bursts of speed that most fish can make to capture prey or to avoid being captured. A hooked bluefin tuna was once "clocked" at 66 km per hour (41 mph); swordfish and marlin are thought to be capable of incredible bursts of speed approaching, or even exceeding, 110 km per hour (68 mph). Such high speeds can be sustained for no more than 1 to 5 seconds.

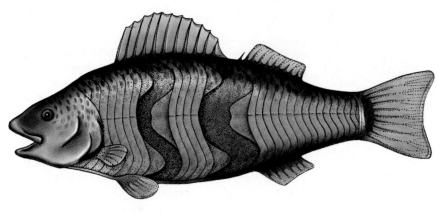

Figure 27-23
Trunk musculature of a teleost fish, partly dissected to show internal arrangement of the muscle bands (myomeres). The myomeres are folded into a complex, nested grouping, an arrangement that favors stronger and more controlled swimming.

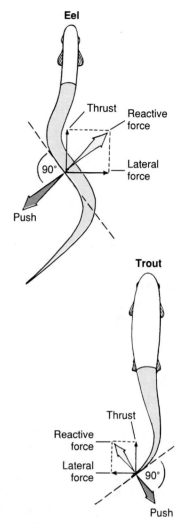

Figure 27-24
Movements of swimming fishes, showing the forces developed by an eel-shaped and spindle-shaped fish.

The propulsive mechanism of a fish is its trunk and tail musculature. The axial, locomotory musculature is composed of zigzag bands, called **myomeres.** The muscle fibers in each myomere are relatively short and connect the tough connective tissue partitions that separate each myomere from the next. On the surface the myomeres take the shape of a **W** lying on its side (Figure 27-23) but internally the bands are complexly folded and nested so that the pull of each myomere extends over several vertebrae. This arrangement produces more power and finer control of movement since many myomeres are involved in bending a given segment of the body.

Understanding how fishes swim can be approached by studying the motion of a very flexible fish such as an eel (Figure 27-24). The movement is serpentine, not unlike that of a snake, with waves of contraction moving backward along the body by alternate contraction of the myomeres on either side. The anterior end of the body bends less than the posterior end, so that each undulation increases in amplitude as it travels along the body. While undulations move backward, the bending of the body pushes laterally against the water, producing a **reactive force** that is directed forward, but at an angle. It can be analyzed as having two components: **thrust,** which is used to overcome drag and propels the fish forward, and **lateral force,** which tends to make the fish's head "yaw," or deviate from the course in the same direction as the tail. This side-to-side head movement is very obvious in a swimming eel or shark, but many fishes have a large, rigid head with enough surface resistance to minimize yaw.

The movement of an eel is reasonably efficient at low speed, but its body shape generates too much frictional drag for rapid swimming. Fishes that swim rapidly, such as trout, are less flexible and limit the body undulations mostly to the caudal region (Figure 27-24). Muscle force generated in the large anterior muscle mass is transferred through tendons to the relatively nonmuscular caudal peduncle and tail where thrust is generated. This form of swimming reaches its highest development in the tunas, whose bodies do not flex at all. Virtually all the thrust is derived from powerful beats of the tail fin (Figure 27-25). Many fast oceanic fishes such as marlin, swordfish, amberjacks, and wahoo have swept-back tail fins shaped much like a sickle. Such fins are the aquatic counterpart of the high-aspect ratio wings of the swiftest birds (p. 584).

Swimming is the most economical form of animal locomotion, largely because aquatic animals are almost perfectly supported by their medium and need expend little energy to overcome the force of gravity. If we compare the energy cost per kilogram of body weight of traveling 1 km by

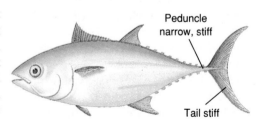

Figure 27-25
Bluefin tuna, showing adaptations for fast swimming. Powerful trunk muscles pull on the slender tail stalk. Since the body does not bend, all of the thrust comes from beats of the stiff, sickle-shaped tail.

different forms of locomotion, we find swimming to cost only 0.39 kcal (salmon) as compared with 1.45 kcal for flying (gull) and 5.43 for walking (ground squirrel). However, part of the unfinished business of biology is understanding how fish and aquatic mammals are able to move through the water while creating almost no turbulence. The secret lies in the way aquatic animals bend their bodies and fins (or flukes) to swim and in the friction-reducing properties of the body surface.

NEUTRAL BUOYANCY AND THE SWIM BLADDER

All fishes are slightly heavier than water because their skeletons and other tissues contain heavy elements that are present only in trace amounts in natural waters. To keep from sinking, sharks must always keep moving forward in the water. The asymmetrical (heterocercal) tail of a shark provides the necessary tail lift as it sweeps to and fro in the water, and the broad head and flat pectoral fins (Figure 27-8) act as angled planes to provide head lift. Sharks are also aided in buoyancy by having very large livers containing a special fatty hydrocarbon called **squalene** that has a density of only 0.86. The liver thus acts like a large sack of buoyant oil that helps to compensate for the shark's heavy body.

By far the most efficient flotation device is a gas-filled space. The **swim bladder** serves this purpose in the bony fishes (Figure 27-26). It arose from the paired lungs of the primitive Devonian bony fishes. Lungs were probably a ubiquitous feature of the Devonian freshwater bony fishes when, as we have seen, the alternating wet and dry climate would have made such an accessory respiratory structure essential for life. Swim bladders are present in most pelagic bony fishes but are absent in tunas, most abyssal fishes, and most bottom dwellers, such as flounders and sculpins.

By adjusting the volume of gas in the swim bladder, a fish can achieve neutral buoyancy and remain suspended indefinitely at any depth with no muscular effort. There are severe technical problems, however. If the fish descends to a greater depth, the swim bladder gas is compressed so that the fish becomes heavier and tends to sink. Gas must be added to

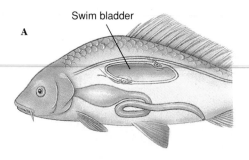

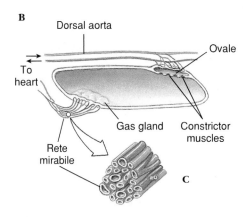

Figure 27-26

A, Swim bladder of a teleost fish. The swim bladder lies in the coelom just beneath the vertebral column. **B,** Gas is secreted into the swim bladder by the gas gland. Gas from the blood is moved into the gas gland by the rete mirabile, a complex array of tightly-packed capillaries that act as a countercurrent multiplier to build up the oxygen concentration. The arrangement of venous and arterial capillaries in the rete is shown in **C.** To release gas during ascent, a muscular valve opens, allowing gas to enter the ovale from which the gas is removed by the circulation.

the bladder to establish a new equilibrium buoyancy. If the fish swims up, the gas in the bladder expands, making the fish lighter. Unless gas is removed, the fish will rise with ever-increasing speed while the bladder continues to expand.

Fishes adjust gas volume in the swim bladder in two ways. The less specialized fishes (trout, for example) have a **pneumatic duct** that connects the swim bladder to the esophagus; these forms must come to the surface and gulp air to charge the bladder and obviously are restricted to relatively shallow depths. More specialized teleosts have lost the pneumatic duct. In these fishes, the gas must originate in the blood and be secreted into the swim bladder. Gas exchange depends

on two highly specialized areas: a **gas gland** that secretes gas into the bladder and a **resorptive area,** or "ovale," that can remove gas from the bladder. The gas gland is supplied by a remarkable network of blood capillaries, called the **rete mirabile** ("marvelous net") that functions as a countercurrent exchange system to trap gases, especially oxygen, and prevent their loss to the circulation (Figure 27-26).

The amazing effectiveness of this device is exemplified by a fish living at a depth of 2400 m (8000 feet). To keep the bladder inflated at that depth, the gas inside (mostly oxygen, but also variable amounts of nitrogen, carbon dioxide, argon, and even some carbon monoxide) must have a pressure exceeding 240 atmospheres, which is much greater than the pressure in a fully charged steel gas cylinder. Yet the oxygen pressure in the fish's blood cannot exceed 0.2 atmosphere—equal to the oxygen pressure at the sea surface.

Physiologists who were at first baffled by the secretion mechanism now understand how it operates. In brief, the gas gland secretes lactic acid, which enters the blood, causing a localized high acidity in the rete mirabile that forces hemoglobin to release its load of oxygen. The capillaries in the rete are arranged so that the released oxygen accumulates in the rete, eventually reaching such a high pressure that the oxygen diffuses into the swim bladder. The final gas pressure attained in the swim bladder depends on the length of the rete capillaries; they are relatively short in fishes living near the surface, but are extremely long in deep-sea fishes.

RESPIRATION

Fish gills are composed of thin filaments covered with a thin epidermal membrane that is folded repeatedly into platelike **lamellae** (Figure 27-27). These are richly supplied with blood vessels. The gills are located inside the pharyngeal cavity and are covered with a movable flap, the **operculum.** This arrangement provides excellent protection to the delicate gill filaments,

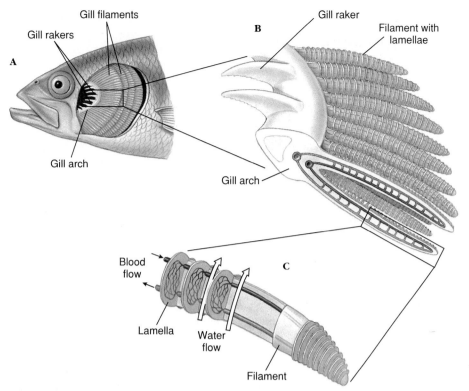

Figure 27-27
Gills of fish. Bony, protective flap covering the gills (operculum) has been removed, **A,** to reveal branchial chamber containing the gills. There are four gill arches on each side, each bearing numerous filaments. A portion of gill arch **(B)** shows gill rakers that project forward to strain out food and debris, and gill filaments that project to the rear. A single gill filament **(C)** is dissected to show the blood capillaries within the platelike lamellae. Direction of water flow (*large arrows*) is opposite the direction of blood flow.

streamlines the body, and makes possible a pumping system for moving water through the mouth, across the gills, and out the operculum. Instead of opercular flaps as in bony fishes, the elasmobranchs have a series of **gill slits** out of which the water flows. In both elasmobranchs and bony fishes the branchial mechanism is arranged to pump water continuously and smoothly over the gills, even though to an observer it appears that fish breathing is pulsatile. The flow of water is opposite to the direction of blood flow (countercurrent flow), the best arrangement for extracting the greatest possible amount of oxygen from the water. Some bony fishes can remove as much as 85% of the oxygen from the water passing over their gills. Very active fishes, such as herring and mackerel, can obtain sufficient water for their high oxygen demands only by continuously swimming forward to force water into the open mouth and across the gills. This is called ram ventilation. Such fish will be

asphyxiated if placed in an aquarium that restricts free swimming movements, even though the water is saturated with oxygen.

A surprising number of fishes can live out of water for varying lengths of time by breathing air. Several devices are employed by different fishes. We already have described the lungs of the lungfishes, *Polypterus,* and the extinct lobe-finned fishes. Freshwater eels often make overland excursions during rainy weather, using the skin as a major respiratory surface. The bowfin, *Amia,* has both gills and a lunglike swim bladder. At low temperatures it uses only its gills, but as the temperature and the fish's activity increase, it breathes mostly air with its swim bladder. The electric eel has degenerate gills and must supplement gill respiration by gulping air through its vascular mouth cavity. One of the best air breathers of all is the Indian climbing perch, which spends most of its time on land near

the water's edge, breathing air through special air chambers above the much-reduced gills.

OSMOTIC REGULATION

Fresh water is an extremely dilute medium with a salt concentration (0.001 to 0.005 gram moles per liter [M]) much below that of the blood of freshwater fishes (0.2 to 0.3 M). Water therefore tends to enter their bodies osmotically, and salt is lost by diffusion outward. Although the scaled and mucous-covered body surface is almost totally impermeable to water, water gain and salt loss do occur across the thin membranes of the gills. Freshwater fishes are **hyperosmotic regulators** that have several defenses against these problems (Figure 27-28). First, the excess water is pumped out by the **mesonephric** kidney (p. 657), which is capable of forming very dilute urine. Second, special **salt-absorbing cells** located in the gill epithelium actively move salt ions, principally sodium and chloride, from the water to the blood. This, together with salt present in the fish's food, replaces diffusive salt loss. These mechanisms are so efficient that a freshwater fish devotes only a small part of its total energy expenditure to keeping itself in osmotic balance.

Perhaps 90% of all bony fishes are restricted to either a freshwater or a seawater habitat because they are incapable of osmotic regulation in the "wrong" habitat. Most freshwater fishes quickly die if placed in seawater, as will marine fishes placed in fresh water. However, some 10% of all teleosts can pass back and forth with ease between both habitats. These **euryhaline fishes** (Gr. *eurys,* broad, + *hals,* salt) are of two types: those such as many flounders, sculpins, and killifish that live in estuaries or certain intertidal areas where the salinity fluctuates throughout the day; and those such as salmon, shad, and eels, that spend part of their life cycle in fresh water and part in seawater.

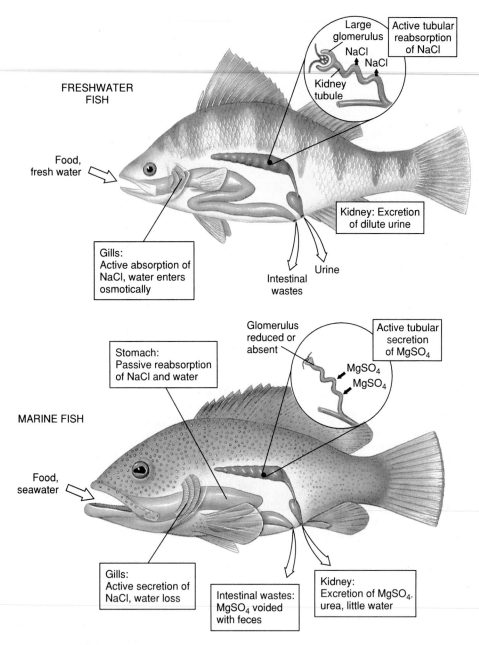

Figure 27-28

Osmotic regulation in freshwater and marine bony fishes. A freshwater fish maintains osmotic and ionic balance in its dilute environment by actively absorbing sodium chloride across the gills (some salt is gained with food). To flush out excess that constantly enters the body, the glomerular kidney produces a dilute urine by reabsorbing sodium chloride. A marine fish must drink seawater to replace water lost osmotically to its salty environment. Sodium chloride and water are absorbed from the stomach. Excess sodium chloride is actively transported outward by the gills. Divalent sea salts, mostly magnesium sulfate, are eliminated with feces and secreted by the tubular kidney.

Marine bony fishes are **hypoosmotic regulators** that encounter a completely different set of problems. Having a much lower blood salt concentration (0.3 to 0.4 M) than the seawater around them (about 1 M), they tend to lose water and gain salt. The marine teleost fish quite literally risks drying out, much like a desert mammal deprived of water. Again, marine bony fishes, like their freshwater counterparts, have evolved an appropriate set of defenses (Figure 27-28). To compensate for water loss, the marine teleost drinks seawater. Although this behavior obviously brings needed water into the body, it is unfortunately accompanied by a great deal of unneeded salt. Un-

wanted salt is disposed in two ways: (1) the major sea salt ions (sodium, chloride, and potassium) are carried by the blood to the gills where they are secreted outward by special **salt-secretory cells;** and (2) the remaining ions, mostly the divalent ions (magnesium, sulfate, and calcium), are left in the intestine and voided with the feces. However, a small but significant fraction of these residual divalent salts in the intestine, some 10% to 40% of the total, penetrates the intestinal mucosa and enters the bloodstream. These ions are excreted by the kidney. Unlike the freshwater fish kidney, which forms its urine by the usual filtration-resorption sequence typical of most vertebrate kidneys (pp. 658 to 663), the marine fish's kidney excretes divalent ions by tubular secretion. Since very little if any filtrate is formed, the glomeruli have lost their importance and disappeared altogether in some marine teleosts. The pipefishes, and the goosefish shown in Figure 27-30, are examples of "aglomerular" marine fishes.

FEEDING BEHAVIOR

For any fish, feeding is one of the main concerns of day-to-day living. Although many a luckless angler would swear otherwise, the fact is that a fish devotes more time and energy to eating, or searching for food to eat, than to anything else. Throughout the long evolution of fishes, there has been unrelenting selective pressure for those adaptations that enable a fish to come out on the better end of the eat-or-be-eaten contest. Certainly the most far-reaching single event was the evolution of jaws. Their possessors were freed from a mud-grubbing or parasitic existence and could adopt a predatory mode of life. Improved means of capturing larger prey demanded stronger muscles, more agile movement, better balance, and improved special senses. More than any other aspect of its life habit, feeding behavior shapes the fish.

Most fishes are **carnivores** that prey on a myriad of animal foods from zooplankton and insect larvae to

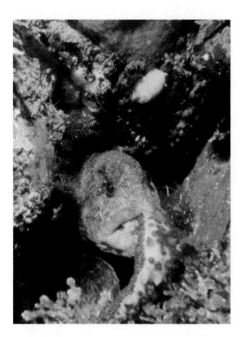

Figure 27-29
Wolf eel, *Anarrhichthys ocellatus,* feeding on a sea cucumber it has captured and pulled to the opening of its den.

Figure 27-30
Goosefish *Lophius piscatorius* awaits its meal. Above its head swings a modified dorsal fin spine ending in a fleshy tentacle that contracts and expands in a convincing wormlike manner. When a fish approaches the alluring bait, the huge mouth opens suddenly, creating a strong inward current that sweeps the prey inside. In a split second all is over.

large vertebrates. Some deep-sea fishes are capable of eating victims nearly twice their own size—an adaptation for life in a world where meals are necessarily infrequent. Most advanced ray-finned fishes cannot masticate their food as we can because doing so would block the current of water across the gills. Some, however, such as the wolf eel (Figure 27-29), have molarlike teeth in the jaws for crushing their prey, which may include hard-bodied crustaceans. Others that do grind their food use powerful pharyngeal teeth in the throat. Most carnivorous fish almost invariably swallow their prey whole, using sharp-pointed teeth in the jaws and on the roof of the mouth to seize their prey. The incompressibility of water makes the task even easier for many large-mouthed predators. When the mouth is opened, a negative pressure is created that sweeps the victim inside (Figure 27-30).

A second group of fishes are **herbivores** that eat plants and algae. Although the plant eaters are relatively few in number, they are crucial intermediates in the food chain, especially in freshwater rivers, lakes, and ponds that contain very little plankton.

The **suspension-feeders** that crop the abundant microorganisms of the sea form a third and diverse group of fishes ranging from fish larvae to basking sharks. However, the most characteristic group of plankton feeders are the herringlike fishes (menhaden, herring, anchovies, capelin, pilchards, and others) that are for the most part **pelagic** (open-sea dwellers) and travel in large schools. Both phytoplankton and the smaller zooplankton are strained from the water with a sievelike device, the gill rakers (Figure 27-27). Because plankton feeders are the most abundant of all fishes, they are important food for numerous larger but less abundant carnivores. Many freshwater fishes also depend on plankton for food.

A fourth group of fishes contains **omnivores** that feed on both plant and animal food. Finally there are the **scavengers** that feed on organic debris (detritus) and the **parasites** that suck the body fluids of other fishes.

Digestion in most fishes follows the vertebrate plan. Except in several fishes that lack stomachs altogether, the food proceeds from stomach to tubular intestine, which tends to be short in carnivores but may be ex-

tremely long and coiled in herbivorous forms. In the herbivorous carp, for example, the intestine may be nine times the body length, an adaptation for the lengthy digestion required for plant carbohydrates. In carnivores, some protein digestion may be initiated in the acid medium of the stomach, but the principal function of the stomach is to store the often large and infrequent meals while awaiting their reception by the intestine.

Digestion and absorption proceed simultaneously in the intestine. A curious feature of ray-finned fishes, especially the teleosts, is the presence of numerous **pyloric ceca** (Figure 27-15) found in no other vertebrate group. Their primary function appears to be fat absorption, although all classes of digestive enzymes (protein-, carbohydrate-, and fat-splitting) are secreted there. They number from two or three to several hundred in some advanced teleost species.

MIGRATION

Eel

For centuries naturalists had been puzzled about the life history of the freshwater eel *Anguilla* (an-gwil′la) (L. eel), a common and commercially important species of coastal streams of the North Atlantic. Eels are **catadromous** (Gr. *kata,* down, + *dromos,* running), meaning that they spend most of their lives in fresh water but migrate to the sea to spawn. Each fall, large numbers of eels were seen swimming down the rivers toward the sea, but no adults ever returned. Each spring countless numbers of young eels, called "elvers" (Figure 27-31), each about the size of a wooden matchstick, appeared in the coastal rivers and began swimming upstream. Beyond the assumption that eels must spawn somewhere at sea, the location of their breeding grounds was completely unknown.

The first clue was provided by two Italian scientists, Grassi and Calandruccio, who in 1896 reported that elvers were not larval eels but rather were relatively advanced juveniles. The true larval eels, they discovered, were tiny,

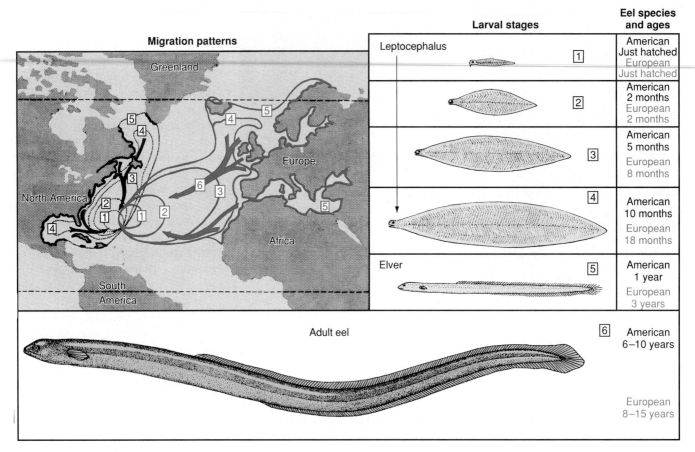

Figure 27-31

Life histories of the European eel, *Anguilla anguilla,* and American eel, *Anguilla rostrata.* Migration patterns of European species are shown in pink. Migration patterns of American species are shown in black. Boxed numbers refer to stages of development. Note that the American eel completes its larval metamorphosis and sea journey in one year. It requires nearly three years for the European eel to complete its much longer journey.

leaf-shaped, completely transparent creatures that bore absolutely no resemblance to an eel. They had been called **leptocephali** (Gr. *leptos,* slender, + *kephale,* head) by early naturalists, who never suspected their true identity. In 1905 Johann Schmidt, supported by the Danish government, began a systematic study of eel biology that he continued until his death in 1933. With the cooperation of captains of commercial vessels plying the Atlantic, thousands of the leptocephali were caught in different areas of the Atlantic with the plankton nets Schmidt supplied them. By noting where larvae in different stages of development were captured, Schmidt and his colleagues eventually reconstructed the spawning migrations.

When the adult eels leave the coastal rivers of Europe and North America, they swim steadily and apparently at great depth for 1 to 2

months until they reach the Sargasso Sea, a vast area of warm oceanic water southeast of Bermuda (Figure 27-31). Here, at depths of 300 m or more, the eels spawn and die. The minute larvae then begin an incredible journey back to the coastal rivers of Europe. Drifting with the Gulf Stream and preyed on constantly by numerous predators, they reach the middle of the Atlantic after 2 years. By the end of the third year they reach the coastal waters of Europe where the leptocephali metamorphose into elvers, with an unmistakable eel-like body form (Figure 27-31). Here the males and females part company; the males remain in the brackish waters of coastal rivers and estuaries while the females continue up the rivers, often traveling hundreds of miles upstream. After 8 to 15 years of growth, the females, now 1 m or more long, return to the sea to join the smaller males;

both return to the ancestral breeding grounds thousands of miles away to complete the life cycle.

Recent enzyme electrophoresis analysis of eel larvae confirmed not only the existence of separate European and American species but also Schmidt's belief that the European and American eels spawn in partially overlapping areas of the Sargasso Sea.

Schmidt found that the American eel (*Anguilla rostrata*) could be distinguished from the European eel (*A. vulgaris*) because it had fewer vertebrae—an average of 107 in the American eel as compared with an average 114 in the European species. Since the American eel is much closer to the North American coastline, it requires only about 8 months to make the journey.

Figure 27-32
Migrating Pacific sockeye salmon.

salmon can navigate on cloudy days and at night, indicating that sun navigation, if used at all, cannot be the salmon's only navigational cue. Fish also (again, like birds) appear able to detect and navigate to the earth's magnetic field. Finally, fishery biologists concede that salmon may not require precise navigational abilities at all, but instead may use ocean currents, temperature gradients, and food availability to reach the general coastal area where "their" river is located. From this point, they would navigate by their imprinted odor map, making correct turns at each stream junction until they reach their natal stream.

Salmon runs in the Pacific Northwest have been devasted by a lethal combination of spawning stream degradation by logging, pollution and, especially, by more than 50 hydroelectric dams which obstruct upstream migration of adult salmon and kill downstream migrants as they pass through the dams' power-generating turbines. In addition, the chain of reservoirs behind the dams, which has converted the Columbia and Snake Rivers into a series of lakes, increases mortality of young salmon migrating downstream by slowing their passage to the sea. The result is that the annual run of wild salmon is today only about 3% of the 10 to 16 million fish that ascended the rivers 150 years ago. While recovery plans have been delayed by the power industry, environmental groups argue that in the long run losing the salmon will be more expensive to the regional economy than making the changes now that will allow salmon stocks to recover.

Homing Salmon

The life history of salmon is nearly as remarkable as that of the eel and certainly has received far more popular attention. Salmon are **anadromous;** that is, they spend their adult lives at sea but return to fresh water to spawn. The Atlantic salmon (*Salmo salar*) and the Pacific salmon (six species of the genus *Oncorhynchus* [on-ko-rink'us]) have this practice, but there are important differences among the seven species. The Atlantic salmon (as well as the closely related steelhead trout) make upstream spawning runs year after year. The six Pacific salmon species (king, sockeye, silver, humpback, chum, and Japanese masu) each make a single spawning run (Figure 27-32), after which they die.

The virtually infallible homing instinct of the Pacific species is legendary: after migrating downstream as a smolt, a sockeye salmon ranges many hundreds of miles over the Pacific for nearly 4 years, grows to 2 to 5 kg in weight, and then returns almost unerringly to spawn in the headwaters of its parent stream. Some straying does occur and is an important means of increasing gene flow and populating new streams.

Experiments by A. D. Hasler and others have shown that homing salmon are guided upstream by the characteristic odor of their parent stream. When the salmon finally reach the spawning beds of their parents (where they themselves were hatched), they spawn and die. The following spring, the newly hatched fry transform into smolts before and during the downstream migration. At this time they are imprinted (p. 763) with the distinctive odor of the stream, which is apparently a mosaic of compounds released by the characteristic vegetation and soil in the watershed of the parent stream. They also seem to imprint on the odors of other streams they pass while migrating downriver and use these odors in reverse sequence as a map during the upriver migration as returning adults.

How do salmon find their way to the mouth of the coastal river from the trackless miles of the open ocean? Salmon move hundreds of miles away from the coast, much too far to be able to detect their parent stream odor. There are experiments suggesting that some migrating fish, like birds, can navigate by orienting to the position of the sun. However, migrant

REPRODUCTION AND GROWTH

In a group as diverse as the fishes, it is no surprise to find extraordinary variations on the basic theme of sexual reproduction. Most fishes favor a simple theme: they are **dioecious,** with **external fertilization** and **external development** of the eggs and embryos. This mode of reproduction is called **oviparous** (meaning

Figure 27-33

Rainbow surfperch *Hypsurus caryi* giving birth. All of the West Coast surfperches (family Embiotocidae) are ovoviviparous.

Figure 27-34

Male banded jawfish *Opistognathus macrognathus* orally brooding its eggs. The male retrieves the female's spawn and incubates the eggs until they hatch. During brief periods when the jawfish is feeding, the eggs are left in the burrow.

"egg-producing"). However, as tropical fish enthusiasts are well aware, the ever-popular guppies and mollies of home aquaria bear their young alive after development in the ovarian cavity of the mother (Figure 27-33). These forms are said to be **ovoviviparous,** meaning "live egg-producing." As described earlier in this chapter (p. 508), some sharks develop a kind of placental attachment through which the young are nourished during gestation. These forms, like placental mammals, are **viviparous** ("alive-producing").

Let us return to the much more common oviparous mode of reproduction. Many marine fishes are extraordinarily profligate egg producers. Males and females come together in great schools and, without mating, release vast numbers of germ cells into the water to drift with the current. Large female cod may release 4 to 6 million eggs at a single spawning. Less than one in a million will survive the numerous perils of the ocean to reach reproductive maturity.

Unlike the minute, buoyant, transparent eggs of pelagic marine teleosts, those of near-shore species are larger, typically yolky, nonbuoyant, and adhesive. On the whole, fishes living in coastal waters where wave action and along-shore currents are prevalent dispose of their eggs in a more conservative manner. Some bury their eggs, many attach them to vegetation, some deposit them in nests, and some even incubate them in their mouths (Figure 27-34). Many coastal species guard their eggs. Intruders expecting an easy meal of eggs may be met with a vivid and often belligerent display by the guard, which is almost always the male.

Freshwater fishes almost invariably produce nonbuoyant eggs. Those, such as perch, that provide no parental care simply scatter their myriads of eggs among weeds or along the bottom. Freshwater fishes that do provide some form of egg care produce fewer, larger eggs that enjoy a better chance for survival.

Elaborate preliminaries to mating are the rule for freshwater fishes. The female Pacific salmon, for example, performs a ritualized mating "dance" with her breeding partner after arriving at the spawning bed in a fast-flowing, gravel-bottomed stream (Figure 27-35). She then turns on her side and scoops out a nest with her tail. As the eggs are laid by the female, they are fertilized by the male (Figure 27-35). After the female covers the eggs with gravel, the exhausted fish dies and drifts downstream.

Soon after the egg of an oviparous species is laid and fertilized, it takes up water and the outer layer hardens. Cleavage follows, and the blastoderm is formed, sitting astride a relatively enormous yolk mass. Soon the yolk mass is enclosed by the developing blastoderm, which then begins to assume a fishlike shape. The fish hatches as a larva carrying a semitransparent sac of yolk which provides its food supply until the mouth and digestive tract have developed. The larva then begins searching for its own food. After a period of growth the larva undergoes a metamorphosis, especially dramatic in many marine species such as the freshwater eel described previously (Figure 27-31). Body shape is refashioned, fin and color patterns change, and the animal becomes a juvenile bearing the unmistakable definitive body form of its species.

Growth is temperature dependent. Consequently, fish living in temperate regions grow rapidly in summer when temperatures are high and food is abundant but nearly stop growing in winter. Annual rings in the scales reflect this seasonal growth (Figure 27-36), a distinctive record of convenience to fishery biologists who wish to determine a fish's age. Unlike birds and mammals, which stop growing after reaching adult size, most fishes after attaining reproductive maturity continue to grow for as long as they live. This may be a selective advantage for the species, since the larger the fish, the more germ cells it produces and the greater its contribution to future generations.

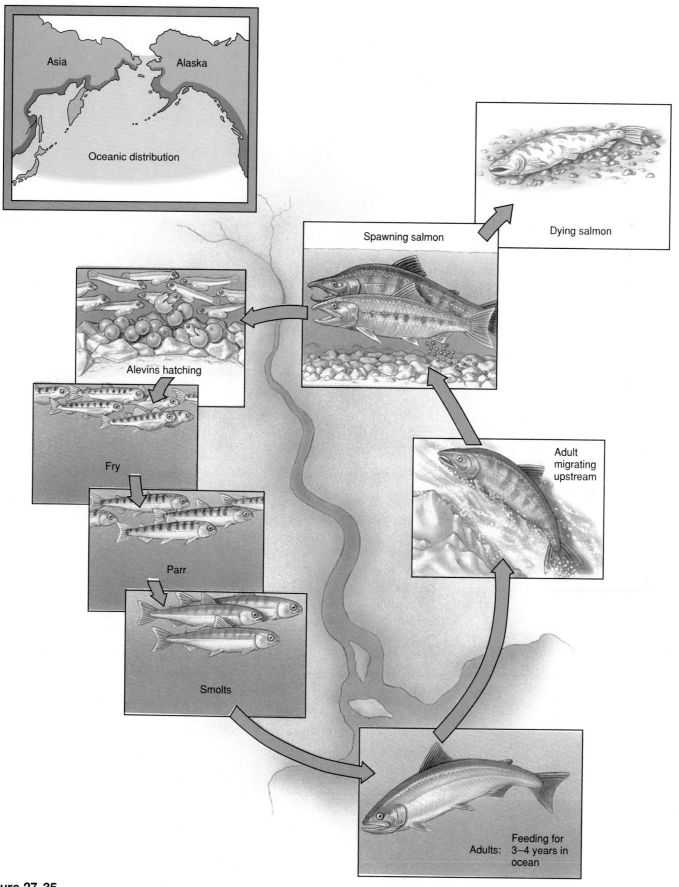

Asia

Alaska

Oceanic distribution

Spawning salmon

Dying salmon

Alevins hatching

Fry

Parr

Smolts

Adult migrating upstream

Adults: Feeding for 3–4 years in ocean

Figure 27-35

Spawning Pacific salmon and development of the eggs and young.

CLASSIFICATION OF LIVING FISHES

The following Linnaean classification of major fish taxa mostly follows that of Nelson (1994). The probable relationships of these traditional groupings together with the major extinct groups of fishes are shown in a cladogram in Figure 27-2. Other schemes of classification have been proposed. Because of the difficulty of determining relationships among the numerous living and fossil species, we can appreciate why fish classification has undergone, and will continue to undergo, continuous revision.

Phylum Chordata

Subphylum Vertebrata

Superclass Agnatha (ag′na-tha) (Gr. *a*, not, + *gnathos*, jaw) **(Cyclostomata).** No jaws; cartilaginous skeleton; paired fins absent; one or two semicircular canals; notochord persistent.

Class Myxini (mik-sy′ny) (Gr. *myxa*, slime): **hagfishes.** Mouth terminal with four pairs of tentacles; buccal funnel absent; nasal sac with duct to pharynx; gill pouches, 5 to 15 pairs; partially hermaphroditic. Examples: *Myxine, Bdellostoma.*

Class Cephalaspidomorphi (sef-a-lass′pe-do-morf′e) (Gr. *kephalē*, head, + *aspidos*, shield, + *morphē*, form) **(Petromyzontes): lampreys.** Mouth suctorial with horny teeth; nasal sac not connected to mouth; gill pouches, seven pairs. Examples: *Petromyzon, Lampetra.*

Superclass Gnathostomata (na′tho-sto′ma-ta) (Gr. *gnathos*, jaw, + *stoma*, mouth). Jaws present; usually paired limbs; three pairs of semicircular canals; notochord persistent or replaced by vertebral centra.

Class Chondrichthyes (kon-drik′thee-eez) (Gr. *chondros*, cartilage, + *ichthys*, fish): **cartilaginous fishes.** Cartilaginous skeleton; teeth not fused to jaws; no swim bladder; intestine with spiral valve.

Subclass Elasmobranchii (e-laz′mo-bran′kee-i) (Gr. *elasmos*, plated, + *branchia*, gills): **sharks, skates, rays.** Placoid scales or no scales; five to seven gill arches and gills in separate clefts along pharynx. Examples: *Squalus, Raja.*

Subclass Holocephali (hol′o-sef′a-li) (Gr. *holos*, entire, + *kephalē*, head): **chimaeras,** or **ghostfish.** Gill slits covered with operculum; jaws with tooth plates; single nasal opening; without scales; accessory clasping organs in male; lateral line an open groove. Examples: *Chimaera, Hydrolagus.*

Class Osteichthyes (os′te-ik′thee-eez) (Gr. *osteon*, bone, + *ichthys*, a fish): **bony fishes.** Body primitively fusiform but variously modified; skeleton mostly ossified; single gill opening on each side covered with operculum; usually swim bladder or lung.

Subclass Actinopterygii (ak′ti-nop-te-rij′ee-i) (Gr. *aktis*, ray, + *pteryx*, fin, wing): **ray-finned fishes.** Paired fins supported by dermal rays and without basal lobed portions; nasal sacs open only to outside. Examples: *Salmo, Perca.*

Subclass Sarcopterygii (sar-cop-te-rij′ee-i) (Gr. *sarkos*, flesh, + *pteryx*, fin, wing): **fleshy-finned fishes.** Heavy bodied; paired fins with sturdy internal skeleton of basic tetrapod type and musculature; muscular lobes at bases of anal and second dorsal fins; diphycercal tail; intestine with spiral valve. Examples: *Latimeria* (coelacanth); *Neoceratodus, Protopterus, Lepidosiren* (lungfishes).

Figure 27-36

Scale growth. Fish scales disclose seasonal changes in growth rate. Growth is interrupted during winter, producing year marks (annuli). Each year's increment in scale growth is a ratio to the annual increase in body length. Otoliths (ear stones) and certain bones can also be used in some species to determine age and growth rate.

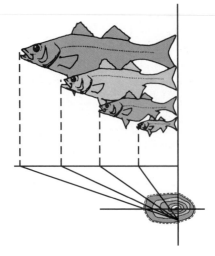

Summary

Fishes are poikilothermic, gill-breathing aquatic vertebrates with fins. They include the oldest vertebrate groups, having originated from an unknown chordate ancestor in the Cambrian period or possibly earlier. Four classes of fishes are recognized. Least derived are the jawless hagfishes (class Myxini) and lampreys (class Cephalaspidomorphi), remnant groups having an eel-like body form without paired fins; a cartilaginous skeleton (although their ancestors, the ostracoderms, had bony skeletons); a notochord that persists throughout life; and a disclike mouth adapted for sucking or biting. All other vertebrates have jaws, a major development in vertebrate evolution. Members of the class Chondrichthyes (sharks, rays, skates, and chimaeras) are a secure group having a cartilaginous skeleton (a degenerative feature), paired fins, excellent sensory equipment, and an active, characteristically predaceous habit. To the fourth class of fishes belong the bony fishes (class Osteichthyes), which may be subdivided into two stems of descent. One

stem is a relic group, the fleshy-finned fishes of the subclass Sarcopterygii, represented today by the lungfishes and the coelacanth. The terrestrial vertebrates arose from within one lineage of this group. The second stem is the ray-finned fishes (subclass Actinopterygii), a huge and diverse modern assemblage containing nearly all of the familiar freshwater and marine fishes.

The modern bony fishes (teleost fishes) have radiated into approximately 24,600 species that reveal an enormous diversity of adaptations, body form, behavior, and habitat preference. Fishes swim by undulatory contractions of the body muscles, which generate thrust (propulsive force) and lateral force. Flexible fishes oscillate the whole body, but in more rapid swimmers the undulations are limited to the caudal region or tail fin alone.

Most pelagic bony fishes achieve neutral buoyancy in water using a gas-filled swim bladder, the most effective gas-secreting device known in the animal

kingdom. The gills of fishes, having efficient countercurrent flow between water and blood, facilitate high rates of oxygen exchange. All fishes but the hagfishes show well-developed osmotic and ionic regulation, achieved principally by the kidneys and the gills.

With the exception of the jawless agnathans, all fishes have jaws that are variously modified for carnivorous, herbivorous, planktivorous, and omnivorous feeding modes.

Many fishes are migratory to some extent, and some, such as freshwater eels and anadromous salmon, make remarkable migrations of great length and precision. Fishes reveal an extraordinary range of sexual reproductive strategies. Most fishes are oviparous, but ovoviviparous and viviparous fishes are not uncommon. The reproductive investment may be in large numbers of germ cells with low survival (many marine fishes) or in fewer germ cells with greater parental care for better survival (freshwater fishes).

Review Questions

1. Give a brief description of the fishes citing characteristics that would distinguish them from all other animals.
2. What characteristics distinguish the hagfishes and lampreys (superclass Agnatha) from all other fishes?
3. Describe feeding behavior in hagfishes and lampreys. How do they differ?
4. Describe the life cycle of the sea lamprey, *Petromyzon marinus,* and the history of its invasion of the Great Lakes.
5. In what ways are sharks well equipped for the predatory life habit?
6. Describe the structure of the lateral line system of sharks and explain its function.
7. Explain how bony fishes differ from sharks and rays in the following systems or features: skeleton, tail shape, scales, buoyancy, respiration, position of mouth, reproduction.
8. Describe the discovery of the living coelacanth. What is the evolutionary significance of the group to which it belongs?

9. Match the ray-finned fishes in the right column with the group to which each belongs in the left column:

    _____ Chondrosteans   a. Perch
    _____ Nonteleost   b. Sturgeon
          neopterygians   c. Gar
    _____ Teleosts   d. Salmon
                           e. Paddlefish
                           f. Bowfin

10. Give three characteristics of modern teleost fishes that distinguish them from the chondrosteans and nonteleost neopterygians.
11. Give the geographical locations of the three surviving genera of lungfishes and explain how they differ in their ability to survive out of water. Which of the three is the least specialized?
12. Compare the swimming movements of the eel with that of the trout, and explain why the latter is more efficient for rapid locomotion.
13. Explain the purpose and function of the swim bladder in teleost fishes. How is gas volume adjusted in the swim bladder?

14. What is meant by "countercurrent flow" as it applies to fish gills?
15. Compare the osmotic problem and the mechanism of osmotic regulation in freshwater and marine bony fishes.
16. Two principal groups of fishes, with respect to feeding behavior, are the carnivores and the suspension-feeders. How are these two groups adapted for their feeding behavior?
17. Describe the life cycle of the European eel. How does the life cycle of the American eel differ from that of the European?
18. How do adult Pacific salmon find their way back to their parent stream to spawn?
19. What mode of reproduction in fishes is described by each of the following terms: oviparous, ovoviviparous, viviparous?
20. Reproduction in marine pelagic fishes and in freshwater fishes is distinctively different. How and why do they differ?

Selected References

See also general references for Part III, p. 626.

Bone, Q., and N. B. Marshall. 1982. Biology of fishes. New York, Chapman & Hall. *Concise, well-written, and well-illustrated primer on the functional process of fishes.*

Conniff, R. 1991. The most disgusting fish in the sea. Audubon **93**(2):100–108 (March). *Recent discoveries shed light on the life history of the enigmatic hagfish that fishermen loathe.*

Horn, M. H., and R. N. Gibson. 1988. Intertidal fishes. Sci. Am. **258**:64–70 (Jan.). *Describes the special adaptations of fishes living in a demanding environment.*

Long, J. A. 1995. The rise of fishes: 500 million years of evolution. Baltimore, The Johns Hopkins University Press. *A lavishly illustrated evolutionary history of fishes.*

Moyle, P. B. 1993. Fish: an enthusiast's guide. Berkeley, University of California Press. *Abbreviated version of the Moyle and Cech ichthyology textbook.*

Moyle, P. B., and J. J. Cech, Jr. 1982. Fishes: an introduction to ichthyology. Englewood Cliffs, New Jersey, Prentice-Hall, Inc. *Textbook written in a lively style and stressing function and ecology rather than morphology; good treatment of the fish groups.*

Nelson, J. S. 1994. Fishes of the world, ed. 3. New York, John Wiley & Sons, Inc. *Authoritative classification of all major groups of fishes.*

Stevens, J. D. (ed.) 1987. Sharks. New York, Facts on File Publications. *Evolution, biology, and behavior of sharks, handsomely illustrated.*

Thomson, K. S. 1991. Living fossil. The story of the coelacanth. New York, W.W. Norton.

Webb, P. W. 1984. Form and function in fish swimming. Sci. Am. **251**:72–82 (July). *Specializations of fish for swimming and analysis of thrust generation.*

28

The Early Tetrapods and Modern Amphibians

Phylum Chordata
Class Amphibia

From Water to Land in Ontogeny and Phylogeny

The chorus of frogs beside a pond on a spring evening heralds one of nature's dramatic events. Masses of frog eggs soon hatch into limbless, gill-breathing, fishlike tadpole larvae. Warmed by the late spring sun, they feed and grow. Then, almost imperceptibly, a remarkable transformation takes place. Hindlegs appear and gradually lengthen. The tail shortens. The larval teeth are lost, and the gills are replaced by lungs. Eyelids develop. The forelegs emerge. In a matter of weeks the aquatic tadpole has completed its metamorphosis to an adult frog.

The evolutionary transition from water to land occurred not in weeks but over millions of years. A lengthy series of alterations cumulatively fitted the vertebrate body plan for life on land. The origin of land vertebrates is no less a remarkable feat for this fact—a feat that incidentally would have a poor chance of succeeding today because well-established competitors make it impossible for a poorly adapted transitional form to gain a foothold.

Amphibians are the only living vertebrates that have a transition from water to land in both their ontogeny and phylogeny. Even after some 350 million years of evolution, few amphibians are completely land adapted; most are quasiterrestrial, hovering between aquatic and land environments. This double life is expressed in their name. Even the amphibians that are best adapted for a terrestrial existence cannot stray far from moist conditions. Many, however, have developed ways to keep their eggs out of open water where the larvae would be exposed to enemies. ■

POSITION IN THE ANIMAL KINGDOM

The amphibians are ectothermic, primitively quadrupedal vertebrates, with glandular skin and dependance on water for their reproduction. They are one of two major groups of living descendants of early Devonian tetrapods, the first vertebrates to evolve the adaptations to breathe, support themselves, move, and pickup airborne sounds and odors on land, while minimizing water loss. The other group is the amniotes, the reptiles, birds, and mammals that completed the movement onto land by evolving adaptations that freed them from their dependence on water for reproduction.

BIOLOGICAL CONTRIBUTIONS

1. **Strong skeletal framework** to support body weight on land, and the **tetrapod leg** with associated shoulder/hip girdle for walking on land.
2. A respiratory system with **lungs** (some modern amphibians are gilled, and some lack both lungs and gills) and paired **internal nostrils** (choanae) which enable breathing through the nose.
3. **Double circulation** with functionally separated pulmonary and systemic circuits and a **three-chambered heart. Pulmonary arteries and veins** supply the lungs and return oxygenated blood to the heart.
4. Ancestral aquatic sensory receptors were modified for life on land. The ear with **tympanic membrane** (eardrum) and **stapes** (columella) for transmitting vibrations to the inner ear is designed to pick up airborne sounds. For vision in air, the cornea rather than the lens became the principal refractive surface for bending light; **eyelids** and **lachrymal glands** evolved to protect and wash the eye. A well-developed **olfactory epithelium** lining the nasal cavity evolved to pick up airborne odors.

Adaptation for life on land is a major theme of the remaining vertebrate groups. These animals form a monophyletic unit known as the **tetrapods.** The amphibians and the amniotes (including reptiles, birds, and mammals) represent the two major extant branches of tetrapod phylogeny. In this chapter, we review what is known about the origins of terrestrial vertebrates and discuss the amphibian lineage in detail. We discuss the major amniote groups in Chapters 29 through 31.

MOVEMENT ONTO LAND

The movement from water to land is perhaps the most dramatic event in animal evolution, because it involves the invasion of a habitat that in many respects is more hazardous for life. Life originated in water. Animals are mostly water in composition, and all cellular activities occur in water. Nevertheless, organisms eventually invaded the land, carrying their watery composition with them. Plants and insects made the transition much earlier than vertebrates, and the pulmonate snails experimented with life on land at approximately the same time that the earliest terrestrial vertebrates evolved. Although the invasion of land required modification of almost every system in the vertebrate body, aquatic and terrestrial vertebrates retain many basic structural and functional similarities. We see the transition between the aquatic and terrestrial vertebrate body plans most clearly today in the many living amphibians that make this transition during their own life histories.

Beyond the obvious difference in water content, there are several important physical differences that animals must accommodate when moving from water to land. These include (1) oxygen content, (2) density, (3) temperature regulation, and (4) habitat diversity. Oxygen is at least 20 times more abundant in air and it diffuses much more rapidly through air than through water. Consequently, terrestrial animals can obtain oxygen far more easily than aquatic ones once they possess the appropriate adaptations, such as lungs. Air, however, has approximately 1000 times less buoyant density than water and is approximately 50 times less viscous. It therefore provides relatively little support against gravity, requiring the terrestrial animal to develop strong limbs and to remodel the skeleton to achieve adequate structural support. Air fluctuates in temperature more readily than water does, and terrestrial environments therefore experience harsh and unpredictable cycles of freezing, thawing, drying, and flooding. Terrestrial animals require behavioral and physiological strategies to protect themselves from thermal extremes; one such important strategy is the homeothermy (regulated constant body temperature) of the birds and mammals.

Despite its hazards, the terrestrial environment offers a great variety of new habitats including coniferous, temperate, and tropical forests, grasslands, deserts, mountains, oceanic islands, and polar regions. The provision of safe shelter for the protection of vulnerable eggs and young may be accomplished much more readily in many of these terrestrial habitats than in aquatic ones.

EARLY EVOLUTION OF TERRESTRIAL VERTEBRATES

DEVONIAN ORIGIN OF THE TETRAPODS

The Devonian period, beginning some 400 million years ago, was a time of mild temperatures and alternating droughts and floods. It was during this period that some primarily aquatic vertebrates evolved two features that would be important for permitting the subsequent evolution for life on land: lungs and limbs.

The Devonian freshwater environment was very unstable. During dry periods, many pools and streams evaporated, water became foul, and the dissolved oxygen disappeared. Only those fishes able to acquire atmospheric oxygen survived such conditions. Gills were unsuitable because in air the filaments collapsed, dried, and quickly lost their function. Virtually all freshwater fishes surviving this period, including the lobe-finned fishes and the lungfishes, had a kind of lung that developed as an outgrowth of the pharynx. It was relatively simple to enhance the efficiency of the air-filled cavity by improving its vascularity with a rich capillary network, and by supplying it with arterial blood from the last (sixth) pair of aortic arches. Oxygenated blood returned directly to the heart by a pulmonary vein to form a complete pulmonary circuit. Thus the **double circulation** characteristic of all tetrapods originated: a systemic circulation serving the body and a pulmonary circulation supplying the lungs.

The vertebrate limbs also arose during the Devonian period. Although fish fins at first appear very different from the jointed limbs of tetrapods, an examination of the bony elements of the paired fins of the lobe-finned fishes shows that they broadly resemble the equivalent limbs of amphibians. In *Eusthenopteron,* a Devonian rhipidistian, we can recognize an upper arm bone (humerus) and two forearm bones (radius and ulna) as well as other elements that can be homologized with the wrist bones of tetrapods (Figure 28-1). *Eusthenopteron* could walk— more accurately flop—along the bottom mud of pools with its fins, since backward and forward movement of the fins was limited to about 20 to 25 degrees. *Acanthostega,* one of the earliest known Devonian amphibians, had well-formed tetrapod legs with clearly formed digits on both fore- and hindlimbs, but the limbs were too weakly constructed to enable the animal to hoist its body off the surface for proper walking on land. *Ichthyostega,* however, with its fully developed shoulder girdle, bulky limb bones, well-developed muscles, and other adaptations for terrestrial life, must

have been able to pull itself onto land, although it is doubtful that it could have walked very well.

Until recently it was thought that the early tetrapods had five fingers and five toes on their hands and feet, the basic pentadactyl plan of most living tetrapods today. However, newly discovered fossils of Devonian tetrapods show that all of them had more than five digits, that is, they were "polydactylous." Only later did the five-digit pattern become stabilized in the different tetrapod lineages.

Movement onto land was clearly a revolution in vertebrate history. How did it come about? A long-accepted scenario developed by Harvard paleontologist Alfred Romer suggested that when freshwater pools of the Devonian evaporated during seasonal droughts, the aquatic vertebrates were forced to move to others that still contained water. The fleshy fins of the sarcopterygians (the living coelacanth and lungfishes, and the extinct "rhipidistians," see pp. 513 through 514) could be adapted as paddles to lever their way across land in search of water. Those with strong fins lived to reproduce. According to this hypothesis, land travel and the gradual development of legs originated as a means for survival in water. This view has changed with the recent discovery of more complete fossils of the earliest known tetrapods. Although *Acanthostega* had tetrapod legs (Figure 28-1), in every other respect it was a fully aquatic animal. A consensus emerging now is that tetrapods evolved their legs underwater and only then, for reasons unknown, began to pull themselves onto land.

As noted above, evidence points to the lobe-finned fishes (rhipidistians) as the closest relatives of the tetrapods; in cladistic terms they contain the sister group of tetrapods (Figures 28-2 and 28-3). Both the rhipidistians and early tetrapods such as *Acanthostega* and *Ichthyostega* shared several characteristics of skull, teeth, and pectoral girdle. *Ichthyostega* (Gr. *ichthys,* fish, + *stegē,* roof, or covering, in reference to the roof of the skull which was shaped like that of a fish)

represents an early offshoot of tetrapod phylogeny that possessed several adaptations, in addition to jointed limbs, that equipped it for life on land. These include a stronger backbone and associated muscles to support the body in air, new muscles to elevate the head, strengthened shoulder and hip girdles, a protective rib cage, a more advanced ear structure for picking up airborne sounds, a foreshortening of the skull, and a lengthening of the snout that improved olfactory powers for detecting dilute airborne odors. Yet *Ichthyostega* still resembled aquatic forms in retaining a tail complete with fin rays and in having opercular (gill) bones.

The bones of *Ichthyostega,* the most thoroughly studied of all early tetrapods, were first discovered on an East Greenland mountainside in 1897 by Swedish scientists looking for three explorers lost two years earlier during an ill-fated attempt to reach the North Pole by hot-air balloon. Later expeditions by Gunnar Säve-Söderberg uncovered skulls of *Ichthyostega* but Säve-Söderberg died, at age 38, before he was able to make a thorough study of the skulls. After Swedish paleontologists returned to the Greenland site where they found the remainder of *Ichthyostega's* skeleton, Erik Jarvik, one of Säve-Söderberg's assistants, assumed the task of examining the skeleton in detail. This became his life's work, resulting in a description of *Ichthyostega* that stands as the most detailed of any Paleozoic tetrapod. Jarvik suffered a crippling stroke at age 88 in 1994, but had by then virtually completed an extensive monograph on *Ichthyostega* which was due to be published in late 1995.

CARBONIFEROUS RADIATION OF THE TETRAPODS

The capricious Devonian period was followed by the Carboniferous period, characterized by a warm, wet climate during which mosses and

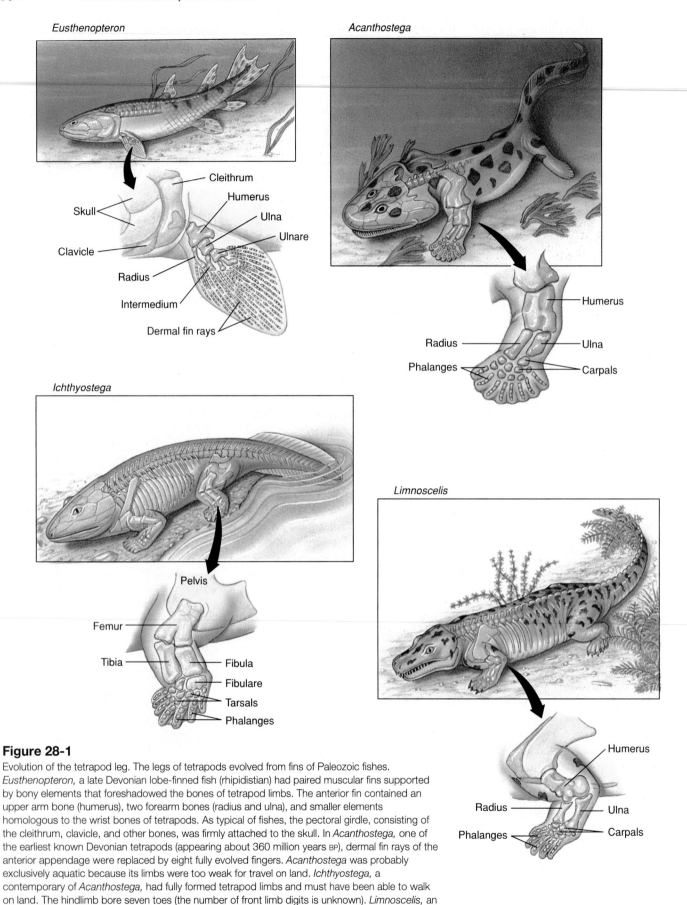

Eusthenopteron

Cleithrum
Skull
Humerus
Ulna
Ulnare
Clavicle
Radius
Intermedium
Dermal fin rays

Acanthostega

Humerus
Radius
Ulna
Phalanges
Carpals

Ichthyostega

Pelvis
Femur
Tibia
Fibula
Fibulare
Tarsals
Phalanges

Limnoscelis

Humerus
Radius
Ulna
Phalanges
Carpals

Figure 28-1

Evolution of the tetrapod leg. The legs of tetrapods evolved from fins of Paleozoic fishes. *Eusthenopteron,* a late Devonian lobe-finned fish (rhipidistian) had paired muscular fins supported by bony elements that foreshadowed the bones of tetrapod limbs. The anterior fin contained an upper arm bone (humerus), two forearm bones (radius and ulna), and smaller elements homologous to the wrist bones of tetrapods. As typical of fishes, the pectoral girdle, consisting of the cleithrum, clavicle, and other bones, was firmly attached to the skull. In *Acanthostega,* one of the earliest known Devonian tetrapods (appearing about 360 million years BP), dermal fin rays of the anterior appendage were replaced by eight fully evolved fingers. *Acanthostega* was probably exclusively aquatic because its limbs were too weak for travel on land. *Ichthyostega,* a contemporary of *Acanthostega,* had fully formed tetrapod limbs and must have been able to walk on land. The hindlimb bore seven toes (the number of front limb digits is unknown). *Limnoscelis,* an anthracosaur amphibian of the Carboniferous (about 300 million years BP) had five digits on both front and hindlimbs, the basic pentadactyl model which became the tetrapod standard.

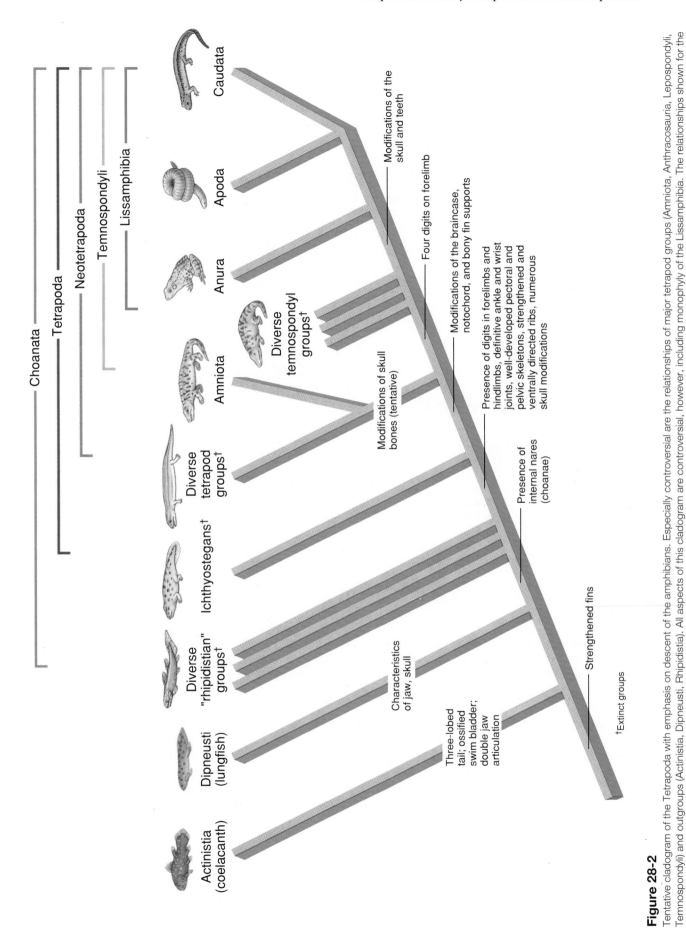

Figure 28-2

Tentative cladogram of the Tetrapoda with emphasis on descent of the amphibians. Especially controversial are the relationships of major tetrapod groups (Amniota, Anthracosauria, Lepospondyli, Temnospondyli) and outgroups (Actinistia, Dipneusti, Rhipidistia). All aspects of this cladogram are controversial, however, including monophyly of the Lissamphibia. The relationships shown for the three groups of Lissamphibia are based on recent molecular evidence.

Source: Modified from E. W. Gaffney in the Bulletin of the Carnegie Museum of Natural History 13:92–105 (1979).

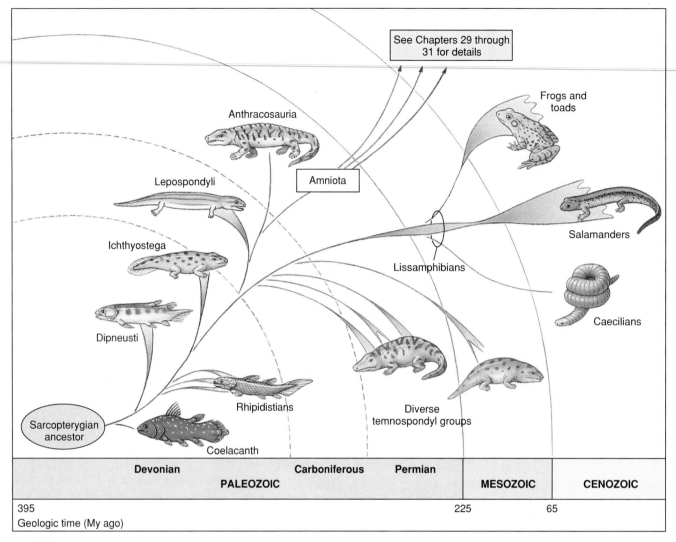

Figure 28-3
Early tetrapod evolution and the descent of amphibians. The tetrapods share most recent common ancestry with the Devonian rhipidistians. The amphibians share most recent common ancestry with the diverse temnospondyls of the Carboniferous and Permian periods of the Paleozoic, and Triassic period of the Mesozoic.

large ferns grew in profusion on a swampy landscape. Tetrapods radiated quickly in this environment to produce a great variety of forms, feeding on the abundance of insects, insect larvae, and aquatic invertebrates available. The evolutionary relationships of the early tetrapod groups are still very controversial. We present a tentative cladogram (see Figure 28-2) that almost certainly will undergo future revision as new data are collected. Several extinct lineages plus the **Lissamphibia,** which contains the modern amphibians, are placed in a group termed the **temnospondyls** (see Figure 28-2). This group is distin-

guished by having generally only four digits on the forelimb rather than the five characteristics of most tetrapods.

The lissamphibians diversified during the Carboniferous to produce the ancestors of the three major groups of amphibians alive today, the **frogs** (Anura or Salientia), **salamanders** (Caudata or Urodela), and **caecilians** (Apoda or Gymnophiona). The early amphibians improved their adaptations for living in water during this period. Their bodies became flatter for moving about in shallow water. Early salamanders developed weak limbs and the tail became better developed as a swimming organ. Even the anurans

(frogs and toads), which are now largely terrestrial as adults, developed specialized hindlimbs with webbed feet better suited for swimming than for movement on land. All amphibians use their porous skin as a primary or accessory breathing organ. This specialization was encouraged by the swampy surroundings of the Carboniferous period but presented serious desiccation problems for life on land.

Two additional generally recognized but nonetheless controversial groupings of Carboniferous and Permian tetrapods, the **lepospondyls** and **anthracosaurs,** are judged on the basis of skull structure to be closer to the amniotes than to the temno-

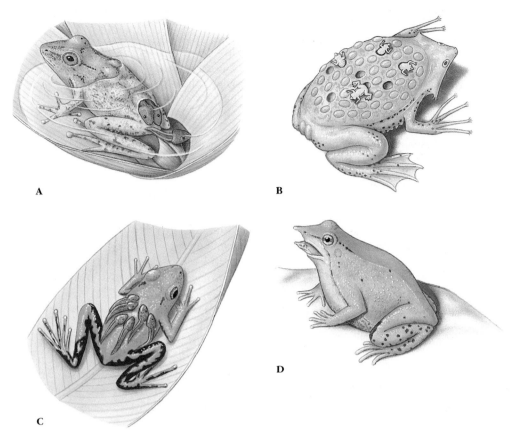

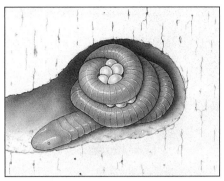

Figure 28-5
Female caecilian coiled around eggs in burrow.

Figure 28-4
Reproductive strategies of anurans. **A,** Female South American pygmy marsupial frog *Flectonotus pygmaeus* carries developing larvae in a dorsal pouch. **B,** Female Surinam frog carries eggs embedded in specialized brooding pouches on the dorsum; froglets emerge and swim away when development is complete. **C,** Male poison arrow frog *Phyllobates bicolor* carries tadpoles adhering to its back. **D,** Tadpoles of a male Darwin's frog *Rhinoderma darwinii* develop into froglets in its vocal pouch. When ready to emerge, a froglet crawls into the parent's mouth, which the parent opens to allow the froglet's escape.

spondyls (see Figure 28-3). Together they form a second major branch of tetrapod phylogeny that will be covered in Chapters 29 through 31.

THE MODERN AMPHIBIANS

The three living amphibian orders comprise more than 3900 species. Most share general adaptations for life on land, including skeletal strengthening and a shifting of special sense priorities from the ancestral lateral line system to the senses of smell and hearing. For this, both the olfactory epithelium and the ear are redesigned to improve sensitivities to airborne odors and sounds.

Nonetheless, most amphibians meet the problems of independent life on land only halfway. In the ancestral life history of amphibians, eggs are aquatic and hatch to produce an aquatic larval form that uses gills for breathing. A metamorphosis follows in which gills are lost and lungs, which are present throughout larval life, are then activated for respiration. Many amphibians retain this general pattern but there are some important exceptions. Some salamanders lack a complete metamorphosis and retain a permanently aquatic, larval morphology throughout life. Others live entirely on land and lack the aquatic larval phase completely. Both of these are evolutionarily derived conditions. Some frogs also have acquired a strictly terrestrial existence by eliminating the aquatic larval stage. Some frogs, salamanders, and caecilians that undergo the complete metamorphic life cycle nonetheless remain in the water as adults rather than moving onto land during their metamorphosis.

Even the most terrestrial amphibians remain dependent on very moist if not aquatic environments. Their skin is thin, and it requires moisture for protection against desiccation in air. An intact frog loses water nearly as rapidly as a skinless frog. Amphibians also require moderately cool environments. Being ectothermic, their body temperature is determined by and varies with the environment, greatly restricting where they can live. This is especially important for reproduction. Eggs are not well protected from desiccation, and they must be shed directly into the water or onto moist terrestrial surfaces. The completely terrestrial amphibians may lay eggs under logs or rocks, in the moist forest floor, in flooded tree holes, in pockets on the mother's back (Figure 28-4), or in folds of the body wall. One species of Australian frog even broods its young in its vocal pouch (see Figure 28-4).

We now highlight the special characteristics of the three major groups of amphibians. We will expand on the coverage of general amphibian features when discussing the groups in which particular features have been studied most extensively. For most features, this will be the frogs.

CAECILIANS: ORDER GYMNOPHIONA (APODA)

The order Gymnophiona contains approximately 160 species of elongate, limbless, burrowing creatures commonly called **caecilians** (Figure 28-5).

CHARACTERISTICS OF MODERN AMPHIBIANS

1. Skeleton mostly bony, with varying numbers of vertebrae; ribs present in some, absent or fused to vertebrae in others; notochord does not persist; exoskeleton absent
2. Body forms vary greatly from an elongated trunk with distinct head, neck, and tail to a compact, depressed body with fused head and trunk and no intervening neck
3. **Limbs usually four (tetrapod),** although some are legless; forelimbs of some much smaller than hindlimbs, in others all limbs small and inadequate; webbed feet often present; no true nails or claws; **forelimb usually with four digits** but sometimes five and sometimes fewer
4. **Skin smooth and moist with many glands,** some of which may be poison glands; pigment cells (chromatophores) common, of considerable variety; no scales, except concealed dermal ones in some

5. Mouth usually large with small teeth in upper or both jaws; two nostrils open into anterior part of mouth cavity
6. Respiration by lungs (absent in some salamanders), skin, and gills in some, either separately or in combination; external gills in the larval form and may persist throughout life in some
7. **Circulation with three-chambered heart,** two atria and one ventricle, and a **double circulation through the heart;** skin abundantly supplied with blood vessels
8. Ectothermal
9. Excretory system of paired mesonephric kidneys; urea main nitrogenous waste
10. Ten pairs of cranial nerves
11. Separate sexes; fertilization mostly internal in salamanders and caecilians, mostly external in frogs and toads; predominantly oviparous, some ovoviviparous or viviparous; metamorphosis usually present; **mesolecithal** (moderately yolky) **eggs with jellylike membrane coverings**

They occur in tropical forests of South America (their principal home), Africa, and Southeast Asia. They possess a long, slender body, small scales in the skin of some, many vertebrae, long ribs, no limbs, and a terminal anus. The eyes are small, and most species are totally blind as adults. Special sensory tentacles occur on the snout. Because they are almost entirely burrowing or aquatic, they are seldom seen by humans. Their food consists mostly of worms and small invertebrates, which they find underground. Fertilization is internal, and the male is provided with a protrusible copulatory organ. The eggs are usually deposited in moist ground near the water. The larvae may be aquatic, or the complete larval development may occur in the egg. In some species the eggs are carefully guarded in folds of the body during their development. Viviparity also is common in some caecilians, with the embryos obtaining nourishment by eating the wall of the oviduct.

SALAMANDERS: ORDER CAUDATA (URODELA)

As its name suggests, the order Caudata (L. *caudatus,* having a tail) consists of about 360 species of tailed amphibians: the salamanders and newts. Salamanders are found in almost all northern temperate regions of the world, and they have great abundance and diversity in North America. Salamanders are found also in the tropical areas of Central and northern South America. Salamanders are typically small; most of the common North American salamanders are less than 15 cm long. Some aquatic forms are considerably longer, and the Japanese giant salamander may exceed 1.5 m in length.

Most salamanders have limbs set at right angles to the body, with forelimbs and hindlimbs of approximately equal size. In some aquatic and burrowing forms, the limbs are rudimentary or absent.

Salamanders are carnivorous both as larvae and adults, preying on worms, small arthropods, and small molluscs. Most eat only things that are moving. Since their food is rich in proteins, they do not store in their bodies great quantities of fat or glycogen. Like all amphibians, they are ectotherms and have a low metabolic rate.

Breeding Behavior

Some salamanders are wholly aquatic throughout their life cycle, but most are metamorphic, having aquatic larvae and terrestrial adults that live in moist places under stones and rotten logs. The eggs of most salamanders are fertilized internally, usually after the female picks up a packet of sperm **(spermatophore)** that previously has been deposited by the male on a leaf or stick (Figure 28-6). Aquatic species lay their eggs in clusters or stringy masses in the water. Their eggs hatch to produce an aquatic larva having external gills and a finlike tail. Completely terrestrial species deposit eggs in small, grapelike clusters under logs or in excavations in soft moist earth, and many species remain to guard the eggs (Figure 28-7). These species have **direct development.** They have bypassed the larval stage and hatch as miniature versions of their parents. The most complex life cycle is observed in some American newts, whose aquatic larvae metamorphose to form terrestrial juveniles that later metamorphose again to produce secondarily aquatic, breeding adults (Figure 28-8). Many newt populations skip the terrestrial "red eft" stage, however, remaining entirely aquatic.

Respiration

Salamanders demonstrate an unusually diverse array of respiratory mechanisms. They share the general amphibian condition of having extensive vascular nets in their skin that serve the respiratory exchange of oxygen and carbon dioxide. At various stages of their life history, salamanders also

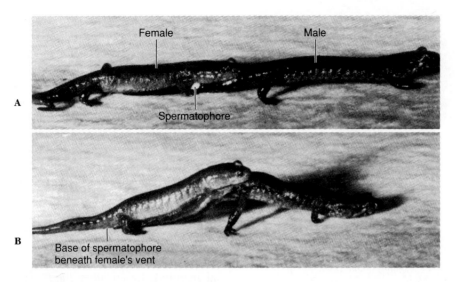

A

Female   Male

Spermatophore

B

Base of spermatophore
beneath female's vent

Figure 28-6
Courtship and sperm transfer in the pygmy salamander, *Desmognathus wrighti.* After judging the
female's receptivity by the presence of her chin on his tail base, the male deposits a spermatophore on
the ground, then moves forward a few paces. **A,** The white mass of the sperm atop a gelatinous base is
visible at the level of the female's forelimb. The male moves ahead, the female following until the
spermatophore is at the level of her vent. **B,** The female has recovered the sperm mass in her vent, while
the male arches his tail, tilting the female upward and presumably facilitating recovery of the sperm mass.

Figure 28-7
Female dusky salamander (*Desmognathus* sp.)
attending eggs. Many salamanders exercise
parental care of the eggs, which includes
rotating the eggs and protecting them from
fungal infections and predation by various
arthropods and other salamanders.

may have external gills, lungs, both,
or neither of these. Salamanders that
have an aquatic larval stage hatch
with gills, but lose them later if a
metamorphosis occurs. Several diverse lineages of salamanders have
evolved permanently aquatic forms
that fail to undergo a complete metamorphosis and retain their gills and
finlike tail throughout life. Lungs, the
most widespread respiratory organ of
terrestrial vertebrates, are present
from birth in the salamanders that
have them, and become active following metamorphosis.

Although we normally associate
lungs with terrestrial organisms and
gills with aquatic ones, salamander
evolution has produced aquatic forms
that breathe primarily with lungs and
terrestrial forms that lack them completely. The amphiumas of the salamander family Amphiumidae have
evolved a completely aquatic life history with a greatly reduced metamorphosis. The amphiumas nonetheless
lose their gills before adulthood and
then breathe primarily by lungs. They
periodically point their nostrils above
the surface of the water to get air.

The amphiumas provide a curious
contrast to many species of the family
Plethodontidae that are entirely terres-

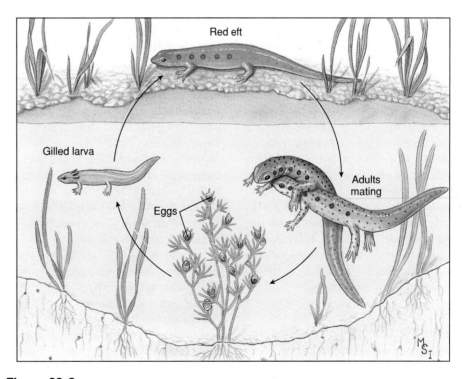

Red eft

Gilled larva

Eggs

Adults
mating

Figure 28-8
Life history of the red-spotted newt, *Notophthalmus viridescens* of the family Salamandridae. In many
habitats the aquatic larva metamorphoses into a brightly colored "red eft" stage, which remains on land
from 1 to 3 years before transforming into a secondarily aquatic adult.

trial but have dispensed with lungs
entirely. This large family contains
more than 220 species including many
of the familiar North American salamanders (see Figures 28-6, 28-7 and
28-9). The efficiency of cutaneous res-

piration is increased by the penetration of a capillary network into the
epidermis or by the thinning of the
epidermis over superficial dermal capillaries. Cutaneous respiration is supplemented by the pumping of air in

Figure 28-9
Longtail salamander *Eurycea longicauda,* a common plethodontid salamander.

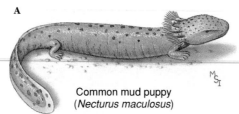

Common mud puppy
(*Necturus maculosus*)

Axolotl
(*Ambystoma mexicanum*)

Figure 28-10
Paedomorphosis in salamanders. **A,** The mud puppy *Necturus* sp. is a permanently gilled (perennibranchiate) aquatic form. **B,** An axolotl (*Ambystoma tigrinum*) may remain permanently gilled, or, should its pond habitat dry up, metamorphose to a terrestrial form that loses its gills and develops lungs.

and out of the mouth where the respiratory gases are exchanged across the vascularized membranes of the buccal (mouth) cavity (buccopharyngeal breathing). The lungless plethodontids may have originated in swift streams where lungs would have been a disadvantage by providing too much buoyancy, and where the water is so cool and well oxygenated that cutaneous respiration alone was sufficient for life. Some plethodontids have aquatic larvae whose gills are lost at metamorphosis. Others retain a permanently larval form with gills throughout life. Many others are completely terrestrial and bear the distinction of being the only vertebrates to have neither lungs nor gills at any stage of their life history. It is odd that the most completely terrestrial lineage of salamanders evolved in a group that completely lacks lungs.

Paedomorphosis

A persistent phylogenetic trend observed in salamander evolution is for descendants to retain into adulthood features that were present only in the pre-adult stages of their ancestors. Some characteristics of the ancestral adult morphology are consequently eliminated. This condition is called **paedomorphosis** (Gr. "child form"; see Chapter 9, p. 162). The most dramatic form of paedomorphosis occurs in those species that become sexually mature while retaining their gills, aquatic life habit, and other larval characteristics. These nonmetamorphic species are said to be **perennibranchiate** ("permanently gilled"). The mud puppies of the genus *Necturus* (Figure 28-10), which live on bottoms of ponds and lakes are an extreme example. These and many other salamanders are obligately perennibranchiate; they have never been observed to metamorphose under any conditions.

Some other species of salamanders that reach sexual maturity with larval morphology but, unlike permanent larvae such as *Necturus,* may metamorphose to terrestrial forms under certain environmental conditions. Good examples are found in *Ambystoma tigrinum* and related species from Mexico and the United States. The gilled individuals are called **axolotls** (Figure 28-10). Their typical habitat consists of small ponds that can disappear through evaporation in dry weather. When this happens, the axolotl metamorphoses to a terrestrial form, losing its gills and developing lungs. It can then travel across the land in search of new sources of water, to which it must return to reproduce. Axolotls are forced to metamorphose artificially when they are treated with the thyroid hormone, thyroxine. Thyroxine is essential for amphibian metamorphosis. The pituitary gland appears not to become fully active in the nonmetamorphosing forms, thereby failing to release the hormone thyrotropin, which is required to stimulate the thyroid gland to produce thyroxine (thyroid function in vertebrates is discussed on pp. 748–749).

Paedomorphosis takes many different forms in different groups of salamanders. It may affect the body as a whole or may be restricted to one or a few specific structures. The amphiumas mentioned previously lose their gills and activate their lungs before maturity, but they retain many general features of the larval body form. Paedomorphosis is important even in the terrestrial plethodontids that never have an aquatic larval stage. We can see the effects of paedomorphosis, for example, in the shape of the hands and feet of the tropical plethodontid genus *Bolitoglossa* (Figure 28-11). The ancestral morphology of *Bolitoglossa* features well-formed digits that grow out from the pad of the hand or foot during development. Some species have enhanced their ability to climb smooth vegetation, such as banana trees, by halting the growth of the digits and retaining throughout life a padlike foot. This padlike foot can produce adhesion and suction to attach the salamander to smooth vertical surfaces, and thereby serves an important adaptive function.

FROGS AND TOADS: ORDER ANURA (SALIENTIA)

The more than 3450 species of frogs and toads that make up the order Anura (Gr. *an,* without, + *oura,* tail) are for most people the most familiar amphibians. The Anura are an old group, known from the Jurassic period, 150 million years ago. Frogs and toads occupy a great variety of habitats. Their

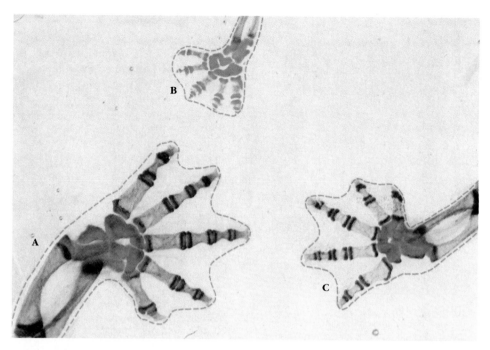

Figure 28-11

The foot structure of representatives of three different species of the tropical plethodontid salamander genus *Bolitoglossa*. These specimens have been treated chemically to clear the skin and muscles and to stain the bone red and cartilage blue. The species having the most fully ossified and distinct digits **(A, C)** live primarily on the forest floor. The species having the padlike foot caused by restricted digital growth **(B)** climbs smooth leaves and stems using the foot surface to produce suction or adhesion for attachment. The padlike foot evolved by paedomorphosis; it was derived evolutionarily by truncating development of the foot to prevent full digital development.

Figure 28-12

Two common North American frogs. **A,** Bullfrog, *Rana catesbeiana,* largest American frog and mainstay of the frog-leg epicurean market (family Ranidae). **B,** Green tree frog *Hyla cinerea,* a common inhabitant of swamps of the southeastern United States (family Hylidae). Note adhesive pads on the feet.

aquatic mode of reproduction and water-permeable skin prevent them from wandering too far from sources of water, however, and their ectothermy bars them from polar and subarctic habitats. The name of the order, Anura, refers to an obvious group characteristic, the absence of tails in adults. Although all pass through a tailed larval stage during development, only the genus *Ascaphus* contains a tail in the adult. Frogs and toads are specialized for jumping, as suggested by the alternative order name, Salientia, which means leaping.

We see in the appearance and life habit of their larvae further distinctions between the Anura and Caudata. The eggs of most frogs hatch into a tadpole ("polliwog"), having a long, finned tail, both internal and external gills, no legs, specialized mouthparts for herbivorous feeding (some tadpoles, and salamander larvae, are carnivorous), and a highly specialized internal anatomy. They look and act altogether differently from adult frogs. The metamorphosis of the frog tadpole to the adult frog is thus a striking transformation. The perennibranchiate condition never occurs in frogs and toads as it does in salamanders.

In addition to their importance in biomedical research and education, frogs have long served the epicurean frog-leg market. Mainstay of this market is the bullfrog, which is in such heavy demand in Europe (especially France) and the United States—the worldwide harvest is an estimated 200 million bullfrogs (about 10,000 metric tons) annually—that its populations have fallen drastically as the result of excessive exploitation and the draining and pollution of wetlands. Most are Asian bullfrogs imported from India and Bangladesh, some 80 million collected each year from rice fields in Bangladesh alone. With so many insect-eating frogs removed from the ecosystem, rice production is threatened from uncontrolled, flourishing insect populations. In the United States, attempts to raise bullfrogs in farms have not been successful, mainly because bullfrogs are voracious eating machines that normally will only accept living prey, such as insects, crayfish, and other frogs.

Frogs and toads are divided into 21 families. The best-known frog families in North America are the Ranidae, which contains most of our familiar frogs (Figure 28-12A), and the Hylidae, the tree frogs (Figure 28-12B). True toads, belonging to the family Bufonidae, have short legs, stout bodies, and thick skins, usually with prominent warts (Figure 28-13). However, the term "toad" is used rather loosely to refer also to more or less terrestrial members of several other families.

The largest anuran is the West African *Conraua goliath,* which is more than 30 cm long from tip of nose to anus (Figure 28-14). This giant eats animals as big as rats and ducks. The smallest frog recorded is *Psyllophryne didactyla,* measuring less than 1 cm in length; it is also the smallest known tetrapod. This tiny

Figure 28-13
American toad *Bufo americanus* (family Bufonidae). This principally nocturnal yet familiar amphibian feeds on large numbers of insect pests and on snails and earthworms. The warty skin contains numerous glands that produce a surprisingly poisonous milky fluid, providing the toad excellent protection from a variety of potential predators.

Figure 28-14
Conraua (*Gigantorana*) *goliath* (family Ranidae) of West Africa, the world's largest frog. This specimen weighed 3.3 kg (approximately 7½ pounds).

Courtesy American Museum of Natural History, Neg. #125617.

What is responsible for the widely reported decline in amphibian, especially frog, populations around the world? Puzzling is the evidence that whereas amphibian populations are falling in various parts of the world, in other areas they are doing well. No single explanation fits all instances of declines. In some instances, population changes are simply random fluctuations caused by periodic droughts and other naturally occurring phenomena. How-ever, several other environmental factors have been implicated in amphib-ian declines: habitat destruction and modification; rises in environmental pollutants such as acid rain, fungicides, herbicides, and industrial chemicals; diseases; and introduction of nonnative predators and competitors. Recently it was shown that depletion of the ozone shield in the stratosphere and the con-sequent increase in ultraviolet radiation reaching the earth's surface caused severe losses in developing embryos of two frog species in the American west. Frog and toad eggs, exposed as they are on the surface of ponds, are especially sensitive to the damaging action of ultraviolet radiation on cellular DNA. Thus one or more explanations do seem to explain certain population declines; in other instances the reasons for the declines are not obvious.

frog, which can be more than covered by a dime, is found in the Brazilian rain forest. The largest American frog is the bullfrog, *Rana catesbeiana* (see Figure 28-12A), which reaches a head and body length of 20 cm.

Habitats and Distribution

Probably the most abundant frogs are the approximately 260 species of the genus *Rana* (Gr. frog), found all over the temperate and tropical regions of the world except in New Zealand, the oceanic islands, and southern South America. They are usually found near water, although some, such as the wood frog *R. sylvatica,* spend most of their time on damp forest floors. The wood frog probably returns to pools only for breeding in early spring. The larger bullfrogs, *R. catesbeiana,* and green frogs, *R. clamitans,* are nearly always found in or near permanent water or swampy regions. The leop-ard frog, *R. pipiens,* has a wider vari-ety of habitats and, with all of its sub-species and forms, is the most widespread of all the North American frogs. This is the species most com-monly used in biology laboratories and for classical electrophysiological research. It has been found in some

form in nearly every state, although sparingly represented along the ex-treme western part of the Pacific coast. It also extends far into northern Canada and as far south as Panama.

Within the range of any species, frogs are often restricted to certain lo-calities (for instance, to specific streams or pools) and may be absent or scarce in similar habitats else-where. The pickerel frog (*R. palustris*) is especially noteworthy in this re-spect because it is known to be abun-dant only in certain localized regions. Recent studies have shown that many populations of frogs worldwide may be suffering declines in numbers and becoming even more patchy than usual in their distributions. The causes of this phenomenon are unknown.

Most of the larger frogs are solitary in their habits except during the breed-ing season. During the breeding period most of them, especially the males, are very noisy. Each male usually takes possession of a particular perch near water, where he may remain for hours or even days, trying to attract a female to that spot. At times frogs are mainly silent, and their presence is not de-tected until they are disturbed. When they enter the water, they dart about swiftly and reach the bottom of the

pool, where they kick up a cloud of muddy water. In swimming, they hold the forelimbs near the body and kick backward with the webbed hindlimbs, which propel them forward. When they come to the surface to breathe, only the head and foreparts are ex-posed and, since they usually take ad-vantage of any protective vegetation, they are difficult to see.

During the winter months most frogs in temperate climates hibernate in the soft mud of the bottoms of pools and streams. Naturally their life processes are at a very low ebb during their hibernation period, and such en-ergy as they need is derived from the glycogen and fat stored in their bodies during the spring and summer months. The more terrestrial frogs, such as tree frogs, hibernate in the humus of the forest floor. They are tolerant of low temperatures, and many actually survive prolonged freezing of as much as 65% of their total body water. Such frost-tolerant frogs prepare for winter by accumulating massive quantities of glucose or glycerol in body fluids, which protects major organs from the normally damaging effects of ice crystal formation.

While native American amphibians continue to disappear as wetlands are drained, an exotic frog introduced into southern California has found the climate quite to its liking. The African clawed frog *Xenopus laevis* (Figure 28-15) is a voracious, aggressive, primarily aquatic frog that is rapidly displacing native frogs and fish from several waterways and is spreading rapidly. The species was introduced into North America in the 1940s when they were used extensively in human pregnancy tests. When more efficient tests appeared in the 1960s, some hospitals simply dumped surplus frogs into nearby streams, where the prolific breeders have become almost indestructible pests. As is so often the case with alien wildlife introductions, benign intentions frequently lead to serious problems.

Adult frogs have numerous enemies, such as snakes, aquatic birds, turtles, raccoons, and humans; fish prey on tadpoles, and only a few tadpoles survive to maturity. Although usually defenseless, many frogs and toads in the tropics and subtropics are aggressive, jumping and biting at predators. Some defend themselves by feigning death. Most anurans can blow up their lungs so that they are difficult to swallow. When disturbed along the margin of a pond or brook, a frog often remains quite still; when it thinks it is detected, it jumps, not always into the water where enemies may be lurking, but into grassy cover on the bank. When held in the hand, a frog may cease its struggles for an instant to put its captor off guard and then leap violently, at the same time voiding its urine. Their best protection is their ability to leap and their use of poison

Figure 28-15

African clawed frog, *Xenopus laevis*. The claws, an unusual feature in frogs, are on the hind feet. This frog has been introduced into California, where it is considered a serious pest.

glands. Bullfrogs in captivity do not hesitate to snap at tormentors and are capable of inflicting painful bites.

Integument and Coloration

The skin of the frog is thin and moist, and it is attached loosely to the body only at certain points. Histologically the skin is composed of two layers: an outer stratified **epidermis** and an inner spongy **dermis** (Figure 28-16). The outer layer of epidermal cells (which is shed periodically when a frog or toad "molts") contains deposits of **keratin,** a tough, fibrous protein that provides a certain measure of protection against abrasion and loss of water from the skin. The more terrestrial amphibians such as toads have especially heavy deposits of keratin. But amphibian keratin is soft, unlike the hard keratin that forms scales, claws, feathers, horns, and hair of amniotes.

The inner layer of the epidermis gives rise to two types of integumentary glands that grow down into the loose dermal tissues below. Small **mucous** glands secrete a protective mucous waterproofing onto the skin surface, and large **serous** glands produce

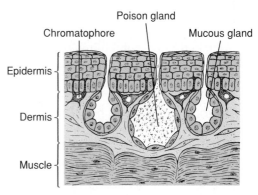

Figure 28-16

Section through frog skin.

a whitish, watery poison that is highly irritating to would-be predators. All amphibians produce a skin poison, but its effectiveness varies from species to species and with different predators. The extremely toxic poison of three species of *Phyllobates,* a genus of small South American dendrobatid frogs, is used by a western Colombian Indian tribe to poison the points of blowgun darts. Most species of the family Dendrobatidae produce toxic skin secretions, some of which are among the most lethal animal secretions known, drop for drop more poisonous even than the venoms of sea snakes or any of the most poisonous arachnids.

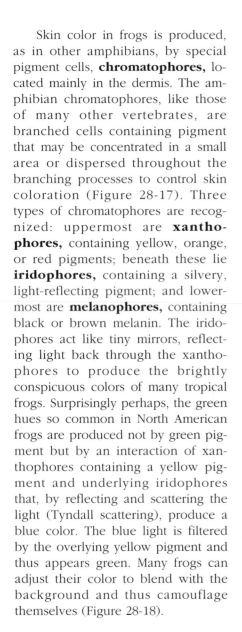

Figure 28-17
Pigment cells (chromatophores). **A,** Pigment dispersed. **B,** Pigment concentrated. The pigment cell does not contract or expand; color effects are produced by streaming of cytoplasm, carrying pigment granules into cell branches for maximum color effect or to the center of the cell for minimum effect. Control over dispersal or concentration of pigment is mostly by light stimuli acting through a pituitary hormone.

Figure 28-18
Cryptic coloration of the gray frog, *Hyla versicolor*. Camouflage is so good that the presence of this frog usually is disclosed only at night by its resonant, flutelike call.

Skin color in frogs is produced, as in other amphibians, by special pigment cells, **chromatophores,** located mainly in the dermis. The amphibian chromatophores, like those of many other vertebrates, are branched cells containing pigment that may be concentrated in a small area or dispersed throughout the branching processes to control skin coloration (Figure 28-17). Three types of chromatophores are recognized: uppermost are **xanthophores,** containing yellow, orange, or red pigments; beneath these lie **iridophores,** containing a silvery, light-reflecting pigment; and lowermost are **melanophores,** containing black or brown melanin. The iridophores act like tiny mirrors, reflecting light back through the xanthophores to produce the brightly conspicuous colors of many tropical frogs. Surprisingly perhaps, the green hues so common in North American frogs are produced not by green pigment but by an interaction of xanthophores containing a yellow pigment and underlying iridophores that, by reflecting and scattering the light (Tyndall scattering), produce a blue color. The blue light is filtered by the overlying yellow pigment and thus appears green. Many frogs can adjust their color to blend with the background and thus camouflage themselves (Figure 28-18).

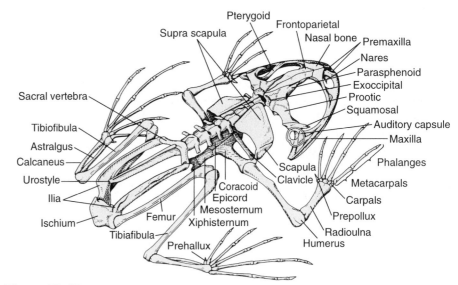

Figure 28-19
Skeleton of the bullfrog, *Rana catesbeiana*.

Skeletal and Muscular Systems

In amphibians, as in other vertebrates, the well-developed **endoskeleton** of bone and cartilage provides a framework for the muscles in movement and protection for the viscera and nervous systems. But movement onto land and the necessity of transforming paddlelike fins into tetrapod legs capable of supporting the body's weight introduced a new set of stress and leverage problems. The changes are most noticeable in the anurans, whose entire musculoskeletal system is specialized for jumping and swimming by simultaneous extensor thrusts of the hindlimbs.

The amphibian vertebral column assumes a new role as a support from which the abdomen is slung and to which the limbs are attached. Since amphibians move with limbs instead of swimming with serial contractions of the trunk musculature, the vertebral column has lost much of the original flexibility characteristic of fishes. Rather it has become a rigid frame for transmitting force from the hindlimbs to the body. The anurans are further specialized by an extreme shortening of the body. Typical frogs have only nine trunk vertebrae and a rodlike **urostyle,** which represents several fused caudal vertebrae (coccyx) (Figure 28-19). The limbless caecilians, which obviously have not shared these specializations for tetrapod locomotion, may have as many as 285 vertebrae.

The frog skull is also vastly altered as compared with its vertebrate ancestors; it is much lighter in weight and more flattened in profile and has fewer bones and less ossification. The front part of the skull, wherein are located the nose, eyes, and brain, is better developed, whereas the back of the skull, which contained the gill apparatus in fishes, is much reduced (see Figure 28-19).

The pattern of bones and muscles in the limbs is the typical tetrapod type. There are three main joints in each limb (hip, knee, and ankle; or shoulder, elbow, and wrist). The foot is typically five-rayed (pentadactyl) and the hand is four-rayed with both foot and hand having several joints in each of the digits (see Figure 28-19). It is a repetitive system that can be plausibly derived from one resembling the bone structure of the rhipidistian lobe-fin, which is distinctly suggestive of the amphibian limb (see Figure 28-1). It is not difficult to imagine how selective pressures through millions of years remodeled ancestral lobe-fins into limbs.

The muscles of the limbs are presumably homologous to the radial muscles that move the fins of fishes up and down, but the muscular arrangement has become so complex in the tetrapod limb that it is no longer possible to see parallels between this and fin musculature. Despite its complexity, we can recognize two major groups of muscles on any limb: an anterior and ventral group that pulls the limb forward and toward the midline (protraction and adduction), and a second set of posterior and dorsal muscles that serves to draw the limb back and away from the body (retraction and abduction).

The trunk musculature, which in fishes is segmentally organized into powerful muscular bands (myomeres, p. 515) for locomotion by lateral flexion, was much modified during amphibian evolution. The dorsal (epaxial) muscles are arranged to support the head and brace the vertebral column. The ventral (hypaxial) muscles of the belly are more developed in amphibians than in fishes, since they must support the viscera in air without the buoying assistance of water.

Respiration and Vocalization

Amphibians use three respiratory surfaces for gas exchange in air: the skin (cutaneous breathing), the mouth (buccal breathing), and the lungs. Frogs and toads show a much greater dependence on lung breathing than do salamanders; nevertheless, the skin continues to serve as an important supplementary avenue for gas exchange in anurans, especially during hibernation in winter. Even under normal conditions when lung breathing predominates, most of the carbon dioxide is lost across the skin while most of the oxygen is taken up across the lungs.

The lungs are supplied by pulmonary arteries (derived from the sixth aortic arches) and blood returns directly to the left atrium by the pulmonary veins. Frog lungs are ovoid, elastic sacs with their inner surfaces divided into a network of septa that are in turn subdivided into small terminal air chambers called alveoli. The alveoli of the frog lung are much larger than those of amniote vertebrates, and consequently the frog lung has a smaller relative surface available for gas exchange: the respiratory surface of the common *Rana pipiens* is about 20 cm^2 per cubic centimeter of air contained, compared with 300 cm^2 for humans. The problem in lung evolution was not the development of a good internal vascular surface, but rather the problem of moving air. A frog is a positive-pressure breather that fills its lungs by forcing air into them; this contrasts with the negative-pressure system of amniotes. The sequence and explanation of breathing in a frog are shown in Figure 28-20. One can easily follow this sequence in a living frog at rest: rhythmical throat movements of mouth breathing may continue some time before flank movements indicate that the lungs are being emptied and refilled.

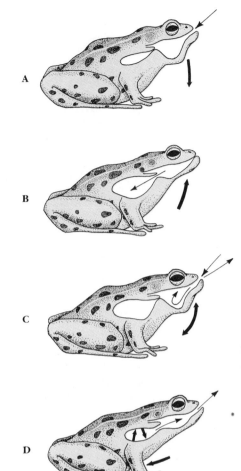

Figure 28-20
Breathing in a frog. Frogs are positive-pressure breathers, that fill their lungs by forcing air into them. **A,** Floor of mouth is lowered, drawing air in through nostrils. **B,** With nostrils closed and glottis open, the frog forces air into its lungs by elevating floor of mouth. **C,** Mouth cavity rhythmically ventilates for a period. **D,** Lungs are emptied by contraction of body-wall musculature and by elastic recoil of lungs.

Both male and female frogs have **vocal cords,** but those of the male are much better developed. They are located in the **larynx,** or voice box. Sound is produced by passing air back and forth over the vocal cords between the lungs and a large pair of sacs (vocal pouches) in the floor of the mouth. The latter also serve as effective resonators in the male. The chief function of the voice is to attract mates. Most species utter characteristic sounds that identify them. Nearly everyone is familiar with the springtime calls of the spring peeper, which produces a high-pitched sound surprisingly strident for such a tiny frog.

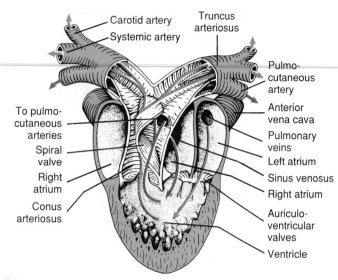

Figure 28-21
Structure of the frog heart. *Red arrows,* oxygenated blood. *Blue arrows,* deoxygenated blood.

Circulation

As in fishes, circulation in amphibians is a closed system of arteries and veins serving a vast peripheral network of capillaries through which blood is forced by the action of a single pressure pump, the heart. The principal changes in circuitry involve the shift from gill to lung breathing. With the elimination of gills, a major obstacle to blood flow was removed from the arterial circuit. But two new problems arose. The first was to provide a blood circuit to the lungs. As we have seen, this problem was solved by converting the sixth aortic arch into pulmonary arteries to serve the lungs and by developing new pulmonary veins for returning oxygenated blood to the heart. The second and evidently more difficult evolutionary problem was to separate the pulmonary circulation from the rest of the body's circulation. Oxygenated blood from the lungs would be selectively sent to the body and deoxygenated venous return from the body would be selectively sent to the lungs. In effect this meant creating a double circulation consisting of separate pulmonary and systemic circuits. Tetrapods solved the problem by evolving a partition down the center of the heart, creating a double pump, one for each circuit. Amphibians and reptiles have made the separation to varying degrees. Birds and mammals have the most completely divided hearts containing two atria and two ventricles.

The frog heart (Figure 28-21) has two separate atria and a single undivided ventricle. Blood from the body (systemic circuit) first enters a large receiving chamber, the sinus venosus, which forces blood into the right atrium. The left atrium receives freshly oxygenated blood from the lungs. Up to this point the deoxygenated blood from the body and oxygenated blood from the lungs are separated. But now both atria contract almost simultaneously, driving both right and left atrial blood into the single undivided **ventricle.** We should expect that complete admixture of the two circuits would happen in the ventricle. In fact there is evidence that in at least some amphibians they remain mostly separated, so that when the ventricle contracts, oxygenated pulmonary blood enters the systemic circuit and deoxygenated systemic blood circulates through the pulmonary circuit. The **spiral valve** in the **conus arteriosus** (see Figure 28-21) may play an important role in maintaining selective distribution. The matter is controversial and has defied complete analysis despite the application of advanced techniques using radiopaque media and high-speed photography.

Feeding and Digestion

Frogs are carnivorous, as are most other adult amphibians, and they feed on insects, spiders, worms, slugs, snails, millipedes, and nearly anything else that moves and is small enough to swallow whole. They snap at moving prey with their protrusible tongue, which is attached to the front of the mouth and is free behind. The free end of the tongue is highly glandular and produces a sticky secretion that adheres to the prey. When teeth are present on the premaxillae, maxillae, and vomers, they are used to prevent escape of prey, not for biting or chewing. The digestive tract is relatively short in adult amphibians, a characteristic of most carnivores, and it produces a variety of enzymes for breaking down proteins, carbohydrates, and fats.

The larval stages of anurans (tadpoles) are usually herbivorous, feeding on pond algae and other vegetable matter; they have a relatively long digestive tract because their bulky food must be submitted to time-consuming fermentation before useful products can be absorbed.

Nervous System and Special Senses

The three fundamental parts of the brain—forebrain (telencephalon), concerned with the sense of smell; midbrain (mesencephalon), concerned with vision; and hindbrain (rhombencephalon), concerned with hearing and balance—have undergone dramatic developmental trends as the vertebrates moved onto land and improved their environmental awareness. In general there is increasing cephalization with emphasis on information processing by the brain and a corresponding loss of indepen-

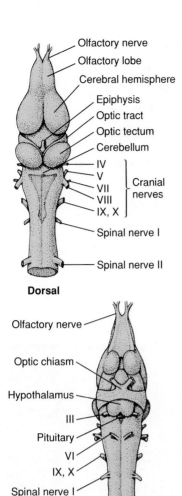

Figure 28-22

Brain of a frog, dorsal and ventral views.

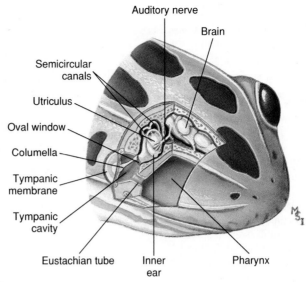

Figure 28-23

Cutaway of frog head showing ear structure. Sound vibrations are transmitted from the tympanic membrane by way of the columella to the inner ear. The eustachian tube allows pressure equilibration between the tympanic cavity and the pharynx.

dence of the spinal ganglia, which are capable of only stereotyped reflexive behavior. Nonetheless, a headless frog preserves an amazing degree of purposive and highly coordinated behavior. With only the spinal cord intact, it maintains normal body posture and can with purposive accuracy raise its leg to wipe from its skin a piece of filter paper soaked in dilute acid. It will even use the opposite leg if the closer leg is held.

The forebrain (Figure 28-22) contains the olfactory center, which assumes greatly increased importance for the detection of dilute airborne odors on land. The sense of smell is in fact one of the dominant special senses in frogs. The remainder of the forebrain, the cerebrum, is of little importance in amphibians. Instead, complex integrative activities of the frog

are located in the midbrain optic lobes. The hindbrain is divided into an anterior cerebellum and a posterior medulla. The cerebellum (see Figure 28-22) is concerned with equilibrium and movement coordination and is not well developed in amphibians, especially in terrestrial species that stay close to the ground and are not noted for dexterity of movement. The cerebellum becomes vastly developed in the fast-moving birds and mammals. The medulla is really the enlarged anterior end of the spinal cord through which pass all sensory neurons except those of vision and smell. Here are located centers for auditory reflexes, respiration, swallowing, and vasomotor control.

The evolution of a semiterrestrial life for the amphibians has necessitated a reordering of sensory receptor priorities on land. The pressure-sensitive lateral line (acousticolateral) system of fishes remains only in the aquatic larvae of amphibians and in a few strictly aquatic adult amphibian species. This system of course can serve no useful purpose on land, since it was designed to detect and localize objects in water by reflected pressure waves. Instead the task of detecting airborne sounds falls on the ear.

The ear of a frog is by amniote standards a simple structure: a middle

ear closed externally by a large **tympanic membrane** (eardrum) and containing a **columella** (stapes) that transmits vibrations to the inner ear (Figure 28-23). The latter contains the **utricle,** from which arise three semicircular canals, and a **saccule** bearing a diverticulum, the **lagena.** The lagena is partly covered with a **tectorial membrane** that in its fine structure is not unlike that of the much more complex mammalian cochlea. In most frogs this structure is sensitive to low-frequency sound energy not greater than 4000 Hz (cycles per second); in the bullfrog the main frequency response is in the 100 to 200 Hz range, which matches the energy of the male frog's low-pitched call.

Vision is the dominant special sense in many amphibians (the mostly blind caecilians are obvious exceptions). Several modifications of the ancestral aquatic eye were required to adapt it for use in air. Lachrymal glands and eyelids evolved to keep the eye moist, wiped free of dust, and shielded from injury. Since the cornea is exposed to air, it is an important refractive surface, removing much of the burden from the lens of bending light rays and focusing the image on the retina. As in the fishes, accommodation (adjusting focus for near and distant objects) is accomplished by moving the lens. But unlike the eyes of most fishes, the amphibian eye at rest is adjusted for distant objects and the lens is moved forward to focus on nearby objects.

Keeping a sharp image on the retina for approaching or receding objects requires accommodation, and this is accomplished in different ways by different vertebrates. The eye of the bony fishes and lampreys is adjusted for near vision; to focus on distant objects, the lens must be moved backward. In amphibians, sharks, and snakes, the relaxed eye is focused on distant objects and the lens is moved *forward* to focus on nearby objects. In birds, mammals, and all reptiles except snakes, the lens accommodates by changing its *curvature* rather than by being moved forward or backward. The resting eye in these forms is adjusted for distant vision, and to focus on nearby objects the lens curvature is increased; that is, the lens is squeezed (or, in some, allowed to relax) into a rounded shape.

The **retina** contains both **rods and cones,** the latter providing frogs with color vision. The iris contains well-developed circular and radial muscles and can rapidly expand or contract the aperture (pupil) to adjust to changing illumination. The upper lid of the eye is fixed, but the lower is folded into a transparent **nictitating membrane** capable of moving across the eye surface (Figure 28-24). Frogs and toads generally possess good vision, a property of crucial importance to animals that rely on quick escape to avoid their numerous predators and on accurate movements to capture rapidly moving prey.

Other sensory receptors include tactile and chemical receptors in the skin, taste buds on the tongue and palate, and a well-developed olfactory epithelium lining the nasal cavity.

Reproduction

Because frogs and toads are ectothermic, they breed, feed, and grow only during the warmer seasons of the year. One of the first drives after the dor-

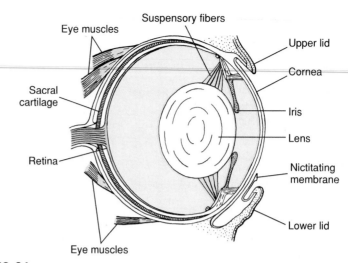

Figure 28-24
Amphibian eye.

mant period is breeding. In the spring males croak and call vociferously to attract females. When their eggs are mature, the females enter the water and are clasped by the males in a process called **amplexus** (Figure 28-25). As the female lays the eggs, the male discharges sperm over the eggs to fertilize them. After fertilization, the jelly layers absorb water and swell. The eggs are laid in large masses, usually anchored to vegetation.

Development of the fertilized egg (zygote) begins almost immediately (Figure 28-26). By repeated division (cleavage) the egg is converted into a hollow ball of cells (blastula). The blastula undergoes gastrulation and then continues to differentiate to form an embryo with a tail bud. At 6 to 9 days, depending on the temperature, a tadpole hatches from the protective jelly coats that had surrounded the original fertilized egg.

At the time of hatching, the tadpole has a distinct head and body with a compressed tail. The mouth is located on the ventral side of the head and is provided with horny jaws for scraping off vegetation from objects for food. Behind the mouth is a ventral adhesive disc for clinging to objects. In front of the mouth are two deep pits, which later develop into the nostrils. Swellings are found on each side of the head, and these later become external gills. There are three pairs of external gills, which later transform into internal

Figure 28-25
A male green frog, *Hyla cinerea,* clasps a larger female during the breeding season in a South Carolina swamp. Clasping (amplexus) is maintained until the female deposits her eggs. Like most tree frogs, these are capable of rapid and marked color changes; the male here, normally green, has darkened during amplexus.

gills that become covered with a flap of skin (the operculum) on each side. On the right side the operculum completely fuses with the body wall, but on the left side a small opening, the spiracle (L. *spiraculum,* air hole) remains, through which water flows after entering the mouth and passing the internal gills. The hindlegs appear first during metamorphosis, whereas the forelimbs are hidden for a time by the folds of the operculum. The tail is resorbed, the intestine becomes much shorter, the mouth undergoes a transformation into the adult condition,

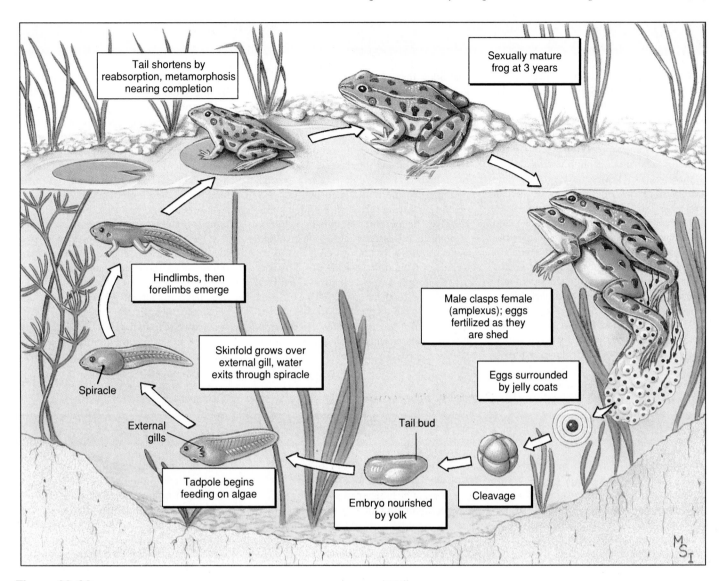

Figure 28-26
Life cycle of a leopard frog.

lungs develop, and the gills are resorbed (Figure 28-26). The leopard frog usually completes its metamorphosis within 3 months; the bullfrog takes 2 or 3 years to complete the process.

Migration of frogs and toads is correlated with their breeding habits. Males usually return to a pond or stream before the females, which they then attract by their calls. Some salamanders are also known to have a strong homing instinct, returning year after year to the same pool for reproduction guided by olfactory cues. The initial stimulus for migration in many cases is attributable to a seasonal cycle in the gonads plus hormonal changes that increase the frogs' sensitivity to temperature and humidity changes.

CLASSIFICATION OF CLASS AMPHIBIA

Order Gymnophiona (jim′no-fy′o-na) (Gr. *gymnos,* naked, + *ophioneos,* of a snake) **(Apoda): caecilians.** Body elongate; limbs and limb girdle absent; mesodermal scales present in skin of some; tail short or absent; 95 to 285 vertebrae; pantropical, 6 families, 34 genera, approximately 160 species.
Order Caudata (caw-dot′uh) (L. *caudatus,* having a tail) **(Urodela): salamanders.** Body with head, trunk, and tail; no scales; usually two pairs of equal limbs; 10 to 60 vertebrae; predominantly holarctic; 9 living families, 62 genera, approximately 360 species.
Order Anura (uh-nur′uh) (Gr. *an,* without, + *oura,* tail) **(Salientia): frogs, toads.** Head and trunk fused; no tail; no scales; two pairs of limbs; large mouth; lungs; 6 to 10 vertebrae including urostyle (coccyx); cosmopolitan, predominantly tropical; 21 living families; 301 genera; approximately 3450 species.

Summary

Amphibians are ectothermic, primitively quadrupedal vertebrates that have glandular skin and that breathe by lungs, gills, or skin. They are the survivors of one of two major branches of tetrapod phylogeny, the other one being represented today by the amniotes. The modern amphibians consist of three major evolutionary groups. The caecilians (order Gymnophiona) are a small tropical group of limbless, elongate forms. The salamanders (order Caudata) are tailed amphibians that have retained the generalized four-legged body plan of their Paleozoic ancestors. The frogs and toads (order Anura) are the largest group of modern amphibians, all of which are specialized for a jumping mode of locomotion.

Most amphibians have a biphasic life cycle that begins with an aquatic larva that later metamorphoses to produce a terrestrial adult that returns to the water to lay eggs. Some frogs, salamanders, and caecilians have evolved direct development that omits the aquatic larval stage and some caecilians have evolved viviparity. Salamanders are unique among amphibians in having evolved several perennibranchiate species that retain a permanently larval morphology throughout life, eliminating the terrestrial phase completely. The perennibranchiate condition is obligate in some species, but in others metamorphosis to a terrestrial form can be induced by the drying of the pond habitat.

Although amphibians have evolved adaptations to the aquatic phase of their life history, the adaptations to their terrestrial existence are particularly noteworthy. Respiratory exchange of gases occurs across the porous skin in all amphibians and is supplemented in most amphibians by lungs. Oddly, the most highly terrestrial salamanders lack lungs whereas some aquatic forms use lungs as their major respiratory structure. Life on land also required strengthening and redirection of skeletal elements, especially the ribs, pectoral and pelvic girdles, and limbs. Derived features of the amphibian auditory and visual systems and associated regions of the brain facilitate sensory perception on land.

Despite their adaptations for terrestrial life, the adults and eggs of all amphibians require cool, moist environments if not actual pools or streams. The eggs and adult skin have no effective protection against very cold, hot, or dry conditions, greatly restricting the adaptive radiation of amphibians to environments that have moderate temperatures and abundant water.

Review Questions

1. As compared with the aquatic habitat, the terrestrial habitat offers both advantages and problems for an animal making the transition from water to land. Summarize how these differences might have influenced the early evolution of tetrapods.
2. Describe the different modes of respiration used in amphibians. What paradox do the amphiumas and terrestrial plethodontids present regarding the association of lungs with life on land?
3. The evolution of the tetrapod leg was one of the most important advances in vertebrate history. Describe the supposed sequence in its evolution.
4. Compare the general life history patterns of salamanders with those of frogs. Which group shows the greater variety of evolutionary changes of the ancestral biphasic amphibian life cycle?
5. Give the literal meaning of the name Gymnophiona. What animals are included in this amphibian order, what do they look like, and where do they live?
6. What is the literal meaning of the order names Caudata and Anura? What major features distinguish the members of these two orders from each other?
7. Describe the breeding behavior of a typical woodland salamander.
8. How has paedomorphosis been important to the evolutionary diversification of the salamanders?
9. Describe the integument of a frog. What is responsible for skin color in frogs?
10. Describe amphibian circulation.
11. Explain how the forebrain, midbrain, hindbrain, and the sensory structures with which each brain division is concerned have developed to meet the sensory requirements for amphibian life on land.
12. Briefly describe the reproductive behavior of frogs. In what important ways do frogs and salamanders differ in their reproduction?

Selected References

See also general references for Part III, p. 626.

Blaustein, A. R., and D. B. Wake. 1995. The puzzle of declining amphibian populations. Sci. Am. **272:**52–57 (Apr.). *Amphibian populations are dwindling in many parts of the world. The causes are multiple, but all derive from human activities.*

Conant, R., and J. T. Collins. 1991. A field guide to reptiles and amphibians, ed. 3. The Peterson field guide series. Boston, Houghton Mifflin Company. *Updated version of a popular field guide; color illustrations and distribution maps for all species.*

del Pino, E. M. 1989. Marsupial frogs. Sci. Am. **260:**110–118 (May). *Several species of tropical frogs incubate their eggs on the female's back, often in a special pouch, and emerge as advanced tadpoles or fully formed froglets.*

Duellman, W. E. 1992. Reproductive strategies of frogs. Sci. Am. **267:**80–87 (July). *Many frogs have evolved improbable reproductive strategies that have permitted colonization of land.*

Duellman, W. E., and L. R. Trueb. 1986. Biology of amphibians. New York, McGraw-Hill Book Company. *Important comprehensive sourcebook of information on amphibians, extensively referenced and illustrated.*

Gibbons, W. 1983. Their blood runs cold: adventures with reptiles and amphibians. University, Alabama, University of Alabama Press. *Delightful account of personal experiences of a herpetologist, filled with engaging stories and interesting facts.*

Halliday, T. R., and K. Adler (eds). 1986. The encyclopedia of reptiles and amphibians. New York, Facts on File, Inc. *Excellent authoritative reference work with high-quality illustrations.*

Hanken, J. 1989. Development and evolution in amphibians. Am. Sci. **77:**336–343 (July–Aug.). *Explains how the diversity in amphibian morphology has been achieved by modifications in development.*

Lewis, S. 1989. Cane toads: an unnatural history. New York, Dolphin/Doubleday. *Based on an amusing and informative film of the same title that describes the introduction of cane toads to Queensland, Australia and the unexpected consequences of their population explosion there. "If Monty Python teamed up with National Geographic, the result would be* Cane Toads.*"*

Moffett, M. W. 1995. Poison-dart frogs: lurid and lethal. National Geographic **187**(5):98–111 (May). *Photographic essay of frogs that can be lethal even to the touch.*

Narins, P. M. 1995. Frog communication. Sci. Am. **273:**78–83 (Aug.). *Frogs employ several strategies to hear and be heard amidst the cacophony of chorusing of many frogs.*

29

The Reptiles

Phylum Chordata
Class Reptilia

Enclosing the Pond

The amphibians, with well-developed legs, redesigned sensory and respiratory systems, and modifications of the postcranial skeleton for supporting the body in air, have made a notable conquest of land. But, with shell-less eggs and often gill-breathing larvae, their development remains hazardously tied to water. The lineage containing reptiles, birds, and mammals developed an egg that could be laid on land. This shelled egg, perhaps more than any other adaptation, unshackled the early reptiles from the aquatic environment by freeing the developmental process from dependence on aquatic or very moist terrestrial environments. In fact, the "pond-dwelling" stages were not eliminated but enclosed within a series of extraembryonic membranes that provided complete support for embryonic development. One membrane, the amnion, encloses a fluid-filled cavity, the "pond," within which the developing embryo floats. Another membranous sac, the allantois, serves both as a respiratory surface and as a chamber for the storage of nitrogenous wastes. Enclosing these membranes is a third membrane, the chorion, through which oxygen and carbon dioxide freely pass. Finally, surrounding and protecting everything is a porous, parchmentlike or leathery shell.

With the last ties to aquatic reproduction severed, conquest of the land by the vertebrates was ensured. The Paleozoic tetrapods that developed this reproductive pattern were the ancestors of a single, monophyletic assemblage called the Amniota, named after the innermost of the three extraembryonic membranes, the amnion. Before the end of the Paleozoic era the amniotes had diverged into multiple lineages that gave rise to all the reptilian groups, the birds, and the mammals. ■

Members of the paraphyletic class Reptilia (rep-til′e-a) (L. *repto,* to creep) include the first truly terrestrial vertebrates. With nearly 7000 species (approximately 300 species in the United States and Canada) occupying a great variety of aquatic and terrestrial habitats, they are diverse and abundant. Nevertheless, reptiles are perhaps remembered best for what they once were, rather than for what they are now. The Age of Reptiles, which lasted for more than 165 million years, saw the appearance of a great radiation of reptilian lineages into a bewildering array of terrestrial and aquatic forms. Among these were the herbivorous and carnivorous dinosaurs, many of huge stature and awesome appearance, that dominated animal life on land. Then, during a mass extinction at the end of Mesozoic era, they suddenly declined. Among the few reptilian lineages to emerge from the Mesozoic extinction are today's reptiles. One of these, the tuatara (*Sphenodon*) of New Zealand, is the sole survivor of a group that otherwise disappeared 100 million years ago. But others, especially the lizards and snakes, have radiated since the Mesozoic extinction into diverse and abundant groups. Under-standing the 300-million-year-old history of reptile life on earth has been complicated by widespread convergent and parallel evolution among the many lineages and by large gaps in the fossil record.

ORIGIN AND ADAPTIVE RADIATION OF REPTILES

As mentioned in the prologue to this chapter, the amniotes are a monophyletic group that evolved in the late Paleozoic. Most paleontologists agree that the amniotes arose from a group of amphibian-like tetrapods, the anthracosaurs, during the early Carboniferous period of the Paleozoic. By the late Carboniferous (approximately 300 million years ago), the amniotes had separated into three lineages. The first lineage, the **anapsids** (Gr. *an,* without, + *apsis,* arch), is characterized by a skull having no temporal opening behind the orbits, the skull behind the orbits being completely roofed with dermal bone (Figure 29-2). This group is represented today only by the turtles. Their morphology is an odd mix of ancestral and derived characters that has scarcely changed at all since the turtles first appeared in the fossil record in the Triassic some 200 million years ago.

The second lineage, the **diapsids** (Gr. *di,* double, + *apsis,* arch), gave rise to all other reptilian groups and to the birds (Figure 29-1). The diapsid skull was characterized by the presence of two temporal openings: one pair located low on the cheeks, and a second pair positioned above the lower pair and separated from them by a bony arch (Figure 29-2). Three subgroups of diapsids appeared. The **lepidosaurs** include the extinct marine ichthyosaurs and all of the modern reptiles with the exception of the turtles and crocodilians. The more derived **archosaurs** comprised the dinosaurs and their relatives, and the living crocodilians and birds. A third, smaller subgroup, the **sauropterygians** included several extinct aquatic groups, the most conspicuous of which were the large, long-necked plesiosaurs.

The third lineage was the **synapsids** (Gr. *syn,* together, + *apsis,* arch), the mammal-like reptiles. The synapsid skull had a single pair of temporal openings located low on the cheeks and bordered by a bony arch (Figure 29-2). The synapsids were the first amniote group to diversify, giving rise first to the pelycosaurs, later to the therapsids, and finally to mammals (Figure 29-1).

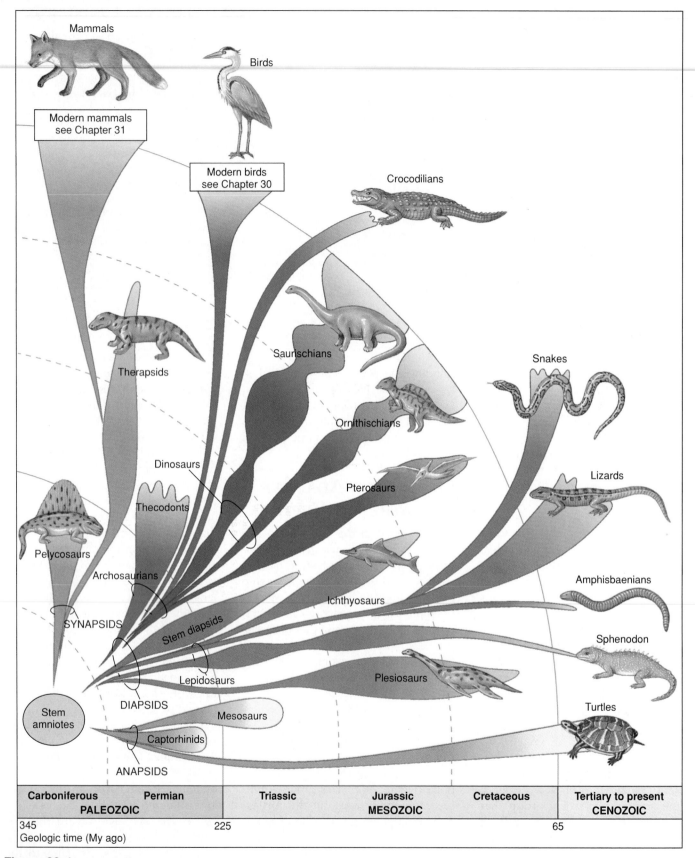

Mammals

Birds

Modern mammals
see Chapter 31

Modern birds
see Chapter 30

Crocodilians

Therapsids

Saurischians

Snakes

Ornithischians

Dinosaurs

Lizards

Thecodonts

Pterosaurs

Pelycosaurs

Archosaurians

Ichthyosaurs

Amphisbaenians

SYNAPSIDS

Stem diapsids

Sphenodon

Lepidosaurs

Plesiosaurs

DIAPSIDS

Stem
amniotes

Mesosaurs

Turtles

Captorhinids

ANAPSIDS

Carboniferous	Permian	Triassic	Jurassic	Cretaceous	Tertiary to present
PALEOZOIC		**MESOZOIC**			**CENOZOIC**

345   225   65
Geologic time (My ago)

Figure 29-1

Evolution of the amniotes. The evolutionary origin of amniotes occurred by the evolution of an amniotic egg that made reproduction on land possible, although this egg may well have developed before the earliest amniotes had ventured far on land. The amniote assemblage, which includes the reptiles, birds, and mammals, evolved from a lineage of small, lizardlike forms known as captorhinids that retained the skull pattern of the early tetrapods. First to diverge from the primitive stock were the mammal-like reptiles, characterized by a skull pattern termed the synapsid condition. All other amniotes, including the birds and all living reptiles except the turtles, have a skull pattern known as diapsid. The turtles have a skull pattern known as anapsid. The great Mesozoic radiation of reptiles may have been caused partly by the increased variety of ecological habitats into which the amniotes could move.

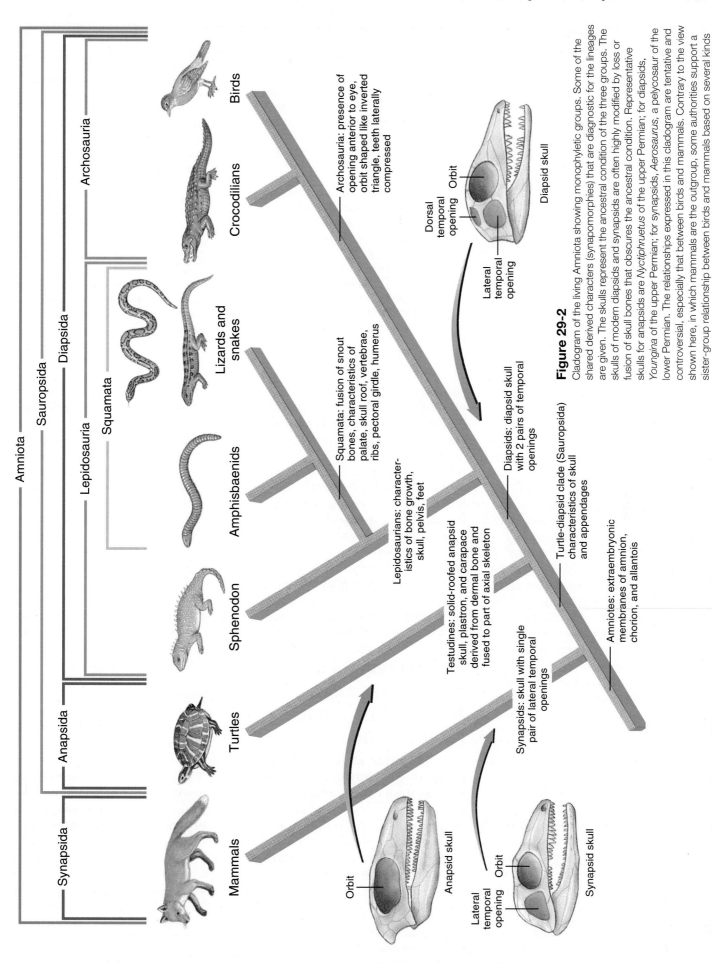

Figure 29-2

Cladogram of the living Amniota showing monophyletic groups. Some of the shared derived characters (synapomorphies) that are diagnostic for the lineages are given. The skulls represent the ancestral condition of the three groups. The skulls of modern diapsids and synapsids are often highly modified by loss or fusion of skull bones that obscures the ancestral condition. Representative skulls for anapsids are *Nyctiphruetus* of the upper Permian; for diapsids, *Youngina* of the upper Permian; for synapsids, *Aerosaurus*, a pelycosaur of the lower Permian. The relationships expressed in this cladogram are tentative and controversial, especially that between birds and mammals. Contrary to the view shown here, in which mammals are the outgroup, some authorities support a sister-group relationship between birds and mammals based on several kinds of molecular and physiological evidences.

Source: Data from F. H. Pough, J. B. Heiser, and W. N. McFarland, Vertebrate Life, 3d edition. Macmillan, New York, 1989.

Labels within the cladogram:

- Archosauria: presence of opening anterior to eye, orbit shaped like inverted triangle, teeth laterally compressed
- Squamata: fusion of snout bones, characteristics of palate, skull roof, vertebrae, ribs, pectoral girdle, humerus
- Lepidosaurians: characteristics of bone growth, skull, pelvis, feet
- Diapsids: diapsid skull with 2 pairs of temporal openings
- Testudines: solid-roofed anapsid skull, plastron, and carapace derived from dermal bone and fused to part of axial skeleton
- Turtle-diapsid clade (Sauropsida) characteristics of skull and appendages
- Amniotes: extraembryonic membranes of amnion, chorion, and allantois
- Synapsids: skull with single pair of lateral temporal openings

Skull labels: Dorsal temporal opening, Orbit, Lateral temporal opening, Diapsid skull; Orbit, Anapsid skull; Lateral temporal opening, Orbit, Synapsid skull

Taxa: Birds, Crocodilians, Lizards and snakes, Amphisbaenids, Sphenodon, Turtles, Mammals

Clade names: Amniota, Sauropsida, Diapsida, Archosauria, Lepidosauria, Squamata, Anapsida, Synapsida

CHANGES IN TRADITIONAL CLASSIFICATION OF REPTILES

With increasing use of cladistic methodology in zoology, and its insistence on hierarchical arrangement of monophyletic groups (see p. 201), important changes have been made in the traditional classification of reptiles. The class Reptilia is no longer recognized by cladists as a valid taxon because it is not monophyletic. As customarily defined, the class Reptilia excludes the birds which descend from the most recent common ancestor of the reptiles. Consequently, the reptiles are a **paraphyletic** group because they do not include all descendants of their most recent common ancestor. Reptiles can be identified only as amniotes that are not birds. This is clearly shown in the phylogenetic tree of the amniotes (Figure 29-1).

An example of this problem is the shared ancestry of birds and crocodilians. Based solely on shared derived characteristics, crocodilians and birds are sister groups; that is they are more recently descended from a common ancestor than either is from any other living reptilian lineage. In other words, birds and crocodilians belong to a monophyletic group apart from other reptiles and, according to the rules of cladism, should be assigned to a clade that separates them from the remaining reptiles. This clade is in fact recognized; it is the Archosauria (Figures 29-1 and 29-2), a grouping that also includes the extinct dinosaurs. Therefore birds should be classified as reptiles. The archosaurs plus their sister group, the lepidosaurs (tuataras, lizards, snakes, and amphisbaenids), comprise a monophyletic group that some taxonomists call the Reptilia. The term Reptilia is thereby redefined to include birds in contrast to its traditional usage. However, evolutionary taxonomists argue that birds represent a novel adaptive zone and grade of organization whereas crocodilians remain within the traditionally recognized reptilian adaptive zone and grade. In this view, the morphological and ecological novelty of birds has been recognized by main-

taining the traditional classification that places the crocodilians in the class Reptilia and birds in the class Aves. Such conflicts of opinion between proponents of the two major competing schools of taxonomy (cladistics and evolutionary taxonomy) have had the healthy effect of forcing zoologists to reevaluate their views of amniote genealogy and how vertebrate classifications should represent genealogy and degree of divergence. In our treatment we retain the class Reptilia because this is still standard taxonomic practice, but we emphasize that this taxonomy is likely to be discontinued.

CHARACTERISTICS OF REPTILES THAT DISTINGUISH THEM FROM AMPHIBIANS

1. Reptiles have tough, dry, scaly skin offering protection against desiccation and physical injury. The skin consists of a thin epidermis, shed periodically, and a much thicker, well-developed **dermis** (Figure 29-3). The dermis is provided with **chromatophores,** the color-bearing cells that give many lizards and snakes their colorful hues. It is

also the layer that, unfortunately for their bearers, is converted into alligator and snakeskin leather, so esteemed for expensive pocketbooks and shoes. The characteristic **scales** of reptiles are formed largely of keratin. They are derived mostly from the epidermis and thus are not homologous to fish scales, which are bony, dermal structures. In some reptiles, such as alligators, the scales remain throughout life, growing gradually to replace wear. In others, such as snakes and lizards, new scales grow beneath the old, which are then shed at intervals. Turtles add new layers of keratin under the old layers of the platelike scutes, which are modified scales. In snakes the old skin (epidermis and scales) is turned inside out when discarded; lizards split out of the old skin leaving it mostly intact and right side out, or it may slough off in pieces.

2. The shelled (amniotic) egg of reptiles contains food and protective membranes for supporting embryonic development on land. Reptiles lay their eggs in sheltered locations on land. The young hatch as lung-breathing juveniles rather than as aquatic larvae. The appearance of the shelled egg (Figure 29-4) widened the

CHARACTERISTICS OF CLASS REPTILIA

1. Body varied in shape, compact in some, elongated in others; **body covered with an exoskeleton of horny epidermal scales** with the addition sometimes of bony dermal plates; **integument with few glands**
2. **Limbs paired, usually with five toes,** and adapted for climbing, running, or paddling; absent in snakes and some lizards
3. Skeleton well ossified; ribs with sternum (sternum absent in snakes) forming a complete thoracic basket; **skull with one occipital condyle**
4. Respiration by lungs; **no gills;** cloaca used for respiration by some; branchial arches in embryonic life
5. Three-chambered heart; **crocodilians with four-chambered heart;** usually one pair of aortic arches; systemic and pulmonary circuits functionally separated
6. Ectothermic; many thermoregulate behaviorally
7. **Metanephric kidney (paired); uric acid main nitrogenous waste**
8. Nervous system with the optic lobes on the dorsal side of brain; **12 pairs of cranial nerves** in addition to nervus terminalis
9. Sexes separate; **fertilization internal**
10. **Eggs covered with calcareous or leathery shells; extraembryonic membranes (amnion, chorion,** and **allantois)** present during embryonic life; **no aquatic larval stages**

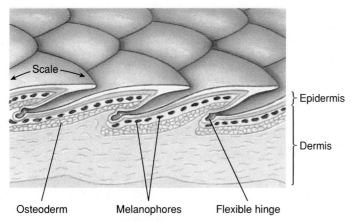

Figure 29-3
Section of the skin of a reptile showing the overlapping epidermal scales.

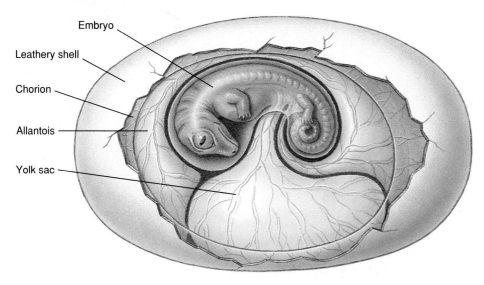

Figure 29-4
Amniotic egg. The embryo develops within the amnion and is cushioned by amniotic fluid. Food is provided by yolk from the yolk sac and metabolic wastes are deposited within the allantois. As development proceeds, the allantois fuses with the chorion, a membrane lying against the inner surface of the shell; both membranes are supplied with blood vessels that assist in the exchange of oxygen and carbon dioxide across the porous shell. Because this kind of egg is an enclosed, self-contained system, it is often called a "cleidoic" egg (Gr. *kleidoun,* to lock in).

division between the evolving amphibians and reptiles and, probably more than any other adaptation, contributed to the evolutionary establishment of reptiles.

3. The reptilian jaws are efficiently designed for applying crushing or gripping force to prey. The jaws of fish and amphibians were designed for quick jaw closure, but once the prey was seized, little static force could be applied. In reptiles jaw muscles became larger, longer, and arranged for much better mechanical advantage.

4. Reptiles have some form of copulatory organ, permitting internal fertilization. Internal fertilization is obviously a requirement for a shelled egg, since the sperm must reach the egg before the egg is enclosed. Sperm from the paired testes are carried by the vasa deferentia to the copulatory organ, which is an evagination of the cloacal wall. The female system consists of paired ovaries and oviducts. The glandular walls of the oviducts secrete albumin (source of amino acids, minerals, and water for the embryo) and shells for the large eggs.

5. Reptiles have a more efficient circulatory system and higher blood pressure than amphibians. In all reptiles the right atrium, which receives unoxygenated blood from the body, is completely partitioned from the left atrium, which receives oxygenated blood from the lungs. In the crocodilians there are two completely separated ventricles as well (Figure 29-5); in other reptiles the ventricle is incompletely separated. Even in reptiles with incomplete separation of the ventricles, flow patterns within the heart prevent admixture of pulmonary (oxygenated) and systemic (unoxygenated) blood; all reptiles therefore have two functionally separate circulations.

6. Reptile lungs are better developed than those of amphibians. Reptiles depend almost exclusively on lungs for gas exchange, supplemented by pharyngeal membrane respiration in some aquatic turtles. Unlike the amphibians, which *force* air into the lungs with mouth muscles, the reptiles *suck* air into the lungs by enlarging the pleural cavity, either by expanding the rib cage (snakes and lizards) or by movement of internal organs (turtles and crocodilians). Reptiles have no muscular diaphragm (found only in mammals). Cutaneous respiration (gas exchange across the skin), so important to amphibians, has been completely abandoned by the reptiles.

7. Reptiles have evolved efficient strategies for water conservation. Although the reptilian kidney is of the advanced metanephros with its own passageway (ureter) to the exterior, it lacks loops of Henle of the more advanced mammalian metanephros, and thus is unable to produce a urine more concentrated than its body fluids. Instead, many reptiles have salt glands located near the nose or eyes (in the tongue of saltwater crocodiles) which secrete a salty fluid that is strongly hyperosmotic to the body fluids. Nitrogenous wastes are excreted as uric acid, rather than urea or ammonia. Uric acid has a low solubility and precipitates out of solution readily, allowing water to be conserved; the urine of many reptiles is a semisolid suspension.

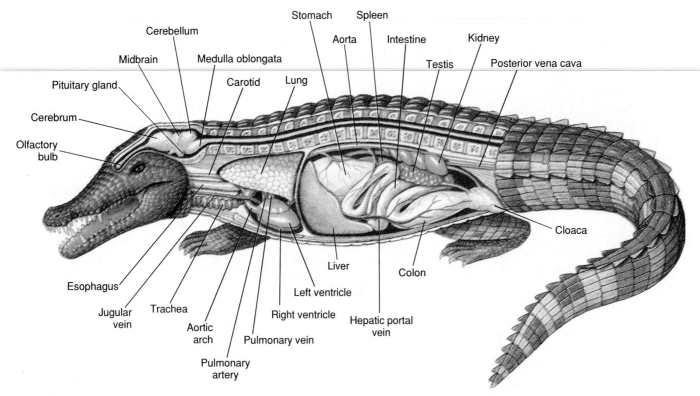

Figure 29-5
Internal structure of a male crocodile.

8. All reptiles, except the limbless members, have better body support than the amphibians and more efficiently designed limbs for travel on land. Nevertheless, most modern reptiles walk with their legs splayed outward and their belly close to the ground. Most dinosaurs, however, (and some modern lizards) walked on upright legs held beneath the body, the best arrangement for rapid movement and for the support of body weight. Many of the dinosaurs walked on powerful hindlimbs alone.

9. The reptilian nervous system is considerably more complex than the amphibian. Although the reptile's brain is small, the cerebrum is increased in size relative to the rest of the brain. The crocodilians have the first true cerebral cortex (neopallium). Central nervous system connections are more advanced, permitting complex kinds of behavior unknown in amphibians. With the exception of hearing, sense organs in general are well developed. Jacobson's organ, a specialized olfactory chamber present

in many tetrapods, is highly developed in lizards and snakes. Odors are carried to the organ by the tongue.

CHARACTERISTICS AND NATURAL HISTORY OF REPTILIAN ORDERS

ANAPSID REPTILES: SUBCLASS ANAPSIDA

Order Testudines (Chelonia): Turtles

Turtles descended from one of the earliest anapsid lineages, probably a group known as the procolophonids of the late Permian, but turtles themselves do not appear in the fossil record until the Upper Triassic, some 200 million years ago. From the Triassic, turtles plodded on to the present with very little change in their early morphology. They are enclosed in shells consisting of dorsal **carapace** (Fr. shell, shield) and a ventral **plastron** (Fr. breastplate). Clumsy and

unlikely as they appear to be within their protective shells, they are nonetheless a varied and ecologically diverse group that seems able to adjust to human presence. The shell is so much a part of the animal that it is fused to thoracic vertebrae and ribs (Figure 29-6). The shell is composed of two layers: an outer horny layer of keratin and an inner layer of bone. New layers of keratin are laid down beneath the old as the turtle grows and ages. Lacking teeth, the turtle jaw is provided with tough, horny plates for gripping food (Figure 29-7).

The terms "turtle," "tortoise," and "terrapin" are applied variously to different members of the turtle order. In North American usage, they are all correctly called turtles. The term "tortoise" is frequently given to land turtles, especially the large forms. British usage of the terms is different: "tortoise" is the inclusive term, whereas "turtle" is applied only to the aquatic members.

One consequence of living in a rigid shell with fused ribs is that a turtle cannot expand its chest to breathe. Turtles solved this problem by employing certain abdominal and pectoral muscles as a "diaphragm." Air is drawn in by contracting limb flank muscles to make the body cavity larger. Exhalation is also active and is accomplished by drawing the shoulder girdle back into the shell, thus compressing the viscera and forcing air out of the lungs. Many water turtles gain enough oxygen by just pumping water in and out of the mouth cavity; this enables them to remain submerged for long periods when inactive. When active they must lung-breathe more frequently.

A turtle's brain, like that of other reptiles, is small, never exceeding 1% of the body weight. The cerebrum, however, is larger than that of an amphibian, and turtles are able to learn a maze about as quickly as a rat. Turtles have both a middle and an inner ear, but perception of sound is poor. Not unexpectedly, therefore, turtles are virtually mute (the biblical "voice of the turtle" refers to the turtledove), although many tortoises utter grunting or roaring sounds during mating (Figure 29-8). Compensating for poor hearing are a good sense of smell, acute vision, and color perception evidently as good as that of humans.

Turtles are oviparous. Fertilization is internal and all turtles, even the marine forms, bury their shelled, amniotic eggs in the ground. Usually considerable care is exercised in constructing the nest, but once the eggs are deposited and covered, the female deserts them. An odd feature of turtle reproduction is that in some turtle families, as in all crocodilians and some lizards, the nest temperature determines the sex of the hatchlings. In turtles, low temperatures during incubation produce males and high temperatures produce females. All of these temperature-dependent reptiles lack sex chromosomes.

The great marine turtles, buoyed by their aquatic environment, may reach 2 m in length and 725 kg in weight. One such heavyweight is the

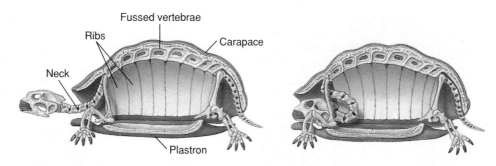

Figure 29-6
Skeleton and shell of a turtle, showing fusion of vertebrae and ribs with the carapace. The long and flexible neck allows the turtle to withdraw its head into its shell for protection.

Figure 29-7
Snapping turtle, *Chelydra serpentina,* showing the absence of teeth. Instead, the jaw edges are covered with a horny plate.

Figure 29-8
Mating Galápagos tortoises. The male has a concave plastron that fits over the highly convex carapace of the female, helping to provide stability during mating. Males utter a roaring sound during mating, the only time they are known to emit vocalizations.

Figure 29-9
Green sea turtle, *Chelonia mydas*. Green turtles are herbivores that subsist on marine grasses and algae. Sea turtles range widely in the oceans, returning to land only to deposit their eggs. Sea turtles are found in all tropical oceans.

Figure 29-10
Alligator snapping turtle *Macroclemys temmincki* of the southeastern United States lies on the bottom, mouth agape, luring fish and other unwary prey by undulating a pink, wormlike protrusion from its tongue. Any prey attempting to eat the bait is instantly captured in powerful jaws.

leatherback. The green turtle (Figure 29-9), so named because of its greenish body fat, may exceed 360 kg, although most individuals of this economically valuable and heavily exploited species seldom live long enough to reach anything approaching this size. Some land tortoises may weigh several hundred kilograms, such as the giant tortoises of the Galápagos Islands that so intrigued Darwin during his visit there in 1835. Most tortoises are rather slow moving; an hour of determined trudging carries a large Galápagos tortoise approximately 300 m (although they may move much more rapidly for short distances). Their low metabolism probably explains their longevity, for some are believed to live more than 150 years.

The shell, like a medieval coat of armor, offers obvious advantages. The head and appendages can be drawn in for protection. The familiar box tortoise (*Terrapene carolina*) has a plastron that is hinged, forming two movable parts that can be pulled up against the carapace so tightly that one can hardly force a knife blade between the shells. Some turtles, such as the large eastern snapping turtle (*Chelydra serpentina*), have reduced shells, making complete withdrawal

for protection quite impossible. Snappers, however, have another formidable defense, as their name implies (Figure 29-7). They are entirely carnivorous, living on fish, frogs, waterfowl, or almost anything that comes within reach of their powerful jaws. The alligator snapper lures unwary fish into its mouth with a "bait" (Figure 29-10). Snappers are wholly aquatic and come ashore only to lay their eggs.

DIAPSID REPTILES: SUBCLASS DIAPSIDA

The diapsid reptiles, that is, reptiles having a skull with two pairs of temporal openings (Figure 29-2), are classified into three lineages (superorders; see the Classification of Amniotes and Living Reptiles on p. 566). The two with living representatives are the superorder Lepidosauria, containing the lizards, snakes, worm lizards, and *Sphenodon;* and the superorder Archosauria, containing the crocodilians.

Order Squamata: Lizards, Snakes, and Worm Lizards

The squamates are the most recent and diverse products of diapsid evolution, making up approximately 95% of all

known living reptiles. Lizards appear in the fossil record as early as the Permian, but they do not begin their radiation until the Cretaceous period of the Mesozoic era when the dinosaurs were at the climax of their radiation. Snakes appeared during the late Cretaceous period, probably from a group of lizards whose descendants include the Gila monster and monitor lizards. Two specializations in particular characterize snakes: extreme elongation of the body and accompanying displacement and rearrangement of internal organs; and specializations for eating large prey. The amphisbaenians (worm lizards), which first appear in the fossil record of the early Cenozoic era, have structural specializations associated with a burrowing habit.

The diapsid skulls of the squamates are modified from the ancestral diapsid condition by the loss of dermal bone ventral and posterior to the lower temporal opening. This has allowed the evolution in most lizards of a mobile skull having movable joints. Such a skull is called a **kinetic skull.** The quadrate, which in other reptiles is fused to the skull, has a joint at its dorsal end, as well as its usual articulation with the lower jaw. In addition, there are joints in the palate and across the roof of the skull that allow the

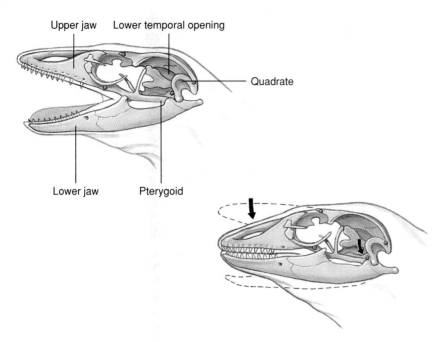

Figure 29-11

Kinetic diapsid skull of a modern lizard (monitor lizard, *Varanus* sp.) showing the joints (indicated by dots) that allow the snout and upper jaw to move on the rest of the skull. The quadrate can move at its dorsal end and ventrally at both the lower jaw and the pterygoid. The front part of the braincase is also flexible, allowing the snout to be raised. Note that the lower temporal opening is very large with no lower border; this modification of the diapsid condition, common in modern lizards, provides space for expansion of large jaw muscles. The upper temporal opening lies dorsal and medial to the postorbital-squamosal arch and is not visible in this drawing.

Figure 29-12

Tokay, *Gekko gecko,* of Southeast Asia has a true voice and is named after the strident repeated *to-kay, to-kay* call.

Figure 29-13

A large male marine iguana, *Amblyrhynchus cristatus,* of the Galápagos Islands, feeding underwater on algae. This is the only marine lizard in the world. It has special salt-removing glands in the eye orbits and long claws that enable it to cling to the bottom while feeding on small red and green algae, its principal diet. It may dive to depths exceeding 10 m (33 feet) and remain submerged more than 30 minutes.

snout to be tilted up (Figure 29-11). The specialized mobility of the skull enables lizards to seize and manipulate their prey; it also increases the effective closing force of the jaw musculature. The skull of snakes is even more kinetic than that of lizards. Such exceptional skull mobility is considered a major factor in the diversification of lizards and snakes.

Suborder Sauria: Lizards.   The lizards are an extremely diverse group, including terrestrial, burrowing, aquatic, arboreal and aerial members. Among the more familiar groups in this varied suborder are the **geckos** (Figure 29-12), small, agile, mostly nocturnal forms with adhesive toe pads that enable them to walk upside down and on vertical surfaces; the **iguanas,** often brightly colored New World lizards with ornamental crests, frills, and throat fans, and a group that includes the remarkable marine iguana of the Galápagos Islands (Figure 29-13); the **skinks,** with elongate bodies and reduced limbs; and the **chameleons,** a group of arboreal lizards, mostly of Africa and Madagascar. The chameleons are entertaining creatures that catch insects with the sticky-tipped tongue that can be flicked accurately and rapidly to a distance greater than the length of their body (Figure 29-14). The great majority of lizards have four limbs and relatively short bodies, but in many the limbs are degenerate, and a few such as the glass lizards (Figure 29-15) are completely limbless.

Most lizards have movable eyelids, whereas a snake's eyes are permanently covered with a transparent cap. Lizards have keen vision for daylight (retinas rich in both cones and rods; see p. 736 for discussion of color vision), although one group, the nocturnal geckos, has retinas composed entirely of rods. Most lizards have an external ear that snakes lack. The inner ear of lizards is variable in structure, but as with other reptiles, hearing does not play an important role in the lives of most lizards. Geckos are exceptions because the males are strongly vocal (to announce territory and discourage other males from approaching), and it is reasonable to assume that they can hear their own vocalizations. Other species of lizards vocalize in defensive behavior.

Many lizards live in the world's hot and arid regions. Since their skin lacks glands, water loss by this avenue is much reduced. They produce a semisolid urine with a high content of crystalline uric acid. This is an excellent mechanism for conserving water and is found in other groups living successfully in arid habitats

THE MESOZOIC WORLD OF DINOSAURS

When in 1841 the English anatomist Richard Owen coined the term *dinosaur* ("terrible lizard") to describe fossil Mesozoic reptiles of gigantic size, only three poorly known dinosaur genera were distinguished. But with new and marvelous fossil discoveries quickly following, by 1887 zoologists were able to distinguish two groups of dinosaurs based on differences in the structure of the pelvic girdles. The Saurischia ("lizard-hipped") had a simple, three-pronged pelvis with the hip bones arranged much as they are in other reptiles. The large bladelike ilium is attached to the backbone by stout ribs. The pubis and ischium extend ventrally and posteriorly respectively, and all three bones meet at the hip socket, a deep opening on the side of the pelvis. The Ornithischia ("bird-hipped") had a somewhat more complex pelvis. The ileum and ischium were arranged similarly in ornithischians and saurischians, but the ornithischian pubis was a narrow, rod-shaped bone with anteriorly and posteriorly directed processes lying alongside the ischium. Oddly, while the ornithischian pelvis, as the name suggests, was similar to that of birds, birds are of the saurischian lineage.

Dinosaurs and their living relatives, the birds, are archosaurs ("ruling lizards"), a group that includes thecodonts (early archosaurs restricted to the Triassic), crocodiles, and pterosaurs (refer to the classification of the amniotes on p. 566). As traditionally recognized, the dinosaurs are a paraphyletic group because they do not include birds which are descended from the most recent common ancestor of dinosaurs.

From among the various archosaurian radiations of the Triassic there emerged a thecodont lineage with limbs drawn under the body to provide an upright posture. This lineage gave rise to the earliest dinosaurs of the Late Triassic. In *Herrerasaurus*, a bipedal dinosaur from Argentina, we see one of the most distinctive characteristics of dinosaurs: walking upright on pillar-like legs, rather than on legs splayed outward as with modern amphibians and reptiles. This arrangement allowed the legs to support the great weight of the body while providing an efficient and rapid stride.

Although their ancestry is unclear, two groups of saurischian dinosaurs have been proposed based on differences in feeding habits and locomotion: the carnivorous and bipedal theropods, and the herbivorous and quadrupedal sauropods (sauropodomorphs). *Coelophysis* was an early theropod with a body form typical of all theropods: powerful hindlegs with three-toed feet; long, heavy counterbalancing tail; slender, grasping forelimbs; flexible neck; and a large head with jaws armed with dagger-like teeth. Large predators such as *Allosaurus,* common during the Jurassic, were replaced by even more massively built carnivores of the Cretaceous, such as *Tyrannosaurus,* which reached a length of 14.5 m (47 ft), stood nearly 6 m high, and weighed more than 7200 kg (8 tons). Not all predatory saurischians were massive; several were swift and nimble, such as *Velociraptor* ("speedy predator") of the Upper Cretaceous.

Herbivorous saurischians, the quadrupedal sauropods, appeared in the Late Triassic. Although early sauropods were small- and medium-sized dinosaurs, those of the Jurassic and Cretaceous attained gigantic proportions, the largest terrestrial vertebrates ever to have lived. *Brachiosaurus* reached 25 m (82 ft) in length and may have weighed in excess of 30,000 kg (33 tons). Even larger sauropods have been discovered; *Supersaurus* was 43 m (140 ft) long. With long necks and long front legs, the sauropods were the first vertebrates adapted to feed on trees. They reached their greatest diversity in the Jurassic and began to decline in overall abundance and diversity during the Cretaceous.

The second group of dinosaurs, the Ornithischia, were all herbivorous. Although more varied, even grotesque, in appearance than saurischians, the ornithischians are united by several derived skeletal features that indicate common ancestry. The huge back-plated *Stegosaurus* of the Jurassic is a well-known example of armored ornithischians which comprised two of the five major groups of ornithischians. Even more shielded with bony plates than the stegosaurs were the heavily built ankylosaurs, "armored tanks" of the dinosaur world. As the Jurassic gave way to the Cretaceous, several groups of unarmored ornithischians appeared, although many bore impressive horns. The steady increase in ornithiscian diversity in the Cretaceous paralleled a concurrent gradual decline in giant sauropods which had flourished in the Jurassic. *Triceratops* is representative of horned dinosaurs that were common in the Upper Cretaceous. Even more prominent in the Upper Cretaceous were the duck-billed dinosaurs (hadrosaurs) which are believed to have lived in large herds. Many hadrosaurs had skulls elaborated with crests that probably functioned as vocal resonators to produce species-specific calls.

Sixty-five million years ago, the last of the Mesozoic dinosaurs became extinct, leaving birds as the only surviving lineage of archosaurs. There is increasingly convincing evidence that the demise of dinosaurs coincided with the impact on earth of a large asteroid that produced devastating worldwide environmental upheaval. We continue to be fascinated by the awe-inspiring, often staggeringly large creatures that dominated the Mesozoic era for 165 million years—an incomprehensibly long period of time. Today, inspired by clues from fossils and footprints from a lost world, scientists continue to piece together the puzzle of how the various dinosaur groups arose, behaved, and diversified.

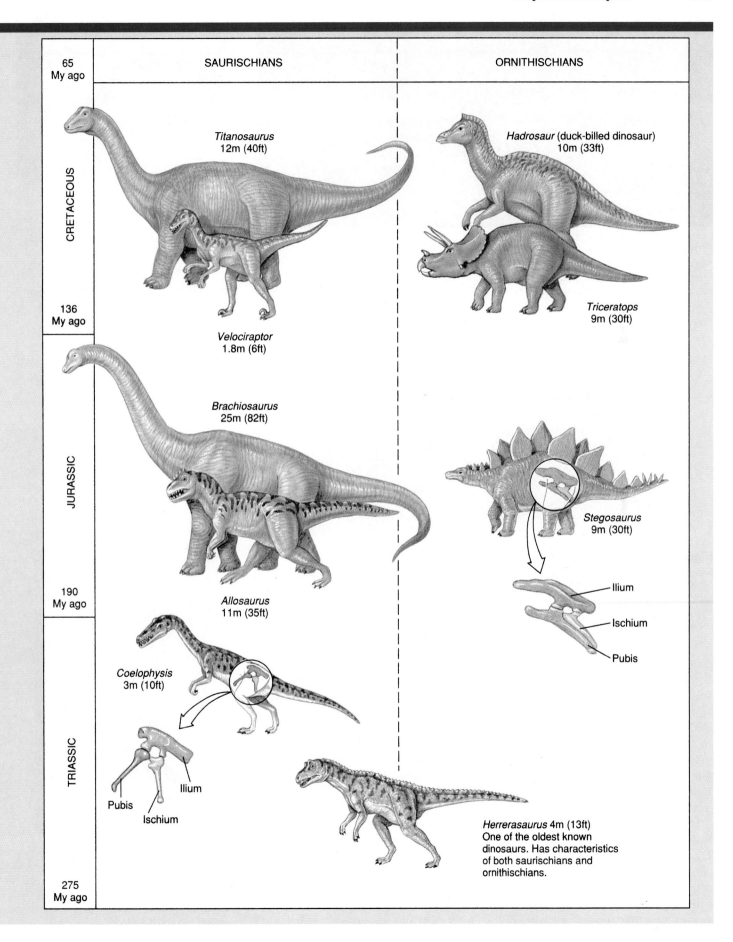

65 My ago

CRETACEOUS

136 My ago

JURASSIC

190 My ago

TRIASSIC

275 My ago

SAURISCHIANS

ORNITHISCHIANS

Titanosaurus
12m (40ft)

Hadrosaur (duck-billed dinosaur)
10m (33ft)

Triceratops
9m (30ft)

Velociraptor
1.8m (6ft)

Brachiosaurus
25m (82ft)

Stegosaurus
9m (30ft)

Ilium

Ischium

Pubis

Allosaurus
11m (35ft)

Coelophysis
3m (10ft)

Ilium

Pubis

Ischium

Herrerasaurus 4m (13ft)
One of the oldest known
dinosaurs. Has characteristics
of both saurischians and
ornithischians.

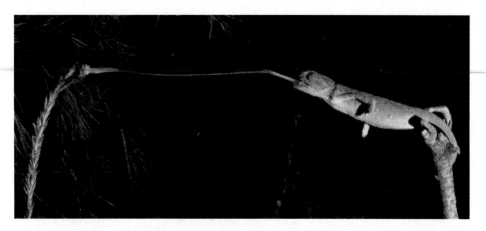

Figure 29-15

A glass lizard, *Ophisaurus* sp., of the southeastern United States. This legless lizard feels stiff and brittle to the touch and has an extremely long, fragile tail that readily fractures when the animal is struck or seized. Most specimens, such as this one, have only a partly regenerated tip to replace a much longer tail previously lost. Glass lizards can be readily distinguished from snakes by the deep, flexible groove running along each side of the body. They feed on worms, insects, spiders, birds' eggs, and small reptiles.

Figure 29-14

A chameleon snares a dragonfly. After cautiously edging close to its target, the chameleon suddenly lunges forward, anchoring its tail and feet to the branch. A split second later, it launches its sticky-tipped, foot-long tongue to trap the prey. The eyes of this common European chameleon (*Chamaeleo chamaeleon*) are swiveled forward to provide binocular vision and excellent depth perception.

Figure 29-16

Gila monster, *Heloderma suspectum,* of southwestern United States desert regions and the related Mexican bearded lizard are the only venomous lizards known. These brightly colored, clumsy-looking lizards feed principally on birds' eggs, nesting birds, mammals, and insects. Unlike poisonous snakes, the Gila monster secretes venom from glands in its lower jaw. The chewing bite is painful to humans but seldom fatal.

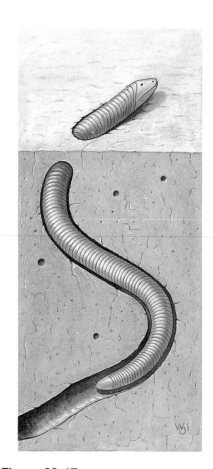

Figure 29-17

A worm lizard of the suborder Amphisbaenia. Worm lizards are burrowing forms with a solidly constructed skull used as a digging tool. The species pictured, *Amphisbaena alba,* is widely distributed in South America.

(birds, insects, and pulmonate snails). Some, such as the Gila monster of the southwestern United States deserts, store fat in their tails, which they use during drought to provide energy and metabolic water (Figure 29-16). The way many lizards keep their body temperature relatively constant by behavioral thermoregulation is described in Chapter 33 (p. 664).

Suborder Amphisbaenia: Worm Lizards. The somewhat inappropriate common name "worm lizards" describes a group of highly specialized, burrowing forms that are neither worms nor true lizards but certainly

are related to the latter. The name of the suborder literally means "double walk," in reference to their peculiar ability to move backward nearly as effectively as forward. They have elongate, cylindrical bodies of nearly uniform diameter, and most lack any trace of external limbs (Figure 29-17). The soft skin is divided into numerous rings, which combined with the absence of visible eyes and ears (both are hidden under skin) make the amphisbaenians look like earthworms. The resemblance, although superficial, is the kind of structural convergence that often occurs when two totally unrelated groups come to

Figure 29-18
Black rat snake, *Elaphe obsoleta obsoleta,* swallowing a chipmunk.

Figure 29-19
Parrot snake, *Leptophis ahaetulla.* The slender body of this Central American tree snake is an adaptation for sliding along branches without weighing them down.

occupy similar habitats. The amphisbaenians have an extensive distribution in South America and tropical Africa. In the United States, one species, *Rhineura florida,* is found in Florida where it is known as the "graveyard snake."

Suborder Serpentes: Snakes. Snakes are entirely limbless and lack both the pectoral and pelvic girdles (the latter persists as a vestige in pythons and boas). The numerous vertebrae of snakes, shorter and wider than those of tetrapods, permit quick lateral undulations through grass and over rough terrain. The ribs increase rigidity of the vertebral column, providing more resistance to lateral stresses. The elevation of the neural spine gives the numerous muscles more leverage.

The highly kinetic skull and feeding apparatus of snakes, which enable them to eat prey several times their own diameter, are perhaps their most remarkable specialization. The two halves of the lower jaw (mandibles) are joined only by muscles and skin, allowing them to spread widely apart. Many of the skull bones are so loosely articulated that the entire skull can flex asymmetrically to accommodate oversized prey (Figure 29-18). Since the snake must keep breathing during the slow process of swallowing, the tracheal opening (glottis) is thrust forward between the two mandibles.

The cornea of the snake's eye is permanently protected with a trans-

Figure 29-20
A timber rattlesnake, *Crotalus horridus,* flicks its tongue to smell its surroundings. Scent particles trapped on the tongue's surface are transferred to Jacobson's organs, olfactory organs in the roof of the mouth. Note the heat-sensitive pit organ between the nostril and eye.

parent membrane called a spectacle, which, together with reduced eyeball mobility, gives snakes the cold, unblinking stare that many people find unnerving. Most snakes have relatively poor vision, with tree-living snakes of the tropical forest being a conspicuous exception (Figure 29-19). Some arboreal snakes possess excellent binocular vision that helps them track prey through branches where scent trails would be impossible to follow.

Snakes have no external or middle ears. This, together with the absence of any obvious response to aerial sounds, led to the widespread opinion that snakes are totally deaf. But snakes do have internal ears, and recent work has shown quite clearly that within a limited range of low frequencies (100 to 700 Hz), hearing in snakes com-

pares favorably with that of most lizards. Snakes are also quite sensitive to vibrations carried in the ground.

Nevertheless, for most snakes it is the chemical senses and not vision and hearing that are employed to hunt their prey. In addition to the usual olfactory areas in the nose, which are not well developed, there are **Jacobson's organs,** a pair of pitlike organs in the roof of the mouth. These are lined with an olfactory epithelium and are richly innervated. The forked tongue, flicking through the air, picks up scent particles and conveys them to the mouth; the tongue is then drawn past Jacobson's organs or the tips of the forked tongue are inserted directly into the organs (Figure 29-20). Information is then transmitted to the brain where scents are identified.

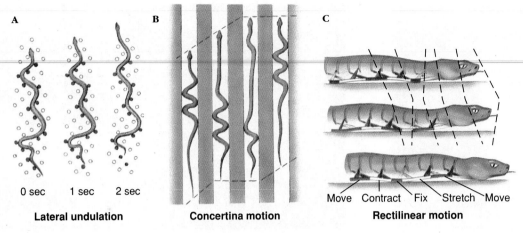

Figure 29-21

Snake locomotion. **A,** Lateral undulation. **B,** Concertina motion. **C,** Rectilinear motion. Refer to text for explanation.

As with lizards, the snake's body is entirely covered with a tough, impervious skin. The hard scales are set together, sometimes overlapping like shingles on a roof, with the skin folded inward between the scales. The skin is not elastic, and when required to stretch, as it must after the snake has enjoyed a hefty meal, it does so in a novel manner. The skin folds are pulled out straight, leaving the scales separated like islands on the skin.

Snakes have evolved several solutions to the obvious problem of movement without legs. The most typical pattern of movement is lateral undulation (Figure 29-21A). Movement follows an **S**-shaped path, with the snake propelling itself by exerting lateral force against surface irregularities. The snake seems to "flow," since the moving loops appear stationary with respect to the ground. Lateral undulatory movement is fast and efficient under most but not all circumstances. **Concertina movement** (Figure 29-21B) enables a snake to move in a narrow passage, as when climbing a tree by using the irregular channels in the bark. The snake extends forward while bracing **S**-shaped loops against the sides of the channel. To advance in a straight line as when stalking prey, many heavy-bodied snakes employ **rectilinear movement.** Two or three sections of the body rest on the ground to support the snake's weight. The intervening sections are lifted free of the ground and pulled forward by muscles (shown in red in Figure 29-21C) that originate on ribs and insert on the ventral skin. Rectilinear movement is a slow but effective way of moving inconspicuously toward prey, even when there are no surface irregularities. **Sidewinding** is a fourth form of movement that enables desert vipers to move with surprising speed across loose, sandy surfaces with minimum surface contact. The sidewinder rattlesnake moves by throwing its body forward in loops with its body lying at an angle of about 60 degrees to its direction of travel.

Snakes of the subfamily Crotalinae within the family Viperidae are called **pit vipers** because of special heat-sensitive pits on their heads, between the nostrils and the eyes (Figures 29-20 and 29-22). All of the best-known North American poisonous snakes are pit vipers, such as the several species of rattlesnakes, the water moccasin, and the copperhead. The pits are supplied with a dense packing of free nerve endings from the fifth cranial nerve. These respond to radiant energy in the long-wave infrared (5000 to 15,000 nm) and are especially sensitive to the heat emitted by the warm-bodied birds and mammals that are their food (infrared wavelengths of about 10,000 nm). Some measurements suggest that pit organs can distinguish temperature

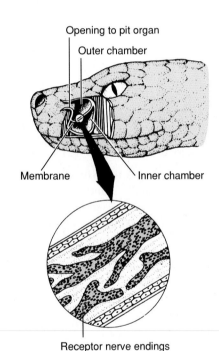

Figure 29-22

Pit organ of rattlesnake, a pit viper. Cutaway shows location of a deep membrane that divides the pit into inner and outer chambers. Heat-sensitive nerve endings are concentrated in the membrane.

differences of only 0.003° C from a radiating surface. Pit vipers use the pit organs to track warm-blooded prey and to aim strikes with great accuracy, as effectively in total darkness as in daylight. Boa constrictors and pythons also have heat receptors, but the anatomy is quite different from that of pit vipers, suggesting that they probably evolved independently.

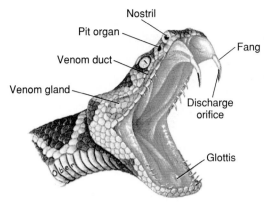

Figure 29-23

Head of rattlesnake showing the venom apparatus. The venom gland, a modified salivary gland, is connected by a duct to the hollow fang.

Nostril

Pit organ

Venom duct

Venom gland

Fang

Discharge orifice

Glottis

Figure 29-24

Nonvenomous African house snake, *Boaedon fuluginosus,* constricting a mouse before swallowing it.

All vipers have a pair of teeth on the maxillary bones modified as fangs. These lie in a membrane sheath when the mouth is closed. When the viper strikes, a special muscle and bone lever system erects the fangs when the mouth opens (Figure 29-23). The fangs are driven into the prey by the thrust of the strike, and venom is injected into the wound through a channel in the fangs. A viper immediately releases its prey after the bite and waits until it is paralyzed or dead. Then the snake swallows the prey whole. The bite of a pit viper can be dangerous to humans, although in many instances the snake injects very little venom when it bites. Approximately 8000 bites but only 12 deaths from pit vipers are reported each year in the United States.

The tropical and subtropical countries are the homes of most species of snakes, both of the venomous and nonvenomous varieties. Even there, less than one-third of snakes are venomous. The nonvenomous snakes kill their prey by constriction (Figure 29-24) or by biting and swallowing. Their diet tends to be restricted, many feeding principally on rodents, whereas others feed on fishes, frogs, and insects. Some African, Indian, and neotropical snakes are egg eaters. Poisonous snakes are usually divided into four groups based on the

type of fangs. Vipers (family Viperidae) have highly developed tubular fangs at the front of the mouth; the group includes the American pit vipers previously mentioned and the Old World true vipers, which lack facial heat-sensing pits. Among the latter are the common European adder and the African puff adder. A second family of poisonous snakes (family Elapidae) has short, permanently erect fangs so that the venom must be injected by repeated bites. In this group are the cobras (Figure 29-25), mambas, coral snakes, and kraits. The highly poisonous sea snakes are usually placed in a third family (Hydrophiidae). The very large family Colubridae, which contains most of the familiar (and nonvenomous) snakes, does include at least two poisonous snakes that have been responsible for human fatalities: the African boomslang and the African twig snake. Both are rear-fanged snakes that normally use their venom to quiet struggling prey.

The saliva of even harmless snakes possesses limited toxic qualities, and it is logical that there was a natural selection for this toxic tendency as snakes evolved. Snake venoms have traditionally been divided into two types. The **neurotoxic** type acts mainly on the nervous system, affecting the optic nerves (causing blindness) or the phrenic nerve of the

Figure 29-25

Yellow cobra, *Naja flava,* of Africa. Cobras erect the front of the body and flatten the neck as a threat display and before attacking. Although the cobra's strike range is limited, all cobras are dangerous because of the extreme toxicity of the venom.

diaphragm (causing paralysis of respiration). The **hemorrhagin** type breaks down the red blood corpuscles and blood vessels and produces extensive hemorrhaging of blood into the tissue spaces. In fact, most snake venoms are complex mixtures of various fractions that attack different organs in specific ways; they seldom can be assigned categorically to one or the other of the traditional types.

The toxicity of a venom is measured by the median lethal dose on laboratory animals (LD_{50}). By this standard the venoms of the Australian tiger snake and some of the sea snakes appear to be the most deadly of poisons drop for drop. However, several larger snakes are more dangerous. The aggressive king cobra, which may exceed 5.5 m in length, is the largest and probably the most dangerous of all poisonous snakes. In India, where snakes come in constant contact with people, some 200,000 snakebites cause more than 9000 deaths each year.

The LD$_{50}$ (median lethal dose) has been the standardized procedure for assaying the toxicity of drugs for chemical safety; it was originally developed in the 1920s by pharmacologists. In practice, small samples of laboratory animals, usually mice, are exposed to a graded series of doses of the drug or toxin. The dose that kills 50% of the animals in the test period is recorded as the LD$_{50}$. Expensive and time consuming, this classical procedure is being replaced by alternative methods that greatly reduce the number of animals needed. Among these alternatives are cytotoxicity tests that evaluate the ability of test substances to kill cells, and toxikinetic procedures that measure the interaction of a drug or toxin with a living system.

Most snakes are **oviparous** (L. *ovum,* egg, + *parere,* to bring forth) species that lay their shelled, elliptical eggs beneath rotten logs, under rocks, or in holes dug in the ground. Most of the remainder, including all the American pit vipers, except the tropical bushmaster, are ovoviviparous (L. *ovum,* egg, + *vivus,* living, + *parere,* to bring forth), giving birth to well-formed young. Very few snakes are viviparous (L. *vivus,* living, + *parere,* to bring forth); in these snakes a primitive placenta forms, permitting the exchange of materials between the embryonic and maternal bloodstreams. Snakes are able to store sperm and can lay several clutches of fertile eggs at long intervals after a single mating.

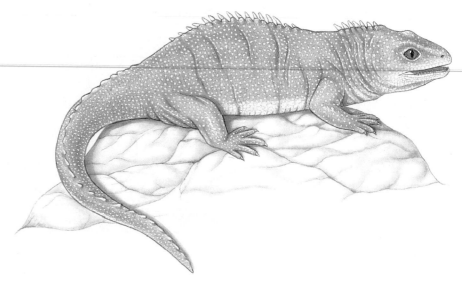

Figure 29-26
Tuatara, *Sphenodon* sp., a living representative of the order Sphenodonta. This "living fossil" reptile has, on top of the head, a well-developed parietal "eye" with retina, lens, and nervous connections to the brain. Although covered with scales, this third eye is sensitive to light. The parietal eye may have been an important sense organ in early reptiles. The tuatara is found today only on certain islands off the coastline of New Zealand.

Order Sphenodonta: The Tuatara

The order Sphenodonta is represented by two living species of the genus *Sphenodon* (Gr. *sphenos,* wedge, + *odontos,* tooth) of New Zealand (Figure 29-26). The tuatara is the sole survivor of the sphenodontid lineage that radiated modestly during the early Mesozoic era but declined toward the end of the Mesozoic. The tuatara was once widespread throughout the two main islands of New Zealand but is now restricted to small islets of Cook Strait and off the northeast coast of North Island where, under protection from the New Zealand government, it should survive.

The tuatara is a lizardlike form 66 cm long or less that lives in burrows often shared with petrels. They are slow-growing animals with a long life; one is recorded to have lived 77 years.

The tuatara has captured the interest of zoologists because of numerous features that are almost identical to those of Mesozoic fossils 200 million years ago. These features include a diapsid skull with two temporal openings bounded by complete arches. Tuataras also bear a well-developed median parietal eye complete with

elements of cornea, lens, and retina (although since it is buried beneath opaque skin this "third eye" can register only changes in light intensity, and its function, if any, remains unknown). In many other respects *Sphenodon* resembles lizards of the early Mesozoic. *Sphenodon* represents one of the slowest rates of evolution known among the vertebrates.

Order Crocodilia: Crocodiles and Alligators

The modern crocodilians are the only surviving reptiles of the archosaurian lineage that gave rise to the great Mesozoic radiation of dinosaurs and their kin and to the birds. Although the modern crocodiles belong to a lineage that began its radiation in the late Cretaceous period, they differ little in structural details from primitive crocodilians of the early Mesozoic. Having remained mostly unchanged for nearly 200 million years, crocodilians face an uncertain future in a world dominated by humans. Modern crocodilians are divided into three families: the alligators and caimans, mostly a New World group; the croc-

A

B

Figure 29-27
Crocodilians. **A,** Nile crocodile (*Crocodylus niloticus*) basking. The fourth tooth of the lower jaw fits *outside* the slender upper jaw; alligators lack this feature. **B,** American alligator (*Alligator mississipiensis*), an increasingly noticeable resident of rivers, bayous, and swamps of the southeastern United States.

odiles, which are widely distributed and include the saltwater crocodile, one of the largest living reptiles; and the gavials, represented by a single species in India and Burma.

All crocodilians have an elongate, robust, well-reinforced skull and massive jaw musculature arranged to provide a wide gape and rapid, powerful closure. Teeth are set in sockets, a type of dentition called **thecodont** that was typical of all archosaurs as well as the earliest birds. Another adaptation, found in no other vertebrate except the mammals, is a complete secondary palate. This innovation allows crocodilians to breathe when the mouth is filled with water or food (or both).

The estuarine crocodile (*Crocodylus porosus*), found in southern Asia, and the Nile crocodile (*C. niloticus;* Figure 29-27A) grow to great size (adults weighing 1000 kg have been reported) and are swift and aggressive. Crocodiles are known to attack animals as large as cattle, deer, and people. Alligators (Figure 29-27B) are less aggressive than crocodiles and certainly far less dangerous to humans. Large alligators are powerful animals nevertheless, and adults have almost no enemies but humans. The chink in their formidable armor is the developmental stages. Nests left unguarded by the mother are almost certain to be discovered and raided by any of several mammals that relish eggs, and the young hatchlings may be devoured by large fish.

Alligators are unusual among reptiles in being able to make definite vocalizations. The male alligator can give loud bellows in the mating season. In the United States, *Alligator mississipi-* *ensis* (Figure 29-27B) is the only species of alligator; *Crocodylus acutus,* restricted to extreme southern Florida, is the only species of crocodile.

Alligators and crocodiles are oviparous. Usually 20 to 50 eggs are laid in a mass of vegetation and guarded by the mother. The mother hears vocalizations from the hatching young and responds by opening the nest to allow the hatchlings to escape. As with many turtles and some lizards, the incubation temperature of the eggs determines the sex ratio of the offspring. However, unlike turtles (p. 555), low nest temperatures produce only females, whereas high nest temperatures produce only males. This results in highly unbalanced sex ratios in some areas. For example, in one study area in Louisiana, female hatchlings outnumbered males five to one.

CLASSIFICATION OF AMNIOTES EMPHASIZING EARLY AMNIOTES AND LIVING REPTILES

The following Linnaean classification is adapted from Carroll (1988)* and agrees with the genealogical relationships of living reptiles shown in Figure 29-2. Extinct groups are indicated by a dagger.

Subclass Anapsida (a-nap'se-duh) (Gr. *an*, without, + *apsis*, arch): **anapsids.** Amniotes having some primitve features, such as a skull with no temporal opening.

Order Captorhinida† (kap-to-rine'uh-duh) (Gr. *kapto*, to seize, + *rhinos*, nose). Amniotes of Carboniferous and early Permian.

Order Testudines (tes-tu'din-eez) (L. *testudo*, tortoise) **(Chelonia): turtles.** Body in a bony case of dorsal carapace and ventral plastron; jaws with horny beaks instead of teeth; vertebrae and ribs fused to overlying carapace; tongue not extensible; neck usually retractable; approximately 330 species.

Subclass Diapsida (di-ap'se-duh) (Gr. *di*, double, + *apsis*, arch): **diapsids.** Amniotes having a skull with two temporal openings.

Superorder Lepidosauria (lep-i-do-sor'ee-uh) (Gr. *lepidos*, scale, + *sauros*, lizard). Diapsid lineage appearing in the Permian; characterized by sprawling posture; no bipedal specializations; diapsid skull often modified by loss of one or both temporal arches.

Order Squamata (skwa-ma'ta) (L. *squamatus*, scaly, + *ata*, characterized by): snakes, lizards, amphisbaenians. Skin of horny epidermal scales or plates, which is shed; quadrate movable; skull kinetic (except amphisbaenians); vertebrae usually concave in front; paired copulatory organs.

Suborder Lacertilia (lay-sur-till'ee-uh) (L. *lacerta*, lizard) **(Sauria): lizards.** Body slender, usually with four limbs; rami of lower jaw fused; eyelids movable; external ear present; this paraphyletic suborder contains approximately 3300 species.

Suborder Amphisbaenia (am'fis-bee'nee-a) (L. *amphis*, double, + *baina*, to walk): worm lizards. Body elongate and of nearly uniform diameter; no legs (except one genus with short front legs); skull bones interlocked for burrowing (not kinetic); limb girdles vestigial; eyes hidden beneath skin; only one lung; approximately 135 species.

Suborder Serpentes (sur-pen'tes) (L. *serpere*, to creep): **snakes.** Body elongate; limbs, ear openings, and middle ear absent; mandibles joined anteriorly by ligaments; eyelids fused into transparent spectacle; tongue forked and protrusible; left lung reduced or absent; approximately 2300 species.

Order Sphenodonta (sfen'o-don'tuh) (Gr. *sphen*, wedge, + *odontos*, tooth) **(Rhynchocephalia).** Primitive diapsid skull; vertebrae biconcave; quadrate immovable; parietal eye present. *Sphenodon* only extant species.

Order Ichthyosauria† (ik'thee-o-sor'ee-uh) (Gr. *ichthys*, fish, + *sauros*, lizard). Mesozoic marine dolphin-shaped diapsids with reduced limbs.

Superorder Sauropterygia† (sor-op-ter-ig'ee-uh) (Gr. *sauros*, lizard, + *pteryginos*, winged). Mesozoic marine reptiles.

Order Plesiosauria† (plees'ee-o-sor'ee-uh) (Gr. *plesios*, near, + *sauros*, lizard). Long-necked Mesozoic marine reptiles with paddlelike limbs.

Superorder Archosauria (ark'uh-sor'ee-uh) (Gr. *archōn*, ruling, + *sauros*, lizard). Advanced diapsids, mostly terrestrial, but some specialized for flight.

Order Thecodontia† (thek'uh-dont'ee-uh) (Gr. *theke*, encased [socket], + *odontos*, tooth). Dominant Triassic archosaurs with teeth set in sockets; bipedal tendency.

Order Crocodilia (crok'uh-dil'ee-uh) (L. *crocodilus*, crocodile): **crocodilians.** Skull elongate and massive; nares terminal; secondary palate present; four-chambered heart; vertebrae usually concave in front; forelimbs usually of five digits; hindlimbs of four digits; quadrate immovable; advanced social behavior; 25 species.

Order Pterosauria† (ter'uh-sor'ee-uh) (Gr. *pteron*, winged, + *sauros*, lizard). Mesozoic archosaurs with membranous wings; extensive radiation.

Order Saurischia (sor-ish'ee-uh) (Gr. *sauros*, lizard, + *ischion*, hip). Mesozoic dinosaurs; bipedal carnivores and quadrupedal herbivores; primitive (reptilian) hip structure.

Suborder Sauropodomorpha† (sor'uh-pod-uh-morf'uh) (Gr. *sauros*, lizard, + *podos*, foot, + *morphē*, form). Herbivorous saurischians including Mesozoic giants such as *Brachiosaurus, Apatosaurus* and *Diplodocus.*

Suborder Theropoda (the-ro'po-duh) (Gr. *thēr*, wild beast, + *podos*, foot). Carnivorous saurischians including huge predators such as *Tyrannosaurus* and small, agile predators such as *Deinonychus* and *Velociraptor.* Birds are descended from this lineage.

Order Ornithischia† (orn'uh-thish'ee-uh) (Gr. *ornis*, bird, + *ischion*, hip). Mesozoic dinosaurs; bipedal and quadrupedal herbivores such as *Stegosaurus* and *Tricerotops;* advanced (birdlike) hip structure.

Subclass Synapsida (sin-ap'si-duh) (Gr. *syn*, together, + *apsis*, arch). Amniotes having skull with one pair of lateral temporal openings; mammal-like reptiles.

Order Pelycosauria† (pel'uh-ko-sor'ee-uh) (Gr. *pelyx*, wooden bowl, + *sauros*, lizard). Carboniferous and Permian synapsids with many primitive amniote characteristics; carnivorous and herbivorous.

Order Therapsida (ther-ap'-si-duh) (Gr. *ther*, wild beast, + *apsis*, arch). Permian and Triassic synapsids with many mammal-like characteristics; both carnivores and herbivores. Living mammals are descended from this lineage.

*Carroll, R. L. 1988. *Vertebrate paleontology and evolution.* New York, W. H. Freeman and Company.

Summary

The reptiles diverged phylogenetically from a group of labyrinthodont amphibians during the late Paleozoic era, some 300 million years ago. Their success as terrestrial vertebrates is attributed in large part to the evolution of the amniotic egg, which, with its three extraembryonic membranes, provided support for full embryonic development within the protection of a shell. Thus the reptiles could lay their eggs on land. Reptiles are also distinguished from amphibians by their dry, scaly skin that limits water loss; more powerful jaws; internal fertilization; and advanced circulatory, respiratory, excretory, and nervous systems. Like amphibians, reptiles are ectotherms, but most exercise considerable behavioral control over their body temperature.

Before the end of the Paleozoic era, the amniotes began a radiation that separated into three lineages: the anapsids, which gave rise to the turtles; the synapsids, a lineage that led to the modern mammals; and the diapsid lineage, which led to all other reptiles and to the birds. The great burst of reptilian radiation during the Mesozoic era produced a worldwide fauna of great diversity.

The turtles (order Testudines) with their distinctive shells have changed little in design since the Triassic period. Turtles are a small group of long-lived terrestrial, semiaquatic, aquatic, and marine species. They lack teeth. All are oviparous and all, including the marine forms, bury their eggs.

The lizards, snakes, and worm lizards (order Squamata) make up 95% of all living reptiles. Lizards (suborder Lacertilia) are a diversified and successful group adapted for walking, running, climbing, swimming, and burrowing. They are distinguished from snakes by typically having two pairs of legs (some species are legless), united lower jaw halves, movable eyelids, external ears, and absence of fangs. Many lizards are well adapted for survival under hot and arid desert conditions.

The worm lizards (suborder Amphisbaenia) are a small tropical group of legless squamates highly adapted for burrowing.

Snakes (suborder Serpentes), in addition to being entirely limbless, are characterized by their elongate bodies and a highly kinetic skull that permits them to swallow whole prey that may be much larger than the snake's own diameter. Most snakes rely on the chemical senses, especially Jacobson's organs, to hunt prey, rather than on weakly developed visual and auditory senses. Two groups of snakes (pit vipers and boids) have unique infrared-sensing organs for tracking warm-bodied prey. Many snakes are venomous.

The tuatara of New Zealand (order Sphenodonta) is a relict species and sole survivor of a group that otherwise disappeared 100 million years ago. It bears several features that are almost identical to those of Mesozoic fossil diapsids.

The crocodiles, alligators, and caimans (order Crocodilia) are the only living reptilian representatives of the archosaurian lineage that gave rise to the extinct dinosaurs and the living birds. Crocodilians have several adaptations for a carnivorous, semi-aquatic life, including a massive skull with powerful jaws, and a secondary palate. They have the most complex social behavior of any reptile.

Review Questions

1. What were the three major amniote radiations of the Mesozoic and from which lineage or lineages did the birds and mammals descend? How could you distinguish among anapsid, diapsid, and synapsid skulls?
2. What changes in egg design allowed the reptiles to lay eggs on land? Why is the egg often called an "amniotic" egg? What are the "amniotes"?
3. Why are the reptiles considered a paraphyletic rather than a monophyletic group?
4. Describe the ways in which the reptiles are more functionally or structurally suited for terrestriality than the amphibians.
5. What are the main characteristics of reptile skin and how would you distinguish it from frog skin?
6. Describe the principal structural features of turtles that would distinguish them from any other reptilian order.
7. How might nest temperature affect egg development in turtles? In crocodilians?
8. What is meant by a "kinetic" skull and what benefit does it confer? How are snakes able to eat such large prey?
9. In what ways are the special senses of snakes similar to those of lizards, and in what ways have they evolved for specialized feeding strategies?
10. How do snakes and crocodilians breathe when their mouths are full of food?
11. What is the function of Jacobson's organ of snakes?
12. What is the function of the "pit" of pit vipers?
13. What is the difference in the structure or location of the fangs of a rattlesnake, a cobra, and an African boomslang?
14. Most snakes are oviparous, but some are ovoviviparous or viviparous. What do these terms mean and what would you have to know to be able to assign a particular snake to one of these reproductive modes?
15. Describe how a snake moves by lateral undulation. Why might this form of locomotion be inefficient on an unstable surface (such as sand) or a surface lacking irregularities? What forms of locomotion would work for a snake under these conditions?
16. Why is the tuatara (*Sphenodon*) of special interest to biologists? Where would you have to go to see one in its natural habitat?
17. From which diapsid lineage have the crocodilians descended? What other major fossil and living vertebrate groups belong to this same lineage? In what structural and behavioral ways are the crocodilians more advanced than other living reptiles?

Selected References

See also general references for Part III, p. 626.

Alexander, R. M. 1991. How dinosaurs ran. Sci. Am. **264:**130–136 (April). *By applying the techniques of modern physics and engineering, a zoologist calculates that the large dinosaurs walked slowly but were capable of a quick run; none required the buoyancy of water for support.*

Alvarez, W., and F. Asaro. 1990. An extraterrestrial impact. Sci. Am. **263:**78–84 (Oct.). *This article and an accompanying article by V. E. Courtillot, "A volcanic eruption," present opposing interpretations of the cause of the Cretaceous mass extinction that led to the demise of the dinosaurs.*

Cogger, H. G., and R. G. Zweifel (eds). 1992. Reptiles and amphibians. New York, Smithmark Publishers, Inc. *This comprehensive, up-to-date, and lavishly illustrated volume was written by some of the best-known herpetologists in the field.*

Crews, D. 1994. Animal sexuality. Sci. Am. **270:**108–114 (Jan.) *The reproductive strategies of reptiles, including nongenetic sex determination, provide insights into the origins and functions of sexuality.*

Gibbons, W. 1983. Their blood runs cold: adventures with reptiles and amphibians. University, Alabama, University of Alabama Press. *Lots of interesting reptile lore in this engaging book.*

Halliday, T. R., and K. Adler (eds). 1986. The encyclopedia of reptiles and amphibians. New York, Facts on File, Inc. *Comprehensive and beautifully illustrated treatment of the reptilian groups with helpful introductory sections on origins and characteristics.*

Lillywhite, H. B. 1988. Snakes, blood circulation and gravity. Sci. Am. **259:**92–98 (Dec.).

Even long snakes are able to maintain blood circulation when the body is extended vertically (head up posture) through special circulatory reflexes that control blood pressure.

Lohmann, K. J. 1992. How sea turtles navigate. Sci. Am. **266:**100–106 (Jan.). *Recent evidence suggests that sea turtles use the earth's magnetic field and the direction of ocean waves to navigate back to their natal beaches to nest.*

Norman, D. 1991. Dinosaur! New York, Prentice-Hall. *Highly readable account of the life and evolution of dinosaurs, with fine illustrations.*

Zug, G. R. 1993. Herpetology: an introductory biology of amphibians and reptiles. New York, Academic Press, Inc. *Introductory college-level textbook.*

30

The Birds

*Phylum Chordata
Class Aves*

Long Trip to a Summer Home

Perhaps it was ordained that birds, having mastered flight, would use this power to make the long and arduous seasonal migrations that have captured human wonder and curiosity. For the advantages of migration are many. Moving between southern wintering regions and northern summer breeding regions enables birds to sustain their intense metabolism with abundant and unfailing sources of food. In the far North long summer days and the abundance of insects combine to provide parents with ample food for rearing their young. Predators of birds are not so abundant in the far North, and the brief once-a-year appearance of vulnerable young birds does not encourage the buildup of predator populations. Migration also vastly increases the amount of space available for breeding and reduces aggressive territorial behavior.

Finally, migration favors homeostasis—the balancing of physiological processes that maintains internal stability—by allowing birds to avoid climatic extremes.

Still, the wonder of the migratory pageant remains, and there is much yet to learn about its mechanisms. What times migration, and what determines that each bird shall store sufficient fuel for the journey? How did the sometimes difficult migratory routes originate, and what cues do birds use in navigation? And what was the origin of this instinctive force to follow the retreat of winter northward? For it is instinct that drives the migratory waves in spring and fall, instinctive blind obedience that carries most birds successfully to their northern nests, while countless others fail and die, winnowed by the ever-challenging environment. ■

POSITION IN THE ANIMAL KINGDOM

The birds are a lineage of endothermic, diapsid amniotes that evolved flight in the Jurassic period of the Mesozoic. Phylogenetically, they are most closely related to certain theropod dinosaurs, a group of bipedal carnivores with birdlike skeletal characteristics. Their closest living relatives are the crocodilians. The morphological characteristics and great uniformity of structure of birds relate almost entirely to the strict demands of flight, and the mobility that flight provides is responsible for the most distinctive aspects of their behavior and ecology.

BIOLOGICAL CONTRIBUTIONS

1. Feathers are unique to the bird lineage and distinguish birds from all other animals. The evolution of feathers was the single most important event leading to the capacity for flight.
2. In addition to feathers, several other essential adaptations contribute to the two prime requirements for flight: increase in power and decrease in weight. These adaptations include forelimbs modified as strong wings, hollow bones, horny bill (rather than heavy jaws and teeth), endothermy, high metabolic rate (six to ten times as high as reptiles of similar weight and body temperature), large hearts and high-pressure circulation, highly efficient respiratory system, keen vision, and excellent neuromuscular coordination.
3. Birds occupy almost every available habitat on the earth's surface and, within the constraints imposed by the requirements for flight, have radiated modestly in body form, especially in bill adaptations.
4. The unparalleled mobility of birds has enabled many to benefit from the advantages of making seasonal and long-distance migrations. Migration enables birds to secure seasonal habitats most beneficial for breeding, finding food, avoiding predators, and reducing interspecific competition.

Of the vertebrates, birds of the class Aves (ay′veez) (L. pl. of *avis,* bird) are the most noticeable, the most melodious, and many think the most beautiful. With more than 9000 species distributed over nearly the entire earth, birds far outnumber all other vertebrates except the fishes. Birds are found in forests and deserts, in mountains and prairies, and on all oceans. Four species are known to have visited the North Pole, and one, a skua, was seen at the South Pole. Some birds live in total blackness in caves, finding their way about by echolocation, and others dive to depths greater than 45 m to prey on aquatic life. The "bee" hummingbird of Cuba, weighing in at only 1.8 g, is the smallest vertebrate endotherm.

The single unique feature that distinguishes birds from other animals is their feathers. If an animal has feathers, it is a bird; if it lacks feathers, it is not a bird. No other vertebrate group bears such an easily recognizable and foolproof identification tag.

There is great uniformity of structure among birds. Despite approximately 150 million years of evolution, during which they proliferated and adapted themselves to specialized ways of life, we have no difficulty recognizing a bird as a bird. In addition to feathers, all birds have forelimbs modified into wings (although they may not be used for flight); all have hindlimbs adapted for walking, swimming, or perching; all have horny beaks; and all lay eggs. The reason for this great structural and functional uniformity is that birds evolved into flying machines. This fact greatly restricts diversity, so much more evident in other vertebrate classes. For example, birds do not begin to approach the diversity seen in their endothermic evolutionary peers, the mammals, a group that includes forms as dissimilar as whale, porcupine, bat, and giraffe.

Birds share with mammals the greatest development of organ systems in the animal kingdom. But a bird's entire anatomy is designed around flight and its perfection. An airborne life for a large vertebrate is a highly demanding evolutionary challenge. A bird must, of course, have wings for support and propulsion. Bones must be light and hollow yet serve as a rigid airframe. The respiratory system must be highly efficient to meet the intense metabolic demands of flight and serve also as a thermoregulatory device to maintain a constant body temperature. A bird must have a rapid and efficient digestive system to process an energy-rich diet; it must have a high metabolic rate; and it must have a high-pressure circulatory system. Above all, birds must have a finely tuned nervous system and acute senses, especially superb vision, to handle the complex problems of headfirst, high-velocity flight.

ORIGIN AND RELATIONSHIPS

Approximately 147 million years ago, a flying animal drowned and settled to the bottom of a shallow marine lagoon in what is now Bavaria, Germany. It was rapidly covered with a fine silt and eventually fossilized. There it remained until discovered in 1861 by a workman splitting slate in a limestone quarry. The fossil was approximately the size of a crow, with a skull not unlike that of modern birds except that the beaklike jaws bore small bony teeth set in sockets like those of reptiles (Figure 30-1). The skeleton was decidedly reptilian with a long bony tail, clawed fingers, and abdominal ribs. It might have been classified as a reptile except that it carried the unmistakable imprint of **feathers,** those marvels of biological engineering that only birds possess. *Archaeopteryx lithographica* (ar-kee-op′ter-ix lith-o-graf′e-ca, Gr., meaning "ancient wing inscribed in stone"), as the fossil was named, was an especially fortunate discovery because the fossil record of birds is disappointingly meager. The finding was also dramatic because it proved beyond reasonable doubt the phylogenetic relatedness of birds and reptiles.

A   B

Figure 30-1

Archaeopteryx, a 147-million-year-old ancestor of modern birds. **A,** Cast of the second and most nearly perfect fossil of *Archaeopteryx,* which was discovered in a Bavarian stone quarry. Six specimens of *Archaeopteryx* have been discovered, the most recent one in 1987. **B,** Reconstruction of *Archaeopteryx.*

A, *Courtesy American Museum of Natural History, Neg. #125065.*

Zoologists had long recognized the similarity of birds and reptiles. The skulls of birds and reptiles abut against the first neck vertebra by a single ball-and-socket joint, the occipital condyle (mammals have two condyles). Birds and reptiles have a single middle ear bone, the stapes (mammals have three middle ear bones). Birds and reptiles have a lower jaw composed of five or six bones, whereas the lower jaw of mammals has one mandibular bone, the dentary. Birds and reptiles excrete their nitrogenous wastes as uric acid whereas mammals excrete theirs as urea. Birds and reptiles lay similar yolked eggs with the early embryo developing on the surface by shallow cleavage divisions.

The distinguished English zoologist Thomas Henry Huxley was so impressed with these and many other anatomical and physiological affinites that he called birds "glorified reptiles" and classified them with a group of dinosaurs called theropods that displayed several birdlike characteristics (Figures 30-2 and 30-3). Theropod dinosaurs share many derived characters with birds, the most obvious of which is the elongate, mobile, S-shaped neck. As shown in the cladogram (Figure 30-3), theropods belong to a lineage of diapsid reptiles, the archosaurians, that includes crocodilians and pterosaurs, as well as the dinosaurs. There is now overwhelming evidence that Huxley was correct: birds' closest phylogenetic affinity is to the theropod dinosaurs. The only anatomical feature required to link bird ancestry with the theropod dinosaurs was feathers, and this was provided by the discovery of *Archaeopteryx.* However, recent fossil discoveries have complicated the picture of bird origins and renewed the debate over which amniote lineage was ancestral to birds.

The discovery in Texas in 1983 of a small Triassic reptile with several birdlike features has opened a heated debate on the origin of birds. Named *Protoavis texensis* ("first bird from Texas"), the provocative fossil is considered by its discoverer to be the earliest known bird fossil, predating *Archaeopteryx* by 75 million years. Other specialists, however, believe the reconstructed fossil is really a small meat-eating theropod dinosaur and question the interpretation of avian-like features of *Protoavis.* For the moment, *Archaeopteryx* remains on its perch as the earliest known bird.

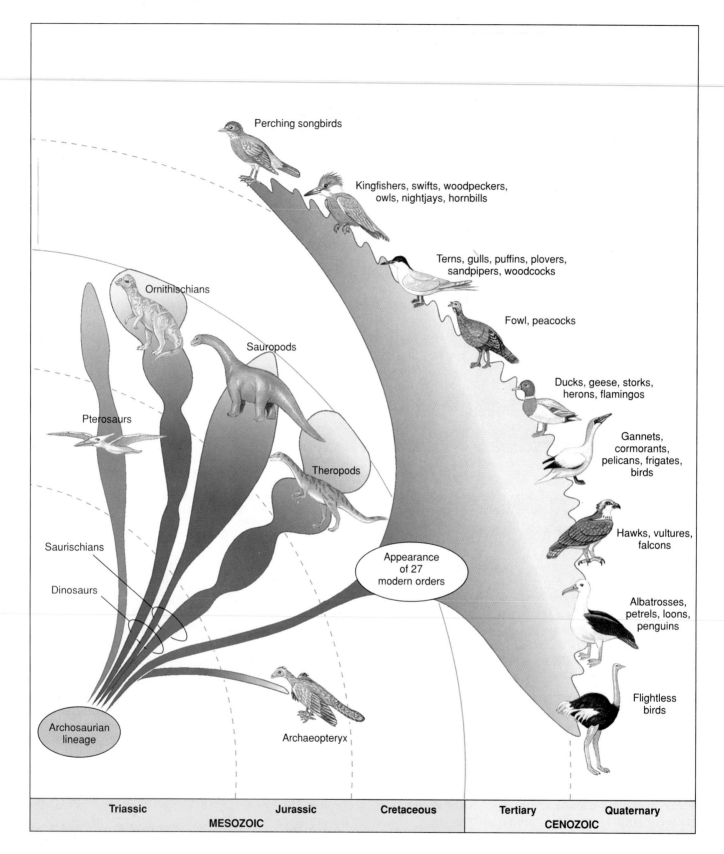

Figure 30-2

Evolution of modern birds. Of 27 living bird orders, 9 of the largest are shown. The earliest known bird, *Archaeopteryx,* lived in the Upper Jurassic, about 147 million years ago. *Archaeopteryx* uniquely shares many specialized aspects of its skeleton with the smaller theropod dinosaurs and is considered to have evolved within the theropod lineage. Evolution of modern bird orders occurred rapidly during the Cretaceous and early Tertiary periods.

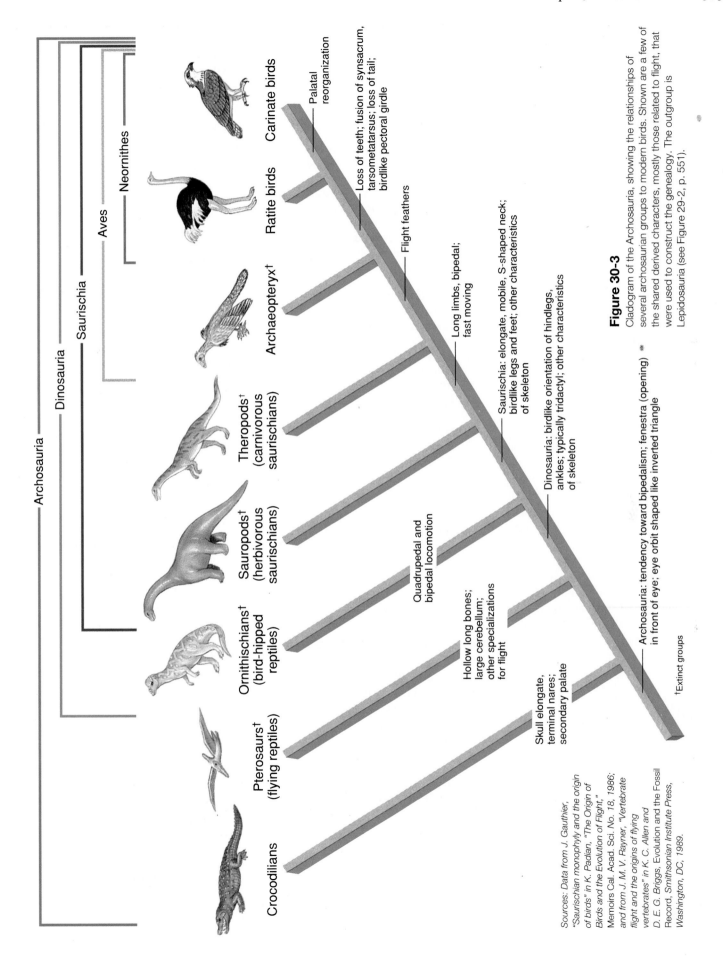

Figure 30-3

Cladogram of the Archosauria, showing the relationships of several archosaurian groups to modern birds. Shown are a few of the shared derived characters, mostly those related to flight, that were used to construct the genealogy. (see Figure 29-2, p. 551).

Sources: Data from J. Gauthier, "Saurischian monophyly and the origin of birds" in K. Padian, "The Origin of Birds and the Evolution of Flight," Memoirs Cal. Acad. Sci. No. 18, 1986; and from J. M. V. Rayner, "Vertebrate flight and the origins of flying vertebrates" in K. C. Allen and D. E. G. Briggs, Evolution and the Fossil Record, Smithsonian Institute Press, Washington, DC, 1989.

Figure 30-4
One of the strangest birds in a strange land, the flightless cormorant, *Nannopterum harrisi*, of the Galápagos Islands dries its wings after a fishing forage. It is a superb swimmer, propelling itself through the water with its feet to catch fish and octopuses. The flightless cormorant is an example of a carinate bird (having a keeled sternum) that has lost the keel and the ability to fly.

Living birds (Neornithes) are divided into two groups: (1) **ratite** (rat′ite) (L. *ratitus,* marked like a raft, from *ratis,* raft), the large flightless ostrichlike birds and the kiwis, which have a flat sternum with poorly developed pectoral muscles, and (2) **carinate** (L. *carina,* keel), the flying birds that have a keeled sternum on which the powerful flight muscles insert. This division originated from the view that the flightless birds (ostrich, emu, kiwi, rhea) represented a separate line of descent that never attained flight. This idea is now completely rejected. The ostrichlike ratites clearly have descended from flying ancestors. Furthermore, not all carinate, or keeled, birds can fly and many of them even lack keels (Figure 30-4). Flightlessness has appeared independently among many groups of birds; the fossil record reveals flightless wrens, pigeons, parrots, cranes, ducks, auks, and even a flightless owl. Penguins are flightless although they use their wings to "fly" through the water (p. 203). Flightlessness has almost always evolved on islands where few terrestrial predators are found. The flightless birds living on continents today are the large ratites (ostrich, rhea, cassowary, emu), which

CHARACTERISTICS OF CLASS AVES

1. Body usually spindle shaped, with four divisions: head, neck, trunk, and tail; **neck disproportionately long** for balancing and food gathering
2. Limbs paired with the **forelimbs usually modified for flying;** posterior pair variously adapted for perching, walking, and swimming; foot with four toes (2 or 3 toes in some)
3. Epidermal **covering of feathers** and **leg scales;** thin integument of epidermis and dermis; no sweat glands; oil or preen gland at root of tail; **pinna of ear rudimentary**
4. **Fully ossified skeleton with air cavities;** skull bones fused with **one occipital condyle;** each jaw covered with a horny sheath, forming a **beak; no teeth;** ribs with strengthening processes; **tail not elongate;** sternum well developed with keel or reduced with no keel; **single bone in middle ear**
5. Nervous system well developed, with brain and 12 pairs of cranial nerves
6. Circulatory system of **four-chambered heart,** with the **right aortic arch persisting;** reduced renal portal system; nucleated red blood cells
7. Endothermic
8. Respiration by slightly expansible lungs, with thin **air sacs** among the visceral organs and skeleton; **syrinx (voice box)** near junction of trachea and bronchi
9. Excretory system of metanephric kidney; ureters open into cloaca; **no bladder;** semisolid urine; uric acid main nitrogenous waste
10. Sexes separate; testes paired, with the vas deferens opening into the cloaca; **females with left ovary and oviduct only;** copulatory organ in ducks, geese, ratites, and a few others
11. Fertilization internal; **amniotic eggs with much yolk and hard calcareous shells;** embryonic membranes in egg during development; **incubation external;** young active at hatching **(precocial)** or helpless and naked **(altricial);** sex determination by females (females heterogametic)

can run fast enough to escape predators. The ostrich can run 70 km (42 miles) per hour, and claims of speeds of 96 km (60 miles) per hour have been made. The evolution and dispersal of flightless birds are discussed on pp. 160 and 787 respectively.

The bodies of flightless birds are dramatically redesigned to remove all of the restrictions of flight. The keel of the sternum is lost, and heavy flight muscles (as much as 17% of the body weight of flying birds), as well as other specialized flight apparatus, disappear. Since body weight is no longer a restriction, flightless birds tend to become large. Several extinct flightless birds were enormous: the giant moas of New Zealand weighed more than 225 kg (500 pounds) and the elephantbird of Madagascar, the largest bird that ever lived, probably weighed nearly 450 kg (about 1000 pounds) and stood nearly 2 m tall.

FORM AND FUNCTION

Just as an airplane must be designed and built according to rigid aerodynamic specifications if it is to fly, so too must birds meet stringent structural requirements if they are to stay airborne. All the special adaptations found in flying birds contribute to two things: more power and less weight. Flight by humans became possible when they developed an internal combustion engine and learned how to reduce the weight-to-power ratio to a critical point. Birds accomplished flight millions of years ago. But birds must do much more than fly. They must feed themselves and convert food into high-energy fuel; they must escape predators; they must be able to repair their own injuries; they must be able to air-condition themselves when overheated and heat themselves when too cool; and, most important of all, they must reproduce themselves.

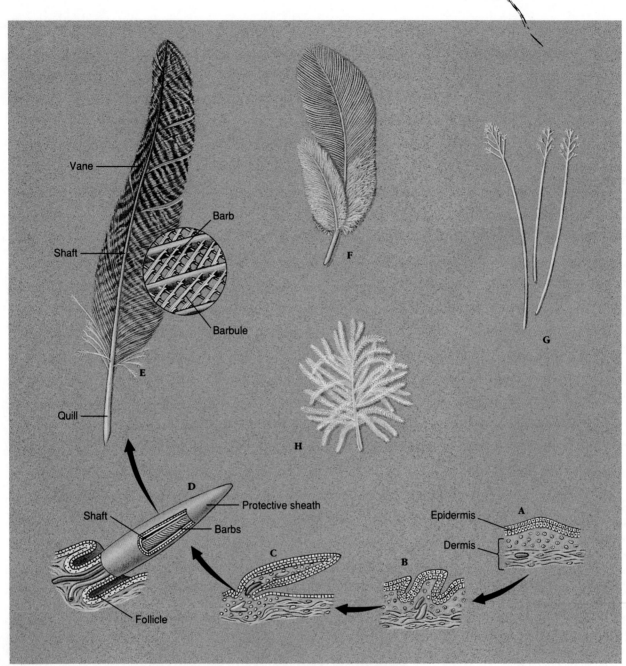

Figure 30-5
Types of bird feathers and their development. **A** to **E,** Successive stages in the development of a vaned, or contour, feather. Growth occurs within a protective sheath, **D,** that splits open when growth is complete, allowing the mature feather to spread flat. **F** to **H,** Other feather varieties, including a pheasant feather with aftershaft, **F,** filoplumes, **G,** and down feathers, **H.**

FEATHERS

A feather is very lightweight, yet it possesses remarkable toughness and tensile strength. The most typical of bird feathers are the **contour feathers,** vaned feathers that cover and streamline the bird's body. A contour feather consists of a hollow **quill,** or calamus, emerging from a skin follicle, and a **shaft,** or rachis, which is a continuation of the quill and bears numerous **barbs** (Figure 30-5). The barbs are arranged in a closely parallel fashion and spread diagonally outward from both sides of the central shaft to form a flat, expansive, webbed surface, the **vane.** There may be several hundred barbs in the vane.

If the feather is examined with a microscope, each barb appears to be a miniature replica of the feather with numerous parallel filaments called **barbules** set in each side of the barb and spreading laterally from it. There may be 600 barbules on each side of a barb, adding up to more than 1 million barbules for the feather. The barbules of one barb overlap the barbules of a neighboring barb in a herringbone pattern and are held together with great tenacity by tiny hooks. Should two adjoining barbs become separated—and considerable force is needed to pull the vane apart—they are instantly zipped together again by drawing the feather through the fingertips. The bird, of course, does this with its bill, and much of a bird's time is occupied with preening to keep its feathers in perfect condition.

Types of Feathers

There are different types of bird feathers, serving different functions. **Contour feathers** (Figure 30-5E) give the bird its outward form and are the type we have already described. Contour feathers that extend beyond the body and are used in flight are called **flight feathers. Down feathers** (Figure 30-5H) are soft tufts hidden beneath the contour feathers. They are soft because their barbules lack hooks. They are especially abundant on the breast and abdomen of water birds and on young quail and grouse and function principally to conserve heat. **Filoplume feathers** (Figure 30-5G) are hairlike, degenerate feathers; each is a weak shaft with a tuft of short barbs at the tip. They are the "hairs" of a plucked fowl. They have no known function. The bristles around the mouths of flycatchers and whippoorwills are probably modified filoplumes. A fourth type of highly modified feather, the **powder-down feather**, is found on herons, bitterns, hawks, and parrots. Tips of these disintegrate as they grow, releasing a talclike powder that helps to waterproof the feathers and give them metallic luster.

Origin and Development

Like the reptiles' scale to which it is homologous, a feather develops from an epidermal elevation overlying a nourishing dermal core (Figure 30-5A). However, rather than flattening like a scale, the feather bud rolls into a cylinder and sinks into the follicle from which it is growing. During growth, pigments (lipochromes and melanin) are added to the epidermal cells. As the feather enlarges and nears the end of its growth, the soft rachis and barbs are transformed into hard structures by the deposition of keratin. The protective sheath splits apart, allowing the feather end to protrude and the barbs to unfold.

Molting

When fully grown, a feather, like mammalian hair, is a dead structure. The shedding, or molting, of feathers is a highly orderly process. Except in penguins, which molt all at once, feathers are discarded gradually to avoid the appearance of bare spots. Flight and tail feathers are lost in exact pairs, one from each side, so that balance is maintained (Figure 30-6). Replacements emerge before the next pair is lost, and most birds can continue to fly unimpaired during the molting period; however, many water birds (ducks, geese, loons, and others) lose all their primary feathers at once and are grounded during the molt. Many prepare for this by moving before molting to isolated bodies of water where they can find food and more easily escape enemies. Nearly all birds molt at least once a year, usually in late summer after the nesting season.

The vivid color of feathers is of two kinds: pigmentary and structural. Red, orange, and yellow feathers are colored by pigments, called lipochromes, deposited in the feather barbules as they are formed. Black, brown, red-brown, and gray colors are from a different pigment, melanin. The blue feathers of the blue jay, indigo buntings, and bluebirds depend not on pigment but on the scattering of shorter wavelengths of light by particles within the feather; these are structural colors. Blue feathers are usually underlain by melanin, which absorbs certain wavelengths, thus intensifying the blue. Such feathers look the same from any angle of view. Green colors are almost always a combination of yellow pigment and blue feather structure. Another kind of structural color is the beautiful iridescent color of many birds, which ranges from red, orange, copper, and gold to green, blue, and violet. Iridescent color is based on interference that causes light waves to reinforce, weaken, or eliminate each other. Iridescent colors may change with the angle of view; the quetzal, for example, looks blue from one angle and green from another. In the animal kingdom, only tropical reef fishes can vie with birds for intensity and vividness of color.

Figure 30-6
Osprey, *Pandion haliaetus* (order Falconiformes), landing on nest. Note alulas (*arrows*). Feathers are molted in sequence in exact pairs so that balance is maintained during flight.

SKELETON

One of the major structural requirements for flight is a light, yet sturdy, skeleton. As compared with the earliest known bird, *Archaeopteryx* (Figure 30-7A), the bones of modern birds are phenomenally light, delicate, and laced with air cavities. Such **pneumatized** bones (Figure 30-8) are nevertheless strong. The skeleton of a frigate bird with a 2.1 m (7-foot) wingspan weighs only 114 grams (4 ounces), less than the weight of all its feathers.

As archosaurs, birds evolved from ancestors with diapsid skulls (p. 549). However, the skulls of modern birds are so highly specialized that it is difficult to see any trace of the original diapsid condition. The bird skull is built lightly and mostly fused into one piece. A pigeon skull weighs only 0.21% of its body weight; by comparison the skull of a rat weighs 1.25% of its body weight. The braincase and orbits are large to accommodate a bulging brain and the large eyes needed for quick motor coordination and superior vision.

In *Archaeopteryx*, both jaws contained teeth set in sockets, an archosaurian characteristic. Modern birds are completely toothless, having instead a horny (keratinous) beak molded around the bony jaws. The mandible is a complex of several bones that hinge on movable bones, the quadrates. The articulation provides a double-jointed action that permits the mouth to gape widely. Most birds have kinetic skulls (kinetic skulls of lizards are described on p. 556). The attachment of the upper jaw to the skull is flexible; this allows the upper jaw to move slightly, thus increasing the gape. In some birds, parrots for example, the upper jaw is especially flexible because it is hinged to the skull.

The vertebral column of birds is highly specialized for flight. Its most distinctive feature is its rigidity. Most of the vertebrae except the **cervicals** (neck vertebrae) are fused together and with the pelvic girdle to form a stiff but light framework to support the legs and provide rigidity for flight. To assist in this rigidity, the ribs are mostly fused with the vertebrae, pectoral girdle, and sternum. Except in the flightless birds, the sternum bears a large, thin keel that provides for the attachment of the powerful flight muscles. Because the sternum was completely missing in *Archaeopteryx* (Figure 30-7B), there was no anchorage for the flight muscles equivalent to that of modern birds. This is one of the principal reasons why *Archaeopteryx* could not have done any strenuous wing-beating. *Archaeopteryx* did, however, have a furcula (wishbone) on which enough pectoral muscle could have attached to permit weak flight.

The bones of the forelimbs are highly modified for flight. They are reduced in number, and several are fused together. Despite these alterations, the bird wing is clearly a rearrangement of the basic vertebrate tetrapod limb from which it arose (p. 530), and all the elements—upper arm, forearm, wrist, and fingers—are represented in modified form (Figure 30-7). The birds' legs have undergone less pronounced modification than the wings, since they are still designed principally for walking, as well as for perching, scratching, food gathering, and occasionally for swimming, as were those of their archosaurian ancestors.

MUSCULAR SYSTEM

The locomotor muscles of the wings are relatively massive to meet the demands of flight. The largest of these is the **pectoralis,** which depresses the wings in flight. Its antagonist is the **supracoracoideus** muscle, which raises the wing (Figure 30-9). Surprisingly, perhaps, this latter muscle is not located on the backbone (anyone who has been served the back of a chicken knows that it offers little meat) but is positioned under the pectoralis on the breast. It is attached by a tendon to the upper side of the humerus of the wing so that it pulls from below by an ingenious "rope-and-pulley" arrangement. Both of these muscles are anchored to the keel. Thus, with the main muscle mass low in the body, aerodynamic stability is improved.

The main leg muscle mass is located in the thigh, surrounding the femur, and a smaller mass lies over the tibiotarsus (shank or "drumstick"). Strong but thin tendons extend downward through sleevelike sheaths to the toes. Consequently the feet are nearly devoid of muscles, explaining the thin, delicate appearance of the bird leg. This arrangement places the main muscle mass near the bird's center of gravity and at the same time allows great agility to the slender, lightweight feet. Since the feet are made up mostly of bone, tendon, and tough, scaly skin, they are highly resistant to damage from freezing. When a bird perches on a branch, an ingenious toe-locking mechanism (Figure 30-10) is activated, which prevents the bird from falling off its perch when asleep. The same mechanism causes the talons of a hawk or owl automatically to sink deeply into its victim as the legs bend under the impact of the strike. The powerful grip of a bird of prey was described by L. Brown*

> When an eagle grips in earnest, one's hand becomes numb, and it is quite impossible to tear it free, or to loosen the grip of the eagle's toes with the other hand. One just has to wait until the bird relents, and while waiting one has ample time to realize that an animal such as a rabbit would be quickly paralyzed, unable to draw breath, and perhaps pierced through and through by the talons in such a clutch.

Birds have lost the long reptilian tail, still fully evident in *Archaeopteryx*, and have substituted a pincushion-like muscle mound into which the tail feathers are rooted. It contains a perplexing array of tiny muscles, as many as 1000 in some species, which control the crucial tail feathers. The most complex muscular system of all is found in the neck of birds; the thin and stringy muscles, elaborately interwoven and subdivided, provide the bird's neck with the ultimate in vertebrate flexibility.

*From Brown, L. 1970. *Eagles,* New York. Arco Publishing.

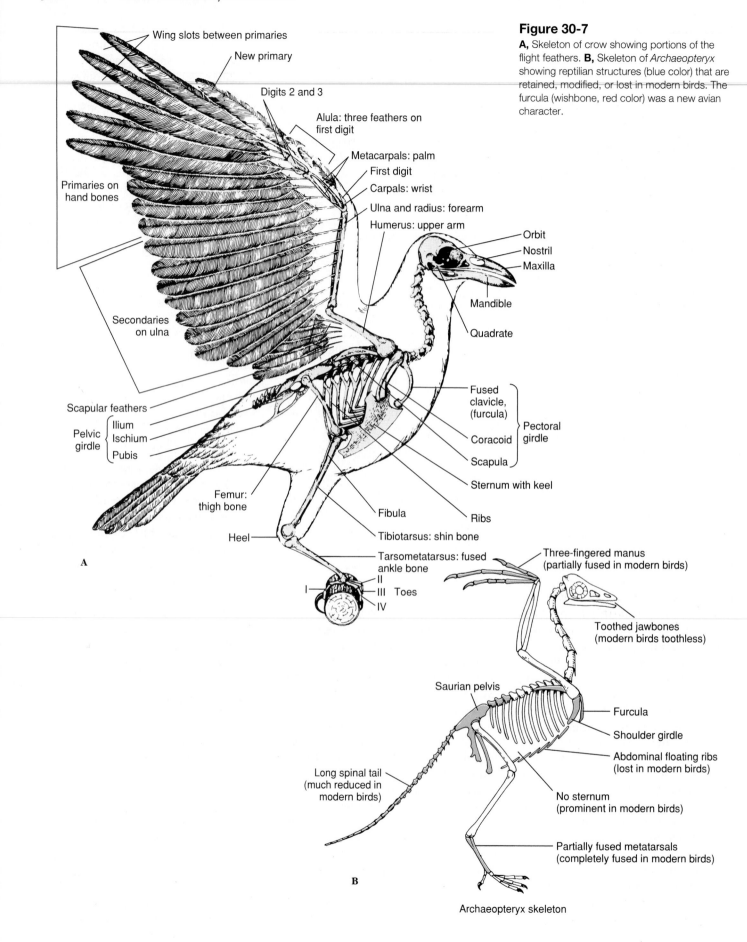

Wing slots between primaries

New primary

Digits 2 and 3

Alula: three feathers on first digit

Metacarpals: palm

First digit

Carpals: wrist

Ulna and radius: forearm

Humerus: upper arm

Primaries on hand bones

Secondaries on ulna

Scapular feathers

Pelvic girdle — Ilium / Ischium / Pubis

Femur: thigh bone

Heel

Orbit

Nostril

Maxilla

Mandible

Quadrate

Fused clavicle, (furcula)

Coracoid

Scapula

Pectoral girdle

Sternum with keel

Fibula

Ribs

Tibiotarsus: shin bone

Tarsometatarsus: fused ankle bone

II / III / IV   Toes

I

A

Three-fingered manus (partially fused in modern birds)

Toothed jawbones (modern birds toothless)

Saurian pelvis

Furcula

Shoulder girdle

Abdominal floating ribs (lost in modern birds)

Long spinal tail (much reduced in modern birds)

No sternum (prominent in modern birds)

Partially fused metatarsals (completely fused in modern birds)

B

Archaeopteryx skeleton

Figure 30-7

A, Skeleton of crow showing portions of the flight feathers. **B,** Skeleton of *Archaeopteryx* showing reptilian structures (blue color) that are retained, modified, or lost in modern birds. The furcula (wishbone, red color) was a new avian character.

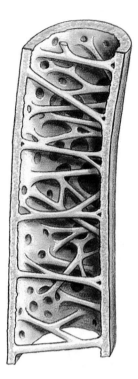

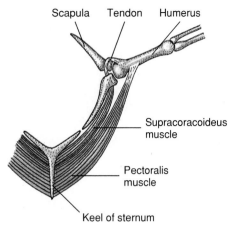

Scapula   Tendon   Humerus

Supracoracoideus
muscle

Pectoralis
muscle

Keel of sternum

Figure 30-9
Flight muscles of a bird are arranged to keep the center of gravity low in the body. Both major flight muscles are anchored on the sternum keel. Contraction of the pectoralis muscle pulls the wing downward. Then, as the pectoralis relaxes, the supracoracoideus muscle contracts and, acting as a pulley system, pulls the wing upward.

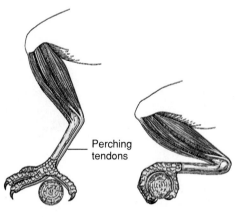

Perching
tendons

Figure 30-10
Perching mechanism of a bird. When a bird settles on a branch, tendons automatically tighten, closing the toes around the perch.

Figure 30-8
Hollow wing bone of a songbird showing the stiffening struts and air spaces that replace bone marrow. Such "pneumatized" bones are remarkably light and strong.

FOOD, FEEDING, AND DIGESTION

Birds have evolved along with food resources in nearly every environment on earth. In their early evolution, most birds were carnivorous, feeding principally on insects. Insects were well established on the earth's surface in both variety and numbers long before birds made their appearance, and they presented an enormously valuable food resource only partly exploited by amphibians and reptiles. With the advantage of flight, birds could hunt insects on the wing and carry their assault to insect refuges mostly inaccessible to their earthbound tetrapod peers. Today, there is a bird to hunt nearly every insect; they probe the soil, search the bark, scrutinize every leaf and twig, and drill into insect galleries hidden in tree trunks.

Other animal foods (worms, molluscs, crustaceans, fish, frogs, reptiles, mammals, as well as other birds) all found their way into the diet of birds. A very large group, nearly one-fifth of

all birds, feeds on nectar. Some birds are omnivores (often termed **euryphagous,** or "wide-eating" species) that will eat whatever is seasonally abundant. However, omnivorous birds must compete with numerous other omnivores for the same broad spectrum of food. Others are specialists (called **stenophagous,** or "narrow-eating" species) that have the pantry to themselves—but at a price. Should the food specialty be reduced or destroyed for some reason (disease, adverse climate, and the like), their very survival may be jeopardized.

The beaks of birds are strongly adapted to specialized food habits—from generalized types such as the strong, pointed beaks of crows, to grotesque, highly specialized ones in flamingoes, hornbills, and toucans (Figure 30-11). The beak of a woodpecker is a straight, hard, chisel-like device. Anchored to a tree trunk with its tail serving as a brace, the woodpecker delivers powerful, rapid blows to build nests or expose the burrows of wood-boring insects. It then uses its long, flexible, barbed tongue to seek out insects in their galleries. The woodpecker's skull is especially thick to absorb shock.

How much do birds eat? By a peculiar twist of reality, the common-

place "to eat like a bird" is supposed to signify a diminutive appetite. Yet birds, because of their intense metabolism, are voracious feeders. Small birds with their high metabolic rate eat relatively more than large birds. This happens because the oxygen consumption increases only about three-fourths as rapidly as body weight. For example, the resting metabolic rate (oxygen consumed per gram of body weight) of a hummingbird is 12 times that of a pigeon and 25 times that of a chicken. A 3 g hummingbird may eat 100% of its body weight in food each day, an 11 g blue tit about 30%, and a 1880 g domestic chicken, 3.4%. Obviously the weight of food consumed also depends on water content of the food, since water has no nutritive value. A 57 g Bohemian waxwing was estimated to eat 170 g of watery *Cotoneaster* berries in one day—three times its body weight! Seed-eaters of equivalent size might eat only 8 g of dry seeds per day.

Birds rapidly process their food with efficient digestive equipment. A shrike can digest a mouse in 3 hours, and berries will pass completely through the digestive tract of a thrush in just 30 minutes. Furthermore, birds utilize a very high percentage of the food they eat. There are no teeth in

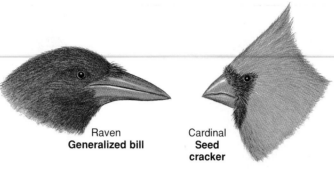

Figure 30-11
Some bills of birds showing variety of adaptations.

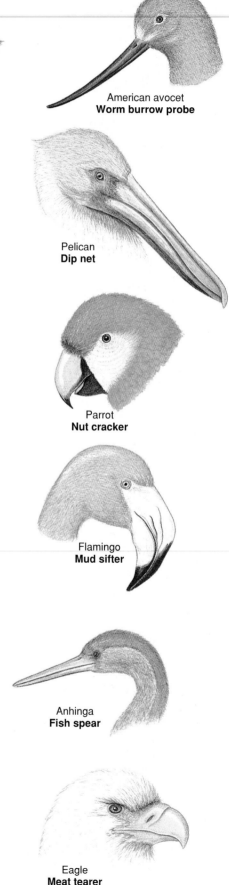

American avocet
Worm burrow probe

Pelican
Dip net

Parrot
Nut cracker

Flamingo
Mud sifter

Anhinga
Fish spear

Eagle
Meat tearer

the mouth, and the poorly developed salivary glands mainly secrete mucus for lubricating the food and the slender, horn-covered tongue. There are few taste buds, although all birds can taste to some extent. Hummingbirds and some others have sticky tongues, and woodpeckers have tongues that are barbed at the end. From the short **pharynx** a relatively long, muscular, elastic **esophagus** extends to the **stomach.** Many birds have an enlargement **(crop)** at the lower end of the esophagus that serves as a storage chamber.

In pigeons, doves, and some parrots the crop not only stores food but also produces milk by the breakdown of epithelial cells of the lining. This "bird milk" is regurgitated by both male and female into the mouth of the young squabs. It has a much higher fat content than cow's milk.

The stomach proper consists of a **proventriculus,** which secretes gastric juice, and the muscular **gizzard,** which is lined with horny plates that serve as millstones for grinding the food. To assist in the grinding process, birds swallow coarse, gritty objects or pebbles, which lodge in the gizzard. Certain birds of prey such as owls form pellets of indigestible materials, for example, bones and fur, in the proventriculus and eject them through the mouth. At the junction of the intestine with the rectum are paired **ceca,** which may be well developed in some birds. Two **bile ducts** from the **gallbladder** or liver and two or three **pancreatic ducts** empty into the duodenum, or first part of the intestine. The **liver** is relatively large and bilobed. The terminal part of the digestive system is the **cloaca,** which also receives the genital ducts and ureters; in young birds the dorsal wall of the cloaca bears the **bursa of Fabricius,** which processes the B lymphocytes that are important in the immune response (p. 676).

CIRCULATORY SYSTEM

The general plan of bird circulation is not greatly different from that of mammals, although their shared derived characteristics were evolved in parallel. The four-chambered heart is large, with strong ventricular walls; thus, birds share with mammals a complete separation of the respiratory and systemic circulations. However, the right aortic arch, instead of the left as in the mammals, leads to the dorsal aorta. The two jugular veins in the neck are connected by a cross vein, an adaptation for shunting the blood from one jugular to the other as the head is turned around. The brachial and pectoral arteries to the wings and breast are unusually large.

The heartbeat is extremely fast, and, as in mammals, there is an inverse relationship between heart rate and body weight. For example, a turkey has a heart rate at rest of about 93 beats per minute, a chicken has 250 beats per minute, and a black-capped chickadee has 500 beats per minute when asleep, which may increase to a phenomenal 1000 beats per minute during exercise. Blood pressure in birds is roughly equivalent to that in mammals of similar size.

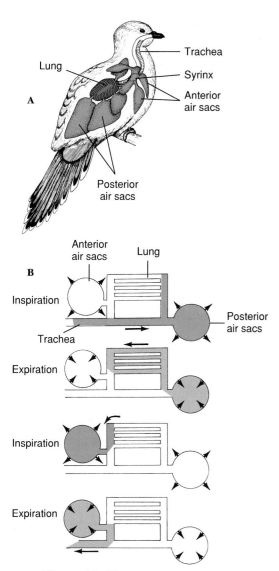

Figure 30-12
Respiratory system of a bird. **A,** Lungs and air sacs. One side of the bilateral air sac system is shown. **B,** Movement of a single volume of air through the bird's respiratory system. Two full respiratory cycles are required to move the air through the system.

Bird's blood contains **nucleated, biconvex erythrocytes.** (Mammals, the only other endothermic vertebrates, have enucleated, biconcave erythrocytes that are somewhat smaller than those of birds.) The **phagocytes,** or mobile ameboid cells of the blood, are very active and efficient in birds in the repair of wounds and in destroying microbes.

RESPIRATORY SYSTEM

The respiratory system of birds differs radically from the lungs of reptiles and mammals and is marvelously adapted for meeting the high metabolic demands of flight. In birds the finest branches of the bronchi, rather than ending in saclike alveoli as in mammals, are developed as tubelike **parabronchi** through which the air flows continuously. Also unique is the extensive system of nine interconnecting **air sacs** that are located in pairs in the thorax and abdomen and even extend by tiny tubes into the centers of the long bones (Figure 30-12). The air sacs connect to the lungs in such a way that perhaps 75% of the inspired air bypasses the lungs and flows directly into the posterior air sacs, which serve as reservoirs for fresh air. On expiration, this oxygenated air is shunted through the lung and collected in the anterior air sacs. From there it flows directly to the outside. The air flow sequence is shown in Figure 30-12. The advantage of such a system is obvious: the lungs receive fresh air during both inspiration and expiration. An almost continuous stream of oxygenated air is passed through a system of richly vascularized parabronchi. Although many details of the bird's respiratory system are not yet understood, it is clearly the most efficient of any vertebrate system.

The remarkable efficiency of the bird respiratory system is emphasized by bar-headed geese that routinely migrate over the Himalayan mountains and have been sighted flying over Mt. Everest (8848 meters or 29,141 feet) under conditions of severe hypoxia. They reach altitudes of 9000 meters in less than a day, without the acclimatization that is absolutely essential for humans even to approach the upper reaches of Mt. Everest.

In addition to performing its principal respiratory function, the air sac system helps cool the bird during vigorous exercise. A pigeon, for example, produces about 27 times more heat when flying than when at rest. The air sacs have numerous diverticula that extend inside the larger pneumatic bones of the pectoral and pelvic girdles, wings, and legs. Because they contain warmed air, they provide considerable buoyancy to the bird.

EXCRETORY SYSTEM

The relatively large paired metanephric kidneys are attached to the dorsal abdominal wall in a depression against the sacral vertebrae and pelvis. The kidney is composed of many thousands of **nephrons,** each consisting of a renal corpuscle and a nephric tubule. Urine is formed in the usual way by glomerular filtration followed by selective modification of the filtrate in the tubule (the details of this sequence are given on pp. 659 to 661). Urine passes by way of **ureters** to the **cloaca.** There is no urinary bladder.

Birds, like reptiles, excrete their nitrogenous wastes as uric acid, rather than urea, an adaptation that originated with the evolution of the shelled egg. In shelled eggs, all excretory products must remain in the eggshell with the growing embryo. If urea were produced, it would quickly accumulate in solution to toxic levels. Uric acid, however, crystallizes out of solution and can be stored harmlessly within the egg shell. Thus from an embryonic necessity was born an adult virtue. Because of uric acid's low solubility, a bird can excrete 1 g of uric acid in only 1.5 to 3 ml of water, whereas a mammal may require 60 ml of water to excrete 1 g of urea. The concentration of uric acid occurs almost entirely in the cloaca, where it is combined with fecal material, and the water is reabsorbed. Thus despite having kidneys that are less effective in true concentrative ability than mammalian kidneys, birds can form urine containing uric acid nearly 3000 times more concentrated than it is in the blood. Even the most effective mammalian kidneys—those of certain desert rodents—can excrete urea only about 25 times the plasma concentration.

Marine birds (also marine turtles) must excrete large loads of salt eaten with their food and in the seawater they drink. Seawater contains about 3% salt and is three times saltier than a bird's body fluids. Yet the bird

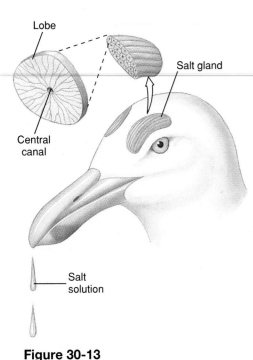

Figure 30-13
Salt glands of a marine bird (gull). One salt gland is located above each eye. Each gland consists of several lobes arranged in parallel. One lobe is shown in cross section, much enlarged. Salt is secreted into many radially arranged tubules, then flows into a central canal that leads into the nose.

kidney cannot concentrate salt in urine above about 0.3%. The problem is solved by special **salt glands,** one located above each eye (Figure 30-13). These glands are capable of excreting a highly concentrated solution of sodium chloride—up to twice the concentration of seawater. The salt solution runs out the internal or external nostrils, giving gulls, petrels, and other sea birds a perpetual runny nose. The size of the salt gland in some birds depends on how much salt the bird takes in its diet. For example, a race of mallard ducks living a semimarine life in Greenland has salt glands 10 times larger than those of ordinary freshwater mallards.

NERVOUS AND SENSORY SYSTEMS

A bird's nervous and sensory system accurately reflects the complex problems of flight and a highly visible existence, in which it must gather food, mate, defend territory, incubate and rear young, and correctly distinguish friend from foe. The brain of a bird

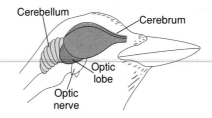

Figure 30-14
Bird brain showing principal divisions.

has well-developed **cerebral hemispheres, cerebellum,** and **midbrain tectum** (optic lobes) (Figure 30-14). The **cerebral cortex**—the portion in mammals that becomes the chief coordinating center—is thin, unfissured, and poorly developed in birds. But the core of the cerebrum, the **corpus striatum,** has enlarged into the principal integrative center of the brain, controlling such activities as eating, singing, flying, and all the complex instinctive reproductive activities. Relatively intelligent birds, such as crows and parrots, have larger cerebral hemispheres than do less intelligent birds such as chickens and pigeons. The **cerebellum** is a crucial coordinating center where muscle-position sense, equilibrium sense, and visual cues are all assembled and used to coordinate movement and balance. The **optic lobes,** laterally bulging structures of the midbrain, form a visual association apparatus comparable to the visual cortex of mammals.

Except in flightless birds, ducks, and vultures, the senses of smell and taste are poorly developed in birds. This deficiency, however, is more than compensated by good hearing and superb vision, the keenest in the animal kingdom. As in mammals, the bird ear consists of three regions: (1) the **external ear,** a sound-conducting canal extending to the eardrum, (2) the **middle ear,** containing a rodlike **columella** that transmits vibrations, and (3) the **inner ear,** where the organ of hearing, the **cochlea,** is located. The bird cochlea is much shorter than the coiled mammalian cochlea, yet birds can hear roughly the same range of sound frequencies as humans. Actually, the

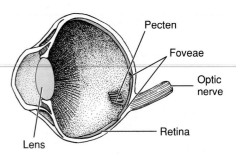

Figure 30-15
Hawk eye has all the structural components of the mammalian eye, plus a peculiar pleated structure, the pecten, believed to provide nourishment to the retina. The extraordinarily keen vision of the hawk is attributed to the extreme density of cone cells in the foveae: 1.5 million per fovea compared to 0.2 million for humans.

bird ear far surpasses that of humans in capacity to distinguish differences in intensities and to respond to rapid fluctuations in pitch.

The bird eye resembles that of other vertebrates in gross structure but is relatively larger, less spherical, and almost immobile; instead of turning their eyes, birds turn their heads with their long and flexible necks to scan the visual field. The light-sensitive **retina** (Figure 30-15) is elaborately equipped with rods (for dim night vision) and cones (for color vision). Cones predominate in day birds, and rods are more numerous in nocturnal birds. A distinctive feature of the bird eye is the **pecten,** a highly vascularized organ attached to the retina near the optic nerve and jutting out into the vitreous humor (Figure 30-15). The pecten is thought to provide nutrients and oxygen to the eye. It may do more, but its function remains largely a mystery.

The position of a bird's eyes in its head is correlated with its life habits. Vegetarians that must avoid predators have eyes placed laterally to give a wide view of the world; predaceous birds such as hawks and owls have eyes directed to the front. In birds of prey and some others, the **fovea,** or region of keenest vision on the retina, is placed in a deep pit, which makes it necessary for the bird to focus exactly on the source. Many birds, moreover, have two foveae on the retina (Figure 30-15): the central one for sharp monocular views and the posterior one for binocular vision. Woodcocks can probably see binocularly both forward and backward. The visual acuity of a hawk is believed to be 8 times that of a human (enabling it to see clearly a crouching rabbit more than a mile away), and an owl's ability to see in dim light is more than 10 times that of a human. Birds have good color vision, especially toward the red end of the spectrum.

Many birds can see into the ultraviolet, enabling them to view environmental features inaccessible to us but accessible to insects (such as flowers with ultraviolet-reflecting "nectar guides" that attract pollinating insects). Several species of ducks, hummingbirds, kingfishers, and passerines (songbirds) can see in the near ultraviolet (UV) down to 370 nm (the human eye filters out ultraviolet light below 400 nm). For what purpose do birds use their UV-sensitivity? Some, such as hummingbirds, may be attracted to nectar-guiding flowers, like insects. But, for the others, the benefit derived from UV-sensitivity is a matter of conjecture.

FLIGHT

What prompted the evolution of flight in birds, the ability to rise free of earthbound concerns, as almost every human has dreamed of doing? The origin of flight was the result of complex selective pressures. The air was a relatively unexploited habitat stocked with flying insects for food. Flight also offered escape from terrestrial predators and opportunity to travel rapidly and widely to establish new breeding areas and to benefit from year-round favorable climate by migrating north and south with the seasons.

The fossil evidence is too meager to provide us with a recorded history of the origin of bird flight, but it must have happened in one of two ways: birds began to fly by climbing to a high place and gliding down, or by flapping their way into the air from the ground. The "ground-up" hypothesis holds that birds were ground-

dwelling runners with primitive wings used to snare insects. With continued enlargement the protowings eventually enabled the running animal to flap its way into the air. The more widely favored "trees-down" hypothesis suggests that birds passed through an arboreal apprenticeship of tree climbing, leaping through the trees, parachuting, gliding, and finally fully powered flight. One thing seems certain: feathers were an absolute requirement for flight. The evolutionary origin of feathers preceded flight; feathers arose for their thermoregulatory role and made possible the subsequent evolution of flight. There is absolutely no support for the idea that bird ancestors were originally membrane-winged flyers, like bats, that later developed feathers.

Bird Wing as a Lift Device

The bird wing is an airfoil that is subject to recognized laws of aerodynamics. It is streamlined in cross section, with a slightly concave lower surface (**cambered**) and with small, tight-fitting feathers where the leading edge meets the air (Figure 30-16). Air slips smoothly over the wing, creating lift with minimum drag. Some lift is produced by positive pressure against the undersurface of the wing. But on the upper side, where the airstream must travel farther and faster over the convex surface, a negative pressure is created that provides more than two-thirds of the total lift.

The lift-to-drag ratio of an airfoil is determined by the angle of tilt (angle of attack) and the airspeed (Figure 30-16). A wing carrying a given load can pass through the air at high speed and small angle of attack or at low speed and larger angle of attack. As speed decreases, lift can be increased by increasing the angle of attack, but drag forces also increase. Finally a point is reached at which the angle of attack becomes too steep; turbulence appears on the upper surface, lift is destroyed, and stalling occurs. Stalling can be delayed or prevented by placing a **wing slot** along the leading edge so that a layer of rapidly moving air is directed across

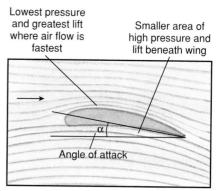
Air flow around wing

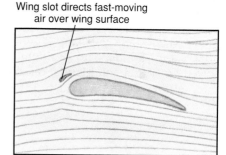
Stalling at low speed

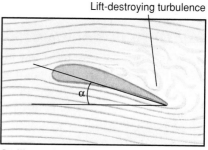
Preventing stall with wing slots

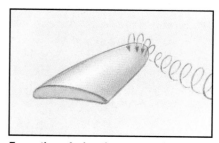
Formation of wing tip vortex

Figure 30-16
Air patterns formed by the airfoil, or wing, moving from right to left. At low speed the angle of attack (α) must increase to maintain lift but this increases the threat of stalling. The upper figures show how low-speed stalling can be prevented with wing slots. Wing tip vortex (*bottom*), a turbulence that tends to develop at high speeds, reduces flight efficiency. The effect is reduced in wings that sweep back and taper to a tip.

the upper wing surface. Wing slots were and still are used in aircraft traveling at a low speed. In birds, two kinds of wing slots have developed: (1) the **alula,** or group of small feathers on the thumb (Figures 30-6 and 30-7), which provides a midwing slot, and (2) **slotting between the primary feathers,** which provides a wing-tip slot. In a number of songbirds, these together provide stall-preventing slots for nearly the entire outer (and aerodynamically more important) half of the wing.

Basic Forms of Bird Wings

Bird wings vary in size and form because the successful exploitation of different habitats has imposed special aerodynamic requirements. Four types of bird wings are easily recognized.*

Elliptical Wings. Birds that must maneuver in forested habitats, such as sparrows, warblers, doves, woodpeckers, and magpies (Figure 30-17A), have elliptical wings. This type has a **low aspect ratio** (ratio of length to average width). The wings of the highly

*Saville, D. B. O. 1957. Adaptive evolution in the avian wing. Evolution **11:**212–224.

maneuverable British Spitfire fighter plane of World War II fame conformed closely to the outline of the sparrow wing. Elliptical wings are highly slotted between the primary feathers; this helps prevent stalling during sharp turns, low-speed flight, and frequent landing and takeoff. Each separated primary feather behaves as a narrow wing with a high angle of attack, providing high lift at low speed. The high maneuverability of the elliptical wing is exemplified by the tiny chickadee, which, if frightened, can change course within 0.03 second.

High-Speed Wings. Birds that feed on the wing, such as swallows, hummingbirds, and swifts, or that make long migrations, such as plovers, sandpipers, terns and gulls, (Figure 30-17B), have wings that sweep back and taper to a slender tip. They are rather flat in section, have a moderately high aspect ratio, and lack the wing-tip slotting characteristic of the preceding group. Sweepback and wide separation of the wing tips reduce "tip vortex," a drag-creating turbulence that tends to develop at wing tips at faster speeds. This type of wing is aerodynamically efficient for high-speed flight but cannot easily keep a bird airborne at low speeds.

The fastest birds, such as sandpipers, clocked at 175 km (109 miles) per hour, belong to this group.

Soaring Wings. The oceanic soaring birds have **high-aspect ratio** wings resembling those of sailplanes. This group includes albatrosses, frigate birds, and gannets (Figure 30-17C). Such long, narrow wings lack wing slots and are adapted for high speed, high lift, and dynamic soaring. They have the highest aerodynamic efficiency of all wings but are less maneuverable than the wide, slotted wings of land soarers. Dynamic soarers have learned how to exploit the highly reliable sea winds, using adjacent air currents of different velocities.

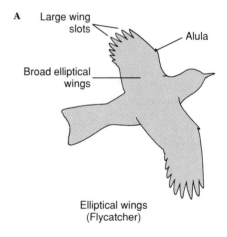

A Large wing slots — Alula

Broad elliptical wings

Elliptical wings (Flycatcher)

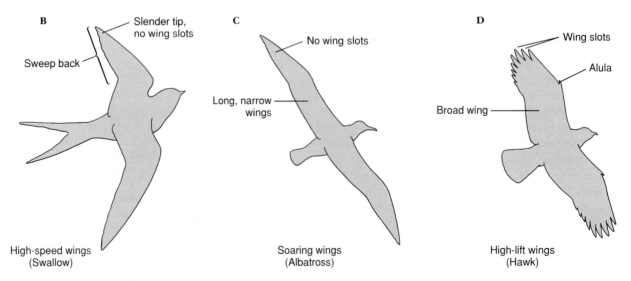

B Slender tip, no wing slots

Sweep back

High-speed wings (Swallow)

C No wing slots

Long, narrow wings

Soaring wings (Albatross)

D Wing slots — Alula

Broad wing

High-lift wings (Hawk)

Figure 30-17
Four basic forms of bird wings.

Figure 30-18
In normal flapping flight of strong fliers like ducks, the wings sweep downward and forward fully extended. Thrust is provided by the primary feathers at the wing tips. To begin the upbeat, the wing is bent, bringing it upward and backward. The wing then extends, ready for the next downbeat.

High-Lift Wings. Vultures, hawks, eagles, owls, and ospreys (Figure 30-17D)—predators that carry heavy loads—have wings with slotting, alulas, and pronounced camber, all of which promote high lift at low speed. Many of these birds are land soarers, with broad, slotted wings that provide the sensitive response and maneuverability required for static soaring in the capricious air currents over land.

Flapping Flight

This basic form of flight is so complex that complete analysis is still not possible—yet young birds fly almost perfectly on their maiden flight. More than a century ago an English zoologist reared swallow fledglings in a space so confining that they could not fully extend their wings. Yet when released at the age when swallows normally fly, they flew immediately and without practice.

Two forces are required for flapping flight: a vertical *lifting* force to support the bird's weight, and a horizontal *thrusting* force to move the bird forward against the resistive forces of friction. Thrust is provided mainly by the primary feathers at the wing tips, while the secondary feathers of the inner wing, which do not move so far or so fast, act as an airfoil, providing mainly lift. Greatest power is applied on the downstroke. The primary feathers are bent upward and twist to a steep angle of attack, biting into the air like a propeller (Figure 30-18). The entire wing (and the bird's body) is pulled forward. On the upstroke, the primary feathers bend in the opposite direction so that their upper surfaces twist into a positive angle of attack to produce thrust, just as the lower surfaces did on the downstroke. A powered upstroke is essential for hovering flight, as in hummingbirds, and is important for fast, steep takeoffs by small birds with elliptical wings.

Migration and Navigation

The advantages of migration were described in the prologue to this chapter. Not all birds migrate, of course, but the majority of North American and European species do, and the biannual journeys of some are truly extraordinary undertakings. Migration is both the greatest adventure and the greatest risk in the life of a migratory bird.

Migration Routes

Most migratory birds have well-established routes trending north and south. Since most birds (and other animals) live in the Northern Hemisphere, where most of the earth's land mass is concentrated, most birds are south-in-winter and north-in-summer migrants. Of the 4000 or more species of migrant birds (a little less than half the total bird species), most breed in the more northern latitudes of the hemisphere; the percentage of migrants in Canada is far higher than the percentage of migrants in Mexico, for example. Some use different routes in the fall and spring (Figure 30-19).

Some, especially certain aquatic species, complete their migratory routes in a very short time. Others, however, make the trip in a leisurely manner, often stopping along the way to feed. Some of the warblers are known to take 50 to 60 days to migrate from their winter quarters in Central America to their summer breeding areas in Canada.

Not all members of a species migrate at the same time; there is a great deal of straggling so that some members do not reach the summer breeding grounds until others are well along with their nesting. Many of the smaller species migrate at night and feed by day; others migrate chiefly in the daytime; and many swimming and wading birds migrate by either day or night. The height at which they fly varies greatly. Migrants tend to fly higher over water than over land and higher at night than during the day.

Many birds are known to follow landmarks, such as rivers and coastlines, but others do not hesitate to fly directly over large bodies of water in their routes. Some birds have very wide migration lanes, whereas others, such as certain sandpipers, are restricted to very narrow ones, keeping well to the coastlines because of their food requirements.

Some species are known for their long-distance migrations. The Arctic tern, greatest globe spanner of all, breeds north of the Arctic Circle and in winter is found in the Antarctic regions. This species is also known to take a circuitous route in migrations from North America, passing over to

Figure 30-19

Migrations of the bobolink and golden plover. The bobolink commutes 22,500 km (14,000 miles) each year between nesting sites in North America and its wintering range in Argentina, a phenomenal feat for such a small bird. Although the breeding range has extended to colonies in western areas, these birds take no shortcuts but adhere to the ancestral seaboard route. The golden plover flies a loop migration, striking out across the Atlantic in its southward autumnal migration but returning in the spring by way of Central America and the Mississippi Valley because ecological conditions are more favorable at that time.

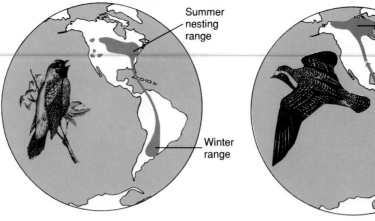

Bobolink   **Golden plover**

the coastlines of Europe and Africa and then to their winter quarters, a trip that may exceed 18,000 km (11,200 miles). Other birds that breed in Alaska follow a more direct line down the Pacific coast of North and South America.

Many small songbirds also make great migration treks (Figure 30-19). Africa is a favorite wintering ground for European birds, and many fly there from Central Asia as well.

STIMULUS FOR MIGRATION

Humans have known for centuries that the onset of the reproductive cycle of birds is closely related to season. Only within the last 60 years, however, has it been proved that the lengthening days of late winter and early spring stimulate the development of the gonads and accumulation of fat—both important internal changes that predispose birds to migrate northward. There is evidence that increasing day length stimulates the anterior lobe of the pituitary into activity. The release of pituitary gonadotropic hormone in turn sets in motion a complex series of physiological and behavioral changes, resulting in gonadal growth, fat deposition, migration, courtship and mating behavior, and care of the young.

DIRECTION FINDING IN MIGRATION

Numerous experiments suggest that most birds navigate chiefly by sight.

Birds recognize topographical landmarks and follow familiar migratory routes—a behavior assisted by flock migration, during which navigational resources and experience of older birds can be pooled. But in addition to visual navigation, birds make use of a variety of orientation cues at their disposal. Birds have a highly accurate innate sense of time. They also have an innate sense of direction; and very recent work adds credence to an old, much debated hypothesis that birds can detect and navigate by the earth's magnetic field. All of these resources are inborn and instinctive, although a bird's navigational abilities may improve with experience.

In the early 1970s W. T. Keeton showed that the flight bearings of homing pigeons were significantly disturbed by magnets attached to the birds' heads, or by minor fluctuations in the geomagnetic field. But until recently the nature and position of a magnetic receptor in pigeons remained a mystery. Deposits of a magnetic substance called magnetite (Fe_3O_4) have been discovered in the neck musculature of pigeons and migratory white-crowned sparrows. If this material were coupled to sensitive muscle receptors, as has been proposed, the structure could serve as a magnetic compass that would enable birds to detect and orient their migrations to the earth's magnetic field.

Experiments by German ornithologists G. Kramer and E. Sauer and American ornithologist S. Emlen demonstrated convincingly that birds can navigate by celestial cues: the sun by day and the stars by night. Using special circular cages, Kramer concluded that birds maintain compass direction by referring to the sun, regardless of the time of day (Figure 30-20). This is called **sun-azimuth orientation** (*azimuth,* compass bearing of the sun). Sauer's and Emlen's ingenious planetarium experiments strongly suggest that some birds, probably many, are able to detect and navigate by the North Star axis around which the constellations appear to rotate.

Some of the remarkable feats of bird navigation still defy rational explanation. Most birds undoubtedly use a combination of environmental and innate cues to migrate. Migration is a rigorous undertaking. The target is often small, and natural selection relentlessly prunes off individuals making errors in migration, leaving only the best navigators to propagate the species.

SOCIAL BEHAVIOR AND REPRODUCTION

The adage says "birds of a feather flock together," and many birds are indeed highly social creatures. Especially during the breeding season, sea birds gather, often in enormous colonies, to nest and rear young (Figure 30-21). Land birds, with some conspicuous

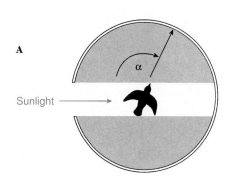

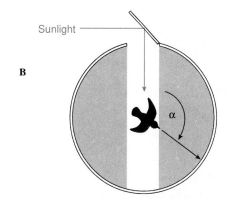

Figure 30-20
Gustav Kramer's experiments with sun-compass navigation in starlings. **A,** In a windowed, circular cage, the bird fluttered to align itself in the direction it would normally follow if it were free. **B,** When the true angle of the sun is deflected with a mirror, the bird maintains the same relative position to the sun. This shows that these birds use the sun as a compass. The bird navigates correctly throughout the day, changing its orientation to the sun as the sun moves across the sky.

Figure 30-21
Part of a colony of Northern gannets, *Morus bassanus,* showing extremely close spacing between pairs in this highly social bird. Order Pelecaniformes.

Figure 30-22
Cooperative feeding behavior by the white pelican, *Pelecanus onocrotalus.* **A,** Pelicans form a horseshoe to drive fish together. **B,** Then they plunge simultaneously to scoop fish in their huge bills. These photographs were taken 2 seconds apart.

exceptions, such as starlings and rooks, tend to be less gregarious than sea birds during breeding and to seek isolation for rearing their brood. But these same species that covet separation from their kind during breeding may aggregate for migration or feeding. Togetherness offers advantages: mutual protection from enemies, greater ease in finding mates, less opportunity for individual straying during migration, and mass huddling for protection against low night temperatures during migration. Certain species, such as pelicans (Figure 30-22), may use highly organized cooperative behavior to feed. At no time are the highly organized social interactions of birds more evident than during the breeding season, as they stake out territorial claims, select mates, build nests, incubate and hatch their eggs, and rear their young.

REPRODUCTIVE SYSTEM

In the male the paired **testes** and accessory ducts are similar to those in many other vertebrates. From the **testes** the **vasa deferentia** run to the cloaca. Before being discharged, the sperm are stored in the **seminal vesicle,** the enlarged distal end of the vas deferens. This seminal vesicle may become so large with stored sperm during the breeding season that it causes a cloacal protuberance. The high body temperature, which tends to inhibit spermatogenesis in the testes, is probably counteracted by the cooling effect of the abdominal air sacs. The testes of birds undergo a great enlargement at the breeding season, as much as 300 fold, and then shrink to tiny bodies afterward. Some birds, including ducks and geese,

have a large, well-developed **copulatory organ** (penis), provided with a groove on its dorsal side for the transfer of sperm. Most birds, however, lack a penis, and copulation is a matter of bringing the cloacal surfaces into contact, usually while the male stands on the female's back (Figure 30-23). Some swifts copulate in flight.

In the female of most birds, only the left ovary and oviduct develop; those on the right dwindle to vestigial structures. Eggs discharged from the ovary are picked up by the expanded end of the oviduct, the **infundibulum** (Figure 30-24). The oviduct runs posteriorly to the cloaca. While the eggs are passing down the oviduct, **albumin,** or egg white, from special glands is added to them; farther down the oviduct, the shell membrane, shell, and shell pigments are also secreted about the egg. Fertilization takes place in the upper oviduct several hours before the layers of albumin, shell membranes, and shell are added. Sperm remain alive in the female oviduct for many days after a single mating. Hen eggs show good fertility for 5 or 6 days after mating, but then fertility drops rapidly.

Figure 30-23
Copulation in birds. In most bird species the male lacks a penis. The male copulates by standing on the back of the female, pressing his cloaca against that of the female, and passing sperm to the female.

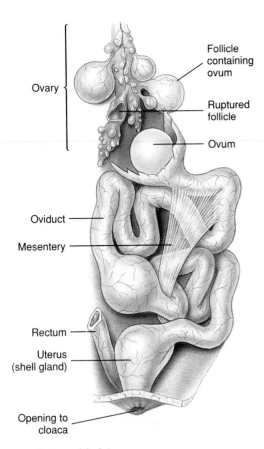

Figure 30-24
Reproductive system of a female bird.

However, the occasional egg will be fertile as long as 30 days after separation of the hen from the rooster.

MATING SYSTEMS

The two most common types of mating systems in animals are **monogamy,** in which an individual mates with only one partner each breeding season, and **polygamy,** in which an individual mates with two or more partners each breeding period. Monogamy is rare in most animal groups, but in birds it is the general rule: more than 90% of the birds are monogamous. In a few bird species such as swans and geese, partners are chosen for life and often remain together throughout the year. Seasonal monogamy is more common, however, in the great majority of migrant birds, which pair up during the breeding season but lead independent lives the rest of the year.

> The term *"polygamy"* ("many marriages") is used when the sex of the individual possessing a plurality of mates is not specified. The most common form of polygamy is polygyny ("many females"), in which a male mates with more than one female. Much less common is polyandry ("many males"), in which a female mates with more than one male per breeding season.

One reason that monogamy is much more common among birds than among mammals is that female birds are not equipped, as mammals are, with a built-in food supply for the young. Accordingly, the ability of the two sexes to provide parental care, especially food for the young, is more equal in birds than in mammals. A female bird will choose a male whose parental investment in their young is apt to be high and avoid a male that has mated with another female. If the male had mated with another female, he could at best divide his time between his two mates and might even devote most of his attention to the alternate mate. Consequently, females enforce monogamy.

Monogamy in birds is also encouraged by the need for the male to secure and defend a territory before he can attract a mate. The male may sing a great deal to announce his presence to females and to discourage rival males from entering his territory. The female wanders from one territory to another, seeking a male with foraging territory that offers the best chances for reproductive success. Usually a male is able to defend an area that provides just enough resources for one nesting female.

The most common form of polygamy in birds, when it occurs, is **polygyny** ("many females"), in which a male mates with more than one female. In many species of grouse, the males gather in a collective display ground, the **lek,** which is divided into individual territories, each vigorously defended by a displaying male (Figure 30-25). There is nothing of value in the lek to the female except the male, and all he can offer are his genes, for only the females care for the young. Usually there are a dominant male and several subordinate males in the lek. Competition among males for females is intense, but the females appear to choose the dominant male for mating because, presumably, social rank correlates with genetic quality.

NESTING AND CARE OF YOUNG

To produce offspring, all birds lay eggs that must be incubated by one or both parents. The eggs of most songbirds (order Passeriformes) require approximately 14 days for hatching; those of ducks and geese require at least twice that long. Most of the duties of incubation fall on the female, although in many instances both parents share the task, and occasionally only the male performs this work.

Most birds build some form of nest in which to rear their young. Some birds simply lay their eggs on the bare ground or rocks and make no pretense of nest building. Others build elaborate nests such as the pendant nests constructed by orioles, the delicate lichen-covered mud nests of hummingbirds (Figure 30-26) and flycatchers, the chimney-shaped mud

Labels on Figure 30-24:
Ovary
Oviduct
Mesentery
Rectum
Uterus (shell gland)
Opening to cloaca
Follicle containing ovum
Ruptured follicle
Ovum

Figure 30-25
Dominant male sage grouse, *Centrocercus urophasianus,* surrounded by several hens that have been attracted by his "booming" display.

Figure 30-26
Broad-tailed hummingbird, *Selasphorus platycercus,* feeding young in its nest of plant down and spider webs and decorated on the outside with lichens. The female builds the nest, incubates the two pea-sized eggs, and rears the young with no assistance from the male. These frail-looking but pugnacious little birds make arduous seasonal migrations between Canada and Mexico.

Altricial
One-day-old meadowlark

Precocial
One-day-old ruffed grouse

Figure 30-27
Comparison of 1-day-old altricial and precocial young. The altricial meadowlark (*left*) is born nearly naked, blind, and helpless. The precocial ruffed grouse (*right*) is covered with down, alert, strong legged, and able to feed itself.

nests of cliff swallows, the floating nests of rednecked grebes, and the huge brush pile nests of Australian brush turkeys. Most birds take considerable pains to conceal their nests from enemies. Woodpeckers, chickadees, bluebirds, and many others place their nests in tree hollows or other cavities; kingfishers excavate tunnels in the banks of streams for their nests; and birds of prey build high in lofty trees or on inaccessible cliffs. Nest parasites such as the brown-headed cowbird and the European cuckoo build no nests at all but simply lay their eggs in the nests of birds smaller than themselves. When the eggs hatch, the foster parents care for the cowbird young which outcompete the host's own hatchlings.

Newly hatched birds are of two types: **precocial** and **altricial.** The precocial young, such as quail, fowl, ducks, and most water birds, are covered with down when hatched and can run or swim as soon as their plumage is dry (Figure 30-27). The altricial ones, on the other hand, are naked and helpless at birth and remain in the nest for a week or more. The young of both types require care from parents for some time after hatching. They must be fed, guarded, and protected against rain and sun. The parents of altricial species must carry food to their young almost constantly, for most young birds will eat more than their weight each day. This enormous food consumption explains the rapid growth of the young and their quick exit from the nest. The food of the young, depending on the species, includes worms, insects, seeds, and fruit. Pigeons and doves

are peculiar in feeding their young "pigeon's milk," a creamy mixture of broken-down cells shed from the lining of the parent's crop.

Nesting success is very low with many birds, especially in altricial species. One investigation several years ago of 170 altricial bird nests reported that only 21% produced at least one young. The annual censusing of birds shows that nesting success is even lower today. Of the many causes of nesting failures, predation by raccoons, skunks, opossums, blue jays, crows, and others, especially in suburban and rural woodlots, and nest parasitism by the brown-headed cowbird are the most important factors. Birds of prey probably have a much higher percentage of reproductive success than songbirds.

BIRD POPULATIONS

Bird populations, like those of other animal groups, vary in size from year to year. Snowy owls, for example, are subject to population cycles that closely follow cycles in their food supply, mainly rodents. Voles, mice, and lemmings in the north have a fairly regular 4-year cycle of abundance; at population peaks, predator populations of foxes, weasels, and buzzards, as well as snowy owls, increase because there is abundant food for rearing their young. After a crash in the rodent population, snowy owls move south,

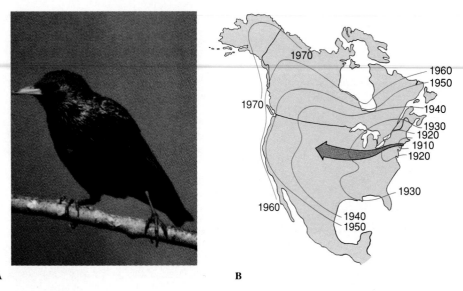

Figure 30-28

A, Starling, *Sturnus vulgaris*. Starlings are omnivorous, eating mostly insects in spring and summer and shifting to wild fruits in the fall. **B,** Colonization of North America by starlings after the introduction of 120 birds into Central Park in New York City in 1890. There are now perhaps 100 million starlings in the United States alone, testimony to the great reproductive potential of birds.

Figure 30-29

Sport-shooting passenger pigeons in Louisiana during the nineteenth century. Relentless sport and market hunting, before the establishment of state and federal hunting regulations, eventually dropped the population too low to sustain colonial breeding. The last passenger pigeon died in captivity in 1914.

seeking alternative food supplies. They occasionally appear in large numbers in southern Canada and the northern United States, where their total absence of fear of humans makes them easy targets for thoughtless hunters.

Occasionally the activities of people bring about spectacular changes in bird distribution. Both starlings (Figure 30-28) and house sparrows have been accidentally or deliberately introduced into numerous countries, and they have become the two most abundant bird species on earth, with the exception of domestic fowl.

Humans also are responsible for the extinction of many bird species. More than 80 species of birds have, since 1695, followed the last dodo to extinction. Many were victims of changes in their habitat or competition with better-adapted species. But several have been hunted to extinction, among them the passenger pigeon, which only a century ago darkened the skies over North America in incredible numbers estimated in the billions (Figure 30-29).

Today, game bird hunting is a well-managed renewable resource in the United States and Canada, and

while hunters kill millions of game birds each year, none of the 74 bird species legally hunted is endangered. Hunting interests, by acquiring large areas of wetlands for migratory bird refuges and sanctuaries, have contributed to the recovery of both game and nongame birds.

Lead poisoning of waterfowl is a side effect of hunting. Before long-delayed federal regulations went into effect in 1991, requiring the use of nonlead shot for all inland and coastal waterfowl hunting, shotguns scattered more than 3000 tons of lead each year in the United States alone. When waterfowl eat the pellets (which they mistake for seeds), the pellets are ground and eroded in their gizzards, facilitating the absorption of lead into the blood. Lead poisoning paralyzes or weakens birds, leading to death by starvation. Today, birds are still dying from ingesting lead shot that has accumulated over the years.

Of particular concern is the recent sharp decline of songbirds in the United States and southern Canada. Amateur birdwatchers and ornithologists have recorded that many songbird species that were abundant as recently as 40 years ago are now suddenly scarce. There are several reasons for the decline. Intensification of agriculture, permitted by the use of herbicides, pesticides, and fertilizers, has deprived ground-nesting birds of fields that were left fallow before the use of these agents. The excessive fragmentation of forests throughout much of the United States has increased exposure of nests of forest-dwelling species to nest predators such as blue jay, raccoons, and opossums, and to nest parasites such as the brown-headed cowbird. The rapid loss of tropical forests—approximately 170,000 square kilometers each year, an area about the size of the state of Washington*—is depriving some 250 species of songbird migrants of their

*Brown, L. R. 1993. State of the world 1993. New York, W. W. Norton & Company.

wintering homes. Of all the long-term threats facing songbird populations, tropical deforestation is the most serious and most intractable to change. If the rate of deforestation accelerates in the next few decades as expected, the world's tropical forests will have disappeared by 2040 (Terborgh, 1992).

Some birds, such as robins, house sparrows, and starlings, can accommodate to these changes, and may even thrive on them. But for most the changes are adverse. Terborgh (1992) warns that unless we take leadership in managing our natural resources wisely we soon could be facing the silent spring that Rachel Carson envisioned 30 years ago.

Figure 30-30
Ostrich *Struthio camelus* of Africa, the largest of all living birds. Order Struthioniformes.

CLASSIFICATION OF LIVING BIRDS OF CLASS AVES*

The class Aves is made up of more than 9600 species distributed among some 27 orders of living birds and a few fossil orders. Very few birds remain to be discovered. Of the 27 orders, four (or five depending on the classification system) are **ratite**, or flightless, birds (ostriches, rheas, cassowaries and emus, and kiwis), although flightlessness is not restricted to these groups. The remaining 23 orders are **carinate** birds (birds with a keeled sternum).

Class Aves (L. *avis,* bird)

Subclass Archaeornithes (Gr. *archaios,* ancient, + *ornis,* bird). Birds of the late Jurassic and early Cretaceous bearing many primitive characteristics. *Archaeopteryx.*

Subclass Neornithes (Gr. *neos,* new, + *ornis,* bird). Extinct and living birds with well-developed sternum and usually with keel; tail reduced; metacarpals and some carpals fused together. Cretaceous to Recent.

Superorder Paleognathae (Gr. *palaios,* ancient, + *gnathos,* jaw). Modern birds with primitive archosaurian palate. Ratites (with unkeeled sternum) and tinamous (with keeled sternum).

Order Struthioniformes (stroo'thi-on-i-for'meez) (L. *struthio,* ostrich, + *forma,* form): **ostrich.** One species, the flightless ostrich of Africa (*Struthio camelus*) (Figure 30-30) is the largest of living birds, with some specimens being 2.4 m tall and weighing 135 kg. The feet are provided with only two toes of unequal size covered with pads, which enable the birds to travel rapidly through sandy country.

Order Rheiformes (re'i-for'meez) (Gr. mythology, *Rhea,* mother of Zeus, + form): **rheas.** Two species of flightless birds restricted to South America; often called the American ostriches.

Order Casuariiformes (kazh'u-ar'ee-i-for'meez) (N.L. *Casuarius,* type genus, + *form*): **cassowaries, emus.** Four species of flightless birds found in Australia, New Guinea, and a few other islands. Some specimens may reach a height of 1.5 m.

Order Apterygiformes (ap'te-rij'i-for'meez) (Gr. *a,* not, + *pteryx,* wing, + form): **kiwis.** Three species of kiwis, flightless birds about the size of domestic fowl, found only in New Zealand. Only the merest vestige of a wing is present. The egg is extremely large for the size of the bird.

Order Tinamiformes (tin-am'i-for'meez) (N.L. *Tinamus,* type genus, + form): **tinamous.** Ground-dwelling, grouselike birds of Central and South America. About 60 species.

continued

*The traditional bird classification given here, called a morphological taxonomy, is based on the careful comparison of shared derived anatomical characters within and between bird groups. A new and still controversial biochemical classification based on degrees of similarity between DNAs of living birds from all over the world is believed by its proponents to represent true phylogenetic relationships much better than the traditional morphological classification. The biochemical taxonomy has produced several astonishing realignments. Most prominent of these is the sweeping revision of the order Ciconiiformes which, as revised, includes penguins, loons, grebes, albatrosses, and birds of prey, all previously placed in separate orders. DNA hybridization studies prove the close relatedness of these groups, whose true genetic affinities are masked by divergent evolution. Biochemical taxonomy, now under review by the American Ornithological Union, is certain to produce significant revision of the traditional taxonomy which has been the standard for more than a century. The biochemical classification reported by Sibley and Ahlquist (1990) is compared to the traditional morphological classification by Proctor and Lynch (1993).

Figure 30-31
Greater flamingos *Phoenicopterus ruber* on an alkaline lake in East Africa. Order Ciconiiformes.

Figure 30-32
Laughing gulls *Larus atricilla* in flight. Order Charadriiformes.

CLASSIFICATION OF LIVING BIRDS OF CLASS AVES—*continued*

Superorder Neognathae (Gr. *neos,* new, + *gnathos,* jaw). Modern birds with flexible palate.

Order Sphenisciformes (sfe-nis′i-for′meez) (Gr. *sphēniskos,* dim. of *sphen,* wedge, from the shortness of the wings, + form): **penguins.** Web-footed marine swimmers of the southern seas from Antarctica to the Galápagos Islands. Although penguins are carinate birds, they use their wings as paddles for swimming rather than for flight. About 17 species.

Order Gaviiformes (gay′vee-i-for′meez) (L. *gavia,* bird, probably sea mew, + form): **loons.** The four species of loons are remarkable swimmers and divers with short legs and heavy bodies. They live exclusively on fish and small aquatic forms. The familiar great northern diver (*Gavia immer*) is found mainly in northern waters of North America and Eurasia.

Order Podicipediformes (pod′i-si-ped′i-for′meez) (L. *podex,* rump, + *pes, pedis,* foot): **grebes.** These are short-legged divers with lobate-webbed toes. The pied-billed grebe (*Podilymbus podiceps*) is a familiar example of this order. Grebes are most common in old ponds where they build their raftlike floating nests. Eighteen species, worldwide distribution.

Order Procellariiformes (pro-sel-lar′ee-i-for′meez) (L. *procella,* tempest, + form): **albatrosses, petrels, fulmars, shearwaters.** All are marine birds with hooked beak and tubular nostrils. In wingspan (more than 3.6 m in some), albatrosses are the largest of flying birds. About 100 species, worldwide distribution.

Order Pelecaniformes (pel-e-can-i-for′meez) (Gr. *pelekan,* pelican, + form): **pelicans, cormorants, gannets, boobies, and so on.** These are colonial fish-eaters with throat pouch and all four toes of each foot included within the web. About 55 species, worldwide distribution, especially in the tropics.

Order Ciconiiformes (si-ko′nee-i-for′meez) (L. *ciconia,* stork, + form): **herons, bitterns, storks, ibises, spoonbills, flamingoes** (Figure 30-31). These are long-necked, long-legged, mostly colonial waders. A familiar eastern North American representative is the great blue heron (*Ardea herodias*), which frequents marshes and ponds. About 90 species, worldwide distribution.

Order Anseriformes (an′ser-i-for′meez) (L. *anser,* goose, + form): **swans, geese, ducks.** The members of this order have broad bills with filtering ridges at their margins, a foot web restricted to the front toes, and a long breastbone with a low keel. About 150 species, worldwide distribution.

Order Falconiformes (fal′ko-ni-for′meez) (L. *falco,* falcon, + form): **eagles, hawks, vultures, falcons, condors, buzzards.** Diurnal birds of prey. All are strong fliers with keen vision. About 270 species, worldwide distribution.

Order Galliformes (gal′li-for′meez) (L. *gallus,* cock, + form): **quail, grouse, pheasants, ptarmigan, turkeys, domestic fowl.** Chickenlike ground-nesting herbivores with strong beaks and heavy feet. The bobwhite quail (*Colinus virginianus*) is found all over the eastern half of the United States. The ruffed grouse (*Bonasa umbellus*) is found in about the same region, but in the woods instead of the open pastures and grain fields, which the bobwhite frequents. About 250 species, worldwide distribution.

Order Gruiformes (groo′i-for′meez) (L. *grus,* crane, + form): **cranes, rails, coots, gallinules.** Prairie and marsh breeders. About 215 species, worldwide distribution.

Order Charadriiformes (ka-rad′ree-i-for′meez) (N.L. *Charadrius,* genus of plovers, + form): **gulls** (Figure 30-32), **oyster catchers, plovers, sandpipers, terns, woodcocks, turnstones, lapwings, snipe, avocets, phalaropes, skuas, skimmers, auks, puffins.** All are shorebirds. They are strong fliers and are usually colonial. About 330 species, worldwide distribution.

Order Columbiformes (co-lum'bi-for'meez) (L. *columba,* dove, + form): **pigeons, doves.** All have short necks, short legs, and a short, slender bill. About 290 species, worldwide distribution.

Order Psittaciformes (sit'ta-si-for'meez) (L. *psittacus,* parrot, + form): **parrots, parakeets.** Birds with hinged and movable upper beak, fleshy tongue. About 320 species, pantropical distribution.

Order Cuculiformes (ku-koo'li-for'meez) (L. *cuculus,* cuckoo, + form): **cuckoos, roadrunners.** The common European cuckoo (*Cuculus canorus*) lays its eggs in the nests of smaller birds, which rear the young cuckoos. The American cuckoos, black billed and yellow billed, usually rear their own young. About 150 species, worldwide distribution.

Order Strigiformes (strij'i-for'meez) (L. *strix,* screech owl, + form): **owls.** Nocturnal predators with large eyes, powerful beaks and feet, and silent flight. About 135 species, worldwide distribution.

Order Caprimulgiformes (kap'ri-mul'ji-for'meez) (L. *caprimulgus,* goatsucker, + form): **goatsuckers, nighthawks, whippoorwills.** Night and twilight feeders with small, weak legs and wide mouths fringed with bristles. The whippoorwills (*Antrostomus vociferus*) are common in the woods of the eastern states, and the nighthawk (*Chordeiles minor*) is often seen and heard in the evening flying around city buildings. About 95 species, worldwide distribution.

Order Apodiformes (up-pod'i-for'meez) (Gr. *apous,* footless, + form): **swifts, hummingbirds.** These are small birds with short legs and rapid wingbeat. The familiar chimney swift (*Chaetura pelagia*) fastens its nest in chimneys by means of saliva. A swift found in China builds a nest of saliva that is used by the Chinese for soup making. Most species of hummingbirds are found in the tropics, but there are 14 species in the United States, of which only one, the ruby-throated hummingbird, is found in the eastern part of the country. About 400 species, worldwide distribution.

Order Coliiformes (ka-ly'i-for'meez) (Gr. *kolios,* green woodpecker, + form): **mousebirds.** Small birds of uncertain relationship. Six species restricted to southern Africa.

Order Trogoniformes (tro-gon'i-for'meez) (Gr. *trōgon,* gnawing, + form): **trogons.** Richly colored, long-tailed birds. About 35 species, pantropical distribution.

Order Coraciiformes (ka-ray'see-i-for'meez or kor'uh-sigh'uh-for'meez) (N.L. *coracii* from Gr. *korakias,* a kind of raven, + form): **kingfishers, hornbills, and so on.** Birds with strong, prominent bills that nest in cavities. In the eastern half of the United States, the belted kingfisher (*Megaceryle alcyon*) is common along most waterways of any size. About 200 species, worldwide distribution.

Order Piciformes (pis'i-for'meez) (L. *picus,* woodpecker, + form): **woodpeckers, toucans, puffbirds, honeyguides.** Birds with highly specialized bills and having two toes extending forward and two backward. All nest in cavities. There are many species of woodpeckers in North America, the most common of which are the flickers and the downy, hairy, red-bellied, redheaded, and yellow-bellied woodpeckers. The largest is the pileated woodpecker, which is usually found in deep and remote woods. Worldwide distribution.

Order Passeriformes (pas'er-i-for'meez) (L. *passer,* sparrow, + form): **perching songbirds** (Figure 30-33). This is the largest order of birds, containing 56 families and 60% of all birds. Most have a highly developed syrinx. Their feet are adapted for perching on thin stems and twigs. The young are altricial. To this order belong many birds with beautiful songs such as the thrushes, warblers, mockingbird, meadowlark, and hosts of others. Others of this order, such as the swallows, magpie, starling, crows, raven, jays, nuthatch, and creeper, have no songs worthy of the name. More than 5000 species, worldwide distribution.

Figure 30-33
Ground finch *Geospiza fuliginosa,* one of the famous Darwin's finches of the Galápagos Islands. Order Passeriformes.

Summary

The more than 9600 species of living birds are egg-laying, endothermic vertebrates covered with feathers and having the forelimbs modified as wings. Birds are closest phylogenetically to the theropods, a group of Mesozoic dinosaurs with several birdlike characteristics. The oldest known fossil bird, *Archaeopteryx* from the Jurassic period of the Mesozoic era, had numerous reptilian characteristics and was almost identical to certain theropod dinosaurs except that it had feathers. It is probably not in the direct lineage leading to modern birds but can be considered a sister group to modern birds.

The adaptations of birds for flight are of two basic kinds: those reducing body weight and those promoting more power for flight. Feathers, the hallmark of birds, are complex derivatives of reptilian scales and combine lightness with strength, water repellency, and high insulative value. Body weight is further reduced by elimination of some bones, fusion of others (to provide rigidity for flight), and the presence in many bones of hollow, air-filled spaces. The light, horny bill, replacing the heavy jaws and teeth of reptiles, serves as both hand and mouth for all birds and is variously adapted for different feeding habits.

Adaptations that provide power for flight include a high metabolic rate and body temperature coupled with an energy-rich diet; a highly efficient respiratory system consisting of a system of air sacs arranged to pass air through the lungs during both inspiration and expiration; powerful flight and leg muscles arranged to place muscle weight near the bird's center of gravity; and an efficient, high-pressure circulation.

Birds have keen eyesight, good hearing, poorly developed sense of smell, and superb coordination for flight. The metanephric kidneys produce uric acid as the principal nitrogenous waste.

Birds fly by applying the same aerodynamic principles as an airplane and using similar equipment: wings for lift and support, a tail for steering and landing control, and wing slots for control at low flight speed. Flightlessness in birds is unusual but has evolved independently in several bird orders, usually on islands where terrestrial predators are absent; all are derived from flying ancestors.

Bird migration refers to regular movements between summer nesting places and wintering regions. Spring migration to the north, where more food is available for nestlings, enhances reproductive success. Many cues are used for direction finding in migration, including innate sense of direction and ability to navigate by the sun, the stars, or the earth's magnetic field.

The highly developed social behavior of birds is manifested in vivid courtship displays, mate selection, territorial behavior, and incubation of eggs and care of the young.

Review Questions

1. Explain the significance of the discovery of *Archaeopteryx*. Why did this fossil prove beyond reasonable doubt that birds share an ancestor with some reptilian groups?
2. Birds are broadly divided into two groups: ratite and carinate. Explain what these terms mean and briefly discuss the appearance of flightlessness in birds.
3. The special adaptations of birds all contribute to two essentials for flight: more power and less weight. Explain how each of the following contributes to one or both of these two essentials: feathers, skeleton, muscle distribution, digestive system, circulatory system, respiratory system, excretory system, reproductive system.
4. How do marine birds rid themselves of excess salt?
5. In what ways are the bird's ears and eyes specialized for the demands of flight?
6. Explain how the bird wing is designed to provide lift. What design features help to prevent stalling at low flight speeds?
7. Describe the four basic forms of bird wings. How does wing shape correlate with bird size and nature of flight (whether powered or soaring)?
8. What are the advantages of seasonal migration for birds?
9. Describe the different navigational resources birds may use in long-distance migration.
10. What are some of the advantages of social aggregation among birds?
11. More than 90% of all bird species are monogamous. Explain why monogamy is so much more common among birds than among mammals.
12. Briefly describe an example of polygyny among birds.
13. Define the terms precocial and altricial as they relate to birds.
14. Offer some examples of how human activities have harmed birds.

Selected References

See also general references for Part III, p. 626.

Burton, R. 1985. Bird behavior. New York, Alfred A. Knopf, Inc. *Well-written and well-illustrated summary of bird behavior.*

Emlen, S. T. 1975. The stellar-orientation system of a migratory bird. Sci. Am. **233:**102-111 (Aug.). *Describes fascinating research with indigo buntings, revealing their ability to navigate by the center of celestial rotation at night.*

Feduccia, A. 1980. The age of birds. Cambridge, Massachusetts, Harvard University Press. *Semipopular but authoritative account of bird evolution. Excellent text and illustrations.*

Norbert, U. M. 1990. Vertebrate flight. New York, Springer-Verlag. *Detailed review of the mechanics, physiology, morphology, ecology, and evolution of flight. Covers bats as well as birds.*

Proctor, N. S., and P. J. Lynch. 1993. Manual of ornithology: avian structure and function. New Haven, Connecticut, Yale University Press.

Sibley, C. G., and J. E. Ahlquist. 1990. Phylogeny and classification of birds: a study in molecular evolution. New Haven, Yale University Press. *A comprehensive application of DNA annealing experiments to the problem of resolving avian phylogeny.*

Terborgh, J. 1992. Why American songbirds are vanishing. Sci. Am. **266:**98–104 (May). *The number of songbirds in the United States has been dropping sharply. The author suggests the reasons why.*

Terres, J. K. 1980. The Audubon Society encyclopedia of North American birds. New York, Alfred A. Knopf, Inc. *Comprehensive, authoritative, and richly illustrated.*

Waldvogel, J. A. 1990. The bird's eye view. Am. Sci. **78:**342–353 (July–Aug.). *Birds possess visual abilities unmatched by humans. So how can we know what they really see?*

Wellnhofer, P. 1990. Archaeopteryx. Sci. Am., **262:**70–77 (May). *Description of perhaps the most important fossil ever discovered.*

Welty, J. C. and L. Baptista. 1988. The life of birds, ed. 4. Philadelphia, Saunders College Publishing. *Among the best of the ornithology texts; lucid style and well illustrated.*

31

The Mammals

Phylum Chordata
Class Mammalia

The Tell-Tale Hair

If Fuzzy Wuzzy, the bear that had no hair (according to the children's rhyme), was truly hairless, he could not have been a mammal or a bear. For hair is as much an unmistakable characteristic of mammals as feathers are of birds. If an animal has hair it is a mammal; if it lacks hair it must be something else. It is true that many aquatic mammals are nearly hairless (whales, for example) but it can usually be found (with a bit of searching) at least in vestigial form somewhere on the body of the adult. Unlike feathers, which evolved from converted reptilian scales, mammalian hair is a completely new epidermal structure. Mammals use their hair for protection from the elements, for protective coloration and concealment, for waterproofing and buoyancy, and for behavioral signaling; they have turned the hairs into sensitive vibrissae on their

snouts and into prickly quills. Perhaps most important of all, mammals use their hair for thermal insulation, which allows them to enjoy the great advantages of homeothermy. Warm-blooded animals in most climates and at sunless times benefit from this natural and controllable protective insulation.

Hair, of course, is only one of several features that together characterize a mammal and help us to understand the mammalian evolutionary achievement. Among these are a highly developed placenta for feeding the embryo; mammary glands for nourishing the newborn; and a surpassingly advanced nervous system that far exceeds in performance that of any other animal group. It is doubtful, however, that even with this winning combination of adaptations, the mammals could have triumphed as they have without their hair. ■

POSITION IN THE ANIMAL KINGDOM

The modern mammals are descendants of the synapsid lineage of amniotes that appeared in the Permian period. The synapsid lineage is characterized in the primitive condition by having a skull with a single temporal opening. Modern mammals are endothermic and homeothermic, have bodies partially or wholly covered with hair, and have mammary glands that secrete milk for the nourishment of the young. These derived characteristics, together with several distinctive skeletal characteristics, a highly developed nervous system, and complex individual and social behavior, distinguish the mammals from all other amniotes. Their genetic plasticity and numerous derived adaptations have enabled mammals to invade almost every environment on earth that supports life.

BIOLOGICAL CONTRIBUTIONS

1. Mammals share with birds both **endothermy** and **homeothermy** which permit a high level of activity at night, and year-round penetration into low temperature habitats denied to ectothermic vertebrates.
2. The **placenta** in placental mammals allows developing young to feed and grow in a protected environment

during the most vulnerable period of their lives. After birth the young continue to feed by suckling from **mammary glands.** A long period of parental care and education allows the young to acquire skills needed for survival.

3. The **specialization of mammalian teeth** for different functions permitted the evolution of many different feeding specializations in mammals. The **secondary palate,** which separates the air passage from the food passage, enables mammals to hold and partially break down food in their mouths without interrupting breathing.
4. The highly evolved brain, especially the large **neocortex,** has bequeathed mammals with a well-developed memory and the capacity to learn rapidly and to respond appropriately to problems they had not previously encountered. The highly elaborated **sense organs** and **special senses,** particularly those of hearing, smell, and touch, contribute an inflow of environmental information that, together with their processing brain centers, provide mammals with a level of environmental awareness and responsiveness unequaled in the animal kingdom.

Mammals, with their highly developed nervous system and numerous ingenious adaptations, occupy almost every environment on earth that supports life. Although not a large group (about 4450 species as compared with more than 9000 species of birds, approximately 24,600 species of fishes, and 800,000 species of insects), the class Mammalia (mam-may′lee-a) (L. *mamma,* breast) is overall the most biologically differentiated group in the animal kingdom. Many potentialities that dwell more or less latently in other vertebrates are highly developed in mammals. Mammals are exceedingly diverse in size, shape, form, and function. They range in size from the recently discovered Kitti's hognosed bat, weighing only 1.5 g, to the whales, some of which exceed 100 tons.

Yet, despite their adaptability and in some instances because of it, mam-

mals have been influenced by the presence of humans more than any other group of animals. We have domesticated numerous mammals for food and clothing, as beasts of burden, and as pets. We use millions of mammals each year in biomedical research. We have introduced alien mammals into new habitats, occasionally with benign results but more frequently with unexpected disaster. Although history provides us with numerous warnings, we continue to overcrop valuable wild stocks of mammals. The whale industry has threatened itself with total collapse by exterminating its own resource—a classic example of self-destruction in the modern world, in which competing segments of an industry are intent only on reaping all they can today as though tomorrow's supply were of no concern whatever. In some cases de-

struction of a valuable mammalian resource has been deliberate, such as the officially sanctioned (and tragically successful) policy during the Indian wars of exterminating the bison to drive the Plains Indians into starvation. Although commercial hunting has declined, the ever-increasing human population with the accompanying destruction of wild habitats has harassed and disfigured the mammalian fauna. Approximately 300 species and subspecies of mammals are considered endangered by the International Union for the Conservation of Nature and Natural Resources (IUCN), including most cetaceans, cats (except domestic cats), otters, and primates (except humans).

An international moratorium on all commercial whaling took effect in 1986. However some countries that objected to the moratorium, notably Japan, are still killing hundreds of whales each year under the guise of "scientific" whaling.

We are becoming increasingly aware that our presence on this planet as the most powerful product of organic evolution makes us responsible for the character of our natural environment. Since our welfare has been and continues to be closely related to that of the other mammals, it is clearly in our interest to preserve the natural environment of which all mammals, ourselves included, are a part. We need to remember that nature can do without us but we cannot exist without nature.

ORIGIN AND EVOLUTION OF MAMMALS

The evolutionary descent of mammals from their earliest amniote ancestors is perhaps the most fully documented transition in vertebrate history. From the fossil record, we can trace the derivation over 150 million years of endothermic, furry mammals from their small, ectothermic, hairless ancestors. Skull structures and especially the teeth are the most abundant fossils, and it is largely from these structures that we can identify the evolutionary descent of the mammals.

The structure of the skull roof permits us to identify three major groups of amniotes that diverged in the Carboniferous period of the Paleozoic era, the **synapsids, anapsids, and diapsids** (p. 549). The synapsid group, which includes the mammals and their ancestors, has a pair of openings in the skull roof for the attachment of jaw muscles (Figure 31-2). This was the first amniote lineage to radiate widely into terrestrial habitats. The anapsid group is characterized by solid skulls and includes the turtles and their ancestors. The diapsids have two pairs of openings in the skull roof (Figure 29-2, p. 551) and this group contains the dinosaurs, lizards, snakes, crocodilians, birds, and their ancestors.

The earliest synapsids radiated extensively into diverse herbivorous and carnivorous forms that are often collectively called **pelycosaurs** (Figures 31-1 and 31-2). These early synapsids were the most common amniotes of the early Permian. The pelycosaurs share a general outward resemblance to lizards, but this resemblance is misleading. The pelycosaurs are not closely related to lizards, which are diapsids, nor are they a monophyletic group. One early synapsid group, the carnivorous sphenacodontines, is evolutionarily most closely related to the **therapsids** (Figure 31-2), the only synapsid group that survived beyond the Paleozoic. In the therapsids we see for the first time an efficient erect gait with upright limbs positioned beneath the body. Since stability was reduced by raising the animal from the ground, the muscular coordination center of the brain, the cerebellum, took on an expanded role. The therapsids radiated into numerous herbivorous and carnivorous forms but most disappeared during the great extinction event at the end of the Permian.

Only the last therapsid subgroup to evolve, the **cynodonts,** survived to enter the Mesozoic. The cynodonts evolved several novel features including a high metabolic rate, which supported a more active life; increased jaw musculature, permitting a stronger bite; several skeletal changes, supporting greater agility; and a secondary bony palate (labeled hard palate in Figure 31-3), enabling the animal to breathe while holding prey in its mouth. The secondary palate would be important to subsequent mammalian evolution by permitting the young to breathe while suckling. Along with the improved biomechanical shift to upright posture in cynodonts, the long bones became more slender and they developed bony processes at the joints for firmer muscle attachment. The number of ribs was reduced, a change that probably improved flexibility of the spinal column. Within the diverse cynodont clade, a small carnivorous group called tritheledontids (Figure 31-2) most closely resembles the mammals, sharing with them several derived features of the skull and teeth.

The earliest mammals of the late Triassic were small mouse- or shrew-sized animals with enlarged cranium, jaws redesigned for a shearing action, and a new type of dentition in which the teeth were replaced only once (deciduous and permanent teeth). This contrasts with the primitive amniote pattern of continual tooth replacement throughout life. The earliest mammals were almost certainly endothermic, although their body temperature would have been rather lower than modern placental mammals. Hair was essential for insulation, and the presence of hair implies that sebaceous and sweat glands must also have evolved at this time to lubricate

CHARACTERISTICS OF CLASS MAMMALIA

1. **Body covered with hair,** but reduced in some
2. **Integument** with **sweat, scent, sebaceous,** and **mammary glands**
3. Skeletal features: skull with **two occipital condyles** and **secondary bony palate;** middle ear with **three ossicles** (malleus, incus, stapes); **seven cervical vertebrae** (except some xenarthrans [edentates] and the manatee); **pelvic bones fused**
4. Mouth with **diphyodont teeth** (milk, or deciduous, teeth replaced by a permanent set of teeth); teeth heterodont in most (varying in structure and function); lower jaw a **single enlarged bone (dentary)**
5. **Movable eyelids** and **fleshy external ears (pinnae)**
6. Four limbs (reduced or absent in some) adapted for many forms of locomotion
7. Circulatory system of a four-chambered heart, **persistent left aorta,** and **nonnucleated, biconcave red blood corpuscles**
8. Respiratory system of lungs with alveoli, and voice box (larynx); **secondary palate** (anterior bony palate and posterior continuation of soft tissue, the soft palate) separates air and food passages (Figure 31-3); **muscular diaphragm** for air exchange separates thoracic and abdominal cavities
9. Excretory system of metanephros kidneys and ureters that usually open into a bladder
10. Brain highly developed, especially **neocerebrum;** 12 pairs of cranial nerves
11. Endothermic and homeothermic
12. Cloaca present only in monotremes (present as shallow cloaca in marsupials)
13. Separate sexes; reproductive organs of a penis, testes (usually in a scrotum), ovaries, oviducts, and vagina
14. Internal fertilization; **eggs develop in a uterus** with **placental attachment** (placenta rudimentary in marsupials and absent in monotremes); **fetal membranes (amnion, chorion, allantois);** sex determination by males (heterogametic)
15. Young nourished by **milk from mammary glands**

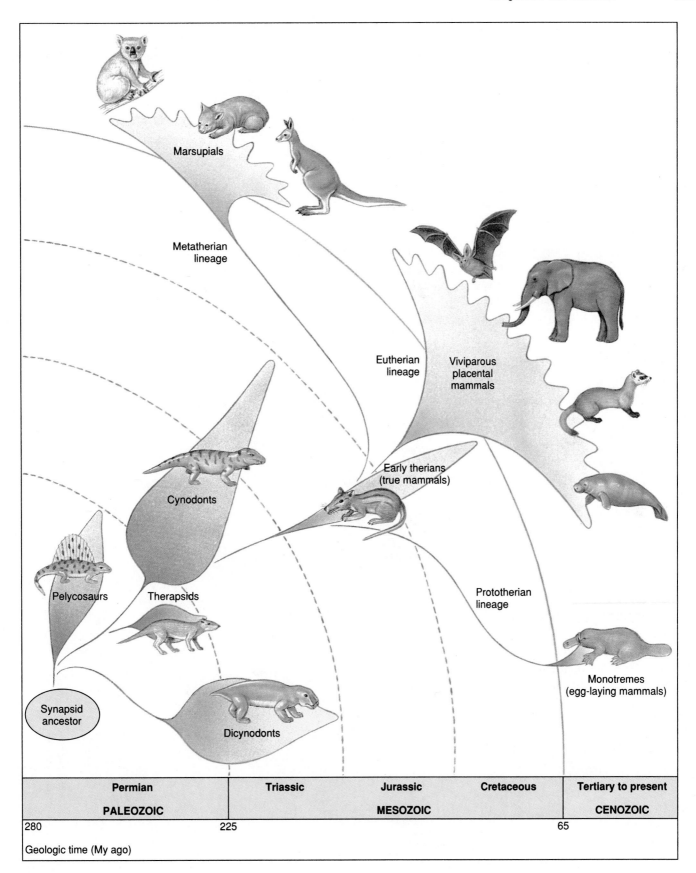

Figure 31-1

Evolution of the major groups of synapsids. The synapsid lineage, characterized by a lateral temporal opening, began with the pelycosaurs, early mammal-like amniotes of the Permian. The pelycosaurs radiated extensively and evolved changes in the jaws, teeth, and body form that presaged several mammalian characteristics. These trends continued in their successors, the therapsids, especially in the cynodonts. One lineage of cynodonts gave rise in the Triassic to the therians, the true mammals. Fossil evidence, as currently interpreted, indicates that all three groups of living mammals—monotremes, marsupials, and placentals—are derived from the same lineage. The great radiation of modern placental orders occurred during the Cretaceous and Tertiary periods.

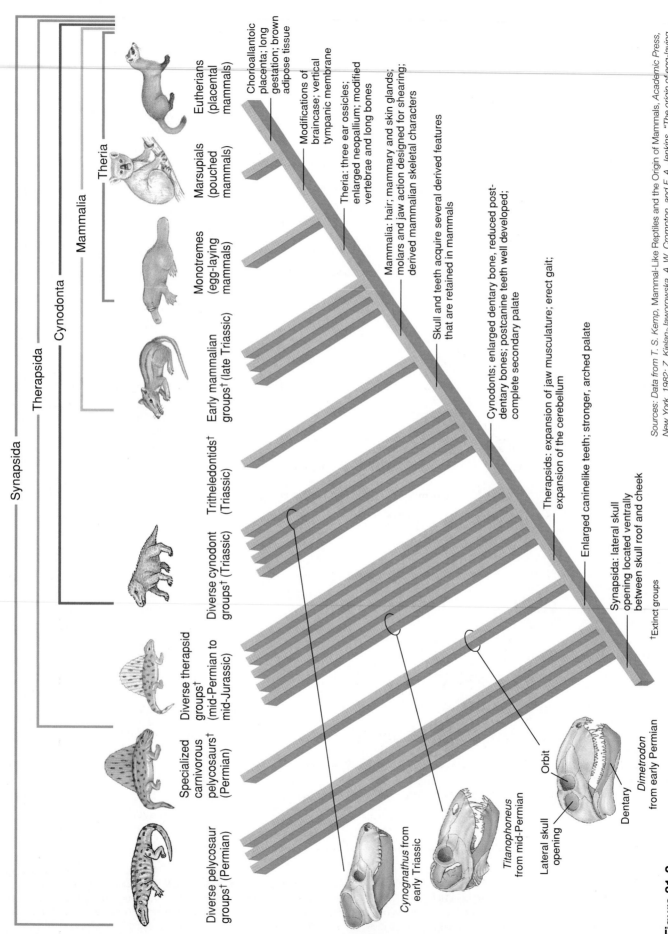

Figure 31-2

Abbreviated cladogram of the synapsids emphasizing the origins of important characteristics of the mammals (shown to the right of the cladogram). Extinct groups are indicated by a dagger. The skulls show the progressive increase in size of the dentary relative to other bones in the lower jaw.

Synapsida

Therapsida

Cynodonta

Mammalia

Theria

Eutherians (placental mammals)

Marsupials (pouched mammals)

Monotremes (egg-laying mammals)

Early mammalian groups† (late Triassic)

Tritheledontids† (Triassic)

Diverse cynodont groups† (Triassic)

Diverse therapsid groups† (mid-Permian to mid-Jurassic)

Specialized carnivorous pelycosaurs† (Permian)

Diverse pelycosaur groups† (Permian)

Chorioallantoic placenta; long gestation; brown adipose tissue

Modifications of braincase; vertical tympanic membrane

Theria: three ear ossicles; enlarged neopallium; modified vertebrae and long bones

Mammalia: hair; mammary and skin glands; molars and jaw action designed for shearing; derived mammalian skeletal characters

Skull and teeth acquire several derived features that are retained in mammals

Cynodonts; enlarged dentary bone, reduced post-dentary bones; postcanine teeth well developed; complete secondary palate

Therapsids: expansion of jaw musculature; erect gait; expansion of the cerebellum

Enlarged caninelike teeth; stronger, arched palate

Synapsida: lateral skull opening located ventrally between skull roof and cheek

†Extinct groups

Cynognathus from early Triassic

Titanophoneus from mid-Permian

Orbit

Dimetrodon from early Permian

Lateral skull opening

Dentary

Sources: Data from T. S. Kemp, Mammal-Like Reptiles and the Origin of Mammals, Academic Press, New York, 1982; Z. Kielan-Jaworowska, A. W. Crompton, and F. A. Jenkins, "The origin of egg-laying mammals" in Nature 326:871–873 (1987); J. Gauthier, A. G. Kluge, and T. Rowe, "Amniote phylogeny and the importance of fossils" in Cladistics 4:105–209 (1988); R. L. Carroll, Vertebrate Paleontology and Evolution, W. H. Freeman, New York, 1988; and F. H. Pough, J. B. Heiser, and W. N. McFarland, Vertebrate Life, 3d edition, Macmillan, New York, 1989.

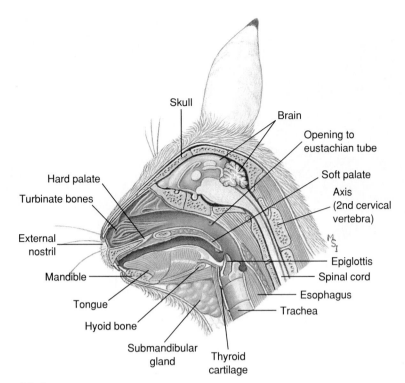

Figure 31-3
Sagittal section through the head of a rabbit.

Labels (clockwise): Skull, Brain, Opening to eustachian tube, Soft palate, Axis (2nd cervical vertebra), Epiglottis, Spinal cord, Esophagus, Trachea, Thyroid cartilage, Submandibular gland, Hyoid bone, Tongue, Mandible, External nostril, Turbinate bones, Hard palate

the hair and control heat loss. The fossil record is silent on the appearance of mammary glands, but they must have evolved before the end of the Triassic. The young of the early mammals probably hatched from eggs in a very immature condition, totally dependent on maternal milk, warmth, and protection. This mode of reproduction occurs today only in the monotremes (echidnas and platypus).

Oddly, the early mammals of the mid-Triassic, having developed nearly all of the novel attributes of modern mammals, had to wait for another 150 million years before they could achieve their great diversity. In the meantime the dinosaurs became diverse and abundant, while all nonmammalian synapsid groups became extinct. But mammals survived, first as shrewlike, probably nocturnal, creatures. Then, in the Cretaceous, especially during the Eocene epoch that began about 54 million years ago, the modern mammals began to expand rapidly. The great Cenozoic radiation of mammals is partly attributed to the numerous environments vacated by the many amniote groups that became extinct at the end of the Cretaceous. The mammalian radiation was almost certainly promoted

by the fact that mammals were agile, endothermic, intelligent, adaptable, and gave birth to living young, which they protected and nourished from their own milk supply, thus dispensing with vulnerable eggs laid in nests.

The class Mammalia includes 21 orders: one order containing the monotremes, one order containing the marsupials, and 19 orders of placentals. A complete classification is on pp. 619 to 623.

STRUCTURAL AND FUNCTIONAL ADAPTATIONS OF MAMMALS

INTEGUMENT AND ITS DERIVATIVES

The mammalian skin and its modifications especially distinguish mammals as a group. As the interface between the animal and its environment, the skin is strongly molded by the animal's way of life. In general the skin is thicker in mammals than in other classes of vertebrates, although as in all vertebrates it is made up of **epidermis** and **dermis** (see Figure 32-1B,

p. 632). Among the mammals the dermis becomes much thicker than the epidermis. The epidermis is relatively thin where it is well protected by hair, but in places that are subject to much contact and use, such as the palms or soles, its outer layers become thick and cornified with keratin.

Hair

Hair is especially characteristic of mammals, although humans are not very hairy creatures, and in whales hair is reduced to only a few sensory bristles on the snout. A hair grows out of a hair follicle that, although an epidermal structure, is sunk into the dermis of the skin (see Figure 32-1B, p. 632). The hair grows continuously by rapid proliferation of cells in the follicle. As the hair shaft is pushed upward, new cells are carried away from their source of nourishment and die, turning into the same dense type of fibrous protein, called **keratin,** that constitutes nails, claws, hooves, and feathers.

Mammals characteristically have two kinds of hair forming the **pelage** (fur coat): (1) dense and soft **underhair** for insulation and (2) coarse and longer **guard hair** for protection against wear and to provide coloration. The underhair traps a layer of insulating air. In aquatic animals, such as the fur seal, otter, and beaver, it is so dense that it is almost impossible to wet. In water the guard hairs become wet and mat down, forming a protective blanket over the underhair (Figure 31-4).

A hair is more than a strand of keratin. It consists of three layers: the medulla or pith in the center of the hair, the cortex with pigment granules next to the medulla, and the outer cuticle composed of imbricated scales. The hair of different mammals shows a considerable range of structure. It may be deficient in cortex, such as the brittle hair of deer, or it may be deficient in medulla, such as the hollow, air-filled hairs of the wolverine. The hairs of rabbits and some others are scaled to interlock when pressed together. Curly hair, such as that of sheep, grows from curved follicles.

Figure 31-4

American beaver, *Castor canadensis,* about to cut down an aspen tree. This second largest rodent (the South American capybara is larger) has a heavy waterproof pelage consisting of long, tough guard hairs overlying the thick, silky underhair so valued in the fur trade. Order Rodentia, family Castoridae.

A

B

Figure 31-5

Snowshoe, or varying, hare, *Lepus americanus* in **A,** brown summer coat and, **B,** white winter coat. In winter, extra hair growth on the hind feet broadens the animal's support in snow. Snowshoe hares are common residents of the taiga (northern coniferous forests) and are an important food for lynxes, foxes, and other carnivores. Population fluctuations of hares and their predators are closely related. Order Lagomorpha.

When a hair reaches a certain length, it stops growing. Normally it remains in the follicle until a new growth starts, whereupon it falls out. In humans, hair is shed and replaced throughout life. But in most mammals, there are periodic molts of the entire coat.

In the simplest cases, such as foxes and seals, the coat is shed once each year during the summer months. Most mammals have two annual molts, one in the spring and one in the fall. The summer coat is always much thinner than the winter coat and in some it may be a different color. Several of the northern mustelid carnivores, such as the weasel, have white winter coats and brown-colored summer coats. It was once believed that the white inner pelage of arctic animals conserves body heat by reducing radiation loss; in fact, dark and white pelages radiate heat equally well. The winter white of arctic animals is simply camouflage in a land of snow. The varying hare of North America has three annual molts: the white winter coat is replaced by a brownish gray summer coat, and this is replaced in autumn by a grayer coat, which is soon shed to reveal the winter white coat beneath (Figure 31-5). The white fur of arctic mammals in winter (leukemism) is not to be con-

fused with albinism, caused by a recessive gene that blocks pigment formation. Albinos have red eyes and pinkish skin, whereas arctic animals in their winter coats have dark eyes and often dark-colored ear tips, noses, and tail tips.

Outside the Arctic, most mammals wear somber colors that are protective. Often the species is marked with "salt-and-pepper" coloration or a disruptive pattern that helps make it inconspicuous in its natural surroundings. Examples are the spots of leopards and fawns and the stripes of tigers. Other mammals, such as skunks, advertise their presence with conspicuous warning coloration.

The hair of mammals has become modified to serve many purposes. The bristles of hogs, the spines of porcupines and their kin, and the vibrissae on the snouts of most mammals are examples. **Vibrissae,** commonly called "whiskers," are really sensory hairs that provide a tactile sense to many mammals. The slightest movement of a vibrissa generates impulses in sensory nerve endings that travel to special sensory areas in the brain. The vibrissae are especially long in nocturnal and burrowing animals.

Porcupines, hedgehogs, echidnas, and a few other mammals have devel-

Figure 31-6

Dogs are frequent victims of the porcupine's impressive armor. Unless removed (usually by a veterinarian) the quills will continue to work their way deeper in the flesh causing great distress and may lead to the victim's death.

oped an effective and dangerous spiny armor. The spines of the common North American porcupine break off at the bases when struck and, aided by backward-pointing hooks on the tips, work deeply into their victims. To assist slow learners, such as dogs, in understanding what they are dealing with, porcupines rattle the spines and prominently display the white markings on the quills toward their tormentors (Figure 31-6).

Horns and Antlers

Three kinds of horns or hornlike substances are found in mammals. **True horns,** found in **ruminants** (for example, cud-chewers such as sheep and cattle), are hollow sheaths of keratinized epidermis that embrace a core of bone arising from the skull. Horns are not normally shed, usually are not branched (although they may be greatly curved), and are found in both sexes (except pronghorn antelope in which they occur only in the male).

Antlers of the deer family are entirely bone when mature. During their annual growth, antlers develop beneath a covering of highly vascular soft skin called **"velvet"** (Figure 31-7). When growth of the antlers is complete just before the breeding season, the blood vessels constrict and the stag tears off the velvet by rubbing the antlers against trees. The antlers are dropped after the breeding season. New buds appear a few months later to herald the next set of antlers. For several years each new pair of antlers is larger and more elaborate than the previous set. The annual growth of antlers places a strain on the mineral metabolism, since during the growing season a large moose or elk must accumulate 50 or more pounds of calcium salts from its vegetable diet.

The **rhinoceros horn** is the third kind of horn. Hairlike keratinized filaments that arise from dermal papillae are cemented together to form a single horn.

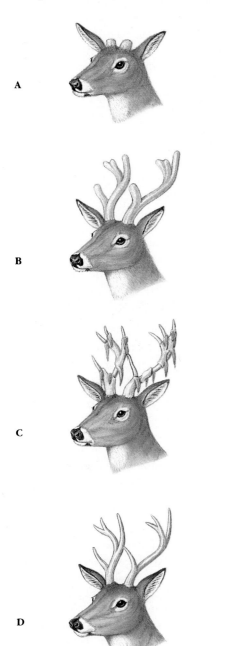

Figure 31-7

Annual growth of buck deer antlers. **A,** Antlers begin growth in late spring, stimulated by pituitary gonadotropins. **B,** The bone grows very rapidly until halted by a rapid rise in testosterone production by the testes. **C,** The skin (velvet) dies and sloughs off. **D,** Testosterone levels peak during the fall breeding season. The antlers are shed in January as testosterone levels subside.

The escalating trade in rhino products—especially rhino horn— during the last three decades, is pushing Asian and African rhinos to the brink of extinction. Rhino horn is valued in China as an agent for reducing fever, and for treating heart, liver, and skin diseases; and in North India as an aphrodisiac. Such supposed medicinal values are totally without pharmacological basis. The principal use of rhino horns, however, is to fashion handles for daggers in the Middle East. Because of their phallic shape, rhino horn daggers are traditional gifts at puberty rites. Between 1969 and 1977, horns from 8000 slaughtered rhinos were imported into North Yemen alone.

Glands

Of all vertebrates, mammals have the greatest variety of integumentary glands. Most fall into one of four classes: sweat, scent, sebaceous, and mammary. All are derivatives of the epidermis.

Sweat glands are simple, tubular, highly coiled glands that occur over much of the body in most mammals. They are not present in other vertebrates. Two kinds of sweat glands may be distinguished: eccrine and apocrine (Figure 31-8). **Eccrine glands** secrete a watery sweat that, when evaporated on the skin's surface, draws heat from the skin and cools it. They occur in hairless regions, especially the foot pads, in most mammals, although in horses and most primates they are scattered all over the body. They are much reduced or absent in rodents, rabbits, whales, and others. **Apocrine glands,** the second type of sweat glands, are larger than eccrine glands and have longer and more winding ducts. Their secretory coil is in the dermis and extends deep into the hypodermis. They always open into the follicle of a hair or where a hair has been. Apocrine glands develop approximately at sexual puberty and are restricted (in the human species) to the axillae (armpits), mons pubis, breasts, external auditory canals, prepuce, scrotum, and a few other places. Their secretion is not watery, like ordinary sweat (eccrine gland), but is a milky, whitish or yellow secretion that dries on the skin to form a plasticlike film. Apocrine glands are not involved in heat regulation, but their activity is correlated with certain aspects of the sex cycle, among other possible functions.

Scent glands are present in nearly all mammals. Their location and functions vary greatly. They are used in communication with members of the same species, to mark territorial boundaries, for warning, or for defense. Scent-producing glands are located in orbital, metatarsal, and interdigital regions (deer); behind the eyes and on the cheek (pica and woodchuck); penis (muskrats, beavers, and many canines); base of the tail (wolves

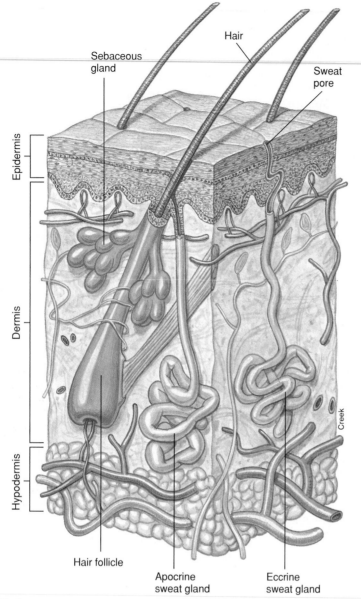

Sebaceous
gland

Hair

Sweat
pore

Epidermis

Dermis

Hypodermis

Hair follicle

Apocrine
sweat gland

Eccrine
sweat gland

Creek

Figure 31-8

Glands in human skin. Sebaceous glands produce sebum, which lubricates the hair and skin. Sweat glands are of two kinds. The more common eccrine glands secrete a watery sweat that cools the skin as it evaporates. Apocrine glands, more limited in body distribution, produce a milky secretion that is not involved in temperature regulation but may play a role in sexual attraction.

and foxes); back of the head (drome-dary); and anal region (skunks, minks, and weasels). The latter, the most odoriferous of all glands, open by ducts into the anus; their secretions can be discharged forcefully for several feet. During the mating season many mammals give off strong scents for attracting the opposite sex. Humans also are endowed with scent glands. But civilization has taught us to dislike our own scent, a concern that has stimulated a lucrative deodorant industry to produce an endless output of soaps and odor-masking compounds.

Sebaceous glands are intimately associated with hair follicles, although some are free and open directly onto the surface. The cellular lining of the gland itself is discharged in the secretory process and must be renewed for further secretion. These gland cells become distended with a fatty accumulation, then die, and are expelled as a greasy mixture called **sebum** into the hair follicle. Called a "polite fat" because it does not turn rancid, it serves as a dressing to keep the skin and hair pliable and glossy. Most mammals have sebaceous

glands all over the body; in humans they are most numerous in the scalp and on the face.

Mammary glands, which provide the name for mammals, are probably modified apocrine glands. Whatever their evolutionary origin, they occur on all female mammals and in a rudimentary form on all male mammals. They develop by the thickening of the epidermis to form a milk line along each side of the abdomen in the embryo. On certain parts of these lines the mammae appear while the intervening parts of the ridge disappear. In the human female the mammary glands begin to increase in size at puberty because of fat accumulation and reach their maximum development in approximately the twentieth year. The breasts (or mammae) undergo additional development during pregnancy. In other mammals the mammae are swollen only periodically when they are distended with milk during pregnancy and subsequent nursing of the young.

FOOD AND FEEDING

Mammals have exploited an enormous variety of food sources; some mammals require highly specialized diets, whereas others are opportunistic feeders that thrive on diversified diets. In all, food habits and physical structure are inextricably linked. A mammal's adaptations for attack and defense and its specializations for finding, capturing, chewing, swallowing, and digesting food all determine a mammal's shape and habits.

Teeth, perhaps more than any other single physical characteristic, reveal the life habit of a mammal (Figure 31-9). All mammals have teeth, except monotremes, anteaters, and certain whales, and their modifications are correlated with what the mammal eats.

As the mammals evolved during the Mesozoic, major changes occurred in the teeth and jaws. Unlike the uniform **homodont** dentition of the reptiles, mammalian teeth became differentiated to perform specialized functions such as cutting, seizing, gnawing, tearing, grinding, or chewing. Teeth differentiated in this manner are called **heterodont.** Typically, the mammalian

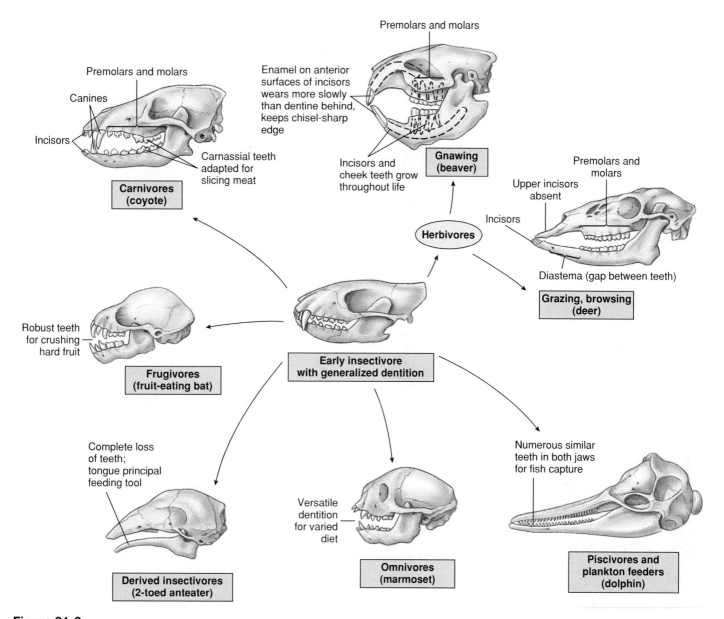

Figure 31-9
Feeding specializations of major trophic groups of eutherian mammals. The early eutherians were insectivores; all other types are descended from them.

dentition is differentiated into four types: **incisors,** with simple crowns and sharp edges, used mainly for snipping or biting; **canines,** with long conical crowns, specialized for piercing; **premolars,** with compressed crowns and one or two cusps, suited for shearing and slicing; and **molars,** with large bodies and variable cusp arrangement, used for crushing and grinding. The primitive tooth formula, which expresses the number of each tooth type in one-half of the upper and lower jaw, was I 3/3, C 1/1, PM 4/4, M 3/3. Members of the order Insectivora (for example, shrews), some omnivores, and carnivores come closest to this primitive pattern (Figure 31-9).

Unlike the reptiles, the mammals no longer continuously replace their teeth throughout their lives. Most mammals grow just two sets of teeth: a temporary set, referred to as **deciduous,** or **milk,** teeth, which is replaced by a permanent set when the skull has grown large enough to accommodate a full set. Only the incisors, canines, and premolars are deciduous; the molars are never replaced and the single permanent set must last a lifetime.

Feeding Specializations

The feeding, or trophic, apparatus of a mammal—the teeth and jaws, tongue, and alimentary canal—are adapted to its particular feeding habits. On the basis of food habits, mammals may be divided among several **trophic groups** (nutritional groups) as shown in Figure 31-9. The three basic trophic groups are insectivores, carnivores, and herbivores, but many other feeding specializations have evolved.

Insectivores are small mammals, usually opportunistic feeders, that feed on a variety of small invertebrates, such as worms and grubs, as well as insects. Examples are shrews, moles, anteaters, and most bats. Since insectivores eat little fibrous vegetable matter that requires prolonged fermentation, their intestinal tract tends to be short (Figure 31-10).

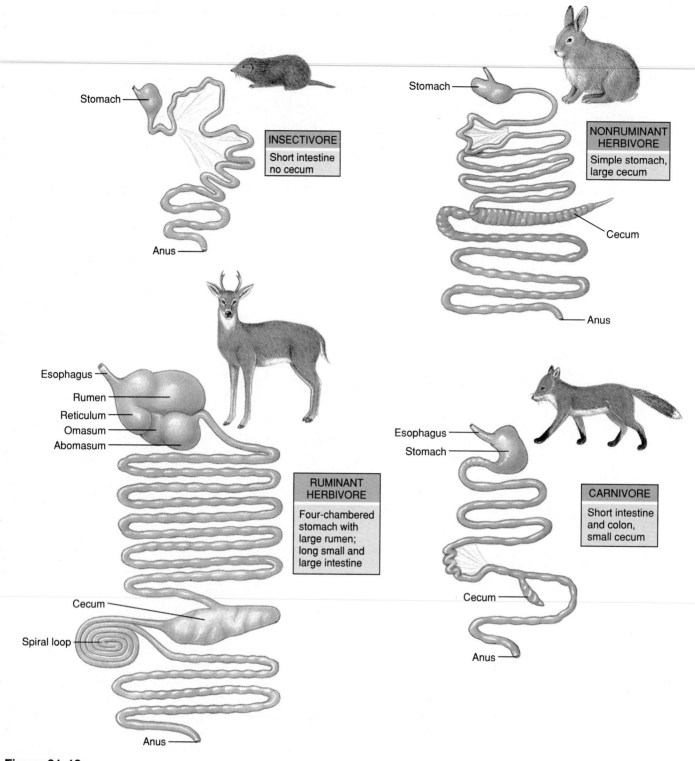

Figure 31-10
Digestive systems of mammals, showing different morphology with different diets.

The insectivorous category is not a sharply distinguished one because carnivores and omnivores often include insects in their diets. Even many rodents, which are considered herbivores, may have a mixed diet of insect larvae, seeds, and fruits.

Herbivorous mammals that feed on grasses and other vegetation form two main groups: the **browsers** and **grazers,** such as the ungulates (hooved mammals including horses, deer, antelope, cattle, sheep, and goats), and the **gnawers,** such as the rodents, and rabbits and hares. In herbivores, the canines are reduced in size or absent, whereas the molars, which are adapted for grinding, are broad and usually high-crowned. Rodents have chisel-sharp incisors that grow throughout life and must be worn away to keep pace with their continual growth (Figure 31-9).

Herbivorous mammals have a number of interesting adaptations for dealing with their fibrous diet of plant food. **Cellulose,** the structural carbohydrate of plants, is a potentially nutritious foodstuff, composed of long chains of glucose units. However, the glucose molecules in cellulose are linked by a type of chemical bond that few enzymes can attack. No vertebrates synthesize cellulose-splitting enzymes. Instead, the herbivorous vertebrates harbor anaerobic bacteria and protozoa in huge fermentation chambers in the gut. These microorganisms break down and metabolize the cellulose, releasing a variety of fatty acids, sugars, and starches that the host animal can absorb and use.

In some herbivores, such as horses and zebras, rabbits and hares, elephants, and many rodents, the gut has a spacious sidepocket, or diverticulum, called a **cecum,** which serves as a fermentation chamber and absorptive area (Figure 31-10). Hares, rabbits, and some rodents often eat their fecal pellets (**coprophagy**), giving the food a second pass through the fermenting action of the intestinal bacteria.

The **ruminants** (cattle, bison, buffalo, goats, antelopes, sheep, deer, giraffes, and okapis) have a huge **four-chambered stomach** (Figure 31-10).

Figure 31-11

Lionesses, *Panthera leo,* eating a wildebeest. Lions stalk prey and then charge suddenly to surprise the victim. They lack stamina for a long chase. Lions gorge themselves with the kill, then sleep and rest for periods as long as one week before eating again. Order Carnivora, family Felidae.

When a ruminant feeds, grass passes down the esophagus to the **rumen,** where it is broken down by bacteria and protozoa and then formed into small balls of cud. At its leisure the ruminant returns the cud to its mouth where the cud is deliberately chewed at length to crush the fiber. Swallowed again, the food returns to the rumen where it is digested by the cellulolytic bacteria and protozoa. The pulp passes to the **reticulum,** then to the **omasum,** where water, soluble food, and microbial products are absorbed. The remainder proceeds to the **abomasum** ("true" acid stomach), where proteolytic enzymes are secreted and normal digestion takes place.

Herbivores in general have large, long digestive tracts and must eat a considerable amount of plant food to survive. A large African elephant weighing 6 tons must consume 135 to 150 kg (300 to 400 pounds) of rough fodder each day to obtain sufficient nourishment for life.

Carnivorous mammals feed mainly on herbivores. This group includes foxes, dogs, weasels, wolverines, fishers, cats, lions, and tigers. Carnivores are well-equipped with biting and piercing teeth and powerful clawed limbs for killing their prey. Since their protein diet is much more easily digested than is the woody food of herbivores, their digestive tract is shorter and the cecum small or absent.

Carnivores organize their feeding into discrete meals rather than feeding continuously (as do most herbivores) and therefore have much more leisure time for play and exploration.

Note that the terms "insectivores" and "carnivores" have two different uses in mammals: to describe diet and to denote specific taxonomic orders of mammals. For example, not all carnivores belong to the order Carnivora (many marsupials, pinnipeds, and cetaceans are carnivorous) and not all members of the order Carnivora are carnivorous. Many are opportunistic feeders and some, such as the panda, are strict vegetarians.

In general, carnivores lead more active—and by human standards more interesting—lives than do the herbivores. Since a carnivore must find and catch its prey, there is a premium on intelligence; many carnivores, such as the cats, are noted for their stealth and cunning in hunting prey (Figure 31-11). This has led to a selection of herbivores capable either of defending themselves or of detecting and escaping carnivores. Thus for the herbivores, there has been a premium on keen senses and agility. Some herbivores, however, survive by virtue of their sheer size (for example, elephants) or by defensive group behavior (for example, muskoxen).

Humans have changed the rules in the carnivore-herbivore contest. Carnivores, despite their intelligence, have suffered much from human presence and have been virtually exterminated in some areas. Small herbivores, on the other hand, with their potent reproductive ability, have consistently defeated our most ingenious efforts to banish them from our environment. The problem of rodent pests in agriculture has intensified; we have removed carnivores, which served as the herbivores' natural population control, but have not been able to devise a suitable substitute.

Omnivorous mammals live on both plants and animals for food. Examples are pigs, raccoons, rats, bears, and most primates (including humans). Many carnivorous forms also eat fruits, berries, and grasses when hard pressed. The fox, which usually feeds on mice, small rodents, and birds, eats frozen apples, beechnuts, and corn when its normal food sources are scarce.

For most mammals, searching for food and eating occupy most of their active life. Seasonal changes in food supplies are considerable in temperate zones. Living may be easy in the summer when food is abundant, but in winter many carnivores must range far and wide to eke out a narrow existence. Some migrate to regions where food is more abundant. Others hibernate and sleep the winter months away.

Many mammals build up food stores during periods of plenty. This habit is most pronounced in rodents, such as squirrels, chipmunks, gophers, and certain mice. All tree squirrels—red, fox, and gray—collect nuts, conifer seeds, and fungi and bury these in caches for winter use. Often each item is hidden in a different place (scatter hoarding) and marked by a scent to assist relocation in the future. Some of the caches of the chipmunks and squirrels may exceed a bushel (Figure 31-12).

Body Weight and Food Consumption

The relationship between body size and metabolic rate was discussed in relation to food consumption of birds

Figure 31-12
Eastern chipmunk, *Tamias striatus,* with cheek pouches stuffed with seeds to be carried to a hidden cache. It will try to store at least a half-bushel of food for the winter. It hibernates but awakens periodically to eat some of its cached food. Order Rodentia, family Sciuridae.

(p. 579). The smaller the animal, the greater is its metabolic rate and the more it must eat relative to its body size. This happens because the metabolic rate of an animal—and therefore the amount of food it must eat to sustain this metabolic rate—varies in rough proportion to the relative surface area rather than to the body weight. Surface area is proportional to approximately 0.7 power of body weight, and the amount of food a mammal (or bird) eats also is roughly proportional to a 0.7 power of its body weight. A 3 g mouse will consume *per gram body weight* five times more food than does a 10 kg dog and about 30 times more food than does a 50,000 kg elephant. One can easily see why small mammals (shrews, bats, and mice) must spend much more time hunting and eating food than do large mammals. The smallest shrews weighing only 2 g may eat more than their body weight each day and will starve to death in a few hours if deprived of food (Figure 31-13). In contrast, a large carnivore can remain fat and healthy with only one meal every few days. The mountain lion is known to kill an average of one deer a week, although it will kill more frequently when game is abundant.

Figure 31-13
The shorttail shrew, *Blarina brevicauda,* eating a grasshopper. This tiny but fierce mammal, with a prodigious appetite for insects, mice, snails, and worms, spends most of its time underground and so is seldom seen by humans. Shrews are believed to resemble the insectivorous ancestors of placental mammals. Order Insectivora, family Soricidae.

MIGRATION

Migration is a much more difficult undertaking for mammals than for birds. Not surprisingly, few mammals make regular seasonal migrations, preferring instead to center their activities in a defined and limited home range. Nevertheless, there are some striking examples of mammalian migrations. More migrators are found in North America than on any other continent.

An example is the barren-ground caribou of Canada and Alaska, which undertakes direct and purposeful mass migrations spanning 160 to 1100 km (100 to 700 miles) twice annually (Figure 31-14). From winter ranges in the boreal forests (taiga), they migrate rapidly in late winter and spring to calving ranges on the barren grounds (tundra). The calves are born in mid-June. As the summer progresses, they are increasingly harassed by warble and nostril flies that bore into their flesh, by mosquitoes that drink their blood (estimated at a liter per caribou each week during the height of the mosquito season), and by wolves that prey on the calves. They move southward in July and August, feeding little along the way. In September they reach the forest and feed there almost continuously on low ground vegetation. Mating (rut) occurs in October.

A

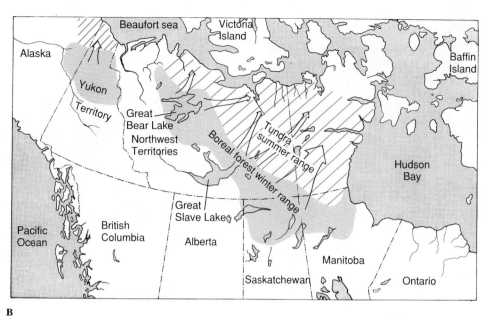

B

Figure 31-14

Barren-ground caribou, *Rangifer tarandus,* of Canada and Alaska. **A,** Adult male caribou in autumn pelage and antlers in velvet. **B,** Summer and winter ranges of some major caribou herds in Canada and Alaska (other herds not shown occur on Baffin Island and in western and central Alaska). The principal spring migration routes are indicated by arrows; routes vary considerably from year to year. The same species is known as reindeer in Europe. Order Artiodactyla, family Cervidae.

The caribou have suffered a drastic decline in numbers since early times when there were several million of them. By 1958 less than 200,000 remained in Canada. The decline has been attributed to several factors, including habitat alteration from exploration and development in the North, but especially to excessive hunting. For example the Western Arctic herd in Alaska exceeded 250,000 caribou in 1970. Following five years of heavy unregulated hunting, a 1976 census revealed only about 65,000 animals left. After restricting hunting, the herd had increased to 140,000 by 1980 and was expected to reach its original population of 250,000 in the 1990s. However, the proposed scheme to open the Arctic National Wildlife Refuge to petroleum development threatens this recovery.

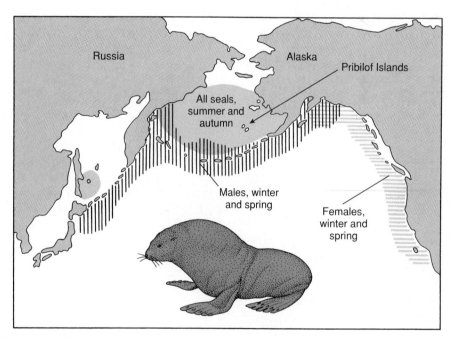

Figure 31-15

Annual migrations of the fur seal, showing the separate wintering grounds of males and females. Both males and females of the larger Pribilof population migrate in early summer to the Pribilof Islands, where the females give birth to their pups and then mate with the males. Order Pinnipedia, family Otariidae.

The plains bison, before its deliberate near extinction by humans, made huge circular migrations to separate summer and winter ranges.

The longest mammal migrations are made by the oceanic seals and whales. One of the most remarkable migrations is that of the fur seal, which breeds on the Pribilof Islands approximately 300 km (185 miles) off the coast of Alaska and north of the Aleutian Islands. From wintering grounds off southern California the females journey as much as 2800 km (1740 miles) across open ocean, arriving in the spring at the Pribilofs where they congregate in enormous numbers (Figure 31-15). The young are born within a few hours or days after arrival of the cows. Then the bulls, having already arrived and established territories, collect

Figure 31-16

Flying squirrel, *Glaucomys sabrinus,* coming in for a landing. Area of undersurface is nearly trebled when gliding skin is spread. Glides of 40 to 50 m are possible. Good maneuverability during flight is achieved by adjusting the position of the gliding skin with special muscles. Flying squirrels are nocturnal and have superb night vision. Order Rodentia, family Sciuridae.

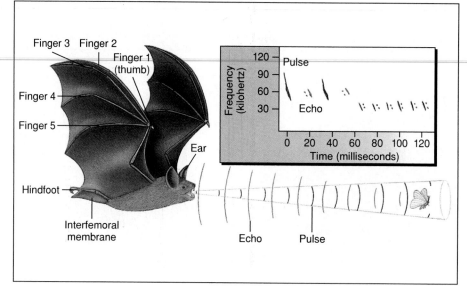

Figure 31-17

Echolocation of an insect by the little brown bat *Myotis lucifugus.* Frequency modulated pulses are directed in a narrow beam from the bat's mouth. As the bat nears its prey, it emits shorter, lower signals at a faster rate. Order Chiroptera.

harems of cows, which they guard with vigilance. After the calves have been nursed for approximately three months, cows and juveniles leave for their long migration southward. The bulls do not follow but remain in the Gulf of Alaska during the winter.

Although we might expect bats, the only winged mammals, to use their gift of flight to migrate, few of them do. Most spend the winter in hibernation. The four species of American bats that do migrate, the red bat, silver-haired bat, hoary bat, and Brazilian free-tailed bat, spend their summers in the northern or western states and their winters in the southern United States or Mexico.

FLIGHT AND ECHOLOCATION

Mammals have not exploited the skies to the same extent that they have the terrestrial and aquatic environments. However, many mammals scamper about in trees with amazing agility; some can glide from tree to tree, and one group, the bats, is capable of full flight. Gliding and flying evolved independently in several groups of mammals, including the marsupials, rodents, flying lemurs, and bats. Anyone who has watched a gibbon perform in a zoo

realizes there is something akin to flight in this primate, too. Among the arboreal squirrels, all of which are nimble acrobats, by far the most efficient is the flying squirrel (Figure 31-16). These forms actually glide rather than fly, using the gliding skin that extends from the sides of the body.

Bats, the only group of flying mammals, are nocturnal and thus occupy a niche left vacant by birds. Their achievement is attributed to two things: flight and the capacity to navigate by echolocation. Together these adaptations enable bats to fly and avoid obstacles in absolute darkness, to locate and catch insects with precision, and to find their way deep into caves (another habitat largely ignored by both mammals and birds) where they sleep away the daytime hours.

Research has been concentrated on members of the family Vespertilionidae, to which most of the common North American bats belong. When in flight, these bats emit short pulses 5 to 10 msec in duration in a narrow directed beam from the mouth or nose (Figure 31-17). Each pulse is frequency modulated; that is, it is highest at the beginning, up to 100,000 hertz (Hz, cycles per second), and sweeps down to perhaps 30,000

Many insectivores (for example, shrews and tenrecs) use echolocation, but it is crudely developed as compared with bats. The toothed whales, however, have a highly developed capacity to locate objects by echolocation. Totally blind sperm whales that are in perfect health have been captured with food in their stomachs. Although the mechanism of sound production and reception remains imperfectly understood, it is thought that low- and high-frequency clicks produced in the sinus passages are focused into a narrow beam by a lens-shaped body in the forehead (the "melon"). Returning echos are channeled through oil-filled sinuses in the lower jaw to the inner ear. Toothed whales can apparently determine the size, shape, speed, distance, direction, and density of objects in the water and know the position of every whale in the pod.

Hz at the end. Sounds of this frequency are ultrasonic to the human ear, which has an upper limit of about 20,000 Hz. When the bat is searching for prey, it produces about 10 pulses per second. If a prey is detected, the rate increases rapidly up to 200 pulses per second in the final phase of approach and capture. The pulses are spaced so that the echo of each is received before the next pulse is emitted, an adaptation that prevents jamming. Since the transmission-to-reception time decreases as the bat approaches an object, it can increase the pulse frequency to obtain more information about the object. The pulse length is also shortened as the bat nears the object. It is interesting that some prey of bats, certain nocturnal moths for example, have evolved ultrasonic detectors used to detect and avoid approaching bats (p. 730).

The external ears of bats are large, like hearing trumpets, and shaped variously in different species. Less is known about the bat's inner ear, but it obviously is capable of receiving the ultrasonic sounds emitted. Biologists believe that bat navigation is so refined that the bat builds up a mental image of its surroundings from echo scanning that is virtually as complete as the visual image from eyes of diurnal animals.

For reasons not fully understood, all bats are nocturnal, even the fruit-eating bats that use vision and olfaction instead of sonar to find their food. The tropics and subtropics have many nectar-feeding bats that are important pollinators for a wide variety of chiropterophilous ("bat-loving") plants. The flowers of these plants open at night, are white or light in color, and emit a musky, bat-like odor that the nectar-feeding bats find attractive.

The famed tropical vampire bat has razor-sharp incisors that it uses to shave away the epidermis of its prey, exposing underlying capillaries. After infusing an anticoagulant to keep the blood flowing, it laps up its meal and stores it in a specially modified stomach.

REPRODUCTION

Reproductive Cycles

Most mammals have definite mating seasons, usually in the winter or spring and timed to coincide with the most favorable time of the year for rearing the

Figure 31-18

African lions *Panthera leo* mating. Lions breed at any season, although predominantly in spring and summer. During the short period a female is receptive, she may mate repeatedly. Three or four cubs are born after gestation of 100 days. Once the mother introduces the cubs into the pride, they are treated with affection by both adult males and females. Cubs go through an 18- to 24-month apprenticeship learning how to hunt and then are frequently driven from the pride to manage themselves. Order Carnivora, family Felidae.

Photo by Kjell Sandved/Visuals Unlimited.

young after birth. Many male mammals are capable of fertile copulation at any time, but the female mating function is restricted to a time during a periodic cycle, known as the **estrous cycle.** The female receives the male only during a relatively brief period known as **estrus,** or heat (Figure 31-18).

The estrous cycle is divided into stages marked by characteristic changes in the ovary, uterus, and vagina. **Proestrus,** or period of preparation, when new ovarian follicles grow, is followed by **estrus,** when mating occurs. Almost simultaneously the ovarian follicles burst, releasing the eggs **(ovulation),** which are fertilized. In all placental mammals the fertilized egg then implants itself in the uterine wall and pregnancy follows. However, should mating and fertilization not occur, estrus is followed by **metestrus,** a period of repair. This stage is followed by **diestrus,** during which the uterus becomes small and anemic. The cycle then repeats itself, beginning with proestrus.

How often females are in heat varies greatly among the different mammals. Animals that have only a single estrus during the breeding sea-

A curious phenomenon that lengthens the gestation period of many mammals is delayed implantation. The blastocyst remains dormant while its implantation in the uterine wall is postponed for periods of a few weeks to several months. For many mammals (for example, bears, seals, weasels, badgers, bats, and many deer) delayed implantation is a device for extending gestation so that the young are born at the time of year that is best for their survival.

son are called **monestrous;** those that have a recurrence of estrus during the breeding season are called **polyestrous.** Dogs, foxes, and bats belong to the first group; field mice and squirrels are all polyestrous as are many mammals living in the more tropical regions of the earth. The Old World monkeys and humans have a somewhat different cycle in which the postovulation period is terminated by **menstruation,** during which the lining of the uterus (endometrium) collapses and is discharged with some blood. This is called a **menstrual cycle** and is described in Chapter 6 (p. 96).

Reproductive Patterns

There are three different patterns of reproduction in mammals. One pattern is represented by the egg-laying (oviparous) mammals, the **monotremes.** The duck-billed platypus has one breeding season each year. The ovulated eggs, usually two, are fertilized in the oviduct. As they continue down the oviduct, various glands add albumin and then a thin, leathery shell to each egg. When laid, the eggs are about the size of a robin's egg. The platypus lays its eggs in a burrow nest where they are incubated for about 12 days. After hatching, the young suck milk from the mother's nipples for a prolonged period. Thus in monotremes there is no gestation (period of pregnancy) and the developing embryo draws on nutrients stored in the egg, much as do the embryos of reptiles and birds. But in common with all other mammals, the monotremes rear their young on milk.

The **marsupials** are pouched, viviparous mammals that exhibit a second pattern of reproduction. Although only the eutherians are referred to as "placental mammals," the marsupials do have a primitive type of placenta, called a choriovitelline, or yolk sac, placenta. The embryo (blastocyst) of a marsupial is at first encapsulated by shell membranes and floats free for several days in the uterine fluid. After "hatching"

from the shell membranes, the embryo does not implant, or "take root" in the uterus as it does in eutherians, but it does erode a shallow depression in the uterine wall in which it lies and absorbs nutrient secretions from the mucosa by way of the vascularized yolk sac. Gestation (the intrauterine period of development) is brief in marsupials, and all marsupials give birth to tiny young that are effectively still embryos, both

anatomically and physiologically. However, early birth is followed by a prolonged interval of lactation and parental care (Figure 31-19).

In red kangaroos (Figure 31-20) the first pregnancy of the season is followed by a 33-day gestation, after which the young (joey) is born, crawls to the pouch without assistance from the mother, and attaches to a nipple. The mother immediately

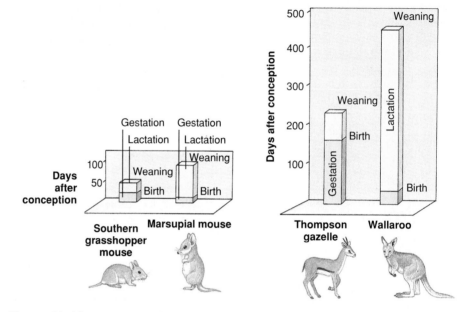

Figure 31-19

Comparison of gestation and lactation periods between matched pairs of ecologically similar species of marsupial and placental mammals. The graph shows that marsupials have shorter intervals of gestation and much longer intervals of lactation than in similar species of placentals.

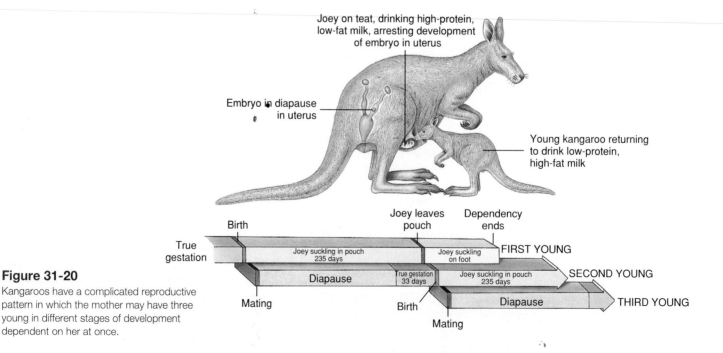

Figure 31-20

Kangaroos have a complicated reproductive pattern in which the mother may have three young in different stages of development dependent on her at once.

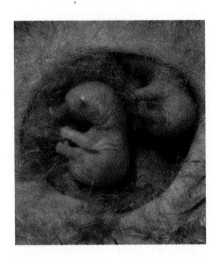

Figure 31-21
Opossums, *Didelphis marsupialis,* 15 days old, fastened to teats in mother's pouch. When born after a gestation period of only 12 days, they are the size of honeybees. They remain attached to the nipples for 50 to 60 days. Order Marsupialia, family Didelphidae.

becomes pregnant again, but the presence of a suckling young in the pouch arrests development of the new embryo in the uterus at about the 100-cell stage. This period of arrest, called embryonic diapause, lasts approximately 235 days during which time the first joey is growing in the pouch. When the joey leaves the pouch, the uterine embryo resumes development and is born about a month later. The mother again becomes pregnant, but because the second joey is suckling, once again development of the new embryo is arrested. Meanwhile, the first joey returns to the pouch from time to time to suckle. At this point the mother has three young of different ages dependent on her: a joey on foot, a joey in the pouch, and a diapause embryo in the uterus. There are variations on this remarkable sequence—not all marsupials have developmental delays like kangaroos, and some do not even have pouches—but in all, the young are born at an extremely early stage of development and undergo prolonged development while dependent on a teat (Figure 31-21).

The third pattern of reproduction is that of the viviparous **placental mammals,** the eutherians. In placen-

tals, the reproductive investment is in prolonged gestation, unlike the marsupials in which the reproductive investment is in prolonged lactation (Figure 31-19). The embryo remains in the mother's uterus, nourished by food supplied through a chorioallantoic type of placenta (described on p. 118), an intimate connection between mother and young. The length of gestation is longer in placentals than marsupials, and in large mammals it is much longer (Figure 31-19). For example, mice have a gestation period of 21 days; rabbits and hares, 30 to 36 days; cats and dogs, 60 days; cattle, 280 days; and elephants, 22 months. But there are important exceptions (nature seldom offers perfect correlations). Baleen whales, the largest mammals, carry their young for only 12 months, while bats, no larger than mice, have gestation periods of 4 to 5 months. The condition of the young at birth also varies. An antelope bears its young well furred, eyes open, and able to run about. Newborn mice, however, are blind, naked, and helpless. We all know how long it takes a human baby to gain its footing. Human growth is in fact slower than that of any other mammal, and this is one of the distinctive attributes that sets us apart from other mammals.

The number of young produced by mammals in a season depends on the mortality rate, which, for some mammals such as mice, may be high at all age levels. Usually, the larger the animal, the smaller the number of young in a litter. Small rodents, which serve as prey for many carnivores, usually produce more than one litter of several young each season. Meadow mice are known to produce as many as 17 litters of four to nine young in a year. Most carnivores have but one litter of three to five young per year. Large mammals, such as elephants and horses, give birth to a single young with each pregnancy. An elephant produces, on average, four calves during her reproductive life of perhaps 50 years.

The renowned fecundity of meadow mice, and the effect of removing the natural predators from rodent populations, is felicitously expressed in this excerpt from Thornton Burgess's **Portrait of a Meadow Mouse**

He's fecund to the nth degree
In fact this really seems to be
His one and only honest claim
To anything approaching fame.
In just twelve months, should all survive,
A million mice would be alive—
His progeny. And this, 'tis clear,
Is quite a record for a year.
Quite unsuspected, night and day
They eat the grass that would be hay.
On any meadow, in a year,
The loss is several tons, I fear.
Yet man, with prejudice for guide,
The checks that nature doth provide
Destroys. The meadow mouse survives
And on stupidity he thrives.

TERRITORY AND HOME RANGE

Many mammals have territories—areas from which individuals of the *same* species are excluded. In fact, many wild mammals, like many people, are basically unfriendly to their own kind, especially so to their own sex during the breeding season. If the mammal dwells in a burrow or den, this area forms the center of its territory. If it has no fixed address, the territory is marked out, usually with the highly developed scent glands described earlier in this chapter. Territories vary greatly in size depending on the size of the animal and its feeding habits. The grizzly bear has a territory of several square miles, which it guards zealously against all other grizzlies.

Mammals usually use natural features of their surroundings in staking their claims. These are marked with secretions from the scent glands or by urinating or defecating. When an intruder knowingly enters another's marked territory, it is immediately placed at a psychological disadvantage. Should a challenge follow, the intruder almost invariably breaks off the encounter in a submissive display

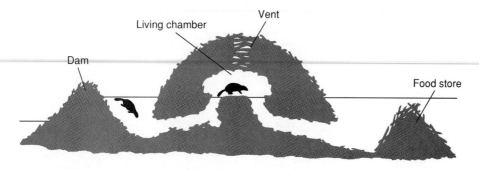

Figure 31-22

Each beaver colony constructs its own lodge in a pond it creates by damming a stream. Each year the mother bears four or five young; when the third litter arrives, the 2-year-olds are driven out of the colony. They will establish new colonies elsewhere. Order Rodentia, family Castoridae.

Figure 31-23

Immature black-tailed prairie dogs, *Cynomys ludovicianus* (order Rodentia), greeting adult. These highly social prairie dwellers are plant eaters that provide an important source of food to many animals. They live in elaborate tunnel systems so closely interwoven that they form "towns" of as many as 1000 individuals. Towns are subdivided into family units, each with one or two males, several females, and their litters. Although prairie dogs display ownership of burrows with territorial calls, they are friendly with inhabitants of adjacent burrows. The name "prairie dog" derives from the sharp, doglike bark they make when danger threatens. Order Rodentia, family Sciuridae.

characteristic for the species. Territoriality and aggressive and submissive displays are described in more detail in Chapter 38 (pp. 765 to 768).

A beaver colony is a family unit, and beavers are among several mammalian species in which the male and female form a strong monogamous bond that lasts a lifetime. Because beavers invest considerable time and energy in constructing a lodge and dam and storing food for winter (Figure 31-22), the family, especially the adult male, vigorously defends its real estate against intruding beavers. Most of the work of building dams and lodges is undertaken by male beavers, but the females help when not occupied with their young.

An interesting exception to the strong territorial nature of most mammals is the prairie dog, which lives in large, friendly communities called prairie dog "towns" (Figure 31-23). When a new litter has been reared, the adults relinquish the old home to the young and move to the edge of the community to establish a new home. Such a practice is totally antithetical to the behavior of most mammals, which drive off the young when they are self-sufficient.

The **home range** of a mammal is a much larger foraging area surrounding a defended territory. Home ranges are not defended in the same way as is a territory; home ranges may, in fact, overlap, producing a neutral zone used by the owners of several territories for seeking food.

MAMMALIAN POPULATIONS

A population of animals includes all the members of a species that share a particular space and potentially interbreed (Chapter 40). All mammals (like other organisms) live in ecological communities, each composed of numerous populations of different animal and plant species. Each species is affected by the activities of other species and by other changes, especially climatic, that occur. Thus populations are always changing in size. Populations of small mammals are lowest before the breeding season and greatest just after the addition of the new members. Beyond these expected changes in population size, mammalian populations may fluctuate from other causes.

Irregular fluctuations are commonly produced by variations in climate, such as unusually cold, hot, or dry weather, or by natural catastrophes, such as fires, hailstorms, and hurricanes. These are **density-independent** causes because they affect a population whether it is crowded or dispersed. However, the most spectacular fluctuations are **density dependent;** that is, they are correlated with population crowding (density-dependent and density-independent causes of growth limitation are discussed on pp. 807–808).

Cycles of abundance are common among many rodent species. One of the best-known examples is the mass migrations of the Scandinavian and arctic North American lemmings following population peaks. Lemmings (Figure 31-24) breed all year, although more in the summer than in the winter. The gestation period is only 21 days; young born at the beginning of the summer are weaned in 14 days and are capable of reproducing by the end of the summer. At the peak of their population density, having devastated the vegetation by tunneling and grazing, they begin long, mass migrations to find new undamaged habitats for food and space. They swim across streams and small lakes as they go but cannot distinguish these from large lakes, rivers, and the sea, in which they drown. Since lemmings are the main diet of many carnivorous mammals and birds, any change in lemming population density affects all their predators as well.

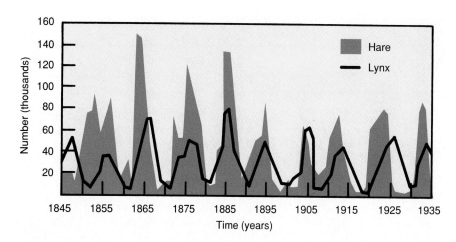

Figure 31-24
Collared lemming, *Dicrostonyx* sp., a small rodent of the far north. Populations of lemmings fluctuate widely. Order Rodentia, family Muridae.

Figure 31-25
Changes in population size of varying hare and lynx in Canada as indicated by pelts received by the Hudson's Bay Company. The abundance of lynx (predator) follows that of the hare (prey).

In his book *The Arctic* (1974. Montreal, Infacor, Ltd.), Canadian naturalist Fred Bruemmer describes the growth of lemming populations in arctic Canada:

"After a population crash one sees few signs of lemmings; there may be only one to every 10 acres. The next year, they are evidently numerous; their runways snake beneath the tundra vegetation, and frequent piles of rice-sized droppings indicate the lemmings fare well. The third year one sees them everywhere. The fourth year, usually the peak year of their cycle, the populations explode. Now more than 150 lemmings may inhabit each acre of land and they honeycomb it with as many as 4000 burrows. Males meet frequently and fight instantly. Males pursue females and mate after a brief but ardent courtship. Everywhere one hears the squeak and chitter of the excited, irritable, crowded animals. At such times they may spill over the land in manic migrations."

The varying hare (snowshoe rabbit) of North America shows 10-year cycles in abundance. The well-known fecundity of rabbits enables them to produce litters of three or four young as many as five times per year. The density may increase to 4000 hares competing for food in each square mile of northern forest. Predators (owls, minks, foxes, and especially lynxes) also increase (Figure 31-25). Then the population crashes precipitously for reasons that have long been a puzzle to scientists. Rabbits die in great numbers, not from lack of food or from an epidemic disease (as was once believed) but evidently from some density-dependent psychogenic cause. As crowding increases, hares become more aggressive, show signs of fear and defense, and stop breeding. The entire population reveals symptoms of pituitary-adrenal gland exhaustion, an endocrine imbalance called "shock disease," which results in death. These dramatic crashes are not well understood. Whatever the causes, population crashes that follow superabundance, although harsh, permit the vegetation to recover, providing the survivors with a much better chance for successful breeding.

HUMANS AND MAMMALS

Some 10,000 years ago, at the time people developed agricultural methods, they also began the domestication of mammals. Dogs were certainly among the first to be domesticated, probably entering voluntarily into their human dependence. The dog is an extremely adaptable and genetically plastic species derived from wolves. Much less genetically variable and certainly less social than dogs is the domestic cat, probably derived from an African race of wildcat. Wildcats look like oversized domestic cats and are still widespread in Africa and Eurasia. The domestication of cattle, buffaloes, sheep, and pigs probably came much later. It is believed that the beasts of burden—horses, camels, oxen, and llamas—probably were subdued by early nomadic peoples. Certain domestic species no longer exist as wild animals, for example, the one-humped dromedary camel of North Africa and the llama and alpaca of South America. All of the truly domestic animals breed in captivity and have become totally dependent on humans; many have been molded by selective breeding to yield characteristics that are desirable for our purposes.

Some mammals hold special positions as "domestic" animals. The elephant has never been truly domesticated because it will seldom breed in captivity. In Asia, adults are captured and submit to a life of toil with astonishing docility. The reindeer of northern Scandinavia are domesticated only in the sense that they are "owned" by nomadic peoples who continue to follow them in their seasonal migrations. The eland of Africa is undergoing experimental domestication in several places. It is placid, gentle, and immune to native diseases and produces excellent meat.

Figure 31-26

Brown rat *Rattus norvegicus*. Living all too successfully beside human habitations, the brown rat not only causes great damage to food stores but also spreads disease, including bubonic plague (a disease, carried by infected fleas, that greatly influenced human history in medieval Europe), typhus, infectious jaundice, *Salmonella* food poisoning, and rabies. Order Rodentia, family Muridae.

Figure 31-27

A prosimian, the Mindanao tarsier, *Tarsius syrichta carbonarius* of Mindanao Island in the Philippines.

The activities of mammals can in some instances conflict with human activities. Rodents and rabbits are capable of inflicting staggering damage to growing crops and stored food (Figure 31-26). We have provided an inviting forage for rodents with our agriculture and convenienced them further by removing their natural predators. Rodents also carry various diseases. Bubonic plague and typhus are carried by house rats. Tularemia, or rabbit fever, is transmitted to humans by the wood tick carried by rabbits, woodchucks, muskrats, and other rodents. Rocky Mountain spotted fever is carried to humans by ticks from ground squirrels and dogs; Lyme disease is transmitted by ticks from white-tailed deer. Trichina worms and tapeworms are acquired by humans who eat the meat of infected hogs, cattle, and other mammals.

In the introduction to this chapter, we alluded to the discouraging exploitation of the whales as one example of our inability to reconcile human needs with the preservation of wildlife. The extermination of a species for commercial gain is so totally indefensible that no debate is required. Once a species is extinct, no amount of scientific or technical ingenuity will bring it back. What has taken millions of years to evolve can be destroyed in a decade of thoughtless exploitation. Many people are concerned with the

awesome impact we have on wildlife, and there is more determination today to reverse a regrettable trend than ever before. If given a chance, mammals will usually make spectacular recoveries from human depredations, as have the sea otter and the saiga antelope, both once in danger of extinction and now numerous.

HUMAN EVOLUTION

Darwin devoted an entire book, *The Descent of Man and Selection in Relation to Sex,* largely to human evolution. The idea that humans shared common descent with apes and other animals was repugnant to the Victorian world, which responded with predictable outrage (Figure 9-14, p. 160). Because there was at the time virtually no fossil evidence linking humans with apes, Darwin built his case mostly on anatomical comparisons between humans and apes. To Darwin, the close resemblances between apes and humans could be explained only by common descent.

The search for fossils, especially for a "missing link" that would provide a connection between apes and humans, began when two skeletons of Neanderthals were collected in the 1880s. Then in 1891, Eugene Dubois discovered the famous Java man (*Homo erectus*). The most spectacular discoveries, however, have been made

in the last three decades in Africa, especially between 1967 and 1977, which American paleoanthropologist Donald C. Johanson calls the "golden decade." During this same period, comparative biochemical studies showed that humans and chimpanzees are as similar genetically as many sibling species. Comparative cytology showed that the chromosomes of humans and apes are homologous. We are no longer searching for a mythical "missing link" to establish the common descent of humans and apes, our closest living relatives.

EVOLUTIONARY RADIATION OF THE PRIMATES

Humans are primates, a fact that even the pre-evolutionist Linnaeus recognized. All primates share certain significant characteristics: grasping fingers on all four limbs, flat fingernails instead of claws, and forward-pointing eyes with binocular vision and excellent depth perception. The details of primate phylogeny are not entirely clear. The following synopsis will highlight the probable relationships of the major primate groups.

The earliest primate was probably a small, nocturnal animal similar in appearance to tree shrews. This ancestral primate stock split into two major lineages, one of which gave rise to the **prosimians,** which include **lemurs, tarsiers** (Figure 31-27), and **lorises;**

A   B

Figure 31-28

Monkeys. **A,** Red-howler monkeys, an example of the New World monkeys. **B,** The olive baboon, an example of the Old World monkeys.

Figure 31-29

The gorilla, an example of the anthropoid apes.

and the other to the **simians,** which include the monkeys (Figure 31-28) and apes (Figure 31-29). Prosimians and many simians are arboreal (tree-dwellers), which is probably the ancestral life-style for both groups. Arboreality probably stimulated the evolution of increased intelligence. Flexible limbs are essential for active animals moving through trees. Grasping hands and feet, in contrast to the clawed feet of squirrels and other rodents, enable the primates to grip limbs, hang from branches, seize food and manipulate it, and, most significantly, use tools. Highly developed sense organs, especially good vision, and proper coordination of limb and finger muscles are essential for an active arboreal life. Of course, sense organs are no better than the brain that processes sensory information. Precise timing, judgment of distance, and alertness require a large cerebral cortex.

The earliest simian fossils appeared in Africa in late Eocene deposits, some 40 million years ago. Many of these primates became day-active rather than nocturnal, making vision the dominant special sense, now enhanced by color vision. The simians comprise three major monophyletic groups; (1) the New World monkeys of South America (ceboids; Figure 31-28A), including the howler monkey, spider monkey, and the tamarin, (2) the Old World monkeys (cercopithecoids), including the baboon (Figure 31-28B), mandrill, and

the colobus monkey, and (3) the anthropoid apes (see Figure 31-29). The Old World monkeys and anthropoid apes (including humans) are sister taxa, and together form the sister group of the New World monkeys. In addition to their geographic separation, Old World monkeys differ from New World monkeys in lacking a grasping tail, and having close-set nostrils, more opposable, grasping thumbs, and more derived teeth. Apes first appear in 25-million-year-old fossils. At this time the woodland savannas were arising in Africa, Europe, and North America. Perhaps motivated by the greater abundance of food on the ground, these apes left the trees and became largely terrestrial.

THE FIRST HOMINIDS

About 8 million years ago during the Miocene epoch the gradual replacement of forests with grasslands in eastern Africa provided an impetus for apes to adapt to an open environment, the savannas. Because of the benefits of standing upright (better view of predators, freeing of hands for using tools and gathering food) emerging hominids gradually evolved upright posture. This important transition was an enormous leap because it required extensive redesigning of the skeleton and muscle attachments.

Evidence of the earliest hominids of this period is remarkably sparse. Although several different early

hominids have been identified, they all disappeared virtually without a trace of their descendants. Not until about 4 million years ago, after a lengthy fossil gap, do the first "near humans" appear. One was the recently discovered *Australopithecus afarensis,* a short, bipedal hominid with a face and brain size resembling those of a chimpanzee. Numerous fossils of this species have now been unearthed, the most celebrated of which was the 40% complete skeleton of a female discovered in 1974 by Donald Johanson and named "Lucy" (Figures 31-30 and 31-31). Many paleoanthropologists believe that *Australopithecus afarensis* represents the ancestral stock of all human and humanlike forms that followed.

EMERGENCE OF *HOMO*, THE TRUE HUMAN

Between 3 and 4 million years ago two quite separate hominid lines emerged that coexisted for at least 2 million years. One was the bipedal *Australopithecus africanus,* the "southern African ape," with a brain size only about one-third as large as that of modern humans. A different line of australopithecines was large and robust (Figure 31-32) and probably approached the size of a gorilla.

Until they became extinct between 1.75 and 1 million years ago, the australopithecines shared the

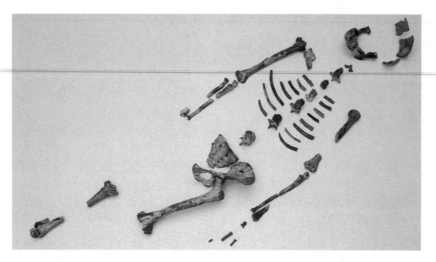

Figure 31-30
Lucy (*Australopithecus afarensis*), the most nearly complete skeleton of an early hominid ever found. Lucy is dated at 2.9 million years old. A nearly complete skull of *A. afarensis* was discovered in 1994.

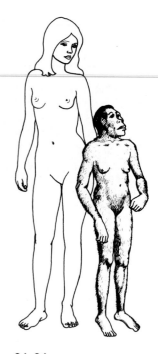

Figure 31-31
A reconstruction of the appearance of Lucy (*right*) compared with a modern human (*left*).

countryside with an advanced, fully erect hominid, *Homo habilis,* the first true human (Figure 31-32). *Homo habilis,* meaning "able man," was more lightly built but larger brained than the australopithecines and unquestionably used stone and bone tools. This species appeared about 2 million years ago and survived for perhaps one-half of a million years.

About 1.5 million years ago *Homo erectus* appeared, probably as a descendant of *Homo habilis. Homo erectus* was a large hominid standing 150 to 170 cm (5 to 5.5 feet) tall, with a low but distinct forehead, strong browridges, and a brain capacity of about 1000 cc (about intermediate between the brain capacity of *Homo habilis* and modern humans) (Figure 31-32). *Homo erectus* was a social species living in tribes of 20 to 50 people. *Homo erectus* had a successful and complex culture and became widespread throughout the tropical and temperate Old World.

Homo Sapiens: Modern Hominids

After the disappearance of *Homo erectus* about 300,000 years ago, subsequent human evolution and the establishment of *Homo sapiens* ("wise man") threaded a complex course.

From among the many early subcultures of *Homo sapiens,* the **Neanderthals** emerged about 130,000 years ago (Figure 31-32). With a brain capacity well within the range of modern humans, the Neanderthals were proficient hunters and tool-users. The Neanderthals were not homogeneous but varied geographically in response to local conditions and the isolation of populations from one another. They dominated the Old World in the late Pleistocene epoch.

About 30,000 years ago the Neanderthals were replaced and quite possibly exterminated by modern humans. The geographical origin of modern humans is obscure. They were tall people with a culture very different from that of the Neanderthals. Implement crafting developed rapidly, and human culture became enriched with aesthetics, artistry, and sophisticated language.

In closing our discussion of human evolution, it is important to note that the recognition of species in *Homo* (and to some degree also in other hominid fossils) is based entirely on morphology. Recognition of three distinct species of *Homo* does not necessarily imply the occurrence of branching speciation in this lineage; it is perhaps equally likely that we are observing phyletic change within a single species through time, and using the species

names only to denote different grades of evolution. There is clearly only a single species of *Homo* alive today.

The Unique Human Position

Biologically, *Homo sapiens* is a product of the same processes that have directed the evolution of every organism from the time of life's origin. Mutation, isolation, genetic drift, and natural selection have operated for us as they have for other animals. Yet we have what no other animal has, a nongenetic cultural evolution that provides a constant feedback between past and future experience. Our symbolic languages, capacities for conceptual thought, knowledge of our history, and abilities to manipulate our environment emerge from this nongenetic cultural endowment. Finally, we owe much of our cultural and intellectual achievements to our arboreal ancestry which bequeathed us with binocular vision, superb visuotactile discrimination, and manipulative skills in the use of our hands. If the horse (with one toe instead of five fingers) had human mental capacity, could it have accomplished what humans have?

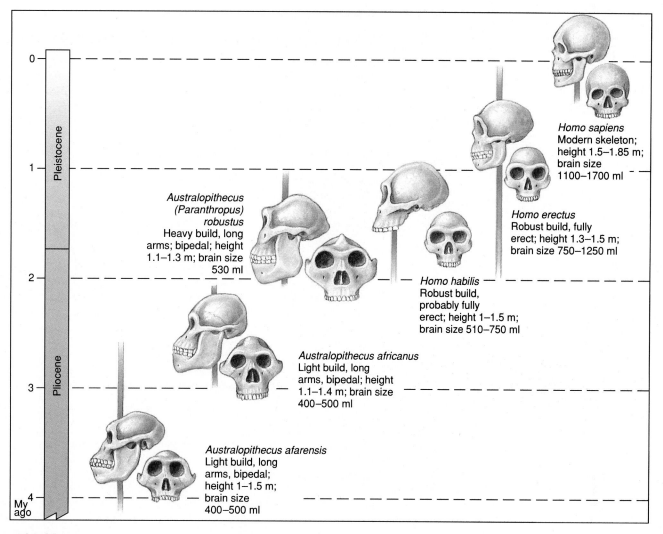

Figure 31-32
Hominid skulls, showing several of the best-known hominid lines preceding modern humans (*Homo sapiens*). The time span of existence for each species, as indicated by the fossil record, is suggested by the vertical red lines.

CLASSIFICATION OF LIVING MAMMALIAN ORDERS*

Class Mammalia
  Subclass Prototheria (pro´to-thir´ee-a) (Gr. *prōtos,* first, + *thēr,* wild animal). Cretaceous and early Cenozoic mammals. Extinct.
  Infraclass Ornithodelphia (or´ni-tho-del´fee-a) (Gr. *ornis,* bird, + *delphys,* womb). Monotreme mammals.
    Order Monotremata (mon´o-tre´ma-tah) (Gr. *monos,* single, + *trēma,* hole): **egg-laying (oviparous) mammals: duck-billed platypus, spiny anteater.** The three species in this order are from Australia, Tasmania, and New Guinea. The most noted member of the order is the duck-billed platypus (*Ornithorhynchus anatinus*). The spiny anteater, or echidna (*Tachyglossus*), has a long, narrow snout adapted for feeding on ants, its chief food.
  Subclass Theria (thir´ee-a) (Gr. *thēr,* wild animal). Extant mammals.
    Infraclass Metatheria (met´a-thir´e-a) (Gr. *meta,* after, + *thēr,* wild animal). Marsupial mammals.

continued

*Based on Nowak, R. M. 1991. Walker's Mammals of the world, ed. 5, Baltimore, The Johns Hopkins University Press.

CLASSIFICATION OF LIVING MAMMALIAN ORDERS—*cont'd*

Order Marsupialia (mar-su′pe-ay′le-a) (Gr. *marsypion,* little pouch): **viviparous pouched mammals: opossums, kangaroos, koalas, Tasmanian wolves, wombats, bandicoots, numbats, and others.** These mammals are characterized by an abdominal pouch, the **marsupium,** in which they rear their young. The young are nourished in the uterus for a short time by way of a yolk-sac placenta. Only the opossum is found in the Americas, but the order is the dominant group of mammals in Australia; 260 species.

Infraclass Eutheria (yu-thir′e-a) (Gr. *eu,* true, + *thēr,* wild animal). The viviparous placental mammals.

Order Insectivora (in-sec-tiv′o-ra) (L. *insectum,* an insect, + *vorare,* to devour): **insect-eating mammals: shrews, hedgehogs, tenrecs, moles.** The principal food is insects. Insectivores, widely distributed over the world except Australia and New Zealand, are small, sharp-snouted animals with primitive characters that spend a great part of their lives underground. The shrews are among the smallest mammals known; 390 species.

Order Macroscelidea (mak-ro-sa-lid′e-a) (Gr. *makros,* large, + *skelos,* leg): **elephant shrews.** These are secretive mammals with long legs, a snoutlike nose adapted for foraging for insects, large eyes, and are widespread in Africa; 15 species.

Order Dermoptera (der-mop′ter-a) (Gr. *derma,* skin, + *pteron,* wing): **flying lemurs.** These are related to the true bats and consist of the single genus *Galeopithecus.* They are found in the Malay peninsula in the East Indies. They are not lemurs (which are primates) and cannot fly in the strict sense of the word, but glide like flying squirrels; two species.

Order Chiroptera (ky-rop′ter-a) (Gr. *cheir,* hand, + *pteron,* wing): **bats.** The wings of bats, the only true flying mammals, are modified forelimbs in which the second to fifth digits are elongated to support a thin integumental membrane for flying. The first digit (thumb) is short with a claw. The common North American forms are the little brown bat (*Myotis*), the free-tailed bat (*Tadarida*), which lives in the Carlsbad Caverns, and the large brown bat (*Eptesicus*). In the Old World tropics the fruit bats, or "flying foxes," (*Pteropus*) are the largest of all bats, with a wingspread of 1.2 to 1.5 m; they live chiefly on fruits; 986 species.

Order Scandentia (skan-dent′e-a) (L. *scandentis,* climbing): **tree shrews.** Tree shrews are small, squirrel-like mammals of the tropical rain forests of southern and southeastern Asia. Despite their name, many are not especially well-adapted for life in trees, and some are almost completely terrestrial; 16 species.

Order Primates (pry-may′teez) (L. *prima,* first): **prosimians, monkeys, apes, humans.** This order stands first in the animal kingdom in brain development, with especially large cerebral hemispheres. Most species are arboreal, apparently derived from tree-dwelling insectivores. The primates represent the end product of a line that branched off early from other mammals and have retained many primitive characteristics. It is believed that their tree-dwelling habits of agility in capturing food or avoiding enemies were largely responsible for their advances in brain structure. As a group they are generalized with five digits (usually provided with flat nails) on both forelimbs and hindlimbs. All except humans have their bodies covered with hair. Forelimbs are often adapted for grasping, as are the hindlimbs sometimes. The group is singularly lacking in claws, scales, horns, and hoofs. There are two suborders;* 233 species.

Suborder Prosimii (pro-sim′ee-i) (Gr. *pro,* before, + *simia,* ape): **lemurs, bush babies, tarsiers, lorises, pottos.** These are arboreal primates, with their second toe provided with a claw and a long nonprehensile tail. They

---

*G. G. Simpson's (1945) division of the primates into prosimian and anthropoid suborders is followed here, but it is no longer recognized by many mammalogists who hold conflicting views on classification, especially at the order and family levels.

look like a cross between squirrels and monkeys. They are found in the forests of Madagascar, Africa, the Malay peninsula, and the Philippines. Their food consists of both plants and small animals.

Suborder Anthropoidea (an'thro-poy'de-a) (Gr. *anthropos,* man): **monkeys, gibbons, apes, humans.** There are three superfamilies.

Superfamily Ceboidea (se-boi'de-a) (Gr. *kebos,* long-tailed monkey): **Platyrhinii.** New World monkeys, characterized by the broad flat nasal septum, nonopposable thumb, prehensile tail, and the absence of ischial callosities and cheek pouches. Familiar members of this superfamily are the capuchin monkey (*Cebus*) of the organ grinder, the spider monkey (*Ateles*), and the howler monkey (*Alouatta*).

Superfamily Cercopithecoidea (sur'ko-pith'e-coy'de-a) (Gr. *kerkos,* tail, + *pithekos,* monkey): **Catarrhini.** Old World monkeys with the external nares close together; many have internal cheek pouches. They never have prehensile tails, there are calloused ischial tuberosities on their buttocks, and their thumbs are opposable. Examples are the savage mandrill (*Cynocephalus*); the rhesus monkey (*Macaca*), widely used in biological investigation; and the proboscis monkey (*Nasalis*).

Superfamily Hominoidea (hom'i-noi'de-a) (L. *homo, hominis,* man). The anthropoid apes and humans make up this superfamily. Their chief characteristics are lack of a tail and lack of cheek pouches. Three families are recognized according to evolutionary taxonomy. The Hylobatidae family includes the gibbons (*Hylobates*). The Pongidae family includes the higher apes: orangutan (*Pongo*), chimpanzee (*Pan*), and gorilla (*Gorilla*). The third family, Hominidae, is represented by a single living species (*Homo sapiens*), modern humans. Humans differ from the members of family Pongidae in being more erect, in having shorter arms and larger thumbs, and in having lighter jaws with smaller front teeth. Most of the apes also have much more prominent ridges over the eyes. Many differences between the human and the anthropoid apes are associated with greater human intelligence, human speech centers in the brain, and the absence of the arboreal habit. Cladistic taxonomy does not recognize the paraphyletic family Pongidae because the most recent common ancestor of family Pongidae is also the ancestor of the family Hominidae (see p. 202 and Figure 11-7).

Order Xenarthra (ze-nar'thra) (Gr. *xenos,* intrusive, + *arthron,* joint) (formerly Edentata [L. *edentatus,* toothless]): **anteaters, armadillos, sloths.** Species of this order are either toothless (anteaters) or have simple, peglike teeth (sloths and armadillos). Most live in South and Central America, although the nine-banded armadillo (*Dasypus novemcinctus*) is common in the southern United States; 30 species.

Order Pholidota (fol'i-do'ta) (Gr. *pholis,* horny scale): **pangolins.** An odd group of mammals whose bodies are covered with overlapping horny scales that have arisen from fused bundles of hair. Their home is in tropical Asia and Africa; seven species.

Order Lagomorpha (lag'o-mor'fa) (Gr. *lagos,* hare, + *morphē,* form): **rabbits, hares, pikas** (Figure 31-33). Lagomorphs have long, constantly growing incisors, like rodents, but unlike rodents, they have an additional pair of peglike incisors growing behind the first pair. All lagomorphs are herbivores with cosmopolitan distribution; 69 species.

Order Rodentia (ro-den'che-a) (L. *rodere,* to gnaw): **gnawing mammals: squirrels** (Figure 31-34), **rats, woodchucks.** The rodents, comprising nearly 40% of all mammalian species, are characterized by two pairs of razor-sharp incisors used for gnawing through the toughest pods and shells

continued

Figure 31-33

A pika, *Ochotona princeps,* atop a rockslide in Alaska. This little rat-sized mammal does not hibernate but prepares for winter by storing dried grasses beneath boulders. Order Lagomorpha.

Figure 31-34

Eastern gray squirrel, *Sciurus carolinensis.* This common resident of Eastern towns and hardwood forests serves as an important reforestation agent by planting numerous nuts that sprout into trees. Order Rodentia, family Sciuridae.

Figure 31-35
Humpback whale, *Megaptera novaeangliae,*
breaching. Among the most acrobatic of whales,
humpbacks appear to breach to stun fish
schools or to communicate information to other
herd members. Order Cetacea, family
Balaenopteridae.

Figure 31-36
Grizzly bear, *Ursus horribilis,* of Alaska. Grizzlies,
once common in the lower 48 states, are now
confined largely to wilderness areas. Order
Carnivora, family Ursidae.

CLASSIFICATION OF LIVING MAMMALIAN ORDERS—*cont'd*

for food. With their impressive reproductive powers, adaptability, and capacity to invade all terrestrial habitats, they are of great ecological significance. Important families of this order are **Sciuridae** (squirrels and woodchucks), **Muridae** (rats and house mice), **Castoridae** (beavers), **Erethizontidae** (porcupines), **Geomyidae** (pocket gophers), and **Cricetidae** (hamsters, deer mice, gerbils, voles, lemmings); 1814 species.

Order Cetacea (see-tay′she-a) (L. *cetus,* whale): **whales** (Figure 31-35), **dolphins, porpoises.** The anterior limbs of cetaceans are modified into broad flippers; the posterior limbs are absent. Some have a fleshy dorsal fin and the tail is divided into transverse fleshy flukes. The nostrils are represented by a single or double blowhole on top of the head. They have no hair except a few hairs on the muzzle, no skin glands except the mammary and those of the eye, no external ear, and small eyes. The order is divided into the **toothed whales** (suborder Odontoceti), represented by dolphins, porpoises, and sperm whales; and the **baleen whales** (suborder Mysticeti) represented by the rorquals, right whales, and gray whales. The baleen whales are generally larger than toothed whales. The blue whale, a rorqual, is the heaviest animal that has ever lived. Rather than teeth, baleen whales have a straining device of whalebone (baleen) attached to the palate, used to filter plankton from the water; 79 species.

Order Carnivora (car-niv′o-ra) (L. *caro,* flesh, + *vorare,* to devour): **flesh-eating mammals: dogs, wolves, cats, bears** (Figure 31-36), **weasels.** All Carnivora except the giant panda have predatory habits, and their teeth are especially adapted for tearing flesh. They are distributed all over the world except in the Australian and Antarctic regions where there are no native forms. Among the more familiar families are **Canidae** (the dog family), consisting of dogs, wolves, foxes, and coyotes; **Felidae** (the cat family), whose members include the domestic cats, tigers, lions, cougars, and lynxes; **Ursidae** (the bear family), made up of bears; **Procyonidae** (raccoons); and **Mustelidae** (the fur-bearing family), containing the martens, skunks, weasels, otters, badgers, minks, and wolverines; 240 species.

Order Pinnipedia (pi-ni-peed′e-a) (L. *pinna,* feather, + *ped,* foot): **sea lions** (Figure 31-37), **seals, and walruses.** The limbs of these aquatic carnivores have been modified as flippers for swimming. Most are saltwater forms, and their food consists mostly of fish; 34 species.

Order Tubulidentata (tu′byu-li-den-ta′ta) (L. *tubulus,* tube, *dens,* tooth): **aardvark.** The name "aardvark" is Dutch for earth pig, a peculiar animal with a piglike body found in Africa; one species.

Order Proboscidea (pro′ba-sid′e-a) (Gr. *proboskis,* elephant's trunk, from *pro,* before, + *boskein,* to feed): **proboscis mammals: elephants.** These,

Figure 31-37
A Galápagos sea lion bull, *Zalophus
californianus,* barks to indicate his territorial
ownership. Order Pinnipedia, family Otariidae.

Figure 31-38
Herd of reindeer, *Rangifer tarandus,* during annual roundup by
Laplanders in northern Sweden. The same species is known as
caribou in North America. Order Artiodactyla, family Cervidae.

the largest of living land animals, have two upper incisors elongated as tusks, and the molar teeth are well developed. The Asiatic or Indian elephant (*Elephas maximus*) has long been domesticated and is trained to do heavy work. The taming of the African elephant (*Loxodonta africana*) is more difficult but was done extensively by the ancient Carthaginians and Romans, who employed them in their armies; two species.

Order Hyracoidea (hy′ra-coi′de-a) (Gr. *hyrax,* shrew): **hyraxes** (coneys). Coneys are herbivores that are restricted to Africa and Syria. They have some resemblance to short-eared rabbits but have teeth like rhinoceroses, with hooves on their toes and pads on their feet. They have four toes on the front feet and three toes on the back; seven species.

Order Sirenia (sy-re′ne-a) (Gr. *seiren,* sea nymph): **sea cows and manatees.** The sirenians are large, clumsy, aquatic mammals with large head, no hindlimbs, and forelimbs modified into flippers. The sea cow (dugong) of tropical coastlines of East Africa, Asia, and Australia and three species of manatees of the Caribbean area and Florida, Amazon River, and West Africa are the only living species. A fifth species, the large Steller's sea cow, was hunted to extinction by humans in the mid-eighteenth century; four species.

Order Perissodactyla (pe-ris′so-dak′ti-la) (Gr. *perissos,* odd, + *dactylos,* toe): **odd-toed hoofed mammals: horses, asses, zebras, tapirs, rhinoceroses.** The odd-toed hoofed mammals have an odd number of toes (one or three), each with a cornified hoof. Both the Perissodactyla and the Artiodactyla are often referred to as **ungulates** (L. *ungula,* hoof), or hoofed mammals, with teeth adapted for chewing. The horse family (Equidae), which also includes asses and zebras, has only one functional toe. Tapirs have a short proboscis formed from the upper lip and nose. The rhinoceros (*Rhinoceros*) includes several species found in Africa and Southeast Asia. All are herbivorous; 17 species.

Order Artiodactyla (ar′te-o-dak′ti-la) (Gr. *artios,* even, + *daktylos,* toe): **even-toed hoofed mammals: swine, camels, deer and their allies** (Figure 31-38), **hippopotamuses, antelopes, cattle, sheep, goats.** Most of these ungulates have two toes, although the hippopotamus and some others have four (Figure 31-39). Each toe is sheathed in a cornified hoof. Many, such as the cow, deer, and sheep have horns or antlers. Many are ruminants, that is, animals that chew the cud. Like Perissodactyla, they are strictly herbivorous. The group is divided into nine living families and many extinct ones and includes some of the most valuable domestic animals. The order is commonly divided into three suborders: the **Suina** (pigs, peccaries, and hippopotamuses), the **Tylopoda** (camels), and the **Ruminantia** (deer, giraffes, sheep, cattle, and so on); 211 species.

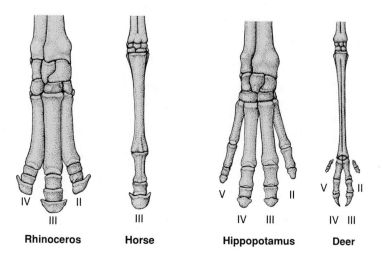

IV   II

III

Rhinoceros

III

Horse

V   II

IV   III

Hippopotamus

V   II

IV III

Deer

Figure 31-39

Odd-toed and even-toed ungulates. The rhinoceros and horse (order Perissodactyla) are odd-toed; the hippopotamus and deer (order Artiodactyla) are even-toed. The lighter, faster mammals run on only one or two toes.

Summary

Mammals are endothermic and homeothermic vertebrates whose bodies are insulated by hair and who nurse their young with milk. The approximately 4450 species of mammals are descended from the synapsid lineage of amniotes that arose in the Carboniferous period of the Paleozoic era. Their evolution can be traced from the pelycosaurs of the Permian period to the therapsids of the late Permian and Triassic periods of the Mesozoic era. One group of the therapsids, the cynodonts, gave rise during the Triassic to the therians, the true mammals. Mammalian evolution was accompanied by the appearance of many important derived characteristics, among these the enlarged brain with greater sensory integration, high metabolic rate, endothermy, and many changes in the skeleton that supported a more active life. Mammals diversified rapidly during the Tertiary period of the Cenozoic era.

Mammals are named for the glandular milk-secreting organs of the female (rudimentary in the male), a unique adaptation which, combined with prolonged parental care, buffers the infants from the demands of foraging for themselves and eases the transition to adulthood. Hair, the integumentary outgrowth that covers most mammals, serves variously for mechanical protection, thermal insulation, protective coloration, and waterproofing. Mammalian skin is rich in glands: sweat glands that function in evaporative cooling, scent glands used in social interactions, and sebaceous glands that secrete lubricating skin oil. All placental mammals have deciduous teeth that are replaced by permanent teeth (diphyodont dentition). The four groups of teeth—incisors, canines, premolars, and molars—may be highly modified in different mammals for specialized feeding tasks, or they may be absent.

The food habits of mammals strongly influence their body form and physiology. Insectivores feed mainly on insects and other small invertebrates. Herbivorous mammals have special adaptations for harboring the intestinal bacteria that break down cellulose of the woody diet, and they have developed adaptations for detecting and escaping predators. Carnivorous mammals feed mainly on herbivores, have a simple digestive tract, and have developed adaptations for a predatory life. Omnivores feed on both plant and animal foods.

Some marine, terrestrial, and aerial mammals migrate; some migrations, such as those of fur seals and caribou, are extensive. Migrations are usually made toward favorable climatic and optimal food and calving conditions, or to bring the sexes together for mating.

Mammals with true flight, the bats, are nocturnal and thus avoid direct competition with birds. Most employ ultrasonic echolocation to navigate and feed in darkness.

The living mammals with the most primitive characters are the egg-laying monotremes of the Australian region. After hatching, the young are nourished with the mother's milk. All other mammals are viviparous. Embryos of marsupials have brief gestation periods, are born underdeveloped, and complete their early growth in the mother's pouch, nourished by milk. The remaining 19 of the 21 orders of mammals are eutherians, mammals that develop an advanced placental attachment between mother and embryos through which the embryos are nourished for a prolonged period.

Mammal populations fluctuate from both density-dependent and density-independent causes and some mammals, particularly rodents, may experience extreme cycles of abundance in population density. The unqualified success of mammals as a group cannot be attributed to greater organ system perfection, but rather to their impressive overall adaptability—the capacity to fit more perfectly in total organization to environmental conditions and thus exploit virtually every habitat on earth.

Darwinian evolutionary principles give us great insight into our own origins. Humans are primates, a mammalian group that descended from a shrewlike ancestor. The common ancestor of all modern primates was arboreal and had grasping fingers and forward-facing eyes capable of binocular vision. Primates radiated over the last 80 million years to form two major lines of descent: the prosimians (lemurs, lorises, and tarsiers) and the simians (monkeys, apes, and hominids). The earliest hominids appeared about 4 million years ago; these were the bipedal australopithecines. These gave rise to, and coexisted with, the species *Homo habilis,* the first user of stone tools. *Homo erectus* appeared about 1.5 million years ago and was eventually replaced by *Homo sapiens* some 300,000 years ago.

Review Questions

1. Describe the evolution of mammals, tracing the synapsid lineage from early amniote ancestors to true mammals. How would you distinguish the skull of a synapsid from that of diapsid?

2. Describe some of the structural and functional adaptations that appeared in the early amniotes that foreshadowed the mammalian body plan. Which mammalian attributes do you think were especially important to the successful radiation of mammals?

3. Hair is believed to have evolved in the therapsids in response as an adaptation for insulation, but modern mammals have adapted hair for several other purposes. Describe these.

4. What is distinctive about each of the following: horns of the ruminants, antlers of the deer family, and the horn of the rhinoceros? Briefly describe the growth cycle of antlers.

5. Describe the location and principal function(s) of each of the following skin glands: sweat glands (of two kinds, eccrine and apocrine), scent glands, sebaceous glands, and mammary glands.

6. Define the terms "diphyodont" and "heterodont" and explain how both terms apply to mammalian dentition.

7. Describe the food habits of each of the following groups: insectivores, herbivores, carnivores, and omnivores. Can you give the common names of some mammals belonging to each group?

8. Most herbivorous mammals depend on cellulose as their main energy source, yet no mammal synthesizes cellulose-splitting enzymes. How are the digestive tracts of mammals specialized for symbiotic digestion of cellulose?

9. Describe the annual migrations of barren-ground caribou and fur seals.

10. Explain what is distinctive about the life habit and mode of navigation in bats.

11. Describe and distinguish the patterns of reproduction in monotremes, marsupials, and placental mammals. What aspects of mammalian reproduction are present in *all* mammals but in no other class of vertebrates?

12. Distinguish between territory and home range in mammals.

13. What is the difference between density-dependent and density-independent causes of fluctuations of the size of mammalian populations?

14. Describe the hare-lynx population cycle, considered a classic example of a prey-predator relationship (Figure 31-25). From your examination of the cycle, can you formulate a hypothesis for the explanation of the oscillations?

15. What do the terms Theria, Metatheria, Eutheria, Monotremata, and Marsupialia literally mean, and what mammals are grouped under each taxon?

16. What anatomical characteristics set the primates apart from other mammals?

17. What role does the fossil named "Lucy" play in the reconstruction of human evolutionary history?

18. In what ways do the genera *Australopithecus* and *Homo,* which coexisted for at least 2 million years, differ?

19. When approximately did the different species of *Homo* appear and how did they differ socially?

20. What major attributes make the human position in animal evolution unique?

Selected References

See also general references for Part III, p. 626.

Chepko-Sade, B. D., and Z. T. Halpin. 1987. Mammalian dispersal patterns: the effects of social structure and population genetics. Chicago, University of Chicago Press. *A summary of discoveries on the population dynamics, social structures, and population genetics of mammals with contributions by many mammalogists.*

Eisenberg, J. F. 1981. The mammalian radiations: an analysis of trends in evolution, adaptation, and behavior. Chicago, University of Chicago Press. *Wide-ranging, authoritative synthesis of mammalian evolution and behavior.*

Grzimek's encyclopedia of mammals. 1990. vol. 1–5. New York, McGraw-Hill Publishing Company. *Valuable source of information on all mammal orders.*

Halstead, L. B. 1978. The evolution of mammals. London, Eurobook, Ltd. *Richly illustrated, semipopular treatment.*

Jones, S., R. Martin, and D. Pilbeam. 1992. Cambridge encyclopedia of human evolution. Cambridge, England, Cambridge University Press. *Comprehensive and informative encyclopedia written for the nonspecialist. Highly readable and highly recommended.*

Kemp, T. S. 1982. Mammal-like reptiles and the origin of mammals. New York, Academic Press, Inc. *Comprehensive synthesis. The final chapter summarizes the earlier chapters and offers a model of evolutionary history of the mammal-like reptiles and primitive mammals.*

Macdonald, D. (ed) 1984. The encyclopedia of mammals. New York, Facts on File Publications. *Coverage of all mammal orders and families, enhanced with fine photographs and color artwork.*

Nowak, R. M. 1991. Walker's mammals of the world, ed. 5. Baltimore, The Johns Hopkins University Press. *The definitive illustrated reference work on mammals, with descriptions of all extant and recently extinct species.*

Preston-Mafham, R., and K. Preston-Mafham. 1992. Primates of the world. New York, Facts on File Publications. *A small "primer" with high quality photographs and serviceable descriptions.*

Rice, J. A. (ed.). 1994. The marvelous mammalian parade. Natural History **103**(4):39–91. *A special multi-authored section on mammalian evolution.*

Rismiller, P. D., and R. S. Seymour. 1991. The echidna. Sci. Am. **294**:96–103 (Feb.). *Recent studies of this fascinating monotreme have revealed many secrets of its natural history and reproduction.*

Savage, R. J. G., and M. R. Long. 1986. Mammal evolution: an illustrated guide. New York, Facts on File Publications. *Profusely illustrated survey of fossil mammals.*

Stringer, C. B. 1990. The emergence of modern humans. Sci. Am. **263**:98–104 (Dec.). *A review of the geographical origins of modern humans.*

Suga, N. 1990. Biosonar and neural computation in bats. Sci. Am. **262**:60–68 (June). *How the bat nervous system processes echolocation signals.*

Vaughan, T. A. 1978. Mammalogy, ed. 2. Philadelphia, W.B. Saunders Company. *Comprehensive treatment of the mammalian characteristics and orders.*

References to Part III

The references below pertain to groups covered in more than one chapter of Part III. They include a number of very valuable field manuals that aid in identification, as well as general texts.

Alexander, R. M. 1981. The chordates, ed. 2. New York, Cambridge University Press. *A general treatment that emphasizes experimental approaches to determining functional basis of chordate structure.*

Alexander, R. M. 1991. Animals. New York, Cambridge University Press. *A somewhat eclectic survey of the animal kingdom which, like Alexander's earlier book on the chordates, emphasizes experimental zoology.*

Barrington, E. J. W. 1965. The biology of the Hemichordata and Protochordata. San Francisco, W. H. Freeman. *This comprehensive treatment remains an essential source of information for three deuterostome groups that share ancestry with the chordates.*

Barrington, E. J. W. 1979. Invertebrate structure and function, ed. 2. New York, John Wiley & Sons, Inc. *Excellent account of function in major invertebrate groups.*

Benton, M. J. 1991. Vertebrate paleontology: biology and evolution. London, Unwin Hyman. *Highly readable introduction to vertebrate paleontology, much less exhaustive in coverage than Robert Carroll's treatment (listed below); suitable for all levels of readers.*

Brusca, R. C., and G. J. Brusca. 1990. Invertebrates. Sunderland, Massachusetts, Sinauer Associates, Inc. *Invertebrate text organized around the bauplan ("body plan") concept—structural range, architectural limits, and functional aspects of a design—for each phylum. Includes cladistic analysis of phylogeny for most groups.*

Carroll, R. L. 1988. Vertebrate paleontology and evolution. New York, W. H. Freeman and Company. *This comprehensive text is the first authoritative, detailed review of vertebrate fossil history to have appeared since Romer's classic* Vertebrate Paleontology. *An invaluable tour de force.*

Conant, R., and J. T. Collins. 1991. Field guide to reptiles and amphibians: Eastern and Central North America. The Peterson Field Guide Series. New York, Houghton Mifflin Company. *Excellent color illustrations and practical range maps are the hallmarks of this popular guide book; includes sections on capturing, transporting, and caring for reptiles and amphibians.*

Conway Morris, S., J. D. George, R. Gibson, and H. M. Platt (eds.). 1985. The origins and relationships of lower invertebrates. Oxford, Clarendon Press. *Technical discussions of phylogenetic relationships among lower invertebrates; essential for serious students of this topic.*

Cox, F. E. G. (ed.). 1993. Modern parasitology, ed. 2. Oxford, England, Blackwell Scientific Publications. *Gives a brief introduction to the biology of various groups of parasites, then good chapters on modern techniques of biochemistry, molecular biology, immunology, chemotherapy, and others, as applied to parasitology.*

Fotheringham, N. 1980. Beachcomber's guide to Gulf Coast marine life. Houston, Texas, Gulf Publishing Company. *Coverage arranged by habitats. No keys, but most common forms that occur near shore can be identified.*

Gosner, K. L. 1979. A field guide to the Atlantic seashore: invertebrates and seaweeds of the Atlantic coast from the Bay of Fundy to Cape Hatteras. The Peterson Field Guide Series. Boston, Houghton Mifflin Company. *A helpful aid for students of the invertebrates found along the northeastern coast of the United States.*

Grzimek, H. C. B. (ed.) 1984. Grzimek's animal life encyclopedia. 13 vols. New York, Van Nostrand Reinhold Company. *Comprehensive overview of the animal kingdom, beautifully illustrated with numerous full-color drawings and photographs, with emphasis on animal behavior. Although heavily weighted toward the higher vertebrates, the set is nevertheless a goldmine of information for zoologists of all persuasions.*

Humann, P. 1992. Reef creature identification. Florida, Caribbean, Bahamas. Jacksonville, Florida, New World Publications. *Excellent field guide to aid identification of Atlantic reef invertebrates except corals.*

Hyman, L. H. 1940–1967. The invertebrates, 6 vols. New York, McGraw-Hill Book Company. *Informative discussions on the phylogenies of most of the invertebrates are treated in this outstanding series of monographs. Volume 1 contains a discussion of the colonial hypothesis of the origin of metazoa, and volume 2 contains a discussion of the origin of bilateral animals, body cavities, and metamerism.*

Kaplan, E. H. 1988. A field guide to southeastern and Caribbean seashores: Cape Hatteras to the Gulf Coast, Florida, and the Caribbean. A Peterson Field Guide Series. Boston, Houghton Mifflin Company. *More than just a field guide, this comprehensive book is filled with information on the biology of seashore animals; complements Gosner's field guide which covers animals north of Cape Hatteras.*

Kozloff, E. N. 1987. Marine invertebrates of the Pacific Northwest. Seattle, University of Washington Press. *Contains keys for many marine groups.*

Kozloff, E. N. 1990. Invertebrates. Philadelphia, Saunders College Publishing. *A good invertebrate text, less exhaustive than Ruppert and Barnes (1994) or Brusca and Brusca (1990). Cladistic analysis not emphasized.*

Lane, R. P., and R. W. Crosskey. 1992. Medically important insects and arachnids. London, Chapman and Hall. *The most up-to-date medical entomology text available.*

Meglitsch, P. A., and F. R. Schram. 1991. Invertebrate zoology, ed. 3. New York, Oxford University Press. *Schram's thorough revision of the older Meglitsch text. Includes cladistic treatments.*

Morris, R. H., D. P. Abbott, and E. C. Haderlie. 1980. Intertidal invertebrates of California. Stanford, Stanford University Press. *An essential reference on the most important invertebrates of the intertidal zone in California. Contains 900 color photographs.*

Nielson, C. 1995. Animal evolution: interrelationships of the living phyla. New York, Oxford University Press. *Cladistic analysis of morphology is used to develop sister-group relationships of the living Metazoa. An advanced but essential reference.*

Norman, D. 1994. Prehistoric life: the rise of the vertebrates. New York, Macmillan. *Semipopular treatment of the evolution of vertebrate life; excellent illustrations.*

Page, L. M., and B. M. Burr. 1991. Field guide to freshwater fishes: North America north of Mexico. New York, Houghton Mifflin.

Parker, S. P. (ed.). 1982. Synopsis and classification of living organisms, 2 vols. New York, McGraw-Hill Book Company. *A comprehensive reference to the classification of living organisms with descriptions of taxa above the generic level. Contains 8200 synoptic articles on the biology of many of the groups.*

Pearse, V., J. Pearse, M. Buchsbaum, and R. Buchsbaum. 1987. Living invertebrates. Palo Alto, California, Blackwell Scientific Publications, and Pacific Grove, California, Boxwood Press. *Readable account of invertebrates; many photographs.*

Pennak, R. W. 1989. Freshwater invertebrates of the United States, ed. 3. New York, John Wiley & Sons, Inc. *Contains keys for identification of freshwater invertebrates, with brief account of each group. Indispensable for freshwater biologists.*

Pough, F. H., J. B. Heiser, and W. N. McFarland. 1989. Vertebrate life, ed. 3. New York, Macmillan Publishing Company. *This* thoroughly rewritten edition of a popular text stresses ancestry and unifying themes of the vertebrates; it is the first vertebrate biology text to adopt a wholly cladistic approach to vertebrate classification.

Radinsky, L. B. 1987. The evolution of vertebrate design. Chicago, University of Chicago Press. *This short book covers vertebrate anatomy, diversity, and evolution, using unusual "block style" drawings that clearly reveal the operating mechanisms under discussion.*

Ricketts, E. F., J. Calvin, and J. W. Hedgpeth. (revised by D. W. Phillips). 1985. Between Pacific tides, ed. 5. Stanford, Stanford University Press. *A revision of a classic work in marine biology. It stresses the habits and habitats of the Pacific coast invertebrates, and the illustrations are revealing. It includes an excellent, annotated systematic index and bibliography.*

Roberts, L. S., and J. Janovy, Jr. 1996. Foundations of parasitology, ed. 5. Dubuque, Wm. C. Brown Publishers. *Highly readable and up-to-date account of parasitic protistans, worms, and arthropods.*

Rogers, E. 1986. Looking at vertebrates: a practical guide to vertebrate adaptations. Essex, England, Longman Group Limited. *Well-illustrated practical textbook of vertebrate form and function; ideal for undergraduates looking for a friendly approach that avoids excessive terminology.*

Ruppert, E. E., and R. D. Barnes. 1994. Invertebrate zoology, ed. 6. Philadelphia, Saunders College Publishing. *Authoritative, detailed coverage of the invertebrate phyla.*

Smith, D. L. 1977. A guide to marine coastal plankton and marine invertebrate larvae. Dubuque, Iowa, Kendall/Hunt Publishing Company. *Valuable manual for identification of marine plankton, which is usually not covered in most field guides.*

Smith, R. I., and J. T. Carlton (eds.). 1975. Light's manual: intertidal invertebrates of the central California coast, ed. 3. Berkeley, University of California Press. *Has keys for identification of intertidal invertebrates from central California.*

Willmer, P. 1990. Invertebrate relationships. Patterns in animal evolution. Cambridge, Cambridge University Press. *Articulate statement of invertebrate phylogeny from a non-cladist. Good account of the polyphyletic hypothesis for the origin of arthropods.*

Young, J. Z. 1981. The life of vertebrates, ed. 3. Oxford, Clarendon Press. *This is the updated, comprehensive classic of vertebrate biology.*

32

Support, Protection, and Movement

Of Grasshoppers and Superman

"A dog," remarked Galileo in the fifteenth century, "could probably carry two or three such dogs upon his back; but I believe that a horse could not carry even one of its own size." Galileo was referring to the principle of scaling, a procedure that allows us to understand the consequences of changing body size. A grasshopper can jump to a height of 50 times the length of its body, yet a man in a standing jump cannot clear an obstacle that is no higher than he is tall. Without an understanding of scaling, this comparison could easily lead us to the erroneous conclusion that there is something very special about insects and their musculature. To the authors of a nineteenth-century entomology text it seemed that "This wonderful strength of insects is doubtless the result of something peculiar in the structure and arrangement of their muscles, and principally their extraordinary power of contraction." But grasshopper muscles are in fact no more powerful than human muscles because *muscles of small and large animals exert the same force per cross-sectional area.* Grasshoppers leap high in proportion to their size because they are small, not because they possess extraordinary muscles.

The authors of this nineteenth-century text further suggested that it was fortunate that higher animals were withheld the powers of insects, for they would surely have "caused the early desolation of the world." More probably, such powers would have led to their own desolation. For earthly mortals would need more than superhuman muscles were they to leap in the proportions of a grasshopper. They would require superhuman tendons, superhuman ligaments, and superhuman bones to withstand the stresses of mighty contractions, not to mention the crushing strains of landing again on earth at terminal velocity. The feats of Superman would be quite impossible were he built of the structural materials available to earthbound animals, rather than of the wondrous materials available to the inhabitants of the mythical planet Krypton. ■

INTEGUMENT AMONG VARIOUS GROUPS OF ANIMALS

The integument is the outer covering of the body, a protective wrapping that includes the skin and all structures derived from or associated with the skin, such as hair, setae, scales, feathers, and horns. In most animals it is tough and pliable, providing mechanical protection against abrasion and puncture and forming an effective barrier against the invasion of bacteria. It may provide moisture proofing against fluid loss or gain. The skin helps protect the underlying cells against the damaging action of the ultraviolet rays of the sun. In addition to being a protective cover, the skin serves a variety of important regulatory functions. For example, in endothermic animals, it is vitally concerned with temperature regulation, since most of the body's heat is lost through the skin; it contains mechanisms that cool the body when it is too hot and slow heat loss when the body is too cold. The skin contains sensory receptors that provide essential information about the immediate environment. It has excretory functions and in some animals respiratory functions as well. Through skin pigmentation the organism can make itself more or less conspicuous. Skin secretions can make the animal sexually attractive or repugnant or provide olfactory cues that influence behavioral interactions between individuals.

INVERTEBRATE INTEGUMENT

Many protozoa have only the delicate cell or plasma membranes for external coverings; others, such as *Paramecium,* have developed a protective pellicle. Most multicellular invertebrates, however, have more complex tissue coverings. The principal covering is a single-layered **epidermis.** Some invertebrates have added a secreted noncellular **cuticle** over the epidermis for additional protection.

The molluscan epidermis is delicate and soft and contains mucous glands, some of which secrete the calcium carbonate of the shell. Cephalopod molluscs (squids and octopuses) have developed a more complex integument, consisting of cuticle, simple epidermis, a layer of connective tissue, a layer of reflecting cells (iridocytes), and a thicker layer of connective tissue.

Arthropods have the most complex of invertebrate integuments, providing not only protection but also skeletal support. The development of a firm exoskeleton and jointed appendages suitable for the attachment of muscles has been a key feature in the extraordinary diversity of this phylum, the largest of animal groups. The arthropod integument consists of a single-layered **epidermis** (also called more precisely **hypodermis**), which secretes a complex cuticle of two zones (Figure 32-1A). The thicker inner zone, the **procuticle,** is composed of protein and chitin (a polysaccharide) laid down in layers (lamellae) much like the veneers of plywood. The outer zone of cuticle, lying on the external surface above the procuticle, is the thin **epicuticle.** The epicuticle is a nonchitinous complex of proteins and lipids that provides a protective moisture-proofing barrier to the integument.

The arthropod cuticle may remain as a tough but soft and flexible layer, as it is in many microcrustaceans and insect larvae. However, it may be hardened by either of two ways. In the decapod crustaceans, for example, crabs and lobsters, the cuticle is stiffened by **calcification,** the deposition of calcium carbonate in the outer layers of the procuticle. In insects hardening occurs when the protein molecules bond together with stabilizing cross-linkages within and between the adjacent lamellae of the procuticle. The result of this process, called **sclerotization,** is the formation of a highly resistant and insoluble protein, **sclerotin.** Arthropod cuticle is one of the toughest materials synthesized by animals; it is strongly resistant to pressure and tearing and can withstand boiling in concentrated alkali, yet it is light, having a specific mass of only 1.3 (1.3 times the mass of water).

When arthropods molt, the epidermal cells first divide by mitosis. Enzymes secreted by the epidermis dissolve most of the procuticle. The digested materials are then absorbed and consequently are not lost to the body. Then in the space beneath the old cuticle a new epicuticle and procuticle are formed. After the old cuticle is shed, the new cuticle is thickened and calcified or sclerotized.

VERTEBRATE INTEGUMENT

The basic plan of the vertebrate integument, as exemplified by human skin (Figure 32-1B), includes a thin, outer stratified epithelial layer, the **epidermis,** derived from ectoderm and an inner, thicker layer, the **dermis,** or true skin, which is of mesodermal origin.

Although the epidermis is thin and appears simple in structure, it gives rise to most derivatives of the integument, such as hair, feathers, claws, and hooves. The dermis contains blood vessels, collagenous fibers, nerves, pigment cells, fat cells, and connective tissue cells called fibroblasts. These elements support, cushion, and nourish its overlying partner, which is devoid of blood vessels.

The epidermis is a stratified squamous epithelium (p. 188) consisting usually of several layers of cells. The basal part is made up of cells that undergo frequent mitosis to renew the layers that lie above. As the outer layers of cells are displaced upward by new generations of cells beneath, an exceedingly tough, fibrous protein called **keratin** accumulates in the interior of the cells. Gradually, keratin replaces all metabolically active cytoplasm. The cell dies and is eventually shed, lifeless and scalelike. Such is the origin of dandruff as well as a significant fraction of household dust. This process is called **keratinization,** and the cell, thus transformed, is said to be **cornified.** Cornified cells, highly resistant to abrasion and water

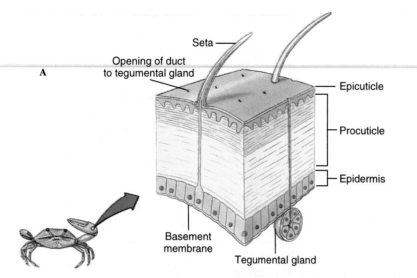

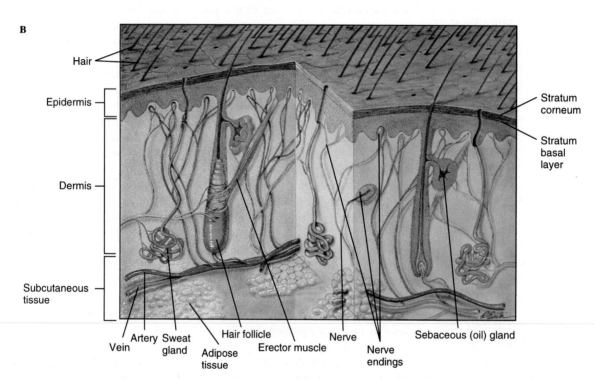

A, Structure of arthropod (crustacean) body wall showing cuticle and epidermis. **B,** Structure of human skin (dermis and epidermis) and hypodermis.

Figure 32-1

diffusion, comprise the **stratum corneum.** This epidermal layer becomes especially thick in areas exposed to persistent pressure or friction, such as calluses and the human palms and soles.

The **dermis,** as already mentioned, mainly serves a supportive role for the epidermis. Nevertheless, true bony structures, where they occur in the integument, are always dermal derivatives. Heavy bony plates were common in ostracoderms and placoderms of the Paleozoic era but

Lizards, snakes, turtles, and crocodilians were among the first to exploit the adaptive possibilities of the remarkably tough protein keratin. The reptilian epidermal scale that develops from keratin is a much lighter and more flexible structure than the bony, dermal scale of fishes, yet it provides excellent protection from abrasion and desiccation. Scales may be overlapping structures, as in snakes and some lizards, or develop into plates, as in turtles and crocodilians.

In birds, keratin found new uses. Feathers, beaks, and claws, as well as scales, are all epidermal structures composed of dense keratin. Mammals continued to capitalize on keratin's virtues by turning it into hair, hooves, claws, and nails. As a result of its keratin content, hair is by far the strongest material in the body. It has a tensile strength comparable to that of rolled aluminum and is nearly twice as strong, weight for weight, as the strongest bone.

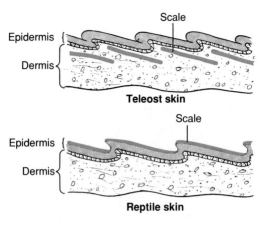

Figure 32-2

Integument of bony fishes and lizards. Bony (teleost) fishes have bony scales from dermis, and lizards have horny scales from epidermis. Dermal scales of fishes are retained throughout life. Since a new growth ring is added to each scale each year, fishery biologists use scales to tell the age of fishes. Epidermal scales of reptiles are shed periodically.

occur in few modern fishes. Scales of contemporary fishes are bony dermal structures that have evolved from the bony armor of the Paleozoic fishes but are much smaller and more flexible. Although of dermal origin, fish scales are intimately associated with the thin, overlying epidermis. In some species the scales protrude through the epidermis, but typically the epidermis forms a continuous sheath that is reflected under the overlapping scales (Figure 32-2). Dermal bone also forms the flat bones of the skull and gives rise to antlers, which are outgrowths of dermal frontal bone.

Animal Coloration

The colors of animals may be vivid and dramatic when serving as important recognition marks or as warning coloration, or they may be subdued or cryptic when used for camouflage. Integumentary color is usually produced by pigments, but in many insects and in some vertebrates, especially birds, certain colors are produced by the physical structure of the surface tissue, which reflects certain light wavelengths and eliminates others. Colors produced this way are called **structural color,** and they are responsible for the most beautifully iridescent and metallic hues to be

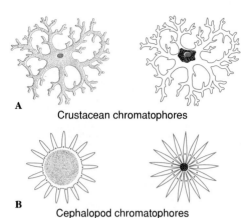

Figure 32-3

Chromatophores. **A,** The crustacean chromatophore showing the pigment dispersed (*left*) and concentrated (*right*). Vertebrate chromatophores are similar. **B,** The cephalopod chromatophore is an elastic capsule surrounded by muscle fibers that, when contracted (*left*), stretch out the capsule to expose the pigment.

found in the animal kingdom. Many butterflies and beetles and a few fishes thus share with birds the distinction of being the earth's most resplendent animals. Certain structural colors of feathers are caused by minute, air-filled spaces or pores that reflect white light (white feathers) or some portion of the spectrum (for example, Tyndall blue coloration produced by scattering of light [see note, p. 576]). Iridescent colors that change hue as the animal's angle shifts with respect to the observer are produced when light is reflected from several layers of thin, transparent film. By phase interference, light waves reinforce, weaken, or eliminate each other to produce some of the purest and most brilliant colors we know.

More common than structural colors in animals are **pigments** (biochromes), an extremely varied group of large molecules that reflect light rays. In crustaceans and ectothermic vertebrates these pigments are contained in large cells with branching processes, called **chromatophores** (Figure 32-3A). The pigment may concentrate in the center of the cell in an aggregate too small to be visible, or it may disperse throughout the cell and its processes, providing maximum display. The chromatophores of cephalopod molluscs are entirely different (Figure 32-3B). Each is a small sac-like cell filled with pigment granules and surrounded by muscle cells that, when contracted, stretch the whole cell out into a pigmented sheet. When the muscles relax, the elastic chromatophore quickly shrinks to a small sphere. With such pigment cells the squids and octopuses can alter their color more rapidly than any other animal.

The most widespread of animal pigments are the **melanins,** a group of black or brown polymers that are responsible for the various earth-colored shades that most animals wear. Yellow and red colors are often caused by **carotenoid** pigments, which are frequently contained within special pigment cells called **xanthophores.** Most vertebrates are incapable of synthesizing their own carotenoid pigments but must obtain them directly or indirectly from plants. Two entirely different classes of pigments called ommochromes and pteridines are usually responsible for the yellow pigments of molluscs and arthropods. Green colors are rare; when they occur, they are usually produced by yellow pigment overlying blue structural color. **Iridophores,** a third type of chromatophore, contain crystals of guanine or some other purine, rather than pigment. They produce a silvery or metallic effect by reflecting light.

By vertebrate standards, mammals are a somber-colored group (p. 602). Most mammals are more or less color blind, a deficiency that is doubtless connected with the lack of bright colors in the group. Exceptions are the brilliantly colored skin patches of some baboons and mandrills. Significantly, primates have color vision and thus can appreciate such eye-catching ornaments. The muted colors of mammals are caused by melanin, which is deposited in growing hair by dermal melanophores.

Injurious Effects of Sunlight

The familiar vulnerability of the human skin to sunburn reminds us of the potentially damaging effects of ultraviolet radiation on protoplasm. Many animals, such as protozoa and flatworms, if exposed to the sun in shallow water are damaged or killed by ultraviolet radiation. Most land animals are protected from such damage by the screening action of special body coverings, for example, the cuticle of arthropods, the scales of reptiles, and the feathers and fur of birds and mammals. Humans, however, are "naked apes" that lack the furry protection of most other mammals. We must depend on thickening of the epidermis **(stratum corneum)** and on epidermal pigmentation for protection. Most ultraviolet radiation is absorbed in the epidermis, but about 10% penetrates the dermis. Damaged cells in both the epidermis and dermis release histamine and other vasodilator substances that cause blood vessel enlargement in the dermis and the characteristic red coloration of sunburn. Light skins suntan through the formation of the pigment **melanin** in the deeper epidermis and by "pigment darkening," that is, the photooxidative blackening of bleached pigment already present in the epidermis. Unfortunately, tanning does not bestow perfect protection. Sunlight still ages the skin prematurely, and tanning itself causes the skin to become dry and leathery. The damaging effects of sunlight accumulate over the years and are responsible for skin cancer, the most common of malignancies among Caucasians. Each year approximately 400,000 cases of skin cancer are reported in the United States alone.

SKELETAL SYSTEMS

Skeletons are supportive systems that provide rigidity to the body, surfaces for muscle attachment, and protection for vulnerable body organs. The familiar bone of the vertebrate skeleton is only one of several kinds of supportive and connective tissues serving various binding and weight-bearing functions, which are described in this discussion.

HYDROSTATIC SKELETONS

Not all skeletons are rigid; many invertebrate groups use their body fluids as an internal hydrostatic skeleton.

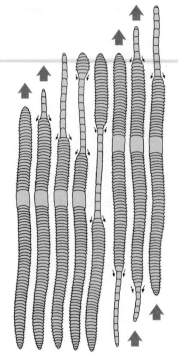

Figure 32-4
How an earthworm moves forward. When circular muscles contract, longitudinal muscles are stretched by internal fluid pressure and the worm elongates. Then, by alternate contraction of longitudinal and circular muscles, a wave of contraction passes from anterior to posterior. Bristlelike setae are extended to anchor the animal and prevent slippage.

The muscles in the body wall of the earthworm, for example, have no firm base for attachment but develop muscular force by contracting against the coelomic fluids, which are enclosed within a limited space and are incompressible, much like the hydraulic brake system of an automobile.

Alternate contractions of the circular and longitudinal muscles of the body wall enable the worm to thin and thicken, setting up backward-moving waves of motion that propel the animal forward (Figure 32-4). Earthworms and other annelids are helped by the septa that separate the body into more or less independent compartments (Figure 18-1, p. 352). An obvious advantage is that if a worm is punctured or even cut into pieces, each part can still develop pressure and move. Worms that lack internal compartments, for example, the lugworm *Arenicola* (Figure 18-5, p. 354), are rendered helpless if the body fluid is lost through a wound.

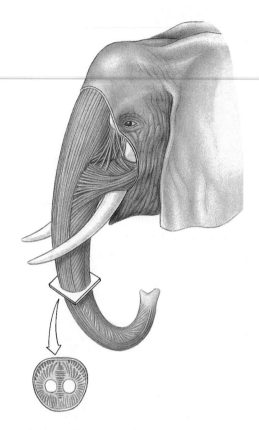

Figure 32-5
Muscular trunk of an elephant, an example of a muscular hydrostat.

There are many examples in the animal kingdom of muscles that not only produce movement but also provide a unique form of skeletal support. The elephant's trunk is an excellent example of a structure that lacks any obvious form of skeletal support, yet is capable of bending, twisting, elongating, and lifting heavy weights (Figure 32-5). The elephant's trunk, the tongues of mammals and reptiles, and the tentacles of cephalopod molluscs are examples of **muscular hydrostats.** Like the hydrostatic skeletons of worms, muscular hydrostats work because they are composed of incompressible tissues that remain at constant volume. The remarkably diverse movements of muscular hydrostats depend on muscles arranged in complex patterns.

RIGID SKELETONS

Rigid skeletons differ from hydrostatic skeletons in one fundamental way: rigid skeletons consist of rigid elements, usually jointed, to which muscles can attach. Muscles can only contract; to be lengthened they must be extended by the pull of an antagonistic set of muscles. Rigid skeletons provide the anchor points required by opposing sets of muscles, such as flexors and extensors.

There are two principal types of rigid skeletons: the **exoskeleton,** typical of molluscs and arthropods, and the **endoskeleton,** characteristic of echinoderms and vertebrates. The invertebrate exoskeleton may be mainly protective, but it may also perform a vital role in locomotion. An exoskeleton may take the form of a shell, a spicule, or a calcareous, proteinaceous, or chitinous plate. It may be rigid, as in molluscs, or jointed and movable, as in arthropods. Unlike the endoskeleton, which grows with the animal, the exoskeleton is often a limiting coat of armor that must be periodically molted to make way for an enlarged replacement (molting in crustaceans is described on p. 390). Some invertebrate exoskeletons, such as the shells of snails and bivalves, grow with the animal.

> The arthropod-type exoskeleton is perhaps a better arrangement for small animals than a vertebrate-type endoskeleton because a hollow cylindrical tube can support much more weight without collapsing than can a solid cylindrical rod of the same material and weight. Arthropods can thus enjoy both protection and structural support from their exoskeleton. But for larger animals the hollow cylinder would be completely impractical. If made thick enough to support the body weight, it would be too heavy to lift; but if kept thin and light, it would be extremely sensitive to buckling or shattering on impact. Finally, can you imagine the sad plight of an animal the size of an elephant when it shed its exoskeleton to molt?

The vertebrate endoskeleton is formed inside the body and is composed of bone and cartilage, which are forms of dense connective tissue. Bone not only supports and protects but is also the major body reservoir for calcium and phosphorus. In amniote vertebrates the red blood cells and certain white blood cells are formed in the bone marrow.

Notochord and Cartilage

The **notochord** (see Figure 26-1, p. 480) is a semirigid supportive axial rod of the protochordates and all vertebrate larvae and embryos. It is composed of large, vacuolated cells and is surrounded by layers of elastic and fibrous sheaths. It is a stiffening device, preserving body shape during locomotion. Except in the jawless vertebrates (lampreys and hagfishes), the notochord is surrounded or replaced by the backbone during embryonic development.

Cartilage is the major skeletal element of some vertebrates. The jawless fishes (for example, lampreys) and the elasmobranchs (sharks, skates, and rays) have purely cartilaginous skeletons, which oddly enough is a derived feature, since their Paleozoic ancestors had bony skeletons. Other vertebrates as adults have principally bony skeletons with some cartilage interspersed. Cartilage is a soft, pliable, characteristically deep-lying tissue. Unlike most connective tissues, which are quite variable in form, cartilage is basically the same wherever it is found. The basic form, **hyaline cartilage,** has a clear, glassy appearance (see Figure 10-6, p. 189). It is composed of cartilage cells **(chondrocytes)** surrounded by firm complex protein gel interlaced with a meshwork of collagenous fibers. Blood vessels are virtually absent—the reason that sports injuries involving cartilage heal poorly. In addition to forming the cartilaginous skeleton of some vertebrates and that of all vertebrate embryos, hyaline cartilage makes up the articulating surfaces of many bone joints of most adult vertebrates and the supporting tracheal, laryngeal, and bronchial rings.

Cartilage similar to hyaline cartilage occurs in some invertebrates, for example in the radula of gastropod molluscs and in the lophophore of brachiopods. The cartilage of cephalopod molluscs is of a special type with long, branching processes that resemble the cells of vertebrate bone.

Bone

Bone is a living tissue that differs from other connective and supportive tissues by having significant deposits of inorganic calcium salts laid down in an extracellular matrix. Its structural organization is such that bone has nearly the tensile strength of cast iron, yet is only one-third as heavy.

Bone is never formed in vacant space but is always laid down by replacement of areas occupied by some form of connective tissue. Most bone develops from cartilage and is called **endochondral** ("within-cartilage") or **replacement bone.** The embryonic cartilage is gradually eroded leaving it extensively honeycombed; bone-forming cells then invade these areas and begin depositing calcium salts around the strandlike remnants of the cartilage. A second type of bone is **intramembranous bone,** which develops directly from sheets of embryonic cells. Dermal bone, mentioned earlier, is a type of intramembranous bone. In tetrapod vertebrates intramembranous bone is restricted to bones of the face and cranium; the remainder of the skeleton is endochondral bone. Whatever its embryonic origin, once bone is fully formed there is no difference in the histological structure: endochondral and intramembranous bone look the same.

Fully formed bone, however, may vary in density. **Cancellous** (or spongy) **bone** consists of an open, interlacing framework of bony tissue, oriented to give maximum strength under the normal stresses and strains that the bone receives. All bone develops first as cancellous bone, but some bones, through further deposition of bone salts, become **compact.** Compact bone is dense, appearing solid to the unaided eye. Both cancellous and compact bone are found in the typical long bones of the body (Figure 32-6).

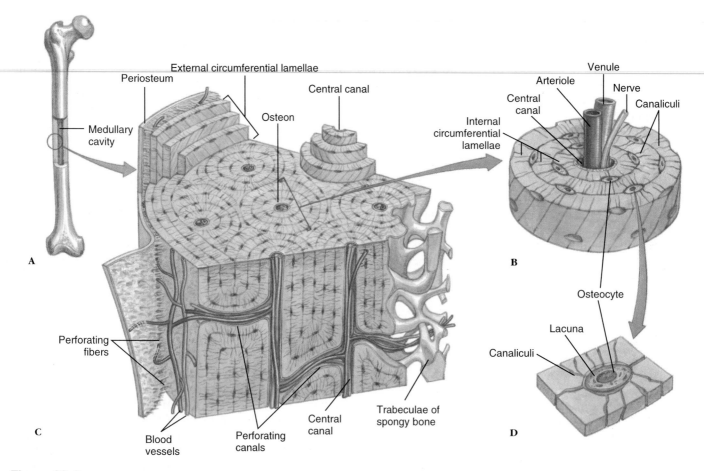

External circumferential lamellae

Periosteum

Central canal

Osteon

Medullary cavity

Venule

Arteriole

Nerve

Central canal

Canaliculi

Internal circumferential lamellae

A

B

Perforating fibers

Osteocyte

Lacuna

Canaliculi

C

Central canal

Trabeculae of spongy bone

Perforating canals

Blood vessels

D

Figure 32-6
Structure of compact bone. **A,** Adult long bone with a cut into the medullary cavity. **B,** Enlarged section showing osteons, the basic histological unit of bone.
C, Enlarged view of an osteon showing the concentric lamellae and the osteocytes (bone cells) arranged within lacunae. **D,** An osteocyte within a lacuna.
Bone cells receive nutrients from the circulatory system via tiny canaliculi that interlace the calcified matrix. Bone cells are known as osteoblasts when they
are building bone, but, in mature bone shown here, they become resting osteocytes. Bone is covered with compact connective tissue called periosteum.

Microscopic Structure of Bone.
Compact bone is composed of a calcified bone matrix arranged in concentric rings. The rings contain cavities **(lacunae)** filled with bone cells **(osteocytes)** which are interconnected by many minute passages **(canaliculi).** These serve to distribute nutrients throughout the bone. This entire organization of lacunae and canaliculi is arranged into an elongated cylinder called an **osteon** (also called **haversian system)** (Figure 32-6). Bone consists of bundles of osteons cemented together and interconnected with blood vessels and nerves. Because of blood vessels and nerves throughout bone, it is living tissue, although nonliving "ground substance" predominates. As a result of its living state, bone breaks can heal, and bone diseases can be as painful as any other tissue disease.

Following menopause, a woman loses 5% to 6% of her bone mass annually, often leading to the disease osteoporosis and increasing the risk of bone fractures. Dietary supplementation with calcium has been advocated to prevent such losses, but even large doses of calcium alone have little effect in slowing demineralization unless accompanied by therapy with the female sex hormone estrogen (because ovarian production of estrogen drops significantly after menopause). Among animals, only humans, especially females, are troubled with osteoporosis, perhaps a consequence of the long postreproductive life of the human species.

Bone growth is a complex restructuring process, involving both its destruction internally by bone-resorbing cells **(osteoclasts)** and its deposition externally by bone-building cells **(osteoblasts).** Both processes occur simultaneously so that the marrow cavity inside grows larger by bone resorption while new bone is laid down outside by bone deposition. Bone growth responds to several hormones, particularly the **parathyroid hormone** from the parathyroid gland, which stimulates bone resorption, and **calcitonin** from the thyroid gland, which inhibits bone resorption. These two hormones, together with a derivative of vitamin D, are responsible for maintaining a constant level of calcium in the blood (p. 750).

Plan of the Vertebrate Skeleton

The vertebrate skeleton is composed of two main divisions: the **axial skeleton,** which includes the skull, vertebral column, sternum, and ribs, and the **appendicular skeleton,** which includes the limbs (or fins or wings) and the pectoral and pelvic girdles (Figure 32-7). Not surprisingly, the skeleton has undergone extensive remodeling in the course of vertebrate evolution. The move from water to land forced dramatic changes in body form. With increased cephalization, the further concentration of brain, sense organs, and food-gathering and respiratory apparatus in the head, the skull became the most intricate portion of the skeleton. Some early fishes had as many as 180 skull bones (a source of frustration to paleontologists) but through loss of some bones and fusion of others, skull bones became greatly reduced in number during the evolution of the tetrapods. Amphibians and lizards have 50 to 95, and mammals, 35 or fewer. Humans have 29.

The vertebral column is the main stiffening axis of the postcranial skeleton. In fishes it serves much the same function as the notochord from which it is derived; that is, it provides points for muscle attachment and prevents telescoping of the body during muscle contraction. With the evolution of the amphibious and terrestrial tetrapods, the vertebrate body was no longer buoyed by the aquatic environment. The vertebral column became structurally adapted to withstand new regional stresses transmitted to the column by the two pairs of appendages. In the amniote tetrapods (reptiles, birds, and mammals), the vertebrae are differentiated into **cervical** (neck), **thoracic** (chest), **lumbar** (back), **sacral** (pelvic), and **caudal** (tail) vertebrae. In birds and also in humans the caudal vertebrae are reduced in number and size, and the sacral vertebrae are fused. The number of vertebrae varies among the different vertebrates. The python seems to lead the list with more than 400. In humans (Figure 32-7) there are 33 in a young child, but in adults 5 are fused to form the **sacrum** and 4 to form the **coccyx.** Besides the sacrum and coccyx, humans have 7 cervical, 12 thoracic, and

5 lumbar vertebrae. The number of cervical vertebrae (7) is constant in nearly all mammals, whether the neck is short as in dolphins, or long as in giraffes.

The first two cervical vertebrae, the **atlas** and the **axis,** are modified to support the skull and permit pivotal movements. The atlas bears the globe of the head much as the mythological Atlas bore the earth on his shoulders. The axis, the second vertebra, permits the head to turn from side to side.

Ribs are long or short skeletal structures that articulate medially with vertebrae and extend into the body wall. Fishes have a pair of ribs for every vertebra; they serve as stiffening elements in the connective tissue septa that separate the muscle segments and thus improve the effectiveness of the muscle contractions. Many fishes have both dorsal and ventral ribs, and some have numerous riblike intermuscular bones as well—all of which increase the difficulty and reduce the pleasure of eating certain kinds of fish. Other vertebrates have a reduced number of ribs, and some, such as the familiar leopard frog, have no ribs at all. Primates other than humans have 13 pairs of ribs; humans have 12 pairs, although approximately 1 person in 20 has a thirteenth pair. In mammals the ribs together form the thoracic basket, which supports the chest wall and prevents collapse of the lungs.

Most vertebrates, fishes included, have paired appendages. All fishes except the agnathans have thin pectoral and pelvic fins that are supported by the pectoral and pelvic girdles, respectively. Tetrapods (except caecilians, snakes, and limbless lizards) have two pairs of **pentadactyl** (five-toed) limbs, also supported by girdles. The pentadactyl limb is similar in all tetrapods, alive and extinct; even when highly modified for various modes of life, the elements are rather easily homologized (the evolution of the pentadactyl limb is illustrated in Figure 28-1, p. 530).

Modifications of the basic pentadactyl limb for life in different envi-

ronments involve the distal elements much more frequently than the proximal, and it is far more common for bones to be lost or fused than for new ones to be added. Horses and their relatives evolved a foot structure for fleetness by elongation of the third toe. In effect, a horse stands on its third fingernail (hoof), much like a ballet dancer standing on the tips of the toes. The bird wing is a good example of distal modification. The bird embryo bears 13 distinct wrist and hand bones (carpals and metacarpals), which are reduced to three in the adult. Most of the finger bones (phalanges) are lost, leaving four bones in three digits (see p. 578). The proximal bones (humerus, radius, and ulna), however, are only slightly modified in the bird wing.

In nearly all tetrapods the pelvic girdle is firmly attached to the axial skeleton, since the greatest locomotory forces transmitted to the body come from the hindlimbs. The pectoral girdle, however, is much more loosely attached to the axial skeleton, providing the forelimbs with greater freedom for manipulative movements.

Effect of Body Size on Bone Stress

As Galileo realized in 1638, the ability of animals' limbs to support a load decreases as animals increase in size (chapter opening essay, p. 630). Imagine two animals, one twice as long as the other, that are proportionally identical. That is, the larger animal is twice as long, twice as wide, and twice as tall as the smaller. The volume (and the weight) of the larger animal will be eight times the volume of the smaller ($2 \times 2 \times 2 = 8$). However, the strength of the larger animal's legs will be only four times the strength of the smaller, because bone, tendon, and muscle strength are proportional to the cross-sectional area. So, as Galileo noted, eight times the weight would have to be carried by only four times the strength. Because the maximum strength of mammalian bone is rather uniform per unit of

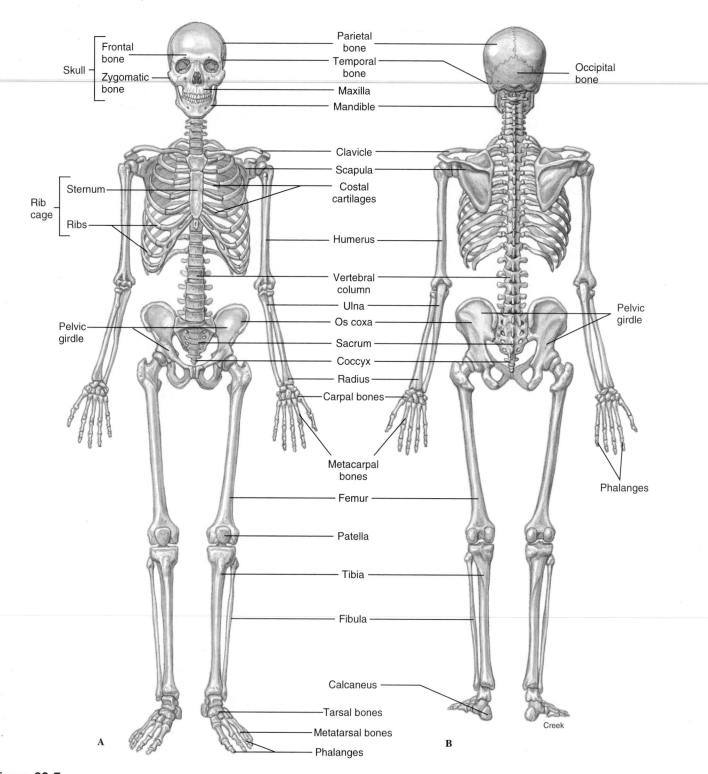

Figure 32-7
Human skeleton. **A,** Ventral view. **B,** Dorsal view. In comparison with other mammals, the human skeleton is a patchwork of primitive and specialized parts. Erect posture, brought about by specialized changes in legs and pelvis, enabled the primitive arrangement of arms and hands (arboreal adaptation of human ancestors) to be used for manipulation of tools. Development of the skull and brain followed as a consequence of the premium natural selection put on dexterity and ability to appraise the environment.

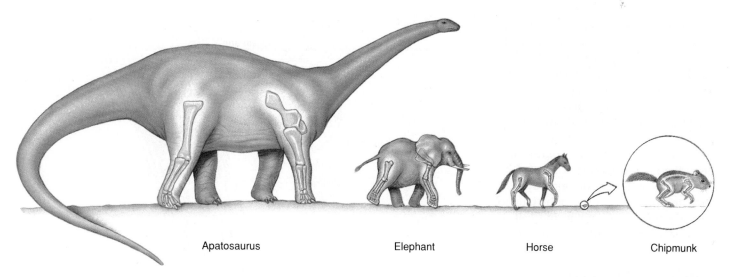

Apatosaurus   Elephant   Horse   Chipmunk

Figure 32-8

Comparison of postures in small and large mammals, showing the effect of scale. Because of its more upright posture, bone stresses in the horse are similar to those in the chipmunk. In mammals larger than horses (above about 300 kg), greatly increased stresses require that bones become exceedingly robust and that the animal lose agility.

cross-sectional area, how can animals become larger without placing unbearable stresses on long limb bones? One obvious solution is to make bones stouter and therefore stronger. However, throughout much of their size range, bone shape in different sized mammals does not change much. Instead, mammals have adapted their limb posture so that stresses are shifted to align with the long axis of the bones, rather than transversely. Small animals the size of a chipmunk run in a crouched limb posture, whereas a large mammal, such as a horse, has adopted an upright posture (Figure 32-8). Bones and muscles are capable of carrying far more weight when aligned more closely with the ground reaction force, as they are in a horse's leg. In this way, peak bone stresses during strenuous activity are no greater for a galloping horse than for a running chipmunk or dog.

For animals larger than horses, further mechanical advantage by changing limb posture is not possible because the limbs are fully upright. Instead, the long bones of an elephant weighing 2.5 metric tons, and those of the enormous dinosaur *Apatosaurus,* weighing an estimated 34 metric tons, are (were) extremely thick and robust (Figure 32-8), pro-

viding the safety factor these massive animals require(d). However, top running speeds of the largest terrestrial mammals decline with increasing size. Nevertheless, recent calculations of bone stresses in dinosaurs suggests that even the largest were capable of considerable agility (Alexander, 1991).

ANIMAL MOVEMENT

Movement is an important characteristic of animals. Animal movement occurs in many forms in animal tissues, ranging from barely discernible streaming of cytoplasm to extensive movements of powerful striated muscles. Most animal movement depends on a single fundamental mechanism: **contractile proteins,** which can change their form to elongate or contract. This contractile machinery is always composed of ultrafine fibrils—fine filaments, striated fibrils, or tubular fibrils (microtubules)—arranged to contract when powered by **ATP.** By far the most important protein contractile system is the **actomyosin system,** composed of two proteins, **actin** and **myosin.** This is an almost universal biomechanical system found from protozoa to vertebrates; it performs a long list of diverse functional roles. Cilia and flagella, however, are composed of

different proteins, and thus are exceptions to the rule. In this discussion we examine the three principal kinds of animal movement: ameboid, ciliary, and muscular.

AMEBOID MOVEMENT

Ameboid movement is a form of movement especially characteristic of amebas and other unicellular forms; it is also found in many wandering cells of metazoans, such as white blood cells, embryonic mesenchyme, and numerous other mobile cells that move through the tissue spaces. Ameboid cells change their shape by sending out and withdrawing **pseudopodia** (false feet) from any point on the cell surface. Beneath the plasmalemma lies a nongranular layer, the gel-like **ectoplasm,** which encloses the more liquid **endoplasm** (see Figure 12-4, p. 217).

Research with a variety of ameboid cells, including the pathogen-fighting phagocytes present in blood, has produced a consensus model to explain pseudopodial extension and ameboid crawling. Optical studies of an ameba in movement suggest the outer layer of ectoplasm surrounds a rather fluid core of endoplasm. Movement depends on actin and other regulatory proteins. According to one hypothesis (Stossel,

1994), as the pseudopod extends, hydrostatic pressure forces actin subunits into the pseudopod where they assemble into a network to form a gel state. At the trailing edge of the gel, where the network disassembles, the freed actin interacts with myosin to create a contractile force that pulls the cell along behind the extending pseudopod. Locomotion is assisted by membrane-adhesion proteins that attach temporarily to the substrate to provide traction, enabling the cell to crawl steadily forward.

CILIARY AND FLAGELLAR MOVEMENT

Cilia are minute, hairlike, motile processes that extend from the surfaces of the cells of many animals. They are a particularly distinctive feature of ciliate protistans, but except for the nematodes in which motile cilia are absent and the arthropods in which they are rare, cilia are found in all major groups of animals. Cilia perform many roles either in moving small organisms such as unicellular ciliates, flagellates, and the ctenophores (Figure 32-10B) through their aquatic environment or in propelling fluids and materials across the epithelial surfaces of larger animals.

Cilia are of remarkably uniform diameter (0.2 to 0.5 μm) wherever they are found. The electron microscope has shown that each cilium contains a peripheral circle of nine double microtubules arranged around two single microtubules in the center (Figure 32-9). (Several exceptions to the 9 + 2 arrangement have been noted; for example, sperm tails of flatworms have but one central microtubule, and sperm tails of a mayfly have no central microtubule.) Each microtubule is composed of a spiral array of protein subunits called **tubulin.** The microtubule doublets around the periphery are connected to each other and to the central pair of microtubules by a complex system of connective elements. Also extending from each doublet is a pair of arms composed of the protein **dynein.** The dynein arms, which act as cross

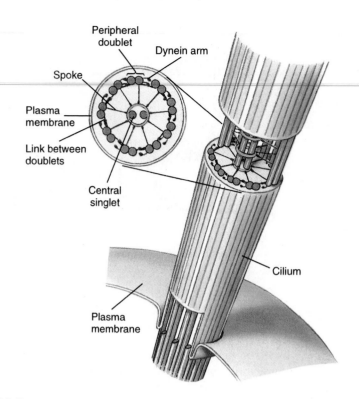

Figure 32-9

Cross section of a cilium showing the microtubules and connecting elements of the 9 + 2 arrangement typical of both cilia and flagella.

bridges between the doublets, operate to produce a sliding force between the microtubules.

A **flagellum** is a whiplike structure longer than a cilium and usually present singly or in small numbers at one end of a cell. They are found in members of flagellate protistans, in animal spermatozoa, and in sponges. The main difference between a cilium and a flagellum is in their beating pattern rather than in their structure, since both look alike internally. A flagellum beats symmetrically with snakelike undulations so that the water is propelled parallel to the long axis of the flagellum. A cilium, in contrast, beats asymmetrically with a fast power stroke in one direction followed by a slow recovery during which the cilium bends as it returns to its original position (Figure 32-10A). The water is propelled parallel to the ciliated surface (Figure 32-10B).

Although the mechanism of ciliary movement is not completely understood, it is known that the microtubules behave as "sliding filaments" that move past one another much like the sliding filaments of vertebrate skeletal muscle

that is described in the next discussion. Several kinds of evidence show that during contraction, the dynein arms link to adjacent microtubules, then swivel and release in repeated cycles. The continuous cycling of hundreds of thousands of dynein arms between the peripheral doublets pulls the doublets past each in a smooth sliding motion. The connecting elements of spokes and links then convert the doublet sliding force into a bending motion of the cilium or flagellum.

MUSCULAR MOVEMENT

Contractile tissue is most highly developed in muscle cells called **fibers.** Although muscle fibers themselves can do work only by contraction and cannot actively lengthen, they can be arranged in so many different configurations and combinations that almost any movement is possible.

Types of Vertebrate Muscle

Vertebrate muscle is broadly classified on the basis of the appearance of

Figure 32-10

A, Flagellum beats in wavelike undulations, propelling water parallel to the main axis of itself. Cilium propels water in direction parallel to the cell surface. **B,** Movement of cilia in comb plates of a ctenophore. Note how the waves of beating comb plates pass down a comb row, opposite the direction of the power stroke of individual cilia. The movement of one comb plate lifts the plate below it and so triggers the next lower plate and so on.

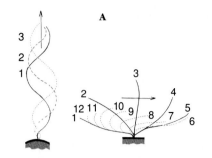

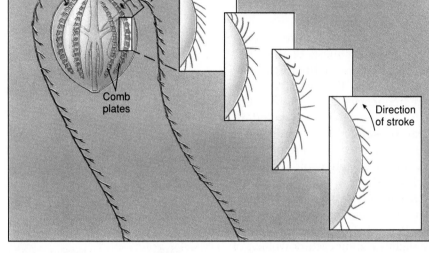

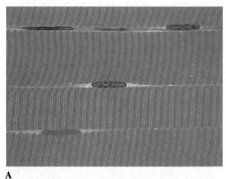

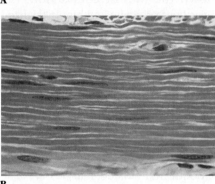

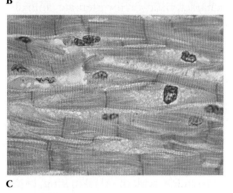

muscle cells (fibers) when viewed with a light microscope. **Striated muscle** appears transversely striped (striated), with alternating dark and light bands (Figure 32-11A). We can recognize two types of striated muscle: **skeletal** and **cardiac muscle.** A third kind of vertebrate muscle is **smooth** (or visceral) **muscle,** which lacks the characteristic alternating bands of the striated type.

Skeletal muscle is typically organized into sturdy, compact bundles or bands (Figure 32-11A). It is called skeletal muscle because it is attached to skeletal elements and is responsible for movements of the trunk, appendages, respiratory organs, eyes, mouthparts, and so on. Skeletal muscle **fibers** are extremely long, cylindrical, multinucleate cells that may reach from one end of the muscle to the other. They are packed into bundles called **fascicles** (L. *fasciculus,* small bundle), which are enclosed by tough connective tissue. The fascicles are in turn grouped into a discrete **muscle** surrounded by a thin connective tissue layer. Most skeletal muscles taper at their ends, where they connect to bones by tendons. Other muscles, such as the ventral abdominal muscles, are flattened sheets.

In most fishes, amphibians, and to some extent lizards and snakes, there is a segmented organization of muscles alternating with the vertebrae. The skeletal muscles of other vertebrates, by splitting, fusion, and shifting, have developed into specialized muscles best suited for manipulating the jointed appendages that have evolved for locomotion on land. Skeletal muscle contracts powerfully and quickly but fatigues more rapidly than does smooth muscle. Skeletal muscle is sometimes called **voluntary muscle** because it is stimulated by motor fibers and is under conscious cerebral control.

Smooth muscle lacks the striations typical of skeletal muscle (Figure 32-11B). The cells are long, tapering strands, each containing a single nucleus. Smooth muscle cells are organized into sheets of muscle circling the walls of the alimentary canal, blood vessels, respiratory passages, and urinary and genital ducts. Smooth muscle is typically slow acting and can maintain prolonged contractions with very little energy expenditure. It is under the control of the autonomic nervous system; thus, unlike skeletal muscle, its contractions are involuntary and unconscious. The principal functions of smooth muscles are to push the material in a tube, such as the intestine, along its way by active contractions or to regulate the diameter of a tube, such as a blood vessel, by sustained contraction.

Figure 32-11

Photomicrographs of types of vertebrate muscle. **A,** Skeletal muscle (human) showing several striated fibers (cells) lying side by side. **B,** Smooth muscle (human) showing absence of striations. Note elongate nuclei in the long fibers. **C,** Cardiac muscle (monkey). Note the vertical bars, called intercalated discs, joining separate fibers end to end.

Cardiac muscle, the seemingly tireless muscle of the vertebrate heart, combines certain characteristics of both skeletal and smooth muscle (Figure 32-11C). It is fast acting and striated like skeletal muscle, but contraction is under involuntary autonomic control like smooth muscle. Actually the autonomic nerves serving the heart can only speed up or slow down the rate of contraction; the heartbeat originates within specialized cardiac muscle, and the heart continues to beat even after all autonomic nerves are severed (heart excitation is described on p. 684). Cardiac muscle is composed of closely opposed, but separate, uninucleate cell fibers.

Types of Invertebrate Muscle

Smooth and striated muscles are also characteristic of invertebrate animals, but there are many variations of both types and even instances in which the structural and functional features of vertebrate smooth and striated muscle are combined. Striated muscle appears in invertebrate groups as diverse as the cnidarians and the arthropods. The thickest muscle fibers known, approximately 3 mm in diameter and 6 cm long, are those of giant barnacles and of Alaska king crabs living along the Pacific coast of North America. Such large muscle cells lend themselves well to physiological studies and are understandably popular with muscle physiologists.

In the limited space available to treat the great diversity of muscle structure and function in the invertebrate assemblage, we have selected for discussion two functional extremes: the specialized adductor muscles of molluscs and the fast flight muscles of insects.

Bivalve molluscan muscles contain fibers of two types. One kind is striated muscle that can contract rapidly, enabling the bivalve to snap shut its valves when disturbed. Scallops use these "fast" muscle fibers to swim in their awkward manner (see Figure 17-24B, p. 335). The second muscle type is smooth muscle, capable of slow, long-lasting contractions. Using these fibers, a bivalve can keep its valves tightly shut for hours or even days. Such adductor muscles use little metabolic energy and receive remarkably few nerve impulses to maintain the activated state. The contracted state has been likened to a "catch mechanism" involving some kind of stable cross-linkage between the contractile proteins within the fiber. However, despite considerable research, there is still much uncertainty about how this adductor mechanism works.

Insect flight muscles are virtually the functional antithesis of the slow, holding muscles of bivalves. The wings of some of the small flies operate at frequencies greater than 1000 beats per second. The so-called **fibrillar muscle,** which contracts at these frequencies—far greater than even the most active of vertebrate muscles—shows unique characteristics. It has very limited extensibility; that is, the wing leverage system is arranged so that the muscles shorten only slightly during each downbeat of the wings. Furthermore, the muscles and wings operate as a rapidly oscillating system in an elastic thorax (see Figure 21-12, p. 411). Since the muscles rebound elastically and are activated by stretch during flight, they receive impulses only periodically rather than one impulse per contraction; one reinforcement impulse for every 20 or 30 contractions is enough to keep the system active. Insect flight muscles are described in more detail in Chapter 21 (p. 410).

Structure of Striated Muscle

As mentioned earlier, striated muscle is so named because of the periodic bands, plainly visible under the light microscope, that pass across the widths of the muscle cells. Each cell, or **fiber,** is a multinucleated tube containing numerous **myofibrils,** packed together and invested by the cell membrane, the **sarcolemma** (Figure 32-12). The myofibril contains

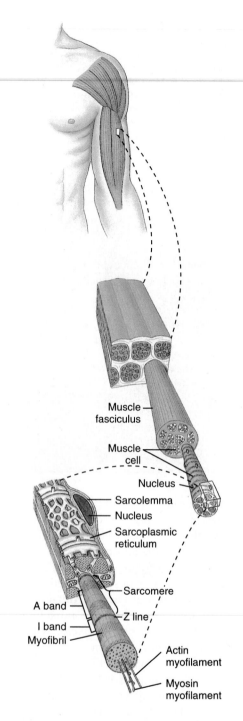

Muscle fasciculus

Muscle cell

Nucleus
Sarcolemma
Nucleus
Sarcoplasmic reticulum

Sarcomere

A band
I band
Myofibril

Z line

Actin myofilament

Myosin myofilament

Figure 32-12

Organization of skeletal muscle from gross to molecular level. A skeletal muscle (*top*) is composed of thousands of multinucleated muscle fibers (*center*), each containing thousands of myofibrils (*bottom*), Each myofibril contains numerous thick (myosin) and thin (actin) filaments that interact to slide past each other during contraction to shorten the muscle. The sarcoplasmic reticulum is a network of tubules surrounding the myofibrils that serves as a communication system for carrying a depolarization to the filaments within the muscle fiber.

two types of **myofilaments:** thick filaments composed of the protein **myosin,** and thin filaments, composed of the protein **actin.** These are the actual contractile proteins of the muscle. The thin filaments are held together by a dense structure called the Z line. The functional unit of the myofibril, the **sarcomere,** extends between successive Z lines. These anatomical relationships are diagramed in Figure 32-12.

Human muscle tissue develops before birth, and a newborn child's complement of skeletal muscle fibers is all that he or she will ever have. But while an adult male weight lifter and a young boy have a similar number of muscle fibers, the weight lifter may be several times the boy's strength because repeated high-intensity, short-duration exercise has induced the synthesis of additional actin and myosin filaments. Each fiber has hypertrophied, becoming larger and stronger. Endurance exercise such as long-distance running produces a very different response. Fibers do not become greatly stronger but develop more mitochondria and myoglobin and become adapted for a high rate of oxidative phosphorylation. These changes, together with the development of more capillaries serving the fibers, lead to increased capacity for long-duration activity.

Each thick filament is made up of myosin molecules packed together in an elongate bundle (Figure 32-13). Each myosin molecule is composed of two polypeptide chains, each having a club-shaped head. Lined up as they are in a bundle to form a thick filament, the double heads of each myosin molecule face outward from the center of the filament. These heads act as the molecular cross bridges that interact with the thin filaments during contraction.

The thin filaments are more complex because they are composed of three different proteins. The backbone of the thin filament is a double strand of the protein actin, twisted into a double helix. Surrounding the actin fila-

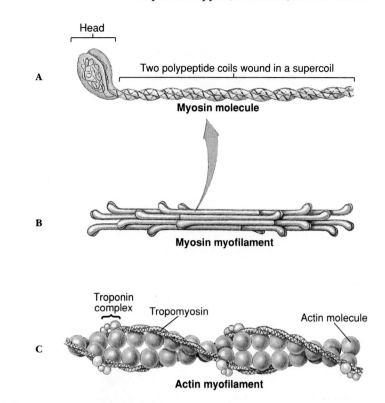

Figure 32-13

Molecular structure of thick and thin myofilaments of skeletal muscle. **A,** The myosin molecule is composed of two polypeptides coiled together and expanded at their ends into a globular head. **B,** The thick myofilament is composed of a bundle of myosin molecules with the globular heads extended outward. **C,** The thin myofilament consists of a double strand of actin surrounded by two tropomyosin strands. A globular protein complex, troponin, occurs in pairs at every seventh actin unit. Troponin is a calcium-dependent switch that controls the interaction between actin and myosin.

ment are two thin strands of another protein, **tropomyosin,** that lie near the grooves between the actin strands. Each tropomyosin strand is itself a double helix as shown in Figure 32-13C.

The third protein of the thin filament is **troponin,** a complex of three globular proteins located at intervals along the filament. Troponin is a calcium-dependent switch that acts as the control point in the contraction process.

Sliding Filament Model of Muscle Contraction

In the 1950s the English physiologists A. F. Huxley and H. E. Huxley independently proposed the **sliding filament model** to explain striated muscle contraction. According to this model, the thick and thin filaments become linked together by molecular cross bridges, which act as levers to pull the filaments past each other. During contraction, the cross bridges on

the thick filaments swing rapidly back and forth, alternately attaching to and releasing from special receptor sites on the thin filaments, and drawing the thin filaments past the thick in a kind of ratchet action. As contraction continues, the Z lines are pulled closer together (Figure 32-14). Thus the sarcomere shortens. Because all of the sarcomere units shorten together, the muscle contracts. Relaxation is a passive process. When the cross bridges between the thick and thin filaments release, the sarcomeres are free to lengthen. This requires some force, which is usually supplied by antagonistic muscles or the force of gravity.

Control of Contraction

Muscle contracts in response to nerve stimulation. If the nerve supply to a muscle is severed, the muscle **atrophies,** or wastes away. Skeletal muscle fibers are innervated by motor neurons whose cell bodies are located in the

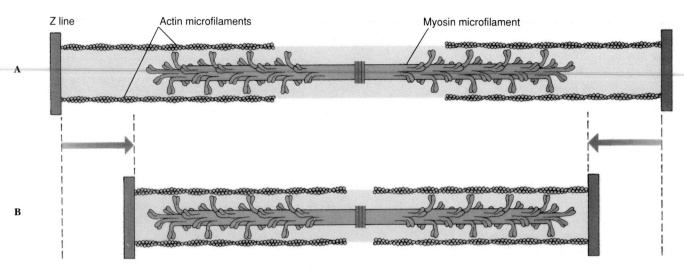

Figure 32-14
Sliding myofilament model, showing how thick and thin myofilaments interact during contraction. **A,** Muscle relaxed. **B,** Muscle contracted.

spinal cord. Each cell body gives rise to a motor axon that leaves the spinal cord to travel by way of a peripheral nerve trunk to a muscle where it branches repeatedly into many terminal branches. Each terminal branch innervates a single muscle fiber. Depending on the type of muscle, a single motor axon may innervate as few as three or four muscle fibers (where very precise control is needed, such as the muscles that control eye movement) or as many as 2000 muscle fibers (where precise control is not required, such as the postural muscles of the back). The motor neuron and all the muscle fibers it innervates is called a **motor unit.** The motor unit is the functional unit of skeletal muscle. When a motor neuron fires, the action potential passes to all the fibers of the motor unit and each is stimulated to contract simultaneously. The total force exerted by a muscle depends on the number of motor units activated. Precise control of movement is achieved by varying the number of motor units activated at any one time. A smooth and steady increase in muscle tension is produced by increasing the number of motor units brought into play; this is called motor unit **recruitment.**

The Myoneural Junction

The place where a motor axon terminates on a muscle fiber is called

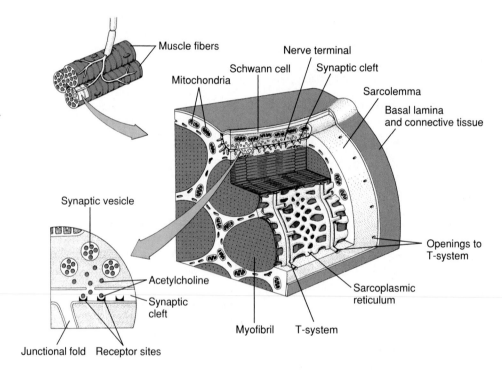

Figure 32-15
Section of vertebrate skeletal muscle showing nerve-muscle synapse (myoneural junction), sarcoplasmic reticulum, and connecting transverse tubules (T-tubule system). Arrival of a nerve impulse at the synapse triggers the release of acetylcholine into synaptic cleft (*inset at left*). The binding of transmitter molecules to receptors generates membrane depolarization. This spreads across the sarcolemma, into the T-tubule system, and to the sarcoplasmic reticulum where the sudden release of calcium sets in motion the contractile machinery of the myofibril.

the **myoneural junction** (Figure 32-15). At the junction is a tiny gap, or **synaptic cleft,** that thinly separates a nerve fiber and muscle fiber. In the vicinity of the junction, the neuron stores a chemical, **acetylcholine,** in minute vesicles known as **synaptic vesicles.** Acetylcholine is released when a nerve impulse reaches a synapse. This substance is a chemical mediator that diffuses across the narrow junction and acts on the muscle fiber membrane to generate an electrical depolarization.

The depolarization spreads rapidly through the muscle fiber, causing it to contract. Thus the synapse is a special chemical bridge that couples together the electrical activities of nerve and muscle fibers.

Built into vertebrate skeletal muscle is an elaborate conduction system that serves to carry the depolarization from the myoneural junction to the densely packed filaments within the fiber. Along the surface of the sarcolemma are numerous invaginations that project as a system of tubules into the muscle fiber. This is called the **T-system** (Figure 32-15). The T-system is continuous with the **sarcoplasmic reticulum,** a system of fluid-filled channels that runs parallel to the myofilaments. The system is ideally arranged for speeding the electrical depolarization from the myoneural junction to the myofilaments within the fiber.

Excitation-Contraction Coupling

How does the electrical depolarization activate the contractile machinery? In the resting, unstimulated muscle, shortening does not occur because the thin tropomyosin strands surrounding the actin myofilaments lie in a position that prevents the myosin heads from attaching to actin. When the muscle is stimulated and the electrical depolarization arrives at the sarcoplasmic reticulum surrounding the fibrils, calcium ions are released (Figure 32-16). Some of the calcium binds to the control protein troponin. Troponin immediately undergoes changes in shape that allow tropomyosin to move out of its blocking position, exposing active sites on the actin myofilaments. The myosin heads then bind to these sites, forming cross bridges between adjacent thick and thin myofilaments. This sets in motion an **attach-pull-release cycle** that occurs in a series of steps as shown in Figure 32-16. The release of bond energy from ATP activates the myosin head, which swings 45 degrees, at the same time releasing a molecule of ADP. This is the power stroke that pulls the actin filament a distance of about 10 nm, and it comes

to an end when another ATP molecule binds to the myosin head, inactivating the site. Thus each cycle requires the expenditure of energy in the form of ATP (Figure 32-16).

Shortening will continue as long as nerve impulses arrive at the myoneural junction and free calcium remains available around the myofilaments. The attach-pull-release cycle can repeat again and again, 50 to 100 times per second, pulling the thick and thin filaments past each other. While the distance each sarcomere can shorten is very small, this distance is multiplied by the thousands of sarcomeres lying end to end in a muscle fiber. Consequently, a strongly contracting muscle may shorten by as much as one-third its resting length.

When stimulation stops, the calcium is quickly pumped back into the sarcoplasmic reticulum. Troponin resumes its original configuration, tropomyosin moves back into its blocking position on actin, and the muscle relaxes.

Energy for Contraction

Adenosine triphosphate (ATP) is the immediate source of energy for muscle, but the amount present will sustain contraction for only a fraction of a second. However, vertebrate muscle contains a much larger reservoir of high-energy phosphate, creatine phosphate. This compound contains even more free bond energy than ATP (p. 66) and thus can readily transfer its bond energy to ADP to form ATP.

Creatine phosphate + ADP → ATP + Creatine

The reserves of creatine phosphate are soon depleted in rapidly contracting muscle and must be restored by the oxidation of carbohydrate. The major store of carbohydrate in muscle is glycogen. In fact, about three-fourths of all the glycogen in the body is stored in muscle (most of the rest is stored in the liver). Glycogen can be readily converted into glucose-6-phosphate, the first stage of glycolysis that leads into mitochondrial respiration and the generation of ATP (p. 71).

The role of creatine phosphate, the high-energy compound of vertebrates that regenerates ATP during muscle contraction, is replaced in most invertebrates by arginine phosphate. Neither creatine phosphate nor arginine phosphate is stored in quantities sufficient to support contractions for more than a few seconds; sustained contractions depend ultimately on the oxidation of fuels: carbohydrates and fats.

If muscular contraction is not too vigorous or too prolonged, glucose can be completely oxidized to carbon dioxide and water by **aerobic metabolism.** During prolonged or heavy exercise, however, the blood flow to the muscles, although greatly increased above the resting level, is insufficient to supply oxygen as rapidly as required for the complete oxidation of glucose. When this happens, the contractile machinery receives its energy largely by **anaerobic glycolysis,** a process that does not require oxygen (p. 71). The ability to take advantage of this anaerobic pathway, although not nearly as efficient as the aerobic one, is of great importance; without it, all forms of heavy muscular exertion would be impossible.

During anaerobic glycolysis, glucose is degraded to lactic acid with the release of energy. This is used to resynthesize creatine phosphate, which in turn passes the energy to ADP for the resynthesis of ATP. Lactic acid accumulates in the muscle and diffuses rapidly into the general circulation. If the muscular exertion continues, the buildup of lactic acid causes enzyme inhibition and fatigue. Thus the anaerobic pathway is a self-limiting one, since continued heavy exertion leads to exhaustion. The muscles incur an **oxygen debt** because the accumulated lactic acid must be oxidized by extra oxygen. After the period of exertion, oxygen consumption remains elevated until all of the lactic acid has been oxidized or resynthesized to glycogen.

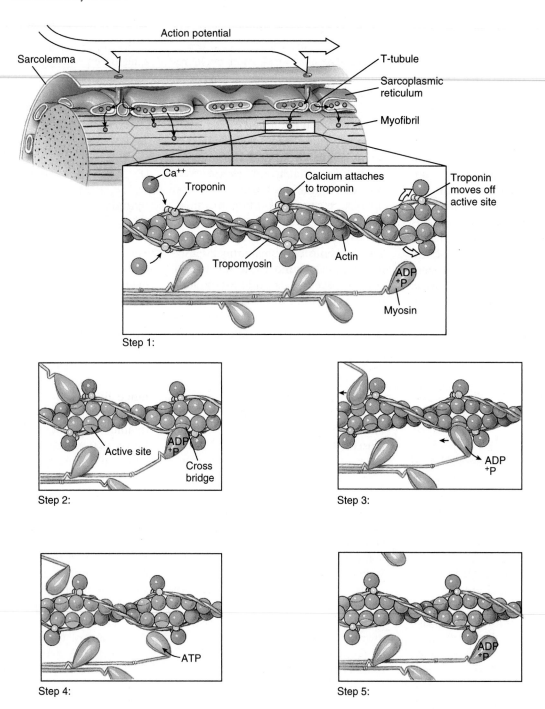

Step 1:

Step 2:

Step 3:

Step 4:

Step 5:

Figure 32-16

Excitation-contraction coupling in vertebrate skeletal muscle. **Step 1:** An action potential spreads along the sarcolemma and is conducted inward to the sarcoplasmic reticulum by way of T tubules (T-tubule system). Calcium ions released from the sarcoplasmic reticulum diffuse rapidly into the myofibrils and bind to troponin molecules on the actin molecule. Troponin molecules are moved away from the active sites. **Step 2:** Myosin cross bridges bind to the exposed active sites. **Step 3:** Using the energy stored in ATP, the myosin head swings toward the center of the sarcomere. ADP and a phosphate group are released. **Step 4:** The myosin head binds another ATP molecule; this frees the myosin head from the active site on actin. **Step 5:** The myosin head splits ATP, retaining the energy released as well as the ADP and the phosphate group. The cycle can now be repeated as long as calcium is present to open active sites on the actin molecules.

Muscle Performance

Fast and Slow Fibers

The skeletal muscles of vertebrates consist of more than one type of fiber. Some muscles contain a high proportion of **slow fibers,** which are specialized for slow, sustained contractions without fatigue. Slow fibers are important in maintaining posture in terrestrial vertebrates. Such muscles are often called **red muscles** because they contain an extensive blood supply, a high density of mitochondria for supplying ATP, and abundant stored myoglobin to supply oxygen reserves, all of which give the muscle a red color. The "dark meat" of a chicken leg is a familiar example.

Other muscles have a high proportion of **fast fibers,** also called twitch fibers, which are capable of fast, powerful contractions. Typically, such muscles cannot maintain their activity for long periods without becoming fatigued, but they are essential for fast movements. Lacking an efficient blood supply and high density of mitochondria and myoglobin, fast muscles may be pale in color and are often called **white muscles.** However, not all fast fibers are alike. Some animals, such as dogs and ungulates (hoofed animals), are capable of running for long periods of time because they have muscles with a high percentage of fast fibers with high oxidative capacity and efficient blood delivery. Such muscles work under aerobic conditions. Members of the cat family, however, have running muscles made up almost entirely of fast fibers that operate anaerobically. During a chase, such muscles build up a substantial oxygen debt that is replenished after the chase. For example, a cheetah after a high-speed chase lasting less than a minute, will pant heavily for 30 to 40 minutes before its oxygen debt is paid off.

Importance of Tendons in Energy Storage

When mammals walk or run, much kinetic energy is stored from step to step as elastic strain energy in the tendons. For example, during running the Achilles tendon is stretched by a combination of the downward force of the body on the foot and the contraction of the calf muscles. The tendon then recoils, extending the foot while the muscle is still contracted, propelling the leg forward (Figure 32-17). An extreme example of this bouncing ball principle is the bounding of a kangaroo, which essentially bounces along on its tendons, utilizing the effect of gravity. This type of movement uses far less energy than would be required if every step relied solely on alternate muscle contraction and relaxation.

There are many examples of elastic storage in the animal kingdom. It is used in the ballistic jumps of grasshoppers and fleas, in the wing hinges of flying insects, in the hinge ligaments of bivalve molluscs, and in the highly elastic large dorsal ligament (ligamentum nuchae) that helps support the head of hoofed mammals.

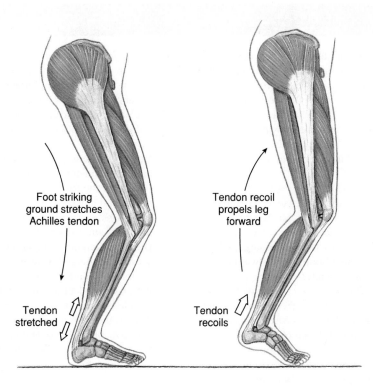

Foot striking
ground stretches
Achilles tendon

Tendon
stretched

Tendon recoil
propels leg
forward

Tendon
recoils

Figure 32-17

Energy storage in the Achilles tendon of the human leg. During running, stretching of the Achilles tendon when the foot strikes the ground stores kinetic energy that is released to propel the leg forward.

Summary

An animal is wrapped in a protective covering, the integument, which may be as simple as the delicate plasma membrane of an ameba or as complex as the skin of a mammal. The arthropod exoskeleton is the most complex of invertebrate integuments, consisting of a two-layered cuticle secreted by a single-layered epidermis. It may be hardened by calcification of sclerotization and must be molted at intervals to permit body growth. Vertebrate integument consists of two layers: the epidermis, which gives rise to various derivatives such as hair, feathers, and claws; and the dermis, which supports and nourishes the epidermis. It also is the origin of bony derivatives such as fish scales and deer antlers.

Integument color is of two kinds: structural color, produced by refraction or scattering of light by particles in the integument, and pigmentary color, produced by pigments that are usually confined to special pigment cells (chromatophores).

Skeletons are supportive systems that may be hydrostatic or rigid. The hydrostatic skeletons of several soft-walled invertebrate groups depend on body-wall muscles that contract against a noncompressible internal fluid of constant volume. In a similar manner, muscular hydrostats, such as the tongue of mammals and reptiles, and the trunk of elephants, rely on muscle bundles arranged in complex patterns to produce movement without either skeletal support or a liquid-filled cavity. Rigid skeletons have evolved with attached muscles that act with the supportive skeleton to produce movement. Arthropods have an external skeleton, which must be shed periodically to make way for an enlarged replacement. The vertebrates developed an internal skeleton, a framework formed of cartilage or bone, that can grow with the animal, while, in the case of bone, additionally serving as a reservoir of calcium and phosphate.

Animal movement, whether in the form of cytoplasmic streaming, ameboid movement, or the contraction of an organized muscle mass, depends on specialized contractile proteins. The most important of these is the actomyosin system, which is usually organized into elongate thick and thin filaments that slide past one another during contraction. When a muscle is stimulated, an electrical depolarization is conducted into the muscle fibers through the sarcoplasmic reticulum, causing the release of calcium. Calcium binds to a protein troponin complex associated with the thin actin filament. This causes tropomyosin to shift out of its blocking position and allows the myosin heads to cross-bridge with the actin filament. Powered by ATP, the myosin heads swivel back and forth to pull the thick and thin filaments past each other. Phosphate bond energy for contraction is supplied by carbohydrate fuels through a storage intermediate, creatine phosphate.

Vertebrate skeletal muscle consists of variable percentages of both slow fibers, used principally for sustained postural contractions, and fast fibers, used in locomotion. Tendons are important in locomotion because the kinetic energy stored in stretched tendons at one stage of a locomotory cycle is released at a subsequent stage.

Review Questions

1. Describe the structure of the arthropod integument, and explain the difference in the way the cuticle is hardened in crustaceans and in insects.
2. Distinguish between epidermis and dermis in vertebrate integument, and describe the structural derivatives of these two layers.
3. What is the difference between structural colors and colors based on pigments? How do the chromatophores of vertebrates and cephalopod molluscs differ in structure and function?
4. Explain how the human skin develops some protection against the damaging effects of ultraviolet solar radiation.
5. Explain what a hydrostatic skeleton is and how it is used in locomotion. What is a muscular hydrostat? Offer examples of both hydrostatic skeleton and muscular hydrostat.
6. What is hyaline cartilage? Compare its distribution and function in lower and higher vertebrates.

7. What is the difference between endochondral and membranous bone? Between spongy and compact bone?
8. Discuss the role of osteoclasts, osteoblasts, parathyroid hormone, and calcitonin in bone growth.
9. Name the major skeletal components included in the axial and appendicular skeleton.
10. Explain why load stresses placed on bones tend to increase as terrestrial animals become larger. What solutions to this problem have evolved that allow animals to become large, while maintaining bone stresses within margins of safety?
11. Describe the interaction of endoplasm and ectoplasm in ameboid movement.
12. Compare the structure and function of a cilium with those of a flagellum.
13. Describe the structural and functional features that distinguish each of the three types of vertebrate muscle.

14. What functional features set molluscan smooth muscle and insect fibrillar muscle apart from any known vertebrate muscle?
15. Explain how skeletal muscle shortens according to the sliding filament hypothesis.
16. Describe in sequence the events in muscle stimulation, explaining the role of each of the following: motor unit, myoneural junction, acetylcholine, sarcoplasmic reticulum, calcium, troponin, and tropomyosin.
17. Describe the immediate and reserve sources of energy for muscle contraction. Under what circumstances is an oxygen debt incurred during muscle contraction?
18. What is the difference between fast and slow fibers in skeletal muscle? What is the importance of tendons in promoting efficiency of locomotion?

Selected References

See also general references for Part IV, p. 774.

Alexander, R. M. 1982. Locomotion in animals. New York, Chapman and Hall. *Concise, fully comparative treatment. Introduced with a discussion of "sources of power" followed by treatment of mechanisms and energetics of locomotion on land, in water, and in the air. Undergraduate level.*

Alexander, R. M. 1991. How dinosaurs ran. Sci. Am. **264**:130–136 (April). *Did the Mesozoic dinosaurs plod sluggishly along or run? The author suggests they may have been formidable running machines.*

Alexander, R. M. 1992. The human machine. New York, Columbia University Press. *Describes all kinds of human movement with the human body viewed as an engineered machine. Well chosen illustrations.*

Caplan, A. J. 1984. Cartilage. Sci. Am. **251**:84–94 (Oct.). *Structure, aging and development of vertebrate cartilage.*

Hadley, N. F. 1986. The arthropod cuticle. Sci. Am. **255**:104–112 (July). *Describes properties of this complex covering that account for much of the adaptive success of arthropods.*

McMahon, T. A. 1984. Muscles, reflexes, and locomotion. Princeton, Princeton University Press. *Comprehensive, ranging from basic muscle mechanics to coordinated motion. Although sprinkled with mathematical models, the text is lucid throughout.*

Nadel, E. R. 1985. Physiological adaptations to aerobic training. Am. Sci. **73**(4):334–343 (July–Aug.). *The studies reported here on energy conversion in muscle were crucial to the training of a pilot for the Daedalus project, the successful world record 119-kilometer flight of a human-powered aircraft in April, 1988 (reported in Am. Sci. July–Aug., 1988).*

Shipman, P., A. Walker, and D. Bichell. 1985. The human skeleton. Cambridge, Massachusetts, Harvard University Press. *Comprehensive view of the human skeleton.*

Spearman, R. I. C. 1973. The integument: a textbook on skin biology. Cambridge, Cambridge University Press. *Comparative treatment, embracing both invertebrates and vertebrates.*

Stossel, T. P. 1994. The machinery of cell crawling. Sci. Am. **271**:54–63 (Sept.). *Cell crawling depends on the orderly assembly and disassembly of an actin protein scaffold.*

33

Homeostasis

Osmotic Regulation, Excretion, and Temperature Regulation

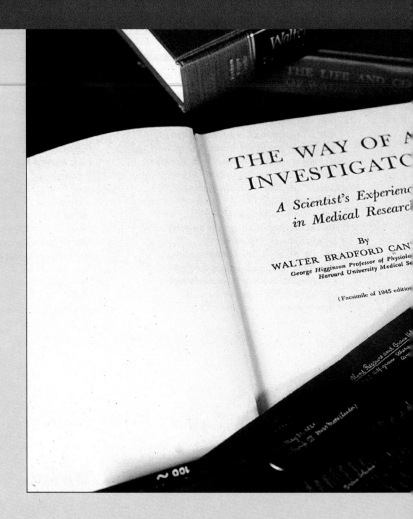

Homeostasis: Birth of a Concept

The tendency toward internal stabilization of the animal body was first recognized by Claude Bernard, great French physiologist of the nineteenth century who, through his studies of blood glucose and liver glycogen, discovered the first internal secretions. Out of a lifetime of study and experimentation gradually grew the principle for which this retiring and lonely man is best remembered, that of the constancy of the internal environment, a principle that in time would pervade physiology and medicine. Years later, at Harvard University, the American physiologist Walter B. Cannon (Figure 33–1) reshaped and restated Bernard's idea. Developed out of his studies of the nervous system and reactions to stress, he described the ceaseless balancing and rebalancing of physiological processes that maintain stability and restore the normal state when it has been disturbed. And

he gave it a name: homeostasis. The term soon flooded the medical literature of the 1930s. Physicians spoke of getting their patients back into homeostasis. Even politicians and sociologists saw what they considered to be deep nonphysiological implications. Cannon enjoyed this broadened application of the concept and later suggested that democracy was the form of government that took the homeostatic middle course. Despite the enduring importance of the homeostasis concept, Cannon never received the Nobel Prize—one of several acknowledged oversights of the Nobel Committee. At the end of his life, Cannon expressed his ideas about scientific research in his autobiography, *The Way of an Investigator*. This engaging book describes the resourceful career of a homespun man whose life embodied the traits that favor successful research. ◼

Figure 33-1

Walter Bradford Cannon (1871 to 1945), Harvard professor of physiology who coined the term "homeostasis" and developed the concept originated by French physiologist Claude Bernard (Figure 34-2, p. 671).

From J. F. Fulton & L. G. Wilson, Selected Readings in the History of Physiology, *1966. Courtesy of Charles C. Thomas, Publisher, Springfield, Illinois.*

The concept of homeostasis, described in the opening essay, permeates all physiological thinking and is the theme of this and the following chapter. Although the concept of homeostasis was first developed from studies with mammals, it applies to single-celled organisms as well as to vertebrates. Potential changes in the internal environment arise from two sources. First, metabolic activities require a constant supply of materials, such as oxygen, nutrients, and salts, that cells withdraw from their surroundings and that must be replaced. Cellular activity also produces waste products that must be disposed. Second, the internal environment responds to changes in the organism's external environment. Changes from either source must be stabilized by the physiological mechanisms of homeostasis.

In more complex metazoans, homeostasis is maintained by coordinated activities of the circulatory, nervous, and endocrine systems, and especially by the organs that serve as sites of exchange with the external environment. These last include the kidneys, lungs or gills, digestive tract, and integument. Through these organs oxygen, foods, minerals, and other constituents of body fluids enter, water is exchanged, heat is lost, and metabolic wastes are eliminated.

We will look first at the problems of controlling the internal fluid environment in aquatic animals. Next we will briefly examine how these problems are solved by terrestrial animals and consider the function of the organs that regulate the internal state. Finally we will look at the strategies that have evolved for living in a world of changing temperatures.

WATER AND OSMOTIC REGULATION

HOW MARINE INVERTEBRATES MEET PROBLEMS OF SALT AND WATER BALANCE

Most marine invertebrates are in osmotic equilibrium with their seawater environment. They have body surfaces that are permeable to salts and water so that their body fluid concentration rises or falls in conformity with changes in concentrations of seawater. Because such animals are incapable of regulating their body fluid osmotic pressure, they are referred to as **osmotic conformers.** Invertebrates living in the open sea are seldom exposed to osmotic fluctuations because the ocean is a highly stable environment. Oceanic invertebrates have, in fact, very limited abilities to withstand osmotic change. If they should be exposed to dilute seawater, they die quickly because their cells cannot tolerate dilution and are helpless to prevent it. These animals are restricted to living in a narrow salinity range and are said to be **stenohaline** (Gr. *stenos,* narrow, + *hals,* salt). An example is the marine spider crab (Figure 33-2).

Conditions along the coasts and in estuaries and river mouths are much less constant than those of the open ocean. Here animals must be able to withstand large and often abrupt changes in salinity as the tides ebb and flow and mix with fresh water draining from rivers. These animals are termed **euryhaline** (Gr.

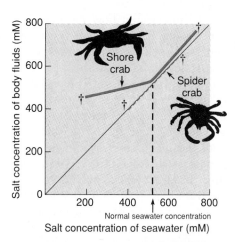

Figure 33-2

Salt concentration of body fluids of two crabs as affected by variations in the seawater concentration. The 45-degree line represents equal concentration between body fluids and seawater. Since the spider crab cannot regulate the salt concentration of its fluids, it conforms to whatever changes happen in the seawater. The shore crab, however, can regulate osmotic concentration of its fluids to some degree because in dilute seawater the shore crab can hold its salt concentration above seawater concentration. For example, when seawater is 200 mM (millimolar), the shore crab's body fluid is approximately 430 mM. Crosses at ends of lines indicate limits of tolerance for each species.

eurys, broad, + *hals,* salt), meaning that they can survive a wide range of salinity change, mainly because they demonstrate varying powers of **osmotic regulation.** For example, the brackish-water shore crab can resist dilution of body fluids by dilute (brackish) seawater (Figure 33-2). Although the concentration of salts in the body fluids falls, it does so less rapidly than the fall in seawater concentration. This crab is a **hyperosmotic regulator** because in a dilute environment it can maintain the salt concentration of its blood above that of the surrounding water.

What is the advantage of hyperosmotic regulation over osmotic conformity, and how is this regulation accomplished? The advantage is that by regulating against excessive dilution, thus protecting the cells from extreme changes, these crabs can live successfully in the physically unstable but biologically rich coastal environment. Their powers of regulation are limited, however, because if the water is highly diluted, their regulation fails and they die.

To understand how the brackish-water shore crab and other coastal invertebrates achieve hyperosmotic regulation, let us examine the problems they face. First, the salt concentration of the internal fluids is greater than in the dilute seawater outside. This causes a steady osmotic influx of water. As with the membrane osmometer containing a salt solution (p. 52), water diffuses inward because it is more concentrated outside than inside. The shore crab is not nearly as permeable as a membrane osmometer—most of its shelled body surface is, in fact, almost impermeable to water—but the thin respiratory surfaces of the gills are highly permeable. Obviously the crab cannot insulate its gills with an impermeable hide and still breathe. The problem is solved by removing the excess water through the action of the kidney (the antennal gland located in the crab's head).

The second problem is salt loss. Again, because the animal is saltier than its environment, it cannot avoid loss of ions by outward diffusion across the gills. Salt is also lost in the urine. This problem is solved by special salt-secreting cells in the gills that actively remove ions from the dilute seawater and move them into the blood, thus maintaining the internal osmotic concentration. This is an **active transport** process that requires energy because ions must be transported against a concentration gradient, that is, from a lower salt concentration (in the dilute seawater) to an already higher one (in the blood).

INVASION OF FRESH WATER

Some 400 million years ago, during the Silurian and Lower Devonian periods, the major groups of jawed fishes began to penetrate into brackish-water estuaries and then gradually into freshwater rivers. Before them lay a new, unexploited habitat already stocked with food in the form of insects and other invertebrates, which had preceded them into fresh water. However, the advantages of this new habitat

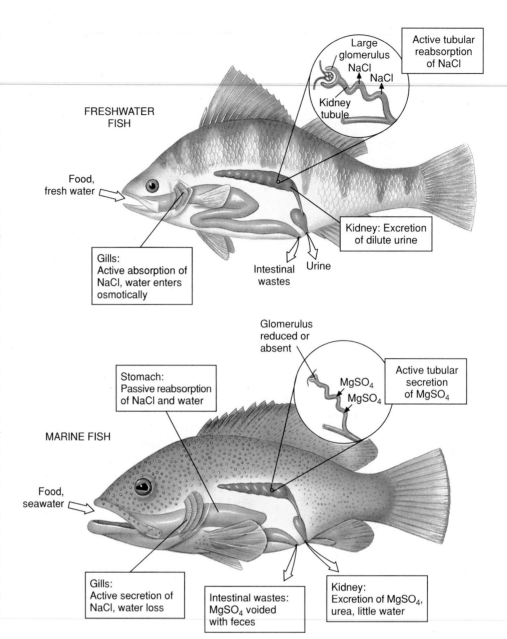

Figure 33-3

Osmotic regulation in freshwater and marine bony fishes. A freshwater fish maintains osmotic and ionic balance in its dilute environment by actively absorbing sodium chloride across the gills (some salt enters with food). To flush out excess water that constantly enters the body, the glomerular kidney produces a dilute urine by reabsorbing sodium chloride. A marine fish must drink seawater to replace water lost osmotically to its salty environment. Sodium chloride and water are absorbed from the stomach. Excess sodium chloride is secreted outward by the gills. Divalent sea salts, mostly magnesium sulfate, are eliminated with feces and secreted by the tubular kidney.

were traded off for a tough physiological challenge: the necessity of developing effective osmotic regulation.

Freshwater animals must keep the salt concentration of their body fluids higher than that of the water in which they live. Water enters their bodies osmotically, and salt is lost by diffusion outward. Their problems are similar to those of the brackish-water crab, but more severe and unremitting. Fresh water is much more dilute than are coastal estuaries, and there is no retreat, no salty sanctuary into

Figure 33-4

Exchange of water and solute in a frog. Water enters the highly permeable skin and is excreted by the kidney. The skin also actively transports ions (sodium chloride) from the environment. The kidney forms a dilute urine by reabsorbing sodium chloride. Urine flows into the urinary bladder, where, during temporary storage, most of the remaining sodium chloride is removed and returned to the blood.

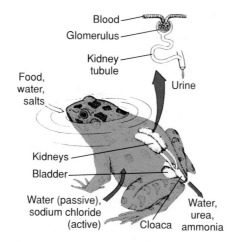

which the freshwater animal can retire for osmotic relief. It must and has become a permanent and highly efficient hyperosmotic regulator.

The scaled and mucus-covered surface of a fish is about as waterproof as any flexible surface can be. In addition, freshwater fishes have several defenses against the problems of water gain and salt loss. First, water that inevitably enters by osmosis across the gills is pumped out by the kidney, which is capable of forming a very dilute urine (Figure 33-3). Second, special salt-absorbing cells located in the gills move salt ions, principally sodium and chloride, from the water to the blood. This, together with salt present in the fish's food, replaces diffusive salt loss. These mechanisms are so efficient that a freshwater fish devotes only a small part of its total energy expenditure to keeping itself in osmotic balance.

Crayfishes, aquatic insect larvae, mussels, and other freshwater animals are also hyperosmotic regulators and face the same hazards as freshwater fishes; they tend to gain too much water and lose too much salt. Not surprisingly, all of these forms solved these problems in the same direct way that fishes did. They excrete the excess water as urine and they actively absorb salt from the water by some salt-transporting mechanism on the body surface.

Amphibians living in water also must compensate for salt loss by ac-

tively absorbing salt from the water (Figure 33-4). They use their skin for this purpose. Physiologists learned some years ago that pieces of frog skin continue to transport sodium and chloride actively for hours when removed and placed in a specially balanced salt solution. Fortunately for biologists, but unfortunately for frogs, these animals were so easily collected and maintained in the laboratory that frog skin became a favorite membrane system for studies of ion-transport phenomena.

RETURN OF FISHES TO THE SEA

The marine bony fishes maintain the salt concentration of their body fluids at approximately one-third that of seawater (body fluids = 0.3 to 0.4 gram mole per liter [M]; seawater = 1 M). They are **hypoosmotic regulators** because their body fluids are substantially more dilute than their seawater environment. Bony fishes living in the oceans today are descendants of earlier freshwater bony fishes that moved back into the sea during the Triassic period approximately 200 million years ago. The return to their ancestral sea was probably prompted by unfavorable climatic conditions on land and the deterioration of freshwater habitats, but we can only guess at the reasons. During the many millions of years that the early freshwater fishes were adapting themselves so well to their environment, they established a

concentration of salt in the body equivalent to approximately one-third that of seawater, thus setting the pattern for all subsequent vertebrate evolution, whether aquatic, terrestrial, or aerial. The ionic composition of vertebrate body fluid is remarkably similar to that of dilute seawater too, a fact that is undoubtedly related to their marine heritage.

> By expressing concentration of salt in seawater or body fluids in molarity, we are saying that the osmotic strength is equivalent to the molar concentration of an ideal solute having the same osmotic strength. In fact, seawater and animal body fluids are not ideal solutions because they contain electrolytes that dissociate in solution. A 1 M solution of sodium chloride (which dissociates in solution) has a much greater osmotic strength than a 1 M solution of glucose, an ideal solute that does not dissociate in solution. Consequently, biologists usually express the osmotic strength of a biological solution in osmolarity rather than in molarity. A 1 osmolar solution exerts the same osmotic pressure as a 1 M solution of a nonelectrolyte.

When some of the freshwater bony fishes of the Triassic period ventured back to the sea, they encountered a new set of problems. Having a much lower internal osmotic concentration than the seawater around them, they lost water and gained salt. Indeed the marine bony fish literally risks drying out, much like a desert mammal deprived of water.

In brief, to compensate for loss of water, the marine teleost drinks seawater (Figure 33-3). This is absorbed from the intestine, and the major sea salt, sodium chloride, is carried by the blood to the gills, where specialized salt-secreting cells transport it back into the surrounding sea. The ions remaining in the intestinal residue, especially magnesium, sulfate, and calcium, are voided with the feces or excreted by the kidney.

In this roundabout way, marine fishes rid themselves of the excess sea salts they have drunk, resulting in a net gain of water. This replaces the water lost by osmosis. Samuel Taylor Coleridge's ancient mariner, surrounded by "water, water, everywhere, nor any drop to drink" would undoubtedly have been tormented even more had he known of the marine fishes' ingenious solution for thirst. A marine fish carefully regulates the amount of seawater it drinks, consuming only enough to replace water loss and no more.

The cartilaginous sharks and rays (elasmobranchs) achieve osmotic balance in a completely different way. This group is almost totally marine. The salt composition of shark's blood is similar to that of the bony fishes, but the blood also carries a large content of organic compounds, especially urea and trimethylamine oxide. Urea is, of course, a metabolic waste that most animals quickly excrete in the urine. The shark kidney, however, conserves urea, causing it to accumulate in the blood. The urea, added to the usual blood electrolytes, raises the osmotic pressure of blood to exceed slightly that of seawater. In this way the sharks and their kin turn an otherwise useless waste material into an asset, eliminating the osmotic problem encountered by the marine bony fishes.

The high concentration of urea in the blood of sharks and their kin—more than 100 times as high as in mammals—could not be tolerated by most other vertebrates. In the latter, such high concentrations of urea disrupt the peptide bonds of proteins, altering protein configuration. Sharks have adapted biochemically to the presence of the urea that permeates all their body fluids, even penetrating freely into the cells. So accommodated are the elasmobranchs to urea that their tissues cannot function without it, and the heart will stop beating in its absence.

How Terrestrial Animals Maintain Salt and Water Balance

The problems of living in an aquatic environment seem small indeed compared with the problems of life on land. Since animal bodies are mostly water, all metabolic activities proceed in water, and life itself originated in water, it would seem that animals were meant to stay in water. Yet many animals, like the plants preceding them, moved onto land, carrying their watery composition with them. Once on land, the terrestrial animals continued their adaptive radiation, solving the threat of desiccation, until they became abundant even in some of the most arid parts of the earth.

Terrestrial animals lose water by evaporation from respiratory and body surfaces, excretion in the urine, and elimination in the feces. Such losses are replaced by water in the food, drinking water if it is available, and formation of **metabolic water** in the cells by oxidation of foodstuffs, especially carbohydrates. Certain insects, such as desert roaches, certain ticks and mites, and the mealworm, are able to absorb water vapor directly from atmospheric air. In some desert rodents, the metabolic water gain may constitute most of the animals' water intake.

Particularly revealing is a comparison of water balance in the human being, a nondesert mammal that drinks water, with that of the kangaroo rat, a desert rodent that may drink no water at all (Table 33-1). The kangaroo rat gains all of its water from its food (90% as metabolic water derived from the oxidation of foodstuffs, 10% as free moisture in the food). Even though humans eat foods with a much higher water content than the dry seeds that make up much of the kangaroo rat's diet, people must still drink half their total water requirement.

The excretion of wastes presents a special problem in water conserva-

Given ample water to drink, humans can tolerate extremely high temperatures while preventing a rise in body temperature. Our ability to keep cool by evaporation was impressively demonstrated more than 200 years ago by a British scientist who remained for 45 minutes in a room heated to 260° F (126° C). A steak he carried in with him was thoroughly cooked, but he remained uninjured and his body temperature did not rise. Sweating rates may exceed 3 liters of water per hour under such conditions and cannot be long tolerated unless the lost water is replaced by drinking. Without water, a human continues to sweat unabatedly until the water deficit exceeds 10% of the body weight, when collapse occurs. With a water deficit of 12% a human is unable to swallow even if offered water, and death occurs when the water deficit reaches about 15% to 20%. Few people can survive more than a day or two in a desert without water. Thus people are not physiologically well adapted for desert climates but prosper there nonetheless by virtue of their technological culture.

tion. The primary end product of protein breakdown is ammonia, a highly toxic material. Fishes can easily excrete ammonia across their gills, since there is an abundance of water to wash it away. Terrestrial insects, reptiles, and birds have no convenient way to rid themselves of toxic ammonia; instead, they convert it into uric acid, a nontoxic, almost insoluble compound. This enables them to excrete a semisolid urine with little water loss. The use of uric acid has another important benefit. Reptiles and birds lay amniotic eggs enclosing the embryos (Figure 29-4, p. 553), their stores of food and water, and whatever wastes that accumulate during embryonic development. By converting ammonia to uric acid, the developing embryo's waste can be precipitated into solid crystals, which are stored harmlessly within the egg until hatching.

Table 33-1	**Water Balance in a Human and a Kangaroo Rat, a Desert Rodent**	
	Human (%)	**Kangaroo Rat (%)**
Gains		
Drinking	48	0
Free water in food	40	10
Metabolic water	12	90
Losses		
Urine	60	25
Evaporation (lungs and skin)	34	70
Feces	6	5

Source: Some data from K. Schmidt-Nielsen, *How Animals Work.* Cambridge University Press, 1972.

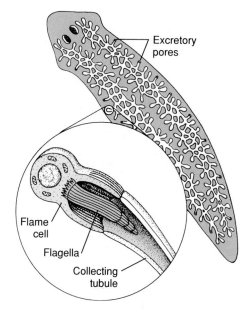

Figure 33-5

Flame cell system of a flatworm. Body fluids collected by flame cells (protonephridia) are passed down a system of ducts to excretory pores on the body surface.

Marine birds and turtles have evolved an effective solution for excreting the large loads of salt eaten with their food. Located above each eye is a special **salt gland** capable of excreting a highly concentrated solution of sodium chloride—up to twice the concentration of seawater. In birds the salt solution runs out the nares (see p. 582). Marine lizards and turtles, like Alice in Wonderland's Mock Turtle, shed their salt gland secretion as salty tears. Salt glands are important accessory organs of salt excretion in these animals because their kidneys cannot produce a concentrated urine, as can mammalian kidneys.

INVERTEBRATE EXCRETORY STRUCTURES

Many protozoa and some freshwater sponges have special excretory organelles called contractile vacuoles. The more complex invertebrates have excretory organs that are basically tubular structures, forming urine by first producing an ultrafiltrate or fluid secretion of the blood. This enters the proximal end of the tubule and is modified continuously as it flows down the tubule. The final product is urine.

CONTRACTILE VACUOLE

The tiny, spherical, intracellular vacuole of protozoa and freshwater sponges is not a true excretory organ, since ammonia and other nitrogenous wastes of metabolism readily leave the cell by direct diffusion across the cell membrane into the surrounding water. The contractile vacuole is an organ of water balance. Because the cytoplasm of freshwater protozoa is considerably saltier than their freshwater environment, they tend to draw water into themselves by osmosis. This excess water is removed by the contractile vacuole. How the contractile vacuole functions in *Amoeba proteus* and other protists has long remained a mystery. The "classical" view, illustrated in Figure 12-7, p. 219, is that excess water and ions from the cytoplasm (such as Na$^+$ and K$^+$) collect in numerous tiny vesicles surrounding the single thin membrane of the contractile vacuole. The ions are then actively reabsorbed, thus recovering the intracellular ions, and a dilute solution is produced which is discharged into the contractile vacuole. As water accumulates within it the vacuole grows and finally collapses, emptying its contents through a pore on the surface, and the cycle is rhythmically repeated. Recent advances in immunofluorescence and electron microscopy* have altered this classical view. Rather than a vacuole surrounded by many independent vesicles, the contractile vacuole system consists of a network of continuous membranous channels that are populated with numerous proton pumps (proton pumps were described in connection with the electron transport chain in Chapter 5, p. 70). Although the mechanism for filling is still not fully understood, the proton pumps apparently create H$^+$ and HCO$^-$ gradients that draw

*Heuser, J., Q. Zhu, and M. Clarke. 1993. Jour. Cell Biol., **121**(6):1311–1327.

water into the vacuole. These ions are excreted along with water when the vacuole empties. Since an isosmotic solution is expelled, this model resolves a problem with the classical view, which required that the contractile vacuole retained water against an osmotic gradient, an almost impossible task for a simple bilayer membrane.

Contractile vacuoles are common in freshwater protozoa, sponges, and radiate animals (such as hydra), but rare or absent in marine forms of these groups, which are isosmotic with seawater and consequently neither lose nor gain too much water.

NEPHRIDIUM

The most common type of invertebrate excretory organ is the nephridium, a tubular structure designed to maintain the appropriate osmotic balance. One of the simplest arrangements is the flame cell system (or **protonephridium**) of the acoelomates (flatworms) and some pseudocoelomates.

In planaria and other flatworms the protonephridial system takes the form of two highly branched duct systems distributed throughout the body (Figure 33-5). Fluid enters the system

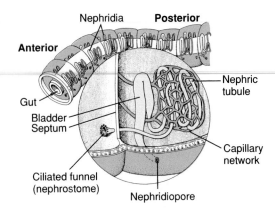

Figure 33-6

Excretory system of an earthworm. Each segment has a pair of large nephridia suspended in a fluid-filled coelom. Each nephridium occupies two segments because the ciliated funnel (nephrostome) drains the segment anterior to the segment containing the rest of the nephridium.

through specialized "flame cells," moves slowly into and down the tubules, and is excreted through pores that open at intervals on the body surface. The rhythmical beat of the flagellar tuft, suggestive of a tiny flickering flame, creates a negative pressure that draws fluid through delicate interdigitations between the flame cell and the tubule cell and drives it into the tubular portion of the system. In the tubule, water and metabolites valuable to the body are recovered by reabsorption, leaving wastes behind to be expelled. Nitrogenous wastes (mainly ammonia) diffuse across the surface of the body.

The extensive branching of the flame cell system is a consequence of the absence of a circulatory system in these acoelomate animals. This is very unlike the condensed kidneys of vertebrates and many invertebrates, which depend on the circulatory system to deliver wastes for excretion.

The protonephridium just described is a **closed** system. The tubules are closed on the inner end and urine is formed from a fluid that must first enter the tubules by being transported across flame cells. A more advanced type of nephridium is the **open,** or "true," nephridium **(metanephridium)** that is found in several of the eucoelomate phyla such as the annelids (Figure

33-6), the molluscs, and several smaller phyla. The metanephridium is more advanced than the protonephridium in two important ways. First, the tubule is open at *both* ends, allowing fluid to be swept into the tubule through a ciliated funnel-like opening, the **nephrostome.** Second, the metanephridium is surrounded by a network of blood vessels that assists in urine formation by reabsorbing water, salts, sugars, amino acids and other valuable materials from the tubular fluid.

Despite these differences, the basic process of urine formation is the same in protonephridia and metanephridia: fluid enters and flows continuously through a tubule where the fluid is selectively modified by (1) withdrawing valuable solutes from it and returning these to the body (reabsorption) and (2) adding waste solutes to it (secretion). The sequence ensures the removal of wastes from the body without the loss of materials valuable to the body. We will see that the kidneys of vertebrates operate in basically the same way.

Arthropod Kidneys

The paired **antennal glands** of crustaceans, located in the ventral part of the head (Figure 33-7), are an advanced design of the basic nephridial organ. However, they lack open nephrostomes. Instead, a filtrate of the blood is formed in the end sac by the hydrostatic pressure within the hemocoel. In the tubular portion of the gland, the filtrate is modified by the selective reabsorption of certain salts and the active secretion of others. Thus crustaceans have excretory organs that are basically vertebrate-like in the functional sequence of urine formation.

Insects and spiders have a unique excretory system consisting of **Malpighian tubules** that operate in conjunction with specialized glands in the wall of the rectum (Figure 33-8). These thin, elastic, blind Malpighian tubules are closed and lack an arterial supply. Urine formation is initiated by the active secretion of salts, largely

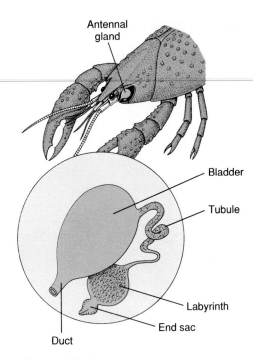

Figure 33-7

Antennal glands of a crayfish. These are filtration kidneys, that is, a filtrate of the blood is formed in the end sac. The filtrate is converted into urine as it passes down the tubule toward the bladder.

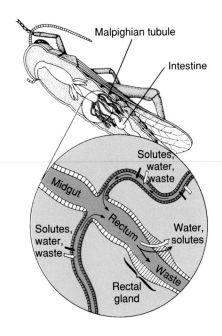

Figure 33-8

Malpighian tubules of insects. Malpighian tubules are located at the juncture of the midgut and hindgut (rectum). Solutes, especially potassium, are actively secreted into the tubules from the surrounding arthropod hemolymph. Water and wastes follow. This fluid drains into the rectum, where solutes and water are actively reabsorbed, leaving wastes to be excreted.

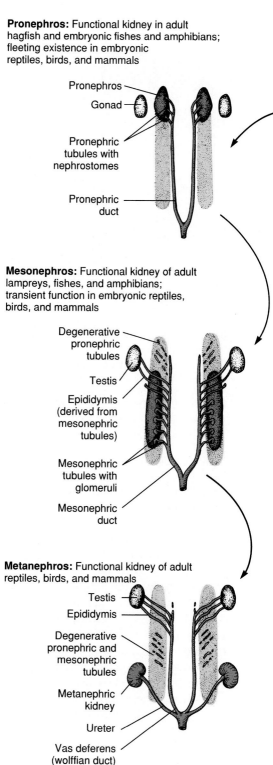

Pronephros: Functional kidney in adult hagfish and embryonic fishes and amphibians; fleeting existence in embryonic reptiles, birds, and mammals

Pronephros
Gonad
Pronephric tubules with nephrostomes
Pronephric duct

Mesonephros: Functional kidney of adult lampreys, fishes, and amphibians; transient function in embryonic reptiles, birds, and mammals

Degenerative pronephric tubules
Testis
Epididymis (derived from mesonephric tubules)
Mesonephric tubules with glomeruli
Mesonephric duct

Metanephros: Functional kidney of adult reptiles, birds, and mammals

Testis
Epididymis
Degenerative pronephric and mesonephric tubules
Metanephric kidney
Ureter
Vas deferens (wolffian duct)

Figure 33-9
Comparative development of male vertebrate kidney. *Red,* functional structures. *Light red,* degenerative or undeveloped parts.

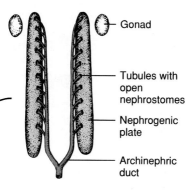

Gonad
Tubules with open nephrostomes
Nephrogenic plate
Archinephric duct

Archinephros: Kidney found in embryo of hagfish; this is the inferred ancestral condition of the vertebrate kidney.

potassium, into the tubules from the hemolymph. This primary secretion of ions creates an osmotic drag that pulls water, solutes, and nitrogenous wastes, especially uric acid, into the tubule. Uric acid enters the upper end of the tubule as soluble potassium urate, which precipitates as insoluble uric acid in the proximal end of the tubule. Once the formative urine drains into the rectum, most of the water and potassium are reabsorbed by specialized rectal glands, leaving behind uric acid and other wastes that are disposed in the feces. The Malpighian tubule excretory system is ideally suited for life in dry environments and has contributed to the adaptive radiation of insects on land.

VERTEBRATE KIDNEY

ANCESTRY AND EMBRYOLOGY

From comparative studies of development it is believed that the kidney of the earliest vertebrates extended the length of the coelomic cavity and was composed of segmentally arranged tubules, each resembling an invertebrate nephridium. Each tubule opened at one end into the coelom by a nephrostome and at the other end into a common **archinephric duct.** This ancestral kidney has been called the **archinephros** ("ancient kidney"), and a segmented kidney very similar to an archinephros is found in the embryos of hagfishes and caecilians (Figure 33-9). Almost from the beginning, the reproductive system, which develops beside the excretory system from the same segmental blocks of trunk mesoderm, made use of the nephric ducts as a convenient conducting system for reproductive products. Thus although the two systems have nothing functionally in common, they are closely associated in their use of common ducts.

Kidneys of living vertebrates developed from this primitive plan. During embryonic development of the amniotic vertebrates, there is a succession of three developmental stages of kidneys: **pronephros, mesonephros,** and **metanephros** (Figure 33-9). Some, but not all, of these stages are observed also in other vertebrate groups. In all vertebrate embryos, the pronephros is the first kidney to appear. It is located anteriorly in the body and becomes part of the persistent kidney only in adult hagfishes. In all other vertebrates it degenerates during development and is replaced by a more centrally located mesonephros. The mesonephros is the functional kidney of embryonic amniotes (reptiles, birds, and mammals), and contributes to the adult kidney (called an opisthonephros) of fishes and amphibians.

The metanephros, characteristic of adult amniotes, is distinguished in several ways from the pronephros and mesonephros. It is more caudally located and it is a much larger, more compact structure containing a very large number of nephric tubules. It is drained by a new duct, the **ureter,** which developed when the old archinephric duct was relinquished to the reproductive system of the male for sperm transport. Thus the three successive kidney types—pronephros, mesonephros, metanephros—succeed each other embryologically, and to some extent phylogenetically, in amniotes.

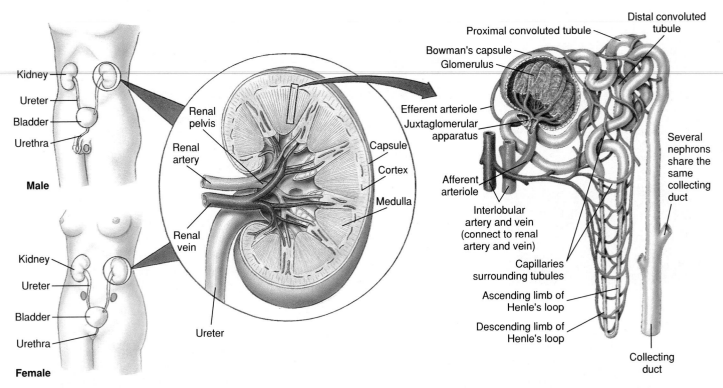

Figure 33-10
Urinary system of humans, with enlargements showing detail of the kidney and a single nephron.

VERTEBRATE KIDNEY FUNCTION

The vertebrate kidney is part of many interlocking mechanisms that maintain homeostasis. The kidney plays a prominent role in this regulatory council because it is the principal organ that regulates the volume and composition of the internal fluid environment. While the vertebrate kidney is commonly described as an excretory organ, the removal of metabolic wastes is incidental to its regulatory function.

The organization of the vertebrate kidney differs somewhat in different groups of vertebrates, but in all the functional unit is a tubule, the **nephron,** and urine is formed by three well-defined physiological processes: **filtration, reabsorption,** and **secretion.** The following discussion focuses mainly on the mammalian kidney, which is the most completely understood regulatory organ.

The two human kidneys are small organs comprising less than 1% of the body weight. Yet they receive a remarkable 20% to 25% of the total cardiac output, some 2000 liters of blood each day. This vast blood flow is channeled to approximately 2 million nephrons,

which make up the bulk of the two kidneys. Each nephron begins with an expanded chamber, the **renal corpuscle,** containing a tuft of capillaries called the **glomerulus** (glo-mer′yoo-lus). Blood pressure in the capillaries forces a protein-free **filtrate** into an expanded chamber known as **Bowman's capsule.** From here the filtrate journeys down a long, twisted **renal tubule** consisting of several segments that perform different functional processes. The filtrate passes first into the **proximal convoluted tubule,** then into a long, thin-walled **loop of Henle,** which drops deep into the inner portion of the kidney (the medulla) before returning to the outer portion (the cortex) where it joins the **distal convoluted tubule.** From the distal tubule the fluid empties into a **collecting duct** which drains into the **renal pelvis.** Here the urine is collected before being carried by the **ureter** to the **urinary bladder.** These anatomical relationships are shown in Figure 33-10.

The urine that enters the renal pelvis is very different from the filtrate produced in the renal corpuscle. During its travels through the renal tubule

and collecting duct, both the composition and concentration of the original filtrate change. Some solutes such as glucose and sodium have been reabsorbed while other materials, such as hydrogen ions and urea have been concentrated in the urine.

The nephron, with its pressure filter and tubule, is intimately associated with the blood circulation (Figure 33-11). Blood from the aorta enters each kidney through a large **renal artery,** which divides into a branching system of smaller arteries. The arterial blood reaches the renal corpuscle through an **afferent arteriole** and leaves by way of an **efferent arteriole.** From the efferent arteriole the blood travels to an extensive capillary network that surrounds and supplies the proximal and distal convoluted tubules and the loop of Henle (Figure 33-10). This capillary network provides a means for the pickup and delivery of materials that are reabsorbed or secreted by the kidney tubules. From these capillaries blood is collected by veins that unite to form the **renal vein.** This vein returns the blood to the vena cava.

Figure 33-11

Scanning electron micrograph of a cast of the microcirculation of the mammalian kidney, showing several glomeruli and associated blood vessels. The Bowman's capsule, which normally surrounds each glomerulus, has been digested away in preparing the cast.

From Tissues and Organs: A Text-Atlas of Scanning Electron Microscopy, *by Richard G. Kessel and Randy H. Kardon, W. H. Freeman and Co.,* © 1979.

GLOMERULAR FILTRATION

Let us now return to the glomerulus, where the process of urine formation begins. The glomerulus acts as a specialized mechanical filter in which a protein-free filtrate of the plasma is driven by the blood pressure across the capillary walls and into the fluid-filled space of Bowman's capsule. Solute molecules small enough to pass the slit pores of the capillary wall are carried through with the water in which they are dissolved. Red blood cells and the plasma proteins, however, are withheld because they are too large to pass the filter (Figure 33-12). The outward pressure forcing water and solutes through the capillary membrane must be sufficient to exceed the colloid osmotic pressure of the blood (created by the plasma proteins that cannot pass the filter; see p. 687) and the resistance to flow down the tubule.

The filtrate now enters the renal tubular system where it will undergo extensive modification before becoming urine. Approximately 180 liters (nearly 50 gallons) of filtrate are

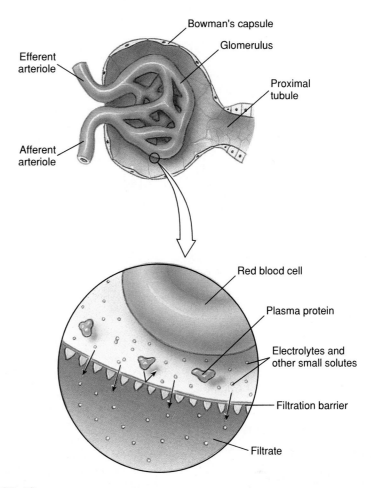

Figure 33-12

Bowman's capsule and glomerulus, showing (*enlargement*) the filtration of fluid through the glomerular capillary membrane. Water, electrolytes, and other small molecules pass the porous filtration barrier, but the plasma proteins are too large to pass the barrier. The filtrate is thus protein free.

formed each day by the human kidneys, a volume many times exceeding the total blood volume. If this volume of water and the valuable nutrients and salts it contains were lost, the body would soon be depleted of these compounds. This does not happen because nearly all of the filtrate is reabsorbed.

The conversion of filtrate into urine involves two processes: (1) modification of the composition of the filtrate through tubular reabsorption and secretion, and (2) changes in the total osmotic concentration of the urine through the regulation of water excretion.

TUBULAR REABSORPTION

Some 60% of the filtrate volume and virtually all of the glucose, amino acids, vitamins and other valuable nutrients are reabsorbed in the proximal convo-luted tubule. Much of this reabsorption is by **active transport,** in which cellular energy is used to transport materials from the tubular fluid to the surrounding capillary network from which they will reenter the blood circulation. Electrolytes such as sodium, potassium, calcium, bicarbonate, and phosphate are reabsorbed by ion pumps which are carrier proteins driven by the hydrolysis of ATP (ion pumps are described on p. 53). Because an essential function of the kidney is to regulate the plasma concentrations of the electrolytes, all are individually reabsorbed by ion pumps specific for each electrolyte. Some are strongly reabsorbed and others weakly reabsorbed, depending on the body's need to conserve each mineral. Some materials are passively reabsorbed. Negatively charged chloride ions, for example, passively accompany the active reabsorption of positively charged sodium

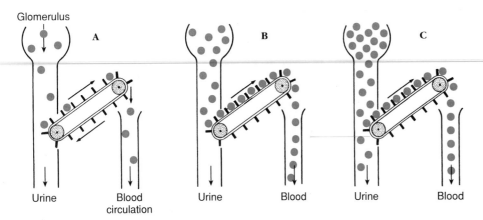

Glomerulus

A   B   C

Urine   Blood circulation   Urine   Blood   Urine   Blood

Figure 33-13

The mechanism for the tubular reabsorption of glucose can be likened to a conveyor belt running at constant speed. **A,** When the concentration of glucose in the filtrate is low, all is reabsorbed. **B,** When the glucose concentration in the filtrate has reached the transport maximum, all carrier sites for glucose are occupied. If the glucose rises further, **C,** as in the disease diabetes mellitus, some glucose escapes the carriers and appears in the urine.

ions in the proximal convoluted tubule. Water, too, is passively withdrawn from the tubule, as it follows osmotically the active reabsorption of solutes.

In the disease diabetes mellitus ("sweet running through"), glucose rises to abnormally high concentrations in the blood plasma (hyperglycemia) because the hormone insulin, which enables the body cells to take up glucose, is deficient. As the blood glucose rises above a normal level of about 100 mg/100 ml of plasma, the concentration of glucose in the filtrate also rises, and more glucose must be reabsorbed by the proximal tubule. Eventually a point is reached (about 300 mg/100 ml of plasma) at which the reabsorptive capacity of the tubular cells is saturated. This is the transport maximum for glucose. Should the plasma glucose continue to rise, glucose spills over into the urine. In untreated diabetes the victim's urine tastes sweet, thirst is unrelenting, and the body wastes away despite a large food intake. In England the disease for centuries was appropriately called the "pissing evil."

For most substances there is an upper limit to the amount of substance that can be reabsorbed. This upper limit is termed the **transport maximum** for that substance. For example, glucose is normally completely reab-

sorbed by the kidney because the transport maximum for the glucose reabsorptive mechanism is poised well above the amount of glucose normally present in the plasma filtrate. Should the plasma glucose concentration exceed this threshold level, as in the disease diabetes mellitus, glucose appears in the urine (Figure 33-13).

Unlike glucose, most electrolytes are excreted in the urine in variable amounts. The reabsorption of sodium, the dominant cation in the plasma, illustrates the flexibility of the reabsorption process. Approximately 600 g of sodium is filtered by the human kidneys every 24 hours. Nearly all of this is reabsorbed, but the exact amount is precisely matched to sodium intake. With a normal sodium intake of 4 g per day, the kidney excretes 4 g and reabsorbs 596 g each day. A person on a low-salt diet of 0.3 g of sodium per day still maintains salt balance because only 0.3 g escapes reabsorption. But with a very high salt intake, much above 20 g a day, the kidney cannot excrete sodium as fast as it enters. The unexcreted sodium chloride holds additional water in the body fluids, and the person begins to gain weight. (The salt intake of the average North American is about 6 to 18 g a day, approximately 20 times more than the body needs, and three times more than is considered acceptable for those predisposed to high blood pressure.)

The human kidney can adapt to excrete large quantities of salt (sodium chloride) under conditions of high salt intake. In societies accustomed to widespread use of foods heavily salted for preservation (for example, salted pork and salt herring) daily intakes may approach or even exceed 100 g. Body weight remains normal under such conditions. However, the acute ingestion of 20–40 g/day by volunteers unadapted to such large intakes of salt caused swelling of tissues, increase in body weight, and some increase in blood pressure.

The final adjustment of filtrate composition is carried out in the distal convoluted tubule. The sodium reabsorbed by the proximal convoluted tubule—some 85% of the total filtered—is obligatory reabsorption, that is, this amount will be reabsorbed independent of sodium intake. In the distal convoluted tubule, however, sodium reabsorption is controlled by **aldosterone,** a steroid hormone from the adrenal gland (p. 752). Aldosterone increases the retention of sodium by the distal tubules and thus decreases the loss of sodium in the urine. The secretion of aldosterone is regulated mainly by the enzyme **renin,** produced by the **juxtaglomerular apparatus,** a complex of cells located in the afferent arteriole at its junction with the glomerulus (Figure 33-10). Renin is released in response to a low blood sodium level or to low blood pressure (which can occur if the blood volume drops too low). Renin then initiates a series of enzymatic events that culminates in the production of **angiotensin,** a blood protein that has several related effects. First, it stimulates the release of aldosterone, which acts in turn to increase sodium reabsorption by the distal tubule. Second, it increases the secretion of **antidiuretic hormone** (vasopressin, discussed later in the chapter), which promotes water conservation by the kidney. Third, it increases the blood pressure. Finally, it stimulates thirst. These actions of angiotensin tend to reverse the circumstances (low blood sodium and low blood pressure and/or blood volume) that triggered the secretion of renin. Sodium and water are conserved, and blood volume and blood pressure are restored to normal.

The flexibility of distal reabsorption of sodium varies considerably in different animals: it is restricted in humans but very broad in many rodents. These differences have appeared because selective pressures during evolution have resulted in rodents adapted for dry environments. They must conserve water and at the same time excrete considerable sodium. Humans, however, were not designed to accommodate the large salt appetites many have. Our closest relatives, the great apes, are vegetarians with an average salt intake of less than 0.5 g a day.

TUBULAR SECRETION

In addition to reabsorbing materials from the plasma filtrate, the kidney tubules are able to secrete certain substances into the tubular fluid. This process, which is the reverse of tubular reabsorption, enables the kidney to build up the urine concentrations of materials to be excreted, such as hydrogen and potassium ions, drugs, and various foreign organic materials. The distal convoluted tubule is the site of most tubular secretion.

In the kidneys of bony marine fishes, reptiles, and birds, tubular secretion is a much more highly developed process than it is in mammalian kidneys. Marine bony fishes actively secrete large amounts of magnesium and sulfate, which are by-products of their mode of osmotic regulation. Reptiles and birds excrete uric acid instead of urea as their major nitrogenous waste. The material is actively secreted by the tubular epithelium. Since uric acid is nearly insoluble, it forms crystals in the urine and requires little water for excretion. Thus the excretion of uric acid is an important adaptation for water conservation.

WATER EXCRETION

The osmotic pressure of the blood is closely regulated by the kidney. When fluid intake is high, the kidney

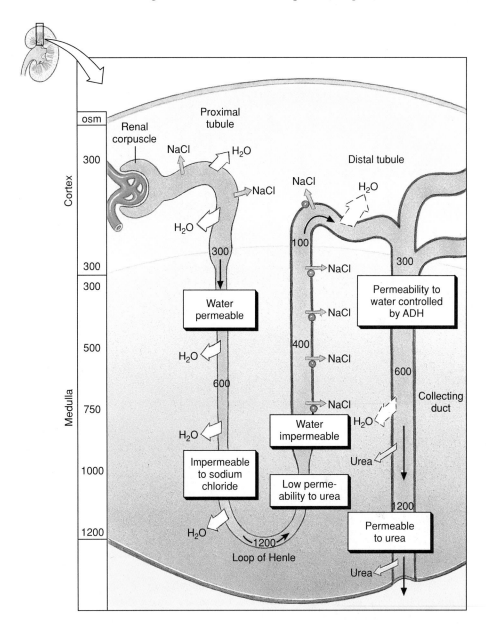

Figure 33-14

Mechanism of urine concentration in mammals. Sodium and chloride are pumped from the ascending limb of the loop of Henle, and water is withdrawn passively from the descending limb, which is impermeable to sodium chloride. Sodium chloride and urea reabsorbed from the collecting duct raise the osmotic concentration in the kidney medulla, creating an osmotic gradient for the controlled reabsorption of water from the collecting duct.

excretes a dilute urine, saving salts and excreting water. When fluid intake is low, the kidney conserves water by forming a concentrated urine. A dehydrated person can concentrate urine to approximately four times blood osmotic concentration. This important ability to concentrate urine enables us to excrete wastes with minimal loss of water.

The capacity of the kidney of mammals and some birds to produce a concentrated urine involves an interac-

tion between the loop of Henle and the collecting ducts. This interplay results in the formation of an osmotic gradient in the kidney, as shown in Figure 33-14. In the cortex, the interstitial fluid is isosmotic with the blood, but deep in the medulla the osmotic concentration is 4 times greater than that of the blood (in rodents and desert mammals that can produce highly concentrated urine the osmotic gradient is much greater than in humans). The high osmotic concentrations in the

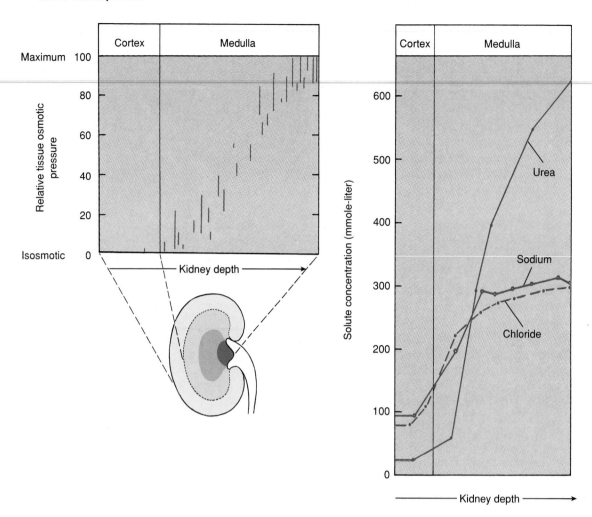

Figure 33-15

Osmotic concentration of tissue fluid in the mammalian kidney. Tissue fluid is isosmotic in the kidney cortex (*to left in diagram*) but osmotic concentration increases continuously through the medulla, reaching a maximum at the papilla where the urine drains into the ureter.

medulla are produced by an exchange of ions in the loop of Henle by **countercurrent multiplication.** "Countercurrent" refers to the opposite directions of fluid movement in the two limbs of the loop of Henle: down in the descending limb and up in the ascending limb. "Multiplication" refers to the increasing osmotic concentration in the medulla that results from ion exchange between the two limbs of the loop.

The functional characteristics of this system are as follows. The descending limb of the loop of Henle is permeable to water but impermeable to solutes. The ascending limb is relatively impermeable to both water and solutes. Sodium chloride is actively transported out of the thick portion of the ascending limb and into the surrounding tissue fluid (Figure 33-14). As

the interstitium surrounding the loop becomes more concentrated with solute, water is passively withdrawn from the descending limb. The tubular fluid in the base of the loop, now more concentrated, moves up the ascending limb, where still more sodium chloride is pumped out. In this way the effect of active ion transport in the ascending limb is multiplied as more water is withdrawn from the descending limb and more concentrated fluid is presented to the ascending limb ion pump. Also contributing significantly to tissue fluid concentration at the bottom of the loop is urea, which is reabsorbed from the collecting duct that lies parallel to the loop (Figures 33-14 and 33-15).

The final adjustment of urine concentration does not occur in the loops of Henle but in the collecting ducts.

Formative urine that enters the distal tubule from the loop of Henle is dilute (because of active salt withdrawal) and is diluted still more by the active reabsorption of more sodium chloride in the distal tubule. The formative urine, low in solutes but carrying urea, now flows down into the collecting duct. Because of the high concentration of solutes surrounding the collecting duct, water is withdrawn from the urine. As the urine becomes more concentrated, urea also diffuses out and adds to the high osmotic pressure in the kidney medulla (Figure 33-15).

The amount of water saved and the final concentration of the urine depend on the permeability of the walls of the collecting duct. This is controlled by the **antidiuretic hormone** (ADH, or vasopressin), which is released by the posterior pituitary gland (neurohypophysis). The release of this hormone is governed in turn by special receptors in the brain that constantly sense the osmotic pressure of the blood. When the blood osmotic pressure increases, as during dehydration, more ADH is released from the pituitary gland. ADH increases the permeability of the collecting duct, probably by expanding the size of pores in the walls of the duct. Then, as the fluid in the collecting duct passes through the hyperosmotic region of the kidney medulla, water diffuses through the pores into the surrounding interstitial fluid and is carried away by the blood circulation (Figure 33-10). The urine loses water and becomes more concentrated. Given this sequence of events for dehydration, it is not difficult to anticipate how the system responds to overhydration: the pituitary stops releasing ADH, the pores in the collecting duct walls close, and a large volume of dilute urine is excreted.

The varying ability of different mammals to form a concentrated urine is closely correlated with the length of the loops of Henle. The beaver, which has no need to conserve water in its aquatic environment, has short loops and can concentrate its urine to only approximately twice that of the blood plasma. Humans, with relatively longer loops, can concentrate urine 4.2 times that of the blood. As we would anticipate, desert mammals have much greater urine concentrating powers. The camel can produce a urine 8 times the plasma concentration, the gerbil 14 times, and the Australian hopping mouse 22 times. In this creature, the greatest urine concentrator of all, the loops of Henle extend to the tip of a long renal papilla that pushes out into the mouth of the ureter.

TEMPERATURE REGULATION

We have seen that a fundamental problem facing an animal is keeping its internal environment in a state that permits normal cell function. Biochemical activities are sensitive to the chemical environment and our discussion thus far has examined how the chemical environment is stabilized. Biochemical reactions are also extremely sensitive to temperature. All enzymes have an optimum temperature; at temperatures away from this optimum, enzyme function is hindered. When the body temperature drops too low, metabolic processes are slowed, reducing the amount of energy the animal can muster for activity and reproduction. If the body temperature rises too high, metabolic reactions become unbalanced and enzymatic activity is impaired or even destroyed. Accordingly animals can function only in a restricted range of temperature, usually between 0° to 40° C. Animals must either find a habitat where they do not have to contend with temperature extremes, or they must develop the means of regulating their body temperature independent of environmental temperature extremes.

A temperature difference of 10° C has become a standard that is used to measure the temperature sensitivity of a biological function. This value, called the Q_{10}, is determined (for temperature intervals of exactly 10° C) simply by dividing the value of a rate function (such as metabolic rate or the rate of an enzymatic reaction) at the higher temperature by the value of the rate function at the lower temperature. In general metabolic reactions have Q_{10} values of about 2.0 to 3.0. Purely physical processes, such as diffusion, have much lower Q_{10} values, usually close to 1.0.

ECTOTHERMY AND ENDOTHERMY

The terms "cold-blooded" and "warm-blooded" have long been used to divide animals into two groups: invertebrates and vertebrates that feel cold to the touch, and those, such as humans, other mammals, and birds, that do not. It is true that the body temperature of mammals and birds is usually (though not always) warmer than the air temperature, but a "cold-blooded" animal is not necessarily cold. Tropical fishes, and insects and reptiles basking in the sun, may have body temperatures equaling or surpassing those of mammals. Conversely, many "warm-blooded" mammals hibernate, allowing their body temperature to approach the freezing point of water. Thus the terms "warm-blooded" and "cold-blooded" are hopelessly subjective and nonspecific but are so firmly entrenched in our vocabulary that most biologists find it easier to accept the usage than to try to change people.

The terms **poikilothermic** (variable body temperature) and **homeothermic** (constant body temperature) are frequently used by zoologists as alternatives to "cold-blooded" and "warm-blooded," respectively. These terms, which refer to variability of body temperature, are more precise and more informative, but they still offer difficulties. For example, deep-sea fishes live in an environment that has no perceptible temperature change. Even though their body temperature is absolutely stable, day in and day out, to call such fishes homeotherms would distort the intended application of the term. Furthermore, among the homeothermic birds and mammals there are many that allow their body temperature to change between day and night, or, as with hibernators, between seasons.

Physiologists prefer yet another way to describe body temperatures, one that reflects the fact that an animal's body temperature is a balance between heat gain and heat loss. All animals produce heat from cellular metabolism, but in most the heat is conducted away as fast as it is produced. In these animals, the **ectotherms**—and the overwhelming majority of animals belong to this group—the body temperature is determined solely by the environment. Many ectotherms exploit their environment behaviorally to select areas of more favorable temperature (such as basking in the sun) but the source of energy used to increase body temperature comes from the environment, not from within the body. Alternatively

Figure 33-16

How a lizard regulates its body temperature behaviorally. In the morning the lizard absorbs the sun's heat through its head while keeping the rest of its body protected from the cool morning air. Later it will emerge to bask. At noon, with its body temperature high, it seeks shade from the hot sun. When the air temperature drops in the late afternoon, it emerges and lies parallel to the sun's rays.

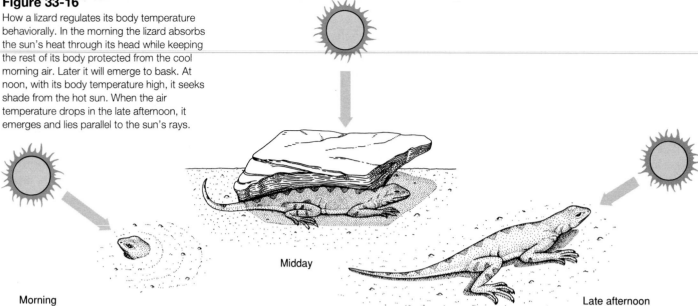

Morning

Midday

Late afternoon

there are some animals that are able to generate and retain enough heat to elevate their own temperature to a high but stable level. Because the source of their body heat is internal, they are called **endotherms.** These favored few in the animal kingdom are the birds and mammals, as well as a few reptiles and fast-swimming fishes, and certain insects that are at least partially endothermic. Endothermy allows birds and mammals to stabilize their internal temperature so that biochemical processes and nervous system functions can proceed at steady high levels of activity. Endotherms can thus remain active in winter and exploit habitats denied to ectotherms.

How Ectotherms Achieve Temperature Independence

Behavioral Adjustments

Although ectotherms cannot control their body temperature physiologically, many are able to regulate their body temperature behaviorally with considerable precision. Ectotherms often have the option of seeking out areas in the environment where the temperature is favorable to their activities. Some ectotherms, such as desert lizards, exploit hour-to-hour changes in solar radiation to keep their body temperature relatively constant (Figure 33-16). In the early morning they

emerge from their burrows and bask in the sun with their bodies flattened to absorb heat. As the day warms, they turn to face the sun to reduce exposure, and raise their bodies from the hot substrate. In the hottest part of the day they may retreat to their burrows. Later they emerge to bask as the sun sinks lower and the air temperature drops.

These behavioral patterns help to maintain a relatively steady body temperature of 36° to 39° C while the air temperature varies between 29° and 44° C. Some lizards can tolerate intense midday heat without shelter. The desert iguana of the southwestern United States prefers a body temperature of 42° C when active and can tolerate a rise to 47° C, a temperature that is lethal to all birds and mammals and most other lizards. The term "cold-blooded" obviously does not apply to these animals!

Metabolic Adjustments

Even without the help of the behavioral adjustments just described, most ectotherms can adjust their metabolic rates to the prevailing temperature so that the intensity of metabolism remains mostly unchanged. This is called **temperature compensation** and it involves complex biochemical and cellular adjustments. These adjustments enable a fish or a salaman-

der, for example, to benefit from almost the same level of activity in both warm and cold environments. Thus, whereas endotherms achieve metabolic homeostasis by regulating their body temperature, ectotherms accomplish much the same by directly regulating their metabolism. This also is a form of homeostasis.

Temperature Regulation in Endotherms

Most mammals have body temperatures between 36° and 38° C, somewhat lower than those of birds, which range between 40° and 42° C. Constant temperature is maintained by a delicate balance between heat production and heat loss—not a simple matter when these animals are alternating between periods of rest and bursts of activity.

Heat is produced by the animal's metabolism. This includes the oxidation of foods, basal cellular metabolism, and muscular contraction. Because much of an endotherm's daily caloric intake is required to generate heat, especially in cold weather, the endotherm must eat more food than an ectotherm of the same size. Heat is lost by radiation, conduction, and convection (air movement) to a cooler environment and by evaporation of water (Figure 33-17). A bird or mammal can control both processes of heat production and heat

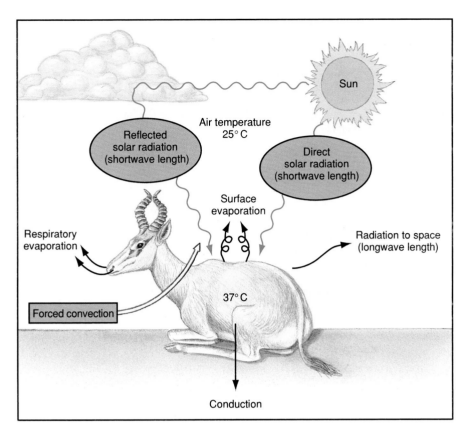

Air temperature
25° C

Reflected
solar radiation
(shortwave length)

Direct
solar radiation
(shortwave length)

Surface
evaporation

Respiratory
evaporation

Radiation to space
(longwave length)

Forced convection

37° C

Conduction

Figure 33-17

Exchange of heat between the animal and its environment on a warm day. Blue arrows indicate sources of net heat gain by the animal (all radiation); black arrows are avenues of net heat loss (evaporative cooling, conduction to the ground, longwave radiation into space, and forced convection by the wind). If the air and ground temperatures were warmer than the animal, the arrows for forced convection, conduction, and radiation would be reversed. Then the animal could lose heat only by evaporative cooling.

loss within rather wide limits. If the animal becomes too cool, it can generate heat by increasing muscular activity (exercise or shivering) and by decreasing heat loss by increasing its insulation. If it becomes too warm, it decreases heat production and increases heat loss. We will examine these processes in the examples that follow.

Adaptations for Hot Environments

Despite the harsh conditions of deserts—intense heat during the day, cold at night, and scarcity of water, vegetation, and cover—many kinds of animals live there successfully. The smaller desert mammals are mostly **fossorial** (living mainly in the ground) or **nocturnal** (active at night). The lower temperature and higher humidity of burrows help to reduce water loss by evaporation. As

explained earlier in this chapter (p. 654), desert animals such as the kangaroo rat and the American desert ground squirrels can, if necessary, derive the water they need from their dry food, drinking no water at all. Such animals produce a highly concentrated urine and form almost completely dry feces.

Large desert ungulates (hooved mammals that chew their cud) obviously cannot escape searing solar radiation by living in burrows. Animals such as camels and desert antelopes (gazelle, oryx, and eland) possess a number of adaptations for coping with heat and dehydration. Those of the eland are shown in Figure 33-18. Mechanisms for controlling water loss and preventing overheating are closely linked. The glossy, pallid color of fur reflects direct sunlight, and fur itself is an excellent insulation that works to resist heat. Heat is lost by convection and

conduction from the underside of the eland where the pelage is very thin. Fat tissue of the eland, an essential food reserve, is concentrated in a single hump on the back, instead of being uniformly distributed under the skin where it would impair loss of heat by radiation. The eland avoids evaporative water loss—the only means an animal has for cooling itself when the environmental temperature is higher than that of the body—by permitting its body temperature to decrease during the cool night and then increase slowly during the day as the body stores heat. Only when the body temperature reaches 41° C must the eland prevent further rise through **evaporative cooling** by sweating and panting. Water is also conserved by producing a concentrated urine and dry feces. All of these adaptations are also found developed to a similar or even greater degree in camels, the most perfectly adapted of all large desert mammals.

Adaptations for Cold Environments

In cold environments mammals use two major mechanisms to maintain homeothermy: (1) **decreased conductance,** that is, reduction of heat loss by increasing the effectiveness of the insulation, and (2) **increased heat production.**

The excellent insulation of the thick pelage of arctic mammals is familiar. In all mammals living in the cold regions of the earth, fur thickness increases in winter, sometimes by as much as 50%. Thick underhair is the principal insulating layer, whereas the longer and more visible guard hair serves as protection against wear and for protective coloration. However, unlike the well-insulated trunk of the body, the body extremities (legs, tail, ears, nose) of arctic mammals are thinly insulated and exposed to rapid cooling. To prevent these parts from becoming major avenues of heat loss, they are allowed to cool to low temperatures, often approaching the freezing point. As warm arterial blood passes into a leg, for example, heat is shunted directly from artery to vein and carried

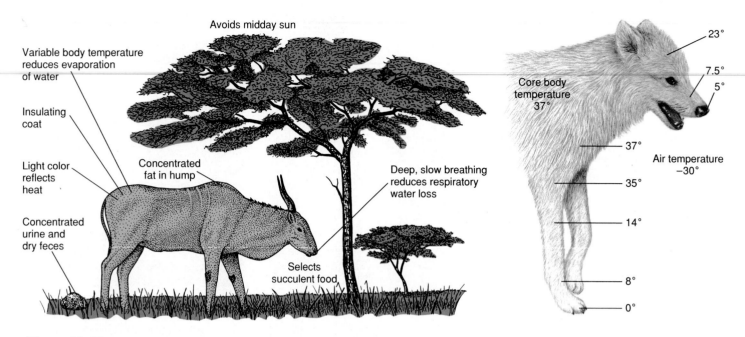

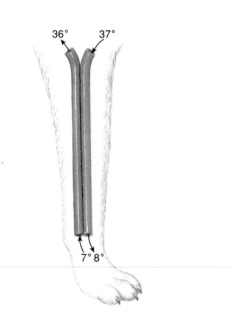

Figure 33-18

Physiological and behavioral adaptations of the common eland for regulating temperature in the hot, arid savanna of central Africa.

Figure 33-19

Countercurrent heat exchange in the leg of an arctic wolf. The upper diagram shows how the extremities cool when the animal is exposed to low air temperatures. The lower diagram depicts a portion of the front leg artery and vein, showing how heat is exchanged between outflowing arterial and inflowing venous blood. Heat is thus shunted back into the body and conserved.

back to the core of the body (Figure 33-19). This shunt prevents the loss of valuable heat through the poorly insulated distal regions of the leg. A consequence of this **peripheral heat exchange system** is that legs and feet must operate at low temperatures. Temperatures of the feet of the arctic fox and barren-ground caribou are just above the freezing point; in fact, the temperature may be below 0° C in the footpads and hooves. To keep feet supple and flexible at such low temperatures, fats in the extremities have very low melting points, perhaps 30° C lower than ordinary body fats.

In severely cold conditions all mammals can produce more heat by **augmented muscular activity** through exercise or shivering. We are all familiar with the effectiveness of both activities. A person can increase heat production as much as 18-fold by violent shivering when maximally stressed by cold. Another source of heat is the increased oxidation of foods, especially from stores of brown adipose tissue (brown fat; see p. 709). This mechanism is called **nonshivering thermogenesis.**

Small mammals the size of lemmings, voles, and mice meet the challenge of cold environments in a different way. Small animals are not as well insulated as large mammals because thickness of fur is limited by the need to maintain mobility. Consequently these forms have exploited the excellent insulating qualities of snow successfully by living under it in runways on the forest floor, where their food also is located. In this **subnivean environment** the temperature seldom drops below −5° C although the air temperature above may fall to −50° C. The snow insulation decreases thermal conductance from small mammals just as thick pelage does for large mammals. Living beneath the snow is really a type of avoidance response to cold.

ADAPTIVE HYPOTHERMIA IN BIRDS AND MAMMALS

Endothermy is energetically expensive. Whereas an ectotherm can survive for weeks in a cold environment without eating, an endotherm must always have energy resources to supply its high metabolic rate. The problem is especially acute for small birds and mammals which, because of their intense metabolism, may have to consume food approaching their own body weight each day to maintain homeothermy (food consumption by birds is related on p. 579, and by mammals on p. 608). It is not surprising then that a few small birds and mammals have evolved ways to abandon homeothermy for periods ranging from a few hours to several months, allowing their body temperature to fall until it approaches or equals the temperature of surrounding air.

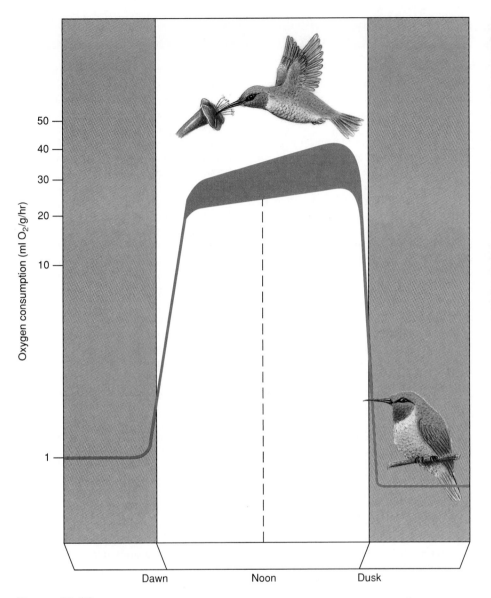

Figure 33-20
Torpor in hummingbirds. Body temperature and oxygen consumption are high when hummingbirds are active during the day but may drop to one-twentieth these levels during periods of food shortage. Torpor vastly lowers demands on the bird's limited energy reserves.

Figure 33-21
Hibernating woodchuck *Marmota monax* (order Rodentia) in den exposed by road-building work sleeps on, unaware of the intrusion. Woodchucks begin hibernating in late September while the weather is still warm and may sleep six months. The animal is rigid and decidedly cold to the touch. Breathing is imperceptible, as slow as one breath every five minutes. Although it appears to be dead, it will awaken if the den temperature drops dangerously low.

Some very small mammals, such as bats, maintain high body temperatures when active but allow their body temperature to drop profoundly when inactive and asleep. This is called **daily torpor,** an adaptive hypothermia that provides enormous saving of energy to small endotherms that are never more than a few hours away from starvation at normal body temperatures. Hummingbirds also may drop their body temperature at night when food supplies are low (Figure 33-20).

Many small and medium-sized mammals in northern temperate regions solve the problem of winter scarcity of food and low temperature by entering a prolonged and controlled state of dormancy: **hibernation.** True hibernators, such as ground squirrels, jumping mice, marmots, and woodchucks (Figure 33-21), prepare for hibernation by building up stores of body fat. Entry into hibernation is gradual. After a series of "test drops" during which body temperature decreases a few degrees and then returns to normal, the animal cools to within a degree or less of the ambient temperature. Metabolism decreases to a fraction of normal. In the ground squirrel, for example, the respiratory rate decreases from a normal rate of 200 per minute to 4 or 5 per minute, and the heart rate from 150 to 5 beats per minute. During arousal the hibernator both shivers violently and employs nonshivering thermogenesis to produce heat.

Some mammals, such as bears, badgers, raccoons, and opossums, enter a state of prolonged sleep in winter with little or no decrease in body temperature. This is not true hibernation. Bears of the northern forest den-up for several months. A bear's heart rate may decrease from 40 to 10 beats per minute, but body temperature remains normal and the bear is awakened if sufficiently disturbed. One intrepid but reckless biologist learned how lightly a bear sleeps when he crawled into a den and attempted to measure the bear's rectal temperature with a thermometer!

Summary

Throughout life, matter and energy pass through the body, potentially disturbing the internal physiological state. Homeostasis, the ability of an organism to maintain internal stability despite such challenges, is a characteristic of all living systems. Homeostasis involves the coordinated activity of several physiological and biochemical mechanisms, and it is possible to relate some major events in animal evolution to increasing internal independence from the consequences of environmental change. In this chapter we have examined two aspects of homeostasis: (1) the varying ability of animals to stabilize the osmotic and chemical composition of the blood, and (2) the capacity of animals to regulate their temperatures in thermally challenging environments.

Most marine invertebrates must either depend on the osmotic stability of the ocean to which they conform, or be able to tolerate wide fluctuations in environmental salinity. Some of the latter show limited powers of osmotic regulation, that is, the capacity to resist internal osmotic change, through the evolution of specialized regulatory organs. All animals living in fresh water are hyperosmotic to their environment and have developed mechanisms for recovering salt from the environment and eliminating excess water that enters the body osmotically.

All vertebrate animals, except the hagfishes, show excellent osmotic homeostasis. The marine bony fishes maintain their body fluids distinctly hypoosmotic to their environment by drinking seawater and physiologically distilling it. The elasmobranchs (sharks and their kin) have adopted a strategy of near-osmotic conformity by retaining urea in the blood.

The kidney is the most important organ for regulating the chemical and osmotic composition of the blood. In all metazoa the kidney is some variation on a basic theme: a tubular structure that forms urine by introducing a fluid secretion or filtrate of the blood or interstitial fluid into a tubule in which it is selectively modified to form urine. Terrestrial vertebrates have especially sophisticated kidneys, since they must be able to regulate closely the water content of the blood by balancing gains and expenditures. The basic excretory unit is the nephron, composed of a glomerulus in which an ultrafiltrate of the blood is formed, and a long nephric tubule in which the formative urine is selectively modified by the tubular epithelium. Water, salts, and other valuable materials are passed by reabsorption to the peritubular circulation, and certain wastes are passed by secretion from the circulation to the tubular urine. All mammals and some birds can produce urine more concentrated than blood by means of a countercurrent multiplier system localized in the loops of Henle, a specialization not found in other vertebrates.

Temperature has a profound effect on the rate of biochemical reactions and, consequently, on the metabolism and activity of all animals. Animals may be classified according to whether body temperature is variable (poikilothermic) or stable (homeothermic), or by the source of body heat, whether external (ectothermic) or internal (endothermic).

Ectotherms partially free themselves from thermal constraints by seeking out habitats with favorable temperatures, by behavioral thermoregulation, or by adjusting their metabolism to the prevailing temperature through biochemical alterations.

The endothermic birds and mammals differ from ectotherms in having a much higher production of metabolic heat and a much lower conductance of heat from the body. They maintain constant body temperature by balancing heat production with loss.

Small mammals in hot environments for the most part escape intense heat and reduce evaporative water loss by burrowing. Large mammals employ several strategies for dealing with direct exposure to heat, including reflective insulation, heat storage by the body, and evaporative cooling.

Endotherms in cold environments maintain body temperature by decreasing heat loss with thickened pelage or plumage, by peripheral cooling, and by increasing heat production through shivering or nonshivering thermogenesis.

Adaptive hypothermia is a strategy used by small mammals and birds to blunt energy demands during periods of inactivity (daily torpor) or periods of prolonged cold and minimal food availability (hibernation).

Review Questions

1. Define homeostasis. Can you suggest why osmotic properties of the blood might be considered a *dynamic* steady state?
2. Distinguish between the following pairs of terms: osmotic conformity and osmotic regulation; stenohaline and euryhaline; hyperosmotic and hypoosmotic.
3. Explain why marine bony fish, but not freshwater bony fish, must drink seawater to maintain osmotic balance.
4. Most marine invertebrates are osmotic conformers. How does their body fluid differ from that of the cartilaginous sharks and rays, which are also in near-osmotic equilibrium with their environment?
5. What strategy does the kangaroo rat use that allows it to exist in the desert without drinking any water?
6. In what animals would you expect to find a salt gland? What is its function?
7. Relate the function of the contractile vacuole to the following experimental observations: to expel an amount of fluid equal in volume to the volume of the animal required 4 to 53 minutes for some freshwater protozoa, and between 2 and 5 hours for some marine species.
8. How does a protonephridium differ structurally and functionally from a true nephridium (metanephridium)? In what ways are they similar?
9. Describe the developmental stages of the kidney in amniotes. How does the developmental sequence for amniotes differ from that of amphibians and fishes?
10. In what ways does the nephridium of an earthworm parallel the human nephron in structure and function?
11. Describe what happens during the following stages in urine formation in the mammalian nephron: filtration, tubular reabsorption, tubular secretion.

12. Explain how the cycling of sodium chloride between the descending and ascending limbs of the loop of Henle in the mammalian kidney, and the special permeability of these tubules, produces high osmotic concentrations in interstitial fluids in the kidney medulla.

13. Explain how the antidiuretic hormone (vasopressin) controls the excretion of water in the mammalian kidney.

14. Define the following terms and comment on the limitations (if any) of each in describing the thermal relationships of animals: poikilothermy, homeothermy, ectothermy, endothermy.

15. Defend the statement: "Both ectotherms and endotherms achieve metabolic homeostasis in unstable thermal environments, but they do so by employing different physiological strategies."

16. Large mammals live successfully in deserts and in the arctic. Describe the different adaptations mammals use to maintain homeothermy in each environment.

17. Explain why it is advantageous for certain small birds and mammals to abandon homeothermy during brief or extended periods of their lives.

Selected References

See also general references for Part IV, p. 774.

Cossins, A. R., and K. Bowler. 1987. Temperature biology of animals. London, Chapman and Hall. *Comprehensive treatment of both ectotherms and endotherms.*

Dantzler, W. H. 1989. Comparative physiology of the vertebrate kidney. *Comprehensive review of vertebrate renal function.*

Hardy, R. N. 1983. Homeostasis, ed. 2. The Institute of Biology's Studies in Biology no. 63, London, Edward Arnold. *Introduces the history of the homeostasis concept; temperature and osmotic regulation are treated in the final chapter.*

Rankin, J. C., and J. Davenport. 1981. Animal osmoregulation. New York, John Wiley & Sons, Inc. *Concise and selective treatment.*

Riegel, J. A. 1972. Comparative physiology of renal excretion. New York, Hafner Publishing Company. *Excellent survey of excretory systems both vertebrate and invertebrate.*

Schmidt-Nielsen, K. 1981. Countercurrent systems in animals. Sci. Am. **244:**118–128 (May). *Explains how countercurrent systems transfer heat, gases, or ions between fluids moving in opposite directions.*

Smith, H. W. 1953. From fish to philosopher. Boston, Little, Brown & Company. *Classic account of vertebrate kidney evolution.*

Storey, K. B., and J. M. Storey. 1990. Frozen and alive. Sci. Am. **263:**92–97 (Dec.). *Explains how many animals have evolved strategies for surviving complete or almost complete freezing during the winter months.*

34

Internal Fluids

*Immunity, Circulation,
and Gas Exchange*

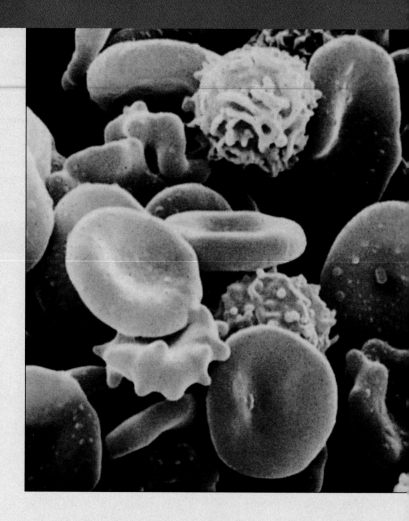

William Harvey's Discovery

Ceaselessly, during a human life, the heart pumps blood through the arteries, capillaries, and veins: about 5 liters per minute, until by the end of a normal life the heart has contracted some 2.5 billion times and pumped 300,000 tons of blood. When the heart stops its contractions, life also ends.

The crucial importance of the heart and its contractions for human life has been known since antiquity, probably almost as long as humans have existed. However, the circuit flow of blood, the notion that the heart pumps blood into arteries through the circulation and receives it back in veins, only became known a few hundred years ago. The first correct description of blood flow by the English physician William Harvey initially received vigorous opposition when published in 1628. Centuries before, the Greek anatomist Galen had taught that air enters the heart from the windpipe and that blood was able to pass from one ventricle to the other through "pores" in the interventricular septum. He also believed that

blood first flowed out of the heart into all vessels, then returned—a kind of ebb and flow of blood. Even though there was almost nothing correct about this concept, it was still doggedly trusted at the time of Harvey's publication. Harvey's conclusions were based on sound experimental evidence. He used a variety of animals for his experiments and chided human anatomists, saying that if only they had acquainted themselves with the anatomy of the lower vertebrates, they would have understood the blood's circuit. By tying ligatures on arteries, he noticed that the region between the heart and ligature swelled up. When veins were tied off, the swelling occurred beyond the ligature. When blood vessels were cut, blood flowed in arteries from the cut end nearest the heart; the reverse happened in veins. By means of such experiments, Harvey discovered the correct scheme of blood circulation, even though he could not see the capillaries that connected the arterial and venous flows. ■

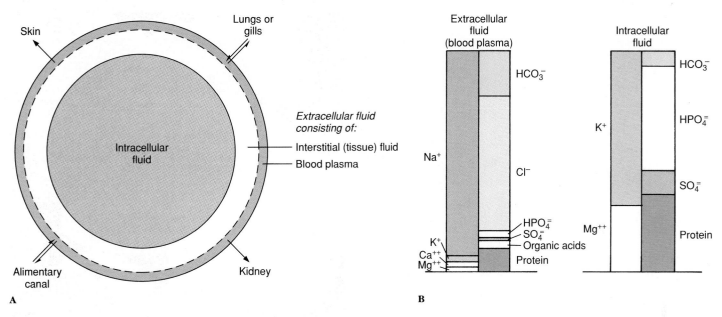

Figure 34-1

Fluid compartments of the body. **A,** All body cells can be represented as belonging to a single large fluid compartment that is completely surrounded and protected by extracellular fluid (*milieu intérieur*). This fluid is further subdivided into plasma and interstitial fluid. All exchanges with the environment occur across the plasma compartment. **B,** Electrolyte composition of extracellular and intracellular fluids. Total equivalent concentration of each major constituent is shown. Equal amounts of anions (negatively charged ions) and cations (positively charged ions) are in each fluid compartment. Note that sodium and chloride, major plasma electrolytes, are virtually absent from intracellular fluid (actually they are present in low concentration). Note the much higher concentration of protein inside the cells.

Single-celled organisms live in direct contact with their environment. They obtain nutrients and oxygen and release wastes directly across the cell surface. These organisms as so small that no special internal system of transport, beyond the normal streaming movements of the cytoplasm, is required. Even some simple multicellular forms, such as sponges, cnidarians, and flatworms, lack the internal complexity and metabolic demands that would require a circulatory system. Most of the other multicellular organisms, because of their size, activity, and complexity, need a specialized circulatory system to transport nutrients and respiratory gases to and from all tissues of the body. In addition to serving these primary transport needs, circulatory systems have acquired additional functions; hormones are moved about, finding their way to target organs where they assist the nervous system to integrate organismal function. Water, electrolytes, and the many other constituents of body fluids are distributed and exchanged between different organs and tissues. An effective response to disease and injury is vastly accelerated by an efficient circulatory system. Homeothermic birds and mammals depend heavily on the blood circulation to conserve or dissipate heat as required for the maintenance of constant body temperature.

INTERNAL FLUID ENVIRONMENT

The body fluid of a single-celled organism is the cellular cytoplasm, a liquid-gel substance in which the various membrane systems and organelles of the cell are suspended. In multicellular animals the body fluids are divided into two main phases, the **intracellular** and the **extracellular.** The intracellular phase (also called intracellular fluid) is the collective fluid inside all the body's cells. The extracellular phase (or fluid) is the fluid outside and surrounding the cells (see Figure 34-1A). Thus the cells, sites of the body's crucial metabolic activities, are bathed by their own aqueous environment, the extracellular fluid that buffers them from the often harsh physical and chemical changes occurring outside the body. The importance of the extracellular fluid was first emphasized by the great French physiologist Claude Bernard (Figure 34-2). In animals having closed circulatory systems (vertebrates, annelids, and a

Figure 34-2

French physiologist Claude Bernard (1813 to 1878), one of the most influential of nineteenth-century physiologists. Bernard believed in the constancy of the *milieu intérieur* ("internal environment"), which is the extracellular fluid bathing the cells. He pointed out that it is through the *milieu intérieur* that foods and wastes and gases are exchanged and through which chemical messengers are distributed. He wrote, "The living organism does not really exist in the external environment (the outside air or water) but in the liquid *milieu intérieur* . . . that bathes the tissue elements."

From J. F. Fulton & L. G. Wilson, Selected Readings in the History of Physiology, *1966. Courtesy of Charles C. Thomas, Publisher, Springfield, Illinois.*

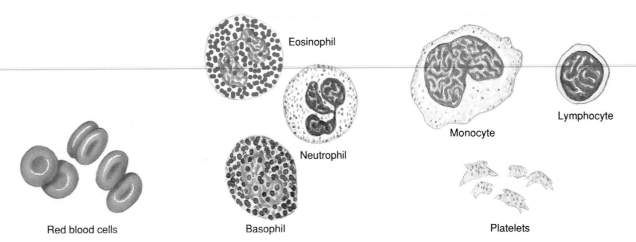

Figure 34-3
Formed elements of human blood. Hemoglobin-containing red blood cells of humans and other mammals lack nuclei, but those of all other vertebrates have nuclei. Various leukocytes provide a wandering system of protection for the body. Platelets participate in the blood's clotting mechanism.

few other invertebrate groups; see p. 343) the extracellular fluid is further subdivided into **blood plasma** and **interstitial (intercellular) fluid** (Figure 34-1A). The blood vessels contain the plasma, whereas the interstitial fluid, or tissue fluid as it is sometimes called, occupies the space surrounding the cells in the body. Nutrients and gases passing between the vascular plasma and the cells must traverse this narrow fluid separation. Interstitial fluid is constantly formed from the plasma by filtration through the capillary walls.

COMPOSITION OF THE BODY FLUIDS

All these fluid spaces—plasma, interstitial, and intracellular—differ from each other in solute composition, but all have one feature in common: they are mostly water. Despite their firm appearance, animals are 70% to 90% water. Humans, for example, are approximately 70% water by weight. Of this, 50% is cell water, 15% is interstitial fluid water, and the remaining 5% is in the blood plasma. It is the plasma space that serves as the pathway of exchange between the cells of the body and the outside world. This exchange of respiratory gases, nutrients, and wastes is accomplished by specialized organs (kidney, lung, gill, alimentary canal), as well as by the skin (Figure 34-1A).

The body fluids contain many inorganic and organic substances in so-

lution. Principal among these are inorganic electrolytes and proteins. **Sodium, chloride,** and **bicarbonate ions** are the chief extracellular electrolytes, whereas **potassium, magnesium,** and **phosphate ions** and **proteins** are the major intracellular electrolytes (Figure 34-1B). These differences are dramatic; they are always maintained despite the continuous flow of materials into and out of the cells of the body. The two subdivisions of the extracellular fluid—plasma and interstitial fluid—have similar compositions except that plasma has more proteins, which are mostly too large to filter through the capillary wall into the interstitial fluid.

COMPOSITION OF BLOOD

Among the invertebrates that lack a circulatory system (such as flatworms and cnidarians) it is not possible to distinguish a true "blood." These forms possess a clear, watery tissue fluid containing some phagocytic cells, a little protein, and a mixture of salts similar to seawater. The "blood" of invertebrates with open circulatory systems is more complex and is often referred to as hemolymph (Gr. *haimo,* blood, + L. *lympha,* water). Invertebrates with closed circulatory systems, on the other hand, maintain a clear separation between blood contained within blood vessels and tissue (interstitial) fluid surrounding the vessels.

In vertebrates, blood is a complex liquid tissue composed of plasma and formed elements, mostly red cells (also called corpuscles), suspended in the plasma. If we separate the red blood corpuscles and other formed elements from the fluid components by centrifugation, we find that blood is approximately 55% plasma and 45% formed elements.

The composition of mammalian blood is as follows:

Plasma
1. Water 90%
2. Dissolved solids, consisting of the plasma proteins (albumin, globulins, fibrinogen), glucose, amino acids, electrolytes, various enzymes, antibodies, hormones, metabolic wastes, and traces of many other organic and inorganic materials
3. Dissolved gases, especially oxygen, carbon dioxide, and nitrogen

Formed elements (Figure 34-3)
1. Red blood cells (erythrocytes), containing hemoglobin for the transport of oxygen and carbon dioxide
2. White blood cells (leukocytes), serving as scavengers particulate matter and as defensive cells
3. Cell fragments (platelets in mammals) or cells (thrombocytes in other vertebrates) that function in blood coagulation

The plasma proteins are a diverse group of large and small proteins that perform numerous functions. The major protein groups are (1) **albumins,** the most abundant group, constituting 60% of the total, which help to keep the plasma in osmotic equilibrium with the cells of the body; (2) **globulins,** a diverse group of high-molecular weight proteins (35% of total) that includes immunoglobulins and various metal-binding proteins; and (3) **fibrinogen,** a very large protein that functions in blood coagulation. Blood **serum** is plasma minus the proteins involved in clot formation (see the following).

Red blood cells, or **erythrocytes,** are present in enormous numbers in the blood, approximately 5.4 billion per milliliter of blood in adult men and 4.8 billion in adult women. In mammals and birds, red cells form continuously from large nucleated **erythroblasts** in the red bone marrow (in other vertebrates the kidneys and spleen are the principal sites of red blood cell production). During erythrocyte formation hemoglobin is synthesized and the cells divide several times. In mammals the nucleus shrinks during development to a small remnant and eventually disappears altogether. Many other characteristics of a typical cell also are lost: ribosomes, mitochondria, and most enzyme systems. What is left is a biconcave disc consisting of a baglike membrane packed with about 280 million molecules of the blood-transporting pigment hemoglobin. Approximately 33% of the erythrocyte by weight is hemoglobin. The biconcave shape (Figure 34-3) is a mammalian innovation that provides a larger surface for gas diffusion than would a flat or spherical shape. All other vertebrates have nucleated erythrocytes that are usually ellipsoidal in shape (Figure 34-4).

The erythrocyte enters the circulation for an average life span of approximately 4 months. During this time it may journey 11,000 km, squeezing repeatedly through the capillaries, which are sometimes so narrow that the erythrocyte must bend to pass through. At last it fragments and is quickly en-

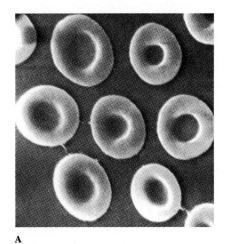
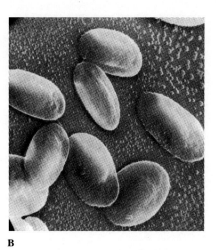

A   B

Figure 34-4

Mammalian and amphibian red blood cells. **A,** The erythrocytes of a gerbil are biconcave discs containing hemoglobin and surrounded by a tough stroma. **B,** The frog erythrocytes are convex discs, each containing a nucleus, which is plainly visible in the scanning electron micrograph as a bulge in the center of each cell. (Magnifications: mammalian erythrocytes, × 6300; frog erythrocytes, × 2400.)

gulfed by large scavenger cells called **macrophages** located in the liver, bone marrow, and spleen. Iron from the hemoglobin is salvaged to be used again; the rest of the heme is converted to **bilirubin,** a bile pigment. It is estimated that the human body produces 10 million erythrocytes and destroys another 10 million every second.

The white blood cells, or **leukocytes,** form a wandering system of protection for the body. In adults they number only approximately 7.5 million per milliliter of blood, a ratio of 1 white cell to 700 red cells. There are several kinds of white blood cells: **granulocytes** (subdivided into **neutrophils, basophils,** and **eosinophils**), and **nongranulocytes,** the lymphocytes and monocytes (Figure 34-3). We will discuss the role of the leukocytes in the body's defense mechanisms later.

HEMOSTASIS: PREVENTION OF BLOOD LOSS

It is essential that animals have ways of preventing the rapid loss of body fluids after an injury. Since blood is flowing and is under considerable hydrostatic pressure, it is especially vulnerable to hemorrhagic loss.

When a vessel is damaged, smooth muscle in the wall of the vessel contracts, which causes the vessel lumen to narrow, sometimes so strongly that

blood flow is completely stopped. This simple but highly effective means of preventing hemorrhage is used by invertebrates and vertebrates alike. Beyond this first defense against blood loss, all vertebrates, as well as some of the larger, active invertebrates with high blood pressures, have special cellular elements and proteins in the blood that are capable of forming plugs, or clots, at the injury site.

In vertebrates **blood coagulation** is the dominant hemostatic defense. Blood clots form as a tangled network of fibers from one of the plasma proteins, **fibrinogen.** The transformation of fibrinogen into a **fibrin** meshwork (Figure 34-5) that entangles blood cells to form a gel-like clot is catalyzed by the enzyme thrombin. Thrombin is normally present in the blood in an inactive form called **prothrombin,** which must be activated for coagulation to occur.

In this process, blood platelets (Figure 34-3) play a vital role. Platelets are formed in red bone marrow from certain large cells that regularly pinch off bits of their cytoplasm; thus they are fragments of cells. There are 150,000 to 300,000 platelets per cubic millimeter of blood. When the normally smooth inner surface of a blood vessel is disrupted, either by a break or by deposits of a cholesterol-lipid material, the platelets rapidly adhere

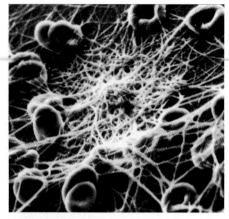

Figure 34-5

Human red blood cells trapped in fibrin clot. Clotting is initiated after tissue damage by the disintegration of platelets in the blood, resulting in a complex series of intravascular reactions that end with the conversion of a plasma protein, fibrinogen, into long, tough, insoluble polymers of fibrin. Fibrin and entangled erythrocytes form the blood clot, which arrests bleeding. An aggregation of platelets probably underlies the raised mass of fibrin in the center.

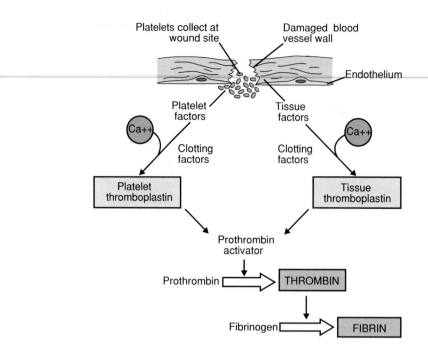

Figure 34-6

Stages in the formation of fibrin.

to the surface and release **thromboplastin** and other clotting factors. These factors, along with factors released from damaged tissue and with calcium ions, initiate the conversion of prothrombin to the active thrombin. The stages in the formation of fibrin are summarized in Figure 34-6.

The catalytic sequence in this scheme is unexpectedly complex, involving a series of plasma protein factors, each normally inactive until activated by a previous factor in the sequence. The sequence behaves like a "cascade" with each reactant in the sequence leading to a large increase in the amount of the next reactant. At least 13 different plasma coagulation factors have been identified. A deficiency of only a single factor can delay or prevent the clotting process. Why has such a complex clotting mechanism evolved? Probably it is necessary to provide a fail-safe system capable of responding to any kind of internal or external hemorrhage that might occur and yet a system that cannot be activated into forming dangerous intravascular clots in the absence of injury.

Several kinds of clotting abnormalities in humans are known. One of these, hemophilia, is a condition characterized by the failure of the blood to clot, so that even insignificant wounds can cause continuous severe bleeding. It is caused by a rare mutation (the condition occurs in about 1 in 10,000 males) on the X sex chromosome, resulting in an inherited lack of one of the platelet factors in males and in homozygous females. Called the "disease of kings," it once ran through several interrelated royal families of Europe, apparently having originated from a mutation in one of Queen Victoria's parents.

Hemophilia is one of the best known cases of sex-linked inheritance in humans (p. 134). Actually two different loci on the X chromosome are involved. Classic hemophilia (hemophilia A) accounts for about 80% of persons with the condition, and the remainder are caused by Christmas disease (hemophilia B). The allele at each locus results in the deficiency of a different platelet factor.

DEFENSE MECHANISMS OF THE BODY

INNATE IMMUNITY

Most animals have one or more mechanisms to protect themselves against invasion of a foreign body or infectious agent. Many of these are nonspecific, either coincidental properties of structures (such as a tough skin or high stomach acidity), or effective against a wide variety of invaders and not directed against any one in particular (mucous linings, phagocytosis, complement [p. 679], and others). These latter mechanisms may be characteristics evolved as adaptations for defense, and in vertebrates they may be dramatically affected by previous immunizing experience with the invading agent.

Phagocytosis

For defense against an invader, the invader first of all must be recognized. The cells in an animal must "know" when a substance does not belong in that animal; they must recognize

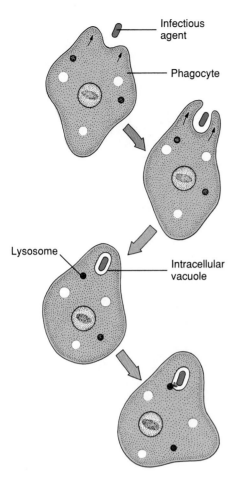

Figure 34-7
Phagocytosis. By pseudopodial movement, the phagocyte engulfs the particle. Lysosomes join with the vacuole containing the agent, pouring their contents (digestive enzymes or lysosomes) into the vacuole to destroy the particle.

"nonself." **Phagocytosis** illustrates nonself recognition, and it is found in almost all metazoa and is a major feeding mechanism among protistans (p. 219). A cell that has this ability is a **phagocyte.** Phagocytosis is a process of engulfment of the invading particle within an invagination of the phagocyte's cell membrane (Figure 34-7). The invagination pinches off and thereby encloses the particle in an intracellular vacuole. Lysosomes empty digestive enzymes into the vacuole to destroy the particle. In metazoan invertebrates the cells performing this function are known as **amebocytes** (or another name, depending on the group of animals). If the particle is too large for phagocytosis, the amebocytes may gather around it and wall it off. In humans there are **fixed** and **mobile**

phagocytes. The fixed phagocytes taken together form the **reticuloendothelial system (RE system)** and are found in the liver, spleen, lymph nodes, and other tissues. The RE system filters out and destroys particles and spent red blood cells from the blood that passes throughout the organs where the cells are located. The mobile phagocytes circulate in the blood and include the granulocytic leukocytes, especially neutrophils, and the monocytes, which become phagocytes (Figure 34-3). When monocytes move into the tissue from the blood, they differentiate into macrophages.

ACQUIRED IMMUNE RESPONSE IN VERTEBRATES

Vertebrates have a specialized system of nonself recognition that results in increased resistance to a *specific* foreign substance or invader after repeated exposures. Any substance that stimulates an immune response is called an **antigen** (Gr. *anti,* against, + *genos,* birth). Antigens may be any of a variety of substances that have a molecular weight greater than 3000, most commonly proteins; they are usually (but not always) foreign to the host. There are two arms of the immune response, known as **humoral** and **cellular.** Humoral immunity is based on **antibodies** (Figure 34-8), which are mostly dissolved in and circulate in the blood, whereas cellular immunity is associated with cell surfaces. There is extensive communication and interaction among the cells of the humoral and cellular responses.

Basis of Self and Nonself Recognition

Nonself recognition is very specific; if tissue from one individual is transplanted into another individual of the same species, the graft will grow for a time and then die as immunity against it arises. Without immunosuppression, tissue grafts can grow successfully only if they are between identical twins or between individuals of highly inbred strains of animals. The molecular basis

for this nonself recognition depends on certain proteins embedded in the cell surface. These proteins are coded by certain genes, now known as the **major histocompatibility complex (MHC)** because they were discovered in tissue graft experiments. The MHC proteins are among the most variable known, and unrelated individuals almost always have different genes. There are two types of MHC proteins: **class I** and **class II.** Class I proteins are found on the surface of virtually all cells, whereas class II MHC proteins are found only on certain cells participating in the immune responses, such as certain lymphocytes and macrophages.

The capability of an immune response develops over a period of time in the early development of the organism. All substances present at the time the capacity develops are recognized as self in later life. Unfortunately, the system of self and nonself recognition sometimes breaks down, and an animal may begin to produce antibodies against some part of its own body. This leads to one of several known *autoimmune diseases,* such as rheumatoid arthritis, multiple sclerosis, and insulin-dependent diabetes mellitus.

Antibodies

Antibodies are proteins called **immunoglobulins.** The basic antibody molecule consists of four polypeptide strands: two identical light chains and two identical heavy chains, held together in a Y-shape by disulfide bonds and hydrogen bonds (Figure 34-8). The amino acid sequence toward the ends of the Y varies in both the heavy and light chains, according to the specific antibody molecule (the **variable region**), and this determines with which antigen the antibody can bind. Each of the ends of the Y forms a cleft that acts as the antigen-binding site (see Figure 34-8), and the specificity of the molecule depends on the shape of the cleft and the properties of the chemical groups that line its walls. The remainder of the antibody is known as the **constant region,** but

Figure 34-8

A, Antibody molecule is composed of two shorter polypeptide chains (light chains) and two longer chains (heavy chains) held together by covalent disulfide bonds. These are further subdivided into variable and constant regions that have independent folding units, or domains, of about 110 amino acids. The folding pattern is more complex than that shown here. Interchain disulfide bonds at the hinge region give the molecule flexibility at that point. The variable domains of both the light and heavy chains have hypervariable ends, which serve as the antigen-binding sites. **B,** Molecular model of antibody molecule.

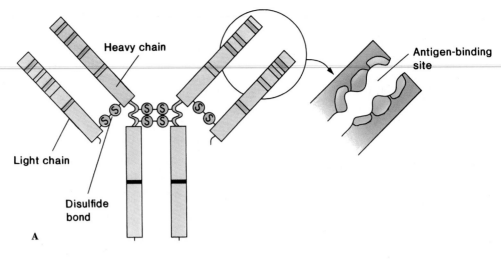

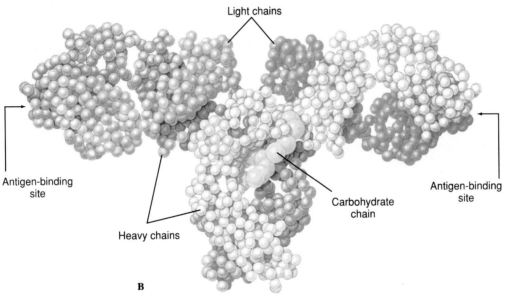

A major problem of immunology is understanding how the mammalian genome could contain the information needed to produce at least a million different antibodies. The answer seems to be that antibody genes occur in pieces, rather than as continuous stretches of DNA, and that the antigen-recognizing sites (variable regions) of the heavy and light chains of the antibody molecules are pieced together from information supplied by separate DNA sequences, which can be shuffled to increase the diversity of the gene products. The immense repertoire of antibodies is achieved in part by complex gene rearrangements and in part by frequent somatic mutations that produce additional variation in protein structure of the variable regions of the heavy and light antibody chains.

the "constant" region also varies to some extent. Variations in the constant region of the heavy chains determine the **class** of the antibodies, abbreviated as **IgM, IgG** (now familiar to many people as "gamma globulin"), **IgA, IgD,** and **IgE.** The class of the antibody controls the role of the antibody in the immune response (for example, whether the antibody is secreted or held on a cell surface), but not the antigen it recognizes.

Cells of the Immune Response

Some of the cells that are important in the immune response, such as granulocytes and macrophages, were mentioned previously. A number of others are mostly encompassed by a category of leukocytes known as **lym-phocytes. B lymphocytes (B cells)** (Figure 34-9) have antibody molecules in their surface and give rise to cells that actively secrete antibodies into the blood. **T lymphocytes (T cells)** have surface receptors that bind antigens, but the receptors are somewhat different in structure from antibodies. A vast number of different kinds of B cells exist in each individual, each kind bearing molecules of antibody on its surface that will bind with one particular antigen, even though that antigen may never have been present in the body previously. There are probably an equally great number of different T cells with receptors for specific antigens.

Lymphocytes are **activated** when they are stimulated to move from their recognition phase, in which they simply bind with a particular antigen, to a phase of proliferation and differentiation. The function of activated cells is to eliminate the antigen. We also speak of activation of cells such as macrophages when they are stimulated to carry out their protective function.

Subsets of T Cells. Communication between cells in the immune response, regulation of the response, and certain effector functions are carried out by different kinds of T cells. Although morphologically similar, subsets of T cells can be distinguished by characteristic proteins in their surface membranes. For example, cells with the protein CD4 (for **c**luster of **d**ifferentiation) are CD4$^+$ and those with CD8 are described as CD8$^+$. A complicated web of interactions, including both CD4$^+$ and CD8$^+$ cells, controls the immune response (Figure 34-10). Some of them (designated T$_H$ for T **h**elper) activate cell-mediated immunity while suppressing the humoral response (T$_H$1 cells), and others (called T$_H$2) activate humoral and suppress cell-mediated immunity.

   Cytotoxic T lymphocytes (CTLs) are CD8$^+$ cells that kill target cells expressing a certain antigen. The CTL binds tightly to the target cell and secretes a protein that causes pores to form in the cell membrane. The target cell then lyses.

Other Cells.   Natural killer cells (NK) are lymphocyte-like cells that can kill virus-infected and tumor cells in the absence of antibody. **Mast cells** are basophil-like cells found in the dermis and other tissues. Their surfaces bear receptors for the constant portions of IgE and IgG.

Cytokines

The 1980s saw rapid advances in our knowledge of how cells of immunity communicate with each other. They do this by means of protein hormones called **cytokines.** Cytokines can produce their effects on the same cells that produce them, on cells nearby, or on cells distant in the body from those that produced the cytokine. A few cytokines important in immune responses follow,

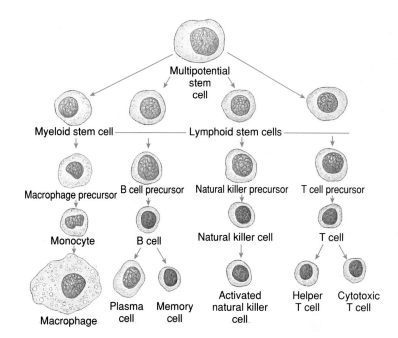

Figure 34-9

Lineage of some cells active in immune response. These cells, as well as red blood cells and other white blood cells are derived from multipotential stem cells in the bone marrow. B cells mature in bone marrow and are released into blood or lymph. Precursors of T cells go through a period in the thymus gland. Precursors of macrophages circulate in blood as monocytes.

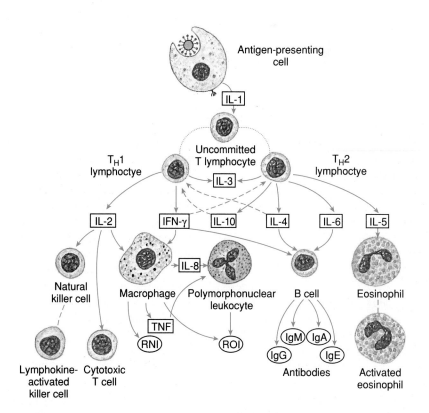

Figure 34-10

Major pathways involved in cell-mediated (T$_H$1) and humoral (T$_H$2) immune responses as mediated by cytokines. Solid arrows indicate positive signals and broken arrows indicate inhibitory signals. Broken lines without arrows indicate path of cellular activation. *IFN-γ,* interferon-γ; *Ig,* immunoglobulin; *IL* interleukin; *T$_H$,* T helper cells; *TNF,* tumor necrosis factor; *RNI* and *ROI,* toxic substances released onto invader.

but for the sake of simplicity, we will not discuss all the cytokines active in the processes shown in Figure 34-10.

1. **Interleukin-1 (IL-1).** The interleukins were originally so-called because they are synthesized by leukocytes and affect leukocytes. We now know that some other kinds of cells can produce interleukins, and interleukins produced by leukocytes can affect other kinds of cells. IL-1 is produced by activated macrophages and mediates the host inflammatory response. It also activates T cells and B cells.

2. **Interleukin-2 (IL-2).** IL-2 is produced by CD4⁺ cells and to a lesser extent by CD8⁺ cells. It is a major growth factor for T and B cells, and it enhances the cytolytic activity of natural killer cells, causing them to become **lymphocyte-activated killer (LAK)** cells.

3. **Interleukin-4 (IL-4).** IL-4 is produced mostly by T_H2 CD4⁺ cells. It is a growth factor for B cells, some CD4⁺ T cells, and mast cells, but it suppresses T_H1 differentiation.

4. **Interleukin-5 (IL-5).** IL-5 is produced by certain CD4⁺ cells and stimulates activation of eosinophils so that they can kill some parasites. It also acts with IL-2 and IL-4 to stimulate growth and differentiation of B cells.

5. **Interleukin-10 (IL-10).** IL-10 is derived from T_H2 CD4⁺ cells, and it inhibits T_H1, CD8⁺, NK, and macrophage cytokine synthesis.

6. **Interferon-γ (IFN-γ).** IFN-γ is produced by some CD4⁺ and almost all CD8⁺ cells. It is a strong macrophage-activating factor, causes a variety of cells to express class II MHC molecules, promotes differentiation of T and B cells, activates neutrophils and NK cells, and activates endothelial cells (see p. 686) to allow lymphocytes to pass through walls of blood vessels.

7. **Tumor necrosis factor (TNF).** Activated macrophages secrete most TNF. It is a major mediator of inflammation. In low concentrations TNF activates endothelial cells, activates granulocytes, and stimulates macrophages and cytokine production (including IL-1, IL-6, and TNF itself). In higher concentrations TNF causes increased synthesis of prostaglandins (p. 747) in the hypothalamus, resulting in fever.

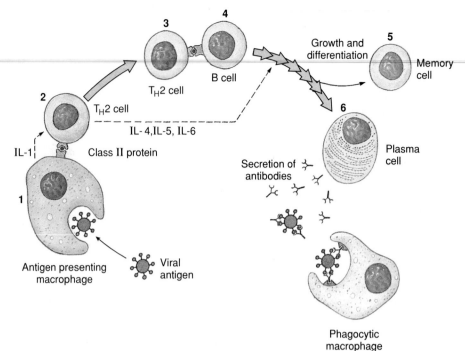

Figure 34-11

Humoral immune response. (1) Macrophage consumes antigen, partially digests it, and displays portions on its surface, along with class II MHC protein, and secretes interleukin-1 (IL-1). (2) T_H2 cell, stimulated by IL-1, recognizes antigen and class II protein on macrophage, is activated, and secretes interleukins 4, 5, and 6 (IL-4, IL-5, IL-6). (3) T_H2 then activates B cell which carries antigen and class II protein on its surface. IL-4, IL-5, and IL-6 stimulate proliferation of B cell line. (4) Activated B cells finally produce many plasma cells that secrete antibody. (5) Some B-cell progeny become memory cells. (6) Antibody produced by plasma cells binds to antigen and stimulates macrophages to consume antigen (opsonization).

Generation of a Humoral Response

When an antigen enters the body, it binds to a specific antibody on the surface of the appropriate B cell, but this is usually insufficient to activate the B cell to multiply. Some of the antigen is taken up by **antigen-presenting cells (APCs),** such as macrophages, that partially digest the antigen. The APCs then incorporate portions of the antigen into their own cell surface (Figures 34-10 and 34-11). The macrophages also secrete IL-1, which activates appropriate T_H2 cells. These T cells recognize the antigen fragments on the surface of the macrophages along with the class II MHC protein on the macrophage surface. (Both the class II protein and the antigen must be present; neither is effective alone.) The T_H2 cells then secrete other interleukins that activate the B cell bearing the same antigen fragment and a class II MHC protein on its surface. The B cell multiplies rapidly and produces many **plasma cells,** which secrete large quantities of antibody for a period of time, then die. Thus if we measure the concentration of the antibody **(titer)** soon after the antigen enters, we can detect little or none. The titer then rises rapidly as the plasma cells secrete antibody, and it may decrease somewhat as they die and the antibody is degraded (Figure 34-12). However, if we give another dose of antigen (the **challenge**), there is no lag, and the antibody titer rises quickly to a higher level than after the first dose. This is the **secondary response,** and it occurs because some of the activated B cells gave rise to long-lived **memory cells.** There are many more memory cells present in the body than the original B lymphocyte with the appropriate antibody on its surface, and they rapidly multiply to produce additional plasma cells.

Functions of Antibody in Host Defense. Antibodies can mediate destruction of an invader (antigen) in a number of ways. A foreign particle, for example, becomes coated with

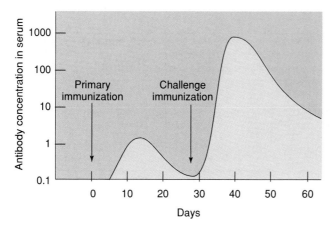

Figure 34-12

Typical immunoglobulin response after primary and challenge immunizations. The secondary response is a result of the large numbers of memory cells produced after the primary B-cell activation.

antibody molecules as their variable regions become bound to it. Macrophages recognize the projecting constant regions and are stimulated to engulf the particle. This is the process of **opsonization.**

Another important process, particularly in the destruction of bacterial cells, is the interaction with **complement.** Complement is a series of 12 enzymes that are activated by bound antibody, and they actually cause cell lysis by destroying the integrity of the cell membrane. Complement may also play a role in opsonization.

The Cell-Mediated Response

Some immune responses involve little, if any, antibody and depend on the action of cells only. In cell-mediated immunity (CMI) an antigenic fragment is also presented by macrophages, but the T_H1 arm of the immune response is activated and the T_H2 arm suppressed. Effector cells are macrophages, granulocytes, cytotoxic T cells, and activated natural killer cells. The specific interaction of lymphocyte and antigen that generates a CMI greatly influences subsequent events in the nonspecific response we call **inflammation.**

Like humoral immunity, CMI shows a secondary response due to large numbers of memory T cells that were produced from the original activation. For example, a second tissue graft (challenge) between the same donor and host will be rejected much more quickly than the first.

Acquired Immune Deficiency Syndrome (AIDS)

AIDS is an extremely serious disease in which the ability to mount an immune response is crippled. The first case was recognized in 1981, and by the year 2000, the World Health Organization estimates 40 million people will have been infected and 10 million will have died.* AIDS patients are continuously plagued by infections with agents (often parasites) that cause insignificant problems in persons with normal immune responses. The disease is caused by a virus (HIV) that preferentially invades and destroys $CD4^+$ lymphocytes. The CD4 surface protein is the major surface receptor for the virus. Normally, $CD4^+$ cells make up 60% to 80% of the T-cell population; in AIDS they can become too rare to be detected. T_H1 cells are relatively more depleted than T_H2 cells, which upsets the balance of immunoregulation and results in persistent B-cell activation.

INFLAMMATION

Inflammation is a vital process in the mobilization of the body defenses against an invading organism or other tissue damage and in the repair of damage thereafter. The inflammatory process is greatly influenced by the prior immunizing experience of the body with the invader and by the duration or persistence of the invader in the body. The processes by which the invader is actually destroyed, however, are themselves nonspecific. Depending on whether the response is primarily cell mediated or humoral,

Many aspects of immunology have been greatly assisted by the discovery of a method for producing stable clones of cells that will produce only one kind of antibody. Such monoclonal antibodies will bind to only *one kind* of antigenic determinant (most proteins bear many different antigenic determinants and thus stimulate the body to produce complex mixtures of antibodies). *Monoclonal* antibodies are made by fusing normal antibody-producing plasma cells with a continuously growing plasma cell line, producing a hybrid of the normal cell with one that can divide indefinitely in culture. This is called a *hybridoma.* Clones are grown from the hybrids, becoming "factories" that produce almost unlimited quantities of one specific antibody. Hybridoma techniques discovered in 1975 have become one of the most important research tools for the immunologist.

Only a few years ago transplantation of organs from one person to another seemed impossible. Then physicians began to transplant kidneys and depress the immune response of the recipient. It was very difficult to immunosuppress the recipient enough that the new organ would not be rejected and at the same time not leave the patient defenseless against infection. Since cyclosporine, a drug derived from a fungus, was discovered not only kidneys, but also hearts, lungs, and livers can be transplanted. Cyclosporine inhibits IL-2 and affects CTLs more than T_HS. It has no effect on other white cells or on healing mechanisms, so that a patient can still mount an immune response but not reject the transplant. However, the patient must continue to take cyclosporine, because if the drug is stopped, the body will recognize the transplanted organ as foreign and reject it.

*World Health Organization. 1994. The current global situation of the HIV/AIDS pandemic. Geneva, Switzerland, Global Programme on AIDS.

Table 34-1	Major Blood Groups							
Blood Type	Genotype	Antigens on Red Blood Cells	Antibodies in Serum	Can Give Blood To	Can Receive Blood From	Frequency in United States (%)		
						Whites	Blacks	Asians
O	O/O	None	Anti-A and anti-B	All	O	45	48	31
A	A/A, A/O	A	Anti-B	A, AB	O, A	41	27	25
B	B/B, B/O	B	Anti-A	B, AB	O, B	10	21	34
AB	A/B	AB	None	AB	All	4	4	10

inflammation may show **delayed type hypersensitivity (DTH)** and **immediate hypersensitivity.**

The DTH reaction is a type of CMI in which the ultimate effectors are activated macrophages. A period of 24 hours or more elapses between the time of antigen introduction and the response, a delay that accounts for the name. The delay occurs because of the time required for the particular T_H1 cells to arrive at the site, become activated, and secrete cytokines. Of the cytokines they secrete, TNF causes leukocytes to adhere to the endothelial cells, first neutrophils, then lymphocytes and monocytes. TNF also causes the endothelial cells to secrete inflammatory cytokines, which increase the mobility of leukocytes and facilitates their passage into the tissues. TNF and IFN-γ also cause the cells of the vessels to change shape, favoring leakage of macromolecules and passage of cells.

As the monocytes pass out of the blood vessels, they become activated macrophages, which are the main effector cells of the DTH. They phagocytize particulate antigen, secrete mediators that promote local inflammation, and secrete cytokines and growth factors that promote healing. If the antigen is not destroyed and removed, its chronic presence leads to deposition of fibrous connective tissue or **fibrosis.**

Immediate hypersensitivity involves exocytosis of their cytoplasmic granules by mast cells in the area. The surfaces of mast cells bear receptors for the constant portions of antibody, especially IgE. Occupation of these sites by antigen-specific antibodies enhances degranulation of the mast cells when the variable portions bind the particular antigen. The granules contain several mediators, such as histamine, that cause dilation of local blood vessels and increased permeability of blood vessels. Escape of blood plasma into the surrounding tissue causes swelling (a condition called **edema**), and engorgement of vessels with blood produces a characteristic redness. Although redness and edema of many immediate hypersensitivity reactions disappear in about an hour, some last 2 to 4 hours. Granulocytes and monocytes accumulate, and the reaction resembles DTH. Immediate hypersensitivity in humans is the basis for allergies and asthma. When the whole body reacts in immediate hypersensitivity, the condition is called **anaphylaxis.** Anaphylaxis can be fatal if not treated rapidly.

Some degree of cell death always occurs, but cell death may not be prominent in minor inflammation. If the cellular debris is confined within a localized area, the pus (spent leukocytes and tissue fluid) may increase in hydrostatic pressure, forming an **abscess.** An area of inflammation that opens out to a skin or mucous surface is an **ulcer.**

BLOOD GROUP ANTIGENS

ABO Blood Types

Blood differs chemically from person to person, and when two different (incompatible) blood types are mixed, **agglutination** (clumping together) of erythrocytes results. The basis of these chemical differences is naturally occurring antigens on the membranes of red blood cells. The best known of these is the ABO blood group. The genes for antigens A and B are inherited as dominant alleles. Thus, as shown in Table 34-1, an individual with, for example, genes *A/A* or *A/O* develops A antigen (blood type A). The presence of a *B* gene produces B antigens (blood type B), and for the genotype *A/B* both A and B antigens develop on the erythrocytes (blood type AB).

There is an odd feature about the ABO system. Normally we would expect that a type A individual would develop antibodies against type B blood only if B cells were introduced into the body. In fact, type A persons always have anti-B antibodies in their blood, even without the prior exposure to type B blood. Similarly, type B individuals carry anti-A antibodies. Type AB blood has neither anti-A nor anti-B antibodies (since if it did, it would destroy its own blood cells), and type O blood has both anti-A and anti-B antibodies.

We see then that the blood group names identify their *antigenic* character. People with type O blood are called universal donors because, lacking antigens, their blood can be infused into a person with any blood type. Even though it contains anti-A and anti-B antibodies, these are so diluted during transfusion that they do not react with A or B antigens in a recipient's blood. In practice, however, clinicians insist on matching blood types to prevent any possibility of incompatibility.

Rh Factor

Karl Landsteiner, an Austrian—later American—physician discovered the ABO blood groups in 1900. In 1940, 10 years after receiving the Nobel Prize, he made still another famous discovery. This was a blood group called the Rh factor, named after the Rhesus monkey, in which it was first found. Approximately 85% of white individuals in the United States have the factor (positive) and the other 15% do not (negative). Rh-positive and Rh-negative bloods are incompatible; shock and even death may follow their mixing when Rh-positive blood is introduced into an Rh-negative person who has been sensitized by an earlier transfusion of Rh-positive blood. Rh incompatibility accounts for a peculiar and often fatal form of anemia of newborn infants called erythroblastosis fetalis. If an Rh-negative mother has an Rh-positive baby (father is Rh-positive) she can become immunized by the fetal blood during the birth process. Anti-Rh antibodies can seep across the placenta during a subsequent pregnancy and agglutinate the fetal blood.

The genetics of the Rh factor are very much more complicated than it was believed when the factor was first discovered. Some authorities think that three genes located close together on the same chromosome are involved, whereas others adhere to a system of one gene with many alleles. In 1968 a revision of the single gene concept listed 37 alleles necessary to account for the phenotypes then known. Furthermore, the frequency of the various alleles varies greatly between whites, Asians, and blacks.

Erythroblastosis fetalis can now be prevented by giving an Rh-negative mother anti-Rh antibodies just after the birth of her first child. These antibodies remain long enough to neutralize any Rh-positive fetal blood cells that may have entered her circulation, thus preventing her own antibody machinery from being stimulated to produce the Rh-positive antibodies. Active, permanent immunity is blocked. The mother must be treated after every subsequent pregnancy (assuming the father is Rh$^+$). If the mother has already developed an immunity, however, the baby may be saved by an immediate, massive transfusion of blood free of antibodies.

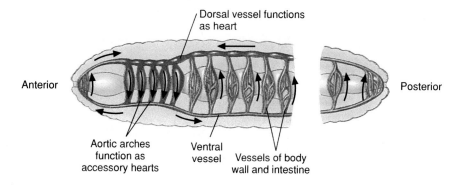

Figure 34-13
Blood flow through the closed vascular system of an earthworm.

CIRCULATION

We pointed out in the opening to this chapter that most animals have evolved mechanisms, in addition to simple diffusion, for transporting materials among various regions of the body. For sponges and radiates the water in which they live provides the medium for transport. Water, propelled by ciliary, flagellar, or body movements, passes through channels or compartments to facilitate the movement of food, respiratory gases, and wastes. True circulatory systems—that is, a system of vessels through which blood moves—become essential with animals so large or so active that diffusional processes alone cannot supply their oxygen needs. An animal's shape obviously is important. The flattened and leaflike acoelomate flatworms, even though many are relatively large animals, have no need for a circulatory system because the distance of any body part from the surface is short; respiratory gases and metabolic wastes transfer by simple diffusion.

A circulatory system having a full complement of components—propulsive organ, arterial distribution system, capillaries, and venous reservoir and return system—is fully recognizable in the annelid worms. In the earthworm (Figure 34-13) there are two main vessels, a dorsal vessel carrying blood toward the head, and a ventral vessel that flows posteriorly, delivering blood to all the body by way of segmental vessels and a dense capillary network. The dorsal vessel drives the blood forward by peristalsis (see p. 703) and thus serves as a heart. Five aortic arches that on each side connect the dorsal and ventral vessels are also contractile and serve as accessory hearts to maintain a steady flow of blood into the ventral vessel. Many of the smaller segmental vessels that deliver blood to tissue capillaries are actively contractile as well. We see then that there is no localized pump pushing the blood through a system of passive tubes; instead the power of contraction is widely distributed throughout the vascular system.

OPEN AND CLOSED CIRCULATIONS

The system just described is a **closed circulation** because the circulating medium, the **blood,** is confined to vessels throughout its journey through the vascular system. Many invertebrates have an **open circulation** in which there are no small blood vessels or capillaries connecting arteries with the veins. In insects and other arthropods, in most molluscs, and in many smaller invertebrate groups blood sinuses, collectively called the **hemocoel,** replace capillary beds found in animals with closed systems. During development of the body cavity in these groups, the blastocoel is not completely obliterated by the expanding mesoderm. This space becomes the hemocoel, which is nothing more than the primary body cavity (persistent blastocoel) through which the blood (also called **hemolymph**) freely circulates (bottom diagrams in Figure 34-14). Since there is no separation of the extracellular

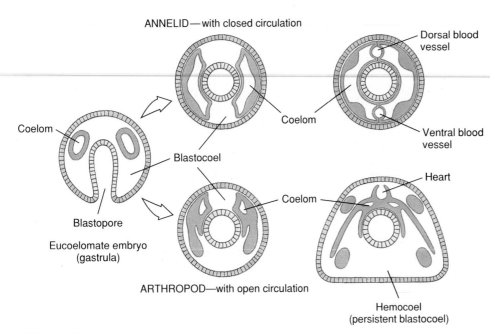

ANNELID—with closed circulation

Coelom

Dorsal blood vessel

Coelom

Ventral blood vessel

Blastocoel

Coelom

Heart

Eucoelomate embryo (gastrula)

Blastopore

ARTHROPOD—with open circulation

Hemocoel (persistent blastocoel)

Figure 34-14

Diagrams showing how open and closed circulatory systems develop. The principal body cavity of arthropods is the persistent blastocoel which becomes the hemocoel; the true coelom remains mostly undeveloped.

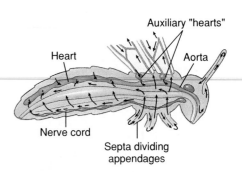

Auxiliary "hearts"

Heart

Aorta

Nerve cord

Septa dividing appendages

Figure 34-15

Circulatory system of an insect. Although the circulatory system is open, the blood is directed through the appendages in channels formed by longitudinal septa. Arrows indicate the course of the circulation.

fluid into blood plasma and lymph (as there is in a closed circulation, p. 687) the blood volume is large and may constitute 20% to 40% of body volume. By contrast, the blood volume in animals with closed circulations (vertebrates, for example) is only about 5% to 10% of body volume.

In arthropods, the heart and all the viscera lie in the hemocoel, bathed by the blood (Figure 34-14). Blood enters the heart through valved openings, the ostia, and the heart's contractions, which resemble a forward-moving peristaltic wave, propel the blood into a limited arterial system. The blood is distributed to the head and other organs, then escapes into the hemocoel. It is routed through the body and appendages by a system of baffles and longitudinal membranes (septa) before returning to the heart. Because the blood pressure is very low in open systems, seldom exceeding 4 to 10 mm Hg, many arthropods have auxiliary hearts or contractile vessels to boost blood flow (Figure 34-15).

During the embryonic development of animals with closed circulatory systems (most annelids, cephalopod molluscs, and all vertebrates) the coelom increases in size to obliterate

the blastocoel and forms a secondary body cavity (top diagrams in Figure 34-14). A system of continuously connected blood vessels develops within the mesoderm. All closed systems have certain features in common. A **heart** pumps the blood into **arteries** that branch and narrow into **arterioles** and then into a vast system of **capillaries.** Blood leaving the capillaries enters **venules** and then **veins** that return the blood to the heart. The capillary walls are thin, permitting rapid rates of transfer of materials between blood and tissues. Closed systems are more suitable for large and active animals because the blood can be moved rapidly to the tissues needing it. In addition, flow to various organs can be readjusted to meet changing needs by varying the diameters of the blood vessels.

Because blood pressures are much higher in closed than in open systems, fluid is constantly filtered across capillary walls into the surrounding tissue spaces. Most of this fluid is drawn back into the capillaries by osmosis (see p. 687). The remainder is recovered by the **lymphatic system** which has evolved in parallel with the high-pressure system of vertebrates.

PLAN OF VERTEBRATE CIRCULATORY SYSTEMS

In vertebrates the principal differences in the blood vascular system involve the gradual separation of the heart into two separate pumps as the vertebrates progressed from aquatic life with gill breathing to fully terrestrial life with lung breathing. These changes are shown in Figure 34-16 which compares the circulation of fish, amphibians, and mammals.

The fish heart contains two main chambers in series, an **atrium** and a **ventricle.** The atrium is preceded by an enlarged chamber, the **sinus venosus,** which collects blood from the venous system to assure a smooth delivery of blood to the heart. Blood makes a single circuit through the fish's vascular system; it is pumped from the heart to the gills, where it is oxygenated, then flows into the dorsal aorta to be distributed to the body organs, and finally returns by veins to the heart. In this circuit the heart must provide sufficient pressure to push the blood through two sequential capillary systems, first that of the gills, and then that of the remainder of the body. The principal disadvantage of the single-circuit system is that the gill capillaries offer so much resistance to blood flow that blood pressures to the body tissues are greatly reduced.

With the evolution of lung breathing and the elimination of the gills between the heart and the aorta, vertebrates developed a high-pressure **double circulation:** a **systemic**

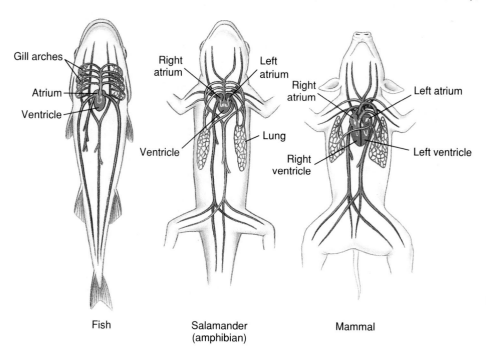

Figure 34-16

Circulatory systems of fish, amphibian, and mammal, showing the evolution of separate systemic and pulmonary circuits in lung-breathing vertebrates.

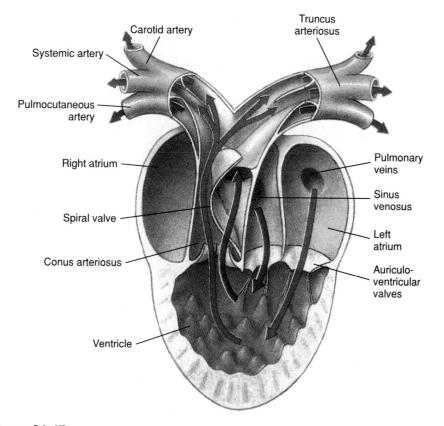

Figure 34-17

Route of blood through a frog heart. Atria are completely separated, and the spiral valve helps to route blood to lungs and systemic circulation.

circuit that provides oxygenated blood to the capillary beds of the body organs; and a **pulmonary circuit** that serves the lungs. We see the beginnings of this major evolutionary change in lungfishes and amphibians. In modern amphibians (frogs, toads, salamanders) the atrium is completely separated by a partition into two atria (Figure 34-17). The right atrium receives venous blood from the body while the left atrium receives oxygenated blood from the lungs. The ventricle is undivided, but venous and arterial blood remain mostly separate by the arrangement of the vessels leaving the heart. Separation of the ventricles is nearly complete in some reptiles (crocodilians) and is completely separate in birds and mammals (Figure 34-18). The systemic and pulmonary circuits are now separate circulations, each served by one half of a dual heart (Figure 34-18).

Mammalian Heart

The four-chambered mammalian heart (Figure 34-18) is a muscular organ located in the thorax and covered by a tough, fibrous sac, the **pericardium.** Blood returning from the lungs collects in the **left atrium,** passes into the **left ventricle,** and is pumped into the body (systemic) circulation. Blood returning from the body flows into the **right atrium,** passes into the **right ventricle,** which pumps it into the lungs. Backflow of blood is prevented by two sets of valves that open and close passively in response to pressure differences between the heart chambers. The **bicuspid** and **tricuspid** valves separate the cavities of the atrium and ventricle in each half of the heart. Where the great arteries, the **pulmonary** from the right ventricle and the **aorta** from the left ventricle, leave the heart, **semilunar valves** prevent backflow into the ventricles.

The contraction of the heart is called **systole** (sis'to-lee), and the relaxation, **diastole** (dy-as'to-lee) (Figure 34-19). The rate of the heartbeat depends on age, sex, and especially exercise. Exercise may increase the **cardiac**

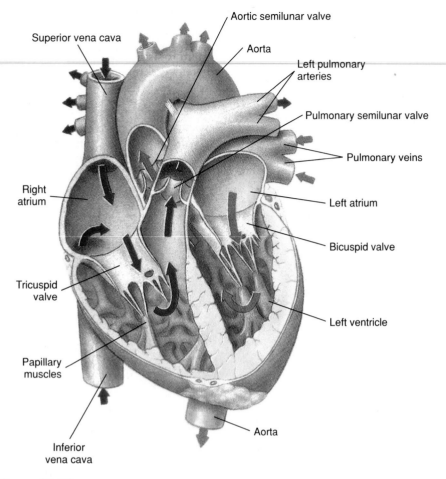

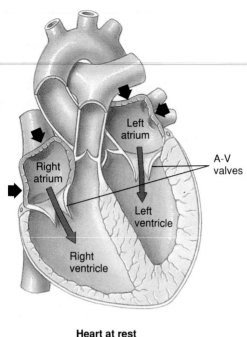

Heart at rest (diastole)

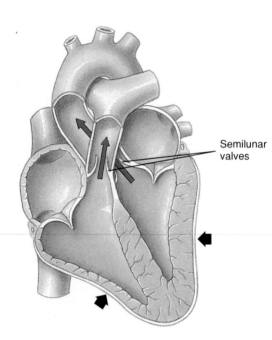

Heart during contraction (systole)

Figure 34-18
Human heart. Deoxygenated blood enters right side of heart and is pumped to the lungs. Oxygenated blood returning from the lungs enters left side of the heart and is pumped to the body. The left ventricular wall is thicker than that of the right ventricle, which needs less muscular force to pump blood into the nearby lungs.

Figure 34-19
Human heart in systole and diastole.

output (volume of blood forced from either ventricle each minute) more than fivefold. Both the heart **rate** and the **stroke volume** increase. Heart rates among vertebrates vary with the general level of metabolism and the body size. The ectothermic codfish has a heart rate of approximately 30 beats per minute; an endothermic rabbit of about the same weight has a rate of 200 beats per minute. Small animals have higher heart rates than do large animals. The heart rate in an elephant is 25 beats per minute, in a human 70 per minute, in a cat 125 per minute, in a mouse 400 per minute, and in the tiny 4 g shrew, the smallest mammal, the heart rate approaches a prodigious 800 beats per minute. We must marvel that the shrew's heart can sustain such a frantic pace throughout this animal's life, brief as it is.

Excitation of the Heart

The vertebrate heart is a muscular pump composed of **cardiac muscle.** Cardiac muscle resembles skeletal muscle—both are types of striated muscle—but the cells are branched and joined end-to-end by junctional complexes to form a complex branching network (see Figure 10-7, p. 190). Unlike skeletal muscle, vertebrate cardiac muscle does not depend on nerve activity to initiate a contraction. Instead, regular contractions are established by specialized cardiac muscle cells, called **pacemaker cells.** In the tetrapod heart it is located in the **sinus node,** a remnant of the sinus venosus in the fishlike ancestor. Electrical activity initiated in the pacemaker spreads over the muscle of the two atria and then, after a slight delay, to the muscle of the ventricles. At this point the electrical activ-

ity is conducted rapidly through the **atrioventricular bundle** to the apex of the ventricle and then continues through specialized fibers (**Purkinje fibers**) up the walls of the ventricles (Figure 34-20). This arrangement allows the contraction to begin at the apex or "tip" of the ventricles and

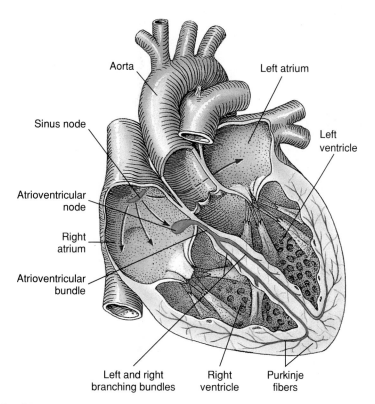

Aorta

Left atrium

Sinus node

Left ventricle

Atrioventricular node

Right atrium

Atrioventricular bundle

Left and right branching bundles

Right ventricle

Purkinje fibers

Figure 34-20

Neuromuscular mechanisms controlling heartbeat. Arrows indicate spread of excitation from the sinus node, across the atria, to the atrioventricular node. Wave of excitation is then conducted very rapidly to ventricular muscle over the specialized conducting bundles and Purkinje fiber system.

spread upward to squeeze out the blood in the most efficient way; it also ensures that both ventricles contract simultaneously. Structural specializations in the Purkinje fibers, such as well-developed intercalated discs and numerous gap junctions facilitate rapid conduction through these fibers.

The **control (cardiac) center** located in the medulla sends out two sets of nerves. Impulses sent along one set, the **vagus** nerves, apply a braking action to the heart rate, and impulses sent along the other set, the **accelerator** nerves, speed it up. Both sets of nerves terminate in the sinus node, thus guiding the activity of the pacemaker.

The cardiac center in turn receives sensory information about a variety of stimuli. Pressure receptors (sensitive to blood pressure) and chemical receptors (sensitive to carbon dioxide and pH) are located at strategic points in the vascular system. The cardiac center uses this information to increase or reduce the heart

rate and cardiac output in response to activity or changes in body position. Feedback mechanisms thus control the heart and keep its activity constantly attuned to needs of the body.

Because the heartbeat is initiated in specialized muscle cells, vertebrate hearts, together with the hearts of molluscs and several other invertebrates, are called **myogenic** ("muscle origin") hearts. Although the nervous system does alter pacemaker activity to slow down or speed up the heart rate, the myogenic heart will beat spontaneously and involuntarily even if completely removed from the body. An isolated turtle or frog heart beats for hours if placed in a balanced salt solution. Some invertebrates, for example decapod crustaceans, have **neurogenic** ("nerve origin") hearts. In these a cardiac ganglion located on the heart serves as pacemaker. If this ganglion is separated from the heart the heart stops beating, even though the ganglion itself remains rhythmically active.

Coronary Circulation

It is no surprise that an organ as active as the heart needs a generous blood supply of its own. The heart muscle of the frog and other amphibians is so thoroughly channeled with spaces between the muscle fibers that the heart's own pumping action squeezes through sufficient oxygenated blood. In birds and mammals, however, the thickness of the heart muscle and its high rate of metabolism require that the heart have its own vascular supply, the **coronary circulation.** The coronary arteries break up into an extensive capillary network surrounding the muscle fibers and provide them with oxygen and nutrients. Heart muscle has an extremely high oxygen demand. Even at rest the heart removes 70% of the oxygen from the blood, in contrast to most other body tissues, which remove only about 25%. Therefore, an increase in the work of the heart must be met by a massive increase in coronary blood flow—up to nine times the resting level during strenuous exercise. Any reduction in coronary circulation due to partial or complete blockage (coronary artery disease) may lead to a heart attack (myocardial infarction) in which heart cells die from lack of oxygen.

Thickening and loss of elasticity in the arteries is known as *arteriosclerosis.* When the arteriosclerosis is caused by fatty deposits of cholesterol in the artery walls, the condition is *atherosclerosis.* Such irregularities in the walls of blood vessels often cause blood to clot around them, forming a *thrombus.* When a bit of the thrombus breaks off and is carried by the blood to lodge elsewhere, it is an *embolus.* If the embolus blocks one of the coronary arteries, the person has a heart attack (a "coronary"). The portion of the heart muscle served by the branch of the coronary artery that is blocked is starved for oxygen. It may be replaced by scar tissue if the person survives.

ARTERIES

All vessels leaving the heart are called arteries whether they carry oxygenated blood (aorta) or deoxygenated blood (pulmonary artery). To withstand high, pounding pressures, arteries are invested with layers of both elastic and tough inelastic connective fibers (Figure 34-21). The elasticity of arteries allows them to yield to the surge of blood leaving the heart during systole and then to compress the fluid column during diastole. This smooths out the blood pressure. Thus the normal arterial pressure in humans varies only between 120 mm Hg (systole) and 80 mm Hg (diastole) (usually expressed as 120/80 or 120 over 80), rather than dropping to zero during diastole as we might expect in a fluid system with an intermittent pump.

As the arteries branch and narrow into **arterioles,** the walls become mostly smooth muscle. Contraction of this muscle narrows the arterioles and reduces the flow of blood. The arterioles thus control the blood flow to body organs, diverting it to where it is most needed. The blood must be pumped with a hydrostatic pressure sufficient to overcome resistance of the narrow passages through which it must flow. Consequently, large animals tend to have higher blood pressure than do small animals.

Blood pressure was first measured in 1733 by Stephan Hales, an English clergyman with unusual inventiveness and curiosity. He tied his mare, which was "to have been killed as unfit for service," on her back and exposed the femoral artery. This he cannulated with a brass tube, connecting it to a tall glass tube with the windpipe of a goose. The use of the windpipe was both imaginative and practical; it gave the apparatus flexibility "to avoid inconveniences that might arise if the mare struggled." The blood rose 8 feet in the glass tube and bobbed up and down with the systolic and diastolic beats of the heart. The weight of the 8-foot column of blood was equal to the blood pressure. We now express

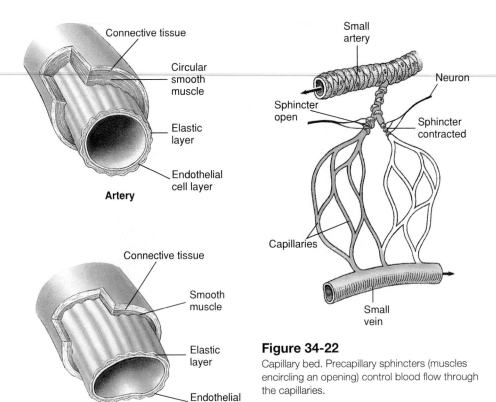

Figure 34-21
Artery and vein, showing layers. Note greater thickness of the muscularis layer (tunica media) in the artery.

Figure 34-22
Capillary bed. Precapillary sphincters (muscles encircling an opening) control blood flow through the capillaries.

this as the height of a column of mercury (Hg), which is 13.6 times heavier than water. Hales' figures, expressed in millimeters of mercury, indicate that he measured a blood pressure of 180 to 200 mm Hg, about normal for a horse.

Today, we measure blood pressure in humans most commonly and easily with an instrument called a **sphygmomanometer.** We inflate a cuff on the upper arm with air to a pressure sufficient to close the arteries in the arm. Holding a stethoscope over the brachial artery (in the crook of the elbow) and slowly releasing the air from the cuff, we can hear the first spurts of blood through the artery as it opens slightly. This is equivalent to the systolic pressure. As the pressure in the cuff decreases, the sound finally disappears as the blood runs smoothly through the artery. The pressure at which the sound disappears is the diastolic pressure.

CAPILLARIES

The Italian Marcello Malpighi was the first to describe the capillaries in 1661, thus confirming the existence of the minute links between the arterial and venous systems that Harvey knew must exist but could not see. Malpighi studied the capillaries of the living frog's lung, which is still one of the simplest and most vivid preparations for demonstrating capillary blood flow.

Capillaries are present in enormous numbers, forming extensive networks in nearly all tissues (Figure 34-22). In muscle there are more than 2000 per square millimeter (1,250,000 per square inch), but not all are open at once. Indeed, perhaps less than 1% are open in resting skeletal muscle. But when the muscle is active, all the capillaries may open to bring oxygen and nutrients to the working muscle fibers and to carry away metabolic wastes.

Capillaries are extremely narrow, averaging about 8 μm in diameter in mammals, which is only slightly wider than the red blood cells that must pass through them. Their walls are formed by a single layer of thin

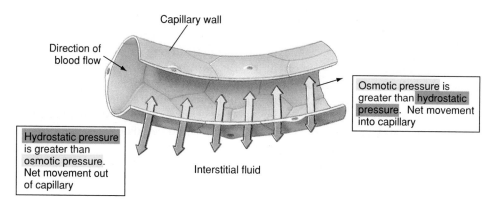

Figure 34-23

Fluid movement across the wall of a capillary. At the arterial end of the capillary, hydrostatic (blood) pressure exceeds colloid osmotic pressure contributed by plasma proteins, and a plasma filtrate is forced out. At the venous end, colloid osmotic pressure exceeds the hydrostatic pressure, and fluid is drawn back in. In this way plasma nutrients are carried out into the interstitial space where they can enter cells, and metabolic end products from the cells are drawn back into the plasma and carried away.

endothelial cells, held together by a delicate basement membrane and connective tissue fibers.

Capillary Exchange

Capillaries are quite permeable to small ions, nutrients, and water. Blood pressure within the capillary tends to force fluids out through the capillary walls and into the surrounding interstitial space (p. 672). Because larger molecules such as plasma proteins cannot pass the capillary wall, an almost protein-free filtrate is forced out. This fluid movement is important in irrigating the interstitial space, in providing tissue cells with oxygen, glucose, amino acids, and other nutrients, and in carrying away metabolic wastes. For capillary exchange to be effective, fluids that leave the capillaries must at some point reenter the circulation. If they did not, fluid would quickly accumulate in tissue spaces, causing edema (p. 680). The delicate balance of fluid exchange across the capillary wall can be accounted for by the two opposing forces of hydrostatic (blood) pressure and osmotic pressure (Figure 34-23).

In the capillary, the blood pressure that pushes water molecules and solutes across the capillary wall is greatest at the start of the capillary and declines along its length as the blood pressure falls (Figure 34-23). Opposing the blood hydrostatic pres-

sure is an osmotic pressure created by the proteins that cannot pass the capillary wall. This **colloid osmotic pressure,** which amounts to about 25 mm Hg in mammalian plasma, tends to draw water back into the capillary from the tissue fluid. The result of these two opposing forces is that water and solutes tend to be filtered out of the arteriolar end of the capillary where the hydrostatic pressure exceeds the osmotic pressure, and to be drawn in again at the venous end where the osmotic pressure exceeds the hydrostatic pressure.

> The actual situation is a bit more complicated because there is a small hydrostatic pressure in the interstitial fluid, and a small amount of protein does leak through the capillary wall. The protein tends to accumulate at the venule end of the capillary, building up a small osmotic pressure there. Although actual calculation of the pressure differences must take into account the interstitial fluid hydrostatic and osmotic pressures, the principle of the capillary fluid shift is as we have presented it.

The amount of fluid filtered across the capillary wall fluctuates greatly among different capillaries. Usually outflow exceeds inflow, and the excess fluid, called **lymph,** remains in

the interstitial spaces between the tissue cells. This excess is picked up and removed by the **lymph capillaries** of the lymphatic system and eventually returned to the circulatory system via larger lymph vessels (see the following text).

VEINS

The venules and veins into which the capillary blood drains for its return journey to the heart are thinner walled, less elastic, and of considerably larger diameter than their corresponding arteries and arterioles (Figure 34-21). Blood pressure in the venous system is low, from approximately 10 mm Hg, where capillaries drain into venules, to approximately zero in the right atrium. Because pressure is so low, the venous return gets assistance from valves in the veins, body muscles surrounding the veins, and the rhythmical action of the lungs. If it were not for these mechanisms, blood might pool in the lower extremities of a standing animal—a very real problem for people who must stand for long periods. Veins that lift blood from the extremities to the heart contain valves that divide the long column of blood into segments. When skeletal muscles contract, as in even slight activity, the veins are squeezed, and the blood within them moves toward the heart because the valves within the veins keep the blood from slipping back. The well-known risk of fainting while standing at stiff attention in hot weather usually can be prevented by deliberately pumping leg muscles. Negative pressure in the thorax created by the inspiratory movement of the lungs also speeds the venous return by sucking the blood up the large vena cava into the heart.

LYMPHATIC SYSTEM

The lymphatic system of vertebrates is an extensive network of thin-walled vessels that arise as blind-ended lymph capillaries in most tissues of the body. These unite to form a tree-like structure of increasingly larger lymph vessels, which finally drain

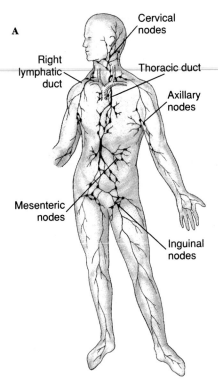

A

Cervical nodes

Right lymphatic duct

Thoracic duct

Axillary nodes

Mesenteric nodes

Inguinal nodes

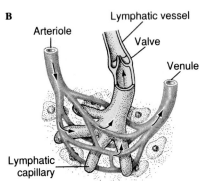

B

Arteriole

Lymphatic vessel

Valve

Venule

Lymphatic capillary

Figure 34-24
Human lymphatic system, showing major vessels, **A,** and a detail of the blood and lymphatic capillaries, **B.**

into veins in the lower neck (Figure 34-24). A principal function of the lymphatic system is to return to the blood the excess fluid (lymph) filtered across capillary walls into the interstitial space. Lymph is similar to plasma but has a much lower concentration of protein. Large molecules, especially fats absorbed from the gut, also reach the circulatory system by way of the lymphatic system. The rate of lymph flow is very low, a minute fraction of the blood flow.

The lymphatic system also plays a central role in the body's defenses. Located at intervals along the lymph vessels are **lymph nodes** (Figure 34-24)

that have several defense-related functions. They are effective filters that remove foreign particles, especially bacteria, that might otherwise enter the general circulation. They are also centers (together with the bone marrow and thymus gland) for the production, maintenance, and distribution of lymphocytes that produce antibodies—essential components of the body's defense mechanisms (p. 676).

RESPIRATION

Energy bound up in food is released by oxidative processes, usually with molecular oxygen as the terminal electron acceptor. Oxygen for this purpose is taken into the body across some respiratory surface. Physiologists find it is convenient to distinguish two separate but interrelated respiratory processes: **cellular respiration,** the oxidative processes that occur within cells (p. 67), and **external respiration,** the exchange of oxygen and carbon dioxide between the organism and the environment. In this section we describe external respiration and the transport of gases from respiratory surfaces to body tissues.

In small organisms such as single-celled protistans, oxygen is acquired and carbon dioxide liberated by direct diffusion across surface membranes. Gas exchange by diffusion alone is possible only for very small organisms less than 1 mm in diameter, where diffusion paths are short and the surface area of the organism is large relative to volume. As animals became larger and evolved a waterproof covering, specialized devices such as lungs and gills evolved that greatly increased the effective surface for gas exchange. But, because gases diffuse so slowly through living tissue, a circulatory system was necessary to distribute the gases to and from the deep tissues of the body. Even these adaptations were inadequate for complex animals with high rates of cellular respiration. The solubility of oxygen in the blood plasma is so low that plasma alone cannot carry enough to support metabolic demands. With the evolution of

special oxygen-transporting blood proteins such as hemoglobin, the oxygen-carrying capacity of the blood increased greatly. Thus what began as a simple and easily satisfied requirement resulted in the evolution of several complex and essential respiratory and circulatory adaptations.

PROBLEMS OF AQUATIC AND AERIAL BREATHING

How an animal respires is determined largely by the nature of its environment. The two great arenas of animal evolution—water and land—are vastly different in their physical characteristics. The most obvious difference is that air contains far more oxygen—at least 20 times more—than does water. For example, water at 5° C fully saturated with air contains approximately 9 ml of oxygen per liter (0.9%); by comparison air contains 209 ml of oxygen per 1000 ml (21%). The density and viscosity of water are approximately 800 and 50 times greater, respectively, than that of air. Furthermore, gas molecules diffuse 10,000 times more rapidly in air than in water. These differences mean that aquatic animals must have evolved very efficient ways of removing oxygen from water. Yet even the most advanced fishes with highly efficient gills and pumping mechanisms may use as much as 20% of their energy just extracting oxygen from water. By comparison, the cost for mammals to breathe is only 1% to 2% of their resting metabolism.

Respiratory surfaces must be thin and always kept wet with a fine film of fluid to allow diffusion of gases across an aqueous phase between the environment and the underlying circulation. This is hardly a problem for aquatic animals, immersed as they are in water, but it is a challenge for air breathers. To keep respiratory membranes moist and protected from injury, air breathers have in general developed invaginations of the body surface and then added pumping mechanisms to move air in and out of the body. The lung is the best example of a successful solution to breathing on land.

In general **evaginations** of the body surface, such as gills, are most suitable for aquatic respiration; **invaginations,** such as lungs and tracheae, are best for air breathing. We can now consider the specific kinds of respiratory organs employed by animals.

RESPIRATORY ORGANS

Cutaneous Respiration

Protozoa, sponges, cnidarians, and many worms respire by direct diffusion of gases between the organism and the environment. We have noted that this kind of **cutaneous respiration** is not adequate when the cellular mass exceeds approximately 1 mm in diameter. However, by greatly increasing the surface of the body relative to its mass, many multicellular animals can supply part or all of their oxygen requirements by direct diffusion. Flatworms are an example of this strategy. Cutaneous respiration frequently supplements gill or lung breathing in larger animals such as amphibians and fishes. For example, an eel can exchange 60% of its oxygen and carbon dioxide through its highly vascular skin. During their winter hibernation, frogs and even turtles exchange all their respiratory gases through the skin while submerged in ponds or springs.

Tracheal Systems

Insects and certain other terrestrial arthropods (centipedes, millipedes, and some spiders) have a highly specialized type of respiratory system, in many respects the simplest, most direct, and most efficient respiratory system found in active animals. It consists of a branching system of tubes **(tracheae)** that extends to all parts of the body (p. 413). The smallest end channels are fluid-filled **tracheoles,** less than 1 μm in diameter, that sink into the plasma membranes of the body cells. Air enters the tracheal system through valvelike openings **(spiracles).** Carbon dioxide diffuses out. Some insects can ventilate the tracheal system with body movements; the familiar telescoping move-

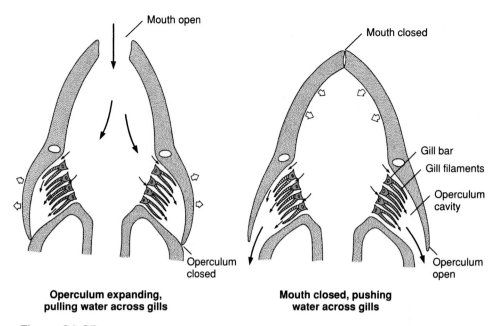

Operculum expanding, pulling water across gills

Mouth closed, pushing water across gills

Figure 34-25

How a fish ventilates its gills. Through the action of two skeletal muscle pumps, one in the mouth cavity, the other in the opercular cavity, water is drawn into the mouth, passes over the gills, and exits through the gill covers (opercular clefts).

ment of the bee abdomen is an example. Because the cells have a direct pipeline to the outside, bringing oxygen in and carrying carbon dioxide out, the insect's respiration is independent of the circulatory system. Consequently, insect blood plays no direct role in oxygen transport.

Gills

Gills of various types are effective respiratory devices for life in water. Gills may be simple **external** extensions of the body surface, such as the **dermal papulae** of sea stars (p. 453) or the **branchial tufts** of marine worms (p. 354) and aquatic amphibians (p. 534). Most efficient are the **internal gills** of fishes (p. 517) and arthropods. Fish gills are thin filamentous structures, richly supplied with blood vessels arranged so that blood flow is opposite to the flow of water across the gills. This arrangement, called **countercurrent flow** (p. 662), provides for the greatest possible extraction of oxygen from water. Water flows over the gills in a steady stream, pulled and pushed by an efficient, two-valved, branchial pump (Figure 34-25). Gill ventilation is often assisted by the fish's forward movement through the water.

Lungs

Gills are unsuitable for life in air because, when removed from the buoying water medium, the gill filaments collapse, dry, and stick together; a fish out of water rapidly asphyxiates despite the abundance of oxygen around it. Consequently air-breathing vertebrates possess lungs, highly vascularized internal cavities. Lungs of a sort are found in certain invertebrates (pulmonate snails, scorpions, some spiders, some small crustaceans), but these structures cannot be very efficiently ventilated.

Lungs that can be ventilated by muscle movements to produce a rhythmic exchange of air are characteristic of terrestrial vertebrates. The most rudimentary vertebrate lungs are those of lungfishes (Dipneusti), which use them to supplement, or even replace, gill respiration during periods of drought. Although of simple construction, the lungfish lung is supplied with a capillary network in its largely unfurrowed walls, a tubelike connection to the pharynx, and a primitive ventilating system for moving air in and out of the lung.

Amphibians have lungs that vary from simple, smooth-walled, baglike

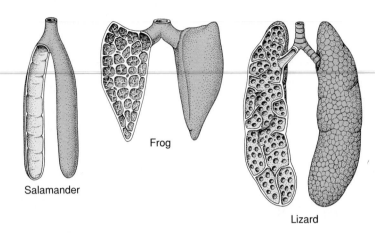

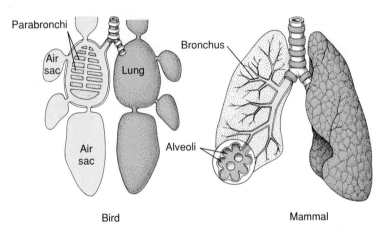

Figure 34-26

Internal structures of lungs among vertebrate groups. Generally, the evolutionary trend has been from simple sacs with little exchange surface between blood and air spaces to complex, lobulated structures, each with complex divisions and extensive exchange surfaces.

lungs of some salamanders to the subdivided lungs of frogs and toads (Figure 34-26). The total surface available for gas exchange is much increased in the lungs of reptiles which are subdivided into numerous interconnecting air sacs. Most elaborate of all is the mammalian lung, a complex of millions of small sacs, called **alveoli** (Figure 34-27), each veiled by a rich vascular network. It has been estimated that human lungs have a total surface area of from 50 to 90 m²—50 times the area of the skin surface—and contain 1000 km of capillaries. A large surface area is essential for the high oxygen uptake required to support the elevated metabolic rate of endothermic mammals.

A disadvantage of the lung is that gas exchange with the blood can take place only in the alveoli, located at the ends of a branching tree of air tubes (trachea, bronchi, and bronchioles [Figure 34-27]). Unlike the efficient one-way flow of water across fish gills, air must enter and exit a lung through the same channel. After exhalation, the air tubes are filled with "used" air from the alveoli which, during the following inhalation, is pulled back into the lungs. The volume of the air in the lung's passageways is called the "dead space." This is air that shuttles back and forth with each breath, adding to the difficulty of properly ventilating the lungs. In fact, lung ventilation in humans is so inefficient that in normal breathing only approximately one-sixth of the air in the lungs is replenished with each inspiration. Even after forced expiration, 20% to 35% of the air remains in the lungs.

In birds, lung efficiency is improved vastly by adding an extensive system of air sacs (Figure 34-26 and p. 581) that serve as air reservoirs during ventilation. On inspiration, some 75% of the incoming air bypasses the lungs to enter the air sacs (gas exchange does not occur here). At expiration some of this fresh air passes directly through the lung passages and eventually into the one-cell thick **air capillaries** where gas exchange takes place. Thus the air capillaries receive nearly fresh air during both inspiration and expiration. The beautifully designed bird lung is the result of selective pressures during the evolution of flight and its high metabolic demands.

Amphibians employ a **positive pressure** action to force air into their lungs, unlike most reptiles, birds, and mammals, which ventilate the lungs by **negative pressure,** that is, sucking air into the lungs by expansion of the thoracic cavity. Frogs ventilate the lungs by first drawing air into the mouth through the **external nares** (nostrils). Then, closing the nares and raising the floor of the mouth, they drive air into the lungs (Figure 34-28). Much of the time, however, frogs rhythmically ventilate only the mouth cavity, a well-vascularized respiratory surface that supplements pulmonary respiration.

STRUCTURE AND FUNCTION OF THE MAMMALIAN RESPIRATORY SYSTEM

Air enters the mammalian respiratory system through the nostrils (external nares), passes through the **nasal chamber,** lined with mucus-secreting epithelium, and then through the **internal nares,** the nasal openings which connect to the **pharynx.** Here, where the pathways of digestion and respiration cross, inhaled air leaves the pharynx by passing into a narrow opening, the **glottis;** food enters the esophagus to pass to the stomach (see Figure 35-11, p. 705). The glottis opens into the **larynx,** or voice box, and then into the **trachea,** or windpipe. The trachea

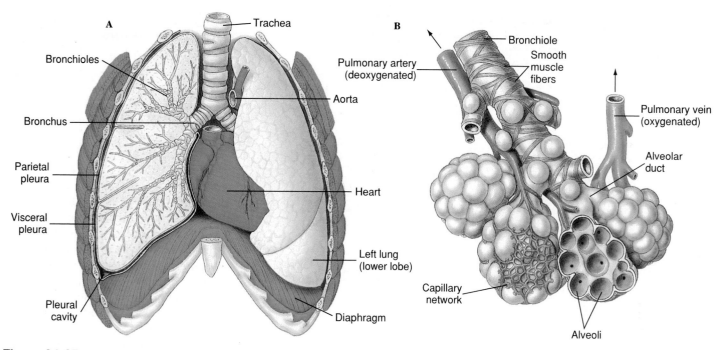

Figure 34-27

A, Lungs of human with right lung shown in section. **B,** Terminal portion of bronchiole showing air sacs with their blood supply. Arrows show direction of blood flow.

branches into two **bronchi,** one to each lung (Figure 34-27). Within the lungs each bronchus divides and subdivides into small tubes **(bronchioles)** that lead via **alveolar ducts** to the air sacs **(alveoli)** (Figure 34-27). The single-layered endothelial walls of the alveoli are thin and moist to facilitate the exchange of gases between the air sacs and the adjacent blood capillaries. Air passageways are lined with both mucus-secreting and ciliated epithelial cells, which play an important role in conditioning the air before it reaches the alveoli. There are partial cartilage rings in the walls of the tracheae, bronchi, and even some of the bronchioles that prevent those structures from collapsing.

In its passage to the air sacs the air undergoes three important changes: (1) it is filtered free from most dust and other foreign substances, (2) it is warmed to body temperature, and (3) it is saturated with moisture.

The lungs consist of a great deal of elastic connective tissue and some muscle. They are covered by a thin layer of tough epithelium known as the **visceral pleura.** A similar layer, the **pari-** **etal pleura,** lines the inner surface of the walls of the chest (Figure 34-27). The two layers of the pleura are in contact and slide over one another as the lungs expand and contract. The "space" between the pleura, called the **pleural cavity,** contains a partial vacuum, which helps keep the lungs expanded to fill the pleural cavity. Therefore no real pleural space exists; the two pleura rub together, lubricated by tissue fluid (lymph). The chest cavity is bounded by the spine, ribs, and breastbone, and floored by the **diaphragm,** a dome-shaped, muscular partition between the chest cavity and abdomen. A muscular diaphragm is found only in mammals.

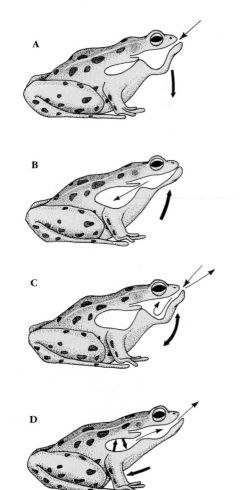

Figure 34-28

Breathing in the frog. The frog, a positive-pressure breather, fills its lungs by forcing air into them. **A,** Floor of mouth is lowered, drawing air in through the nostrils. **B,** With nostrils closed and glottis open, the frog forces air into lungs by elevating the floor of mouth. **C,** Mouth cavity is ventilated rhythmically for a period. **D,** Lungs are emptied by contraction of body-wall musculature and by elastic recoil of lungs.

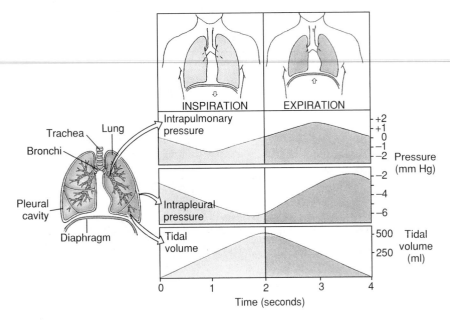

Figure 34-29
Mechanism of breathing in humans.

Mechanism of Breathing

The chest cavity is an air-tight chamber. In **inspiration** the ribs are pulled upward, the diaphragm is contracted and flattened, and the chest cavity is enlarged (Figure 34-29). The resultant increase in volume of the chest cavity causes the air pressure in the lungs to fall below atmospheric pressure: air rushes in through the air passageways to equalize the pressure. Normal **expiration** is a less active process than inspiration. When the muscles relax, the ribs and diaphragm return to their original position, and the chest cavity size decreases, the elastic lungs deflate, and air exits (Figure 34-29).

Coordination of Breathing

Respiration must adjust itself to changing requirements of the body for oxygen. Breathing is normally involuntary and automatic but can come under voluntary control. Normal, quiet breathing is regulated by neurons centered in the medulla of the brain. These neurons spontaneously produce rhythmical bursts that lead to contraction of the diaphragm and the intercostal muscles between the ribs. The rhythm and depth of breathing are precisely regu-

lated by the amount of carbon dioxide in the blood. Exercise raises the carbon dioxide level, and the breathing rate increases. Actually, the stimulatory effects of carbon dioxide are due in part to the increase in blood hydrogen ion concentration, as shown by the following reaction:

$$CO_2 + H_2O \rightleftarrows H_2CO_3 \rightleftarrows H^+ + HCO_3^-$$

Thus, an increase in hydrogen ions accompanies any increase in carbon dioxide and both strongly stimulate the respiratory center in the medulla, leading to increased rate and depth of respiration.

It is well known that swimmers can remain submerged much longer if they vigorously hyperventilate first to blow off carbon dioxide from the lungs. This delays the overpowering urge to surface and breathe. The practice is dangerous because blood oxygen is depleted just as rapidly as without prior hyperventilation, and the swimmer may lose consciousness when the oxygen supply to the brain drops below a critical point. Several documented drownings among swimmers attempting long underwater swimming records have been caused by this practice.

Gaseous Exchange in Lungs and Body Tissues

Diffusion of oxygen and carbon dioxide takes place in accordance with the laws of physical diffusion; that is, the gases pass from regions of higher concentration to those of lower concentration. The air we breathe is not a single gas, but a mixture of gases: about 79% nitrogen, 20.9% oxygen, and small amounts of water vapor and carbon dioxide. The total atmospheric pressure at sea level, 760 mm Hg, represents the combined pressures of all the gases present in air, with each contributing to the total pressure in proportion to its relative abundance. We express the contribution of each gas in a mixture such as air as its **partial pressure.** For example, since air is 20.9% oxygen, the partial pressure of oxygen is 20.9 × 760, which is equal to about 159 mm Hg for dry air. (In fact, atmospheric air is never completely dry, and the varying amount of water vapor present exerts a pressure in proportion to its concentration, like other gases.)

As soon as air enters the respiratory tract, the partial pressures of atmospheric gases change. The differences between inspired air, expired air, and alveolar air are given in Table 34-2 and shown in Figure 34-30. Inspired air becomes saturated with water vapor as it travels through the air-filled passageways toward the alveoli. When inspired air reaches the alveoli and mixes with residual air remaining from the previous respiratory cycle, the oxygen content drops and the carbon dioxide content rises. Upon expiration, air from the alveoli mixes with air in the dead air space to produce still a different mixture (Table 34-2).

Because the partial pressure of oxygen in the lung alveoli is greater (100 mm Hg pressure) than it is in venous blood of lung capillaries (40 mm Hg pressure), oxygen diffuses into the lung capillaries. In a similar manner the carbon dioxide in the blood of the lung capillaries has a higher concentration (46 mm Hg) than has this same gas in the lung alveoli (40 mm Hg), so that carbon dioxide diffuses from the blood in the alveoli.

Table 34-2	Partial Pressures and Gas Concentrations in Air and Body Fluids			
	Nitrogen (N$_2$)	**Oxygen (O$_2$)**	**Carbon Dioxide (CO$_2$)**	**Water Vapor (H$_2$O)**
Inspired air (dry)	600 (79%)	159 (20.9%)	0.2 (0.03%)	—
Alveolar air (saturated)	573 (75.4%)	100 (13.2%)	40 (5.2%)	47 (6.2%)
Expired air (saturated)	569 (74.8%)	116 (15.3%)	28 (3.7%)	47 (6.2%)
Arterial blood	573	100	40	
Peripheral tissues	573	30	50	
Venous blood	573	40	46	

Note: Values expressed in millimeters of mercury (mm Hg). Percentages indicate proportion of total atmospheric pressure at sea level (760 mm Hg). Inspired air is shown as dry, although atmospheric air always contains variable amounts of water. If, for example, atmospheric air at 20° C were half saturated (relative humidity 50%), the partial pressures and percentages would be N$_2$ 593.5 (78.1%); O$_2$ 157 (20.6%); CO$_2$ 0.2 (0.03%); and H$_2$O 8.75 (1.1%).

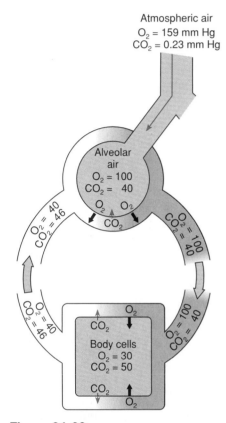

Figure 34-30
Exchange of respiratory gases in lungs and tissue cells. Numbers present partial pressures in millimeters of mercury (mm Hg).

In the tissues respiratory gases also move according to their concentration gradients (Figure 34-30). The concentration of oxygen in the blood (100 mm Hg pressure) is greater than in the tissues (0 to 30 mm Hg pressure), and the carbon dioxide concentration in the tissues (45 to 68 mm Hg pressure) is greater than that in blood (40 mm Hg pressure). The gases in each case diffuse from a high to a low concentration.

Transport of Gases in Blood

In some invertebrates the respiratory gases are simply carried, dissolved in the body fluids. However, the solubility of oxygen is so low in water that it is adequate only for animals with low rates of metabolism. For example, only approximately 1% of a human's oxygen requirement can be transported in this way. Consequently in many invertebrates and in virtually all vertebrates, nearly all of the oxygen and a significant amount of carbon dioxide are transported by special colored proteins, or **respiratory pigments,** in the blood. In most animals (all vertebrates) these respiratory pigments are packaged into blood cells.

Because of the weight of water, the hydrostatic pressure increases the equivalent of 1 atmosphere for every 10 m of depth in seawater, and the pressure of the air supplied to a diver must be increased correspondingly so that it can be drawn into the lungs. Under the increased pressure, additional air dissolves in the blood, the amount depending on the depth and time at depth of the dive. If the diver ascends slowly, the gas comes out of solution imperceptibly and is breathed out from the lungs. However, if the ascent is too rapid, the air comes out of solution and forms bubbles in the blood and other tissues, a condition known as *decompression sickness* or *the bends.* The result is painful and, if severe, can cause paralysis or death.

Sickle cell anemia is an incurable, inherited condition (p. 146) in which a single amino acid in normal hemoglobin (HbA) is substituted. The ability of sickle cell hemoglobin (HbS) to carry oxygen is severely impaired, and the erythrocytes tend to crumple during periods of oxygen stress (for example, during exercise). Capillaries become clogged with the misshapen red cells; the affected area is very painful, and the tissue may die. About 1 in 10 black Americans carry the trait (heterozygous). Heterozygotes do not have sickle cell anemia and live normal lives, but if both parents are heterozygous, each of their offspring has a 25% chance of inheriting the disease.

The most widespread respiratory pigment in the animal kingdom is **hemoglobin,** a red, iron-containing protein present in all vertebrates and many invertebrates. Each molecule of hemoglobin is made up of 5% **heme,** an iron-containing compound giving the red color to blood, and 95% **globin,** a colorless protein. The heme portion of the hemoglobin has a great affinity for oxygen; each gram of hemoglobin can carry a maximum of approximately 1.3 ml of oxygen. Because there are approximately 15 g of hemoglobin in each 100 ml of blood, fully oxygenated blood contains approximately 20 ml of oxygen per 100 ml. Of course, for hemoglobin to be of value to the body it must hold oxygen in a loose, reversible chemical combination so that it can be released to the tissues.

Although hemoglobin is the only vertebrate respiratory pigment, several other respiratory pigments are known among the invertebrates. *Hemocyanin,* a blue, copper-containing protein, is present in the crustaceans and most molluscs. Among other pigments is *chlorocruorin* (klor-a-cru´-o-rin), a green-colored, iron-containing pigment found in four families of polychaete tube worms. Its structure and oxygen-carrying capacity are very similar to those of hemoglobin, but it is carried free in the plasma rather than being enclosed in blood corpuscles. *Hemerythrin* is a red pigment found in some polychaete worms. Although it contains iron, this metal is not present in a heme group (despite the name of the pigment!), and its oxygen-carrying capacity is poor compared to hemoglobin.

The actual amount of oxygen that combines with hemoglobin depends on the oxygen partial pressure surrounding the blood cells. When the oxygen concentration is high, as it is in the capillaries of the lung alveoli, hemoglobin binds oxygen; in the tissues where the prevailing oxygen partial pressure is low, hemoglobin releases its stored oxygen reserves.

Unfortunately for humans and many other animals, hemoglobin has an affinity for carbon monoxide that is about 200 times greater than its affinity for oxygen. Consequently, even when carbon monoxide is present in the atmosphere at lower concentrations than oxygen, it tends to displace oxygen from hemoglobin to form a stable compound called carboxyhemoglobin. Air containing only 0.2% carbon monoxide may be fatal. Because of their higher respiratory rate, children and small animals are poisoned more rapidly than adults. Carbon monoxide is becoming an atmospheric contaminant of ever-increasing proportions as the world's population and industrialization continue to increase rapidly.

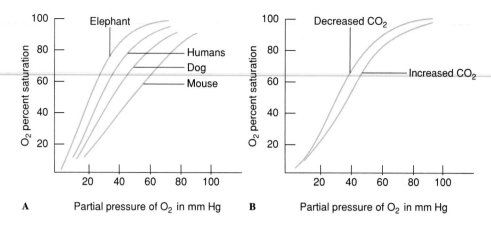

Figure 34-31

Hemoglobin saturation curves. Curves show how amount of oxygen that can bind to hemoglobin is related to oxygen partial pressure. **A,** Small animals have blood that gives up oxygen more readily than does the blood of large animals. **B,** Hemoglobin is also sensitive to carbon dioxide partial pressure. As carbon dioxide enters blood from the tissues, it shifts the curve to the right, decreasing affinity of hemoglobin for oxygen. Thus the hemoglobin unloads more oxygen in the tissues where carbon dioxide concentration is higher.

The relationship of carrying capacity to surrounding oxygen concentration can be expressed by **hemoglobin saturation curves** (also called oxygen dissociation curves [Figure 34-31]). As these curves show, the lower the surrounding oxygen tension, the greater the quantity of oxygen released. This important characteristic of hemoglobin allows more oxygen to be released to those tissues which need it most (those having the lowest partial pressure of oxygen).

Another characteristic facilitating the release of oxygen to the tissues is the sensitivity of oxyhemoglobin (hemoglobin with bound oxygen) to carbon dioxide. Carbon dioxide shifts the hemoglobin saturation curve to the right (Figure 34-31B), a phenomenon called the **Bohr effect** after the Danish scientist who first described it. As carbon dioxide enters the blood from the respiring tissues, it encourages the release of additional oxygen from the hemoglobin. The opposite event occurs in the lungs; as carbon dioxide diffuses from the venous blood into the alveolar space, the hemoglobin saturation curve shifts back to the left, allowing more oxygen to be loaded onto hemoglobin.

The same blood that transports oxygen to the tissues from the lungs must carry carbon dioxide back to the lungs on its return trip. However, unlike oxygen that is transported almost exclu-

sively in combination with hemoglobin, carbon dioxide is transported in three major forms. A small fraction of the carbon dioxide, only about 7%, is carried as the physically dissolved gas in the plasma. The remainder diffuses into the red blood cells. In the red blood cells, most of the carbon dioxide, approximately 70%, is converted to carbonic acid through the action of the enzyme carbonic anhydrase. Carbonic acid immediately dissociates into hydrogen ion and bicarbonate ion. We can summarize the entire reaction as follows:

$$CO_2 + H_2O \xrightleftharpoons{\text{carbonic anhydrase}} H_2CO_3 \rightleftharpoons H^+ + HCO_3^-$$

The hydrogen ion is buffered by several buffer systems in the blood, thus preventing a severe decrease in blood pH. The bicarbonate ion remains in solution in the plasma and red blood cell water since, unlike carbon dioxide, bicarbonate is extremely soluble.

Another fraction of the carbon dioxide, approximately 23%, combines reversibly with hemoglobin. Carbon dioxide does not combine with the heme group but with amino groups of several amino acids to form a compound called carbaminohemoglobin.

All of these reactions are reversible. When the venous blood reaches the lungs, carbon dioxide diffuses out of the red blood cells and into the alveolar air.

Summary

Fluid in the body is mostly water, but it has many substances dissolved in it, including electrolytes and proteins. Vertebrate blood consists of the fluid plasma and the formed elements, including red and white blood cells and platelets. Red blood cells contain the oxygen-carrying pigment hemoglobin. White blood cells are important defensive elements. Platelets are vital in the process of clotting, necessary to prevent excess blood loss when blood vessels are damaged. They release a series of factors that activate prothrombin to thrombin, an enzyme that causes fibrinogen to be changed to the gel form, fibrin.

Most animals have mechanisms to protect themselves against invading agents. Phagocytosis plays a critical role in defense in most multicellular animals and is an important feeding mechanism in many protistans. In vertebrates phagocytes can be fixed (reticuloendothelial system) or mobile (neutrophils and macrophages).

Vertebrate immune systems respond to any of a multitude of *specific* antigens, and the response increases with repeated exposures to the antigen. Certain proteins are the basis of self-nonself recognition in the surface of all cells in the body. These are encoded in genes of the major histocompatibility complex (MHC). Cells important in the immune response are phagocytes (granulocytes, macrophages) and lymphocytes. T lymphocytes (T cells) have receptors for particular antigens in their surface, and B lymphocytes (B cells) have antibodies in their surface and give rise to cells that secrete antibodies (plasma cells). The various subtypes of T cells are distinguished by their functions and by specific proteins in their surface, for example, T helper cells (T_H) and cytotoxic T lymphocytes. The T_H1 cells mediate the cellular arm of the immune response, and T_H2 mediate the humoral (antibody) arm. When one arm is activated, the other tends to be down regulated. Cells of the immune system communicate with each other and with other cells in the body by means of cytokines, such as the various interleukins, interferon-γ, and tumor necrosis factor.

An immune response is initiated when an antigen is taken up by a macrophage, and the macrophage presents parts of the antigen to T helper cells. In a humoral response, specific T_H2 cells stimulate specific B lymphocytes to proliferate, and finally, plasma cells are produced that manufacture antibody to the specific antigen. Some of the activated B cells become memory cells, which are responsible for the secondary response. Among other actions, antibodies help destroy invading substances by opsonization and by lysis via complement. Inflammation is a nonspecific process that is greatly influenced by the body's prior immunizing experience to a particular antigen. Delayed type hypersensitivity is a type of cell-mediated immunity, and immediate hypersensitivity depends on antibodies, particularly IgE on the surfaces of mast cells.

The CD4 surface protein on T_H cells is a receptor for HIV, and destruction of $CD4^+$ cells accounts for much of the damage to the immune system in AIDS.

People have genetically determined antigens in the surfaces of their red blood cells (ABO blood groups and others); blood types must be compatible in transfusions, or the transfused blood will be agglutinated by antibodies in the recipient.

In open circulatory systems, such as those of arthropods and most molluscs, the blood escapes from arteries into a hemocoel which is a primary body cavity derived from the blastocoel. In closed circulatory systems, such as those of annelids, vertebrates, and cephalopod molluscs, the heart pumps blood into arteries, then into arterioles of smaller diameter, through the bed of fine capillaries, through venules, and finally through the veins, which lead back to the heart. In fishes, which have a two-chambered heart with a single atrium and a single ventricle, the blood is pumped to the gills and then directly to the systemic capillaries throughout the body without first returning to the heart. With the evolution of lungs, vertebrates developed a double circulation consisting of a systemic circuit serving the body, and a pulmonary circuit serving the lungs. To be fully efficient, this change required partitioning of both the atrium and ventricle to form a double pump; partitioning was begun in the lungfishes and amphibians which have two atria but an undivided ventricle, and completed in birds and mammals, which have four-chambered hearts.

One-way flow of blood during the heart's contraction (systole) and relaxation (diastole) is assured by valves between the atria and the ventricles and between the ventricles and the pulmonary arteries and the aorta. Although the heart can beat spontaneously, its rate is controlled by nerves from the central nervous system. The heart muscle uses a great deal of oxygen and has a well-developed coronary blood circulation. The walls of arteries are thicker than those of veins, and the connective tissue in the walls of arteries allows them to expand during systole and contract during diastole. Normal arterial blood pressure (hydrostatic) of humans in systole is 120 mm Hg and in diastole, 80 mm Hg. Because the capillary walls are permeable to water, a protein-free filtrate crosses capillary walls, its movement determined by the balance between opposing forces of hydrostatic and protein osmotic pressure. Tissue fluid (lymph) that does not reenter the capillary system is collected by the lymphatic system and returned to the blood by lymph ducts.

Very small animals can depend on diffusion between the external environment and their tissues or cytoplasm for transport of respiratory gases, but larger animals require specialized organs, such as gills, tracheae, or lungs, for this function. Gills and lungs provide an increased surface area for exchange of respiratory gases between the blood and the environment. Many animals have special respiratory pigments and other mechanisms to help transport oxygen and carbon dioxide in the blood. The most widespread respiratory pigment in the animal kingdom, hemoglobin, has a high affinity for oxygen at high oxygen concentrations but releases it at lower concentrations. Vertebrate hemoglobin, which is packaged in red blood cells, combines readily with oxygen in gills or lungs, then releases it in respiring body tissues where the oxygen partial pressure is low. Carbon dioxide is carried from the tissues to the lungs in the blood as the bicarbonate ion, in combination with hemoglobin, and as the dissolved gas.

Review Questions

1. Name the chief intracellular electrolytes and the chief extracellular electrolytes.
2. What is the fate of spent erythrocytes in the body?
3. Outline or briefly describe the sequence of events that leads to blood coagulation.
4. Phagocytosis is an important defense mechanism in most animals. How are phagocytes classified? Name two kinds of cells that are phagocytic.
5. What is the molecular basis of self and nonself recognition?
6. What is the difference between T cells and B cells?
7. What is a cytokine? What are some functions of cytokines?
8. Outline the sequence of events in a humoral immune response from the introduction of antigen to the production of antibody.
9. Define the following: plasma cell, secondary response, memory cell, complement, opsonization, titer, challenge, cytokine, lymphokine, natural killer cell, interleukin-2.
10. In general, what are consequences of activation of the T_H1 arm of the immune response? Activation of the T_H2 arm?
11. Distinguish between class I and class II MHC proteins.

12. Describe a typical inflammatory response.
13. Distinguish between delayed type hypersensitivity and immediate hypersensitivity.
14. What is a major mechanism by which HIV damages the immune system in AIDS?
15. Give the genotypes of each of the following blood types: A, B, O, AB. What happens when a person with type A gives blood to a person with type B? With type AB? With type O?
16. What is the difference between an open and a closed circulatory system? What is the origin of a hemocoel such as that of an arthropod?
17. Place the following in correct order to describe the circuit of blood through the vascular system of a fish: ventricle, gill capillaries, sinus venosus, body tissue capillaries, atrium, dorsal aorta.
18. Trace the flow of blood through the heart of a mammal, naming the four chambers, their valves, and explaining where the blood entering each atrium comes from, and where the blood leaving each ventricle goes.
19. Explain the origin and conduction of the excitation that leads to a heart contraction. Why is the vertebrate heart said to be a myogenic heart?
20. Define the terms systole and diastole.

21. Explain the movement of fluid across the walls of the capillaries. How does the balance of hydrostatic pressure and colloid osmotic pressure determine the direction of net fluid flow?
22. Provide a brief description of the lymphatic system. What are its principle functions?
23. What is an advantage of a fish's gills for breathing in water and a disadvantage for breathing on land?
24. Trace the route of inspired air in humans from the nostrils to the smallest chamber of the lungs. What is the "dead air space" of a mammalian lung and how does it affect the partial pressure of oxygen reaching the alveoli?
25. How does a frog ventilate its lung? Contrast an amphibian's positive pressure breathing with a mammal's negative pressure breathing.
26. Describe the tracheal system of insects. What is the advantage of such a system for a small animal?
27. What is the role of carbon dioxide in the control of the rate and depth of breathing of a mammal?
28. Explain how oxygen is carried in the blood, including specifically the role of hemoglobin. Answer the same question with regard to carbon dioxide transport.

Selected References

See also general references for Part IV, p. 774.

Abbas, A. K., A. R. Lichtman, and J. S. Pober. 1994. Cellular and molecular immunology, ed. 2. Philadelphia, Saunders. *Good account of current immunology.*

von Boehmer, H., and P. Kisielow. 1991. How the immune system learns about self. Sci. Am. **265:**74–81 (Oct.). *Part of the mechanism by which the immune system tolerates "self" antigens may be by a mechanism known as clonal deletion.*

Feder, M. E., and W. W. Burggren. 1985. Skin breathing in vertebrates. Sci. Am. **253:**126–142 (Nov.). *In many amphibians and reptiles the skin supplements and may even replace the work of gills and lungs.*

Gallo, R. C., and L. Montagnier. 1988. AIDS in 1988. Sci. Am. **259:**40–48 (Oct.). *Authors are the American and French scientists who independently discovered the AIDS virus. Lead article in an issue of* Scientific American *entirely devoted to the status and knowledge of AIDS.*

Golde, D. W. 1991. The stem cell. Sci. Am. **265:**86–93 (Dec.). *Undifferentiated cells in the bone marrow give rise to white and red blood cells, macrophages, and platelets.*

Grey, H. M., A. Sette, and S. Buus. 1989. How T cells see antigen. Sci. Am. **261:**56–64 (Nov.). *Describes processing and presenting of antigen to T cells by macrophages.*

Lawn, R. M., and G. A. Vehar. 1986. The molecular genetics of hemophilia. Sci. Am. **254:**48–54 (Mar.). *The gene coding for the factor in the clotting cascade in the most common form of hemophilia has been isolated and cloned with recombinant DNA techniques. Prospects are good for a safe (virus-free), abundant source of the factor for treating hemophiliacs.*

Lillywhite, H. B. 1988. Snakes, blood circulation and gravity. Sci. Am. **259:**92–98 (Dec.). *How a snake's vascular system is designed to counter the effects of gravity.*

Perutz, M. F. 1978. Hemoglobin structure and respiratory transport. Sci. Am. **240:**92–125 (Dec.). *Hemoglobin transports oxygen and carbon dioxide between the lungs and tissues by clicking back and forth between two structures. Perutz and J. C. Kendrew won the Nobel Prize in 1962 for discovering the structure of hemoglobin.*

Randall, D. J., W. W. Burggren, A. P. Farrell, and M. S. Haswell. 1981. The evolution of air breathing in vertebrates. Cambridge, England, Cambridge University Press. *Traces the physiology of air breathing from aquatic ancestors.*

Robinson, T. F., S. M. Factor, and E. H. Sonnenblick. 1986. The heart as a suction pump. Sci. Am. **254:**84–91 (June). *Suggests that filling of heart in diastole is aided by elastic recoil of energy from systole.*

Sinha, A. A., M. T. G. Lopez, and H. O. McDevitt. 1990. Autoimmune diseases: the failure of self tolerance. Science **248:**1380–1388. *Current status of autoimmune disease research, which has focused on the complex formed by MHC proteins, antigens, and T cell receptors.*

Weiss, R. 1994. Of myths and mischief. Discover **15**(12):36–42. *Debunks the astonishing myths that have grown up around AIDS, including the most pernicious of all, that HIV is not the cause.*

Zucker, M. B. 1980. The functioning of the blood platelets. Sci. Am. **242:**86–103 (June). *The small blood elements that act to stop blood flow from a wound also perform complex roles in health and disease.*

35

Digestion and Nutrition

A Consuming Cornucopia

Sir Walter Raleigh observed that the difference between a rich man and a poor man is that the former eats when he pleases while the latter eats when he can get it. In today's crowded world, with 90 million people added each year to the world's population of 5.6 billion, the separation between the well-fed affluent and the hungry and malnourished poor reminds us that time has not diminished the shrewdness of Sir Walter's remark. Unlike the affluent for whom food acquisition requires only the selection of prepackaged foods at a well-stocked supermarket, the world's poor can appreciate that for them, as for the rest of the animal kingdom, procuring food is fundamental to survival. For most animals, eating is the main business of living.

Potential food is everywhere and little remains unexploited. Animals bite, chew, nibble, crush, graze, browse, shred, rasp, filter, engulf, enmesh, suck, and soak up foods of incredible variety. What an animal eats and how it eats it profoundly affects an animal's feeding specialization, its behavior, its physiology, and its internal and external anatomy—in short, both its body form and its role in the web of life. The endless evolutionary jostling between predator and prey has provided compromise adaptations for eating and adaptations for avoiding being eaten. By whatever means food may be secured, there is far less variation among animals in the subsequent digestive simplification of foods. Vertebrates and invertebrates alike use similar digestive enzymes. Even more uniform are the final biochemical pathways for nutrient use and energy transformation. The nourishment of animals is like a cornucopia in which the food flows in rather than out. A great diversity of foods procured by countless feeding adaptations streams into the mouth of the horn, is simplified, and finally applied to the common purpose of survival and reproduction. ■

All organisms require energy to maintain their highly ordered and complex structure. This energy is chemical bond energy that is released by transforming complex compounds acquired from the organism's environment into simpler ones.

The ultimate source of energy for life on earth is the sun. Sunlight is captured by chlorophyll molecules in green plants, which transform a portion of this energy into chemical bond energy (food energy). Green plants are **autotrophic** organisms; they require only inorganic compounds absorbed from their surroundings to provide the raw material for synthesis and growth. Most autotrophic organisms are the chlorophyll-bearing **phototrophs,** although some, the chemosynthetic bacteria, are **chemotrophs;** they gain energy from inorganic chemical reactions.

Almost all animals are **heterotrophic organisms** that depend on already synthesized organic compounds of plants and other animals to obtain the materials they will use for growth, maintenance, and reproduction of their kind. Since the food of animals, normally the complex tissues of other organisms, is usually too bulky to be absorbed directly by cells, it must be broken down, or digested, into soluble molecules that are small enough to be used.

Animals may be divided into a number of categories on the basis of dietary habits. **Herbivorous** animals feed mainly on plant life. **Carnivorous** animals feed mainly on herbivores and other carnivores. **Omnivorous** forms eat both plants and animals.

The ingestion of foods and their simplification by digestion are only initial steps in nutrition. Foods reduced by digestion to soluble, molecular form are **absorbed** into the circulatory system and **transported** to the tissues of the body. There they are **assimilated** into the structure of cells. Oxygen is also transported by blood to the tissues, where food products are **oxidized,** or burned to yield energy and heat. Food not immediately used is **stored** for future use. Wastes produced by oxidation must be **excreted.** Food products unsuitable for digestion are **egested** in the form of feces.

In this chapter we will first examine the feeding adaptations of animals. Next we will discuss digestion and absorption of food. We will close with a consideration of nutritional requirements of animals.

FEEDING MECHANISMS

Few animals can absorb nutrients directly from their external environments. Some blood parasites and intestinal parasites may derive all their nourishment as primary organic molecules by surface absorption. Some aquatic invertebrates may soak up part of their nutritional needs directly from the water. Most animals, however, must work for their meals. The specializations that have evolved for food procurement are almost as numerous as are the species of animals. In this brief discussion we consider some of the major food-gathering devices.

FEEDING ON PARTICULATE MATTER

Drifting microscopic particles are found in the upper hundred meters of the ocean. Most of this multitude is **plankton,** organisms too small to do anything but drift with the ocean's currents. The rest is organic debris, the disintegrating remains of dead plants and animals. Although this oceanic swarm of plankton forms a rich life domain, it is unevenly distributed. The heaviest plankton growth occurs in estuaries and areas of upwelling, where there is an abundant nutrient supply. It is consumed by numerous larger animals, invertebrates and vertebrates, using a variety of feeding mechanisms.

One of the most important and widely employed methods for feeding to have evolved is **suspension feeding** (Figure 35-1). The majority of suspension feeders use ciliated surfaces to produce currents that draw drifting food particles into their mouths. Most suspension-feeding invertebrates, such as tube-dwelling polychaete worms, bivalve molluscs, hemichordates, and most protochordates, entrap the particulate food on mucous sheets that convey the food into the digestive tract. Others, such as fairy shrimps, water fleas, and barnacles, use sweeping movements of their setae-fringed legs to create water currents and entrap food, which is transferred to the mouth. In the freshwater developmental stages of certain insect orders, the organisms use fanlike arrangements of setae or spin silk nets to entrap food.

Suspension feeding has evolved frequently as a secondary modification among representatives of groups that are primarily selective feeders. Examples are many of the microcrustaceans, fishes such as herring, menhaden, and basking sharks, certain birds such as the flamingo, and the largest of all animals, baleen (whalebone) whales. The vital importance of one component of plankton, the diatoms, in supporting a great pyramid of suspension-feeding animals is stressed by N. J. Berrill:[*]

> A humpback whale . . . needs a ton of herring in its stomach to feel comfortably full—as many as five thousand individual fish. Each herring, in turn, may well have 6000 or 7000 small crustaceans in its own stomach, each of which contains as many as 130,000 diatoms. In other words, some 400 billion yellow-green diatoms sustain a single medium-sized whale for a few hours at most.

Another type of particulate feeding exploits deposits of disintegrated organic material (detritus) that accumulates on and in the substratum; this is called **deposit feeding.** Some deposit feeders, such as many annelids and some hemichordates, simply pass the substrate through their bodies, removing from it whatever provides nourishment. Others, such as scaphopod molluscs,

---

[*]Berrill, N. J. 1958. You and the universe. New York, Dodd, Mead & Co.

Figure 35-1

Some suspension feeders and their feeding mechanisms.

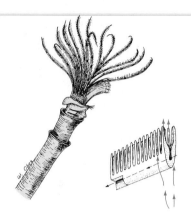

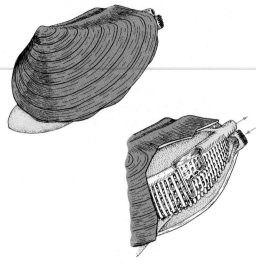

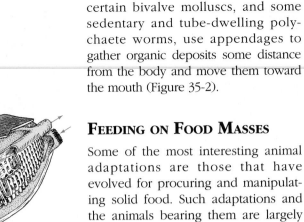

A, Marine fan worms (class Polychaeta, phylum Annelida) have a crown of tentacles. Numerous cilia on the edges of the tentacles draw water (*solid arrows*) between pinnules where food particles are entrapped in mucus; particles are then carried down a "gutter" in the center of the tentacle to the mouth (*broken arrows*).

B, Bivalve molluscs (class Bivalvia, phylum Mollusca) use their gills as feeding devices, as well as for respiration. Water currents created by cilia on the gills carry food particles into the incurrent siphon and between slits in the gills where they are entangled in a mucous sheet covering the gill surface. Particles are then transported by ciliated food grooves to the mouth (not shown). Arrows indicate direction of water movement.

certain bivalve molluscs, and some sedentary and tube-dwelling polychaete worms, use appendages to gather organic deposits some distance from the body and move them toward the mouth (Figure 35-2).

FEEDING ON FOOD MASSES

Some of the most interesting animal adaptations are those that have evolved for procuring and manipulating solid food. Such adaptations and the animals bearing them are largely shaped by what the animal eats.

Predators must be able to locate, capture, hold, and swallow prey. Most

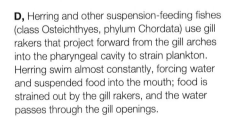

Gills

Gill rakers

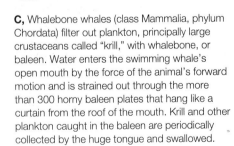

C, Whalebone whales (class Mammalia, phylum Chordata) filter out plankton, principally large crustaceans called "krill," with whalebone, or baleen. Water enters the swimming whale's open mouth by the force of the animal's forward motion and is strained out through the more than 300 horny baleen plates that hang like a curtain from the roof of the mouth. Krill and other plankton caught in the baleen are periodically collected by the huge tongue and swallowed.

D, Herring and other suspension-feeding fishes (class Osteichthyes, phylum Chordata) use gill rakers that project forward from the gill arches into the pharyngeal cavity to strain plankton. Herring swim almost constantly, forcing water and suspended food into the mouth; food is strained out by the gill rakers, and the water passes through the gill openings.

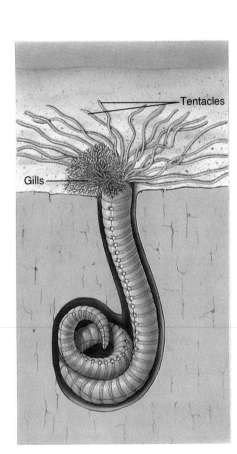

Tentacles

Gills

Figure 35-2

The annelid *Amphitrite* is a deposit feeder that lives in a mucus-lined burrow and extends long feeding tentacles in all directions across the surface. Food trapped on mucus is conveyed along the tentacles to the mouth.

carnivorous animals simply seize food and swallow it intact, although some employ toxins that paralyze or kill prey at the time of capture. Although no true teeth appear among the invertebrates, many have beaks or toothlike structures for biting and holding. A familiar example is the carnivorous polychaete *Nereis,* which possesses a muscular pharynx armed with chitinous jaws that can be everted with great speed to seize prey (Figure 18-3A, p. 353). Once a capture is made, the pharynx is retracted and the prey swallowed. The teeth of fish, amphibians, and reptiles are used principally to grip the prey and prevent its escape until it can be swallowed whole. Snakes and some fishes can swallow enormous meals. This, together with the absence of limbs, is associated with some striking feeding adaptations in these groups: recurved teeth for seizing and holding prey and distensible jaws and stomachs to accommodate their large and infrequent meals (Figure 35-3). Birds lack teeth, but the bills are often provided with serrated edges or the upper bill is hooked for seizing and tearing apart prey.

Many invertebrates are able to reduce food size by shredding devices (such as the shredding mouthparts of many crustaceans) or by tearing devices (such as the beaklike jaws of the cephalopod molluscs). Insects have three pairs of appendages on their heads that serve variously as jaws, chitinous teeth, chisels, tongues, or sucking tubes. Usually the first pair serves as crushing teeth; the second as grasping jaws; and the third, as a probing and tasting tongue.

True mastication, that is, the chewing of food as opposed to tearing or crushing, is found only among mammals. Mammals usually have four different types of teeth, each adapted for specific functions. **Incisors** are designed for biting, cutting, and stripping; **canines** are for seizing, piercing, and tearing; **premolars** and **molars,** at the back of the jaw, are for grinding and crushing (Figure 35-4). This basic pattern is often greatly modified in

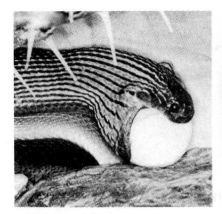

Figure 35-3
This African egg-eating snake, *Dasypeltis,* subsists entirely on hard-shelled birds' eggs, which it swallows whole. Its special adaptations are reduced size and number of teeth, enormously expansible jaw provided with elastic ligaments, and teethlike vertebral spurs that puncture the shell. Shortly after the second photograph was taken, the snake punctured and collapsed the egg, swallowed its contents, and regurgitated the crushed shell.

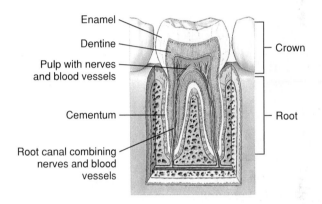

Figure 35-4
Structure of human molar tooth. The tooth is built of three layers of calcified tissue covering: enamel, which is 98% mineral and the hardest material in the body; dentine, which composes the mass of the tooth and is approximately 75% mineral; and cementum, which forms a thin covering over the dentine in the root of the tooth and is very similar to dense bone in composition. The pulp cavity contains loose connective tissue, blood vessels, nerves, and tooth-building cells.

animals having specialized food habits (Figure 35-5; see also Figure 31-9, p. 605). Herbivores have suppressed canines but well-developed molars with enamel ridges for grinding. The well-developed, self-sharpening incisors of rodents grow throughout life and must be worn away by gnawing to keep pace with growth. Some teeth have become so highly modified that they are no longer useful for biting or chewing food. An elephant's tusk (Figure 35-6) is a modified upper incisor used for defense, attack, and rooting, and the male wild boar has modified canines that are used

as weapons. Many feeding specializations of mammals are described on pp. 604–608.

Herbivorous, or plant-eating, animals have evolved special devices for crushing and cutting plant material. Some invertebrates have scraping mouthparts, such as the radula of snails (Figure 17-3, p. 324). Insects such as locusts have grinding and cutting mandibles; herbivorous mammals such as horses and cattle use wide, corrugated molars for grinding. All these mechanisms disrupt the tough cellulose cell wall to accelerate its digestion by intestinal microorganisms, as well as to

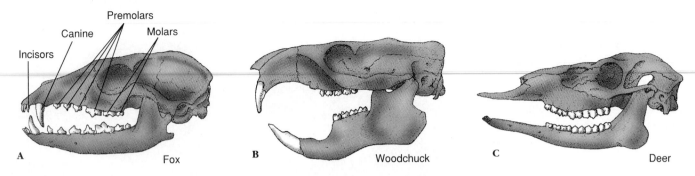

Figure 35-5

Mammalian dentition. **A,** Teeth of gray fox, a carnivore, showing the four types of teeth; **B,** Woodchuck, a rodent, has chisel-like incisors that continue to grow throughout life to replace wear; **C,** White-tailed deer, a browsing ungulate, with flat molars bearing complex ridges suited for grinding.

Figure 35-6

An African elephant loosening soil from a salt lick with its tusk. Elephants use their powerful modified incisors in many ways in the search for food and water: plowing the ground for roots, prying apart branches to reach the edible cambium, and drilling into dry riverbeds for water.

release the cell contents for direct enzymatic breakdown. Thus herbivores are able to digest food that carnivores cannot, and in doing so, convert plant material into protein for consumption by carnivores and omnivores.

FEEDING ON FLUIDS

Fluid feeding is especially characteristic of parasites, but it is practiced among many free-living forms as well. Some internal parasites (endoparasites) simply absorb the nutrient surrounding them, unwittingly provided by the host. Others bite and rasp off host tissue, suck blood, and feed on the contents of the host's intestine. External parasites (ectoparasites) such as leeches, lampreys,

parasitic crustaceans, and insects use a variety of efficient piercing and sucking mouthparts to feed on blood or other body fluid. Unfortunately for humans and other warm-blooded animals, the ubiquitous mosquito excels in its bloodsucking habit. Alighting gently, the mosquito sets about puncturing its prey with an array of six needlelike mouthparts (Figure 21-16, p. 413). One of these is used to inject an anticoagulant saliva (responsible for the irritating itch that follows the "bite" and serving as a vector for microorganisms causing malaria, yellow fever, encephalitis, and other diseases); another mouthpart is a channel through which the blood is sucked. It is of little comfort that only the female dines on blood.

DIGESTION

In the process of digestion, which means literally "carrying asunder," organic foods are mechanically and chemically broken into small units for absorption. Although food solids consist principally of carbohydrates, proteins, and fats, the very components that make up the body of the consumer, these components must first be reduced to their simplest molecular units and dissolved before they can be assimilated. Each animal reassembles some of these digested and absorbed units into organic compounds of the animal's own unique pattern. Cannibalism confers no special metabolic benefit; victims of an animal's own kind are digested just as thoroughly as food composed of another species.

In protozoa and sponges digestion is entirely **intracellular** (Figure 35-7). The food particle is enclosed within a food vacuole by phagocytosis (see p. 53). Digestive enzymes are added and the products of digestion, the simple sugars, amino acids, and other molecules, are absorbed into the cell cytoplasm where they may be used directly or, in the case of multicellular animals, may be transferred to other cells. Food wastes are simply extruded from the cell.

There are important limitations to intracellular digestion. Only particles small enough to be phagocytized can be accepted, and every cell must be capable of secreting all of the necessary enzymes, and of absorbing the products into the cytoplasm. These limitations were resolved with the evolution of an **alimentary system** in which **extracellular** digestion of large food masses could take place. In extracellular digestion certain cells lining the **lumen** (cavity) of the alimentary canal specialize in forming various digestive secretions, whereas others function largely, or entirely, in absorption. Many of the simpler metazoans, such as radiates, turbellarian flatworms, and ribbon worms (nemerteans), practice both intracellular and extracellular digestion. With the evolution of greater complexity and the appearance of complete mouth-to-anus alimentary systems, extracellular digestion became emphasized, together with increasing regional specialization of the digestive tract. For arthropods and vertebrates, digestion is almost entirely extracellular. The

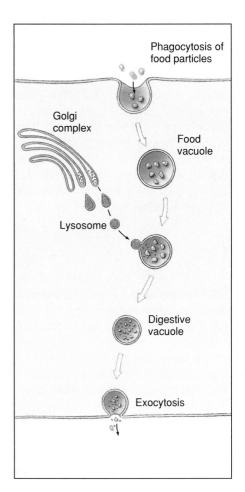

Figure 35-7
Intracellular digestion. Lysosomes containing digestive enzymes (lysozymes) are produced within the cell, possibly by the Golgi complex. Lysosomes fuse with food vacuoles and release enzymes that digest the enclosed food. Usable products of digestion are absorbed into the cytoplasm, and indigestible wastes are expelled.

ingested food is exposed to various mechanical, chemical, and bacterial treatments, to different acidic and alkaline phases, and to digestive juices that are added at appropriate stages as the food passes through the alimentary canal.

ACTION OF DIGESTIVE ENZYMES

Mechanical processes of cutting and grinding by teeth and muscular mixing by the intestinal tract are important in digestion. However, the reduction of foods to small, absorbable units relies principally on chemical breakdown by **enzymes,** discussed in Chapter 5 (p. 64). The digestive enzymes are **hydrolytic** enzymes, or **hydrolases,** so called because food

molecules are split by the process of **hydrolysis;** that is, the breaking of a chemical bond by adding the components of water across it:

$$R{-}R + H_2O \xrightarrow[\text{enzyme}]{\text{Digestive}} R{-}OH + H{-}R$$

In this general enzymatic reaction, R—R represents a food molecule that is split into two products, R—OH and R—H. Usually these reaction products must in turn be split repeatedly before the original molecule is reduced to its numerous subunits. Proteins, for example, are composed of hundreds, or even thousands, of interlinked amino acids, which must be completely separated before the individual amino acids can be absorbed. Similarly, carbohydrates must be reduced to simple sugars. Fats (lipids) are reduced to molecules of glycerol and fatty acids, although some fats, unlike proteins and carbohydrates, may be absorbed without first being completely hydrolyzed. There are specific enzymes for each class of organic compounds. These enzymes are located in specific regions of the alimentary canal in an "enzyme chain," in which one enzyme may complete what another has started. The product then moves posteriorly for still further hydrolysis.

MOTILITY IN THE ALIMENTARY CANAL

Food is moved through the digestive tract by **cilia** or by specialized **musculature,** and often by both. Movement is usually by cilia in the acoelomate and pseudocoelomate metazoa that lack the mesodermally derived gut musculature of true coelomates. Cilia move intestinal fluids and materials also in the eucoelomate bivalve molluscs in which the coelom is weakly developed. In animals with well-developed coeloms, the gut is usually lined with two opposing layers of muscle: a longitudinal layer, in which the smooth muscle fibers run parallel with the length of the gut, and a circular layer, in which the muscle fibers embrace the circumference of

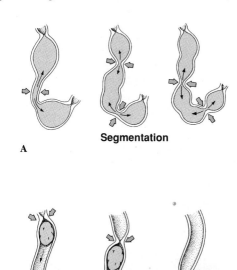

Figure 35-8
Movement of intestinal contents by segmentation and peristalsis. **A,** Segmentational movements of food showing how constrictions squeeze the food back and forth, mixing it with enzymes. The sequential mixing movements occur at about 1-second intervals. **B,** Peristaltic movement, showing how food is propelled forward by a traveling wave of contraction.

the gut. The most characteristic gut movement is **segmentation,** the alternate constriction of rings of smooth muscle of the intestine that constantly divide and squeeze the contents back and forth (Figure 35-8A). Walter B. Cannon of homeostasis fame (p. 651), while still a medical student at Harvard in 1900, was the first to use X rays to watch segmentation in experimental animals that had been fed suspensions of barium sulfate. Segmentation serves to mix food but does not move it through the gut. Another kind of muscular action, called **peristalsis,** sweeps the food down the gut with waves of contraction of circular muscle (Figure 35-8B).

ORGANIZATION AND REGIONAL FUNCTION OF THE ALIMENTARY CANAL

The metazoan alimentary canal can be divided into five major regions: (1) reception, (2) conduction and storage, (3) grinding and early digestion,

RECEPTION:
Mouth parts, tongue, salivary glands

CONDUCTION:
Esophagus

STORAGE:
Stomach
Crop (birds only)

GRINDING:
Gizzard (birds only)
AND
EARLY DIGESTION:
Stomach (acid)

TERMINAL DIGESTION AND ABSORPTION:
Small intestine (alkaline)

WATER ABSORPTION AND CONCENTRATION OF SOLIDS:
Large intestine

DEFECATION

Figure 35-9
Generalized vertebrate digestive tract showing the functions of regions.

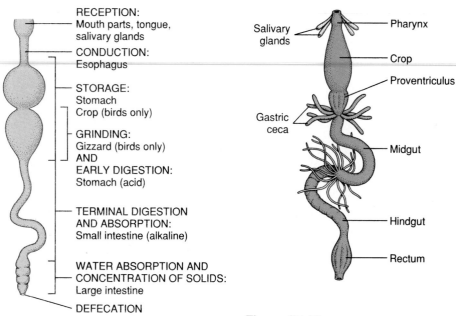

Salivary glands — Pharynx
Crop
Proventriculus
Gastric ceca
Midgut
Hindgut
Rectum

Figure 35-10
Insect digestive system (cockroach). The proventriculus is a gizzard containing chitinous teeth for grinding food.

(4) terminal digestion and absorption, and (5) water absorption and concentration of solids. Food progresses from one region to the next, allowing digestion to proceed in sequential stages (Figures 35-9 and 35-10).

RECEIVING REGION

The first region of the alimentary canal consists of devices for feeding and swallowing. These include the mouthparts (for example mandibles, jaws, teeth, radula, bills), the **buccal cavity** and muscular **pharynx.** Most metazoans other than the filter feeders have **salivary glands** (buccal glands) that produce lubricating secretions containing mucus to assist swallowing (Figure 35-10). Salivary glands often have other specialized functions such as secretion of toxic enzymes for quieting struggling prey and secretion of salivary enzymes to begin digestion. The salivary secretion of the leech, for example, is a complex mixture containing an anesthetic substance (making its bite nearly painless) and several enzymes that prevent blood coagulation and increase blood flow by dilating veins and dissolving the tissue cement that binds cells together.

Salivary **amylase** is a carbohydrate-splitting enzyme that begins the hydrolysis of plant and animal starches. It is found only in certain herbivorous molluscs, some insects, and in primate mammals, including humans. Starches are long polymers of glucose. Salivary amylase does not completely hydrolyze starch, but breaks it down mostly into two-glucose fragments called **maltose.** Some free glucose, as well as longer fragments of starch, is also produced. When the food mass (bolus) is swallowed, salivary amylase continues to act for some time, digesting perhaps half of the starch before the enzyme is inactivated by the acidic environment of the stomach. Further starch digestion resumes beyond the stomach in the intestine.

The tongue is a vertebrate innovation, usually attached to the floor of the mouth, that assists in food manipulation and swallowing. It may be used for other purposes, however, such as food capture (for example, chameleons, woodpeckers, anteaters) or as an olfactory sensor (many lizards and snakes).

In humans, swallowing begins with the tongue pushing the moistened food toward the pharynx. The nasal cavity is reflexively closed by raising the soft palate. As the food slides into the pharynx, the epiglottis tips down over the windpipe, nearly closing it (Figure 35-11). Some particles of food may enter the opening of the windpipe but are prevented from going farther by contraction of laryngeal muscles. Once in the esophagus, the food is forced smoothly toward the stomach by peristaltic contraction of esophageal muscles.

CONDUCTION AND STORAGE REGION

The **esophagus** of vertebrates and many invertebrates serves to transfer food to the digestive region. In many animals the esophagus is expanded into a **crop** (Figure 35-10), used for food storage before digestion. Among vertebrates, only birds have a crop. This serves to store and soften food (grain, for example) before it passes to the stomach, or to allow mild fermentation of food before it is regurgitated to feed nestlings.

REGION OF GRINDING AND EARLY DIGESTION

In most vertebrates, and in some invertebrates, the **stomach** provides for initial digestion as well as for storage and mixing of food with digestive juices. Mechanical breakdown of food, especially plant food with its tough cellulose cell walls, is often continued in herbivorous animals by grinding and crushing devices in the stomach. The muscular **gizzard** of terrestrial oligochaete worms, many arthropods, and birds, is assisted by stones and grit swallowed along with food (annelids and birds) or by hardened linings (for example, the chitinous teeth of the insect proventriculus [Figure 35-10], and the calcareous teeth of the gastric mill of crustaceans).

Digestive diverticula—blind tubules or pouches arising from the main passage—often supplement the stomach of many invertebrates. They are usually lined with a multipurpose epithelium having cells specialized for secreting

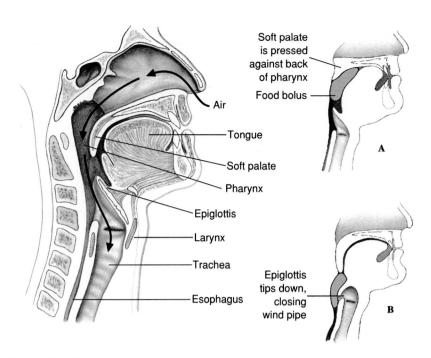

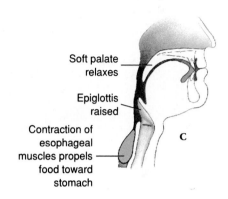

Figure 35-11
Sequential events of swallowing in humans.

mucus or digestive enzymes, or absorption or storage. Examples include the ceca of polychaete annelids, digestive glands of bivalve molluscs, the hepatopancreas of crustaceans, and the pyloric ceca of sea stars.

Herbivorous vertebrates have evolved several strategies for exploiting cellulose-splitting microorganisms to get maximal nutrition from plant food. Despite its abundance on earth, the woody cellulose that encloses plant cells can be broken down only by an enzyme, cellulase, that has limited distribution in the living world. No metazoan animals can produce intestinal cellulase for the direct digestion of cellulose. However many herbivorous metazoans harbor microorganisms (bacteria and protozoa) in their gut that do produce cellulase. These ferment cellulose under the anaerobic conditions of the gut, producing fatty acids and sugars that the herbivore can use. While the ultimate fermentation machine is the multichambered stomach of the cud-chewing ruminants described on p. 607, many other animals harbor microorganisms in other parts of the gut, such as the intestine proper or the cecum.

The stomach of carnivorous and omnivorous vertebrates is typically a U-shaped muscular tube provided with glands that produce a proteolytic enzyme and a strong acid, the latter an adaptation that probably arose for killing prey and checking bacterial activity. When food reaches the stomach, the **cardiac sphincter** opens reflexively to allow the food to enter, then closes to prevent regurgitation back into the esophagus. In humans, gentle peristaltic waves pass over the filled stomach at the rate of approximately three each minute. Churning is most vigorous at the intestinal end where food is steadily released into the **duodenum,** first region of the intestine. Approximately 2 liters of **gastric juice** in humans are secreted each day by deep, tubular glands in the stomach wall. Two types of cells line these glands: **chief cells,** which secrete **pepsin,** and **parietal cells,** which secrete **hydrochloric acid.** Pepsin is a **protease** (protein-splitting enzyme) that acts only in an acid medium (pH 1.6 to 2.4). It is a highly specific enzyme that splits large proteins by preferentially breaking down certain peptide bonds scattered along the peptide chain of the protein molecule. Although pepsin, because of its specificity, cannot completely degrade proteins, it effectively hydrolyzes them into smaller polypeptides. Digestion of protein is completed in the intestine by other proteases that together can split all peptide bonds. Pepsin is present in the stomachs of nearly all vertebrates.

That the stomach mucosa is not digested by its own powerful acid secretions is a result of another gastric secretion, mucin, a highly viscous organic compound that coats and protects the mucosa from both chemical and mechanical injury. We should note that despite the popular misconception of an "acid stomach" being unhealthy, a notion carefully nourished in advertising, stomach acidity is normal and essential. Sometimes, however, the protective mucous coating fails. This is often associated with an infection from a bacterium (*Helicobacter pylori*) that secretes toxins that cause inflammation of the stomach's lining. This may lead to a stomach ulcer.

Rennin (not to be confused with renin, an enzyme produced by the kidney, p. 660) is a milk-curdling enzyme found in the stomach of ruminant mammals. It is probably widely distributed among other mammals. By clotting and precipitating milk proteins, it

Figure 35-12

Dr. William Beaumont at Fort Mackinac, Michigan Territory, collecting gastric juice from Alexis St. Martin.

Courtesy of Wyeth-Ayerst Laboratories.

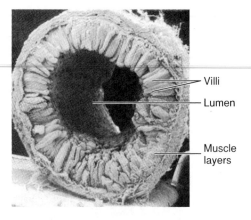

Figure 35-13

Scanning electron micrograph of a rat intestine showing the numerous fingerlike villi that project into the lumen and vastly increase the effective absorptive and secretory surface of the intestine. (× 21)

From Tissues and Organs: A Text-Atlas of Scanning Electron Microscopy, *by Richard G. Kessel and Randy H. Kardon, W. H. Freeman and Co., © 1979.*

slows the movement of milk through the stomach. Rennin extracted from the stomachs of calves is used in making cheese. Human infants, lacking rennin, digest milk proteins with acidic pepsin, just as adults do.

The secretion of the gastric juices is intermittent. Although a small volume of gastric juice is secreted continuously, even during prolonged periods of starvation, secretion is normally increased by the sight and smell of food, presence of food in the stomach, and emotional states such as anxiety and anger.

The most unique and classic investigation in the field of digestion was made by U.S. Army surgeon William Beaumont during the years 1825 to 1833. His subject was a young, hard-living French-Canadian voyageur named Alexis St. Martin, who in 1822 accidentally shot himself in the abdomen with a musket, the blast "blowing off integuments and muscles of the size of a man's hand, fracturing and carrying away the anterior half of the sixth rib, fracturing the fifth, lacerating the lower portion of the left lobe of the lungs, the diaphragm, and perforating the stomach." Miraculously the wound healed, but a permanent opening, or fistula, was formed that permitted Beaumont to see directly into the stomach (Figure 35-12). St. Martin became a

permanent, although temperamental, patient in Beaumont's care, which included food and housing. Over a period of 8 years, Beaumont was able to observe and record how the lining of the stomach changed under different psychological and physiological conditions, how foods changed during digestion, the effect of emotional states on stomach motility, and many other facts about the digestive process of his famous patient.

Region of Terminal Digestion and Absorption: The Intestine

The importance of the intestine varies widely among animal groups. In invertebrates that have extensive digestive diverticula in which food is broken down and phagocytized, the intestine may serve only as a pathway for conducting wastes out of the body. In other invertebrates with simple stomachs, and in all vertebrates, the intestine is equipped for both digestion and absorption.

Devices for increasing the internal surface area of the intestine are highly developed in vertebrates, but are generally absent among invertebrates. Perhaps the most direct way to increase the absorptive surface of the

gut is to increase its length. Coiling of the intestine is common among all vertebrate groups and reaches its highest development in mammals, in which the length of the intestine may exceed eight times the length of the body. Although a coiled intestine is rare among invertebrates, other strategies for increasing surface sometimes occur. For example, the **typhlosole** of terrestrial oligochaete worms (see Figure 18-12C, p. 358), an inward folding of the dorsal intestinal wall that runs the full length of the intestine, effectively increases internal surface area of the gut in a narrow body lacking space for a coiled intestine.

Lampreys and sharks have longitudinal or spiral folds in their intestines. Other vertebrates have developed elaborate folds and minute fingerlike projections called **villi,** that give the inner surface of fresh intestinal tissue the appearance of velvet (Figure 35-13). The electron microscope reveals that each cell lining the intestinal cavity additionally is bordered by hundreds of short, delicate processes called **microvilli** (Figure 35-14C and D). These processes, together with larger villi and intestinal folds, may increase the internal surface area of the intestine more than a million times as compared to a

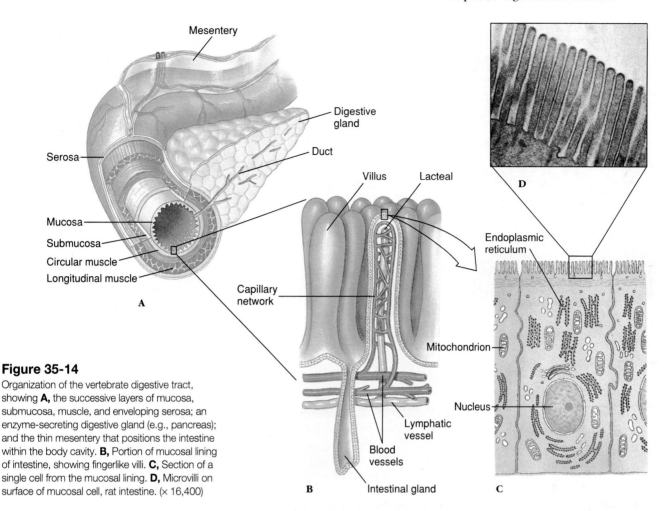

Figure 35-14
Organization of the vertebrate digestive tract, showing **A,** the successive layers of mucosa, submucosa, muscle, and enveloping serosa; an enzyme-secreting digestive gland (e.g., pancreas); and the thin mesentery that positions the intestine within the body cavity. **B,** Portion of mucosal lining of intestine, showing fingerlike villi. **C,** Section of a single cell from the mucosal lining. **D,** Microvilli on surface of mucosal cell, rat intestine. (× 16,400)

smooth cylinder of the same diameter. The absorption of food molecules is enormously facilitated as a result.

Digestion in the Vertebrate Small Intestine

Food is released into the small intestine through the **pyloric sphincter,** which relaxes at intervals to allow entry of acidic stomach contents into the initial segment of the small intestine, the **duodenum.** Two secretions are poured into this region: **pancreatic juice** and **bile** (Figure 35-15). Both of these secretions have a high bicarbonate content, especially pancreatic juice, which effectively neutralizes gastric acid, raising the pH of the liquefied food mass, now called **chyme,** from 1.5 to 7 as it enters the duodenum. This change in pH is essential because all the intestinal enzymes are effective only in a neutral or slightly alkaline medium.

Cells of the intestinal mucosa, like those of the stomach mucosa, are subjected to considerable wear and are constantly undergoing replacement. Cells deep in the crypt between adjacent villi divide rapidly and migrate up the villus. In mammals the cells reach the tip of the villus in about two days. There they are shed, along with their membrane enzymes, into the lumen at the rate of some 17 billion a day along the length of the human intestine. Before they are shed, however, these cells differentiate into absorptive cells that transport nutrients into the network of blood and lymph vessels, once digestion is complete.

Pancreatic Enzymes. The pancreatic secretion of vertebrates contains several enzymes of major importance in digestion (Figure 35-15). Two powerful proteases, **trypsin** and **chymotrypsin,** continue enzymatic digestion of proteins begun by pepsin, which is now inactivated by the alkalinity of the intestine. Trypsin and chymotrypsin, like pepsin, are highly specific proteases that split apart peptide bonds deep inside the protein molecule. The hydrolysis of the peptide linkage may be shown as:

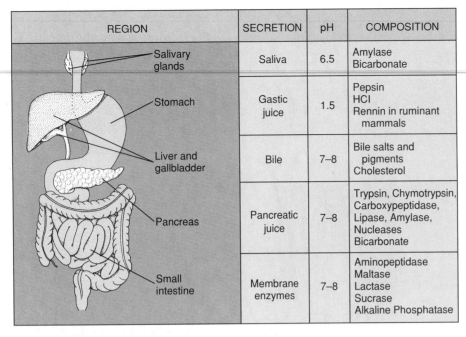

REGION	SECRETION	pH	COMPOSITION
Salivary glands	Saliva	6.5	Amylase Bicarbonate
Stomach	Gastic juice	1.5	Pepsin HCl Rennin in ruminant mammals
Liver and gallbladder	Bile	7–8	Bile salts and pigments Cholesterol
Pancreas	Pancreatic juice	7–8	Trypsin, Chymotrypsin, Carboxypeptidase, Lipase, Amylase, Nucleases Bicarbonate
Small intestine	Membrane enzymes	7–8	Aminopeptidase Maltase Lactase Sucrase Alkaline Phosphatase

Figure 35-15
Secretions of the mammalian alimentary canal with the principal components and the pH of each secretion.

Pancreatic juice also contains **carboxypeptidase,** which splits amino acids off the carboxyl ends of polypeptides; **pancreatic lipase,** which hydrolyzes fats into fatty acids and glycerol; **pancreatic amylase,** a starch-splitting enzyme identical to salivary amylase in its action; and **nucleases,** which degrade RNA and DNA to nucleotides.

Membrane Enzymes. The cells lining the intestine have digestive enzymes embedded in their surface membrane that continue digestion of carbohydrates, proteins, and phosphate compounds (Figure 35-15). These enzymes of the microvillus membrane (Figure 35-14D) include **aminopeptidase** that splits terminal amino acids from the amino end of short peptides, and several **disaccharidases,** enzymes that split 12-carbon sugar molecules into 6-carbon units. The disaccharidases include **maltase,** which splits maltose into two molecules of glucose; **sucrase,** which splits sucrose to fructose and glucose; and **lactase,** which breaks down lactose (milk sugar) into glucose and galactose. Also present is **alkaline phosphatase,** an enzyme that attacks a variety of phosphate compounds.

Bile. The liver secretes bile into the **bile duct,** which drains into the upper intestine (duodenum). Between meals

Although milk is the universal food of newborn mammals and one of the most complete human foods, many adult humans cannot digest milk because they are deficient in lactase, the enzyme that hydrolyzes lactose (milk sugar). Lactose intolerance is genetically determined. It is characterized by abdominal bloating, cramps, flatulence, and watery diarrhea, all appearing within 30 to 90 minutes after ingesting milk or its unfermented by-products. (Fermented dairy products, such as yogurt and cheese, create no intolerance problems.)

Northern Europeans and their descendants, which include the majority of North American whites, are most tolerant of milk. Many other ethnic groups are generally intolerant to lactose, including the Japanese, Chinese, Jews in Israel, Eskimos, South American Indians, and most African blacks. Only about 30% of North American blacks are tolerant; those who are tolerant are mostly descendants of slaves brought from east and central Africa where dairying is traditional and tolerance to lactose is high.

the bile collects in the **gallbladder,** an expansible storage sac that releases bile when stimulated by the presence of fatty food in the duodenum. Bile contains water, bile salts, and pig-

ments, but no enzymes. The **bile salts** (mainly sodium taurocholate and sodium glycocholate) are essential for digestion of fats. Fats, because of their tendency to remain in large, water-resistant globules, are especially resistant to enzymatic digestion. Bile salts reduce the surface tension of fat globules, allowing the churning action of the intestine to break up fats into tiny droplets. With the total surface exposure of fat particles greatly increased, fat-splitting lipases are able to reach and hydrolyze the triglyceride molecules. The yellow-green color of bile is produced by **bile pigments,** breakdown products of hemoglobin from worn-out red blood cells. The bile pigments also give the feces its characteristic color.

Bile production is only one of the liver's many functions. This highly versatile organ is a storehouse for glycogen, production center for the plasma proteins, site of protein synthesis and detoxification of protein wastes, site for destruction of worn red blood cells, and center for metabolism of fat, amino acids, and carbohydrates.

Absorption

Little food is absorbed in the stomach because digestion is still incomplete and because of limited surface exposure. Some materials, however, such as drugs and alcohol, are absorbed mostly there, which contributes to their rapid action. Most digested food is absorbed from the small intestine where the numerous finger-shaped villi provide an enormous surface area through which materials can pass from the intestinal lumen into the circulation.

Carbohydrates are absorbed almost exclusively as simple sugars (monosaccharides, for example, glucose, fructose, and galactose) because the intestine is virtually impermeable to polysaccharides. Proteins are absorbed principally as their amino acid subunits, although small amounts of small proteins or peptide fragments sometimes may be absorbed. Simple sugars and amino acids are transferred across the intestinal epithelium by both passive and active processes.

Immediately after a meal these materials are in such high concentration in the gut that they readily diffuse into the blood, where their concentration is initially lower. However, if absorption were passive only, we would expect transfer to cease as soon as the concentrations of a substance became equal on both sides of the intestinal epithelium. This would permit valuable nutrients to be lost in the feces. In fact, very little is lost because passive transfer is supplemented by an **active transport** mechanism (p. 53) located in the epithelial cells that transfers food molecules into the blood. Materials thus are moved *against* their concentration gradient, a process requiring the expenditure of energy. Although not all food products are actively transported, those that are, such as glucose, galactose, and most of the amino acids, are handled by transport mechanisms that are specific for each kind of molecule.

As already described, fat droplets are emulsified by bile salts and then digested by pancreatic lipase. Triglycerides are broken into fatty acids and monoglycerides, which are absorbed by simple diffusion. However, free fatty acids never enter the blood. Instead, during their passage through the intestinal epithelial cells, the fatty acids are resynthesized into triglycerides that pass out of the cells and into **lacteals** (Figure 35-14B). From the lacteals, fat droplets enter the lymph system (Figure 34-24, p. 688) and eventually pass into the blood through the thoracic duct. After a fatty meal, even a peanut butter sandwich, the presence of numerous fat droplets in the blood imparts a milky appearance to the blood plasma.

REGION OF WATER ABSORPTION AND CONCENTRATION OF SOLIDS

In the large intestine the indigestible remnants of digestion are consolidated by reabsorption of water to form solid or semisolid feces for removal from the body by **defecation.** Reabsorption of water is of special significance in insects, especially those living in dry environments, which must (and do) conserve nearly all water entering the rectum. Specialized **rectal glands** absorb water and ions as needed, leaving behind fecal pellets that are almost completely dry. In reptiles and birds, which also produce nearly dry feces, most of the water is reabsorbed in the cloaca. A white pastelike feces is formed containing both indigestible food wastes and uric acid.

The colon of humans contains enormous numbers of bacteria, which first enter the sterile colon of the newborn infant with its food. In adults approximately one-third of the dry weight of feces is bacteria; these include harmless bacilli as well as cocci that can cause serious illness should they escape into the abdomen or bloodstream. Normally the body's defenses prevent invasion of such bacteria. The bacteria break down organic wastes in the feces and provide some nutritional benefit by synthesizing certain vitamins (vitamin K and small quantities of some of the B vitamins), which are absorbed by the body.

REGULATION OF FOOD INTAKE

Most animals unconsciously adjust intake of food to balance energy expenditure. If energy expenditure is increased by greater physical activity, more food is consumed. Most vertebrates, from fish to mammals, eat for calories rather than bulk because, if the diet is diluted with fiber, they respond by eating more. Similarly, intake is adjusted downward following a period of several days when caloric intake is too high.

Intake of food is regulated in large part by a "hunger" center located in the hypothalamus of the brain. The level of sugar in the blood has an important influence on this center because hunger coincides with decreasing levels of glucose. While most animals seem able to stabilize their weight at normal levels with ease, many humans cannot. It is becoming clear that many obese people do not eat more food than thin people. Rather they have a reduced capacity to burn off excess calories by "nonshivering thermogenesis" (p. 666). Placental mammals are unique in having a dark adipose tissue called **brown fat,** specialized for the generation of heat. Newborn mammals, including human infants, have much more brown fat than adults. In human infants brown fat is located in the chest, upper back, and near the kidneys. The abundant mitochondria in brown fat contains a membrane protein called **thermogenin** that acts to uncouple oxidative phosphorylation (p. 70). In people of average weight an increased caloric intake induces brown fat to dissipate excess energy as heat through the uncoupling action of thermogenin. This is referred to as "diet-induced thermogenesis." In many people tending toward obesity, this capacity is diminished because they have less brown fat or because their brown fat does not respond to hypothalamic signals as it does in people of average weight. There are other reasons for obesity in addition to the fact that many people simply eat too much. Fat stores are supervised by the hypothalamus, which is set at a point that may be higher or lower than the norm. A high setpoint can be lowered somewhat by exercise, but as dieters are painfully aware, the body defends its fat stores with remarkable tenacity.

The body contains two kinds of adipose tissue that perform completely different functions. White adipose tissue, which comprises the bulk of body fat, is adapted for the storage of fat derived mainly from surplus fats and carbohydrates in the diet. It is distributed throughout the body, particularly in the deep layers of the skin. Brown adipose tissue is highly specialized for mediating nonshivering thermogenesis rather than for the storage of fat. It is brown because it is packed with mitochondria containing large quantities of iron-bearing cytochrome molecules. In ordinary body cells, ATP is generated by the flow of electrons down the respiratory chain (p. 70). This ATP is then used to power various cellular processes. In brown fat cells heat is generated instead of ATP. Thermogenesis is activated by the sympathetic nervous system, which responds to signals from the hypothalamus.

NUTRITIONAL REQUIREMENTS

The food of animals must include **carbohydrates, proteins, fats, water, mineral salts,** and **vitamins.** Carbohydrates and fats are required as fuels for energy and for the synthesis of various substances and structures. Proteins (actually the amino acids of which they are composed) are needed for the synthesis of specific proteins and other nitrogen-containing compounds. Water is required as the solvent for body chemistry and as a major component of all the fluids of the body. Inorganic salts are required as the anions and cations of body fluids and tissues and form important structural and physiological components throughout the body. Vitamins are accessory factors from food that are often built into the structure of many enzymes.

A vitamin is a relatively simple organic compound that is not a carbohydrate, fat, protein, or mineral and that is required in very small amounts in the diet for some specific cellular function. Vitamins are not sources of energy but are often associated with the activity of important enzymes that serve vital metabolic roles. Plants and many microorganisms synthesize all the organic compounds they need; animals, however, have lost certain synthetic abilities during their long evolution and depend ultimately on plants to supply these compounds. Vitamins therefore represent synthetic gaps in the metabolic machinery of animals.

Vitamins are usually classified as fat soluble (soluble in fat solvents such as ether) or water soluble. The water-soluble vitamins include the B complex and vitamin C (Table 35-1). Vitamins of the B complex, so grouped because the original B vitamin was subsequently found to consist of several distinct molecules, tend to be found together in nature. Almost all animals, vertebrate and invertebrate, require the B vitamins; they are "universal" vitamins. The dietary need for vitamin C and the fat-soluble vitamins A, D, E, and K is mostly restricted to the vertebrates,

although some are required by certain invertebrates. Even within groups of close relationship, requirements for vitamins are relative, not absolute. A rabbit does not require vitamin C, but guinea pigs and humans do. Some songbirds require vitamin A, but others do not.

The recognition years ago that many human diseases and those of domesticated animals were caused by or associated with dietary deficiencies led biologists to search for specific nutrients that would prevent such diseases. These studies eventually yielded a list of **essential nutrients** for human beings and other animal species studied. Essential nutrients are those needed for normal growth and maintenance and that *must* be supplied in the diet. In other words, it is "essential" that these nutrients be in the diet because the animal cannot synthesize them from other dietary constituents. Nearly 30 organic compounds (amino acids and vitamins) and 21 elements have been established as essential for humans (Table 35-1). Considering that the body contains thousands of different organic compounds, the list in Table 35-1 is remarkably short. Animal cells have marvelous powers of synthesis, enabling them to build compounds of enormous variety and complexity from a small, select group of raw materials.

In the average diet of North Americans approximately 50% of the total calories (energy content) comes from carbohydrates and 40% comes from lipids. Proteins, essential as they are for structural needs, supply only a little more than 10% of the total calories of the average diet of North Americans. Carbohydrates are widely consumed because they are more abundant and cheaper than proteins or lipids. Actually humans and many other animals can subsist on diets devoid of carbohydrates, provided sufficient total calories and essential nutrients are present. Eskimos, before the decline of their native culture, lived on a diet that was high in fat and protein and very low in carbohydrate.

Lipids are needed principally to provide energy. However, at least

Table 35-1	Human Nutrient Requirements

Amino Acids

Phenylalanine	Methionine
Lysine	Cystine
Isoleucine	Tryptophan
Leucine	Threonine
Valine	

Polyunsaturated Fatty Acids

Arachidonic
Linoleic
Linolenic

Water-Soluble Vitamins

Thiamine (B_1)	Folacin
Riboflavin (B_2)	Vitamin B_{12}
Niacin	Biotin
Pyridoxine (B_6)	Choline
Pantothenic acid	Ascorbic acid (C)

Fat-Soluble Vitamins

A, D, E, and K

Minerals

Calcium	Silicon
Phosphorus	Vanadium
Sulfur	Tin
Potassium	Nickel
Chlorine	Selenium
Sodium	Manganese
Magnesium	Iodine
Iron	Molybdenum
Fluorine	Chromium
Zinc	Cobalt
Copper	

Adapted from "The Requirements of Human Nutrition" by Nevin S. Scrimshaw and Vernon R. Young. Copyright © September 1976 by Scientific American, Inc. All rights reserved. Adapted by permission.

three fatty acids are essential for humans because they cannot be synthesized. Much interest and research have been devoted to lipids in our diets because of the association between fatty diets and the disease **atherosclerosis.** The matter is complex, but evidence suggests that atherosclerosis may occur when the diet is high in saturated lipids (lipids with no double bonds in the carbon chains of the fatty acids) but low in polyunsaturated lipids (two or more double bonds in the carbon chains).

Atherosclerosis (Gr. *atheroma,* tumor containing gruel-like matter, + *sclerosis,* to harden) is a degenerative disease in which fatty substances are deposited in the lining of arteries, resulting in narrowing of the passage and eventual hardening and loss of elasticity.

Proteins are expensive foods and limited in the diet. Proteins, of course, are not themselves the essential nutrients but rather contain essential amino acids. Of the 20 amino acids commonly found in proteins, 9 and possibly 11 are essential to humans (Table 35-1). The rest can be synthesized from other amino acids. Generally, animal proteins have more of the essential amino acids than do proteins of plant origin. All 9 of the essential amino acids must be present simultaneously in the diet for protein synthesis. If one or more is missing, the use of the other amino acids will be reduced proportionately; they cannot be stored and are broken down for energy. Thus heavy reliance on a single plant source will inevitably lead to protein deficiency. This problem can be corrected if two kinds of plant proteins having complementary strengths in essential amino acids are ingested together. For example, a balanced protein diet can be prepared by mixing wheat flour, which is deficient only in lysine, with a legume (peas or beans), which is a good source of lysine but deficient in methionine and cysteine. Each plant complements the other by having adequate amounts of those amino acids that are deficient in the other.

Because animal proteins are rich in essential amino acids, they are in great demand in all countries. North Americans eat far more animal proteins than do Asians and Africans. In 1989 the annual per capita consumption of red meat was 76 kg in the United States, 27 kg in Japan, 12 kg in Egypt, and 1 kg in India.* The high consumption of meat in North America and Europe carries the price of a high death rate from the so-called diseases of affluence: heart disease, stroke, and certain kinds of cancer.

Undernourishment and malnourishment rank as two of the world's oldest problems and remain major health problems today, afflicting an eighth of the human population. Growing children and pregnant and lactating women are especially vulnerable to the

*Brown, L. R. 1991. State of the world 1991. New York, Worldwatch Institute/W. W. Norton & Company, p. 159.

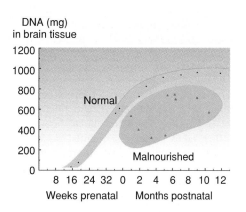

DNA (mg) in brain tissue

Figure 35-16

Effect of early malnutrition on cell number (measured as total DNA content) in the human brain. This graph shows that malnourished infants (*purple oval*) have far fewer brain cells than do normal infants (*green growth curve*).

devastating effects of malnutrition. Cell proliferation and growth in the human brain are most rapid in the terminal months of gestation and the first year after birth. Adequate protein for neuron development is a requirement during this critical time to prevent neurological dysfunction. The brains of children who die of protein malnutrition during the first year of life have 15% to 20% fewer brain cells than those of normal children (Figure 35-16). Malnourished children who survive this period suffer permanent brain damage and cannot be helped by later corrective treatment (Figure 35-17).

Two different types of severe food deficiency are recognized: marasmus, general undernourishment from a diet low in both calories and protein, and kwashiorkor, protein malnourishment from a diet adequate in calories but deficient in protein. Marasmus (Gr. *marasmos,* to waste away) is common in infants weaned too early and placed on low-calorie–low-protein diets; these children are listless, and their bodies waste away. Kwashiorkor is a West African word describing a disease a child gets when displaced from the breast by a newborn sibling. This disease is characterized by retarded growth, anemia, weak muscles, a bloated body with typical pot belly, acute diarrhea, susceptibility to infection, and high mortality.

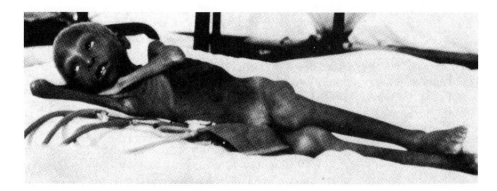

Figure 35-17

Biafran refugee child suffering severe malnutrition.

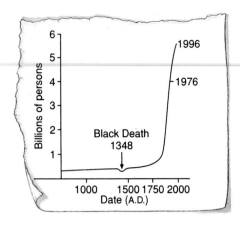

Figure 35-18

Portion of a graph for human population growth since A.D. 800, as it appeared in the 1979 edition of this book when the earth's population had passed 4 billion three years earlier, and updated to show the 1996 figure of approximately 5.6 billion.

The major cause of the world's precarious food situation is recent rapid population growth. The world population was 2 billion in 1930, reached 3 billion in 1960, passed 5.5 billion in 1992 (Figure 35-18), and is expected to reach 8.9 billion by the year 2030,* several years ahead of earlier estimates. Approximately 90 million people are added each year. The equivalent of the total U.S. population of 250 million people is added to the world every 33 months. Yet, as the demand for food increases, the world each year loses billions of tons of topsoil and trillions of gallons of groundwater needed to grow food crops. In the view of many, the exploding human population is a major force driving the global environmental crisis.

*Population Division, United Nations, New York, 1993.

Summary

Autotrophic organisms (mostly green plants), using inorganic compounds as raw materials, capture the energy of sunlight through photosynthesis and produce complex organic molecules. Heterotrophic organisms (bacteria, fungi, and animals) use the organic compounds synthesized by plants, and chemical bond energy stored therein for their own nutritional and energy needs.

A large group of animals with very different levels of complexity feed by filtering out minute organisms and other particulate matter from the water. Others feed on organic detritus deposited in the substrate. Selective feeders, on the other hand, have evolved mechanisms for manipulating larger food masses, including various devices for seizing, scraping, boring, tearing, biting, and chewing. Fluid feeding is characteristic of endoparasites, which may absorb food across the general body surface, and of ectoparasites, herbivores, and predators that have developed specialized mouthparts for piercing and sucking.

Digestion is the process of breaking down food mechanically and chemically into molecular subunits for absorption. Digestion is intracellular in the protozoa and sponges. In more complex metazoans it is supplemented, and finally replaced entirely, by extracellular digestion, which takes place in sequential stages in a tubular cavity, the alimentary canal. Food is received in the mouth and mixed with lubricating saliva, then passed down the esophagus to regions where the food may be stored (crop), or ground (gizzard), or acidified and subjected to early digestion (vertebrate stomach). Among vertebrates, most digestion occurs in the small intestine. Enzymes from the pancreas and membrane enzymes embedded in intestinal mucosal cells hydrolyze proteins, carbohydrates, fats, nucleic acids, and various phosphate compounds. The liver secretes bile, containing salts that emulsify fats. Once foods are digested, their products are absorbed as molecular subunits (monosaccharides, amino acids, and fatty acids) into the blood or lymph vessels of the villi of the small intestine. The large intestine (colon) serves mainly to absorb water and minerals from the food wastes as they pass through it. It also contains symbiotic bacteria that produce certain vitamins.

Most animals balance food intake with energy expenditure. Food intake is regulated primarily by a hunger center located in the hypothalamus. In mammals, should caloric intake exceed requirements for energy, the excess calories normally are dissipated as heat in specialized brown fat tissue. A deficiency in this response is one cause of human obesity.

All animals require a balanced diet containing both fuels (mainly carbohydrates and lipids) and structural and functional components (proteins, minerals, and vitamins). For every multicellular animal, certain amino acids, lipids, vitamins, and minerals are "essential" dietary factors that cannot be produced by the animal's own synthetic machinery. Animal proteins are better-balanced sources of amino acids than are plant proteins, which tend to lack one or more essential amino acids. Undernourishment and protein malnourishment are among the world's major health problems, afflicting millions of people.

Review Questions

1. Distinguish between the following pairs of terms: autotrophic and heterotrophic; phototrophic and chemotrophic; herbivores and carnivores; omnivores and insectivores.
2. Suspension feeding is one of the most important methods of feeding among animals. Explain the characteristics, advantages, and limitations of suspension feeding, and name three different groups of animals that are suspension feeders.
3. An animal's feeding adaptations are an integral part of an animal's behavior and usually shape the appearance of the animal itself. Discuss the contrasting feeding adaptations of carnivores and herbivores.
4. Explain how food is propelled through the digestive tract.
5. Compare intracellular with extracellular digestion and explain the advantages of the latter over intracellular digestion.
6. What structural modifications vastly increase the internal surface area of the intestine, and why is this large surface area important?
7. Trace the digestion and final absorption of a carbohydrate (starch) in the vertebrate gut, naming the carbohydrate-splitting enzymes, where they are found, the breakdown products of starch digestion, and in what form they are finally absorbed.
8. As in question 7, trace the digestion and final absorption of a protein.
9. Explain how fats are emulsified, digested, and absorbed in the vertebrate gut.
10. Explain the phrase "diet-induced thermogenesis" and relate it to the problem of obesity in some people.
11. Name the basic classes of foods that serve mainly as (1) fuels and as (2) structural and functional components.
12. Define a vitamin. What are the water-soluble and the fat-soluble vitamins?
13. Explain what is meant by the term "essential nutrients."
14. Explain the difference between saturated and unsaturated lipids, and comment on the current interest in these compounds as they relate to human health.
15. What is meant by "protein complementarity" among plant foods?

Selected References

See also general references for Part IV, p. 774.

Blaser, M. J. 1996. The bacteria behind ulcers. Sci. Am. **274:**104–107 (Jan.). *We now know that most cases of stomach ulcers are caused by acid-loving microbes. At least one-third of the human population are infected although most do not become ill.*

Carr, D. E. 1971. The deadly feast of life. Garden City, New York, Doubleday & Company. *What and how animals eat told with insight and wit.*

Doyle, J. 1985. Altered harvest: agriculture, genetics, and the fate of the world's food supply. New York, Viking Penguin, Inc. *Examines the politics of the agricultural revolution and the environmental and biological costs of the American food production system.*

Griggs, B. 1986. The food factor. New York, Viking Penguin, Inc. *Packed with facts on nutrition and eating habits with an international perspective and emphasis on food's relation to disease.*

Jennings, J. B. 1973. Feeding, digestion and assimilation in animals, ed. 2. New York, St. Martin's Press, Inc. *A general, comparative approach. Excellent account of feeding mechanisms in animals.*

Magee, D. F. and A. F. Dalley, II. 1986. Digestion and the structure and function of the gut. Basel, Switzerland, S. Karger AG. *Comprehensive treatment of mammalian (mostly human) digestion.*

Milton, K. 1993. Diet and primate evolution. Sci. Am. **269:**86–93 (Aug.). *Studies with primates suggest that modern human diets often diverge greatly from those to which the human body may be adapted.*

Moog, F. 1981. The lining of the small intestine. Sci. Am. **245:**154–176 (Nov.). *Describes how the mucosal cells actively process foods.*

Owen, J. 1980. Feeding strategy. Chicago, University of Chicago Press. *Well-written and generously illustrated book from the series "Survival in the Wild."*

Sanderson, S. L., and R. Wassersug. 1990. Suspension-feeding vertebrates. Sci. Am. **262:**96–101 (Mar.). *A variety of vertebrates, some enormous in size, eat by filtering out small organisms from massive amounts of water passed through a feeding apparatus.*

Stevens, C. E. 1988. Comparative physiology of the vertebrate digestive system. New York, Cambridge University Press. *Lucid and balanced treatment of anatomical characteristics of vertebrate digestive systems and the physiology and biochemistry of food digestion.*

Weindrach, R. 1996. Caloric restriction and aging. Sci. Am. **274:**46–52 (Jan.). *Organisms from single-celled protists to mammals live longer on well-balanced but low-calorie diets. The potential benefits for humans are examined.*

36

Nervous Coordination

Nervous System and Sense Organs

The Private World of the Senses

By any measure, people enjoy a rich sensory world. We are continually assailed by information from the senses of vision, hearing, taste, olfaction, and touch. These classic five senses are supplemented by sensory inputs of cold, warmth, vibration, and pain, as well as by information from numerous internal sensory receptors that operate silently and automatically to help keep our interior domain working smoothly. It is our senses that provide us with our impression of the environment. Yet the world our senses perceive is uniquely human. We share this exclusive world with no other animal, nor can we venture into the sensory world of any other animal except as an abstraction through our imagination.

The idea that each animal enjoys an unshared sensory world was first conceived by Jakob von Uexküll, a seldom cited German biologist of the early part of this century. Von Uexküll asks us to try to enter the world of a tick through our imagination, supplemented by what we know of tick biology.

It is a world of temperature, of light and dark, and of a single odor, that of butyric acid, a chemical common to all mammals. Insensible to all other stimuli, the tick clambers up a blade of grass to wait, for years if necessary, for the cues that will betray the presence of her prey. Later, swollen with blood, she drops to the earth, lays her eggs, and dies. The tick's impoverished sensory world, devoid of sensory luxuries and fine-tuned by natural selection for the world she will encounter, has ensured her single goal, reproduction.

A bird and a bat may share for a moment precisely the same environment. The worlds of their perceptions, however, are vastly different, structured by the limitations of the sensory windows each employs and by the brain that garners and processes what it needs for survival. For one it is a world dominated by vision; for the other, echolocation. The world of each is alien to the other, just as their worlds are to us. ■

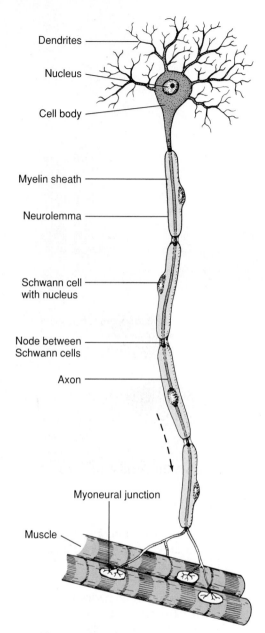

Figure 36-1
Structure of a motor (efferent) neuron.

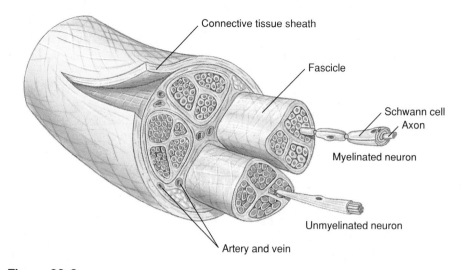

Figure 36-2
Structure of a nerve showing nerve fibers surrounded by various layers of connective tissue. Such a nerve may contain thousands of both efferent and afferent fibers.

The nervous system originated in a fundamental property of life: **irritability,** the ability to respond to environmental stimuli (Chapter 1, p. 9). The response may be simple, such as a protozoan moving to avoid a noxious substance, or quite complex, such as a vertebrate animal responding to elaborate signals of courtship. A protistan receives and responds to a stimulus, all within the confines of a single cell. The evolution of multicellularity and more complex levels of animal organization required increasingly complex mechanisms for com-

munication between cells and organs. This is accomplished by two principal means: **neural** and **hormonal.** Relatively rapid communication is by neural mechanisms and involves propagated electrochemical changes in cell membranes. The basic plan of the nervous system is to code information and to transmit and process it for appropriate action. These functions will be examined in this chapter. Relatively less rapid or long-term adjustments in animals are governed by hormonal mechanisms, the subject of the next chapter.

THE NEURON: FUNCTIONAL UNIT OF THE NERVOUS SYSTEM

The **neuron** is a cell body with all its processes. Although neurons assume many shapes, depending on their function and location, a typical kind is shown diagrammatically in Figure 36-1. From the nucleated body extend **processes** of two types. All but the simplest nerve cells have one or more cytoplasmic **dendrites.** As the name dendrite suggests (Gr. *dendron,* tree), these are often profusely branched. They are the nerve cell's receptive apparatus, often receiving information from several different sources at once. Some of these inputs are excitatory, others inhibitory.

From the nucleated cell body extends a single **axon** (Gr. *axon,* axle), often a long fiber (meters in length in the largest mammals), relatively uniform in diameter, that typically carries impulses away from the cell body. In vertebrates and some complex invertebrates, the axon is usually covered with an insulating sheath.

Neurons are commonly classified as **afferent,** or sensory; **efferent,** or motor; and **interneurons,** which are neither sensory nor motor but connect neurons with the other neurons. Afferent and efferent neurons lie mostly outside the central nervous system (brain and nerve cord) while interneurons, which in humans make up 99% of all nerve cells in the body, lie mostly within the central nervous system. Afferent neurons are connected to **receptors,** which function to convert some environmental stimuli into nerve impulses, which are carried by the afferent neurons into the central nervous system. Here impulses may be perceived as conscious sensation. Impulses also move to efferent neurons, which carry them out by the peripheral system to **effectors,** such as muscles or glands.

In vertebrates, nerve processes (usually axons) usually are bundled together in a well-formed wrapping of connective tissue to form a **nerve** (Figure 36-2). The cell bodies of these nerve processes are located either in

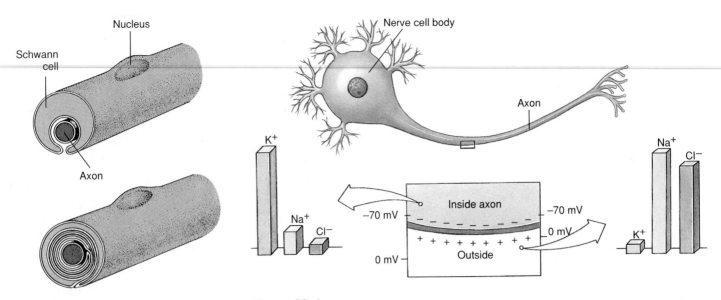

Figure 36-3

Development of the myelin sheath. The Schwann cell grows around the axon, then rotates around it, enclosing the axon in a tight, multilayered sheath. The myelin sheath insulates the nerve axon and facilitates the transmission of nerve impulses.

Figure 36-4

Ionic composition inside and outside a resting nerve cell. An active sodium-potassium exchange pump located in the cell membrane drives sodium to the outside, keeping its concentration low inside. Potassium concentration is high inside. Although the membrane is "leaky" to potassium, this ion is held inside by the repelling positive charge outside the membrane.

the central nervous system or in **ganglia,** which are discrete bundles of nerve cell bodies located outside the central nervous system.

Surrounding the neurons are non-nervous **neuroglial cells** (often simply called "glial" cells) that have a special relationship to the nerve cells. Neuroglial cells are extremely numerous in the vertebrate brain, where they outnumber nerve cells 10 to 1 and may make up almost half the volume of the brain. Some glial cells form intimate insulating sheaths of lipid-containing myelin around nerve fibers. Vertebrate peripheral nerves are often enclosed by **myelin,** an insulating sheath laid down in concentric rings by special glial cells called **Schwann cells** (Figure 36-3). Other functional roles of the glial cells are only now becoming known. Certain glial cells, called astrocytes because of their radiating, starlike shape, serve as a scaffold during brain development, enabling migrating neurons to find their way to their destination from points of origin. Astrocytes are also essential for the regenerative process that follows brain injury. Unfortunately, astrocytes also participate in several diseases of the nervous system, including parkinsonism and multiple sclerosis.

NATURE OF THE NERVE IMPULSE

The **nerve impulse** is the chemical-electrical message of nerves, the common functional denominator of all nervous system activity. Despite the incredible complexity of the nervous system of many animals, nerve impulses are basically alike in all nerves and in all animals. An impulse is an "all-or-none" phenomenon; either the fiber is conducting an impulse, or it is not. Because all impulses are alike, the only way a nerve fiber can vary its effect on the tissue it innervates is by changing the frequency of impulse conduction. Frequency change is the language of a nerve fiber. A fiber may conduct no impulses at all or very few per second up to a maximum approaching 1000 per second. The higher the frequency (or rate) of conduction, the greater is the level of excitation.

The Resting Potential

Membranes of nerve cells, like all cellular membranes, have a special permeability that creates ionic imbalances. The interstitial fluid surrounding nerve cells contains relatively high concentrations of sodium (Na$^+$) and chloride (Cl$^-$) ions, but a low concentration of potassium ions (K$^+$). Inside the neuron, the ratio is reversed: the K$^+$ concentration is high but the Na$^+$ and Cl$^-$ concentrations are low (Figure 36-4; see also Figure 34-1B, p. 671). These differences are pronounced; there is approximately 10 times more Na$^+$ outside than in and 25 to 30 times more K$^+$ inside than out.

When at rest, the membrane of a nerve cell is selectively permeable to K$^+$, which can pass the membrane through special potassium channels in the membrane. The permeability to Na$^+$ and Cl$^-$ is nearly zero because these channels are closed in the resting membrane. Potassium ions tend to diffuse outward through the membrane, following the gradient of potassium concentration. Because chloride ions cannot follow, however, an excess of positively charged potassium ions accumulates outside the cell. This creates a charged membrane that is positive outside and negative inside. Very quickly the positive charge outside reaches a level that prevents any more K$^+$ from diffusing out of the axon (because like charges repel each other). Now the resting membrane is at equilibrium, with a

resting membrane potential that exactly balances the concentration gradient that forces the K⁺ out. The resting potential is usually -70 mV (millivolts), with the inside of the membrane negative to the outside.

The Action Potential

The nerve impulse is a rapidly moving change in electrical potential called the **action potential** (Figure 36-5). It is a very rapid and brief depolarization of the membrane of the nerve fiber. In most nerve fibers, the action potential does not simply return the membrane potential to zero but instead overshoots zero. In other words, the membrane potential reverses for an instant so that the outside becomes negative as compared with the inside. Then, as the action potential moves ahead, the membrane returns to its normal resting potential ready to conduct another impulse. The entire event occupies approximately a millisecond. Perhaps the most significant property of the nerve impulse is that it is self-propagating; that is, once started the impulse moves ahead automatically, much like the burning of a fuse.

What causes the reversal of polarity in the cell membrane during passage of an action potential? We have seen that the resting potential depends on the high membrane permeability (leakiness) to K⁺, some 50 to 70 times greater than the permeability to Na⁺. When the action potential arrives at a given point, Na⁺ channels suddenly open, permitting a flood of Na⁺ to diffuse into the axon from the outside. Actually only a very minute amount of Na⁺ moves across the membrane—less than one millionth of the Na⁺ outside— but this sudden rush of positive ions wipes out the local membrane resting potential. The membrane is **depolarized,** creating an electrical "hole." Potassium ions, finding their electrical barrier gone, begin to move out. Then, as the action potential passes, the membrane quickly regains its resting properties. It becomes once again practically impermeable to Na⁺ and the outward movement of K⁺ is checked.

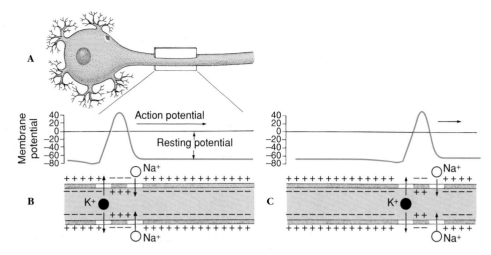

Figure 36-5

Conduction of the action potential of a nerve impulse. The impulse originates in the cell body of the neuron **A** and moves toward the right. **B** and **C** show the electrical event and associated changes in localized membrane permeability to sodium and potassium. The position of the action potential in **C** is shown about 4 milliseconds after **B**. When the impulse arrives at a point, sodium gates are opened, allowing sodium ions to rush in. Sodium inflow reverses the membrane polarity, making the inner surface of the axon positive and the outside negative. Sodium gates then close and potassium gates open. Potassium ions can now penetrate the membrane and restore the normal resting potential.

The rising phase of the action potential is associated with the rapid influx (inward movement) of Na⁺ (Figure 36-5). When the action potential reaches its peak, the Na⁺ permeability is restored to normal, and K⁺ permeability briefly increases above the resting level. This causes the action potential to drop rapidly toward the resting membrane level.

The Sodium Pump

The resting cell membrane has a very low permeability to Na⁺. Nevertheless some Na⁺ leaks across, even in the resting condition. When the axon is active, Na⁺ flows inward with each passing impulse. If not removed, the accumulation of Na⁺ inside the axon would cause the resting potential of the fiber to decay. This is prevented by **sodium pumps,** each a complex of protein subunits embedded in the plasma membrane of the axon (see Figure 4-19, p. 54). Each sodium pump uses the energy stored in ATP to transport sodium from the inside to the outside of the membrane. The sodium pump in nerve axons, as in many other cell membranes, also moves K⁺ into the axon while it is moving Na⁺ out. Thus it is a **sodium-potassium exchange pump** that

helps to restore the ion gradients of both Na⁺ and K⁺. Recently it was discovered that the astrocytes (mentioned earlier) help to maintain the correct balance of ions surrounding neurons by sweeping away excess potassium produced during neuronal activity.

HIGH-SPEED CONDUCTION

Although the ionic and electrical events associated with action potentials are much the same throughout the animal kingdom, this is not true for the speed at which action potentials move down nerve axons. Conduction velocities vary enormously from nerve to nerve and from animal to animal—from as slow as 0.1 m/sec in sea anemones to as fast as 120 m/sec in some mammalian motor axons. In most invertebrates, speed of conduction is closely related to the diameter of the axon. Small axons conduct slowly because internal resistance to current flow is high. Where fast conduction velocities are important for quick response, such as in locomotion to capture prey or to avoid capture, axon diameters are larger. The giant axon of the squid is nearly 1 mm in diameter and carries impulses 10 times faster than ordinary fibers in the same animal. The squid's

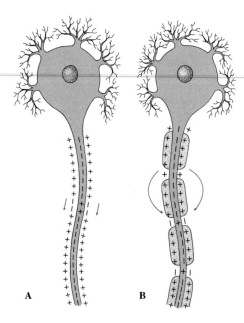

Figure 36-6

Impulse conduction in unmyelinated and myelinated fibers. In unmyelinated fibers **A,** the action potential spreads continuously along the entire length of the axon. In myelinated fibers **B,** the action potential leaps from node to node, bypassing the insulated portions of the fiber. This is saltatory conduction, which is much faster than continuous conduction.

giant axon innervates the animal's mantle musculature and is used for powerful mantle contractions when the animal swims by jet propulsion. Similar giant axons enable earthworms, which are normally slow-moving animals, to withdraw almost instantaneously into their burrows when startled.

Vertebrates have achieved high-conduction velocities in a different way, by a cooperative relationship between axons and the investing layers of myelin laid down by the Schwann cells described earlier. The insulating myelin sheaths are interrupted at intervals by nodes (called **nodes of Ranvier**) where the surface of the axon is exposed to the fluid surrounding the nerve. In these **myelinated fibers** the action potential does not move as a continuous wave of depolarization down the fiber, since this is prevented by the myelin insulation, but instead leaps from node to node (Figure 36-6). The ion pumps and channels that move ions

across the membrane are concentrated in each node. Once the action potential starts down the axon, the depolarization of the first node sets up an electrical current that stretches out to the neighboring node, causing it to depolarize and trigger an action potential. Thus the action potential leaps from node to node, a kind of conduction called **saltatory** (L. *salto,* to dance, leap). The gain in efficiency as compared with invertebrate nonmyelinated fibers is impressive. For example, a frog myelinated axon only 12 μm in diameter conducts nerve impulses at the same speed as a squid axon 350 μm in diameter.

Some invertebrates including prawns and insects also have fast fibers invested with multiple layers of a myelin-like substance that is interrupted at intervals much like the myelinated fibers of vertebrates. Conduction rates, though not as fast as vertebrate saltatory conduction, are much faster than unmyelinated fibers of the same diameter in other invertebrates.

SYNAPSES: JUNCTION POINTS BETWEEN NERVES

When an action potential passes down an axon to its terminal, it must cross a small gap, the synapse (Gr. *synapsis,* contact, union), separating it from another neuron or an effector organ. Two distinct kinds of synapses are known: electrical and chemical.

Electrical synapses, although much less common than chemical synapses, have been demonstrated in several invertebrate groups and are probably rather common in the nervous systems of many vertebrates. Electrical synapses are points at which ionic currents flow directly across a narrow **gap junction** (see Figure 4-15, p. 50) from one neuron to another.

Electrical synapses show no time lag and consequently are important for escape reactions.

Much more complex than electrical synapses are **chemical synapses.** These contain packets of specialized chemicals called **neurotransmitters.** Neurons bringing impulses toward chemical synapses are called **presynaptic neurons;** those carrying impulses away are **postsynaptic** neurons. At the synapse, the membranes are separated by a narrow gap, the **synaptic cleft,** having a width of approximately 20 nm.

The axon of most neurons divides at its end into many branches, each of which bears a synaptic knob that sits on the dendrites or cell body of the next neuron (Figure 36-7). The axon terminations of several neurons may almost cover a nerve cell body and its dendrites with thousands of synapses. Because a single impulse coming down a nerve axon splays out into many branches and synaptic endings on the next nerve cell, many impulses converge on the cell body at one instant.

The 20 nm fluid-filled gap between presynaptic and postsynaptic membranes prevents impulses from spreading directly to the postsynaptic neuron. Instead the synaptic knobs secrete a specific transmitter, usually **acetylcholine,** that communicates chemically with the postsynaptic cell. Inside the synaptic knobs are numerous tiny vesicles, each containing several thousand molecules of acetylcholine. Additional acetylcholine is also present in the cytoplasm of the synaptic knobs. Evidence suggests that when an impulse arrives at the terminal knob a sequence of events as portrayed in Figure 36-8 occurs. The action potential opens protein channels in the presynaptic membrane, allowing a pulse of acetylcholine to diffuse across the gap in a fraction of a millisecond and bind briefly to receptor molecules on the postsynaptic membrane. This creates a voltage change in the postsynaptic membrane. Whether the voltage change is large enough to trigger a postsynaptic potential depends on how many acetylcholine molecules are released and how many channels are opened. The acetylcholine is rapidly destroyed by the enzyme **acetylcholinesterase.** This is important because, if not inactivated in this way, the transmitter would continue to stimulate indefinitely. The organophosphate insecticides (such as malathion) and certain military nerve gases are poisonous for precisely this reason; that is, they block acetylcholinesterase. The final step in the sequence is the resynthesis of acetylcholine and its storage in vesicles, ready to respond to another impulse.

Several different chemical neurotransmitters have been identified in both vertebrate and invertebrate nervous systems. Some, such as acetylcholine and norepinephrine, depolarize the postsynaptic membrane; these are **excitatory synapses.** Other neurotransmitters, such as gamma aminobutyric acid (GABA), hyperpolarize the postsynaptic membrane; this tends to stabilize the membrane against depolarization. These are **inhibitory synapses.** Most nerve cells in the central nervous system have both excitatory and inhibitory synapses among the hundreds or thousands of synaptic knobs on the dendrites and cell body of each nerve cell.

It is the net balance of all excitatory and inhibitory inputs received by a postsynaptic cell that determines whether or not it will generate an action potential. If many excitatory impulses are received at one time, they may reduce the membrane potential enough in the postsynaptic membrane to elicit an action potential. Inhibitory impulses, however, stabilize the postsynaptic membrane, making it less likely that an action potential will be generated. The synapse is of great functional importance because it is a crucial part of the decision-making equipment of the central nervous system. In the synapse information is modulated from one nerve to the next.

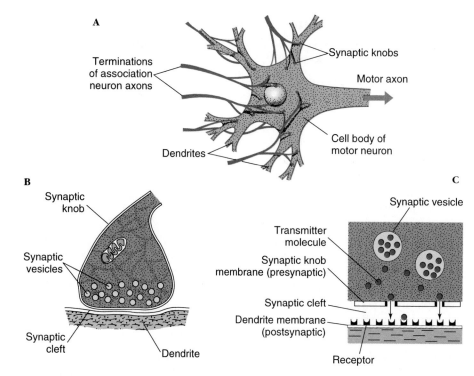

Figure 36-7

Transmission of impulses across nerve synapses. **A,** The cell body of a motor nerve is shown with the terminations of interneurons. Each termination ends in a synaptic knob; thousands of synaptic knobs may rest on a single nerve cell body and its dendrites. **B,** A synaptic knob enlarged 60 times more than in **A.** An impulse traveling down the axon causes protein channels in the presynaptic membrane to open, releasing neurotransmitter molecules into the cleft. **C,** Diagram of a synaptic cleft at the ultrastructural level. Neurotransmitter molecules move rapidly across the gap to bind briefly with receptor molecules in the postsynaptic membrane. This produces a change in the potential of the postsynaptic membrane.

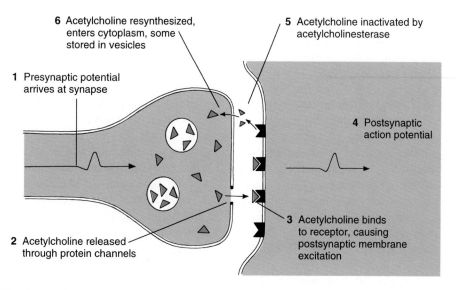

Figure 36-8

Sequence of events in synaptic transmission.

EVOLUTION OF THE NERVOUS SYSTEM

INVERTEBRATES: DEVELOPMENT OF CENTRALIZED NERVOUS SYSTEMS

The various metazoan phyla reveal a progressive increase in complexity of nervous systems that probably reflects in a general way the stages in the evolution of the nervous system. The simplest pattern of invertebrate nervous systems is the **nerve net** of radiate animals, such as sea anemones, jellyfish, hydra, and comb jellies (Figure 36-9A). The nerve net is a quantum leap in complexity beyond sensory systems of the protozoa, which lack nerves. The nerve net forms an extensive network that is found in and under the epidermis over all the body. An impulse starting in one part of this net is conducted in all directions, since synapses in most radiates do not restrict transmission to one-way movement, as they do in more complex animals. There are no differentiated sensory, motor, or connector components in the strict meaning of those terms. Branches of the nerve net connect to receptors in the epidermis and to epithelial cells that have contractile properties. Most responses tend to be generalized, yet many are astonishingly complex for so simple a nervous system. It is interesting that this type of nervous system is found among vertebrates in the form of nerve plexuses located, for example, in the intestinal wall, where they govern generalized intestinal movements such as peristalsis and segmentation.

Bilateral nervous systems, first seen in the flatworms, represent a distinct increase in complexity over the nerve net of radiate animals. Flatworms have two anterior ganglia of nerve cells from which two main nerve trunks run posteriorly, with lateral branches extending to the various parts of the body (Figure 36-9B). This is the simplest nervous system showing differentiation into a **peripheral nervous system** (a communication

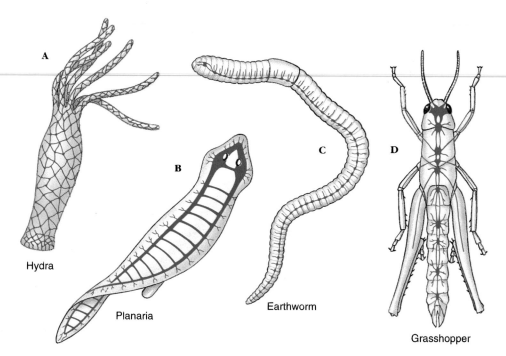

Hydra

Planaria

Earthworm

Grasshopper

Figure 36-9

Invertebrate nervous systems. **A,** The nerve net of radiates, the simplest neural organization. **B,** The flatworm system, the simplest linear-type nervous system of two nerve trunks connected to a complex neuronal network. **C,** The annelid nervous system, organized into a bilobed brain and ventral cord with segmental ganglia. **D,** Arthropod nervous system with large ganglia and with more elaborate sense organs.

network extending to all parts of the body) and a **central nervous system,** which coordinates everything. More complex invertebrates have a more centralized nervous system, two longitudinal nerve cords fused (although still recognizable) and many ganglia present. The elaborate nervous system of annelids consists of a bilobed brain, a double nerve cord with ganglia, and distinctive **afferent** (sensory) and **efferent** (motor) neurons (Figure 36-9C). The segmental ganglia serve as relay stations for coordinating regional activity.

The basic plan of the molluscan nervous system is a series of three pairs of well-defined ganglia, but in the cephalopods (such as octopus and squid), the ganglia have burgeoned into textured nervous centers of great complexity, such as those of the octopus, which contain more than 160 million cells. Sense organs, too, are highly developed. Consequently, the complexity of cephalopod behavior far outstrips that of any other invertebrate.

The basic plan of the arthropod nervous system (Figure 36-9D) resembles that of annelids, but the ganglia are larger and sense organs are much better developed. Social behavior is often elaborate, particularly in the hymenopteran insects (bees, wasps, and ants), and most arthropods are capable of considerable manipulation of their environment. Despite the complexity of much insect behavior, insects are nevertheless reflex-bound animals incapable of involved learned behavior principally because of their small size.

VERTEBRATES: FRUITION OF ENCEPHALIZATION

The basic plan of the vertebrate nervous system is a hollow, *dorsal* nerve cord terminating anteriorly in a large ganglionic mass, the brain. This pattern contrasts with the nerve cord of bilateral invertebrates, which is solid and ventral to the alimentary canal. By far the most important trend in the evolution of vertebrate nervous sys-

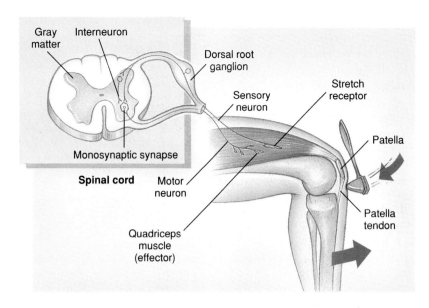

Figure 36-11

The knee-jerk reflex, a simple reflex arc. Sudden pressure on the patellar ligament stretches muscles in the upper leg. Impulses generated in stretch receptors are conducted over afferent (sensory) fibers to the spinal cord and relayed by an interneuron to an efferent (motor) nerve cell body. Impulses pass over efferent axons to the leg muscles (effectors), stimulating them to contract.

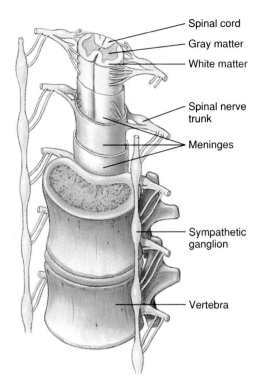

Figure 36-10

Human spinal cord and its protection. Two vertebrae show the position of the spinal cord, emerging spinal nerve trunks, and the sympathetic trunk. The cord is wrapped by three layers of membrane (meninges) between two of which lies a protective bath of cerebrospinal fluid.

tems is the great elaboration of the size, configuration, and functional capacity of the brain, a process called **encephalization.** Vertebrate encephalization has brought to full fruition several functional capabilities including fast responses, great capacity for storage of information, and enhanced complexity and flexibility of behavior. Another consequence of encephalization is the ability to form associations between past, present, and (at least in humans) future events.

The Spinal Cord

The **brain** and **spinal cord** compose the central nervous system. During early embryonic development, the spinal cord and brain begin as an ectodermal neural groove, which by folding and enlarging becomes a long, hollow neural tube. The cephalic end enlarges into the brain vesicles, and the rest becomes the spinal cord. Un-

like any invertebrate nerve cord, the segmental nerves of the spinal cord (31 pairs in humans) are separated into dorsal sensory roots and ventral motor roots. The sensory nerve cell bodies are gathered together into dorsal root (spinal) ganglia. Both dorsal (sensory) and ventral (motor) roots meet beyond the spinal cord to form a mixed spinal nerve (Figure 36-10).

The spinal cord is enclosed by the vertebral canal and is additionally wrapped in three layers of membranes called **meninges** (men-in'jeez; Gr. *meningos,* membrane). In cross section the cord shows two zones (Figure 36-10). The inner zone of gray matter, resembling in shape the wings of a butterfly, contains the cell bodies of motor neurons and interconnecting interneurons (described in the following text). The outer zone of white matter contains bundles of axons and dendrites linking different levels of the cord with each other and with the brain.

The Reflex Arc

Many neurons work in groups called **reflex arcs.** There must be at least two neurons in a reflex arc, but usu-

ally there are more. The parts of a typical reflex arc (as, for example, in the well-known "knee-jerk" reflex, Figure 36-11) are (1) a **receptor,** a sense organ in the skin, muscle, or other organ; (2) an **afferent,** or sensory, neuron, which carries impulses toward the central nervous system; (3) the **central nervous system,** where synaptic connections are made between the sensory neurons and the interneurons; (4) the **efferent,** or motor, neuron, which makes a synaptic connection with the interneuron and carries impulses from the central nervous system; and (5) the **effector,** by which the animal responds to environmental changes. Examples of effectors are muscles, glands, ciliated cells, nematocysts of the radiate animals, electric organs of fish, and certain pigmented cells called chromatophores.

A reflex arc in its simplest form consists of only two neurons—a sensory (afferent) neuron and a motor (efferent) neuron. Usually, however, interneurons are interposed between sensory and motor neurons (Figure 36-11). Interneurons may connect afferent and efferent neurons on the same side of the spinal cord or on opposite sides, or it may connect them on different levels of the spinal cord, either on the same or opposite sides. In almost any reflex act a number of reflex arcs are involved. For instance, a single afferent neuron may make synaptic connections with many efferent neurons. In a similar way an efferent neuron may receive impulses from many afferent neurons.

A **reflex act** is the response to a stimulus acting over a reflex arc. It is involuntary, meaning that it is not under the control of the will. Many vital processes of the body, such as control of breathing, heartbeat, diameter of blood vessels, and sweat gland secretion are reflex acts. Some reflex acts are innate; others are acquired through learning.

The Brain

Unlike the spinal cord, which has changed little in structure during vertebrate evolution, the brain has changed dramatically. From the primitive linear brain of fishes and amphibians, it has expanded into the deeply fissured, enormously intricate brain of mammals (Figure 36-12). It reaches its greatest complexity in the human brain, which contains some 35 billion nerve cells, each of which may receive information from tens of thousands of synapses at one time. The ratio between the weight of the brain and that of the spinal cord affords a fair criterion of an animal's intelligence. In fish and amphibians this ratio is approximately 1:1; in humans the ratio is 55:1—in other words, the brain is 55 times heavier than the spinal cord. Although the human brain is not the largest (the sperm whale's brain is seven times heavier) nor the most convoluted (that of the porpoise is even more wrinkled), it is by all odds the best in overall performance. This "great ravelled knot," as the British physiologist Sir Charles Sherrington called the human brain, in fact may be so complex that it will never be able to understand its own function.

The brain of the early vertebrate fishes had three principal divisions: a

Although the large size of their brain undoubtedly makes humans the wisest of animals, it is apparent that they can do without much of it and still remain wise. Brain scans of persons with hydrocephalus (enlargement of the head as a result of pressure disturbances that cause the brain ventricles to enlarge many times their normal size) show that although many of them are functionally disabled, others are nearly normal. The cranium of one person with hydrocephalus was nearly filled with cerebrospinal fluid and the only remaining cerebral cortex was a thin layer of tissue, 1 mm thick, pressed against the cranium. Yet this young man, with only 5% of his brain, had achieved first-class honors in mathematics at a British university and was socially normal. This and other similarly dramatic observations suggest that there is enormous redundancy and spare capacity in corticocerebral function. It also suggests that the deep structures of the brain, which are relatively spared in hydrocephalus, may perform functions once believed to be performed solely by the cortex.

forebrain, the **prosencephalon;** a midbrain, the **mesencephalon;** and a hindbrain, the **rhombencephalon** (Figure 36-13). Each of the three parts was concerned with one or more of the special senses: the forebrain with the sense of smell, the midbrain with vision, and the hindbrain with hearing and balance. These primitive but very fundamental concerns of the brain have been in some instances amplified and in others reduced or overshadowed during continued evolution as sensory priorities were shaped by the animal's habitat and way of life.

Hindbrain. The **medulla,** the most posterior division of the brain, is really a conical continuation of the spinal cord. The medulla, together with the more anterior midbrain, constitutes the "brain stem," an area that controls numerous vital and largely subconscious

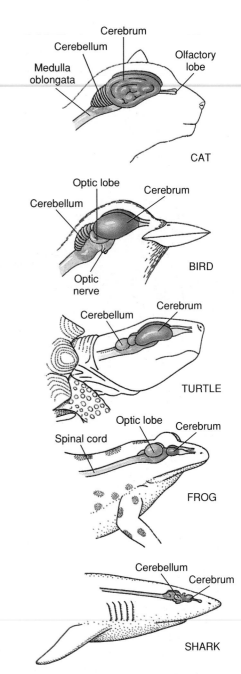

Figure 36-12
Evolution of the vertebrate brain. Note the progressive increase in size of the cerebrum. The cerebellum, concerned with equilibrium and motor coordination, is largest in animals whose balance and precise motor movements are well developed (fishes, birds, and mammals).

activities such as heartbeat, respiration, vascular tone, and swallowing. The **pons,** also a part of the hindbrain, is a thick bundle of fibers that carry impulses from one side of the cerebellum to the other, and to higher brain centers.

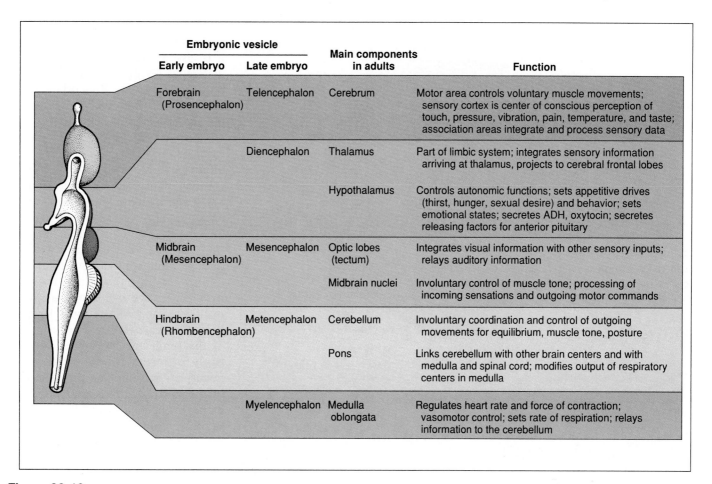

Embryonic vesicle		Main components in adults	Function
Early embryo	**Late embryo**		
Forebrain (Prosencephalon)	Telencephalon	Cerebrum	Motor area controls voluntary muscle movements; sensory cortex is center of conscious perception of touch, pressure, vibration, pain, temperature, and taste; association areas integrate and process sensory data
	Diencephalon	Thalamus	Part of limbic system; integrates sensory information arriving at thalamus, projects to cerebral frontal lobes
		Hypothalamus	Controls autonomic functions; sets appetitive drives (thirst, hunger, sexual desire) and behavior; sets emotional states; secretes ADH, oxytocin; secretes releasing factors for anterior pituitary
Midbrain (Mesencephalon)	Mesencephalon	Optic lobes (tectum)	Integrates visual information with other sensory inputs; relays auditory information
		Midbrain nuclei	Involuntary control of muscle tone; processing of incoming sensations and outgoing motor commands
Hindbrain (Rhombencephalon)	Metencephalon	Cerebellum	Involuntary coordination and control of outgoing movements for equilibrium, muscle tone, posture
		Pons	Links cerebellum with other brain centers and with medulla and spinal cord; modifies output of respiratory centers in medulla
	Myelencephalon	Medulla oblongata	Regulates heart rate and force of contraction; vasomotor control; sets rate of respiration; relays information to the cerebellum

Figure 36-13

Divisions of the vertebrate brain.

The **cerebellum,** lying dorsal to the medulla, controls equilibrium, posture, and movement (Figure 36-14). Its development is directly correlated with the animal's mode of locomotion, agility of limb movement, and balance. It is usually weakly developed in amphibians and reptiles, forms that live close to the ground, and well developed in the more agile bony fishes. It reaches its apogee in birds and mammals in which it is greatly expanded and folded. The cerebellum does not initiate movement but operates as a precision error-control center, or servo-mechanism, that programs a movement initiated somewhere else, such as the motor cortex. Primates and especially humans, who possess a manual dexterity far surpassing that of other animals, have the most complex cerebellum. Movements of the hands and fingers may involve the cerebellar coordina-tion of the simultaneous contraction and relaxation of hundreds of individual muscles.

Midbrain. The midbrain consists mainly of the **tectum** (including the optic lobes), which contains nuclei that serve as centers for visual and auditory reflexes. (In neurophysiological usage a *nucleus* is a small aggregation of nerve cell bodies within the central nervous system.) The midbrain has undergone little evolutionary change in structure among vertebrates but has changed markedly in function. It mediates the most complex behavior of fishes and amphibians. Such integrative functions were gradually assumed by the forebrain in amniotes. In mammals the midbrain is mainly a reflex center for eye muscles and a relay and analysis center for auditory information.

Forebrain. Just anterior to the midbrain lie the **thalamus** and **hypothalamus,** the most posterior elements of the forebrain. The egg-shaped thalamus is a major relay station that analyzes and passes sensory information to higher brain centers. In the hypothalamus are several "housekeeping" centers that regulate body temperature, water balance, appetite, and thirst—all functions concerned with the maintenance of internal constancy (homeostasis) Neurosecretory cells located in the hypothalamus produce several pituitary-regulating neurohormones (described in Chapter 37). The hypothalamus also contains centers for diverse emotions such as pleasure, aggression, and sexual arousal.

The anterior portion of the forebrain, the cerebrum (Figure 36-14), can be divided into two anatomically distinct areas, the **paleocortex** and

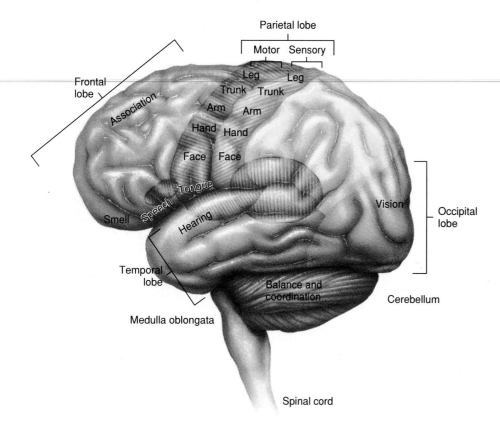

Figure 36-14
External view of the human brain, showing lobes of the cerebrum and localization of major function of the cerebrum and cerebellum.

neocortex. Originally concerned with smell, it became well developed in the advanced fishes and early terrestrial vertebrates, which depend on this special sense. In mammals and especially in primates the paleocortex is a deep-lying area called the rhinencephalon ("nose brain"), because many of its functions depend on olfaction. Better known as the **limbic system,** it mediates several species-specific behaviors that relate to fulfilling needs such as feeding and sex.

Although a late arrival in vertebrate evolution, the neocortex completely overshadows the paleocortex and has become so expanded that it envelops much of the forebrain and all of the midbrain (Figure 36-14). Almost all of the integrative activities primitively assigned to the midbrain were transferred to the neocortex, or cerebral cortex as it is usually called.

Functions in the cerebrum have been localized by direct stimulation of exposed brains of people and experimental animals, postmortem examination of persons suffering from various lesions, and surgical removal of specific brain areas in experimental animals. The cortex contains discrete motor and sensory areas (Figures 36-14 and 36-15) as well as large "silent" regions, called **association areas,** concerned with memory, judgment, reasoning, and other integrative functions. These regions are not directly connected to sense organs or muscles.

Thus in mammals, and especially in humans, separate parts of the brain mediate conscious and unconscious functions. The unconscious mind, all of the brain except the cerebral cortex, governs numerous vital functions that are removed from conscious control: respiration, blood pressure, heart rate, hunger, thirst, temperature balance, salt balance, sexual drive, and basic (sometimes irrational) emotions. It is also a complex endocrine gland that regulates the body's subservient endocrine system. The other, conscious mind, the cerebral cortex, is the site of higher mental activities (for example, planning and reasoning), memory, and integration of sensory information. Memory appears to transcend all parts of the brain rather than being a property of any particular part of the brain as was once believed.

The right and left hemispheres of the cerebral cortex are bridged through the corpus callosum, a neural connection through which the two hemispheres are able to transfer information and coordinate mental activities. In humans, the two hemispheres of the brain are specialized for entirely different functions: the left hemisphere (controlling the right side of the body) for language development, mathematical and learning capabilities, and sequential thought processes; and the right hemisphere (controlling the left side of the body) for spatial, musical, artistic, intuitive, and perceptual activities. It has been known for a long time that even extensive damage to the right hemisphere may cause varying degrees of left-sided paralysis but has little effect on intellect and speech. Conversely, damage to the left hemisphere usually causes loss of speech and may have disastrous effects on intellect. Since these differences in brain symmetry and function exist at birth, they appear to be inborn rather than the result of developmental or environmental effects as once believed.

Hemispheric specialization has long been considered a unique human trait, but was recently discovered in the brains of songbirds in which one side of the brain is specialized for song production.

The Peripheral Nervous System

The peripheral nervous system includes all nervous tissue outside of

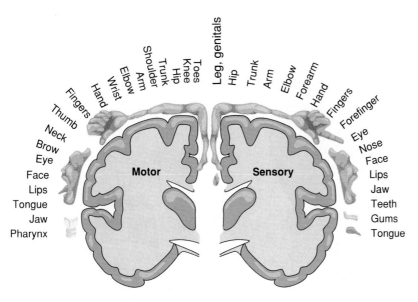

Figure 36-15

Arrangement of sensory and motor cortices. Localization of sensory terminations from different parts of the body are shown at left; origin of descending motor pathways are shown at right. The motor cortex lies in front of the sensory cortex, so the two are not superimposed. These maps grew out of the 1930s work of Canadian neurosurgeon Wilder Penfield. Recent research shows that the motor cortex is not as orderly as the map suggests; rather there is a more diffuse correspondence between cortical areas and the areas of the body they control.

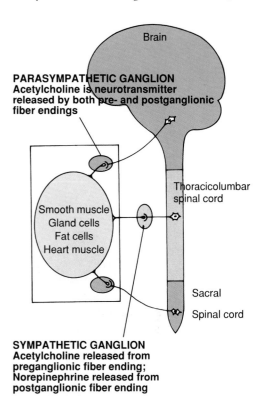

Figure 36-16

General organization of the autonomic nervous system.

the central nervous system. It consists of two functional divisions: the **afferent division** that brings sensory information to the central nervous system, and the **efferent division** that conveys motor commands to muscles and glands. The efferent division consists of two components, (1) the **somatic nervous system,** which supplies the skeletal muscle, and (2) the **autonomic nervous system,** which supplies smooth muscle, cardiac muscle, and glands.

Autonomic Nervous System. The autonomic system governs the involuntary, internal functions of the body that do not ordinarily affect consciousness, such as the movements of the alimentary canal and heart, the contraction of the smooth muscle of the blood vessels, urinary bladder, iris of the eye, and others, plus the secretions of various glands.

Autonomic nerves originate in the brain or spinal cord as do the nerves of the somatic nervous system, but unlike the latter, the autonomic fibers consist of not one but two motor neurons.

They synapse once after leaving the cord and before arriving at the effector organ. These synapses are located outside the spinal cord in ganglia. Fibers passing from the cord to the ganglia are called preganglionic autonomic fibers; those passing from the ganglia to the effector organs are called postganglionic fibers. These relationships are illustrated in Figure 36-16.

Subdivisions of the autonomic system are the **parasympathetic** and the **sympathetic** systems. Most organs in the body are innervated by both sympathetic and parasympathetic fibers, and their actions are antagonistic (Figure 36-17). If one fiber speeds up an activity, the other slows it down. However, neither kind of nerve is exclusively excitatory or inhibitory. For example, parasympathetic fibers inhibit heartbeat but excite peristaltic movements of the intestine; sympathetic fibers increase heartbeat but slow down peristaltic movement.

The parasympathetic system consists of motor neurons, some of which emerge from the brain stem by certain

cranial nerves and others of which emerge from the sacral (pelvic) region of the spinal cord (Figures 36-16 and 36-17). In the sympathetic division the nerve cell bodies of all the preganglionic fibers are located in the thoracic and upper lumbar areas of the spinal cord. Their fibers pass out through the ventral roots of the spinal nerves, separate from these, and go to the sympathetic ganglia (Figure 36-17), which are paired and form a chain on each side of the spinal column.

All preganglionic fibers, whether sympathetic or parasympathetic, release acetylcholine at the synapse with the postganglionic cells. However, the parasympathetic postganglionic fibers release acetylcholine at their endings, whereas the sympathetic postganglionic fibers with few exceptions release norepinephrine (also called noradrenaline). This difference is another important characteristic distinguishing the two parts of the autonomic nervous system.

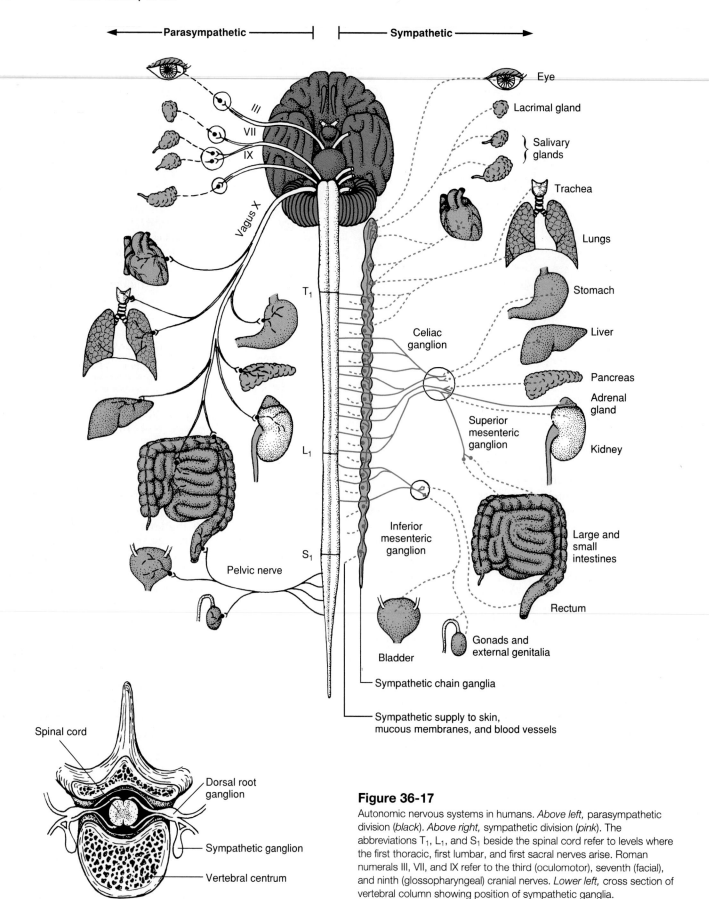

Parasympathetic   **Sympathetic**

Eye

Lacrimal gland

Salivary glands

Trachea

Lungs

Stomach

Liver

Pancreas

Adrenal gland

Kidney

Large and small intestines

Rectum

Gonads and external genitalia

III

VII

IX

Vagus X

T_1

L_1

S_1

Celiac ganglion

Superior mesenteric ganglion

Inferior mesenteric ganglion

Pelvic nerve

Bladder

Sympathetic chain ganglia

Sympathetic supply to skin, mucous membranes, and blood vessels

Spinal cord

Dorsal root ganglion

Sympathetic ganglion

Vertebral centrum

Figure 36-17

Autonomic nervous systems in humans. *Above left,* parasympathetic division (*black*). *Above right,* sympathetic division (*pink*). The abbreviations T_1, L_1, and S_1 beside the spinal cord refer to levels where the first thoracic, first lumbar, and first sacral nerves arise. Roman numerals III, VII, and IX refer to the third (oculomotor), seventh (facial), and ninth (glossopharyngeal) cranial nerves. *Lower left,* cross section of vertebral column showing position of sympathetic ganglia.

As a general rule the parasympathetic division is active under resting conditions when such functions as eating, digestion, urination, and other vegetative activities are emphasized. The sympathetic division is active under conditions of physical activity and stress. Under such conditions the heart rate increases, blood vessels to the skeletal muscles dilate, blood vessels in the viscera constrict, activity of the intestinal tract decreases, and metabolic rate increases. The importance of these responses in emergency reactions (sometimes referred to as the fight-flight response) are described in the next chapter (p. 753). It should be noted, however, that the sympathetic division is also active during resting conditions in maintaining normal blood pressure and body temperature.

SENSE ORGANS

Animals require a constant inflow of information from the environment to regulate their lives. Sense organs are specialized receptors designed for detecting environmental status and change. An animal's sense organs are its first level of environmental perception; they are channels for bringing information to the brain.

A **stimulus** is some form of energy—electrical, mechanical, chemical, or radiant. The task of the sense organ is to transform the energy form of the stimulus it receives into nerve impulses, the common language of the nervous system. In a very real sense, then, sense organs are biological transducers. A microphone, for example, is a transducer that converts mechanical (sound) energy into electrical energy. Like the microphone that is sensitive only to sound, sense organs are, as a rule, specific for one kind of stimulus. Thus eyes respond only to light, ears to sound, pressure receptors to pressure, and chemoreceptors to chemical molecules. But again, all of these different forms of energy are converted into nerve impulses.

Since all nerve impulses are qualitatively alike, how do animals perceive and distinguish the different sensations of varying stimuli? The answer is that the real perception of sensation is done in localized regions of the brain, where each sensory organ has its own hookup. Impulses arriving at a particular sensory area of the brain can be interpreted in only one way. For example, pressure on the eye causes us to see "stars" or other visual patterns; the mechanical distortion of the eye initiates impulses in the optic nerve fibers that are perceived as light sensations. Although such an operation probably could never be done, a deliberate surgical switching of optic and auditory nerves would cause the recipient literally to see thunder and hear lightning!

CLASSIFICATION OF RECEPTORS

Receptors are traditionally classified on the basis of their location. Those near the external surface, called **exteroceptors,** keep the animal informed about the external environment. Internal parts of the body are provided with **interoceptors,** which pick up stimuli from the internal organs. Muscles, tendons, and joints have **proprioceptors,** which are sensitive to changes in the tension of muscles and provide the organism with a sense of body position. Sometimes receptors are classified by the form of energy to which the receptors respond, such as **chemical, mechanical, light,** or **thermal.**

CHEMORECEPTION

Chemoreception is the oldest and most universal sense in the animal kingdom. It probably guides the behavior of animals more than any other sense. Protozoa use **contact chemical receptors** to locate food and adequately oxygenated water and to avoid harmful substances. These receptors elicit an orientation behavior, called **chemotaxis,** toward or away from the chemical source. Most metazoans have specialized **distance chemical receptors.** These are often developed to a remarkable degree of sensitivity. Distance chemoreception, usually referred to as a sense of smell or olfaction, guides feeding behavior, location and selection of sexual mates, territorial and trail marking, and alarm reactions of numerous animals.

The social insects and some other animals, including mammals, produce species-specific compounds, called **pheromones,** which constitute a highly developed chemical language. Pheromones are a diverse group of organic compounds that an animal releases to affect the physiology or behavior of another individual of the same species. Ants, for example, are walking batteries of glands (Figure 36-18) that produce numerous chemical signals. These include releaser pheromones, such as alarm and trail pheromones, and primer pheromones, which alter the endocrine and reproductive systems of different castes in the colony. Insects bear a variety of chemoreceptors on the surface of the body for sensing specific pheromones, as well as other, nonspecific odors.

In all vertebrates and in insects as well, the senses of **taste** and **smell** are clearly distinguishable. Although there are similarities between taste and smell receptors, in general the

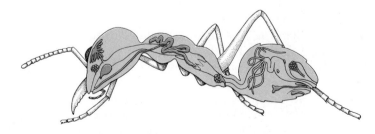

Figure 36-18
Pheromone-producing glands of an ant.

sense of taste is more restricted in response and is less sensitive than the sense of smell. Nervous centers for taste and smell are also located in different parts of the brain.

In vertebrates, taste receptors are found in the mouth cavity and especially on the tongue (Figure 36-19), where they provide a means for judging foods before they are swallowed. A **taste bud** consists of a cluster of several receptor cells surrounded by supporting cells; it is provided with a small external pore through which the slender tips of the sensory cells project. Chemicals being tasted apparently combine with specific receptor sites on the microvilli of the receptor cells. Because they are subject to the wear and tear of abrasive foods, taste buds have a short life (5 to 10 days in mammals) and are continually being replaced.

The four basic taste sensations possessed by humans—sour, salty, bitter, and sweet—are each attributable to a different kind of taste bud. The tastes for salty and sweet are found mainly on the tip of the tongue, bitter at the base of the tongue, and sour along the sides of the tongue. Of these, the bitter taste is by far the most sensitive, because it provides early warning against potentially dangerous substances, many of which are bitter.

The sense of smell is more complex than taste, and until very recently odor research has lagged behind other areas of sensory physiology. Although the olfactory sense is a primal sense for many animals, used for the identification of food, sexual mates, and predators, olfaction is most highly developed in mammals. Even humans, although a species not celebrated for detecting smells, can discriminate perhaps 10,000 different odors. The human nose can detect 1/25 of one-millionth of 1 mg of mercaptan, the odoriferous substance of the skunk. Even so, our olfactory abilities compare poorly with those of other mammals that rely on olfaction for survival. A

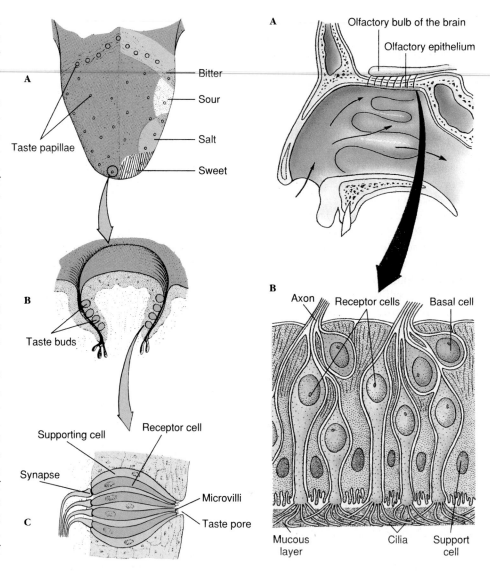

Figure 36-19
Taste receptors. **A,** Surface of human tongue showing regions of maximum sensitivity to the four primary taste sensations. **B,** Position of taste buds on a taste papilla. **C,** Structure of a taste bud.

Figure 36-20
Human olfactory epithelium. **A,** The epithelium is a patch of tissue positioned in the roof of the nasal cavity. **B,** It is composed of supporting cells, basal cells, and olfactory receptor cells with cilia protruding from their free ends.

dog explores new surroundings with its nose much as we do with our eyes. A dog's nose is justifiably renowned; with some odorous sources a dog's nose is at least a million times more sensitive than ours. Dogs are assisted in their proficiency by having a nose located close to the ground where odors from passing creatures tend to linger.

The olfactory endings are located in a special epithelium covered with a

thin film of mucus, positioned deep in the nasal cavity (Figure 36-20). Within the epithelium lie millions of olfactory neurons, each with several hairlike cilia protruding from the free end. Odor molecules entering the nose bind to receptor proteins located in the cilia; this binding generates an electrical signal that travels along axons to the olfactory bulb of the brain. From here odor information is sent to the olfactory cortex where odors are analyzed. Odor information is then projected to higher brain centers where they effect emotions, thoughts, and behavior.

Recently, using the techniques of gene cloning and molecular hybridization (p. 145), researchers discovered a large family of genes in mammals (including humans) that appears to code for odor reception. Each of the approximately 1000 genes discovered encodes a separate type of odor receptor. Since mammals can detect at least 10,000 different odors, each receptor must respond to several odor molecules, and each odor molecule must bind with several types of receptors, each of which responds to a part of the molecule's structure. Brain mapping techniques have shown that each olfactory neuron projects to a characteristic location on the olfactory bulb, providing a two-dimensional map that identifies which receptors have been activated in the nose. Projected to the brain, this information is recognized as a unique scent.

Because the flavor of food depends on odors reaching the olfactory epithelium through the throat passage, taste and smell are easily confused. All the various "tastes" other than the four basic ones (sweet, sour, bitter, salty) are really the results of flavor molecules reaching the olfactory epithelium in this manner. Food loses its appeal during a common cold because a stuffy nose blocks odors rising from the mouth.

MECHANORECEPTION

Mechanoreceptors are sensitive to quantitative forces such as touch, pressure, stretching, sound, vibration, and gravity—in short, they respond to motion. Animals require a steady flow of information from mechanoreceptors in order to interact with their environment, feed themselves, maintain normal posture, and to walk, swim, or fly.

Touch and Pain

The **pacinian corpuscle,** a relatively large mechanoreceptor that registers deep touch and pressure in mam-

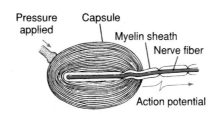 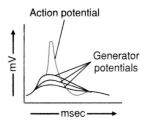

Figure 36-21

Response of pacinian corpuscle to applied pressure. Progressively stronger pressure produces stronger receptor potentials. When the threshold stimulus is reached, an all-or-none action potential is generated in the afferent nerve fiber.

malian skin, illustrates the general properties of mechanoreceptors. They are common in the deep layers of the skin, connective tissue surrounding muscles and tendons, and the abdominal mesenteries. Each corpuscle consists of a nerve terminus surrounded by a capsule of numerous, concentric, onionlike layers of connective tissue (Figure 36-21). Pressure at any point on the capsule distorts the nerve ending, producing a graded **receptor potential.** This is a local flow of electric current. Progressively stronger stimuli lead to correspondingly stronger receptor potentials until a **threshold current** is produced; this initiates an action potential in the sensory nerve fibers. Stronger stimuli will produce a burst of action potentials. However, if the pressure applied is sustained, the corpuscle quickly adjusts to the new shape and no longer responds. This is referred to as **adaptation** (not to be confused with the evolutionary meaning of this term [Chapter 9]) and is characteristic of many kinds of touch receptors, which are admirably suited to detecting a sudden mechanical change but readily adapt to new conditions. We are aware of new pressures when we put on our shoes in the morning, but we are glad not to be reminded all day long that we have them on.

Invertebrates, especially the insects, have many kinds of receptors sensitive to touch. Such receptors are well endowed with tactile hairs sensitive to both touch and vibrations. Superficial touch receptors of vertebrates are distributed all over the body but tend to be concentrated in areas especially important for exploring and interpreting

the environment. In most vertebrates these areas are on the face and extremities of the limb. Of the more than half million separate sensitive spots on the surface of the human body, most are found on the lips, tongue, and fingertips (Figure 36-15). Many touch receptors are bare nerve-fiber terminals, but there is an assortment of other kinds of receptors of varying shapes and sizes. Each hair follicle is crowded with receptors that are sensitive to touch.

Pain receptors are relatively unspecialized nerve fiber endings that respond to a variety of stimuli signaling possible or real damage to tissues. It is still uncertain whether pain fibers respond directly to injury or indirectly to some substance such as histamine, which is released by damaged cells.

Pain is a distress call from the body signaling some noxious stimulus or internal disorder. Although there is no cortical pain center, discrete areas have been located in the brain stem where pain messages from the periphery terminate. These areas contain two kinds of small peptides, endorphins and enkephalins, that have morphinelike or opiumlike activity. When released, they bind with specific opiate receptors in the midbrain. They are the body's own analgesics.

Just as pain is a sign of danger, sensory pleasure is a sign of a stimulus useful to the subject. Pleasure depends on the internal state of the animal and is judged with reference to homeostasis and some physiological set point.

Lateral Line System of Fishes

The lateral line is a distant touch reception system for detecting wave vibrations and currents in water. The receptor cells, called **neuromasts,** are located on the body surface in aquatic amphibians and some fishes, but in many fishes they are located within canals running beneath the epidermis; these canals open at intervals to the surface (Figure 36-22). Each neuromast is a collection of hair cells with the sensory hairs embedded in a gelatinous, wedge-shaped mass known as a **cupula.** The cupula projects into the center of the lateral line canal so that it bends in response to any disturbance of water on the body surface. The lateral line system is one of the principal sensory systems that guide fishes in their movements and in the location of predators, prey, and social partners (p. 507).

Hearing

The ear is a specialized receptor for detecting sound waves in the surrounding environment. Because sound communication and reception are an integral part of the lives of terrestrial vertebrates, we may be surprised to discover that most invertebrates inhabit a silent world. Only certain arthropod groups—crustaceans, spiders, and insects—have developed true sound-receptor organs. Even among the insects, only the locusts, cicadas, crickets, grasshoppers, and most moths possess ears, and these are of simple design: a pair of air pockets, each enclosed by a tympanic membrane that passes sound vibrations to sensory cells. Despite their spartan construction, insect ears are beautifully designed to detect the sound of a potential mate or a rival male.

Especially interesting are the ultrasonic detectors of certain nocturnal moths. These have evolved specifically to detect approaching bats and thus lessen the moth's chance of becoming a bat's evening meal (echolocation in bats is described on p. 610). Each moth ear possesses just two receptors (Figure 36-23). One of these, known as the A_1 receptor, will respond to the ultrasonic

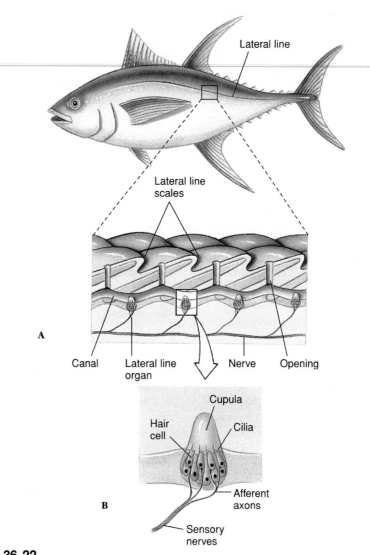

Figure 36-22

Lateral line system. **A,** Lateral line of a bony fish with both exposed and hidden neuromasts. **B,** Structure of a neuromast (lateral line organ).

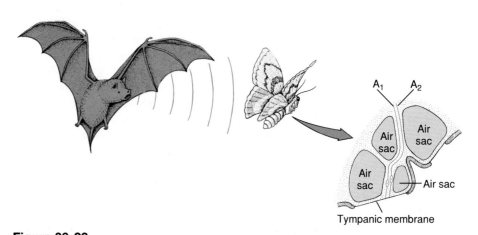

Figure 36-23

Ear of a moth used to detect approaching bats. See text for explanation.

cries of a bat that is still too far away to detect the moth. As the bat approaches and its cries increase in intensity, the receptor fires more rapidly, informing the moth that the bat is coming nearer. Since the moth has two ears, its nervous system can determine the bat's position by comparing firing rates from the two ears. The moth's strategy is to fly away before the bat detects it. But if the bat continues its approach, the second (A_2) receptor in each ear, which responds only to high-intensity sounds, will fire. The moth responds immediately with an evasive maneuver, usually making a power dive to a bush or the ground where it is safe because the bat cannot distinguish the moth's echo from those of the surroundings.

In its evolution, the vertebrate ear originated as a balance organ, the **labyrinth.** In all jawed vertebrates, from fishes to mammals, the labyrinth has a similar structure, consisting of two small chambers called the **saccule** and the **utricle,** and three **semicircular canals.** In fish the base of the saccule is extended into a tiny pocket (the **lagena**) that, during the evolution of the vertebrates, developed into the hearing receptor of tetrapods. With continued elaboration and elongation in the birds and mammals, the fingerlike lagena evolved into the **cochlea.**

The human ear (Figure 36-24) is representative of mammalian ears. The outer, or external, ear collects the sound waves and funnels them through the **auditory canal** to the eardrum or **tympanic membrane** lying next to the middle ear. The middle ear is an air-filled chamber containing a remarkable chain of three tiny bones, or ossicles, known as the **malleus** (hammer), **incus** (anvil), and **stapes** (stirrup), named because of their fancied resemblance to these objects. These bones conduct the sound waves across the middle ear (Figure 36-24B). The bridge of bones is so arranged that the force of sound waves pushing against the tympanic membrane is amplified as much as 90 times where the stapes contacts the **oval window** of the inner ear. Muscles attached to the middle ear bones contract when the ear receives very loud noises, providing the inner ear some protection from damage. The middle ear connects with the pharynx by means of the **eustachian tube,** which permits pressure equalization on both sides of the tympanic membrane.

The origin of the three tiny bones of the mammalian middle ear—the malleus, incus, and stapes—is one of the most extraordinary and well-documented transitions in vertebrate evolution. Amphibians, reptiles, and birds have a single rodlike ear ossicle, the stapes (also called the columella), which originated as a jaw support (the hyomandibular) as seen in fishes (see Figure 26-16, p. 495). With the evolution of the earliest tetrapods, the braincase became firmly sutured to the skull, and the hyomandibular, no longer needed to brace the jaw, became converted into the stapes. In a similar way, the two additional ear ossicles of the mammalian middle ear—the malleus and incus—originated from parts of the jaw of the early vertebrates. The quadrate bone of the reptilian upper jaw became the incus, and the articular bone of the lower jaw became the malleus. The homology of reptilian jaw bones to mammalian ear bones is clearly documented in fossil record and in the embryological development of mammals.

Within the inner ear is the organ of hearing, the **cochlea** (Gr. *cochlea,* snail's shell), which is coiled in mammals, making two and one-half turns in humans (Figure 36-24B). The cochlea is divided longitudinally into three tubular canals running parallel with one another. This relationship is indicated in Figure 36-25, in which the cochlea is shown stretched out. These canals become progressively smaller from the base of the cochlea to the apex. One of these canals is called the **vestibular canal;** its base is closed by the oval window. The **tympanic canal,** which is in communication with the vestibular canal at the tip of the cochlea, has its base closed by the **round window.** Between these two canals is the **cochlear duct,** which contains the **organ of Corti,** the actual sensory apparatus. Within the organ of Corti are fine rows of hair cells that run lengthwise from the base to the tip of the cochlea (Figure 36-24C). There are at least 24,000 hair cells in the human ear. The 80-100 "hairs" on each cell are actually microvilli and a single large cilium which project into the endolymph of the cochlear canal. Each cell is connected with neurons of the auditory nerve. The hair cells rest on the **basilar membrane,** which separates the tympanic canal and cochlear duct, and they are covered by the **tectorial membrane,** lying directly above them.

When a sound wave strikes the ear, its energy is transmitted through the ossicles of the middle ear to the oval window, which oscillates back and forth, driving the fluid of the vestibular and tympanic canals before it. Because these fluids are noncompressible, an inward movement of the oval window produces a corresponding outward movement of the round window. The fluid oscillations also cause the basilar membrane with its hair cells to vibrate simultaneously.

According to the **place hypothesis of pitch discrimination** formulated by Georg von Békésy, different areas of the basilar membrane respond to different frequencies; that is, for every sound frequency, there is a specific "place" on the basilar membrane where the hair cells respond to that frequency. The initial displacement of the basilar membrane starts a wave traveling down the membrane, much as flipping a rope at one end starts a wave moving down the rope (Figure 36-26). The displacement wave increases in amplitude as it moves from the oval window toward the apex of the cochlea, reaching a maximum at the region of the basilar membrane where the natural frequency of the membrane corresponds to the sound frequency. Here, the membrane vibrates with such ease that the energy of the traveling wave is completely dissipated. Hair cells in that region are stimulated and the impulses conveyed to the fibers of the auditory nerve. Those impulses that are carried by

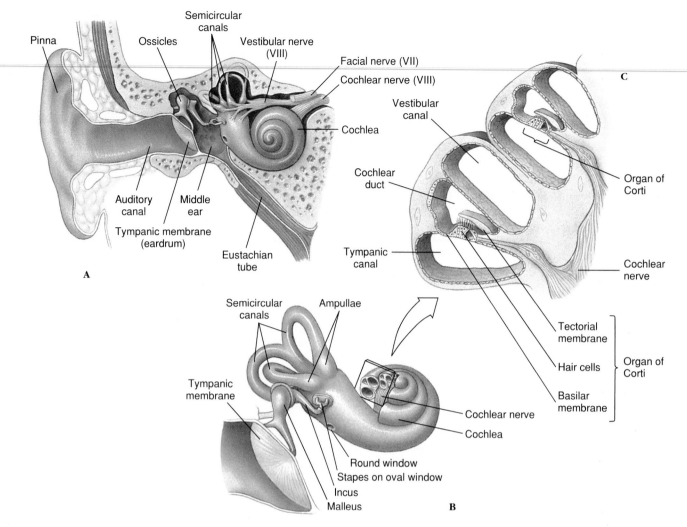

Figure 36-24

Human ear. **A,** Longitudinal section showing external, middle, and inner ear. **B,** Enlargement of middle ear and inner ear. The cochlea of the inner ear has been opened to show the arrangement of canals within, **C,** Enlarged cross section of cochlea showing the organ of Corti.

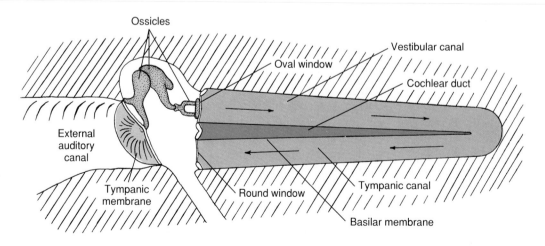

Figure 36-25

Mammalian ear as it would appear with the cochlea stretched out. Sound waves transmitted to the oval window produce vibration waves that travel down the basilar membrane. High-frequency vibrations cause the membrane to resonate at the end near the oval window before dying; low-frequency tones travel farther down the basilar membrane.

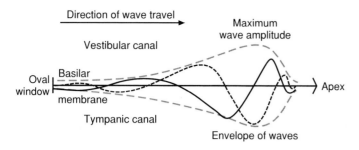

Figure 36-26
Traveling waves along the basilar membrane. The oval window is at left, and the cochlear apex at right. The two wave formations (*solid* and *dashed lines*) occur at separate instants of time. The curves in color represent the extreme displacements of the membrane because of traveling waves.

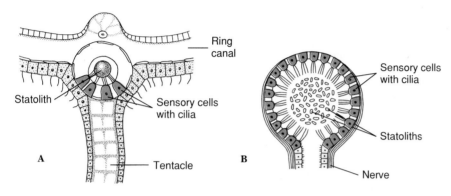

Figure 36-27
Types of statocysts, static balance organs of invertebrates. **A,** Statocyst of the medusa of the hydrozoan *Obelia.* **B,** Statocyst of the bivalve mollusc *Pecten.*

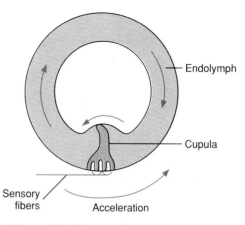

Figure 36-28
Semicircular canal, showing bending of cupula on the ampulla during angular acceleration.

certain fibers of the auditory nerve are interpreted by the hearing center as particular tones. The **loudness** of a tone depends on the number of hair cells stimulated, whereas the **timbre,** or quality, of a tone is produced by the pattern of the hair cells stimulated by sympathetic vibration. This latter characteristic of tone enables us to distinguish between different human voices and different musical instruments, although the notes in each case may be of the same pitch and loudness.

Sense of Equilibrium

In invertebrates, specialized sense organs for monitoring gravity and low-frequency vibrations often appear as **statocysts.** Each is a simple sac lined with hair cells and containing a heavy calcareous structure, the **statolith** (Figure 36-27). The delicate, hairlike filaments of the sensory cells are activated by the shifting position of the statolith when the animal changes po-

sition. Statocysts are found in many invertebrate phyla from radiates to arthropods. They are all built on similar principles.

In vertebrates, the organ of equilibrium is the **labyrinth.** It consists of two small chambers **(saccule** and **utricle)** and three **semicircular canals** (Figure 36-24B). The utricle and saccule are static balance organs that, like invertebrate statocysts, give information about the position of the head or body with respect to the force of gravity. As the head is tilted in one direction or another, stony accretions press on different groups of hair cells; these send nerve impulses to the brain, which interprets this information with reference to head position.

The semicircular canals of vertebrates are designed to respond to **rotational acceleration** and are relatively insensitive to linear acceleration. The three semicircular canals are at right angles to each other, one for each axis of rotation. They are filled with fluid

(endolymph), and within each canal is a bulblike enlargement, the **ampulla,** which contains hair cells. The hair cells are embedded in a gelatinous membrane, the **cupula,** which projects into the fluid. When the head rotates, fluid in the canal at first tends not to move because of inertia. Since the cupula is attached, its free end is pulled in the direction opposite to the direction of rotation (Figure 36-28). Bending of the cupula distorts and excites the hair cells embedded in it, and this stimulation increases the discharge rate over the afferent nerve fibers leading from the ampulla to the brain. This produces the sensation of rotation. Since the three canals of each ear are in different planes, acceleration in any direction stimulates at least one ampulla.

PHOTORECEPTION: VISION

Light-sensitive receptors are called **photoreceptors.** These receptors range all the way from simple light-sensitive cells scattered randomly on the body surface of many invertebrates (dermal light sense) to the exquisitely developed camera-type eye of vertebrates. Eyespots of astonishingly advanced organization appear even in some protozoa. That of the dinoflagellate *Nematodinium* bears a lens, a light gathering chamber, and a photoreceptive pigment cup—all developed within a single-celled organism (Figure 36-29). The dermal light receptors of many invertebrates are of

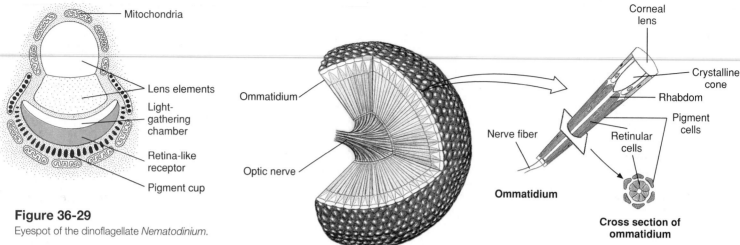

Figure 36-29
Eyespot of the dinoflagellate *Nematodinium*.

Compound eye

Figure 36-30
Compound eye of an insect. A single ommatidium is shown enlarged at right.

much simpler design. They are far less sensitive than optic receptors, but they are important in locomotory orientation, pigment distribution in chromatophores, photoperiodic adjustment of reproductive cycles, and other behavior changes.

More highly organized eyes, many capable of excellent image formation, are based on one or the other of two different principles: a single-lens, camera-type eye such as those of cephalopod molluscs and vertebrates; or a multifaceted (compound) eye as in arthropods. Arthropod **compound eyes** are composed of many independent visual units called **ommatidia** (Figure 36-30). The eye of a bee contains about 15,000 of these units, each of which views a separate narrow sector of the visual field. Such eyes form a mosaic of images of varying brightness from the separate units. Resolution (that is, the ability to see objects sharply) is poor as compared with that of a vertebrate eye. A fruit fly, for example, must be closer than 3 cm to see another fruit fly as anything but a single spot. However, the compound eye is especially well suited to detecting motion, as anyone knows who has tried to swat a fly.

The eyes of certain annelids, molluscs, and all vertebrates are built like a camera—or rather we should say that a camera is modeled somewhat after the vertebrate eye. The camera-type eye contains in the front a light-tight chamber and lens system, which

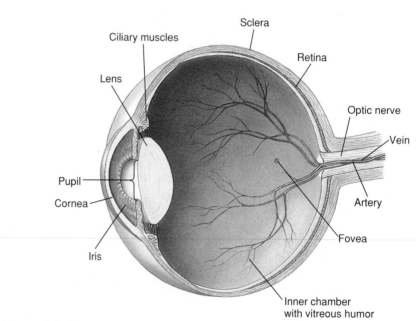

Figure 36-31
Structure of the human eye.

focuses an image of the visual field on a light-sensitive surface (the retina) in the back (Figure 36-31). Because eyes and cameras are based on the same laws of optics, we can wear eyeglasses to correct optical defects.

The spherical eyeball is built of three layers: (1) a tough outer white **sclera** that provides support and protection, (2) middle **choroid coat,** containing blood vessels for nourishment, and (3) light-sensitive **retina** (Figure 36-31). The **cornea** is a transparent anterior modification of the sclera. A circular, pigmented curtain, the **iris,** regulates the size of the light opening, the **pupil.** Just behind the iris is the **lens,** a transparent, elastic oval disc that, with the aid of **ciliary muscles,** can alter the curvature of

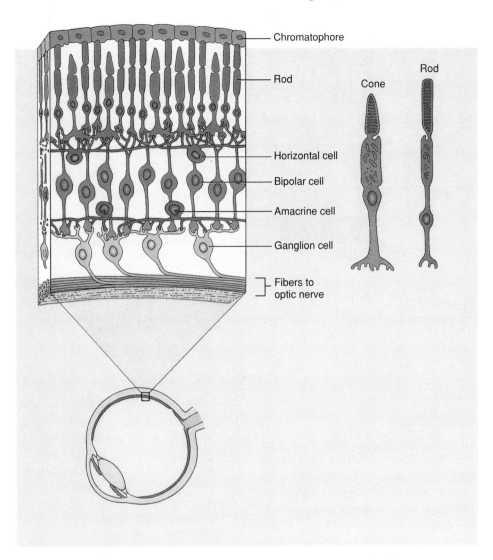

Chromatophore

Rod

Cone

Rod

Horizontal cell

Bipolar cell

Amacrine cell

Ganglion cell

Fibers to optic nerve

Figure 36-32
Structure of the primate retina, showing the organization of intermediate neurons that connect the photoreceptor cells to the ganglion cells of the optic nerve.

The **fovea centralis,** the region of keenest vision, is located in the center of the retina, in direct line with the center of the lens and cornea. It contains only cones, a vertebrate specialization for diurnal (daytime) vision. The acuity of an animal's eyes depends on the density of cones in the fovea. The human fovea and that of a lion contain approximately 150,000 cones per square millimeter. But many water and field birds have up to 1 million cones per square millimeter. Their eyes are as good as our eyes would be if aided by eight-power binoculars.

At the peripheral parts of the retina only rods are found. Rods are high-sensitivity receptors for dim light. At night, the cone-filled fovea is unresponsive to low levels of light and we become functionally color blind ("at night all cats are gray"). Under nocturnal conditions, the position of greatest visual acuity is not at the center of the fovea but at its edge. Thus it is easier to see a dim star at night by looking slightly to one side of it.

One of the several marvels of the vertebrate eye is its capacity to compress the enormous range of light intensities presented to it into a narrow range that can be handled by the optic nerve fibers. Light intensity between a sunny noon and starlight differs more than 10 billion to 1. Rods quickly saturate with high light intensity, but the cones do not; they shift their operating range with changing ambient light intensity so that a high-contrast image is perceived over a broad range of light conditions. This is made possible by complex interactions among the network of nerve cells that lie between the cones and the ganglion cells that generate the retinal output to the brain.

Chemistry of Vision

Each rod contains a light-sensitive pigment known as **rhodopsin.** Each rhodopsin molecule consists of a large protein, **opsin,** which behaves as an enzyme, and a small carotenoid molecule, **retinal,** a derivative of vitamin A. When a quantum of light strikes a rod

the lens and bend the rays to focus them on the retina. In terrestrial vertebrates the cornea actually does most of the bending of light rays, whereas the lens adjusts focus for near and far objects. Between the cornea and the lens is the **outer chamber** filled with watery **aqueous humor;** between the lens and the retina is the much larger **inner chamber** filled with viscous **vitreous humor.**

The retina is composed of several cell layers (Figure 36-32). The outermost layer, closest to the sclera, consists of pigment cells. Adjacent to this layer are the photoreceptors, the **rods** and **cones.** Approximately 125 million rods and 1 million cones are present in each human eye. Cones are primarily concerned with color vision in ample light; rods, with colorless vision in dim light. Next is a network of **intermediate neurons** (bipolar, horizontal, and amacrine cells) that process and relay visual information from the photoreceptors to the **ganglion cells** whose axons form the optic nerve. The network permits much convergence, especially for rods. Information from several hundred rods may converge on a single ganglion cell, an adaptation that greatly increases the effectiveness of rods in dim light. Cones show very little convergence. By coordinating activities between different ganglion cells, and adjusting the sensitivities of bipolar cells, the horizontal and amacrine cells improve overall contrast and quality of the visual image.

and is absorbed by the rhodopsin molecule, retinal is isomerized, that is, the shape of the molecule is changed. This triggers the enzymatic activity of opsin, which sets in motion a biochemical sequence of several steps. This complex sequence, the details of which have only recently been worked out (Stryer, 1987), behaves as an excitatory cascade that vastly amplifies the energy of a single photon to generate a nerve impulse in the rod.

The amount of intact rhodopsin in the retina depends on the intensity of light reaching the eye. The dark-adapted eye contains much rhodopsin and is very sensitive to weak light. Conversely in the light-adapted eye, most of the rhodopsin is broken down into retinal and opsin. It takes approximately half an hour for the light-adapted eye to fully accommodate to darkness, while the rhodopsin level is gradually built up.

The composition of the cone pigments is not yet completely clear, but they probably are similar to rhodopsin, containing retinal combined with a special protein, **cone opsin.** Cones function to perceive color and require 50 to 100 times more light for stimulation than do rods. Consequently, night vision is almost totally rod vision. Unlike humans, who have both day and night vision, some vertebrates specialize for one or the other. Strictly nocturnal animals, such as bats and owls, have pure rod retinas. Purely diurnal forms, such as the common gray squirrel and some birds, have only cones. They are, of course, virtually blind at night.

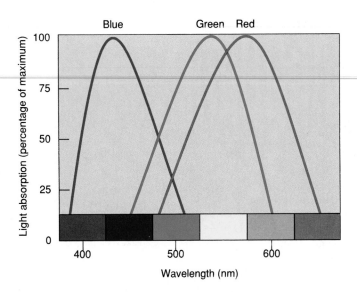

Figure 36-33
The absorption spectrum of human vision. Three types of cones absorb maximally at 430 nm (blue cones), 540 nm (green cones), and 575 nm (red cones).

Color Vision

In 1802 the English physician and physicist Thomas Young speculated that we see color by the relative excitation of three kinds of photoreceptors: one each for red, green, and blue. It took many years, but in the 1960s Young's prescient hypothesis was eventually supported through the combined work of several groups of researchers. The blue cones absorb the most light at 430 nm, the green cones at 540 nm, and the red cones at 575 nm (Figure 36-33). Colors are perceived by comparing the levels of excitation of the three different kinds of cones. For example, a light having a wavelength of 530 nm would excite the green cones 95%, the red cones about 70%, and the blue ones not at all. This comparison is made both in nerve circuits in the retina and in the visual cortex of the brain, and the brain interprets this combination as green.

Color vision is present in some members of all vertebrate groups with the possible exception of the amphibians. Bony fishes and birds have particularly good color vision. Surprisingly, most mammals are color blind; exceptions are primates and a few other species such as squirrels.

Summary

The nervous system is a rapid communication system that interacts continuously with the endocrine system in the control and coordination of body function. The basic unit of nervous integration in all animals is the neuron, a highly specialized cell designed to conduct self-propagating impulses, called action potentials, to other cells. Action potentials are transmitted from one nerve to another across synapses which may be either electrical or chemical. The thin gap between nerves at chemical synapses is bridged by a chemical transmitter, usually acetylcholine, which is released from the synaptic knob.

The simplest organization of neurons into a system is the nerve net of cnidarians, basically a plexus of nerve cells that, with additions, is the basis of the nervous systems of several invertebrate phyla. With the appearance of ganglia (nerve centers) in the bilateral flatworms, nervous systems differentiated into central and peripheral divisions. In vertebrates, the central nervous system consists of the brain and spinal cord. Fishes and amphibians have a three-part linear brain, whereas in mammals, the cerebral cortex has become a vastly enlarged multicomponent structure that has assumed the most important integrative activities of the nervous system. It completely overshadows the ancient brain, which is consigned to the role of relay center and to serving numerous unconscious but nonetheless vital functions such as breathing, blood pressure, and heart rate.

In humans the left cerebral hemisphere is usually specialized for language and mathematical skills while the right hemisphere is specialized for visual-spatial and musical skills.

The peripheral nervous system connects the central nervous system to receptors and effector organs. It is divided broadly into an afferent system, which conducts sensory signals to the central nervous system, and an efferent system, which conveys motor impulses to effector organs. The autonomic nervous system is a motor system with its own separate set of fibers. It is subdivided into anatomically distinct sympathetic and parasympathetic systems, each of which sends fibers to most body organs. Generally the sympathetic system governs excitatory activities and the parasympathetic system governs maintenance and restoration of body resources.

Sensory organs are receptors designed especially to respond to internal or environmental change. The most primitive and ubiquitous sense is chemoreception. Chemoreceptors may be contact receptors such as the vertebrate sense of taste, or distance receptors such as smell, which detects airborne molecules. Most hypotheses of olfaction postulate some kind of molecular integration between odorant and receptor; the resulting signals are expressed as a spatial pattern that is interpreted by the brain as a particular odor.

The receptors for touch, pain, equilibrium, and hearing are all mechanical force receptors. Touch and pain receptors are characteristically simple structures, but hearing and equilibrium are highly specialized senses based on special hair cells that respond to mechanical deformation. Sound waves received by the ear are mechanically amplified and transmitted to the inner ear where different areas of the cochlea respond to different sound frequencies. Equilibrium receptors, also located in the inner ear, consist of two saclike static balance organs and three semicircular canals that detect movement.

Vision receptors (photoreceptors) are associated with special pigment molecules that photochemically decompose in the presence of light and, in doing so, trigger nerve impulses in optic fibers. The advanced compound eye of arthropods is especially well suited to detecting motion in the visual field. Vertebrates have a camera eye with focusing optics. The photoreceptor cells of the retina are of two kinds: rods, designed for high sensitivity with dim light, and cones, designed for color vision in daylight. Cones predominate in the fovea centralis of the human eye, the area of keenest vision. Rods are more abundant in the peripheral areas of the retina.

Review Questions

1. Define the following terms: neuron, axon, dendrite, myelin sheath, afferent neuron, efferent neuron, association neuron.
2. What are glial cells in the nervous system and what functions do they perform?
3. Explain how the permeability properties of a nerve fiber give rise to the resting (equilibrium) potential of the fiber. What is the importance of the sodium pump in maintaining the resting potential?
4. What ionic and electrical changes occur during the passage of an action potential along a nerve fiber?
5. Explain the different ways in which invertebrates and vertebrates have achieved high velocities for the conduction of nerve impulses. Can you suggest why the invertebrate solution would not be suitable for the homeothermic birds and mammals?
6. Describe the microstructure of a chemical synapse. Summarize what happens when an action potential arrives at a synapse.
7. Describe the cnidarian (radiate) nervous system. How is the tendency toward centralization of the nervous system manifested in flatworms, annelids, molluscs, and arthropods?
8. How does the vertebrate spinal cord differ morphologically from the nerve cord of invertebrates?
9. Name the components of a typical reflex arc. What is the difference between a reflex arc and a reflex act?
10. Name the major functions associated with the following brain structures: medulla, cerebellum, tectum, thalamus, hypothalamus, cerebrum, limbic system.
11. What functional activities are associated with the left and the right hemispheres of the cerebral cortex?
12. What is the autonomic nervous system and what activities does it perform that distinguish it from the central nervous system? Why can the autonomic nervous system be described as a "two-neuron" system?
13. Give the meaning of the statement, "The idea that all sense organs behave as biological transducers is a uniting concept in sensory physiology."
14. Chemoreception in vertebrates and insects is mediated through the clearly distinguishable senses of taste and smell. Contrast these two senses in humans in terms of anatomical location and nature of the receptors and sensitivity to chemical molecules.
15. Explain how the ultrasonic detectors of certain nocturnal moths are adapted to help them escape an approaching bat.
16. Outline the place theory of pitch discrimination as an explanation of the human ear's ability to distinguish between sounds of different frequencies.
17. Explain how the semicircular canals of the ear are designed to detect rotation of the head in any directional plane.
18. Explain what happens when light strikes a dark-adapted rod that leads to the generation of a nerve impulse. What is the difference between rods and cones in their sensitivity to light?

Selected References

See also general references for Part IV, p.774.

Axel, R. 1995. The molecular logic of smell. Sci. Am. **273**:154–159 (Oct.). *Recent research has revealed a surprisingly large family of genes that code for odor molecules. This and other findings help to illuminate how the nose and brain may perceive scents.*

Bullock, T. H., R. Orkland, and A. Grinnell. 1977. Introduction to nervous systems. San Francisco, W. H. Freeman and Company. *Excellent comparative treatment.*

Dunant, Y., and M. Israel, 1985. The release of acetylcholine. Sci. Am. **252**:58–66 (Apr.). *Recent studies have altered prevailing views of the events at a synapse during impulse transmission.*

Freeman, W. J. 1991. The physiology of perception. Sci. Am. **264**:78–85 (Feb.) *How the brain transforms sensory messages almost instantly into conscious perceptions.*

Hudspeth, A. J. 1983. The hair cells of the inner ear. Sci. Am. **248**:54–64 (Jan.). *How these biological transducers work.*

Jacobson, M. 1993. Foundations of neuroscience. New York, Plenum Press. *The historical development of neuroscience and its outstanding personages—and the dangers of hero worship of individual neuroscientists.*

Nathan, P. 1982. The nervous system, ed. 2. Oxford, England, Oxford University Press. *One of the best of several semipopular accounts of the nervous system.*

Nathans, J. 1989. The genes for color vision. Sci. Am. **260**:42–49. (Feb.). *The recent isolation of genes that encode color-detecting proteins of the human eye provide clues about the evolution of color vision.*

Snyder, S. H. 1985. The molecular basis of communication between cells. Sci. Am. **253**:132–141 (Oct.). *Describes the different actions of neurotransmitters.*

Stebbins, W. C. 1983. The acoustic sense of animals. Cambridge, Massachusetts, Harvard University Press. *Broadly comparative introduction to the physics, physiology, natural history, and evolution of hearing.*

Stevens, C. F. 1979. The neuron. Sci. Am. **241**:54–65 (Sept.). *Its function is detailed.*

Stryer, L. 1987. The molecules of visual excitation. Sci. Am. **257**:42–50 (July). *Describes the cascade of molecular events following light absorption by a rod cell that leads to a nerve signal.*

Werblin, F. S. 1973. The control of sensitivity in the retina. Sci. Am. **228**:70-79 (Jan.). *Studies of neuron interactions in the retina help to explain its versatility over wide-ranging light conditions.*

37

Chemical Coordination

Endocrine System

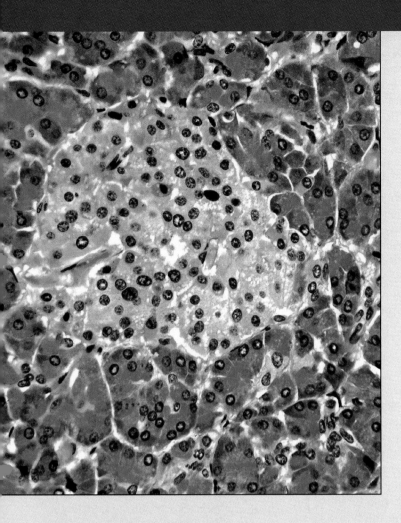

The Crucial Experiment

The birth date of endocrinology as a science is usually given as 1902, the year two English physiologists, W. H. Bayliss and E. H. Starling (Figure 37-1), demonstrated the action of a hormone in a classic experiment that is still considered a model in the use of the scientific method. Bayliss and Starling were interested in determining how the pancreas secreted its digestive juice into the small intestine at the proper time of the digestive process. Only one year earlier, the Russian physiologist Ivan Pavlov had demonstrated that the smell, taste, and thought of food provokes the release of gastric and pancreatic juice. Pavlov suggested that acidic food entering the intestine triggered a nervous reflex that released pancreatic juice. To test this hypothesis, Bayliss and Starling cut away all of the nerves serving a tied-off loop of the small intestine of an anesthetized dog, leaving the isolated loop connected to the body only by its circulation. Injecting acid into the nerveless loop, they saw a pronounced flow of

pancreatic juice. Clearly, Pavlov was wrong. Rather than a nervous reflex, some chemical messenger had circulated from the intestine to the pancreas, causing the pancreas to secrete. Yet acid itself could not be the factor because it had no effect when injected directly into the circulation.

Bayliss and Starling then designed the crucial experiment that was to usher in the new science of endocrinology. Suspecting that the chemical messenger originated in the mucosal lining of the intestine, they next prepared an extract of scrapings from the mucosa, injected it into the dog's circulation, and were rewarded with an abundant flow of pancreatic juice. They named the messenger present in the intestinal mucosa *secretin*. Later Starling coined the term *hormone* to describe all such chemical messengers, since he correctly surmised that secretin was only the first of many hormones awaiting discovery. ■

A

B

Figure 37-1

Founders of endocrinology. **A,** Sir William H. Bayliss (1860 to 1924). **B,** Ernest H. Starling (1866 to 1927).

From J. F. Fulton & L. G. Wilson, Selected Readings in the History of Physiology, *1966. Courtesy of Charles C. Thomas, Publisher, Springfield, Illinois.*

The endocrine system, the second great integrative system controlling the body's activities, communicates by chemical messengers called **hormones** (Gr. *hormōn,* to excite). Hormones are chemical compounds that are released into the blood in small amounts and transported by the circulatory system throughout the body to distant **target cells** where they initiate physiological responses.

Many hormones are secreted by **endocrine glands,** small, well-vascularized *ductless* glands composed of groups of cells arranged in cords or plates. Since the endocrine glands have no ducts, their only connection with the rest of the body is by the bloodstream; they must capture their raw materials from the extensive blood supply they receive and secrete their finished hormonal products into it. **Exocrine glands,** in contrast, are provided with ducts for discharging their secretions onto a free surface. Examples of exocrine glands are sweat glands and sebaceous glands of skin, salivary glands, and the various enzyme-secreting glands lining the walls of the stomach and intestine.

The classic definitions of hormones and endocrine glands given above, like so many other generalizations in biology, may have to be altered as new information appears. Some hormones, such as certain neurosecretions, may never enter the general circulation at all. Furthermore, there is good evidence that many hormones, such as insulin, are synthesized in minute amounts in a variety of nonendocrine tissues (nerve cells, for example), and some, such as cytokines, are secreted by cells of the immune system (p. 677). Such hormones may function as local **tissue factors,** substances that stimulate cell growth or some biochemical process. Most hormones, however, are blood borne and therefore diffuse into every tissue space in the body. This is quite unlike the discrete action of the nervous system with its network of cablelike nerve fibers that selectively send messages to specific points.

Compared with the nervous system, the endocrine system is slow acting because of the time required for a hormone to reach the appropriate tissue, cross the capillary endothelium, and diffuse through tissue fluid to, and sometimes into, cells. The minimum response time is seconds and may be much longer. Hormonal responses in general are long lasting (minutes to days) whereas those under nervous control are short-term (milliseconds to minutes). We expect to find endocrine control where a sustained effect is required, as in many metabolic and growth processes, or where some concentration or rate of secretion must be maintained at a particular level. Despite such differences, the nervous and endocrine systems function as a single, interdependent system. There is no sharp separation between the two. Endocrine glands often receive directions from the brain. Conversely, several hormones act on the nervous system and may significantly affect many kinds of animal behavior.

All hormones are low-level signals. Even when an endocrine gland is secreting maximally, the hormone is so greatly diluted by the large volume of blood it enters that its plasma concentration seldom exceeds 10^{-9} M (or one billionth of a 1 M concentration). Some target cells respond to plasma concentrations of hormone as low as 10^{-12} M. Since hormones have far-reaching and often powerful influences on cells, it is evident that their effects are vastly amplified at the cellular level.

MECHANISMS OF HORMONE ACTION

The widespread distribution of hormones in the body makes it possible for certain hormones, such as the growth hormone of the pituitary gland, to affect most, if not all, cells during specific stages of cellular differentiation. Other hormones produce highly specific responses only in certain target cells and at certain times. Such specificity is made possible by **receptor molecules** on or in the target cells. A hormone will engage only those cells that display the receptor that, by virtue of its specific molecular shape, will bind with the hormone molecule. Other cells are insensitive to the hormone's presence because they lack the specific receptors. Hormones act through two kinds of receptors: **membrane-bound receptors** and **nuclear receptors.**

MEMBRANE-BOUND RECEPTORS AND THE SECOND MESSENGER CONCEPT

Many hormones, such as most amino acid derivatives, and the peptide hormones that are too large to pass cell membranes, bind to receptor sites present on the surface of target cell membranes. The combination of hormone and receptor forms a complex that triggers the release of a molecule from the inner surface of the membrane.

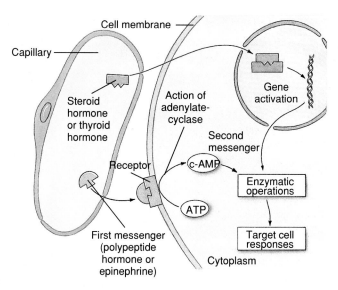

Figure 37-2

Mechanisms of hormone action. Peptide hormones and epinephrine act through cyclic AMP. The combination of hormone with a membrane receptor stimulates the enzyme adenylate cyclase to catalyze the formation of cyclic AMP (second messenger). Steroid hormones and thyroid hormones penetrate the cell membrane to combine with a nuclear receptor that activates gene transcription.

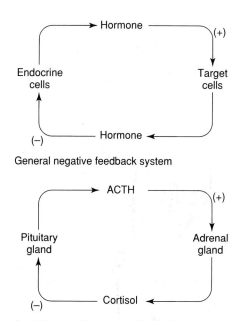

General negative feedback system

Specific example of a negative feedback system

Figure 37-3

Negative feedback systems.

The hormone thus behaves as a **first messenger** that causes the release of a **second messenger** in the cytoplasm. At least six different molecules have been identified as second messengers, but the most important is **cyclic AMP.** Cyclic AMP is formed when the binding of hormone to receptor activates the enzyme **adenylate cyclase** in the target cell membrane. This enzyme converts ATP to cyclic AMP (the second messenger), which then acts to modify the direction and rate of cytoplasmic reactions (Figure 37-2). Since many molecules of cyclic AMP may be manufactured after a single hormone molecule has been bound, the message is amplified, perhaps many thousands of times.

Cyclic AMP mediates the actions of many peptide hormones, including parathyroid hormone, glucagon, adrenocorticotropic hormone (ACTH), thyrotropic hormone (TSH), melanophore-stimulating hormone (MSH), and vasopressin. It also mediates the action of epinephrine (also called adrenaline), an amino acid derivative.

NUCLEAR RECEPTORS

Unlike the peptide hormones and epinephrine, which are much too large to pass through cell membranes, the **steroid hormones** (for example, estrogen, testosterone, and aldosterone), are lipid-soluble molecules that readily diffuse through cell membranes. Once inside the cytoplasm, steroid hormones bind selectively to receptor molecules found in the nucleus of the target cells. The hormone-receptor complex then activates specific genes. As a result, gene transcription is increased, and messenger RNA molecules are synthesized on specific sequences of DNA. Moving from the nucleus into the cytoplasm, the newly formed mRNA initiates the formation of key enzymes, thus setting in motion the hormone's observed effect (Figure 37-2). Thyroid hormones and the insect-molting hormone, ecdysone, also act through nuclear receptors.

As compared with the peptide hormones that act *indirectly* through second messenger systems, steroids have a *direct* effect on protein synthesis because they combine with a nuclear receptor that activates specific genes.

HOW SECRETION RATES OF HORMONES ARE CONTROLLED

Hormones influence cellular functions by altering the rates of many different biochemical processes. Many affect enzyme activity and thus alter cellular metabolism, some change membrane permeability, some regulate the synthesis of cellular proteins, and some stimulate the release of hormones from other endocrine glands. Because these are all dynamic processes that must adapt to changing metabolic demands, they must be regulated, not merely activated, by the appropriate hormones. This regulation is achieved by precisely controlled release of a hormone into the blood. However, the concentration of a hormone in the plasma depends on two factors: its rate of secretion and the rate at which it is inactivated and removed from the circulation. Consequently, if secretion is to be correctly controlled, an endocrine gland requires information about the level of its own hormone(s) in the plasma.

Many hormones, especially those of the pituitary gland, are controlled by negative feedback systems that operate between the glands secreting the hormones and the target cells (Figure 37-3). A feedback pattern is one in which the output is constantly compared with a set point, like a thermostat. For example, ACTH, secreted by the pituitary, stimulates the adrenal gland (the target cells) to secrete cortisol. As the cortisol level in the plasma rises, it acts on, or "feeds back" on, the pituitary gland to inhibit the release of ACTH. Thus any deviation from the set point (a specific plasma level of cortisol) leads to corrective action in the opposite direction (Figure 37-3). Such a negative feedback system is highly effective in preventing extreme oscillations in hormonal output. However, hormonal feedback systems are more complex than a rigid "closed-loop" system such as the thermostat that controls the central heating system in a house, because hormonal feedback may be altered by input from the nervous system or by metabolites or other hormones.

INVERTEBRATE HORMONES

In many metazoan phyla, the principal source of hormones is **neurosecretory cells,** specialized nerve cells capable of synthesizing and secreting hormones. Their products, called neurosecretions or neurosecretory hormones, are discharged directly into the circulation. Neurosecretion is an ancient physiological activity. Because it serves as a crucial link between the nervous and endocrine systems, it is believed that hormones first evolved as nerve cell secretions. Later, nonnervous endocrine glands appeared, especially among the vertebrates, but remained chemically linked to the nervous system by the neurosecretory hormones.

Neurosecretory hormones occur in all metazoan groups. The most extensively studied neurosecretory process is the control of development and metamorphosis of insects. In insects, as in other arthropods, growth is a series of steps in which the rigid, nonexpansible exoskeleton is periodically discarded and replaced with a new, larger one. Most insects undergo a process of metamorphosis (p. 417), in which a series of juvenile stages, each requiring the formation of a new exoskeleton, end with a molt. In some orders the

Juvenile hormone of silkworm

Molting hormone (α-ecdysone) of silkworm

change to the adult form is gradual. In others the adult is separated from the larval stages by a quiescent form, the pupa, and the change to the adult is abrupt. Hormonal control of both types is the same.

Insect physiologists have discovered that molting and metamorphosis are controlled by the interaction of two hormones, one favoring growth and the differentiation of adult structures and the other favoring the retention of juvenile structures. These two hormones are the **molting hormone** (also called **ecdysone** [ek'duh-sone; ek-die'sone]), produced by the prothoracic gland, and the **juvenile hormone,** produced by the corpora allata (Figure 37-4). The structure of both hormones has been determined. Extraction from 1000 kg (about 1 ton) of silkworm pupae was required to show that the molting hormone is a steroid. The juvenile hormone has an entirely different structure.

The molting hormone is under the control of **brain hormone** (also called ecdysiotropin or prothoracicotropic hormone). This hormone is a polypeptide (molecular weight about 5000) that is produced by neurosecretory cells of the brain. It is transported by axons to the corpora allata where it is stored. At intervals during juvenile growth, the release of brain hormone into the blood stimulates the prothoracic gland to secrete molting hormone. Molting hormone appears to act directly on the chromosomes to set in motion the changes resulting in

a molt. The molting hormone favors the development of adult structures. It is held in check, however, by the juvenile hormone, which favors the development of juvenile characteristics. During juvenile life the juvenile hormone predominates and each molt yields another larger juvenile (Figure 37-4). Finally the output of juvenile hormone decreases, allowing the final metamorphosis to the adult stage.

The precise location of brain hormone in the brain of pupal tobacco hornworms was revealed by N. Agui by delicate microdissection. Using a human eyebrow hair, he was able to isolate the single cell in each brain hemisphere that contained brain hormone activity. Thus only two cells, each about 20 μm in diameter, produce this insect's total supply of ecdysiotropin. In an age when sophisticated instrumentation has removed much of the tedium (and some of the creativity) from research, it is refreshing to learn that certain biological mysteries succumb only to skillful use of the human hand.

Chemists have synthesized several potent analogs of the juvenile hormone, which hold great promise as insecticides. Minute quantities of these synthetic analogs induce abnormal final molts or prolong or block development. Unlike the usual chemical insecticides, they are highly specific and ecologically benign.

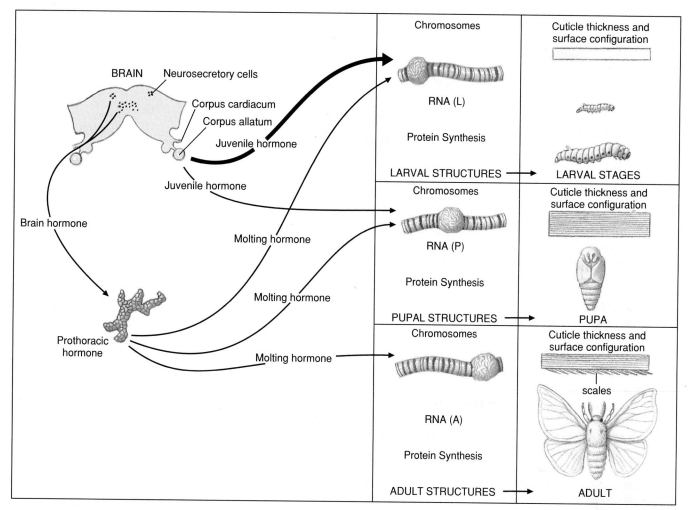

Figure 37-4

Endocrine control of molting in a moth, typical of insects having complete metamorphosis. Many moths mate in the spring or summer, and eggs soon hatch into the first of several larval stages, called instars. After the final larval molt, the last and largest larva (caterpillar) spins a cocoon in which it pupates. The pupa overwinters, and an adult emerges in the spring to start a new generation. Juvenile hormone and molting hormone interact to control molting and pupation. Many genes are activated during metamorphosis, as seen by the puffing of chromosomes (center column). Puffs form in sequence during successive molts. Changes in cuticle thickness and surface characteristics are shown at right.

Vertebrate Endocrine Glands and Hormones

In the remainder of this chapter we describe some of the best understood and most important of the vertebrate hormones. While the following discussion is limited principally to a brief overview of mammalian hormonal mechanisms (since laboratory mammals and humans have always been the objects of the most intensive research), we will point out some important differences in the functional roles of hormones among the different vertebrate groups.

Hormones of the Pituitary Gland, Hypothalamus, and Pineal Gland

The pituitary gland, or **hypophysis,** is a small gland (0.5 g in humans) lying in a well-protected position between the roof of the mouth and the floor of the brain (Figure 37-5). It is a two-part gland having a double embryological origin. The **anterior pituitary** (adenohypophysis) is derived embryologically from the roof of the mouth. The **posterior pituitary** (neurohypophysis) arises from a ventral portion of the brain, the **hypothalamus,** and is connected to it by a stalk, the **infundibulum.** Although

the anterior pituitary lacks any anatomical connection to the brain, it is functionally connected to it by a special portal circulatory system. A portal circulation is one that delivers blood from one capillary bed to another (Figures 37-5 and 37-6).

Anterior Pituitary

The anterior pituitary consists of an **anterior lobe** (pars distalis) and an **intermediate lobe** (pars intermedia) as shown in Figure 37-5. The anterior pituitary produces seven hormones whose functions have been clearly established. All but one are released by the anterior lobe.

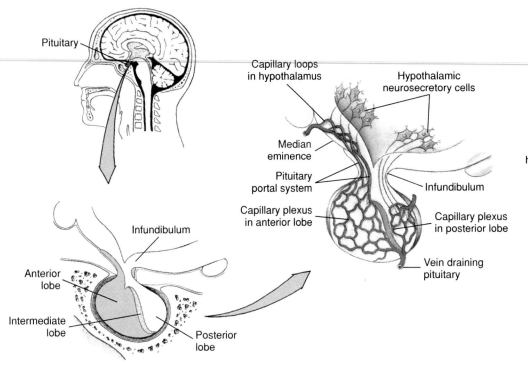

Figure 37-5

Human pituitary gland. The posterior lobe is connected directly to the hypothalamus by neurosecretory fibers. The anterior lobe is indirectly connected to the hypothalamus by a portal circulation (shown in red) beginning in the base of the hypothalamus and ending in the anterior pituitary.

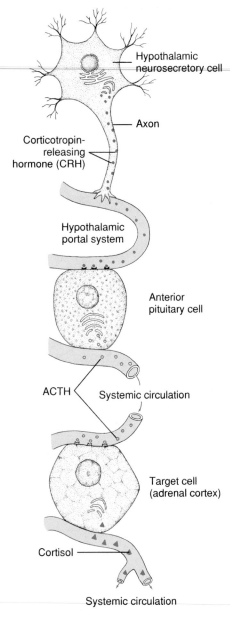

Figure 37-6

Relationship of hypothalamic, pituitary, and target-gland hormones. The hormone sequence controlling the release of cortisol from the adrenal cortex is used as an example.

Four hormones of the anterior pituitary are **tropic hormones** (pronounced trōpic, from the Greek *tropē,* to turn toward) that regulate other endocrine glands (Table 37-1). The **thyroid-stimulating hormone (TSH)** or **thyrotropin,** stimulates the production of thyroid hormones by the thyroid gland. Two of the tropic hormones are commonly called **gonadotropins** because they act on the gonads (ovary of the female, testis of the male). These are the **follicle-stimulating hormone (FSH)** and the **luteinizing hormone (LH).** FSH promotes egg production and the secretion of estrogen in females. In males FSH supports sperm production. LH induces ovulation and the secretion of the female sex hormones progesterone and estrogen. In males LH promotes the production of male sex hormones. LH is often called the interstitial cell stimulating hormone (ICSH) in males, but it is the same hormone chemically. The fourth tropic hormone, **adrenocortiotropic hormone**

(ACTH), increases the production and secretion of steroid hormones from the adrenal cortex.

Prolactin and the structurally related growth hormone **(GH)** are proteins. Prolactin is essential for preparing the mammary glands for lactation; after birth it is required for production of milk. Prolactin has also been implicated in parental behavior in a wide variety of vertebrates. Growth hormone (also called somatotropin) performs a vital role in governing body growth through its stimulatory effect on cellular mitosis and on synthesis of messenger RNA and protein, especially in new tissue of young vertebrates. If produced in excess, the growth hormone causes giantism. A deficiency of this hormone in the human child leads to dwarfism. Unlike the tropic hormones, prolactin acts directly on their target tissues rather than through other hormones. Growth hormone acts primarily through a polypeptide hormone, somatomedin, produced by the liver.

The only anterior pituitary hormone produced by the intermediate lobe (Figure 37-5) is the **melanophore-stimulating hormone (MSH).** In cartilaginous and bony fishes, amphibians, and reptiles, MSH is a direct-acting hormone that promotes dispersion of the pigment melanin within melanophores, causing darkening of the skin. In birds and mammals, MSH is produced by cells in the anterior

Table 37-1	Hormones of the Vertebrate Pituitary			
	Hormone	**Chemical Nature**	**Principal Action**	**Hypothalamic Controls**
Adenohypophysis Anterior lobe	Thyroid-stimulating hormone (TSH)	Glycoprotein	Stimulates thyroid hormone synthesis and secretion	TSH-releasing hormone (TRH)
	Follicle-stimulating hormone (FSH)	Glycoprotein	Female: follicle maturation and estrogen synthesis Male: stimulates sperm production	Gonadotropin-releasing hormone (GnRH)[1]
	Luteinizing hormone (LH, ICSH)	Glycoprotein	Female: stimulates ovulation, corpus luteum formation, estrogen and progesterone synthesis Male: testosterone secretion	Gonadotropin-releasing hormone (GnRH)[1]
	Prolactin (PL)	Protein	Mammary gland growth, milk synthesis in mammals; various reproductive and nonreproductive functions in lower vertebrates	Prolactin release-inhibiting factor (PIF) and (in birds) Prolactin releasing factor (PRF)
	Growth hormone (GH) (Somatotropin)	Protein	Stimulates growth	Growth hormone release-inhibiting hormone (Somatostatin) and Growth hormone releasing hormone (Somatocrinin)
	Adrenocorticotropic hormone (ACTH)	Polypeptide	Stimulates steroid hormone synthesis by adrenal cortex	Corticotropin-releasing hormone (CRH)
Intermediate lobe[2]	Melanophore-stimulating hormone (MSH)	Polypeptide	Pigment dispersion in melanophores of ectotherms; function unclear in endotherms.	MSH release-inhibiting factor (MIF)
Neurohypophysis Posterior lobe	Oxytocin	Octapeptide	In mammals stimulates milk ejection and uterine contractions	
	Antidiuretic hormone (Vasopressin)	Octapeptide	In mammals increases water reabsorption by kidney	
	Vasotocin	Octapeptide	Present in all vertebrate classes; in tetrapods increases water reabsorption	
	Mesotocin	Octapeptide	In lungfish, amphibians, reptiles decreases water reabsorption by kidney	

[1]One GnRH hormone regulates both FSH and LH according to recent experimental evidence.
[2]Birds and some mammals lack an intermediate lobe. In these forms, MSH is produced by the anterior lobe.

pituitary rather than the intermediate lobe (birds and some mammals lack an intermediate lobe altogether), but its physiological function remains unclear. MSH appears unrelated to pigmentation in the endotherms, although it will cause darkening of the skin in humans if injected into the circulation. Until recently, many endocrinologists thought MSH was a vestigial hormone, but interest has been rekindled by studies showing that it enhances memory and growth of the mammalian fetus. MSH and ACTH are derived from a precursor molecule called pro-opiomelanocortin that is transcribed and translated from a single gene.

Hypothalamus and Neurosecretion

Because of the strategic importance of the pituitary in influencing most of the hormonal activities in the body, the pituitary was once called the "master gland." This description is not appropriate, however, because the tropic hormones are regulated by a higher council, the neurosecretory centers of the hypothalamus. The hypothalamus is itself under the ultimate control of the brain. The hypothalamus contains groups of neurosecretory cells, which are specialized in giant nerve cells (Figure 37-6). These cells manufacture polypeptide hormones, called **releasing hormones or release inhibiting hormones** (or "factors"), which then travel down the nerve fibers to their endings in the median eminence. Here they enter a capillary network to complete their journey to the anterior pituitary by way of the pituitary portal system. The hypothalamic hormones then stimulate or inhibit the release of the various anterior pituitary hormones. Several hypothalamic releasing hormones have been discovered since the demonstration in 1955 of a corticotropin-releasing hormone (Table 37-1). There are one or more releasing hormones regulating each of the seven pituitary hormones. Several of the releasing hormones have now been isolated in pure state and characterized chemically (the identification and action of some of the hypothalamic hormones listed in Table 37-1 is still tentative). All are peptides.

Posterior Pituitary

The hypothalamus is also the source of two hormones of the posterior lobe of the pituitary (Table 37-1). They are formed in neurosecretory cells in the hypothalamus, then transported down the infundibular stalk and into the posterior lobe, ending in proximity to blood capillaries, which the hormones enter when released (see Figure 37-5). In a sense the posterior lobe is not a true endocrine gland, but a storage and release center for hormones man-ufactured entirely in the hypothalamus. The two posterior lobe hormones of mammals, oxytocin and vasopressin, are chemically very much alike. Both are polypeptides consisting of eight amino acids and are called octapeptides (Figure 37-7). These hormones are among the fastest acting hormones, since they are capable of producing a response within seconds of their release from the posterior lobe.

Oxytocin has two important specialized reproductive functions in adult female mammals. It stimulates contraction of uterine smooth muscles during parturition (birth of the young). In clinical practice, oxytocin is used to induce delivery during a difficult labor and to prevent uterine hemorrhage after birth. The second action of oxytocin is that of milk ejection by the mammary glands in response to suckling. Although present, oxytocin has no known function in the male.

Vasopressin, the second posterior lobe hormone, acts on the collecting duct of the kidney to increase water reabsorption and thus restrict urine flow, as already described on p. 662. It is therefore often called the **antidiuretic hormone.** Vasopressin has a second, weaker effect of increasing the blood pressure through its generalized constrictor effect on the smooth muscles of the arterioles. Although the name "vasopressin" unfortunately suggests that vasoconstriction is the hormone's major effect, it is probably of little physiological importance, except perhaps to help sustain blood pressure during a severe hemorrhage.

All jawed vertebrates secrete two posterior lobe hormones that are quite similar to those of mammals. All are octapeptides, but there is some variation in structure because of amino acid substitutions in three of eight amino acid positions in the molecule.

Of all the posterior lobe hormones, **vasotocin** (Table 37-1) has the widest phylogenetic distribution and is believed to be the parent hormone from which the other octapeptides evolved. It is found in all vertebrate classes except mammals. It is a water-balance hormone in amphibians, espe-

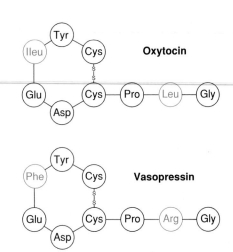

Figure 37-7

Posterior lobe hormones of mammals. Both oxytocin and vasopressin consist of eight amino acids (the two sulfur-linked cysteine molecules are considered a single amino acid, cystine). Oxytocin and vasopressin are identical except for amino acid substitutions in the blue positions. The abbreviations represent amino acids.

cially toads, in which it acts to conserve water by (1) increasing permeability of the skin (to promote water absorption from the environment), (2) stimulating water reabsorption from the urinary bladder, and (3) decreasing urine flow. The action of vasotocin is best understood in amphibians, but it appears to play some water-conserving role in birds and reptiles as well.

Pineal Gland

In all vertebrates the dorsal part of the brain, the diencephalon (Figure 36–13, p. 723), gives rise to a sac-like evagination called the pineal complex which lies just beneath the skull in a mid-line position. In ectothermic vertebrates the pineal complex contains glandular tissue and a photoreceptive sensory organ involved in pigmentation responses and in light-dark biological rhythms. In lampreys, many amphibians, lizards, and the tuatara (*Sphenodon,* p. 564), the median photoreceptive organ is so well developed, containing structures analogous to the lens and cornea of lateral eyes, that it is often called a third eye. In birds and mammals, the pineal complex has evolved into an entirely glandular structure called the pineal gland. The pineal gland produces the hor-

Structure of melatonin

mone **melatonin.** Melatonin secretion is strongly affected by exposure to light. Its production is lowest during daylight hours and highest at night. In nonmammalian vertebrates, the pineal gland is responsible for maintaining **circadian rhythms**—self-generated (endogenous) rhythms that are about 24 hours in length. A circadian rhythm serves as a biological clock for many physiological processes that follow a regular pattern.

In mammals, the pineal has lost most of the capacity to pace circadian rhythms. Instead, an area of the hypothalamus called the suprachiasmatic nucleus has become the primary circadian pacemaker in mammals, although the pineal gland still produces melatonin nightly and serves to reinforce the circadian rhythm of the suprachiasmatic nucleus. In mammals in which seasonal rhythms in reproduction are regulated by **photoperiod,** the pineal hormone melatonin plays a critical role in regulating gonadal activity. In long-day breeders, such as ferrets, hamsters, and deer mice, the reduced light stimulation with shortening day length in autumn increases melatonin secretion. Melatonin, by inhibiting the synthesis and secretion of the gonadotropin-releasing hormone (GnRH), suppresses reproductive activity during the winter months. The lengthening days in the spring have the opposite effect and reproductive activities are resumed. Short-day breeders, such as white-tailed deer, silver fox, spotted skunk, and sheep, are stimulated by reduced day length in the fall; in these, increasing melatonin levels in the fall stimulates, rather than inhibits, reproductive activity.

Only recently the pineal gland has been shown to produce subtle and incompletely understood effects on circadian and annual rhythms in nonphotoperiodic mammals (such as humans). For example, melatonin secretion has been linked to a sleeping and eating disorder in humans known as seasonal affective disorder (SAD). Some people living in northern latitudes, where day lengths are short in winter and when melatonin production is elevated, become depressed in winter, sleep long periods, and may go on eating binges. Often this wintertime disorder can be treated by exposure to sun lamps with full-spectrum light; such exposure depresses melatonin secretion by the pineal gland.

NONENDOCRINE HORMONES

Brain Neuropeptides

The blurred distinction between the endocrine and nervous systems is nowhere more evident than in the nervous system, where a growing list of hormonelike neuropeptides have been discovered in central and peripheral nervous systems of vertebrates and invertebrates. In mammals, approximately 40 neuropeptides (short chains of amino acids) have been located using immunological labeling with fluorescent antibodies that can be visualized in histological sections under the microscope. Many are known to lead double lives—to be capable of behaving both as hormones, carrying signals from gland cells to their targets, and as neurotransmitters, relaying signals between nerve cells. For example, both oxytocin and vasopressin have been discovered at widespread sites in the brain by immunochemical methods. Related to this discovery is the fascinating observation that people and experimental animals injected with minute quantities of vasopressin experience enhanced learning and improved memory. This effect of vasopressin in brain tissue is unrelated to its well-known antidiuretic function in the kidney (p. 662). Several hormones, such as gastrin and cholecystokinin (p. 754) (which long had been supposed to function only in the gastrointestinal tract), have been discovered in the cerebral cortex and hippocampus. In addition to its gastrointestinal actions, cholecystokinin is known to function in the control of feeding and satiety and may serve other functions as a brain neuroregulator.

The radioimmunoassay technique developed by Solomon Berson and Rosalyn Yalow about 1960 has revolutionized endocrinology and neurochemistry. First, antibodies to the hormone of interest (insulin, for example) are prepared by injecting guinea pigs or rabbits with the hormone. Then, a fixed amount of radioactively labeled insulin and unlabeled insulin antibodies is mixed with the sample of blood plasma to be measured. The native insulin in the blood plasma and the radioactive insulin compete for antibodies. The more insulin there is in the sample, the less radioactive insulin will bind to the antibodies. Bound and unbound insulin are then separated, and their radioactivities are measured together with those of appropriate standards to determine the amount of insulin present in the blood sample. The method is so incredibly sensitive that it can measure the equivalent of a cube of sugar dissolved in one of the Great Lakes.

Among the dramatic developments in this field was the discovery in 1975 of the endorphins and enkephalins, neuropeptides that bind with opiate receptors and influence perception of pleasure and pain (see note on p. 729). The endorphins and enkephalins are found also in brain circuits that modulate several other functions unrelated to pleasure and pain, such as control of blood pressure, body temperature, and body movement. Even more intriguing, the endorphins are derived from the same prohormone that gives rise to the anterior pituitary hormones ACTH and MSH. The discovery in the brain of a complex family of compounds whose functions and interrelationships are not yet clear has spawned an active area of biomedical research.

Prostaglandins

Prostaglandins are a family of long-chain unsaturated fatty acids that were discovered in seminal fluid in the 1930s. At first they were thought to be produced by the prostate gland (hence the name) but have now been found

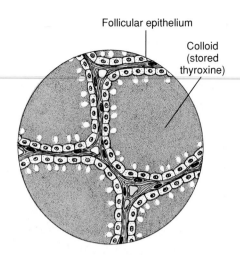

Structure of a typical prostaglandin

in virtually all mammalian tissues. Prostaglandins act as local hormones that have diverse actions on many different tissues, making generalizations about their effects difficult. Many of their effects, however, involve smooth muscle. In some tissues prostaglandins regulate vasodilation or vasoconstriction by their often antagonistic action on the smooth muscle in the walls of blood vessels. They are known to stimulate contraction of uterine smooth muscle during childbirth. There is also evidence that overproduction of uterine prostaglandins is responsible for the symptoms of painful menstruation (dysmenorrhea) experienced by many women. Several inhibitors of prostaglandins that provide relief from these symptoms have now been approved. Among other actions of prostaglandins is their intensification of pain in damaged tissues, mediation of the inflammatory response, and involvement in fever. In many instances the physiological significance of the numerous reported effects of prostaglandins is unknown.

Cytokines

For some years we have known that cells of the immune system somehow communicated with each other and that this communication was crucial to the immune response. Now we understand that polypeptide hormones called cytokines (p. 677) mediate communication between the cells of immunity. Cytokines can affect the cells that secrete them, affect nearby cells, and like other hormones, they can affect cells in distant locations. Their target cells bear specific receptors for the cytokine bound to the surface membrane. Cytokines coordi-

nate a complex network, with some target cells being activated, stimulated to divide and often to secrete their own cytokines. The same cytokine that activates some cells may suppress division of other target cells.

HORMONES OF METABOLISM

Another important group of hormones adjusts the delicate balance of metabolic activities. The rates of chemical reactions within cells are often regulated by long sequences of enzymes. Although such sequences are complex, each step in a pathway is mostly self-regulating as long as the equilibrium between substrate, enzyme, and product remains stable. However, hormones may alter the activity of crucial enzymes in a metabolic process, thereby accelerating or inhibiting the entire process. It should be emphasized that hormones never initiate enzymatic processes but alter their rate, speeding them up or slowing them down. The most important hormones of metabolism are those of the thyroid, parathyroid, adrenal glands, and pancreas.

Thyroid Hormones

Two hormones, **thyroxine and triiodothyronine** (T_4 and T_3, respectively) are secreted by the thyroid gland. This large endocrine gland is located in the neck of all vertebrates. The thyroid is composed of thousands of tiny spherelike units, called follicles, where thyroid hormone is synthesized, stored, and released into the bloodstream as needed. The size of the follicles, and the amount of stored thyroxine and triiodothyronine they contain, depends on the activity of the gland (Figure 37-8).

One of the unique characteristics of the thyroid is its high concentration of **iodine;** in most animals this single gland contains well over half the body store of iodine. The epithelial cells of the thyroid follicles actively trap iodine from the blood and combine it with the amino acid tyrosine, creating the two thyroid hormones. Each molecule

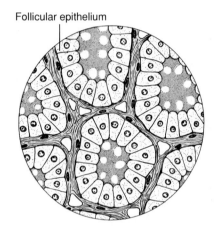

Inactive follicles

Active follicles

Figure 37-8

Appearance of thyroid gland follicles viewed through the microscope (approximately ×350). When inactive, the follicles are distended with colloid, the storage form of thyroid hormone, and the epithelial cells are flattened. When active, the colloid disappears as thyroid hormone is secreted into the circulation, and the epithelial cells become greatly enlarged.

Thyroxine

of thyroxine contains four atoms of iodine. Triiodothyronine has three instead of four iodine atoms. Thyroxine is formed in much greater amounts than triiodothyronine, but in many animals triiodothyronine is the more

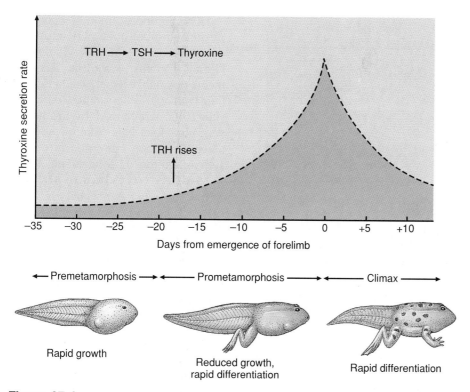

← Premetamorphosis → ← Prometamorphosis → ← Climax →

Rapid growth

Reduced growth, rapid differentiation

Rapid differentiation

Figure 37-9

Effect of thyroxine on growth and metamorphosis of a frog. The release of TRH from the hypothalamus at the end of premetamorphosis sets in motion the hormonal changes (increased TSH and thyroid hormone) leading to metamorphosis. Thyroxine levels are maximal at the time the forelimbs emerge.

physiologically active hormone. Both hormones have two important actions. One is to promote normal growth and development of the nervous system of growing animals. The other is to stimulate the metabolic rate.

Undersecretion of thyroid hormones in fish, birds, and mammals dramatically impairs growth, especially of the nervous system. The human **cretin,** a mentally retarded dwarf, is the result of thyroid malfunction from a very early age. Conversely, the oversecretion of thyroid hormones causes precocious development in all vertebrates, but its effect is particularly prominent in fish and amphibians. In frogs and toads, the transformation from aquatic herbivorous tadpole without lungs or legs to semiterrestrial or terrestrial carnivorous adult with lungs and four legs, occurs when the thyroid gland becomes active at the end of larval development. Stimulated by a rise in the thyroxine level of the blood, metamorphosis and climax occur (Figure 37-9). Growth of the frog after metamorphosis is directed by the growth hormone.

In birds and mammals, the control of oxygen consumption and heat production is the best known action of the thyroid hormones. The thyroid maintains metabolic activity of homeotherms (birds and mammals) at a normal level. Oversecretion of the thyroid hormones will speed up body processes as much as 50%, resulting in irritability, nervousness, fast heart rate, intolerance of warm environments, and loss of body weight despite increased appetite. Undersecretion of the thyroid hormones slows metabolic activities, which can cause loss of mental alertness, slowing of the heart rate, muscular weakness, increased sensitivity to cold, and weight gain. An important function of the thyroid gland is to promote adaptation to cold environments by increasing heat production. The thyroid hormones stimulate cells to produce more heat and store less chemical energy (ATP); in other words, thyroxine *reduces* the efficiency of cellular oxidative phosphorylation (p. 70). Consequently many cold-adapted mammals have

heartier appetites and eat more food in winter than in summer although their activity is about the same in both seasons. In winter, a larger portion of the food is being converted directly into body-warming heat.

The synthesis and release of the thyroid hormones are governed by **thyrotropic hormone** (TSH) from the anterior pituitary gland (Table 37-1). TSH is in turn regulated by the thyrotropin-releasing hormone (TRH) of the hypothalamus. As noted earlier, TRH is part of a higher regulatory council that controls the tropic hormones of the anterior pituitary. TSH controls thyroid activity through an excellent example of negative feedback. If thyroid hormone level in the blood decreases, more TSH is released, which returns the level to normal. Should the thyroid hormone level rise too high, it acts on the anterior pituitary to inhibit TSH release. With declining TSH output, the thyroid is less stimulated and the level of thyroid hormone in the blood returns to normal. Such a system is obviously very effective in damping oscillations in hormonal output by the target gland. It can be overridden, however, by neural stimuli, such as exposure to cold, which can directly stimulate increased release of TRH and thus TSH.

Some years ago, a condition called goiter was common among people living in the Great Lakes region of the United States and Canada, as well as some other parts of the world, such as the Swiss Alps. This type of goiter is an enlargement of the thyroid gland caused by a deficiency of iodine in the food and water. By striving to produce thyroid hormone with not enough iodine available, the gland hypertrophies, sometimes so much that the entire neck region becomes swollen (Figure 37-10). Goiter caused by iodine deficiency is seldom seen in North America because of the widespread use of iodized salt. However, it is estimated that even today 200 million people worldwide experience varying degrees of goiter, mostly in high mountains of Latin America, Europe, and Asia.

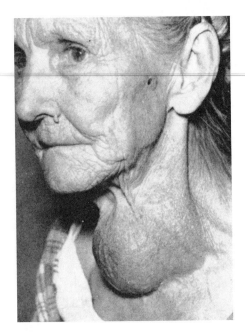

Figure 37-10
A large goiter caused by iodine deficiency. By enlarging enormously, the thyroid gland can extract enough iodine from the blood to synthesize the body's requirement for thyroid hormone.

Hormonal Regulation of Calcium Metabolism

Closely associated with the thyroid gland and in some animals buried within it are the **parathyroid glands.** These tiny glands occur as two pairs in humans but vary in number and position in other vertebrates. They were discovered at the end of the nineteenth century when the fatal effects of "thyroidectomy" were traced to the unknowing removal of the parathyroid glands together with the thyroid gland. In birds and mammals, including humans, removal of the parathyroid glands causes the level of calcium in the blood to decrease rapidly. This leads to a serious increase in nervous system excitability, severe muscular spasms and tetany, and finally death.

Before considering how hormones maintain calcium homeostasis, it is helpful to summarize mineral metabolism in bone, a densely packed storehouse of both calcium and phosphorus. Bone contains approximately 98% of the calcium and 80% of the phosphorus in the body. Although bone is second only to teeth as the most durable material in the body (as evi-

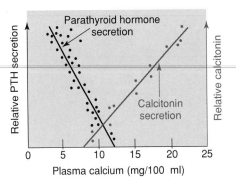

Figure 37-11
How rate of secretion of parathyroid hormone (PTH) and calcitonin respond to changes in blood calcium level in a mammal.

denced by the survival of fossil bones for millions of years) it is in a state of constant turnover in the living body. Bone-building cells **(osteoblasts)** synthesize the organic fibers of the bone matrix which later become mineralized with a form of calcium phosphate called hydroxyapatite. Bone-resorbing cells **(osteoclasts),** are giant cells that dissolve the bony matrix, releasing calcium and phosphate into the blood. These opposing activities allow bone constantly to remodel itself, especially in the growing animal, for structural improvements to counter new mechanical stresses on the body. They additionally provide a vast and accessible reservoir of minerals that can be withdrawn as needed for general cellular requirements.

The level of calcium in the blood is maintained by the action of three hormones which coordinate the absorption, storage, and excretion of calcium ions. If the blood calcium should decrease slightly, the parathyroid gland increases its secretion of **parathyroid hormone (PTH).** This stimulates the osteoclasts to dissolve bone adjacent to these cells, thus releasing calcium and phosphate into the bloodstream and returning the blood calcium level to normal. Parathyroid hormone also decreases the rate of calcium excretion by the kidney and increases production of the hormone 1,25-dihydroxy vitamin D (see the following text). The parathyroid hormone level varies inversely with blood calcium level, as shown in Figure 37-11.

The second hormone involved in calcium metabolism in all tetrapods is derived from vitamin D. Vitamin D, like all vitamins, is a dietary requirement. But unlike other vitamins, vitamin D may also be synthesized in the skin from a precursor by irradiation with ultraviolet light from the sun. Vitamin D is then converted in a two-step oxidation to a hormonal form, **1,25-dihydroxyvitamin D.** This steroid hormone is essential for active calcium absorption by the gut (Figure 37-12). It also promotes the synthesis of a protein that transports calcium in the blood. The production of 1,25-dihydroxyvitamin D is stimulated by low plasma phosphate as well as by an increase in PTH secretion.

In humans, a deficiency of vitamin D causes rickets, a disease characterized by low blood calcium and weak, poorly calcified bones that tend to bend under postural and gravitational stresses. Rickets has been called a disease of northern winters, when sunlight is minimal. It was once common in the smoke-darkened cities of England and Europe.

The third calcium-regulating hormone, **calcitonin,** is secreted by specialized cells (C cells) in the thyroid gland of mammals and in the ultimobranchial gland of other vertebrates. Calcitonin is released in response to elevated levels of calcium in the blood. It rapidly suppresses calcium withdrawal from bone, decreases the intestinal absorption of calcium, and increases the excretion of calcium by the kidneys. Calcitonin thus protects the body against a dangerous increase in the level of calcium in the blood, just as parathyroid hormone protects it from a dangerous decrease in blood calcium (Figure 37-12). Calcitonin has been identified in all vertebrate groups but, except in mammals, its functional role is uncertain.

Hormones of the Adrenal Cortex

The mammalian adrenal gland is a double gland composed of two unrelated types of glandular tissue: an outer region of adrenocortical cells, or

cortex, and an inner region of specialized cells, the **medulla** (Figure 37-13). In nonmammalian vertebrates the homologs of the adrenocortical and medullary cells are organized quite differently; they may be intermixed or distinct, but never arranged in a cortex-medulla relationship as in mammals.

At least 30 different compounds have been isolated from adrenocortical tissue, all of them closely related lipoidal compounds known as steroids. Only a few of these compounds are true steroid hormones; most are various intermediates in the synthesis of steroid hormones from **cholesterol** (Figure 37-14). The corticosteroid hormones are commonly classified into two groups, according to their function: glucocorticoids and mineralocorticoids.

Glucocorticoids, such as **cortisol** (see Figure 37-14) and **corticosterone,** are concerned with food metabolism, inflammation, and stress. They promote the synthesis of glucose from compounds other than carbohydrates, particularly amino acids and fats. This process is called **gluconeogenesis.** The overall effect is to increase the

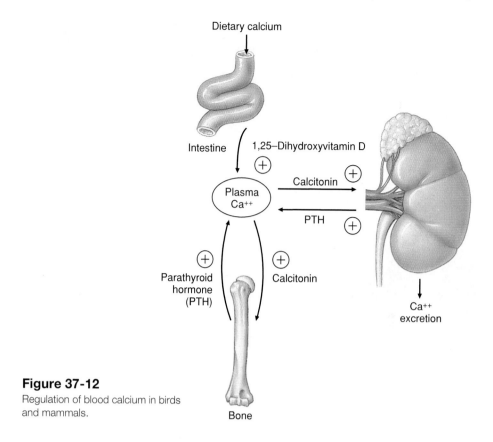

Figure 37-12

Regulation of blood calcium in birds and mammals.

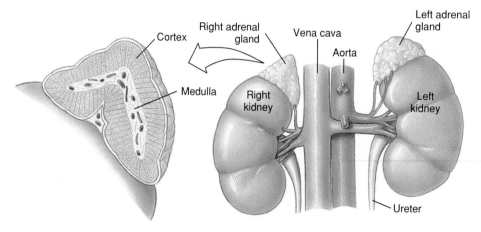

Figure 37-13

Paired adrenal glands of humans, showing gross structure and position on the upper poles of the kidneys. Steroid hormones are produced by the cortex. The sympathetic hormones epinephrine and norepinephrine are produced by the medulla.

The adrenal steroid hormones, especially the glucocorticoids, are remarkably effective in relieving the *symptoms* of rheumatoid arthritis, allergies, and various disorders of the connective tissue, skin, and blood. Following the report in 1948 by P. S. Hench and his colleagues at the Mayo Clinic that cortisone dramatically relieved the pain and crippling effects of advanced arthritis, the steroid hormones were hailed by the media as "wonder drugs." Optimism was soon dimmed, however, when it became apparent that severe side effects always attended long-term administration of the antiinflammatory steroids. Steroid therapy lulls the adrenal cortex into inactivity and may permanently impair the body's capacity to produce its own steroids. Today steroid therapy is applied with caution, because it is realized that the inflammatory response is a necessary part of the body's defenses.

level of glucose in the blood to provide a quick energy source for muscle and nervous tissue. The glucocorticoids are also important in diminishing the immune response to various inflammatory conditions. Because several diseases of humans are inflammatory diseases (for example, allergies, hypersensitivity, and rheumatoid arthritis), these corticosteroids have important medical applications.

The synthesis and secretion of the glucocorticoids is controlled principally by ACTH of the anterior pituitary (see Figure 37-6). ACTH is also controlled by the corticotropin-releasing hormone (CRH) of the hypothalamus (Table 37-1). As with pituitary control of the thyroid, a negative feedback relationship exists between CRH, ACTH, and the adrenal cortex. An increase in the release of glucocorticoids

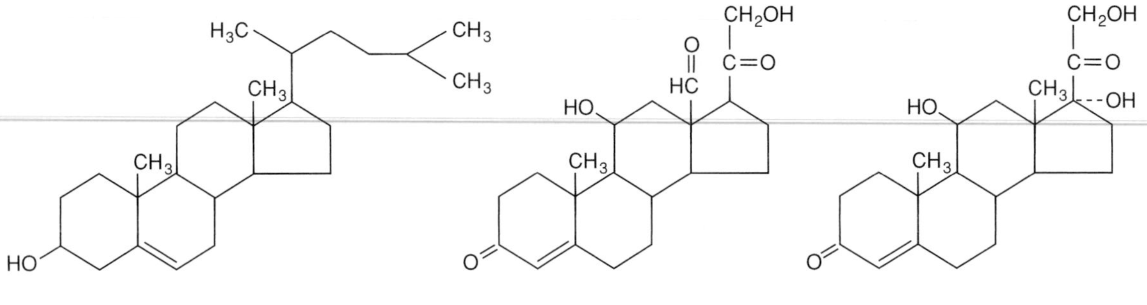

Figure 37-14

Hormones of the adrenal cortex. Cortisol (a glucocorticoid) and aldosterone (a mineralocorticoid) are two of several steroid hormones synthesized from cholesterol in the adrenal cortex.

suppresses the output of CRH and ACTH; the resulting decline in the blood level of CRH and ACTH then feeds back to the adrenal cortex to inhibit the further release of the glucocorticoids. An opposite cycle of events happens if the blood level of glucocorticoids drops: ACTH output increases which in turn stimulates the secretion of glucocorticoids. CRH is known to mediate stressful stimuli through the adrenal axis.

Mineralocorticoids, the second group of corticosteroids, are those that regulate salt balance. **Aldosterone** (see Figure 37-14) is by far the most important steroid of this group. Aldosterone promotes the tubular reabsorption of sodium and chloride and the tubular excretion of potassium by the kidney. Since sodium usually is in short supply in the diet of many animals and potassium is in excess, the mineralocorticoids play vital roles in preserving the correct balance of blood electrolytes. The salt-regulating action of aldosterone is controlled principally by the renin-angiotensin system, described on p. 660.

The mineralocorticoids *oppose* the antiinflammatory effect of the glucocorticoids. In other words, they promote the *inflammatory* defense of the body to various noxious stimuli. Although these opposing actions of the corticosteroids seem self-defeating, they actually are not. They are necessary to maintain readiness of the body's defenses for any stress or threat of disease, yet prevent these defenses from becoming so powerful that they turn against the body's own tissues.

The adrenocortical tissue also produces androgens (Gr. *andros,* man, + *genesis,* origin), which, as the name implies, are similar in effect to the male sex hormone, testosterone. The adrenal androgens promote some of the changes that occur just before puberty in the human male and female. The recent development of so-called **anabolic steroids,** synthetic hormones related to testosterone, has led to widespread abuse of steroids among athletes. These

The use of anabolic (tissue-building) steroids by athletes became major news following Ben Johnson's drug-fueled win of the 100-meter race at the 1988 Olympics. Despite almost universal condemnation by Olympic, medical, and college sports authorities, an unscientific and clandestine program of experimentation with anabolic steroids has become popular with many amateur and professional athletes in many countries. There may be 3 million regular anabolic steroid users in the United States alone. Most anabolic steroids are purchased illegally through a black market with annual sales of some $400 million. The extensive use of anabolic steroids by Olympic athletes has been documented by Robert Voy who served as Chief Medical Officer for the United States Olympic Committee from 1985 to 1989, when he quit in frustration over a crisis which he believes is destroying the Olympic ideal (Voy, R., 1991, *Drugs, sport, and politics,* Leisure Press).

synthetics (and testosterone) cause hypertrophy of skeletal muscle and may improve performance that depends on strength. Unfortunately they also have serious side effects, including testicular atrophy, periods of irritability, abnormal liver function, and cardiovascular disease.

Hormones of the Adrenal Medulla

The adrenal medullary cells secrete two structurally similar hormones: **epinephrine** (adrenaline) and **norepinephrine** (noradrenaline). The adrenal medulla is derived embryologically from the same tissue that gives rise to the postganglionic sympathetic neurons of the autonomic nervous system (p. 725). Norepinephrine serves as a neurotransmitter at the endings of the sympathetic nerve fibers. Thus functionally, as well as embryologically, the adrenal medulla can be considered a very large sympathetic nerve ending.

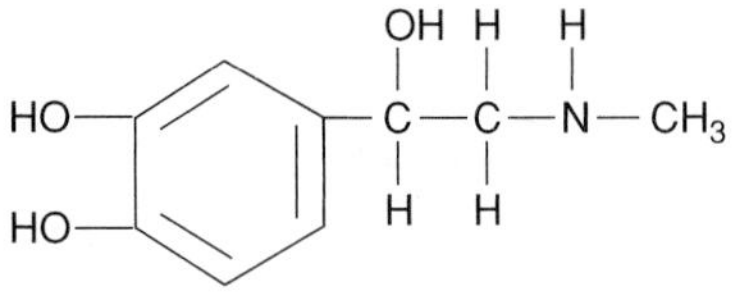

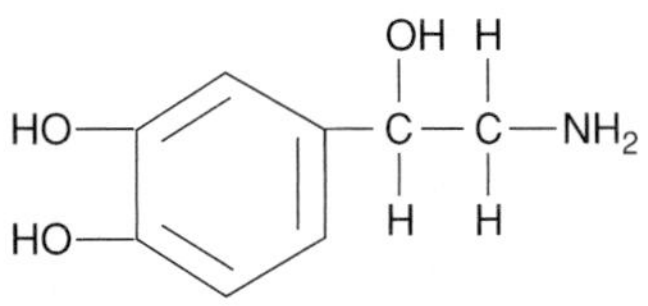

It is not surprising then that the adrenal medullary hormones and the sympathetic nervous system have the same general effects on the body. These effects center on responses to emergencies, such as fear and strong emotional states, flight from danger, fighting, pain, lack of oxygen, and blood loss. Walter B. Cannon, of homeostasis fame (p. 651), termed these "fight or flight" responses that are appropriate for survival. We are all familiar with the increased heart rate, tightening of the stomach, dry mouth, trembling muscles, general feeling of anxiety, and increased awareness that attends sudden fright or other strong emotional states. These effects are attributable to increased activity of the sympathetic nervous system and to the rapid release into the blood of epinephrine from the adrenal medulla.

Epinephrine and norepinephrine have many other effects of which we are not as aware, including constriction of the arterioles (which, together with the increased heart rate, increases the blood pressure), mobilization of liver glycogen and fat stores to release glucose and fatty acids for energy, increased oxygen consumption and heat production, hastening of blood coagulation, and inhibition of the gastrointestinal tract. All of these changes prepare the body for emergencies or stressful conditions.

Insulin from the Islet Cells of the Pancreas

The pancreas is both an exocrine and an endocrine organ (Figure 37-15). The *exocrine* portion produces pancreatic juice, a mixture of digestive enzymes and bicarbonate ions that is conveyed by a duct to the digestive tract. Scattered within the extensive exocrine portion of the pancreas are numerous small islets of tissue, called **islets of Langerhans** (see Figure 37-15). This *endocrine* portion of the gland makes up only 1% to 2% of the total weight of the pancreas. The islets are without ducts and secrete their hormones directly into blood vessels that extend throughout the pancreas.

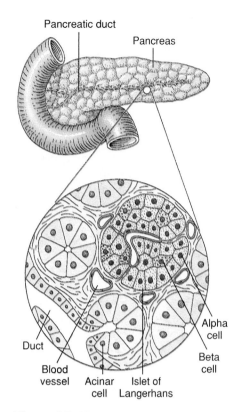

Pancreatic duct

Pancreas

Alpha cell

Beta cell

Duct

Blood vessel   Acinar cell   Islet of Langerhans

Figure 37-15

The pancreas is composed of two kinds of glandular tissue: exocrine acinar cells that secrete digestive juices which enter the intestine through the pancreatic duct, and the endocrine islets of Langerhans. The islets of Langerhans secrete the hormones insulin and glucagon directly into the blood circulation.

Two polypeptide hormones are secreted by different cell types within the islets: **insulin,** produced by the **beta cells,** and **glucagon,** produced by the **alpha cells.** Insulin and glucagon have antagonistic actions of great importance in the metabolism of carbohydrates and fats. Insulin is essential for the use of blood glucose by cells, especially skeletal muscle cells. Insulin promotes the entry of glucose into body cells through its action on a glucose transporter molecule found in cell membranes. Without insulin, body cells cannot use glucose. The level of glucose in the blood rises to abnormally high levels (hyperglycemia) to exceed the transport maximum of the kidney (p. 660), and sugar (glucose) appears in the urine. Lack of insulin also inhibits the uptake of amino acids by skeletal muscle, and fats and muscle are broken down to provide energy. The body cells starve while the urine abounds in the very substance the body craves. The disease, called diabetes mellitus, afflicts nearly 5% of the human population in varying degrees of severity. If left untreated, it can lead to severe damage to kidneys, eyes, and blood vessels, and greatly shorten life expectancy.

In 1982, insulin became the first hormone produced by genetic engineering (recombinant DNA technology, p. 145) to be marketed for human use. Recombinant insulin has the exact structure of human insulin and therefore will not stimulate an immune response, which has often been a problem for diabetics receiving insulin purified from pig or cow pancreas.

The first extraction of insulin in 1921 by two Canadians, Frederick Banting and Charles Best, was one of the most dramatic and important events in the history of medicine. Many years earlier two German scientists, J. Von Mering and O. Minkowski, discovered that surgical removal of the pancreas of dogs invariably caused severe symptoms of diabetes, resulting in the animal's death within a few weeks. Many attempts were made to isolate the diabetes preventive factor, but all failed because powerful protein-splitting digestive enzymes in the exocrine portion of the pancreas destroyed the hormone during extraction procedures. Following a hunch, Banting, in collaboration with Best and his physiology professor J. J. R. Macleod, tied off the pancreatic ducts of several dogs. This caused the exocrine portion of the gland with its hormone-destroying enzyme to degenerate but left the islets' tissues healthy, since they were independently served by their own blood supply. Banting and Best then successfully extracted insulin from these glands. Injected into another dog, the insulin immediately lowered the level of sugar in the blood (Figure 37-16). Their experiment paved the way for the commercial extraction of insulin from slaughterhouse animals. It meant that millions of people with diabetes, previously doomed to invalidism or death, could look forward to more normal lives.

Figure 37-16

Charles H. Best and Sir Frederick Banting in 1921 with the first dog to be kept alive by insulin.

From J. F. Fulton & L. G. Wilson, Selected Readings in the History of Physiology, *1966. Courtesy of Charles C. Thomas, Publisher, Springfield, Illinois.*

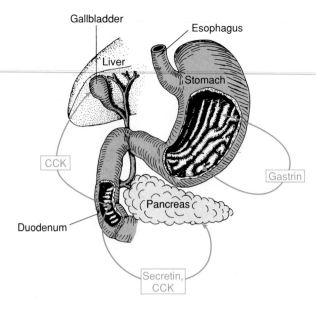

Figure 37-17

Hormones of digestion. Arrows show source and target of three gastrointestinal hormones. CCK = cholecystokinin.

Glucagon, the second hormone of the pancreas, has several effects on carbohydrate and fat metabolism that are opposite to the effects of insulin. For example, glucagon raises the blood glucose level (by converting liver glycogen to glucose), whereas insulin lowers blood glucose. Glucagon and insulin do not have the same effects in all vertebrates, and in some, glucagon is lacking altogether. Glucagon is an example of a hormone that operates through the cyclic AMP second-messenger system.

HORMONES OF DIGESTION

The digestive process is coordinated by a family of hormones produced by the body's most diffuse endocrine system, the gastrointestinal tract. These hormones are examples of the many substances produced by the vertebrate body that have hormonal function, yet are not necessarily produced by discrete endocrine glands. Because of their diffuse origins the gastrointestinal (GI) hormones have been difficult to isolate and study and only recently have they been researched in depth.

Among the principal GI hormones are gastrin, cholecystokinin (CCK), and secretin (Figure 37-17). **Gastrin** is a small polypeptide hormone produced by endocrine cells in the pyloric portion of the stomach. Gastrin is secreted when protein food enters the stomach. Its main actions are to stimulate hydrochloric acid secretion and to increase gastric motility. Gastrin is an unusual hormone in that it exerts its action on the same organ from which it is secreted. **CCK** is also a polypeptide hormone, and it has a striking structural resemblance to gastrin, suggesting that the two arose by duplication of ancestral genes. CCK has at least three distinct functions. It stimulates gallbladder contraction and thus increases the flow of bile salts into the intestine; it stimulates an enzyme-rich secretion from the pancreas; and it acts on the brain to contribute a feeling of well-being after a meal. The principal action of **secretin,** the first hormone to be discovered (see the opening essay on p. 739), is to stimulate the release of an alkaline pancreatic fluid that neutralizes stomach acid as it enters the intestine. It also aids fat digestion by inhibiting gastric motility and increasing bile production.

Several other GI hormones have been isolated recently and their structure determined. All are peptides. It is now well established that several peptide hormones are present in both the GI tract and in the central nervous system. One of these is CCK, which has been found in high concentrations in the cerebral cortex of mammals. By providing a feeling of satiety after eating (mentioned previously) it may play some role in regulating appetite. Several other GI peptides appear to play neurotransmitter roles in the brain. This unexpected versatility has served to broaden our concept of hormones as molecules capable of functioning in several different ways.

Summary

Hormones are chemical messengers synthesized by special endocrine cells and other cells and transported by the blood to target cells where they affect cell function by altering specific biochemical processes. Specificity of response is ensured by the presence on or in the target cells of protein receptors that bind only selected hormones. Hormone effects are vastly amplified in the target cells by acting through one of two basic mechanisms. Many hormones, including epinephrine, glucagon, vasopressin, and some hormones of the anterior pituitary, cause production of a "second messenger," such as cyclic AMP, that relays the hormone's message from a surface receptor to the cell's biochemical machinery. Steroid hormones and the thyroid hormones operate through nuclear receptors. A hormone-receptor complex is formed that induces protein synthesis by setting gene transcription in motion.

Most invertebrate hormones are products of neurosecretory cells. The best understood invertebrate endocrine system is that controlling molting and metamorphosis in insects. A juvenile insect grows by passing through a series of molts under the control of two hormones, one (the molting hormone) favoring molting to an adult and the other (the juvenile hormone) favoring retention of juvenile characteristics. The molting hormone is under the control of a neurosecretory hormone (ecdysiotropin) from the brain.

The vertebrate endocrine system is orchestrated by the pituitary gland. The anterior lobe of the pituitary produces seven well-characterized hormones. Four of these are tropic hormones that regulate subservient endocrine glands: thyrotropic hormone (TSH), which controls the secretion of thyroid hormones; adrenocorticotropic hormone (ACTH), which stimulates the release of steroid hormones by the adrenal cortex; and follicle-stimulating hormone (FSH) and luteinizing hormone (LH), which act on the ovaries and testes. Three direct-acting hormones are (1) prolactin, which plays several diverse roles, including the stimulation of milk production, (2) growth hormone that governs body growth; and (3) melanophore-stimulating hormone (MSH), which controls melanophore dispersion in ectothermic vertebrates. The release of all of the anterior lobe hormones is regulated in part by hypothalamic neurosecretory products called releasing hormones (or release-inhibiting hormones). The hypothalamus also produces two neurosecretory hormones, which are stored in and released from the posterior lobe of the pituitary. In mammals these two hormones are oxytocin, which stimulates milk production, and vasopressin (antidiuretic hormone), which acts on the kidney to restrict urine production. In amphibians, reptiles, and birds, vasotocin replaces vasopressin as the water-balance hormone.

The pineal gland, a derivative of the pineal complex of the diencephalon of the brain, produces the hormone melatonin. In many vertebrates melatonin, which is strongly affected by light-dark cycles, maintains circadian rhythms. In mammals that become reproductively active when day lengths increase, and possibly in other vertebrates, melatonin regulates seasonal reproductive cycles.

The recent application of ultrasensitive radioimmunochemical techniques has revealed many neuropeptides in the brain, several of which behave as neurotransmitters in the brain but as hormones elsewhere in the body. The classical definition of a hormone has been modified to include other chemical messengers, such as prostaglandins and cytokines, that originate in sources other than clearly defined endocrine glands.

Several hormones play important roles in regulating cellular metabolic activities. The two thyroid hormones, thyroxine and triiodothyronine, control growth, development of the nervous system, and cellular metabolism. Calcium metabolism is regulated principally by three hormones: parathyroid hormone from the parathyroid glands, a hormonal derivative of vitamin D, 1,25-dihydroxyvitamin D, and calcitonin from the thyroid gland. Parathyroid hormone and 1,25-dihydroxyvitamin D increase plasma calcium levels; calcitonin decreases plasma calcium levels.

The principal steroid hormones of the adrenal cortex are glucocorticoids, which stimulate formation of glucose from nonglucose sources, and mineralocorticoids, which regulate blood electrolyte balance. The adrenal medulla is the source of epinephrine and norepinephrine, which have many effects, including assisting the sympathetic nervous system in emergency responses.

Sugar metabolism is regulated by the antagonistic action of two pancreatic hormones. Insulin is needed for cellular use of blood glucose and the uptake of amino acids in muscle. Glucagon opposes the action of insulin.

Several gastrointestinal hormones coordinate digestive functions. They include gastrin, which stimulates acid secretion by the stomach; CCK, which stimulates gallbladder and pancreatic secretion; and secretin, which stimulates bicarbonate secretion from the pancreas and inhibits gastric motility.

Review Questions

1. Outline the famous experiment of Bayliss and Starling that marks the birth of endocrinology. What might their *hypothesis* have been?
2. Provide definitions for the following: hormone, endocrine gland, exocrine gland, hormone receptor molecule.
3. Hormone receptor molecules are the key to understanding the specificity of hormone action on target cells. Describe and distinguish between receptors located on the cell surface and those located in the nucleus of target cells. Name two hormones whose action is mediated through each type of receptor.
4. What is the importance of feedback systems in the control of hormonal output? Offer an example of a hormonal feedback pattern.
5. Explain how the three hormones involved in insect growth—molting hormone, juvenile hormone, and brain hormone—interact in molting and metamorphosis.
6. Name seven hormones produced by the anterior pituitary gland. Why are four of these seven hormones called "tropic hormones"? Explain how the secretion of the anterior pituitary hormones is controlled by neurosecretory cells in the hypothalamus.
7. Describe the chemical nature and function of two posterior lobe hormones, oxytocin and vasopressin. What is distinctive about the way these neurosecretory hormones are secreted as compared with the neurosecretory release hormones that control the anterior pituitary hormones?
8. What is the evolutionary origin of the pineal gland of birds and mammals? Explain the role of the pineal hormone melatonin in regulating seasonal reproductive rhythms in some mammals. Does melatonin have any function in humans?
9. What are endorphins and enkephalins? What are prostaglandins?
10. What function have the recently described hormones called cytokines in the body?
11. What are the two most important functions of the thyroid hormones?
12. Explain how you would interpret the graph in Figure 37-11 to show that PTH and calcitonin act in a complementary way to control the blood calcium level.
13. Describe the principal functions of the two major groups of adrenal cortical steroids, the glucocorticoids and the mineralocorticoids. To what extent do these names provide clues to their function?
14. Where are the hormones epinephrine and norepinephrine produced and what is their relationship to the sympathetic nervous system and its response to emergencies?
15. Explain the actions of the hormones of the islets of Langerhans on the level of glucose in the blood. What is the consequence of insulin insufficiency as in the disease diabetes mellitus?
16. Name three hormones of the gastrointestinal tract and explain how they assist in the coordination of gastrointestinal function.

Selected References

See also general references for Part IV, p. 774.

Bentley, P. J. 1982. Comparative vertebrate endocrinology, ed. 2. Cambridge, Cambridge University Press. *Undergraduate text with good evolutionary perspective.*

Bohlander, F. F. 1989. Molecular endocrinology. New York, Academic Press. *Excellent synthesis of a fast-moving field.*

Chester-Jones, I., P. M. Ingleton, and J. G. Phillips (eds.). 1987. Fundamentals of comparative vertebrate endocrinology. New York, Plenum Press. *Contributed chapters.*

Gorbman, A., W. W. Dickhoff, S. R. Vigna, N. B. Clark, and C. Ralph. 1983. Comparative endocrinology. New York, John Wiley & Sons. *Authoritative text, although the comparative treatment is limited to vertebrates.*

Hadley, M. E. 1992. Endocrinology, ed 3. Englewood Cliffs, New Jersey, Prentice-Hall, Inc. *Undergraduate level textbook in vertebrate endocrinology.*

Laufer, H., and G. H. Downer (eds.). 1988. Endocrinology of selected invertebrate types. New York, Alan R. Liss. *Eighteen contributed chapters and an introductory chapter on the comparative endocrinology of invertebrates, radiates through echinoderms.*

Lienhard, G. E., J. W. Slot, D. E. James, and M. M. Mueckler. 1992. How cells absorb glucose. Sci. Am. **266:**89–91 (Jan.). *How insulin regulates the function of a special transporter molecule that moves glucose across cell membranes.*

Snyder, S. H. 1985. The molecular basis of communication between cells. Sci. Am. **253:** 132–141 (Oct.). *Review of vertebrate hormonal communication.*

38

Animal Behavior

The Lengthening Shadow of One Man

For as long as people have walked the earth, their lives have been touched by, indeed interwoven with, the lives of other animals. They hunted and fished them, domesticated them, ate them and were eaten by them, made pets of them, revered them, hated and feared them, immortalized them in art, song, and verse, fought them, and loved them. The very survival of ancient people depended on knowledge of wild animals. To stalk animals, people had to know their habits and behaviors. As the hunting societies of primitive people gave way to agricultural civilizations, an awareness was retained of the interrelationship with other animals, and the need to understand their behaviors increased.

This is still evident today. Zoos attract more visitors than ever before; wildlife television shows are increasingly popular; game-watching safaris to Africa constitute a thriving enterprise; and millions of pet animals share the cities with us—more than a half million pet dogs live in New York City alone. Although people have always been interested in the behavior of animals, the science of animal behavior is a

newcomer to biology. Charles Darwin, with the uncanny insight of genius, prepared for the reception of animal behavior by showing how natural selection would favor specialized behavioral patterns for survival. Darwin's pioneering book, *The Expression of the Emotions of Man and Animals,* published in 1872, mapped out a strategy for behavioral research still in use today. However science in 1872 was unprepared for Darwin's central insight that behavioral patterns, no less than bodily structures, are selected for and have evolutionary histories. Another 60 years would pass before such concepts would begin to flourish within behavioral science.

It was Ralph Waldo Emerson who said that an institution is the lengthening shadow of one man. For Charles Darwin the shadow is long indeed, for he brought into being entire fields of knowledge, such as evolution, ecology, and finally, after a long gestation, animal behavior. Above all, he altered the way we think about ourselves, the earth we inhabit, and the animals that share it with us. ■

In 1973, the Nobel Prize in Physiology or Medicine was awarded to three pioneering zoologists, Karl von Frisch, Konrad Lorenz, and Niko Tinbergen (Figure 38-1). The citation stated that these three were the principal architects of the new science of **ethology,**

A

B

C

Figure 38-1

Pioneers of the science of ethology. **A,** Konrad Lorenz (1903 to 1989). **B,** Karl von Frisch (1886 to 1982). **C,** Niko Tinbergen (1907 to 1988).

the scientific study of animal behavior, particularly under natural conditions. It was the first time any contributor to the behavioral sciences was so honored, and it meant that the discipline of animal behavior, which really has its roots in the work of Charles Darwin, had arrived.

THE SCIENCE OF ANIMAL BEHAVIOR

Behavioral biologists have traditionally asked two kinds of questions about behavior: *how* animals behave and *why* they behave as they do. "How" questions are concerned with immediate or **proximate causation.** For example, a biologist might wish to explain the singing of a male white-throated sparrow in the spring in terms of hormonal or neural mechanisms. Such physiological causes of behavior, that is, the mechanisms the animal uses in its behavior, are proximate factors. Alternatively, a biologist might ask what function singing serves the sparrow, and then seek to understand those events in the ancestry of birds that led to springtime singing. These are "why" questions that focus on **ultimate causation,** that is, the evolutionary origin and purpose of a behavior. These are really independent approaches to behavior, because understanding *how* the sparrow sings does not depend on what function singing serves, and vice versa. Students of animal behavior consider this distinction significant. Studies of proximate and ultimate causation are both important, but each may be of limited value in understanding the other.

The study of animal behavior has arisen from several different historical roots, and there is no universally accepted term for the whole subject. Today we can recognize three different experimental approaches to the study of animal behavior: comparative psychology, ethology, and sociobiology. **Comparative psychology** emerged from efforts to find general laws of behavior that would apply to many species, and preferably to humans as well. Early research that depended

heavily on inference was later replaced by replicable experimental approaches that concentrated on a few species, particularly white rats, pigeons, dogs, and occasionally primates. Following criticisms that the discipline lacked an evolutionary perspective and focused too narrowly on the white rat as a model for other organisms, many comparative psychologists developed more truly comparative investigations, some of these conducted in the field.

The aim of the second approach, **ethology,** has been to describe the behavior of an animal in its *natural habitat.* Most ethologists have been zoologists. Their laboratory has been the out-of-doors, and early ethologists gathered their data by field observation. They also conducted experiments, often with nature providing the variables, but increasingly ethologists have manipulated the variables for their own purposes by using animal models, playing recordings of animal vocalizations and altering the habitat. Modern ethologists also conduct many experiments in the laboratory where they can test their predictions under closely controlled conditions. However, ethologists usually take pains to compare laboratory observations with observations of free-ranging animals in undisturbed natural environments.

Ethology emphasizes the importance of ultimate factors affecting behavior. One of the great contributions of von Frisch, Lorenz, and Tinbergen was to demonstrate that behavioral traits are measurable entities like anatomical or physiological traits. This was to become the central theme of ethology: behavioral traits can be isolated and measured and they have evolutionary histories.

Sociobiology, the ethological study of social behavior, originated with the 1975 publication of E. O. Wilson's *Sociobiology: The New Synthesis.* Wilson describes social behavior as reciprocal communication of a cooperative nature (transcending mere sexual activity) that permits a group of organisms of the same species to become organized in a cooperative manner. In a complex system of social interactions, individuals are highly dependent on others for their daily living. While social behavior appears in many groups of animals, Wilson identified four "pinnacles" of complex social behavior. These are (1) colonial invertebrates, such as the Portuguese man-of-war (p. 262), which is a tightly-knit composite of individual organisms; (2) social insects, such as ants, bees, and termites, which have developed sophisticated systems of communication; (3) nonhuman mammals, such as dolphins, elephants, and some primates, which have highly developed social systems; and (4) humans.

The inclusion by Wilson of human behavior in sociobiology, and his references to the genetic foundation of many human social behaviors, has been strongly criticized. Complex systems of human social interactions, including religion, economic systems, and such objectionable characteristics as racism, sexism, and war, are emergent properties (p. 6) of human culture and its history. Is it meaningful to search for a specific genetic basis or justification for such phenomena? Many would answer "no," and look instead to the field of sociology, rather than sociobiology, to help us understand the complex, emergent properties of human societies.

DESCRIBING BEHAVIOR: PRINCIPLES OF CLASSICAL ETHOLOGY

The ethologists, through step-by-step analysis of the behavior of animals in nature, focused on the relatively invariant components of behavior. From such studies emerged several concepts that were first popularized in Tinbergen's influential book, *The Study of Instinct* (1951).

The basic concepts of ethology can be approached by considering the egg-retrieval response of the greylag goose (Figure 38-2), described by Lorenz and Tinbergen in a famous paper published in 1938. If Lorenz and Tinbergen presented a female greylag goose with an egg a short distance from her nest, she would rise,

Figure 38-2
Egg-rolling behavior of the greylag goose (*Anser anser*).

extend her neck until the bill was just over the egg and then bend her neck, pulling the egg carefully into the nest.

Although this behavior appeared to be intelligent, Tinbergen and Lorenz noticed that if they removed the egg once the goose had begun her retrieval, or if the egg being retrieved slipped away and rolled down the outer slope of the nest, the goose would continue the retrieval movement without the egg until she was again settled comfortably on her nest. Then, seeing that the egg had not been retrieved, she would begin the egg-rolling pattern all over again.

Thus the bird performed the egg-rolling behavior as if it were a program that, once initiated, had to run to completion. Lorenz and Tinbergen viewed egg-retrieval as a "fixed" pattern of behavior: a motor pattern that is mostly invariable in its performance. A behavior of this type, carried out in an orderly, predictable sequence is often called **stereotypical behavior.** Of course, stereotyped behavior may not be performed identically on all occasions. But it should be recognizable, even when performed inappropriately. Further experiments by Tinbergen disclosed that the greylag goose was not particularly discriminating about what she retrieved. Almost any smooth and rounded object placed outside the nest would trigger the egg-rolling behavior; even a small toy dog and a large yellow balloon were dutifully retrieved.

But once the goose settled down on such objects, they obviously did not feel right and she discarded them.

Lorenz and Tinbergen realized that the presence of the egg outside the nest must act as a stimulus, or trigger, that released the egg-retrieval behavior. Lorenz termed the triggering stimulus a **releaser**; a simple feature in the environment that would trigger a certain innate behavior. Or, because the animal usually responded to some specific aspect of the releaser (sound, shape, or color, for example) the effective stimulus was called a **sign stimulus.** Ethologists have described hundreds of examples of sign stimuli. In every case the response is highly predictable. For example, the alarm call of adult herring gulls always releases a crouching freeze-response in the chicks. Or, to cite an example given in an earlier chapter (p. 730), certain nocturnal moths take evasive maneuvers or drop to the ground when they hear the ultrasonic cries of bats that feed on them; most other sounds do not release this response.

These examples illustrate the predictable and programmed nature of much animal behavior. This is even more evident when stereotyped behavior is released inappropriately. In the spring the male three-spined stickleback selects a territory that it defends vigorously against other males. The underside of the male becomes bright red, and the approach

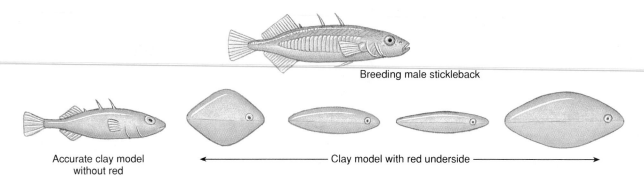

Breeding male stickleback

Accurate clay model without red

← ——————— Clay model with red underside ———————→

Figure 38-3

Stickleback models used to study territorial behavior. The carefully made model of the stickleback (*left*), without a red belly, is attacked much less frequently by a territorial male stickleback than the four simple red-bellied models.

Figure 38-4

Two models of the English robin. The bundle of red feathers is attacked by male robins, whereas the stuffed juvenile bird (*right*) without a red breast is ignored.

of another red-bellied male will release a threat posture or even an aggressive attack. Tinbergen's suspicion that the red belly of the male served as a releaser for aggression was reinforced when a passing red postal truck evoked attacks from the males in his aquarium. Tinbergen then carried out experiments using a series of models, which he presented to the males. He found that they vigorously attacked any model bearing a red stripe, even a plump lump of wax with a red underside. Yet a carefully made model that closely resembled a male stickleback but lacked the red belly was ignored (Figure 38-3). Tinbergen discovered other examples of stereotyped behavior released by simple sign stimuli. Male English robins furiously attacked a bundle of red feathers placed in their territory but ignored a stuffed juvenile robin without the red feathers (Figure 38-4).

We have seen in the examples above that there are costs to programmed behavior because it may lead to improper responses. Fortunately for red-bellied sticklebacks and red-breasted English robins, their aggressive response toward red works appropriately most of the time because red objects are uncommon in the worlds of these animals. But why don't these and other animals simply *reason* out the correct response rather than relying on mostly automatic responses? Under conditions that are relatively consistent and predictable, automatic preprogrammed responses may be most efficient. Even if they can or could, thinking about or learn-ing the correct response may take too much time. Releasers have the advantage of focusing the animal's attention on the relevant signal, and the release of a preprogrammed stereotyped behavior will enable an animal to respond rapidly when speed may be essential for survival.

CONTROL OF BEHAVIOR

From the beginning, the mostly invariable and predictable nature of stereotyped behavior suggested to ethologists that they were dealing with inherited, or **innate,** behavior. Many kinds of preprogrammed behavior appear suddenly in animals and are indistinguishable from similar behavior performed by older, experienced individuals. Orb-weaving spiders build their webs without practice, and male crickets court females without lessons from more experienced crickets or by learning from trial and error. To such behaviors the term innate, or instinctive, has been applied. Such words suggest that these kinds of behavioral patterns are absolutely committed and will develop in the same way regardless of environment. This idea, called instinct theory, has fallen out of fashion with behavioral scientists because it cannot be shown that a behavior develops independently of experience. Critics of instinct theory argue that all forms of behavior depend on an interaction of the organism and environment, beginning with the fertilized egg. Genes code for proteins and not directly for behavior. Even with web-building spiders and courting crickets, the environment is bound to have some influence. Given a different environment in which to develop, the resulting behavior may be different.

Nevertheless, it seems incontestable that many complex sequences of behavior in invertebrate animals are largely invariate in their execution, are not learned, and appear to be programmed by a set of rules. It is easy to understand why programmed behavior is important for survival, especially for animals that never know their parents. They must be equipped to respond to the world immediately and correctly as soon as they emerge into it. It is also evident that more complex animals with longer lives and with parental care or other opportunities for social interactions may improve or change their behavior by learning.

THE GENETICS OF BEHAVIOR

The hereditary transmission of most innate behaviors is complex. However, there are a few examples of behavioral differences within species that show simple Mendelian transmission from parents to offspring. Perhaps the most convincing example is the inheritance of hygienic behavior in bees. Honey bees are susceptible to a bacterial disease, American foul-

brood (*Bacillus larvae*). A bee larva that catches foulbrood dies. If the bees remove the dead larvae from the hive they reduce the chance of the infection spreading.

Some strains of bees, called "hygienic," uncap hive cells containing rotting larvae and carry them out of the hive. W. C. Rothenbuhler found that there are two components to this behavior: first the removal of cell caps, and second the removal of the larvae. Hygienic bees have homozygous recessive genotypes for two different genes. Uncapping behavior is performed by individuals homozygous for a recessive allele, *u*, at one gene, and removal behavior is performed by individuals homozygous for a recessive allele, *r,* at a second gene. When Rothenbuhler crossed hygienic bees (*u/u r/r*) with a nonhygienic strain (*U/U R/R*), he found that all the hybrids (*U/u R/r*) were nonhygienic. Thus only workers having both genes in the homozygous recessive condition show the complete behavior. Next, Rothenbuhler performed a "backcross" between the hybrids and the hygienic parental strain. As we should expect if hygienic behavior is transmitted by allelic variation at two genes, four different kinds of bees resulted. Approximately one-quarter of

the bees were homozygous recessive for both *u* and *r* and showed the complete behavior: they both uncapped the cells and removed the bees. Another quarter of the offspring (*u/u R/r* or *u/u R/R*) uncapped but did not remove dead bees. Another quarter (*U/u r/r* or *U/U r/r*) did not uncap, but would remove the larvae if another worker uncapped the cells. Workers that were homozygous or heterozygous for the dominant allele at both genes (*U/u R/r*) would not perform either part of the cleaning behavior (Figure 38-5). The results showed clearly that each component of the cleaning behavior was associated with one, independently segregating gene.

Most inherited behaviors do not show simple segregation and independence; instead, hybrids of subspecies or species commonly show intermediate or confused behavior. A classic study of the effect of cross-breeding on behavior was carried out by W. C. Dilger on nest-building behavior in different species of lovebirds. Lovebirds are small parrots of the genus *Agapornis* (Figure 38-6). Each species has its own method of courtship and technique for carrying nesting material. Fischer's lovebirds (*A. personata fischeri*) cut long strips

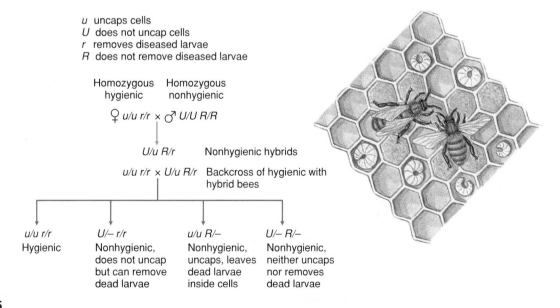

u uncaps cells
U does not uncap cells
r removes diseased larvae
R does not remove diseased larvae

Homozygous   Homozygous
hygienic   nonhygienic

♀ *u/u r/r* × ♂ *U/U R/R*

U/u R/r   Nonhygienic hybrids

u/u r/r × *U/u R/r*   Backcross of hygienic with hybrid bees

u/u r/r   *U/– r/r*   *u/u R/–*   *U/– R/–*
Hygienic   Nonhygienic,   Nonhygienic,   Nonhygienic,
                 does not uncap   uncaps, leaves   neither uncaps
                 but can remove   dead larvae   nor removes
                 dead larvae   inside cells   dead larvae

Figure 38-5
The genetics of hygienic behavior in honeybees, as demonstrated by W. C. Rothenbuhler. The results are explained by assuming that there are two independently assorting genes, one associated with uncapping cells containing diseased larvae, and the other associated with removing the larvae from the cells. See text for further explanation.

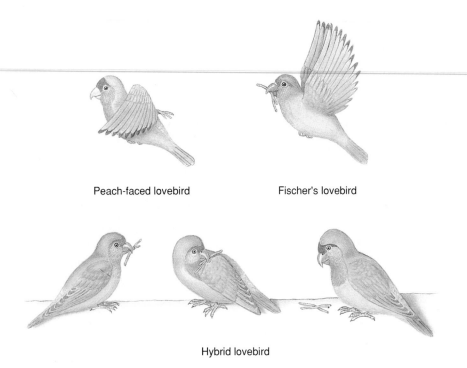

Peach-faced lovebird

Fischer's lovebird

Hybrid lovebird

Figure 38-6

Confused behavior in hybrid lovebirds (*Agapornis* sp.). The peach-faced lovebird carries nest-building material tucked into its feathers; Fischer's lovebird carries nest-building material in its beak. The hybrids attempted both carrying methods, neither method accomplished successfully.

Figure 38-7

The sea hare *Aplysia* sp., an opisthobranch gastropod used in many neurophysiological and behavioral studies.

of nesting material from vegetation, then carry this to the nest, one strip at a time. Peach-faced lovebirds (*A. roseicollis*) carry several strips of torn nesting material at one time by tucking them into the feathers of the lower back and rump. Dilger, who was able to cross the two species successfully, found that the hybrids displayed a confused conflict between the tendency to carry material in the feathers (inherited from the peach-faced lovebirds) and the tendency to carry material in the bill (inherited from Fischer's lovebird) (see Figure 38-6). Both feather-tucking and bill-carrying were attempted but neither was performed correctly. The hybrids had inherited a behavior that was intermediate between that of the parents. With experience the hybrids improved their carrying ability by tending to carry the material in their bills, like Fischer's lovebird.

LEARNING AND THE DIVERSITY OF BEHAVIOR

Another aspect of behavior is learning, which we can define as the mod-

ification of behavior through experience. An excellent model system for the study of learning processes has been the marine opisthobranch snail *Aplysia* (Figure 38-7), the subject of intense experimentation by E. R. Kandel and his associates. The gills of *Aplysia* are partly covered by the mantle cavity and open to the outside by a siphon (Figure 38-8). If one prods the siphon, *Aplysia* withdraws its siphon and gills and folds them up in the mantle cavity. This simple protective response, called the gill withdrawal reflex, will be repeated when *Aplysia* extends its siphon again. But if the siphon is touched repeatedly, *Aplysia* decreases the gill-withdrawal response and finally comes to ignore the stimulus altogether. This is a widespread form of learning called **habituation.** If now *Aplysia* is given a noxious stimulus (for example, an electric shock) to the head at the same time the siphon is touched, it becomes **sensitized** to the stimulus and withdraws its gills as completely as it did before habituation occurred. Sensitization, then, can reverse any previous habituation.

The mechanisms of habituation and sensitization in *Aplysia* are known because this is one of the rare instances in which the nervous pathways involved have been completely revealed. Receptors in the siphon are connected through sensory neurons (black pathways in Figure 38-8) to motor neurons (blue pathway in Figure 38-8) that control the gill-withdrawal muscles and muscles of the mantle cavity. Kandel found that repeated stimulation of the siphon caused a decline in the release of synaptic transmitter from the sensory neurons. The sensory neurons continue to fire when the siphon is probed but, with less neurotransmitter being released into the synapse, the system becomes less responsive.

Sensitization requires the action of a different kind of neuron called a facilitating interneuron. These interneurons make connections between sensory neurons in the animal's head and the motor neurons that control the gill and mantle muscles (see Figure 38-8). When the sensory neurons in the head are stimulated by an electric shock, they fire the facilitating interneurons, which end on the synaptic terminals of the sensory neurons (red pathways in Figure 38-8). These endings in turn cause an *increase* in the amount of transmitter released by the siphon sensory neurons. This increases the state of

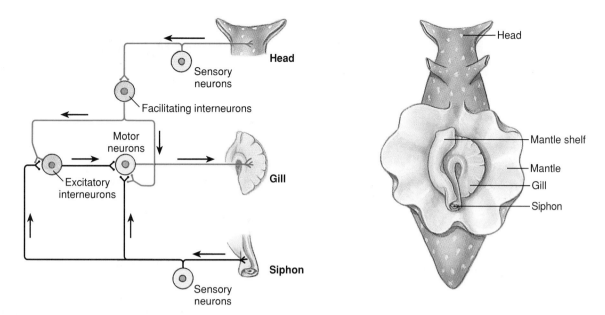

Figure 38-8
Neural circuitry concerned with habituation and sensitization of the gill-withdrawal reflex in the marine snail *Aplysia.* See text for explanation.

excitation in the excitatory interneurons and motor neurons leading to the gill and mantle muscles. The motor neurons now fire more readily than before. The system is now sensitized because any stimulus to the siphon will produce a strong gill-withdrawal response.

The studies of habituation and sensitization on *Aplysia* by Kandel and his colleagues represent the most complete account yet made of the neural and molecular mechanisms involved with learning. Kandel's studies indicate that the strengthening or weakening of the gill-withdrawal reflex involves changes in transmitter levels in existing synapses. However, we know that some cases of more complex kinds of learning involve the formation of new neural pathways and connections, as well as changes in existing circuits.

Imprinting

Another kind of learned behavior is **imprinting,** the imposition of a stable behavior pattern in a young animal by exposure to particular stimuli during a critical period in the animal's development. As soon as a newly hatched gosling or duckling is strong enough to walk, it follows its mother away from the nest. After it has followed the mother for some time it follows no other animal (Figure 38-9). But, if the eggs are hatched in an incubator or if the mother is separated from the eggs as they hatch, the goslings follow the first large object they see. As they grow, the young geese prefer the artificial "mother" to anything else, including their true mother. The goslings are said to be imprinted on the artificial mother.

Imprinting was observed at least as early as the first century A.D. when the Roman naturalist Pliny the Elder wrote of "a goose which followed Lacydes as faithfully as a dog." Konrad Lorenz was the first to study imprinting objectively and systematically. When Lorenz hand-reared goslings, they formed an immediate and permanent attachment to him and waddled after him wherever he went (see Figure 38-1A). They could no longer be induced to follow their own mother or another human being. Lorenz found that the imprinting period is confined to a brief sensitive period in the individual's early life and that once established the imprinted bond is mostly retained for life.

What imprinting shows is that the brain of the goose (or the brain of numerous other birds and mammals that show imprinting-like behavior) ac-

Figure 38-9
Canada goose, *Branta canadensis,* with her imprinted young.

commodates the imprinting experience. Natural selection favors the evolution of a brain that imprints in this way, in which following the mother and obeying her commands are important for survival. The fact that a gosling can be made to imprint to a mechanical toy duck or a person under artificial conditions is a cost to the system that can be tolerated because goslings seldom encounter these stimuli in their natural environment. The disadvantages of the system's simplicity are outweighed by the advantages of its reliability.

We will cite one final example to complete our consideration of learning. The males of many species of birds

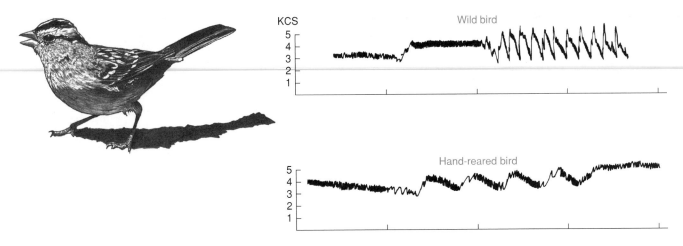

KCS
Wild bird

Hand-reared bird

Figure 38-10
Sound spectrograms of songs of white-crowned sparrows, *Zonotrichia leucophrys. Top,* natural songs of wild bird; *bottom,* abnormal song of isolated bird.

have characteristic territorial songs that identify the singers to the other birds and announce territorial rights to other males of that species. Like many other songbirds, the male white-crowned sparrow must learn the song of its species by hearing the song of its father. If the sparrow is hand-reared in acoustic isolation in the laboratory, it develops an abnormal song (Figure 38-10). But if the isolated bird is allowed to hear recordings of normal white-crowned sparrow songs during a critical period of 10 to 50 days after hatching, it learns to sing normally. It even imitates the local dialect it hears.

It might appear from this result that characteristics of the song are determined by learning alone. However, if during the critical learning period, the isolated male white-crowned sparrow is played a recording of another species of sparrow, even a closely related one, it does not learn the song. It learns only the song appropriate to its own species. Thus although the song must be learned, the brain is constrained to recognize and to learn vocalizations produced by males of its species alone. Learning the wrong song would result in behavioral chaos, and natural selection favors a system that eliminates such errors.

SOCIAL BEHAVIOR

When we think of "social" animals we are likely to think of highly structured honeybee colonies, herds of antelope

grazing on the African plains (Figure 38-11), schools of herring, or flocks of starlings. But social behavior of animals *of the same species* living together is by no means limited to such obvious examples in which individuals influence one another.

In the broad sense, any kind of interaction resulting from the response of one animal to another of the same species represents social behavior. Even a pair of rival males squaring off for a fight over the possession of a female is a social interaction, despite our perceptual bias as people that might encourage us to label it antisocial. Social aggregations are only one kind of social behavior, and indeed not all aggregations of animals are social.

Clouds of moths attracted to a light at night, barnacles attracted to a common float, or trout gathering in the coolest pool of a stream are groupings of animals responding to environmental signals. Social aggregations, on the other hand, depend on signals from the animals themselves. They remain together and do things together by influencing one another.

Not all animals showing sociality are social to the same degree. While all sexually reproducing species must at least cooperate enough to achieve fertilization, among some animals breeding is about the only adult sociality to occur. Alternatively, swans, geese, albatrosses, and beavers, to name just a few, form strong monoga-

Figure 38-11
Mixed herd of topi and common zebra grazing on the savanna of tropical Africa.

mous bonds that last a lifetime. The most persistent social bonds usually form between mothers and their young, and these bonds for bird and mammals usually terminate at fledging or weaning.

ADVANTAGES OF SOCIALITY

Living together may be beneficial in many ways. One obvious benefit for social aggregations is defense, both passive and active, from predators. Musk-oxen that form a passive defensive circle when threatened by a wolf pack are much less vulnerable than an individual facing the wolves alone.

As an example of active defense, a breeding colony of gulls, alerted by the alarm calls of a few, attack predators

Figure 38-12

An infant yellow baboon (*Papio cyanocephalus*) "jockey rides" its mother. Later, as the infant is weaned, the mother-infant bond weakens and the infant will be refused rides.

en masse, this is certain to discourage a predator more effectively than individual attacks. Members of a town of prairie dogs, although divided into social units called coteries, cooperate by warning each other with a special bark when danger threatens. Thus every individual in a social organization benefits from the eyes, ears, and noses of all other members of the group.

Sociality offers several benefits to animal reproduction. It facilitates encounters between males and females, which, for solitary animals, may consume much time and energy. Sociality also helps synchronize reproductive behavior through the mutual stimulation that individuals have on one another. Among colonial birds the sounds and displays of courting individuals set in motion prereproductive endocrine changes in other individuals. Because there is more social stimulation, large colonies of gulls produce more young per nest than do small colonies. Furthermore, the parental care that social animals provide their offspring increases survival of the brood (Figure 38-12). Social living provides opportunities for individuals to give aid and to share food with young other than their own. Such interactions within a social network have resulted in some intricate cooperative behavior among parents, their young, and their kin.

Of the many other advantages of social organization noted by ethologists, we will mention only a few in this brief treatment: cooperation in hunting for food; huddling for mutual protection from severe weather; opportunities for division of labor, which is especially well developed in the social insects with their caste systems; and the potential for learning and transmitting useful information through the society. This last advantage of social organization is discussed in the following example.

On the Galápagos Islands, hunting by humans in the last century had so greatly thinned the giant tortoise population on one island that the few surviving males and females seldom, if ever, met. Lichens grew on the females' backs because there were no males to scrub them off during mating! Research personnel saved the tortoise from the inevitable extinction by collecting them together in a pen, where they began to reproduce.

Observers of a seminatural colony of macaque monkeys in Japan recount an interesting example of acquiring and passing tradition in a society. The macaques were provisioned with sweet potatoes and wheat at a feeding station on the beach of an island colony. One day a young female named Imo was observed washing the sand off a sweet potato in seawater. The behavior was quickly imitated by Imo's playmates and later by Imo's mother. Still later when the young members of the troop became mothers they waded into the sea to wash their potatoes; their offspring imitated them without hesitation. The tradition was firmly established in the troop (Figure 38-13).

Some years later, Imo, an adult, discovered that she could separate wheat from sand by tossing a handful of sandy wheat in the water; allowing the sand to sink, she could scoop up the floating wheat to eat. Again, within a few years, wheat-sifting became a tradition in the troop.

Figure 38-13

Japanese macaque washing sweet potatoes. The tradition began when a young female named Imo began washing sand from the potatoes before eating them. Younger members of the troop quickly imitated the behavior.

Imo's peers and social inferiors copied her innovations most readily. The adult males, her superiors in the social hierarchy, would not adopt the practice but continued laboriously to pick wet sand grains off their sweet potatoes and scout the beach for single grains of wheat.

Social living also has some disadvantages as compared with a solitary existence for some animals. Species that survive by camouflage from potential predators profit by being dispersed. Large predators benefit from a solitary existence for a different reason, their requirement for a large supply of prey. Thus there is no overriding adaptive advantage to sociality that inevitably selects against the solitary way of life. It depends on the ecological situation.

AGGRESSION AND DOMINANCE

Many animal species are social because of the numerous benefits that sociality offers. This requires cooperation. At the same time animals, like governments, tend to look out for their own interests. In short, they are in competition with one another because of limitations in the common resources that all require for life. Animals may compete for food, water, sexual mates, or shelter when such requirements are limited in quantity and are therefore worth a fight.

Much of what animals do to resolve competition is called **aggression,** which we may define as an offensive physical action, or threat, to force others to abandon something they own or might attain. Many ethologists consider aggression part of a somewhat more inclusive interaction called **agonistic** (Gr. contest) behavior, referring to any activity related to fighting, whether it be aggression, defense, submission, or retreat.

Contrary to the widely held notion that aggressive behavior aims at the destruction or at least defeat of an opponent, most aggressive encounters are ritualized duels that lack the atmosphere of violence that we usually associate with fighting. Many species possess specialized weapons such as teeth, beaks, claws, or horns that are used for protection from, or predation on, other species. Although potentially dangerous, such weapons are seldom used in any effective way against members *of their own species.*

Animal aggression within the species seldom results in injury or death because animals have evolved many symbolic aggressive displays that carry mutually understood meanings. Fights over mates, food, or territory become ritualized jousts rather than bloody, no-holds-barred battles. When fiddler crabs spar for territory, their large claws usually are only slightly opened. Even in the most intense fighting when the claws are used, the crabs grasp each other in a way that prevents reciprocal injury. Rival male poisonous snakes engage in stylized bouts by winding themselves together; each attempts to butt the other's head with its own until one becomes so fatigued that it retreats. The rivals never bite each other. Many species of fish contest territorial boundaries with lateral display threats, the males puffing themselves to look as threatening as possible. The encounter is usually settled when either animal perceives itself obviously inferior, folds up its fins, and swims off. Rival giraffes engage in largely symbolic "necking" matches in which two males standing side by side wrap and unwrap their necks around each other (Figure 38-14). Neither uses its potentially lethal hooves on the other, and neither is injured.

The animals fight as though programmed by rules that prevent serious injury. Fights between rival bighorn rams are spectacular to watch, and the sound of clashing horns may be heard for hundreds of meters (Figure 38-15). But the skull is so well protected by the massive horns that injury occurs only by accident. Nevertheless, despite these constraints, aggressive encounters can on occasion be true fights to the death. If African male elephants are unable to resolve dominance conflicts painlessly with ritual postures, they may resort to incredibly violent battles, with each trying to plunge its tusks into the most vulnerable parts of the opponent's body.

More commonly, however, the loser of a ritualized encounter may simply run away, or signal defeat by a specialized subordination ritual. If it becomes evident to him that he is going to lose anyway, he is better off communicating his submission as quickly as possible and avoiding the cost of a real thrashing. Such submissive displays that signal the end of a fight may be almost the opposite of threat displays (Figure 38-16). In his book *Expression of the Emotions in Man and Animals* (1872), Charles Darwin described the seemingly opposite nature of threat and appeasement displays as the "principle of antithesis." The principle remains accepted by ethologists today.

Figure 38-14

Male Masai giraffes, *Giraffa camelopardalis,* fight for social dominance. Such fights are largely symbolic, seldom resulting in injury.

Figure 38-15

Male bighorn sheep, *Ovis canadensis,* fight for social dominance during the breeding season.

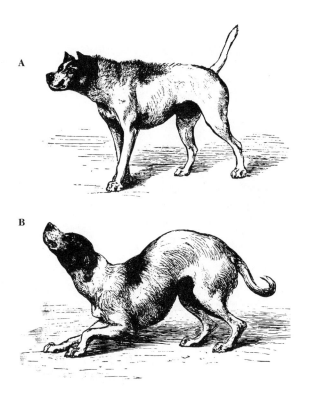

Figure 38-16
Darwin's principle of antithesis as exemplified by the postures of dogs. **A,** A dog approaches another dog with hostile, aggressive intentions. **B,** The same dog is in a humble and conciliatory state of mind. The signals of aggressive display have been reversed.

Sometimes the space defended moves with the individual. This individual distance, as it is called, can be observed as the spacing between swallows or pigeons on a wire, in gulls lined up on the beach, or in people lined up for a bus.

The winner of an aggressive competition is dominant to the loser, the subordinate. For the victor, dominance means enhanced access to all the contested resources that contribute to reproductive success: food, mates, and territory. In a social species, dominance interactions often take the form of a dominance hierarchy. One animal at the top wins encounters with all other members in the social group; the second in rank wins all but those with the top-ranking individual.

Such a simple, ordered hierarchy was first observed in chickens by Schjelderup-Ebbe, who called the hierarchy a "peck-order." Once social ranking is established, actual pecking diminishes and is replaced by threats, bluffs, and bows. Top hens and cocks get unquestioned access to feed and water, dusting areas, and the roost. The system works because it reduces the social tensions that would constantly surface if animals had to fight all the time over social position.

The subordinates in any social order are the expendables of the social group. They almost never get a chance to reproduce, and when times get difficult they are the first to die. During times of food scarcity, the death of the weaker members helps to protect the resource for the stronger members. Rather than sharing food, the population excess is sacrificed. This is not viewed by contemporary behaviorists as resulting from some direct, purposeful process ensuring "good for the species"; it is a *consequence* of the individual advantage that the stronger, dominant individuals possess during such circumstances.

TERRITORIALITY

Territorial ownership is another facet of sociality in animal populations. A territory is a fixed area from which intruders *of the same species* are excluded. This involves defending the area from intruders and spending long periods of time being conspicuous on the site. Territorial defense has been observed in numerous animals: insects, crustaceans, other invertebrates, fishes, amphibians, lizards, birds, and mammals, including humans.

Territoriality is generally an alternative to dominance behavior, although both systems may be observed operating in the same species. A territorial system may work well when the population is low, but it may break down with increasing population density and be replaced with dominance hierarchies with all animals occupying the same space.

Like every other competitive endeavor, territoriality carries both costs and advantages. It is beneficial when it ensures access to limited resources, *unless* the territorial boundaries cannot be maintained with little effort. The presumed benefits of a territory are, in fact, numerous: uncontested access to a foraging area; enhanced attractiveness to females thus reducing the problems of pair-bonding, mating, and rearing the young; reduced disease transmission; reduced vulnerability to predators. But the advantages of holding a territory begin to wane if the individual must spend most of the time in boundary disputes with neighbors.

Most of the time and energy required for territoriality are expended when the territory is first established. Once the boundaries are located they tend to be respected, and aggressive behavior diminishes as territorial neighbors come to recognize each other. Indeed, neighbors may look so peaceful that an observer who was not present when the territories were established may conclude (incorrectly) that the animals are not territorial. A "beachmaster" sea lion (that is, a dominant male with a harem) seldom quarrels with his neighbors who have their own territories to defend. However, he must be constantly vigilant against bachelor bulls who challenge the beachmaster for harem privileges.

Of all the vertebrate classes, birds are the most conspicuously territorial. Most male songbirds establish territories in the early spring and defend these vigorously against all males of the same species during spring and summer when mating and nesting are at their height. A male song sparrow, for example, has a territory of approximately three-fourths of an acre. In any given area, the number of song sparrows remains approximately the same year after year. The population remains stable because the young occupy territories of adults that die or are killed. Any surplus in the song sparrow population is excluded from territories and thus not able to mate or nest.

Sea birds such as gulls, gannets, boobies, and albatrosses occupy colonies that are divided into very small territories just large enough for nesting (Figure 38-17). The territories of these birds cannot include their fishing grounds, since they all forage in the sea where the food is always shifting in location and shared by all.

Territorial behavior is not as prominent with mammals as it is with birds. Mammals are less mobile than birds, and this makes it more difficult for them to patrol a territory for trespassers. Instead, many mammals have **home ranges** (p. 613). A home range is the total area an individual traverses in its activities. It is not an exclusive, defended preserve but overlaps with the home ranges of other individuals of the same species.

For example, the home ranges of baboon troops overlap extensively, although a small part of each range becomes the recognized territory of each troop for its exclusive use. Home ranges may shift considerably with the seasons. A baboon troop may have to shift to a new range during the dry season to obtain water and better grass. Elephants, before their movements were restricted by humans, made long seasonal migrations across the African savanna to new feeding ranges. However, the home ranges established for each season are remarkably consistent in size.

Figure 38-17
Gannet nesting colony. Note precise spacing of nests with each occupant just beyond pecking distance of its neighbors.

ANIMAL COMMUNICATION

Only through communication can one animal influence the behavior of another. Compared with the enormous communicative potential of human speech, however, nonhuman communication is severely restricted. Animals may communicate by sounds, scents, touch, and movement. Indeed any sensory channel may be used, and in this sense animal communication has richness and variety.

Unlike our language, which is composed of words with definite meanings that may be rearranged to generate an almost infinite array of new meanings and images, communication of other animals consists of a limited repertoire of signals. Typically, each signal conveys one and only one message. These messages cannot be divided or rearranged to construct *new kinds* of information. A single message from the sender may, however, contain several bits of relevant information for the receiver.

The song of a cricket announces to an unfertilized female the species

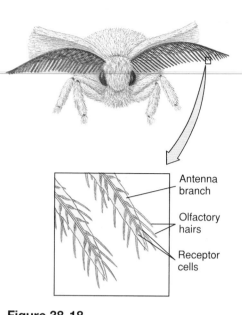

Figure 38-18
Large antennae of the male silkworm moth *Bombyx mori;* these are especially sensitive to the sex attractant (pheromone) released by the female moth.

Antenna branch

Olfactory hairs

Receptor cells

of the sender (males of different species have different songs), his sex (only males sing), his location (source of the song), and social status (only a male able to defend the area around his burrow sings from one location). This is all crucial information to the female and accomplishes a biological function. But there is no way for the male to alter his song to provide additional information concerning food, predators, or habitat, which might improve his mate's chances of survival and thus enhance his own fitness.

Chemical Sex Attraction in Moths

Mate attraction in silkworm moths illustrates an extreme case of stereotyped, single message communication that has evolved to serve a single biological function: mating. Virgin female silkworm moths have special glands that produce a chemical sex attractant to which the males are sensitive. Adult males smell with their large bushy antennae, covered with thousands of sensory hairs that function as receptors (Figure 38-18). Most of these receptors are sensitive to the chemical the dance consists of three components: (1) a circle with a diameter about three times the length of the bee, (2) a straight run while waggling the abdomen from side to side and emitting a pulsed, low-frequency sound, and (3) another circle, turning in the opposite direction from the first. This dance is repeated many times with the circling alternating clockwise and counterclockwise.

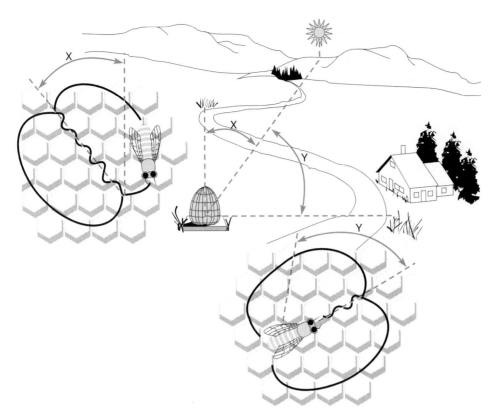

Figure 38-19
Waggle dance of the honeybee used to communicate both the direction and distance of a food source. The straight run of the waggle dance indicates direction according to the position of the sun (angles *X* and *Y*).

attractant **bombykol** (a complex alcohol named after the silkworm *Bombyx mori*) and to nothing else.

To attract the male, the female merely sits quietly and emits a minute amount of bombykol, which is carried downwind. When a few molecules reach the male's antennae, he is stimulated to fly upwind in search of the female. His search is at first random, but, when by chance he approaches within a few hundred yards of the female, he encounters a concentration gradient of the attractant. Guided by the gradient, he flies toward the female, finds her, and copulates with her.

In this example of chemical communication the attractant bombykol, a pheromone (p. 727), serves as a signal to bring the sexes together. Its effectiveness is ensured because natural selection favors the evolution of males with antennal receptors sensitive enough to detect the attractant at great distances (several miles). Males with a genotype that produces a less sensitive sen-

sory system fail to locate a female and thus are reproductively eliminated from the population.

Language of the Bees

One of the most sophisticated and complex of all nonhuman communication systems is the symbolic language of bees. Honeybees are able to communicate the location of food resources when these sources are too distant to be located easily by individual bees. Communication is done by dances, which are mainly of two forms. The form having the most communicative richness is the **waggle dance** (Figure 38-19). Bees most commonly execute these dances when a forager has returned from the rich source, carrying either nectar in her stomach or pollen grains packed in basketlike spaces formed by hairs on her legs. The waggle dance is roughly in the pattern of a figure eight made against the vertical surface on the comb inside the hive. One cycle of

The significance of the bee dances was discovered in 1943 by German zoologist Karl von Frisch, one of the recipients of the 1973 Nobel Prize. Despite detailed and extensive experiments by von Frisch and others that supported his original interpretations of the bee dances, the experiments have been criticized, especially by American biologist Adrian Wenner, who suggests that the correlation between dance symbolism and food location is accidental. He argued that foraging bees bring back odors characteristic of the food source, and that recruits are stimulated by dance to search for flowers bearing those odors. Wenner, with P. H. Wells, has reviewed his studies as a scientific autobiography and polemic (*Anatomy of a controversy: the question of a language among bees,* 1990, Columbia University Press). Wenner and Wells' assertions have generated strong controversy and have stimulated more rigorously controlled research on the bee dances. Recently, researchers have constructed a robot bee that can be moved through the waggle dance while producing the dance song with vibrating metal "wings." When operated in a hive, the computer-directed robot successfully recruited attending bees to visit preselected food dishes outside the hive that had never been visited previously. These experiments provide convincing evidence that the bee dances do communicate both direction and distance information to foraging bees.

The straight, waggle run is the important informational component of the dance. Waggle dances are performed almost always in clear weather, and the direction of the straight run is related to the position of the sun. If

Figure 38-20
A pair of Galápagos blue-footed boobies, *Sula nebouxii,* display to each other. The male (*right*) is sky pointing; the female (*left*) is parading. Such vivid, stereotyped, communicative displays serve to maintain reciprocal stimulation and cooperative behavior during courtship, mating, nesting, and care of the young.

the forager has located food directly toward the sun, she will make her waggle run straight upward over the vertical surface of the comb. If food was located 60 degrees to the right of the sun, her waggle run is 60 degrees to the right of vertical. We see then that the waggle run points at the same angle relative to the vertical as the food is located relative to the sun.

Distance of the food source is also coded into bee dances. If the food is close to the hive (less than 50 m), the forager employs a simpler dance called the **round dance.** The forager simply turns a complete clockwise circle, then turns, and completes a counterclockwise circle, a performance that is repeated many times. Other workers cluster around the scout and become stimulated by the dance as well as by the odor of nectar and pollen grains from flowers she has visited. The recruits then fly out and search in all directions but do not stray far. The round dance carries the message that food is to be found in the vicinity of the hive.

If the food source is farther away, the round dances become waggle dances, which provide both distance and directional information. The tempo of the waggle dance is inversely related to the food's distance. If the food is about 100 m away, each figure-eight cycle lasts about 1.25 seconds; if 1000 m away, it lasts about 3 seconds; and if about 8 km (5 miles) away, it lasts 8 seconds. When food is plentiful, the bees may not dance at all. But when food is scarce, the dancing becomes intense, and the other workers cluster around the returning scouts and follow them through the dance patterns.

Communication by Displays

A display is a kind of behavior or series of behaviors that serves a communicative purpose. The release of sex attractant by the female moth and the dances of bees just described are examples of displays; so are the alarm calls of herring gulls, song of the white-crowned sparrow, courtship dance of the sage grouse, and "eyespots" on the hind wings of certain moths that are quickly exposed to startle potential predators. A true display is a behavior pattern that has been modified through evolution to make it increasingly effective in serving a communicative function. This process is called **ritualization.** Through ritualization, simple movements or traits become more intensive, conspicuous, or precise, and acquire the function of a signal. The result of such intensification is to reduce the possibility of misunderstanding.

The elaborate pair-bonding behavior of the blue-footed boobies (Figure 38-20) exemplifies this point. These displays are performed with maximum intensity when the birds come together after a period of separation. The male at right in the illustration is sky pointing: the head and tail are pointed skyward and the wings are swiveled forward in a seemingly impossible position to display their glossy upper surfaces to the female. This is accompanied by a high-piping whistle. The female at left, for her part, is parading. She steps with exaggerated slow deliberation, lifting each brilliant blue foot in turn, as though holding it aloft momentarily for the male to admire. Such highly personalized displays, performed with droll solemnity, appear comical, even inane to the observer. Indeed the boobies, whose name is derived from the Spanish word "bobo" meaning clown, presumably were so designated for their amusing antics.

Needless to say, for the birds amusement plays no part in the ceremonies. The exaggerated nature of the displays ensures that the message is not missed or misunderstood. Such displays are essential to establish and maintain a strong pair bond between male and female. This requirement also explains the repetitious nature of the displays that follow one another throughout courtship and until laying of eggs. Redundancy of displays maintains a state of mutual stimulation between male and female, ensuring the degree of cooperation necessary for copulation and subsequent incubation and care of the young. A sexually aroused male has little success with an indifferent female.

Even highly organized communicative behavior does not imply an animal's intelligent appreciation of its purpose. Sometimes apparently intelligent behavior misfires, and such incidents often emphasize its stereotyped nature. The following excerpt contains a perfect example of the automatic release of inappropriate behavior in a gannet colony:*

> A male of an old pair flew into his nest. Normally, he would bite his mate on the head with some violence and then go through a long and complicated meeting ceremony, an ecstatic display confined to members of a pair. Unfortunately, the female had caught her lower mandible in a loop of fish netting that was firmly anchored in the structure of the nest. Every time she tried to raise her head to perform the meeting ceremony with her mate she merely succeeded in opening her upper mandible whilst the lower remained fixed in the netting. So she apparently threatened the male with widely gaping beak and he immediately responded by attacking her. With each attack she lowered and turned away her bill (the way in which a female gannet appeases an aggressive male). At once the male stopped biting her and she again turned to greet him but simply repeated the beak-opening and drew another attack. And so it went on despite the fact that these two birds had been mated for years and that the netting, the cause of all the trouble, was clearly visible.

Communication between Humans and Other Animals

One uncertainty in studies of animal communication is understanding what sensory channel an animal is using. The signals may be visual displays, odors, vocalizations, tactile vibrations, or electrical currents (as, for example, among certain fishes). Even more difficult is establishing two-way communication between animals and humans, since the investigator must translate meanings into symbols that the animal can understand. Furthermore, people are poor social partners for most other animals. However, in the 1970s a female chimpanzee named Washoe was taught to employ *and combine* gestures, using words from the American Sign Language for the deaf, much as people use spoken words. The discovery that manual gestures and expressive motions were much more appropriate than vocalizations in communicating with apes was considered a major breakthrough in behavioral research. Since Washoe, sign-language studies have been extended to other chimpanzees and to gorillas; several have acquired "vocabularies" of several hundred reliable signs, some invented by the apes themselves.

> Whether apes are able to combine signals (that is, "words") to produce sentences—an ability that is crucial to true language—cannot yet be answered conclusively. Furthermore, although apes use signals for requests, commands, and pleas, they do not appear able to make statements aimed at changing the behavior or beliefs of the other party. Critics of the first ape-language studies charged that most of the signed utterances of apes were prompted by the teacher. They argued that the investigators had fallen victim to the "Clever Hans" error, after a horse prodigy of the turn of the century that appeared to do sums in his head and tap out answers with his hoof, but in fact was responding to subtle and inadvertent cues from the trainer. More recent studies appear to have dealt satisfactorily with this criticism by isolating the ape from the teacher during the training session.

The animal behaviorist Irven DeVore has reported how choosing the proper channel for dialogue can have more than academic interest:*

One day on the savanna I was away from my truck watching a baboon troop when a young juvenile came and picked up my binoculars. I knew if the glasses disappeared into the troop they'd be lost, so I grabbed them back. The juvenile screamed. Immediately every adult male in the troop rushed at me—I realized what a cornered leopard must feel like. The truck was 30 or 40 feet away. I had to face the males. I started smacking my lips very loudly, a gesture that says as strongly as a baboon can, "I mean you no harm." The males came charging up, growling, snarling, showing their teeth. Right in front of me they halted, cocked their heads to one side—and started lip-smacking back to me. They lip-smacked. I lip-smacked, "I mean you no harm." "I mean *you* no harm." It was, in retrospect, a marvelous conversation. But while my lips talked baboon, my feet edged toward the truck until I could leap inside and close the door.

The study of animal communication has made great strides in recent years, buoyed by the assimilation of a wealth of facts and information about communication in many species. The animal world is filled with communication. In recognizing that reasoning and insight are not required for effective, highly organized behavior, we should not conclude that other animals are, as Descartes proclaimed in the seventeenth century, nothing more than machines. Although the gannet in the example given earlier lacked the "intelligence" to free his mate by purposefully disentangling her beak from the netting, he was capable of appropriately making thousands of strategic choices during his lifetime: how to find and hold a mate, where to build a nest and how to defend it, how to locate and catch evasive marine food, and what to do when the environment changes. All this and more requires endless behavioral adjustments to new situations. Conceivably this might be accomplished by a machine, but only by one of staggering complexity.

*Nelson, B. 1968. Galápagos: islands of birds. London, Longmans, Green & Company, Ltd. Reprinted by permission of William Morrow & Company.

*DeVore, Irven. The marvels of animal behavior. 1972. Washington, D.C., National Geographic Society.

Summary

Animal behavior has emerged as a scientific discipline from three different experimental approaches. Comparative psychology emphasizes the identification of mechanisms controlling behavior, using relatively few species, with the intent that these might have wide applicability among animals. Ethology is the study of the behavior, both innate and learned, of animals in their natural habitats. Ethologists have shown that behavioral traits have evolutionary histories and evolve by natural selection. Sociobiology aims to understand how and why social behavior in animals has evolved. Both ethology and sociobiology distinguish between studies that focus on the mechanisms of behavior (proximate causation) and those that focus on function or evolution of behavior (ultimate causation).

Students of animal behavior have observed and cataloged many behavioral patterns of animals that are highly predictable and almost invariable in performance. Often these patterns are triggered, or "released," by specific, and usually simple, environmental stimuli, called sign stimuli. Although such formalized behaviors may be released inappropriately at times, they are efficient and enable the animal to respond rapidly. The development of behavioral patterns depends on an interaction between the organism and the environment in which the animal lives. For this reason, behavioral scientists prefer not to describe behaviors—those that are largely invariable in their performance—as "instinctive" or "innate."

Behavior may be modified by learning through experience. Two simple kinds of learning behavior are habituation, which is the reduction or elimination of a behavioral response in the absence of any reward or punishment; and sensitization, in which a repeated stimulus increases the strength of a behavioral response. The gill-withdrawal reflex of the marine mollusc *Aplysia* is described as a protective response that can be modified experimentally to show either habituation or sensitization. The modification of the alarm response of herring gull chicks is another example of habituation. Another form of learning is imprinting, the lasting recognition bond that forms early in life between the young of many social animals and their mothers.

Social behavior is the behavior of a species when the members interact with one another. In social organizations, animals tend to remain together, communicate with each other, and usually resist intrusions by "outsiders." The advantages of sociality include cooperative defense from predators, cooperative searching for food, improved reproductive performance and parental care of the young, and transmission of useful information through the society. Because social animals compete with one another for resources (such as food, sexual mates, and shelter), conflicts are often resolved by a form of overt hostility called aggression. Most aggressive encounters between conspecifics are stylized bouts involving more bluff than intent to injure or kill. Dominance hierarchies, in which a priority of access to common resources is established by aggression, are common in social organizations. Territoriality is an alternative to dominance. A territory is a defended area from which intruders of the same species are excluded.

Communication, often considered the essence of social organization, is the means by which animals influence the behavior of other animals, using sounds, scents, visual displays, touch, or other sensory signals. As compared with the richness of human language, animals communicate with a very limited repertoire of signals. One of the most famous examples of animal communication is that of the symbolic dances of honeybees. Birds communicate by calls and songs and, especially, by visual displays. By ritualization, simple movements have evolved to become conspicuous signals having definite meanings.

Review Questions

1. How do the experimental approaches of comparative psychology and ethology differ? Comment on the aims and methods employed by each.
2. The egg-retrieval behavior of greylag geese is an excellent example of a highly predictable behavior. Interpret this behavior within the framework of classic ethology, using these terms: releaser, sign stimulus, and stereotyped behavior. Interpret the territorial defense behavior of male three-spined sticklebacks in the same context.
3. The idea that behavior must be *either* innate or learned has been called the "nature versus nurture" controversy. What reasons are there for believing that such a strict dichotomy does not exist?
4. Two kinds of simple learning are habituation and imprinting. Distinguish between these two types of learning, and offer an example of each.
5. Some strains of bees show hygienic behavior by uncapping cells containing larvae infected with a bacterial disease called foulbrood and removing the dead larvae from the hive. What is the evidence that this behavior is transmitted by two independently segregating genes?
6. Discuss the advantages of sociality for animals. If social living has so many advantages, why do many animals successfully live alone?
7. Suggest why aggression, which might seem counterproductive, exists among social animals.
8. What is the selective advantage, to the probable winner as well as the loser, of ritualized aggression over unrestrained fighting?
9. Of what use is a territory to an animal, and how is a territory established and kept? What is the difference between territory and home range?
10. Comment on the limitations of animal communication as compared with those of human communication.
11. The dance language used by returning forager honeybees to specify the location of food is a remarkable example of complex communication among "simple" animals. How is direction and distance information coded into the waggle dance of the bees? What is the purpose of the round dance?
12. What is meant by "ritualization" in display communication? What is the adaptive significance of ritualization?
13. Early efforts by humans to communicate vocally with chimpanzees were almost total failures. Recently, however, researchers have learned how to communicate successfully with apes. How was this done?

Selected References

Alcock, J. 1993. Animal behavior: an evolutionary approach, ed. 5. Sunderland, Massachusetts, Sinauer Associates, Inc. *Clearly written and well-illustrated discussion of the genetics, physiology, ecology, and history of behavior in an evolutionary perspective.*

Attenborough, D. 1990. The trials of life: a natural history of animal behavior. Boston, Little, Brown and Company. *Superb photographs and flowing text describes the life cycles of organisms, often focusing on unusual and exciting patterns of behavior.*

Drickamer, L. C., and S. H. Vessey. 1992. Animal behavior: mechanisms, ecology and evolution, ed. 3. Dubuque, William C. Brown. *Comprehensive, with helpful discussions on the methods and experimental approaches used to answer behavioral questions.*

Gould, J. L., and C. G. Gould. 1994. The animal mind. New York, Scientific American Library. *Attractively illustrated, engagingly written exploration of animal behavior and the efforts of researchers to determine what happens inside the minds of animals.*

Greenspan, R. J. 1995. Understanding the genetic construction of behavior. Sci. Am. **272:**72–78 (Apr.). *Studies of courtship and mating in fruit flies indicates that behavior is regulated by many multipurpose genes, each of which handles diverse responsibilities in the body.*

Grier, J. W., and T. Burk. 1992. Biology of animal behavior. St. Louis, Mosby-Year Book, Inc. *Popular text with broad coverage of ethology, comparative psychology, and neurobiology.*

Huber, F., and J. Thorson. 1985. Cricket auditory communication. Sci. Am. **253:**60–68 (Dec.). *How female crickets respond to the mating song of the male, and the brain neuronal machinery underlying the response.*

Kirchner, W. H., and W. F. Towne. 1994. The sensory basis of the honeybee's dance language. Sci. Am. **270:**74–80 (June). *Experiments with a robotic honeybee that can dance and emit sounds similar to living bees show conclusively that the dance language successfully recruits foragers to food outside the hive.*

Lorenz, K. Z. 1952. King Solomon's ring. New York, Thomas Y. Crowell Company, Inc. *One of the most delightful books ever written about the behavior of animals.*

Manning, A., and M. S. Dawkins. 1992. An introduction to animal behaviour, ed. 4. Cambridge, England, Cambridge University Press. *Survey of animal behavior, drawing from ethology, physiology, and comparative psychology.*

Preston-Mafham, R. and K. Preston-Mafham. 1993. The encyclopedia of land invertebrate behavior. Cambridge, Massachusetts, The MIT Press. *Numerous examples of fascinating invertebrate behavior in a series of informative and beautifully illustrated essays. Highly recommended.*

Ridley, M. 1995. Animal behavior: an introduction to behavioral mechanisms, development, and ecology, ed. 2. Oxford, Blackwell Scientific Publications. *The principles of animal behavior presented with well-chosen examples and clear illustrations.*

Savage-Rumbaugh, E. S. 1986. Ape language: from conditioned response to symbol. New York, Columbia University Press. *Details the author's studies as well as the general area of ape language.*

Slater, P. J. B. 1987. Encyclopedia of animal behavior. Encyclopedia of Animals series. New York, Facts on File, Inc. *Hardly an "encyclopedia" but the coverage is extensive and the illustrations excellent.*

References to Part IV

The books listed in the following selection are mainly textbooks covering wide areas of physiology. They vary considerably in depth and in the level of background in biology and chemistry required of the reader for a full understanding, as indicated in the annotations.

Alexander, R. M. (ed.). 1987). The encyclopedia of animal biology. New York, Facts on File, Inc. *Semipopular account answering common questions about animal function. Beautifully illustrated.*

Annual Review of Physiology. 1939–present. Palo Alto, California, Annual Reviews, Inc. *Contributed review articles selected from all major disciplines of physiology.*

Berne, R. M., and M. N. Levy. 1993. Physiology. St. Louis, Mosby Year Book. *Comprehensive general and mammalian (mostly human) physiology.*

Eckert, R., D. Randall, and G. Augustine. 1988. Animal physiology, ed. 3. *Special emphasis on general principles and on membrane, neural, and sensory physiology.*

Guyton, A. C., and J. E. Hall. 1996. Textbook of medical physiology, ed. 9. Philadelphia, W. B. Saunders Company. *A detailed but readable treatment of medical physiology.*

Hoar, W. S. 1983. General and comparative physiology, ed. 3. Englewood Cliffs, New Jersey, Prentice-Hall, Inc. *Balanced synopsis of comparative physiology.*

Marshall, P. T., and G. M. Hughes. 1981. Physiology of mammals and other vertebrates, ed. 2. Cambridge, England, Cambridge University Press. *Undergraduate comparative physiology; clearly presented and illustrated organ-system approach.*

Martini, F. 1995. Fundamentals of anatomy and physiology, ed. 3. Englewood Cliffs, New Jersey, Prentice-Hall. *College-level human anatomy and physiology; comprehensive coverage.*

Miller, J. 1978. The body in question. New York, Random House, Inc. *Selective, historically grounded view of human body function, based on a television series.*

Prosser, C. L. (ed.). 1991. Comparative animal physiology, ed. 4. Philadelphia. W. B. Saunders Company. *Advanced treatise.*

Prosser, C. L. 1986. Adaptational biology: molecules to organisms. New York, John Wiley & Sons. *A wide-ranging synthesis of general biological principles and the interrelationships between disciplines, from molecular to whole-organism biology. Advanced.*

Schmidt-Nielsen, K. 1990. Animal physiology: adaptation and environment, ed. 4. New York, Cambridge University Press. *Clearly written, selective treatment of comparative physiology, emphasizing physiological adaptations to the environment.*

Seeley, R. R., T. D. Stephens, and P. Tate. 1992. Anatomy & physiology, ed. 2. St. Louis, Mosby-Year Book, Inc. *Readable and well-illustrated treatment of human anatomy and physiology.*

Smith, A. 1986. The body. New York, Viking Penguin, Inc. *A fascinating store of facts about the human body.*

Vander, A. J., J. H. Sherman, and D. S. Luciano. 1994. Human physiology: the mechanisms of the body function, ed. 5. New York, McGraw-Hill Book Company. *Excellent intermediate-level human physiology text.*

Withers, P. C. 1992. Comparative animal physiology. New York, Saunders College Publishing.

Young, J. Z. 1981. The life of vertebrates, ed. 3. Oxford, England, Oxford University Press. *The comprehensive treatment and trenchant writing style of this classic have been retained in this updating, which has increased emphasis on physiology, ecology, and behavior.*

The Animal and Its Environment

39

The Biosphere and Animal Distribution

Spaceship Earth

All life is confined to a thin veneer of the earth called the **biosphere.** From the first remarkable photographs of earth taken from the *Apollo* spacecraft, revealing a beautiful blue and white globe lying against the limitless backdrop of space, viewers were struck and perhaps humbled by our isolation and insignificance in the enormity of the universe. The phrase "spaceship earth" became a part of the vocabulary, and the realization evolved that all the resources we will ever have for sustaining life are restricted to a thin layer of land and sea and a narrow veil of atmosphere above it. We could better appreciate just how thin the biosphere is if we could shrink the earth and all of its dimensions to a 1 m sphere. We would no longer perceive vertical dimensions on the earth's surface. The highest mountains would fail to penetrate a thin coat of paint applied to our shrunken earth; a fingernail's scratch on the surface would exceed the depth of the ocean's deepest trenches.

The earth's biosphere and the organisms in it have evolved together. In the continuous interchange between organism and environment, both have been altered, and a favorable relationship preserved. The earth's biosphere, with its living and nonliving components, is not a static thing but has undergone an evolution in every way as dramatic as the evolution of the animal kingdom. Today the biosphere is changing rapidly under the impact of humans, one of the greatest agents of biotic disturbance the earth has ever known. Only the historical bombardment of the earth by asteroids has produced a greater disturbance of the earth's biota. ■

In a universe of billions of stars, our earth is a small planet circling an ordinary star. Thousands of other stars are like our sun with planetary systems that conceivably could support life. Yet, of all these, our planet is the only one that we *know* supports life. Until proven otherwise, the earth is unique, a true wonder of an infinite universe.

What makes earth an especially fit environment for life? Most biologists would agree that foremost is the presence of liquid water on the earth's surface. Water, with its many extraordinary physical properties, provided the medium for the origin of life and bestowed on earth a moderate climate suitable for life's continued evolution. Many other properties of earth make it optimal for life. Among these are a steady supply of light and heat from an unfailing sun; a suitable range of temperature for life, neither too hot nor too cold; a supply of the major and minor elements required by living matter; and a gravity force strong enough to hold an extensive gaseous atmosphere.

The many properties that make the earth wonderfully suitable for life were first recognized and examined in detail by Lawrence J. Henderson (1878 to 1942) in his book *The Fitness of the Environment,* published in 1913. The profound insights of this distinguished Harvard biochemist and physiologist were remarkable, appearing as they did before ecology had become a science. His insightful understanding of reciprocity between organism and environment has become a principle that underlies all ecological science. Henderson's book deserves a broader appreciation than it has received; it is, for example, seldom mentioned in ecology textbooks.

The organism and its environment share a reciprocal relationship. The environment is modified by organisms, and populations of organisms are modified by the evolutionary process to adapt them to the environment and its changes. As an open system, an animal is forever receiving and giving off materials and energy. The building materials for life are obtained from the physical environment, either directly by producers such as green plants or indirectly by consumers that return inorganic substances to the environment by excretion or by the decay and disintegration of their bodies.

The living form is a transient link that is built of environmental materials, which are then returned to the environment to be used again in the re-creation of new life. Life, death, decay, and re-creation have been the cycle of existence since life began.

The primitive earth of 3.5 billion years ago, barren, stormy, and volcanic with a reducing atmosphere of ammonia, methane, and water, was wonderfully fit for the prebiotic syntheses that led to life's beginnings. Yet, it was totally unsuited, indeed lethal, for the kinds of living organisms that inhabit the earth today, just as early forms of life could not survive in our present environment. The appearance of free oxygen in the atmosphere, produced largely if not almost entirely by life, is an example of the reciprocity between organism and environment. Although oxygen was at first poisonous to early forms of life, its gradual accumulation from photosynthesis over the ages forced protective biochemical alterations to appear that led eventually to complete dependence on oxygen by most organisms. As living organisms adapt and evolve, they act on and produce changes in their environment. In so doing they must themselves change.

DISTRIBUTION OF LIFE ON EARTH

BIOSPHERE AND ITS SUBDIVISIONS

The biosphere as usually defined is the thin outer layer of the earth capable of supporting life. It is probably best viewed as a global system that includes all life on earth and the physical environments in which living organisms exist and interact. The nonliving subdivisions of the biosphere include the lithosphere, hydrosphere, and atmosphere.

The **lithosphere** is the rocky material of the earth's outer shell and is the ultimate source of all mineral elements required by living organisms. The **hydrosphere** is the water on or near the earth's surface, and it extends into the lithosphere and the atmosphere. Water is distributed over the earth by a global hydrological cycle of evaporation, precipitation, and runoff. Five-sixths of the evaporation is from the ocean, and more water is evaporated from the ocean than is returned to it by precipitation. Oceanic evaporation therefore provides much of the rainfall that supports life on land. The gaseous component of the biosphere, the **atmosphere,** extends to some 3500 km above the surface of the earth, but all life is confined to the lowest 8 to 15 km (troposphere). The screening layer in the atmosphere of oxygen-ozone is concentrated mostly between 20 and 25 km. The main gases present in the troposphere are (by volume) nitrogen, 78%; oxygen, 21%; argon, 0.93%; carbon dioxide, 0.03%; and variable amounts of water vapor.

Atmospheric oxygen has originated almost entirely from photosynthesis. As discussed in Chapter 3, the primitive earth contained a reducing atmosphere devoid of oxygen. When oxygen-evolving photosynthesis appeared about 3 billion years ago, oxygen gradually began to accumulate in the atmosphere. It is believed that by the mid-Paleozoic era, some 400 million years BP, the oxygen concentration had reached its present level of about 21%. Since then, oxygen consumption by animals and plants has approximately equaled oxygen production. The present surplus of free oxygen in the atmosphere resulted from the fossilization of plants before they could decay or be consumed by animals. As these vast stores of fossil fuels are burned by our industrialized civilization, the oxygen surplus that accumulated over the ages conceivably could be depleted. Fortunately, this is unlikely for two reasons: (1) most of the total fossilized carbon is in the form of noncombustible shales and rocks, and

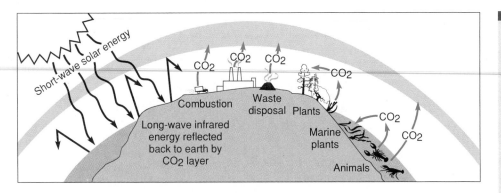

Figure 39-1

"Greenhouse effect." Carbon dioxide and water vapor in the atmosphere are transparent to sunlight but absorb heat energy reradiated from the earth, leading to warming of atmospheric air.

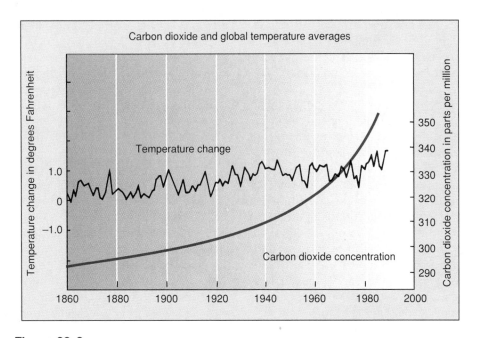

Figure 39-2

Rise in global atmospheric carbon dioxide and global temperature averages for the past 140 years. Data points before 1958 come from analysis of air trapped in bubbles in glacial ice from sites around the world. Atmospheric carbon dioxide has climbed steadily for more than a century while the earth's temperature has followed a more erratic upward trend.

The concern over the long-range effects of increasing atmospheric carbon dioxide, primarily from the burning of fossil fuel, stems not from mere conjecture. Atmospheric carbon dioxide increased from about 280 parts per million (ppm) before the Industrial Revolution to an average 335 ppm today and is increasing at a rate of 1.3 ppm per year. It is expected to exceed 600 ppm in the next century. In the past century the global temperature has increased 0.4° C, and it will have increased 3° C when the carbon dioxide has doubled in the next century, a warming of unprecedented magnitude. This is expected to shift climatic zones profoundly, creating drought conditions in the American wheatlands and wetter conditions along the coasts. Melting of the earth's ice sheets is another probability, with flooding of coastal lowlands throughout the world. Whatever the outcome, a fascinating, unplanned global experiment is underway.

TERRESTRIAL ENVIRONMENTS: BIOMES

A biome is a major biotic unit bearing a characteristic and easily recognized array of plant life. Botanists long ago recognized that the terrestrial environment of the earth could be divided into large units having a distinctive vegetation, such as forests, prairies, and deserts. Animal distribution has always been more difficult to map, because plant and animal distributions do not exactly coincide. Over time zoogeographers came to accept plant distributions as the basic biotic units and recognized the biomes as distinctive combinations of plants and animals. A biome is therefore identified by its dominant plant formation (Figure 39-3), but, since animals depend on plants, each biome supports a characteristic fauna.

Each biome is distinctive, but its borders are not. Anyone who has traveled across North America knows that plant communities grade into one another over broad areas. The moist deciduous forests of the Appalachians

(2) the oxygen reserves in the atmosphere and in the oceans are so enormous that the supply could last thousands of years even if all photosynthetic replenishment suddenly were to cease.

There is concern, however, that the rapid input of carbon dioxide into the atmosphere from the burning of fossil fuels may significantly affect the earth's heat budget. Much of the sun's short-wave light energy absorbed by the earth's surface reradiates as longer-wave infrared heat energy (Figure 39-1). Materials in the atmosphere, especially carbon diox-ide and water vapor, impede this heat loss and allow the atmosphere to warm up. This is called the "greenhouse effect," since the atmosphere acts to trap reradiated heat from the earth in much the same way the glass of a greenhouse traps the heat reradiated by the plants and soil inside. While the greenhouse effect provides conditions essential for all life on earth, there is concern that the gradual accumulation of carbon dioxide could lead to an increase in the temperature of the biosphere as a whole (Figure 39-2).

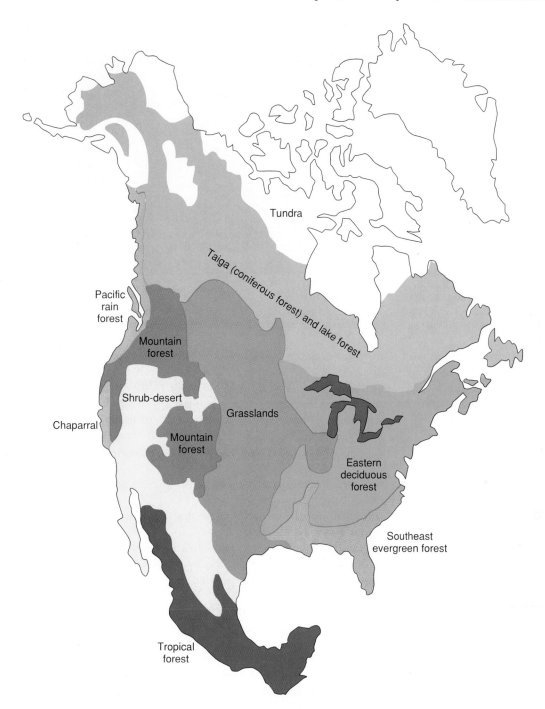

Figure 39-3

Major biomes of North America. Boundaries between biomes are not distinct as shown but grade into one another over broad areas.

give way gradually to the drier oak forests of the upper Mississippi Valley, and then to oak woodlands with grassy understory. This yields to tall and mixed prairies (now corn and wheatlands), then to desert grasslands, and finally to desert shrublands. The indistinct boundaries where the dominant plants of adjacent biomes are mixed together form an almost continuous gradient called an **ecocline.** Thus biomes

are in some sense abstractions, a convenient way for us to organize our concepts about different communities. Nevertheless, anyone can distinguish a grassland, deciduous forest, coniferous forest, or shrub desert by the dominant plants in each. And we can make reasonable assumptions about the kinds of animals that live in each biome.

Distinctiveness of a biome is determined mainly by climate, the char-

acteristic pattern of rainfall and temperature of each region and the solar radiation it receives. Global variation in climate arises from the uneven heating of the atmosphere by the sun. Because of the lower angle of the sun's rays striking the higher latitudes, atmospheric heating is less there than at the equator (Figure 39-4).

The principal terrestrial biomes are temperate deciduous forest, temperate

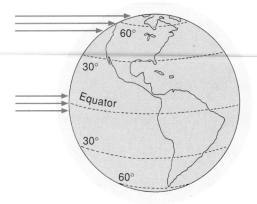

Figure 39-4

The earth's climate is determined by differential solar radiation between the higher latitudes and equator. Solar energy is spread across a much larger, slanting surface area at high latitudes than is an equivalent amount of energy at the equator.

Figure 39-5

A bull moose browses on dwarf birch in the coniferous forest biome. Note shedding of antler skin ("velvet"), signifying that antler growth is complete and that breeding season is approaching.

coniferous forest, tropical forest, grassland, tundra, and desert. In this brief survey, we will refer especially to the biomes of North America and will consider predominant features of each.

Temperate Deciduous Forest

The temperate deciduous forest, best developed in eastern North America, encompasses several forest types that change gradually from the northeast to the south. Deciduous, broad-leaved trees such as oak, maple, and beech that shed their leaves in winter predominate. Seasonal aspects are better defined in this biome than in any other. The deciduous habit is an adaptation of dormancy for low-energy levels from the sun in winter and freezing winter temperatures. In summer, the relatively dense forests form a closed canopy that creates deep shade below. Consequently there has been a selection for understory plants that grow rapidly in the spring and flower early before the canopy develops. The mean annual precipitation is relatively high (75 to 125 cm, or 30 to 50 inches) and rain falls periodically throughout the year.

Animal communities in deciduous forests respond to seasonal change in various ways. Some, such as the insect-eating warblers, migrate. Others, such as the woodchuck, hibernate during the winter months. Others that are unable to escape survive by using the available food (for example, deer) or stored food supplies (for example, squirrels). Hunting and habitat loss have eliminated virtually all the large carnivores that once roamed the eastern forests, such as mountain lions, bobcats, and wolves. Deer, on the other hand, thrive in second-growth forests under the protection of strict hunting management. Insect and invertebrate communities are abundant in deciduous forests because decaying logs and forest floor litter provide excellent shelter.

The heavy exploitation of the deciduous forests of North America began in the seventeenth century and reached a peak in the nineteenth century. Logging removed nearly all of the once-magnificent stands of temperate hardwoods. With the opening of the prairie for agriculture, many eastern farms were abandoned and allowed to return gradually to deciduous forests.

Coniferous Forest

In North America the coniferous forests form a broad, continuous, continent-wide belt stretching across Canada and Alaska, and south through the Rocky Mountains into Mexico. It continues across northern Eurasia, making it one of the largest plant formations on earth. It is dominated by evergreens—pine, fir, spruce, and cedar—which are adapted to withstand freezing and take full advantage of short summer growing seasons. The conical trees with their flexible branches shed snow easily. The northern area is the **boreal** (northern) **forest,** often called **taiga** (a Russian word, pronounced "tie-ga"). The taiga is dominated by white and black spruce, balsam, subalpine fir, larch, and birch. In the central region of North America, the taiga merges into **lake forest,** dominated by white pine, red pine, and eastern hemlock. However, most of this forest was destroyed by exploitive logging and was replaced by shrubby second growth that still characterizes much of Michigan, Wisconsin, southern Ontario, and Minnesota today. The large **southern evergreen forests** occupy much of the southeastern United States. The last old growth coniferous forests of the Pacific northwest are rapidly falling to commercial logging.

Mammals of the boreal and lake coniferous forests are deer, moose (Figure 39-5), elk, snowshoe hare, a variety of rodents, carnivores such as wolves, foxes, wolverines, lynxes, weasels, and martins, and the omnivorous bears. They are adapted physiologically or behaviorally for long, cold, snowy winters. Common birds

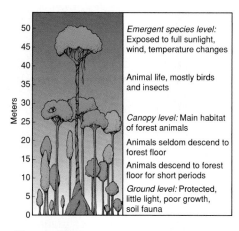

Figure 39-6

Profile of tropical forest, showing stratification of animal and plant life into six strata. The animal biomass is small compared with the biomass of the trees.

are chickadees, nuthatches, warblers, and jays. One bird, the red crossbill, has a beak specialized for picking seeds from cones. Mosquitoes and flies are pests to both animals and humans in this biome. Southern coniferous forests lack many mammals found in the north, but they have more snakes, lizards, and amphibians.

Tropical Forest

The worldwide equatorial belt of tropical forests is an area of high rainfall, high humidity, relatively high and constant temperatures, and little seasonal variation in day length. These conditions have nurtured luxurious, uninterrupted growth that reaches its greatest intensity in the rain forests. In sharp contrast to temperate deciduous forests, dominated as they are by relatively few tree species, tropical forests contain thousands of species, none of which is dominant. A single hectare typically contains 50 to 70 tree species as compared with 10 to 20 tree species in an equivalent area of hardwood forest in the eastern United States. Climbing plants and epiphytes are common among the trunks and limbs. A distinctive feature of tropical forests is the stratification of life into six, and occasionally as many as eight, feeding strata (Figure 39-6).

Insectivorous birds and bats occupy the air above the canopy; below it birds, fruit bats, and mammals feed

on leaves and fruit. In the middle zones are arboreal mammals (such as monkeys and tree sloths), numerous birds, insectivorous bats, insects, and amphibians. A middle zone of climbing animals, such as squirrels and civets, range up and down the trunks, feeding from all strata. On the ground are large mammals lacking climbing ability, such as the large rodents of South America (for example, capybara, paca, and agouti) and members of the pig family. Finally, a mixed group of small insectivorous, carnivorous, and herbivorous animals searches the litter and lower tree trunks for food. No other biome can match the tropical forest in incredible variety of animal species. Food webs are intricate and notoriously difficult for ecologists to unravel.

The tropical forests, especially the enormous expanse centered in the Amazon Basin, are the most seriously threatened of forest ecosystems. Large areas are being cleared for agriculture by "slash-and-burn" methods, but, because of low soil fertility, farms are soon abandoned. It may seem paradoxical that a biome as luxuriant as the tropical forest should have poor soil. This occurs because nutrients released by decomposition are rapidly recycled by plants, leaving no reservoir of humus. In many areas, once the plants are removed, the soil rapidly becomes a hard, bricklike crust called **laterite.** Tropical plants cannot recolonize such areas.

Each month an area of undisturbed tropical forest the size of El Salvador is converted to other uses. Unlike the forest clearing in the temperate zones, which made possible sustained, productive agriculture, tropical soils quickly become depleted, forcing farmers to move on to clear more forest. Another pressure on tropical forests is multinational timber companies that own and cut large tracts of timber to make furniture for developed countries. Cattle ranchers also clear huge forest tracts to raise beef that is later sold to North America's fast-food chains. As one specialist remarked, "Everyone's hand is on the chain saw."

Figure 39-7

Bison grazing on a short-grass prairie.

Grassland

The North American prairie biome is one of the most extensive grasslands in the world, extending from the Rocky Mountain edge on the west to the eastern deciduous forest on the east, and from northern Mexico in the south to the Canadian provinces of Alberta, Saskatchewan, and Manitoba in the north. The original grassland associations of plants and animals have been almost completely destroyed by humans. The prairies today have been transformed into the most productive agricultural region in the world, dominated by monocultures of cereal grains. In grazing lands virtually all the major native grasses have been replaced by alien species. Vast areas of Arizona and New Mexico have been converted from lush grasslands to parched desert by more than a century of livestock overgrazing. Of the once dominant herbivore, bison, (Figure 39-7) very few survive, but jackrabbits, prairie dogs, ground squirrels, and antelope remain. Mammalian predators include coyotes, ferrets, and badgers, although, of these, only coyotes are common.

Tundra

The tundra is characteristic of severe, cold climatic regions, especially the treeless Arctic regions and high mountaintops. Plant life must adapt itself to a short growing season of about 60 days

Figure 39-8
A large male caribou on the Alaskan tundra. The gregarious caribou travel in large herds, feeding in summer on grasses, dwarf willow, and birch, but in winter almost exclusively on lichen.

and to a soil that remains frozen for most of the year. Most tundra regions are covered with bogs, marshes, ponds, and a spongy mat of decayed vegetation, although high tundras may be covered only with lichens and grasses. Despite the thin soil and short growing season, the vegetation of dwarf woody plants, grasses, sedges, and lichens may be quite profuse. The plants of the alpine tundra of high mountains, such as the Rockies and Sierra Nevadas, differ from the Arctic tundra in some respects. Characteristic animals of the Arctic tundra are the lemming, caribou (Figure 39-8), musk-ox, arctic fox, arctic hare, ptarmigan, and (during the summer) many migratory birds.

Desert

Deserts are arid regions where rainfall is low (less than 25 cm a year), and water evaporation is high. The North American desert is of two parts, the hot deserts of the southwest (Mohave, Sonoran, and Chihuahuan) and the cool, high desert in the rain shadow of the High Sierras and the Cascade mountains. Desert plants, such as thorny shrubs and cacti, have reduced foliage, drought-resistant seeds, and other adaptations for conserving water. Many large desert animals have

developed remarkable anatomical and physiological adaptations for keeping cool and conserving water (p. 665). Most smaller animals avoid the most severe conditions by living in burrows or developing nocturnal habits. Mammals found there include the mule deer, peccary, cottontail, jackrabbit, kangaroo rat, and ground squirrel. Typical birds are the roadrunner, cactus wren, turkey vulture, and burrowing owl. Reptiles are numerous, and a few species of toads are common. Arthropods include a great variety of insects and arachnids.

> Deserts are expanding rapidly. Between 1882 and 1952 the area of the earth's land surface occupied by desert increased from an estimated 9.4 to 23.3%. Since 1965, 650,000 km² of grazing land was added to the Sahara Desert of Africa, the largest desert on earth, because of an extended drought combined with overgrazing by livestock.

AQUATIC ENVIRONMENTS

Inland Waters

Inland waters are divided broadly into running-water, or **lotic** (L. *lotus,* action

of washing) habitats, and standing-water, or **lentic** (L. *lentus,* slow) habitats. Lotic habitats follow a gradient from mountain brooks to streams and rivers. Brooks and streams with a high velocity of water flow are high in dissolved oxygen as a result of their turbulence. Energy input is chiefly in the form of organic detritus washed from adjacent terrestrial areas. More slowly moving rivers have less dissolved oxygen and more floating algae and plants. Their fauna is tolerant of lower oxygen concentration.

Lentic habitats, such as ponds and lakes, tend to have still lower concentrations of oxygen, particularly in the deeper areas. Animals living on the bottom or on submerged vegetation **(benthos)** include snails and mussels, crustaceans, and a wide variety of insects. Many swimming forms, called **nekton,** are found in lakes and larger ponds. Depending on the nutrients available, there may be a large contingent of small floating or weakly swimming plants and animals **(plankton).** Ponds and lakes have short lifespans—a few hundred to many thousands of years depending on size and rate of sedimentation—and undergo great physical change as they age. The Great Lakes of North America that occupy depressions gouged out by the glacial advances of the Pleistocene epoch became ice free about 5000 years ago.

> A striking exception to the short lifetimes of most lakes is Lake Baikal in southern Siberia. This enormous lake, 1741 m deep (more than 1 mile), is by far the oldest lake in the world, dating from at least the Paleocene— more than 60 million years BP. The speciation of sculpins in Lake Baikal is illustrated in Figure 9-20, p. 164.

Many freshwater habitats have been severely damaged by human pollution such as the dumping of toxic industrial wastes and enormous quantities of sewage. Of the Great Lakes,

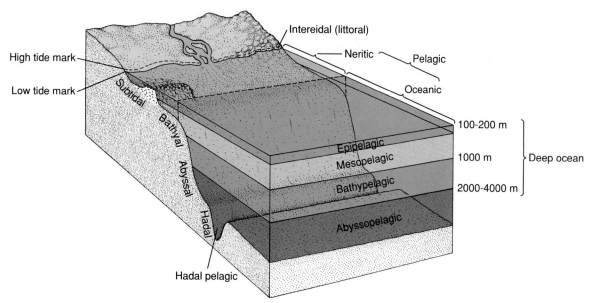

Figure 39-9
Major marine zones.

Oceans

By almost any measure, the oceans represent by far the largest portion of the earth's biosphere. They cover 71% of the earth's surface to an average depth of 3.75 km (2.3 miles), with their greatest depths reaching to more than 11.5 km (7.2 miles) below sea level. The marine world is relatively uniform as compared with land, and in many respects it is less demanding on life forms. However, the evident monotony of the ocean's surface belies the variety of life below. The oceans are the cradle of life, and this is reflected by the variety of organisms living there—more than 200,000 species of protists, plants, and animals. The vast majority of these forms, about 98%, live on the seabed **(benthic);** only 2% live freely in the open ocean

(pelagic). Of the benthic forms, most occur in the intertidal zone or shallow depths of the oceans. Less than 1% live in the deep ocean below 2000 m.

The most productive areas are concentrated along continental margins and a few areas where the waters are enriched by organic nutrients and debris lifted by upwelling currents into the sunlit, or **photic,** zone, where photosynthetic activity occurs. With certain notable exceptions (see box in Chapter 40, p. 797), all life below the photic zone must be supported by the light "rain" of organic particles from above.

The life of the ocean is divided into regions, or provinces, each with its own distinctive life forms (Figure 39-9). The **littoral,** or **intertidal,** zone, where sea and land meet, is paradoxically both the harshest and the richest of all marine environments. It, as well as the animals living there, is subjected to pounding surf, sun, wind, rain, extreme temperature fluctuations, erosion, and sedimentation. Yet because of the diversity of available habitats and the bounteous supply of nutrients, animals such as barnacles, snails, chitons, limpets, mussels, sea urchins, and many others

flourish there. Below the littoral zone is the **sublittoral,** or **subtidal,** zone, which is always submerged. It also supports a rich variety of animal life, as well as forests of brown algae.

An **estuary** is a semienclosed transition zone where fresh water flows into the sea. Despite an unstable salinity caused by the variable entry of fresh water, the estuary is a nutrient-rich habitat that supports a diverse fauna.

The **neritic,** or shallow water, zone surrounds the continents and extends to the edge of the continental shelf—approximately to a depth of 200 m (Figure 39-9). This zone is more productive than the open ocean because it benefits from nutrients delivered by rivers and by upwelling at the edge of the continental shelf. Algal growth is prolific, which in turn supports a diverse animal life, including most of the world's fisheries.

Areas of **upwelling,** although small and restricted to a few regions, are vital sources of nutrient renewal for the surface photic zone. Some of the world's most productive fisheries are—or were—centered on upwelling regions. Before its collapse in 1972, the Peruvian anchovy fishery that

---

Lake Erie has been the most seriously affected by the inflow of large amounts of nitrates and phosphates. These nutrients fertilize the lake, creating huge blooms of algae that die and sink to the bottom to decompose and rob the lake of oxygen. As a result, all levels of aquatic life are adversely affected.

depended on the Peru Current provided 22% of all the fishes caught in the world! Earlier, the California sardine fishery and the Japanese herring fishery, both fisheries of upwelling regions, were intensively harvested to the point of collapse and have never recovered.

The vast open ocean is known as the **pelagic** realm (Figure 39-10). Despite its size (comprising 90% of the total oceanic area), the pelagic realm is relatively impoverished biologically because, as organisms die, they sink out of the photic zone, carrying nutrients into the bathypelagic zone where they are immobilized. However, in areas of upwelling and where oceanic currents converge, nutrients are replenished and productivity may be high. The enormously productive polar seas are an example. Before their populations were overexploited by humans, the baleen whales probably consumed around 77 million tons of Antarctic krill (a shrimplike animal, p. 398) per year, far more than the entire catch of all the fishes, crustaceans, and molluscs taken by all the world's fishing fleet in any single year. The enormous krill population was sustained by phytoplankton, the base of the food chain, which in turn flourished because of the abundance of nutrients in the Antarctic sea.

Below the surface, or **epipelagic,** layers of the pelagic realm are the great ocean depths, characterized by enormous pressure, perpetual darkness, and a constant temperature near 0° C. It remained a world unknown to humans until recently, when baited cameras, bathyscaphs, and deep-water trawls have been lowered to view and sample the ocean bottom. There are several distinct habitats in the ocean depths (Figure 39-10). The **mesopelagic** is the "twilight zone," which receives dim light and supports a varied community of animals. Below the mesopelagic is a world of

perpetual darkness, divided into three depth zones as shown in Figure 39-10: bathypelagic, abyssopelagic, and hadopelagic. Deep-sea forms depend on that meager portion of the gentle rain of organic debris from above that escapes consumption by organisms in the water column. On the sea floor exists the benthos, represented by sea anemones, sea urchins, crustaceans, polychaete worms, and fishes—indeed nearly all major metazoan groups. Most are deposit feeders characterized by very slow growth (because of the scarcity of food) and long lives.

Recently self-contained benthic communities of animals that are completely independent of solar energy and the rain of organic debris from above were discovered adjacent to vents of hot water issuing from rifts in the ocean floor (see box in Chapter 40, p. 797).

ANIMAL DISTRIBUTION (ZOOGEOGRAPHY)

The study of zoogeography tries to explain why animals are distributed as they are, their patterns of dispersal, and the factors responsible for their dispersal. With the exception of the human species, which is able to live nearly anywhere on earth, and creatures such as the house mouse and the cockroach, which share human habitations, the different kinds of animals typically occupy limited geographic areas. It is not always easy to explain why animals are distributed as they are, since similar habitats on separate continents may be occupied by quite different kinds of animals. A particular species may be absent from a region that supports similar animals because of barriers that prevent it from getting there or because established populations of other animals prevent it from colonizing.

Thus there are good reasons why animals are found where they are (or

are not found where one thinks they ought to be), and we would like to discover these reasons. Usually this means studying the past. The fossil record plainly shows that animals once flourished in regions from which they are now absent. Extinction has played a major role, but many groups left descendants that migrated to other regions and survived. For example, the ancestors of camels originated in North America, where their fossils are found, but spread by way of Alaska to Eurasia and Africa where they are represented today by true camels; and to South America during the Pleistocene epoch, where their descendants survive as llamas, alpacas, guanacos and vicuñas. (The Pleistocene began about 1.7 million years BP and ended about 11 thousand years BP; see the geological time table on the back inside-cover.) Then camels became extinct in North America about 10,000 years BP at the close of the Ice Age. Thus the history of an animal species or its ancestor must be known before one can understand why it is where it is. The earth's surface is undergoing constant change. Many areas that are now land were once covered with seas; fertile plains may be claimed by advancing desert; impassable mountain barriers may arise where none existed before; or inhospitable ice fields may retreat before a warmer climate to be replaced by forests. Geological change has been responsible for much of the alteration in animal (and plant) distribution and has been a powerful influence in shaping organic evolution.

DISJUNCT DISTRIBUTIONS

A major problem for zoogeographers is to explain the numerous instances of discontinuous or **disjunct distributions:** closely related species living in widely separated areas of a continent, or even the world (Figure 39-11). How could a group of animals

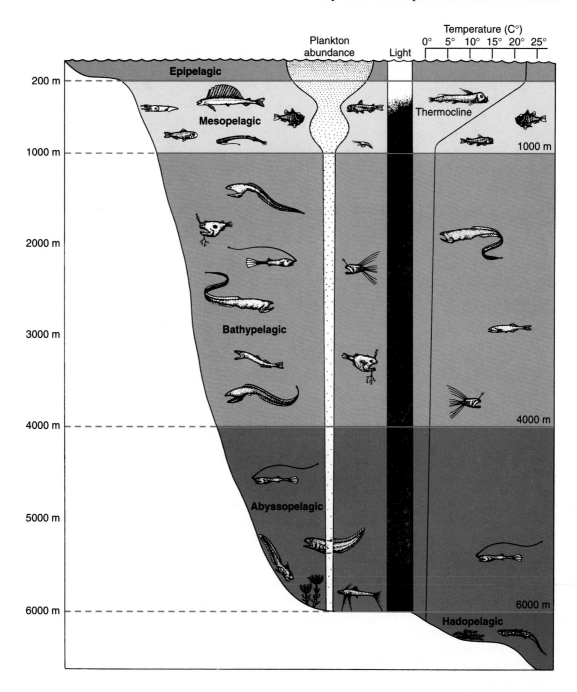

Figure 39-10

Life of the pelagic zones. Each zone supports a distinct community of organisms. Animals in zones below the mesopelagic depend on the meager rain of food that sinks out of the epipelagic and mesopelagic zones.

Figure 39-11
Disjunct distributions in North America. **A,** Moles of the family Talpidae probably entered North America across the Bering land bridge that once joined North America and Asia during the Tertiary period. Eastern and western populations are now separated by the Rocky Mountains. **B,** Gopher turtles of the genus *Gopherus* are now separated into three fully isolated populations.

become so dispersed geographically? There are two possible ways for a disjunct distribution to arise. Either a population moves from its place of origin to a new location **(dispersal),** traversing intervening territory that is unsuited for long-term colonization, or the environment changes, breaking a once continuously-distributed species into geographically separated populations **(vicariance).** Vicariance may involve climatic changes that contract and fragment the areas of habitat favorable for a species, or it may involve physical movement of landmasses or waterways that carry different populations of a species away from each other.

DISTRIBUTION BY DISPERSAL

By dispersal, animals spread into new localities from their places of origin. Dispersal involves *emigration* from one region and *immigration* into another. Dispersal is a *one-way,* outward movement that must be distinguished from *periodic* movement back and forth between two localities, such as the seasonal migration of many birds. Dispersing animals may move actively under their own power, or may be passively dispersed by wind, by floating or rafting on rivers, lakes, or the sea, or by hitching rides on other animals. Animals are expected to expand their geographic distributions in this manner across all favorable habitat that is accessible to them. For example, as the last Pleistocene glaciers retreated northward, habitats favorable for many temperate species became available on formerly glaciated territory in North America, Europe, and Asia. Species that originated immediately south of the glaciated territory prior to glacial retreat then expanded northward as new habitats appeared. Because the reproductive rate of animal populations is great, there is a continuous pressure on the populations to expand across all favorable habitats.

Dispersal easily explains the movement of animal populations into favorable habitats that are geographically adjacent to their places of origin. This produces an expanded but geographically continuous distribution. Can dispersal also explain the origins of geographically disjunct distributions? For example, the flightless ratite birds (Figure 39-12) inhabit disjunct landmasses primarily of the Southern Hemisphere including Africa, Australia, Madagascar, New Guinea, New Zealand, and South America. These landmasses are separated from each other by ocean, a very strong barrier to ratite dispersal. To explain this distribution by dispersal, one must postulate a **center of origin** from which the group dispersed to reach all of the widely separated landmasses on which it is now found. Because the ratites do not fly, the dispersalist hypothesis requires intermittent, passive rafting of individuals across the ocean. Is this hypothesis reasonable?

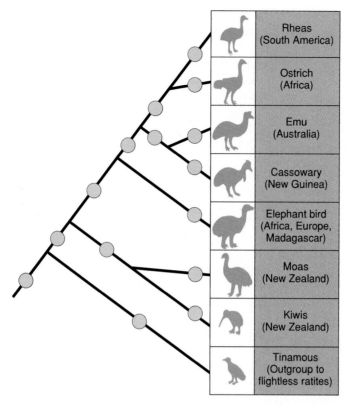

Figure 39-12

The phylogenetic relationships inferred for ratite birds (Chapter 9, pp. 160–161). Vicariance biogeography proposes that these ratite species descended from an ancestral species that was widespread in the Southern Hemisphere when Africa, Australia, Madagascar, New Guinea, New Zealand, and South America were connected. By moving apart, these landmasses fragmented both themselves and the ratite populations they contained. If the vicariance hypothesis is correct, the phylogenetic branching sequence inferred for the allopatric ratite species reflects the sequence by which their landmasses broke apart from each other. This hypothesis is tested by looking for similar phylogenetic patterns in other groups of animals and plants whose ancestral populations would have been fragmented by the same geological events. The widespread geographic distribution of the elephant bird suggests that it has dispersed following the fragmentation of landmasses.

We know from studies of the Galápagos Islands and Hawaii (Chapter 9) that occasional, long-distance dispersal of terrestrial animals and plants across oceans does occur. This is the only way that terrestrial animals could colonize islands produced by oceanic volcanoes. For the ratites and many other discontinuously distributed animals, however, there is an alternative to the hypothesis that disjunct distribution was produced by dispersal over unfavorable habitat. This is the hypothesis of vicariance (L. *vicarius,* a substitute).

DISTRIBUTION BY VICARIANCE

Disjunct distributions of animals may be created by physical changes in the environment that cause formerly con-tinuous habitats to become disjunct. Areas that were once joined may become separated by barriers that are effectively impenetrable for many animals that inhabit them. The study of fragmentation of biotas in this manner is called **vicariance biogeography.** At the species level, "vicariance" is often used as a synonym of "allopatry," which is simply a distribution of populations in geographically separated areas (Chapter 9). Lava flows from a volcano may cause a formerly continuous forest to become separated into geographically discontinuous patches, thereby breaking many species of plants and animals into geographically isolated populations.

Perhaps the most dramatic vicariant phenomenon in the earth's history is continental drift, through which a once continuous landmass was sequentially broken into continents and islands separated by ocean (see the following text). All terrestrial and freshwater animal species that had spread across the initially continuous landmass became sequentially fragmented into many populations on different continents and islands separated by ocean. Vicariance by continental drift theory gives us another hypothesis for explaining the disjunct distribution of ratite birds; they may descend from an ancestral species that was widespread in the Southern Hemisphere when Africa, Australia, Madagascar, New Guinea, New Zealand, and South America were in closer contact than they are today. When these landmasses moved apart across the ocean, the ancestral ratite species would have fragmented into disjunct populations that evolved independently, producing the diversity of forms that we observe today.

How do we test hypotheses of vicariance biogeography? Reconstructing the past histories of environmental changes might seem impossible, but we do have a very powerful method for testing such hypotheses, and it is based on the systematic methods presented in Chapter 11.

Suppose that the different ratite species evolved allopatrically as continental drifting sequentially broke their terrestrial environment into isolated pieces. If we construct a cladogram or phylogenetic tree of the ratites as shown in Figure 39-12, the earliest divergence should correspond to the first vicariant event that fragmented their common ancestral species. All subsequent branching events on the tree should correspond sequentially to the subsequent vicariant events that fragmented the major ratite lineages further. Our tree hypothetically reconstructs the history of vicariant events for the group. If we erase the names of the species from the terminal branches of the tree and replace them with the geographic areas in which each species is found, we have a hypothesis of the sequential separation of the different geographic areas. We can test this vicariant hypothesis further by identifying other groups of terrestrial organisms that have different species in each of the same geographic areas as the ratites. If our hypothesis is correct, these groups were fragmented geographically by the same vicariant events that fragmented the ratites. We therefore predict that the cladogram or phylogenetic trees constructed for species in the other groups will show the same branching pattern as the ratite tree when we replace the species names with those of the areas they inhabit. If this is confirmed, we have a **general area cladogram** that depicts the history of fragmentation of the different geographic areas studied. This general area cladogram can be investigated further using geological and climatic studies.

In many groups of organisms, it is likely that both vicariant and dispersal events have contributed to the evolution of disjunct distributional patterns. The methods of vicariance biogeography will be very useful for finding such cases. Indeed, the ratite cladogram is not just a simple grouping of birds that inhabit nearby areas. We can ask whether any branches on a cladogram representing a particular group of species are inconsistent with the general area cladogram for the geographic areas that the species inhabit. Suppose that the cladogram for a particular taxon is consistent with the area cladogram except for the placement of a single branch. We explain most of the geographic disjunctions within the taxon by vicariance but look for dispersal to explain the single branch that is not compatible with the general area cladogram. In this way, we can focus our study of dispersal on the specific cases in which it is most likely to have occurred.

CONTINENTAL DRIFT THEORY

It is no accident that the current enthusiasm for vicariance biogeography coincides with the recent acceptance of the continental drift theory by geologists. The continental drift theory is not new (it was proposed in 1912 by the German meteorologist Alfred Wegener), but it remained controversial and largely neglected until the newly proposed theory of **plate tectonics** provided a mechanism to account for drifting continents (unfortunately, Wegener did not live to see his hypothesis accepted). According to the theory of plate tectonics (tectonics means "deforming movement"), the earth's surface is composed of 6 to 10 rocky plates, about 100 km thick, that shift about on a more malleable underlying layer. Wegener proposed that the earth's continents had been drifting like rafts following the breakup of a single great landmass called Pangaea ("all land"). According to recent workers who have considerably revised Wegener's dating, this occurred some 200 million years ago. Two great supercontinents were formed: a northern Laurasia and a southern Gondwana, separated from each other by the Tethys Sea (Figure 39-13). At the end of the Jurassic period, some 135 million years ago, the supercontinents began to fragment and drift apart. Laurasia split into North America, most of Eurasia, and Greenland. Gondwana split into South America, Africa, Madagascar, Arabia, India, Australia, and Antarctica. This theory is supported by the appearance of fit between the continents, by airborne paleomagnetic surveys, by seismographic studies, by the presence of mid-ocean ridges where the tectonic plates are born, and by a wealth of biological data.

Continental drift explains several otherwise puzzling distributions of animals, such as the similarity of invertebrate fossils in Africa and South America, as well as certain similarities in present-day fauna at the same latitudes on the two continents. However, the continents have been separated for all of the Cenozoic era and probably for much of the Mesozoic era as well, much too long to explain the distributions of some modern organisms such as placental mammals. Continental drift theory is, nevertheless, enormously useful in explaining interconnections between flora and fauna of the past.

The present distribution of marsupial mammals is an excellent example of the influence of continental breakup. Marsupials appeared in the Middle Cretaceous period, about 100 million years BP, probably in South America. Because South America was at that time connected to Australia through Antarctica (which was then much warmer than it is today), the marsupials spread through all three continents. They also moved into North America, but there they encountered the placental mammals, which had dispersed to that continent from Asia. The marsupials evidently could not coexist with placentals, and so became extinct in North America. (North American marsupials today, the opossums, are relatively recent arrivals from South America.) The placentals followed the marsupials into South America, but by that time the marsupials had expanded and were too firmly established to be driven into extinction. In the meantime, about 50 million years BP, Australia drifted apart from Antarctica, barring entrance to the placentals. Australia remained in isolation, allowing the marsupials to diversify into the present rich and varied fauna.

Temporary land bridges also have been important pathways of dispersal. An important and well-established land bridge that no longer exists connected Asia and North America across the Bering Strait. It was across this corridor that the placentals moved from Asia into North America.

Today a land bridge connects North and South America at the Isthmus of Panama. But from the mid-Eocene epoch (50 million years BP) to the end of the Pliocene epoch (3 million years BP), the two continents were completely separated by water. During this long period, the major groups of mammals evolved in distinctive directions on each continent. When the land bridge was reestablished at the end of the Pliocene epoch, a tide of mammals began to flow in both directions (Figure 39-14). This has been called the

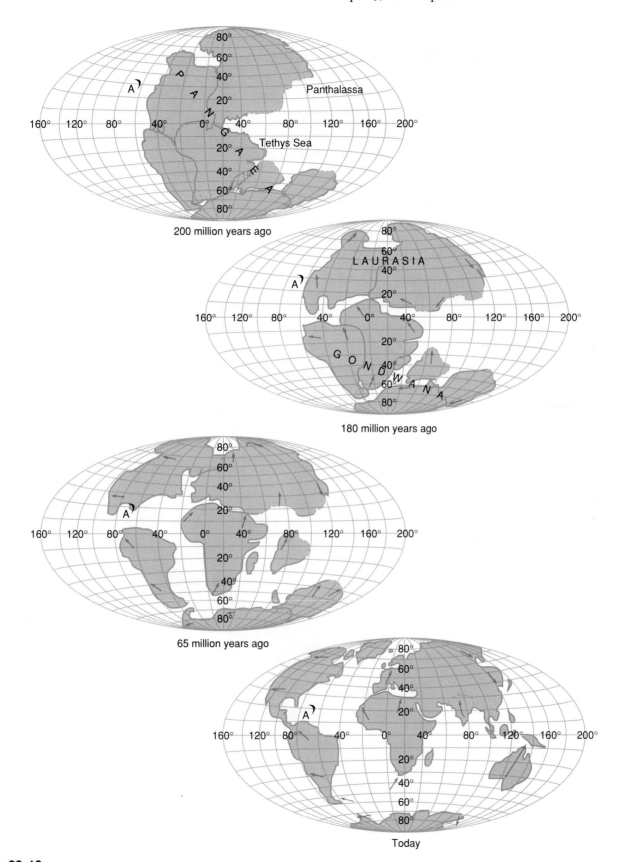

200 million years ago

180 million years ago

65 million years ago

Today

Figure 39-13

Hypothesized drift of continents over the past 200 million years from an original single landmass to their present positions. The universal landmass Pangaea first separated into two supercontinents (Laurasia and Gondwana). These later broke up into smaller continents. The arrows indicate vector movements of the continents. The black crescent labeled A is a modern geographical reference point representing the Antilles arc in the West Indies.

Adapted from "The breakup of Pangaea" by Robert S. Dietz and John C. Holden. Copyright © October 1970 by Scientific American, Inc. All rights reserved. Adapted by permission.

Figure 39-14

The Great American Interchange. The Isthmus of Panama emerged approximately 3 million years ago, permitting the extensive interchange of many families of mammals. At top are representatives of 38 South American genera that walked north across the isthmus. At bottom are representatives of 47 North American genera that migrated to South America. The North American immigrants diversified rapidly after entering South America. South American immigrants to North America diversified little and most became extinct.

"Great American Interchange," one of the most important minglings of distinct continental faunas in the earth's history. For a period both continents gained in mammalian diversity, but the extinction of large numbers of mammals on both continents soon followed. North American carnivores such as raccoons, weasels, foxes, dogs, cats (including sabercats), and bears began preying on South American mammals, which previously had evolved in an environment free of carnivores. Other North American invaders included hoofed mammals (horses, tapirs, peccaries, llamas, deer, antelopes, and mastodonts), rabbits, and several families of rodents. These displaced many South American residents occupying similar habitats. Today nearly half of the South American mammals are descendants of recent North American invaders. Only a few South American invaders survived in North America: porcupines, armadillos, and opossums. Several other South American groups, including giant ground sloths, glyptodonts, anteaters, giant aquatic capybaras, toxodonts (rhino-sized plant eaters), and giant armadillos, entered North America but subsequently became extinct there.

Summary

The biosphere is the thin life-containing blanket surrounding the earth. The presence of life on earth is possible because numerous conditions for life are fulfilled on this planet. These include a steady supply of energy from the sun, presence of water, a suitable range of temperatures, the correct proportion of major and minor elements, and the screening of lethal ultraviolet radiation by atmospheric ozone. The earth's environment and living organisms have evolved together, each deeply marking the other.

The biosphere comprises the lithosphere, the earth's rocky shell; the hydrosphere, the global distribution of water; and the atmosphere, the blanket of gas surrounding the earth.

The earth's terrestrial environment is composed of biomes that bear a distinctive array of plant life and associated animal life. The eastern deciduous forest is characterized by distinct seasons and autumn leaf fall. North of the deciduous forest is the coniferous forest, which in its northern range is called taiga, an area dominated by needle-leafed trees adapted for heavy snowfall. Animals of the taiga are adapted for long, snowy winters.

The tropical forest is the richest biome, characterized in part by a great diversity of plant species and the vertical stratification of animal habitats. Most tropical forest soils rapidly deteriorate when the forest is removed.

The most modified biome is the grassland, or prairie, which has been largely converted to agriculture and grazing. The tundra biome of the far north and the desert biome are both severe environments for animal life, but they are populated nevertheless with organisms that have evolved appropriate adaptations.

Freshwater habitats include rivers and streams (lotic habitats) and ponds and lakes (lentic habitats). All are geologically ephemeral habitats that are strongly influenced by nutrient input.

The oceans occupy 71% of the earth's surface. The photic, or sunlit, zone supports photosynthetic activity by phytoplankton. The rain of nutrients from the photic zone supports the great diversity of life below on the sea bed (benthos). The littoral, or tidal, zone is biologically rich but physically harsh. The neritic, or shallow-water, zone overlying the continental shelf is the locus of the world's great fisheries, which are especially productive in areas of upwelling where nutrients are constantly renewed. The open ocean, or pelagic zone, occupies most of the ocean's area but has low biological productivity.

Zoogeography is the study of animal distribution on earth. Animals have become distributed by dispersal, the spread of populations from the centers of origin, and by vicariance, the separation of populations by barriers. Continental drift, now strongly supported by plate tectonic theory, helps explain how animal groups become geographically separated so that evolutionary diversification can occur. It also explains how certain groups, such as marsupial mammals, can become isolated from others. Temporary land bridges have also served as important pathways for animal dispersal.

Review Questions

1. What are the special conditions on earth that make this planet especially fit for life?
2. What is the justification for saying that the earth and the life on it have evolved together and that each has deeply influenced the other?
3. What is the biosphere? How would you distinguish between the following subdivisions of the biosphere: lithosphere, hydrosphere, atmosphere?
4. What is the origin of oxygen on earth? What would happen to the earth's supply of oxygen if photosynthesis were suddenly to cease?
5. What is the evidence that increasing carbon dioxide levels in the atmosphere are responsible for the increase in the "greenhouse effect"?
6. What is a biome? Briefly describe six examples of biomes.
7. What are some very productive marine environments, and why are they so productive?
8. What is the source of nutrients for animals living in the deep-sea habitat?
9. What are some reasons why a species may be absent from a habitat or region to which it should adapt well?
10. Define and distinguish between the alternative explanations for disjunct distributions among animals: dispersal and vicariance.
11. Who first proposed the continental drift theory and what finally convinced geologists that the theory was correct?
12. In what way does the continental drift theory help explain the present distribution of marsupial mammals on earth?
13. What was the Great American Interchange, when did it occur, and what were the results?

Selected References

Cox, C. B., and P. D. Moore. 1993. Biogeography: an ecological and evolutionary approach, ed. 5. Boston, Blackwell Scientific Publications. *Highly readable account with a strong ecological emphasis.*

Dietz, R. S., and J. C. Holden. 1970. The breakup of Pangaea. Sci. Am. **223**:30–41 (Oct.). *The sequence of continental drift since the early Mesozoic era is mapped.*

Henderson, L. J. 1913. The fitness of the environment. New York, Macmillan, Inc. *This short but influential book, one of the great classics of biological literature, explains how conditions on our planet made life possible.*

Marshall, L. G. 1988. Land mammals and the Great American Interchange. Am. Sci. **76**:380–388 (July-Aug.). *Mammalian faunas of North and South America, having developed in isolation for millions of years, were suddenly allowed to intermingle when the Panamanian land bridge emerged 3 million years ago.*

Mielke, H. 1989. Patterns of life: biogeography of a changing world. Boston, Unwin Hyman. *Comprehensive textbook of biogeography, unique in embracing an extensive treatment of human impact on the biosphere.*

Pielou, E. C. 1979. Biogeography. New York, John Wiley & Sons, Inc. *A landmark publication in the field with a superb analysis of contemporary thinking in biogeography. Highly recommended.*

White, R. M. 1990. The great climate debate. Sci. Am. **263**:36–43 (July). *Greenhouse warming and the prospect of global warming is the subject of scientific and political controversy. Can greenhouse warming be avoided?*

Wiley, E. O. 1988. Vicariance biogeography. Ann. Rev. Ecol. Systemat. **19**:271–290. *A review of the science of vicariance biogeography.*

40

Animal Ecology

Every Species Has Its Niche

The lavish richness of the earth's biomass is organized into a hierarchy of interacting units: the individual organism, the population, the community, and finally the ecosystem, that most bewilderingly complex of all natural systems. Central to ecological study is the habitat, the spatial location where an animal lives. What an animals does in its habitat, its profession as it were, is its niche: how it gets its food, how it arranges for its reproductive perpetuity—in short, how it survives and stays

adapted in the Darwinian sense. The niche is a product of evolution and once it is established, no other species in the community can evolve to exploit exactly the same resources. This illustrates the "competitive exclusion principle": no two species will occupy exactly the same niche. Different species are therefore able to form an ecological community in which each has a different role in their shared environment.

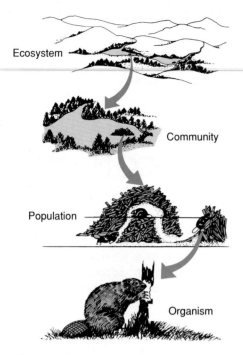

Figure 40-1
Relatonships between ecosystem, community, population, and organism.

Living things within the biosphere may be examined at several different levels of ecological organization. The most inclusive level of organization below the biosphere is the **ecosystem** (Figure 40-1). An ecosystem is a complex, self-sustaining natural system of which living organisms are a part, together with the nonliving components. All the interactions that bind the living **(biotic)** and nonliving **(abiotic)** components together are included in the ecosystem. A particular ecosystem is usually identified by the investigator studying it; it may be quite large, such as a grassland, a forest, a lake, or even an ocean, or it may be more restricted, such as a riverbank or tree-hole. Whatever the size or the biological structures it contains, certain characteristics of any ecosystem can be described. The sun's energy is fixed by plants and then transferred to consumer and decomposers. Nutrients are cycled and recycled through the various living components of the ecosystem. No ecosystem is ever completely closed, however. There is always some flow of resources and organisms into and out of an ecosystem.

The next level of organization is the **community,** an assemblage of living organisms sharing the same environment and having a certain distinctive unity (Figure 40-1). Communities comprise the living elements of an ecosystem. Like ecosystems, communities may be large or small, ranging from the coniferous forest community that may span a continent to the inhabitants of a rotting log community or the community of microorganisms living in a human's large intestine. The elements of a community are closely interdependent.

The **population,** the next lower level of organization, is a group of organisms of the *same species* sharing a particular space. The size of a population is quite variable. It may be defined by obvious boundaries (such as the population of a species of fish in a lake) or it may extend across a large geographic area. A local population of interbreeding members is called a **deme;** they share a **gene pool** and thus are a distinct genetic unit. Every community is composed of several populations, including those of plants, animals, and microorganisms. Energy and nutrients flow through a population. Its size is regulated by its relationships to other populations in the community and by the abiotic characteristics of the ecosystem in which it is found.

At the base of the ecological hierarchy within the biosphere is the **organism** itself. It is the living expression of the species. Each organism responds to its environment. To understand why animals are distributed as they are, ecologists must examine the varied mechanisms that animals use to compensate for environmental stresses and alterations.

Ecologists have, in fact, become increasingly interested in the physiological and behavioral mechanisms of animals. Both are intrinsic to the animal-habitat interrelationship. For example, the success of certain endothermic species (such as birds and mammals) under extreme temperature conditions such as in the Arctic or in a desert depends on near-perfect balance between heat production and heat loss and between appropriate insulation and special heat exchangers. Other species succeed in these situations by escaping the most extreme conditions by migration, hibernation, or torpidity. Insects, fishes, and other ectotherms (animals whose body temperature is dependent on heat in the environment) compensate for temperature change by altering biochemical and cellular processes involving enzymes, lipid organization, and the neuroendocrine system (p. 664). Thus the physiological capacities with which the animal is endowed permit it to live under changing and often adverse environmental conditions. Physiological studies are necessary to answer the "how" questions of ecology.

The animal's behavioral responses are also part of the animal-habitat interaction and of interest to the ecologist. Behavioral responses are important for obtaining food, finding shelter, escaping enemies and unfavorable environments, finding a mate, courting, and caring for the young. Those that improve adaptability to the environment assist in survival and the evolution of the species.

The term **ecology,** coined in the last century by the German zoologist Ernst Haeckel, is derived from the Greek *oikos,* meaning "house" or "place to live." Haeckel called ecology the "relation of the animal to its organic as well as inorganic environment." Although we no longer restrict ecology to animals alone, Haeckel's definition is still basically sound.

The term **environment** is often used in reference to the organism's immediate surroundings, but it is not always clear whether it is meant to include the living, as well as the nonliving, surroundings. To the ecologist it certainly includes both. Ultimately the environment consists of everything in the universe external to the organism. It is important to emphasize that the organism actively determines and modifies many aspects of its environment through its behavior and use of environmental resources. As George Gaylord Simpson (see Figure 11-5, p. 202) has noted, the distinction between organism and environment is overemphasized in many ecological studies; it is more important to concentrate on the complex interrelationship in which organism and environment are not really separable.

> Not infrequently the word "ecology" is misused as a synonym for environment, which often makes biologists wince. As people concerned about the environment, we can be environmentalists; a person engaged in the scientific study of the relationship of organisms and their environment is an ecologist. He or she is usually an environmentalist too, but environment is not the same as ecology.

The ecologist may choose to focus on any level of organization within the biosphere. **Ecosystem analysis** is interdisciplinary, incorporating physics, chemistry, and other sciences to comprehend the factors determining the distribution and abundance of organisms. **Community ecology** is of more restricted scope; it is possible to focus on the interactions of a few species and to study energy transfers in detail. **Population biology** stresses genetics, evolution, seasonal changes, and other phenomena affecting the dynamics of populations. Some ecologists study the organism itself **(organismic biology)** to see how it responds to the environment, hour by hour, day by day; such studies have become physiological and behavioral. All contribute to ecological understanding. None stands alone.

ECOSYSTEM ECOLOGY

The abiotic component of an ecosystem can be characterized by its physical parameters such as temperature, moisture, light, and altitude and by its chemical features, which include various essential nutrients. These characteristics establish the basic quality of the ecosystem.

The biotic component—the populations of plants, animals, and microorganisms that form the communities of the ecosystem—may be categorized into producers, consumers, and decomposers. The **producers,** algae, green plants, and cyanobacteria, are **autotrophs,** which use the energy of the sun to synthesize sugars from carbon dioxide by photosynthesis. This energy is made available to the **consumers** and **decomposers.** They are **heterotrophs,** which exploit the self-nourishing autotrophs by converting organic compounds of plants into compounds required for their own growth and activity. Consumers include herbivores, carnivores, omnivores, and parasites. Decomposers are consumers that perform the final breakdown of complex organic materials of dead organisms or their fecal material into simple inorganic constituents.

In this discussion we consider the flow of energy through the ecosystem, which involves the concepts of productivity and the food chain, trophic levels, biogeochemical cycling, and limiting factors within the environment.

SOLAR RADIATION AND PHOTOSYNTHESIS

Virtually all life depends on the energy of the sun. The sun releases energy produced by the nuclear transmutation of hydrogen to helium. Solar radiation is received at the earth's surface in wavelengths of approximately 280 to 13,500 nm (nanometers; 1000 nm = 1 micrometer). Ultraviolet radiation with wavelengths of less than 280 nm is cut off sharply by the ozone layer in the upper atmosphere (Figure 40-2). Radiation with wavelengths greater than 760 nm is long-wave infrared radiation that heats the atmosphere, warms the earth, and produces currents of air and water. The most important part of solar radiation is the wavelengths between 310 and 760 nm;

this is the portion that we call **visible light** because of its effect on the human retina. It is also the range that drives all important photobiological processes, including photosynthesis, photochemical effects, phototropism (orientation of plants with respect to light), and animal vision.

Photosynthesis involves the storage of a part of the sun's energy as potential energy in organic molecules. Light striking a green plant is absorbed by chlorophyll, sending low-energy electrons into a higher energy level. These excited electrons drop back to a ground state in approximately 10^{-7} second, but in this brief interval their energy is channeled into a sequence of energy-yielding reactions. Part of the energy is used to synthesize ATP; the remainder causes the reduction of pyridine nucleotides (NADP$^+$). Both ATP and reduced NADP are then used to synthesize sugars from carbon dioxide and water. Photosynthesis in the individual leaf begins at low light intensity and increases at first linearly; that is, the rate of photosynthesis increases as light intensity increases until it reaches a maximum. The leaf achieves its highest rate of photosynthesis at only approximately one-tenth the intensity of full sunlight.

The amount of solar energy that reaches the earth's atmosphere is estimated at 15.3×10^8 gram calories per square meter each year (g cal/m^2/yr). Much of this energy is dissipated by dust particles or consumed in the evaporation of water. Only a small fraction drives the photosynthetic conversion of carbon dioxide to carbohydrates. Calculated on an annual or growing season basis, the photosynthetic efficiency of land area is approximately 0.3% and that of the ocean approximately 0.13%. These estimates are low because they are based on the total energy available for the year rather than on the growing season alone. During brief intervals of very active growth, plants may store a maximum of 19% of the available light energy.

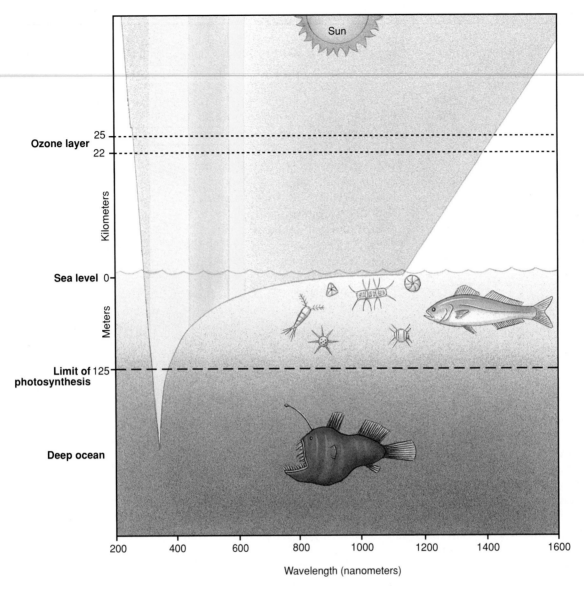

Figure 40-2
Narrowing of the spectrum and intensity of sunlight by absorption and attenuation in atmosphere and seawater.

PRODUCTION AND THE FOOD CHAIN

The energy accumulated by plants in the photosynthesis is called **production.** Because it is the first step in the input of energy into the ecosystem, the *rate* of energy storage (that is, energy storage per unit of time) by plants is known as **primary productivity.** Primary productivity can be expressed as units of energy stored per unit of area for a given time period (for example, joules/m²/day) or dry organic matter stored per unit of area for a given time period (for example, kg/ha/day). The total rate of energy storage, the **gross productivity,** is not entirely available for growth because plants also use energy for maintenance and reproduction. When this energy consumption, or plant **respiration,** is subtracted from the gross productivity, the **net primary productivity** remains. Plant growth results in the accumulation of plant **biomass.** Biomass (also referred to as "standing crop") is usually expressed either as the *weight* of dry organic matter per unit of area (for example, kg/m²) or as the amount of stored food *energy* per unit of area at any given time (for example, joules/m²). Biomass thus differs from productivity, which is the *rate* at which organic matter is formed by photosynthesis. It is possible, therefore, for a system to be highly productive but, because it is rapidly grazed by consumers, to have a low biomass.

The gross productivity of different ecosystems depends on the availability of nutrients, the limitations of temperature, and the availability of moisture. Highly productive ecosystems include intertidal communities, floodplain forests, swamps and marshes, estuaries, coral reefs, and certain crop ecosystems (for example, rice and sugar cane). In such systems net productivity can exceed 14,000 g/m²/year

LIFE WITHOUT THE SUN

For many years, it was believed that all animals depended directly or indirectly on primary production from solar energy. However, in 1977 and 1979, dense communities of animals were discovered living on the sea floor adjacent to vents of hot water issuing from the rifts (Galápagos Rift and East Pacific Rise) where tectonic plates on the sea floor are slowly spreading apart. These communities (see figure) included several species of molluscs, some crabs, polychaete worms, enteropneusts (acorn worms), and giant pogonophoran worms. The temperature of seawater above and immediately around vents is 7° to 23° C where it is heated by basaltic intrusions, whereas the surrounding normal seawater is 2° C.

It was discovered that the producers in the vent communities are chemoautotrophic bacteria that derive energy from the oxidation of the large amounts of hydrogen sulfide in the vent water and fix carbon dioxide into organic carbon. Some of the animals in the vent communities, for example, the bivalve molluscs, are filter feeders that ingest the bacteria. Others, such as the giant pogonophoran tubeworms (see p. 434), which lack mouths and digestive tracts, harbor colonies of symbiotic bacteria in their tissues and use the organic carbon that these bacteria synthesize.

Thus the deep-sea vent communities are self-contained, closed systems that depend entirely on energy issuing from the earth's interior. Each is a miniature ecosystem, separate from other known ecosystems which depend on solar energy and photosynthesis.

A population of giant pogonophoran tubeworms grows in dense profusion near a Galápagos Rift thermal vent, photographed at 2800 m (about 9000 feet) from the deep submersible *Alvin*. Also visible in the photograph are mussels and crabs.

of dry organic matter (intertidal kelp and sea palm communities). Less productive (1000 to 2000 g/m²/yr) ecosystems include most temperate forests, most agricultural crops, lakes and streams, and grasslands. Least productive (70 to 200 g/m²/yr) ecosystems are tundra and alpine regions, deserts, and the open ocean—the latter a surprise to many people who mistakenly view the ocean as a rich and unharvested source of food for all. Extreme desert, rocky, and icy regions have virtually zero productivity.

The way in which the net primary productivity of plants in a community supports the remaining life in the community is best seen in the **food chain.** Plants are eaten by consumers, which are themselves consumed by other consumers, in a series of steps. Food chains are descriptions of the way energy flows through the ecosystem. A diagram of a food chain shows arrows leading from one species to another, meaning that the first species is food for the second. But the first may be food for several other organisms as well. Seldom, in fact, does one organism live exclusively on another. Although some food chains are simple and short, for example, the one in which the whale feeds mainly on plankton, it is more common for several food chains to be interwoven into a complex **food web** (Figure 40-3).

The concepts of food chains and ecological pyramids were invented and first explained in 1923 by Charles Elton, a young ecologist at Oxford University. Working for a summer on a treeless arctic island, Elton watched the arctic foxes as they roamed, noting what they ate and, in turn, what their prey had eaten, until he was able to trace the complex cycling of nitrogen in food throughout the animal community. Elton realized that life in a food chain comes in discrete sizes, because each form had evolved to be much bigger than the thing it eats. He thus explained the common observation that large animals are rare while small animals are common.

Despite their complexity, food chains tend to follow an identifiable pattern. Green plants or algae, the base of the food chain, are eaten by grazing **herbivores,** which convert the stored energy into animal tissue. This is the base of the **grazing food chain.** Herbivores may be eaten by small **carnivores** and these by large carnivores.

There may be two or three, sometimes even four, levels of carnivores. At the end of the chain are the **top carnivores,** which, lacking predators, decompose after death, replenishing the soil with nutrients for plants that start the chain. At each level, beginning with plants, are parasites, which qualify as herbivores or carnivores depending on whether they are parasites of plants or animals. They usually deprive their hosts of insignificant amounts of energy, but other detrimental effects may be significant. Sometimes the parasites have parasites.

There are numerous examples of food chains. In the forest, for instance, many small insects (primary consumers) feed on plants (producers). A smaller number of spiders and carnivorous insects (secondary consumers) prey on small insects; still fewer insectivorous birds (tertiary consumers) live on the spiders and carnivorous insects; and finally one or two hawks (quaternary or top consumers) prey on the insectivorous birds.

The decomposers have traditionally been considered the final step in the herbivore-carnivore food chain, since they reduce organic matter into nutrients that become available again to the producers. Ecologists now recognize

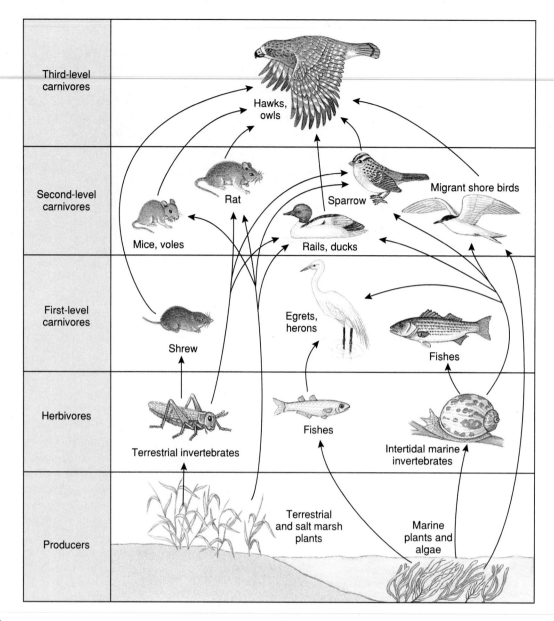

Figure 40-3
Midwinter food web in *Salicornia* salt marsh of San Francisco Bay area.

that the decomposers comprise their own distinct **detritus food chain,** consisting of **detritus feeders,** such as earthworms, mites, millipedes, crabs, aquatic worms, and molluscs, and **microorganisms,** such as bacteria and fungi. Dead organic matter, such as fallen leaves or dead animals, is decomposed and used by fungi, bacteria, and protozoa. Detritus feeders then eat the microorganisms as well as much of the dead organic matter directly. They are in turn eaten by small carnivores; the detritus food chain thus leads up into grazing food chains where it is important in recycling nutrients into grazing food chains.

TROPHIC LEVELS

Each step in the food chain may be considered a **trophic level** (meaning feeding level, from the Greek *trophé,* food). At each transfer to the next trophic level, 80% to 90% of the available energy is lost as heat because most of the fuel obtained in food is burned to do the work of staying alive. Only the 10% to 20% of food calories remaining can be used to build bodies. Since only about 10% of the available energy is passed to the next trophic level, the number of steps in the chain is usually limited to four or five. Therefore the number of

top consumers that can be supported by a given biomass of plants depends on the length of the chain.

Humans, who occupy a position at the end of the chain, may eat the grain that fixes the sun's energy; this very short chain represents an efficient use of the potential energy. Humans also may eat beef from animals that eat grass that fixes the sun's energy. The addition of a trophic level decreases the available energy by an order of 10. In other words, it requires 10 times as much plant biomass to feed humans as meat eaters as to feed humans as grain eaters. Let us consider the person who eats the bass that eats the sunfish that eats the zooplankton that eats the phytoplankton that fixes the sun's energy. The 10-fold loss of energy occurring at each trophic level in this five-step chain requires that the pond must produce 5 tons of phytoplankton biomass for a person to gain a pound by eating bass. If the human population depended on bass for survival, we would quickly exhaust this resource.

These figures must be considered as we look to the sea for food. The productivity of the oceans is very low and limited largely to regions of up-welling where nutrients are brought up and made available to the phyto-plankton producers, and to estuaries, marshes, and reefs. Such areas oc-cupy only a small part of the ocean. The rest is a watery void.

Ocean fisheries supply 18% of the world's protein, but most of this is used to supplement livestock and poultry feed. If we remember the rule of 10-to-1 loss in energy with each transfer of material between trophic levels, then the use of fish as food for livestock rather than as food for humans is poor utilization of a valuable resource in a protein-deficient world. Of the fishes that we do eat, the preference is for species such as flounder, tuna, and halibut, which are three or four steps up the food chain. Every 125 g of tuna requires 1 metric ton of phy-toplankton food to produce. If hu-mans are to derive greater benefit from the oceans as a food source in the future, we must eat more of the fishes that are at lower trophic lev-els. Such fishes are unfortunately less acceptable to humans as food.

When we examine the food chain in terms of biomass at each level, it is apparent that we can construct eco-logical pyramids of numbers or of bio-mass. A pyramid of numbers (Figure 40-4A), also known as an Eltonian pyramid (after the British ecologist Charles Elton, who first devised the scheme), depicts the numbers of indi-vidual organisms that are transferred between each trophic level. This pyra-mid provides a vivid impression of the

Steve Stinson and Roanoke Times and World-News

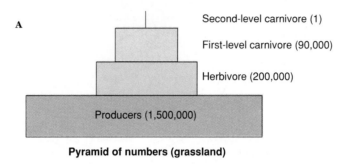

Pyramid of numbers (grassland)

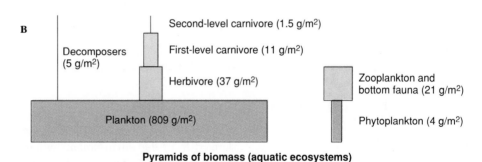

Pyramids of biomass (aquatic ecosystems)

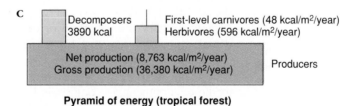

Pyramid of energy (tropical forest)

Figure 40-4

Ecological pyramids of numbers, biomass, and energy. Pyramids are generalized, since the area within each trophic level is not scaled proportionally to quantitative differences in units given.

great difference in numbers of organ-isms involved in each step of the chain, and supports the familiar obser-vation that large predatory animals are much rarer than the small animals on which they feed. However, a pyramid of numbers does not indicate the ac-tual weight of organisms at each level.

More instructive are pyramids of biomass (Figure 40-4B), which depict the total bulk, or "standing crop," of

organisms at each trophic level. Such pyramids usually slope upward be-cause mass and energy are lost at each transfer. However, in some aquatic ecosystems in which the pro-ducers are the algae, which have short life spans and rapid turnover rates, the pyramid is inverted. This happens be-cause the algae can tolerate heavy ex-ploitation by the zooplankton con-sumers. Therefore the base of the

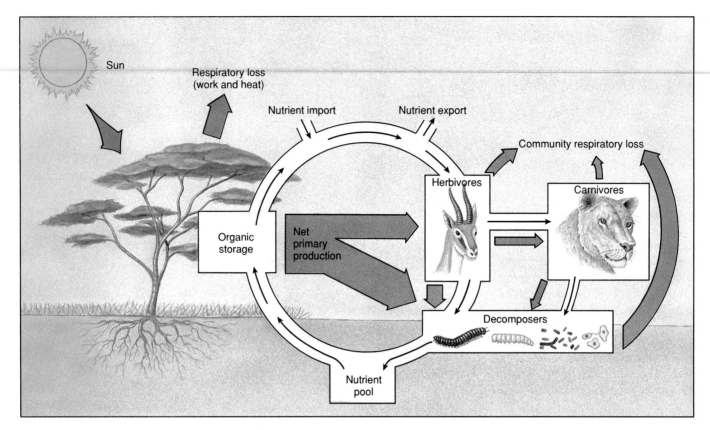

Figure 40-5
Nutrient cycles and energy flow in a terrestrial ecosystem. Note that nutrients are recycled, whereas energy flow (*red*) is one way.

pyramid (biomass of phytoplankton) is smaller than the biomass of zooplankton it supports. We could liken this inverted pyramid to a person who weighs far more than the food in a refrigerator, but who can be sustained from the refrigerator because the food is constantly replenished.

A third type of pyramid is the pyramid of energy, which shows rate of energy flow between levels (Figure 40-4C). An energy pyramid is never inverted because less energy is transferred from each level than was put into it. A pyramid of energy gives the best overall picture of community structure because it is based on production. In the example above, the phytoplankton *production* exceeds that of zooplankton production, although the biomass of phytoplankton is less than the biomass of zooplankton (because of heavy grazing by the zooplankton consumers).

NUTRIENT CYCLES

All of the elements essential for life are derived from the environment, where they are present in the air, soil, rocks, and water. When plants and animals die and their bodies decay, or when organic substances are burned or oxidized, the elements and inorganic compounds essential for life processes (nutrients) are released and returned to the environment. Decomposers fulfill an essential role in this process by feeding on the remains of plants and animals and on fecal material. The result is that nutrients flow in a perpetual cycle between the biotic and abiotic components of the ecosystem.

Nutrient cycles are often called **biogeochemical cycles** because they involve exchanges between living organisms (bio-) and the rocks, air, and water of the earth's crust (geo-). Geochemistry is the discipline that deals

with the chemical composition of the earth and the exchange of elements therein.

Nutrient cycles and energy flow are closely interrelated, since both influence the abundance of organisms in an ecosystem. However, unlike nutrients, which recirculate, energy flow follows one direction; it does not follow a cycle because it is lost as heat as it is used. The continuous input of energy from the sun keeps nutrients flowing and the ecosystem functioning. This interrelationship is depicted in Figure 40-5. Among the most important biogeochemical cycles are those of carbon and nitrogen.

Carbon Cycle

Carbon is the basic constituent of organic compounds and living tissue and is required by plants for the fixation of energy by photosynthesis. It is hardly necessary to emphasize the total dependence of life on the availability of this element. Carbon circulates between carbon dioxide (CO_2 gas in the atmosphere and living organisms through assimilation and respiration, and between the oceans, lakes, and streams. The oceans contain an enormous reserve of carbon dioxide (50 times as much CO_2 as contained in the atmosphere) and are an important repository for CO_2 released by respiration of the world's biota and by human activities. Carbon dioxide is also withdrawn into long-term reserves of fossil fuel deposits (humus and peat and finally coal and oil) (Figure 40-6).

The cycling of carbon parallels and is linked to the flow of energy that begins with the fixation of energy during photosynthetic production. Plants synthesize glucose, a 6-carbon compound, from CO_2 that is withdrawn from the atmosphere; they then use this sugar to build higher carbohydrates, especially cellulose, the main structural carbohydrate of plants. Plants require 1.6 kg of CO_2 from the atmosphere for each kilogram of cellulose produced. Carbon dioxide is a rare gas, present in the atmosphere today at a concentration of only 0.03%, in contrast to the relative abundance of atmospheric oxygen, 21% of air.

Two aspects of the low concentration of CO_2 require emphasis. The first is that CO_2 availability limits energy fixation by plants. Physiologists have shown that, if the atmospheric CO_2 is increased by 10%, plant photosynthesis increases 5% to 8%. Therefore carbon dioxide may be a limiting resource for land plants when growth is not restricted by other potentially limiting factors, such as nutrients, water, or temperature.

The second point is that there has been a tremendous increase in the consumption of fossil fuels by humans in the last two decades. More than 10 billion tons of CO_2 enters the atmosphere each year from industrial and agricultural activities; of this amount approximately 80% comes from burning fossil fuels and 20% from the clearing of forests. The CO_2 concentration

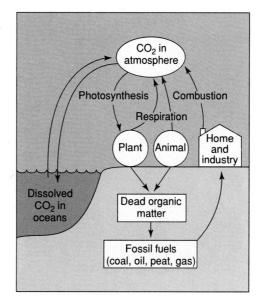

Figure 40-6
Carbon cycle showing circulation of carbon as CO_2 gas between living components of environment and long-term storage as fossil fuels.

in the atmosphere has been rising ever since measurements began (Figure 39-2, p. 778) leading to concern that even small increases in CO_2 which traps radiated heat, may increase the temperature of the earth's biosphere ("greenhouse effect," p. 778). To what extent has the consumption of fossil fuels contributed to this rise? Ecologists have been able to answer this question by determining the ratios of different isotopes of carbon in air, water, fossil fuels, and the biota. These analyses show that the increase in atmospheric CO_2 comes both from the burning of fossil fuels and from the clearing of forests, with oceans absorbing only about one half the extra CO_2.

Nitrogen Cycle

Like carbon, nitrogen is a basic and essential constituent of living material, particularly in proteins and nucleic acids. Despite the high concentration of nitrogen in the atmosphere (78% of air), it is almost totally unavailable to living organisms in its gaseous state (N_2). The nitrogen cycle converts N_2 into a chemical form that living organisms can use. This conversion is called **nitrogen fixation.** Some atmospheric N_2 is fixed by

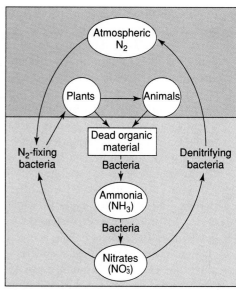

Figure 40-7
Nitrogen cycle showing circulation of nitrogen between organisms and through environment (*brown*). Microorganisms responsible for key conversions are indicated in circles and boxes.

lightning, which produces ammonia and nitrates that are carried to earth by rain and snow. But at least 10 times as much N_2 is biologically fixed by bacteria and cyanobacteria (Figure 40-7). These **nitrogen fixers** are the essential link between the accessible inorganic nitrogen pool and all other organisms.

Most important in terrestrial systems are bacteria associated with legumes (members of the pea family), whose nitrogenous contribution to soil enrichment is well known. The old agricultural practice of allowing fields to lie fallow (without crops) for a season every few years enabled N_2 fixers to replenish the element in a usable form. Aerobic rhizobia bacteria produce nodules on the roots of legumes in which molecular N_2 is converted to ammonia and nitrates, which plants can use to build protein. Plant proteins are transferred to consumers, which build their own proteins from the amino acids supplied. Plants' and animals' waste products and their ultimate decomposition provide for the return of organic nitrogen to the substrate. Then decomposers break down proteins, freeing ammonia, which is used by bacteria in a series of steps as shown in Figure 40-7.

Finally nitrate (NO_3^-) is produced, which may be used once again by plants or may be degraded to inorganic nitrogen by denitrifying bacteria and returned to the atmosphere. Nitrates also are carried by runoff to streams and lakes and eventually to the sea. A cycle similar to this terrestrial cycle occurs in aquatic ecosystems except that there is a steady loss of nitrogen to deep-sea sediments.

The nitrogen cycle is a near-perfect, self-regulating cycle in which losses in one phase are balanced by gains in another. There is little absolute change in nitrogen in the biosphere as a whole. However, human activities have caused steady losses of soil nitrogen by slowing natural addition of organic nitrogen through current agricultural practice and by the harvesting of timber. The latter causes an especially heavy outflow, from both timber removal and soil disturbance.

COMMUNITIES

Communities represent the most tangible concept in ecology. We are all familiar with the differences between forest, grassland, desert, and salt marsh, and we have little trouble picturing, at least in a general way, the kinds of plant and animal communities associated with each of these ecosystems. Communities comprise the *biotic* portion of the ecosystem; each consists of a certain combination of species that forms a functional unit. Although communities are sometimes difficult to define because the assemblages of species within similar communities are not always the same, communities do exist, and all possess a number of attributes. It is beyond the scope of this book to discuss all aspects of form and function of communities. In this discussion we examine two important principles that operate in community organization.

ECOLOGICAL DOMINANCE AND KEYSTONE SPECIES

Biological communities are typically dominated by a single species or lim-

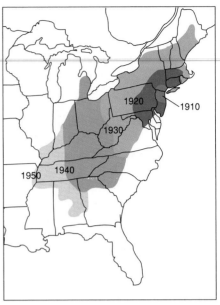

Figure 40-8

American chestnut tree *Castanea dentata,* and the geographic spread of the chestnut blight following its accidental introduction into New York City in 1900.

ited group of species that greatly influence the nature of the local environment. All other species in that community must adapt to conditions created by the dominants. Communities often are named for the dominant organisms, for example, black spruce forest, beech-maple forest, oyster community, and coral reef. It is not always easy to specify just what constitutes a dominant organism. It may be the most numerous, the largest, or the most productive, or it may in some other manner exert the greatest influence on the rest of the community. In a woodland a tree obviously means more to the community than a poison ivy plant (although a casual visitor may carry away a more lasting impression of the poison ivy).

Dominant species are often so because they occupy space that might otherwise be occupied by other species. When a dominant species is eliminated for some reason, the community changes. The American chestnut once dominated large regions of the eastern United States, where it made up more than 40% of the overstory trees in climax deciduous forests. After the invasion in 1900 of the chestnut blight (a fungus), which was apparently brought into New York City on nursery stock from Asia,

the chestnut was eliminated from its entire range within 50 years (Figure 40-8). While the disappearance of a tree of enormous commercial value was a great loss to the human community, the forest community quickly accommodated to the change. The chestnut was replaced by oak, hickory, beech, poplar, and red maple, and chestnut-oak forests became oak and oak-hickory forests.

For the American chestnut, to be dominant did not mean to be essential. The basic structure of eastern hardwood forests was not significantly altered. Sometimes, however, the disappearance of a single species can transform the fundamental framework of a community. Such species are called **keystone species.** For example, in 1983, a massive wave of death swept through the Caribbean populations of the sea urchin *Diadema antillarum,* destroying more than 95% of the animals. The immediate effect was on the algal community, which, no longer grazed by the urchins, changed from a thin mat to a thick canopy of different algal composition. Both productivity and diversity on the coral reefs declined. *Diadema antillarum* was clearly a keystone species for Caribbean coral reef communities. On the West Coast, the sea star *Pisaster*

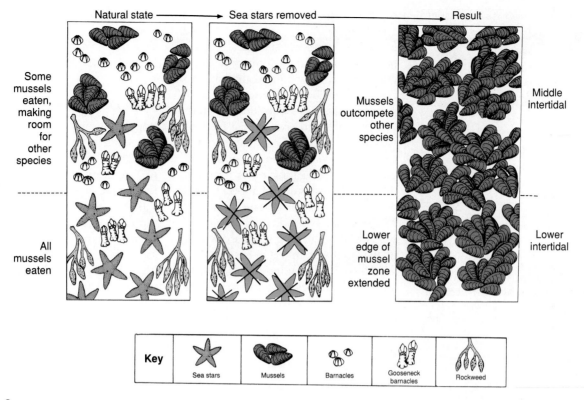

Key: Sea stars — Mussels — Barnacles — Gooseneck barnacles — Rockweed

Natural state ——→ Sea stars removed ——————→ Result

Some mussels eaten, making room for other species

All mussels eaten

Mussels outcompete other species

Lower edge of mussel zone extended

Middle intertidal

Lower intertidal

Figure 40-9
The experimental removal of a keystone species, the predatory sea star *Pisaster ochraceus,* from an intertidal community completely changes the structure of the community. With their principal predator missing, mussels form dense beds by out-competing and replacing other intertidal species.

ochraceus is a keystone species. Sea stars are a major predator of the mussel *Mytilus californianus.* When the sea stars were experimentally removed from a patch of Washington state coastline, the mussels expanded, taking over all the space previously occupied by some 25 species of other invertebrates and algae (Figure 40-9). Keystone species have been identified in other communities. Not all communities include keystone species, but in those that do, keystone species are critical to community structure.

THE CONCEPT OF ECOLOGICAL NICHE

An animal's position in the environment is characterized by more than the habitat in which we expect to find it; it also has a "profession," a special role in life that distinguishes it from all other species. This is its **niche,** which Elton defined as the animal's place in the biotic environment, that is, what it does and its relation to its food and its enemies. The entire set of conditions under which an animal population can live and replace itself is called its **fundamental niche.** The actual set of conditions under which an animal population exists is called its **realized niche,** which is a subset of the fundamental niche. The fundamental niche represents a biological potential of the population whereas the realized niche is the outcome of an interaction of that potential with a given set of abiotic and biotic environmental conditions.

It is an accepted rule that populations of two species cannot occupy the same niche at the same time and place; this is the **principle of competitive exclusion** that we mentioned in the prologue to this chapter. If the populations of two species did occupy the same niche, they would be in direct competition for exactly the same food and space. Should this happen, one species would have to change its realized niche, move to a different habitat, or face extinction. Closely related species sharing the same habitat might be expected to have similar niches and be in danger of direct competition.

One way closely related species coexist is by partitioning the resources among them. For example, in parts of eastern and southern Africa the two species of rhinoceros, the black and the white, share the same habitat. However, the black rhinoceros is a browser, feeding on leaves and woody plants, whereas the white rhinoceros is a grazer, eating grasses and herbs. They do not compete for the same food, and consequently they occupy distinct ecological niches.

In his classic study of Darwin's finches on the Galápagos Islands, the English ornithologist David Lack noticed that bill sizes of these birds depended on whether they occurred together on the same island (Figure 40-10). On the islands Daphne and Los Hermanos where *Geospiza fuliginosa* and *G. fortis* occur separately, and are therefore not in competition with each other, beak sizes are nearly identical. But on the island Santa Cruz, where both *G. fuliginosa* and *G. fortis* coexist, their bill sizes do not overlap. This suggests resource partitioning, since

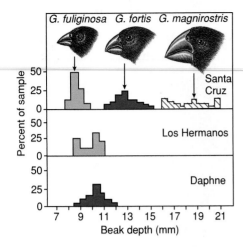

Figure 40-10

Displacement of beak sizes in Darwin's finches from the Galápagos Islands. Beak depths are given for the ground finches *Geospiza fuliginosa* and *G. fortis* where they occur together (sympatric) on Santa Cruz Island and where they occur alone on the islands Daphne and Los Hermanos. *G. magnirostris* is another large ground finch that lives on Santa Cruz.

bill size determines the size of seeds selected for food. Recent work by American ornithologists Peter and Rosemary Grant and their coworkers has confirmed what Lack suspected: *G. fuliginosa* with its smaller bill selects smaller seeds than does *G. fortis* with its larger bill. This is an example of the work of evolution in the past. Where the two species coexisted, competition between them forced displacement of the bill sizes so that competition ceased. The absence of competition that we see today has been described as "the ghost of competition past."

The niche concept is the basis for understanding how communities are structured, with special reference to competitive interactions among closely related species. The role of competition in structuring populations remains controversial, however, as does the concept of niche.

THE CONCEPT OF GUILD

The examples just given show how closely related species that live together live in ways that avoid competition. The concept of **guild** provides another approach to understanding the role of competition in structuring communities.

A guild is a group of species that makes a living in the same way. Just as a guild in medieval times constituted a brotherhood of men plying a common trade, a guild of species are those that share a common livelihood. Species that compete for food form one kind of guild. For example, bird species that feed on nectar form a guild, those that fly out to catch insects on the wing form another guild, and seed-eating birds form another. A classical study of a food-limited guild was Robert MacArthur's study of foraging behavior of five species of wood warblers nesting in spruce woods in the northeastern United States. The warblers were strikingly sensitive to small differences in habitat (Figure 40-11). One species searched for prey only on the outer branches of spruce crowns, another used the top 60% of the tree's outer and inner branches, although not next to the trunk, another concentrated on the inner branches closer to the trunk, another used the midsection from the periphery to the trunk, and still another foraged in the bottom 20% of the tree. These results suggested that a bird's niche was defined by structural differences in habitat. They also show that several species that resemble each other and are potential competitors can coexist using an array of microhabitats that exist within a common habitat. The concept of guild allows us to suggest, for example, why tropical forests support a much greater bird diversity than temperate habitats. Because of their structural complexity and year-round production of resources, tropical forests support more guilds than do temperate forests, and each guild contains more species than are found in temperate forest guilds.

POPULATIONS

A population is a group of organisms belonging to the same species that shares a particular space. Whether the population is gray squirrels in an eastern woods or bluegill sunfish in a farm fishpond, it bears a number of attributes unique to the group. A population shares a common gene pool; it has a certain density, birth rate, death rate, age ratio, and reproductive potential; and it grows and differentiates much like the individual organisms of which it is composed.

POPULATION INTERACTIONS

Populations in a community may affect each other in various ways, although in certain cases there may be no apparent interaction (neutral effect). One interaction, **competition,** has already been mentioned. Although it is true that two populations cannot occupy the same niche at the same time and place, some degree of competition often may occur in natural populations. Because natural selection favors decreased competition, this often leads to diversification of niches, as shown by two species of Galápagos finches described previously. Another adaptation may be that two populations use similar resources but at different times of the year. One population may be excluded by a superior competitor from a habitat it could otherwise occupy. When two such populations occupy adjacent habitats, competition may be intense along the border between them.

Another interaction is the **predator-prey** relationship, in which individuals of a predator population kill and eat individuals of a prey population. Often a population has more than one species of predator, and predators commonly have more than one species of prey. Predators are usually large relative to their prey, and they usually consume several to many prey animals in their lifetime. A conceptually related interaction is the **parasite-host** relationship. Both parasites and predators feed at the expense of their host-prey. However, parasites are small relative to their hosts, and parasites have only one host—at least at a given stage in their life cycle. In most cases, parasites do not kill their hosts because, if the host dies, the parasite perishes with it. Most parasites are **symbiotic;** that is, they live in **(endoparasites)** or on **(ectoparasites)** the body of their host. Organisms such as mosquitoes

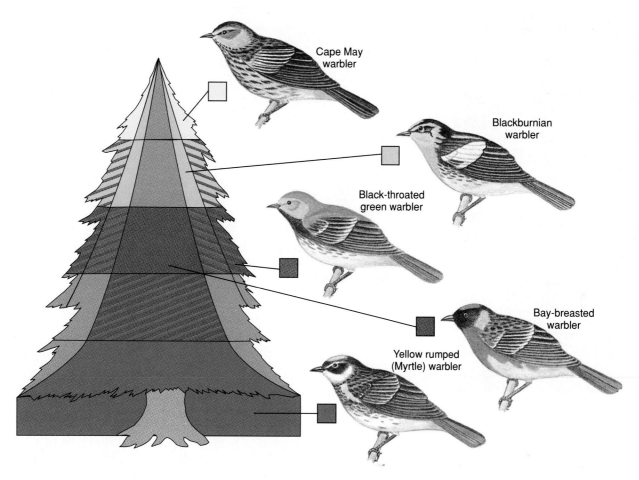

Figure 40-11

Distribution of foraging effort among five species of wood warblers in a northeastern spruce forest. The warblers form a feeding guild.

and biting flies are intermediate between parasites and predators as we have defined them; they are sometimes referred to as **micropredators** or **intermittent parasites.**

Most people are aware that lions, tigers, and wolves are predators, but the world of invertebrates also includes numerous predaceous animals. These range from protozoa, jellyfish and their relatives, and various worms to predaceous insects, sea stars, and many others.

Another type of interaction that is sometimes symbiotic is **commensalism,** in which the commensal enjoys some benefit from the relationship but does not harm the host. A classic example of commensalism is the association of pilot fishes and remoras with sharks. These fish get the "crumbs" left over when the host makes its kill, but we now know that some remoras also feed on ectoparasites of the sharks. This is an example of commensalism grading into **mutualism,** an interaction in which both species benefit (Figure 40-12). An example of mutualism is that of the termite and the protozoa living in its gut. The protozoa can digest the wood eaten by the termite, although the termite cannot, and the termite lives on the waste products of the protozoan metabolism. In return, the protozoa gain a congenial place to live and a food supply. Mutualism has been recognized in recent years as much more prevalent in the animal kingdom than previously believed. Many cases of mutualistic interactions are symbiotic, but many are not. Some investigators prefer to distinguish **protocooperation** from mutualism. Protocooperation describes mutually beneficial interactions that are not physiologically necessary to the survival of the partners.

GROWTH OF POPULATIONS

Living organisms have reproductive potential beyond that required for replacement of their numbers, some far in excess of such requirement. Some female insects lay thousands of eggs, field mice can produce as many as 17 litters of four to seven young each year, and a single female codfish may spawn 6 million eggs per season. It has been calculated that a bacterium dividing three times per hour would produce a colony a foot deep over the entire earth in 1½ days, and the colony would be over our heads 1 hour later. Unrestricted geometric (or exponential) growth can happen only in an environment with bounteous resources and no competition. This kind of growth, shown by the steeply rising curve in Figure 40-13, is called the **intrinsic growth rate** of a population and is referred to by the symbol **r.** Of course, the growth potential of a

Figure 40-12
Among the many examples of mutualism that abound in nature is the whistling thorn acacia of the African savanna and the ants that make their homes in the acacia's swollen galls. The acacia provides both protection for the ants' larvae (*lower photograph of opened gall*) and honeylike secretions used by the ants as food. In turn, the ants protect the tree from herbivores by swarming out as soon as the tree is touched. Giraffes, however, which love the tender acacia leaves, seem immune to the ants' fiery bites.

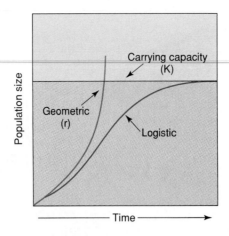

Figure 40-13
Growth of populations, showing geometric (exponential) growth of a species in an unlimited environment, and logistic growth in a limited environment.

population can never proceed unchecked. We are in no danger of being buried in bacteria, or codfish, or any other organism because as the population becomes more crowded, individuals compete with each other for limited resources, especially food and space. Eventually the population ceases to grow larger. Most populations tend to reach a certain density and then fluctuate about that level. What keeps populations in check? How is the "balance of nature" explained? Or is nature ever really "in balance"?

If we plot the numbers of individuals on a graph against units of time, the population describes a growth history that is sigmoid in shape (Figure 40-13). In the beginning, with ample food and space, the starting population breeds and grows as fast as its reproductive potential allows. We see this kind of growth—the intrinsic growth rate—as early summer plankton blooms in lakes, eruptions of insect pests, and growth of weeds in old fields. But, as resources become depleted and the upper population limit is approached, growth slows and finally stops altogether. This limit, the maximum density the environment can support, is called the **carrying capacity** of the environment and is represented by the symbol **K.** The simple sigmoid curve has been demonstrated repeatedly in the laboratory with bacteria, protozoa, and some insects. The mathematical form of the growth curve, described by the **logistic equation,** is explained in the box on p. 807.

When populations reach the carrying capacity of the environment, they may remain at this level, or they may fluctuate above and below the upper limit. For example, when sheep were introduced on the island of Tasmania around 1800, growth of their populations was represented by a sigmoid curve with a small overshoot, followed by mild oscillations around a final population size of 1,700,000 sheep (Figure 40-14A). A similar pattern but with larger fluctuations was recorded for a population of ring-necked pheasants introduced on an island in Ontario, Canada (Figure 40-14B). Some populations oscillate cyclically with large year-to-year changes in abundance. One of the best-known examples is the cyclic oscillation in the abundance of lynx and snowshoe hare in which peaks of abundance alternate with spectacular "crashes" (see Figure 31-25, p. 615). One important reason for such oscillations is that the carrying capacity of the environment itself changes. Nature is seldom "in balance," at least not for very long, because of weather fluctuations, spread of diseases, changes in predation patterns, and other disturbances.

The growth of the human population was slow for a very long time. For most of their evolutionary history, humans were hunters and gatherers who depended on and were limited by the natural productivity of the environment. The first surge in population, from 150,000 to 5 million, coincided with the development of toolmaking. With the development of agriculture, the carrying capacity of the environment increased, and the population grew steadily from 5 million around 8000 B.C., when agriculture was introduced, to 16 million around 4000 B.C. Despite the toll taken by terrible famines, disease, and war, the population reached 500 million by 1650. With the coming of the Industrial Revolution in Europe and England in the eighteenth century, followed by a medical revolution, discovery of new lands for colonization, and better agriculture practices, the carrying capacity of the earth for humans increased dramatically. The population doubled to 1 billion around 1850. It doubled again to 2 billion by 1930, to 4 billion in 1976, passed 5.5 billion in 1992, and is expected to reach 9 billion by the year 2025. Thus the growth has been exponential and remains high (Figure 40-14C).

Unlike other animals, which can do little to increase the carrying capacity of their environment, humans have

EXPONENTIAL AND LOGISTIC GROWTH

The sigmoid growth curve (see Figure 40-13) can be described by a simple model called the **logistic equation.** The slope at any point on the growth curve is the *growth rate,* that is, how rapidly the population size is changing with time. If **N** represents the number of organisms and **t** the time, we can, in the language of calculus, express growth as an *instantaneous rate:*

dN/dt = the rate of change in the number of organisms per time at a particular instant in time.

When populations are growing in an environment of unlimited resources (unlimited food and space, and no competition from other organisms), growth is limited only by the inherent capacity of the population to reproduce itself. Under these ideal conditions growth is expressed by the symbol **r,** which is defined as the intrinsic rate of population growth per capita. The index **r** is actually the difference between the birth rate and death rate per individual in the population at any instant. The growth rate of the population as a whole is then:

$$\frac{d\text{N}}{d\text{t}} = r\text{N}$$

This expression describes the rapid, **exponential growth** illustrated by the

early upward-curving portion of the sigmoid growth curve (see Figure 40-13). But growth rate for populations in the real world slows as the upper limit is approached, and eventually stops altogether. At this point **N** has reached its maximum density; that is, the space being studied has become "saturated" with animals. This limit is called the **carrying capacity** of the environment and is expressed by the symbol **K.** The sigmoid population growth curve can now be described by the logistic equation, which is written as follows:

$$\frac{d\text{N}}{d\text{t}} = r\text{N}\left(\frac{\text{K} - \text{N}}{\text{K}}\right)$$

This equation states that the rate of increase per unit of time (dN/dt) = rate of growth per capita (**r**) × population size (**N**) × unutilized freedom for growth ([**K − N**]/**K**). One can see from the equation that when the population approaches the carrying capacity, that is, when **K − N** approaches 0, dN/dt also approaches 0 and the curve will flatten. It is not uncommon for populations to overshoot the carrying capacity of the environment so that **N** exceeds **K.** When this happens, the population runs out of some resource (usually food or shelter). The rate of growth, dN/dt, then becomes negative and the population must decline.

HOW POPULATION GROWTH IS CURBED

What determines the number of animals in a natural population? For a laboratory culture of animals the answer seems fairly clear. They grow until they reach the carrying capacity of the environment, at which point growth is suppressed by competition for the limited resources of space and food. Thus the forces that limit growth arise from within the population; that is, they are **density-dependent** mechanisms caused by crowding. Food and space limitations are commonly observed density-dependent factors, but others, such as diseases and predation, may operate as well.

Transmision of infectious **disease** is usually much higher under crowded conditions. Epidemics may sweep through crowded populations as did the terrible bubonic plague (Black Death) through the crowded, dirty cities of Europe in the fourteenth century. Parasitism also causes more mortality than would prevail at lower densities. **Predation** increases when prey becomes abundant and easy to catch, and larger populations of prey are often followed by an increase in the predator populations. **Shelter** for suitable nest sites and as refuge from bad weather becomes less available with crowding.

The continuous competition for food and space to live brings forth yet another density-dependent force: **stress.** Although the physiology of stress is not thoroughly understood, there is ample evidence that, when certain natural populations become crowded—such as populations of lemmings, voles, and snowshoe hares—a neuroendocrine imbalance involving the pituitary and adrenal glands appears. Growth is suppressed. Reproduction fails, and individuals become irritable and aggressive. Under such conditions snowshoe hares suffer from a lethal "shock disease" that results in a population crash.

done so repeatedly. Unfortunately, when the carrying capacity is increased, humans respond as would any animal by increasing their population. Because the earth's resources are finite, the time inevitably will come when the carrying capacity can be extended no further. The question is whether humans will be able to anticipate this limit and check growth of their population in time or whether we will overshoot the resources and experience a crash. Indeed, for millions of people of the Third World time already has run out. The rapid growth of the human population is not something that thinking people view with optimism.

Recent surveys provide hope that the growth of the human population is slackening. Between 1970 and 1995 the annual growth rate decreased from 1.9% to 1.6%. At 1.6%, it will take 43.3 years for the world population to double rather than 36.5 years at the higher annual growth rate figure. The decrease is credited to better family planning. Despite the drop in growth rate, the greatest surge in population lies ahead, with a projected three billion people added within the next three decades, the most rapid increase ever in human numbers.

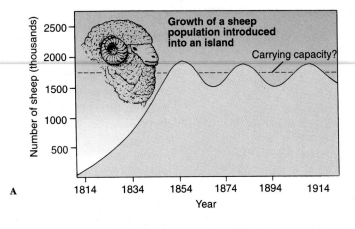

A

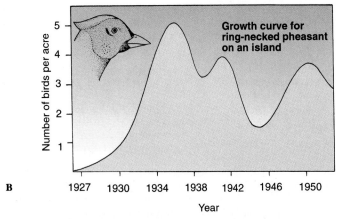

B

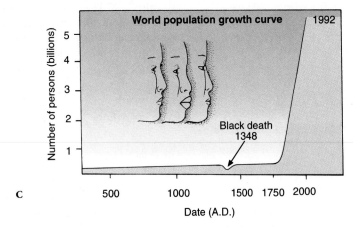

C

Figure 40-14

Growth curves for sheep **A,** ring-necked pheasant **B,** and world human populations **C** throughout history. Note that while the sheep population on an island is stable because of human control of the population, the ring-necked pheasant population oscillates greatly, probably because of large changes in carrying capacity. Where would you place the carrying capacity for the human population?

Overcrowding may also cause **emigration** away from the birth area. Overcrowded mice increase their locomotor activity and begin to explore new areas. In the case of lemmings, overpopulation produces the famous mass "marches" recorded at intervals in Scandinavia (p. 614). When songbirds become overcrowded, the less successful become "vagabond" birds that are forced out into less preferred habitats. Emigration is one major force resulting in the colonization of new habitats. The rapid dispersal of the European starling throughout the United States and southern Canada, following its introduction into New York City in 1890, is an excellent example (p. 590).

Not all of the forces that limit growth are density dependent. Extreme changes in the weather or unusually cold, hot, wet, or dry weather is a **density independent** hazard of varying severity to animal populations. Local populations of insects are sometimes pushed to the point of extinction by a severe winter. Hurricanes and volcanic eruptions may destroy entire populations. Hailstorms have been known to kill most of the young of wading-birds. Prairie grass and forest fires, droughts, and floods all take their toll on animal populations.

Thus population numbers are influenced by factors generated within the population and by forces from without, and usually no single mechanisms can account fully for how growth is curbed in a given population. Natural populations are controlled by density-dependent and density-independent forces. Humans are the greatest force of all. By altering habitats of animals we change the balances that set the old limits to abundance. Our activities can increase or exterminate whole populations of animals.

Summary

The most inclusive level of environmental organization below the biosphere is the ecosystem, which includes interacting biotic and abiotic components. The biotic components are communities of various species of organisms, each species being represented by a population, and each population being the sum of the individual organisms that share a gene pool.

Energy flows into the ecosystem as sunlight. Nearly all life depends on the conversion of that energy into organic compounds by the fixation of carbon dioxide in green plants. The only animal communities known that are completely independent of solar energy are recently discovered communities associated with thermal vents of the deep sea.

Food chains of producers (photosynthetic organisms) and consumers (herbivores and carnivores) may be recognized, but in nature, complex food webs exist. Ecological pyramids can be constructed to show the decrease in numbers of individuals, biomass, and energy at each level of the food chain. Nutrients, such as carbon, nitrogen, and other elements, are cycled and recycled in the ecosystem. Each community is usually dominated by populations of one or a few species that greatly influence the nature of the local environment. Each population of a species has its own niche, or the role it plays in the ecosystem. If the populations of several species exploit the environment in a similar way, they may be considered members of a guild. Populations in the community may have no interaction (neutral), or they may interact as competitors, predators and prey, parasites and hosts, commensals, mutuals, or protocooperators.

Populations introduced into an environment where there is an upper limit to resources usually grow in a manner that can be described by a sigmoid curve; at first, when food and space are not limiting, the growth is exponential, and then the curve levels off as the carrying capacity of the environment is approached. The curve can be expressed mathematically as the logistic equation. This equation takes into account the intrinsic rate of increase of the population and the carrying capacity of the environment. Density-dependent mechanisms for controlling populations include competition, disease, and predation. Density-independent mechanisms include extreme weather, abrupt environmental change, and hazardous conditions for the progeny with accompanying high mortality.

Review Questions

1. The term ecology is derived from the Greek meaning "house" or "place to live." However, as used by scientists, the term "ecology" is not the same as "environment." What is the distinction between the terms?
2. How would you distinguish between ecosystem, community, and population?
3. Only about one-half of the solar energy reaching the outer limit of the atmosphere penetrates to the earth's surface. What happens to solar radiation as it passes through the atmosphere?
4. Recently discovered animal communities surrounding deep-sea thermal vents apparently exist in total independence of solar energy. How can this be possible?
5. Define *production* as the word is used in ecology. What is the distinction between production, primary productivity, gross productivity, and net primary productivity? What is the distinction between productivity and biomass (or standing crop)?
6. What is a food chain? How does a food chain differ from a food web?
7. Distinguish between producers, herbivores, carnivores, and detritus feeders in a food chain.
8. What is a trophic level, and how does it relate to a food chain?
9. How is it possible to have an inverted pyramid of biomass in which the consumers have a greater biomass than the producers? Can you think of an example of an inverted pyramid of *numbers* in which there are, for example, more herbivores than plants on which they feed?
10. The pyramid of energy has been offered as an example of the second law of thermodynamics (p. 63). Why?
11. What pools of CO_2 are involved in the carbon cycle? What is the significance of the carbon cycle?
12. What biological and nonbiological processes are involved in nitrogen fixation?
13. Dominance among species of plants sometimes has been called "space capture." Is this an acceptable description? Why or why not?
14. Define the niche concept. How does the "realized niche" of a population differ from its "fundamental niche"? How does the concept of niche differ from the concept of guild?
15. Define predation. How does the predator-prey relationship differ from the parasite-host relationship?
16. What are some types of symbiotic interactions among animals? Is mutualism always a symbiotic relationship?
17. Contrast exponential and logistic growth of a population. Under what conditions might you expect a population to exhibit exponential growth? Why cannot exponential growth be perpetuated indefinitely?
18. Growth of a population may be hindered by either density-dependent or density-independent mechanisms. Define and contrast these two mechanisms. Offer examples of how growth of the human population might be curbed by either agent.

Selected References

Berner, R. A., and A. C. Lasaga. 1989. Modeling the geochemical carbon cycle. Sci. Am. **260:**74–81 (Mar.). *Geochemical processes that transfer carbon among land, sea, and atmosphere follow long-term cycles that may have warmed the earth in the past through the greenhouse effect. Humans have now short-circuited the long-term cycle.*

Bertness, M. D. 1992. The ecology of a New England salt marsh. Am. Sci. **80**(3):260–268 (May–June). *Animals and plants of this harsh environment must both compete and cooperate to survive.*

Brooks, D. R., and D. A. McLennan. 1991. Phylogeny, ecology and behavior. Chicago, University of Chicago Press. *A discussion of how systematic methods can be used to enhance the study of ecology and animal behavior.*

Colinvaux, P. 1993. Ecology 2. New York, John Wiley & Sons. *Comprehensive college textbook.*

Kates, R. W. 1994. Sustaining life on the earth. Sci. Am. **271:**114–122 (Oct.). *Major technological revolutions—toolmaking, agriculture, and manufacturing—have triggered geometric growth in the human population. The author asks whether we can learn enough about biological, physical, and social reality to fashion a future that our planet can sustain.*

Krebs, C. J. 1993. Ecology: the experimental analysis of distribution and abundance, ed. 4. New York, Harper & Row, Publishers. *Important treatment of population ecology.*

Moore, P. D. (ed.). 1987. The encyclopedia of animal ecology. New York, Facts on File Publications. *Although ecological*

principles are sparsely treated in this book, the world's major ecosystems are surveyed with extensive text and beautifully illustrated with photographs, paintings, and helpful diagrams.

Pianka, E. R. 1993. Evolutionary ecology, ed. 5. New York, Harper & Row, Publishers. *An introduction to ecology written from an evolutionary perspective.*

Smith, R. L. 1995. Ecology and field biology, ed. 5. HarperCollins, Publishers. *Clearly written, well-illustrated general ecology text.*

Tunnicliffe, V. 1992. Hydrothermal-vent communities of the deep sea. Am. Sci. **80**(4):336–349 (July–Aug.). *At hot vents along mid-ocean ridges, nuclear and chemical energy make possible exotic ecosystems that have evolved in near-total isolation.*

APPENDIX A
Development of Zoology

Certain key discoveries have greatly influenced progress in the study of zoology. This appendix presents some of the major landmarks in the development of biology and the individuals whose names are commonly associated with them. It is very difficult to appraise the historical development of any field of study. One investigator often gets the credit for an important discovery when many others should share in the prestige. No one individual has a monopoly on ideas; advances in science are built on the work of many minds. Especially in recent years, important discoveries have resulted from the efforts of teams of investigators. Furthermore, because knowledge of a topic accrues over a period of time, it is often difficult to decide in which year the "discovery" occurred.

In this brief outline the student may be able to see some of the relationships that exist between one discovery and another. The discoveries are not completely isolated, as they may sometimes appear. One may note also that fundamental discoveries in a particular branch of biology tend to be grouped fairly close together chronologically because that particular interest may have dominated the thought of biological investigators at that time.

The reader may wish to locate important discoveries in a given area without reading the entire list. Such an effort may be made easier by use of the following list of key symbols. Any such system of classification suffers some limitations, however. Many basic discoveries are not easily categorized because they have implications in wide areas. The list of categories must be short, or it becomes too cumbersome. Nevertheless, we hope that the classification will be useful.

☐ Taxonomy, systematics, and environmental fields (behavior, ecology, and the like)
○ Tissue- and organ-level anatomy and physiology
△ Cell biology
∗ Developmental biology
‡ Genetics
× Evolution and paleontology
† Of *particular* medical significance

ORIGINS OF BASIC CONCEPTS AND KEY DISCOVERIES IN ZOOLOGY

○ **circa 450–370 B.C.:** *Hippocrates. Establishment of the first biomedical tradition.*

This Greek physician developed an extensive body of anatomical and physiological knowledge that formed the basis of the revival of research in anatomy and physiology in the Renaissance. He is best remembered as the "Father of Medicine" who devised the code of medical ethics now administered as an oath to beginning medical practitioners.

384–322 B.C.: *Aristotle. The foundation of zoology as a science.*

Although this pioneer zoologist and philosopher cannot be appraised by modern standards, there is scarcely a major subdivision of zoology to which he did not make some contribution. However, Aristotle was more the philosopher and poet than scientist, and much of his biological writing is riddled with erroneous opinion.

○ **130–200 A.D.:** *Galen. Development of anatomy and physiology.*

This Roman investigator has been praised for his clear concept of scientific methods and blamed for passing down certain glaring errors that persisted for centuries. His influence was so great that for centuries students considered him the final authority on anatomical and physiological subjects.

1347: *William of Occam. Occam's razor.*

This principle of science has received its name from the fact that it is supposed to cut out unnecessary and irrelevant hypotheses in the explanation of phenomena. The gist of the principle is that, of several possible explanations, the one that is simplest, has the fewest assumptions, and is most consistent with the data at hand is the most probable.

○ **1543:** *Vesalius, Andreas. First modern interpretation of anatomical structures.*

With his insight into fundamental structure, Vesalius ushered in the dawn of modern biological investigation. Many aspects of his interpretation of anatomy are just now beginning to be appreciated.

○ **1616–1628:** *Harvey, William. First accurate description of blood circulation.*

Harvey's classic demonstration of blood circulation was the key experiment that laid the foundation of modern physiology. He explained bodily processes in physical terms, cleared away much of the mystical interpretation, and gave an auspicious start to experimental physiology.

○ **1627:** *Aselli, G. First demonstration of lacteal vessels.*

This discovery, coming at the same time as Harvey's great work, supplemented the discovery of circulation.

○ **1649:** *Descartes, René. Early concept of reflex action.*

Descartes postulated that impulses originating at the receptors of the body were carried to the central nervous system where they activated muscles and glands by what he called "reflection."

∗ **1651:** *Harvey, William. Aphorism of Harvey:* Omne vivum ex ovo (*all life from the egg*).

Although Harvey's work as an embryologist is overshadowed by his demonstration of blood circulation, his *De Generatione Animalium,* published in 1651, contains many sound observations on embryological processes.

○ **1652:** *Bartholin, Thomas. Discovery of lymphatic system.*

The significance of the thoracic duct in its relation to the circulation was determined in this investigation.

○ **1658:** *Swammerdam, Jan. Description of red blood corpuscles.*

This discovery, together with Swammerdam's observations on the valves of the lymphatics and the alterations in shape of muscles during contraction, represented early advancements in the microscopical study of bodily structures.

○ **1661:** *Malpighi, Marcello. Demonstration of capillary circulation.*

By demonstrating the capillaries in the lung of a frog, Malpighi was able to complete the scheme of blood circulation. Harvey never saw capillaries and thus never included them in his description.

△ **1665:** *Hooke, Robert. Discovery of the cell.*

Hooke's investigations were made with cork, and the term "cell" fits cork much better than it does animal cells, but by tradition the misnomer has stuck.

△ **1672:** *de Graaf, R. Description of ovarian follicles.*

De Graaf's name is given to the mature ovarian follicle, but he believed that the follicles were the actual ova, an error later corrected by von Baer.

△□ **1675–1680:** *van Leeuwenhoek, Anthony. Discovery of protozoa.*

The discoveries of this eccentric Dutch microscopist revealed a whole new world of biology.

□ **1693:** *Ray, J. Concept of species.*

Although Ray's work on classification was later overshadowed by that of Linnaeus, Ray was really the first to apply the species concept to a particular kind of organism and to point out the variations that exist among the members of a species.

○ **1733:** *Hales, Stephen. First measurement of blood pressure.*

This was further proof that the bodily processes could be measured quantitatively—more than a century after Harvey's momentous demonstration.

✳ **1745:** *Bonnet, Charles. Discovery of natural parthenogenesis.*

Although somewhat unusual in nature, this phenomenon has yielded much information about meiosis and other cytological problems.

□ **1758:** *Linnaeus, Carolus. Development of binomial nomenclature system of taxonomy.*

So important is this work in taxonomy that 1758 is regarded as the starting point in the determination of the generic and specific names of animals. Besides the value of his binomial system, Linnaeus gave taxonomists a valuable working model of conciseness and clearness that has never been surpassed.

✳ **1759:** *Wolff, Kaspar F. Embryological theory of epigenesis.*

This embryologist, the greatest before von Baer, did much to overthrow the preformation theory then in vogue and, despite many shortcomings, laid the basis for the modern interpretation of embryology.

○ **1760:** *Hunter, John. Development of comparative investigations of animal structure.*

This vigorous eighteenth-century anatomist gave a powerful impetus not only to anatomical observations but also to the establishment of natural history museums.

□ **1763:** *Adanson, M. Concept of empirical taxonomy.*

This botanist proposed a scheme of classification that grouped individuals into taxa according to shared characteristics. A species would have the maximum number of shared characteristics according to this scheme. The concept has been revived recently by exponents of numerical taxonomy. It lacks the evaluation of the evolutionary concept and has been criticized on this account.

‡ **1763:** *Kölreuter, J.G. Discovery of quantitative inheritance (multiple genes).*

Kölreuter, a pioneer in plant hybridization, found that certain plant hybrids had characteristics more or less intermediate between the parents in the F_1 generation, but in the F_2 there were many gradations from one extreme to the other. An explanation was not forthcoming until after Mendel's laws were discovered.

□ **1768–1779:** *Cook, James. Influence of geographical exploration on development of biology.*

This famous sea captain made possible a greater range of biological knowledge because of the able naturalists whom he took on his voyages of discovery.

△ **1772:** *Priestley, J., and J. Ingenhousz. Concept of photosynthesis.*

These investigators first pointed out some of the major aspects of this important phenomenon, such as the use of light energy for converting carbon dioxide and water into released oxygen and retained carbon.

△ **1774:** *Priestley, J. Discovery of oxygen.*

The discovery of this element is of great biological interest because it helped in determining the nature of oxidation and the exact role of respiration in organisms.

△ **1778:** *Lavoisier, Antoine L. Nature of animal respiration demonstrated.*

A basis for the chemical interpretation of the life process was given a great impetus by the careful quantitative studies of the changes during breathing made by this great investigator. His work also meant the final overthrow of the mystical phlogiston hypothesis that had held sway for so long.

□ **1781:** *Abildgaard, P. First experimental life cycle of a tapeworm.*

Life cycles of parasites may be very complicated, involving several hosts. That a parasitic worm could require more than one host was a revolutionary concept in Abildgaard's day and was widely disbelieved until the work of Küchenmeister over 60 years later.

✕ **1791:** *Smith, William. Correlation between fossils and geological strata.*

By observing that certain types of fossils were peculiar to particular strata, Smith was able to work out a method for estimating geological age. He laid the basis of stratigraphic geology.

○ **1792:** *Galvani, L. Animal electricity.*

The lively controversy between Galvani and Volta over the twitching of frog legs led to extensive investigation of precise methods of measuring the various electrical phenomena of animals.

✕ **1796:** *Cuvier, Georges. Development of vertebrate paleontology.*

Cuvier compared the structure of fossil forms with that of living ones and concluded that there had been a succession of organisms that had become extinct and were succeeded by the creation of new ones. To account for this extinction, Cuvier held to the theory of catastrophism, or the simultaneous extinction of animal populations by natural cataclysms.

✕ **1801:** *de Lamarck, J.B. Evolutionary concept of use and disuse.*

Lamarck gave the first clear-cut expression of a hypothesis to account for organic evolution. His assumption was that acquired characteristics were inherited; modern evolutionists have refuted this part of the hypothesis.

○ **1802:** *Young, T. Hypothesis of trichromatic color vision.*

Young's hypothesis suggested that the retina contained three kinds of light-sensitive substances, each having a maximum sensitivity in a different region of the spectrum and each being transmitted separately to the brain. The three substances combined produced the colors of the environment. The three pigments responsible are located in three kinds of cones. Young's explanation has been modified in certain details by other investigators.

○ **1811:** *Bell, C., and F. Magendie. Discovery of the functions of dorsal and ventral roots of spinal nerves.*

This demonstration was a starting point for an anatomical and functional investigation of the most complex system in the body.

✳ **1817:** *Pander, C.H. First description of three germ layers.*

The description of the three germ layers was first made on the chick, and later the concept was extended by von Baer to include all vertebrates.

✳ **1827:** *von Baer, Karl. Discovery of mammalian ovum.*

The very tiny ova of mammals escaped de Graaf's eyes, but von Baer brought mammalian reproduction into line with that of other animals by detecting the ova and their true relation to the follicles.

△ **1828:** *Brown, Robert. Brownian movement first described.*

This interesting phenomenon is characteristic of living matter and sheds some light on the structure of cells.

☐ **1828:** *Thompson, J.V. Nature of plankton.*

Thompson's collections of these small forms with a tow net, together with his published descriptions, are the first records of the vast community of planktonic animals. He was also the first to work out the true nature of barnacles.

△ **1828:** *Wöhler, F. First to synthesize an organic compound.*

Wöhler succeeded in making urea (a compound formed in the body) from the inorganic substance ammonium cyanate; thus, $NH_4OCN \rightarrow NH_2CONH_2$. This success in producing an organic substance synthetically was the stimulus that resulted in the preparation of thousands of compounds by others.

✳✕ **1830:** *von Baer, Karl. Biogenetic law formulated.*

Von Baer's conception of this law was conservative and sounder in its implication than has been the case with many other biologists (Haeckel, for instance). Von Baer stated that embryos of higher and lower forms resemble each other more the earlier they are compared in their development, and *not* that the embryos of higher forms resemble the adults of lower organisms.

✕ **1830:** *Lyell, C. Modern concept of geology.*

The influence of this concept not only did away with the catastrophic theory but also gave a logical interpretation of fossil life and the correlation between the formation of rock strata and the animal life that existed at the time these formations were laid down.

△ **1831:** *Brown, Robert. First description of cell nucleus.*

Others had seen nuclei, but Brown was the first to name the structure and to regard the nucleus as a general phenomenon. This description was an important preliminary to the formation of the cell theory a few years later, for Schleiden acknowledged the importance of the nucleus in the development of the cell concept.

○ **1833:** *Hall, Marshall. Concept of reflex action.*

Hall described the method by which a stimulus can produce a response independently of sensation or volition and coined the term "reflex action." It remained for the outstanding work of Sir Charles Sherrington in the twentieth century to explain much of the complex nature of reflexes.

†○ **1835:** *Bassi, Agostino. First demonstration of a microorganism as an infective agent.*

Bassi's discovery that a certain disease of silkworms was caused by a small fungus represents the beginning of the germ theory of disease that was to prove so fruitful in the hands of Pasteur and other able investigators.

△ **1835:** *Dujardin, Felix. Description of living matter (protoplasm).*

Dujardin associated the jellylike substance that he found in protozoa and that he called "sarcode" with the life process. This substance was later called "protoplasm," and the sarcode idea may be considered a significant landmark in the development of the protoplasm concept.

†○ **1835:** *Owen, Richard. Discovery of* Trichinella.

This versatile investigator is chiefly remembered for his researches in anatomy, but his discovery of this very common parasite in humans is an important landmark in the history of parasitology.

△○ **1838:** *von Liebig, Justus. Foundation of biochemistry.*

The idea that vital activity could be explained by chemicophysical factors was an important concept for biological investigators studying the nature of life.

△ **1838–1839:** *Schleiden, M.J., and T. Schwann. Formulation of the cell theory.*

The cell doctrine, with its basic idea that all plants and animals were made up of similar units, represents one of the truly great landmarks in biological progress. Many biologists believe the work of Schleiden and Schwann has been rated much higher than it deserves in light of what others had done to develop the cell

concept. R. Dutrochet (1824) and H. von Mohl (1831) described all tissues as being composed of cells.

△ **1838:** *Mulder, G.J. Concept of the nature of proteins.*

Mulder proposed the name "protéine," later "protein," for the basic constituents of protoplasmic materials.

△ **1839–1846:** *Purkinje, J.E., and Hugo von Mohl. Concept of protoplasm established.*

Purkinje proposed the name "protoplasm" for living matter, and von Mohl did extensive work on its nature, but it remained for Max Schultze (1861) to give a clear-cut concept of the relations of protoplasm to cells and its essential unity in all organisms.

☐ **1839:** *Verhulst, P.F. Logistical theory of population growth.*

According to this theory, animal populations have a slow initial growth rate that gradually speeds up until it reaches a maximum and then slows down to a state of equilibrium. By plotting the logarithm of the total number of individuals against time, an S-shaped curve results that is somewhat similar for all populations.

☐○ **1840:** *von Liebig, J., and F.F. Blackman. The law of minimum requirements.*

Von Liebig interpreted plant growth as being dependent on essential requirements that are present in minimum quantity; some substances, even in minute quantities, were needed for normal growth. Later Blackman discovered that the rate of photosynthesis is restricted by the factor that operates at a limiting intensity (for example, light or temperature). The concept has been modified in various ways since it was formulated. We now know that a factor interaction other than the minimum ones plays a part.

○ **1840:** *Müller, J. Theory of specific nerve energies.*

This theory states that the kind of sensation experienced depends on the nature of the sense organ with which the stimulated nerve is connected. The optic nerve, for instance, conveys the impression of vision however it is stimulated.

○ **1842:** *Bowman, William. Histological structure of the nephron (kidney unit).*

Bowman's accurate description of the nephron afforded physiologists an opportunity to attack the problem of how the kidney separates waste from the blood.

✳ **1842:** *Steenstrup, Johann J.S. Alternation of generations described.*

Metagenesis, or alternation of sexual and asexual reproduction in the life cycle, exists in many animals and plants. The concept had been introduced before this time by L.A. de Chamisso (1819).

× **1843:** *Owen, Richard. Concepts of homology and analogy.*

Homology, as commonly understood, refers to similarity in embryonic origin and development, whereas analogy is the likeness between two organs in their functioning. Owen's concept of homology was merely that of the same organ in different animals under all varieties of form and function.

○ **1844:** *Ludwig, C. Filtration theory of renal excretion.*

About the same time that Bowman described the renal corpuscle, Ludwig showed that this unit functions as a passive filter and that the filtrate that passes through it from the blood carries into the renal tubules the waste products that are concentrated by the resorption of water as the filtrate moves down the tubules. Other investigators, Cushney, Starling, and Richards, have confirmed and extended his observations.

○△ **1845:** *von Helmholtz, H., and J.R. von Mayer. The formulation of the law of conservation of energy.*

This landmark in human thinking showed that in any system, living or nonliving, the power to perform mechanical work always tends to decrease unless energy is added from without. Physiological investigation could now advance on the hypothesis that the living organism is an energy machine and obeys the law of energetics.

‡ **1848:** *Hofmeister, W. Discovery of chromosomes.*

This investigator made sketches of bodies, later known to be chromosomes, in the nuclei of pollen mother cells (*Tradescantia*). Schneider further described these elements in 1873, and Waldeyer named them in 1888.

□ **1848:** *von Siebold, Carl T.E. Establishment of the status of protozoa.*

Von Siebold emphasized the unicellular nature of protozoa, fitted them into the recently developed cell theory, and established them as the basic phylum of the animal kingdom.

○ **1851:** *Bernard, Claude. Discovery of the vasomotor system.*

Bernard showed how the amount of blood distributed to the various tissues by the small arterioles was regulated by vaso-motor nerves of the sympathetic nervous system, as he demonstrated in the ear of a white rabbit.

✱**1851:** *Waller, A.V. Importance of the nucleus in regeneration.*

When nerve fibers are cut, the parts of the fibers peripheral to the cut degenerate in a characteristic fashion. This wallerian degeneration enables one to trace the course of fibers through the nervous system. In the regeneration process the nerve cell body (with its nucleus) is necessary for the downgrowth of the fiber from the proximal segment.

○ **1852:** *von Helmholtz, Herman. Determination of rate of nervous impulse.*

This landmark in nerve physiology showed that the phenomena of nervous activity could be measured experimentally and expressed numerically.

○△ **1852:** *von Kölliker, Albrecht. Establishment of histology as a science.*

Many histological structures were described with marvelous insight by this great investigator, starting with the publication of the first text in histology (*Handbuch der Gewebelehre*).

○ **1852:** *Stannius, H.F. Stannius's experiment on the heart.*

By tying a ligature as a constriction between the sinus venosus and the atrium in the frog and also one around the atrioventricular groove, Stannius was able to demonstrate that the muscle tissues of the atria and ventricles have independent and spontaneous rhythms. His observations also indicated that the sinus is the pacemaker of the heartbeat.

†□ **1852:** *Küchenmeister, G.F.H. Relationship of "bladder worms" to adult tapeworms.*

This was the first demonstration that "bladder worms" (cysticerci) were the juveniles of taeniid tapeworms. It was followed rapidly by numerous other life-cycle studies on cestodes.

✱ **1854:** *Newport, G. Discovery of the entrance of spermatozoon into a frog's egg.*

This was a significant step in cellular embryology, although its real meaning was not revealed until the concept of fertilization as the union of two pronuclei was formulated about 20 years later (Hertwig, 1875).

†○ **1855:** *Addison, T. Discovery of adrenal disease.*

The importance of the adrenals in maintaining normal body functions was shown when Addison described a syndrome of general disorders associated with the pathology of this gland.

× **1856:** *Discovery of Neanderthal fossil hominid* (Homo sapiens).

Many specimens of this type of fossil hominid have been discovered (Europe, Asia, Africa) since the first one was found near Düsseldorf, Germany. Their culture was Mousterian (100,000 to 40,000 years BP), and although they were of short stature, their cranial capacity was as large as or larger than that of modern humans.

○ **1857:** *Bernard, Claude. Discovery of formation of glycogen by the liver.*

Bernard's demonstration that the liver forms glycogen from substances brought to it by the blood showed that the body can build up complex substances as well as tear them down.

□ **1858:** *Sclater, P.L. Distribution of animals on basis of zoological regions.*

This was the first serious attempt to study the geographical distribution of organisms, a study that eventually led to the present science of zoogeography. A.R. Wallace worked along similar lines.

✱△ **1858:** *Virchow, Rudolf. Aphorism of Virchow:* Omnis cellula e cellula (*every cell from a cell*).

△ **1858:** *Virchow, Rudolf. Formation of the concept of disease from the viewpoint of cell structure.*

He laid the basis of modern pathology by stressing the role of the cell in diseased tissue.

× **1859:** *Darwin, Charles. The concept of natural selection as a factor in evolution.*

Although Darwin did not originate the concept of organic evolution, no one has been more influential in the development of evolutionary thought. The publication of *The Origin of Species* represents the greatest and most influential single landmark in the history of biology.

○ **1860:** *Bernard, Claude. Concept of the constancy of the internal environment (homeostasis).*

Bernard's realization that there are mechanisms in living organisms that tend to maintain internal conditions within relatively narrow limits, despite external change, is one of the most valuable generalizations in physiology.

✱ **1860:** *Pasteur, Louis. Aphorism of Pasteur:* Omne vivum e vivo (*every living thing from the living*).

✱ **1860:** *Pasteur, Louis. Refutation of spontaneous generation.*

Pasteur's experiment with the open S-shaped flask proved conclusively that fermentation or putrefaction resulted from microbes, thus ending the long-standing controversy regarding spontaneous generation.

☐ **1860:** *Wallace, A.R. The Wallace line of faunal delimitation.*

As originally proposed by Wallace, there was a sharp boundary between the Australian and Oriental faunal regions so that a geographical line drawn between certain islands of the Malay Archipelago, through the Makassar Strait between Borneo and Celebes, and between the Philippines and the Sanghir Islands separated two distinct and contrasting zoological regions. On one side Australian forms predominated; on the other, Oriental ones. The validity of this division has been questioned by zoogeographers in the light of more extensive knowledge of the faunas of the two regions.

△ **1861:** *Graham, Thomas. Colloidal states of matter.*

This work on colloidal solutions has resulted in one of the most fruitful concepts regarding protoplasmic systems.

☐ **1862:** *Bates, H.W. Concept of mimicry.*

Mimicry refers to the advantageous resemblance of one species to another for protection against predators. It involves palatable or edible species that imitate dangerous or inedible species that employ warning coloration against potential enemies.

☐ **1864:** *Haeckel, Ernst. Modern zoological classification.*

The broad features of zoological classification as we know it today were outlined by Haeckel and others, especially R.R. Lankester, in the third quarter of the nineteenth century. B. Hatschek is another zoologist who deserves much credit for the modern scheme, which is constantly undergoing revision. Other schemes of classification in the early nineteenth century did much to resolve the difficult problem of classification, such as those of Cuvier, de Lamarck, Leuckart, Ehrenberg, Vogt, Gegenbauer, and Schimkivitch.

△ **1864:** *Schultze, Max. Protoplasmic bridges between cells.*

The connection of one cell to another by means of protoplasmic bridges has been demonstrated in both plants and animals. Modern techniques, including electron microscopy, have shown that many cells communicate by gap junctions, which have channels capable of passing molecules of up to 1000 daltons in molecular weight.

‡ **1866:** *Haeckel, Ernst. Nuclear control of inheritance.*

At the time the hypothesis was formed, the view that the nucleus transmitted the inheritance of an animal had no evidence to substantiate it.

☐ **1866:** *Kovalevski, A. Taxonomic position of tunicates.*

The great Russian embryologist showed in the early stages of development the similarity between amphioxus and the tunicates and how the latter could be considered a branch of the phylum Chordata.

‡ **1866:** *Mendel, Gregor. Formulation of the first two laws of heredity.*

These first clear-cut statements of inheritance made possible an analysis of hereditary patterns with mathematical precision. Rediscovered in 1900, his paper confirmed the experimental data geneticists had found at that time. Since that time Mendelian genetics has been considered the chief cornerstone of hereditary investigation.

○ **1866:** *Schultze, Max. Histological analysis of the retina.*

Schultze's fundamental discovery that the retina contained two types of visual cells—rods and cones—helped to explain the physiological differences of vision in high and low intensities of light.

×* **1867:** *Kovalevski, A. Germ layers of invertebrates.*

The concept of the primary germ layers laid down by Pander and von Baer was extended to invertebrates by this investigator. He found that the same three germ layers arose in the same fashion as those in vertebrates. Thus an important embryological unity was established for the animal kingdom.

○ **1869:** *Langerhans, P. Discovery of islet cells in pancreas.*

These islets were first found in the rabbit, and since their discovery they have been one of the most investigated tissues in animals. The hypothesis that they have come from the transformation of pancreatic acinus cells has now given way to the belief that they originated from the embryonic tubules of the pancreatic duct.

△ **1869:** *Miescher, F. Isolation of nucleoprotein.*

From pus cells, a pathological product, Miescher was able to demonstrate in the nuclei certain phosphorus-rich substances, nucleic acids, which are bound to proteins to form nucleoproteins. In recent years these complex molecules have been the focal point of significant biochemical

investigations on the chemical properties of genes, with the wider implications of a better understanding of growth, heredity, and evolution.

☐ **1871:** *Quetelet, Lambert A. Foundations of biometry.*

The applications of statistics to biological problems is called biometry. By working out the distribution curve of the height of soldiers, Quetelet showed biologists how the systematic study of the relationships of numerical data could become a powerful tool for analyzing data in evolution, genetics, and other biological fields.

☐ **1872–1876:** Challenger *expedition.*

Not only did this expedition establish the science of oceanography, but a vast amount of material was collected that greatly extended the knowledge of the variety and range of animal life.

☐ **1872:** *Dohrn, Anton. Establishment of Naples Biological Station.*

The establishment of this famous station marked the first of the great biological stations, and it rapidly attained international importance as a center of biological investigation.

○ **1872:** *Ludwig, C., and E.F.W. Pflüger. Gas exchange of the blood.*

By means of the mercurial blood pump, these investigators separated the gases from the blood and thereby threw much light on the nature of gas exchange and the place where oxidation occurs (in the tissues).

×☐ **1874:** *Haeckel, Ernst H. Taxonomic position of phylum Chordata.*

The great German evolutionist based many of his conclusions on the work of the Russian embryologist Kovalevski, who in 1866 showed that the tunicates as well as amphioxus had vertebrate affinities.

× **1874:** *Haeckel, Ernst H. Gastrea hypothesis of metazoan ancestry.*

According to this hypothesis, the hypothetical ancestor of all the Metazoa consisted of two layers (ectoderm and endoderm) similar to the gastrula stage in embryonic development, and the endoderm arose as an invagination of the blastula. Thus the diploblastic stage of ontogeny was to be considered as the repetition of this ancestral form. This concept in modified form has had wide acceptance.

‡* **1875:** *Hertwig, Oskar. Concept of fertilization as the conjugation of two sex cells.*

The fusion of the pronuclei of the two gametes in the process of fertilization paved the way for the concept that the nuclei contained the hereditary factors and that both maternal and paternal factors are brought together in the zygote.

△ **1875:** *Strasburger, Eduard. Description of mitotic cell division.*

The accurate description of the processes of cell division that Strasburger made in plants represents a great pioneer work in the rapid development of cytology during the last quarter of the nineteenth century.

△ **1876:** *Pasteur, Louis. The Pasteur effect.*

In his studies of fermentation processes in wine making, Pasteur discovered that cells consumed much more glucose and produced much more lactic acid in the absence of oxygen than in its presence. It is now known that the effect occurs in many different kinds of cells and is a consequence of the lower energy yield of glycolysis compared with oxidative phosphorylation.

○ **1877:** *Pfeffer, Wilhelm. Concept of osmosis and osmotic pressure.*

Pfeffer's experiments on osmotic pressure and the determination of the pressure in different concentrations laid the foundation for an understanding of a general phenomenon in all organisms.

○ **1878:** *Balfour, Francis M. Relationship of the adrenal medulla to the sympathetic nervous system.*

By showing that the adrenal medulla has the same origin as the sympathetic nervous system, Balfour really laid the foundation for the interesting concept of the similarity in the action of epinephrine and sympathetic nervous mediation.

○△ **1878:** *Kühne, Willy. Nature of enzymes.*

The study of the action of chemical catalysts has steadily increased with biological advancement and now represents one of the most fundamental aspects of biochemistry.

△ **1879:** *Flemming, Walther. Chromatin described and named.*

Flemming shares with a few others (especially Strasburger, 1875) the description of the details of mitotic cell division. The part of the nucleus that stains deeply he called **chromatin** (colored), which gives rise to the chromosomes.

✳ **1879:** *Fol, Hermann. Penetration of ovum by a spermatozoon described.*

Fol was the first to describe a thin cone-like body extending outward from the egg to meet the sperm. Compare Dan's acrosome reaction (1954).

△ **1879:** *Kossel, Albrecht. Isolation of nucleoprotein.*

Nucleoproteins were isolated in the heads of fish sperm; they make up the major part of chromatin. They are combinations of proteins with nucleic acids, and this study was one of the first of the investigations of a topic that interests biochemists at the present time.

Nobel Laureate (1910).

† **1880:** *Laveran, C.L.A. Protozoa as pathogenic agents.*

This French investigator first demonstrated that the causative organism for malaria is a protozoan. This discovery led to other investigations that revealed the role the protozoa play in causing diseases such as sleeping sickness and kala-azar. It remained for Sir Ronald Ross to discover the role of mosquitos in spreading malaria.

Nobel Laureate (1907).

○ **1880:** *Ringer, S. Influence of blood ions on heart contraction.*

The pioneer work of this investigator determined the inorganic ions necessary for contraction of frog hearts and made possible an evaluation of heart metabolism and the replacement of body fluids.

△ **1882:** *Flemming, Walther. First accurate counts of nuclear filaments (chromosomes) made.*

△ **1882:** *Flemming, Walther. Mitosis and spireme named.*

†△ **1882:** *Metchnikoff, Élie. Role of phagocytosis in immunity.*

The theory that microbes are ingested and destroyed by certain white corpuscles (phagocytes) shares with the theory of chemical bodies (antibodies) the chief explanation for the body's natural immunity. Metchnikoff first studied phagocytosis in the cells of starfish and water fleas (*Daphnia*).

Nobel Laureate (1908).

△ **1882:** *Strasburger, Eduard. Cytoplasm and nucleoplasm named.*

□ **1882–1924:** *The* Albatross *of the Fish Commission.*

The *Albatross,* under the direction of the U.S. Fish Commission, was second only to the famed *Challenger* in advancing scientific knowledge about oceanography.

○△ **1883:** *Golgi, Camillo, and R. Cajal. Silver nitrate technique for nervous elements.*

The development and refinement of this technique gave a completely new picture of the intricate relationships of neurons. Modifications of this method have given valuable information concerning the cellular element—the Golgi apparatus.

Nobel Laureates (1906).

✳ **1883:** *Hertwig, Oskar. Origin of term "mesenchyme."*

This important tissue may arise embryonically from all three germ layers but arises chiefly from the mesoderm. It is a protoplasmic network whose meshes are filled with a fluid intercellular substance. Mesenchyme gives rise to a great variety of tissues, and both its cells and intercellular substance may be variously modified; its most common derivative is connective tissue.

†□ **1883:** *Leuckart, Rudolph, and A.P. Thomas. Life history of sheep liver flukes.*

This discovery represents the first time a complete life cycle was worked out for a digenetic trematode. Working independently, these investigators found that the trematode used a snail intermediate host.

‡ **1883:** *Roux, Wilhelm. Allocation of hereditary functions to chromosomes.*

This hypothesis could not have been much more than a guess when Roux made it, but how fruitful was the idea in the light of the enormous amount of evidence since accumulated!

△ **1884:** *Flemming, W., E. Strasburger, and E. van Beneden. Demonstration that nuclear filaments (chromosomes) double in number by longitudinal division.*

This concept represented a further step in understanding the precise process in indirect cell division.

○ **1884:** *Rubner, Max. Quantitative determinations of the energy value of foods.*

Although Liebig and others had estimated calorie values of foods, the investigations of Rubner put their determinations on a sound basis. His work made possible a scientific explanation for metabolism and a basis for the study of comparative nutrition.

‡ **1885:** *Hertwig, Oskar, and E. Strasburger. Concept of the nucleus as the basis of heredity.*

The development of this idea occurred before Mendel's laws of heredity were rediscovered in 1900, but it anticipated the important role the nucleus with its chromosomes was to assume in hereditary transmission.

‡ **1885:** *Rabl, Karl. Concept of the individuality of the chromosomes.*

The view that the chromosomes retain their individuality through all stages of the cell cycle is accepted by all cytologists, and much evidence has accumulated in support of the hypothesis. However, it has been virtually impossible to demonstrate the chromosome's individuality through all stages.

*   **1885:** *Roux, Wilhelm. Mosaic theory of development.*

In the early development of the frog's egg, Roux showed that the determinants for differentiations were segregated in the early cleavage stages and that each cell or groups of cells would form only certain parts of the developing embryo (mosaic or determinate development). Later, other investigators showed that in many forms blastomeres, separated early, would give rise to whole embryos (indeterminate development).

*‡   **1885:** *Weismann, August. Formulation of germ plasm theory.*

Weismann's great theory of the germ plasm stresses the idea that there are two types of protoplasm—germ plasm, which gives rise to the reproductive cells or gametes, and somatoplasm, which furnishes all the other cells. Germ plasm, according to the theory, is continuous from generation to generation, whereas the somatoplasm dies with each generation and does not influence the germ plasm.

1886: *Establishment of Woods Hole Biological Station in Massachusetts.*

This station is by all odds the greatest center of its kind in the world. Many of the biologists in America have studied there, and the station has increasingly attracted investigators from many other countries as well. The influence of Woods Hole on the progress of biology cannot be overestimated.

△   **1887:** *Fischer, Emil. Structural patterns of proteins.*

The importance of proteins in biological systems has made their study the central theme of much modern biochemical work.

Nobel Laureate (1902).

*   **1887:** *Haeckel, Ernst H. Concept of organic form and symmetry.*

Symmetry refers to the spatial relations and arrangements of parts in such a way as to form geometrical designs. Although many others before Haeckel's time had studied and described types of animal form, Haeckel gave us our present concepts of organic symmetry, as revealed in his monograph on radiolarians collected on the *Challenger* expedition.

‡   **1889:** *Hertwig, R., and E. Maupas. True nature of conjugation in paramecium.*

The process of conjugation was described (even by van Leeuwenhoek) many times and its sexual significance interpreted, but these two investigators independently showed the details of the divisions before conjugation and the mutual exchange of the micronuclei during the process.

†○   **1889:** *von Mering, J., and O. Minkowski. Effect of pancreatectomy.*

The classic experiment of removing the pancreas stimulated research that led to the isolation of the pancreatic hormone insulin by Banting (1922).

†   **1890:** *Smith, Theobald. Role of an arthropod in disease transmission.*

The transmission of the sporozoan *Babesia*, which is the active agent in causing Texas cattle fever, by the tick *Boophilus*, represented the first demonstration of the role of an arthropod as a vector of disease.

*   **1891:** *Driesch, Hans. Discovery of totipotent cleavage.*

The discovery that each of the first several blastomeres, if separated from each other in the early cleavage of the sea urchin embryo, would develop into a complete embryo stimulated investigation on totipotent and other types of development.

×   **1890–1891:** *Dubois, Eugene. Discovery of the fossil human* Pithecanthropus erectus (*now* Homo erectus).

Although not the first fossil human to be found, the java man represents one of the first significant primitive humans that have been discovered.

×   **1893:** *Dollo, I. Concept of the irreversibility of evolution.*

In general, the overall evolutionary process is one-way and irreversible insofar as a whole complex genetic system is concerned, although back mutations and restricted variations may occur over and over.

○   **1893:** *His, Wilhelm. Anatomy and physiology of the atrioventricular node and bundle.*

The specialized conducting tissue of the heart has given rise to many investigations, not the least of which were those of His, whose name was given to the intricate system of branches that are reflected over the inner surface of the ventricles.

*   **1894:** *Driesch, Hans. Constancy of nuclear potentiality.*

Driesch's view was that all nuclei of an organism were equipotential but that the activity of nuclei varied as tissues differentiated. More recent studies have emphasized the crucial importance of the cytoplasm of the embryo in determining expression of the nuclear genome.

□   **1894:** *Merriam, C.H. Concept of life zones in North America.*

This scheme is based on temperature criteria and the importance of temperature in the distribution of plants and animals. According to this concept, animals and plants are restricted in their northward distribution by the total quantity of heat during the season of growth and reproduction, and their southward distribution is restricted by the mean temperature during the hottest part of the year.

†□   **1895:** *Bruce, D. Life cycle of protozoan blood parasite* (Trypanosoma).

The relation of this parasite to the tsetse fly and to wild and domestic animal infection in Africa is an early demonstration of the role of arthropods as vectors of disease.

†○   **1895:** *Roentgen, W. Discovery of x-rays.*

This great discovery was quickly followed by its application in the interpretation of bodily structures and processes and represents one of the great tools in biological research.

Nobel Laureate (1901).

×   **1896:** *Baldwin, J.M. Baldwin evolutionary effect.*

It is the belief that genetic selection of genotypes will be channeled or canalized in the same direction as the adaptive modifications that were formerly nonhereditary. It is now understood that the *capacity* to respond to environmental conditions is itself hereditary.

×   **1896:** *Russian Hydrographic Survey. Biology of Lake Baikal.*

This lake in Siberia is more than 800 km long and 80 km wide and has an extreme depth of more than a mile (the deepest lake in the world). Its unique fauna is a striking example of evolutionary results from long-continued isolation. Almost all of the species in certain groups are endemic (found nowhere else). This remarkable fauna represents the survival of ancient freshwater animals that have become extinct in surrounding areas.

○   **1897:** *Abel, J.J., and A.C. Crawford. Isolation of the first hormone (epinephrine).*

The purification and isolation of one of the active principles of the adrenal medulla led to discovery of its chemical nature and naming by J. Takamine (1901) and its synthesis by F. Stolz (1904).

△○   **1897:** *Buchner, E. Discovery of zymase.*

Buchner's discovery that an enzyme (a nonliving substance) manufactured by yeast cells was responsible for fermentation resolved many problems that had baffled Pasteur and other investigators. Zymase is now known to consist of a number of enzymes.

Nobel Laureate (1907).

× **1897:** *Canadian Geological Survey. Dinosaur fauna of Alberta, Canada.*

The discovery and exploration of the Upper Cretaceous fossil beds along the Red Deer River in Alberta revealed one of the richest dinosaur faunal finds known.

†○ **1897:** *Eijkman, Christian. Discovery of the cause of a dietary deficiency disease.*

Eijkman's pioneer work on the causes of beriberi led to the isolation of the antineuritic vitamin (thiamine). This work may be called the key discovery that resulted in the development of the important vitamin concept.

Nobel Laureate (1929).

†□ **1897:** *Ross, Ronald. Life history of malarial parasite* (Plasmodium).

This notable achievement represents a great landmark in the field of parasitology and the climax of the work of many investigators on the problem.

Nobel Laureate (1902).

△○ **1897:** *Sherrington, C.S. Concept of the synapse in the nervous system.*

If the nervous system is composed of discrete units, or neurons, functional connections must exist between these units. Sherrington showed how individual nerve cells could exert integrative influences on other nerve cells by graded excitatory or inhibitory synaptic actions. The electron microscope has in recent years added much to our knowledge of synaptic structure.

Nobel Laureate (1932).

△ **1898:** *Benda, C., and C. Golgi. Discovery of mitochondria and the Golgi apparatus.*

These interesting cytoplasmic inclusions were actually seen by various observers before this date, but they were not named in 1898, and the real study of them began at this time. W. Flemming and R. Altmann first demonstrated mitochondria. The Golgi apparatus was demonstrated by V. St. George (1867) and G. Platner (1855), but Camillo Golgi, with his silver nitrate impregnation method, gave the first clear description of the apparatus in nerve cells. Mitochondria are now known to play an important role in cell respiration and energy production.

× **1898:** *Osborn, Henry F. Concept of adaptive radiation in evolution.*

This concept states that, starting from a common ancestral type, many different forms of evolutionary adaptations may occur. In this way evolutionary divergence can take place, and the occupation of a variety of ecological niches is made possible, according to the adaptive nature of the invading species.

× **1900:** *Chamberlain, T.C. Hypothesis of the freshwater origin of vertebrates.*

The evidence for this hypothesis as first proposed was based mainly on the discovery of early vertebrate fossils in sediments thought to be of freshwater origin, such as Old Red Sandstone. The fossil evidence has been reevaluated by recent investigators, who interpret it to support a marine origin of the vertebrates.

‡ **1900:** *Correns, Karl E., Erick Tschermak von Seysenegg, and Hugo De Vries. Rediscovery of Mendel's laws of heredity.*

In their genetic experiments on plants these three investigators independently obtained results similar to Mendel's, and in their survey of the literature they found that Mendel had published his now-famous laws in 1866. A few years later, W. Bateson and others found that the same laws applied to animals.

× **1900:** *Andrews, C.F. Discovery of fossil beds in the Fayum Depression region of Egypt.*

Here Andrews and others discovered numerous Eocene and Oligocene fossils of protomonkeys believed to be ancestral to Old World monkeys and thus in the lineage leading to hominids.

†○ **1900:** *Landsteiner, Karl. Discovery of blood groups.*

This fundamental discovery made possible successful blood transfusions and initiated intense work on the biochemistry of blood.

Nobel Laureate (1930).

✳ **1900:** *Loeb, Jacques. Discovery of artificial parthenogenesis.*

The possibility of getting eggs that normally undergo fertilization to develop by chemical and mechanical methods has been accomplished in a number of different animals, from sea urchin and frog eggs (Loeb, 1900) to the rabbit egg (Pincus, 1936). The phenomenon has been a useful tool in studying the biology of fertilization.

‡ **1901:** *Montgomery, T.H. Homologous pairing of maternal and paternal chromosomes in the zygote.*

Montgomery showed that in synapsis before reduction division, each pair is made up of a maternal and a paternal chromosome. This phenomenon, verified a year later by W.S. Sutton, is of fundamental importance in the segregation of hereditary factors (genes).

×‡ **1901:** *De Vries, Hugo. Mutation theory of evolution.*

De Vries concluded from his study of the evening primrose *Oenothera lamarckiana* that new characteristics appear suddenly and are inheritable. Although the variations in *Oenothera* were probably not mutations at all, since many of them represented hybrid combinations, the evidence that mutations are the ultimate source of all new variation has steadily mounted.

‡ **1902:** *McClung, Clarence E. Discovery of sex chromosomes.*

The discovery in the grasshopper that a certain chromosome (X) had a synaptic mate (Y) different in appearance or else lacked a mate altogether gave rise to the theory that certain chromosomes determined sex. H. Henking actually discovered the X chromosome in 1891.

○ **1903:** *Bayliss, William M., and Ernest H. Starling. Action of the hormone secretin.*

The demonstration of the action of secretin, a hormone released from the mucosa of the stomach, marked the real birth of the science of endocrinology. It was the first unequivocal proof that physiological functions could be chemically integrated without participation of the nervous system.

‡ **1903:** *Boveri, Theodor, and W.S. Sutton. Parallelism between chromosome behavior and Mendelian segregation.*

This concept states that synaptic mates in meiosis correspond to the Mendelian alternative characters and that the formula of character inheritance of Mendel could be explained by the behavior of the chromosomes during maturation. This is therefore a cytological demonstration of Mendelian genetics.

‡ **1903:** *Sutton, W.S. Constitution of the diploid group of chromosomes.*

Sutton related Mendelian transmission of inherited characters to cytology by showing that the diploid group of chromosomes is made up of two chromosomes of each recognizable size, one member of which is paternal and the other maternal in origin.

○ **1904:** *Cannon, Walter B. Mechanics of digestion observed by means of x-ray films.*

The clever application of x-rays to a study of the movements and other aspects of the digestive system has revealed an enormous amount of information on the physiology of the alimentary canal. Cannon first used this technique in 1898.

○ **1904:** *Carlson, Anton J. Pacemaking activity of neurogenic hearts.*

Heartbeats may originate in muscle (myogenic hearts) or in ganglion cells (neurogenic hearts). Carlson showed that in the arthropod *Limulus* the pacemaker was located in certain ganglia on the dorsal surface of the heart and that experimental alterations of these ganglia by temperature or other means altered the heart rate. Neurogenic hearts are restricted to certain invertebrates.

□ **1904:** *Jennings, Herbert S. Behavior patterns in protozoa.*

The careful investigations of this lifelong student of the behavior of these lower organisms led to concepts such as the trial-and-error behavior and many of our important beliefs about the various forms of tropisms and taxes.

□ **1904:** *Nuttall, George H.F. Serological relationships of animals.*

This method of determining animal relationships is striking evidence of evolution. It has been used in recent years to help establish the taxonomic position of animals.

○ **1905:** *Haldane, John B.S., and John G. Priestley. Role of carbon dioxide in the regulation of breathing.*

By their clever technique of obtaining samples of air from the lung alveoli, these investigators showed how the constancy of carbon dioxide concentration in the alveoli and its relation to the concentration in the blood were the chief regulators of the mechanism of respiration.

‡ **1906:** *Bateson, William, and Reginald C. Punnett. Discovery of linkage of hereditary units.*

Although Bateson and Punnett first discovered linkage in sweet peas, it was Morgan and associates who correctly interpreted this great genetic concept. All seven pairs of Mendel's alternative characteristics were on separate chromosomes, which simplified the problem.

○ **1906:** *Einthoven, W. Mechanism of the electrocardiogram.*

The invention of the string galvanometer (1903) by Einthoven supplied a precise tool for measuring the bioelectrical activity of the heart and quickly led to the electrocardiogram, which gives accurate information about disturbances of the heart's rhythm.

Nobel Laureate (1924).

○ **1906:** *Hopkins, Frederick G. Analysis of dietary deficiency.*

Hopkins tried to explain dietary deficiency by a biochemical investigation of the lack of essential amino acids in the diet, an approach that has led to many important investigations in nutritional requirements.

Nobel Laureate (1929).

△ **1906:** *Tswett, M. Principle of chromatography.*

This is the separation of chemical components in a mixture by differential migration of materials according to structural properties within a special porous sorptive medium. It was refined and became a widely used method 30 or 40 years later (Martin and Synge, 1941). The technique was first used by Tswett to separate pigments of plants.

† **1906–1913:** *Gorgas, William C. Control of malaria and yellow fever.*

By intelligent and industrious application of the knowledge that mosquitoes carry malaria and yellow fever, Gorgas saved 71,000 lives and made possible the construction of the Panama Canal.

* **1907:** *Boveri, Theodor. Qualitative differences of chromosomes.*

Boveri showed in his classic experiment with sea urchin eggs that chromosomes have qualitatively different effects on development. He found that only those cells that had one of each kind of chromosome developed into larvae; those cells that did not have representatives of each kind of chromosome failed to develop.

○△ **1907:** *Harrison, Ross. Tissue culture technique.*

The culturing of living tissues in vitro, that is, outside the body, has given biologists an important tool for studying tissue structure and growth. Harrison overcame great technical difficulties to successfully culture living tissues in suitable media.

○ **1907:** *Hopkins, Frederick G. Relationship of lactic acid to muscular contraction.*

Hopkins showed that, after being formed in muscular contraction, a part of the lactic acid is oxidized to furnish energy for the resynthesis of the remaining lactic acid into glycogen. This discovery did much to clarify part of the metabolic reactions involved in the complicated process of muscular contraction.

Nobel Laureate (1929).

* **1907:** *Wilson, H.V. Reorganization of sponge cells.*

In this classic experiment, Wilson showed that the disaggregation of sponges, by squeezing them through fine silk bolting cloth so that they are sepa-

rated into minute cell clumps, resulted in the surviving cells coming together and organizing themselves into small sponges when they were in seawater.

‡ **1908:** *Garrod, A.E. Discovery that gene products are proteins.*

Certain hereditary disorders are caused by enzyme deficiencies wherein the mutations of a single gene may be responsible for controlling the specificity of a particular enzyme in certain disorders. Garrod's work was largely ignored until 1940.

×‡ **1908:** *Hardy, Godfrey H., and W. Weinberg, Hardy-Weinberg population formula.*

This important theorem states that, in the absence of factors (such as mutation and selection) causing change in gene frequency, the frequency of a particular gene in any large population will reach an equilibrium in one generation and thereafter will remain stable regardless of whether the genes are dominant or recessive. Its mathematical expression forms the basis for the calculations of population genetics.

○△ **1909:** *Arrhenius, S., and S.P.L. Sörensen. Dissociation theory.*

The sensitivity of most biological systems to acid and alkaline conditions has made pH values of the utmost importance in biological research.

Arrhenius, Nobel Laureate (1903).

‡ **1909:** *Castle, William E., and J.C. Philips. The inviolability of germ cells to somatic cell influences.*

That germ cells are relatively free from somatic cell influences was shown by the substitution of a black guinea pig ovary in a white guinea pig, which gave rise to black offspring when mated with a black male.

× **1909:** *Johannsen, W.L. Limitations of natural selection on pure lines.*

This investigator found that when a hereditary group of characteristics becomes genetically homogeneous, natural selection cannot change the genetic constitution with regard to these characteristics. He showed that selection could not itself create new genotypes but was effective only in isolating genotypes already present in the group; it therefore could not effect evolutionary changes directly.

†□ **1909:** *Nicolle, Charles J.H. Body louse as vector of typhus fever.*

The demonstration that typhus fever was transmitted from patient to patient by the bite of the body louse paved the way for the control of this dreaded epidemic disorder by using DDT to delouse populations.

Nobel Laureate (1928).

○ **1910:** *Dale, Henry H. Nature of histamine.*

Dale and colleagues found that an extract from ergot had the properties of histamine (β-imidazolyl ethylamine), which can be produced synthetically by splitting off carbon dioxide from the amino acid histidine. It is also a constituent of all tissue cells, from which it may be released by injuries or other causes. The pronounced effect of histamine in dilating small blood vessels, contracting smooth muscle, and stimulating glands has caused it to be associated with many physiological phenomena, such as anaphylaxis, shock, and allergies.

Nobel Laureate (1936).

† **1910:** *Ehrlich, Paul. Chemotherapy in the treatment of disease.*

The discovery of salvarsan as a cure for syphilis represents the first great discovery in this field. Another was the dye sulfanilamide, discovered by Domagk in 1935.

Nobel Laureate (1908).

□ **1910:** *Heinroth, O. Concept of imprinting as a type of behavior.*

This is a special type of learning that is demonstrated by birds (and possibly other animals). It is based on the observation that in many species of birds the hatchlings are attracted to the first large object they see and thereafter will follow that object to the exclusion of all others.

‡ **1910:** *Morgan, Thomas H. Discovery of sex linkage.*

Morgan and colleagues discovered that the results of a cross between a white-eyed male and a red-eyed female in *Drosophila* were different from those obtained from the reciprocal cross of a red-eyed male with a white-eyed female. This was a crucial experiment because it showed for the first time that how a trait behaved in heredity depended on the sex of the parent, in contrast to most Mendelian characters, which behave genetically the same way whether introduced by a male or female parent.

‡ **1910–1920:** *Morgan, Thomas H. Establishment of the theory of the gene.*

The extensive work of Morgan and associates on the localization of hereditary factors (by genetic experiments) on the chromosomes of the fruit fly (*Drosophila*) represents some of the most significant work ever performed in the field of heredity.

Nobel Laureate (1933).

□ **1910:** *Murray, J., and J. Hjort. Deep-sea expedition of the* Michael Sars.

Of the many expeditions for exploring the depths of the oceans, the *Michael Sars* expedition, made in the North Atlantic regions, must rank among the foremost. The expedition yielded an immense amount of information about deep-sea animals, as well as important data regarding the ecological patterns of animal distribution in the sea.

□ **1910:** *Pavlov, Ivan P. Concept of the conditioned reflex.*

The idea that acquired reflexes play an important role in the nervous reaction patterns of animals has greatly influenced the development of modern psychology.

Nobel Laureate (1904; for earlier work on the physiology of digestion).

* **1911:** *Child, Charles M. Axial gradient hypothesis.*

This hypothesis attempts to explain the pattern of metabolism from the standpoint of localized regional differences along the axes of organisms. The differences in the metabolic rate of different areas have made possible an understanding of certain aspects of regeneration, development, and growth.

× **1911:** *Cuénot, L. The preadaptation concept.*

Preadaptation refers to a morphological or physiological characteristic that has been selected for (arisen) in one environment, but that is coincidentally adaptive in a new environment. This favorable conjunction of characteristics and suitable environment is considered to be an important factor in progressive evolution (opportunistic evolution).

† **1911:** *Funk, Casimer. Vitamin hypothesis.*

Vitamin deficiency diseases are those in which effects can be definitely traced to the lack of some essential constituent of the diet. Thus beriberi is caused by an insufficient amount of thiamine, scurvy by a lack of vitamin C, and so on.

* **1911:** *Harvey, E.B. Cortical changes in the egg during fertilization.*

In the mature egg, cortical granules gather at the surface of the egg, but on activation of the egg these granules, beginning at the point of sperm contact, disappear in a wavelike manner around the egg. These granules are supposed to release material during their breakdown that helps form the ensuing fertilization membrane.

△ **1911:** *Rutherford, Ernest. Concept of the atomic nucleus.*

This cornerstone of modern physics must be of equal interest to the biologist in the light of the rapid advances of molecular biology. Rutherford also discovered (1920) the proton, one of the charged particles in the atomic nucleus.

Nobel Laureate (1908).

× **1911:** *Walcott, Charles D. Discovery of Burgess shale fossils.*

The discovery of a great assemblage of beautifully preserved invertebrates in the Burgess shale of British Columbia and their careful study by an American paleontologist represent a landmark in the fossil record of invertebrates. These fossils date from the middle Cambrian age and include the striking *Aysheaia,* which has a resemblance to the extant *Peripatus.*

○* **1912:** *Gudernatsch, J.F. Role of the thyroid gland in the metamorphosis of frogs.*

This investigator found that the removal of the thyroid gland of tadpoles prevented metamorphosis into frogs, and also that the feeding of thyroid extracts to tadpoles induced precocious metamorphosis. In 1919 W.W. Swingle showed that the presence and absence of inorganic iodine would produce the same results.

□ **1912:** *Wegener, Alfred L. Concept of continental drift.*

This theory postulates that the continents were originally joined together in one or two large masses that gradually broke up during geological time, and the fragments drifted apart to form the current landmasses. The theory has been revived recently on the basis of surveys of paleomagnetism, seismographic studies, tectonic plate studies, and a wealth of biological evidence, for example, the finding of amphibian fossils in Antarctic regions.

△ **1913:** *Michaelis, L., and M. Menten. Enzyme-substrate complex.*

On theoretical and mathematical grounds these investigators showed that an enzyme forms an intermediate compound with its substrate (the enzyme-substrate complex) that subsequently decomposes to release the free enzyme and the reaction products. D. Keilin of Cambridge University and B. Chance of the University of Pennsylvania later demonstrated the presence of such a complex by color changes and measurements of its rate of formation and breakdown that agree with theoretical predictions.

× **1913:** *Reck, H. Discovery of Olduvai Gorge fossil deposits.*

This region in East Africa has yielded an immense number of early mammalian fossils as well as the tools of Stone Age humans, such as stone axes. Among the interesting fossils discovered were elephants with lower jaw tusks, horses with three toes, the odd ungulate (chalicothere) with claws on the toes, and, more recently, several hominid fossils (see Leakey, Mary D., 1959).

☐ **1913:** *Shelford, Victor E. Law of ecological tolerance.*

This law states that the potential success of an organism in a specific environment depends on how it can adjust within the range of its toleration to the various factors to which the organism is exposed.

‡ **1913:** *Sturtevant, A.H. Formation of first chromosome map.*

By the method of crossover percentages, it has been possible to locate the genes in their relative positions on chromosomes—one of the most fruitful discoveries in genetics, for it led to the extensive mapping of the chromosomes in *Drosophila.*

○ **1913:** *Tashiro, Shiro. Metabolic activity of propagated nerve impulse.*

The detection of slight increases in carbon dioxide production in stimulated nerves, as compared with inactive ones, was evidence that conduction in nerves is a chemical change. Later (1926), A.V. Hill was able to measure the heat given off during the passage of an impulse. Increased oxygen consumption and other metabolic changes have also been measured in excited nerves.

○ **1914:** *Kendall, Edward C. Isolation of thyroxine.*

The isolation of thyroxine in crystalline form was a landmark in endocrinology. It was artificially synthesized by Harington in 1927.

* **1914:** *Lillie, Frank R. Role of fertilizin in fertilization.*

According to this hypothesis, the jelly coat of eggs contains a substance, fertilizin, now known to be an acid mucopolysaccharide, which combines with the antifertilizin on the surface of sperm and causes the sperm to clump together.

‡ **1914:** *Shull, George H. Concept of heterosis.*

When two standardized strains or races are crossed, the resulting hybrid generation may be noticeably superior to both parents as shown by greater vigor, vitality, and resistance to unfavorable environmental conditions. First worked out in corn (maize), such hybrid vigor may also be manifested by other kinds of hybrids. Although its exact nature is still obscure, the phenomenon may be caused by the bringing together in the hybrid of many dominant genes of growth and vigor that were scattered among the two

inbred parents, or it may be caused by the complementary reinforcing action of genes when brought together.

‡ **1916:** *Bridges, Calvin B. Discovery of nondisjunction.*

Bridges explained an aberrant genetic result by a suggested formula that later he was able to confirm by cytological examination. A pair of chromosomes failed to disjoin at the reduction division so that both chromosomes passed into the same cell. It was definite proof that genes are located on the chromosomes.

☐ **1917:** *Grinnell, Joseph. Concept of the ecological niche.*

The niche is the role of an animal in relation to all of the resources in its environment, both physical and biological. C. Elton has done much to develop the concept.

○ **1918:** *Krogh, August. Regulation of the motor mechanism of capillaries.*

The mechanism of the differential distribution of blood to the various tissues has posed many problems from the days of Harvey and Malpighi, but Krogh showed that capillaries were not merely passive in this distribution but that they had the power to contract or dilate actively, according to the needs of the tissues. Both nervous and chemical controls are involved.

Nobel Laureate (1920).

☐ **1918:** *Loeb, Jacques. Forced movements, tropisms, and animal conduct.*

Regarded as the founder of a mechanist "school," Loeb opposed the generally anthropomorphic and teleological interpretation of animal behavior. While Loeb carried his mechanistic doctrine to an extreme, he did inspire more rigorous, experimental approaches in animal behavior studies.

○ **1918:** *Starling, Ernest H. The law of the heart.*

Within physiological limits, the more the ventricles are filled with incoming blood, the greater is the force of their contraction at systolé. This is an adaptive mechanism for supplying more blood to tissues when it is needed. It is now recognized that only the hearts of lower vertebrates obey Starling's law; the intact heart of mammals is regulated principally by nervous and hormonal controls.

1919: *Aston, Francis W. Discovery of isotopes.*

Radioactive isotopes have proved of the utmost value in biological research because of the possibility of tracing the course of various elements in living organisms.

Nobel Laureate (1922).

△ **1920:** *Herzog, R.O., and W. Jancke. Development of x-ray diffractometry.*

When a parallel beam of x-rays is passed through crystallized biological material, the rays are spread and an image of the diffracted pattern is recorded (by rings, spots, and the like). By measurements and mathematical calculations information about structure can be obtained.

△ **1921:** *Hopkins, Frederick G. Isolation of glutathione.*

The discovery of this sulfur-containing compound gave a great impetus to the study of the complicated nature of cellular oxidation and metabolism.

Nobel Laureate (1929).

○ **1921:** *Langley, John N. Concept of the functional autonomic nervous system.*

Langley's concept of the functional aspects of the autonomic system dealt mainly with the mammalian type, but the fundamental principles of the system have been applied with modification to other groups as well. Two divisions—sympathetic and parasympathetic—are recognized in the functional interpretation of excitation and inhibition in the antagonistic nature of the two divisions.

○ **1921:** *Loewi, Otto, and H.H. Dale. Isolation of acetylcholine.*

This key demonstration led to the neurohumoral concept of the transmission of nerve impulses to muscles.

Nobel Laureates (1936).

○ **1921:** *Richards, A.N. Collection and analysis of glomerular filtrate of the kidney.*

This experiment was direct evidence of the role of the glomeruli as mechanical filters of cell-free and protein-free fluid from the blood and was striking confirmation of the Ludwig-Cushny theory of kidney excretion.

* **1921:** *Spemann, Hans. Organizer concept in embryology.*

Certain parts of the developing embryo known as organizers induce specific developmental patterns in responding tissue. While many aspects of induction remain obscure, Spemann's work was important in showing that development is a sequence of programmed stages and that each stage is necessary for the next.

Nobel Laureate (1935).

†○ **1921, 1922:** *Banting, Frederick, C.H. Best, and J.J.R. Macleod. Extraction of insulin.*

The great success of this hormone in relieving a distressful disease, diabetes mellitus, and the dramatic way in which active extracts were obtained have made the isolation of this hormone the best known in the field of endocrinology.

Banting and Macleod, Nobel Laureates (1923).

○ **1922:** *Erlanger, Joseph, and H.S. Gasser. Differential conduction of nerve impulses.*

By using the cathode-ray oscillograph, these investigators found that there were several different types of mammalian nerve fibers that could be distinguished structurally and that had different rates of conducting nervous impulses according to the thickness of the nerve sheaths (most rapid in the thicker ones).

Nobel Laureates (1944).

○ **1922:** *Kopeč, S. Concept of neurosecretory systems.*

This concept has emerged from the work of many investigators. R. Cajalin (1899) discovered that the nerve fibers to the neurohypophysis are extensions of the neurons of the hypothalamus. Kopeč found that the substance responsible for metamorphosis in the larva of moths originated in the brain where certain cerebral ganglia served as glands of internal secretion. Kopeč's pioneering studies were largely ignored by vertebrate endocrinologists until the 1950s, when through the studies of B. Scharrer, E. Scharrer, B. Hanström, G.W. Harris, and others it was recognized that vertebrates possess a neurosecretory system (hypothalamo-hypophyseal) analogous to invertebrate systems.

□ **1922:** *Schjelderup-Ebbe, T. Social dominance-subordinance hierarchies.*

This observer found certain types of social hierarchies among birds in which higher-ranking individuals could peck those of lower rank without being pecked in return. Those of the first rank dominated those of the second rank, who dominated those of the third, and so on. Such an organization, once formed, may be permanent.

□ **1922:** *Schmidt, Johann. Life history of the freshwater eel.*

The long, patient work of this oceanographer in solving the mystery of eel migration from the freshwater streams of Europe to their spawning grounds in the Sargasso Sea near Bermuda represents one of the most romantic achievements in natural history.

△ **1923:** *de Hevesy, George. First isotopic tracer method.*

Tracer methodology has proved especially useful in biochemistry and physiology. For instance, it has been possible by the use of these labeled units to determine the fate of the particular molecule in all steps of a metabolic process and the nature of many enzymatic reactions. The exact locations of many elements in the body have been traced by this method.

Nobel Laureate (1943).

△ **1923:** *Warburg, Otto. Manometric methods for studying metabolism of living cells.*

The Warburg apparatus has been useful in measuring gas exchange and other metabolic processes of living tissues. It proved of great value in the study of enzymatic reactions in living systems and was a standard tool in many biochemical laboratories for many years.

Nobel Laureate (1931).

○ **1924:** *Cleveland, L.R. Symbiotic relationships between termites and intestinal flagellates.*

This study was made on one of the most remarkable examples of mutualism known in the animal kingdom. Equally important were the observations this investigator and others made of the symbiosis between the wood-roach (*Cryptocercus*) and its intestinal protozoa.

○ **1924:** *Houssay, Bernardo A. Role of the pituitary gland in regulation of carbohydrate metabolism.*

This investigator showed that, when a dog had been made diabetic by pancreatectomy, the resulting hyperglycemia and glucosuria could be abolished by removing the anterior pituitary gland.

Nobel Laureate (1947).

× **1925:** *Dart, Raymond. Discovery of* Australopithecus africanus.

This important fossil, once referred to as the "missing link," bore many human characteristics but had a brain capacity only slightly greater than the higher apes. Most authorities place this hominid on or near the main branch of human ancestry.

△ **1925:** *Mast, S.O. Nature of ameboid movement.*

By studying the reversible sol-gel transformation in the protoplasm of an ameba, Mast not only gave a logical interpretation of ameboid movement but also initiated many fruitful concepts about the contractile nature of cytoplasmic gel systems, such as the furrowing movements (cytokinesis) in cell divisions.

†○ **1925:** *Minot, G.R., M.W.P. Murphy, and G.H. Whipple. Liver treatment of pernicious anemia.*

These investigators discovered that the feeding of raw liver had a pronounced effect in the treatment of pernicious anemia (a serious blood disorder). Much later investigation by numerous workers has led to some understandings of the antipernicious factor, vitamin B_{12}, whose chemical name is cyanocobalamin. This complex vitamin contains, among other components, porphyrin that has a cobalt atom instead of iron or magnesium at its center.

Nobel Laureate (1934).

□ **1925:** *Rowan, William. Photoperiodism hypothesis of bird migration.*

Rowan demonstrated that birds subjected to artificially increased day length in winter increased the size of their gonads and showed a striking tendency to migrate out of season. Other workers, stimulated by Rowan's experiments, showed that naturally increasing day length in spring acts through the brain neurosecretory system to set in motion the migratory disposition in birds.

△ **1926:** *Sumner, James B. Isolation of enzyme urease.*

The isolation of the first enzyme in crystalline form was a key discovery and was followed by others that have helped unravel the complex nature of these important biological substances.

Nobel Laureate (1946).

○ **1927:** *Bozler, E. Analysis of nerve net components.*

Bozler's demonstration that the nerve net of cnidarians was made up of separate cells and contained synaptic junctions resolved the old problem of whether or not the plexus in this group of animals was an actual network.

○ **1927:** *Eggleton, P., G.P. Eggleton, C.H. Fiske, and Y. Subbarow. Role of phosphagen (phosphocreatine) in muscular contraction.*

The demonstration that phosphagen is broken down during muscular contraction and then is resynthesized during recovery gave an entirely new concept of the initial energy necessary for the contraction process. Confirmation of this discovery received a great impetus from the discovery of E. Lungsgaard (1930) that muscles poisoned with monoiodoacetic acid, which inhibits the production of lactic acid from glycogen, would still contract and that the amount of phosphocreatine broken down was proportional to the energy liberated. It is now known that phosphocreatine is a high-energy phosphate storage compound, readily donating its phosphate group to ADP to form ATP.

○ **1927:** *Heymans, Corneille. Role of carotid and aortic reflexes in respiratory control.*

The carotid sinus and aortic areas contain pressoreceptors and chemoreceptors, the former responding to mechanical stimulation, such as blood pressure, and the latter to oxygen lack. When the pressoreceptors are stimulated, respiration is inhibited; when the chemoreceptors are stimulated, the respiratory rate is increased. (A much more potent stimulator of respiration, however, is the effect of carbon dioxide on the respiratory center of the brain.)

Nobel Laureate (1938).

‡ **1927:** *Muller, Hermann J. Artificial induction of mutations.*

By subjecting fruit flies (*Drosophila*) to mild doses of x-rays, Muller found that the rate of mutation could be increased 150 times over the normal rate.

Nobel Laureate (1946).

× **1927:** *Stensiö, E.A. Appraisal of the cephalaspid (ostracoderm) fish fossil.*

The replacement of amphioxus as a prototype of vertebrate ancestry by the ammocoete lamprey larva has to a great extent been the result of this careful fossil reconstruction. It is generally believed that living Agnatha (lampreys and hagfishes) and these ancient forms are descended from the same common ancestor.

× **1928:** *Garstang, Walter. Hypothesis of the ascidian ancestry of chordates.*

According to this hypothesis, primitive chordates were sessile, filter-feeding marine organisms very similar to present-day ascidians. The actively swimming prevertebrate was considered a later stage in chordate evolution. The tadpole ascidian larva, with its basic organization of a vertebrate, had evolved within the group by progressive evolution and by neoteny became sexually mature, ceased to metamorphose into a sessile, mature ascidian, and became the ancestral vertebrate.

‡△ **1928:** *Griffith, F. Discovery of the transforming principle (DNA) in bacteria (genetic transduction).*

By injecting living nonencapsulated bacteria and dead encapsulated bacteria of the *Pneumococcus* strain into mice, it was found that the former acquired the ability to grow a capsule and that this ability was transmitted to succeeding generations. This active agent or transforming principle was isolated from the encapsulated type by other workers later and was found to consist of DNA.

F. Sanfelice (1893) had actually found the same principle when he discovered that nonpathogenic bacilli grown in a culture medium containing the metabolic products of true tetanus bacilli would also produce toxins and would do so for many generations.

△○ **1928:** *Weiland, H., and A. Windaus. Structure of the cholesterol molecule.*

Sterol chemistry has been a focal point in the investigation of such biological products as vitamins, sex hormones, and cortisone. Cholesterol is the precursor of bile acids and steroid hormones in animals.

Nobel Laureates (1927, 1928).

○ **1929:** *Berger, Hans. Demonstration of brain waves.*

The science of electroencephalography, or the electrical recording of brain activity, has revealed much about both the healthy and the diseased brain.

○ **1929:** *Butenandt, A., and E.A. Doisy. Isolation of estrone.*

This discovery was the first isolation of a sex hormone and was arrived at independently by these two investigators. Estrone was found to be the urinary and transformed product of estradiol, the actual hormone. The male hormone testosterone was synthesized by Butenandt and L. Ruzicka in 1931. The second female hormone, progesterone, was isolated from the corpora lutea of sow ovaries in 1934.

Doisy, Nobel Laureate (1943).

† **1929:** *Fleming, Alexander. Discovery of penicillin.*

The chance discovery of this drug from molds and its development by H. Florey a few years later gave us the first of a notable line of antibiotics that have revolutionized medicine.

Nobel Laureate (1945).

○ **1929:** *Heymans, Corneille. Discovery of the role of the carotid sinus and bodies in regulating the respiratory center and arterial blood pressure.*

Heymans found that the carotid sinus contains pressure-sensitive receptors (baroreceptors) that, acting through a nervous reflex to the medulla, help to keep the blood pressure stabilized. He also discovered oxygen-sensitive chemoreceptors in the carotid and aortic bodies. These are the origin of a reflex that responds to a drop in blood oxygen tension by increasing respiratory rate and blood pressure.

Nobel Laureate (1938).

△ **1929:** *Lohmann, K., C. Fiske, and Y. Subbarow. Discovery of ATP.*

The discovery of ATP (adenosine triphosphate) was the culmination of a long search for the direct source of energy in biochemical reactions of many varieties, such as muscular contraction, vitamin action, and many enzymatic systems.

× **1930:** *Fisher R.A. Statistical analysis of evolutionary variations.*

With Sewall Wright and J.B.S. Haldane, Fisher analyzed mathematically the interrelationships of the factors of mutation rates, population sizes, selection values, and others in the evolutionary process. Although many of their theories are in the empirical stage, evolutionists in general agree that they have great significance in evolutionary interpretation.

‡ **1931:** *Stern, C., H. Creighton, and B. McClintock. Cytological demonstration of crossing over.*

Proof that crossing over in genes is correlated with exchange of material by homologous chromosomes was independently obtained by Stern in *Drosophila* and by Creighton and McClintock in corn. By using crosses of strains that had homologous chromosomes distinguishable individually, they definitely demonstrated cytologically that genetic crossing over was accompanied by chromosomal exchange.

○□ **1932:** *Bethe, A. Concept of the ectohormone (pheromone).*

An organism secretes these substances to the outside of the body, where they affect the physiology or behavior of another individual of the same species. They have been demonstrated in insects, where they may function in trail making, sex attraction, development control, and the like.

× **1932:** *Danish scientific expedition. Discovery of fossil amphibians (ichthyostegids).*

These fossils were found in the upper Devonian sediments in eastern Greenland and appear to be intermediate between lobe-finned rhipidistians (*Osteolepis*) and early amphibians. They are among the oldest known tetrapods that can be considered amphibians. Many of their characters show primitive amphibian conditions.

× **1932:** *Lewis, G. Edward. Discovery of* Ramapithecus *fossil.*

Ramapithecus brevirostris was a small humanlike ape of the Miocene and the earliest known hominid. Since Lewis's initial find in India, other fossil fragments of the genus have been discovered in East Africa, Greece, Turkey, and Hungary.

×‡ **1932:** *Wright, Sewall. Genetic drift as a factor in evolution.*

In small populations the Hardy-Weinberg formula of gene frequency may not apply because of "sampling error." If some genotypes are by chance underrepresented in matings in a small population, gene frequencies in the population may change.

○ **1933:** *Goldblatt, M., and U.S. von Euler. Discovery of prostaglandins.*

These fatty acid derivative compounds have been isolated from many mammalian tissues (such as seminal plasma, pancreas, seminal vesicle, brain, and kidney). They have a variety of hormonelike and pharmacological actions, such as stimulating smooth muscle contractions and relaxation, lowering blood pressure, and inhibiting enzymes and hormones.

× **1933, 1938:** *Haldane, John B.S., and A.I. Oparin. Heterotroph hypothesis of the origin of life.*

This hypothesis is based on the idea that life was generated from nonliving matter under the conditions that existed before the appearance of life and that have not been duplicated since. The hypothesis stresses the idea that living systems at present make it impossible for any incipient life to gain a foothold as primordial life was able to do.

‡ **1933:** *Painter, T.S., E. Heitz, and H. Bauer. Rediscovery of giant salivary chromosomes.*

These interesting chromosomes were first described by Balbiani in 1881, but their true significance was not realized until these investigators rediscovered them. It has been possible in a large measure to establish the chromosome theory of inheritance by comparing the actual cytological chromosome maps of salivary chromosomes with the linkage maps obtained by genetic experimentation.

○ **1933:** *Wald, George. Discovery of vitamin A in the retina.*

The discovery that vitamin A is a part of the visual purple molecule of the rods not only gave a better understanding of an important vitamin but also showed how night blindness can occur when there is a deficiency of this vitamin in the diet.

Nobel Laureate (1967).

○ **1934:** *Dam, Carl P.H., and E.A. Doisy. Identification of vitamin K.*

The isolation and synthesis of this vitamin are important not merely because of the vitamin's practical value in certain forms of hemorrhage but also because of the light they throw on the physiological mechanism of blood clotting.

Nobel Laureate (1943).

△ **1934–1935:** *Danielli, J.F., and H. Davson. Concept of the cell (plasma) membrane.*

Danielli proposed a hypothesis that cell membranes consist of two layers of lipid molecules surrounded on the inner and outer surfaces by a layer of protein molecules. The electron microscope reveals the plasma membrane of a thickness of 7.5 to 10 nm, consisting of two dark (protein) membranes (2.5 to 3 nm thick) separated by a light (lipid) interval membrane of 2.5 to 3 nm thickness. Recent studies confirm the Danielli-Davson model, while demonstrating a greater dynamic role for the peripheral proteins.

○ **1934:** *Wigglesworth, V.B. Role of the corpus allatum gland in insect metamorphosis.*

This small gland lies close to the brain of an insect, and it has been shown that during the larval stage this gland secretes a juvenile hormone that causes the larval characteristics to be retained. Metamorphosis and maturation occur as the gland secretes less and less of the hormone. Removal of the gland causes the larva to undergo precocious metamorphosis; grafting the gland into a mature larva will cause the latter to grow into a giant larval form. The gland was first described by A. Nabert in 1913.

○ **1935:** *Hanstöm, B. Discovery of the X-organ in crustaceans.*

This organ, together with the related sinus gland, constitutes an anatomical complex that has proved of great interest in understanding crustacean endocrinology. The neurosecretory cells in the X-organ (part of the brain) produce a molt-inhibiting hormone that is stored in the sinus gland of the eyestalk, while a molting hormone is produced in the Y-organ. The interrelations of these two hormones control the molting process.

○ **1935–1936:** *Kendall, Edward C., and P.S. Hench. Discovery of cortisone.*

Kendall had first isolated this substance, which he called compound E, from the adrenal glands. Its final stages were prepared by Hench later and involved a long tedious chemical process. A hormone that controls the synthesis and release of cortisone and similar steroids was isolated in 1943 from the pituitary and was called adrenocorticotropic hormone (ACTH).

Nobel Laureates (1950).

△ **1935:** *Stanley, Wendell M. Isolation of a virus in crystalline form.*

This achievement of isolating a virus (tobacco mosaic) is noteworthy not only in giving information about these small agents responsible for many diseases but also in affording much speculation on the differences between the living and the nonliving.

Nobel Laureate (1946).

○ **1936:** *Young, J.Z. Demonstration of giant fibers in squid.*

These giant fibers are formed by the fusion of the axons of many neurons whose cell bodies are found in a ganglion near the head. Each fiber is really a tube, up to 1 mm in diameter, consisting of an external sheath filled with liquid axoplasm. Because of their size, they have been a very valuable system in neurophysiology.

△ **1937:** *Findley, G.W.M., and F.O. MacCullum. Discovery of interferon.*

These protein substances, produced by cells in response to viruses and other foreign substances, selectively inhibit virus replication and are one of the body's important defenses against viral infection. They also play a role in regulation of the immune response.

△ **1937:** *Krebs, Hans A. Krebs cycle (citric acid, tricarboxylic acid).*

In a brilliant example of reasoning and deduction, Krebs showed the existence of a cycle of reactions central to aerobic energy production. In the cycle, two carbon fragments from carbohydrate, fatty acids, or amino acids are oxidized to carbon dioxide. The electrons derived from the oxidation reactions in the cycle are passed through a further series of reactions (electron transport system), in which energy is captured in ATP, and finally to oxygen as the final electron acceptor to produce water.

Nobel Laureate (1953).

× **1937:** *Sonneborn, T.M. Discovery of mating types in paramecium.*

Sonneborn's discovery that only individuals of complementary physiological classes (mating types) would conjugate opened up a new era of protozoan investigation that promises to shed new light on the problems of species concept and evolution.

△○ **1937:** *König, P., and A. Tiselius. Development of electrophoresis.*

Electrophoresis is an extremely valuable method of separating proteins in a solution and, after numerous refinements, is currently in wide use. It depends on the differential migration of proteins on an inert carrier when an electrical field is applied.

□ **1938:** *Remane, A. Discovery of the new phylum Gnathostomulida.*

This marine phylum was first described by P. Ax in 1956. The members of the phylum are small wormlike forms (about 0.5 mm long) and show great diversity among the different species. One of their major characteristics is their complicated jaws provided with teeth. They live in sandy substrata and are worldwide in distribution.

○ **1938:** *Schoenheimer, R. Use of radioactive isotopes to demonstrate synthesis of bodily constituents.*

By labeling amino acids, fats, carbohydrates, and the like with radioactive isotopes, it was possible to show how these were incorporated into the various constituents of the body. Such experiments demonstrated that parts of the cell were constantly being synthesized and broken down and that the body must be considered a dynamic equilibrium.

□ **1938:** *Skinner, B.F. Measurement of motivation in animal behavior.*

Skinner worked out a technique for measuring the rewarding effect of a stimulus, or the effects of learning on voluntary behavior. His experimental animals (rats) were placed in a special box (Skinner's box) containing a lever that the animal could manipulate. When the rat pressed the lever, small pellets of food would or would not be released, according to the experimental conditions.

○△ **1938:** *Svedberg, Theodor. Development of ultracentrifuge.*

In biological and medical investigation this instrument has been widely used for the purification of substances, the determination of particle sizes in colloidal systems, the relative densities of materials in living cells, the production of abnormal development, and the study of many problems concerned with electrolytes.

Nobel Laureate (1926).

○ **1939:** *Brown, Frank A., Jr., and O. Cunningham. Demonstration of molt-inhibiting hormone in eyestalk of crustaceans.*

Although C. Zeleny (1905) and others had shown that eyestalk removal shortened the intermolt period in crustaceans, Brown and Cunningham were the first to present evidence to explain the effect as being caused by a molt-inhibiting hormone present in the sinus gland.

× **1939:** *Discovery of coelacanth fishes.*

The collection of a living specimen of this ancient fish (*Latimeria*), followed later by other specimens, has brought about a complete reappraisal of this "living fossil" with reference to its ancestry of the amphibians and land forms.

✳ **1939:** *Hörstadius, S. Analysis of the basic pattern of regulative and mosaic eggs in development.*

The masterful work of this investigator has done much to resolve the differences in the early development of regulative eggs (in which each of the early blastomeres can give rise to a whole embryo) and mosaic eggs (in which isolated blastomeres produce only fragments of an embryo; also called determinate cleavage). Regulative eggs were shown to have two kinds of substances, both of which were necessary in proper ratios to produce normal embryos. Each of the early blastomeres has this proper ratio and thus can develop into a complete embryo; in mosaic eggs the regulative power is restricted to a much earlier timescale in development (before cleavage), and thus each isolated blastomere will give rise only to a fragment.

×□ **1939:** *Huxley, Julian. Concept of the cline in evolutionary variation.*

This concept refers to the gradual and continuous variation in character over an extensive area because of adjustments to changing conditions. This idea of character gradients has proved a very fruitful one in the analysis of the mechanism of evolutionary processes, for such a variability helps to explain the initial stages in the transformation of species.

†‡ **1940:** *Landsteiner, Karl, and A.S. Wiener. Discovery of Rh blood factor.*

Not only was a knowledge of the Rh factor of importance in solving a fatal disease of infancy, but it has also yielded a great deal of information about relationships between human races.

Landsteiner, Nobel Laureate (1930).

△‡ **1941:** *Beadle, George W., and E.L. Tatum. Biochemical mutation.*

By subjecting the bread mold *Neurospora* to x-ray irradiation, they found that genes responsible for the synthesis of certain vitamins and amino acids were inactivated (mutated) so that a strain of this mold carrying the mutant genes could no longer grow unless these particular vitamins and amino acids were added to the medium on which the mold was growing. This discovery—that alterations in various genes resulted in loss of a corresponding biosynthetic enzyme—first revealed the precise way in which genes and enzymes are related.

Nobel Laureates (1958).

○ **1941:** *Cori, Carl F., and Gerty T. Cori. Lactic acid metabolic cycle.*

The regeneration of muscle glycogen reserves in mammals involves the passage of lactic acid from the muscles through the blood to the liver, the conversion of lactic acid there to glycogen, the production of

blood glucose from the liver glycogen, and the synthesis of muscle glycogen from the blood glucose.

Nobel Laureates (1947).

△○ **1941:** *Martin, A.J.P., and R.L.M. Synge. Development of partition chromatography.*

This is a powerful method for separation of chemically similar substances in mixtures. It depends on differential solubility of a compound in stationary and moving phases of solvents. Variants of the method include column, paper, and thin-layer chromatography.

○ **1941:** *Szent-Györgyi, Albert. Role of ATP in muscular contraction.*

The demonstration showing that muscles get their energy for contraction from ATP (adenosine triphosphate) has done much to explain many aspects of muscle physiology.

Nobel Laureate (1937).

✳ **1942:** *McClean, D, and I.M. Rowlands. Discovery of the enzyme hyaluronidase in mammalian sperm.*

This enzyme dissolves the cement substance of the follicle cells that surround the mammalian egg and facilitates the passage of the sperm to the egg. This discovery not only aided in resolving some of the difficult problems of the fertilization process but also offers a logical explanation of cases of infertility in which too few sperm may not carry enough of the enzyme to afford a passage through the inhibiting follicle cells.

△ **1943:** *Claude, A. Isolation of cell constituents.*

By differential centrifugation, Claude found it possible to separate, in relatively pure form, particulate components, such as mitochondria, microsomes, and nuclei. These investigations led immediately to a more precise knowledge of the chemical nature of these cell constituents and aided the elucidation of the structure and physiology of the mitochondria—one of the great triumphs in the biochemistry of the cell.

Nobel Laureate (1974).

✳ **1943:** *Holtfreter, J. Tissue synthesis from dissociated cells.*

By dissociating the cells of embryonic tissues of amphibians (by dissolving with enzymes or other agents the intercellular cement that holds the cells together) and heaping them in a mass, he found that the cells in time coalesced and formed the type of tissue from which they had come. This is an application to vertebrates of Wilson's discovery with sponge cells. It showed that, by some mechanism, cells could recognize other cells of the same type.

‡ **1943:** *Sonneborn, T.M. Extranuclear inheritance.*

The view that in paramecia cytoplasmic determiners (plasmagenes) that are self-reproducing and capable of mutation can produce genetic variability has thrown additional light on the role of the cytoplasm in hereditary patterns.

△‡ **1944:** *Avery, O.T.C., C.M. MacLeod, and M. McCarty. Agent responsible for bacterial transformation.*

These workers were able to show that the bacterial transformation of nonencapsulated bacteria to encapsulated cells was really attributable to the DNA fraction of debris from disrupted encapsulated cells to which the nonencapsulated cells were exposed. This key demonstration showed for the first time that nucleic acids and not proteins were the basic material of heredity, and initiated the study of molecular genetics. However, the profound significance of Avery's conclusions was only gradually recognized.

△ **1945:** *Cori, Carl F. Hormone influence on enzyme activity.*

The delicate balance that insulin and the diabetogenic hormone of the pituitary exercise over the activity of the enzyme hexokinase in carbohydrate metabolism has opened up a whole new field of the regulative action of hormones on enzymes.

Nobel Laureate (1947).

□ **1945:** *Griffin, D., and R. Galambos. Development of the concept of echolocation.*

Echolocation refers to a type of perception of objects at a distance by which echoes of sound are reflected back from obstacles and detected acoustically. These investigators found that bats generated their own ultrasonic sounds that were reflected back to their own ears so that they were able to avoid obstacles in their flight without the aid of vision. Their work climaxed an interesting series of experiments inaugurated as early as 1793 by Spallanzani, who believed that bats avoided obstacles in the dark by reflection of sound waves to their ears. Others who laid the groundwork for the novel concept were C. Jurine (1794), who proved that ears were the all-important organs in perception; H.S. Maxin (1912), who advanced the idea that the bat made use of low-frequency sounds inaudible to human ears; and H. Hartridge (1920), who proposed the hypothesis that bats emitted sounds of high frequencies and short wavelengths (ultrasonic sounds).

△ **1945:** *Lipmann, Fritz A. Discovery of coenzyme A.*

The discovery of this important coenzyme made possible a better understanding of the breaking down of fatty acid chains and furnished important further knowledge of the reactions in the tricarboxylic acid cycle.

△ **1945:** *Porter, Keith R. Description of the endoplasmic reticulum.*

The endoplasmic reticulum is a very complex cytoplasmic structure consisting of a lacelike network of irregular anastomosing tubules and vesicular expansions within the cytoplasmic matrix. Associated with the reticulum complex are small, dense granules of ribonucleoprotein and other granules known as microsomes that are fragments of the endoplasmic reticulum. The reticulum complex plays an important role in the synthesis of proteins.

‡ **1946:** *Lederberg, J., and E.L. Tatum. Sexual recombination in bacteria.*

These investigators found that two different strains of bacteria (*Escherichia coli*) could undergo conjugation and exchange genetic material, thereby producing a hereditary strain with characteristics of the two parent strains. W. Hayes (1952) found that recombination still occurred after one parent strain was killed.

Nobel Laureates (1958).

× **1946:** *Libby, Willard F. Radiocarbon dating of fossils.*

The radiocarbon age determination is based on the fact that carbon 14 in the dead organism disintegrates at the rate of one half in 5560 years, one half of the remainder in the next 5560 years, and so on. This is based on the assumption that the isotope is mixed equally through all living matter and that the cosmic rays (which form the isotopes) have not varied much in periods of many thousands of years. The limitation of the method is around 40,000 years.

Nobel Laureate (1960).

× **1946:** *White, E.I. Discovery of the primitive chordate fossil* Jaymoytius.

The discovery of this fossil in deposits of Silurian rock in Scotland threw some light on the early ancestry of vertebrates, specifically the origin of the living agnatha (hag fishes and lampreys). More recent finds of other anapsid fossils, such as *Mayomyzon*, are strikingly similar to the living lampreys.

○ **1947:** *Holtz, P. Discovery of norepinephrine (noradrenaline).*

This hormone (vasoconstriction effects) has since been found in most verte-

brates and shares with epinephrine (metabolic effects) the functions of the chromaffin part of the adrenal gland.

× **1947:** *Sprigg, R.C. Discovery of Precambrian Ediacara fossil bed.*

The discovery of a rich deposit of Precambrian fossils in the Ediacara Hills of South Australia has been of particular interest because the scarcity of such fossils in the past has given rise to vague and uncertain explanations about Precambrian life. It was all the more remarkable that the fossils discovered were those of soft-bodied forms, such as jellyfish, soft corals, and segmented worms, including the amazing *Spriggina,* which shows relationship to the trilobites. The fossils are prearthropod but not preannelid.

△ **1947:** *Szent-Györgyi, Albert. Concept of the contractile substance actinomyosin.*

This protein complex consists of the two components, actin and myosin, and is considered the source of muscular contraction when triggered by ATP. Neither actin nor myosin will contract singly.

Nobel Laureate (1937).

‡ **1947:** *McClintock, Barbara. Concept of mobile genetic elements.*

Studying the pattern of certain mutations in maize, McClintock deduced that it must be caused by movement of genetic elements ("jumping genes") within the genome. Her observations were ignored or discounted for years until mobile genetic elements were discovered in bacteria in the 1960s, then in many kinds of eukaryotes in the 1970s. We now know that transposable genes help to account for the huge diversity of antibodies in vertebrates.

Nobel Laureate (1983).

□ **1948:** *von Frisch, Karl. Communication mechanisms of honeybees.*

After many years of patient work, von Frisch deciphered the meaning of bee "dances," behavior patterns of individual bees returning to the hive. The dances communicate information to the other bees regarding location of food and water, location and suitability of sites for a new hive, and other information.

Nobel Laureate (1973).

□ **1948:** *Hess, Walter R. Localization of instinctive impulse patterns in the brain.*

By inserting electrodes through the skull, fixing them in position, and allowing such holders to heal in place, it was possible to study the brain of an animal in its ordinary activities. When the rat could automatically and at will stimulate itself by pressing a lever, it did so frequently when the electrode was inserted in the hypothalamus region of the brain, indicating a pleasure center. In this way, by placing electrodes at different centers, rats can be made to gratify such drives as thirst, sex, and hunger.

Nobel Laureate (1949).

△ **1948:** *Hogeboom, G.H., W.C. Schneider, and G.E. Palade. Separation of mitochondria from the cell.*

This was an important discovery in unraveling the amazing enzymatic activity of the mitochondria in the tricarboxylic acid cycle. The role mitochondria play in the energy transfer of the cell has earned for these rod-shaped bodies the appellation "powerhouse" of the cell.

Palade, Nobel Laureate (1974).

△ **1949:** *de Duve, C. Discovery of the lysosome organelle.*

These small particles were first identified chemically, and later (1955) morphologically with the electron microscope. They contain the enzymes for digestion of materials taken into the cell by pinocytosis and phagocytosis and for the digestion of the cell's own cytoplasm. When the cell dies, their enzymes are also released and digest the cell (autolysis).

Nobel Laureate (1974).

† **1949:** *Enders, J.F., F.C. Robbins, and T.H. Weller. Cell culture of animal viruses.*

These investigators found that poliomyelitis virus could be grown in ordinary tissue cultures of nonnervous tissue instead of being restricted to host systems of laboratory animals or embryonated chick eggs.

Nobel Laureates (1954).

○ **1949:** *von Euler, U.S. Role of norepinephrine as transmitter.*

Von Euler isolated and identified norepinephrine as the neurotransmitter in the sympathetic nervous system (1940). Later he isolated and characterized norepinephrine storage granules in nerves and described how the substance was taken up, stored, and released by them.

Nobel Laureate (1970).

‡ **1949:** *Pauling, Linus. Genic control of protein structure.*

Pauling and his colleagues demonstrated a direct connection between specific chemical differences in protein molecules and alterations in genotypes. Making use of the hemoglobin of patients with sickle cell anemia (which is caused by a homozygous condition of an abnormal gene), Pauling was able by the method of electrophoresis to show a considerable difference in the behavior of this hemoglobin in an electric field compared with that from a heterozygote or from a normal person.

Nobel Laureate (1954).

○ **1949:** *Selye, Hans. Concept of the stress syndrome.*

In 1937 Selye began his experiments, which led to what he called the "general adaptation syndrome." This involved the chain reactions of many hormones, such as cortisone and ACTH, in meeting stress conditions faced by an organism. Whenever the stress experience exceeds the limitations of these body defenses, serious degenerative disorders may result.

△ **1950:** *Caspersson, T., and J. Brachet. Biosynthesis of proteins.*

Investigations to solve the vital problem of protein synthesis from free amino acids have been under way since Emil Fischer's (1887) outstanding work on protein structure. Many competent research workers, such as Bergmann, Lipmann, and Schoenheimer, have made contributions to an understanding of protein synthesis, but Caspersson and Brachet were the first to point out the significant role of ribonucleic acid (RNA) in the process—and most biochemical investigations of the problem since that time have been directed along this line.

△ **1950:** *Chargaff, Erwin. Base composition of DNA.*

The discovery that the amount of purines was equal to the amount of pyrimidines in DNA, the amount of adenine was equal to that of thymine, and the amount of cytosine was equal to that of guanine paved the way for the DNA model of Watson and Crick. The two major functions of DNA are replication and hereditary information storage.

□ **1950:** *Lorenz, Konrad. Development of ethology.*

Lorenz is regarded as the founder of ethology, a biological approach to analyzing behavior, emphasizing a comparative method and stressing innate factors in behavior development. His classic studies on the greylag goose first came to the attention of English-speaking scientists during the 1950s, though the work itself commenced in the 1930s.

Nobel Laureate (1973).

□ **1950:** *von Frisch, Karl. Discovery of polarized light perception by honeybees.*

Honeybees are able to detect, and navigate to, polarized sunlight reaching the earth's surface, using specialized retinular cells in a specific but small area of the compound eyes. Subsequently it has been found that some crustaceans, cephalopod molluscs, and teleost fishes, as well as many other terrestrial arthropods, can orient by polarized light.

△ **1950s:** *Fischer, Edmond, and Edwin Krebs. Role of phosphorylation in enzyme activation.*

Fischer and Krebs showed that glycogen phosphorylase was activated and inactivated by the reversible addition of a phosphate group. We now know that phosphorylation controls the activities of hundreds of enzymes, regulating functions as diverse as hormonal responses, muscle contraction, immune responses, and cell growth and division.

Nobel Laureates (1992).

† **1950s–1970s:** *Elion, Gertrude B., and George H. Hitchings. Design of purine antimetabolites for use in chemotherapy.*

Certain analogs of purines can act as antagonists of nucleic acid metabolism in rapidly dividing cells or viruses. This has led to treatments of acute leukemia, gout, herpes-virus infections, and prevention of organ transplant rejection.

Nobel Laureates (1988).

* **1950s–1970s:** *Levi-Montalcini, Rita, and Stanley Cohen. Discovery of tissue-specific growth factors.*

Levi-Montalcini and Cohen discovered and characterized nerve growth factor and epidermal growth factor, which are necessary for the growth, development, and maintenance of nerve cells and epidermal cells, respectively. Since then, numerous additional growth factors that act on many cell types have been discovered. Some oncogenes are derived from genes for growth factors or growth-factor receptors.

Nobel Laureates (1986).

* **1952:** *Briggs, Robert, and Thomas J. King. Demonstration of possible differentiated nuclear genotypes.*

The belief that all cells of a particular organism have the same genetic endowment has been questioned as the result of the work of these investigators who transplanted nuclei of different ages and sources from blastulas and early gastrulas into enucleated zygotes and observed variable and abnormal development.

☐ **1952:** *Kramer, G. Orientation of birds to positional changes of sun.*

Kramer showed that birds (starlings and pigeons) can be trained to find food in accordance with the position of the sun. He found that the general orientation of the birds shifted at a rate (when exposed to a constant artificial sun) that could be predicted on the basis of the birds' correcting for the normal rotation of the earth. Birds were able to orient themselves in a definite direction with references to the sun, whether the light of the sun reached them directly or was reflected by mirrors. They were capable also of finding food at any time of day, indicating an ability to compensate for the sun's motion across the sky.

△ **1952:** *Palade, G.E. Analysis of the fine structure of the mitochondrion.*

The important role of mitochondria in the enzymatic systems and cellular metabolism has focused much investigation on the structure of these cytoplasmic inclusions. Each mitochondrion is bounded by two membranes; the outer is smooth and the inner is thrown into small folds or cristae that project into a homogeneous matrix in the interior. Some modifications of this pattern are found.

Nobel Laureate (1974).

△ **1952:** *Zinder, N., and J. Lederberg. Discovery of the transduction principle.*

Transduction is the transfer of DNA from one bacterial cell to another by means of a phage. It occurs when an infective phage picks up from its disintegrated host a small fragment of the host's DNA and carries it to a new host where it becomes a part of the genetic equipment of the new bacterial cell.

Lederberg, Nobel Laureate (1958).

‡△ **1953:** *Crick, Francis H.C., James D. Watson, and Maurice H.F. Wilkins. Chemical structure of DNA.*

Based on knowledge of the chemical nature of DNA and studies of x-ray diffraction by Wilkins, Crick and Watson formulated the hypothesis that the DNA molecule was made up of two chains twisted around each other in a helical structure with pairs of nitrogenous bases projecting toward each other—adenine opposite thymine and guanine opposite cytosine. Genes are considered to be segments of these molecules with the sequence of bases coding for the amino acids in a protein. Each of the complementary strands acts as a model or template to form a new

strand before cell division. The hypothesis has been widely accepted and is now amply confirmed.

Nobel Laureates (1962).

△ **1953:** *Palade, G.D. Description of cytoplasmic ribosomes.*

The description of ribonucleic acid-rich granules (usually on the endoplasmic reticulum) represents one of the key links in the unraveling of the mechanism of protein synthesis. The ribosomes (about 25 nm in diameter) serve as the site of protein synthesis; the number of ribosomes (polysomes) involved in the synthesis of a protein depends on the length of messenger RNA and the protein being synthesized.

Nobel Laureate (1974).

✕ **1953:** *Urey, Harold, and Stanley Miller. Demonstration of the possible primordia of life.*

By exposing a mixture of water vapor, ammonia, methane, and hydrogen gas to electrical discharge (to simulate lightning) for several days, these investigators found that several complex organic substances, such as the amino acids glycine and alanine, were formed when the water vapor was condensed into water. This demonstration offered a very plausible hypothesis to explain how the early beginnings of life substances could have started by the formation of organic substances from inorganic ones.

○† **1953:** *Sperry, R.W. Independence of brain hemispheric function.*

By severing the bundle of nerve fibers separating the two cortical hemispheres, Sperry developed the "split-brain" technique, which he used to show that the two brain halves govern two sets of activities. He also contributed much to developmental neurobiology.

Nobel Laureate (1981).

△ **1954:** *Del Castillo, J., and B. Katz. Impulse transmission at nerve junction.*

These researchers demonstrated that the chemical transmitter substance is released from the presynaptic terminal in discrete multimolecular packets, or quanta, each containing several thousand molecules.

Katz, Nobel Laureate (1970).

✳ **1954:** *Dan, J.C. Acrosome reaction.*

In echinoderms, annelids, and molluscs Dan showed that the acrosome region of the spermatozoon forms a filament and releases an unknown substance at the time of fertilization. Evidence suggests that the filament is associated with the formation of the fertilization cone. Other observers had described similar filaments before Dan

made his detailed descriptions. The filament (1 to 75 μm long) may play an important role in the entrance of the sperm into the cytoplasm.

○ **1954:** *Du Vigneaud, V. Synthesis of pituitary hormones.*

This investigator isolated the posterior pituitary hormones oxytocin and vasopressin. Both were found to be polypeptides, and oxytocin was the first polypeptide hormone to be produced artificially. Oxytocin contracts the uterus during childbirth and releases the mother's milk; vasopressin raises blood pressure and decreases urine production.

○ **1954:** *Huxley, H.E., A.F. Huxley, and J. Hanson. Sliding filament model of muscular contraction.*

By means of electron microscopic studies and x-ray diffraction studies, these investigators showed that the proteins actin and myosin were found as separate filaments that apparently produced contraction by a sliding reaction in the presence of ATP. The concept is widely accepted.

A.F. Huxley, Nobel Laureate (1963).

△ **1954:** *Sanger, Frederick. Structure of the insulin molecule.*

Insulin is the important hormone used in the treatment of diabetes. It was the first protein for which a complete amino acid sequence was known. The molecule was found to be made up of 17 different amino acids in 51 amino acid units. Although one of the smallest proteins, its formula contains 777 atoms.

Nobel Laureate (1958).

△ **1955–1957:** *Kornberg, Arthur, and S. Ochoa. Synthesis of nucleic acids outside of cells (in vitro).*

By mixing the enzyme polymerase, extracted from the bacterium *Escherichia coli*, with a mixture of nucleotides and a tiny amount of DNA, Kornberg was able to produce synthetic DNA. Ochoa obtained the synthesis of RNA in a similar manner by using the enzyme polynucleotide phosphorylase from the bacterium *Azotobacter vinelandii*. This significant work reveals more insight into the mechanism of nucleic acid duplication in the cell.

Nobel Laureates (1959).

✳ **1955:** *Kettlewell, H.B.D. Natural selection in action: industrial melanism in moths.*

Using field techniques, Kettlewell was able to demonstrate a selective advantage of dark (melanic) mutations in a barklike cryptic moth species in industrial areas of Great Britain, where the three trunks had been darkened by soot. This demonstration provided an evolutionary explanation for the increasing incidence of melanism in cryptic moths that has been noted throughout the industrialized regions of the world and constitutes a classic example of the effects of a strong selective pressure.

○ **1956:** *von Békésy, Georg. The traveling wave hypothesis of hearing.*

Helmholtz (1868) had proposed the resonance hypothesis of hearing on the basis that each cross fiber of the basilar membrane, which increases in width from the base to the apex of the cochlea, resonates at a different frequency; von Békésy showed that a traveling wave of vibration is set up in the basilar membrane and reaches a maximal vibration in the part of the membrane appropriate for that frequency.

Nobel Laureate (1961).

△ **1956:** *Borsook, H., and P.C. Zamecnik. Site of protein synthesis.*

By injecting radioactive amino acids into an animal, they found that the ribosome of the endoplasmic reticulum is the place where proteins are formed.

‡ **1956:** *Ingram, V.M. Nature of a mutation.*

By tracing the change in one amino acid unit out of more than 300 units that make up the protein hemoglobin, Ingram was able to pinpoint the difference between normal hemoglobin and the mutant form of hemoglobin that causes sickle cell anemia.

△ **1956:** *Sutherland, Earl W., and T.W. Rall. Discovery of cyclic AMP.*

An intracellular mediating agent (cyclic adenosine-3′, 5′-monophosphate, or cyclic AMP) is found in all living tissues, and changes in cyclic AMP levels (by effects of hormones) cause the hormones to produce different target effects depending on the type of cell in which they are found. Cyclic AMP is formed by a metabolic reaction in which the enzyme adenyl cyclase converts ATP to cyclic AMP.

Sutherland, Nobel Laureate (1971).

‡ **1956:** *Tjio, J.H., and A. Levan. Revision of human chromosome count.*

The time-honored number of chromosomes in humans, 48 (diploid), was found by careful cytological technique to be 46 instead.

△ **1957:** *Calvin, Melvin. Chemical pathways in photosynthesis.*

By using radioactive carbon 14, Calvin and colleagues were able to analyze step by step the incorporation of carbon dioxide and the identity of each intermediate product involved in the formation of carbohydrates and proteins by plants.

Nobel Laureate (1961).

△ **1957:** *Holley, Robert W. The role of transfer RNA in protein synthesis.*

Nucleotides of transfer RNAs differ from each other only in their bases. Holley also devised methods that precisely established the transfer RNAs used in transfer of certain amino acids to the site of protein synthesis.

Nobel Laureate (1968).

□ **1957:** *Ivanov, A.V. Analysis of phylum Pogonophora (beard worms).*

Specimens of this phylum were collected in 1900 and represent one of the most recent phyla to be discovered and evaluated in the animal kingdom. Collections have been made in the waters of Indonesia, the Okhotsk Sea, the Bering Sea, and the Pacific Ocean. They are found mostly in the abyssal depths. Thought originally to belong to the Deuterostomia division, more recent evidence indicates that they are protostomes. At present 80 species divided into two orders have been described.

△ **1957:** *Perutz, Max F., and J.C. Kendrew. Structure of hemoglobin.*

The mapping of a complex globular protein molecule of 600 amino acids and 10,000 atoms arranged in a three-dimensional pattern represented one of the great triumphs in biochemistry. Myoglobin of muscle, which acts as a storehouse for oxygen and contains only one heme group instead of four (hemoglobin), was found to contain 150 amino acids.

Nobel Laureates (1962).

□ **1957:** *Sauer, E. Celestial navigation by birds.*

By subjecting Old World warblers to various synthetic night skies of star settings in a planetarium, Sauer was able to demonstrate that the birds made use of the stars to guide them in their migrations.

○ **1958:** *Lerner, A.B. Discovery of melatonin in the pineal glands.*

Lerner and associates discovered that the production of melatonin in the pineal gland is increased in darkness and decreased in the light. Activity of the hormone appears important in regulation of gonadal functions affected by photoperiod, and its discovery represented a breakthrough in understanding the function of the pineal gland.

△ **1958:** *Meselson, M., and F.W. Stahl. Confirmation "in vivo" of the duplicating mechanism in DNA.*

This was really confirmation of the self-copying of DNA in accordance with Watson and Crick's scheme of the structure of DNA. These investigators found that, after producing a culture of bacterial cells labeled with heavy nitrogen 15 and then transferring these bacteria to a cultural medium of light nitrogen 14, the resulting bacteria had a DNA density intermediate between heavy and light, as would be expected on the basis of the Watson-Crick hypothesis.

○ **1959:** *Burnet, F. Macfarlane. Clonal selection hypothesis of immunity.*

One of the most puzzling aspects of the acquired immune response has been how to account genetically for the ability to generate specific antibodies against the enormous variety of potential antigens. N.K. Jerne suggested that the information necessary to generate the antibodies was present in the host before exposure to the antigens. Burnet proposed that we have a large variety of antibody-producing cells, each present in such a small number that its product cannot be detected. Upon exposure to a particular antigen, a clone of cells that can make antibody to that antigen is "selected" and multiplies rapidly, thus increasing antibody to a detectable level. A corollary to the theory is that antigens present when the organism is embryonic or very young are recognized as "self," and the immune system does not respond. This was confirmed by P. Medawar.

Burnet and Medawar, Nobel Laureates (1960); Jerne, Nobel Laureate (1984).

‡ **1959:** *Ford, C.E., P.A. Jacobs, and J.H. Tjio. Chromosomal basis of sex determination in humans.*

By discovering that certain genetic defects were associated with an abnormal somatic chromosomal constitution, it was possible to determine that male-determining genes in humans were located on the Y chromosome. Thus a combination of XXY (47 instead of the normal 46 diploid number) produced sterile males (Klinefelter's syndrome), and those with XO combinations (45 diploid number) gave rise to Turner's syndrome or immature females.

× **1959:** *Leakey, Mary D. Discovery of* Australopithecus (Zinjanthropus) boisei *fossil hominid.*

This fossil is a robust form within the small-brained, large-jawed genus *Australopithecus,* first discovered by Dart (1925). Its status is at present controversial: some believe it to be a species distinct from *A. africanus,* whereas others believe that *A. africanus* and *A. boisei* are females and males within a single, polytypic species.

○ **1959:** *Butenandt, A.F.J. Chemical identification of a pheromone.*

Butenandt and associates chemically analyzed the first pheromone—the sex attractant substance of silk moths (*Bombyx mori*). Named bombykol, it is a doubly unsaturated fatty alcohol with 16 carbon atoms. The chemical nature of numerous other pheromones has since been determined.

△† **1950s–present:** *Bennacerraf, B., J. Dausset, and G. Snell. Genetics and function of the major histocompatibility complex (MHC).*

This complex of genes codes for cell surface antigens. These antigens are critically important in various interactions between cells, such as rejection of transplanted tissues and control of the immune response. Much work remains to be done before we completely understand the role of the MHC in immunity.

Nobel Laureates (1980).

Nobel Laureates (1986).

○ **1959–1960:** *Yalow, Rosalyn, and S.A. Berson. Development of radioimmunoassay.*

The development of this technique made it possible for the first time to measure minute quantities of a hormone directly in a complex mixture of proteins such as serum. A radioactively labeled hormone is mixed with a specific antibody to the hormone, together with a sample of the unknown. The labeled compound competes with the unlabeled in the formation of the antigen-antibody complex, and the amount of unknown is obtained by comparison with a standard curve prepared with known amounts of unlabeled hormone. The technique was originally described for insulin, but it has been rapidly extended to many other hormones and has revolutionized endocrinology.

Yalow, Nobel Laureate (1977).

□ **1960:** *Tinbergen, Niko. Development of the concept of specific searching images in predators.*

Tinbergen's studies of songbird predation in Dutch pinewoods helped to lay the foundation for the science of ethology.

Nobel Laureate (1973).

△ **1960:** *Hurwitz, J., A. Stevens, and S. Weiss. Enzymatic synthesis of messenger RNA.*

The exciting knowledge of the coding system of DNA and its translation to protein synthesis (ribosomes) was further elucidated when it was discovered that an enzyme, RNA polymerase, was responsible for the synthesis of RNA from a template pattern of DNA.

△ **1960:** *Jacob, François, and Jacques Monod. The operon hypothesis.*

The operon hypothesis is a postulated model of how enzyme synthesis is regulated in prokaryotes. The model proposes that refinement in regulation involves an inducible system for allowing structural genes to synthesize needed enzymes and a repressible system that cuts off the synthesis of unneeded enzymes.

Nobel Laureates (1965).

△ **1960:** *Strell, M., and R.B. Woodward. Synthesis of chlorophyll a.*

Strell and Woodward with the aid of many co-workers finally solved this problem, which had been the goal of organic chemists for generations.

Woodward, Nobel Laureate (1965).

† **1960s:** *Thomas, E. Donnall, and Joseph Murray. First successful organ and tissue transplants between unrelated donors.*

Previously, organ transplantation between donors other than identical twins had been impossible because of immune tissue rejection. Discovery of methods to suppress immune rejection and tissue typing based on major histocompatibility complex (to find suitable donors) vastly increased successful transplantation rate.

Nobel Laureates (1990).

△ **1961:** *Hurwitz, J., A. Stevens, and S.B. Weiss. Confirmation of messenger RNA.*

Messenger RNA transcribes directly the genetic message of the nuclear DNA and moves to the cytoplasm, where it becomes associated with a number of ribosomes or submicroscopic particles containing protein and nonspecific structural RNA. Here the messenger RNA molecules serve as templates against which amino acids are arranged in the sequence corresponding to the coded instructions carried by messenger RNA.

‡△ **1961:** *Jacob, François, and Jacques Monod. The role of messenger RNA in the genetic code.*

The transmission of information from the DNA code in the genes to the ribosomes represents an important step in the unraveling of the genetic code, and these investigators proposed certain deductions for confirmation of the hypothesis, which in general has been established by many researchers.

Nobel Laureates (1965).

○ **1961:** *Miller, J.F.A. Function of the thymus gland.*

Long known as a transitory organ that persists during the early growth period of animals, the thymus is now recognized as a vital processing center in the development of certain lymphocytes (T cells). These cells are very important in the immune response of vertebrates.

△ **1961:** *Mitchell, Peter. The chemiosmotic-coupling hypothesis of ATP formation.*

Although it is known that the oxidation of food molecules in the cell results in net synthesis of ATP from ADP and inorganic phosphate, the molecular mechanism to drive the reaction is still unclear. Current evidence supports the chemiosmotic hypothesis, which suggests that the energy derived from electron transport serves to pump hydrogen ions across the inner mitochondrial membrane, actively setting up an electrochemical gradient. The gradient of hydrogen ions is then coupled with an ATPase complex in the inner membrane, providing the free energy to form the high-energy phosphate bond.

‡△ **1961:** *Nirenberg, Marshall W., and J.H. Matthaei. Deciphering the genetic code.*

By adding a synthetic RNA composed entirely of uracil nucleotides to a mixture of amino acids, these investigators obtained a polypeptide made up solely of a single amino acid, phenylalanine. On the basis of a triplet code, it was concluded that the RNA code word for phenylaline was UUU and its DNA complement was AAA. This was the beginning of decoding.

Nirenberg, Nobel Laureate (1968).

○ **1962:** *Copp, Harold. Discovery of calcitonin.*

This peptide hormone is produced by the ultimobranchial glands (lower vertebrates) or their embryological derivatives, the parafollicular cells of the thyroid gland (mammals). It lowers the blood calcium and increases calcium uptake by bone cells, thus antagonizing the action of parathormone.

△‡ **1962–1972:** *Arber, W., H.P. Smith, and D. Nathans. Discovery, isolation, and characterization of restriction endonucleases.*

Use of these bacterial enzymes has been essential in the recent explosion of knowledge in molecular genetics and recombinant DNA technology. Arber is given credit for predicting the existence of such enzymes, Smith for isolating the first restriction endonuclease and describing its reaction, and Nathans for applying the enzymes in studies of gene regulation and organization.

Nobel Laureates (1978).

× **1964:** *Hamilton, W.D. Concepts of kin selection and inclusive fitness.*

The problem of neuter castes in social insects had destroyed Lamarckian and baffled Darwinian evolutionary explanations. Hamilton, however, demonstrated that the peculiar mode of reproduction in most social insects having neuter castes (haplodiploidy, or males haploid and females diploid) resulted in a situation in which sisters would share more common genes with one another than with their own progeny. Hence, helping to rear "kin" (sisters) could be more adaptive from the point of view of one's genes (inclusive fitness) than rearing one's own young (individual fitness). These ideas have since been extended to social behaviors of many species—vertebrate as well as invertebrate—and provide a fundamental core of the emerging field of sociobiology.

× **1964:** *Hoyer, B.H., B.J. McCarthy, and E.T. Bolton. Phylogeny and DNA sequence.*

These investigators presented evidence that certain homologies exist among the polynucleotide sequences in the DNA of such different forms as fishes and humans. Such sequences may represent genes that have been retained with little change throughout vertebrate history. Possible phenotypic expressions may be bilateral symmetry, notochord, hemoglobin, and so on.

△‡ **1966:** *Khorana, H.G. Proof of code assignments in the genetic code.*

By using alternating codons (CUC and UCU) in an artificial RNA chain, Khorana was able to synthesize a polypeptide of alternating amino acids (leucine and serine) for which these codons respectively stood.

Nobel Laureate (1968).

△ **1967:** *Katz, Bernard, and R. Miledi. Calcium entry at nervous synapses.*

These investigators proposed that the arrival of an action potential at a presynaptic terminal causes calcium influx that facilitates binding of synaptic vesicles with the presynaptic membrane.

□ **1967:** *MacArthur, Robert H., and E.O. Wilson. Theoretical ecology.*

Mathematical models in conjunction with field studies form the basis of theoretical ecology. This is well shown in the *Theory of Island Biogeography* by MacArthur and Wilson. By the methods

found in their book and the investigations of many other researchers, it has been possible to determine the equilibrium of species, the number of extinctions, and other factors on islands. This viewpoint has been a revelation in ecological studies.

△‡ **1967:** *Ptashne, M. Isolation of first repressor.*

Repressors are protein substances supposedly formed by regulatory genes; they function by preventing a structural gene from making its product when not needed by the cells.

□ **1968:** *Goodall, Jane. Behavior of free-living chimpanzees.*

A long-term complete study of primate social behavior was done in the field, providing an impetus and stimulus for a large number of similar studies on other primates.

△† **1960s–1970s:** *Bergström, S., B. Samuelsson, and J. Vane. Characterization of prostaglandins.*

Prostaglandins are chemical transmitters of intracellular and intercellular signals. They are involved in a variety of physiological and pathological functions. Bergström is credited with isolating prostaglandins and determining their structures; Samuelsson determined their biosynthesis and metabolism; and Vane found that vascular endothelium produces a prostaglandin (prostacyclin) that inhibits platelet aggregation.

Nobel Laureates (1982).

△ **1960s–1970s:** *Gilman, Alfred, and Martin Rodbell. Discovery of G proteins.*

Many chemical signals (for example, hormones, growth factors, neurotransmitters) bind to specific receptors on the cell surface. The cellular effect of many such signals occurs when cyclic AMP is synthesized. Transduction of the signal is mediated by a protein that requires GTP for its action (G protein).

Nobel Laureates (1994).

○ **1960s–1970s:** *Hubel, D.H., and T.N. Wiesel. Understanding of stereoscopic vision.*

Analysis of processing of images in eyes and brain opened a new field in vision physiology.

Nobel Laureates (1981).

△ **1970:** *Temin, H.M., and D. Baltimore. Demonstration of DNA synthesis from an RNA template.*

Many viruses carry RNA rather than DNA as a genetic material, and it was not known how the RNA virus could enter the host cell and induce the host to synthesize

more RNA virus particles. Temin and Baltimore independently found an RNA-dependent DNA polymerase in RNA viruses; thus the RNA virus reproduces by entering a host cell and producing a DNA copy from the RNA template; then the host cell machinery makes new virus particles from the DNA template.

Nobel Laureates (1975).

○△ **1970:** *Edelman, Gerald M, and R.R. Porter. The structure of gamma globulin.*

By using myeloma tumors (which contain pure immunoglobulin proteins) the investigators were able to work out after many years a complete analysis of the large gamma globulin molecule, which is made up of 1320 amino acids and 19,996 atoms, with a molecular weight of 150,000.

Nobel Laureates (1972).

‡* **1970s–1980s:** *Tonegawa, Susumu. Mechanisms of antibody diversity.*

The immune system of the body can make enormous numbers of different antibodies against the multitude of antigens it encounters. How to account genetically for antibody diversity has been a major challenge in immunology. A system of somatic mutations, multiple copies of gene segments, and gene rearrangements all play a role in generation of antibody diversity.

Nobel Laureate (1987).

△ **1971:** *Berg, P., D. Jackson, and R. Symons. Recombinant DNA.*

A DNA molecule from both a bacterial virus and an animal tumor virus were opened with a restriction endonuclease, then spliced together. This was the first recombinant DNA from two different organisms.

Berg, Nobel Laureate (1980).

△○ **1971:** *Cheung, W.Y. Discovery of calmodulin.*

The calcium-binding protein interacts reversibly with intracellular calcium to form a complex that regulates a broad spectrum of cellular activities.

† **1971–1973:** *Cormack, A.M., and G.N. Hounsfield. Invention of computer-assisted tomography (CAT scanner).*

This x-ray diagnostic technique produces clear images of internal body structures. A basic problem was how to obtain accurate measurements of the x-ray attenuation coefficient for all points in the x-ray section being examined. Cormack and Hounsfield each developed mathematical solutions to the problem.

Nobel Laureates (1979).

× **1972:** *Gould, Stephen J., and N. Eldridge. Concept of punctuated equilibrium.*

This hypothesis proposes that species remain virtually unchanged for long periods of time until a sudden spurt of rapid evolution produces a distinct species. Since the hypothesis is an alternative to more orthodox Darwinian gradualism, it has produced lively debate among evolutionists.

○△ **1972:** *Jerne, Niels K. Hypothesis of anti-idiotype regulation of the immune system.*

After the discovery by J. Oudin and H. Kunkel that antibodies themselves bear sites (idiotypes) that stimulate production of other antibodies against them (anti-idiotypes), Jerne proposed that regulation of the immune system was mediated by a complex network of antibodies specific for other antibodies. Binding of antibodies to each other could be stimulatory or suppressive. There is considerable evidence to support this.

Nobel Laureate (1984).

△ **1972:** *Singer, S.J., and G.L. Nicolson. Fluid-mosaic model of biological membranes.*

The plasma membrane of cells contains a bimolecular leaflet of lipids with the surface interrupted by proteins. Some proteins, called extrinsic proteins, are attached to the lipid surface, whereas others, called intrinsic proteins, penetrate the bilayer and may completely span the membrane. This model remains the most widely accepted model of membrane structure.

○ **1972:** *Woodward, R.B., and A. Eschenmoser. The synthesis of vitamin B₁₂.*

Vitamin B_{12} was the last vitamin to be synthesized. This complex molecule is not made up of polymers; new methods of organic chemistry were required before it could be synthesized. It contains the metal ion cobalt.

☐ **1973:** *The Endangered Species Act.*

This omnibus legislation, created in 1966 and greatly strengthened in 1973, is the first United States domestic law concerned exclusively with wildlife and enacted for purely altruistic motives. Its strength resides in that portion of the act specifically prohibiting federally funded projects from jeopardizing endangered species or their habitats. Some 400 species of plants and animals have been placed on the endangered list.

△ **1974:** *Brown, M., and J. Goldstein. Discovery of the LDL receptor on cell surfaces.*

Cholesterol, which is an essential component of cell membranes, is taken up by body cells in association with low density lipoproteins (LDL). Discovery that the presence of a receptor for LDL on the cell surface was essential for this process had important implications for understanding of endocytosis of other macromolecules. Brown and Goldstein also described the intracellular regulation of cholesterol metabolism, and they cloned the gene for the LDL receptor and thus were able to work out the molecular structure of the receptor.

Nobel Laureates (1985).

△ **1975:** *Miller, J., and P. Lu. Alteration of amino acid sequence.*

By inserting known amino acids to replace naturally occurring amino acids in the lac repressor protein, these investigators have been able to determine which amino acids are necessary for the various functions of the lac repressor.

△† **1975:** *Milstein, Cesar, and G. Köhler. Hybridoma technique and monoclonal antibodies.*

Hybridomas are cells formed by hybridizing myeloma cells (from immune system tumors) with lymphocytes. Each hybridoma produces a clone of identical daughter cells, all manufacturing identical antibodies. Monoclonal antibodies are of great value in research, are useful diagnostic tools, and have great potential in disease therapy.

Nobel Laureates (1984).

△ **1975:** *Sanger, J.W. Chromosome migration on spindle fibers.*

The proteins (actin and myosin) that contract muscles were found to be involved in the migration of chromosomes along the spindle fibers. Actin bundles are present on chromosomal spindle fibers.

△ **1975–1977:** *Gilbert, W., A. Maxam, and F. Sanger. Methods for determining base sequence in DNA.*

Practical methods for determining the base sequence in DNA have great implications for molecular genetics and DNA technology.

Gilbert and Sanger, Nobel Laureates (1980).

△ **1976–1980:** *Neher, Erwin, and Bert Sakmann. Development of the patch clamp.*

The "patch clamp" is a technique for measuring movements of ions through very small areas of cell membrane. A patch of cell membrane is sucked onto or into a tiny pipette of 0.5 μm diameter, and current changes caused by individual ion channels can be monitored. This technique has revolutionized neuroscience and cell biology.

Nobel Laureates (1991).

‡‡ **1976:** *Bishop, J. Michael, and Harold E. Varmus. Discovery of oncogenes.*

Genes that cause cancer (oncogenes) are derived from normal genes (proto-oncogenes) that have functional roles in healthy cells.

Nobel Laureates (1989).

*‡△ **1977:** *Roberts, Richard, Philip Sharp, and others. Discovery of introns and exons in DNA.*

Roberts and Sharp and others in their research groups showed that genes are often broken up by lengthy tracts of DNA that do not specify protein structure. The coding sections came to be called exons, and intervening noncoding stretches are introns.

Nobel Laureates (1993).

*‡ **1978:** *Lewis, Edward B. Discovery of homeotic genes.*

Lewis identified a series of control genes in *Drosophila* that seems to regulate the activity of other genes. Later known as homeotic genes, we now know that they are highly conserved and that they play similar functions in a wide variety of organisms.

Nobel Laureate (with C. Nüsslein-Volhard and E. Wieschaus) (1995).

1979: *Alvarez, W., L.W. Alvarez, F. Asaro, and H.V. Michel. Asteroid impacts cause mass extinctions observed in the fossil record.*

Anomalous levels of iridium at the boundary of Cretaceous/Tertiary rocks suggest that impacts of asteroids on the earth stimulated the mass extinction observed there. Subsequent studies of shocked quartz, soot, and spherules at the Cretaceous/Tertiary boundary throughout the world upheld predictions of the asteroid impact hypothesis. The asteroid impacts would have sent debris into the atmosphere, producing darkness and cold temperatures and eventually acid rain, wildfires, and a greenhouse effect, thereby endangering many species. The asteroid impact hypothesis may account for other mass extinctions in the fossil record, but this remains controversial.

*‡ **1980:** *Nüsslein-Volhard, Christiane, and Eric Wieschaus. Discovery of genes that control activation of homeotic genes.*

Groups of genes they named "gap," "pair-rule," and "polarity" genes were discovered in *Drosophila* that control the homeotic genes. Like the homeotic genes, they are highly conserved. Thousands of researchers are now analyzing these genes and gene families in mice, chickens, zebra fish, humans, and other organisms.

Nobel Laureates (with E.B. Lewis) (1995).

✳‡△ **1980s:** *Smith, Michael. Site-directed mutagenesis.*

Smith worked out a technique whereby a researcher could produce a DNA with a specific mutation at a particular site coding for a certain amino acid, thus producing a customized protein. This technique is used by virtually everyone in protein engineering and molecular biology.

Nobel Laureate (with K.B. Mullis) (1993).

△ **1981:** *Complete sequencing of human mitochondrial DNA.*

This landmark research by a group at England's Medical Research Council revealed that the human mitochondrial genome contains 16,569 base pairs. Packed into the genome with remarkable economy are the genes for making two ribosomal RNAs, 22 transfer RNAs, and 13 assorted proteins.

× **1982–1986:** *Vrba, Elisabeth, Stephen J. Gould, Niles Eldredge, and others. The hierarchical expansion of Darwinian evolutionary theory.*

The causal explanation of evolutionary change is expanded to include selective processes acting at different levels of biological complexity (genic, organismal, species) and different scales of evolutionary time. The phenomenon of sorting (differential survival and reproduction of varying individuals) is separated conceptually from selection, a cause of sorting based on the interaction between varying individuals (genes, organisms, species) and their respective environments. This discovery refines and expands our understanding of evolutionary processes.

△✳ **1982–1984:** *Altman, Sidney, and Thomas R. Cech. Discovery of RNA catalysis.*

Earlier dogma held that proteins were responsible for all enzymatic activity. Discovering that RNA introns could be edited out of precursor mRNA and that some RNA had catalytic properties in the absence of protein has led to the belief that most or all RNA splicing is carried out by catalytic RNA (ribozymes).

Nobel Laureates (1989).

□△‡× **1986:** *Mullis, Kary B. The polymerase chain reaction.*

This biochemical technique can amplify selected genes in vitro from a sample of genomic DNA. It greatly improves our ability to obtain DNA sequences for genes of interest, even from degraded DNA samples. Applications include analysis of DNA sequences preserved in fossil organisms and DNA obtained from blood samples for forensic studies.

Nobel Laureate (with M. Smith) (1993).

BOOKS AND PUBLICATIONS THAT HAVE GREATLY INFLUENCED DEVELOPMENT OF ZOOLOGY

Aristotle. 336–323 B.C. De anima, Historia animalium, De partibus animalium, *and* De generatione animalium. *These biological works of the Greek thinker have exerted an enormous influence on biological thinking for centuries.*

Vesalius, Andreas. 1543. De fabrica corporis humani. *This work is the foundation of modern anatomy and represents a break with the Galen tradition. His representations of some anatomical subjects, such as muscles, have never been surpassed. Moreover, he treated anatomy as a living whole, a viewpoint adopted by most present-day anatomists.*

Fabricius of Aquapendente. 1600–1621. De formato foetu *and* De formatione ovi pulli. *This was the first illustrated work on embryology and may be said to be the beginning of the modern study of development.*

Harvey, William. 1628. Essay on the motion of the heart and the blood. *This great work represents one of the first accurate explanations in physical terms of an important physiological process. It initiated an experimental method of observation that gave an impetus to research in all fields of biology.*

Descartes, René. 1637. Discourse on method. *This philosophical essay gave a great stimulus to a mechanistic interpretation of biological phenomena.*

Buffon, Georges. 1749–1804. Histoire naturelle. *This extensive work of many volumes collected together natural history facts in a popular and pleasing style. It had a great influence in stimulating a study of nature. Many eminent biological thinkers, such as Erasmus, Darwin and Lamarck, were influenced by its generalizations.*

Linnaeus, Carolus. 1758. Systema naturae. *This work lays the basis of the classification of animals and plants. With few modifications, the taxonomic principles outlined therein have been universally adopted by biologists.*

Wolff, Caspar Friedrich. 1759. Theoria generationis. *The theory of epigenesis was here set forth for the first time in opposition to the preformation theory of development so widely held before Wolff's work.*

von Haller, Albrecht. 1760. Elementa physiologiae. *An extensive summary of various aspects of physiology that greatly influenced physiological thinking for many years. Some of the basic concepts laid down therein are still considered valid, especially those on the nervous system.*

Malthus, Thomas R. 1798. Essay on population. *This work stimulated evolutionary thinking among such people as Darwin and Wallace.*

de Lamarck, Jean Baptiste. 1809. Philosophie zoologique. *This publication was of great importance in focusing the attention of biologists on the problem of the role of the environment as a factor in evolution. Lamarck's belief that all species came from other species represented one of the first clear-cut statements on the mutability of species, even though his theory of use and disuse is not accepted by biologists today.*

Cuvier, Georges. 1817. Le règne animal. *A comprehensive biological work that dealt with classification and a comparative study of animal structures. Its plates are still of value, but the general plan of the work was marred by a disbelief in evolution and a faith in the doctrine of geological catastrophes. The book, however, exerted an enormous influence on contemporary zoological thought.*

von Baer, Karl Ernst. 1828–1837. Entwickelungsgeschichte der Thiere. *In this important work are laid down the fundamental principles of germ layer formation and the similarity of corresponding stages in the development of embryos that have proved to be the foundation studies of modern embryology.*

Audubon, John J. 1827–1938. The birds of America. *The most famous of all ornithological works, it has served as the model for all monographs dealing with a specific group of animals. The plates are the work of a master artist.*

Lyell, Charles. 1830–1833. Principles of geology. *This great work exerted a profound influence on biological thinking, for it did away with the theory of catastrophism and prepared the way for an evolutionary interpretation of fossils and the forms that arose from them.*

Beaumont, William. 1833. Experiments and observations on the gastric juice and the physiology of digestion. *Beaumont's observations on various functions of the stomach and digestion were accurate and thorough. This classic work paved the way for the brilliant investigations of Pavlov, Cannon, and Carlson of later generations.*

Müller, Johannes. 1834–1840. Handbook of physiology. *The principles set down in this work by perhaps the greatest nineteenth century physiologist has set the pattern for the development of the science of physiology.*

Darwin, Charles. 1839. Journal of researches (Voyage of the *Beagle*). *This book reveals the training and development of the* naturalist and the material that led to the formulation of Darwin's concept of organic evolution.

Schwann, Theodor. 1839. Mikroskopische Untersuchungen über die Uebereinstimmung in der Struktur und dem Wachstum der Thiere und Pflanzen. *The basic principles concerning the cell doctrine are presented in this classic work.*

Kölliker, Albrecht. 1852. Mikroskopische Anatomie. *This was the first textbook in histology and contains contributions of the greatest importance in this field. Many of the histological descriptions Kölliker made have never needed correction. In many of his biological views he was far ahead of his time.*

Maury, Matthew F. 1855. The physical geography of the sea. *This work has often been called the first textbook on oceanography. This pioneer treatise stressed the integration of such knowledge as was then available about tides, winds, currents, depths, circulation, and such matters. Maury's work represents a real starting point in the fascinating study of the oceans and has had a great influence in stimulating investigations in this field.*

Virchow, Rudolf. 1858. Die Cellularpathologie. *In this work Virchow made the first clear distinction between normal and diseased tissues and demonstrated the real nature of pathological cells. The work also represents the death knell of the old humoral pathology, which had held sway for so long.*

Darwin, Charles. 1859. On the origin of species. *The most influential book ever published in biology. Although built around the theme that natural selection is the most important factor in evolution, the enormous influence of the book can be attributed to the extensive array of evolutionary evidence it presented. It also stimulated constructive thinking on a subject that had been vague and confusing before Darwin's time.*

Marsh, George P. 1864. Man and nature: physical geography as modified by human action. *A work that had an early and important influence on the conservation movement in America.*

Mendel, Gregor. 1866. Versuche über Pflanzenhybriden. *Careful, controlled pollination technique and statistical analysis gave a scientific explanation that has influenced all geneticists after the "rediscovery" in 1900 of Mendel's classic paper on the two basic laws of inheritance.*

Owen, Richard. 1866. Anatomy and physiology of the vertebrates. *This work contains an enormous amount of personal observation on the structure and physiology of animals, and some of the basic concepts of structure and function, such as homolog and analog, are here defined for the first time.*

Brehm, Alfred E. 1869. Tierleben. *The many editions of this work over many years have indicated its importance as a general natural history.*

Bronn, Heinrich G. (editor). 1873 to present. Klassen and Ordnungen des Tier-Reiches. *This great work is made up of exhaustive treatises on the various groups of animals by numerous authorities. Its growth extends over many years, and it is one of the most valuable works ever published in zoology.*

Balfour, Francis M. 1880. Comparative embryology. *This is a comprehensive summary of embryological work on both vertebrates and invertebrates up to the time it was published. It is often considered the beginning of modern embryology.*

Semper, Karl. 1881. Animal life as affected by the natural conditions of existence. *This work first pointed out the modern ecological point of view and laid the basis for many ecological concepts of existence that have proved important in the further development of this field of study.*

Bütschli, Otto. 1889. Protozoen (Bronn's Klassen und Ordnungen des Tier-Reichs). *This monograph has been of the utmost importance to students of protozoa. No other work on a like scale has ever been produced in this field of study.*

von Hertwig, Richard. 1892. Lehrbuch der Zoologie. *This text has proved to be an invaluable source of material for many generations of zoologists. Its illustrations have been widely used in other textbooks.*

Weismann, August. 1892. Das Keimplasma. *Weismann predicted from purely theoretical considerations the necessity of meiosis or reduction of the chromosomes in the germ cell cycle, a postulate that was quickly confirmed cytologically by others.*

Hertwig, Oskar. 1893. Zelle und Gewebe. *In this work a clear distinction is made between histology as the science of tissues and cytology as the science of cell structure and function. Cytology as a study in its own right dates from this time.*

Korschelt, E., and K. Heider. 1893. Lehrbuch der vergleichende Entwicklungsgeschichte der wirbellosen Thiere, 4 vols. *A treatise that has been a valuable tool for all workers in the difficult field of invertebrate embryology.*

Wilson, Edmund B. 1896. The cell in development and heredity. *This and subsequent editions represented the most outstanding work of its kind in the English language. Its influence in directing the development of cytogenetics cannot be overestimated, and in summarizing the many investigations in cytology, the book has served as one of the most useful tools in the field.*

Pavlov, Ivan. 1897. Le travail des glandes digestives. *This work, a landmark in the study of the digestive system, describes many of Pavlov's now classic experiments, such as the gastric pouch technique and the rate of gastric secretions.*

De Vries, Hugo. 1901. Die Mutationstheorie. *The belief that evolution is the result of sudden changes or mutations is advanced by one who is commonly credited with the initiation of this line of investigation into the causes of evolution.*

Sherrington, Charles. 1906. The integrative action of the nervous system. *The basic concepts of neurophysiology elaborated in this book, especially the concept of the integrative action of the nervous system, remains the basis of modern neurophysiology.*

Garrod, Archibald. 1909. Inborn errors of metabolism. *This pioneer book showed that certain congenital diseases were caused by defective genes that failed to produce the proper enzymes for normal functioning. It laid the basis for biochemical genetics, which later received a great impetus from the work of Beadle and Tatum.*

Henderson, Lawrence J. 1913. The fitness of the environment. *This book pointed out in a specific way the reciprocity that exists between living and nonliving nature and how organic matter is fitted to the inorganic environment. It has exerted a considerable influence on the study of ecological aspects of adaptation.*

Shelford, Victor E. 1913. Animal communities in temperate America. *This work was a pioneer in the field of biotic community ecology.*

Bayliss, William M. 1915. Principles of genetic physiology. *If a classic book must meet the requirements of masterly analysis and synthesis of what is known in a particular discipline, then this great work must be called one.*

Matthew, W.D. 1915. Climate and evolution. *Matthew, in contrast to Wegener, viewed the positions of continents as permanent. He explained distribution of plants and animals by dispersal between continents across land bridges, such as the Bering bridge between Asia and Alaska and the Panamanian bridge between Central and South America. Matthew's ideas dominated biogeographical thinking until the revival of Wegener's hypotheses in the 1960s and 1970s.*

Morgan, Thomas H., A.H. Sturtevant, C.B. Bridges, and H.J. Muller. 1915. The mechanism of Mendelian heredity. *This book gave an analysis and synthesis of Mendelian inheritance as formulated from the epoch-making investigations of the authors. This classic work stands as a cornerstone of our modern interpretation of heredity.*

Wegener, Alfred. 1915. The origin of continents and oceans. *In the first (German) edition of this book, Wegener developed the idea of continental drift. Wegener's interpretation of continental history was out of favor for many decades but was substantiated by geophysical work (plate tectonics) in the 1960s and 1970s. Biogeographers now invoke Wegenerian hypotheses to explain distribution of many groups of plants and animals.*

Doflein, F. 1916. Lehrbuch der Protozoenkunde, ed. 6 (revised by E. Reichenow, 1949). *A standard treatise on protozoa. Its many editions have proved helpful to all workers in this field.*

Thompson, D'Arcy W. 1917. Growth and form. *In this pioneering work the author attempted to reduce the great diversity of life into common themes and designs.*

Kukenthal, W., and T. Krumbach. 1923. Handbuch der Zoologie. *An extensive treatise on zoology that covers all phyla. The work has been an invaluable tool for all zoologists who are interested in the study of a particular group.*

Fisher, Ronald A. 1930. Genetical basis of natural selection. *This work has exerted an enormous influence on the newer synthesis of evolutionary mechanisms that emerged in the 1930s.*

Dobzhansky, Theodosius. 1937. Genetics and the origin of species. *The vast change in the explanation of the mechanism of evolution, which emerged about 1930, is well analyzed in this work by a master evolutionist. Other syntheses of this new biological approach to the evolutionary problems have appeared since this work was published, but none of them has surpassed the clarity and fine integration of Dobzhansky's work.*

Spemann, Hans. 1938. Embryonic development and induction. *In this work the author summarizes his pioneer investigations that have proved so fruitful in experimental embryology.*

Hyman, Libbie H. 1940. The invertebrates: Protozoa through Ctenophora. 1951. Platyhelminthes and Rhynchocoela, The acoelomate Bilateria. 1951. Acanthocephala, Aschelminthes, and Entoprocta, the pseudocoelomate Bilateria. 1955. Echinodermata, the coelomate Bilateria. 1959. Smaller coelomate groups: Chaetognatha, Hemichordata, Pogonophora, Phoronida, Ectoprocta, Brachiopoda, Sipunculida. The coelomate Bilateria. 1967. Mollusca I. *This series of volumes is a monumental work by a single author and is the only such coverage of the invertebrates in the English language. It is a benchmark in completeness, accuracy, and incisive analysis. It remains an absolutely essential reference for all teachers and investigators in invertebrate zoology.*

Schrödinger, Erwin. 1945. What is life? *The emphasis placed on the physical explanation of life and the popularization of the idea of a chemical genetic "codescript" provided a new point of view of biological phenomena so well expressed in the current molecular biological revolution.*

Lack, David. 1947. Darwin's finches. *This classic, written several years after Lack's study visit to the Galápagos Islands, introduced competition theory into animal ecology, stressed the importance of ecological isolation in speciation, and provided a cogent model for adaptive radiation.*

Grassé, P.P. (editor). 1948. Traité de zoologie. *This is a series of many treatises by various specialists on both invertebrates and vertebrates. Since it is relatively recent and comprehensive, the work is valuable to all students who desire detailed information on the various animal groups.*

Allee, W.C., A.E. Emerson, O. Park, T. Park, and K.P. Schmidt. 1949. Principles of animal ecology. *The basic ecological principles laid down in this comprehensive work are a landmark in the field.*

Blum, H.F. 1951. Time's arrow and evolution. *In this thoughtful book, Blum explores the relation between the second law of thermodynamics ("time's arrow") and organic evolution and recognizes that mutation and natural selection have been restricted to certain channels in accordance with the law, even though these two factors appear to controvert the principle of pointing the direction of events in time.*

Tinbergen, Niko. 1951. The study of instinct. *The classic statement of the approaches and theoretical interpretations of early ethologists, especially K. Lorenz, is presented. This work brought ethology to the attention of American behavioral scientists, generating both controversy and collaboration.*

Crick, Francis H.C., and James D. Watson. 1953. Genetic implications of the structure of deoxyribonucleic acid. *The solution of the structure of the DNA molecule has served as the cornerstone for an explanation of genetic replication and control of the cell's attributes and functions.*

Simpson, George G. 1953. The major features of evolution. *In a comprehensive synthesis of modern evolutionary theory, the author draws on evidence from paleontology, population genetics, and systematics. This book, along with Simpson's earlier* Tempo and mode in evolution, *have greatly influenced our thinking about the mechanisms of evolution within the framework of natural selection.*

Burnet, F. MacFarlane. 1959. The clonal selection theory of acquired immunity. *This was an important new theory to explain the specificity of the acquired immune response and has to a great extent replaced the template theory formulated by Haurowitz in 1930.*

Carson, Rachel L. 1962. Silent spring. *Though a polemic addressed to a lay audience, this book was highly valuable in increasing public awareness of the danger to the ecosystem of indiscriminate use of pesticides.*

Mayr, Ernst. 1963. Animal species and evolution. *This lucid and thoroughly documented landmark in evolutionary study helped to clarify and integrate the emerging synthetic theory of evolution.*

Watson, James D. 1965. Molecular biology of the gene. *With perhaps the clearest statement of what molecular biology is,*

this important book introduced the field to the new generation of investigators in molecular genetics.

Williams, George C. 1966. Adaptation and natural selection. *This scholarly and insightful work, with its well-reasoned explanation of adaptation and its convincing defense of natural selection, was highly influential in clarifying the role of natural selection in evolutionary change.*

MacArthur, R.H. 1972. Geographical ecology. *One of the most important theoretical ecology books to have appeared in the 1970s.*

Gurdon, John B. 1974. The control of gene expression in animal development. *Gurdon developed an experimental system, extended from the pioneering nuclear transplantation experiments of Briggs and King, that enabled him to transfer genetic information from one animal to another. His studies, summarized in this influential work, are interpreted to mean that nuclei of embryonic cells do not undergo alterations or loss of genetic material during development.*

Lewontin, Richard C. 1974. The genetic basis of evolutionary change. *This book promoted the study of protein polymorphisms to measure the amount of genetic variation present in natural populations. The development of the alternative theories of neutral polymorphism versus balancing natural selection is covered in detail. The methods described in this book for the first time made possible studies on a large scale of population genetics in natural populations of animals.*

Wilson, Edward O. 1975. Sociobiology: the new synthesis. *The rapidly developing fields of behavioral biology, behavioral ecology, and population genetics are reviewed. Wilson attempts to provide the basis for a new biological science, utilizing concepts developed in these heretofore disparate fields, devoted to analyzing social behaviors at all phylogenetic levels.*

Gould, Stephen Jay. 1977. Ontogeny and phylogeny. *This book reexamined the ancient problem of relationship between organismal development and evolution and revitalized this field. Numerous studies of heterochrony—the evolutionary changes in the timing of organismal development—followed publication of this book and provided most of our present knowledge about the evolution of morphology and life histories in animals.*

APPENDIX B
Basic Structure of Matter

For the convenience of students who have not had a course in basic chemistry, or for those who wish to review the material, we are providing a brief introduction here.

ELEMENTS AND ATOMS

All matter is composed of **elements,** which are substances that cannot be subdivided further by ordinary chemical reactions. Only 92 elements occur naturally, but the elements may be combined by chemical bonds into a vast number of different compounds. The elements are designated by one or two letters derived from their Latin or English names (Table B-1). The elements are composed of discrete units called **atoms,** which are the smallest components into which an element can be subdivided by normal chemical means. Combination of the atoms of an element with each other or with those of other elements by chemical bonds creates **molecules.** When molecules are composed of the atoms of two or more different kinds of elements, they are a **compound.**

In a chemical formula, the symbol for an element stands for one atom of the element, with additional atoms indicated by appropriately placed numbers. Thus atmospheric nitrogen is N_2 (each molecule is composed of two atoms of nitrogen), and water is H_2O (two atoms of hydrogen and one of oxygen in each molecule), and so on.

SUBATOMIC PARTICLES

Each atom is composed of subatomic particles, of these there are three with which we need concern ourselves: protons, neutrons, and electrons. Every atom consists of a positively charged nucleus surrounded by a negatively charged system of electrons (Figure B-1). The nucleus, containing most of the atom's mass, is made up of protons and neutrons clustered together in a very small volume. These two particles have about the same mass, each being about 2000 times heavier than an electron. The protons bear positive charges, and the neutrons are uncharged (neutral). Although the number of protons in the nucleus is the same as the number of electrons around the nucleus, the number of neutrons may vary. For every positively charged proton in the nucleus, there is a negatively charged electron; the total charge of the atom is thus neutral.

The **atomic number** of an element is equal to the number of protons in the nucleus, whereas the **atomic mass** is nearly equal to the number of protons plus the number of neutrons (explanation of why atomic mass is not exactly equal to protons plus neutrons can be found in any introductory chemistry text). The mass of the electrons may be neglected because it is only 1/1836 that of a proton or neutron.

ISOTOPES

It is possible for two atoms of the same element to have the same number of protons in their nuclei but have a different number of neutrons. Such different forms, having the same number of protons but different

Table B-1	Some of the Most Important Elements in Living Organisms		
Element	**Symbol**	**Atomic Number**	**Approximate Atomic Weight**
Carbon	C	6	12
Oxygen	O	8	16
Hydrogen	H	1	1
Nitrogen	N	7	14
Phosphorus	P	15	31
Sodium	Na	11	23
Sulfur	S	16	32
Chlorine	Cl	17	35
Potassium	K	19	39
Calcium	Ca	20	40
Iron	Fe	26	56
Iodine	I	53	127

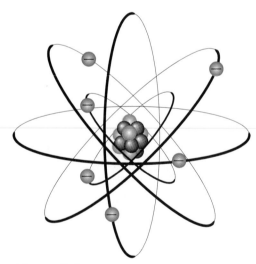

Figure B-1

Structure of an atom. Planetary system of negatively charged electrons around a dense nucleus of positively charged protons and uncharged neutrons.

atomic masses, are called **isotopes.** For example, the predominant form of hydrogen in nature has 1 proton and no neutron (^1H) (Figure B-2). Another form (deuterium [^2H]) has 1 proton and 1 neutron. Tritium (^3H) has 1 proton and 2 neutrons. Some isotopes

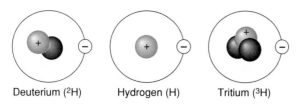

Deuterium (²H)   Hydrogen (H)   Tritium (³H)

Figure B-2

Three isotopes of hydrogen. Of the three isotopes, hydrogen 1 makes up about 99.98% of all hydrogen, and deuterium (heavy hydrogen) makes up about 0.02%. Tritium is radioactive and is found only in traces in water. Numbers indicate approximate atomic weights. Most elements are mixtures of isotopes. Some elements (for example, tin) have as many as 10 isotopes.

are unstable, undergoing a spontaneous disintegration with the emission of one or more of three types of particles, or rays: gamma rays (a form of electromagnetic radiation), beta rays (electrons), and alpha rays (positively charged helium nuclei stripped of their electrons). These unstable isotopes are said to be **radioactive.** Using radioisotopes, biologists are able to trace movements of elements and tagged compounds through organisms. Our present understanding of metabolic pathways in animals and plants is in large part the result of this powerful analytical tool. Among the commonly used radioisotopes are carbon 14 (^{14}C), tritium, and phosphorus 32 (^{32}P).

ELECTRON "SHELLS" OF ATOMS

According to Niels Bohr's planetary model of the atom, the electrons revolve around the nucleus of an atom in circular orbits of precise energy and size. All of the orbits of any one energy and size compose an electron shell. This simplified picture of the atom has been greatly modified by more recent experimental evidence; definite electron pathways are no longer hypothesized, and an electron shell is more vaguely understood as a thick region of space around the nucleus rather than a narrow shell of a particular radius out in space.

However, the old planetary model with the idea of electronic shells is still useful in interpreting chemical phenomena. The number of concentric shells required to contain an element's electrons varies with the element. Each shell can hold a maximum number of electrons. The first shell next to the atomic nucleus can hold a maximum of 2 electrons, and the second shell can hold 8; other shells also have a maximum number, but no atom can have more than 8 electrons in its outermost shell. Inner shells are filled first, and if there are not enough electrons to fill all the shells, the outer shell is left incomplete. Hydrogen has 1 proton in its

nucleus and 1 electron in its single orbit but no neutron. Since its shell can hold 2 electrons, it has an incomplete shell. Helium has 2 electrons in its single shell, and its nucleus is made up of 2 protons and 2 neutrons. Since the 2-electron arrangement in helium's shell is the maximum number for this shell, the shell is closed and precludes all chemical activity. There is no known compound of helium. Neon is another inert (chemically inactive) gas because its outer shell contains 8 electrons, the maximum number (Figure B-3). However, stable compounds of xenon (an inert gas) with fluorine and oxygen are formed under special conditions. Oxygen has an atomic number of 8. Its 8 electrons are arranged with 2 in the first shell and 6 in the second shell (Figure B-3). It is active chemically, forming compounds with almost all elements except inert gases.

CHEMICAL BONDS

As we noted above, atoms joined to each other by chemical bonds form molecules, and atoms of each element form molecules with each other or with atoms of other elements in particular ways, depending on the number of electrons in their outer orbits.

IONIC BONDS

Elements react in such a way as to gain a stable configuration of electrons in their outer shells. The number of electrons in the outer shell varies from 0 to 8. With either 0 or 8 in this shell, the element is chemically inactive. When there are fewer than 8 electrons in the outer shell, the atom will tend to lose or gain electrons to have an outer shell of 8. This will give the atom a net electrical charge, and the atom is now an **ion.** Atoms with 1 to 3 electrons in the outer shell tend to lose them to other atoms and to become positively charged ions because of the excess pro-

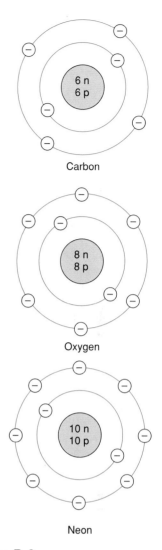

Carbon

Oxygen

Neon

Figure B-3

Electron shells of three common atoms. Since no atom can have more than 8 electrons in its outermost shell and two electrons in its innermost shell, neon is chemically inactive. However, the second shells of carbon and oxygen, with 4 and 6 electrons, respectively, are open so that these elements are electronically unstable and react chemically whenever appropriate atoms come into contact. Chemical properties of atoms are determined by their outermost electron shells.

tons in the nucleus. Atoms with 5 to 7 electrons in the outer orbit tend to gain electrons from other atoms and to become negatively charged ions because of the greater number of electrons than protons. Positive and negative ions tend to unite.

Every atom has a tendency to complete its outer shell to increase its stability in the presence of other atoms. Let us examine how two atoms with incomplete outer shells, sodium and chlorine, can interact to fill their outer shells. Sodium, with 11 electrons, has 2 electrons in its

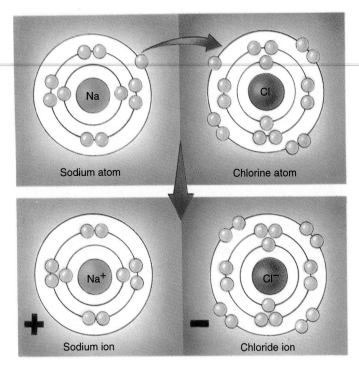

Figure B-4

Ionic bond. When one atom of sodium and one of chlorine react to form a molecule, a single electron in the outer shell of sodium is transferred to the outer shell of chlorine. This causes the outer or second shell (third shell is now empty) of sodium to have 8 electrons and also chlorine to have 8 electrons in its outer or third shell. The compound thus formed is sodium chloride (NaCl). By losing 1 electron, sodium becomes a positive ion, and by gaining 1 electron, chlorine becomes a negative ion (chloride). This ionic bond is the strong electrostatic force acting between positively and negatively charged ions.

first shell, 8 in its second shell, and only 1 in the third shell. The third shell is highly incomplete; if this third-shell electron were lost, the second shell would be the outermost shell and would produce a stable atom. Chlorine, with 17 electrons, has 2 in the first shell, 8 in the second, and 7 in the incomplete third shell. Chlorine must gain an electron to fill the outer shell and become a stable atom. Clearly, the transfer of the third-shell sodium electron to the incomplete chlorine third shell would yield simultaneous stability to both atoms.

Sodium, now with 11 protons but only 10 electrons, becomes electropositive (Na^+). In gaining an electron from sodium, chlorine contains 18 electrons but only 17 protons and thus becomes an electronegative chloride ion (Cl^-). Since unlike charges attract, a strong electrostatic force, called an **ionic bond** (Figure B-4), is formed. The ionic compound formed, sodium chloride, can be represented in electron dot notation ("fly-speck formulas") as:

$$Na \cdot + \cdot \overset{\cdot \cdot}{\underset{\cdot \cdot}{Cl}} : \longrightarrow Na^+ + (: \overset{\cdot \cdot}{\underset{\cdot \cdot}{Cl}} :)^-$$

The number of dots shows the number of electrons present in the outer shell of the atom: 7 in the case of the neutral chlorine atom and 8 for the chloride ion; 1 in the case of the neutral sodium atom and none for the sodium ion.

If an element with 2 electrons in its outer shell, such as calcium, reacts with chlorine, it must give them both up, one to each of two chlorine atoms, and calcium becomes doubly positive:

$$Ca : + 2 \cdot \overset{\cdot \cdot}{\underset{\cdot \cdot}{Cl}} : \longrightarrow Ca^{2+} + 2(: \overset{\cdot \cdot}{\underset{\cdot \cdot}{Cl}} :)^-$$

Processes that involve the **loss of electrons** are called **oxidation** reactions; those that involve the **gain of electrons** are **reduction** reactions. Since oxidation and reduction always occur simultaneously, each of these processes is really a "half-reaction." The entire reaction is called an **oxidation-reduction** reaction, or simply a **redox** reaction. The terminology is confusing because oxidation-reduction reactions involve electron transfers, rather than (necessarily) any reaction with oxygen. However, it is easier to learn the system than to try to change accepted usage.

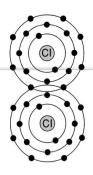

Figure B-5

Covalent bond. Each chlorine atom has 7 electrons in its outer shell, and by sharing one pair of electrons, each atom acquires a complete outer shell of 8 electrons, thus forming a molecule of chlorine (Cl_2).

COVALENT BONDS

Stability can also be achieved when two atoms share electrons. Let us again consider the chlorine atom, which, as we have seen, has an incomplete 7-electron outer shell. Stability is attained by gaining an electron. One way this can be done is for two chlorine atoms to *share* one pair of electrons (Figure B-5). To do this, the two chlorine atoms must *overlap* their third shells so that the electrons in these shells can now spread themselves over both atoms, thereby completing the filling of both shells. Many other elements can form covalent (or electron-pair) bonds. Examples are hydrogen (H_2):

$$H \cdot + H \cdot \longrightarrow H : H$$

and oxygen (O_2):

$$\overset{\cdot \cdot}{\underset{\cdot \cdot}{O}} : + : \overset{\cdot \cdot}{\underset{\cdot \cdot}{O}} \longrightarrow \overset{\cdot \cdot}{\underset{\cdot \cdot}{O}} : : \overset{\cdot \cdot}{\underset{\cdot \cdot}{O}}$$

In this case oxygen must share two pairs of electrons to achieve stability. Each atom now has 8 electrons available to its outer shell, the stable number.

Covalent bonds are of great significance to living systems, since the major elements of living matter (carbon, oxygen, nitrogen, hydrogen) almost always share electrons in strong covalent bonds. The stability of these bonds is essential to the integrity of DNA and other macromolecules, which, if easily dissociated, would result in biological disorder.

The outer shell of carbon contains 4 electrons. This element is endowed with great potential for forming a variety of atomic configurations with itself and other molecules. It can, for example, share its electrons with hydrogen to form methane

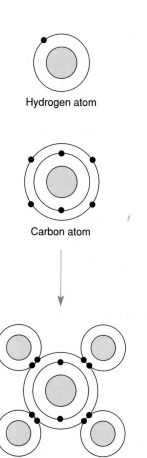

Hydrogen atom

Carbon atom

Methane (CH₄)

Figure B-6

In methane the four hydrogen atoms each share an electron with a carbon atom. They are arranged symmetrically around the carbon atom and form a pyramid-shaped tetrahedron in which each of the hydrogen atoms is equally distant from the others.

(Figure B-6). Carbon now achieves stability with 8 electrons, and each hydrogen atom becomes stable with two electrons. Carbon can also bond with itself (and hydrogen) to form, for example, ethane:

$$H:C:C:H \quad \text{or} \quad H—C—C—H$$

Carbon also forms covalent bonds with oxygen:

$$·C· + 2 O: → O::C::O$$

This is a "double-bond" configuration usually written as $O = C = O$. Carbon can even form triple bonds as, for example, in acetylene:

$$H:C:::C:H \text{ or } H—C≡C—H$$

The significant aspect of each of these molecules is that each carbon gains a share in 4 electrons from atoms nearby,

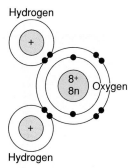

Figure B-7

Molecular structure of water. The two hydrogen atoms bonded covalently to an oxygen atom are arranged at an angle of about 105 degrees to each other. Since the electrical charge is not symmetrical, the molecule is polar with positively and negatively charged ends.

thus attaining the stability of 8 electrons. The sharing may occur between carbon and other elements or other carbon atoms, and in many instances the 8-electron stability is achieved by means of multiple bonds.

These examples only begin to illustrate the amazing versatility of carbon. It is a part of virtually all compounds comprising living substance, and without carbon, life as we know it would not exist.

HYDROGEN BONDS

Hydrogen bonds are described as "weak" bonds because they require little energy to break. They do not form by transfer or sharing of electrons, but result from unequal charge distribution on a molecule, so that the molecule is polar. For example, the two hydrogen atoms that share electrons with an oxygen atom to form water (H_2O) are not 180 degrees away from each other around the oxygen, but form an angle of about 105 degrees (Figure B-7). Thus the side of the molecule away from the hydrogen atoms is more negative, and the hydrogen side is more positive (contrast the methane molecule [Figure B-6], in which the equidistant placement of hydrogen atoms cancels out charge displacements). The electrostatic attraction between the electropositive part of one molecule forms a hydrogen bond with the electronegative part of an adjacent molecule. The ability of water molecules to form hydrogen bonds with each other (Figure B-8) accounts for many unusual properties of this unique substance (p. 22). Hydrogen bonds are important in the formation and function of other biologically active substances, such as proteins and nucleic acids (pp. 26 through 28).

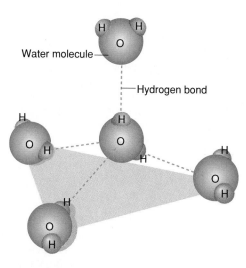

Water molecule

Hydrogen bond

Figure B-8

Geometry of water molecules. Each water molecule is linked by hydrogen bonds (dashed lines) to four other water molecules. If imaginary lines are used to connect the divergent oxygen atoms, a tetrahedron is obtained. In ice, the individual tetrahedrons associate to form an open lattice structure.

ACIDS, BASES, AND SALTS

The hydrogen ion (H^+) is one of the most important ions in living organisms. The hydrogen atom contains a single electron. When this electron is completely transferred to another atom (not just shared with another atom as in the covalent bonds with carbon), only the hydrogen nucleus with its positive proton remains. Any molecule that dissociates in solution and gives rise to a hydrogen ion is an **acid.** An acid is classified as strong or weak, depending on the extent to which the acid molecule dissociates in solution. Examples of strong acids that dissociate completely in water are hydrochloric acid ($HCl → H^+ + Cl^-$) and nitric acid ($HNO_3 → H^+ + NO_3^-$). Weak acids, such as carbonic acid ($H_2CO_3 → H^+ + HCO_3^-$), dissociate only slightly. A solution of carbonic acid is mostly undissociated carbonic acid molecules with only a small number of bicarbonate (HCO_3^-) and hydrogen ions (H^+) present.

A **base** contains negative ions called hydroxide ions and may be defined as a molecule or ion that will accept a proton (hydrogen ion). Bases are produced when compounds containing them are dissolved in water. Sodium hydroxide (NaOH) is a strong base because it will dissociate completely in water into sodium (Na^+) and hydroxide (OH^-) ions. Among the characteristics of bases is their ability to combine with hydrogen ions, thus decreasing the concentration of the hydrogen ions. Like acids, bases vary in the extent to which they dissociate in aqueous solutions into hydroxide ions.

A **salt** is a compound resulting from the chemical interaction of an acid and a base. Common salt, sodium chloride (NaCl), is formed by the interaction of hydrochloric acid (HCl) and sodium hydroxide (NaOH). In water the HCl is dissociated into H^+ and Cl^- ions. The hydrogen and hydroxide ions combine to form water (H_2O), and the sodium and chloride ions remain as a dissolved form of salt (Na^+Cl^-):

$$H^+Cl^- + Na^+OH^- \rightarrow Na^+Cl^- + H_2O$$

Acid   Base   Salt

Organic acids are usually characterized by having in their molecule the carboxyl group ($-COOH$). They are weak acids because a relatively small proportion of the H^+ reversibly dissociates from the carboxyl:

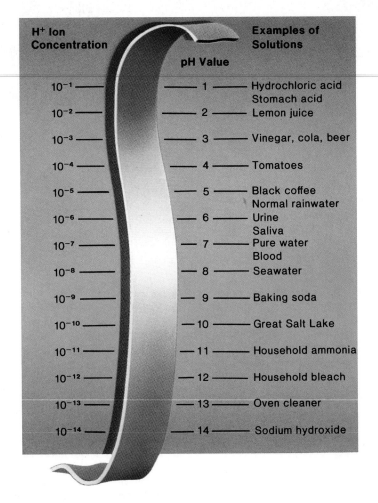

R refers to an atomic grouping unique to the molecule. Some common organic acids are acetic, citric, formic, lactic, and oxalic.

HYDROGEN ION CONCENTRATION (pH)

Solutions are classified as acid, basic, or neutral according to the proportion of hydrogen (H^+) and hydroxide (OH^-) ions they possess. In acid solutions there is an excess of hydrogen ions; in alkaline, or basic, solutions the hydroxide ion is more common; and in neutral solutions both hydrogen and hydroxide ions are present in equal numbers.

To express the acidity or alkalinity of a substance, a logarithmic scale, a type of mathematical shorthand, is employed that uses the numbers 1 to 14. This is the pH, defined as:

$$pH = \log_{10} \frac{1}{[H]}$$

or

$$pH = -\log_{10}[H^+]$$

Thus pH is the negative logarithm of the hydrogen ion concentration in moles per liter. In other words, when the hydrogen ion concentration is expressed exponentially, pH is the exponent, but with the *opposite* sign; if $[H^+] = 10^{-2}$, then pH = $-(-2) = +2$. Unfortunately, pH can be a confusing concept because, as the $[H^+]$ decreases, the pH increases. Numbers below 7 indicate an acid range, and numbers above 7 indicate alkalinity (Figure B-9). The number 7 indicates neutrality, that is, the presence of equal numbers of H^+ and OH^- ions. According to this logarithmic

scale, a pH of 3 is 10 times more acid than one of 4; a pH of 9 is 10 times more alkaline than one of 8.

BUFFER ACTION

The hydrogen ion concentration in the extracellular fluids must be regulated so that metabolic reactions within the cell will not be adversely affected by a constantly changing hydrogen ion concentration, to which they are extremely sensitive. A change in pH of only 0.2 from the normal mammalian blood pH of about 7.35 can cause serious metabolic disturbances. To maintain pH within physiological limits, there are certain substances in cells and organisms that tend to compensate for any change in the pH when acids or alkalies are produced in metabolic reactions or are added to the body fluids. These are called **buffers.** A buffer is a mixture of slightly ionized weak acid and its completely ionized salt. In such a system, added H^+ combines with the anion of the salt to form undissociated acid, and added

OH^- combines with H^+ from the weak acid molecule to form water. The most important buffers in blood and other extracellular fluids are the bicarbonates and phosphates, and organic molecules such as amino acids and proteins are important buffers within cells. The bicarbonate buffer system consists of carbonic acid (H_2CO_3, a weak acid) and its salt, sodium bicarbonate ($NaHCO_3$). Sodium bicarbonate is strongly ionized into sodium ions (Na^+) and bicarbonate ions (HCO_3^-). When a strong acid (for example, HCl) is added to the fluid, the H^+ ions of the dissociated acid will react with the bicarbonate ion (HCO_3^-) to form a very weak acid, carbonic acid, which dissociates only slightly. Thus the H^+ ions from the HCl are removed and the pH is little altered. When a strong base (for example, NaOH) is added to the fluid, the OH^- ions of the strong base will react with the carbonic acid by removing H^+ ions from the H_2CO_3, to form water and bicarbonate ions. Again the H^+ ion concentration in solution is little altered and the pH remains nearly unchanged.

Figure B-9

The pH scale. A pH of 7 is considered neutral. Values below 7 are acidic, and the lower the value, the more acidic the solution. Values above 7 are basic or alkaline, and the higher the value, the more basic the solution. Representative fluids with approximate pH values are listed.

GLOSSARY

This glossary lists definitions, pronunciations, and derivations of the most important recurrent technical terms, units, and names (excluding taxa) used in the text.

A

abiotic (ā′bī-äd′ik) (Gr. *a*, without, + *biōtos*, life, livable). Characterized by the absence of life.

abomasum (ab′ō-mā′səm) (L. *ab*, from, + *omasum*, paunch). Fourth and last chamber of the stomach of ruminant mammals.

aboral (ab-o′rəl) (L. *ab*, from, + *os*, mouth). A region of an animal opposite the mouth.

abscess (ab′ses) (L. *abscessus*, a going away). Dead cells and tissue fluid confined in a localized area, causing swelling.

acanthodians (a′kan-thō′dē-əns) (Gr. *akantha*, prickly, thorny). A group of the earliest known true jawed fishes from Lower Silurian to Lower Permian.

acanthor (ə-kan′thor) (Gr. *akantha*, spine or thorn, + *or*). First larval form of acanthocephalans in the intermediate host.

acclimatization (ə-klī′mə-də-zā-shən) (L. *ad*, to, + Gr. *klima*, climate). Gradual physiological adaptation in response to relatively long-lasting environmental changes.

acetabulum (as′ə-tab′ū-ləm) (L. a little saucer for vinegar). True sucker, especially in flukes and leeches; the socket in the hip bone that receives the thigh bone.

acicula (ə-sik′ū-lə) (L. *acicula*, a small needle). Needlelike supporting bristle in parapodia of some polychaetes.

acid. A molecule that dissociates in solution to produce a hydrogen ion (H⁺).

acinus (as′ə-nəs), pl. **acini** (as′ə-ni) (L. grape). A small lobe of a compound gland or a saclike cavity at the termination of a passage.

acoelomate (ā-sēl′ə-māt′) (Gr. *a*, not, + *koilōma*, cavity). Without a coelom, as in flatworms and proboscis worms.

acontium (ə-kän′chē-əm), pl. **acontia** (Gr. *akontion*, dart). Threadlike structure bearing nematocysts located on mesentery of sea anemone.

acrocentric (ak′rō-sen′trək) (gr. *akros*, tip, + *kentron*, center). Chromosome with centromere near the end.

acron (a′crän) (Gr. *akron*, mountaintop, fr. *akros*, tip). Preoral region of an insect.

actin (Gr. *aktis*, ray). A protein in the contractile tissue that forms the thin myofilaments of striated muscle.

actinotroch (ək-tin′ə-trōk) (Gr. *aktis*, ray, beam, + *trochos*, wheel). Larval form found in Phoronida.

active transport. Mediated transport in which a permease transports a molecule across a cell membrane against a concentration gradient; requires expenditure of energy; contrast with **facilitated diffusion.**

adaptation (L. *adaptatus*, fitted). An anatomical structure, physiological process, or behavioral trait that evolved by natural selection and improves an organism's ability to survive and leave descendants.

adaptive radiation. Evolutionary diversification that produces numerous ecologically disparate lineages from a single ancestral one, especially when this diversification occurs within a short interval of geological time.

adaptive value. Degree to which a characteristic helps an organism to survive and reproduce or lends greater fitness in its environment; selective advantage.

adaptive zone. A characteristic reaction and mutual relationship between environment and organism ("way of life") demonstrated by a group of evolutionarily related organisms.

adductor (ə-duk′tər) (L. *ad*, to, + *ducere*, to lead). A muscle that draws a part toward a median axis, or a muscle that draws the two valves of a mollusc shell together.

adenine (ad′nēn, ad′ə-nēn) (Gr. *adēn*, gland, + *ine*, suffix). A purine base; component of nucleotides and nucleic acids.

adenosine (ə-den′ə-sen) **(di-, tri) phosphate** (ADP and ATP). A nucleotide composed of adenine, ribose sugar, and two (ADP) or three (ATP) phosphate units; ATP is an energy-rich compound that, with ADP, serves as a phosphate bond-energy transfer system in cells.

adipose (ad′ə-pōs) (L. *adeps*, fat). Fatty tissue; fatty.

adrenaline (ə-dren′ə-lən) (L. *ad*, to, + *renalis*, pertaining to kidneys). A hormone produced by the adrenal, or suprarenal, gland; epinephrine.

adsorption (ad-sorp′shən) (L. *ad*, to, + *sorbeo*, to absorb). The adhesion of molecules to solid bodies.

aerobic (a-rō′bik) (Gr. *aēr*, air, + *bios*, life). Oxygen-dependent form of respiration.

afferent (af′ə-rənt) (L. *ad*, to, + *ferre*, to bear). Adjective meaning leading or bearing toward some organ, for example, nerves conducting impulses toward the brain or blood vessels carrying blood toward an organ; opposed to **efferent.**

aggression (ə-gres′hən) (L. *aggressus*, attack). An offensive action or procedure.

agonistic behavior (Gr. *agōnistēs*, combatant). An offensive action or threat directed toward another organism.

alate (ā′lāt) (L. *alatus*, wing). Winged.

albumin (al-byū′mən) (L. *albumen*, white of egg). Any of a large class of simple proteins that are important constituents of vertebrate blood plasma and tissue fluids and also present in milk, whites of eggs, and other animal substances.

alimentary (al′ə-men′tə-rē) (L. *alimentum*, food, nourishment). Having to do with nutrition or nourishment.

allantois (ə-lan′tois) (Gr. *allas*, sausage, + *eidos*, form). One of the extraembryonic membranes of the amniotes that functions in respiration and excretion in birds and reptiles and plays an important role in the development of the placenta in most mammals.

allele (ə-lēl′) (Gr. *allēlōn*, of one another). Alternative forms of genes coding for the same trait; situated at the same locus in homologous chromosomes.

allograft (a′lō-graft) (Gr. *allos*, other, + graft). A piece of tissue or an organ transferred from one individual to another individual of the same species, not identical twins; homograft.

allometry (ə-lom′ə-trē) (Gr. *allos*, other, + *metry*, measure). Relative growth of a part in relation to the whole organism.

allopatric (Gr. *allos*, other, + *patra*, native land). In separate and mutually exclusive geographical regions.

---

bat / āpe / ärmadillo / herring / fēmale / finch / līce / crocodile / crōw / duck / ūnicorn / ə indicates unaccented vowel sound "uh" as in mammal, fishes, cardinal, heron, vulture / stress as in bi-ol′o-gy, bi′o-log′i-cal

alpha-helix (Gr. *alpha,* first, + L. *helix,* spiral). Literally the first spiral arrangement of the genetic DNA molecule; regular coiled arrangement of polypeptide chain in proteins; secondary structure of proteins.

altricial (al-tri′shəl) (L. *altrices,* nourishers). Referring to young animals (especially birds) having the young hatched in an immature, dependent condition.

alula (al′yə-lə) (L. dim. of *ala,* wing). The first digit or thumb of a bird's wing, much reduced in size.

alveolus (al-vē′ə-ləs) (L. dim. of *alveus,* cavity, hollow). A small cavity or pit, such as a microscopic air sac of the lungs, terminal part of an alveolar gland, or bony socket of a tooth.

ambulacra (am′byə-lak′rə) (L. *ambulare,* to walk). In echinoderms, radiating grooves where podia of water-vascular system characteristically project to outside.

amebocyte (ə-mē′bə-sīt) (Gr. *amoibē,* change, + *kytos,* hollow vessel). Cell in metazoan invertebrate, often functioning in defense against invading particles.

ameboid (ə-mē′boid) (Gr. *amoibē,* change, + *oid,* like). Ameba-like in putting forth pseudopodia.

amictic (ə-mik′tic) (Gr. *a,* without, + *miktos,* mixed or blended). Pertaining to female rotifers, which produce only diploid eggs that cannot be fertilized, or to the eggs produced by such females. Compare with **mictic.**

amino acid (ə-mē′nō) (amine, an organic compound). An organic acid with an amino group (—NH$_2$). Makes up the structure of proteins.

amitosis (ā′mī-tō′səs) (Gr. *a,* not, + *mitos,* thread). A form of cell division in which mitotic nuclear changes do not occur; cleavage without separation of daughter chromosomes.

amniocentesis (am′nē-ō-sin-tē′səs) (Gr. *amnion,* membrane around the fetus, + *centes,* puncture). Procedure for withdrawing a sample of fluid around the developing embryo for examination of chromosomes in the embryonic cells and other tests.

amnion (am′nē-än) (Gr. *amnion,* membrane around the fetus). The innermost of the extraembryonic membranes forming a fluid-filled sac around the embryo in amniotes.

amniote (am′nē-ōt). Having an amnion; as a noun, an animal that develops an amnion in embryonic life, that is, reptiles, birds, and mammals.

amphiblastula (am′fə-blas′chə-lə) (Gr. *amphi,* on both sides, + *blastos,* germ, + L. *ula,* small). Free-swimming larval stage of certain marine sponges; blastula-like but with only the cells of the animal pole flagellated; those of the vegetal pole unflagellated.

amphid (am′fəd) (Gr. *amphidea,* anything that is bound around). One of a pair of anterior sense organs in certain nematodes.

amphipathic (am-fi-pa′thək) (Gr. *amphi,* on both sides, + *pathos,* suffering, passion). Adjective to describe a molecule with one part soluble in water (polar) and another part insoluble in water (nonpolar).

amplexus (am-plek′səs) (L. embrace). The copulatory embrace of frogs or toads.

ampulla (am-pūl′ə) (L. flask). Membranous vesicle; dilation at one end of each semicircular canal containing sensory epithelium; muscular vesicle above tube foot in water-vascular system of echinoderms.

amylase (am′ə-lās′) (L. *amylum,* starch, + *ase,* suffix meaning enzyme). An enzyme that breaks down starch into smaller units.

anabolism (ə-na′bə-li′zəm) (Gr. *ana,* up, + *bol,* to throw, + *ism,* suffix meaning state of condition). Constructive metabolism.

anadromous (an-ad′rə-məs) (Gr. *anadromos,* running upward). Refers to fishes that migrate up streams from the sea to spawn.

anaerobic (an′ə-rō′bik) (Gr. *an,* not, + *aēr,* air, + *bios,* life). Not dependent on oxygen for respiration.

analogy (L. *analogus,* ratio). Similarity of function but not of origin.

anaphylaxis (an′ə-fə-lax′əs) (Gr. *ana-,* up, + *phylax,* guard). A systemic (whole body) immediate hypersensitivity reaction.

anapsid (ə-nap′səd) (Gr. *an,* without, + *apsis,* arch). Amniotes in which the skull lacks temporal openings, with turtles the only living representatives.

anastomosis (ə-nas′tə-mō′səs) (Gr. *ana,* again, + *stoma,* mouth). A union of two or more blood vessels, fibers, or other structures to form a branching network.

androgen (an′drə-jən) (Gr. *anēr, andros,* man, + *genēs,* born). Any of a group of vertebrate male sex hormones.

androgenic gland (an′drō-jen′ək) (Gr. *anēr,* male, + *gennaein,* to produce). Gland in Crustacea that causes development of male characteristics.

aneuploidy (an′ū-ploid′ē) (Gr. *an,* without, not, + *eu,* good, well, + *ploid,* multiple of). Loss or gain of a chromosome, cells of the organism have one fewer than normal chromosome number, or one extra chromosome, for example, trisomy 21 (Down syndrome).

angiotensin (an′jē-o-ten′sən) (Gr. *angeion,* vessel, + L. *tensio,* to stretch). Blood protein formed from the interaction of renin and a liver protein, causing increased blood pressure and stimulating release of aldosterone and ADH.

Angstrom (after Ångström, Swedish physicist). A unit of one ten-millionth of a millimeter (one ten-thousandth of a micrometer); it is represented by the symbol Å.

anhydrase (an-hī′drās) (Gr. *an,* not, + *hydōr,* water, + *ase,* enzyme suffix). An enzyme involved in the removal of water from a compound. Carbonic anhydrase promotes the conversion of carbonic acid into water and carbon dioxide.

anisogametes (an′īs-ō-gam′ēts) (Gr. *anisos,* unequal, + *gametēs,* spouse). Gametes of a species that differ in form or size.

anlage (än′lä-gə) (Ger. laying out, foundation). Rudimentary form; primordium.

annulus (an′yəl-əs) (L. ring). Any ringlike structure, such as superficial rings on leeches.

antenna (L. sail yard). A sensory appendage on the head of arthropods, or the second pair of the two such pairs of structures in crustaceans.

antennal gland. Excretory gland of Crustacea located in the antennal metamere.

anterior (L. comparative of *ante,* before). The head end of an organism, or (as an adjective) toward that end.

anthracosaurs (an-thrak′ə-sors) (Gr. *anthrax,* coal, carbon, + *sauros,* lizard). A group of Paleozoic labyrinthodont amphibians.

antibodies (an′tē-bod′ēz). Proteins (immunoglobulins) in cell surfaces and dissolved in blood, capable of combining with the antigens that stimulated their production.

anticodon (an′tī-kō′don). A sequence of three nucleotides in transfer RNA that is complementary to a codon in messenger RNA.

antigen (an′ti-jən). Any substance capable of stimulating an immune response, most often a protein.

aperture (ap′ər-chər) (L. *apertura* from *aperire,* to uncover). An opening; the opening into the first whorl of a gastropod shell.

apex (ā′peks) (L. summit). Highest or uppermost point; the lower pointed end of the heart.

apical (ā′pə-kl) (L. *apex,* tip). Pertaining to the tip or apex.

apical complex. A certain combination of organelles found in the protozoan phylum Apicomplexa.

apocrine (ap′ə-krən) (Gr. *apo,* away, + *krinein,* to separate). Applies to a type of mammalian sweat gland that produces a viscous secretion by breaking off a part of the cytoplasm of secreting cells.

apoptosis (a′pə-tō′səs) (Gr. *apo-,* prefix meaning away from, + *ptōsis,* a falling). Genetically determined cell death, "programmed" cell death.

bat / āpe / ärmadillo / herring / fēmale / finch / līce / crocodile / crōw / duck / ūnicorn / ə indicates unaccented vowel sound "uh" as in mammal, fishes, cardinal, heron, vulture / stress as in bi-ol′o-gy, bi′o-log′i-cal

apopyle (ap'ə-pīl) (Gr. *apo,* away from, + *pylē,* gate). In sponges, opening of the radial canal into the spongocoel.

appendicular (L. *ad,* to, + *pendere,* to hang). Pertaining to appendages; pertaining to vermiform appendix.

arboreal (är-bōr'ē-al) (L. *arbor,* tree). Living in trees.

archaeocytes (ärk'ē-ō-sites) (Gr. *archaios,* beginning, + *kytos,* hollow vessel). Ameboid cells of varied function in sponges.

archenteron (ärk-en'tə-rän) (Gr. *archē,* beginning, + *enteron,* gut). The main cavity of an embryo in the gastrula stage; it is lined with endoderm and represents the future digestive cavity.

archinephros (ärk'ē-nəf'rōs) (Gr. *archaois,* ancient, + *nephros,* kidney). Ancestral vertebrate kidney, existing today only in the embryo of hagfishes.

archosaur (är'kə-sor) (Gr. *archōn,* ruling, + *sauros,* lizard). Advanced diapsid vertebrates, a group that includes the living crocodiles and the extinct pterosaurs and dinosaurs.

areolar (a-rē'ə-ler) (L. *areola,* small space). A small area, such as spaces between fibers of connective tissue.

arginine phosphate. Phosphate storage compound (phosphagen) found in many invertebrates and used to regenerate stores of ATP.

Aristotle's lantern. Masticating apparatus of some sea urchins.

arteriole (är-tir'ē-ōl) (L. *arteria,* artery). A small arterial branch that delivers blood to a capillary network.

artery (ärt'ə-rē) (L. *arteria,* artery). A blood vessel that carries blood away from the heart and toward a peripheral cavity.

artiodactyl (är'ti-o-dak'təl) (Gr. *artios,* even, + *daktylos,* toe). One of an order of mammals with two or four digits on each foot.

asconoid (Gr. *askos,* bladder). Simplest form of sponges, with canals leading directly from the outside to the interior.

asexual. Without distinct sexual organs; not involving formation of gametes.

assimilation (L. *assimilatio,* bringing into conformity). Absorption and building up of digested nutriments into complex organic protoplasmic materials.

atherosclerosis (a'thə-rō-sklə-rō'səs) (Gr. *athērōma,* tumor full of gruel-like material, + *sklērōs,* hard). Disease characterized by fatty plaques forming in the inner lining of arteries.

atoke (ā'tōk) (Gr. *a,* without, + *tokos,* offspring). Anterior, nonreproductive part of a marine polychaete, as distinct from the posterior, reproductive part (epitoke) during the breeding season.

atoll (ə-tol') (Maldivian, *atolu*). A coral reef or island surrounding a lagoon.

atom. The smallest unit of an element, composed of a dense nucleus of protons and (usually) neutrons surrounded by a system of electrons.

ATP. Adenosine triphosphate. In biochemistry, an ester of adenosine and triphosphoric acid.

atrium (ā'trē-əm) (L. *atrium,* vestibule). One of the chambers of the heart; also, the tympanic cavity of the ear; also, the large cavity containing the pharynx in tunicates and cephalochordates.

auricle (aw'ri-kl) (L. *auricula,* dim. of *auris,* ear). One of the less muscular chambers of the heart; atrium; the external ear, or pinna; any earlike lobe or process.

auricularia (ə-rik'u-lar'e-ə) (L. *auricula,* a small ear). A type of larva found in Holothuroidea.

autogamy (aw-täg'ə-me) (Gr. *autos,* self, + *gamos,* marriage). Condition in which the gametic nuclei produced by meiosis fuse within the same organism that produced them to restore the diploid number.

autosome (aw'tə-sōm) (Gr. *autos,* self, + *sōma,* body). Any chromosome that is not a sex chromosome.

autotomy (aw-täd'ə-mē) (Gr. *autos,* self, + *tomos,* a cutting). The breaking off of a part of the body by the organism itself.

autotroph (aw'tə-trōf) (Gr. *autos,* self, + *trophos,* feeder). An organism that makes its organic nutrients from inorganic raw materials.

autotrophic nutrition (Gr. *autos,* self, + *trophia,* denoting nutrition). Nutrition characterized by the ability to use simple inorganic substances for the synthesis of more complex organic compounds, as in green plants and some bacteria.

avicularium (L. *avicula,* small bird, + *aria,* like or connected with). Modified zooid that is attached to the surface of the major zooid in Ectoprocta and resembles a bird's beak.

axial (L. *axis,* axle). Relating to the axis, or stem; on or along the axis.

axocoel (ak'sə-cēl) (Gr. *axon,* an axle, + *koilos,* hollow). The most anterior of three coelomic spaces that appear during larval echinoderm development.

axolotl (ak'sə-lot'l) (Nahuatl, *atl,* water, + *xolotl,* doll, servant, spirit). Larval stage of any of several species of the genus *Ambystoma* (such as *Ambystoma tigrinum*) exhibiting neotenic reproduction.

axon (ak'sän) (Gr. *axōn*). Elongate extension of a neuron that conducts impulses away from the cell body and toward the synaptic terminals.

axoneme (aks'ə-nēm) (L. *axis,* axle + Gr. *nēma,* thread). The microtubules in a cilium or flagellum, usually arranged as a circlet of nine pairs enclosing one central pair; also, the microtubules of an axopodium.

axopodium (ak'sə-pō'di-um) (Gr. *axon,* an axis, + *podion,* small foot). Long, slender, more or less permanent pseudopodium found in certain sarcodine protozoa. (Also **axopod.**)

B

B cell. A type of lymphocyte that is most important in the humoral immune response.

barrier reef. A coral reef that runs approximately parallel to the shore and is separated from the shore by a lagoon.

basal body. Also known as kinetosome and blepharoplast, a cylinder of nine triplets of microtubules found basal to a flagellum or cilium; same structure as a centriole.

base. A molecule that dissociates in solution to produce a hydroxide ion.

basis, basipodite (bā'səs, bā-si'pə-dīt) (Gr. *basis,* base, + *pous, podos,* foot). The distal or second joint of the protopod of a crustacean appendage.

bathypelagic (bath'ə-pe-laj'ik) (Gr. *bathys,* deep, + *pelagos,* open sea). Relating to or inhabiting the deep sea.

benthos (ben'thäs) (Gr. depth of the sea). Organisms that live along the bottom of the seas and lakes; adj., **benthic.** Also, the bottom itself.

bilirubin (bil'ə-ru-bən) (L. *bilis,* bile, + *rubeo,* to be red). A breakdown product of the heme group of hemoglobin, excreted in the bile.

binary fission. A mode of asexual reproduction in which the animal splits into two approximately equal offspring.

biogenesis (bī'ō-jen'ə-səs) (Gr. *bios,* life, + *genesis,* birth). The doctrine that life originates only from preexisting life.

biological species concept. A reproductive community of populations (reproductively isolated from others) that occupies a specific niche in nature.

bioluminescence. Method of light production by living organisms in which usually certain proteins (luciferins), in the presence of oxygen and an enzyme (luciferase), are converted to oxyluciferins with the liberation of light.

biomass (Gr. *bios,* life, + *maza,* lump or mass). The weight of total living organisms or of a species population per unit of area.

biome (bī'ōm) (Gr. *bios,* life, + *ōma,* abstract group suffix). Complex of plant and animal communities characterized by climatic and soil conditions; the largest ecological unit.

biosphere (bī'os,* life, + *sphaira,* globe). That part of earth containing living organisms.

biotic (bī-äd'ik) (Gr. *biōtos,* life, livable). Of or relating to life.

bipinnaria (L. *bi,* double, + *pinna,* wing, + *aria,* like or connected with). Free-swimming, ciliated, bilateral larva of the asteroid echinoderms; develops into the brachiolaria larva.

biramous (bī-rām′əs) (L. *bi*, double, + *ramus*, a branch). Adjective describing appendages with two distinct branches, contrasted with uniramous, unbranched.

bivalent (bī-vāl′ənt) (L. *bi*, double, + *valen*, strength, worth). The pairs of homologous chromosomes at synapsis in the first meiotic division, a tetrad.

blastocoel (blas′tə-sēl) (Gr. *blastos*, germ, + *koilos*, hollow). Cavity of the blastula.

blastocyst (blast′ō-sist) (Gr. *blastos*, germ, + *kystis*, bladder). Mammalian embryo in the blastula stage.

blastomere (Gr. *blastos*, germ, + *meros*, part). An early cleavage cell.

blastopore (Gr. *blastos*, germ, + *poros*, passage, pore). External opening of the archenteron in the gastrula.

blastula (Gr. *blastos*, germ, + L. *ula*, dim.). Early embryological stage of many animals; consists of a hollow mass of cells.

blending. See **polygenic inheritance.**

blepharoplast (blə-fä′rə-plast) (Gr. *blepharon*, eyelid, + *plastos*, formed). See **basal body.**

blood plasma. The liquid, noncellular fraction of blood, including dissolved substances.

blood type. Characteristic of human blood given by the particular antigens on the membranes of the erythrocytes, genetically determined, causing agglutination when incompatible groups are mixed; the blood types are designated A, B, O, AB, Rh negative, Rh positive, and others.

Bohr effect. A characteristic of hemoglobin that causes it to dissociate from oxygen in greater degree at higher concentrations of carbon dioxide.

boreal (bōr′ē-əl) (L. *boreas*, north wind). Relating to a northern biotic area characterized by a predominance of coniferous forests and tundra.

B.P. Before the present.

brachial (brak′ē-əl) (L. *brachium*, forearm). Referring to the arm.

brachiolaria (brak′ē-ō-lār′ē-ə) (L. *brachiola*, little arm, + *aria*, pertaining to). This asteroid larva develops from the bipinnaria larva and has three preoral holdfast processes.

brain hormone. See **ecdysiotropin.**

branchial (brank′ē-əl) (Gr. *branchia*, gills). Referring to gills.

bronchiole (brän′kē-ōl) (Gr. *bronchion*, dim. of *bronchos*, windpipe). Small, thin-walled branch of the bronchus.

bronchus (brän′kəs) pl. **bronchi** (Gr. *bronchos*, windpipe). Either of two primary divisions of the trachea that lead to the right and left lung.

brown fat. Mitochondria-rich, heat-generating adipose tissue of endothermic vertebrates.

buccal (buk′əl) (L. *bucca*, cheek). Referring to the mouth cavity.

budding. Reproduction in which the offspring arises as an outgrowth from the parent and is initially smaller than the parent. Failure of the offspring to separate from the parent leads to colony formation.

buffer. Any substance or chemical compound that tends to keep pH levels constant when acids or bases are added.

bursa pl. **bursae** (M.L. *bursa*, pouch, purse made of skin). A sac-like cavity. In ophiuroid echinoderms, pouches opening at bases of arms and functioning in respiration and reproduction (genitorespiratory bursae).

C

calciferous glands (kal-si′fə-rəs). Glands in an earthworm that secrete calcium ions into the gut.

calorie (kal′ə-rē) (L. *calere*, to be warm). Unit of heat defined as the amount of heat required to heat 1 g of water from 14.5 to 15.5° C; 1 cal = 4.184 joules in the International System of Units.

calyx (kā′liks) (L. bud cup of a flower). Any of various cup-shaped zoological structures.

cancellous (kan′səl-əs) (L. *cancelli*, latticework, + *osus*, full of). Having a spongy or porous structure.

capitulum (ka-pi′tə-ləm) (L. small head). Term applied to small, headlike structures of various organisms, including projection from body of ticks and mites carrying mouthparts.

captacula (kap-tak′ū-lə) (L. *captare*, to lie in wait for). Tentacles extending from head of scaphopod molluscs, used in feeding.

carapace (kar′ə-pās) (F. from Sp. *carapacho*, shell). Shieldlike plate covering the cephalothorax of certain crustaceans; dorsal part of the shell of a turtle.

carbohydrate (L. *carbo*, charcoal, + Gr. *hydōr*, water). Compounds of carbon, hydrogen, and oxygen having the generalized formula $(CH_2O)_n$; aldehyde or ketone derivatives of polyhydric alcohols, with hydrogen and oxygen atoms attached in a 2:1 ratio.

carboxyl (kär-bäk′səl) (carbon + oxygen + yl, chemical radical suffix). The acid group of organic molecules —COOH.

cardiac (kär′dē-ak) (Gr. *kardia*, heart). Belonging or relating to the heart.

carinate (kar′ə-nāt) (L. *carina*, keel). Having a keel, in particular the flying birds with a keeled sternum for the insertion of flight muscles.

carnivore (kar′nə-vōr′) (L. *carnivorous*, flesh eating). One of the flesh-eating mammals of the order Carnivora. Also, any organism that eats animals. Adj., **carnivorous.**

carotene (kär′ə-tēn) (L. *carota*, carrot, + *ene*, unsaturated straight-chain hydrocarbons). A red, orange, or yellow pigment belonging to the group of carotenoids; precursor of vitamin A.

carrying capacity. The maximum number of individuals that can persist under specified environmental conditions.

cartilage (L. *cartilago;* akin to L. *cratis*, wickerwork). A translucent elastic tissue that makes up most of the skeleton of embryos, very young vertebrates, and adult cartilaginous fishes, such as sharks and rays; in higher forms much of it is converted into bone.

caste (kast) (L. *castus*, pure, separated). One of the polymorphic forms within an insect society, each caste having its specific duties, as queen, worker, soldier, and so on.

catabolism (Gr. *kata*, downward, + *bol*, to throw, + *ism*, suffix meaning state of condition). Destructive metabolism; process in which complex molecules are reduced to simpler ones.

catadromous (kə-tad′rə-məs) (Gr. *kata*, down, + *dromos*, a running). Refers to fishes that migrate from fresh water to the ocean to spawn.

catalyst (kad′ə-ləst) (Gr. *kata*, down, + *lysis*, a loosening). A substance that accelerates a chemical reaction but does not become a part of the end product.

caudal (käd′l) (L. *cauda*, tail). Constituting, belonging to, or relating to a tail.

caveolae (ka-vē′ə-lē) (L. *cavea*, a cave, + dim. suffix). The invaginated vesicles and pits in potocytosis.

cDNA. See **complementary DNA.**

cecum, caecum (sē′kəm) (L. *caecus*, blind). A blind pouch at the beginning of the large intestine; any similar pouch.

cell-mediated immune response. Immune response involving cell surfaces only, not antibody production, specifically the T_H1 arm of the immune response. Contrast **humoral immune response.**

cellulose (sel′ū-lōs) (L. *cella*, small room). Chief polysaccharide constituent of the cell wall of green plants and some fungi; an insoluble carbohydrate $(C_6H_{10}O_5)_n$ that is converted into glucose by hydrolysis.

centriole (sen′trē-ol) (Gr. *kentron*, center of a circle, + L. *ola*, small). A minute cytoplasmic organelle usually found in the centrosome and considered to be the active division center of the animal cell; organizes spindle fibers during mitosis and meiosis. Same structure as basal body or kinetosome.

centrolecithal (sen′tro-les′ə-thəl) (Gr. *kentron*, center, + *lekithos*, yolk, + Eng. *al*, adjective). Pertaining to an insect egg with the yolk concentrated in the center.

centromere (sen′trə-mir) (Gr. *kentron*, center, + *meros*, part). A localized

---

bat / āpe / ärmadillo / herring / fēmale / finch / līce / crocodile / crōw / duck / ūnicorn / ə indicates unaccented vowel sound "uh" as in mammal, fishes, cardinal, heron, vulture / stress as in bi-ol′o-gy, bi′o-log′i-cal

constriction in a characteristic position on a given chromosome, bearing the kinetochore.

centrosome (sen′trə-sōm) (Gr. *kentron,* center, + *sōma,* body). Minute body in cytoplasm of many plant and animal cells that contains one or two centrioles and is the center of dynamic activity in mitosis.

cephalization (sef′ə-li-zā-shən) (Gr. *kephale,* head). The process by which specialization, particularly of the sensory organs and appendages, become localized in the head end of animals.

cephalothorax (sef′ə-lä-thō′raks) (Gr. *kephale,* head, + thorax). A body division found in many Arachnida and higher Crustacea, in which the head is fused with some or all of the thoracic segments.

cercaria (ser-kar′ē-ə) (Gr. *kerkos,* tail, + L. *aria,* like or connected with). Tadpolelike larva of trematodes (flukes).

cervical (sər′və-kəl) (L. *cervix,* neck). Relating to a neck.

character. (kar′ik-tər). A component of phenotype (including specific molecular, morphological, behavioral or other features) used by systematists to diagnose species or higher taxa, or to evaluate phylogenetic relationships among different species or higher taxa, or relationships among populations within a species.

charging. In protein synthesis, a reaction catalyzed by tRNA synthetase, in which an amino acid is attached to its particular tRNA molecule.

chelicera (kə-lis′ə-rə) pl. **chelicerae** (Gr. *chēlē,* claw, + *keras,* horn). One of a pair of the most anterior head appendages on the members of the subphylum Chelicerata.

chelipeds (kēl′ə-peds) (Gr. *chēlē,* claw, + L. *pes,* foot). Pincerlike first pair of legs in most decapod crustaceans; specialized for seizing and crushing.

chemoautotroph (ke′mō-aw′tō-trōf) (Gr. *chemeia,* transmutation, + *autos,* self, + *trophos,* feeder). An organism utilizing inorganic compounds as a source of energy.

chemotaxis (kē′mō-tak′səs) (Gr. *chēmeia,* an infusion, + *taxō > tassō,* to put in order). Orientation movement of cells or organisms in response to a chemical stimulus.

chemotroph (kem′ə-trōf) (Gr. *chēmeia,* an infusion, + *tropē,* to turn). An organism that derives nourishment from inorganic substances without using chlorophyll.

chiasma (kī-az′mə), pl. **chiasmata** (Gr. cross). An intersection or crossing, as of nerves; a connection point between homologous chromatids where crossing over has occurred at synapsis.

chitin (kī′tən) (Fr. *chitine,* from Gr. *chitōn,* tunic). A horny substance that forms part of the cuticle of arthropods and is found sparingly in certain other invertebrates; a nitrogenous polysaccharide insoluble in water, alcohol, dilute acids, and digestive juices of most animals.

chlorocruorin (klō′rō-kroo′ə-rən) (Gr. *chlōros,* light green, + L. *cruor,* blood). A greenish iron-containing respiratory pigment dissolved in the blood plasma of certain marine polychaetes.

chlorogogen cells (klōr′ə-gog-ən) (Gr. *chlōros,* light green, + *agōgos,* a leading, a guide). Modified peritoneal cells, greenish or brownish, clustered around the digestive tract of certain annelids; apparently they aid in elimination of nitrogenous wastes and in food transport.

chlorophyll (klō′rə-fil) (Gr. *chlōros,* light green, + *phyllōn,* leaf). Green pigment found in plants and in some animals; necessary for photosynthesis.

chloroplast (klō′rə-plast) (Gr. *chlōros,* light green, + *plastos,* molded). A plastid containing chlorophyll and usually other pigments, found in cytoplasm of plant cells.

choanocyte (kō-an′ə-sīt) (Gr. *choanē,* funnel, + *kytos,* hollow vessel). One of the flagellate collar cells that line cavities and canals of sponges.

cholinergic (kōl′i-nər′jik) (Gr. *chōle,* bile, + *ergon,* work). Type of nerve fiber that releases acetylcholine from axon terminal.

chorion (kō′rē-on) (Gr. *chorion,* skin). The outer of the double membrane that surrounds the embryo of reptiles, birds, and mammals; in mammals it contributes to the placenta.

choroid (kōr′oid) (Gr. *chorion,* skin, + *eidos,* form). Delicate, highly vascular membrane; in vertebrate eye; the layer between the retina and sclera.

chromatid (krō′mə-tid) (Gr. *chromato,* from *chrōma,* color, + L. *id,* feminine stem for particle of specified kind). A replicated chromosome joined to its sister chromatid by the centromere; separates and becomes daughter chromosome at anaphase of mitosis or anaphase of the second meiotic division.

chromatin (krō′mə-tin) (Gr. *chrōma,* color). The nucleoprotein material of a chromosome; the hereditary material containing DNA.

chromatophore (krō-mat′ə-fōr) (Gr. *chrōma,* color, + *pherein,* to bear). Pigment cell, usually in the dermis, in which usually the pigment can be dispersed or concentrated.

chromomere (krō′mō-mir) (Gr. *chrōma,* color, + *meros,* part). One of the chromatin granules of characteristic size on the chromosome; may be identical with a gene or a cluster of genes.

chromonema (krō-mə-nē′mə) (Gr. *chrōma,* color, + *nēma,* thread). A convoluted thread in prophase of mitosis or the central thread in a chromosome.

chromoplast (krō′mə-plast) (Gr. *chrōma,* color, + *plastos,* molded). A plastid containing pigment.

chromosome (krō′mə-sōm) (Gr. *chrōma,* color, + *sōma,* body). A complex body, spherical or rod shaped, that arises from the nuclear network during mitosis, splits longitudinally, and carries a part of the organism's genetic information as genes composed of DNA.

chrysalis (kris′ə-lis) (L. from Gr. *chrysos,* gold). The pupal stage of a butterfly.

chyme (kīm) (Gr. *chymos,* juice). Semifluid mass of partly digested food in stomach and small intestine as digestion proceeds.

cilium (sil′i-əm), pl. **cilia** (L. eyelid). A hairlike, vibratile organelle process found on many animal cells. Cilia may be used in moving particles along the cell surface or, in ciliate protozoans, for locomotion.

cinclides (sing′klid-əs), sing. **cinclis** (sing′kləs) (Gr. *kinklis,* latticed gate or partition). Small pores in the external body wall of sea anemones for extrusion of acontia.

circadian (sər-kād′ē-ən) (L. *circa,* around, + *dies,* day). Occurring at a period of approximately 24 hours.

cirrus (sir′əs) (L. curl). A hairlike tuft on an insect appendage; locomotor organelle of fused cilia; male copulatory organ of some invertebrates.

cisternae (sis-ter′nē) (L. *cista,* box). Space between membranes of the endoplasmic reticulum within cells.

cistron (sis′trən) (L. *cista,* box). A series of codons in DNA that code for an entire polypeptide chain.

clade (klād) (Gr. *klados,* branch). A taxon or other group consisting of an ancestral species and all of its descendants, forming a distinct branch on a phylogenetic tree.

cladistics (klad-is′-təks) (Gr. *klados,* branch, sprout). A system of arranging taxa by analysis of evolutionarily derived characteristics so that the arrangement will reflect phylogenetic relationships.

cladogram (klād′ə-gr m) (Gr. *klados,* branch, + *gramma,* letter). A branching diagram showing the pattern of sharing of evolutionarily derive 1 characters among species or higher taxa.

clathrin (kla′thrən) (L. *chathri,* latticework). A protein forming a lattice structure lining the invaginated pits during receptor-mediated endocytosis.

cleavage (O.E. *cleofan,* to cut). Process of nuclear and cell division in animal zygote.

climax (klī′maks) (Gr. *klimax,* ladder). Stage of relative stability attained by a community of organisms, often the culminating development of a natural succession. Also, orgasm.

climax community (Gr. *klimax,* ladder, staircase, climax). A self-perpetuating, more-or-less stable community of organisms that continues as long as environmental conditions under which it developed prevail.

cline (klīn) (Gr. *klinein,* slope, bend). A pattern of directional, often gradual, change in phenotype or genotype of populations across a geographic transect.

clitellum (klī-tel′əm) (L. *clitellae*, packsaddle). Thickened saddlelike portion of certain midbody segments of many oligochaetes and leeches.

cloaca (klō-ā′kə) (L. sewer). Posterior chamber of digestive tract in many vertebrates, receiving feces and urogenital products. In certain invertebrates, a terminal portion of digestive tract that serves also as respiratory, excretory, or reproductive tract.

clone (klōn) (Gr. *klōn*, twig). All descendants derived by asexual reproduction from a single individual.

cnidoblast (nī′də-blast) (Gr. *knidē*, nettle, + *blastos*, germ). See **cnidocyte.**

cnidocil (nī′də-sil) (Gr. *knidē*, nettle, + L. *cilium*, hair). Modified cilium on nematocyst-bearing cnidocytes in cnidarians; triggers nematocyst.

cnidocyte (nī′də-sīt) (Gr. *knidē*, nettle, + *kytos*, hollow vessel). Modified interstitial cell that holds the nematocyst; during development of the nematocyst, the cnidocyte is a cnidoblast.

coacervate (kō′ə-sər′vət) (L. *coacervatus*, to heap up). An aggregate of colloidal droplets held together by electrostatic forces.

coagulation (kō-ag′ū-lā-shən). Process in which a series of enzymes are activated, resulting in clotting of blood.

cochlea (kōk′lēə) (L. snail, from Gr. *kochlos*, a shellfish). A tubular cavity of the inner ear containing the essential organs of hearing; occurs in crocodiles, birds, and mammals; spirally coiled in mammals.

cocoon (kə-kun′) (Fr. *cocon*, shell). Protective covering of a resting or developmental stage, sometimes used to refer to both the covering and its contents; for example, the cocoon of a moth or the protective covering for the developing embryos in some annelids.

codominance. See **intermediate inheritance.**

codon (kō′dän) (L. code, + on). In messenger RNA a sequence of three adjacent nucleotides that codes for one amino acid.

coelenteron (sē-len′tər-on) (Gr. *koilos*, hollow, + *enteron*, intestine). Internal cavity of a cnidarian; gastrovascular cavity; archenteron.

coelom (sē′lōm) (Gr. *koilōma*, cavity). The body cavity in triploblastic animals, lined with mesodermal peritoneum.

coelomocyte (sē′lō′mə-sīt) (Gr. *koilōma*, cavity, + *kytos*, hollow vessel). Another name for amebocyte; primitive or undifferentiated cell of the coelom and the water-vascular system.

coelomoduct (sē-lō′mə-dukt) (Gr. *koilos*, hollow, + L. *ductus*, a leading). A duct that carries gametes or excretory products (or both) from the coelom to the exterior.

coenecium, coenoecium (sə-nēs[h]′ē-əm) (Gr. *koinos*, common, + *oikion*, house). The common secreted investment of an ectoproct colony; may be chitinous, gelatinous, or calcareous.

coenenchyme (sēn′ən-kīm) (Gr. *koinos*, shared in common, + *enchyma*, something poured in). Extensive mesogleal tissue between the polyps of an alcyonarian (phylum Cnidaria) colony.

coenocytic (sē-nə-sit′ik) (Gr. *koinos*, common, + *kytos*, hollow vessel). A tissue in which the nuclei are not separated by cell membranes; syncytial.

coenosarc (sē′nə-särk) (Gr. *koinos*, shared in common, + *sarkos*, flesh). The inner, living part of hydrocauli in hydroids.

coenzyme (kō-en′zīm) (L. prefix, *co*, with, + Gr. *enzymos*, leavened, from *en*, in, + *zymē*, leaven). A required substance in the activation of an enzyme; a prosthetic or nonprotein constituent of an enzyme.

collagen (käl′ə-jən) (Gr. *kolla*, glue, + *genos*, descent). A tough, fibrous protein occurring in vertebrates as the chief constituent of collagenous connective tissue; also occurs in invertebrates, for example, the cuticle of nematodes.

collenchyme (käl′ən-kīm) (Gr. *kolla*, glue, + *enchyma*, infusion). A gelatinous mesenchyme containing undifferentiated cells; found in cnidarians and ctenophores.

collencyte (käl′lən-sīt) (Gr. *kolla*, glue, + *en*, in, + *kytos*, hollow vessel). A type of cell in sponges that is star shaped and apparently contractile.

colloblast (käl′ə-blast) (Gr. *kolla*, glue, + *blastos*, germ). A glue-secreting cell on the tentacles of ctenophores.

colloid (kä′loid) (Gr. *kolla*, glue, + *eidos*, form). A two-phase system in which particles of one phase are suspended in the second phase.

columella (kä′lə-mel′ə) (L. small column). Central pillar in gastropod shells.

comb plate. One of the plates of fused cilia that are arranged in rows for ctenophore locomotion.

commensalism (kə-men′səl-iz′əm) (L. *cum*, together with, + *mensa*, table). A relationship in which one individual lives close to or on another and benefits, and the host is unaffected; often symbiotic.

community (L. *communitas*, community, fellowship). An assemblage of organisms that are associated in a common environment and interact with each other in a self-sustaining and self-regulating relation.

competition. Some degree of overlap in ecological niches of two populations in the same community, such that both depend on the same food source, shelter, or other resources, and negatively affect each other's survival.

complement. Collective name for a series of enzymes and activators in the blood, some of which may bind to antibody and may lead to rupture of a foreign cell.

complementary DNA (cDNA). DNA prepared by transcribing the base sequence from mRNA into DNA by reverse transcriptase; also called **copy DNA.**

compound. A substance whose molecules are composed of atoms of two or more elements.

condensation reaction. A chemical reaction in which reactant molecules are combined by the removal of a water molecule (a hydrogen from one and a hydroxyl from the other reactant).

condyle (kän′dīl) (Gr. *kondylos*, bump). A process on a bone used for articulation.

conjugation (kon′ju-ga′shən) (L. *conjugare*, to yoke together). Temporary union of two ciliate protozoa while they are exchanging chromatin material and undergoing nuclear phenomena resulting in binary fission. Also, formation of cytoplasmic bridges between bacteria for transfer of plasmids.

conspecific (L. *com*, together, + *species*). A member of the same species.

contractile vacuole. A clear fluid-filled cell vacuole in protozoa and a few lower metazoa; takes up water and releases it to the outside in a cyclical manner, for osmoregulation and some excretion.

control. That part of a scientific experiment to which the experimental variable is not applied but which is similar to the experimental group in all other respects.

coprophagy (kə-prä′fə-jē) (Gr. *kopros*, dung, + *phagein*, to eat). Feeding on dung or excrement as a normal behavior among animals; reinjestion of feces.

copulation (Fr. from L. *copulare*, to couple). Sexual union to facilitate the reception of sperm by the female.

copy DNA. See **complementary DNA.**

coralline algae. Algae that precipitate calcium carbonate in their tissues; important contributors to coral reef mass.

corium (kō′re-um) (L. *corium*, leather). The deep layer of the skin; dermis.

cornea (kor′nē-ə) (L. *corneus*, horny). The outer transparent coat of the eye.

corneum (kor′nē-əm) (L. *corneus*, horny). Epithelial layer of dead, keratinized cells. Stratum corneum.

cornified (kor′nə-fīd) (L. *corneus*, horny). Adjective for conversion of epithelial cells into nonliving, keratinized cells.

corona (kə-rō′nə) (L. crown). Head or upper portion of a structure; ciliated disc on anterior end of rotifers.

corpora allata (kor′pə-rə əl-la′tə) (L. *corpus*, body, + *allatum*, aided). Endocrine glands in insects that produce juvenile hormone.

corpora cardiaca (kor′pə-rə kar-dī′ə-cə) (L. *corpus*, body, + Gr. *kardiakos*, belonging to the heart). Paired organs behind the brain of insects, serve as storage and release organs for brain hormone.

cortex (kor′teks) (L. bark). The outer layer of a structure.

covalent bond. A chemical bond in which electrons are shared between atoms.

coxa, coxopodite (kox′ə, kəx-ä′pə-dīt) (L. *coxa,* hip, + Gr. *pous, podos,* foot). The proximal joint of an insect or arachnid leg; in crustaceans, the proximal joint of the protopod.

creatine phosphate. High-energy phosphate compound found in the muscle of vertebrates and some invertebrates, used to regenerate stores of ATP.

cretin (krēt′n) (Fr. *crétin,* [dialect], fr. L. *christianus,* Christian, to indicate idiots so afflicted were also human). A human with severe mental, somatic, and sexual retardation resulting from hypothyroidism during early stages of development.

crista (kris′ta), pl. **cristae** (L. *crista,* crest). A crest or ridge on a body organ or organelle; a platelike projection formed by the inner membrane of mitochondrion.

crossing over. Exchange of parts of nonsister chromatids at synapsis in the first meiotic division.

cryptobiotic (Gr. *kryptos,* hidden, + *biŏticus,* pertaining to life). Living in concealment; refers to insects and other animals that live in secluded situations, such as underground or in wood; also tardigrades and some nematodes, rotifers, and others that survive harsh environmental conditions by assuming for a time a state of very low metabolism.

ctenidia (te-ni′dē-ə) (Gr. *kteis,* comb). Comb-like structures, especially gills of molluscs; also applied to comb plates of Ctenophora.

ctenoid scales (ten′oid) (Gr. *kteis, ktenos,* comb). Thin, overlapping dermal scales of the more advanced fishes; exposed posterior margins have fine, toothlike spines.

cupula (kū′pū-lə) (L. little tub). Small inverted cup-like structure housing another structure; gelatinous matrix covering hair cells in lateral line and equilibrium organs.

cuticle (kū′ti-kəl) (L. *cutis,* skin). A protective, noncellular, organic layer secreted by the external epithelium (hypodermis) of many invertebrates. In higher animals the term refers to the epidermis or outer skin.

cyanobacteria (sī-an-ō-bak-ter′ē-ə) (Gr. *kyanos,* a dark-blue substance, + *bakterion,* dim. of *baktron,* a staff). Photosynthetic prokaryotes, also called blue-green algae, cyanophytes.

cyanophyte (sī-an′ō-fīt) (Gr. *kyanos,* a dark-blue substance, + *phyton,* plant). A cyanobacterium, blue-green alga.

cyclin. A protein important in the control of the cell division cycle and mitosis.

cycloid scales (sī′-kloid) (Gr. *kyklos,* circle). Thin, overlapping dermal scales of the more primitive fishes; posterior margins are smooth.

cydippid larva (sī-dip′pid) (Gr. *kydippe,* mythological Athenian maiden). Free-swimming larva of most ctenophores; superficially similar to the adult.

cynodonts (sin′ə-dänts) (Gr. *kynodōn,* canine tooth). A group of mammal-like carnivorous synapsids of the Upper Permian and Triassic.

cyrtocyte (ser′tō-sīt) (Gr. *kyrtē,* a fish basket, cage, + *kytos,* hollow vessel). A protonephridial cell with a single flagellum enclosed in a cylinder of cytoplasmic rods.

cystacanth (sis′tə-kanth) (Gr. *kystis,* bladder, pouch, + *akantha,* thorn). Juvenile stage of an acanthocephalan that is infective to the definitive host.

cysticercoid (sis′tə-ser′koəd) (Gr. *kystis,* bladder, + *kerkos,* tail, + *eidos,* form). A type of juvenile tapeworm composed of a solid-bodied cyst containing an invaginated scolex; contrast with **cysticercus.**

cysticercus (sis′tə-ser′kəs) (Gr. *kystis,* bladder, + *kerkos,* tail). A type of juvenile tapeworm in which an invaginated and introverted scolex is contained in a fluid-filled bladder; contrast with **cysticercoid.**

cystid (sis′tid) (Gr. *kystis,* bladder). In an ectoproct, the dead secreted outer parts plus the adherent underlying living layers.

cytochrome (sī′tə-krōm) (Gr. *kytos,* hollow vessel, + *chrŏma,* color). Several iron-containing pigments that serve as electron carriers in aerobic respiration.

cytokine (sī′tə-kīn) (Gr. *kytos,* hollow vessel, + *kinein,* to move). A molecule secreted by an activated or stimulated cell, for example, macrophages, that causes physiological changes in certain other cells.

cytokinesis (sī′tə-kin-ē′sis) (Gr. *kytos,* hollow vessel, + *kinesis,* movement). Division of the cytoplasm of a cell.

cytopharynx (Gr. *kytos,* hollow vessel, + *pharynx,* throat). Short tubular gullet in ciliate protozoa.

cytoplasm (sī′tə-plasm) (Gr. *kytos,* hollow vessel, + *plasma,* mold). The living matter of the cell, excluding the nucleus.

cytoproct (sī′tə-prokt) (Gr. *kytos,* hollow vessel, + *prŏktos,* anus). Site on a protozoan where undigestible matter is expelled.

cytopyge (sī′tə-pīj) (Gr. *kytos,* hollow vessel, + *pyge,* rump or buttocks). In some protozoa, localized site for expulsion of wastes.

cytosol (sī′tə-sol) (Gr. *kytos,* hollow vessel, + L. *sol,* from *solutus,* to loosen). Unstructured portion of the cytoplasm in which the organelles are bathed.

cytosome (sī′tə-sōm) (Gr. *kytos,* hollow vessel, + *sŏma,* body). The cell body inside the plasma membrane.

cytostome (sī′tə-stōm) (Gr. *kytos,* hollow vessel, + *stoma,* mouth). The cell mouth in many protozoa.

cytotoxic T cells (Gr. *kytos,* hollow vessel, + toxin). A special T cell activated during cell-mediated immune responses that recognizes and destroys virus-infected cells.

D

dactylozooid (dak-til′ə-zō-id) (Gr. *dakos,* bite, sting, + *tylos,* knob, + *zŏon,* animal). A polyp of a colonial hydroid specialized for defense or killing food.

Darwinism. Theory of evolution emphasizing common descent of all living organisms, gradual change, multiplication of species and natural selection.

data sing. **datum** (Gr. *dateomai,* to divide, cut in pieces). The results in a scientific experiment, or descriptive observations, upon which a conclusion is based.

deciduous (də-sij′ə-wəs) (L. *decidere,* to fall off). Shed or falling off at end of a growing period.

deduction (L. *deductus,* led apart, split, separated). Reasoning from the general to the particular, that is, from given premises to their necessary conclusion.

definitive host. The host in which sexual reproduction of a symbiont takes place; if no sexual reproduction, then the host in which the symbiont becomes mature and reproduces; contrast **intermediate host.**

delayed type hypersensitivity. Inflammatory reaction based primarily on cell-mediated immunity.

deme (dēm) (Gr. populace). A local population of closely related animals.

demography (də-mäg′rə-fē) (Gr. *demos,* people, + *graphy*). The properties of the rate of growth and the age structure of populations.

dendrite (den′drīt) (Gr. *dendron,* tree). Any of nerve cell processes that conduct impulses toward the cell body.

deoxyribonucleic acid (DNA). The genetic material of all organisms, characteristically organized into linear sequences of genes.

deoxyribose (dē-ok′sē-rī′bōs) (L. *deoxy,* loss of oxygen, + *ribose,* a pentose sugar). A 5-carbon sugar having 1 oxygen atom less than ribose; a component of deoxyribonucleic acid (DNA).

dermal (Gr. *derma,* skin). Pertaining to the skin; cutaneous.

dermis. The inner, sensitive mesodermal layer of skin; corium.

desmosome (dez′mə-sōm) (Gr. *desmos,* bond, + *sŏma,* body). Buttonlike plaque serving as an intercellular connection.

determinate cleavage. The type of cleavage, usually spiral, in which the fate of the blastomeres is determined very early in development; mosaic cleavage.

detritus (də-trī′tus) (L. that which is rubbed or worn away). Any fine particulate debris of organic or inorganic origin.

Deuterostomia (dū'də-rō-stō'mē-ə) (Gr. *deuteros,* second, secondary, + *stoma,* mouth). A group of higher phyla in which cleavage is indeterminate (regulative) and primitively radial. The endomesoderm is enterocoelous, and the mouth is derived away from the blastopore. Includes Echinodermata, Chordata, and a number of minor phyla. Compare with Protostomia.

dextral (dex'trəl) (L. *dexter,* right-handed). Pertaining to the right; in gastropods, shell is dextral if opening is to right of columella when held with spire up and facing observer.

diapause (dī'ə-pawz) (Gr. *diapausis,* pause). A period of arrested development in the life cycle of insects and certain other animals in which physiological activity is very low and the animal is highly resistant to unfavorable external conditions.

diapsids (dī-ap'səds) (Gr. *di,* two, + *apsis,* arch). Amniotes in which the skull bears two pairs of temporal openings; includes reptiles (except turtles) and birds.

diastole (dī-as'tə-lē) (Gr. *diastolē,* dilation). Passive relaxation and expansion of the heart during which the chambers are filled with blood.

diffusion (L. *diffusus,* dispersion). The movement of particles or molecules from area of high concentration of the particles or molecules to area of lower concentration.

digitigrade (dij'ə-də-grād) (L. *digitus,* finger, toe, + *gradus,* step, degree). Walking on the digits with the posterior part of the foot raised; compare plantigrade.

dihybrid (dī-hī'brəd) (Gr. *dis,* twice, + L. *hibrida,* mixed offspring). A hybrid whose parents differ in two distinct characters; an offspring having two different alleles at two different loci, for example, *A/a B/b.*

dimorphism (dī-mor'fizm) (Gr. *di,* two, + *morphē,* form). Existence within a species of two distinct forms according to color, sex, size, organ structure, and so on. Occurrence of two kinds of zooids in a colonial organism.

dioecious (dī-ē'shəs) (Gr. *di,* two, + *oikos,* house). Having male and female organs in separate individuals.

diphycercal (dif'i-ser'kəl) (Gr. *diphyēs,* two-fold, + *kerkos,* tail). A tail that tapers to a point, as in lungfishes; vertebral column extends to tip without upturning.

diphyodont (di'fi-ə-dänt) (Gr. *diphyēs,* two-fold, + *odous,* tooth). Having deciduous and permanent sets of teeth successively.

diploblastic (di'plə-blas'tək) (Gr. *diploos,* double, + *blastos,* bud). Organism with two germ layers, endoderm and ectoderm.

diploid (dip'loid) (Gr. *diploos,* double, + *eidos,* form). Having the somatic (double, or 2N) number of chromosomes or twice the number characteristic of a gamete of a given species.

disaccharides (dī-sak'ə-rīds) (Gr. *dis,* twice, + L. *saccharum,* sugar). A class of sugars (such as lactose, maltose, and sucrose) that yield two monosaccharides on hydrolysis.

distal (dis'təl). Farther from the center of the body than a reference point.

DNA. See **deoxyribonucleic acid.**

dominance hierarchy. A social ranking, formed through agonistic behavior, in which individuals are associated with each other so that some have greater access to resources than do others.

dominant. An allele that is expressed regardless of the nature of the corresponding allele on the homologous chromosome.

dorsal (dor'səl) (L. *dorsum,* back). Toward the back, or upper surface, of an animal.

Down syndrome. A congenital syndrome including mental retardation, caused by the cells in a person's body having an extra chromosome 21; also called trisomy 21.

dual-gland adhesive organ. Organs in the epidermis of most turbellarians, with three cell types; viscid and releasing gland cells and anchor cells.

duodenum (dū-ə-dēn'əm) (L. *duodeni,* twelve each, fr. its length, about 12 fingers' width). The first and shortest portion of the small intestine lying between the pyloric end of the stomach and the jejunum.

dyad (dī'əd) (Gr. *dyas,* two). One of the groups of two chromosomes formed by the division of a tetrad during the first meiotic division.

E

eccrine (ek'rən) (Gr. *ek,* out of, + *krinein,* to separate). Applies to a type of mammalian sweat gland that produces a watery secretion.

ecdysiotropin (ek-dē-zē-o-tro'pən) (Gr. *ekdysis,* to strip off, escape, + *tropos,* a turn, change). Hormone secreted in brain of insects that stimulates prothoracic gland to secrete molting hormone. Prothoracicotropic hormone; brain hormone.

ecdysis (ek'də-sis) (Gr. *ekdysis,* to strip off, escape). Shedding of outer cuticular layer; molting, as in insects or crustaceans.

ecdysone (ek-dī'sōn) (Gr. *ekdysis,* to strip off). Molting hormone of arthropods, stimulates growth and ecdysis, produced by prothoracic glands in insects and Y organs in crustaceans.

ecocline (ek'ō-klīn) (Gr. *oikos,* home, + *klino,* to slope, recline). The gradient between adjacent biomes; a gradient of environmental conditions.

ecology (Gr. *oikos,* house, + *logos,* discourse). Part of biology that deals with the relationship between organisms and their environment.

ecosystem (ek'ō-sis-təm) (eco[logy] from Gr. *oikos,* house, + system). An ecological unit consisting of both the biotic communities and the nonliving (abiotic) environment, which interact to produce a stable system.

ecotone (ek'ō-tōn) (eco[logy] from Gr. *oikos,* home, + *tonos,* stress). The transition zone between two adjacent communities.

ectoderm (ek'tō-derm) (Gr. *ektos,* outside, + *derma,* skin). Outer layer of cells of an early embryo (gastrula stage); one of the germ layers, also sometimes used to include tissues derived from ectoderm.

ectognathous (ek'tə-nā'thəs) (Gr. *ektos,* outside, without, + *gnathos,* jaw). Derived character of most insects; mandibles and maxillae not in pouches.

ectolecithal (ek'tō-les'ə-thəl) (Gr. *ektos,* outside, + *lekithos,* yolk). Yolk for nutrition of the embryo contributed by cells that are separate from the egg cell and are combined with the zygote by envelopment within the eggshell.

ectoneural (ek'tə-nu-rəl) (Gr. *ektos,* outside, without, + *neuron,* nerve). Oral (chief) nervous system in echinoderms.

ectoplasm (ek'tō-plazm) (Gr. *ektos,* outside, + *plasma,* form). The cortex of a cell or that part of cytoplasm just under the cell surface; contrasts with **endoplasm.**

ectothermic (ek'tō-therm'ic) (Gr. *ektos,* outside, + *thermē,* heat). Having a variable body temperature derived from heat acquired from the environment; contrasts with **endothermic.**

edema (ē-dē'mə) (Gr. *oidēma,* swelling). Escape of fluid from blood into interstitial space, causing swelling.

effector (L. *efficere,* bring to pass). An organ, tissue, or cell that becomes active in response to stimulation.

efferent (ef'ə-rənt) (L. *ex,* out, + *ferre,* to bear). Leading or conveying away from some organ, for example, nerve impulses conducted away from the brain, or blood conveyed away from an organ; contrasts with **afferent.**

egestion (ē-jes'chən) (L. *egestus,* to discharge). Act of casting out indigestible or waste matter from the body by any normal route.

electron. A subatomic particle with a negative charge and a mass of 9.1066×10^{-28} gram.

eleocyte (el'ē-ə-sīt) (Gr. *elaion,* oil, + *kytos,* hollow vessel). Fat-containing cells in annelids that originate from the chlorogogen tissue.

elephantiasis (el-ə-fən-tī'ə-səs). Disfiguring condition caused by chronic infection with filarial worms *Wuchereria bancrofti* and *Brugia malayi.*

embryogenesis (em'brē-ō-jen'ə-səs) (Gr. *embryon,* embryo, + *genesis,* origin). The origin and development of the embryo; embryogeny.

---

bat / āpe / ärmadillo / herring / fēmale / finch / līce / crocodile / crōw / duck / ūnicorn / ə indicates unaccented vowel sound "uh" as in mammal, fishes, cardinal, heron, vulture / stress as in bi-ol'o-gy, bi'o-log'i-cal

emergence (L. *e*, out, + *mergere*, to plunge). The appearance of properties in a biological system (at the molecular, cellular, organismal, or species levels) that cannot be deduced from knowledge of the component parts taken separately or in partial combinations; such properties are termed **emergent properties.**

emigrate (L. *emigrare*, to move out). To move *from* one area to another to take up residence.

emulsion (ə-məl′shən) (L. *emulsus*, milked out). A colloidal system in which both phases are liquids.

endemic (en-dem′ik) (Gr. *en*, in, + *demos*, populace). Peculiar to a certain region or country; native to a restricted area; not introduced.

endergonic (en-dər-gän′ik) (Gr. *endon*, within, + *ergon*, work). Used in reference to a chemical reaction that requires energy; energy absorbing.

endite (en′dīt) (Gr. *endon*, within). Medial process on an arthropod limb.

endochondral (en′dō-kän′drōl) (Gr. *endon*, within, + *chondros*, cartilage). Occurring with the substance of cartilage, especially bone formation.

endocrine (en′də-krən) (Gr. *endon*, within, + *krinein*, to separate). Refers to a gland that is without a duct and that releases its product directly into the blood or lymph.

endocytosis (en′dō-sī-tō′səs) (Gr. *endon*, within, + *kytos*, hollow vessel). The engulfment of matter by phagocytosis, potocytosis, receptor-mediated endocytosis, and by bulk-phase (nonspecific) endocytosis.

endoderm (en′də-dərm) (Gr. *endon*, within, + *derma*, skin). Innermost germ layer of an embryo, forming the primitive gut; also may refer to tissues derived from endoderm.

endognathous (en′də-nā-thəs) (Gr. *endon*, within, + *gnathous*, jaw). Ancestral character in insects, found in orders Diplura, Collembola, and Protura, in which the mandibles and maxillae are located in pouches.

endolecithal (en′də-les′ə-thəl) (Gr. *endon*, within, + *lekithos*, yolk). Yolk for nutrition of the embryo incorporated into the egg cell itself.

endolymph (en′də-limf) (Gr. *endon*, within, + *lympha*, water). Fluid that fills most of the membranous labyrinth of the vertebrate ear.

endometrium (en′də-mē′trē-əm) (Gr. *endon*, within, + *mētra*, womb). The mucous membrane lining the uterus.

endoplasm (en′də-pla-zm) (Gr. *endon*, within, + *plasma*, mold or form). The portion of cytoplasm that immediately surrounds the nucleus.

endoplasmic reticulum. A complex of membranes within a cell; may bear ribosomes (rough) or not (smooth).

endopod, endopodite (en′də-päd, en-dop′ə-dīt) (Gr. *endon*, within, + *pous, podos*, foot). Medial branch of a biramous crustacean appendage.

endopterygote (en′dəp-ter′i-gōt) (Gr. *endon*, within, + *pteron*, feather, wing). Insect in which the wing buds develop internally; has holometabolous metamorphosis.

endorphin (en-dor′fin) (contraction of endogenous morphine). Group of opiate-like brain neuropeptides that modulate pain perception and are implicated in many other functions.

endoskeleton (Gr. *endon*, within, + *skeletos*, hard). A skeleton or supporting framework within the living tissues of an organism; contrasts with **exoskeleton.**

endosome (en′də-sōm) (Gr. *endon*, within, + *sōma*, body). Nucleolus in nucleus of some protozoa that retains its identity through mitosis.

endostyle (en′də-stīl) (Gr. *endon*, within, + *stylos*, a pillar). Ciliated groove(s) in the floor of the pharynx of tunicates, cephalochordates, and larval jawless fishes useful for accumulating and moving food particles to the stomach.

endothelium (en-də-thē′lē-əm) (Gr. *endon*, within, + *thēlē*, nipple). Squamous epithelium lining internal body cavities such as heart and blood vessels. Adj., **endothelial.**

endothermic (en′də-therm′ic) (Gr. *endon*, within, + *thermē*, heat). Having a body temperature determined by heat derived from the animal's own oxidative metabolism; contrasts with **ectothermic.**

enkephalin (en-kef′ə-lin) (Gr. *endon*, within, + *kephale*, head). Group of small brain neuropeptides with opiate-like qualities.

enterocoel (en′tər-ō-sēl′) (Gr. *enteron*, gut, + *koilos*, hollow). A type of coelom formed by the outpouching of a mesodermal sac from the endoderm of the primitive gut.

enterocoelic mesoderm formation. Embryonic formation of mesoderm by a pouchlike outfolding from the archenteron, which then expands and obliterates the blastocoel, thus forming a large cavity, the coelom, lined with mesoderm.

enterocoelomate (en′ter-ō-sēl′ō-māte) (Gr. *enteron*, gut, + *koilōma*, cavity, + Eng. *ate*, state of). An animal having an enterocoel, such as an echinoderm or a vertebrate.

enteron (en′tə-rän) (Gr. *intestine*). The digestive cavity.

entomology (en′tə-mol′ə-jē) (Gr. *entoma*, an insect, + *logos*, discourse). Study of insects.

entozoic (en-tə-zō′ic) (Gr. *entos*, within, + *zōon*, animal). Living within another animal; internally parasitic (chiefly parasitic worms).

entropy (en′trə-pē) (Gr. *en*, in, on, + *tropos*, turn, change in manner). A quantity that is the measure of energy in a system not available for doing work.

enzyme (en′zīm) (Gr. *enzymos*, leavened, from *en*, in, + *zyme*, leaven). A substance, produced by living cells, that is capable of speeding up specific chemical transformations, such as hydrolysis, oxidation, or reduction, but is unaltered itself in the process; a biological catalyst.

eocytes (ē′ə-sīts) (Gr. *ēōs*, the dawn, + *kytos*, hollow vessel). A group of prokaryotes currently classified among the Archaebacteria but possibly a sister group of eukaryotes.

ephyra (ef′ə-rə) (Gr. *Ephyra*, Greek city). Refers to castlelike appearance. Medusa bud from a scyphozoan polyp.

epidermis (ep′ə-dər′məs) (Gr. *epi*, on, upon, + *derma*, skin). The outer, nonvascular layer of skin of ectodermal origin; in invertebrates, a single layer of ectodermal epithelium.

epididymis (ep′ə-did′ə-məs) (Gr. *epi*, on, upon, + *didymos*, testicle). Part of the sperm duct that is coiled and lying near the testis.

epigenesis (ep′ə-jen′ə-sis) (Gr. *epi*, on, upon, + *genesis*, birth). The embryological (and generally accepted) view that an embryo is a new creation that develops and differentiates step by step from an initial stage; the progressive production of new parts that were nonexistent as such in the original zygote.

epigenetics (ep′ə-je-net′iks) (Gr. *epi*, on, upon, + *genesis*, birth). Study of the relationship between genotype and phenotype as mediated by developmental processes.

epipod, epipodite (ep′ē-päd, e-pip′ə-dīt) (Gr. *epi*, on, upon, + *pous, podos*, foot). A lateral process on the protopod of a crustacean appendage, often modified as a gill.

epistasis (e-pis′tə-səs) (Gr. *epi*, on, upon, + *stasis*, standing). Prevention of expression of an allele at one locus by an allele at another locus.

epistome (ep′i-stōm) (Gr. *epi*, on, upon, + *stoma*, mouth). Flap over the mouth in some lophophorates bearing the protocoel.

epithelium (ep′ə-thē′lē-əm) (Gr. *epi*, on, upon, + *thēlē*, nipple). A cellular tissue covering a free surface or lining a tube or cavity.

epitoke (ep′ə-tōk) (Gr. *epitokos*, fruitful). Posterior part of a marine polychaete when swollen with developing gonads during the breeding season; contrast with **atoke.**

erythroblastosis fetalis (ə-rith′rə-blas-tō′səs fə-tal′əs) (Gr. *erythros*, red, + *blastos*, germ, + *osis*, a disease; L. *fetalis*, relating to a fetus). A disease of newborn infants caused when Rh-negative mothers develop antibodies against the Rh-positive blood of the fetus. See **blood type.**

erythrocyte (ə-rith′rə-sīt) (Gr. *erythros*, red, + *kytos*, hollow vessel). Red blood cell; has hemoglobin to carry oxygen from lungs or gills to tissues; during formation in mammals, erythrocytes lose their nuclei, those of other vertebrates retain the nuclei.

esthete (es-thēt′) (Gr. *esthēs*, a garment). Light sensory receptor on a shell of a chiton (phylum Mollusca).

estrus (es′trəs) (L. *oestrus*, gadfly, frenzy). The period of heat, or rut, especially of the female during ovulation of the egg. Associated with maximum sexual receptivity.

estuary (es′chə-we′rē) (L. *aestuarium*, estuary). An arm of the sea where the tide meets the current of a freshwater drainage.

ethology (e-thäl′-ə-jē) (Gr. *ethos*, character, + *logos*, discourse). The study of animal behavior in natural environments.

euchromatin (ū′krō-mə-tən) (Gr. *eu*, good, well, + *chrōma*, color). Part of the chromatin that takes up stain less than heterochromatin, contains active genes.

eukaryotic, eucaryotic (ū′ka-rē-ot′ik) (Gr. *eu*, good, true, + *karyon*, nut, kernel). Organisms whose cells characteristically contain a membrane-bound nucleus or nuclei; contrasts with **prokaryotic.**

euploidy (ū′ploid′ē) (Gr. *eu*, good, well, + *ploid*, multiple of). Change in chromosome number from one generation to the next in which there is an addition or deletion of a complete set of chromosomes in the progeny; the most common type is polyploidy.

euryhaline (ū′-rə-hā′līn) (Gr. *eurys*, broad, + *hals*, salt). Able to tolerate wide ranges of saltwater concentrations.

euryphagous (yə-rif′ə-gəs) (Gr. *eurys*, broad, + *phagein*, to eat). Eating a large variety of foods.

eurytopic (ū-rə-täp′ik) (Gr. *eurys*, broad, + *topos*, place). Refers to an organism with a wide environmental range.

eutely (ū′te-lē) (Gr. *euteia*, thrift). Condition of a body composed of a constant number of cells or nuclei in all adult members of a species, as in rotifers, acanthocephalans, and nematodes.

evagination (ē-vaj′ə-nā′shən) (L. *e*, out, + *vagina*, sheath). An outpocketing from a hollow structure.

evolution (L. *evolvere*, to unfold). Organic evolution encompasses all changes in the characteristics and diversity of life on earth throughout its history.

evolutionary duration. The length of time that a species or higher taxon exists in geological time.

evolutionary species concept. A single lineage of ancestral-descendant populations that maintains its identity from other such lineages and has its own evolutionary tendencies and historical fate; differs from the biological species concept by explicitly including a time dimension and including asexual lineages.

evolutionary taxonomy. A system of classification, formalized by George Gaylord Simpson, that groups species into Linnean higher taxa representing a hierarchy of distinct adaptive zones; such taxa may be monophyletic or paraphyletic but not polyphyletic.

excision repair. Means by which cells are able to repair certain kinds of damage (dimerized pyrimidines) in their DNA.

exergonic (ek′sər-gän′ik) (Gr. *exō*, outside of, + *ergon*, work). An energy-yielding reaction.

exite (ex′īt) (Gr. *exō*, outside). Process from lateral side of an arthropod limb.

exocrine (ek′sə-krən) (Gr. *exō*, outside, + *krinein*, to separate). A type of gland that releases its secretion through a duct; contrasts with **endocrine.**

exocytosis (eks′ə-sī-tō′səs) (Gr. *exo*, outside, + *kytos*, hollow vessel). Transport of a substance from inside a cell to the outside.

exon (ex′ən) (Gr. *exō*, outside). Part of the mRNA as transcribed from the DNA that contains a portion of the information necessary for final gene product.

exopod, exopodite (ex′ə-päd, ex-äp′ə-dīt) (Gr. *exō*, outside, + *pous, podos*, foot). Lateral branch of a biramous crustacean appendage.

exopterygote (ek′səp-ter′i-gōt) (Gr. *exō*, without, + *pteron*, feather, wing). Insect in which the wing buds develop externally during nymphal instars; has hemimetabolous metamorphosis.

exoskeleton (ek′sō-skel′ə-tən) (Gr. *exō*, outside, + *skeletos*, hard). A supporting structure secreted by ectoderm or epidermis; external, not enveloped by living tissue, as opposed to **endoskeleton.**

experiment (L. *experiri*, to try). A trial made to support or disprove a hypothesis.

exteroceptor (ek′stər-ō-sep′tər) (L. *exter*, outward, + *capere*, to take). A sense organ excited by stimuli from the external world.

F

facilitated diffusion. Mediated transport in which a permease makes possible diffusion of a molecule across a cell membrane in the direction of a concentration gradient; contrast with **active transport.**

FAD. Abbreviation for flavine adenine dinucleotide, an electron acceptor in the respiratory chain.

fascicle (fas′ə-kəl) (L. *fasciculus*, small bundle). A small bundle, usually referring to a collection of muscle fibers or nerve axons.

fatty acid. Any of a series of saturated organic acids having the general formula $C_nH_{2n}O_2$, occurs in natural fats of animals and plants.

fermentation (L. *fermentum*, ferment). Enzymatic transformation, without oxygen, of organic substrates, especially carbohydrates, yielding products such as alcohols, acids, and carbon dioxide.

fiber (L. *fibra*, thread). A fiberlike cell or strand of protoplasmic material produced or secreted by a cell and lying outside the cell.

fibril (L. *fibra*, thread). A strand of protoplasm produced by a cell and lying within the cell.

fibrillar (fī′brə-lər) (L. *fibrilla*, small fiber). Composed of or pertaining to fibrils or fibers.

fibrin. Protein that forms a meshwork, trapping erythrocytes, to become blood clot. Precursor is fibrinogen.

fibrosis (fī-brō′səs). Deposition of fibrous connective tissue in a localized site, during process of tissue repair or to wall off a source of antigen.

filipodium (fi′li-pō′de-əm) (L. *filum*, thread, + Gr. *pous, podos*, a foot). A type of pseudopodium that is very slender and may branch but does not rejoin to form a mesh.

filter feeding. Any feeding process by which particulate food is filtered from water in which it is suspended.

fission (L. *fissio*, a splitting). Asexual reproduction by a division of the body into two or more parts.

fitness. Degree of adjustment and suitability for a particular environment. Genetic fitness is relative contribution of one genetically distinct organism to the next generation; organisms with high genetic fitness are naturally selected and become prevalent in a population.

flagellum (flə-jel′em) pl. **flagella** (L. a whip). Whiplike organelle of locomotion.

flame cell. Specialized hollow excretory or osmoregulatory structure of one or several small cells containing a tuft of flagella (the "flame") and situated at the end of a minute tubule; connected tubules ultimately open to the outside. See **solenocyte, protonephridium.**

fluke (O.E. *flōc*, flatfish). A member of class Trematoda or class Monogenea. Also, certain of the flatfishes (order Pleuronectiformes).

FMN. Abbreviation for flavin mononucleotide, the prosthetic group of a protein (flavoprotein) and a carrier in the electron transport chain in respiration.

food vacuole. A digestive organelle in the cell.

foraminiferan (for′əm-i-nif′-ər-ən) (L. *foramin*, hole, perforation, + *fero*, to bear). A member of the class Granuloreticulosea (phylum Sarcomastigophora) bearing a test with many openings.

fossil (fos′əl). Any remains or impression of an organism from a past geological age that has been preserved by natural processes, usually by mineralization in the earth's crust.

fossorial (fä-sōr′ē-əl) (L. *fossor*, digger). Characterized by digging or burrowing.

bat / āpe / ärmadillo / herring / fēmale / finch / līce / crocodile / crōw / duck / ūnicorn / ə indicates unaccented vowel sound "uh" as in mammal, fishes, cardinal, heron, vulture / stress as in bi-ol′o-gy, bi′o-log′i-cal

fouling. Contamination of feeding or respiratory areas of an organism by excrement, sediment, or other matter. Also, accumulation of sessile marine organisms on the hull of a boat or ship so as to impede its progress through the water.

founder event. Establishment of a new population by a small number of individuals (sometimes a single female carrying fertile eggs) that disperse from their parental population to a new location geographically isolated from the parental population.

fovea (fō′vē-ə) (L. small pit). A small pit or depression; especially the fovea centralis, a small rodless pit in the retina of some vertebrates, a point of acute vision.

free energy. The energy available for doing work in a chemical system.

frontal plane. A plane parallel to the main axis of the body and at right angles to the sagittal plane.

fusiform (fū′zə-form) (L. *fusus,* spindle, + *forma,* shape). Spindle shaped; tapering toward each end.

G

gamete (ga′mēt, gə-mēt′) (Gr. *gamos,* marriage). A mature haploid sex cell; usually, male and female gametes can be distinguished. An egg or a sperm.

gametic meiosis. Meiosis that occurs during formation of the gametes, as in humans and other metazoa.

gametocyte (gə-mēt′ə-sīt) (Gr. *gametēs,* spouse, + *kytos,* hollow vessel). The mother cell of a gamete, that is, immature gamete.

ganglion (gang′lē-ən) pl. **ganglia** (Gr. little tumor). An aggregation of nerve tissue containing nerve cells.

ganoid scales (ga′noid) (Gr. *ganos,* brightness). Thick, bony, rhombic scales of some primitive bony fishes; not overlapping.

gap junction. An area of tiny canals communicating the cytoplasm between two cells.

gastrodermis (gas′tro-dər′mis) (Gr. *gastēr,* stomach, + *derma,* skin). Lining of the digestive cavity of cnidarians.

gastrolith (gas′trə-lith) (Gr. *gastēr,* stomach, + *lithos,* stone). Calcareous body in the wall of the cardiac stomach of crayfish and other Malacostraca, preceding the molt.

gastrovascular cavity (Gr. *gastēr,* stomach, + L. *vasculum,* small vessel). Body cavity in certain lower invertebrates that functions in both digestion and circulation and has a single opening serving as both mouth and anus.

gastrozooid (gas′trə-zō-id) (Gr. *gastēr,* stomach, + *zōon,* animal). The feeding polyp of a hydroid, a hydranth.

gastrula (gas′trə-lə) (Gr. *gastēr,* stomach, + L. *ula,* dim.). Embryonic stage, usually cap or sac shaped, with walls of two layers of cells surrounding a cavity (archenteron) with one opening (blastopore).

gastrulation (gas′trə-lā′shən) (Gr. *gastēr,* stomach). Process by which an early metazoan embryo becomes a gastrula, acquiring first two and then three layers of cells.

gel (jel) (from gelatin, from L. *gelare,* to freeze). That state of a colloidal system in which the solid particles form the continuous phase and the fluid medium the discontinuous phase.

gemmule (je′mūl) (L. *gemma,* bud, + *ula,* dim.). Asexual, cystlike reproductive unit in freshwater sponges; formed in summer or autumn and capable of overwintering.

gene (Gr. *genos,* descent). A nucleic acid sequence (usually DNA) that encodes a functional polypeptide or RNA sequence.

gene pool. A collection of all of the alleles of all of the genes in a population.

genetic drift. Random change in allelic frequencies in a population occurring by chance. In small populations, genetic variation at a locus may be lost by chance fixation or a single allelic variant.

genome (jē′nōm) (Gr. *genos,* offspring, + *ōma,* abstract group). All the DNA in a haploid set of chromosomes (nuclear genome), organelle (mitochondrial genome, chloroplast genome) or virus (viral genome, which in some viruses consists of RNA rather than DNA).

genotype (jēn′ō-tīp) (Gr. *genos,* offspring, + *typos,* form). The genetic constitution, expressed and latent, of an organism; the total set of genes present in the cells of an organism; contrasts with **phenotype.**

genus (jē-nus), pl. **genera** (L. race). A group of related species with taxonomic rank between family and species.

germ layer. In the animal embryo, one of three basic layers (ectoderm, endoderm, mesoderm) from which the various organs and tissues arise in the multicellular animal.

germ plasm. Cell lineages giving rise to the germ cells of a multicellular organism, as distinct from the somatoplasm.

gestation (jes-tā′shən) (L. *gestare,* to bear). The period in which offspring are carried in the uterus.

globulins (glo′bū-lənz) (L. *globus,* a globe, ball, + *-ulus,* ending denoting tendency). A large group of compact proteins with high molecular weight; includes immunoglobulins (antibodies).

glochidium (glō-kid′e-əm) (Gr. *glochis,* point, + *idion,* dim.). Bivalved larval stage of freshwater mussels.

glomerulus (glä-mer′u-ləs) (L. *glomus,* ball). A tuft of capillaries projecting into a renal corpuscle in a kidney. Also, a small spongy mass of tissue in the proboscis of hemichordates, presumed to have an excretory function. Also, a concentration of nerve fibers situated in the olfactory bulb.

gluconeogenesis (glū-cō-nē-ō-gən′ə-səs) (Gr. *glykys,* sweet, + *neos,* new, + *genesis,* origin). Synthesis of glucose from protein or lipid precursors.

glycogen (glī′kə-jən) (Gr. *glykys,* sweet, + *genēs,* produced). A polysaccharide constituting the principal form in which carbohydrate is stored in animals; animal starch.

glycolysis (glī-kol′ə-səs) (Gr. *glykys,* sweet, + *lysis,* a loosening). Enzymatic breakdown of glucose (especially) or glycogen into phosphate derivatives with release of energy.

gnathobase (nāth′ə-bās′) (Gr. *gnathos,* jaw, + base). A median basic process on certain appendages in some arthropods, usually for biting or crushing food.

gnathostomes (nath′ə-stōms) (Gr. *gnathos,* jaw, + *stoma,* mouth). Vertebrates with jaws.

Golgi complex (gōl′jē) (after Golgi, Italian histologist). An organelle in cells that serves as a collecting and packaging center for secretory products.

gonad (gō′nad) (N.L. *gonas,* primary sex organ). An organ that produces gametes (ovary in the female and testis in the male).

gonangium (gō-nan′jē-əm) (N.L. *gonas,* primary sex organ, + *angeion,* dim. of vessel). Reproductive zooid of hydroid colony (Cnidaria).

gonoduct (Gr. *gonos,* seed, progeny, + duct). Duct leading from a gonad to the exterior.

gonopore (gän′ə-pōr) (Gr. *gonos,* seed, progeny, + *poros,* an opening). A genital pore found in many invertebrates.

grade (L. *gradus,* step). A level of organismal complexity or adaptive zone characteristic of a group of evolutionarily related organisms.

gradualism (graj′ə-wal-iz′əm). A component of Darwin's evolutionary theory postulating that evolution occurs by the temporal accumulation of small, incremental changes, usually across very long periods of geological time; it opposes claims that evolution can occur by large, discontinuous or macromutational changes.

granulocytes (gran′ū-lə-sīts) (L. *granulus,* small grain, + Gr. *kytos,* hollow vessel). White blood cells (neutrophils, eosinophils, and basophils) bearing "granules" (vacuoles) in their cytoplasm that stain deeply.

green gland. Excretory gland of certain Crustacea; the antennal gland.

gregarious (L. *grex,* herd). Living in groups or flocks.

guanine (gwä′nēn) (Sp. from Quechua, *huanu,* dung). A white crystalline purine base, $C_5H_5N_5O$, occurring in various animal tissues and in guano and other animal excrements.

guild (gild) (M.E. *gilde,* payment, tribute). In ecology, a group of species that exploit the same class of environment in a similar way.

gynandromorph (ji-nan′drə-mawrf) (Gr. *gyn*, female, + *andr*, male, + *morphē*, form). An abnormal individual exhibiting characteristics of both sexes in different parts of the body; for example the left side of a bilateral organism may show characteristics of one sex and the right side those of the other sex.

gynocophoric canal (gī′nə-kə-fōr′ik) (Gr. *gynē*, woman, + *pherein*, to carry). Groove in male schistosomes (certain trematodes) that carries the female.

H

habitat (L. *habitare*, to dwell). The place where an organism normally lives or where individuals of a population live.

habituation. A kind of learning in which continued exposure to the same stimulus produces diminishing responses.

halter (hal′tər), pl. **halteres** (hal-ti′rēz) (Gr. leap). In Diptera, small club-shaped structure on each side of the metathorax representing the hindwings; believed to be sense organs for balancing; also called balancer.

haplodiploidy (Gr. *haploos*, single, + *diploos*, double, + *eidos*, form). Reproduction in which haploid males are produced parthenogenetically, and diploid females are from fertilized eggs.

haploid (Gr. *haploos*, single). The reduced, or N, number of chromosomes, typical of gametes, as opposed to the diploid, or 2N, number found in somatic cells. In certain groups, mature organisms may have a haploid number of chromosomes.

Hardy-Weinberg equilibrium. Mathematical demonstration that the Mendelian hereditary process does not change the populational frequencies of alleles or genotypes across generations, and that changes in allelic or genotypic frequencies requires factors such as natural selection, genetic drift in finite populations, recurring mutation, migration of individuals among populations, and nonrandom mating.

hectocotylus (hek-tə-kät′ə-ləs) (Gr. *hekaton*, hundred, + *kotylē*, cup). Specialized, and sometimes autonomous, arm that serves as a male copulatory organ in cephalopods.

hemal system (hē′məl) (Gr. *haima*, blood). System of small vessels in echinoderms; function unknown.

hemerythrin (hē′mə-rith′rin) (Gr. *haima*, blood, + *erythros*, red). A red, iron-containing respiratory pigment found in the blood of some polychaetes, sipunculids, priapulids, and brachiopods.

hemimetabolous (he′mi-mə-ta′bə-ləs) (Gr. *hēmi*, half, + *metabolē*, change). Refers to gradual metamorphosis during development of insects, without a pupal stage.

hemocoel (hēm′ə-sēl) (Gr. *haima*, blood, + *koiloma*, cavity). Major body space in arthropods replacing the coelom, contains the blood (hemolymph).

hemoglobin (Gr. *haima*, blood, + L. *globulus*, globule). An iron-containing respiratory pigment occurring in vertebrate red blood cells and in blood plasma of many invertebrates; a compound of an iron porphyrin heme and globin proteins.

hemolymph (hē′mə-limf) (Gr. *haima*, blood, + L. *lympha*, water). Fluid in the coelom or hemocoel of some invertebrates that represents the blood and lymph of vertebrates.

hemozoin (hē-mə-zo′ən) (Gr. *haima*, blood, + *zōon*, an animal). Insoluble digestion product of malaria parasites produced from hemoglobin.

hepatic (hə-pat′ik) (Gr. *hēpatikos*, of the liver). Pertaining to the liver.

herbivore ([h]ərb′ə-vōr′) (L. *herba*, green crop, + *vorare*, to devour). Any organism subsisting on plants. Adj., **herbivorous.**

heredity (L. *heres*, heir). The faithful transmission of biological traits from parents to their offspring.

hermaphrodite (hə[r]-maf′rə-dīt) (Gr. *hermaphroditos*, containing both sexes; from Greek mythology, Hermaphroditos, son of Hermes and Aphrodite). An organism with both male and female functional reproductive organs. **Hermaphroditism** may refer to an aberration in unisexual animals; **monoecy** implies that this is the normal condition for the species.

hermatypic (hər-mə-ti′pik) (Gr. *herma*, reef, + *typos*, pattern). Relating to reef-forming corals.

heterocercal (het′ər-o-sər′kəl) (Gr. *heteros*, different, + *kerkos*, tail). In some fishes, a tail with the upper lobe larger than the lower, and the end of the vertebral column somewhat upturned in the upper lobe, as in sharks.

heterochromatin (he′tə-rō-krōm′ə-tən) (Gr. *heteros*, different, + *chrōma*, color). Chromatin that stains intensely and appears to represent inactive genetic areas.

heterochrony (hed′ə-rō-krōn-y) (Gr. *heteros*, different, + *chronos*, time). Evolutionary change in the relative time of appearance or rate of development of characteristics from ancestor to descendant.

heterodont (hed′ə-ro-dänt) (Gr. *heteros*, different, + *odous*, tooth). Having teeth differentiated into incisors, canines, and molars for different purposes.

heterotroph (hət′ə-rō-träf) (Gr. *heteros*, different, + *trophos*, feeder). An organism that obtains both organic and inorganic raw materials from the environment in order to live; includes most animals and those plants that do not carry on photosynthesis.

heterozygote (het′ə-rō-zī′gōt) (Gr. *heteros*, different, + *zygōtos*, yoked). An organism in which homologous chromosomes contain different allelic forms (often dominant and recessive) of a locus; derived from a zygote formed by union of gametes of dissimilar allelic constitution.

hexamerous (hek-sam′ər-əs) (Gr. *hex*, six, + *meros*, part). Six parts, specifically, symmetry based on six or multiples thereof.

hibernation (L. *hibernus*, wintry). Condition, especially of mammals, of passing the winter in a torpid state in which the body temperature drops nearly to freezing and the metabolism drops close to zero.

hierarchical system. A scheme arranging organisms into a series of taxa of increasing inclusiveness, as illustrated by Linnean classification.

histogenesis (his-tō-jen′ə-sis) (Gr. *histos*, tissue, + *genesis*, descent). Formation and development of tissue.

histone (hi′stōn) (Gr. *histos*, tissue). Any of several simple proteins found in cell nuclei and complexed at one time or another with DNA. Histones yield a high proportion of basic amino acids on hydrolysis; characteristic of eukaryotes.

holoblastic cleavage (Gr. *holo*, whole, + *blastos*, germ). Complete and approximately equal division of cells in early embryo. Found in mammals, amphioxus, and many aquatic invertebrates that have eggs with a small amount of yolk.

holometabolous (hō′lō-mə-ta′bə-ləs) (Gr. *holo*, complete, + *metabolē*, change). Complete metamorphosis during development.

holophytic nutrition (hōl′ō-fit′ik) (Gr. *holo*, whole, + *phyt*, plant). Occurs in green plants and certain protozoa and involves synthesis of carbohydrates from carbon dioxide and water in the presence of light, chlorophyll, and certain enzymes.

holozoic nutrition (hōl′ō-zō′ik) (Gr. *holo*, whole, + *zoikos*, of animals). Type of nutrition involving ingestion of liquid or solid organic food particles.

homeobox (hō′mē-ō-box) (Gr. *homoios*, like, resembling, + L. *buxus*, boxtree [used in the sense of enclosed, contained]). A highly conserved 180-base pair sequence found in regulatory sequences of protein-coding genes that regulate development.

homeostasis (hō′mē-ō-stā′sis) (Gr. *homeo*, alike, + *stasis*, state or standing). Maintenance of an internal steady state by means of self-regulation.

homeothermic (hō′mē-ō-thər′mik) (Gr. *homeo*, alike, + *thermē*, heat). Having a nearly uniform body temperature, regulated independent of the environmental temperature; "warm blooded."

homeotic genes (hō-mē-ät′ik) (Gr. *homoios*, like, resembling). Genes, identified through mutations, that give developmental identity to specific body segments.

home range. The area over which an animal ranges in its activities. Unlike territories, home ranges are not defended.

hominid (häm′ə-nid) (L. *homo, hominis,* man). A member of the family Hominidae, now represented by one living species, *Homo sapiens.*

hominoid (häm′ə-noid). Relating to the Hominoidea, a superfamily of primates to which the great apes and humans are assigned.

homocercal (hō′mə-ser′kəl) (Gr. *homos,* same, common, + *kerkos,* tail). A tail with the upper and lower lobes symmetrical and the vertebral column ending near the middle of the base, as in most teleost fishes.

homodont (hō′mō-dänt) (Gr. *homos,* same, + *odous,* tooth). Having all teeth similar in form.

homograft. See **allograft.**

homology (hō-mäl′ə-jē) (Gr. *homologos,* agreeing). Similarity of parts or organs of different organisms caused by evolutionary derivation from a corresponding part or organ in a remote ancestor, and usually having a similar embryonic origin. May also refer to a matching pair of chromosomes. Serial homology is the correspondence in the same individual of repeated structures having the same origin and development, such as the appendages of arthropods. Adj., **homologous.**

homoplasy (hō′mō′plā′sē). Phenotypic similarity among characteristics of different species or populations (including molecular, morphological, behavioral or other features) that does not accurately represent patterns of common evolutionary descent (= nonhomologous similarity); it is produced by evolutionary parallelism, convergence and/or reversal, and is revealed by incongruence among different characters on a cladogram or phylogenetic tree.

homozygote (hō-mə-zī′gōt) (Gr. *homos,* same, + *zygotos,* yoked). An organism in which the pair of alleles for a trait is composed of the same genes (either dominant or recessive but not both). Adj., **homozygous.**

humoral (hū′mər-əl) (L. *humor,* a fluid). Pertaining to an endocrine secretion.

humoral immune response. Immune response involving production of antibodies, specifically the T_H2 arm of the immune response. Contrast **cell-mediated immune response.**

hyaline (hī′ə-lən) (Gr. *hyalos,* glass). Adj., glassy, translucent. Noun, a clear, glassy, structureless material occurring, for example, in cartilage, vitreous body, mucin, and glycogen.

hybridoma (hī-brid-ō′mah) (contraction of hybrid + myeloma). Fused product of a normal and a myeloma (cancer) cell, which has some of the characteristics of the normal cell.

hydatid cyst (hī-da′təd) (Gr. *hydatis,* watery vesicle). A type of cyst formed by juveniles of certain tapeworms (*Echinococcus*) in their vertebrate hosts.

hydranth (hī′dranth) (Gr. *hydōr,* water, + *anthos,* flower). Nutritive zooid of hydroid colony.

hydrocaulus (hī′drə-kä′ləs) (Gr. *hydōr,* water, + *kaulos,* stem of a plant). Stalks or "stems" of a hydroid colony, the parts between the hydrorhiza and the hydranths.

hydrocoel (hī′-drə-sēl) (*Gr.* hydōr, water, + *koilos,* hollow). Second or middle coelomic compartment in echinoderms; left hydrocoel gives rise to water vascular system.

hydrogen bond. A relatively weak chemical bond resulting from unequal charge distribution within molecules, in which a hydrogen atom covalently bonded to another atom is attracted to the electronegative portion of another molecule.

hydroid. The polyp form of a cnidarian as distinguished from the medusa form. Any cnidarian of the class Hydrozoa, order Hydroida.

hydrolysis (Gr. *hydōr,* water, + *lysis,* a loosening). The decomposition of a chemical compound by the addition of water; the splitting of a molecule into its groupings so that the split products acquire hydrogen and hydroxyl groups.

hydrorhiza (hī′drə-rī′zə) (Gr. *hydōr,* water, + *rhiza,* a root). Rootlike stolon that attaches a hydroid to its substrate.

hydrosphere (Gr. *hydōr,* water, + *sphaira,* ball, sphere). Aqueous envelope of the earth.

hydrostatic pressure. The pressure exerted by a fluid (gas or liquid), defined as force per unit area. For example, the hydrostatic pressure of one atmosphere (1 atm) is 14.7 lb/in^2.

hydrostatic skeleton. A mass of fluid or plastic parenchyma enclosed within a muscular wall to provide the support necessary for antagonistic muscle action; for example, parenchyma in acoelomates and perivisceral fluids in pseudocoelomates serve as hydrostatic skeletons.

hydrothermal vent. A submarine hot spring; seawater seeping through the sea bottom is heated by magma and expelled back into the sea through the hydrothermal vent.

hydroxyl (hydrogen + oxygen, + yl). Containing an OH⁻ group, a negatively charged ion formed by alkalies in water.

hyomandibular (hī-ō-mən-dib′yə-lər) (Gr. *hyoeides* [shaped like the Gr. letter upsilon ϒ, + *eidos,* form], + L. *mandere,* to chew). Bone derived from the hyoid gill arch, forming part of articulation of the lower jaw of fishes, and forming the stapes of the ear of amniotic vertebrates.

hyperosmotic (Gr. *hyper,* over, + *ōsmos,* impulse). Refers to a solution whose osmotic pressure is greater than that of another solution to which it is compared; contains a greater concentration of dissolved particles and gains water through a selectively permeable membrane from a solution containing fewer particles; contrasts with **hypoosmotic.**

hyperparasitism (hī′pər-par′ə-sid-iz-əm) (Gr. *hyper,* over, + *para,* beside, + *sitos,* food). Parasitism of a parasite by another parasite.

hypertrophy (hī-pər′trə-fē) (Gr. *hyper,* over, + *trophē,* nourishment). Abnormal increase in size of a part or organ.

hypodermis (hī′pə-dər′mis) (Gr. *hypo,* under, + L. *dermis,* skin). The cellular layer lying beneath and secreting the cuticle of annelids, arthropods, and certain other invertebrates.

hypoosmotic (Gr. *hypo,* under, + *ōsmos,* impulse). Refers to a solution whose osmotic pressure is less than that of another solution with which it is compared or taken as a standard; contains a lesser concentration of dissolved particles and loses water during osmosis; contrasts with **hyperosmotic.**

hypophysis (hī-pof′ə-sis) (Gr. *hypo,* under, + *physis,* growth). Pituitary body.

hypostome (hī′pə-stōm) (Gr. *hypo,* under, + *stoma,* mouth). Name applied to structure in various invertebrates (such as mites and ticks), located at posterior or ventral area of mouth.

hypothalamus (hī-pō-thal′ə-mis) (Gr. *hypo,* under, + *thalamos,* inner chamber). A ventral part of the forebrain beneath the thalamus; one of the centers of the autonomic nervous system.

hypothesis (Gr. *hypothesis,* foundation, supposition). A statement or proposition that can be tested by experiment.

hypothetico-deductive (Gr. *hypotithenai,* to suppose, + L. *deducere,* to lead). Scientific process of making a conjecture and then seeking empirical tests that potentially lead to its rejection.

I

imago (ə-mā′gō). The adult and sexually mature insect.

immediate hypersensitivity. Inflammatory reaction based primarily on humoral immunity.

immunoglobulin (im′yə-nə-glä′byə-lən) (L. *immunis,* free, + *globus,* globe). Any of a group of plasma proteins, produced by plasma cells, that participates in the immune response by combining with the antigen that stimulated its production. Antibody.

imprinting (im′print-ing) (L. *imprimere,* to impress, imprint). Rapid and usually stable learning pattern appearing early in the life of a member of a social species and involving recognition of its own species; may involve attraction to the first moving object seen.

inbreeding. The tendency among members of a population to mate preferentially with close relatives.

incomplete dominance. See **intermediate inheritance.**

incus (in′kəs) (L. *incus,* anvil). The middle of a chain of three bones of the mammalian middle ear.

indeterminate cleavage. A type of embryonic development in which the fate of the blastomeres is not determined very early as to tissues or organs, for example, in echinoderms and vertebrates; regulative cleavage.

indigenous (ən-dij′ə-nəs) (L. *indigena,* native). Pertains to organisms that are native to a particular region; not introduced.

induction (L. *inducere, inductum,* to lead). Reasoning from the particular to the general, that is, deriving a general statement (hypothesis) based on individual observations. In embryology, the alteration of cell fates as the result of interaction with neighboring cells.

inductor (in-duk′ter) (L. *inducere,* to introduce, lead in). In embryology, a tissue or organ that causes the differentiation of another tissue or organ.

inflammation (in′fləm-mā′shən) (L. *inflammare,* from *flamma,* flame). The complicated physiological process in mobilization of body defenses against foreign substances and infectious agents and repair of damage from such agents.

infraciliature (in-fra-sil′e-ə-tər) (L. *infra,* below, + *cilia,* eyelashes). The organelles just below the cilia in ciliate protozoa.

infundibulum (in′fun-dib′u-ləm) (L. funnel). Stalk of the neurohypophysis linking the pituitary to the diencephalon.

innate (i-nāt′) (L. *innatus,* inborn). A characteristic based partly or wholly on genetic or epigenetic constitution.

instar (inz′tär) (L. form). Stage in the life of an insect or other arthropod between molts.

instinct (L. *instinctus,* impelled). Stereotyped, predictable, genetically programmed behavior. Learning may or may not be involved.

integument (ən-teg′ū-mənt) (L. *integumentum,* covering). An external covering or enveloping layer.

intercellular (in-tər-sel′yə-lər) (L. *inter,* among, + *cellula,* chamber). Occurring between body cells.

interferons. Several cytokines encoded by different genes, important in mediation of natural immunity and inflammation.

interleukin-1. A cytokine produced by macrophages that stimulates T helper lymphocytes.

interleukin-2. A lymphokine produced by T helper lymphocytes that leads to proliferation of T helper cells and other T lymphocytes.

intermediary meiosis. Meiosis that occurs neither during gamete formation nor immediately after zygote formation, resulting in both haploid and diploid generations, such as in foraminiferan protozoa.

intermediate host. A host in which some development of a symbiont occurs, but in which maturation and sexual reproduction do not take place.

intermediate inheritance. Neither of alternate alleles of a gene are completely dominant, and heterozygote shows a condition intermediate between or different from homozygotes for each allele.

interstitial (in-tər-sti′shəl) (L. *inter,* among, + *sistere,* to stand). Situated in the interstices or spaces between structures such as cells, organs, or grains of sand.

intracellular (in-tra-sel′yə-lər) (L. *intra,* inside, + *cellula,* chamber). Occurring within a body cell or within body cells.

intrinsic growth rate. Exponential growth rate of a population, that is, the difference between the density-independent components of the birth and death rates of a natural population with stable age distribution.

intron (in′trän) (L. *intra,* within). Portion of mRNA as transcribed from DNA that will not form part of mature mRNA, and therefore does not encode an amino-acid sequence in the protein product.

introvert (L. *intro,* inward, + *vertere,* to turn). The anterior narrow portion that can be withdrawn (introverted) into the trunk of a sipunculid worm.

invagination (in-vaj′ə-nā′shən) (L. *in,* in, + *vagina,* sheath). An infolding of a layer of tissue to form a sac-like structure.

inversion (L. *invertere,* to turn upside down). A turning inward or inside out, as in embryogenesis of sponges; also, reversal in order of genes or reversal of a chromosome segment.

ion. An atom or group of atoms with a net positive or negative electrical charge because of the loss or gain of electrons.

ionic bond. A chemical bond formed by transfer of one or more electrons from one atom to another; characteristic of salts.

iridophore (ī-rid′ə-fōr) (Gr. *iris,* rainbow, or iris of eye). Iridescent or silvery chromatophores containing crystals or plates of guanine or other purine.

irritability (L. *irritare,* to provoke). A general property of all organisms involving the ability to respond to stimuli or changes in the environment.

isogametes (īs′o-gam′ēts) (Gr. *isos,* equal, + *gametēs,* spouse). Gametes of a species in which gametes of both sexes are alike in size and appearance.

isolecithal (ī′sə-les′ə-thəl) (Gr. *isos,* equal, + *lekithos,* yolk, + *al*). Pertaining to a zygote (or ovum) with yolk evenly distributed. Homolecithal.

isosmotic. A liquid having the same osmotic pressure as another, reference liquid.

isotonic (Gr. *isos,* equal, + *tonikos,* tension). Pertaining to solutions having the same or equal osmotic pressure; isosmotic.

isotope (Gr. *isos,* equal, + *topos,* place). One of several different forms (species) of a chemical element, differing from each other in atomic mass but not in atomic number.

J

juvenile hormone. Hormone produced by the corpora allata of insects; among its effects are maintenance of larval or nymphal characteristics during development.

juxtaglomerular apparatus (jək′stə-glä-mer′yə-lər) (L. *juxta,* close to, + *glomus,* ball). Complex of sensory cells located in the afferent arteriole adjacent to the glomerulus and a loop of the distal tubule, which produces the enzyme renin.

K

kentrogon (ken′trə-gən) (Gr. *kentron,* a point, spine, + *gonos,* progeny, generation). A larva of the cirripede order Rhizocephala (subphylum Crustacea) that functions to inject the parasite cells into the host hemocoel.

keratin (ker′ə-tən) (Gr. *kera,* horn, + *in,* suffix of proteins). A scleroprotein found in epidermal tissues and modified into hard structures such as horns, hair, and nails.

keystone species. A species (typically a predator) whose removal leads to reduced species diversity within the community.

kinesis (kə-nē′səs) (Gr. *kinēsis,* movement). Movements by an organism in random directions in response to stimulus.

kinetochore (kī-nēt′ə-kōr) (Gr. *kinein,* to move, + *choris,* asunder, apart). A disc of proteins located on the centromere, specialized to interact with the spindle fibers during mitosis.

kinetodesma (kə-nē′tə-dez′mə). pl. **kinetodesmata** (Gr. *kinein,* to move, + *desma,* bond). Fibril arising from the kinetosome of a cilium in a ciliate protozoan, and passing along the kinetosomes of cilia in that same row.

kinetosome (kən-ēt′ə-sōm) (Gr. *kinētos,* moving, + *sōma,* body). The self-duplicating granule at the base of the flagellum or cilium; similar to centriole, also called basal body or blepharoplast.

bat / āpe / ärmadillo / herring / fēmale / finch / līce / crocodile / crōw / duck / ūnicorn / ə indicates unaccented vowel sound "uh" as in mammal, fishes, cardinal, heron, vulture / stress as in bi-ol′o-gy, bi′o-log′i-cal

kinety (kə-nē′tē) (Gr. *kinein,* to move). All the kinetosomes and kinetodesmata of a row of cilia.

kinin (kī′nin) (Gr. *kinein,* to move, + *in,* suffix of hormones). A type of local hormone that is released near its site of origin; also called parahormone or tissue hormone.

K-selection (from the K term in the logistic equation). Natural selection under conditions that favor survival when populations are controlled primarily by density-dependent factors.

kwashiorkor (kwash-ē-or′kər) (from Ghana). Malnutrition caused by diet high in carbohydrate and extremely low in protein.

L

labium (lā′bē-əm) (L. a lip). The lower lip of the insect formed by fusion of the second pair of maxillae.

labrum (lā′brəm) (L. a lip). The upper lip of insects and crustaceans situated above or in front of the mandibles; also refers to the outer lip of a gastropod shell.

labyrinth (L. *labyrinthus,* labyrinth). Vertebrate internal ear, composed of a series of fluid-filled sacs and tubules (membranous labyrinth) suspended within bone cavities (osseous labyrinth).

labyrinthodont (lab′ə-rin′thə-dänt) (Gr. *labyrinthos,* labyrinth, + *odous, odontos,* tooth). A group of Paleozoic amphibians containing the temnospondyls and the anthracosaurs.

lachrymal (lak′rə-məl) (L. *lacrimia,* tear). Secreting or relating to tears.

lacteal (lak′te-əl) (L. *lacteus,* of milk). Noun, one of the lymph vessels in the villus of the intestine. Adj., relating to milk.

lacuna (lə-kū′nə), pl. **lacunae** (L. pit, cavity). A sinus; a space between cells; a cavity in cartilage or bone.

lagena (lə-jē′nə) (L. large flask). Portion of the primitive ear in which sound is translated into nerve impulses; evolutionary beginning of cochlea.

Lamarckism. Hypothesis, as expounded by Jean Baptiste de Lamarck, of evolution by the acquisition during an organism's lifetime of characteristics that are transmitted directly to offspring.

lamella (lə-mel′ə) (L. dim. of *lamina,* plate). One of the two plates forming a gill in a bivalve mollusc. One of the thin layers of bone laid concentrically around an osteon (Haversian canal). Any thin, platelike structure.

lappets. Lobes around the margin of scyphozoan medusae (phylum Cnidaria).

larva (lar′və), pl. **larvae** (L. a ghost). An immature stage that is quite different from the adult.

larynx (lar′inks) (Gr., the larynx, gullet). Modified upper portion of respiratory tract of air-breathing vertebrates, bounded by the glottis above and the trachea below; voice box; adj., **laryngeal** (lə-rin′j(ē)əl), relating to the larynx.

lateral (L. *latus,* the side, flank). Of or pertaining to the side of an animal; a *bilateral* animal has two sides.

laterite (lad′ə-rīt) (L. *later,* brick). Group of hard, red soils from topical areas that show intense weathering and leaching of bases and silica, leaving aluminum hydroxides and iron oxides; adj. **lateritic.**

lek (lek) (Sw. play, game). An area where animals assemble for communal courtship display and mating.

lemniscus (lem-nis′kəs) (L. ribbon). One of a pair of internal projections of the epidermis from the neck region of Acanthocephala, which functions in fluid control in the protrusion and invagination of the proboscis.

lentic (len′tik) (L. *lentus,* slow). Of or relating to standing water such as swamp, pond, or lake.

lepidosaurs (lep′ə-dō-sors) (L. *lepidos,* scale, + *sauros,* lizard). A lineage of diapsid reptiles that appeared in the Permian and that includes the modern snakes, lizards, amphisbaenids, and tuataras, and the extinct ichthyosaurs.

leptospondyls (lep′ə-spänd′ls) (Gr. *lepos,* scale, + *spondylos,* vertebra). A group of Paleozoic amphibians distinguished by the possession of spool-shaped vertebral centra.

leptocephalus (lep′tə-sef′ə-ləs) pl. **leptocephali** (Gr. *leptos,* thin, + *kephalē,* head). Transparent, ribbonlike migratory larva of the European or American eel.

leukemism (lū′kə-mi-zəm) (Gr. *leukos,* white, + *ismos,* condition of). Presence of white pelage or plumage in animals with normally pigmented eyes and skin.

leukocyte (lū′kə-sīt) (Gr. *leukos,* white, + *kytos,* hollow vessel). Any of several kinds of white blood cells (for example, granulocytes, lymphocytes, monocytes), so-called because they bear no hemoglobin, as do red blood cells.

library. In molecular biology, a set of clones containing recombinant DNA. Obtained from and representing the genome of the organism.

ligament (lig′ə-mənt) (L. *ligamentum,* bandage). A tough, dense band of connective tissue connecting one bone to another.

ligand (lī′gənd) (L. *ligo,* to bind). A molecule that specifically binds to a receptor; for example, a hormone (ligand) binds specifically to its receptor on the cell surface.

limax form (lī′məx) (L. *limax,* slug). Form of pseudopodial movement in which entire organism moves without extending a discrete pseudopodium.

lipase (lī′pās) (Gr. *lipos,* fat, + *ase,* enzyme suffix). An enzyme that accelerates the hydrolysis or synthesis of fats.

lipid, lipoid (li′pid) (Gr. *lipos,* fat). Certain fatlike substances, often containing other groups such as phosphoric acid; lipids combine with proteins and carbohydrates to form principal structural components of cells.

lithosphere (lith′ə-sfir) (Gr. *lithos,* rock, + *sphaira,* ball). The rocky component of the earth's surface layers.

littoral (lit′ə-rəl) (L. *litoralis,* seashore). Adj., pertaining to the shore. Noun, that portion of the sea floor between the extent of high and low tides, intertidal; in lakes, the shallow part from the shore to the lakeward limit of aquatic plants.

lobopodium (lō′bə-pō′de-əm) (Gr. *lobos,* lobe, + *pous, podos,* foot). Blunt, lobelike pseudopodium.

locus (lō′kəs), pl. **loci** (lō′sī) (L. place). Position of a gene in a chromosome.

logistic equation. A mathematical expression describing an idealized sigmoid curve of population growth.

lophocyte (lō′fə-sīt) (Gr. *lophos,* crest, + *kytos,* hollow vessel). Type of sponge amebocyte that secretes bundles of fibrils.

lophophore (lōf′ə-fōr) (Gr. *lophos,* crest, + *phoros,* bearing). Tentacle-bearing ridge or arm within which is an extension of the coelomic cavity in lophophorate animals (ectoprocts, brachiopods, and phoronids).

lorica (lo′rə-kə) (L. corselet). Protective external case found in some protozoa, rotifers, and others.

lotic (lo′tik) (L. *lotus,* action of washing or bathing). Of or pertaining to running water, such as a brook or river.

lumbar (lum′bär) (L. *lumbus,* loin). Relating to or near the loins or lower back.

lumen (lū′mən) (L. light). The cavity of a tube or organ.

lymph (limf) (L. *lympha,* water). The interstitial (intercellular) fluid in the body, also the fluid in the lymphatic system.

lymphocyte (lim′fō-sīt) (L. *lympha,* water, goddess of water, + Gr. *kytos,* hollow vessel). Cell in blood and lymph that has central role in immune responses. See **T cell** and **B cell.**

lymphokine (limf′ə-kīn) (L. *lympha,* water, + Gr. *kinein,* to move). A molecule secreted by an activated or stimulated lymphocyte that causes physiological changes in certain other cells.

lysosome (lī′sə-sōm) (Gr. *lysis,* loosing, + *sōma,* body). Intracellular organelle consisting of a membrane enclosing several digestive enzymes that are released when the lysosome ruptures.

M

macroevolution (L. *makros*, long, large, + *evolvere*, to unfold). Evolutionary change on a grand scale, encompassing the origin of novel designs, evolutionary trends, adaptive radiation, and mass extinction.

macrogamete (mak′rə-gam′ēt) (Gr. *makros*, long, large, + *gamos*, marriage). The larger of the two gamete types in a heterogametic organism, considered the female gamete.

macromere (mak′rə-mer′) (Gr. *makros*, long, large, + *meros*, part). The largest size class of blastomeres in a cleaving embryo when the blastomeres differ in size from one another.

macromolecule. A very large molecule, such as a protein, polysaccharide, or nucleic acid.

macronucleus (ma′krō-nū′klē-əs) (Gr. *makros*, long, large, + *nucleus*, kernel). The larger of the two kinds of nuclei in ciliate protozoa; controls all cell functions except reproduction.

macrophage (mak′rə-fāj) (Gr. *makros*, long, large, + *phagō*, to eat). A phagocytic cell type in vertebrates that performs crucial functions in the immune response and inflammation, such as presenting antigenic epitopes to T cells and producing several cytokines.

madreporite (ma′drə-pōr′īt) (Fr. *madrépore*, reef-building coral, + *ite*, suffix for some body parts). Sievelike structure that is the intake for the water-vascular system of echinoderms.

major histocompatibility complex (MHC). Complex of genes coding for proteins inserted in the cell membrane; the proteins are the basis of self-nonself recognition by the immune system.

malacostracan (mal′ə-käs′trə-kən) (Gr. *malako*, soft, + *ostracon*, shell). Any member of the crustacean subclass Malacostraca, which includes both aquatic and terrestrial forms of crabs, lobsters, shrimps, pillbugs, sand fleas, and others.

malleus (mal′ē-əs) (L. hammer). The ossicle attached to the tympanum in middle ears of mammals.

malpighian tubules (mal-pig′ē-ən) (Marcello Malpighi, Italian anatomist, 1628–1694). Blind tubules opening into the hindgut of nearly all insects and some myriapods and arachnids, and functioning primarily as excretory organs.

mantle. Soft extension of the body wall in certain invertebrates, for example, brachiopods and molluscs, which usually secretes a shell; thin body wall of tunicates.

manubrium (man-ū′bri-əm) (L. handle). The portion projecting from the oral side of a jellyfish medusa, bearing the mouth; oral cone; presternum or anterior part of sternum; handle-like part of malleus of ear.

marasmus (mə-raz′məs) (Gr. *marasmos*, to waste away). Malnutrition, especially of infants, caused by a diet deficient in both calories and protein.

marsupial (mär-sū′pē-əl) (Gr. *marsypion*, little pouch). One of the pouched mammals of the subclass Metatheria.

mastax (mas′təx) (Gr. jaws). Pharyngeal mill of rotifers.

matrix (mā′triks) (L. *mater*, mother). The intercellular substance of a tissue, or that part of a tissue into which an organ or process is set.

maturation (L. *maturus*, ripe). The process of ripening; the final stages in the preparation of gametes for fertilization.

maxilla (mak-sil′ə) (L. dim. of *mala*, jaw). One of the upper jawbones in vertebrates; one of the head appendages in arthropods.

maxilliped (mak-sil′ə-ped) (L. *maxilla*, jaw, + *pes*, foot). One of the pairs of head appendages located just posterior to the maxilla in crustaceans, a thoracic appendage that has become incorporated into the feeding mouthparts.

medial (mē′dē-əl). Situated, or occurring, in the middle.

mediated transport. Transport of a substance across a cell membrane mediated by a carrier molecule in the membrane.

medulla (mə-dul′ə) (L. marrow). The inner portion of an organ in contrast to the cortex or outer portion. Also, hindbrain.

medusa (mə-dū′sə) (Gr. mythology, female monster with snake-entwined hair). A jellyfish, or the free-swimming stage in the life cycle of cnidarians.

Mehlis' gland (me′ləs). Glands of uncertain function surrounding the ootype of trematodes and cestodes.

meiofauna (mī′ō-faw-nə) (Gr. *meion*, smaller, + L. *faunus*, god of the woods). Small invertebrates found in the interstices between sand grains.

meiosis (mī-ō′səs) (Gr. from *mieoun*, to make small). The nuclear changes by means of which the chromosomes are reduced from the diploid to the haploid number; in animals, usually occurs in the last two divisions in the formation of the mature egg or sperm.

melanin (mel′ə-nin) (Gr. *melas*, black). Black or dark-brown pigment found in plant or animal structures.

melanophore (mel′ə-nə-fōr, mə-lan′ə-fōr) (Gr. *melania*, blackness, + *pherein*, to bear). Black or brown chromatophore containing melanin.

memory cells. Population of long-lived B lymphocytes remaining after initial immune response that provides for the secondary response.

meninges (mə-nin′jez), sing. **meninx** (Gr. *mēninx*, membrane). Any of three membranes (arachnoid, dura mater, pia mater) that envelop the vertebrate brain and spinal cord. Also, solid connective tissue sheath enclosing the central nervous system of some vertebrates.

menopause (men′ō-pawz) (Gr. *men*, month, + *pauein*, to cease). In the human female, that time of life when ovulation ceases; cessation of the menstrual cycle.

menstruation (men′stroo-ā′shən) (L. *menstrua*, the menses, from *mensis*, month). The discharge of blood and uterine tissue from the vagina at the end of a menstrual cycle.

meroblastic (mer-ə-blas′tik) (Gr. *meros*, part, + *blastos*, germ). Partial cleavage occurring in zygotes having a large amount of yolk at the vegetal pole; cleavage restricted to a small area on the surface of the egg.

merozoite (me′rə-zō′it) (Gr. *meros*, part, + *zōon*, animal). A very small trophozoite at the stage just after cytokinesis has been completed in multiple fission of a protozoan.

mesenchyme (me′zən-kīm) (Gr. *mesos*, middle, + *enchyma*, infusion). Embryonic connective tissue; irregular or amebocytic cells often embedded in gelatinous matrix.

mesentery (mes′ən-ter′ē) (L. *mesenterium*, mesentery). Peritoneal fold serving to hold the viscera in position.

mesocoel (mez′ō-sēl) (Gr. *mesos*, middle, + *koilos*, hollow). Middle body coelomic compartment in some deuterostomes, anterior in lophophorates, corresponds to hydrocoel in echinoderms.

mesoderm (me′zə-dərm) (Gr. *mesos*, middle, + *derma*, skin). The third germ layer, formed in the gastrula between the ectoderm and endoderm; gives rise to connective tissues, muscle, urogenital and vascular systems, and the peritoneum.

mesoglea (mez′ō-glē′ə) (Gr. *mesos*, middle, + *glia*, glue). The layer of jellylike or cement material between the epidermis and gastrodermis in cnidarians and ctenophores; also may refer to jellylike matrix between epithelial layers in sponges.

mesohyl (me′sə-hil) (Gr. *mesos*, middle, + *hylē*, a wood). Gelatinous matrix surrounding sponge cells; mesoglea, mesenchyme.

mesolecithal (me′zō-ləs′ə-thəl) (Gr. *mesos*, middle, + *lekithos*, yolk). Pertaining to a zygote (or ovum) having a moderate amount of yolk concentrated in the vegetal pole.

mesonephros (me-zō-nef′rōs) (Gr. *mesos*, middle, + *nephros*, kidney). The middle of three pairs of embryonic renal organs in vertebrates. Functional kidney of fishes and amphibians; its collecting duct is a Wolffian duct. Adj., **mesonephric.**

mesosome (mez′ə-sōm) (Gr. *mesos*, middle, + *sōma*, body). The portion of the body in lophophorates and some deuterostomes that contains the mesocoel.

messenger RNA (mRNA). A form of ribonucleic acid that carries genetic information from the gene to the ribosome, where it determines the order of amino acids as a polypeptide is formed.

metabolism (Gr. *metabolē,* change). A group of processes that includes digestion, production of energy (respiration), and synthesis of molecules and structures by organisms; the sum of the constructive (anabolic) and destructive (catabolic) processes.

metacentric (me′tə-sen′trək) (Gr. *meta,* between, among, after, + *kentron,* center). Chromosome with centromere at or near the middle.

metacercaria (me′tə-sər-ka′rē-ə) (Gr. *meta,* between, among, after, + *kerkos,* tail, + L. *aria,* connected with). Fluke juvenile (cercaria) that has lost its tail and has become encysted.

metacoel (met′ə-sēl) (Gr. *meta,* between, among, after, + *koilos,* hollow). Posterior coelomic compartment in some deuterostomes and lophophorates; corresponds to somatocoel in echinoderms.

metamere (met′ə-mēr) (Gr. *meta,* after, + *meros,* part). A repeated body unit along the longitudinal axis of an animal; a somite, or segment.

metamerism (mə-ta′mə-ri′zəm) (Gr. *meta,* between, among, after, + *meros,* part). Condition of being made up of serially repeated parts (metameres); serial segmentation.

metamorphosis (Gr. *meta,* between, among, after, + *morphē,* form, + *osis,* state of). Sharp change in form during postembryonic development, for example, tadpole to frog or larval insect to adult.

metanephridium (me′tə-nə-fri′di-əm) (Gr. *meta,* between, among, after, + *nephros,* kidney). A type of tubular nephridium with the inner open end draining the coelom and the outer open end discharging to the exterior.

metanephros (me′tə-ne′fräs) (Gr. *meta,* between, among, after, + *nephros,* kidney). Embryonic renal organs of vertebrates arising behind the mesonephros; the functional kidney of reptiles, birds, and mammals. It is drained from a ureter.

metasome (met′ə-som) (Gr. *meta,* after, behind, + *sōma,* body). The portion of the body in lophophorates and some deuterostomes that contains the metacoel.

metazoa (met-ə-zō′ə) *Gr.* meta, after, + *zōon,* animal). Multicellular animals.

MHC. See **major histocompatibility complex.**

microevolution (mī-krō-ev-ə-lü′shən). (L. *mikros,* small, + *evolvere,* to unfold). A change in the gene pool of a population across generations.

microfilament (mī′krō-fil′ə-mənt) (Gr. *mikros,* small, + L. *filum,* a thread). A thin, linear structure in cells; of actin in muscle cells and others.

microfilariae (mīk′rə-fil-ar′ē-ē) (Gr. *mikros,* small, + L. *filum,* a thread). Partially developed juveniles borne alive by filarial worms (phylum Nematoda).

microgamete (mīk′rə-gam′et) (Gr. *mikros,* small, + *gamos,* marriage). The smaller of the two gamete types in a heterogametic organism, considered the male gamete.

micromere (mīk′rə-mer′) (Gr. *mikros,* small, + *meros,* part). The smallest size class of blastomeres in a cleaving embryo when the blastomeres differ in size from one another.

micron (μ) (mī′krän) (Gr. neuter of *mikros,* small). One one-thousandth of a millimeter; about 1/25,000 of an inch. Now largely replaced by micrometer (μm).

microneme (mī′krə-nēm) (Gr. *mikros,* small, + *nēma,* thread). One of the types of structures composing the apical complex in the phylum Apicomplexa, slender and elongate, leading to the anterior and thought to function in host cell penetration.

micronucleus. A small nucleus found in ciliate protozoa; controls the reproductive functions of these organisms.

micropyle (mīk′rə-pīl) (Gr. *mikros,* small, + *pileos,* a cap). The small opening through which the cells emerge from a gemmule (phylum Porifera).

microthrix. See **microvillus.**

microtubule (Gr. *mikros,* small, + L. *tubule,* pipe). A long, tubular cytoskeletal element with an outside diameter of 20 to 27 μm. Microtubules influence cell shape and play important roles during cell division.

microvillus (Gr. *mikros,* small, + L. *villus,* shaggy hair). Narrow, cylindrical cytoplasmic projection from epithelial cells; microvilli form the brush border of several types of epithelial cells. Also, microvilli with unusual structure cover the surface of cestode tegument (also called **microthrix** [pl. **microtriches**]).

mictic (mik′tik) (Gr. *miktos,* mixed or blended). Pertaining to haploid egg of rotifers or the females that lay such eggs.

mineralocorticoids (min(ə)rəl-ō-kord′ə-koids) (M. E. *minerale,* ore, + L. *cortex,* bark, + *oid,* suffix denoting likeness of form). Hormones of the adrenal cortex, especially aldosterone, that regulate salt balance.

miracidium (mīr′ə-sid′ē-əm) (Gr. *meirakidion,* youthful person). A minute ciliated larval stage in the life of flukes.

mitochondrion (mīd′ə-kän′drē-ən) (Gr. *mitos,* a thread, + *chondrion,* dim. of *chondros,* corn, grain). An organelle in the cell in which aerobic metabolism takes place.

mitosis (mī-tō′səs) (Gr. *mitos,* thread, + *osis,* state of). Nuclear division in which there is an equal qualitative and quantitative division of the chromosomal material between the two resulting nuclei; ordinary cell division.

molecule. A configuration of atomic nuclei and electrons bound together by chemical bonds.

monocyte (mon′ə-sīt) (Gr. *monos,* single, + *kytos,* hollow vessel). A type of leukocyte that becomes a phagocytic cell (macrophage) after moving into tissues.

monoecious (mə-nē′shəs) (Gr. *monos,* single, + *oikos,* house). Having both male and female gonads in the same organism; hermaphroditic.

monogamy (mə-näg′ə-mē) adj. **monogamous** (Gr. *monos,* single, + *gamos,* marriage). The condition of having a single mate at any one time.

monohybrid (Gr. *monos,* single, + L. *hybrida,* mongrel). A hybrid offspring of parents different in one specified character.

monomer (mä′nə-mər) (Gr. *monos,* single, + *meros,* part). A molecule of simple structure, but capable of linking with others to form polymers.

monophyly (män′ə-fī-lē) (Gr. *monos,* single, + *phyle,* tribe). The condition that a taxon or other group of organisms contains the most recent common ancestor of the group and all of its descendants; contrasts with **polyphyly** and **paraphyly.**

monosaccharide (män′nə-sa′kə-rīd) (Gr. *monos,* one, + *sakcharon,* sugar, from Sanskrit *sarkarā,* gravel, sugar). A simple sugar that cannot be decomposed into smaller sugar molecules; the most common are pentoses (such as ribose) and hexoses (such as glucose).

monozoic (mo′nə-zō′ik) (Gr. *monos,* single, + *zōon,* animal). Tapeworms with a single proglottid, do not undergo strobilation to form chain of proglottids.

morphogenesis (mor′fə-je′nə-səs) (Gr. *morphē,* form, + *genesis,* origin). Development of the architectural features of organisms; formation and differentiation of tissues and organs.

morphology (Gr. *morphē,* form, + L. *logia,* study, from Gr. *logos,* work). The science of structure. Includes cytology, the study of cell structure; histology, the study of tissue structure; and anatomy, the study of gross structure.

morula (mär′u-lə) (L. *morum,* mulberry, + *ula,* dim.). Solid ball of cells in early stage of embryonic development.

mosaic cleavage. Embryonic development characterized by independent differentiation of each part of the embryo; determinate cleavage.

mucin (mū′sən) (L. *mucus,* nasal mucus). Any of a group of glycoproteins secreted by certain cells, especially those of salivary glands.

mucus (mū′kəs) (L. *mucus,* nasal mucus). Viscid, slippery secretion rich in mucins produced by secretory cells such as those in mucous membranes. Adj., **mucous.**

Müller's larva.   Free-swimming ciliated larva that resembles a modified ctenophore, characteristic of certain marine polyclad turbellarians.

multiple fission.   A mode of asexual reproduction in some protistans in which the nuclei divide more than once before cytokinesis occurs.

mutation   (mū-tā′shən) (L. *mutare,* to change). A stable and abrupt change of a gene; the heritable modification of a characteristic.

mutualism   (mū′chə-wə-li′zəm) (L. *mutuus,* lent, borrowed, reciprocal). A type of interaction in which two different species derive benefit from their association and in which the association is necessary to both; often symbiotic.

myelin   (mī′ə-lən) (Gr. *myelos,* marrow). A fatty material forming the medullary sheath of nerve fibers.

myocyte   (mī′ə-sīt) (Gr. *mys,* muscle, + *kytos,* hollow vessel). Contractile cell (pinacocyte) in sponges.

myofibril   (Gr. *mys,* muscle, + L. dim. of *fibra,* fiber). A contractile filament within muscle or muscle fiber.

myogenic   (mī′o-jen′ik) (Gr. *mys,* muscle, + N.L., *genic,* giving rise to). Originating in muscle, such as heartbeat arising in vertebrate cardiac muscle because of inherent rhythmical properties of muscle rather than because of neural stimuli.

myomere   (mī′ə-mer) (Gr. *mys,* muscle, + *meros,* part). A muscle segment of successive segmental trunk musculature.

myosin   (mī′ə-sin) (Gr. *mys,* muscle, + *in,* suffix, belonging to). A large protein of contractile tissue that forms the thick myofilaments of striated muscle. During contraction it combines with actin to form actomyosin.

myotome   (mī′ə-tōm) (Gr. *mys,* muscle, + *tomos,* cutting). That part of a somite destined to form muscles; the muscle group innervated by a single spinal nerve.

N

nacre   (nā′kər) (F. mother-of-pearl). Innermost lustrous layer of mollusc shell, secreted by mantle epithelium. Adj., **nacreous.**

NAD.   Abbreviation of nicotinamide adenine dinucleotide, an electron acceptor or donor in many metabolic reactions.

nares   (na′rēz), sing. **naris** (L. nostrils). Openings into the nasal cavity, both internally and externally, in the head of a vertebrate.

natural killer cells.   Lymphocyte-like cells that can kill virus-infected cells and tumor cells in the absence of antibody.

natural selection.   A nonrandom reproduction of varying organisms in a population that results in the survival of those best adapted to their environment and elimination of those less well adapted; leads to evolutionary change if the variation is heritable.

nauplius   (naw′plē-əs) (L. a kind of shellfish). A free-swimming microscopic larval stage of certain crustaceans, with three pairs of appendages (antennules, antennae, and mandibles) and median eye. Characteristic of ostracods, copepods, barnacles, and some others.

nekton   (nek′tən) (Gr. neuter of *nēktos,* swimming). Term for actively swimming organisms, essentially independent of wave and current action. Compare with **plankton.**

nematocyst   (ne-mad′ə-sist′) (Gr. *nēma,* thread, + *kystis,* bladder). Stinging organelle of cnidarians.

neo-Darwinism   (nē′ō′ där′wə-niz′əm). A modified version of Darwin's evolutionary theory that eliminates elements of the Lamarckian inheritance of acquired characteristics and pangenesis that were present in Darwin's formulation; this theory originated with August Weissmann in the late nineteenth century and, after incorporating Mendelian genetic principles, has become the currently favored version of Darwinian evolutionary theory.

neopterygian   (nē-äp′tə-rij′ē-ən) (Gr. *neos,* new, + *pteryx,* fin). Any of a large group of bony fishes that includes most modern species.

neotenine.   See **juvenile hormone.**

neoteny   (nē′ə-tē′nē, nē-ot′ə-nē) (Gr. *neos,* new, + *teinein,* to extend). An evolutionary process by which organismal development is retarded relative to sexual maturation; produces a descendant that reaches sexual maturity while retaining a morphology characteristic of the preadult or larval stage of an ancestor.

nephridiopore   (nə-frid′ē-ə-pōr) (Gr. *nephros,* kidneys, + *porus,* pore). An external excretory opening in invertebrates.

nephridium   (nə-frid′ē-əm) (Gr. *nephridios,* of the kidney). One of the segmentally arranged, paired excretory tubules of many invertebrates, notably the annelids. In a broad sense, any tubule specialized for excretion and/or osmoregulation; with an external opening and with or without an internal opening.

nephron   (ne′frän) (Gr. *nephros,* kidney). Functional unit of kidney structure of vertebrates, consisting of a Bowman's capsule, an enclosed glomerulus, and the attached uriniferous tubule.

nephrostome   (nef′rə-stōm) (Gr. *nephros,* kidney, + *stoma,* mouth). Ciliated, funnel-shaped opening of a nephridium.

neritic   (nə-rid′ik) (Gr. *nērites,* a mussel). Portion of the sea overlying the continental shelf, specifically from the subtidal zone to a depth of 200 m.

nested hierarchy.   A pattern in which species are ordered into a series of increasingly more inclusive clades according to the taxonomic distribution of synapomorphies.

neurogenic   (nū-rä-jen′ik) (Gr. *neuron,* nerve, + N.L. *genic,* give rise to). Originating in nervous tissue, as does the rhythmical beat of some arthropod hearts.

neuroglia   (nū-räg′le-ə) (Gr. *neuron,* nerve, + *glia,* glue). Tissue supporting and filling the spaces between the nerve cells of the central nervous system.

neurolemma   (nū-rə-lem′ə) (Gr. *neuron,* nerve, + *lemma,* skin). Delicate nucleated outer sheath of a nerve cell; sheath of Schwann.

neuromast   (Gr. *neuron,* sinew, nerve, + *mastos,* knoll). Cluster of sense cells on or near the surface of a fish or amphibian that is sensitive to vibratory stimuli and water.

neuron   (Gr. nerve). A nerve cell.

neuropodium   (nū′rə-pō′de-əm) (Gr. *neuron,* nerve, + *pous, podos,* foot). Lobe of parapodium nearer the ventral side in polychaete annelids.

neurosecretory cell   (nu′rō-sə-krēd′ə-rē). Any cell (neuron) of the nervous system that produces a hormone.

neutron.   A subatomic particle lacking an electrical charge and having a mass 1839 times that of an electron and found in the nucleus of atoms.

niche.   The role of an organism in an ecological community; its unique way of life and its relationship to other biotic and abiotic factors.

nictitating membrane   (nik′tə-tā-ting) (L. *nicto,* to wink). Third eyelid, a transparent membrane of birds and many reptiles and mammals, that can be pulled across the eye.

nitrogen fixation   (Gr. *nitron,* soda, + *gen,* producing). Reduction of molecular nitrogen to ammonia by some bacteria and cyanobacteria, often followed by **nitrification,** the oxidation of ammonia to nitrites and nitrates by other bacteria.

nondisjunction.   Failure of a pair of homologous chromosomes to separate during meiosis, leading to one gamete with n + 1 chromosomes (see **trisomy**) and another gamete with n − 1 chromosomes.

notochord   (nōd′ə-kord′) (Gr. *nōtos,* back, + *chorda,* cord). An elongated cellular cord, enclosed in a sheath, which forms the primitive axial skeleton of chordate embryos and adult cephalochordates.

bat / āpe / ärmadillo / herring / fēmale / finch / līce / crocodile / crōw / duck / ūnicorn / ə indicates unaccented vowel sound "uh" as in mammal, fishes, cardinal, heron, vulture / stress as in bi-ol′o-gy, bi′o-log′i-cal

notopodium (nō′tə-pō′de-əm) (Gr. *nōtos*, back, + *pous, podos*, foot). Lobe of parapodium nearer the dorsal side in polychaete annelids.

nucleic acid (nu′klē′ik) (L. *nucleus*, kernel). One of a class of molecules composed of joined nucleotides; chief types are deoxyribonucleic acid (DNA), found in cell nuclei (chromosomes) and mitochondria, and ribonucleic acid (RNA), found both in cell nuclei (chromosomes and nucleoli) and in cytoplasmic ribosomes.

nucleoid (nu′klē-oid) (L. *nucleus*, kernel, + *oid*, like). The region in a prokaryotic cell where the chromosome is found.

nucleolus (nu-klē′ə-ləs) (dim. of L. *nucleus*, kernel). A deeply staining body within the nucleus of a cell and containing RNA; nucleoli are specialized portions of certain chromosomes that carry multiple copies of the information to synthesize ribosomal RNA.

nucleoplasm (nu′klē-ə-plazm′) (L. *nucleus*, kernel, + Gr. *plasma*, mold). Protoplasm of nucleus, as distinguished from cytoplasm.

nucleoprotein. A molecule composed of nucleic acid and protein; occurs in the nucleus and cytoplasm of all cells.

nucleosome (nu′klē-ə-som) (L. *nucleus*, kernel, + *sōma*, body). A repeating subunit of chromatin in which one and three-quarter turns of the double-helical DNA are wound around eight molecules of histones.

nucleotide (nu′klē-ə-tīd). A molecule consisting of phosphate, 5-carbon sugar (ribose or deoxyribose), and a purine or a pyrimidine; the purines are adenine and guanine, and the pyrimidines are cytosine, thymine, and uracil.

nucleus (nū′klē-əs) (L. *nucleus*, a little nut, the kernel). The organelle in eukaryotes that contains the chromatin and which is bounded by a double membrane (nuclear envelope).

nuptial flight (nup′shəl). The mating flight of insects, especially that of the queen with male or males.

nurse cells. Single cells or layers of cells surrounding or adjacent to other cells or structures for which the nurse cells provide nutrient or other molecules (for example, for insect oocytes or *Trichinella* spp. juveniles).

nymph (L. *nympha*, nymph, bride). An immature stage (following hatching) of a hemimetabolous insect that lacks a pupal stage.

O

ocellus (ō-sel′əs) (L. dim. of *oculus*, eye). A simple eye or eyespot in many types of invertebrates.

octomerous (ok-tom′ər-əs) (Gr. *oct*, eight, + *meros*, part). Eight parts, specifically, symmetry based on eight.

odontophore (ō-don′tə-fōr′) (Gr. *odous*, tooth, + *pherein*, to carry). Tooth-bearing organ in molluscs, including the radula, radular sac, muscles, and cartilages.

olfactory (äl-fakt′(ə)-rē) (L. *olor*, smell, + *factus*, to bring about). Pertaining to the sense of smell.

omasum (ō-mā′səm) (L. paunch). The third compartment of the stomach of a ruminant mammal.

ommatidium (ä′mə-tid′ē-əm) (Gr. *omma*, eye, + *idium*, small). One of the optical units of the compound eye of arthropods.

omnivore (äm′nə-vōr) (L. *omnis*, all, + *vorare*, to devour). An animal that uses a variety of animal and plant material in its diet.

oncogene (än′kə-jen) (Gr. *onkos*, protuberance, tumor, + *genos*, descent). Any of a number of genes that are associated with neoplastic growth (cancer). The gene in its benign state, either inactivated or carrying on its normal role, is a **proto-oncogene.**

oncomiracidium (än′kō-mīr′ə-sid′ē-əm) (Gr. *onkos*, barb, hook, + *meirakidion*, youthful person). A ciliated larva of a monogenetic trematode.

oncosphere (än′kəs-fər) (Gr. *onkinos*, a hook, + *sphaira*, ball). Rounded larva common to all cestodes, bears hooks.

ontogeny (än-tä′jə-nē) (Gr. *ontos*, being, + *geneia*, act of being born, from *genēs*, born). The course of development of an individual from egg to senescence.

oocyst (ō′ə-sist) (Gr. *ōion*, egg, + *kystis*, bladder). Cyst formed around zygote of malaria and related organisms.

oocyte (ō′ə-sīt) (Gr. *ōion*, egg, + *kytos*, hollow). Stage in formation of ovum, just preceding first meiotic division (primary oocyte) or just following first meiotic division (secondary oocyte).

ooecium (ō-ēs′ē-əm) (Gr. *ōion*, egg, + *oikos*, house, + L. *ium*, from). Brood pouch; compartment for developing embryos in ectoprocts.

oogenesis (ō-ə-jen′ə-səs) (Gr. *ōion*, egg, + *genesis*, descent). Formation, development, and maturation of a female gamete or ovum.

oogonium (ō′ə-gōn′ē-əm) (Gr. *ōion*, egg, + *gonos*, offspring). A cell that, by continued division, gives rise to oocytes; an ovum in a primary follicle immediately before the beginning of maturation.

ookinete (ō-ə-ki′nēt) (Gr. *ōion*, egg, + *kinein*, to move). The motile zygote of malarial parasites.

ootid (ō-ə-tid′) (Gr. *ōion*, egg, + *idion*, dim.). Stage of formation of ovum after second meiotic division following expulsion of second polar body.

ootype (ō′ə-tīp) (Gr. *ōion*, egg, + *typos*, mold). Part of oviduct in flatworms that receives ducts from vitelline glands and Mehlis' gland.

operculum (ō-per′kū-ləm) (L. cover). The gill cover in bony fishes; horny plate in some snails.

operon (äp′ə-rän). A genetic unit consisting of a cluster of genes under the control of other genes, found in prokaryotes.

ophthalmic (äf-thal′mik) (Gr. *ophthalamos*, an eye). Pertaining to the eye.

opisthaptor (ä′pəs-thap′tər) (Gr. *opisthen*, behind, + *haptein*, to fasten). Posterior attachment organ of a monogenetic trematode.

opisthosoma (ō-pis′thə-sō′mə) (Gr. *opisthe*, behind, + *sōma*, body). Posterior body region in arachnids and pogonophorans.

opsonization (op′sən-i-zā′shən) (Gr. *opsonein*, to buy victuals, to cater). The facilitation of phagocytosis of foreign particles by phagocytes in the blood or tissues, mediated by antibody bound to the particles.

organelle (Gr. *organon*, tool, organ, + L. *ella*, dim.). Specialized part of a cell; literally, a small organ that performs functions analogous to organs of multicellular animals.

organizer (or′gan-ī-zer) (Gr. *organos*, fashioning). Area of an embryo that directs subsequent development of other parts.

orthogenesis (ōr′thō-jen′ə-səs). A unidirectional trend in the evolutionary history of a lineage as revealed by the fossil record; also, a now discredited, anti-Darwinian evolutionary theory, popular around 1900, postulating that genetic momentum forced lineages to evolve in a predestined linear direction that was independent of external factors and often led to decline and extinction.

osculum (os′kū-ləm) (L. *osculum*, a little mouth). Excurrent opening in a sponge.

osmole. Molecular weight of a solute, in grams, divided by the number of ions or particles into which it dissociates in solution. Adj., **osmolar.**

osmoregulation. Maintenance of proper internal salt and water concentrations in a cell or in the body of a living organism, active regulation of internal osmotic pressure.

osmosis (oz-mō′sis) (Gr. *ōsmos*, act of pushing, impulse). The flow of solvent (usually water) through a semipermeable membrane.

osmotic potential. Osmotic pressure.

osmotroph (oz′mə-trōf) (Gr. *ōsmos*, a thrusting, impulse, + *trophē*, to eat). A heterotrophic organism that absorbs dissolved nutrients.

osphradium (äs-frā′dē-əm) (Gr. *osphradion*, small bouquet, dim. of *osphra*, smell). A sense organ in aquatic snails and bivalves that tests incoming water.

ossicles (L. *ossiculum*, small bone). Small separate pieces of echinoderm endoskeleton. Also, tiny bones of the middle ear of vertebrates.

osteoblast (os′tē-ō-blast) (Gr. *osteon*, bone, + *blastos*, bud). A bone-forming cell.

osteoclast (os′tē-ō-oclast) (Gr. *osteon*, bone, + *klan*, to break). A large, multinucleate cell that functions in bone dissolution.

osteocyte (os′tē-ə-sīt) (Gr. *osteon*, bone, + *kytos*, hollow). A bone cell that is characteristic of adult bone, has developed from an osteoblast, and is isolated in a lacuna of the bone substance.

osteon (ōs′tē-än) (Gr. bone). Unit of bone structure; Haversian system.

osteostracans (os-tē-äs′trə-kəns) (Gr. *osteon*, bone, + *ostrakon*, shell). A group of Paleozoic (Upper Silurian to Upper Devonian) agnathans belonging to the order Cephalaspidiformes.

ostium (L. door). Opening.

otolith (ōd′əl-ith′) (Gr. *ous, otos,* ear, + *lithos,* stone). Calcareous concretions in the membranous labyrinth of the inner ear of lower vertebrates, or in the auditory organ of certain invertebrates.

outgroup. In phylogenetic systematic studies, a species or group of species closely related to but not included in a taxon whose phylogeny is being studied, and used to polarize variation of characters and to root the phylogenetic tree.

oviger (ō′vi-jər) (L. *ovum*, egg, + *gerere*, to bear). Leg that carries eggs in pycnogonids.

oviparity (ō′və-pa′rəd-ē) (L. *ovum*, egg, + *parere*, to bring forth). Reproduction in which eggs are released by the female; development of offspring occurs outside the maternal body. Adj., **oviparous** (ō-vip′ə-rəs).

ovipositor (ō′və-päz′əd-ər) (L. *ovum*, egg, + *positor*, builder, placer, + *or*, suffix denoting agent or doer). In many female insects a structure at the posterior end of the abdomen for laying eggs.

ovoviviparity (ō′vo-vī-və-par′ə-dē) (L. *ovum*, egg, + *vivere*, to live, + *parere*, to bring forth). Reproduction in which eggs develop within the maternal body without additional nourishment from the parent and hatch within the parent, or immediately after laying. Adj., **ovoviviparous** (ō′vo-vī-vip′ə-rəs).

ovum (L. *ovum*, egg). Mature female germ cell (egg).

oxidation (äk′sə-dā-shən) (Fr. *oxider,* to oxidize, from Gr. *oxys*, sharp, + *ation*). The loss of an electron by an atom or molecule; sometimes addition of oxygen chemically to a substance. Opposite of reduction, in which an electron is accepted by an atom or molecule.

oxidative phosphorylation (äk′sə-dād′iv fäs′fər-i-lā′shən). The conversion of inorganic phosphate to energy-rich phosphate of ATP, involving electron transport through a respiratory chain to molecular oxygen.

P

p53 protein. A tumor suppressor protein with critical functions in normal cells. A mutation in the gene that encodes it, *p53,* can result in loss of control over cell division and thus cancer.

paedogenesis (pē-dō-jen′ə-sis) (Gr. *pais,* child, + *genēs,* born). Reproduction by immature or larval animals caused by acceleration of maturation. Progenesis.

paedomorphosis (pē-dō-mor′fə-səs) (Gr. *pais,* child, + *morphē,* form). Displacement of ancestral juvenile features to later stages of the ontogeny of descendants.

pair bond. An affiliation between an adult male and an adult female for reproduction. Characteristic of monogamous species.

pallium (pal′e-əm) (L. mantle). Mantle of a mollusc or brachiopod.

pangenesis (pan-jen′ə-sis) (Gr. *pan,* all, + *genesis,* descent). Darwin's hypothesis that hereditary characteristics are carried by individual body cells that produce particles that collect in the germ cells.

papilla (pə-pil′ə) (L. nipple). A small nipplelike projection. A vascular process that nourishes the root of a hair, feather, or developing tooth.

papula (pa′pū-lə) (L. pimple). Respiratory processes on skin of sea stars; also, pustules on skin.

parabiosis (pa′rə-bī-ō′sis) (Gr. *para,* beside, + *biosis,* mode of life). The fusion of two individuals, resulting in mutual physiological intimacy.

paramylon bodies (par′ə-mī-lən) (Gr. *para,* beside, + *mylos,* mill, grinder). Organelles containing the starch-like substance paramylon; in some algae and flagellates.

paraphyly (par′ə-fī-lē) (Gr. *para,* beside, + *phyle,* tribe). The condition that a taxon or other group of organisms contains the most recent common ancestor of all members of the group but excludes some descendants of that ancestor; contrasts with **monophyly** and **polyphyly.**

parapodium (pa′rə-pō′dē-əm) (Gr. *para,* beside, + *pous, podos,* foot). One of the paired lateral processes on each side of most segments in polychaete annelids; variously modified for locomotion, respiration, or feeding.

parasitism (par′ə-sīd′izəm) (Gr. *parasitos,* from *para,* beside, + *sitos,* food). The condition of an organism living in or on another organism (host) at whose expense the parasite is maintained; destructive symbiosis.

parasympathetic (par′ə-sim-pə-thed′ik) (Gr. *para,* beside, + *sympathes,* sympathetic, from *syn,* with, + *pathos,* feeling). One of the subdivisions of the autonomic nervous system, whose fibers originate in the brain and in anterior and posterior parts of the spinal cord.

parenchyma (pə-ren′kə-mə) (Gr. anything poured in beside). In lower animals, a spongy mass of vacuolated mesenchyme cells filling spaces between viscera, muscles, or epithelia; in some, cell bodies of muscle cells. Also, the specialized tissue of an organ as distinguished from the supporting connective tissue.

parenchymula (pa′rən-kī′mū-lə) (Gr. *para,* beside, + *enchyma,* infusion). Flagellated, solid-bodied larva of some sponges.

parietal (pä-rī-ə-təl) (L. *paries,* wall). Something next to, or forming part of, a wall of a structure.

parthenogenesis (pär′thə-nō-gen′ə-sis) (Gr. *parthenos,* virgin, + L. from Gr. *genesis,* origin). Unisexual reproduction involving the production of young by females not fertilized by males; common in rotifers, cladocerans, aphids, bees, ants, and wasps. A parthenogenetic egg may be diploid or haploid.

pathogenic (path′ə-jen′ik) (Gr. *pathos,* disease, + N.L. *genic,* giving rise to). Producing or capable of producing disease.

PCR. See **polymerase chain reaction.**

peck order. A hierarchy of social privilege in a flock of birds.

pecten (L. comb). Any of several types of comblike structures on various organisms, for example, a pigmented, vascular, and comblike process that projects into the vitreous humor from the retina at a point of entrance of the optic nerve in the eyes of all birds and many reptiles.

pectines (pek′tīnz) (L. comb, pl. of **pecten**). Sensory appendage on abdomens of scorpions.

pectoral (pek′tə-rəl) (L. *pectoralis,* from *pectus,* the breast). Of or pertaining to the breast or chest; to the pectoral girdle; or to a pair of horny shields of the plastron of certain turtles.

pedalium (pə-dal′ē-əm) (L. *pedalis,* of or belonging to the foot). Flattened blade at the base of the tentacles in cubozoan medusae (Cnidaria).

pedal laceration. Asexual reproduction found in sea anemones, a form of fission.

pedicel (ped′ə-sel) (L. *pediculus,* little foot). A small or short stalk or stem. In insects, the second segment of an antenna or the waist of an ant.

pedicellaria (ped′ə-sə-lar′ē-ə) (L. *pediculus,* little foot, + *aria,* like or connected with). One of many minute pincerlike organs on the surface of certain echinoderms.

pedipalps (ped′ə-palps′) (L. *pes, pedis,* foot, + *palpus,* stroking, caress). Second pair of appendages of arachnids.

pedogenesis See **paedogenesis.**

peduncle (pē′dun-kəl) (L. *pedunculus*, dim. of *pes*, foot). A stalk. Also, a band of white matter joining different parts of the brain.

pelage (pel′ij) (Fr. fur). Hairy covering of mammals.

pelagic (pə-laj′ik) (Gr. *pelagos*, the open sea). Pertaining to the open ocean.

pellicle (pel′ə-kəl) (L. *pellicula*, dim. of *pellis*, skin). Thin, translucent, secreted envelope covering many protozoa.

pelvic (pel′vik) (L. *pelvis*, a basin). Situated at or near the pelvis, as applied to girdle, cavity, fins, and limbs.

pelycosaur (pel′ə-kō-sor) (Gr. *pelyx*, basin, + *sauros*, lizard). Any of a group of carnivorous Permian synapsids distinguished by powerful jaws, stabbing teeth, and a large skin-covered sail on the back.

pentadactyl (pen-tə-dak′təl) (Gr. *pente*, five, + *daktylos*, finger). With five digits, or five fingerlike parts, to the hand or foot.

pentamerous symmetry (pen-tam′ər-əs) (Gr. *pente*, five, + *meros*, part). A radial symmetry based on five or multiples thereof.

peptidase (pep′tə-dās) (Gr. *peptein*, to digest, + *ase*, enzyme suffix). An enzyme that breaks down simple peptides, releasing amino acids.

peptide bond. A bond that binds amino acids together into a polypeptide chain, formed by removing an OH from the carboxyl group of one amino acid and an H from the amino group of another to form an amide group —CO—NH—.

perennibranchiate (pə-ren′ə-brank′ē-āt) (L. *perennis*, throughout the year, + Gr. *branchia*, gills). Having permanent gills, relating especially to certain paedomorphic salamanders.

pericardium (pə-ri-kär′dē-əm) (Gr. *peri*, around, + *kardia*, heart). Area around heart; membrane around heart.

periostracum (pe-rē-äs′trə-kəm) (Gr. *peri*, around, + *ostrakon*, shell). Outer horny layer of a mollusc shell.

peripheral (pə-ri′fər-əl) (Gr. *peripherein*, to move around). Structure or location distant from center, near outer boundaries.

periproct (per′ə-präkt) (Gr. *peri*, around, + *prōktos*, anus). Region of aboral plates around the anus of echinoids.

perisarc (per′ə-särk) (Gr. *peri*, around, + *sarx*, flesh). Sheath covering the stalk and branches of a hydroid.

perissodactyl (pə-ris′ə-dak′təl) (Gr. *perissos*, odd, + *daktylos*, finger, toe). Pertaining to an order of ungulate mammals with an odd number of digits.

peristalsis (per′ə-stal′səs) (Gr. *peristaltikos*, compressing around). The series of alternate relaxations and contractions that serve to force food through the alimentary canal.

peristomium (per′ə-stō′mē-əm) (Gr. *peri*, around, + *stoma*, mouth). Foremost true segment of an annelid; it bears the mouth.

peritoneum (per′ə-tə-nē′əm) (Gr. *peritonaios*, stretched around). The membrane that lines the coelom and covers the coelomic viscera.

permease. A transporter molecule; a molecule in the cell membrane that makes it possible for another molecule (to which the membrane is not otherwise permeable) to be transported across the membrane, that is, mediated transport.

petaloids (pe′tə-loids) (Gr. *petalon*, leaf, + *eidos*, form). Describes flowerlike arrangement of respiratory podia in irregular sea urchins.

pH (*potential* of *hydrogen*). A symbol referring to the relative concentration of hydrogen ions in a solution; pH values are from 0 to 14, and the lower the value, the more acid or hydrogen ions in the solution. Equal to the negative logarithm of the hydrogen ion concentration.

phagocyte (fag′ə-sīt) (Gr. *phagein*, to eat, + *kytos*, hollow vessel). Any cell that engulfs and devours microorganisms or other particles.

phagocytosis (fag′ə-sī-tō-səs) (Gr. *phagein*, to eat, + *kytos*, hollow vessel). The engulfment of a particle by a phagocyte or a protozoan.

phagosome (fa′gə-sōm) (Gr. *phagein*, to eat, + *sōma*, body). Membrane-bound vesicle in cytoplasm containing food material engulfed by phagocytosis.

phagotroph (fag′ə-trōf) (Gr. *phagein*, to eat, + *trophē*, food). A heterotrophic organism that ingests solid particles for food.

pharynx (far′inks), pl. **pharynges** (Gr. *pharynx*, gullet). The part of the digestive tract between the mouth cavity and the esophagus that, in vertebrates, is common to both digestive and respiratory tracts. In cephalochordates the gill slits open from it.

phasmid (faz′mid) (Gr. *phasma*, apparition, phantom, + *id*). One of a pair of glands or sensory structures found in the posterior end of certain nematodes.

phenetic (fə-ne′tik) (Gr. *phaneros*, visible, evident). Refers to the use of a criterion of overall similarity to classify organisms into taxa; contrasts with classifications based explicitly on a reconstruction of phylogeny.

phenotype (fē′nə-tīp) (Gr. *phainein*, to show). The visible or expressed characteristics of an organism, controlled by the genotype, but not all genes in the genotype are expressed.

phenotypic gradualism. The hypothesis that new traits, even those that are strikingly different from ancestral ones, evolve by a long series of small, incremental steps.

pheromone (fer′ə-mōn) (Gr. *pherein*, to carry, + *hormōn*, exciting, stirring up). Chemical substance released by one organism that influences the behavior or physiological processes of another organism.

phosphagen (fäs′fə-jən) (phosphate + gen). A term for creatine phosphate and arginine phosphate, which store and may be sources of high-energy phosphate bonds.

phosphatide (fäs′fə-tīd′) (phosphate + ide). A lipid with phosphorus, such as lecithin. A complex phosphoric ester lipid, such as lecithin, found in all cells. Phospholipid.

phosphorylation (fäs′fə-rə-lā′shən). The addition of a phosphate group, that is, —PO_3, to a compound.

photoautotroph (fōt-ō-aw′-tō-trōf) (Gr. *photōs*, light, + *autos*, self, + *trophos*, feeder). An organism requiring light as a source of energy for making organic nutrients from inorganic raw materials.

photosynthesis (fōt-ō-sin′thə-sis) (Gr. *phōs*, light, + *synthesis*, action or putting together). The synthesis of carbohydrates from carbon dioxide and water in chlorophyll-containing cells exposed to light.

phototaxis (fōt-ō-tak′sis) (Gr. *phōs*, light, + *taxis*, arranging, order). A taxis in which light is the orienting stimulus. An involuntary tendency for an organism to turn toward (positive) or away from (negative) light.

phototrophs (fōt′-ō-trōfs) (Gr. *phōs*, *phōtos*, light, + *trophē*, nourishment). Organisms capable of using CO_2 in the presence of light as a source of metabolic energy.

phyletic gradualism. A model of evolution in which morphological evolutionary change is continuous and incremental and occurs mainly within unbranched species or lineages over long periods of geological time; contrasts with **punctuated equilibrium.**

phyllopodium (fī′lə-pō′dē-əm) (Gr. *phyllon*, leaf, + *pous*, *podos*, foot). Leaflike swimming appendage of branchiopod crustaceans.

phylogenetic species concept. An irreducible (basal) cluster of organisms, diagnosably distinct from other such clusters, and within which there is a parental pattern of ancestry and descent.

phylogenetic systematics. See **cladistics.**

phylogeny (fī-läj′ə-nē) (Gr. *phylon*, tribe, race, + *geneia*, origin). The origin and diversification of any taxon, or the evolutionary history of its origin and diversification, usually presented in the form of a dendrogram.

phylum (fī′ləm), pl. **phyla** (N.L. from Gr. *phylon*, race, tribe). A chief category, between kingdom and class, of taxonomic classifications into which are grouped organisms of common descent that share a fundamental pattern of organization.

physiology (L. *physiologia*, natural science). A branch of biology dealing with the organic processes and phenomena of an organism or any of its parts or of a particular bodily process.

phytoflagellates (fī-tə-fla′jə-lāts). Members of the class Phytomastigophorea, plantlike flagellates.

phytophagous (fī-täf′ə-gəs) (Gr. *phyton,* plant, + *phagein,* to eat). Organisms that feed on plants.

pilidium (pī-lid′ē-əm) (Gr. *pilidion,* dim. of *pilos,* felt cap). Free-swimming, hat-shaped larva of nemertine worms.

pinacocyte (pin′ə-kō-sīt′) (Gr. *pinax,* tablet, + *kytos,* hollow vessel). Flattened cells composing dermal epithelium in sponges.

pinacoderm (pə-nak′ə-dərm) (Gr. *pinax,* plank, tablet, + *derma,* skin). The layer of pinacocytes in sponges.

pinna (pin′ə) (L. feather, sharp point). The external ear. Also a feather, wing, or fin or similar part.

pinocytosis (pin′o-sī-tō′sis, pīn′o-sī-to′sis) (Gr. *pinein,* to drink, + *kytos,* hollow vessel, + *osis,* condition). Taking up of fluid by endocytosis; cell drinking.

placenta (plə-sen′tə) (L. flat cake). The vascular structure, embryonic and maternal, through which the embryo and fetus are nourished while in the uterus.

placode (pla′kōd) (Gr. *plakos,* flat round plate). Localized, plate-like thickening of vertebrate head ectoderm from which a specialized structure develops; such structures include eye lens, special sense organs, and certain neurons.

placoderms (plak′ə-dərm) (Gr. *plax,* plate, + *derma,* skin). A group of heavily armored jawed fishes of the Lower Devonian to Lower Carboniferous.

placoid scale (pla′koid) (Gr. *plax, plakos,* tablet, plate). Type of scale found in cartilaginous fishes, with basal plate of dentin embedded in the skin and a backward-pointing spine tipped with enamel.

plankton (plank′tən) (Gr. neuter of *planktos,* wandering). The passively floating animal and plant life of a body of water; compares with **nekton.**

plantigrade (plan′tə-grād′) (L. *planta,* sole, + *gradus,* step, degree). Pertaining to animals that walk on the whole surface of the foot (for example, humans and bears); compares with **digitigrade.**

planula (plan′yə-lə) (N.L. dim. from L. *planus,* flat). Free-swimming, ciliated larval type of cnidarians; usually flattened and ovoid, with an outer layer of ectodermal cells and an inner mass of endodermal cells.

planuloid ancestor (plan′yə-loid) (L. *planus,* flat, + Gr. *eidos,* form). Hypothetical form representing ancestor of Cnidaria and Platyhelminthes.

plasma cell (plaz′mə) (Gr. *plasma,* a form, mold). A descendant cell of a B cell, functions to secrete antibodies.

plasmalemma (plaz′mə-lem-ə) (Gr. *plasma,* a form, mold, + *lemma,* rind, sheath). The cell membrane.

plasma membrane (plaz′mə) (Gr. *plasma,* a form, mold). A living, external, limiting, protoplasmic structure that functions to regulate exchange of nutrients across the cell surface.

plasmid (plaz′məd) (Gr. *plasma,* a form, mold). A small circle of DNA that may be carried by a bacterium in addition to its genomic DNA.

plasmodium (plaz-mō′dē-əm) (Gr. *plasma,* a form, mold, + *eidos,* form). Multinucleate ameboid mass, syncytial.

plastid (plas′təd) (Gr. *plast,* formed, molded, + L. *id,* feminine stem for particle of specified kind). A membranous organelle in plant cells functioning in photosynthesis and/or nutrient storage, for example, chloroplast.

plastron (plast′trən) (Fr. *plastron,* breast plate). Ventral body shield of turtles; structure in corresponding position in certain arthropods; thin film of gas retained by epicuticle hairs of aquatic insects.

platelet (plāt′lət) (Gr. dim. of *plattus,* flat). A tiny, incomplete cell in the blood that releases substances initiating blood clotting.

pleiotropic (plī-ə-trō′pic) (Gr. *pleiōn,* more, + *tropos,* to turn). Pertaining to a gene producing more than one effect; affecting multiple phenotypic characteristics.

pleopod (plē′ə-päd) (Gr. *plein,* to sail, + *pous, podos,* foot). One of the swimming appendages on the abdomen of a crustacean.

plesiomorphic (plē′sē-ə-mōr′fik). An ancestral condition of a variable character.

pleura (plu′rə) (Gr. side, rib). The membrane that lines each half of the thorax and covers the lungs.

plexus (plek′səs) (L. network, braid). A network, especially of nerves or blood vessels.

pluteus (plü′dē-əs), pl. **plutei** (L. *pluteus,* movable shed, reading desk). Echinoid or ophiuroid larva with elongated processes like the supports of a desk; originally called "painter's easel larva."

pneumostome (nū′mə-stōm) (Gr. *pneuma,* breathing, + *stoma,* mouth). The opening of the mantle cavity (lung) of pulmonate gastropods to the outside.

podium (pō′de-əm) (Gr. *pous, podos,* foot). A footlike structure, for example, the tube foot of echinoderms.

poikilothermic (poi-ki′lə-thər′mik) (Gr. *poikilos,* variable, + thermal). Pertaining to animals whose body temperature is variable and fluctuates with that of the environment; cold blooded; compares with **ectothermic.**

polarity (Gr. *polos,* axis). In systematics, the ordering of alternative states of a taxonomic character from ancestral to successively derived conditions in an evolutionary transformation series. In developmental biology, the tendency for the axis of an ovum to orient corresponding to the axis of the mother. Also, condition of having opposite poles; differential distribution of gradation along an axis.

polarization (L. *polaris,* polar, + Gr. *iz,* make). The arrangement of positive electrical charges on one side of a surface membrane and negative electrical charges on the other side (in nerves and muscles).

Polian vesicles (pō′le-ən) (from G. S. Poli, Italian naturalist). Vesicles opening into ring canal in most asteroids and holothuroids.

polyandry (pol′ē-an′drē) (Gr. *polys,* many, + *anēr,* man). Condition of having more than one male mate at one time.

polygamy (pə-lig′ə-mē)) (Gr. *polys,* many, + *gamos,* marriage). Condition of having more than one mate at a time.

polygenic inheritance. Inheritance of traits influenced by multiple alleles; traits show continuous variation between extremes; offspring are usually intermediate between the two parents; also known as **blending** and **quantitative inheritance.**

polygyny (pə-lij′ə-nē) (Gr. *polys,* many, + *gynē,* woman). Condition of having more than one female mate at one time.

polymer (pä′lə-mər) (Gr. *polys,* many, + *meros,* part). A chemical compound composed of repeated structural units called monomers.

polymerase chain reaction (PCR). A technique for preparing large quantities of DNA from tiny samples, making it easy to clone a specific gene as long as part of the sequence of the gene is known.

polymerization (pə-lim′ər-ə-zā′shən). The process of forming a polymer or polymeric compound.

polymorphism (pä′lē-mor′fi-zəm) (Gr. *polys,* many, + *morphē,* form). The presence in a species of more than one structural type of individual.

polynucleotide (poly + nucleotide). A nucleotide of many mononucleotides combined.

polyp (pä′lip) (Gr. *polypous,* many-footed). Individual of the phylum Cnidaria, generally adapted for attachment to the substratum at the aboral end, often form colonies.

polypeptide (pä-lē-pep′tīd) (Gr. *polys,* many, + *peptein,* to digest). A molecule consisting of many joined amino acids, not as complex as a protein.

polyphyly (päl′ē-fī′lē) (Gr. *polys,* many, + *phylon,* tribe). The condition that a taxon or other group of organisms does not contain the most recent common ancestor of all members of the group, implying that it has multiple evolutionary origins; such groups are not valid as formal taxa and are recognized as such only through error. Contrasts with **monophyly** and **paraphyly.**

polyphyodont (pä′lē-fī′ə-dänt) (Gr. *polyphyes,* manifold, + *odous,* tooth). Having several sets of teeth in succession.

polypide (pä′li-pīd) (L. *polypus,* polyp). An individual or zooid in a colony, specifically in ectoprocts, which has a lophophore, digestive tract, muscles, and nerve centers.

polyploid (pä′lə-ploid′) (Gr. *polys,* many, + *ploidy,* number of chromosomes). An organism possessing more than two full homologous sets of chromosomes.

polysaccharide (pä′lē-sak′ə-rid, -rīd). (Gr. *polys,* many, + *sakcharon,* sugar, from Sanskrit *sarkarā,* gravel, sugar). A carbohydrate composed of many monosaccharide units, for example, glycogen, starch, and cellulose.

polysome (polyribosome) (Gr. *polys,* many, + *sōma,* body). Two or more ribosomes connected by a molecule of messenger RNA.

polytene chromosomes (pä′li-tēn) (Gr. *polys,* many, + *tainia,* band). Chromosomes in the somatic cells of some insects in which the chromatin replicates repeatedly without undergoing mitosis.

polyzoic (pä′lē-zō′ik) (Gr. *polys,* many, + *zōon,* animal). A tapeworm forming a strobila of several to many proglottids; also, a colony of many zooids.

pongid (pän′jəd) (L. *Pongo,* type genus of orangutan). Of or relating to the primate family Pongidae, comprising the anthropoid apes (gorillas, chimpanzees, gibbons, orangutans).

population (L. *populus,* people). A group of organisms of the same species inhabiting a specific geographical locality.

populational gradualism. The observation that new genetic variants become established in a population by increasing their frequencies across generations incrementally, initially from one or a few individuals and eventually characterizing a majority of the population.

porocyte (pō′rə-sīt) (Gr. *porus,* passage, pore, + *kytos,* hollow vessel). Type of cell found in asconoid sponges through which water enters the spongocoel.

portal system (L. *porta,* gate). System of large veins beginning and ending with a bed of capillaries; for example, hepatic portal and renal portal system in vertebrates.

posterior (L. latter). Situated at or toward the rear of the body; situated toward the back; in human anatomy the upright posture makes posterior and dorsal identical.

potocytosis (pä′tə-sī-tō′səs) (Gr. *potos,* a drinking, + *kytos,* hollow vessel). Endocytosis of certain small molecules and ions bound to specific receptors limited to

small areas on the cell surface. The areas of the receptors are invaginated and pinch off to form tiny vesicles. See **caveolae.**

preadaptation. The possession of a trait that coincidentally predisposes an organism for survival in an environment different from those encountered in its evolutionary history.

prebiotic synthesis. The chemical synthesis that occurred before the emergence of life.

precocial (prē-kō′shəl) (L. *praecoquere,* to ripen beforehand). Referring (especially) to birds whose young are covered with down and are able to run about when newly hatched.

predaceous, predacious (prē-dā′shəs) (L. *praedator,* a plunderer, *praeda,* prey). Living by killing and consuming other animals; predatory.

predator (pred′ə-tər) (L. *praedator,* a plunderer, *praeda,* prey). An organism that preys on other organisms for its food.

prehensile (prē-hen′səl) (L. *prehendere,* to seize). Adapted for grasping.

primary bilateral symmetry. Usually applied to a radially symmetrical organism descended from a bilateral ancestor and developing from a bilaterally symmetrical larva.

primary radial symmetry. Usually applied to a radially symmetrical organism that did not have a bilateral ancestor or larva, in contrast to a secondarily radial organism.

primate (prī-māt) (L. *primus,* first). Any mammal of the order Primates, which includes the tarsiers, lemurs, marmosets, monkeys, apes, and humans.

primitive (L. *primus,* first). Primordial; ancient; little evolved; said of characteristics closely approximating those possessed by early ancestral types.

proboscis (prō-bäs′əs) (Gr. *pro,* before, + *boskein,* feed). A snout or trunk. Also, tubular sucking or feeding organ with the mouth at the end as in planarians, leeches, and insects. Also, the sensory and defensive organ at the anterior end of certain invertebrates.

producers (L. *producere,* to bring forth). Organisms, such as plants, able to produce their own food from inorganic substances.

production. In ecology, the energy accumulated by an organism that becomes incorporated into new biomass.

progesterone (prō-jes′tə-rōn′) (L. *pro,* before, + *gestare,* to carry). Hormone secreted by the corpus luteum and the placenta; prepares the uterus for the fertilized egg and maintains the capacity of the uterus to hold the embryo and fetus.

proglottid (prō-gläd′əd) (Gr. *proglōttis,* tongue tip, from *pro,* before, + *glōtta,* tongue, + *id,* suffix). Portion of a tapeworm containing a set of reproductive organs; usually corresponds to a segment.

prohormone (prō′hor-mōn) (Gr. *pro,* before, + *hormaein,* to excite). A precursor of a hormone, especially a peptide hormone.

prokaryotic, procaryotic (pro-kar′ē-ät′ik) (Gr. *pro,* before, + *karyon,* kernel, nut). Not having a membrane-bound nucleus or nuclei. Prokaryotic cells characterize the bacteria and cyanobacteria.

promoter. A region of DNA to which the RNA polymerase must have access for transcription of a structural gene to begin.

pronephros (prō-nef′rəs) (Gr. *pro,* before, + *nephros,* kidney). Most anterior of three pairs of embryonic renal organs of vertebrates, functional only in adult hagfishes and larval fishes and amphibians, and vestigial in mammalian embryos. Adj., **pronephric.**

proprioceptor (prō′prē-ə-sep′tər) (L. *proprius,* own, particular, + receptor). Sensory receptor located deep within the tissues, especially muscles, tendons, and joints, that is responsive to changes in muscle stretch, body position, and movement.

prosimian (prō-sim′ē-ən) (Gr. *pro,* before, + L. *simia,* ape). Any member of a group of arboreal primates including lemurs, tarsiers, and lorises, but excluding monkeys, apes, and humans.

prosoma (prō-sōm′ə) (Gr. *pro,* before, + *sōma,* body). Anterior part of an invertebrate in which primitive segmentation is not visible; fused head and thorax of arthropod; cephalothorax.

prosopyle (präs′ə-pīl) (Gr. *prosō,* forward, + *pyle,* gate). Connections between the incurrent and radial canals in some sponges.

prostaglandins (präs′tə-glan′dəns). A family of fatty-acid hormones, originally discovered in semen, known to have powerful effects on smooth muscle, nerves, circulation, and reproductive organs.

prostomium (prō-stōm′ē-əm) (Gr. *protos,* first, + *stoma,* mouth, + *-idion,* dim. ending). Anterior closure of a metameric animal, anterior to the mouth.

protandrous (prō-tan′drəs) (Gr. *prōtos,* first, + *anēr,* male). Condition of hermaphroditic animals and plants in which male organs and their products appear before the corresponding female organs and products, thus preventing self-fertilization.

protease (prō′tē-ās) (Gr. *protein,* + *ase,* enzyme). An enzyme that digests proteins; includes proteinases and peptidases.

protein (prō′tēn, prō′tē-ən) (Gr. *protein,* from *proteios,* primary). A macromolecule of carbon, hydrogen, oxygen, and nitrogen and sometimes sulfur and phosphorus; composed of chains of amino acids joined by peptide bonds; present in all cells.

prothoracic glands. Glands in the prothorax of insects that secrete the hormone ecdysone.

prothoracicotropic hormone. See **ecdysiotropin.**

prothrombin (prō-thräm′bən) (Gr. *pro,* before, + *thrombos,* clot). A constituent of blood plasma that is changed to thrombin by a catalytic sequence that includes thromboplastin, calcium, and plasma globulins; involved in blood clotting.

protist (prō′tist) (Gr. *protos,* first). A member of the kingdom Protista, generally considered to include the protozoa and eukaryotic algae.

protocoel (prō′tə-sēl) (Gr. *protos,* first, + *koilos,* hollow). The anterior coelomic compartment in some deuterostomes, corresponds to the axocoel in echinoderms.

protocooperation. A mutually beneficial interaction between organisms in which the interaction is not physiologically necessary to the survival of either.

proton. A subatomic particle with a positive electrical charge and having a mass of 1836 times that of an electron; found in the nucleus of atoms.

protonephridium (prō′tə-nə-frid′ē-əm) (Gr. *protos,* first, + *nephros,* kidney). Primitive osmoregulatory or excretory organ consisting of a tubule terminating internally with flame bulb or solenocyte; the unit of a flame bulb system.

proto-oncogene. See **oncogene.**

protoplasm (prō′tə-plazm) (Gr. *protos,* first, + *plasma,* form). Organized living substance; cytoplasm and nucleoplasm of the cell.

protopod, protopodite (prō′tə-päd, prō-top′ə-dīt) (Gr. *protos,* first, + *pous, podos,* foot). Basal portion of crustacean appendage, containing coxa and basis.

Protostomia (prō′tə-stō′mē-ə) (Gr. *protos,* first, + *stoma,* mouth). A group of phyla in which cleavage is determinate, the coelom (in coelomate forms) is formed by proliferation of mesodermal bands (schizocoelic formation), the mesoderm is formed from a particular blastomere (called 4d), and the mouth is derived from or near the blastopore. Includes the Annelida, Arthropoda, Mollusca, and a number of minor phyla. Compares with **Deuterostomia.**

proventriculus (prō′ven-trik′ū-ləs) (L. *pro,* before, + *ventriculum,* ventricle). In birds the glandular stomach between the crop and gizzard. In insects, a muscular dilation of foregut armed internally with chitinous teeth.

proximal (L. *proximus,* nearest). Situated toward or near the point of attachment; opposite of distal, distant.

proximate cause (L. *proximus,* nearest, + *causa*). The factors that underlie the functioning of a biological system at a particular place and time, including those responsible for metabolic, physiological, and behavioral functions at the molecular, cellular, organs, and population levels.

pseudocoel (sū′do-sēl) (Gr. *pseudēs,* false, + *koilōma,* cavity). A body cavity not lined with peritoneum and not a part of the blood or digestive systems, embryonically derived from the blastocoel.

pseudopodium (sū′də-pō′dē-əm) (Gr. *pseudēs,* false, + *podion,* small foot, + *eidos,* form). A temporary cytoplasmic protrusion extended out from a protozoan or ameboid cell, and serving for locomotion or for taking up food.

puff. Strands of DNA spread apart at certain locations on giant chromosomes of some flies where that DNA is being transcribed.

pulmonary (pul′mən-ner-ē) (L. *pulmo,* lung, + *aria,* suffix denoting connected to). Relating to or associated with lungs.

punctuated equilibrium. A model of evolution in which morphological evolutionary change is discontinuous, being associated primarily with discrete, geologically instantaneous events of speciation leading to phylogenetic branching; morphological evolutionary stasis characterizes species between episodes of speciation; contrasts with **phyletic gradualism.**

pupa (pū′pə) (L. girl, doll, puppet). Inactive quiescent stage of the holometabolous insects. It follows the larval stages and precedes the adult stage.

purine (pū′rēn) (L. *purus,* pure, + *urina,* urine). Organic base with carbon and nitrogen atoms in two interlocking rings. The parent substance of adenine, guanine, and other naturally occurring bases.

pygidium (pī-jid′e-əm) (Gr. *pygē,* rump, buttocks, + *-idion,* dim. ending). Posterior closure of a metameric animal, bearing the anus.

pyrimidine (pī-rim′ə-dēn) (alter. of pyridine, from Gr. *pyr,* fire, + *id,* adj. suffix, + *ine*). An organic base composed of a single ring of carbon and nitrogen atoms; parent substance of several bases found in nucleic acids.

Q

quantitative inheritance. See **polygenic inheritance.**

queen. In entomology, the single fully developed female in a colony of social insects such as bees, ants, and termites, distinguished from workers, nonreproductive females, and soldiers.

R

radial canals. Canals along the ambulacra radiating from the ring canal of echinoderms; also choanocyte-lined canals in syconoid sponges.

radial cleavage. Embryonic development in which early cleavage planes are symmetrical to the polar axis, each blastomere of one tier lying directly above the corresponding blastomere of the next layer; indeterminate cleavage.

radial symmetry. A morphological condition in which the parts of an animal are arranged concentrically around an oral-aboral axis, and more than one imaginary plane through this axis yields halves that are mirror images of each other.

radiolarian (rā′dē-ə-la′rē-ən) (L. *radius,* ray, spoke of a wheel, + *Lar,* tutelary god of house and field). Members of the classes Acantharea, Phaeodarea, and Polycystinea (phylum Sarcomastigophora) with actinopodia and beautiful tests.

radioles (rā′dē-ōlz) (L. *radius,* ray, spoke of a wheel). Featherlike processes from the head of many tubicolous polychaete worms (phylum Annelida), used primarily for feeding.

radula (re′jə-lə) (L. scraper). Rasping tongue found in most molluscs.

Ras protein. A protein that initiates a cascade of reactions leading to cell division when a growth factor is bound to the cell surface. The gene encoding Ras becomes an oncogene when a mutation produces a form of Ras protein that initiates the cascade even in the absence of the growth factor.

ratite (ra′tīt) (L. *ratis,* raft). Referring to birds having an unkeeled sternum; compares with **carinate.**

recapitulation. Summing up or repeating; hypothesis that an individual repeats its phylogenetic history in its development.

receptor-mediated endocytosis. Endocytosis of large molecules, which are bound to surface receptors in clathrin-coated pits.

recessive. An allele that must be homozygous for the allele to be expressed.

recombinant DNA. DNA from two different species, such as a virus and a mammal, combined into a single molecule.

redia (rē′dē-ə), pl. **rediae** (rē′dē-ē) (from Redi, Italian biologist). A larval stage in the life cycle of flukes; it is produced by a sporocyst larva, and in turn gives rise to many cercariae.

reduction. In chemistry, the gain of an electron by an atom or molecule of a substance; also the addition of hydrogen to, or the removal of oxygen from, a substance.

regulative development. Progressive determination and restriction of initially totipotent embryonic material.

releaser (L. *relaxare,* to unloose). Simple stimulus that elicits an innate behavior pattern.

bat / āpe / ärmadillo / herring / fēmale / finch / līce / crocodile / crōw / duck / ūnicorn / ə indicates unaccented vowel sound "uh" as in mammal, fishes, cardinal, heron, vulture / stress as in bi-ol′o-gy, bi′o-log′i-cal

renin (rē′nin) (L. *ren*, kidney). An enzyme produced by the kidney juxtaglomerular apparatus that initiates changes leading to increased blood pressure and increased sodium reabsorption.

rennin (re′nən) (M.E. *renne*, to run). A milk-clotting endopeptidase secreted by the stomach of some young mammals, including bovine calves and human infants.

replication (L. *replicatio*, a folding back). In genetics, the duplication of one or more DNA molecules from the preexisting molecule.

reproductive barrier (L. *re* + *producere*, to lead forward; M.F. *barriere*, bar). The factors that prevent one sexually propagating population from interbreeding and exchanging genes with another population.

repugnatorial glands (L. *repugnare*, to resist). Glands secreting a noxious substance for defense or offense, for example, as in the millipedes.

respiration (L. *respiratio*, breathing). Gaseous interchange between an organism and its surrounding medium. In the cell, the release of energy by the oxidation of food molecules.

restriction endonuclease. An enzyme that cleaves a DNA molecule at a particular base sequence.

rete mirabile (rē′tē mə-rab′ə-lē) (L. wonderful net). A network of small blood vessels so arranged that the incoming blood runs countercurrent to the outgoing blood and thus makes possible efficient exchange between the two bloodstreams. Such a mechanism serves to maintain the high concentration of gases in the fish swim bladder.

reticular (rə-tik′ū-lər) (L. *reticulum*, small net). Resembling a net in appearance or structure.

reticuloendothelial system (rə-tic′ū-lō-en-dō-thēl′i-əl) (L. *reticulum*, dim. of net, + Gr. *endon*, within, + *thele*, nipple). The fixed phagocytic cells in the tissues, especially the liver, lymph nodes, spleen, and others; also called RE system.

reticulopodia (rə-tik′ū-lə-pō′dē-ə) (L. *retiulum*, dim. of *rete*, net, + *podos*, *pous*, foot). Pseudopodia that branch and rejoin extensively.

retina (ret′nə, ret′ən-ə) (L. *rete*, net). The posterior sensory membrane of the eye that receives images.

rhabdite (rab′dit) (Gr. *rhabdos*, rod). Rodlike structures in the cells of the epidermis or underlying parenchyma in certain turbellarians. They are discharged in mucous secretions.

rheoreceptor (rē′ə-rē-cep′tər) (Gr. *rheos*, a flowing, + receptor). A sensory organ of aquatic animals that responds to water current.

rhinophore (rī′nə-fōr) (Gr. *rhis*, nose, + *pherein*, to carry). Chemoreceptive tentacles in some molluscs (opisthobranch gastropods).

rhopalium (rō-pā′lē-əm) (N.L. from Gr. *rhopalon*, a club). One of the marginal, club-shaped sense organs of certain jellyfishes; tentaculocyst.

rhoptries (rōp′trēz) (Gr. *rhopalon*, club, + *tryō*, to rub, wear out). Club-shaped bodies in Apicomplexa composing one of the structures of the apical complex; open at anterior and apparently functioning in penetration of host cell.

rhynchocoel (ring′kō-sēl) (Gr. *rhynchos*, snout, + *koilos*, hollow). In nemerteans, the dorsal tubular cavity that contains the inverted proboscis. It has no opening to the outside.

ribosome (rī′bə-sōm). Subcellular structure composed of protein and ribonucleic acid. May be free in the cytoplasm or attached to the membranes of the endoplasmic reticulum; functions in protein synthesis.

ritualization. In ethology, the evolutionary modification, usually intensification, of a behavior pattern to serve communication.

RNA. Ribonucleic acid, of which there are several different kinds, such as messenger RNA, ribosomal RNA, and transfer RNA (mRNA, rRNA, tRNA).

RNA world. Hypothetical stage in the evolution of life on earth in which both catalysis and replication were performed by RNA, not protein enzymes and DNA.

rostellum (räs-tel′ləm) (L. small beak). Projecting structure on scolex of tapeworm, often with hooks.

rostrum (räs′trəm) (L. ship's beak). A snoutlike projection on the head.

rumen (rū′mən) (L. cud). The large first compartment of the stomach of ruminant mammals.

ruminant (rūm′ə-nənt) (L. *ruminare*, to chew the cud). Cud-chewing artiodactyl mammals with a complex four-chambered stomach.

S

saccule (sa′kūl) (L. *sacculus*, small bag). Small chamber of the membranous labyrinth of the inner ear.

sacrum Adj. **sacral** (sā′krəm, sā′krəl) (L. *sacer*, sacred). Bone formed by fused vertebrae to which pelvic girdle is attached; pertaining to the sacrum.

sagittal (saj′ə-dəl) (L. *sagitta*, arrow). Pertaining to the median anteroposterior plane that divides a bilaterally symmetrical organism into right and left halves.

salt (L. *sal*, salt). The reaction product of an acid and a base; dissociates in water solution to negative and positive ions, but not H⁺ or OH⁻.

saprophagous (sə-präf′ə-gəs) (Gr. *sapros*, rotten, + *phagos*, from *phagein*, to eat). Feeding on decaying matter; saprobic; saprozoic.

saprophyte (sap′rə-fīt) (Gr. *sapros*, rotten, + *phyton*, plant). A plant living on dead or decaying organic matter.

saprozoic nutrition (sap-rə-zō′ik) (Gr. *sapros*, rotten, + *zōon*, animal). Animal nutrition by absorption of dissolved salts and simple organic nutrients from surrounding medium; also refers to feeding on decaying matter.

sarcolemma (sär′kə-lem′ə) (Gr. *sarx*, flesh, + *lemma*, rind). The thin, noncellular sheath that encloses a striated muscle fiber.

sarcomere (sär′kə-mir) (Gr. *sarx*, flesh, + *meros*, part). Transverse segment of striated muscle believed to be the fundamental contractile unit.

sarcoplasm (sär′kə-plaz′əm) (Gr. *sarx*, flesh, + *plasma*, mold). The clear, semifluid cytoplasm between the fibrils of muscle fibers.

sauropterygians (so-räp′tə-rij′ē-əns) (Gr. *sauros*, lizard, + *pteryginos*, winged). Mesozoic marine reptiles.

schizocoel (skiz′ō-sēl) (Gr. *schizo*, from *schizein*, to split, + *koilōma*, cavity). A coelom formed by the splitting of embryonic mesoderm. Noun, **schizocoelomate,** an animal with a schizocoel, such as an arthropod or mollusc. Adj., **schizocoelous.**

schizocoelous mesoderm formation (skiz′ō-sēl-ləs). Embryonic formation of the mesoderm as cords of cells between ectoderm and endoderm; splitting of these cords results in the coelomic space.

schizogony (skə-zä′gə-nē) (Gr. *schizein*, to split, + *gonos*, seed). Multiple asexual fission.

sclerite (skler′it) (Gr. *skleros*, hard). A hard chitinous or calcareous plate or spicule; one of the plates making up the exoskeleton of arthropods, especially insects.

scleroblast (skler′ə-blast) (Gr. *sklēros*, hard, + *blastos*, germ). An amebocyte specialized to secrete a spicule, found in sponges.

sclerocyte (skler′ə-sīt) (Gr. *sklēros*, hard, + *kytos*, hollow vessel). An amebocyte in sponges that secretes spicules.

sclerotic (skler-äd′ik) (Gr. *sklēros*, hard). Pertaining to the tough outer coat of the eyeball.

sclerotization (sklėr′ə-tə-zā′shən). Process of hardening of the cuticle of arthropods by the formation of stabilizing cross linkages between peptide chains of adjacent protein molecules.

scolex (skō′leks) (Gr. *skōlēx*, worm, grub). The holdfast, or so-called head, of a tapeworm; bears suckers and, in some, hooks, and posterior to it new proglottids are differentiated.

scrotum (skrō′təm) (L. bag). The pouch that contains the testes in most mammals.

scyphistoma (sī-fis′tə-mə) (Gr. *skyphos*, cup, + *stoma*, mouth). A stage in the development of scyphozoan jellyfish just after the larva becomes attached, the polyp form of a scyphozoan.

sebaceous (sə-bāsh′əs) (L. *sebaceus*, made of tallow). A type of mammalian epidermal gland that produces a fatty substance.

sedentary (sed′ən-ter-ē). Stationary, sitting, inactive; staying in one place.

selectively permeable. Permeable to small particles, such as water and certain inorganic ions, but not to larger molecules.

seminiferous (sem-ə-nif′rəs) (L. *semen*, semen, + *ferre*, to bear). Pertains to the tubules that produce or carry semen in the testes.

semipermeable (L. *semi*, half, + *permeabilis*, capable of being passed through). Permeable to small particles, such as water and certain inorganic ions, but not to larger molecules.

sensillum, pl. **sensilla** (sin-si′ləm) (L. *sensus*, sense). A small sense organ, especially in the arthropods.

septum, pl. **septa** (L. fence). A wall between two cavities.

serial homology. See **homology.**

serosa (sə-rō′sə) (N.L. from L. *serum*, serum). The outer embryonic membrane of birds and reptiles; chorion. Also, the peritoneal lining of the body cavity.

serotonin (sir′ə-tōn′ən) (L. *serum*, serum). A phenolic amine, found in the serum of clotted blood and in many other tissues, that possesses several poorly understood metabolic, vascular, and neural functions; 5-hydroxytryptamine.

serous (sir′əs) (L. *serum*, serum). Watery, resembling serum; applied to glands, tissue, cells, fluid.

serum (sir′əm) (L. whey, serum). The liquid that separates from the blood after coagulation; blood plasma from which fibrinogen has been removed. Also, the clear portion of a biological fluid separated from its particulate elements.

sessile (ses′əl) (L. *sessilis*, low, dwarf). Attached at the base; fixed to one spot, not able to move about.

seta (sīd′ə), pl. **setae** (sē′tē) (L. bristle). A needlelike chitinous structure of the integument of annelids, arthropods, and others.

sex chromosomes. Chromosomes that determine gender of an animal. They may bear a few or many other genes.

sibling species. Reproductively isolated species that are so similar morphologically that they are difficult or impossible to distinguish using morphological characters.

bat / āpe / ärmadillo / herring / fēmale / finch / līce / crocodile / crōw / duck / ūnicorn / ə indicates unaccented vowel sound "uh" as in mammal, fishes, cardinal, heron, vulture / stress as in bi-ol′o-gy, bī′o-log′i-cal

sickle cell anemia. A condition that causes the red blood cells to collapse (sickle) under oxygen stress. The condition becomes manifest when an individual is homozygous for the gene for hemoglobin-S (HbS).

siliceous (sə-li′shəs) (L. *silex*, flint). Containing silica.

simian (sim′ē-ən) (L. *simia*, ape). Pertaining to monkeys or apes.

sinistral (si′nə-strəl, sə-ni′stral) (L. *sinister*, left). Pertaining to the left; in gastropods, shell is sinistral if opening is to left of columella when held with spire up and facing observer.

sinus (sī′nəs) (L. curve). A cavity or space in tissues or in bone.

siphonoglyph (sī′fän′ə-glif′) (Gr. *siphōn*, reed, tube, siphon, + *glyphē*, carving). Ciliated furrow in the gullet of sea anemones.

siphuncle (sī′fun-kəl) (L. *siphunculus*, small tube). Cord of tissue running through the shell of a nautiloid, connecting all chambers with body of animal.

sister group. The relationship between a pair of species or higher taxa that are each other's closest phylogenetic relatives.

sociobiology. Ethological study of social behavior in humans or other animals.

solenia (sō-len′ē-ə) (Gr. *sōlēn*, pipe). Channels through the coenenchyme connecting the polyps in an alcyonarian colony (phylum Cnidaria).

solenocyte (sō-len′ə-sīt) (Gr. *sōlēn*, pipe, + *kytos*, hollow vessel). Special type of flame bulb in which the bulb bears a flagellum instead of a tuft of flagella. See **flame cell, protonephridium.**

soma (sō′mə) (Gr. body). The whole of an organism except the germ cells (germ plasm).

somatic (sō-mat′ik) (Gr. *sōma*, body). Refers to the body, for example, somatic cells in contrast to germ cells.

somatocoel (sə-mat′ə-sēl) (Gr. *sōma*, the body, + *koilos*, hollow). Posterior coelomic compartment of echinoderms; left somatocoel gives rise to oral coelom, and right somatocoel becomes aboral coelom.

somatoplasm (sō′mə-də-pla′zm) (Gr. *sōma*, body, + *plasma*, anything formed). The living matter that makes up the mass of the body as distinguished from germ plasm, which makes up the reproductive cells. The protoplasm of body cells.

somite (sō′mīt) (Gr. *soma*, body). One of the blocklike masses of mesoderm arranged segmentally (metamerically) in a longitudinal series beside the neural tube of the embryo; metamere.

sorting. Differential survival and reproduction among varying individuals; often confused with natural selection which is one possible cause of sorting.

speciation (spē′sē-ā′shən) (L. *species*, kind). The evolutionary process or event by which new species arise.

species (spē′shez, spē′sēz) sing. and pl. (L. particular kind). A group of interbreeding individuals of common ancestry that are reproductively isolated from all other such groups; a taxonomic unit ranking below a genus and designated by a binomen consisting of its genus and the species name.

spermatheca (spər′mə-thē′kə) (Gr. *sperma*, seed, + *thēkē*, case). A sac in the female reproductive organs for the reception and storage of sperm.

spermatid (spər′mə-təd) (Gr. *sperma*, seed, + *eidos*, form). A growth stage of a male reproductive cell arising by division of a secondary spermatocyte; gives rise to a spermatozoon.

spermatocyte (spər-mad′ə-sīt) (Gr. *sperma*, seed, + *kytos*, hollow vessel). A growth stage of a male reproductive cell; gives rise to a spermatid.

spermatogenesis (spər-mad′ə-jen′-ə-səs) (Gr. *sperma*, seed, + *genesis*, origin). Formation and maturation of spermatozoa.

spermatogonium (spər′mad-ə-gō′nē-əm) (Gr. *sperma*, seed, + *gonē*, offspring). Precursor of mature male reproductive cell; gives rise directly to a spermatocyte.

spermatophore (spər-mad′ə-fōr′) (Gr. *sperma*, *spermatos*, seed, + *pherein*, to bear). Capsule or packet enclosing sperm, produced by males of several invertebrate groups and a few vertebrates.

sphincter (sfingk′tər) (Gr. *sphinkter*, band, sphincter, from *sphingein*, to bind tight). A ring-shaped muscle capable of closing a tubular opening by constriction.

spicule (spi′kyul) (L. dim. *spica*, point). One of the minute calcareous or siliceous skeletal bodies found in sponges, radiolarians, soft corals, and sea cucumbers.

spiracle (spi′rə-kəl) (L. *spiraculum*, from *spirare*, to breathe). External opening of a trachea in arthropods. One of a pair of openings on the head of elasmobranchs for passage of water. Exhalent aperture of tadpole gill chamber.

spiral cleavage. A type of embryonic cleavage in which cleavage planes are diagonal to the polar axis and unequal cells are produced by the alternate clockwise and counterclockwise cleavage around the axis of polarity; determinate cleavage.

spongin (spun′jin) (L. *spongia*, sponge). Fibrous, collagenous material making up the skeletal network of horny sponges.

spongioblast (spun′je-o-blast) (Gr. *spongos*, sponge, + *blastos*, bud). Cell in a sponge that secretes spongin, a protein.

spongocoel (spun′jō-sēl) (Gr. *spongos*, sponge, + *koilos*, hollow). Central cavity in sponges.

spongocyte (spun′jō-sīt) (Gr. *spongos*, sponge, + *kytos*, hollow vessel). A cell in sponges that secretes spongin.

sporocyst (spō′rə-sist) (Gr. *sporos*, seed, + *kystis*, pouch). A larval stage in the life cycle of flukes; it originates from a miracidium.

sporogony (spor-äg′ə-nē) (Gr. *sporos*, seed, + *gonos*, birth). Multiple fission to produce sporozoites after zygote formation.

sporozoite (spō′rə-zō′īt) (Gr. *sporos*, seed, + *zōon*, animal, + *ite*, suffix for body part). A stage in the life history of many sporozoan protozoa; released from oocysts.

squalene (skwā′lēn) (L. *squalus*, a kind of fish). A liquid acyclic triterpene hydrocarbon found especially in the liver oil of sharks.

squamous epithelium (skwā′məs) (L. *squama*, scale, + *osus*, full of). Simple epithelium of flat, nucleated cells.

stapes (stā′pēz) (L. stirrup). Stirrup-shaped innermost bone of the middle ear.

statoblast (stad′ə-blast) (Gr. *statos*, standing, fixed, + *blastos*, germ). Biconvex capsule containing germinative cells and produced by most freshwater ectoprocts by asexual budding. Under favorable conditions it germinates to give rise to new zooid.

statocyst (Gr. *statos*, standing, + *kystis*, bladder). Sense organ of equilibrium; a fluid-filled cellular cyst containing one or more granules (statoliths) used to sense direction of gravity.

statolith (Gr. *statos*, standing, + *lithos*, stone). Small calcareous body resting on tufts of cilia in the statocyst.

stenohaline (sten-ə-hā′līn, -lən) (Gr. *stenos*, narrow, + *hals*, salt). Pertaining to aquatic organisms that have restricted tolerance to changes in environmental saltwater concentration.

stenophagous (stə-näf′ə-gəs) (Gr. *stenos*, narrow, + *phagein*, to eat). Eating few kinds of foods.

stenotopic (sten-ə-tä′pik) (Gr. *stenos*, narrow, + *topos*, place). Refers to an organism with a narrow range of adaptability to environmental change; having a restricted environmental distribution.

stereogastrula (ste′rē-ə-gas′trə-lə) (Gr. *stereos*, solid, + *gastēr*, stomach, + L. *ula*, dim.). A solid type of gastrula, such as the planula of cnidarians.

stereom (ster′ē-ōm) (Gr. *stereos*, solid, hard, firm). Meshwork structure of endoskeletal ossicles of echinoderms.

stereotyped behavior. A pattern of behavior repeated with little variation in performance.

sternum (ster′nəm) (L. breastbone). Ventral plate of an arthropod body segment; breastbone of vertebrates.

sterol (ste′rōl), steroid (ste′roid) (Gr. *stereos*, solid, + L. *ol*, from *oleum*, oil). One of a class of organic compounds containing a molecular skeleton of four fused carbon rings; it includes cholesterol, sex hormones, adrenocortical hormones, and vitamin D.

stigma (Gr. *stigma*, mark, tatoo mark). Eyespot in certain protozoa. Spiracle of certain terrestrial arthropods.

stolon (stō′lən) (L. *stolō, stolonis*, a shoot, or sucker of a plant). A rootlike extension of the body wall giving rise to buds that may develop into new zooids, thus forming a compound animal in which the zooids remain united by the stolon. Found in some colonial anthozoans, hydrozoans, ectoprocts, and ascidians.

stoma (stō′mə) (Gr. mouth). A mouthlike opening.

stomochord (stō′mə-kord) (Gr. *stoma*, mouth, + *chordē*, cord). Anterior evagination of the dorsal wall of the buccal cavity into the proboscis of hemichordates; the buccal diverticulum.

strobila (strō′bə-lə) (Gr. *strobilē*, lint plug like a pine cone [*strobilos*]). A stage in the development of the scyphozoan jellyfish. Also, the chain of proglottids of a tapeworm.

strobilation (strō′bə-lā′shən) (Gr. *strobilos*, a pine cone). Repeated, linear budding of individuals, as in scyphozoans (phylum Cnidaria), or sets of reproductive organs, as in tapeworms (phylum Platyhelminthes).

stroma (strō′mə) (Gr. *strōma*, bedding). Supporting connective tissue framework of an animal organ; filmy framework of red blood corpuscles and certain cells.

structural gene. A gene carrying the information to construct a protein.

subnivean (səb-ni′vē-ən) (L. *sub*, under, below, + *nivis*, snow). Applied to environments beneath snow, in which snow insulates against a colder atmospheric temperature.

substrate. The substance upon which an enzyme acts; also, a base or foundation (substratum); and the substance or base on which an organism grows.

sycon (sī′kon) (Gr. *sykon*, fig). A type of canal system in certain sponges. Sometimes called syconoid.

symbiosis (sim′bī-ōs′əs, sim′bē-ōs′əs) (Gr. *syn*, with, + *bios*, life). The living together of two different species in an intimate relationship. Symbiont always benefits; host may benefit, may be unaffected, or may be harmed (mutualism, commensalism, and parasitism).

sympatric (sim′pa′-trik) (Gr. *syn*, with, + *patra*, native land). Having the same or overlapping regions of geographical distribution. Noun, **sympatry.**

symplesiomorphy (sim-plē′sē-ə-mōr′fē). Sharing among species of ancestral characteristics, not indicative that the species comprise a monophyletic group.

synapomorphy (sin-ap′o-mor′fē) (Gr. *syn*, together with, + *apo*, of, + *morphe*, form). Shared, evolutionarily derived character states that are used to recover patterns of common descent among two or more species.

synapse (si′naps, si-naps′) (Gr. *synapsis*, contact, union). The place at which a nerve impulse passes between neuron processes, typically from an axon of one nerve cell to a dendrite of another nerve cell.

synapsids (si-nap′sədz) (Gr. *synapsis*, contact, union). An amniote lineage comprising the mammals and the ancestral mammal-like reptiles, having a skull with a single pair of temporal openings.

synapsis (si-nap′səs) (Gr. *synapsis*, contact, union). The time when the pairs of homologous chromosomes lie alongside each other in the first meiotic division.

synaptonemal complex (sin-ap′tə-nē′məl) (Gr. *synapsis*, a joining together, + *nēma*, thread). The structure that holds homologous chromosomes together during synapsis in prophase of meiosis I.

syncytium (sən-sish′e-əm) adj. **syncytial** (Gr. *syn*, with, + *kytos*, hollow). A multinucleated cell.

syndrome (sin′drōm) (Gr. *syn*, with, + *dramein*, to run). A group of symptoms characteristic of a particular disease or abnormality.

syngamy (sin′gə-mē) (Gr. *syn*, with, + *gamos*, marriage). Fertilization of one gamete with another individual gamete to form a zygote, found in most animals with sexual reproduction.

synkaryon (sin-ker′e-on) (Gr. *syn*, with, + *karyon*, nucleus). Zygote nucleus resulting from fusion of pronuclei.

syrinx (sir′inks) (Gr. shepherd's pipe). The vocal organ of birds located at the base of the trachea.

systematics (sis-tə-mat′iks). Science of classification and reconstruction of phylogeny.

systole (sis′tə-lē) (Gr. *systolē*, drawing together). Contraction of heart.

T

T cell. A type of lymphocyte important in cellular immune response and in regulation of most immune responses.

tactile (tak′til) (L. *tactilis*, able to be touched, from *tangere*, to touch). Pertaining to touch.

tagma, pl. tagmata (Gr. *tagma*, arrangement, order, row). A compound body section of an arthropod resulting from embryonic fusion of two or more segments; for example, head, thorax, abdomen.

tagmatization, tagmosis. Organization of the arthropod body into tagmata.

taiga (tī′gä) (Russ.). Habitat zone characterized by large tracts of coniferous forests, long, cold winters, and short summers; most typical in Canada and Siberia.

tantulus (tan′tə-ləs) (Gr. *tantulus*, so small). Larva of a tantulocaridan (subphylum Crustacea).

taxis (tak′sis), pl. **taxes** (Gr. *taxis*, arrangement). An orientation movement by a (usually) simple organism in response to an environmental stimulus.

taxon (tak′son), pl. **taxa** (Gr. *taxis,* arrangement). Any taxonomic group or entity.

taxonomy (tak-sän′ə-mi) (Gr. *taxis,* arrangement, + *nomos,* law). Study of the principles of scientific classification; systematic ordering and naming of organisms.

tectum (tek′təm) (L. roof). A rooflike structure, for example, dorsal part of capitulum in ticks and mites.

tegmen (teg′mən) (L. *tegmen,* a cover). External epithelium of crinoids (phylum Echinodermata).

tegument (teg′ū-ment) (L. *tegumentum,* from *tegere,* to cover). An integument: specifically external covering in cestodes and trematodes, formerly believed to be a cuticle.

telencephalon (tel′en-sef′ə-lon) (Gr. *telos,* end, + *encephalon,* brain). The most anterior vesicle of the brain; the anterior-most subdivision of the prosencephalon that becomes the cerebrum and associated structures.

teleology (tel′ē-äl′ə-jē) (Gr. *telos,* end, + L. *logia,* study of, from Gr. *logos,* word). The philosophical view that natural events are goal directed and are preordained, as opposed to the scientific view of mechanical determinism.

telocentric (tē′lō-sen′trək) (Gr. *telos,* end, + *kentron,* center). Chromosome with centromere at the end.

telolecithal (te-lō-les′ə-thəl) (Gr. *telos,* end, + *lekithos,* yolk, + *al*). Having the yolk concentrated at one end of an egg.

telson (tel′sən) (Gr. *telson,* extremity). Posterior projection of the last body segment in many crustaceans.

temnospondyls (tem-nō-spän′dəls) (Gr. *temnō,* to cut, + *spondylos,* vertebra). A large lineage of amphibians that extended from the Carboniferous to the Triassic.

template (tem′plət). A pattern or mold guiding the formation of a duplicate; often used with reference to gene duplication.

tendon (ten′dən) (L. *tendo,* tendon). Fibrous band connecting muscle to bone or other movable structure.

tentaculocyst (ten-tak′u-lō-sist) (L. *tentaculum,* feeler, + *kystis,* pouch). One of the sense organs along the margin of medusae; a rhopalium.

tergum (ter′gəm) (L. back). Dorsal part of an arthropod body segment.

territory (L. *territorium,* from *terra,* earth). A restricted area preempted by an animal or pair of animals, usually for breeding purposes, and guarded from other individuals of the same species.

test (L. *testa,* shell). A shell or hardened outer covering.

tetrad (te′trad) (Gr. *tetras,* four). Group of two pairs of chromatids at synapsis and resulting from the replication of paired homologous chromosomes; the bivalent.

tetrapods (te′trə-päds) (Gr. *tetras,* four, + *pous, podos,* foot). Four-footed vertebrates; the group includes amphibians, reptiles, birds, and mammals.

thecodonts (thēk′ə-dänts) (Gr. *thēkē,* box, + *odontos,* tooth). A large assemblage of Triassic archosaurian diapsids of the order Thecodontia and characterized by having teeth set in sockets.

therapsids (thə-rap′sidz) (Gr. *theraps,* an attendant). Extinct Mesozoic mammal-like reptiles from which true mammals evolved.

thermocline (thər′mō-klīn) (Gr. *thermē,* heat, + *klinein,* to swerve). Layer of water separating upper warmer and lighter water from lower colder and heavier water in a lake or sea; a stratum of abrupt change in water temperature.

thoracic (thō-ra′sək) (L. *thōrax,* chest). Pertaining to the thorax or chest.

thrombin. Enzyme catalyzing fibrinogen transformation into fibrin. Percursor is **prothrombin.**

Tiedemann's bodies (tēd′ə-mənz) (from F. Tiedemann, German anatomist). Four or five pairs of pouchlike bodies attached to the ring canal of sea stars, apparently functioning in production of coelomocytes.

tight junction. Region of actual fusion of cell membranes between two adjacent cells.

tissue (ti′shu) (M.E. *tissu,* tissue). An aggregation of cells, usually of the same kind, organized to perform a common function.

titer (tī′tər) (Fr. *titrer,* to titrate). Concentration of a substance in a solution as determined by titration.

tornaria (tor-na′rē-ə) (L. *tornare,* to turn). A free-swimming larva of enteropneusts that rotates as it swims; resembles somewhat the bipinnaria larva of echinoderms.

torsion (L. *torquere,* to twist). A twisting phenomenon in gastropod development that alters the position of the visceral and pallial organs by 180 degrees.

toxicyst (tox′i-sist) (Gr. *toxikon,* poison, + *kystis,* bladder). Structures possessed by predatory ciliate protozoa, which on stimulation expel a poison to subdue the prey.

trabecular net (trə-bek′ū-lər) (L. *trabecula,* a small beam). Network of living tissue formed by pseudopodia of amebocytes in Hexactinellida (phylum Porifera).

trachea (trā′kē-ə) (M.L. windpipe). The windpipe. Also, any of the air tubes of insects.

transcription. Formation of messenger RNA from the coded DNA.

transduction. Condition in which bacterial DNA (and the genetic characteristics it bears) is transferred from one bacterium to another by the agent of viral infection.

transfer RNA (tRNA). A form of RNA of about 70 or 80 nucleotides, which are adapter molecules in the synthesis of proteins. A specific amino acid molecule is carried by transfer RNA to a ribosome-messenger RNA complex for incorporation into a polypeptide.

transformation. Condition in which DNA in the environment of bacteria somehow penetrates them and is incorporated into their genetic complement, so that their progeny inherit the genetic characters so acquired.

translation (L. a transferring). The process in which the genetic information present in messenger RNA is used to direct the order of specific amino acids during protein synthesis.

transporter. See **permease.**

transverse plane (L. *transversus,* across). A plane or section that lies or passes across a body or structure.

trichinosis (trik-ən-o′səs). Disease caused by infection with the nematode *Trichinella spiralis.*

trichocyst (trik′ə-sist) (Gr. *thrix,* hair, + *kystis,* bladder). Sac-like protrusible organelle in the ectoplasm of ciliates, which discharges as a threadlike weapon of defense.

triglyceride (trī-glis′ə-rīd) (Gr. *tria,* three, + *glykys,* sweet, + *ide,* suffix denoting compound). A triester of glycerol with one, two, or three acids.

trimerous (trī′mə-rəs) (Gr. *treis,* three, + *meros,* a part). Body in three main divisions, as in lophophorates and some deuterostomes.

tripartite (trī-par′tīt). See **trimerous.**

triploblastic (trip′lō-blas′tik) (Gr. *triploos,* triple, + *blastos,* germ). Pertaining to metazoa in which the embryo has three primary germ layers—ectoderm, mesoderm, and endoderm.

trisomy 21. See **Down syndrome.**

trochophore (trōk′ə-fōr) (Gr. *trochos,* wheel, + *pherein,* to bear). A free-swimming ciliated marine larva characteristic of most molluscs and certain ectoprocts, brachiopods, and marine worms; an ovoid or pyriform body with preoral circlet of cilia and sometimes a secondary circlet behind the mouth.

trophallaxis (trōf′ə-lak′səs) (Gr. *trophē,* food, + *allaxis,* barter, exchange). Exchange of food between young and adults, especially certain social insects.

trophi (trō′fi) (Gr. *trophos,* one who feeds). Jaw-like structures in the mastax of rotifers.

trophic (trō′fək) (Gr. *trophē,* food). Pertaining to feeding and nutrition.

trophoblast (trōf′ə-blast) (Gr. *trephein,* to nourish, + *blastos,* germ). Outer ectodermal nutritive layer of blastodermic vesicle; in mammals it is part of the chorion and attaches to the uterine wall.

trophosome (trof′ə-sōm) (Gr. *trophē,* food, + *sōma,* body). Organ in poganophorans bearing mutualistic bacteria, derived from midgut.

bat / āpe / ärmadillo / herring / fēmale / finch / līce / crocodile / crōw / duck / ūnicorn / ə indicates unaccented vowel sound "uh" as in mammal, fishes, cardinal, heron, vulture / stress as in bi-ol′o-gy, bi′o-log′i-cal

trophozoite (trŏf′ə-zō′ĭt) (Gr. *trophē*, food, + *zōon*, animal). Adult stage in the life cycle of a protozoan in which it is actively absorbing nourishment.

tropic (trä′pic) (Gr. *tropē*, to turn toward). Related to the tropics (tropical); in endocrinology, a hormone that influences the action of another hormone or endocrine gland (usually pronounced trō′pic).

tropomyosin (trōp′ə-mī′ə-sən) (Gr. *tropos*, turn, + *mys*, muscle). Low-molecular weight protein surrounding the actin filaments of striated muscle.

troponin (trə-pōn′in). Complex of globular proteins positioned at intervals along the actin filament of skeletal muscle; thought to serve as a calcium-dependent switch in muscle contraction.

tube feet (podia). Numerous small, muscular, fluid-filled tubes projecting from body of echinoderms; part of water-vascular system; used in locomotion, clinging, food handling, and respiration.

tubercle (tū′bər-kəl) (L. *tuberculum,* small hump). Small protuberance, knob, or swelling.

tubulin (tū′bū-lən) (L. *tubulus,* small tube, + *in,* belonging to). Globular protein forming the hollow cylinder of microtubules.

tumor necrosis factor. A cytokine, the most important source of which is macrophages, that is a major mediator of inflammation.

tundra (tun′drə) (Russ. from Lapp, *tundar,* hill). Terrestrial habitat zone, located between taiga and polar regions; characterized by absence of trees, short growing season, and mostly frozen soil during much of the year.

tunic (L. *tunica,* tunic, coat). In tunicates, a cuticular, cellulose-containing covering of the body secreted by the underlying body wall.

tympanic (tim-pan′ik) (Gr. *tympanon,* drum). Relating to the tympanum that separates the outer and middle ear (eardrum).

type specimen. A specimen deposited in a museum that formally defines the name of the species that it represents.

typhlosole (tif′lə-sōl′) (Gr. *typhlos,* blind, + *sōlēn,* channel, pipe). A longitudinal fold projecting into the intestine in certain invertebrates such as the earthworm.

typology (tī-päl′ə-jē) (L. *typus,* image). A classification of organisms in which members of a taxon are perceived to share intrinsic, essential properties, and variation among organisms is regarded as uninteresting and unimportant.

U

ulcer (ul-sər) (L. *ulcus,* ulcer). An abscess that opens through the skin or a mucous surface.

ultimate cause (L. *ultimatus,* last, + *causa*). The evolutionary factors responsible for the origin, state of being, or purpose of a biological system.

umbilical (L. *umbilicus,* navel). Refers to the navel, or umbilical cord.

umbo (um′bō), pl. **umbones** (əm-bō′nēz) (L. boss of a shield). One of the prominences on either side of the hinge region in a bivalve mollusc shell. Also, the "beak" of a brachiopod shell.

ungulate (un′gū-lət) (L. *ungula,* hoof). Hooved. Noun, any hooved mammal.

uniformitarianism (ū′nə-fōr′mə-ter′ē-ə-niz′əm). Methodological assumptions that the laws of chemistry and physics have remained constant throughout the history of the earth, and that past geological events occurred by processes that can be observed today.

ureter (ūr′ə-tər) (Gr. *ouētēr,* ureter). Duct carrying urine from kidney to bladder.

urethra (ū-rē′thrə) (Gr. *ourethra,* urethra). The tube from the urinary bladder to the exterior in both sexes.

uropod (ū′rə-pod) (Gr.. *oura,* tail, + *pous, podos,* foot). Posteriormost appendage of many crustaceans.

utricle (ū′trə-kəl) (L. *utriculus,* little bag). That part of the inner ear containing the receptors for dynamic body balance; the semicircular canals lead from and to the utricle.

V

vacuole (vak′yə-wōl) (L. *vacuus,* empty, + Fr. *ole,* dim.). A membrane-bounded, fluid-filled space in a cell.

valence (vā′ləns) (L. *valere,* to have power). Degree of combining power of an element as expressed by the number of atoms of hydrogen (or its equivalent) that the element can hold (if negative) or displace in a reaction (if positive). The oxidation state of an element in a compound. The number of electrons gained, shared, or lost by an atom when forming a bond with one or more other atoms.

valve (L. *valva,* leaf of a double door). One of the two shells of a typical bivalve mollusc or brachiopod.

variation (L. *varius,* various). Differences among individuals of a group or species that cannot be ascribed to age, sex, or position in the life cycle.

vector (L. a bearer, carrier, from *vehere, vectum,* to carry). Any agent that carries and transmits pathogenic microorganisms from one host to another host. Also, in molecular biology, an agent such as bacteriophage or plasmid that carries recombinant DNA.

veins (vānz) (L. *vena,* a vein). Blood vessels that carry blood toward the heart; in insects, fine extensions of the tracheal system that support the wings.

velarium (və-la′rē-əm) (L. *velum,* veil, covering). Shelf-like extension of the subumbrella edge in cubozoans (phylum Cnidaria).

veliger (vēl′ə-jər, vel-) (L. *velum,* veil, covering). Larval form of certain molluscs; develops from the trochophore and has the beginning of a foot, mantle, shell, and so on.

velum (vē′ləm) (L. veil, covering). A membrane on the subumbrella surface of jellyfish of class Hydrozoa. Also, a ciliated swimming organ of the veliger larva.

ventral (ven′trəl) (L. *venter,* belly). Situated on the lower or abdominal surface.

venule (ven′ūl) (L. *venula,* dim. of *vena,* vein). Small vessel conducting blood from capillaries to vein; small vein of insect wing.

vermiform (ver′mə-form) (L. *vermis,* worm, + *forma,* shape). Adjective to describe any wormlike organism; an adult (nematogen) rhombozoan (phylum Mesozoa).

vestige (ves′tij) (L. *vestigium,* footprint). A rudimentary organ that may have been well developed in some ancestor or in the embryo.

vibrissa (vī-bris′ə), pl. **vibrissae** (L. nostril-hair). Stiff hairs that grow from the nostrils or other parts of the face of many mammals and that serve as tactile organs; "whiskers."

vicariance (vī-kar′ē-ənts) (L. *vicarius,* a substitute). Geographical separation of populations, especially as imposed by discontinuities in the physical environment that fragmented populations that were formerly geographically continuous.

villus (vil′əs), pl. **villi** (L. tuft of hair). A small fingerlike, vascular process on the wall of the small intestine. Also one of the branching, vascular processes on the embryonic portion of the placenta.

virus (vī′rəs) (L. slimy liquid, poison). A submicroscopic noncellular particle composed of a nucleoprotein core and a protein shell; parasitic; will grow and reproduce in a host cell.

viscera (vis′ər-ə) (L. pl. of *viscus*, internal organ). Internal organs in the body cavity.

visceral (vis′ər-əl). Pertaining to viscera.

vitalism (L. *vita*, life). The view that natural processes are controlled by supernatural forces and cannot be explained through the laws of physics and chemistry alone, as opposed to mechanism.

vitamin (L. *vita*, life, + *amine*, from former supposed chemical origin). An organic substance required in small amounts for normal metabolic function; must be supplied in the diet or by intestinal flora because the organism cannot synthesize it.

vitellaria (vi′təl-lar′e-ə) (L. *vitellus*, yolk of an egg). Structures in many flatworms that produce vitelline cells, that is, cells that provide eggshell material and nutrient for the embryo.

vitelline gland. See **vitellaria.**

vitelline membrane (və-tel′ən, vī′təl-ən) (L. *vitellus*, yolk of an egg). The noncellular membrane that encloses the egg cell.

bat / āpe / ärmadillo / herring / fēmale / finch / līce / crocodile / crōw / duck / ūnicorn / ə indicates unaccented vowel sound "uh" as in mammal, fishes, cardinal, heron, vulture / stress as in bi-ol′o-gy, bi′o-log′i-cal

viviparity (vī′və-par′ə-dē) (L. *vivus*, alive, + *parere*, to bring forth). Reproduction in which eggs develop within the female body, with nutritional aid of maternal parent as in therian mammals, many reptiles, and some fishes; offspring are born as juveniles. Adj., **viviparous** (vī-vip′ə-rəs).

W

water-vascular system. System of fluid-filled closed tubes and ducts peculiar to echinoderms; used to move tentacles and tube feet that serve variously for clinging, food handling, locomotion, and respiration.

X

xanthophore (zan′thə-fōr) (Gr. *xanthos*, yellow, + *pherein*, to bear). A chromatophore containing yellow pigment.

X-organ. Neurosecretory organ in eyestalk of crustaceans that secretes molt-inhibiting hormone.

Y

Y-organ. Gland in the antennal or maxillary segment of some crustaceans that secretes molting hormone.

Z

zoecium, zooecium (zō-ē′shē-əm) (Gr. *zōon*, animal, + *oikos*, house). Cuticular sheath or shell of Ectoprocta.

zoochlorella (zō′ə-klōr-el′ə) (Gr. *zōon*, life, + *Chlorella*). Any of various minute green algae (usually *Chlorella*) that live symbiotically within the cytoplasm of some protozoa and other invertebrates.

zooflagellates (zō′ə-fla′jə-lāts). Members of the Zoomastigophora, the animal-like flagellates (phylum Sarcomastigophora).

zooid (zō-id) (Gr. *zōon*, life). An individual member of a colony of animals, such as colonial cnidarians and ectoprocts.

zooxanthella (zo′ə-zan-thəl′ə) (Gr. *zōon*, animal, + *xanthos*, yellow). A minute dinoflagellate alga living in the tissues of many types of marine invertebrates.

zygote (Gr. *zygōtos*, yoked). The fertilized egg.

zygotic meiosis. Meiosis that takes place within the first few divisions after zygote formation; thus all stages in the life cycle other than the zygote are haploid.

CREDITS

Photos

Part Openers

1: Cleveland P. Hickman, Jr.; **2:** © Tom Tietz/Tony Stone Images; **3:** Larry S. Roberts; **4:** © James Martin/Tony Stone Images; **5:** Cleveland P. Hickman, Jr.

Chapter 1

Opener: Cleveland P. Hickman, Jr.; **1.1a:** © Dave B. Fleetham/Visuals Unlimited; **1.1b:** © Steve McCutcheon/Visuals Unlimited; **1.1c:** © Peter Ziminski/Visuals Unlimited; **1.1d:** © Link/Visuals Unlimited; **1.1e:** © T. E. Adams/Visuals Unlimited; **1.2a:** Courtesy of IMB U.K. Scientific Centre; **1.3:** © John D. Cunningham/Visuals Unlimited; **1.4:** © David M. Phillips/Visuals Unlimited; **1.5a:** N. P. Salzman; **1.5b:** © Ed Reschke; **1.5c:** © Ken Highfill/Photo Researchers, Inc.; **1.5, Bottom Left:** Larry S. Roberts; **1.5, Top Right:** © William Ober; **1.6:** © A. C. Barrington Brown/Photo Researchers, Inc.; **1.7a:** © M. Abbey/Visuals Unlimited; **1.7b:** © S. Dalton/National Audubon Society Collection/Photo Researchers, Inc.; **1.8 a,b:** © D. Kline/Visuals Unlimited; **1.12 a,b:** © Michael Tweedie/Photo Researchers, Inc.; **1.13:** Courtesy American Museum of Natural History, Neg. #326668; **1.17 a,b:** Courtesy Gregor Mendel Museum, Brno, Czechoslovakia; **1.20:** © Carolina Biological Supply/Phototake; **p. 18:** Foundations For Biomedical Research

Chapter 2

Opener: Larry S. Roberts; **2.3:** © G. I. Bernard/Animals Animals/Earth Scenes

Chapter 3

Opener: NASA; **3.1:** The Bettman Archive; **3.4b:** Courtesy Kevin Walsh, U.S.C.D.; **3.5:** Courtesy R. M. Syren and S. W. Fox, Institute of Molecular Evolution/University of Miami, Coral Gables, Florida; **3.6:** Cleveland P. Hickman, Jr.

Chapter 4

Opener: © Bill Ober; **4.1a:** © John D. Cunningham/Visuals Unlimited; **4.1b:** From C. R. Morgan and R. A. Jersild, Jr., 1970. *Anat. Rec.* 166:575–586.; **4.5:** Courtesy A. Wayne Vogl; **4.6:** Courtesy Susumo Ito; **4.7:** Courtesy G. E. Palade, University of California School of

Medicine; **4.8b:** Courtesy Richard Rodewald; **4.9b, 4.11b:** Courtesy of Charles Flickinger; **4.12:** Courtesy A. Wayne Vogl; **4.13:** © K. G. Murti/Visuals Unlimited; **4.14b:** Courtesy Kent McDonald; **4.25 a–e:** © Times Mirror Higher Education Group, Inc./Kingsley Stern, photographer

Chapter 5

Opener: © Gary W. Carter/Visuals Unlimited

Chapter 6

Opener: © Francis Leroy, Biocosmos/SPL/Photo Researchers, Inc.; **6.3:** © Robert Humbert/Biological Photo Service; **6.8:** From R. G. Kessel and R. H. Kardon, *Tissues and Organs: A Text-Atlas of Scanning Electron Microscopy,* 1979, W. H. Freeman and Co.

Chapter 7

Opener: Courtesy MBL Archive, Woods Hole Oceanographic Institution; **7.5:** Courtesy G. Schatten; **7.17:** © F. R. Turner/Biological Photo Service

Chapter 8

Opener: Photograph courtesy of Calgene, Inc.; **8.1b:** Courtesy Gregor Mendel Museum, Brno, Czechoslovakia; **8.6:** Courtesy M. L. Barr/H. W. Barr; **8.8a:** © Peter J. Bryant/Biological Photo Service

Chapter 9

Opener: © John N. A. Lott/Biological Photo Service; **9.1a,** Neg. #326662, **9.1b, 9.2, 9.3:** Courtesy The Natural History Museum, London; **9.5a:** © Bridgeman/Art Resource; **9.5b:** © Stock Montage; **9.6, 9.7:** Cleveland P. Hickman, Jr.; **9.8a:** © Ken Lucas/Biological Photo Service; **9.8b:** © A. J. Copley/Visuals Unlimited; **9.8c:** © Roberta Hess Poinar; **9.8d:** Courtesy G. O. Poinar, University of California at Berkeley; **9.9a:** Courtesy W. Boehm; **9.10:** Cleveland P. Hickman, Jr.; **9.14:** Courtesy Library of Congress; **9.18:** Courtesy M. K. Kelley, courtesy of Harvard University Press; **9.22b:** Cleveland P. Hickman, Jr.; **9.23:** Courtesy of Storrs Agricultural Experiment Station, University of Connecticut at Storrs; **9.26:** Photo by Fritz Goro; **9.28:** Courtesy Timothy W. Ransom/Biological Photo Service; **9.29:** © S. Krasemann/Photo Researchers, Inc.; **9.30b:** Courtesy Dr. Robert

K. Selander; **9.34:** Courtesy of the Canada Center for Remote Sensing, Energy, Mines, and Resources, Canada

Chapter 10

Opener: From *Animals; A Pictorial Archive from 19th Century Sources* selected by Jim Hartner. 1979 Dove Publications, NY; **10.4 Top Left, 10.4 Middle Left, 10.4 Bottom Left, 10.5 Top Left, 10.5 Bottom Left, 10.6a, 10.6b:** © E. Reschke; **10.6c:** Cleveland P. Hickman, Jr.; **10.6d, 10.7a, 10.7b, 10.7c:** E. Reschke

Chapter 11

Opener: Cleveland P. Hickman, Jr.; **11.1:** Courtesy Library of Congress; **11.5:** Courtesy American Museum of Natural History, Neg. #334101; **11.6a:** © M. Coe/OSF/Animals Animals/Earth Scenes; **11.6b:** © D. Allen/OSF/Animals Animals/Earth Scenes; **11.8:** Courtesy of Dr. George W. Byers, University of Kansas; **11.9:** © Kjell Sandved

Chapter 12

Opener: © M. Abbey/Visuals Unlimited; **12.2:** Courtesy L. Tetley; **12.3b:** Courtesy Dr. Ian R. Gibbons; **12.5:** © M. Abbey/Visuals Unlimited; **12.6a:** Courtesy L. Evans Roth; **12.15a:** © Manfred Kage/Peter Arnold; **12.15b:** © A. M. Siegelman/Visuals Unlimited; **12.17a, 12.17b, 12.17c:** © John Shaw/Tom Stack and Associates; **12.18:** Courtesy J. and M. Cachon. From Lee, J. J., S. H. Hutner, and E. C. Bovee (editors). 1985. *An Illustrated Guide to the Protozoa,* Society of Protozoologists, Allen Press, Lawrence, KS; **12.23:** © Carolina Biological Supply/Phototake

Chapter 13

Opener: Larry S. Roberts; **13.6, 13.8, 13.14a, 13.14b, 13.14c:** Larry S. Roberts

Chapter 14

Opener: Larry S. Roberts; **14.1a, 14.5:** © R. Harbo; **14.6:** © Carolina Biological Supply/Phototake; **14.8:** © Cabisco/Visuals Unlimited; **14.11:** © D. W. Gotshall; **14.14a, 14.14b:** Larry S. Roberts; **14.15:** C. Lane; **14.16, 14.17:** © R. Harbo; **14.19:** © D. P. Wilson/Frank Lane Picture Agency Lmtd.; **14.21:** © D. W. Gotshall; **14.22a:** © J. L. Rotman; **14.22b, 14.24a, 14.24b:** © R. Harbo; **14.25:** © R. F. Myers/Visuals Unlimited;

14.26a: Cleveland P. Hickman, Jr.; **14.26b, 14.26c:** Larry S. Roberts; **14.28:** © R. Wallace/ Visuals Unlimited; **14.29a, 14.31a, 14.31b:** Larry S. Roberts; **14.31c:** © W. C. Ober; **14.32:** © Kjell Sandved; **14.33b:** Cleveland P. Hickman, Jr.; **14.34a:** © J. L. Rotman; **14.34b:** © Kjell Sandved

Chapter 15

Opener: © A. Kerstitch/Visuals Unlimited; **15.2:** © Cabisco/Visuals Unlimited; **15.4, 15.12a:** Larry S. Roberts; **15.13:** R. E. Kuntz, From H. Zaiman *A Pictorial Presentation of Parasites*; **15.14:** © Arthur M. Seigelman/Visuals Unlimited; **15.19:** © Cabisco/Visuals Unlimited; **15.20:** Larry S. Roberts; **15.22:** Cleveland P. Hickman, Jr.

Chapter 16

Opener: Courtesy D. Despommier/From H. Zaiman *A Pictorial Presentation*; **16.12a:** Frances M. Hickman; **16.12b:** G. W. Kelley, Jr./From H. Zaiman *A Pictorial Presentation of Parasites*; **16.13:** E. Pike/From H. Zaiman *A Pictorial Presentation of Parasites*; **16.14:** H. Zaiman/From *A Pictorial Presentation of Parasites*; **16.15a:** © R. Calentine/Visuals Unlimited; **16.15b:** Courtesy H. Zaiman/ From *A Pictorial Presentation of Parasites*; **16.16:** Contributed by E. L. Schiller, AFIP; **16.17:** Larry S. Roberts

Chapter 17

Opener: Larry S. Roberts; **17.1a, 17.1b, 17.1c:** © R. Harbo; **17.1d:** © D. W. Gothshall; **17.1e:** © Fred Bavendam/Peter Arnold, Inc.; **17.3b:** Larry S. Roberts; **17.7:** © Kjell Sandved; **17.10:** © R. Harbo; **17.15a, 17.15b:** © D. W. Gotshall; **17.16a, 17.16b:** © A. Kerstitch; **17.19a:** © R. Harbo; **17.19b:** © Tom Phillipp; **17.20a:** © Kjell Sandved; **17.20b:** © R. Harbo; **17.21a, 17.21b:** Cleveland P. Hickman Jr.; **17.22:** © Tom Phillipp; **17.23a:** Larry S. Roberts; **17.23b:** Cleveland P. Hickman, Jr.; **17.24a:** © R. Harbo; **17.24b:** © D. P. Wilson/Frank Lane Picture Agency Lmtd.; **17.25, 17.27a, 17.27b:** Larry S. Roberts; **17.28a:** © R. Harbo; **17.34b:** Richard J. Neves; **17.35:** Larry S. Roberts; **17.36a:** Courtesy M. Butschler, Vancouver Public Aquarium; **17.37:** Larry S. Roberts; **17.38a:** © Dave Fleetham/Tom Stack & Associates

Chapter 18

Opener: Larry S. Roberts; **18.2a, 18.2b:** Larry S. Roberts; **18.7:** General Biological Supply; **18.8:** © S. Elems/Visuals Unlimited; **18.9:** Larry S. Roberts; **18.11:** © W. C. Jorgensen/Visuals Unlimited; **18.17:** © G. L. Twiest/Visuals Unlimited; **18.20:** Photograph by T. Branning; **18.22:** Cleveland P. Hickman, Jr.

Chapter 19

Opener: © A. J. Copley/Visuals Unlimited; **19.1a, 19.1b:** © A. J. Copley/Visuals Unlimited; **19.7, 19.8, 19.9, 19.10a, 19.10b:** © J. H. Gerard/Nature Press; **19.11a:**

© J. Alcock/Visuals Unlimited; **19.11b:** Cleveland P. Hickman, Jr.; **19.12a, 19.12b:** © J. H. Gerard/Nature Press; **19.13, 19.14:** Larry S. Roberts; **19.15:** © D. S. Snyder/Visuals Unlimited; **19.16:** © A. M. Siegelman/Visuals Unlimited; **19.17:** © John D. Cunningham/ Visuals Unlimited

Chapter 20

Opener: © T. E. Adams/Visuals Unlimited; **20.22a:** Cleveland P. Hickman, Jr.; **20.22b:** © R. Harbo; **20.24a:** Cleveland P. Hickman, Jr.; **20.25:** Larry S. Roberts; **20.26a:** © R. Harbo; **20.26b, 20.26c:** © Kjell Sandved; **20.28a:** Cleveland P. Hickman, Jr.; **20.28b:** © R. Harbo; **20.28c:** Cleveland P. Hickman, Jr.; **20.28d:** Larry S. Roberts; **20.28e:** © J. H. Gerard/Nature Press; **20.29:** Larry S. Roberts

Chapter 21

21.1a: © L. A. Wenner/Visuals Unlimited; **21.2a:** Cleveland P. Hickman, Jr.; **21.3b:** © Dan Kline/Visuals Unlimited; **21.4b:** Cleveland P. Hickman, Jr.; **21.7a, 21.7b:** © Ron West/Nature Photography; **21.9a, 21.9b:** © Kjell Sandved; **21.10:** Cleveland P. Hickman, Jr.; **21.11:** © J. H. Gerard/Nature Press; **21.14:** © John D. Cunningham/Visuals Unlimited; **21.15a:** Cleveland P. Hickman, Jr.; **21.15b:** © J. H. Gerard/Nature Press; **21.20a:** © Kjell Sandved; **21.20b:** Cleveland P. Hickman, Jr.; **21.22a, 21.22b:** © Robert Brons/Biological Photo Service; **21.23:** © J. H. Gerard/Nature Press; **21.25a:** Cleveland P. Hickman, Jr.; **21.25b:** © J. H. Gerard/Nature Press; **21.25c:** © Carolina Biological Supply/Phototake; **21.26a, 21.26b:** © J. H. Gerard/Nature Press; **21.27a, 21.27b, 21.27c:** © Kjell Sandved; **21.28:** © J. H. Gerard/Nature Press; **21.29:** Courtesy J. E. Lloyd; **21.30:** K. Lorenzen © 1979. Educational Images; **21.31a:** © J. H. Gerard/Nature Press; **21.31b, 21.32a:** © Kjell Sandved; **21.32b:** Larry S. Roberts; **21.33, 21.34a:** © L. L. Rue, III; **21.34b, 21.34c:** © J. H. Gerard/Nature Press; **21.35a, 21.35b:** © Kjell Sandved; **21.35c:** Cleveland P. Hickman, Jr.; **21.35d:** © Kjell Sandved

Chapter 22

Opener: Larry S. Roberts; **22.6:** Courtesy J. F. Grassle, Woods Hole Oceanographic Institution; **22.10:** Courtesy J. Ubelaker.; **22.13:** Courtesy D. R. Nelson; **22.15:** Courtesy D. R. Nelson; **22.16:** From R. M. Sayre, *Trans. Am. Microsc.* 88:266–274, 1969.

Chapter 23

Opener: Larry S. Roberts; **23.3a, 23.3b:** Larry S. Roberts; **23.4:** © Ken Lucas/Biological Photo Service; **23.5a, 23.5b:** © Robert Brons/ Biological Photo Service; **23.6:** Larry S. Roberts

Chapter 24

Opener: © Ken Lucas/Visuals Unlimited; **24.1a, 24.1b:** © R. Harbo; **24.1c:** Larry S. Roberts; **24.1d, 24.4f, 24.5a:** © R. Harbo;

24.5b: © D. W. Gotshall; **24.6a, 24.6b:** © J. L. Rotman; **24.8:** Larry S. Roberts; **24.11a:** © R. Harbo; **24.11b:** © D. W. Gotshall; **24.14a, 24.14b, 24.15:** © R. Harbo; **24.16a:** © A. Kerstitch; **24.16b, 24.16c:** © R. Harbo; **24.16d:** © W. C. Ober; **24.16e:** © Kjell Sandved; **24.17a, 24.17b:** © A. Kerstitch; **24.18a, 24.18b:** Larry S. Roberts; **24.21a, 24.21b, 24.21c, 24.24a, 24.24b:** © R. Harbo; **24.24c:** Larry S. Roberts; **24.26:** D. W. Gotshall

Chapter 25

Opener: © Charles Wyttenbach, Univ. of Kansas/Biological Photo Service; **25.1b:** Thuesen, E. V., and R. Bieri, 1987. Canad. J. Zool. 65:181–187

Chapter 26

Opener: © Heather Angel; **26.4:** Courtesy of R. P. S. Jeffries, The Natural History Museum, London; **26.6:** Cleveland P. Hickman, Jr.; **26.8:** © D. B. Fleetham/Visuals Unlimited; **26.10b:** Cleveland P. Hickman, Jr.

Chapter 27

Opener: Jonathan Green; **27.9:** © J. R. Rotman; **27.20a, 27.20b:** Courtesy John G. Shedd Aquarium/Patrice Ceisel; **27.29:** © D. W. Gotshall; **27.30:** © J. L. Rotman; **27.32:** © Will Troyer/Visuals Unlimited; **27.33:** © D. W. Gotshall; **27.34:** Courtesy F. McConnaughey

Chapter 28

Opener: Cleveland P. Hickman, Jr.; **28.6a, 28.6b:** Courtesy L. Houck; **28.9:** Cleveland P. Hickman, Jr.; **28.11:** Allan Larson; **28.12a:** © Ken Lucas/Biological Photo Service; **28.12b, 28.13:** Cleveland P. Hickman, Jr.; **28.14:** Courtesy American Museum of Natural History, Neg. #125617; **28.15, 28.18, 28.25:** Cleveland P. Hickman, Jr.

Chapter 29

Opener: Courtesy Jessie Cohen/National Zoological Park; **29.7, 29.8:** Cleveland P. Hickman,; **29.9:** Jonathan Green; **29.12:** © J. H. Gerard/Nature Press; **29.13:** Cleveland P. Hickman, Jr.; **29.14:** © J. Andrada; **29.15, 29.16, 29.18:** © L. L. Rue, III; **29.19:** Cleveland P. Hickman, Jr.; **29.20, 29.25:** © L. L. Rue, III; **29.27a, 29.27b:** Cleveland P. Hickman, Jr.

Chapter 30

Opener: © William J. Weber/Visuals Unlimited; **30.1a:** Courtesy American Museum of Natural History, Neg. #125065; **30.4:** Cleveland P. Hickman, Jr.; **30.6:** © J. L. McAlonay/Visuals Unlimited; **30.21:** © D. Poe/Visuals Unlimited; **30.22, 30.25, 30.26, 30.28a:** © L. L. Rue, III; **30.29:** Courtesy Culver Pictures; **30.30, 30.31, 30.32, 30.33:** Cleveland P. Hickman, Jr.

Chapter 31

Opener: © L. L. Rue, III; **31.4, 31.5a, 31.5b:** © L. L. Rue, III; **31.6:** R. E. Treat; **31.11:** © L. L. Rue, III; **31.12, 31.13:** © J. Gerlach/

Visuals Unlimited; **31.14a:** Cleveland P. Hickman, Jr.; **31.16:** © S. Maslowski/Visuals Unlimited; **31.18:** © Kjell Sandved/Visuals Unlimited; **31.21:** © L. L. Rue, III; **31.23:** © M. H. Tierney, Jr./Visuals Unlimited; **31.24:** © G. Herben/Visuals Unlimited; **31.26:** © L. L. Rue, III; **31.27, 31.28a:** Courtesy San Diego Zoo; **31.28b:** Cleveland P. Hickman, Jr.; **31.29:** © J. McDonald/Visuals Unlimited; **31.30:** © John Reader; **31.33, 31.34:** Cleveland P. Hickman, Jr.; **31.35:** © Bill Ober; **31.36, 31.37, 31.38:** Cleveland P. Hickman, Jr.

Chapter 32

Opener: © Stephen Dalton/Photo Researchers, Inc.; **32.11a:** © G. W. Willis M.D./Biological Photo Service; **32.11b:** © E. Reschke; **32.11c:** © G. W. Willis M.D./Biological Photo Service

Chapter 33

Opener: Cleveland P. Hickman, Jr.; **33.1:** From J. F. Fulton and L. G. Wilson, *Selected Readings in the History of Physiology,* 1966. Courtesy of Charles C. Thomas, Publisher, Springfield, IL.; **33.11:** From R. G. Kessel and R. H. Kardon, *Tissues and Organs: A Text-Atlas of Scanning Electron Microscopy,* 1979 W. H. Freeman and Co.; **33.21:** © L. L. Rue, III

Chapter 34

Opener: © David M. Phillips/Visuals Unlimited; **34.2:** From J. F. Fulton and L. G. Wilson, *Selected Readings in the History of Physiology,* 1966. Courtesy of Charles C. Thomas, Publisher, Springfield, Il.; **34.4a,** Graziadei; **34.4b:** Courtesy P. P. C. Graziadei; **34.5:** Courtesy N. F. Rodman

Chapter 35

Opener: Cleveland P. Hickman, Jr.; **35.3:** Courtesy Carl Gans; **35.6:** Cleveland P. Hickman, Jr.; **35.12:** Courtesy of Wyeth-Ayerst Laboratories; **35.13:** From R. G. Kessel and R. H. Kardon, *Tissues and Organs: A Text-Atlas of Scanning Electron Microscopy,* 1979 W. H. Freeman and Co.; **35.14d:** J. D. Berlin; **35.17:** Hospital Tribune 8:1, 1974.

Chapter 36

Opener: © D. H. Ellis/Visuals Unlimited

Chapter 37

Opener: © Ed Reschke; **37.1a, 37.1b:** From J. F. Fulton and L. G. Wilson, *Selected Readings in the History of Physiology,* 1966. Courtesy of Charles C. Thomas, Publisher, Springfield, IL.; **37.10:** From J. A. Prior, et al., *Physical Diagnosis,* 1981 Mosby-Year Book, Inc.; **37.16:** From J. F. Fulton and L. G. Wilson, *Selected Readings in the History of Physiology,* 1966. Courtesy of Charles C. Thomas, Publisher, Springfield, IL.

Chapter 38

Opener: Cleveland P. Hickman, Jr.; **38.1a:** Thomas McAvoy, Life Magazine © 1955. Time Inc.; **38.1b:** Courtesy of W. S. Hoar; **38.1c:** Courtesy Lary Shaffer; **38.7:** Cleveland P.

Hickman, Jr.; **38.9:** © L. L. Rue, III; **38.11, 38.12:** Cleveland P. Hickman Jr.; **38.14, 38.15:** © L. L. Rue, III; **38.17:** Cleveland P. Hickman, Jr.

Chapter 39

Opener: NASA; **39.5:** Cleveland P. Hickman, Jr.; **39.7:** © William J. Weber/Visuals Unlimited; **39.8:** Cleveland P. Hickman, Jr.

Chapter 40

Opener: Cleveland P. Hickman, Jr.; **Box 40.1:** © D. Foster/WHOI/Visuals Unlimited; **40.12a, 40.12b:** Cleveland P. Hickman, Jr.

Line Art and Text

Chapter 1

1.12: Source: After P. M. Brakefield, Industrial melanism: Do we have the answers?, *Trends in Ecology and Evolution* 2:117–122, 1987; **1.14:** Source: From S. Gould, *Ontogeny and phylogeny.* Harvard University Press, 1977; **1.15:** Source: After W. Bock, *Evolution.* 24:704–722, 1970.

Chapter 2

2.2: From Peter H. Raven and George B. Johnson, *Biology,* 4th edition. Copyright © 1996 Times Mirror Higher Education Group, Inc., Dubuque, Iowa. Reprinted by permission. All Rights Reserved; **2.4, 2.14:** From Peter H. Raven and George B. Johnson, *Understanding Biology,* 3d edition. Copyright © 1995 Times Mirror Higher Education Group, Inc., Dubuque, Iowa. Reprinted by permission. All Rights Reserved.

Chapter 3

3.8: From Peter H. Raven and George B. Johnson, *Biology,* 4th edition. Copyright © 1996 Times Mirror Higher Education Group, Inc., Dubuque, Iowa. Reprinted by permission. All Rights Reserved.

Chapter 4

4.4, 4.8a, 4.9a, 4.11a, 4.17, 4.18a: From Peter H. Raven and George B. Johnson, *Understanding Biology,* 3d edition. Copyright © 1995 Times Mirror Higher Education Group, Inc., Dubuque, Iowa. Reprinted by permission. All Rights Reserved; **4.19:** From Peter H. Raven and George B. Johnson, *Biology,* 3d edition. Copyright © 1992 Mosby-Year Book. Reprinted by permission of Times Mirror Higher Education Group, Inc., Dubuque, Iowa. All Rights Reserved. **4.22:** Source: After A. W. Murray and M. W. Kirschner, *Sci. Am.* 264:56–63, March 1991; **4.23:** Source: After J. Darnell, H. Lodish, and D. Baltimore, *Molecular cell biology.* Scientific American Books, New York, 1986.

Chapter 5

5.1: From Peter H. Raven and George B. Johnson, *Understanding Biology,* 3d edition. Copyright © 1995 Times Mirror Higher Education Group, Inc., Dubuque, Iowa. Reprinted by permission. All Rights Reserved; **5.3:** From Peter

H. Raven and George B. Johnson, *Biology,* 4th edition. Copyright © 1996 Times Mirror Higher Education Group, Inc., Dubuque, Iowa. Reprinted by permission. All Rights Reserved; **5.6:** From Peter H. Raven and George B. Johnson, *Understanding Biology,* 3d edition. Copyright © 1995 Times Mirror Higher Education Group, Inc., Dubuque, Iowa. Reprinted by permission. All Rights Reserved; **5.8:** From Peter H. Raven and George B. Johnson, *Biology,* 4th edition. Copyright © 1996 Times Mirror Higher Education Group, Inc., Dubuque, Iowa. Reprinted by permission. All Rights Reserved.

Chapter 6

6.4: Source: After C. J. Cole, Unisexual lizards, *Sci. Am.* 250:94–100, January 1984; **6.17:** Source: After A. Ulmann, G. Teutsch, and D. Philbert, RU 486, *Sci. Am.* 262:42–48, June 1990; **6.19:** From Peter H. Raven and George B. Johnson, *Biology,* 4th edition. Copyright © 1996 Times Mirror Higher Education Group, Inc., Dubuque, Iowa. Reprinted by permission. All Rights Reserved; **6.20:** Source: After J. Langman, *Medical embryology,* 4th ed. Williams & Wilkins, Baltimore, 1981.

Chapter 7

7.1: Source: From N. Hartsoeker, *Essai de dioprique,* 1694. **7.3:** Source: After D. Epel, The program of fertilization, *Sci. Am.* 237:128–138, November 1977; **7.10:** Source: After S. F. Gilbert, *Developmental Biology,* 4th ed. Sinauer Associates, Sunderland, MA, 1994, and other sources; **7.11:** Source: After L. W. Browder, C. A. Erickson, and W. R. Jeffrey, *Developmental biology.* Saunders College Publishing, Philadelphia, 1991; **7.14:** Source: After S. F. Gilbert, *Developmental Biology,* 3d ed., Sinauer Associates, Sunderland, MA, 1991, and T. J. King, Nuclear transplantation in amphibia, *Meth. Cell Physiol.* 2:1–36, 1966; **7.15:** Source: After R. G. McKinnell, *Cloning: Nuclear transplantation in amphibia.* University of Minnesota Press, Minneapolis, 1978; **7.18:** William McGinnis; adapted by the studio of Wood Ronsaville Harlin, Inc., for Howard Hughes Medical Institute as published in *From Egg to Adult* © 1992. Reprinted by permission; **7.19:** Source: After E. M. De Robertis, O. Guillermo, and C. V. E. Wright, Homeobox genes and the vertebrate body plan, *Sci. Am.* 263:46–52, July 1990; **7.21:** Source: After B. M. Patten, The first heart beats and the beginning of embryonic circulation, *Am. Sci.* 39:225–243, April 1951; **7.22:** Source: C. P. Hickman Jr., The larval development of the sand sole, Paralichthys melanostictus, Washington State Fisheries Research Papers 2:38–47, 1959; **7.32:** Source: After D. M. Raup and J. J. Sepkoski Jr., Mass extinctions in the marine fossil record, *Science* 215:1502–1504, 1982.

Chapter 8

8.16: From Etkin W. 1973. A representation of the structure of DNA. Figure. *BioScience* 23:653. © 1973 American Institute of Biological Sciences. Reprinted by permission; **8.18:** Source:

After P. Chambon, *Sci. Am.* 244:60–71, May 1981; **8.20:** From Peter H. Raven and George B. Johnson, *Understanding Biology,* 3d edition. Copyright © 1995 Times Mirror Higher Education Group, Inc., Dubuque, Iowa. Reprinted by permission. All Rights Reserved.

Chapter 9

9.4: Source: After A. Moorehead, *Darwin and the Beagle.* Harper & Row, New York, 1969; **9.12:** Source: From J. J. Sepkoski Jr., *Paleobiology* 7:36–53, 1981; **9.13:** From Peter H. Raven and George B. Johnson, *Biology,* 4th edition. Copyright © 1996 Times Mirror Higher Education Group, Inc., Dubuque, Iowa. Reprinted by permission. All Rights Reserved; **9.15:** Source: After J. Cracraft, *Ibis* 116:294–521, 1974; **9.17:** Source: After A. Dumeril, *Ann. Sci. Nat. Zool.* 7:229–254, 1867; **9.19:** Source: After R. Highton and S. A. Henry, *Evol. Biol.* 4:211–256, 1970; **9.20:** Source: After D. N. Taliev, *Sculpins of Baikal (Cottoidei).* Acad. Sci. USSR Moscow, 1955; **9.21:** Source: After P. R. Grant, Speciation and adaptive radiation of Darwin's finches, *American Scientist* 69:653–663, 1981; **9.22a:** From Peter H. Raven and George B. Johnson, *Biology,* 4th edition. Copyright © 1996 Times Mirror Higher Education Group, Inc., Dubuque, Iowa. Reprinted by permission. All Rights Reserved; **9.27:** Source: After A. E. Mourant, *The distribution of human blood.* Ryerson Press, Toronto, 1954; **9.30a:** Source: After P. W. Hedrick, *Population biology.* Jones and Bartlett, Boston, 1984; **9.33:** Source: After E. S. Vrba, *Living Fossils,* ed. by N. Eldredge and S. M. Stanley. Springer Verlag, New York, 1983.

Chapter 10

10.1: Source: After J. T. Bonner, *The Evolution of Complexity.* Princeton University Press, 1988; **10.2:** Source: After C. R. Taylor, K. Schmidt-Nielsen, and J. L. Raab, Scaling of energetic cost of running to body size in animals, *American Journal of Physiology* 219(4):1106, October 1970; **10.8:** From Kent M. Van De Graaff and Stuart Ira Fox, *Concepts of Human Anatomy & Physiology,* 4th edition. Copyright © 1995 Times Mirror Higher Education Group, Inc., Dubuque, Iowa. Reprinted by permission. All Rights Reserved.

Chapter 11

11.3: Source: After E. O. Wiley, *Phylogenetics.* John Wiley & Sons, New York, 1981; **11.10:** Source: After Th. Dobzhansky, *A Century of Darwin,* edited by S. A. Barnett. Harvard University Press, Cambridge, 1958.

Chapter 12

12.1: Source: After J. Lasman in *Journal of Protozoology* 24:244–248, 1977; **12.3a:** From Peter H. Raven and George B. Johnson, *Biology,* 4th edition. Copyright © 1996 Times Mirror Higher Education Group, Inc., Dubuque, Iowa. Reprinted by permission. All Rights Reserved.

Chapter 13

13.1: Source: After E. A. Lapan and H. Morowitz, *Sci. Am.* 227:94–101, December 1972; **13.3:** Source: After K. G. Grell, *Z. Morph. Tiere* 73:297–314, 1972; **13.4, 13.9:** From Peter H. Raven and George B. Johnson, *Biology,* 4th edition. Copyright © 1996 Times Mirror Higher Education Group, Inc., Dubuque, Iowa. Reprinted by permission. All Rights Reserved;

Chapter 15

15.9: Source: After G. D. Schmidt and L. S. Roberts, *Foundations of Parasitology,* 4th ed. Mosby-Year Book, St. Louis, 1989; **15.15:** Source: From J. F. Mueller and H. J. Van Cleave, *Roosevelt wildlife annals,* 1932; **15.16:** Source: After D. J. Morseth in *Journal of Parasitology* 53:492–500, 1967; **15.26:** From W. E. Sterrer, Systematics and evolution within the Gnathostomulida, *System. Zool.* 21:151, 1972. Reprinted by permission.

Chapter 16

Extract, page 301: Source: From N. A. Cobb, *Yearbook of the United States Department of Agriculture,* 1914, p. 472; **16.2a:** From M. Voigt and W. Koste, *Rotaria, die Radertiere Mitteleuropas,* 2d ed. Borntraeger, Berlin, 1978; **16.2b:** From W. T. Edmonson, editor, *Ward and Whipple's freshwater biology,* 2d ed. John Wiley and Sons, New York, 1959; **16.2c&d:** From A. Ruttner-Kolisko, *Das Zooplankton der Binnengewasser* 26 (suppl.):1, 1974; **16.6:** From Synopsis and classification of living organisms, edited by S. P. Parker. Copyright © 1982 McGraw-Hill, Inc. Reprinted by permission; **16.8:** Source: After R. M. Kristensen, Loricifera, a new phylum with Aschelminthes characters from the meiobenthos, *Zeitsch. Zool. Syst. Evol.* 21:163, 1983; **16.21:** Source: After C. Con, Kamptozoa, *Klassen und Ordnungen des Tier-Reichs,* vol. 4, part 2, edited by H. G. Bronn. Akademische Verlagsgesselschaft, Leipzig, 1936.

Chapter 18

18.6: Source: From P. Fauvel, Annelides polychetes. Reproduction, *Traite de Zoologie,* vol. 5, part 1, edited by P. P. Grasse. Masson et Cie, Paris, 1959. Modified from W. M. Woodworth, 1907.

Chapter 19

19.3, 19.4: From Synopsis and classification of living organisms, edited by S. P. Parker. Copyright © 1982 McGraw-Hill, Inc. Reprinted by permission.

Chapter 20

20.8: Source: After G. B. Moment, *General zoology.* Houghton Mifflin, Boston, 1967; **20.20:** Source: After G. A. Boxshall and R. J. Lincoln, *J. Crust. Biol.* 3:1–16, 1983.

Chapter 21

Extract, page 404: Source: From the *New York Times,* 20 April 1988; **21.16, 21.18a:** From Peter H. Raven and George B. Johnson, *Biology,* 4th

edition. Copyright © 1996 Times Mirror Higher Education Group, Inc., Dubuque, Iowa. Reprinted by permission. All Rights Reserved.

Chapter 23

23.8b: Source: After W. D. Russell-Hunter, *A Biology of Higher Invertebrates.* Macmillan, New York, 1969.

Chapter 24

24.4: Source: Courtesy of Tim Doyle.

Chapter 25

25.3: Source: After W. D. Russell-Hunter, *A Biology of Higher Invertebrates.* Macmillan, New York, 1969.

Chapter 26

Poem, page 479: Reprinted by permission of Alpha Music Inc; **26.11:** Source: After S. J. Gould, *Wonderful Life.* W. W. Norton, New York, 1989; **26.16:** Source: After R. Zangerl and M. E. Williams, *Paleontology* 18:333–341, 1975.

Chapter 27

27.3b: Source: After R. Conniff, *Audubon,* March 1991; **27.3c:** Source: After F. H. Pough et al., *Vertebrate life,* Macmillan, 1989; and D. Jensen, *Sci. Am.* 214(2):82–90, 1966; **27.3d:** Source: After D. Jensen, *Sci. Am.* 214(2):82–90, 1966; **27.22:** From Peter Castro and Michael E. Huber, *Marine Biology.* Copyright © 1992 Mosby-Year Book. Reprinted by permission of Times Mirror Higher Education Group, Inc., Dubuque, Iowa. All Rights Reserved; **27.24:** From F. H. Pough, J. B. Heiser, and W. N. McFarland, *Vertebrate life,* 3d edition. Macmillan, 1989; **27.27:** Source: After A. N. Baker, F. W. E. Row, and H. E. S. Clark, A new class of Echinodermata from New Zealand, *Nature* 321:862–864, 1986; **27.28:** Source: After D. Webster and M. Webster *Comparative vertebrate morphology.* Academic Press, New York, 1974.

Chapter 28

28.5: Source: After W. E. Duellman and L. Trueb, *Biology of Amphibians.* McGraw Hill, New York, 1986; **28.19:** Source: After M. S. Gordon et al., *Animal function: Principles and adaptations.* Macmillan, New York, 1968.

Chapter 29

29.11: Source: After R. M. Alexander, *The chordates.* Cambridge University Press, England, 1975.

Chapter 30

Extract, page 577: Source: From L. Brown, *Eagles.* Arco Publishing, New York, 1970; **30.1:** From Peter H. Raven and George B. Johnson, *Understanding Biology,* 3d edition. Copyright © 1995 Times Mirror Higher Education Group, Inc., Dubuque, Iowa. Reprinted by permission. All Rights Reserved; **30.7b:** Source: After P. Wellenhofer, Archaeopteryx, *Sci. Am.* 262:70–77, May 1990; **30.28:** Source: After S. R. Johnson and I. T. McCowan, Thermal adaptation as a factor affecting colonizing success of introduced Sturnidae (Aves) in North America, *Canadian Journal of Zoology* 52:1559–1576, 1974; **30.12b:** Source: After K. Schmidt-Nielsen, *Animal Physiology*, 4th edition. Cambridge University Press, 1990.

Chapter 31

31.1: Source: From R. L. Carroll, *Vertebrate paleontology and evolution*. W. H. Freeman and Co., New York, 1988; **31.3:** Source: After J. Z. Young, *The Life of Mammals*. Oxford University Press, Oxford, 1975; **31.8:** From Kent M. Van De Graaff and Stuart Ira Fox, *Concepts of Human Anatomy & Physiology*, 4th edition. Copyright © 1995 Times Mirror Higher Education Group, Inc., Dubuque, Iowa. Reprinted by permission. All Rights Reserved; **31.10:** Source: After E. Rogers, *Looking at vertebrates*. Longman Group, Essex, England, 1986; **31.17:** Source: After N. Suga, Biosonar and neural computation in bats, *Sci. Am.* 262:60–68, June 1990; **31.19:** Source: After J. A. Lillegraven et al., The origin of eutherian mammals, *Biol. Jour. Linn. Soc.* 32:281–336, 1987; **31.20:** Source: After C. R. Austin and R. V. Short, editors, *Reproduction in Mammals: Volume 4, Reproductive Patterns.* Cambridge University Press, New York, 1972.

Chapter 32

32.1b: From Peter H. Raven and George B. Johnson, *Understanding Biology,* 3d edition. Copyright © 1995 Times Mirror Higher Education Group, Inc., Dubuque, Iowa. Reprinted by permission. All Rights Reserved; **32.6, 32.7:** From Kent M. Van De Graaff and Stuart Ira Fox, *Concepts of Human Anatomy & Physiology,* 4th edition. Copyright © 1995 Times Mirror Higher Education Group, Inc., Dubuque, Iowa. Reprinted by permission. All Rights Reserved; **32.8:** Source: After A. A. Biewener, Mammalian terrestrial locomotion and size, *BioScience* 39(11):776–783, 1989; and R. M. Alexander, How dinosaurs ran, *Sci. Am.* 264:130–136, April 1991; **32.10a:** Source: After M. A. Sleigh, *The Biology of Cilia and Flagella.* Pergamon Press, Oxford, 1962; **32.10b:** Source: After M. A. Sleigh and D. I. Barlow, Metachronism and control of locomotion in animals with many propulsive structures, *Aspects of animal movement,* ed. by H. Y. Elder and E. R. Trueman, Cambridge University Press, 1980; **32.14:** From Peter H. Raven and George B. Johnson, *Understanding Biology,* 3d edition. Copyright © 1995 Times Mirror Higher Education Group, Inc., Dubuque, Iowa. Reprinted by permission. All Rights Reserved; **32.17:** Source: After R. Eckert, D. Randall, and G. Augustine, *Animal physiology,* 3d ed. W. H. Freeman and Company, New York, 1988.

Chapter 33

33.3: Source: After D. Webster and M. Webster *Comparative vertebrate morphology.* Academic Press, New York, 1974; **33.4:** Source: After D. Webster and M. Webster *Comparative vertebrate morphology.* Academic Press, New York, 1974; **33.13:** Source: After R. F. Pitts, *Physiology of the kidney and body fluids*, 3d ed.

Mosby-Year Book, St. Louis, 1974; **33.15a:** Source: After H. Wirz, B. Hargitay, and W. Kuhn, *Helv. Physiol. Acta* 9:196–207, 1951; **33.15b:** Source: After K. J. Ullrich and K. H. Jarausch, *Pflugers Archiv.* 262:537–550, 1956; **33.20:** Source: After R. C. Lasiewski, *Physiol. Zool.* 36:122–140, 1963.

Chapter 34

34.8a: From Peter H. Raven and George B. Johnson, *Understanding Biology,* 3d edition. Copyright © 1995 Times Mirror Higher Education Group, Inc., Dubuque, Iowa. Reprinted by permission. All Rights Reserved; **34.8b:** From Peter H. Raven and George B. Johnson, *Biology,* 4th edition. Copyright © 1996 Times Mirror Higher Education Group, Inc., Dubuque, Iowa. Reprinted by permission. All Rights Reserved; **34.28:** Source: After M. S. Gordon et al., *Animal function: Principles and adaptations.* Macmillan, New York, 1968.

Chapter 35

Extract, page 699: Source: From N. J. Berrill, *You and the universe.* Dodd, Mead & Co., New York, 1958; **35.11:** From Peter H. Raven and George B. Johnson, *Biology,* 4th edition. Copyright © 1996 Times Mirror Higher Education Group, Inc., Dubuque, Iowa. Reprinted by permission. All Rights Reserved; **35.15:** Source: After R. Eckert and D. Randall, *Animal physiology,* 2d ed. W. H. Freeman and Company, New York, 1983; **35.16:** Source: After M. Winick, *Malnutrition and brain development.* Oxford University Press, New York, 1976.

Chapter 36

36.11: From Peter H. Raven and George B. Johnson, *Biology,* 3d edition. Copyright © 1992 Mosby-Year Book. Reprinted by permission of Times Mirror Higher Education Group, Inc., Dubuque, Iowa. All Rights Reserved; **36.14:** From Kent M. Van De Graaff and Stuart Ira Fox, *Concepts of Human Anatomy & Physiology,* 4th edition. Copyright © 1995 Times Mirror Higher Education Group, Inc., Dubuque, Iowa. Reprinted by permission. All Rights Reserved; **36.15:** From Peter H. Raven and George B. Johnson, *Understanding Biology,* 3d edition. Copyright © 1995 Times Mirror Higher Education Group, Inc., Dubuque, Iowa. Reprinted by permission. All Rights Reserved; **36.18:** Source: After E. O. Wilson and W. H. Bossert, *Res. Prog. Horm. Res.* 19:673–716, 1963; **36.29:** Source: After F. Lenci and G. Colombetti, *Photoreception and sensory transduction in aneural organisms.* Plenum Press, New York, 1980; **36.31:** From Peter H. Raven and George B. Johnson, *Biology,* 4th edition. Copyright © 1996 Times Mirror Higher Education Group, Inc., Dubuque, Iowa. Reprinted by permission. All Rights Reserved; **36.32, 36.33:** From Peter H. Raven and George B. Johnson, *Understanding Biology,* 2d edition. Copyright © 1991 Mosby-Year Book. Reprinted by permission of Times Mirror Higher Education Group, Inc., Dubuque, Iowa. All Rights Reserved.

Chapter 37

37.11: Source: After D. H. Copp, *J. Endocrinol.* 43:137–161, 1969; **37.9:** Source: After P. J. Bentley, *Comparative vertebrate endocrinology,* 2d ed. Cambridge University Press, 1982.

Chapter 38

Extract, page 771: Source: From B. Nelson, *Galapagos: Islands of birds.* Longmans, Green & Company, London, 1968; **Extract, page 771:** Source: From Irven DeVore, *The marvels of animal behavior.* National Geographic Society, Washington, DC, 1972; **38.2:** Source: After K. Lorenz and N. Tinbergen, *Zeit. Tierpsychol.* 2:1–29, 1938; **38.4:** From N. Tinbergen, *The study of instinct,* Oxford University Press, Oxford, England, 1951; modified from D. Lack, *The life of the robin,* Cambridge University Press, Cambridge, England, 1943; **38.5:** Source: After N. Rothenbuhler, Behavior genetics of nest cleaning honey bees. IV. Responses of F1 and backcross generations to disease-killed brood, *Am. Zool.* 4:111–123, 1964; **38.6:** Source: After W. C. Dilger, The behavior of lovebirds, *Sci. Am.* 206:89–98, January 1962; **38.10:** Source: After J. Alcock, *Animal behavior: An evolutionary approach,* 3d ed., Sinauer Associates, Sunderland, MA, 1984, from a photograph by Masakasu Konishi; **38.16:** From C. Darwin, *Expression of the emotions in man and animals.* Appleton and Co., New York, 1872.

Chapter 39

39.1: From Peter Castro and Michael E. Huber, *Marine Biology.* Copyright © 1992 Mosby-Year Book. Reprinted by permission of Times Mirror Higher Education Group, Inc., Dubuque, Iowa. All Rights Reserved; **39.2:** From W. M. Washington, Where's the heat?, *Natural History,* March 1990; **39.9, 39.10:** From Peter Castro and Michael E. Huber, *Marine Biology.* Copyright © 1992 Mosby-Year Book. Reprinted by permission of Times Mirror Higher Education Group, Inc., Dubuque, Iowa. All Rights Reserved; **39.11a:** Source: After W. H. Burt, *Zoogeography,* ed. by C. L. Hubbs. AAAS Pub. No. 51, Washington, D.C., 1958; **39.11b:** Source: After W. F. Blair, *Zoogeography,* ed. by C. L. Hubbs. AAAS Pub. No. 51, Washington, D.C., 1958; **39.12:** Source: After J. Cracraft, *Ibis* 116:294–521, 1974; **39.14:** Drawings by Marlene Hill Werner, reprinted courtesy of Larry G. Marshall.

Chapter 40

40.4: Source: Data from R. L. Smith, *Ecology and field biology,* 3d edition, Harper and Row, New York, 1980; and E. P. Odum, *Fundamentals of ecology,* 3d ed., W. B. Saunders, Philadelphia, 1971; **40.8:** Source: Map courtesy of Agricultural Research Service; **40.9:** From Peter Castro and Michael E. Huber, *Marine Biology.* Copyright © 1992 Mosby-Year Book. Reprinted by permission of Times Mirror Higher Education Group, Inc., Dubuque, Iowa. All Rights Reserved; **40.10:** Source: Data from D. Lack, *Darwin's finches.* Cambridge University Press, 1947.

ORIGIN OF LIFE AND GEOLOGIC TIME TABLE

Millions of years ago	Bacteria Cyanobacteria Protista Fungi Plants Animals	Oxygen in atmosphere	Events in evolution of life
Present day			Cambrian "explosion"
1000			Appearance of multicellular organisms
2000			Appearance of oxygen in atmosphere
			Photosynthesis
3000			Anaerobic metabolism
			First living systems
			Biopolymers: proteins, polysaccharides, nucleic acids
4000			Chemical evolution
			Early reducing atmosphere of methane, ammonia, water, nitrogen, carbon dioxide
			Origin of solar system
5000			